ENGLISH–SPANISH, SPANISH–ENGLISH

ELECTRICAL
AND COMPUTER
ENGINEERING
DICTIONARY

DICCIONARIO DE INGENIERÍA
ELÉCTRICA Y DE COMPUTADORAS
INGLÉS/ESPAÑOL, ESPAÑOL/INGLÉS

By the same author

Wiley's English-Spanish and Spanish-English Legal Dictionary

Wiley's English-Spanish Spanish-English Dictionary of Psychology and Psychiatry

Por el mismo autor

Diccionario jurídico inglés-español y español-inglés Wiley

Diccionario de psicología y psiquiatría inglés-español español-inglés Wiley

ENGLISH–SPANISH, SPANISH–ENGLISH

ELECTRICAL
AND COMPUTER
ENGINEERING
DICTIONARY

DICCIONARIO DE INGENIERÍA
ELÉCTRICA Y DE COMPUTADORAS
INGLÉS/ESPAÑOL, ESPAÑOL/INGLÉS

Steven M. Kaplan
Lexicographer

Editorial Advisory Board:

Professor Ramón Pallás-Areny
Universitat Politècnica de Catalunya
Barcelona, Spain

Dr. Daniel Malacara
Centro de Investigaciones en Optica A.C.
Leon, México

Sr. Carlos Urbina Pacheco
Torreón, Coahulla, México

Professor Sergio Verdú
Princeton University

A Wiley-Interscience Publication
JOHN WILEY & SONS, INC.
New York   Chichester   Brisbane   Toronto   Singapore

This publication is designed to provide accurate and
authoritative information in regard to the subject
matter covered. It is sold with the understanding that
the publisher is not engaged in rendering any professional
services. If expert assistance is required, the services of a
competent professional person should be sought.

Library of Congress Cataloging in Publication Data:

Kaplan, Steven M.
English-Spanish, Spanish-English electrical and computer
engineering dictionary = Diccionario de ingeniería eléctrica y de
computadoras inglés-español, español-inglés/ Steven M.
Kaplan, lexicographer
      p.   cm.
      "A Wiley-Interscience publication."
      ISBN 0-471-01037-5 (cloth : alk. paper)
      1. Electric engineering—Dictionaries.   2. Computer engineering—
Dictionaries.   3. English language—Dictionaries—Spanish.
4. Electric engineering—Dictionaries—Spanish.   5. Computer
engineering—Dictionaries—Spanish.   6. Spanish language—
Dictionaries—English.   I. Title.
TK9.K37   1995
603—dc20   95-42946
                                                                CIP

Printed in the United States of America

10 9 8 7 6 5 4 3 2 1

PREFACE AND NOTES ON THE USE OF THIS DICTIONARY

There are no special rules for the use of this dictionary. The user simply locates the desired entry and obtains the equivalent. The primary goals for this dictionary were accuracy, clarity, comprehensiveness, currency, and ease of use. Some entries from areas that have diminishing importance, such as telegraphy are included, as they still have a significant presence in the literature of these fields. The dictionary has over 95,000 entries. Since the dictionary is already very large, certain measures have been taken to control size without reducing accuracy. One has been to choose a typesize which is small enough to provide for an affordable volume which can be handled, yet of sufficient size to permit reasonably comfortable reading. Another has been to use one of the two equally accurate forms in Spanish for seven concepts that appear throughout the book. They are:

> "*tensión*" and "*voltaje*" for the English concept of *voltage*
> "*puerta*" and "*compuerta*" for the English concept of *gate*
> "*resistor*" and "*resistencia*" for the English concept of *resistor*
> "*pulso*" and "*impulso*" for the English concept of *pulse*
> "*arrollamiento*" and "*devanado*" for the English concept of *winding*
> "*conmutador*" and "*interruptor*" for the English concept of *switch*
> "*bucle*" and "*lazo*" for the English concept of *loop*

In each case the former is used throughout the dictionary, instead of both, which would unnecessarily add many pages to the text. Many variables were factored in to arrive at which term to use. For example, even though "*tensión*" and "*voltaje*" are equally accurate equivalents for *voltage*, "*tensión*" is used instead of "*voltaje*" since the literature reflects that a greater portion of the Spanish-speaking world uses "*tensión*." "*Voltaje*" and "*devanado*," and the others in this list do appear alphabetically in the Spanish section, and as an additional equivalent in the English section.

Regionalisms are beyond the scope of this dictionary. Regionalisms, as their name implies, apply only to small sections of the English or Spanish speaking worlds. Inclusion of regionalisms would add considerable bulk to the dictionary and would not serve the needs of the vast majority of users. The dictionary provides equivalents which should be clear to those versed in the language and specifically in these fields.

Series of entries that start with the same word include many of the phrases that are most often encountered in the literature. As a group, these phrases will include some that are needed by one user and others by another, and so on. In addition, many of the most looked-up entries are of this type, and the user may look one up to be sure of using an accurate equivalent. This saves the user the time of locating the components of such entries individually, having to put them together, and still having the doubt of whether the result is accurate. Each of these phrasal entries is provided as a separate entry for simplicity of use.

This dictionary is designed to cater to specific needs. A quality general bilingual dictionary should be consulted for words and phrases that are not specifically within the electrical and computer engineering fields. If explanations for terms are desired, monolingual dictionaries in these fields should provide them.

I am very grateful to John Wiley Senior Editor George J. Telecki who provided the idea for this dictionary. His encouragement and understanding from the start, combined with his locating eminent editorial advisors helped provide for the best possible dictionary. I am also appreciative of editorial assistant Rose Leo Kish who in many ways made sure that everything continued along smoothly.

The editorial advisors, Carlos Urbina Pacheco, Ramón Pallás-Areny, Daniel Malacara, and Sergio Verdú played a key role in helping to obtain a book of the highest possible quality.

Steven M. Kaplan

Miami, Florida

PREFACIO Y NOTAS SOBRE EL USO DE ESTE DICCIONARIO

No hay reglas especiales para el uso de este diccionario. El usuario sencillamente ubica la voz de entrada deseada y obtiene el equivalente. Las metas primarias para este diccionario fueron la precisión, claridad, abarcamiento, actualidad, y facilidad de uso. Algunas voces de entradas de áreas de importancia decreciente, tales como la telegrafía se incluyen, pues todavía tienen una presencia significativa en la literatura de estos campos. El diccionario tiene más de 95,000 voces de entradas. Ya que el diccionario ya es muy grande, se han tomado ciertas medidas para controlar el tamaño sin disminuir la precisión. Una ha sido escoger un tamaño de letras el cual es lo suficientemente pequeño para producir un tomo manejable y al alcance, pero a la vez de tamaño suficiente para permitir lectura razonablemente cómoda. Otra ha sido usar una de las dos formas de igual precisión en español para siete conceptos que aparecen a través del libro. Ellos son:

> *tensión* y *voltaje* para el concepto de inglés de *"voltage"*
> *puerta* y *compuerta* para el concepto de inglés de *"gate"*
> *resistor* y *resistencia* para el concepto de inglés de *"resistor"*
> *pulso* e *impulso* para el concepto de inglés de *"pulse"*
> *arrollamiento* y *devanado* para el concepto de inglés de *"winding"*
> *conmutador* e *interruptor* para el concepto de inglés de *"switch"*
> *bucle* y *lazo* para el concepto de inglés de *"loop"*

En cada caso se ha usado el primero a través del diccionario, en vez de ambos, lo cual añadiría innecesariamente muchas páginas al texto. Muchas variables se factorizaron para llegar a cual término usar. Por ejemplo, aunque *tensión* y *voltaje* son equivalentes igual de precisos para *"voltage"*, se usa *tensión* en vez de *voltaje* ya que la literatura refleja que una mayor porción del mundo habla hispana usa *tensión*. *Voltaje* y *devanado*, y los otros en esta lista sí aparecen alfabéticamente en la sección de español, y como un equivalente adicional en la sección de inglés.

Los regionalismos están más allá del alcance de este diccionario. Los regionalismos, como sus nombre indica, aplican sólo a secciones pequeñas de los mundos de habla inglesa o habla hispana. El incluir regionalismos añadiría tamaño considerable al diccionario y no serviría las necesidades de la vasta mayoría de usuarios. El diccionario provee equivalentes los cuales deberían ser claros para aquellos que se desenvuelven en el idioma y específicamente en estos campos.

Series de voces de entradas que empiezan con la misma palabra incluyen muchas de las frases que se encuentran más a menudo en la literatura.   Como grupo, estas frases incluyen algunas que necesita un usuario y otras que necesita otro, y así por el estilo. En adición, muchas de las voces de entradas que más se buscan son de esta índole, y el usuario puede buscar una para tener la certidumbre de usar un equivalente preciso. Esto le economiza al usuario el tiempo de buscar individualmente los componentes de dichas voces de entradas, el tener que unirlas, y todavía tener la duda si el resultado es preciso.   Cada una de estas entradas en frase se provee como voz de entrada separada para la simplicidad de uso.

Este diccionario está diseñado para servir necesidades específicas.   Un diccionario general bilingüe de calidad se debe consultar para palabras y frases que no estén específicamente dentro de los campos de la ingeniería eléctrica y de computadoras.   Si se desean explicaciones para términos, diccionarios monolingües en estos campos las deben proveer.

Le estoy muy agradecido al editor de John Wiley George J. Telecki quien proveyó la idea para este diccionario.   Su apoyo y comprensión desde el comienzo, combinados con su localización de asesores editoriales eminentes ayudaron a proveer el mejor diccionario posible.   También le estoy agradecido al asistente de editor Rose Leo Kish quien en muchas maneras se aseguró de que todo marchara sobre ruedas.

Los asesores editoriales, Carlos Urbina Pacheco, Ramón Pallás-Areny, Daniel Malacara, y Sergio Verdú desempeñaron un papel clave en ayudar a obtener un libro de la mejor calidad posible.

Steven M. Kaplan

Miami, Florida

A address dirección A
A and not B gate puerta A y no B
A-B method método A-B
A-B test prueba A-B
A battery batería A
A board tablero A
A channel canal A
A contact contacto A
a-d converter convertidor a-d
a-d encoder codificador a-d
A-digit selector selector de dígito A
A display indicador visual A
A-law ley A
A operator operador A
A position posición A
A power supply fuente de alimentación A
A station estación A
A supply fuente A
A traffic tráfico A
A voltage tensión A
AB amplifier amplificador AB
AB power pack bloque de alimentación AB
abac ábaco
abacus ábaco
abampere abampere
abampere per square centimeter abampere por
    centímetro cuadrado
abandon abandonar
abandoned abandonado
abandoned call attempt intento de llamada
    abandonado
abandonment abandono
Abbe condenser condensador de Abbe
abbreviate abreviar
abbreviated abreviado
abbreviated address calling llamada de dirección
    abreviada
abbreviated dialing marcación abreviada
abbreviated form of message forma abreviada de
    mensaje
abbreviated installation instalación abreviada
abbreviated selection selección abreviada
abbreviation abreviación
abbreviator abreviador
abcoulomb abcoulomb
abcoulomb per square centimeter abcoulomb por
    centímetro cuadrado
abend final anormal
abend recovery program programa de recuperación
    de final anormal
aberration aberración
abfarad abfarad
abhenry abhenry
ability habilidad
abmho abmho
abnormal anormal
abnormal absorption absorción anormal
abnormal cathode fall caída catódica anormal
abnormal condition condición anormal
abnormal decay decaimiento anormal
abnormal dissipation disipación anormal

abnormal end final anormal
abnormal end of task final de tarea anormal
abnormal exit salida anormal
abnormal glow luminiscencia anormal
abnormal glow discharge descarga luminiscente
    anormal
abnormal oscillation oscilación anormal
abnormal propagation propagación anormal
abnormal radiation radiación anormal
abnormal reflections reflexiones anormales
abnormal statement sentencia anormal
abnormal termination terminación anormal
abnormal triggering disparo anormal
abohm abohm
abohm centimeter abohm centímetro
abort abortar
abort sequence secuencia de abortar
aborted cycle ciclo abortado
aborted job tarea abortada
abrasion abrasión
abrasion machine máquina de abrasión
abrasion resistance resistencia a la abrasión
abrupt junction unión abrupta
abscissa abscisa
absence ausencia
absent ausente
absiemens absiemens
absolute absoluto
absolute accuracy exactitud absoluta
absolute address dirección absoluta
absolute addressing direccionamiento absoluto
absolute altimeter altímetro absoluto
absolute altitude altitud absoluta
absolute ampere ampere absoluto
absolute apparatus aparato absoluto
absolute assembler ensamblador absoluto
absolute block bloqueo absoluto
absolute cell reference referencia de celda absoluta
absolute chronology cronología absoluta
absolute code código absoluto
absolute coding codificación absoluta
absolute command mando absoluto
absolute constant constante absoluta
absolute current level nivel de corriente absoluto
absolute data datos absolutos
absolute delay retardo absoluto
absolute digital position transducer transductor de
    posición digital absoluto
absolute dimension dimensión absoluta
absolute efficiency eficiencia absoluta
absolute electrical units unidades eléctricas
    absolutas
absolute electrode potential potencial de electrodo
    absoluto
absolute electrometer electrómetro absoluto
absolute element elemento absoluto
absolute encoder system sistema codificador
    absoluto
absolute error error absoluto
absolute expression expresión absoluta
absolute frequency frecuencia absoluta
absolute gain ganancia absoluta
absolute gain of antenna ganancia absoluta de
    antena
absolute galvanometer galvanómetro absoluto
absolute humidity humedad absoluta
absolute instruction instrucción absoluta
absolute language lenguaje absoluto
absolute level nivel absoluto
absolute load module módulo de carga absoluta
absolute loader cargador absoluto
absolute luminance threshold umbral de luminancia
    absoluta

absolute machine code código de máquina absoluto
absolute magnitude magnitud absoluta
absolute maximum rating valor máximo absoluto
absolute maximum supply voltage tensión de alimentación máxima absoluta
absolute measurement medida absoluta
absolute measurement of current medida absoluta de corriente
absolute measurement of voltage medida absoluta de tensión
absolute method of measurement método de medida absoluta
absolute minimum resistance resistencia mínima absoluta
absolute ohm ohm absoluto
absolute origin origen absoluto
absolute peak pico absoluto
absolute Peltier coefficient coeficiente de Peltier absoluto
absolute permeability permeabilidad absoluta
absolute permissive block bloqueo permisivo absoluto
absolute permittivity permitividad absoluta
absolute photocathode spectral response respuesta espectral de fotocátodo absoluta
absolute pitch tono absoluto
absolute potential potencial absoluto
absolute power potencia absoluta
absolute power level nivel de potencia absoluta
absolute pressure presión absoluta
absolute pressure transducer transductor de presión absoluta
absolute priority prioridad absoluta
absolute program programa absoluto
absolute programming programación absoluta
absolute readout lectura absoluta
absolute refractory state estado refractario absoluto
absolute scale escala absoluta
absolute Seebeck coefficient coeficiente de Seebeck absoluto
absolute spectral response respuesta espectral absoluta
absolute speed drop caída de velocidad absoluta
absolute speed rise subida de velocidad absoluta
absolute speed variation variación de velocidad absoluta
absolute stability estabilidad absoluta
absolute steady-state deviation desviación de estado estable absoluta
absolute stop parada absoluta
absolute stop signal señal de parada absoluta
absolute system sistema absoluto
absolute system deviation desviación de sistema absoluto
absolute system of units sistema absoluto de unidades
absolute temperature temperatura absoluta
absolute temperature scale escala de temperatura absoluta
absolute term término absoluto
absolute threshold umbral absoluto
absolute tolerance tolerancia absoluta
absolute transient deviation desviación transitoria absoluta
absolute unit unidad absoluta
absolute value valor absoluto
absolute value circuit circuito de valor absoluto
absolute value computer computadora de valor absoluto
absolute value device dispositivo de valor absoluto
absolute value representation representación de valor absoluto
absolute voltage level nivel de tensión absoluto

absolute voltage rise aumento de tensión absoluto
absolute voltage variation variación de tensión absoluta
absolute wavemeter ondámetro absoluto
absolute zero cero absoluto
absorb absorber
absorbance absorbancia
absorbed absorbido
absorbed dose dosis absorbida
absorbed dose index índice de dosis absorbida
absorbed power potencia absorbida
absorbed voltage tensión absorbida
absorbed wave onda absorbida
absorber absorbedor
absorber circuit circuito absorbedor
absorber circuit factor factor de circuito absorbedor
absorber diode diodo absorbedor
absorbing absorbente
absorbing filter filtro absorbente
absorbing Markov chain model modelo de cadena de Markov absorbente
absorbing material material absorbente
absorbing medium medio absorbente
absorbing state estado absorbente
absorptance absortancia
absorptiometer absorciómetro
absorptiometry absorciometría
absorption absorción
absorption band banda de absorción
absorption chromatography cromatografía de absorción
absorption circuit circuito de absorción
absorption coefficient coeficiente de absorción
absorption control control de absorción
absorption cross section sección eficaz de absorción
absorption current corriente de absorción
absorption curve curva de absorción
absorption dynamometer dinamómetro de absorción
absorption edge banda de absorción
absorption energy energía de absorción
absorption fading desvanecimiento por absorción
absorption frequency band banda de frecuencia de absorción
absorption frequency meter frecuencímetro de absorción
absorption limiting frequency frecuencia limitante por absorción
absorption line línea de absorción
absorption loss pérdida por absorción
absorption marker marcador de absorción
absorption modulation modulación por absorción
absorption of charged particles absorción de partículas cargadas
absorption of energy absorción de energía
absorption of waves absorción de ondas
absorption spectrum espectro de absorción
absorption trap trampa de absorción
absorption wavemeter ondámetro de absorción
absorptive attenuator atenuador de absorción
absorptive loss pérdida de absorción
absorptivity absortividad
abstract automata theory teoría de autómatas abstracta
abstract code código abstracto
abstract data type tipo de datos abstracto
abstract machine máquina abstracta
abstract quality calidad abstracta
abstract symbol símbolo abstracto
abstraction abstracción
abtesla abtesla
abundance abundancia
abvolt abvolt
abvolt per centimeter abvolt por centímetro

abwatt abwatt
abweber abweber
academic simulation simulación académica
accelerate acelerar
accelerated acelerado
accelerated life test prueba de duración acelerada
accelerated service test prueba de servicio acelerado
accelerated test prueba acelerada
accelerating acelerador
accelerating anode ánodo acelerador
accelerating conductor conductor acelerador
accelerating electrode electrodo acelerador
accelerating field campo acelerador
accelerating force fuerza aceleradora
accelerating grid rejilla aceleradora
accelerating or decelerating device dispositivo
    acelerador o decelerador
accelerating potential potencial acelerador
accelerating relay relé acelerador
accelerating resistance resistencia aceleradora
accelerating time tiempo de aceleración
accelerating torque par acelerador
accelerating voltage tensión aceleradora
acceleration aceleración
acceleration constant constante de aceleración
acceleration error constant constante de error de
    aceleración
acceleration factor factor de aceleración
acceleration indicator indicador de aceleración
acceleration-insensitive drift rate tasa de
    deslizamiento insensible a la aceleración
acceleration potential potencial de aceleración
acceleration-sensitive drift rate tasa de
    deslizamiento sensible a la aceleración
acceleration space espacio de aceleración
acceleration switch conmutador de aceleración
acceleration time tiempo de aceleración
acceleration torque par de aceleración
acceleration voltage tensión de aceleración
accelerator acelerador
accelerator board tablero acelerador
accelerator grid rejilla aceleradora
accelerator of electrons acelerador de electrones
accelerograph acelerógrafo
accelerometer acelerómetro
accentuated acentuado
accentuation acentuación
accentuator acentuador
accentuator circuit circuito acentuador
accept aceptar
acceptable aceptable
acceptable energized background noise level nivel
    de ruido de fondo energizado aceptable
acceptable-environmental-range test prueba de
    intervalo ambiental aceptable
acceptable quality level nivel de calidad aceptable
acceptable source fuente aceptable
acceptable terminal partial discharge level nivel de
    descarga parcial terminal aceptable
acceptance angle ángulo de aceptación
acceptance criteria criterios de aceptación
acceptance sampling plan plan de muestreo de
    aceptación
acceptance test prueba de aceptación
accepted test prueba aceptada
accepting aceptante
accepting semiconductor semiconductor aceptante
accepting station estación aceptante
acceptor aceptor
acceptor atom átomo aceptor
acceptor circuit circuito aceptor
acceptor impurity impureza aceptora
acceptor level nivel aceptor

acceptor material material aceptor
acceptor semiconductor semiconductor aceptor
access acceso
access arm brazo de acceso
access barred acceso obstaculizado
access capability capacidad de acceso
access code código de acceso
access conflict conflicto de acceso
access control control de acceso
access-control field campo de control de acceso
access-control function función de control de acceso
access-control mechanism mecanismo de control de
    acceso
access-control register registro de control de acceso
access-control words palabras de control de acceso
access coupler acoplador de acceso
access cycle ciclo de acceso
access environment ambiente de acceso
access fitting accesorio de acceso
access level nivel de acceso
access line línea de acceso
access macro macro de acceso
access matrix matriz de acceso
access mechanism mecanismo de acceso
access method método de acceso
access mode modo de acceso
access name nombre de acceso
access path trayectoria de acceso
access permission permiso de acceso
access priority prioridad de acceso
access procedure procedimiento de acceso
access right derecho de acceso
access selector selector de acceso
access technique técnica de acceso
access time tiempo de acceso
access type tipo de acceso
access unit unidad de acceso
access value valor de acceso
accessibility accesibilidad
accessible accesible
accessible terminal terminal accesible
accessible voltage drop caída de tensión accesible
accessory accesorio
accessory equipment equipo accesorio
accessory outlet tomacorriente accesorio
accessory probe sonda accesoria
accessory receptacle receptáculo accesorio
accessory unit unidad accesoria
accidental error error accidental
accidental ground tierra accidental
accidental loss pérdida accidental
accidental triggering disparo accidental
accidental voltage transfer transferencia de tensión
    accidental
accommodation acomodación
accompanying audio channel canal de audio
    acompañante
accordion acordeón
accounting computer computadora de contabilidad
accounting machine máquina de contabilidad
accounting routine rutina de contabilidad
accumulate acumular
accumulated acumulado
accumulated down time tiempo muerto acumulado
accumulated error error acumulado
accumulated jitter fluctuación acumulada
accumulated service years años de servicio
    acumulados
accumulated total total acumulado
accumulated value valor acumulado
accumulating counter contador acumulante
accumulating reproducer reproductor acumulante
accumulating stimulus estímulo acumulante

accumulation acumulación
accumulative acumulativo
accumulator acumulador
accumulator address dirección de acumulador
accumulator battery batería de acumuladores
accumulator function función de acumulador
accumulator jump instruction instrucción de salto de acumulador
accumulator register registro de acumulador
accumulator shift instruction instrucción de desplazamiento de acumulador
accumulator substation subestación de acumuladores
accumulator transfer instruction instrucción de transferencia de acumulador
accuracy exactitud, precisión
accuracy class clase de exactitud
accuracy class for metering clase de exactitud para medición
accuracy control character carácter de control de exactitud
accuracy control system sistema de control de exactitud
accuracy rating clasificación de exactitud
accuracy test prueba de exactitud
accurate exacto, preciso
acetate acetato
acetate base base de acetato
acetate disk disco de acetato
acetate tape cinta de acetato
achieved reliability confiabilidad alcanzada
achromatic acromático
achromatic lens lente acromática
achromatic locus lugar acromático
achromatic scale escala acromática
achromatic stimulus estímulo acromático
acid ácido
acid cell pila de ácido
acid-resistant resistente al ácido
acidometer acidómetro
acknowledge character carácter de reconocimiento
acknowledging device dispositivo confirmante
acknowledging switch conmutador confirmante
acknowledging whistle silbato confirmante
acknowledgment reconocimiento
acknowledgment character carácter de reconocimiento
acknowledgment signal señal de reconocimiento
acknowledgment signal unit unidad de señal de reconocimiento
aclinic line línea aclínica
acorn tube tubo de bellota
acoustic acústico
acoustic absorption absorción acústica
acoustic absorption loss pérdida por absorción acústica
acoustic absorptivity absortividad acústica
acoustic admittance admitancia acústica
acoustic amplifier amplificador acústico
acoustic attenuation constant constante de atenuación acústica
acoustic branch rama acústica
acoustic bridge puente acústico
acoustic burglar alarm alarma contra ladrones acústica
acoustic capacitance capacitancia acústica
acoustic clarifier clarificador acústico
acoustic communication comunicación acústica
acoustic compensator compensador acústico
acoustic compliance docilidad acústica
acoustic concentration concentración acústica
acoustic conductivity conductividad acústica
acoustic correction corrección acústica

acoustic coupler acoplador acústico
acoustic coupling acoplamiento acústico
acoustic damping amortiguamiento acústico
acoustic delay retardo acústico
acoustic delay line línea de retardo acústica
acoustic depth finder buscador de profundidad acústico
acoustic detector detector acústico
acoustic device dispositivo acústico
acoustic diffraction difracción acústica
acoustic dispersion dispersión acústica
acoustic distortion distorsión acústica
acoustic elasticity elasticidad acústica
acoustic electric transducer transductor eléctrico acústico
acoustic enclosure caja acústica
acoustic feedback retroalimentación acústica
acoustic filter filtro acústico
acoustic flowmeter flujómetro acústico
acoustic frequency frecuencia acústica
acoustic frequency response respuesta de frecuencia acústica
acoustic generator generador acústico
acoustic grating retículo acústico
acoustic-gravity wave onda gravitacional acústica
acoustic homing system sistema de guía acústico
acoustic horn bocina acústica
acoustic howl aullido acústico
acoustic impedance impedancia acústica
acoustic impedance meter medidor de impedancia acústica
acoustic impulse impulso acústico
acoustic inductance inductancia acústica
acoustic inertance inertancia acústica
acoustic inhibition inhibición acústica
acoustic input entrada acústica
acoustic intensity intensidad acústica
acoustic interferometer interferómetro acústico
acoustic labyrinth laberinto acústico
acoustic lens lente acústica
acoustic line línea acústica
acoustic load carga acústica
acoustic mass masa acústica
acoustic material material acústico
acoustic memory memoria acústica
acoustic mine mina acústica
acoustic mirage espejismo acústico
acoustic mode modo acústico
acoustic noise ruido acústico
acoustic ohm ohm acústico
acoustic oscillation oscilación acústica
acoustic output salida acústica
acoustic phase coefficient coeficiente de fase acústica
acoustic phase constant constante de fase acústica
acoustic phase inverter inversor de fase acústico
acoustic phonograph fonógrafo acústico
acoustic phonon fonón acústico
acoustic pickup fonocaptor acústico
acoustic power potencia acústica
acoustic power factor factor de potencia acústica
acoustic pressure presión acústica
acoustic propagation propagación acústica
acoustic radiating element elemento radiante acústico
acoustic radiation radiación acústica
acoustic radiation pressure presión de radiación acústica
acoustic radiator radiador acústico
acoustic radiometer radiómetro acústico
acoustic reactance reactancia acústica
acoustic reading lectura acústica
acoustic receiver receptor acústico

acoustic recorder registrador acústico
acoustic reflection reflexión acústica
acoustic reflectivity reflectividad acústica
acoustic refraction refracción acústica
acoustic regeneration regeneración acústica
acoustic resistance resistencia acústica
acoustic resonance resonancia acústica
acoustic resonator resonador acústico
acoustic scattering dispersión acústica
acoustic shock choque acústico
acoustic signal señal acústica
acoustic stimulus estímulo acústico
acoustic storage almacenamiento acústico
acoustic suspension suspensión acústica
acoustic system sistema acústico
acoustic telegraph telégrafo acústico
acoustic transducer transductor acústico
acoustic transmission transmisión acústica
acoustic transmission system sistema de transmisión
　　acústica
acoustic transmissivity transmisividad acústica
acoustic treatment tratamiento acústico
acoustic wave onda acústica
acoustic-wave amplifier amplificador de ondas
　　acústicas
acoustic-wave filter filtro de ondas acústicas
acoustical acústico
acoustically tunable optical filter filtro óptico
　　sintonizable acústicamente
acoustics acústica
acoustimeter acustímetro
acoustoelectric acustoeléctrico
acoustoelectric amplifier amplificador
　　acustoeléctrico
acoustoelectric effect efecto acustoeléctrico
acoustoelectronics acustoelectrónica
acoustooptic acustoóptico
acoustooptic device dispositivo acustoóptico
acoustooptic effect efecto acustoóptico
acoustooptical acustoóptico
acoustooptical cell celda acustoóptica
acquisition adquisición
acquisition and control system sistema de
　　adquisición y control
acquisition and tracking radar radar de adquisición
　　y rastreo
acquisition phase fase de adquisición
acquisition probability probabilidad de adquisición
acquisition radar radar de adquisición
acquisition time tiempo de adquisición
acronym acrónimo
across-the-line starter arrancador directo
acrylic resin resina acrílica
actinic actínico
actinic rays rayos actínicos
actinism actinismo
actinium actinio
actinodielectric actinodieléctrico
actinoelectric effect efecto actinoeléctrico
actinoelectricity actinoelectricidad
actinometer actinómetro
action acción
action chart diagrama de acción
action code código de acción
action current corriente de acción
action entries entradas de acción
action indicator indicador de acción
action period periodo de acción
action portion porción de acción
action potential potencial de acción
action routine rutina de acción
action spike punta de acción
activate activar

activated activado
activated cathode cátodo activado
activated filament filamento activado
activation activación
activation analysis análisis por activación
activation curve curva de activación
activation detector detector de activación
activation of block activación de bloque
activation of emitter activación de emisor
activation polarization polarización de activación
activation record registro de activación
activation time tiempo de activación
activator activador
active activo
active antenna antena activa
active antenna array red de antenas activas
active area área activa
active arm brazo activo
active balance equilibrio activo
active cell celda activa
active communications satellite satélite de
　　comunicaciones activo
active component componente activo
active component of current componente activo de
　　corriente
active computer computadora activa
active conductor conductor activo
active contact contacto activo
active control system sistema de control activo
active current corriente activa
active-current compensator compensador de
　　corriente activa
active data dictionary diccionario de datos activo
active database base de datos activa
active decoder descodificador activo
active device dispositivo activo
active dimension dimensión activa
active dissemination diseminación activa
active electric network red eléctrica activa
active electrode electrodo activo
active electronic countermeasures contramedidas
　　electrónicas activas
active element elemento activo
active energy energía activa
active energy meter medidor de energía activa
active file archivo activo
active filter filtro activo
active guidance guiado activo
active-high signal señal alta activa
active homing guiado activo
active impedance impedancia activa
active index índice activo
active infrared detection detección infrarroja activa
active ingredient ingrediente activo
active jamming interferencia intencional activa
active laser medium medio de láser activo
active length longitud activa
active line línea activa
active link enlace activo
active load carga activa
active logic lógica activa
active-low signal señal baja activa
active maintenance time tiempo de mantenimiento
　　activo
active mass masa activa
active master file archivo maestro activo
active material material activo
active modulator modulador activo
active monitor monitor activo
active network red activa
active page página activa
active parameter parámetro activo
active-passive network red activa-pasiva

active potential potencial activo
active power potencia activa
active power meter medidor de potencia activa
active power relay relé de potencia activa
active pressure presión activa
active preventive maintenance time tiempo de
  mantenimiento preventivo activo
active program programa activo
active record registro activo
active redundancy redundancia activa
active reflection coefficient coeficiente de reflexión
  activa
active region región activa
active repair time tiempo de reparación activa
active repeater repetidor activo
active satellite satélite activo
active sonar sonar activo
active state estado activo
active station estación activa
active store memoria activa
active substrate substrato activo
active system sistema activo
active tracking system sistema de seguimiento activo
active transducer transductor activo
active voltage tensión activa
active wait espera activa
active window ventana activa
activity actividad
activity-based simulation simulación basada en
  actividad
activity content contenido de actividad
activity curve curva de actividad
activity factor factor de actividad
activity file archivo de actividad
activity level nivel de actividad
activity loading carga de actividad
activity meter medidor de actividad
activity queue cola de actividad
activity ratio razón de actividad
activity response respuesta de actividad
actual address dirección real
actual argument argumento real
actual cathode cátodo real
actual circuit circuito real
actual code código real
actual data transfer rate velocidad de transferencia
  de datos real
actual decimal point punto decimal real
actual efficiency eficiencia real
actual execution ejecución real
actual generation generación real
actual ground tierra real
actual height altura real
actual instruction instrucción real
actual key llave real
actual parameter parámetro real
actual power potencia real
actual rhythm ritmo real
actual time tiempo real
actual transfer rate velocidad de transferencia real
actual value valor real
actuating current corriente de accionamiento
actuating device dispositivo de accionamiento
actuating force fuerza de accionamiento
actuating mechanism mecanismo de accionamiento
actuating quantity cantidad de accionamiento
actuating signal señal de accionamiento
actuating system sistema de accionamiento
actuating time tiempo de accionamiento
actuating voltage tensión de accionamiento
actuation actuación
actuation device dispositivo de actuación
actuation time tiempo de actuación

actuator actuador
actuator switch conmutador actuador
actuator valve válvula actuadora
acutance acutancia
acute exposure exposición aguda
acyclic feeding alimentación acíclica
acyclic machine máquina acíclica
ad hoc ad hoc
adapt adaptar
adaptability adaptabilidad
adaptation adaptación
adaptation data datos de adaptación
adaptation matrix matriz de adaptación
adaptation parameter parámetro de adaptación
adapter adaptador
adapter circuit circuito adaptador
adapter head cabeza adaptadora
adapter kit juego de adaptadores
adapter plate placa adaptadora
adapter plug clavija adaptadora
adapting adaptante
adapting connector conector adaptante
adaptive adaptivo
adaptive antenna system sistema de antenas
  adaptivo
adaptive coding codificación adaptiva
adaptive color shift desplazamiento de color
  adaptivo
adaptive colorimetric shift desplazamiento
  colorimétrico adaptivo
adaptive communication comunicación adaptiva
adaptive control control adaptivo
adaptive control system sistema de control adaptivo
adaptive equalization igualación adaptiva
adaptive maintenance mantenimiento adaptivo
adaptive prediction predicción adaptiva
adaptive system sistema adaptivo
adaptivity adaptividad
adaptor adaptador
Adcock antenna antena de Adcock
Adcock direction-finder radiogoniómetro de Adcock
Adcock radio range radiofaro direccional de Adcock
Adcock system sistema de Adcock
add-and-subtract relay relé de suma y resta
add control control de suma
add instruction instrucción de suma
add mode modo de suma
add on añadir
add operation operación de suma
add output salida de suma
add record registro de suma
add-subtract time tiempo de suma y resta
add time tiempo de suma
add transaction transacción de suma
added filter filtro adicional
addend sumando
addend register registro de sumando
adder sumador
adder-accumulator sumador-acumulador
adder circuit circuito sumador
adder-subtracter sumador-restador
adding amplifier amplificador sumador
adding circuit circuito sumador
adding machine máquina sumadora
adding network red sumadora
addition suma, adición
addition agent agente de suma
addition identity identidad de suma
addition record registro de suma
addition time tiempo de suma
additional adicional
additional character carácter adicional
additional hardware equipo físico adicional

additional read station estación de lectura adicional
additional space espacio adicional
additional tone tono adicional
additional twisting torcido adicional
additive aditivo
additive color color aditivo
additive color filter filtro de color aditivo
additive color system sistema de colores aditivos
additive colorimeter colorímetro aditivo
additive filter filtro aditivo
additive mixture mezcla aditiva
additive primaries primarios aditivos
additive process proceso aditivo
additron aditrón
address dirección
address assignment asignación de dirección
address bus bus de dirección
address calculation sort clasificación de cálculos de dirección
address code código de dirección
address code switch conmutador de código de dirección
address comparator comparador de dirección
address computation cómputo de dirección
address constant constante de dirección
address conversion conversión de dirección
address cycle ciclo de dirección
address decoder descodificador de dirección
address device dispositivo de direcciones
address field campo de dirección
address file archivo de dirección
address format formato de dirección
address generation generación de dirección
address identification identificación de dirección
address indicator indicador de dirección
address modification modificación de dirección
address modifier modificador de dirección
address-only cycle ciclo de dirección solamente
address-only transaction transacción de dirección solamente
address part parte de dirección
address portion porción de dirección
address reference referencia de dirección
address register registro de dirección
address selection selección de dirección
address signal señal de dirección
address space espacio de dirección
address substitution sustitución de dirección
address table sorting clasificación de tabla de direcciones
address transfer transferencia de dirección
address translation traducción de dirección
addressable direccionable
addressable memory memoria direccionable
addressable point punto direccionable
addressable register registro direccionable
addressed direccionado
addressed board tablero direccionado
addressed direct access acceso directo direccionado
addressed memory memoria direccionada
addressed transmission transmisión direccionada
addressing direccionamiento
addressing capacity capacidad de direccionamiento
addressing exception excepción de direccionamiento
addressing level nivel de direccionamiento
addressing machine máquina de direccionamiento
addressing mode modo de direccionamiento
addressing system sistema de direccionamiento
addressless sin dirección
addressless instruction instrucción sin dirección
adequate performance funcionamiento adecuado
adhesion adhesión
adhesion coefficient coeficiente de adhesión

adhesive tape cinta adhesiva
adhesive-tape resistor resistor de cinta adhesiva
adiabatic adiabático
adiabatic damping amortiguamiento adiabático
adiabatic demagnetization desmagnetización adiabática
adiabatic process proceso adiabático
adiactinic adiactínico
adjacency adyacencia
adjacent adyacente
adjacent audio channel canal de audio adyacente
adjacent carrier portador adyacente
adjacent channel canal adyacente
adjacent-channel attenuation atenuación de canal adyacente
adjacent-channel interference interferencia de canal adyacente
adjacent-channel operation operación de canal adyacente
adjacent-channel selectivity selectividad de canal adyacente
adjacent-channel selectivity and desensitization selectividad y desensibilización de canal adyacente
adjacent exchange central adyacente
adjacent frequency frecuencia adyacente
adjacent sound channel canal de sonido adyacente
adjacent video carrier portadora de video adyacente
adjoining adyacente
adjoint system sistema adjunto
adjust ajustar
adjust line mode modo de ajuste de líneas
adjustable ajustable
adjustable capacitor capacitor ajustable
adjustable center-tap resistor resistor de derivación central ajustable
adjustable component componente ajustable
adjustable condenser condensador ajustable
adjustable constant-speed motor motor de velocidad constante ajustable
adjustable frequency frecuencia ajustable
adjustable gain ganancia ajustable
adjustable instrument instrumento ajustable
adjustable point coma ajustable, punto ajustable
adjustable resistor resistor ajustable
adjustable-speed motor motor de velocidad ajustable
adjustable thermostatic switch conmutador termostático ajustable
adjustable transformer transformador ajustable
adjustable varying-speed motor motor de velocidad variable ajustable
adjustable varying-voltage control control de tensión variable ajustable
adjustable voltage tensión ajustable
adjustable voltage control control de tensión ajustable
adjustable voltage divider divisor de tensión ajustable
adjusted circuit circuito ajustado
adjusted decibels decibeles ajustados
adjusted speed velocidad ajustada
adjustment ajuste
adjustment accuracy exactitud de ajuste
adjustment device dispositivo de ajuste
adjustment frequency frecuencia de ajuste
adjustment parameter parámetro de ajuste
adjustment tolerance tolerancia de ajuste
administrative downtime tiempo muerto administrativo
admittance admitancia, admisión
admittance matrix matriz de admitancia
admittance parameters parámetros de admitancia
adsorbent adsorbente

adsorption adsorción
advance ball bola de avance
advance feed tape cinta de alimentación de avance
advance information información de avance
advance wire hilo de avance
advanced license licencia avanzada
advanced potential potencial avanzado
advanced radar tracking system sistema de
    seguimiento de radar avanzado
adverse water conditions condiciones de agua
    adversas
adverse weather clima adverso
aeolight lámpara de cátodo frío
aerial antena
aerial adapter adaptador de antena
aerial amplifier amplificador de antena
aerial array red de antenas
aerial attenuator atenuador de antena
aerial beacon aerofaro
aerial cable cable de antena, cable aéreo
aerial cable line línea de cable aéreo
aerial capacitor capacitor de antena
aerial circuit circuito de antena
aerial circuit-breaker cortacircuito de antena
aerial coil bobina de antena
aerial condenser condensador de antena
aerial contact line línea de contacto aéreo
aerial coupler acoplador de antena
aerial crossing cruce aéreo
aerial current corriente de antena
aerial damping amortiguamiento de antena
aerial device dispositivo de antena
aerial directivity directividad de antena
aerial element elemento de antena
aerial feed alimentación de antena
aerial feed impedance impedancia de alimentación
    de antena
aerial feeder alimentador de antena
aerial gain ganancia de antena
aerial-ground circuit circuito antena-tierra
aerial impedance impedancia de antena
aerial inductance inductancia de antena
aerial inductance coil bobina de inductancia de
    antena
aerial loading coil bobina de carga de antena
aerial noise ruido de antena
aerial power potencia de antena
aerial radiation radiación de antena
aerial radiation resistance resistencia de radiación
    de antena
aerial reactance reactancia de antena
aerial resistance resistencia de antena
aerial tuning coil bobina de sintonización de antena
aerial tuning condenser condensador de
    sintonización de antena
aerial tuning inductance inductancia de
    sintonización de antena
aerodrome beacon radiofaro de aeródromo
aerodrome control radar radar de control de
    aeródromo
aerodrome control radio station estación de radio
    de control de aeródromo
aerodrome control service servicio de control de
    aeródromo
aerodrome control tower torre de control de
    aeródromo
aerodrome hazard beacon radiofaro de peligro de
    aeródromo
aerodynamic aerodinámico
aerodynamic sound sonido aerodinámico
aerodynamics aerodinámica
aerogenerator aerogenerador
aerogram aerograma

aeromagnetic aeromagnético
aerometeorograph aerometeorógrafo
aerometer aerómetro
aerometry aerometría
aeromobile service servicio aeromóvil
aeronautical administrative message mensaje
    administrativo aeronáutico
aeronautical beacon radiofaro aeronáutico
aeronautical broadcasting service servicio de
    radiodifusión aeronáutico
aeronautical electronics electrónica aeronáutica
aeronautical emergency frequency frecuencia de
    emergencia aeronáutica
aeronautical fixed service servicio fijo aeronáutico
aeronautical fixed service station estación de
    servicio fijo aeronáutico
aeronautical fixed station estación fija aeronáutica
aeronautical fixed telecommunications network red
    de telecomunicaciones fija aeronáutica
aeronautical fixed telecommunications service
    servicio de telecomunicaciones fijo aeronáutico
aeronautical ground light luz de superficie
    aeronáutica
aeronautical ground radio station estación de radio
    de superficie aeronáutica
aeronautical ground station estación de superficie
    aeronáutica
aeronautical information información aeronáutica
aeronautical light luz aeronáutica
aeronautical marker-beacon station estación de
    radiobaliza aeronáutica
aeronautical mobile service servicio móvil
    aeronáutico
aeronautical mobile telecommunications service
    servicio de telecomunicaciones aeronáuticas móvil
aeronautical radio service servicio de radio
    aeronáutico
aeronautical radio station estación de radio
    aeronáutica
aeronautical radionavigation radionavegación
    aeronáutica
aeronautical radionavigation service servicio de
    radionavegación aeronáutico
aeronautical station estación aeronáutica
aeronautical telecommunication service servicio de
    telecomunicación aeronáutica
aeronautical telecommunication station estación de
    telecomunicación aeronáutica
aeronautical telecommunications
    telecomunicaciones aeronáuticas
aerophare aerofaro
aerophase aerofase
aerophone aerófono
aerophysics aerofísica
aeroplane flutter titilación de aeroplano
aerosol aerosol
aerosol development desarrollo de aerosol
afferent aferente
aftereffect efecto posterior
afterglow luminiscencia persistente
afterimage imagen persistente
afterpulse impulso inducido
agendum call card tarjeta de llamada de agenda
agent agente
agent error error de agente
agent status estado de agente
aggregate agregado
aggregate baud rate velocidad en bauds agregada
aggregate function función agregada
aggregate signal señal agregada
aggressive carbon dioxide dióxido de carbono
    agresivo
aging envejecimiento

aging factor factor de envejecimiento
aging mechanism mecanismo de envejecimiento
aging of insulator envejecimiento de aislador
agitator agitador
agonic line línea agónica
aid to navigation ayuda de navegación
aided tracking seguimiento ayudado
aiming radar radar de puntería
air battery batería de aire
air-blast circuit-breaker cortacircuito de chorro de
  aire
air capacitor capacitor de aire
air cell pila de aire
air circuit circuito de aire
air circuit-breaker cortacircuito de aire
air column columna de aire
air compressor compresor de aire
air conditioning acondicionamiento de aire
air conduction conducción de aire
air contamination contaminación de aire
air contamination indicator indicador de
  contaminación de aire
air contamination meter medidor de contaminación
  de aire
air contamination monitor monitor de
  contaminación de aire
air cooled enfriado por aire
air-cooled component componente enfriado por aire
air-cooled transistor transistor enfriado por aire
air-cooled tube tubo enfriado por aire
air cooling enfriamiento por aire
air-core cable cable de núcleo de aire
air-core coil bobina de núcleo de aire
air-core inductance inductancia de núcleo de aire
air-core transformer transformador de núcleo de
  aire
air-dielectric coax cable coaxial de dieléctrico de
  aire
air duct conducto de aire
air environment ambiente de aire
air equivalent radiation dose dosis de radiación
  equivalente en aire
air flow flujo de aire
air gap espacio de aire, entrehierro
air-gap line línea de espacio de aire
air-gap surge arrester protector de sobretensión de
  espacio de aire
air-gap surge protector protector de sobretensión de
  espacio de aire
air-gap technique técnica de espacio de aire
air-ground communication comunicación aire-tierra
air-ground control radio station estación de radio
  de control aire-tierra
air-ground frequency frecuencia aire-tierra
air-ground radio frequency frecuencia de radio
  aire-tierra
air-insulated cable cable aislado por aire
air-insulated line línea aislada por aire
air intake toma de aire
air line línea de aire
air mass masa de aire
air mobile service servicio móvil aéreo
air-position indicator indicador de posición en el
  aire
air-route surveillance radar radar de vigilancia de
  rutas aéreas
air-space cable cable con espacio de aire
air-space coax cable coaxial con espacio de aire
air surveillance radar radar de vigilancia aérea
air switch conmutador de aire
air-to-air aire a aire
air-to-air communication comunicación de aire a
  aire

air-to-ground aire a tierra
air-to-ground communication comunicación de aire
  a tierra
air-to-ground radio frequency frecuencia de radio
  de aire a tierra
air-to-surface aire a superficie
air-traffic control control de tráfico aéreo
air-traffic control radar radar de control de tráfico
  aéreo
air-traffic control safety beacon radiofaro de
  seguridad de control de tráfico aéreo
air ventilation ventilación por aire
air-zinc cell pila de aire-cinc
airborne beacon faro aerotransportado
airborne control system sistema de control
  aerotransportado
airborne electronic equipment equipo electrónico
  aerotransportado
airborne equipment equipo aerotransportado
airborne intercept radar radar de intercepción
  aerotransportado
airborne noise ruido aéreo
airborne radar radar aerotransportado
airborne radar platform plataforma de radar
  aerotransportada
airborne radar responder respondedor de radar
  aerotransportado
airborne radioactivity radiactividad aérea
airborne receiver receptor aerotransportado
airborne replier respondedor aerotransportado
airborne search radar radar de búsqueda
  aerotransportado
airborne sensor sensor aerotransportado
airborne surveillance radar radar de vigilancia
  aerotransportado
airborne transmitter transmisor aerotransportado
airborne weather radar radar meteorológico
  aerotransportado
aircraft antenna antena de aeronave
aircraft bonding conexión general de aeronave
aircraft compass compás de aeronave
aircraft contactor contactor de aeronave
aircraft equipment equipo de aeronave
aircraft flutter titilación de aeronave
aircraft light luz de aeronave
aircraft radar radar de aeronave
aircraft station estación de aeronave
aircraft transmitter transmisor de aeronave
aircraft warning advertencia de aeronave
aircraft warning service servicio de advertencia de
  aeronave
airfield radar radar de campo de aviación
airflow corriente de aire
airplane effect efecto de avión
airplane flutter titilación de avión
airplane insulator aislador de avión
airplane receiver receptor de avión
airplane transmitter transmisor de avión
airport beacon radiofaro de aeropuerto
airport control station estación de control de
  aeropuerto
airport danger beacon radiofaro de peligro de
  aeropuerto
airport radar radar de aeropuerto
airport runway beacon radiofaro de pista de
  aeropuerto
airport surface-detection equipment equipo de
  detección de superficie de aeropuerto
airport surveillance radar radar de vigilancia de
  aeropuerto
airport traffic control control de tráfico de
  aeropuerto
airport traffic control tower torre de control de

tráfico de aeropuerto
airspeed indicator indicador de velocidad del aire
airtight hermético
airwaves ondas aéreas
airway beacon radiofaro de aerovía
airway communication station estación de
comunicación de aerovía
alarm alarma
alarm bell timbre de alarma
alarm checking revisión de alarma
alarm circuit circuito de alarma
alarm condition condición de alarma
alarm contact contacto de alarma
alarm device dispositivo de alarma
alarm function función de alarma
alarm indication signal señal de indicación de
alarma
alarm indicator indicador de alarma
alarm lamp lámpara de alarma
alarm relay relé de alarma
alarm sending envío de alarma
alarm signal señal de alarma
alarm summary printout impresión de resumen de
alarmas
alarm switch conmutador de alarma
alarm tone tono de alarma
albedo albedo
albedograph albedógrafo
Alexanderson alternator alternador de Alexanderson
Alexanderson antenna antena de Alexanderson
Alford antenna antena de Alford
Alford loop antenna antena de cuadro de Alford
Alfven velocity velocidad de Alfven
Alfven wave onda de Alfven
algebraic adder sumador algebraico
algebraic coding function función de codificación
algebraica
algebraic expression expresión algebraica
algebraic function función algebraica
algebraic language lenguaje algebraico
algebraic manipulation manipulación algebraica
algebraic operation operación algebraica
algebraic sign signo algebraico
algebraic sum suma algebraica
algorithm algoritmo
algorithm analysis análisis de algoritmos
algorithmic algorítmico
algorithmic language lenguaje algorítmico
algorithmic routine rutina algorítmica
algorithmic translation traducción algorítmica
alias alias
align alinear
aligned-grid tube tubo de rejillas alineadas
alignment alineación
alignment chart diagrama de alineación
alignment jitter fluctuación de alineación
alignment kit juego de alineación
alignment pin terminal de alineación
alignment tool herramienta de alineación
alive vivo
alkali alcalino
alkali metal metal alcalino
alkaline accumulator acumulador alcalino
alkaline battery batería alcalina
alkaline cell pila alcalina
alkaline cleaning limpieza alcalina
alkaline earth tierra alcalina
alkaline-earth metal metal alcalinotérreo
alkaline secondary battery batería secundaria
alcalina
alkaline storage battery acumulador alcalino
all-band antenna antena de toda banda
all-electric todo eléctrico

all-electronic todo electrónico
all-or-nothing todo o nada
all-or-nothing relay relé todo o nada
all-pass filter filtro pasatodo
all-pass function función pasatodo
all-pass network red pasatodo
all-pass transducer transductor pasatodo
all-relay system sistema automático de relés
all-wave de toda onda
all-wave antenna antena de toda onda
all-wave generator generador de toda onda
all-wave oscillator oscilador de toda onda
all-wave receiver receptor de toda onda
all-wave signal receiver receptor de señales de toda
onda
all-weather de todo clima
all-weather distribution distribución en todo clima
alligator clip pinza cocodrilo
allocate asignar
allocated baseline línea base asignada
allocated channel canal asignado
allocated configuration identification identificación
de configuración asignada
allocated logical storage almacenamiento lógico
asignado
allocated physical storage almacenamiento físico
asignado
allocated-use circuit circuito de uso asignado
allocated variable variable asignada
allocation asignación
allocation factor factor de asignación
allocation of frequencies asignación de frecuencias
allocation recovery recuperación de asignación
allocation scheme esquema de asignación
allocation table tabla de asignación
allocation technique técnica de asignación
allochromatic alocromático
allot asignar
allot frequencies asignar frecuencias
alloter switch conmutador de asignación
allotropic alotrópico
allowable continuous current corriente continua
permisible
allowable distortion distorsión permisible
allowable load carga permisible
allowable record configuration configuración de
registro permisible
allowable stress estrés permisible
allowable tolerance tolerancia permisible
allowable voltage tensión permisible
allowance tolerancia
allowed band banda permitida
allowed energy band banda de energía permitida
alloy aleación
alloy deposition deposición de aleación
alloy-diffused transistor transistor de difusión de
aleación
alloy diode diodo de aleación
alloy junction unión de aleación
alloy-junction diode diodo de unión de aleación
alloy-junction transistor transistor de unión de
aleación
alloy plate placa de aleación
alloy transistor transistor de aleación
alloyed aleado
alloyed steel acero aleado
alloyed transistor transistor aleado
alpha alfa
alpha cutoff frequency frecuencia de corte alfa
alpha decay desintegración alfa
alpha detector detector alfa
alpha disintegration desintegración alfa
alpha emitter emisor alfa

alpha particle partícula alfa
alpha radiation radiación alfa
alpha rays rayos alfa
alpha system sistema alfa
alphabetic character carácter alfabético
alphabetic character set conjunto de caracteres alfabéticos
alphabetic code código alfabético
alphabetic coding codificación alfabética
alphabetic data datos alfabéticos
alphabetic data code código de datos alfabéticos
alphabetic field campo alfabético
alphabetic keyboard teclado alfabético
alphabetic list lista alfabética
alphabetic-numeric alfabético-numérico
alphabetic shift cambio alfabético
alphabetic signal señal alfabética
alphabetic string secuencia alfabética
alphabetic string word palabra de secuencia alfabética
alphabetic telegraphy telegrafía alfabética
alphageometric alfageométrico
alphanumeric alfanumérico
alphanumeric character carácter alfanumérico
alphanumeric character set conjunto de caracteres alfanuméricos
alphanumeric code código alfanumérico
alphanumeric coding codificación alfanumérica
alphanumeric conversion conversión alfanumérica
alphanumeric data datos alfanuméricos
alphanumeric data code código de datos alfanuméricos
alphanumeric display device dispositivo de visualización alfanumérica
alphanumeric field campo alfanumérico
alphanumeric information información alfanumérica
alphanumeric keyboard teclado alfanumérico
alphanumeric notation notación alfanumérica
alphanumeric optical reader lectora óptica alfanumérica
alphanumeric reader lectora alfanumérica
alphanumeric readout lectura alfanumérica
alphanumeric sorting clasificación alfanumérica
alphaphotographic alfafotográfico
alphatron alfatrón
alter alterar
alterable alterable
alterable memory memoria alterable
alteration alteración
alternate alternativo
alternate channel canal alternativo
alternate-channel interference interferencia de canal alternativo
alternate character set conjunto de caracteres alternativos
alternate description descripción alternativa
alternate digit inversion inversión de dígito alternativo
alternate display visualización alternativa
alternate frequency frecuencia alternativa
alternate function key tecla de función alternativa
alternate function switch conmutación de función alternativa
alternate index índice alternativo
alternate instruction instrucción alternativa
alternate key tecla alternativa
alternate-mark inversion code código de inversión de marcas alternativas
alternate-mark inversion signal señal de inversión de marcas alternativas
alternate mode modo alternativo
alternate name nombre alternativo
alternate operation operación alternativa

alternate power source fuente de alimentación alternativa
alternate program programa alternativo
alternate-program device dispositivo de programa alternativo
alternate recovery recuperación alternativa
alternate route ruta alternativa
alternate routine rutina alternativa
alternate routing encaminamiento alternativo
alternate selection selección alternativa
alternating-charge characteristic característica de carga alterna
alternating component componente alterno
alternating-current corriente alterna
alternating-current analog computer computadora analógica de corriente alterna
alternating-current automatic block bloqueo automático de corriente alterna
alternating-current base current corriente de base de corriente alterna
alternating-current base resistance resistencia de base de corriente alterna
alternating-current base voltage tensión de base de corriente alterna
alternating-current bias polarización de corriente alterna
alternating-current breakdown voltage tensión de ruptura de corriente alterna
alternating-current bridge puente de corriente alterna
alternating-current cable cable de corriente alterna
alternating-current cathode current corriente de cátodo de corriente alterna
alternating-current cathode resistance resistencia de cátodo de corriente alterna
alternating-current cathode voltage tensión de cátodo de corriente alterna
alternating-current circuit circuito de corriente alterna
alternating-current circuit-breaker cortacircuito de corriente alterna
alternating-current collector current corriente de colector de corriente alterna
alternating-current collector resistance resistencia de colector de corriente alterna
alternating-current collector voltage tensión de colector de corriente alterna
alternating-current commutator machine máquina de colector de corriente alterna
alternating-current commutator motor motor de colector de corriente alterna
alternating-current component componente de corriente alterna
alternating-current continuous wave onda continua de corriente alterna
alternating-current controller controlador de corriente alterna
alternating-current conversion factor factor de conversión de corriente alterna
alternating-current converter convertidor de corriente alterna
alternating-current-coupled flip-flop basculador acoplado de corriente alterna
alternating-current coupling acoplamiento de corriente alterna
alternating-current dialing marcación por corriente alterna
alternating-current-direct-current corriente alterna-corriente continua
alternating-current-direct-current motor motor de corriente alterna-corriente continua
alternating-current-direct-current receiver receptor de corriente alterna-corriente continua

alternating-current-direct-current ringing llamada de corriente alterna-corriente continua

alternating-current directional overcurrent relay relé de sobrecorriente direccional de corriente alterna

alternating-current distribution distribución de corriente alterna

alternating-current drain current corriente de drenador de corriente alterna

alternating-current drain resistance resistencia de drenador de corriente alterna

alternating-current drain voltage tensión de drenador de corriente alterna

alternating-current electric locomotive locomotora eléctrica de corriente alterna

alternating-current electrical field strength intensidad de campo eléctrico de corriente alterna

alternating-current electrical field strength meter medidor de intensidad de campo eléctrico de corriente alterna

alternating-current electrolytic capacitor capacitor electrolítico de corriente alterna

alternating-current electromagnet electroimán de corriente alterna

alternating-current emitter current corriente de emisor de corriente alterna

alternating-current emitter resistance resistencia de emisor de corriente alterna

alternating-current emitter voltage tensión de emisor de corriente alterna

alternating-current equipment equipo de corriente alterna

alternating-current erase borrado por corriente alterna

alternating-current erasing borrado por corriente alterna

alternating-current erasing head cabeza de borrado de corriente alterna

alternating-current filter filtro de corriente alterna

alternating-current gate voltage tensión de puerta de corriente alterna

alternating-current generator generador de corriente alterna

alternating-current grid voltage tensión de rejilla de corriente alterna

alternating-current hum zumbido de corriente alterna

alternating-current interruption interrupción de corriente alterna

alternating-current line línea de corriente alterna

alternating-current line filter filtro de línea de corriente alterna

alternating-current line voltage tensión de línea de corriente alterna

alternating-current magnetic bias polarización magnética de corriente alterna

alternating-current magnetic biasing polarización magnética por corriente alterna

alternating-current measurement medida de corriente alterna

alternating-current meter medidor de corriente alterna

alternating-current motor motor de corriente alterna

alternating-current noise ruido de corriente alterna

alternating-current noise immunity inmunidad al ruido de corriente alterna

alternating-current outlet salida de corriente alterna

alternating-current plate current corriente de placa de corriente alterna

alternating-current plate resistance resistencia de placa de corriente alterna

alternating-current plate voltage tensión de placa de corriente alterna

alternating-current power potencia de corriente alterna

alternating-current power-line fields campos de líneas de potencia de corriente alterna

alternating-current power supply fuente de alimentación de corriente alterna

alternating-current pulse impulso de corriente alterna

alternating-current receiver receptor de corriente alterna

alternating-current reclosing relay relé de reconexión de corriente alterna

alternating-current relay relé de corriente alterna

alternating-current resistance resistencia de corriente alterna

alternating-current saturable reactor reactor saturable de corriente alterna

alternating-current selection selección de corriente alterna

alternating-current signaling system sistema de señalización por corriente alterna

alternating-current source fuente de corriente alterna

alternating-current source current fuente de corriente alterna

alternating-current source voltage fuente de tensión alterna

alternating-current standby power potencia de reserva de corriente alterna

alternating-current switch conmutador de corriente alterna

alternating-current system sistema de corriente alterna

alternating-current test prueba de corriente alterna

alternating-current time overcurrent relay relé de sobrecorriente de tiempo de corriente alterna

alternating-current transducer transductor de corriente alterna

alternating-current transmission transmisión de corriente alterna

alternating-current voltage tensión de corriente alterna

alternating electromagnetic field campo electromagnético alterno

alternating electromotive force fuerza electromotriz alterna

alternating field campo alterno

alternating function función alterna

alternating gradient gradiente alterno

alternating light luz alterna

alternating operation operación alterna

alternating voltage tensión alterna

alternation alternación

alternative alternativa

alternative denial negación alternativa

alternative frequency frecuencia alternativa

alternative route ruta alternativa

alternative routing encaminamiento alternativo

alternative sector sector alternativo

alternative system console consola de sistema alternativo

alternative test method método de prueba alternativo

alternator alternador

alternator-rectifier exciter excitador de alternador-rectificador

alternator transmitter transmisor de alternador

altimeter altímetro

altimeter station estación de altímetro

altitude altitud

altitude delay retardo de altitud

altitude signal señal de altitud

alumina alúmina

aluminium aluminio

aluminized aluminizado
aluminized cathode-ray tube tubo de rayos catódicos aluminizado
aluminized screen pantalla aluminizada
aluminized tube tubo aluminizado
aluminum aluminio
aluminum antimonide antimoniuro de aluminio
aluminum arrester protector de aluminio
aluminum-cell arrester protector de células de aluminio
aluminum conductor conductor de aluminio
aluminum sheath vaina de aluminio
amalgam amalgama
amateur band banda de aficionados
amateur radio radio de aficionados
amber light luz ámbar
ambient ambiente
ambient air aire ambiente
ambient air temperature temperatura de aire ambiente
ambient background fondo ambiente
ambient compensation compensación de ambiente
ambient conditions condiciones ambiente
ambient effects efectos del ambiente
ambient humidity humedad ambiente
ambient level nivel ambiente
ambient light luz ambiente
ambient-light filter filtro de luz ambiente
ambient noise ruido ambiente
ambient operating temperature temperatura de operación ambiente
ambient operating-temperature range intervalo de temperatura de operación ambiente
ambient pressure presión ambiente
ambient radiation radiación ambiente
ambient radio noise ruido de radio ambiente
ambient temperature temperatura ambiente
ambient-temperature range intervalo de temperatura ambiente
ambient-temperature time constant constante de tiempo de temperatura ambiente
ambient thermostatic switch conmutador termostático ambiente
ambiguity ambigüedad
ambiguity error error de ambigüedad
ambiguity function función de ambigüedad
ambiguous ambiguo
ambiguous count cuenta ambigua
ambiguous reference referencia ambigua
ambiguousness ambigüedad
ambipolar ambipolar
ambipolar diffusion difusión ambipolar
americium americio
ammeter amperímetro
ammeter shunt derivación de amperímetro
ammonium chloride cloruro de amonio
amorphous amorfo
amorphous semiconductor semiconductor amorfo
amorphous substance sustancia amorfa
amortisseur amortiguador
amortisseur bar barra amortiguadora
amortisseur winding arrollamiento amortiguador
amount indicator indicador de cantidad
amount of error cantidad de error
amount of modulation cantidad de modulación
amount of radiation cantidad de radiación
ampacity ampacidad
amperage amperaje
ampere ampere
Ampere's law ley de Ampere
Ampere's rule regla de Ampere
Ampere's theorem teorema de Ampere
ampere-foot ampere-pie

ampere-hour ampere-hora
ampere-hour capacity capacidad en ampere-horas
ampere-hour efficiency eficiencia en ampere-horas
ampere-hour meter medidor de ampere-horas
Ampere-Laplace theorem teorema de Ampere-Laplace
ampere-minute ampere-minuto
ampere per meter ampere por metro
ampere-turn ampere-vuelta
ampere-turn coefficient coeficiente de ampere-vueltas
ampliation ampliación
amplidyne amplidino
amplification amplificación
amplification band banda de amplificación
amplification current corriente de amplificación
amplification factor factor de amplificación
amplification power potencia de amplificación
amplification stage etapa de amplificación
amplification tube tubo de amplificación
amplified amplificado
amplified back bias contrapolarización amplificada
amplified signal señal amplificada
amplified spontaneous emission emisión espontanea amplificada
amplifier amplificador
amplifier bandwidth ancho de banda de amplificador
amplifier chain cadena de amplificadores
amplifier circuit circuito amplificador
amplifier class rating clasificación de clase de amplificador
amplifier diode diodo amplificador
amplifier distortion distorsión de amplificador
amplifier gain ganancia de amplificador
amplifier ground tierra de amplificador
amplifier input entrada de amplificador
amplifier noise ruido de amplificador
amplifier nonlinearity no linealidad de amplificador
amplifier output salida de amplificador
amplifier pentode pentodo de amplificador
amplifier power potencia de amplificador
amplifier-rectifier amplificador-rectificador
amplifier response respuesta de amplificador
amplifier tube tubo amplificador
amplifier unit unidad amplificadora
amplify amplificar
amplifying amplificador, amplificante
amplifying circuit circuito amplificador
amplifying delay line línea de retardo amplificadora
amplifying detector detector amplificador
amplifying equipment equipo amplificador
amplifying power potencia amplificadora
amplifying repeater repetidor amplificador
amplifying tube tubo amplificador
amplifying valve válvula amplificadora
amplifying winding arrollamiento amplificador
amplitron amplitrón
amplitude amplitud
amplitude-amplitude characteristic característica amplitud-amplitud
amplitude-amplitude distortion distorsión amplitud-amplitud
amplitude analyzer analizador de amplitud
amplitude change cambio de amplitud
amplitude characteristic característica de amplitud
amplitude comparison comparación de amplitudes
amplitude-comparison circuit circuito de comparación de amplitudes
amplitude-comparison monopulse monoimpulso comparador de amplitudes
amplitude compression compresión de amplitud
amplitude control control de amplitud
amplitude demodulation desmodulación de amplitud

amplitude density distribution distribución de densidad de amplitud
amplitude detector detector de amplitud
amplitude discriminator discriminador de amplitud
amplitude distortion distorsión de amplitud
amplitude distortion factor factor de distorsión de amplitud
amplitude distribution distribución de amplitud
amplitude distribution analysis análisis de distribución de amplitud
amplitude distribution meter medidor de distribución de amplitud
amplitude error error de amplitud
amplitude factor factor de amplitud
amplitude fading desvanecimiento de amplitud
amplitude-frequency characteristic característica amplitud-frecuencia
amplitude-frequency distortion distorsión amplitud-frecuencia
amplitude-frequency response respuesta amplitud-frecuencia
amplitude gate puerta de amplitud
amplitude jitter fluctuación de amplitud
amplitude level nivel de amplitud
amplitude limiter limitador de amplitud
amplitude-limiter circuit circuito limitador de amplitud
amplitude-limiter stage etapa limitadora de amplitud
amplitude limiting limitación de amplitud
amplitude-limiting circuit circuito limitador de amplitud
amplitude modulated modulado en amplitud
amplitude-modulated generator generador modulado en amplitud
amplitude-modulated signal señal modulada en amplitud
amplitude-modulated signal generator generador de señal modulada en amplitud
amplitude-modulated transmission transmisión modulada en amplitud
amplitude-modulated transmitter transmisor modulado en amplitud
amplitude-modulated wave onda modulada en amplitud
amplitude-modulation modulación de amplitud
amplitude-modulation broadcasting radiodifusión de modulación de amplitud
amplitude-modulation compression compresión de modulación de amplitud
amplitude-modulation detector detector de modulación de amplitud
amplitude-modulation distortion distorsión de modulación de amplitud
amplitude-modulation factor factor de modulación de amplitud
amplitude-modulation-frequency-modulation receiver receptor de modulación de amplitud-modulación de frecuencia
amplitude-modulation-frequency-modulation transmitter transmisor de modulación de amplitud-modulación de frecuencia
amplitude-modulation-frequency-modulation tuner sintonizador de modulación de amplitud-modulación de frecuencia
amplitude-modulation noise ruido de modulación de amplitud
amplitude-modulation noise level nivel de ruido de modulación de amplitud
amplitude modulator modulador de amplitud
amplitude noise ruido de amplitud
amplitude of noise amplitud de ruido
amplitude permeability permeabilidad de amplitud
amplitude range intervalo de amplitud

amplitude ratio razón de amplitud
amplitude reference level nivel de referencia de amplitud
amplitude resonance resonancia de amplitud
amplitude response respuesta de amplitud
amplitude selection selección de amplitud
amplitude selector selector de amplitud
amplitude separation separación de amplitud
amplitude separator separador de amplitud
amplitude shift desplazamiento de amplitud
amplitude spectrum espectro de amplitud
amplitude suppression ratio razón de supresión de amplitud
amplitude variation variación de amplitud
amplitude-versus-frequency distortion distorsión de frecuencia contra amplitud
anacoustic anacústico
anacoustic zone zona anacústica
analog analógico
analog adder sumador analógico
analog adder-subtracter sumador-restador analógico
analog amplifier amplificador analógico
analog and digital data datos analógicos y digitales
analog carrier transmission transmisión de portadora analógica
analog channel canal analógico
analog circuit circuito analógico
analog communications comunicaciones analógicas
analog comparator comparador analógico
analog computer computadora analógica
analog controller controlador analógico
analog data datos analógicos
analog demodulation desmodulación analógica
analog detector detector analógico
analog device dispositivo analógico
analog differentiator diferenciador analógico
analog-digital analógico-digital
analog-digital converter convertidor analógico-digital
analog display visualizador analógico
analog divider divisor analógico
analog electronics electrónica analógica
analog frequency meter frecuencímetro analógico
analog function función analógica
analog gate puerta analógica
analog indicator indicador analógico
analog information información analógica
analog input entrada analógica
analog input card tarjeta de entrada analógica
analog input-output entrada-salida analógica
analog integrated circuit circuito integrado analógico
analog integration integración analógica
analog integrator integrador analógico
analog inverting adder sumador inversor analógico
analog machine máquina analógica
analog meter medidor analógico
analog model modelo analógico
analog multiplexer multiplexor analógico
analog multiplier multiplicador analógico
analog network red analógica
analog output salida analógica
analog output card tarjeta de salida analógica
analog quantity cantidad analógica
analog record registro analógico
analog recorder registrador analógico
analog repeater repetidor analógico
analog representation representación analógica
analog sample muestra analógica
analog sampling muestreo analógico
analog signal señal analógica
analog simulation simulación analógica
analog subtracter restador analógico

analog summer sumador analógico
analog switch conmutador analógico
analog switching conmutación analógica
analog system sistema analógico
analog telemetering telemedición analógica
analog-to-digital analógico a digital
analog-to-digital conversion conversión analógica a digital
analog-to-digital converter convertidor analógico a digital
analog-to-frequency converter convertidor analógico a frecuencia
analog transmission transmisión analógica
analog variable variable analógica
analog voltage comparator comparador de tensión analógico
analogical analógico
analogous análogo
analogous pole polo análogo
analysis análisis
analysis block bloque de análisis
analysis by absorption análisis por absorción
analysis instrumentation instrumentación de análisis
analysis phase fase de análisis
analysis routine rutina de análisis
analysis time tiempo de análisis
analytic analítico
analytical analítico
analyzer analizador
analyzing analizador
analyzing spectrometer espectrómetro analizador
anastigmatic anastigmático
anastigmatic lens lente anastigmática
anchor ancla, armadura
anchor light luz de anclaje
anchor point punto de anclaje
anchor rod barra de anclaje
anchorage anclaje
ancillary auxiliar
ancillary circuit circuito auxiliar
ancillary equipment equipo auxiliar
ancillary logic lógica auxiliar
ancillary position posición auxiliar
AND circuit circuito Y
AND element elemento Y
AND function función Y
AND gate puerta Y
AND relationship relación Y
Anderson bridge puente de Anderson
anechoic anecoico
anechoic chamber cámara anecoica
anechoic enclosure caja anecoica
anechoic room sala anecoica
anechoic studio estudio anecoico
anelectric aneléctrico
anelectrotonus anelectrotono
anemograph anemógrafo
anemometer anemómetro
aneroid altimeter altímetro aneroide
angel eco parásito
angel echo eco parásito
angle ángulo
angle fluctuation fluctuación de ángulo
angle jamming interferencia intencional de ángulo
angle modulation modulación de ángulo
angle noise ruido de ángulo
angle of advance ángulo de avance
angle of approach lights luces de ángulo de aproximación
angle of arrival ángulo de llegada
angle of attack ángulo de ataque
angle of azimuth ángulo de acimut
angle of beam ángulo de haz

angle of climb ángulo de subida
angle of collimation ángulo de colimación
angle of conduction ángulo de conducción
angle of convergence ángulo de convergencia
angle of current flow ángulo de flujo de corriente
angle of cut ángulo de corte
angle of declination ángulo de declinación
angle of deflection ángulo de deflexión
angle of departure ángulo de salida
angle of depression ángulo de depresión
angle of descent ángulo de descenso
angle of deviation ángulo de desviación
angle of dispersement ángulo de dispersión
angle of dispersion ángulo de dispersión
angle of divergence ángulo de divergencia
angle of elevation ángulo de elevación
angle of extinction ángulo de extinción
angle of field ángulo de campo
angle of flow ángulo de flujo
angle of ignition ángulo de ignición
angle of incidence ángulo de incidencia
angle of lag ángulo de retardo
angle of lead ángulo de avance
angle of overlap ángulo de solape
angle of phase difference ángulo de diferencia de fase
angle of polarization ángulo de polarización
angle of protection ángulo de protección
angle of radiation ángulo de radiación
angle of reflection ángulo de reflexión
angle of refraction ángulo de refracción
angle of retard ángulo de retardo
angle of rotation ángulo de rotación
angle of scattering ángulo de dispersión
angle optimum bunching agrupamiento óptimo angular
angle tower torre angular
angle tracking seguimiento angular
angle tracking noise ruido de seguimiento angular
angstrom angstrom
angstrom unit unidad angstrom
angular angular
angular acceleration aceleración angular
angular-acceleration sensitivity sensibilidad de aceleración angular
angular accelerometer acelerómetro angular
angular accuracy exactitud angular
angular aperture abertura angular
angular bearing rumbo angular
angular carrier frequency frecuencia portadora angular
angular correlation correlación angular
angular dependence dependencia angular
angular deviation desviación angular
angular deviation loss pérdida de desviación angular
angular deviation sensitivity sensibilidad de desviación angular
angular difference diferencia angular
angular displacement desplazamiento angular
angular distance distancia angular
angular distribution distribución angular
angular frequency frecuencia angular
angular height altura angular
angular length longitud angular
angular mechanical impedance impedancia mecánica angular
angular misalignment loss pérdida por desalineación angular
angular modulation modulación angular
angular momentum momento angular
angular phase difference diferencia de fase angular
angular position posición angular
angular rate tasa angular

angular reading equipment equipo de lectura angular
angular resolution resolución angular
angular resolver resolvedor angular
angular scintillation centelleo angular
angular separation separación angular
angular spacing espaciado angular
angular variation variación angular
angular velocity velocidad angular
angular velocity sensitivity sensibilidad de velocidad angular
angular vibration sensitivity sensibilidad de vibración angular
angular width anchura angular
anharmonic anarmónico
anharmonic oscillator oscilador anarmónico
anhysteresis anhistéresis
anhysteretic magnetization magnetización anhisterética
anhysteretic state estado anhisterético
anion anión
aniseikon aniseicón
anisotropic anisotrópico
anisotropic astigmatism astigmatismo anisotrópico
anisotropic medium medio anisotrópico
anisotropic radiator radiador anisotrópico
anisotropic substrate substrato anisotrópico
anneal recocer
annealed recocido
annealed copper cobre recocido
annealed copper wire hilo de cobre recocido
annealed laminations laminaciones recocidas
annealed wire hilo recocido
annealing recocido
annotation anotación
annual cycle ciclo anual
annual load factor factor de carga anual
annular anular
annular conductor conductor anular
annular effect efecto anular
annular electromagnet electroimán anular
annular resonator resonador anular
annular transistor transistor anular
annulling anulador
annulling network red anuladora
annunciation relay relé de anunciación
annunciator anunciador
annunciator relay relé anunciador
annunciator system sistema anunciador
anodal anodal
anode ánodo
anode angle ángulo anódico
anode arm brazo anódico
anode balancing coil bobina equilibradora anódica
anode battery batería anódica
anode bend curva anódica
anode breakdown voltage tensión de ruptura anódica
anode bridge puente anódico
anode butt residuo anódico
anode-bypass capacitor capacitor de desacoplo anódico
anode-cathode capacitance capacitancia ánodo-cátodo
anode characteristic característica anódica
anode circuit circuito anódico
anode circuit-breaker cortacircuito anódico
anode circuit detector detector de circuito anódico
anode circuit efficiency eficiencia de circuito anódico
anode cleaning limpieza anódica
anode conductance conductancia anódica
anode connector conector anódico
anode converter convertidor anódico

anode cooling enfriamiento anódico
anode corrosion corrosión anódica
anode corrosion efficiency eficiencia de corrosión anódica
anode current corriente anódica
anode current characteristic característica de corriente anódica
anode current modulation modulación de corriente anódica
anode dark space espacio obscuro anódico
anode detection detección anódica
anode detector detector anódico
anode differential resistance resistencia diferencial anódica
anode dissipation disipación anódica
anode drop caída anódica
anode effect efecto anódico
anode efficiency eficiencia anódica
anode face cara anódica
anode fall caída anódica
anode firing disparo anódico
anode follower seguidor anódico
anode-follower amplifier amplificador de seguidor anódico
anode glow luminiscencia anódica
anode hum zumbido anódico
anode impedance impedancia anódica
anode input power potencia de entrada anódica
anode layer capa anódica
anode load carga anódica
anode load impedance impedancia de carga anódica
anode modulation modulación anódica
anode neutralization neutralización anódica
anode paralleling reactor reactor en paralelo anódico
anode plate placa anódica
anode polarization polarización anódica
anode potential potencial anódico
anode potential fall caída de potencial anódico
anode power potencia anódica
anode power input entrada de potencia anódica
anode power supply fuente de alimentación anódica
anode power supply voltage tensión de fuente de alimentación anódica
anode pulsing pulsación anódica
anode radiator radiador anódico
anode-ray current corriente de rayos anódicos
anode rays rayos anódicos
anode region región anódica
anode resistance resistencia anódica
anode resonator resonador anódico
anode saturation saturación anódica
anode scrap residuo anódico
anode sensitivity sensibilidad anódica
anode sheath vaina anódica
anode strap banda anódica
anode supply fuente anódica
anode supply voltage tensión de alimentación anódica
anode support soporte anódico
anode terminal terminal anódica
anode thermal capacity capacidad térmica anódica
anode-to-cathode ánodo a cátodo
anode-to-cathode characteristic característica de ánodo a cátodo
anode-to-cathode voltage tensión de ánodo a cátodo
anode voltage tensión anódica
anode voltage drop caída de tensión anódica
anodic anódico
anodic area área anódica
anodic oxidation oxidación anódica
anodic passivation pasivación anódica
anodic polarization polarización anódica

anodic protection protección anódica
anodic reaction reacción anódica
anodic solution solución anódica
anodically anódicamente
anodization anodización
anodize anodizar
anodized anodizado
anodizing anodización
anolyte anólito
anomalistic anomalístico
anomalistic period periodo anomalístico
anomalous anómalo
anomalous dispersion dispersión anómala
anomalous propagation propagación anómala
anomalous Zeeman effect efecto de Zeeman
    anómalo
anomaly anomalía
anoptic anóptico
anotron anotrón
anoxemia toximeter toxímetro de anoxemia
answer cord cordón de contestación
answer lamp lámpara de contestación
answer signal señal de contestación
answerback contestación
answerback code código de contestación
answerback unit unidad de contestación
answered call llamada contestada
answering contestación, respuesta
answering circuit circuito de contestación
answering delay retardo de contestación
answering device dispositivo de contestación
answering equipment equipo de contestación
answering interval intervalo de contestación
answering jack jack de contestación
answering line línea de contestación
answering machine máquina de contestación
answering time tiempo de contestación
answering wave onda de contestación
antenna antena
antenna adapter adaptador de antena
antenna ammeter amperímetro de antena
antenna amplifier amplificador de antena
antenna array red de antenas
antenna attenuator atenuador de antena
antenna bandwidth ancho de banda de antena
antenna beamwidth anchura de haz de antena
antenna booster amplificador de antena
antenna cable cable de antena
antenna capacitor capacitor de antena
antenna circuit circuito de antena
antenna circuit-breaker cortacircuito de antena
antenna coil bobina de antena
antenna coincidence coincidencia de antena
antenna condenser condensador de antena
antenna connection conexión de antena
antenna connector conector de antena
antenna correction factor factor de corrección de
    antena
antenna coupler acoplador de antena
antenna coupling acoplamiento de antena
antenna coupling circuit circuito de acoplamiento de
    antena
antenna coupling condenser condensador de
    acoplamiento de antena
antenna coupling unit unidad de acoplamiento de
    antena
antenna cross-section sección transversal de antena
antenna crosstalk diafonía entre antenas
antenna current corriente de antena
antenna detector detector de antena
antenna diplexer diplexor de antena
antenna directivity directividad de antena
antenna directivity diagram diagrama de

directividad de antena
antenna director director de antena
antenna discrimination discriminación de antena
antenna distribution unit unidad de distribución de
    antena
antenna duplexer duplexor de antena
antenna effect efecto de antena
antenna effective area área efectiva de antena
antenna efficiency eficiencia de antena
antenna element elemento de antena
antenna eliminator eliminador de antena
antenna energy energía de antena
antenna excitation excitación de antena
antenna factor factor de antena
antenna feed alimentación de antena
antenna feed impedance impedancia de alimentación
    de antena
antenna feeder alimentador de antena
antenna feeding alimentación de antena
antenna field campo de antena
antenna front-to-back ratio eficacia direccional de
    antena
antenna gain ganancia de antena
antenna ground system sistema de tierra de antena
antenna height altura de antena
antenna illumination efficiency eficiencia de
    iluminación de antena
antenna impedance impedancia de antena
antenna impedance transformer transformador de
    impedancia de antena
antenna-induced potential potencial inducido por
    antena
antenna inductance inductancia de antena
antenna input entrada de antena
antenna insulator aislador de antena
antenna lens lente de antena
antenna loading carga de antena
antenna loading coil bobina de carga de antena
antenna lobe lóbulo de antena
antenna loss pérdida de antena
antenna loss damping amortiguamiento de pérdida
    de antena
antenna matching adaptación de antena
antenna-matching circuit circuito de adaptación de
    antena
antenna model modelo de antena
antenna multicoupler multiacoplador de antena
antenna noise ruido de antena
antenna pair par de antenas
antenna park parque de antenas
antenna pattern patrón de antena
antenna polarization polarización de antena
antenna power potencia en antena
antenna power gain ganancia de potencia en antena
antenna preamplifier preamplificador de antena
antenna radiation radiación de antena
antenna radiation damping amortiguamiento de
    radiación de antena
antenna radiator radiador de antena
antenna range intervalo de antena
antenna reactance reactancia de antena
antenna reflector reflector de antena
antenna region región de antena
antenna relay relé de antena
antenna resistance resistencia de antena
antenna resonance resonancia de antena
antenna resonant frequency frecuencia resonante de
    antena
antenna scanner explorador de antena
antenna stage etapa de antena
antenna switch conmutador de antena
antenna system sistema de antena
antenna temperature temperatura de antena

antenna terminals terminales de antena
antenna tuner sintonizador de antena
antenna tuning sintonización de antena
antenna tuning coil bobina de sintonización de antena
antenna tuning condenser condensador de sintonización de antena
antenna tuning inductance inductancia de sintonización de antena
antenna tuning unit unidad de sintonización de antena
antenna wire hilo de antena
antennafier antena-amplificador
anticapacitance switch conmutador anticapacitivo
anticathode anticátodo
anticipation anticipación
anticipation mode modo de anticipación
anticipatory anticipatorio
anticipatory buffering almacenamiento intermedio anticipatorio
anticipatory paging paginación anticipatoria
anticlutter contra ecos parásitos
anticlutter circuit circuito contra ecos parásitos
anticlutter gain control control de ganancia contra ecos parásitos
anticoincidence anticoincidencia
anticoincidence circuit circuito anticoincidencia
anticoincidence gate puerta anticoincidencia
anticoincidence operation operación anticoincidencia
anticoincidence selector selector anticoincidencia
anticoincidence unit unidad anticoincidencia
anticollision anticolisión
anticollision device dispositivo anticolisión
anticollision light luz anticolisión
anticollision radar radar anticolisión
anticorrosion anticorrosión
anticorrosive anticorrosivo
anticyclotron anticiclotrón
antidamping antiamortiguamiento
antifading antidesvanecimiento
antifading antenna antena antidesvanecimiento
antiferroelectric antiferroeléctrico
antiferroelectric chain cadena antiferroeléctrica
antiferroelectric crystal cristal antiferroeléctrico
antiferroelectric material material antiferroeléctrico
antiferroelectric substance sustancia antiferroeléctrica
antiferromagnetic antiferromagnético
antiferromagnetic material material antiferromagnético
antiferromagnetic resonance resonancia antiferromagnética
antiferromagnetic substance sustancia antiferromagnética
antiferromagnetic susceptibility susceptibilidad antiferromagnética
antiferromagnetism antiferromagnetismo
antihunt antifluctuación
antihunt circuit circuito antifluctuación
antihunt device dispositivo antifluctuación
antihunt transformer transformador antifluctuación
antihysteresis antihistéresis
antihysteresis grid rejilla antihistéresis
antiinduction antiinducción
antiinductive antiinductivo
antiinterference antiinterferencia
antiinterference antenna antena antiinterferencia
antiinterference condenser condensador antiinterferencia
antijamming contra interferencia intencional
antijamming array red contra interferencia intencional
antijamming device dispositivo contra interferencia

intencional
antilogarithm antilogaritmo
antilogous antílogo
antilogous pole polo antílogo
antiluminescent antiluminiscente
antimagnetic antimagnético
antimagnetic alloy aleación antimagnética
antimatter antimateria
antimicrophonic antimicrofónico
antimony antimonio
antineutrino antineutrino
antineutron antineutrón
antinode antinodo
antinoise antirruido
antinoise carrier-operated circuit circuito operado por portadora antirruido
antinoise material material antirruido
antinoise microphone micrófono antirruido
antinucleon antinucleón
antioxidant antioxidante
antiparallel antiparalelo
antiparallel connection conexión antiparalela
antiparallel coupling acoplamiento antiparalelo
antiparticle antipartícula
antiphase antifase
antiplugging protection protección contra frenado de contramarcha
antipodal antípoda
antipolar antipolar
antipole antipolo
antiproton antiprotón
antiquark antiquark
antiradar antirradar
antiradiation antirradiación
antireflection antirreflexión
antireflection coating revestimiento antirreflexión
antiresonance antirresonancia
antiresonant antirresonante
antiresonant circuit circuito antirresonante
antiresonant frequency frecuencia antirresonante
antiresonant line línea antirresonante
antiresonant winding arrollamiento antirresonante
antiscatter antidispersión
antiscatter grid rejilla antidispersión
antisidetone contra efectos locales
antisidetone device dispositivo contra efectos locales
antisidetone equipment equipo contra efectos locales
antisidetone induction coil bobina de inducción contra efectos locales
antisidetone telephone set aparato telefónico con dispositivo contra efectos locales
antiskating control control antipatinaje
antiskating device dispositivo antipatinaje
antistatic antiestático
antistatic agent agente antiestático
antistatic antenna antena antiestática
antistickoff voltage tensión desplazadora de cero falso
antitransmit-receive box caja contra transmisión-recepción
antitransmit-receive switch conmutador contra transmisión-recepción
antitransmit-receive tube tubo contra transmisión-recepción
antivibrator antivibrador
antivoice-operated transmission transmisión con bloqueo por voz
aperiodic aperiódico
aperiodic antenna antena aperiódica
aperiodic circuit circuito aperiódico
aperiodic current corriente aperiódica
aperiodic damping amortiguamiento aperiódico
aperiodic discharge descarga aperiódica

aperiodic element elemento aperiódico
aperiodic function función aperiódica
aperiodic instrument instrumento aperiódico
aperiodic phenomenon fenómeno aperiódico
aperiodic time constant constante de tiempo
    aperiódica
aperiodic variation variación aperiódica
aperture abertura, apertura
aperture angle ángulo de abertura
aperture antenna antena de abertura
aperture blockage bloqueo de abertura
aperture compensation compensación de abertura
aperture compensator compensador de abertura
aperture core núcleo de abertura
aperture correction corrección de abertura
aperture distortion distorsión de abertura
aperture efficiency eficiencia de abertura
aperture equalization igualación de abertura
aperture illumination iluminación de abertura
aperture lens lente de abertura
aperture mask máscara de abertura
aperture plane plano de abertura
aperture plate placa de abertura
aperture ratio razón de abertura
aperture time tiempo de abertura
apochromatic apocromático
apochromatic condenser condensador apocromático
apochromatism apocromatismo
apogee apogeo
apostilb apostilbio
apparatus aparato
apparatus insulator aislador de aparato
apparatus insulator unit unidad de aislador de
    aparato
apparatus termination terminación de aparato
apparatus thermal device dispositivo térmico de
    aparato
apparent aparente
apparent altitude altitud aparente
apparent bearing rumbo aparente
apparent brightness brillo aparente
apparent candlepower potencia lumínica aparente
apparent charge carga aparente
apparent dead time tiempo muerto aparente
apparent definition definición aparente
apparent discharge magnitude magnitud de
    descarga aparente
apparent energy energía aparente
apparent error error aparente
apparent horizon horizonte aparente
apparent inductance inductancia aparente
apparent internal resistance resistencia interna
    aparente
apparent magnitude magnitud aparente
apparent motion moción aparente
apparent output power potencia de salida aparente
apparent permeability permeabilidad aparente
apparent position posición aparente
apparent power potencia aparente
apparent power loss pérdida de potencia aparente
apparent power meter medidor de potencia aparente
apparent radiated power potencia radiada aparente
apparent reflectance reflectancia aparente
apparent sag caída aparente
apparent separation separación aparente
apparent time constant constante de tiempo aparente
apparent vertical vertical aparente
apparent visual angle ángulo visual aparente
apparent watts watts aparentes
appearance potential potencial de aparición
applause meter medidor de aplausos
Applegate diagram diagrama de Applegate
Appleton layer capa de Appleton

appliance aparato eléctrico
appliance branch circuit circuito de derivación de
    aparato eléctrico
appliance impedance impedancia de aparato
    eléctrico
appliance load carga de aparato eléctrico
appliance outlet tomacorriente de aparato eléctrico
appliance wire hilo de aparato eléctrico
applicability aplicabilidad
application aplicación
application audit auditoría de aplicación
application development language lenguaje de
    desarrollo de aplicaciones
application development system sistema de
    desarrollo de aplicaciones
application diagram diagrama de aplicación
application factor factor de aplicación
application generator generador de aplicación
application integrity integridad de aplicación
application-oriented language lenguaje orientado a
    aplicaciones
application plan plan de aplicación
application point punto de aplicación
application processing program programa de
    procesamiento de aplicación
application program programa de aplicación
application programming language lenguaje de
    programación de aplicaciones
application routine rutina de aplicación
application schematic diagram diagrama
    esquemático de aplicación
application service servicio de aplicación
application software programas de aplicación
application structure estructura de aplicación
application support soporte de aplicación
application valve válvula de aplicación
applicative order orden aplicativo
applicator aplicador
applied aplicado
applied-potential test prueba de potencial aplicado
applied power potencia aplicada
applied shock choque aplicado
applied voltage tensión aplicada
applied voltage test prueba de tensión aplicada
applique circuit circuito de adaptación
approach aproximación
approach area área de aproximación
approach beacon radiofaro de aproximación
approach circuit circuito de aproximación
approach control control de aproximación
approach control radar radar de control de
    aproximación
approach control service servicio de control de
    aproximación
approach indicator indicador de aproximación
approach-light beacon radiofaro de luz de
    aproximación
approach lighting alumbrado de aproximación
approach-lighting relay relé de alumbrado de
    aproximación
approach-lighting system sistema de alumbrado de
    aproximación
approach lights luces de aproximación
approach locking bloqueo de aproximación
approach locking device dispositivo de bloqueo de
    aproximación
approach navigation navegación de aproximación
approach path trayectoria de aproximación
approach signal señal de aproximación
approach speed velocidad de aproximación
approach zone zona de aproximación
approval test prueba de aprobación
approved aprobado

approximate data datos aproximados
approximate value valor aproximado
approximation aproximación
aqua pura agua pura
aquavox aquavox
arbiter árbitro
arbitrary arbitrario
arbitrary constant constante arbitraria
arbitrary function filter filtro de función arbitraria
arbitrary sequence computer computadora de
  secuencia arbitraria
arbitrary zero cero arbitrario
arbitration arbitraje
arbitration bus bus de arbitraje
arbitration contest competencia de arbitraje
arbitration cycle ciclo de arbitraje
arbitration operation operación de arbitraje
arbitration timing control control de tiempo de
  arbitraje
arc arco
arc angle ángulo de arco
arc cathode cátodo de arco
arc chute conducto de arco
arc control control de arco
arc control device dispositivo de control de arco
arc converter convertidor de arco
arc cosecant cosecante de arco
arc cosine coseno de arco
arc cotangent cotangente de arco
arc crater cráter de arco
arc current corriente de arco
arc discharge descarga de arco
arc-discharge rectifier rectificador de descarga de
  arco
arc-discharge tube tubo de descarga de arco
arc drop caída de arco
arc-drop loss pérdida de caída de arco
arc-drop voltage tensión de caída de arco
arc extinction extinción de arco
arc failure falla por arco
arc function función de arco
arc furnace horno de arco
arc gate puerta de arco
arc heating calentamiento por arco
arc lamp lámpara de arco
arc length longitud de arco
arc loss pérdida de arco
arc maintenance mantenimiento de arco
arc maintenance device dispositivo de
  mantenimiento de arco
arc minute minuto de arco
arc of contact arco de contacto
arc oscillation oscilación de arco
arc oscillator oscilador de arco
arc path trayectoria de arco
arc rectifier rectificador de arco
arc resistance resistencia de arco
arc secant secante de arco
arc second segundo de arco
arc sensor sensor de arco
arc sine seno de arco
arc spectrum espectro de arco
arc stability estabilidad de arco
arc suppression supresión de arco
arc-suppressor diode diodo supresor de arco
arc switch conmutador de arco
arc tangent tangente de arco
arc transmitter transmisor de arco
arc tube tubo de arco
arc voltage tensión de arco
arc welding soldadura por arco
arc-welding electrode electrodo de soldadura por
  arco

arcback retroarco
architectural arquitectónico
architectural design diseño arquitectónico
architecture arquitectura
archival archival
archival database base de datos archivados
archival storage almacenamiento de datos archivados
archived archivado
arcing formación de arcos
arcing contact contacto de formación de arcos
arcing horn electrodos de guarda
arcing ring anillo de guarda
arcing time tiempo de formación de arcos
arcover salto de arco
arcover voltage tensión de salto de arco
arcthrough falla de aislamiento
area área
area code código de área
area control error error de control de área
area control radar radar de control de área
area frequency-response characteristic
  característica de respuesta de frecuencia de área
area monitor monitor de área
area monitoring monitoreo de área
area network red de área
area redistribution redistribución de área
area search búsqueda de área
area supplementary control control suplementario
  de área
area traffic control control de tráfico de área
argon argón
argon glow lamp lámpara luminiscente de argón
argon laser láser de argón
argon tube tubo de argón
argument argumento
argument separator separador de argumentos
arithmetic aritmética
arithmetic address dirección aritmética
arithmetic and logic unit unidad aritmética y lógica
arithmetic check comprobación aritmética
arithmetic circuit circuito aritmético
arithmetic constant constante aritmética
arithmetic conversion conversión aritmética
arithmetic element elemento aritmético
arithmetic expression expresión aritmética
arithmetic function función aritmética
arithmetic instruction instrucción aritmética
arithmetic logic unit unidad aritmética-lógica
arithmetic logical operation operación
  aritmética-lógica
arithmetic mean media aritmética
arithmetic mode modo aritmético
arithmetic operation operación aritmética
arithmetic overflow desbordamiento aritmético
arithmetic point coma aritmética, punto aritmético
arithmetic progression progresión aritmética
arithmetic reactive factor factor reactivo aritmético
arithmetic register registro aritmético
arithmetic relation relación aritmética
arithmetic section sección aritmética
arithmetic shift desplazamiento aritmético
arithmetic sum suma aritmética
arithmetic symmetry simetría aritmética
arithmetic underflow desbordamiento negativo
  aritmético
arithmetic unit unidad aritmética
arithmetical aritmético
arm brazo
armature armadura, inducido
armature band banda del inducido
armature band insulation aislamiento de banda del
  inducido
armature bar barra del inducido

armature coil bobina del inducido
armature core núcleo del inducido
armature duct conducto del inducido
armature gap entrehierro del inducido
armature reactance reactancia del inducido
armature reaction reacción del inducido
armature relay relé de armadura
armature resistance resistencia del inducido
armature spider estrella del inducido
armature terminal terminal del inducido
armature travel recorrido de armadura
armature voltage tensión del inducido
armature voltage control control de tensión del
   inducido
armature winding arrollamiento del inducido
armor armadura
armor clamp abrazadera de armadura
armored cable cable blindado
Armstrong frequency-modulation system sistema
   de modulación de frecuencia de Armstrong
Armstrong modulation modulación de Armstrong
Armstrong oscillator oscilador de Armstrong
Armstrong superheterodyne circuit circuito
   superheterodino de Armstrong
Armstrong superregenerative circuit circuito
   superregenerativo de Armstrong
arrangement arreglo, ordenación
array red, red de antenas
array control control de red
array device dispositivo de red
array element elemento de red de antenas
array factor factor de red de antenas
array field campo de red
array subsystem subsistema de red
arrester protector
arrester discharge capacity capacidad de descarga
   de protector
arrester disconnector desconector de protector
arrester ground tierra de protector
arrowhead punta de flecha
arsenic arsénico
articulation articulación
articulation equivalent equivalente de articulación
articulation reduction reducción de articulación
articulation test prueba de articulación
artifact artefacto
artificial artificial
artificial acoustic target blanco acústico artificial
artificial antenna antena artificial
artificial delay line línea de retardo artificial
artificial dielectric dieléctrico artificial
artificial dielectric lens lente dieléctrica artificial
artificial ear oído artificial
artificial earth tierra artificial
artificial echo eco artificial
artificial extension line línea de extensión artificial
artificial eye ojo artificial
artificial fading desvanecimiento artificial
artificial ground tierra artificial
artificial hand mano artificial
artificial horizon horizonte artificial
artificial intelligence inteligencia artificial
artificial ionization ionización artificial
artificial language lenguaje artificial
artificial line línea artificial
artificial load carga artificial
artificial mouth boca artificial
artificial piezoelectric crystal cristal piezoeléctrico
   artificial
artificial pupil pupila artificial
artificial radioactivity radiactividad artificial
artificial satellite satélite artificial
artificial traffic tráfico artificial

artificial transmission line línea de transmisión
   artificial
artificial voice voz artificial
artificially artificialmente
artificially induced inducido artificialmente
artwork diseño
asbestos asbesto
ascending ascendente
ascending node nodo ascendente
ascending sort clasificación ascendente
askarel askarel
aspect aspecto
aspect ratio razón de aspecto
asperities asperezas
assay ensayo
assemble ensamblar
assemble-and-go ensamblar y ejecutar
assembled ensamblado
assembled origin origen ensamblado
assembler ensamblador
assembler code código ensamblador
assembler instruction instrucción de ensamblador
assembler language lenguaje ensamblador
assembler program programa ensamblador
assembling phase fase de ensamblaje
assembling time tiempo de ensamblaje
assembly ensamblaje
assembly code código de ensamblaje
assembly language lenguaje de ensamblaje
assembly language instruction instrucción de
   lenguaje de ensamblaje
assembly language processor procesador de
   lenguaje de ensamblaje
assembly language program programa de lenguaje
   de ensamblaje
assembly listing listado de ensamblaje
assembly machine máquina de ensamblaje
assembly phase fase de ensamblaje
assembly process proceso de ensamblaje
assembly program programa de ensamblaje
assembly routine rutina de ensamblaje
assembly system sistema de ensamblaje
assembly technique técnica de ensamblaje
assembly time tiempo de ensamblaje
assembly unit unidad de ensamblaje
assert afirmar
asserted afirmado
assertion aserción
assessed reliability confiabilidad evaluada
assign asignar
assign frequencies asignar frecuencias
assignable asignable
assignable frequency frecuencia asignable
assigned asignado
assigned error error asignado
assigned frequency frecuencia asignada
assigned indexing indexación asignada
assigned value valor asignado
assignment asignación
assignment indexing indexación de asignación
assignment of frequencies asignación de frecuencias
assignment statement sentencia de asignación
assistance call llamada de asistencia
assisted asistido
assisted operation operación asistida
associated asociado
associated circuits circuitos asociados
associated electrical apparatus aparato eléctrico
   asociado
associated equipment equipo asociado
associated structural parts partes estructurales
   asociadas
association asociación

associative asociativo
associative lookup búsqueda asociativa
associative memory memoria asociativa
associative storage almacenamiento asociativo
associative storage register registro de
    almacenamiento asociativo
assumed binary point coma binaria implícita, punto
    binario implícito
assumed decimal point coma decimal implícita,
    punto decimal implícito
assumed option opción implícita
assumed position posición implícita
assumed radix point coma base implícita, punto base
    implícito
assured access protocol protocolo de acceso seguro
astable astable
astable circuit circuito astable
astable multivibrator multivibrador astable
astatic astático
astatic control control astático
astatic element elemento astático
astatic galvanometer galvanómetro astático
astatic magnetic system sistema magnético astático
astatic measuring instrument instrumento de
    medición astático
astatic microphone micrófono astático
astatic pair par astático
astatine astatino
astigmatic astigmático
astigmatism astigmatismo
Aston dark space espacio obscuro de Aston
Aston mass spectrometer espectrómetro de masa de
    Aston
astrionics astriónica
astrocompass astrocompás
astrodynamics astrodinámica
astronomical position posición astronómica
astronomical unit unidad astronómica
astronomical unit of distance unidad de distancia
    astronómica
astrotracker astroseguidor
asymmetric asimétrico
asymmetrical asimétrico
asymmetrical amplitude modulation modulación de
    amplitud asimétrica
asymmetrical cell célula asimétrica
asymmetrical conduction conducción asimétrica
asymmetrical conductivity conductividad asimétrica
asymmetrical control control asimétrico
asymmetrical current corriente asimétrica
asymmetrical deflection deflexión asimétrica
asymmetrical device dispositivo asimétrico
asymmetrical distortion distorsión asimétrica
asymmetrical distribution distribución asimétrica
asymmetrical effect efecto asimétrico
asymmetrical input entrada asimétrica
asymmetrical input-output entrada-salida asimétrica
asymmetrical multivibrator multivibrador
    asimétrico
asymmetrical output salida asimétrica
asymmetrical phase control control de fase
    asimétrico
asymmetrical reflector reflector asimétrico
asymmetrical refractor refractor asimétrico
asymmetrical resonance curve curva de resonancia
    asimétrica
asymmetrical sideband banda lateral asimétrica
asymmetrical sideband transmission transmisión de
    banda lateral asimétrica
asymmetrical terminal voltage tensión de terminal
    asimétrica
asymmetrical wave onda asimétrica
asymmetry asimetría

asymmetry control control de asimetría
asymptote asíntota
asymptotic asintótico
asymptotic availability disponibilidad asintótica
asymptotic breakdown voltage tensión de ruptura
    asintótica
asymptotic expression expresión asintótica
asynchronous asincrónico, asíncrono
asynchronous circuit circuito asincrónico
asynchronous communication comunicación
    asincrónica
asynchronous computer computadora asincrónica
asynchronous condenser condensador asincrónico
asynchronous data transfer transferencia de datos
    asincrónica
asynchronous data transmission transmisión de
    datos asincrónica
asynchronous device dispositivo asincrónico
asynchronous entry point punto de entrada
    asincrónico
asynchronous exit salida asincrónica
asynchronous flow flujo asincrónico
asynchronous impedance impedancia asincrónica
asynchronous input entrada asincrónica
asynchronous machine máquina asincrónica
asynchronous memory memoria asincrónica
asynchronous mode modo asincrónico
asynchronous modem módem asincrónico
asynchronous motor motor asincrónico
asynchronous multiplex múltiplex asincrónico
asynchronous multiplexer multiplexor asincrónico
asynchronous operation operación asincrónica
asynchronous procedure procedimiento asincrónico
asynchronous processing procesamiento asincrónico
asynchronous reactance reactancia asincrónica
asynchronous resistance resistencia asincrónica
asynchronous starting arranque asincrónico
asynchronous terminal terminal asincrónico
asynchronous terminal emulator emulador de
    terminal asincrónica
asynchronous torque par asincrónico
asynchronous transmission transmisión asincrónica
asynchronous vibrator vibrador asincrónico
asyndetic asindético
athermancy atermancia
atmosphere atmósfera
atmospheric atmosférico
atmospheric absorption absorción atmosférica
atmospheric absorption noise ruido de absorción
    atmosférica
atmospheric bending flexión atmosférica
atmospheric condition monitor monitor de
    condición atmosférica
atmospheric contamination contaminación
    atmosférica
atmospheric duct conducto atmosférico
atmospheric electric disturbance perturbación
    eléctrica atmosférica
atmospheric electricity electricidad atmosférica
atmospheric humidity humedad atmosférica
atmospheric interference interferencia atmosférica
atmospheric ionization ionización atmosférica
atmospheric noise ruido atmosférico
atmospheric paths trayectorias atmosféricas
atmospheric pressure presión atmosférica
atmospheric radio duct conducto de radio
    atmosférico
atmospheric radio wave onda de radio atmosférica
atmospheric radio window ventana de radio
    atmosférica
atmospheric radioactivity radiactividad atmosférica
atmospheric reflection reflexión atmosférica
atmospheric refraction refracción atmosférica

atmospheric scatter dispersión atmosférica
atmospheric transmissivity transmisividad
  atmosférica
atmospherical atmosférico
atmospherics atmosféricos
atom átomo
atomic atómico
atomic absorption absorción atómica
atomic absorption coefficient coeficiente de
  absorción atómica
atomic absorption spectrophotometer
  espectrofotómetro de absorción atómica
atomic amplifier amplificador atómico
atomic battery batería atómica
atomic charge carga atómica
atomic clock reloj atómico
atomic condition condición atómica
atomic data element elemento de datos atómico
atomic disintegration desintegración atómica
atomic electricity electricidad atómica
atomic energy energía atómica
atomic energy level nivel de energía atómica
atomic fission fisión atómica
atomic frequency frecuencia atómica
atomic fusion fusión atómica
atomic mass masa atómica
atomic mass unit unidad de masa atómica
atomic migration migración atómica
atomic number número atómico
atomic photoelectric effect efecto fotoeléctrico
  atómico
atomic power potencia atómica
atomic radiation radiación atómica
atomic ratio razón atómica
atomic reactor reactor atómico
atomic resonance resonancia atómica
atomic scattering dispersión atómica
atomic spectrum espectro atómico
atomic structure estructura atómica
atomic theory teoría atómica
atomic time tiempo atómico
atomic type tipo atómico
atomic unit of energy unidad de energía atómica
atomic weight peso atómico
atomic weight unit unidad de peso atómico
atomically atómicamente
atomicity atomicidad
atomics atómica
atomistics atomística
attachment aditamento, conexión
attachment plug clavija de conexión
attachment point punto de conexión
attachment unit interface interfaz de unidad de
  conexión
attack ataque
attack time tiempo de ataque
attemperator atemperador
attempt intento
attention atención
attention cycle ciclo de atención
attention display indicación visual de atención
attention signal señal de atención
attenuate atenuar
attenuated atenuado
attenuated band banda atenuada
attenuated field campo atenuado
attenuating atenuante
attenuating coupler acoplador atenuante
attenuation atenuación
attenuation accuracy exactitud de atenuación
attenuation band banda de atenuación
attenuation box caja de atenuación
attenuation characteristic característica de

atenuación
attenuation coefficient coeficiente de atenuación
attenuation compensation compensación de
  atenuación
attenuation compensator compensador de
  atenuación
attenuation constant constante de atenuación
attenuation curve curva de atenuación
attenuation distortion distorsión de atenuación
attenuation equalization igualación de atenuación
attenuation equalizer ecualizador de atenuación
attenuation equivalent equivalente de atenuación
attenuation factor factor de atenuación
attenuation-frequency characteristic característica
  de frecuencia de atenuación
attenuation-frequency distortion distorsión de
  frecuencia de atenuación
attenuation measurement medida de atenuación
attenuation network red de atenuación
attenuation ratio razón de atenuación
attenuation vector vector de atenuación
attenuator atenuador
attenuator equalizer ecualizador de atenuación
attenuator system sistema de atenuación
attenuator tube tubo de atenuación
attitude actitud, orientación
attitude control control de actitud
attitude gyroelectric indicator indicador
  giroeléctrico de actitud
attofarad attofarad
attraction atracción
attractive potential potencial de atracción
attribute atributo
attribute byte byte de atributo
attribute character carácter de atributo
attribute code código de atributo
attribute data element elemento de datos de atributo
audar audar
audibility audibilidad
audibility curve curva de audibilidad
audible audible
audible alarm alarma audible
audible busy signal señal de ocupado audible
audible cab indicator indicador de cabina audible
audible call llamada audible
audible frequency frecuencia audible
audible indication indicación audible
audible noise ruido audible
audible range intervalo audible
audible signal señal audible
audible signal device dispositivo de señal audible
audible sound sonido audible
audible spectrum espectro audible
audible test prueba audible
audible tone tono audible
audio audio
audio amplification amplificación de audio
audio amplifier amplificador de audio
audio analyzer analizador de audio
audio band banda de audio
audio cable cable de audio
audio channel canal de audio
audio circuit circuito de audio
audio communication line línea de comunicación de
  audio
audio component componente de audio
audio control control de audio
audio converter convertidor de audio
audio detector detector de audio
audio distortion distorsión de audio
audio frequency frecuencia de audio
audio-frequency amplifier amplificador de
  audiofrecuencia

audio-frequency band banda de audiofrecuencia
audio-frequency bandwidth ancho de banda de audiofrecuencia
audio-frequency distortion distorsión de audiofrecuencia
audio-frequency feedback retroalimentación de audiofrecuencia
audio-frequency filter filtro de audiofrecuencia
audio-frequency harmonic armónica de audiofrecuencia
audio-frequency harmonic distortion distorsión armónica de audiofrecuencia
audio-frequency meter medidor de audiofrecuencias
audio-frequency noise ruido de audiofrecuencia
audio-frequency oscillator oscilador de audiofrecuencia
audio-frequency peak limiter limitador de picos de audiofrecuencia
audio-frequency range intervalo de audiofrecuencia
audio-frequency response respuesta de audiofrecuencia
audio-frequency shift desplazamiento de audiofrecuencia
audio-frequency-shift modulator modulador de desplazamiento de audiofrecuencia
audio-frequency signal señal de audiofrecuencia
audio-frequency spectrum espectro de audiofrecuencia
audio-frequency tone tono de audiofrecuencia
audio-frequency transformer transformador de audiofrecuencia
audio-frequency transistor transistor de audiofrecuencia
audio-frequency voltage tensión de audiofrecuencia
audio-frequency voltmeter vóltmetro de audiofrecuencia
audio generator generador de audio
audio howl aullido de audio
audio image imagen de audio
audio input power potencia de entrada de audio
audio-level meter medidor de nivel de audio
audio limiter limitador de audio
audio line línea de audio
audio mixer mezclador de audio
audio oscillator oscilador de audio
audio output salida de audio
audio output power potencia de salida de audio
audio peak pico de audio
audio peak limiter limitador de picos de audio
audio power potencia de audio
audio power output salida de potencia de audio
audio response respuesta de audio
audio response unit unidad de respuesta de audio
audio signal señal de audio
audio signal generator generador de señal de audio
audio spectrum espectro de audio
audio squelch silenciador de audio
audio subcarrier subportadora de audio
audio system sistema de audio
audio terminal terminal de audio
audio-tone channel canal de tono de audio
audio transformer transformador de audio
audio transmitter transmisor de audio
audio-visual audiovisual
audio-visual system sistema audiovisual
audio voltage tensión de audio
audio wire hilo de audio
audioconference audioconferencia
audiogram audiograma
audiologist audiólogo
audiology audiología
audiometer audiómetro
audiometric audiométrico

audiometrist audiometrista
audiometry audiometría
audion audión
audiotape audiocinta
audit auditoría
audit trail pista de auditoría
auditory auditivo
auditory inhibition inhibición auditiva
auditory mirage espejismo auditivo
auditory sensation area área de sensación auditiva
augend cosumando
augend register registro de cosumando
Auger effect efecto de Auger
Auger electron electrón de Auger
augment aumentar
augmented carrier portadora aumentada
augmented operation code código de operación aumentado
augmenter aumentador
aural aural
aural harmonic armónica aural
aural radio range radiofaro direccional aural
aural reception recepción aural
aural signal señal aural
aural transmitter transmisor aural
aurora aurora
auroral auroral
auroral absorption absorción auroral
auroral attenuation atenuación auroral
auroral effects efectos aurorales
auroral flutter titilación auroral
auroral hiss siseo auroral
auroral interference interferencia auroral
auroral opening abertura auroral
auroral oval óvalo auroral
auroral propagation propagación auroral
auroral reflection reflexión auroral
auroral zone zona auroral
austenitic austenítico
authorized autorizado
authorized bandwidth ancho de banda autorizado
authorized channel canal autorizado
authorized operator operador autorizado
authorized path trayectoria autorizada
authorized program programa autorizado
authorized state estado autorizado
authorized user usuario autorizado
autoalarm autoalarma
autobias autopolarización
autocapacity autocapacidad
autocoder autocodificador
autocondensation autocondensación
autoconduction autoconducción
autocontrol autocontrol
autocontrol device dispositivo de autocontrol
autocorrection autocorrección
autocorrelation autocorrelación
autocorrelation coefficient coeficiente de autocorrelación
autocorrelation function función de autocorrelación
autodial automarcado
autodyne autodino
autodyne circuit circuito autodino
autodyne detector detector autodino
autodyne oscillator oscilador autodino
autodyne receiver receptor autodino
autodyne reception recepción autodina
autoelectric autoeléctrico
autoelectric emission emisión autoeléctrica
autoelectrolysis autoelectrólisis
autoelectronic effect efecto autoelectrónico
autoerection autoerección
autoexcitation autoexcitación

autoexcited autoexcitado
autoexciter autoexcitador
autoheterodyne autoheterodino
autoignition autoignición
autoinduction autoinducción
autoinductive autoinductivo
autoionization autoionización
automata theory teoría de autómatas
automate automatizar
automated automatizado
automated assembly equipment equipo de
    ensamblaje automatizado
automated audit auditoría automatizada
automated communication comunicación
    automatizada
automated data medium medio de datos
    automatizado
automated data processing procesamiento de datos
    automatizado
automated design tool herramienta de diseño
    automatizado
automated dictionary diccionario automatizado
automated equipment equipo automatizado
automated glossary glosario automatizado
automated language processing procesamiento de
    lenguaje automatizado
automated lexicon léxico automatizado
automated logic diagram diagrama lógico
    automatizado
automated office oficina automatizada
automated teller machine máquina de cajero
    automatizado
automated test case generator generador de casos
    de prueba automatizado
automated test data generator generador de datos
    de prueba automatizado
automated test generator generador de pruebas
    automatizado
automated thesaurus tesauro automatizado
automated verification system sistema de
    verificación automatizado
automated verification tools herramientas de
    verificación automatizadas
automatic automático
automatic abstract resumen automático
automatic acceleration aceleración automática
automatic alarm alarma automática
automatic alarm equipment equipo de alarma
    automática
automatic amplitude control control de amplitud
    automático
automatic answering contestación automática
automatic approach control control de
    aproximación automático
automatic area área automática
automatic audible alarm alarma audible automática
automatic back bias contrapolarización automática
automatic background control control de fondo
    automático
automatic bass compensation compensación de
    bajos automático
automatic bass control control de bajos automático
automatic bias polarización automática
automatic block signal system sistema de señal de
    bloqueo automático
automatic block system sistema de bloqueo
    automático
automatic brightness control control de brillo
    automático
automatic cab signal señal de cabina automática
automatic cab signal system sistema de señal de
    cabina automática
automatic calculator calculadora automática

automatic calendar calendario automático
automatic call llamada automática
automatic call distributor distribuidor de llamadas
    automático
automatic calling unit unidad de llamada automática
automatic capacitor capacitor automático
automatic carriage carro automático
automatic central central automática
automatic check comprobación automática
automatic chrominance control control de
    crominancia automática
automatic circuit circuito automático
automatic circuit-breaker cortacircuito automático
automatic circuit closer cierracircuito automático
automatic code código automático
automatic coding codificación automática
automatic combustion control control de
    combustión automático
automatic computer computadora automática
automatic computing computación automática
automatic connection conexión automática
automatic constant constante automática
automatic contrast control control de contraste
    automático
automatic control control automático
automatic control system sistema de control
    automático
automatic controller controlador automático
automatic crossover transición automática
automatic current limiter limitador de corriente
    automático
automatic current regulator regulador de corriente
    automático
automatic cutout desconexión automática
automatic data acquisition adquisición de datos
    automática
automatic data processing procesamiento de datos
    automático
automatic data processing equipment equipo de
    procesamiento de datos automático
automatic data processing system sistema de
    procesamiento de datos automático
automatic data switch conmutador de datos
    automático
automatic data transmission transmisión de datos
    automática
automatic degausser desmagnetizador automático
automatic dialing unit unidad de marcación
    automática
automatic dictionary diccionario automático
automatic digital computer computadora digital
    automática
automatic digital network red digital automática
automatic direct-control telecommunications system
    sistema de telecomunicaciones de control directo
    automático
automatic direction-finder radiogoniómetro
    automático
automatic dispatching system sistema de despacho
    automático
automatic distribution distribución automática
automatic document feeder alimentador de
    documentos automático
automatic duplication duplicación automática
automatic elevation measurement medida de
    elevación automática
automatic equipment equipo automático
automatic error correction corrección de errores
    automática
automatic error detection detección de errores
    automática
automatic exchange central automática
automatic extraction turbine turbina de extracción

automática

automatic fading desvanecimiento automático

automatic feed alimentación automática

automatic fire alarm alarma de incendios automática

automatic fire-alarm system sistema de alarma de incendio automática

automatic fire detector detector de incendios automático

automatic flight control control de vuelo automático

automatic flight control system sistema de control de vuelo automático

automatic focusing enfoque automático

automatic frequency control control de frecuencia automática

automatic frequency correction corrección de frecuencia automática

automatic function función automática

automatic gain control control de ganancia automático

automatic gain control voltage tensión de control de ganancia automático

automatic gate bias polarización de puerta automática

automatic generation control control de generación automática

automatic grid bias polarización de rejilla automática

automatic height control control de altura automática

automatic hold retención automática

automatic hyphenation separación con guiones automática

automatic identification identificación automática

automatic index índice automático

automatic indicator indicador automático

automatic information processing procesamiento de información automático

automatic installation instalación automática

automatic intercept intercepción automática

automatic interlocking enclavamiento automático

automatic interrupt interrupción automática

automatic interruption interrupción automática

automatic level compensation compensación de nivel automático

automatic level control control de nivel automático

automatic level regulation regulación de nivel automática

automatic limiter limitador automático

automatic limiting control control de limitación automático

automatic line adjust ajuste de línea automático

automatic line sectionalizer seccionador de línea automático

automatic line selection selección de línea automática

automatic load carga automática

automatic load compensation compensación de carga automática

automatic load control control de carga automático

automatic load limitation limitación de carga automática

automatic loader cargador automático

automatic long-distance service servicio de larga distancia automático

automatic machine control equipment equipo de control de máquina automático

automatic maintenance mantenimiento automático

automatic message accounting contabilidad de mensajes automática

automatic mode of operation modo automático de operación

automatic modulation control control de modulación automático

automatic monitoring equipment equipo de monitoreo automático

automatic monitoring of system monitoreo automático de sistema

automatic multiband antenna antena multibanda automática

automatic noise limiter limitador de ruido automático

automatic noise limiting limitación de ruido automática

automatic noise suppression supresión de ruido automática

automatic number identification identificación de número automática

automatic numbering numeración automática

automatic opening abertura automática

automatic operation operación automática

automatic outage interrupción automática

automatic overload protection protección contra sobrecargas automática

automatic pagination paginación automática

automatic peak limiter limitador de picos automático

automatic phase control control de fase automático

automatic pilot piloto automático

automatic-pilot servo motor servomotor de piloto automático

automatic polarity polaridad automática

automatic potentiometer potenciómetro automático

automatic processing procesamiento automático

automatic program programa automático

automatic program control control de programa automático

automatic program interrupt interrupción de programa automática

automatic programming programación automática

automatic protection protección automática

automatic protective device dispositivo protector automático

automatic quality control control de calidad automático

automatic radar equipment equipo de radar automático

automatic radio compass radiocompás automático

automatic radio transmitter radiotransmisor automático

automatic recovery program programa de recuperación automática

automatic regulation regulación automática

automatic regulator regulador automático

automatic relay retransmisión automática

automatic repeat key tecla de repetición automática

automatic repeater repetidor automático

automatic repeater station estación repetidora automática

automatic repetition repetición automática

automatic reset reposición automática

automatic retransmission retransmisión automática

automatic retransmitter retransmisor automático

automatic reversing inversión automática

automatic route selection selección de ruta automática

automatic routine rutina automática

automatic scanning exploración automática

automatic scanning receiver receptor de exploración automática

automatic search búsqueda automática

automatic selectivity control control de selectividad automático

automatic send-receive emisión-recepción automática

automatic send-receive set aparato de emisión-recepción automático

automatic send-receive unit unidad de emisión-recepción automática
automatic sensitivity control control de sensibilidad automático
automatic sequencing secuenciamiento automático
automatic sequential operation operación secuencial automática
automatic service servicio automático
automatic short-circuit protection protección de cortocircuito automática
automatic short-circuiter cortocircuitador automático
automatic shutoff desconexión automática
automatic signal señal automática
automatic signaling señalización automática
automatic smoke alarm system sistema de alarma de humo automático
automatic speed adjustment ajuste de velocidad automático
automatic starter arrancador automático
automatic starting arranque automático
automatic station estación automática
automatic stop parada automática
automatic storage almacenamiento automático
automatic substation subestación automática
automatic switch conmutador automático
automatic switch center centro de conmutadores automáticos
automatic switchboard cuadro conmutador automático
automatic switching conmutación automática
automatic switching center centro de conmutación automático
automatic switching control control de conmutación automático
automatic switching control equipment equipo de control de conmutación automático
automatic switching equipment equipo de conmutación automático
automatic switching system sistema de conmutación automático
automatic synchronizer sincronizador automático
automatic system sistema automático
automatic telecommunications exchange central de telecomunicaciones automática
automatic telecommunications system sistema de telecomunicaciones automático
automatic telegraphy telegrafía automática
automatic test prueba automática
automatic test equipment equipo de prueba automático
automatic threshold variation variación de umbral automática
automatic time corrector corrector de tiempo automático
automatic time switch conmutador de tiempo automático
automatic track follower seguidor de trayectoria automática
automatic tracking seguimiento automático
automatic tracking antenna antena de seguimiento automático
automatic train control control de tren automático
automatic train control application aplicación de control de tren automático
automatic transfer transferencia automática
automatic transfer equipment equipo de transferencia automático
automatic transfer switch conmutador de transferencia automático
automatic transformer control control de transformador automático
automatic transformer control equipment equipo

de control de transformador automático
automatic transit tránsito automático
automatic transmission transmisión automática
automatic transmitter transmisor automático
automatic triggering disparo automático
automatic tripping abertura automática
automatic tuning sintonización automática
automatic variable variable automática
automatic verifier verificador automático
automatic voice operation operación por voz automática
automatic voltage regulator regulador de tensión automático
automatic volume control control de volumen automático
automatic volume recognition reconocimiento de volumen automático
automatic zero cero automático
automatically automáticamente
automatically controlled controlado automáticamente
automatically programmed tools herramientas programadas automáticamente
automatically regulated regulado automáticamente
automaticity automaticidad
automatics automática
automation automatización
automatization automatización
automatize automatizar
automaton autómata
automonitor automonitor
autonavigator autonavegante
autopilot autopiloto
autopilot coupler acoplador de autopiloto
autoplex autoplex
autoradar plot comparador cartográfico
autoradiography autorradiografía
autoregulation autorregulación
autorotation autorrotación
autostarter autoarrancador
autosyn autosincrónico, autosíncrono
autosynchronous autosincrónico
autotracking seguimiento automático
autotransducter autotransductor
autotransformer autotransformador
autotransformer starter autotransformador de arranque
autotransformer starting arranque por autotransformador
autoverification autoverificación
auxiliary auxiliar
auxiliary actuator actuador auxiliar
auxiliary anode ánodo auxiliar
auxiliary arc arco auxiliar
auxiliary arm brazo auxiliar
auxiliary branch rama auxiliar
auxiliary building edificio auxiliar
auxiliary burden carga auxiliar
auxiliary carrier portadora auxiliar
auxiliary circuit circuito auxiliar
auxiliary circuit-breaker cortacircuito auxiliar
auxiliary compartment compartimiento auxiliar
auxiliary compressor compresor auxiliar
auxiliary computer system sistema de computadora auxiliar
auxiliary contacts contactos auxiliares
auxiliary converter convertidor auxiliar
auxiliary data datos auxiliares
auxiliary device dispositivo auxiliar
auxiliary electrode electrodo auxiliar
auxiliary equipment equipo auxiliar
auxiliary file archivo auxiliar
auxiliary function función auxiliar

auxiliary generator generador auxiliar
auxiliary line línea auxiliar
auxiliary memory memoria auxiliar
auxiliary motor motor auxiliar
auxiliary operation operación auxiliar
auxiliary power potencia auxiliar
auxiliary power supply fuente de alimentación
   auxiliar
auxiliary power supply line línea de fuente de
   alimentación auxiliar
auxiliary power transformer transformador de
   potencia auxiliar
auxiliary processor procesador auxiliar
auxiliary program programa auxiliar
auxiliary receiver receptor auxiliar
auxiliary register registro auxiliar
auxiliary relay relé auxiliar
auxiliary route ruta auxiliar
auxiliary routine rutina auxiliar
auxiliary secondary winding arrollamiento
   secundario auxiliar
auxiliary station estación auxiliar
auxiliary station line filter filtro de línea de estación
   auxiliar
auxiliary storage almacenamiento auxiliar
auxiliary storage unit unidad de almacenamiento
   auxiliar
auxiliary switch conmutador auxiliar
auxiliary switchboard cuadro conmutador auxiliar
auxiliary switching conmutación auxiliar
auxiliary switching unit unidad de conmutación
   auxiliar
auxiliary transformer transformador auxiliar
auxiliary transmitter transmisor auxiliar
auxiliary winding arrollamiento auxiliar
availability disponibilidad
availability factor factor de disponibilidad
availability model modelo de disponibilidad
availability ratio razón de disponibilidad
available disponible
available accuracy exactitud disponible
available capacity capacidad disponible
available conversion gain ganancia de conversión
   disponible
available conversion power potencia de conversión
   disponible
available current corriente disponible
available file space espacio de archivo disponible
available gain ganancia disponible
available generation generación disponible
available hours horas disponibles
available line línea disponible
available memory memoria disponible
available power potencia disponible
available power efficiency eficiencia de potencia
   disponible
available power gain ganancia de potencia
   disponible
available power response respuesta de potencia
   disponible
available short-circuit current corriente de
   cortocircuito disponible
available signal-to-noise ratio razón de señal a ruido
   disponible
available state estado disponible
available time tiempo disponible
avalanche avalancha
avalanche breakdown ruptura por avalancha
avalanche characteristic característica de avalancha
avalanche conduction conducción por avalancha
avalanche current corriente por avalancha
avalanche device dispositivo de avalancha
avalanche diode diodo de avalancha

avalanche effect efecto de avalancha
avalanche impedance impedancia de avalancha
avalanche noise ruido de avalancha
avalanche photodiode fotodiodo de avalancha
avalanche protection protección de avalancha
avalanche rectifier rectificador de avalancha
avalanche transistor transistor de avalancha
average absolute burst magnitude magnitud de
   ráfaga absoluta media
average absolute pulse amplitude amplitud de
   impulso absoluto medio
average absolute value valor absoluto medio
average access time tiempo de acceso medio
average active power potencia activa media
average amplitude amplitud media
average brightness brillo medio
average calculation operation operación de cálculo
   medio
average carrier level nivel de portadora medio
average charge carga media
average current corriente media
average data transfer rate velocidad de
   transferencia de datos media
average delay retardo medio
average detector detector medio
average deviation desviación media
average electrode current corriente de electrodo
   media
average inside air temperature temperatura del aire
   interior media
average latency latencia media
average level nivel medio
average life vida media
average luminance luminancia media
average mutual information información mutua
   media
average noise ruido medio
average noise factor factor de ruido medio
average noise figure cifra de ruido media
average noise temperature temperatura de ruido
   media
average operation time tiempo de operación medio
average output current corriente de salida media
average picture level nivel de imagen medio
average power potencia media
average power output salida de potencia media
average pulse amplitude amplitud de impulso media
average quality calidad media
average queue time tiempo de cola medio
average rectified current corriente rectificada media
average speed velocidad media
average traffic tráfico medio
average transfer rate velocidad de transferencia
   media
average transformation rate velocidad de
   transformación media
average transmission rate velocidad de transmisión
   media
average value valor medio
average value of power valor medio de potencia
average voice level nivel de voz medio
average voltage tensión media
average voltage of electrode tensión media de
   electrodo
aversive shock choque aversivo
aviation channel canal de aviación
aviation electronics electrónica de aviación
aviation instrument instrumento de aviación
aviation instrumentation instrumentación de
   aviación
aviation radiocommunication radiocomunicación de
   aviación
Avogadro's constant constante de Avogadro

axial axial
axial aberration aberración axial
axial flow flujo axial
axial incidence incidencia axial
axial leads terminales axiales
axial mode modo axial
axial projection proyección axial
axial propagation constant constante de propagación axial
axial quadrupole cuadripolo axial
axial ratio razón axial
axial ray rayo axial
axial response respuesta axial
axial sensitivity sensibilidad axial
axiality axialidad
axiotron axiotrón
axis eje
axis inversion inversión de ejes
axis of abscissas eje de abscisas
axis of imaginaries eje de imaginarias
axis of ordinates eje de ordenadas
axis of reals eje de reales
axle bearing cojinete de eje
Ayrton-Mather galvanometer shunt derivación de galvanómetro de Ayrton-Mather
Ayrton-Perry winding arrollamiento de Ayrton-Perry
Ayrton shunt derivación de Ayrton
azel azel
azel display indicador azel
azel indicator indicador azel
azimuth acimut, azimut
azimuth adjustment ajuste de acimut
azimuth alignment alineación de acimut
azimuth blanking extinción de acimut
azimuth control control de acimut
azimuth deviation desviación de acimut
azimuth discrimination discriminación de acimut
azimuth gain reduction reducción de ganancia de acimut
azimuth loss pérdida de acimut
azimuth rate velocidad de acimut
azimuth resolution resolución de acimut
azimuth stabilization estabilización de acimut
azimuthal acimutal
azuza azuza

B

B battery batería B
B channel canal B
B display indicador visual B
B operator operador B
B position posición B
B power supply fuente de alimentación B
B register registro B
B station estación B
B supply fuente B
B voltage tensión B
babble murmullo
back ampere-turn amperevuelta antagonista
back beam haz posterior
back bias contrapolarización, retropolarización
back conductance conductancia inversa
back conduction conducción inversa
back-connected fuse fusible conectado por atrás
back-connected switch conmutador conectado por atrás
back contact contacto de reposo
back coupling retroacoplamiento
back current corriente inversa
back diode diodo inverso
back echo eco reflejado
back electrode electrodo posterior
back electromotive force fuerza contraelectromotriz
back emission emisión inversa
back-end processor procesador posterior
back focal length longitud focal posterior
back gate puerta posterior
back lobe lóbulo posterior
back porch escalón posterior
back radiation radiación posterior
back reference referencia posterior
back resistance resistencia inversa
back scatter retrodispersión
back-shunt circuit circuito de contraderivación
back-shunt keying tecleado equilibrado
back stop tope posterior
back tabulation tabulación inversa
back-to-back adosado
back-to-back circuit circuito en oposición
back-to-back connection conexión en oposición
back-to-back devices dispositivos en oposición
back-to-back method método de oposición
back-to-back repeater repetidor desmodulador
back-to-back testing pruebas en oposición
back voltage contratensión
back wave onda posterior
back-wave radiation radiación de ondas posteriores
backbone circuit circuito principal
backbone system sistema principal
backemission retroemisión
background fondo
background activity actividad de fondo
background area área de fondo
background brightness brillo de fondo
background brightness control control de brillo de fondo
background color color de fondo
background communication comunicación de fondo

background control control de fondo
background count cuenta de fondo
background count rate ritmo de cuenta de fondo
background effect efecto de fondo
background image imagen de fondo
background ionization voltage tensión de ionización
   de fondo
background job tarea de fondo
background level nivel de fondo
background monitor monitor de fondo
background noise ruido de fondo
background noise level nivel de ruido de fondo
background printing impresión de fondo
background processing procesamiento de fondo
background program programa de fondo
background radiation radiación de fondo
background reflectance reflectancia de fondo
background region región de fondo
background response respuesta de fondo
background sound sonido de fondo
backing respaldo
backing file archivo de respaldo
backing plate placa de apoyo
backing storage almacenamiento de respaldo
backlash juego, contrapresión
backplane tablero de conectores de circuitos
   impresos
backplane interface logic lógica de interfaz de
   tablero de conectores de circuitos impresos
backplate placa posterior
backpressure contrapresión
backscatter retrodispersión
backscatter coefficient coeficiente de retrodispersión
backscatter factor factor de retrodispersión
backscatter field campo de retrodispersión
backscattering retrodispersión
backscattering coefficient coeficiente de
   retrodispersión
backscattering cross-section sección transversal de
   retrodispersión
backscattering technique técnica de retrodispersión
backspace code código de retroceso
backspace key tecla de retroceso
backspace mechanism mecanismo de retroceso
backswing retroimpulso
backup respaldo, reserva, equipo de reserva, retorno
   de potencia
backup arrangement arreglo de respaldo
backup battery batería de respaldo
backup computer computadora de respaldo
backup controller controlador de respaldo
backup copy copia de respaldo
backup diskette disquete de respaldo
backup file archivo de respaldo
backup light luz de respaldo
backup procedure procedimiento de respaldo
backup protection protección de respaldo
backup relay relé de respaldo
backup station estación de respaldo
backup storage almacenamiento de respaldo
backup system sistema de respaldo
backup tape cinta de respaldo
backup unit unidad de respaldo
backwall muro posterior
backward channel canal inverso
backward conductance conductancia inversa
backward diode diodo inverso
backward direction dirección inversa
backward file recovery recuperación de archivo
   inversa
backward frequency frecuencia inversa
backward processing procesamiento inverso
backward reference referencia inversa

backward resistance resistencia inversa
backward search búsqueda inversa
backward signal señal inversa
backward sort clasificación inversa
backward-wave magnetron magnetrón de ondas
   inversas
backward-wave oscillator oscilador de ondas
   inversas
backward-wave tube tubo de ondas inversas
backward waves ondas inversas, ondas regresivas
backwound coil bobina superpuesta
bactericidal bactericida
bactericidal exposure exposición bactericida
bactericidal flux flujo bactericida
bactericidal lamp lámpara bactericida
bactericidal radiation radiación bactericida
bad definition mala definición
bad sector sector defectuoso
bad track pista defectuosa
baffle bafle, pantalla acústica, pantalla de desviación
bakelite baquelita
bakelite tube tubo de baquelita
bakelized baquelizado
balance balance, equilibrio
balance attenuation atenuación de equilibrio
balance check comprobación de equilibrio
balance coil bobina de equilibrio
balance control control de equilibrio
balance electrometer electrómetro de balanza
balance equation ecuación de equilibrio
balance impedance impedancia de equilibrio
balance indicator indicador de equilibrio
balance ratio razón de equilibrio
balance relay relé de equilibrio
balance test prueba de equilibrio
balance to ground equilibrio respecto a tierra
balance wire hilo de equilibrio
balanced equilibrado
balanced aggregate signal señal agregada
   equilibrada
balanced amplifier amplificador equilibrado
balanced antenna antena equilibrada
balanced antenna input entrada de antena
   equilibrada
balanced antenna system sistema de antenas
   equilibrado
balanced bridge puente equilibrado
balanced bridge transition transición de puente
   equilibrado
balanced capacitance capacitancia equilibrada
balanced capacitive coupling acoplamiento
   capacitivo equilibrado
balanced circuit circuito equilibrado
balanced code código equilibrado
balanced conditions condiciones equilibradas
balanced converter convertidor equilibrado
balanced currents corrientes equilibradas
balanced data link enlace de datos equilibrado
balanced demodulator desmodulador equilibrado
balanced detector detector equilibrado
balanced discriminator discriminador equilibrado
balanced duplexer duplexor equilibrado
balanced electronic voltmeter vóltmetro electrónico
   equilibrado
balanced error error equilibrado
balanced filter filtro equilibrado
balanced impedance impedancia equilibrada
balanced input entrada equilibrada
balanced line línea equilibrada
balanced line system sistema de línea equilibrada
balanced load carga equilibrada
balanced loop cuadro equilibrado, lazo equilibrado
balanced loop antenna antena de cuadro equilibrada

balanced low-pass filter filtro pasabajos equilibrado
balanced magnetic amplifier amplificador magnético equilibrado
balanced merge fusión equilibrada
balanced metallic circuit circuito metálico equilibrado
balanced method método de cero, método de equilibrio
balanced mixer mezclador equilibrado
balanced modulation modulación equilibrada
balanced modulator modulador equilibrado
balanced multivibrator multivibrador equilibrado
balanced network red equilibrada
balanced open line línea abierta equilibrada
balanced oscillator oscilador equilibrado
balanced output salida equilibrada
balanced pair par equilibrado
balanced phase detector detector de fase equilibrado
balanced polyphase source fuente polifásica equilibrada
balanced polyphase system sistema polifásico equilibrado
balanced probe sonda equilibrada
balanced sound sonido equilibrado
balanced state estado equilibrado
balanced station estación equilibrada
balanced system sistema equilibrado
balanced telephone line línea telefónica equilibrada
balanced termination terminación equilibrada
balanced-to-unbalanced transformer transformador de simétrico a asimétrico
balanced transformer transformador equilibrado
balanced transmission line línea de transmisión equilibrada
balanced-unbalanced transformer transformador simétrico-asimétrico
balanced voltages tensiones equilibradas
balancer equilibrador
balancer transformer transformador equilibrador
balancing equilibrador
balancing action acción equilibradora
balancing antenna antena equilibradora
balancing attenuator atenuador equilibrador
balancing battery batería equilibradora
balancing bus bus equilibrador
balancing capacitor capacitor equilibrador
balancing circuit circuito equilibrador
balancing coil bobina equilibradora
balancing condenser condensador equilibrador
balancing flux flujo equilibrador
balancing force fuerza equilibradora
balancing impedance impedancia equilibradora
balancing line línea equilibradora
balancing load carga equilibradora
balancing network red equilibradora
balancing resistance resistencia equilibradora
balancing speed velocidad equilibradora
balancing unit unidad equilibradora
ballast estabilizador, reactancia auxiliar, bobina de inductancia, balasto
ballast lamp lámpara estabilizadora
ballast resistance resistencia estabilizadora
ballast resistor resistor estabilizador
ballast tube tubo estabilizador
ballast valve válvula estabilizadora
ballistic balística
ballistic circuit-breaker cortacircuito balístico
ballistic factor factor balístico
ballistic galvanometer galvanómetro balístico
ballistic measurement medida balística
ballistic missile misil balístico
ballistics balística
balloon antenna antena de globo

balun balún
banana jack jack banana
banana pin terminal banana
banana plug clavija banana
banana plug terminal terminal de clavija banana
band banda
band amplification amplificación de banda
band amplifier amplificador de banda
band articulation articulación de banda
band attenuation atenuación de banda
band center centro de banda
band compressor compresor de banda
band edge límite de banda
band edge energy energía de límite de banda
band-elimination filter filtro de eliminación de banda
band filter filtro de banda
band-limiting filter filtro limitador de banda
band number número de banda
band of frequencies banda de frecuencias
band of regulated voltage banda de tensión regulada
band overlap solape de banda
band pressure presión de banda
band pressure level nivel de presión de banda
band printer impresora de banda
band rejection rechazo de banda
band-rejection filter filtro de rechazo de banda
band repeater repetidor de banda
band selector selector de banda
band sound pressure level nivel de presión sonora de banda
band spectrum espectro de bandas
band splitter divisor de banda
band splitting división de banda
band suppression supresión de banda
band-suppression filter filtro de supresión de banda
band switch conmutador de banda
band switching conmutación de banda
band theory teoría de banda
bandpass paso de banda
bandpass amplifier amplificador de paso de banda
bandpass circuit circuito de paso de banda
bandpass coupling acoplamiento de paso de banda
bandpass filter filtro de paso de banda
bandpass response respuesta de paso de banda
bandpass tuning sintonización de paso de banda
bandset capacitor capacitor de fijación de banda
bandset dial cuadrante de fijación de banda
bandspread ensanche de banda
bandspread tuning capacitor capacitor de sintonización de ensanche de banda
bandspread tuning control control de sintonización de ensanche de banda
bandspreader ensanchador de banda
bandspreading ensanchamiento de banda
bandstop rechazo de banda
bandstop filter filtro de rechazo de banda
bandswitch conmutador de banda
bandswitching conmutación de banda
bandwidth ancho de banda
bandwidth adjuster ajustador de ancho de banda
bandwidth balancing mechanism mecanismo equilibrador de ancho de banda
bandwidth compression compresión de ancho de banda
bandwidth control control de ancho de banda
bandwidth ratio razón de ancho de banda
bank banco, serie
banked transformers transformadores en paralelo
banked winding arrollamiento superpuesto
bantam tube tubo miniatura
bar barra
bar code código de barras

bar code optical scanner explorador óptico de código de barras
bar code reader-sorter lectora-clasificadora de código de barras
bar code scanner explorador de código de barras
bar generator generador de barras
bar graph gráfica de barras
bar-graph meter medidor de gráfica de barras
bar insulation aislamiento de barra
bar magnet imán de barra
bar pattern patrón de barras
bar printer impresor de barras
bar test pattern patrón de prueba de barras
bare conductor conductor desnudo
bare copper wire hilo de cobre desnudo
bare electrode electrodo desnudo
bare lamp lámpara desnuda
bare resistance wire hilo resistivo desnudo
bare wire alambre desnudo, hilo desnudo
barium bario
barium-strontium oxide óxido de bario-estroncio
barium titanate titanato de bario
Barkhausen criterion criterio de Barkhausen
Barkhausen criterion for oscillators criterio de Barkhausen para osciladores
Barkhausen effect efecto de Barkhausen
Barkhausen eliminator eliminador de Barkhausen
Barkhausen interference interferencia de Barkhausen
Barkhausen jumps saltos de Barkhausen
Barkhausen-Kurz oscillation oscilación de Barkhausen-Kurz
Barkhausen-Kurz oscillator oscilador de Barkhausen-Kurz
Barkhausen oscillation oscilación de Barkhausen
Barkhausen oscillator oscilador de Barkhausen
Barkhausen tube tubo de Barkhausen
barn barn
Barnett effect efecto de Barnett
Barnett method método de Barnett
barograph barógrafo
barometer barómetro
barometer effect efecto de barómetro
barometric barométrico
barometric altimeter altímetro barométrico
barometric gradient gradiente barométrico
barometric pressure presión barométrica
barometric switch conmutador barométrico
baroresistor barorresistor
barostat barostato
baroswitch baroconmutador
barothermograph barotermógrafo
barrage jamming interferencia intencional multibanda
barrel distortion distorsión en barril
barrel winding arrollamiento en barril
barretter resistencia estabilizadora, bolómetro
barretter mount montura de bolómetro
barrier barrera
barrier balance equilibrio de barrera
barrier capacitance capacitancia de barrera
barrier cell célula de barrera
barrier diode diodo de barrera
barrier frequency frecuencia de barrera
barrier grid rejilla de barrera
barrier height altura de barrera
barrier layer capa de barrera
barrier-layer capacitance capacitancia de capa de barrera
barrier-layer cell célula de capa de barrera
barrier-layer resistance resistencia de capa de barrera
barrier potential potencial de barrera

barrier region región de barrera
barrier resistance resistencia de barrera
barrier strip tira de barrera
barrier transistor transistor de barrera
barrier voltage tensión de barrera
baryon barión
basal basal
base base
base address dirección de base
base address recorder registrador de dirección de base
base address register registro de dirección de base
base ambient temperature temperatura ambiente base
base apparent power potencia aparente base
base bias polarización base
base cluster grupo de base
base current corriente de base
base electrode electrodo de base
base element elemento base
base feed alimentación de base
base film película de base
base frequency frecuencia de base
base group grupo de base
base impedance impedancia de base
base insulator aislador de base
base item artículo de base
base level nivel de base
base light luz de base
base limiter limitador de base
base limiting limitación de base
base line línea de base
base line break interrupción de línea de base
base line data datos de línea de base
base load carga base
base load control control de carga base
base-loaded antenna antena de carga de base
base location localización de base
base material material de base
base memory memoria de base
base metal metal de base
base number número de base
base pin terminal de base
base plate placa de base
base potential potencial de base
base quantity cantidad de base
base receiver receptor de base
base region región de base
base register registro de base
base resistance resistencia de base
base resistivity resistividad de base
base resistor resistor de base
base speed velocidad base
base spreading resistance resistencia de dispersión de base
base station estación de base
base tabulation tabulación de base
base terminal terminal de base
base unit unidad de base
base value valor de base
base version versión de base
base voltage tensión de base
baseband banda base
baseband amplifier amplificador de banda base
baseband bridge puente de banda base
baseband channel canal de banda base
baseband channel capacity capacidad de canal de banda base
baseband data datos de banda base
baseband frequency frecuencia de banda base
baseband frequency response respuesta de frecuencia de banda base
baseband impedance impedancia de banda base

baseband input entrada de banda base
baseband modem módem de banda base
baseband regulator regulador de banda base
baseband response function función de respuesta de banda base
baseband signal señal de banda base
baseband switch conmutador de banda base
baseband transfer function función de transferencia de banda base
baseboard tablero de base
basegroup grupo de base
basegroup modem módem de grupo de base
baseline línea base
baseline delay retardo de línea base
baseline direction dirección de línea base
baseline restoration restauración de línea base
baseline stabilizer estabilizador de línea base
basic básico
basic access method método de acceso básico
basic alternating voltage tensión alterna básica
basic amplifier amplificador básico
basic assembler ensamblador básico
basic band banda básica
basic bandwidth ancho de banda básico
basic cell celda básica
basic circuit circuito básico
basic coding codificación básica
basic control control básico
basic control element elemento de control básico
basic control mode modo de control básico
basic controller controlador básico
basic counter contador básico
basic current range intervalo de corriente básico
basic cycle ciclo básico
basic data exchange intercambio de datos básicos
basic device dispositivo básico
basic element elemento básico
basic frequency frecuencia básica
basic functions funciones básicas
basic group grupo básico
basic information unit unidad de información básica
basic input-output system sistema de entrada-salida básico
basic instruction instrucción básica
basic language lenguaje básico
basic load carga básica
basic loop bucle básico
basic metallic rectifier rectificador metálico básico
basic mode modo básico
basic monitor monitor básico
basic network red básica
basic noise ruido básico
basic operating system sistema operativo básico
basic part parte básica
basic power amplifier amplificador de potencia básico
basic protection protección básica
basic radiance radiancia básica
basic reference standard patrón de referencia básico
basic repetition frequency frecuencia de repetición básica
basic repetition rate velocidad de repetición básica
basic research investigación básica
basic storage almacenamiento básico
basic supergroup supergrupo básico
basic switch connection conexión de conmutador básica
basic transmission loss pérdida de transmisión básica
basic transmission unit unidad de transmisión básica
basic tuner sintonizador básico
basic voltage range intervalo de tensión básica
basis base

basket-weave coil bobina en hilado canastilla
bass bajos
bass attenuator atenuador de bajos
bass boost refuerzo de bajos
bass booster reforzador de bajos
bass compensation compensación de bajos
bass compensator compensador de bajos
bass control control de bajos
bass correction corrección de bajos
bass corrector corrector de bajos
bass loudspeaker altavoz de bajos
bass-reflex enclosure caja reflectora de bajos
bass-reflex speaker altavoz reflector de bajos
bass-resonant frequency frecuencia resonante de bajos
bass suppression supresión de bajos
bat-handle switch conmutador de palanca
bat-wing antenna antena en mariposa
batch lote
batch compiler compilador por lotes
batch control control por lotes
batch fabrication process proceso de fabricación por lotes
batch file archivo de serie de instrucciones
batch processing procesamiento por lotes
batch processing mode modo de procesamiento por lotes
batch processing program programa de procesamiento por lotes
batch production producción por lotes
batch sort clasificación por lotes
batch testing pruebas por lotes
bath voltage tensión de baño
bathtub capacitor capacitor en bañera
bathtub condenser condensador en bañera
bathtub curve curva en bañera
bathypitometer batipitómetro
bathythermograph batitermógrafo
Batten system sistema de Batten
battery batería
battery acid ácido de batería
battery alarm alarma de batería
battery backup system sistema de reserva de batería
battery base base de batería
battery bell timbre de batería
battery bias polarización de batería
battery capacity capacidad de batería
battery cell celda de batería
battery charge carga de batería
battery charge indicator indicador de carga de batería
battery charger cargador de baterías
battery charging carga de baterías
battery charging circuit circuito de carga de baterías
battery charging generator generador de carga de baterías
battery charging rectifier rectificador de carga de baterías
battery charging relay relé de carga de baterías
battery charging resistor resistor de carga de baterías
battery circuit circuito de batería
battery clip pinza de batería
battery current corriente de batería
battery-current regulation regulación de corriente de batería
battery duty cycle ciclo de trabajo de batería
battery element elemento de batería
battery eliminator eliminador de batería
battery-fed alimentado por batería
battery life vida de batería
battery main switch conmutador principal de batería
battery-operated operado por batería

battery power supply fuente de alimentación de batería
battery powered alimentado por batería
battery rack bastidor de baterías, soporte de baterías
battery receiver receptor de batería
battery stand estante de baterías
battery supply suministro de batería
battery-supply circuit circuito de suministro de batería
battery-supply coil bobina de suministro de batería
battery-supply conductor conductor de suministro de batería
battery-supply relay relé de suministro de batería
battery test prueba de batería
battery tube tubo de batería
battery voltage tensión de batería
battery wire hilo de batería
baud baudio
baud rate velocidad de bauds
baud rate generator generador de velocidad de bauds
Baudot code código de Baudot
Baudot system sistema de Baudot
Baxandall tone control control de tonos de Baxandall
bay sección, lugar
bayonet base base de bayoneta
bayonet cap casquillo de bayoneta
bayonet insulator aislador de bayoneta
bayonet mount montura de bayoneta
bayonet socket zócalo de bayoneta
bazooka bazuca
beacon radiofaro, faro, baliza
beacon delay retardo de radiofaro
beacon equation ecuación de radiofaro
beacon receiver receptor de radiofaro
beacon reception recepción de radiofaro
beacon station estación de radiofaro
beacon transmitter transmisor de radiofaro
beaconage balizamiento
beaconing balizamiento
bead perla
bead thermistor termistor de cuenta
beaded coax cable coaxial perlado
beaded support soporte perlado
beam haz
beam accelerating anode ánodo acelerador de haz
beam accelerating voltage tensión aceleradora de haz
beam aligner alineador de haz
beam alignment alineación de haz
beam angle ángulo de haz
beam antenna antena de haz
beam axis eje de haz
beam bender desviador de haz
beam bending desviación de haz
beam convergence convergencia de haces
beam coupling acoplamiento de haz
beam coupling coefficient coeficiente de acoplamiento de haz
beam crossover cruce de haz
beam current corriente de haz
beam cutoff extinción de haz
beam deflection deflexión de haz
beam-deflection tube tubo de deflexión de haz
beam deflector deflector de haz
beam defocusing desenfoque de haz
beam diameter diámetro de haz
beam divergence divergencia de haz
beam effect efecto de haz
beam efficiency eficiencia de haz
beam electrode electrodo de haz
beam energy energía de haz

beam expander expansor de haz
beam extinction supresión de haz
beam finder buscador de haz
beam former formador de haz
beam-forming plate placa de formación de haz
beam-forming tetrode tetrodo de formación de haz
beam hole abertura de haz
beam intermodulation intermodulación de haces
beam jitter fluctuación de haz
beam locater ubicador de haz
beam magnet imán de convergencia de haces
beam modulation modulación de haz
beam modulation percentage porcentaje de modulación de haz
beam multiplier tube tubo multiplicador de haz
beam noise ruido de haz
beam of electrons haz de electrones
beam of neutrons haz de neutrones
beam parametric amplifier amplificador paramétrico de haz
beam pattern patrón de haz
beam pointing apuntamiento de haz
beam position posición de haz
beam power potencia de haz
beam power tube tubo de potencia de haces
beam reactor reactor de haz
beam resonator resonador de haz
beam-rider guidance guiado por haz
beam shaping formación de haz
beam signal señal de haz
beam solid angle ángulo sólido de haz
beam splitter divisor de haz
beam splitter camera cámara tricroma
beam splitting división de haz
beam spread dispersión de haz
beam suppression supresión de haz
beam switching conmutación de haz
beam-switching tube tubo de conmutación de haz
beam system sistema de haz
beam tetrode tetrodo de haz
beam transadmittance transadmitancia de haz
beam transmission transmisión de haz
beam transmission line línea de transmisión de haz
beam tube tubo de haz
beam velocity velocidad de haz
beam waveguide guíaondas de haz
beamwidth anchura de haz
beamwidth error error de anchura de haz
beamwidth of antenna anchura de haz de antena
bearing rumbo, soporte, cojinete
bearing accuracy exactitud de rumbo
bearing resolution resolución de rumbo
bearing sensitivity sensibilidad de rumbo
bearing transmission transmisión de rumbo
bearing transmission unit unidad de transmisión de rumbo
bearing transmitter transmisor de rumbo
beat pulsación, batido
beat frequency frecuencia de batido
beat-frequency generator generador de frecuencia de batido
beat-frequency oscillator oscilador de frecuencia de batido
beat-frequency output salida de frecuencia de batido
beat indicator indicador de batido
beat method método de batido
beat note nota de batido
beat note detector detector de nota de batido
beat oscillator oscilador de batido
beat reception recepción heterodina
beat tone tono de batido
beating batido
beavertail beam haz en cola de castor

beavertail beam antenna antena de haz en cola de castor
becquerel bequerel
Becquerel effect efecto de Becquerel
bedspring antenna antena direccional de reflector plano
beep tono breve
beeper dispositivo emisor de tonos breves, controlador de avión sin piloto
Beer's law ley de Beer
beginning of file comienzo de archivo
beginning of tape comienzo de cinta
bel bel
bell timbre
bell character carácter de timbre
bell-shaped curve curva acampanada
bell switch conmutador de timbre
bell transformer transformador de timbre
bell wire hilo de timbre
Bellini-Tosi antenna antena de Bellini-Tosi
Bellini-Tosi direction-finder radiogoniómetro de Bellini-Tosi
belt drive transmisión por correa
belt generator generador de correa
bench test prueba de banco
benchmark prueba de referencia
benchmark problem problema de prueba de referencia
benchmark program programa de prueba de referencia
benchmark routine rutina de prueba de referencia
bend codo, curvado
bend amplitude amplitud de codo
bend coupling acoplamiento curvo
bend frequency frecuencia de codo
bending effect efecto de flexión
Benedick effect efecto de Benedick
benito benito
bent antenna antena doblada
bent gun cañón doblado
berkelium berquelio
beryllia berilia
beryllium bronze bronce al berilio
beryllium copper cobre al berilio
beryllium oxide óxido de berilio
Bessel function función de Bessel
beta beta
beta activity actividad beta
beta circuit circuito beta
beta cutoff frequency frecuencia de corte beta
beta decay desintegración beta
beta disintegration desintegración beta
beta emitter emisor beta
beta function función beta
beta particle partícula beta
beta radiation radiación beta
beta ray spectrometer espectrómetro de rayos beta
beta ray spectrum espectro de rayos beta
beta rays rayos beta
beta test prueba beta
beta-to-alpha conversion conversión de beta a alfa
beta transformation transformación beta
beta wave onda beta
betatron betatrón
bevatron bevatrón
Beverage antenna antena de Beverage
beyond-the-horizon circuit circuito transhorizonte
beyond-the-horizon communication comunicación transhorizonte
beyond-the-horizon propagation propagación transhorizonte
bezel bisel
biamplifier system sistema biamplificador

bias polarización, sesgo
bias battery batería de polarización
bias cell pila de polarización
bias compensator compensador de polarización
bias control control de polarización
bias current corriente de polarización
bias current drift deslizamiento de corriente de polarización
bias distortion distorsión de polarización
bias filter filtro de polarización
bias frequency frecuencia de polarización
bias modulation modulación de polarización
bias oscillator oscilador de polarización
bias point punto de polarización
bias potential potencial de polarización
bias power potencia de polarización
bias resistor resistor de polarización
bias source fuente de polarización
bias spectrum espectro de polarización
bias stabilization estabilización de polarización
bias supply fuente de polarización
bias telegraph distortion distorsión telegráfica de polarización
bias voltage tensión de polarización
bias winding arrollamiento de polarización
biased polarizado
biased amplifier amplificador polarizado
biased detector detector polarizado
biased diode diodo polarizado
biased rectifier rectificador polarizado
biased relay relé polarizado
biasing battery batería de polarización
biasing cell pila de polarización
biasing method método de polarización
biasing potential potencial de polarización
biasing resistance resistencia de polarización
biasing resistor resistor de polarización
biaxial biaxial
biaxial loudspeaker altavoz biaxial
biaxial test prueba biaxial
bichromate cell pila de bicromato
biconcave bicóncavo
biconditional bicondicional
biconditional gate puerta bicondicional
biconditional operation operación bicondicional
biconic bicónico
biconical bicónico
biconical antenna antena bicónica
biconical horn antenna antena de bocina bicónica
biconical reflectance reflectancia bicónica
biconical transmittance transmitancia bicónica
biconvex biconvexo
bidirectional bidireccional
bidirectional antenna antena bidireccional
bidirectional buffer almacenamiento intermedio bidireccional
bidirectional bus bus bidireccional
bidirectional bus driver controlador de bus bidireccional
bidirectional clamping circuit circuito de fijación de nivel bidireccional
bidirectional clipping circuit circuito recortador bidireccional
bidirectional counter contador bidireccional
bidirectional coupler acoplador bidireccional
bidirectional current corriente bidireccional
bidirectional data bus bus de datos bidireccional
bidirectional flow flujo bidireccional
bidirectional loudspeaker altavoz bidireccional
bidirectional microphone micrófono bidireccional
bidirectional operation operación bidireccional
bidirectional pin terminal bidireccional
bidirectional printer impresora bidireccional

bidirectional printing impresión bidireccional
bidirectional pulse impulso bidireccional
bidirectional reflectance reflectancia bidireccional
bidirectional relay relé bidireccional
bidirectional signal señal bidireccional
bidirectional signal line línea de señales
  bidireccional
bidirectional thyristor tiristor bidireccional
bidirectional transducer transductor bidireccional
bidirectional transistor transistor bidireccional
bidirectional transmittance transmitancia
  bidireccional
bidirectional triode thyristor tiristor triódico
  bidireccional
bifilar bifilar
bifilar cable cable bifilar
bifilar electrometer electrómetro bifilar
bifilar inductor inductor bifilar
bifilar oscillograph oscilógrafo bifilar
bifilar resistance resistencia bifilar
bifilar resistor resistor bifilar
bifilar suspension suspensión bifilar
bifilar transformer transformador bifilar
bifilar winding arrollamiento bifilar
bifurcated bifurcado
bifurcated contact contacto bifurcado
bifurcated feeder alimentador bifurcado
bifurcation bifurcación
bigrid birrejilla
bihemispherical reflectance reflectancia
  bihemisférica
bihemispherical transmittance transmitancia
  bihemisférica
bilateral bilateral
bilateral amplifier amplificador bilateral
bilateral antenna antena bilateral
bilateral circuit circuito bilateral
bilateral conductivity conductividad bilateral
bilateral control control bilateral
bilateral element elemento bilateral
bilateral impedance impedancia bilateral
bilateral instrument instrumento bilateral
bilateral network red bilateral
bilateral scanning exploración bilateral
bilateral switching conmutación bilateral
bilateral transducer transductor bilateral
billboard antenna antena de reflector plano
billing error error de facturación
billing error probability probabilidad de error de
  facturación
billing report informe de facturación
billing tape cinta de facturación
bimetal aleación bimetálica
bimetallic bimetálico
bimetallic element elemento bimetálico
bimetallic instrument instrumento bimetálico
bimetallic relay relé bimetálico
bimetallic switch conmutador bimetálico
bimetallic thermometer termómetro bimetálico
bimetallic thermostat termostato bimetálico
bimetallic wire hilo bimetálico
bimodal bimodal
bimorph bimorfo
bimorphous bimorfo
bimorphous cell célula bimorfa
binary binario
binary adder sumador binario
binary algebra álgebra binaria
binary alloy aleación binaria
binary arithmetic aritmética binaria
binary arithmetic operation operación aritmética
  binaria
binary automatic computer computadora automática

binaria
binary base base binaria
binary Boolean operation operación de Boole
  binaria
binary capacity capacidad binaria
binary card tarjeta binaria
binary cell celda binaria
binary chain cadena binaria
binary channel canal binario
binary character carácter binario
binary character set conjunto de caracteres binarios
binary circuit circuito binario
binary code código binario
binary-coded de codificación binaria, codificado en
  binario
binary-coded address dirección codificada en
  binario
binary-coded character carácter codificado en
  binario
binary-coded decimal decimal codificado en binario
binary-coded decimal character code código de
  caracteres decimales codificado en binario
binary-coded decimal character set conjunto de
  caracteres decimales codificado en binario
binary-coded decimal code código decimal
  codificado en binario
binary-coded decimal digit dígito decimal codificado
  en binario
binary-coded decimal notation notación decimal
  codificada en binario
binary-coded decimal number número decimal
  codificado en binario
binary-coded decimal system sistema decimal
  codificado en binario
binary-coded digit dígito codificado en binario
binary-coded notation notación codificada en binario
binary-coded octal octal codificado en binario
binary-coded octal notation notación octal
  codificada en binario
binary-coded set conjunto codificado en binario
binary coding codificación binaria
binary column columna binaria
binary component componente binaria
binary configuration configuración binaria
binary converter convertidor binario
binary core núcleo binario
binary counter contador binario
binary data datos binarios
binary-decimal binario-decimal
binary-decimal code código binario-decimal
binary-decimal conversion relay relé de conversión
  binario-decimal
binary decoder descodificador binario
binary device dispositivo binario
binary digit dígito binario
binary digit character carácter de dígito binario
binary digit string secuencia de dígitos binarios
binary divider divisor binario
binary division división binaria
binary element elemento binario
binary element string secuencia de elementos
  binarios
binary equivalent equivalente binario
binary file archivo binario
binary format formato binario
binary function función binaria
binary group grupo binario
binary image imagen binaria
binary image data datos de imagen binarios
binary incremental representation representación
  incremental binaria
binary information información binaria
binary input entrada binaria

binary insertion point coma de inserción binaria, punto de inserción binario
binary item artículo binario
binary language lenguaje binario
binary logic lógica binaria
binary logic element elemento de lógica binario
binary logic system sistema de lógica binario
binary mode modo binario
binary modulation modulación binaria
binary notation notación binaria
binary number número binario
binary number system sistema numérico binario
binary numeral numeral binario
binary numeration system sistema de numeración binario
binary one uno binario
binary operand operando binario
binary operation operación binaria
binary operator operador binario
binary output salida binaria
binary point coma binaria, punto binario
binary position posición binaria
binary preset switch conmutador de preestablecimiento binario
binary quantity cantidad binaria
binary relation relación binaria
binary relay relé binario
binary representation representación binaria
binary scale escala binaria
binary search búsqueda binaria
binary sequence secuencia binaria
binary signal señal binaria
binary sort clasificación binaria
binary stage etapa binaria
binary storage element elemento de almacenamiento binario
binary system sistema binario
binary-to-decimal conversion conversión de binario a decimal
binary-to-decimal converter convertidor de binario a decimal
binary-to-hexadecimal conversion conversión de binario a hexadecimal
binary-to-octal conversion conversión de binario a octal
binary tree árbol binario
binary unit unidad binaria
binary variable variable binaria
binary word palabra binaria
binary zero cero binario
binaural binaural
binaural effect efecto binaural
binaural sound sonido binaural
binder ligador
binding force fuerza ligante
binding post terminal de sujeción
binding screw tornillo de sujeción
binistor binistor
binocular binocular
binode binodo
binomial binomio
binomial antenna array red de antenas binomia
binomial array red binomia
binomial theorem teorema binomio
biochemical cell célula bioquímica
bioelectric bioeléctrico
bioelectric circuit circuito bioeléctrico
bioelectrical bioeléctrico
bioelectricity bioelectricidad
bioelectronics bioelectrónica
bioengineering bioingeniería
bionics biónica
biooelectrogenesis bioelectrogénesis

Biot-Savart Law ley de Biot-Savart
biotelemetry biotelemetría
biotron biotrón
bipartition bipartición
bipartition angle ángulo de bipartición
biphase bifásico
biphase compensation compensación bifásica
biphase current corriente bifásica
biphase half-wave rectifier rectificador de media onda bifásica
bipolar bipolar
bipolar amplifier amplificador bipolar
bipolar circuit circuito bipolar
bipolar code código bipolar
bipolar code violation violación de código bipolar
bipolar coding codificación bipolar
bipolar decoder descodificador bipolar
bipolar device dispositivo bipolar
bipolar electrode electrodo bipolar
bipolar electrode system sistema de electrodos bipolares
bipolar fuse fusible bipolar
bipolar generator generador bipolar
bipolar input entrada bipolar
bipolar integrated circuit circuito integrado bipolar
bipolar line línea bipolar
bipolar machine máquina bipolar
bipolar operation operación bipolar
bipolar pulse impulso bipolar
bipolar receiver receptor bipolar
bipolar semiconductor semiconductor bipolar
bipolar signal señal bipolar
bipolar transistor transistor bipolar
bipolar transmission transmisión bipolar
bipolar video video bipolar
bipolar violation violación bipolar
bipolar voltmeter vóltmetro bipolar
bipolar winding arrollamiento bipolar
bipolarity bipolaridad
biquinary biquinario
biquinary code código biquinario
biquinary notation notación biquinaria
biquinary number número biquinario
biquinary numeration system sistema de numeración biquinario
birdie silbido
birefringence birrefrigencia
Birmingham wire gage calibre de alambres de Birmingham
bisector bisector
bismuth bismuto
bismuth cathode cátodo de bismuto
bismuth fluxmeter flujómetro de bismuto
bismuth thermocouple termopar de bismuto
bispheric biesférico
bispherical biesférico
bistable biestable
bistable amplifier amplificador biestable
bistable circuit circuito biestable
bistable component componente biestable
bistable device dispositivo biestable
bistable element elemento biestable
bistable logic function función lógica biestable
bistable multivibrator multivibrador biestable
bistable operation operación biestable
bistable relaxation circuit circuito de relajación biestable
bistable relay relé biestable
bistable switching circuit circuito conmutador biestable
bistable trigger circuit circuito de disparo biestable
bistable unit unidad biestable
bistate de dos estados

bistatic biestático
bistatic cross-section sección transversal biestática
bistatic radar radar biestático
bisymmetric bisimétrico
bit bit
bit check comprobación de bit
bit combination combinación de bits
bit configuration configuración de bits
bit density densidad de bits
bit error error de bit
bit error probability probabilidad de error de bit
bit error rate tasa de errores de bits
bit-error ratio razón de errores a bits
bit location localización de bit
bit manipulation manipulación de bit
bit map mapa de bits
bit memory organization organización de memoria
    de bits
bit parity paridad de bit
bit pattern patrón de bits
bit position posición de bit
bit rate velocidad de bits
bit retention time tiempo de retención de bits
bit sampling muestreo de bits
bit slice rebanada de bits
bit storage almacenamiento de bits
bit stream corriente de bits, flujo de bits
bit stream transmission transmisión de corriente de
    bits, transmisión de flujo de bits
bit string secuencia de bits
bit time tiempo de bit
bit transfer rate velocidad de transferencia de bits
bitonic bitónico
bits per inch bits por pulgada
bits-second bits-segundo
bivalent bivalente
Bjerknes equation ecuación de Bjerknes
black after white negro tras blanco
black annealed wire hilo recocido en negro
black box caja negra
black-box concept concepto de caja negra
black-box model modelo de caja negra
black compression compresión de negro
black level nivel de negro
black level alignment alineación del nivel de negro
black level control control de nivel de negro
black level stability estabilidad de nivel de negro
black recording registro negro
black saturation saturación de negro
black signal señal negra
black spot punto negro
black transmission transmisión negra
blackbody cuerpo negro
blackbody radiation radiación de cuerpo negro
blackout apagón, supresión de haz
blackout signal señal de supresión de haz
blackout voltage tensión de supresión de haz
blank en blanco, cristal en blanco
blank area área en blanco
blank card tarjeta en blanco
blank cell celda en blanco
blank character carácter en blanco
blank column columna en blanco
blank diskette disquete en blanco
blank level nivel de borrado
blank line línea en blanco
blank medium medio en blanco
blank segment segmento en blanco
blank tape cinta en blanco
blank wire hilo en blanco
blanked picture signal señal de imagen borrada
blanketing cubrimiento
blanking extinción, borrado

blanking amplifier amplificador de extinción
blanking circuit circuito de extinción
blanking generator generador de extinción
blanking interval intervalo de extinción
blanking level nivel de extinción
blanking pedestal pedestal de extinción
blanking period periodo de extinción
blanking pulse impulso de extinción
blanking signal señal de extinción
blanking time tiempo de extinción
blasting sobrecarga
bleaching blanqueo
bleeder resistor de drenaje
bleeder current corriente de drenaje
bleeder divider divisor de drenaje
bleeder power potencia de drenaje
bleeder resistance resistencia de drenaje
bleeder resistor resistor de drenaje
bleeder temperature temperatura de drenaje
blemish imperfección
blind approach aproximación ciega
blind approach beacon radiofaro de aproximación
    ciega
blind approach beacon system sistema de radiofaro
    de aproximación ciega
blind flight vuelo ciego
blind landing aterrizaje ciego
blind navigation navegación ciega
blind reception recepción ciega
blind sector sector ciego
blind speed velocidad ciega
blind spot punto ciego
blind study estudio ciego
blind takeoff despegue ciego
blind transmission transmisión ciega
blind zone zona ciega
blinking parpadeo
blinking lamp lámpara parpadeante
blinking light luz parpadeante
blip señal visual
blip-scan ratio razón de señal visual a exploración
Bloch equations ecuaciones de Bloch
Bloch functions funciones de Bloch
Bloch theorem teorema de Bloch
Bloch wall pared de Bloch
block bloque
block address dirección de bloque
block allocation asignación de bloque
block chaining encadenamiento de bloque
block character carácter de bloque
block check comprobación por bloques
block check character carácter de comprobación por
    bloques
block code código de bloque
block coding codificación por bloques
block copy copia de bloque
block definition definición de bloque
block delete eliminación de bloque
block diagram diagrama en bloques
block error error de bloque
block error rate tasa de errores en bloques
block-error ratio razón error-bloques
block format formato de bloque
block gap espacio de bloques
block handler manipulador de bloques
block handling routine rutina de manejo de bloques
block indicator indicador de bloque
block input processing procesamiento de entrada de
    bloques
block length longitud de bloque
block mark marca de bloque
block mode modo de bloque
block move movimiento de bloque

block movement movimiento de bloque
block multiplexer multiplexor de bloque
block multiplexer mode modo de multiplexor de bloque
block operation operación de bloque
block output salida de bloque
block parity paridad de bloque
block post puesto de bloqueo
block prefix prefijo de bloque
block processor procesador de bloque
block protection protección de bloque
block read cycle ciclo de lectura de bloque
block read transaction transacción de lectura de bloque
block separator separador de bloques
block serial number número de serie de bloque
block signal señal de bloqueo
block size tamaño de bloque
block sort clasificación de bloques
block station estación de bloque
block structure estructura de bloque
block synchronization sincronización de bloques
block system sistema de bloqueo
block transfer transferencia de bloque
block transfer function función de transferencia de bloques
block write cycle ciclo de escritura de bloque
block write transaction transacción de escritura de bloque
blocked bloqueado
blocked dial tone tono de marcar bloqueado
blocked grid rejilla bloqueada
blocked impedance impedancia bloqueada
blocked list lista bloqueada
blocked oscillator oscilador bloqueado
blocked record registro bloqueado
blocked resistance resistencia bloqueada
blockette subdivisión de bloque
blocking bloqueo
blocking action acción de bloqueo
blocking battery batería de bloqueo
blocking capacitor capacitor de bloqueo
blocking characteristic característica de bloqueo
blocking circuit circuito de bloqueo
blocking condenser condensador de bloqueo
blocking contact contacto de bloqueo
blocking device dispositivo de bloqueo
blocking factor factor de bloqueo
blocking filter filtro de bloqueo
blocking interference interferencia de bloqueo
blocking interval intervalo de bloqueo
blocking layer capa de bloqueo
blocking magnet imán de bloqueo
blocking mode modo de bloqueo
blocking oscillator oscilador de bloqueo
blocking oscillator circuit circuito de oscilador de bloqueo
blocking oscillator driver excitador de oscilador de bloqueo
blocking oscillator synchronization sincronización de oscilador de bloqueo
blocking period periodo de bloqueo
blocking pulse impulso de bloqueo
blocking ratio razón de bloqueo
blocking relay relé de bloqueo
blocking resistance resistencia de bloqueo
blocking signal señal de bloqueo
blocking system sistema de bloqueo
blocking transformer transformador de bloqueo
blocking voltage tensión de bloqueo
blooming floración
blooper receptor que radia señales
blow fundir

blowback ampliación
blower soplador
blowing fusión, soplo
blowing current corriente de fusión
blowing point punto de fusión
blown fundido
blown fuse fusible fundido
blowout extinción, fusión
blowout coil bobina de extinción
blowout magnet imán de extinción
blue amplifier amplificador de azul
blue beam haz de azul
blue beam corrector corrector del haz de azul
blue box caja azul
blue filter filtro azul
blue gain control control de ganancia de azul
blue glow luminiscencia azul
blue gun cañón de azul
blue restorer restaurador de azul
blue signal señal de azul
blue video voltage tensión de video de azul
blueprint copia heliográfica
blur desenfoque, perturbación
blur factor factor de perturbación
blur level nivel de perturbación
blurred desenfocado
blurring desenfoque
board tablero, cuadro, placa
board signal señal de tablero
bobbin carrete
bobbin core núcleo de carrete
bobbin height altura de carrete
Bode curve curva de Bode
Bode diagram diagrama de Bode
Bode plot gráfica de Bode
body-antenna effect efecto cuerpo-antena
body capacitance capacitancia del cuerpo
body effect efecto del cuerpo
body electrode electrodo conectado al cuerpo
body temperature temperatura de cuerpo
bogey valor medio, eco falso, eco no identificado
bogie valor medio, eco falso, eco no identificado
Bohr magneton magnetón de Bohr
Bohr radius radio de Bohr
boilerplate text texto estandarizado almacenado
boiling ebullición
bolometer bolómetro
bolometer bridge puente de bolómetro
bolometer element elemento de bolómetro
bolometer mount montura de bolómetro
bolometer smoke detector detector de humo de bolómetro
bolometer unit unidad de bolómetro
bolometric bolométrico
bolometric bias polarización bolométrica
bolometric brightness brillo bolométrico
bolometric detector detector bolométrico
bolometric instrument instrumento bolométrico
bolometric power meter medidor de potencia bolométrico
bolometric technique técnica bolométrica
Boltzmann's constant constante de Boltzmann
Boltzmann's principle principio de Boltzmann
bolus bolus
bolus material material de bolus
bombardment bombardeo
bombardment conductivity conductividad por bombardeo
bond enlace, adherencia
bond strength fuerza de adherencia
bonded enlazado, adherido
bonded-barrier transistor transistor de base aleada
bonded motor motor adherido

bonding ligamento, conexión, adhesión
bone-conduction conducción ósea
bone-conduction receiver receptor de conducción ósea
bone-conduction transducer transductor de conducción ósea
book capacitor capacitor de libro
Boolean algebra álgebra de Boole
Boolean approach acercamiento de Boole
Boolean calculator calculadora de Boole
Boolean calculus cálculo de Boole
Boolean data datos de Boole
Boolean equation ecuación de Boole
Boolean expression expresión de Boole
Boolean function función de Boole
Boolean logic lógica de Boole
Boolean operation operación de Boole
Boolean operator operador de Boole
Boolean search búsqueda de Boole
Boolean value valor de Boole
Boolean variable variable de Boole
boom jirafa, reverberación, mástil
boost refuerzo
boost capacitor capacitor de refuerzo
boost charge carga de refuerzo
booster reforzador
booster amplifier amplificador de refuerzo
booster battery batería de refuerzo
booster coil bobina de arranque
booster diode diodo de refuerzo
booster gain ganancia de reforzador
booster transformer transformador auxiliar de retorno
boot manguito aislador, cubierta aislante
bootstrap autocarga, rutina de carga
bootstrap card tarjeta de rutina de carga
bootstrap circuit circuito autoelevador
bootstrap driver excitador autoelevador
bootstrap input program programa de entrada de rutina de carga
bootstrap loader cargador de rutina de carga
bootstrap routine rutina de carga
border borde
border delineation delineación de borde
border detection detección de borde
boresighting alineación óptica
boric acid ácido bórico
bornite bornita
boron boro
both-way bidireccional
both-way circuit circuito bidireccional
both-way communication comunicación bidireccional
both-way operation operación bidireccional
both-way repeater repetidor bidireccional
bottom antenna antena inferior
bottom channel canal inferior
bottom contact contacto inferior
bottom electrode electrodo inferior
bottom surface superficie inferior
bottom-up design diseño de abajo-arriba
bounce rebote
bounce of contacts rebote de contactos
bound charge carga latente
bound electron electrón ligado
boundary límite
boundary alignment alineación de límite
boundary condition condición de límite
boundary defect defecto de límite
boundary light luz de límite
boundary marker marcador de límite
boundary marker beacon radiobaliza de límite
boundary potential potencial de límite

boundary value valor límite
boundary wave onda de límite
bow-tie antenna antena de corbata de lazo
bow-tie test prueba de corbata de lazo
box antenna antena en caja
box diagram diagrama de caja
Boys radiomicrometer radiomicrómetro de Boys
bracket soporte
Bradley detector detector de Bradley
Bragg's equation ecuación de Bragg
Bragg's law ley de Bragg
Bragg angle ángulo de Bragg
Bragg curve curva de Bragg
Bragg diffraction difracción de Bragg
Bragg reflection reflexión de Bragg
Bragg region región de Bragg
Bragg scattering dispersión de Bragg
Bragg spectrometer espectrómetro de Bragg
braid trenza
braided cable cable trenzado
braided wire alambre trenzado
brain waves ondas cerebrales
brake control control de freno
brake light luz de freno
brake test prueba de freno
brake type tipo de freno
braking frenaje, frenado
braking characteristic característica de frenaje
braking circuit circuito de frenaje
braking couple par de frenaje
braking magnet imán de frenaje
braking resistor resistor de frenaje
braking response respuesta de frenaje
braking system sistema de frenaje
braking test prueba de frenaje
braking torque par de frenaje
branch rama, derivación
branch address dirección de derivación
branch cable cable de derivación
branch circuit circuito de derivación
branch conductor conductor de derivación
branch current corriente de derivación
branch impedance impedancia de derivación
branch instruction instrucción de derivación
branch joint conexión de derivación
branch line línea de derivación
branch point nodo
branch voltage tensión de derivación
branched winding arrollamiento ramificado
branching ramificación, derivación
branching filter filtro de ramificación
branching network red de ramificación
brass latón
Braun electroscope electroscopio de Braun
Braun tube tubo de Braun
brazing soldadura fuerte
breadboard base de montaje temporal, montaje de pruebas
breadboard circuit circuito de montaje temporal
breadboard construction construcción de montaje temporal
breadboard model modelo de montaje temporal
break ruptura, abertura, interrupción
break-before-make contacts contactos de abrir antes de cerrar
break contact contacto de ruptura
break current corriente de ruptura
break impulse impulso de ruptura
break-in interrumpir
break jack jack de ruptura
break period periodo de abertura
break spark chispa de ruptura
break time tiempo de abertura

breakaway separación, arranque
breakaway force fuerza de arranque
breakaway friction fricción de arranque
breakaway torque par de arranque
breakdown falla, fallo, ruptura, disrupción
breakdown current corriente de ruptura
breakdown diode diodo de avalancha
breakdown impedance impedancia de avalancha
breakdown resistance resistencia de ruptura
breakdown strength rigidez dieléctrica
breakdown torque par máximo
breakdown voltage tensión de ruptura
breaker ruptor, cortacircuito
breaking ruptura
breaking capacity capacidad de ruptura
breaking current corriente de ruptura
breaking load carga de ruptura
breakover transición conductiva
breakover current corriente de transición conductiva
breakover point punto de transición conductiva
breakover voltage tensión de transición conductiva
breakpoint punto de ruptura, punto de interrupción
breakpoint instruction instrucción de interrupción
breakpoint program programa de interrupción
breakpoint switch conmutador de interrupción
breathing respiración
bremsstrahlung bremsstrahlung
Brewster's angle ángulo de Brewster
Brewster's law ley de Brewster
bridge puente
bridge amplifier amplificador de puente
bridge balance equilibrio de puente
bridge balance control control de equilibrio de puente
bridge circuit circuito en puente
bridge-connected amplifier amplificador de conexión en puente
bridge connection conexión en puente
bridge contact contacto de puente
bridge control control de puente
bridge current corriente de puente
bridge detector detector de puente
bridge diplexer diplexor de puente
bridge duplex dúplex de puente
bridge feedback retroalimentación en puente
bridge fuse fusible de puente
bridge generator generador de puente
bridge indicator indicador de puente
bridge input circuit circuito de entrada de puente
bridge limiter limitador de puente
bridge network red en puente
bridge oscillator oscilador de puente
bridge ratio razón de puente
bridge rectifier rectificador en puente
bridge rectifier circuit circuito rectificador en puente
bridge source fuente de puente
bridge transformer transformador de puente
bridge transition transición de puente
bridge-type impedance meter medidor de impedancia de tipo de puente
bridge-type meter medidor de tipo de puente
bridge-type oscillator oscilador de tipo de puente
bridge-type power meter medidor de potencia de tipo de puente
bridged-T network red en T puenteada
bridging puenteo
bridging amplifier amplificador de puente
bridging connection conexión en puente
bridging coupler acoplador de puente
bridging gain ganancia de puenteo
bridging loss pérdida de puenteo
bridging resistance resistencia en paralelo

Bridgman effect efecto de Bridgman
Bridgman relation relación de Bridgman
bright echo eco brillante
bright emitter cathode cátodo emisor brillante
bright filament tube tubo de filamento brillante
brightness brillo
brightness component componente de brillo
brightness contrast contraste de brillo
brightness control control de brillo
brightness distortion distorsión de brillo
brightness of surface brillo de superficie
brightness ratio razón de brillos
brightness signal señal de brillo
brightness temperature temperatura de brillo
brilliance brillantez, brillo
brilliance control control de brillantez
brilliance modulation modulación de brillantez
brilliancy brillantez
Brillouin function función de Brillouin
Brillouin zone zona de Brillouin
British thermal unit unidad térmica británica
broad impulse impulso ancho
broad response respuesta ancha
broad tuning sintonización ancha
broadband banda ancha
broadband amplification amplificación de banda ancha
broadband amplifier amplificador de banda ancha
broadband antenna antena de banda ancha
broadband channel canal de banda ancha
broadband coaxial cable transmission transmisión por cable coaxial de banda ancha
broadband device dispositivo de banda ancha
broadband electrical noise ruido eléctrico de banda ancha
broadband emission emisión de banda ancha
broadband interference interferencia de banda ancha
broadband noise ruido de banda ancha
broadband radio noise ruido de radio de banda ancha
broadband response spectrum espectro de respuesta de banda ancha
broadband signal señal de banda ancha
broadband system sistema de banda ancha
broadband transmission transmisión de banda ancha
broadband tuning sintonización de banda ancha
broadcast radiodifusión, emisión
broadcast address dirección de radiodifusión
broadcast band banda de radiodifusión
broadcast center centro de radiodifusión
broadcast channel canal de radiodifusión
broadcast data set conjunto de datos de radiodifusión
broadcast interference interferencia de radiodifusión
broadcast message mensaje emitido
broadcast receiver receptor de radiodifusión
broadcast service servicio de radiodifusión
broadcast station estación de radiodifusión
broadcast transmission transmisión de radiodifusión
broadcast transmitter transmisor de radiodifusión
broadcaster radiodifusor
broadcasting radiodifusión
broadside antenna antena de radiación transversal
broadside array red de radiación transversal
Broca galvanometer galvanómetro de Broca
broken circuit circuito abierto
broken tape indicator indicador de cinta rota
bromine bromo
Bronson resistance resistencia de Bronson
bronze bronce
bronze conductor conductor de bronce
bronzewelded soldado con bronce

bronzewelding soldadura con bronce
Brooks variable inductometer inductómetro variable de Brooks
Brown and Sharpe gage calibre de Brown y Sharpe
brownian movement movimiento browniano
brownout reducción de tensión de línea
Bruce antenna antena de Bruce
brush escobilla
brush box caja de escobillas
brush discharge descarga en penacho
brush holder portaescobillas
brush holder arm brazo de portaescobillas
brush holder support soporte de portaescobillas
brush set conjunto de escobillas
brush shift decalaje de escobillas
brush station estación de escobillas
brush voltage tensión de escobillas
brushless sin escobillas
brushless machine máquina sin escobillas
brushless synchronous machine máquina sincrónica sin escobillas
brute force fuerza bruta
brute-force filter filtro no calculado
brute supply alimentación no calculada
bubble chart diagrama en burbujas
bubble memory memoria de burbujas
bubble memory operation operación de memoria de burbujas
bubble sort clasificación por burbujas
bucket compartimiento
bucking oposición
bucking circuit circuito de compensación
bucking coil bobina compensadora
bucking voltage tensión de compensación
buckling alabeo
buffer amortiguador, almacenador intermedio, memoria intermedia, separador
buffer amplifier amplificador separador
buffer battery batería compensadora
buffer capacitor capacitor amortiguador
buffer circuit circuito de almacenador intermedio, circuito separador
buffer condenser condensador amortiguador
buffer delay retardo de memoria intermedia
buffer depletion agotamiento de memoria intermedia
buffer limiter limitador separador
buffer memory memoria intermedia
buffer memory register registro de memoria intermedia
buffer register registro de memoria intermedia
buffer stage etapa separadora
buffer storage almacenamiento intermedio
buffer store almacenamiento intermedio
buffer unit unidad de memoria intermedia
buffering amortiguamiento, almacenamiento intermedio, separación
bug error de programa, parásito, indicador de posición, defecto
bug free sin errores de programa
building-block technique técnica de bloques de construcción
buildup acumulación, aumento
buildup factor factor de acumulación
built-in incorporado
built-in antenna antena incorporada
built-in automatic check comprobación automática incorporada
built-in check comprobación incorporada
built-in font tipo de letra incorporada
built-in function función incorporada
built-in overvoltage protection protección contra sobretensiones incorporada
built-in resistor resistor incorporado

built-in routine rutina incorporada
built-in simulation simulación incorporada
built-in simulator simulador incorporado
built-in test prueba incorporada
built-in test equipment equipo de prueba incorporado
built-in transformer transformador incorporado
bulb ampolla, bombilla
bulge protuberancia
bulk volumen
bulk effect efecto masivo
bulk eraser borrador masivo
bulk memory memoria masiva
bulk storage almacenamiento masivo
bulk storage device dispositivo de almacenamiento masivo
bulk storage unit unidad de almacenamiento masivo
bunched cables cables agrupados
bunched circuit circuito agrupado
bunched electron beam haz de electrones agrupados
buncher resonador agrupador, agrupador
buncher grid rejilla agrupadora
buncher resonator resonador de cavidad de entrada
bunching agrupamiento
bunching angle ángulo de agrupamiento
bunching device dispositivo de agrupamiento
bunching frequency frecuencia de agrupamiento
bunching of electrons agrupamiento de electrones
bunching section sección de agrupamiento
bunching time tiempo de agrupamiento
Bunet's formula fórmula de Bunet
Bunsen cell pila de Bunsen
Bunsen photometer fotómetro de Bunsen
burden carga
burden impedance impedancia de carga
burden regulation regulación de carga
buried cable cable enterrado
burn remanencia de imagen, quemadura
burn-in estabilización, asentamiento
burn-in period periodo de estabilización
burning quemadura
burnisher bruñidor
burnishing bruñido
burnout quemadura, falla
burst incremento repentino, ráfaga
burst duration duración de ráfaga
burst error error por ráfaga
burst gate puerta de sincronización cromática
burst magnitude magnitud de ráfaga
burst transmission transmisión por ráfagas
bus bus, canal, barra colectora
bus arbitration arbitraje de bus
bus circuit circuito de bus
bus conflict conflicto de bus
bus controller controlador de bus
bus cycle ciclo de bus
bus driver controlador de bus
bus interface interfaz de bus
bus line línea de bus
bus master controlador de bus
bus network red de bus
bus operation operación de bus
bus priority structure estructura de prioridades de bus
bus structure estructura de bus
bus system sistema de bus
bus timer temporizador de bus
busbar barra colectora
busbar insulator aislador de barra colectora
busbar relay relé de barra colectora
bushing pasamuros, aislador de entrada
bushing insulator aislador pasamuros
bushing transformer transformador pasamuros

business machine máquina de oficina
business oriented computer computadora orientada a negocios
busy hour hora cargada, hora ocupada
busy line línea cargada, línea ocupada
busy state estado de ocupado
busy test prueba de ocupado
busy time tiempo de ocupado
busy tone tono de ocupado
busy verification verificación de ocupado
Butler oscillator oscilador de Butler
butt welding soldadura a tope
butterfly capacitor capacitor de mariposa
butterfly circuit circuito de mariposa
butterfly condenser condensador de mariposa
butterfly curve curva en mariposa
butterfly oscillator oscilador de mariposa
butterfly resonator resonador de mariposa
butterfly valve válvula de mariposa
Butterworth filter filtro de Butterworth
Butterworth function función de Butterworth
button botón
button capacitor capacitor tipo botón
button microphone micrófono de botón
buzz zumbido
buzzer zumbador
bypass derivación
bypass anode ánodo de derivación
bypass capacitor capacitor de derivación
bypass condenser condensador de derivación
bypass current corriente de derivación
bypass device dispositivo de derivación
bypass filter filtro de derivación
bypass line línea de derivación
bysynchronous bisíncrono
bysynchronous transmission transmisión bisíncrona
byte byte
byte address dirección de byte
byte manipulation manipulación de bytes
byte multiplexing multiplexión de bytes
byte space espacio de byte
bytes per inch bytes por pulgada

C

cabinet caja, gabinete
cable cable
cable accessories accesorios de cables
cable adapter adaptador de cable
cable address dirección cablegráfica
cable armor armadura de cable
cable assembly ensamblaje de cable
cable attenuation atenuación de cable
cable box caja de cable
cable bracket soporte de cable
cable capacitance capacitancia de cable
cable channel canal de cable
cable circuit circuito de cable
cable clamp abrazadera de cable
cable code código de cable
cable communication comunicación por cable
cable compensation compensación por cable
cable complement complemento de cable
cable conductor conductor de cable
cable connection conexión de cable
cable connector conector de cable
cable core alma de cable
cable correction corrección de cable
cable creepage corrimiento de cable
cable delay circuit circuito de retardo por cable
cable distribution point punto de distribución de cable
cable drum tambor de cable
cable duct conducto de cable, conducto portacables
cable end punta de cable
cable entrance entrada de cable
cable entry entrada de cable
cable equalizer ecualizador de cable
cable filler relleno de cable
cable form forma de cable
cable hanger grapa de cable
cable head cabeza de cable
cable insulator aislador de cable
cable jacket chaqueta de cable
cable joint junta de cable
cable laying tendido de cables
cable link enlace por cable
cable loss pérdida de cable
cable Morse code código de Morse para cable
cable network red de cable
cable parameter parámetro de cable
cable penetration penetración de cable
cable pillar pilar de cables
cable post poste de cable
cable powered alimentado por cable
cable rack bastidor de cables
cable ring anillo para cables
cable route ruta de cable
cable run tendido de cable
cable segment segmento de cable
cable separator separador de cable
cable shackle cubierta de cable
cable sheath vaina de cable
cable-sheath insulator aislador de vaina de cable
cable shield blindaje de cable
cable shoe terminal de cable

cable splice empalme de cable
cable splicer empalmador de cables
cable splicing empalme de cable
cable stripper desforrador de cables
cable supporting structure estructura de soporte de
  cables
cable system sistema de cables
cable television televisión por cable
cable terminal terminal de cables
cable terminal box caja de terminal de cables
cable terminator terminador de cables
cable test prueba de cable
cable transfer transferencia por cable
cable tray bandeja de cables
cable tray system sistema de bandeja de cables
cable type tipo de cable
cabled cableado
cablegram cablegrama
cabling cableado
cabling diagram diagrama de cableado
cache memory memoria inmediata
cadmium cadmio
cadmium cell pila de cadmio
cadmium electrode electrodo de cadmio
cadmium lamp lámpara de cadmio
cadmium-plated cadmiado
cadmium-plating cadmiado
cadmium standard cell pila patrón de cadmio
cadmium-sulfide cell célula de sulfuro de cadmio
cadmium-sulfide crystal rectifier rectificador de
  cristal de sulfuro de cadmio
cage jaula
cage antenna antena en jaula
calcium calcio
calculate calcular
calculated address dirección calculada
calculated field campo calculado
calculated value valor calculado
calculating operation operación de cálculo
calculating speed velocidad de cálculo
calculating time tiempo de cálculo
calculating unit unidad de cálculo
calculation specifications especificaciones de cálculo
calculator calculadora
calculator chip chip calculador
calculator mode modo calculador
calculograph calculógrafo
calendar of conversion calendario de conversión
calibrate calibrar
calibrated calibrado
calibrated capacitance capacitancia calibrada
calibrated capacitor capacitor calibrado
calibrated cell pila calibrada
calibrated condenser condensador calibrado
calibrated frequency frecuencia calibrada
calibrated instrument instrumento calibrado
calibrated leak fuga calibrada
calibrated measurement medida calibrada
calibrated meter medidor calibrado
calibrated scale escala calibrada
calibrated signal señal calibrada
calibrated sweep barrido calibrado
calibrating calibrador
calibrating bridge puente calibrador
calibrating circuit circuito calibrador
calibrating marker marca calibradora
calibrating tape cinta calibradora
calibration calibración
calibration accuracy exactitud de calibración
calibration certificate certificado de calibración
calibration circle círculo de calibración
calibration constancy constancia de calibración
calibration constant constante de calibración

calibration curve curva de calibración
calibration cycle ciclo de calibración
calibration error error de calibración
calibration factor factor de calibración
calibration indicator indicador de calibración
calibration interval intervalo de calibración
calibration laboratory laboratorio de calibración
calibration level nivel de calibración
calibration marker marcador de calibración
calibration oscillator oscilador de calibración
calibration procedure procedimiento de calibración
calibration scale escala de calibración
calibration signal señal de calibración
calibration standard patrón de calibración
calibration test prueba de calibración
calibration voltage tensión de calibración
calibration wavelength longitud de onda de
  calibración
calibrator calibrador
calibrator crystal cristal calibrador
californium californio
call llamada
call acceptance aceptación de llamada
call-accepted signal señal de aceptación de llamada
call announcer anunciador de llamada
call attempt intento de llamada
call attempts per hour intentos de llamada por hora
call by name llamar por nombre
call by reference llamar por referencia
call by value llamar por valor
call by variable llamar por variable
call capacity capacidad de llamadas
call channel canal de llamada
call circuit circuito de llamada
call control control de llamadas
call-control character carácter de control de
  llamadas
call-control procedure procedimiento de control de
  llamadas
call-directing code código de direccionamiento de
  llamada
call direction code código de dirección de llamada
call distributor distribuidor de llamada
call duration duración de llamada
call finder buscador de llamada
call forwarding traslado de llamadas
call graph gráfico de llamadas
call identifier identificador de llamadas
call-indicating device dispositivo indicador de
  llamadas
call indicator indicador de llamadas
call instruction instrucción de llamada
call letters letras de llamada
call meter medidor de llamadas
call-not-accepted signal señal de no aceptación de
  llamada
call number número de llamada
call processing procesamiento de llamadas
call progress signal señal de progreso de llamada
call rate tasa de llamadas, tarifa de llamada
call relay relé de llamada
call request signal señal de solicitud de llamada
call response respuesta de llamada
call routine rutina de llamada
call signal señal de llamada
call subroutine subrutina de llamada
call switch conmutador de llamada
call trace rastreo de llamada
call waiting espera de llamadas
called line línea llamada
called party parte llamada
called program programa llamado
calling amplifier amplificador de llamada

calling channel canal de llamada
calling circuit circuito de llamada
calling device dispositivo de llamadas
calling frequency frecuencia de llamada
calling line línea de llamada
calling list lista de llamadas
calling macro macro de llamada
calling program programa de llamada
calling relay relé de llamada
calling repeater circuit circuito repetidor de llamada
calling sequence secuencia de llamada
calling signal señal de llamada
calling test prueba de llamada
calling wave onda de llamada
calomel calomel
calomel electrode electrodo de calomel
calomel half-cell semicelda de calomel
calometric calométrico
calometric power meter medidor de potencia
    calométrico
calorescence calorescencia
calorescent calorescente
calorimeter calorímetro
calorimetric calorimétrico
calorimetric power potencia calorimétrica
Calzecchi-Onesti effect efecto de Calzecchi-Onesti
cambric cambray
camera amplifier amplificador de cámara
camera cable cable de cámara
camera chain cadena de cámara
camera control monitor monitor de control de
    cámara
camera control unit unidad de control de cámara
camera monitor monitor de cámara
camera preamplifier preamplificador de cámara
camera signal señal de cámara
camera storage tube tubo de almacenamiento de
    cámara
camera tube tubo de cámara
camp-on puesta en espera
Campbell's formula fórmula de Campbell
Campbell bridge puente de Campbell
Campbell-Colpitts bridge puente de
    Campbell-Colpitts
canal rays rayos canales
cancel character carácter de cancelación
cancel indicator indicador de cancelación
cancel key tecla de cancelación
canceling signal señal de cancelación
cancellation cancelación
cancellation circuit circuito de cancelación
cancellation control control de cancelación
cancellation ratio razón de cancelación
candela candela
candle bujía
candlepower potencia lumínica de bujía
candoluminescence candoluminiscencia
cannibalization canibalización
capability margin margen de capacidad
capacimeter capacímetro
capacitance capacitancia, capacidad
capacitance altimeter altímetro de capacitancia
capacitance box caja de capacitancias
capacitance bridge puente de capacitancias
capacitance coupling acoplamiento por capacitancia
capacitance current corriente de capacitancia
capacitance diode diodo de capacitancia
capacitance divider divisor por capacitancia
capacitance feedback retroalimentación de
    capacitancia
capacitance filter filtro de capacitancia
capacitance meter medidor de capacitancia
capacitance ratio razón de capacitancias

capacitance relay relé de capacitancia
capacitance unbalance desequilibrio de capacitancias
capacitive capacitivo
capacitive accelerometer acelerómetro capacitivo
capacitive amplifier amplificador capacitivo
capacitive antenna antena capacitiva
capacitive attenuator atenuador capacitivo
capacitive circuit circuito capacitivo
capacitive component componente capacitivo
capacitive connection conexión capacitiva
capacitive coupling acoplamiento capacitivo
capacitive current corriente capacitiva
capacitive diaphragm diafragma capacitivo
capacitive diode diodo capacitivo
capacitive divider divisor capacitivo
capacitive division división capacitiva
capacitive feedback retroalimentación capacitiva
capacitive load carga capacitiva
capacitive potentiometer potenciómetro capacitivo
capacitive probe sonda capacitiva
capacitive reactance reactancia capacitiva
capacitive transducer transductor capacitivo
capacitive tuning sintonización capacitiva
capacitive tuning screw tornillo de sintonización
    capacitiva
capacitive voltage divider divisor de tensión
    capacitivo
capacitive welding soldadura capacitiva
capacitive window ventana capacitiva
capacitor capacitor, condensador
capacitor amplifier amplificador de capacitor
capacitor antenna antena de capacitancia
capacitor bank batería de capacitores
capacitor box caja de capacitores
capacitor checker comprobador de capacitores
capacitor constant constante de capacitor
capacitor decade década de capacitores
capacitor element elemento de capacitor
capacitor equipment equipo de capacitor
capacitor filter filtro de capacitor
capacitor group grupo de capacitores
capacitor-input filter filtro de entrada capacitiva
capacitor installation instalación de capacitor
capacitor leakage fuga de capacitor
capacitor loudspeaker altavoz de capacitor
capacitor microphone micrófono de condensador
capacitor motor motor de condensador
capacitor pickup fonocaptor electrostático
capacitor plate placa de capacitor
capacitor platform plataforma de capacitores
capacitor rack bastidor de capacitores
capacitor read only storage almacenamiento sólo de
    lectura de capacitor
capacitor-resistor unit unidad de capacitor-resistor
capacitor segment segmento de capacitor
capacitor storage almacenamiento por capacitor
capacitor switch conmutador de capacitor
capacitor tester probador de capacitores
capacitor transducer transductor de capacitor
capacitor unit unidad de capacitor
capacitor voltage transformer transformador de
    tensión por capacitor
capacitron capacitrón
capacity capacidad
capacity bridge puente de capacidad
capacity cell célula de capacidad
capacity coupling acoplamiento capacitivo
capacity deviation desviación de capacidad
capacity emergency emergencia de capacidad
capacity factor factor de capacidad
capacity limits límites de capacidad
capacity meter medidor de capacidad
capacity-operated relay relé operado por capacidad

capacity reactance reactancia de capacidad
capacity resources recursos de capacidad
capacity retention retención de capacidad
capacity test prueba de capacidad
capacity unbalance desequilibrio de capacidad
capillary electrometer electrómetro capilar
capristor capristor
capstan cabrestante
capstan shaft eje de cabrestante
captive oscillator oscilador cautivo
capture captura
capture area área de captura
capture coefficient coeficiente de captura
capture effect efecto de captura
capture efficiency eficiencia de captura
capture probability probabilidad de captura
capture ratio razón de captura
capture spot punto de captura
caractron caractrón
carbon carbón
carbon anode ánodo de carbón
carbon arc arco de carbón
carbon-arc lamp lámpara de arco de carbón
carbon brush escobilla de carbón
carbon button cápsula de carbón
carbon-consuming cell pila de consumo de carbón
carbon contact contacto de carbón
carbon-contact pickup fonocaptor de contacto de
    carbón
carbon control control de carbón
carbon dioxide laser láser de dióxido de carbono
carbon dust polvo de carbón
carbon electrode electrodo de carbón
carbon filament filamento de carbón
carbon-filament lamp lámpara de filamento de
    carbón
carbon-film resistor resistor de película de carbón
carbon granules granalla de carbón
carbon-granules microphone micrófono de granalla
    de carbón
carbon-graphite brush escobilla de carbón-grafito
carbon hydrophone hidrófono de carbón
carbon lamp lámpara de carbón
carbon microphone micrófono de carbón
carbon pickup fonocaptor de carbón
carbon-pile regulator regulador de placas de carbón
carbon-pile voltage regulator regulador de tensión
    de placas de carbón
carbon potentiometer potenciómetro de carbón
carbon resistor resistor de carbón
carbon rheostat reóstato de carbón
carbon tetrachloride fuse fusible de tetracloruro de
    carbono
carbon tip punta de carbón
carbon transmitter transmisor de carbón
carbon-zinc cell pila de carbón-cinc
carbonize carbonizar
carbonized carbonizado
carbonized anode ánodo carbonizado
carbonized cloth tela carbonizada
carbonized filament filamento carbonizado
carbonized plate placa carbonizada
carborundum carborundo
carborundum detector detector de carborundo
carcinotron carcinotrón
card tarjeta, ficha
card back dorso de tarjeta
card bed cama de tarjetas
card box caja de tarjetas
card capacity capacidad de tarjeta
card channel canal de tarjetas
card code código de tarjeta
card collator colacionador de tarjetas

card column columna de tarjeta
card-controlled calculator calculadora controlada
    por tarjeta
card-controlled tape punch perforadora de tarjetas
    controlada por tarjeta
card counter contador de tarjetas
card cycle ciclo de tarjeta
card data format formato de datos de tarjeta
card design diseño de tarjeta
card edge borde de tarjeta
card encoder codificador de tarjetas
card face cara de tarjeta
card feed alimentador de tarjetas
card feed attachment dispositivo alimentador de
    tarjetas
card feeding alimentación de tarjetas
card field campo de tarjeta
card file archivo de tarjetas
card fluff pelusa de tarjetas
card format formato de tarjeta
card guide guía de tarjetas
card handling manejo de tarjetas
card hopper almacén de alimentación de tarjetas
card image imagen de tarjeta
card-image file archivo de imágenes de tarjeta
card indexing indexación de tarjetas
card input entrada de tarjetas
card interpreter interpretador de tarjetas
card jamming atascamiento de tarjetas
card lever palanca de tarjetas
card loader cargador de tarjetas
card machine máquina de tarjetas
card matching apareamiento de tarjetas
card module módulo de tarjetas
card-operated equipment equipo operado por
    tarjetas
card operating system sistema de operación de
    tarjetas
card-oriented computer computadora orientada a
    tarjetas
card output salida de tarjetas
card path trayectoria de tarjetas
card printing impresión de tarjetas
card program programa de tarjetas
card-programmed programado por tarjetas
card puller extractor de tarjetas
card punch perforadora de tarjetas
card-punch buffer almacenador intermedio de
    perforadora de tarjetas
card-punch unit unidad de perforadora de tarjetas
card-punching rate velocidad de perforación de
    tarjetas
card reader lectora de tarjetas
card reader unit unidad lectora de tarjetas
card reading station estación de lectura de tarjetas
card receiver receptor de tarjetas
card record registro de tarjeta
card reproducer reproductor de tarjetas
card reversing device dispositivo de inversión de
    tarjetas
card row fila de tarjetas
card sensing detección de tarjetas
card set conjunto de tarjetas
card slot ranura de tarjetas
card sorter clasificadora de tarjetas
card sorting machine máquina clasificadora de
    tarjetas
card speed velocidad de tarjetas
card stacker acumulador de tarjetas
card system sistema de tarjetas
card system format formato de sistema de tarjetas
card tester probador de tarjetas
card-to-card tarjeta a tarjeta

card-to-disk tarjeta a disco
card-to-disk conversion conversión de tarjeta a disco
card-to-disk converter convertidor de tarjeta a disco
card-to-magnetic tape converter convertidor de tarjeta a cinta magnética
card-to-tape tarjeta a cinta
card-to-tape conversion conversión de tarjeta a cinta
card-to-tape converter convertidor de tarjeta a cinta
card track pista de tarjetas
card tray bandeja de tarjetas
card type tipo de tarjeta
card verification verificación de tarjetas
card verifier verificador de tarjetas
card verifying verificación de tarjetas
card wreck atascamiento de tarjetas
cardiac monitor monitor cardíaco
cardiac pacemaker marcapasos cardíaco
cardiac stimulator estimulador cardíaco
cardiogram cardiograma
cardiograph cardiógrafo
cardioid cardioide
cardioid diagram diagrama cardioide
cardioid microphone micrófono cardioide
cardioid pattern patrón cardioide
cardioid reception recepción en cardioide
cardiotachometer cardiotacómetro
cards per minute tarjetas por minuto
Carey-Foster bridge puente de Carey-Foster
Carey-Foster mutual inductance bridge puente de inductancia mutua de Carey-Foster
carnauba wax cera de carnauba
Carnot theorem teorema de Carnot
carriage control control de carro
carriage return retorno de carro
carriage return signal señal de retorno de carro
carrier portadora, portador
carrier amplifier amplificador de portadora
carrier amplitude amplitud de portadora
carrier-amplitude regulation regulación de amplitud de portadora
carrier beat batido de portadora
carrier beating batido de portadora
carrier channel canal de portadora
carrier choke choque de portadora
carrier chrominance signal señal de crominancia de portadora
carrier circuit circuito de portadora
carrier color signal señal de color de portadora
carrier communication comunicación por portadora
carrier-communication equipment equipo de comunicación por portadora
carrier component componente de portadora
carrier concentration concentración de portadores
carrier conditioner acondicionador de portadora
carrier control control de portadora
carrier-controlled modulation modulación controlada por portadora
carrier current corriente de portadora
carrier-current channel canal de corriente portadora
carrier-current communication comunicación por corriente portadora
carrier-current control control por corriente portadora
carrier-current protection protección por corriente portadora
carrier-current receiver receptor de corriente portadora
carrier-current relay relé de corriente portadora
carrier-current telegraphy telegrafía por corriente portadora
carrier-current telephone circuit circuito telefónico por corriente portadora
carrier-current transmitter transmisor de corriente portadora

carrier density densidad de portadores
carrier deviation desviación de portadora
carrier-deviation meter medidor de desviación de portadora
carrier dispersion dispersión de portadora
carrier elimination eliminación de portadora
carrier-elimination filter filtro de eliminación de portadora
carrier equipment equipo de portadora
carrier filter filtro de portadora
carrier frequency frecuencia de portadora
carrier-frequency pulse impulso de frecuencia portadora
carrier-frequency range intervalo de frecuencia portadora
carrier-frequency shift desplazamiento de frecuencia portadora
carrier-frequency spacing espaciado de frecuencias portadoras
carrier-frequency stability estabilidad de frecuencia portadora
carrier-frequency stereo disk disco estereofónico de frecuencia portadora
carrier group grupo de portadoras
carrier hum zumbido de portadora
carrier injection inyección de portadores
carrier leak fuga de portadora
carrier level nivel de portadora
carrier lifetime vida de portadores
carrier line línea de portadora
carrier loading carga de portadora
carrier mobility movilidad de portadores
carrier modulation modulación de portadora
carrier multiplexing system sistema de multiplexión por portadora
carrier noise ruido de portadora
carrier noise level nivel de ruido de portadora
carrier-operated relay relé operado por portadora
carrier oscillator oscilador de portadora
carrier output salida de portadora
carrier population población de portadores
carrier power potencia de portadora
carrier power output salida de potencia de portadora
carrier primary flow flujo primario de portadores
carrier recovery recuperación de portadora
carrier repeater repetidor de portadora
carrier sense sentido de portadora
carrier shift desplazamiento de portadora
carrier shift indicator indicador de desplazamiento de portadora
carrier signal señal de portadora
carrier signaling señalización de portadora
carrier spacing espaciado de portadoras
carrier storage almacenamiento de portadores
carrier storage time tiempo de almacenamiento de portadores
carrier subgroup subgrupo de portadoras
carrier supply fuente de portadores
carrier suppression supresión de portadora
carrier swing oscilación de portadora
carrier synchronization sincronización de portadora
carrier system sistema de portadora
carrier telegraphy telegrafía por portadora
carrier telephone channel canal telefónico de portadora
carrier telephone circuit circuito telefónico de portadora
carrier telephony telefonía por portadora
carrier terminal terminal de portadora
carrier terminal equipment equipo de terminal de portadora
carrier-to-noise ratio razón de portadora a ruido

carrier transfer filter filtro de transferencia de portadora
carrier transmission transmisión por portadora
carrier transport transporte de portadores
carrier velocity velocidad de portadores
carrier voltage tensión de portadora
carrier wave onda portadora
carrier-wave supply fuente de ondas portadoras
carrier-wave telecommunication telecomunicación por ondas portadoras
carrier-wave telephone terminal terminal telefónica de onda portadora
carrier-wave terminal terminal de onda portadora
carrier-wave transmission transmisión por ondas portadoras
carry digit dígito de arrastre, dígito de acarreo
carry feedback retroalimentación de arrastre
carry register registro de arrastre
carry time tiempo de arrastre
carrying capacity capacidad de corriente
Cartesian coordinate system sistema de coordenadas cartesianas
Cartesian coordinates coordenadas cartesianas
Cartesian plane plano cartesiano
cartridge cartucho
cartridge diode diodo de cartucho
cartridge drive unidad de cartucho
cartridge drive system sistema de unidad de cartucho
cartridge fuse fusible de cartucho
cartridge size tamaño de cartucho
cascade cascada
cascade amplifier amplificador en cascada
cascade capacitor capacitor en cascada
cascade carry arrastre en cascada
cascade connection conexión en cascada
cascade control control en cascada
cascade control system sistema de control en cascada
cascade converter convertidor en cascada
cascade image tube tubo de imagen en cascada
cascade junction unión en cascada
cascade limiter limitador en cascada
cascade network red en cascada
cascade noise ruido en cascada
cascade rectifier circuit circuito de rectificadores en cascada
cascade refrigeration system sistema de refrigeración en cascada
cascade set conjunto en cascada
cascade sort clasificación en cascada
cascade tube tubo en cascada
cascade voltage doubler doblador de tensión en cascada
cascaded amplifier amplificador en cascada
cascading conexión en cascada
cascode cascodo
cascode amplifier amplificador cascodo
cascode circuit circuito cascodo
cascode tuner sintonizador cascodo
case caja
case capacitance capacitancia de caja
case temperature temperatura de caja
Cassegrain antenna antena de Cassegrain
Cassegrain feed alimentación de Cassegrain
Cassegrain reflector antenna antena con reflector de Cassegrain
cassette casete
cassette drive unidad de casete
cassette interface interfaz de casete
cassette recorder magnetófono, grabadora de casete, registrador de casete
cassette tape cinta de casete

cassette unit unidad de casete
castor oil aceite de castor
cat's whisker electrodo en bigote de gato
catacoustics catacústica
catadioptric catadióptrico
catalog file archivo de catálogo
cataloged data set conjunto de datos catalogados
cataloged procedure procedimiento catalogado
catalysis catálisis
catalyst catalizador
catalytic catalítico
catalytic agent agente catalítico
cataphoresis cataforesis
catastrophic error error catastrófico
catastrophic failure falla catastrófica
catcher diode diodo limitador
catcher space espacio de captación
category storage almacenamiento por categorías
catelectrotonus catelectrotono
catenary aerial cable cable aéreo catenario
catenary contact line línea de contacto catenaria
catenary suspension suspensión catenaria
cathamplifier catamplificador
cathautograph catautógrafo
cathetron catetrón
cathode cátodo
cathode activity actividad catódica
cathode-anode capacitance capacitancia cátodo-ánodo
cathode beam haz catódico
cathode bias polarización catódica
cathode-bias resistor resistor de polarización catódica
cathode bombardment bombardeo catódico
cathode border límite catódico
cathode bypass desvío de cátodo
cathode bypass capacitor capacitor de desvío de cátodo
cathode cleaning limpieza catódica
cathode coating revestimiento catódico
cathode-coating impedance impedancia de revestimiento catódico
cathode compensation compensación catódica
cathode contamination contaminación catódica
cathode corrosion corrosión catódica
cathode-coupled amplifier amplificador acoplado por cátodo
cathode-coupled multivibrator multivibrador acoplado por cátodo
cathode-coupled oscillator oscilador acoplado por cátodo
cathode-coupled phantastron fantastrón acoplado por cátodo
cathode coupling acoplamiento por cátodo
cathode current corriente catódica
cathode-current density densidad de corriente catódica
cathode dark region región obscura catódica
cathode dark space espacio obscuro catódico
cathode deposit depósito catódico
cathode diameter diámetro catódico
cathode disintegration desintegración catódica
cathode drop caída catódica
cathode efficiency eficiencia catódica
cathode element elemento catódico
cathode emission emisión catódica
cathode evaporation evaporación catódica
cathode face cara catódica
cathode fall caída catódica
cathode fall of potential caída de potencial catódica
cathode-filament capacitance capacitancia cátodo-filamento
cathode follower seguidor catódico

cathode glow luminosidad catódica
cathode grid rejilla catódica
cathode-grid capacitance capacitancia cátodo-rejilla
cathode heating rate velocidad de calentamiento catódico
cathode heating time tiempo de calentamiento catódico
cathode hum zumbido catódico
cathode impedance impedancia catódica
cathode injection inyección catódica
cathode-interface capacitance capacitancia de interfaz catódica
cathode-interface impedance impedancia de interfaz catódica
cathode-interface resistance resistencia de interfaz catódica
cathode keying manipulación por cátodo
cathode layer capa catódica
cathode-loaded amplifier amplificador cargado por el cátodo
cathode loading factor factor de carga catódica
cathode luminescence luminiscencia catódica
cathode luminous sensitivity sensibilidad luminosa catódica
cathode-modulated amplifier amplificador modulado por el cátodo
cathode modulation modulación por cátodo
cathode neutralization neutralización catódica
cathode poisoning envenenamiento catódico
cathode potential fall caída de potencial catódica
cathode preheating time tiempo de precalentamiento catódico
cathode protection protección de cátodo
cathode pulse modulation modulación de impulsos por cátodo
cathode-ray current corriente de rayos catódicos
cathode-ray direction-finder radiogoniómetro de rayos catódicos
cathode-ray display visualizador de rayos catódicos
cathode-ray furnace horno de rayos catódicos
cathode-ray indicator indicador de rayos catódicos
cathode-ray lamp lámpara de rayos catódicos
cathode-ray oscillograph oscilógrafo de rayos catódicos
cathode-ray oscilloscope osciloscopio de rayos catódicos
cathode-ray oscilloscope tube tubo de osciloscopio de rayos catódicos
cathode-ray screen pantalla de rayos catódicos
cathode-ray storage almacenamiento de rayos catódicos
cathode-ray storage tube tubo de almacenamiento por rayos catódicos
cathode-ray television tube tubo de televisión de rayos catódicos
cathode-ray tube tubo de rayos catódicos
cathode-ray tube display unit unidad visualizadora de tubo de rayos catódicos
cathode-ray tube function key tecla de función de tubo de rayos catódicos
cathode-ray tube memory memoria de tubo de rayos catódicos
cathode-ray tube projector proyector de tubo de rayos catódicos
cathode-ray tube screen pantalla de tubo de rayos catódicos
cathode-ray tube storage almacenamiento de tubo de rayos catódicos
cathode-ray tube terminal terminal de tubo de rayos catódicos
cathode-ray tuning indicator indicador de sintonización de rayos catódicos
cathode-ray voltmeter vóltmetro de rayos catódicos

cathode-rays rayos catódicos
cathode regeneration regeneración catódica
cathode region región catódica
cathode resistor resistor catódico
cathode screen pantalla catódica
cathode sleeve manguito catódico
cathode spot punto catódico
cathode sputtering pulverización catódica
cathode terminal terminal catódica
cathode voltage tensión catódica
cathodic catódico
cathodic amplifier amplificador catódico
cathodic bombardment bombardeo catódico
cathodic cleaning limpieza catódica
cathodic corrosion corrosión catódica
cathodic evaporation evaporación catódica
cathodic polarization polarización catódica
cathodic protection protección catódica
cathodic reaction reacción catódica
cathodofluorescence catodofluorescencia
cathodograph catodógrafo
cathodoluminescence catodoluminiscencia
cathodophosphorescence catodofosforescencia
cathodyne catodino
catholyte católito
cation catión
catwhisker electrodo en bigote de gato
caustic-soda cell pila de soda cáustica
caustic-soda electrolyte electrólito de soda cáustica
cavitation cavitación
cavitation damage daño por cavitación
cavitation noise ruido de cavitación
cavity cavidad
cavity coupling acoplamiento de cavidad
cavity filter filtro de cavidad
cavity frequency meter frecuencímetro de cavidad
cavity impedance impedancia de cavidad
cavity magnetron magnetrón de cavidades
cavity maser máser de cavidad
cavity oscillator oscilador de cavidades
cavity radiation radiación de cavidad
cavity ratio razón de cavidad
cavity resonance resonancia de cavidad
cavity resonator resonador de cavidad
cavity-type diode amplifier amplificador de diodo de tipo de cavidad
cavity-type resonator resonador de tipo de cavidad
cavity wavemeter ondámetro de cavidad
ceiling máximo
ceiling current corriente máxima
ceiling voltage tensión máxima
cell célula, celda, pila
cell address dirección de celda
cell array red de celdas
cell cavity cavidad de célula
cell constant constante de celda
cell definition definición de celda
cell format formato de celda
cell protection protección de celda
cell reference referencia de celda
cell-type tube tubo de tipo de célula
cell voltage tensión de celda
cellular telephone teléfono celular
celluloid celuloide
cellulose acetate acetato de celulosa
cellulose acetate disk disco de acetato de celulosa
cellulose acetate tape cinta de acetato de celulosa
cellulose nitrate nitrato de celulosa
cellulose nitrate disk disco de nitrato de celulosa
Celsius scale escala Celsius
center expansion expansión central
center feed alimentación central
center-feed antenna antena de alimentación central

center-feed linear antenna antena lineal de alimentación central
center frequency frecuencia central
center-frequency delay retardo de frecuencia central
center-loaded antenna antena cargada por el centro
center loading carga central
center of beam centro de haz
center of channel centro de canal
center of distribution centro de distribución
center of gravity centro de gravedad
center of mass centro de masa
center of radiation centro de radiación
center-stable relay relé estable en posición central
center-tap derivación central
center-tap keying manipulación por derivación central
center-tap modulation modulación por derivación central
center tapped con derivación central
center-tapped coil bobina con derivación central
center-tapped filament filamento con derivación central
center-tapped inductor inductor con derivación central
center-tapped potentiometer potenciómetro con derivación central
center-tapped resistor resistor con derivación central
center-tapped transformer transformador con derivación central
center-tapped winding arrollamiento con derivación central
center-zero meter medidor con cero central
centering circuit circuito de centrado
centering control control de centrado
centering current corriente de centrado
centering current unit unidad de corriente de centrado
centering potentiometer potenciómetro de centrado
centerline línea central
centi centi
centigrade scale escala centígrada
centimeter-gram-second system sistema cegesimal
centimetric waves ondas centimétricas
centipoise centipoise
centisecond centisegundo
central alarm station estación de alarmas central
central battery batería central
central-battery signaling señalización por batería central
central computer computadora central
central control control central
central control room sala de control central
central control unit unidad de control central
central database base de datos central
central distributed system sistema distribuido central
central information file archivo de información central
central marker marcador central
central mass storage almacenamiento masivo central
central memory memoria central
central processing unit unidad de procesamiento central
central processing unit element elemento de unidad de procesamiento central
central processing unit expander expansor de unidad de procesamiento central
central processing unit operation operación de unidad de procesamiento central
central processing unit processor procesador de unidad de procesamiento central
central processing unit time tiempo de unidad de procesamiento central

central processor procesador central
central-processor limited limitado por procesador central
central processor organization organización del procesador central
central processor unit unidad del procesador central
central ray rayo central
central rectifier system sistema de rectificación central
central services module módulo de servicios central
central station estación central
central station equipment equipo de estación central
central storage almacenamiento central
central subsystem subsistema central
central terminal terminal central
central terminal unit unidad terminal central
central unit unidad central
centralized computer network red de computadoras centralizada
centralized control control centralizado
centralized data processing procesamiento de datos centralizado
centralized network red centralizada
centralized network configuration configuración de red centralizada
centralized power source fuente de energía centralizada
centralized processing procesamiento centralizado
centralized system sistema centralizado
centralized traffic control control de tráfico centralizado
centralograph centralógrafo
centrifugal force fuerza centrífuga
centrifugal relay relé centrífugo
centrifugal switch conmutador centrífugo
centrifugation centrifugación
centrifugation potential potencial de centrifugación
centripetal force fuerza centrípeta
ceramic amplifier amplificador cerámico
ceramic-based microcircuit microcircuito basado en cerámica
ceramic capacitor capacitor cerámico
ceramic cartridge cartucho cerámico
ceramic chip chip cerámico
ceramic condenser condensador cerámico
ceramic core núcleo cerámico
ceramic crystal transducer transductor de cristal cerámico
ceramic dielectric dieléctrico cerámico
ceramic dielectric capacitor capacitor de dieléctrico cerámico
ceramic diode diodo cerámico
ceramic disk disco cerámico
ceramic element elemento cerámico
ceramic element microphone micrófono con elemento cerámico
ceramic ferroelectric ferroeléctrico cerámico
ceramic filter filtro cerámico
ceramic insulator aislador cerámico
ceramic magnet imán cerámico
ceramic microphone micrófono cerámico
ceramic phonographic cartridge cartucho fonográfico cerámico
ceramic pickup fonocaptor cerámico
ceramic piezoelement piezoelemento cerámico
ceramic relay relé cerámico
ceramic resistor resistor cerámico
ceramic-to-metal seal junta hermética de cerámica a metal
ceramic transducer transductor cerámico
ceramic tube tubo cerámico
Cerenkov radiation radiación de Cerenkov

ceresin wax cera de ceresina
cerium cerio
cermet cermet
cermet resistor resistor cermet
certification accuracy exactitud de certificación
certification tests pruebas de certificación
certified design diseño certificado
certified tape cinta certificada
certified unit unidad certificada
cesium cesio
cesium-beam tube tubo de haz de cesio
cesium cell célula de cesio
cesium diode diodo de cesio
cesium electron tube tubo electrónico de cesio
cesium hollow cathode cátodo hueco de cesio
cesium phototube fototubo de cesio
cesium thermionic converter convertidor
    termoiónico de cesio
cesium-vapor lamp lámpara de vapor de cesio
cesium-vapor rectifier rectificador de vapor de cesio
chad residuos de perforación
chad box caja de residuos de perforación
chad duct conducto de residuos de perforación
chadded tape cinta con residuos de perforación
chadless tape cinta con perforación parcial
chaff tiras metálicas antirradar
chain amplifier amplificador en cadena
chain broadcasting radiodifusión en cadena
chain calculation cálculo en cadena
chain code código de cadena
chain error error de cadena
chain link record registro de enlace de cadena
chain list lista en cadena
chain of transmitters cadena de transmisores
chain printer impresora en cadena
chain printer unit unidad impresora en cadena
chain radar system sistema de radar en cadena
chain reaction reacción en cadena
chain switch conmutador de cadena
chained files archivos encadenados
chained list lista encadenada
chained-list search búsqueda de lista encadenada
chained procedures procedimientos encadenados
chained program programa encadenado
chained record registro encadenado
chained sector sector encadenado
chaining search búsqueda en cadena
change bit bit de cambio
change character carácter de cambio
change detection detección de cambios
change direction protocol protocolo de cambio de
    dirección
change dump vaciado de cambios
change file archivo de cambios
change of control cambio de control
change record registro de cambios
change sign signo de cambio
change tape cinta de cambios
changeover contact contacto de conmutación
changeover frequency frecuencia de transición
changeover time tiempo de conmutación
changer cambiador
channel canal
channel access acceso de canal
channel activity comparator comparador de
    actividad de canal
channel adapter adaptador de canales
channel address dirección de canal
channel address word palabra de dirección de canal
channel allocation asignación de canales
channel analyzer analizador de canales
channel assignment asignación de canales
channel balance equilibrio de canales

channel balance control control de equilibrio de
    canales
channel bandwidth ancho de banda de canal
channel bank banco de canales
channel busy tone tono de ocupado de canal
channel calibration calibración de canal
channel capacity capacidad de canal
channel carrier portadora de canal
channel-carrier frequency frecuencia de portadora
    de canal
channel-carrier generator generador de portadora
    de canal
channel-carrier oscillator oscilador de portadora de
    canal
channel center frequency frecuencia central de canal
channel check comprobación de canal
channel combiner combinador de canales
channel command mando de canal
channel command word palabra de mando de canal
channel connector conector de canales
channel control block bloque de control de canal
channel control field campo de control de canal
channel control unit unidad de control de canal
channel control word palabra de control de canal
channel controller controlador de canales
channel crystal cristal de canal
channel demodulator desmodulador de canal
channel designator designador de canal
channel distribution distribución de canales
channel effect efecto de canal
channel entry entrada de canal, selección de canal
channel equipment equipo de canal
channel filter filtro de canal
channel frequency frecuencia de canal
channel group grupo de canales
channel handover traspaso de canal
channel indicator indicador de canal
channel input-output command mando de
    entrada-salida de canal
channel lights luces de canal
channel loading carga de canal
channel manager administrador de canal
channel mixer mezclador de canales
channel modulation modulación de canal
channel modulator modulador de canal
channel multiplexing multiplexión de canales
channel of approach canal de acercamiento
channel overload sobrecarga de canal
channel pass filter filtro de paso de canal
channel program programa de canal
channel program block bloque de programa de canal
channel program translation traducción de
    programa de canal
channel pulse impulso de canal
channel quality indicator indicador de calidad de
    canal
channel reliability confiabilidad de canal
channel request solicitud de canal
channel restorer restaurador de canales
channel reversal inversión de canales
channel sampling rate velocidad de muestreo de
    canal
channel select signal señal de selección de canal
channel select time tiempo de selección de canal
channel selector selector de canales
channel selector unit unidad selectora de canales
channel sensitivity sensibilidad de canal
channel separation separación de canales
channel set conjunto de canales
channel shift cambio de canal
channel shifter cambiador de canal
channel spacing espaciado de canales
channel status condition condición de estado de

canal
channel status field campo de estado de canal
channel status routine rutina de estado de canal
channel status table tabla de estado de canal
channel status word palabra de estado de canal
channel supergroup supergrupo de canales
channel synchronizer sincronizador de canales
channel terminal equipment equipo terminal de canal
channel-to-channel adapter adaptador de canal a canal
channel-to-channel connection conexión de canal a canal
channel unit unidad de canal
channel-utilization index índice de utilización de canal
channel waiting queue cola de espera de canales
channel wave onda de canal
channel width anchura de canal
channel work area área de trabajo de canal
channeling canalización
channeling equipment equipo de canalización
channelizing canalización
Chaoul tube tubo de Chaoul
Chapman region región de Chapman
chapron winding arrollamiento chaprón
character carácter
character address dirección de carácter
character adjustment ajuste de carácter
character arrangement table tabla de arreglo de caracteres
character attribute atributo de carácter
character-based program programa basado en caracteres
character boundary límite de carácter
character box caja de caracteres
character cell celda de carácter
character check comprobación de caracteres
character code código de carácter
character coded codificado en caracteres
character constant constante de caracteres
character counter contador de caracteres
character crowding apiñamiento de caracteres
character cycle ciclo de carácter
character cylinder cilindro de caracteres
character data datos de caracteres
character density densidad de caracteres
character design diseño de carácter
character device dispositivo de caracteres
character display visualización de caracteres
character display device dispositivo de visualización de caracteres
character distortion distorsión de carácter
character edge borde de carácter
character element elemento de carácter
character emitter emisor de caracteres
character-error rate tasa de errores de caracteres
character-error ratio razón de errores de caracteres
character expansion expansión de carácter
character expression expresión de caracteres
character field campo de caracteres
character font tipo de carácter
character form forma de carácter
character generation generación de caracteres
character generator generador de caracteres
character graphics gráficos de caracteres
character group grupo de caracteres
character identifier identificador de caracteres
character impact printer impresora de impacto de caracteres
character increment incremento de carácter
character interval intervalo de caracteres
character key tecla de carácter

character map mapa de caracteres
character matrix matriz de carácter
character mode modo de carácter
character modifier modificador de carácter
character operator operador de caracteres
character-oriented orientado a caracteres
character-oriented computer computadora orientada a caracteres
character-oriented display visualización orientada a caracteres
character-oriented protocol protocolo orientado a caracteres
character parity paridad de caracteres
character pattern patrón de caracteres
character period periodo de caracteres
character phase fase de caracteres
character position posición de caracteres
character printer impresor de caracteres
character properties propiedades de caracteres
character reader lectora de caracteres
character reading lectura de caracteres
character recognition reconocimiento de caracteres
character recognition device dispositivo de reconocimiento de caracteres
character reference point punto de referencia de caracteres
character relation relación de caracteres
character row fila de caracteres
character sensing detección de caracteres
character set conjunto de caracteres
character signal señal de carácter
character size control control de tamaño de caracteres
character space espacio de carácter
character spacing espaciado de caracteres
character string secuencia de caracteres
character subset subconjunto de caracteres
character table tabla de caracteres
character transfer rate velocidad de transferencia de caracteres
character type tipo de carácter
character variable variable de carácter
character wheel rueda de caracteres
characteristic característica
characteristic acoustic impedance impedancia acústica característica
characteristic angle ángulo característico
characteristic anode voltage tensión anódica característica
characteristic band banda característica
characteristic boundary límite característico
characteristic conductivity conductividad característica
characteristic constant constante característica
characteristic curve curva característica
characteristic data datos característicos
characteristic distortion distorsión característica
characteristic energy energía característica
characteristic equation ecuación característica
characteristic frequency frecuencia característica
characteristic impedance impedancia característica
characteristic induction inducción característica
characteristic insertion loss pérdida de inserción característica
characteristic modulation modulación característica
characteristic overflow desbordamiento característico
characteristic quantity cantidad característica
characteristic radiation radiación característica
characteristic series serie característica
characteristic sign signo característico
characteristic temperature temperatura característica

characteristic underflow desbordamiento negativo característico
characteristic wave onda característica
characteristic wave impedance impedancia de onda característica
characters per inch caracteres por pulgada
characters per minute caracteres por minuto
characters per second caracteres por segundo
charge carga
charge acceptance aceptación de carga
charge amplifier amplificador de carga
charge carrier portador de carga
charge-coupled device dispositivo acoplado por cargas
charge-coupled device memory dispositivo de memoria acoplado por cargas
charge density densidad de carga
charge discriminator discriminador de carga
charge efficiency eficiencia de carga
charge factor factor de carga
charge holding retención de carga
charge indicating device dispositivo indicador de carga
charge indicator indicador de carga
charge multiplication multiplicación de carga
charge neutralization neutralización de carga
charge neutralizer neutralizador de carga
charge of electron carga de electrón
charge retention retención de carga
charge source fuente de carga
charge storage almacenamiento de carga
charge-storage tube tubo de almacenamiento de carga
charge switch conmutador de carga
charge time constant constante de tiempo de carga
charge-to-mass carga a masa
charge transfer transferencia de carga
charge-transfer device dispositivo de transferencia de carga
charge-transfer spectrum espectro de transferencia de carga
charge voltage tensión de carga
charged battery batería cargada
charged particle partícula cargada
charger cargador
charging carga
charging board tablero de carga
charging choke choque de carga
charging circuit circuito de carga
charging coefficient coeficiente de carga
charging current corriente de carga
charging panel panel de carga
charging period periodo de carga
charging rate velocidad de carga
charging resistance resistencia de carga
charging resistor resistor de carga
charging time tiempo de carga
charging unit unidad de carga
charging voltage tensión de carga
chart diagrama
chart format formato de diagrama
chart layout trazado de diagrama
chassis chasis
chassis ground masa de chasis
chassis holder portachasis
chatter vibración
cheater cord cable prolongador
Chebyshev array red de Chebyshev
Chebyshev filter filtro de Chebyshev
check comprobación
check beam haz de comprobación
check bit bit de comprobación
check box caja de comprobación

check bus bus de comprobación
check card tarjeta de comprobación
check character carácter de comprobación
check digit dígito de comprobación
check field campo de comprobación
check indicator indicador de comprobación
check indicator instruction instrucción de indicador de comprobación
check key clave de comprobación
check length longitud de comprobación
check multiple múltiple de comprobación
check number número de comprobación
check position posición de comprobación
check problem problema de comprobación
check punch perforación de comprobación
check reading lectura de comprobación
check redundancy redundancia de comprobación
check register registro de comprobación
check routine rutina de comprobación
check row fila de comprobación
check sample muestra de comprobación
check sorter clasificadora de comprobación
check sum suma de comprobación
check symbol símbolo de comprobación
check total total de comprobación
check window ventana de comprobación
check word palabra de comprobación
checking circuit circuito de comprobación
checking circuitry circuitería de comprobación
checking code código de comprobación
checking device dispositivo de comprobación
checking factor factor de comprobación
checking program programa de comprobación
checking routine rutina de comprobación
checklist lista de comprobación
checkout comprobación
checkout equipment equipo de comprobación
checkout routine rutina de comprobación
checkout time tiempo de comprobación
checkpoint punto de comprobación
checkpoint data set conjunto de datos de punto de comprobación
checkpoint dump vaciado de puntos de comprobación
checkpoint entry especificación de puntos de comprobación
checkpoint file archivo de puntos de comprobación
checkpoint record registro de puntos de comprobación
checkpoint recovery recuperación de puntos de comprobación
checkpoint register registro de puntos de comprobación
checkpoint restart reiniciación de puntos de comprobación
checkpoint routine rutina de puntos de comprobación
checkpoint start iniciación de puntos de comprobación
checkpoint tape cinta de puntos de comprobación
checksum suma de comprobación
cheese antenna antena en queso
chemical corrosion corrosión química
chemical deposition deposición química
chemical detector detector químico
chemical effect efecto químico
chemical energy energía química
chemical equivalent equivalente químico
chemical erosion erosión química
chemical exchange intercambio químico
chemical-film dielectric dieléctrico de película química
chemical light luz química

chemical load carga química
chemical passivation pasivación química
chemical passivity pasividad química
chemical protector protector químico
chemical rectifier rectificador químico
chemical reduction reducción química
chemical resistor resistor químico
chemical switch conmutador químico
Cherenkov coupler acoplador de Cherenkov
Child's law ley de Child
Child-Langmuir equation ecuación de
    Child-Langmuir
chip chip, pastilla
chip architecture arquitectura de chip
chip box caja de residuos de perforaciones
chip capacitor capacitor en chip
chip carrier portachip
chip circuit circuito de chip
chip enable habilitación de chip
chip enable input entrada de habilitación de chip
chip materials materiales de chip
chip microprocessor microprocesador de chip
chip positioner posicionador de chips
chip reduction reducción de chip
chip resistor resistor en chip
chip select selección de chip
chip size tamaño de chip
chip system sistema de chips
chip tray bandeja de residuos de perforaciones
Chireix-Mesny antenna antena de Chireix-Mesny
chirp chirrido
chirp modulator modulador de chirrido
chirp radar radar de chirrido
chirping chirrido
Chladni's plates placas de Chladni
chlorine cloro
choke choque
choke coil bobina de choque
choke-coil modulation modulación de bobina de
    choque
choke coupling acoplamiento por choque
choke-input filter filtro de choque de entrada
choke joint junta de choque
choke piston pistón de choque
chopped display indicador visual conmutado
chopped impulse impulso seccionado
chopped-impulse voltage tensión de impulso
    seccionado
chopped mode modo seccionado
chopped signal señal seccionada
chopper interruptor periódico, seccionador,
    descrestador
chopper amplifier amplificador de interrupción
    periódica
chopper converter convertidor de interrupción
    periódica
chopper stabilization estabilización por interrupción
    periódica
chopper-stabilized amplifier amplificador
    estabilizado por interrupción periódica
chopper transistor transistor de interrupción
    periódica
chopping interrupción, descrestación, seccionado
chopping frequency frecuencia de interrupción
chord cuerda
Christiansen antenna antena de Christiansen
chroma croma
chroma amplifier amplificador de croma
chroma circuit circuito de croma
chroma control control de croma
chroma detector detector de croma
chroma oscillator oscilador de croma
chromatic cromático

chromatic aberration aberración cromática
chromatic adaptation adaptación cromática
chromatic amplifier amplificador cromático
chromatic color color cromático
chromatic correction corrección cromática
chromatic dispersion dispersión cromática
chromatic distortion distorsión cromática
chromatic fidelity fidelidad cromática
chromatic lens lente cromática
chromatic picture tube tubo de imagen cromático
chromatic spectrum espectro cromático
chromaticity cromaticidad
chromaticity coefficient coeficiente de cromaticidad
chromaticity coordinate coordenada de cromaticidad
chromaticity diagram diagrama de cromaticidad
chromaticity flicker parpadeo de cromaticidad
chromaticity information información de
    cromaticidad
chromaticity signal señal de cromaticidad
chromatron cromatrón
chrome-plated cromado
chromel cromel
chrominance crominancia
chrominance amplifier amplificador de crominancia
chrominance amplitude amplitud de crominancia
chrominance band banda de crominancia
chrominance cancellation cancelación de
    crominancia
chrominance carrier portadora de crominancia
chrominance-carrier reference referencia de
    portadora de crominancia
chrominance channel canal de crominancia
chrominance-channel bandwidth ancho de banda de
    canal de crominancia
chrominance circuit circuito de crominancia
chrominance component componente de
    crominancia
chrominance demodulator desmodulador de
    crominancia
chrominance gain control control de ganancia de
    crominancia
chrominance information información de
    crominancia
chrominance mixer mezclador de crominancia
chrominance modulator modulador de crominancia
chrominance primary primario de crominancia
chrominance signal señal de crominancia
chrominance signal component componente de señal
    de crominancia
chrominance subcarrier subportadora de
    crominancia
chrominance-subcarrier demodulator
    desmodulador de subportadora de crominancia
chrominance-subcarrier modulator modulador de
    subportadora de crominancia
chrominance-subcarrier oscillator oscilador de
    subportadora de crominancia
chrominance-subcarrier reference referencia de
    subportadora de crominancia
chrominance-subcarrier regenerator regenerador
    de subportadora de crominancia
chrominance tube tubo de crominancia
chrominance video signals videoseñales de
    crominancia
chromium cromo
chromoscope cromoscopio
chronistor cronistor
chronograph cronógrafo
chronometer cronómetro
chronopher cronófero
chronophotometer cronofotómetro
chronorelay cronorrelé
chronoscope cronoscopio

chronotron cronotrón
chute conducto
circle of confusion círculo de confusión
circle of declination círculo de declinación
circuit circuito
circuit algebra álgebra de circuitos
circuit analysis análisis de circuito
circuit analyzer analizador de circuitos
circuit arrangement arreglo de circuitos
circuit balance equilibrio de circuito
circuit board tablero de circuitos
circuit-breaker cortacircuito, disyuntor
circuit-breaker compartment compartimiento de
   cortacircuito
circuit-breaker grouping agrupamiento de
   cortacircuito
circuit burden carga de circuito
circuit busy hour hora ocupada de circuito, hora
   pico de circuito
circuit capacitance capacitancia de circuito
circuit capacity capacidad de circuito
circuit card tarjeta de circuitos
circuit closing cierre de circuito
circuit component componente de circuito
circuit conditioning acondicionamiento de circuito
circuit configuration configuración de circuito
circuit control panel panel de control de circuito
circuit controller controlador de circuitos
circuit design diseño de circuitos
circuit development desarrollo de circuitos
circuit diagram diagrama de circuito
circuit dropout interrupción de circuito
circuit efficiency eficiencia de circuito
circuit element elemento de circuito
circuit emulator emulador de circuito
circuit engineer ingeniero de circuitos
circuit equivalent equivalente de circuito
circuit fabrication fabricación de circuitos
circuit fault falla de circuito
circuit function función de circuito
circuit-gap admittance admitancia de espacio de
   circuito
circuit group grupo de circuitos
circuit-group busy hour hora pico de grupo de
   circuitos
circuit hole agujero de circuito
circuit identification identificación de circuito
circuit identification number número de
   identificación de circuito
circuit interrupter interruptor de circuito
circuit layout trazado de circuito
circuit load carga de circuito
circuit logic lógica de circuitos
circuit loss pérdidas de circuito
circuit malfunction falla de circuito
circuit-malfunction analysis análisis de falla de
   circuito·
circuit model modelo de circuito
circuit noise ruido de circuito
circuit-noise level nivel de ruido de circuito
circuit-noise meter medidor de ruido de circuito
circuit overload sobrecarga de circuito
circuit parameter parámetro de circuito
circuit properties propiedades de circuito
circuit protection protección de circuito
circuit reliability confiabilidad de circuito
circuit schematic esquema de circuito
circuit serial number número de serie de circuito
circuit simplification simplificación de circuito
circuit stability estabilidad de circuito
circuit switch conmutador de circuito
circuit switching conmutación de circuito
circuit synthesis síntesis de circuitos

circuit test prueba de circuito
circuit tester probador de circuitos
circuit theory teoría de circuitos
circuit voltage tensión de circuito
circuital circuital
circuitron circuitrón
circuitry circuitería
circular angle ángulo circular
circular antenna antena circular
circular-aperture array red de abertura circular
circular array red circular
circular baffle bafle circular
circular beam haz circular
circular-beam multiplier multiplicador de haz
   circular
circular coil bobina circular
circular current corriente circular
circular electric mode modo eléctrico circular
circular electric wave onda eléctrica circular
circular frequency frecuencia circular
circular function función circular
circular grating rejilla circular
circular guide guía circular
circular horn bocina circular
circular-horn radiator radiador de bocina circular
circular list lista circular
circular magnet imán circular
circular magnetic field campo magnético circular
circular magnetic wave onda magnética circular
circular mil mil circular
circular polarization polarización circular
circular-polarization duplexer duplexor de
   polarización circular
circular queue cola circular
circular reference referencia circular
circular scan exploración circular
circular scanning exploración circular
circular shift desplazamiento circular
circular sweep barrido circular
circular sweep generation generación de barrido
   circular
circular trace trazo circular
circular waveguide guíaondas circular
circulating circulatorio, circulante
circulating current corriente circulatoria
circulating memory memoria circulatoria
circulating register registro circulatorio
circulating storage almacenamiento circulatorio
circulation of electrolyte circulación de electrólito
circulator circulador
citizen's band banda ciudadana
cladding revestimiento
cladding diameter diámetro de revestimiento
cladding mode modo de revestimiento
cladding ray rayo de revestimiento
clamp abrazadera
clamp tube tubo fijador de nivel
clamper fijador de nivel
clamping fijación de nivel, sujeción
clamping circuit circuito de fijación de nivel
clamping diode diodo de fijación de nivel
Clapp-Gouriet oscillator oscilador de Clapp-Gouriet
Clapp oscillator oscilador de Clapp
clapper badajo
Clark cell pila de Clark
class A amplifier amplificador clase A
class A component componente clase A
class A modulation modulación clase A
class A modulator modulador clase A
class A operation operación clase A
class A oscilloscope osciloscopio clase A
class A signal area área de señales clase A
class A station estación clase A

class AB amplifier amplificador clase AB
class AB modulator modulador clase AB
class B amplifier amplificador clase B
class B modulation modulación clase B
class B modulator modulador clase B
class B operation operación clase B
class B oscilloscope osciloscopio clase B
class B station estación clase B
class condition condición de clase
class name nombre de clase
class of insulation clase de aislamiento
class of service clase de servicio
classical electron radius radio de electrón clásico
classification light luz de clasificación
Clausius-Mosotti equation ecuación de
  Clausius-Mosotti
Clausius-Mosotti-Lorentz-Lorenz equation
  ecuación de Clausius-Mosotti-Lorentz-Lorenz
clean room sala limpia
cleanup process proceso de limpieza
clear area área libre
clear band banda libre
clear channel canal libre
clear device dispositivo libre
clear display visualizador borrado
clear input entrada libre
clear memory borrar la memoria
clear screen pantalla borrada
clear signal señal abierta
clear storage register borrar los registros de
  almacenamiento
clear the line liberar la línea
clearance between open contacts distancia entre
  contactos abiertos
clearance between poles espacio entre polos
clearing device dispositivo de borrar
clearing prefix prefijo de borrar
click method método de chasquidos
climate chamber cámara climática
climatometer climatómetro
climb space espacio para trepar
clip pinza, abrazadera
clipboard tablilla
clipper recortador
clipper amplifier amplificador recortador
clipper circuit circuito recortador
clipper limiter limitador recortador
clipper tube tubo recortador
clipping recortamiento, recorte, limitación
clipping amplifier amplificador de limitación
clipping circuit circuito de limitación
clipping level nivel de limitación
clipping point punto de limitación
clipping time tiempo de limitación
clock accuracy exactitud de reloj
clock-calendar board tablero de reloj-calendario
clock comparator comparador de reloj
clock counter contador de reloj
clock cycle ciclo de reloj
clock frequency frecuencia de reloj
clock generator generador de reloj
clock module módulo de reloj
clock oscillator oscilador de reloj
clock pulse impulso de reloj
clock pulse generator generador de impulsos de
  reloj
clock rate frecuencia de reloj
clock signal señal de reloj
clock signal generator generador de señal de reloj
clock speed velocidad de reloj
clock time tiempo de reloj
clocked flip-flop basculador temporizado
clockwise dextrorso, dextrógiro

clockwise capacitor capacitor dextrorso
clockwise-polarized wave onda con polarización
  dextrorsa
close coupling acoplamiento cerrado
close-spaced array red apiñada
close-talk microphone micrófono para hablar de
  cerca
closed array red cerrada
closed bus system sistema de bus cerrado
closed circuit circuito cerrado
closed-circuit alarm device dispositivo de alarma de
  circuito cerrado
closed-circuit alarm system sistema de alarma de
  circuito cerrado
closed-circuit arrangement arreglo en circuito
  cerrado
closed-circuit cell pila de circuito cerrado
closed-circuit communication comunicación en
  circuito cerrado
closed-circuit cooling enfriamiento en circuito
  cerrado
closed-circuit current corriente en circuito cerrado
closed-circuit signaling señalización de circuito
  cerrado
closed-circuit television televisión por circuito
  cerrado
closed-circuit voltage tensión en circuito cerrado
closed core núcleo cerrado
closed-core choke choque de núcleo cerrado
closed-core inductor inductor de núcleo cerrado
closed-core transformer transformador de núcleo
  cerrado
closed cycle ciclo cerrado
closed-cycle fuel cell pila de combustible de ciclo
  cerrado
closed-cycle system sistema de ciclo cerrado
closed file archivo cerrado
closed loop bucle cerrado
closed-loop control control de bucle cerrado
closed-loop control circuit circuito de control de
  bucle cerrado
closed-loop control system sistema de control de
  bucle cerrado
closed-loop gain ganancia de bucle cerrado
closed-loop input impedance impedancia de entrada
  de bucle cerrado
closed-loop output impedance impedancia de salida
  de bucle cerrado
closed-loop program programa de bucle cerrado
closed-loop system sistema de bucle cerrado
closed-loop voltage gain ganancia de tensión de
  bucle cerrado
closed magnetic circuit circuito magnético cerrado
closed routine rutina cerrada
closed subroutine subrutina cerrada
closed system sistema cerrado
closed waveguide guíaondas cerrado
closedown cierre
closing cierre
closing action acción de cierre
closing coil bobina de cierre
closing relay relé de cierre
closing spark chispa de cierre
closing time tiempo de cierre
closure cierre
cloud pulse impulso de nube
cloverleaf antenna antena en trébol
cluster grupo
cluster control unit unidad de control de grupo
cluster controller controlador de grupo
cluster entry especificación de grupo
cluster interface interfaz de grupo
cluster station estación de grupo

cluster workstations estaciones de trabajo en grupo
clustered devices dispositivos agrupados
clustering agrupamiento
clustering index índice de agrupamiento
clutch embrague
clutch disengaging desembrague
clutch point punto de embrague
clutter ecos parásitos
clutter attenuation atenuación de ecos parásitos
clutter eliminator eliminador de ecos parásitos
clutter filter filtro de ecos parásitos
clutter reflectivity reflectividad de ecos parásitos
clutter suppression supresión de ecos parásitos
coactive maintenance mantenimiento coactivo
coarse adjustment ajuste grueso
coarse control control grueso
coarse tuning sintonización gruesa
coast station estación costera
coastal refraction refracción costera
coated cathode cátodo revestido
coated electrode electrodo revestido
coated filament filamento revestido
coated wire hilo revestido
coating revestimiento
coating activation activación de revestimiento
coating conductivity conductividad de revestimiento
coating thickness grosor de revestimiento
coax coaxial, coaxil
coaxial antenna antena coaxial
coaxial attenuator atenuador coaxial
coaxial bolometer bolómetro coaxial
coaxial-bolometer mount montura de bolómetro
    coaxial
coaxial cable cable coaxial
coaxial-cable interface interfaz de cable coaxial
coaxial-cable network red de cables coaxiales
coaxial-cable section sección de cable coaxial
coaxial-cable segment segmento de cable coaxial
coaxial capacitor capacitor coaxial
coaxial cavity cavidad coaxial
coaxial circuit circuito coaxial
coaxial conductor conductor coaxial
coaxial connector conector coaxial
coaxial control station estación de control coaxial
coaxial controls controles coaxiales
coaxial-cylinder magnetron magnetrón de cilindros
    coaxiales
coaxial detector detector coaxial
coaxial dielectric dieléctrico coaxial
coaxial diode diodo coaxial
coaxial-dipole antenna antena dipolo coaxial
coaxial directional coupler acoplador direccional
    coaxial
coaxial electrode system sistema de electrodos
    coaxiales
coaxial feeder alimentador coaxial
coaxial filter filtro coaxial
coaxial hybrid híbrido coaxial
coaxial input entrada coaxial
coaxial isolator aislador coaxial
coaxial jack jack coaxial
coaxial line línea coaxial
coaxial-line circuit circuito de línea coaxial
coaxial-line frequency meter frecuencímetro de
    línea coaxial
coaxial-line oscillator oscilador de línea coaxial
coaxial-line resonator resonador de línea coaxial
coaxial-line tube tubo de línea coaxial
coaxial loudspeaker altavoz coaxial
coaxial pair par coaxial
coaxial plug clavija coaxial
coaxial receptacle receptáculo coaxial
coaxial relay relé coaxial

coaxial speaker altavoz coaxial
coaxial stub adaptador coaxial
coaxial switch conmutador coaxial
coaxial switching conmutación coaxial
coaxial terminal terminal coaxial
coaxial transistor transistor coaxial
coaxial transmission line línea de transmisión
    coaxial
coaxial tuner sintonizador coaxial
coaxial wattmeter wáttmetro coaxial
coaxial wavemeter ondámetro coaxial
cobalt cobalto
cobalt anode ánodo de cobalto
cobinotron cobinotrón
cochannel canal común
cochannel broadcasting radiodifusión por canal
    común
cochannel interference interferencia de canal común
Cockcroft-Walton accelerator acelerador de
    Cockcroft-Walton
codan codán
code código
code address dirección de código
code area área de código
code audit auditoría de código
code bar barra de código
code-bar switch conmutador de barras de código
code beacon radiofaro de código
code block bloque de código
code character carácter de código
code checking time tiempo de comprobación de
    código
code combination combinación de códigos
code conversion conversión de código
code converter convertidor de código
code delay retardo de código
code dependent system sistema dependiente de
    códigos
code digit dígito de código
code-directing characters caracteres de
    direccionamiento de código
code disk disco de códigos
code element unit unidad de elementos de código
code elements elementos de código
code emitter emisor de código
code generator generador de código
code group grupo de códigos
code holes perforaciones de código
code independent system sistema independiente de
    códigos
code key clave de código
code letter letra de código
code level nivel de código
code lever palanca de código
code light luz de código
code line index índice de líneas de código
code lines líneas de código
code machine máquina de código
code parameter parámetro de códigos
code pattern patrón de código
code point punto de código
code position posición de código
code reader lectora de códigos
code receiver receptor de códigos
code ringing llamada en código
code segment segmento de códigos
code selector selector de código
code set conjunto de códigos
code sheet hoja de códigos
code sign signo de código
code speed velocidad de códigos
code symbol símbolo de código
code system sistema de códigos

code table tabla de códigos
code transition level nivel de transición de código
code translation traducción de código
code translator traductor de códigos
code transmitter transmisor de códigos
code value valor de código
code violation violación de código
code word palabra de código
codec codificador-descodificador
coded codificado
coded arithmetic data datos aritméticos codificados
coded call llamada codificada
coded character carácter codificado
coded character set conjunto de caracteres
  codificado
coded decimal decimal codificado
coded-decimal digit dígito decimal codificado
coded-decimal notation notación decimal codificada
coded-decimal sequence secuencia decimal
  codificada
coded element set conjunto de elementos codificados
coded font tipo de letra codificado
coded graphics gráficos codificados
coded image imagen codificada
coded mark inversion inversión de marca codificada
coded notation notación codificada
coded number número codificado
coded passive reflector reflector pasivo codificado
coded program programa codificado
coded representation representación codificada
coded sequence secuencia codificada
coded set conjunto codificado
coded signal señal codificada
coded stop parada codificada
coded tape cinta codificada
codeposition codeposición
coder codificador
coder circuit circuito codificador
coder-decoder codificador-descodificador
codification codificación
coding codificación
coding check comprobación de codificación
coding circuit circuito de codificación
coding error error de codificación
coding line línea de codificación
coding process proceso de codificación
coding pulse impulso de codificación
coding scheme esquema de codificación
coding sequence secuencia de codificación
coding sheet hoja de codificación
coding signal señal de codificación
coding system sistema de codificación
codistor codistor
coefficient coeficiente
coefficient of absorption coeficiente de absorción
coefficient of attenuation coeficiente de atenuación
coefficient of coupling coeficiente de acoplamiento
coefficient of diffusion coeficiente de difusión
coefficient of directivity coeficiente de directividad
coefficient of exploitation coeficiente de explotación
coefficient of harmonic distortion coeficiente de
  distorsión armónica
coefficient of linear absorption coeficiente de
  absorción lineal
coefficient of linear diffusion coeficiente de difusión
  lineal
coefficient of mutual induction coeficiente de
  inducción mutua
coefficient of occupation coeficiente de ocupación
coefficient of performance coeficiente de
  rendimiento
coefficient of reflection coeficiente de reflexión
coefficient of selective absorption coeficiente de

absorción selectiva
coefficient of self-induction coeficiente de
  autoinducción
coefficient of utilization coeficiente de utilización
coefficient of variation coeficiente de variación
coercive field campo coercitivo
coercive force fuerza coercitiva
coercivity coercitividad
cogeneration cogeneración
cohered video video coherente
coherence coherencia
coherence area área de coherencia
coherence distance distancia de coherencia
coherence length longitud de coherencia
coherence time tiempo de coherencia
coherency coherencia
coherent coherente
coherent area área coherente
coherent bundle grupo coherente
coherent carrier portadora coherente
coherent demodulation desmodulación coherente
coherent detection detección coherente
coherent detector detector coherente
coherent echo eco coherente
coherent electroluminescent device dispositivo
  electroluminiscente coherente
coherent gate puerta coherente
coherent integration integración coherente
coherent light luz coherente
coherent-light detection detección de luz coherente
coherent-light oscillator oscilador de luz coherente
coherent-light radar radar de luz coherente
coherent memory filter filtro de memoria coherente
coherent oscillations oscilaciones coherentes
coherent oscillator oscilador coherente
coherent-pulse operation operación de impulsos
  coherentes
coherent-pulse radar radar de impulsos coherentes
coherent pulses impulsos coherentes
coherent radiation radiación coherente
coherent reference referencia coherente
coherent scattering dispersión coherente
coherent signal señal coherente
coherent signal processing procesamiento de señal
  coherente
coherent system sistema coherente
coherent transponder transpondedor coherente
coherent video video coherente
coherer cohesor
cohesion cohesión
coil bobina
coil antenna antena en bobina
coil box caja de bobinas
coil brace plataforma de bobinas
coil comparator comparador de bobinas
coil constant constante de bobina
coil dissipation disipación de bobina
coil form forma de bobina, soporte de bobina
coil ignition ignición por bobina
coil insulation aislamiento de bobina
coil loading carga por bobinas
coil magnification factor factor de magnificación de
  bobina
coil neutralization neutralización por bobina
coil-neutralization circuit circuito de neutralización
  por bobina
coil resistance resistencia de bobina
coil serving revestimiento de bobina
coil set conjunto de bobinas
coil side lado de bobina
coil signal señal de bobina
coil spacing espaciado de bobinas
coil tap toma de bobina

coincidence coincidencia
coincidence amplifier amplificador de coincidencia
coincidence channel canal de coincidencia
coincidence circuit circuito de coincidencia
coincidence correction corrección de coincidencia
coincidence counter contador de coincidencia
coincidence counting conteo de coincidencia
coincidence currents corrientes coincidentes
coincidence detector detector de coincidencia
coincidence effect efecto de coincidencia
coincidence error error de coincidencia
coincidence factor factor de coincidencia
coincidence frequency frecuencia de coincidencia
coincidence gate puerta de coincidencia
coincidence loss pérdida por coincidencia
coincidence magnet imán de coincidencia
coincidence mixer mezclador de coincidencia
coincidence multiplier multiplicador de coincidencia
coincidence selector selector de coincidencia
coincident-current selection selección por corrientes
    coincidentes
coincident signal elements elementos de señal
    coincidentes
cold antenna antena fría
cold boot arranque en frío, rutina de carga fría
cold cathode cátodo frío
cold-cathode arc arco de cátodo frío
cold-cathode counter tube tubo contador de cátodo
    frío
cold-cathode diode diodo de cátodo frío
cold-cathode discharge descarga de cátodo frío
cold-cathode discharge lamp lámpara de descarga
    de cátodo frío
cold-cathode electron emitter emisor de electrones
    de cátodo frío
cold-cathode emission emisión de cátodo frío
cold-cathode gas tube tubo de gas de cátodo frío
cold-cathode ionization gage medidor de ionización
    de cátodo frío
cold-cathode lamp lámpara de cátodo frío
cold-cathode rectifier rectificador de cátodo frío
cold-cathode tube tubo de cátodo frío
cold chamber cámara fría
cold connection conexión fría
cold electrode electrodo frío
cold emission emisión fría
cold flow flujo frío
cold junction soldadura fría
cold light luz fría
cold neutron neutrón frío
cold pressure welding soldadura por presión fría
cold resistance resistencia fría
cold restart reiniciación fría
cold spot punto frío
cold start arranque frío
cold start iniciación fría
cold weld soldadura fría
colidar colidar
collate intercalar, colacionar
collation intercalación
collation file archivo de intercalación
collation operation operación de intercalación
collation sequence secuencia de intercalación
collation tape cinta de intercalación
collator intercaladora, colacionador
collecting bar barra colectora
collecting electrode electrodo colector
collecting grid rejilla colectora
collection efficiency eficiencia de recolección
collective antenna antena colectiva
collective-call sign señal de llamada colectiva
collective transmission transmisión colectiva
collector colector

collector anode ánodo colector
collector-base barrier barrera colector-base
collector-base capacitance capacitancia colector-base
collector-base circuit circuito colector-base
collector-base conductance conductancia
    colector-base
collector-base diode diodo colector-base
collector capacitance capacitancia de colector
collector current corriente de colector
collector cutoff current corriente de corte de
    colector
collector dissipation disipación de colector
collector dot punto colector
collector efficiency eficiencia de colector
collector electrode electrodo colector
collector family familia de colectores
collector grid rejilla colectora
collector junction unión de colector
collector mesh malla de colector
collector modulation modulación de colector
collector multiplication multiplicación de colector
collector plate placa colectora
collector positioning posicionamiento de colector
collector resistance resistencia de colector
collector ring anillo colector
collector saturation current corriente de saturación
    de colector
collector shoe patín
collector transition capacitance capacitancia de
    transición de colector
collector voltage tensión de colector
collimate colimar
collimated colimado
collimated rays rayos colimados
collimation colimación
collimation equipment equipo de colimación
collimator colimador
collinear colineal
collinear antenna antena colineal
collinear array red colineal
collinear dipole antenna antena dipolo colineal
collinear element elemento colineal
Collins coupler acoplador de Collins
collision-avoidance system sistema para evitar
    colisiones, sistema anticolisiones
collision detection detección de colisiones
collision warning device dispositivo de aviso de
    colisiones
collision warning radar radar de aviso de colisiones
collodion colodión
colloidal graphite grafito coloidal
colloidal particle partícula coloidal
color aberration aberración de colores
color amplifier amplificador de colores
color balance equilibrio de colores
color bar barra de colores
color-bar generator generador de barras de colores
color-bar pattern patrón de barras de colores
color-bar test pattern patrón de prueba de barras de
    colores
color blending mezcla de colores
color breakup descomposición de colores
color brilliance control control de brillantez de
    colores
color broadcasting radiodifusión de colores
color burst ráfaga de color
color carrier portadora de colores
color-carrier reference referencia de portadora de
    colores
color cell célula de colores
color center centro de color
color channel canal de colores
color code código de colores

color coded codificado por colores
color coder codificador de colores
color coding codificación de colores
color comparator comparador de colores
color comparison comparación de colores
color component componente de color
color contamination contaminación de colores
color control control de colores
color convergence convergencia de colores
color-convergence control control de convergencia de colores
color converter convertidor de colores
color-coordinate transformation transformación de coordenadas de color
color correction corrección de color
color-correction factor factor de corrección de color
color-correction filter filtro de corrección de color
color decoder descodificador de color
color detector detector de colores
color-difference signal señal de diferencia de color
color dilution dilución de color
color discrimination discriminación de colores
color disk disco de colores
color dot punto de color
color edging coloración de borde
color encoder codificador de colores
color equation ecuación de colores
color equipment equipo de colores
color fidelity fidelidad de colores
color field trama de color
color-field corrector corrector de trama de color
color filter filtro de colores
color flicker parpadeo de colores
color fringing franjas de colores marginales
color gamut gama de colores
color gate puerta de color
color generator generador de colores
color graphics gráficos en colores
color graphics adapter adaptador de gráficos en colores
color impurity impureza de color
color information información de colores
color intensity intensidad de color
color killer supresor de colores
color-killer circuit circuito supresor de colores
color-killer stage etapa de supresor de colores
color-killer tube tubo supresor de colores
color kinescope cinescopio de colores
color level nivel de color
color match equilibrio de colores
color meter colorímetro
color mixer mezclador de colores
color mixture mezcla de colores
color monitor monitor de colores
color-origination equipment equipo de originación de colores
color oscillator oscilador de colores
color overload sobresaturación de colores
color phase fase de color
color phase detector detector de fase de color
color picture screen pantalla de imagen de colores
color picture signal señal de imagen de colores
color picture tube tubo de imagen de colores
color plane plano de colores
color primaries colores primarios
color printer impresora de colores
color purity pureza de colores
color-purity adjustment ajuste de pureza de colores
color-purity magnet imán de pureza de colores
color quality calidad de colores
color range gama de colores
color receiver receptor de colores
color registration superimposición de colores

color rendering index índice de concordancia de color percibido
color response respuesta de color
color sampling rate frecuencia de muestreo de colores
color sampling sequence secuencia de muestreo de colores
color saturation saturación de color
color screen pantalla de colores
color sensitivity sensibilidad de colores
color separation separación de colores
color sequence secuencia de colores
color sidebands bandas laterales de colores
color signal señal de colores
color space espacio de colores
color spectrum espectro de colores
color splitter divisor de colores
color standardization normalización de colores
color stimulus estímulo de color
color subcarrier subportadora de colores
color-subcarrier frequency frecuencia de subportadora de colores
color-subcarrier monitor monitor de subportadora de colores
color-subcarrier oscillator oscilador de subportadora de colores
color-subcarrier reference referencia de subportadora de colores
color switching conmutación de colores
color synchronization sincronización de colores
color-synchronization system sistema de sincronización de colores
color synchronizer sincronizador de colores
color synchronizing circuit circuito de sincronización de colores
color synchronizing signal señal de sincronización de colores
color system sistema de colores
color table tabla de colores
color television televisión a colores, televisión en colores
color-television picture tube tubo de imagen de televisión a colores
color-television receiver receptor de televisión a colores
color-television signal señal de televisión a colores
color-television standards patrones de televisión a colores
color temperature temperatura de color
color temperature meter medidor de temperatura de color
color transmission transmisión de colores
color transmitter transmisor de colores
color triad triada de colores
color triangle triángulo de colores
color video signal videoseñal de colores
color weather radar radar meteorológico de colores
colored filter filtro coloreado
colorimeter colorímetro
colorimetric colorimétrico
colorimetric characteristic característica colorimétrica
colorimetric distortion distorsión colorimétrica
colorimetric filter filtro colorimétrico
colorimetric purity pureza colorimétrica
colorimetric shift cambio colorimétrico
colorimetric system sistema colorimétrico
colorimetry colorimetría
Colpitts oscillator oscilador de Colpitts
columbium columbio
column columna
column binary card tarjeta binaria en columnas
column binary code código binario en columnas

column binary mode modo binario en columnas
column indicator indicador de columna
column punch perforadora de columnas
column separator separador de columnas
column sort clasificación de columna
column speaker altavoz de columna
coma lobe lóbulo de coma
comb amplifier amplificador en peine
comb antenna antena en peine
comb filter filtro en peine
combination combinación
combination bridge puente combinado
combination cable cable combinado
combination detector detector combinado
combination microphone micrófono combinado
combination of punches combinación de
    perforaciones
combination rubber tape cinta de goma combinada
combination speaker altavoz combinado
combination starter arrancador combinado
combination tone tono combinado
combination tube tubo combinado
combinational combinacional
combinational circuit circuito combinacional
combinational function función combinacional
combinational logic lógica combinacional
combinational logic element elemento de lógica
    combinacional
combinational logic function función de lógica
    combinacional
combinatorial combinatorio
combinatorial logic lógica combinatoria
combined access acceso combinado
combined condition condición combinada
combined file archivo combinado
combined head cabeza combinada
combined network red combinada
combined reactance reactancia combinada
combined read-write head cabeza de
    lectura-escritura combinada
combined telephone set aparato telefónico
    combinado
combiner combinador
combiner circuit circuito combinador
combiner panel panel combinador
combining circuit circuito combinador
combining network red combinadora
combining unit unidad combinadora
command mando, comando, mandato, orden
command address dirección de mando
command area área de mandos
command chain cadena de mandos
command chaining encadenamiento de mandos
command character carácter de mando
command code código de mando
command control control de mandos
command control program programa de control de
    mandos
command definition definición de mando
command display visualización de mandos
command display station estación de visualización
    de mandos
command driven accionado por mandos
command-driven program programa accionado por
    mandos
command field campo de mandos
command file archivo de mandos
command file processor procesador de archivo de
    mandos
command function key tecla de función de mandos
command generator generador de mandos
command installation instalación de mandos
command interrupt mode modo de interrupción de

mandos
command key tecla de mandos
command key indicator indicador de tecla de
    mandos
command language lenguaje de mandos
command line línea de mandos
command line operating system sistema de
    operación de línea de mandos
command link enlace de mandos
command list lista de mandos
command menu menú de mandos
command mode modo de mandos
command name nombre de mando
command network red de mandos
command procedure procedimiento de mandos
command processing procesamiento de mandos
command processing program programa de
    procesamiento de mandos
command processor procesador de mandos
command pulse impulso de mando
command reader lectora de mandos
command recognition character carácter de
    reconocimiento de mando
command reference referencia de mandos
command retry reintento de mando
command scan exploración de mando
command signal señal de mando
command statement sentencia de mando
command transfer transferencia de mandos
comment field campo de comentarios
comment line línea de comentarios
comment statement sentencia de comentarios
commercial device dispositivo comercial
commercial electric power energía eléctrica
    comercial
commercial language lenguaje comercial
commercial receiver receptor comercial
commercial telephone circuit circuito telefónico
    comercial
common area área común
common-base amplifier amplificador de base común
common-base circuit circuito de base común
common-base connection conexión en base común
common battery batería común
common-battery supply fuente de batería común
common-battery system sistema de batería común
common block bloque común
common buffer amortiguador común
common-capacitor coupling acoplamiento por
    capacitor común
common-cathode circuit circuito de cátodo común
common channel canal común
common-channel interference interferencia de canal
    común
common-channel operation operación de canal
    común
common-collector amplifier amplificador de colector
    común
common-collector circuit circuito de colector común
common-collector connection conexión de colector
    común
common-component coupling acoplamiento por
    componente común
common control control común
common control section sección de control común
common coupling acoplamiento común
common-emitter circuit circuito de emisor común
common-emitter connection conexión de emisor
    común
common equipment equipo común
common field campo común
common-frequency broadcasting radiodifusión por
    frecuencia común

common-gate amplifier amplificador de puerta común
common-gate circuit circuito de puerta común
common-grid circuit circuito de rejilla común
common ground tierra común
common hardware equipo físico común
common impedance impedancia común
common-impedance coupling acoplamiento por impedancia común
common-inductor coupling acoplamiento por inductor común
common language lenguaje común
common logarithm logaritmo común
common logic lógica común
common machine language lenguaje de máquina común
common mode modo común
common-mode conversion conversión de modo común
common-mode gain ganancia de modo común
common-mode input circuit circuito de entrada de modo común
common-mode input impedance impedancia de entrada de modo común
common-mode input signal señal de entrada de modo común
common-mode interface interfaz de modo común
common-mode interference interferencia de modo común
common-mode noise ruido de modo común
common-mode rejection rechazo de modo común
common-mode rejection ratio razón de rechazo de modo común
common-mode rejection voltage tensión de rechazo de modo común
common-mode signal señal de modo común
common-mode voltage tensión de modo común
common-mode voltage amplification amplificación de tensión de modo común
common-mode voltage gain ganancia de tensión de modo común
common-mode voltage range intervalo de tensión de modo común
common-mode voltage rejection rechazo de tensión de modo común
common negative negativo común
common network red común
common pool memoria común
common program programa común
common-resistor coupling acoplamiento por resistor común
common return retorno común
common segment segmento común
common service area área de servicio común
common software programas comunes
common-source amplifier amplificador de fuente común
common-source circuit circuito de fuente común
common storage almacenamiento común
common storage area área de almacenamiento común
common supply alimentación común
common system sistema común
common working frequency frecuencia de trabajo común
communication comunicación
communication adapter adaptador de comunicaciones
communication area área de comunicaciones
communication band banda de comunicación
communication card tarjeta de comunicaciones
communication center centro de comunicaciones
communication channel canal de comunicación

communication chart diagrama de comunicaciones
communication circuit circuito de comunicación
communication conductor conductor de comunicación
communication control control de comunicaciones
communication control character carácter de control de comunicaciones
communication control unit unidad de control de comunicaciones
communication controller controlador de comunicaciones
communication device dispositivo de comunicaciones
communication file archivo de comunicaciones
communication file definition definición de archivo de comunicaciones
communication identifier identificador de comunicación
communication interface interfaz de comunicaciones
communication interface circuit circuito de interfaz de comunicaciones
communication interrupt control program programa de control de interrupción de comunicaciones
communication line adapter adaptador de línea de comunicaciones
communication lines líneas de comunicaciones
communication link enlace de comunicaciones
communication medium medio de comunicación
communication network management administración de red de comunicaciones
communication parameters parámetros de comunicaciones
communication parameters list lista de parámetros de comunicaciones
communication port puerto de comunicaciones
communication processor procesador de comunicaciones
communication queue cola de comunicaciones
communication region región de comunicación
communication reliability confiabilidad de comunicación
communication routing encaminamiento de comunicaciones
communication server servidor de comunicaciones
communication services servicios de comunicaciones
communication system sistema de comunicaciones
communication terminal terminal de comunicaciones
communication tower torre de comunicaciones
communication traffic tráfico de comunicaciones
communications buffer memoria intermedia de comunicaciones
communications cable cable de comunicaciones
communications computer computadora de comunicaciones
communications control device dispositivo de control de comunicaciones
communications control program programa de control de comunicaciones
communications control station estación de control de comunicaciones
communications facilities instalaciones de comunicaciones
communications interface equipment equipo de interfaz de comunicaciones
communications monitor monitor de comunicaciones
communications network red de comunicaciones
communications-oriented computer computadora orientada a comunicaciones
communications processing procesamiento de comunicaciones

communications program programa de comunicaciones
communications protocol protocolo de comunicaciones
communications receiver receptor de comunicaciones
communications recording registro de comunicaciones
communications satellite satélite de comunicaciones
communications software programas de comunicaciones
communications terminal terminal de comunicaciones
communications test set conjunto de prueba de comunicaciones
communications word palabra de comunicaciones
communications zone indicator indicador de zona de comunicaciones
community-antenna television televisión por antena comunitaria
community television system sistema de televisión comunitaria
commutating angle ángulo de conmutación
commutating capacitor capacitor de conmutación
commutating field campo de conmutación
commutating group grupo de conmutación
commutating impedance impedancia de conmutación
commutating period periodo de conmutación
commutating pole polo de conmutación
commutating reactance reactancia de conmutación
commutating rectifier rectificador de conmutación
commutating resistance resistencia de conmutación
commutating voltage tensión de conmutación
commutating winding arrollamiento de conmutación
commutation conmutación
commutation elements elementos de conmutación
commutation factor factor de conmutación
commutation failure falla de conmutación
commutation frequency frecuencia de conmutación
commutation interval intervalo de conmutación
commutator conmutador, colector
commutator bar delga de colector
commutator brush escobilla de colector
commutator induction motor motor de inducción de colector
commutator insulation aislamiento de colector
commutator motor motor de colector
commutator retaining ring anillo de retención de colector
commutator ripple ondulación de colector
commutator riser unión de colector
commutator segment segmento de colector
commutator sleeve manguito de colector
commutator surface superficie de colector
compact disk disco compacto
compact system sistema compacto
compacting compactación
compaction compactación
compander compresor-expansor
companding compresión-expansión
companding law ley de compresión-expansión
companion keyboard teclado complementario
companion loudspeaker altavoz complementario
comparative comparativo
comparative sort clasificación comparativa
comparative test prueba comparativa
comparator comparador
comparator check comprobación de comparador
comparator circuit circuito comparador
compare data comparar datos
compare instruction instrucción de comparar
comparing unit unidad de comparación
comparison amplifier amplificador de comparación

comparison bridge puente de comparación
comparison definition definición de comparación
comparison electrode electrodo de comparación
comparison equipment equipo de comparación
comparison error error de comparación
comparison lamp lámpara de comparación
comparison measurement medida por comparación
comparison operation operación de comparación
comparison operator operador de comparación
comparison receiver receptor de comparación
comparison scale escala de comparación
comparison standard patrón de comparación
comparison station estación de comparación
comparison surface superficie de comparación
comparison test prueba de comparación
comparison value valor de comparación
compass compás, brújula
compass bearing rumbo por compás
compatibility compatibilidad
compatibility circuit circuito de compatibilidad
compatibility mode modo de compatibilidad
compatibility test prueba de compatibilidad
compatible connector conector compatible
compatible hardware equipo físico compatible
compatible integrated circuit circuito integrado compatible
compatible monolithic integrated circuit circuito integrado monolítico compatible
compatible software programas compatibles
compatible system sistema compatible
compensated amplifier amplificador compensado
compensated current transformer transformador de corriente compensado
compensated diode detector detector de diodo compensado
compensated-impurity resistor resistor con impurezas compensadas
compensated ionization chamber cámara de ionización compensada
compensated-loop direction-finder radiogoniómetro de cuadro compensado
compensated motor motor compensado
compensated preamplifier preamplificador compensado
compensated scan exploración compensada
compensated semiconductor semiconductor compensado
compensated video amplifier videoamplificador compensado
compensated volume control control de volumen compensado
compensated winding arrollamiento compensado
compensating adjustment ajuste compensador
compensating capacitor capacitor compensador
compensating charge carga compensadora
compensating circuit circuito compensador
compensating coil bobina compensadora
compensating current corriente compensadora
compensating diode diodo compensador
compensating element elemento compensador
compensating error error compensador
compensating filter filtro compensador
compensating magnet imán compensador
compensating network red compensadora
compensating resistor resistor compensador
compensating voltage tensión compensadora
compensating wave onda compensadora
compensating winding arrollamiento compensador
compensation compensación
compensation filter filtro de compensación
compensation resistor resistor de compensación
compensation signal señal de compensación
compensation theorem teorema de compensación

compensator compensador
compensator starter arrancador compensador
compensatory compensatorio
compilation compilación
compilation time tiempo de compilación
compilation unit unidad de compilación
compilator compilador
compile compilar
compile-and-go compilar y ejecutar
compile duration duración de compilación
compile phase fase de compilación
compile routine rutina de compilación
compile time tiempo de compilación
compiler compilador
compiler-based system sistema basado en el
   compilador
compiler code código de compilador
compiler diagnostics diagnósticos de compilador
compiler generator generador de compilador
compiler language lenguaje de compilador
compiler options opciones de compilador
compiler program programa de compilador
compiler system sistema de compilador
compiling compilación
compiling computer computadora de compilación
compiling duration duración de compilación
compiling phase fase de compilación
compiling program programa de compilación
compiling routine rutina de compilación
compiling time tiempo de compilación
complement complemento
complement base base de complemento
complement form forma de complemento
complement procedure procedimiento de
   complemento
complementary complementario
complementary antennas antenas complementarias
complementary colors colores complementarios
complementary commutation conmutación
   complementaria
complementary flip-flop basculador complementario
complementary function función complementaria
complementary metal-oxide semiconductor
   semiconductor metal-óxido complementario
complementary operation operación complementaria
complementary operator operador complementario
complementary pair par complementario
complementary pushpull circuit circuito en
   contrafase complementario
complementary rectifier rectificador
   complementario
complementary symmetry simetría complementaria
complementary-symmetry circuit circuito de
   simetría complementaria
complementary-symmetry device dispositivo de
   simetría complementaria
complementary-transistor amplifier amplificador de
   transistores complementarios
complementary-transistor pair par de transistores
   complementarios
complementary transistors transistores
   complementarios
complementary wave onda complementaria
complementary wavelength longitud de onda
   complementaria
complementation complementación
complementer complementador
complementing circuit circuito complementario
complementing mode modo complementario
complete binary tree árbol binario completo
complete carry arrastre completo
complete circuit circuito completo
complete connection conexión completa

complete diffusion difusión completa
complete impulse impulso completo
complete instruction instrucción completa
complete modulation modulación completa
complete operation operación completa
complete routine rutina completa
complete self-excitation autoexcitación completa
complete suppression supresión completa
completed call llamada completada
completion code código de terminación
completion message mensaje de terminación
complex bipolar bipolar complejo
complex cathode cátodo complejo
complex condition condición compleja
complex conductivity conductividad compleja
complex data datos complejos
complex display indicador visual complejo
complex function función compleja
complex impedance impedancia compleja
complex notation notación compleja
complex number número complejo
complex operator operador complejo
complex periodic wave onda periódica compleja
complex permeability permeabilidad compleja
complex permittivity permitividad compleja
complex plane plano complejo
complex quantity cantidad compleja
complex radar target blanco de radar complejo
complex radiation radiación compleja
complex reflector reflector complejo
complex relative attenuation atenuación relativa
   compleja
complex signal señal compleja
complex text texto complejo
complex tone tono complejo
complex variable variable compleja
complex waveform forma de onda compleja
complex waves ondas complejas
complexity complejidad
compliance docilidad, elasticidad
compole compolo
component componente
component address dirección de componente
component aging envejecimiento de componentes
component assembly ensamblaje de componentes
component density densidad de componentes
component error error de componente
component failure falla de componente
component failure rate tasa de fallas de
   componentes
component function función de componente
component layout distribución de componentes
component stress estrés de componentes
component symbol símbolo de componente
composite compuesto
composite attenuation atenuación compuesta
composite cable cable compuesto
composite card tarjeta compuesta
composite circuit circuito compuesto
composite color monitor monitor de colores
   compuesto
composite color signal señal de colores compuesta
composite conductor conductor compuesto
composite console consola compuesta
composite controlling voltage tensión de control
   compuesta
composite current corriente compuesta
composite electrode electrodo compuesto
composite file archivo compuesto
composite filter filtro compuesto
composite gain ganancia compuesta
composite key clave compuesta
composite level nivel compuesto

composite link enlace compuesto
composite loudspeaker altavoz compuesto
composite module módulo compuesto
composite operator operador compuesto
composite picture signal videoseñal compuesta
composite plate depósito compuesto
composite pulse impulso compuesto
composite signal señal compuesta
composite symbol símbolo compuesto
composite synchronization signal señal de sincronización compuesta
composite television signal señal de televisión compuesta
composite transducer transductor compuesto
composite tube tubo compuesto
composite type tipo compuesto
composite video video compuesto
composite video signal videoseñal compuesta
composite voltage tensión compuesta
composite wave filter filtro de ondas compuesto
composite waveform forma de onda compuesta
composition capacitor capacitor de composición
composition resistor resistor de composición
compound compuesto
compound brush escobilla compuesta
compound circuit circuito compuesto
compound condition condición compuesta
compound connection conexión compuesta
compound cryosar criosar compuesto
compound excitation excitación compuesta
compound filter filtro compuesto
compound generator generador compuesto
compound horn bocina compuesta
compound lens lente compuesta
compound logic lógica compuesta
compound logical element elemento lógico compuesto
compound modulation modulación compuesta
compound motor motor compuesto
compound resonator resonador compuesto
compound signal señal compuesta
compound statement sentencia compuesta
compound switching conmutación compuesta
compound transistor transistor compuesto
compound winding arrollamiento compuesto
compress comprimir
compressed-air capacitor capacitor de aire comprimido
compressed-air speaker altavoz de aire comprimido
compressed encoding codificación comprimida
compressed file archivo comprimido
compressed mode modo comprimido
compressed response respuesta comprimida
compressed video video comprimido
compression compresión
compression cone cono de compresión
compression factor factor de compresión
compression phase fase de compresión
compression ratio razón de compresión
compression wave onda de compresión
compressor compresor
compressor amplifier amplificador compresor
compressor network red compresora
Compton absorption absorción de Compton
Compton diffusion difusión de Compton
Compton effect efecto de Compton
Compton meter medidor de Compton
Compton scattering dispersión de Compton
Compton shift cambio de Compton
compulsory equipment equipo compulsorio
computability computabilidad
computation computación
computational computacional

computational ability capacidad computacional
computational error error computacional
computational load carga computacional
computational power potencia computacional
computational stability estabilidad computacional
compute computar
compute mode modo de computar
computed computarizado
computed tomography tomografía computarizada
computer computadora, ordenador
computer-aided design diseño asistido por computadora
computer-aided engineering ingeniería asistida por computadora
computer-aided inspection inspección asistida por computadora
computer-aided instruction instrucción asistida por computadora
computer-aided manufacturing manufactura asistida por computadora
computer-aided production producción asistida por computadora
computer antibody anticuerpo de computadora
computer application aplicación de computadora
computer architecture arquitectura de computadora
computer assisted asistido por computadora
computer-assisted diagnosis diagnóstico asistido por computadora
computer-assisted instruction instrucción asistida por computadora
computer-assisted tomography tomografía asistida por computadora
computer backed respaldado por computadora
computer based basado en computadora
computer-based instruction instrucción basada en computadora
computer-based learning aprendizaje basado en computadora
computer-based training entrenamiento basado en computadora
computer capacity capacidad de computadora
computer center centro de computadoras
computer circuits circuitos de computadora
computer code código de computadora
computer communications comunicaciones por computadora
computer-communications system sistema de comunicaciones por computadora
computer compatible compatible con computadora
computer component componente de computadora
computer configuration configuración de computadora
computer console consola de computadora
computer control control por computadora
computer control console consola de control por computadora
computer control mode modo de control por computadora
computer control panel panel de control por computadora
computer controlled controlado por computadora
computer controlled network red controlada por computadora
computer data datos de computadora
computer dependent program programa dependiente de computadora
computer design diseño de computadora
computer diode diodo de computadora
computer directed dirigido por computadora
computer driven accionado por computadora
computer efficiency eficiencia de computadora
computer engineer ingeniero de computadoras
computer engineering ingeniería de computadoras

computer equipment equipo de computadora
computer file archivo de computadora
computer game juego de computadora
computer generated generado por computadora
computer graphics gráficos de computadora
computer image processing procesamiento de imágenes por computadora
computer independent language lenguaje independiente de computadoras
computer information system sistema de información de computadoras
computer input entrada de computadora
computer instruction instrucción de computadora
computer instruction code código de instrucciones de computadora
computer instruction set conjunto de instrucciones de computadora
computer integrated manufacturing manufactura integrada por computadora
computer interface unit unidad de interfaz de computadora
computer language lenguaje de computadora
computer learning aprendizaje por computadora
computer-limited limitado por computadora
computer-limited operation operación limitada por computadora
computer log diario de computadora
computer machinery maquinaria de computadora
computer managed administrado por computadora
computer memory memoria de computadora
computer meter medidor de computadora
computer network red de computadora
computer network components componentes de red de computadoras
computer numerical control control numérico de computadoras
computer operation operación de computadora
computer operator operador de computadora
computer-oriented language lenguaje orientado a computadoras
computer output salida de computadora
computer power potencia de computadora
computer processing procesamiento de computadora
computer processing cycle ciclo de procesamiento de computadora
computer program programa de computadora
computer programmer programador de computadora
computer programming programación de computadora
computer programming language lenguaje de programación de computadora
computer resource allocation asignación de recursos de computadora
computer ribbon cinta de computadora
computer room sala de computadoras
computer routine rutina de computadora
computer section sección de computadoras
computer security seguridad de computadoras
computer services servicio de computadoras
computer simulated simulado por computadora
computer simulation simulación por computadora
computer simulator simulador de computadora
computer storage almacenamiento de computadora
computer storage tube tubo de almacenamiento de computadora
computer stored almacenado por computadora
computer stored data datos almacenados en computadora
computer system sistema de computadora
computer tape cinta de computadora
computer technician técnico de computadora
computer-television interface interfaz

computadora-televisión
computer terminal terminal de computadora
computer time tiempo de computadora
computer virus virus de computadora
computer word palabra de computadora
computerized computarizado
computerized axial tomography tomografía axial computarizada
computerized data datos computarizados
computerized file archivo computarizado
computerized mail correo computarizado
computerized numerical control control numérico computarizado
computerized report informe computarizado
computing computación
computing amplifier amplificador computador
computing efficiency eficiencia de cómputo
computing element elemento computador
computing equipment equipo computador
computing link enlace computador
computing machine máquina computadora
computing network red de cómputo
computing power potencia de cómputo
computing speed velocidad de cómputo
comraz comraz
concatenate concatenar
concatenated concatenado
concatenated data set conjunto de datos concatenados
concatenated field campo concatenado
concatenation concatenación
concealed wiring alambrado oculto
concentrated-arc lamp lámpara de arco concentrado
concentrated beam haz concentrado
concentrated inductance inductancia concentrada
concentrated winding arrollamiento concentrado
concentration cell pila de concentración
concentration electrode electrodo de concentración
concentration gradient gradiente de concentración
concentration line línea de concentración
concentration polarization polarización por concentración
concentration position posición de concentración
concentrator concentrador
concentric concéntrico
concentric cable cable concéntrico
concentric capacitor capacitor concéntrico
concentric conductor conductor concéntrico
concentric control control concéntrico
concentric electrode system sistema de electrodos concéntricos
concentric grid rejilla concéntrica
concentric jack jack concéntrico
concentric line línea concéntrica
concentric-line oscillator oscilador de línea concéntrica
concentric pair par concéntrico
concentric plug clavija concéntrica
concentric receptacle receptáculo concéntrico
concentric resonator resonador concéntrico
concentric transmission line línea de transmisión concéntrica
concentric winding arrollamiento concéntrico
concentricity concentricidad
concept phase fase de concepto
conceptual design diseño conceptual
concordance concordancia
concordance file archivo de concordancia
concurrency concurrencia
concurrency logic lógica de concurrencia
concurrency management administración de concurrencia
concurrent concurrente

concurrent computer computadora concurrente
concurrent control control concurrente
concurrent control system sistema de control concurrente
concurrent conversion conversión concurrente
concurrent execution ejecución concurrente
concurrent jobs tareas concurrentes
concurrent operating control control operativo concurrente
concurrent operation operación concurrente
concurrent processes procesos concurrentes
concurrent processing procesamiento concurrente
concurrent program execution ejecución de programas concurrentes
concurrent programming programación concurrente
concurrent programming language lenguaje de programación concurrente
concurrent programs programas concurrentes
concurrent reorganization reorganización concurrente
concurrent tasks tareas concurrentes
condensation condensación
condensation coefficient coeficiente de condensación
condensation temperature temperatura de condensación
condensed instructions instrucciones condensadas
condenser condensador
condenser bank banco de condensadores
condenser checker comprobador de condensador
condenser coupling acoplamiento por condensador
condenser electroscope electroscopio de condensador
condenser leakage fuga de condensador
condenser loudspeaker altavoz de condensador
condenser microphone micrófono de condensador
condenser pickup fonocaptor de condensador
condenser storage almacenamiento en condensador
condensing routine rutina condensadora
condition code código de condición
condition entry especificación de condición
condition indicator indicador de condición
condition name nombre de condición
condition value valor de condición
conditional condicional
conditional assembly ensamblaje condicional
conditional branch rama condicional
conditional branch instruction instrucción de rama condicional
conditional breakpoint interrupción condicional
conditional breakpoint instruction instrucción de interrupción condicional
conditional call llamada condicional
conditional capture captura condicional
conditional code código condicional
conditional construction construcción condicional
conditional control control condicional
conditional control structure estructura de control condicional
conditional control transfer transferencia de control condicional
conditional dump vaciado condicional
conditional entry entrada condicional
conditional entry code código de entrada condicional
conditional expression expresión condicional
conditional instruction instrucción condicional
conditional jump salto condicional
conditional operation operación condicional
conditional operator operador condicional
conditional restart reiniciación condicional
conditional signal señal condicional
conditional stability estabilidad condicional
conditional statement sentencia condicional
conditional stop parada condicional

conditional stop instruction instrucción de parada condicional
conditional test prueba condicional
conditional transfer transferencia condicional
conditional transfer of control transferencia condicional de control
conditional variable variable condicional
conditioned carrier portadora condicionada
conditioning acondicionamiento
conductance conductancia
conductance meter conductómetro
conductance relay relé de conductancia
conducting band banda de conducción
conducting coating revestimiento conductor
conducting direction dirección de conducción
conducting element elemento de conducción
conducting interval intervalo de conducción
conducting layer capa de conducción
conducting material material conductor
conducting medium medio conductor
conducting paint pintura conductora
conducting period periodo de conducción
conducting plane plano de conducción
conducting salt sal conductora
conduction conducción
conduction angle ángulo de conducción
conduction band banda de conducción
conduction cooling enfriamiento por conducción
conduction current corriente de conducción
conduction-current modulation modulación de corriente de conducción
conduction electron electrón de conducción
conduction error error de conducción
conduction field campo de conducción
conduction interval intervalo de conducción
conduction ratio razón de conducción
conductive conductor
conductive cell célula conductora
conductive coating revestimiento conductor
conductive coupling acoplamiento conductor
conductive gasket empaque conductor
conductive material material conductor
conductive medium medio conductor
conductive paint pintura conductora
conductive pattern patrón conductor
conductive pencil lápiz conductor
conductive plastic plástico conductor
conductivity conductividad
conductivity bridge puente de conductividad
conductivity cell célula de conductividad
conductivity chamber cámara de conductividad
conductivity meter medidor de conductividad
conductivity modulation modulación de conductividad
conductivity-modulation transistor transistor de modulación de conductividad
conductor conductor
conductor density densidad de conductores
conductor flat cable cable plano de conductores
conductor insulation aislamiento de conductor
conductor loss pérdida por conductor
conductor system sistema de conductores
conductor temperature temperatura de conductor
conduit conducto
conduit system sistema de conductos
cone antenna antena cónica
cone of nulls cono de radiación nula
cone of protection cono de protección
cone of silence cono de silencio
cone resonance resonancia de cono
cone speaker altavoz de cono
cónference call llamada de conferencia
conference circuit circuito de conferencia

conference connection conexión de conferencia
conference repeater repetidor de conferencia
confidence check comprobación de confianza
confidence factor factor de confianza
confidence interval intervalo de confianza
confidence level nivel de confianza
confidence limits límites de confianza
confidence test prueba de confianza
configurable configurable
configuration configuración
configuration control control de configuración
configuration dependent dependiente de configuración
configuration factor factor de configuración
configuration file archivo de configuración
configuration image imagen de configuración
configuration independent independiente de configuración
configuration section sección de configuración
configuration state estado de configuración
confined beam haz confinado
confined electron flow flujo de electrones confinado
confocal confocal
confocal resonator resonador confocal
conformal antenna antena conforme
conformal array red conforme
conformance conformidad
conformance test prueba de conformidad
confusion region región de confusión
conical antenna antena cónica
conical array red cónica
conical beam haz cónico
conical horn bocina cónica
conical-horn antenna antena de bocina cónica
conical monopole antenna antena monopolar cónica
conical scanning exploración cónica
conical sweep barrido cónico
conical wave onda cónica
conjugate branches ramas conjugadas
conjugate bridge puente conjugado
conjugate conductors conductores conjugados
conjugate impedances impedancias conjugadas
connect conectar
connect time tiempo de conexión
connected load carga conectada
connected position posición conectada
connected system sistema conectado
connecting cable cable de conexión
connecting circuit circuito de conexión
connecting cord cordón de conexión
connecting device dispositivo de conexión
connecting diagram diagrama de conexión
connecting plug clavija de conexión
connecting relay relé de conexión
connecting terminal terminal de conexión
connecting wire hilo de conexión
connection conexión
connection assembly ensamblaje de conexión
connection bar barra de conexión
connection box caja de conexiones
connection contact contacto de conexión
connection diagram diagrama de conexiones
connection identifier indentificador de conexión
connection insulation aislamiento de conexión
connection number número de conexión
connection plate placa de conexión
connection terminal terminal de conexión
connection time tiempo de conexión
connectionless sin conexiones
connectionless service servicio sin conexiones
connectivity conectividad
connectivity platform plataforma de conectividad
connector conector

connector interface interfaz de conector
connector style estilo de conector
connector switch conmutador de conector
connector variant variante de conector
conoscope conoscopio
consecutive data set conjunto de datos consecutivos
consecutive numbering numeración consecutiva
consecutive operation operación consecutiva
consecutive organization organización consecutiva
consecutive processing procesamiento consecutivo
consequent poles polos consecuentes
conservation of energy conservación de energía
conservation of radiance conservación de radiancia
conservative specification especificación conservadora
consistency check comprobación de consistencia
console consola
console display visualización de consola
console function función de consola
console keyboard teclado de consola
console printer impresora de consola
console switch conmutador de consola
consonance consonancia
constant constante
constant-amplitude carrier portadora de amplitud constante
constant-amplitude recording registro de amplitud constante
constant area área constante
constant bandwidth ancho de banda constante
constant-conductance network red de conductancia constante
constant current corriente constante
constant-current characteristic característica de corriente constante
constant-current charge carga de corriente constante
constant-current curve curva de corriente constante
constant-current filter filtro de corriente constante
constant-current generator generador de corriente constante
constant-current governor regulador de corriente constante
constant-current modulation modulación de corriente constante
constant-current motor motor de corriente constante
constant-current power supply fuente de alimentación de corriente constante
constant-current regulation regulación de corriente constante
constant-current regulator regulador de corriente constante
constant-current response respuesta de corriente constante
constant-current retention retención de corriente constante
constant-current source fuente de corriente constante
constant-current supply fuente de corriente constante
constant-current system sistema de corriente constante
constant-current transformer transformador de corriente constante
constant data datos constantes
constant-delay discriminator discriminador de retardo constante
constant efficiency eficiencia constante
constant factor factor constante
constant field campo constante
constant-frequency control control por frecuencia constante
constant-gain amplifier amplificador de ganancia constante

constant-impedance line línea de impedancia constante
constant-impedance multiplexer multiplexor de impedancia constante
constant-impedance system sistema de impedancia constante
constant-k filter filtro de k constante
constant-k network red de k constante
constant-luminance signal señal de luminancia constante
constant-luminance system sistema de luminancia constante
constant-luminance transmission transmisión de luminancia constante
constant multiplier multiplicador constante
constant-potential regulator regulador de potencial constante
constant-power dissipation line línea de disipación de potencia constante
constant-power motor motor de potencia constante
constant-ratio code código de razón constante
constant-resistance network red de resistencia constante
constant-resistance structure estructura de resistencia constante
constant-speed motor motor de velocidad constante
constant-tone generator generador de tono constante
constant-torque motor motor de par constante
constant-velocity recorder grabadora de velocidad constante
constant-velocity recording grabación a velocidad constante
constant voltage tensión constante
constant-voltage charge carga de tensión constante
constant-voltage generator generador de tensión constante
constant-voltage line línea de tensión constante
constant-voltage rectifier rectificador de tensión constante
constant-voltage regulation regulación de tensión constante
constant-voltage regulator regulador de tensión constante
constant-voltage source fuente de tensión constante
constant-voltage system sistema de tensión constante
constant-voltage transformer transformador de tensión constante
constantan constantán
construction test prueba de construcción
consumer reliability risk riesgo de confiabilidad de consumidor
contact contacto
contact alignment alineación de contactos
contact arc arco de contacto
contact area área de contacto
contact arrangement arreglo de contactos
contact bias polarización de contacto
contact blade cuchilla de contacto
contact block bloque de contacto
contact bounce rebote de contacto
contact breaker interruptor de contacto
contact button botón de contacto
contact chatter vibración de contacto
contact circuit circuito de contacto
contact clip pinza de contacto
contact closure cierre de contacto
contact combination combinación de contactos
contact conductor conductor de contacto
contact corrosion corrosión de contacto
contact detector detector de contacto
contact drop caída de contacto
contact effect efecto de contacto
contact electricity electricidad de contacto

contact electrode electrodo de contacto
contact electromotive force fuerza electromotriz de contacto
contact element elemento de contacto
contact encoder codificador de contacto
contact erosion erosión de contacto
contact film película de contacto
contact finger dedo de contacto
contact follow acompañamiento de contacto
contact force fuerza de contacto
contact gap espacio de contacto
contact input entrada de contacto
contact line línea de contacto
contact-making meter medidor de contacto
contact material material de contactos
contact mechanism mecanismo de contacto
contact microphone micrófono de contacto
contact miss pérdida de contacto
contact modulator modulador de contacto
contact noise ruido de contacto
contact pin terminal de contacto
contact piston pistón de contacto
contact plating depósito por contacto
contact potential potencial de contacto
contact-potential barrier barrera de potencial de contacto
contact-potential bias polarización por potencial de contacto
contact-potential difference diferencia de potencial de contacto
contact-potential energy barrier barrera de energía de potencial de contacto
contact pressure presión de contacto
contact protection protección de contacto
contact protector protector de contacto
contact rail carril de contacto
contact rectifier rectificador de contacto
contact resistance resistencia de contacto
contact sense module módulo de detección de contactos
contact separating force fuerza separadora de contacto
contact separation separación de contactos
contact size tamaño de contacto
contact strip regleta de contacto
contact surface superficie de contacto
contact switch conmutador de contacto
contact travel distancia de recorrido de contacto
contact tube tubo de contacto
contact unit unidad de contacto
contact voltage tensión de contacto
contact voltmeter vóltmetro de contacto
contact wire hilo de contacto
contactless sin contactos
contactor contactor
contactor button botón contactor
contactor tube tubo contactor
contaminant contaminante
contaminated material material contaminado
contamination contaminación
content-addressed storage almacenamiento direccionado por contenido
contextual declaration declaración contextual
contiguous contiguo
contiguous alphabet alfabeto contiguo
contiguous area área contigua
contiguous data structure estructura de datos contiguos
continental code código continental
contingency interrupt interrupción de contingencia
contingency procedure procedimiento de contingencia
continuation card tarjeta de continuación

continuation column columna de continuación
continuation file archivo de continuación
continuation indicator indicador de continuación
continuation line línea de continuación
continuity continuidad
continuity check comprobación de continuidad
continuity test prueba de continuidad
continuity tester probador de continuidad
continuous air monitor monitor de aire continuo
continuous amplifier amplificador continuo
continuous audible signal señal audible continua
continuous cards tarjetas continuas
continuous carrier portadora continua
continuous circuit circuito continuo
continuous commercial service servicio comercial continuo
continuous comparator comparador continuo
continuous control control continuo
continuous-control system sistema de control continuo
continuous corona corona continua
continuous coverage cobertura continua
continuous cumulative demand demanda acumulativa continua
continuous current corriente continua
continuous-current test prueba de corriente continua
continuous duty servicio continuo
continuous-duty circuit circuito de servicio continuo
continuous-duty current corriente de servicio continuo
continuous-duty current rating clasificación de corriente de servicio continuo
continuous-duty rating clasificación de servicio continuo
continuous electrode electrodo continuo
continuous error error continuo
continuous flow flujo continuo
continuous function función continua
continuous lighting alumbrado continuo
continuous load carga continua
continuous load rating clasificación de carga continua
continuous loading carga continua
continuous-loop operation operación de bucle continuo
continuous mark marca continua
continuous memory memoria continua
continuous mode modo continuo
continuous network red continua
continuous noise ruido continuo
continuous operation operación continua
continuous oscillation oscilación continua
continuous path operation operación de trayectoria continua
continuous power potencia continua
continuous processing procesamiento continuo
continuous pulse impulso continuo
continuous radiation radiación continua
continuous rating régimen continuo
continuous recorder registrador continuo
continuous service servicio continuo
continuous signal señal continua
continuous simulation simulación continua
continuous spectrum espectro continuo
continuous system sistema continuo
continuous test prueba continua
continuous tone tono continuo
continuous tone image imagen de tono continuo
continuous tuner sintonizador continuo
continuous variable variable continua
continuous wave onda continua
continuous-wave filter filtro de onda continua
continuous-wave laser láser de onda continua

continuous-wave magnetron magnetrón de ondas continuas
continuous-wave monitor monitor de onda continua
continuous-wave oscillator oscilador de onda continua
continuous-wave radar radar de onda continua
continuous-wave reference signal señal de referencia de onda continua
continuous-wave tracking system sistema de seguimiento de onda continua
continuous-wave transmitter transmisor de ondas continuas
continuous welding soldadura continua
continuous X-rays rayos X continuos
continuously adjustable continuamente ajustable
continuously-adjustable inductor inductor continuamente ajustable
continuously-adjustable transformer transformador continuamente ajustable
contour contorno
contradirectional timing sincronización contradireccional
contrast contraste
contrast control control de contraste
contrast enhancement realzado de contraste
contrast gradient gradiente de contraste
contrast hue tinte de contraste
contrast photometer fotómetro de contraste
contrast potentiometer potenciómetro de contraste
contrast range intervalo de contraste
contrast ratio razón de contraste
contrast reduction reducción de contraste
contrast sensitivity sensibilidad de contraste
control accuracy exactitud de control
control action acción de control
control agent agente de control
control apparatus aparato de control
control application aplicación de control
control area área de control
control battery batería de control
control block bloque de control
control board tablero de control
control box caja de control
control-break control-interrupción
control brush escobilla de control
control bus bus de control
control byte byte de control
control cable cable de control
control card tarjeta de control
control card specifications especificaciones de tarjeta de control
control center centro de control
control change cambio de control
control character carácter de control
control characteristic característica de control
control chart diagrama de control
control circuit circuito de control
control-circuit transformer transformador de circuito de control
control-circuit voltage tensión de circuito de control
control code código de control
control command mando de control
control compartment compartimiento de control
control computer computadora de control
control conductor conductor de control
control console consola de control
control contact contacto de control
control counter contador de control
control coupling acoplamiento de control
control cubicle cubículo de control
control current corriente de control
control cycle ciclo de control
control data datos de control

control data field campo de datos de control
control desk pupitre de control
control device dispositivo de control
control electrode electrodo de control
control element elemento de control
control equipment equipo de control
control field campo de control
control field line línea de campo de control
control flow flujo de control
control flux flujo de control
control format formato de control
control function función de control
control generator generador de control
control grid rejilla de control
control-grid bias polarización de rejilla de control
control-grid current corriente de rejilla de control
control-grid injection inyección por rejilla de control
control-grid modulation modulación por rejilla de control
control group grupo de control
control hole perforación de control
control inductor inductor de control
control information información de control
control instruction instrucción de control
control-instruction register registro de instrucción de control
control interval intervalo de control
control interval access acceso de intervalo de control
control interval split división de intervalo de control
control key tecla de control
control language lenguaje de control
control language program programa de lenguaje de control
control language variable variable de lenguaje de control
control level nivel de control
control level indicator indicador de nivel de control
control light luz de control
control line línea de control
control logic lógica de control
control loop bucle de control
control loudspeaker altavoz de control
control machine máquina de control
control magnet imán de control
control mark marca de control
control mechanism mecanismo de control
control memory memoria de control
control method método de control
control microprogram microprograma de control
control mode modo de control
control module módulo de control
control module area área de módulo de control
control number número de control
control operation operación de control
control operator operador de control
control output salida de control
control output module módulo de salida de control
control panel panel de control
control path trayectoria de control
control point punto de control
control position posición de control
control potentiometer potenciómetro de control
control precision precisión de control
control procedure procedimiento de control
control program programa de control
control punching perforación de control
control quantity cantidad de control
control radar radar de control
control range intervalo de control
control ratio razón de control
control record registro de control
control record card tarjeta de registro de control
control rectifier rectificador de control

control region región de control
control register registro de control
control relationship relación de control
control relay relé de control
control room sala de control
control routine rutina de control
control routine interrupt interrupción de rutina de control
control scheme esquema de control
control section sección de control
control sequence secuencia de control
control sequencer secuenciador de control
control sheet hoja de control
control signal señal de control
control software programas de control
control specification especificación de control
control standards patrones de control
control statement sentencia de control
control statement analyzer analizador de sentencias de control
control station estación de control
control stop parada de control
control storage almacenamiento de control
control storage processor procesador de almacenamiento de control
control structure estructura de control
control switch conmutador de control
control switching control control de conmutación de control
control system sistema de control
control table mesa de control
control tape cinta de control
control technique técnica de control
control terminal terminal de control
control total total de control
control tower torre de control
control track pista de control
control transfer transferencia de control
control transfer instruction instrucción de transferencia de control
control transfer statement sentencia de transferencia de control
control transformer transformador de control
control tube tubo de control
control unit unidad de control
control voltage tensión de control
control-voltage amplifier amplificador de tensión de control
control winding arrollamiento de control
control word palabra de control
control zone zona de control
controllable controlable
controllable load carga controlable
controlled access acceso controlado
controlled airspace espacio aéreo controlado
controlled area área controlada
controlled availability disponibilidad controlada
controlled avalanche avalancha controlada
controlled-avalanche device dispositivo de avalancha controlada
controlled-avalanche diode diodo de avalancha controlada
controlled-avalanche rectifier rectificador de avalancha controlada
controlled carrier portadora controlada
controlled-carrier modulation modulación de portadora controlada
controlled-carrier modulator modulador de portadora controlada
controlled-carrier transmission transmisión de portadora controlada
controlled circuit circuito controlado
controlled condition condición controlada

controlled-coupling transformer transformador de acoplamiento controlado
controlled electric drainage drenaje eléctrico controlado
controlled environment ambiente controlado
controlled flight vuelo controlado
controlled list lista controlada
controlled medium medio controlado
controlled memory memoria controlada
controlled output salida controlada
controlled overvoltage test prueba de sobretensión controlada
controlled parameter parámetro controlado
controlled rectifier rectificador controlado
controlled sender emisor controlado
controlled system sistema controlado
controlled variable variable controlada
controlled voltage tensión controlada
controlled warm-up time tiempo de calentamiento controlado
controller controlador
controller card tarjeta de controlador
controller configuration configuración de controlador
controller current corriente de controlador
controller data datos de controlador
controller equipment equipo de controlador
controller function función de controlador
controller information información de controlador
controller information field campo de información de controlador
controller section sección de controlador
controller storage almacenamiento de controlador
controller unit unidad de controlador
controlling controlador
controlling circuit circuito controlador
controlling electrode electrodo controlador
controlling element elemento controlador
controlling exchange central controladora
controlling factor factor controlador
controlling file archivo controlador
controlling instrument instrumento controlador
controlling position posición controladora
controlling recorder registrador controlador
controlling section sección controladora
controlling system sistema controlador
convection convección
convection cooling enfriamiento por convección
convection current corriente de convección
convection-current modulation modulación por corriente de convección
convective convectivo
convective diffusion difusión convectiva
convective discharge descarga convectiva
convectron convectrón
convenience outlet tomacorriente de conveniencia, toma de corriente
convenience receptacle receptáculo de conveniencia, enchufe de corriente
conventional current corriente convencional
conventional equipment equipo convencional
conventional programming programación convencional
convergence convergencia
convergence adjustment ajuste de convergencia
convergence amplifier amplificador de convergencia
convergence checker comprobador de convergencia
convergence circuit circuito de convergencia
convergence coil bobina de convergencia
convergence control control de convergencia
convergence correction corrección de convergencia
convergence electrode electrodo de convergencia
convergence factor factor de convergencia

convergence frequency frecuencia de convergencia
convergence function función de convergencia
convergence line línea de convergencia
convergence magnet imán de convergencia
convergence phase control control de fase de convergencia
convergence plane plano de convergencia
convergence ratio razón de convergencia
convergence surface superficie de convergencia
convergence zone zona de convergencia
convergent convergente
convergent beam haz convergente
converging convergente
converging lens lente convergente
converging rays rayos convergentes
conversational compiler compilador conversacional
conversational file archivo conversacional
conversational interaction interacción conversacional
conversational language lenguaje conversacional
conversational mode modo conversacional
conversational operation operación conversacional
conversational processing procesamiento conversacional
conversational programming programación conversacional
conversational system sistema conversacional
conversational terminal terminal conversacional
converse piezoelectric effect efecto piezoeléctrico converso
conversion conversión
conversion code código de conversión
conversion conductance conductancia de conversión
conversion detector detector de conversión
conversion device dispositivo de conversión
conversion efficiency eficiencia de conversión
conversion equipment equipo de conversión
conversion exciter excitador de conversión
conversion factor factor de conversión
conversion gain ganancia de conversión
conversion-gain ratio razón de ganancia de conversión
conversion loss pérdida de conversión
conversion mode modo de conversión
conversion process proceso de conversión
conversion program programa de conversión
conversion rate velocidad de conversión
conversion ratio razón de conversión
conversion routine rutina de conversión
conversion table tabla de conversión
conversion time tiempo de conversión
conversion transconductance transconductancia de conversión
conversion transducer transductor de conversión
conversion transformer transformador de conversión
conversion voltage gain ganancia de tensión de conversión
convert convertir
converted command mando convertido
converter convertidor
converter noise ruido de convertidor
converter system sistema convertidor
converter tube tubo convertidor
converter unit unidad convertidora
convexo-concave convexo-concavo
Conwell-Weisskopf equation ecuación de Conwell-Weisskopf
coolant refrigerante
coolant distribution unit unidad de distribución de refrigerante
cooled infrared detector detector infrarrojo enfriado
Coolidge tube tubo de Coolidge

Coolidge X-ray tube tubo de rayos X de Coolidge
cooling enfriamiento
cooling fan ventilador de enfriamiento
cooling fins aletas de enfriamiento
cooling medium medio de enfriamiento
cooling period periodo de enfriamiento
cooling system sistema de enfriamiento
cooling time tiempo de enfriamiento
cooperative network red cooperativa
coordinate data receiver receptor de datos coordenados
coordinate data transmitter transmisor de datos coordenados
coordinate system sistema de coordenadas
coordination complex complejo de coordinación
coordination distance distancia de coordinación
coordination of insulation coordinación de aislamiento
coplanar electrodes electrodos coplanares
coplanar-grid tube tubo de rejillas coplanares
coplanar loop cuadro coplanar
copolarization copolarización
copper cobre
copper brush escobilla de cobre
copper cable cable de cobre
copper-clad steel acero recubierto de cobre
copper-clad wire hilo recubierto de cobre
copper-constantan thermocouple termopar cobre-constantán
copper loss pérdidas en el cobre
copper-oxide diode diodo de óxido de cobre
copper-oxide modulator modulador de óxido de cobre
copper-oxide photocell fotocélula de óxido de cobre
copper-oxide photovoltaic cell célula fotovoltaica de óxido de cobre
copper-oxide rectifier rectificador de óxido de cobre
copper-sulfide rectifier rectificador de sulfuro de cobre
copper tubing tubería de cobre
copper winding arrollamiento de cobre
coprocessor coprocesador
copy check verificación de copia
copy control control de copia
copy cycle ciclo de copiado
copy function función de copiado
copy modification modificación de copia
copy operation operación de copiado
copy protection protección contra copia
Corbino disk disco de Corbino
Corbino effect efecto de Corbino
cord cordón
cord circuit circuito de cordón
cord fastener sujetador de cordón
cord terminal terminal de cordón
cordless sin cordón
cordless plug clavija sin cordón
cordless switch conmutador sin cordón
cordless switchboard cuadro conmutador sin cordones
cordless switching conmutación sin cordones
core núcleo, alma
core allocation asignación de núcleo
core array red de núcleos
core dump vaciado de memoria de núcleos
core dump program programa de vaciado de memoria de núcleos
core dump routine rutina de vaciado de memoria de núcleos
core end plate placa terminal de núcleo
core image imagen de memoria de núcleos
core image buffer almacenador intermedio de imagen de memoria de núcleos

core load carga de memoria de núcleos
core loss pérdidas en el núcleo
core matrix matriz de núcleos
core memory memoria de núcleos
core memory resident residente en memoria de núcleos
core plane plano de núcleos
core saturation saturación de núcleo
core storage almacenamiento de núcleos
core storage unit unidad de almacenamiento de núcleos
core transformer transformador de núcleo
core-type induction heater calentador de inducción de tipo de núcleo
core wrapping envoltura de núcleo
coreless sin núcleo
coreless induction heater calentador de inducción sin núcleo
corner antenna antena angular
corner diffraction difracción de esquinas
corner effect efecto de esquinas
corner frequency frecuencia de esquina
corner reflection reflexión angular
corner-reflection antenna antena de reflexión angular
corner reflector reflector angular
corner-reflector antenna antena con reflector angular
corona corona
corona current corriente de corona
corona detector detector de corona
corona discharge descarga en corona
corona effect efecto de corona
corona extinction voltage tensión de extinción de corona
corona inception voltage tensión inicial de corona
corona loss pérdidas por corona
corona shield pantalla anticorona
corona stabilizer estabilizador de corona
corona voltmeter vóltmetro de efecto corona
coroutine corutina
correcting current corriente correctora
correcting mechanism mecanismo corrector
correcting network red correctora
correcting signal señal correctora
correction corrección
correction factor factor de corrección
correction field campo de corrección
correction key tecla de corrección
correction program programa de corrección
correction rate velocidad de corrección
correction time tiempo de corrección
corrective correctivo
corrective action acción correctiva
corrective feedback retroalimentación correctiva
corrective filter filtro correctivo
corrective maintenance mantenimiento correctivo
corrective maintenance time tiempo de mantenimiento correctivo
corrective network red correctiva
corrector corrector
corrector circuit circuito corrector
correlation correlación
correlation coefficient coeficiente de correlación
correlation detection detección de correlación
correlation detector detector de correlación
correlation distance distancia de correlación
correlation function función de correlación
correlation identification identificación de correlación
correlation matrix matriz de correlación
correlation name nombre de correlación
correlation-type receiver receptor de tipo de

correlación
correlator correlacionador
correspondence center centro de correspondencia
corrosion corrosión
corrosion-resistant resistente a la corrosión
corrosion-resistant metal metal resistente a la
corrosión
corrugated-surface antenna antena de superficie
corrugada
corrupted file archivo viciado
corruption corrupción
cosecant cosecante
cosecant antenna antena de cosecante
cosecant-squared antenna antena de cosecante
cuadrada
cosecant-squared beam haz de cosecante cuadrada
cosecant-squared beam antenna antena de haz de
cosecante cuadrada
cosecant-squared pattern patrón de cosecante
cuadrada
cosine coseno
cosine emission law ley de emisión del coseno
cosine law ley del coseno
cosine wave onda cosenoidal
cosine winding arrollamiento cosenoidal
cosmic noise ruido cósmico
cosmic radiation radiación cósmica
cosmic rays rayos cósmicos
cosmotron cosmotrón
cotangent cotangente
Cotton-Mouton effect efecto de Cotton-Mouton
Cottrell hardening endurecimiento de Cottrell
Cottrell process proceso de Cottrell
coulomb coulombio
Coulomb's law ley de Coulomb
Coulomb attraction atracción de Coulomb
Coulomb field campo de Coulomb
Coulomb force fuerza de Coulomb
Coulomb interactions interacciones de Coulomb
Coulomb potential potencial de Coulomb
Coulomb repulsion repulsión de Coulomb
coulombmeter coulómbmetro
coulometer coulómetro
count cuenta
count up cuenta progresiva
countdown cuenta regresiva
counter contador
counter binary binario contador
counter capacity capacidad de contador
counter circuit circuito contador
counter entry entrada de contador
counter exit salida de contador
counter measurement medida de contador
counter overvoltage sobretensión de contador
counter position posición de contador
counter tube tubo contador
counterbalance contrapeso
counterclockwise sinistrórsum, en sentido contrario
de las manecillas del reloj
counterclockwise-polarized wave onda polarizada
sinistrórsum
counterelectrode contraelectrodo
counterelectromotive cell pila contraelectromotriz
counterelectromotive force fuerza
contraelectromotriz
countergate contrapuerta
counterpoise contrapeso
countervoltage contratensión
counting circuit circuito de conteo
counting efficiency eficiencia de conteo
counting interval intervalo de conteo
counting ionization chamber cámara de ionización
de conteo

counting loop bucle de conteo
counting mechanism mecanismo de conteo
counting operation operación de conteo
counting rate tasa de conteo
counting relay relé de conteo
counting sequence secuencia de conteo
counting time tiempo de conteo
couple par
coupled antennas antenas acopladas
coupled circuits circuitos acoplados
coupled field vectors vectores de campo acoplados
coupled impedance impedancia acoplada
coupled oscillators osciladores acoplados
coupled transistors transistores acoplados
coupler acoplador
coupling acoplamiento
coupling aperture abertura de acoplamiento
coupling capacitance capacitancia de acoplamiento
coupling capacitor capacitor de acoplamiento
coupling circuit circuito de acoplamiento
coupling coefficient coeficiente de acoplamiento
coupling coil bobina de acoplamiento
coupling component componente de acoplamiento
coupling condenser condensador de acoplamiento
coupling constant constante de acoplamiento
coupling diode diodo de acoplamiento
coupling-dissipation factor factor de disipación de
acoplamiento
coupling efficiency eficiencia de acoplamiento
coupling element elemento de acoplamiento
coupling factor factor de acoplamiento
coupling hole abertura de acoplamiento
coupling impedance impedancia de acoplamiento
coupling loop bucle de acoplamiento
coupling loss pérdida por acoplamiento
coupling mechanism mecanismo de acoplamiento
coupling network red de acoplamiento
coupling probe sonda de acoplamiento
coupling ratio razón de acoplamiento
coupling resistance resistencia de acoplamiento
coupling resistor resistor de acoplamiento
coupling slot ranura de acoplamiento
coupling switch conmutador de acoplamiento
coupling tap toma de acoplamiento
coupling transformer transformador de
acoplamiento
coupling unit unidad de acoplamiento
course-indicating beacon radiofaro indicador de
rumbo
course indicator indicador de rumbo
course line línea de rumbo
course-line computer computadora de línea de
rumbo
course-line deviation desviación de línea de rumbo
course-line deviation indicator indicador de
desviación de línea de rumbo
course selector selector de rumbo
course sensitivity sensibilidad de rumbo
cover plate placa de cubierta
coverage cobertura
coverage area área de cobertura
coverage diagram diagrama de cobertura
covered conductor conductor recubierto
covered wire hilo recubierto
cracked-carbon resistor resistor de carbón
craqueado
cradle-guard red de protección
crash falla completa
crater lamp lámpara de cráter
crazing agrietamiento
creepage corrimiento
crest factor factor de cresta
crest value valor de cresta

crest voltage tensión de cresta
crest voltmeter vóltmetro de cresta
crevice corrosion corrosión en hendiduras
crimp contact contacto por presión
crisscross neutralization neutralización entrecruzada
crisscross rectifier circuit circuito rectificador
  entrecruzado
critical absorption wavelength longitud de onda de
  absorción crítica
critical angle ángulo crítico
critical anode voltage tensión de ánodo crítica
critical area área crítica
critical band banda crítica
critical bandwidth ancho de banda crítico
critical build-up resistance resistencia de cebado
  crítica
critical build-up speed velocidad de cebado crítica
critical characteristic característica crítica
critical controlling current corriente de control
  crítica
critical coupling acoplamiento crítico
critical coupling factor factor de acoplamiento
  crítico
critical current corriente crítica
critical damping amortiguamiento crítico
critical dimension dimensión crítica
critical failure falla crítica
critical field campo crítico
critical frequency frecuencia crítica
critical grid current corriente de rejilla crítica
critical grid voltage tensión de rejilla crítica
critical high-power level nivel de potencia alta
  crítico
critical humidity humedad crítica
critical inductance inductancia crítica
critical induction inducción crítica
critical load carga crítica
critical magnetic field campo magnético crítico
critical path trayectoria crítica
critical path method método de trayectoria crítica
critical potential potencial crítico
critical process control control de procesos críticos
critical resistance resistencia crítica
critical section sección crítica
critical software programas críticos
critical speed velocidad crítica
critical system sistema crítico
critical temperature temperatura crítica
critical tuning sintonización crítica
critical voltage tensión crítica
critical wavelength longitud de onda crítica
crocodile clip pinza cocodrilo
Crookes dark space espacio obscuro de Crookes
Crookes radiometer radiómetro de Crookes
Crookes tube tubo de Crookes
Crosby circuit circuito de Crosby
Crosby exciter excitador de Crosby
cross antenna antena en cruz
cross assembler ensamblador cruzado
cross assembler-compiler ensamblador-compilador
  cruzado
cross assembler program programa de ensamblador
  cruzado
cross bearings marcaciones cruzadas
cross-channel communication comunicación de
  canal cruzado
cross-checking comprobación cruzada
cross-color diafotía cromática
cross compiler compilador cruzado
cross configuration configuración cruzada
cross-connected neutralization neutralización por
  conexión cruzada
cross-connection conexión cruzada

cross control control cruzado
cross-control circuit circuito de control cruzado
cross-correlation correlación cruzada
cross-coupled multivibrator multivibrador
  interacoplado
cross-coupling interacoplamiento
cross-current corriente cruzada
cross flux flujo cruzado
cross-modulation modulación cruzada
cross-modulation distortion meter medidor de
  distorsión de modulación cruzada
cross-modulation factor factor de modulación
  cruzada
cross-modulation noise ruido de modulación cruzada
cross-neutralization neutralización cruzada
cross-neutralized circuit circuito con neutralización
  cruzada
cross-polarization polarización cruzada
cross-polarization discrimination discriminación de
  polarización cruzada
cross product producto vectorial
cross program programa cruzado
cross programming programación cruzada
cross programming system sistema de programación
  cruzada
cross-protection protección cruzada
cross reference referencia cruzada
cross reference generator generador de referencias
  cruzadas
cross-section sección transversal
cross-sectional area área de sección transversal
cross validation validación cruzada
crossband banda cruzada
crossband communication comunicación de banda
  cruzada
crossband operation operación de banda cruzada
crossband transponder transpondedor de banda
  cruzada
crossbar barra cruzada
crossbar equipment equipo de barras cruzadas
crossbar selector selector de barras cruzadas
crossbar switch conmutador de barras cruzadas
crossbar switching system sistema de conmutación
  de barras cruzadas
crossbar system sistema de barras cruzadas
crossbar transformer transformador de barras
  cruzadas
crossed-field amplifier amplificador de campos
  cruzados
crossed-field device dispositivo de campos cruzados
crossed-field interaction interacción de campos
  cruzados
crossed-field multiplier multiplicador de campos
  cruzados
crossed-field multiplier phototube fototubo
  multiplicador de campos cruzados
crossed-pointer indicator indicador de agujas
  cruzadas
crossed variable variable cruzada
crosshatch generator generador de trama de imagen
crosshatch pattern patrón de trama de imagen
crossing point punto de cruce
crossing structure estructura de cruce
crossover punto de cruce, cruce
crossover circuit circuito de cruce
crossover distortion distorsión de cruce
crossover frequency frecuencia de cruce
crossover gate puerta de cruce
crossover loss pérdidas de cruce
crossover network red divisora
crossover point punto de cruce
crossover region región de cruce
crossover time tiempo de cruce

crosspoint punto de cruce
crosspoint reed relay relé de laminilla de punto de cruce
crosspoint switch conmutador de punto de cruce
crosstalk diafonía
crosstalk attenuation atenuación de diafonía
crosstalk coupling acoplamiento por diafonía
crosstalk index índice de diafonía
crosstalk level nivel de diafonía
crosstalk loss pérdidas por diafonía
crosstalk measurement medida de diafonía
crosstalk meter medidor de diafonía
crosstalk noise ruido de diafonía
crosstalk suppression filter filtro de supresión de diafonía
crosstalk unit unidad de diafonía
crowbar acción de palanca
cryoelectronics crioelectrónica
cryogen criógeno
cryogenic criogénico
cryogenic computer computadora criogénica
cryogenic conductor conductor criogénico
cryogenic device dispositivo criogénico
cryogenic electronics electrónica criogénica
cryogenic element elemento criogénico
cryogenic film película criogénica
cryogenic liquid líquido criogénico
cryogenic memory memoria criogénica
cryogenic motor motor criogénico
cryogenic storage almacenamiento criogénico
cryogenic temperature temperatura criogénica
cryogenic transformer transformador criogénico
cryogenics criogenia
cryometer criómetro
cryoresistance criorresistencia
cryosar criosar
cryoscope crioscopio
cryoscopic crioscópico
cryosistor criosistor
cryostat criostato
cryotron criotrón
cryotronics criotrónica
cryptographic criptográfico
cryptographic algorithm algoritmo criptográfico
cryptographic communication comunicación criptográfica
cryptographic equipment equipo criptográfico
cryptographic key clave criptográfica
cryptographic master key clave maestra criptográfica
cryptographic technique técnica criptográfica
crystal cristal
crystal activity actividad de cristal
crystal amplifier amplificador de cristal
crystal analysis análisis de cristal
crystal audio receiver receptor de audio de cristal
crystal axes ejes de cristal
crystal-calibrated calibrado por cristal
crystal calibrator calibrador de cristal
crystal capacitor capacitor de cristal
crystal cartridge cartucho de cristal
crystal control control por cristal
crystal-controlled converter convertidor controlado por cristal
crystal-controlled master oscillator oscilador maestro controlado por cristal
crystal-controlled oscillator oscilador controlado por cristal
crystal-controlled receiver receptor controlado por cristal
crystal-controlled transmitter transmisor controlado por cristal
crystal-controlled tuning sintonización controlada

por cristal
crystal counter contador de cristal
crystal current corriente de cristal
crystal cut corte de cristal
crystal detector detector de cristal
crystal-determined frequency frecuencia determinada por cristal
crystal diffraction difracción de cristal
crystal diode diodo de cristal
crystal-diode detector detector de diodo de cristal
crystal-diode mixer mezclador de diodo de cristal
crystal-diode probe sonda de diodo de cristal
crystal dynamics dinámica de cristal
crystal earphone audífono de cristal
crystal filter filtro de cristal
crystal frequency frecuencia de cristal
crystal-frequency changer cambiador de frecuencia de cristal
crystal headphone audífono de cristal
crystal holder portacristal
crystal hydrophone hidrófono de cristal
crystal imperfection imperfección de cristal
crystal impurity impureza de cristal
crystal laser láser de cristal
crystal lattice red cristalina
crystal-lattice filter filtro de red cristalina
crystal loudspeaker altavoz de cristal
crystal marker marcador de cristal
crystal meter medidor de cristal
crystal microphone micrófono de cristal
crystal mixer mezclador de cristal
crystal operation operación de cristal
crystal oscillator oscilador de cristal
crystal oven horno para cristal
crystal photocell fotocélula de cristal
crystal pickup fonocaptor de cristal
crystal plate placa de cristal
crystal probe sonda de cristal
crystal pulling extracción de cristal
crystal radio radio de cristal
crystal receiver receptor de cristal
crystal rectifier rectificador de cristal
crystal resistor resistor de cristal
crystal resonator resonador de cristal
crystal sensor sensor de cristal
crystal set receptor de cristal
crystal slab placa de cristal
crystal socket zócalo de cristal
crystal spectrograph espectrógrafo de cristal
crystal spectrometer espectrómetro de cristal
crystal-stabilized transmitter transmisor estabilizado por cristal
crystal system sistema cristalino
crystal tester probador de cristal
crystal tetrode tetrodo de cristal
crystal transducer transductor de cristal
crystal triode triodo de cristal
crystal unit unidad de cristal
crystal vibrator vibrador de cristal
crystal video receiver receptor de video de cristal
crystal video rectifier rectificador de video de cristal
crystal voltmeter vóltmetro de cristal
crystalline anisotropy anisotropía cristalina
crystalline field campo cristalino
crystalline material material cristalino
crystallogram cristalograma
crystallography cristalografía
crystallomagnetic cristalomagnético
cubical antenna antena cúbica
Cuccia coupler acoplador de Cuccia
cue circuit circuito de señalización
cumulative demand demanda acumulativa
cumulative-demand meter medidor de demanda

acumulativa
cumulative-demand register registro de demanda
  acumulativa
cumulative dose dosis acumulativa
cumulative error error acumulativo
cumulative ionization ionización acumulativa
curie curie
Curie's law ley de Curie
Curie point punto de Curie
Curie temperature temperatura de Curie
Curie-Weiss law ley de Curie-Weiss
Curie-Weiss temperature temperatura de
  Curie-Weiss
curium curio
current corriente
current address dirección corriente
current amplification amplificación de corriente
current-amplification factor factor de amplificación
  de corriente
current amplifier amplificador de corriente
current antinode antinodo de corriente
current attenuation atenuación de corriente
current balance equilibrio de corriente
current-balance ratio razón de equilibrio de
  corriente
current-balance relay relé de equilibrio de corriente
current-balance switch conmutador de equilibrio de
  corriente
current bias polarización de corriente
current carrier portador de corriente
current-carrying capacity capacidad de corriente
current cell celda corriente
current cell indicator indicador de celda corriente
current circuit circuito de corriente
current coil bobina de contador
current collector colector de corriente
current comparator comparador de corrientes
current compensator compensador de corriente
current consumption consumo de corriente
current-controlled amplifier amplificador
  controlado por corriente
current density densidad de corriente
current detector detector de corriente
current device dispositivo corriente, dispositivo en
  uso
current distribution distribución de corriente
current divider divisor de corriente
current drain consumo de corriente
current drive unidad corriente, unidad en uso
current echo eco de corriente
current efficiency eficiencia de corriente
current error error de corriente
current-fed antenna antena alimentada por corriente
current feed alimentación de corriente
current feedback retroalimentación de corriente
current feedback circuit circuito de
  retroalimentación de corriente
current flow flujo de corriente
current gain ganancia de corriente
current generator generador de corriente
current governor regulador de corriente
current hogging acaparamiento de corriente
current instruction instrucción corriente, instrucción
  en ejecución
current instruction counter contador de instrucción
  corriente, contador de instrucción en ejecución
current instruction register registro de instrucción
  corriente, registro de instrucción en ejecución
current integrator integrador de corriente
current intensity intensidad de corriente
current interrupter interruptor de corriente
current lag retardo de corriente
current lead adelanto de corriente

current limit límite de corriente
current limiter limitador de corriente
current limiting limitación de corriente
current-limiting coil bobina limitadora de corriente
current-limiting device dispositivo limitador de
  corriente
current-limiting fuse fusible limitador de corriente
current-limiting fuse unit unidad de fusible
  limitadora de corriente
current-limiting resistor resistor limitador de
  corriente
current line línea corriente, línea en uso
current loop vientre de corriente, bucle corriente,
  bucle de corriente
current magnitude magnitud de corriente
current margin margen de corriente
current measurement medida de corriente
current meter medidor de corriente
current-meter operation operación de medidor de
  corriente
current-mode logic lógica en modo de corriente
current node nodo de corriente
current noise ruido de corriente
current output salida de corriente
current period periodo corriente, periodo actual
current position posición corriente, posición actual
current priority level nivel de prioridad corriente,
  nivel de prioridad actual
current probe sonda de corriente
current pulse impulso de corriente
current rating corriente nominal
current ratio razón de corrientes
current record registro corriente, registro actual
current rectifier rectificador de corriente
current regulation regulación de corriente
current regulator regulador de corriente
current-regulator tube tubo regulador de corriente
current relay relé de corriente
current reverser inversor de corriente
current ripple ondulación de corriente
current saturation saturación de corriente
current sensing muestreo de corriente
current-sensing resistor resistor de muestreo de
  corriente
current sensitivity sensibilidad de corriente
current spreading dispersión de corriente
current stabilizer estabilizador de corriente
current state estado corriente, estado actual
current status estado corriente, estado actual
current strength intensidad de corriente
current supply alimentación de corriente
current surge sobrecorriente
current tap toma de corriente
current terminals terminales de corriente
current test prueba de corriente
current transformer transformador de corriente
current vector vector de corriente
current-voltage feedback retroalimentación
  corriente-tensión
current-voltage relation relación corriente-tensión
current wave onda de corriente
cursor cursor
cursor control control de cursor
cursor key tecla de cursor
cursor movement movimiento de cursor
curtain array red en cortina
curtain reflector reflector en cortina
cushioned socket portalámpara antivibratoria
custom designed diseñado a la medida
custom made hecho a la medida
Cutler antenna antena de Cutler
cutoff corte, cortacircuito
cutoff attenuator atenuador de corte

cutoff bias polarización de corte
cutoff characteristic característica de corte
cutoff current corriente de corte
cutoff field campo de corte
cutoff frequency frecuencia de corte
cutoff limiting limitación de corte
cutoff parabola parábola de corte
cutoff point punto de corte
cutoff potential potencial de corte
cutoff receiver receptor de corte
cutoff region región de corte
cutoff relay relé de corte
cutoff switch conmutador de corte
cutoff tube tubo de corte
cutoff voltage tensión de corte
cutoff waveguide guíaondas de corte
cutoff wavelength longitud de onda de corte
cutout cortacircuito, conmutador, interruptor, fusible
cutout element elemento de corte
cutout point punto de corte
cutter cortador, fonoincisor, grabador
cybernetics cibernética
cycle ciclo
cycle availability disponibilidad de ciclos
cycle count cuenta de ciclos
cycle counter contador de ciclos
cycle index índice de ciclos
cycle index counter contador de índice de ciclos
cycle life vida de ciclos
cycle shift desplazamiento de ciclos
cycle stealing robo de ciclos
cycle termination terminación de ciclo
cycle time tiempo de ciclo
cycle timer sincronizador de ciclos
cycles per second ciclos por segundo
cyclic cíclico
cyclic binary code código binario cíclico
cyclic check comprobación cíclica
cyclic code código cíclico
cyclic control signal señal de control cíclica
cyclic decimal code código decimal cíclico
cyclic distortion distorsión cíclica
cyclic feed alimentación cíclica
cyclic irregularity irregularidad cíclica
cyclic magnetization magnetización cíclica
cyclic memory memoria cíclica
cyclic noise ruido cíclico
cyclic redundancy redundancia cíclica
cyclic redundancy check comprobación de redundancia cíclica
cyclic redundancy code código de redundancia cíclica
cyclic search búsqueda cíclica
cyclic shift desplazamiento cíclico
cyclic storage almacenamiento cíclico
cyclic storage access acceso al almacenamiento cíclico
cyclic voltage variation variación de tensión cíclica
cyclically magnetized magnetizado cíclicamente
cycling ciclado
cyclogram ciclograma
cycloidotron cicloidotrón
cyclophon ciclofón
cyclotron ciclotrón
cyclotron beam haz ciclotrónico
cyclotron frequency frecuencia ciclotrónica
cyclotron radiation radiación ciclotrónica
cyclotron wave onda ciclotrónica
cylinder boundary límite de cilindro
cylinder magnet imán en cilindro
cylinder overflow area área de desbordamiento de cilindro
cylinder record availability disponibilidad de registros de cilindro
cylindrical antenna antena cilíndrica
cylindrical array red cilíndrica
cylindrical capacitor capacitor cilíndrico
cylindrical cathode cátodo cilíndrico
cylindrical cavity cavidad cilíndrica
cylindrical-cavity resonator resonador de cavidad cilíndrica
cylindrical contour contorno cilíndrico
cylindrical inductor inductor cilíndrico
cylindrical magnet imán cilíndrico
cylindrical reflector reflector cilíndrico
cylindrical wave onda cilíndrica
cylindrical waveguide guíaondas cilíndrico
cylindrical winding arrollamiento cilíndrico
Czochralski method método de Czochralski
Czochralski technique técnica de Czochralski

D

D'Arsonval current corriente de D'Arsonval
D'Arsonval galvanometer galvanómetro de
  D'Arsonval
D'Arsonval meter medidor de D'Arsonval
D'Arsonval movement mecanismo de D'Arsonval
daisy chain cadena en margarita
daisy chain bus bus de cadena en margarita
daisy chain interrupt interrupción de cadena en
  margarita
daisy chain structure estructura de cadena en
  margarita
daisy wheel rueda en margarita
daisy wheel printer impresora de rueda en margarita
damage assessment evaluación de daños
Damon effect efecto de Damon
damped amortiguado
damped electric oscillations oscilaciones eléctricas
  amortiguadas
damped filter filtro amortiguado
damped galvanometer galvanómetro amortiguado
damped harmonic system sistema armónico
  amortiguado
damped impedance impedancia amortiguada
damped loudspeaker altavoz amortiguado
damped meter medidor amortiguado
damped natural frequency frecuencia natural
  amortiguada
damped oscillation oscilación amortiguada
damped speaker altavoz amortiguado
damped vibration vibración amortiguada
damped wave onda amortiguada
damped-wave decay decaimiento de ondas
  amortiguadas
dampen amortiguar
damper amortiguador
damper diode diodo amortiguador
damper tube tubo amortiguador
damper winding arrollamiento amortiguador
damping amortiguamiento
damping action acción de amortiguamiento
damping circuit circuito amortiguador
damping coefficient coeficiente de amortiguamiento
damping constant constante de amortiguamiento
damping control control de amortiguamiento
damping diode diodo de amortiguamiento
damping factor factor de amortiguamiento
damping fluid fluido amortiguador
damping magnet imán amortiguador
damping material material amortiguador
damping mechanism mecanismo amortiguador
damping ratio razón de amortiguamiento
damping resistance resistencia amortiguadora
damping resistor resistor amortiguador
damping signal señal amortiguadora
damping torque par amortiguador
damping tube tubo amortiguador
danger signal señal de peligro
Daniell cell pila Daniell
daraf daraf
dark conduction conducción en obscuridad
dark current corriente en obscuridad

dark-current pulses impulsos de corriente en
  obscuridad
dark discharge descarga en obscuridad
dark heater calentador en obscuridad
dark resistance resistencia en obscuridad
dark space espacio obscuro
dark spot punto obscuro, mancha obscura
dark-spot signal señal de punto obscuro
dark-trace tube tubo de pantalla absorbente, tubo de
  trazo obscuro
darkening obscurecimiento
Darlington amplifier amplificador de Darlington
dash raya
dash-dot telegraphy telegrafía de rayas y puntos
dashpot amortiguador
dashpot relay relé de amortiguador
data datos
data abstraction abstracción de datos
data access arrangement arreglo de acceso a datos
data access program programa de acceso a datos
data access protocol protocolo de acceso a datos
data access register registro de acceso a datos
data acquisition adquisición de datos
data-acquisition and control adquisición de datos y
  control
data-acquisition control system sistema de control
  de adquisición de datos
data-acquisition equipment equipo de adquisición de
  datos
data-acquisition module módulo de adquisición de
  datos
data-acquisition system sistema de adquisición de
  datos
data-acquisition terminal terminal de adquisición de
  datos
data adapter adaptador de datos
data adapter unit unidad adaptadora de datos
data address dirección de datos
data aggregate agregado de datos
data analysis análisis de datos
data analysis display unit unidad de visualización de
  análisis de datos
data area área de datos
data attribute atributo de datos
data bank banco de datos
data base base de datos
data bit bit de datos
data block bloque de datos
data breakpoint punto de interrupción de datos
data buffer memoria intermedia de datos
data bus bus de datos
data bus enable habilitación de bus de datos
data capture captura de datos
data card tarjeta de datos
data carrier portador de datos
data carrier storage almacenamiento de portador de
  datos
data cartridge cartucho de datos
data cell celda de datos
data center centro de datos
data chain cadena de datos
data chaining encadenamiento de datos
data channel canal de datos
data channel control control de canal de datos
data channel multiplexer multiplexor de canal de
  datos
data character carácter de datos
data character set conjunto de caracteres de datos
data characteristic característica de datos
data check comprobación de datos
data circuit circuito de datos
data class clase de datos
data code código de datos

data code conversion conversión de código de datos
data collection recopilación de datos
data collection equipment equipo de recopilación de datos
data collection station estación de recopilación de datos
data collection terminal terminal de recopilación de datos
data collector recopilador de datos
data communication comunicación de datos
data communications buffer memoria intermedia de comunicaciones de datos
data communications channel canal de comunicaciones de datos
data communications control unit unidad de control de comunicaciones de datos
data communications equipment equipo de comunicaciones de datos
data communications feature característica de comunicaciones de datos
data communications interface interfaz de comunicaciones de datos
data communications line línea de comunicaciones de datos
data communications network red de comunicaciones de datos
data communications processing procesamiento de comunicaciones de datos
data communications processor procesador de comunicaciones de datos
data communications system sistema de comunicaciones de datos
data communications terminal terminal de comunicaciones de datos
data compaction compactación de datos
data complementation complementación de datos
data component componente de datos
data compression compresión de datos
data concentration concentración de datos
data concentrator concentrador de datos
data connection conexión de datos
data constant constante de datos
data contamination contaminación de datos
data control control de datos
data control block bloque de control de datos
data control section sección de control de datos
data control unit unidad de control de datos
data control word palabra de control de datos
data conversion conversión de datos
data conversion line línea de conversión de datos
data converter convertidor de datos
data corruption corrupción de datos
data count field campo de cuenta de datos
data coupler acoplador de datos
data coupling acoplamiento de datos
data cycle ciclo de datos
data definition definición de datos
data definition language lenguaje de definición de datos
data definition name nombre de definición de datos
data definition statement sentencia de definición de datos
data deletion borrado de datos
data delimiter delimitador de datos
data density densidad de datos
data description descripción de datos
data description entry entrada de descripción de datos
data description language lenguaje de descripción de datos
data description specifications especificaciones de descripción de datos
data description table tabla de descripción de datos

data descriptor descriptor de datos
data design diseño de datos
data dictionary diccionario de datos
data directory directorio de datos
data disk disco de datos
data display visualización de datos
data display station estación de visualización de datos
data display system sistema de visualización de datos
data dissemination diseminación de datos
data distribution system sistema de distribución de datos
data division división de datos
data domain dominio de datos
data element elemento de datos
data encoder codificador de datos
data encoding codificación de datos
data encoding device dispositivo de codificación de datos
data encryption codificación de datos
data encryption key clave de codificación de datos
data entry entrada de datos
data entry device dispositivo de entrada de datos
data entry form forma de entrada de datos
data entry keyboard teclado de entrada de datos
data entry panel panel de entrada de datos
data error error de datos
data evaluation evaluación de datos
data exchange intercambio de datos
data field campo de datos
data file archivo de datos
data file tape cinta de archivo de datos
data flow flujo de datos
data flow-chart diagrama de flujo de datos
data flow-diagram diagrama de flujo de datos
data format formato de datos
data gathering recopilación de datos
data gathering system sistema de recopilación de datos
data generated generado por datos
data generator generador de datos
data group grupo de datos
data handling manejo de datos
data-handling capacity capacidad de manejo de datos
data-handling equipment equipo de manejo de datos
data-handling system sistema de manejo de datos
data hierarchy jerarquía de datos
data import importación de datos
data independence independencia de datos
data input entrada de datos
data input bus bus de entrada de datos
data input device dispositivo de entrada de datos
data input station estación de entrada de datos
data insertion inserción de datos
data instruction instrucción de datos
data integrity integridad de datos
data interchange intercambio de datos
data interchange code código de intercambio de datos
data interchange format formato de intercambio de datos
data inventory inventario de datos
data item artículo de datos
data layout arreglo de datos
data leakage fuga de datos
data level nivel de datos
data library biblioteca de datos
data line línea de datos
data line monitor monitor de línea de datos
data link enlace de datos
data link adapter adaptador de enlace de datos

data link control control de enlace de datos
data link control character carácter de control de enlace de datos
data link control layer capa de control de enlace de datos
data link control protocol protocolo de control de enlace de datos
data link escape character carácter de escape de enlace de datos
data link layer capa de enlace de datos
data link level nivel de enlace de datos
data link protocol protocolo de enlace de datos
data list lista de datos
data logger registrador de datos
data logging registración de datos
data-logging equipment equipo de registración de datos
data maintenance mantenimiento de datos
data management administración de datos
data management system sistema de administración de datos
data manipulation manipulación de datos
data manipulation language lenguaje de manipulación de datos
data matrix matriz de datos
data medium medio de datos
data migration migración de datos
data mode modo de datos
data model modelo de datos
data modem módem de datos
data modification modificación de datos
data module módulo de datos
data movement movimiento de datos
data movement time tiempo de movimiento de datos
data multiplexer multiplexor de datos
data name nombre de datos
data network red de datos
data organization organización de datos
data origination originación de datos
data output salida de datos
data path trayectoria de datos
data pattern patrón de datos
data pickup captor de datos, captura de datos
data playback reproducción de datos
data plotter trazador de datos
data pointer indicador de datos
data preparation preparación de datos
data preparation device dispositivo de preparación de datos
data preparation equipment equipo de preparación de datos
data printout impresión de datos
data privacy privacidad de datos
data processing procesamiento de datos
data-processing card tarjeta de procesamiento de datos
data-processing center centro de procesamiento de datos
data-processing cycle ciclo de procesamiento de datos
data-processing equipment equipo de procesamiento de datos
data-processing field campo de procesamiento de datos
data-processing machine máquina de procesamiento de datos
data-processing management administración de procesamiento de datos
data-processing node nodo de procesamiento de datos
data-processing sequence secuencia de procesamiento de datos
data-processing standards patrones de procesamiento de datos
data-processing station estación de procesamiento de datos
data-processing step paso de procesamiento de datos
data-processing system sistema de procesamiento de datos
data-processing system security seguridad de sistema de procesamiento de datos
data-processing task tarea de procesamiento de datos
data-processing technology tecnología de procesamiento de datos
data-processing terminal terminal de procesamiento de datos
data processor procesador de datos
data protection protección de datos
data purification purificación de datos
data quality calidad de datos
data rate velocidad de datos
data receiver receptor de datos
data reception recepción de datos
data-reception system sistema de recepción de datos
data record registro de datos
data record format formato de registro de datos
data recorder registrador de datos
data recording registro de datos
data recording control control de registro de datos
data recording device dispositivo de registro de datos
data recording medium medio de registro de datos
data reduction reducción de datos
data-reduction system sistema de reducción de datos
data redundancy redundancia de datos
data register registro de datos
data reliability confiabilidad de datos
data representation representación de datos
data resource recurso de datos
data retrieval recuperación de datos
data retrieval system sistema de recuperación de datos
data routing encaminamiento de datos
data routing system sistema de encaminamiento de datos
data safety seguridad de datos
data security seguridad de datos
data segment segmento de datos
data selection selección de datos
data selector selector de datos
data selector-multiplexer selector-multiplexor de datos
data set conjunto de datos
data set control block bloque de control de conjunto de datos
data set deletion borrado de conjunto de datos
data set equipment equipo de conjunto de datos
data set group grupo de conjunto de datos
data set identification identificación de conjunto de datos
data set label etiqueta de conjunto de datos
data set name nombre de conjunto de datos
data set organization organización de conjunto de datos
data set profile perfil de conjunto de datos
data set security seguridad de conjunto de datos
data set size tamaño de conjunto de datos
data sheet hoja de datos
data signal señal de datos
data source fuente de datos
data space espacio de datos
data specification especificación de datos
data speed velocidad de datos
data stabilization estabilización de datos
data statement sentencia de datos

data station estación de datos
data storage almacenamiento de datos
data storage device dispositivo de almacenamiento de datos
data storage medium medio de almacenamiento de datos
data storage technique técnica de almacenamiento de datos
data stream corriente de datos
data structure estructura de datos
data-structure diagram diagrama de estructura de datos
data sublanguage sublenguaje de datos
data synchronizer sincronizador de datos
data system sistema de datos
data tape cinta de datos
data terminal terminal de datos
data-terminal equipment equipo de terminal de datos
data-terminal unit unidad de terminal de datos
data track pista de datos
data traffic tráfico de datos
data transcription transcripción de datos
data transducer transductor de datos
data transfer transferencia de datos
data-transfer bus bus de transferencia de datos
data-transfer controller controlador de transferencia de datos
data-transfer mode modo de transferencia de datos
data-transfer operation operación de transferencia de datos
data-transfer phase fase de transferencia de datos
data-transfer rate velocidad de transferencia de datos
data-transfer register registro de transferencia de datos
data-transfer speed velocidad de transferencia de datos
data transformation transformación de datos
data translation traducción de datos
data transmission transmisión de datos
data-transmission control unit unidad de control de transmisión de datos
data-transmission equipment equipo de transmisión de datos
data-transmission line línea de transmisión de datos
data transmission rate velocidad de transmisión de datos
data-transmission system sistema de transmisión de datos
data-transmission terminal terminal de transmisión de datos
data-transmission unit unidad de transmisión de datos
data transmitter transmisor de datos
data type tipo de datos
data unit unidad de datos
data validation validación de datos
data validation operation operación de validación de datos
data validity validez de datos
data value valor de datos
data verification verificación de datos
data volatility volatilidad de datos
data words palabras de datos
database base de datos
database access method método de acceso de base de datos
database administration administración de base de datos
database command language lenguaje de mandos de base de datos
database component componente de base de datos

database creation creación de base de datos
database definition definición de base de datos
database descriptor descriptor de base de datos
database design diseño de base de datos
database environment ambiente de base de datos
database extract extracto de base de datos
database files archivos de base de datos
database integrity integridad de base de datos
database machine máquina de base de datos
database management administración de base de datos
database management program programa de administración de base de datos
database management system sistema de administración de base de datos
database manipulation manipulación de base de datos
database organization organización de base de datos
database position posición de base de datos
database processing procesamiento de base de datos
database record registro de base de datos
database reorganization reorganización de base de datos
database security seguridad de base de datos
database segment segmento de base de datos
database server servidor de base de datos
database size tamaño de base de datos
database structure estructura de base de datos
database sublanguage sublenguaje de base de datos
database system sistema de base de datos
datacom comunicación de datos
Davisson coordinates coordenadas de Davisson
davit pescante
day-night sound level nivel de sonido diurno-nocturno
daylight effect efecto diurno
daylight factor factor de luz solar
daylight lamp lámpara de luz solar
daylight range alcance diurno
de Broglie's equation ecuación de de Broglie
de Broglie waves ondas de de Broglie
deaccentuator desacentuador
deactivate desactivar
deactivation desactivación
deactivation date fecha de desactivación
deactuating pressure presión desactuante
dead area área muerta
dead band banda muerta
dead-band rating clasificación de banda muerta
dead battery batería muerta
dead break rompimiento muerto
dead-break connector conector de rompimiento muerto
dead-center position posición de centro exacto
dead circuit circuito muerto
dead earth tierra perfecta
dead end extremo muerto
dead-end station estación terminal
dead-end tower torre terminal
dead file archivo muerto
dead front frente muerto
dead-front mounting montura de frente muerto
dead-front panel panel de frente muerto
dead-front switch conmutador de frente muerto
dead ground tierra perfecta
dead interval intervalo muerto
dead layer capa muerta
dead-layer thickness grosor de capa muerta
dead level nivel muerto
dead line línea muerta
dead load carga muerta
dead period periodo muerto
dead-reckoning navigation navegación a la estima

dead room sala anecoica
dead sector sector muerto
dead short cortocircuito total
dead space espacio muerto
dead spot punto muerto
dead stretch estiramiento muerto
dead time tiempo muerto
dead-time correction corrección de tiempo muerto
dead zone zona muerta
dead zone unit unidad de zona muerta
deadbeat aperiódico
deadbeat ammeter amperímetro aperiódico
deadbeat discharge descarga aperiódica
deadbeat galvanometer galvanómetro aperiódico
deadbeat instrument instrumento aperiódico
deadbeat meter medidor aperiódico
deadlock punto muerto
deallocate desasignar
deallocation desasignación
debatable time tiempo debatible
deblock desbloquear
debounced switch conmutador sin rebotes
debug depurar
debugging depuración
debugging aid routine rutina de ayuda en depuración
debugging line línea de depuración
debugging period periodo de depuración
debugging phase fase de depuración
debugging program programa de depuración
debugging routine rutina de depuración
debugging section sección de depuración
debugging statement sentencia de depuración
debunching desagrupamiento
debye debye
Debye effect efecto de Debye
Debye equation ecuación de Debye
Debye frequency frecuencia de Debye
Debye-Jauncey scattering dispersión de Debye-Jauncey
Debye length longitud de Debye
Debye potentials potenciales de Debye
Debye-Scherrer method método de Debye-Scherrer
Debye specific heat calor específico de Debye
Debye temperature temperatura de Debye
Debye unit unidad de Debye
Debye-Waller factor factor de Debye-Waller
deca deca
decade amplifier amplificador de décadas
decade attenuator atenuador de décadas
decade band banda de décadas
decade box caja de décadas
decade bridge puente de décadas
decade capacitance box caja de décadas de capacitancias
decade capacitor capacitor de décadas
decade counter contador de décadas
decade divider divisor de décadas
decade inductance box caja de décadas de inductancias
decade inductor inductor de décadas
decade oscillator oscilador de décadas
decade resistance box caja de décadas de resistencias
decade resistor resistor de décadas
decade scaler escala decádica
decametric waves ondas decamétricas
decatron decatrón
decay decaimiento, disminución, desintegración
decay characteristic característica de persistencia
decay coefficient coeficiente de decaimiento
decay constant constante de desintegración
decay curve curva de disminución
decay factor factor de decaimiento

decay rate rapidez de decaimiento
decay time tiempo de decaimiento
decca decca
decca navigation navegación decca
decca navigation system sistema de navegación decca
decca radar radar decca
decca system sistema decca
decelerated electron electrón desacelerado
decelerating electrode electrodo desacelerador
decelerating element elemento desacelerador
decelerating relay relé desacelerador
deceleration desaceleración
deceleration time tiempo de desaceleración
decelerometer decelerómetro
decentralized computer network red de computadoras descentralizada
decentralized data processing procesamiento de datos descentralizado
decentralized processing procesamiento descentralizado
deci deci
decibel decibel, decibelio
decibel meter decibelímetro
decibel scale escala decibelimétrica
decibels above one kilowatt decibeles sobre un kilowatt
decibels above one volt decibeles sobre un volt
decibels above one watt decibeles sobre un watt
decibels above reference noise decibeles sobre ruido de referencia
decigram decigramo
decile decila
deciliter decilitro
decimal decimal
decimal alignment alineación decimal
decimal arithmetic operation operación aritmética decimal
decimal attenuator atenuador decimal
decimal-binary switch conmutador decimal-binario
decimal candle bujía decimal
decimal code código decimal
decimal-coded digit dígito codificado en decimal
decimal comma coma decimal
decimal digit dígito decimal
decimal digit position posición de dígito decimal
decimal encoder codificador decimal
decimal equivalent equivalente decimal
decimal form forma decimal
decimal fraction fracción decimal
decimal marker marcador decimal
decimal metric circuit circuito métrico decimal
decimal notation notación decimal
decimal number system sistema numérico decimal
decimal overflow desbordamiento decimal
decimal place lugar decimal
decimal point coma decimal, punto decimal
decimal point alignment alineación de coma decimal, alineación de punto decimal
decimal pulse impulso decimal
decimal representation representación decimal
decimal symbol símbolo decimal
decimal system sistema decimal
decimal tabulation tabulación decimal
decimal-to-binary conversion conversión de decimal a binario
decimal-to-hexadecimal conversion conversión de decimal a hexadecimal
decimetric band banda decimétrica
decimetric waves ondas decimétricas
decineper decineper
decipher descifrar
decision box caja de decisión

decision circuit circuito de decisión
decision criterion criterio de decisión
decision element gate puerta de elementos de decisión
decision elements elementos de decisión
decision feedback system sistema de retroalimentación de decisión
decision flow-chart diagrama de flujo de decisión
decision gate puerta de decisión
decision instruction instrucción de decisión
decision level nivel de decisión
decision mechanism mecanismo de decisión
decision procedure procedimiento de decisión
decision rule regla de decisión
decision structure estructura de decisión
decision support soporte de decisión
decision-support system sistema de soporte de decisiones
decision table tabla de decisión
decision theory teoría de decisión
decision threshold umbral de decisión
decision tree árbol de decisión
decision value valor de decisión
deck plataforma
deck arrangement arreglo de plataforma
declarative language lenguaje declarativo
declarative macro instruction macroinstrucción declarativa
declarative operation operación declarativa
declarative statement sentencia declarativa
declination declinación
declinometer declinómetro
decodable descifrable
decode descodificar
decoded descodificado
decoder descodificador
decoder chip chip descodificador
decoder circuit circuito descodificador
decoder-demultiplexer descodificador-desmultiplexor
decoder unit unidad descodificadora
decoding descodificación
decoding circuit circuito de descodificación
decoding network red de descodificación
decoding oscillator oscilador de descodificación
decoding routine rutina de descodificación
decoding signal señal de descodificación
decometer decómetro
decommutation desconmutación
decommutator desconmutador
decompile descompilar
decompiler descompilador
decomposition descomposición
decomposition potential potencial de descomposición
decomposition voltage tensión de descomposición
decompression descompresión
decouple desacoplar
decoupler desacoplador
decoupling desacoplamiento
decoupling capacitor capacitor de desacoplamiento
decoupling circuit circuito de desacoplamiento
decoupling component componente de desacoplamiento
decoupling condenser condensador de desacoplamiento
decoupling filter filtro de desacoplamiento
decoupling network red de desacoplamiento
decoupling resistance resistencia de desacoplamiento
decoupling resistor resistor de desacoplamiento
decrement decremento
decrement factor factor de decremento
decremeter decrémetro

decremeter capacitor capacitor de decrémetro
decrypt descifrar
decryption descifrado
dectra dectra
dectra system sistema dectra
dedicated cable cable dedicado
dedicated channel canal dedicado
dedicated circuit circuito dedicado
dedicated computer computadora dedicada
dedicated data set conjunto de datos dedicado
dedicated device dispositivo dedicado
dedicated file server servidor de archivos dedicado
dedicated line línea dedicada
dedicated machine control control de máquina dedicado
dedicated register registro dedicado
dedicated service servicio dedicado
dedicated storage almacenamiento dedicado
dedicated system sistema dedicado
deductive logic lógica deductiva
dee line soporte de electrodos en D
deemphasis desacentuación
deemphasis amplifier amplificador de desacentuación
deemphasis circuit circuito de desacentuación
deemphasis device dispositivo de desacentuación
deemphasis network red de desacentuación
deenergize desenergizar
deenergized desenergizado
deep discharge descarga profunda
deep field campo profundo
deexcitation desexcitación
default attribute atributo por omisión
default drive unidad por omisión
default extension extensión por omisión
default initial value valor inicial por omisión
default option opción por omisión
default parameter parámetro por omisión
default program programa por omisión
default value valor por omisión
defect defecto, agujero
defect conduction conducción por agujeros
defective defectuoso
defective fuse fusible defectuoso
defective track pista defectuosa
deferred address dirección diferida
deferred addressing direccionamiento diferido
deferred constant constante diferida
deferred entry entrada diferida
deferred exit salida diferida
deferred maintenance mantenimiento diferido
deferred printing impresión diferida
deferred processing procesamiento diferido
deferred restart reiniciación diferida
deferred update actualización diferida
defibrillation desfibrilación
defibrillator desfibrilador
defined zero cero definido
definite-purpose circuit-breaker cortacircuito de uso definido
definite-purpose component componente de uso definido
definite-purpose controller controlador de uso definido
definite-purpose motor motor de uso definido
definite time tiempo definido
definite-time delay retardo de tiempo definido
definite-time relay relé de tiempo definido
definition definición
definition chart imagen de prueba de definición
definition converter convertidor de definición
definition pattern patrón de definición
deflecting coil bobina de deflexión, bobina

desviadora
deflecting current corriente de deflexión
deflecting electrode electrodo de deflexión
deflecting field campo de deflexión
deflecting force fuerza de deflexión
deflecting plate placa de deflexión
deflecting potential potencial de deflexión
deflecting torque par de deflexión
deflecting voltage tensión de deflexión
deflecting yoke yugo de deflexión
deflection deflexión, desviación
deflection angle ángulo de deflexión
deflection axis eje de deflexión
deflection center centro de deflexión
deflection characteristic característica de deflexión
deflection chassis chasis de deflexión
deflection circuit circuito de deflexión
deflection coefficient coeficiente de deflexión
deflection coil bobina de deflexión
deflection defocusing desenfoque por deflexión
deflection electrode electrodo de deflexión
deflection factor factor de deflexión
deflection generator generador de deflexión
deflection oscillator oscilador de deflexión
deflection plane plano de deflexión
deflection plate placa de deflexión
deflection polarity polaridad de deflexión
deflection sensibility sensibilidad de deflexión
deflection sensitivity sensibilidad de deflexión
deflection system sistema de deflexión
deflection valve válvula de deflexión
deflection voltage tensión de deflexión
deflection yoke yugo de deflexión
deflector deflector
deflector circuit circuito de deflexión
deflector coil bobina de deflexión
deflector plate placa de deflexión
defocusing desenfoque
deformation potential potencial de deformación
degassing desgasificación
degauss desmagnetizar
degausser desmagnetizador
degaussing desmagnetización
degaussing cable cable de desmagnetización
degaussing circuit circuito de desmagnetización
degaussing coil bobina de desmagnetización
degaussing control control de desmagnetización
degaussing generator generador de
 desmagnetización
degeneracy degeneración
degenerate amplifier amplificador degenerado
degenerate conduction band banda de conducción
 degenerada
degenerate mode modo degenerado
degenerate parametric amplifier amplificador
 paramétrico degenerado
degenerate semiconductor semiconductor
 degenerado
degeneration degeneración
degenerative degenerativo
degenerative amplifier amplificador degenerativo
degenerative bridge circuit circuito de puente
 degenerativo
degenerative feedback retroalimentación
 degenerativa
degenerative resistor resistor degenerativo
degradation degradación
degradation factor factor de degradación
degraded backup reserva degradada
degraded operation operación degradada
degraded voltage condition condición de tensión
 degradada
degree grado

degree of accuracy grado de exactitud
degree of asymmetry grado de asimetría
degree of coherence grado de coherencia
degree of current rectification grado de
 rectificación de corriente
degree of damping grado de amortiguamiento
degree of dissociation grado de disociación
degree of distortion grado de distorsión
degree of vacuum grado de vacío
degree of vibration grado de vibración
degree of voltage rectification grado de rectificación
 de tensión
degrees of freedom grados de libertad
dehumidification deshumidificación
dehumidifier deshumidificador
deinstall desinstalar
deionization desionización
deionization grid rejilla de desionización
deionization potential potencial de desionización
deionization rate velocidad de desionización
deionization time tiempo de desionización
deionization voltage tensión de desionización
deionize desionizar
deionizing grid rejilla de desionización
deka deca
dekatron decatrón
dekatron tube tubo decatrón
delamination deslaminación
delay retardo
delay action acción de retardo
delay angle ángulo de retardo
delay cable cable de retardo
delay characteristic característica de retardo
delay circuit circuito de retardo
delay coincidence circuit circuito de coincidencia de
 retardo
delay counter contador de retardo
delay curve curva de retardo
delay device dispositivo de retardo
delay dispersion dispersión de retardo
delay distortion distorsión de retardo
delay element elemento de retardo
delay equalizer ecualizador de retardo
delay lens lente de retardo
delay line línea de retardo
delay-line memory memoria de línea de retardo
delay-line memory unit unidad de memoria de línea
 de retardo
delay-line register registro de línea de retardo
delay-line storage almacenamiento de línea de
 retardo
delay medium medio de retardo
delay multivibrator multivibrador de retardo
delay network red de retardo
delay programming programación de retardo
delay pulsing pulsación de retardo
delay relay relé de retardo
delay screen pantalla de retardo
delay switch conmutador de retardo
delay switching system sistema de conmutación de
 retardo
delay time tiempo de retardo
delay timer temporizador de retardo
delay unit unidad de retardo
delayed access acceso retardado
delayed action acción retardada
delayed application aplicación retardada
delayed automatic gain control control de ganancia
 automática retardada
delayed break ruptura retardada
delayed call llamada retardada
delayed close cierre retardado
delayed closure cierre retardado

delayed coincidence coincidencia retardada
delayed contacts contactos retardados
delayed detection detección retardada
delayed make cierre retardado
delayed open abertura retardada
delayed output salida retardada
delayed potential potencial retardado
delayed-pulse generator generador de impulsos retardados
delayed relay relé retardado
delayed release desenganche retardado
delayed repeater repetidor retardado
delayed repeater satellite satélite repetidor retardado
delayed response respuesta retardada
delayed scanning exploración retardada
delayed sweep barrido retardado
delayed test prueba retardada
delayed updating actualización retardada
delaying retardante
delaying sweep barrido retardante
delete eliminar, borrar
delete character carácter de borrado
delete function función de borrado
delete transaction transacción de borrado
deletion record registro de borrado
delimiter delimitador
deliquescent delicuescente
deliquescent material material delicuescente
Dellinger effect efecto de Dellinger
delta delta
delta antenna antena en delta
delta circuit circuito en delta
delta connection conexión en delta
delta grouping agrupamiento en delta
delta-matched antenna antena de adaptación en delta
delta matching transformer transformador de adaptación en delta
delta modulation modulación en delta
delta network red en delta
delta pulse-code modulation modulación de impulsos codificados en delta
delta rays rayos delta
delta-sigma modulation modulación delta-sigma
delta voltage tensión en delta
delta waves ondas delta
deltic deltic
deltic method método deltic
demagnetization desmagnetización
demagnetization curve curva de desmagnetización
demagnetization effect efecto de desmagnetización
demagnetize desmagnetizar
demagnetizer desmagnetizador
demagnetizing desmagnetizante
demagnetizing current corriente desmagnetizante
demagnetizing factor factor desmagnetizante
demagnetizing field campo desmagnetizante
demagnetizing force fuerza desmagnetizante
demand charge carga por demanda
demand constant constante de demanda
demand deviation desviación de demanda
demand factor factor de demanda
demand failure rate tasa de fallas por demanda
demand interval intervalo de demanda
demand limiter limitador de demanda
demand meter medidor de demanda
demand processing procesamiento inmediato
demand rate tasa de demanda
demand register registro de demanda
demarcation demarcación
demarcation current corriente de demarcación
demarcation level nivel de demarcación
demarcation potential potencial de demarcación
demarcation strip franja de demarcación

Dember effect efecto de Dember
demineralization desmineralización
demodularization desmodularización
demodulated desmodulado
demodulated signal señal desmodulada
demodulated voltage tensión desmodulada
demodulation desmodulación
demodulator desmodulador
demodulator probe sonda desmoduladora
demodulator rectifier rectificador desmodulador
demodulator tube tubo desmodulador
demonstration program programa de demostración
demountable desmontable
demountable tube tubo desmontable
demultiplexer desmultiplexor
demultiplexing desmultiplexor
demultiplexing circuit circuito desmultiplexor
denary denario
denary band banda denaria
dendrite dendrita
dendritic dendrítico
dendritic growth crecimiento dendrítico
denominator denominador
dense binary code código binario denso
densitometer densitómetro
density densidad
density coefficient coeficiente de densidad
density modulation modulación de densidad
density of electrons densidad de electrones
denuder separador
dependable operation operación confiable
dependent contact contacto dependiente
dependent job tarea dependiente
dependent linearity linealidad dependiente
dependent node nodo dependiente
dependent segment segmento dependiente
dependent station estación dependiente
dependent variable variable dependiente
deplete agotar
depletion-enhancement mode modo de agotamiento-enriquecimiento
depletion field-effect-transistor transistor de efecto de campo de agotamiento
depletion layer capa de agotamiento
depletion-layer capacitance capacitancia de capa de agotamiento
depletion-layer rectification rectificación de capa de agotamiento
depletion-layer transistor transistor de capa de agotamiento
depletion mode modo de agotamiento
depletion region región de agotamiento
depletion voltage tensión de agotamiento
depolarization despolarización
depolarization factor factor de despolarización
depolarization field campo de despolarización
depolarizer despolarizador
depolarizing despolarizante, despolarizador
depolarizing agent agente despolarizante
depolarizing mix mezcla despolarizante
deposited carbon resistor resistor de carbón depositado
deposited metal metal depositado
deposition deposición
deposition speed velocidad de deposición
deposition time tiempo de deposición
deposition voltage tensión de deposición
depth indicator indicador de profundidad
depth of cut profundidad de corte
depth of discharge profundidad de descarga
depth of heating profundidad de calentamiento
depth of modulation profundidad de modulación
depth of penetration profundidad de penetración

dequeue eliminar de una cola
derate reducir
derating reducción
derating curve curva de reducción
derating factor factor de reducción
derivative derivada, derivado
derivative action acción derivada
derivative control control derivado
derivative function función derivada
derived center channel canal central derivado
derived data datos derivados
derived field campo derivado
derived function función derivada
derived pulse impulso derivado
derived relation relación derivada
derived units unidades derivadas
Dershem electrometer electrómetro de Dershem
desaturation desaturación
DeSauty's bridge puente de DeSauty
descending order orden descendiente
descending sequence secuencia descendiente
descending sort clasificación descendiente
descriptive model modelo descriptivo
descriptive parameter parámetro descriptivo
desensitization desensibilización
desensitize desensibilizar
desiccant desecante
design diseño
design analysis análisis de diseño
design analyzer analizador de diseño
design automation automatización de diseño
design-center rating clasificación nominal
design compatibility compatibilidad de diseños
design defect defecto de diseño
design description descripción de diseño
design element elemento de diseño
design factor factor de diseño
design feature característica de diseño
design improvement mejora de diseño
design inspection inspección de diseño
design language lenguaje de diseño
design level nivel de diseño
design life vida de diseño
design limitation limitación de diseño
design limits límites de diseño
design margin margen de diseño
design-maximum rating clasificación máxima
design phase fase de diseño
design pressure presión de diseño
design program programa de diseño
design requirement requisito de diseño
design review revisión de diseño
design specifications especificaciones de diseño
design tests pruebas de diseño
design unit unidad de diseño
design verification verificación de diseño
design voltage tensión de diseño
designation designación
designation hole perforación de designación
designation number número de designación
desired polarization polarización deseada
desired signal señal deseada
desired value valor deseado
deskstand telephone teléfono de columna
desktop computer computadora de escritorio
desktop microcomputer microcomputadora de escritorio
desolder desoldar
dessicator desecador
destination destino
destination address dirección de destino
destination code código de destino
destination field campo de destino

destination file archivo de destino
destination indicator indicador de destino
destination queue cola de destino
destination register registro de destino
destination station estación de destino
destination system sistema de destino
destination terminal terminal de destino
Destriau effect efecto de Destriau
destruction point punto de destrucción
destructive addition suma destructiva
destructive breakdown ruptura destructiva
destructive cursor cursor destructivo
destructive interference interferencia destructiva
destructive operation operación destructiva
destructive reading lectura destructiva
destructive readout lectura destructiva
destructive test prueba destructiva
detachable desconectable
detachable cable cable desconectable
detachable keyboard teclado desconectable
detail constant constante de detalle
detail contrast contraste de detalle
detail file archivo de detalle
detail record registro de detalle
detailed detallado
detailed design diseño detallado
detectability detectabilidad
detectability factor factor de detectabilidad
detectable detectable
detectable element elemento detectable
detectable leak fuga detectable
detectable segment segmento detectable
detected error error detectado
detecting detector
detecting element elemento detector
detecting probe sonda detectora
detection detección
detection channel canal de detección
detection efficiency eficiencia de detección
detection limit límite de detección
detection probability probabilidad de detección
detection radar radar de detección
detection threshold umbral de detección
detectivity detectividad
detectophone detectófono
detector detector
detector amplifier amplificador detector
detector balanced bias polarización equilibrada de detector
detector bias polarización de detector
detector cell célula detectora
detector circuit circuito detector
detector diode diodo detector
detector probe sonda detectora
detector stage etapa detectora
detector tube tubo detector
detent retén
detent disk disco de retén
detent mechanism mecanismo de retén
determinant determinante
detune desintonizar
detuning desintonización
deupdating desactualización
deuterium deuterio
deuterium discharge tube tubo de descarga de deuterio
development methodology metodología de desarrollo
development specification especificación de desarrollo
development system sistema de desarrollo
development technique técnica de desarrollo
development time tiempo de desarrollo
deviation desviación

deviation absorption absorción por desviación
deviation distortion distorsión de desviación
deviation factor factor de desviación
deviation meter desviómetro, medidor de desviación
deviation monitor monitor de desviación
deviation ratio razón de desviación
deviation sensitivity sensibilidad de desviación
deviation system sistema de desviación
device dispositivo, equipo
device address dirección de dispositivo
device class clase de dispositivo
device code código de dispositivo
device complexity complejidad de dispositivo
device control block bloque de control de dispositivo
device control character carácter de control de
    dispositivo
device control unit unidad de control de dispositivo
device dependence dependencia de dispositivo
device dependent dependiente de dispositivo
device driver unidad de dispositivo, programa de
    operación de dispositivo
device emulation emulación de dispositivo
device field campo de dispositivo
device independence independencia de dispositivo
device independent independiente de dispositivo
device input queue cola de entrada de dispositivo
device name nombre de dispositivo
device number número de dispositivo
device parameter parámetro de dispositivo
device priority prioridad de dispositivo
device processor procesador de dispositivo
device selection character carácter de selección de
    dispositivo
device selection check comprobación de selección de
    dispositivo
device space espacio de dispositivo
device status estado de dispositivo
device type tipo de dispositivo
dew point punto de rocío
Dewar flask frasco de Dewar
diac diac
diacritical current corriente diacrítica
diactor diactor
diagnostic program programa de diagnóstico
diagnostic programming programación de
    diagnóstico
diagnostic routine rutina de diagnóstico
diagnostic software programas de diagnóstico
diagnostic test prueba de diagnóstico
diagnostic unit unidad de diagnóstico
diagnotor diagnotor
diagram diagrama
diagrammatic diagramático
dial dial, cuadrante, disco
dial attenuator atenuador de cuadrante
dial cable cable de cuadrante
dial cord cordón de cuadrante
dial impulse impulso de disco
dial lamp lámpara de cuadrante
dial light luz de cuadrante
dial pulse impulso de disco
dial recorder registrador de cuadrante
dial resolution resolución de cuadrante
dial selector selector de cuadrante
dial telephone teléfono de disco
dial tone tono de marcar, señal de línea
dial-tone delay retardo de tono de marcar
dial-up marcar
dialer marcador
dialing marcación, marcado, discado
diamagnetic diamagnético
diamagnetic effect efecto diamagnético
diamagnetic material material diamagnético

diamagnetic substance sustancia diamagnética
diamagnetic susceptibility susceptibilidad
    diamagnética
diamagnetism diamagnetismo
diamagnetometer diamagnetómetro
diametral diametral
diametral voltage tensión diametral
diamond antenna antena en rombo
diamond stylus aguja de diamante
diamond winding arrollamiento en rombo
diapason diapasón
diaphragm diafragma
diathermic diatérmico
diathermy diatermia
diathermy interference interferencia por diatermia
diatomic diatómico
dibit dibit
dichotomizing search búsqueda dicotómica
dichotomy dicotomía
dichroic filter filtro dicroico
dichroic mirror espejo dicroico
dichroism dicroismo
dichromate cell pila de bicromato
dicing cortado en dados
Dicke radiometer radiómetro de Dicke
die dado, troquel, pastilla
die cast fundido a presión
dielectric dieléctrico
dielectric absorption absorción dieléctrica
dielectric amplifier amplificador dieléctrico
dielectric antenna antena dieléctrica
dielectric breakdown ruptura dieléctrica
dielectric breakdown voltage tensión de ruptura
    dieléctrica
dielectric capacity capacidad dieléctrica
dielectric circuit circuito dieléctrico
dielectric coating revestimiento dieléctrico
dielectric coefficient coeficiente dieléctrico
dielectric constant constante dieléctrica
dielectric core núcleo dieléctrico
dielectric crystal cristal dieléctrico
dielectric current corriente dieléctrica
dielectric diode diodo dieléctrico
dielectric dispersion dispersión dieléctrica
dielectric displacement desplazamiento dieléctrico
dielectric dissipation disipación dieléctrica
dielectric dissipation factor factor de disipación
    dieléctrica
dielectric fatigue fatiga de dieléctrico
dielectric film película dieléctrica
dielectric flux flujo dieléctrico
dielectric flux density densidad de flujo dieléctrico
dielectric gas gas dieléctrico
dielectric guide guía dieléctrica
dielectric heater calentador dieléctrico
dielectric heating calentamiento dieléctrico
dielectric hysteresis histéresis dieléctrica
dielectric hysteresis loss pérdida por histéresis
    dieléctrica
dielectric isolation aislamiento dieléctrico
dielectric layer capa dieléctrica
dielectric lens lente dieléctrica
dielectric-lens antenna antena de lente dieléctrica
dielectric loss pérdida dieléctrica
dielectric loss angle ángulo de pérdida dieléctrica
dielectric loss factor factor de pérdida dieléctrica
dielectric matching plate placa de adaptación
    dieléctrica
dielectric material material dieléctrico
dielectric mirror espejo dieléctrico
dielectric permittivity permitividad dieléctrica
dielectric phase angle ángulo de fase de dieléctrico
dielectric polarization polarización dieléctrica

dielectric power factor factor de potencia de dieléctrico
dielectric radiator radiador dieléctrico
dielectric rating clasificación de dieléctrico
dielectric reactance reactancia dieléctrica
dielectric relaxation relajación dieléctrica
dielectric resistance resistencia dieléctrica
dielectric rigidity rigidez dieléctrica
dielectric rod varilla dieléctrica
dielectric-rod antenna antena de varilla dieléctrica
dielectric-rod waveguide guíaondas de varilla dieléctrica
dielectric separator separador dieléctrico
dielectric strain esfuerzo dieléctrico
dielectric strength rigidez dieléctrica
dielectric stress estrés dieléctrico
dielectric susceptibility susceptibilidad dieléctrica
dielectric test prueba dieléctrica
dielectric viscosity viscosidad dieléctrica
dielectric waveguide guíaondas dieléctrico
dielectric wire hilo dieléctrico
dielectricity dielectricidad
dielectrometer dielectrómetro
difference amplifier amplificador de diferencia
difference band banda de diferencia
difference channel canal de diferencia
difference detector detector de diferencia
difference frequency frecuencia de diferencia
difference limen limen de diferencia
difference of potential diferencia de potencial
difference pattern patrón de diferencia
difference signal señal de diferencia
difference tone tono de diferencia
differential diferencial
differential absorption ratio razón de absorción diferencial
differential amplifier amplificador diferencial
differential amplitude amplitud diferencial
differential analyzer analizador diferencial
differential angle ángulo diferencial
differential anode conductance conductancia de ánodo diferencial
differential anode resistance resistencia de ánodo diferencial
differential calculus cálculo diferencial
differential capacitance capacitancia diferencial
differential capacitance characteristic característica de capacitancia diferencial
differential capacitance voltage tensión de capacitancia diferencial
differential capacitor capacitor diferencial
differential coefficient coeficiente diferencial
differential coil bobina diferencial
differential comparator comparador diferencial
differential compound motor motor compuesto diferencial
differential condenser condensador diferencial
differential control control diferencial
differential control current corriente de control diferencial
differential control voltage tensión de control diferencial
differential cooling enfriamiento diferencial
differential current corriente diferencial
differential delay retardo diferencial
differential discriminator discriminador diferencial
differential distortion distorsión diferencial
differential Doppler frequency frecuencia de Doppler diferencial
differential dump vaciado diferencial
differential duplex dúplex diferencial
differential duplex installation instalación dúplex diferencial

differential duplex system sistema dúplex diferencial
differential equation ecuación diferencial
differential excitation excitación diferencial
differential flutter titilación diferencial
differential frequency meter frecuencímetro diferencial
differential gain ganancia diferencial
differential gain control control de ganancia diferencial
differential gain control circuit circuito de control de ganancia diferencial
differential galvanometer galvanómetro diferencial
differential gap intervalo diferencial, espacio diferencial
differential generator generador diferencial
differential governor regulador diferencial
differential heating calentamiento diferencial
differential impedance impedancia diferencial
differential-inductance transducer transductor de inductancia diferencial
differential induction coil bobina de inducción diferencial
differential input entrada diferencial
differential-input amplifier amplificador de entrada diferencial
differential-input capacitance capacitancia de entrada diferencial
differential-input impedance impedancia de entrada diferencial
differential-input measurement medida de entrada diferencial
differential-input rating clasificación de entrada diferencial
differential-input resistance resistencia de entrada diferencial
differential-input signal señal de entrada diferencial
differential-input voltage tensión de entrada diferencial
differential instrument instrumento diferencial
differential ionization chamber cámara de ionización diferencial
differential memory dump vaciado de memoria diferencial
differential microphone micrófono diferencial
differential-mode attenuation atenuación de modo diferencial
differential-mode delay retardo de modo diferencial
differential-mode gain ganancia de modo diferencial
differential-mode input entrada de modo diferencial
differential-mode interference interferencia de modo diferencial
differential-mode radio noise ruido de radio de modo diferencial
differential-mode rejection rechazo de modo diferencial
differential-mode signal señal de modo diferencial
differential-mode voltage tensión de modo diferencial
differential modulation modulación diferencial
differential motor motor diferencial
differential multiplexer multiplexor diferencial
differential nonlinearity no linealidad diferencial
differential output current corriente de salida diferencial
differential permeability permeabilidad diferencial
differential permittivity permitividad diferencial
differential phase fase diferencial
differential phase shift desplazamiento de fase diferencial
differential position posición diferencial
differential pressure presión diferencial
differential-pressure transducer transductor de presión diferencial

differential protection protección diferencial
differential protective relay relé de protección
　diferencial
differential protective system sistema de protección
　diferencial
differential receiver receptor diferencial
differential regulator regulador diferencial
differential relay relé diferencial
differential resistance resistencia diferencial
differential selsyn selsyn diferencial
differential signal señal diferencial
differential stage etapa diferencial
differential synchro sincro diferencial
differential threshold umbral diferencial
differential transducer transductor diferencial
differential transformer transformador diferencial
differential transmitter transmisor diferencial
differential unbalance desequilibrio diferencial
differential voltage tensión diferencial
differential voltage gain ganancia de tensión
　diferencial
differential voltmeter vóltmetro diferencial
differential winding arrollamiento diferencial
differentiate diferenciar
differentiated diferenciado
differentiating diferenciador
differentiating amplifier amplificador diferenciador
differentiating circuit circuito diferenciador
differentiating network red diferenciadora
differentiator diferenciador
diffracted wave onda difractada
diffraction difracción
diffraction angle ángulo de difracción
diffraction camera cámara de difracción
diffraction delay line línea de retardo de difracción
diffraction efficiency eficiencia de difracción
diffraction field campo de difracción
diffraction grating rejilla de difracción
diffraction loss pérdida por difracción
diffraction pattern patrón de difracción
diffraction phenomenon fenómeno de difracción
diffraction propagation propagación por difracción
diffraction region región de difracción
diffraction spectrum espectro de difracción
diffraction zone zona de difracción
diffractograph difractógrafo
diffractometer difractómetro
diffractor difractor
diffuse difuso
diffuse boundary límite difuso
diffuse field campo difuso
diffuse-field response respuesta de campo difuso
diffuse-field sensitivity sensibilidad de campo difuso
diffuse phase transition transición de fase difusa
diffuse reference electrode electrodo de referencia
　difuso
diffuse reflection reflexión difusa
diffuse reflection factor factor de reflexión difusa
diffuse reflectivity reflectividad difusa
diffuse refraction refracción difusa
diffuse sound sonido difuso
diffuse sound field campo de sonido difuso
diffuse transmission transmisión difusa
diffuse transmission factor factor de transmisión
　difusa
diffused-alloy transistor transistor de aleación difusa
diffused-base transistor transistor de base difusa
diffused device dispositivo por difusión
diffused diode diodo difuso
diffused junction unión por difusión
diffused-junction rectifier rectificador de unión por
　difusión
diffused-junction transistor transistor de unión por

　difusión
diffused-planar transistor transistor planar por
　difusión
diffused resistor resistor por difusión
diffused sound sonido difuso
diffused transistor transistor por difusión
diffusion difusión
diffusion bonding unión por difusión
diffusion by reflection difusión por reflexión
diffusion by transmission difusión por transmisión
diffusion capacitance capacitancia de difusión
diffusion coefficient coeficiente de difusión
diffusion constant constante de difusión
diffusion current corriente de difusión
diffusion distance distancia de difusión
diffusion layer capa de difusión
diffusion length longitud de difusión
diffusion process proceso de difusión
diffusion pump bomba de difusión
diffusion theory teoría de difusión
diffusion time tiempo de difusión
diffusion transistor transistor de difusión
digit dígito
digit absorption absorción de dígitos
digit compression compresión de dígitos
digit current corriente de dígito
digit delay device dispositivo de retardo de dígito
digit delay element elemento de retardo de dígito
digit emitter emisor de dígitos
digit filter filtro de dígito
digit key tecla de dígito
digit period periodo de dígito
digit place lugar de dígito
digit plane plano de dígito
digit position posición de dígito
digit pulse pulso de dígito
digit row fila de dígitos
digit selector selector de dígitos
digit time tiempo de dígito
digit transfer transferencia de dígitos
digit-transfer bus bus de transferencia de dígitos
digit-transfer track pista de transferencia de dígitos
digital digital
digital analyzer analizador digital
digital barometer barómetro digital
digital bit rate velocidad de bits digital
digital bus bus digital
digital capacitance meter medidor de capacitancia
　digital
digital capacity capacidad digital
digital circuit circuito digital
digital clock reloj digital
digital communications comunicaciones digitales
digital communications system sistema de
　comunicaciones digitales
digital comparator comparador digital
digital compression compresión digital
digital computer computadora digital
digital computer storage almacenamiento de
　computadora digital
digital control control digital
digital control system sistema de control digital
digital controller controlador digital
digital converter convertidor digital
digital counter contador digital
digital data datos digitales
digital-data cable cable de datos digitales
digital-data circuit circuito de datos digitales
digital-data link enlace de datos digitales
digital-data processor procesador de datos digitales
digital-data system sistema de datos digitales
digital-data transmission transmisión de datos
　digitales

digital device   dispositivo digital
digital differential analyzer   analizador diferencial digital
digital display   visualización digital
digital display area   área de visualización digital
digital display unit   unidad de visualización digital
digital divider   divisor digital
digital electrometer   electrómetro digital
digital electronics   electrónica digital
digital entry   entrada digital
digital equipment   equipo digital
digital filter   filtro digital
digital frequency meter   frecuencímetro digital
digital image   imagen digital
digital increment   incremento digital
digital indicator   indicador digital
digital information   información digital
digital information display   visualización de información digital
digital information transmission   transmisión de información digital
digital input   entrada digital
digital integrated circuit   circuito integrado digital
digital integrator   integrador digital
digital interface   interfaz digital
digital keyboard   teclado digital
digital line link   enlace de línea digital
digital line section   sección de línea digital
digital link   enlace digital
digital logic   lógica digital
digital logic element   elemento lógico digital
digital logic module   módulo lógico digital
digital milliwatt   miliwatt digital
digital monitor   monitor digital
digital multimeter   multímetro digital
digital multiplex   múltiplex digital
digital multiplexing   multiplexión digital
digital multiplier   multiplicador digital
digital output   salida digital
digital output module   módulo de salida digital
digital panel meter   medidor de panel digital
digital path   trayectoria digital
digital phase shifter   desfasador digital
digital photometer   fotómetro digital
digital potentiometer   potenciómetro digital
digital power meter   medidor de potencia digital
digital processor   procesador digital
digital quantity   cantidad digital
digital reading   lectura digital
digital readout   lectura digital
digital recorder   registrador digital
digital recording   registro digital
digital reference signal   señal de referencia digital
digital repeater   repetidor digital
digital representation   representación digital
digital rotary transducer   transductor rotativo digital
digital section   sección digital
digital signal   señal digital
digital signal processing   procesamiento de señales digitales
digital simulation   simulación digital
digital sort   clasificación digital
digital sound   sonido digital
digital subset   subconjunto digital
digital subtracter   restador digital
digital switch   conmutador digital
digital switching   conmutación digital
digital telemeter   telemedidor digital
digital television   televisión digital
digital temperature indicator   indicador de temperatura digital
digital termination system   sistema de terminación digital

digital thermometer   termómetro digital
digital-to-analog conversion   conversión de digital a analógico
digital-to-analog converter   convertidor de digital a analógico
digital transducer   transductor digital
digital transmission   transmisión digital
digital voltmeter   vóltmetro digital
digital voltmeter-potentiometer   vóltmetro-potenciómetro digital
digital wattmeter   wáttmetro digital
digitalize   digitalizar
digitization   digitización
digitize   digitalizar
digitron   digitrón
diheptal base   base diheptal
dilatometer   dilatómetro
dimensional   dimensional
dimensional analysis   análisis dimensional
dimensional ratio   razón dimensional
dimensional stability   estabilidad dimensional
dimensional system   sistema dimensional
dimensionless   adimensional
dimensionless parameter   parámetro adimensional
dimensionless quantity   cantidad adimensional
dimmer   reductor de intensidad
diode   diodo
diode action   acción de diodo
diode amplifier   amplificador de diodo
diode array   red de diodos
diode bias   polarización de diodo
diode capacitance   capacitancia de diodo
diode capacitor   capacitor de diodo
diode characteristic   característica de diodo
diode checker   comprobador de diodo
diode clipper   recortador de diodo
diode converter   convertidor de diodo
diode current   corriente de diodo
diode current meter   medidor de corriente de diodo
diode damper   amortiguador de diodo
diode demodulator   desmodulador de diodo
diode detector   detector de diodo
diode electronic switch   conmutador electrónico de diodo
diode equation   ecuación de diodo
diode equivalent   equivalente de diodo
diode feedback rectifier   rectificador por retroalimentación de diodo
diode function generator   generador de funciones de diodo
diode fuse   fusible de diodo
diode gate   puerta de diodo
diode-heptode   diodo-heptodo
diode holder   portadiodo
diode impedance   impedancia de diodo
diode lamp   lámpara de diodo
diode laser   láser de diodo
diode light source   fuente de luz de diodo
diode limiter   limitador de diodo
diode load   carga de diodo
diode load resistance   resistencia de carga de diodo
diode load resistor   resistor de carga de diodo
diode logic   lógica de diodos
diode matrix   matriz de diodos
diode mixer   mezclador de diodo
diode modulator   modulador de diodo
diode noise limiter   limitador de ruido de diodos
diode oscillator   oscilador de diodo
diode pack   bloque de diodos
diode peak detector   detector de picos de diodo
diode peak voltmeter   vóltmetro de picos de diodo
diode-pentode   diodo-pentodo
diode probe   sonda de diodo

diode product detector detector de productos de diodo
diode pulse amplifier amplificador de impulsos de diodo
diode recovery time tiempo de recuperación de diodo
diode rectification rectificación por diodo
diode rectifier rectificador de diodo
diode resistor resistor de diodo
diode storage almacenamiento de diodo
diode storage time tiempo de almacenamiento de diodo
diode switch conmutador de diodo
diode temperature stabilization estabilización de temperatura de diodo
diode tester probador de diodo
diode-tetrode diodo-tetrodo
diode transducer transductor de diodo
diode transistor transistor diodo
diode-transistor logic lógica diodo-transistor
diode-triode diodo-triodo
diode tube tubo diodo
diode varactor diodo varactor
diode varistor diodo varistor
diode voltmeter vóltmetro de diodo
diodemeter diodímetro
diopter dioptría
dioptoscope dioptoscopio
dioptral dioptral
dioptrometer dioptrómetro
diotron diotrón
dip coating revestimiento por inmersión
dip encapsulation encapsulación por inmersión
dip impregnation impregnación por inmersión
dip plating depósito por inmersión
dip soldering soldadura por inmersión
diplex díplex
diplex operation operación díplex
diplex radio transmission radiotransmisión díplex
diplex receiving device dispositivo de recepción díplex
diplex reception recepción díplex
diplex system sistema díplex
diplex transmission transmisión díplex
diplexer diplexor
diplexing diplexión
dipolar dipolar
dipolarization dipolarización
dipole dipolo
dipole antenna antena dipolo
dipole electric field campo eléctrico de dipolo
dipole magnetic field campo magnético de dipolo
dipole moment momento dipolar
dipole radiator radiador dipolo
dipped component componente inmerso
dipping inmersión
dipping electrode electrodo de inmersión
dipping needle aguja de inmersión
direct access acceso directo
direct-access device dispositivo de acceso directo
direct-access method método de acceso directo
direct-access mode modo de acceso directo
direct-access queue cola de acceso directo
direct-access storage almacenamiento de acceso directo
direct-access storage device dispositivo de almacenamiento de acceso directo
direct-access unit unidad de acceso directo
direct-acting recorder registrador de accionamiento directo
direct-acting recording instrument instrumento de registro de accionamiento directo
direct-acting valve válvula de accionamiento directo

direct activation activación directa
direct address dirección directa
direct addressing direccionamiento directo
direct allocation asignación directa
direct alternating-current converter convertidor de corriente alterna directo
direct-axis component componente longitudinal
direct-axis subtransient impedance impedancia subtransitoria longitudinal
direct-axis subtransient reactance reactancia subtransitoria longitudinal
direct-axis subtransient voltage tensión subtransitoria longitudinal
direct-axis synchronous impedance impedancia sincrónica longitudinal
direct-axis synchronous reactance reactancia sincrónica longitudinal
direct-axis transient impedance impedancia transitoria longitudinal
direct-axis transient reactance reactancia transitoria longitudinal
direct-axis transient voltage tensión transitoria longitudinal
direct call llamada directa
direct capacitance capacitancia directa
direct chaining encadenamiento directo
direct circuit circuito directo
direct code código directo
direct coding codificación directa
direct communication comunicación directa
direct commutation conmutación directa
direct component componente directo
direct control control directo
direct control feature característica de control directo
direct conversion conversión directa
direct-conversion receiver receptor de conversión directa
direct-coupled amplifier amplificador de acoplamiento directo
direct-coupled attenuation atenuación de acoplamiento directo
direct-coupled transistor logic lógica de transistores de acoplamiento directo
direct coupling acoplamiento directo
direct-coupling amplifier amplificador de acoplamiento directo
direct-current corriente continua
direct-current-alternating-current converter convertidor corriente continua-corriente alterna
direct-current amplifier amplificador de corriente continua
direct-current analog computer computadora analógica de corriente continua
direct-current automatic block bloqueo automático de corriente continua
direct-current balance equilibrio de corriente continua
direct-current balancer equilibrador de corriente continua
direct-current base current corriente de base de corriente continua
direct-current base resistance resistencia de base de corriente continua
direct-current base voltage tensión de base de corriente continua
direct-current bell timbre de corriente continua
direct-current breakdown voltage tensión de ruptura de corriente continua
direct-current bus bus de corriente continua
direct-current cathode current corriente de cátodo de corriente continua
direct-current cathode resistance resistencia de

cátodo de corriente continua

direct-current cathode voltage tensión de cátodo de corriente continua

direct-current centering centrado por corriente continua

direct-current circuit circuito de corriente continua

direct-current circuit-breaker cortacircuito de corriente continua

direct-current collector current corriente de colector de corriente continua

direct-current collector resistance resistencia de colector de corriente continua

direct-current collector voltage tensión de colector de corriente continua

direct-current compensator compensador de corriente continua

direct-current component componente de corriente continua

direct-current connection conexión de corriente continua

direct-current converter convertidor de corriente continua

direct-current coupling acoplamiento de corriente continua

direct-current discharge descarga de corriente continua

direct-current distribution distribución de corriente continua

direct-current drain current corriente de consumo de corriente continua

direct-current drain resistance resistencia de consumo de corriente continua

direct-current drain voltage tensión de consumo de corriente continua

direct-current dump vaciado de corriente continua

direct-current electric locomotive locomotora eléctrica de corriente continua

direct-current emitter current corriente de emisor de corriente continua

direct-current emitter resistance resistencia de emisor de corriente continua

direct-current emitter voltage tensión de emisor de corriente continua

direct-current equipment equipo de corriente continua

direct-current erase borrado por corriente continua

direct-current erasing borrado por corriente continua

direct-current erasing head cabeza de borrado por corriente continua

direct-current filter filtro de corriente continua

direct-current gate current corriente de puerta de corriente continua

direct-current gate resistance resistencia de puerta de corriente continua

direct-current gate voltage tensión de puerta de corriente continua

direct-current generator generador de corriente continua

direct-current grid bias polarización de rejilla de corriente continua

direct-current grid current corriente de rejilla de corriente continua

direct-current grid voltage tensión de rejilla de corriente continua

direct-current input current corriente de entrada de corriente continua

direct-current input power potencia de entrada de corriente continua

direct-current inverter inversor de corriente continua

direct-current leakage fuga de corriente continua

direct-current leakage current corriente de fuga de

corriente continua

direct-current level nivel de corriente continua

direct-current load carga de corriente continua

direct-current loop signaling señalización por bucle de corriente continua

direct-current magnetic biasing polarización magnética de corriente continua

direct-current motor motor de corriente continua

direct-current motor control control de motor de corriente continua

direct-current noise ruido de corriente continua

direct-current noise margin margen de ruido de corriente continua

direct-current operated relay relé operado por corriente continua

direct-current operating point punto de operación de corriente continua

direct-current oscilloscope osciloscopio de corriente continua

direct-current overcurrent relay relé de sobrecorriente de corriente continua

direct-current overvoltage relay relé de sobretensión de corriente continua

direct-current picture transmission transmisión de imagen de corriente continua

direct-current plate current corriente de placa de corriente continua

direct-current plate resistance resistencia de placa de corriente continua

direct-current plate voltage tensión de placa de corriente continua

direct-current polarizing voltage tensión polarizante de corriente continua

direct-current positioning posicionamiento por corriente continua

direct-current power potencia de corriente continua

direct-current power supply fuente de alimentación de corriente continua

direct-current probe sonda de corriente continua

direct-current receiver receptor de corriente continua

direct-current reference referencia de corriente continua

direct-current reinsertion reinserción de corriente continua

direct-current relay relé de corriente continua

direct-current resistance resistencia en corriente continua

direct-current resistivity resistividad de corriente continua

direct-current restoration restauración de corriente continua

direct-current restorer restaurador de corriente continua

direct-current restorer diode diodo restaurador de corriente continua

direct-current restoring circuit circuito restaurador de corriente continua

direct-current signaling señalización por corriente continua

direct-current source fuente de corriente continua

direct-current source current corriente de fuente de corriente continua

direct-current source resistance resistencia de fuente de corriente continua

direct-current source voltage tensión de fuente de corriente continua

direct-current supply voltage tensión de alimentación de corriente continua

direct-current suppressor voltage tensión supresora de corriente continua

direct-current tachometer tacómetro de corriente continua

direct-current telegraphy telegrafía por corriente continua
direct-current threshold umbral de corriente continua
direct-current thyratron tiratrón de corriente continua
direct-current transducer transductor de corriente continua
direct-current transformer transformador de corriente continua
direct-current transmission transmisión de corriente continua
direct-current tuning voltage tensión de sintonización de corriente continua
direct-current undercurrent relay relé de corriente mínima de corriente continua
direct-current undervoltage relay relé de tensión mínima de corriente continua
direct-current voltage tensión de corriente continua
direct-current voltmeter vóltmetro de corriente continua
direct cycle ciclo directo
direct data entry entrada de datos directa
direct data set conjunto de datos directo
direct data transmission transmisión de datos directa
direct deactivation desactivación directa
direct demand demanda directa
direct digital control control digital directo
direct display visualización directa
direct display unit unidad de visualización directa
direct-distance dialing marcación de distancia directa
direct drive accionamiento directo
direct-drive machine máquina de accionamiento directo
direct-drive motor motor de accionamiento directo
direct-drive torque motor motor de par de accionamiento directo
direct-drive tuning sintonización por accionamiento directo
direct-drive tuning system sistema de sintonización por accionamiento directo
direct effect efecto directo
direct electric drainage drenaje eléctrico directo
direct electromotive force fuerza electromotriz directa
direct excitation excitación directa
direct feeder alimentador directo
direct file archivo directo
direct frequency modulation modulación de frecuencia directa
direct grid bias polarización de rejilla directa
direct ground tierra directa
direct heating calentamiento directo
direct illumination iluminación directa
direct impulse impulso directo
direct induced current corriente inducida directa
direct input entrada directa
direct-input circuit circuito de entrada directa
direct-insert routine rutina de inserción directa
direct-insert subroutine subrutina de inserción directa
direct instruction instrucción directa
direct interelectrode capacitance capacitancia interelectródica directa
direct interlocking enclavamiento directo
direct international circuit circuito internacional directo
direct light luz directa
direct lighting alumbrado directo
direct line línea directa
direct measurement medida directa
direct memory access acceso a memoria directo

direct memory access controller controlador de acceso a memoria directo
direct memory access transfer transferencia de acceso a memoria directo
direct method método directo
direct noise amplifier amplificador de ruido directo
direct numerical control control numérico directo
direct operation operación directa
direct output salida directa
direct pickup captación directa
direct piezoelectric effect efecto piezoeléctrico directo
direct piezoelectricity piezoelectricidad directa
direct playback reproducción directa
direct processing procesamiento directo
direct pulse impulso directo
direct radiation radiación directa
direct radiation monitor monitor de radiación directa
direct-radiator loudspeaker altavoz de radiador directo
direct ratio razón directa
direct ray rayo directo
direct reading lectura directa
direct-reading digital instrument instrumento digital de lectura directa
direct-reading frequency meter frecuencímetro de lectura directa
direct-reading indicator indicador de lectura directa
direct-reading instrument instrumento de lectura directa
direct-reading meter medidor de lectura directa
direct recording registro directo
direct-recording instrument instrumento de registro directo
direct relation relación directa
direct resistance coupling acoplamiento por resistencia directa
direct scanning exploración directa
direct selection selección directa
direct short cortocircuito directo
direct sound wave onda sonora directa
direct storage almacenamiento directo
direct storage access acceso de almacenamiento directo
direct stroke descarga directa
direct-stroke protection protección contra descarga directa
direct substitution sustitución directa
direct synthesizer sintetizador directo
direct test prueba directa
direct transit tránsito directo
direct-transit circuit circuito de tránsito directo
direct transmission transmisión directa
direct viewing visión directa
direct-viewing tube tubo de visión directa
direct vision visión directa
direct voltage tensión continua
direct wave onda directa
direct Wiedemann effect efecto de Wiedemann directo
direction dirección
direction angle ángulo de dirección
direction finder radiogoniómetro
direction-finder antenna antena radiogoniométrica
direction-finder deviation desviación radiogoniométrica
direction-finder sensitivity sensibilidad radiogoniométrica
direction finding radiogoniometría
direction indicator indicador de dirección
direction indicator lamp lámpara indicadora de dirección

direction of current flow dirección de flujo de corriente
direction of polarization dirección de polarización
direction of propagation dirección de propagación
direction of rotation dirección de rotación
direction of transmission dirección de transmisión
direction rectifier rectificador de dirección
direction semaphore semáforo de dirección
direction signal señal de dirección
directional direccional
directional antenna antena direccional
directional array red direccional
directional baffle bafle direccional
directional beam haz direccional
directional characteristic característica direccional
directional circuit-breaker cortacircuito direccional
directional control control direccional
directional counter contador direccional
directional coupler acoplador direccional
directional current protection protección de corriente direccional
directional diagram diagrama direccional
directional diode diodo direccional
directional discrimination discriminación direccional
directional effect efecto direccional
directional element elemento direccional
directional emissivity emisividad direccional
directional filter filtro direccional
directional gain ganancia direccional, índice de direccionalidad
directional gyroscope giroscopio direccional
directional homing guiado direccional
directional hydrophone hidrófono direccional
directional interlocking enclavamiento direccional
directional lighting alumbrado direccional
directional lobe lóbulo direccional
directional microphone micrófono direccional
directional operation operación direccional
directional overcurrent protection protección de sobrecorriente direccional
directional overcurrent relay relé de sobrecorriente direccional
directional pattern patrón direccional
directional phase changer desfasador direccional
directional phase shifter desfasador direccional
directional power protection protección de potencia direccional
directional power relay relé de potencia direccional
directional reception recepción direccional
directional relay relé direccional
directional response respuesta direccional
directional response pattern patrón de respuesta direccional
directional selectivity selectividad direccional
directional separation filter filtro de separación direccional
directional transmission transmisión direccional
directional transmitter transmisor direccional
directive directivo
directive antenna antena directiva
directive beacon radiofaro directivo
directive efficiency eficiencia directiva
directive feed alimentación directiva
directive gain ganancia directiva
directive properties propiedades directivas
directive radiation radiación directiva
directivity directividad
directivity angle ángulo de directividad
directivity diagram diagrama de directividad
directivity factor factor de directividad
directivity index índice de directividad
directivity pattern patrón de directividad

directivity signal señal de directividad
directly coupled acoplado directamente
directly grounded conectado a tierra directamente
directly heated cathode cátodo calentado directamente
directly heated thermistor termistor calentado directamente
directly heated thermocouple termopar calentado directamente
directly measurable medible directamente
directly operated operado directamente
director element elemento director
director system sistema director
directory key clave de directorio
directory structure estructura de directorio
directrix directriz
disable desactivar
disassemble desensamblar
disassembler desensamblador
disaster continue routine rutina de continuar tras desastre
disc disco
discharge descarga
discharge capacity capacidad de descarga
discharge circuit circuito de descarga
discharge coil bobina de descarga
discharge counter contador de descargas
discharge current corriente de descarga
discharge current limiting device dispositivo limitador de corriente de descarga
discharge detector detector de descarga
discharge device dispositivo de descarga
discharge energy energía de descarga
discharge-energy test prueba de energía de descarga
discharge extinction voltage tensión de extinción de descarga
discharge gap espacio de descarga
discharge inception voltage tensión de comienzo de descarga
discharge indicator indicador de descarga
discharge key llave de descarga
discharge lamp lámpara de descarga
discharge phenomena fenómenos de descarga
discharge potential potencial de descarga
discharge probe sonda de descarga
discharge rate régimen de descarga
discharge resistor resistor de descarga
discharge switch conmutador de descarga
discharge time constant constante de tiempo de descarga
discharge tube tubo de descarga
discharge valve válvula de descarga
discharge voltage tensión de descarga
discharger descargador
discharging current corriente de descarga
discharging rate régimen de descarga
discharging resistor resistor de descarga
discomfort glare resplandor incomodante
discone antenna antena discocónica
disconnect desconectar
disconnect signal señal de desconexión
disconnect switch seccionador, conmutador de desconexión
disconnectable device dispositivo desconectable
disconnected position posición desconectada
disconnecter desconectador
disconnecting desconectador
disconnecting device dispositivo desconectador
disconnecting fuse fusible desconectador
disconnecting plug clavija desconectadora
disconnecting switch seccionador, conmutador de desconexión
disconnection desconexión

disconnector desconectador
discontinuity discontinuidad
discontinuity capacitance capacitancia de discontinuidad
discontinuous discontinuo
discontinuous amplifier amplificador discontinuo
discontinuous function función discontinua
discrete discreto
discrete capacitor capacitor discreto
discrete circuit circuito discreto
discrete comparator comparador discreto
discrete component componente discreto
discrete-component microcircuit microcircuito de componentes discretos
discrete device dispositivo discreto
discrete discontinuity discontinuidad discreta
discrete electric impulse impulso eléctrico discreto
discrete element elemento discreto
discrete frequency frecuencia discreta
discrete inductor inductor discreto
discrete information source fuente de información discreta
discrete level nivel discreto
discrete model modelo discreto
discrete output salida discreta
discrete part parte discreta
discrete programming programación discreta
discrete pulse impulso discreto
discrete quantity cantidad discreta
discrete representation representación discreta
discrete resistor resistor discreto
discrete sampling muestreo discreto
discrete signal señal discreta
discrete simulation simulación discreta
discrete thin-film component componente de película delgada discreto
discrete tone tono discreto
discrete type tipo discreto
discrete unit unidad discreta
discrete variable variable discreta
discriminating discriminante
discriminating element elemento discriminante
discriminating relay relé discriminante
discriminating repeater repetidor discriminante
discriminating selector selector discriminante
discrimination discriminación
discrimination factor factor de discriminación
discriminator discriminador
discriminator amplifier amplificador discriminador
discriminator circuit circuito discriminador
discriminator stage etapa discriminadora
discriminator transformer transformador discriminador
discriminator tuner sintonizador discriminador
discriminator tuning device dispositivo sintonizador discriminador
dish antenna antena parabólica
disintegration desintegración
disintegration voltage tensión de desintegración
disintegrator desintegrador
disjunction disyunción
disk disco
disk access time tiempo de acceso a disco
disk adapter adaptador de disco
disk allocation table tabla de asignación de disco
disk anode ánodo de disco
disk antenna antena de disco
disk attenuator atenuador de disco
disk block bloque de discos
disk buffer memoria intermedia de disco
disk cache memoria inmediata de disco
disk capacitor capacitor de disco
disk capacity capacidad de disco

disk cartridge cartucho de disco
disk cathode cátodo de disco
disk ceramic capacitor capacitor cerámico de disco
disk coil bobina en disco
disk control unit unidad de control de disco
disk controller controlador de disco
disk crash falla completa de disco
disk device dispositivo de disco
disk drive unidad de disco
disk drive system sistema de unidad de disco
disk dump vaciado de disco, volcado de disco
disk duplication duplicación de disco
disk dynamo dínamo de disco
disk file archivo de disco
disk generator generador de disco
disk initialization program programa de inicialización de disco
disk interface interfaz de disco
disk memory memoria de disco
disk operating system sistema operativo de disco
disk operating system booting rutina de carga de sistema operativo de disco
disk optimizer optimizador de disco
disk optimizing optimización de disco
disk organization organización de disco
disk pack conjunto de discos
disk partition partición de disco
disk programming system sistema de programación de disco
disk recorder registrador de disco
disk recording registro en disco
disk recording unit unidad de registro en disco
disk rectifier rectificador de disco
disk resident system sistema residente en disco
disk resistor resistor de disco
disk scanner explorador de disco
disk-seal tube tubo faro
disk sector sector de disco
disk server servidor de disco
disk signal señal de disco
disk sort clasificación de disco
disk sort program programa de clasificación de disco
disk space espacio de disco
disk speed velocidad de disco
disk storage almacenamiento en disco
disk storage controller controlador de almacenamiento en disco
disk storage file archivo de almacenamiento en disco
disk storage module módulo de almacenamiento en disco
disk storage subsystem subsistema de almacenamiento en disco
disk storage unit unidad de almacenamiento en disco
disk subsystem subsistema de disco
disk system sistema de disco
disk system format formato de sistema de disco
disk thermistor termistor en disco
disk-type resistor resistor de tipo de disco
disk unit unidad de disco
disk varistor varistor en disco
disk volume volumen de disco
disk winding arrollamiento en disco
diskette disquete, disco flexible
diskette compatibility compatibilidad de disquete
diskette drive unidad de disquete
dislocation dislocación
dispenser cathode cátodo compensado
dispersibility dispersibilidad
dispersion dispersión
dispersion coefficient coeficiente de dispersión
dispersion current corriente de dispersión
dispersion diagram diagrama de dispersión

dispersion matrix matriz de dispersión
dispersion meter dispersímetro
dispersion pattern patrón de dispersión
dispersion photometer fotómetro de dispersión
dispersive dispersivo
dispersive bandwidth ancho de banda dispersivo
dispersive dielectric dieléctrico dispersivo
dispersive electrode electrodo dispersivo
dispersive medium medio dispersivo
dispersive power poder dispersivo
dispersive reference electrode electrodo de
  referencia dispersivo
displaced charge carga desplazada
displacement desplazamiento
displacement angle ángulo de desplazamiento
displacement antiresonance antirresonancia de
  desplazamiento
displacement byte byte de desplazamiento
displacement cell pila de desplazamiento
displacement current corriente de desplazamiento
displacement flux density densidad de flujo de
  desplazamiento
displacement measurement medida de
  desplazamiento
displacement meter medidor de desplazamiento
displacement transducer transductor de
  desplazamiento
display visualizador, visualización, presentación
  visual, indicador visual, pantalla
display adapter adaptador de visualización,
  adaptador de visualizador
display attribute atributo de visualización
display background fondo de visualización
display buffer memoria intermedia de visualización
display buffer memory memoria intermedia de
  visualización
display column columna de visualización
display command mando de visualización
display component componente de visualización
display console consola de visualización
display control control de visualización
display cursor cursor de visualización
display cycle ciclo de visualización
display cycle time tiempo de ciclo de visualización
display device dispositivo de visualización,
  visualizador
display element elemento de visualización
display emulation emulación de visualización
display field campo de visualización
display file archivo de visualización
display format formato de visualización
display group grupo de visualización
display image imagen de visualización
display instruction instrucción de visualización
display light luz de visualizador
display line línea de visualizador
display list lista de visualización
display loss pérdida de visualización
display mode modo de visualización
display module módulo de visualización
display monitor monitor de visualización
display option opción de visualización
display panel panel de visualización
display position posición de visualización
display random access memory memoria de acceso
  aleatorio de visualización
display reference level nivel de referencia de
  visualización
display register registro de visualización
display resolution resolución de visualización
display screen pantalla de visualización
display segment segmento de visualización
display space espacio de visualización

display station estación de visualización
display station field campo de estación de
  visualización
display-storage tube tubo de almacenamiento de
  visualización
display surface superficie de visualización
display switch conmutador de visualización
display terminal terminal de visualizador
display time tiempo de visualización
display tolerance tolerancia de visualización
display tube tubo de visualizador
display type tipo de visualización
display unit unidad de visualización
display visibility visibilidad de visualización
display window ventana de visualización
displayed part parte presentada
disposable component componente desechable
disruption disrupción
disruption spark chispa de disrupción
disruptive disruptivo
disruptive critical voltage tensión crítica disruptiva
disruptive discharge descarga disruptiva
disruptive discharge voltage tensión de descarga
  disruptiva
disruptive force fuerza disruptiva
disruptive voltage tensión disruptiva
dissector disector
dissector tube tubo disector
dissipation disipación
dissipation constant constante de disipación
dissipation electrode electrodo de disipación
dissipation factor factor de disipación
dissipation line línea de disipación
dissipation rate velocidad de disipación
dissipative disipativo
dissipative material material disipativo
dissipative system sistema disipativo
dissipator disipador
dissociation disociación
dissonance disonancia
dissymmetrical disimétrico
dissymmetrical network red disimétrica
dissymmetrical transducer transductor disimétrico
distance dialing marcación de distancia
distance mark marca de distancia
distance protection protección de distancia
distance relay relé de distancia
distance resolution resolución de distancia
distant field campo distante
distant source fuente distante
distorted distorsionado
distorted current corriente distorsionada
distorted signal señal distorsionada
distorted sine wave onda sinusoidal distorsionada
distortion distorsión
distortion analyzer analizador de distorsión
distortion bridge puente de distorsión
distortion correction corrección de distorsión
distortion delay retardo por distorsión
distortion factor factor de distorsión
distortion-limited operation operación limitada por
  distorsión
distortion meter distorsiómetro
distortion tolerance tolerancia de distorsión
distortionless sin distorsión
distortionless modulation modulación sin distorsión
distortionless variable attenuator atenuador variable
  sin distorsión
distress call llamada de socorro
distress communication comunicación de socorro
distress frequency frecuencia de socorro
distress receiver receptor de socorro
distress signal señal de socorro

distress wave onda de socorro
distributed distribuido
distributed amplifier amplificador distribuido
distributed capacitance capacitancia distribuida
distributed capacity capacidad distribuida
distributed circuit circuito distribuido
distributed component componente distribuido
distributed constant constante distribuida
distributed control network red de control
   distribuida
distributed data processing procesamiento de datos
   distribuido
distributed data processing network red de
   procesamiento de datos distribuida
distributed database base de datos distribuida
distributed design diseño distribuido
distributed element elemento distribuido
distributed-element circuit circuito de elementos
   distribuidos
distributed emission emisión distribuida
distributed-emission photodiode fotodiodo de
   emisión distribuida
distributed free space espacio libre distribuido
distributed function función distribuida
distributed inductance inductancia distribuida
distributed intelligence inteligencia distribuida
distributed load carga distribuida
distributed network red distribuida
distributed numerical control control numérico
   distribuido
distributed paramagnetic amplifier amplificador
   paramagnético distribuido
distributed parameter parámetro distribuido
distributed-parameter network red de parámetros
   distribuidos
distributed pole polo distribuido
distributed-pole motor motor de polos distribuidos
distributed processing procesamiento distribuido
distributed processing system sistema de
   procesamiento distribuido
distributed resistance resistencia distribuida
distributed system sistema distribuido
distributed winding arrollamiento distribuido
distributing distribuidor
distributing amplifier amplificador distribuidor
distributing block bloque distribuidor
distributing board cuadro distribuidor
distributing cable cable distribuidor
distributing finder buscador distribuidor
distributing frame repartidor
distributing panel panel distribuidor
distribution distribución
distribution amplifier amplificador de distribución
distribution board cuadro de distribución
distribution box caja de distribución
distribution cable cable de distribución
distribution center centro de distribución
distribution chamber cámara de distribución
distribution coefficient coeficiente de distribución
distribution control control de distribución
distribution diskette disquete de distribución
distribution entry entrada de distribución
distribution factor factor de distribución
distribution frame repartidor
distribution function función de distribución
distribution line línea de distribución
distribution list lista de distribución
distribution network red de distribución
distribution panel panel de distribución
distribution pattern patrón de distribución
distribution relay relé de distribución
distribution request solicitud de distribución
distribution sort clasificación de distribución

distribution station estación de distribución
distribution substation subestación de distribución
distribution switchboard cuadro conmutador de
   distribución
distribution system sistema de distribución
distribution tape cinta de distribución
distribution transformer transformador de
   distribución
distribution zone zona de distribución
distributor distribuidor
distributor box caja distribuidora
disturbance perturbación
dither vibración
dithering vibración
diurnal variation variación diurna
diurnal wave onda diurna
divergence divergencia
divergence coefficient coeficiente de divergencia
divergence loss pérdida por divergencia
divergent divergente
divergent beam haz divergente
divergent series serie divergente
diverging divergente
diverging lens lente divergente
diverging wave onda divergente
diversity diversidad
diversity comparator comparador de diversidad
diversity factor factor de diversidad
diversity gain ganancia por diversidad
diversity radar radar en diversidad
diversity receiver receptor en diversidad
diversity reception recepción en diversidad
diversity selector selector de diversidad
diversity signal señal de diversidad
diversity system sistema en diversidad
diversity terminal terminal en diversidad
diverter desviador
divided-carrier modulation modulación de portadora
   dividida
divided circuit circuito derivado
divider divisor
divider bridge puente divisor
divider probe sonda divisora
dividing arm brazo divisor
dividing network red divisora
division subroutine subrutina de división
Dobrowolsky generator generador de Dobrowolsky
document assembly ensamblaje de documento
document chaining encadenamiento de documentos
document comparison comparación de documentos
document conversion conversión de documento
document design diseño de documento
document drive unidad de documento
document processing procesamiento de documentos
document reader lectora de documentos
document retrieval recuperación de documentos
document sorter clasificadora de documentos
document sorting clasificación de documentos
documentation documentación
documentation level nivel de documentación
doghouse perrera, casa de sintonía
Doherty amplifier amplificador de Doherty
Doherty linear amplifier amplificador lineal de
   Doherty
Dolezalek electrometer electrómetro de Dolezalek
domain dominio
domain operator operador de dominio
domain rotation rotación de dominio
domestic electronics electrónica doméstica
domestic induction heater calentador por inducción
   doméstico
dominant mode modo dominante
dominant series serie dominante

dominant term término dominante
dominant wave onda dominante
dominant wavelength longitud de onda dominante
donor donador
donor impurity impureza donadora
donor level nivel donador
donor semiconductor semiconductor donador
doorknob capacitor capacitor en pomo
doorknob tube tubo en pomo
dopant dopante
dope dopar
doped junction unión dopada
doped semiconductor semiconductor dopado
doping dopado
doping agent agente de dopado
doping compensation compensación de dopado
doping gas gas de dopado
doping level nivel de dopado
doping method método de dopado
Doppler broadening ensanchamiento de Doppler
Doppler cabinet caja de Doppler
Doppler displacement desplazamiento de Doppler
Doppler effect efecto de Doppler
Doppler enclosure caja de Doppler
Doppler filter filtro de Doppler
Doppler frequency frecuencia de Doppler
Doppler laser láser de Doppler
Doppler navigation system sistema de navegación de Doppler
Doppler navigator navegador de Doppler
Doppler radar radar de Doppler
Doppler shift desplazamiento de Doppler
Doppler signal señal de Doppler
Doppler system sistema de Doppler
Doppler tracking seguimiento de Doppler
doran doran
dormant state estado latente
dosage meter dosímetro
dot punto
dot-and-dash telegraphy telegrafía de puntos y rayas
dot cycle ciclo de punto
dot frequency frecuencia de puntos
dot generator generador de puntos
dot interlacing entrelazado de puntos
dot matrix matriz de puntos
dot-matrix display visualizador de matriz de puntos
dot-matrix printer impresora de matriz de puntos
dot pattern patrón de puntos
double-anode diode diodo de ánodo doble
double armature doble armadura
double-base diode diodo de doble base
double-base junction transistor transistor de unión de doble base
double-base tetrode tetrodo de doble base
double-base transistor transistor de doble base
double-beam cathode ray tube tubo de rayos catódicos de doble haz
double-beam oscillograph oscilógrafo de doble haz
double-beam oscilloscope osciloscopio de doble haz
double-beam radar radar de doble haz
double-bounce signal señal de doble rebote
double-break contact contacto de doble ruptura
double-break switch conmutador de doble ruptura
double bridge puente doble
double buffering almacenamiento intermedio doble
double bus system sistema de doble bus
double-button microphone micrófono de doble botón
double card tarjeta doble
double-channel duplex dúplex de doble canal
double-channel simplex símplex de doble canal
double-circuit system sistema de circuito doble
double-circuit tuning sintonización por circuito

doble
double connection doble conexión
double conversion doble conversión
double-conversion receiver receptor de doble conversión
double-conversion superheterodyne superheterodino de doble conversión
double-current generator generador de doble corriente
double density doble densidad
double density recording registro de doble densidad
double-diamond antenna antena de doble diamante
double-diffused transistor transistor de difusión doble
double diode doble diodo
double-diode demodulator desmodulador de doble diodo
double-diode limiter limitador de doble diodo
double-diode rectifier rectificador de doble diodo
double-doped transistor transistor de doble dopado
double-ended cord cordón con dos clavijas
double error doble error
double-faced tape cinta de dos caras
double feed doble alimentación
double filament doble filamento
double-hump effect efecto de doble pico
double-hump resonance resonancia de doble pico
double-hump resonance curve curva de resonancia de doble pico
double-hump response respuesta de doble pico
double-hump wave onda de doble pico
double image doble imagen
double insulation doble aislamiento
double layer doble capa
double length doble longitud
double limiter doble limitador
double local oscillator oscilador local de dos señales
double-make contacts contactos de doble cierre
double-make switch conmutador de doble cierre
double modulation doble modulación
double picture doble imagen
double-pole bipolar
double-pole head cabeza bipolar
double-precision doble precisión
double-precision addition adición de doble precisión
double-precision hardware equipo físico de doble precisión
double-precision number número de doble precisión
double-precision system sistema de doble precisión
double processing doble procesamiento
double processing system sistema de doble procesamiento
double processor doble procesador
double pulse doble impulso
double-pulse reading lectura de doble impulso
double-pulse wave onda de doble impulso
double punch doble perforación
double punching doble perforación
double reaction doble reacción
double regeneration doble regeneración
double register doble registro
double response doble respuesta
double screen doble pantalla
double sheath doble revestimiento
double shield doble blindaje
double sideband doble banda lateral
double-sideband modulation modulación de doble banda lateral
double-sideband system sistema de doble banda lateral
double-sideband transmission transmisión de doble banda lateral
double-sideband transmitter transmisor de doble

banda lateral
double-sided de dos lados
double-sided diskette disquete de dos lados
double signal doble señal
double-signal receiver receptor de doble señal
double-stream amplifier amplificador de dos
  trayectorias
double-surface transistor transistor de doble
  superficie
double-throw de dos posiciones
double-throw circuit-breaker cortacircuito de dos
  posiciones
double-throw contact contacto de dos posiciones
double-throw switch conmutador de dos posiciones
double triode doble triodo
double-tuned amplifier amplificador de doble
  sintonía
double-tuned circuit circuito de doble sintonía
double-tuned detector detector de doble sintonía
double winding doble arrollamiento
double winding generator generador de doble
  arrollamiento
doubler doblador
doubler circuit circuito doblador
doubler stage etapa dobladora
doublet doblete
doublet antenna antena de doblete
doubling doblamiento
doughnut antenna antena toroidal
doughnut coil bobina toroidal
Dow oscillator oscilador de Dow
dowel espiga
down-conversion conversión descendente
down-converter convertidor descendente
down time tiempo muerto
downlink señal de bajada
downlink beamwidth ancho de haz de señal de
  bajada
downlink frequency frecuencia de señal de bajada
download transferir, descargar
downward compatibility compatibilidad descendente
downward compression compresión descendente
downward heterodyning heterodinaje descendente
downward modulation modulación descendente
drag angle ángulo de arrastre
drag cup taza de arrastre
drag-cup motor motor de taza de arrastre
drag magnet imán de arrastre
dragging arrastramiento
drain consumo
drain current corriente de consumo
drainage equipment equipo de drenaje
drainage unit unidad de drenaje
dress arreglo
drift deslizamiento, deriva
drift angle ángulo de deriva
drift bar barra de deriva
drift compensation compensación de deslizamiento
drift correction corrección de deriva
drift-correction angle ángulo de corrección de
  deriva
drift current corriente de deslizamiento
drift electrode electrodo de corrimiento
drift field campo de arrastre
drift-field transistor transistor de campo de arrastre
drift mobility movilidad de deslizamiento
drift motion moción de deriva
drift rate tasa de deslizamiento
drift space espacio de corrimiento
drift speed velocidad de arrastre
drift stabilization estabilización de deslizamiento
drift transistor transistor de campo interno
drift tunnel túnel de corrimiento

drift velocity velocidad de deslizamiento
drift voltage deslizamiento de tensión
drip loop lazo de goteo
dripproof a prueba de goteo
dripproof construction construcción a prueba de
  goteo
dripproof enclosure caja a prueba de goteo
dripproof machine máquina a prueba de goteo
dripproof motor motor a prueba de goteo
driptight hermético contra el goteo
drive impulso, accionamiento, excitación, unidad
drive belt correa de transmisión
drive circuit circuito de excitación
drive control control de excitación
drive designator designador de unidad
drive letter letra de unidad
drive number número de unidad
drive pattern patrón de interferencia
drive pin espiga antideslizante
drive power potencia de excitación
drive pulse impulso de mando
drive transformer transformador de excitación
driven antenna antena excitada
driven element elemento excitado
driven multivibrator multivibrador sincronizado
driven section sección excitada
driver excitador, controlador, altavoz
driver amplifier amplificador excitador
driver circuit circuito excitador
driver coil bobina excitadora
driver element elemento de excitación
driver impedance impedancia de etapa excitadora
driver inductance inductancia de etapa excitadora
driver module módulo de controlador
driver resistance resistencia de etapa excitadora
driver software programas de controlador
driver stage etapa excitadora
driver transformer transformador de etapa
  excitadora
driver tube tubo de etapa excitadora
driving circuit circuito de excitación
driving current corriente de excitación
driving grid rejilla de mando
driving-point admittance admitancia de entrada
driving-point impedance impedancia de entrada
driving power potencia de excitación
driving pulse impulso de excitación
driving signal señal de excitación
driving voltage tensión de excitación
drone avión radioguiado
drop bajada
drop bar barra de descarga
drop bracket soporte vertical
drop cable cable de bajada
drop circuit circuito de bajada
drop-in creación no intencionada de bits
drop indicator indicador por bajada
drop level nivel de bajada
drop line línea de bajada
drop test prueba de bajada
drop wire hilo de acometida
dropout desprendimiento, desaccionamiento, falla de
  registro, omisión no intencionada de bits
dropout current corriente de desaccionamiento
dropout power potencia de desaccionamiento
dropout ratio razón de desaccionamiento
dropout time tiempo de desaccionamiento
dropout value valor de desprendimiento
dropout voltage tensión de desaccionamiento
dropping resistor resistor de caída
drum tambor, cilindro
drum capacitor capacitor de cilindro
drum controller controlador de tambor

drum dial cuadrante de tambor
drum dump vaciado de tambor
drum factor factor de tambor
drum magnet imán de cilindro
drum mark marca de tambor
drum memory memoria de tambor
drum parity paridad de tambor
drum printer impresora de tambor
drum programmer programador de tambor
drum receiver receptor de tambor
drum recorder registrador de tambor
drum resistor resistor en cilindro
drum scanner explorador de cilindro
drum speed velocidad de tambor
drum storage almacenamiento de tambor
drum switch conmutador de tambor
drum transmitter transmisor de tambor
drum-type controller controlador de tambor
drum-type memory memoria de tambor
drum varistor varistor de cilindro
drum winding arrollamiento en tambor
dry seco
dry battery batería de pilas secas
dry cell pila seca
dry-cell battery batería de pilas secas
dry circuit circuito seco
dry-contact rectifier rectificador de contacto seco
dry contacts contactos secos
dry-disk rectifier rectificador de disco seco
dry electrolytic capacitor capacitor electrolítico seco
dry flashover voltage tensión de arqueo en seco
dry friction fricción seca
dry pile pila seca
dry-plate rectifier rectificador de placa seca
dry rectifier rectificador seco
dry-reed relay relé de laminillas secas
dry-reed switch conmutador de laminillas secas
dry run operación preliminar de prueba
dry shelf life vida de almacenamiento seco
dry transformer transformador seco
dry-type transformer transformador de tipo seco
dual-beam cathode ray tube tubo de rayos catódicos de doble haz
dual-beam oscilloscope osciloscopio de doble haz
dual-beam system sistema de doble haz
dual bus doble bus
dual capacitor doble capacitor
dual card tarjeta doble
dual-channel amplifier amplificador de dos canales
dual-channel audio amplifier amplificador de audio de dos canales
dual-channel controller controlador de dos canales
dual-channel oscilloscope osciloscopio de dos canales
dual-channel sound sonido de dos canales
dual-channel system sistema de dos canales
dual coaxial cable cable coaxial doble
dual coding doble codificación
dual-coil relay relé de dos bobinas
dual-cone speaker altavoz de dos conos
dual contact doble contacto
dual control doble control
dual-control heptode heptodo de doble control
dual-control selector selector de doble control
dual controller doble controlador
dual conversation doble conversación
dual-conversion receiver receptor de doble conversión
dual density doble densidad
dual density diskette disquete de doble densidad
dual diode doble diodo
dual diode-dual triode doble diodo-doble triodo
dual diversity doble diversidad

dual-diversity receiver receptor en doble diversidad
dual-diversity reception recepción en doble diversidad
dual-diversity system sistema en doble diversidad
dual-emitter transistor transistor de doble emisor
dual-frequency calibrator calibrador de dos frecuencias
dual-frequency induction heater calentador por inducción de dos frecuencias
dual gate doble puerta
dual intensity doble intensidad
dual local oscillator oscilador local de dos señales
dual meter medidor doble
dual modulation doble modulación
dual network doble red
dual operation operación doble
dual pentode doble pentodo
dual pickup doble fonocaptor
dual-polarized antenna antena de doble polarización
dual potentiometer doble potenciómetro
dual processor doble procesador
dual processor system sistema de doble procesador
dual recording doble registro
dual resistor doble resistor
dual speed doble velocidad
dual storage doble almacenamiento
dual switchboard cuadro conmutador doble
dual trace doble trazo
dual-trace oscilloscope osciloscopio de doble trazo
dual-trace recorder registrador de doble trazo
dual triode doble triodo
dual use de doble uso
dual-voltage coil bobina de dos tensiones
duality dualidad
dubbing mezcla de sonidos, rerregistro, regrabación
duct conducto
duct effect efecto de conducto
duct entrance entrada de conducto
duct sealing cierre hermético de conductos
duct system sistema de conducto
ductilimeter ductilímetro
ductility ductilidad
ductility factor factor de ductilidad
ducting canalización
Duddell arc arco de Duddell
Duerdoth's stability margin margen de estabilidad de Duerdoth
dull emitter emisor débil
dull-emitter filament filamento de emisión débil
dummy ficticio, inoperante, simulado
dummy activity actividad ficticia
dummy address dirección ficticia
dummy antenna antena ficticia
dummy coil bobina ficticia
dummy component componente ficticio
dummy control section sección de control ficticia
dummy data datos ficticios
dummy data set conjunto de datos ficticios
dummy device dispositivo ficticio
dummy device assignment asignación de dispositivo ficticio
dummy entry entrada ficticia
dummy fuse fusible ficticio
dummy instruction instrucción ficticia
dummy line línea ficticia
dummy load carga ficticia
dummy-load rheostat reóstato de carga ficticia
dummy module módulo ficticio
dummy panel panel ficticio
dummy parameter parámetro ficticio
dummy resistor resistor ficticio
dummy section sección ficticia
dummy variable variable ficticia

dump vaciado, vuelco, volcado
dump and restart vaciado y reiniciación
dump check comprobación de vaciado
dump point punto de vaciado
dump subroutine subrutina de vaciado
dumping vaciado, volcado
dumping resistor resistor de descarga
duodecal socket zócalo duodecal
duodecimal duodecimal
duodecimal notation notación duodecimal
duodecimal number system sistema numérico
  duodecimal
duodiode duodiodo
duodiode-pentode duodiodo-pentodo
duodiode-tetrode duodiodo-tetrodo
duodiode-triode duodiodo-triodo
duolateral coil bobina duolateral
duoplasmatron duoplasmatrón
duopole duopolo
duosexadecimal duosexadecimal
duotricenary duotricenario
duotriode duotriodo
duplex dúplex
duplex apparatus aparato dúplex
duplex artificial line línea artificial equilibradora
duplex cable cable dúplex
duplex cavity cavidad dúplex
duplex channel canal dúplex
duplex channeling canalización dúplex
duplex circuit circuito dúplex
duplex communication comunicación dúplex
duplex computer system sistema de computadoras
  dúplex
duplex console consola dúplex
duplex diode diodo dúplex
duplex line línea dúplex
duplex operation operación en dúplex
duplex printing impresión dúplex
duplex switchboard cuadro conmutador dúplex
duplex system sistema dúplex
duplex transmission transmisión dúplex
duplex tube tubo dúplex
duplex winding arrollamiento dúplex
duplexer duplexor
duplexer filter filtro duplexor
duplexing duplexión
duplicate line línea duplicada
duplicate record registro duplicado
duplicate service servicio duplicado
duplication check comprobación de duplicación
duplication factor factor de duplicación
duplicator duplicador
dural dural
duralumin duraluminio
duration control control de duración
duration time tiempo de duración
durometer durómetro
dust collector recolector de polvo
dust core núcleo de polvo
dust cover cubierta contra polvo
dust precipitator precipitador de polvo
dust resistant resistente al polvo
dustproof a prueba de polvo
dusttight hermético contra el polvo
dusttight enclosure caja hermética contra el polvo
duty cycle ciclo de trabajo
duty factor factor de trabajo
duty ratio razón de trabajo
dyadic operation operación diádica
dyadic operator operador diádico
dyna dina
dynameter dinámetro
dynamic dinámico

dynamic acceleration aceleración dinámica
dynamic access acceso dinámico
dynamic accuracy exactitud dinámica
dynamic address dirección dinámica
dynamic allocation asignación dinámica
dynamic amplifier amplificador dinámico
dynamic analysis análisis dinámico
dynamic analyzer analizador dinámico
dynamic area área dinámica
dynamic balancing equilibrio dinámico
dynamic base current corriente de base dinámica
dynamic base resistance resistencia de base
  dinámica
dynamic base voltage tensión de base dinámica
dynamic behavior conducta dinámica
dynamic bias control control de polarización
  dinámico
dynamic braking frenaje dinámico
dynamic buffering almacenamiento intermedio
  dinámico
dynamic calibration calibración dinámica
dynamic cathode current corriente de cátodo
  dinámica
dynamic cathode resistance resistencia de cátodo
  dinámica
dynamic cathode voltage tensión de cátodo dinámica
dynamic cell celda dinámica
dynamic characteristic característica dinámica
dynamic check comprobación dinámica
dynamic circuit circuito dinámico
dynamic collector current corriente de colector
  dinámica
dynamic collector resistance resistencia de colector
  dinámica
dynamic collector voltage tensión de colector
  dinámica
dynamic contact resistance resistencia de contacto
  dinámica
dynamic control control dinámico
dynamic control function función de control
  dinámico
dynamic convergence convergencia dinámica
dynamic-convergence circuit circuito de
  convergencia dinámica
dynamic-convergence signal señal de convergencia
  dinámica
dynamic-convergence voltage tensión de
  convergencia dinámica
dynamic cooling enfriamiento dinámico
dynamic curve curva dinámica
dynamic damper amortiguador dinámico
dynamic debugging depuración dinámica
dynamic decay decaimiento dinámico
dynamic demonstrator demostrador dinámico
dynamic deviation desviación dinámica
dynamic device dispositivo dinámico
dynamic device reconfiguration reconfiguración de
  dispositivo dinámico
dynamic diode tester probador de diodo dinámico
dynamic directional stability estabilidad direccional
  dinámica
dynamic drain current corriente de consumo
  dinámica
dynamic drain resistance resistencia de consumo
  dinámica
dynamic drain voltage tensión de consumo dinámica
dynamic dump vaciado dinámico
dynamic electric field campo eléctrico dinámico
dynamic electricity electricidad dinámica
dynamic electrode potential potencial de electrodo
  dinámico
dynamic emitter current corriente de emisor
  dinámica

dynamic emitter resistance resistencia de emisor dinámica
dynamic emitter voltage tensión de emisor dinámica
dynamic equilibrium equilibrio dinámico
dynamic error error dinámico
dynamic factor factor dinámico
dynamic flip-flop basculador dinámico
dynamic focus enfoque dinámico
dynamic focusing enfoque dinámico
dynamic gain ganancia dinámica
dynamic gate voltage tensión de puerta dinámica
dynamic grid voltage tensión de rejilla dinámica
dynamic heating calentamiento dinámico
dynamic impedance impedancia dinámica
dynamic input capacitance capacitancia de entrada dinámica
dynamic limiter limitador dinámico
dynamic link enlace dinámico
dynamic load carga dinámica
dynamic loudspeaker altavoz dinámico
dynamic luminous sensitivity sensibilidad luminosa dinámica
dynamic magnetic field campo magnético dinámico
dynamic memory memoria dinámica
dynamic memory allocation asignación de memoria dinámica
dynamic microphone micrófono dinámico
dynamic model modelo dinámico
dynamic noise limiter limitador de ruido dinámico
dynamic noise suppressor supresor de ruido dinámico
dynamic output impedance impedancia de salida dinámica
dynamic parameter parámetro dinámico
dynamic partition partición dinámica
dynamic pickup fonocaptor dinámico
dynamic plate current corriente de placa dinámica
dynamic plate impedance impedancia de placa dinámica
dynamic plate resistance resistencia de placa dinámica
dynamic plate voltage tensión de placa dinámica
dynamic printout impresión dinámica
dynamic problem checking comprobación de problemas dinámica
dynamic program loading carga de programa dinámica
dynamic programming programación dinámica
dynamic quality calidad dinámica
dynamic quantity magnitud dinámica
dynamic radiation test prueba de radiación dinámica
dynamic random access memory memoria de acceso aleatorio dinámica
dynamic range margen dinámico
dynamic ratio razón dinámica
dynamic reconfiguration reconfiguración dinámica
dynamic regulation regulación dinámica
dynamic regulator regulador dinámico
dynamic relocation relocalización dinámica
dynamic relocation program programa de relocalización dinámica
dynamic reproducer reproductor dinámico
dynamic resistance resistencia dinámica
dynamic resource allocation asignación de recursos dinámica
dynamic response respuesta dinámica
dynamic restructuring reestructuración dinámica
dynamic sensitivity sensibilidad dinámica
dynamic sequential control control secuencial dinámico
dynamic spatial reconstructor reconstructor espacial dinámico
dynamic speaker altavoz dinámico

dynamic stability estabilidad dinámica
dynamic stop parada dinámica
dynamic storage almacenamiento dinámico
dynamic storage allocation asignación de almacenamiento dinámica
dynamic storage area área de almacenamiento dinámica
dynamic strain esfuerzo dinámico
dynamic subroutine subrutina dinámica
dynamic support system sistema de soporte dinámico
dynamic suspension suspensión dinámica
dynamic test prueba dinámica
dynamic transfer characteristic característica de transferencia dinámica
dynamic transistor tester probador de transistores dinámico
dynamic tube tester probador de tubos dinámico
dynamic unbalance desequilibrio dinámico
dynamic wavemeter ondámetro dinámico
dynamics dinámica
dynamo dínamo
dynamoelectric dinamoeléctrico
dynamoelectric amplifier amplificador dinamoeléctrico
dynamoelectric machinery maquinaria dinamoeléctrica
dynamometer dinamómetro
dynamometer ammeter amperímetro dinamométrico
dynamometer instrument instrumento dinamométrico
dynamometer multiplier multiplicador dinamométrico
dynamometer test prueba dinamométrica
dynamometer voltmeter vóltmetro dinamométrico
dynamometer wattmeter wáttmetro dinamométrico
dynamophone dinamófono
dynamotor dinamotor
dynatron dinatrón
dynatron effect efecto dinatrón
dynatron oscillator oscilador dinatrón
dyne dina
dyne-centimeter dina-centímetro
dynistor dinistor
dynode dinodo
dysprosium disprosio

E

E core núcleo en E
early-failure period periodo de fallas tempranas
earphone audífono
earphone coupler acoplador de audífono
earpiece auricular
earth tierra
earth's magnetic field campo magnético de la tierra
earth capacity capacidad a tierra
earth circuit circuito de tierra
earth connection conexión a tierra
earth currents corrientes de tierra
earth impedance impedancia de tierra
earth inductor inductor de tierra
earth leakage fuga a tierra
earth-leakage current corriente de fuga a tierra
earth plate placa de tierra
earth potential potencial de tierra
earth resistance resistencia de tierra
earth resistivity resistividad de tierra
earth resonance resonancia de la tierra
earth return retorno por tierra
earth-return circuit circuito con retorno por tierra
earth-return phantom circuit circuito fantasma con
    retorno por tierra
earth-return system sistema con retorno por tierra
earth station estación terrestre
earth system sistema de tierra
earth terminal terminal de tierra
earth wire hilo de tierra
earthed conectado a tierra
earthed circuit circuito conectado a tierra
earthed neutral neutro conectado a tierra
earthing puesta a tierra
earthing circuit circuito de puesta a tierra
earthing relay relé de puesta a tierra
earthing switch conmutador de puesta a tierra
earthing terminal terminal de puesta a tierra
ebiconductivity conductividad por bombardeo
    electrónico
ebonite ebonita
ebonite spacer espaciador de ebonita
ebonite tube tubo de ebonita
eccentric excéntrico
eccentric circle círculo excéntrico
eccentric groove surco excéntrico
eccentricity excentricidad
Eccles-Jordan circuit circuito de Eccles-Jordan
Eccles-Jordan flip-flop basculador de Eccles-Jordan
Eccles-Jordan multivibrator multivibrador de
    Eccles-Jordan
echelon escalón
echelon antenna antena escalonada
echelon grating rejilla escalonada
echelon telegraphy telegrafía escalonada
echo eco
echo amplitude amplitud de eco
echo area área de eco
echo attenuation atenuación de eco
echo blip señal visual de eco
echo box caja de ecos
echo cancellation cancelación de ecos

echo chamber cámara ecoica
echo check comprobación por eco
echo current corriente reflejada
echo detection detección por ecos
echo eliminator eliminador de eco
echo frequency frecuencia de eco
echo image imagen fantasma
echo intensifier intensificador de eco
echo interference interferencia por eco
echo level nivel de eco
echo-location localización por eco
echo matching apareamiento de ecos
echo meter ecómetro
echo mode modo de eco
echo path trayectoria de eco
echo phenomenon fenómeno de eco
echo power potencia de eco
echo pulse impulso de eco
echo range alcance de eco
echo ranging localización por ecos
echo record registro de eco
echo selector selector de eco
echo signal señal de eco
echo sounder ecosonador
echo sounding ecosondeo
echo sounding system sistema de ecosondeo
echo splitting división de ecos
echo suppression supresión de ecos
echo supressor supresor de eco
echo voltage tensión de eco
echo wave onda reflejada
echoencephalograph ecoencefalógrafo
echoencephalography ecoencefalografía
echogram ecograma
echograph ecógrafo
echoless sin ecos
echoless chamber cámara anecoica
echometer ecómetro
echometry ecometría
echoscope ecoscopio
eclipse effect efecto de eclipse
economic dispatch despacho económico
economic dispatch control control de despacho
    económico
economy energy energía de economía
economy power potencia de economía
eddy current corriente parásita
eddy-current braking frenaje por corrientes
    parásitas
eddy-current device dispositivo de corrientes
    parásitas
eddy-current heating calentamiento por corrientes
    parásitas
eddy-current loss pérdida por corrientes parásitas
eddy-current tachometer tacómetro de corrientes
    parásitas
edge control control de bordes
edge detection detección de borde
edge effect efecto de borde
edge enhancement realzado de bordes
edge image imagen de borde
edge-lighted iluminado por los bordes
edge-lighted dial cuadrante iluminado por los bordes
edge-lighted scale escala iluminada por los bordes
edge lighting iluminación por los bordes
edge linking enlace de bordes
edge perforated perforado en los bordes
edge pixel pixel de borde
edge-punched card tarjeta perforada en los bordes
edging coloración extraña en bordes
Edison accumulator acumulador de Edison
Edison base casquillo de rosca
Edison battery batería de Edison

Edison cell pila de Edison
Edison distribution system sistema de distribución de Edison
Edison effect efecto de Edison
Edison storage battery acumulador de Edison
edit editar
edit code código de edición
edit command mando de edición
edit mask máscara de edición
edit mode modo de edición
edit program programa de edición
edit routine rutina de edición
edit statement sentencia de edición
edit word palabra de edición
editing edición
editing character carácter de edición
editing mask máscara de edición
editing statement sentencia de edición
editing symbol símbolo de edición
editing terminal terminal de edición
effective acoustic center centro acústico efectivo
effective actuation time tiempo de actuación efectivo
effective address dirección efectiva
effective ampere ampere efectivo
effective antenna height altura de antena efectiva
effective antenna length longitud de antena efectiva
effective antenna resistance resistencia de antena efectiva
effective aperture abertura efectiva
effective area área efectiva
effective band banda efectiva
effective bandwidth ancho de banda efectivo
effective bunching angle ángulo de agrupamiento efectivo
effective byte location localización de byte efectiva
effective capacitance capacitancia efectiva
effective capacity capacidad efectiva
effective center centro efectivo
effective center of mass centro de masa efectivo
effective conductivity conductividad efectiva
effective current corriente efectiva
effective cutoff frequency frecuencia de corte efectiva
effective data transfer rate velocidad de transferencia de datos efectiva
effective dead time tiempo muerto efectivo
effective dielectric constant constante dieléctrica efectiva
effective echo area área de eco efectiva
effective efficiency eficiencia efectiva
effective electric angle ángulo eléctrico efectivo
effective energy energía efectiva
effective field intensity intensidad de campo efectiva
effective force fuerza efectiva
effective gap capacitance capacitancia de espacio efectiva
effective height altura efectiva
effective inductance inductancia efectiva
effective input admittance admitancia de entrada efectiva
effective input capacitance capacitancia de entrada efectiva
effective input impedance impedancia de entrada efectiva
effective instruction instrucción efectiva
effective isolation aislamiento efectivo
effective language lenguaje efectivo
effective load resistance resistencia de carga efectiva
effective margin margen efectivo
effective noise factor factor de ruido efectivo
effective noise temperature temperatura de ruido efectiva
effective output admittance admitancia de salida efectiva
effective output capacitance capacitancia de salida efectiva
effective output impedance impedancia de salida efectiva
effective parallel capacitance capacitancia en paralelo efectiva
effective parallel resistance resistencia en paralelo efectiva
effective percentage of modulation porcentaje efectivo de modulación
effective phase angle ángulo de fase efectivo
effective power potencia efectiva
effective printing rate velocidad de impresión efectiva
effective radiated power potencia radiada efectiva
effective radiation radiación efectiva
effective range alcance efectivo
effective reactance reactancia efectiva
effective relative permeability permeabilidad relativa efectiva
effective resistance resistencia efectiva
effective resistivity resistividad efectiva
effective selectivity selectividad efectiva
effective series inductance inductancia en serie efectiva
effective series resistance resistencia en serie efectiva
effective signal radiated señal radiada efectiva
effective sound pressure presión sonora efectiva
effective speed velocidad efectiva
effective temperature temperatura efectiva
effective thermal resistance resistencia térmica efectiva
effective time tiempo efectivo
effective transfer rate velocidad de transferencia efectiva
effective transmission transmisión efectiva
effective transmission rate velocidad de transmisión efectiva
effective transmission speed velocidad de transmisión efectiva
effective value valor efectivo
effective voltage tensión efectiva
effective wavelength longitud de onda efectiva
effectively bonded adherido efectivamente
effectively grounded conectado a tierra efectivamente
efferent eferente
efficiency eficiencia, rendimiento
efficiency curve curva de eficiencia
efficiency diode diodo de recuperación
efficiency factor factor de eficiencia
efficiency of rectification eficiencia de rectificación
efficiency test prueba de eficiencia
effluve efluvio
eight-bit alphanumeric code código alfanumérico de ocho bits
eight-bit decimal code código decimal de ocho bits
eight-hour rating clasificación de ocho horas
eight-level code código de ocho niveles
Einstein's law ley de Einstein
Einstein-de Haas effect efecto de Einstein-de Haas
Einstein photoelectric equation ecuación fotoeléctrica de Einstein
einsteinium einstenio
Einthoven string galvanometer galvanómetro de hilo de Einthoven
elapsed time tiempo transcurrido
elapsed-time indicator indicador de tiempo transcurrido
elapsed-time meter medidor de tiempo transcurrido
elapsed-time printout impresión de tiempo

transcurrido
elastance elastancia
elastic buffer memoria intermedia elástica
elastic limit límite elástico
elastic wave onda elástica
elasticity elasticidad
elastivity elastividad
elastomer elastómero
elastomeric insulation aislamiento elastomérico
elastoresistance elastorresistencia
electra electra
electret electreto
electret microphone micrófono de electreto
electret recorder registrador de electreto
electric eléctrico
electric absorption absorción eléctrica
electric accounting machine máquina de contabilidad eléctrica
electric alarm alarma eléctrica
electric anesthesia anestesia eléctrica
electric annealing recocido eléctrico
electric appliance aparato eléctrico
electric arc arco eléctrico
electric arc furnace horno de arco eléctrico
electric arc welding soldadura por arco eléctrico
electric axis eje eléctrico
electric background noise ruido de fondo eléctrico
electric balance equilibrio eléctrico
electric bell campana eléctrica
electric bias polarización eléctrica
electric bistoury bisturí eléctrico
electric boresight visor eléctrico
electric brake freno eléctrico
electric braking frenaje eléctrico
electric brazing soldadura fuerte eléctrica
electric breakdown ruptura eléctrica
electric breakdown voltage tensión de ruptura eléctrica
electric bridge puente eléctrico
electric bus ómnibus eléctrico
electric buzzer zumbador eléctrico
electric cable cable eléctrico
electric calculator calculadora eléctrica
electric capacitance capacitancia eléctrica
electric cell célula eléctrica
electric center centro eléctrico
electric charge carga eléctrica
electric chronograph cronógrafo eléctrico
electric chronometer cronómetro eléctrico
electric circuit circuito eléctrico
electric clock reloj eléctrico
electric communication comunicación eléctrica
electric component componente eléctrico
electric conducting path trayectoria de conducción eléctrica
electric conduction conducción eléctrica
electric conductivity conductividad eléctrica
electric connection conexión eléctrica
electric connector conector eléctrico
electric constant constante eléctrica
electric contact contacto eléctrico
electric control control eléctrico
electric control apparatus aparato de control eléctrico
electric controller controlador eléctrico
electric convection convección eléctrica
electric cooling enfriamiento eléctrico
electric counter contador eléctrico
electric coupler acoplador eléctrico
electric coupling acoplamiento eléctrico
electric current corriente eléctrica
electric current density densidad de corriente eléctrica

electric damping amortiguamiento eléctrico
electric delay retardo eléctrico
electric delay line línea de retardo eléctrica
electric density densidad eléctrica
electric deposition deposición eléctrica
electric depth finder buscador de profundidad eléctrico
electric dipole dipolo eléctrico
electric discharge descarga eléctrica
electric-discharge lamp lámpara de descarga eléctrica
electric-discharge machining maquinado por descarga eléctrica
electric-discharge tube tubo de descarga eléctrica
electric discontinuity discontinuidad eléctrica
electric disintegration desintegración eléctrica
electric dispersion dispersión eléctrica
electric displacement desplazamiento eléctrico
electric-displacement density densidad de desplazamiento eléctrico
electric dissipation disipación eléctrica
electric domain dominio eléctrico
electric double layer doble capa eléctrica
electric doublet doblete eléctrico
electric drainage drenaje eléctrico
electric drive accionamiento eléctrico
electric dust precipitator precipitador de polvo eléctrico
electric dynamometer dinamómetro eléctrico
electric echo eco eléctrico
electric efficiency eficiencia eléctrica
electric elasticity elasticidad eléctrica
electric elevator elevador eléctrico
electric energy energía eléctrica
electric equipment equipo eléctrico
electric escalator escalera eléctrica
electric eye ojo eléctrico
electric fence cerca eléctrica
electric fidelity fidelidad eléctrica
electric field campo eléctrico
electric-field component componente de campo eléctrico
electric-field induction inducción de campo eléctrico
electric-field intensity intensidad de campo eléctrico
electric-field line línea de campo eléctrico
electric-field pickup captor de campo eléctrico
electric-field strength intensidad de campo eléctrico
electric-field strength meter medidor de intensidad de campo eléctrico
electric-field vector vector de campo eléctrico
electric filament lamp lámpara de filamento eléctrico
electric filter filtro eléctrico
electric fluid fluido eléctrico
electric flux flujo eléctrico
electric-flux density densidad de flujo eléctrico
electric-flux lines líneas de flujo eléctrico
electric focusing enfoque eléctrico
electric force fuerza eléctrica
electric furnace horno eléctrico
electric generator generador eléctrico
electric glow luminiscencia eléctrica
electric governor regulador eléctrico
electric heat accumulator acumulador de calor eléctrico
electric heater calentador eléctrico
electric heating calefacción eléctrica
electric home appliance aparato electrodoméstico
electric horn bocina eléctrica
electric hygrometer higrómetro eléctrico
electric hysteresis histéresis eléctrica
electric image imagen eléctrica
electric impulse impulso eléctrico
electric induction inducción eléctrica

electric inertia inercia eléctrica
electric influence influencia eléctrica
electric installation instalación eléctrica
electric insulator aislador eléctrico
electric interconnection interconexión eléctrica
electric interference interferencia eléctrica
electric interlocking enclavamiento eléctrico
electric lamp lámpara eléctrica
electric leakage fuga eléctrica
electric length longitud eléctrica
electric light luz eléctrica
electric lighting alumbrado eléctrico
electric line línea eléctrica
electric lines of force líneas de fuerza eléctrica
electric lock cerradura eléctrica
electric locomotive locomotora eléctrica
electric loss pérdida eléctrica
electric machine máquina eléctrica
electric mechanism mecanismo eléctrico
electric meter medidor eléctrico
electric mirror espejo eléctrico
electric mode modo eléctrico
electric modulation modulación eléctrica
electric moment momento eléctrico
electric motor motor eléctrico
electric motor controller controlador de motor
    eléctrico
electric needle aguja eléctrica
electric network red eléctrica
electric noise ruido eléctrico
electric-noise signal señal de ruido eléctrico
electric-noise suppressor supresor de ruido eléctrico
electric operation operación eléctrica
electric oscillations oscilaciones eléctricas
electric plant planta eléctrica
electric polarization polarización eléctrica
electric potential potencial eléctrico
electric potential difference diferencia de potencial
    eléctrico
electric power potencia eléctrica
electric power distribution panel panel de
    distribución de potencia eléctrica
electric precipitation precipitación eléctrica
electric precipitator precipitador eléctrico
electric pressure presión eléctrica
electric probe sonda eléctrica
electric propulsion propulsión eléctrica
electric propulsion system sistema de propulsión
    eléctrica
electric pulsation pulsación eléctrica
electric pulse impulso eléctrico
electric pyrometer pirómetro eléctrico
electric radiation radiación eléctrica
electric radiator radiador eléctrico
electric recording registro eléctrico
electric reset reposición eléctrica
electric reset relay relé de reposición eléctrica
electric residue residuo eléctrico
electric resistance resistencia eléctrica
electric resistivity resistividad eléctrica
electric rigidity rigidez eléctrica
electric rings anillos eléctricos
electric scanning exploración eléctrica
electric screen pantalla eléctrica
electric sensitivity sensibilidad eléctrica
electric shield pantalla eléctrica
electric shock choque eléctrico
electric signal señal eléctrica
electric signal storage tube tubo de almacenamiento
    de señales eléctricas
electric sleep sueño eléctrico
electric soldering iron hierro de soldar eléctrico
electric spark chispa eléctrica

electric-spark erosion erosión por chispa eléctrica
electric-spark machining maquinado por chispa
    eléctrica
electric sparking chispeo eléctrico
electric steel acero eléctrico
electric stopwatch cronómetro eléctrico
electric storage almacenamiento eléctrico
electric storage subsystem subsistema de
    almacenamiento eléctrico
electric storm tormenta eléctrica
electric strain esfuerzo eléctrico
electric-strain gage deformímetro eléctrico
electric strength rigidez eléctrica
electric stress estrés eléctrico
electric stroboscope estroboscopio eléctrico
electric supply suministro eléctrico
electric-supply equipment equipo de suministro
    eléctrico
electric-supply lines líneas de suministro eléctrico
electric-supply station estación de suministro
    eléctrico
electric surface density densidad superficial eléctrica
electric susceptibility susceptibilidad eléctrica
electric system sistema eléctrico
electric tachometer tacómetro eléctrico
electric tape cinta eléctrica
electric telecommunication telecomunicación
    eléctrica
electric telegraphy telegrafía eléctrica
electric telemeter electrotelémetro
electric telemetering electrotelemedición
electric terminal terminal eléctrica
electric thermometer termómetro eléctrico
electric thermostat termostato eléctrico
electric timer temporizador eléctrico
electric tool herramienta eléctrica
electric traction tracción eléctrica
electric transcription transcripción eléctrica
electric transducer transductor eléctrico
electric tuning sintonización eléctrica
electric vector vector eléctrico
electric voltage tensión eléctrica
electric volume density densidad volumétrica
    eléctrica
electric watch reloj eléctrico
electric wave onda eléctrica
electric-wave filter filtro de ondas eléctricas
electric welding soldadura eléctrica
electric wind viento eléctrico
electric wiring alambrado eléctrico
electrical absorption absorción eléctrica
electrical anesthesia anestesia eléctrica
electrical angle ángulo eléctrico
electrical arc arco eléctrico
electrical assembler ensamblador eléctrico
electrical attraction atracción eléctrica
electrical axis eje eléctrico
electrical bandspread ensanche de banda eléctrico
electrical bias polarización eléctrica
electrical boresight visor eléctrico
electrical breakdown ruptura eléctrica
electrical capacitance capacitancia eléctrica
electrical center centro eléctrico
electrical centering centrado eléctrico
electrical code código eléctrico
electrical conduction conducción eléctrica
electrical conductivity conductividad eléctrica
electrical constant constante eléctrico
electrical contact contacto eléctrico
electrical control control eléctrico
electrical coupler acoplador eléctrico
electrical coupling acoplamiento eléctrico
electrical current corriente eléctrica

electrical degree grado eléctrico
electrical depression depresión eléctrica
electrical diagram diagrama eléctrico
electrical discharge descarga eléctrica
electrical-discharge machining maquinado por descargas eléctricas
electrical distance distancia eléctrica
electrical doublet doblete eléctrico
electrical drainage drenaje eléctrico
electrical efficiency eficiencia eléctrica
electrical element elemento eléctrico
electrical energy energía eléctrica
electrical engineer ingeniero electricista
electrical equipment equipo eléctrico
electrical erosion erosión eléctrica
electrical excitability excitabilidad eléctrica
electrical filter filtro eléctrico
electrical firing disparo eléctrico
electrical forming formación eléctrica
electrical gearing engranaje eléctrico
electrical glass vidrio eléctrico
electrical impulse impulso eléctrico
electrical inertia inercia eléctrica
electrical initiation iniciación eléctrica
electrical installation instalación eléctrica
electrical instrument instrumento eléctrico
electrical insulating material material aislante eléctrico
electrical insulation system sistema de aislamiento eléctrico
electrical integrator integrador eléctrico
electrical interchangeability intercambialidad eléctrica
electrical interference interferencia eléctrica
electrical interlocking enclavamiento eléctrico
electrical length longitud eléctrica
electrical load carga eléctrica
electrical measurement medida eléctrica
electrical network red eléctrica
electrical neutrality neutralidad eléctrica
electrical noise ruido eléctrico
electrical operation operación eléctrica
electrical outlet tomacorriente eléctrico, toma de corriente eléctrica
electrical overtravel sobrecarrera eléctrica
electrical parameter parámetro eléctrico
electrical polarity polaridad eléctrica
electrical power potencia eléctrica
electrical radian radián eléctrico
electrical reproduction reproducción eléctrica
electrical repulsion repulsión eléctrica
electrical reserve reserva eléctrica
electrical reset reposición eléctrica
electrical resistance resistencia eléctrica
electrical resistivity resistividad eléctrica
electrical resolver resolutor eléctrico
electrical scan exploración eléctrica
electrical sensitivity sensibilidad eléctrica
electrical sheet hoja eléctrica
electrical specification especificación eléctrica
electrical steel acero eléctrico
electrical stimulation estimulación eléctrica
electrical stress estrés eléctrico
electrical suppressor supresor eléctrico
electrical system sistema eléctrico
electrical tape cinta eléctrica
electrical taste sabor eléctrico
electrical transcription transcripción eléctrica
electrical transducer transductor eléctrico
electrical unit unidad eléctrica
electrical wavelength longitud de onda eléctrica
electrical welding soldadura eléctrica
electrical window ventana eléctrica

electrical zero cero eléctrico
electrically alterable alterable eléctricamente
electrically connected conectado eléctricamente
electrically driven accionado eléctricamente
electrically erasable borrable eléctricamente
electrically heated calentado eléctricamente
electrically operated operado eléctricamente
electrically operated valve válvula operada eléctricamente
electrically tuned oscillator oscilador sintonizado eléctricamente
electrically variable capacitor capacitor variable eléctricamente
electrically variable inductor inductor variable eléctricamente
electrically variable resistor resistor variable eléctricamente
electricity electricidad
electricity meter contador de electricidad
electrification electrificación
electrification time tiempo de electrificación
electrified electrificado
electrified railway ferrocarril electrificado
electrified track pista electrificada
electrify electrificar
electroacoustic electroacústico
electroacoustic amplifier amplificador electroacústico
electroacoustic apparatus aparato electroacústico
electroacoustic device dispositivo electroacústico
electroacoustic extensometer extensómetro electroacústico
electroacoustic resonance resonancia electroacústica
electroacoustic transducer transductor electroacústico
electroacoustical electroacústico
electroacoustics electroacústica
electroaffinity electroafinidad
electroanalysis electroanálisis
electroanalyzer electroanalizador
electroanesthesia electroanestesia
electroatomic electroatómico
electroballistics electrobalística
electrobath baño electrolítico
electrobiology electrobiología
electrobioscopy electrobioscopía
electrocapillarity electrocapilaridad
electrocardiogram electrocardiograma
electrocardiograph electrocardiógrafo
electrocardiography electrocardiografía
electrocardiophonograph electrocardiofonógrafo
electrocatalysis electrocatálisis
electrocautery electrocauterización
electrochemical electroquímico
electrochemical cell célula electroquímica
electrochemical corrosion corrosión electroquímica
electrochemical deterioration deterioro electroquímico
electrochemical diffused collector transistor transistor de colector difuso electroquímico
electrochemical diffusion difusión electroquímica
electrochemical equivalent equivalente electroquímico
electrochemical hardening endurecimiento electroquímico
electrochemical measurement medida electroquímica
electrochemical migration migración electroquímica
electrochemical oxidation oxidación electroquímica
electrochemical passivation pasivación electroquímica
electrochemical polarization polarización electroquímica

electrochemical recording registro electroquímico
electrochemical rectifier rectificador electroquímico
electrochemical reduction reducción electroquímica
electrochemical series serie electroquímica
electrochemical switch conmutador electroquímico
electrochemical telegraphy telegrafía electroquímica
electrochemical transducer transductor electroquímico
electrochemical valve válvula electroquímica
electrochemistry electroquímica
electrochromic display visualizador electrocrómico
electrochronometer electrocronómetro
electrocoagulation electrocoagulación
electrocooling electroenfriamiento
electrocorticogram electrocorticograma
electroculture electrocultivo
electrocute electrocutar
electrocution electrocución
electrocybernetics electrocibernética
electrode electrodo
electrode admittance admitancia de electrodo
electrode bias polarización de electrodo
electrode capacitance capacitancia de electrodo
electrode characteristic característica de electrodo
electrode conductance conductancia de electrodo
electrode control control de electrodo
electrode current corriente de electrodo
electrode dark current corriente en obscuridad de electrodo
electrode dissipation disipación de electrodo
electrode drop caída de electrodo
electrode efficiency eficiencia de electrodo
electrode holder portaelectrodo
electrode impedance impedancia de electrodo
electrode insulator aislador de electrodo
electrode inverse current corriente inversa de electrodo
electrode potential potencial de electrodo
electrode radiator radiador de electrodo
electrode reactance reactancia de electrodo
electrode resistance resistencia de electrodo
electrode spacing espaciado de electrodos
electrode support soporte de electrodo
electrode susceptance susceptancia de electrodo
electrode voltage tensión de electrodo
electrodeless discharge descarga sin electrodos
electrodeless tube tubo sin electrodos
electrodeposit electrodepósito
electrodeposition electrodeposición
electrodermal electrodérmico
electrodermal effect efecto electrodérmico
electrodermal reaction reacción electrodérmica
electrodermogram electrodermograma
electrodermography electrodermografía
electrodesiccation electrodesecación
electrodiagnosis electrodiagnóstico
electrodissolution electrodisolución
electrodynamic electrodinámico
electrodynamic braking frenaje electrodinámico
electrodynamic compressor compresor electrodinámico
electrodynamic force fuerza electrodinámica
electrodynamic instrument instrumento electrodinámico
electrodynamic loudspeaker altavoz electrodinámico
electrodynamic microphone micrófono electrodinámico
electrodynamic relay relé electrodinámico
electrodynamics electrodinámica
electrodynamometer electrodinamómetro
electroencephalogram electroencefalograma
electroencephalograph electroencefalógrafo
electroencephaloscope electroencefaloscopio

electroerosion electroerosión
electroextraction electroextracción
electroform electroformar
electroformed electroformado
electroforming electroformación
electrogalvanic electrogalvánico
electrogalvanize electrogalvanizar
electrogalvanized electrogalvanizado
electrograph electrógrafo
electrographic electrográfico
electrographic recording registro electrográfico
electrographite electrografito
electrohydraulic electrohidráulico
electrohydraulic servomechanism servomecanismo electrohidráulico
electrohydraulic servovalve servoválvula electrohidráulica
electrohydraulic system sistema electrohidráulico
electrojet electrochorro
electrokinetic electrocinético
electrokinetic energy energía electrocinética
electrokinetic hydrophone hidrófono electrocinético
electrokinetic potential potencial electrocinético
electrokinetics electrocinética
electroluminescence electroluminiscencia
electroluminescent electroluminiscente
electroluminescent diode diodo electroluminiscente
electroluminescent display visualizador electroluminiscente
electroluminescent display device dispositivo de visualización electroluminiscente
electroluminescent lamp lámpara electroluminiscente
electroluminescent panel panel electroluminiscente
electroluminescent screen pantalla electroluminiscente
electrolysis electrólisis
electrolyte electrólito
electrolytic electrolítico
electrolytic action acción electrolítica
electrolytic capacitor capacitor electrolítico
electrolytic cathode cátodo electrolítico
electrolytic cell celda electrolítica
electrolytic cleaning limpieza electrolítica
electrolytic condenser condensador electrolítico
electrolytic conduction conducción electrolítica
electrolytic conductivity conductividad electrolítica
electrolytic copper cobre electrolítico
electrolytic corrosion corrosión electrolítica
electrolytic current meter medidor de corriente electrolítico
electrolytic decomposition descomposición electrolítica
electrolytic deposition deposición electrolítica
electrolytic detector detector electrolítico
electrolytic diaphragm diafragma electrolítico
electrolytic dissociation disociación electrolítica
electrolytic excess voltage sobretensión electrolítica
electrolytic gas gas electrolítico
electrolytic interrupter interruptor electrolítico
electrolytic ionization ionización electrolítica
electrolytic iron hierro electrolítico
electrolytic metallization metalización electrolítica
electrolytic meter medidor electrolítico
electrolytic oxidation oxidación electrolítica
electrolytic polarization polarización electrolítica
electrolytic polishing pulido electrolítico
electrolytic potential potencial electrolítico
electrolytic recorder registrador electrolítico
electrolytic recording registro electrolítico
electrolytic rectifier rectificador electrolítico
electrolytic reduction reducción electrolítica
electrolytic refining refinado electrolítico

electrolytic resistor resistor electrolítico
electrolytic solution solución electrolítica
electrolytic switch conmutador electrolítico
electrolytic tank tanque electrolítico
electrolytic valve válvula electrolítica
electrolyze electrolizar
electrolyzer electrolizador
electromagnet electroimán
electromagnetic electromagnético
electromagnetic adherence adherencia electromagnética
electromagnetic ammeter amperímetro electromagnético
electromagnetic anechoic chamber cámara anecoica electromagnética
electromagnetic attraction atracción electromagnética
electromagnetic bias polarización electromagnética
electromagnetic brake freno electromagnético
electromagnetic braking frenaje electromagnético
electromagnetic cathode-ray tube tubo de rayos catódicos electromagnético
electromagnetic clock reloj electromagnético
electromagnetic clutch embrague electromagnético
electromagnetic communications comunicaciones electromagnéticas
electromagnetic compatibility compatibilidad electromagnética
electromagnetic complex complejo electromagnético
electromagnetic component componente electromagnético
electromagnetic constant constante electromagnética
electromagnetic coupling acoplamiento electromagnético
electromagnetic crack detector detector de grietas electromagnético
electromagnetic cylinder cilindro electromagnético
electromagnetic damping amortiguamiento electromagnético
electromagnetic deflection deflexión electromagnética
electromagnetic deflection coil bobina de deflexión electromagnética
electromagnetic deflection tube tubo de deflexión electromagnética
electromagnetic delay line línea de retardo electromagnética
electromagnetic device dispositivo electromagnético
electromagnetic disturbance perturbación electromagnética
electromagnetic energy energía electromagnética
electromagnetic energy conversion conversión de energía electromagnética
electromagnetic environment ambiente electromagnético
electromagnetic field campo electromagnético
electromagnetic field induction inducción de campo electromagnético
electromagnetic field intensity intensidad de campo electromagnético
electromagnetic flowmeter flujómetro electromagnético
electromagnetic flux flujo electromagnético
electromagnetic focusing enfoque electromagnético
electromagnetic force fuerza electromagnética
electromagnetic frequency spectrum espectro de frecuencias electromagnético
electromagnetic horn bocina electromagnética
electromagnetic induction inducción electromagnética
electromagnetic inertia inercia electromagnética
electromagnetic instrument instrumento electromagnético

electromagnetic interaction interacción electromagnética
electromagnetic interference interferencia electromagnética
electromagnetic lens lente electromagnética
electromagnetic loudspeaker altavoz electromagnético
electromagnetic mass masa electromagnética
electromagnetic microphone micrófono electromagnético
electromagnetic mirror espejo electromagnético
electromagnetic noise ruido electromagnético
electromagnetic oscillograph oscilógrafo electromagnético
electromagnetic oscilloscope osciloscopio electromagnético
electromagnetic pickup fonocaptor electromagnético
electromagnetic pressure presión electromagnética
electromagnetic propagation propagación electromagnética
electromagnetic pulse impulso electromagnético
electromagnetic pump bomba electromagnética
electromagnetic radiation radiación electromagnética
electromagnetic reaction reacción electromagnética
electromagnetic receiver receptor electromagnético
electromagnetic relay relé electromagnético
electromagnetic release desenganche electromagnético
electromagnetic remote control control remoto electromagnético
electromagnetic repulsion repulsión electromagnética
electromagnetic safety mechanism mecanismo de seguridad electromagnético
electromagnetic scanning exploración electromagnética
electromagnetic screen pantalla electromagnética
electromagnetic seismograph sismógrafo electromagnético
electromagnetic separation separación electromagnética
electromagnetic separator separador electromagnético
electromagnetic shield pantalla electromagnética
electromagnetic signal señal electromagnética
electromagnetic speaker altavoz electromagnético
electromagnetic spectrum espectro electromagnético
electromagnetic stress estrés electromagnético
electromagnetic susceptibility susceptibilidad electromagnética
electromagnetic switch conmutador electromagnético
electromagnetic system sistema electromagnético
electromagnetic theory of light teoría de la luz electromagnética
electromagnetic torquemeter torsiómetro electromagnético, medidor electromagnético de par
electromagnetic transducer transductor electromagnético
electromagnetic transmitter transmisor electromagnético
electromagnetic tube tubo electromagnético
electromagnetic unit unidad electromagnética
electromagnetic vibrator vibrador electromagnético
electromagnetic viscometer viscosímetro electromagnético
electromagnetic waves ondas electromagnéticas
electromagnetic yoke yugo electromagnético
electromagnetically electromagnéticamente
electromagnetically driven accionado electromagnéticamente
electromagnetically generated pulses impulsos

generados electromagnéticamente
electromagnetics electromagnética
electromagnetism electromagnetismo
electromagnetizer electromagnetizador
electromechanic electromecánico
electromechanical electromecánico
electromechanical amplifier amplificador
    electromecánico
electromechanical bell campana electromecánica
electromechanical brake freno electromecánico
electromechanical commutator conmutador
    electromecánico
electromechanical computer computadora
    electromecánica
electromechanical counter contador electromecánico
electromechanical coupling factor factor de
    acoplamiento electromecánico
electromechanical device dispositivo
    electromecánico
electromechanical energy energía electromecánica
electromechanical filter filtro electromecánico
electromechanical flip-flop basculador
    electromecánico
electromechanical force factor coeficiente de
    acoplamiento electromecánico
electromechanical frequency meter frecuencímetro
    electromecánico
electromechanical interlocking enclavamiento
    electromecánico
electromechanical manipulator manipulador
    electromecánico
electromechanical modulator modulador
    electromecánico
electromechanical oscillator oscilador
    electromecánico
electromechanical oscilloscope osciloscopio
    electromecánico
electromechanical recorder registrador
    electromecánico
electromechanical recording registro
    electromecánico
electromechanical rectifier rectificador
    electromecánico
electromechanical register registro electromecánico
electromechanical relay relé electromecánico
electromechanical repeater repetidor
    electromecánico
electromechanical switch conmutador
    electromecánico
electromechanical switching system sistema de
    conmutación electromecánico
electromechanical timer temporizador
    electromecánico
electromechanical transducer transductor
    electromecánico
electromechanical valve válvula electromecánica
electromechanics electromecánica
electromedical apparatus aparato electromédico
electromedical equipment equipo electromédico
electrometallurgy electrometalurgia
electrometer electrómetro
electrometer amplifier amplificador electrométrico
electrometer circuit circuito electrométrico
electrometer triode triodo electrométrico
electrometer tube tubo electrométrico
electrometry electrometría
electromigration electromigración
electromotion electromoción
electromotive electromotriz
electromotive force fuerza electromotriz
electromotive series serie electromotriz
electromotor electromotor
electromyogram electromiograma

electromyograph electromiógrafo
electromyographic electromiográfico
electromyographic potential potencial
    electromiográfico
electromyography electromiografía
electron electrón
electron accelerator acelerador electrónico
electron avalanche avalancha electrónica
electron ballistics balística electrónica
electron band banda electrónica
electron beam haz electrónico
electron-beam deflection system sistema de
    deflexión de haz electrónico
electron-beam fabrication generación de haz
    electrónico
electron-beam focusing enfoque de haz electrónico
electron-beam generator generador de haz
    electrónico
electron-beam heating calentamiento por haz
    electrónico
electron-beam induced conductivity conductividad
    inducida de haz electrónico
electron-beam instrument instrumento de haz
    electrónico
electron-beam intensity intensidad de haz
    electrónico
electron-beam loading carga de haz electrónico
electron-beam machining maquinado por haz
    electrónico
electron-beam magnetometer magnetómetro de haz
    electrónico
electron-beam recording registro por haz
    electrónico
electron-beam tube tubo de haz electrónico
electron-beam welding soldadura por haz electrónico
electron bombardment bombardeo electrónico
electron bunching agrupamiento de electrones
electron camera cámara electrónica
electron capture detector detector de captura
    electrónica
electron charge carga electrónica
electron-charge density densidad de carga
    electrónica
electron cloud nube electrónica
electron collection recolección electrónica
electron collection chamber cámara de recolección
    electrónica
electron collection surface superficie de recolección
    electrónica
electron collection time tiempo de recolección
    electrónica
electron collector colector electrónico
electron collision colisión electrónica
electron concentration concentración electrónica
electron conduction conducción electrónica
electron conductor conductor electrónico
electron-coupled multivibrator multivibrador de
    acoplamiento electrónico
electron-coupled oscillator oscilador de
    acoplamiento electrónico
electron coupler acoplador electrónico
electron coupling acoplamiento electrónico
electron crystallography cristalografía electrónica
electron current corriente electrónica
electron density densidad electrónica
electron device dispositivo electrónico
electron diffraction difracción electrónica
electron diffraction chamber cámara de difracción
    electrónica
electron discharge tube tubo de descarga electrónica
electron drift flujo electrónico
electron drift velocity velocidad de flujo electrónico
electron emission emisión electrónica

electron flow flujo electrónico
electron flux flujo electrónico
electron focusing enfoque electrónico
electron gas gas electrónico
electron gun cañón electrónico
electron-hole pair par electrón-hueco
electron image imagen electrónica
electron image tube tubo de imagen electrónica
electron impact impacto electrónico
electron injection inyección de electrones
electron injector inyector de electrones
electron jet chorro electrónico
electron lens lente electrónica
electron linear accelerator acelerador lineal electrónico
electron magnetic moment momento magnético electrónico
electron magnetic resonance resonancia magnética electrónica
electron mass masa de electrón
electron microanalyzer microanalizador electrónico
electron microprobe microsonda electrónica
electron microscope microscopio electrónico
electron microscopy microscopía electrónica
electron mirror espejo electrónico
electron motion moción de electrón
electron multiplication multiplicación electrónica
electron multiplier multiplicador electrónico
electron multiplier phototube fototubo multiplicador electrónico
electron multiplier tube tubo multiplicador electrónico
electron optics óptica electrónica
electron orbit órbita electrónica
electron oscillator oscilador electrónico
electron pair par de electrones
electron paramagnetic resonance resonancia paramagnética electrónica
electron physics física electrónica
electron probe sonda electrónica
electron-probe microanalysis microanálisis por sonda electrónica
electron-ray indicator tube tubo indicador catódico
electron relay relé electrónico
electron resonance resonancia electrónica
electron rest mass masa en reposo de electrón
electron sheath vaina de electrones, cubierta de electrones
electron shell capa electrónica
electron source fuente de electrones
electron specific charge carga específica de electrón
electron spin espín de electrón
electron spin resonance resonancia del espín electrónico
electron stream corriente electrónica
electron-stream potential potencial de corriente electrónica
electron-stream transmission efficiency eficiencia de transmisión de corriente electrónica
electron synchrotron sincrotrón electrónico
electron telescope telescopio electrónico
electron trajectory trayectoria electrónica
electron transfer transferencia electrónica
electron transit time tiempo de tránsito de electrón
electron tube tubo electrónico
electron-tube amplifier amplificador de tubos electrónicos
electron-tube coupler acoplador de tubo electrónico
electron-tube generator generador de tubo electrónico
electron velocity velocidad de electrón
electron-volt electrón-volt
electron wave onda electrónica

electron-wave tube tubo de ondas electrónicas
electronarcosis electronarcosis
electronegative electronegativo
electronegative potential potencial electronegativo
electronegativity electronegatividad
electronic electrónico
electronic accelerator acelerador electrónico
electronic accounting machine máquina de contabilidad electrónica
electronic adder sumador electrónico
electronic affinity afinidad electrónica
electronic altimeter altímetro electrónico
electronic amplification amplificación electrónica
electronic amplifier amplificador electrónico
electronic amplifying tube tubo amplificador electrónico
electronic analog accelerator acelerador analógico electrónico
electronic analog computer computadora analógica electrónica
electronic analog simulator simulador analógico electrónico
electronic autopilot autopiloto electrónico
electronic balance equilibrio electrónico
electronic balancer equilibrador electrónico
electronic beam haz electrónico
electronic bell campana electrónica
electronic board tablero electrónico
electronic brain cerebro electrónico
electronic breadboard montaje temporal electrónico, montaje preliminar electrónico
electronic bug problema electrónico indeterminado, manipulador electrónico
electronic bulletin board tablero de anuncios electrónico
electronic buzzer zumbador electrónico
electronic calculating punch perforadora calculadora electrónica
electronic calculator calculadora electrónica
electronic carillon carillón electrónico
electronic cash register caja registradora electrónica
electronic charge carga electrónica
electronic chronoscope cronoscopio electrónico
electronic circuit circuito electrónico
electronic circuitry circuitería electrónica
electronic clock reloj electrónico
electronic commutator conmutador electrónico
electronic component componente electrónico
electronic computer computadora electrónica
electronic computer memory memoria de computadora electrónica
electronic condenser condensador electrónico
electronic conduction conducción electrónica
electronic conductivity conductividad electrónica
electronic contactor contactor electrónico
electronic control control electrónico
electronic control circuit circuito de control electrónico
electronic control cubicle cubículo de control electrónico
electronic control loop bucle de control electrónico
electronic controller controlador electrónico
electronic counter contador electrónico
electronic counting circuit circuito contador electrónico
electronic coupling acoplamiento electrónico
electronic crossover separador electrónico de frecuencias, cruce electrónico
electronic current corriente electrónica
electronic current meter medidor de corriente electrónica
electronic data processing procesamiento de datos electrónico

electronic data processing center centro de procesamiento de datos electrónico
electronic data processing equipment equipo de procesamiento de datos electrónico
electronic data processing machine máquina de procesamiento de datos electrónico
electronic data processing system sistema de procesamiento de datos electrónico
electronic data processor procesador de datos electrónico
electronic decade counter contador de décadas electrónico
electronic densitometer densitómetro electrónico
electronic device dispositivo electrónico
electronic differential analyzer analizador diferencial electrónico
electronic differentiator diferenciador electrónico
electronic digit emitter emisor de dígitos electrónico
electronic digital computer computadora digital electrónica
electronic distance measurement medida de distancia electrónica
electronic distributor distribuidor electrónico
electronic divider divisor electrónico
electronic dust precipitator precipitador de polvo electrónico
electronic efficiency eficiencia electrónica
electronic emission emisión electrónica
electronic equilibrium equilibrio electrónico
electronic equipment equipo electrónico
electronic eye ojo electrónico
electronic field campo electrónico
electronic filter filtro electrónico
electronic flash destello electrónico
electronic flash lamp lámpara de destello electrónico
electronic flash tube tubo de destellos electrónicos
electronic flasher destellador electrónico
electronic flow flujo electrónico
electronic flowmeter flujómetro electrónico
electronic frequency meter frecuencímetro electrónico
electronic frequency synthesizer sintetizador de frecuencias electrónico
electronic-gap admittance admitancia de espacio electrónica
electronic gas gas electrónico
electronic gate puerta electrónica
electronic generator generador electrónico
electronic gyroscopic repeater repetidor giroscópico electrónico
electronic heater calentador electrónico
electronic heating calentamiento electrónico
electronic hum zumbido electrónico
electronic hygrometer higrómetro electrónico
electronic ignition ignición electrónica
electronic induction inducción electrónica
electronic information processing procesamiento de información electrónico
electronic instrument instrumento electrónico
electronic instrumentation instrumentación electrónica
electronic integrator integrador electrónico
electronic intelligence inteligencia electrónica
electronic interference interferencia electrónica
electronic inverter inversor electrónico
electronic jamming interferencia intencional electrónica
electronic key manipulador electrónico
electronic keying manipulación electrónica
electronic light amplifier amplificador de luz electrónico
electronic line scanning exploración de líneas electrónica

electronic locator localizador electrónico
electronic lock cerradura electrónica
electronic magnification magnificación electrónica
electronic mail correo electrónico
electronic mail network red de correo electrónico
electronic malfunction falla electrónica
electronic matrix matriz electrónica
electronic measurement medida electrónica
electronic memory memoria electrónica
electronic memory tube tubo de memoria electrónica
electronic microammeter microamperímetro electrónico
electronic micrometer micrómetro electrónico
electronic microphone micrófono electrónico
electronic microscope microscopio electrónico
electronic microvoltmeter microvóltmetro electrónico
electronic milliammeter miliamperímetro electrónico
electronic millivoltmeter milivóltmetro electrónico
electronic mirror espejo electrónico
electronic mobility movilidad electrónica
electronic modulation modulación electrónica
electronic multimeter multímetro electrónico
electronic multiplier multiplicador electrónico
electronic music música electrónica
electronic navigation navegación electrónica
electronic optics óptica electrónica
electronic orbit órbita electrónica
electronic oscillation oscilación electrónica
electronic packaging empaquetado electrónico
electronic paramagnetic resonance resonancia paramagnética electrónica
electronic part parte electrónica
electronic passband filter filtro pasabanda electrónico
electronic phase meter fasímetro electrónico
electronic photometer fotómetro electrónico
electronic pickup fonocaptor electrónico
electronic picoammeter picoamperímetro electrónico
electronic picovoltmeter picovóltmetro electrónico
electronic plotter trazador electrónico
electronic position indicator indicador de posición electrónico
electronic potentiometer potenciómetro electrónico
electronic power converter convertidor de potencia electrónico
electronic power supply fuente de alimentación electrónica
electronic precipitator precipitador electrónico
electronic printed circuit circuito impreso electrónico
electronic product producto electrónico
electronic profilometer perfilómetro electrónico
electronic pulse impulso electrónico
electronic reader lector electrónico
electronic recording registro electrónico
electronic rectifier rectificador electrónico
electronic register registro electrónico
electronic regulation regulación electrónica
electronic regulator regulador electrónico
electronic relay relé electrónico
electronic repeater repetidor electrónico
electronic resistor resistor electrónico
electronic scanning exploración electrónica
electronic sextant sextante electrónico
electronic spectroscope espectroscopio electrónico
electronic spin espín electrónico
electronic stethoscope estetoscopio electrónico
electronic stimulator estimulador electrónico
electronic storage almacenamiento electrónico
electronic storage register registro de almacenamiento electrónico
electronic storage tube tubo de almacenamiento

electrónico
electronic stroboscope estroboscopio electrónico
electronic subtracter restador electrónico
electronic surge sobretensión electrónica
electronic switch conmutador electrónico
electronic switching conmutación electrónica
electronic switching circuit circuito de conmutación electrónica
electronic switching system sistema de conmutación electrónica
electronic tachometer tacómetro electrónico
electronic technique técnica electrónica
electronic temperature temperatura electrónica
electronic temperature recorder registrador de temperatura electrónico
electronic temperature regulator regulador de temperatura electrónico
electronic thermal conductivity conductividad térmica electrónica
electronic thermometer termómetro electrónico
electronic thermoregulator termorregulador electrónico
electronic timer temporizador electrónico
electronic transformer transformador electrónico
electronic trigger circuit circuito de disparo electrónico
electronic tube tubo electrónico
electronic tube rectifier rectificador de tubo electrónico
electronic tuning sintonización electrónica
electronic tuning hysteresis histéresis de sintonización electrónica
electronic tuning range intervalo de sintonización electrónica
electronic tuning sensitivity sensibilidad de sintonización electrónica
electronic valve válvula electrónica
electronic variable attenuator atenuador variable electrónico
electronic video recorder registrador de video electrónico, grabadora de video electrónica
electronic video recording registro de video electrónico, grabación de video electrónica
electronic voltage regulation regulación de tensión electrónica
electronic voltage regulator regulador de tensión electrónico
electronic voltage stabilizer estabilizador de tensión electrónico
electronic voltmeter vóltmetro electrónico
electronic watch reloj electrónico
electronic wattmeter wáttmetro electrónico
electronically electrónicamente
electronically controlled controlado electrónicamente
electronically controlled speed velocidad controlada electrónicamente
electronically regulated regulado electrónicamente
electronically regulated power supply fuente de alimentación regulada electrónicamente
electronics electrónica
electrooculogram electrooculograma
electrooculography electrooculografía
electrooptic effect efecto electroóptico
electrooptical electroóptico
electrooptical frequency meter frecuencímetro electroóptico
electrooptical transistor transistor electroóptico
electrooptics electroóptica
electroosmosis electroósmosis
electroosmotic electroosmótico
electroosmotic potential potencial electroosmótico
electrophilic electrofílico
electrophone electrófono

electrophonic electrofónico
electrophonic effect efecto electrofónico
electrophony electrofonía
electrophoresis electroforesis
electrophoresis equipment equipo de electroforesis
electrophoretic electroforético
electrophoretic deposition deposición electroforética
electrophoretic potential potencial electroforético
electrophorus electróforo
electrophotographic electrofotográfico
electrophotographic process proceso electrofotográfico
electrophotography electrofotografía
electrophotometer electrofotómetro
electrophysics electrofísica
electrophysiology electrofisiología
electroplate galvanoplastiar
electroplated galvanoplastiado
electroplating galvanoplastia
electropneumatic electroneumático
electropneumatic brake freno electroneumático
electropneumatic control control electroneumático
electropneumatic controller controlador electroneumático
electropneumatic interlocking enclavamiento electroneumático
electropneumatic regulator regulador electroneumático
electropneumatic relay relé electroneumático
electropneumatic servomotor servomotor electroneumático
electropneumatic signal señal electroneumática
electropneumatic signaling señalización electroneumática
electropneumatic valve válvula electroneumática
electropolar electropolar
electropolishing electropulido
electropositive electropositivo
electropositive potential potencial electropositivo
electroreduction electrorreducción
electrorefine electrorrefinar
electrorefining electrorrefinado
electroresistive electrorresistivo
electroresistive effect efecto electrorresistivo
electroretinogram electrorretinograma
electroscope electroscopio
electrosection electrosección
electrosensitive electrosensible
electrosensitive paper papel electrosensible
electrosensitive printer impresora electrosensible
electrosensitive recording registro electrosensible
electrosensitive tape cinta electrosensible
electroshock electrochoque
electrostatic electrostático
electrostatic accelerator acelerador electrostático
electrostatic actuator actuador electrostático
electrostatic adhesion adhesión electrostática
electrostatic amplifier amplificador electrostático
electrostatic attraction atracción electrostática
electrostatic capacitance capacitancia electrostática
electrostatic capacitor capacitor electrostático
electrostatic capacity capacidad electrostática
electrostatic cathode-ray tube tubo de rayos catódicos electrostático
electrostatic charge carga electrostática
electrostatic component componente electrostático
electrostatic constant constante electrostática
electrostatic convergence convergencia electrostática
electrostatic cooling enfriamiento electrostático
electrostatic copier copiadora electrostática
electrostatic corona corona electrostática
electrostatic coupling acoplamiento electrostático
electrostatic deflection deflexión electrostática

electrostatic deflection sensitivity sensibilidad de deflexión electrostática
electrostatic deflection tube tubo de deflexión electrostática
electrostatic discharge descarga electrostática
electrostatic electrometer electrómetro electrostático
electrostatic electron microscope microscopio electrónico electrostático
electrostatic electrophotography electrofotografía electrostática
electrostatic energy energía electrostática
electrostatic error error electrostático
electrostatic field campo electrostático
electrostatic field intensity intensidad de campo electrostático
electrostatic filter filtro electrostático
electrostatic flux flujo electrostático
electrostatic focusing enfoque electrostático
electrostatic galvanometer galvanómetro electrostático
electrostatic generator generador electrostático
electrostatic gyroscope giroscopio electrostático
electrostatic headphone audífono electrostático
electrostatic hysteresis histéresis electrostática
electrostatic image imagen electrostática
electrostatic induction inducción electrostática
electrostatic influence influencia electrostática
electrostatic instrument instrumento electrostático
electrostatic lens lente electrostática
electrostatic line of force línea de fuerza electrostática
electrostatic loudspeaker altavoz electrostático
electrostatic machine máquina electrostática
electrostatic memory memoria electrostática
electrostatic memory tube tubo de memoria electrostático
electrostatic microphone micrófono electrostático
electrostatic microscope microscopio electrostático
electrostatic particle precipitation precipitación de partículas electrostática
electrostatic phase shifter desfasador electrostático
electrostatic photocopier fotocopiadora electrostática
electrostatic photomultiplier fotomultiplicador electrostático
electrostatic picture tube tubo de imagen electrostático
electrostatic plotter trazador electrostático
electrostatic potential potencial electrostático
electrostatic precipitation precipitación electrostática
electrostatic precipitator precipitador electrostático
electrostatic pressure presión electrostática
electrostatic printer impresora electrostática
electrostatic process proceso electrostático
electrostatic receiver receptor electrostático
electrostatic recording registro electrostático
electrostatic relay relé electrostático
electrostatic scanning exploración electrostática
electrostatic screen pantalla electrostática
electrostatic separation separación electrostática
electrostatic separator separador electrostático
electrostatic series serie electrostática
electrostatic shield pantalla electrostática
electrostatic speaker altavoz electrostático
electrostatic storage almacenamiento electrostático
electrostatic strain esfuerzo electrostático
electrostatic stress estrés electrostático
electrostatic system sistema electrostático
electrostatic transducer transductor electrostático
electrostatic tweeter altavoz de agudos electrostático
electrostatic unit unidad electrostática
electrostatic voltmeter vóltmetro electrostático
electrostatic wattmeter wáttmetro electrostático
electrostatic wave onda electrostática

electrostatically electrostáticamente
electrostatics electrostática
electrostatography electroestatografía
electrostenolysis electroestenólisis
electrostimulation electroestimulación
electrostriction electroestricción
electrostrictive electroestrictivo
electrostrictive ceramic cerámica electroestrictiva
electrostrictive relay relé electroestrictivo
electrostrictive transducer transductor electroestrictivo
electrosurgery electrocirugía
electrosynthesis electrosíntesis
electrotape electrocinta
electrotaxis electrotaxia
electrotechnology electrotecnología
electrotellurograph electrotelurógrafo
electrotherapeutics electroterapéutica
electrotherapy electroterapia
electrothermal electrotérmico
electrothermal efficiency eficiencia electrotérmica
electrothermal expansion element elemento de expansión electrotérmico
electrothermal instrument instrumento electrotérmico
electrothermal printer impresora electrotérmica
electrothermal recorder registrador electrotérmico
electrothermal recording registro electrotérmico
electrothermal relay relé electrotérmico
electrothermic electrotérmico
electrothermic device dispositivo electrotérmico
electrothermic element elemento electrotérmico
electrothermic instrument instrumento electrotérmico
electrothermic power indicator indicador de potencia electrotérmico
electrothermic power meter medidor de potencia electrotérmico
electrothermic unit unidad electrotérmica
electrothermics electrotermia
electrotitration electrotitración
electrotonic electrotónico
electrotonic wave onda electrotónica
electrotonus electrotono
electrotropism electrotropismo
electrovalence electrovalencia
electrowelding electrosoldadura
electrowinning extracción electrolítica
element elemento
element address dirección de elemento
element error rate tasa de errores de elementos
elemental area área elemental
elemental charge carga elemental
elemental semiconductor semiconductor elemental
elementary charge carga elemental
elementary circuit circuito elemental
elementary data element elemento de datos elemental
elementary diagram diagrama elemental
elementary item artículo elemental
elementary particle partícula elemental
elevated antenna antena elevada
elevation angle ángulo de elevación
elevator control control de elevador
elevator signal-transfer switch conmutador de transferencia de señal de elevador
eliminator eliminador
ell ele
elliptic loudspeaker altavoz elíptico
elliptical elíptico
elliptical field campo elíptico
elliptical function función elíptica
elliptical polarization polarización elíptica

elliptical waveguide guíaondas elíptico
elliptically polarized wave onda de polarización
  elíptica
ellipticity elipticidad
elongation elongación
emanation emanación
embedded empotrado, interconstruído
embedded command mando empotrado
embedded system sistema empotrado
embossed-foil printed circuit circuito impreso sobre
  hoja metálica en relieve
embossed-groove recording registro por surcos en
  relieve
embrittlement fragilización
emergency antenna antena de emergencia
emergency battery batería de emergencia
emergency beacon radiofaro de emergencia
emergency cell elemento de emergencia
emergency channel canal de emergencia
emergency circuit circuito de emergencia
emergency communication comunicación de
  emergencia
emergency coupling acoplamiento de emergencia
emergency equipment equipo de emergencia
emergency fan ventilador de emergencia
emergency frequency frecuencia de emergencia
emergency lighting alumbrado de emergencia
emergency machine máquina de emergencia
emergency maintenance mantenimiento de
  emergencia
emergency maintenance time tiempo de
  mantenimiento de emergencia
emergency power supply fuente de alimentación de
  emergencia
emergency radio channel canal de radio de
  emergencia
emergency radio transmitter transmisor de radio de
  emergencia
emergency rating clasificación de emergencia
emergency receiver receptor de emergencia
emergency repair reparación de emergencia
emergency route ruta de emergencia
emergency service servicio de emergencia
emergency stop parada de emergencia
emergency switch conmutador de emergencia
emergency switchboard cuadro conmutador de
  emergencia
emergency system sistema de emergencia
emergency transfer switch conmutador de
  transferencia de emergencia
emergency transmitter transmisor de emergencia
emergency voltage limit límite de tensión de
  emergencia
emission emisión
emission band banda de emisión
emission capability capacidad de emisión
emission characteristic característica de emisión
emission chart diagrama de emisión
emission code código de emisión
emission current corriente de emisión
emission efficiency eficiencia de emisión
emission frequency frecuencia de emisión
emission limitation limitación por emisión
emission lines líneas de emisión
emission map mapa de emisión
emission meter medidor de emisión
emission mode modo de emisión
emission power potencia de emisión
emission saturation saturación de emisión
emission spectrum espectro de emisión
emission tester probador de emisión
emission threshold umbral de emisión
emission type tipo de emisión

emission velocity velocidad de emisión
emission wavelength longitud de onda de emisión
emissive emisivo
emissive material material emisivo
emissive power poder emisivo
emissivity emisividad
emit emitir
emitron emitrón
emittance emitancia
emitted electron electrón emitido
emitter emisor
emitter-base circuit circuito emisor-base
emitter-base junction unión emisor-base
emitter-base voltage tensión emisor-base
emitter bias polarización de emisor
emitter-coupled logic circuit circuito lógico
  acoplado por emisor
emitter-coupled transistor logic circuit circuito
  lógico de transistores acoplado por emisor
emitter current corriente de emisor
emitter cutoff current corriente de corte de emisor
emitter degeneration degeneración por emisor
emitter efficiency eficiencia de emisor
emitter electrode electrodo emisor
emitter follower seguidor de emisor
emitter impedance impedancia de emisor
emitter junction unión de emisor
emitter resistance resistencia de emisor
emitter saturation voltage tensión de saturación de
  emisor
emitter stabilization estabilización de emisor
emitter station estación emisora
emitter voltage tensión de emisor
emitting power poder emisor
emitting source fuente emisora
emphasizer acentuador
empirical empírico
empirical curve curva empírica
empirical design diseño empírico
empirical method método empírico
empirical probability probabilidad empírica
empirical propagation model modelo de
  propagación empírico
empty level nivel vacío
empty medium medio vacío
empty string secuencia vacía
empty tape cinta vacía
emulate emular
emulation emulación
emulator emulador
emulator circuit circuito emulador
emulator program programa emulador
enable habilitar, activar
enable pulse impulso de habilitación
enabled habilitado, activado
enabled condition condición habilitada
enabled module módulo habilitado
enabling gate puerta habilitante
enabling pulse impulso habilitante
enabling signal señal habilitante
enameled cable cable esmaltado
enameled wire hilo esmaltado
encapsulant encapsulante
encapsulated encapsulado
encapsulated circuit circuito encapsulado
encapsulated coil bobina encapsulada
encapsulated component componente encapsulado
encapsulated element elemento encapsulado
encapsulated module módulo encapsulado
encapsulating encapsulador
encapsulating material material encapsulador
encapsulation encapsulación
encephalogram encefalograma

encephalography encefalografía
encipher cifrar
enciphered facsimile facsímil cifrado
enclosed arc arco encerrado
enclosed capacitor capacitor encerrado
enclosed fuse fusible encerrado
enclosed relay relé encerrado
enclosed switch conmutador encerrado
enclosed switchboard cuadro conmutador encerrado
enclosure caja
enclosure resonance resonancia de caja
encode codificar
encoded data datos codificados
encoded image imagen codificada
encoder codificador
encoder interface interfaz de codificador
encoding codificación
encoding circuit circuito de codificación
encrypt cifrar
end-around carry transporte cíclico, acarreo cíclico
end-around shift desplazamiento cíclico
end cell elemento de regulación
end connection conexión terminal
end device dispositivo terminal
end effect efecto de terminación
end-fed antenna antena alimentada por el extremo
end feed alimentación por el extremo
end-fire antenna antena de radiación longitudinal
end-fire array red de radiación longitudinal
end instrument instrumento terminal
end item artículo final
end mark marca de fin
end-of-address fin de dirección
end-of-area fin de área
end-of-block fin de bloque
end-of-block signal señal de fin de bloque
end-of-card fin de tarjeta
end-of-data fin de datos
end-of-data mark marca de fin de datos
end-of-field fin de campo
end-of-field mark marca de fin de campo
end-of-file fin de archivo
end-of-file indicator indicador de fin de archivo
end-of-file mark marca de fin de archivo
end-of-file routine rutina de fin de archivo
end-of-job fin de trabajo
end-of-job card tarjeta de fin de trabajo
end-of-job control card tarjeta de control de fin de
    trabajo
end-of-line fin de línea
end-of-message fin de mensaje
end-of-message character carácter de fin de mensaje
end-of-page indicator indicador de fin de página
end-of-procedure fin de procedimiento
end-of-program fin de programa
end-of-record fin de registro
end-of-record word palabra de fin de registro
end-of-tape fin de cinta
end-of-tape mark marca de fin de cinta
end-of-tape routine rutina de fin de cinta
end-of-tape sensor sensor de fin de cinta
end-of-text fin de texto
end-of-text character carácter de fin de texto
end-of-transmission fin de transmisión
end-of-transmission character carácter de fin de
    transmisión
end plate placa terminal
end-plate magnetron magnetrón de placas terminales
end point punto final
end-point control control de punto final
end position posición final
end section sección terminal
end shield pantalla terminal

end spaces espacios finales
end use uso final
end user usuario final
endless loop bucle sin fin
endodyne endodino
endodyne circuit circuito endodino
endodyne reception recepción endodina
endoergic endoérgico
endoradiosonde endorradiosonda
endothermic endotérmico
endothermic reaction reacción endotérmica
endurance limit límite de durabilidad
endurance ratio razón de durabilidad
endurance test prueba de durabilidad
energetic energético
energize energizar, excitar
energized energizado, excitado
energized anode ánodo energizado
energized dipole dipolo energizado
energizer energizador, excitador
energy energía
energy absorption absorción de energía
energy absorption coefficient coeficiente de
    absorción de energía
energy amplification amplificación de energía
energy balance equilibrio energético
energy band banda de energía
energy-band diagram diagrama de banda de energía
energy-band structure estructura de banda de
    energía
energy capacity capacidad de energía
energy cell pila de energía
energy component componente energético
energy consumption consumo de energía
energy control center centro de control de energía
energy conversion conversión de energía
energy conversion device dispositivo de conversión
    de energía
energy conversion unit unidad de conversión de
    energía
energy converter convertidor de energía
energy density densidad de energía
energy diagram diagrama energético
energy distribution distribución de energía
energy efficiency eficiencia energética
energy exchange intercambio de energía
energy flux flujo de energía
energy flux density densidad de flujo de energía
energy gain ganancia de energía
energy gap salto de energía
energy gap separación energética
energy gradient gradiente de energía
energy level nivel de energía
energy-level diagram diagrama de niveles de energía
energy loss pérdida de energía
energy product producto de energía
energy-product curve curva de producto de energía
energy production producción de energía
energy ratio razón de energía
energy redistribution redistribución de energía
energy resolution resolución de energía
energy selector selector de energía
energy spectrum espectro energético
energy state estado energético
energy storage almacenamiento de energía
energy-storage capacitor capacitor de
    almacenamiento de energía
energy-storage device dispositivo de almacenamiento
    de energía
energy-storage system sistema de almacenamiento
    de energía
energy transfer transferencia de energía
energy transfer rate velocidad de transferencia de

energía
energy transformation transformación de energía
energy unit unidad de energía
engine starter arrancador de motor
enhance realzar
enhanced-carrier demodulation desmodulación de portadora realzada
enhancement type transistor transistor de tipo de acrecentamiento, transistor de tipo de enriquecimiento
enqueue añadir a una cola
enter mode modo de entrada
entity attribute atributo de entidad
entity type tipo de entidad
entrance cable cable de acometida
entrance switch conmutador de entrada
entropy entropía
entry entrada, especificación
entry address dirección de entrada
entry block bloque de entrada
entry call llamada de entrada
entry condition condición de entrada
entry constant constante de entrada
entry data datos de entrada
entry instruction instrucción de entrada
entry level nivel de entrada
entry line línea de entrada
entry name nombre de entrada
entry point punto de entrada
entry queue cola de entrada
entry screen pantalla de entrada
entry sequence secuencia de entrada
entry symbol símbolo de entrada
entry time tiempo de entrada
entry value valor de entrada
entry variable variable de entrada
envelope envolvente
envelope analyzer analizador de envolvente
envelope delay retardo de envolvente
envelope-delay distortion distorsión por retardo de envolvente
envelope demodulator desmodulador de envolvente
envelope detector detector de envolvente
envelope distortion distorsión de envolvente
environmental chamber cámara ambiental
environmental coefficient coeficiente ambiental
environmental conditions condiciones ambientales
environmental control control ambiental
environmental extremes extremos ambientales
environmental factors factores ambientales
environmental hazards peligros ambientales
environmental impact impacto ambiental
environmental radio noise ruido de radio ambiental
environmental simulation simulación ambiental
environmental specifications especificaciones ambientales
environmental stability estabilidad ambiental
environmental temperature temperatura ambiental
environmental test prueba ambiental
environmental test chamber cámara de pruebas ambiental
environmentally sealed sellado ambientalmente
environmentproof a prueba del ambiente
ephapse efapsis
epitaxial epitaxial
epitaxial base transistor transistor de base epitaxial
epitaxial deposition deposición epitaxial
epitaxial device dispositivo epitaxial
epitaxial field-effect-transistor transistor de efecto de campo epitaxial
epitaxial film película epitaxial
epitaxial growth crecimiento epitaxial
epitaxial-growth mesa transistor transistor de

meseta de crecimiento epitaxial
epitaxial-growth process proceso de crecimiento epitaxial
epitaxial junction unión epitaxial
epitaxial layer capa epitaxial
epitaxial mesa transistor transistor de meseta epitaxial
epitaxial planar transistor transistor planar epitaxial
epitaxial process proceso epitaxial
epitaxial processing technique técnica de procesamiento epitaxial
epitaxial thyristor tiristor epitaxial
epitaxial transistor transistor epitaxial
epitaxic epitáxico
epitaxic growth crecimiento epitáxico
epitaxy epitaxia
epoxy resin resina epóxica
epsilon épsilon
equal alternations alternaciones iguales
equal-energy source fuente de energía constante
equal-energy spectrum espectro de energía constante
equal-loudness curve curva de intensidad sonora equivalente
equality circuit circuito de igualdad
equalization ecualización, igualación
equalization circuit circuito de ecualización
equalization control control de ecualización
equalization curve curva de ecualización
equalization voltage tensión de ecualización
equalizer ecualizador
equalizing battery batería ecualizadora
equalizing charge carga ecualizadora
equalizing coil bobina ecualizadora
equalizing connection conexión ecualizadora
equalizing current corriente ecualizadora
equalizing line línea ecualizadora
equalizing network red ecualizadora
equalizing preamplifier preamplificador ecualizador
equalizing pressure presión ecualizadora
equalizing pulses impulsos ecualizadores
equalizing resistor resistor ecualizador
equalizing voltage tensión ecualizadora
equation solver resolvedor de ecuaciones
equatorial orbit órbita ecuatorial
equilibrium equilibrio
equilibrium carrier portador en equilibrio
equilibrium condition condición de equilibrio
equilibrium conductivity conductividad de equilibrio
equilibrium electrode potential potencial de electrodo de equilibrio
equilibrium energy energía de equilibrio
equilibrium length longitud de equilibrio
equilibrium method método de equilibrio
equilibrium mode modo de equilibrio
equilibrium mode simulator simulador de modo de equilibrio
equilibrium potential potencial de equilibrio
equilibrium reaction potential potencial de reacción de equilibrio
equilibrium state estado de equilibrio
equilibrium voltage tensión de equilibrio
equiphase equifase
equiphase surface superficie equifase
equiphase zone zona equifase
equipment equipo
equipment chain cadena de equipo
equipment compatibility compatibilidad de equipos
equipment design diseño de equipo
equipment failure falla de equipo
equipment ground tierra de equipo
equipment life vida de equipo
equipment noise ruido de equipo
equipment number número de equipo

equipment panel panel de equipo
equipment qualification calificación de equipo
equipment system sistema de equipos
equipment test prueba de equipo
equipment unit unidad de equipo
equipotential equipotencial
equipotential cathode cátodo equipotencial
equipotential connection conexión equipotencial
equipotential line línea equipotencial
equipotential space espacio equipotencial
equipotential surface superficie equipotencial
equipotential winding arrollamiento equipotencial
equisignal equiseñal
equisignal beacon radiofaro equiseñal
equisignal localizer localizador equiseñal
equisignal radio range beacon radiofaro direccional
  equiseñal
equisignal sector sector equiseñal
equisignal surface superficie equiseñal
equisignal zone zona equiseñal
equivalence equivalencia
equivalence element elemento de equivalencia
equivalence operation operación de equivalencia
equivalent equivalente
equivalent absorbing power poder de absorción
  equivalente
equivalent absorption absorción equivalente
equivalent acoustics acústica equivalente
equivalent binary digits dígitos binarios equivalentes
equivalent bit rate velocidad de bits equivalente
equivalent capacitance capacitancia equivalente
equivalent circuit circuito equivalente
equivalent component density densidad de
  componentes equivalente
equivalent concentration concentración equivalente
equivalent conductance conductancia equivalente
equivalent conductivity conductividad equivalente
equivalent contrast contraste equivalente
equivalent dark-current input entrada de corriente
  en obscuridad equivalente
equivalent decrement decremento equivalente
equivalent delay line línea de retardo equivalente
equivalent delta delta equivalente
equivalent differential input capacitance
  capacitancia de entrada diferencial equivalente
equivalent differential input impedance impedancia
  de entrada diferencial equivalente
equivalent differential input resistance resistencia
  de entrada diferencial equivalente
equivalent diode diodo equivalente
equivalent diode voltage tensión de diodo
  equivalente
equivalent disturbing current corriente
  perturbadora equivalente
equivalent disturbing voltage tensión perturbadora
  equivalente
equivalent equations ecuaciones equivalentes
equivalent faults fallas equivalentes
equivalent field luminescence luminiscencia de
  campo equivalente
equivalent four-wire system sistema de cuatro hilos
  equivalente
equivalent grid voltage tensión de rejilla equivalente
equivalent height altura equivalente
equivalent hours horas equivalentes
equivalent impedance impedancia equivalente
equivalent inductance inductancia equivalente
equivalent loudness intensidad sonora equivalente
equivalent network red equivalente
equivalent noise bandwidth ancho de banda de ruido
  equivalente
equivalent noise conductance conductancia de ruido
  equivalente

equivalent noise current corriente de ruido
  equivalente
equivalent noise input entrada de ruido equivalente
equivalent noise resistance resistencia de ruido
  equivalente
equivalent noise temperature temperatura de ruido
  equivalente
equivalent optics óptica equivalente
equivalent oscillating system sistema oscilante
  equivalente
equivalent permeability permeabilidad equivalente
equivalent plate voltage tensión de placa equivalente
equivalent reactance reactancia equivalente
equivalent resistance resistencia equivalente
equivalent separation separación equivalente
equivalent series circuit circuito en serie equivalente
equivalent series resistance resistencia en serie
  equivalente
equivalent sine wave onda sinusoidal equivalente
equivalent sine wave power potencia de onda
  sinusoidal equivalente
equivalent source fuente equivalente
equivalent time tiempo equivalente
erasable borrable
erasable memory memoria borrable
erasable storage almacenador borrable
erase borrar
erase button botón de borrado
erase character carácter de borrado
erase circuit circuito de borrado
erase current corriente de borrado
erase generator generador de borrado
erase head cabeza de borrado
erase magnet imán de borrado
erase oscillator oscilador de borrado
erase signal señal de borrado
eraser borrador
erasing current corriente de borrado
erasing cycle ciclo de borrado
erasing field campo de borrado
erasing frequency frecuencia de borrado
erasing head cabeza de borrado
erasing key tecla de borrado
erasing rate tasa de borrado
erasing speed velocidad de borrado
erasing time tiempo de borrado
erasing voltage tensión de borrado
erasing winding arrollamiento de borrado
erasure borrado
erbium erbio
erg erg
ergograph ergógrafo
ergometer ergómetro
erlang erlang
erosion erosión
erratic errático
erratic noise ruido errático
erroneous erróneo
erroneous bit bit erróneo
erroneous block bloque erróneo
error error
error amplifier amplificador de error
error analysis análisis de error
error analysis routine rutina de análisis de error
error burst incremento repentino de errores
error card tarjeta de error
error category categoría de error
error character carácter de error
error checking comprobación de errores
error-checking code código comprobador de errores
error checking routine rutina de comprobación de
  errores
error code código de error

error coefficient coeficiente de error
error condition condición de error
error condition statement sentencia de condición de error
error constant constante de error
error control control de error
error control character carácter de control de error
error control software programas de control de errores
error-correcting code código corrector de errores
error-correcting routine rutina correctora de errores
error-correcting system sistema corrector de errores
error-correcting telegraph telégrafo corrector de errores
error correction corrección de errores
error-correction code código de corrección de errores
error-correction routine rutina de corrección de errores
error corrector corrector de errores
error count cuenta de errores
error current corriente de error
error curve curva de error
error-detecting code código detector de errores
error-detecting routine rutina detectora de errores
error-detecting system sistema detector de errores
error detection detección de errores
error-detection circuit circuito de detección de errores
error-detection code código de detección de errores
error-detection routine rutina de detección de errores
error detector detector de errores
error diagnostics diagnósticos de errores
error dump vaciado de errores
error file archivo de errores
error-free sin error
error-free second segundo sin error
error indicator indicador de errores
error interrupt interrupción de error
error light luz de error
error list lista de errores
error message mensaje de error
error model modelo de error
error percentage porcentaje de error
error prediction predicción de errores
error prediction model modelo de predicción de errores
error-prone propenso a errores
error range intervalo de errores
error rate tasa de errores
error ratio razón de errores
error record registro de errores
error recording registro de errores
error recovery recuperación de errores
error recovery routine rutina de recuperación de errores
error report informe de errores
error routine rutina de error
error seeding siembra de errores
error signal señal de error
error signal key tecla de señal de error
error stop parada por error
error tape cinta de errores
error tolerance tolerancia de errores
error transfer function función de transferencia de error
error voltage tensión de error
error volume analysis análisis de volumen de errores
errored second segundo con errores
errorless sin error
errorproof a prueba de error

escape character carácter de escape
escape code código de escape
escape key tecla de escape
escape ratio razón de escape
escapement escape
escutcheon placa decorativa
Esnault-Pelterie formula fórmula de Esnault-Pelterie
essential electrical systems sistemas eléctricos esenciales
estimated life vida estimada
estimated maximum load carga máxima estimada
eta eta
etchant grabador químico
etched circuit circuito grabado
etched printed circuit circuito impreso grabado
etching grabación química
Ettinghausen effect efecto de Ettinghausen
eureka eureka
europium europio
eutectic eutéctico
eutectic alloy aleación eutéctica
eutectic bond enlace eutéctico
evacuation evacuación
evaluation period periodo de evaluación
evanescent field campo evanescente
evanescent mode modo evanescente
evanescent wave onda evanescente
evanescent waveguide guíaondas evanescente
evaporation evaporación
even harmonic armónica par
even line línea par
even-line field campo de líneas pares
even parity paridad par
even parity check comprobación de paridad par
event control block bloque de control de evento
event counter contador de eventos
exalted carrier portadora incrementada
exalted-carrier receiver receptor de portadora incrementada
exalted-carrier reception recepción de portadora incrementada
except gate puerta de excepción
exception line línea de excepción
exception message mensaje de excepción
exception report informe de excepciones
exception response respuesta de excepción
excess carrier portador excedente
excess charge carga excedente
excess conduction conducción por electrones excedentes
excess current sobrecorriente
excess electron electrón excedente
excess fifty exceso de cincuenta
excess-fifty code código de exceso de cincuenta
excess meter medidor de excedente
excess minority carrier portador minoritario excedente
excess noise ruido excedente
excess noise factor factor de ruido excedente
excess sound pressure presión sonora excedente
excess-three code código de exceso de tres
excess voltage sobretensión
exchange intercambio, central
exchange area zona de central
exchange buffering almacenamiento intermedio de intercambio
exchange equipment equipo de central
exchange file archivo de intercambio
excitability excitabilidad
excitability curve curva de excitabilidad
excitation excitación
excitation anode ánodo de excitación
excitation band banda de excitación

excitation coefficient coeficiente de excitación
excitation control system sistema de control de excitación
excitation current corriente de excitación
excitation current ratio razón de corriente de excitación
excitation energy energía de excitación
excitation equipment equipo de excitación
excitation frequency frecuencia de excitación
excitation losses pérdidas por excitación
excitation mode modo de excitación
excitation potential potencial de excitación
excitation probability probabilidad de excitación
excitation purity pureza de excitación
excitation response respuesta de excitación
excitation rheostat reóstato de excitación
excitation state estado de excitación
excitation system sistema de excitación
excitation system rated voltage tensión nominal de sistema de excitación
excitation system stability estabilidad de sistema de excitación
excitation system stabilizer estabilizador de sistema de excitación
excitation system voltage response respuesta de tensión de sistema de excitación
excitation voltage tensión de excitación
excitation wave onda de excitación
excitation winding arrollamiento de excitación
excitator excitador
excited-field loudspeaker altavoz de campo excitado
exciter excitador
exciter ceiling voltage tensión máxima excitadora
exciter circuit circuito excitador
exciter field campo excitador
exciter-field circuit circuito de campo excitador
exciter-field current corriente de campo excitador
exciter-field winding arrollamiento de campo excitador
exciter lamp lámpara excitadora
exciter relay relé excitador
exciter unit unidad excitadora
exciting anode ánodo excitador
exciting coil bobina excitadora
exciting current corriente excitadora
exciting field campo excitador
exciting power potencia excitadora
exciting voltage tensión excitadora
exciton excitón
excitron excitrón
exclusive band banda exclusiva
exclusive mode modo excluyente
exclusive NOR NO-O excluyente
exclusive OR O excluyente
exclusive use uso exclusivo
excursion excursión
executable program programa ejecutable
execution ejecución
execution cycle ciclo de ejecución
execution efficiency eficiencia de ejecución
execution instruction instrucción de ejecución
execution phase fase de ejecución
execution routine rutina de ejecución
execution sequence secuencia de ejecución
execution time tiempo de ejecución
executive control control ejecutivo
executive control logic lógica de control ejecutivo
executive control program programa de control ejecutivo
executive control statement sentencia de control ejecutivo
executive control system sistema de control ejecutivo

executive cycle ciclo ejecutivo
executive diagnostic diagnóstico ejecutivo
executive dump vaciado ejecutivo
executive instruction instrucción ejecutiva
executive language lenguaje ejecutivo
executive program programa ejecutivo
executive routine rutina ejecutiva
executive system control control de sistema ejecutivo
executive system routine rutina de sistema ejecutivo
exit salida
exit cycle ciclo de salida
exit macro instruction macroinstrucción de salida
exit point punto de salida
exit program programa de salida
exit routine rutina de salida
exothermic exotérmico
exothermic reaction reacción exotérmica
expandable expansible
expandable gate puerta expansible
expanded display visualización ensanchada
expanded memory memoria expandida
expanded mode modo expandido
expanded position indicator indicador de posición ensanchado
expanded scale escala ensanchada
expanded-scale meter medidor de escala ensanchada
expanded storage almacenamiento ensanchado
expanded sweep barrido ensanchado
expander expansor
expander board tablero expansor
expansibility expansibilidad
expansion expansión
expansion board tablero de expansión
expansion bus bus de expansión
expansion card tarjeta de expansión
expansion chamber cámara de expansión
expansion interface interfaz de expansión
expansion ratio razón de expansión
expansion slot ranura de expansión
expansion time tiempo de expansión
expansion unit unidad de expansión
expendable desechable
experiment experimento
experimental experimental
experimental chassis chasis experimental
experimental model modelo experimental
experimental service servicio experimental
experimental station estación experimental
explicit address dirección explícita
explicit function función explícita
explicit route ruta explícita
exploring coil bobina exploradora
exploring electrode electrodo explorador
explosion chamber cámara de explosión
explosionproof apparatus aparato a prueba de explosiones
explosionproof device dispositivo a prueba de explosiones
explosionproof enclosure caja a prueba de explosiones
explosionproof fuse fusible a prueba de explosiones
explosionproof loudspeaker altavoz a prueba de explosiones
exponent overflow desbordamiento de exponente
exponent underflow desbordamiento negativo de exponente
exponential exponencial
exponential amplifier amplificador exponencial
exponential curve curva exponencial
exponential damping amortiguamiento exponencial
exponential decay decaimiento exponencial
exponential decrease reducción exponencial

exponential equation ecuación exponencial
exponential function función exponencial
exponential horn bocina exponencial
exponential increase aumento exponencial
exponential line línea exponencial
exponential quantity cantidad exponencial
exponential rheostat reóstato exponencial
exponential series serie exponencial
exponential sweep barrido exponencial
exponential transmission line línea de transmisión
    exponencial
exponential waveform forma de onda exponencial
export exportar
exposed expuesto
exposed installation instalación expuesta
exposed lamp lámpara expuesta
exposure meter exposímetro
exposure period periodo de exposición
exposure time tiempo de exposición
expulsion fuse fusible de expulsión
expulsion fuse unit unidad de fusible de expulsión
extended address dirección extendida
extended addressing direccionamiento extendido
extended binary tree árbol binario extendido
extended capability capacidad extendida
extended character set conjunto de caracteres
    extendido
extended cutoff tube tubo de corte extendido
extended delta connection conexión en delta
    extendida
extended graphic character set conjunto de
    caracteres gráficos extendido
extended memory memoria extendida
extended network red extendida
extended operating system sistema operativo
    extendido
extended operator control control de operador
    extendido
extended planned outage interrupción planificada
    extendida
extended precision precisión extendida
extended response byte byte de respuesta extendido
extended response field campo de respuesta
    extendido
extended route ruta extendida
extended segment segmento extendido
extended source fuente extendida
extensible coupling acoplamiento extensible
extensimeter extensímetro
extension board tablero de extensión
extension cable cable de extensión
extension character carácter de extensión
extension cord cordón de extensión
extension line línea de extensión
extension loudspeaker altavoz de extensión
extension register registro de extensión
extensometer extensómetro
external cable cable externo
external call llamada externa
external capacitor capacitor externo
external cavity cavidad externa
external characteristic característica externa
external circuit circuito externo
external command mando externo
external connection conexión externa
external connector conector externo
external control controles externos
external critical damping resistance resistencia de
    amortiguamiento crítico externa
external cross-modulation modulación cruzada
    externa
external damping device dispositivo de
    amortiguamiento externo

external data datos externos
external data file archivo de datos externo
external data item dato externo
external data record registro de datos externo
external delay retardo externo
external device code código de dispositivo externo
external device status estado de dispositivo externo
external domain dominio externo
external error error externo
external extension extensión externa
external feedback retroalimentación externa
external field campo externo
external field influence influencia de campo externo
external function función externa
external hard disk disco duro externo
external impedance impedancia externa
external indicator indicador externo
external insulation aislamiento externo
external interruption interrupción externa
external line línea externa
external load carga externa
external loudspeaker altavoz externo
external memory memoria externa
external modem módem externo
external modulation modulación externa
external noise ruido externo
external power loss pérdida de potencia externa
external power supply fuente de alimentación
    externa
external priority prioridad externa
external procedure procedimiento externo
external register registro externo
external reset reposición externa
external resistance resistencia externa
external signal señal externa
external speaker altavoz externo
external storage almacenamiento externo
external switch conmutador externo
external symbol símbolo externo
external temperature temperatura externa
externally operable operable externamente
externally programmable programable externamente
externally programmed programado externamente
externally ventilated machine máquina ventilada
    externamente
extinction potential potencial de extinción
extinction voltage tensión de extinción
extra-high voltage tensión extra alta
extra-high voltage system sistema de tensión extra
    alta
extracameral extracameral
extract extraer
extract instruction instrucción de extracción
extractor extractor
extracurrent extracorriente
extraneous component componente extraño
extraneous emission emisión extraña
extraneous impulse noise ruido de impulsos extraños
extraneous mechanical noise ruido mecánico
    extraño
extraneous noise ruido extraño
extraneous response respuesta extraña
extraneous signal señal extraña
extraordinary load carga extraordinaria
extraordinary ray rayo extraordinario
extraordinary wave onda extraordinaria
extreme extremo
extremely high frequency frecuencia
    extremadamente alta
extremely low frequency frecuencia extremadamente
    baja
extrinsic extrínseco
extrinsic conductance conductancia extrínseca

extrinsic conduction conducción extrínseca
extrinsic property propiedad extrínseca
extrinsic semiconductor semiconductor extrínseco
extrinsic transconductance transconductancia
  extrínseca
extrusion extrusión
eyepiece ocular

F

fabrication tolerance tolerancia de fabricación
Fabry-Perot interferometer interferómetro de
  Fabry-Perot
face-down feed alimentación boca abajo
faceplate fondo, placa frontal
faceplate controller controlador de disco
faceplate rheostat reóstato de placa frontal
facilitation facilitación
facilities instalaciones
facom facom
facsimile facsímil
facsimile bandwidth ancho de banda de facsímil
facsimile broadcast station estación de radiodifusión
  de facsímil
facsimile density densidad de facsímil
facsimile equipment equipo de facsímil
facsimile machine máquina de facsímil
facsimile network red de facsímil
facsimile receiver receptor de facsímil
facsimile recorder registrador de facsímil
facsimile reproducer reproductor de facsímil
facsimile signal señal de facsímil
facsimile-signal level nivel de señal de facsímil
facsimile standards patrones de facsímil
facsimile system sistema de facsímil
facsimile telegraphy telegrafía en facsímil
facsimile terminal terminal de facsímil
facsimile transient fluctuación transitoria de facsímil
facsimile transmission transmisión de facsímil
facsimile transmitter transmisor de facsímil
factor factor
factor of influence factor de influencia
factor of merit factor de mérito
factor of quality factor de calidad
factor of safety factor de seguridad
factor storage almacenamiento de factores
factor storage entry entrada de almacenamiento de
  factores
factorial factorial
factoring factorización
fade desvanecimiento
fade in aumento progresivo
fade out desaparición progresiva
fader desvanecedor
fader circuit circuito desvanecedor
fader control control desvanecedor
fading desvanecimiento
fading allowance tolerancia de desvanecimiento
fading area área de desvanecimiento
fading circuit circuito de desvanecimiento
fading depth profundidad de desvanecimiento
fading loss pérdida por desvanecimiento
fading machine máquina de desvanecimiento
fading margin margen de desvanecimiento
fading range margen de desvanecimiento
fading rate tasa de desvanecimientos
fading spectrum espectro de desvanecimiento
Fahnestock clip pinza de Fahnestock
Fahrenheit scale escala Fahrenheit
failsafe protegido contra daños por fallas
failsafe feature característica de protección contra

daños por fallas
failsafe operation operación protegida contra daños
   por fallas
failsoft falla que no impide funcionamiento
failsoft system sistema que permite funcionamiento
   tras falla
failure falla, fallo
failure analysis análisis de falla
failure category categoría de falla
failure cause causa de falla
failure criteria criterios de falla
failure data datos de falla
failure distribution distribución de fallas
failure indicator indicador de falla
failure light luz de falla
failure logging registro de fallas
failure mechanism mecanismo de falla
failure mode modo de falla
failure prediction predicción de fallas
failure rate tasa de fallas
failure rate level nivel de tasa de fallas
failure ratio razón de fallas
failure recovery recuperación tras falla
failure unit unidad de falla
fair weather buen clima
fair weather distribution distribución de buen clima
faithful modulation modulación fiel
faithful reproduction reproducción fiel
fall of potential caída de potencial
fall-off caída
fall-off time tiempo de caída
fall time tiempo de caída
fall time of pulse tiempo de caída de impulso
fallback circuit circuito de reserva
fallback procedure procedimiento de reserva
falling characteristic característica descendente
falling edge borde descendente
falling-off caída
false add suma falsa
false alarm falsa alarma
false-alarm probability probabilidad de falsa alarma
false-alarm time tiempo entre falsas alarmas
false code código falso
false-code check comprobación de código falso
false course rumbo falso
false echo falso eco
false error error falso
false identification identificación falsa
false operation operación falsa
false-operation probability probabilidad de
   operación falsa
false precision precisión falsa
false retrieval recuperación falsa
false signal señal falsa
false tripping abertura falsa
family of curves familia de curvas
family of frequencies familia de frecuencias
fan antenna antena en abanico
fan beam haz en abanico
fan marker marcador en abanico
fan-marker beacon radiobaliza en abanico
fan-shaped antenna antena en forma de abanico
fanin convergencia de entrada, capacidad de entrada
fanin circuit circuito de convergencia de entrada
fanned beam haz en abanico
fanned-beam antenna antena de haz en abanico
fanout divergencia de salida, capacidad de salida
fanout circuit circuito de divergencia de salida
far-end crosstalk telediafonía
far-end crosstalk attenuation atenuación
   telediafónica
far-end crosstalk deviation desviación telediafónica
far-end crosstalk isolation aislamiento telediafónico

far-end crosstalk reduction reducción telediafónica
far field campo lejano
far-field diffraction pattern patrón de difracción de
   campo lejano
far-field pattern patrón de campo lejano
far-field radiation pattern patrón de radiación de
   campo lejano
far-field region región de campo lejano
far infrared infrarrojo lejano
far infrared maser máser de infrarrojo lejano
far ultraviolet ultravioleta lejano
far zone zona lejana
farad farad
faraday faraday
Faraday's law ley de Faraday
Faraday's law of induction ley de Faraday de
   inducción
Faraday's laws of electrolysis leyes de Faraday de
   electrólisis
Faraday cage jaula de Faraday
Faraday constant constante de Faraday
Faraday cylinder cilindro de Faraday
Faraday dark space espacio obscuro de Faraday
Faraday effect efecto de Faraday
Faraday electrostatic screen pantalla electrostática
   de Faraday
Faraday rotation rotación de Faraday
Faraday rotation device dispositivo de rotación de
   Faraday
Faraday rotator rotador de Faraday
Faraday screen pantalla de Faraday
Faraday shield jaula de Faraday
Faraday tube tubo de Faraday
faradic farádico
faradic current corriente farádica
faradic rectification rectificación farádica
faradimeter faradímetro
faradism faradismo
faradization faradización
faradmeter faradímetro
Farnsworth detector tube tubo detector de
   Farnsworth
Farnsworth iconoscope iconoscopio de Farnsworth
Farnsworth multiplier multiplicador de Farnsworth
fast access acceso rápido
fast-access storage almacenamiento de acceso rápido
fast-acting de acción rápida
fast-acting circuit-breaker cortacircuito de acción
   rápida
fast-acting relay relé de acción rápida
fast diode diodo rápido
fast forward avance rápido
fast forward system sistema de avance rápido
fast groove surco rápido
fast-operate relay relé de funcionamiento rápido
fast-release relay relé de desenganche rápido
fast reverse retroceso rápido
fast time tiempo rápido
fast-time constant constante de tiempo corto
fast wave onda rápida
fastener fijador, sujetador
fatal error error fatal
fathometer fatómetro
fatigue fatiga
fatigue breakage ruptura por fatiga
fatigue effect efecto de fatiga
fatigue limit límite de fatiga
fatigue of metal fatiga de metal
fatigue resistance resistencia a la fatiga
fault falla, fallo
fault alarm alarma de fallas
fault-alarm circuit circuito de alarma de fallas
fault alarm system sistema de alarma de fallas

fault alarm tone tono de alarma de fallas
fault analysis análisis de falla
fault category categoría de falla
fault circuit circuito de falla
fault code código de falla
fault coder codificador de fallas
fault current corriente de falla
fault decoder descodificador de fallas
fault-detecting system sistema detector de fallas
fault detection detección de fallas
fault detector detector de fallas
fault display indicación visual de fallas
fault display equipment equipo de indicación visual de fallas
fault display panel panel de indicación visual de fallas
fault electrode current corriente de electrodo anormal
fault encoder codificador de fallas
fault finder localizador de fallas
fault finding localización de fallas
fault-finding test prueba de localización de fallas
fault ground tierra por falla
fault-indicating indicador de fallas
fault-indicating equipment equipo indicador de fallas
fault indication indicación de fallas
fault indicator indicador de fallas
fault-indicator panel panel indicador de fallas
fault-initiating switch conmutador iniciador de fallas
fault isolation aislamiento de falla
fault isolation analysis routine rutina de análisis de aislamiento de falla
fault location localización de fallas
fault-location program programa de localización de fallas
fault-location test prueba de localización de fallas
fault masking enmascaramiento de falla
fault relay relé de falla
fault reporting indicación de fallas
fault-reporting circuit circuito indicador de fallas
fault-reporting equipment equipo indicador de fallas
fault-reporting system sistema indicador de fallas
fault-reporting tone tono indicador de fallas
fault seeding siembra de fallas
fault signal señal de falla
fault signaling señalización de fallas
fault supervision supervisión de fallas
fault symptom síntoma de falla
fault time tiempo de falla
fault tolerance tolerancia a fallas
fault tolerant tolerante a fallas
fault tone tono de fallas
fault transmitter transmisor de fallas
faulty defectuoso
faulty circuit circuito defectuoso
faulty connection conexión defectuosa
faulty design diseño defectuoso
faulty hardware equipo físico defectuoso
faulty insulation aislamiento defectuoso
faulty line línea defectuosa
faulty part parte defectuosa
faulty reading lectura defectuosa
faulty signal señal defectuosa
faulty software programas defectuosos
faulty station estación defectuosa
Faure plate placa de Faure
fax fax
feature característica
feed alimentación
feed belt correa de alimentación
feed check comprobación de alimentación
feed circuit circuito de alimentación

feed clutch embrague de alimentación
feed current corriente de alimentación
feed cycle ciclo de alimentación
feed failure falla de alimentación
feed gap abertura de alimentación
feed hole perforación de alimentación
feed hopper almacén de alimentación
feed horn bocina de alimentación
feed index índice de alimentación
feed line línea de alimentación
feed mechanism mecanismo de alimentación
feed motor motor de alimentación
feed pitch paso de alimentación
feed point punto de alimentación
feed pump bomba de alimentación
feed rate tasa de alimentación
feed reel carrete de alimentación
feed stability estabilidad de alimentación
feed system sistema de alimentación
feed track pista de alimentación
feed tray bandeja de alimentación
feed unit unidad de alimentación
feed valve válvula de alimentación
feed wire hilo de alimentación
feedback retroalimentación, retroacción
feedback admittance admitancia de retroalimentación
feedback amplifier amplificador de retroalimentación
feedback attenuation atenuación de retroalimentación
feedback bridge puente de retroalimentación
feedback capacitance capacitancia de retroalimentación
feedback circuit circuito de retroalimentación
feedback coil bobina de retroalimentación
feedback control control por retroalimentación
feedback control loop bucle de control por retroalimentación
feedback control signal señal de control por retroalimentación
feedback control system sistema de control por retroalimentación
feedback coupling acoplamiento por retroalimentación
feedback cutter fonoincisor con retroalimentación
feedback element elemento de retroalimentación
feedback factor factor de retroalimentación
feedback impedance impedancia de retroalimentación
feedback information información de retroalimentación
feedback input current corriente de entrada de retroalimentación
feedback loop bucle de retroalimentación
feedback oscillator oscilador de retroalimentación
feedback path trayectoria de retroalimentación
feedback percentage porcentaje de retroalimentación
feedback ratio razón de retroalimentación
feedback ratio razón de retroalimentación
feedback regulator regulador de retroalimentación
feedback resistance resistencia de retroalimentación
feedback signal señal de retroalimentación
feedback stabilization estabilización por retroalimentación
feedback system sistema con retroalimentación
feedback transfer function función de transferencia de retroalimentación
feedback voltage tensión de retroalimentación
feedback winding arrollamiento de retroalimentación
feeder alimentador
feeder brush escobilla colectora
feeder cable cable alimentador

feeder delay noise ruido por retardo de alimentador
feeder distortion distorsión de alimentador
feeder line línea alimentadora
feeder loss pérdida por alimentador
feeder relay relé alimentador
feeder voltage regulator regulador de tensión de alimentador
feeding alimentación
feeding belt correa de alimentación
feeding bridge puente de alimentación
feeding cable cable de alimentación
feeding mechanism mecanismo de alimentación
feeding point punto de alimentación
feedthrough pasante
feedthrough capacitor capacitor pasante
feedthrough insulator aislador pasante
feedthrough terminal terminal pasante
feeler palpador
female plug clavija hembra
femto femto
femtoampere femtoampere
femtofarad femtofarad
femtovolt femtovolt
Fermat's principle principio de Fermat
fermi fermi
Fermi-Dirac distribution function función de distribución de Fermi-Dirac
Fermi level nivel de Fermi
fermium fermio
fernico fernico
ferretic ferrético
ferric induction inducción férrica
ferric oxide óxido férrico
ferrielectric ferrieléctrico
ferrielectricity ferrieléctricidad
ferrimagnetic ferrimagnético
ferrimagnetic amplifier amplificador ferrimagnético
ferrimagnetic circuit circuito ferrimagnético
ferrimagnetic limiter limitador ferrimagnético
ferrimagnetic material material ferrimagnético
ferrimagnetism ferrimagnetismo
ferristor ferristor
ferrite ferrita
ferrite antenna antena de ferrita
ferrite circulator circulador de ferrita
ferrite core núcleo de ferrita
ferrite-core antenna coil bobina de antena con núcleo de ferrita
ferrite-core memory memoria de núcleos de ferrita
ferrite-core memory unit unidad de memoria de núcleos de ferrita
ferrite-core storage almacenamiento de núcleos de ferrita
ferrite device dispositivo de ferrita
ferrite filter filtro de ferrita
ferrite isolator aislador de ferrita
ferrite junction device dispositivo de unión de ferrita
ferrite limiter limitador de ferrita
ferrite load isolator aislador de carga de ferrita
ferrite memory memoria de ferrita
ferrite ring anillo de ferrita
ferrite rod varilla de ferrita
ferrite-rod antenna antena de varilla de ferrita
ferrite rotator rotador de ferrita
ferrite separator separador de ferrita
ferrite switch conmutador de ferrita
ferrite toroid toroide de ferrita
ferrite-tuned oscillator oscilador sintonizado por ferrita
ferritic ferrítico
ferroacoustic ferroacústico
ferrodynamic ferrodinámico
ferrodynamic instrument instrumento ferrodinámico

ferrodynamic relay relé ferrodinámico
ferroelastic ferroelástico
ferroelastic crystal cristal ferroelástico
ferroelectric ferroeléctrico
ferroelectric amplifier amplificador ferroeléctrico
ferroelectric axis eje ferroeléctrico
ferroelectric capacitor capacitor ferroeléctrico
ferroelectric ceramic cerámica ferroeléctrica
ferroelectric converter convertidor ferroeléctrico
ferroelectric crystal cristal ferroeléctrico
ferroelectric Curie point punto de Curie ferroeléctrico
ferroelectric Curie temperature temperatura de Curie ferroeléctrica
ferroelectric dielectric dieléctrico ferroeléctrico
ferroelectric domain dominio ferroeléctrico
ferroelectric flip-flop basculador ferroeléctrico
ferroelectric material material ferroeléctrico
ferroelectric polymer polímero ferroeléctrico
ferroelectric scanning exploración ferroeléctrica
ferroelectricity ferroelectricidad
ferroinductance ferroinductancia
ferroinductor ferroinductor
ferromagnetic ferromagnético
ferromagnetic amplifier amplificador ferromagnético
ferromagnetic component componente ferromagnético
ferromagnetic core núcleo ferromagnético
ferromagnetic domain dominio ferromagnético
ferromagnetic electron electrón ferromagnético
ferromagnetic material material ferromagnético
ferromagnetic resonance resonancia ferromagnética
ferromagnetic spinel espinela ferromagnética
ferromagnetic substance sustancia ferromagnética
ferromagnetic tape cinta ferromagnética
ferromagnetics ferromagnética
ferromagnetism ferromagnetismo
ferrometer ferrómetro
ferroresonance ferrorresonancia
ferroresonant ferrorresonante
ferroresonant circuit circuito ferrorresonante
ferroresonant counter contador ferrorresonante
ferroresonant flip-flop basculador ferrorresonante
ferroresonant ring anillo ferrorresonante
ferroresonant shift register registro de desplazamiento ferrorresonante
ferroresonant voltage regulation regulación de tensión ferrorresonante
ferroresonant voltage regulator regulador de tensión ferrorresonante
ferroresonator ferrorresonador
ferrospinel ferroespinela
ferrous ferroso
ferrous oxide óxido ferroso
ferrule manguito de empalme, casquillo
fetch buscar
fetch cycle ciclo de búsqueda
fetch instruction instrucción de búsqueda
fetch program programa de búsqueda
fetch protection protección contra búsqueda
fetch routine rutina de búsqueda
fiber fibra
fiber bandwidth ancho de banda de fibra
fiber buffer amortiguador de fibra
fiber bundle paquete de fibras
fiber electrometer electrómetro de hilo
fiber insulation aislamiento de fibra
fiber metallurgy metalurgia de fibras
fiber needle aguja de fibra
fiber-optic fibroóptico, de fibra óptica, por fibra óptica
fiber-optic bandwidth ancho de banda fibroóptica

fiber-optic bit rate velocidad de bits fibroóptica
fiber-optic bundle paquete fibroóptico
fiber-optic cable cable fibroóptico, cable de fibras
  ópticas
fiber-optic capacity capacidad fibroóptica
fiber-optic communication comunicación
  fibroóptica, comunicación por fibras ópticas
fiber-optic component componente fibroóptico
fiber-optic coupling acoplamiento fibroóptico
fiber-optic dispersion dispersión fibroóptica
fiber-optic distortion distorsión fibroóptica
fiber-optic local module módulo local fibroóptico
fiber-optic probe sonda fibroóptica
fiber-optic scrambler codificador fibroóptico
fiber-optic transceiver transceptor fibroóptico
fiber optics fibroóptica, óptica de fibras
fiber stylus estilete de fibra
fiberglass fibra de vidrio
fiberglass insulation aislamiento de fibra de vidrio
fiberoptics fibroóptica, óptica de fibras
fiberscope fibroscopio
Fibonacci number número de Fibonacci
Fibonacci search búsqueda de Fibonacci
Fibonacci series serie de Fibonacci
fibrillation fibrilación
fictitious ficticio
fictitious earth tierra ficticia
fictitious power potencia ficticia
fictitious voltage generator generador de tensión
  ficticio
fidelity fidelidad
fidelity check comprobación de fidelidad
field campo
field accelerating relay relé acelerador de campo
field accumulation acumulación de campo
field amplitude amplitud de campo
field application relay relé de aplicación de campo
field area área de campo
field attribute atributo de campo
field-by-field processing procesamiento de campo
  por campo
field circuit-breaker cortacircuito de excitación de
  campo
field coil bobina de campo
field comparison comparación de campos
field contact voltage tensión de contacto de campo
field control control de campo
field data datos de campo
field data code código de datos de campo
field definition definición de campo
field density densidad de campo
field depth profundidad de campo
field description specifications especificaciones de
  descripción de campo
field descriptor descriptor de campo
field detector detector de campo
field direction dirección de campo
field discharge descarga de campo
field discharge protection protección de descarga de
  campo
field displacement desplazamiento de campo
field distortion distorsión de campo
field distribution distribución de campo
field disturbance perturbación de campo
field disturbance sensor sensor de perturbación de
  campo
field effect efecto de campo
field-effect controlled controlado por efecto de
  campo
field-effect device dispositivo de efecto de campo
field-effect diode diodo de efecto de campo
field-effect modulation modulación por efecto de
  campo

field-effect tetrode tetrodo de efecto de campo
field-effect transistor transistor de efecto de campo
field-effect tube tubo de efecto de campo
field-effect varistor varistor de efecto de campo
field emission emisión por campo
field-emission microscope microscopio de emisión
  de campo
field-enhanced photoelectric emission emisión
  fotoeléctrica realzada por campo
field-enhanced secondary emission emisión
  secundaria realzada por campo
field equation ecuación de campo
field erasing borrado de campo
field-excitation control control de excitación de
  campo
field-excitation current corriente de excitación de
  campo
field-failure protection protección de falla de campo
field-failure relay relé de falla de campo
field flattener aplanador de campo
field-flattener lens lente aplanadora de campo
field forcing forzado de campo
field forcing relay relé de forzado de campo
field-free emission current corriente de emisión de
  campo nulo
field frequency frecuencia de campo
field-frequency control control de frecuencia de
  campo
field gradient gradiente de campo
field indicator indicador de campo
field initialization inicialización de campo
field intensity intensidad de campo
field-intensity contour contorno de intensidad de
  campo
field-intensity measurement medida de intensidad de
  campo
field-intensity meter medidor de intensidad de
  campo
field-intensity pattern patrón de intensidad de campo
field-ion emission emisión iónica de campo
field-ion microscope microscopio iónico de campo
field ionization ionización de campo
field length longitud de campo
field length type tipo de longitud de campo
field level nivel de campo
field level specification especificación de nivel de
  campo
field line línea de campo
field magnet imán de campo
field marker marcador de campo
field measurement medida de campo
field mesh malla de campo
field meter medidor de campo
field name nombre de campo
field-neutralizing coil bobina neutralizadora de
  campo
field-neutralizing magnet imán neutralizador de
  campo
field of force campo de fuerza
field overflow desbordamiento de campo
field particle partícula de campo
field pattern patrón de campo
field period periodo de campo
field pole polo de campo
field protection protección de campo
field protective relay relé protector de campo
field pulse impulso de campo
field ratio razón de campo
field reference field campo de referencia de campo
field regulator regulador de campo
field relay relé de campo
field-reliability test prueba de confiabilidad de
  campo

field resistance resistencia de campo
field resistor resistor de campo
field rheostat reóstato de campo
field scanning exploración de campo
field-scanning device dispositivo de exploración de campo
field selection selección de campo
field separator separador de campo
field-sequential system sistema de secuencia de campos
field strength intensidad de campo
field-strength calculation cálculo de intensidad de campo
field-strength contour contorno de intensidad de campo
field-strength distribution distribución de intensidad de campo
field-strength measurement medida de intensidad de campo
field-strength meter medidor de intensidad de campo
field-strength pattern patrón de intensidad de campo
field-strength recorder registrador de intensidad de campo
field-strength recording registro de intensidad de campo
field string secuencia de campo
field suppressor supresor de campo
field sweep barrido de campo
field switch conmutador de campo
field test prueba práctica
field transfer transferencia de campo
field tube tubo de campo
field uniformity uniformidad de campo
field variation variación de campo
field voltage tensión de campo
field winding arrollamiento de campo
fieldistor fildistor
fifth generation computer computadora de quinta generación
fifth generation language lenguaje de quinta generación
figurative constant constante figurativa
figure-of-eight reception recepción con diagrama en ocho
figure of merit factor de mérito
filament filamento
filament activity test prueba de actividad de filamento
filament battery batería de filamento
filament choke choque de filamento
filament circuit circuito de filamento
filament coil bobina de filamento
filament cold resistance resistencia de filamento frío
filament current corriente de filamento
filament current consumption consumo de corriente de filamento
filament current regulator regulador de corriente de filamento
filament electrometer electrómetro de filamento
filament emission emisión de filamento
filament heating current corriente de calentamiento de filamento
filament hot resistance resistencia de filamento caliente
filament hum zumbido de filamento
filament lag retardo de filamento
filament lamp lámpara de filamento
filament lead conexión de filamento
filament modulation effect efecto de modulación de filamento
filament power potencia de filamento
filament power supply fuente de alimentación de filamento

filament reactivation reactivación de filamento
filament resistance resistencia de filamento
filament return retorno de filamento
filament rheostat reóstato de filamento
filament saturation saturación de filamento
filament switch conmutador de filamento
filament transformer transformador de filamento
filament-type bolometer bolómetro de tipo de filamento
filament-type tube tubo de tipo de filamento
filament voltage tensión de filamento
filament winding arrollamiento de filamento
filamentary filamentario
filamentary cathode cátodo filamentario
file archivo
file access mode modo de acceso de archivo
file addressing direccionamiento de archivo
file allocation table tabla de asignación de archivo
file attribute atributo de archivo
file batching agrupamiento de archivos
file beginning comienzo de archivo
file blocking agrupamiento de archivos
file card tarjeta de archivo
file cleanup limpieza de archivo
file closing cierre de archivo
file compression compresión de archivo
file concatenation concatenación de archivos
file condensation condensación de archivo
file conditioning acondicionamiento de archivo
file connector conector de archivo
file consolidation consolidación de archivos
file constant constante de archivo
file control control de archivo
file control block bloque de control de archivo
file control table tabla de control de archivo
file conversion conversión de archivo
file conversion program programa de conversión de archivo
file deletion borrado de archivo
file description descripción de archivo
file description entry entrada de descripción de archivo
file description specifications especificaciones de descripción de archivo
file design diseño de archivo
file directory directorio de archivo
file error error de archivo
file expression expresión de archivo
file feed alimentador de archivos
file format formato de archivo
file fragmentation fragmentación de archivo
file gap espacio entre archivos
file handler manipulador de archivos
file-handling routine rutina de manipulación de archivo
file identification identificación de archivo
file identifier identificador de archivo
file label etiqueta de archivo
file layout arreglo de archivo
file link enlace de archivos
file maintenance mantenimiento de archivos
file maintenance protocol protocolo de mantenimiento de archivos
file management administración de archivos
file management program programa de administración de archivos
file management system sistema de administración de archivos
file manager administrador de archivos
file manipulator manipulador de archivos
file map mapa de archivo
file mark marca de archivo
file name nombre de archivo

file name extension extensión de nombre de archivo
file opening apertura de archivo
file-opening routine rutina de apertura de archivo
file organization organización de archivo
file-oriented programming programación orientada a archivos
file parameter parámetro de archivo
file position indicator indicador de posición de archivo
file print impresión de archivo
file privilege privilegio de archivo
file processing procesamiento de archivos
file protection protección de archivos
file protection device dispositivo de protección de archivos
file protection ring anillo de protección de archivos
file reconstitution reconstitución de archivo
file reconstruction reconstrucción de archivo
file reconstruction procedure procedimiento de reconstrucción de archivo
file recovery recuperación de archivo
file reference referencia de archivo
file scan exploración de archivo
file search búsqueda de archivo
file search device dispositivo de búsqueda de archivos
file section sección de archivo
file security seguridad de archivos
file separator separador de archivo
file separator character carácter de separación de archivos
file sequence number número de secuencia de archivo
file server servidor de archivo
file set conjunto de archivos
file sharing compartimiento de archivos
file size tamaño de archivo
file storage almacenamiento de archivo
file structure estructura de archivo
file table tabla de archivos
file tape cinta de archivos
file transfer transferencia de archivo
file transfer function función de transferencia de archivo
file transfer protocol protocolo de transferencia de archivo
file type tipo de archivo
file unit unidad de archivo
file updating actualización de archivo
file variable variable de archivo
file version versión de archivo
fill character carácter de relleno
fill-in station estación de relleno
filled band banda llena
filled tape cinta llena
filler relleno
filler bit bit de relleno
filler character carácter de relleno
filler metal metal de relleno
filling relleno
filling character carácter de relleno
film película
film badge dosímetro de película
film bolometer bolómetro de película
film capacitor capacitor de película
film chain cadena de películas
film coefficient coeficiente de película
film deposition deposición de película
film dosimeter dosímetro de película
film integrated circuit circuito integrado de películas
film reader lector de película
film recorder registrador en película
film recording registro en película

film reproducer reproductor de película
film resistor resistor de película
film scanning exploración de película
film speed velocidad de película
filter filtro
filter attenuation atenuación por filtro
filter attenuation band banda de atenuación por filtro
filter capacitor capacitor de filtro
filter center centro de filtro
filter choke choque de filtro
filter circuit circuito de filtro
filter coil bobina de filtro
filter condenser condensador de filtro
filter cutoff corte de filtro
filter discrimination discriminación de filtro
filter echo eco de filtro
filter effectiveness efectividad de filtro
filter factor factor de filtro
filter impedance compensator compensador de impedancia de filtro
filter inductor inductor de filtro
filter network red de filtros
filter panel panel de filtro
filter passband banda de paso de filtro
filter section sección de filtro
filter selectivity selectividad de filtro
filter slot ranura de filtro
filter stop band banda de supresión por filtro
filter transmission band banda de transmisión de filtro
filtered filtrado
filtered output salida filtrada
filtering filtrado
filtering action acción filtrante
filtering circuit circuito filtrante
filtering network red filtrante
filtration filtración
filtron filtrón
fin aleta
fin antenna antena en aleta
final actuation time tiempo de actuación final
final adjustment ajuste final
final amplifier amplificador final
final amplifier stage etapa amplificadora final
final amplifier tube tubo amplificador final
final anode ánodo final
final condition condición final
final contact pressure presión de contacto final
final control element elemento de control final
final group selector selector de grupo final
final power amplifier amplificador de potencia final
final selector selector final
final stage etapa final
final state estado final
final tube tubo final
final value valor final
final voltage tensión final
final wrap envoltura final
find and replace encontrar y reemplazar
finder buscador
finder switch conmutador buscador
fine adjustment ajuste fino
fine-adjustment screw tornillo de ajuste fino
fine-chrominance primary primario de crominancia fina
fine control control fino
fine-control element elemento de control fino
fine focusing enfoque fino
fine frequency control control de frecuencia fino
fine groove surco fino
fine-groove record disco de surco fino
fine index índice fino

fine-mesh grid rejilla de malla fina
fine structure estructura fina
fine-structure constant constante de estructura fina
fine tuning sintonización fina
fine-tuning control control de sintonización fina
fine-tuning vernier nonio de sintonización fina
finish lead hilo externo
finishing acabado
finishing rate régimen de terminación de carga
finite finito
finite clipping limitación finita
finite increment incremento finito
finite series serie finita
finite transmission line línea de transmisión finita
finsen finsen
fire alarm alarma de fuego
fire-alarm signal señal de alarma de fuego
fire-alarm system sistema de alarma de fuego
fire detector detector de fuego
fire-extinguisher system sistema extintor de fuego
fire protection protección contra fuego
fire-resistance rating clasificación de resistencia a
  fuego
fire-resistant resistente al fuego
fireproof a prueba de fuego, incombustible
fireproof material material a prueba de fuego
firing disparo, cebado
firing angle ángulo de disparo
firing circuit circuito de disparo
firing point punto de disparo
firing potential potencial de cebado
firing voltage tensión de cebado
firm power potencia disponible
firmware programas inalterables
first anode primer ánodo
first audio stage primera etapa de audio
first card primera tarjeta
first detector primer detector
first digit primer dígito
first dynode primer dinodo
first Fresnel region primera región de Fresnel
first Fresnel zone primera zona de Fresnel
first generation computer computadora de primera
  generación
first generation language lenguaje de primera
  generación
first harmonic primera armónica
first-in-first-out primero en entrar-primero en salir
first law of thermodynamics primera ley de
  termodinámica
first-level address dirección de primer nivel
first-level message mensaje de primer nivel
first-level storage almacenamiento de primer nivel
first line find búsqueda de primera línea
first local oscillator primer oscilador local
first-order effect efecto de primer orden
first-order multiplexer multiplexor de primer orden
first reading primera lectura
first reading station estación de primera lectura
first selector primer selector
first stage primera etapa
first Townsend discharge primera descarga de
  Townsend
first transaction primera transacción
first transition primera transición
first transition duration duración de primera
  transición
fishbone antenna antena en espina de pescado
fishing wire culebra
fishpaper papel de pescado
fitting accesorio
five-channel code código de cinco canales
five-electrode tube tubo de cinco electrodos

five-element code código de cinco elementos
five-element tube tubo de cinco elementos
five-layer device dispositivo de cinco capas
five-level code código de cinco niveles
five-unit code código de cinco unidades
five wire system sistema de cinco hilos
fixed antenna antena fija
fixed area área fija
fixed armature inducido fijo
fixed attenuator atenuador fijo
fixed audio tone tono de audio fijo
fixed beacon radiofaro fijo
fixed bias polarización fija
fixed binary data datos binarios fijos
fixed block bloque fijo
fixed block file archivo de bloque fijo
fixed block format formato de bloque fijo
fixed block length longitud de bloque fijo
fixed brush escobilla fija
fixed capacitor capacitor fijo
fixed channel canal fijo
fixed circuit circuito fijo
fixed component componente fijo
fixed composition resistor resistor de composición
  fija
fixed condenser condensador fijo
fixed connector conector fijo
fixed contact contacto fijo
fixed control panel panel de control fijo
fixed crystal cristal fijo
fixed-crystal detector detector de cristal fijo
fixed cycle ciclo fijo
fixed-cycle operation operación de ciclo fijo
fixed data datos fijos
fixed decimal decimal fijo
fixed-decimal arithmetic aritmética de decimal fijo
fixed-decimal data datos de decimal fijo
fixed-diode detector detector de diodo fijo
fixed direction-finder radiogoniómetro fijo
fixed disk disco fijo
fixed disk drive unidad de disco fijo
fixed disk storage almacenamiento de disco fijo
fixed drive unidad fija
fixed echo eco fijo
fixed equipment equipo fijo
fixed error error fijo
fixed feed alimentación fija
fixed field campo fijo
fixed-field addressing direccionamiento de campo
  fijo
fixed-field coding codificación de campo fijo
fixed file archivo fijo
fixed file attribute atributo de archivo fijo
fixed focus foco fijo
fixed-form coding codificación de forma fija
fixed format formato fijo
fixed-format message mensaje de formato fijo
fixed frequency frecuencia fija
fixed-frequency amplifier amplificador de
  frecuencia fija
fixed-frequency band filter filtro de banda de
  frecuencia fija
fixed-frequency oscillator oscilador de frecuencia
  fija
fixed-frequency receiver receptor de frecuencia fija
fixed-frequency transmitter transmisor de
  frecuencia fija
fixed function función fija
fixed-function generator generador de función fija
fixed-function terminal terminal de función fija
fixed-function work station estación de trabajo de
  función fija
fixed inductor inductor fijo

fixed instruction instrucción fija
fixed-instruction computer computadora de instrucciones fijas
fixed length longitud fija
fixed-length file record registro de archivo de longitud fija
fixed-length operation operación de longitud fija
fixed-length record registro de longitud fija
fixed-length record system sistema de registro de longitud fija
fixed-length string secuencia de longitud fija
fixed-length word palabra de longitud fija
fixed light luz fija
fixed logic lógica fija
fixed memory memoria fija
fixed multiplier multiplicador fijo
fixed oscillator oscilador fijo
fixed parabolic equalizer ecualizador parabólico fijo
fixed parameter parámetro fijo
fixed-parameter system sistema de parámetros fijos
fixed partition división fija
fixed pitch tono fijo
fixed-pitch instrument instrumento de tono fijo
fixed point coma fija, punto fijo
fixed-point arithmetic aritmética de coma fija, aritmética de punto fijo
fixed-point binary data datos binarios de coma fija, datos binarios de punto fijo
fixed-point calculation cálculo de coma fija, cálculo de punto fijo
fixed-point constant constante de coma fija, constante de punto fijo
fixed-point data datos de coma fija, datos de punto fijo
fixed-point format formato de coma fija, formato de punto fijo
fixed-point notation notación de coma fija, notación de punto fijo
fixed-point number número de coma fija, número de punto fijo
fixed-point operation operación de coma fija, operación de punto fijo
fixed-point representation representación de coma fija, representación de punto fijo
fixed-point representation system sistema de representación de coma fija, sistema de representación de punto fijo
fixed-point system sistema de coma fija, sistema de punto fijo
fixed position posición fija
fixed program programa fijo
fixed-program computer computadora de programa fijo
fixed-radix notation notación de base fija
fixed-radix numbering system sistema de numeración de base fija
fixed-radix numeration numeración de base fija
fixed ratio razón fija
fixed ratio code código de razón fija
fixed routine rutina fija
fixed sequential format formato secuencial fijo
fixed service servicio fijo
fixed-service system sistema de servicio fijo
fixed signal señal fija
fixed-size record registro de tamaño fijo
fixed spacing separación fija
fixed station estación fija
fixed-step potentiometer potenciómetro de pasos fijos
fixed storage almacenamiento fijo
fixed-storage area área de almacenamiento fija
fixed transmitter transmisor fijo

fixed variable variable fija
fixed-voltage generator generador de tensión fija
fixed-voltage winding arrollamiento de tensión fija
fixed word length longitud de palabra fija
fixture aparato
fixture wire hilo de aparato
flag bit bit indicador
flag byte byte indicador
flame alarm alarma de llama
flame arc lamp lámpara de arco de llama
flame attenuation atenuación por llama
flame control control de llama
flame cutting oxicorte
flame detector detector de llama
flame-failure control control de combustible por falla de llama
flame lamp lámpara de llama
flame microphone micrófono de llama
flame photometer fotómetro de llama
flame-resistant ignífugo
flame-resistant cable cable ignífugo
flame safety electrode electrodo de seguridad de llama
flame spectrometer espectrómetro de llama
flame spectrophotometer espectrofotómetro de llama
flameproof antideflagrante
flameproof apparatus aparato antideflagrante
flameproof machine máquina antideflagrante
flameproof terminal box caja de terminales antideflagrante
flameproof wire hilo antideflagrante
flammability inflamabilidad
flammable inflamable
flange brida
flange connector conector de bridas
flange coupling acoplamiento por bridas
flange joint junta de bridas
flange ring aro de brida
flanged valve válvula de bridas
flanking flanqueo
flanking effect efecto de flanqueo
flap attenuator atenuador de aleta
flare bocina, destello
flare angle cambio de diámetro de guía de ondas
flare factor factor de abocinado
flare spot mancha superluminosa
flaring abocinado
flaring constant constante de abocinado
flaring test prueba de abocinado
flash destello, impulso
flash arc arco de destello
flash delay retardo de destello
flash-delay unit unidad de retardo de destello
flash generator generador de destellos
flash plating galvanoplastia rápida
flash point punto de inflamabilidad
flash radiography radiografía instantánea
flash test prueba instantánea
flash tube tubo de destellos
flashback voltage tensión inversa de ionización
flashbulb lámpara de destello
flasher destellador
flasher light-emitting diode diodo emisor de luz destellador
flasher relay relé destellador
flashing destello
flashing apparatus aparato de destellos
flashing beacon radiofaro de destellos
flashing lamp lámpara de destellos
flashing light luz de destellos
flashing rate frecuencia de destellos
flashing signal señal de destellos

flashing test prueba de destellos
flashing voltage tensión de destello
flashlamp lámpara de destello
flashlight linterna
flashover contorneo, arqueo
flashover voltage tensión de contorneamiento, tensión disruptiva
flashtube tubo de destellos
flat baffle bafle plano
flat cable cable plano
flat coil bobina plana
flat electrode electrodo plano
flat fading desvanecimiento plano
flat field campo plano
flat file archivo plano
flat flexible cable cable flexible plano
flat frequency characteristic característica de frecuencia plana
flat frequency response respuesta de frecuencia plana
flat gain ganancia plana
flat-gain amplifier amplificador de ganancia plana
flat line línea no resonante
flat loss pérdida plana
flat loudspeaker altavoz plano
flat pack conjunto plano
flat-panel display visualizador de panel plano
flat plate placa plana
flat-plate antenna antena de placa plana
flat-plate collector colector de placa plana
flat response respuesta plana
flat-response characteristic característica de respuesta plana
flat-response curve curva de respuesta plana
flat screen pantalla plana
flat sheet hoja plana
flat top tope plano, parte plana
flat-top antenna antena de tope plano
flat-top response respuesta plana
flat-topping recortamiento
flat tube tubo plano
flat tuning sintonización plana
flat-type relay relé de tipo plano
flat-type resistor resistor de tipo plano
flat wire hilo plano
flat zone zona plana
flattened aplanado
flattened curve curva aplanada
flattening aplanamiento
flaw imperfección
flaw detector detector de imperfecciones
Fleming's generator rule regla de generadores de Fleming
Fleming's left-hand rule regla de la mano izquierda de Fleming
Fleming's right-hand rule regla de la mano derecha de Fleming
Fleming's rule regla de Fleming
Fleming-Kennelly law ley de Fleming-Kennelly
Fleming valve válvula de Fleming
Fletcher-Munson curves curvas Fletcher-Munson
Fletcher-Munson effect efecto de Fletcher-Munson
Flewelling circuit circuito de Flewelling
flex life longevidad ante flexiones
flexibility flexibilidad
flexible flexible
flexible cable cable flexible
flexible-cable connector conector de cable flexible
flexible coaxial cable cable coaxial flexible
flexible conduit conducto flexible
flexible connection conexión flexible
flexible connector conector flexible
flexible contact contacto flexible

flexible cord cordón flexible
flexible coupler acoplador flexible
flexible coupling acoplamiento flexible
flexible disk disco flexible
flexible disk drive unidad de disco flexible
flexible equipment equipo flexible
flexible flat cable cable plano flexible
flexible interruption system sistema de interrupción flexible
flexible lead hilo flexible
flexible metal conduit conducto de metal flexible
flexible mounting montura flexible
flexible pipe tubo flexible
flexible resistor resistor flexible
flexible rubber conduit conducto de goma flexible
flexible shaft eje flexible
flexible tube tubo flexible
flexible tubing tubería flexible
flexible waveguide guíaondas flexible
flexion flexión
flexion point emission current corriente de emisión de punto de flexión
flexode flexodo
flexural noise ruido por flexión
flicker parpadeo
flicker effect efecto de parpadeo
flicker frequency frecuencia de parpadeo
flicker fusion frequency frecuencia de fusión de parpadeo
flicker noise ruido de parpadeo
flicker photometer fotómetro de parpadeo
flicker threshold umbral de parpadeo
flickering parpadeo
flight control control de vuelo
flight indicator indicador de vuelo
flight information información de vuelo
flight instrument instrumento de vuelo
flight path trayectoria de vuelo
flight-path computer computadora de trayectoria de vuelo
flight-path deviation desviación de trayectoria de vuelo
flight-path deviation indicator indicador de desviación de trayectoria de vuelo
flight-path deviation meter medidor de desviación de trayectoria de vuelo
flight recorder registrador de vuelo
flight simulator simulador de vuelo
flight test prueba de vuelo
Flinders bar barrita de Flinders
flint glass cristal de roca
flip chip chip reversible
flip coil bobina exploradora
flip-flop basculador
flip-flop circuit circuito basculador
flip-flop device dispositivo basculador
flip-flop relay relé biestable
float switch conmutador de flotador
floating action acción flotante
floating address dirección flotante
floating balance equilibrio flotante
floating battery batería flotante
floating beacon faro flotante
floating carrier portadora flotante
floating-carrier modulation modulación de portadora flotante
floating-carrier system sistema de portadora flotante
floating character carácter flotante
floating charge carga flotante
floating circuit circuito flotante
floating control control flotante
floating controller controlador flotante
floating decimal decimal flotante

floating-decimal arithmetic aritmética de decimal flotante
floating gate puerta flotante
floating grid rejilla flotante
floating head cabeza flotante
floating input entrada flotante
floating-input measurement medida de entrada flotante
floating instrument instrumento flotante
floating junction unión flotante
floating-junction transistor transistor de unión flotante
floating neutral neutro flotante
floating output salida flotante
floating paraphase amplifier amplificador parafásico flotante
floating paraphase inverter inversor parafásico flotante
floating point coma flotante, punto flotante
floating-point arithmetic aritmética de coma flotante, aritmética de punto flotante
floating-point calculation cálculo de coma flotante, cálculo de punto flotante
floating-point coefficient coeficiente de coma flotante, coeficiente de punto flotante
floating-point computer computadora de coma flotante, computadora de punto flotante
floating-point constant constante de coma flotante, constante de punto flotante
floating-point data datos de coma flotante, datos de punto flotante
floating-point format formato de coma flotante, formato de punto flotante
floating-point item artículo de coma flotante, artículo de punto flotante
floating-point notation notación de coma flotante, notación de punto flotante
floating-point number número de coma flotante, número de punto flotante
floating-point operation operación de coma flotante, operación de punto flotante
floating-point processor procesador de coma flotante, procesador de punto flotante
floating-point radix base de coma flotante, base de punto flotante
floating-point register registro de coma flotante, registro de punto flotante
floating-point representation representación de coma flotante, representación de punto flotante
floating-point representation system sistema de representación de coma flotante, sistema de representación de punto flotante
floating-point routine rutina de coma flotante, rutina de punto flotante
floating-point system sistema de coma flotante, sistema de punto flotante
floating potential potencial flotante
floating probe sonda flotante
floating resistor resistor flotante
floating source fuente flotante
floating station estación flotante
floating string secuencia flotante
floating symbolic address dirección simbólica flotante
floating zero cero flotante
floating zone zona flotante
floating-zone technique técnica de zona flotante
flood gun cañón de inundación
flooded cable cable inundado
floppy disk disco flexible
floppy-disk drive unidad de disco flexible
floppy-disk sector sector de disco flexible
floppy-disk system sistema de discos flexibles

floppy-disk track pista de disco flexible
flotation fluid fluido de flotación
flow flujo
flow area área de flujo
flow chart organigrama, diagrama de flujo
flow control control de flujo
flow diagram organigrama, diagrama de flujo
flow direction dirección de flujo
flow line línea de flujo
flow of control flujo de control
flow path trayectoria de flujo
flow rate velocidad de flujo
flow relay relé de flujo
flow soldering soldadura por ola
flow switch conmutador de flujo
flowchart organigrama
flowmeter flujómetro
flowmetering flujometría
fluctuating fluctuante
fluctuating current corriente fluctuante
fluctuating power potencia fluctuante
fluctuating voltage tensión fluctuante
fluctuation fluctuación
fluctuation loss pérdida por fluctuación
fluctuation noise ruido de fluctuación
fluctuation time tiempo de fluctuación
fluctuation voltage tensión de fluctuación
fluid fluido
fluid damping amortiguamiento por fluido
fluid-flow alarm alarma de flujo de fluido
fluid-flow control control de flujo de fluido
fluid-flow gage medidor de flujo de fluido
fluid-flow indicator indicador de flujo de fluido
fluid-flow meter medidor de flujo de fluido
fluid-flow switch conmutador de flujo de fluido
fluid-level control control de nivel de fluido
fluid-level gage medidor de nivel de fluido
fluid-level indicator indicador de nivel de fluido
fluid-level meter medidor de nivel de fluido
fluid-level switch conmutador de nivel de fluido
fluid-pressure control control de presión de fluido
fluid-pressure gage medidor de presión de fluido
fluid-pressure indicator indicador de presión de fluido
fluid-pressure meter medidor de presión de fluido
fluid-pressure switch conmutador de presión de fluido
fluidics fluídica
fluorescence fluorescencia
fluorescence spectroscopy espectroscopia de fluorescencia
fluorescent fluorescente
fluorescent lamp lámpara fluorescente
fluorescent material material fluorescente
fluorescent radiation radiación fluorescente
fluorescent screen pantalla fluorescente
fluorescent tube tubo fluorescente
fluorescent X-rays rayos X fluorescentes
fluorimeter fluorímetro
fluorimetry fluorimetría
fluorine fluor
fluorocarbon resin resina fluorocarbúrica
fluorographic recording registro fluorográfico
fluorometer fluorómetro
fluorometry fluorometría
fluorophotometer fluorofotómetro
fluoroscope fluoroscopio
fluoroscopic fluoroscópico
fluoroscopic image intensifier intensificador de imagen fluoroscópica
fluoroscopic screen pantalla fluoroscópica
fluoroscopy fluoroscopia
flush ras

flush outlet tomacorriente embutido, toma de corriente empotrada
flush receptacle receptáculo embutido, receptáculo empotrado
flutter titilación, centelleo
flutter echo eco múltiple
flutter rate velocidad de titilación
flux flujo
flux density densidad de flujo
flux distribution distribución de flujo
flux gate puerta de flujo
flux gradient gradiente de flujo
flux guide guía de flujo
flux leakage fuga de flujo
flux lines líneas de flujo
flux linkage acoplamiento inductivo, encadenamiento de flujo
flux pattern patrón de flujo
flux refraction refracción de flujo
flux tube tubo de flujo
fluxgraph flujógrafo
fluxmeter flujómetro
flyback retorno
flyback checker comprobador de retorno
flyback line línea de retorno
flyback period periodo de retorno
flyback power supply fuente de alimentación de retorno
flyback rectifier rectificador de retorno
flyback time tiempo de retorno
flyback transformer transformador de retorno
flyback velocity velocidad de retorno
flyback wave onda de retorno
flying head cabeza móvil
flying spot punto móvil
flying-spot microscope microscopio de punto móvil
flying-spot scanner explorador de punto móvil
flying-spot scanning exploración por punto móvil
flying-spot tube tubo de punto móvil
flywheel circuit circuito de volante
flywheel effect efecto de volante
flywheel rotor rotor de volante
flywheel-stored energy energía almacenada en volante
flywheel synchronization sincronización de volante
flywheel tuning sintonización de volante
focal focal
focal aberration aberración focal
focal distance distancia focal
focal length distancia focal
focal plane plano focal
focal point punto focal
focal spot punto focal
focal time tiempo focal
focometer focómetro
focus foco, enfoque
focus anode ánodo de enfoque
focus coil bobina de enfoque
focus control control de enfoque
focus current regulator regulador de corriente de enfoque
focus-defocus mode modo de enfoque-desenfoque
focus grid rejilla de enfoque
focus inductor concentrador
focus rectifier rectificador de enfoque
focus screen pantalla de enfoque
focusing enfoque
focusing anode ánodo de enfoque
focusing coil bobina de enfoque
focusing control control de enfoque
focusing cup cúpula de enfoque
focusing current corriente de enfoque
focusing device dispositivo de enfoque

focusing drift deslizamiento de enfoque
focusing electrode electrodo de enfoque
focusing field campo de enfoque
focusing grid rejilla de enfoque
focusing index índice de enfoque
focusing lens lente de enfoque
focusing magnet imán de enfoque
focusing magnetic field campo magnético de enfoque
focusing range margen de enfoque
focusing scale escala de enfoque
focusing screen pantalla de enfoque
focusing structure estructura de enfoque
focusing system sistema de enfoque
focusing tube tubo de enfoque
focusing voltage tensión de enfoque
foil hoja metálica, lámina metálica
foil capacitor capacitor de hojas metálicas
foil conductor conductor de hoja metálica
foil electrode electrodo de hoja metálica
foil electroscope electroscopio de hoja metálica
folded cavity cavidad doblada
folded dipole dipolo doblado
folded-dipole antenna antena de dipolo doblado
folded-dipole array red de dipolos doblados
folded-dipole director director de dipolos doblados
folded-dipole reflector reflector de dipolo doblado
folded doublet doblete doblado
folded filament filamento doblado
folded horn bocina doblada
folded-horn enclosure caja de bocina doblada
folded monopole monopolo doblado
folded-monopole antenna antena de monopolo doblado
folded pattern patrón doblado
folded vertical radiator radiador vertical doblado
foldover distorsión por imágenes superpuestas
follow current corriente subsiguiente
follow-on current corriente subsiguiente
follower seguidor
follower circuit circuito seguidor
following blacks borde negro tras imagen blanca
following whites borde blanco tras imagen negra
font tipo de letras
font cartridge cartucho de tipo de letras
font change cambio de tipo de letras
font characteristic característica de tipo de letras
font data set conjunto de datos de tipo de letras
font family familia de tipo de letras
font reticle retículo de tipo de letras
font section sección de tipo de letras
font set conjunto de tipo de letras
foolproof a prueba de impericia
foot-candle bujía-pie
foot-lambert lambert-pie
foot-pound libra-pie
foot-pound-second system sistema de pie-libra-segundo
foot switch conmutador de pie
footing base
forbidden band banda prohibida
forbidden character carácter prohibido
forbidden character code código de caracteres prohibidos
forbidden combination combinación prohibida
forbidden-combination check comprobación de combinación prohibida
forbidden digit dígito prohibido
forbidden-digit check comprobación de dígito prohibido
forbidden energy band banda de energía prohibida
forbidden reaction reacción prohibida
forbidden reflection reflexión prohibida
forbidden region región prohibida

forbidden zone zona prohibida
force fuerza
force constant constante de fuerza
force diagram diagrama de fuerza
force differential diferencia de fuerzas
force factor factor de fuerza
force field campo de fuerza
force-free field campo libre de fuerza
force pump bomba impelente
forced-air cooled enfriado por aire forzado
forced-air cooling enfriamiento por aire forzado
forced coding codificación forzada
forced control field campo de control forzado
forced drainage drenaje forzado
forced electric drainage drenaje eléctrico forzado
forced frequency frecuencia forzada
forced interruption interrupción forzada
forced oscillations oscilaciones forzadas
forced outage interrupción forzada
forced outage duration duración de interrupción forzada
forced outage hours horas de interrupción forzada
forced release liberación forzada
forced response respuesta forzada
forced sequence secuencia forzada
forced-ventilated machine máquina de ventilación forzada
foreground area área de prioridad
foreground job tarea de prioridad
foreground processing procesamiento de prioridad
foreground program programa de prioridad
foreground task tarea de prioridad
foreign body cuerpo extraño
fork amplifier amplificador de diapasón
fork-controlled controlado por diapasón
fork frequency frecuencia de diapasón
fork oscillator oscilador de diapasón
fork tone modulation modulación de tono por diapasón
forked circuit circuito bifurcado
forked working operación bifurcada
form distortion distorsión de forma
form factor factor de forma
form-feed character carácter de alimentación de formulario
form-feed key tecla de alimentación de formulario
form-feed stop key tecla de paro de alimentación de formulario
formal classification clasificación formal
formal language lenguaje formal
formal logic lógica formal
formal parameter parámetro formal
formal specification especificación formal
formal test prueba formal
formaldehyde formaldehído
formant formante
formant filter filtro formante
format formato
format character carácter de formato
format character set conjunto de caracteres de formato
format check comprobación de formato
format classification clasificación de formato
format control control de formato
format control word palabra de control de formato
format conversion conversión de formato
format-conversion language lenguaje de conversión de formato
format definition definición de formato
format description descripción de formato
format detail detalle de formato
format error error de formato
format file archivo de formato

format identification field campo de identificación de formato
format item artículo de formato
format line línea de formato
format mode modo de formato
format record registro de formato
format selection selección de formato
format selector selector de formato
format storage almacenamiento de formato
format test prueba de formato
format type tipo de formato
formation formación
formation voltage tensión de formación
formatless sin formato
formatted formateado
formatted diskette disquete formateado
formatted display visualización formateada
formatted dump vaciado formateado
formatted image imagen formateada
formatted information información formateada
formatted input-output entrada-salida formateada
formatted program programa formateado
formatted record registro formateado
forming formación
forsterite forsterita
fortuitous conductor conductor fortuito
fortuitous distortion distorsión fortuita
forward-acting regulator regulador de accionamiento progresivo
forward anode voltage tensión de ánodo directa
forward-backward counter contador progresivo-regresivo
forward bias polarización directa
forward breakover transición conductiva directa
forward breakover voltage tensión de transición conductiva directa
forward characteristic característica directa
forward conductance conductancia directa
forward conduction conducción directa
forward current corriente directa
forward current density densidad de corriente directa
forward current-transfer ratio razón de transferencia de corriente directa
forward direction dirección conductora
forward drop caída directa
forward gain ganancia directa
forward loss pérdida directa
forward path trayectoria de progresión
forward power loss pérdida de potencia directa
forward recovery recuperación directa
forward recovery time tiempo de recuperación directa
forward resistance resistencia directa
forward scan exploración directa
forward scatter dispersión directa
forward-scatter equipment equipo de dispersión directa
forward-scatter installation instalación de dispersión directa
forward-scatter propagation propagación por dispersión directa
forward scattering dispersión directa
forward signal señal directa
forward signaling señalización directa
forward sweep barrido directo
forward transadmittance transadmitancia directa
forward transconductance transconductancia directa
forward transfer transferencia directa
forward transfer current ratio razón de corriente de transferencia directa
forward transfer function función de transferencia directa

forward transimpedance transimpedancia directa
forward voltage tensión directa
forward voltage drop caída de tensión directa
forward wave onda directa
fosdic fosdic
Foster-Seeley detector detector de Foster-Seeley
Foster-Seeley discriminator discriminador de Foster-Seeley
Foster theorem teorema de Foster
Foucault currents corrientes de Foucault
Foucault effect efecto de Foucault
Foucault interrupter interruptor de Foucault
foul electrolyte electrólito impuro
four-address code código de cuatro direcciones
four-address computer computadora de cuatro direcciones
four-address instruction instrucción de cuatro direcciones
four-channel sound system sistema de sonido de cuatro canales
four-course radio range radiofaro direccional de cuatro vías
four-electrode tube tubo de cuatro electrodos
four-electrode valve válvula de cuatro electrodos
four-element bridge puente de cuatro elementos
four-element tube tubo de cuatro elementos
four-frequency diplex díplex de cuatro frecuencias
four-layer diode diodo de cuatro capas
four-layer transistor transistor de cuatro capas
four-level laser láser de cuatro niveles
four-phase system sistema de cuatro fases
four-plus-one address dirección de cuatro más uno
four-pole tetrapolar
four-pole switch conmutador tetrapolar
four-terminal device dispositivo de cuatro terminales
four-terminal network red de cuatro terminales
four-track recording registro en cuatro pistas
four-track tape cinta de cuatro pistas
four-wire amplifier amplificador de cuatro hilos
four-wire channel canal de cuatro hilos
four-wire circuit circuito de cuatro hilos
four-wire repeater repetidor de cuatro hilos
four-wire system sistema de cuatro hilos
Fourier analysis análisis de Fourier
Fourier series serie de Fourier
Fourier spectrum espectro de Fourier
fourth generation computer computadora de cuarta generación
fourth generation language lenguaje de cuarta generación
fractional binary binario fraccional
fractional digit dígito fraccional
fractional exponent exponente fraccional
fractional fixed point coma fija fraccional, punto fijo fraccional
fractional gain ganancia fraccional
fractional horsepower caballo de fuerza fraccional
fractional part parte fraccional
fragment fragmento
fragmentation fragmentación
fragmentation index índice de fragmentación
Frahm frequency meter frecuencímetro de Frahm
frame cuadro, encuadre
frame alignment alineación de cuadro
frame-alignment signal señal de alineación de cuadro
frame antenna antena de cuadro
frame direction-finder radiogoniómetro de cuadro
frame distortion distorsión de cuadro
frame frequency frecuencia de cuadro
frame grid rejilla de cuadro
frame-grid tube tubo de rejilla de cuadro
frame of reference sistema de referencia

frame period periodo de cuadro
frame rate frecuencia de cuadro
frame repetition rate frecuencia de repetición de cuadro
frame synchronization sincronización de cuadro
frame-synchronizing signal señal sincronizadora de cuadro
framework armazón, marco
framing encuadre
framing bit bit de encuadre
framing coil bobina de encuadre
framing control control de encuadre
framing magnet imán de encuadre
framing pulse impulso de encuadre
framing signal señal de encuadre
francium francio
franklin franklin
Franklin antenna antena de Franklin
Franklin oscillator oscilador de Franklin
Fraunhofer diffraction difracción de Fraunhofer
Fraunhofer region región de Fraunhofer
frayed cable cable deshilachado
free capacitance capacitancia libre
free carrier portador libre
free charge carga libre
free cyanide cianuro libre
free electron electrón libre
free field campo libre
free-field acoustic response respuesta acústica de campo libre
free-field current response respuesta de corriente de campo libre
free-field current sensitivity sensibilidad de corriente de campo libre
free-field emission emisión de campo libre
free-field microphone micrófono de campo libre
free-field response respuesta de campo libre
free-field voltage response respuesta de tensión de campo libre
free-field voltage sensitivity sensibilidad de tensión de campo libre
free-floating grid rejilla flotante libre
free format formato libre
free grid rejilla libre
free gyro giroscopio libre
free impedance impedancia libre
free line línea libre
free magnetic pole polo magnético libre
free magnetism magnetismo libre
free motion moción libre
free network red libre
free oscillations oscilaciones libres
free path trayectoria libre
free progressive wave onda progresiva libre
free reel carrete libre
free routing encaminamiento libre
free-running frequency frecuencia propia
free-running multivibrator multivibrador libre
free space espacio libre
free-space antenna antena en espacio libre
free-space attenuation atenuación en espacio libre
free-space characteristic impedance impedancia característica en espacio libre
free-space diagram diagrama de espacio libre
free-space field intensity intensidad de campo en espacio libre
free-space loss pérdida en espacio libre
free-space path trayectoria en espacio libre
free-space pattern patrón de espacio libre
free-space permeability permeabilidad en espacio libre
free-space power gain ganancia de potencia en espacio libre

free-space propagation propagación en espacio libre
free-space radiation pattern patrón de radiación en espacio libre
free-space transmission transmisión en espacio libre
free-space wavelength longitud de onda en espacio libre
free speed velocidad libre
free storage almacenamiento libre
free vibration vibración libre
freezing level nivel de congelación
freezing point punto de congelación
Fremodyne detector detector de Fremodyne
freon freón
frequency frecuencia
frequency accuracy exactitud de frecuencia
frequency adapter adaptador de frecuencia
frequency adjustment panel panel de ajuste de frecuencia
frequency agility agilidad de frecuencia
frequency allocation asignación de frecuencias
frequency allotment asignación de frecuencias
frequency allotted frecuencia asignada
frequency amplifier amplificador de frecuencia
frequency analysis análisis de frecuencias
frequency assignment asignación de frecuencias
frequency band banda de frecuencias
frequency-band number número de banda de frecuencias
frequency bandwidth ancho de banda de frecuencias
frequency bias polarización de frecuencia
frequency bridge puente de frecuencias
frequency calibration calibración de frecuencia
frequency calibrator calibrador de frecuencias
frequency change cambio de frecuencia
frequency-change signaling señalización por cambios de frecuencia
frequency changer cambiador de frecuencia
frequency-changer crystal cristal cambiador de frecuencia
frequency-changer heptode heptodo cambiador de frecuencia
frequency-changer modulator modulador cambiador de frecuencia
frequency-changer oscillator oscilador cambiador de frecuencia
frequency-changer stage etapa cambiadora de frecuencia
frequency-changer triode triodo cambiador de frecuencia
frequency-changer unit unidad cambiadora de frecuencia
frequency channel canal de frecuencias
frequency characteristic característica de frecuencia
frequency check comprobación de frecuencia
frequency comparator comparador de frecuencias
frequency comparison comparación de frecuencias
frequency-compensated attenuator atenuador de compensación de frecuencia
frequency-compensating circuit circuito compensador de frecuencia
frequency compensation compensación de frecuencia
frequency component componente de frecuencia
frequency condition condición de frecuencia
frequency constant constante de frecuencia
frequency contour contorno de frecuencia
frequency control control de frecuencia
frequency conversion conversión de frecuencia
frequency converter convertidor de frecuencia
frequency-converting network red convertidora de frecuencia
frequency coordination coordinación de frecuencias
frequency-correcting circuit circuito corrector de

frecuencia
frequency correction corrección de frecuencia
frequency counter contador de frecuencia
frequency coverage cobertura de frecuencias
frequency cutoff frecuencia de corte
frequency demodulation desmodulación de frecuencia
frequency demultiplication desmultiplicación de frecuencia
frequency departure variación de frecuencia
frequency dependence dependencia de frecuencia
frequency-dependent dependiente de frecuencia
frequency detector detector de frecuencia
frequency-determining determinante de frecuencia
frequency-determining component componente determinante de frecuencia
frequency-determining element elemento determinante de frecuencia
frequency-determining network red determinante de frecuencia
frequency deviation desviación de frecuencia
frequency-deviation control control de desviación de frecuencia
frequency-deviation meter medidor de desviación de frecuencia
frequency dial cuadrante de frecuencias
frequency difference diferencia de frecuencias
frequency discrimination discriminación de frecuencia
frequency discriminator discriminador de frecuencia
frequency displacement desplazamiento de frecuencia
frequency distortion distorsión de frecuencia
frequency-distortion curve curva de distorsión de frecuencia
frequency distribution distribución de frecuencias
frequency-distribution curve curva de distribución de frecuencias
frequency divergence divergencia de frecuencia
frequency diversity diversidad de frecuencia
frequency divider divisor de frecuencia
frequency-dividing multivibrator multivibrador divisor de frecuencia
frequency-dividing network red divisora de frecuencia
frequency-division multiplex múltiplex por división de frecuencia
frequency-division multiplexing multiplexión por división de frecuencia
frequency-division multiplexing system sistema de multiplexión por división de frecuencia
frequency doubler doblador de frecuencia
frequency-doubler stage etapa dobladora de frecuencia
frequency doubling doblamiento de frecuencia
frequency drift deslizamiento de frecuencia
frequency error error de frecuencia
frequency excursion excursión de frecuencia
frequency filter filtro de frecuencias
frequency flexibility flexibilidad de frecuencia
frequency frogging cruzamiento de frecuencias
frequency gap separación entre frecuencias
frequency generator generador de frecuencias
frequency hysteresis histéresis de frecuencia
frequency-independent independiente de frecuencia
frequency-independent antenna antena independiente de frecuencia
frequency indicator indicador de frecuencia
frequency influence influencia por frecuencia
frequency interlace entrelazado de frecuencias
frequency interleaving entrelazado de frecuencias
frequency inverter inversor de frecuencia
frequency jitter fluctuación de frecuencia

frequency keying manipulación de frecuencia
frequency limit límite de frecuencia
frequency linearity linealidad de frecuencia
frequency loading carga de frecuencia
frequency measurement medida de frecuencia
frequency measurer medidor de frecuencia
frequency-measuring device dispositivo de medida de frecuencia
frequency meter frecuencímetro
frequency-modulated broadcasting radiodifusión por modulación de frecuencia
frequency-modulated cyclotron ciclotrón de modulación de frecuencia
frequency-modulated pickup fonocaptor de modulación de frecuencia
frequency-modulated radar radar de modulación de frecuencia
frequency-modulated transmitter transmisor de modulación de frecuencia
frequency-modulation modulación de frecuencia
frequency-modulation-amplitude-modulation multiplier multiplicador de modulación de frecuencia y modulación de amplitud
frequency-modulation-amplitude-modulation tuner sintonizador de modulación de frecuencia y modulación de amplitud
frequency-modulation broadcast band banda de radiodifusión por modulación de frecuencia
frequency-modulation broadcast channel canal de radiodifusión por modulación de frecuencia
frequency-modulation broadcast station estación de radiodifusión por modulación de frecuencia
frequency-modulation carrier portadora de modulación de frecuencia
frequency-modulation detector detector de modulación de frecuencia
frequency-modulation deviation desviación de modulación de frecuencia
frequency-modulation discriminator discriminador de modulación de frecuencia
frequency-modulation feedback retroalimentación de modulación de frecuencia
frequency-modulation-frequency-modulation modulación de frecuencia-modulación de frecuencia
frequency-modulation limiter limitador de modulación de frecuencia
frequency-modulation link enlace de modulación de frecuencia
frequency-modulation noise ruido de modulación de frecuencia
frequency-modulation-phase-modulation modulación de frecuencia-modulación de fase
frequency-modulation radar radar de modulación de frecuencia
frequency-modulation receiver receptor de modulación de frecuencia
frequency-modulation reception recepción de modulación de frecuencia
frequency-modulation recording registro de modulación de frecuencia, grabación por modulación de frecuencia
frequency-modulation repeater repetidor de modulación de frecuencia
frequency-modulation stereo estéreo por modulación de frecuencia
frequency-modulation transmitter transmisor de modulación de frecuencia
frequency-modulation tuner sintonizador de modulación de frecuencia
frequency modulator modulador de frecuencia
frequency monitor monitor de frecuencia
frequency monitoring monitoreo de frecuencia

frequency multiplex múltiplex por frecuencia
frequency multiplication multiplicación de frecuencia
frequency multiplication stage etapa de multiplicación de frecuencia
frequency multiplier multiplicador de frecuencia
frequency-multiplier klystron klistrón multiplicador de frecuencia
frequency-multiplier stage etapa multiplicadora de frecuencia
frequency-multiplying multiplicador de frecuencia
frequency-multiplying amplifier amplificador multiplicador de frecuencia
frequency-multiplying circuit circuito multiplicador de frecuencia
frequency-multiplying transformer transformador multiplicador de frecuencia
frequency of occurrence frecuencia de ocurrencia
frequency offset desplazamiento de frecuencia
frequency overlap solape de frecuencias
frequency pairing pareo de frecuencias
frequency protection protección de frecuencia
frequency pulling corrimiento adelantado de frecuencia
frequency pulsing pulsación de frecuencia
frequency pushing corrimiento atrasado de frecuencia
frequency quadrupler cuadruplicador de frecuencia
frequency quintupler quintuplicador de frecuencia
frequency range intervalo de frecuencias
frequency ratio razón de frecuencias
frequency-ratio counter contador de razón de frecuencias
frequency-ratio measurement medida de razón de frecuencias
frequency-ratio meter medidor de razón de frecuencias
frequency record registro de frecuencia
frequency regulation regulación de frecuencia
frequency regulator regulador de frecuencia
frequency rejection rechazo de frecuencia
frequency relay relé de frecuencia
frequency resolution resolución de frecuencia
frequency response respuesta de frecuencia
frequency-response analysis análisis de respuesta de frecuencia
frequency-response characteristic característica de respuesta de frecuencia
frequency-response curve curva de respuesta de frecuencia
frequency-response equalization ecualización de respuesta de frecuencia
frequency-response function función de respuesta de frecuencia
frequency-response measurement medida de respuesta de frecuencia
frequency-response recorder registrador de respuesta de frecuencia
frequency-response test prueba de respuesta de frecuencia
frequency run recorrido de frecuencias
frequency scanning exploración de frecuencia
frequency selection selección de frecuencia
frequency-selection sensor sensor de selección de frecuencia
frequency-selective amplifier amplificador selectivo a la frecuencia
frequency-selective fading desvanecimiento selectivo a la frecuencia
frequency-selective relay relé selectivo a la frecuencia
frequency-selective ringing llamada selectiva a la frecuencia

frequency-selective switching circuit circuito conmutador selectivo a la frecuencia
frequency-selective voltmeter vóltmetro selectivo a la frecuencia
frequency selectivity selectividad de frecuencia
frequency selector selector de frecuencias
frequency-sensitive sensible a la frecuencia
frequency-sensitive bridge puente sensible a la frecuencia
frequency-sensitive measurement medida sensible a la frecuencia
frequency-sensitive relay relé sensible a la frecuencia
frequency sensitivity sensibilidad de frecuencia
frequency separation separación entre frecuencias
frequency separator separador de frecuencias
frequency sharing compartimiento de frecuencias
frequency shift desplazamiento de frecuencia
frequency-shift amount cantidad de desplazamiento de frecuencia
frequency-shift keying manipulación por desplazamiento de frecuencia
frequency-shift modulation modulación por desplazamiento de frecuencia
frequency-shift Morse code código de Morse por desplazamiento de frecuencia
frequency-shift signaling señalización por desplazamiento de frecuencia
frequency-shift telegraphy telegrafía por desplazamiento de frecuencia
frequency-shift transmission transmisión por desplazamiento de frecuencia
frequency shifter desplazador de frecuencia
frequency sliding deslizamiento de frecuencia
frequency span intervalo de frecuencias
frequency spectrograph espectrógrafo de frecuencia
frequency spectrum espectro de frecuencias
frequency spread dispersión de frecuencias
frequency stability estabilidad de frecuencia
frequency stabilization estabilización de frecuencia
frequency-stabilizing voltage tensión estabilizante de frecuencia
frequency-stable oscillator oscilador de frecuencia estable
frequency staggering escalonamiento de frecuencias
frequency standard patrón de frecuencia
frequency subband subbanda de frecuencia
frequency sum suma de frecuencias
frequency sweep barrido de frecuencia
frequency-sweep band banda de barrido de frecuencia
frequency-sweep range intervalo de barrido de frecuencia
frequency swing oscilación de frecuencia
frequency synthesizer sintetizador de frecuencias
frequency table tabla de frecuencias
frequency threshold umbral de frecuencia
frequency-to-voltage converter convertidor de frecuencia a tensión
frequency tolerance tolerancia de frecuencia
frequency transfer device dispositivo de transferencia de frecuencia
frequency transfer unit unidad de transferencia de frecuencia
frequency transformation transformación de frecuencia
frequency transformer transformador de frecuencia
frequency translation traslación de frecuencia
frequency trebling triplicación de frecuencia
frequency tripler triplicador de frecuencia
frequency-type telemeter telemedidor de tipo de frecuencia
frequency variation variación de frecuencia

frequency-variation method método de variación de frecuencia
frequency vernier nonio de frecuencia
fresnel fresnel
Fresnel clearance espacio de Fresnel
Fresnel contour contorno de Fresnel
Fresnel diffraction difracción de Fresnel
Fresnel diffraction pattern patrón de difracción de Fresnel
Fresnel ellipse elipse de Fresnel
Fresnel lens lente de Fresnel
Fresnel lens antenna antena de lente de Fresnel
Fresnel number número de Fresnel
Fresnel pattern patrón de Fresnel
Fresnel reflection reflexión de Fresnel
Fresnel-reflection method método de reflexión de Fresnel
Fresnel region región de Fresnel
Fresnel zone zona de Fresnel
friction fricción, rozamiento
friction coefficient coeficiente de fricción
friction factor factor de fricción
friction feed alimentación por fricción
friction tape cinta de fricción
frictional electricity electricidad por fricción
frictional error error por fricción
frictional loss pérdida por fricción
fringe area área marginal
fringe capacitance capacitancia marginal
fringe effect efecto marginal
fringing efecto marginal
fritting fritado, desmenuzado
frogging cruzamiento
front contact contacto de trabajo
front electrode electrodo frontal
front end sección de entrada
front feed alimentación frontal
front loading carga frontal
front panel panel frontal
front-panel control control de panel frontal
front-panel switch conmutador de panel frontal
front porch umbral anterior, escalón frontal
front projection proyección frontal
front-surface mirror espejo de superficie reflectora frontal
front-to-back ratio eficacia direccional
front-to-rear ratio eficacia direccional
frost alarm alarma de congelación
frosted lamp lámpara mateada
Fuchs antenna antena de Fuchs
fuel cell pila de combustible
fuel-cell system sistema de pilas de combustible
fuel economy economía de combustible
fuel-flow alarm alarma de flujo de combustible
fuel-flow control control de flujo de combustible
fuel-flow gage medidor de flujo de combustible
fuel-flow indicator indicador de flujo de combustible
fuel-flow meter medidor de flujo de combustible
fuel-flow switch conmutador de flujo de combustible
fuel gage medidor de combustible
fuel-gage calibrator calibrador de medidor de combustible
fuel meter medidor de combustible
fuel-pressure indicator indicador de presión de combustible
fuel-pressure meter medidor de presión de combustible
fulguration fulguración
full adder sumador completo
full availability disponibilidad completa
full backup reserva completa
full binary tree árbol binario completo
full bridge puente completo

full carrier portadora completa
full deflection deflexión completa
full duplex dúplex completo
full-duplex system sistema dúplex completo
full echo suppressor supresor de eco completo
full excursion excursión completa
full-impulse wave onda de choque completa
full load plena carga
full-load current corriente de plena carga
full-load power potencia de plena carga
full-load voltage tensión de plena carga
full-load wattage wattaje de plena carga
full power plena potencia
full-power frequency response respuesta de
    frecuencia de plena potencia
full precision precisión completa
full radiator radiador completo
full-range speaker altavoz de gama completa
full-range transducer transductor de gama completa
full rating plena carga
full scale valor de límite de escala, valor de fondo de
    escala
full-scale accuracy exactitud de límite de escala
full-scale current corriente de límite de escala
full-scale deflection deflexión de límite de escala
full-scale error error de límite de escala
full-scale frequency frecuencia de límite de escala
full-scale reading lectura de límite de escala
full-scale sensitivity sensibilidad de límite de escala
full-scale value valor de límite de escala
full screen pantalla completa
full-screen mode modo de pantalla completa
full-screen panel panel de pantalla completa
full-screen processing procesamiento de pantalla
    completa
full speed velocidad completa
full track pista llena
full-track head cabeza de pista completa
full-track recording registro de pista completa
full-wave antenna antena de onda completa
full-wave detector detector de onda completa
full-wave doubler doblador de onda completa
full-wave loop antenna antena de cuadro de onda
    completa
full-wave rectification rectificación de onda
    completa
full-wave rectifier rectificador de onda completa
full-wave rectifier circuit circuito rectificador de
    onda completa
full-wave rectifier tube tubo rectificador de onda
    completa
full-wave vibrator vibrador de onda completa
full-wave voltage doubler doblador de tensión de
    onda completa
full width at half maximum anchura completa a
    media altura
fully automatic completamente automático
fully automatic circuit circuito completamente
    automático
fully automatic operation operación completamente
    automática
fully connected network red completamente
    conectada
fully energized completamente energizado
fully equalized completamente ecualizado
fully inverted file archivo completamente invertido
function función
function character carácter de función
function check comprobación de función
function code código de función
function digit dígito de función
function generator generador de funciones
function hole perforación de función

function identifier identificador de función
function key tecla de función
function library biblioteca de funciones
function mode modo de función
function procedure procedimiento de función
function program programa de función
function punch perforación de función
function relay relé de función
function signal señal de función
function subprogram subprograma de función
function switch conmutador de función
function table tabla de funciones
functional funcional
functional address dirección funcional
functional block bloque funcional
functional character carácter funcional
functional chart organigrama, diagrama funcional
functional check comprobación funcional
functional code código funcional
functional component componente funcional
functional configuration configuración funcional
functional decomposition descomposición funcional
functional dependency dependencia funcional
functional design diseño funcional
functional designation designación funcional
functional device dispositivo funcional
functional diagram diagrama funcional
functional electronic block bloque electrónico
    funcional
functional element elemento funcional
functional interface interfaz funcional
functional language lenguaje funcional
functional macro instruction macroinstrucción
    funcional
functional message mensaje funcional
functional module módulo funcional
functional part parte funcional
functional programming programación funcional
functional requirement requisito funcional
functional section sección funcional
functional specification especificación funcional
functional subsystem subsistema funcional
functional symbol símbolo funcional
functional test prueba funcional
functional unit unidad funcional
functor functor
fundamental fundamental
fundamental band banda fundamental
fundamental carrier portadora fundamental
fundamental category categoría fundamental
fundamental circuit circuito fundamental
fundamental component componente fundamental
fundamental efficiency eficiencia fundamental
fundamental frequency frecuencia fundamental
fundamental-frequency component componente de
    frecuencia fundamental
fundamental-frequency energy energía de
    frecuencia fundamental
fundamental group grupo fundamental
fundamental harmonic armónica fundamental
fundamental mode modo fundamental
fundamental operation operación fundamental
fundamental resonance resonancia fundamental
fundamental suppression supresión fundamental
fundamental swing oscilación fundamental
fundamental tone tono fundamental
fundamental units unidades fundamentales
fundamental wave onda fundamental
fundamental wavelength longitud de onda
    fundamental
funnel antenna antena en embudo
fuse fusible
fuse alarm alarma de fusible

fuse base base de fusible
fuse board tablero de fusibles
fuse box caja de fusibles
fuse carrier portafusible
fuse cartridge cartucho de fusible
fuse characteristic característica de fusible
fuse clip pinza de fusible
fuse connector conector de fusible
fuse current rating corriente nominal de fusible
fuse cutout cortacircuito de fusible
fuse element elemento fusible
fuse failure falla de fusible
fuse filler relleno de fusible
fuse holder portafusible
fuse insulator aislador de fusible
fuse light luz de fusible
fuse link elemento fusible
fuse melting time tiempo de fusión de fusible
fuse panel panel de fusibles
fuse puller sacafusibles
fuse rating clasificación de fusible
fuse tester probador de fusible
fuse tongs tenazas de fusibles
fuse unit unidad de fusible
fuse wire hilo fusible
fused fundido, protegido por fusible
fused electrolyte electrólito fundido
fused-electrolyte cell pila de electrólito fundido
fused junction unión por fusión
fused-junction diode diodo de unión por fusión
fused-junction transistor transistor de unión por fusión
fused quartz cuarzo fundido
fused-quartz lamp lámpara de cuarzo fundido
fused semiconductor semiconductor fundido
fuseholder portafusible
fusible fusible
fusible alloy aleación fusible
fusible cutout cortacircuito fusible
fusible resistor resistor fusible
fusible wire hilo fusible
fusing current corriente de fusión
fusion fusión
fusion frequency frecuencia de fusión
fusion welding soldadura por fusión
future labels etiquetas futuras

G

gadolinium gadolinio
gage medidor, calibre, calibrador
gage electrometer electrómetro calibrador
gaging medición, calibraje
gaging circuit circuito calibrador
gain ganancia
gain adjustment ajuste de ganancia
gain-bandwidth factor factor ganancia-ancho de banda
gain-bandwidth product producto ganancia-ancho de banda
gain check comprobación de ganancia
gain checker comprobador de ganancia
gain circuit circuito de ganancia
gain control control de ganancia
gain-crossover frequency frecuencia de ganancia cruzada
gain equalization control control de ecualización de ganancia
gain equalizer ecualizador de ganancia
gain equation ecuación de ganancia
gain-frequency characteristic característica ganancia-frecuencia
gain function función de ganancia
gain integrator integrador de ganancia
gain-level linearity linealidad de nivel de ganancia
gain linearity linealidad de ganancia
gain-loss ganancia-pérdida
gain margin margen de ganancia
gain measurement medida de ganancia
gain-measuring set gananciómetro
gain meter medidor de ganancia
gain of antenna ganancia de antena
gain ratio razón de ganancia
gain reduction reducción de ganancia
gain regulator regulador de ganancia
gain sensitivity control control de sensibilidad de ganancia
gain set gananciómetro
gain stability estabilidad de ganancia
gain stage etapa de ganancia
gain temperature coefficient coeficiente de temperatura de ganancia
gain turndown reducción de ganancia
galactic magnetic field campo magnético galáctico
galactic noise ruido galáctico
galactic radio noise ruido de radio galáctico
galactic radio waves ondas de radio galácticas
galaxy noise ruido galáctico
galaxy radio signals señales de radio galácticas
galena galena
galena crystal cristal de galena
galena detector detector de galena
gallium galio
gallium antimonide antimoniuro de galio
gallium arsenide arseniuro de galio
gallium-arsenide diode diodo de arseniuro de galio
gallium-arsenide laser láser de arseniuro de galio
gallium-arsenide varactor varactor de arseniuro de galio
gallium-based alloy aleación basada en galio

gallium phosphide fosfuro de galio
gallium-phosphide diode diodo de fosfuro de galio
Galton whistle silbato de Galton
galvanic galvánico
galvanic action acción galvánica
galvanic anode ánodo galvánico
galvanic bath baño galvánico
galvanic cell pila galvánica
galvanic corrosion corrosión galvánica
galvanic couple par galvánico
galvanic current corriente galvánica
galvanic deterioration deterioro galvánico
galvanic electricity electricidad galvánica
galvanic metallization metalización galvánica
galvanic pile pila galvánica
galvanic series serie galvánica
galvanic skin response respuesta de la piel galvánica
galvanic taste sabor galvánico
galvanism galvanismo
galvanization galvanización
galvanize galvanizar
galvanized iron hierro galvanizado
galvanizing galvanización
galvanomagnetic galvanomagnético
galvanomagnetic effect efecto galvanomagnético
galvanometer galvanómetro
galvanometer coil bobina de galvanómetro
galvanometer constant constante de galvanómetro
galvanometer driver excitador de galvanómetro
galvanometer lamp lámpara de galvanómetro
galvanometer modulator modulador galvanométrico
galvanometer recorder registrador galvanométrico
galvanometer relay relé galvanométrico
galvanometer sensitivity sensibilidad de
  galvanómetro
galvanometer shunt derivación de galvanómetro
galvanometric galvanométrico
galvanometry galvanometría
galvanoplastics galvanoplastia
galvanoplasty galvanoplastia
galvanoscope galvanoscopio
galvanotaxis galvanotaxia
galvanotherapy galvanoterapia
galvanotropism galvanotropismo
galvatron galvatrón
game port puerto de juegos
gamma gamma
gamma amplifier amplificador de gamma
gamma characteristics características gamma
gamma correction corrección de gamma
gamma decay desintegración gamma
gamma distribution distribución gamma
gamma emitter emisor gamma
gamma ferric oxide óxido férrico gamma
gamma function función gamma
gamma match adaptador gamma
gamma radiation radiación gamma
gamma-ray absorption absorción de rayos gamma
gamma-ray altimeter altímetro de rayos gamma
gamma-ray camera cámara de rayos gamma
gamma-ray counter contador de rayos gamma
gamma-ray emission rate tasa de emisión de rayos
  gamma
gamma-ray probe sonda de rayos gamma
gamma-ray resolution resolución de rayos gamma
gamma-ray source fuente de rayos gamma
gamma-ray spectrometer espectrómetro de rayos
  gamma
gamma-ray spectrum espectro de rayos gamma
gamma rays rayos gamma
gamma spectroscopy espectroscopia gamma
gammameter gammámetro
gamut gama

gang acoplar mecánicamente
gang assembly ensamblaje en tándem
gang capacitor capacitor tándem
gang control control en tándem
gang punch multiperforadora
gang punching multiperforación
gang switch conmutadores acoplados
ganged acoplado mecánicamente, en tándem
ganged capacitors capacitores acoplados
  mecánicamente
ganged circuits circuitos en tándem
ganged potentiometers potenciómetros acoplados
  mecánicamente
ganged rheostats reóstatos acoplados mecánicamente
ganged switches conmutadores acoplados
  mecánicamente
ganged tuning sintonización en tándem
ganging acoplamiento mecánico
gap espacio, abertura, intervalo, entrehierro,
  separación
gap admittance admitancia de espacio
gap arrester protector con espacios
gap character carácter de intervalo
gap coding codificación por intervalos
gap depth profundidad del entrehierro
gap digit dígito de intervalo
gap effect efecto del entrehierro
gap energy energía del entrehierro
gap filling relleno de espacios
gap length longitud del entrehierro
gap loss pérdida por entrehierro
gap reluctance reluctancia del entrehierro
gap width anchura del entrehierro
gapless sin espacios
garbage basura, residuo
garbage collection compactación de memoria
garble hacer ininteligible
garbled ininteligible
garbled message mensaje ininteligible
garnet granate
garnet maser máser de granate
gas gas
gas-actuated relay relé accionado por gas
gas admittance valve válvula de admisión de gas
gas admixture ratio razón de mezcla de gases
gas adsorption adsorción de gas
gas amplification amplificación de gas
gas-amplification factor factor de amplificación de
  gas
gas barrier insulation aislamiento de barrera de gas
gas breakdown ruptura de gas
gas capacitor capacitor de gas
gas cell pila de gas
gas cleanup limpieza de gas
gas current corriente de gas
gas density densidad de gas
gas detector detector de gases
gas dielectric capacitor capacitor de dieléctrico
  gaseoso
gas diode diodo de gas
gas discharge descarga gaseosa
gas-discharge display visualizador de descarga
  gaseosa
gas-discharge lamp lámpara de descarga gaseosa
gas-discharge tube tubo de descarga gaseosa
gas escape escape de gas
gas-filled cable cable de gas
gas-filled counter tube tubo contador de gas
gas-filled diode diodo de gas
gas-filled filament lamp lámpara de filamento de gas
gas-filled joint junta de gas
gas-filled lamp lámpara de gas
gas-filled photocell fotocélula de gas

gas-filled rectifier rectificador de gas
gas-filled relay relé de gas
gas-filled tetrode tetrodo de gas
gas-filled thyratron tiratrón de gas
gas-filled triode triodo de gas
gas-filled tube tubo de gas
gas flow flujo de gas
gas-flow alarm alarma de flujo de gas
gas-flow control control de flujo de gas
gas-flow counter contador de flujo de gas
gas-flow counter tube tubo contador de flujo de gas
gas-flow error error de flujo de gas
gas-flow gage medidor de flujo de gas
gas-flow indicator indicador de flujo de gas
gas-flow meter medidor de flujo de gas
gas-flow switch conmutador de flujo de gas
gas focusing enfoque por gas
gas grooves surcos por gas
gas-insulated substation subestación aislada por gas
gas-insulated transformer transformador aislado por gas
gas laser láser de gas
gas leak fuga de gas
gas-leak detection equipment equipo de detección de fugas de gas
gas-leak indicator indicador de fugas de gas
gas leakage fuga de gas
gas magnification magnificación por gas
gas maser máser de gas
gas multiplication multiplicación de gas
gas multiplication factor factor de multiplicación de gas
gas noise ruido por gas
gas panel panel de gas
gas phototube fototubo de gas
gas-plasma display visualizador de plasma gaseoso
gas pressure presión de gas
gas-pressure alarm alarma de presión de gas
gas-pressure control control de presión de gas
gas-pressure gage medidor de presión de gas
gas-pressure indicator indicador de presión de gas
gas-pressure meter medidor de presión de gas
gas-pressure relay relé de presión de gas
gas ratio razón de vacío
gas relay relé de gas
gas reservoir reserva de gas
gas sensor sensor de gas
gas sniffer detector de gases
gas tetrode tetrodo de gas
gas triode triodo de gas
gas tube tubo de gas
gas-tube lightning arrester pararrayos con tubo de gas
gas-tube oscillator oscilador con tubo de gas
gas valve válvula de gas
gas volume volumen de gas
gas X-ray tube tubo de rayos X de gas
gaseous gaseoso
gaseous annealing recocido gaseoso
gaseous conduction conducción gaseosa
gaseous dielectric dieléctrico gaseoso
gaseous-dielectric cable cable de dieléctrico gaseoso
gaseous diffusion difusión gaseosa
gaseous discharge descarga gaseosa
gaseous medium medio gaseoso
gaseous mixture mezcla gaseosa
gaseous phototube fototubo gaseoso
gaseous rectifier rectificador gaseoso
gaseous tube tubo gaseoso
gaseous-tube generator generador de tubo gaseoso
gaseous voltage regulator regulador de tensión gaseosa
gasket junta

gasproof a prueba de gas
gasproof machine máquina a prueba de gas
gassing generación de gas, gasificación
gassy tube tubo con gas residual
gastight hermético contra gas
gaston gaston
gate puerta, compuerta, impulso de mando
gate array arreglo de puertas
gate-array chip chip de arreglo de puertas
gate capacitance capacitancia de puerta
gate circuit circuito de puerta
gate closer cerrador de puerta
gate control control de puerta
gate-controlled controlado por puerta
gate-controlled delay time tiempo de retardo controlado por puerta
gate-controlled rise time tiempo de subida controlado por puerta
gate-controlled switch conmutador controlado por puerta
gate current corriente de puerta
gate detector detector de puerta
gate electrode electrodo de puerta
gate element elemento de puerta
gate generator generador de impulsos de desbloqueo
gate impedance impedancia de puerta
gate leakage current corriente inversa de puerta
gate of entry puerta de entrada
gate power dissipation disipación de potencia de puerta
gate protective action acción protectora de puerta
gate pulse impulso de puerta
gate recovery time tiempo de recuperación de puerta
gate reverse current corriente inversa de puerta
gate signal señal de puerta
gate suppression supresión de puerta
gate switch conmutador de puerta
gate terminal terminal de puerta
gate timing sincronización de puerta
gate trigger current corriente de disparo de puerta
gate trigger diode diodo de disparo de puerta
gate trigger voltage tensión de disparo de puerta
gate tube tubo de puerta
gate turn-off current corriente de interrupción de puerta
gate turn-off switch conmutador controlado por puerta
gate turn-off voltage tensión de interrupción de puerta
gate valve válvula de puerta
gate voltage tensión de puerta
gate winding arrollamiento de puerta
gated amplifier amplificador controlado por puerta
gated beam haz controlado
gated-beam detector detector de haz controlado
gated-beam tube tubo de haz controlado
gated bidirectional thyristor tiristor bidireccional controlado
gated demodulator desmodulador controlado
gated flip-flop basculador controlado
gated oscillator oscilador controlado
gated signal señal controlada
gated sweep barrido controlado
gateway trayectoria de acceso
gating desbloqueo periódico, selección de señal
gating circuit circuito intermitente
gating pulse impulso de selección
gauge medidor, calibre, calibrador
gauging medición, calibraje
gausitron gausitrón
gauss gauss
Gauss' law ley de Gauss
Gauss' theorem teorema de Gauss

gaussage gausaje
Gaussian beam haz de Gauss
Gaussian curve curva de Gauss
Gaussian distribution distribución de Gauss
Gaussian filter filtro de Gauss
Gaussian function función de Gauss
Gaussian noise ruido de Gauss
Gaussian-noise generator generador de ruido de
  Gauss
Gaussian pulse impulso de Gauss
Gaussian random noise ruido aleatorio de Gauss
Gaussian response respuesta de Gauss
Gaussian system sistema de Gauss
Gaussian waveform forma de onda de Gauss
Gaussian well pozo de Gauss
gaussmeter gausiómetro
gear equipo, dispositivo, engranaje
gear motor motor con engranaje reductor
gear ratio razón de engranaje
gear-wheel pattern patrón de rueda dentada
Gee Gee
Gee indicator indicador Gee
Gee receiver receptor Gee
Gee system sistema Gee
Geiger counter contador de Geiger
Geiger-Mueller counter contador de Geiger-Mueller
Geiger-Mueller counter tube tubo contador de
  Geiger-Mueller
Geiger-Mueller detector detector de Geiger-Mueller
Geiger-Mueller region región de Geiger-Mueller
Geiger-Mueller threshold umbral de Geiger-Mueller
Geiger-Mueller tube tubo de Geiger-Mueller
Geiger region región de Geiger
Geiger threshold umbral de Geiger
Geiger tube tubo de Geiger
Geissler discharge descarga de Geissler
Geissler tube tubo de Geissler
gel gel
genemotor genemotor
general coverage cobertura general
general-coverage receiver receptor de cobertura
  general
general diffused lighting alumbrado difuso general
general format formato general
general ground tierra general
general-information radar radar de información
  general
general lighting alumbrado general
general processing level nivel de procesamiento
  general
general program programa general
general purpose de uso general
general-purpose branch circuit circuito de
  derivación de uso general
general-purpose bridge puente de uso general
general-purpose circuit-breaker cortacircuito de uso
  general
general-purpose component componente de uso
  general
general-purpose computer computadora de uso
  general
general-purpose controller controlador de uso
  general
general-purpose data concentrator concentrador de
  datos de uso general
general-purpose digital computer computadora
  digital de uso general
general-purpose diode diodo de uso general
general-purpose enclosure caja de uso general
general-purpose function generator generador de
  funciones de uso general
general-purpose induction motor motor de
  inducción de uso general

general-purpose interface interfaz de uso general
general-purpose library biblioteca de uso general
general-purpose loader cargador de uso general
general-purpose motor motor de uso general
general-purpose operating program programa
  operativo de uso general
general-purpose operating system sistema operativo
  de uso general
general-purpose processor procesador de uso
  general
general-purpose program programa de uso general
general-purpose radar radar de uso general
general-purpose register registro de uso general
general-purpose relay relé de uso general
general-purpose storage almacenamiento de uso
  general
general-purpose system sistema de uso general
general-purpose tester probador de uso general
general-purpose transformer transformador de uso
  general
general-purpose transistor transistor de uso general
general queue cola general
general reference file archivo de referencia general
general register registro general
general routine rutina general
general search búsqueda general
general service servicio general
general storage almacenamiento general
general-storage assignment asignación de
  almacenamiento general
general-storage entry entrada de almacenamiento
  general
general-use switch conmutador de uso general
general-use volume volumen de uso general
generalized generalizado
generalized entity entidad generalizada
generalized impedance impedancia generalizada
generalized impedance converter convertidor de
  impedancia generalizada
generalized property propiedad generalizada
generalized quantity magnitud generalizada
generalized routine rutina generalizada
generalized sequential access method método de
  acceso secuencial generalizado
generalized sort clasificación generalizada
generalized sort-merge program programa de
  clasificación-fusión generalizado
generate generar
generated address dirección generada
generated data set conjunto de datos generado
generated error error generado
generated noise ruido generado
generated voltage tensión generada
generating electric field meter medidor de campo
  eléctrico generador
generating function función generadora
generating magnetometer magnetómetro generador
generating plant planta eléctrica, central eléctrica
generating program programa generador
generating routine rutina generadora
generating set planta eléctrica
generating station estación generadora, central
  generadora
generating-station efficiency eficiencia de estación
  generadora
generating-station reserve reserva de estación
  generadora
generating unit unidad generadora
generating voltmeter vóltmetro generador
generation generación
generation data group grupo de datos de generación
generation data set conjunto de datos de generación
generation number número de generación

generation parameter parámetro de generación
generation phase fase de generación
generation program programa de generación
generation rate velocidad de generación
generation routine rutina de generación
generator generador
generator armature inducido de generador
generator distortion factor factor de distorsión de generador
generator drive adapter adaptador de impulsor de generador
generator efficiency eficiencia de generador
generator field accelerating relay relé acelerador de campo de generador
generator field control control de campo de generador
generator field decelerating relay relé desacelerador de campo de generador
generator language lenguaje de generador
generator-motor generador-motor
generator noise ruido de generador
generator program programa generador
generator regulator regulador de generador
generic genérico
generic card set conjunto de cartas genérico
generic connection assembly ensamblaje de conexión genérico
generic data element elemento de datos genérico
generic design diseño genérico
generic equipment equipo genérico
generic font name nombre de diseño de letra genérico
generic mode modo genérico
generic name nombre genérico
generic program unit unidad de programa genérica
generic unit unidad genérica
geoalert geoalerta
geocentric geocéntrico
geodesic geodésico
geodesic lens antenna antena de lente geodésica
geodesic satellite satélite geodésico
geodimeter geodímetro
geoelectric geoeléctrico
geoelectric effect efecto geoeléctrico
geoelectric response respuesta geoeléctrica
geoelectric survey levantamiento geoeléctrico
geoelectricity geoelectricidad
geographical address dirección geográfica
geographical address control control de dirección geográfica
geographical addressing direccionamiento geográfico
geomagnetic geomagnético
geomagnetic activity actividad geomagnética
geomagnetic effect efecto geomagnético
geomagnetism geomagnetismo
geometric amplification factor factor de amplificación geométrico
geometric capacitance capacitancia geométrica
geometric correction corrección geométrica
geometric distortion distorsión geométrica
geometric energy flux flujo de energía geométrico
geometric error error geométrico
geometric factor factor geométrico
geometric fidelity fidelidad geométrica
geometric mean media geométrica
geometric optics óptica geométrica
geometric progression progresión geométrica
geometric symmetry simetría geométrica
geometry geometría
geostationary orbit órbita geoestacionaria
geostationary satellite satélite geoestacionario
geothermal geotérmico

geothermal power station central eléctrica geotérmica
geotropism geotropismo
German silver plata alemana
germanium germanio
germanium alloy-junction transistor transistor de unión de aleación de germanio
germanium crystal cristal de germanio
germanium-crystal detector detector de cristal de germanio
germanium-crystal diode diodo de germanio
germanium diode diodo de germanio
germanium-diode detection detección por diodo de germanio
germanium dioxide dióxido de germanio
germanium junction diode diodo de unión de germanio
germanium junction photodiode fotodiodo de unión de germanio
germanium junction photoelectric cell célula fotoeléctrica de unión de germanio
germanium NPN transistor transistor NPN de germanio
germanium photocell fotocélula de germanio
germanium photodiode fotodiodo de germanio
germanium PN junction unión PN de germanio
germanium power rectifier rectificador de potencia de germanio
germanium rectifier rectificador de germanio
germanium transistor transistor de germanio
germanium wafer oblea de germanio
get operation operación de obtención
getter rarefactor
ghost fantasma, imagen fantasma
ghost beacon radiofaro fantasma
ghost circuit circuito fantasma
ghost image imagen fantasma
ghost mode modo fantasma
ghost pulse impulso fantasma
ghost signal señal fantasma
giant-pulse laser láser de impulsos gigantes
Gibb's phenomenon fenómeno de Gibbs
giga giga
gigabit gigabit
gigabyte gigabyte
gigacycle gigaciclo
gigaelectronvolt gigaelectronvolt
gigahertz gigahertz
gigameter gigámetro
gigaohm gigaohm
gigawatt gigawatt
gilbert gilbert
gilberts per centimeter gilberts por centímetro
Gill-Morrell oscillator oscilador de Gill-Morrell
gimbal balancín
Giorgi system sistema de Giorgi
Giorgi units unidades de Giorgi
glare resplandor
glass vidrio
glass amplifier amplificador de vidrio
glass-bonded mica mica adherida con vidrio
glass bulb ampolla de vidrio
glass-bulb rectifier rectificador de ampolla de vidrio
glass capacitor capacitor de vidrio
glass cartridge cartucho de vidrio
glass-cartridge fuse fusible de cartucho de vidrio
glass dielectric capacitor capacitor de dieléctrico de vidrio
glass diode diodo de vidrio
glass dosimeter dosímetro de vidrio
glass electrode electrodo de vidrio
glass envelope envolvente de vidrio
glass fiber fibra de vidrio

glass half-cell semicelda de vidrio
glass insulator aislador de vidrio
glass-metal joint junta vidrio-metal
glass-metal seal cierre vidrio-metal
glass plate placa de vidrio
glass plate capacitor capacitor de placa de vidrio
glass resistor resistor de vidrio
glass seal cierre de vidrio
glass tube tubo de vidrio
glass wall pared de vidrio
glassine papel cristal
glassivation pasivación con vidrio
glide angle ángulo de planeo
glide path trayectoria de planeo
glide-path angle ángulo de trayectoria de planeo
glide-path equipment equipo de trayectoria de planeo
glide-path indicator indicador de trayectoria de planeo
glide-path landing aterrizaje radioguiado
glide-path landing system sistema de aterrizajes radioguiado
glide-path localizer localizador de trayectoria de planeo
glide-path receiver receptor de trayectoria de planeo
glide-path sector sector de trayectoria de planeo
glide-path station estación de trayectoria de planeo
glide-path transmitter transmisor de trayectoria de planeo
glide slope inclinación de planeo
glide-slope angle ángulo de inclinación de planeo
glide-slope deviation desviación de inclinación de planeo
glide-slope facility instalación de inclinación de planeo
glide-slope receiver receptor de inclinación de planeo
glide-slope sector sector de inclinación de planeo
glide-slope transmitter transmisor de inclinación de planeo
glint parpadeo
glitter parpadeo
global global
global area área global
global backup reserva global
global change cambio global
global character carácter global
global code código global
global control section sección de control global
global data datos globales
global format formato global
global name nombre global
global object objeto global
global operation operación global
global parameter parámetro global
global processor procesador global
global replace reemplazo global
global search búsqueda global
global search and replace búsqueda y reemplazo global
global storage almacenamiento global
global variable variable global
global variable symbol símbolo de variable global
global zone zona global
gloss brillo
gloss factor factor de brillo
glossmeter brillancímetro, medidor de brillo
glow luminiscencia
glow current corriente de luminiscencia
glow discharge descarga luminiscente
glow-discharge cathode cátodo de descarga luminiscente
glow-discharge lamp lámpara de descarga

luminiscente
glow-discharge microphone micrófono de descarga luminiscente
glow-discharge rectifier rectificador de descarga luminiscente
glow-discharge region región de descarga luminiscente
glow-discharge tube tubo de descarga luminiscente
glow-discharge voltage regulator regulador de tensión de descarga luminiscente
glow factor factor de luminiscencia
glow lamp lámpara luminiscente
glow potential potencial de luminiscencia
glow switch conmutador por luminiscencia
glow-transfer counter tube tubo contador de transferencia luminiscente
glow tube tubo luminiscente
glow voltage tensión de luminiscencia
glucinium glucinio
go-ahead sequence secuencia de proseguir
go-and-return channel canal de ida y vuelta
go channel canal de ida
go to ir a
goggles gafas de protección
Golay cell celda de Golay
gold oro
gold-bonded diode diodo adherido con oro
gold coating revestimiento de oro
gold-doped germanium germanio dopado con oro
gold doping dopado con oro
gold leaf hoja de oro
gold-leaf electroscope electroscopio de hojas de oro
gold-plated dorado
gold-plated contacts contactos dorados
gold-plated grid rejilla dorada
gold plating dorado
Goldschmidt alternator alternador de Goldschmidt
goniograph goniógrafo
goniometer goniómetro
goniometry goniometría
goniophotometer goniofotómetro
googol googol
googolplex googolplex
Goto circuit circuito de Goto
Goto pair par de Goto
governing system sistema regulador
governor regulador
governor-controlled speed velocidad controlada por regulador
graceful degradation funcionamiento tras falla
graceful exit salida sin percances
gradation gradación
grade grado
grade of service grado de servicio
graded graduado
graded-base transistor transistor de base graduada
graded brush escobilla graduada
graded filter filtro graduado
graded-index profile perfil de índice graduado
graded insulation aislamiento graduado
graded junction unión graduada
graded-junction transistor transistor de unión graduada
graded potentiometer potenciómetro graduado
gradient gradiente
gradient meter gradientímetro, medidor de gradiente
gradient microphone micrófono de gradiente
grading gradación, múltiple parcial
grading group subgrupo
gradiometer gradiómetro
gradual gradual
gradual control control gradual
gradual failure falla gradual

graduated graduado
Graetz bridge puente de Graetz
grain grano
grain boundary límite intergranular
grain-oriented steel acero de grano orientado
graininess granosidad
gram gramo
gram-atom átomo gramo
gram-atomic weight peso atómico-gramo
gram-calorie caloría-gramo
gram-centimeter centímetro-gramo
gram-equivalent equivalente-gramo
gram-molecule molécula-gramo
Gramme armature inducido de Gramme
Gramme ring anillo de Gramme
gramophone gramófono
gramophone amplifier amplificador gramofónico
gramophone record disco gramofónico
grandfather cycle ciclo abuelo
grandfather file archivo abuelo
grandfather tape cinta abuelo
granular granular
granular carbon carbón granular
granular microphone micrófono granular
granularity granulosidad
graph gráfica, gráfico
graph plotter trazador de gráficos
graph screen pantalla cuadriculada
graphecon graficón
graphic gráfico
graphic character carácter gráfico
graphic character modification modificación de
  caracteres gráficos
graphic data output area área de salida de datos
  gráficos
graphic data processing procesamiento de datos
  gráficos
graphic data set conjunto de datos gráficos
graphic data structure estructura de datos gráficos
graphic display visualización gráfica
graphic display mode modo de visualización gráfica
graphic display program programa de visualización
  gráfica
graphic display resolution resolución de
  visualización gráfica
graphic display system sistema de visualización
  gráfica
graphic display terminal terminal de visualización
  gráfica
graphic display unit unidad de visualización gráfica
graphic documentation documentación gráfica
graphic input device dispositivo de entrada gráfica
graphic input-output entrada-salida gráfica
graphic instrumentation instrumentación gráfica
graphic job processing procesamiento de trabajo
  gráfico
graphic job processor procesador de trabajo gráfico
graphic output salida gráfica
graphic output device dispositivo de salida gráfica
graphic output terminal terminal de salida gráfica
graphic panel panel gráfico
graphic processing unit unidad de procesamiento
  gráfico
graphic recorder registrador gráfico
graphic solution solución gráfica
graphic string secuencia gráfica
graphic subset subconjunto gráfico
graphic symbol símbolo gráfico
graphic terminal terminal gráfica
graphical gráfico
graphical analysis análisis gráfico
graphical display device dispositivo de visualización
  gráfica

graphical input device dispositivo de entrada gráfico
graphical model modelo gráfico
graphics gráficos
graphics access method método de acceso de
  gráficos
graphics adapter adaptador de gráficos
graphics applications aplicaciones de gráficos
graphics card tarjeta de gráficos
graphics data file archivo de datos gráficos
graphics display visualización de gráficos
graphics field campo de gráficos
graphics file format formato de archivo de gráficos
graphics hierarchy jerarquía de gráficos
graphics input entrada de gráficos
graphics-input hardware equipo físico de entrada de
  gráficos
graphics language lenguaje de gráficos
graphics light pen lápiz óptico de gráficos
graphics mode modo de gráficos
graphics output salida de gráficos
graphics-output hardware equipo físico de salida de
  gráficos
graphics plotting trazado de gráficos
graphics printer impresora de gráficos
graphics processor procesador de gráficos
graphics program programa de gráficos
graphics resolution resolución de gráficos
graphics segment segmento de gráficos
graphics software programas de gráficos
graphics symbol set conjunto de símbolos gráficos
graphics system sistema de gráficos
graphics text texto de gráficos
graphite grafito
graphite anode ánodo de grafito
graphite brush escobilla de grafito
graphite electrode electrodo de grafito
graphite-line resistor resistor de línea de grafito
graphite resistor resistor de grafito
graphitization grafitización
graphophone grafófono
grasshopper fuse fusible con alarma
grate magnet imán de parrilla
graticule retículo
graticule pattern patrón reticulado
grating rejilla, retículo
grating converter convertidor reticular
grating generator generador de rejilla
grating reflector reflector de rejilla
Gratz rectifier rectificador de Gratz
gravitational constant constante gravitacional
gravitational energy energía gravitacional
gravitational field campo gravitacional
gravity gravedad
gravity cell pila de gravedad
gravity gradient stabilization estabilización por
  gradiente de gravedad
gravity switch conmutador de gravedad
gravity wave onda gravitacional
gray body cuerpo gris
Gray code código de Gray
gray level nivel del gris
gray scale escala de grises
gray scale images imágenes en escala de grises
gray tin estaño gris
grazing angle ángulo rasante
grazing loss pérdida por efecto rasante
grazing path trayectoria rasante
greaseproof a prueba de grasa
great circle círculo máximo
great circle bearing rumbo de círculo máximo
green amplifier amplificador de verde
green filter filtro de verde
green gain control control de ganancia de verde

green gun cañón de verde
green information información de verde
green laser láser de luz verde
green restorer restaurador de verde
green signal señal de verde
green video voltage tensión del video de verde
grenz rays rayos límite
grenz tube tubo de rayos límite
grid rejilla, red de distribución
grid-anode capacitance capacitancia rejilla-ánodo
grid-anode characteristic característica rejilla-ánodo
grid battery batería de rejilla
grid bias polarización de rejilla
grid-bias battery batería de polarización de rejilla
grid-bias cell pila de polarización de rejilla
grid-bias circuit circuito de polarización de rejilla
grid-bias detector detector de polarización de rejilla
grid-bias modulation modulación por polarización
de rejilla
grid-bias resistor resistor de polarización de rejilla
grid-bias shift desplazamiento de polarización de
rejilla
grid-bias voltage tensión de polarización de rejilla
grid-block keying manipulación por bloqueo de
rejilla
grid blocking bloqueo de rejilla
grid cap casquete de rejilla
grid capacitance capacitancia de rejilla
grid capacitor capacitor de rejilla
grid-cathode capacitance capacitancia rejilla-cátodo
grid characteristic característica de rejilla
grid circuit circuito de rejilla
grid-circuit alignment alineamiento de circuito de
rejilla
grid-circuit impedance impedancia de circuito de
rejilla
grid-circuit tester probador de circuito de rejilla
grid clip pinza de rejilla
grid coil bobina de rejilla
grid condenser condensador de rejilla
grid conductance conductancia de rejilla
grid connecting ring anillo de conexión de rejilla
grid constraint constreñimiento de rejilla
grid contact potential potencial de contacto de rejilla
grid control control por rejilla
grid-control circuit circuito de control por rejilla
grid-controlled rectifier rectificador de control por
rejilla
grid-controlled tube tubo de control por rejilla
grid current corriente de rejilla
grid-current characteristic característica de
corriente de rejilla
grid-current indicator indicador de corriente de
rejilla
grid cutoff voltage tensión de corte de rejilla
grid detection detección por rejilla
grid-dip meter ondámetro de absorción
grid dissipation disipación de rejilla
grid drive excitación de rejilla
grid-drive characteristic característica de excitación
de rejilla
grid-drive detector detector de excitación de rejilla
grid-driving power potencia de excitación de rejilla
grid electrode electrodo de rejilla
grid emission emisión de rejilla
grid excitation excitación de rejilla
grid family of curves familia de curvas de rejilla
grid-filament capacity capacidad rejilla-filamento
grid filling relleno de rejilla
grid-glow tube tubo de rejilla luminiscente
grid impedance impedancia de rejilla
grid-input impedance impedancia de entrada de
rejilla

grid ionization chamber cámara de ionización de
rejilla
grid keying manipulación por rejilla
grid leak fuga de rejilla
grid-leak bias polarización por fuga de rejilla
grid-leak detector detector de fuga de rejilla
grid-leak resistance resistencia de fuga de rejilla
grid-leak resistor resistor de fuga de rejilla
grid limiter limitador de rejilla
grid limiting limitación por rejilla
grid-load resistance resistencia de carga de rejilla
grid-load resistor resistor de carga de rejilla
grid loading effect efecto de carga de rejilla
grid locking bloqueo de rejilla
grid mesh malla de rejilla
grid milliammeter miliamperímetro de rejilla
grid-modulated amplifier amplificador modulado
por rejilla
grid modulation modulación por rejilla
grid modulator modulador por rejilla
grid neutralization neutralización de rejilla
grid phase shift desplazamiento de fase de rejilla
grid pin patilla de rejilla
grid pitch paso de rejilla
grid-plate capacitance capacitancia rejilla-placa
grid-plate capacity capacidad rejilla-placa
grid-plate characteristic característica rejilla-placa
grid-plate transconductance transconductancia
rejilla-placa
grid polarization voltage tensión de polarización de
rejilla
grid-pool tube tubo de cátodo de mercurio con rejilla
grid potential potencial de rejilla
grid potentiometer potenciómetro de rejilla
grid power loss pérdida de potencia de rejilla
grid-pulse modulation modulación por impulsos de
rejilla
grid pulsing pulsación por rejilla
grid rectification rectificación por rejilla
grid resistance resistencia de rejilla
grid resistor resistor de rejilla
grid return retorno de rejilla
grid-return lead conexión de retorno de rejilla
grid rheostat reóstato de rejilla
grid-screen capacitance capacitancia rejilla-pantalla
grid-screen capacity capacidad rejilla-pantalla
grid signal señal de rejilla
grid stopper freno de rejilla
grid stopper resistor resistor de freno de rejilla
grid structure estructura de rejilla
grid suppressor supresor de rejilla
grid swing oscilación de rejilla
grid tank tanque de rejilla
grid tank capacitance capacitancia de tanque de
rejilla
grid tank inductance inductancia de tanque de rejilla
grid tank voltage tensión de tanque de rejilla
grid-to-cathode capacitance capacitancia de rejilla a
cátodo
grid-to-plate capacitance capacitancia de rejilla a
placa
grid-to-plate transconductance transconductancia de
rejilla a placa
grid-to-screen capacitance capacitancia de rejilla a
pantalla
grid transformer transformador de rejilla
grid tuning sintonización por rejilla
grid voltage tensión de rejilla
grid voltage supply fuente de tensión de rejilla
grid waveform forma de onda de rejilla
grid wire hilo de rejilla
gridistor gridistor
grille rejilla

grommet arandela aislante
groove surco
groove angle ángulo de surco
groove deformation deformación de surco
groove depth profundidad de surco
groove diameter diámetro de surco
groove shape forma de surco
groove speed velocidad de surco
groove wall pared de surco
groove width anchura de surco
gross actual generation generación real bruta
gross available capacity capacidad disponible bruta
gross generation generación bruta
gross information content contenido de información bruto
gross maximum capacity capacidad máxima bruta
gross maximum generation generación máxima bruta
ground tierra, conexión a tierra, masa
ground absorption absorción de tierra
ground antenna antena en tierra
ground-based radar radar basado en tierra
ground bus bus de puesta a tierra
ground cable cable de puesta a tierra
ground circuit circuito de tierra
ground clamp abrazadera de tierra
ground clip pinza de puesta a tierra
ground clutter ecos parásitos de tierra
ground conductivity conductividad de tierra
ground conductor conductor de tierra
ground conduit conducto de tierra
ground connection conexión a tierra
ground constant constante de tierra
ground contact contacto a tierra
ground-contact indicator indicador de contacto a tierra
ground control control desde tierra
ground-controlled approach aproximación controlada desde tierra
ground-controlled approach system sistema de aproximación controlada desde tierra
ground current corriente de tierra
ground data equipment equipo de datos terrestre
ground detector detector de tierra
ground dielectric constant constante dieléctrica de tierra
ground effect efecto de tierra
ground efficiency eficiencia de tierra
ground electrode electrodo de tierra
ground environment ambiente de tierra
ground equalizer coil bobina ecualizadora de tierra
ground equalizer inductors inductores ecualizadores de tierra
ground equipment equipo de tierra
ground fault pérdida de conexión a tierra, tierra accidental
ground-fault interrupter interruptor accionado por pérdida de conexión a tierra
ground-ground communication comunicación tierra-tierra
ground indication indicación de tierra
ground indicator indicador de tierra
ground insulation aislamiento de tierra
ground lead conexión de tierra
ground leak fuga a tierra
ground level nivel fundamental
ground loop bucle de tierra
ground mat tapete de tierra
ground noise ruido de fondo
ground outlet tomacorriente con conexión a tierra, toma de corriente con conexión a tierra
ground plane plano de tierra
ground-plane antenna antena con plano de tierra

ground plate placa de tierra
ground-position indicator indicador de posición relativa
ground potential potencial de tierra
ground protection protección de tierra
ground radar station estación de radar terrestre
ground reference station estación de referencia terrestre
ground-reflected wave onda reflejada por tierra
ground reflection reflexión terrestre
ground relay relé de tierra
ground relay protection protección por relé de tierra
ground resistance resistencia de tierra
ground return retorno a tierra, eco terrestre
ground-return circuit circuito de retorno a tierra
ground rod varilla de tierra
ground speed velocidad relativa a la tierra
ground state estado fundamental
ground station estación terrestre
ground support equipment equipo de apoyo terrestre
ground surveillance radar radar de vigilancia terrestre
ground switch conmutador de tierra
ground system sistema de tierra
ground terminal terminal de tierra
ground-to-air communication comunicación de tierra al aire
ground-to-ground tierra a tierra
ground-to-ground communication comunicación de tierra a tierra
ground wave onda terrestre
ground wire hilo de tierra
grounded conectado a tierra, conectado a masa
grounded-anode amplifier amplificador con ánodo a masa
grounded antenna antena con conexión a tierra
grounded-base amplifier amplificador con base a masa
grounded-base circuit circuito con base a masa
grounded-base configuration configuración con base a masa
grounded-base connection conexión con base a masa
grounded-cathode amplifier amplificador con cátodo a masa
grounded-cathode circuit circuito con cátodo a masa
grounded-cathode connection conexión con cátodo a masa
grounded circuit circuito puesto a tierra, circuito puesto a masa
grounded-collector amplifier amplificador con colector a masa
grounded-collector circuit circuito con colector a masa
grounded-collector configuration configuración con colector a masa
grounded-collector connection conexión con colector a masa
grounded conductor conductor puesto a tierra
grounded-emitter amplifier amplificador con emisor a masa
grounded-emitter circuit circuito con emisor a masa
grounded-emitter configuration configuración con emisor a masa
grounded-emitter connection conexión con emisor a masa
grounded-gate amplifier amplificador con puerta a masa
grounded-gate circuit circuito con puerta a masa
grounded-grid amplifier amplificador con rejilla a masa
grounded-grid circuit circuito con rejilla a masa
grounded-grid configuration configuración con

rejilla a masa
grounded-grid oscillator oscilador con rejilla a masa
grounded-grid triode triodo con rejilla a masa
grounded-grid triode circuit circuito de triodo con rejilla a masa
grounded-grid triode mixer mezclador de triodo con rejilla a masa
grounded outlet tomacorriente con conexión a tierra
grounded parts partes con conexión a tierra
grounded-plate amplifier amplificador con placa a masa
grounded-source amplifier amplificador con fuente a masa
grounded-source circuit circuito con fuente a masa
grounded system sistema con conexión a tierra
grounding puesta a tierra, puesta a masa
grounding cable cable de puesta a tierra
grounding circuit circuito de puesta a tierra
grounding clamp abrazadera de puesta a tierra
grounding conductor conductor de puesta a tierra
grounding connection conexión de puesta a tierra
grounding contact contacto de puesta a tierra
grounding cord cordón de puesta a tierra
grounding device dispositivo de puesta a tierra
grounding electrode electrodo de puesta a tierra
grounding outlet tomacorriente con conexión a tierra
grounding plate placa de tierra
grounding point punto de puesta a tierra
grounding rod varilla de tierra
grounding switch conmutador de puesta a tierra
grounding terminal terminal de puesta a tierra
grounding transformer transformador de puesta a tierra
group grupo
group access acceso de grupo
group address dirección de grupo
group address message mensaje de dirección de grupo
group addressing direccionamiento de grupo
group allocation asignación de grupos
group amplifier amplificador de grupo
group busy señal de grupo ocupado
group-busy signal señal de grupo ocupado
group-busy tone tono de grupo ocupado
group calculation cálculo de grupo
group carrier portadora de grupo
group carrier generator generador de portadora de grupo
group center centro de grupo
group code código de grupo
group control control de grupo
group control station estación de control de grupo
group delay retardo de grupo
group-delay characteristic característica de retardo de grupo
group-delay distortion distorsión de retardo de grupo
group-delay equalizer ecualizador de retardo de grupo
group-delay noise ruido de retardo de grupo
group-delay time tiempo de retardo de grupo
group demodulator desmodulador de grupo
group distribution frame repartidor de grupos
group-engaged tone tono de grupo ocupado
group frequency frecuencia de grupo
group identification identificación de grupo
group index índice de grupo
group indication indicación de grupo
group item artículo de grupo
group link enlace de grupos
group loop bucle de grupo
group mark marca de grupo
group marker marcador de grupo

group modem módem de grupo
group modulating equipment equipo de modulación de grupo
group modulation modulación de grupo
group modulation equipment equipo de modulación de grupo
group modulator modulador de grupo
group name nombre de grupo
group number número de grupo
group occupancy meter medidor de ocupación de grupo
group operation operación en grupo
group printing impresión de grupo
group processing procesamiento de grupo
group section sección de grupo
group selection selección de grupo
group selector selector de grupo
group separator separador de grupos
group separator character carácter separador de grupos
group sorting clasificación de grupo
group sorting device dispositivo de clasificación de grupo
group transfer point punto de transferencia de grupo
group velocity velocidad de grupo
grouped-frequency operation operación por frecuencias agrupadas
grouped positions posiciones agrupadas
grouped records registros agrupados
grouping agrupamiento
grouping center centro de agrupamiento
grouping exchange central de agrupamiento
grouping factor factor de agrupamiento
grouping network red de agrupamiento
grout lechada
Grove cell pila de Grove
growler probador de inducidos
grown-diffused transistor transistor de capas de difusión
grown junction unión por crecimiento
grown-junction diode diodo de unión por crecimiento
grown-junction photocell fotocélula de unión por crecimiento
grown-junction transistor transistor de unión por crecimiento
guard band banda de guarda
guard bit bit de guarda
guard circle círculo de guarda
guard circuit circuito de guarda
guard circuitry circuitería de guarda
guard digit dígito de guarda
guard electrode electrodo de guarda
guard interval intervalo de guarda
guard plate placa de guarda
guard position posición de guarda
guard potential potencial de guarda
guard relay relé de guarda
guard ring anillo de guarda
guard shield blindaje de guarda
guard signal señal de guarda
guard terminal terminal de guarda
guard tone tono de guarda
guard tube tubo de guarda
guard wire hilo de guarda
guarded input entrada con guarda
Gudden-Pohl effect efecto de Gudden-Pohl
Gudden-Pohl law ley de Gudden-Pohl
guidance guiado, guianza
guidance equipment equipo de guiado
guidance system sistema de guiado
guide guía
guide edge borde de guía

guide elbow codo de guía
guide field campo de guía
guide flange brida de guía
guide junction unión de guía
guide line línea de guía
guide margin margen de guía
guide pin patilla de guía
guide plate placa de guía
guide rod varilla de guía
guide-wave impedance impedancia de onda de guía
guide wavelength longitud de onda en la guía
guided guiado
guided electromagnetic wave onda electromagnética
    guiada
guided propagation propagación guiada
guided ray rayo guiado
guided wave onda guiada
Guillemin effect efecto de Guillemin
Guillemin line línea de Guillemin
guillotine attenuator atenuador de guillotina
guillotine capacitor capacitor de guillotina
Gunn diode diodo de Gunn
Gunn effect efecto de Gunn
Gunn-effect circuit circuito de efecto de Gunn
Gunn-effect oscillator oscilador de efecto de Gunn
Gunn oscillations oscilaciones de Gunn
Gunn oscillator oscilador de Gunn
gutta-percha gutapercha
guy anchor anclaje de viento
guy wire riosta
guyed antenna mast mástil de antena arriostrado
guyed pole poste arriostrado
guyed tower torre arriostrada
guying arriostramiento
gyrator girador
gyro giroscopio
gyrocompass girocompás
gyrofrequency girofrecuencia
gyromagnetic giromagnético
gyromagnetic compass compás giromagnético
gyromagnetic coupler acoplador giromagnético
gyromagnetic effect efecto giromagnético
gyromagnetic filter filtro giromagnético
gyromagnetic frequency frecuencia giromagnética
gyromagnetic limiter limitador giromagnético
gyromagnetic material material giromagnético
gyromagnetic ratio razón giromagnética
gyromagnetic resonance resonancia giromagnética
gyropilot giropiloto
gyroscope giroscopio
gyroscopic giroscópico
gyroscopic compass compás giroscópico
gyroscopic effect efecto giroscópico
gyrostabilized platform plataforma giroestabilizada
gyrostat giróstato
gyrostatic girostático
gyrostatic compass compás girostático
gyrotron girotrón
gyrotropic girotrópico

hafnium hafnio
hair hygrometer higrómetro de cabello
hair-trigger altamente sensible
hair wire hilo muy fino
hairline línea fina
hairpin antenna antena en horquilla
hairpin coil bobina en horquilla
hairpin coupling coil bobina de acoplamiento en
    horquilla
hairpin pickup coil bobina captadora en horquilla
hairspring pelo
halation halación, halo
half-add semisuma
half-adder semisumador
half-amplitude duration duración de media amplitud
half-amplitude recovery time tiempo de
    recuperación de media amplitud
half-bridge semipuente
half-carry semiacarreo
half-cell semicelda, semipila
half-cycle semiciclo
half-cycle impulse magnetizer magnetizador de
    impulso de semiciclo
half-cycle magnetizer magnetizador de semiciclo
half-dipole semidipolo
half-duplex semidúplex
half-duplex channel canal semidúplex
half-duplex circuit circuito semidúplex
half-duplex operation operación semidúplex
half-duplex repeater repetidor semidúplex
half-duplex service servicio semidúplex
half-duplex system sistema semidúplex
half-duplex transmission transmisión semidúplex
half-frequency semifrecuencia
half-height device dispositivo de media altura
half-height drive unidad de media altura
half-life media vida
half-load media carga
half-nut media tuerca
half-period semiperiodo
half-power beamwidth anchura de haz de media
    potencia
half-power frequencies frecuencias de media
    potencia
half-power point punto de media potencia
half-power width anchura de media potencia
half-step semitono
half-subtracter semirrestador
half tap toma media
half-time emitter emisor de medio tiempo
half-track recorder registrador de media pista
half-track tape cinta de media pista
half-wave media onda
half-wave antenna antena de media onda
half-wave dipole dipolo de media onda
half-wave doubler doblador de media onda
half-wave feeder alimentador de media onda
half-wave line línea de media onda
half-wave loop antenna antena de cuadro de media
    onda
half-wave plate placa de media onda

half-wave radiator radiador de media onda
half-wave rectification rectificación de media onda
half-wave rectifier rectificador de media onda
half-wave transmission line línea de transmisión de media onda
half-wave vibrator vibrador de media onda
half-wave voltage doubler doblador de tensión de media onda
half-width semianchura
half-word media palabra
halftone semitono
halftone characteristic característica de semitono
halide haluro
halide crystal cristal haloideo
Hall analog multiplier multiplicador analógico de Hall
Hall angle ángulo de Hall
Hall coefficient coeficiente de Hall
Hall constant constante de Hall
Hall effect efecto de Hall
Hall-effect device dispositivo de efecto de Hall
Hall-effect fluxmeter flujómetro de efecto de Hall
Hall-effect generator generador por efecto de Hall
Hall-effect modulator modulador por efecto de Hall
Hall-effect multiplier multiplicador por efecto de Hall
Hall-effect probe sonda de efecto de Hall
Hall field campo de Hall
Hall generator generador de Hall
Hall mobility movilidad de Hall
Hall modulator modulador de Hall
Hall multiplier multiplicador de Hall
Hall network red de Hall
Hall plate placa de Hall
Hall probe sonda de Hall
Hall terminals terminales de Hall
Hall voltage tensión de Hall
Hallwacks effect efecto de Hallwacks
halo halo
halogen halógeno
halogen counter contador de halógeno
halogen-quenched counter tube tubo contador autoextintor de halógeno
Halstead system sistema de Halstead
halt parada
halt condition condición de parada
halt indicator indicador de parada
halt instruction instrucción de parada
halt mode modo de parada
Hamilton's principle principio de Hamilton
hammer martillo
Hamming code código de Hamming
Hamming distance distancia de Hamming
hand burnishing bruñido a mano
hand capacitance capacitancia de mano
hand effect efecto de mano
hand feeding alimentación a mano
hand generator generador de mano
hand key manipulador
hand lamp lámpara de mano
hand-operated device dispositivo operado a mano
hand operation operación a mano
hand punch perforadora de mano
hand radar radar de mano
hand receiver receptor de mano
hand reset reposición manual
hand-reset relay relé de reposición manual
hand rule regla de la mano
hand telephone set aparato microtelefónico
hand-type pointer indicador de tipo de mano
hand winding arrollamiento a mano
hand wiring alambrado a mano
handie-talkie radioteléfono portátil

handling device dispositivo de manejo
handover traspaso
handset microteléfono
handset amplifier amplificador de microteléfono
handset earphone audífono de microteléfono
handshake intercambio de impulsos de sincronización
handshake interface interfaz de intercambio de impulsos de sincronización
handshaking intercambio de impulsos de sincronización
handwriting reader lectora de escritura
hang-up prevention prevención de paradas imprevistas
hang-up signal señal de colgado
hangover persistencia de bajas
hangup parada imprevista
harbor control radar radar de control de puerto
hard clipping recortamiento duro
hard copy copia impresa
hard-copy device dispositivo de copia impresa
hard-copy interface interfaz de copia impresa
hard-copy output salida de copia impresa
hard-copy terminal terminal de copia impresa
hard cosmic ray rayo cósmico duro
hard disk disco duro
hard-disk backup program programa de reserva de disco duro
hard-disk cartridge cartucho de disco duro
hard-disk controller controlador de disco duro
hard-disk cylinder cilindro de disco duro
hard-disk system sistema de disco duro
hard-drawn wire hilo estirado en duro
hard failure falla de equipo, falla completa
hard image imagen dura
hard magnetic material material fuertemente magnético
hard rubber goma dura
hard solder soldante duro
hard superconductor superconductor duro
hard tube tubo duro
hard vacuum alto vacío
hard-vacuum tube tubo de alto vacío
hard wood madera dura
hard X-rays rayos X duros
hardness dureza
hardness tester durómetro
hardware equipo físico, hardware
hardware address dirección de equipo físico
hardware assembler ensamblador de equipo físico
hardware availability ratio razón de disponibilidad de equipo físico
hardware check comprobación de equipo físico
hardware component componente de equipo físico
hardware configuration configuración de equipo físico
hardware configuration item artículo de configuración de equipo físico
hardware description language lenguaje de descripción de equipo físico
hardware design language lenguaje de diseño de equipo físico
hardware device dispositivo de equipo físico
hardware error error de equipo físico
hardware error recovery recuperación de error de equipo físico
hardware failure falla de equipo físico
hardware interrupt interrupción de equipo físico
hardware language lenguaje de equipo físico
hardware logic diagram diagrama de lógica de equipo físico
hardware monitor monitor de equipo físico
hardware recovery recuperación de equipo físico

hardware reset reposición de equipo físico
hardware resources recursos de equipo físico
Harkness circuit circuito de Harkness
harmful interference interferencia dañina
harmful radiation radiación dañina
harmonic armónico, armónica
harmonic absorber eliminador armónico
harmonic accentuation acentuación armónica
harmonic accentuator acentuador armónico
harmonic amplifier amplificador armónico
harmonic analysis análisis armónico
harmonic analyzer analizador armónico
harmonic antenna antena armónica
harmonic attenuation atenuación armónica
harmonic attenuator atenuador armónico
harmonic component componente armónico
harmonic composition composición armónica
harmonic content contenido armónico
harmonic conversion transducer transductor de
  conversión armónico
harmonic current corriente armónica
harmonic detector detector armónico
harmonic distortion distorsión armónica
harmonic-distortion meter medidor de distorsión
  armónica
harmonic-distortion percentage porcentaje de
  distorsión armónica
harmonic distribution distribución armónica
harmonic elimination eliminación armónica
harmonic eliminator eliminador armónico
harmonic emission emisión armónica
harmonic excitation excitación armónica
harmonic factor factor armónico
harmonic filter filtro armónico
harmonic filtering filtrado armónico
harmonic flow field campo de flujo armónico
harmonic frequency frecuencia armónica
harmonic function función armónica
harmonic generator generador armónico
harmonic intensification intensificación armónica
harmonic intensifier intensificador armónico
harmonic intensity intensidad armónica
harmonic interference interferencia armónica
harmonic leakage power potencia de fuga armónica
harmonic mean media armónica
harmonic motion moción armónica
harmonic oscillation oscilación armónica
harmonic oscillator oscilador armónico
harmonic percentage porcentaje armónico
harmonic producer productor armónico
harmonic progression progresión armónica
harmonic ratio razón armónica
harmonic reducer reductor armónico
harmonic reduction reducción armónica
harmonic resonance resonancia armónica
harmonic response respuesta armónica
harmonic ringing llamada armónica
harmonic series serie armónica
harmonic series of tones serie armónica de tonos
harmonic suppression supresión armónica
harmonic suppressor supresor armónico
harmonic telegraphy telegrafía armónica
harmonic telephone ringer timbre telefónico
  armónico
harmonic tolerance tolerancia armónica
harmonic transformer transformador armónico
harmonic wave onda armónica
harmonically related bands bandas relacionadas
  armónicamente
harness grupo de conductores
harp antenna antena en arpa
hartley hartley
Hartley oscillator oscilador de Hartley

hash chasquidos, ecos parásitos
hash filter filtro de chasquidos
hash total total de comprobación
Hay bridge puente de Hay
hazard peligro
hazard beacon radiofaro de peligro
hazardous peligroso
hazardous area área peligrosa
hazardous electrical condition condición eléctrica
  peligrosa
hazardous environment ambiente peligroso
hazardous material material peligroso
hazardous substance sustancia peligrosa
Hazeltine circuit circuito de Hazeltine
Hazeltine neutralization neutralización de Hazeltine
Hazeltine neutralizing circuit circuito neutralizador
  de Hazeltine
head cabeza
head alignment alineación de cabeza
head amplifier amplificador de cabeza
head assembly ensamblaje de cabeza, cabezal
head-cleaning device dispositivo limpiador de cabeza
head-cleaning diskette disquete limpiador de cabeza
head crash choque de cabeza
head degausser desmagnetizador de cabeza
head demagnetizer desmagnetizador de cabeza
head gap entrehierro de cabeza
head positioning posicionamiento de cabeza
head receiver audífono de cabeza
head slot ranura de cabeza
head stack unidad de cabezas múltiples
head-to-tape contact contacto de cabeza a cinta
header canal transversal, cabecera
header block bloque de cabecera
header card tarjeta de cabecera
header label etiqueta de cabecera
header record registro de cabecera
header segment segmento de cabecera
header specifications especificaciones de cabecera
header table tabla de cabecera
heading rumbo
headphone audífono
headphone adapter adaptador de audífono
headphone amplifier amplificador de audífono
headphone receiver receptor de audífono
headroom altura libre
headset audífonos
hearing aid audífono
hearing-aid battery batería de audífono
heart fibrillation fibrilación cardiaca
heat absorption absorción de calor
heat aging degeneración por calor
heat balance equilibrio térmico
heat capacity capacidad térmica
heat coil bobina térmica
heat conductivity conductividad térmica
heat-conductivity gage medidor de conductividad
  térmica
heat detector detector de calor
heat dissipation disipación térmica
heat dissipator disipador térmico
heat drift deriva térmica
heat effect efecto térmico
heat efficiency eficiencia térmica
heat energy energía térmica
heat exchange intercambio de calor
heat exchanger intercambiador de calor
heat-exchanger circuit circuito intercambiador de
  calor
heat-exchanger cooling system sistema de
  enfriamiento de intercambiador de calor
heat flow flujo térmico
heat gradient gradiente de calor

heat-insulating termoaislante
heat insulation termoaislamiento
heat loss pérdida de calor, pérdida por calor
heat of emission calor de emisión
heat of fusion calor de fusión
heat of reaction calor de reacción
heat of vaporization calor de vaporización
heat pump bomba de calor
heat radiator radiador de calor
heat rate eficiencia térmica
heat relay relé térmico
heat remover extractor de calor
heat resistance resistencia al calor
heat-resistant glass vidrio resistente al calor
heat-sensitive termosensible, sensible al calor
heat-sensitive resistor resistor termosensible
heat-sensitive switch conmutador termosensible
heat shield pantalla térmica
heat-shrink tubing tubería termocontráctil
heat sink disipador térmico
heat source fuente de calor
heat spot punto de calor
heat transfer transferencia de calor
heat treatment tratamiento térmico
heat unit unidad térmica
heat wave onda térmica
heater calefactor
heater bias polarización de calefactor
heater circuit circuito calefactor
heater current corriente de calefactor
heater-current consumption consumo de corriente
   de calefactor
heater hum zumbido de calefactor
heater power potencia de calefactor
heater power supply fuente de alimentación de
   calefactor
heater string serie de calefactores
heater temperature temperatura de calefactor
heater transformer transformador de calefactor
heater voltage tensión de calefactor
heater-voltage coefficient coeficiente de tensión de
   calefactor
heater wattage wattaje de calefactor
heating calentamiento
heating cable cable de calentamiento
heating cell célula de calentamiento
heating chamber cámara de calentamiento
heating cycle ciclo de calentamiento
heating depth profundidad de calentamiento
heating effect efecto de calentamiento
heating element elemento de calentamiento
heating limit límite de calentamiento
heating of anode calentamiento de ánodo
heating of grid calentamiento de rejilla
heating pattern patrón de calentamiento
heating power potencia de calentamiento
heating resistor resistor de calentamiento
heating station estación de calentamiento
heating surface superficie de calentamiento
heating system sistema de calentamiento
heating time tiempo de calentamiento
heating unit unidad de calentamiento
heatproof a prueba de calor
heatsink disipador térmico
Heaviside-Campbell bridge puente de
   Heaviside-Campbell
Heaviside-Hertz equations ecuaciones de
   Heaviside-Hertz
Heaviside-Kennelly layer capa de
   Heaviside-Kennelly
Heaviside layer capa de Heaviside
Heaviside-Lorentz electromagnetic system sistema
   electromagnético de Heaviside-Lorentz

Heaviside-Lorentz system sistema de
   Heaviside-Lorentz
Heaviside mutual-inductance bridge puente de
   inductancia mutua de Heaviside
Heaviside units unidades de Heaviside
heavy-duty contact contacto reforzado
heavy-duty relay relé reforzado
heavy electron electrón pesado
heavy hydrogen hidrógeno pesado
heavy isotope isótopo pesado
heavy metal metal pesado
heavy water agua pesada
hecto hecto
hectoampere hectoampere
hectometer hectómetro
hectometric hectométrico
hectometric wave onda hectométrica
hectowatt hectowatt
heel effect efecto de talón
heelpiece culata
Hefner candle bujía de Hefner
Hefner lamp lámpara de Hefner
Hefner unit unidad de Hefner
height control control de altura
height effect efecto de altura
height finder radar telealtímetro
height-finder radar radar telealtímetro
height gain ganancia por altura
height marker marcador de altura
height of antenna altura de antena
Heil oscillator oscilador de Heil
Heil tube tubo de Heil
Heisenberg uncertainty principle principio de
   incertidumbre de Heisenberg
Heising modulation modulación de Heising
helical antenna antena helicoidal
helical element elemento helicoidal
helical grid rejilla helicoidal
helical line línea helicoidal
helical path trayectoria helicoidal
helical plate placa helicoidal
helical potentiometer potenciómetro helicoidal
helical scan exploración helicoidal
helical scanning exploración helicoidal
helical sweep barrido helicoidal
helical transmission line línea de transmisión
   helicoidal
helicone helicono
helioelectric cell célula helioeléctrica
helionics heliónica
heliostat heliostato
helitron helitrón
helitron oscillator oscilador helitrón
helium helio
helium lamp lámpara de helio
helium mass spectrometer espectrómetro de masa
   de helio
helium-neon laser láser helio-neón
helium spectrometer espectrómetro de helio
helix hélice
helix antenna antena helicoidal
helix circuit circuito helicoidal
helix current corriente helicoidal
helix line línea helicoidal
helix recorder registrador helicoidal
helix waveguide guíaondas helicoidal
Helmholtz coil bobina de Helmholtz
Helmholtz double layer capa doble de Helmholtz
Helmholtz pendulum péndulo de Helmholtz
Helmholtz resonator resonador de Helmholtz
help key tecla de ayuda
help panel panel de ayuda
hemimorphic hemimórfico

hemispherical hemisférico
henry henry
heptode heptodo
hermaphrodite coaxial connector conector coaxial hermafrodita
hermaphrodite connector conector hermafrodita
hermaphroditic plug clavija hermafrodítica
hermetic hermético
hermetic motor motor hermético
hermetic refrigerant refrigerante hermético
hermetic seal cierre hermético
hermetical hermético
hermetically sealed sellado herméticamente
hermetically-sealed relay relé sellado herméticamente
Heroult furnace horno de Heroult
herringbone pattern patrón en espina de pescado
Herschel-Quincke tube tubo de Herschel-Quincke
hertz hertz
Hertz antenna antena de Hertz
Hertz dipole dipolo de Hertz
Hertz effect efecto de Hertz
Hertz oscillator oscilador de Hertz
Hertz radiator radiador de Hertz
Hertz vector vector de Hertz
Hertzian antenna antena hertziana
Hertzian beam haz hertziano
Hertzian cable cable hertziano
Hertzian oscillator oscilador hertziano
Hertzian radiation radiación hertziana
Hertzian spectroscopy espectroscopia hertziana
Hertzian vector vector hertziano
Hertzian waves ondas hertzianas
hesitation hesitación
heterochromatic heterocromático
heterodyne heterodino
heterodyne conversion transducer transductor de conversión heterodino
heterodyne detection detección heterodina
heterodyne detector detector heterodino
heterodyne eliminator eliminador heterodino
heterodyne filter filtro heterodino
heterodyne frequency frecuencia heterodina
heterodyne frequency meter frecuencímetro heterodino
heterodyne harmonic analyzer analizador armónico heterodino
heterodyne interference interferencia heterodina
heterodyne method método heterodino
heterodyne mixer mezclador heterodino
heterodyne modulator modulador heterodino
heterodyne oscillator oscilador heterodino
heterodyne receiver receptor heterodino
heterodyne reception recepción heterodina
heterodyne repeater repetidor heterodino
heterodyne-type frequency meter frecuencímetro de tipo heterodino
heterodyne wave analyzer analizador de ondas heterodino
heterodyne wavemeter ondámetro heterodino
heterodyne whistle silbido heterodino
heterodyning heterodinaje
heterogeneity heterogeneidad
heterogenous heterogéneo
heterogenous radiation radiación heterogénea
heteropolar heteropolar
heteropolar alternator alternador heteropolar
heteropolar generator generador heteropolar
heteropolar machine máquina heteropolar
heterostatic method método heterostático
heuristic heurístico
heuristic approach acercamiento heurístico
heuristic program programa heurístico

heuristic programming programación heurística
heuristic routine rutina heurística
Heusler alloys aleación de Heusler
hexadecimal hexadecimal
hexadecimal constant constante hexadecimal
hexadecimal digit dígito hexadecimal
hexadecimal notation notación hexadecimal
hexadecimal number número hexadecimal
hexadecimal number system sistema numérico hexadecimal
hexadecimal numeral número hexadecimal
hexadecimal numeration system sistema de numeración hexadecimal
hexadecimal point coma hexadecimal, punto hexadecimal
hexadecimal system sistema hexadecimal
hexadecimal-to-decimal conversion conversión de hexadecimal a decimal
hexode hexodo
hickey casquillo conectador
hidden code código oculto
hidden field campo oculto
hidden line línea oculta
hidden surface superficie oculta
hierarchical jerárquico
hierarchical database base de datos jerárquica
hierarchical database management system sistema de administración de base de datos jerárquica
hierarchical decomposition descomposición jerárquica
hierarchical direct-access method método de acceso directo jerárquico
hierarchical direct organization organización directa jerárquica
hierarchical file system sistema de archivos jerárquico
hierarchical model modelo jerárquico
hierarchical network red jerárquica
hierarchical sequence secuencia jerárquica
hierarchical sequential-access method método de acceso secuencial jerárquico
hierarchical sequential organization organización secuencial jerárquica
hierarchical storage almacenamiento jerárquico
hierarchical structure estructura jerárquica
hierarchy jerarquía
high address dirección alta
high-amperage current corriente de alto amperaje
high band banda alta
high bit rate alta velocidad de bits
high boost refuerzo de altas
high conductivity alta conductividad
high contrast alto contraste
high-contrast image imagen de alto contraste
high current alta corriente
high definition alta definición
high-definition image imagen de alta definición
high-definition radar radar de alta definición
high-definition television televisión de alta definición
high density alta densidad
high-density diskette disquete de alta densidad
high-density tape cinta de alta densidad
high-efficiency amplifier amplificador de alta eficiencia
high-efficiency grid modulation modulación por rejilla de alta eficiencia
high-efficiency linear amplifier amplificador lineal de alta eficiencia
high-efficiency modulation modulación de alta eficiencia
high-energy material material de alta energía
high-energy particle partícula de alta energía
high fidelity alta fidelidad

high-fidelity amplifier amplificador de alta fidelidad
high-fidelity equipment equipo de alta fidelidad
high-fidelity headphones audífonos de alta fidelidad
high-fidelity receiver receptor de alta fidelidad
high-fidelity recording registro de alta fidelidad
high-fidelity signal señal de alta fidelidad
high filter filtro de altas
high frequency alta frecuencia
high-frequency alternator alternador de alta frecuencia
high-frequency amplifier amplificador de alta frecuencia
high-frequency antenna antena de alta frecuencia
high-frequency bias polarización de alta frecuencia
high-frequency biasing polarización de alta frecuencia
high-frequency broadcast band banda de radiodifusión de alta frecuencia
high-frequency broadcast station estación de radiodifusión de alta frecuencia
high-frequency broadcast transmitter transmisor de radiodifusión de alta frecuencia
high-frequency broadcasting radiodifusión de alta frecuencia
high-frequency choke choque de alta frecuencia
high-frequency compensation compensación de alta frecuencia
high-frequency condenser microphone micrófono de condensador de alta frecuencia
high-frequency converter convertidor de alta frecuencia
high-frequency crystal cristal de alta frecuencia
high-frequency current corriente de alta frecuencia
high-frequency direction finder radiogoniómetro de alta frecuencia
high-frequency driver altavoz de alta frecuencia
high-frequency furnace horno de alta frecuencia
high-frequency heating calentamiento por alta frecuencia
high-frequency loudspeaker altavoz de alta frecuencia
high-frequency oscillator oscilador de alta frecuencia
high-frequency radar radar de alta frecuencia
high-frequency resistance resistencia en alta frecuencia
high-frequency response respuesta en alta frecuencia
high-frequency speaker altavoz de alta frecuencia
high-frequency transformer transformador de alta frecuencia
high-frequency transmitter transmisor de alta frecuencia
high-frequency trimmer condensador de ajuste de altas frecuencias
high-frequency unit unidad de alta frecuencia
high-gain amplifier amplificador de alta ganancia
high-gain antenna antena de alta ganancia
high-gain loop bucle de alta ganancia
high-gamma tube tubo de alta gamma
high impedance alta impedancia
high-impedance output salida de alta impedancia
high-impedance relay relé de alta impedancia
high-impedance rotor rotor de alta impedancia
high-impedance triode triodo de alta impedancia
high-impedance voltmeter vóltmetro de alta impedancia
high-information delta modulation modulación en delta de alta información
high intensity alta intensidad
high-intensity discharge lamp lámpara de descarga de alta intensidad
high-intensity light luz de alta intensidad
high level alto nivel

high-level compiler compilador de alto nivel
high-level detector detector de alto nivel
high-level format formato de alto nivel
high-level input current corriente de entrada de alto nivel
high-level language lenguaje de alto nivel
high-level message mensaje de alto nivel
high-level modulation modulación de alto nivel
high-level multiplexer multiplexor de alto nivel
high-level output current corriente de salida de alto nivel
high-level programming language lenguaje de programación de alto nivel
high-level recovery recuperación de alto nivel
high-mu tube tubo de alta mu
high-mu valve válvula de alta mu
high order orden alto
high-order digit dígito de orden alto
high-order language lenguaje de orden alto
high-order position posición de orden alto
high-pass filter filtro pasaaltos
high-pass filter method método de filtro pasaaltos
high-performance de alto rendimiento
high-performance channel canal de alto rendimiento
high-performance equipment equipo de alto rendimiento
high-performance navigation system sistema de navegación de alto rendimiento
high-potential alto potencial
high-potential test prueba de alto potencial
high-power rectifier rectificador de alta potencia
high-power triode triodo de alta potencia
high-power tube tubo de alta potencia
high-power valve válvula de alta potencia
high pressure alta presión
high-pressure contact contacto de alta presión
high-pressure sodium lamp lámpara de sodio de alta presión
high-pressure vacuum pump bomba de vacío de alta presión
high-purity germanium germanio de alta pureza
high-rate battery charger cargador de batería de alto régimen
high-rate charge carga de alto régimen
high-reactance rotor rotor de alta reactancia
high-reactance transformer transformador de alta reactancia
high-reliability electric line línea eléctrica de alta confiabilidad
high-reliability relay relé de alta confiabilidad
high-reliability tube tubo de alta confiabilidad
high resistance alta resistencia
high-resistance joint junta de alta resistencia
high-resistance probe sonda de alta resistencia
high-resistance rotor rotor de alta resistencia
high-resistance short cortocircuito de alta resistencia
high-resistance voltmeter vóltmetro de alta resistencia
high resolution alta resolución
high-resolution graphics gráficos de alta resolución
high-resolution power supply fuente de alimentación de alta resolución
high-resolution radar radar de alta resolución
high sensitivity alta sensibilidad
high speed alta velocidad
high-speed buffer almacenamiento intermedio de alta velocidad
high-speed bus bus de alta velocidad
high-speed carry acarreo de alta velocidad
high-speed circuit-breaker cortocircuito de alta velocidad
high-speed computer computadora de alta velocidad
high-speed data transfer transferencia de datos de

alta velocidad
high-speed diode diodo de alta velocidad
high-speed electron beam haz electrónico de alta
velocidad
high-speed excitation system sistema de excitación
de alta velocidad
high-speed feed alimentación de alta velocidad
high-speed flip-flop basculador de alta velocidad
high-speed grounding switch conmutador de puesta
a tierra de alta velocidad
high-speed line línea de alta velocidad
high-speed logic lógica de alta velocidad
high-speed memory memoria de alta velocidad
high-speed oscillography oscilografía de alta
velocidad
high-speed oscilloscope osciloscopio de alta
velocidad
high-speed particle partícula de alta velocidad
high-speed plotter trazador de alta velocidad
high-speed printer impresora de alta velocidad
high-speed processor procesador de alta velocidad
high-speed punch perforadora de alta velocidad
high-speed reader lectora de alta velocidad
high-speed regulator regulador de alta velocidad
high-speed relay relé de alta velocidad
high-speed scan exploración de alta velocidad
high-speed storage almacenamiento de alta velocidad
high-speed switch conmutador de alta velocidad
high-speed switching transistor transistor de
conmutación de alta velocidad
high-speed telegraph transmission transmisión
telegráfica de alta velocidad
high-speed terminal terminal de alta velocidad
high-speed transistor transistor de alta velocidad
high temperature alta temperatura
high tension alta tensión
high-tension battery batería de alta tensión
high-tension cable cable de alta tensión
high-tension coil bobina de alta tensión
high-tension current corriente de alta tensión
high-tension line línea de alta tensión
high-threshold logic lógica de alto umbral
high-usage trunk troncal de alto uso
high-usage trunk group grupo de enlace de alto uso
high vacuum alto vacío
high-vacuum gage medidor de alto vacío
high-vacuum phototube fototubo de alto vacío
high-vacuum rectifier rectificador de alto vacío
high-vacuum thermocouple termopar de alto vacío
high-vacuum tube tubo de alto vacío
high-vacuum valve válvula de alto vacío
high-velocity scanning exploración de alta velocidad
high voltage alta tensión
high-voltage apparatus aparato de alta tensión
high-voltage battery batería de alta tensión
high-voltage capacity capacidad de alta tensión
high-voltage cover sonda de alta tensión
high-voltage feed alimentación de alta tensión
high-voltage power supply fuente de alimentación de
alta tensión
high-voltage probe sonda de alta tensión
high-voltage rectifier rectificador de alta tensión
high-voltage regulator regulador de alta tensión
high-voltage relay relé de alta tensión
high-voltage system sistema de alta tensión
high-voltage test prueba de alta tensión
high-voltage time test prueba de tiempo de alta
tensión
high-voltage transformer transformador de alta
tensión
high-voltage winding arrollamiento de alta tensión
higher-frequency component componente de
frecuencia superior

higher layer capa superior
higher level nivel superior
higher-level indicator indicador de nivel superior
higher-level language lenguaje de nivel superior
higher-order language lenguaje de orden superior
higher-order mode modo de orden superior
highlight área brillante
highly directional antenna antena altamente
direccional
highly sensitive receiver receptor altamente sensible
highway trayectoria principal
highway crossing signal señal de cruce de ferrocarril
hill-and-dale recording registro vertical
hinged-iron ammeter amperímetro de hierro
articulado
hinged-iron instrument instrumento de hierro
articulado
hiss siseo
histogram histograma
hit acierto, perturbación momentánea, coincidencia
de dos pulsos
hit ratio razón de aciertos
Hittorf dark space espacio obscuro de Hittorf
Hittorf tube tubo de Hittorf
hodoscope hodoscopio
Hoeppner connection conexión de Hoeppner
Hoffmann electrometer electrómetro de Hoffmann
hoghorn bocina curvada
hoghorn antenna antena de bocina
hoistway trayectoria de ascensor
hold sostenimiento, detención, mantenimiento,
retención
hold circuit circuito de retención
hold-closed mechanism mecanismo para mantener
cerrado
hold code código de retención
hold condition condición de retención
hold control control de retención
hold current corriente de sostenimiento
hold electrode electrodo de contacto permanente
hold mode modo de mantenimiento
hold queue cola de retención
hold time tiempo de retención
holding anode ánodo mantenedor
holding beam haz de sostenimiento
holding coil bobina de sostenimiento
holding control control de retención
holding current corriente de sostenimiento
holding gun cañón de mantenimiento
holding magnet imán de retención
holding relay relé de retención
holding time tiempo de retención
holding torque par de retención
holding winding arrollamiento de mantenimiento
hole hueco, abertura, agujero, perforación
hole capture captura de huecos
hole conduction conducción por huecos
hole configuration configuración de huecos
hole current corriente de huecos
hole density densidad de huecos
hole-electron pair par hueco-electrón
hole flow flujo de huecos
hole injection inyección de huecos
hole injector inyector de huecos
hole mobility movilidad de huecos
hole pattern patrón de huecos
hole site lugar de perforación
hole storage acumulación de huecos
hole-storage factor factor de acumulación de huecos
hole trap trampa de huecos
Hollerith code código de Hollerith
Hollerith equipment equipo de Hollerith
hollow anode ánodo hueco

hollow beam haz hueco
hollow cathode cátodo hueco
hollow-cathode tube tubo de cátodo hueco
hollow conductor conductor hueco
hollow core núcleo hueco
hollow-core conductor conductor de núcleo hueco
hollow cylinder cilindro hueco
hollow electrode electrodo hueco
holmium holmio
holocamera holocámara
hologram holograma
holographic holográfico
holographic memory memoria holográfica
holographic plate placa holográfica
holographic recorder registrador holográfico
holographic storage almacenamiento holográfico
holography holografía
Holt amplifier amplificador de Holt
Holtz tube tubo de Holtz
home appliance aparato doméstico
homeostasis homeostasis
hometaxial-base transistor transistor de base
　　homotaxial
homing guiado, recalada, reposición
homing antenna antena buscadora
homing beacon radiofaro direccional
homing device dispositivo guiador
homing relay relé de reposición
homing station estación guiadora
homochromatic homocromático
homochromatic gain ganancia homocromática
homodyne homodino
homodyne detection detección homodina
homodyne receiver receptor homodino
homodyne reception recepción homodina
homogeneity homogeneidad
homogeneous homogéneo
homogeneous cladding revestimiento homogéneo
homogeneous line broadening ensanchamiento de
　　línea homogéneo
homogeneous magnetic field campo magnético
　　homogéneo
homogeneous radiation radiación homogénea
homogeneous redundancy redundancia homogénea
homogeneous series serie homogénea
homogeneous tape cinta homogénea
homomorphism homomorfismo
homopolar homopolar
homopolar alternator alternador homopolar
homopolar field impedance impedancia de campo
　　homopolar
homopolar generator generador homopolar
homopolar induction inducción homopolar
homopolar magnet imán homopolar
honeycomb coil bobina en panal
honeycomb winding arrollamiento en panal
hook-on instrument instrumento de medida con
　　pinza
hook switch conmutador de gancho
hook transistor transistor de gancho
Hooke's law ley de Hooke
hookup conexión, esquema de montaje
hookup wire hilo de conexión
hoot stop parada con señal auditiva
hop salto
hopper almacén de alimentación
horizon horizonte
horizontal horizontal
horizontal adjustment ajuste horizontal
horizontal advance avance horizontal
horizontal amplification amplificación horizontal
horizontal amplifier amplificador horizontal
horizontal angle of deviation ángulo horizontal de

　　desviación
horizontal axis eje horizontal
horizontal beam haz horizontal
horizontal beamwidth anchura de haz horizontal
horizontal blanking extinción horizontal
horizontal-blanking pulse impulso de extinción
　　horizontal
horizontal-blanking signal señal de extinción
　　horizontal
horizontal centering centrado horizontal
horizontal centering control control de centrado
　　horizontal
horizontal channel canal horizontal
horizontal component componente horizontal
horizontal convergence convergencia horizontal
horizontal-convergence control control de
　　convergencia horizontal
horizontal coordinates coordenadas horizontales
horizontal definition definición horizontal
horizontal deflection deflexión horizontal
horizontal-deflection coils bobinas de deflexión
　　horizontal
horizontal-deflection electrodes electrodos de
　　deflexión horizontal
horizontal-deflection oscillator oscilador de
　　deflexión horizontal
horizontal-deflection plates placas de deflexión
　　horizontal
horizontal dipole dipolo horizontal
horizontal directivity directividad horizontal
horizontal-directivity diagram diagrama de
　　directividad horizontal
horizontal-discharge tube tubo de descarga
　　horizontal
horizontal doublet doblete horizontal
horizontal-drive control control de excitación
　　horizontal
horizontal dynamic convergence convergencia
　　dinámica horizontal
horizontal dynamic focus enfoque dinámico
　　horizontal
horizontal feed alimentación horizontal
horizontal field strength intensidad de campo
　　horizontal
horizontal field-strength diagram diagrama de
　　intensidad de campo horizontal
horizontal flyback retorno horizontal
horizontal frequency frecuencia horizontal
horizontal frequency response respuesta de
　　frecuencia horizontal
horizontal gain ganancia horizontal
horizontal-gain control control de ganancia
　　horizontal
horizontal hold sincronismo horizontal
horizontal-hold control control de sincronismo
　　horizontal
horizontal-hum bars barras de zumbido horizontales
horizontal input entrada horizontal
horizontal instability inestabilidad horizontal
horizontal-line frequency frecuencia de líneas
　　horizontales
horizontal linearity linealidad horizontal
horizontal-linearity adjustment ajuste de linealidad
　　horizontal
horizontal-linearity control control de linealidad
　　horizontal
horizontal lock enganche horizontal
horizontal-lock circuit circuito de enganche
　　horizontal
horizontal machine máquina horizontal
horizontal microinstruction microinstrucción
　　horizontal
horizontal multivibrator multivibrador horizontal

horizontal oscillator oscilador horizontal
horizontal-output stage etapa de salida horizontal
horizontal-output transformer transformador de
  salida horizontal
horizontal-output tube tubo de salida horizontal
horizontal parity check comprobación de paridad
  horizontal
horizontal pattern patrón horizontal
horizontal polarization polarización horizontal
horizontal-positioning control control de
  posicionamiento horizontal
horizontal pulse impulso horizontal
horizontal quantity cantidad horizontal
horizontal radiation pattern patrón de radiación
  horizontal
horizontal recording registro horizontal
horizontal-repetition rate frecuencia de exploración
  horizontal
horizontal resolution resolución horizontal
horizontal retrace retorno horizontal
horizontal-retrace blanking extinción de retorno
  horizontal
horizontal-retrace line línea de retorno horizontal
horizontal scale escala horizontal
horizontal scan barrido horizontal
horizontal scanning exploración horizontal
horizontal-scanning frequency frecuencia de
  exploración horizontal
horizontal-scanning generator generador de
  exploración horizontal
horizontal sensitivity sensibilidad horizontal
horizontal signal señal horizontal
horizontal sweep barrido horizontal
horizontal-sweep amplifier amplificador de barrido
  horizontal
horizontal-sweep circuit circuito de barrido
  horizontal
horizontal-sweep frequency frecuencia de barrido
  horizontal
horizontal-sweep generator generador de barrido
  horizontal
horizontal-sweep rate frecuencia de barrido
  horizontal
horizontal-sweep transformer transformador de
  barrido horizontal
horizontal-sync discriminator discriminador de
  sincronismo horizontal
horizontal-sync pulse impulso de sincronismo
  horizontal
horizontal synchronization sincronización horizontal
horizontal wave onda horizontal
horizontal waveform forma de onda horizontal
horizontal-width control control de anchura
  horizontal
horizontally horizontalmente
horizontally-polarized wave onda polarizada
  horizontalmente
horn bocina, cuerno
horn antenna antena de bocina
horn aperture abertura de bocina
horn arrester pararrayos de cuernos
horn loudspeaker altavoz de bocina
horn mouth boca de bocina
horn reflector antenna antena de reflector de bocina
horn speaker altavoz de bocina
horn throat garganta de bocina
horsepower caballo de fuerza
horsepower-hour caballo de fuerza-hora
horseshoe coil bobina en herradura
horseshoe magnet imán en herradura
host computer computadora primaria
host machine máquina principal
host processor procesador principal

host system sistema principal
hot caliente, activo
hot-air soldering soldadura por aire caliente
hot carrier portador activo
hot-carrier diode diodo de portadores activos
hot cathode cátodo caliente
hot-cathode discharge lamp lámpara de descarga de
  cátodo caliente
hot-cathode lamp lámpara de cátodo caliente
hot-cathode rectifier rectificador de cátodo caliente
hot-cathode tube tubo de cátodo caliente
hot-cathode X-ray tube tubo de rayos X de cátodo
  caliente
hot electrode electrodo caliente
hot filament filamento caliente
hot-filament gas detector detector de gases de
  filamento caliente
hot junction unión caliente
hot line línea energizada
hot-pen recorder registrador de pluma caliente
hot resistance resistencia caliente
hot spark chispa caliente
hot spot punto caliente
hot-strip ammeter amperímetro de franja caliente
hot-wire ammeter amperímetro de hilo caliente
hot-wire anemometer anemómetro de hilo caliente
hot-wire detector detector de hilo caliente
hot-wire flowmeter flujómetro de hilo caliente
hot-wire instrument instrumento de hilo caliente
hot-wire meter medidor de hilo caliente
hot-wire microphone micrófono de hilo caliente
hot-wire relay relé de hilo caliente
hot-wire transducer transductor de hilo caliente
hotplate placa para calentar
hottest-spot temperature temperatura del punto más
  caliente
hour-discharge rate régimen de descarga por hora
hour meter medidor de horas
house cable cable de casa
housekeeping check comprobación de preparación
housekeeping command mando de preparación
housekeeping instruction instrucción de preparación
housekeeping operation operación de preparación
housekeeping program programa de preparación
housekeeping routine rutina de preparación
housing caja, envuelta
howl aullido
howler aullador
howler circuit circuito de aullador
hue tinte, matiz
hue control control de tintes
hue range gama de tintes
Hull diagram diagrama de Hull
hum zumbido
hum bar barra de zumbido
hum bucking reducción de zumbido
hum-bucking coil bobina de reducción de zumbido
hum field campo de zumbido
hum-free libre de zumbido
hum-free relay relé libre de zumbido
hum frequency frecuencia de zumbido
hum interference interferencia de zumbido
hum jitter fluctuación por zumbido
hum level nivel de zumbido
hum modulation modulación de zumbido
hum voltage tensión de zumbido
human engineering ingeniería humana
human factors factores humanos
humidification humidificación
humidimeter humidímetro
humidity humedad
humidity detector detector de humedad
humidity indicator indicador de humedad

humidity meter humidímetro
humidity-proof a prueba de humedad
humidity protection protección contra humedad
humidity sensor sensor de humedad
hunting oscilación, penduleo, exploración
Huygens' principle principio de Huygens
Huygens' sources fuentes de Huygens
hybrid híbrido
hybrid active circuit circuito activo híbrido
hybrid balance equilibrio híbrido
hybrid circuit circuito híbrido
hybrid coil bobina híbrida
hybrid computer computadora híbrida
hybrid computer system sistema de computadora
    híbrido
hybrid coupler acoplador híbrido
hybrid coupling acoplamiento híbrido
hybrid electromagnetic wave onda electromagnética
    híbrida
hybrid integrated circuit circuito integrado híbrido
hybrid interface interfaz híbrida
hybrid junction unión híbrida
hybrid microcircuit microcircuito híbrido
hybrid microstructure microestructura híbrida
hybrid mode modo híbrido
hybrid network red híbrida
hybrid panel panel híbrido
hybrid parameter parámetro híbrido
hybrid reflectometer reflectómetro híbrido
hybrid ring anillo híbrido
hybrid set conjunto híbrido
hybrid simulation simulación híbrida
hybrid system sistema híbrido
hybrid tee T híbrida
hybrid thin-film circuit circuito de película delgada
    híbrido
hybrid transformer transformador híbrido
hybrid transmission transmisión híbrida
hybrid wave onda híbrida
hydraulic circuit circuito hidráulico
hydroacoustic hidroacústico
hydroacoustic transducer transductor hidroacústico
hydroacoustics hidroacústica
hydrodynamic hidrodinámico
hydrodynamic oscillator oscilador hidrodinámico
hydrodynamic pressure presión hidrodinámica
hydroelectric hidroeléctrico
hydroelectric generating station estación generadora
    hidroeléctrica
hydroelectric machine máquina hidroeléctrica
hydroelectric power potencia hidroeléctrica
hydroelectric power station central hidroeléctrica
hydroelectric station estación hidroeléctrica
hydrogen hidrógeno
hydrogen atmosphere atmósfera de hidrógeno
hydrogen atom átomo de hidrógeno
hydrogen electrode electrodo de hidrógeno
hydrogen-ion concentration concentración de iones
    de hidrógeno
hydrogen lamp lámpara de hidrógeno
hydrogen line línea de hidrógeno
hydrogen maser máser de hidrógeno
hydrogen thyratron tiratrón de hidrógeno
hydrokinetic hidrocinético
hydrolysis hidrólisis
hydromagnetic hidromagnético
hydromagnetic capacitor capacitor hidromagnético
hydromagnetic condenser condensador
    hidromagnético
hydromagnetic oscillation oscilación hidromagnética
hydromagnetic wave onda hidromagnética
hydromagnetics hidromagnética
hydrometer hidrómetro

hydrophone hidrófono
hydrostatic pressure presión hidrostática
hygristor higristor
hygrograph higrógrafo
hygrometer higrómetro
hygroscopic material material higroscópico
hygrostat higrostato
hygrothermograph higrotermógrafo
hyperacoustical zone zona hiperacústica
hyperacoustics hiperacústica
hyperbolic cosecant cosecante hiperbólica
hyperbolic cosine coseno hiperbólico
hyperbolic cotangent cotangente hiperbólica
hyperbolic error error hiperbólico
hyperbolic-field multiplier multiplicador de campo
    hiperbólico
hyperbolic function función hiperbólica
hyperbolic horn bocina hiperbólica
hyperbolic logarithm logaritmo hiperbólico
hyperbolic navigation navegación hiperbólica
hyperbolic navigation system sistema de navegación
    hiperbólica
hyperbolic radian radián hiperbólico
hyperbolic secant secante hiperbólica
hyperbolic sine seno hiperbólico
hyperbolic tangent tangente hiperbólica
hyperdirective antenna antena hiperdirectiva
hyperfocal distance distancia hiperfocal
hyperfrequency hiperfrecuencia
hyperfrequency circuit circuito de hiperfrecuencias
hyperfrequency tube tubo de hiperfrecuencias
hyperfrequency waves ondas de hiperfrecuencia
hyperharmonic series serie hiperarmónica
hyperon hiperón
hyperpolarization hiperpolarización
hypersonic hipersónico
hypersonic speed velocidad hipersónica
hypersound hipersonido
hypothesis hipótesis
hypothetical reference circuit circuito de referencia
    hipotético
hypsogram hipsograma
hypsometer hipsómetro
hysteresigram histeresigrama
hysteresigraph histeresígrafo
hysteresimeter histeresímetro
hysteresis histéresis
hysteresis brake freno por histéresis
hysteresis clutch embrague por histéresis
hysteresis coefficient coeficiente de histéresis
hysteresis curve curva de histéresis
hysteresis distortion distorsión por histéresis
hysteresis energy energía de histéresis
hysteresis error error por histéresis
hysteresis factor factor de histéresis
hysteresis heat calor por histéresis
hysteresis heater calentador por histéresis
hysteresis heating calentamiento por histéresis
hysteresis loop bucle de histéresis
hysteresis loss pérdida por histéresis
hysteresis meter histeresímetro
hysteresis motor motor de histéresis
hysteresis region región de histéresis
hysteresis torque par de histéresis
hysteresis voltage tensión de histéresis
hysteresiscope histeresiscopio

I

iconoscope iconoscopio
iconoscope camera cámara de iconoscopio
iconoscopic iconoscópico
iconotron iconotrón
ideal ideal
ideal bunching agrupamiento ideal
ideal capacitor capacitor ideal
ideal component componente ideal
ideal conductor conductor ideal
ideal crystal cristal ideal
ideal dielectric dieléctrico ideal
ideal efficiency eficiencia ideal
ideal filter filtro ideal
ideal gas gas ideal
ideal junction unión ideal
ideal magnetization magnetización ideal
ideal noise diode diodo de ruido ideal
ideal permeability permeabilidad ideal
ideal rectifier rectificador ideal
ideal site lugar ideal
ideal transducer transductor ideal
ideal transformer transformador ideal
ideal value valor ideal
ideal voltage source fuente de tensión ideal
idealized idealizado
idealized system sistema idealizado
identification identificación
identification beacon radiofaro de identificación
identification card tarjeta de identificación
identification-card reader lectora de tarjetas de
    identificación
identification division división de identificación
identification group grupo de identificación
identification label etiqueta de identificación
identification light luz de identificación
identification mark marca de identificación
identification panel panel de identificación
identification phase fase de identificación
identification sequence secuencia de identificación
identification sign signo de identificación
identification system sistema de identificación
identified identificado
identifier identificador
identifier word palabra identificadora
identity element elemento de identidad
identity operation operación de identidad
identity unit unidad de identidad
ideographic ideográfico
ideographic character carácter ideográfico
ideographic character set conjunto de caracteres
    ideográficos
idiochromatic idiocromático
idioelectric idioeléctrico
idiostatic method método idiostático
idle vacante, libre, en reposo
idle channel canal vacante
idle channel code código de canal vacante
idle channel noise ruido de canal vacante
idle character carácter de relleno
idle circuit circuito en reposo
idle current corriente reactiva

idle line línea libre
idle period periodo de reposo
idle power potencia reactiva
idle-power regulation regulación de potencia
    reactiva
idle state estado de reposo
idle time tiempo de reposo
idler circuit circuito complementario
idler frequency frecuencia complementaria
idler wheel rueda de transferencia
idling operación en espera
idling circuit circuito complementario
idling current corriente en espera
idling frequency frecuencia complementaria
idling power potencia en espera
idling voltage tensión en espera
IF-AND-ONLY-IF SI-Y-SOLO-SI
IF-AND-ONLY-IF element elemento SI-Y-SOLO-SI
IF-AND-ONLY-IF gate puerta SI-Y-SOLO-SI
IF-AND-ONLY-IF operation operación
    SI-Y-SOLO-SI
IF expression expresión SI
IF instruction instrucción SI
IF-THEN SI-ENTONCES
IF-THEN element elemento SI-ENTONCES
IF-THEN operation operación SI-ENTONCES
igniter ignitor, encendedor
ignition ignición, encendido
ignition cable cable de ignición
ignition circuit circuito de ignición
ignition coil bobina de ignición
ignition control control de ignición
ignition delay retardo de ignición
ignition failure falla de ignición
ignition interference interferencia por ignición
ignition lag retardo de ignición
ignition magneto magneto de ignición
ignition noise ruido de ignición
ignition point punto de ignición
ignition potential potencial de ignición
ignition rectifier rectificador de ignición
ignition reserve reserva de ignición
ignition switch conmutador de ignición
ignition system sistema de ignición
ignition transformer transformador de ignición
ignition voltage tensión de ignición
ignitor ignitor, encendedor
ignitor discharge descarga de encendedor
ignitor electrode electrodo de encendido
ignitor firing time tiempo de encendido de
    encendedor
ignitor interaction interacción de encendedor
ignitor leakage resistance resistencia de fuga de
    encendedor
ignitor oscillation oscilación de encendedor
ignitor voltage drop caída de tensión de encendedor
ignitron ignitrón
ignore character carácter de cancelación
Ilgner system sistema de Ilgner
illegal address dirección ilegal
illegal character carácter ilegal
illegal code código ilegal
illegal entry entrada ilegal
illegal instruction instrucción ilegal
illegal operation operación ilegal
illuminance iluminancia
illuminant iluminante
illuminate iluminar
illuminated iluminado
illuminated instrument instrumento iluminado
illuminated push-button pulsador iluminado
illuminated switch conmutador iluminado
illumination iluminación

illumination control control de iluminación
illumination meter luxómetro
illumination unit unidad de iluminación
illuminometer iluminómetro
illustrative diagram diagrama ilustrativo
image imagen
image admittance admitancia imagen
image amplifier amplificador de imagen
image analysis análisis de imagen
image antenna antena imagen
image area área de imagen
image attenuation atenuación imagen
image attenuation coefficient coeficiente de
    atenuación imagen
image attenuation constant constante de atenuación
    imagen
image attenuation factor factor de atenuación
    imagen
image block bloque de imagen
image brilliance brillantez de imagen
image burn imagen fija producida por quemadura,
    imagen retenida
image carrier portadora de imagen
image cell celda de imagen
image component componente de imagen
image compression compresión de imagen
image contrast contraste de imagen
image converter convertidor de imagen
image-converter panel panel convertidor de imagen
image-converter tube tubo convertidor de imagen
image data datos de imagen
image detail detalle de imagen
image dissection disección de imagen
image dissector disector de imagen
image-dissector tube tubo disector de imagen
image drift deslizamiento de imagen
image effect efecto de imagen
image element elemento de imagen
image enhancement realzado de imagen
image-enhancement system sistema de realzado de
    imagen
image field campo de imagen
image framing encuadre de imagen
image frequency frecuencia imagen
image iconoscope iconoscopio de imagen,
    supericonoscopio
image impedance impedancia imagen
image intensification intensificación de imagen
image intensifier intensificador de imagen
image-intensifier panel panel intensificador de
    imagen
image-intensifier tube tubo intensificador de imagen
image interference interferencia imagen
image-interference ratio razón de interferencia
    imagen
image mode modo de imagen
image monitor monitor de imagen
image orthicon orticón de imagen, superorticón
image-orthicon camera cámara con superorticón
image parameters parámetros de imagen
image pattern patrón de imagen
image phase fase imagen
image phase constant constante de fase imagen
image phase factor factor de fase imagen
image processing procesamiento de imagen
image processor procesador de imagen
image ratio razón imagen
image reconstruction reconstrucción de imagen
image reconstructor reconstructor de imagen
image regeneration regeneración de imagen
image rejection rechazo de imagen
image reproducer reproductor de imagen
image response respuesta imagen

image restoration restauración de imagen
image retention retención de imagen
image segmentation segmentación de imagen
image sensor sensor de imagen
image sharpness nitidez de imagen
image shift desplazamiento de imagen
image signal señal imagen
image-storage device dispositivo de almacenamiento
    de imagen
image-storage tube tubo de almacenamiento de
    imagen
image table tabla de imagen
image transfer transferencia de imagen
image transfer coefficient coeficiente de
    transferencia de imagen
image transfer constant constante de transferencia
    de imagen
image transmission transmisión de imagen
image tube tubo de imagen
imaginary axis eje imaginario
imaginary component componente imaginario
imaginary line línea imaginaria
imaginary number número imaginario
imaginary part parte imaginaria
imitation imitación
immediate inmediato
immediate access acceso inmediato
immediate-access storage almacenamiento de acceso
    inmediato
immediate address dirección inmediata
immediate addressing direccionamiento inmediato
immediate cancel cancelación inmediata
immediate command mando inmediato
immediate control control inmediato
immediate data datos inmediatos
immediate instruction instrucción inmediata
immediate maintenance mantenimiento inmediato
immediate operand operando inmediato
immediate processing procesamiento inmediato
immediate request solicitud inmediata
immediate-request mode modo de solicitud
    inmediata
immediate response respuesta inmediata
immediate-response mode modo de respuesta
    inmediata
immediate ringing llamada inmediata
immediate signal señal inmediata
immediate task tarea inmediata
immediate value valor inmediato
immersion heater calentador por inmersión
immersion plating depósito por inmersión
immersion thermostat termostato de inmersión
immersionproof a prueba de inmersión
immittance immitancia
immittance bridge puente de immitancias
immittance comparator comparador de immitancias
immittance function función de immitancia
immittance matrix matriz de immitancia
immunity inmunidad
impact excitation excitación por impacto
impact fluorescence fluorescencia por impacto
impact ionization ionización por impacto
impact noise ruido por impacto
impact printer impresora por impactos
impact strength resistencia a impactos
impact stress estrés de impacto
impact test prueba de impacto
impactproof a prueba de impactos
impaired insulation aislamiento deteriorado
impairment deterioro
impedance impedancia
impedance adapter adaptador de impedancia
impedance angle ángulo de impedancia

impedance arm brazo de impedancia
impedance bond conexión de impedancia
impedance branch rama de impedancia
impedance bridge puente de impedancia
impedance-bridge assembly ensamblaje de puente de impedancia
impedance characteristic característica de impedancia
impedance coil bobina de impedancia
impedance comparator comparador de impedancias
impedance compensator compensador de impedancia
impedance component componente de impedancia
impedance converter convertidor de impedancia
impedance corrector corrector de impedancia
impedance-coupled amplifier amplificador acoplado por impedancia
impedance-coupled stage etapa acoplada por impedancia
impedance coupling acoplamiento por impedancia
impedance drop caída por impedancia
impedance factor factor de impedancia
impedance frequency-meter frecuencímetro de impedancia
impedance function función de impedancia
impedance heating calentamiento por impedancia
impedance inverter inversor de impedancia
impedance irregularity irregularidad de impedancia
impedance level nivel de impedancia
impedance magnetometer magnetómetro de impedancia
impedance match adaptación de impedancias
impedance matching adaptación de impedancias
impedance-matching device dispositivo de adaptación de impedancias
impedance-matching filter filtro de adaptación de impedancias
impedance-matching network red de adaptación de impedancias
impedance-matching ratio razón de adaptación de impedancias
impedance-matching transformer transformador de adaptación de impedancias
impedance measurement medida de impedancia
impedance meter impedancímetro
impedance mismatch desadaptación de impedancias
impedance protection protección de impedancia
impedance range intervalo de impedancias
impedance ratio razón de impedancias
impedance relay relé de impedancia
impedance transformation transformación de impedancia
impedance transformer transformador de impedancia
impedance triangle triángulo de impedancia
impedance variation variación de impedancia
impedance vector vector impedancia
impedance voltage tensión de impedancia
impedor impedor
imperative macro instruction macroinstrucción imperativa
imperative statement sentencia imperativa
imperfect imperfecto
imperfect conductor conductor imperfecto
imperfect debugging depuración imperfecta
imperfect dielectric dieléctrico imperfecto
imperfect emission emisión imperfecta
imperfect polarization polarización imperfecta
imperfection imperfección
implementation implementación
implementation design diseño de implementación
implementation phase fase de implementación
implementation requirement requisito de implementación

implicit implícito
implicit address dirección implícita
implicit addressing direccionamiento implícito
implicit computation computación implícita
implicit declaration declaración implícita
implicit function función implícita
implied address dirección implícita
implied addressing direccionamiento implícito
implied binary point coma binaria implícita, punto binario implícito
implied decimal point coma decimal implícita, punto decimal implícito
implied radix point coma de la base implícita, punto de la base implícita
implosion implosión
import importar
imprecision imprecisión
impregnant impregnante
impregnated impregnado
impregnated carbon carbón impregnado
impregnated cathode cátodo impregnado
impregnated coil bobina impregnada
impregnated insulation aislamiento impregnado
impregnated tape cinta impregnada
impregnation impregnación
impressed voltage tensión aplicada
improper character carácter impropio
improper ferroelectric ferroeléctrico impropio
improvement threshold umbral de mejora
impulse impulso
impulse accumulator acumulador de impulsos
impulse bandwidth ancho de banda de impulso
impulse circuit circuito de impulsos
impulse code código de impulsos
impulse counter contador de impulsos
impulse-difference relay relé de diferencia de impulsos
impulse emitter emisor de impulsos
impulse excitation excitación por impulsos
impulse frequency frecuencia de impulsos
impulse function función de impulsos
impulse generator generador de impulsos
impulse noise ruido de impulsos
impulse period periodo de impulsos
impulse ratio razón de impulso
impulse recorder registrador de impulsos
impulse recording registro de impulsos
impulse regeneration regeneración de impulsos
impulse regenerator regenerador de impulsos
impulse relay relé de impulsos
impulse repeater repetidor de impulsos
impulse repetition repetición de impulsos
impulse selector selector de impulsos
impulse separator separador de impulsos
impulse speed velocidad de impulsos
impulse test prueba de impulsos
impulse transformer transformador de impulsos
impulse transmission transmisión de impulsos
impulse voltage tensión de impulso
impulse wave onda de impulsos
impulsing impulsión
impulsion impulsión
impulsive impulsivo
impulsive admittance admitancia impulsiva
impulsive discharge descarga impulsiva
impulsive force fuerza impulsiva
impulsive load carga impulsiva
impulsive noise ruido impulsivo
impurity impureza
impurity agent agente impuro
impurity atom átomo de impureza
impurity center centro de impurezas

impurity concentration concentración de impurezas
impurity density densidad de impurezas
impurity element elemento impuro
impurity interaction interacción de impurezas
impurity ion ion de impureza
impurity level nivel de impurezas
impurity material material impuro
impurity scattering dispersión de impurezas
impurity semiconductor semiconductor de
    impurezas
in circuit en circuito
in-circuit emulator emulador en circuito
in-circuit tester probador en circuito
in contact en contacto
in line en línea, en serie
in opposition en oposición
in parallel en paralelo
in phase en fase
in-phase carrier portadora en fase
in-phase component componente en fase
in-phase current corriente en fase
in-phase feedback retroalimentación en fase
in-phase recording registro en fase
in-phase signal señal en fase
in-phase voltage tensión en fase
in series en serie
inaccuracy inexactitud
inaccurate inexacto
inactive inactivo
inactive area área inactiva
inactive block bloque inactivo
inactive character carácter inactivo
inactive file archivo inactivo
inactive leg rama inactiva
inactive lines líneas inactivas
inactive link enlace inactivo
inactive mass storage almacenamiento masivo
    inactivo
inactive modules módulos inactivos
inactive node nodo inactivo
inactive program programa inactivo
inactive record registro inactivo
inactive region región inactiva
inactive state estado inactivo
inactive station estación inactiva
inactive time tiempo inactivo
inactive window ventana inactiva
inaudible frequency frecuencia inaudible
inbound circuit circuito de llegada
inbound data datos de llegada
incandescence incandescencia
incandescent incandescente
incandescent bulb bombilla incandescente
incandescent cathode cátodo incandescente
incandescent electric lamp lámpara eléctrica
    incandescente
incandescent filament lamp lámpara de filamento
    incandescente
incandescent lamp lámpara incandescente
incandescent source fuente incandescente
inception voltage tensión inicial
inch-pound libra-pulgada
inching avance por cierres sucesivos
inching device dispositivo de avance lento
inching relay relé de avance lento
incidence incidencia
incidence angle ángulo de incidencia
incident incidente
incident beam haz incidente
incident field intensity intensidad de campo incidente
incident field strength intensidad de campo incidente
incident light luz incidente
incident power potencia incidente

incident pulse impulso incidente
incident ray rayo incidente
incident record registro incidente
incident sound sonido incidente
incident wave onda incidente
incidental incidental
incidental amplitude modulation modulación de
    amplitud incidental
incidental characteristic característica incidental
incidental frequency modulation modulación de
    frecuencia incidental
incidental programming programación incidental
incidental radiator radiador incidental
incidental time tiempo incidental
incipient failure falla incipiente
inclination inclinación
inclinometer inclinómetro
inclusion inclusión
inclusive OR operation operación O incluyente
inclusive reference referencia incluyente
inclusive segments segmentos incluyentes
incoherency incoherencia
incoherent incoherente
incoherent detection detección incoherente
incoherent detector detector incoherente
incoherent light luz incoherente
incoherent oscillations oscilaciones incoherentes
incoherent radiation radiación incoherente
incoherent scattering dispersión incoherente
incoherent waves ondas incoherentes
incoming de llegada
incoming channel canal de llegada
incoming circuit circuito de llegada
incoming current corriente de llegada
incoming data datos de llegada
incoming feeder alimentador de llegada
incoming group grupo de llegada
incoming internal traffic tráfico interno de llegada
incoming line línea de llegada
incoming positions posiciones de llegada
incoming record registro de llegada
incoming selector selector de llegada
incoming traffic tráfico de llegada
incoming trunk enlace de llegada
incomplete incompleto
incomplete diffusion difusión incompleta
incomplete program programa incompleto
incomplete routine rutina incompleta
incomplete suppression supresión incompleta
incompletely incompletamente
inconsistency inconsistencia
inconsistent inconsistente
incorrect modulation modulación incorrecta
incorrect operation operación incorrecta
incorrect relay operation operación de relé
    incorrecta
increment incremento
incremental incremental
incremental backup reserva incremental
incremental binary representation representación
    binaria incremental
incremental compaction compactación incremental
incremental compiler compilador incremental
incremental computer computadora incremental
incremental data datos incrementales
incremental development desarrollo incremental
incremental digital recorder registrador digital
    incremental
incremental dimension dimensión incremental
incremental display visualizador incremental
incremental execution ejecución incremental
incremental feed alimentación incremental
incremental frequency change cambio de frecuencia

incremental
incremental frequency control control de frecuencia incremental
incremental gain ganancia incremental
incremental hysteresis loss pérdida por histéresis incremental
incremental impedance impedancia incremental
incremental impulse impulso incremental
incremental inductance inductancia incremental
incremental induction inducción incremental
incremental integrator integrador incremental
incremental movement movimiento incremental
incremental permeability permeabilidad incremental
incremental plotter trazador incremental
incremental programming programación incremental
incremental representation representación incremental
incremental resistance resistencia incremental
incremental sensitivity sensibilidad incremental
incremental spacing espaciado incremental
incremental sweep barrido incremental
incremental system sistema incremental
incremental tuning sintonización incremental
independent components componentes independientes
independent contact contacto independiente
independent data communication comunicación de datos independiente
independent excitation excitación independiente
independent failure falla independiente
independent feeder alimentador independiente
independent mode modo independiente
independent operation operación independiente
independent overflow area área de desbordamiento independiente
independent sideband banda lateral independiente
independent switch conmutador independiente
independent tracks pistas independientes
independent transformer transformador independiente
independent variable variable independiente
independent voltage control control de tensión independiente
index índice
index card tarjeta de índice
index character carácter de índice
index code código de índice
index counter contador índice
index cycle ciclo de índice
index entry entrada de índice
index factor factor índice
index file archivo índice
index level nivel de índice
index mark marca de índice
index of modulation índice de modulación
index of refraction índice de refracción
index point punto índice
index position posición de índice
index profile perfil de índice
index record registro índice
index register registro índice
index space espacio de índice
index word palabra índice
indexation indexación
indexed indexado
indexed access acceso indexado
indexed address dirección indexada
indexed addressing direccionamiento indexado
indexed data set conjunto de datos indexados
indexed file archivo indexado
indexed organization organización indexada
indexed segment segmento indexado

indexed sequential access acceso secuencial indexado
indexed sequential access method método de acceso secuencial indexado
indexed sequential file archivo secuencial indexado
indexed sequential organization organización secuencial indexada
indexing indexación
indexing segment segmento de indexación
indexing system sistema de indexación
indexing technique técnica de indexación
indicated indicado
indicated angle ángulo indicado
indicated bearing rumbo indicado
indicated horsepower caballos de fuerza indicados
indicated power potencia indicada
indicated value valor indicado
indicating circuit circuito indicador
indicating controller controlador indicador
indicating device dispositivo indicador
indicating fuse fusible indicador
indicating instrument instrumento indicador
indicating lamp lámpara indicadora
indicating relay relé indicador
indicating scale escala indicadora
indicating unit unidad indicadora
indicating wattmeter wáttmetro indicador
indicator indicador
indicator board cuadro indicador
indicator lamp lámpara indicadora
indicator light luz indicadora
indicator-light panel panel de luces indicadoras
indicator panel panel indicador
indicator probe sonda indicadora
indicator tube tubo indicador
indicator unit unidad indicadora
indicial indicial
indicial admittance admitancia indicial
indicial response respuesta indicial
indirect indirecto
indirect activation activación indirecta
indirect address dirección indirecta
indirect addressing direccionamiento indirecto
indirect backscatter retrodispersión indirecta
indirect communication comunicación indirecta
indirect commutation conmutación indirecta
indirect component componente indirecto
indirect control control indirecto
indirect coupling acoplamiento indirecto
indirect energy loss pérdida de energía indirecta
indirect entry entrada indirecta
indirect excitation excitación indirecta
indirect frequency modulation modulación de frecuencia indirecta
indirect glare resplandor indirecto
indirect ground tierra indirecta
indirect heating calentamiento indirecto
indirect illumination iluminación indirecta
indirect instruction instrucción indirecta
indirect interlocking enclavamiento indirecto
indirect light luz indirecta
indirect lighting alumbrado indirecto
indirect measurement medida indirecta
indirect mode modo indirecto
indirect operation operación indirecta
indirect piezoelectricity piezoelectricidad indirecta
indirect ray rayo indirecto
indirect scanning exploración indirecta
indirect stroke descarga indirecta
indirect-stroke protection protección contra descarga indirecta
indirect wave onda indirecta
indirectly indirectamente

indirectly controlled controlado indirectamente
indirectly controlled variable variable controlada indirectamente
indirectly fed antenna antena alimentada indirectamente
indirectly grounded conectado a tierra indirectamente
indirectly heated cathode cátodo calentado indirectamente
indirectly heated thermistor termistor calentado indirectamente
indirectly heated thermocouple termopar calentado indirectamente
indirectly heated tube tubo calentado indirectamente
indirectly heated valve válvula calentada indirectamente
indium indio
indium antimonide antimoniuro de indio
indium arsenide arseniuro de indio
indium phosphide fosfuro de indio
indium transistor transistor de indio
individual address dirección individual
individual control control individual
individual distortion distorsión individual
individual drive accionamiento individual
individual field campo individual
individual lighting alumbrado individual
individual line línea individual
individual protection protección individual
individual transmission transmisión individual
individual trunk enlace individual
indoor antenna antena interior
indoor apparatus aparato interior
indoor arrester protector interior
indoor electrical equipment equipo eléctrico interior
indoor installation instalación interior
indoor loop antenna antena de cuadro interior
indoor radiation radiación interior
indoor termination terminación interior
indoor transformer transformador interior
induced inducido
induced charge carga inducida
induced current corriente inducida
induced electromotive force fuerza electromotriz inducida
induced emission emisión inducida
induced environment ambiente inducido
induced failure falla inducida
induced field current corriente de campo inducida
induced hum zumbido inducido
induced jitter fluctuación inducida
induced noise ruido inducido
induced polarization polarización inducida
induced radioactivity radiactividad inducida
induced voltage tensión inducida
inducing current corriente inductora
inductance inductancia
inductance box caja de inductancias
inductance bridge puente de inductancias
inductance-capacitance inductancia-capacitancia
inductance-capacitance bridge puente de inductancia-capacitancia
inductance-capacitance filter filtro de inductancia-capacitancia
inductance-capacitance meter medidor de inductancia-capacitancia
inductance-capacitance-resistance inductancia-capacitancia-resistencia
inductance-capacitance-resistance bridge puente de inductancia-capacitancia-resistencia
inductance coil bobina de inductancia
inductance device dispositivo de inductancia
inductance factor factor de inductancia

inductance filter filtro de inductancia
inductance meter inductancímetro
inductance standard patrón de inductancia
inductance switch conmutador de inductancia
inductance tube tubo de inductancia
inductance-tube modulator modulador de tubo de inductancia
induction inducción
induction brazing soldadura fuerte por inducción
induction bridge puente de inducción
induction coil bobina de inducción
induction compass compás de inducción
induction coupling acoplamiento por inducción
induction density densidad de inducción
induction effect efecto de inducción
induction factor factor de inducción
induction field campo de inducción
induction flux flujo de inducción
induction force fuerza de inducción
induction frequency converter convertidor de frecuencia de inducción
induction furnace horno de inducción
induction generator generador de inducción
induction heater calentador por inducción
induction heating calentamiento por inducción
induction instrument instrumento de inducción
induction loss pérdidas por inducción
induction loudspeaker altavoz de inducción
induction machine máquina de inducción
induction meter medidor de inducción
induction modulator modulador de inducción
induction motor motor de inducción
induction noise ruido de inducción
induction oscillator oscilador de inducción
induction potentiometer potenciómetro de inducción
induction regulator regulador de inducción
induction relay relé de inducción
induction seismograph sismógrafo de inducción
induction speaker altavoz de inducción
induction system sistema de inducción
induction voltage regulator regulador de tensión por inducción
induction wattmeter wáttmetro de inducción
induction welding soldadura por inducción
induction zone zona de inducción
inductive inductivo
inductive capacitor capacitor inductivo
inductive capacity capacidad inductiva
inductive circuit circuito inductivo
inductive component componente inductivo
inductive coordination coordinación inductiva
inductive coupling acoplamiento inductivo
inductive effect efecto inductivo
inductive feedback retroalimentación inductiva
inductive heater calentador inductivo
inductive heating calentamiento inductivo
inductive hum zumbido inductivo
inductive impedance impedancia inductiva
inductive influence influencia inductiva
inductive interference interferencia inductiva
inductive kick sobretensión inductiva
inductive load carga inductiva
inductive logic lógica inductiva
inductive microphone micrófono inductivo
inductive neutralization neutralización inductiva
inductive reactance reactancia inductiva
inductive resistance resistencia inductiva
inductive resistor resistor inductivo
inductive switching conmutación inductiva
inductive transducer transductor inductivo
inductive tuning sintonización inductiva
inductive winding arrollamiento inductivo
inductive window ventana inductiva

inductively inductivamente
inductivity inductividad
inductometer inductómetro
inductophone inductófono
inductor inductor
inductor alternator alternador inductor
inductor coil bobina inductora
inductor microphone micrófono inductor
industrial control control industrial
industrial electronic equipment equipo electrónico industrial
industrial electronics electrónica industrial
industrial interference interferencia industrial
industrial radiography radiografía industrial
industrial tube tubo industrial
ineffective time tiempo ineficaz
inert gas gas inerte
inertance inertancia
inertia inercia
inertia relay relé de inercia
inertia switch conmutador de inercia
inertial inercial
inertial effect efecto inercial
inertial guidance guiado inercial
inertial navigation equipment equipo de navegación inercial
inertial navigation system sistema de navegación inercial
infection infección
infiltration infiltración
infinite infinito
infinite attenuation atenuación infinita
infinite-attenuation frequency frecuencia de atenuación infinita
infinite baffle bafle infinito
infinite capacity capacidad infinita
infinite frequency frecuencia infinita
infinite impedance impedancia infinita
infinite-impedance detector detector de impedancia infinita
infinite line línea infinita
infinite loop bucle infinito
infinite resolution resolución infinita
infinite series serie infinita
infinite transmission line línea de transmisión infinita
infinitesimal infinitesimal
infinitesimal dipole dipolo infinitesimal
infinity infinito
infinity plug clavija de infinidad
infix notation notación infija
infix operator operador infijo
inflection point punto de inflexión
inflection-point current corriente de punto de inflexión
inflection-point emission current corriente de emisión de punto de inflexión
inflection-point voltage tensión de punto de inflexión
influence influencia
influence machine máquina de influencia
influenced conductor conductor influenciado
informatics informática
information información
information analysis análisis de información
information bandwidth ancho de banda de información
information bits bits de información
information center centro de información
information channel canal de información
information circuit circuito de información
information communication system sistema de comunicación de información
information content contenido de información

information demand demanda de información
information display visualización de información
information environment ambiente de información
information feedback retroalimentación de información
information-feedback system sistema de retroalimentación de información
information field campo de información
information flow flujo de información
information flow analysis análisis de flujo de información
information flow control control de flujo de información
information gate puerta de información
information gathering recopilación de información
information graphics gráficos de información
information handling manejo de información
information-handling system sistema de manejo de información
information interchange intercambio de información
information line línea de información
information link enlace de información
information-management system sistema de administración de información
information medium medio de información
information network red de información
information overload sobrecarga de información
information path trayectoria de información
information processing procesamiento de información
information-processing center centro de procesamiento de información
information-processing machine máquina de procesamiento de información
information-processing system sistema de procesamiento de información
information pulse impulso de información
information rate velocidad de información
information request solicitud de información
information requirement requisito de información
information resource management administración de recursos de información
information retrieval recuperación de información
information-retrieval system sistema de recuperación de información
information separator separador de información
information service servicio de información
information source fuente de información
information speed velocidad de información
information storage almacenamiento de información
information-storage capacity capacidad de almacenamiento de información
information-storage device dispositivo de almacenamiento de información
information-storage medium medio de almacenamiento de información
information-storage tube tubo de almacenamiento de información
information system sistema de información
information technology tecnología de información
information technology equipment equipo de tecnología de información
information traffic tráfico de información
information transfer transferencia de información
information transmission transmisión de información
information-transmission system sistema de transmisión de información
information word palabra de información
infra infra
infraacoustic infraacústico
infradyne infradino
infradyne receiver receptor infradino

infradyne reception recepción infradina
infralow frequency frecuencia infrabaja
infrared infrarrojo
infrared absorption spectrum espectro de absorción
 de infrarrojo
infrared beacon radiofaro de infrarrojo
infrared camera cámara de infrarrojo
infrared communication comunicación por
 infrarrojo
infrared detector detector de infrarrojo
infrared diode diodo de infrarrojo
infrared emitter emisor de infrarrojo
infrared energy energía infrarroja
infrared fiber optics fibroóptica de infrarrojo
infrared guidance guiado por infrarrojo
infrared heat calor infrarrojo
infrared homing guiado por infrarrojo
infrared illuminator iluminador de infrarrojo
infrared image converter convertidor de imágenes
 de infrarrojo
infrared lamp lámpara de infrarrojos
infrared light luz infrarroja
infrared maser máser de infrarrojo
infrared microscope microscopio de infrarrojo
infrared motion detector detector de moción por
 infrarrojos
infrared optics óptica de infrarrojo
infrared photography fotografía con infrarrojo
infrared polarizer polarizador de infrarrojo
infrared radiation radiación infrarroja
infrared rays rayos infrarrojos
infrared receiver receptor de infrarrojo
infrared source fuente de infrarrojo
infrared spectroscopy espectroscopia infrarroja
infrared spectrum espectro infrarrojo
infrared transmitter transmisor de infrarrojo
infrared waves ondas infrarrojas
infrared window ventana de infrarrojo
infrasonic infrasónico
infrasonic frequency frecuencia infrasónica
infrasonics infrasónica
infrasound infrasonido
inharmonic distortion distorsión inharmónica
inherent inherente
inherent component componente inherente
inherent distortion distorsión inherente
inherent error error inherente
inherent feedback retroalimentación inherente
inherent filter filtro inherente
inherent filtration filtración inherente
inherent interference interferencia inherente
inherent memory memoria inherente
inherent noise ruido inherente
inherent stability estabilidad inherente
inherent storage almacenamiento inherente
inherited error error heredado
inhibit inhibir
inhibit gate puerta de inhibición
inhibit pulse impulso de inhibición
inhibit signal señal de inhibición
inhibiting input entrada inhibidora
inhibiting signal señal inhibidora
inhibition inhibición
inhibition gate puerta de inhibición
inhibitor inhibidor
inhomogeneous inhomogéneo
initial approach aproximación inicial
initial charge carga inicial
initial condition condición inicial
initial drain corriente inicial
initial electron current corriente electrónica inicial
initial failure falla inicial
initial instructions instrucciones iniciales

initial inverse voltage tensión inversa inicial
initial ionizing event evento ionizante inicial
initial magnetization curve curva de magnetización
 inicial
initial order orden inicial
initial permeability permeabilidad inicial
initial position posición inicial
initial procedure procedimiento inicial
initial program load carga de programa inicial
initial program loader cargador de programa inicial
initial ray rayo inicial
initial reversible capacitance capacitancia reversible
 inicial
initial signal señal inicial
initial state estado inicial
initial susceptibility susceptibilidad inicial
initial tension tensión inicial
initial test prueba inicial
initial value valor inicial
initial velocity velocidad inicial
initial-velocity current corriente de velocidad inicial
initial voltage tensión inicial
initialization inicialización
initialize inicializar
initiate iniciar
injection inyección
injection efficiency eficiencia de inyección
injection grid rejilla de inyección
injection gun cañón de inyección
injection laser láser de inyección
injection luminescent diode diodo luminiscente de
 inyección
injector inyector
injector electrode electrodo inyector
injector grid rejilla inyectora
ink bleed sangrado de tinta
ink-jet galvanometer galvanómetro de chorro de
 tinta
ink-jet printer impresora de chorro de tinta
ink-mist recorder registrador de vapor de tinta
ink recorder registrador de tinta
ink-vapor recorder registrador de vapor de tinta
inline en línea, en serie
inline code código en línea
inline coding codificación en línea
inline data processing procesamiento de datos en
 línea
inline processing procesamiento en línea
inline readout lectura en línea
inline sequence secuencia en línea
inline subminiature tube tubo subminiatura en línea
inline subroutine subrutina en línea
inline tuning sintonización en línea
inner interno
inner conductor conductor interno
inner connector conector interno
inner contact contacto interno
inner jacket envoltura interna
inner macro instruction macroinstrucción interna
inner marker beacon radiobaliza interna
inner memory memoria interna
inner panel panel interno
inner program programa interno
inoperative inoperante
inorganic inorgánico
inorganic electrode electrodo inorgánico
inorganic electrolyte electrólito inorgánico
input entrada
input activity actividad de entrada
input admittance admitancia de entrada
input area área de entrada
input bias current corriente de polarización de
 entrada

input block bloque de entrada
input-bound limitado por entrada
input buffer memoria intermedia de entrada
input capacitance capacitancia de entrada
input capacitor capacitor de entrada
input capacity capacidad de entrada
input card tarjeta de entrada
input cavity cavidad de entrada
input channel canal de entrada
input choke choque de entrada
input circuit circuito de entrada
input conductance conductancia de entrada
input converter convertidor de entrada
input-converter module módulo de convertidor de
    entrada
input coupling capacitor capacitor de acoplamiento
    de entrada
input coupling transformer transformador de
    acoplamiento de entrada
input current corriente de entrada
input data datos de entrada
input data set conjunto de datos de entrada
input data validation validación de datos de entrada
input deck paquete de entrada
input device dispositivo de entrada
input document documento de entrada
input electrode electrodo de entrada
input element elemento de entrada
input equipment equipo de entrada
input error voltage tensión de error de entrada
input factor factor de entrada
input feeder alimentador de entrada
input field campo de entrada
input file archivo de entrada
input filter filtro de entrada
input format formato de entrada
input frequency frecuencia de entrada
input gap espacio de entrada
input hopper almacén de alimentación de entrada
input impedance impedancia de entrada
input instruction code código de instrucción de
    entrada
input job queue cola de trabajos de entrada
input job stream corriente de trabajos de entrada
input keyboard teclado de entrada
input language lenguaje de entrada
input level nivel de entrada
input-level control control de nivel de entrada
input-limited limitado por entrada
input line línea de entrada
input loading carga de entrada
input medium medio de entrada
input message mensaje de entrada
input mode modo de entrada
input module módulo de entrada
input noise current corriente de ruido de entrada
input noise voltage tensión de ruido de entrada
input offset voltage tensión de desequilibrio de
    entrada
input-output entrada-salida
input-output activity actividad de entrada-salida
input-output area área de entrada-salida
input-output bandwidth ancho de banda de
    entrada-salida
input-output block bloque de entrada-salida
input-output board tablero de entrada-salida
input-output bound limitado por entrada-salida
input-output buffer memoria intermedia de
    entrada-salida
input-output bus bus de entrada-salida
input-output channel canal de entrada-salida
input-output channel selection selección de canal de
    entrada-salida

input-output characteristic característica de
    entrada-salida
input-output control control de entrada-salida
input-output control module módulo de control de
    entrada-salida
input-output control program programa de control
    de entrada-salida
input-output control system sistema de control de
    entrada-salida
input-output control unit unidad de control de
    entrada-salida
input-output controller controlador de entrada-salida
input-output coupling acoplamiento de
    entrada-salida
input-output device dispositivo de entrada-salida
input-output diskette slot ranura de disquete de
    entrada-salida
input-output equipment equipo de entrada-salida
input-output file archivo de entrada-salida
input-output instruction instrucción de
    entrada-salida
input-output interface interfaz de entrada-salida
input-output interface operation operación de
    interfaz de entrada-salida
input-output interrupt interrupción de entrada-salida
input-output interruption interrupción de
    entrada-salida
input-output isolation aislamiento de entrada-salida
input-output limited limitado por entrada-salida
input-output map mapa de entrada-salida
input-output medium medio de entrada-salida
input-output memory address dirección de memoria
    de entrada-salida
input-output mode modo de entrada-salida
input-output model modelo de entrada-salida
input-output operation operación de entrada-salida
input-output operation mode modo de operación de
    entrada-salida
input-output order orden de entrada-salida
input-output overlap solape de entrada-salida
input-output poles polos de entrada-salida
input-output port puerto de entrada-salida
input-output procedure procedimiento de
    entrada-salida
input-output processor procesador de entrada-salida
input-output register registro de entrada-salida
input-output routine rutina de entrada-salida
input-output scan exploración de entrada-salida
input-output section sección de entrada-salida
input-output slot ranura de entrada-salida
input-output statement sentencia de entrada-salida
input-output status estado de entrada-salida
input-output status response repuesta de estado de
    entrada-salida
input-output storage almacenamiento de
    entrada-salida
input-output switching conmutación de
    entrada-salida
input-output symbol símbolo de entrada-salida
input-output system sistema de entrada-salida
input-output table tabla de entrada-salida
input-output traffic control control de tráfico de
    entrada-salida
input-output unit unidad de entrada-salida
input-output voltage differential diferencial de
    tensión de entrada-salida
input peripheral periférico de entrada
input potentiometer potenciómetro de entrada
input power potencia de entrada
input procedure procedimiento de entrada
input process proceso de entrada
input-process-output entrada-procesamiento-salida
input program programa de entrada

input protection protección de entrada
input pulse impulso de entrada
input queue cola de entrada
input reader lectora de entrada
input record registro de entrada
input recorder registrador de entrada
input reference referencia de entrada
input register registro de entrada
input request solicitud de entrada
input resistance resistencia de entrada
input resistor resistor de entrada
input resonator resonador de entrada
input-resonator grid rejilla de resonador de entrada
input routine rutina de entrada
input section sección de entrada
input sensitivity sensibilidad de entrada
input signal señal de entrada
input specifications especificaciones de entrada
input speed velocidad de entrada
input stage etapa de entrada
input state estado de entrada
input station estación de entrada
input storage almacenamiento de entrada
input stream corriente de entrada
input-stream control control de corriente de entrada
input subsystem subsistema de entrada
input system sistema de entrada
input tape cinta de entrada
input terminal terminal de entrada
input time constant constante de tiempo de entrada
input track pista de entrada
input transformer transformador de entrada
input translator traductor de entrada
input tube tubo de entrada
input uncertainty incertidumbre de entrada
input unit unidad de entrada
input validation validación de entrada
input voltage tensión de entrada
input winding arrollamiento de entrada
input work queue cola de trabajos de entrada
inquiry consulta
inquiry display terminal terminal de visualización
    de consulta
inquiry message mensaje de consulta
inquiry mode modo de consulta
inquiry processing procesamiento de consulta
inquiry program programa de consulta
inquiry-reply consulta-respuesta
inquiry-response consulta-respuesta
inquiry-response operation operación de
    consulta-respuesta
inquiry-response terminal terminal de
    consulta-respuesta
inquiry session sesión de consulta
inquiry station estación de consulta
inquiry terminal terminal de consulta
inquiry unit unidad de consulta
inrush irrupción
inrush current corriente de irrupción
insert core núcleo de inserción
insert earphone audífono de inserción
insert mode modo de inserción
insert tube tubo de inserción
inserted subroutine subrutina insertada
insertion inserción
insertion attenuation atenuación por inserción
insertion character carácter de inserción
insertion gain ganancia por inserción
insertion length longitud de inserción
insertion loss pérdida por inserción
insertion phase shift desfase de inserción
insertion point punto de inserción
insertion resistance resistencia de inserción

insertion sort clasificación de inserción
insertion track pista de inserción
inside antenna antena interior
inside cable cable interior
inside conductor conductor interior
inside diameter diámetro interior
inside lead hilo interior
inside spider araña interior
inside transformer transformador interior
inspection inspección
inspection window ventanilla de inspección
inspectoscope inspectoscopio
instability inestabilidad
installation instalación
installation cycle ciclo de instalación
installation date fecha de instalación
installation diagram diagrama de instalación
installation diskette disquete de instalación
installation instructions instrucciones de instalación
installation program programa de instalación
installation routine rutina de instalación
installation test prueba de instalación
installation time tiempo de instalación
installation verification procedure procedimiento de
    verificación de instalación
installed capacity capacidad instalada
installed life vida instalada
installed reserve reserva instalada
instant instante
instantaneous instantáneo
instantaneous access acceso instantáneo
instantaneous acoustic power potencia acústica
    instantánea
instantaneous amplitude amplitud instantánea
instantaneous automatic gain control control de
    ganancia automático instantáneo
instantaneous automatic volume control control de
    volumen automático instantáneo
instantaneous companding compresión-expansión
    instantánea
instantaneous contacts contactos instantáneos
instantaneous current corriente instantánea
instantaneous data transfer transferencia de datos
    instantánea
instantaneous demand demanda instantánea
instantaneous disk disco instantáneo
instantaneous forward current corriente directa
    instantánea
instantaneous forward voltage tensión directa
    instantánea
instantaneous frequency frecuencia instantánea
instantaneous frequency correlation correlación de
    frecuencia instantánea
instantaneous load factor factor de carga instantáneo
instantaneous multiplier multiplicador instantáneo
instantaneous overcurrent sobrecorriente
    instantánea
instantaneous overcurrent relay relé de
    sobrecorriente instantánea
instantaneous peak power potencia pico instantánea
instantaneous power potencia instantánea
instantaneous power output salida de potencia
    instantánea
instantaneous recording registro instantáneo
instantaneous relay relé instantáneo
instantaneous release desenganche instantáneo
instantaneous reset reposición instantánea
instantaneous reverse current corriente inversa
    instantánea
instantaneous reverse voltage tensión inversa
    instantánea
instantaneous sample muestra instantánea
instantaneous sampling muestreo instantáneo

instantaneous sound power potencia acústica instantánea
instantaneous sound pressure presión acústica instantánea
instantaneous speech power potencia vocal instantánea
instantaneous storage almacenamiento instantáneo
instantaneous value valor instantáneo
instantaneous voltage tensión instantánea
instantaneous voltage value valor de tensión instantáneo
instantaneously instantáneamente
instruction instrucción
instruction address dirección de instrucción
instruction address register registro de dirección de instrucción
instruction area área de instrucción
instruction card tarjeta de instrucción
instruction character carácter de instrucción
instruction code código de instrucción
instruction control control de instrucciones
instruction control unit unidad de control de instrucciones
instruction counter contador de instrucciones
instruction cycle ciclo de instrucción
instruction decoder descodificador de instrucciones
instruction execution logic lógica de ejecución de instrucción
instruction field campo de instrucción
instruction format formato de instrucción
instruction length longitud de instrucción
instruction length code código de longitud de instrucción
instruction mix mezcla de instrucciones
instruction modification modificación de instrucción
instruction modifier modificador de instrucción
instruction register registro de instrucción
instruction set conjunto de instrucciones
instruction statement sentencia de instrucción
instruction storage almacenamiento de instrucciones
instruction tape cinta de instrucciones
instruction time tiempo de instrucción
instruction word palabra de instrucción
instrument instrumento
instrument accuracy exactitud de instrumento
instrument amplifier amplificador de instrumento
instrument approach aproximación por instrumentos
instrument-approach system sistema de aproximación por instrumentos
instrument board tablero de instrumentos
instrument calibration calibración de instrumento
instrument chopper interruptor periódico de instrumento
instrument error error de instrumento
instrument flight vuelo por instrumentos
instrument fuse fusible de instrumento
instrument installation instalación de instrumento
instrument lamp lámpara de instrumentos
instrument landing aterrizaje por instrumentos
instrument-landing antenna antena de aterrizaje por instrumentos
instrument-landing control panel panel de control de aterrizaje por instrumentos
instrument-landing station estación de aterrizaje por instrumentos
instrument-landing system sistema de aterrizaje por instrumentos
instrument-landing system indicator indicador de sistema de aterrizaje por instrumentos
instrument layout arreglo de instrumentos, distribución de instrumentos
instrument light luz de instrumentos
instrument lighting alumbrado de instrumentos

instrument mounting montaje de instrumentos
instrument multiplier multiplicador de instrumento
instrument panel tablero de instrumentos
instrument panel light luz de tablero de instrumentos
instrument preamplifier preamplificador de instrumento
instrument range alcance de instrumento
instrument reading lectura de instrumento
instrument relay relé de instrumento
instrument resistance resistencia de instrumento
instrument room sala de instrumentos
instrument sensitivity sensibilidad de instrumento
instrument shunt derivación de instrumento
instrument switch conmutador de instrumento
instrument transformer transformador de medidor
instrument-type relay relé de tipo de instrumento
instrumental instrumental
instrumental error error instrumental
instrumental input entrada instrumental
instrumentation instrumentación
instrumentation amplifier amplificador de instrumentación
instrumentation radar radar de instrumentación
insulant aislante
insulate aislar
insulated aislado
insulated cabinet caja aislada
insulated cable cable aislado
insulated carbon resistor resistor de carbón aislado
insulated clip pinza aislada
insulated conductor conductor aislado
insulated coupling acoplamiento aislado
insulated device dispositivo aislado
insulated enclosure caja aislada
insulated grid rejilla aislada
insulated joint junta aislante
insulated neutral neutro aislado
insulated probe sonda aislada
insulated resistor resistor aislado
insulated return retorno aislado
insulated-return system sistema de retorno aislado
insulated section sección aislada
insulated system sistema aislado
insulated tool herramienta aislada
insulated wire hilo aislado
insulating aislante
insulating base base aislante
insulating capacity capacidad aislante
insulating clamp abrazadera aislante
insulating device dispositivo aislante
insulating gloves guantes aislantes
insulating joint junta aislante
insulating mat tapete aislante
insulating material material aislante
insulating oil aceite aislante
insulating sheet hoja aislante
insulating sleeve manguito aislante
insulating stool taburete aislante
insulating tape cinta aislante
insulating tool herramienta aislante
insulating transformer transformador aislante
insulating varnish barniz aislante
insulation aislamiento
insulation breakdown ruptura de aislamiento
insulation breakdown current corriente de ruptura de aislamiento
insulation class clase de aislamiento
insulation failure falla de aislamiento
insulation fault falla de aislamiento
insulation indicator indicador de aislamiento
insulation level nivel de aislamiento
insulation measurement medida de aislamiento

insulation power factor factor de potencia de aislamiento
insulation rating clasificación de aislamiento
insulation resistance resistencia de aislamiento
insulation system sistema de aislamiento
insulation test prueba de aislamiento
insulation voltage tensión de aislamiento
insulator aislador
insulator arcover salto de arco de aislador
insulator bracket soporte de aislador
insulator loss pérdida por aislador
insulator-type transformer transformador de tipo aislador
integer entero
integral integral
integral action acción integral
integral calculus cálculo integral
integral control control integral
integral control action acción de control integral
integral control programming programación de control integral
integral controller controlador integral
integral coupling acoplamiento integral
integral multiple múltiplo integral
integral number número entero
integrand integrando
integrate integrar
integrated integrado
integrated alarm system sistema de alarma integrado
integrated amplifier amplificador integrado
integrated antenna system sistema de antena integrado
integrated capacitor capacitor integrado
integrated circuit circuito integrado
integrated-circuit array red de circuitos integrados
integrated-circuit card tarjeta de circuitos integrados
integrated-circuit doping dopado de circuito integrado
integrated-circuit electrical contact contacto eléctrico de circuito integrado
integrated communications adapter adaptador de comunicaciones integrado
integrated component componente integrado
integrated computer system sistema de computadora integrado
integrated configuration configuración integrada
integrated data processing procesamiento de datos integrado
integrated database base de datos integrada
integrated demand demanda integrada
integrated-demand meter medidor de demanda integrada
integrated digital network red digital integrada
integrated electronics electrónica integrada
integrated emulator emulador integrado
integrated emulator program programa de emulador integrado
integrated equipment component componente de equipo integrado
integrated file adapter adaptador de archivos integrado
integrated flux flujo integrado
integrated heating system sistema de calentamiento integrado
integrated modem módem integrado
integrated monolithic circuit circuito monolítico integrado
integrated optical circuit circuito óptico integrado
integrated program programa integrado
integrated radiation radiación integrada
integrated resistor resistor integrado
integrated software programas integrados
integrated storage control control de

almacenamiento integrado
integrated system sistema integrado
integrating integrador
integrating amplifier amplificador integrador
integrating apparatus aparato integrador
integrating circuit circuito integrador
integrating dosimeter dosímetro integrador
integrating effect efecto integrador
integrating filter filtro integrador
integrating frequency meter frecuencímetro integrador
integrating galvanometer galvanómetro integrador
integrating instrument instrumento integrador
integrating mechanism mecanismo integrador
integrating meter medidor integrador
integrating network red integradora
integrating photometer fotómetro integrador
integrating preamplifier preamplificador integrador
integrating relay relé integrador
integrating timer temporizador integrador
integrating unit unidad integradora
integration integración
integration level nivel de integración
integration test prueba de integración
integration voltage tensión de integración
integrator integrador
integrator capacitor capacitor integrador
integrator circuit circuito integrador
integrator switch conmutador integrador
integrity integridad
intelligent controller controlador inteligente
intelligent printer impresora inteligente
intelligent terminal terminal inteligente
intelligibility inteligibilidad
intelligible inteligible
intensification intensificación
intensified intensificado
intensified field campo intensificado
intensifier intensificador
intensifier electrode electrodo intensificador
intensifying intensificador
intensifying ring anillo intensificador
intensifying screen pantalla intensificadora
intensimeter intensímetro
intensitometer intensitómetro
intensity intensidad
intensity amplifier amplificador de intensidad
intensity control control de intensidad
intensity distribution distribución de intensidad
intensity distribution curve curva de distribución de intensidad
intensity level nivel de intensidad
intensity modulation modulación de intensidad
intensity of current intensidad de corriente
intensity of field intensidad de campo
intensity of illumination intensidad de iluminación
intensity of magnetization intensidad de magnetización
intensity of radiation intensidad de radiación
intensity of traffic intensidad de tráfico
interaction interacción
interaction circuit circuito de interacción
interaction crosstalk diafonía de interacción
interaction factor factor de interacción
interaction gap espacio de interacción
interaction impedance impedancia de interacción
interaction loss pérdida por interacción
interaction space espacio de interacción
interactive interactivo
interactive compiler compilador interactivo
interactive computing computación interactiva
interactive debugging depuración interactiva
interactive debugging system sistema de depuración

interactiva
interactive display visualizador interactivo
interactive electrical systems sistemas eléctricos interactivos
interactive environment ambiente interactivo
interactive graphics gráficos interactivos
interactive graphics system sistema de gráficos interactivos
interactive job trabajo interactivo
interactive language lenguaje interactivo
interactive mode modo interactivo
interactive multiprocessor system sistema de multiprocesador interactivo
interactive operation operación interactiva
interactive partition partición interactiva
interactive processes procesos interactivos
interactive processing procesamiento interactivo
interactive program programa interactivo
interactive simulator simulador interactivo
interactive subsystem subsistema interactivo
interactive system sistema interactivo
interactive terminal terminal interactiva
interbase resistance resistencia entre bases
interbase voltage tensión entre bases
interblock gap espacio entre bloques
interblock space espacio entre bloques
intercarrier interportadora
intercarrier method método de interportadora
intercarrier noise suppression supresión de ruido de interportadora
intercarrier receiver receptor de interportadora
intercarrier sound system sistema de sonido por interportadora
intercarrier system sistema de interportadora
intercepted station estación interceptada
interception circuit circuito de intercepción
interchange intercambio
interchange code código de intercambio
interchange point punto de intercambio
interchangeability intercambialidad
interchangeable intercambiable
interchangeable control panel panel de control intercambiable
interchangeable unit unidad intercambiable
interchannel correlation correlación entre canales
interchannel crosstalk diafonía entre canales
interchannel interference interferencia entre canales
interchannel noise suppressor supresor de ruido entre canales
intercharacter increment incremento entre caracteres
intercharacter space espacio entre caracteres
intercharacter spacing espaciado entre caracteres
intercom intercomunicador
intercommunicating system sistema de intercomunicación
intercommunication intercomunicación
intercommunication system sistema de intercomunicación
intercommunicator intercomunicador
interconnect interconectar
interconnected interconectado
interconnected network red interconectada
interconnected system sistema interconectado
interconnecting circuit circuito interconectador
interconnection interconexión
interconnection device dispositivo de interconexión
interconnection diagram diagrama de interconexiones
interconnection line línea de interconexión
interconnection pattern patrón de interconexiones
interconnector interconectador
intercontinental circuit circuito intercontinental

intercontinental traffic tráfico intercontinental
intercrystalline intercristalino
intercrystalline disintegration desintegración intercristalina
interdendritic corrosion corrosión interdendrítica
interdigit interval intervalo entre dígitos
interdigit time tiempo entre dígitos
interdigital interdigital
interdigital delay line línea de retardo interdigital
interdigital magnetron magnetrón interdigital
interdigital resonator resonador interdigital
interdigital tube tubo interdigital
interelectrode capacitance capacitancia interelectródica
interelectrode capacity capacidad interelectródica
interelectrode coupling acoplamiento interelectródico
interelectrode insulation aislamiento interelectródico
interelectrode space espacio interelectródico
interelectrode transadmittance transadmitancia interelectródica
interelectrode transconductance transconductancia interelectródica
interelectrode transit time tiempo de tránsito interelectródico
interelement capacitance capacitancia interelemento
interface interfaz
interface adapter adaptador de interfaz
interface bus bus de interfaz
interface card tarjeta de interfaz
interface channel canal de interfaz
interface circuit circuito de interfaz
interface circuitry circuitería de interfaz
interface control control de interfaz
interface control check comprobación de control de interfaz
interface converter convertidor de interfaz
interface debugging depuración de interfaz
interface effect efecto de interfaz
interface module módulo de interfaz
interface processor procesador de interfaz
interface requirement requisito de interfaz
interface resistance resistencia de interfaz
interface routine rutina de interfaz
interface specification especificación de interfaz
interface standard patrón de interfaz
interface system sistema de interfaz
interface unit unidad de interfaz
interfacial interfacial
interfacial connection conexión interfacial
interference interferencia
interference analyzer analizador de interferencia
interference area área de interferencia
interference attenuator atenuador de interferencia
interference band banda de interferencia
interference control control de interferencia
interference eliminator eliminador de interferencia
interference factor factor de interferencia
interference field campo de interferencia
interference field strength intensidad de campo de interferencia
interference filter filtro de interferencia
interference generator generador de interferencia
interference inverter inversor de interferencia
interference level nivel de interferencia
interference measurement medida de interferencia
interference pattern patrón de interferencia
interference power potencia de interferencia
interference pulse impulso de interferencia
interference spectrum espectro de interferencia
interference suppression supresión de interferencia
interference suppressor supresor de interferencia
interference threshold umbral de interferencia

interference trap trampa de interferencia
interference unit unidad de interferencia
interfering interferente
interfering electric signal señal eléctrica interferente
interfering field campo interferente
interfering frequency frecuencia interferente
interfering signal señal interferente
interfering voltage tensión interferente
interfering wave onda interferente
interferometer interferómetro
interfix interrelación
intergranular corrosion corrosión intergranular
interharmonic beats batidos interharmónicos
interim storage almacenamiento temporal
interlace entrelazar
interlace factor factor de entrelazado
interlaced entrelazado
interlaced field campo entrelazado
interlaced memory memoria entrelazada
interlaced scanning exploración entrelazada
interlacing entrelazamiento
interleave intercalar
interleaved memory memoria intercalada
interleaved windings arrollamientos intercalados
interleaving intercalación
interlock enclavamiento, bloqueo
interlock circuit circuito de enclavamiento
interlock device dispositivo de enclavamiento
interlock relay relé de enclavamiento
interlock switch conmutador de enclavamiento
interlock system sistema de enclavamiento
interlocking enclavamiento
interlocking device dispositivo de enclavamiento
interlocking relay relé de enclavamiento
interlocking signals señales de enclavamiento
interlocking system sistema de enclavamiento
intermediate amplifier amplificador intermedio
intermediate block character carácter de bloque intermedio
intermediate cable cable intermedio
intermediate center centro intermedio
intermediate circuit circuito intermedio
intermediate control control intermedio
intermediate-control change cambio de control intermedio
intermediate-control cycle ciclo de control intermedio
intermediate-control data datos de control intermedio
intermediate-control field campo de control intermedio
intermediate-control relay relé de control intermedio
intermediate cycle ciclo intermedio
intermediate distributing frame repartidor intermedio
intermediate echo suppressor supresor de eco intermedio
intermediate equipment equipo intermedio
intermediate frequency frecuencia intermedia
intermediate-frequency amplification amplificación de frecuencia intermedia
intermediate-frequency amplifier amplificador de frecuencia intermedia
intermediate-frequency carrier portadora de frecuencia intermedia
intermediate-frequency channel canal de frecuencia intermedia
intermediate-frequency converter convertidor de frecuencia intermedia
intermediate-frequency harmonic interference interferencia armónica de frecuencia intermedia
intermediate-frequency interconnection interconexión de frecuencia intermedia
intermediate-frequency interference interferencia de frecuencia intermedia
intermediate-frequency interference ratio razón de interferencia de frecuencia intermedia
intermediate-frequency response ratio razón de respuesta de frecuencia intermedia
intermediate-frequency selectivity selectividad de frecuencia intermedia
intermediate-frequency signal señal de frecuencia intermedia
intermediate-frequency stage etapa de frecuencia intermedia
intermediate-frequency transformer transformador de frecuencia intermedia
intermediate horizon horizonte intermedio
intermediate keyboard teclado intermedio
intermediate language lenguaje intermedio
intermediate layer capa intermedia
intermediate maintenance mantenimiento intermedio
intermediate memory memoria intermedia
intermediate memory storage almacenamiento de memoria intermedia
intermediate node nodo intermedio
intermediate power amplifier amplificador de potencia intermedio
intermediate range alcance intermedio
intermediate repeater repetidor intermedio
intermediate result resultado intermedio
intermediate routing function función de enrutamiento intermedia
intermediate routing node nodo de enrutamiento intermedio
intermediate section sección intermedia
intermediate state estado intermedio
intermediate station estación intermedia
intermediate storage almacenamiento intermedio
intermediate store almacenamiento intermedio
intermediate subcarrier subportadora intermedia
intermediate switching conmutación intermedia
intermediate system sistema intermedio
intermediate total total intermedio
intermetallic intermetálico
intermetallic alloy aleación intermetálica
intermetallic compound compuesto intermetálico
intermetallic contact contacto intermetálico
intermittent intermitente
intermittent action acción intermitente
intermittent condition condición intermitente
intermittent control control intermitente
intermittent current corriente intermitente
intermittent defect defecto intermitente
intermittent direct current corriente continua intermitente
intermittent duty servicio intermitente
intermittent-duty rating clasificación de servicio intermitente
intermittent-duty relay relé de servicio intermitente
intermittent error error intermitente
intermittent failure falla intermitente
intermittent fault falla intermitente
intermittent interference interferencia intermitente
intermittent light luz intermitente
intermittent operation operación intermitente
intermittent oscillation oscilación intermitente
intermittent periodic duty servicio periódico intermitente
intermittent periodic service servicio periódico intermitente
intermittent rating clasificación de servicio intermitente
intermittent service servicio intermitente
intermittent-service area área de servicio

intermitente
intermittent signal señal intermitente
intermittent test prueba intermitente
intermittent tone tono intermitente
intermodulate intermodular
intermodulation intermodulación
intermodulation analyzer analizador de
 intermodulación
intermodulation coefficient coeficiente de
 intermodulación
intermodulation distortion distorsión de
 intermodulación
intermodulation-distortion analyzer analizador de
 distorsión de intermodulación
intermodulation-distortion percentage porcentaje de
 distorsión de intermodulación
intermodulation-distortion ratio razón de distorsión
 de intermodulación
intermodulation frequency frecuencia de
 intermodulación
intermodulation interference interferencia de
 intermodulación
intermodulation level nivel de intermodulación
intermodulation meter medidor de intermodulación
intermodulation noise ruido de intermodulación
intermodulation products productos de
 intermodulación
intermolecular intermolecular
internal interno
internal absorptance absortancia interna
internal address field campo de dirección interna
internal amplification amplificación interna
internal arithmetic aritmética interna
internal audit auditoría interna
internal block bloque interno
internal blocking bloqueo interno
internal bus bus interno
internal cable cable interno
internal calibration calibración interna
internal capacity capacidad interna
internal clock reloj interno
internal code código interno
internal command mando interno
internal congestion congestión interna
internal connection conexión interna
internal connector conector interno
internal control control interno
internal control system sistema de control interno
internal conversion conversión interna
internal correction voltage tensión de corrección
 interna
internal data datos internos
internal data file archivo de datos internos
internal data item artículo de datos internos
internal data model modelo de datos internos
internal data record registro de datos internos
internal data representation representación de datos
 internos
internal detector of temperature detector interno de
 temperatura
internal electric current corriente eléctrica interna
internal energy energía interna
internal field campo interno
internal file connector conector de archivo interno
internal format formato interno
internal frequency reference referencia de
 frecuencia interna
internal hard disk disco duro interno
internal heating calentamiento interno
internal impedance impedancia interna
internal input impedance impedancia de entrada
 interna
internal input resistance resistencia de entrada

interna
internal insulation aislamiento interno
internal interruption interrupción interna
internal line línea interna
internal load carga interna
internal logic state estado lógico interno
internal memory memoria interna
internal model modelo interno
internal modem módem interno
internal modulation modulación interna
internal noise ruido interno
internal object objeto interno
internal output circuit circuito de salida interno
internal output impedance impedancia de salida
 interna
internal output resistance resistencia de salida
 interna
internal oxidation oxidación interna
internal power loss pérdida de potencia interna
internal power supply fuente de alimentación interna
internal procedure procedimiento interno
internal processing procesamiento interno
internal reader lectora interna
internal record registro interno
internal reference voltage tensión de referencia
 interna
internal reflection reflexión interna
internal resistance resistencia interna
internal schema esquema interno
internal shield blindaje interno
internal sort clasificación interna
internal storage almacenamiento interno
internal stress estrés interno
internal test prueba interna
internal traffic tráfico interno
internal transmittance transmitancia interna
internal transport instruction instrucción de
 transporte interno
internal triggering disparo interno
internal variable variable interna
international ampere ampere internacional
international broadcast station estación de
 radiodifusión internacional
international call llamada internacional
international call sign distintivo de llamada
 internacional
international candle bujía internacional
international circuit circuito internacional
international code código internacional
international connection conexión internacional
international coulomb coulomb internacional
international direct-distance dialing marcación de
 distancia directa internacional
international distance dialing marcación de
 distancia internacional
international electrical system sistema eléctrico
 internacional
international electrical units unidades eléctricas
 internacionales
international farad farad internacional
international henry henry internacional
international joule joule internacional
international line línea internacional
international Morse code código de Morse
 internacional
international number número internacional
international ohm ohm internacional
international phonetic alphabet alfabeto fonético
 internacional
international radio silence silencio de radio
 internacional
international second segundo internacional
international standard patrón internacional

international switching center centro de
    conmutación internacional
international system sistema internacional
International System of Units Sistema Internacional
    de Unidades
international telegraph circuit circuito telegráfico
    internacional
international telephone circuit circuito telefónico
    internacional
international telephone line línea telefónica
    internacional
international telephone service servicio telefónico
    internacional
international telex télex internacional
international traffic tráfico internacional
international unit unidad internacional
international volt volt internacional
international watt watt internacional
interphase tension tensión interfásica
interphase transformer transformador interfásico
interphone interfono
interphone amplifier amplificador de interfono
interphone connection conexión de interfono
interphone system sistema de interfono
interpolate interpolar
interpolation interpolación
interpolation function función de interpolación
interpolator interpolador
interpole interpolo
interpreter interpretador
interpreter code código interpretador
interpreter program programa interpretador
interpreter routine rutina interpretadora
interpretive interpretativo
interpretive code código interpretativo
interpretive execution ejecución interpretativa
interpretive language lenguaje interpretativo
interpretive program programa interpretativo
interpretive routine rutina interpretativa
interrecord gap espacio entre registros
interresonance interresonancia
interrogating pulse impulso interrogante
interrogation interrogación
interrogation antenna antena de interrogación
interrogation frequency frecuencia de interrogación
interrogation pulse impulso de interrogación
interrogation signal señal de interrogación
interrogator interrogador
interrogator-responder interrogador-respondedor
interrupt interrumpir
interrupt device dispositivo de interrupción
interrupt handling manejo de interrupción
interrupt identification identificación de
    interrupción
interrupt input line línea de entrada de interrupción
interrupt latency latencia de interrupción
interrupt level nivel de interrupción
interrupt level subroutine subrutina de nivel de
    interrupción
interrupt mode modo de interrupción
interrupt operation operación de interrupción
interrupt priority prioridad de interrupción
interrupt processing routine rutina de
    procesamiento de interrupción
interrupt register registro de interrupción
interrupt request solicitud de interrupción
interrupt request block bloque de solicitud de
    interrupción
interrupt response time tiempo de respuesta de
    interrupción
interrupt service device dispositivo de servicio de
    interrupción
interrupt service routine rutina de servicio de

    interrupción
interrupt signal señal de interrupción
interrupt signal feedback retroalimentación de señal
    de interrupción
interrupt system sistema de interrupción
interruptable interrumpible
interruptable state estado interrumpible
interrupted interrumpido
interrupted circuit circuito interrumpido
interrupted continuous wave onda continua
    interrumpida
interrupted direct current corriente continua
    interrumpida
interrupted ringing llamada interrumpida
interrupted waves ondas interrumpidas
interrupter interruptor
interrupter contacts contactos interruptores
interrupter frequency frecuencia de interruptor
interrupting capacity capacidad interruptora
interrupting current corriente interruptora
interruption interrupción
interruption code código de interrupción
interruption duration duración de interrupción
interruption-duration index índice de duración de
    interrupción
interruption frequency frecuencia de interrupciones
interruption-frequency index índice de frecuencia
    de interrupciones
intersatellite communication comunicación entre
    satélites
intersection intersección
interstage capacitor capacitor entre etapas
interstage coupling acoplamiento entre etapas
interstage diode diodo entre etapas
interstage punching perforación entre posiciones
    usuales
interstage shielding blindaje entre etapas
interstage transformer transformador entre etapas
interstation noise ruido entre estaciones
interstation noise suppression supresión de ruido
    entre estaciones
intersymbol interference interferencia entre
    símbolos
intersystem communications comunicaciones entre
    sistemas
intersystem electromagnetic compatibility
    compatibilidad electromagnética entre sistemas
intertie interconexión
intertripping disparo interdependiente
interval intervalo
interval meter medidor de intervalo
interval mode modo de intervalo
interval timer temporizador de intervalo
intervalometer intervalómetro
intervalve circuit circuito intervalvular
intervalve coupling acoplamiento intervalvular
interword gap espacio entre palabras
interword space espacio entre palabras
interword spacing espaciado entre palabras
intoxication tester probador de intoxicación
intramodal distortion distorsión intramodal
intrinsic intrínseco
intrinsic availability disponibilidad intrínseca
intrinsic-barrier diode diodo de barrera intrínseca
intrinsic-barrier transistor transistor de barrera
    intrínseca
intrinsic brightness brillo intrínseco
intrinsic brilliance brillantez intrínseca
intrinsic characteristic característica intrínseca
intrinsic coercive force fuerza coercitiva intrínseca
intrinsic coercivity coercitividad intrínseca
intrinsic concentration concentración intrínseca
intrinsic conduction conducción intrínseca

intrinsic conductivity conductividad intrínseca
intrinsic control control intrínseco
intrinsic electric strength rigidez eléctrica intrínseca
intrinsic flux flujo intrínseco
intrinsic flux density densidad de flujo intrínseco
intrinsic function función intrínseca
intrinsic hue tinte intrínseco
intrinsic impedance impedancia intrínseca
intrinsic induction inducción intrínseca
intrinsic mobility movilidad intrínseca
intrinsic noise ruido intrínseco
intrinsic permeability permeabilidad intrínseca
intrinsic property propiedad intrínseca
intrinsic region región intrínseca
intrinsic-region transistor transistor de región
  intrínseca
intrinsic resistance resistencia intrínseca
intrinsic resistivity resistividad intrínseca
intrinsic semiconductor semiconductor intrínseco
intrinsic temperature temperatura intrínseca
intrinsic temperature range intervalo de
  temperatura intrínseca
intrinsic transconductance transconductancia
  intrínseca
intrusion alarm alarma contra intrusión
intrusion sensor sensor de intrusión
invalid address dirección no válida
invalid character carácter no válido
invalid code código no válido
invalid instruction instrucción no válida
invalid key tecla no válida
invar invar
inverse inverso
inverse amplification factor factor de amplificación
  inverso
inverse anode voltage tensión anódica inversa
inverse beta beta inversa
inverse bias polarización inversa
inverse characteristics características inversas
inverse conduction conducción inversa
inverse current corriente inversa
inverse current relay relé de corriente inversa
inverse direction dirección inversa
inverse electrical characteristics características
  eléctricas inversas
inverse electrode current corriente de electrodo
  inversa
inverse feedback retroalimentación inversa
inverse-feedback amplifier amplificador de
  retroalimentación inversa
inverse-feedback filter filtro de retroalimentación
  inversa
inverse function función inversa
inverse grid current corriente de rejilla inversa
inverse grid potential potencial de rejilla inverso
inverse grid voltage tensión de rejilla inversa
inverse impedance impedancia inversa
inverse leakage fuga inversa
inverse limiter limitador inverso
inverse network red inversa
inverse peak voltage tensión de pico inversa
inverse photoelectric effect efecto fotoeléctrico
  inverso
inverse piezoelectric effect efecto piezoeléctrico
  inverso
inverse ratio razón inversa
inverse resistance resistencia inversa
inverse resonance resonancia inversa
inverse-square law ley inversa de cuadrados
inverse suppressor supresor de onda inversa
inverse time-lag relay relé de retardo inverso
inverse trigonometric function función
  trigonométrica inversa

inverse video video inverso
inverse voltage tensión inversa
inverse Wiedeman effect efecto de Wiedemann
  inverso
inversely inversamente
inversion inversión
inversion point punto de inversión
inversion ratio razón de inversión
invert invertir
inverted invertido
inverted amplification factor factor de amplificación
  invertida
inverted amplifier amplificador invertido
inverted channel canal invertido
inverted exponential horn bocina exponencial
  invertida
inverted file archivo invertido
inverted input entrada invertida
inverted sideband banda lateral invertida
inverted structure estructura invertida
inverted tube tubo invertido
inverter inversor
inverting adder sumador inversor
inverting connection conexión inversora
inverting input entrada inversora
invisible failure falla invisible
involuntary interrupt interrupción involuntaria
inward traffic tráfico de llegada
iodine yodo
ion ion
ion accelerating voltage tensión aceleradora iónica
ion activity actividad iónica
ion avalanche avalancha iónica
ion beam haz iónico
ion-beam probe sonda de haz iónico
ion burn mancha iónica
ion charge carga iónica
ion-charge density densidad de carga iónica
ion cluster grupo iónico
ion-collection time tiempo de recolección iónica
ion concentration concentración iónica
ion conduction current corriente de conducción
  iónica
ion convection current corriente de convección
  iónica
ion counter contador iónico
ion current corriente iónica
ion deflection deflexión iónica
ion density densidad iónica
ion exchange intercambio iónico
ion-exchange technique técnica de intercambio
  iónico
ion flow flujo iónico
ion gun cañón iónico
ion implantation implantación iónica
ion laser láser iónico
ion migration migración iónica
ion mobility movilidad iónica
ion noise ruido iónico
ion pump bomba iónica
ion ray rayo iónico
ion sensor sensor iónico
ion sheath vaina iónica
ion source fuente iónica
ion spot mancha iónica
ion trap trampa iónica
ion-trap magnet imán de trampa iónica
ionic iónico
ionic beam haz iónico
ionic binding forces fuerzas ligantes iónicas
ionic bond enlace iónico
ionic charge carga iónica
ionic conduction conducción iónica

ionic crystal cristal iónico
ionic current corriente iónica
ionic diffusion difusión iónica
ionic focusing enfoque iónico
ionic-heated cathode cátodo de calentamiento iónico
ionic-heated cathode tube tubo con cátodo de calentamiento iónico
ionic loudspeaker altavoz iónico
ionic microphone micrófono iónico
ionic mobility movilidad iónica
ionic polarizability polarizabilidad iónica
ionic semiconductor semiconductor iónico
ionic strength fuerza iónica
ionic switch conmutador iónico
ionic valve válvula iónica
ionization ionización
ionization chamber cámara de ionización
ionization chamber detector detector de cámara de ionización
ionization coefficient coeficiente de ionización
ionization current corriente de ionización
ionization density densidad de ionización
ionization effect efecto de ionización
ionization energy energía de ionización
ionization factor factor de ionización
ionization gage medidor de ionización
ionization-gage tube tubo medidor de ionización
ionization instrument instrumento de ionización
ionization loss pérdida por ionización
ionization measurement medida de ionización
ionization noise ruido por ionización
ionization path trayectoria de ionización
ionization potential potencial de ionización
ionization pressure presión de ionización
ionization pulse impulso de ionización
ionization rate velocidad de ionización
ionization resistance resistencia de ionización
ionization smoke detector detector de humo de ionización
ionization spectrometer espectrómetro de ionización
ionization time tiempo de ionización
ionization transducer transductor de ionización
ionization vacuum gage vacuómetro de ionización
ionization voltage tensión de ionización
ionize ionizar
ionized ionizado
ionized gas gas ionizado
ionized layer capa ionizada
ionized liquid líquido ionizado
ionizing ionizante
ionizing event evento ionizante
ionizing radiation radiación ionizante
ionogen ionógeno
ionogram ionograma
ionometer ionómetro
ionophone ionófono
ionosonde ionosonda
ionosphere ionósfera
ionospheric ionosférico
ionospheric absorption absorción ionosférica
ionospheric current corriente ionosférica
ionospheric defocusing desenfoque ionosférico
ionospheric disturbance perturbación ionosférica
ionospheric error error ionosférico
ionospheric focusing enfoque ionosférico
ionospheric height error error de altura ionosférico
ionospheric irregularities irregularidades ionosféricas
ionospheric layers capas ionosféricas
ionospheric prediction predicción ionosférica
ionospheric propagation propagación ionosférica
ionospheric ray rayo ionosférico
ionospheric reflection reflexión ionosférica

ionospheric region región ionosférica
ionospheric scatter dispersión ionosférica
ionospheric storm tormenta ionosférica
ionospheric wave onda ionosférica
iraser iraser
iridescence iridiscencia
iridium iridio
iris iris
iron hierro
iron-cobalt alloy aleación hierro-cobalto
iron-constantan thermocouple termopar hierro-constantán
iron core núcleo de hierro
iron-core coil bobina con núcleo de hierro
iron-core transformer transformador con núcleo de hierro
iron dust polvo de hierro
iron loss pérdida en el hierro
iron magnet imán de hierro
iron-nickel alloy aleación hierro-níquel
iron oxide óxido de hierro
iron pyrite pirita de hierro
iron-vane meter medidor de paleta de hierro
ironclad acorazado
ironclad electromagnet electroimán acorazado
ironclad plate placa acorazada
irradiance irradiancia
irradiate irradiar
irradiated irradiado
irradiation irradiación, radiación
irrational number número irracional
irrecoverable error error irrecuperable
irregular irregular
irregularity irregularidad
irreversible irreversible
irreversible electrolytic process proceso electrolítico irreversible
irreversible energy loss pérdida de energía irreversible
irreversible flux loss pérdida de flujo irreversible
irreversible process proceso irreversible
isinglass colapez, mica
island effect efecto de islote
isocandela curve curva isocandela
isochromatic isocromático
isochrone isocrona
isochronism isocronismo
isochronous isócrono
isochronous circuits circuitos isócronos
isochronous modulation modulación isócrona
isochronous scanning exploración isócrona
isochronous transmission transmisión isócrona
isoclinic line línea isoclínica
isodose isodosis
isodynamic line línea isodinámica
isoelectric isoeléctrico
isoelectric point punto isoeléctrico
isogonal isogono
isogonic line línea isogónica
isolantite isolantita
isolate aislar
isolated aislado
isolated amplifier amplificador aislado
isolated conductor conductor aislado
isolated digital output module módulo de salida digital aislado
isolated input entrada aislada
isolated input-output module módulo de entrada-salida aislado
isolated location localización aislada
isolated network red aislada
isolated-neutral system sistema con neutro aislado
isolated power system sistema de alimentación

aislado
isolated speaker altavoz aislado
isolating aislante
isolating amplifier amplificador aislante
isolating capacitor capacitor aislante
isolating circuit circuito aislante
isolating diode diodo aislante
isolating resistor resistor aislante
isolating switch conmutador aislante
isolating tape cinta aislante
isolating transformer transformador aislante
isolation aislamiento
isolation amplifier amplificador de aislamiento
isolation capacitor capacitor de aislamiento
isolation device dispositivo de aislamiento
isolation diode diodo de aislamiento
isolation filter filtro de aislamiento
isolation level nivel de aislamiento
isolation network red de aislamiento
isolation resistor resistor de aislamiento
isolation transformer transformador de aislamiento
isolator aislador, seccionador
isolith isolito
isolux isolux
isolux line línea isolux
isomagnetic isomagnético
isomagnetic line línea isomagnética
isomer isómero
isoplanar isoplanar
isothermal isotérmico
isotope isótopo
isotope effect efecto isotópico
isotope shift desplazamiento isotópico
isotopic isotópico
isotopic effect efecto isotópico
isotron isotrón
isotronics isotrónica
isotropic isotrópico
isotropic antenna antena isotrópica
isotropic detector detector isotrópico
isotropic dielectric dieléctrico isotrópico
isotropic gain ganancia isotrópica
isotropic radiator radiador isotrópico
isotropic scattering dispersión isotrópica
isotropic source fuente isotrópica
item artículo, ítem, elemento
item file archivo de artículos
iteration iteración
iteration factor factor de iteración
iteration structure estructura de iteración
iterative iterativo
iterative algorithm algoritmo iterativo
iterative attenuation atenuación iterativa
iterative attenuation factor factor de atenuación
  iterativa
iterative calculation cálculo iterativo
iterative coding codificación iterativa
iterative configuration configuración iterativa
iterative filter filtro iterativo
iterative impedance impedancia iterativa
iterative loop bucle iterativo
iterative operation operación iterativa
iterative process proceso iterativo
iterative program programa iterativo
iterative routine rutina iterativa
iterative sequence secuencia iterativa
iterative transfer constant constante de transferencia
  iterativa

J

jack jack, conjuntor
jack base base de jack
jack box caja de jacks
jack connector conector de jack
jack-ended junction unión terminada en jack
jack field campo de jacks
jack mounting montura de jacks
jack panel panel de jacks
jack space espacio de jacks
jack spacer espaciador de jacks
jack strip regleta de jacks
jacket chaqueta, forro, cubierta
Jacob's law ley de Jacob
jag distorsión por desincronización
jam atascamiento
jam circuit circuito contra atascamientos
jammer perturbador
jammer band banda perturbadora
jammer emission emisión perturbadora
jamming interferencia intencional, perturbación
  intencional, atascamiento
jamming device dispositivo de interferencia
  intencional
jamming effectiveness efectividad de interferencia
  intencional
jamming resistance resistencia a interferencia
  intencional
jamproof a prueba de atascamientos
Jansky noise ruido de Jansky
jar recipiente
jet chorro
jewel bearing cojinete de joya
jig soporte
jitter fluctuación
jitter-free signal señal libre de fluctuaciones
jitter tolerance tolerancia a fluctuaciones
job trabajo
job batch lote de trabajos
job card tarjeta de trabajo
job class clase de trabajo
job code código de trabajo
job control control de trabajo
job control block bloque de control de trabajos
job control card tarjeta de control de trabajos
job control file archivo de control de trabajos
job control information información de control de
  trabajos
job control language lenguaje de control de trabajos
job control processor procesador de control de
  trabajos
job control program programa de control de
  trabajos
job control statement sentencia de control de
  trabajos
job control table tabla de control de trabajos
job definition definición de trabajo
job description descripción de trabajo
job dividing división de trabajo
job-entry subsystem subsistema de entrada de
  trabajos
job flow flujo de trabajo

job flow control control de flujo de trabajo
job function función de trabajo
job information station estación de información de trabajos
job initialization inicialización de trabajo
job initialization processing level nivel de procesamiento de inicialización de trabajo
job-input device dispositivo de entrada de trabajos
job-input file archivo de entrada de trabajos
job-input stream corriente de entrada de trabajos
job-level field campo de nivel de trabajos
job library biblioteca de trabajos
job management administración de trabajos
job message queue cola de mensajes de trabajo
job name nombre de trabajo
job number número de trabajo
job-oriented orientado a trabajos
job-oriented terminal terminal orientada a trabajos
job output salida de trabajos
job-output device dispositivo de salida de trabajos
job-output stream corriente de salida de trabajos
job priority prioridad de trabajo
job processing procesamiento de trabajos
job-processing control control de procesamiento de trabajos
job queue cola de trabajos
job recorder registrador de trabajos
job seeker buscador de trabajos
job sequence secuencia de trabajo
job set conjunto de trabajos
job set control control de conjunto de trabajos
job set generator generador de conjunto de trabajos
job set program programa de conjunto de trabajos
job stack pila de trabajos
job stacking apilamiento de trabajos
job statement sentencia de trabajo
job step paso de trabajo
job step initiation iniciación de paso de trabajo
job step task tarea de paso de trabajo
job stream corriente de trabajos
job stream processor procesador de corriente de trabajos
job table tabla de trabajos
jogging avance por cierres sucesivos
Johnson counter contador de Johnson
Johnson curve curva de Johnson
Johnson-Lark-Horowitz effect efecto de Johnson-Lark-Horowitz
Johnson-Lark-Horowitz formula fórmula de Johnson-Lark-Horowitz
Johnson noise ruido de Johnson
joint junta, empalme
joint access acceso en común
joint box caja de junta
joint circuit circuito en común
joint communications comunicaciones en común
joint insulation aislamiento de junta
joint use uso en común
jointer empalmador
jointing sleeve manguito de empalme
Joly transformer transformador de Joly
Jones plug clavija de Jones
Josephson effect efecto de Josephson
Joshi effect efecto de Joshi
joule joule
Joule's law ley de Joule
Joule calorimeter calorímetro de Joule
Joule constant constante de Joule
Joule effect efecto de Joule
Joule heat calor de Joule
Joule heat gradient gradiente de calor de Joule
Joule heat loss pérdida de calor de Joule
Joule loss pérdida por efecto de Joule

joulemeter medidor en joules
joystick palanca de mando
judder distorsión por movimientos de equipo
jump salto
jump instruction instrucción de salto
jump line línea de salto
jump operation operación de salto
jump routine rutina de salto
jumper puente, cable de puente
jumper cable cable de puente
jumper connection conexión de puente
jumper wire hilo de puente
junction unión, punto de unión, enlace
junction area área de unión
junction barrier barrera de unión
junction battery batería de unión
junction box caja de unión
junction cable cable de unión
junction capacitance capacitancia de unión
junction capacitor capacitor de unión
junction circuit circuito de unión
junction circulator circulador de unión
junction diagram diagrama de uniones
junction diode diodo de unión
junction effect efecto de unión
junction field-effect-transistor transistor de efecto de campo de unión
junction fieldistor fildistor de unión
junction filter filtro de unión
junction group grupo de uniones
junction hunter buscador de uniones
junction laser láser de unión
junction light source fuente de luz de unión
junction line línea de unión
junction loss pérdida de unión
junction network red de uniones
junction panel panel de uniones
junction photocell fotocélula de unión
junction photodiode fotodiodo de unión
junction phototransistor fototransistor de unión
junction plane plano de unión
junction point punto de unión
junction pole poste de unión
junction rectifier rectificador de unión
junction resistance resistencia de unión
junction station estación de unión
junction temperature temperatura de unión
junction tetrode tetrodo de uniones
junction transistor transistor de uniones
junction tube tubo de unión
junctor conjuntor
justification routine rutina de justificación
jute yute
jute-protected cable cable forrado con yute
juxtaposed elements elementos yuxtapuestos
juxtaposed images imágenes yuxtapuestas

K

kallirotron kalirotrón
kallitron oscillator oscilador kalitrón
Kalman filter filtro de Kalman
kappa kappa
Karlous cell célula de Karlous
Karlous lens system sistema de lentes de Karlous
Karlous system sistema de Karlous
Karnaugh map mapa de Karnaugh
keep-alive anode ánodo auxiliar
keep-alive circuit circuito de ionización residual, circuito de mantenimiento
keep-alive discharge descarga de mantenimiento
keep-alive electrode electrodo de excitación, electrodo de entretenimiento
keep-alive voltage tensión de sostenimiento
keeper armadura de imán
Kelvin's law ley de Kelvin
Kelvin absolute electrometer electrómetro absoluto de Kelvin
Kelvin balance balanza de Kelvin
Kelvin bridge puente de Kelvin
Kelvin contacts contactos de Kelvin
Kelvin double bridge puente doble de Kelvin
Kelvin effect efecto de Kelvin
Kelvin electrometer electrómetro de Kelvin
Kelvin scale escala de Kelvin
Kelvin voltmeter vóltmetro de Kelvin
Kendall effect efecto de Kendall
Kennelly's law ley de Kennelly
Kennelly-Heaviside layer capa de Kennelly-Heaviside
Kennelly layer capa de Kennelly
kenopliotron kenopliotrón
kenotron kenotrón
keraunophone keraunófono
kernel núcleo
Kerr cell célula de Kerr
Kerr effect efecto de Kerr
Kerr electrooptical effect efecto electroóptico de Kerr
Kerr electrooptical shutter obturador electroóptico de Kerr
Kerr magnetooptical effect efecto magnetoóptico de Kerr
Kerr shutter obturador de Kerr
Kew magnetometer magnetómetro de Kew
key llave, clave, tecla, manipulador
key assignment asignación de teclado
key character carácter clave
key chirps chirridos de manipulación
key-click filter filtro de chasquidos de manipulación
key clicks chasquidos de manipulación
key data datos claves
key field campo clave
key file archivo clave
key punch perforadora de teclado
key station estación de control
key-to-disk unit unidad de teclado a disco
key-to-tape unit unidad de teclado a cinta
key tube tubo de manipulación
key-up position posición de reposo

keyboard teclado
keyboard actuator actuador de teclado
keyboard buffer memoria intermedia de teclado
keyboard computer computadora de teclado
keyboard control key tecla de control de teclado
keyboard device dispositivo de teclado
keyboard encoder codificador de teclado
keyboard entry entrada por teclado
keyboard function key tecla de función de teclado
keyboard keyer manipulador de teclado
keyboard layout distribución de teclado
keyboard lockout bloqueo de teclado
keyboard lockup bloqueo de teclado
keyboard macro macro de teclado
keyboard mechanism mecanismo de teclado
keyboard mode modo de teclado
keyboard operated operado por teclado
keyboard perforator perforadora de teclado
keyboard processor procesador de teclado
keyboard punch perforadora de teclado
keyboard terminal terminal de teclado
keyboard transmitter transmisor de teclado
keyed carrier portadora manipulada
keyed continuous wave onda continua manipulada
keyed damped wave onda amortiguada manipulada
keyed rainbow generator generador de arco iris
keyed wave onda manipulada
keyer manipulador
keying manipulación, tecleado
keying chirps chirridos de manipulación
keying circuit circuito de manipulación
keying cycle ciclo de manipulación
keying device dispositivo de manipulación
keying filter filtro de chasquidos de manipulación
keying frequency frecuencia de manipulación
keying head cabeza de manipulación
keying interval intervalo de manipulación
keying monitor monitor de manipulación
keying pause pausa de manipulación
keying rate velocidad de manipulación
keying relay relé de manipulación
keying sideband banda lateral de manipulación
keying speed velocidad de manipulación
keying test prueba de manipulación
keying transients fluctuaciones transitorias de manipulación
keying tube tubo de manipulación
keying wave onda de manipulación
keyless ringing llamada sin llave
keypunch perforadora
keypunch error rate tasa de errores de perforación
keyshelf estante de llaves
keystone distortion distorsión trapecial
keystone effect efecto trapecial
keystoning distorsión trapecial
keyway ranura posicionadora
keyword palabra clave
keyword in context palabra clave en contexto
keyword out of context palabra clave fuera de contexto
keyword parameter parámetro de palabra clave
kickback contragolpe, retorno
kickback power supply fuente de alimentación de retorno
Kikuchi lines líneas de Kikuchi
killer supresor, impureza
killer circuit circuito supresor
kilo kilo
kiloampere kiloampere
kilobaud kilobaud
kilobit kilobit
kilobyte kilobyte
kilocalorie kilocaloría

kilocurie kilocurie
kilocycle kilociclo
kiloelectronvolt kiloelectrónvolt
kilogauss kilogauss
kilogram kilogramo
kilohenry kilohenry
kilohertz kilohertz
kiloline kilolínea
kilolumen kilolumen
kilomaxwell kilomaxwell
kilometer kilómetro
kilometric kilométrico
kilometric wave onda kilométrica
kilooersted kilooersted
kiloohm kiloohm
kiloroentgen kilorroentgen
kilosecond kilosegundo
kilovar kilovar
kilovar-hour kilovar-hora
kilovolt kilovolt
kilovolt-ampere kilovolt-ampere
kilovoltmeter kilovóltmetro
kilowatt kilowatt
kilowatt-hour kilowatt-hora
kilowatt-hour meter watthorímetro
kinematic cinemático
kinematograph cinematógrafo
kinescope cinescopio
kinescope display visualizador cinescópico
kinescope recorder registrador cinescópico
kinescope recording registro cinescópico
kinescope transcription transcripción cinescópica
kinescope tube tubo cinescópico
kinescopic cinescópico
kinescopic recording registro cinescópico
kinetic cinético
kinetic depth effect efecto de profundidad cinético
kinetic energy energía cinética
kinetograph cinetógrafo
kinetoscope cinetoscopio
King oscillator oscilador de King
Kipp relay relé de Kipp
Kirchhoff's current law ley de corrientes de
    Kirchhoff
Kirchhoff's equation ecuación de Kirchhoff
Kirchhoff's first law primera ley de Kirchhoff
Kirchhoff's law ley de Kirchhoff
Kirchhoff's radiation law ley de radiación de
    Kirchhoff
Kirchhoff's second law segunda ley de Kirchhoff
Kirchhoff's voltage law ley de tensiones de
    Kirchhoff
kit juego
kite-supported antenna antena apoyada por cometa
Klipsch enclosure caja de Klipsch
Klipsch horn bocina de Klipsch
klydonogram clidonograma
klydonograph clidonógrafo
klystron klistrón
klystron amplifier amplificador klistrón
klystron frequency multiplier multiplicador de
    frecuencia klistrón
klystron generator generador klistrón
klystron harmonic generator generador armónico
    klistrón
klystron oscillator oscilador klistrón
klystron repeater repetidor klistrón
knife-blade fuse fusible de cuchilla
knife-contact switch conmutador de contacto de
    cuchilla
knife fuse fusible de cuchilla
knife lightning arrester pararrayos de cuchilla
knife switch conmutador de cuchilla

knob botón
knob insulator aislador de botón
knockout agujero ciego
Koch resistance resistencia de Koch
Kolster decremeter decrémetro de Kolster
Kooman antenna antena de Kooman
kovar kovar
Kozanowski oscillator oscilador de Kozanowski
kraft paper papel kraft
Kramer system sistema de Kramer
krarupization krarupización
Kraus antenna antena de Kraus
kryptol criptol
krypton criptón
krypton discharge tube tubo de descarga de criptón
krypton ion laser láser iónico de criptón
Kundt's law ley de Kundt
Kundt effect efecto de Kundt
Kundt tube tubo de Kundt
kurtosis curtosis
kymograph quimógrafo

L

label etiqueta
label alignment alineación de etiqueta
label area área de etiqueta
label checking comprobación de etiquetas
label group grupo de etiquetas
label identifier identificador de etiqueta
label information area área de información de etiquetas
label information cylinder cilindro de información de etiquetas
label processing procesamiento de etiquetas
label processing routine rutina de procesamiento de etiquetas
label record registro de etiqueta
label set conjunto de etiquetas
labile lábil
labile oscillator oscilador lábil
laboratory laboratorio
laboratory conditions condiciones de laboratorio
laboratory environment ambiente de laboratorio
laboratory measurement medida de laboratorio
laboratory measuring equipment equipo de medida de laboratorio
laboratory oscillator oscilador de laboratorio
laboratory oscilloscope osciloscopio de laboratorio
laboratory power supply fuente de alimentación de laboratorio
laboratory simulation simulación de laboratorio
laboratory standard patrón de laboratorio
laboratory system sistema de laboratorio
laboratory test prueba de laboratorio
labyrinth baffle bafle de laberinto
labyrinth loudspeaker altavoz de laberinto
labyrinth speaker altavoz de laberinto
lace perforación en cadena
laced wiring alambrado atado
lacing perforación en cadena
lacquer disk disco de laca
lacquer original original en laca
lacquer recording registro en laca
ladar ladar
ladder attenuator atenuador en escalera
ladder filter filtro en escalera
ladder network red en escalera
ladder-type attenuator atenuador en escalera
ladder-type filter filtro en escalera
lag retardo, persistencia de imagen
lag network red de retardo
lag variable variable de retardo
lagged-demand meter medidor de demanda con retardo
lagging current corriente retardada
lagging load carga retardada
lagging phase fase retardada
Lalande cell pila Lalande
Lamb wave onda de Lamb
lambda lambda
lambda wave onda lambda
lambert lambert
Lambert's cosine law ley del coseno de Lambert
Lambert's law ley de Lambert

Lambert's law of illumination ley de Lambert de iluminación
Lambertian radiator radiador de Lambert
Lambertian surface superficie de Lambert
laminated laminado
laminated antenna antena laminada
laminated brush escobilla laminada
laminated contact contacto laminado
laminated core núcleo laminado
laminated disk disco laminado
laminated ferromagnetic core núcleo ferromagnético laminado
laminated iron core núcleo de hierro laminado
laminated magnet imán laminado
laminated pole polo laminado
laminated record disco laminado
laminated sheet hoja laminada
lamination laminación
laminography laminografía
lamp lámpara
lamp amplifier amplificador de lámpara
lamp bank banco de lámparas
lamp-bank resistor resistor de banco de lámparas
lamp base casquillo de lámpara
lamp bulb bombilla de lámpara
lamp cap casquillo de lámpara
lamp cord cordón de lámpara, cordón eléctrico
lamp dimmer reductor de intensidad de lámpara
lamp driver excitador de lámpara
lamp extractor extractor de lámparas
lamp-holder portalámpara
lamp jack jack de lámpara
lamp mount portalámpara
lamp mounting portalámpara
lamp panel panel de lámparas
lamp regulator regulador de lámpara
lamp resistance resistencia de lámpara
lamp socket portalámpara
lampblack negro de humo
lampholder portalámpara
Lampkin oscillator oscilador de Lampkin
lanac lanac
lanac beacon radiofaro lanac
lanac directional antenna antena direccional lanac
land cable cable terrestre
land mobile service servicio móvil terrestre
land mobile station estación móvil terrestre
land path trayectoria terrestre
land return retorno terrestre
land station estación terrestre
Landau damping amortiguamiento de Landau
Landau formula fórmula de Landau
Landau length longitud de Landau
landing beacon radiofaro de aterrizaje
landing beam haz de aterrizaje
landing direction indicator indicador de dirección de aterrizaje
landing direction light luz de dirección de aterrizaje
landing instrument instrumento de aterrizaje
landing light luz de aterrizaje
landing path trayectoria de aterrizaje
landing signal señal de aterrizaje
landing zone zona de aterrizaje
landline línea terrestre
landline circuit circuito por línea terrestre
landline relay retransmisión por líneas terrestres
landline system sistema de líneas terrestres
landmark beacon faro de referencia
landscape orientation orientación horizontal
Langevin function función de Langevin
Langevin ion ion de Langevin
Langevin-Pauli formula fórmula de Langevin-Pauli
Langevin theory of diamagnetism teoría de

diamagnetismo de Langevin
Langmuir's equation ecuación de Langmuir
Langmuir's law ley de Langmuir
Langmuir dark space espacio obscuro de Langmuir
language lenguaje
language character set conjunto de caracteres de lenguaje
language code código de lenguaje
language conversion program programa de conversión de lenguaje
language converter convertidor de lenguaje
language interpreter interpretador de lenguaje
language name nombre de lenguaje
language processor procesador de lenguaje
language standard patrón de lenguaje
language statement sentencia de lenguaje
language subset subconjunto de lenguaje
language subsystem subsistema de lenguaje
language translation traducción de lenguaje
language translation program programa de traducción de lenguaje
language translator traductor de lenguaje
lanthanum lantanio
lap joint junta de solape
lap winding arrollamiento imbricado
lapel microphone micrófono de solapa
Laplace's law ley de Laplace
Laplace equation ecuación de Laplace
Laplace transform transformación de Laplace
lapping recubrimiento, lapidado
lapse lapso
large-angle scattering dispersión en gran ángulo
large-angle tube tubo de gran ángulo
large capacity gran capacidad
large-capacity cable cable de gran capacidad
large-capacity core storage almacenamiento de núcleos de gran capacidad
large-capacity storage almacenamiento de gran capacidad
large-scale integrated circuit circuito integrado en gran escala
large-scale integration integración en gran escala
large signal señal fuerte, señal grande
large-signal analysis análisis de señales fuertes
large-signal characteristics características de señales fuertes
large-signal component componente de señales fuertes
large-signal equivalent circuit circuito equivalente de señales fuertes
large-signal input entrada de señal fuerte
large-signal operation operación de señales fuertes
large-signal power gain ganancia de potencia de señal fuerte
Larmor frequency frecuencia de Larmor
Larmor orbit órbita de Larmor
Larmor precession precesión de Larmor
Larmor precession frequency frecuencia de precesión de Larmor
Larsen effect efecto de Larsen
laryngaphone laringófono
laryngophone laringófono
laser láser
laser altimeter altímetro láser
laser beam haz láser
laser-beam communication comunicación por haz láser
laser burst ráfaga láser, descarga láser
laser camera cámara láser
laser capacitor capacitor láser
laser cavity cavidad láser
laser cutting corte láser
laser diode diodo láser

laser disk disco láser
laser Doppler radar radar de Doppler láser
laser drill taladro láser
laser flash tube tubo de destellos láser
laser medium medio láser
laser memory memoria láser
laser optical videodisk system sistema de videodisco óptico láser
laser photocoagulator fotocoagulador láser
laser print head cabeza impresora láser
laser printer impresora láser
laser radar radar láser
laser storage almacenamiento láser
laser surgery cirugía láser
laser system sistema láser
laser-welded soldado con láser
laser welding soldadura láser
lashed cable cable con suspensión continua
lashing suspensión continua
lasing acción láser
last-line release desconexión por la última línea
latch cerrojo
latching current corriente de enganche
latching relay relé de enclavamiento
latching thermal switch conmutador térmico de enganche
late contacts contactos tardíos
latency latencia
latency time tiempo de latencia
latent latente
latent electronic image imagen electrónica latente
latent energy energía latente
latent image imagen latente
latent period periodo latente
lateral lateral
lateral adjustment knob botón de ajuste lateral
lateral cable cable lateral
lateral chromatic aberration aberración cromática lateral
lateral compliance docilidad lateral
lateral compression compresión lateral
lateral conductor conductor lateral
lateral-correction magnet imán de corrección lateral
lateral deviation desviación lateral
lateral displacement desplazamiento lateral
lateral image movement movimiento de imagen lateral
lateral insulator aislador lateral
lateral inversion inversión lateral
lateral load carga lateral
lateral magnet imán lateral
lateral motion moción lateral
lateral parity paridad lateral
lateral parity check comprobación de paridad lateral
lateral profile perfil lateral
lateral projection proyección lateral
lateral recording registro lateral
lateral redundancy check comprobación de redundancia lateral
lateral reproducer reproductor lateral
lateral stability estabilidad lateral
lateral transcription transcripción lateral
lateral wave onda lateral
lateral width anchura lateral
lateral winding arrollamiento lateral
latitude effect efecto de latitud
Latour alternator alternador de Latour
lattice celosía, red
lattice amplifier amplificador en celosía
lattice energy energía de red
lattice filter filtro en celosía
lattice imperfection imperfección de red
lattice network red en celosía

lattice section sección en celosía
lattice structure estructura en celosía
lattice winding arrollamiento en celosía
Laue diagram diagrama de Laue
Laue equations ecuaciones de Laue
Laue pattern patrón de Laue
launching alimentación, lanzamiento, emisión
launching unit unidad de alimentación
Laurent polarimeter polarímetro de Laurent
Lauritsen electroscope electroscopio de Lauritsen
law of averages ley de promedios
law of charges ley de cargas
law of electric charges ley de cargas eléctricas
law of electromagnetic induction ley de inducción electromagnética
law of electromagnetic systems ley de sistemas electromagnéticos
law of electrostatic attraction ley de atracción electrostática
law of electrostatic repulsion ley de repulsión electrostática
law of induced current ley de corrientes inducidas
law of induction ley de inducción
law of inverse squares ley inversa de cuadrados
law of lenses ley de lentes
law of magnetism ley de magnetismo
law of normal distribution ley de distribución normal
law of Pfeffer ley de Pfeffer
law of radiation ley de radiación
law of reflection ley de reflexión
law of refraction ley de refracción
Lawrence accelerator acelerador de Lawrence
Lawrence tube tubo de Lawrence
lawrencium laurencio
layer capa
layer-to-layer adhesion adhesión de capa a capa
layer-to-layer signal transfer transferencia de señal de capa a capa
layer-to-layer transfer transferencia de capa a capa
layer-to-layer voltage tensión de capa a capa
layer winding arrollamiento en capas
layerwound coil bobina arrollada en capas
layout arreglo, distribución, diagrama
layout of equipment arreglo de equipo
lazy-H antenna antena de doble dipolo
lead hilo de conexión, conexión, conductor, avance, plomo
lead accumulator acumulador de plomo
lead-acid battery batería de plomo-ácido
lead-acid cell celda de plomo-ácido
lead-antimony battery batería de plomo-antimonio
lead bath baño de plomo
lead-bath furnace horno de baño de plomo
lead battery batería de plomo
lead-calcium battery batería de plomo-calcio
lead capacitance capacitancia de conexión
lead castle castillo de plomo
lead cell celda de plomo
lead chloride cloruro de plomo
lead-coated revestido de plomo
lead connection conexión de plomo
lead-covered recubierto de plomo
lead-covered cable cable recubierto de plomo
lead-covered wire hilo recubierto de plomo
lead dioxide dióxido de plomo
lead dioxide primary cell pila primaria de dióxido de plomo
lead equivalent equivalente de plomo
lead glass vidrio de plomo
lead-in bajada, conductor de entrada
lead-in groove surco de entrada
lead-in insulator aislador pasamuros

lead-in spiral espiral de entrada
lead-in wire hilo de entrada
lead inductance inductancia de conexiones
lead-out groove surco de salida
lead-over groove surco entre registros
lead oxide óxido de plomo
lead polarity polaridad de conexiones
lead rubber caucho al plomo
lead-rubber apron mandil de caucho al plomo
lead-rubber gloves guantes de caucho al plomo
lead screw tornillo de avance, tornillo de guía
lead selenide seleniuro de plomo
lead selenide photocell fotocélula de seleniuro de plomo
lead sheath vaina de plomo
lead shielding blindaje de plomo
lead sleeve manguito de plomo
lead storage battery acumulador de plomo
lead sulfate sulfato de plomo
lead sulfide sulfuro de plomo
lead sulfide photoconductive cell célula fotoconductora de sulfuro de plomo
lead telluride telururo de plomo
lead telluride cell célula de telururo de plomo
lead time tiempo de avance
lead-zinc storage battery acumulador de plomo-cinc
lead zirconate-titanate zirconato-titanato de plomo
lead zirconate-titanate ceramic cerámica de zirconato-titanato de plomo
leader guía, cinta de arrastre, cabecera
leader cable cable guía
leader-cable system sistema de cable guía
leader card tarjeta de cabecera
leader record registro de cabecera
leader tape cinta de arrastre
leading blacks borde negro ante imagen blanca
leading current corriente adelantada
leading echo eco adelantado
leading edge borde de entrada, borde delantero
leading-edge pulse time tiempo de borde delantero de impulso
leading ghost fantasma adelantado
leading-in cable cable de entrada
leading-in insulator aislador de entrada
leading-in tube tubo de entrada
leading-in wire hilo de entrada
leading load carga adelantada
leading-through insulator aislador pasamuros
leading voltage tensión adelantada
leading whites borde blanco ante imagen negra
leading zero cero a la izquierda
leading zero suppression supresión de ceros a la izquierda
leadover groove surco entre registros
leaf electroscope electroscopio de hojas
leak fuga
leak conductance conductancia de fugas
leak current corriente de fugas
leak detection detección de fugas
leak detector detector de fugas
leak locator localizador de fugas
leakage fuga
leakage coefficient coeficiente de fuga
leakage conductance conductancia de fuga
leakage current corriente de fuga
leakage detector detector de fuga
leakage distance distancia de fuga
leakage factor factor de fuga
leakage finder buscador de fugas
leakage flow flujo de fuga
leakage flux flujo de fuga
leakage indicator indicador de fugas
leakage inductance inductancia de fuga

leakage path trayectoria de fuga
leakage power potencia de fuga
leakage radiation radiación de fuga
leakage reactance reactancia de fuga
leakage resistance resistencia de fuga
leakage test prueba de fugas
leakage voltage tensión de fuga
leakance perditancia
leakproof a prueba de fugas
leaky con fugas
leaky capacitor capacitor con fugas
leaky condenser condensador con fugas
leaky-grid rectification rectificación por fugas de
    rejilla
leaky line línea con fugas
leaky pipe tubo con fugas
leaky ray rayo con fugas
leaky tube tubo con fugas
leaky-wave antenna antena de onda con fugas
leaky waveguide guíaondas con fugas
leapfrog test prueba por saltos
leapfrogging avance por saltos
learning program programa de aprendizaje
leased channel canal arrendado
leased circuit circuito arrendado
leased line línea arrendada
leased network red arrendada
least significant menos significativo
least significant bit bit menos significativo
least significant character carácter menos
    significativo
least significant digit dígito menos significativo
Leblanc connection conexión de Leblanc
Leblanc system sistema de Leblanc
Lecher bridge puente de Lecher
Lecher frame cuadro de Lecher
Lecher lines líneas de Lecher
Lecher oscillator oscilador de Lecher
Lecher wires hilos de Lecher
Leclanche cell pila de Leclanche
Leduc current corriente de Leduc
Leduc effect efecto de Leduc
left-hand polarized wave onda con polarización
    levógira
left-hand rule regla de la mano izquierda
left-hand taper aumento de resistencia a la izquierda
left-hand thread rosca izquierda
left shift desplazamiento a la izquierda
left zero control control de ceros a la izquierda
leg patilla, rama
Lenard rays rayos de Lenard
Lenard tube tubo de Lenard
length longitud
length of break longitud de ruptura
length of cable longitud de cable
length of scanning line longitud de línea de
    exploración
length of travel longitud de recorrido
lens lente
lens antenna antena de lente
lens aperture abertura de lente
lens disk disco de lentes
lens drum tambor de lentes
lens multiplication factor factor de multiplicación de
    lente
lens speed abertura relativa
lenticular lenticular
Lenz's law ley de Lenz
Lepel discharger descargador de Lepel
less than menor que
level nivel
level above threshold nivel sobre el umbral
level clipper recortador de nivel

level compensator compensador de nivel
level control control de nivel
level control table tabla de control de nivel
level coordination coordinación de niveles
level detector detector de nivel
level diagram diagrama de niveles
level gage medidor de nivel
level indicator indicador de nivel
level line línea de nivel
level monitoring monitoreo de nivel
level number número de nivel
level of access nivel de acceso
level of addressing nivel de direccionamiento
level of confidence nivel de confianza
level of modulation nivel de modulación
level of priority nivel de prioridad
level recorder registrador de nivel
level regulation regulación de nivel
level surface superficie de nivel
level switch conmutador de nivel
level table tabla de niveles
level translator traductor de nivel
leveling nivelación
lever-actuated accionado por palanca
lever key llave de palanca
lever switch conmutador de palanca
Lewis antenna antena de Lewis
lexical analyzer analizador de léxico
Leyden bottle botella de Leyden
Leyden jar botella de Leyden
liberator tank tanque de liberación
library biblioteca
library backup system sistema de respaldo de
    biblioteca
library block bloque de biblioteca
library character set conjunto de caracteres de
    biblioteca
library control sector sector de control de biblioteca
library file archivo de biblioteca
library list lista de bibliotecas
library macro definition macrodefinición de
    biblioteca
library maintenance program programa de
    mantenimiento de biblioteca
library name nombre de biblioteca
library program programa de biblioteca
library routine rutina de biblioteca
library software programas de biblioteca
library subroutine subrutina de biblioteca
library tape cinta de biblioteca
Lichtenberg figure figura de Lichtenberg
Lichtenberg figure camera cámara de figura de
    Lichtenberg
lid tapa
lidar lidar
lie detector detector de mentiras
life cycle ciclo de vida
life-cycle phase fase de ciclo de vida
life expectancy expectativa de vida
life test prueba de duración
lifter levantador, armadura
lifting magnet electroimán elevador
light luz
light absorption absorción de luz
light-activated switch conmutador activado por luz
light-actuated device dispositivo accionado por luz
light adaptation adaptación a luz
light amplifier amplificador de luz
light-amplifying device dispositivo amplificador de
    luz
light beacon baliza luminosa
light-beam haz luminoso
light-beam communication comunicación por haz

luminoso
light-beam galvanometer galvanómetro de haz luminoso
light-beam localizer localizador de haz luminoso
light-beam meter medidor de haz luminoso
light-beam oscillograph oscilógrafo de haz luminoso
light-beam pickup captor de haz luminoso
light-beam pointer indicador de haz luminoso
light-beam receiver receptor de haz luminoso
light-beam recorder registrador de haz luminoso
light-beam transmitter transmisor de haz luminoso
light-carrier injection inyección de portadora luminosa
light center centro de luz
light chopper interruptor periódico de luz
light control control de luz
light-controlled oscillator oscilador controlado por luz
light-dependent resistor resistor dependiente de luz
light diffraction difracción de luz
light diffuser difusor de luz
light dimmer reductor de intensidad de luz
light emission emisión de luz
light emitter emisor de luz
light-emitting diode diodo emisor de luz
light-emitting film película emisora de luz
light filter filtro de luz
light flash destello de luz
light flasher destellador de luz
light flux flujo de luz
light-flux level nivel de flujo de luz
light fluxmeter flujómetro de luz
light guide guía de luz
light gun cañón de luz
light impulse impulso de luz
light-induced electricity electricidad inducida por luz
light load carga liviana
light meter fotómetro
light-microsecond microsegundo-luz
light modulation modulación de luz
light modulator modulador de luz
light negative fotorresistente
light-operated relay relé operado por luz
light-operated switch conmutador operado por luz
light pattern patrón de luz
light pen lápiz óptico
light-pen detection detección de lápiz óptico
light-pen operation operación de lápiz óptico
light-pen system sistema de lápiz óptico
light pipe conducto de luz
light positive fotoconductor
light quantum cuanto de luz
light ray rayo luminoso
light-ray oscilloscope osciloscopio de rayo luminoso
light reflection reflexión de luz
light relay fotorrelé
light scattering dispersión de luz
light-sensitive sensible a la luz
light-sensitive cathode cátodo sensible a la luz
light-sensitive detector detector sensible a la luz
light-sensitive diode diodo sensible a la luz
light-sensitive film película sensible a la luz
light-sensitive material material sensible a la luz
light-sensitive resistor resistor sensible a la luz
light-sensitive sensor sensor sensible a la luz
light-sensitive surface superficie sensible a la luz
light-sensitive tube tubo sensible a la luz
light sensor sensor de luz
light source fuente de luz
light spectrum espectro de luz
light spot punto luminoso
light-spot scanner explorador de punto luminoso móvil

light standard patrón de luz
light traffic tráfico liviano
light valve fotoválvula
light wave onda luminosa
light-wave telephony telefonía por onda luminosa
lighted switch conmutador iluminado
lighthouse tube tubo tipo faro
lighthouse-tube oscillator oscilador de tubo tipo faro
lighting alumbrado
lighting circuit circuito de alumbrado
lighting effectiveness factor factor de efectividad de alumbrado
lighting equipment equipo de alumbrado
lighting facilities instalaciones de alumbrado
lighting feeder alimentador de alumbrado
lighting load carga de alumbrado
lighting outlet tomacorriente de alumbrado
lighting switch conmutador de alumbrado
lighting system sistema de alumbrado
lighting transformer transformador de alumbrado
lightning rayo
lightning arrester pararrayos
lightning conductor conductor de pararrayos
lightning detector detector de rayos
lightning overvoltage sobretensión por rayos
lightning protection protección contra rayos
lightning protective device dispositivo protector contra rayos
lightning rod barra pararrayos
lightning strike descarga atmosférica
lightning stroke descarga de rayo
lightning surge sobretensión por rayo
lightning switch conmutador de antena-tierra
lightningproof a prueba de rayos
like poles polos similares
Lilienfeld amplifier amplificador de Lilienfeld
Lilienfeld effect efecto de Lilienfeld
Lilienfeld tube tubo de Lilienfeld
limen limen
limit límite
limit check comprobación de límite
limit control control de límite
limit load carga límite
limit-load factor factor de carga límite
limit of error límite de error
limit of resolution límite de resolución
limit rays rayos límite
limit switch conmutador de límite
limit test prueba de límites
limited limitado
limited availability disponibilidad limitada
limited capacity capacidad limitada
limited input-output entrada-salida limitada
limited integrator integrador limitado
limited signal señal limitada
limited stability estabilidad limitada
limiter limitador
limiter circuit circuito limitador
limiting limitador
limiting ambient temperature temperatura ambiente límite
limiting amplifier amplificador limitador
limiting aperture abertura limitadora
limiting circuit circuito limitador
limiting coil bobina limitadora
limiting condition condición limitadora
limiting current corriente límite
limiting device dispositivo limitador
limiting diode diodo limitador
limiting error error límite
limiting factor factor limitador
limiting filter filtro limitador
limiting frequency frecuencia límite

limiting operation operación limitadora
limiting polarization polarización limitadora
limiting resistor resistor limitador
limiting resolution resolución límite
limiting stage etapa limitadora
limiting temperature temperatura límite
limiting tube tubo limitador
limiting value valor límite
lin-log receiver receptor lineal-logarítmico
linac linac
Lindemann electrometer electrómetro de Lindemann
Lindemann glass vidrio de Lindemann
line línea
line adapter adaptador de línea
line-adjusting circuit circuito ajustador de línea
line amplifier amplificador de línea
line amplifier-rectifier amplificador-rectificador de línea
line amplitude amplitud de línea
line-amplitude control control de amplitud de línea
line balance equilibrio de línea
line-balance converter convertidor de equilibrio de línea
line battery batería de línea
line bias polarización de línea
line blanking extinción de línea
line-blanking interval intervalo de extinción de línea
line blocked bloqueo
line breadth anchura de línea
line break ruptura de línea
line breaker interruptor de línea
line broadcasting radiodifusión por línea
line brush escobilla de línea
line buffer almacenador intermedio de línea
line busy tone tono de línea ocupada
line carrier portadora en línea
line character carácter de línea
line characteristic distortion distorsión característica de línea
line charging current corriente de carga de línea
line choking coil bobina de choque
line circuit circuito de línea
line circuit-breaker cortacircuito de línea
line clear trayectoria libre
line code código de línea
line command mando de línea
line communication comunicación por línea
line concentration concentración de líneas
line concentrator concentrador de líneas
line conditioning condicionamiento de línea
line conductor conductor de línea
line connection conexión de línea
line connector conector de línea
line constant constante de línea
line control control de línea
line-control block bloque de control de línea
line-control characters caracteres de control de línea
line-control unit unidad de control de línea
line converter convertidor de línea
line coordination coordinación de línea
line cord cordón de alimentación
line-cord resistor resistor de cordón de alimentación
line corrector corrector de línea
line counter contador de líneas
line coupling acoplamiento de línea
line current corriente de línea
line delay retardo de línea
line deletion character carácter de borrado de línea
line density densidad de línea
line description descripción de línea
line detection detección de línea
line device dispositivo de línea

line diffuser difusor de línea
line display visualizador de línea
line distortion distorsión de línea
line divider divisor de línea
line drawing dibujo de líneas
line driver excitador de línea
line drop caída de línea
line-drop compensator compensador de caída de línea
line elongation elongación de línea
line-end symbol símbolo de fin de línea
line equalization ecualización de línea
line equalizer ecualizador de línea
line equipment equipo de línea
line failure rate tasa de fallas de línea
line fault falla de línea
line feed alimentación de línea
line-feed character carácter de alimentación de línea
line-feed code código de alimentación de línea
line-feed key tecla de alimentación de línea
line feeder alimentador de línea
line filter filtro de línea
line finder buscador de línea
line finding localización de línea
line-finding function función de localización de línea
line flicker parpadeo de línea
line flyback retorno de línea
line-flyback pulse impulso de retorno de línea
line-focus tube tubo de enfoque de línea
line free desbloqueado
line frequency frecuencia de líneas
line-frequency distortion distorsión de frecuencia de líneas
line-frequency generator generador de frecuencia de línea
line-frequency regulation regulación de frecuencia de línea
line fuse fusible de línea
line generator generador de línea
line group grupo de línea
line group selector selector de grupo de línea
line height altura de línea
line hydrophone hidrófono de línea
line impedance impedancia de línea
line integral integral de línea
line-intensifying pulse impulso intensificador de línea
line interface base base de interfaz de línea
line interlace entrelazado de líneas
line interlacing entrelazado de líneas
line interrupter interruptor de línea
line jack jack de línea
line-jump scanning exploración por saltos de líneas
line key tecla de línea
line lamp lámpara de línea
line leakage fuga de línea
line length longitud de línea
line lengthener prolongador de línea
line level nivel de línea
line link enlace de línea
line load carga de línea
line-load control control de carga de línea
line lockout bloqueo de línea
line loop bucle de línea
line loss pérdida de línea
line maintenance mantenimiento de línea
line management administración de líneas
line-matching transformer transformador de adaptación de línea
line microphone micrófono de línea
line mode modo de línea
line-mode switching conmutación de modo de línea
line noise ruido de línea

line number número de línea
line-number access acceso de número de línea
line of centers línea de centros
line of code línea de código
line of communication línea de comunicación
line of flux línea de flujo
line of force línea de fuerza
line of induction línea de inducción
line of magnetic force línea de fuerza magnética
line of position línea de posición
line of propagation línea de propagación
line-of-sight línea visual
line-of-sight circuit circuito de línea visual
line-of-sight distance distancia de línea visual
line-of-sight network red de línea visual
line-of-sight path trayectoria de línea visual
line-of-sight radio link radioenlace de línea visual
line-of-sight range alcance de línea visual
line-of-sight system sistema de línea visual
line output salida de línea
line-output pentode pentodo de salida de línea
line-output transformer transformador de salida de línea
line pad atenuador de línea
line parameters parámetros de línea
line period periodo de línea
line pilot piloto de línea
line-pilot frequency frecuencia piloto de línea
line plot trazado de línea
line plug clavija de línea
line post poste de línea
line power potencia de línea
line power supply fuente de alimentación de línea
line printer impresora de línea
line printer controller controlador de impresora de línea
line printing impresión por línea
line-protection equipment equipo de protección de línea
line pulse impulso de línea
line pulser pulsador de línea
line pulsing pulsación de línea
line radio radiodifusión por línea
line reactance reactancia de línea
line receiver receptor de línea
line redundancy redundancia de línea
line redundancy level nivel de redundancia de línea
line reference referencia de línea
line reflection reflexión de línea
line regulation regulación de línea
line relay relé de línea
line repeater repetidor de línea
line response mode modo de respuesta de línea
line return retorno de línea
line scan exploración de líneas
line scanning exploración de líneas
line-scanning frequency frecuencia de exploración de líneas
line-scanning oscillator oscilador de exploración de líneas
line section sección de línea
line segment segmento de línea
line selection selección de línea
line selector selector de línea
line-sequential system sistema de secuencia de líneas
line side lado de línea
line signal señal de línea
line source fuente de línea
line spacing espaciado de líneas
line spectrum espectro de líneas
line speed velocidad de línea
line splitter divisor de línea
line splitting división de línea

line stabilization estabilización de línea
line-stabilized oscillator oscilador estabilizado por línea
line stretcher extensor de línea
line style estilo de línea
line surge sobretensión de línea
line sweep circuit circuito de barrido de línea
line switch conmutador de línea
line switching conmutación de líneas
line synchronization sincronización de líneas
line-synchronizing pulse impulso sincronizador de línea
line-synchronizing signal señal sincronizadora de línea
line terminal terminal de línea
line-to-line voltage tensión entre fases
line transducer transductor de línea
line transformer transformador de línea
line translation traducción de líneas
line transmission transmisión por línea
line trigger disparador de línea
line triggering disparo de línea
line type tipo de línea
line unit unidad de línea
line voltage tensión de línea
line-voltage connection conexión de tensión de línea
line-voltage drop caída de tensión de línea
line-voltage regulator regulador de tensión de línea
line-voltage variations variaciones de tensión de línea
line waveform forma de onda de línea
line width anchura de línea
line wire hilo de línea
lineal lineal
linear lineal
linear absorption coefficient coeficiente de absorción lineal
linear acceleration aceleración lineal
linear accelerator acelerador lineal
linear accelerometer acelerómetro lineal
linear actuator actuador lineal
linear adder sumador lineal
linear amplification amplificación lineal
linear amplifier amplificador lineal
linear amplitude response respuesta de amplitud lineal
linear analog computer computadora analógica lineal
linear antenna antena lineal
linear antenna array red de antenas lineal
linear array red lineal
linear attenuation coefficient coeficiente de atenuación lineal
linear automatic regulation regulación automática lineal
linear backward-wave oscillator oscilador de ondas regresivas lineal
linear beam tube tubo de haz lineal
linear bilateral impedance impedancia bilateral lineal
linear calibration calibración lineal
linear characteristic característica lineal
linear charge density densidad de carga lineal
linear circuit circuito lineal
linear clipping recorte lineal
linear computer computadora lineal
linear configuration configuración lineal
linear control control lineal
linear counter contador lineal
linear crosstalk diafonía lineal
linear data structure estructura de datos lineal
linear decrement decremento lineal
linear delay unit unidad de retardo lineal

linear density densidad lineal
linear detection detección lineal
linear detector detector lineal
linear differential transformer transformador diferencial lineal
linear direct-current amplifier amplificador de corriente continua lineal
linear discontinuity discontinuidad lineal
linear dispersion dispersión lineal
linear distortion distorsión lineal
linear drift deriva lineal
linear electrical constants constantes eléctricas lineales
linear electrical parameters parámetros eléctricos lineales
linear electron accelerator acelerador electrónico lineal
linear element elemento lineal
linear energy transfer transferencia de energía lineal
linear equation ecuación lineal
linear feedback control control con retroalimentación lineal
linear feedback control system sistema de control con retroalimentación lineal
linear filter filtro lineal
linear flow flujo lineal
linear function función lineal
linear gate puerta lineal
linear induction motor motor de inducción lineal
linear input logic lógica de entrada lineal
linear insulation aislamiento lineal
linear integrated circuit circuito integrado lineal
linear interpolation interpolación lineal
linear ion accelerator acelerador iónico lineal
linear ion density densidad iónica lineal
linear ionization ionización lineal
linear lamp lámpara lineal
linear light luz lineal
linear list lista lineal
linear magnetostriction magnetoestricción lineal
linear magnetron magnetrón lineal
linear magnification magnificación lineal
linear measure medida lineal
linear modulation modulación lineal
linear modulator modulador lineal
linear motion transducer transductor de moción lineal
linear motor motor lineal
linear network red lineal
linear oscillator oscilador lineal
linear polarization polarización lineal
linear potentiometer potenciómetro lineal
linear power amplifier amplificador de potencia lineal
linear power pentode pentodo de potencia lineal
linear prediction predicción lineal
linear program programa lineal
linear programming programación lineal
linear programming control language lenguaje de control de programación lineal
linear programming system sistema de programación lineal
linear pulse amplifier amplificador de impulsos lineal
linear rectification rectificación lineal
linear rectifier rectificador lineal
linear regulator regulador lineal
linear representation representación lineal
linear resistance resistencia lineal
linear resistor resistor lineal
linear response respuesta lineal
linear sawtooth waveform onda en diente de sierra lineal

linear scale escala lineal
linear scan exploración lineal
linear scanning exploración lineal
linear search búsqueda lineal
linear-slope equalization ecualización de pendiente lineal
linear sort clasificación lineal
linear structure estructura lineal
linear sweep barrido lineal
linear sweep voltage tensión de barrido lineal
linear system sistema lineal
linear taper variación lineal
linear time base base de tiempo lineal
linear transducer transductor lineal
linear transformer transformador lineal
linear varying parameter parámetro con variación lineal
linear varying-parameter network red de parámetros con variación lineal
linear vibration vibración lineal
linear voltage-frequency characteristic característica tensión-frecuencia lineal
linearity linealidad
linearity adjustment ajuste de linealidad
linearity checker comprobador de linealidad
linearity control control de linealidad
linearity distortion distorsión de linealidad
linearity error error de linealidad
linearity of response linealidad de respuesta
linearity of signal linealidad de señal
linearity pattern generator generador de patrón de linealidad
linearity region región de linealidad
linearity sector sector de linealidad
linearizable linealizable
linearizable quadripole cuadripolo linealizable
linearize linealizar
linearly linealmente
linearly modulated voltage tensión modulada linealmente
linearly polarized polarizado linealmente
linearly polarized field campo polarizado linealmente
linearly polarized mode modo polarizado linealmente
linearly polarized sound wave onda sonora polarizada linealmente
linearly polarized wave onda polarizada linealmente
liner revestidor, revestimiento, forro
lines per minute líneas por minuto
lining revestimiento, forro
link enlace, eslabón
link-access procedure procedimiento de acceso de enlace
link address dirección de enlace
link-attached conectado por enlace
link-attached station estación conectada por enlaces
link-attached terminal terminal conectada por enlaces
link attribute atributo de enlace
link box caja de seccionamiento
link circuit circuito de enlace
link connection conexión de enlace
link control control de enlace
link coupling acoplamiento de eslabón
link escape character carácter de escape de enlace
link field campo de enlace
link fuse fusible de hilo descubierto
link group grupo de enlace
link header cabecera de enlace
link indicator indicador de enlace
link management administración de enlaces
link protocol protocolo de enlace

link register registro de enlace
link station estación de enlace
link transmitter transmisor de enlace
link word palabra de enlace
linkage enlace, encadenamiento
linkage field campo de enlace
linkage instruction instrucción de enlace
linked circuit-breaker cortacircuito enlazado
linked format formato enlazado
linked linear list lista lineal enlazada
linked list lista enlazada
linked subroutine subrutina enlazada
linked switch conmutador enlazado
linking enlace, encadenamiento
linotron linotrón
lip microphone micrófono de labio
liquid líquido
liquid absorption absorción de líquido
liquid capacitor capacitor líquido
liquid cathode cátodo líquido
liquid cell celda líquida
liquid conductor conductor líquido
liquid-cooled enfriado por líquido
liquid cooling enfriamiento por líquido
liquid counter tube tubo contador para líquidos
liquid crystal cristal líquido
liquid-crystal display visualizador de cristal líquido
liquid detector detector de líquido
liquid dielectric dieléctrico líquido
liquid-filled capacitor capacitor lleno de líquido
liquid-filled transformer transformador lleno de
    líquido
liquid-flow alarm alarma de flujo de líquido
liquid-flow control control de flujo de líquido
liquid-flow counter tube tubo contador de flujo de
    líquido
liquid-flow gage medidor de flujo de líquido
liquid-flow indicator indicador de flujo de líquido
liquid-flow meter medidor de flujo de líquido
liquid-flow switch conmutador de flujo de líquido
liquid fuse fusible líquido
liquid fuse unit unidad de fusible líquido
liquid-impregnated capacitor capacitor impregnado
    en líquido
liquid-jet oscillograph oscilógrafo de chorro de
    líquido
liquid junction unión líquida
liquid-junction potential potencial de unión de
    líquidos
liquid laser láser líquido
liquid-level alarm alarma de nivel de líquido
liquid-level control control de nivel de líquido
liquid-level gage medidor de nivel de líquido
liquid-level indicator indicador de nivel de líquido
liquid-level meter medidor de nivel de líquido
liquid-level recorder registrador de nivel de líquido
liquid-level sensor sensor de nivel de líquido
liquid-level switch conmutador de nivel de líquido
liquid load carga líquida
liquid-pressure alarm alarma de presión de líquido
liquid-pressure control control de presión de líquido
liquid-pressure gage medidor de presión de líquido
liquid-pressure indicator indicador de presión de
    líquido
liquid-pressure meter medidor de presión de líquido
liquid-pressure switch conmutador de presión de
    líquido
liquid rectifier rectificador líquido
liquid resistor resistor líquido
liquid rheostat reóstato líquido
liquid semiconductor semiconductor líquido
liquid switch conmutador líquido
Lissajous figure figura de Lissajous

Lissajous pattern patrón de Lissajous
list lista
list element elemento de lista
list processing procesamiento de listas
list-processing language lenguaje de procesamiento
    de listas
list sorting clasificación de lista
list structure estructura de lista
listing listado
liter litro
literal constant constante literal
literal operands operandos literales
lithium litio
lithium battery batería de litio
lithium cell pila de litio
lithium-drifted detector detector por deriva de litio
lithium ferrite-core memory memoria de núcleos de
    ferrita lítica
litz wire hilo litz
Litzendraht wire hilo de Litzendraht
live vivo, activo
live circuit circuito vivo
live load carga viva
live-metal part parte de metal vivo
live part parte viva
live wire hilo vivo
load carga
load address dirección de carga
load admittance admitancia de carga
load-and-go cargar y ejecutar
load-and-store cargar y almacenar
load angle ángulo de carga
load balance equilibrio de carga
load-break connector conector de ruptura de carga
load-break switch conmutador de ruptura de carga
load-break test prueba de ruptura de carga
load capacitance capacitancia de carga
load capacity capacidad de carga
load card tarjeta de carga
load cell célula de carga
load center centro de carga
load characteristic característica de carga
load circuit circuito de carga
load-circuit efficiency eficiencia de circuito de carga
load-circuit power input entrada de potencia de
    circuito de carga
load coil bobina de carga
load control control de carga
load controller controlador de carga
load current corriente de carga
load curve curva de carga
load-disconnect relay relé de desconexión de carga
load dispatcher despachador de carga
load distribution distribución de carga
load diversity diversidad de cargas
load division división de carga
load-duration curve curva de duración de carga
load end extremo de carga
load factor factor de carga
load flicker parpadeo de carga
load-frequency characteristic característica
    carga-frecuencia
load group grupo de carga
load impedance impedancia de carga
load-impedance diagram diagrama de impedancia de
    carga
load-indicating resistor resistor indicador de carga
load instruction instrucción de carga
load-interrupter switch conmutador interruptor de
    carga
load isolator aislador de carga
load key tecla de carga
load life vida de carga

load line línea de carga
load list lista de carga
load losses pérdidas por carga
load map mapa de carga
load matching adaptación de carga
load-matching network red de adaptación de carga
load-matching switch conmutador de adaptación de
    carga
load mode modo de carga
load module módulo de carga
load point punto de carga
load power potencia de carga
load profile perfil de carga
load program programa de carga
load regulation regulación de carga
load regulator regulador de carga
load resistance resistencia de carga
load resistor resistor de carga
load restoration restauración de carga
load rheostat reóstato de carga
load saturation curve curva de saturación de carga
load stabilizer estabilizador de carga
load switch conmutador de carga
load time tiempo de carga
load torque par de carga
load transfer switch conmutador de transferencia de
    carga
load voltage tensión de carga
load-voltage stabilization estabilización de tensión
    de carga
load wattage wattaje de carga
loaded cargado
loaded antenna antena cargada
loaded cable cable cargado
loaded circuit circuito cargado
loaded impedance impedancia cargada
loaded line línea cargada
loaded pair par cargado
loader cargador
loader routine rutina de cargador
loading carga
loading capacity capacidad de carga
loading coil bobina de carga
loading-coil box caja de bobinas de carga
loading-coil case caja de bobinas de carga
loading-coil pot caja de bobinas de carga
loading-coil spacing espaciado de bobinas de carga
loading cycle ciclo de carga
loading disk disco de carga
loading error error de carga
loading factor factor de carga
loading inductance inductancia de carga
loading operation operación de carga
loading platform plataforma de carga
loading point punto de carga
loading procedure procedimiento de carga
loading process proceso de carga
loading program programa de carga
loading resistor resistor de carga
loading routine rutina de carga
loading terminal terminal de carga
loading tray bandeja de carga
loading unit unidad de carga
loading zone zona de carga
lobe lóbulo
lobe switch conmutador de lóbulo
lobe switching conmutación de lóbulo
lobe width anchura de lóbulo
lobing lobulación
local action acción local
local area área local
local area network red de área local
local backup respaldo local

local battery batería local
local-battery exchange central de batería local
local-battery inductor coil bobina inductora de
    batería local
local-battery switchboard cuadro conmutador de
    batería local
local-battery system sistema de batería local
local-battery telephone set aparato telefónico de
    batería local
local broadcast station estación de radiodifusión
    local
local cable cable local
local call llamada local
local cell pila local
local central office oficina central local
local channel canal local
local clock reloj local
local compaction compactación local
local connection conexión local
local control control local
local current corriente local
local data datos locales
local device controller controlador de dispositivo
    local
local echo eco local
local exchange central local
local fault falla local
local fault display unit unidad indicadora de falla
    local
local feedback retroalimentación local
local jack jack local
local junction unión local
local-junction circuit circuito de unión local
local lighting alumbrado local
local line línea local
local loop bucle local
local memory memoria local
local mode modo local
local network red local
local network control program programa de control
    de red local
local noise ruido local
local oscillator oscilador local
local-oscillator frequency check comprobación de
    frecuencia de oscilador local
local-oscillator radiation radiación de oscilador local
local-oscillator tube tubo de oscilador local
local processor procesador local
local program programa local
local reception recepción local
local repeater repetidor local
local sensitivity sensibilidad local
local service servicio local
local session identification identificación de sesión
    local
local source fuente local
local station estación local
local storage almacenamiento local
local system library biblioteca de sistema local
local system queue area área de cola de sistema
    local
local terminal terminal local
local transaction transacción local
local transmission transmisión local
local trunk enlace local
local update procedure procedimiento de
    actualización local
local variable variable local
local work station estación de trabajo local
localization localización
localized localizado
localizer localizador
localizer beam haz localizador

localizer receiver receptor localizador
localizer sector sector localizador
localizer transmitter transmisor localizador
location localización, ubicación
location area área de localización
location counter contador de localizaciones
location indicator indicador de localización
location monitor monitor de localización
location symbol símbolo de localización
locator localizador
locator device dispositivo localizador
lock-down retención en reposo
lock-in amplifier amplificador de enganche
lock-in base base de enganche
lock-in range intervalo de enganche
lock-in relay relé de enclavamiento
lock-in technique técnica de enganche
lock-in tube tubo de enganche
lock mode modo de bloqueo
lock-out bloqueo, cierre
lock-up relay relé de retención
locked file archivo cerrado
locked groove surco cerrado
locked oscillator oscilador sincronizado
locked-oscillator detector detector de oscilador
  sincronizado
locked-oscillator operation operación de oscilador
  sincronizado
locked record registro cerrado
locked-up bloqueado
locked-up keyboard teclado bloqueado
locking bloqueo, cierre, sincronización
locking device dispositivo de bloqueo
locking-out relay relé de bloqueo
locking relay relé de bloqueo
lockout bloqueo, cierre
lockout amplifier amplificador de bloqueo
lockout-free libre de bloqueos
lockout gate puerta de bloqueo
lockout mechanism mecanismo de bloqueo
lockout module módulo de bloqueo
lockout relay relé de bloqueo
lockout tube tubo de bloqueo
lockout voltage tensión de bloqueo
lockover circuit circuito biestable
lockup retención en trabajo
lodar lodar
lodestone piedra imán
Loftin-White circuit circuito de Loftin-White
log logaritmo, registro
log in entrar al sistema, iniciar una sesión
log off salir del sistema, terminar una sesión
log on entrar al sistema, iniciar una sesión
log out salir del sistema, terminar una sesión
log-periodic antenna antena periódica logarítmica
log receiver receptor logarítmico
log response respuesta logarítmica
log-response amplifier amplificador de respuesta
  logarítmica
logarithm logaritmo
logarithmic logarítmico
logarithmic amplifier amplificador logarítmico
logarithmic attenuator atenuador logarítmico
logarithmic circuit circuito logarítmico
logarithmic curve curva logarítmica
logarithmic decrement decremento logarítmico
logarithmic direct-current amplifier amplificador
  de corriente continua logarítmico
logarithmic equation ecuación logarítmica
logarithmic frequency scale escala de frecuencias
  logarítmica
logarithmic graph gráfica logarítmica
logarithmic horn bocina logarítmica

logarithmic-law capacitor capacitor de ley
  logarítmica
logarithmic mean media logarítmica
logarithmic meter medidor logarítmico
logarithmic motor motor logarítmico
logarithmic periodic antenna antena periódica
  logarítmica
logarithmic photometer fotómetro logarítmico
logarithmic potentiometer potenciómetro
  logarítmico
logarithmic pulse amplifier amplificador de
  impulsos logarítmico
logarithmic receiver receptor logarítmico
logarithmic response respuesta logarítmica
logarithmic scale escala logarítmica
logarithmic-scale meter medidor de escala
  logarítmica
logarithmic search búsqueda logarítmica
logarithmic series serie logarítmica
logarithmic threshold umbral logarítmico
logarithmic unit unidad logarítmica
logarithmic voltmeter vóltmetro logarítmico
logarithmically logarítmicamente
logic lógica
logic analysis análisis lógico
logic analyzer analizador lógico
logic array red lógica
logic board tablero lógico
logic card tarjeta lógica
logic chart diagrama lógico
logic circuit circuito lógico
logic comparator comparador lógico
logic comparison comparación lógica
logic connective operador lógico
logic decision decisión lógica
logic design diseño lógico
logic device dispositivo lógico
logic diagram diagrama lógico
logic difference diferencia lógica
logic diode diodo lógico
logic element elemento lógico
logic error error lógico
logic file archivo lógico
logic flow chart organigrama lógico
logic function función lógica
logic gate puerta lógica
logic instruction instrucción lógica
logic level nivel lógico
logic map mapa lógico
logic multiplication multiplicación lógica
logic network red lógica
logic operation operación lógica
logic operator operador lógico
logic probe sonda lógica
logic product producto lógico
logic programming programación lógica
logic programming language lenguaje de
  programación lógica
logic relay relé lógico
logic shift desplazamiento lógico
logic signal señal lógica
logic state estado lógico
logic sum suma lógica
logic swing oscilación lógica
logic symbol símbolo lógico
logic unit unidad lógica
logical lógico
logical access level nivel de acceso lógico
logical address dirección lógica
logical analyzer analizador lógico
logical block bloque lógico
logical card tarjeta lógica
logical channel canal lógico

logical circuit circuito lógico
logical cohesion cohesión lógica
logical comparison comparación lógica
logical constant constante lógica
logical data design diseño de datos lógico
logical data model modelo de datos lógico
logical data structure estructura de datos lógica
logical data transfer transferencia de datos lógica
logical database base de datos lógica
logical decision decisión lógica
logical design diseño lógico
logical destination destino lógico
logical device address dirección de dispositivo lógico
logical diagram diagrama lógico
logical difference diferencia lógica
logical element elemento lógico
logical equivalence equivalencia lógica
logical error error lógico
logical escape symbol símbolo de escape lógico
logical expression expresión lógica
logical file archivo lógico
logical flow chart organigrama lógico
logical format formato lógico
logical function función lógica
logical gate puerta lógica
logical-gate circuit circuito de puerta lógica
logical group grupo lógico
logical group number número de grupo lógico
logical indicator indicador lógico
logical input device dispositivo de entrada lógico
logical instruction instrucción lógica
logical line línea lógica
logical link enlace lógico
logical link control control de enlace lógico
logical link control procedure procedimiento de
control de enlace lógico
logical link control protocol protocolo de control de
enlace lógico
logical link path trayectoria de enlace lógico
logical message mensaje lógico
logical monitor monitor lógico
logical network red lógica
logical number número lógico
logical operation operación lógica
logical operator operador lógico
logical output device dispositivo de salida lógica
logical page página lógica
logical product producto lógico
logical record registro lógico
logical relation relación lógica
logical representation representación lógica
logical schema esquema lógico
logical segment segmento lógico
logical shift desplazamiento lógico
logical storage almacenamiento lógico
logical storage address dirección de almacenamiento
lógico
logical structure estructura lógica
logical sum suma lógica
logical symbol símbolo lógico
logical terminal terminal lógica
logical type tipo lógico
logical unit unidad lógica
logical unit block bloque de unidad lógica
logical unit name nombre de unidad lógica
logical unit number número de unidad lógica
logical value valor lógico
logical variable variable lógica
logon logonio
loktal loktal
loktal base base loktal
loktal socket zócalo loktal
London equation ecuación de London

London superconductivity equation ecuación de
superconductividad de London
long distance larga distancia
long-distance broadcasting radiodifusión a larga
distancia
long-distance call llamada a larga distancia
long-distance circuit circuito a larga distancia
long-distance communication comunicación a larga
distancia
long-distance connection conexión a larga distancia
long-distance dialing marcación a larga distancia
long-distance reception recepción a larga distancia
long-distance selection selección a larga distancia
long-distance telecommunication circuit circuito de
telecomunicación a larga distancia
long-distance traffic tráfico a larga distancia
long-distance transmission transmisión a larga
distancia
long duration larga duración
long-duration echo eco de larga duración
long format formato largo
long-life de vida larga
long-life tube tubo de vida larga
long line línea larga
long-line effect efecto de línea larga
long-period fading desvanecimiento de largo periodo
long-period fluctuation fluctuación de largo periodo
long-persistence screen pantalla de larga persistencia
long-pull magnet electroimán de largo recorrido
long-range de largo alcance
long-range electrostatic effect efecto electrostático
de largo alcance
long-range force fuerza de largo alcance
long-range navigation navegación de largo alcance
long-range navigation system sistema de navegación
de largo alcance
long-range radar radar de largo alcance
long-range reception recepción de largo alcance
long-shunt connection conexión de derivación larga
long-tail pair pareja acoplada por el cátodo
long-tailed pair pareja acoplada por el cátodo
long-term a largo plazo
long-term drift deslizamiento a largo plazo
long-term fading desvanecimiento a largo plazo
long-term instability inestabilidad a largo plazo
long-term stability estabilidad a largo plazo
long-term storage almacenamiento a largo plazo
long-throw de recorrido largo
long-throw suspension suspensión de recorrido largo
long-travel cone cono de recorrido largo
long wave onda larga
long-wave band banda de ondas largas
long-wave broadcasting station estación de
radiodifusión de ondas largas
long-wave diathermy diatermia de ondas largas
long-wave propagation propagación de ondas largas
long-wave radar radar de ondas largas
long-wave spectrum espectro de ondas largas
long-wire antenna antena de hilo largo
longevity longevidad
longitudinal longitudinal
longitudinal attenuation atenuación longitudinal
longitudinal balance equilibrio longitudinal
longitudinal carrier cable cable portador
longitudinal
longitudinal check comprobación longitudinal
longitudinal check character carácter de
comprobación longitudinal
longitudinal chromatic aberration aberración
cromática longitudinal
longitudinal circuit circuito longitudinal
longitudinal coil bobina longitudinal
longitudinal current corriente longitudinal

longitudinal delay line línea de retardo longitudinal
longitudinal differential protection protección diferencial longitudinal
longitudinal effect efecto longitudinal
longitudinal electromotive force fuerza electromotriz longitudinal
longitudinal heating calentamiento longitudinal
longitudinal impedance impedancia longitudinal
longitudinal interference interferencia longitudinal
longitudinal magnetic field campo magnético longitudinal
longitudinal magnetization magnetización longitudinal
longitudinal magnification magnificación longitudinal
longitudinal mode modo longitudinal
longitudinal oscillation oscilación longitudinal
longitudinal overvoltage sobretensión longitudinal
longitudinal parity paridad longitudinal
longitudinal parity check comprobación de paridad longitudinal
longitudinal perforation perforación longitudinal
longitudinal profile perfil longitudinal
longitudinal recording head cabeza de registro longitudinal, cabeza de grabación longitudinal
longitudinal redundancy redundancia longitudinal
longitudinal redundancy check comprobación de redundancia longitudinal
longitudinal redundancy check character carácter de comprobación de redundancia longitudinal
longitudinal resonance resonancia longitudinal
longitudinal signal señal longitudinal
longitudinal slot ranura longitudinal
longitudinal spherical aberration aberración esférica longitudinal
longitudinal stability estabilidad longitudinal
longitudinal surge variación repentina longitudinal
longitudinal voltage tensión longitudinal
longitudinal wave onda longitudinal
loom conducto flexible
loop bucle, ciclo, cuadro, lazo, antinodo, anillo
loop adapter adaptador de bucle
loop antenna antena de cuadro
loop body cuerpo de bucle
loop checking comprobación por bucle
loop circuit circuito de bucle
loop code código de bucle
loop configuration configuración en bucle
loop connection conexión en bucle
loop control control de bucle
loop-control statement sentencia de control de bucle
loop-control structure estructura de control de bucle
loop-control variable variable de control de bucle
loop counter contador de bucles
loop current corriente en bucle
loop-delay unit unidad de retardo de bucle
loop dialing marcación en bucle
loop-difference signal señal diferencial de bucle
loop direction finder radiogoniómetro de cuadro
loop element elemento de bucle
loop error error de bucle
loop-error signal señal de error de bucle
loop feature característica de bucle
loop-feedback signal señal de retroalimentación de bucle
loop filter filtro de bucle
loop gain ganancia de bucle
loop galvanometer galvanómetro de bucle
loop initialization inicialización de bucle
loop-input signal señal de entrada al bucle
loop loss pérdida de bucle
loop margin margen de bucle
loop modification modificación de bucle

loop network red en bucle
loop operation operación de bucle
loop-output signal señal de salida de bucle
loop program programa en bucle
loop resistance resistencia de bucle
loop resistor resistor de bucle
loop-return signal señal de retorno en bucle
loop signaling señalización en bucle
loop structure estructura de bucle
loop system sistema en bucle
loop test prueba en bucle
loop-transfer function función de transferencia de bucle
loop-transfer ratio razón de transferencia de bucle
loop transmission transmisión en bucle
loop transmittance transmitancia de bucle
loop update actualización de bucle
loopback test prueba en bucle
looped circuit circuito en bucle
looped filament filamento en bucle
looping operación en bucle
loopstick antenna antena de varilla
loose cathode deposit depósito catódico flojo
loose connection conexión floja
loose coupler acoplador flojo
loose coupling acoplamiento flojo
lorac lorac
loran lorán
loran antenna antena de lorán
loran guidance guiado por lorán
loran indicator indicador de lorán
loran receiver receptor de lorán
loran signal señal de lorán
loran station estación de lorán
loran transmitter transmisor de lorán
Lorentz condition condición de Lorentz
Lorentz field campo de Lorentz
Lorentz force fuerza de Lorentz
Lorentz force equation ecuación de fuerza de Lorentz
Lorentz instrument-landing system sistema de aterrizaje por instrumentos de Lorentz
Lorentz law ley de Lorentz
Lorentz-Lorentz law ley de Lorentz-Lorentz
Lorentz number número de Lorentz
loss pérdida
loss angle ángulo de pérdidas
loss current corriente de pérdidas
loss factor factor de pérdidas
loss-free libre de pérdidas
loss function función de pérdidas
loss index índice de pérdidas
loss meter medidor de pérdidas
loss modulation modulación de pérdidas
loss of information pérdida de información
loss of power pérdida de potencia
loss of signal pérdida de señal
loss resistance resistencia a pérdidas
loss tangent tangente de pérdidas
losser amortiguador
losser circuit circuito amortiguado
Lossev effect efecto de Lossev
lossless sin pérdidas
lossless antenna antena sin pérdidas
lossless encoding codificación sin pérdidas
lossless line línea sin pérdidas
lossless network red sin pérdidas
lossless termination terminación sin pérdidas
lossy line línea con pérdidas
lossy medium medio con pérdidas
lossy termination terminación con pérdidas
lossy waveguide guíaondas con pérdidas
lost call llamada perdida

loudness intensidad sonora, sonoridad
loudness compensation compensación de intensidad sonora
loudness compensator compensador de intensidad sonora
loudness control control de intensidad sonora
loudness gain ganancia de intensidad sonora
loudness level nivel de intensidad sonora
loudness rating clasificación de intensidad sonora
loudspeaker altavoz
loudspeaker amplifier amplificador de altavoz
loudspeaker baffle bafle de altavoz
loudspeaker cabinet caja de altavoz
loudspeaker capacitor capacitor de altavoz
loudspeaker condenser condensador de altavoz
loudspeaker cone cono de altavoz
loudspeaker control control de altavoz
loudspeaker damping amortiguamiento de altavoz
loudspeaker diaphragm diafragma de altavoz
loudspeaker dividing network red divisora de altavoz
loudspeaker efficiency eficiencia de altavoz
loudspeaker enclosure caja de altavoz
loudspeaker housing caja de altavoz
loudspeaker impedance impedancia de altavoz
loudspeaker line línea de altavoz
loudspeaker magnet imán de altavoz
loudspeaker matching transformer transformador de adaptación de altavoz
loudspeaker mechanism mecanismo de altavoz
loudspeaker-microphone altavoz-micrófono
loudspeaker motor motor de altavoz
loudspeaker phasing enfasamiento de altavoz
loudspeaker polarity polaridad de altavoz
loudspeaker relay relé de altavoz
loudspeaker-reversal switch conmutador de inversión de altavoces
loudspeaker system sistema de altavoces
loudspeaker voice coil bobina móvil de altavoz
loudspeaker volume control control de volumen de altavoz
low activity baja actividad
low band banda baja
low-brightness fluorescent lamp lámpara fluorescente de bajo brillo
low-capacitance contacts contactos de baja capacitancia
low-capacitance probe sonda de baja capacitancia
low-capacitance tube tubo de baja capacitancia
low-capacity cable cable de baja capacidad
low consumption bajo consumo
low current corriente débil
low definition baja definición
low-definition image imagen de baja definición
low-definition television televisión de baja definición
low-density tape cinta de baja densidad
low distortion baja distorsión
low-distortion amplifier amplificador de baja distorsión
low-distortion limiting amplifier amplificador limitador de baja distorsión
low drain bajo consumo
low-drift amplifier amplificador de baja deriva
low energy baja energía
low-energy circuit circuito de baja energía
low-energy neutron neutrón de baja energía
low-energy particle partícula de baja energía
low-energy power circuit circuito de potencia de baja energía
low-field loss pérdida de campo débil
low filter filtro de bajos
low frequency baja frecuencia
low-frequency amplifier amplificador de baja frecuencia

low-frequency choke choque de baja frecuencia
low-frequency compensation compensación de baja frecuencia
low-frequency direction finder radiogoniómetro de baja frecuencia
low-frequency distortion distorsión en baja frecuencia
low-frequency driver altavoz de baja frecuencia
low-frequency furnace horno de baja frecuencia
low-frequency impedance corrector corrector de impedancia de baja frecuencia
low-frequency induction furnace horno por inducción de baja frecuencia
low-frequency induction heater calentador por inducción de baja frecuencia
low-frequency induction heating calentamiento por inducción de baja frecuencia
low-frequency loran lorán de baja frecuencia
low-frequency loudspeaker altavoz de baja frecuencia
low-frequency padder pader de baja frecuencia
low-frequency response respuesta en baja frecuencia
low-frequency side lado de baja frecuencia
low-frequency signaling señalización de baja frecuencia
low impedance baja impedancia
low-impedance electrode electrodo de baja impedancia
low-impedance head cabeza de baja impedancia
low-impedance magnetic microphone micrófono magnético de baja impedancia
low-impedance pickup captor de baja impedancia
low-impedance triode triodo de baja impedancia
low intensity baja intensidad
low-intensity light luz de baja intensidad
low level bajo nivel
low-level code código de bajo nivel
low-level contact contacto de bajo nivel
low-level crystal cristal de bajo nivel
low-level format formato de bajo nivel
low-level grid modulation modulación por rejilla de bajo nivel
low-level input current corriente de entrada de bajo nivel
low-level language lenguaje de bajo nivel
low-level logic lógica de bajo nivel
low-level modulation modulación de bajo nivel
low-level output current corriente de salida de bajo nivel
low-level programming language lenguaje de programación de bajo nivel
low-level signal señal de bajo nivel
low load carga baja
low-loss de pocas pérdidas
low-loss ceramic cerámica de pocas pérdidas
low-loss dielectric material material dieléctrico de pocas pérdidas
low-loss insulator aislador de pocas pérdidas
low-loss line línea de pocas pérdidas
low-loss material material de pocas pérdidas
low-mu tube tubo de bajo mu
low noise bajo ruido
low-noise amplifier amplificador de bajo ruido
low-noise receiver receptor de bajo ruido
low-noise tube tubo de bajo ruido
low order orden bajo
low-order bit bit de orden bajo
low-order position posición de orden bajo
low-pass acoustic filter filtro acústico pasabajos
low-pass cell célula pasabajos
low-pass filter filtro pasabajos
low-pass filter circuit circuito de filtro pasabajos

low power baja potencia
low-power amplifier amplificador de baja potencia
low-power circuit circuito de baja potencia
low-power modulation modulación de baja potencia
low-power transmitter transmisor de baja potencia
low pressure baja presión
low-pressure mercury lamp lámpara de mercurio de baja presión
low-pressure sodium lamp lámpara de sodio de baja presión
low-pressure vacuum pump bomba de vacío de baja presión
low priority baja prioridad
low rate bajo régimen
low-rate discharge descarga de bajo régimen
low resistance baja resistencia
low resolution baja resolución
low-resolution graphics gráficos de baja resolución
low speed baja velocidad
low-speed electron beam haz electrónico de baja velocidad
low-speed logic lógica de baja velocidad
low-speed storage almacenamiento de baja velocidad
low-speed terminal terminal de baja velocidad
low-speed transmission transmisión de baja velocidad
low-striking voltage tensión de cebado baja
low temperature baja temperatura
low-temperature fluorescent lamp lámpara fluorescente de baja temperatura
low tension baja tensión
low vacuum bajo vacío
low-velocity scanning exploración de baja velocidad
low voltage baja tensión
low-voltage bar barra de baja tensión
low-voltage battery batería de baja tensión
low-voltage current corriente de baja tensión
low-voltage power cable cable de alimentación de baja tensión
low-voltage rectifier rectificador de baja tensión
low-voltage system sistema de baja tensión
low-voltage transformer transformador de baja tensión
low-wattage bajo wattaje
lower contacts contactos inferiores
lower critical field campo crítico inferior
lower curtate parte inferior de tarjeta perforada
lower frequency frecuencia inferior
lower limit límite inferior
lower sideband banda lateral inferior
lower sideband converter convertidor de banda lateral inferior
lowest effective power potencia efectiva mínima
lowest usable frequency frecuencia utilizable más baja
lowest useful frequency frecuencia útil más baja
lug terminal, lengüeta de conexión
lug splice empalme de aletas
Lukasiewicz notation notación de Lukasiewicz
lumen lumen
lumen-hour lumen-hora
lumen-meter lumen-metro
lumen per watt lumens por watt
lumen-second lumen-segundo
luminaire luminaria
luminaire efficiency eficiencia de luminaria
luminance luminancia
luminance amplifier amplificador de luminancia
luminance channel canal de luminancia
luminance coefficient coeficiente de luminancia
luminance component componente de luminancia
luminance contrast contraste de luminancia
luminance demodulator desmodulador de luminancia
luminance-difference threshold umbral diferencial de luminancia
luminance factor factor de luminancia
luminance flicker parpadeo de luminancia
luminance function función de luminancia
luminance information información de luminancia
luminance primary primario de luminancia
luminance ratio razón de luminancia
luminance signal señal de luminancia
luminance temperature temperatura de luminancia
luminance threshold umbral de luminancia
luminescence luminiscencia
luminescence efficiency eficiencia de luminiscencia
luminescence threshold umbral de luminiscencia
luminescent luminiscente
luminescent cell célula luminiscente
luminescent discharge descarga luminiscente
luminescent material material luminiscente
luminescent screen pantalla luminiscente
luminescent-screen tube tubo de pantalla luminiscente
luminophor luminóforo
luminosity luminosidad
luminosity coefficients coeficientes de luminosidad
luminosity factor factor de luminosidad
luminous luminoso
luminous area área luminosa
luminous beam haz luminoso
luminous coefficient coeficiente luminoso
luminous density densidad luminosa
luminous dial cuadrante luminoso
luminous discharge descarga luminosa
luminous-discharge lamp lámpara de descarga luminosa
luminous edge borde luminoso
luminous efficiency eficiencia luminosa
luminous electrode spot mancha electródica luminosa
luminous element elemento luminoso
luminous emittance emitancia luminosa
luminous energy energía luminosa
luminous flux flujo luminoso
luminous flux density densidad de flujo luminoso
luminous heat calor luminoso
luminous intensity intensidad luminosa
luminous level nivel luminoso
luminous output salida luminosa
luminous power potencia luminosa
luminous reflectance reflectancia luminosa
luminous ring anillo luminoso
luminous screen pantalla luminosa
luminous sensitivity sensibilidad luminosa
luminous signal señal luminosa
luminous spectrum espectro luminoso
luminous transmittance transmitancia luminosa
lumistor lumistor
lumped capacitor capacitor concentrado
lumped capacity capacidad concentrada
lumped circuit circuito concentrado
lumped component componente concentrado
lumped constant constante concentrada
lumped-constant circuit circuito de constantes concentradas
lumped-constant filter filtro de constantes concentradas
lumped delay line línea de retardo concentrada
lumped element elemento concentrado
lumped impedance impedancia concentrada
lumped inductance inductancia concentrada
lumped inductor inductor concentrado
lumped parameter parámetro concentrado
lumped resistor resistor concentrado

lumped voltage　tensión concentrada
Luneberg lens　lente de Luneberg
Luneberg lens antenna　antena de lente de Luneberg
lutetium　lutecio
lux　lux
lux-second　lux-segundo
Luxemberg effect　efecto de Luxemberg
luxmeter　luxómetro
Lysholm grid　rejilla de Lysholm

machine acceptable　aceptable por máquina
machine address　dirección de máquina
machine check　comprobación de máquina
machine-check handler　manipulador de
　comprobación de máquina
machine-check indicator　indicador de comprobación
　de máquina
machine-check interruption　interrupción de
　comprobación de máquina
machine code　código de máquina
machine-code instruction　instrucción de código de
　máquina
machine coding　codificación de máquina
machine configuration　configuración de máquina
machine control　control de máquina
machine control electronics　electrónica de control
　de máquina
machine current　corriente de máquina
machine cycle　ciclo de máquina
machine-dependent　dependiente de máquina
machine equation　ecuación de máquina
machine error　error de máquina
machine execution priority　prioridad de ejecución
　de máquina
machine failure　falla de máquina
machine fault　falla de máquina
machine file　archivo de máquina
machine function　función de máquina
machine halt　parada de máquina
machine hardware　equipo físico de máquina
machine identification code　código de identificación
　de máquina
machine impedance　impedancia de máquina
machine independence　independencia de máquina
machine independent　independiente de máquina
machine-independent language　lenguaje
　independiente de máquina
machine information　información de máquina
machine information code　código de información de
　máquina
machine instruction　instrucción de máquina
machine instruction code　código de instrucción de
　máquina
machine instruction set　conjunto de instrucciones de
　máquina
machine intelligence　inteligencia de máquina
machine interface　interfaz de máquina
machine interruption　interrupción de máquina
machine-interruption code　código de interrupción
　de máquina
machine language　lenguaje de máquina
machine-language code　código de lenguaje de
　máquina
machine-language programming　programación de
　lenguaje de máquina
machine learning　aprendizaje de máquina
machine load　carga de máquina
machine log　diario de máquina
machine logic　lógica de máquina
machine logic design　diseño lógico de máquina
machine operation　operación de máquina

machine-oriented language lenguaje orientado a
    máquina
machine-oriented programming system sistema de
    programación orientado a máquina
machine pass pasada de máquina
machine performance rendimiento de máquina
machine processable procesable por máquina
machine program programa de máquina
machine programming programación de máquina
machine readable legible por máquina
machine-readable data datos legibles por máquina
machine-readable information información legible
    por máquina
machine-readable medium medio legible por
    máquina
machine recognizable reconocible por máquina
machine ringing llamada automática
machine room sala de máquinas
machine run pasada de máquina
machine-sensible detectable por máquina
machine-sensible information información
    detectable por máquina
machine state estado de máquina
machine storage almacenamiento de máquina
machine time tiempo de máquina
machine units unidades de máquina
machine-usable medium medio ultilizable por
    máquina
machine utilization utilización de máquina
machine variable variable de máquina
machine voltage tensión de máquina
machine winding arrollamiento de máquina
machine word palabra de máquina
machine-word length longitud de palabra de
    máquina
machine zero cero de máquina
machined maquinado
machining maquinado
Maclaurin series serie de Maclaurin
macro macro, macroinstrucción
macro assembler macroensamblador
macro assembly macroensamblaje
macro-assembly language lenguaje de
    macroensamblaje
macro-assembly program programa de
    macroensamblaje
macro call llamada de macro
macro command macromando
macro definition macrodefinición
macro expansion macroexpansión
macro flowchart macroorganigrama
macro-generating program programa
    macrogenerador
macro generation macrogeneración
macro generator macrogenerador
macro instruction macroinstrucción
macro language macrolenguaje
macro library biblioteca de macroinstrucciones
macro operation macrooperación
macro processing macroprocesamiento
macro processor macroprocesador
macro program macroprograma
macro programming macroprogramación
macro prototype prototipo de macro
macro statement macrosentencia
macroassembler macroensamblador
macroassembly macroensamblaje
macrocode macrocódigo
macrocoding macrocodificación
macrocommand macromando
macrodefinition macrodefinición
macroexpansion macroexpansión
macroflowchart macroorganigrama

macrogeneration macrogeneración
macrogenerator macrogenerador
macroinstruction macroinstrucción
macroinstruction design diseño de macroinstrucción
macrolanguage macrolenguaje
macrooperation macrooperación
macroprocessing macroprocesamiento
macroprocessor macroprocesador
macroprogram macroprograma
macroprogramming macroprogramación
macroscopic macroscópico
macrosonics macrosónica
macrostatement macrosentencia
macrowave macroonda
madistor madistor
magazine almacén, almacén de alimentación,
    cartucho
magic eye ojo mágico
magic-eye tuning sintonización por ojo mágico
magic tee T mágica
magnal base base magnal
magnal socket zócalo magnal
magnalium magnalio
magnesium magnesio
magnesium anode ánodo de magnesio
magnesium cell pila de magnesio
magnesium-copper-sulfide rectifier rectificador de
    magnesio-sulfuro de cobre
magnesium-flouride phosphor fósforo de fluoruro
    de magnesio
magnesium-oxide emitter emisor de óxido de
    magnesio
magnesium-silicate phosphor fósforo de silicato de
    magnesio
magnesium-tungstate phosphor fósforo de tungstato
    de magnesio
magnet imán, electroimán
magnet armature armadura de imán
magnet assembly ensamblaje de imán
magnet battery batería de imanes
magnet carrier portaimán
magnet case caja de imán
magnet coil bobina de electroimán
magnet core núcleo de electroimán
magnet gap entrehierro de imán
magnet keeper armadura de imán
magnet meter medidor de imán
magnet motor motor de imán
magnet polarity polaridad de imán
magnet protector armadura de imán
magnet steel acero para imanes
magnet system sistema de imanes
magnet tester probador de imanes
magnet wire hilo para electroimanes
magnet yoke yugo de imán
magnetic magnético
magnetic activity actividad magnética
magnetic aftereffect efecto posterior magnético
magnetic aging envejecimiento magnético
magnetic air-gap entrehierro
magnetic amplification amplificación magnética
magnetic amplifier amplificador magnético
magnetic analysis análisis magnético
magnetic analyzer analizador magnético
magnetic anisotropy anisotropía magnética
magnetic anomaly detector detector de anomalías
    magnéticas
magnetic antenna antena magnética
magnetic armature armadura magnética
magnetic-armature loudspeaker altavoz de
    armadura magnética
magnetic-armature microphone micrófono de
    armadura magnética

magnetic assembly conjunto magnético
magnetic associative memory memoria asociativa
  magnética
magnetic attraction atracción magnética
magnetic axis eje magnético
magnetic azimuth acimut magnético
magnetic balance equilibrio magnético
magnetic band banda magnética
magnetic bar barra magnética
magnetic barrier barrera magnética
magnetic bearing rumbo magnético
magnetic bias polarización magnética
magnetic biasing polarización magnética
magnetic-biasing coil bobina de polarización
  magnética
magnetic blow extinción magnética
magnetic blower extintor magnético
magnetic blowout extinción magnética
magnetic-blowout circuit-breaker cortacircuito de
  extinción magnética
magnetic bottle botella magnética
magnetic brake freno magnético
magnetic braking frenado magnético
magnetic bridge puente magnético
magnetic bubble memory memoria de burbujas
  magnéticas
magnetic bubbles burbujas magnéticas
magnetic capacity capacidad magnética
magnetic card tarjeta magnética
magnetic card file archivo de tarjetas magnéticas
magnetic card reader lectora de tarjetas magnéticas
magnetic card storage almacenamiento de tarjetas
  magnéticas
magnetic cartridge fonocaptor magnético, cápsula
  magnética
magnetic cassette casete magnético
magnetic cell célula magnética
magnetic centering centrado magnético
magnetic character carácter magnético
magnetic character reader lectora de caracteres
  magnéticos
magnetic character recognition reconocimiento de
  caracteres magnéticos
magnetic character sorter clasificador de caracteres
  magnéticos
magnetic charge carga magnética
magnetic charge density densidad de carga
  magnética
magnetic chuck mandril magnético
magnetic circuit circuito magnético
magnetic circuit-breaker cortacircuito magnético
magnetic clutch embrague magnético
magnetic coating revestimiento magnético
magnetic code código magnético
magnetic coil bobina magnética
magnetic compass compás magnético
magnetic compensator compensador magnético
magnetic component componente magnético
magnetic conduction current corriente de
  conducción magnética
magnetic conductivity conductividad magnética
magnetic constant constante magnética
magnetic contactor contactor magnético
magnetic control relay relé de control magnético
magnetic controller controlador magnético
magnetic convergence convergencia magnética
magnetic core núcleo magnético
magnetic-core memory memoria de núcleos
  magnéticos
magnetic-core plane plano de núcleos magnéticos
magnetic-core register registro de núcleos
  magnéticos
magnetic-core storage almacenamiento de núcleos

magnéticos
magnetic-core storage plane plano de
  almacenamiento de núcleos magnéticos
magnetic-core storage unit unidad de
  almacenamiento de núcleos magnéticos
magnetic-core unit unidad de núcleos magnéticos
magnetic couple par magnético
magnetic coupling acoplamiento magnético
magnetic course rumbo magnético
magnetic crack detection detección de grietas
  magnética
magnetic crack detector detector de grietas
  magnético
magnetic creep arrastre magnético
magnetic creeping arrastre magnético
magnetic current corriente magnética
magnetic current density densidad de corriente
  magnética
magnetic curve curva magnética
magnetic cutter grabador magnético
magnetic cycle ciclo magnético
magnetic damper amortiguador magnético
magnetic damping amortiguamiento magnético
magnetic data-storage system sistema de
  almacenamiento de datos magnético
magnetic declination declinación magnética
magnetic deflecting coil bobina de deflexión
  magnética
magnetic deflection deflexión magnética
magnetic delay retardo magnético
magnetic delay line línea de retardo magnética
magnetic demodulation desmodulación magnética
magnetic density densidad magnética
magnetic detector detector magnético
magnetic deviation desviación magnética
magnetic device dispositivo magnético
magnetic dip inclinación magnética
magnetic dipole dipolo magnético
magnetic dipole moment momento dipolar
  magnético
magnetic direction-finder goniómetro magnético
magnetic direction-indicator indicador de dirección
  magnético
magnetic discontinuity discontinuidad magnética
magnetic discriminator discriminador magnético
magnetic disk disco magnético
magnetic-disk file archivo de disco magnético
magnetic-disk layout distribución de disco
  magnético
magnetic-disk memory memoria de disco magnético
magnetic-disk storage almacenamiento de disco
  magnético
magnetic-disk storage unit unidad de
  almacenamiento de disco magnético
magnetic-disk unit unidad de disco magnético
magnetic displacement desplazamiento magnético
magnetic disturbance perturbación magnética
magnetic domain dominio magnético
magnetic doublet doblete magnético
magnetic drag arrastre magnético
magnetic-drag coupling acoplamiento de arrastre
  magnético
magnetic-drag tachometer tacómetro de arrastre
  magnético
magnetic drive arrastre magnético
magnetic drum tambor magnético
magnetic-drum computer computadora de tambor
  magnético
magnetic-drum delay system sistema de retardo de
  tambor magnético
magnetic-drum memory memoria de tambor
  magnético
magnetic-drum separator separador de tambor

magnético

magnetic-drum storage almacenamiento de tambor magnético

magnetic-drum storage device dispositivo de almacenamiento de tambor magnético

magnetic-drum storage unit unidad de almacenamiento de tambor magnético

magnetic-drum unit unidad de tambor magnético

magnetic dry-reed switch conmutador de laminillas secas magnético

magnetic electron microscope microscopio electrónico magnético

magnetic electron multiplier multiplicador electrónico magnético

magnetic elongation elongación magnética

magnetic energy energía magnética

magnetic energy density densidad de energía magnética

magnetic energy storage almacenamiento de energía magnética

magnetic equator ecuador magnético

magnetic explorer explorador magnético

magnetic feedback retroalimentación magnética

magnetic field campo magnético

magnetic-field coil bobina de campo magnético

magnetic-field component componente de campo magnético

magnetic-field expeller expulsor de campo magnético

magnetic-field gradient gradiente de campo magnético

magnetic-field induction inducción de campo magnético

magnetic-field intensity intensidad de campo magnético

magnetic-field interference interferencia de campo magnético

magnetic-field line línea de campo magnético

magnetic-field penetration penetración de campo magnético

magnetic-field strength intensidad de campo magnético

magnetic film película magnética

magnetic-film file archivo de película magnética

magnetic-film memory memoria de película magnética

magnetic-film storage almacenamiento de película magnética

magnetic flaw detector detector de imperfecciones magnético

magnetic flip-flop basculador magnético

magnetic flowmeter flujómetro magnético

magnetic fluid fluido magnético

magnetic flux flujo magnético

magnetic flux density densidad de flujo magnético

magnetic flux leakage fuga de flujo magnético

magnetic focusing enfoque magnético

magnetic-focusing coil bobina de enfoque magnético

magnetic force fuerza magnética

magnetic friction fricción magnética

magnetic friction clutch embrague de fricción magnética

magnetic gap entrehierro

magnetic gate puerta magnética

magnetic grate parrilla magnética

magnetic gripper sujetador magnético

magnetic hardness dureza magnética

magnetic head cabeza magnética

magnetic heading rumbo magnético

magnetic holder sujetador magnético

magnetic hum zumbido magnético

magnetic hysteresis histéresis magnética

magnetic hysteresis loop bucle de histéresis

magnética

magnetic hysteresis loss pérdida por histéresis magnética

magnetic impulse impulso magnético

magnetic impulser impulsor magnético

magnetic inclination inclinación magnética

magnetic induction inducción magnética

magnetic-induction gyroscope giroscopio de inducción magnética

magnetic ink tinta magnética

magnetic ink character carácter de tinta magnética

magnetic ink character reader lectora de caracteres de tinta magnética

magnetic ink character recognition reconocimiento de caracteres de tinta magnética

magnetic instability inestabilidad magnética

magnetic integrated circuit circuito integrado magnético

magnetic intensity intensidad magnética

magnetic inverter inversor magnético

magnetic iron ore magnetita

magnetic iron oxide óxido de hierro magnético

magnetic isotope separator separador de isótopos magnético

magnetic keyboard teclado magnético

magnetic latching enganche magnético

magnetic latching relay relé de enganche magnético

magnetic leakage dispersión magnética

magnetic lens lente magnética

magnetic line línea magnética

magnetic line of force línea de fuerza magnética

magnetic linkage enlace magnético

magnetic load carga magnética

magnetic locking bloqueo magnético

magnetic logic circuit circuito lógico magnético

magnetic logic element elemento lógico magnético

magnetic loss pérdida magnética

magnetic loss angle ángulo de pérdida magnética

magnetic loss factor factor de pérdida magnética

magnetic loudspeaker altavoz magnético

magnetic material material magnético

magnetic measurement medida magnética

magnetic medium medio magnético

magnetic memory memoria magnética

magnetic memory computer computadora de memoria magnética

magnetic memory plate placa de memoria magnética

magnetic mercury switch conmutador de mercurio magnético

magnetic meridian meridiano magnético

magnetic microphone micrófono magnético

magnetic microscope microscopio magnético

magnetic mirror espejo magnético

magnetic modulator modulador magnético

magnetic moment momento magnético

magnetic monopole monopolo magnético

magnetic needle aguja magnética

magnetic north norte magnético

magnetic ore magnetita

magnetic oxide óxido magnético

magnetic particle partícula magnética

magnetic path trayectoria magnética

magnetic permeability permeabilidad magnética

magnetic phase modulator modulador de fase magnético

magnetic pickup fonocaptor magnético

magnetic pickup cartridge fonocaptor magnético

magnetic pickup head fonocaptor magnético

magnetic plated wire hilo con capa magnética

magnetic polarity polaridad magnética

magnetic polarizability polarizabilidad magnética

magnetic polarization polarización magnética

magnetic pole polo magnético

magnetic pole strength intensidad de polo magnético
magnetic potential potencial magnético
magnetic potential difference diferencia de potencial magnético
magnetic potentiometer potenciómetro magnético
magnetic powder polvo magnético
magnetic powder-coated tape cinta revestida de polvo magnético
magnetic powder core núcleo de polvo magnético
magnetic power potencia magnética
magnetic power factor factor de potencia magnética
magnetic pressure presión magnética
magnetic printer impresora magnética
magnetic printing impresión magnética
magnetic probe sonda magnética
magnetic pulley polea magnética
magnetic pump bomba magnética
magnetic reaction analyzer analizador de reacción magnético
magnetic reader lectora magnética
magnetic reading head cabeza de lectura magnética
magnetic readout lectura magnética
magnetic recorder registrador magnético
magnetic recording registro magnético
magnetic recording disk disco de registro magnético
magnetic recording head cabeza de registro magnético
magnetic recording medium medio de registro magnético
magnetic recording reproducer reproductor de registro magnético
magnetic recording tape cinta de registro magnético
magnetic recording wire hilo de registro magnético
magnetic reflector reflector magnético
magnetic regulator regulador magnético
magnetic relay relé magnético
magnetic reluctance reluctancia magnética
magnetic remanence remanencia magnética
magnetic reproducer reproductor magnético
magnetic reproducing head cabeza reproductora magnética
magnetic repulsion repulsión magnética
magnetic resistance resistencia magnética
magnetic resistivity resistividad magnética
magnetic resonance resonancia magnética
magnetic resonance spectrum espectro de resonancia magnética
magnetic reversal inversión magnética
magnetic rigidity rigidez magnética
magnetic rod varilla magnética
magnetic rotation rotación magnética
magnetic saturation saturación magnética
magnetic scan exploración magnética
magnetic screen pantalla magnética
magnetic-screen effect efecto de pantalla magnética
magnetic sensitivity sensibilidad magnética
magnetic separation separación magnética
magnetic separator separador magnético
magnetic sheet hoja magnética
magnetic shell capa magnética
magnetic shield pantalla magnética
magnetic shielding blindaje magnético
magnetic shift register registro de desplazamiento magnético
magnetic shunt derivación magnética
magnetic sound recording registro de sonido magnético
magnetic sound system sistema de sonido magnético
magnetic sound track pista de sonido magnética
magnetic south sur magnético
magnetic speaker altavoz magnético
magnetic spectrograph espectrógrafo magnético
magnetic spectrometer espectrómetro magnético

magnetic starter arrancador magnético
magnetic state estado magnético
magnetic storage almacenamiento magnético
magnetic storage device dispositivo de almacenamiento magnético
magnetic storage medium medio de almacenamiento magnético
magnetic storage unit unidad de almacenamiento magnético
magnetic store almacenamiento magnético
magnetic storm tormenta magnética
magnetic strain deformación magnética
magnetic strain energy energía de deformación magnética
magnetic strength fuerza magnética
magnetic stress estrés magnético
magnetic strip franja magnética
magnetic strip card tarjeta con franja magnética
magnetic surface superficie magnética
magnetic susceptibility susceptibilidad magnética
magnetic sweep barrido magnético
magnetic switch conmutador magnético
magnetic system sistema magnético
magnetic tachometer tacómetro magnético
magnetic tape cinta magnética
magnetic-tape adapter adaptador de cinta magnética
magnetic-tape cartridge cartucho de cinta magnética
magnetic-tape cassette recorder magnetófono
magnetic-tape code código de cinta magnética
magnetic-tape computer computadora de cinta magnética
magnetic-tape control unit unidad de control de cinta magnética
magnetic-tape controller controlador de cinta magnética
magnetic-tape converter convertidor de cinta magnética
magnetic-tape core núcleo de cinta magnética
magnetic-tape deck unidad de cinta magnética, mecanismo de arrastre de cinta magnética
magnetic-tape drive unidad de cinta magnética, mecanismo de arrastre de cinta magnética
magnetic-tape encoder codificador de cinta magnética
magnetic-tape eraser borrador de cinta magnética
magnetic-tape file archivo en cinta magnética
magnetic-tape handler mecanismo de arrastre de cinta magnética
magnetic-tape head cabeza de cinta magnética
magnetic-tape library biblioteca de cintas magnéticas
magnetic-tape memory memoria en cinta magnética
magnetic-tape parity paridad de cinta magnética
magnetic-tape reader lectora de cinta magnética
magnetic-tape recorder registrador en cinta magnética
magnetic-tape recording registro en cinta magnética
magnetic-tape recording head cabeza de registro en cinta magnética
magnetic-tape recording unit unidad de registro en cinta magnética
magnetic-tape reproducer reproductor de cinta magnética
magnetic-tape station estación de cinta magnética
magnetic-tape storage almacenamiento de cinta magnética
magnetic-tape subsystem subsistema de cinta magnética
magnetic-tape system sistema de cinta magnética
magnetic-tape tester probador de cinta magnética
magnetic-tape unit unidad de cinta magnética
magnetic temperature compensator compensador de temperatura magnético

magnetic test coil bobina de muestreo magnético
magnetic thick film película magnética gruesa
magnetic thickness measurer medidor de grosor
    magnético
magnetic thin film película magnética delgada
magnetic time relay relé de tiempo magnético
magnetic track pista magnética, rumbo magnético
magnetic transducer transductor magnético
magnetic transfer transferencia magnética
magnetic tuning sintonización magnética
magnetic-vane meter medidor de paleta magnética
magnetic variation variación magnética
magnetic vector vector magnético
magnetic vector potential potencial vectorial
    magnético
magnetic viscosity viscosidad magnética
magnetic whirl remolino magnético
magnetic wire hilo magnético
magnetic wire storage almacenamiento de hilo
    magnético
magnetic yoke yugo magnético
magnetically magnéticamente
magnetically biased polarizado magnéticamente
magnetically damped instrument instrumento
    amortiguado magnéticamente
magnetics magnética
magnetism magnetismo
magnetite magnetita
magnetizability magnetizabilidad
magnetizable magnetizable
magnetization magnetización
magnetization by induction magnetización por
    inducción
magnetization curve curva de magnetización
magnetization cycle ciclo de magnetización
magnetize magnetizar
magnetized magnetizado
magnetized head cabeza magnetizada
magnetized ink character carácter de tinta
    magnetizada
magnetizer magnetizador
magnetizing magnetizante
magnetizing coil bobina magnetizante
magnetizing current corriente magnetizante
magnetizing field campo magnetizante
magnetizing force fuerza magnetizante
magnetizing head cabeza magnetizante
magnetizing inductance inductancia magnetizante
magneto magneto
magneto ignition ignición por magneto
magneto ringing llamada por magneto
magneto system sistema por magneto
magneto telephone teléfono de magneto
magneto telephone set aparato telefónico de magneto
magnetoacoustic magnetoacústico
magnetoacoustic method método magnetoacústico
magnetoacoustic wave onda magnetoacústica
magnetoacoustics magnetoacústica
magnetocardiogram magnetocardiograma
magnetocardiograph magnetocardiógrafo
magnetochemistry magnetoquímica
magnetoconductivity magnetoconductividad
magnetocrystalline magnetocristalino
magnetodiode magnetodiodo
magnetodynamic magnetodinámico
magnetodynamic flow flujo magnetodinámico
magnetodynamic head cabeza magnetodinámica
magnetoelastic magnetoelástico
magnetoelastic effect efecto magnetoelástico
magnetoelastic energy energía magnetoelástica
magnetoelectric magnetoeléctrico
magnetoelectric effect efecto magnetoeléctrico
magnetoelectric generator generador

magnetoeléctrico
magnetoelectric induction inducción
    magnetoeléctrica
magnetoelectric relay relé magnetoeléctrico
magnetoelectric transducer transductor
    magnetoeléctrico
magnetofluidmechanic magnetofluidomecánico
magnetofluidmechanics magnetofluidomecánica
magnetofluidynamic magnetofluidodinámico
magnetofluidynamics magnetofluidodinámica
magnetogram magnetograma
magnetograph magnetógrafo
magnetohydrodynamic magnetohidrodinámico
magnetohydrodynamic converter convertidor
    magnetohidrodinámico
magnetohydrodynamic effect efecto
    magnetohidrodinámico
magnetohydrodynamic equilibrium equilibrio
    magnetohidrodinámico
magnetohydrodynamic generator generador
    magnetohidrodinámico
magnetohydrodynamic gyroscope giroscopio
    magnetohidrodinámico
magnetohydrodynamic instability inestabilidad
    magnetohidrodinámica
magnetohydrodynamic power energía
    magnetohidrodinámica
magnetohydrodynamic power generation
    generación de energía magnetohidrodinámica
magnetohydrodynamic power generator generador
    de energía magnetohidrodinámico
magnetohydrodynamic shock wave onda de choque
    magnetohidrodinámica
magnetohydrodynamic system sistema
    magnetohidrodinámico
magnetohydrodynamic wave onda
    magnetohidrodinámica
magnetohydrodynamics magnetohidrodinámica
magnetoionic magnetoiónico
magnetoionic duct conducto magnetoiónico
magnetoionic environment ambiente magnetoiónico
magnetoionic Faraday rotation rotación de Faraday
    magnetoiónica
magnetoionic medium medio magnetoiónico
magnetoionic polarization polarización
    magnetoiónica
magnetoionic wave component onda componente
    magnetoiónica
magnetoionics magnetoiónica
magnetomechanical magnetomecánico
magnetomechanical damping amortiguación
    magnetomecánica
magnetomechanical factor factor magnetomecánico
magnetomechanical hysteresis histéresis
    magnetomecánica
magnetomechanical ratio razón magnetomecánica
magnetometer magnetómetro
magnetomotive magnetomotriz
magnetomotive force fuerza magnetomotriz
magnetomotive force component componente de
    fuerza magnetomotriz
magneton magnetón
magnetooptical magnetoóptico
magnetooptical effect efecto magnetoóptico
magnetooptical laser láser magnetoóptico
magnetooptical rotation rotación magnetoóptica
magnetooptics magnetoóptica
magnetopause magnetopausa
magnetophone magnetófono
magnetoplasmadynamic magnetoplasmadinámico
magnetoplasmadynamic generator generador
    magnetoplasmadinámico
magnetoplasmadynamics magnetoplasmadinámica

magnetoresistance magnetorresistencia
magnetoresistive magnetorresistivo
magnetoresistive coefficient coeficiente
  magnetorresistivo
magnetoresistive effect efecto magnetorresistivo
magnetoresistive metal metal magnetorresistivo
magnetoresistive ratio razón magnetorresistiva
magnetoresistivity magnetorresistividad
magnetoresistor magnetorresistor
magnetoscope magnetoscopio
magnetoscopic magnetoscópico
magnetoscopic inspection inspección
  magnetoscópica
magnetoscopy magnetoscopía
magnetosphere magnetosfera
magnetostatic magnetostático
magnetostatic field campo magnetostático
magnetostatics magnetostática
magnetostriction magnetoestricción,
  magnetostricción
magnetostriction effect efecto de magnetoestricción
magnetostriction filter filtro de magnetoestricción
magnetostriction hydrophone hidrófono de
  magnetoestricción
magnetostriction loudspeaker altavoz de
  magnetoestricción
magnetostriction magnetometer magnetómetro de
  magnetoestricción
magnetostriction microphone micrófono de
  magnetoestricción
magnetostriction oscillator oscilador de
  magnetoestricción
magnetostriction speaker altavoz de
  magnetoestricción
magnetostrictive magnetoestrictivo
magnetostrictive ácoustic delay line línea de retardo
  acústica magnetoestrictiva
magnetostrictive contraction contracción
  magnetoestrictiva
magnetostrictive delay line línea de retardo
  magnetoestrictiva
magnetostrictive delay-line storage almacenamiento
  de línea de retardo magnetoestrictiva
magnetostrictive effect efecto magnetoestrictivo
magnetostrictive expansion expansión
  magnetoestrictiva
magnetostrictive film transducer transductor de
  película magnetoestrictiva
magnetostrictive filter filtro magnetoestrictivo
magnetostrictive loudspeaker altavoz
  magnetoestrictivo
magnetostrictive microphone micrófono
  magnetoestrictivo
magnetostrictive oscillator oscilador
  magnetoestrictivo
magnetostrictive relay relé magnetoestrictivo
magnetostrictive resonator resonador
  magnetoestrictivo
magnetostrictive transceiver transceptor
  magnetoestrictivo
magnetostrictive transducer transductor
  magnetoestrictivo
magnetostrictive ultrasonic vibrator vibrador
  ultrasónico magnetoestrictivo
magnetostrictive vibration vibración
  magnetoestrictiva
magnetostrictor magnetoestrictor
magnetothermal magnetotérmico
magnetothermal circuit-breaker cortacircuito
  magnetotérmico
magnetothermal resistance resistencia
  magnetotérmica
magnetothermoelastic magnetotermoelástico

magnetothermoelasticity magnetotermoelasticidad
magnetothermoelectric magnetotermoeléctrico
magnetothermoelectric effect efecto
  magnetotermoeléctrico
magnetotorsion magnetotorsión
magnetotransistor magnetotransistor
magnetron magnetrón
magnetron arcing formación de arco en magnetrón
magnetron effect efecto magnetrón
magnetron oscillator oscilador magnetrón
magnetron rectifier rectificador magnetrón
magnetron tube tubo magnetrón
magnettor magnetor
magnification magnificación, aumento
magnification circuit circuito resonador
magnification factor factor de aumento, factor de
  sobretensión
magnifying power potencia de magnificación
magnistor magnistor
magnitude magnitud
magnitude reference line línea de referencia de
  magnitud
magtape cinta magnética
mailing list program programa de lista de
  direcciones
main anode ánodo principal
main bang impulso piloto, impulso principal
main beam haz principal
main cable cable principal
main capacitance capacitancia principal
main carrier portadora principal
main carrier cable cable portador principal
main channel canal principal
main circuit circuito principal
main circuit-breaker cortacircuito principal
main command menu menú de mandos principal
main computer computadora principal
main contacts contactos principales
main control control principal
main control room sala de control principal
main control switch conmutador de control principal
main control unit unidad de control principal
main disconnect switch seccionador principal
main distributing frame repartidor principal
main distribution frame repartidor principal
main electrode electrodo principal
main equipment equipo principal
main exchange central principal
main exciter excitador principal
main feedback loop bucle de retroalimentación
  principal
main file archivo principal
main file tape cinta de archivo principal
main frame unidad central, unidad de procesamiento
  central, maxicomputadora
main gap intervalo principal
main generator generador principal
main ground tierra principal
main job tarea principal
main keyboard teclado principal
main lead conductor principal
main line línea principal
main line switch conmutador de línea principal
main lobe lóbulo principal
main memory memoria principal
main menu menú principal
main network red principal
main network address dirección de red principal
main path trayectoria principal
main phase fase principal
main pole polo principal
main power input entrada de potencia principal
main power switch conmutador de alimentación

principal
main program programa principal
main protection protección principal
main receiver receptor principal
main reflector reflector principal
main repeater repetidor principal
main repeating section sección de repetición principal
main route ruta principal
main routine rutina principal
main scanner explorador principal
main section sección principal
main segment segmento principal
main sideband banda lateral principal
main station estación principal
main storage almacenamiento principal
main storage database base de datos de almacenamiento principal
main storage partition partición de almacenamiento principal
main storage processor procesador de almacenamiento principal
main storage region región de almacenamiento principal
main store almacenamiento principal
main switch conmutador principal
main switchboard cuadro conmutador principal
main system sistema principal
main task tarea principal
main transformer transformador principal
main transmitter transmisor principal
main unit unidad principal
main winding arrollamiento principal
mainframe unidad central, unidad de procesamiento central, maxicomputadora
mainframe computer maxicomputadora
mains líneas principales, red eléctrica, red
mains antenna antena de red
mains cable cable de entrada de red
mains failure falla de red
mains frequency frecuencia de red
mains hum zumbido de red
mains noise ruido de red
mains supply alimentación por red
mains voltage regulator regulador de tensión de red
maintained contact contacto mantenido
maintenance mantenimiento
maintenance access processor procesador de acceso de mantenimiento
maintenance analysis procedure procedimiento de análisis de mantenimiento
maintenance control control de mantenimiento
maintenance control panel panel de control de mantenimiento
maintenance cycle ciclo de mantenimiento
maintenance file archivo de mantenimiento
maintenance-free libre de mantenimiento
maintenance instructions instrucciones de mantenimiento
maintenance interval intervalo de mantenimiento
maintenance level nivel de mantenimiento
maintenance log diario de mantenimiento
maintenance mode modo de mantenimiento
maintenance of equipment mantenimiento de equipo
maintenance of service mantenimiento de servicio
maintenance panel panel de mantenimiento
maintenance plan plan de mantenimiento
maintenance processor procesador de mantenimiento
maintenance program programa de mantenimiento
maintenance routine rutina de mantenimiento
maintenance schedule programa de mantenimiento
maintenance service servicio de mantenimiento

maintenance test prueba de mantenimiento
maintenance time tiempo de mantenimiento
major control data datos de control principal
major control field campo de control principal
major cycle ciclo principal
major defect defecto principal
major face cara principal
major failure falla principal
major lobe lóbulo principal
major loop bucle principal
major node nodo principal
major repair reparación principal
major sort clasificación principal
major task tarea principal
majority carrier portador mayoritario
majority decision gate compuerta de decisión mayoritaria
majority element elemento mayoritario
majority emitter emisor mayoritario
majority logic lógica mayoritaria
make cierre
make-and-break ratio razón de cierre-abertura
make-before-break cierre antes de abertura
make-before-break contacts contactos de cerrar antes de abrir
make contact contacto de cierre
make impulse impulso de cierre
make position posición de cierre
make pulse impulso de cierre
make time tiempo de cierre
make timer temporizador de cierre
making capacity capacidad de cierre
making current corriente de cierre
making-current test prueba de corriente de cierre
malfunction mal funcionamiento, falla, fallo, avería
malfunction alert aviso de mal funcionamiento
malfunction report informe de mal funcionamiento
malfunction routine rutina de mal funcionamiento
malleability maleabilidad
Malter effect efecto de Malter
management system sistema de administración
manganese manganeso
manganese-dioxide depolarizer despolarizador de dióxido de manganeso
manganin manganina
manganin wire hilo de manganina
manipulate manipular
manipulation manipulación
manipulator manipulador
manometer manómetro
Mansbridge condenser condensador de Mansbridge
manual manual
manual arc welding soldadura por arco manual
manual block system sistema de bloqueo manual
manual central office oficina central manual
manual command mando manual
manual contrast control control de contraste manual
manual control control manual
manual control panel panel de control manual
manual control system sistema de control manual
manual controller controlador manual
manual cutout cortacircuito manual, conmutador manual
manual data input entrada de datos manual
manual dialing marcación manual
manual direction-finder radiogoniómetro manual
manual entry entrada manual
manual equipment equipo manual
manual exchange central manual
manual feed alimentación manual
manual fire-alarm signal señal de alarma de incendios manual
manual function función manual

manual hold   retención manual
manual indicator   indicador manual
manual input   entrada manual
manual input register   registro de entrada manual
manual input unit   unidad de entrada manual
manual mode   modo manual
manual mode operation   operación en modo manual
manual operation   operación manual
manual perforator   perforador manual
manual position   posición manual
manual potentiometer   potenciómetro manual
manual preset   preajuste manual
manual regulation   regulación manual
manual reset   reposición manual
manual-reset relay   relé de reposición manual
manual retransmission   retransmisión manual
manual ringing   llamada manual
manual selection   selección manual
manual storage   almacenamiento manual
manual storage switch   conmutador de
    almacenamiento manual
manual switch   conmutador manual
manual switchboard   cuadro conmutador manual
manual switching   conmutación manual
manual system   sistema manual
manual telecommunications system   sistema de
    telecomunicaciones manual
manual telegraphy   telegrafía manual
manual telephone   teléfono manual
manual telephone set   aparato telefónico manual
manual telephone system   sistema telefónico manual
manual test equipment   equipo de prueba manual
manual transmission   transmisión manual
manual tuning   sintonización manual
manual volume control   control de volumen manual
manual welding   soldadura manual
manually   manualmente
manually operated   operado manualmente
manufacturing tolerance   tolerancia de manufactura
Marconi antenna   antena de Marconi
Marconi effect   efecto de Marconi
margin   margen
margin of allowance   margen de tolerancia
margin of commutation   margen de conmutación
margin of power   margen de potencia
margin of stability   margen de estabilidad
margin-punched card   tarjeta perforada en los
    márgenes
marginal   marginal
marginal check   comprobación marginal
marginal current   corriente marginal
marginal fault   falla marginal
marginal-punched card   tarjeta perforada en los
    márgenes
marginal relay   relé marginal
marginal stability   estabilidad marginal
marginal test   prueba marginal
marginal voltage check   comprobación de tensión
    marginal
marine broadcast station   estación de radiodifusión
    marítima
marine electronic equipment   equipo electrónico
    marítimo
marine radio   radio marítimo
marine radiobeacon   radiofaro marítimo
marine radiobeacon station   estación de radiofaro
    marítimo
maritime radionavigation service   servicio de
    radionavegación marítima
maritime wave   onda marítima
mark   marca
mark density   densidad de marcas
mark reader   lectora de marcas

mark reading   lectura de marcas
mark scanning   exploración de marcas
mark sensing   detección de marcas
mark-sensing device   dispositivo de detección de
    marcas
mark-space ratio   razón marca-espacio
marker   marca, marcador, radiobaliza
marker adder   sumador de marcas
marker antenna   antena de radiobaliza
marker beacon   radiobaliza
marker-beacon receiver   receptor de radiobaliza
marker calibrator   calibrador marcador
marker frequency   frecuencia marcadora
marker generator   generador marcador
marker light   luz marcadora
marker pip   impulso marcador
marker pulse   impulso marcador
marker radiobeacon   radiobaliza
marker receiver   receptor de radiobaliza
marker signal   señal marcadora
marker thread   hilo marcador
marking bias   polarización de trabajo
marking condition   condición de trabajo
marking current   corriente de trabajo
marking element   elemento de trabajo
marking frequency   frecuencia de trabajo
marking impulse   impulso de trabajo
marking interval   intervalo de trabajo
marking pulse   impulso de trabajo
marking signal   señal de trabajo
marking switching   conmutación de trabajo
marking tone   tono de trabajo
marking wave   onda de trabajo
Markov chain   cadena de Markov
Marx generator   generador de Marx
maser   máser
maser amplifier   amplificador máser
mask   máscara
masking   enmascaramiento
masking effect   efecto de enmascaramiento
masking level   nivel de enmascaramiento
masking measurement   medida de enmascaramiento
masking sound   sonido de enmascaramiento
masking voltage   tensión de enmascaramiento
mass absorption coefficient   coeficiente de absorción
    de masa
mass analyzer   analizador de masa
mass attenuation coefficient   coeficiente de
    atenuación de masa
mass data   datos masivos
mass-impregnated and drained insulation
    aislamiento impregnado y escurrido
mass-impregnated nondrained insulation
    aislamiento impregnado sin escurrido
mass memory   memoria masiva
mass number   número de masa
mass production   producción en masa
mass reactance   reactancia de masa
mass resistivity   resistividad de masa
mass spectrograph   espectrógrafo de masas
mass spectrometer   espectrómetro de masa
mass spectrometry   espectrometría de masa
mass spectroscope   espectroscopio de masa
mass spectroscopy   espectroscopia de masa
mass spectrum   espectro de masa
mass storage   almacenamiento masivo
mass storage control   control de almacenamiento
    masivo
mass storage device   dispositivo de almacenamiento
    masivo
mass storage file   archivo de almacenamiento masivo
mass storage volume   volumen de almacenamiento
    masivo

mass unit unidad de masa
mast antenna antena de mástil
master address dirección maestra
master address space espacio de dirección maestra
master antenna antena colectiva
master antenna system sistema de antena colectiva
master brightness control control de brillo maestro
master card tarjeta maestra
master cartridge cartucho maestro
master chip chip maestro
master clock reloj maestro
master clock frequency frecuencia de reloj maestro
master compass compás magistral, compás maestro,
    brújula maestra
master computer computadora maestra
master console consola maestra
master contactor contactor maestro
master control control maestro
master control board tablero de control maestro
master control code código de control maestro
master control desk pupitre de control maestro
master control monitor monitor de control maestro
master control panel panel de control maestro
master control position posición de control maestro
master control room sala de control maestro
master control routine rutina de control maestro
master controller controlador maestro
master data datos maestros
master data file archivo de datos maestros
master drawing dibujo maestro
master driver excitador maestro
master element elemento maestro
master file archivo maestro
master file record registro de archivo maestro
master file tape cinta de archivo maestro
master gage calibre maestro
master gain control control de ganancia maestro
master gyrocompass girocompás magistral,
    girocompás maestro
master image imagen maestra
master index índice maestro
master information información maestra
master input file archivo de entrada maestro
master instruction tape cinta de instrucciones
    maestra
master library biblioteca maestra
master library tape cinta de biblioteca maestra
master mode modo maestro
master monitor monitor maestro
master node nodo maestro
master oscillator oscilador maestro
master oscillator-power amplifier oscilador
    maestro-amplificador de potencia
master output file archivo de salida maestro
master pattern patrón matriz
master processor procesador maestro
master program programa maestro
master program file archivo de programas maestros
master radar radar maestro
master record registro maestro
master recording monitor monitor de registro
    maestro
master regulator regulador maestro
master relay relé maestro
master routine rutina maestra
master sound receiver receptor de sonido maestro
master station estación maestra
master switch conmutador maestro
master synchronization pulse impulso de
    sincronización maestro
master synchronizer sincronizador maestro
master tape cinta maestra
master television antenna antena de televisión

maestra
master terminal terminal maestra
master transmitter transmisor maestro
master unit unidad maestra
master valve válvula maestra
master workstation estación de trabajo maestra
mastergroup grupo maestro
mastergroup demodulator desmodulador de grupo
    maestro
mastergroup modulator modulador de grupo
    maestro
mastergroup section sección de grupo maestro
match adaptación, comparación, equilibrio,
    apareamiento
matched components componentes apareados
matched doublets dobletes apareados
matched electron tubes tubos electrónicos apareados
matched filter filtro adaptado
matched impedance impedancia adaptada
matched-impedance antenna antena de impedancia
    adaptada
matched-impedance attenuator atenuador de
    impedancia adaptada
matched junction unión adaptada
matched line línea adaptada
matched load carga adaptada
matched microphones micrófonos apareados
matched pair par apareado
matched termination terminación adaptada
matched transmission line línea de transmisión
    adaptada
matched waveguide guíaondas adaptado
matching adaptación, comparación, equilibrio,
    apareamiento
matching device dispositivo de adaptación
matching diaphragm diafragma de adaptación
matching error error de adaptación
matching impedance impedancia de adaptación
matching network red de adaptación
matching pillar varilla de adaptación
matching pole varilla de adaptación
matching quadripole cuadripolo de adaptación
matching resistor resistor de adaptación
matching section sección de adaptación
matching strip franja de adaptación
matching stub adaptador de impedancia, línea de
    adaptación
matching transformer transformador de adaptación
mathematical analysis análisis matemático
mathematical check comprobación matemática
mathematical control mode modo de control
    matemático
mathematical function función matemática
mathematical logic lógica matemática
mathematical model modelo matemático
mathematical parameter parámetro matemático
mathematical programming programación
    matemática
mathematical programming system sistema de
    programación matemática
mathematical simulation simulación matemática
mathematical subroutine subrutina matemática
mathematical symbol símbolo matemático
matrix matriz
matrix analysis análisis matricial
matrix character carácter matricial
matrix circuit circuito matricial
matrix connection conexión matricial
matrix equation ecuación matricial
matrix memory memoria matricial
matrix network red matricial
matrix notation notación matricial
matrix operation operación matricial

matrix printer impresora de matriz de puntos
matrix printing impresión por matrices
matrix store almacenamiento matricial
matrix unit unidad matricial
matte surface superficie mate
matter materia
Matteucci effect efecto de Matteucci
mavar mavar
maxima and minima máximos y mínimos
maximal máximo
maximal flatness amplifier amplificador de aplanamiento máximo
maximal signal strength intensidad de señal máxima
maximum máximo
maximum access time tiempo máximo de acceso
maximum amplitude amplitud máxima
maximum anode dissipation disipación máxima anódica
maximum applied load carga máxima aplicada
maximum available gain ganancia máxima disponible
maximum available power gain ganancia máxima de potencia disponible
maximum average power potencia máxima media
maximum average power output salida máxima de potencia media
maximum baseband frequency frecuencia máxima de banda base
maximum capability capacidad máxima
maximum capacity capacidad máxima
maximum carrier amplitude amplitud máxima de portadora
maximum continuous power potencia máxima continua
maximum continuous rating régimen máximo continuo
maximum control current corriente máxima de control
maximum current corriente máxima
maximum current relay relé de corriente máxima
maximum deflection deflexión máxima
maximum demand demanda máxima
maximum-demand indicator indicador de demanda máxima
maximum-demand recorder registrador de demanda máxima
maximum design voltage tensión máxima de diseño
maximum deviation desviación máxima
maximum-deviation sensitivity sensibilidad de desviación máxima
maximum discharge current corriente máxima de descarga
maximum distortion distorsión máxima
maximum duration duración máxima
maximum effective aperture abertura máxima efectiva
maximum equivalent conductance conductancia máxima equivalente
maximum equivalent inductance inductancia máxima equivalente
maximum excursion excursión máxima
maximum exposure temperature temperatura máxima de exposición
maximum file size tamaño máximo de archivo
maximum frequency deviation desviación máxima de frecuencia
maximum frequency swing oscilación máxima de frecuencia
maximum generation generación máxima
maximum keying frequency frecuencia máxima de manipulación
maximum line length longitud máxima de línea
maximum load carga máxima

maximum modulating frequency frecuencia máxima de modulación
maximum modulation frequency frecuencia máxima de modulación
maximum observed frequency frecuencia máxima observada
maximum operating pressure presión máxima de operación
maximum operating voltage tensión máxima de operación
maximum output salida máxima
maximum output voltage tensión máxima de salida
maximum peak inverse voltage tensión inversa máxima de pico
maximum peak plate current corriente de placa máxima de pico
maximum peak reverse voltage tensión inversa máxima de pico
maximum percentage modulation porcentaje máximo de modulación
maximum permeability permeabilidad máxima
maximum permissible voltage drop caída de tensión máxima permisible
maximum power potencia máxima
maximum power output salida máxima de potencia
maximum power transfer theorem teorema de transferencia máxima de potencia
maximum rating clasificación máxima
maximum record level nivel máximo de registro
maximum response speed velocidad máxima de respuesta
maximum retention time tiempo máximo de retención
maximum safe input power potencia máxima de entrada segura
maximum sensitivity sensibilidad máxima
maximum signal level nivel máximo de señal
maximum signal power output salida máxima de potencia de señal
maximum sound pressure presión máxima acústica
maximum specific consumption consumo máximo específico
maximum speed velocidad máxima
maximum storage time tiempo máximo de almacenamiento
maximum system deviation desviación máxima de sistema
maximum system voltage tensión máxima de sistema
maximum undistorted output salida máxima sin distorsión
maximum undistorted power output salida máxima de potencia sin distorsión
maximum usable frequency frecuencia máxima utilizable
maximum usable frequency factor factor de frecuencia máxima utilizable
maximum usable power potencia máxima utilizable
maximum useful output salida máxima útil
maximum value valor máximo
maximum voltage tensión máxima
maximum wattage wattaje máximo
maxwell maxwell
Maxwell's equation ecuación de Maxwell
Maxwell's law ley de Maxwell
Maxwell's rule regla de Maxwell
Maxwell bridge puente de Maxwell
Maxwell coil bobina de Maxwell
Maxwell inductance bridge puente de inductancias de Maxwell
Maxwell mutual-inductance bridge puente de inductancia mutua de Maxwell
maxwell-turn maxwell-vuelta
McLeod gage medidor de McLeod

McNally tube tubo de McNally
McProud test prueba de McProud
Meacham oscillator oscilador de Meacham
mean media, medio
mean accuracy exactitud media
mean amplitude amplitud media
mean carrier frequency frecuencia de portadora media
mean charge carga media
mean-charge characteristic característica de carga media
mean delay time tiempo de retardo medio
mean deviation desviación media
mean down time tiempo muerto medio
mean electron energy energía electrónica media
mean electron velocity velocidad electrónica media
mean energy energía media
mean energy production producción de energía media
mean error error medio
mean force fuerza media
mean free path recorrido libre medio
mean horizontal intensity intensidad horizontal media
mean information content contenido de información medio
mean instrument reading lectura de instrumento media
mean ionization energy energía de ionización media
mean life vida media
mean occupancy ocupación media
mean path recorrido medio
mean power potencia media
mean pulse time tiempo de impulso medio
mean repair time tiempo de reparación medio
mean stopping time tiempo de parada medio
mean temperature temperatura media
mean temperature difference diferencia de temperatura media
mean time between failures tiempo medio entre fallas
mean time between outages tiempo medio entre interrupciones
mean time to failure tiempo medio hasta falla
mean time to first failure tiempo medio hasta la primera falla
mean time to repair tiempo medio hasta reparación
mean value valor medio
mean value indicator indicador de valor medio
mean velocity velocidad media
mean waiting time tiempo de espera medio
measurable medible
measurand mensurando
measure of information medida de información
measured current corriente medida
measured curve curva medida
measured feedback retroalimentación medida
measured service servicio medido
measured signal señal medida
measured temperature temperatura medida
measured value valor medido
measured variable variable medida
measurement medida
measurement component componente de medida
measurement device dispositivo de medida
measurement energy energía de medida
measurement equipment equipo de medida
measurement error error de medida
measurement mechanism mecanismo de medida
measurement procedure procedimiento de medida
measurement range gama de medida
measurement standard patrón de medida
measurement system sistema de medida

measurement terminals terminales de medida
measurement uncertainty incertidumbre de medida
measurement voltage divider divisor de tensión de medida
measuring accuracy exactitud de medida
measuring amplifier amplificador de medida
measuring antenna antena de medida
measuring bridge puente de medida
measuring circuit circuito de medida
measuring delay time tiempo de retardo de medida
measuring device dispositivo de medida
measuring diode diodo de medida
measuring equipment equipo de medida
measuring frequency frecuencia de medida
measuring head cabeza de medida
measuring instrument instrumento de medida
measuring loop bucle de medida
measuring microphone micrófono de medida
measuring microscope microscopio de medida
measuring modulator modulador de medida
measuring point punto de medida
measuring potentiometer potenciómetro de medida
measuring range intervalo de medida
measuring receiver receptor de medida
measuring relay relé de medida
measuring signal señal de medida
measuring spark-gap chispómetro, medidor de chispa
measuring station estación de medida
measuring system sistema de medida
measuring technique técnica de medida
measuring transductor transductor de medida
measuring unit unidad de medida
mechanical mecánico
mechanical-action relay relé de acción mecánica
mechanical adjustment ajuste mecánico
mechanical admittance admitancia mecánica
mechanical axis eje mecánico
mechanical bandspread ensanche de banda mecánico
mechanical bias polarización mecánica
mechanical centering centrado mecánico
mechanical compliance docilidad mecánica
mechanical computer computadora mecánica
mechanical connection conexión mecánica
mechanical contactor contactor mecánico
mechanical damping amortiguamiento mecánico
mechanical data processing procesamiento de datos mecánico
mechanical differential diferencial mecánico
mechanical drive accionamiento mecánico
mechanical efficiency eficiencia mecánica
mechanical electromagnetic pump bomba electromagnética mecánica
mechanical electrometer electrómetro mecánico
mechanical electroplating galvanoplastia mecánica
mechanical energy energía mecánica
mechanical equivalent of heat equivalente mecánico de calor
mechanical equivalent of light equivalente mecánico de luz
mechanical filter filtro mecánico
mechanical force fuerza mecánica
mechanical impedance impedancia mecánica
mechanical instability inestabilidad mecánica
mechanical interchangeability intercambialidad mecánica
mechanical interface interfaz mecánica
mechanical joint junta mecánica
mechanical life vida mecánica
mechanical limit límite mecánico
mechanical load carga mecánica
mechanical loss pérdida mecánica

mechanical mass masa mecánica
mechanical modulator modulador mecánico
mechanical noise ruido mecánico
mechanical operation operación mecánica
mechanical oscillograph oscilógrafo mecánico
mechanical overtravel sobrecarrera mecánica
mechanical phenomenon fenómeno mecánico
mechanical phonograph fonógrafo mecánico
mechanical polarization polarización mecánica
mechanical power potencia mecánica
mechanical reactance reactancia mecánica
mechanical recorder registrador mecánico
mechanical recording registro mecánico
mechanical recording head cabeza de registro
    mecánica
mechanical rectification rectificación mecánica
mechanical rectifier rectificador mecánico
mechanical register registro mecánico
mechanical relay relé mecánico
mechanical reproducer reproductor mecánico
mechanical resistance resistencia mecánica
mechanical scanner explorador mecánico
mechanical scanning exploración mecánica
mechanical shock choque mecánico
mechanical stability estabilidad mecánica
mechanical stress estrés mecánico
mechanical sweep barrido mecánico
mechanical switching device dispositivo de
    conmutación mecánico
mechanical time constant constante de tiempo
    mecánica
mechanical tuning sintonización mecánica
mechanical tuning range margen de sintonización
    mecánica
mechanical wave filter filtro de ondas mecánico
mechanical zero cero mecánico
mechanically mecánicamente
mechanically produced producido mecánicamente
mechanically timed relay relé temporizado
    mecánicamente
mechanics mecánica
mechanism mecanismo
mechanoelectric mecanoeléctrico
mechanoelectric transducer transductor
    mecanoeléctrico
mechanoelectronic mecanoeléctrico
median mediana
medium medio
medium access control control de acceso de medio
medium access control procedure procedimiento de
    control de acceso de medio
medium band banda media
medium frequency frecuencia media
medium-frequency transmitter transmisor de
    frecuencia media
medium-gain amplifier amplificador de ganancia
    media
medium-impedance triode triodo de impedancia
    media
medium of propagation medio de propagación
medium-range radar radar de alcance medio
medium-scale integration integración de escala
    media
medium speed velocidad media
medium tension tensión media
medium voltage tensión media
medium-voltage system sistema de tensión media
medium wave onda media
medium-wave antenna antena de onda media
medium-wave band banda de onda media
medium-wave frequency frecuencia de onda media
medium-wave spectrum espectro de onda media
medium-wave transmitter transmisor de onda media

medium wavelength longitud de onda media
mega mega
megabar megabar
megabit megabit
megabyte megabyte
megacurie megacurie
megacycle megaciclo
megacycles per second megaciclos por segundo
megadyne megadina
megaelectron-volt megaelectrón-volt
megahertz megahertz
megalumen megalumen
megampere megampere
megaphone megáfono
megaroentgen megarroentgen
megaton megatonelada
megatron megatrón
megavar megavar
megavolt megavolt
megavolt-ampere megavolt-ampere
megawatt megawatt
megawatt-hour megawatt-hora
megger megóhmetro
megohm megohm
megohm-farads megohm-farads
megohm-microfarads megohm-microfarads
megohmmeter megóhmetro
Meissner circuit circuito de Meissner
Meissner effect efecto de Meissner
Meissner oscillator oscilador de Meissner
mel mel
meltback process proceso de refusión
meltback transistor transistor de refusión
melting point punto de fusión
melting time tiempo de fusión
membrane keyboard teclado de membrana
membrane potential potencial de membrana
membrane switch conmutador de membrana
memistor memistor
memory memoria
memory action acción de memoria
memory address dirección de memoria
memory address counter contador de direcciones de
    memoria
memory address register registro de direcciones de
    memoria
memory address space espacio de direcciones de
    memoria
memory addressing direccionamiento de memoria
memory addressing mode modo de
    direccionamiento de memoria
memory allocation asignación de memoria
memory allocation manager administrador de
    asignación de memoria
memory area área de memoria
memory array red de memoria
memory bank banco de memoria
memory block bloque de memoria
memory board tablero de memoria
memory buffer register registro de almacenamiento
    intermedio de memoria
memory bus bus de memoria
memory capacity capacidad de memoria
memory cell célula de memoria
memory chip chip de memoria
memory circuit circuito de memoria
memory code disk disco de código de memoria
memory compaction compactación de memoria
memory controller controlador de memoria
memory core núcleo de memoria
memory cycle ciclo de memoria
memory data register registro de datos de memoria
memory-dependent dependiente de memoria

memory device dispositivo de memoria
memory diagram diagrama de memoria
memory disk disco de memoria
memory drum tambor de memoria
memory dump vaciado de memoria
memory error error de memoria
memory exchange intercambio de memoria
memory expansion expansión de memoria
memory expansion module módulo de expansión de memoria
memory guard guarda de memoria
memory hierarchy jerarquía de memoria
memory latency time tiempo de latencia de memoria
memory layout distribución de memoria
memory location localización de memoria
memory management administración de memoria
memory map mapa de memoria
memory mechanism mecanismo de memoria
memory module módulo de memoria
memory organization organización de memoria
memory overflow desbordamiento de memoria
memory page página de memoria
memory parity paridad de memoria
memory port puerto de memoria
memory power potencia de memoria
memory printout impresión de memoria
memory process proceso de memoria
memory protection protección de memoria
memory protection device dispositivo de protección de memoria
memory protection option opción de protección de memoria
memory register registro de memoria
memory relay relé con memoria
memory resident residente en memoria
memory resident program programa residente en memoria
memory retention retención en memoria
memory scan exploración de memoria
memory scan option opción de exploración de memoria
memory size tamaño de memoria
memory space espacio de memoria
memory stage etapa de memoria
memory storage almacenamiento de memoria
memory structural units unidades estructurales de memoria
memory system sistema de memoria
memory tape cinta de memoria
memory tube tubo de memoria
memory type tipo de memoria
memory unit unidad de memoria
memory word palabra de memoria
memoscope memoscopio
memotron memotrón
mendelevium mendelevio
mensuration mensuración
mensuration system sistema de mensuración
menu menú
menu area área de menú
menu display visualización de menú
menu-driven accionado por menú
menu-driven program programa accionado por menú
menu-driven software programas accionados por menú
menu screen pantalla de menú
mercury mercurio
mercury arc arco de mercurio
mercury-arc converter convertidor de arco de mercurio
mercury-arc inverter inversor de arco de mercurio
mercury-arc rectifier rectificador de arco de mercurio

mercury battery batería de mercurio
mercury capsule cápsula de mercurio
mercury cathode cátodo de mercurio
mercury cell pila de mercurio
mercury-contact relay relé de contactos de mercurio
mercury delay line línea de retardo de mercurio
mercury diffusion pump bomba de difusión de mercurio
mercury discharge lamp lámpara de vapor de mercurio
mercury displacement relay relé de desplazamiento de mercurio
mercury fluorescent lamp lámpara fluorescente de mercurio
mercury-jet commutation conmutación por chorro de mercurio
mercury-jet commutator conmutador de chorro de mercurio
mercury-jet switch conmutador de chorro de mercurio
mercury lamp lámpara de mercurio
mercury memory memoria de mercurio
mercury oxide cell pila de óxido de mercurio
mercury-pool cathode cátodo de baño de mercurio
mercury-pool cathode tube tubo de cátodo de baño de mercurio
mercury-pool rectifier rectificador de baño de mercurio
mercury pump bomba de mercurio
mercury rectifier rectificador de mercurio
mercury relay relé de mercurio
mercury storage almacenamiento de mercurio
mercury switch conmutador de mercurio
mercury tank tanque de mercurio
mercury thermostat termostato de mercurio
mercury vapor vapor de mercurio
mercury-vapor lamp lámpara de vapor de mercurio
mercury-vapor rectifier rectificador de vapor de mercurio
mercury-vapor thyratron tiratrón de vapor de mercurio
mercury-vapor tube tubo de vapor de mercurio
mercury-vapor valve válvula de vapor de mercurio
mercury-wetted contact relay relé de contactos humedecidos por mercurio
mercury-wetted reed relay relé de laminillas humedecidas por mercurio
mercury-wetted relay relé de contactos humedecidos por mercurio
merge fusión
merge file archivo de fusión
merge generator generador de fusión
merge level nivel de fusión
merge pass pasada de fusión
merge routine rutina de fusión
merge-sort fusión-clasificación
meridional ray rayo meridional
mesa mesa
mesa diode diodo mesa
mesa transistor transistor mesa
mesh malla
mesh circuit circuito de malla
mesh connection conexión de malla
mesh current corriente de malla
mesh equations ecuaciones de mallas
mesh network red de malla
mesh resistance resistencia de malla
mesh voltage tensión de malla
Mesny circuit circuito de Mesny
mesochronous mesócrono
message mensaje
message accumulation acumulación de mensajes

message authentication code código de autenticación de mensajes
message blocking bloqueo de mensajes
message buffering almacenamiento intermedio de mensajes
message channel canal de mensajes
message circuit circuito de mensajes
message concentrator concentrador de mensajes
message control control de mensajes
message control program programa de control de mensajes
message data datos de mensajes
message data set conjunto de datos de mensajes
message delete borrado de mensajes
message delete option opción de borrado de mensajes
message description descripción de mensajes
message-driven program programa accionado por mensajes
message element elemento de mensaje
message exchange intercambio de mensajes
message exchange unit unidad de intercambio de mensajes
message feedback retroalimentación de mensajes
message field campo de mensajes
message file archivo de mensajes
message format formato de mensajes
message handler manipulador de mensajes
message handling manejo de mensajes
message identification identificación de mensajes
message identification code código de identificación de mensajes
message identifier identificador de mensaje
message indicator indicador de mensaje
message interception intercepción de mensajes
message line línea de mensajes
message logging registro de mensajes
message priority prioridad de mensajes
message processing procesamiento de mensajes
message processing program programa de procesamiento de mensajes
message processing region región de procesamiento de mensajes
message queue cola de mensajes
message recovery recuperación de mensajes
message recovery point punto de recuperación de mensajes
message retransmission retransmisión de mensajes
message retrieval recuperación de mensajes
message routine rutina de mensajes
message source fuente de mensajes
message space espacio de mensajes
message switch conmutador de mensajes
message switching conmutación de mensajes
message switching center centro de conmutación de mensajes
message switching system sistema de conmutación de mensajes
message transfer rate velocidad de transferencia de mensajes
message transmission transmisión de mensajes
metacompiler metacompilador
metadata metadata
metadyne metadina
metal metal
metal antenna antena metálica
metal-arc welding soldadura por arco metálico
metal backing recubrimiento metálico interior
metal-base transistor transistor de base metálica
metal brush escobilla metálica
metal-ceramic metalocerámico
metal-ceramic construction construcción metalocerámica

metal-ceramic technique técnica metalocerámica
metal-clad blindado
metal-clad switch conmutador blindado
metal-clad switchgear conmutador blindado
metal-coated metalizado
metal-coated bulb ampolla metalizada
metal-cone tube tubo de cono metálico
metal deposition deposición metálica
metal electrode electrodo metálico
metal-enclosed apparatus aparato con cubierta metálica
metal-enclosed equipment equipo con cubierta metálica
metal-enclosed switch conmutador con cubierta metálica
metal-filament lamp lámpara de filamento metálico
metal film película metálica
metal-film diode diodo de película metálica
metal-film resistor resistor de película metálica
metal finder buscador de metales
metal fog neblina metálica
metal foil hoja metálica
metal lens lente metálica
metal locator ubicador de metales
metal master negativo original, original metálico
metal mist neblina metálica
metal negative negativo original, negativo metálico
metal-oxide film resistor resistor de película de metal-óxido
metal-oxide resistor resistor de metal-óxido
metal-oxide-semiconductor semiconductor metal-óxido
metal-oxide-semiconductor transistor transistor de semiconductor metal-óxido
metal-oxide-silicon silicio metal-óxido
metal-oxide-silicon capacitor capacitor de silicio metal-óxido
metal-oxide varistor varistor de metal-óxido
metal-plate rectifier rectificador de placa metálica
metal rectifier rectificador metálico
metal tube tubo metálico
metal valve válvula metálica
metal vapor vapor metálico
metal-vapor lamp lámpara de vapor metálico
metalanguage metalenguaje
metallic antenna lens lente de antena metálica
metallic binding forces fuerzas ligantes metálicas
metallic circuit circuito metálico
metallic crystal cristal metálico
metallic currents corrientes metálicas
metallic-disk rectifier rectificador de discos metálicos
metallic-film resistor resistor de película metálica
metallic insulator aislador metálico
metallic lens lente metálica
metallic noise ruido metálico
metallic rectifier rectificador metálico
metallic resistor resistor metálico
metallization metalización
metallize metalizar
metallized metalizado
metallized brush escobilla metalizada
metallized capacitor capacitor metalizado
metallized lamp lámpara metalizada
metallized-paper capacitor capacitor de papel metalizado
metallized-polycarbonate capacitor capacitor de policarbonato metalizado
metallized resistor resistor metalizado
metallized screen pantalla metalizada
metallized-screen tube tubo con pantalla metalizada
metallizing metalización
metalloid metaloide

metamer metámero
metascope metascopio
meteor-burst communications comunicaciones por estelas meteóricas
meteor-burst signals señales por estelas meteóricas
meteor echo eco meteórico
meteor ionization trail estela de ionización meteórica
meteor scatter dispersión meteórica
meteor trail estela meteórica
meteor-trail reflections reflexiones de estelas meteóricas
meteoric meteórico
meteoric ionization ionización meteórica
meteorograph meteorógrafo
meteorological meteorológico
meteorological conditions condiciones meteorológicas
meteorological information información meteorológica
meteorological instrumentation instrumentación meteorológica
meteorological station estación meteorológica
meter medidor, contador, metro
meter-ampere metro-ampere
meter amplifier amplificador de medidor
meter armature inducido de medidor
meter bracket soporte de medidor
meter-candle metro-bujía
meter case caja de medidor
meter circuit circuito de medidor
meter correction factor factor de corrección de medidor
meter cover tapa de medidor
meter display indicador visual de medidor
meter face cara de medidor
meter faceplate placa frontal de medidor
meter frame soporte de medidor
meter indication indicación de medidor
meter-kilogram-second metro-kilogramo-segundo
meter-kilogram-second system sistema metro-kilogramo-segundo
meter lamp lámpara de medidor
meter mechanism mecanismo de medidor
meter movement mecanismo de medidor
meter panel panel de medidor
meter protector protector de medidor
meter rating clasificación de medidor
meter reading lectura de medidor
meter readout lectura de medidor
meter rectifier rectificador de medidor
meter relay relé medidor
meter resistance resistencia de medidor
meter scale escala de medidor
meter scale factor factor de escala de medidor
meter sensitivity sensibilidad de medidor
meter shunt derivación de medidor
meter speed velocidad de medidor
meter support soporte de medidor
meter switch conmutador de medidor
meter time tiempo de medidor
meter torque par de medidor
meter-type modulator modulador de tipo de medidor
meter-type relay relé de tipo de medidor
meter wire hilo de medidor
metered circuit circuito medido
metered line línea medida
metered service servicio medido
metering medición
metering technique técnica de medición
method of control método de control
method of operation método de operación
metric métrico

metric band banda métrica
metric system sistema métrico
metric-wave broadcasting radiodifusión por ondas métricas
metric-wave measurement medida por ondas métricas
metric-wave propagation propagación de ondas métricas
metric waves ondas métricas
metrology metrología
metron metronio
metron capacity capacidad en metronios
metron density densidad en metronios
metronome metrónomo
mho mho
mica mica
mica capacitor capacitor de mica
mica condenser condensador de mica
mica dielectric capacitor capacitor de dieléctrico de mica
mica plate placa de mica
micanite micanita
micarta micarta
Michelson interferometer interferómetro de Michelson
micro micro
microalloy diffused transistor transistor de difusión microaleado
microalloy transistor transistor microaleado
microammeter microamperímetro
microammeter meter microamperímetro
microampere microampere
microanalyzer microanalizador
microarchitecture microarquitectura
microassembler microensamblador
microbalance microbalanza
microband microbanda
microbar microbar
microbarn microbarn
microbarograph microbarógrafo
microbeam microhaz
microcard microtarjeta
microchannel microcanal
microchannel card tarjeta de microcanal
microchip microchip
microcircuit microcircuito
microcircuit isolation aislamiento de microcircuitos
microcircuit wafer oblea de microcircuitos
microcircuitry microcircuitería
microcode microcódigo
microcode assembler ensamblador de microcódigo
microcode instruction instrucción de microcódigo
microcode instruction set conjunto de instrucciones de microcódigo
microcoding microcodificación
microcoding device dispositivo de microcodificación
microcomponent microcomponente
microcomputer microcomputadora
microcomputer addressing mode modo de direccionamiento de microcomputadora
microcomputer analyzer analizador de microcomputadora
microcomputer architecture arquitectura de microcomputadora
microcomputer card tarjeta de microcomputadora
microcomputer central processing unit unidad de procesamiento central de microcomputadora
microcomputer chip chip de microcomputadora
microcomputer classification clasificación de microcomputadora
microcomputer component componente de microcomputadora
microcomputer control panel panel de control de

microcomputadora
microcomputer control system sistema de control de microcomputadora
microcomputer disk operating system sistema operativo de disco de microcomputadora
microcomputer element elemento de microcomputadora
microcomputer execution cycle ciclo de ejecución de microcomputadora
microcomputer function función de microcomputadora
microcomputer input-output architecture arquitectura de entrada-salida de microcomputadora
microcomputer instrument instrumento de microcomputadora
microcomputer machine control control de máquina de microcomputadora
microcomputer system sistema de microcomputadora
microcomputer terminal terminal de microcomputadora
microcomputer timing module módulo de temporización de microcomputadora
microcontrol microcontrol
microcontrol microprogramming microprogramación de microcontrol
microcontroller microcontrolador
microcontroller interface interfaz de microcontrolador
microcontroller simulator simulador de microcontrolador
microcrystal microcristal
microcurie microcurie
microcycle microciclo
microdensitometer microdensitómetro
microdisk microdisco
microdiskette microdisquete
microelectric microeléctrico
microelectric device dispositivo microeléctrico
microelectrode microelectrodo
microelectrolysis microelectrólisis
microelectronic microelectrónico
microelectronic circuit circuito microelectrónico
microelectronic circuitry circuitería microelectrónica
microelectronic device dispositivo microelectrónico
microelectronic module módulo microelectrónico
microelectronics microelectrónica
microelectrophoresis microelectroforesis
microelectroscope microelectroscopio
microelement microelemento
microelement wafer oblea de microelemento
microenergy microenergía
microenergy logic circuit circuito lógico de microenergía
microfarad microfarad
microfaradmeter microfaradímetro
microfiche microficha
microfiche reader lectora de microfichas
microfilm micropelícula
microfilm cartridge cartucho de micropelícula
microfilm printer impresora de micropelícula
microflash microdestello
microflash lamp lámpara de microdestellos
microfloppy disk microdisco flexible
microform microforma
microform reader lectora de microformas
microfriction potentiometer potenciómetro de microfricción
microfunction microfunción
microgalvanometer microgalvanómetro
microgauss microgauss

microgram microgramo
micrographics micrográficos
micrography micrografía
microgroove microsurco
microgroove disk disco microsurco
microgroove record disco microsurco
microhenry microhenry
microhm microhm
microhmmeter microóhmetro
microimage microimagen
microinch micropulgada
microinstruction microinstrucción
microinstruction display visualización de microinstrucciones
microinstruction sequence secuencia de microinstrucciones
microlamp microlámpara
microleak microfuga
microliter microlitro
microlock microenganche
micrologic micrológica
micromachining micromaquinado
micromanipulator micromanipulador
micromesh micromalla
micrometer micrómetro
micrometric micrométrico
micrometry micrometría
micromho micromho
micromhometer micromhómetro
micromicro micromicro, pico
microminiature microminiatura
microminiature component componente microminiatura
microminiature lamp lámpara microminiatura
microminiature tube tubo microminiatura
micromodule micromódulo
micron micrón
micronics micrónica
microoperation microoperación
microoscillograph microoscilógrafo
microphenomenon microfenómeno
microphone micrófono
microphone adapter adaptador de micrófono
microphone amplifier amplificador de micrófono
microphone button botón de micrófono
microphone cable cable de micrófono
microphone cartridge cápsula de micrófono
microphone center centro de micrófono
microphone head cabeza de micrófono
microphone hiss siseo de micrófono
microphone hum zumbido de micrófono
microphone input entrada de micrófono
microphone jack jack de micrófono
microphone noise ruido de micrófono
microphone oscillator oscilador de micrófono
microphone preamplifier preamplificador de micrófono
microphone signal señal de micrófono
microphone test prueba de micrófono
microphone tester probador de micrófonos
microphone transformer transformador de micrófono
microphone transmitter transmisor de micrófono
microphonic microfónico
microphonic contact contacto microfónico
microphonic detector detector microfónico
microphonic effect efecto microfónico
microphonic noise ruido microfónico
microphonic resonance resonancia microfónica
microphonic transformer transformador microfónico
microphonic tube tubo microfónico
microphonics microfónica

microphonism microfonismo
microphonograph microfonógrafo
microphonoscope microfonoscopio
microphony microfonía
microphotograph microfotografía
microphotography microfotografía
microphotometer microfotómetro
microphysics microfísica
micropower micropotencia
micropower circuit circuito de micropotencia
microprobe microsonda
microprocessing microprocesamiento
microprocessing system sistema de microprocesamiento
microprocessing unit unidad de microprocesamiento
microprocessor microprocesador
microprocessor analyzer analizador de microprocesador
microprocessor architecture arquitectura de microprocesador
microprocessor assembler ensamblador de microprocesador
microprocessor assembler simulator simulador de ensamblador de microprocesador
microprocessor automation automatización de microprocesador
microprocessor cache memory memoria inmediata de microprocesador
microprocessor card tarjeta de microprocesador
microprocessor chip chip de microprocesador
microprocessor chip characteristics características de chip de microprocesador
microprocessor classification clasificación de microprocesador
microprocessor compiler compilador de microprocesador
microprocessor component componente de microprocesador
microprocessor configuration configuración de microprocesador
microprocessor controller controlador de microprocesador
microprocessor debugging program programa de depuración de microprocesador
microprocessor development system sistema de desarrollo de microprocesador
microprocessor instruction instrucción de microprocesador
microprocessor instruction set conjunto de instrucciones de microprocesador
microprocessor instrument instrumento de microprocesador
microprocessor intelligence inteligencia de microprocesador
microprocessor language assembler ensamblador de lenguaje de microprocesador
microprocessor memory interface interfaz de memoria de microprocesador
microprocessor modem módem de microprocesador
microprocessor monitor monitor de microprocesador
microprocessor register registro de microprocesador
microprocessor system sistema de microprocesador
microprocessor system analyzer analizador de sistema de microprocesador
microprocessor system organization organización de sistemas de microprocesador
microprocessor terminal terminal de microprocesador
microprogram microprograma
microprogram assembly language lenguaje de ensamblaje de microprograma
microprogram control control de microprograma

microprogram control function función de control de microprograma
microprogram control section sección de control de microprograma
microprogram control unit unidad de control de microprograma
microprogram debugging depuración de microprograma
microprogram development desarrollo de microprograma
microprogram display visualización de microprograma
microprogram emulation emulación de microprograma
microprogram field campo de microprograma
microprogram instruction set conjunto de instrucciones de microprograma
microprogram machine instructions instrucciones de máquina de microprograma
microprogrammable microprogramable
microprogrammable computer computadora microprogramable
microprogrammable instruction instrucción microprogramable
microprogrammable processor procesador microprogramable
microprogrammed microprogramado
microprogrammed control control microprogramado
microprogrammed processor procesador microprogramado
microprogrammed subroutine subrutina microprogramada
microprogramming microprogramación
microprogramming techniques técnicas de microprogramación
micropulsation micropulsación
microradar microrradar
microradiography microrradiografía
microradiometer microrradiómetro
microreflectometer microrreflectómetro
microroentgen microrroentgen
microroutine microrrutina
microscale microescala, microbalanza
microscope microscopio
microscopic microscópico
microscopic examination examen microscópico
microscopic state estado microscópico
microscopy microscopía
microsecond microsegundo
microsiemens microsiemens
microspacing microespaciamiento
microspark microchispa
microspectrophotometer microespectrofotómetro
microstrip microbanda
microstructure microestructura
microswitch microconmutador
microsyn microsin
microsystem microsistema
microsystems electronics electrónica de microsistemas
microtelephone microteléfono
microtron microtrón
microvolt microvolt
microvoltage microtensión
microvolter fuente de microtensiones
microvoltmeter microvóltmetro
microvolts per meter microvolts por metro
microwafer microoblea
microwatt microwatt
microwattage microwattaje
microwattmeter microwáttmetro
microwave microonda

microwave absorber absorbedor de microondas
microwave acoustics acústica de microondas
microwave antenna antena de microondas
microwave attenuator atenuador de microondas
microwave beacon radiofaro de microondas
microwave beam haz de microondas
microwave cavity cavidad de microondas
microwave cavity resonator resonador de cavidad de microondas
microwave channel canal de microondas
microwave circulator circulador de microondas
microwave coupler acoplador de microondas
microwave delay line línea de retardo de microondas
microwave direction finder radiogoniómetro de microondas
microwave discriminator discriminador de microondas
microwave dish antena parabólica de microondas
microwave early warning aviso temprano por microondas
microwave equipment equipo de microondas
microwave filter filtro de microondas
microwave frequencies hiperfrecuencias
microwave generation generación de microondas
microwave gyrator girador de microondas
microwave integrated circuit circuito integrado de microondas
microwave interferometer interferómetro de microondas
microwave interferometry interferometría de microondas
microwave lens lente de microondas
microwave link enlace por microondas
microwave-link system sistema de enlace por microondas
microwave logic circuit circuito lógico de microondas
microwave mirror espejo de microondas
microwave mixer mezclador de microondas
microwave multiplier phototube fototubo multiplicador de microondas
microwave noise generator generador de ruido de microondas
microwave optical system sistema óptico de microondas
microwave oscillator oscilador de microondas
microwave oven horno de microondas
microwave phototube fototubo de microondas
microwave radio link radioenlace por microondas
microwave radio relay radioenlace por microondas
microwave radio relay system sistema de radioenlace por microondas
microwave radiometer radiómetro de microondas
microwave radiometry radiometría de microondas
microwave range gama de microondas
microwave receiver receptor de microondas
microwave refractometer refractómetro de microondas
microwave region región de microondas
microwave relay radioenlace por microondas, relé de microondas
microwave relay link radioenlace por microondas
microwave repeater repetidor de microondas
microwave resonator resonador de microondas
microwave route ruta de microondas
microwave signal señal de microondas
microwave signal generator generador de señales de microondas
microwave sound channel canal de sonido de microondas
microwave spectroscopy espectroscopia de microondas
microwave spectrum espectro de microondas

microwave station estación de microondas
microwave system sistema de microondas
microwave terminal terminal de microondas
microwave tower torre de microondas
microwave transistor transistor de microondas
microwave transmission line línea de transmisión de microondas
microwave transmission path trayectoria de transmisión de microondas
microwave transmitter transmisor de microondas
microwave tube tubo de microondas
microwave turbulence turbulencia de microondas
microwave ultrasonics ultrasónica de hiperfrecuencias
midband centro de banda
midband frequency frecuencia de centro de banda
middle marker beacon radiobaliza intermedia
middle-range loudspeaker altavoz de registro medio
midfrequency frecuencia de centro de banda
midicomputer midicomputadora
midposition contact contacto de posición neutra
midrange speaker altavoz de registro medio
Mie scattering dispersión de Mie
migration migración
Miller amplifier amplificador de Miller
Miller bridge puente de Miller
Miller capacitance capacitancia de Miller
Miller circuit circuito de Miller
Miller effect efecto de Miller
Miller-effect amplifier amplificador de efecto de Miller
Miller integrator integrador de Miller
Miller oscillator oscilador de Miller
milli mili
milliammeter miliamperímetro
milliampere miliampere
milliampere-second miliampere-segundo
millibar milibar
millibarn milibarn
millicurie milicurie
millifarad milifarad
milligauss miligauss
milligram miligramo
millihenry milihenry
millilambert mililambert
milliliter mililitro
millimaxwell milimaxwell
millimeter milímetro
millimeter band banda milimétrica
millimeter wave onda milimétrica
millimeter-wave amplifier amplificador de ondas milimétricas
millimeter-wave tube tubo de ondas milimétricas
millimetric milimétrico
millimetric radar radar milimétrico
millimetric wave onda milimétrica
millimicro milimicro
millimole milimol
milliohm miliohm
milliohmmeter milióhmetro
millions of instructions per second millones de instrucciones por segundo
milliphot milifot
milliradian milirradián
milliroentgen milirroentgen
millirutherford milirutherford
millisecond milisegundo
millivolt milivolt
millivolt-ampere milivolt-ampere
millivolt potentiometer potenciómetro de milivolts
millivoltmeter milivóltmetro
millivolts per meter milivolts por metro
milliwatt miliwatt

milliwattmeter miliwáttmetro
mineral mineral
mineral detector detector de mineral
mineral oil aceite mineral
mineral-oil capacitor capacitor de aceite mineral
miniassembler miniensamblador
miniassembler program programa miniensamblador
miniature miniatura
miniature brush escobilla miniatura
miniature compander compresor-expansor miniatura
miniature component componente miniatura
miniature lamp lámpara miniatura
miniature microphone micrófono miniatura
miniature thyratron tiratrón miniatura
miniature tube tubo miniatura
miniature valve válvula miniatura
miniaturization miniaturización
miniaturize miniaturizar
miniaturized miniaturizado
miniaturized circuit circuito miniaturizado
miniaturized electronic equipment equipo
    electrónico miniaturizado
minicalculator minicalculadora
minicomputer minicomputadora
minicomputer concentrator concentrador de
    minicomputadora
minicomputer interface interfaz de minicomputadora
minicomputer system sistema de minicomputadora
minicomputer terminal terminal de
    minicomputadora
minidisk minidisco
minifloppy minidisco flexible
minigroove record disco minisurco
minimization minimización
minimum mínimo
minimum access acceso mínimo
minimum-access code código de acceso mínimo
minimum-access programming programación de
    acceso mínimo
minimum-access routine rutina de acceso mínimo
minimum amplitude amplitud mínima
minimum blowing current corriente de fusión
    mínima
minimum capacitance capacitancia mínima
minimum circuit-breaker cortacircuito de mínimo
minimum configuration configuración mínima
minimum contact contacto de mínimo
minimum current corriente mínima
minimum cutout cortacircuito de mínimo
minimum delay code código de retardo mínimo
minimum delay coding codificación de retardo
    mínimo
minimum delay programming programación de
    retardo mínimo
minimum demand demanda mínima
minimum detectable change cambio mínimo
    detectable
minimum detectable leak fuga mínima detectable
minimum detectable power potencia mínima
    detectable
minimum detectable signal señal mínima detectable
minimum discernible signal señal mínima
    discernible
minimum distance code código de distancia mínima
minimum equivalent equivalente mínimo
minimum firing power potencia mínima de disparo
minimum flashover voltage tensión mínima
    disruptiva
minimum fusing current corriente mínima de fusión
minimum illumination iluminación mínima
minimum interval intervalo mínimo
minimum ionization ionización mínima
minimum latency latencia mínima

minimum latency code código de latencia mínima
minimum latency coding codificación de latencia
    mínima
minimum latency programming programación de
    latencia mínima
minimum latency routine rutina de latencia mínima
minimum maintenance mantenimiento mínimo
minimum-make corrector corrector de cierre
    mínimo
minimum net loss pérdida neta mínima
minimum operate current corriente mínima de
    operación
minimum pause pausa mínima
minimum-phase network red de fase mínima
minimum resistance resistencia mínima
minimum run time tiempo mínimo de ejecución
minimum sampling frequency frecuencia mínima de
    muestreo
minimum signal level nivel de señal mínimo
minimum starting voltage tensión mínima de
    arranque
minimum-strength signal señal de intensidad
    mínima
minimum usable time tiempo mínimo utilizable
minimum wavelength longitud mínima de onda
minimum working net loss pérdida neta mínima de
    trabajo
miniperipheral miniperiférico
miniscope miniscopio
miniswitch miniconmutador
minitrack minipista
minometer minómetro
minor adjustment ajuste menor
minor arc arco menor
minor class field campo de clase menor
minor control control menor
minor control data datos de control menor
minor control field campo de control menor
minor cycle ciclo menor
minor face cara menor
minor failure falla menor
minor lobe lóbulo menor
minor loop bucle menor
minor node nodo menor
minor structure estructura menor
minor total total menor
minority carrier portador minoritario
minority-carrier injection inyección de portadores
    minoritarios
minority emitter emisor minoritario
Minter stereophonic system sistema estereofónico
    de Minter
minuend minuendo
minus charge carga negativa
minute minuto
mirror espejo
mirror antenna antena de espejo
mirror drum tambor de espejos
mirror effect efecto de espejo
mirror galvanometer galvanómetro de espejo
mirror-galvanometer oscillograph oscilógrafo de
    galvanómetro de espejo
mirror image imagen de espejo
mirror instrument instrumento de espejo
mirror interferometer interferómetro de espejo
mirror oscillograph oscilógrafo de espejo
mirror reading lectura por espejo
mirror reflection reflexión de espejo
mirror-reflection echo eco por reflexión de espejo
mirror scale escala de espejo
mirror-type meter medidor de tipo de espejo
mirror wheel rueda de espejos
misaligned head cabeza desalineada

misalignment desalineación
miscellaneous function función miscelánea
miscellaneous information información miscelánea
miscellaneous interception intercepción miscelánea
misfeed alimentación errónea
misfire falla de encendido
misleading precision precisión engañosa
misleading signal señal engañosa
mismatch desadaptación
mismatch factor factor de desadaptación
mismatch loss pérdida por desadaptación
mismatched impedances impedancias desadaptadas
misplace colocar mal
misprint error de impresión
misread error de lectura
missing-pulse factor factor de impulsos ausentes
mistor mistor
mix mezcla
mixed base base mixta
mixed-base notation notación de base mixta
mixed-base number número de base mixta
mixed-base numeration system sistema de
  numeración de base mixta
mixed-channel output salida de canales mixtos
mixed circuit circuito mixto
mixed coupling acoplamiento mixto
mixed data set conjunto de datos mixtos
mixed data string secuencia de datos mixtos
mixed environment ambiente mixto
mixed highs altas mixtas
mixed-light lamp lámpara de luz mixta
mixed list lista mixta
mixed modulation modulación mixta
mixed notation notación mixta
mixed number número mixto
mixed polarization polarización mixta
mixed polyelectrode potential potencial de
  polielectrodo mixto
mixed radix base mixta
mixed-radix notation notación de base mixta
mixed receiver receptor mixto
mixed reflection reflexión mixta
mixed route ruta mixta
mixed semiconductor semiconductor mixto
mixed service servicio mixto
mixed state estado mixto
mixed sweep barrido mixto
mixed traffic tráfico mixto
mixed transmission transmisión mixta
mixer mezclador
mixer amplifier amplificador mezclador
mixer circuit circuito mezclador
mixer coil bobina mezcladora
mixer condenser condensador mezclador
mixer crystal cristal mezclador
mixer diode diodo mezclador
mixer hexode hexodo mezclador
mixer klystron klistrón mezclador
mixer noise ruido de mezclador
mixer oscillator oscilador mezclador
mixer preamplifier preamplificador mezclador
mixer rectifier rectificador mezclador
mixer stage etapa mezcladora
mixer tube tubo mezclador
mixer valve válvula mezcladora
mixing amplifier amplificador mezclador
mixing circuit circuito mezclador
mixing condenser condensador mezclador
mixing device dispositivo mezclador
mixing microphone micrófono mezclador
mixing tube tubo mezclador
mixture mezcla
mnemonic mnemónico

mnemonic address dirección mnemónica
mnemonic address code código de dirección
  mnemónica
mnemonic code código mnemónico
mnemonic language lenguaje mnemónico
mnemonic name nombre mnemónico
mnemonic notation notación mnemónica
mnemonic operation code código de operación
  mnemónica
mnemonic symbol símbolo mnemónico
mobile móvil
mobile antenna antena móvil
mobile communications comunicaciones móviles
mobile communications equipment equipo de
  comunicaciones móviles
mobile communications system sistema de
  comunicaciones móviles
mobile radio service servicio de radio móvil
mobile receiver receptor móvil
mobile relay station estación retransmisora móvil
mobile service servicio móvil
mobile station estación móvil
mobile substation subestación móvil
mobile system sistema móvil
mobile telemetering telemedición móvil
mobile telephone system sistema telefónico móvil
mobile transmitter transmisor móvil
mobile unit unidad móvil
mobile unit substation subestación de unidad móvil
mobility movilidad
modal modal
modal analysis análisis modal
modal direction dirección modal
modal distance distancia modal
modal noise ruido modal
modal point punto modal
mode modo
mode change cambio de modo
mode change character carácter de cambio de modo
mode changer cambiador de modo
mode conversion conversión de modo
mode conversion loss pérdida por conversión de
  modo
mode coupling acoplamiento de modos
mode field campo de modo
mode filter filtro de modos
mode indicator indicador de modo
mode jump salto de modo
mode name nombre de modo
mode number número de modo
mode of emission modo de emisión
mode of failure modo de falla
mode of operation modo de operación
mode of oscillation modo de oscilación
mode of propagation modo de propagación
mode of reception modo de recepción
mode of resonance modo de resonancia
mode of vibration modo de vibración
mode purity pureza de modo
mode selector selector de modo
mode separation separación de modos
mode shift desplazamiento de modo
mode suppressor supresor de modo
mode switch conmutador de modo
mode transformer transformador de modo
mode width anchura de modo
model modelo
model statement sentencia modelo
modem módem
modem check comprobación de módem
modem check control control de comprobación de
  módem
modem communication comunicación por módem

modem control device dispositivo de control de módem
modem eliminator eliminador de módem
modem multiplexer multiplexor de módem
modem splitter divisor de módem
modem synchronization sincronización de módem
modem tester probador de módem
moderator moderador
modification modificación
modification level nivel de modificación
modification loop bucle de modificación
modified modificado
modified data datos modificados
modified index of refraction índice de refracción modificado
modified refractive index índice de refracción modificado
modifier modificador
modifier register registro modificador
modify modificar
moding falla por impureza de modo
modor modor
modular modular
modular circuit circuito modular
modular coding codificación modular
modular component componente modular
modular connector conector modular
modular construction construcción modular
modular converter convertidor modular
modular decomposition descomposición modular
modular design diseño modular
modular junction panel panel de uniones modular
modular processing procesamiento modular
modular program programa modular
modular programming programación modular
modular system sistema modular
modular technique técnica modular
modularity modularidad
modularity level nivel de modularidad
modularization modularización
modulate modular
modulated modulado
modulated amplification stage etapa de amplificación modulada
modulated amplifier amplificador modulado
modulated beam haz modulado
modulated-beam photoelectric system sistema fotoeléctrico de haz modulado
modulated carrier portadora modulada
modulated carrier wave onda portadora modulada
modulated color subcarrier subportadora de colores modulada
modulated continuous wave onda continua modulada
modulated current corriente modulada
modulated-current electroplating galvanoplastia por corriente modulada
modulated electron beam haz electrónico modulado
modulated emission emisión modulada
modulated ion beam haz iónico modulado
modulated light luz modulada
modulated light beam haz luminoso modulado
modulated light ray rayo luminoso modulado
modulated oscillation oscilación modulada
modulated oscillator oscilador modulado
modulated output stage etapa de salida modulada
modulated power potencia modulada
modulated pulse impulso modulado
modulated scattering dispersión modulada
modulated signal generator generador de señales moduladas
modulated stage etapa modulada
modulated telegraphy telegrafía modulada

modulated wave onda modulada
modulating modulador
modulating amplifier amplificador modulador
modulating anode ánodo modulador
modulating-anode klystron klistrón de ánodo modulador
modulating circuit circuito modulador
modulating current corriente moduladora
modulating electrode electrodo modulador
modulating frequency frecuencia moduladora
modulating power potencia moduladora
modulating power supply fuente de alimentación moduladora
modulating signal señal moduladora
modulating valve válvula moduladora
modulating wave onda moduladora
modulation modulación
modulation bandwidth ancho de banda de modulación
modulation bars barras de modulación
modulation capability capacidad de modulación
modulation characteristic característica de modulación
modulation checker comprobador de modulación
modulation code código de modulación
modulation coefficient coeficiente de modulación
modulation coherence coherencia de modulación
modulation condition condición de modulación
modulation contrast contraste de modulación
modulation cycle ciclo de modulación
modulation depth profundidad de modulación
modulation distortion distorsión de modulación
modulation element elemento de modulación
modulation-eliminator bridge puente eliminador de modulación
modulation envelope envolvente de modulación
modulation factor factor de modulación
modulation fidelity fidelidad de modulación
modulation frequency frecuencia de modulación
modulation-frequency range gama de frecuencias de modulación
modulation index índice de modulación
modulation level nivel de modulación
modulation limiter limitador de modulación
modulation limiting limitación de modulación
modulation linearity linealidad de modulación
modulation meter medidor de modulación
modulation monitor monitor de modulación
modulation noise ruido de modulación
modulation parameter parámetro de modulación
modulation peak pico de modulación
modulation percentage porcentaje de modulación
modulation rate rapidez de modulación
modulation ratio razón de modulación
modulation rise aumento de modulación
modulation section sección de modulación
modulation sideband banda lateral de modulación
modulation signal señal de modulación
modulation suppression supresión de modulación
modulation threshold umbral de modulación
modulation transformer transformador de modulación
modulation wave onda de modulación
modulator modulador
modulator-amplifier modulador-amplificador
modulator choke choque de modulador
modulator choke coil bobina de choque de modulador
modulator crystal cristal modulador
modulator-demodulator modulador-desmodulador
modulator driver excitador de modulador
modulator glow tube tubo luminiscente modulador
modulator grid rejilla moduladora

modulator power potencia moduladora
modulator relay relé modulador
modulator stage etapa moduladora
modulator tube tubo modulador
modulator valve válvula moduladora
module módulo
module check comprobación de módulo
module test prueba de módulo
modulo N check comprobación módulo N
modulometer modulómetro
modulus módulo
modulus of elasticity módulo de elasticidad
mogul base casquillo goliat
moire muaré
moist húmedo
moisture humedad
moisture absorption absorción de humedad
moisture content contenido de humedad
moisture indicator indicador de humedad
moisture meter humidímetro
moisture-proof a prueba de humedad
moisture-proof connector conector a prueba de
  humedad
moisture resistance resistencia a la humedad
moisture-resistant resistente a la humedad
moisture-resistant element elemento resistente a la
  humedad
moisture trap trampa de humedad
moistureproof a prueba de humedad, hidrófugo
moistureproof connector conector a prueba de
  humedad
moistureproof cord cordón a prueba de humedad
moistureproof cover cubierta a prueba de humedad
molality molalidad
molar conductance conductancia molar
molar solution solución molar
molarity molaridad
mold molde
molded moldeado
molded capacitor capacitor moldeado
molded case caja moldeada
molded ceramic capacitor capacitor cerámico
  moldeado
molded coil bobina moldeada
molded component componente moldeado
molded electrolytic capacitor capacitor electrolítico
  moldeado
molded flat capacitor capacitor plano moldeado
molded glass capacitor capacitor de vidrio moldeado
molded-in moldeado en la pieza
molded inductor inductor moldeado
molded mica capacitor capacitor de mica moldeado
molded paper capacitor capacitor de papel
  moldeado
molded porcelain capacitor capacitor de porcelana
  moldeada
molded resistor resistor moldeado
molded transistor transistor moldeado
molded tube capacitor capacitor de tubo moldeado
molding moldeado
molectronics electrónica molecular
molecular molecular
molecular amplifier amplificador molecular
molecular beam haz molecular
molecular-beam frequency standard patrón de
  frecuencia de haz molecular
molecular-beam oscillator oscilador de haz
  molecular
molecular circuit circuito molecular
molecular concentration concentración molecular
molecular conductance conductancia molecular
molecular conductivity conductividad molecular
molecular density densidad molecular

molecular device dispositivo molecular
molecular diagram diagrama molecular
molecular dilution dilución molecular
molecular disassociation disociación molecular
molecular electronics electrónica molecular
molecular energy level nivel de energía molecular
molecular flow flujo molecular
molecular flux flujo molecular
molecular free path trayectoria libre molecular
molecular frequency standard patrón de frecuencia
  molecular
molecular friction fricción molecular
molecular functional block bloque funcional
  molecular
molecular ion beam haz iónico molecular
molecular ion collision colisión iónica molecular
molecular lattice red molecular
molecular level nivel molecular
molecular mass masa molecular
molecular microwave amplifier amplificador de
  microondas molecular
molecular microwave spectroscopy espectroscopia
  por microondas molecular
molecular oscillator oscilador molecular
molecular repulsion repulsión molecular
molecular resistivity resistividad molecular
molecular rotation rotación molecular
molecular spectrum espectro molecular
molecular structure estructura molecular
molecular technique técnica molecular
molecular theory of magnetism teoría molecular del
  magnetismo
molecular velocity velocidad molecular
molecular vibration vibración molecular
molecular weight peso molecular
molecule molécula
moleculoy moleculoy
molybdenum molibdeno
moment momento
momentary action acción momentánea
momentary-action switch conmutador de acción
  momentánea
momentary contact contacto momentáneo
momentary-contact action acción de contacto
  momentáneo
momentary-contact switch conmutador de contacto
  momentáneo
momentary current corriente momentánea
momentary disturbance perturbación momentánea
momentary drainage drenaje momentáneo
momentary flow flujo momentáneo
momentary interruption interrupción momentánea
momentary load carga momentánea
momentary output salida momentánea
momentary overload sobrecarga momentánea
momentary-overload relay relé de sobrecarga
  momentánea
momentary push-button pulsador momentáneo
momentary rating clasificación momentánea
momentary surge sobretensión momentánea
momentary switching conmutación momentánea
monadic monádico
monadic Boolean operator operador de Boole
  monádico
monadic operation operación monádica
monadic operator operador monádico
monatomic monoatómico
monatomic layer capa monoatómica
monaural monoaural
monaural amplifier amplificador monoaural
monaural preamplifier preamplificador monoaural
monaural recorder registrador monoaural
monaural sound system sistema de sonido

monoaural
monaxial monoaxial
monel metal metal monel
moniscope moniscopio
monitor monitor
monitor amplifier amplificador monitor
monitor antenna antena monitora
monitor console consola monitora
monitor desk pupitre monitor
monitor earphone audífono monitor
monitor head cabeza monitora
monitor ionization chamber cámara de ionización monitora
monitor line línea monitora
monitor loudspeaker altavoz monitor
monitor meter medidor monitor
monitor mode modo monitor
monitor panel panel monitor
monitor program programa monitor
monitor receiver receptor monitor
monitor routine rutina monitora
monitor set aparato monitor
monitor speaker altavoz monitor
monitor system sistema monitor
monitor terminal terminal monitora
monitor unit unidad monitora
monitoring monitoreo, monitorización
monitoring alarm alarma de monitoreo
monitoring amplifier amplificador de monitoreo
monitoring antenna antena de monitoreo
monitoring channel canal de monitoreo
monitoring circuit circuito de monitoreo
monitoring desk pupitre de monitoreo
monitoring device dispositivo de monitoreo
monitoring earphone audífono de monitoreo
monitoring element elemento de monitoreo
monitoring key llave de monitoreo
monitoring loudspeaker altavoz de monitoreo
monitoring meter medidor de monitoreo
monitoring oscilloscope osciloscopio de monitoreo
monitoring panel panel de monitoreo
monitoring program programa de monitoreo
monitoring receiver receptor de monitoreo
monitoring station estación de monitoreo
monitoring system sistema de monitoreo
monitoring tube tubo de monitoreo
monkey chatter modulación cruzada
monoanodic monoanódico
monoanodic rectifier rectificador monoanódico
monocell monocélula
monochromatic monocromático
monochromatic aberration aberración monocromática
monochromatic component componente monocromático
monochromatic emissivity emisividad monocromática
monochromatic filter filtro monocromático
monochromatic illuminator iluminador monocromático
monochromatic light luz monocromática
monochromatic picture imagen monocromática
monochromatic radiation radiación monocromática
monochromatic ray rayo monocromático
monochromatic sensitivity sensibilidad monocromática
monochromatic spectrum espectro monocromático
monochromaticity monocromaticidad
monochromator monocromador
monochrome monocromo
monochrome channel canal monocromo
monochrome display adapter adaptador de visualizador monocromo

monochrome display device dispositivo de visualización monocroma
monochrome frequency frecuencia monocroma
monochrome image imagen monocroma
monochrome information información monocroma
monochrome monitor monitor monocromo
monochrome picture imagen monocroma
monochrome picture tube tubo de imagen monocroma
monochrome receiver receptor monocromo
monochrome reception recepción monocroma
monochrome set aparato monocromo
monochrome signal señal monocroma
monochrome television televisión monocroma
monochrome transmission transmisión monocroma
monochrome video signal videoseñal monocroma
monochromic monocromo
monoclinic monoclínico
monoclinic crystal cristal monoclínico
monoclinic system sistema monoclínico
monocrystal monocristal
monocrystalline material material monocristalino
monocrystallization monocristalización
monode monodo
monoenergetic monoenergético
monoenergetic beam haz monoenergético
monoenergetic electrons electrones monoenergéticos
monoenergetic ion beam haz iónico monoenergético
monoenergetic radiation radiación monoenergética
monofier monofier
monogroove stereophonic recording registro estereofónico monosurco
monoimpulse monoimpulso
monokinetic monocinético
monolayer monocapa
monolithic monolítico
monolithic capacitor capacitor monolítico
monolithic chip chip monolítico
monolithic circuit circuito monolítico
monolithic conduit conducto monolítico
monolithic integrated circuit circuito integrado monolítico
monolithic storage almacenamiento monolítico
monolithic technology tecnología monolítica
monometallic monometálico
monomial monomio
monomolecular monomolecular
monomolecular film película monomolecular
monomolecular layer capa monomolecular
monophonic monofónico
monophonic receiver receptor monofónico
monophonic reception recepción monofónica
monophonic record disco monofónico
monophonic recorder registrador monofónico
monophonic sound system sistema de sonido monofónico
monophonic system sistema monofónico
monophony monofonía
monopiece monopieza
monopolar monopolar
monopolar electrode electrodo monopolar
monopolar electrode system sistema de electrodos monopolar
monopole monopolo
monopole antenna antena monopolo
monoprogramming monoprogramación
monopulse monoimpulso
monopulse radar radar de monoimpulsos
monorange speaker altavoz de gama completa
monoscope monoscopio
monoscopic monoscópico
monoscopic camera cámara monoscópica
monoscopic generator generador monoscópico

monospacing monoespaciamiento
monostable monoestable
monostable blocking oscillator oscilador de bloqueo monoestable
monostable circuit circuito monoestable
monostable device dispositivo monoestable
monostable multivibrator multivibrador monoestable
monostable timer temporizador monoestable
monostable trigger circuit circuito de disparo monoestable
monostatic monostático
monostatic radar radar monostático
monostatic reflectivity reflectividad monostática
monostatic reflector reflector monostático
monotonicity monotonicidad
monotron monotrón
monovalent monovalente
moon bounce rebote lunar
Moore code código de Moore
Moore lamp lámpara de Moore
Morse circuit circuito de Morse
Morse code código de Morse
Morse code signal señal de código de Morse
Morse receiver receptor de Morse
Morse relay relé de Morse
Morse set aparato de Morse
Morse signal señal de Morse
Morse telegraph telégrafo de Morse
Morse telegraphy telegrafía de Morse
mosaic mosaico
mosaic electrode electrodo de mosaico
mosaic screen pantalla mosaico
mosaic telegraphy telegrafía mosaico
Moseley's law ley de Moseley
Mossbauer effect efecto de Mossbauer
most significant bit bit más significativo
most significant character carácter más significativo
most significant digit dígito más significativo
motherboard tablero matriz, placa madre
motion moción
motion detector detector de moción
motion frequency frecuencia natural
motion indicator indicador de moción
motion-picture pickup transductor cinematográfico
motion-picture projector proyector cinematográfico
motional electric field campo eléctrico cinético
motional energy energía cinética
motional field campo cinético
motional impedance impedancia cinética
motive power fuerza motriz
motive unit unidad motriz
motor motor
motor amplifier amplificador de motor
motor armature inducido de motor
motor-board bloque magnetofónico, tablero de motor
motor brush escobilla de motor
motor capacitor capacitor de motor
motor condenser condensador de motor
motor control control de motor
motor control circuit circuito de control de motor
motor control mechanism mecanismo de control de motor
motor control switch conmutador de control de motor
motor control unit unidad de control de motor
motor converter convertidor de motor
motor-driven accionado por motor
motor-driven actuator actuador accionado por motor
motor-driven relay relé accionado por motor
motor effect efecto de motor
motor efficiency eficiencia de motor

motor field campo de motor
motor-field control control de campo de motor
motor-field regulator regulador de campo de motor
motor-generator motor-generador
motor-generator set grupo convertidor
motor meter medidor de motor
motor-operated operado por motor
motor-operated control control operado por motor
motor-run capacitor capacitor de funcionamiento de motor
motor-speed control control de velocidad de motor
motor-start capacitor capacitor de arranque de motor
motor-start relay relé de arranque de motor
motor-starter arrancador de motor
motor switch conmutador de motor
motor torque par motor
motor-type relay relé de tipo de motor
motor unit unidad de motor
motor ventilator ventilador de motor
motorboard bloque magnetofónico, tablero de motor
motorboating ruido de motor fuera de borda, ruido de canoa
motorized motorizado
motorized keyboard teclado motorizado
motorized switch conmutador motorizado
Mott scattering formula fórmula de dispersión de Mott
mount montura, soporte
mountain effect efecto de montaña
mounting montaje
mounting assembly ensamblaje de montaje, conjunto de montaje
mounting block bloque de montaje
mounting board tablero de montaje
mounting bracket soporte de montaje
mounting clip abrazadera de montaje
mounting diagram diagrama de montaje
mounting frame marco de montaje
mounting hardware equipo físico de montaje
mounting hole agujero de montaje
mounting layout distribución de montaje
mounting plate placa de montaje
mounting position posición de montaje
mounting shelf estante de montaje
mounting stand estante de montaje
mounting strip regleta de montaje
mounting tray bandeja de montaje
mouse ratón
mouthpiece boquilla, bocina
movable móvil
movable-anode tube tubo de ánodo móvil
movable antenna antena móvil
movable-coil galvanometer galvanómetro de bobina móvil
movable-coil meter medidor de bobina móvil
movable-coil modulator modulador de bobina móvil
movable contact contacto móvil
movable-disk relay relé de disco móvil
movable element elemento móvil
movable-head disk disco de cabeza móvil
movable-head disk unit unidad de disco de cabeza móvil
movable-iron meter medidor de hierro móvil
movement movimiento, mecanismo
moving-armature loudspeaker altavoz de armadura móvil
moving beam haz móvil
moving coil bobina móvil
moving-coil ammeter amperímetro de bobina móvil
moving-coil galvanometer galvanómetro de bobina móvil
moving-coil hydrophone hidrófono de bobina móvil

moving-coil instrument instrumento de bobina móvil
moving-coil loudspeaker altavoz de bobina móvil
moving-coil mechanism mecanismo de bobina móvil
moving-coil meter medidor de bobina móvil
moving-coil microphone micrófono de bobina móvil
moving-coil motor motor de bobina móvil
moving-coil movement mecanismo de bobina móvil
moving-coil pickup fonocaptor de bobina móvil
moving-coil regulator regulador de bobina móvil
moving-coil relay relé de bobina móvil
moving-coil speaker altavoz de bobina móvil
moving-coil voltmeter vóltmetro de bobina móvil
moving conductor conductor móvil
moving-conductor hydrophone hidrófono de
   conductor móvil
moving-conductor loudspeaker altavoz de conductor
   móvil
moving-conductor microphone micrófono de
   conductor móvil
moving-conductor receiver receptor de conductor
   móvil
moving contact contacto móvil
moving element elemento móvil
moving field campo móvil
moving grid rejilla móvil
moving-head disk disco de cabeza móvil
moving iron hierro móvil
moving-iron ammeter amperímetro de hierro móvil
moving-iron instrument instrumento de hierro móvil
moving-iron loudspeaker altavoz electromagnético
moving-iron meter medidor de hierro móvil
moving-iron microphone micrófono
   electromagnético
moving-iron receiver receptor de hierro móvil
moving-iron voltmeter vóltmetro de hierro móvil
moving load carga móvil
moving magnet imán móvil
moving-magnet galvanometer galvanómetro de imán
   móvil
moving-magnet instrument instrumento de imán
   móvil
moving-magnet magnetometer magnetómetro de
   imán móvil
moving mirror espejo móvil
moving-mirror oscillograph oscilógrafo de espejo
   móvil
moving needle aguja móvil
moving-needle galvanometer galvanómetro de aguja
   móvil
moving probe sonda móvil
moving unit unidad móvil
moving vane paleta móvil
moving-vane meter medidor de paleta móvil
moving wave onda móvil
mu mu
mu circuit circuito mu
mu factor factor mu
Muller tube tubo de Muller
multiaccess multiacceso
multiaccess system sistema multiacceso
multiaddress multidirección
multiaddress instruction instrucción multidirección
multiaddress message mensaje multidirección
multianode multiánodo
multianode rectifier rectificador multiánodo
multianode tube tubo multiánodo
multiaperture multiabertura
multiaperture device dispositivo multiabertura
multiar multiar
multiband multibanda
multiband amplifier amplificador multibanda
multiband antenna antena multibanda
multiband device dispositivo multibanda

multiband oscilloscope osciloscopio multibanda
multiband receiver receptor multibanda
multibeam multihaz
multibeam antenna antena multihaz
multibeam cathode-ray oscilloscope osciloscopio de
   rayos catódicos multihaz
multibeam oscilloscope osciloscopio multihaz
multibeam tube tubo multihaz
multibreak circuit-breaker cortacircuito
   multiruptura
multicathode multicátodo
multicathode counter tube tubo contador
   multicátodo
multicathode tube tubo multicátodo
multicavity multicavidad
multicavity magnetron magnetrón multicavidad
multicellular multicelular
multicellular horn bocina multicelular
multicellular loudspeaker altavoz multicelular
multichannel multicanal
multichannel amplifier amplificador multicanal
multichannel analyzer analizador multicanal
multichannel antenna antena multicanal
multichannel carrier portadora multicanal
multichannel circuit circuito multicanal
multichannel coaxial cable cable coaxial multicanal
multichannel counter contador multicanal
multichannel crystal oscillator oscilador de cristal
   multicanal
multichannel equipment equipo multicanal
multichannel field-effect transistor transistor de
   efecto de campo multicanal
multichannel link enlace multicanal
multichannel loudspeaker altavoz multicanal
multichannel operation operación multicanal
multichannel oscillograph oscilógrafo multicanal
multichannel potentiometer potenciómetro
   multicanal
multichannel pulse analyzer analizador de impulsos
   multicanal
multichannel radio-frequency tuner sintonizador de
   radiofrecuencia multicanal
multichannel radio link radioenlace multicanal
multichannel radio relay radioenlace multicanal
multichannel radio relay system sistema de
   radioenlace multicanal
multichannel radio transmitter radiotransmisor
   multicanal
multichannel receiver receptor multicanal
multichannel remote control control remoto
   multicanal
multichannel repeater repetidor multicanal
multichannel signal señal multicanal
multichannel spectrograph espectrógrafo multicanal
multichannel stereophonic system sistema
   estereofónico multicanal
multichannel stereophony estereofonía multicanal
multichannel system sistema multicanal
multichannel telegraphy telegrafía multicanal
multichannel teletype teletipo multicanal
multichannel television televisión multicanal
multichannel terminal terminal multicanal
multichannel transmitter transmisor multicanal
multichip multichip
multichip circuit circuito multichip
multichip integrated circuit circuito integrado
   multichip
multichrome multicromo
multicircuit multicircuito
multicoil relay relé multibobina
multicomponent multicomponente
multicomponent signal señal multicomponente
multicomputer multicomputadora

multicomputer system sistema multicomputadora
multicon multicón
multiconductor multiconductor
multiconductor cable cable multiconductor
multicontact multicontacto
multicontact connector conector multicontacto
multicontact switch conmutador multicontacto
multicoupler multiacoplador
multicurrent multicorriente
multicycle multiciclo
multidimensional multidimensional
multidimensional coding system sistema de
　codificación multidimensional
multidimensional magnetic memory memoria
　magnética multidimensional
multidimensional modulation modulación
　multidimensional
multidimensional system sistema multidimensional
multidirectional multidireccional
multidirectional radio transmission
　radiotransmisión multidireccional
multielectrode multielectrodo
multielectrode tube tubo multielectrodo
multielectrode valve válvula multielectrodo
multielement multielemento
multielement antenna antena multielemento
multielement detector detector multielemento
multielement tube tubo multielemento
multiemitter multiemisor
multiemitter transistor transistor multiemisor
multifiber multifibra
multifiber cable cable multifibra
multifiber joint junta multifibra
multifilament multifilamento
multifilament wire hilo multifilamento
multifile multiarchivo
multifile sorting clasificación multiarchivo
multifile tape cinta multiarchivo
multiflow multiflujo
multifocal multifocal
multifrequency multifrecuencia
multifrequency antenna antena multifrecuencia
multifrequency code código multifrecuencia
multifrequency dipole dipolo multifrecuencia
multifrequency heterodyne generator generador
　heterodino multifrecuencia
multifrequency pulse impulso multifrecuencia
multifrequency pulsing current corriente pulsante
　multifrecuencia
multifrequency signaling señalización
　multifrecuencia
multifrequency system sistema multifrecuencia
multifrequency test prueba multifrecuencia
multifrequency transmitter transmisor
　multifrecuencia
multifunction multifunción
multifunction board tablero multifunción
multifunction module módulo multifunción
multifunction terminal terminal multifunción
multigenerational tape cinta multigeneracional
multigrid tube tubo multirrejilla
multigrounded con múltiples conexiones a tierra
multigrounded line línea con múltiples conexiones a
　tierra
multigroup multigrupo
multigun multicañón
multigun cathode-ray tube tubo de rayos catódicos
　multicañón
multigun tube tubo multicañón
multihole multiagujero, multiperforación
multihole coupler acoplador multiagujero
multihole directional coupler acoplador direccional
　multiagujero

multihop propagation propagación por saltos
　múltiples
multihop propagation system sistema de
　propagación por saltos múltiples
multihop transmission transmisión por saltos
　múltiples
multijunction multiunión
multijunction device dispositivo multiunión
multilayer multicapa
multilayer board tablero multicapa
multilayer circuit circuito multicapa
multilayer coil bobina multicapa
multilayer dielectric reflector reflector dieléctrico
　multicapa
multilayer filter filtro multicapa
multilayer winding arrollamiento multicapa
multileaf multihoja
multilevel multinivel
multilevel access acceso multinivel
multilevel action acción multinivel
multilevel address dirección multinivel
multilevel addressing direccionamiento multinivel
multilevel circuit circuito multinivel
multilevel information channel canal de información
　multinivel
multilevel security seguridad multinivel
multilevel sort clasificación multinivel
multilevel storage almacenamiento multinivel
multiline multilínea
multiline controller controlador multilínea
multiline microprocessor microprocesador
　multilínea
multilink multienlace
multiloop multibucle
multimesh multimalla
multimesh filter filtro multimalla
multimesh network red multimalla
multimeter multímetro
multimodal multimodal
multimode multimodo
multimode distortion distorsión multimodo
multimode fiber fibra multimodo
multimode horn bocina multimodo
multimode interferometer interferómetro multimodo
multimode laser láser multimodo
multimode operation operación multimodo
multimode resonator resonador multimodo
multimode waveguide guíaondas multimodo
multioperational multioperacional
multioutlet multitomacorriente, toma de corriente
　múltiple
multioutput multisalida
multipactor multipactor
multipactor rectifier rectificador multipactor
multipair multipar
multipair cable cable de pares múltiples
multipass multipasada
multipath cancellation cancelación por trayectorias
　múltiples
multipath delay retardo por trayectorias múltiples
multipath distortion distorsión por trayectorias
　múltiples
multipath effect efecto de trayectorias múltiples
multipath error error de trayectorias múltiples
multipath propagation propagación por trayectorias
　múltiples
multipath reception recepción por trayectorias
　múltiples
multipath transmission transmisión por trayectorias
　múltiples
multiphase multifásico
multiphase alternator alternador multifásico
multiphase current corriente multifásica

multiphase generator generador multifásico
multiphase program programa multifásico
multiphase system sistema multifásico
multiple múltiple
multiple access acceso múltiple
multiple-access computer computadora de acceso
     múltiple
multiple-access modulation modulación de acceso
     múltiple
multiple-access network red de acceso múltiple
multiple-access satellite system sistema de satélite
     de acceso múltiple
multiple-access system sistema de acceso múltiple
multiple-access virtual machine máquina virtual de
     acceso múltiple
multiple address dirección múltiple
multiple-address code código de direcciones
     múltiples
multiple-address instruction instrucción de
     direcciones múltiples
multiple-address message mensaje de direcciones
     múltiples
multiple antenna antena múltiple
multiple apparatus aparato múltiple
multiple assembly ensamblaje múltiple
multiple assignment asignación múltiple
multiple-bay antenna antena de secciones múltiples
multiple-beam klystron klistrón de haces múltiples
multiple break ruptura múltiple
multiple-break contacts contactos de ruptura
     múltiple
multiple bus structure estructura de bus múltiple
multiple call llamada múltiple
multiple-cavity magnetron magnetrón de cavidades
     múltiples
multiple chamber cámara múltiple
multiple channel canal múltiple
multiple-channel amplifier amplificador multicanal
multiple-channel oscilloscope osciloscopio
     multicanal
multiple-channel system sistema multicanal
multiple-chip circuit circuito multichip
multiple circuit circuito múltiple
multiple-coil relay relé de bobinas múltiples
multiple computer system sistema de computadoras
     múltiples
multiple conductor conductor múltiple
multiple-conductor cable cable de conductores
     múltiples
multiple connection conexión múltiple
multiple connector conector múltiple
multiple console consola múltiple
multiple contact contacto múltiple
multiple-contact relay relé de contactos múltiples
multiple-contact switch conmutador de contactos
     múltiples
multiple control control múltiple
multiple converter convertidor múltiple
multiple course rumbo múltiple
multiple coverage cobertura múltiple
multiple crosstalk diafonía múltiple
multiple-current generator generador de corrientes
     múltiples
multiple digit selection selección de dígitos múltiples
multiple duct conducto múltiple
multiple echo eco múltiple
multiple electrode electrodo múltiple
multiple-electrode welding soldadura por electrodos
     múltiples
multiple electrometer electrómetro múltiple
multiple error error múltiple
multiple excitation excitación múltiple
multiple-feedback amplifier amplificador de

     retroalimentación múltiple
multiple feeder alimentador múltiple
multiple field campo múltiple
multiple file archivo múltiple
multiple-file tape cinta de archivo múltiple
multiple-grid tube tubo de rejillas múltiples
multiple-grid valve válvula de rejillas múltiples
multiple-hop transmission transmisión por saltos
     múltiples
multiple image imagen múltiple
multiple integral integral múltiple
multiple interrupt interrupción múltiple
multiple ionization ionización múltiple
multiple jack jack múltiple
multiple-job processing procesamiento de trabajos
     múltiples
multiple joint junta múltiple
multiple-length number número de longitud múltiple
multiple-line printing impresión de líneas múltiples
multiple-line telephone teléfono de líneas múltiples
multiple-loop feedback system sistema de
     retroalimentación de bucle múltiple
multiple-loudspeaker system sistema de altavoces
     múltiples
multiple-marker generator generador de marcas
     múltiples
multiple message mode modo de mensajes múltiples
multiple metering medición múltiple
multiple modulation modulación múltiple
multiple monitor system sistema de monitores
     múltiples
multiple parallel winding arrollamiento paralelo
     múltiple
multiple-pass printing impresión por pasadas
     múltiples
multiple path trayectoria múltiple
multiple plug clavija múltiple
multiple precision precisión múltiple
multiple-precision arithmetic aritmética de precisión
     múltiple
multiple process proceso múltiple
multiple process operating system sistema operativo
     de procesos múltiples
multiple programming programación múltiple
multiple punching perforación múltiple
multiple-purpose de funciones múltiples
multiple-purpose meter medidor de funciones
     múltiples
multiple-purpose tester probador de funciones
     múltiples
multiple-ratio transformer transformador de
     razones múltiples
multiple reception recepción múltiple
multiple rectifier rectificador múltiple
multiple rectifier circuit circuito rectificador
     múltiple
multiple-reflection echo eco de reflexiones múltiples
multiple-reflection losses pérdidas por reflexiones
     múltiples
multiple resonance resonancia múltiple
multiple-resonant line línea resonante múltiple
multiple rhombic antenna antena rómbica múltiple
multiple routing encaminamiento múltiple
multiple scanning exploración múltiple
multiple scattering dispersión múltiple
multiple selection selección múltiple
multiple sound track pista de sonido múltiple
multiple speakers altavoces múltiples
multiple-speed motor motor de velocidades
     múltiples
multiple-spot scanning exploración por puntos
     múltiples
multiple-spot welder soldadora de puntos múltiples

multiple-spot welding soldadura por puntos múltiples
multiple string processing procesamiento de secuencias múltiples
multiple switch conmutador múltiple
multiple switchboard cuadro conmutador múltiple
multiple switching conmutación múltiple
multiple system sistema múltiple
multiple-tariff meter medidor de tarifas múltiples
multiple terminal terminal múltiple
multiple-track recording registro en pistas múltiples
multiple transformer transformador múltiple
multiple transmission transmisión múltiple
multiple transmitter transmisor múltiple
multiple tube tubo múltiple
multiple tuner sintonizador múltiple
multiple tuning sintonización múltiple
multiple-twin cable cable de pares gemelos
multiple-twin quad cuadrete de dos cables de pares gemelos
multiple-unit antenna antena de unidades múltiples
multiple-unit capacitor capacitor de unidades múltiples
multiple-unit condenser condensador de unidades múltiples
multiple-unit control control de unidades múltiples
multiple-unit semiconductor device dispositivo semiconductor de unidades múltiples
multiple-unit steerable antenna antena orientable de unidades múltiples
multiple-unit tube tubo de unidades múltiples
multiple-unit valve válvula de unidades múltiples
multiple-use de usos múltiples
multiple-use card tarjeta de usos múltiples
multiple-user system sistema de usuarios múltiples
multiple winding arrollamiento múltiple
multiple-wire antenna antena de hilos múltiples
multiplex múltiplex
multiplex adapter adaptador múltiplex
multiplex apparatus aparato múltiplex
multiplex channel canal múltiplex
multiplex communication comunicación múltiplex
multiplex data station estación de datos múltiplex
multiplex data terminal terminal de datos múltiplex
multiplex device dispositivo múltiplex
multiplex equipment equipo múltiplex
multiplex interface interfaz múltiplex
multiplex link enlace múltiplex
multiplex mode modo múltiplex
multiplex modulation modulación múltiplex
multiplex operation operación múltiplex
multiplex stereophonic broadcasting system sistema de radiodifusión estereofónica múltiplex
multiplex system sistema múltiplex
multiplex telegraphy telegrafía múltiplex
multiplex telephony telefonía múltiplex
multiplex transmission transmisión múltiplex
multiplex transmission system sistema de transmisión múltiplex
multiplex transmitter transmisor múltiplex
multiplex winding arrollamiento múltiplex
multiplexed system sistema múltiplex
multiplexer multiplexor
multiplexing multiplexión
multiplicand multiplicando
multiplication device dispositivo de multiplicación
multiplication factor factor de multiplicación
multiplication point punto de multiplicación
multiplication shift desplazamiento de multiplicación
multiplication time tiempo de multiplicación
multiplier multiplicador
multiplier amplifier amplificador multiplicador
multiplier factor factor multiplicador
multiplier photocell fotocélula multiplicadora

multiplier phototube fototubo multiplicador
multiplier potentiometer potenciómetro multiplicador
multiplier prefix prefijo multiplicador
multiplier probe sonda multiplicadora
multiplier quotient cociente multiplicador
multiplier-quotient register registro de cociente multiplicador
multiplier register registro multiplicador
multiplier resistance resistencia multiplicadora
multiplier stage etapa multiplicadora
multiplier tube tubo multiplicador
multiplier unit unidad multiplicadora
multiply-divide instruction instrucción de multiplicar-dividir
multiply operation operación de multiplicación
multiplying multiplicador
multiplying medium medio multiplicador
multiplying power poder multiplicador
multiplying punch perforadora multiplicadora
multiplying servo servomotor multiplicador
multiplying system sistema multiplicador
multiplying winding arrollamiento multiplicador
multipoint multipunto
multipoint circuit circuito multipunto
multipoint connection conexión multipunto
multipoint line línea multipunto
multipoint link enlace multipunto
multipoint network red multipunto
multipoint switch conmutador multipunto
multipoint welder soldadora multipunto
multipolar multipolar
multipolar cutout cortacircuito multipolar
multipolar fuse fusible multipolar
multipolar generator generador multipolar
multipolar machine máquina multipolar
multipolar motor motor multipolar
multipolar switch conmutador multipolar
multipole multipolar
multipole breaker cortacircuito multipolar
multipole moment momento multipolar
multipole switch conmutador multipolar
multiport multipuerto, multipuerta
multiport component componente multipuerto
multiport memory memoria multipuerto
multiport network red multipuerto
multiposition multiposición
multiposition action acción multiposición
multiposition relay relé multiposición
multiposition selector switch conmutador selector multiposición
multiposition switch conmutador multiposición
multiprecision multiprecisión
multiprocess multiproceso
multiprocessing multiprocesamiento
multiprocessing operation operación de multiprocesamiento
multiprocessing organization organización de multiprocesamiento
multiprocessing system sistema de multiprocesamiento
multiprocessor multiprocesador
multiprocessor system sistema multiprocesador
multiprogram multiprograma
multiprogram computer computadora multiprograma
multiprogramming multiprogramación
multiprogramming executive system sistema ejecutivo de multiprogramación
multiprogramming interrupt interrupción de multiprogramación
multiprogramming priority prioridad de multiprogramación

multiprogramming system   sistema de multiprogramación
multipunching   multiperforación
multipurpose   de funciones múltiples
multipurpose meter   medidor de funciones múltiples
multipurpose probe   sonda de funciones múltiples
multipurpose tester   probador de funciones múltiples
multipurpose tube   tubo de funciones múltiples
multirange   multiescala, multigama
multirange ammeter   amperímetro multiescala
multirange amplifier   amplificador multigama
multirange instrument   instrumento multiescala
multirange meter   medidor multiescala
multirange preamplifier   preamplificador multigama
multirange voltmeter   vóltmetro multiescala
multisection   multisección
multisection capacitor   capacitor multisección
multisection coil   bobina multisección
multisection filter   filtro multisección
multisegment   multisegmento
multisegment magnetron   magnetrón multisegmento
multisensor   multisensor
multisequence   multisecuencia
multiskip transmission   transmisión por saltos múltiples
multispeed   multivelocidad
multispeed induction motor   motor de inducción multivelocidad
multispiral disk   disco multiespiral
multistable   multiestable
multistable circuit   circuito multiestable
multistable switching circuit   circuito de conmutación multiestable
multistage   multietapa
multistage amplifier   amplificador multietapa
multistage device   dispositivo multietapa
multistage feedback   retroalimentación multietapa
multistage magnetic amplifier   amplificador magnético multietapa
multistage oscillation   oscilación multietapa
multistage process   proceso multietapa
multistage transmitter   transmisor multietapa
multistage tube   tubo multietapa
multistage X-ray tube   tubo de rayos X multietapa
multistandard receiver   receptor multipatrón
multistate   multiestado
multistate circuit   circuito multiestado
multistation   multiestación
multistep   multiescalonado
multistrand   multitrenza
multistrand cable   cable multitrenza
multistrand wire   hilo multitrenza
multiswitch   multiconmutador
multiswitching   multiconmutación
multisync monitor   monitor de multisincronización
multisystem   multisistema
multisystem mode   modo de multisistema
multisystem network   red de multisistema
multitap transformer   transformador multitoma
multitap winding   arrollamiento multitoma
multitape   multicinta
multitape memory   memoria multicinta
multitask   multitarea
multitask operation   operación multitarea
multitask processing   procesamiento multitarea
multitasking   multitarea
multitasking operating system   sistema operativo multitarea
multitasking system   sistema multitarea
multiterminal   multiterminal
multitester   multiprobador
multitone   multitono
multitrace   multitrazo

multitrace recorder   registrador multitrazo
multitrace recording   registro multitrazo
multitrack   multipista
multitrack head   cabeza multipista
multitrack magnetic system   sistema magnético multipista
multitrack recording   registro multipista
multitrack recording system   sistema de registro multipista
multitron   multitrón
multitube   multitubo
multiturn   multivuelta
multiturn potentiometer   potenciómetro multivuelta
multiunit   multiunidad
multiunit system   sistema multiunidad
multiunit tube   tubo multiunidad
multiuser   multiusuario
multiuser system   sistema multiusuario
multivalent   multivalente
multivariable   multivariable
multivelocity   multivelocidad
multivelocity electron beam   haz electrónico multivelocidad
multivibrator   multivibrador
multivibrator circuit   circuito multivibrador
multivoltage   multitensión
multivolume   multivolumen
multivolume file   archivo multivolumen
multiwire   multihilo, multialambre
multiwire antenna   antena multihilo
multiwire branch circuit   circuito de derivación multihilo
multiwire connector   conector multihilo
multiwire doublet   doblete multihilo
multiwire doublet antenna   antena de doblete multihilo
multiwire element   elemento multihilo
mumetal   mumetal
Munsell color system   sistema de colores de Munsell
Munsell system   sistema de Munsell
Munsell value   valor de Munsell
Muntz metal   metal de Muntz
Murray loop test   prueba de bucle de Murray
mush area   área de perturbación
mushroom lamp   lámpara en hongo
mushroom loudspeaker   altavoz en hongo
mushroom valve   válvula en hongo
mushroom ventilator   ventilador en hongo
music chip   chip de música
music power   potencia musical
music power test   prueba de potencia musical
music synthesizer   sintetizador musical
musical circuit   circuito musical
musical echo   eco musical
musical frequency   frecuencia musical
musical level   nivel musical
musical tone   tono musical
mute antenna   antena ficticia
mutilated   mutilado
muting circuit   circuito silenciador
muting device   dispositivo silenciador
muting switch   conmutador silenciador
muting system   sistema silenciador
mutual capacitance   capacitancia mutua
mutual-capacitance attenuator   atenuador de capacitancia mutua
mutual capacity   capacidad mutua
mutual characteristic   característica mutua
mutual circuit   circuito mutuo
mutual coil   bobina mutua
mutual conductance   conductancia mutua
mutual-conductance measurement   medida de conductancia mutua

mutual-conductance meter medidor de conductancia mutua
mutual coupling acoplamiento mutuo
mutual impedance impedancia mutua
mutual inductance inductancia mutua
mutual-inductance attenuator atenuador de inductancia mutua
mutual-inductance bridge puente de inductancia mutua
mutual-inductance coupling acoplamiento de inductancia mutua
mutual-inductance factor factor de inductancia mutua
mutual-inductance transducer transductor de inductancia mutua
mutual induction inducción mutua
mutual inductor inductor mutuo
mutual information información mutua
mutual interaction interacción mutua
mutual interference interferencia mutua
mutual reactance reactancia mutua
mutual resistance resistencia mutua
mutual synchronization sincronización mutua
mutually mutuamente
mutually coupled circuits circuitos mutuamente acoplados
mutually coupled coils bobinas mutuamente acopladas
myoelectric mioeléctrico
myoelectric signals señales mioeléctricas
myoelectricity mioelectricidad
myriameter miriámetro
myriametric waves ondas miriamétricas

N

n-level logic lógica de n niveles
n-n junction unión n-n
n-plus-one address dirección de n más uno
N-type area área tipo N
N-type base base tipo N
N-type capacitor capacitor tipo N
N-type conduction conducción tipo N
N-type conductivity conductividad tipo N
N-type emitter emisor tipo N
N-type germanium germanio tipo N
N-type material material tipo N
N-type semiconductor semiconductor tipo N
N-type silicon silicio tipo N
N-type transistor transistor tipo N
Nagaoka's constant constante de Nagaoka
naked wire hilo desnudo
nameplate rating potencia nominal
NAND circuit circuito NO-Y
NAND element elemento NO-Y
NAND flip-flop basculador NO-Y
NAND gate puerta NO-Y
NAND operation operación NO-Y
nano nano
nanoampere nanoampere
nanocircuit nanocircuito
nanocomputer nanocomputadora
nanofarad nanofarad
nanohenry nanohenry
nanoinstruction nanoinstrucción
nanosecond nanosegundo
nanosecond circuit circuito de nanosegundo
nanostore nanoalmacenamiento
nanovolt nanovolt
nanovoltmeter nanovóltmetro
nanowatt nanowatt
nanowattmeter nanowáttmetro
Naperian base base neperiana
Naperian logarithm logaritmo neperiano
narrow-angle diffusion difusión de ángulo angosto
narrow band banda angosta
narrow bandwidth banda angosta
narrow-sector recorder registrador de sector angosto
narrowband banda angosta
narrowband amplifier amplificador de banda angosta
narrowband analyzer analizador de banda angosta
narrowband axis eje de banda angosta
narrowband channel canal de banda angosta
narrowband detector detector de banda angosta
narrowband filter filtro de banda angosta
narrowband frequency-modulation modulación de frecuencia de banda angosta
narrowband interference interferencia de banda angosta
narrowband limiter limitador de banda angosta
narrowband radio noise ruido de radio de banda angosta
narrowband receiver receptor de banda angosta
narrowband response respuesta de banda angosta
narrowband response spectrum espectro de

respuesta de banda angosta
narrowband television televisión de banda angosta
narrowband voice modulation modulación de voz de banda angosta
national number número nacional
national numbering numeración nacional
national numbering plan plan de numeración nacional
national standard patrón nacional
nationwide call llamada nacional
nationwide network red nacional
native code código nativo
native compiler compilador nativo
native mode modo nativo
natural antenna frequency frecuencia de antena natural
natural bandwidth ancho de banda natural
natural binary binario natural
natural capacitance capacitancia natural
natural convection convección natural
natural disintegration desintegración natural
natural electricity electricidad natural
natural environment ambiente natural
natural excitation excitación natural
natural frequency frecuencia natural
natural function generator generador de función natural
natural horizon horizonte natural
natural inductance inductancia natural
natural interference interferencia natural
natural logarithm logaritmo natural
natural magnet imán natural
natural magnetism magnetismo natural
natural mode modo natural
natural model modelo natural
natural movement movimiento natural
natural noise ruido natural
natural number número natural
natural oscillation oscilación natural
natural period periodo natural
natural piezoelectric crystal cristal piezoeléctrico natural
natural process proceso natural
natural radiation radiación natural
natural radio noise ruido de radio natural
natural resonance resonancia natural
natural resonant frequency frecuencia resonante natural
natural sequence secuencia natural
natural stability estabilidad natural
natural stability limit límite de estabilidad natural
natural wavelength longitud de onda natural
navaglobe navaglobo
navar navar
navigation aid ayuda de navegación
navigation beacon faro de navegación
navigation computer computadora de navegación
navigation instrument instrumento de navegación
navigation lamp lámpara de navegación
navigation light luz de navegación
navigation radar radar de navegación
navigation receiver receptor de navegación
navigation satellite satélite de navegación
navigational aid ayuda navegacional
navigational beacon faro navegacional
navigational computer computadora navegacional
navigational electronics electrónica navegacional
navigational instrument instrumento navegacional
navigational radar radar navegacional
navigational satellite satélite navegacional
near echo eco próximo
near-end crosstalk paradiafonía
near-end crosstalk attenuation atenuación paradiafónica
near-end crosstalk deviation desviación paradiafónica
near-end crosstalk isolation aislamiento paradiafónico
near field campo próximo
near-field diffraction pattern patrón de difracción de campo próximo
near-field pattern patrón de campo próximo
near-field radiation pattern patrón de radiación de campo próximo
near-field region región de campo próximo
near-field scanning exploración de campo próximo
near infrared infrarrojo próximo
near infrared region región de infrarrojo próximo
near region región próxima
near ultraviolet region región de ultravioleta próxima
near zone zona próxima
necessary bandwidth ancho de banda necesario
necessary condition condición necesaria
necessary operation operación necesaria
neck cuello
needle aguja
needle chatter sonido de aguja
needle check comprobación por aguja
needle deviation desviación de aguja
needle drag fricción de aguja
needle electrode electrodo de aguja
needle force fuerza de aguja
needle gap distancia entre puntas
needle holder portaaguja
needle memory memoria de agujas
needle noise ruido de aguja
needle pointer indicador de aguja
needle pressure presión de aguja
needle probe sonda de aguja
needle scratch ruido de aguja
needle sort clasificación por aguja
needle talk sonido de aguja
needle telegraph telégrafo de aguja
needle test point punto de prueba de aguja
needle-tip probe sonda de punta de aguja
needle valve válvula de aguja
negate negar
negated negado
negated condition condición negada
negation negación
negation circuit circuito de negación
negation element elemento de negación
negation gate puerta de negación
negative negativo
negative acceleration aceleración negativa
negative acknowledgment character carácter de reconocimiento negativo
negative afterimage imagen persistente negativa
negative amplitude modulation modulación de amplitud negativa
negative angle ángulo negativo
negative answer respuesta negativa
negative bar barra negativa
negative bias polarización negativa
negative booster reductor de tensión
negative capacitance capacitancia negativa
negative charge carga negativa
negative compliance docilidad negativa
negative conductance conductancia negativa
negative conductor conductor negativo
negative coupling acoplamiento negativo
negative crystal cristal negativo
negative dispersion dispersión negativa
negative distortion distorsión negativa
negative electricity electricidad negativa

negative electrification electrificación negativa
negative electrode electrodo negativo
negative electron electrón negativo
negative exponent exponente negativo
negative feedback retroalimentación negativa
negative-feedback amplifier amplificador de
  retroalimentación negativa
negative-feedback arrangement arreglo con
  retroalimentación negativa
negative-feedback factor factor de retroalimentación
  negativa
negative function función negativa
negative gain ganancia negativa
negative-gain amplifier amplificador de ganancia
  negativa
negative ghost imagen fantasma negativa
negative ghost image imagen fantasma negativa
negative glow luminiscencia negativa
negative grid rejilla negativa
negative-grid bias polarización de rejilla negativa
negative-grid generator generador de rejilla negativa
negative-grid oscillator oscilador de rejilla negativa
negative-grid region región de rejilla negativa
negative-grid voltage tensión de rejilla negativa
negative image imagen negativa
negative impedance impedancia negativa
negative impedance booster reforzador de
  impedancia negativa
negative impedance converter convertidor de
  impedancia negativa
negative impedance oscillator oscilador de
  impedancia negativa
negative impedance repeater repetidor de
  impedancia negativa
negative indication indicación negativa
negative indicator indicador negativo
negative inductance inductancia negativa
negative ion ion negativo
negative-ion generator generador de iones negativos
negative lead conductor negativo
negative light modulation modulación de luz
  negativa
negative line línea negativa
negative loading carga negativa
negative logarithm logaritmo negativo
negative logic lógica negativa
negative logic conversion conversión de lógica
  negativa
negative-mass amplifier amplificador de masa
  negativa
negative matrix matriz negativa
negative measurement error error de medida
  negativo
negative modulation modulación negativa
negative modulation factor factor de modulación
  negativa
negative modulation peak pico de modulación
  negativa
negative number número negativo
negative peak pico negativo
negative-peak rectifier rectificador de pico negativo
negative-peak voltmeter vóltmetro de pico negativo
negative phase fase negativa
negative phase-sequence impedance impedancia de
  secuencia de fase negativa
negative phase-sequence reactance reactancia de
  secuencia de fase negativa
negative phase-sequence relay relé de secuencia de
  fase negativa
negative phase-sequence resistance resistencia de
  secuencia de fase negativa
negative picture imagen negativa
negative picture modulation modulación de imagen

negativa
negative picture phase fase de imagen negativa
negative plate placa negativa
negative polarity polaridad negativa
negative-polarity signal señal de polaridad negativa
negative pole polo negativo
negative potential potencial negativo
negative proton protón negativo
negative pulse impulso negativo
negative reactance reactancia negativa
negative reaction reacción negativa
negative region región negativa
negative resistance resistencia negativa
negative-resistance amplifier amplificador de
  resistencia negativa
negative-resistance coefficient coeficiente de
  resistencia negativa
negative-resistance device dispositivo de resistencia
  negativa
negative-resistance diode diodo de resistencia
  negativa
negative-resistance effect efecto de resistencia
  negativa
negative-resistance element elemento de resistencia
  negativa
negative-resistance magnetron magnetrón de
  resistencia negativa
negative-resistance oscillator oscilador de
  resistencia negativa
negative-resistance region región de resistencia
  negativa
negative-resistance repeater repetidor de resistencia
  negativa
negative resistor resistor negativo
negative response respuesta negativa
negative-sequence component componente de
  secuencia negativa
negative-sequence reactance reactancia de secuencia
  negativa
negative-sequence resistance resistencia de
  secuencia negativa
negative side lado negativo
negative temperature temperatura negativa
negative temperature coefficient coeficiente de
  temperatura negativo
negative terminal terminal negativa
negative torque par negativo
negative transconductance transconductancia
  negativa
negative-transconductance oscillator oscilador de
  transconductancia negativa
negative-transconductance tube tubo de
  transconductancia negativa
negative transition transición negativa
negative transmission transmisión negativa
negative voltage feedback retroalimentación de
  tensión negativa
negative voltage jump salto de tensión negativo
negative wire hilo negativo
negatively negativamente
negatively biased polarizado negativamente
negator negador
negatron negatrón
nematic nemático
nematic crystal cristal nemático
nematic-crystal display visualizador de cristal
  nemático
nematic liquid líquido nemático
neodymium neodimio
neodymium amplifier amplificador de neodimio
neodymium-glass laser láser de vidrio de neodimio
neodymium laser láser de neodimio
neon neón

neon bulb lámpara de neón, bombilla de neón
neon-bulb flip-flop basculador de lámparas de neón
neon-bulb gate puerta de lámpara de neón
neon-bulb indicator indicador de lámpara de neón
neon-bulb logic lógica de lámparas de neón
neon-bulb memory memoria de lámparas de neón
neon-bulb multivibrator multivibrador de lámparas de neón
neon-bulb oscillator oscilador de lámpara de neón
neon-bulb overmodulation indicator indicador de sobremodulación de lámpara de neón
neon-bulb overvoltage indicator indicador de sobretensión de lámpara de neón
neon-bulb peak indicator indicador de picos de lámpara de neón
neon-bulb ring counter contador en anillo de lámparas de neón
neon-bulb storage almacenamiento de lámparas de neón
neon-bulb stroboscope estroboscopio de lámpara de neón
neon-bulb voltage regulator regulador de tensión de lámpara de neón
neon-bulb volume indicator indicador de volumen de lámparas de neón
neon circuit tester probador de circuito de neón
neon glow lamp lámpara luminiscente de neón
neon-helium laser láser de neón-helio
neon indicator indicador de neón
neon indicator tube tubo indicador de neón
neon lamp lámpara de neón
neon-lamp readout dispositivo de lectura de lámpara de neón
neon light luz de neón
neon-light indicator indicador de luz de neón
neon oscillator oscilador de neón
neon stabilizer estabilizador de neón
neon tester probador de neón
neon triode triodo de neón
neon tube tubo de neón
neoprene neopreno
neotron neotrón
neper neper
nephelometer nefelómetro
nephelometry nefelometría
neptunium neptunio
Nernst bridge puente de Nernst
Nernst effect efecto de Nernst
Nernst-Ettinghausen effect efecto de Nernst-Ettinghausen
Nernst lamp lámpara de Nernst
nesistor nesistor
nest anidar
nested anidado
nested block bloque anidado
nested command mando anidado
nested loop bucle anidado
nested procedure procedimiento anidado
nested subroutine subrutina anidada
nesting anidamiento
nesting loop bucle de anidamiento
net authentication autenticación de red
net capability capacidad neta
net capacitance capacitancia neta
net component componente neto
net current corriente neta
net echo loss pérdida de eco neta
net energy gain ganancia de energía neta
net gain ganancia neta
net generation generación neta
net impedance impedancia neta
net inductance inductancia neta
net information content contenido de información

neto
net load carga neta
net load capability capacidad de carga neta
net loop gain ganancia de bucle neta
net loss pérdida neta
net margin margen neto
net power potencia neta
net radiometer radiómetro neto
net reactance reactancia neta
net resistance resistencia neta
net system energy energía de sistema neta
net voltage tensión neta
network red
network address dirección de red
network administrator administrador de red
network analysis análisis de red
network analyzer analizador de red
network application aplicación de red
network architecture arquitectura de red
network bridge puente de red
network buffer memoria intermedia de red
network calculator calculadora de red
network chart esquema de red
network communications circuits circuitos de comunicaciones por red
network configuration table tabla de configuración de red
network constant constante de red
network control control de red
network control mode modo de control de red
network control processor procesador de control de red
network control program programa de control de red
network control program station estación de programa de control de red
network database base de datos de red
network database management administración de base de datos de red
network database management system sistema de administración de base de datos de red
network failure falla de red
network feeder alimentador de red
network filter filtro de red
network flow flujo de red
network function función de red
network geometry geometría de redes
network interface interfaz de red
network interface card tarjeta de interfaz de red
network interface controller controlador de interfaz de red
network load carga de red
network load analysis análisis de carga de red
network management administración de red
network management process proceso de administración de red
network master relay relé maestro de red
network model modelo de red
network node nodo de red
network of circuits red de circuitos
network of lines red de líneas
network operating system sistema operativo de red
network operations center centro de operaciones de red
network operator operador de red
network-operator console consola de operador de red
network parameter parámetro de red
network path trayectoria de red
network protector protector de red
network redundancy redundancia de red
network relay relé de red
network resource recurso de red

network security seguridad de red
network server servidor de red
network service servicio de red
network stabilization estabilización de red
network station estación de red
network structure estructura de red
network synthesis síntesis de red
network terminal terminal de red
network theory teoría de redes
network topology topología de redes
network transfer function función de transferencia de red
network unbalance desequilibrio de red
network version versión de red
Neumann's law ley de Neumann
neuristor neuristor
neuroelectricity neuroelectricidad
neutral neutro, conductor neutro
neutral absorber absorbedor neutro
neutral-anode magnetron magnetrón de ánodo neutro
neutral circuit circuito neutro
neutral conductor conductor neutro
neutral diffuser difusor neutro
neutral filter filtro neutro
neutral ground tierra neutra
neutral line línea neutra
neutral plane plano neutro
neutral point punto neutro
neutral position posición neutra
neutral relay relé neutro
neutral stability estabilidad neutra
neutral state estado neutro
neutral terminal terminal neutra
neutral transmission transmisión neutra
neutral wire hilo neutro
neutral zone zona neutra
neutralization neutralización
neutralization indicator indicador de neutralización
neutralize neutralizar
neutralized neutralizado
neutralized amplifier amplificador neutralizado
neutralizing neutralizante
neutralizing capacitor capacitor neutralizante
neutralizing circuit circuito neutralizante
neutralizing coil bobina neutralizante
neutralizing condenser condensador neutralizante
neutralizing voltage tensión neutralizante
neutrodyne neutrodino
neutrodyne receiver receptor neutrodino
neutron neutrón
neutron beam haz neutrónico
neutron crystallography cristalografía neutrónica
neutron current corriente neutrónica
neutron energy energía neutrónica
neutron flux flujo neutrónico
neutron spectrometer espectrómetro neutrónico
neutronic neutrónico
neutronics neutrónica
newton newton
Newton's laws of motion leyes de moción de Newton
nexus nexo
nibble nibble
Nichols radiometer radiómetro de Nichols
nichrome nicromo
nickel níquel
nickel-cadmium níquel-cadmio
nickel-cadmium accumulator acumulador de níquel-cadmio
nickel-cadmium battery batería de níquel-cadmio
nickel-cadmium cell celda de níquel-cadmio
nickel-cadmium storage battery acumulador de níquel-cadmio

nickel delay line línea de retardo de níquel
nickel-iron níquel-hierro
nickel-iron accumulator acumulador de níquel-hierro
nickel-iron battery batería de níquel-hierro
nickel-oxide diode diodo de óxido de níquel
nickel-plating niquelado
nickel silver plata níquel
Nicol prism prisma de Nicol
night alarm alarma nocturna
night effect efecto nocturno
night-effect error error por efecto nocturno
night error error nocturno
night frequency frecuencia nocturna
night range alcance nocturno
nines complement complemento a nueve
ninety-column card tarjeta de noventa columnas
ninety six-column card tarjeta de noventa y seis columnas
niobium niobio
Nipkow disk disco de Nipkow
nit nit
nitrocellulose nitrocelulosa
nitrogen nitrógeno
nitrogen lamp lámpara de nitrógeno
no-address instruction instrucción sin dirección
no-bias relay relé sin polarización
no-break power source fuente de energía sin interrupción
no-break power supply fuente de alimentación sin interrupción
no-break power unit unidad de energía sin interrupción
no-break system sistema sin interrupción
no-crosstalk amplifier amplificador sin diafonía
no-delay service servicio sin espera
no-load adjustment ajuste sin carga
no-load admittance admitancia sin carga
no-load characteristic característica sin carga
no-load current corriente sin carga
no-load losses pérdidas sin carga
no-load operation operación sin carga
no-load reading lectura sin carga
no-load speed velocidad sin carga
no-load voltage tensión sin carga
no-load work trabajo sin carga
no-noise amplifier amplificador sin ruido
no-op instruction instrucción de no operación
no-operation command mando de no operación
no-operation instruction instrucción de no operación
no parity sin paridad
no-signal current corriente sin señal
no-wait operation operación sin espera
nobelium nobelio
Nobili's rings anillos de Nobili
noble noble
noble gas gas noble
noble metal metal noble
noctovision noctovisión
nodal nodal
nodal analysis análisis nodal
nodal plane plano nodal
nodal point punto nodal
node nodo
node identification identificación de nodo
node identifier identificador de nodo
node of current nodo de corriente
node of pressure nodo de presión
node of voltage nodo de tensión
node operator operador de nodo
node type tipo de nodo
node voltage tensión de nodo

Nodon valve válvula de Nodon
nodules nódulos
noise ruido
noise abatement reducción de ruido
noise abatement procedure procedimiento de reducción de ruido
noise acceptance aceptación de ruido
noise advantage ventaja sobre ruido
noise amplifier amplificador de ruido
noise amplifier-rectifier amplificador-rectificador de ruido
noise amplitude distribution distribución de amplitud de ruido
noise analysis análisis de ruido
noise analyzer analizador de ruido
noise area área de ruido
noise audiogram audiograma de ruido
noise audiometer audiómetro de ruido
noise background fondo de ruido
noise-balancing circuit circuito equilibrador de ruido
noise-balancing system sistema equilibrador de ruido
noise bandwidth ancho de banda de ruido
noise-cancelling microphone micrófono de cancelación de ruido
noise capability capacidad de ruido
noise clipper recortador de ruido
noise clipping recorte de ruido
noise constant constante de ruido
noise control control de ruido
noise current corriente de ruido
noise-current generator generador de corriente de ruido
noise deadening amortiguación de ruido
noise degradation degradación por ruido
noise detector detector de ruido
noise diffusion difusión de ruido
noise diode diodo de ruido
noise elimination eliminación de ruido
noise eliminator eliminador de ruido
noise emission emisión de ruido
noise equivalent bandwidth ancho de banda equivalente de ruido
noise equivalent input entrada equivalente de ruido
noise equivalent power potencia equivalente de ruido
noise factor factor de ruido
noise-factor meter medidor de factor de ruido
noise failure falla por ruido
noise field campo de ruido
noise field intensity intensidad de campo de ruido
noise figure cifra de ruido
noise filter filtro de ruido
noise-free libre de ruido
noise generator generador de ruido
noise-generator diode diodo generador de ruido
noise-generator tube tubo generador de ruido
noise grade nivel de ruido
noise immunity inmunidad al ruido
noise index índice de ruido
noise-induced inducido por ruido
noise insulation aislamiento de ruido
noise-insulation factor factor de aislamiento de ruido
noise-intensity meter medidor de intensidad de ruido
noise interference interferencia de ruido
noise killer supresor de ruido
noise level nivel de ruido
noise-level meter medidor de nivel de ruido
noise-level test prueba de nivel de ruido
noise limiter limitador de ruido
noise limiting limitación de ruido
noise load carga de ruido
noise-making productor de ruido
noise-making circuit circuito productor de ruido
noise margin margen de ruido

noise measurement medida de ruido
noise-measuring set aparato medidor de ruido
noise meter medidor de ruido
noise-operated operado por ruido
noise output salida de ruido
noise power potencia de ruido
noise-power ratio razón de potencias de ruido
noise pressure presión de ruido
noise pressure equivalent equivalente de presión de ruido
noise producer productor de ruido
noise pulse impulso de ruido
noise quieting reducción de ruido
noise rating clasificación de ruido
noise ratio razón de ruido
noise rectifier rectificador de ruido
noise reducer reductor de ruido
noise-reducing antenna antena reductora de ruido
noise reduction reducción de ruido
noise-reduction amplifier amplificador de reducción de ruido
noise rejection rechazo de ruido
noise residue residuo de ruido
noise sensing detección de ruido
noise signal señal de ruido
noise signal generator generador de señal de ruido
noise silencer silenciador de ruido
noise source fuente de ruido
noise spectrum espectro de ruido
noise spike pico de ruido
noise squelch silenciador de ruido
noise storm tormenta de ruido
noise suppression supresión de ruido
noise-suppression circuit circuito de supresión de ruido
noise-suppression relay relé de supresión de ruido
noise supressor supresor de ruido
noise temperature temperatura de ruido
noise test prueba de ruido
noise transmission effect efecto de transmisión de ruido
noise transmission impairment deterioro por transmisión de ruido
noise unit unidad de ruido
noise voltage tensión de ruido
noise-voltage generator generador de tensión de ruido
noise wave onda de ruido
noiseless sin ruido
noiseless operation operación sin ruido
noiseless recording registro sin ruido
noiseproof a prueba de ruido
noisy ruidoso
noisy channel canal ruidoso
noisy circuit circuito ruidoso
noisy control control ruidoso
noisy digit dígito ruidoso
noisy line línea ruidosa
noisy mode modo ruidoso
noisy operation operación ruidosa
nomenclature nomenclatura
nomenclature standard patrón de nomenclatura
nominal nominal
nominal antenna impedance impedancia de antena nominal
nominal apparent power potencia aparente nominal
nominal band banda nominal
nominal bandwidth ancho de banda nominal
nominal battery voltage tensión de batería nominal
nominal black negro nominal
nominal capacitance capacitancia nominal
nominal circuit voltage tensión de circuito nominal
nominal consumption consumo nominal

nominal control current corriente de control nominal
nominal current corriente nominal
nominal cutoff frequency frecuencia de corte nominal
nominal data datos nominales
nominal discharge current corriente de descarga nominal
nominal frequency frecuencia nominal
nominal impedance impedancia nominal
nominal input entrada nominal
nominal input power potencia de entrada nominal
nominal level nivel nominal
nominal line width anchura de línea nominal
nominal load carga nominal
nominal margin margen nominal
nominal output salida nominal
nominal power potencia nominal
nominal power factor factor de potencia nominal
nominal power rating potencia nominal
nominal rate tasa nominal
nominal rating clasificación nominal
nominal ratio razón nominal
nominal resistance resistencia nominal
nominal service servicio nominal
nominal speed velocidad nominal
nominal system voltage tensión de sistema nominal
nominal tolerance tolerancia nominal
nominal value valor nominal
nominal voltage tensión nominal
nominal white blanco nominal
nominal working voltage tensión de trabajo nominal
nonaddressable no direccionable
nonadjustable no ajustable
nonamplifying no amplificador
nonamplifying detector detector no amplificador
nonarithmetic no aritmético
nonarithmetic shift desplazamiento no aritmético
nonaudible no audible, inaudible
nonautomatic no automático
nonautomatic operation operación no automática
nonblank no en blanco
nonblinking meter medidor no oscilante
nonblocking system sistema sin bloqueos
nonbridging sin puenteo, no ponteante
nonbridging contact contacto sin puenteo
nonbridging switching conmutación sin puenteo
nonbuffered sin memoria intermedia
noncancellable no cancelable
nonchargeable battery batería no recargable
noncoded no codificado
noncoded graphics gráficos no codificados
noncoherent no coherente, incoherente
noncoherent integration integración no coherente
noncoherent radiation radiación no coherente
noncoherent signal señal no coherente
noncompensated no compensado
noncompensated amplifier amplificador no compensado
nonconducting no conductor
nonconducting diode diodo no conductor
nonconducting material material no conductor
nonconductive no conductor
nonconductor no conductor
noncontact sin contacto
noncontact temperature measurement medida de temperatura sin contacto
noncontinuous no continuo, discontinuo
noncontinuous electrode electrodo no continuo
noncontinuous enclosure caja no continua
noncorrosive no corrosivo
noncorrosive alloy aleación no corrosiva
noncorrosive flux flujo no corrosivo

noncorrosive paste pasta no corrosiva
noncritical no crítico
noncritical dimension dimensión no crítica
noncritical failure falla no crítica
noncrystalline no cristalino
nondelay sin retardo
nondelay fuse fusible sin retardo
nondemodulating no desmodulante
nondestructive no destructivo
nondestructive addition suma no destructiva
nondestructive cursor cursor no destructivo
nondestructive memory memoria no destructiva
nondestructive operation operación no destructiva
nondestructive read lectura no destructiva
nondestructive reading lectura no destructiva
nondestructive readout lectura no destructiva
nondestructive test prueba no destructiva
nondestructive tester probador no destructivo
nondeviated sin desviación
nondeviated absorption absorción sin desviación
nondirectional no direccional
nondirectional antenna antena no direccional
nondirectional beacon radiofaro no direccional
nondirectional microphone micrófono no direccional
nondirectional radio beacon radiofaro no direccional
nondirective no directivo
nondirective antenna antena no directiva
nondispersive no dispersivo
nondispersive waves ondas no dispersivas
nondissipative no disipativo
nondissipative load carga no disipativa
nondissipative network red no disipativa
nondissipative stub adaptador no disipativo
nonelastic no elástico
nonelectric no eléctrico
nonelectrical no eléctrico
nonelectrified no electrificado
nonelectrolyte no electrólito
nonelectronic no electrónico
nonelectronic meter medidor no electrónico
nonembedded no empotrado
nonembedded command mando no empotrado
nonequivalence circuit circuito de no equivalencia
nonequivalence element elemento de no equivalencia
nonequivalence operation operación de no equivalencia
nonequivalent no equivalente
nonerasable no borrable
nonerasable storage almacenamiento no borrable
nonessential no esencial
nonessential circuit circuito no esencial
nonessential service servicio no esencial
nonexecutable no ejecutable
nonexecutable statement sentencia no ejecutable
nonferrous no ferroso, no férrico
nonferrous metal metal no ferroso, metal no férrico
nonferrous shield blindaje no ferroso, blindaje no férrico
nonflammable no inflamable, ininflamable
nonharmonic frequency frecuencia no armónica
nonharmonic oscillations oscilaciones no armónicas
nonhomogeneous no homogéneo, inhomogéneo
nonhygroscopic no higroscópico
nonidentity operation operación de no identidad
noninduced no inducido
noninduced current corriente no inducida
noninductive no inductivo
noninductive capacitor capacitor no inductivo
noninductive circuit circuito no inductivo
noninductive coil bobina no inductiva
noninductive condenser condensador no inductivo

noninductive load carga no inductiva
noninductive resistance resistencia no inductiva
noninductive resistor resistor no inductivo
noninductive shunt derivación no inductiva
noninductive winding arrollamiento no inductivo
noninsulated no aislado
noninsulated resistor resistor no aislado
noninterchangeable no intercambiable
noninterchangeable fuse fusible no intercambiable
noninterlaced no entrelazado
noninterlaced system sistema no entrelazado
noninverting connection conexión no inversora
noninverting input entrada no inversora
noninverting parametric device dispositivo
    paramétrico no inversor
nonionic no iónico
nonionizing no ionizante
nonionizing particle partícula no ionizante
nonionizing radiation radiación no ionizante
nonlinear no lineal
nonlinear amplifier amplificador no lineal
nonlinear bridge puente no lineal
nonlinear capacitor capacitor no lineal
nonlinear characteristic característica no lineal
nonlinear circuit circuito no lineal
nonlinear coil bobina no lineal
nonlinear correlation correlación no lineal
nonlinear data structure estructura de datos no
    lineal
nonlinear detection detección no lineal
nonlinear detector detector no lineal
nonlinear dielectric dieléctrico no lineal
nonlinear distortion distorsión no lineal
nonlinear distortion coefficient coeficiente de
    distorsión no lineal
nonlinear distortion factor factor de distorsión no
    lineal
nonlinear element elemento no lineal
nonlinear filtering filtrado no lineal
nonlinear function función no lineal
nonlinear ideal capacitor capacitor ideal no lineal
nonlinear inductor inductor no lineal
nonlinear network red no lineal
nonlinear operator operador no lineal
nonlinear optimization optimización no lineal
nonlinear parameter parámetro no lineal
nonlinear potentiometer potenciómetro no lineal
nonlinear programming programación no lineal
nonlinear resistance resistencia no lineal
nonlinear resistor resistor no lineal
nonlinear response respuesta no lineal
nonlinear scale escala no lineal
nonlinear scanning exploración no lineal
nonlinear scattering dispersión no lineal
nonlinear system sistema no lineal
nonlinearity no linealidad
nonlinearity error error de no linealidad
nonlinearity index índice de no linealidad
nonloaded no cargado
nonloaded cable cable no cargado
nonloaded line línea no cargada
nonloaded pair par no cargado
nonlocal no local
nonlocalized no localizado
nonlossy sin pérdidas
nonlossy cable cable sin pérdidas
nonmagnetic no magnético
nonmagnetic alloy aleación no magnética
nonmagnetic base base no magnética
nonmagnetic compass compás no magnético
nonmagnetic recording medium medio de registro
    no magnético
nonmathematical no matemático

nonmechanical no mecánico
nonmechanical switching device dispositivo de
    conmutación no mecánica
nonmetal no metal
nonmetallic no metálico
nonmetallic conduction conducción no metálica
nonnumeric no numérico
nonnumeric character carácter no numérico
nonnumeric data processing procesamiento de datos
    no numérico
nonnumeric item artículo no numérico
nonnumeric programming programación no
    numérica
nonnumerical carácter no numérico
nonohmic no óhmico
nonohmic response respuesta no óhmica
nonoperable no operable
nonoperating no operativo
nonoperational no operacional
nonoperational mode modo no operacional
nonoscillating no oscilante
nonoscillating detector detector no oscilante
nonoscillatory no oscilatorio
nonoverlapping no solapante
nonpaired no pareado
nonpaired data datos no pareados
nonparallel no paralelo
nonparametric no paramétrico
nonperiodic no periódico
nonperiodic vibration vibración no periódica
nonplanar no planar
nonplanar network red no planar
nonpolar no polar
nonpolar crystal cristal no polar
nonpolar relay relé no polar
nonpolarizable no polarizable
nonpolarizable electrode electrodo no polarizable
nonpolarizable reference electrode electrodo de
    referencia no polarizable
nonpolarized no polarizado
nonpolarized electrolytic capacitor capacitor
    electrolítico no polarizado
nonpolarized relay relé no polarizado
nonprint character carácter de no impresión
nonprint code código de no impresión
nonprint impulse impulso de no impresión
nonprint instruction instrucción de no impresión
nonprivileged no privilegiado
nonprivileged instruction instrucción no privilegiada
nonproductive no productivo, improductivo
nonproductive task tarea no productiva
nonprogrammed no programado
nonprogrammed halt parada no programada
nonradioactive no radiactivo
nonreactive no reactivo
nonreactive circuit circuito no reactivo
nonreactive load carga no reactiva
nonreactive resistance resistencia no reactiva
nonrechargeable no recargable
nonrechargeable battery batería no recargable
nonreciprocal no recíproco
nonreciprocal component componente no recíproco
nonreciprocal device dispositivo no recíproco
nonreciprocal gain ganancia no recíproca
nonreciprocal loss pérdida no recíproca
nonreciprocal microwave component componente
    de microondas no recíproco
nonreciprocity no reciprocidad
nonrecoverable no recuperable, irrecuperable
nonrecoverable error error no recuperable
nonrecoverable transaction transacción no
    recuperable
nonrecurrent no recurrente

nonrecurrent sweep barrido no recurrente
nonrecurring no recurrente
nonrecurring pulse impulso no recurrente
nonreflecting no reflejante
nonreflecting attenuator atenuador no reflejante
nonregenerative no regenerativo
nonregenerative detector detector no regenerativo
nonregenerative receiver receptor no regenerativo
nonremovable no removible
nonremovable disk disco no removible
nonrenewable no renovable
nonrenewable fuse unit unidad de fusible no
    renovable
nonrepetitive no repetitivo
nonrepetitive sweep barrido no repetitivo
nonreproducing codes códigos de no reproducción
nonreset sin reposición
nonreset timer temporizador sin reposición eléctrica
nonresident no residente
nonresident portion porción no residente
nonresident program programa no residente
nonresident routine rutina no residente
nonresonant no resonante
nonresonant antenna antena no resonante
nonresonant line línea no resonante
nonresonant load carga no resonante
nonresonant transmission line línea de transmisión
    no resonante
nonresonating no resonante
nonresonating antenna antena no resonante
nonreturn-to-zero sin retorno a cero
nonreturn-to-zero recording registro sin retorno a
    cero
nonreversible no reversible, irreversible
nonreversible connector conector no reversible
nonreversible plug clavija no reversible
nonsaturable no saturable
nonsaturable amplifier amplificador no saturable
nonsaturated no saturado
nonsaturated color color no saturado
nonsaturated logic lógica no saturada
nonselective no selectivo
nonselective absorber absorbedor no selectivo
nonselective diffuser difusor no selectivo
nonselective radiator radiador no selectivo
nonself-maintained no automantenido
nonself-synchronizing no autosincronizante
nonseparable no separable, inseparable
nonsequential no secuencial
nonsequential computer computadora no secuencial
nonsequential queue cola no secuencial
nonsequential scanning exploración no secuencial
nonshorting sin cortocircuito
nonshorting switch conmutador sin cortocircuito
nonsimultaneous no simultáneo
nonsimultaneous transmission transmisión no
    simultánea
nonsinusoidal no sinusoidal
nonsinusoidal wave onda no sinusoidal
nonspecific no específico
nonspecular no especular
nonspecular reflection reflexión no especular
nonstationary no estacionario
nonstorage sin almacenamiento
nonstorage display visualizador sin almacenamiento
nonstructural no estructural
nonsymmetrical no simétrico, asimétrico
nonsymmetrical wave onda no simétrica
nonsynchronized no sincronizado
nonsynchronized scanning exploración no
    sincronizada
nonsynchronous no sincrónico, asincrónico,
    asíncrono

nonsynchronous network red no sincrónica
nonsynchronous operation operación no sincrónica
nonsynchronous receiver receptor no sincrónico
nonsynchronous recorder registrador no sincrónico
nonsynchronous satellite satélite no sincrónico
nonsynchronous switch conmutador no sincrónico
nonsynchronous switching conmutación no
    sincrónica
nonsynchronous vibrator vibrador no sincrónico
nontext message mensaje sin texto
nonthermal no térmico
nonthermal radiation radiación no térmica
nonuniform no uniforme
nonuniform field campo no uniforme
nonventilated no ventilado
nonvolatile no volátil
nonvolatile memory memoria no volátil
nonvolatile storage almacenamiento no volátil
NOR circuit circuito NO-O
NOR element elemento NO-O
NOR gate puerta NO-O
NOR operation operación NO-O
normal normal
normal band banda normal
normal binary binario normal
normal calomel electrode electrodo de calomel
    normal
normal cathode fall caída catódica normal
normal channel canal normal
normal circuit circuito normal
normal concentration concentración normal
normal condition condición normal
normal contact contacto normal
normal control control normal
normal control field campo de control normal
normal current corriente normal
normal curve curva normal
normal cut corte normal
normal dilution dilución normal
normal dimension dimensión normal
normal distribution distribución normal
normal-distribution curve curva de distribución
    normal
normal electrode electrodo normal
normal energy level nivel de energía normal
normal failure period periodo de fallas normal
normal flow flujo normal
normal frequency frecuencia normal
normal glow luminiscencia normal
normal glow discharge descarga luminiscente
    normal
normal hydrogen electrode electrodo de hidrógeno
    normal
normal impedance impedancia normal
normal induction inducción normal
normal-induction curve curva de inducción normal
normal install instalación normal
normal joint unión normal
normal lamp lámpara normal
normal linearity linealidad normal
normal load carga normal
normal magnetization magnetización normal
normal margin margen normal
normal mode modo normal
normal-mode noise ruido de modo normal
normal mode of radiation modo normal de radiación
normal mode of vibration modo normal de
    vibración
normal-mode rejection rechazo de modo normal
normal number número normal
normal operating level nivel de operación normal
normal operation operación normal
normal overload sobrecarga normal

normal permeability permeabilidad normal
normal polarization polarización normal
normal position posición normal
normal power output salida de potencia normal
normal pressure presión normal
normal priority prioridad normal
normal program execution ejecución de programa normal
normal propagation propagación normal
normal radiation radiación normal
normal range alcance normal
normal rating clasificación normal
normal reaction reacción normal
normal recording level nivel de registro normal
normal response respuesta normal
normal response mode modo de respuesta normal
normal restart reiniciación normal
normal route ruta normal
normal sequence secuencia normal
normal solution solución normal
normal speed velocidad normal
normal state estado normal
normal stress estrés normal
normal temperature temperatura normal
normal threshold umbral normal
normal vibration vibración normal
normal voltage limit límite de tensión normal
normal weather clima normal
normalization normalización
normalization routine rutina de normalización
normalization signal señal de normalización
normalize normalizar
normalized normalizado
normalized admittance admitancia normalizada
normalized directivity directividad normalizada
normalized form forma normalizada
normalized frequency frecuencia normalizada
normalized impedance impedancia normalizada
normalized pattern patrón normalizado
normalized response respuesta normalizada
normalizing normalización
normally normalmente
normally closed normalmente cerrado
normally closed contact contacto normalmente cerrado
normally open normalmente abierto
normally open contact contacto normalmente abierto
north magnetic pole polo norte magnético
north pole polo norte
Norton's equivalent equivalente de Norton
Norton's theorem teorema de Norton
Norton transformation transformación de Norton
NOT-AND circuit circuito NO-Y
NOT-AND element elemento NO-Y
NOT-AND gate puerta NO-Y
NOT-AND operation operación NO-Y
NOT circuit circuito NO
NOT element elemento NO
NOT gate puerta NO
NOT logic lógica NO
not ready no listo
not ready condition condición de no listo
not recoverable no recuperable
notation notación
notch muesca, punto, entalla
notch amplifier amplificador con filtro de muesca
notch antenna antena de ranura
notch coding codificación por muesca
notch diplexer diplexor de muesca
notch filter filtro de muesca
notch network red de muesca
notch sweep barrido de muesca
notched card tarjeta muescada

notching ratio finura de regulación
noval noval
noval base base noval
noval tube tubo noval
novar novar
noy noy
NP junction unión NP
NP junction transistor transistor de unión NP
NP0 capacitor capacitor NP0
NPIN junction transistor transistor de uniones NPIN
NPIN transistor transistor NPIN
NPIP transistor transistor NPIP
NPN junction transistor transistor de uniones NPN
NPN transistor transistor NPN
NPNP device dispositivo NPNP
NPNP transistor transistor NPNP
nu nu
nuclear nuclear
nuclear battery batería nuclear
nuclear charge carga nuclear
nuclear clock reloj nuclear
nuclear electricity electricidad nuclear
nuclear energy energía nuclear
nuclear force fuerza nuclear
nuclear induction inducción nuclear
nuclear magnetic moment momento magnético nuclear
nuclear magnetic resonance resonancia magnética nuclear
nuclear magneton magnetón nuclear
nuclear reaction reacción nuclear
nucleonics nucleónica
nucleus núcleo
null nulo
null balance equilibrio a cero
null-balance potentiometer potenciómetro de equilibrio a cero
null-balance system sistema de equilibrio a cero
null bridge puente de cero
null character carácter nulo
null character string secuencia de caracteres nulos
null current corriente nula
null cycle ciclo nulo
null data datos nulos
null detection detección de cero
null detector detector de cero
null field campo nulo
null frequency frecuencia nula
null indication indicación de cero
null indicator indicador de cero
null instruction instrucción nula
null line línea nula
null matrix matriz nula
null meter medidor de cero
null method método de cero
null modem módem nulo
null name nombre nulo
null point punto nulo
null representation representación nula
null-seeking relay relé buscador de cero
null set conjunto nulo
null statement sentencia nula
null string secuencia nula
null transaction transacción nula
null value valor nulo
null voltage tensión nula
number base base numérica
number crunching cómputos en masa
number field campo numérico
number notation notación numérica
number of layers número de capas
number of scanning lines número de líneas de

exploración
number representation representación numérica
number representation system sistema de
   representación numérica
number system sistema numérico
number system base base de sistema numérico
numbering numeración
numbering device dispositivo de numeración
numbering diagram diagrama de numeración
numbering machine máquina de numeración
numbering plan plan de numeración
numbering system sistema de numeración
numbering transmitter transmisor de numeración
numeral numeral
numeralization numeralización
numeration numeración
numeration system sistema de numeración
numerator numerador
numeric numérico
numeric analysis análisis numérico
numeric bit data datos de bits numéricos
numeric character carácter numérico
numeric computer computadora numérica
numeric constant constante numérica
numeric control control numérico
numeric coprocessor coprocesador numérico
numeric data datos numéricos
numeric database base de datos numéricos
numeric element elemento numérico
numeric field campo numérico
numeric information información numérica
numeric keyboard teclado numérico
numeric operator operador numérico
numeric position posición numérica
numeric processing procesamiento numérico
numeric processor procesador numérico
numeric punch perforación numérica
numeric representation representación numérica
numeric sequence secuencia numérica
numeric shift desplazamiento numérico
numeric sort clasificación numérica
numeric tape cinta numérica
numeric value valor numérico
numerical numérico
numerical analysis análisis numérico
numerical aperture abertura numérica
numerical bit data datos de bits numéricos
numerical character carácter numérico
numerical character set conjunto de caracteres
   numéricos
numerical character subset subconjunto de
   caracteres numéricos
numerical code código numérico
numerical coding codificación numérica
numerical computer computadora numérica
numerical constant constante numérica
numerical control control numérico
numerical control system sistema de control
   numérico
numerical control tape cinta de control numérico
numerical coprocessor coprocesador numérico
numerical data datos numéricos
numerical data code código de datos numéricos
numerical database base de datos numéricos
numerical display visualización numérica
numerical element elemento numérico
numerical field campo numérico
numerical form forma numérica
numerical format formato numérico
numerical index índice numérico
numerical information información numérica
numerical integration integración numérica
numerical item artículo numérico

numerical keyboard teclado numérico
numerical keypad teclado numérico
numerical machine máquina numérica
numerical model modelo numérico
numerical operator operador numérico
numerical position posición numérica
numerical processing procesamiento numérico
numerical processor procesador numérico
numerical processor chip chip de procesador
   numérico
numerical punch perforación numérica
numerical punching perforación numérica
numerical ratio razón numérica
numerical readout lectura numérica
numerical recording registro numérico
numerical reliability confiabilidad numérica
numerical representation representación numérica
numerical selection selección numérica
numerical sequence secuencia numérica
numerical shift desplazamiento numérico
numerical shift mode modo de desplazamiento
   numérico
numerical sort clasificación numérica
numerical tape cinta numérica
numerical value valor numérico
numerical verifier verificador numérico
numerical word palabra numérica
numerically numéricamente
nutation nutación
nuvistor nuvistor
nylon nilón
Nyquist criterion criterio de Nyquist
Nyquist diagram diagrama de Nyquist
Nyquist formula fórmula de Nyquist
Nyquist frequency frecuencia de Nyquist
Nyquist interval intervalo de Nyquist
Nyquist noise rule regla de ruido de Nyquist
Nyquist rate velocidad de Nyquist
Nyquist theorem teorema de Nyquist

object code　código objeto
object computer　computadora objeto
object computer entry　entrada de computadora
　　objeto
object configuration　configuración objeto
object file　archivo objeto
object format　formato objeto
object language　lenguaje objeto
object machine　máquina objeto
object module　módulo objeto
object-oriented design　diseño orientado a objetos
object-oriented language　lenguaje orientado a
　　objetos
object phase　fase objeto
object point　punto objeto
object program　programa objeto
object routine　rutina objeto
objective lens　objetivo
objective loudness rating　clasificación de intensidad
　　sonora objetiva
objective measurement　medida objetiva
objective noise meter　medidor de ruido objetivo
objective photometer　fotómetro objetivo
oblique　oblicuo
oblique incidence　incidencia oblicua
oblique-incidence probing　sondeo con incidencia
　　oblicua
oblique-incidence transmission　transmisión con
　　incidencia oblicua
oblique-incidence wave　onda de incidencia oblicua
oblique projection　proyección oblicua
oblique view　vista oblicua
oblique winding　arrollamiento oblicuo
obliquity　oblicuidad
oboe　oboe
oboe system　sistema oboe
obscure zone　zona obscura
observable　observable
observable error　error observable
observable temperature　temperatura observable
observation　observación
observation aperture　apertura de observación
observation desk　pupitre de observación
observation device　dispositivo de observación
observation error　error de observación
observation mirror　espejo de observación
observation time　tiempo de observación
observation window　ventanilla de observación
observational　observacional
observed bearing　rumbo aparente
observed reliability　confiabilidad observada
obsolescence　obsolescencia
obsolescence-free　sin obsolescencia
obsolescence-prone　con tendencia a la obsolescencia
obsolete　obsoleto
obstacle gain　ganancia por obstáculo
obstruction light　luz de obstrucción
occluded gas　gas ocluido
occultation　ocultación
occulting light　luz de ocultación
occupation　ocupación

occupation efficiency　eficiencia de ocupación
occupied band　banda ocupada
occupied bandwidth　ancho de banda ocupado
occurrence　ocurrencia
octal　octal
octal base　base octal
octal-base relay　relé de base octal
octal-base tube　tubo de base octal
octal digit　dígito octal
octal notation　notación octal
octal number　número octal
octal number system　sistema numérico octal
octal numeral　número octal
octal numeration system　sistema de numeración
　　octal
octal plug　clavija octal
octal point　coma octal, punto octal
octal socket　zócalo octal
octal system　sistema octal
octal-to-binary conversion　conversión de octal a
　　binario
octal-to-decimal conversion　conversión de octal a
　　decimal
octal tube　tubo octal
octantal error　error octantal
octave　octava
octave band　banda de octava
octave-band amplifier　amplificador de banda de
　　octava
octave-band analysis　análisis de banda de octava
octave-band noise analyzer　analizador de ruido de
　　banda de octava
octave-band oscillator　oscilador de banda de octava
octave-band pressure level　nivel de presión de
　　banda de octava
octave filter　filtro de octava
octave pressure level　nivel de presión de octava
octet　octeto
octode　octodo
octode tube　tubo octodo
octonary　octonario
octonary signal　señal octonaria
octonary system　sistema octonario
ocular prism　prisma ocular
odd channel　canal impar
odd-even bit　bit de paridad
odd-even check　comprobación de paridad
odd-even interleaving　intercalación de paridad
odd harmonic　armónica impar
odd-harmonic distortion　distorsión por armónicas
　　impares
odd-harmonic intensification　intensificación de
　　armónicas impares
odd line　línea impar
odd-line field　campo de líneas impares
odd-line interlace　entrelazado de líneas impares
odd-line interlacing　entrelazamiento de líneas
　　impares
odd-numbered line　línea impar
odd parity　paridad impar
odd-parity bit　bit de paridad impar
odd-parity check　comprobación de paridad impar
odd scanning field　campo de exploración impar
odograph　odógrafo
odometer　odómetro
oersted　oersted
off-axis mode　modo fuera de eje
off-balance　desequilibrado
off-balance bridge　puente desequilibrado
off-center display　visualización descentrada
off-center-fed antenna　antena alimentada fuera de
　　centro
off-center feed　alimentación fuera de centro

off-circuit fuera de circuito
off-delay retardo de desconexión
off-frequency fuera de frecuencia
off-frequency signal señal fuera de frecuencia
off-hook descolgado
off-hook condition condición de descolgado
off isolation aislamiento de desconexión
off-limit fuera del límite
off-line fuera de línea, indirecto
off-line application aplicación fuera de línea
off-line correction corrección fuera de línea
off-line device dispositivo fuera de línea
off-line diagnostic program programa de diagnóstico fuera de línea
off-line equipment equipo fuera de línea
off-line function función fuera de línea
off-line mode modo fuera de línea
off-line operation operación fuera de línea
off-line processing procesamiento fuera de línea
off-line storage almacenamiento fuera de línea
off-line system sistema fuera de línea
off-line test prueba fuera de línea
off-line unit unidad fuera de línea
off-on operation operación de desconexión y conexión
off-peak period periodo de carga reducida
off period periodo de desconexión
off position posición de desconexión
off-punch perforación desalineada
off-scale fuera de escala
off state estado de desconexión
off-state voltage tensión en estado de desconexión
off the air fuera del aire
off-time tiempo de desconexión, tiempo de inactividad
off tune fuera de sintonía
offhook descolgado
office alarm alarma de oficina
office selector selector de oficina
offline fuera de línea, indirecto
offset desequilibrio, desviación, compensación, línea secundaria
offset angle ángulo de excentricidad
offset channel canal desplazado
offset current corriente de desplazamiento
offset diode diodo de compensación
offset frequency frecuencia desplazada
offset oscillator oscilador desplazado
offset stacker dispositivo de apilamiento desplazado
offset stacking apilamiento desplazado
offset-stacking device dispositivo de apilamiento desplazado
offset voltage tensión de desequilibrio
offset zero cero desplazado
ohm ohm
Ohm's law ley de Ohm
ohm-centimeter ohm-centímetro
ohm-mile ohm-milla
ohmage ohmiaje
ohmic óhmico
ohmic component componente óhmico
ohmic contact contacto óhmico
ohmic drop caída óhmica
ohmic heating calentamiento óhmico
ohmic load carga óhmica
ohmic loss pérdida óhmica
ohmic overvoltage sobretensión óhmica
ohmic region región óhmica
ohmic resistance resistencia óhmica
ohmic resistance test prueba de resistencia óhmica
ohmic response respuesta óhmica
ohmic value valor óhmico
ohmic voltage component componente óhmico de

tensión
ohmmeter óhmetro
ohms per square ohms por cuadrado
ohms per volt ohms por volt
ohms-per-volt rating clasificación de ohms por volt
oil bath baño de aceite
oil capacitor capacitor de aceite
oil circuit-breaker cortacircuito en aceite
oil condenser condensador de aceite
oil-cooled enfriado por aceite
oil-cooled transformer transformador enfriado por aceite
oil-cooled tube tubo enfriado por aceite
oil dielectric dieléctrico de aceite
oil-diffusion pump bomba de difusión de aceite
oil-filled cable cable lleno de aceite
oil-filled capacitor capacitor lleno de aceite
oil-filled circuit-breaker cortacircuito lleno de aceite
oil-filled transformer transformador lleno de aceite
oil-fuse cutout cortacircuito de fusible en aceite
oil heater calentador de aceite
oil-immersed apparatus aparato bañado en aceite
oil-immersed circuit-breaker cortacircuito bañado en aceite
oil-immersed transformer transformador bañado en aceite
oil-immersed tube tubo bañado en aceite
oil-impregnated impregnado en aceite
oil-impregnated capacitor capacitor impregnado en aceite
oil-impregnated paper capacitor capacitor de papel impregnado en aceite
oil-insulated transformer transformador aislado por aceite
oil insulator aislador de aceite
oil-level gage indicador de nivel de aceite
oil pressure presión de aceite
oil-pressure alarm alarma de presión de aceite
oil-pressure gage indicador de presión de aceite
oil pump bomba de aceite
oil switch conmutador en aceite
oiled paper papel aceitado
oiltight hermético contra aceite
omega omega
omegatron omegatrón
omit function función de omisión
omit line línea de omisión
omitted card tarjeta omitida
omnibearing rumbo por radiofaro omnidireccional
omnibearing indicator indicador acimutal automático
omnibus channel canal ómnibus
omnibus telegraph system sistema telegráfico ómnibus
omnidirectional omnidireccional
omnidirectional antenna antena omnidireccional
omnidirectional antenna array red de antenas omnidireccionales
omnidirectional beacon radiofaro omnidireccional
omnidirectional gain ganancia omnidireccional
omnidirectional hydrophone hidrófono omnidireccional
omnidirectional microphone micrófono omnidireccional
omnidirectional radio beacon radiofaro omnidireccional
omnidirectional radio range radiofaro omnidireccional
omnidirectional range radiofaro omnidireccional
omnidirectional range station estación de radiofaro omnidireccional
omnidirectional transmitter transmisor omnidireccional

omnidirective omnidirectivo
omnidirective antenna antena omnidirectiva
omnigraph omnígrafo
omnirange radiofaro omnidireccional
on air en el aire
on-call channel canal asignado sin exclusividad
on-course signal señal de en rumbo
on-delay retardo de conexión
on-demand system sistema a demanda
on-hook colgado
on-hook condition condición de colgado
on-hook signal señal de colgado
on-line en línea, directo
on-line application aplicación en línea
on-line backup respaldo en línea
on-line batch processing procesamiento por lotes en línea
on-line central file archivo central en línea
on-line computer computadora en línea
on-line computer system sistema de computadora en línea
on-line data processing procesamiento de datos en línea
on-line data reduction reducción de datos en línea
on-line database base de datos en línea
on-line debugging depuración en línea
on-line device dispositivo en línea
on-line diagnostics diagnósticos en línea
on-line equipment equipo en línea
on-line help ayuda en línea
on-line information información en línea
on-line information service servicio de información en línea
on-line mode modo en línea
on-line operation operación en línea
on-line plotter trazador en línea
on-line process proceso en línea
on-line processing procesamiento en línea
on-line program programa en línea
on-line software programas en línea
on-line storage almacenamiento en línea
on-line system sistema en línea
on-line test prueba en línea
on-line test program programa de prueba en línea
on-line test system sistema de prueba en línea
on-line unit unidad en línea
on-line work trabajo en línea
on-off action acción todo o nada
on-off circuit circuito de conexión y desconexión
on-off control control de conexión y desconexión
on-off cycle ciclo de conexión y desconexión
on-off device dispositivo de conexión y desconexión
on-off keying manipulación todo o nada
on-off operation operación de conexión y desconexión
on-off ratio razón de conexión-desconexión
on-off switch conmutador de conexión-desconexión, conmutador de encendido
on-off telegraphy telegrafía de todo o nada
on-peak period periodo de carga aumentada
on period periodo de conexión
on position posición de conexión
on state estado de conexión
on-state voltage tensión de estado de conexión
on station en estación
on the air en el aire
on-the-fly al vuelo
on-the-fly printer impresora al vuelo
on-the-fly printing impresión al vuelo
on-time tiempo de conexión, tiempo activo
ondograph ondógrafo
ondoscope ondoscopio
ondulator ondulador

one-address code código de una dirección
one-address computer computadora de una dirección
one-address instruction instrucción de una dirección
one-address message mensaje de una dirección
one-chip computer computadora de un chip
one condition condición uno
one-cycle multivibrator multivibrador de ciclo único
one-digit adder sumador de un dígito
one-digit subtracter restador de un dígito
one-dimensional de una dimensión, unidimensional
one-directional de una dirección, unidireccional
one-directional meter medidor de una dirección
one-element gate puerta de un elemento
one-element rotary antenna antena rotativa de un elemento
one-fluid cell pila de electrólito único
one-for-one assembler ensamblador de uno por uno
one-for-one compiler compilador de uno por uno
one-frequency de una frecuencia
one-frequency radiocompass radiocompás de una frecuencia
one gate puerta uno
one-hour duty servicio de una hora
one-input terminal terminal de entrada única
one-level address dirección de nivel único
one-level code código de nivel único
one-level memory memoria de nivel único
one-level signal señal de nivel único
one-level storage almacenamiento de nivel único
one-level subroutine subrutina de nivel único
one output salida uno
one-output signal señal de salida única
one-output terminal terminal de salida única
one-pass assembler ensamblador de una pasada
one-pass compiler compilador de una pasada
one-phase monofásico, de una fase
one-phase alternator alternador monofásico
one-phase current corriente monofásica
one-plus-one address dirección de uno más uno
one-plus-one instruction instrucción de uno más uno
one-point control control de un punto
one-pole monopolo, de un polo
one-position winding arrollamiento de un piso
one-shot circuit circuito monoestable
one-shot logic lógica monoestable
one-shot multivibrator multivibrador monoestable
one-sided de un lado, unilateral
one-signal de una señal
one-stage de una etapa
one state estado uno
one-step operation operación de un paso
one-third-octave band banda de un tercio de octava
one-time fuse fusible de un solo uso
one-to-one assembler ensamblador de uno a uno
one-to-one ratio razón de uno a uno
one-to-one transformer transformador de uno a uno
one-to-one translator traductor de uno a uno
one-to-zero ratio razón de salida uno a salida cero
one-unit call llamada de una unidad
one-way unidireccional, de una dirección
one-way channel canal unidireccional
one-way circuit circuito unidireccional
one-way communication comunicación unidireccional
one-way conduction conducción unidireccional
one-way connection conexión unidireccional
one-way mobile system sistema móvil unidireccional
one-way radio radio unidireccional
one-way repeater repetidor unidireccional
one-way switch conmutador unidireccional
one-way transmission transmisión unidireccional
ones complement complemento a uno
online en línea, directo

opacimeter opacímetro
opacity opacidad
opal lamp lámpara opalina
opaque opaco
opaque body cuerpo opaco
opaque plasma plasma opaco
opaque screen pantalla opaca
opdar opdar
open abierto, cortado
open-air ionization chamber cámara de ionización de aire libre
open antenna antena abierta
open-arc lamp lámpara de arco abierto
open-back cabinet caja abierta por atrás
open-bus system sistema de bus abierto
open-center display visualizador de centro abierto
open-chassis construction construcción de chasis abierto
open circuit circuito abierto
open-circuit admittance admitancia en circuito abierto
open-circuit alarm device dispositivo de alarma de circuito abierto
open-circuit alarm system sistema de alarma de circuito abierto
open-circuit breakdown voltage tensión de ruptura en circuito abierto
open-circuit-breaker cortacircuito abierto
open-circuit cell pila de circuito abierto
open-circuit characteristic característica en circuito abierto
open-circuit current corriente en circuito abierto
open-circuit impedance impedancia en circuito abierto
open-circuit inductance inductancia en circuito abierto
open-circuit jack jack de circuito abierto
open-circuit line línea en circuito abierto
open-circuit monitor monitor de circuito abierto
open-circuit parameter parámetro en circuito abierto
open-circuit plug clavija de circuito abierto
open-circuit potential potencial en circuito abierto
open-circuit reactance reactancia en circuito abierto
open-circuit resistance resistencia en circuito abierto
open-circuit signaling señalización de circuito abierto
open-circuit termination terminación de circuito abierto
open-circuit test prueba de circuito abierto
open-circuit transition transición en circuito abierto
open-circuit voltage tensión en circuito abierto
open-circuit voltage gain ganancia de tensión en circuito abierto
open-circuited line línea en circuito abierto
open code código abierto
open collector colector abierto
open conductor conductor abierto
open core núcleo abierto
open cutout cortacircuito abierto
open cycle ciclo abierto
open-delta connection conexión en delta abierta
open-ended system sistema abierto
open feeder alimentador abierto
open framework armazón abierto
open fuse fusible abierto
open-fuse cutout cortacircuito de fusible abierto
open generator generador abierto
open grid rejilla abierta
open-grid connection conexión de rejilla abierta
open line línea abierta
open-line alarm alarma de línea abierta
open loop bucle abierto
open-loop bandwidth ancho de banda en bucle abierto

open-loop control control en bucle abierto
open-loop control system sistema de control en bucle abierto
open-loop gain ganancia en bucle abierto
open-loop input impedance impedancia de entrada en bucle abierto
open-loop measurement medida en bucle abierto
open-loop output impedance impedancia de salida en bucle abierto
open-loop output resistance resistencia de salida en bucle abierto
open-loop servomechanism servomecanismo de bucle abierto
open-loop system sistema de bucle abierto
open-loop test prueba de bucle abierto
open-loop transfer function función de transferencia en bucle abierto
open-loop voltage gain ganancia de tensión en bucle abierto
open machine máquina abierta
open magnetic circuit circuito magnético abierto
open mode modo abierto
open motor motor abierto
open network red abierta
open path trayectoria abierta
open phase fase abierta
open-phase protection protección contra fase abierta
open-phase relay relé de fase abierta
open plug clavija abierta
open position posición abierta
open relay relé abierto
open resonator resonador abierto
open routine rutina abierta
open subroutine subrutina abierta
open system sistema abierto
open-system architecture arquitectura de sistema abierto
open temperature pickup captor de temperatura abierta
open transmission line línea de transmisión abierta
open-type machine máquina abierta
open-type relay relé abierto
open window ventana abierta
open wire hilo aéreo, línea aérea
open-wire feeder alimentador de línea aérea
open-wire line línea aérea
open-wire loop bucle de línea aérea
open-wire transmission line línea de transmisión aérea
opening abertura
opening of a file abertura de un archivo
opening time tiempo de abertura
operable operable
operable equipment equipo operable
operable time tiempo operable
operand operando
operand field campo operando
operand matrix matriz operando
operate current corriente de operación
operate delay retardo de operación
operate interval intervalo de operación
operate lag retardo de operación
operate level nivel de operación
operate mode modo de operación
operate position posición de operación
operate time tiempo de operación
operate voltage tensión de operación
operating ambient temperature temperatura ambiente de operación
operating angle ángulo de operación
operating area área de operación
operating band banda de operación

operating bias polarización de operación
operating channel canal de operación
operating characteristic característica de operación
operating circuit circuito de operación
operating code código de operación
operating coil bobina de trabajo
operating conditions condiciones de operación
operating console consola de operación
operating contact contacto de trabajo
operating control control de operación
operating current corriente de operación
operating cycle ciclo de operación
operating data datos de operación
operating device dispositivo de operación
operating diskette disquete operativo
operating duty trabajo de operación
operating environment ambiente operativo
operating feature característica de operación
operating frequency frecuencia de operación
operating-frequency range gama de frecuencias de operación
operating instructions instrucciones de operación
operating interface interfaz de operación
operating life vida de operación
operating line línea de operación
operating mechanism mecanismo de operación
operating mode modo de operación
operating overload sobrecarga de operación
operating panel panel de operación
operating parameters parámetros de operación
operating period periodo de operación
operating point punto de operación
operating-point shift desplazamiento del punto de operación
operating position posición de operación
operating power potencia de operación
operating procedure procedimiento de operación
operating range alcance de operación
operating ratio razón de operación
operating reserve reserva de operación
operating sensitivity sensibilidad de operación
operating sequence secuencia de operación
operating speed velocidad de operación
operating station estación de operación
operating system sistema operativo
operating temperature temperatura de operación
operating-temperature limit límite de temperatura de operación
operating-temperature range intervalo de temperatura de operación
operating time tiempo de operación
operating unit unidad de operación
operating voltage tensión de operación
operating-voltage range margen de tensión de operación
operation operación, funcionamiento
operation analysis análisis de operación
operation characteristics características de operación
operation code código de operación
operation-code field campo de código de operación
operation-code register registro de código de operación
operation control control de operación
operation control language lenguaje de control de operación
operation control statement sentencia de control de operación
operation cycle ciclo de operación
operation decoder descodificador de operación
operation duration duración de operación
operation exception excepción de operación
operation expression expresión de operación

operation factor factor de operación
operation field campo de operación
operation grouping agrupamiento de operaciones
operation indicator indicador de operación
operation influence influencia de operación
operation mode modo de operación
operation monitor monitor de operaciones
operation number número de operación
operation part parte de operación
operation procedure procedimiento de operación
operation sequence secuencia de operaciones
operation time tiempo de operación
operational operacional
operational amplifier amplificador operacional
operational character carácter operacional
operational control control operacional
operational control data datos de control operacional
operational differential amplifier amplificador diferencial operacional
operational environment ambiente operacional
operational feature característica operacional
operational flexibility flexibilidad operacional
operational gain ganancia operacional
operational impedance impedancia operacional
operational inductance inductancia operacional
operational maintenance mantenimiento operacional
operational method método operacional
operational mode modo operacional
operational power supply fuente de alimentación operacional
operational procedure procedimiento operacional
operational program programa operacional
operational programming programación operacional
operational readiness disponibilidad operacional
operational relay relé operacional
operational reliability confiabilidad operacional
operational technique técnica operacional
operational test prueba operacional
operational time tiempo operacional
operational unit unidad operacional
operational wavelength longitud de onda operacional
operations per minute operaciones por minuto
operations queue cola de operaciones
operator operador
operator access code código de acceso de operador
operator code código de operador
operator command mando de operador
operator console consola de operador
operator console panel panel de consola de operador
operator control panel panel de control de operador
operator control station estacón de control de operador
operator-oriented machine máquina orientada al operador
operator station estación de operador
ophitron ofitrón
opposed currents corrientes opuestas
opposed forces fuerzas opuestas
opposed windings arrollamientos opuestos
opposing voltage tensión de oposición
opposing winding arrollamiento de oposición
opposite phase fase opuesta
opposite pole polo opuesto
opposite sequence secuencia opuesta
opposition oposición
opposition method método de oposición
optic óptico
optical óptico
optical activity actividad óptica
optical ammeter amperímetro óptico
optical amplification amplificación óptica
optical amplifier amplificador óptico

optical angle ángulo óptico
optical axis eje óptico
optical axis angle ángulo de eje óptico
optical bandwidth ancho de banda óptico
optical bar code código de barras ópticas
optical bar code reader lectora de códigos de barras
  ópticas
optical bench banco óptico
optical cable cable óptico
optical cable assembly ensamblaje de cable óptico
optical cavity cavidad óptica
optical cavity maser máser de cavidad óptica
optical center centro óptico
optical character carácter óptico
optical character reader lectora de caracteres óptica
optical character recognition reconocimiento de
  caracteres ópticos
optical character sensing detección de caracteres
  ópticos
optical combiner combinador óptico
optical communications comunicaciones ópticas
optical comparator comparador óptico
optical computing computación óptica
optical conductor conductor óptico
optical connector conector óptico
optical coupler acoplador óptico
optical coupling acoplamiento óptico
optical data bus bus de datos óptico
optical data links enlaces de datos ópticos
optical density densidad óptica
optical depth profundidad óptica
optical detector detector óptico
optical diffraction difracción óptica
optical disk disco óptico
optical document reader lectora de documentos
  óptica
optical fiber fibra óptica
optical-fiber bundle paquete de fibras ópticas
optical-fiber cable cable de fibra óptica
optical filter filtro óptico
optical filtering filtrado óptico
optical frequency range gama de frecuencias ópticas
optical heterodyning heterodinaje óptico
optical horizon horizonte óptico
optical image imagen óptica
optical indicator indicador óptico
optical isolator aislador óptico
optical landing system sistema de aterrizaje óptico
optical laser disk disco láser óptico
optical lens lente óptica
optical lever amplificador óptico
optical link enlace óptico
optical mark marca óptica
optical mark reader lectora de marcas óptica
optical mark reading lectura de marcas óptica
optical mark recognition reconocimiento de marcas
  óptico
optical maser máser óptico
optical memory memoria óptica
optical mirror espejo óptico
optical mode modo óptico
optical multiplexer multiplexor óptico
optical multiplexing multiplexión óptica
optical multiplexing system sistema de multiplexión
  óptica
optical path trayectoria óptica
optical path length longitud de trayectoria óptica
optical pattern patrón óptico
optical photon fotón óptico
optical playback reproducción óptica
optical power potencia óptica
optical printer impresora óptica
optical probe sonda óptica

optical pumping bombeo óptico
optical pyrometer pirómetro óptico
optical radar radar óptico
optical range alcance óptico
optical reader lectora óptica
optical recognition device dispositivo de
  reconocimiento óptico
optical recording registro óptico
optical reduction reducción óptica
optical reflection reflexión óptica
optical-reflection factor factor de reflexión óptica
optical relay relé óptico
optical repeater repetidor óptico
optical resolution resolución óptica
optical resonance resonancia óptica
optical scanner explorador óptico
optical scanning exploración óptica
optical scanning station estación de exploración
  óptica
optical sound recorder registrador de sonido óptico
optical sound recording registro de sonido óptico
optical sound reproducer reproductor de sonido
  óptico
optical sound track pista de sonido óptica
optical spectrometer espectrómetro óptico
optical spectrum espectro óptico
optical speed velocidad óptica
optical switch conmutador óptico
optical system sistema óptico
optical tachometer tacómetro óptico
optical telecommunication telecomunicación óptica
optical thermometer termómetro óptico
optical thickness grosor óptico
optical track pista óptica
optical transmission transmisión óptica
optical wand varita óptica
optical waveguide guíaondas óptico
optical waveguide connector conector de guíaondas
  óptico
optical waveguide coupler acoplador de guíaondas
  óptico
optical waveguide termination terminación de
  guíaondas óptico
optically ópticamente
optically active ópticamente activo
optically coupled ópticamente acoplado
optically flat ópticamente plano
optics óptica
optimal óptimo
optimization optimización
optimize optimizar
optimized optimizado
optimizer optimizador
optimum óptimo
optimum angle ángulo óptimo
optimum angle of incidence ángulo óptimo de
  incidencia
optimum angle of radiation ángulo óptimo de
  radiación
optimum bias polarización óptima
optimum bunching agrupamiento óptimo
optimum code código óptimo
optimum coding codificación óptima
optimum coupling acoplamiento óptimo
optimum current corriente óptima
optimum damping amortiguamiento óptimo
optimum frequency frecuencia óptima
optimum load carga óptima
optimum load impedance impedancia de carga
  óptima
optimum performance rendimiento óptimo
optimum plate load carga de placa óptima
optimum program programa óptimo

optimum programming programación óptima
optimum reliability confiabilidad óptima
optimum routine rutina óptima
optimum traffic frequency frecuencia de tráfico óptima
optimum working frequency frecuencia de trabajo óptima
option code código de opción
option field campo de opciones
optional opcional
optional device dispositivo opcional
optional halt parada opcional
optional halt instruction instrucción de parada opcional
optional stop parada opcional
optional stop instruction instrucción de parada opcional
optional task tarea opcional
optocoupler optoacoplador
optoelectric optoeléctrico
optoelectric coupler acoplador optoeléctrico
optoelectronic optoelectrónico
optoelectronic amplifier amplificador optoelectrónico
optoelectronic device dispositivo optoelectrónico
optoelectronic element elemento optoelectrónico
optoelectronic integrated circuit circuito integrado optoelectrónico
optoelectronic relay relé optoelectrónico
optoelectronic transistor transistor optoelectrónico
optoelectronics optoelectrónica
optoisolator optoaislador
optophone optófono
OR circuit circuito O
OR element elemento O
OR gate puerta O
OR logic lógica O
OR operation operación O
OR operator operador O
OR relationship relación O
orange peel superficie en piel de naranja
orbit órbita
orbital orbital
orbital-beam tube tubo de haz orbital
orbital electron electrón orbital
orbital energy energía orbital
orbital multiplier multiplicador orbital
orbital period periodo orbital
orbital velocity velocidad orbital
order orden
order card tarjeta de orden
order code código de orden
order entry entrada de orden
order format formato de orden
order of calls orden de llamadas
order of interference orden de interferencia
order of logic orden de lógica
order of magnitude orden de magnitud
order of priority orden de prioridad
order of reflection orden de reflexión
order processing procesamiento de orden
order tone tono de orden
order wire línea de órdenes
order-wire circuit circuito de línea de órdenes
orderly halt parada ordenada
ordinal data datos ordinales
ordinary ordinario
ordinary binary binario ordinario
ordinary component componente ordinario
ordinary differential equation ecuación diferencial ordinaria
ordinary ray rayo ordinario
ordinary wave onda ordinaria

ordinary-wave component componente de onda ordinaria
ordinate ordenada
organic semiconductor semiconductor orgánico
orientation orientación
orientation angle ángulo de orientación
orientation control control de orientación
orientation indicator indicador de orientación
orientation range intervalo de orientación
orientational orientacional
orientational axis eje orientacional
oriented orientado
oriented ferrite ferrita orientada
orifice orificio
origin origen
origin distortion distorsión de origen
original address dirección original
original data datos originales
original lacquer original en laca
original master negativo original
originating de origen
originating circuit circuito de origen
originating point punto de origen
originating station estación de origen
originating tape cinta de origen
originating terminal terminal de origen
originating unit unidad de origen
orioscope orioscopio
orthicon orticón
orthicon camera cámara de orticón
orthiconoscope orticonoscopio
orthocode ortocódigo
orthocore ortonúcleo
orthogonal axes ejes ortogonales
orthogonal mode modo ortogonal
orthogonal projection proyección ortogonal
orthogonality ortogonalidad
orthorhombic ortorrómbico
orthorhombic system sistema ortorrómbico
osciducer osciductor
oscillate oscilar
oscillating oscilante
oscillating arc arco oscilante
oscillating circuit circuito oscilante
oscillating-circuit inductance inductancia de circuito oscilante
oscillating coil bobina oscilante
oscillating crystal cristal oscilante
oscillating-crystal method método del cristal oscilante
oscillating current corriente oscilante
oscillating detector detector oscilante
oscillating diode diodo oscilante
oscillating discharge descarga oscilante
oscillating electric field campo eléctrico oscilante
oscillating klystron klistrón oscilante
oscillating meter medidor oscilante
oscillating rod varilla oscilante
oscillating sort clasificación oscilante
oscillating transducer transductor oscilante
oscillating transformer transformador oscilante
oscillating voltage tensión oscilante
oscillating wire hilo oscilante
oscillation oscilación
oscillation absorber absorbedor de oscilaciones
oscillation amplitude amplitud de oscilación
oscillation constant constante de oscilación
oscillation control control de oscilación
oscillation damping amortiguamiento de oscilaciones
oscillation detector detector de oscilaciones
oscillation efficiency eficiencia de oscilación
oscillation frequency frecuencia de oscilación
oscillation generator generador de oscilaciones

oscillation mode modo de oscilación
oscillation number número de oscilación
oscillation test prueba de oscilación
oscillation time tiempo de oscilación
oscillation transformer transformador de oscilación
oscillator oscilador
oscillator capacitor capacitor oscilador
oscillator circuit circuito oscilador
oscillator coil bobina de oscilador
oscillator-converter oscilador-convertidor
oscillator-detector oscilador-detector
oscillator-doubler oscilador-doblador
oscillator drift deslizamiento de frecuencia de
    oscilador
oscillator frequency frecuencia de oscilador
oscillator grid rejilla osciladora
oscillator harmonic interference interferencia
    armónica de oscilador
oscillator interference interferencia de oscilador
oscillator klystron klistrón oscilador
oscillator-mixer oscilador-mezclador
oscillator-mixer-detector
    oscilador-mezclador-detector
oscillator mode modo de oscilador
oscillator-multiplier oscilador-multiplicador
oscillator padder pader de oscilador
oscillator power supply fuente de alimentación de
    oscilador
oscillator radiation radiación de oscilador
oscillator stabilization estabilización de oscilador
oscillator synchronization sincronización de
    oscilador
oscillator transistor transistor de oscilador
oscillator trimmer condensador de ajuste de
    oscilador
oscillator triode triodo oscilador
oscillator tube tubo oscilador
oscillator tuning sintonización de oscilador
oscillator unit unidad osciladora
oscillator valve válvula osciladora
oscillator wavelength longitud de onda de oscilador
oscillatory oscilante
oscillatory circuit circuito oscilante
oscillatory current corriente oscilante
oscillatory discharge descarga oscilante
oscillatory electromotive force fuerza electromotriz
    oscilante
oscillatory scanning exploración oscilante
oscillatory surge sobretensión oscilante
oscillatory transient fluctuación transitoria oscilante
oscillistor oscilistor
oscillogram oscilograma
oscillograph oscilógrafo
oscillograph galvanometer galvanómetro
    oscilográfico
oscillograph recorder registrador oscilográfico
oscillograph tube tubo de oscilógrafo
oscillographic oscilográfico
oscillographic display visualización oscilográfica
oscillographic record registro oscilográfico
oscillographic recorder registrador oscilográfico
oscillography oscilografía
oscillometer oscilómetro
oscilloscope osciloscopio
oscilloscope camera cámara osciloscópica
oscilloscope differential amplifier amplificador
    diferencial de osciloscopio
oscilloscope display visualizador osciloscópico
oscilloscope recording registro osciloscópico
oscilloscope screen pantalla osciloscópica
oscilloscope trace trazo osciloscópico
oscilloscope tube tubo de osciloscopio
oscilloscopic osciloscópico

oscilloscopic display visualizador osciloscópico
osmiridium osmiridio
osmium osmio
osmosis ósmosis
osmotic osmótico
osmotic coefficient coeficiente osmótico
osmotic pressure presión osmótica
osophone osófono
osteophone osteófono
Ostwald electrode electrodo de Ostwald
Oudin current corriente de Oudin
Oudin resonator resonador de Oudin
out-band signaling señalización fuera de banda
out of balance fuera de equilibrio, desequilibrado
out of band fuera de banda
out-of-band assignment asignación fuera de banda
out-of-band radiation radiación fuera de banda
out-of-band signal señal fuera de banda
out-of-band signaling señalización fuera de banda
out of contact fuera de contacto
out of order fuera de servicio
out-of-phase desfasado
out-of-phase amplifier amplificador desfasado
out-of-phase component componente desfasado
out-of-phase current corriente desfasada
out-of-phase recording registro desfasado
out-of-phase signal señal desfasada
out-of-phase voltage tensión desfasada
out of service fuera de servicio
out-of-service time tiempo fuera de servicio
out of step fuera de sincronismo
out-of-step protection protección contra pérdida de
    sincronismo
out-of-tolerance value valor fuera de tolerancia
out of use fuera de uso
outage interrupción, interrupción de servicio
outage duration duración de interrupción
outage initiation iniciación de interrupción
outage rate tasa de interrupciones
outage state estado de interrupción
outage time tiempo de interrupción
outage type tipo de interrupción
outboard components componentes externos
outdoor exterior
outdoor antenna antena exterior
outdoor apparatus aparato exterior
outdoor arrester protector exterior
outdoor circuit-breaker cortacircuito exterior
outdoor disconnecting switch seccionador exterior
outdoor electrical equipment equipo eléctrico
    exterior
outdoor enclosure caja exterior
outdoor substation subestación exterior
outdoor termination terminación exterior
outdoor transformer transformador exterior
outer brush escobilla exterior
outer coating revestimiento exterior
outer conductor conductor exterior
outer electrode electrodo exterior
outer grid rejilla exterior
outer-grid injection inyección por rejilla exterior
outer macroinstruction macroinstrucción exterior
outer marker radiobaliza exterior
outer marker beacon radiobaliza exterior
outer modulation modulación exterior
outer program programa exterior
outer winding arrollamiento exterior
outgassing fuga de gases
outgoing de salida
outgoing access acceso de salida
outgoing call llamada de salida
outgoing channel canal de salida
outgoing circuit circuito de salida

outgoing circuit-breaker cortacircuito de salida
outgoing current corriente de salida
outgoing exchange central de salida
outgoing feeder alimentador de salida
outgoing group grupo de salida
outgoing junction unión de salida
outgoing line línea de salida
outgoing message mensaje de salida
outgoing position posición de salida
outgoing selector selector de salida
outgoing signal señal de salida
outgoing traffic tráfico de salida
outgoing trunk enlace de salida
outgoing wave onda de salida
outlet tomacorriente, salida
outlet box caja de salida
outphasing desfasamiento, desfase
outphasing modulation modulación por
   desfasamiento
output salida, señal de salida, potencia de salida
output admittance admitancia de salida
output amplifier amplificador de salida
output angle ángulo de salida
output antenna antena de salida
output area área de salida
output attenuator atenuador de salida
output axis eje de salida
output block bloque de salida
output blocking capacitor capacitor de bloqueo de
   salida
output blocking factor factor de bloqueo de salida
output-bound limitado por salida
output buffer memoria intermedia de salida
output bus driver controlador de bus de salida
output capability capacidad de salida
output capacitance capacitancia de salida
output capacitive loading carga capacitiva de salida
output capacitor capacitor de salida
output capacity capacidad de salida
output cavity cavidad de salida
output channel canal de salida
output choke choque de salida
output circuit circuito de salida
output-circuit distortion distorsión de circuito de
   salida
output coefficient coeficiente de salida
output conductance conductancia de salida
output contact contacto de salida
output control control de salida
output coupling capacitor capacitor de acoplamiento
   de salida
output coupling loop bucle de acoplamiento de salida
output coupling transformer transformador de
   acoplamiento de salida
output current corriente de salida
output curve curva de salida
output data datos de salida
output data set conjunto de datos de salida
output device dispositivo de salida
output efficiency eficiencia de salida
output electrode electrodo de salida
output enable habilitación de salida
output energy energía de salida
output equipment equipo de salida
output error voltage tensión de error de salida
output factor factor de salida
output feedback retroalimentación de salida
output feeder alimentador de salida
output field campo de salida
output file archivo de salida
output filter filtro de salida
output format formato de salida
output frequency frecuencia de salida

output gap espacio de salida
output governor regulador de salida
output hopper almacén de alimentación de salida
output impedance impedancia de salida
output impulse amplitude amplitud de impulso de
   salida
output indicator indicador de salida
output jack jack de salida
output job queue cola de trabajos de salida
output job stream corriente de trabajos de salida
output language lenguaje de salida
output leakage current corriente de fuga de salida
output level nivel de salida
output limit límite de salida
output limiter limitador de salida
output limiting limitación de salida
output line línea de salida
output load carga de salida
output load current corriente de carga de salida
output medium medio de salida
output meter medidor de salida
output-meter adapter adaptador de medidor de
   salida
output mode modo de salida
output module módulo de salida
output monitor monitor de salida
output multiplexer multiplexor de salida
output network red de salida
output pentode pentodo de salida
output peripheral periférico de salida
output port puerto de salida
output power potencia de salida
output-power meter medidor de potencia de salida
output printer impresora de salida
output priority prioridad de salida
output procedure procedimiento de salida
output process proceso de salida
output program programa de salida
output pulse impulso de salida
output-pulse amplitude amplitud de impulso de
   salida
output-pulse duration duración de impulso de salida
output punch perforadora de salida
output queue cola de salida
output reactance reactancia de salida
output reference axis eje de referencia de salida
output register registro de salida
output register buffer memoria intermedia de
   registro de salida
output regulator regulador de salida
output request petición de salida
output resistance resistencia de salida
output resonator resonador de salida
output routine rutina de salida
output section sección de salida
output signal señal de salida
output source current fuente de corriente de salida
output specifications especificaciones de salida
output speed velocidad de salida
output stage etapa de salida
output state estado de salida
output stream corriente de salida
output stream control control de corriente de salida
output subsystem subsistema de salida
output tape cinta de salida
output terminal terminal de salida
output test prueba de salida
output time constant constante de tiempo de salida
output transformer transformador de salida
output transistor transistor de salida
output triode triodo de salida
output tube tubo de salida
output unit unidad de salida

output variable variable de salida
output video signal videoseñal de salida
output voltage tensión de salida
output voltage regulation regulación de tensión de salida
output voltage stabilization estabilización de tensión de salida
output voltage swing oscilación de tensión de salida
output voltmeter vóltmetro de salida
output wave onda de salida
output waveform forma de onda de salida
output winding arrollamiento de salida
output window ventana de salida
output work queue cola de trabajos de salida
outside exterior
outside antenna antena exterior
outside cable cable exterior
outside conductor conductor exterior
outside diameter diámetro exterior
outside foil hoja exterior
outside lead hilo externo
outside transformer transformador exterior
outward board cuadro de salida
outward pulse impulso de salida
outward traffic tráfico de salida
outward trunk troncal de salida
oval cathode cátodo ovalado
oval grid rejilla ovalada
oval loudspeaker altavoz ovalado
over-the-horizon communication comunicación transhorizonte
over-the-horizon link enlace transhorizonte
over-the-horizon radar radar transhorizonte
over-the-horizon signal señal transhorizonte
over-the-horizon transmission transmisión transhorizonte
overall amplification amplificación total
overall attenuation atenuación total
overall efficiency eficiencia global
overall electrical efficiency eficiencia eléctrica global
overall feedback retroalimentación global
overall gain ganancia total
overall loss pérdida total
overall loudness intensidad sonora aparente
overall noise factor factor de ruido global
overall noise level nivel de ruido global
overall power efficiency eficiencia de potencia total
overall power gain ganancia de potencia total
overall response curve curva de respuesta global
overall selectivity selectividad global
overall volume volumen global
overbias sobrepolarización
overbunching sobreagrupamiento
overcharge sobrecargar
overcompounded excitation excitación hipercompuesta
overcompounded generator generador hipercompuesto
overcoupled sobreacoplado
overcoupled transformer transformador sobreacoplado
overcoupling sobreacoplamiento
overcurrent sobrecorriente
overcurrent circuit-breaker cortacircuito de sobrecorriente
overcurrent device dispositivo de sobrecorriente
overcurrent factor factor de sobrecorriente
overcurrent protection protección contra sobrecorriente
overcurrent protective device dispositivo protector contra sobrecorriente
overcurrent relay relé de sobrecorriente

overcutting sobrecorte
overdamped sobreamortiguado
overdamped galvanometer galvanómetro sobreamortiguado
overdamping sobreamortiguamiento
overdesign sobrediseñar
overdrive sobreexcitación
overdriven sobreexcitado
overdriven amplifier amplificador sobreexcitado
overdriven unit unidad sobreexcitada
overexcitation sobreexcitación
overexcited sobreexcitado
overexpose sobreexponer
overexposure sobreexposición
overflow desbordamiento, rebose
overflow area área de desbordamiento
overflow bucket compartimiento de desbordamiento
overflow channel canal de desbordamiento
overflow check comprobación de desbordamiento
overflow condition condición de desbordamiento
overflow data datos de desbordamiento
overflow digit dígito de desbordamiento
overflow error error por desbordamiento
overflow exception excepción de desbordamiento
overflow field campo de desbordamiento
overflow indicator indicador de desbordamiento
overflow line línea de desbordamiento
overflow link enlace de desbordamiento
overflow loss pérdida por desbordamiento
overflow meter medidor de desbordamiento
overflow operation operación de desbordamiento
overflow position posición de desbordamiento
overflow record registro de desbordamiento
overflow register registro de desbordamiento
overflow route ruta de desbordamiento
overflow skip salto por desbordamiento
overflow storage almacenamiento de desbordamiento
overflow tape cinta de desbordamiento
overflow track pista de desbordamiento
overflow traffic tráfico de desbordamiento
overflow valve válvula de desbordamiento
overflux sobreflujo
overfrequency sobrefrecuencia
overfrequency relay relé de sobrefrecuencia
overhead cable cable aéreo
overhead conductor conductor aéreo
overhead line línea aérea
overhead wire línea aérea
overland cable cable terrestre
overland route ruta terrestre
overlap solape, ángulo de recubrimiento
overlap angle ángulo de recubrimiento
overlapping solapamiento, recubrimiento
overlapping contacts contactos con solapamiento
overlay capa superpuesta, superposición, recubrimiento
overlay path trayectoria de recubrimiento
overlay program programa de recubrimiento
overlay segment segmento de recubrimiento
overlay transistor transistor de capa superpuesta
overload sobrecarga
overload capacity capacidad de sobrecarga
overload characteristic característica de sobrecarga
overload circuit-breaker cortacircuito de sobrecarga
overload current corriente de sobrecarga
overload cutout cortacircuito de sobrecarga
overload distortion distorsión de sobrecarga
overload factor factor de sobrecarga
overload indication indicación de sobrecarga
overload indicator indicador de sobrecarga
overload indicator lamp lámpara indicadora de sobrecarga
overload level nivel de sobrecarga

overload margin margen de sobrecarga
overload point punto de sobrecarga
overload protection protección contra sobrecarga
overload protective device dispositivo de protección contra sobrecarga
overload pulse impulso de sobrecarga
overload recovery time tiempo de recuperación tras sobrecarga
overload relay relé de sobrecarga
overload release desconexión por sobrecarga
overload time tiempo de sobrecarga
overloaded sobrecargado
overloaded amplifier amplificador sobrecargado
overloaded circuit circuito sobrecargado
overloaded oscillator oscilador sobrecargado
overmodulate sobremodular
overmodulation sobremodulación
overmodulation alarm alarma de sobremodulación
overmodulation indicator indicador de sobremodulación
overpotential sobrepotencial
overpower sobrepotencia
overpower relay relé de sobrepotencia
overpressure sobrepresión
overprint sobreimprimir
overprinting sobreimpresión
overpunch sobreperforación
overpunching sobreperforación
overscanning sobreexploración
overshoot sobredisparo, sobreimpulso, sobreelongación, sobremodulación, sobrealcance
overshoot distortion distorsión de sobreelongación
overspeed sobrevelocidad
overspeed protection protección contra sobrevelocidad
overspeed test prueba de sobrevelocidad
overswing sobreoscilación
overtemperature sobretemperatura
overtemperature alarm circuit circuito de alarma de sobretemperatura
overtemperature detector detector de sobretemperatura
overtemperature indicator indicador de sobretemperatura
overtemperature protection protección contra sobretemperatura
overthrow sobrerrecorrido
overthrow distortion distorsión por sobrerrecorrido
overtone sobretono
overtone crystal cristal de sobretono
overtone oscillator oscilador de sobretono
overtravel sobrerrecorrido
overtravel limit límite de sobrerrecorrido
overtravel switch conmutador de sobrerrecorrido
overvoltage sobretensión
overvoltage circuit-breaker cortacircuito de sobretensión
overvoltage connection conexión de sobretensión
overvoltage cutout cortacircuito de sobretensión
overvoltage protection protección contra sobretensión
overvoltage protection device dispositivo de protección contra sobretensión
overvoltage protection level nivel de protección contra sobretensión
overvoltage relay relé de sobretensión
overvoltage suppressor supresor de sobretensión
overvoltage test prueba de sobretensión
overwrite sobreescribir
Ovshinsky effect efecto de Ovshinsky
Owen bridge puente de Owen
own coding codificación propia
oxidant oxidante

oxidation oxidación
oxidation-reduction potential potencial de oxidación-reducción
oxidation-resistant resistente a la oxidación
oxide óxido
oxide buildup acumulación de óxido
oxide cathode cátodo revestido de óxido
oxide-cathode tube tubo con cátodo revestido de óxido
oxide-cathode valve válvula con cátodo revestido de óxido
oxide-coated revestido de óxido
oxide-coated cathode cátodo revestido de óxido
oxide-coated emitter emisor revestido de óxido
oxide-coated filament filamento revestido de óxido
oxide-coated tape cinta revestida de óxido
oxide coating revestimiento de óxido
oxide dispersion dispersión de óxido
oxide emitter emisor de óxido
oxide film película de óxido
oxide-film capacitor capacitor de película de óxido
oxide ratio razón de óxido
oxide rectifier rectificador de óxido
oxidize oxidar
oxidizing oxidante
oximeter oxímetro
oxy-arc cutting oxicorte por arco
oxygen oxígeno
oxygen analyzer analizador de oxígeno
ozone ozono

P

P-type area área tipo P
P-type base base tipo P
P-type collector colector tipo P
P-type conduction conducción tipo P
P-type conductivity conductividad tipo P
P-type emitter emisor tipo P
P-type germanium germanio tipo P
P-type germanium junction unión de germanio tipo P
P-type material material tipo P
P-type semiconductor semiconductor tipo P
P-type silicon silicio tipo P
pacemaker marcapasos
pack paquete, conjunto
package paquete, conjunto, cubierta
packaged empaquetado, integrado, encapsulado
packaged magnetron magnetrón integrado
packaged system sistema integrado
packaging empaquetado, integración
packaging density densidad de información, densidad de componentes
packed empaquetado
packed binary data datos binarios empaquetados
packed data datos empaquetados
packed data field campo de datos empaquetados
packed decimal decimal empaquetado
packed decimal data datos decimales empaquetados
packed decimal format formato decimal empaquetado
packed decimal item artículo decimal empaquetado
packed field campo empaquetado
packed format formato empaquetado
packed key clave empaquetada
packed numerical field campo numérico empaquetado
packet paquete
packet assembler-disassembler ensamblador-desensamblador de paquete
packet communication comunicación de paquetes
packet data link interface interfaz de enlace de paquetes de datos
packet data link processor procesador de enlace de paquetes de datos
packet length longitud de paquete
packet level nivel de paquete
packet level interface interfaz de nivel de paquete
packet mode operation operación de modo de paquetes
packet mode terminal terminal de modo de paquetes
packet sequencing secuenciamiento de paquetes
packet switching conmutación de paquetes
packet switching data network red de datos de conmutación de paquetes
packet switching network red de conmutación de paquetes
packet transmission transmisión de paquetes
packing empaquetamiento, empaque, compactación
packing density densidad de información, densidad de componentes
packing factor factor de empaquetamiento, densidad de información, densidad de componentes

packing fraction fracción de empaque
pacor pacor
pad atenuador fijo, área terminal, encadenamiento
pad character carácter de relleno
padar padar
padder pader
padding byte byte de relleno
padding capacitor capacitor de compensación
padding character carácter de relleno
padding condenser condensador de compensación
padding device dispositivo de compensación
padding inductance inductancia de compensación
padding record registro de relleno
paddle paleta
paddle-handle switch conmutador de palanca en paleta
page addressing direccionamiento de página
page buffer almacenamiento intermedio de página
page-change hole perforación de cambio de página
page control control de página
page-control block bloque de control de página
page data datos de página
page data set conjunto de datos de página
page definition definición de página
page depth control control de profundidad de página
page description language lenguaje de descripción de página
Page effect efecto de Page
page-end character carácter de final de página
page exit salida de página
page fault falla de página
page frame cuadro de página
page input-output entrada-salida de página
page key clave de página
page layout program programa de distribución de página
page mode modo de página
page number número de página
page orientation orientación de página
page printer impresora de página
page protection protección de página
page reader lectora de páginas
page segment segmento de página
page skip salto de página
page teleprinter teleimpresora de página
page wait espera de página
pageable paginable
pageable dynamic area área dinámica paginable
pageable partition partición paginable
pageable region región paginable
pager dispositivo emisor de tonos breves
pagination paginación
paging paginación, búsqueda
paging algorithm algoritmo de paginación
paging area área de paginación
paging device dispositivo de paginación
paging rate velocidad de paginación
paging system sistema de paginación
paging technique técnica de paginación
pair par
pair attenuation coefficient coeficiente de atenuación de pares
pair production producción de pares
pair production absorption absorción de producción de pares
paired bars barras apareadas
paired brushes escobillas apareadas
paired cable cable de pares
paired data datos apareados
pairing apareamiento
paleomagnetism paleomagnetismo
palladium paladio
Palmer scan exploración de Palmer

pancake coil   bobina plana
pancake loudspeaker   altavoz plano
pancake motor   motor plano
panchromatic   pancromático
panchromatic film   película pancromática
panchromatic plate   placa pancromática
panel   panel, bastidor
panel control   control de panel
panel data set   conjunto de datos de panel
panel definition program   programa de definición de panel
panel efficiency   eficiencia de panel
panel extension   extensión de panel
panel illumination   iluminación de panel
panel lamp   lámpara de panel
panel layout   distribución de panel
panel light   luz de panel
panel meter   medidor de panel
panel mounting   montaje de panel
panel number   número de panel
panel switch   conmutador de panel
panel system   sistema de panel
panoramic   panorámico
panoramic adapter   adaptador panorámico
panoramic analyzer   analizador panorámico
panoramic attenuator   atenuador panorámico
panoramic camera   cámara panorámica
panoramic control   control panorámico
panoramic display   visualización panorámica
panoramic display device   dispositivo de visualización panorámica
panoramic indicator   indicador panorámico
panoramic monitor   monitor panorámico
panoramic potentiometer   potenciómetro panorámico
panoramic presentation   presentación panorámica
panoramic radar   radar panorámico
panoramic receiver   receptor panorámico
panoramic reception   recepción panorámica
panoramic sound   sonido panorámico
pantograph   pantógrafo
pantography   pantografía
paper advance mechanism   mecanismo de avance de papel
paper capacitor   capacitor de papel
paper capacity   capacidad de papel
paper card   tarjeta de papel
paper channel   canal de papel
paper condenser   condensador de papel
paper control   control de papel
paper-control tape   cinta de control de papel
paper-covered wire   hilo recubierto de papel
paper cutter   cortadora de papel
paper deflector   desviador de papel
paper extraction   extracción de papel
paper extractor   extractor de papel
paper feed   alimentación de papel
paper-feed control   control de alimentación de papel
paper-feed guide   guía de alimentación de papel
paper-feed mechanism   mecanismo de alimentación de papel
paper-feed roller   rodillo de alimentación de papel
paper-feed tray   bandeja de alimentación de papel
paper guide   guía de papel
paper handler   manejador de papel
paper handling   manejo de papel
paper holder   sujetador de papel
paper injection   inyección de papel
paper injector   inyector de papel
paper injector-extractor   inyector-extractor de papel
paper-insulated cable   cable aislado por papel
paper-insulated wire   hilo aislado por papel
paper mask   máscara de papel
paper path   trayectoria de papel

paper release   liberación de papel
paper sensor   sensor de papel
paper skip   salto de papel
paper speed   velocidad de papel
paper tape   cinta de papel
paper-tape channel   canal de cinta de papel
paper-tape code   código de cinta de papel
paper-tape controller   controlador de cinta de papel
paper-tape input   entrada de cinta de papel
paper-tape loop   bucle de cinta de papel
paper-tape memory   memoria de cinta de papel
paper-tape perforator   perforadora de cinta de papel
paper-tape punch   perforadora de cinta de papel
paper-tape reader   lectora de cinta de papel
paper-tape reproducer   reproductor de cinta de papel
paper-tape-to-card converter   convertidor de cinta de papel a tarjetas
paper-tape verification   verificación de cinta de papel
paper-tape verifier   verificador de cinta de papel
paper track   pista de papel
paper tractor   tractor de papel
parabola   parábola
parabolic   parabólico
parabolic antenna   antena parabólica
parabolic detection   detección parabólica
parabolic microphone   micrófono parabólico
parabolic mirror   espejo parabólico
parabolic profile   perfil parabólico
parabolic pulse   impulso parabólico
parabolic reflector   reflector parabólico
parabolic-reflector antenna   antena de reflector parabólico
parabolic-reflector microphone   micrófono de reflector parabólico
parabolic waveform   forma de onda parabólica
paraboloid   paraboloide
paraboloid reflector   reflector paraboloidal
paraboloidal   paraboloidal
paraboloidal antenna   antena paraboloidal
paraboloidal reflector   reflector paraboloidal
paraboloidal-reflector antenna   antena de reflector paraboloidal
paraelastic crystal   cristal paraelástico
paraelectric   paraeléctrico
paraelectric phase   fase paraeléctrica
paraelectric region   región paraeléctrica
paraffin   parafina
parageometrical optics   óptica parageométrica
paragraph mark   marca de párrafo
paragutta   paraguta
paragutta insulation   aislamiento de paraguta
parallax   paralaje
parallax compensation   compensación de paralaje
parallax correction   corrección de paralaje
parallax error   error de paralaje
parallax-free   libre de paralaje
parallel   paralelo
parallel access   acceso en paralelo
parallel adder   sumador en paralelo
parallel addition   suma en paralelo
parallel antenna tuning   sintonización de antena en paralelo
parallel arithmetic   aritmética paralela
parallel arithmetic unit   unidad aritmética paralela
parallel arrangement   arreglo en paralelo
parallel bandspread   ensanche de banda paralelo
parallel beam   haz paralelo
parallel binary logic   lógica binaria en paralelo
parallel capacitance   capacitancia en paralelo
parallel capacitor   capacitor en paralelo
parallel cathode heating   calentamiento catódico paralelo
parallel character transmission   transmisión de

caracteres paralela
parallel circuit circuito en paralelo
parallel columns columnas paralelas
parallel-component amplifier amplificador de componentes en paralelo
parallel-component oscillator oscilador de componentes en paralelo
parallel components componentes en paralelo
parallel computer computadora en paralelo
parallel-connected conectado en paralelo
parallel-connected capacitors capacitores conectados en paralelo
parallel-connected generators generadores conectados en paralelo
parallel connection conexión en paralelo
parallel connector conector paralelo
parallel conversion conversión en paralelo
parallel conversion system sistema de conversión en paralelo
parallel cut corte paralelo
parallel-cut crystal cristal de corte paralelo
parallel data medium medio de datos en paralelo
parallel digital computer computadora digital paralela
parallel distribution distribución en paralelo
parallel elements elementos en paralelo
parallel entry entrada en paralelo
parallel equalizer ecualizador en paralelo
parallel equivalent circuit circuito equivalente paralelo
parallel feed alimentación en paralelo
parallel-feed system sistema de alimentación en paralelo
parallel feeder alimentador en paralelo
parallel feeding alimentación en paralelo
parallel flow flujo paralelo
parallel-flow electron gun cañón electrónico de flujo paralelo
parallel full-adder sumador completo en paralelo
parallel full-subtracter restador completo en paralelo
parallel gate circuit circuito de puerta en paralelo
parallel half-adder semisumador en paralelo
parallel half-subtracter semirrestador en paralelo
parallel impedance impedancia en paralelo
parallel-impedance components componentes de impedancia en paralelo
parallel inductance inductancia en paralelo
parallel inductors inductores en paralelo
parallel input-output entrada-salida en paralelo
parallel input-output card tarjeta de entrada-salida en paralelo
parallel interface interfaz paralela
parallel limiter limitador en paralelo
parallel link enlace paralelo
parallel magnetic pulse amplifier amplificador de impulsos magnético en paralelo
parallel memory memoria en paralelo
parallel-network oscillator oscilador de red en paralelo
parallel operation operación en paralelo
parallel output salida en paralelo
parallel partitioning división en paralelo
parallel-plane triode triodo de planos paralelos
parallel-plane waveguide guíaondas de planos paralelos
parallel-plate capacitor capacitor de placas paralelas
parallel-plate lens lente de placas paralelas
parallel-plate oscillator oscilador de placas paralelas
parallel-plate transmission system sistema de transmisión de placas paralelas
parallel-plate waveguide guíaondas de placas paralelas

parallel poll configuration configuración de sondeo paralelo
parallel poll logic lógica de sondeo paralelo
parallel polling sondeo paralelo
parallel printer impresora en paralelo
parallel printing impresión en paralelo
parallel processing procesamiento en paralelo
parallel processing system sistema de procesamiento en paralelo
parallel processor procesador en paralelo
parallel processor architecture arquitectura de procesador en paralelo
parallel programming programación paralela
parallel pumping bombeo en paralelo
parallel punching perforación en paralelo
parallel reading lectura en paralelo
parallel rectifier rectificador en paralelo
parallel-rectifier circuit circuito de rectificadores en paralelo
parallel resistance resistencia en paralelo
parallel-resistance bridge puente de resistencia en paralelo
parallel resistors resistores en paralelo
parallel resonance resonancia en paralelo
parallel-resonant circuit circuito resonante en paralelo
parallel-rod oscillator oscilador de varillas paralelas
parallel-rod tuning sintonización por varillas paralelas
parallel run ejecución en paralelo
parallel run test prueba de ejecución en paralelo
parallel running funcionamiento en paralelo
parallel search búsqueda en paralelo
parallel-search memory memoria de búsqueda en paralelo
parallel-search storage almacenamiento de búsqueda en paralelo
parallel-serial paralelo-serie
parallel-series capacitors capacitores en paralelo-serie
parallel-series circuit circuito en paralelo-serie
parallel sessions sesiones en paralelo
parallel slots ranuras paralelas
parallel storage almacenamiento en paralelo
parallel surfaces superficies paralelas
parallel system sistema en paralelo
parallel-T amplifier amplificador T en paralelo
parallel-T network red T en paralelo
parallel-T oscillator oscilador T en paralelo
parallel terminal terminal en paralelo
parallel-to-serial converter convertidor de paralelo a serie
parallel transducer transductor de acoplamiento en paralelo
parallel transfer transferencia en paralelo
parallel transmission transmisión en paralelo
parallel-tube amplifier amplificador de tubos en paralelo
parallel-tuned sintonizado en paralelo
parallel winding arrollamiento en paralelo
parallel-wire antenna antena de hilos paralelos
parallel-wire line línea de hilos paralelos
parallel-wire resonator resonador de hilos paralelos
paralleling acoplamiento en paralelo, conexión en paralelo
paralleling device dispositivo de acoplamiento en paralelo
paralleling equipment equipo de acoplamiento en paralelo
paralysis circuit circuito de parálisis
paralysis time tiempo de parálisis
paramagnetic paramagnético
paramagnetic absorption absorción paramagnética

paramagnetic amplifier amplificador paramagnético
paramagnetic crystal cristal paramagnético
paramagnetic element elemento paramagnético
paramagnetic material material paramagnético
paramagnetic resonance resonancia paramagnética
paramagnetic substance sustancia paramagnética
paramagnetism paramagnetismo
parameter parámetro
parameter association asociación de parámetros
parameter card tarjeta de parámetro
parameter descriptor descriptor de parámetro
parameter driven accionado por parámetros
parameter list lista de parámetros
parameter medium medio de parámetros
parameter statement sentencia de parámetro
parameter substitution sustitución de parámetro
parameter test prueba de parámetro
parameter word palabra de parámetro
parametric paramétrico
parametric amplification amplificación paramétrica
parametric amplifier amplificador paramétrico
parametric conversion conversión paramétrica
parametric converter convertidor paramétrico
parametric device dispositivo paramétrico
parametric diode diodo paramétrico
parametric-down converter subconvertidor
  paramétrico
parametric excitation excitación paramétrica
parametric frequency converter convertidor de
  frecuencia paramétrico
parametric modulation modulación paramétrica
parametric modulator modulador paramétrico
parametric motor motor paramétrico
parametric oscillator oscilador paramétrico
parametric potentiometer potenciómetro
  paramétrico
parametric programming programación paramétrica
parametric receiver receptor paramétrico
parametric resonance resonancia paramétrica
parametric subharmonic oscillator oscilador
  subarmónico paramétrico
parametric subroutine subrutina paramétrica
parametric-up converter hiperconvertidor
  paramétrico
parametric variation variación paramétrica
parametron parametrón
paramistor paramistor
paraphase amplifier amplificador parafásico
paraphase coupling acoplamiento parafásico
paraphase inverter inversor parafásico
parasitic parásito
parasitic antenna antena parásita
parasitic array red parásita
parasitic capacitance capacitancia parásita
parasitic capacity capacidad parásita
parasitic component componente parásito
parasitic coupling acoplamiento parásito
parasitic current corriente parásita
parasitic disturbance perturbación parásita
parasitic echo eco parásito
parasitic effect efecto parásito
parasitic element elemento parásito
parasitic-element directive antenna antena directiva
  de elementos parásitos
parasitic eliminator eliminador de parásitos
parasitic emission emisión parásita
parasitic inductance inductancia parásita
parasitic induction inducción parásita
parasitic noise ruido parásito
parasitic oscillation oscilación parásita
parasitic reflector reflector parásito
parasitic resistance resistencia parásita
parasitic signal señal parásita

parasitic suppressor supresor de parásitos
paraxial paraxial
paraxial ray rayo paraxial
parfocal parfocal
parity paridad
parity bit bit de paridad
parity check comprobación de paridad
parity checking comprobación de paridad
parity conservation conservación de paridad
parity digit dígito de paridad
parity error error de paridad
parity flag señalizador de paridad
parity generator-checker generador-verificador de
  paridad
parity interrupt interrupción de paridad
parity-line circuit circuito de línea de paridad
parity-line structure estructura de línea de paridad
parity violation violación de paridad
parse analizar sintácticamente
parser analizador sintáctico
parsing análisis sintáctico
parsing procedure procedimiento de análisis
  sintáctico
part parte, pieza
part failure falla de parte, falla de pieza
partial parcial
partial arithmetic aritmética parcial
partial automatic control control automático parcial
partial carry acarreo parcial
partial completion terminación parcial
partial correctness corrección parcial
partial derivative derivada parcial
partial differential equation ecuación diferencial
  parcial
partial differentiation diferenciación parcial
partial directivity directividad parcial
partial discharge descarga parcial
partial dispersion dispersión parcial
partial drive pulse impulso de mando parcial
partial failure falla parcial
partial gain ganancia parcial
partial integration integración parcial
partial multiple múltiple parcial
partial network red parcial
partial node nodo parcial
partial page página parcial
partial pitch paso parcial
partial plating depósito parcial, recubrimiento
  parcial
partial pressure presión parcial
partial product producto parcial
partial read lectura parcial
partial-read pulse impulso de lectura parcial
partial response respuesta parcial
partial result resultado parcial
partial-select output salida de selección parcial
partial sum suma parcial
partial word palabra parcial
partial-write pulse impulso de escritura parcial
partially parcialmente
partially enclosed apparatus aparato parcialmente
  encerrado
partially energized relay relé parcialmente
  energizado
partially inverted file archivo parcialmente invertido
partially occupied band banda parcialmente ocupada
partially selected cell celda parcialmente
  seleccionada
partially suppressed sideband banda lateral
  parcialmente suprimida
particle partícula
particle accelerator acelerador de partículas
particle Coulomb interactions interacciones de

Coulomb entre partículas
particle current density densidad de corriente de partículas
particle density densidad de partículas
particle detector detector de partículas
particle diffusion difusión de partículas
particle displacement desplazamiento de partículas
particle emission emisión de partículas
particle emitter emisor de partículas
particle energy energía de partícula
particle flux flujo de partículas
particle-flux density densidad de flujo de partículas
particle fluxmeter flujómetro de partículas
particle orientation orientación de partículas
particle shape forma de partículas
particle size tamaño de partículas
particle velocity velocidad de partículas
particle wave onda de partículas
partition partición
partition area área de partición
partition control control de partición
partition control descriptor descriptor de control de partición
partition control table tabla de control de partición
partition identifier identificador de partición
partition noise ruido de partición
partitioned access acceso dividido
partitioned access method método de acceso dividido
partitioned data set conjunto de datos divididos
partitioned data set compression compresión de conjuntos de datos divididos
partitioned file archivo dividido
partitioned mode modo dividido
partitioned organization organización dividida
partitioning partición
parts family familia de partes
party line línea compartida
party-line circuit circuito de línea compartida
party-line driver excitador de línea compartida
party-line equipment equipo de línea compartida
party-line operation operación de línea compartida
parylene parileno
parylene capacitor capacitor de parileno
pascal pascal
Paschen's law ley de Paschen
Paschen-Back effect efecto de Paschen-Back
pass element elemento de paso
pass time tiempo de paso
passband banda pasante
passband filter filtro de banda pasante
passband response respuesta de banda pasante
passband ripple ondulación de banda pasante
passband width anchura de banda pasante
passing contact contacto de paso
passing frequency frecuencia pasante
passivate pasivar
passivated pasivado
passivated iron hierro pasivado
passivated oil aceite pasivado
passivated surface superficie pasivada
passivated transistor transistor pasivado
passivating pasivación
passivation pasivación
passive pasivo
passive antenna antena pasiva
passive base base pasiva
passive chromium oxide óxido de cromo pasivo
passive circuit circuito pasivo
passive communications satellite satélite de comunicaciones pasivo
passive component componente pasivo
passive contact contacto pasivo

passive corner reflector reflector angular pasivo
passive decoder descodificador pasivo
passive detection detección pasiva
passive device dispositivo pasivo
passive dipole dipolo pasivo
passive electric network red eléctrica pasiva
passive electrode electrodo pasivo
passive element elemento pasivo
passive equivalent circuit circuito equivalente pasivo
passive film película pasiva
passive graphics gráficos pasivos
passive iron hierro pasivo
passive limiter limitador pasivo
passive mixer mezclador pasivo
passive mode modo pasivo
passive modulator modulador pasivo
passive network red pasiva
passive nonlinear element elemento no lineal pasivo
passive radar radar pasivo
passive radiator radiador pasivo
passive radio beacon radiofaro pasivo
passive reflector reflector pasivo
passive relay relé pasivo
passive repeater repetidor pasivo
passive resistance resistencia pasiva
passive satellite satélite pasivo
passive screen pantalla pasiva
passive sonar sonar pasivo
passive stainless steel acero inoxidable pasivo
passive station estación pasiva
passive substrate sustrato pasivo
passive system sistema pasivo
passive test prueba pasiva
passive tracking system sistema de seguimiento pasivo
passive transducer transductor pasivo
passivity pasividad
password contraseña
password protection protección de contraseña
paste pasta
paste solder soldante en pasta
patch conexión temporal, corrección
patch area área de corrección
patch bay conjunto de paneles de conexiones
patch cable cable de conexión temporal
patch cord cordón de conexión temporal
patch panel panel de conexiones, panel de conmutación
patch routine rutina de corrección
patchboard tablero de conexiones
patching conexión temporal
patching cord cordón de conexión temporal
patching panel panel de conexiones, panel de conmutación
patching resistor resistor de conexión temporal
path trayectoria, vía, ruta
path analysis análisis de trayectoria
path attenuation atenuación de trayectoria
path condition condición de trayectoria
path control control de trayectoria
path control network red de control de trayectoria
path distortion noise ruido de propagación
path information unit unidad de información de trayectoria
path length longitud de trayectoria
path loss pérdida de trayectoria
path name nombre de trayectoria
path profile perfil de trayectoria
path selector selector de trayectoria
path switch conmutador de trayectoria
path test prueba de trayectoria
path transmittance transmitancia de trayectoria
patina pátina

pattern patrón, diagrama, imagen, disposición
pattern class clase de patrón
pattern classification clasificación de patrones
pattern generator generador de imagen patrón
pattern handling manejo de patrón
pattern-handling statement sentencia de manejo de patrón
pattern identification identificación de patrón
pattern multiplication multiplicación de patrones
pattern recognition reconocimiento de patrón
pause pausa
pause control control de pausa
pause instruction instrucción de pausa
peak pico, punta, cresta
peak amplifier amplificador de pico
peak amplitude amplitud de pico
peak anode current corriente pico de ánodo
peak anode voltage tensión pico de ánodo
peak black pico del negro
peak blocked voltage tensión pico bloqueada
peak carrier amplitude amplitud pico de portadora
peak cathode current corriente pico de cátodo
peak chopper recortador de picos
peak clipper recortador de picos
peak clipping recorte de picos
peak current corriente pico
peak data transfer rate velocidad pico de transferencia de datos
peak deflection deflexión pico
peak detector detector de pico
peak discharge current corriente pico de descarga
peak distortion distorsión máxima
peak electrode current corriente pico de electrodo
peak emission capability capacidad pico de emisión
peak envelope power potencia pico de envolvente
peak factor factor de pico
peak field strength intensidad máxima de campo
peak figure cifra máxima
peak filter filtro de pico
peak flux density densidad máxima de flujo
peak forward anode voltage tensión pico de ánodo directo
peak forward drop caída directa de pico
peak frequency deviation desviación pico de frecuencia
peak hours horas de consumo máximo
peak indicator indicador de pico
peak intensity intensidad pico
peak inverse anode voltage tensión pico inversa de ánodo
peak inverse voltage tensión pico inversa
peak level nivel pico
peak limiter limitador de picos
peak limiting limitación de picos
peak-limiting device dispositivo limitador de picos
peak load carga máxima
peak magnetizing force fuerza magnetizante máxima
peak making current corriente pico de establecimiento
peak modulating voltage tensión modulante pico
peak period periodo de tráfico máximo
peak plate current corriente pico de placa
peak plate voltage tensión pico de placa
peak point punto pico
peak power potencia pico
peak power output potencia pico de salida
peak probe sonda de pico
peak pulse amplitude amplitud pico de impulso
peak pulse power potencia pico de impulso
peak radiated power potencia pico de radiación
peak radiofrequency output potencia pico de radiofrecuencia
peak recurrent forward current corriente directa

pico recurrente
peak response respuesta máxima
peak reverse voltage tensión pico inversa
peak separation separación de picos
peak signal level nivel pico de señal
peak signal power potencia pico de señal
peak sound pressure presión sonora máxima
peak speed velocidad máxima
peak-to-peak amplitude amplitud de pico a pico
peak-to-peak excursion excursión de pico a pico
peak-to-peak frequency excursion excursión de frecuencia de pico a pico
peak-to-peak meter medidor de pico a pico
peak-to-peak rectifier rectificador de pico a pico
peak-to-peak swing oscilación de pico a pico
peak-to-peak value valor de pico a pico
peak-to-peak voltage tensión de pico a pico
peak-to-peak voltmeter vóltmetro de pico a pico
peak torque par máximo
peak transfer rate velocidad pico de transferencia
peak value valor pico
peak voltage tensión máxima
peak voltmeter vóltmetro de pico
peak white pico del blanco
peaker circuit circuito diferenciador
peaking corrección, agudización
peaking circuit circuito corrector
peaking coil bobina correctora
peaking control control corrector
peaking network red correctora
peaking resistor resistor corrector
peaking transformer transformador de generador de impulsos
pedal circuit circuito de pedal
pedestal pedestal
pedestal level nivel de pedestal
peg espiga, tapón
pellet holder portapastilla
pellet resistor resistor de pastilla
Peltier coefficient coeficiente de Peltier
Peltier effect efecto de Peltier
Peltier electromotive force fuerza electromotriz de Peltier
Peltier heat calor de Peltier
Peltier junction unión de Peltier
pen plotter trazador de pluma
pen recorder registrador de pluma
pencil beam haz en lápiz, haz filiforme
pencil-beam antenna antena de haz en lápiz
pencil tube tubo tipo lápiz
pendant colgante
pendant station estación colgante
pendant switch conmutador colgante
pendant telephone teléfono colgante
pendulum péndulo
pendulum meter medidor de péndulo
penetrability penetrabilidad
penetrant penetrante
penetrating penetrante
penetrating frequency frecuencia penetrante
penetrating power poder de penetración
penetrating power of radiation poder de penetración de radiación
penetrating radiation radiación penetrante
penetrating rays rayos penetrantes
penetration depth profundidad de penetración
penetration electrode electrodo de penetración
penetration factor factor de penetración
penetration frequency frecuencia de penetración
penetration probability probabilidad de penetración
penetration rate velocidad de penetración
penetration voltage tensión de penetración
penetrometer penetrómetro

penetron penetrón
Penning discharge descarga de Penning
Penning ionization gage medidor de ionización de
  Penning
pentagrid pentarrejilla
pentagrid converter convertidor pentarrejilla
pentagrid converter tube tubo convertidor
  pentarrejilla
pentagrid detector detector pentarrejilla
pentagrid mixer mezclador pentarrejilla
pentagrid tube tubo pentarrejilla
pentatron pentatrón
pentavalent element elemento pentavalente
pentode pentodo
pentode field-effect transistor transistor pentodo de
  efecto de campo
pentode gun cañón pentodo
pentode section sección pentodo
pentode transistor transistor pentodo
percent porcentaje
percent absorption porcentaje de absorción
percent distortion porcentaje de distorsión
percent harmonic distortion porcentaje de distorsión
  armónica
percent modulation porcentaje de modulación
percent-modulation meter medidor de porcentaje de
  modulación
percent ripple porcentaje de ondulación
percent synchronization porcentaje de
  sincronización
percentage porcentaje
percentage coupling porcentaje de acoplamiento
percentage error porcentaje de error
percentage loss porcentaje de pérdidas
percentage modulation porcentaje de modulación
percentage-modulation meter medidor de porcentaje
  de modulación
percentage of modulation porcentaje de modulación
percentage synchronization porcentaje de
  sincronización
percentage uncertainty porcentaje de incertidumbre
perceptron perceptrón
percussion welding soldadura por percusión
percussive welding soldadura percusiva
perfect conductor conductor perfecto
perfect crystal cristal perfecto
perfect dielectric dieléctrico perfecto
perfect diffuser difusor perfecto
perfect diffusion difusión perfecta
perfect emission emisión perfecta
perfect gas gas perfecto
perfect modulation modulación perfecta
perfect optical system sistema óptico perfecto
perfect polarization polarización perfecta
perfect transformer transformador perfecto
perfilometer perfilómetro
perforated perforado
perforated board panel perforado
perforated tape cinta perforada
perforated-tape code código de cinta perforada
perforated-tape reader lectora de cinta perforada
perforated-tape retransmitter retransmisor de cinta
  perforada
perforated-tape transmission transmisión de cinta
  perforada
perforating machine máquina perforadora
perforating mechanism mecanismo de perforación
perforation perforación
perforation device dispositivo de perforación
perforation rate velocidad de perforación
perforator perforador
perforator-checker perforador-comprobador
performance rendimiento, funcionamiento,
  comportamiento
performance capability capacidad de rendimiento
performance characteristic característica de
  rendimiento
performance chart diagrama de rendimiento
performance curve curva de rendimiento
performance evaluation evaluación de rendimiento
performance factor factor de rendimiento
performance feature característica de rendimiento
performance impairment deterioro de rendimiento
performance index índice de rendimiento
performance measurement medida de rendimiento
performance monitor monitor de rendimiento
performance pattern patrón de rendimiento
performance requirement requisito de rendimiento
performance specification especificación de
  rendimiento
performance test prueba de rendimiento
performance verification verificación de
  rendimiento
perikon detector detector perikón
period periodo
period counter contador de periodos
period definition definición de periodo
period measurement medida de periodos
period meter medidor de periodos
period of duty periodo de trabajo
period of excitation periodo de excitación
period of oscillation periodo de oscilación
period of validity periodo de validez
periodic periódico
periodic antenna antena periódica
periodic beam haz periódico
periodic check comprobación periódica
periodic circuit circuito periódico
periodic current corriente periódica
periodic curve curva periódica
periodic damping amortiguamiento periódico
periodic decimal decimal periódico
periodic deviation desviación periódica
periodic disturbance perturbación periódica
periodic duty servicio periódico
periodic electromagnetic wave onda
  electromagnética periódica
periodic filter filtro periódico
periodic flashes destellos periódicos
periodic focusing enfoque periódico
periodic function función periódica
periodic light luz periódica
periodic line línea periódica
periodic magnetic structure estructura magnética
  periódica
periodic monitoring monitoreo periódico
periodic noise ruido periódico
periodic oscillation oscilación periódica
periodic permanent magnet imán permanente
  periódico
periodic pulsation pulsación periódica
periodic pulse train tren de impulsos periódico
periodic quantity magnitud periódica
periodic rating clasificación periódica
periodic resonance resonancia periódica
periodic service servicio periódico
periodic signal señal periódica
periodic table tabla periódica
periodic test prueba periódica
periodic undulation ondulación periódica
periodic vibration vibración periódica
periodic wave onda periódica
periodic waveguide guíaondas periódico
periodically periódicamente
periodicity periodicidad
peripheral periférico

peripheral assignment table tabla de asignación de periféricos
peripheral-bound limitado por periférico
peripheral buffer memoria intermedia periférica
peripheral compatibility compatibilidad de periférico
peripheral component componente periférico
peripheral computer computadora periférica
peripheral control control periférico
peripheral control unit unidad de control periférico
peripheral controller controlador de periféricos
peripheral device dispositivo periférico
peripheral driver controlador de periférico
peripheral electron electrón periférico
peripheral equipment equipo periférico
peripheral interface adapter adaptador de interfaz de periférico
peripheral interface channel canal de interfaz de periférico
peripheral interface module módulo de interfaz de periférico
peripheral interrupt interrupción de periférico
peripheral-limited limitado por periférico
peripheral link enlace periférico
peripheral logical unit unidad lógica periférica
peripheral node nodo periférico
peripheral operation operación periférica
peripheral physical unit unidad física de periférico
peripheral power supply fuente de alimentación de periférico
peripheral processor procesador periférico
peripheral program programa de periférico
peripheral slot ranura de periférico
peripheral subsystem subsistema periférico
peripheral transfer transferencia periférica
peripheral unit unidad periférica
periscopic periscópico
periscopic antenna antena periscópica
peristaltic peristáltico
peristaltic induction inducción peristáltica
permalloy permaloy
permamagnetic speaker altavoz permamagnético
permanence permanencia
permanent permanente
permanent action acción permanente
permanent connection conexión permanente
permanent data datos permanentes
permanent deformation deformación permanente
permanent dynamic storage almacenamiento dinámico permanente
permanent echo eco permanente
permanent elongation elongación permanente
permanent error error permanente
permanent error indicator indicador de error permanente
permanent fault falla permanente
permanent fixed circuit circuito fijo permanente
permanent lamp lámpara permanente
permanent magnet imán permanente
permanent-magnet centering centrado por imán permanente
permanent-magnet crack grieta de imán permanente
permanent-magnet dynamic loudspeaker altavoz dinámico de imán permanente
permanent-magnet erase borrado por imán permanente
permanent-magnet erase head cabeza de borrado por imán permanente
permanent-magnet erasing head cabeza de borrado de imán permanente
permanent-magnet focusing enfoque por imanes permanentes
permanent-magnet galvanometer galvanómetro de

imán permanente
permanent-magnet generator generador de imán permanente
permanent-magnet instrument instrumento de imán permanente
permanent-magnet loudspeaker altavoz de imán permanente
permanent-magnet machine máquina magnetoeléctrica
permanent-magnet material material para imanes permanentes
permanent-magnet meter medidor de imán permanente
permanent-magnet motor magnetomotor
permanent-magnet relay relé de imán permanente
permanent-magnet speaker altavoz de imán permanente
permanent memory memoria permanente
permanent-memory computer computadora de memoria permanente
permanent read-write error error de lectura-escritura permanente
permanent record registro permanente
permanent set deformación permanente
permanent signal señal permanente
permanent storage almacenamiento permanente
permanent stretch alargamiento permanente
permanent virtual circuit circuito virtual permanente
permanently permanentemente
permanently connected conectado permanentemente
permanently earthed conectado a tierra permanentemente
permanently resident residente permanentemente
permanently stored almacenado permanentemente
permatron permatrón
permeability permeabilidad
permeability coefficient coeficiente de permeabilidad
permeability tuning sintonización por permeabilidad
permeable permeable
permeameter permeámetro
permeance permeancia
permendur permendur
perminvar perminvar
permissible permisible
permissible interference interferencia permisible
permissible load carga permisible
permissible signal distortion distorsión de señal permisible
permissible voltage drop caída de tensión permisible
permissive permisivo
permissive block bloque permisivo
permissive control control permisivo
permissive control device dispositivo de control permisivo
permissive relay relé permisivo
permittivity permitividad
permutation permutación
permutation code código de permutación
permutation index índice de permutación
peroxide of lead peróxido de plomo
perpendicular perpendicular
perpendicular magnetization magnetización perpendicular
perpendicular polarization polarización perpendicular
perpendicular recording registro perpendicular
perpetual perpetuo
persistence persistencia
persistence characteristic característica de persistencia
persistence screen pantalla de persistencia
persistent persistente

persistent current corriente persistente
persistent-image device dispositivo de imagen
  persistente
persistent magnetic field campo magnético
  persistente
persistent oscillations oscilaciones persistentes
persistor persistor
persistron persistrón
personal code código personal
personal computer computadora personal
personal computing computación personal
personal document documento personal
personal identification number número de
  identificación personal
personal minicomputer minicomputadora personal
perturbation perturbación
perturbed perturbado
perturbed electric field campo eléctrico perturbado
perturbed field campo perturbado
perturbed magnetic field campo magnético
  perturbado
perveance perveancia
peta peta
pH pH
pH indicator indicador de pH
pH measuring equipment equipo de medida de pH
pH meter medidor de pH
pH recorder registrador de pH
pH value valor de pH
phanotron fanotrón
phantastron fantastrón
phantom fantasma
phantom antenna antena fantasma
phantom channel canal fantasma
phantom circuit circuito fantasma
phantom circuit repeating coil bobina repetidora de
  circuito fantasma
phantom coil bobina fantasma
phantom group grupo fantasma
phantom line línea fantasma
phantom network red fantasma
phantom output salida fantasma
phantom repeating coil bobina repetidora fantasma
phantom signal señal fantasma
phantom target blanco fantasma
phantom telegraph circuit circuito telegráfico
  fantasma
phantophone fantófono
phase fase
phase adjustment ajuste de fase
phase advance avance de fase
phase-advance circuit circuito de avance de fase
phase advancer adelantador de fase
phase alternation alternación de fase
phase-alternation line system sistema de alternación
  de fase por línea
phase ammeter amperímetro de fase
phase amplifier amplificador de fase
phase-amplitude distortion distorsión fase-amplitud
phase angle ángulo de fase, desfase
phase-angle error error de desfase
phase-angle measurement medida de desfase
phase-angle voltmeter vóltmetro de desfase
phase balance equilibrio de fases
phase-balance relay relé de equilibrio de fases
phase-balance voltage tensión de equilibrio de fases
phase balancing coil bobina equilibradora de fases
phase balancing network red equilibradora de fases
phase bandwidth ancho de banda de fase
phase bar barra de fase
phase break sección de separación
phase bunching agrupamiento en fase
phase-by-phase fase por fase

phase center centro de fase
phase center of array centro de fase de red
phase change cambio de fase, desfase
phase-change coefficient coeficiente de desfase
phase changer desfasador
phase characteristic característica de fase
phase coherence coherencia de fase
phase coincidence coincidencia de fases
phase comparison comparación de fase
phase-comparison localizer localizador de
  comparación de fase
phase-comparison protection protección por
  comparación de fase
phase-comparison radar radar de comparación de
  fase
phase-comparison tracking system sistema de
  seguimiento por comparación de fase
phase-compensating component componente
  compensador de fase
phase-compensating network red compensadora de
  fase
phase compensation compensación de fase
phase-compensation network red de compensación
  de fase
phase compensator compensador de fase
phase compressor compresor de fase
phase condition condición de fase
phase conductor conductor de fase
phase constant constante de fase
phase contrast contraste de fase
phase-contrast apparatus aparato de contraste de
  fase
phase control control de fase
phase-control factor factor de regulación
phase-controlled controlado por fase
phase-controlled rectifier rectificador controlado por
  fase
phase-controlled thyratron tiratrón controlado por
  fase
phase converter convertidor de fase
phase-corrected horn bocina con corrección de fase
phase-corrected reflector reflector con corrección
  de fase
phase-correcting network red correctora de fase
phase correction corrección de fase
phase corrector corrector de fase
phase current corriente de fase
phase delay retardo de fase
phase-delay distortion distorsión por retardo de fase
phase-delay error error de retardo de fase
phase-delay time tiempo de retardo de fase
phase detector detector de fase
phase deviation desviación de fase
phase diagram diagrama de fase
phase difference diferencia de fase, desfase
phase-difference indicator indicador de desfase
phase-difference recording registro por desfase
phase-discriminating rectifier rectificador
  discriminador de fase
phase discriminator discriminador de fase
phase displacement desplazamiento de fase, desfase
phase-displacement error error de desfase
phase distortion distorsión de fase
phase-distortion coefficient coeficiente de distorsión
  de fase
phase-distortion measurement medida de distorsión
  de fase
phase distribution distribución de fase
phase division división de fase
phase-division multiplex múltiplex por división de
  fase
phase effect efecto de fase
phase encoding codificación de fase

phase equalizer ecualizador de fase
phase-equalizing network red ecualizadora de fase
phase error error de fase
phase factor factor de fase
phase-frequency distortion distorsión de fase
phase front frente de fase
phase grouping agrupamiento de fases
phase indicator indicador de fase
phase interlocking enclavamiento de fase
phase inversion inversión de fase
phase-inversion cabinet caja de inversión de fase
phase-inversion circuit circuito de inversión de fase
phase-inversion modulation modulación por inversión de fase
phase inverter inversor de fase
phase-inverter tube tubo inversor de fase
phase-inverter valve válvula inversora de fase
phase-isolated transformer transformador de fase aislada
phase jitter fluctuación de fase
phase lag retardo de fase
phase-lag angle ángulo de retardo de fase
phase-lag corrector corrector de retardo de fase
phase lead avance de fase
phase leading avance de fase
phase length longitud de fase
phase library biblioteca de fases
phase linearity linealidad de fase
phase lock enganche de fase
phase-locked circuit circuito de enganche de fase
phase-locked demodulator desmodulador de enganche de fase
phase-locked detector detector de enganche de fase
phase-locked loop bucle de enganche de fase
phase-locked oscillator oscilador de enganche de fase
phase-locked receiver receptor de enganche de fase
phase-locked servosystem servosistema de enganche de fase
phase-locked subharmonic oscillator oscilador subarmónico de enganche de fase
phase logic lógica de fase
phase margin margen de fase
phase measurement medida de fase
phase measurer fasímetro
phase meter fasímetro
phase modifier modificador de fase
phase-modulated modulado en fase
phase-modulated carrier portadora modulada en fase
phase-modulated oscillation oscilación modulada en fase
phase-modulated transmitter transmisor modulado en fase
phase-modulated wave onda modulada en fase
phase modulation modulación de fase
phase-modulation recording registro por modulación de fase
phase-modulation transmitter transmisor de modulación de fase
phase-modulation wave onda de modulación de fase
phase modulator modulador de fase
phase monitor monitor de fase
phase monitoring monitoreo de fase
phase multiplier multiplicador de fase
phase-multiplying transformer transformador multiplicador de fase
phase noise ruido de fase
phase nonlinearity no linealidad de fase
phase opposition oposición de fase
phase pattern patrón de fase
phase precorrection precorrección de fase
phase-propagation ratio razón de propagación de

fase
phase-protective device dispositivo protector de fase
phase quadrature cuadratura de fase
phase recorder registrador de fase
phase recovery time tiempo de recuperación de fase
phase regulator regulador de fase
phase relation relación de fase
phase relationship relación de fase
phase resistance resistencia de fase
phase resolution resolución de fase
phase resonance resonancia de fase
phase response respuesta de fase
phase-response characteristic característica de respuesta de fase
phase reversal inversión de fase
phase-reversal protection protección contra inversión de fase
phase-reversal relay relé de inversión de fase
phase-reversal switch conmutador de inversión de fase
phase-reversing unit unidad inversora de fase
phase rotation rotación de fase
phase-rotation relay relé de rotación de fase
phase-rotation system sistema de rotación de fase
phase-sensitive sensible a la fase
phase-sensitive amplifier amplificador sensible a la fase
phase-sensitive demodulator desmodulador sensible a la fase
phase-sensitive detector detector sensible a la fase
phase sequence secuencia de fases
phase-sequence indicator indicador de secuencia de fases
phase-sequence relay relé de secuencia de fases
phase-sequence reversal inversión de secuencia de fases
phase-sequence test prueba de secuencia de fases
phase-sequence voltage relay relé de tensión de secuencia de fases
phase shift desplazamiento de fase, desfase
phase-shift bridge puente de desfase
phase-shift circuit circuito de desfase
phase-shift control control por desfase
phase-shift discriminator discriminador de desfase
phase-shift discriminator circuit circuito discriminador de desfase
phase-shift feedback circuit circuito de retroalimentación de desfase
phase-shift keying manipulación por desfase
phase-shift microphone micrófono de desfase
phase-shift modulation modulación por desfase
phase-shift network red de desfase
phase-shift oscillator oscilador de desfase
phase-shift standard patrón de desfase
phase-shift telegraph system sistema telegráfico por desfase
phase-shift tone control control de tono por desfase
phase-shifter desfasador
phase-shifting autotransformer autotransformador desfasador
phase-shifting bridge puente desfasador
phase-shifting capacitor capacitor desfasador
phase-shifting circuit circuito desfasador
phase-shifting device dispositivo desfasador
phase-shifting network red desfasadora
phase-shifting transformer transformador desfasador
phase-shifting unit unidad desfasadora
phase simulator simulador de fase
phase spectrum espectro de fase
phase splitter divisor de fase
phase splitting división de fase
phase-splitting circuit circuito divisor de fase

phase-splitting device dispositivo divisor de fase
phase stability estabilidad de fase
phase swinging oscilación de fase
phase-synchronized sincronizado en fase
phase-synchronized transmitter transmisor sincronizado en fase
phase terminal terminal de fase
phase tracking seguimiento de fase
phase transformation transformación de fase
phase transformer transformador de fase
phase transition transición de fase
phase velocity velocidad de fase
phase voltage tensión de fase
phase voltmeter vóltmetro de fase
phase wave onda de fase
phase winding arrollamiento de fase
phased array red en fase
phasemeter fasímetro
phaser enfasador, ajustador de fase
phasing ajuste de fase, enfasamiento
phasing actuator actuador de enfasamiento
phasing channel canal de enfasamiento
phasing equipment equipo de enfasamiento
phasing line línea de enfasamiento
phasing method método de enfasamiento
phasing network red de enfasamiento
phasing signal señal de enfasamiento
phasing time tiempo de enfasamiento
phasing transformer transformador de enfasamiento
phasitron fasitrón
phasor fasor
phenolic fenólico
phenolic insulant aislante fenólico
phenolic insulator aislante fenólico
phenolic material material fenólico
phenolic plastic plástico fenólico
phenolic resin resina fenólica
phenomenon fenómeno
phi phi
Phillips gage medidor de Phillips
Phillips gate puerta de Phillips
phon fonio
phone teléfono, audífono
phone jack jack de audífono, jack de teléfono
phone net red telefónica
phone patch conexión temporal de teléfono
phone plug clavija de audífono, clavija de teléfono
phone reception recepción telefónica
phone transmitter transmisor telefónico
phonic motor motor fónico
phonic wheel rueda fónica
phonics fónica
phono adapter adaptador fonográfico
phono cartridge cápsula fonocaptora
phono equalizer ecualizador fonográfico
phono jack jack fonográfico
phono lead conexión fonográfica
phono panel panel fonográfico
phono pickup captor fonográfico
phono plug clavija fonográfica
phonocardiogram fonocardiograma
phonocardiograph fonocardiógrafo
phonogram fonograma
phonograph fonógrafo
phonograph adapter adaptador fonográfico
phonograph amplifier amplificador fonográfico
phonograph cartridge cápsula fonocaptora
phonograph connection conexión fonográfica
phonograph disk disco fonográfico
phonograph equalizer ecualizador fonográfico
phonograph input entrada fonográfica
phonograph needle aguja fonográfica
phonograph oscillator oscilador fonográfico

phonograph pickup captor fonográfico
phonograph-pickup amplifier amplificador de captor fonográfico
phonograph record disco fonográfico
phonograph recorder registrador fonográfico
phonograph recording registro fonográfico
phonograph reproducer reproductor fonográfico
phonograph reproduction reproducción fonográfica
phonograph stylus aguja fonográfica
phonometer fonómetro
phonometry fonometría
phonon fonón
phonon amplifier amplificador fonónico
phonon laser láser fonónico
phonon maser máser fonónico
phonon scattering dispersión fonónica
phonoreception fonorrecepción
phonorecord fonodisco
phosphor fósforo
phosphor bronze bronce fosforoso
phosphor dot punto fosforescente
phosphorescence fosforescencia
phosphorescent fosforescente
phosphorescent glow luminiscencia fosforescente
phosphorescent paper papel fosforescente
phosphorescent screen pantalla fosforescente
phosphorescing fosforescente
phosphorogen fosforógeno
phosphoroscope fosforoscopio
phosphorous fosforoso
phosphorus fósforo
phot fotio
photicon foticón
photo peak fotopico
photo reader lector fotográfico
photoactive fotoactivo
photoactuator fotoactuador
photocatalyst fotocatalizador
photocathode fotocátodo
photocathode response respuesta de fotocátodo
photocathode sensitivity sensibilidad de fotocátodo
photocell fotocélula
photocell amplifier amplificador de fotocélula
photocell card reader lectora de tarjetas fotoeléctrica
photocell tape reader lectora de cinta fotoeléctrica
photochemical fotoquímico
photochemical activity actividad fotoquímica
photochemical cell célula fotoquímica
photochemical effect efecto fotoquímico
photochemical reaction reacción fotoquímica
photochemistry fotoquímica
photochopper fotointerruptor periódico
photochromic fotocrómico
photochromic compound compuesto fotocrómico
photochromic glass vidrio fotocrómico
photochromic panel panel fotocrómico
photocolorimeter fotocolorímetro
photocomposition fotocomposición
photoconductance fotoconductancia
photoconducting fotoconductor
photoconducting cell célula fotoconductora
photoconducting crystal cristal fotoconductor
photoconducting surface superficie fotoconductora
photoconduction fotoconducción
photoconductive fotoconductivo, fotoconductor
photoconductive camera tube tubo de cámara fotoconductivo
photoconductive cell célula fotoconductiva
photoconductive chopper interruptor periódico fotoconductivo
photoconductive detector detector fotoconductivo
photoconductive effect efecto fotoconductivo
photoconductive infrared detector detector

infrarrojo fotoconductivo
photoconductive material material fotoconductivo
photoconductivity fotoconductividad
photoconductor fotoconductor
photoconductor component componente
fotoconductor
photoconductor effect efecto fotoconductor
photocopier fotocopiadora
photocurrent fotocorriente
photodecomposition fotodescomposición
photodensitometer fotodensitómetro
photodetector fotodetector
photodevice fotodispositivo
photodielectric fotodieléctrico
photodielectric effect efecto fotodieléctrico
photodiffusion fotodifusión
photodiffusion effect efecto de fotodifusión
photodiode fotodiodo
photodiode parametric amplifier amplificador
paramétrico fotodiódico
photodisintegration fotodesintegración
photodissociation fotodisociación
photodosimetry fotodosimetría
photoeffect fotoefecto
photoelastic fotoelástico
photoelasticimeter fotoelasticímetro
photoelasticimetry fotoelasticimetría
photoelasticity fotoelasticidad
photoelectric fotoeléctrico
photoelectric absorption absorción fotoeléctrica
photoelectric alarm alarma fotoeléctrica
photoelectric amplifier amplificador fotoeléctrico
photoelectric attenuation coefficient coeficiente de
atenuación fotoeléctrico
photoelectric autocollimator autocolimador
fotoeléctrico
photoelectric card reader lectora de tarjetas
fotoeléctrica
photoelectric cathode cátodo fotoeléctrico
photoelectric cell célula fotoeléctrica, fotocélula
photoelectric-cell amplifier amplificador de célula
fotoeléctrica
photoelectric-cell pyrometer pirómetro de célula
fotoeléctrica
photoelectric-cell reflector reflector de célula
fotoeléctrica
photoelectric character reader lector de caracteres
fotoeléctrico
photoelectric color comparator comparador de
colores fotoeléctrico
photoelectric colorimeter colorímetro fotoeléctrico
photoelectric colorimetry colorimetría fotoeléctrica
photoelectric conductivity conductividad
fotoeléctrica
photoelectric constant constante fotoeléctrica
photoelectric control control fotoeléctrico
photoelectric conversion process proceso de
conversión fotoeléctrica
photoelectric counter contador fotoeléctrico
photoelectric current corriente fotoeléctrica
photoelectric densitometer densitómetro
fotoeléctrico
photoelectric density meter densitómetro
fotoeléctrico
photoelectric detector detector fotoeléctrico
photoelectric device dispositivo fotoeléctrico
photoelectric directional counter contador
direccional fotoeléctrico
photoelectric disintegration desintegración
fotoeléctrica
photoelectric door opener abridor de puerta
fotoeléctrico
photoelectric effect efecto fotoeléctrico

photoelectric efficiency eficiencia fotoeléctrica
photoelectric electron-multiplier tube tubo
multiplicador fotoeléctrico
photoelectric emission emisión fotoeléctrica
photoelectric eye ojo fotoeléctrico
photoelectric fatigue fatiga fotoeléctrica
photoelectric field-effect-transistor transistor de
efecto de campo fotoeléctrico
photoelectric flame detector detector de llama
fotoeléctrico
photoelectric flaw detector detector de
imperfecciones fotoeléctrico
photoelectric glossmeter brillancímetro fotoeléctrico
photoelectric illumination control system sistema de
control de iluminación fotoeléctrico
photoelectric inspection inspección fotoeléctrica
photoelectric lighting control control de alumbrado
fotoeléctrico
photoelectric lighting controller controlador de
alumbrado fotoeléctrico
photoelectric loop control control de bucle
fotoeléctrico
photoelectric material material fotoeléctrico
photoelectric microscope microscopio fotoeléctrico
photoelectric multiplier multiplicador fotoeléctrico
photoelectric opacimeter opacímetro fotoeléctrico
photoelectric peak pico fotoeléctrico
photoelectric phonograph pickup cápsula
fonocaptora fotoeléctrica
photoelectric photometer fotómetro fotoeléctrico
photoelectric photometry fotometría fotoeléctrica
photoelectric pickup captor fotoeléctrico
photoelectric pinhole detector detector de agujeros
diminutos fotoeléctrico
photoelectric potentiometer potenciómetro
fotoeléctrico
photoelectric probe sonda fotoeléctrica
photoelectric pulse generator generador de impulsos
fotoeléctrico
photoelectric pyrometer pirómetro fotoeléctrico
photoelectric reader lectora fotoeléctrica
photoelectric receiver receptor fotoeléctrico
photoelectric receptor receptor fotoeléctrico
photoelectric recorder registrador fotoeléctrico
photoelectric reflection meter reflectómetro
fotoeléctrico
photoelectric reflectometer reflectómetro
fotoeléctrico
photoelectric register control control de registro
fotoeléctrico
photoelectric relay relé fotoeléctrico
photoelectric scanner explorador fotoeléctrico
photoelectric sensitivity sensibilidad fotoeléctrica
photoelectric signal señal fotoeléctrica
photoelectric smoke alarm alarma de humo
fotoeléctrica
photoelectric smoke detector detector de humo
fotoeléctrico
photoelectric sorter clasificador fotoeléctrico
photoelectric spectrum espectro fotoeléctrico
photoelectric system sistema fotoeléctrico
photoelectric tape reader lectora de cinta
fotoeléctrica
photoelectric threshold umbral fotoeléctrico
photoelectric timer temporizador fotoeléctrico
photoelectric transducer transductor fotoeléctrico
photoelectric tube tubo fotoeléctrico
photoelectric valve válvula fotoeléctrica
photoelectric voltage tensión fotoeléctrica
photoelectric wattmeter wáttmetro fotoeléctrico
photoelectric yield rendimiento fotoeléctrico
photoelectrical fotoeléctrico
photoelectrically fotoeléctricamente

photoelectricity fotoelectricidad
photoelectroluminescence fotoelectroluminiscencia
photoelectromagnetic fotoelectromagnético
photoelectromagnetic effect efecto fotoelectromagnético
photoelectromotive fotoelectromotriz
photoelectromotive force fuerza fotoelectromotriz
photoelectron fotoelectrón
photoelectronic fotoelectrónico
photoelectronic device dispositivo fotoelectrónico
photoelectronic relay relé fotoelectrónico
photoelectronic storage tube tubo de almacenamiento fotoelectrónico
photoelectronics fotoelectrónica
photoemission fotoemisión
photoemissive fotoemisivo
photoemissive camera tube tubo de cámara fotoemisivo
photoemissive cell célula fotoemisiva
photoemissive effect efecto fotoemisivo
photoemissive material material fotoemisivo
photoemissive tube tubo fotoemisivo
photoemitter fotoemisor
photoemitter cathode cátodo fotoemisor
photoengraved fotograbado
photoengraved circuit circuito fotograbado
photofabrication fotofabricación
photoflash fotodestello
photoflash lamp lámpara de fotodestellos
photoflash tube tubo de fotodestellos
photofluorescence fotofluorescencia
photofluorograph fotofluorógrafo
photofluorography fotofluorografía
photoformer fotoformador
photogalvanic fotogalvánico
photogalvanic cell célula fotogalvánica
photogenerator fotogenerador
photoglow tube tubo fotoluminiscente
photogoniometer fotogoniómetro
photogram fotograma
photograph reception recepción de fotografía
photograph transmission transmisión de fotografía
photographic fotográfico
photographic exposure meter exposímetro fotográfico
photographic image imagen fotográfica
photographic reception recepción fotográfica
photographic recorder registrador fotográfico
photographic recording registro fotográfico
photographic routine rutina fotográfica
photographic sound recorder registrador de sonido fotográfico
photographic sound recording registro de sonido fotográfico
photographic sound reproducer reproductor de sonido fotográfico
photographic storage almacenamiento fotográfico
photographic transmission transmisión fotográfica
photoionization fotoionización
photoisolator fotoaislador
photojunction fotounión
photojunction cell célula de fotounión
photolithographic fotolitográfico
photolithographic process proceso fotolitográfico
photoluminescence fotoluminiscencia
photolysis fotólisis
photomagnetic fotomagnético
photomagnetic effect efecto fotomagnético
photomagnetoelectric fotomagnetoeléctrico
photomagnetoelectric effect efecto fotomagnetoeléctrico
photomask fotomáscara
photometer fotómetro

photometer head cabeza fotométrica
photometer unit unidad fotométrica
photometric fotométrico
photometric brightness brillo fotométrico
photometric integrator integrador fotométrico
photometry fotometría
photomicrograph fotomicrografía
photomicrography fotomicrografía
photomixer fotomezclador
photomosaic fotomosaico
photomultiplier fotomultiplicador
photomultiplier cell célula fotomultiplicadora
photomultiplier tube tubo fotomultiplicador
photon fotón
photon absorption absorción fotónica
photon coupling acoplamiento fotónico
photon emission emisión fotónica
photon-emission curve curva de emisión fotónica
photon-emission spectrum espectro de emisión fotónica
photon energy energía fotónica
photon flux flujo fotónico
photon radiation radiación fotónica
photonegative fotonegativo
photoneutron fotoneutrón
photonuclear fotonuclear
photonuclear effect efecto fotonuclear
photooptic fotóptica
photooptic memory memoria fotoóptica
photooxidation fotooxidación
photooxidation cell célula de fotooxidación
photoparametric fotoparamétrico
photoparametric amplifier amplificador fotoparamétrico
photophone fotófono
photophoresis fotoforesis
photopic fotópico
photopositive fotopositivo
photoproton fotoprotón
photoreduction fotorreducción
photoreduction cell célula de fotorreducción
photorelay fotorrelé
photoresistance fotorresistencia
photoresistant fotorresistente
photoresistive fotorresistivo
photoresistive cell célula fotorresistiva
photoresistive material material fotorresistivo
photoresistivity fotorresistividad
photoresistor fotorresistor
photoresistor multiplier multiplicador de fotorresistor
photoscope fotoscopio
photosensitive fotosensible
photosensitive device dispositivo fotosensible
photosensitive electronic device dispositivo electrónico fotosensible
photosensitive junction unión fotosensible
photosensitive material material fotosensible
photosensitive pigment pigmento fotosensible
photosensitive powder polvo fotosensible
photosensitive recording registro fotosensible
photosensitive screen pantalla fotosensible
photosensitivity fotosensibilidad
photosensor fotosensor
photosphere fotosfera
photostimulation fotoestimulación
photosurface fotosuperficie
photoswitch fotoconmutador
phototelegram fototelegrama
phototelegraph fototelégrafo
phototelegraphic fototelegráfico
phototelegraphy fototelegrafía
phototelephony fototelefonía

phototimer fototemporizador
phototransducer fototransductor
phototransistor fototransistor
phototube fototubo
phototube relay relé fotoeléctrico
photovalve fotoválvula
photovaristor fotovaristor
photovoltage fototensión
photovoltaic fotovoltaico
photovoltaic cell célula fotovoltaica
photovoltaic detector detector fotovoltaico
photovoltaic effect efecto fotovoltaico
photovoltaic material material fotovoltaico
photovoltaic module módulo fotovoltaico
photovoltaic panel panel fotovoltaico
photovoltaic power system sistema de alimentación
　fotovoltaico
photovoltaic response respuesta fotovoltaica
photran fotran
physical access level nivel de acceso físico
physical block bloque físico
physical capacity capacidad física
physical circuit circuito físico
physical colorimeter colorímetro físico
physical colorimetry colorimetría física
physical connection conexión física
physical conversion conversión física
physical data independence independencia de datos
　física
physical data structure estructura de datos física
physical database base de datos física
physical database record registro de base de datos
　física
physical device address dirección de dispositivo
　física
physical drive unidad física
physical electronics electrónica física
physical file archivo físico
physical format formato físico
physical input-output address dirección de
　entrada-salida física
physical interface interfaz física
physical layer capa física
physical line línea física
physical link enlace físico
physical magnitude magnitud física
physical map mapa físico
physical measurement medida física
physical message mensaje físico
physical optics óptica física
physical output device dispositivo de salida físico
physical page página física
physical phenomenon fenómeno físico
physical photometer fotómetro físico
physical photometry fotometría física
physical property propiedad física
physical quantity cantidad física
physical receptor receptor físico
physical record registro físico
physical requirement requisito físico
physical segment segmento físico
physical storage almacenamiento físico
physical storage organization organización de
　almacenamiento físico
physical structure estructura física
physical terminal terminal física
physical transmission transmisión física
physical unit unidad física
physical unit block bloque de unidad física
physical unit control point punto de control de
　unidad física
phystron fistrón
pi pi

pi mode modo pi
pi network red en pi
pi section sección pi
pi-section filter filtro de sección pi
pickling decapado
pickling bath baño de decapado
pickling inhibitor inhibidor de decapado
pickling solution solución de decapado
pickoff transductor
pickup fonocaptor, captor, lector, transductor, valor
　de puesta en trabajo, escobilla
pickup arm brazo de fonocaptor
pickup cartridge cápsula fonocaptora
pickup characteristic característica de fonocaptor
pickup coil bobina de captación
pickup current corriente de puesta en trabajo
pickup factor factor de captación
pickup microphone micrófono de captación
pickup needle aguja fonocaptora
pickup probe sonda de captación
pickup ratio razón de captación
pickup value valor de puesta en trabajo
pickup voltage tensión de puesta en trabajo
pico pico
picoammeter picoamperímetro
picoampere picoampere
picocomputer picocomputadora
picocoulomb picocoulomb
picocurie picocurie
picofarad picofarad
picohenry picohenry
picosecond picosegundo
picovolt picovolt
picowatt picowatt
pictorial pictórico
pictorial diagram diagrama pictórico
pictorial wiring diagram diagrama de alambrado
　pictórico
picture amplifier amplificador de imagen
picture amplitude amplitud de imagen
picture analysis análisis de imagen
picture black nivel del negro
picture brightness brillo de imagen
picture carrier portadora de imagen
picture-carrier amplifier amplificador de portadora
　de imagen
picture-carrier frequency frecuencia de portadora
　de imagen
picture-carrier monitor monitor de portadora de
　imagen
picture centering centrado de imagen
picture-centering control control de centrado de
　imagen
picture channel canal de imagen
picture chrominance crominancia de imagen
picture component componente de imagen
picture compression compresión de imagen
picture contrast contraste de imagen
picture-contrast control control de contraste de
　imagen
picture control control de imagen
picture definition definición de imagen
picture detail detalle de imagen
picture detector detector de imagen
picture-detector diode diodo detector de imagen
picture diagram diagrama pictórico
picture dot punto de imagen
picture edge borde de imagen
picture element elemento de imagen
picture fading desvanecimiento de imagen
picture filter filtro de imagen
picture fine adjustment ajuste fino de imagen
picture frequency frecuencia de imagen

picture frequency band banda de frecuencias de imagen
picture generator generador de imagen
picture height altura de imagen
picture-height control control de altura de imagen
picture information información de imagen
picture intermediate frequency frecuencia intermedia de imagen
picture inversion inversión de imagen
picture jitter fluctuación de imagen
picture jump salto vertical de imagen
picture line amplifier amplificador de línea de imagen
picture line frequency frecuencia de líneas de imagen
picture line standard patrón de líneas de imagen
picture linearity linealidad de imagen
picture lock fijación de imagen
picture mixer mezclador de imagen
picture modulation modulación de imagen
picture-modulation component componente de modulación de imagen
picture monitor monitor de imagen
picture output salida de imagen
picture period periodo de imagen
picture point punto de imagen
picture potentiometer potenciómetro de imagen
picture predistortion predistorsión de imagen
picture processing procesamiento de imagen
picture quality calidad de imagen
picture ratio razón de imagen
picture receiver receptor de imagen
picture reception recepción de imagen
picture reconstruction reconstrucción de imagen
picture record registro de imagen
picture repetition rate frecuencia de repetición de imagen
picture reproduction reproducción de imagen
picture resolution resolución de imagen
picture roll corrimiento vertical de imagen
picture rotation rotación de imagen
picture-rotation control control de rotación de imagen
picture signal señal de imagen
picture-signal amplifier amplificador de señal de imagen
picture-signal input entrada de señal de imagen
picture-signal modulation modulación de señal de imagen
picture-signal polarity polaridad de señal de imagen
picture size tamaño de imagen
picture slip deslizamiento vertical de imagen
picture source fuente de imagen
picture storage tube tubo de almacenamiento de imagen
picture synchronization sincronización de imagen
picture synchronizing sincronización de imagen
picture-synchronizing pulse impulso de sincronización de imagen
picture synthesis síntesis de imagen
picture tearing desgarre de imagen
picture telegraphy fototelegrafía
picture tone frecuencia portadora de imagen
picture transformer transformador de imagen
picture transmission transmisión de imagen
picture transmitter transmisor de imagen
picture tube tubo de imagen
picture-tube brightener abrillantador de tubo de imagen
picture weave salto horizontal de imagen
picture white nivel del blanco
picture width anchura de imagen
picture-width control control de anchura de imagen

pie winding arrollamiento en bobinas planas
Pierce gun cañón de Pierce
Pierce oscillator oscilador de Pierce
piezo piezo
piezocrystal piezocristal
piezodielectric piezodieléctrico
piezoelectric piezoeléctrico
piezoelectric accelerometer acelerómetro piezoeléctrico
piezoelectric activity actividad piezoeléctrica
piezoelectric axis eje piezoeléctrico
piezoelectric ceramic cerámica piezoeléctrica
piezoelectric ceramics cerámica piezoeléctrica
piezoelectric constant constante piezoeléctrica
piezoelectric crystal cristal piezoeléctrico
piezoelectric detector detector piezoeléctrico
piezoelectric device dispositivo piezoeléctrico
piezoelectric earphone audífono piezoeléctrico
piezoelectric effect efecto piezoeléctrico
piezoelectric filter filtro piezoeléctrico
piezoelectric gage galga piezoeléctrica
piezoelectric hydrophone hidrófono piezoeléctrico
piezoelectric ignition ignición piezoeléctrica
piezoelectric indicator indicador piezoeléctrico
piezoelectric loudspeaker altavoz piezoeléctrico
piezoelectric microphone micrófono piezoeléctrico
piezoelectric oscillator oscilador piezoeléctrico
piezoelectric pickup fonocaptor piezoeléctrico
piezoelectric pressure gage manómetro piezoeléctrico
piezoelectric probe sonda piezoeléctrica
piezoelectric quartz cuarzo piezoeléctrico
piezoelectric receiver receptor piezoeléctrico
piezoelectric resonator resonador piezoeléctrico
piezoelectric seismograph sismógrafo piezoeléctrico
piezoelectric sensor sensor piezoeléctrico
piezoelectric transducer transductor piezoeléctrico
piezoelectric tweeter altavoz de agudos piezoeléctrico
piezoelectric vibrator vibrador piezoeléctrico
piezoelectricity piezoelectricidad
piezometer piezómetro
piezoresistance piezorresistencia
piezoresistive piezorresistivo
piezoresistive effect efecto piezorresistivo
piezoresistivity piezorresistividad
piezothermal piezotérmico
piezotropism piezotropismo
pigeons palomas
piggyback control control en cascada
pigtail cable de conexión flexible, rabo de cerdo
pile pila
pillbox line línea de placas paralelas
pilot piloto
pilot amplifier amplificador piloto
pilot automatic exchange central automática piloto
pilot autooscillator autooscilador piloto
pilot bell timbre piloto
pilot card tarjeta piloto
pilot carrier portadora piloto
pilot cell celda piloto
pilot channel canal piloto
pilot circuit circuito piloto
pilot controller controlador piloto
pilot exciter excitador piloto
pilot frequency frecuencia piloto
pilot-frequency carrier portadora de frecuencia piloto
pilot-frequency generator generador de frecuencia piloto
pilot fuse fusible piloto
pilot hole agujero piloto
pilot lamp lámpara piloto

pilot light luz piloto
pilot line línea piloto
pilot loudspeaker altavoz piloto
pilot method método piloto
pilot model modelo piloto
pilot oscillator oscilador piloto
pilot production producción piloto
pilot protection protección por piloto
pilot pulse impulso piloto
pilot reference referencia piloto
pilot regulator regulador piloto
pilot relay relé piloto
pilot subcarrier subportadora piloto
pilot system sistema piloto
pilot tape cinta piloto
pilot test prueba piloto
pilot tone tono piloto
pilot-tone detector detector de tono piloto
pilot-tone generator generador de tono piloto
pilot tube tubo piloto
pilot wave onda piloto
pilot wire hilo piloto
pilot-wire protection protección por hilo piloto
pin patilla, perno, espiga, alfiler
pin circle círculo de espigas
pin connection conexión de espiga
PIN diode diodo PIN
pin feed alimentación por espigas
pin feed hole perforación de alimentación por
    espigas
pin insulator aislador de terminal
pin jack jack de patilla
PIN photodiode fotodiodo PIN
pin plug clavija de patilla
pin straightener enderezador de patillas,
    enderezador de espigas
pinch salto de aguja, constricción, pellizco
pinch current corriente de constricción
pinch effect efecto de constricción
pinch magnetic field campo magnético de
    constricción
pinchoff estricción, constricción
pinchoff diode diodo de estricción
pinchoff voltage tensión de estricción
pincushion distorsión en acerico
pincushion distortion distorsión en acerico
pine-tree antenna antena en pino
pine-tree array red en pino
ping impulso
pinhole agujero diminuto, agujero de espiga
pinhole detector detector de agujeros diminutos
pink noise ruido rosado
pip pip, impulso, cresta
Pirani gage medidor de Pirani
piston pistón
piston action acción de pistón
piston attenuator atenuador de pistón
piston type attenuator atenuador de tipo de pistón
pistonphone pistonófono
pit hoyo, depresión
pitch paso, espaciamiento, tono
pitch control control de paso
pitch factor factor de paso
pitch of winding paso de arrollamiento
pitch ratio razón de paso
Pitot tube tubo de Pitot
pits picaduras
pivot pivote
pivot arm brazo de pivote
pivot-mounted montado sobre pivote
pixel pixel
pixel depth profundidad de pixel
pixel scan exploración de pixel

placement algorithm algoritmo de colocación
plain connector conector plano
plain coupling acoplador plano
plan-position indicator indicador de posición en el
    plano
planar planar
planar antenna antena planar
planar array red planar
planar dielectric waveguide guíaondas dieléctrico
    planar
planar diffusion difusión planar
planar diode diodo planar
planar epitaxial passivated diode diodo epitaxial
    pasivado planar
planar epitaxial passivated transistor transistor
    epitaxial pasivado planar
planar epitaxial transistor transistor epitaxial planar
planar integrated circuit circuito integrado planar
planar junction unión planar
planar-junction diode diodo de unión planar
planar-junction transistor transistor de uniones
    planares
planar network red planar
planar photodiode fotodiodo planar
planar process proceso planar
planar silicon transistor transistor de silicio planar
planar transistor transistor planar
planar transmission line línea de transmisión planar
planar triode triodo planar
planar tube tubo planar
Planck's constant constante de Planck
Planck's law ley de Planck
Planck's radiation formula fórmula de radiación de
    Planck
Planck's radiation law ley de radiación de Planck
Planck-Einstein equation ecuación de
    Planck-Einstein
Planckian color color de Planck
Planckian locus lugar de Planck
Planckian radiator radiador de Planck
plane plano
plane antenna antena en hoja
plane aperture abertura plana
plane condenser condensador plano
plane electromagnetic wave onda electromagnética
    plana
plane of incidence plano de incidencia
plane of polarization plano de polarización
plane of projection plano de proyección
plane of propagation plano de propagación
plane of rotation plano de rotación
plane of vibration plano de vibración
plane polarization polarización plana
plane-polarized polarizado en un plano
plane-polarized light luz polarizada en un plano
plane-polarized sound wave onda sonora polarizada
    en un plano
plane-polarized wave onda polarizada en un plano
plane-progressive wave onda progresiva plana
plane reflector reflector plano
plane-reflector antenna antena con reflector plano
plane source fuente plana
plane wave onda plana
planetary electron electrón planetario
planigraph planígrafo
planigraphy planigrafía
planimeter planímetro
planned outage interrupción planificada
planoconcave planocóncavo
planoconvex planoconvexo
plant factor factor de utilización
Plante plate placa de Plante
plasma plasma

plasma accelerator acelerador de plasma
plasma anodization anodización de plasma
plasma balance balance de plasma
plasma beam haz de plasma
plasma current corriente de plasma
plasma density densidad de plasma
plasma diode diodo de plasma
plasma display visualizador de plasma
plasma display panel panel de visualización de plasma
plasma dynamics dinámica de plasma
plasma filament filamento de plasma
plasma frequency frecuencia de plasma
plasma gun cañón de plasma
plasma layer capa de plasma
plasma oscillation oscilación de plasma
plasma pressure presión de plasma
plasma radiation radiación de plasma
plasma resonance resonancia de plasma
plasma sheath vaina de plasma
plasma state estado de plasma
plasma thermocouple termopar de plasma
plasma wave onda de plasma
plasmatic plasmático
plasmatron plasmatrón
plasmoid plasmoide
plasmon plasmón
plasmotron plasmotrón
plastic plástico
plastic capacitor capacitor de plástico
plastic case caja de plástico
plastic effect efecto plástico
plastic-encapsulated encapsulado en plástico
plastic-encapsulated component componente encapsulado en plástico
plastic encapsulation encapsulación plástica
plastic-impregnated impregnado de plástico
plastic-insulated aislado con plástico
plastic-insulated cable cable aislado con plástico
plastic insulation aislamiento con plástico
plastic integrated circuit circuito integrado plástico
plastic mask máscara de plástico
plastic sheet hoja de plástico
plastic state estado plástico
plastic wave onda plástica
plasticity plasticidad
plasticizer plastificante
plate placa
plate battery batería de placa
plate bypass capacitor capacitor de desacoplo de placa
plate bypass condenser condensador de desacoplo de placa
plate capacitance capacitancia de placa
plate-cathode capacitance capacitancia placa-cátodo
plate-cathode capacity capacidad placa-cátodo
plate characteristic característica de placa
plate circuit circuito de placa
plate-circuit efficiency eficiencia de circuito de placa
plate-circuit relay relé de circuito de placa
plate coil bobina de placa
plate condenser condensador de placa
plate conductance conductancia de placa
plate-coupled multivibrator multivibrador de acoplamiento de placa
plate current corriente de placa
plate-current cutoff corte de corriente de placa
plate-current detection detección de corriente de placa
plate-current meter medidor de corriente de placa
plate-current modulation modulación de corriente de placa
plate-current relay relé de corriente de placa

plate-current saturation saturación de corriente de placa
plate-current shift desplazamiento de corriente de placa
plate detection detección de placa
plate detector detector de placa
plate dissipation disipación de placa
plate edge borde de placa
plate efficiency eficiencia de placa
plate efficiency factor factor de eficiencia de placa
plate electrode electrodo de placa
plate-filament capacitance capacitancia placa-filamento
plate-filament capacity capacidad placa-filamento
plate-grid capacitance capacitancia placa-rejilla
plate-grid capacity capacidad placa-rejilla
plate impedance impedancia de placa
plate input entrada de placa
plate input power potencia de entrada de placa
plate keying manipulación por placa
plate load carga de placa
plate load impedance impedancia de carga de placa
plate load resistance resistencia de carga de placa
plate magnet imán de placa
plate meter medidor de placa
plate-modulated modulado en placa
plate-modulated amplifier amplificador modulado en placa
plate modulation modulación de placa
plate neutralization neutralización de placa
plate potential potencial de placa
plate power potencia de placa
plate power input entrada de potencia de placa
plate power output salida de potencia de placa
plate power supply fuente de alimentación de placa
plate pulse modulation modulación de impulsos por placa
plate relay relé de placa
plate resistance resistencia de placa
plate resistor resistor de placa
plate saturation saturación de placa
plate-screen capacitance capacitancia placa-pantalla
plate-screen modulation modulación placa-pantalla
plate spacing espaciado de placas
plate supply alimentación de placa
plate supply voltage tensión de alimentación de placa
plate tank tanque de placa
plate tank capacitance capacitancia de tanque de placa
plate tank coil bobina de tanque de placa
plate tank inductance inductancia de tanque de placa
plate tank voltage tensión de tanque de placa
plate-to-cathode capacitor capacitor de placa a cátodo
plate-to-filament voltage tensión de placa a filamento
plate-to-grid capacitance capacitancia de placa a rejilla
plate-to-grid capacity capacidad de placa a rejilla
plate-to-plate impedance impedancia de placa a placa
plate transformer transformador de placa
plate tuning sintonización de placa
plate tuning capacitance capacitancia de sintonización de placa
plate tuning condenser condensador de sintonización de placa
plate tuning inductance inductancia de sintonización de placa
plate valve válvula de placa
plate vibrator vibrador de placa
plate voltage tensión de placa
plate-voltage regulator regulador de tensión de placa

plateau characteristic característica de meseta
plateau length longitud de meseta
plateau of potential meseta de potencial
plateau relative slope pendiente relativa de meseta
plateau slope pendiente de meseta
plated depositado, enchapado, electrodepositado, galvanoplasteado, metalizado
plated circuit circuito impreso por depósito
plated printed circuit circuito impreso por depósito
platen rodillo, platina
platform independence independencia de plataforma
platform stabilization estabilización de plataforma
plating depósito, enchapado, electrodeposición, galvanoplastia, metalización
plating bath baño de galvanoplastia
plating solution solución de galvanoplastia
plating tank tanque de galvanoplastia
platinite platinita
platinization platinado
platinize platinizar
platinized platinado
platinized quartz cuarzo platinado
platinizing platinado
platinode platinodo
platinoid platinoide
platinotron platinotrón
platinum platino
platinum-clad platinado
platinum-clad copper cobre platinado
platinum-clad electrode electrodo platinado
platinum-cobalt magnet imán de platino-cobalto
platinum-cobalt permanent magnet imán permanente de platino-cobalto
platinum contacts contactos de platino
platinum-plate platinar
platinum-plated platinado
platinum-plating platinado
platinum resistance resistencia de platino
platinum-resistance thermometer termómetro de resistencia de platino
platinum-ruthenium emitter emisor de platino-rutenio
platinum-tellurium thermocouple termopar platino-telurio
platinum-tipped con puntas platinadas
platinum wire hilo de platino
playback lectura, reproducción
playback amplifier amplificador de reproducción
playback channel canal de reproducción
playback characteristic característica de lectura
playback gap entrehierro de reproducción
playback head cabeza lectora, cabeza de reproducción
playback loss pérdida de lectura
playback preamplifier preamplificador de reproducción
playback stylus aguja de reproducción
playback system sistema de reproducción
playback unit unidad de reproducción
player reproductor, tocadiscos, tocacintas, lector
playing area área de registro
playing time tiempo de registro, tiempo de reproducción
pliodynatron pliodinatrón
pliotron pliotrón
plot gráfica, trazado
plotter trazador, graficador
plotting trazado
plotting head cabeza de trazado
plotting machine máquina de trazado
plotting speed velocidad de trazado
plug clavija, enchufe
plug adapter adaptador de clavija

plug-and-jack connection conexión de clavija y jack
plug board cuadro de clavijas
plug connection conexión de clavija
plug connector conector de clavija
plug contact contacto de clavija
plug cutout cortacircuito de tapón
plug fuse fusible de tapón
plug gage calibrador de tapón
plug-in enchufable
plug-in amplifier amplificador enchufable
plug-in capacitor capacitor enchufable
plug-in chassis chasis enchufable
plug-in circuit circuito enchufable
plug-in coil bobina enchufable
plug-in component componente enchufable
plug-in condenser condensador enchufable
plug-in connection conexión enchufable
plug-in connector conector enchufable
plug-in contact contacto enchufable
plug-in device dispositivo enchufable
plug-in fuse fusible enchufable
plug-in head assembly ensamblaje de cabeza enchufable
plug-in lamp lámpara enchufable
plug-in meter medidor enchufable
plug-in module módulo enchufable
plug-in outlet tomacorriente de clavija
plug-in piezoelectric crystal cristal piezoeléctrico enchufable
plug-in relay relé enchufable
plug-in resistor resistor enchufable
plug-in resistor resistor enchufable
plug-in transformer transformador enchufable
plug-in unit unidad enchufable
plug receptacle tomacorriente de clavija
plug seat asiento de clavija
plug sleeve base de clavija
plug switch conmutador de clavija
pluggable enchufable
plugging frenado por contracorriente, bloqueo
plugging device dispositivo de bloqueo
plugging-up device dispositivo de bloqueo
plumbicon plumbicón
plunger pulsador, contacto de presión, pistón
plutonium plutonio
PN barrier barrera PN
PN boundary límite PN
PN hook gancho PN
PN junction unión PN
PN-junction diode diodo de unión PN
PN-junction photocell fotocélula de unión PN
PN-junction rectifier rectificador de unión PN
PN-junction transistor transistor de unión PN
PN rectifier rectificador PN
pneumatic neumático
pneumatic circuit circuito neumático
pneumatic computer computadora neumática
pneumatic control control neumático
pneumatic detector detector neumático
pneumatic logic lógica neumática
pneumatic loudspeaker altavoz neumático
pneumatic receptor receptor neumático
pneumatic relay relé neumático
pneumatic switch conmutador neumático
pneumatic timer temporizador neumático
PNIN transistor transistor PNIN
PNIP transistor transistor PNIP
PNP diffused-junction transistor transistor de unión por difusión PNP
PNP junction unión PNP
PNP junction transistor transistor de unión PNP
PNP phototransistor fototransistor PNP
PNP tetrode tetrodo PNP

PNP transistor   transistor PNP
PNPN device   dispositivo PNPN
PNPN diode   diodo PNPN
PNPN transistor   transistor PNPN
Pockels effect   efecto de Pockels
pocket ammeter   amperímetro de bolsillo
pocket meter   medidor de bolsillo
Poincare sphere   esfera de Poincare
point   punto, punta
point-by-point   punto por punto
point-by-point measurement   medida punto por
   punto
point charge   carga puntual
point contact   contacto de punta
point-contact diode   diodo de contacto de punta
point-contact junction   unión de contacto de punta
point-contact photodiode   fotodiodo de contacto de
   punta
point-contact phototransistor   fototransistor de
   contacto de punta
point-contact rectifier   rectificador de contacto de
   punta
point-contact transistor   transistor de contactos de
   punta
point-contact transistor tetrode   tetrodo transistor de
   contactos de punta
point counter   contador de punta
point counter tube   tubo contador de punta
point detection   detección de aguja
point effect   efecto de punta
point electrode   electrodo de punta
point group   grupo puntual
point identification   identificación de puntos
point impedance   impedancia puntual
point-junction transistor   transistor de punta y
   uniones
point light   luz puntual
point location   localización de puntos, localización de
   comas
point machine   motor de aguja
point of communication   punto de comunicación
point of connection   punto de conexión
point of contact   punto de contacto
point of origin   punto de origen
point of phase   punto de fase
point of sale   punto de venta
point of saturation   punto de saturación
point source   fuente puntual
point-to-point circuit   circuito de punto a punto
point-to-point communication   comunicación de
   punto a punto
point-to-point line   línea de punto a punto
point-to-point link   enlace de punto a punto
point-to-point network   red de punto a punto
point-to-point scanning   exploración de punto a punto
point-to-point station   estación de punto a punto
point-to-point transmission   transmisión de punto a
   punto
point-to-point wiring   alambrado de punto a punto
point transistor   transistor de puntas
pointer   indicador, aguja
pointer deflection   deflexión de aguja
pointer frequency meter   frecuencímetro de aguja
pointer galvanometer   galvanómetro de aguja
pointer instrument   instrumento de aguja
pointer reading   lectura de aguja
pointer-type meter   medidor de tipo de aguja
poise   poise
Poisson's equation   ecuación de Poisson
polar   polar
polar arc   arco polar
polar axis   eje polar
polar circuit   circuito polar

polar comparator   comparador polar
polar coordinates   coordenadas polares
polar curve   curva polar
polar detector   detector polar
polar diagram   diagrama polar
polar molecule   molécula polar
polar orbit   órbita polar
polar reaction   reacción polar
polar relay   relé polarizado
polar transmission   transmisión polar
polarimeter   polarímetro
polarimetry   polarimetría
polariscope   polariscopio
polarity   polaridad
polarity finder   buscador de polaridad
polarity indicator   indicador de polaridad
polarity-protection diode   diodo de protección de
   polaridad
polarity-reversal detector   detector de inversión de
   polaridad
polarity-reversal switch   conmutador de inversión de
   polaridad
polarity reverser   inversor de polaridad, inversor de
   polos
polarity reversing   inversión de polaridad, inversión
   de polos
polarity-sensitive relay   relé sensible a la polaridad
polarity switch   conmutador de polaridad
polarity trap   trampa de polaridad
polarizability   polarizabilidad
polarizable   polarizable
polarization   polarización
polarization apparatus   aparato de polarización
polarization capacitance   capacitancia de
   polarización
polarization changer   cambiador de polarización
polarization charge   carga de polarización
polarization current   corriente de polarización
polarization cycle   ciclo de polarización
polarization discrimination   discriminación de
   polarización
polarization diversity   diversidad de polarización
polarization effect   efecto de polarización
polarization efficiency   eficiencia de polarización
polarization error   error de polarización
polarization fading   desvanecimiento por
   polarización
polarization index   índice de polarización
polarization interferometer   interferómetro de
   polarización
polarization modulation   modulación por
   polarización
polarization pattern   patrón de polarización
polarization phenomenon   fenómeno de polarización
polarization photometer   fotómetro de polarización
polarization plane   plano de polarización
polarization potential   potencial de polarización
polarization ratio   razón de polarización
polarization reactance   reactancia de polarización
polarization receiving factor   factor de recepción de
   polarización
polarization resistance   resistencia de polarización
polarization reversal   inversión de polarización
polarization unit vector   vector unitario de
   polarización
polarization voltage   tensión de polarización
polarize   polarizar
polarized   polarizado
polarized ammeter   amperímetro polarizado
polarized beam   haz polarizado
polarized bell   timbre polarizado
polarized capacitor   capacitor polarizado
polarized differential relay   relé diferencial

polarizado
polarized electric drainage drenaje eléctrico polarizado
polarized electrolytic capacitor capacitor electrolítico polarizado
polarized electromagnet electroimán polarizado
polarized field frequency relay relé de frecuencia de campo polarizado
polarized grid rejilla polarizada
polarized infrared radiation radiación infrarroja polarizada
polarized light luz polarizada
polarized meter medidor polarizado
polarized plug clavija polarizada
polarized proton beam haz protónico polarizado
polarized radiation radiación polarizada
polarized receptacle receptáculo polarizado
polarized relay relé polarizado
polarized ringer timbre polarizado
polarized wave onda polarizada
polarized X-rays rayos X polarizados
polarizer polarizador
polarizing polarizante, polarizador
polarizing battery batería polarizante
polarizing current corriente polarizante
polarizing filter filtro polarizante
polarizing flux flujo polarizante
polarizing interference interferencia polarizante
polarizing prism prisma polarizante
polarizing slot ranura polarizante
polarizing voltage tensión polarizante
polarogram polarograma
polarograph polarógrafo
polarographic polarográfico
polarographic analysis análisis polarográfico
polarography polarografía
polaroid polaroid
polaroid filter filtro polaroid
pole polo, poste
pole arc arco polar
pole changer cambiador de polos
pole face cara polar
pole finder buscapolos
pole gap entrehierro
pole horn cuerno polar
pole indicator indicador de polos
pole piece pieza polar
pole pitch paso polar
pole reverser inversor de polos
pole shoe zapata polar
pole tip extremo polar
pole winding arrollamiento polar
polling sondeo, interrogación
polling characters caracteres de sondeo
polling delay retardo de sondeo
polling interval intervalo de sondeo
polling list lista de sondeo
polling loop bucle de sondeo
polling program programa de sondeo
polling routine rutina de sondeo
polonium polonio
polyanode polianódico
polyanode rectifier rectificador polianódico
polyatomic poliatómico
polycarbonate policarbonato
polycarbonate condenser condensador de policarbonato
polychloroprene policloropreno
polychloroprene-insulated cable cable aislado con policloropreno
polychlorotrifluoroethylene resin resina de policlorotrifluoroetileno
polychromatic policromático

polychromatic radiation radiación policromática
polychromator policromador
polychrome policromo
polychrome picture imagen policroma
polycrystalline material material policristalino
polydirectional polidireccional
polydirectional microphone micrófono polidireccional
polyelectrode polielectrodo
polyelectrolyte polielectrólito
polyenergetic polienergético
polyester poliéster
polyester base base de poliéster
polyester film película de poliéster
polyester plastic plástico de poliéster
polyethylene polietileno
polyethylene disk disco de polietileno
polyethylene-insulated cable cable aislado con polietileno
polygonal poligonal
polygonal coil bobina poligonal
polygraph polígrafo
polyiron polihierro
polyline polilínea
polymer polímero
polymeric polimérico
polymerization polimerización
polymerize polimerizar
polyode poliodo
polyolefin poliolefina
polyoptic polióptico
polyoptic sealing sellado polióptico
polyphase polifásico
polyphase alternator alternador polifásico
polyphase circuit circuito polifásico
polyphase commutator motor motor de colector polifásico
polyphase compensating winding arrollamiento compensador polifásico
polyphase converter convertidor polifásico
polyphase current corriente polifásica
polyphase generator generador polifásico
polyphase induction machine máquina de inducción polifásica
polyphase induction motor motor de inducción polifásico
polyphase machine máquina polifásica
polyphase meter medidor polifásico
polyphase motor motor polifásico
polyphase oscillator oscilador polifásico
polyphase power potencia polifásica
polyphase rectifier rectificador polifásico
polyphase rectifier circuit circuito rectificador polifásico
polyphase selectivity selectividad polifásica
polyphase shunt motor motor en derivación polifásico
polyphase sort clasificación polifásica
polyphase synchronous generator generador síncrono polifásico
polyphase synchronous motor motor síncrono polifásico
polyphase system sistema polifásico
polyphase torque converter convertidor de par polifásico
polyphase transformer transformador polifásico
polyphotal polifotal
polyphote polifoto
polyplexer poliplexor
polypropylene polipropileno
polyrod antenna antena dieléctrica de varillas
polysilicon polisilicio
polystyrene poliestireno

polystyrene capacitor capacitor de poliestireno
polystyrol poliestirol
polystyrol capacitor capacitor de poliestirol
polystyrol condenser condensador de poliestirol
polytetrafluoroethylene politetrafluoretileno
polytetrafluoroethylene resin resina de politetrafluoretileno
polythene politeno
polythene dielectric dieléctrico de politeno
polythene-insulated aislado con politeno
polythene insulator aislador de politeno
polyurethane poliuretano
polyvalence polivalencia
polyvalent polivalente
polyvinyl polivinilo
polyvinyl acetate acetato de polivinilo
polyvinyl chloride cloruro de polivinilo
pool cathode cátodo líquido
pool-cathode tube tubo de cátodo líquido
pool rectifier rectificador de cátodo líquido
pool tube tubo de cátodo líquido
popcorn noise ruido de palomitas de maíz
porcelain porcelana
porcelain base base de porcelana
porcelain capacitor capacitor de porcelana
porcelain enamel esmalte de porcelana
porcelain insulator aislador de porcelana
porcelain reflector reflector de porcelana
porcelain sleeve manguito de porcelana
porcelain socket zócalo de porcelana
porcelain tube tubo de porcelana
porch umbral
porous poroso
porous pot vaso poroso
port puerta, puerto, acceso
port controller controlador de puerto
port selector selector de puerto
portability portabilidad
portable portátil
portable antenna antena portátil
portable apparatus aparato portátil
portable appliance aparato eléctrico portátil
portable battery batería portátil
portable cell elemento portátil
portable computer computadora portátil
portable computer system sistema de computadora portátil
portable data medium medio de datos portátil
portable generator generador portátil
portable lamp lámpara portátil
portable lighting alumbrado portátil
portable mobile station estación móvil portátil
portable mobile unit unidad móvil portátil
portable radio radio portátil
portable receiver receptor portátil
portable recorder registrador portátil, grabadora portátil
portable station estación portátil
portable storage almacenamiento portátil
portable test unit unidad de prueba portátil
portable tester probador portátil
portable transmitter transmisor portátil
portable unit unidad portátil
portable voltmeter vóltmetro portátil
portrait orientation orientación vertical
position posición
position angle ángulo de posición
position code código de posición
position control control de posición
position-control system sistema de control de posición
position-control transducer transductor de control de posición

position-dependent dependiente de posición
position feedback retroalimentación de posición
position fixing fijación de posición
position grouping agrupación de posiciones
position indicator indicador de posición
position lights luces de posición
position line línea de posición
position meter medidor de posición
position-sensitive sensible a la posición
position-sensitive detector detector sensible a la posición
position sensor sensor de posición
position storage almacenamiento de posición
position switch conmutador de posición
position transducer transductor de posición
positional posicional, de posición
positional checking comprobación posicional
positional error error posicional
positional notation notación posicional
positional number system sistema numérico posicional
positional operand operando posicional
positional parameter parámetro posicional
positional representation representación posicional
positional representation system sistema de representación posicional
positional servomechanism servomecanismo posicional
positioner posicionador
positioning posicionamiento
positioning arm brazo de posicionamiento
positioning control system sistema de control de posicionamiento
positioning device dispositivo de posicionamiento
positioning magnet imán de posicionamiento
positioning mechanism mecanismo de posicionamiento
positioning time tiempo de posicionamiento
positive positivo
positive action acción positiva
positive amplitude modulation modulación de amplitud positiva
positive-and-negative booster elevador-reductor
positive angle ángulo positivo
positive area área positiva
positive bias polarización positiva
positive booster elevador
positive brush escobilla positiva
positive busbar barra colectora positiva
positive charge carga positiva
positive column columna positiva
positive conductance conductancia positiva
positive conductor conductor positivo
positive contact contacto positivo
positive control mando directo
positive crystal cristal positivo
positive dispersion dispersión positiva
positive displacement desplazamiento positivo
positive distortion distorsión positiva
positive drive transmisión directa
positive electricity electricidad positiva
positive electrification electrificación positiva
positive electrode electrodo positivo
positive electron electrón positivo
positive exponent exponente positivo
positive feedback retroalimentación positiva
positive-feedback amplifier amplificador de retroalimentación positiva
positive-feedback circuit circuito de retroalimentación positiva
positive frequency modulation modulación de frecuencia positiva
positive function función positiva

positive ghost fantasma positivo
positive glow luminiscencia positiva
positive grid rejilla positiva
positive-grid oscillator oscilador de rejilla positiva
positive-grid oscillator tube tubo de oscilador de rejilla positiva
positive hole hueco positivo
positive image imagen positiva
positive indicator indicador positivo
positive input entrada positiva
positive ion ion positivo
positive-ion accelerator acelerador de iones positivos
positive-ion emission emisión de iones positivos
positive-ion sheath vaina de iones positivos
positive-ion trapping captura de iones positivos
positive lead conductor positivo
positive light modulation modulación positiva
positive line línea positiva
positive logic lógica positiva
positive logic conversion conversión de lógica positiva
positive magnetostriction magnetoestricción positiva
positive matrix matriz positiva
positive measurement error error de medida positivo
positive modulation modulación positiva
positive modulation factor factor de modulación positiva
positive-negative booster elevador-reductor
positive number número positivo
positive peak pico positivo
positive-peak clipper recortador de picos positivos
positive-peak modulation modulación de picos positivos
positive-peak voltmeter vóltmetro de pico positivo
positive phase fase positiva
positive phase-sequence reactance reactancia de secuencia de fase positiva
positive phase-sequence relay relé de secuencia de fase positiva
positive phase-sequence resistance resistencia de secuencia de fase positiva
positive picture imagen positiva
positive picture modulation modulación de imagen positiva
positive picture phase fase de imagen positiva
positive plate placa positiva
positive polarity polaridad positiva
positive-polarity signal señal de polaridad positiva
positive pole polo positivo
positive potential potencial positivo
positive pressure presión positiva
positive pulse impulso positivo
positive ray rayo positivo
positive-ray analysis análisis por rayos positivos
positive-ray current corriente de rayos positivos
positive resistance resistencia positiva
positive resistor resistor positivo
positive response respuesta positiva
positive-sequence impedance impedancia de secuencia positiva
positive-sequence resistance resistencia de secuencia positiva
positive side lado positivo
positive synchronization sincronización positiva
positive temperature coefficient coeficiente de temperatura positivo
positive terminal terminal positiva
positive transition transición positiva
positive transmission transmisión positiva
positive valence valencia positiva
positive value valor positivo
positive voltage tensión positiva

positive wave onda positiva
positive wire hilo positivo
positive zero cero positivo
positively positivamente
positively charged cargado positivamente
positively ionized ionizado positivamente
positron positrón
positron emission emisión de positrones
positron emitter emisor de positrones
possible capacity capacidad posible
post poste, borne, patilla
post-accelerating anode ánodo postacelerador
post-accelerating electrode electrodo postacelerador
post acceleration postaceleración
post-acceleration tube tubo de postaceleración
post-acceleration voltage tensión de postaceleración
post-deflection accelerating electrode electrodo acelerador postdeflexión
post-deflection acceleration aceleración postdeflexión
post-deflection focus enfoque postdeflexión
post-echo posteco
post-equalization postecualización
post-mortem dump vaciado post mortem
post-mortem program programa post mortem
post-mortem routine rutina post mortem
post office box caja de puente
post-processing postprocesamiento
post-processor postprocesador
post-synchronization postsincronización
potassium potasio
potassium chloride cloruro de potasio
potassium cyanide cianuro de potasio
potential potencial, tensión
potential attenuator atenuador de potencial
potential barrier barrera de potencial
potential coil bobina de tensión, bobina en derivación
potential curve curva de potencial
potential diagram diagrama de potencial
potential difference diferencia de potencial
potential distribution distribución de potencial
potential divider divisor de potencial
potential drop caída de potencial
potential energy energía potencial
potential-energy curve curva de energía potencial
potential fall caída de potencial
potential function función potencial
potential fuse fusible de potencial
potential galvanometer galvanómetro de potencial
potential gradient gradiente de potencial
potential minimum mínimo de potencial
potential peak pico de potencial
potential plateau meseta de potencial
potential probe sonda de potencial
potential profile perfil de potencial
potential scattering dispersión de potencial
potential transformer transformador de potencial
potential trough depresión de potencial
potential value valor de potencial
potential variation variación de potencial
potentiometer potenciómetro
potentiometer chain cadena potenciométrica
potentiometer circuit circuito potenciométrico
potentiometer control control potenciométrico
potentiometer divider divisor potenciométrico
potentiometer indicator indicador potenciométrico
potentiometer method método potenciométrico
potentiometer network red potenciométrica
potentiometer rheostat reóstato potenciométrico
potentiometer slider cursor de potenciómetro
potentiometer-type resistor resistor potenciométrico
potentiometer-type rheostat reóstato potenciométrico

potentiometer-type transducer transductor potenciométrico
potentiometric potenciométrico
potentiometric analysis análisis potenciométrico
potentiometric transducer transductor potenciométrico
potentiometric voltmeter vóltmetro potenciométrico
potentiometrically potenciométricamente
potentiostat potenciostato
Potier's coefficient of equivalence coeficiente de equivalencia de Potier
Potier's diagram diagrama de Potier
Potier's electromotive force fuerza electromotriz de Potier
Potier's method método de Potier
Potier's reactance reactancia de Potier
potted circuit circuito encapsulado
potted component componente encapsulado
Potter horn bocina de Potter
Potter multivibrator multivibrador de Potter
Potter oscillator oscilador de Potter
potting encapsulación, encapsulado
potting material material de encapsulación
Poulsen arc arco de Poulsen
poundal poundal
pounds per square inch libras por pulgada cuadrada
powder metallurgy pulvimetalurgia
powder pattern patrón de polvo
powdered pulverizado
powdered-iron core núcleo de hierro pulverizado
powdered magnetic alloy aleación magnética pulverizada
powdered metal metal pulverizado
power potencia, energía, aumento
power-actuated accionado por potencia
power amplification amplificación de potencia
power amplifier amplificador de potencia
power-amplifier device dispositivo amplificador de potencia
power-amplifier stage etapa amplificadora de potencia
power-amplifier tube tubo amplificador de potencia
power-amplifier unit unidad amplificadora de potencia
power-amplifying circuit circuito amplificador de potencia
power attenuation atenuación de potencia
power available potencia disponible
power bandwidth ancho de banda de potencia
power busbar barra colectora de potencia
power cable cable de alimentación
power capability capacidad de potencia
power capacitor capacitor de potencia
power capacity capacidad de potencia
power circuit circuito de potencia
power coefficient coeficiente de potencia
power component componente de potencia
power conductor conductor de alimentación
power connector conector de alimentación
power consumption consumo de potencia
power contact contacto de potencia
power control control de potencia
power control cable cable de control de potencia
power control center centro de control de potencia
power control panel panel de control de potencia
power control relay relé de control de potencia
power control unit unidad de control de potencia
power converter convertidor de potencia
power cord cordón de alimentación
power coupler acoplador de potencia
power current corriente de alimentación, corriente fuerte
power curve curva de potencia

power density densidad de potencia
power detection detección de potencia
power detector detector de potencia
power deviation desviación de potencia
power device dispositivo de potencia
power difference diferencia de potencia
power diode diodo de potencia
power-direction relay relé de dirección de potencia
power-directional relay relé direccional de potencia
power-dissipating disipador de potencia
power-dissipating resistor resistor disipador de potencia
power dissipation disipación de potencia
power-distributing bar barra distribuidora de potencia
power-distributing transformer transformador distribuidor de potencia
power distribution distribución de potencia
power-distribution box caja de distribución de potencia
power-distribution panel panel de distribución de potencia
power-distribution unit unidad de distribución de potencia
power distributor distribuidor de potencia
power divider divisor de potencia
power drain consumo de potencia
power drop caída de potencia
power equalizer ecualizador de potencia
power factor factor de potencia
power-factor capacitor capacitor de factor de potencia
power-factor correction corrección de factor de potencia
power-factor improvement mejora de factor de potencia
power-factor indicator indicador de factor de potencia
power-factor meter medidor de factor de potencia
power-factor regulator regulador de factor de potencia
power-factor relay relé de factor de potencia
power failure falla de alimentación
power-failure alarm alarma de falla de alimentación
power-failure indicator indicador de falla de alimentación
power-failure interrupt interrupción por falla de alimentación
power-failure warning signal señal de aviso de falla de alimentación
power feed alimentación de energía
power filter filtro de potencia
power flow flujo de potencia
power flux flujo de potencia
power flux density densidad de flujo de potencia
power frame marco de potencia
power frequency frecuencia de red
power gain ganancia de potencia
power-gain cutoff frequency frecuencia de corte de ganancia de potencia
power-handling capacity capacidad máxima de potencia
power impulse impulso de potencia
power input potencia de entrada
power inverter inversor de potencia
power klystron klistrón de potencia
power lead línea de alimentación
power level nivel de potencia
power-level calibration calibración de nivel de potencia
power-level difference diferencia de nivel de potencia
power-level indicator indicador de nivel de potencia

power leveler   nivelador de potencia
power limit   límite de potencia
power limiter   limitador de potencia
power-limiting   limitador de potencia
power-limiting reactance   reactancia limitadora de potencia
power line   línea de energía, red eléctrica
power-line carrier   portadora en línea de energía
power-line communication   comunicación por línea de energía
power-line filter   filtro de línea de energía
power-line frequency   frecuencia de línea de energía
power-line monitor   monitor de línea de energía
power loss   pérdida de potencia
power-loss factor   factor de pérdida de potencia
power magnification   magnificación de potencia
power measurement   medida de potencia
power measurement device   dispositivo de medida de potencia
power meter   medidor de potencia, contador
power modulation   modulación de potencia
power-modulation factor   factor de modulación de potencia
power monitor   monitor de potencia
power network   red de energía
power of resolution   poder de resolución
power operation   operación por servomotor
power oscillator   oscilador de potencia
power outage   interrupción de alimentación
power outlet   tomacorriente
power output   potencia de salida
power-output capability   capacidad de potencia de salida
power-output meter   medidor de potencia de salida
power-output tube   tubo de potencia de salida
power pack   unidad de alimentación
power panel   panel de alimentación
power pattern   patrón de potencia
power peak   pico de potencia
power-peak limitation   limitación de picos de potencia
power pentode   pentodo de potencia
power plant   planta eléctrica
power plug   tomacorriente
power point   punto de potencia
power protection   protección de potencia
power pulse   impulso de potencia
power range   gama de potencia
power rating   potencia nominal
power rating test   prueba de potencia nominal
power ratio   razón de potencia
power reactor   reactor de potencia
power receptacle   tomacorriente
power rectification   rectificación de potencia
power rectifier   rectificador de potencia
power rectifying valve   válvula rectificadora de potencia
power regulator   regulador de potencia
power relay   relé de potencia
power requirements   requisitos de alimentación
power reserve   reserva de potencia
power resistor   resistor de potencia
power ringing   llamada de potencia
power selector   selector de potencia
power sensibility   sensibilidad de potencia
power series   serie de potencia
power signal box   puesto de mando
power signaling   señalización de potencia
power source   fuente de alimentación
power spectrum   espectro de potencia
power stage   etapa de potencia
power station   planta eléctrica
power supply   fuente de alimentación

power-supply circuit   circuito de fuente de alimentación
power-supply efficiency   eficiencia de fuente de alimentación
power-supply filter   filtro de fuente de alimentación
power-supply hum   zumbido de fuente de alimentación
power-supply line   línea de fuente de alimentación
power-supply panel   panel de alimentación
power-supply transformer   transformador de alimentación
power-supply unit   unidad de alimentación
power-supply variation   variación de fuente de alimentación
power switch   conmutador de alimentación
power switchboard   cuadro de distribución de fuerza
power switchgroup   combinador de potencia
power system   sistema de alimentación
power system stabilizer   estabilizador de sistema de alimentación
power takeoff   toma de potencia
power tetrode   tetrodo de potencia
power transfer   transferencia de potencia
power transfer relay   relé de transferencia de potencia
power transfer theorem   teorema de transferencia de potencia
power transformation   transformación de potencia
power transformer   transformador de potencia
power transistor   transistor de potencia
power transmission   transmisión de energía
power transmission line   línea de transmisión de energía
power triode   triodo de potencia
power tube   tubo de potencia
power unit   unidad de potencia
power winding   arrollamiento de potencia
powerhouse   planta eléctrica
Poynting vector   vector de Poynting
PP junction   unión PP
practical component   componente práctico
practical electrical units   unidades eléctricas prácticas
practical electromagnetic system   sistema electromagnético práctico
practical system   sistema práctico
practical units   unidades prácticas
pragilbert   pragilbert
pragilberts per weber   pragilberts por weber
praoersted   praoersted
praseodymium   praseodimio
preallocation   preasignación
preamplification   preamplificación
preamplification transformer   transformador de preamplificación
preamplifier   preamplificador
preamplifier stage   etapa preamplificadora
preamplifier tube   tubo preamplificador
preamplifying   preamplificador
preamplifying valve   válvula preamplificadora
preanalysis   preanálisis
preassembled   preensamblado
preassembly   preensamblado
preassembly time   tiempo de preensamblado
preassigned   preasignado
preassigned multiple access   acceso múltiple preasignado
precalibrated   precalibrado
precession   precesión
precession resonance   resonancia de precesión
prechecking   precomprobación
precipitability   precipitabilidad
precipitable   precipitable

precipitation precipitación
precipitation static estática de precipitación
precipitator precipitador
precipitron precipitrón
precise preciso
precision precisión
precision adjustment ajuste de precisión
precision approach aproximación de precisión
precision approach lighting system sistema de
    alumbrado de aproximación de precisión
precision approach radar radar de aproximación de
    precisión
precision device dispositivo de precisión
precision error error de precisión
precision frequency meter frecuencímetro de
    precisión
precision instrument instrumento de precisión
precision lamp lámpara de precisión
precision network red de precisión
precision potentiometer potenciómetro de precisión
precision radar radar de precisión
precision recording registro de precisión
precision resistor resistor de precisión
precision sweep barrido de precisión
precoating prerrevestimiento
precoded precodificado
precompilation precompilación
precompiler precompilador
precompiling precompilación
precondition precondición
preconditioning precondicionamiento
preconduction preconducción
preconduction current corriente de preconducción
preconfigured preconfigurado
preconfigured system definition definición de
    sistema preconfigurado
precontrol precontrol
precorrection precorrección
precursor precursor
precut precortado
precut wire hilo precortado
predefined predefinido
predefined function función predefinida
predefined message mensaje predefinido
predefined process proceso predefinido
predetection predetección
predetermined predeterminado
predetermined code código predeterminado
predetermined counter contador predeterminado
predetermined sequence secuencia predeterminada
predictive control control predictivo
predissociation predisociación
predistortion predistorsión
Preece's formula fórmula de Preece
preecho preeco
preemphasis preénfasis
preemphasis filter filtro de preénfasis
preemphasis network red de preénfasis
preequalization preecualización
preexecution preejecución
prefabricated prefabricado
prefabricated circuit circuito prefabricado
prefabricated unit unidad prefabricada
prefabricated wiring alambrado prefabricado
prefabrication prefabricación
preferential preferencial
preferential flow flujo preferencial
preferential oxidation oxidación preferencial
preferred preferido
preferred numbers números preferidos
preferred orientation orientación preferida
preferred power supply fuente de alimentación
    preferida

preferred tube types tipos de tubos preferidos
preferred values valores preferidos
preferred values of components valores preferidos
    de componentes
prefix prefijo
prefix notation notación por prefijos
prefocus base base prefoco
prefocus lamp lámpara prefoco
prefocused preenfocado
prefocusing preenfoque
preformed preformado
preformed cable cable preformado
preformed winding arrollamiento sobre horma
pregroup pregrupo
preheat precalentar
preheater precalentador
preheating precalentamiento
preheating time tiempo de precalentamiento
preignition preignición
preimpregnate preimpregnar
preimpregnated preimpregnado
preimpregnated insulation aislamiento
    preimpregnado
preindication preindicación
preionization preionización
preionizing preionizante
preliminary preliminar
preliminary amplifier amplificador preliminar
preliminary design diseño preliminar
preliminary information información preliminar
preload precargar
preloaded precargado
premodulation premodulación
premodulation amplifier amplificador de
    premodulación
prenumbering prenumeración
preoperative error error preoperativo
preoscillation preoscilación
preoscillation current corriente de preoscilación
preprinted preimpreso
preprinting preimpresión
preprocess preprocesar
preprocessed preprocesado
preprocessed display visualización preprocesada
preprocessed macro program programa de macros
    preprocesado
preprocessing preprocesamiento
preprocessor preprocesador
preprocessor statement sentencia de preprocesador
preproduction preproducción
preprogrammed preprogramado
prepunch preperforar
prepunched preperforado
prepunched card tarjeta preperforada
prepunched master card tarjeta maestra
    preperforada
preread head cabeza de prelectura
prerecord prerregistrar
prerecorded pregrabado, prerregistrado
prerecorded data medium medio de datos
    prerregistrado
prerecorded disk disco pregrabado
prerecorded tape cinta pregrabada
presducer presductor
preselected preseleccionado
preselected reference voltage tensión de referencia
    preseleccionada
preselection preselección
preselection stage etapa de preselección
preselector preselector
presentation presentación
presentation medium medio de presentación
preset preajustado, preajustable

preset button botón preajustado, botón preajustable
preset controller controlador preajustado, controlador preajustable
preset counter contador preajustado, contador preajustable
preset element elemento preajustado, elemento preajustable
preset parameter parámetro preajustado, parámetro preajustable
preset switch conmutador preajustado, conmutador preajustable
press button botón de presión
press-button lock cierre de botón de presión
press-to-talk button botón de oprimir para hablar
press-to-talk intercom intercomunicador de oprimir para hablar
press-to-talk microphone micrófono de oprimir para hablar
press-to-talk switch conmutador de oprimir para hablar
pressed powder polvo compactado
pressed powder printed circuit circuito impreso de polvo compactado
pressing switch conmutador de presión
pressure presión, tensión
pressure-actuated accionado por presión
pressure-actuated switch conmutador accionado por presión
pressure adjustment ajuste de presión
pressure amplitude amplitud de presión
pressure cable cable a presión
pressure change cambio de presión
pressure coefficient coeficiente de presión
pressure connector conector de presión
pressure contact contacto a presión
pressure-contact switch conmutador de contacto a presión
pressure detector detector de presión
pressure difference diferencia de presión
pressure differential presión diferencial
pressure-differential switch conmutador de presión diferencial
pressure distribution distribución de presión
pressure drop caída de presión, caída de tensión
pressure equalizer ecualizador de presión
pressure equilibrium equilibrio de presión
pressure-feed alimentación a presión
pressure gradient gradiente de presión
pressure-gradient microphone micrófono de gradiente de presión
pressure hydrophone hidrófono de presión
pressure indicator indicador de presión
pressure level nivel de presión
pressure limit límite de presión
pressure loss pérdida de presión
pressure meter medidor de presión
pressure microphone micrófono de presión
pressure-operated operado por presión
pressure pad almohadilla de presión
pressure pickup captor de presión
pressure plate placa de presión
pressure regulation regulación de presión
pressure regulator regulador de presión
pressure relay relé de presión
pressure-relief device dispositivo de alivio de presión
pressure resonance resonancia de presión
pressure response respuesta a presión, respuesta en presión
pressure roller rodillo de presión
pressure sensitive sensible a presión
pressure-sensitive adhesive adhesivo sensible a presión

pressure-sensitive keyboard teclado sensible a presión
pressure-sensitive pen pluma sensible a presión
pressure sensitivity sensibilidad a presión, rendimiento en presión
pressure switch conmutador de presión
pressure test prueba de presión, prueba de tensión
pressure transducer transductor de presión
pressure welding soldadura por presión
pressurestat presostato
pressurization presurización
pressurize presurizar
pressurized presurizado
pressurized transmission line línea de transmisión presurizada
prestorage prealmacenamiento
prestore prealmacenar
presumptive address dirección supuesta
presumptive instruction instrucción supuesta
presumptive loss pérdida supuesta
pretersonic pretersónico
pretunable presintonizable
pretune presintonizar
pretuned presintonizado
pretuned frequency frecuencia presintonizada
pretuned receiver receptor presintonizado
pretuned stage etapa presintonizada
preventive maintenance mantenimiento preventivo
preventive maintenance test prueba de mantenimiento preventivo
preventive maintenance time tiempo de mantenimiento preventivo
preventive service servicio preventivo
prewired precableado
prewired circuit circuito precableado
primary primario
primary address dirección primaria
primary application aplicación primaria
primary application block bloque de aplicación primaria
primary application program programa de aplicación primaria
primary assignment asignación primaria
primary battery batería primaria
primary block bloque primario
primary breakdown disrupción primaria
primary brush escobilla primaria
primary capacitance capacitancia primaria
primary capacitor capacitor primario
primary card tarjeta primaria
primary carrier flow flujo de portadores primario
primary cell célula primaria
primary center centro primario
primary change cambio primario
primary character carácter primario
primary circuit circuito primario
primary cluster grupo primario
primary coating revestimiento primario
primary coil bobina primaria
primary-color filter filtro de color primario
primary-color signal señal de color primario
primary-color unit unidad de color primario
primary colors colores primarios
primary console consola primaria
primary control control primario
primary-control program programa de control primario
primary current corriente primaria
primary-current ratio razón de corriente primaria
primary detecting element elemento de detección primario
primary detector detector primario
primary device dispositivo primario

primary disconnect switch seccionador primario
primary disconnecting switch seccionador primario
primary display sequence secuencia de visualización primaria
primary distribution system sistema de distribución primario
primary electron electrón primario
primary emission emisión primaria
primary emission current corriente de emisión primaria
primary entry point punto de entrada primario
primary extended route ruta extendida primaria
primary factor factor primario
primary failure falla primaria
primary fault falla primaria
primary feed alimentación primaria
primary feedback retroalimentación primaria
primary feeder alimentador primario
primary field campo primario
primary file archivo primario
primary filter filtro primario
primary flow flujo primario
primary frequency standard patrón de frecuencia primaria
primary function función primaria
primary grid emission emisión de rejilla primaria
primary group grupo primario
primary-group modulation modulación de grupo primario
primary hue tinte primario
primary impedance impedancia primaria
primary inductance inductancia primaria
primary initialization inicialización primaria
primary input entrada primaria
primary instruction instrucción primaria
primary ion ion primario
primary ionization ionización primaria
primary keying manipulación primaria
primary library biblioteca primaria
primary light source fuente de luz primaria
primary line línea primaria
primary-line switch conmutador de línea primaria
primary link enlace primario
primary link station estación de enlace primario
primary logical unit unidad lógica primaria
primary memory memoria primaria
primary network red primaria
primary pattern patrón primario
primary power potencia primaria
primary processing procesamiento primario
primary-processing unit unidad de procesamiento primario
primary program programa primario
primary protection protección primaria
primary radar radar primario
primary radiation radiación primaria
primary radiator radiador primario
primary reading brush escobilla de lectura primaria
primary reference standard patrón de referencia primario
primary register registro primario
primary regulator regulador primario
primary relay relé primario
primary request solicitud primaria
primary resistance resistencia primaria
primary return code código de retorno primario
primary route ruta primaria
primary segment segmento primario
primary selector selector primario
primary sequence secuencia primaria
primary service area área de servicio primario
primary source fuente primaria
primary space allocation asignación de espacio

primaria
primary spectrum espectro primario
primary standard patrón primario
primary station estación primaria
primary storage almacenamiento primario
primary task tarea primaria
primary track pista primaria
primary voltage tensión primaria
primary volume volumen primario
primary winding arrollamiento primario
primary zone zona primaria
prime cebar, sensibilizar
prime mover motor primario, fuente de energía primaria
primer cebador
priming cebado, sensibilización
priming illumination iluminación de cebado
priming speed velocidad de cebado
priming system sistema de cebado
principal axis eje principal
principal channel canal principal
principal clock reloj principal
principal focus foco principal
principal mode modo principal
principal path trayectoria principal
principal plane plano principal
principal point punto principal
principal ray rayo principal
principal register registro principal
principal section sección principal
principal stress estrés principal
principal value valor principal
print imprimir
print amplifier amplificador de impresión
print area área de impresión
print-bound limitado por impresión
print buffer almacenamiento intermedio de impresión
print chain cadena de impresión
print check comprobación de impresión
print code código de impresión
print contrast contraste de impresión
print contrast ratio razón de contraste de impresión
print contrast signal señal de contraste de impresión
print control control de impresión
print-control character carácter de control de impresión
print-control device dispositivo de control de impresión
print-control switch conmutador de control de impresión
print data set conjunto de datos de impresión
print density densidad de impresión
print direction dirección de impresión
print drum tambor de impresión
print element elemento de impresión
print engine motor de impresión
print format formato de impresión
print hammer martillo de impresión
print head cabeza de impresión
print intercept routine rutina de intercepción de impresión
print line línea de impresión
print mechanism mecanismo de impresión
print out imprimir
print position posición de impresión
print quality calidad de impresión
print queue cola de impresión
print selection selección de impresión
print speed velocidad de impresión
print statement sentencia de impresión
print station estación de impresión
print storage almacenamiento de impresión

print suppression supresión de impresión
print-through transferencia magnética
print unit unidad de impresión
print-unit capacity capacidad de unidad de impresión
print wheel rueda de impresión
printed impreso
printed capacitor capacitor impreso
printed circuit circuito impreso
printed-circuit board tablero de circuitos impresos
printed-circuit card tarjeta de circuitos impresos
printed-circuit configuration configuración de circuitos impresos
printed-circuit connector conector de circuitos impresos
printed-circuit design diseño de circuitos impresos
printed-circuit generator generador de circuitos impresos
printed-circuit lamp lámpara de circuito impreso
printed-circuit motor motor de circuito impreso
printed-circuit panel panel de circuitos impresos
printed-circuit relay relé de circuito impreso
printed-circuit switch conmutador de circuito impreso
printed circuitry circuitería impresa
printed communication comunicación impresa
printed component componente impreso
printed-component board tablero de componentes impresos
printed conductor conductor impreso
printed contact contacto impreso
printed data datos impresos
printed electronic circuit circuito electrónico impreso
printed element elemento impreso
printed inductor inductor impreso
printed information información impresa
printed resistor resistor impreso
printed switch conmutador impreso
printed tape cinta impresa
printed transmission line línea de transmisión impresa
printed wiring alambrado impreso
printed-wiring armature inducido de alambrado impreso
printed-wiring board tablero de alambrado impreso
printed-wiring card tarjeta de alambrado impreso
printer impresora
printer backup respaldo de impresora
printer-bound limitado por impresora
printer command mando de impresora
printer control control de impresora
printer-control unit unidad de control de impresora
printer controller controlador de impresora
printer description descripción de impresora
printer file archivo de impresora
printer font tipo de letra de impresora
printer input-output device dispositivo de entrada-salida de impresora
printer operation operación de impresora
printer output salida de impresora
printer port puerto de impresora
printer record registro de impresora
printer server servidor de impresora
printer tape cinta de impresora
printing impresión
printing apparatus aparato impresor
printing arm brazo impresor
printing calculator calculadora impresora
printing carriage carro impresor
printing device dispositivo impresor
printing equipment equipo impresor
printing format formato de impresión

printing hammer martillo impresor
printing key tecla de impresión
printing mechanism mecanismo impresor
printing perforator perforador impresor
printing rate velocidad de impresión
printing reperforator reperforador impresor
printing speed velocidad de impresión
printing station estación de impresión
printing telegraph telégrafo impresor
printing unit unidad impresora
printout impresión
priority prioridad
priority circuit circuito de prioridad
priority control control de prioridad
priority indicator indicador de prioridad
priority interrupt interrupción de prioridad
priority interrupt controller controlador de interrupción de prioridad
priority interrupt controller unit unidad de controlador de interrupción de prioridad
priority interrupt level nivel de interrupción de prioridad
priority interrupt module módulo de interrupción de prioridad
priority interrupt system sistema de interrupción de prioridad
priority level nivel de prioridad
priority list lista de prioridad
priority mode modo de prioridad
priority order orden de prioridad
priority processing procesamiento por prioridad
priority program programa de prioridad
priority queue cola de prioridad
priority rules reglas de prioridad
priority scheduler programador de prioridades
priority scheduling programación de prioridades
priority scheme esquema de prioridad
priority sequence secuencia de prioridades
priority signal señal de prioridad
priority wavelength longitud de onda de prioridad
prism prisma
prism antenna antena en prisma
prismatic prismático
privacy code código de privacidad
privacy equipment equipo de privacidad
privacy protection protección de privacidad
privacy switch conmutador de privacidad
privacy system sistema de privacidad
private automatic branch exchange central automática privada conectada a red pública
private automatic exchange central automática privada
private branch exchange central privada conectada a red pública
private circuit circuito privado
private code código privado
private data file archivo de datos privados
private exchange central privada
private file archivo privado
private library biblioteca privada
private line línea privada
private manual branch exchange central manual privada conectada a red pública
private manual exchange central manual privada
private network red privada
private queue cola privada
private receiving station estación receptora privada
private service servicio privado
private station estación privada
private telephone exchange central telefónica privada
private telephone network red telefónica privada
private telephone station estación telefónica privada

private wire circuito privado, hilo de prueba
private-wire circuit circuito privado
private-wire connection conexión por circuito privado
private-wire service servicio por circuito privado
privileged instruction instrucción privilegiada
privileged mode modo privilegiado
privileged operation operación privilegiada
probability probabilidad
probability amplitude amplitud de probabilidad
probability curve curva de probabilidad
probability density densidad de probabilidad
probability multiplier multiplicador de probabilidad
probability of collision probabilidad de colisión
probability of failure probabilidad de falla
probability of ionization probabilidad de ionización
probable error error probable
probable value valor probable
probe sonda, indagación
probe coil bobina de exploración
probe contacts contactos de exploración
probe microphone micrófono sonda
probe thermistor termistor sonda
probe thermocouple termopar sonda
probe transformer transformador sonda
probing sondeo
problem analysis análisis de problema
problem check comprobación de problema
problem description descripción de problema
problem determination determinación de problema
problem diagnosis diagnóstico de problema
problem identification identificación de problema
problem language lenguaje de problemas
problem-oriented language lenguaje orientado a problemas
problem register registro de problemas
problem solving solución de problemas
problem-solving language lenguaje de solución de problemas
problem-solving program programa de solución de problemas
problem statement sentencia de un problema
problem tape cinta de problemas
procedural error error de procedimiento
procedural language lenguaje de procedimientos
procedural statement sentencia de procedimiento
procedural step paso de procedimiento
procedural test prueba de procedimiento
procedure procedimiento
procedure analysis análisis de procedimientos
procedure block bloque de procedimiento
procedure call llamada de procedimiento
procedure chaining encadenamiento de procedimientos
procedure command mando de procedimiento
procedure control control de procedimiento
procedure control instruction instrucción de control de procedimiento
procedure definition definición de procedimiento
procedure level nivel de procedimiento
procedure memory memoria de procedimientos
procedure message mensaje de procedimiento
procedure name nombre de procedimiento
procedure-oriented language lenguaje orientado a procedimientos
procedure statement sentencia de procedimiento
procedure step paso de procedimiento
procedure subprogram subprograma de procedimientos
proceed-to-dial signal señal de invitación a marcar
proceed-to-send signal señal de invitación a transmitir
proceed-to-transmit signal señal de invitación a

transmitir
process proceso
process access group grupo de acceso de proceso
process analysis análisis de proceso
process assembly ensamblado de proceso
process automation automatización de procesos
process-bound limitado por proceso
process computer system sistema de computadora de procesos
process control control de procesos
process-control block bloque de control de procesos
process-control compiler compilador de control de procesos
process-control computer computadora de control de procesos
process-control equipment equipo de control de procesos
process-control language lenguaje de control de procesos
process-control loop bucle de control de procesos
process-control panel panel de control de procesos
process-control peripheral equipment equipo periférico de control de procesos
process-control system sistema de control de procesos
process controller controlador de procesos
process conversion conversión de procesos
process development desarrollo de procesos
process entry entrada de proceso
process flowchart organigrama de proceso
process-interface system sistema de interfaz de proceso
process interrupt interrupción de proceso
process-limited limitado por proceso
process loop bucle de proceso
process monitoring monitoreo de procesos
process-oriented language lenguaje orientado a procesos
process supervisory program programa supervisor de procesos
process time tiempo de proceso
processed circuit circuito procesado
processing procesamiento
processing amplifier amplificador de procesamiento
processing and control element elemento de procesamiento y control
processing area área de procesamiento
processing capability capacidad de procesamiento
processing center centro de procesamiento
processing cycle ciclo de procesamiento
processing efficiency eficiencia de procesamiento
processing flexibility flexibilidad de procesamiento
processing level nivel de procesamiento
processing medium medio de procesamiento
processing mode modo de procesamiento
processing of data procesamiento de datos
processing overlap solape de procesamiento
processing power potencia de procesamiento
processing program programa de procesamiento
processing requirements requisitos de procesamiento
processing section sección de procesamiento
processing sequence secuencia de procesamiento
processing site lugar de procesamiento
processing specifications especificaciones de procesamiento
processing speed velocidad de procesamiento
processing step paso de procesamiento
processing symbol símbolo de procesamiento
processing system sistema de procesamiento
processing time tiempo de procesamiento
processing unit unidad de procesamiento
processor procesador

processor address dirección de procesador
processor address space espacio de dirección de procesador
processor-bound limitado por procesador
processor check comprobación de procesador
processor configuration configuración de procesador
processor controller controlador de procesador
processor demand demanda de procesador
processor-dependent interrupt interrupción dependiente de procesador
processor-error interrupt interrupción por error de procesador
processor-evaluation module módulo de evaluación de procesador
processor input-output channel canal de entrada-salida de procesador
processor instruction instrucción de procesador
processor interface module módulo de interfaz de procesador
processor interrupt interrupción de procesador
processor-limited limitado por procesador
processor module módulo de procesador
processor status word palabra de estado de procesador
processor transfer time tiempo de transferencia de procesador
processor unit unidad de procesador
prod punta de contacto
product demodulator desmodulador de producto
product detection detección de producto
product detector detector de producto
product development program programa de desarrollo de productos
product modulator modulador de producto
product relay relé de producto
product specification especificación de producto
production producción
production automation automatización de producción
production automation computer computadora de automatización de producción
production automation system sistema de automatización de producción
production control control de producción
production language lenguaje de producción
production model modelo de producción
production program programa de producción
production test prueba de producción
production time tiempo de producción
production unit unidad de producción
productive task tarea productiva
profile diagram diagrama de perfil
program programa
program access key tecla de acceso de programa
program activation vector vector de activación de programa
program address counter contador de direcciones de programa
program address register registro de direcciones de programa
program advance key tecla de avance de programa
program amplifier amplificador de programa
program analyzer analizador de programa
program architecture arquitectura de programa
program area área de programa
program area block bloque de área de programa
program assembly ensamblaje de programa
program bank banco de programas
program block bloque de programa
program bug error de programa
program card tarjeta de programa
program chaining encadenamiento de programas
program channel canal de programa

program check comprobación de programa
program circuit circuito de programa
program coding codificación de programa
program compatibility compatibilidad de programas
program compilation compilación de programa
program compiler compilador de programas
program control control de programa
program-control card tarjeta de control de programa
program-control data datos de control de programa
program-control transfer transferencia de control de programa
program-control unit unidad de control de programa
program-controlled controlado por programa
program-controlled input-output entrada-salida controlada por programa
program-controlled interrupt interrupción controlada por programa
program controller controlador de programa
program conversion conversión de programa
program counter contador de programa
program coupling acoplamiento de programa
program cycle ciclo de programa
program data datos de programa
program debugging depuración de programa
program definition definición de programa
program description descripción de programa
program design diseño de programa
program design language lenguaje de diseño de programa
program development desarrollo de programa
program-development system sistema de desarrollo de programas
program-development time tiempo de desarrollo de programas
program device dispositivo de programa
program disk disco de programa
program drum tambor de programa
program entry entrada de programa
program error error de programa
program-error diagnosis diagnóstico de errores de programa
program-error interrupt interrupción por error de programa
program evaluation evaluación de programa
program event evento de programa
program exception excepción de programa
program execution ejecución de programa
program-execution monitor monitor de ejecución de programa
program exit salida de programa
program failure falla de programa
program file archivo de programa
program flow flujo de programa
program flowchart organigrama de programa
program function función de programa
program function key tecla de función de programa
program generation generación de programa
program generation system sistema de generación de programas
program generator generador de programas
program identification identificación de programa
program identification code código de identificación de programa
program indicator code código indicador de programa
program initiator iniciador de programa
program instruction instrucción de programa
program interface interfaz de programa
program interrupt interrupción de programa
program interruption interrupción de programa
program interruption element elemento de interrupción de programa
program isolation aislamiento de programa

program language lenguaje de programa
program level nivel de programa
program level indicator indicador de nivel de programa
program library biblioteca de programas
program line línea de programa
program link enlace de programas
program load carga de programa
program loader cargador de programa
program logic lógica de programa
program loop bucle de programa
program maintenance mantenimiento de programa
program mask máscara de programa
program master control control maestro de programa
program mode modo de programa
program modification modificación de programa
program module módulo de programa
program monitor monitor de programa
program name nombre de programa
program operator operador de programa
program package paquete de programas
program parameter parámetro de programa
program position posición de programa
program priority prioridad de programa
program profile perfil de programa
program protection protección de programa
program recovery table tabla de recuperación de programa
program reference table tabla de referencia de programa
program register registro de programa
program relocation relocalización de programa
program repeat repetición de programa
program repeater repetidor de programa
program run ejecución de programa
program scheme esquema de programa
program security seguridad de programa
program segment segmento de programa
program segmentation segmentación de programa
program-selected terminal terminal seleccionada por programa
program selection selección de programa
program selector selector de programa
program-selector switch conmutador selector de programa
program-sensitive error error sensible al programa
program-sensitive fault falla sensible al programa
program sequence secuencia de programa
program setup preparación de programa, configuración de programa
program sharing compartimiento de programas
program signal señal de programa
program simulator simulador de programa
program specification especificación de programa
program start comienzo de programa
program statement sentencia de programa
program status estado de programa
program status register registro de estado de programa
program status word palabra de estado de programa
program step paso de programa
program stop parada de programa
program stop instruction instrucción de parada de programa
program storage almacenamiento de programas
program storage unit unidad de almacenamiento de programas
program structure estructura de programa
program support soporte de programa
program switching conmutación de programas
program-switching panel panel de conmutación de programas

program synchronization sincronización de programas
program synthesis síntesis de programa
program tape cinta de programa
program test prueba de programa
program test cycle ciclo de prueba de programa
program test system sistema de prueba de programa
program test time tiempo de prueba de programa
program time tiempo de programa
program timer temporizador de programa
program-to-program message mensaje de programa a programa
program translation traducción de programa
program transmission transmisión de programa
program-transmission circuit circuito de transmisión de programas
program transmitter transmisor de programas
program unit unidad de programa
program validation validación de programa
program variable variable de programa
programmable programable
programmable calculator calculadora programable
programmable clock reloj programable
programmable communications interface interfaz de comunicaciones programable
programmable controller controlador programable
programmable controller capability capacidad de controlador programable
programmable controller input entrada de controlador programable
programmable digital computer computadora digital programable
programmable digital logic lógica digital programable
programmable equipment equipo programable
programmable function key tecla de función programable
programmable input-output channel canal de entrada-salida programable
programmable input-output device dispositivo de entrada-salida programable
programmable interval timer temporizador de intervalos programable
programmable logic lógica programable
programmable logic device dispositivo lógico programable
programmable logic system sistema lógico programable
programmable memory memoria programable
programmable peripheral interface interfaz de periférico programable
programmable protected field campo protegido programable
programmable read-only memory memoria sólo de lectura programable
programmable terminal terminal programable
programmable work station estación de trabajo programable
programmatics programática
programmed programado
programmed acceleration aceleración programada
programmed algorithm algoritmo programado
programmed check comprobación programada
programmed control control programado
programmed data transfer transferencia de datos programada
programmed dump vaciado programado
programmed halt parada programada
programmed input-output entrada-salida programada
programmed input-output address dirección de entrada-salida programada
programmed input-output instruction instrucción

de entrada-salida programada
programmed input-output operation operación de entrada-salida programada
programmed instruction instrucción programada
programmed keyboard teclado programado
programmed logic lógica programada
programmed mode modo programado
programmed search búsqueda programada
programmed stop parada programada
programmed switch conmutador programado
programmed symbols símbolos programados
programmer programador
programmer control panel panel de control de programador
programmer subsystem subsistema de programador
programming programación
programming computer computadora de programación
programming control panel panel de control de programación
programming error error de programación
programming flowchart organigrama de programación
programming instructions instrucciones de programación
programming language lenguaje de programación
programming module módulo de programación
programming speed velocidad de programación
programming statement sentencia de programación
programming system sistema de programación
programming tape cinta de programación
programming unit unidad de programación
progression progresión
progressive progresivo
progressive action acción progresiva
progressive interconnection interconexión progresiva
progressive interlace entrelazamiento progresivo
progressive overflow desbordamiento progresivo
progressive phase shift desfase progresivo
progressive scanning exploración progresiva
progressive total total progresivo
progressive wave onda progresiva
progressive-wave antenna antena de ondas progresivas
project control control de proyecto
project control system sistema de control de proyectos
project management administración de proyecto
project management system sistema de administración de proyectos
projection cathode-ray tube tubo de rayos catódicos de proyección
projection comparator comparador de proyección
projection densitometer densitómetro de proyección
projection frequency frecuencia de proyección
projection head cabeza de proyección
projection lamp lámpara de proyección
projection lens lente de proyección
projection optics óptica de proyección
projection receiver receptor de proyección
projection television televisión de proyección
projection television receiver receptor de televisión de proyección
projection tube tubo de proyección
projector proyector
projector arc lamp lámpara de arco de proyección
projector efficiency eficiencia de proyector
projector head cabeza de proyector
projector lens lente de proyección
prolonger cable cable prolongador
promethium prometio
promethium cell pila de prometio

prong espiga, terminal
proof prueba, comprobación
proof control control de prueba
proof factor factor de prueba
proof plane plano de prueba
proof pressure presión de prueba, tensión de prueba
proof total total de prueba
propagated propagado
propagated error error propagado
propagated potential potencial propagado
propagated response respuesta propagada
propagating medium medio de propagación
propagating mode modo de propagación
propagation propagación
propagation characteristic característica de propagación
propagation coefficient coeficiente de propagación
propagation constant constante de propagación
propagation curve curva de propagación
propagation delay retardo de propagación
propagation delay time tiempo de retardo de propagación
propagation distortion analyzer analizador de distorsión de propagación
propagation disturbance perturbación de propagación
propagation error error de propagación
propagation factor factor de propagación
propagation loss pérdida de propagación
propagation medium medio de propagación
propagation mode modo de propagación
propagation path trayectoria de propagación
propagation ratio razón de propagación
propagation reliability confiabilidad de propagación
propagation test prueba de propagación
propagation time tiempo de propagación
propagation velocity velocidad de propagación
proportion proporción
proportional proporcional
proportional action acción proporcional
proportional-action control control de acción proporcional
proportional-action factor factor de acción proporcional
proportional amplifier amplificador proporcional
proportional band banda proporcional
proportional control control proporcional
proportional-control servo servo de control proporcional
proportional-control system sistema de control proporcional
proportional controller controlador proporcional
proportional counter contador proporcional
proportional counter tube tubo contador proporcional
proportional current corriente proporcional
proportional direct current corriente continua proporcional
proportional gain ganancia proporcional
proportional ionization ionización proporcional
proportional ionization chamber cámara de ionización proporcional
proportional loss pérdida proporcional
proportional positioning posicionamiento proporcional
proportional region región proporcional
proportional response respuesta proporcional
proportionality proporcionalidad
proportionate proporcionado
proportioning proporcionamiento
prospective current corriente propia
protactinium protactinio
protectant protector

protected protegido
protected apparatus aparato protegido
protected area área protegida
protected dynamic storage almacenamiento dinámico protegido
protected enclosure caja protegida
protected field campo protegido
protected location localización protegida
protected machine máquina protegida
protected mode modo protegido
protected queue cola protegida
protected queue area área de cola protegida
protected storage almacenamiento protegido
protected zone zona protegida
protecting protector
protecting element elemento protector
protecting lamp lámpara protectora
protecting plate placa protectora
protection protección, dispositivo de protección
protection channel canal de protección
protection character carácter de protección
protection current corriente de protección
protection device dispositivo de protección
protection distance distancia de protección
protection element elemento de protección
protection glass vidrio de protección
protection ground relay relé de protección de puesta a tierra
protection ratio razón de protección
protection system sistema de protección
protective protector
protective action acción protectora
protective barrier barrera protectora
protective bias polarización protectora
protective cable cable protector
protective capacitor capacitor protector
protective channel canal protector
protective circuit circuito protector
protective circuit-breaker cortacircuito protector
protective coating revestimiento protector
protective cover cubierta protectora
protective device dispositivo protector
protective fuse fusible protector
protective gap espacio protector
protective glass vidrio protector
protective grounding puesta a tierra protectora
protective horn bocina protectora
protective margin margen protector
protective material material protector
protective reactance coil bobina de reactancia protectora
protective relay relé protector
protective resistance resistencia protectora
protective resistor resistor protector
protective screen pantalla protectora
protective signaling señalización protectora
protective sleeve manguito protector
protective system sistema protector
protector protector
protector ground puesta a tierra protectora
protector tube tubo protector
protocol protocolo
protocol function función de protocolo
protocol level nivel de protocolo
proton protón
proton accelerator acelerador protónico
proton gun cañón protónico
proton magnetic resonance resonancia magnética protónica
proton magnetometer magnetómetro protónico
proton microscope microscopio protónico
proton synchrotron sincrotrón protónico
prototype prototipo

prototype debugging depuración de prototipo
prototype test prueba de prototipo
proustite proustita
provisional provisional
provisional specification especificación provisional
proximity proximidad
proximity alarm alarma de proximidad
proximity detector detector de proximidad
proximity effect efecto de proximidad
proximity fuse fusible de proximidad
proximity influence influencia de proximidad
proximity switch conmutador de proximidad
pseudo-Brewster angle seudoángulo de Brewster
pseudobinary seudobinario
pseudocarrier seudoportadora
pseudoclock seudorreloj
pseudocode seudocódigo
pseudodynamic memory memoria seudodinámica
pseudoerror seudoerror
pseudoerror detection detección de seudoerrores
pseudoerror detector detector de seudoerrores
pseudoinstruction seudoinstrucción
pseudolanguage seudolenguaje
pseudonoise seudorruido
pseudooperation seudooperación
pseudorandom seudoaleatorio
pseudorandom noise ruido seudoaleatorio
pseudorandom numbers números seudoaleatorios
pseudoregister seudorregistro
pseudostereophonic seudoestereofónico
pseudostereophonic effect efecto seudoestereofónico
pseudotext seudotexto
pseudotimer seudotemporizador
pseudovariable seudovariable
psophometer sofómetro
psophometric sofométrico
psophometric noise ruido sofométrico
psophometric power potencia sofométrica
psophometric voltage tensión sofométrica
psophometric weight peso sofométrico
psophometrically sofométricamente
public-address amplifier amplificador megafónico
public-address speaker altavoz megafónico
public-address system sistema megafónico
public line línea pública
public network red pública
public station estación pública
public telephone teléfono público
public telephone network red telefónica pública
pull-in current corriente de conexión
pull-in time tiempo de conexión
pull-in voltage tensión de conexión
pulling arrastre
pulling effect efecto de arrastre
pulling figure factor de arrastre
pullup current corriente de conexión
pulsate pulsar
pulsating pulsante, pulsatorio
pulsating current corriente pulsante
pulsating direct current corriente continua pulsante
pulsating magnetic field campo magnético pulsante
pulsating quantity cantidad pulsante
pulsating voltage tensión pulsante
pulsating wave onda pulsante
pulsation pulsación
pulsation frequency frecuencia de pulsación
pulsation period periodo de pulsación
pulsation welding soldadura por pulsaciones
pulse impulso, pulso
pulse amplifier amplificador de impulsos
pulse amplitude amplitud de impulsos
pulse-amplitude analyzer analizador de amplitud de impulsos

pulse-amplitude discriminator discriminador de amplitud de impulsos
pulse-amplitude modulation modulación de amplitud de impulsos
pulse-amplitude selector selector de amplitud de impulsos
pulse analyzer analizador de impulsos
pulse bandwidth ancho de banda de impulso
pulse broadening ensanchamiento de impulso
pulse carrier portadora de impulsos
pulse chopper interruptor periódico de impulsos
pulse circuit circuito de impulsos
pulse clipper recortador de impulsos
pulse clipping recortamiento de impulsos
pulse clock reloj de impulsos
pulse code código de impulsos
pulse-code modulation modulación por impulsos codificados
pulse coder codificador de impulsos
pulse coincidence coincidencia de impulsos
pulse communicating system sistema de comunicación por impulsos
pulse communication system sistema de comunicación por impulsos
pulse compression compresión de impulso
pulse correction corrección de impulso
pulse corrector corrector de impulso
pulse count conteo de impulsos
pulse-count discriminator discriminador por conteo de impulsos
pulse-count modulation modulación por conteo de impulsos
pulse counter contador de impulsos
pulse-counter frequency meter frecuencímetro contador de impulsos
pulse counting conteo de impulsos
pulse-counting spectrophotometer espectrofotómetro de conteo de impulsos
pulse-counting system sistema de conteo de impulsos
pulse damping amortiguamiento de impulsos
pulse-damping diode diodo amortiguador de impulsos
pulse decay caída de impulso
pulse decay time tiempo de caída de impulso
pulse decoder descodificador de impulsos
pulse decoding descodificación de impulsos
pulse delay retardo de impulsos
pulse-delay circuit circuito de retardo de impulsos
pulse-delay network red de retardo de impulsos
pulse detector detector de impulsos
pulse dialing marcación por impulsos
pulse discrimination discriminación de impulsos
pulse discriminator discriminador de impulsos
pulse distortion distorsión de impulso
pulse distributing box caja distribuidora de impulsos
pulse distribution distribución de impulsos
pulse-distribution amplifier amplificador de distribución de impulsos
pulse distributor distribuidor de impulsos
pulse-divider circuit circuito divisor de impulsos
pulse Doppler radar radar de Doppler de impulsos
pulse droop inclinación del techo de impulso
pulse duplication duplicación de impulsos
pulse duration duración de impulso
pulse-duration discriminator discriminador de duración de impulsos
pulse-duration error error por duración de impulsos
pulse-duration modulation modulación por duración de impulsos
pulse duty factor factor de trabajo de impulsos
pulse echometer ecómetro de impulsos
pulse edge flanco de impulso
pulse electronic multiplier multiplicador electrónico

de impulsos
pulse emission emisión de impulsos
pulse energy energía de impulsos
pulse envelope envolvente de impulsos
pulse equalizer ecualizador de impulsos
pulse fall time tiempo de caída de impulso
pulse flat techo de impulso
pulse former formador de impulsos
pulse forming formación de impulsos
pulse-forming circuit circuito formador de impulsos
pulse-forming line línea formadora de impulsos
pulse-forming network red formadora de impulsos
pulse frequency frecuencia de impulsos
pulse-frequency divider divisor de frecuencia de impulsos
pulse-frequency modulation modulación de frecuencia de impulsos
pulse-frequency repeater repetidor de frecuencia de impulsos
pulse-frequency spectrum espectro de frecuencias de impulsos
pulse-generating circuit circuito generador de impulsos
pulse generation generación de impulsos
pulse generator generador de impulsos
pulse group grupo de impulsos
pulse height altura de impulso
pulse-height analysis análisis de altura de impulsos
pulse-height analyzer analizador de altura de impulsos
pulse-height detector detector de altura de impulsos
pulse-height discriminator discriminador de altura de impulsos
pulse-height resolution resolución de altura de impulsos
pulse-height selector selector de altura de impulsos
pulse initiator iniciador de impulsos
pulse integration integración de impulsos
pulse integrator integrador de impulsos
pulse interlacing entrelazado de impulsos
pulse interleaving entrelazado de impulsos
pulse interrogation interrogación por impulsos
pulse interval intervalo de impulsos
pulse-interval meter medidor de intervalo de impulsos
pulse-interval modulation modulación de intervalo de impulsos
pulse inverter inversor de impulsos
pulse-ionization chamber cámara de ionización de impulsos
pulse jitter fluctuación de impulsos
pulse length duración de impulso
pulse-length discriminator discriminador de duración de impulsos
pulse-length modulation modulación por duración de impulsos
pulse limiting limitación de impulsos
pulse line línea de impulsos
pulse magnetization magnetización por impulsos
pulse mixing mezcla de impulsos
pulse-mixing circuit circuito mezclador de impulsos
pulse mode modo de impulsos
pulse-mode multiplex múltiplex por modos de impulsos
pulse-modulated modulado por impulsos
pulse-modulated oscillator oscilador modulado por impulsos
pulse-modulated transmission transmisión modulada por impulsos
pulse-modulated wave onda modulada por impulsos
pulse modulation modulación de impulsos
pulse-modulation multiplex múltiplex por modulación de impulsos

pulse-modulation radio link radioenlace por modulación de impulsos
pulse-modulation recording registro por modulación de impulsos
pulse modulator modulador de impulsos
pulse navigation system sistema de navegación por impulsos
pulse noise ruido de impulsos
pulse number número de impulsos
pulse operation operación de impulsos
pulse oscillator oscilador de impulsos
pulse pair par de impulsos
pulse phase fase de un impulso
pulse position posición de impulso
pulse-position modulation modulación por posición de impulsos
pulse power potencia de impulso
pulse radar radar de impulsos
pulse rate frecuencia de repetición de impulsos
pulse ratio razón de impulso
pulse receiver receptor de impulsos
pulse recorder registrador de impulsos
pulse reflection reflexión de impulsos
pulse regeneration regeneración de impulsos
pulse regenerator regenerador de impulsos
pulse repeater repetidor de impulsos
pulse repeating repetición de impulsos
pulse repetition repetición de impulsos
pulse repetition frequency frecuencia de repetición de impulsos
pulse repetition interval intervalo de repetición de impulsos
pulse repetition period periodo de repetición de impulsos
pulse repetition rate frecuencia de repetición de impulsos
pulse repetition time tiempo de repetición de impulsos
pulse reply respuesta de impulso
pulse resolution resolución de impulsos
pulse response respuesta de impulsos
pulse rise time tiempo de subida de impulso
pulse selecting selección de impulsos
pulse selection selección de impulsos
pulse-selection circuit circuito de selección de impulsos
pulse selector selector de impulsos
pulse sensor sensor de impulsos
pulse separation separación de impulsos
pulse sequence secuencia de impulsos
pulse shape forma de impulso
pulse shaper formador de impulsos
pulse shaping formación de impulsos
pulse-shaping circuit circuito formador de impulsos
pulse shortening acortamiento de impulsos
pulse signal señal de impulsos
pulse slope pendiente de impulso
pulse-slope modulation modulación de pendiente de impulso
pulse spacing espaciado de impulsos
pulse-spacing modulation modulación de espaciado de impulsos
pulse spectrum espectro de impulso
pulse spectrum bandwidth ancho de banda de espectro de un impulso
pulse spike impulso parásito
pulse-spike amplitude amplitud de un impulso parásito
pulse start inicio de impulso
pulse stop parada de impulso
pulse stream corriente de impulsos
pulse stretcher ensanchador de impulsos
pulse stretching ensanchamiento de impulsos

pulse superposition superposición de impulsos
pulse switch conmutador de impulsos
pulse switching conmutación de impulsos
pulse-switching circuit circuito de conmutación de impulsos
pulse synthesizer sintetizador de impulsos
pulse tilt inclinación del techo de impulso
pulse time tiempo de impulso
pulse-time analysis análisis de tiempo de impulsos
pulse-time modulation modulación por tiempo de impulsos
pulse-time ratio razón de tiempo de impulsos
pulse timing temporización de impulsos
pulse-timing circuit circuito de temporización de impulsos
pulse top parte superior de impulso
pulse trace trazo de impulso
pulse train tren de impulsos
pulse-train generator generador de trenes de impulsos
pulse-train spectrum espectro de un tren de impulsos
pulse transformer transformador de impulsos
pulse transmission transmisión de impulsos
pulse transmission system sistema de transmisión por impulsos
pulse transmitter transmisor de impulsos
pulse trigger circuit circuito disparador de impulsos
pulse valley valle de impulso
pulse waveform forma de onda de impulso
pulse width anchura de impulso
pulse-width distortion distorsión de anchura de impulso
pulse-width modulation modulación en anchura de impulsos
pulsed pulsado
pulsed beacon radiofaro pulsado
pulsed beam haz pulsado
pulsed carrier portadora pulsada
pulsed cavitation cavitación pulsada
pulsed discharge descarga pulsada
pulsed emission emisión pulsada
pulsed klystron klistrón pulsado
pulsed laser láser pulsado
pulsed light luz pulsada
pulsed light source fuente de luz pulsada
pulsed magnetic field campo magnético pulsado
pulsed magnetron magnetrón pulsado
pulsed maser máser pulsado
pulsed oscillation oscilación pulsada
pulsed oscillator oscilador pulsado
pulsed output salida pulsada
pulsed-output laser láser de salida pulsada
pulsed particle accelerator acelerador de partículas pulsado
pulsed power potencia pulsada
pulsed ruby laser láser de rubí pulsado
pulsed ruby maser máser de rubí pulsado
pulsed signal señal pulsada
pulsed source fuente pulsada
pulsed transmitter transmisor pulsado
pulser pulsador
pulses per second impulsos por segundo
pulsing pulsación
pulsing circuit circuito pulsante
pulsing modulator modulador pulsante
pulsing relay relé pulsante
pulsing system sistema pulsante
pulsing wave onda pulsante
pump bomba
pump efficiency eficiencia de bomba
pumped tube tubo con bombeo continuo
pumped valve válvula con bombeo continuo

pumping bombeo
pumping band banda de bombeo
pumping circuit circuito de bombeo
pumping cycle ciclo de bombeo
pumping frequency frecuencia de bombeo
pumping radiation radiación de bombeo
pumping voltage tensión de bombeo
pumpless rectifier rectificador sin bomba
punch perforar
punch area área de perforación
punch block bloque de perforación
punch brush escobilla de perforación
punch card tarjeta perforable, tarjeta perforada
punch check comprobación de perforación
punch code código de perforación
punch column columna de perforación
punch emitter emisor de perforación
punch feed alimentador de perforadora
punch head cabeza de perforación
punch hole agujero de perforación
punch magnet electroimán de perforación
punch-magnet circuit circuito de electroimán de perforación
punch mechanism mecanismo de perforación
punch position posición de perforación
punch relay relé de perforación
punch selector selector de perforación
punch station estación de perforación
punch suppression supresión de perforación
punch tape cinta de perforación
punch tape code código de cinta de perforación
punch unit unidad de perforación
punched perforado
punched card tarjeta perforada
punched-card calculator calculadora de tarjetas perforadas
punched-card computer computadora de tarjetas perforadas
punched-card electronic computer computadora electrónica de tarjetas perforadas
punched-card equipment equipo de tarjetas perforadas
punched-card field campo de tarjeta perforada
punched-card file archivo de tarjetas perforadas
punched-card format formato de tarjeta perforada
punched-card interpreter interpretador de tarjetas perforadas
punched-card machine máquina de tarjetas perforadas
punched-card programming programación por tarjetas perforadas
punched-card reader lectora de tarjetas perforadas
punched-card sorter clasificadora de tarjetas perforadas
punched-card verifier verificador de tarjetas perforadas
punched path trayectoria perforada
punched tape cinta perforada
punched-tape code código de cinta perforada
punched-tape handler manipulador de cinta perforada
punched-tape program programa de cinta perforada
punched-tape reader lectora de cinta perforada
punched-tape recorder registrador de cinta perforada
puncher perforadora
punching perforación
punching apparatus aparato de perforación
punching block bloque de perforación
punching error error de perforación
punching field campo de perforación
punching instruction instrucción de perforación
punching key tecla de perforación

punching machine máquina de perforación
punching mechanism mecanismo de perforación
punching position posición de perforación
punching rate velocidad de perforación
punching relay relé de perforación
punching section sección de perforación
punching station estación de perforación
punching track pista de perforación
punching unit unidad de perforación
punchless sin perforaciones
puncture voltage tensión de perforación
Pupin coil bobina de Pupin
Pupin system sistema de Pupin
pure binary binario puro
pure continuous waves ondas continuas puras
pure emitter emisor puro
pure generator generador puro
pure inductance inductancia pura
pure oscillation oscilación pura
pure resistance resistencia pura
pure sound sonido puro
pure spectral color color espectral puro
pure tone tono puro
pure-tone component componente de tono puro
pure wave onda pura
purification purificación
purification of electrolyte purificación de electrólito
purifier purificador
purify purificar
purifying purificación
purity pureza
purity adjustment ajuste de pureza
purity coil bobina de pureza
purity control control de pureza
purity magnet imán de pureza
purple plague plaga púrpura
push-button pulsador
push-button actuated accionado por pulsador
push-button board tablero de pulsadores
push-button box caja de pulsadores
push-button circuit-breaker cortacircuito de pulsador
push-button control control por pulsador
push-button controlled controlado por pulsador
push-button current selector selector de corrientes por pulsador
push-button microphone micrófono de pulsador
push-button operated operado por pulsador
push-button operation operación por pulsador
push-button relay relé de pulsador
push-button selection selección por pulsador
push-button switch conmutador de pulsador
push-button switching conmutación por pulsador
push-button tuner sintonizador de pulsador
push-button tuning sintonización por pulsador
push-button tuning system sistema de sintonización por pulsador
push-down desplazamiento descendente
push-down list lista de desplazamiento descendente
push-down queue cola de desplazamiento descendente
push-down store almacenamiento de desplazamiento descendente
push-pull amplification amplificación en contrafase
push-pull amplifier amplificador en contrafase
push-pull amplifier circuit circuito amplificador en contrafase
push-pull arrangement arreglo en contrafase
push-pull carbon microphone micrófono de carbón en contrafase
push-pull cascode cascodo en contrafase
push-pull circuit circuito en contrafase
push-pull connection conexión en contrafase

push-pull currents corrientes en contrafase
push-pull detection circuit circuito detector en
   contrafase
push-pull detector detector en contrafase
push-pull input entrada en contrafase
push-pull magnetic amplifier amplificador
   magnético en contrafase
push-pull microphone micrófono en contrafase
push-pull mixer mezclador en contrafase
push-pull mixing stage etapa mezcladora en
   contrafase
push-pull modulator modulador en contrafase
push-pull operation operación en contrafase
push-pull oscillator oscilador en contrafase
push-pull-parallel amplifier amplificador de
   elementos en paralelo en contrafase
push-pull-parallel circuit circuito de elementos en
   paralelo en contrafase
push-pull-parallel oscillator oscilador de elementos
   en paralelo en contrafase
push-pull repeater repetidor en contrafase
push-pull transformer transformador en contrafase
push-pull transistor amplifier amplificador de
   transistores en contrafase
push-pull voltages tensiones en contrafase
push-push amplifier amplificador en
   contrafase-paralelo
push-push circuit circuito en contrafase-paralelo
push switch conmutador de pulsador
push-to-listen switch conmutador de oprimir para
   escuchar
push-to-talk switch conmutador de oprimir para
   hablar
push-up list lista ascendente
push-up storage almacenamiento ascendente
pushbutton pulsador
pushbutton actuated accionado por pulsador
pushbutton board tablero de pulsadores
pushbutton box caja de pulsadores
pushbutton circuit-breaker cortacircuito de pulsador
pushbutton control control por pulsador
pushbutton controlled controlado por pulsador
pushbutton current selector selector de corrientes
   por pulsador
pushbutton microphone micrófono de pulsador
pushbutton operated operado por pulsador
pushbutton operation operación por pulsador
pushbutton relay relé de pulsador
pushbutton selection selección por pulsador
pushbutton switch conmutador de pulsador
pushbutton switching conmutación por pulsador
pushbutton tuner sintonizador de pulsador
pushbutton tuning sintonización por pulsador
pushbutton tuning system sistema de sintonización
   por pulsador
pushdown desplazamiento descendente
pushdown list lista de desplazamiento descendente
pushdown queue cola de desplazamiento
   descendente
pushdown store almacenamiento de desplazamiento
   descendente
pushing figure índice de corrimiento atrasado
pushup list lista ascendente
pushup storage almacenamiento ascendente
put to earth conectar a tierra
pylon antenna antena de cilindro ranurado
pyramid antenna antena en pirámide
pyramidal piramidal
pyramidal horn antenna antena de bocina piramidal
pyramidal horn radiator radiador de bocina
   piramidal
pyrheliometer pirheliómetro
pyroconductivity piroconductividad

pyroelectric piroeléctrico
pyroelectric effect efecto piroeléctrico
pyroelectric lamp lámpara piroeléctrica
pyroelectric material material piroeléctrico
pyroelectric transducer transductor piroeléctrico
pyroelectricity piroelectricidad
pyroelectrolysis piroelectrólisis
pyrolysis pirólisis
pyromagnetic piromagnético
pyromagnetic effect efecto piromagnético
pyromagnetic motor motor piromagnético
pyromagnetism piromagnetismo
pyrometer pirómetro
pyrometer head cabeza pirométrica
pyrometer probe sonda pirométrica
pyrometer tube tubo pirométrico
pyrometric pirométrico
pyrometric circuit circuito pirométrico
pyrometric switch conmutador pirométrico
pyron detector detector pirón
Pythagorean scale escala de Pitágoras

Q

Q address dirección Q
Q adjustment ajuste Q
Q band banda Q
Q bar barra Q
Q channel canal Q
Q chrominance signal señal de crominancia Q
Q code código Q
Q demodulator desmodulador Q
Q factor factor Q
Q meter medidor de Q
Q multiplier multiplicador de Q
Q point punto Q
Q signal señal Q
Q switching conmutación de Q
Q test prueba Q
Q vector vector Q
Q wave onda Q
quad cuadrete
quad cable cable de cuadretes
quadded cable cable de cuadretes
quadradar cuadradar
quadrant cuadrante
quadrant antenna antena de cuadrante
quadrant electrometer electrómetro de cuadrantes
quadrantal cuadrantal
quadrantal altitude altitud cuadrantal
quadrantal antenna antena cuadrantal
quadrantal deviation desviación cuadrantal
quadrantal error error cuadrantal
quadrantal error correction corrección de error cuadrantal
quadrantal frequency frecuencia cuadrantal
quadraphonic cuadrafónico, cuadrifónico
quadraphonic decoder descodificador cuadrafónico
quadraphonic sound sonido cuadrafónico
quadraphonic system sistema cuadrafónico
quadraphonics cuadrafonía
quadraphony cuadrafonía
quadratic component componente cuadrática
quadratic detector detector cuadrático
quadratic equation ecuación cuadrática
quadratic formula fórmula cuadrática
quadratic programming programación cuadrática
quadrature cuadratura
quadrature amplifier amplificador en cuadratura
quadrature amplitude modulation modulación de amplitud en cuadratura
quadrature axis eje en cuadratura
quadrature-axis component componente transversal
quadrature-axis subtransient impedance impedancia subtransitoria transversal
quadrature-axis synchronous impedance impedancia sincrónica transversal
quadrature-axis transient impedance impedancia transitoria transversal
quadrature channel canal en cuadratura
quadrature coil bobina en cuadratura
quadrature component componente en cuadratura
quadrature current corriente en cuadratura
quadrature detector detector en cuadratura
quadrature-fed alimentado en cuadratura

quadrature field campo en cuadratura
quadrature flux flujo en cuadratura
quadrature grid rejilla en cuadratura
quadrature magnetizing current corriente magnetizante en cuadratura
quadrature modulation modulación en cuadratura
quadrature network red en cuadratura
quadrature-phase subcarrier signal señal de subportadora en cuadratura de fase
quadrature portion porción en cuadratura
quadrature reactance reactancia en cuadratura
quadrature signal señal en cuadratura
quadrature stage etapa en cuadratura
quadrature suppression supresión en cuadratura
quadrature transformer transformador en cuadratura
quadrature tube tubo en cuadratura
quadrature voltage tensión en cuadratura
quadricorrelator cuadricorrelador
quadripartite cuatripartito
quadriphonic cuadrifónico
quadripole cuadripolo
quadripole propagation factor factor de propagación de cuadripolo
quadripole radiation radiación de cuadripolo
quadrivalent tetravalente, cuatrivalente
quadruple diversity diversidad cuádruple
quadruple-diversity receiver receptor de diversidad cuádruple
quadruple-diversity system sistema de diversidad cuádruple
quadruple phantom circuit circuito fantasma cuádruple
quadruple register registro cuádruple
quadruple scanning exploración cuádruple
quadruple system sistema cuádruple
quadrupler cuadruplicador
quadruplex cuádruplex
quadruplex circuit circuito cuádruplex
quadruplex system sistema cuádruplex
quadruplex telegraphy telegrafía cuádruplex
quadruplexer cuadruplexor
quadrupole cuadrupolo
quadrupole coupling constant constante de acoplamiento de cuadripolo
quadrupole radiation radiación de cuadripolo
qualification calificación
qualified life vida calificada
qualimeter cualímetro
qualitative cualitativo
qualitative analysis análisis cualitativo
qualitative test prueba cualitativa
quality calidad, cualidad
quality assurance comprobación de calidad
quality control control de calidad
quality diagnostic diagnóstico de calidad
quality factor factor de calidad
quality-factor meter medidor de factor de calidad
quality figure cifra de calidad
quality index índice de calidad
quanta cuantos
quantimeter cuantímetro
quantitative cuantitativo
quantitative analysis análisis cuantitativo
quantitative detector detector cuantitativo
quantitative measurement medida cuantitativa
quantitative test prueba cuantitativa
quantitatively cuantitativamente
quantity cantidad, magnitud
quantity meter amperhorímetro, medidor de cantidad
quantity of electricity cantidad de electricidad
quantity of illumination cantidad de iluminación

quantity of information cantidad de información
quantity of light cantidad de luz
quantity of radiation cantidad de radiación
quantization cuantificación
quantization distortion distorsión de cuantificación
quantization error error de cuantificación
quantization level nivel de cuantificación
quantization noise ruido de cuantificación
quantize cuantificar
quantized cuantificado
quantized electromagnetic field campo
   electromagnético cuantificado
quantized field campo cuantificado
quantized interaction interacción cuantificada
quantized pulse modulation modulación por
   impulsos cuantificados
quantized signal señal cuantificada
quantized system sistema cuantificado
quantizer cuantificador
quantizing cuantificación
quantizing encoder codificador de cuantificación
quantizing error error de cuantificación
quantizing loss pérdida de cuantificación
quantizing noise ruido de cuantificación
quantizing operation operación de cuantificación
quantometer cuantómetro
quantum cuanto
quantum efficiency eficiencia cuántica
quantum electrodynamics electrodinámica cuántica
quantum electronics electrónica cuántica
quantum equivalence equivalencia cuántica
quantum equivalence principle principio de
   equivalencia cuántica
quantum interference interferencia cuántica
quantum jump salto cuántico
quantum level nivel cuántico
quantum limit límite cuántico
quantum mechanics mecánica cuántica
quantum noise ruido cuántico
quantum number número cuántico
quantum of energy cuanto de energía
quantum of light cuanto de luz
quantum of radiation cuanto de radiación
quantum state estado cuántico
quantum statistics estadística cuántica
quantum theory teoría cuántica
quantum theory of magnetism teoría cuántica del
   magnetismo
quantum transition transición cuántica
quantum voltage tensión cuántica
quantum yield rendimiento cuántico
quark quark
quarter channel cuarto de canal
quarter-phase bifásico
quarter-phase current corriente bifásica
quarter-phase network red bifásica
quarter-phase system sistema bifásico
quarter wave cuarto de onda
quarter-wave antenna antena de cuarto de onda
quarter-wave attenuator atenuador de cuarto de
   onda
quarter-wave balun balún de cuarto de onda
quarter-wave dipole dipolo de cuarto de onda
quarter-wave filter filtro de cuarto de onda
quarter-wave line línea de cuarto de onda
quarter-wave monopole monopolo de cuarto de onda
quarter-wave plate placa de cuarto de onda
quarter-wave radiator radiador de cuarto de onda
quarter-wave receiving antenna antena receptora de
   cuarto de onda
quarter-wave resonance resonancia de cuarto de
   onda
quarter-wave resonant frequency frecuencia

   resonante de cuarto de onda
quarter-wave resonant line línea resonante de cuarto
   de onda
quarter-wave stub adaptador de cuarto de onda
quarter-wave termination terminación en cuarto de
   onda
quarter-wave transformer transformador en cuarto
   de onda
quarter-wave transmission line línea de transmisión
   de cuarto de onda
quarter-wave tuner sintonizador de cuarto de onda
quarter wavelength cuarto de longitud de onda
quarter-wavelength line línea de cuarto de longitud
   de onda
quartic equation ecuación cuártica
quartz cuarzo
quartz bar barra de cuarzo
quartz block bloque de cuarzo
quartz clock reloj de cuarzo
quartz-controlled controlado por cuarzo
quartz-controlled carrier portadora controlada por
   cuarzo
quartz-controlled generator generador controlado
   por cuarzo
quartz-controlled oscillation oscilación controlada
   por cuarzo
quartz crystal cristal de cuarzo
quartz-crystal calibrator calibrador de cristal de
   cuarzo
quartz-crystal oscillator oscilador de cristal de
   cuarzo
quartz-crystal resonator resonador de cristal de
   cuarzo
quartz delay line línea de retardo de cuarzo
quartz-fiber electrometer electrómetro de fibra de
   cuarzo
quartz-fiber electroscope electroscopio de fibra de
   cuarzo
quartz-halogen lamp lámpara de cuarzo-halógeno
quartz-iodine lamp lámpara de cuarzo-yodo
quartz lamp lámpara de cuarzo
quartz lens lente de cuarzo
quartz master oscillator oscilador maestro de cuarzo
quartz monochromator monocromador de cuarzo
quartz oscillator oscilador de cuarzo
quartz plate placa de cuarzo
quartz-plate resonator resonador de placa de cuarzo
quartz prism prisma de cuarzo
quartz resonator resonador de cuarzo
quartz thermometer termómetro de cuarzo
quartz vibrator vibrador de cuarzo
quasar cuasar
quasi cuasi
quasi-analog signal señal cuasianalógica
quasi-bistable circuit circuito cuasibiestable
quasi-conductor cuasiconductor
quasi-dielectric cuasidieléctrico
quasi-dynamic memory memoria cuasidinámica
quasi-elastic cuasielástico
quasi-harmonic vibration vibración cuasiarmónica
quasi-impulsive interference interferencia
   cuasiimpulsiva
quasi-impulsive noise ruido cuasiimpulsivo
quasi-instruction cuasiinstrucción
quasi-language cuasilenguaje
quasi-linear feedback system sistema de
   retroalimentación cuasilineal
quasi-linear sweep barrido cuasilineal
quasi-monostable circuit circuito cuasimonoestable
quasi-negative cuasinegativo
quasi-optical cuasióptico
quasi-optical propagation propagación cuasióptica
quasi-optical range alcance cuasióptico

quasi-optical wave onda cuasióptica
quasi-peak detector detector de cuasipicos
quasi-peak level nivel de cuasipico
quasi-peak receiver receptor de cuasipico
quasi-peak value valor de cuasipico
quasi-positive cuasipositivo
quasi-random cuasialeatorio
quasi-random access memory memoria de acceso cuasialeatorio
quasi-random signal señal cuasialeatoria
quasi-rectangular wave onda cuasirectangular
quasi-single sideband banda lateral cuasiúnica
quasi-square wave onda cuasicuadrada
quasi-stable state estado cuasiestable
quasi-stationary cuasiestacionario
quasi-stellar radio source radiofuente cuasiestelar
quaternary cuaternario
quaternary code código cuaternario
quench extinguir, apagar, enfriar rápidamente, detener abruptamente, templar
quench agent agente extintor, capacitor de extinción
quench capacitor capacitor de extinción
quench circuit circuito de extinción
quench coil bobina de extinción
quench condenser condensador de extinción
quench frequency frecuencia de supresión
quench oscillator oscilador de extinción
quench resistance resistencia de extinción
quench resistor resistor de extinción
quench voltage tensión de extinción
quenched extinguido, amortiguado, apagado
quenched gap descargador de chispa amortiguada, espinterómetro de chispa amortiguada
quenched spark chispa amortiguada
quenched spark-gap descargador de chispa amortiguada, espinterómetro de chispa amortiguada
quencher extintor, amortiguador, apagador
quenching agent agente de extinción
quenching capacitor capacitor de extinción
quenching circuit circuito de extinción
quenching coil bobina de extinción
quenching condenser condensador de extinción
quenching frequency frecuencia de extinción
quenching oscillator oscilador de extinción
quenching resistance resistencia de extinción
quenching resistor resistor de extinción
quenching voltage tensión de extinción
query pregunta
queue cola
queue access method método de acceso por colas
queue control block bloque de control de colas
queue element elemento de cola
queue file archivo de cola
queue management administración de cola
queue manager administrador de cola
queue place lugar en una cola
queued access method método de acceso por colas
queuing puesta en cola
queuing list lista de puesta en cola
queuing theory teoría de colas
quibinary code código quibinario
quick-access de acceso rápido
quick-access storage almacenamiento de acceso rápido
quick-acting de acción rápida
quick-acting relay relé de acción rápida
quick-acting switch conmutador de acción rápida
quick-break de ruptura rápida
quick-break contactor contactor de ruptura rápida
quick-break fuse fusible de ruptura rápida
quick-break switch conmutador de ruptura rápida
quick charge carga rápida

quick closedown cierre rápido
quick-configuration device dispositivo de configuración rápida
quick-connect de conexión rápida
quick-connecting de conexión rápida
quick-connecting device dispositivo de conexión rápida
quick-disconnect de desconexión rápida
quick-disconnect terminal terminal de desconexión rápida
quick-heating de calentamiento rápido
quick-heating filament filamento de calentamiento rápido
quick-insert de inserción rápida
quick-make de cierre rápido
quick-make-and-break switch conmutador de cierre y abertura rápida
quick-make switch conmutador de cierre rápido
quick-opening de abertura rápida
quick-opening switch conmutador de abertura rápida
quick-release coupling acoplamiento de desenganche rápido
quick-release relay relé de desenganche rápido
quick-response recorder registrador de respuesta rápida
quick sort clasificación rápida
quick start inicio rápido
quicksilver azogue
quiescent quiescente, en reposo, inactivo
quiescent antenna antena ficticia
quiescent carrier portadora quiescente
quiescent-carrier modulation modulación de portadora quiescente
quiescent-carrier system sistema de portadora quiescente
quiescent-carrier telephony telefonía de portadora quiescente
quiescent-carrier transmission transmisión de portadora quiescente
quiescent circuit circuito quiescente
quiescent component componente quiescente
quiescent condition condición de reposo
quiescent current corriente de reposo
quiescent-current compensation compensación de corriente de reposo
quiescent point punto de reposo
quiescent power consumption consumo de energía en reposo
quiescent power dissipation disipación de energía en reposo
quiescent push-pull amplifier amplificador en contrafase quiescente
quiescent state estado de reposo
quiescent value valor de reposo
quiet automatic gain control control de ganancia automático silencioso
quiet automatic volume control control de volumen automático silencioso
quiet battery batería silenciosa
quiet channel canal silencioso
quiet circuit circuito silencioso
quiet tuning sintonización silenciosa
quieting silenciamiento
quieting level nivel de silenciamiento
quieting sensitivity sensibilidad de silenciamiento
quieting system sistema de silenciamiento
quinary quinario
quinary code código quinario
quinary counter contador quinario
quinary group grupo quinario
quinhydrone electrode electrodo de quinhidrona
quinhydrone half-cell semicelda de quinhidrona
quintupler quintuplicador

quintuplet quintuplete
quotient expansion expansión de cociente
quotient meter cocientímetro
quotient relay relé de cociente

rabbit-ear antenna antena en orejas de conejo
race carrera
raceway conducto de cables, conducto eléctrico, canaleta
rack bastidor, estante, soporte
rack adapter adaptador de bastidor
rack assembly conjunto de bastidor
rack layout arreglo de bastidor
rack-mount de montaje en bastidor
rack-mounted montado en bastidor
rack-mounted equipment equipo montado en bastidor
rack-mounted unit unidad montada en bastidor
rack mounting montaje en bastidor
rack panel panel de bastidor
racon racon
rad rad
radan radan
radar radar
radar altimeter altímetro de radar
radar altitude altitud por radar
radar antenna antena de radar
radar approach aproximación por radar
radar area área de radar
radar astronomy astronomía de radar
radar band banda de radar
radar beacon radiofaro de radar
radar beam haz de radar
radar bearing rumbo por radar
radar blind spot punto ciego de radar
radar calibration calibración de radar
radar camera cámara de radar
radar chronometer cronómetro de radar
radar chronometry cronometría de radar
radar clutter ecos parásitos de radar
radar console consola de radar
radar control control por radar
radar-control area área de control por radar
radar-controlled controlado por radar
radar coverage cobertura de radar
radar cross-section sección eficaz de radar
radar data datos de radar
radar data display visualización de datos de radar
radar detection detección de radar
radar detector detector de radar
radar display visualizador de radar
radar display unit unidad de visualización de radar
radar disturbance perturbación de radar
radar dome cúpula de radar
radar echo eco de radar
radar element elemento de radar
radar equation ecuación de radar
radar equipment equipo de radar
radar-equipped equipado con radar
radar frequency band banda de frecuencias de radar
radar-guided guiado por radar
radar heading rumbo por radar
radar homing guiado por radar
radar horizon horizonte de radar
radar identification identificación por radar
radar image imagen de radar

radar indicator indicador de radar
radar information información de radar
radar installation instalación de radar
radar jamming interferencia intencional de radar
radar klystron klistrón de radar
radar marker marcador de radar
radar mirage espejismo de radar
radar modulator modulador de radar
radar monitor monitor de radar
radar monitoring monitoreo por radar
radar navigation navegación por radar
radar navigation system sistema de navegación por radar
radar network red de radar
radar noise ruido de radar
radar-operated operado por radar
radar performance figure cifra de rendimiento de radar
radar photograph fotografía de radar
radar pip impulso de radar
radar pulse impulso de radar
radar-pulse modulator modulador de impulsos de radar
radar-pulse repeater repetidor de impulsos de radar
radar radiation radiación de radar
radar range alcance de radar
radar receiver receptor de radar
radar reflection reflexión de radar
radar reflector reflector de radar
radar relay relé de radar
radar repeater repetidor de radar
radar resolution resolución de radar
radar responder respondedor de radar
radar response respuesta de radar
radar scan exploración de radar
radar scanner explorador de radar
radar screen pantalla de radar
radar set equipo de radar
radar shadow sombra de radar
radar signal señal de radar
radar signal recorder registrador de señal de radar
radar signal recording registro de señal de radar
radar speed measurement medición de velocidad por radar
radar station estación de radar
radar synchronizer sincronizador de radar
radar system sistema de radar
radar telescope telescopio de radar
radar tower torre de radar
radar-tracked seguido por radar
radar tracking system sistema de seguimiento por radar
radar transmitter transmisor de radar
radar transponder transpondedor de radar
radar unit unidad de radar
radar warning advertencia por radar
radar wave onda de radar
radarized radarizado
radarscope radariscopio
radarscope display visualizador de radar
radechon radecón
radiac radiac
radiac computer computadora de radiac
radiac detector detector radiac
radial radial
radial beam haz radial
radial-beam tube tubo de haz radial
radial circuit circuito radial
radial conductor conductor radial
radial convergence convergencia radial
radial diffusion difusión radial
radial-diffusion coefficient coeficiente de difusión radial

radial feeder alimentador radial
radial grating retículo radial
radial ground tierra radial
radial ground system sistema de tierra radial
radial lead conexión radial
radial line línea radial
radial network red radial
radial oscillation oscilación radial
radial pattern patrón radial
radial power factor factor de potencia radial
radial slot ranura radial
radial-slot rotor rotor de ranuras radiales
radial sweep barrido radial
radial system sistema radial
radial transfer transferencia radial
radial transmission line línea de transmisión radial
radial wire hilo radial
radian radián
radian frequency frecuencia de oscilación
radiance radiancia
radiance factor factor de radiancia
radiance temperature temperatura de radiancia
radiancy radiancia
radiant radiante
radiant absorptance absortancia radiante
radiant density densidad radiante
radiant efficiency eficiencia radiante
radiant element elemento radiante
radiant emittance emitancia radiante
radiant energy energía radiante
radiant energy density densidad de energía radiante
radiant excitance excitancia radiante
radiant exposure exposición radiante
radiant field campo radiante
radiant flux flujo radiante
radiant flux density densidad de flujo radiante
radiant gain ganancia radiante
radiant heat calor radiante
radiant heater calentador radiante
radiant heating calefacción radiante
radiant heating system sistema de calefacción radiante
radiant intensity intensidad radiante
radiant power potencia radiante
radiant sensitivity sensibilidad radiante
radiate radiar
radiated radiado
radiated energy energía radiada
radiated interference interferencia radiada
radiated noise ruido radiado
radiated power potencia radiada
radiated power output salida de potencia radiada
radiated radio noise ruido de radio radiado
radiated signal señal radiada
radiated spectrum espectro radiado
radiatics radiática
radiating radiante
radiating antenna antena radiante
radiating area área radiante
radiating circuit circuito radiante
radiating curtain cortina radiante
radiating doublet doblete radiante
radiating element elemento radiante
radiating guide guíaondas radiante
radiating slot ranura radiante
radiating surface superficie radiante
radiating tower torre radiante
radiation radiación
radiation absorber absorbedor de radiación
radiation absorption absorción de radiación
radiation angle ángulo de radiación
radiation beam haz de radiación
radiation belt cinturón de radiación

radiation cone cono de radiación
radiation constant constante de radiación
radiation cooling enfriamiento por radiación
radiation counter contador de radiación
radiation-counter tube tubo contador de radiación
radiation danger peligro de radiación
radiation danger zone zona de peligro de radiación
radiation density densidad de radiación
radiation-density constant constante de densidad de
  radiación
radiation detector detector de radiación
radiation diagram diagrama de radiación
radiation effect efecto de radiación
radiation efficiency eficiencia de radiación
radiation energy energía de radiación
radiation exposure exposición a radiación
radiation field campo de radiación
radiation flux flujo de radiación
radiation hazard riesgo de radiación
radiation height altura de radiación
radiation impedance impedancia de radiación
radiation indicator indicador de radiación
radiation intensity intensidad de radiación
radiation ionization ionización por radiación
radiation level nivel de radiación
radiation lobe lóbulo de radiación
radiation loss pérdida por radiación
radiation measurement medida de radiación
radiation mode modo de radiación
radiation monitor monitor de radiación
radiation pattern patrón de radiación
radiation potential potencial de radiación
radiation power potencia de radiación
radiation pressure presión de radiación
radiation pyrometer pirómetro de radiación
radiation resistance resistencia de radiación
radiation-resistant resistente a radiación
radiation-sensitive sensible a radiación
radiation sensor sensor de radiación
radiation sphere esfera de radiación
radiation system sistema de radiación
radiation temperature temperatura de radiación
radiation thermometer termómetro de radiación
radiation trapping atrapamiento de radiación
radiation zone zona de radiación
radiative radiativo
radiative relaxation time tiempo de relajación
  radiativa
radiator radiador
radio radio
radio aid radioayuda
radio altimeter radioaltímetro
radio altitude altitud por radio
radio amplification radioamplificación
radio antenna antena de radio
radio-astronomical radioastronómico
radio astronomy radioastronomía
radio atmosphere radioatmósfera
radio attenuation radioatenuación
radio beacon radiofaro
radio-beacon antenna antena de radiofaro
radio-beacon receiver receptor de radiofaro
radio-beacon station estación de radiofaro
radio-beacon system sistema de radiofaros
radio beam haz de radio
radio broadcast radiodifusión
radio broadcast station estación de radiodifusión
radio broadcasting radiodifusión
radio buoy radioboya
radio carrier radioportadora
radio cast radiodifusión
radio central central de radio
radio channel radiocanal

radio chronometer radiocronómetro
radio circuit circuito de radio
radio communication radiocomunicación
radio compass radiocompás
radio-compass indicator indicador de radiocompás
radio-compass magnetic indicator indicador
  magnético de radiocompás
radio component componente de radio
radio control radiocontrol
radio-control circuit circuito de radiocontrol
radio-controlled radiocontrolado
radio course curso de radio
radio detection radiodetección
radio direction-finder radiogoniómetro
radio direction-finding radiogoniometría
radio-dispatching radiodespacho
radio-dispatching system sistema de radiodespacho
radio disturbance perturbación de radio
radio Doppler radiolocalización por Doppler
radio echo radioeco
radio-echo method método de radioeco
radio-electronic circuit circuito radioelectrónico
radio-electronics radioelectrónica
radio emanation radioemanación
radio emission radioemisión
radio equipment equipo de radio
radio facilities instalaciones de radio
radio facsimile radiofacsímil
radio-facsimile transmission transmisión de
  radiofacsímil
radio fadeout desvanecimiento de radio
radio field intensity intensidad de campo de radio
radio field strength intensidad de campo de radio
radio field strength meter medidor de intensidad de
  campo de radio
radio filter filtro de radio
radio frequency radiofrecuencia
radio-frequency alternator alternador de
  radiofrecuencia
radio-frequency amplification amplificación de
  radiofrecuencia
radio-frequency amplifier amplificador de
  radiofrecuencia
radio-frequency amplifier noise ruido de
  amplificador de radiofrecuencia
radio-frequency amplifier stage etapa amplificadora
  de radiofrecuencia
radio-frequency attenuator atenuador de
  radiofrecuencia
radio-frequency bandwidth ancho de banda de
  radiofrecuencia
radio-frequency booster reforzador de
  radiofrecuencia
radio-frequency bridge puente de radiofrecuencia
radio-frequency cable cable de radiofrecuencia
radio-frequency carrier portadora de
  radiofrecuencia
radio-frequency choke bobina de inducción de
  radiofrecuencia
radio-frequency coil bobina de radiofrecuencia
radio-frequency component componente de
  radiofrecuencia
radio-frequency contactor contactor de
  radiofrecuencia
radio-frequency converter convertidor de
  radiofrecuencia
radio-frequency current corriente de
  radiofrecuencia
radio-frequency energy energía de radiofrecuencia
radio-frequency envelope indicator indicador de
  envolvente de radiofrecuencia
radio-frequency filter filtro de radiofrecuencia
radio-frequency gain ganancia en radiofrecuencia

radio-frequency generator generador de radiofrecuencia

radio-frequency glow discharge descarga luminiscente de radiofrecuencia

radio-frequency grid current corriente de rejilla de radiofrecuencia

radio-frequency harmonic armónica de radiofrecuencia

radio-frequency head cabeza de radiofrecuencia

radio-frequency heating calentamiento por radiofrecuencia

radio-frequency indicator indicador de radiofrecuencia

radio-frequency interference interferencia de radiofrecuencia

radio-frequency intermodulation distortion distorsión de intermodulación de radiofrecuencia

radio-frequency leak fuga de radiofrecuencia

radio-frequency line línea de radiofrecuencia

radio-frequency link enlace de radiofrecuencia

radio-frequency load carga de radiofrecuencia

radio-frequency meter medidor de radiofrecuencia

radio-frequency monitor monitor de radiofrecuencia

radio-frequency oscillator oscilador de radiofrecuencia

radio-frequency output salida de radiofrecuencia

radio-frequency plasma torch soplete de plasma de radiofrecuencia

radio-frequency power potencia de radiofrecuencia

radio-frequency power amplifier amplificador de potencia de radiofrecuencia

radio-frequency power output salida de potencia de radiofrecuencia

radio-frequency power supply fuente de alimentación de radiofrecuencia

radio-frequency preselector preselector de radiofrecuencia

radio-frequency probe sonda de radiofrecuencia

radio-frequency pulse impulso de radiofrecuencia

radio-frequency radiation radiación de radiofrecuencia

radio-frequency record registro de radiofrecuencia

radio-frequency resistance resistencia en radiofrecuencia

radio-frequency response respuesta en radiofrecuencia

radio-frequency selectivity selectividad de radiofrecuencia

radio-frequency signal señal de radiofrecuencia

radio-frequency signal generator generador de señales de radiofrecuencia

radio-frequency spectrometer espectrómetro de radiofrecuencia

radio-frequency spectrum espectro de radiofrecuencias

radio-frequency stage etapa de radiofrecuencia

radio-frequency suppressor supresor de radiofrecuencia

radio-frequency transformer transformador de radiofrecuencia

radio-frequency transistor transistor de radiofrecuencia

radio-frequency transmission line línea de transmisión de radiofrecuencia

radio-frequency tube tubo de radiofrecuencia

radio galaxy radiogalaxia

radio guidance radioguiado

radio homing radioguiado

radio horizon radiohorizonte

radio interference radiointerferencia

radio interference suppression supresión de radiointerferencia

radio interferometer radiointerferómetro

radio link radioenlace

radio-link circuit circuito de radioenlace

radio-link system sistema de radioenlace

radio-link transmitter transmisor de radioenlace

radio loop antena de cuadro

radio magnetic indicator indicador radiomagnético

radio marker beacon radiobaliza de posición

radio mast mástil de antena

radio mesh radiomalla

radio microphone radiomicrófono

radio monitor monitor de radio

radio multiplexing multiplexión de radio

radio navigation radionavegación

radio net red de radio

radio network red de radio

radio noise radiorruido, ruido de radio

radio noise filter filtro de radiorruido

radio noise level nivel de radiorruido

radio noise storm tormenta de radiorruido

radio-operated radiooperado

radio-optical distance distancia radióptica

radio-phonograph radiofonógrafo

radio propagation radiopropagación

radio-propagation path trayectoria de radiopropagación

radio prospecting radioprospección

radio proximity radioproximidad

radio range radiofaro direccional

radio-range antenna antena de radiofaro direccional

radio-range beacon radiofaro direccional

radio-range beam haz de radiofaro direccional

radio-range monitor monitor de radiofaro direccional

radio-range station estación de radiofaro direccional

radio-range transmitter transmisor de radiofaro direccional

radio receiver radiorreceptor

radio reception radiorrecepción

radio recorder radiorregistrador

radio reflector radiorreflector

radio relay radiorrelé, reemisión de radio

radio-relay channel canal de radiorrelé

radio-relay circuit circuito de radiorrelé

radio-relay equipment equipo de radiorrelé

radio-relay network red de radiorrelés

radio-relay route ruta de radiorrelés

radio-relay station estación de radiorrelé

radio-relay system sistema de radiorrelé

radio repeater repetidor de radio

radio repeater station estación repetidora de radio

radio section sección de radio

radio sender radioemisor

radio set equipo de radio

radio shield radioblindaje

radio signal radioseñal

radio source radiofuente

radio spectroscope radioespectroscopio

radio spectrum radioespectro, espectro de radio

radio spectrum analyzer analizador de radioespectro

radio star radioestrella

radio station radioestación

radio system radiosistema

radio telegram radiotelegrama

radio telegraph radiotelégrafo

radio telegraphy radiotelegrafía

radio telescope radiotelescopio

radio tower torre de antena

radio transceiver radiotransceptor

radio transmission radiotransmisión

radio transmitter radiotransmisor

radio-transparent radiotransparente

radio-transparent material material radiotransparente

radio tube tubo de radio
radio tuner sintonizador de radio
radio valve válvula de radio
radio wave onda de radio
radio-wave propagation propagación de ondas de radio
radio-wave scattering dispersión de ondas de radio
radio window ventana de radio
radioacoustic radioacústico
radioacoustics radioacústica
radioactive radiactivo, radioactivo
radioactive deposit depósito radiactivo
radioactive element elemento radiactivo
radioactive emanation emanación radiactiva
radioactive isotope isótopo radiactivo
radioactive material material radiactivo
radioactive source fuente radiactiva
radioactive tracer trazador radiactivo
radioactive transducer transductor radiactivo
radioactivity radiactividad
radioactivity absorber absorbedor de radiactividad
radioactivity counter contador de radiactividad
radioactivity detection detección de radiactividad
radioactivity detector detector de radiactividad
radioactivity monitor monitor de radiactividad
radioactivity standard patrón de radiactividad
radiobeacon radiofaro
radiobroadcast radiodifusión
radiobroadcasting radiodifusión
radiocast radiodifusión
radiocommunication radiocomunicación
radiocommunication circuit circuito de radiocomunicación
radioconductor radioconductor
radiocontrol radiocontrol
radiocrystallography radiocristalografía
radiodetection radiodetección
radiodiffusion radiodifusión
radiodistribution radiodistribución
radiodistribution system sistema de radiodistribución
radioelectric radioeléctrico
radioelectric wave onda radioeléctrica
radioelectricity radioelectricidad
radioelement radioelemento
radiofacsimile radiofacsímil
radiofrequency radiofrecuencia
radiogenic radiogénico
radiogenic heat calor radiogénico
radiogoniometer radiogoniómetro
radiogoniometry radiogoniometría
radiogram radiograma
radiograph radiografía
radiographic radiográfico
radiographic film película radiográfica
radiographically radiográficamente
radiography radiografía
radioguidance radioguiado
radiointerferometer radiointerferómetro
radioisotope radioisótopo
radiolocation radiolocalización
radiolocation station estación de radiolocalización
radiolocator radiolocalizador
radiological radiológico
radiological indicator indicador radiológico
radiological instrument instrumento radiológico
radiological monitor monitor radiológico
radiology radiología
radiolucency radiolucencia
radiolucent radiolucente
radioluminescence radioluminiscencia
radiometallography radiometalografía
radiometeorograph radiometeorógrafo

radiometeorology radiometeorología
radiometer radiómetro
radiometric radiométrico
radiometry radiometría
radionavigation radionavegación
radionavigation mobile station estación móvil de radionavegación
radionavigation system sistema de radionavegación
radionics radiónica
radioopacity radioopacidad
radiopaque radioopaco
radiophone radiófono
radiophonic radiofónico
radiophonograph radiofonógrafo
radiophony radiofonía
radiophoto radiofoto
radiophotogram radiofotograma
radiophotography radiofotografía
radiophotoluminescence radiofotoluminiscencia
radioprobe radiosonda
radioreceiver radiorreceptor
radioreception radiorrecepción
radioresistance radiorresistencia
radioresonance radiorresonancia
radioscope radioscopio
radioscopy radioscopía
radiosensitive radiosensible
radiosensitivity radiosensibilidad
radiosonde radiosonda
radiosonde receiver receptor de radiosonda
radiosonde recorder registrador de radiosonda
radiosonde sensor sensor de radiosonda
radiosonde transmitter transmisor de radiosonda
radiosonic radiosónico
radiosounding radiosondeo
radiotelecommunication radiotelecomunicación
radiotelegram radiotelegrama
radiotelegraph radiotelégrafo
radiotelegraph channel canal radiotelegráfico
radiotelegraph circuit circuito radiotelegráfico
radiotelegraph communication comunicación radiotelegráfica
radiotelegraph receiver receptor radiotelegráfico
radiotelegraph reception recepción radiotelegráfica
radiotelegraph transmitter transmisor radiotelegráfico
radiotelegraphic radiotelegráfico
radiotelegraphy radiotelegrafía
radiotelemetering radiotelemedición
radiotelemetry radiotelemetría
radiotelephone radioteléfono
radiotelephone channel canal radiotelefónico
radiotelephone circuit circuito radiotelefónico
radiotelephone emission emisión radiotelefónica
radiotelephone facilities instalaciones radiotelefónicas
radiotelephone link enlace radiotelefónico
radiotelephone transmitter transmisor radiotelefónico
radiotelephonic radiotelefónico
radiotelephony radiotelefonía
radioteleprinter radioteleimpresora
radioteletype radioteletipo
radioteletypewriter radioteleimpresor
radiotelex radiotélex
radiothermics radiotermia
radiotracer radiotrazador
radiotransmission radiotransmisión
radiotransparent radiotransparente
radiotransparent material material radiotransparente
radium radio
radius radio

radix base
radix complement complemento de base
radix-minus-one complement complemento de base menos uno
radix notation notación de base
radix number número de base
radix numeration numeración de base
radix numeration system sistema de numeración de base
radix point coma de separación, punto de separación
radix search búsqueda de base
radix sort clasificación por base
radix sorting clasificación por base
radome radomo
radon radón
radux radux
rail bond conexión eléctrica de carriles
rail return retorno por carril
railings empalizada
railway signaling señalización ferroviaria
rain attenuation atenuación por lluvia
rain clutter ecos parásitos por lluvia
rainband banda de lluvia
rainbow generator generador de arco iris
rainbow pattern patrón de arco iris
rainproof a prueba de lluvia
raintight hermético contra la lluvia
ramac ramac
Raman effect efecto de Raman
Raman-Nath region región de Raman-Nath
Raman scattering dispersión de Raman
Raman spectrometer espectrómetro de Raman
Raman spectroscopy espectroscopia de Raman
Raman spectrum espectro de Raman
ramark ramark
ramp rampa
ramp generator generador de rampa
Ramsauer effect efecto de Ramsauer
random access acceso aleatorio
random-access device dispositivo de acceso aleatorio
random-access file archivo de acceso aleatorio
random-access memory memoria de acceso aleatorio
random-access memory accounting contabilidad con memoria de acceso aleatorio
random-access memory card tarjeta de memoria de acceso aleatorio
random-access memory circuit circuito de memoria de acceso aleatorio
random-access memory disk disco de memoria de acceso aleatorio
random-access memory loader cargador de memoria de acceso aleatorio
random-access memory refresh refresco de memoria de acceso aleatorio
random-access memory storage almacenamiento de memoria de acceso aleatorio
random-access memory subroutine subrutina de memoria de acceso aleatorio
random-access memory system sistema de memoria de acceso aleatorio
random-access memory terminal terminal de memoria de acceso aleatorio
random-access memory test prueba de memoria de acceso aleatorio
random-access method método de acceso aleatorio
random-access programming programación de acceso aleatorio
random-access storage almacenamiento de acceso aleatorio
random-access storage device dispositivo de almacenamiento de acceso aleatorio
random-access store almacenamiento de acceso aleatorio

random-access time tiempo de acceso aleatorio
random channel canal aleatorio
random check comprobación aleatoria
random coding codificación aleatoria
random coincidence coincidencia aleatoria
random digit dígito aleatorio
random distribution distribución aleatoria
random disturbance perturbación aleatoria
random drift deslizamiento aleatorio
random drift rate tasa de deslizamiento aleatorio
random electrical noise ruido eléctrico aleatorio
random electrostatic field campo electrostático aleatorio
random emission emisión aleatoria
random error error aleatorio
random event evento aleatorio
random failure falla aleatoria
random feed alimentación aleatoria
random file archivo aleatorio
random fluctuations fluctuaciones aleatorias
random incidence incidencia aleatoria
random-incidence microphone micrófono de incidencia aleatoria
random-incidence response respuesta de incidencia aleatoria
random inputs entradas aleatorias
random inquiry consulta aleatoria
random interference interferencia aleatoria
random linear filter filtro lineal aleatorio
random loading carga aleatoria
random logic lógica aleatoria
random-logic design diseño de lógica aleatoria
random-logic device dispositivo de lógica aleatoria
random medium medio aleatorio
random noise ruido aleatorio
random-noise bandwidth ancho de banda de ruido aleatorio
random-noise correction corrección de ruido aleatorio
random-noise generator generador de ruido aleatorio
random-noise input entrada de ruido aleatorio
random number número aleatorio
random-number generation generación de números aleatorios
random-number generator generador de números aleatorios
random-number sequence secuencia de números aleatorios
random output salida aleatoria
random probing sondeo aleatorio
random processing procesamiento aleatorio
random pulsing pulsación aleatoria
random response respuesta aleatoria
random sample muestra aleatoria
random sampling muestreo aleatorio
random scan exploración aleatoria
random-scan display device dispositivo de visualización de exploración aleatoria
random scattering dispersión aleatoria
random sensitivity sensibilidad aleatoria
random separation separación aleatoria
random sequence secuencia aleatoria
random sequential access acceso secuencial aleatorio
random signal señal aleatoria
random signal generator generador de señales aleatorias
random surface superficie aleatoria
random variable variable aleatoria
random variation variación aleatoria
random vibration vibración aleatoria
random winding arrollamiento aleatorio

randomization aleatorización
randomize aleatorizar
randomized aleatorizado
randomly aleatoriamente
randomly arranged arreglado aleatoriamente
randomly distributed distribuido aleatoriamente
randomness aleatoriedad
range alcance, gama, margen, distancia, escala
range ambiguity ambigüedad de distancia
range-amplitude display indicador de
    distancia-amplitud
range-bearing display indicador de distancia-rumbo
range check comprobación de margen
range control control de distancia
range corrector corrector de distancia
range discrimination discriminación de distancia
range-energy relation relación alcance-energía
range error error de distancia
range extension extensión de distancia
range-extension unit unidad de extensión de
    distancia
range-height indicator indicador de distancia-altura
range-height indicator display visualización de
    indicador de distancia-altura
range lights luces de límites
range mark marca de distancia
range marker marcador de distancia
range-marker generator generador de marcas de
    distancia
range measurement medida de distancia
range multiplier multiplicador de alcance
range of brightness gama de brillo
range of colors gama de colores
range of detection alcance de detección
range of frequencies gama de frecuencias
range of measurement gama de medida
range of station alcance de estación
range of temperature gama de temperaturas
range of values gama de valores
range of wavelengths gama de longitudes de onda
range resolution resolución de distancia
range ring anillo de distancia
range selector selector de distancia, selector de
    alcance, selector de banda, selector de escala
range signal señal de distancia
range switch conmutador de distancia, conmutador
    de alcance, conmutador de banda, conmutador de
    escala
range switching conmutación de distancia,
    conmutación de alcance, conmutación de banda,
    conmutación de escala
range transmitter transmisor de distancia
range unit unidad de distancia
Rankine scale escala Rankine
rapcon rapcon
rapid-access de acceso rápido
rapid-access loop bucle de acceso rápido
rapid-access memory memoria de acceso rápido
rapid-access storage almacenamiento de acceso
    rápido
rapid action acción rápida
rapid-action switch conmutador de acción rápida
rapid charging carga rápida
rapid memory memoria rápida
rapid printer impresora rápida
rapid pulse impulso rápido
rapid response respuesta rápida
rapid rise subida rápida
rapid scanning exploración rápida
rapid-scanning motor motor de exploración rápida
rapid service servicio rápido
rapid-start fluorescent lamp lámpara fluorescente
    de encendido rápido

rapid-start lamp lámpara de encendido rápido
rapidity rapidez
rapidity of modulation rapidez de modulación
rare earth tierra rara
rare gas gas raro
rare-gas tube tubo de gas raro
rarefaction rarefacción
rarefied rarificado
raser ráser
raster trama, cuadrícula
raster burn quemadura de trama
raster count cuenta de exploración
raster display device dispositivo de visualización de
    exploración
raster graphics gráficos de exploración
raster grid trama de exploración
raster image processor procesador de imagen de
    exploración
raster pattern patrón de exploración
raster pattern generator generador de patrones de
    exploración
raster scan exploración de trama
raster unit unidad de exploración
ratchet relay relé de trinquete
rate tasa, velocidad, régimen, razón
rate action acción derivada
rate control control derivado
rate effect efecto de transición
rate-grown junction unión por crecimiento variable
rate-grown transistor transistor por crecimiento
    variable
rate gyro giroscopio de velocidades angulares
rate gyroscope giroscopio de velocidades angulares
rate of change tasa de cambio
rate-of-change relay relé de tasa de cambio
rate of charge velocidad de carga
rate of decay velocidad de extinción
rate of deposition velocidad de deposición
rate of diffusion velocidad de difusión
rate of fall velocidad de caída
rate of flow velocidad de flujo, régimen de descarga
rate of growth ritmo de crecimiento
rate of illumination frecuencia de iluminación
rate of interrogation ritmo de interrogación
rate of modulation velocidad de modulación
rate of punching velocidad de perforación
rate of reaction velocidad de reacción
rate of reading velocidad de lectura
rate of rise velocidad de subida
rate-of-rise detector detector de velocidad de subida
rate-of-rise limiter limitador de velocidad de subida
rate-of-rise suppressor supresor de velocidad de
    subida
rate of transmission velocidad de transmisión
rate time tiempo de acción derivada
rated accuracy exactitud nominal
rated apparent efficiency eficiencia aparente
    nominal
rated burden régimen nominal
rated capacity capacidad nominal
rated characteristic característica nominal
rated circuit voltage tensión de circuito nominal
rated contact current corriente de contactos nominal
rated continuous current corriente continua nominal
rated coverage cobertura nominal
rated current corriente nominal
rated direct current corriente continua nominal
rated dissipation disipación nominal
rated duty servicio nominal
rated engine speed velocidad nominal de motor
rated field current corriente de campo nominal
rated field voltage tensión de campo nominal
rated frequency frecuencia nominal

rated frequency deviation desviación de frecuencia nominal
rated horsepower potencia nominal
rated impedance impedancia nominal
rated input entrada nominal
rated input power potencia de entrada nominal
rated internal pressure presión interna nominal
rated life vida nominal
rated line current corriente de línea nominal
rated line voltage tensión de línea nominal
rated load carga nominal
rated making capacity capacidad de cierre nominal
rated operating range alcance de operación nominal
rated output salida nominal
rated output capacity capacidad de salida nominal
rated output current corriente de salida nominal
rated output frequency frecuencia de salida nominal
rated output power potencia de salida nominal
rated output voltage tensión de salida nominal
rated pressure presión nominal, tensión nominal
rated primary current corriente primaria nominal
rated primary voltage tensión primaria nominal
rated quantity cantidad nominal
rated secondary current corriente secundaria nominal
rated secondary voltage tensión secundaria nominal
rated short-circuit capacity capacidad de cortocircuito nominal
rated short-circuit current corriente de cortocircuito nominal
rated speed velocidad nominal
rated system deviation desviación de sistema nominal
rated system frequency frecuencia de sistema nominal
rated system voltage tensión de sistema nominal
rated thermal current corriente térmica nominal
rated torque par nominal
rated value valor nominal
rated voltage tensión nominal
rated working voltage tensión de trabajo nominal
rating clasificación, régimen, régimen nominal, potencia nominal, capacidad nominal, valor nominal
rating plate placa indicadora
ratio razón, relación
ratio-arm bridge puente de brazos de razón
ratio arms brazos de razón
ratio between channels razón entre canales
ratio box caja de razones
ratio calculator calculadora de razón
ratio calibration calibración de razón
ratio control control de razón
ratio detector detector de razón
ratio discriminator discriminador de razón
ratio discriminator circuit circuito discriminador de razón
ratio error error de razón
ratio meter medidor de razón
ratio of attenuation razón de atenuación
ratio of magnification razón de magnificación
ratio of transformation razón de transformación
ratiometer logómetro, medidor de razón
rational racional
rational number número racional
rationalized system sistema racionalizado
rationalized unit unidad racionalizada
raw data datos sin procesar
raw tape cinta en blanco
rawin radioviento
ray rayo
ray beam haz de rayos
ray-control electrode electrodo de control de rayo

ray locking bloqueo de rayo
ray of light rayo de luz
raydist raydist
rayl rayl
Rayleigh-Carson theorem teorema de Rayleigh-Carson
Rayleigh criterion criterio de Rayleigh
Rayleigh cycle ciclo de Rayleigh
Rayleigh density function función de densidad de Rayleigh
Rayleigh disk disco de Rayleigh
Rayleigh distribution distribución de Rayleigh
Rayleigh-Jeans equation ecuación de Rayleigh-Jeans
Rayleigh-Jeans law ley de Rayleigh-Jeans
Rayleigh law ley de Rayleigh
Rayleigh line línea de Rayleigh
Rayleigh number número de Rayleigh
Rayleigh reciprocity theorem teorema de reciprocidad de Rayleigh
Rayleigh scatter dispersión de Rayleigh
Rayleigh scattering dispersión de Rayleigh
Rayleigh surface wave onda de superficie de Rayleigh
Rayleigh wave onda de Rayleigh
reach alcance
reach factor factor de alcance
reachthrough voltage tensión de penetración
reacquisition time tiempo de readquisición
reactance reactancia
reactance amplifier amplificador de reactancia
reactance coil bobina de reactancia
reactance diode diodo de reactancia
reactance drop caída de reactancia
reactance-drop compensation compensación de caída de reactancia
reactance factor factor de reactancia
reactance frequency multiplier multiplicador de frecuencia por reactancia
reactance function función de reactancia
reactance meter reactancímetro
reactance modulator modulador de reactancia
reactance protection protección de reactancia
reactance relay relé de reactancia
reactance-resistance ratio razón de reactancia-resistencia
reactance transistor transistor de reactancia
reactance tube tubo de reactancia
reactance tube modulator modulador de tubo de reactancia
reactance valve válvula de reactancia
reactance voltage tensión de reactancia
reactatron reactatrón
reaction reacción
reaction alternator alternador de reacción
reaction cavity cavidad de reacción
reaction circuit circuito de reacción
reaction coil bobina de reacción
reaction condenser condensador de reacción
reaction coupling acoplamiento de reacción
reaction degree grado de reacción
reaction distance distancia de reacción
reaction energy energía de reacción
reaction formula fórmula de reacción
reaction heat calor de reacción
reaction indicator indicador de reacción
reaction inhibitor inhibidor de reacción
reaction motor motor de reacción
reaction power potencia de reacción
reaction scanning exploración por reacción
reaction stress estrés de reacción
reaction suppressor supresor de reacción
reaction system sistema de reacción
reaction time tiempo de reacción

reaction-time meter medidor de tiempo de reacción
reaction turbine turbina de reacción
reactivate reactivar
reactivation reactivación
reactivator reactivador
reactive reactivo
reactive anode ánodo reactivo
reactive attenuator atenuador reactivo
reactive balance equilibrio reactivo
reactive circuit circuito reactivo
reactive component componente reactivo
reactive current corriente reactiva
reactive damping amortiguamiento reactivo
reactive energy energía reactiva
reactive factor factor reactivo
reactive-factor meter medidor de factor reactivo
reactive kilovolt-ampere kilovolt-ampere reactivo
reactive load carga reactiva
reactive power potencia reactiva
reactive-power relay relé de potencia reactiva
reactive volt-ampere volt-ampere reactivo
reactive voltage tensión reactiva
reactivity reactividad
reactor reactor
read leer, indicar
read access time tiempo de acceso de lectura
read-after-write check comprobación de lectura tras
    escritura
read-ahead queue cola de lectura adelantada
read-and-punch code código de lectura y
    perforación
read-and-write memory memoria de lectura y
    escritura
read board tablero de lectura
read cell celda de lectura
read command mando de lectura
read cycle ciclo de lectura
read-cycle time tiempo de ciclo de lectura
read data check comprobación de datos de lectura
read diode diodo de lectura
read error error de lectura
read head cabeza de lectura
read-only sólo de lectura
read-only access acceso sólo de lectura
read-only attribute atributo sólo de lectura
read-only disk disco sólo de lectura
read-only memory memoria sólo de lectura
read-only memory error error de memoria sólo de
    lectura
read-only memory programmer programador de
    memoria sólo de lectura
read-only memory simulator simulador de memoria
    sólo de lectura
read-only memory terminal terminal de memoria
    sólo de lectura
read-only storage almacenamiento sólo de lectura
read-out leer
read output salida de lectura
read output signal señal de salida de lectura
read path trayectoria de lectura
read-protect proteger contra lectura
read protection protección contra lectura
read pulse impulso de lectura
read rate velocidad de lectura
read screen pantalla de lectura
read system sistema de lectura
read time tiempo de lectura
read unit unidad de lectura
read-unit feed alimentación de unidad de lectura
read wire hilo de lectura
read-write lectura-escritura
read-write access acceso de lectura-escritura
read-write access mode modo de acceso de

lectura-escritura
read-write channel canal de lectura-escritura
read-write check comprobación de lectura-escritura
read-write counter contador de lectura-escritura
read-write cycle ciclo de lectura-escritura
read-write cycle time tiempo de ciclo de
    lectura-escritura
read-write file archivo de lectura-escritura
read-write head cabeza de lectura-escritura
read-write memory memoria de lectura-escritura
read-write protection protección de lectura-escritura
readability legibilidad
readability scale escala de legibilidad
readable legible
readable signal señal legible
reader lector, lectora
reader-interpreter lectora-interpretadora
reader-printer lectora-impresora
reader-sorter lectora-clasificadora
reading lectura
reading access time tiempo de acceso de lectura
reading accuracy exactitud de lectura
reading brush escobilla de lectura
reading channel canal de lectura
reading circuit circuito de lectura
reading head cabeza de lectura
reading machine máquina de lectura
reading pulse impulso de lectura
reading rate velocidad de lectura
reading speed velocidad de lectura
reading station estación de lectura
reading task tarea de lectura
reading time tiempo de lectura
reading track pista de lectura
reading unit unidad de lectura
reading velocity velocidad de lectura
readjust reajustar
readout lectura
readout accuracy exactitud de lectura
readout device dispositivo de lectura
readout dial cuadrante de lectura
readout equipment equipo de lectura
readout error error de lectura
readout indicator indicador de lectura
readout lamp lámpara de lectura
readout resolution resolución de lectura
readout signal señal de lectura
ready listo, disponible
ready condition condición de disponibilidad
ready status estado de disponibilidad
real address dirección real
real axis eje real
real circuit circuito real
real component componente real
real constant constante real
real data datos reales
real focus foco real
real horizon horizonte real
real image imagen real
real image mode modo de imagen real
real line línea real
real memory memoria real
real mode modo real
real number número real
real part parte real
real power potencia real
real storage almacenamiento real
real storage address dirección de almacenamiento
    real
real time tiempo real
real-time application aplicación en tiempo real
real-time batch processing procesamiento por lotes
    en tiempo real

real-time channel canal en tiempo real
real-time clock reloj de tiempo real
real-time clock module módulo de reloj de tiempo real
real-time computer computadora en tiempo real
real-time control control en tiempo real
real-time control system sistema de control en tiempo real
real-time data datos en tiempo real
real-time debugging program programa de depuración en tiempo real
real-time digital system sistema digital en tiempo real
real-time image generation generación de imagen en tiempo real
real-time information system sistema de información en tiempo real
real-time input entrada en tiempo real
real-time input-output entrada-salida en tiempo real
real-time inquiry consulta en tiempo real
real-time interface interfaz en tiempo real
real-time mode modo en tiempo real
real-time monitor monitor en tiempo real
real-time monitoring monitoreo en tiempo real
real-time operating system sistema operativo en tiempo real
real-time operation operación en tiempo real
real-time output salida en tiempo real
real-time printout impresión en tiempo real
real-time processing procesamiento en tiempo real
real-time relative addressing direccionamiento relativo en tiempo real
real-time simulation simulación en tiempo real
real-time switching conmutación en tiempo real
real-time system sistema en tiempo real
real-time transmission transmisión en tiempo real
real value valor real
real variable variable real
real zero cero real
reallocate reasignar
reallocation reasignación
rear feed alimentación posterior
rear scanning exploración posterior
reassemble reensamblar
reassembly reensamblaje
reassign reasignar
reassignment reasignación
Reaumur scale escala Reaumur
rebalance reequilibrar
rebecca equipment equipo rebecca
rebecca-eureka beacon radiofaro rebecca-eureka
rebecca-eureka equipment equipo rebecca-eureka
rebecca-eureka system sistema rebecca-eureka
rebecca system sistema rebecca
rebroadcast rerradiodifusión, reemisión
rebroadcasting rerradiodifusión, reemisión
recalculate recalcular
recalculation recálculo
recalculation method método de recálculo
recalescence recalescencia
recalescent point punto de recalescencia
recalibrate recalibrar
recalibrated recalibrado
recall signal señal de llamada
receive chain cadena de recepción
receive electromagnet electroimán de recepción
receive frequency frecuencia de recepción
receive loop bucle de recepción
receive-only de sólo recepción
receive wave onda de recepción
received power potencia recibida
receiver receptor
receiver bandwidth ancho de banda de receptor

receiver cabinet caja de receptor
receiver card tarjeta de receptor
receiver-exciter receptor-excitador
receiver gating desbloqueo periódico de receptor
receiver isolation aislamiento de receptor
receiver lockout system sistema de bloqueo de receptor
receiver maximum sensitivity sensibilidad máxima de receptor
receiver-modulator receptor-modulador
receiver muting silenciamiento de receptor
receiver-muting circuit circuito silenciador de receptor
receiver noise ruido de receptor
receiver noise figure cifra de ruido de receptor
receiver output salida de receptor
receiver output circuit circuito de salida de receptor
receiver primaries primarios de receptor
receiver radiation radiación de receptor
receiver register registro de receptor
receiver response respuesta de receptor
receiver response time tiempo de respuesta de receptor
receiver selectivity selectividad de receptor
receiver sensitivity sensibilidad de receptor
receiver set aparato receptor
receiver station estación receptora
receiver test prueba de receptor
receiver-transmitter receptor-transmisor
receiver tuning sintonización de receptor
receiving amplifier amplificador receptor
receiving antenna antena receptora
receiving bandpass filter filtro de paso de banda de recepción
receiving bandwidth ancho de banda de recepción
receiving baseband banda base de recepción
receiving baseband amplifier amplificador de banda base de recepción
receiving center centro receptor
receiving circuit circuito receptor
receiving console consola receptora
receiving distributor distribuidor receptor
receiving electrode electrodo receptor
receiving end extremo receptor
receiving equipment equipo receptor
receiving field campo receptor
receiving filter filtro de recepción
receiving intensity intensidad de recepción
receiving level nivel de recepción
receiving location localización de recepción
receiving loop bucle de recepción
receiving-loop loss pérdida de bucle de recepción
receiving net red receptora
receiving pair par de recepción
receiving perforator perforador de recepción
receiving polarization polarización de recepción
receiving position posición de recepción
receiving relay relé de recepción
receiving reperforator reperforador de recepción
receiving set equipo receptor
receiving station estación receptora
receiving system sistema receptor
receiving teletype teletipo receptor
receiving terminal terminal receptora
receiving unit unidad receptora
receptacle receptáculo
receptacle circuit circuito de receptáculo
receptacle outlet toma de receptáculo
reception recepción
reception area área de recepción
reception level nivel de recepción
reception mode modo de recepción
receptivity receptividad

receptor receptor
recessed luminaire luminaria empotrada
recharge recarga
rechargeable recargable
rechargeable battery batería recargable
rechargeable nickel-cadmium battery batería de níquel-cadmio recargable
reciprocal recíproco
reciprocal action acción recíproca
reciprocal bearing rumbo recíproco
reciprocal device dispositivo recíproco
reciprocal extinction extinción recíproca
reciprocal impedances impedancias recíprocas
reciprocal linear dispersion dispersión lineal recíproca
reciprocal networks redes recíprocas
reciprocal ohm ohm recíproco
reciprocal transducer transductor recíproco
reciprocating grid rejilla oscilante
reciprocating pump bomba oscilante
reciprocation reciprocación
reciprocity reciprocidad
reciprocity coefficient coeficiente de reciprocidad
reciprocity method método de reciprocidad
reciprocity principle principio de reciprocidad
reciprocity relation relación de reciprocidad
reciprocity theorem teorema de reciprocidad
reclosing reconexión, recierre
reclosing contact contacto de reconexión
reclosing device dispositivo de reconexión
reclosing fuse cutout cortacircuito de fusible restablecedor
reclosing interval intervalo de reconexión
reclosing relay relé de reconexión
reclosing switch conmutador de reconexión
reclosing time tiempo de reconexión
reclosure reconexión
reclosure switch conmutador de reconexión
recode recodificar
recoding recodificación
recognition reconocimiento
recognition device dispositivo de reconocimiento
recognizable reconocible
recognizable signal señal reconocible
recoil electron electrón de rechazo
recombination recombinación
recombination center centro de recombinación
recombination coefficient coeficiente de recombinación
recombination current corriente de recombinación
recombination rate velocidad de recombinación
recombination velocity velocidad de recombinación
recompile recompilar
reconfiguration reconfiguración
reconstruction reconstrucción
reconversion reconversión
reconvert reconvertir
record registro, disco
record address dirección de registro
record address file archivo de dirección de registro
record advance avance de registro
record area área de registro
record block bloque de registros
record circuit circuito de registro
record code código de registro
record compensator compensador de discos
record condition condición de registro, condición de grabación
record count cuenta de registros
record description descripción de registro
record format formato de registro
record format descriptor descriptor de formato de registro

record gap espacio entre registros
record head cabeza de registro
record key clave de registro
record length longitud de registro
record level nivel de registro
record-level control control de nivel de registro
record-level indicator indicador de nivel de registro
record manager administrador de registros
record medium medio de registro
record mode modo de registro
record name nombre de registro
record number número de registro
record player tocadiscos
record relay relé de registro
record separator separador de registros
record size tamaño de registro
record tape cinta de registro, cinta de grabación
record type tipo de registro
recordable registrable
recorded disk disco grabado, disco registrado
recorded information información registrada
recorded magnetic tape cinta magnética grabada, cinta magnética registrada
recorded tape cinta grabada, cinta registrada
recorded value valor registrado
recorder registrador, grabador
recorder-controller registrador-controlador
recorder head cabeza de registro
recorder pen pluma de registro
recorder tape cinta de registro
recording registro, grabación
recording accelerometer acelerómetro de registro
recording ammeter amperímetro de registro
recording amplifier amplificador de registro
recording apparatus aparato de registro
recording area área de registro
recording bridge puente de registro
recording channel canal de registro
recording characteristic característica de registro
recording circuit circuito de registro
recording controller controlador de registro
recording curve curva de registro
recording decibel meter decibelímetro de registro
recording demand meter medidor de demanda de registro
recording density densidad de registro
recording digital voltmeter vóltmetro digital de registro
recording disk disco para grabación, disco para registro
recording equipment equipo de registro
recording filter filtro de registro
recording frequencies frecuencias de registro
recording frequency-meter frecuencímetro de registro
recording galvanometer galvanómetro de registro
recording gap entrehierro de registro
recording head cabeza de registro
recording head assembly ensamblaje de cabeza de registro, ensamblaje de cabeza grabadora
recording instrument instrumento de registro
recording ionometer ionómetro de registro
recording lamp lámpara de registro
recording level nivel de registro
recording-level indicator indicador de nivel de registro, indicador de nivel de grabación
recording loss pérdida de grabación, pérdida de registro
recording mechanism mecanismo de registro
recording medium medio de registro
recording meter medidor de registro
recording mode modo de registro
recording needle aguja de grabación, aguja de

registro
recording pen pluma de registro
recording-playback head cabeza de registro-lectura, cabeza de grabación-lectura
recording position posición de registro
recording process proceso de registro, proceso de grabación
recording regulator regulador de registro
recording-reproducing head cabeza de registro-reproducción, cabeza de grabación-reproducción
recording speed velocidad de registro
recording spot punto de registro
recording stylus aguja de grabación, aguja de registro
recording surface superficie de registro
recording tape cinta para grabación, cinta para registro
recording time tiempo de registro
recording trace trazo de registro
recording track pista de registro
recording trunk enlace de anotaciones
recording unit unidad de grabación, unidad de registro
recording voltmeter vóltmetro de registro
recording wattmeter wáttmetro de registro
recoverable abend final anormal recuperable
recoverable error error recuperable
recovery recuperación
recovery current corriente de recuperación
recovery function función de recuperación
recovery phase fase de recuperación
recovery point punto de recuperación
recovery procedure procedimiento de recuperación
recovery program programa de recuperación
recovery rate velocidad de recuperación
recovery ratio razón de recuperación
recovery routine rutina de recuperación
recovery time tiempo de recuperación
recovery-time constant constante de tiempo de recuperación
recovery value valor de recuperación
recovery voltage tensión de recuperación
rectangular array red rectangular
rectangular cathode cátodo rectangular
rectangular cavity cavidad rectangular
rectangular coordinates coordenadas rectangulares
rectangular horn bocina rectangular
rectangular-horn antenna antena de bocina rectangular
rectangular-horn radiator radiador de bocina rectangular
rectangular impulse impulso rectangular
rectangular loop bucle rectangular
rectangular plate placa rectangular
rectangular pulse impulso rectangular
rectangular-pulse generation generación de impulsos rectangulares
rectangular-pulse modulation modulación de impulsos rectangulares
rectangular scan exploración rectangular
rectangular signal señal rectangular
rectangular tube tubo rectangular
rectangular voltage tensión rectangular
rectangular wave onda rectangular
rectangular-wave generator generador de onda rectangular
rectangular waveguide guíaondas rectangular
rectangular wire hilo rectangular
rectifiable rectificable
rectification rectificación
rectification efficiency eficiencia de rectificación
rectification factor factor de rectificación

rectification failure falla de rectificación
rectification ratio razón de rectificación
rectified rectificado
rectified alternating current corriente alterna rectificada
rectified output salida rectificada
rectified signal señal rectificada
rectified tension tensión rectificada
rectified value valor rectificado
rectified voltage tensión rectificada
rectifier rectificador
rectifier-amplifier voltmeter vóltmetro rectificador-amplificador
rectifier anode ánodo rectificador
rectifier bridge puente rectificador
rectifier cathode cátodo rectificador
rectifier cell célula rectificadora
rectifier delay bias polarización de retardo de rectificador
rectifier diode diodo rectificador
rectifier disk disco rectificador
rectifier element elemento rectificador
rectifier filter filtro rectificador
rectifier forward current corriente directa de rectificador
rectifier instrument instrumento con rectificador
rectifier junction unión rectificadora
rectifier noise ruido de rectificador
rectifier photocell fotocélula rectificadora
rectifier probe sonda rectificadora
rectifier relay relé de rectificador
rectifier stack pila de rectificadores
rectifier substation subestación de rectificadores
rectifier transformer transformador de rectificador
rectifier tube tubo rectificador
rectifier unit unidad rectificadora
rectify rectificar
rectifying rectificador
rectifying action acción rectificadora
rectifying barrier barrera rectificadora
rectifying circuit circuito rectificador
rectifying detector detector rectificador
rectifying element elemento rectificador
rectifying junction unión rectificadora
rectifying tube tubo rectificador
rectifying valve válvula rectificadora
rectigon rectigón
rectilineal rectilíneo
rectilineal compliance docilidad rectilínea
rectilinear rectilíneo
rectilinear scan exploración rectilínea
rectilinear scanning exploración rectilínea
recurrence recurrencia
recurrence frequency frecuencia de recurrencia
recurrence period periodo de recurrencia
recurrence rate frecuencia de recurrencia
recurrent recurrente
recurrent code código recurrente
recurrent network red recurrente
recurrent period periodo recurrente
recurrent phenomenon fenómeno recurrente
recurrent sweep barrido recurrente
recursion recursión, recurrencia
recursive recursivo, recurrente
recursive algorithm algoritmo recursivo
recursive function función recursiva
recursive procedure procedimiento recursivo
recursive process proceso recursivo
recursive routine rutina recursiva
recursive subroutine subrutina recursiva
recycle timer temporizador de reciclado
recycling reciclado
red amplifier amplificador de rojo

red component componente rojo
red filter filtro de rojo
red-green-blue rojo-verde-azul
red gun cañón de rojo
red signal señal de rojo
red video voltage tensión del video de rojo
rediffusion redifusión
rediffusion channel canal de redifusión
redirect redirigir
redistribution redistribución
redox cell célula de reducción-oxidación
redox potential potencial de reducción-oxidación
redox system sistema de reducción-oxidación
reduce reducir
reduced band banda reducida
reduced carrier portadora reducida
reduced power potencia reducida
reduced speed velocidad reducida
reduced voltage tensión reducida
reduced-voltage switch conmutador de tensión
    reducida
reducer reductor
reducible reducible
reduction reducción
redundancy redundancia
redundancy check comprobación de redundancia
redundancy-check bit bit de comprobación de
    redundancia
redundancy-check character carácter de
    comprobación de redundancia
redundancy factor factor de redundancia
redundancy parity check comprobación de paridad
    por redundancia
redundant redundante
redundant bit bit redundante
redundant character carácter redundante
redundant check comprobación redundante
redundant circuit circuito redundante
redundant code código redundante
redundant digit dígito redundante
redundant equipment equipo redundante
redundant processor procesador redundante
redundant recording registro redundante
redundant system sistema redundante
reed lengüeta, laminilla
reed relay relé de laminillas
reed switch conmutador de laminilla
reed-type switch conmutador de tipo de laminilla
reel carrete
reel-to-reel tape deck mecanismo de arrastre de
    carretes
reel-to-reel tape recorder registrador de carretes,
    reproductor de carretes
reenter reentrar
reenterable reentrable
reenterable program programa reentrable
reenterable routine rutina reentrable
reentrant reentrante
reentrant beam haz reentrante
reentrant cavity cavidad reentrante
reentrant circuit circuito reentrante
reentrant code código reentrante
reentrant oscillator oscilador reentrante
reentrant program programa reentrante
reentrant routine rutina reentrante
reentrant subroutine subrutina reentrante
reentrant winding arrollamiento cerrado
reentry reentrada
reentry point punto de reentrada
reference referencia
reference accuracy exactitud de referencia
reference address dirección de referencia
reference amplifier amplificador de referencia

reference angle ángulo de referencia
reference antenna antena de referencia
reference apparatus aparato de referencia
reference artificial ear oído artificial de referencia
reference axis eje de referencia
reference baseline línea base de referencia
reference bias current corriente de polarización de
    referencia
reference bit bit de referencia
reference black level nivel del negro de referencia
reference block bloque de referencia
reference card tarjeta de referencia
reference cavity cavidad de referencia
reference center centro de referencia
reference channel canal de referencia
reference chromaticity cromaticidad de referencia
reference circuit circuito de referencia
reference color color de referencia
reference coupling acoplamiento de referencia
reference current corriente de referencia
reference current range intervalo de corriente de
    referencia
reference database base de datos de referencia
reference deflection deflexión de referencia
reference designation designación de referencia
reference diode diodo de referencia
reference dipole dipolo de referencia
reference direction dirección de referencia
reference distance distancia de referencia
reference edge borde de referencia
reference electrode electrodo de referencia
reference equivalent equivalente de referencia
reference excursion excursión de referencia
reference field campo de referencia
reference format formato de referencia
reference frequency frecuencia de referencia
reference frequency-meter frecuencímetro de
    referencia
reference generator generador de referencia
reference grid rejilla de referencia
reference indicator indicador de referencia
reference input entrada de referencia
reference input signal señal de entrada de referencia
reference instruction instrucción de referencia
reference instrument instrumento de referencia
reference intensity intensidad de referencia
reference lamp lámpara de referencia
reference language lenguaje de referencia
reference level nivel de referencia
reference line línea de referencia
reference mark marca de referencia
reference modulation modulación de referencia
reference noise ruido de referencia
reference number número de referencia
reference operating conditions condiciones de
    operación de referencia
reference oscillator oscilador de referencia
reference parameter parámetro de referencia
reference pilot piloto de referencia
reference plane plano de referencia
reference point punto de referencia
reference power supply fuente de alimentación de
    referencia
reference pressure presión de referencia
reference printer impresora de referencia
reference pulse pair par de impulsos de referencia
reference quantity cantidad de referencia
reference radius radio de referencia
reference recording registro de referencia
reference sensitivity sensibilidad de referencia
reference signal señal de referencia
reference source fuente de referencia
reference standard patrón de referencia

reference surface superficie de referencia
reference system sistema de referencia
reference tape cinta de referencia
reference temperature temperatura de referencia
reference time tiempo de referencia
reference tone tono de referencia
reference tube tubo de referencia
reference value valor de referencia
reference variable variable de referencia
reference voltage tensión de referencia
reference-voltage circuit circuito de tensión de
    referencia
reference-voltage source fuente de tensión de
    referencia
reference volume volumen de referencia
reference waveguide guíaondas de referencia
reference white level nivel del blanco de referencia
reflectance reflectancia
reflectance factor factor de reflectancia
reflectance spectrophotometer espectrofotómetro de
    reflectancia
reflected reflejado
reflected beam haz reflejado
reflected binary code código binario reflejado
reflected code código reflejado
reflected current corriente reflejada
reflected electromagnetic field campo
    electromagnético reflejado
reflected electron electrón reflejado
reflected energy energía reflejada
reflected flow flujo reflejado
reflected glare resplandor reflejado
reflected heat calor reflejado
reflected impedance impedancia reflejada
reflected light luz reflejada
reflected power potencia reflejada
reflected-power meter medidor de potencia reflejada
reflected ray rayo reflejado
reflected resistance resistencia reflejada
reflected signal señal reflejada
reflected wave onda reflejada
reflecting reflector
reflecting electrode electrodo reflector
reflecting galvanometer galvanómetro reflector
reflecting goniometer goniómetro reflector
reflecting grating rejilla reflectora
reflecting screen pantalla reflectora
reflecting telescope telescopio reflector
reflection reflexión
reflection altimeter altímetro de reflexión
reflection coefficient coeficiente de reflexión
reflection-coefficient meter medidor de coeficiente
    de reflexión
reflection color tube tubo de imagen en color de
    reflexión
reflection effect efecto de reflexión
reflection error error de reflexión
reflection factor factor de reflexión
reflection gain ganancia por reflexión
reflection interval intervalo de reflexión
reflection law ley de reflexión
reflection loss pérdida por reflexión
reflection meter reflectómetro
reflection mode modo de reflexión
reflection modulation modulación de reflexión
reflection plane plano de reflexión
reflectionless sin reflexión
reflectionless transmission line línea de transmisión
    sin reflexión
reflectionless waveguide guíaondas sin reflexión
reflective code código reflejante
reflective element elemento reflector
reflective surface superficie reflectora

reflectivity reflectividad
reflectogram reflectograma
reflectograph reflectógrafo
reflectometer reflectómetro
reflector reflector
reflector antenna antena con reflector
reflector curtain cortina reflectora
reflector electrode electrodo reflector
reflector element elemento reflector
reflector lamp lámpara con reflector
reflector satellite satélite reflector
reflector screen pantalla reflectora
reflector space espacio de reflexión
reflector voltage tensión de reflector
reflex amplification amplificación refleja
reflex amplifier amplificador reflejo
reflex baffle bafle reflejo
reflex bunching agrupamiento por reflejo
reflex cabinet caja refleja
reflex circuit circuito reflejo
reflex coefficient coeficiente reflejo
reflex enclosure caja refleja
reflex klystron klistrón reflejo
reflex loading carga refleja
reflex loudspeaker altavoz reflejo
reflex oscillator oscilador reflejo
reflex transconductance transconductancia refleja
reforming reformación, reformado
refractance refractancia
refracted refractado
refracted light luz refractada
refracted ray rayo refractado
refracted-ray method método de rayos refractados
refracted wave onda refractada
refraction refracción
refraction error error por refracción
refraction index índice de refracción
refraction law ley de refracción
refraction loss pérdida por refracción
refractive coefficient coeficiente de refracción
refractive index índice de refracción
refractive index contrast contraste de índice de
    refracción
refractive index gradient gradiente de índice de
    refracción
refractive index profile perfil de índice de refracción
refractive modulus módulo de refracción
refractive power poder de refracción
refractivity refractividad
refractivity profile perfil de refractividad
refractometer refractómetro
refractometric refractométrico
refractor refractor
refractory refractario
refresh refrescar
refresh buffer almacenamiento intermedio para
    refrescar
refresh circuitry circuitería de refresco
refresh cycle ciclo de refresco
refresh display device dispositivo de visualización
    con refresco
refresh memory memoria para refrescar
refresh rate frecuencia de refresco
refrigerant refrigerante
regenerate regenerar
regeneration regeneración
regeneration circuit circuito de regeneración
regeneration control control de regeneración
regeneration of electrolyte regeneración de
    electrólito
regeneration period periodo de regeneración
regenerative regenerativo
regenerative amplification amplificación

regenerativa
regenerative amplifier amplificador regenerativo
regenerative braking frenado regenerativo
regenerative circuit circuito regenerativo
regenerative coupling acoplamiento regenerativo
regenerative detector detector regenerativo
regenerative divider divisor regenerativo
regenerative feedback retroalimentación
    regenerativa
regenerative memory memoria regenerativa
regenerative modulator modulador regenerativo
regenerative reading lectura regenerativa
regenerative receiver receptor regenerativo
regenerative reception recepción regenerativa
regenerative repeater repetidor regenerativo
regenerative storage almacenamiento regenerativo
regenerator regenerador
region of anode región anódica
region of cathode región catódica
region of operation región de operación
region partitioning división de regiones
regional regional
regional address dirección regional
regional center centro regional
regional channel canal regional
regional interconnection interconexión regional
register registro, contador
register address dirección de registro
register addressing direccionamiento de registros
register arrangement arreglo de registros
register capacity capacidad de registro
register constant constante de contador
register control control de registro
register designator designador de registro
register display unit unidad de visualización de
    registros
register finder buscador de registro
register length longitud de registro
register mark marca de registro
register ratio razón de contador
register reading lectura de registro
register select selección de registro
register selector selector de registro
register transfer transferencia de registro
register-transfer language lenguaje de transferencia
    de registros
register-transfer module módulo de transferencia de
    registros
registration registro
registration error error de registro
registration of colors registro de colores
regression testing prueba de regresión
regular binary binario regular
regular channel canal regular
regular command mando regular
regular reflectance reflectancia regular
regular reflection reflexión regular
regular reflection factor factor de reflexión regular
regular refraction refracción regular
regular station estación regular
regular transmission transmisión regular
regular transmittance transmitancia regular
regulated circuit circuito regulado
regulated frequency frecuencia regulada
regulated power supply fuente de alimentación
    regulada
regulated station estación regulada
regulated supply fuente regulada
regulated voltage tensión regulada
regulating regulador
regulating apparatus aparato regulador
regulating circuit circuito regulador
regulating coil bobina reguladora

regulating device dispositivo regulador
regulating element elemento regulador
regulating grid rejilla reguladora
regulating inductor inductor regulador
regulating pilot piloto regulador
regulating relay relé regulador
regulating servo servo regulador
regulating system sistema regulador
regulating transformer transformador regulador
regulating unit unidad reguladora
regulating winding arrollamiento regulador
regulation regulación
regulation curve curva de regulación
regulator regulador
regulator circuit circuito regulador
regulator diode diodo regulador
regulator triode triodo regulador
regulator tube tubo regulador
reignition reignición, reencendido
Reinartz crystal oscillator oscilador de cristal de
    Reinartz
reinforced plastic plástico reforzado
reinitialization reinicialización, reinicio
reinitialize reinicializar, reiniciar
reinitiate reiniciar
reinsertion reinserción
reinsertion current corriente de reinserción
reinsertion voltage tensión de reinserción
reject character carácter de rechazo
rejection rechazo
rejection band banda de rechazo
rejection circuit circuito de rechazo
rejection filter filtro de rechazo
rejection notch muesca de rechazo
rejector rechazador
rejector circuit circuito rechazador
rel rel
relation character carácter de relación
relation condition condición de relación
relational relacional
relational checking comprobación relacional
relational database base de datos relacional
relational database management administración de
    base de datos relacional
relational expression expresión relacional
relational file archivo relacional
relational language lenguaje relacional
relational model modelo relacional
relational structure estructura relacional
relative relativo
relative accuracy exactitud relativa
relative address dirección relativa
relative addressing direccionamiento relativo
relative addressing mode modo de direccionamiento
    relativo
relative aperture abertura relativa
relative attenuation atenuación relativa
relative bearing rumbo relativo
relative block number número de bloque relativo
relative byte address dirección de byte relativo
relative calibration calibración relativa
relative cell reference referencia de celda relativa
relative code código relativo
relative coding codificación relativa
relative command mando relativo
relative conductance conductancia relativa
relative contrast sensitivity sensibilidad de contraste
    relativa
relative control control relativo
relative current level nivel de corriente relativo
relative damping amortiguamiento relativo
relative data datos relativos
relative data set conjunto de datos relativos

relative delay retardo relativo
relative dielectric constant constante dieléctrica relativa
relative directive gain ganancia directiva relativa
relative drift deriva relativa
relative effectiveness efectividad relativa
relative efficiency eficiencia relativa
relative energy energía relativa
relative equivalent equivalente relativo
relative error error relativo
relative file archivo relativo
relative frequency frecuencia relativa
relative gain ganancia relativa
relative harmonic content contenido armónico relativo
relative humidity humedad relativa
relative instruction instrucción relativa
relative intensity intensidad relativa
relative key clave relativa
relative level nivel relativo
relative luminance luminancia relativa
relative luminosity luminosidad relativa
relative luminosity factor factor de luminosidad relativa
relative luminous efficiency eficiencia luminosa relativa
relative magnitude magnitud relativa
relative order orden relativo
relative organization organización relativa
relative partial gain ganancia parcial relativa
relative pattern patrón relativo
relative permeability permeabilidad relativa
relative permittivity permitividad relativa
relative positioning posicionamiento relativo
relative power potencia relativa
relative power gain ganancia de potencia relativa
relative power level nivel de potencia relativa
relative pressure presión relativa
relative record registro relativo
relative-record number número de registro relativo
relative redundancy redundancia relativa
relative refractive index índice de refracción relativa
relative regulation regulación relativa
relative response respuesta relativa
relative sensibility sensibilidad relativa
relative sequential access method método de acceso secuencial relativo
relative spectral energy distribution distribución de energía espectral relativa
relative speed velocidad relativa
relative speed variation variación de velocidad relativa
relative temperature index índice de temperatura relativa
relative track pista relativa
relative transmission level nivel de transmisión relativo
relative uncertainty incertidumbre relativa
relative value valor relativo
relative velocity velocidad relativa
relative voltage drop caída de tensión relativa
relative voltage level nivel de tensión relativa
relative voltage rise subida de tensión relativa
relative voltage variation variación de tensión relativa
relaxation relajación
relaxation circuit circuito de relajación
relaxation cycle ciclo de relajación
relaxation frequency frecuencia de relajación
relaxation inverter inversor de relajación
relaxation oscillator oscilador de relajación
relaxation time tiempo de relajación
relay relé, relevador

relay actuation time tiempo de actuación de relé
relay actuator actuador de relé
relay adjustment ajuste de relé
relay amplifier amplificador relé
relay antenna antena de relé
relay armature armadura de relé
relay-armature ratio razón de armadura de relé
relay-armature travel recorrido de armadura de relé
relay bias polarización de relé
relay box caja de relés
relay calculator calculadora de relés
relay chain cadena de relés
relay circuit circuito de relé
relay coil bobina de relé
relay-coil resistance resistencia de bobina de relé
relay computer computadora de relés
relay contact contacto de relé
relay-contact bounce rebote de contacto de relé
relay-contact chatter vibración de contacto de relé
relay-contact combination combinación de contactos de relé
relay-contact gap entrehierro de contactos de relé
relay-contact separation separación de contactos de relé
relay-controlled controlado por relé
relay core núcleo de relé
relay critical voltage tensión crítica de relé
relay driver excitador de relé
relay duty cycle ciclo de trabajo de relé
relay electric bias polarización eléctrica de relé
relay element elemento de relé
relay fritting fritado de relé
relay group grupo de relés
relay heater calentador de relé
relay hum zumbido de relé
relay impedance impedancia de relé
relay leakage flux flujo de fuga de relé
relay magnet electroimán de relé
relay magnetic bias polarización magnética de relé
relay mechanical bias polarización mecánica de relé
relay mounting montaje de relé
relay-operated operado por relé
relay operation operación de relé
relay panel panel de relés
relay rack bastidor de relés
relay rating clasificación de relé
relay recovery time tiempo de recuperación de relé
relay satellite satélite relé
relay sensitivity sensibilidad de relé
relay set conjunto de relés
relay station estación relé
relay switch conmutador relé
relay system sistema de relés
relay transmitter transmisor relé
relay tube tubo relé
release liberación, desconexión, desenganche
release coil bobina de liberación
release current corriente de liberación
release delay retardo de liberación
release factor factor de liberación
release force fuerza de liberación
release key tecla de liberación
release magnet electroimán de liberación
release mechanism mecanismo de liberación
release signal señal de liberación
release time tiempo de liberación, tiempo de emisión
releasing current corriente de liberación
releasing magnet electroimán de liberación
releasing mechanism mecanismo de liberación
releasing position posición de liberación
reliability confiabilidad
reliability allocation asignación de confiabilidad
reliability assessment evaluación de confiabilidad

reliability control control de confiabilidad
reliability data datos de confiabilidad
reliability evaluation evaluación de confiabilidad
reliability factor factor de confiabilidad
reliability growth crecimiento de confiabilidad
reliability index índice de confiabilidad
reliability model modelo de confiabilidad
reliability test prueba de confiabilidad
relieving anode ánodo de descarga
reload recargar
reloadable recargable
relocatable reubicable
relocatable address dirección reubicable
relocatable area área reubicable
relocatable assembler ensamblador reubicable
relocatable binary format formato binario
    reubicable
relocatable code código reubicable
relocatable coding codificación reubicable
relocatable expression expresión reubicable
relocatable machine code código de máquina
    reubicable
relocatable module módulo reubicable
relocatable program programa reubicable
relocatable program loader cargador de programa
    reubicable
relocatable subroutine subrutina reubicable
relocate reubicar
relocating assembler ensamblador reubicador
relocating loader cargador reubicador
relocation relocalización, reubicación
relocation constant constante de relocalización
relocation factor factor de relocalización
reluctance reluctancia
reluctance factor factor de reluctancia
reluctance microphone micrófono de reluctancia
    variable
reluctance motor motor de reluctancia
reluctance tuning sintonización por reluctancia
reluctive reluctivo
reluctive transduction transducción reluctiva
reluctivity reluctividad
reluctometer reluctómetro
remagnetizer remagnetizador
remanence remanencia
remanent remanente
remanent charge carga remanente
remanent flux density densidad de flujo remanente
remanent induction inducción remanente
remanent magnetization magnetización remanente
remanent polarization polarización remanente
remodulation remodulación
remodulator remodulador
remote remoto
remote access acceso remoto
remote-access data processing procesamiento de
    datos de acceso remoto
remote alarm alarma remota
remote alarm indication indicación de alarma
    remota
remote amplifier amplificador remoto
remote assistance asistencia remota
remote backup respaldo remoto
remote calculator calculadora remota
remote compass compás remoto
remote computer computadora remota
remote computer system sistema de computadora
    remoto
remote computer system language lenguaje de
    sistema de computadora remoto
remote computing computación remota
remote-computing system sistema de computación
    remota

remote-computing system language lenguaje de
    sistema de computación remota
remote concentrator concentrador remoto
remote console consola remota
remote control control remoto, telemando
remote-control circuit circuito de control remoto
remote-control code código de control remoto
remote-control equipment equipo de control remoto
remote-control exchange central de control remoto
remote-control language lenguaje de control remoto
remote-control manipulator manipulador de control
    remoto
remote-control monitoring system sistema de
    monitoreo de control remoto
remote-control operation operación por control
    remoto
remote-control receiver receptor de control remoto
remote-control signal señal de control remoto
remote-control transmitter transmisor de control
    remoto
remote-control unit unidad de control remoto
remote-controlled controlado remotamente,
    telemandado
remote-controlled station estación controlada
    remotamente
remote cutoff corte remoto
remote-cutoff grid rejilla de corte remoto
remote-cutoff tube tubo de corte remoto
remote data processing procesamiento de datos
    remoto
remote data station estación de datos remota
remote data terminal terminal de datos remota
remote debugging depuración remota
remote detection detección remota
remote device dispositivo remoto
remote display visualización remota
remote-display unit unidad de visualización remota
remote error sensing detección de error remota
remote gain control control de ganancia remoto
remote indication indicación remota
remote inquiry consulta remota
remote job processing procesamiento de trabajo
    remoto
remote key clave remota
remote line línea remota
remote magnetic sensor sensor magnético remoto
remote maintenance mantenimiento remoto
remote manipulation manipulación remota
remote manipulator manipulador remoto
remote meter medidor remoto
remote-metered medido remotamente, telemedido
remote monitor monitor remoto
remote multiplexing system sistema de multiplexión
    remoto
remote network control program programa de
    control de red remoto
remote operation operación remota
remote peripheral periférico remoto
remote processing procesamiento remoto
remote processor procesador remoto
remote program loader cargador de programa
    remoto
remote reading lectura remota
remote-reading indicator indicador de lectura
    remota
remote-reading thermometer termómetro de lectura
    remota
remote readout lectura remota
remote real-time terminal terminal en tiempo real
    remota
remote receiver receptor remoto
remote-receiver station estación receptora remota
remote recording registro remoto

remote sensing detección remota
remote signaling señalización remota
remote station estación remota
remote switching conmutación remota
remote terminal terminal remota
remote-terminal access method método de acceso de terminal remota
remote test prueba remota
remote transmitter station estación transmisora remota
remote transmitting station estación transmisora remota
remote tuning sintonización remota
remote work station estación de trabajo remota
remotely remotamente
remotely controlled controlado remotamente
remotely-controlled station estación controlada remotamente
remotely operated operado remotamente
remotely-operated antenna antena operada remotamente
removable removible
removable disk disco removible
removable-disk storage unit unidad de almacenamiento de disco removible
removable drive unidad removible
removable element elemento removible
removable storage media medios de almacenamiento removibles
remove remover
renewable renovable
renewable fuse fusible renovable
renewable part parte renovable
reorganization reorganización
rep rep
repair reparación
repair time tiempo de reparación
repair urgency urgencia de reparación
repairability reparabilidad
repeat character carácter de repetición
repeat function función de repetición
repeat key tecla de repetición
repeatability repetibilidad
repeatability error error de repetibilidad
repeatable repetible
repeated repetido
repeated call llamada repetida
repeated signal señal repetida
repeatedly repetidamente
repeater repetidor
repeater alarm alarma de repetidor
repeater-alarm circuit circuito de alarma de repetidor
repeater circuit circuito de repetidor
repeater distribution frame distribuidor de repetidores
repeater gain ganancia de repetidor
repeater insertion inserción de repetidor
repeater lamp lámpara de repetidor
repeater rack bastidor de repetidores
repeater station estación repetidora
repeater transmitter transmisor repetidor
repeater tube tubo repetidor
repeater unit unidad repetidora
repeating coil bobina repetidora
repeating-coil rack bastidor de bobinas repetidoras
repeating decimal decimal repetitivo
repeating decimal number número decimal repetitivo
repeating field campo repetitivo
repeating flash tube lámpara de destellos repetitiva
repeating group grupo repetitivo
repeating relay relé repetidor

repeating selector selector repetidor
repeating station estación repetidora
repeller repeledor, reflector
repelling force fuerza repulsiva
reperforator reperforador
repetition repetición
repetition equivalent equivalente de repetición
repetition factor factor de repetición
repetition frequency frecuencia de repetición
repetition instruction instrucción de repetición
repetition rate frecuencia de repetición
repetitive repetitivo
repetitive addition suma repetitiva
repetitive addressing direccionamiento repetitivo
repetitive analog computer computadora analógica repetitiva
repetitive check comprobación repetitiva
repetitive checking comprobación repetitiva
repetitive differential analyzer analizador diferencial repetitivo
repetitive error error repetitivo
repetitive event evento repetitivo
repetitive operation operación repetitiva
repetitive peak pico repetitivo
repetitive peak inverse voltage tensión inversa de pico repetitiva
repetitive phenomenon fenómeno repetitivo
repetitive sweep barrido repetitivo
replacement reemplazo
replacement cartridge cartucho de reemplazo
replacement part parte de reemplazo
replacement tube tubo de reemplazo
replacement unit unidad de reemplazo
replay respuesta, reproducción
replay amplifier amplificador de reproducción
replay chain cadena de reproducción
replay channel canal de reproducción
replier respondedor
reply frequency frecuencia de respuesta
reply phase fase de respuesta
reply pulse impulso de respuesta
report informe, reporte
report file archivo de informe
report generation generación de informe
report generator generador de informes
representation system sistema de representación
representational model modelo representacional
representative representativo
representative calculating time tiempo de cálculo representativo
reprocessing reprocesamiento
reproduce reproducir
reproduce head cabeza de reproducción
reproducer reproductor
reproducibility reproducibilidad
reproducible reproducible
reproducible test prueba reproducible
reproducible value valor reproducible
reproducing reproducción
reproducing apparatus aparato de reproducción
reproducing characteristic característica de reproducción
reproducing head cabeza de reproducción
reproducing stylus aguja de reproducción
reproduction reproducción
reproduction channel canal de reproducción
reproduction code código de reproducción
reproduction loss pérdida de reproducción
reproduction ratio razón de reproducción
reproduction speed velocidad de reproducción
reprogram reprogramar
reprogramming reprogramación
repulsion repulsión

repulsion-induction motor motor de repulsión-inducción
repulsion motor motor de repulsión
repulsion-start induction motor motor de inducción de arranque por repulsión
repulsion-start motor motor de arranque por repulsión
repulsive repulsivo
repulsiveness repulsividad
repunch reperforar
repunching reperforación
request solicitud
request phase fase de solicitud
request unit unidad de solicitud
requester solicitante
requesting solicitante
requesting agent agente solicitante
requesting program programa solicitante
requesting terminal terminal solicitante
required input entrada requerida
required output salida requerida
required reserve reserva requerida
required space espacio requerido
required time tiempo requerido
requirement requisito
requirements analysis análisis de requisitos
requirements inspection inspección de requisitos
requirements phase fase de requisitos
requirements review revisión de requisitos
requirements specification especificación de requisitos
requirements verification verificación de requisitos
reradiate rerradiar
reradiation rerradiación
rerecord rerregistrar, regrabar
rerecording rerregistro, regrabación
rerecording amplifier amplificador de rerregistro, amplificador de regrabación
rerecording compensator compensador de rerregistro, compensador de regrabación
rerecording system sistema de rerregistro, sistema de regrabación
reroute reencaminar
rerouting reencaminamiento
rerun repetición, reejecución, reanudación
rerun point punto de repetición, punto de reejecución, punto de reanudación
rerun routine rutina de repetición, rutina de reejecución, rutina de reanudación
rescan reexplorar
reserve reserva
reserve reserva
reserve bars barras de reserva
reserve battery batería de reserva
reserve cell pila de reserva
reserve circuit circuito de reserva
reserve equipment equipo de reserva
reserve group grupo de reserva
reserve installation instalación de reserva
reserve pair par de reserva
reserve protection protección de reserva
reserve section sección de reserva
reserve set aparato de reserva
reserved reservado
reserved area área reservada
reserved unit unidad reservada
reserved word palabra reservada
reset reposición, puesta a cero, reconectador, restablecedor
reset action acción de reposición, acción de corrección proporcional
reset button botón de reposición
reset check comprobación de reposición

reset circuit circuito de reposición
reset command mando de reposición
reset contact contacto de reposición
reset contactor contactor de reposición
reset control control de reposición
reset current corriente de reposición
reset device dispositivo de reposición
reset interval intervalo de reposición
reset key tecla de reposición
reset lever palanca de reposición
reset position posición de reposición
reset pulse impulso de reposición
reset rate tasa de reposición
reset switch conmutador de reposición
reset terminal terminal de reposición
reset time tiempo de reposición
reset timer temporizador de reposición
reset voltage tensión de reposición
resettability reposicionabilidad
resettable reposicionable
resettable register registro reposicionable
resetting reposición, puesta a cero, reconexión, restablecimiento
resetting device dispositivo de reposición
resetting half-cycle semiciclo de reposición
resetting interval intervalo de reposición
resetting pulse impulso de reposición
resetting time tiempo de reposición, tiempo de retroceso
resident residente
resident access method método de acceso residente
resident compiler compilador residente
resident control program programa de control residente
resident file archivo residente
resident load module módulo de carga residente
resident loader cargador residente
resident macroassembler macroensamblador residente
resident module módulo residente
resident program programa residente
resident program storage almacenamiento de programa residente
resident reader lectora residente
resident routine rutina residente
resident storage almacenamiento residente
resident system sistema residente
residual residual
residual activity actividad residual
residual amplitude modulation modulación de amplitud residual
residual band banda residual
residual bandwidth ancho de banda residual
residual capacitance capacitancia residual
residual charge carga residual
residual control control residual
residual-control program programa de control residual
residual current corriente residual
residual deflection deflexión residual
residual deviation desviación residual
residual discharge descarga residual
residual dispersion dispersión residual
residual energy energía residual
residual error error residual
residual-error check comprobación de error residual
residual-error rate tasa de errores residuales
residual-error ratio razón de errores residuales
residual excitation excitación residual
residual field campo residual
residual flux flujo residual
residual flux density densidad de flujo residual
residual frequency instability inestabilidad de

frecuencia residual
residual frequency modulation modulación de frecuencia residual
residual gas gas residual
residual hum zumbido residual
residual impedance impedancia residual
residual impulse impulso residual
residual inductance inductancia residual
residual induction inducción residual
residual ion ion residual
residual ionization ionización residual
residual loss pérdida residual
residual magnetic induction inducción magnética residual
residual magnetism magnetismo residual
residual magnetization magnetización residual
residual modulation modulación residual
residual noise ruido residual
residual plate placa antirremanente
residual radioactivity radiactividad residual
residual resistance resistencia residual
residual resistivity resistividad residual
residual response respuesta residual
residual ripple ondulación residual
residual spacing error error de espaciado residual
residual stop tope antirremanente
residual stress estrés residual
residual stud placa antirremanente
residual unbalance desequilibrio residual
residual value valor residual
residual vapor vapor residual
residual voltage tensión residual
residue check comprobación de residuo
resilience resiliencia
resin resina
resin-encapsulated encapsulado en resina
resin-encapsulated circuit circuito encapsulado en resina
resin-encapsulated component componente encapsulado en resina
resinous resinoso
resistance resistencia
resistance adapter adaptador de resistencia
resistance attenuator atenuador de resistencia
resistance balance equilibrio de resistencias
resistance box caja de resistencias
resistance braking frenado por resistencia
resistance brazing soldadura fuerte por resistencia
resistance bridge puente de resistencias
resistance-capacitance de resistencia-capacitancia
resistance-capacitance arrangement arreglo de resistencia-capacitancia
resistance-capacitance bridge puente de resistencia-capacitancia
resistance-capacitance characteristic característica de resistencia-capacitancia
resistance-capacitance circuit circuito de resistencia-capacitancia
resistance-capacitance constant constante de resistencia-capacitancia
resistance-capacitance-coupled amplifier amplificador acoplado por resistencia-capacitancia
resistance-capacitance coupling acoplamiento por resistencia-capacitancia
resistance-capacitance differentiator diferenciador de resistencia-capacitancia
resistance-capacitance divider divisor de resistencia-capacitancia
resistance-capacitance filter filtro de resistencia-capacitancia
resistance-capacitance generator generador de resistencia-capacitancia
resistance-capacitance-inductance

resistencia-capacitancia-inductancia
resistance-capacitance network red de resistencia-capacitancia
resistance-capacitance oscillator oscilador de resistencia-capacitancia
resistance-capacitance phase-shifter desfasador de resistencia-capacitancia
resistance-capacitance time constant constante de tiempo de resistencia-capacitancia
resistance-capacitance tuning sintonización por resistencia-capacitancia
resistance coefficient coeficiente de resistencia
resistance coil bobina de resistencia
resistance component componente de resistencia
resistance contact contacto de resistencia
resistance control control por resistencia
resistance-coupled acoplado por resistencia
resistance-coupled amplifier amplificador acoplado por resistencia
resistance coupling acoplamiento por resistencia
resistance drop caída por resistencia
resistance element elemento de resistencia
resistance frame cuadro de resistencia
resistance furnace horno de resistencia
resistance-grounded puesto a tierra por resistencia
resistance heat calor por resistencia
resistance heater calentador de resistencia
resistance heating calentamiento por resistencia
resistance-inductance de resistencia-inductancia
resistance-inductance bridge puente de resistencia-inductancia
resistance-inductance circuit circuito de resistencia-inductancia
resistance-inductance phase-shifter desfasador de resistencia-inductancia
resistance instrument instrumento de resistencia
resistance lamp lámpara de resistencia
resistance loss pérdida por resistencia
resistance magnetometer magnetómetro de resistencia
resistance material material con resistencia
resistance measurement medida de resistencia
resistance metal metal con resistencia
resistance noise ruido de resistencia
resistance pad atenuador de resistencias
resistance protection protección de resistencia
resistance pyrometer pirómetro de resistencia
resistance range gama de resistencias
resistance ratio razón de resistencias
resistance reduction reducción de resistencia
resistance-reduction factor factor de reducción de resistencia
resistance regulation regulación por resistencia
resistance relay relé de resistencia
resistance shunt derivación de resistencia
resistance soldering soldadura por resistencia
resistance-stabilized estabilizado por resistencia
resistance-stabilized oscillator oscilador estabilizado por resistencia
resistance standard patrón de resistencia
resistance-start motor motor de arranque con resistencia
resistance strain gage extensímetro de resistencia
resistance strip regleta de resistencia
resistance substitution box caja de sustitución de resistencias
resistance temperature detector detector de temperatura por resistencia
resistance termination terminación de resistencia
resistance test prueba de resistencia
resistance thermometer termómetro de resistencia
resistance-thermometer bridge puente de termómetro de resistencia

resistance-thermometer detector detector de termómetro de resistencia
resistance thermometry termometría de resistencia
resistance to earth resistencia a tierra
resistance transducer transductor de resistencia
resistance tube tubo de resistencia
resistance-tuned sintonizado por resistencia
resistance-tuned oscillator oscilador sintonizado por resistencia
resistance tuning sintonización por resistencia
resistance unit unidad de resistencia
resistance value valor de resistencia
resistance variation variación de resistencia
resistance welding soldadura por resistencia
resistance welding electrode electrodo de soldadura por resistencia
resistance welding equipment equipo de soldadura por resistencia
resistance wire hilo de resistencia
resistance-wire sensor sensor de hilo de resistencia
resistant resistente
resistive resistivo
resistive adapter adaptador resistivo
resistive attenuator atenuador resistivo
resistive bidirectional coupler acoplador bidireccional resistivo
resistive circuit circuito resistivo
resistive component componente resistivo
resistive conductor conductor resistivo
resistive coupling acoplamiento resistivo
resistive current corriente resistiva
resistive current-limiting device dispositivo limitador de corriente resistivo
resistive cutoff frequency frecuencia de corte resistivo
resistive divider divisor resistivo
resistive element elemento resistivo
resistive impedance impedancia resistiva
resistive load carga resistiva
resistive loss pérdida resistiva
resistive network red resistiva
resistive temperature sensor sensor de temperatura resistivo
resistive transducer transductor resistivo
resistive transduction transducción resistiva
resistive unbalance desequilibrio resistivo
resistive vane paleta resistiva
resistive voltage tensión resistiva
resistive Wheatstone bridge puente de Wheatstone resistivo
resistivity resistividad
resistivity measurement medida de resistividad
resistivity probe sonda de resistividad
resistor resistor, resistencia
resistor-capacitor-transistor logic lógica de resistor-capacitor-transistor
resistor-capacitor unit unidad de resistor-capacitor
resistor color code código de colores de resistores
resistor core núcleo de resistor
resistor-coupled circuit circuito acoplado por resistor
resistor element elemento resistor
resistor fuse resistor fusible
resistor network red de resistores
resistor strip regleta de resistores
resistor tester probador de resistores
resistor-transistor resistor-transistor
resistor-transistor logic lógica de resistor-transistor
resistors in parallel resistores en paralelo
resistors in parallel-series resistores en paralelo-serie
resistors in series resistores en serie
resistors in series-parallel resistores en

serie-paralelo
resnatron resnatrón
resolution resolución, definición
resolution capability capacidad de resolución
resolution chart imagen de prueba de definición
resolution elements elementos de resolución
resolution error error de resolución
resolution pattern imagen de prueba de definición
resolution phase fase de resolución
resolution ratio razón de definición
resolution sensitivity sensibilidad de resolución
resolution test chart imagen de prueba de definición
resolution time tiempo de resolución
resolve resolver
resolver resolvedor, resolutor
resolving capability capacidad de resolución
resolving power poder de resolución, poder de definición
resolving time tiempo de resolución
resolving-time correction corrección de tiempo de resolución
resonance resonancia
resonance absorption absorción de resonancia
resonance amplifier amplificador de resonancia
resonance amplitude amplitud de resonancia
resonance box caja de resonancia
resonance bridge puente de resonancia
resonance capacitor-transformer capacitor-transformador de resonancia
resonance characteristic característica de resonancia
resonance circuit circuito de resonancia
resonance curve curva de resonancia
resonance effect efecto de resonancia
resonance energy energía de resonancia
resonance energy band banda de energía de resonancia
resonance fluorescence fluorescencia de resonancia
resonance flux flujo de resonancia
resonance frequency frecuencia de resonancia
resonance frequency-meter frecuencímetro de resonancia
resonance indicator indicador de resonancia
resonance isolator aislador de resonancia
resonance lamp lámpara de resonancia
resonance level nivel de resonancia
resonance line línea de resonancia
resonance meter medidor de resonancia
resonance method método de resonancia
resonance mode modo de resonancia
resonance oscillatory circuit circuito oscilante de resonancia
resonance peak pico de resonancia
resonance radiation radiación de resonancia
resonance resistance resistencia de resonancia
resonance scattering dispersión de resonancia
resonance spectrum espectro de resonancia
resonance state estado de resonancia
resonance tube tubo de resonancia
resonance vibration vibración de resonancia
resonant resonante
resonant acoustic panel panel acústico resonante
resonant antenna antena resonante
resonant capacitor capacitor resonante
resonant-capacitor magnetron magnetrón de capacitor resonante
resonant-capacitor maser máser de capacitor resonante
resonant cavity cavidad resonante
resonant-cavity magnetron magnetrón de cavidad resonante
resonant-cavity wavemeter ondámetro de cavidad resonante
resonant chamber cámara resonante

resonant circuit circuito resonante
resonant-circuit coupling acoplamiento por circuito resonante
resonant-circuit frequency indicator indicador de frecuencia de circuito resonante
resonant depolarization despolarización resonante
resonant diaphragm diafragma resonante
resonant dipole dipolo resonante
resonant element elemento resonante
resonant extraction extracción resonante
resonant feeder alimentador resonante
resonant filter filtro resonante
resonant frequency frecuencia resonante
resonant gap entrehierro resonante
resonant-gate transistor transistor de puerta resonante
resonant iris iris resonante
resonant line línea resonante
resonant-line amplifier amplificador de línea resonante
resonant-line circuit circuito de línea resonante
resonant-line oscillator oscilador de línea resonante
resonant-line tuner sintonizador de líneas resonantes
resonant mode modo resonante
resonant-mode filter filtro de modo resonante
resonant oscillation oscilación resonante
resonant relay relé resonante
resonant shunt derivación resonante
resonant transmission line línea de transmisión resonante
resonant-voltage rise subida de tensión resonante
resonant window ventana resonante
resonate resonar
resonating resonante
resonating cavity cavidad resonante
resonating window ventana resonante
resonator resonador
resonator current corriente de resonador
resonator grid rejilla de resonador
resonator mode modo de resonador
resonistor resonistor
resort reclasificar
resorting reclasificación
resource access security seguridad de acceso a recursos
resource allocation asignación de recursos
resource allocation processor procesador de asignación de recursos
resource management administración de recursos
resource security seguridad de recursos
resource security file archivo de seguridad de recursos
responder respondedor
responder beacon radiofaro respondedor
response respuesta
response characteristic característica de respuesta
response curve curva de respuesta
response duration duración de respuesta
response frequency frecuencia de respuesta
response function función de respuesta
response linearity linealidad de respuesta
response mode modo de respuesta
response range gama de respuesta
response rate velocidad de respuesta
response shape forma de respuesta
response signal señal de respuesta
response spectrum espectro de respuesta
response speed velocidad de respuesta
response time tiempo de respuesta
responser respondedor
responsivity responsividad
rest contact contacto de reposo
rest current corriente de reposo

rest electromotive force fuerza electromotriz de reposo
rest energy energía en reposo
rest position posición de reposo
rest potential potencial en reposo
restart reiniciar, reanudar
restart address dirección de reinicio, dirección de reanudación
restart instruction instrucción de reinicio, instrucción de reanudación
restart point punto de reinicio, punto de reanudación
restart routine rutina de reinicio, rutina de reanudación
resting contact contacto de reposo
resting frequency frecuencia de reposo
resting potential potencial de reposo
resting state estado de reposo
restitution restitución
restitution delay retardo de restitución
restitution distortion distorsión de restitución
restoration restauración, reposición
restoration circuit circuito de restauración
restore restaurar, reponer
restored wave onda restaurada
restorer restaurador
restoring circuit circuito de restauración
restoring force fuerza de restauración
restoring time tiempo de restauración
restoring torque par de restauración
restriking reencendido, restablecimiento
restriking voltage tensión de reencendido, tensión transitoria de restablecimiento
resultant resultante
resultant color shift desplazamiento total
resultant tone tono resultante
resultant voltage tensión resultante
resulting error error resultante
resume reanudar
reswitch reconmutar
reswitching reconmutación
retained image imagen retenida
retaining coil bobina de retención
retaining plate placa de retención
retaining ring anillo de retención
retaining spring resorte de retención
retardation retardo
retardation coil bobina de retardo
retardation effect efecto de retardo
retardation mechanism mecanismo de retardo
retarded retardado
retarded closing cierre retardado
retarded field campo retardado
retarded opening abertura retardada
retarded potential potencial retardado
retarded relay relé retardado
retarded spark chispa retardada
retarder retardador
retarding retardador
retarding electrode electrodo retardador
retarding field campo retardador
retarding-field detector detector de campo retardador
retarding-field oscillator oscilador de campo retardador
retarding-field tube tubo de campo retardador
retarding force fuerza retardadora
retarding magnet imán retardador
retarding torque par retardador
retention retención
retention characteristic característica de retención
retention circuit circuito de retención
retention failure falla de retención
retention longevity longevidad de retención

retention period periodo de retención
retention time tiempo de retención
retentivity retentividad
reticle retículo
reticulated reticulado
retrace retorno, retrazo
retrace blanking extinción durante retorno
retrace ghost fantasma durante retorno
retrace interval intervalo de retorno
retrace line línea de retorno
retrace period periodo de retorno
retrace ratio razón de retorno
retrace suppression circuit circuito de supresión de retorno
retrace time tiempo de retorno
retractable retractable
retransmission retransmisión
retransmission installation instalación de retransmisión
retransmitter retransmisor
retransmitting installation instalación de retransmisión
retrievable recuperable
retrieval recuperación
retrieval code código de recuperación
retrieval time tiempo de recuperación
retrieve recuperar
retroaction retroacción, reacción
retroactive retroactivo, de reacción
retroactive amplifier amplificador con reacción
retroactively retroactivamente
retrodirective retrodirectivo
retrodirective antenna antena retrodirectiva
retrodirective reflector reflector retrodirectivo
retrodispersion retrodispersión
retrofit retromodificación
retrograde retrógrado
retrograde orbit órbita retrógrada
retrograde rays rayos retrógrados
retroreflecting retrorreflejante
retroreflecting material material retrorreflejante
retroreflecting surface superficie retrorreflejante
retroreflection retrorreflexión
retroreflector retrorreflector
retry period periodo de reintento
retune resintonizar
retuning resintonización
return retorno, vuelta
return address dirección de retorno
return beam haz de retorno
return cable cable de retorno
return channel canal de retorno
return character carácter de retorno
return circuit circuito de retorno
return code código de retorno
return conductor conductor de retorno
return connection conexión de retorno
return current corriente de retorno
return-current coefficient coeficiente de corriente de retorno
return difference diferencia de retorno
return electron electrón de rechazo
return feeder alimentador de retorno
return instruction instrucción de retorno
return interval intervalo de retorno
return lead hilo de retorno
return light luz de retorno
return line línea de retorno
return loss pérdida de retorno
return medium medio de retorno
return path trayectoria de retorno
return period periodo de retorno
return point punto de retorno

return ratio razón de retorno
return routine rutina de retorno
return signal señal de retorno
return-signal intensity intensidad de señal de retorno
return time tiempo de retorno
return to zero retorno a cero
return trace trazo de retorno
return value valor de retorno
return voltage tensión de retorno
return wave onda de retorno
return wire hilo de retorno
reusable reusable
reusable disk storage almacenamiento de disco reusable
reusable file archivo reusable
reusable program programa reusable
reusable routine rutina reusable
reusable subroutine subrutina reusable
reverberant reverberante
reverberant sound sonido reverberante
reverberate reverberar
reverberation reverberación
reverberation absorption coefficient coeficiente de absorción de reverberación
reverberation absorption factor factor de absorción de reverberación
reverberation chamber cámara de reverberación
reverberation-controlled controlado por reverberación
reverberation period periodo de reverberación
reverberation reflection coefficient coeficiente de reflexión de reverberación
reverberation reflection factor factor de reflexión de reverberación
reverberation response respuesta de reverberación
reverberation room sala de reverberación
reverberation system sistema de reverberación
reverberation time tiempo de reverberación
reverberation transmission coefficient coeficiente de transmisión de reverberación
reverberation transmission factor factor de transmisión de reverberación
reverberation unit unidad de reverberación
reverberator reverberador
reverberatory reverberatorio
reversal inversión
reversal of current inversión de corriente
reversal of polarity inversión de polaridad
reversal point punto de inversión
reverse action acción inversa
reverse arm brazo inverso
reverse avalanche avalancha inversa
reverse bias polarización inversa
reverse-biased junction unión con polarización inversa
reverse-blocking triode-thyristor tiristor triodo de bloqueo inverso
reverse breakdown ruptura inversa
reverse-breakdown current corriente de ruptura inversa
reverse-breakdown voltage tensión de ruptura inversa
reverse channel canal inverso
reverse characteristic característica inversa
reverse compatibility compatibilidad inversa
reverse conduction conducción inversa
reverse contact contacto inverso
reverse coupler acoplador inverso
reverse current corriente inversa, contracorriente
reverse-current circuit-breaker cortacircuito de corriente inversa
reverse-current coil bobina de corriente inversa
reverse-current cutout cortacircuito de corriente

inversa
reverse-current protection protección contra
   corriente inversa
reverse-current relay relé de corriente inversa
reverse detection detección inversa
reverse direction dirección inversa
reverse emission emisión inversa
reverse entry entrada inversa
reverse feedback retroalimentación inversa
reverse-feedback amplifier amplificador de
   retroalimentación inversa
reverse gate current corriente de puerta inversa
reverse gate voltage tensión de puerta inversa
reverse grid current corriente de rejilla inversa
reverse grid potential potencial de rejilla inverso
reverse grid voltage tensión de rejilla inversa
reverse image imagen inversa
reverse impulse impulso inverso
reverse motion moción inversa
reverse order orden inverso
reverse period periodo inverso
reverse polarity polaridad inversa
reverse-polarity protection protección contra
   polaridad inversa
reverse position posición inversa
reverse power potencia inversa
reverse-power relay relé de potencia inversa
reverse reaction reacción inversa
reverse reading lectura inversa
reverse recovery interval intervalo de recuperación
   inversa
reverse recovery time tiempo de recuperación
   inversa
reverse resistance resistencia inversa
reverse scan exploración inversa
reverse search búsqueda inversa
reverse transadmittance transadmitancia inversa
reverse transimpedance transimpedancia inversa
reverse voltage tensión inversa
reverse wave onda inversa
reversed invertido
reversed contrast contraste invertido
reversed flow flujo invertido
reversed image imagen invertida
reversed picture imagen invertida
reversed polarity polaridad invertida
reversed-polarity rectifier rectificador de polaridad
   invertida
reverser inversor
reversibility reversibilidad
reversible reversible
reversible booster elevador-reductor
reversible capacitance capacitancia reversible
reversible capacitance characteristic característica
   de capacitancia reversible
reversible cartridge diode diodo de cartucho
   reversible
reversible cell pila reversible
reversible circuit circuito reversible
reversible counter contador reversible
reversible cycle ciclo reversible
reversible electrolytic process proceso electrolítico
   reversible
reversible execution ejecución reversible
reversible magnetic process proceso magnético
   reversible
reversible motor motor reversible
reversible path trayectoria reversible
reversible permeability permeabilidad reversible
reversible permittivity permitividad reversible
reversible process proceso reversible
reversible rotation rotación reversible
reversible susceptibility susceptibilidad reversible

reversible transducer transductor reversible
reversing inversión
reversing contact contacto inversor
reversing contactor contactor inversor
reversing key manipulador inversor
reversing prism prisma inversor
reversing switch conmutador inversor
reversion reversión, inversión
revert revertir, invertir
revertive impulse impulso inverso
revertive pulse impulso inverso
revertive-pulse system sistema de impulsos inversos
revolution revolución
revolving antenna antena giratoria
revolving armature inducido giratorio
revolving-armature generator generador de
   inducido giratorio
revolving beacon radiofaro giratorio
revolving field campo giratorio
revolving-field alternator alternador de campo
   giratorio
revolving-field generator generador de campo
   giratorio
revolving light luz giratoria
revolving radiobeacon radiofaro giratorio
rewind rebobinar, rearrollar
rewind control control de rebobinado
rewinder rebobinador
rewinding rebobinado
rewinding speed velocidad de rebobinado
rewire realambrar
rewireable realambrable
rewrite reescribir
rhenium renio
rheoelectric reoeléctrico
rheoelectricity reoelectricidad
rheograph reógrafo
rheographic reográfico
rheophore reóforo
rheostat reóstato
rheostatic reostático
rheostatic braking frenado reostático
rheostatic control control reostático
rheostatic regulator regulador reostático
rheostatic starting arranque reostático
rheostatically reostáticamente
rheostatically controlled controlado reostáticamente
rheostriction reoestricción
rheotaxial reotaxial
rheotaxial film película reotaxial
rheotaxial growth crecimiento reotaxial
rheotome reotomo
rheotropic reotrópico
rho rho
rho-theta navigation navegación rho-theta
rho-theta navigation system sistema de navegación
   rho-theta
rho-theta system sistema rho-theta
rhodium rodio
rhodium coating revestimiento de rodio
rhodium-plated rodiado
rhodium-plated contact contacto rodiado
rhombic rómbico
rhombic antenna antena rómbica
rhumbatron rumbatrón
rhythmic rítmico
rhythmic light luz rítmica
rhythmic pulsing pulsación rítmica
ribbon cinta
ribbon cable cable de cinta
ribbon cartridge cartucho de cinta
ribbon control control de cinta
ribbon element elemento de cinta

ribbon feed alimentación de cinta
ribbon-feed mechanism mecanismo de alimentación de cinta
ribbon feeding alimentación de cinta
ribbon filament filamento de cinta
ribbon loudspeaker altavoz de cinta
ribbon microphone micrófono de cinta
ribbon transducer transductor de cinta
Rice neutralization neutralización de Rice
Rice neutralizing circuit circuito neutralizador de Rice
Richardson's equation ecuación de Richardson
Richardson-Dushman equation ecuación de Richardson-Dushman
Richardson effect efecto de Richardson
ride gain ajuste de volumen continuo
ridge resalte, costilla
ridge waveguide guíaondas acanalado
Rieke chart diagrama de Rieke
Rieke diagram diagrama de Rieke
right-hand polarized wave onda polarizada a la derecha
right-hand rule regla de la mano derecha
right-hand taper mayoría de resistencia a la derecha
Right-Leduc effect efecto de Right-Leduc
right shift desplazamiento a la derecha
rigid rígido
rigid disk disco rígido
rigid fastening fijación rígida
rigid support soporte rígido
rigid suspension suspensión rígida
rim drive arrastre por borde
ring anillo
ring anode ánodo en anillo
ring antenna antena en anillo
ring armature inducido en anillo
ring-back señal de llamada
ring-back signal señal de llamada
ring bridge puente en anillo
ring bus barra colectora en anillo, bus en anillo
ring circuit circuito en anillo, anillo
ring connection conexión en anillo
ring counter contador en anillo
ring current corriente en anillo
ring-down llamada manual
ring filter filtro en anillo
ring head cabeza en anillo
ring magnet imán en anillo
ring method método de anillo
ring mode filter filtro de modo en anillo
ring modulator modulador en anillo
ring network red en anillo
ring oscillator oscilador en anillo
ring rheostat reóstato en anillo
ring shift desplazamiento en anillo
ring switch conmutador de anillo
ring system sistema en anillo
ring time tiempo de oscilación parásita
ring winding arrollamiento en anillo
ringback señal de llamada
ringback signal señal de llamada
ringdown llamada manual
ringed network red de anillos
ringer timbre, llamador, señalizador
ringing llamada, oscilaciones transitorias, oscilaciones amortiguadas, sonido de timbre
ringing battery batería de llamada
ringing circuit circuito de llamada
ringing current corriente de llamada
ringing-current circuit circuito de corriente de llamada
ringing frequency frecuencia de llamada
ringing impulses impulsos de llamada

ringing periodicity periodicidad de llamada
ringing pilot piloto de llamada
ringing relay relé de llamada
ringing repeater repetidor de llamada
ringing signal señal de llamada
ringing time tiempo de oscilación parásita
ringing tone tono de llamada
riometer riómetro
ripple ondulación
ripple amplitude amplitud de ondulación
ripple component componente de ondulación
ripple current corriente de ondulación
ripple factor factor de ondulación
ripple filter filtro de ondulación
ripple frequency frecuencia de ondulación
ripple generator generador de ondulación
ripple percentage porcentaje de ondulación
ripple ratio razón de ondulación
ripple voltage tensión de ondulación
rise time tiempo de subida
rising characteristic característica ascendente
rising edge borde ascendente
rising-sun magnetron magnetrón de sol naciente
rising-sun resonator resonador de sol naciente
roaming rastreo, seguimiento
Robinson-Adcock direction-finder radiogoniómetro de Robinson-Adcock
Robinson bridge puente de Robinson
Robinson direction-finder radiogoniómetro de Robinson
robot robot
robotic robótico
robotic system sistema robótico
robotics robótica
robotization robotización
robustness robustez
Rochelle salt sal de Rochelle
Rochelle-salt crystal cristal de sal de Rochelle
Rochelle-salt crystal microphone micrófono de cristal de sal de Rochelle
rocker arm balancín
rocker switch conmutador de balancín
rod varilla, barra
rod antenna antena de varilla
rod gap espacio entre varillas
rod magnet imán en varilla
rod reflector reflector de varillas
rod wavemeter ondámetro de varilla
Roebel transposition transposición de Roebel
roentgen roentgen
roentgen meter roentgenómetro
Roentgen ray rayo Roentgen
roentgenogram roentgenograma
roentgenography roentgenografía
roentgenometer roentgenómetro
roentgenoscope roentgenoscopio
roentgenoscopy roentgenoscopía
Roget spiral espiral de Roget
roll rollo, bobina, corrimiento vertical
rollback repetición, reejecución, reanudación
rolling rodamiento, arrollamiento, corrimiento vertical
rolling contact contacto rodante
rolloff caída progresiva
rolloff characteristic característica de caída progresiva
Romex cable cable de Romex
roof mount montura de techo
rooftop antenna antena de techo
room acoustics acústica de sala
room coefficient coeficiente de sala
room index índice de local
room noise ruido de sala, ruido ambiente

room resonance resonancia de sala
room temperature temperatura de sala, temperatura ambiente
root raíz
root directory directorio raíz
root mean square raíz cuadrada de media de cuadrados
root-mean-square amplitude amplitud eficaz
root-mean-square current corriente eficaz
root-mean-square deviation desviación media cuadrática
root-mean-square load carga eficaz
root-mean-square magnitude magnitud eficaz
root-mean-square rating potencia eficaz
root-mean-square sound pressure presión sonora eficaz
root-mean-square value valor eficaz
root-mean-square velocity velocidad eficaz
root-mean-square voltage tensión eficaz
root record registro raíz
rosette roseta, rosetón
rosette plate placa de rosetas
rosin resina, pez griega
rosin connection conexión de resina
rosin-core solder soldadura con núcleo de resina
rosin joint junta de resina
rotameter rotámetro
rotary rotativo, rotatorio
rotary action acción rotativa
rotary-action relay relé de acción rotativa
rotary actuator actuador rotativo
rotary amplifier amplificador rotativo
rotary antenna antena rotativa
rotary attenuator atenuador rotativo
rotary beam haz rotativo
rotary-beam antenna antena de haz rotativo
rotary capacitor capacitor rotativo
rotary condenser condensador rotativo
rotary control control rotativo
rotary converter convertidor rotativo
rotary coupler acoplador rotativo
rotary dialing marcación rotativa
rotary discharger descargador rotativo
rotary dispersion dispersión rotativa
rotary drum tambor rotativo
rotary electrostatic generator generador electrostático rotativo
rotary encoder codificador rotativo
rotary field campo rotativo
rotary generator generador rotativo
rotary interrupter interruptor rotativo
rotary inverter inversor rotativo
rotary joint junta rotativa
rotary machine máquina rotativa
rotary magnetic disk disco magnético rotativo
rotary mast mástil rotativo
rotary motion moción rotativa
rotary-motion sensor sensor de moción rotativo
rotary phase changer desfasador rotativo
rotary phase converter convertidor de fase rotativo
rotary phase shifter desfasador rotativo
rotary positioning posicionamiento rotativo
rotary relay relé rotativo
rotary repeater repetidor rotativo
rotary selector switch conmutador de selección rotativo
rotary solenoid solenoide rotativo
rotary solenoid relay relé de solenoide rotativo
rotary spark-gap descargador rotativo
rotary stepping relay relé de pasos rotativo
rotary stepping switch conmutador de pasos rotativo
rotary switch conmutador rotativo
rotary system sistema rotativo

rotary transformer transformador rotativo
rotary voltmeter vóltmetro rotativo
rotatable giratorio
rotatable antenna antena giratoria
rotatable loop cuadro giratorio
rotatable-loop antenna antena de cuadro giratorio
rotatable phase-adjusting transformer transformador ajustador de fase giratorio
rotatable transformer transformador giratorio
rotate rotar
rotating rotativo
rotating amplifier amplificador rotativo
rotating anode ánodo rotativo
rotating-anode tube tubo de ánodo rotativo
rotating antenna antena rotativa
rotating beacon radiofaro rotativo
rotating beam haz rotativo
rotating coil bobina rotativa
rotating coupler acoplador rotativo
rotating crystal cristal rotativo
rotating direction-finder radiogoniómetro rotativo
rotating disk disco rotativo
rotating electric generator generador eléctrico rotativo
rotating element elemento rotativo
rotating feeder alimentador rotativo
rotating field campo rotativo
rotating-field antenna antena de campo rotativo
rotating grating rejilla rotativa
rotating head cabeza rotativa
rotating interrupter interruptor rotativo
rotating joint junta rotativa
rotating loop cuadro rotativo
rotating-loop antenna antena de cuadro rotativo
rotating-loop direction-finder radiogoniómetro de cuadro rotativo
rotating machine máquina rotativa
rotating magnet imán rotativo
rotating magnetic amplifier amplificador magnético rotativo
rotating memory memoria rotativa
rotating mirror espejo rotativo
rotating plasma plasma rotativo
rotating pole polo rotativo
rotating prism prisma rotativo
rotating radio beacon radiofaro rotativo
rotating rectifier rectificador rotativo
rotating scanner explorador rotativo
rotating spark-gap descargador rotativo
rotating turntable plato rotativo
rotation rotación
rotation angle ángulo de rotación
rotation isolator aislador de rotación
rotation spectrum espectro de rotación
rotational rotacional
rotational analyzer analizador rotacional
rotational band banda rotacional
rotational energy energía rotacional
rotational field campo rotacional
rotational hysteresis histéresis rotacional
rotational velocity velocidad rotacional
rotational wave onda rotacional
rotative rotativo
rotatomotive rotatomotriz
rotator rotador
rotatory rotatorio
rotatory power potencia rotatoria
rotor rotor
rotor coil bobina rotatoria
rotor current corriente rotórica
rotor diameter diámetro del rotor
rotor plate placa móvil
rotor reactance reactancia rotórica

rotor resistance resistencia rotórica
rotor rheostat reóstato rotórico
rotor winding arrollamiento del rotor
round-off redondear
round-off error error de redondeo
round-up redondear hacia arriba
rounded number número redondeado
rounding redondeamiento, redondeo
rounding error error de redondeo
rounding routine rutina de redondeo
Rousseau diagram diagrama de Rousseau
route ruta, vía, línea
routine rutina
routine maintenance mantenimiento rutinario
routine maintenance time tiempo de mantenimiento rutinario
routine test prueba rutinaria
routing encaminamiento, enrutamiento
routing channel canal de encaminamiento
routing code código de encaminamiento
routing data datos de encaminamiento
routing indicator indicador de encaminamiento
routing pattern patrón de encaminamiento
routing step paso de encaminamiento
row fila, hilera
row binary code código binario en fila
row pitch paso entre filas
rubber caucho, goma
rubber-covered cubierto de caucho
rubber-covered cable cable cubierto de caucho
rubber-covered wire hilo cubierto de caucho
rubber insulation aislamiento de caucho
rubber tape cinta de caucho
rubidium rubidio
ruby laser láser de rubí
ruby maser máser de rubí
Ruhmkorff coil bobina de Ruhmkorff
rumble ronquido, ruido mecánico, ruido de motor
rumble filter filtro contra ronquido, filtro contra ruido mecánico
run pasada, ejecución, ciclo, recorrido
run chart diagrama de ejecución
run indicator indicador de ejecución
run time tiempo de ejecución
run time routine rutina de tiempo de ejecución
run unit unidad de ejecución
runaway embalamiento
running light luz de funcionamiento
running time tiempo de funcionamiento
running-time meter medidor de tiempo de funcionamiento
running voltage tensión de funcionamiento
runway localizer localizador de pista
runway localizer beacon radiofaro localizador de pista
rupture ruptura
rural automatic exchange central automática rural
rural line línea rural
rural network red rural
rush hour hora pico
rust-resistant resistente a la herrumbre
ruthenium rutenio
rutherford rutherford
ryotron riotrón

S

sabin sabine
Sabine's law ley de Sabine
Sabine absorption absorción de Sabine
Sabine coefficient coeficiente de Sabine
saddle abrazadera
safe let-go level nivel seguro de soltar
safe load carga segura
safe value valor seguro
safe working space espacio de trabajo seguro
safe working voltage tensión de trabajo segura
safety seguridad
safety arc arco de seguridad
safety class clasificación de seguridad
safety communication comunicación de seguridad
safety control control de seguridad
safety cutout cortacircuito de seguridad
safety device dispositivo de seguridad
safety factor factor de seguridad
safety factor for dropout factor de seguridad para puesta en reposo
safety factor for holding factor de seguridad para mantenimiento
safety factor for pickup factor de seguridad para puesta en trabajo
safety function función de seguridad
safety fuse fusible de seguridad
safety glass vidrio de seguridad
safety ground tierra de seguridad
safety lamp lámpara de seguridad
safety margin margen de seguridad
safety mechanism mecanismo de seguridad
safety outlet tomacorriente puesto a tierra
safety ring anillo de seguridad
safety service servicio de seguridad
safety signal señal de seguridad
safety switch conmutador de seguridad
safety system sistema de seguridad
safety thermostat termostato de seguridad
safety traffic tráfico de seguridad
safety window ventana de seguridad
sag flecha, caída
sagittal plane plano sagital
Sagnac effect efecto de Sagnac
sal ammoniac sal de amoniaco
sal ammoniac cell pila de sal de amoniaco
salient pole polo saliente
Salisbury chamber cámara de Salisbury
Salisbury darkbox caja obscura de Salisbury
salt-spray test prueba de rocío salino
samarium samario
sample muestra
sample and hold muestreo y retención
sample-and-hold amplifier amplificador de muestreo y retención
sample-and-hold circuit circuito de muestreo y retención
sample-and-hold device dispositivo de muestreo y retención
sample size tamaño de muestra
sampled data datos muestreados
sampling muestreo, muestra, conmutación de colores

sampling action　acción de muestreo
sampling circuit　circuito de muestreo
sampling control　control de muestreo
sampling control system　sistema de control de muestreo
sampling controller　controlador de muestreo
sampling element　elemento de muestreo
sampling frequency　frecuencia de muestreo
sampling gate　puerta de muestreo
sampling interval　intervalo de muestreo
sampling order　orden de muestreo
sampling oscilloscope　osciloscopio de muestreo
sampling period　periodo de muestreo
sampling process　proceso de muestreo
sampling rate　frecuencia de muestreo
sampling speed　velocidad de muestreo
sampling switch　conmutador de muestreo
sampling system　sistema de muestreo
sampling theory　teoría del muestreo
sampling voltmeter　vóltmetro de muestreo
sampling window　ventana de muestreo
sanafant　sanafán
sanatron　sanatrón
sand load　carga de arena
sandwich windings　arrollamientos alternados
sapphire needle　aguja de zafiro
sapphire stylus　aguja de zafiro
sapphire substrate　sustrato de zafiro
satellite　satélite
satellite antenna　antena de satélite
satellite communication　comunicación por satélite
satellite computer　computadora satélite
satellite exchange　central satélite
satellite navigation　navegación con satélite
satellite orbit　órbita de satélite
satellite processor　procesador satélite
satellite station　estación satélite
satellite system　sistema de satélites
satellite transmission　transmisión por satélite
satellite transmitter　transmisor satélite
saturable　saturable
saturable capacitor　capacitor saturable
saturable choke　choque saturable
saturable core　núcleo saturable
saturable-core magnetometer　magnetómetro de núcleo saturable
saturable-core oscillator　oscilador de núcleo saturable
saturable-core reactor　reactor de núcleo saturable
saturable magnetic core　núcleo magnético saturable
saturable reactor　reactor saturable
saturable transformer　transformador saturable
saturant　saturante
saturate　saturar
saturated　saturado
saturated color　color saturado
saturated core　núcleo saturado
saturated diode　diodo saturado
saturated operation　operación saturada
saturated solution　solución saturada
saturated vapor　vapor saturado
saturating　saturante
saturating current　corriente saturante
saturating signal　señal saturante
saturation　saturación
saturation absorption　absorción de saturación
saturation characteristic　característica de saturación
saturation control　control de saturación
saturation current　corriente de saturación
saturation curve　curva de saturación
saturation effect　efecto de saturación
saturation factor　factor de saturación
saturation flux　flujo de saturación

saturation flux density　densidad de flujo de saturación
saturation inductance　inductancia de saturación
saturation induction　inducción de saturación
saturation level　nivel de saturación
saturation limiting　limitación por saturación
saturation magnetization　magnetización de saturación
saturation magnetostriction　magnetoestricción de saturación
saturation noise　ruido de saturación
saturation point　punto de saturación
saturation polarization　polarización de saturación
saturation reactance　reactancia de saturación
saturation region　región de saturación
saturation resistance　resistencia de saturación
saturation signal　señal de saturación
saturation state　estado de saturación
saturation value　valor de saturación
saturation voltage　tensión de saturación
saturator　saturador
save　almacenar, conservar
save area　área de almacenaje
sawtooth　diente de sierra
sawtooth amplifier　amplificador de diente de sierra
sawtooth arrester　descargador en diente de sierra
sawtooth current　corriente en diente de sierra
sawtooth generator　generador de ondas en diente de sierra
sawtooth oscillation　oscilación en diente de sierra
sawtooth pulse　impulso en diente de sierra
sawtooth scanning　exploración en diente de sierra
sawtooth sweep　barrido en diente de sierra
sawtooth voltage　tensión en diente de sierra
sawtooth wave　onda en diente de sierra
scalar　escalar
scalar expression　expresión escalar
scalar field　campo escalar
scalar function　función escalar
scalar product　producto escalar
scalar quantity　magnitud escalar
scalar ratio　razón escalar
scalar variable　variable escalar
scale　escala
scale coefficient　coeficiente de escala
scale correction　corrección de escala
scale division　división de escala
scale down　reducir a escala
scale expansion　expansión de escala
scale factor　factor de escala
scale-factor adjustment　ajuste de factor de escala
scale-factor error　error de factor de escala
scale-factor tolerance　tolerancia de factor de escala
scale length　longitud de escala
scale line　línea de escala
scale mark　marca de escala
scale model　modelo a escala
scale multiplier　multiplicador de escala
scale-of-eight circuit　circuito de escala de ocho
scale-of-ten counter　contador decimal
scale of turbulence　escala de turbulencia
scale-of-two circuit　circuito de escala de dos
scale-of-two counter　contador binario
scale range　intervalo de escala
scale selector　selector de escala
scale span　intervalo de escala
scale switch　conmutador de escala
scaler　escala, escalímetro
scaling　desmultiplicación, escalamiento, ajuste
scaling circuit　circuito escalador
scaling factor　factor de escala
scaling ratio　razón de escala
scalloping　festoneado

scan exploración, barrido, trazo
scan amplifier amplificador de exploración
scan angle ángulo de exploración
scan antenna antena de exploración
scan aperture abertura de exploración
scan area área de exploración
scan axis eje de exploración
scan band banda de exploración
scan beam haz de exploración
scan coil bobina de exploración
scan control control de exploración
scan conversion conversión de exploración
scan converter convertidor de exploración
scan-converter tube tubo convertidor de exploración
scan cycle ciclo de exploración
scan density densidad de exploración
scan design diseño de exploración
scan disk disco de exploración
scan distortion distorsión de exploración
scan drum tambor de exploración
scan equipment equipo de exploración
scan field campo de exploración
scan frequency frecuencia de exploración
scan generator generador de exploración
scan head cabeza de exploración
scan hole agujero de exploración
scan instruction logic lógica de instrucción de exploración
scan lamp lámpara de exploración
scan limit límite de exploración
scan line línea de exploración
scan-line frequency frecuencia de líneas de exploración
scan-line length longitud de línea de exploración
scan linearity linealidad de exploración
scan loss pérdida por exploración
scan method método de exploración
scan path trayectoria de exploración
scan pattern patrón de exploración
scan period periodo de exploración
scan pitch paso de exploración
scan point punto de exploración
scan rate velocidad de exploración
scan resolution resolución de exploración
scan sector sector de exploración
scan sensitivity sensibilidad de exploración
scan sequence secuencia de exploración
scan slot ranura de exploración
scan speed velocidad de exploración
scan spot punto de exploración
scan stage etapa de exploración
scan system sistema de exploración
scan time tiempo de exploración
scan unit unidad de exploración
scan velocity velocidad de exploración
scan voltage tensión de exploración
scandium escandio
scanistor escanistor
scanned image imagen explorada
scanner explorador
scanner amplifier amplificador de explorador
scanner unit unidad exploradora
scanning exploración, barrido
scanning amplifier amplificador de exploración
scanning angle ángulo de exploración
scanning antenna antena de exploración
scanning aperture abertura de exploración
scanning area área de exploración
scanning axis eje de exploración
scanning band banda de exploración
scanning beam haz de exploración
scanning circuit circuito de exploración
scanning coil bobina de exploración

scanning control control de exploración
scanning conversion conversión de exploración
scanning converter convertidor de exploración
scanning-converter tube tubo convertidor de exploración
scanning cycle ciclo de exploración
scanning density densidad de exploración
scanning design diseño de exploración
scanning disk disco de exploración
scanning distortion distorsión de exploración
scanning drum tambor de exploración
scanning electron microscope microscopio electrónico de barrido
scanning electron microscopy microscopía electrónica de barrido
scanning equipment equipo de exploración
scanning field campo de exploración
scanning frequency frecuencia de exploración
scanning generator generador de exploración
scanning head cabeza de exploración
scanning hole agujero de exploración
scanning instruction logic lógica de instrucción de exploración
scanning lamp lámpara de exploración
scanning limit límite de exploración
scanning line línea de exploración
scanning-line frequency frecuencia de líneas de exploración
scanning-line length longitud de línea de exploración
scanning linearity linealidad de exploración
scanning loss pérdida por exploración
scanning method método de exploración
scanning microscopy microscopía de barrido
scanning pattern patrón de exploración
scanning period periodo de exploración
scanning pitch paso de exploración
scanning point punto de exploración
scanning rate velocidad de exploración
scanning receiver receptor de exploración
scanning resolution resolución de exploración
scanning sector sector de exploración
scanning sensitivity sensibilidad de exploración
scanning sequence secuencia de exploración
scanning slot ranura de exploración
scanning sonar sonar de exploración
scanning speed velocidad de exploración
scanning spot punto de exploración
scanning stage etapa de exploración
scanning system sistema de exploración
scanning time tiempo de exploración
scanning unit unidad de exploración
scanning velocity velocidad de exploración
scanning voltage tensión de exploración
scanning yoke yugo de exploración
scansion escansión
scatter dispersión, esparcimiento
scatter angle ángulo de dispersión
scatter band banda de dispersión
scatter propagation propagación por dispersión
scatter read lectura dispersa
scatter transmission transmisión por dispersión
scattered disperso, difuso
scattered beam haz disperso
scattered electrons electrones dispersos
scattered field campo difuso
scattered light luz difusa
scattered wave onda dispersa
scattering dispersión, esparcimiento
scattering amplitude amplitud de dispersión
scattering angle ángulo de dispersión
scattering aperture abertura de dispersión
scattering area área de dispersión
scattering attenuation coefficient coeficiente de

atenuación de dispersión
scattering coefficient coeficiente de dispersión
scattering communication comunicación por
dispersión
scattering cone cono de dispersión
scattering cross-section sección transversal de
dispersión
scattering curve curva de dispersión
scattering factor factor de dispersión
scattering frequency frecuencia de dispersión
scattering loss pérdida por dispersión
scattering matrix matriz de dispersión
scattering of electrons dispersión de electrones
scattering of light dispersión de luz
scattering surface superficie de dispersión
scatterometer dispersiómetro
schedule programa, horario
scheduled down time tiempo muerto programado
scheduled frequency frecuencia programada
scheduled interruption interrupción programada
scheduled maintenance mantenimiento programado
scheduled operation operación programada
scheduled outage interrupción programada
scheduled outage rate tasa de interrupciones
programadas
scheduled service servicio programado
schematic esquemático
schematic circuit diagram diagrama de circuito
esquemático
schematic diagram diagrama esquemático
schematic drawing dibujo esquemático
schematic symbol símbolo esquemático
Scherbius machine máquina de Scherbius
Scherbius system sistema de Scherbius
Schering bridge puente de Schering
Schlieren method método de Schlieren
Schmidt optical system sistema óptico de Schmidt
Schmidt optics óptica de Schmidt
Schmidt system sistema de Schmidt
Schmitt circuit circuito de Schmitt
Schmitt limiter limitador de Schmitt
Schmitt phase inverter inversor de fase de Schmitt
Schmitt trigger disparador de Schmitt
Schottky barrier diode diodo de barrera de Schottky
Schottky diode diodo de Schottky
Schottky effect efecto de Schottky
Schottky emission emisión de Schottky
Schottky logic lógica de Schottky
Schottky noise ruido de Schottky
Schrage motor motor de Schrage
Schuler tube tubo de Schuler
Schumann region región de Schumann
scientific compiler compilador científico
scientific language lenguaje científico
scientific notation notación científica
scintillating centelleante
scintillating crystal cristal centelleante
scintillating material material centelleante
scintillation centelleo
scintillation conversion efficiency eficiencia de
conversión de centelleo
scintillation counter contador de centelleos
scintillation-counter head cabeza de contador de
centelleos
scintillation decay time tiempo de decaimiento de
centelleo
scintillation detector detector de centelleos
scintillation duration duración de centelleo
scintillation error error por centelleo
scintillation meter medidor de centelleo
scintillation rise time tiempo de subida de centelleo
scintillation spectrometer espectrómetro de
centelleos

scintillation spectrometry espectrometría de
centelleos
scintillator centelleador
scintillator conversion efficiency eficiencia de
conversión de centelleador
scintillator crystal cristal centelleador
scintillator material material centelleador
scissoring recortado
scotopic vision visión escotópica
Scott connection conexión de Scott
Scott oscillator oscilador de Scott
Scott system sistema de Scott
Scott transformer transformador de Scott
scrambled signal señal codificada
scrambled speech conversación codificada
scrambler codificador
scrambler circuit circuito codificador
scratch rayadura, rasguño, raya
scratch file archivo temporal de tarea
scratch filter filtro de ruido de aguja
scratchpad area área de apuntes
scratchpad memory memoria de apuntes
screen pantalla, monitor
screen address dirección de pantalla
screen amplification factor factor de amplificación
de pantalla
screen angle ángulo de pantalla
screen attribute byte byte de atributo de pantalla
screen battery batería de pantalla
screen brightness brillo de pantalla
screen buffer almacenamiento intermedio de pantalla
screen control control de pantalla
screen current corriente de pantalla
screen decoupling desacoplamiento de pantalla
screen design aid ayuda de diseño de pantalla
screen dissipation disipación de pantalla
screen dropping resistor resistor reductor de tensión
de pantalla
screen dump vaciado de pantalla
screen effect efecto de pantalla
screen efficiency eficiencia de pantalla
screen factor factor de pantalla
screen font tipo de letra de pantalla
screen generator generador de pantalla
screen grid rejilla-pantalla
screen-grid current corriente de rejilla-pantalla
screen-grid modulation modulación por
rejilla-pantalla
screen-grid potential potencial de rejilla-pantalla
screen-grid voltage tensión de rejilla-pantalla
screen illumination iluminación de pantalla
screen image imagen de pantalla
screen luminance luminancia de pantalla
screen material material de pantalla
screen modulation modulación de pantalla
screen potential potencial de pantalla
screen read lectura de pantalla
screen regeneration regeneración de pantalla
screen resistance resistencia de pantalla
screen resistor resistor de pantalla
screen saver protector de pantalla
screen size tamaño de pantalla
screen sweep barrido de pantalla
screen transconductance transconductancia de
pantalla
screen type tipo de pantalla
screen update actualización de pantalla
screen voltage tensión de pantalla
screened antenna antena apantallada
screened cable cable apantallado
screened circuit circuito apantallado
screened ignition ignición apantallada
screened pair par apantallado

screening constant constante de apantallado
screening factor factor de apantallado
screw terminal terminal de tornillo
scribing grabado
scroll desplazar, desplazar hacia arriba, desplazar hacia abajo
scrolling desplazamiento, desplazamiento hacia arriba, desplazamiento hacia abajo
sea clutter ecos parásitos de mar
sea return ecos reflejos de mar
seal cierre, junta hermética
sealable equipment equipo sellable
sealed sellado
sealed beam haz sellado
sealed circuit circuito sellado
sealed contact contacto sellado
sealed end extremo sellado
sealed meter medidor sellado
sealed rectifier rectificador sellado
sealed relay relé sellado
sealed tube tubo sellado
sealing sellado, cierre hermético, obturación
sealing compound compuesto para sellado
sealing current corriente de cierre
sealing material material para sellado
sealing voltage tensión de cierre
seamed tube tubo con costura
seamed tubing tubería con costura
seamless tube tubo sin costura
seamless tubing tubería sin costura
search búsqueda, exploración
search algorithm algoritmo de búsqueda
search and replace búsqueda y reemplazo
search card tarjeta de búsqueda
search coil bobina de exploración
search condition condición de exploración
search criterion criterio de búsqueda
search cycle ciclo de búsqueda
search field campo de búsqueda
search frequency frecuencia de exploración
search memory memoria de búsqueda
search oscillator oscilador de exploración
search probe sonda de exploración
search procedure procedimiento de exploración, procedimiento de búsqueda
search radar radar de exploración
search range alcance de exploración
search string secuencia de búsqueda
search time tiempo de búsqueda
search tree árbol de búsqueda
searcher buscador, explorador
searchlight proyector
seasonal estacional
seasonal diversity diversidad estacional
seasonal effect efecto estacional
seasonal factor factor estacional
seasonal static estática estacional
seating time tiempo de asentamiento
secam secam
secant secante
second segundo
second anode segundo ánodo
second-anode potential potencial de segundo ánodo
second-channel attenuation atenuación de segundo canal
second-channel interference interferencia de segundo canal
second-degree equation ecuación de segundo grado
second detector segundo detector
second generation segunda generación
second generation computer computadora de segunda generación
second generation language lenguaje de segunda generación

second generation tape cinta de segunda generación
second group segundo grupo
second harmonic segunda armónica
second-harmonic distortion distorsión de segunda armónica
second law of thermodynamics segunda ley de termodinámica
second-level address dirección de segundo nivel
second-level addressing direccionamiento de segundo nivel
second-level directory directorio de segundo nivel
second-level message mensaje de segundo nivel
second-level statement sentencia de segundo nivel
second-level storage almacenamiento de segundo nivel
second-level subroutine subrutina de segundo nivel
second-level system sistema de segundo nivel
second-order de segundo orden
second-order distortion distorsión de segundo orden
second-order effect efecto de segundo orden
second Townsend discharge segunda descarga de Townsend
secondary secundario
secondary access method método de acceso secundario
secondary address dirección secundaria
secondary address cycle ciclo de dirección secundaria
secondary application aplicación secundaria
secondary area área secundaria
secondary axis eje secundario
secondary battery batería secundaria
secondary breakdown ruptura secundaria
secondary brush escobilla secundaria
secondary calibration calibración secundaria
secondary capacitance capacitancia secundaria
secondary capacitor capacitor secundario
secondary cell pila secundaria
secondary channel canal secundario
secondary circuit circuito secundario
secondary clock reloj secundario
secondary coil bobina secundaria
secondary color color secundario
secondary console consola secundaria
secondary constant constante secundaria
secondary current corriente secundaria
secondary data datos secundarios
secondary data set conjunto de datos secundarios
secondary destination destino secundario
secondary device dispositivo secundario
secondary distribution distribución secundaria
secondary distribution system sistema de distribución secundaria
secondary electric shock choque eléctrico secundario
secondary electron electrón secundario
secondary-electron conduction conducción de electrones secundarios
secondary-electron multiplier multiplicador de electrones secundarios
secondary emission emisión secundaria
secondary-emission characteristic característica de emisión secundaria
secondary-emission coefficient coeficiente de emisión secundaria
secondary-emission conductivity conductividad de emisión secundaria
secondary-emission multiplier multiplicador de emisión secundaria
secondary-emission photocell fotocélula de emisión secundaria
secondary-emission ratio razón de emisión

secundaria
secondary-emission tube tubo de emisión secundaria
secondary emitter emisor secundario
secondary energy energía secundaria
secondary entry point punto de entrada secundario
secondary failure falla secundaria
secondary fault falla secundaria
secondary feed alimentación secundaria
secondary file archivo secundario
secondary filter filtro secundario
secondary finder buscador secundario
secondary flow flujo secundario
secondary frequency frecuencia secundaria
secondary frequency standard patrón de frecuencia
    secundaria
secondary fuse fusible secundario
secondary grid emission emisión de rejilla
    secundaria
secondary group grupo secundario
secondary impedance impedancia secundaria
secondary index índice secundario
secondary inductance inductancia secundaria
secondary input entrada secundaria
secondary ion ion secundario
secondary ionization ionización secundaria
secondary key clave secundaria
secondary level nivel secundario
secondary-level address dirección de nivel
    secundario
secondary light source fuente de luz secundaria
secondary line línea secundaria
secondary link station estación de enlace secundaria
secondary lobe lóbulo secundario
secondary-lobe suppression supresión de lóbulos
    secundarios
secondary logical unit unidad lógica secundaria
secondary memory memoria secundaria
secondary network red secundaria
secondary outage interrupción secundaria
secondary parameter parámetro secundario
secondary pattern patrón secundario
secondary power potencia secundaria
secondary power supply fuente de alimentación
    secundaria
secondary processing sequence secuencia de
    procesamiento secundario
secondary program programa secundario
secondary radar radar secundario
secondary radiation radiación secundaria
secondary radiator radiador secundario
secondary rays rayos secundarios
secondary reaction reacción secundaria
secondary reading brush escobilla de lectura
    secundaria
secondary relay relé secundario
secondary request solicitud secundaria
secondary resistance resistencia secundaria
secondary route ruta secundaria
secondary segment segmento secundario
secondary service area área de servicio secundario
secondary source fuente secundaria
secondary space allocation asignación de espacio
    secundario
secondary standard patrón secundario
secondary station estación secundaria
secondary storage almacenamiento secundario
secondary storage device dispositivo de
    almacenamiento secundario
secondary storage medium medio de
    almacenamiento secundario
secondary task tarea secundaria
secondary tone tono secundario
secondary transmitter transmisor secundario

secondary turns vueltas secundarias
secondary utilization factor factor de utilización
    secundaria
secondary voltage tensión secundaria
secondary winding arrollamiento secundario
secondary X-rays rayos X secundarios
section sección, tramo
section circuit-breaker cortacircuito de sección
section insulator aislador de sección
section name nombre de sección
section number número de sección
section scanning exploración de sección
section side lado de sección
section switch seccionador
sectional seccionado, seccional
sectionalized antenna antena seccionada
sectionalized linear antenna antena lineal seccionada
sectionalized winding arrollamiento seccionado
sectioning seccionamiento
sector sector
sector cable cable de conductores de sector
sector display visualización de sector
sector scanning exploración por sectores
sectoral sectorial
sectoral horn bocina sectorial
sectoral horn antenna antena de bocina sectorial
secular effect efecto secular
secular variation variación secular
security seguridad
security control control de seguridad
security level nivel de seguridad
security maintenance mantenimiento de seguridad
security program programa de seguridad
security system sistema de seguridad
sediment sedimento, depósito
sedimentation sedimentación
sedimentation potential potencial de sedimentación
Seebeck coefficient coeficiente de Seebeck
Seebeck effect efecto de Seebeck
Seebeck electromotive force fuerza electromotriz de
    Seebeck
seed crystal cristal semilla
seek buscar
seek access time tiempo de acceso de búsqueda
seek area área de búsqueda
seek time tiempo de búsqueda
segment segmento
segment addressing direccionamiento de segmento
segment attribute atributo de segmento
segment extender extensor de segmento
segment mark marca de segmento
segment number número de segmento
segment search argument argumento de búsqueda
    de segmento
segment type tipo de segmento
segmental segmental
segmental conductor conductor segmental
segmental meter medidor segmental
segmental voltmeter vóltmetro segmental
segmentation segmentación
segmentation unit unidad de segmentación
segmented segmentado
segmented program programa segmentado
segmenting segmentación
Seignette salt sal de Seignette
seismogram sismograma
seismograph sismógrafo
seismometer sismómetro
seismophone sismófono
seismoscope sismoscopio
seizing signal señal de toma
seizure signal señal de toma
selcal selcal

select seleccionar
selectable seleccionable
selectable element elemento seleccionable
selectance selectancia
selected seleccionado
selected cell celda seleccionada
selected frequency frecuencia seleccionada
selected mode modo seleccionado
selecting circuit circuito de selección
selecting commutator conmutador de selección
selecting device dispositivo de selección
selecting mechanism mecanismo de selección
selection selección
selection card tarjeta de selección
selection check comprobación de selección
selection circuit circuito de selección
selection code código de selección
selection control control de selección
selection counter contador de selección
selection field campo de selección
selection instruction instrucción de selección
selection level nivel de selección
selection menu menú de selección
selection priority prioridad de selección
selection ratio razón de selección
selection signal señal de selección
selection sort clasificación de selección
selection stage etapa de selección
selection structure estructura de selección
selection switch conmutador de selección
selective selectivo
selective absorption absorción selectiva
selective amplifier amplificador selectivo
selective calling llamada selectiva
selective-calling code código de llamada selectiva
selective corrosion corrosión selectiva
selective detector detector selectivo
selective diffuser difusor selectivo
selective diffusion difusión selectiva
selective digit emitter emisor de dígitos selectivo
selective dump vaciado selectivo
selective electrodeposition electrodeposición selectiva
selective emission emisión selectiva
selective emitter emisor selectivo
selective erase borrado selectivo
selective-erase head cabeza de borrado selectivo
selective fading desvanecimiento selectivo
selective filter filtro selectivo
selective information información selectiva
selective inspection inspección selectiva
selective interference interferencia selectiva
selective jamming interferencia intencional selectiva
selective localization localización selectiva
selective network red selectiva
selective opening abertura selectiva
selective polarization polarización selectiva
selective protection protección selectiva
selective radiation radiación selectiva
selective radiator radiador selectivo
selective receiver receptor selectivo
selective reflection reflexión selectiva
selective relay relé selectivo
selective release desenganche selectivo
selective resonance resonancia selectiva
selective response curve curva de respuesta selectiva
selective ringing llamada selectiva
selective sequence secuencia selectiva
selective squelch silenciador selectivo
selective system sistema selectivo
selective trace program programa de análisis selectivo
selective transmission transmisión selectiva

selective tuning sintonización selectiva
selectively selectivamente
selectivity selectividad
selectivity characteristic característica de selectividad
selectivity control control de selectividad
selectivity curve curva de selectividad
selector selector
selector channel canal selector
selector circuit circuito selector
selector coil bobina selectora
selector magnet imán selector
selector mode modo selector
selector plug clavija de selector
selector pulse impulso selector
selector relay relé selector
selector switch conmutador selector
selector tube tubo selector
selectron selectrón
selenium selenio
selenium amplifier amplificador de selenio
selenium cell célula de selenio
selenium diode diodo de selenio
selenium dioxide dióxido de selenio
selenium photocell fotocélula de selenio
selenium photovoltaic cell célula fotovoltaica de selenio
selenium rectifier rectificador de selenio
selenium relay relé de selenio
self-absorption autoabsorción
self-absorption curve curva de autoabsorción
self-adapting autoadaptivo
self-adapting program programa autoadaptivo
self-adaptive autoadaptivo
self-adaptive computer computadora autoadaptiva
self-adjustable autoajustable
self-aligning autoalineante
self-aligning bearing cojinete autoalineable
self-baking electrode electrodo de autococción
self-balanced autoequilibrado
self-balanced bridge puente autoequilibrado
self-balancing autoequilibrador
self-balancing recorder registrador autoequilibrador
self-bias autopolarización
self-bias resistor resistor de autopolarización
self-biased autopolarizado
self-biased grid rejilla autopolarizada
self-biasing autopolarizante
self-calibrating autocalibrador
self-calibrating instrument instrumento autocalibrador
self-calibration autocalibración
self-capacitance autocapacitancia
self-centering autocentrador
self-check autocomprobación
self-checking autocomprobación
self-checking code código de autocomprobación
self-checking digit dígito de autocomprobación
self-checking instrument instrumento con autocomprobación
self-checking number número de autocomprobación
self-checking routine rutina de autocomprobación
self-cleaning contacts contactos autolimpiantes
self-closing autocerrador
self-compensated autocompensado
self-complementing code código con autocomplementación
self-consistent autoconsistente
self-constriction autoconstricción
self-contained autónomo, incorporado, completo
self-contained control control autónomo
self-contained device dispositivo autónomo
self-contained instrument instrumento autónomo

self-contained mobile unit unidad móvil autónoma
self-contained navigation aid ayuda de navegación autónoma
self-contained power supply fuente de alimentación autónoma
self-contained station estación autónoma
self-control autocontrol
self-controlled autocontrolado
self-controlled oscillator oscilador autocontrolado
self-cooled autoenfriado
self-cooled transformer transformador autoenfriado
self-cooling autoenfriamiento
self-correcting autocorrector
self-correcting code código autocorrector
self-damping autoamortiguamiento
self-diagnostic autodiagnósitco
self-discharge autodescarga
self-energy energía propia
self-equalizing autoecualización
self-erasing autoborrado
self-excitation autoexcitación
self-excited autoexcitado
self-excited alternator alternador autoexcitado
self-excited generator generador autoexcitado
self-excited machine máquina autoexcitada
self-excited oscillator oscilador autoexcitado
self-excited transmitter transmisor autoexcitado
self-exciting autoexcitador
self-extinguishing autoextintor
self-extinguishing circuit circuito autoextintor
self-extinguishing thyratron tiratrón autoextintor
self-feeding autoalimentador
self-focus picture tube tubo de imagen autoenfocador
self-focusing picture tube tubo de imagen autoenfocador
self-generating photocell fotocélula autogeneradora
self-generating transducer transductor autogenerador
self-generative autogenerativo
self-governing autorregulador
self-healing capacitor capacitor autorregenerativo
self-heated autocalentado
self-heated thermocouple termopar autocalentado
self-heating autocalentable
self-heating thermistor termistor autocalentable
self-ignition autoignición, autoencendido
self-impedance autoimpedancia
self-inductance autoinductancia
self-induction autoinducción
self-inductive autoinductivo
self-inductor autoinductor
self-interrupted autointerrumpido
self-interrupted circuit circuito autointerrumpido
self-interrupter autointerruptor
self-latching autoenganchador
self-latching relay relé autoenganchador
self-latching switch conmutador autoenganchador
self-limiting autolimitador
self-loading autocarga
self-loading routine rutina de autocarga
self-lubricating autolubricante
self-lubrication autolubricación
self-luminous autoluminoso
self-maintained automantenido
self-maintained discharge descarga automantenida
self-modification automodificación
self-modulated automodulado
self-modulated amplifier amplificador automodulado
self-modulated oscillator oscilador automodulado
self-operated autooperado, automático
self-organizing autoorganizador
self-orienting autoorientador

self-orienting mechanism mecanismo autoorientador
self-oscillating autooscilante
self-oscillation autooscilación
self-oscillator autooscilador
self-powered autoalimentado
self-powered control control autoalimentado
self-powered device dispositivo autoalimentado
self-powered station estación autoalimentada
self-propagating autopropagante
self-propagation autopropagación
self-propelled autopropulsado
self-protected autoprotegido
self-protected tube tubo autoprotegido
self-protected winding arrollamiento autoprotegido
self-protection autoprotección
self-pulse autoimpulso
self-pulse modulation modulación por autoimpulso
self-pulsing autopulsante
self-pulsing blocking oscillator oscilador de bloqueo autopulsante
self-quenched counter tube tubo contador autoextintor
self-quenching oscillator oscilador autoextintor
self-recording autorregistro
self-recording instrument instrumento autorregistrador
self-rectification autorrectificación
self-rectifier autorrectificador
self-rectifying autorrectificador
self-rectifying device dispositivo autorrectificador
self-rectifying tube tubo autorrectificador
self-rectifying vibrator vibrador autorrectificador
self-regulating autorregulación
self-regulating recorder registrador autorregulador
self-regulation autorregulación
self-reset autorreposición, autorrestablecimiento
self-reset relay relé con autorreposición
self-resetting autorreposición
self-resetting circuit-breaker cortacircuito con autorreposición
self-resetting loop bucle con autorreposición
self-resetting relay relé con autorreposición
self-resistance autorresistencia
self-resonance autorresonancia
self-resonant autorresonante
self-resonant circuit circuito autorresonante
self-resonant frequency frecuencia autorresonante
self-reversal autoinversión
self-rotation autorrotación
self-saturating autosaturante
self-saturating circuit circuito autosaturante
self-saturation autosaturación
self-scattering autodispersión
self-screening autoapantallamiento, autoblindaje
self-sealing autosellador
self-shielding autoblindaje
self-soldering autosoldadura
self-stabilization autoestabilización
self-stabilizing autoestabilizador
self-starting autoarranque
self-starting motor motor con autoarranque
self-starting synchronous motor motor sincrónico con autoarranque
self-sustained autosostenido
self-sustained discharge descarga autosostenida
self-sustained oscillations oscilaciones autosostenidas
self-synchronizing autosincronización
self-synchronous autosincrónico
self-synchronous device dispositivo autosincrónico
self-synchronous instrument instrumento autosincrónico
self-test autoprueba, autocomprobación

self-tuned autosintonizado
self-tuning autosintonización
self-ventilated autoventilado
self-ventilated machine máquina autoventilada
self-ventilated motor motor autoventilado
self-ventilation autoventilación
self-wiping contacts contactos autolimpiantes
selsyn selsyn
selsyn generator generador selsyn
selsyn motor motor selsyn
selsyn system sistema selsyn
selsyn transmitter transmisor selsyn
semantic semántico
semantic error error semántico
semantics semántica
semaphore semáforo
semator semator
semiactive semiactivo
semiactive guidance guiado semiactivo
semiactive repeater repetidor semiactivo
semiadjustable semiajustable
semiadjustable control control semiajustable
semiattended machine máquina semiatendida
semiautomatic semiautomático
semiautomatic controller controlador
    semiautomático
semiautomatic electroplating galvanoplastia
    semiautomática
semiautomatic exchange central semiautomática
semiautomatic installation instalación
    semiautomática
semiautomatic machine máquina semiautomática
semiautomatic operation operación semiautomática
semiautomatic plating depósito semiautomático
semiautomatic ringing llamada semiautomática
semiautomatic signal señal semiautomática
semiautomatic starter arrancador semiautomático
semiautomatic station estación semiautomática
semiautomatic substation subestación
    semiautomática
semiautomatic switching conmutación
    semiautomática
semiautomatic switching system sistema de
    conmutación semiautomática
semiautomatic system sistema semiautomático
semiautomatic telephone system sistema telefónico
    semiautomático
semiautomatic test equipment equipo de prueba
    semiautomático
semicircular error error semicircular
semicompiled semicompilado
semiconductible semiconductible
semiconducting semiconductor
semiconducting ceramic cerámica semiconductora
semiconducting element elemento semiconductor
semiconducting material material semiconductor
semiconducting region región semiconductora
semiconducting tape cinta semiconductora
semiconductor semiconductor
semiconductor amplifier amplificador
    semiconductor
semiconductor chip chip semiconductor
semiconductor contact contacto semiconductor
semiconductor-controlled rectifier rectificador
    controlado por semiconductor
semiconductor detector detector semiconductor
semiconductor device dispositivo semiconductor
semiconductor diode diodo semiconductor
semiconductor doping dopado de semiconductor
semiconductor frequency changer cambiador de
    frecuencia de semiconductor
semiconductor integrated circuit circuito integrado
    semiconductor

semiconductor junction unión semiconductora
semiconductor laser láser de semiconductor
semiconductor maser máser de semiconductor
semiconductor material material semiconductor
semiconductor memory memoria semiconductora
semiconductor-metal junction unión
    semiconductor-metal
semiconductor microphone micrófono
    semiconductor
semiconductor optical maser máser óptico de
    semiconductor
semiconductor power converter convertidor de
    potencia de semiconductor
semiconductor rectifier rectificador de
    semiconductor
semiconductor relay relé de semiconductor
semiconductor storage almacenamiento por
    semiconductor
semiconductor switch conmutador de semiconductor
semiconductorized semiconductorizado
semicontinuous semicontinuo
semicrystalline semicristalino
semicycle semiciclo
semidirect semidirecto
semidirect lighting alumbrado semidirecto
semidirectional semidireccional
semidirectional microphone micrófono
    semidireccional
semiduplex semidúplex
semiduplex circuit circuito semidúplex
semiduplex communication comunicación
    semidúplex
semiduplex operation operación semidúplex
semienclosed semiencerrado, semicerrado
semiindirect lighting luz semiindirecta
semilogarithmic semilogarítmico
semimagnetic controller controlador semimagnético
semimechanical semimecánico
semimechanical system sistema semimecánico
semimetal semimetal
semimetallic semimetálico
semiportable semiportátil
semiproportional semiproporcional
semiprotected semiprotegido
semiprotected machine máquina semiprotegida
semiprotected motor motor semiprotegido
semiremote control control semirremoto
semiresonant semirresonante
semiresonant line línea semirresonante
semiresonant transmission line línea de transmisión
    semirresonante
semirigid semirrígido
semiselective semiselectivo
semiselective ringing llamada semiselectiva
semistable semiestable
semitone semitono
semitransparent semitransparente
semitransparent photocathode fotocátodo
    semitransparente
semiuniversal semiuniversal
senary senario
send-receive relay relé de emisión-recepción
send-receive switch conmutador de
    emisión-recepción
sender emisor, manipulador
sending transmisión, emisión
sending amplifier amplificador de transmisión
sending antenna antena de transmisión
sending end extremo de transmisión
sending-end impedance impedancia del extremo de
    transmisión
sending field campo de transmisión
sending filter filtro de transmisión

sending installation instalación de transmisión
sending oscillator oscilador de transmisión
sending set aparato de transmisión
sending stage etapa de transmisión
sending station estación de transmisión
sense sentido, detección
sense antenna antena de sentido
sense determination determinación de sentido
sense finder indicador de sentido
sense finding indicación de sentido
sense switch conmutador de sentido
sense terminal terminal de detección
sensibility sensibilidad
sensing detección, lectura
sensing circuit circuito sensor
sensing device dispositivo sensor, dispositivo de lectura
sensing element elemento sensor
sensing mark marca sensora
sensing station estación de lectura
sensing unit unidad sensora
sensing wire hilo de lectura
sensitive sensible
sensitive device dispositivo sensible
sensitive element elemento sensible
sensitive layer capa sensible
sensitive lining revestimiento sensible
sensitive magnetic relay relé magnético sensible
sensitive plate placa sensible
sensitive receiver receptor sensible
sensitive relay relé sensible
sensitive switch conmutador sensible
sensitiveness sensibilidad
sensitivity sensibilidad
sensitivity adjustment ajuste de sensibilidad
sensitivity analysis análisis de sensibilidad
sensitivity coefficient coeficiente de sensibilidad
sensitivity control control de sensibilidad
sensitivity decrease reducción de sensibilidad
sensitivity level nivel de sensibilidad
sensitivity loss pérdida de sensibilidad
sensitivity ratio razón de sensibilidad
sensitivity regulator regulador de sensibilidad
sensitivity test prueba de sensibilidad
sensitivity-time control control de tiempo de sensibilidad
sensitization sensibilización
sensitize sensibilizar
sensitized sensibilizado
sensitized surface superficie sensibilizada
sensitizer sensibilizador
sensitizing sensibilización
sensitometer sensitómetro
sensor sensor
sensor-based basado en sensores
sensor-based computer computadora basada en sensores
sensor-control system sistema de control de sensores
separability separabilidad
separable separable
separate separado
separate compilation compilación separada
separate excitation excitación separada
separately separadamente
separately compiled program programa compilado separadamente
separately excited excitado separadamente
separately excited device dispositivo excitado separadamente
separately excited machine máquina excitada separadamente
separating separador
separating amplifier amplificador separador

separating capacitor capacitor separador
separating character carácter separador
separating filter filtro separador
separating tube tubo separador
separation separación
separation circuit circuito de separación
separation distance distancia de separación
separation filter filtro de separación
separation loss pérdida por separación
separation of frequencies separación de frecuencias
separation panel panel de separación
separator separador
separator character carácter separador
separator circuit circuito separador
separator diaphragm diafragma separador
separator tube tubo separador
septanary septenario
septate coaxial cavity cavidad coaxial tabicada
septate waveguide guíaondas tabicado
septendecimal septendecimal
septet septeto
septum septo, tabique
sequence secuencia
sequence break ruptura de secuencia
sequence brush escobilla de secuencia
sequence characteristic característica de secuencia
sequence check comprobación de secuencia
sequence checking comprobación de secuencia
sequence-checking routine rutina de comprobación de secuencia
sequence control control de secuencia
sequence-control register registro de control de secuencia
sequence-control structure estructura de control de secuencia
sequence-control unit unidad de control de secuencia
sequence controller controlador de secuencia
sequence counter contador de secuencias
sequence error error de secuencia
sequence field campo de secuencia
sequence information información de secuencia
sequence number número de secuencia
sequence of operations secuencia de operaciones
sequence of statements secuencia de sentencias
sequence processor procesador de secuencias
sequence programmer programador de secuencias
sequence register registro de secuencias
sequence relay relé de secuencia
sequence set conjunto de secuencias
sequence signal señal de secuencia
sequence specifications especificaciones de secuencia
sequence structure estructura de secuencia
sequence switch conmutador de secuencia
sequence symbol símbolo de secuencia
sequence table tabla de secuencia
sequence timer temporizador de secuencia
sequence unit unidad de secuencia
sequencer secuenciador
sequential secuencial
sequential access acceso secuencial
sequential-access memory memoria de acceso secuencial
sequential-access method método de acceso secuencial
sequential-access storage almacenamiento de acceso secuencial
sequential alarm alarma secuencial
sequential algorithm algoritmo secuencial
sequential analysis análisis secuencial
sequential analyzer analizador secuencial
sequential batch processing procesamiento por lotes secuencial

sequential circuit circuito secuencial
sequential coding codificación secuencial
sequential-coding network red de codificación
    secuencial
sequential cohesion cohesión secuencial
sequential color television televisión a colores
    secuencial
sequential commutation conmutación secuencial
sequential computer computadora secuencial
sequential control control secuencial
sequential-control line línea de control secuencial
sequential data access acceso a datos secuencial
sequential data structure estructura de datos
    secuencial
sequential decoding descodificación secuencial
sequential device dispositivo secuencial
sequential discharge tube tubo de descarga
    secuencial
sequential element elemento secuencial
sequential file archivo secuencial
sequential interlace entrelazado secuencial
sequential interlocking enclavamiento secuencial
sequential list lista secuencial
sequential logic lógica secuencial
sequential logic element elemento de lógica
    secuencial
sequential logic system sistema de lógica secuencial
sequential logical function función lógica secuencial
sequential machine máquina secuencial
sequential memory memoria secuencial
sequential monitoring monitoreo secuencial
sequential operation operación secuencial
sequential organization organización secuencial
sequential output salida secuencial
sequential processing procesamiento secuencial
sequential program programa secuencial
sequential programming programación secuencial
sequential queue cola secuencial
sequential relay relé secuencial
sequential relay circuit circuito de relé secuencial
sequential sampling muestreo secuencial
sequential scanner explorador secuencial
sequential scanning exploración secuencial
sequential search búsqueda secuencial
sequential signal señal secuencial
sequential starter arrancador secuencial
sequential storage almacenamiento secuencial
sequential-storage device dispositivo de
    almacenamiento secuencial
sequential switch conmutador secuencial
sequential switching conmutación secuencial
sequential synchronization sincronización secuencial
sequential system sistema secuencial
sequential timer temporizador secuencial
sequential transfer transferencia secuencial
sequential transmission transmisión secuencial
sequentially secuencialmente
sequentially operated operado secuencialmente
serdes serdes
serial serial, en serie
serial access acceso en serie
serial-access memory memoria de acceso en serie
serial-access storage almacenamiento de acceso en
    serie
serial-access system sistema de acceso en serie
serial adder sumador en serie
serial addition suma en serie
serial arithmetic aritmética en serie
serial-arithmetic unit unidad de aritmética en serie
serial bit bit en serie
serial by bit en serie por bits
serial by character en serie por caracteres
serial by word en serie por palabras

serial code código en serie
serial communication comunicación en serie
serial computer computadora en serie
serial connection conexión en serie
serial data datos en serie
serial decoding descodificación en serie
serial digital adder sumador digital en serie
serial digital computer computadora digital en serie
serial feed alimentación en serie
serial feeding alimentación en serie
serial full-adder sumador completo en serie
serial full-subtracter restador completo en serie
serial half-adder semisumador en serie
serial half-subtracter semirrestador en serie
serial input-output entrada-salida en serie
serial input-output channel canal de entrada-salida
    en serie
serial input-output controller controlador de
    entrada-salida en serie
serial input-output port puerto de entrada-salida en
    serie
serial interface interfaz en serie
serial-interface card tarjeta de interfaz en serie
serial memory memoria en serie
serial-memory system sistema de memoria en serie
serial mouse ratón en serie
serial number número en serie
serial numbering numeración en serie
serial numbering station estación de numeración en
    serie
serial operation operación en serie
serial output salida en serie
serial-parallel serie-paralelo
serial port puerto serie
serial printer impresora en serie
serial processing procesamiento en serie
serial programming programación en serie
serial punching perforación en serie
serial reader lectora en serie
serial reading lectura en serie
serial search búsqueda en serie
serial sort clasificación en serie
serial storage almacenamiento en serie
serial terminal terminal en serie
serial-to-parallel converter convertidor de serie a
    paralelo
serial transfer transferencia en serie
serial transmission transmisión en serie
serial word palabra en serie
serializer-deserializer serializador-deserializador
serially en serie
serially numbered numerado en serie
series serie
series-aiding en serie aditiva
series arrangement arreglo en serie
series capacitance capacitancia en serie
series capacitor capacitor en serie
series circuit circuito en serie
series coil bobina en serie
series compensation compensación en serie
series condenser condensador en serie
series-connected conectado en serie
series connection conexión en serie
series detector detector en serie
series dropping resistor resistor de caída en serie
series dynamo dínamo en serie
series elements elementos en serie
series equalizer ecualizador en serie
series excitation excitación en serie
series exciter excitador en serie
series-fed amplifier amplificador alimentado en serie
series-fed oscillator oscilador alimentado en serie
series-fed vertical antenna antena vertical

alimentada en serie
series feed alimentación en serie
series-feed system sistema de alimentación en serie
series feedback retroalimentación en serie
series field campo por arrollamiento en serie
series filter filtro en serie
series generator generador en serie
series heater calentador en serie
series impedance impedancia en serie
series inductance inductancia en serie
series inductors inductores en serie
series lamp lámpara en serie
series limiter limitador en serie
series loading carga en serie
series magnetic circuits circuitos magnéticos en serie
series modulation modulación en serie
series modulator modulador en serie
series motor motor en serie
series operation operación en serie
series-opposing en serie opositiva
series-parallel serie-paralelo
series-parallel arrangement arreglo en serie-paralelo
series-parallel capacitors capacitores en serie-paralelo
series-parallel circuit circuito en serie-paralelo
series-parallel connection conexión en serie-paralelo
series-parallel control control en serie-paralelo
series-parallel inductors inductores en serie-paralelo
series-parallel network red en serie-paralelo
series-parallel resistors resistores en serie-paralelo
series-parallel switch conmutador en serie-paralelo, combinador
series-parallel winding arrollamiento en serie-paralelo
series-peaking coil bobina correctora en serie
series reactor reactor en serie
series rectifier circuit circuito rectificador en serie
series regulator regulador en serie
series relay relé en serie
series resistance resistencia en serie
series resistors resistores en serie
series resonance resonancia en serie
series-resonance bridge puente de resonancia en serie
series-resonant circuit circuito resonante en serie
series-shunt serie-derivación
series-shunt compensation compensación en serie-derivación
series-shunt network red en serie-derivación
series system sistema en serie
series tee junction unión en T serie
series transducer transductor de acoplamiento en serie
series transformer transformador en serie
series-tuned circuit circuito sintonizado en serie
series welding soldadura en serie
series winding arrollamiento en serie
series-wound arrollado en serie
series-wound generator generador arrollado en serie
series-wound motor motor arrollado en serie
serrasoid modulator modulador serrasoide
serrated pulse impulso aserrado
serrated rotor plate placa de rotor con muescas
serrated vertical pulse impulso vertical aserrado
serrated vertical synchronizing pulse impulso de sincronismo vertical aserrado
serrodyne serrodino
server servidor
service servicio, acometida
service area área de servicio
service band banda de servicio

service bits bits de servicio
service box caja de acometida
service cable cable de acometida
service capacity capacidad de servicio
service channel canal de servicio
service-channel bridge puente de canal de servicio
service-channel demodulator desmodulador de canal de servicio
service-channel modulator modulador de canal de servicio
service circuit circuito de servicio
service class clase de servicio
service code código de servicio
service conditions condiciones de servicio
service conductor conductor de acometida
service connection conexión de servicio, acometida
service connector conector de servicio
service corrosion corrosión por servicio
service current corriente de servicio
service distortion distorsión de servicio
service drop colgante de acometida
service entrance acometida
service environment ambiente de servicio
service equipment equipo de servicio
service factor factor de servicio
service fuse fusible de acometida
service hours horas de servicio
service level nivel de servicio
service life vida de servicio
service line línea de servicio, acometida
service load carga de servicio
service meter contador de servicio
service mode modo de servicio
service modem módem de servicio
service observation observación de servicio
service oscillator oscilador de servicio
service panel panel de servicio
service period periodo de servicio
service program programa de servicio
service range alcance de servicio
service rating clasificación de servicio
service requirement requisito de servicio
service routine rutina de servicio
service set aparato de servicio
service switch conmutador de servicio
service test prueba de servicio
service time tiempo de servicio
service unit unidad de servicio
service voltage tensión de servicio, tensión de acometida
serviceability ratio razón de utilidad
serving revestimiento
servo altimeter servoaltímetro
servo amplifier servoamplificador
servo analyzer servoanalizador
servo assistance servoayuda
servo assisted servoayudado
servo circuit servocircuito
servo control servocontrol
servo controlled servocontrolado
servo controller servocontrolador
servo cylinder servocilindro
servo driven servoaccionado
servo governor servorregulador
servo integrator servointegrador
servo loop servobucle
servo manipulator servomanipulador
servo mechanism servomecanismo
servo modulator servomodulador
servo motor servomotor
servo motorized servomotorizado
servo multiplier servomultiplicador
servo-operated servooperado

servo oscillation oscilación de servomecanismo
servo potentiometer servopotenciómetro
servo probe servosonda
servo regulator servorregulador
servo system servosistema
servo valve servoválvula
servoactuated servoaccionado
servoactuator servoactuador
servoaltimeter servoaltímetro
servoamplifier servoamplificador
servoanalyzer servoanalizador
servoassistance servoayuda
servoassisted servoayudado
servocircuit servocircuito
servocontrol servocontrol
servocontrolled servocontrolado
servocontroller servocontrolador
servocylinder servocilindro
servodriven servoaccionado
servogovernor servorregulador
servointegrator servorintegrador
servoloop servobucle
servomanipulator servomanipulador
servomecanics servomecánica
servomechanism servomecanismo
servomodulator servomodulador
servomotor servomotor
servomotor amplifier amplificador de servomotor
servomotorized servomotorizado
servomultiplier servomultiplicador
servooperated servooperado
servooscillation oscilación de servomecanismo
servopotentiometer servopotenciómetro
servoprobe servosonda
servoregulator servorregulador
servosystem servosistema
servovalve servoválvula
session sesión
session activation activación de sesión
session activation request solicitud de activación de sesión
session control control de sesión
session deactivation desactivación de sesión
session deactivation request solicitud de desactivación de sesión
session end fin de sesión
session establishment establecimiento de sesión
session establishment request solicitud de establecimiento de sesión
session initiation iniciación de sesión, inicio de sesión
session initiation request solicitud de iniciación de sesión
session level nivel de sesión
session parameter parámetro de sesión
session termination terminación de sesión
session termination request solicitud de terminación de sesión
set conjunto, aparato
set analyzer aparato analizador
set noise ruido de aparato
set point punto fijado
set pulse impulso de disposición, impulso de definición
set terminal terminal de disposición, terminal de definición
set-up preparar, configurar, montar
set-up circuit circuito de preparación
set-up impulse impulso de preparación
set-up time tiempo de preparación, tiempo de montura
set value valor fijado
setting ajuste, posición, calibración, disposición

setting index índice de regulación
setting knob botón de regulación
setting point punto de regulación
setting range límites de regulación
settling estabilización, establecimiento
settling time tiempo de estabilización, tiempo de establecimiento
setup montaje, ajuste inicial, preparación, fijación de nivel
setup adjustment ajuste inicial
setup circuit circuito de preparación
setup control control inicial
setup hours horas de preparación
setup procedure procedimiento de preparación
setup time tiempo de preparación
seven-bit byte byte de siete bits
sexadecimal sexadecimal
sexadecimal digit dígito sexadecimal
sexadecimal notation notación sexadecimal
sexadecimal number número sexadecimal
sexadecimal number system sistema numérico sexadecimal
sexagesimal sexagesimal
sexagesimal number system sistema numérico sexagesimal
sextet sexteto
shaded pole polo sombreado
shaded-pole motor motor de polo sombreado
shaded-pole relay relé de polo sombreado
shading sombreado, sombra
shading coil bobina de sombra
shading compensation compensación de sombra
shading-compensation signal señal de compensación de sombra
shading correction corrección de sombra
shading ring anillo de sombra
shading signal señal de compensación de sombra
shadow angle ángulo de sombra
shadow area área de sombra
shadow attenuation atenuación de sombra
shadow effect efecto de sombra
shadow factor factor de sombra
shadow loss pérdida por sombra
shadow mask máscara de sombra
shadow photometer fotómetro de sombra
shadow region región de sombra
shadow-tuning indicator indicador de sintonía por sombra
shadow zone zona de sombra
shaft eje, barra, haz
shaft angle ángulo de eje
shaft-angle encoder codificador de ángulo de eje
shaft contact contacto de eje
shaft current corriente de eje
shaft position posición de eje
shaft-position encoder codificador de posición de eje
shaft-position indicator indicador de posición de eje
shake table mesa de sacudidas, mesa vibratoria
shake test prueba de sacudidas
shakeproof a prueba de sacudidas
shaker sacudidor, vibrador
Shannon's theorem teorema de Shannon
shape factor factor de forma
shaped beam haz conformado
shaped-beam antenna antena de haz conformado
shaped-beam tube tubo de haz conformado
shaped conductor conductor conformado
shaper formador
shaping formación, perfilado
shaping circuit circuito formador
shaping network red correctora
shared access acceso compartido
shared-access path trayectoria de acceso compartido

shared band banda compartida
shared channel canal compartido
shared control control compartido
shared database base de datos compartida
shared device dispositivo compartido
shared environment ambiente compartido
shared executive system sistema ejecutivo compartido
shared file archivo compartido
shared files system sistema de archivos compartidos
shared information system sistema de información compartida
shared line línea compartida
shared logic lógica compartida
shared memory memoria compartida
shared partition partición compartida
shared resources recursos compartidos
shared routine rutina compartida
shared segment segmento compartido
shared service servicio compartido
shared storage almacenamiento compartido
shared systems sistemas compartidos
shared use uso compartido
shared virtual area área virtual compartida
shared wave onda compartida
sharing compartimiento, compartición
sharing of frequencies compartimiento de frecuencias
sharing rules reglas de compartimiento
sharp agudo, nítido
sharp bandpass filter filtro de paso de banda agudo
sharp cutoff corte agudo
sharp-cutoff filter filtro de corte agudo
sharp-cutoff pentode pentodo de corte agudo
sharp-cutoff tube tubo de corte agudo
sharp image imagen nítida
sharp pulse impulso agudo
sharp tuning sintonización fina
sharpness agudeza, nitidez
sharpness limit límite de nitidez
shaving afeitado
shear wave onda rotacional
sheath vaina, cubierta, revestimiento
sheath temperature temperatura de vaina
sheathing envainado, revestimiento
sheet detector detector de hojas
sheet feeder alimentador de hojas
shelf corrosion corrosión por almacenamiento
shelf depreciation depreciación por almacenamiento
shelf life vida en almacenamiento
shelf temperature temperatura de almacenamiento
shelf test prueba de almacenamiento
shell capa, cubierta, cuba, casco
shell circuit resonador bivalvo
shell-type transformer transformador acorazado
Shepherd tube tubo de Shepherd
shield blindaje, pantalla
shield effectiveness efectividad de blindaje
shield grid rejilla de apantallamiento
shield percentage porcentaje de apantallamiento
shielded blindado, apantallado, protegido
shielded building edificio apantallado
shielded cable cable apantallado
shielded conductor cable cable apantallado
shielded electromagnet electroimán apantallado
shielded electroscope electroscopio apantallado
shielded ignition ignición blindada
shielded joint junta blindada
shielded line línea blindada
shielded loop cuadro apantallado
shielded-loop antenna antena de cuadro apantallado
shielded pair par apantallado
shielded probe sonda apantallada

shielded transmission line línea de transmisión apantallada
shielded wire hilo apantallado
shielded X-ray tube tubo de rayos X blindado
shielding blindaje, apantallamiento
shielding effect efecto de apantallamiento
shielding factor factor de apantallamiento
shift desplazamiento, desviación, cambio
shift angle ángulo de desplazamiento
shift byte byte de desplazamiento
shift character carácter de desplazamiento
shift control control de desplazamiento
shift in frequency desplazamiento de frecuencia
shift instruction instrucción de desplazamiento
shift key tecla de cambio
shift pulse impulso de desplazamiento
shift register registro de desplazamiento
shift unit unidad de desplazamiento
shifting desplazamiento, desviación, cambio
shifting element elemento de desplazamiento
ship station estación de barco
ship-to-ship communication comunicación de barco a barco
ship-to-shore communication comunicación de barco a tierra
shipboard antenna antena de barco
shipboard equipment equipo de barco
shock choque, sacudida
shock absorber amortiguador
shock excitation excitación por choque
shock-excited oscillator oscilador excitado por choque
shock hazard riesgo de choque eléctrico
shock-resistant resistente a choques
shock-resistant relay relé resistente a choques
shock test prueba de choque
shock wave onda de choque
Shockley diode diodo de Schockley
shockproof a prueba de choques
shockproof electrical apparatus aparato eléctrico a prueba de choques
shoe zapata
shop-assembled ensamblado en taller
shop test prueba de taller
shoran shorán
shore effect difracción costera
shore station estación costera
shore-to-ship communication comunicación de tierra a barco
short cortocircuito
short block bloque corto
short break interrupción corta
short-circuit cortocircuito
short-circuit admittance admitancia en cortocircuito
short-circuit brake freno de cortocircuito
short-circuit breakdown voltage tensión de ruptura en cortocircuito
short-circuit characteristic característica en cortocircuito
short-circuit conductance conductancia en cortocircuito
short-circuit current corriente de cortocircuito
short-circuit detector detector de cortocircuito
short-circuit feedback admittance admitancia de retroalimentación en cortocircuito
short-circuit finder detector de cortocircuito
short-circuit forward admittance admitancia directa en cortocircuito
short-circuit impedance impedancia en cortocircuito
short-circuit inductance inductancia en cortocircuito
short-circuit input admittance admitancia de entrada en cortocircuito
short-circuit input capacitance capacitancia de

entrada en cortocircuito
short-circuit output admittance admitancia de salida en cortocircuito
short-circuit output capacitance capacitancia de salida en cortocircuito
short-circuit parameter parámetro en cortocircuito
short-circuit protection protección contra cortocircuitos
short-circuit ratio razón de cortocircuito
short-circuit reactance reactancia en cortocircuito
short-circuit resistance resistencia en cortocircuito
short-circuit response respuesta en cortocircuito
short-circuit response curve curva de respuesta en cortocircuito
short-circuit stability estabilidad en cortocircuito
short-circuit test prueba en cortocircuito
short-circuit transadmittance transadmitancia en cortocircuito
short-circuit voltage tensión en cortocircuito
short-circuit welding soldadura por cortocircuito
short-circuited cortocircuitado
short-circuited line línea cortocircuitada
short-circuiter cortocircuitador
short-circuiting cortocircuitador
short-circuiting bar barra cortocircuitadora
short-circuiting device dispositivo cortocircuitador
short-circuiting section sección cortocircuitadora
short-circuiting switch conmutador cortocircuitador
short-circuitproof a prueba de cortocircuitos
short code código corto
short dipole dipolo corto
short-distance circuit circuito de corta distancia
short-duration current pulse impulso de corriente de corta duración
short-duration pulse impulso de corta duración
short-duration voltage pulse impulso de tensión de corta duración
short format formato corto
short indicator indicador de cortocircuito
short instruction instrucción corta
short-instruction format formato de instrucción corta
short interval intervalo corto
short period periodo corto
short-pitch winding arrollamiento de paso corto
short pulse impulso corto
short range corto alcance
short-range navigation system sistema de navegación de corto alcance
short-range radar radar de corto alcance
short-range radiobeacon radiofaro de corto alcance
short response time tiempo corto de respuesta
short rise time tiempo corto de subida
short skip salto corto
short-skip communication comunicación por saltos cortos
short term corto plazo
short-term stability estabilidad a corto plazo
short-time current corriente de corto tiempo
short-time duty servicio de corto tiempo
short-time effect efecto de corto tiempo
short-time overload sobrecarga de corto tiempo
short-time rating capacidad de carga de corto tiempo
short-time stability estabilidad de corto tiempo
short-wave onda corta
short-wave adapter adaptador de ondas cortas
short-wave antenna antena de ondas cortas
short-wave converter convertidor de ondas cortas
short-wave directional antenna antena direccional de ondas cortas
short-wave oscillator oscilador de ondas cortas
short-wave range gama de ondas cortas
short-wave receiver receptor de ondas cortas

short-wave spectrum espectro de ondas cortas
short-wave transmitter transmisor de ondas cortas
shortable amplifier amplificador cortocircuitable
shorted cortocircuitado
shortening capacitor capacitor de acortamiento
shortening condenser condensador de acortamiento
shorting cortocircuitado
shorting bar barra cortocircuitadora
shorting contact contacto cortocircuitador
shorting link enlace cortocircuitador
shorting plug clavija cortocircuitadora
shorting relay relé cortocircuitador
shorting segment segmento cortocircuitador
shorting switch conmutador cortocircuitador
shortwave onda corta
shortwave adapter adaptador de ondas cortas
shortwave antenna antena de ondas cortas
shortwave converter convertidor de ondas cortas
shortwave directional antenna antena direccional de ondas cortas
shortwave oscillator oscilador de ondas cortas
shortwave range gama de ondas cortas
shortwave receiver receptor de ondas cortas
shortwave spectrum espectro de ondas cortas
shortwave transmitter transmisor de ondas cortas
shot effect efecto de granalla
shot-effect noise ruido de efecto de granalla
shot noise ruido de granalla
shrink encogimiento
shrink tubing tubería termocontráctil
shrinkage encogimiento
shunt derivación, shunt
shunt apparatus aparato en derivación
shunt capacitance capacitancia en derivación
shunt capacitor capacitor en derivación
shunt capacity capacidad en derivación
shunt circuit circuito en derivación
shunt coil bobina en derivación
shunt compensation compensación en derivación
shunt condenser condensador en derivación
shunt-connected conectado en derivación
shunt connection conexión en derivación
shunt detector detector en derivación
shunt dynamo dínamo en derivación
shunt efficiency diode diodo de recuperación en derivación
shunt excitation excitación en derivación
shunt-excited excitado en derivación
shunt-fed alimentado en derivación
shunt-fed oscillator oscilador alimentado en derivación
shunt-fed vertical antenna antena vertical alimentada en derivación
shunt feed alimentación en derivación
shunt feedback retroalimentación en derivación
shunt field campo en derivación
shunt-field coil bobina de campo en derivación
shunt-field relay relé de campo en derivación
shunt filter filtro en derivación
shunt generator generador en derivación
shunt impedance impedancia en derivación
shunt leads conductores en derivación
shunt limiter limitador en derivación
shunt line línea en derivación
shunt loading carga en derivación
shunt modulator modulador en derivación
shunt motor motor en derivación
shunt peaking coil bobina correctora en derivación
shunt reactor reactor en derivación
shunt rectifier circuit circuito rectificador en derivación
shunt regulator regulador en derivación
shunt resistance resistencia en derivación

shunt resistor resistor en derivación
shunt-series circuit circuito en derivación-serie
shunt switch conmutador en derivación
shunt-tee junction unión en T paralelo
shunt transformer transformador en derivación
shunt transistor transistor en derivación
shunt transition transición por derivación
shunt tuning sintonización en derivación
shunt winding arrollamiento en derivación
shunt-wound arrollado en derivación
shunt-wound generator generador arrollado en derivación
shunt-wound machine máquina arrollada en derivación
shunt-wound motor motor arrollado en derivación
shunted en derivación, en paralelo, shuntado
shunting derivación, shuntado
shunting capacitance capacitancia en derivación
shunting capacitor capacitor en derivación
shunting condenser condensador en derivación
shunting effect efecto de derivación
shunting resistor resistor en derivación
shunting switch conmutador en derivación
shutter obturador
side armature armadura lateral
side-armature relay relé de armadura lateral
side cap casquillo lateral
side circuit circuito lateral
side component componente lateral
side contact contacto lateral
side echo eco lateral
side emission emisión lateral
side frequency frecuencia lateral
side lobe lóbulo lateral
side-lobe echo eco por lóbulos laterales
side-lobe suppression supresión de lóbulos laterales
side panel panel lateral
side-stable relay relé de dos posiciones
side thrust fuerza lateral
sideband banda lateral
sideband attenuation atenuación de banda lateral
sideband component componente de banda lateral
sideband cutoff corte de banda lateral
sideband cutting eliminación de banda lateral
sideband filter filtro de bandas laterales
sideband frequency frecuencia de banda lateral
sideband interference interferencia de banda lateral
sideband power potencia de bandas laterales
sideband selector selector de bandas laterales
sideband splatter emisiones espurias de banda lateral
sideband suppression supresión de banda lateral
sidelobe lóbulo lateral
sidelobe suppression supresión de lóbulos laterales
sideplate placa lateral
sidetone tono local
sidetone effect efecto local
sidetone telephone teléfono con efecto local
sidetone telephone set aparato telefónico con efecto local
siemens siemens
Siemens' electrodynamometer electrodinamómetro de Siemens
sievert sievert
Sievert chamber cámara de Sievert
sigma sigma
sigmatron sigmatrón
sign signo, señal
sign bit bit de signo
sign character carácter de signo
sign-check indicator indicador de comprobación de signo
sign code código de signo
sign control control de signo

sign conversion conversión de signo
sign digit dígito de signo
sign field campo de signo
sign position posición del signo
signal señal, signo
signal aggregate agregado de señales
signal amplitude amplitud de señal
signal analyzer analizador de señal
signal area área de señal
signal aspect aspecto de señal
signal attenuation atenuación de señal
signal back light luz posterior de señal
signal bias polarización de señal
signal booster reforzador de señal
signal channel canal de señal
signal circuit circuito de señal
signal comparator comparador de señales
signal comparison comparación de señales
signal component componente de señal
signal conditioner acondicionador de señal
signal conditioning acondicionamiento de señal
signal contrast contraste de señal
signal control control de señal
signal conversion conversión de señal
signal-conversion equipment equipo de conversión de señal
signal converter convertidor de señal
signal current corriente de señal
signal decay time tiempo de decaimiento de señal
signal delay retardo de señal
signal diode diodo de señales
signal distance distancia entre señales
signal distortion distorsión de señal
signal distributing distribución de señales
signal duration duración de señal
signal electrode electrodo de señal
signal element elemento de señal
signal envelope envolvente de señal
signal-envelope shape forma de envolvente de señal
signal fading desvanecimiento de señal
signal-flow analysis análisis de flujos de señales
signal-flow diagram diagrama de flujos de señales
signal flux flujo de señal
signal form forma de señal
signal frequency frecuencia de señal
signal frequency amplifier amplificador de frecuencia de señal
signal frequency shift desplazamiento de frecuencia de señal
signal gain ganancia de señal
signal gate puerta de señal
signal generator generador de señales
signal-generator mechanism mecanismo generador de señales
signal identifier identificador de señal
signal-image ratio razón señal-imagen
signal indication indicación de señal
signal injection inyección de señal
signal-injection grid rejilla de inyección de señales
signal-injection test prueba de inyección de señal
signal injector inyector de señal
signal input entrada de señal
signal integration integración de señales
signal intensity intensidad de señal
signal interlocking enclavamiento de señal
signal inversion inversión de señal
signal inverter inversor de señal
signal lamp lámpara de señal
signal level nivel de señal
signal-level indicator indicador de nivel de señal
signal light luz de señal
signal limiter limitador de señal
signal line línea de señal

signal-line current corriente de línea de señal
signal loss pérdida de señal
signal mixer mezclador de señales
signal-noise ratio razón de señal-ruido
signal-operated operado por señales
signal operation operación por señales
signal output salida de señal
signal output current corriente de salida de señal
signal output electrode electrodo de salida de señal
signal overlapping solapamiento de señales
signal parameter parámetro de señal
signal path trayectoria de señal
signal peak pico de señal
signal phase fase de señal
signal plate placa de señal
signal power potencia de señal
signal processing procesamiento de señal
signal processor procesador de señal
signal pulse impulso de señal
signal purity pureza de señal
signal range alcance de señal
signal receiver receptor de señales
signal recording registro de señal
signal recovery recuperación de señal
signal rectification rectificación de señal
signal rectifier rectificador de señales
signal regeneration regeneración de señal
signal regenerator regenerador de señal
signal relay relé de señal
signal reliability confiabilidad de señal
signal resistance resistencia de señal
signal rise time tiempo de subida de señal
signal selector selector de señales
signal separator separador de señales
signal-shaping amplifier amplificador conformador
    de señal
signal-shaping network red conformadora de señal
signal shifter oscilador de frecuencia variable
signal source fuente de señales
signal splitter divisor de señal
signal storage almacenamiento de señales
signal strength intensidad de señal
signal-strength meter medidor de intensidad de señal
signal synthesizer sintetizador de señales
signal threshold umbral de señal
signal time delay retardo de tiempo de señal
signal-to-crosstalk ratio razón de señal a diafonía
signal-to-distortion ratio razón de señal a distorsión
signal-to-image ratio razón de señal a imagen
signal-to-interference ratio razón de señal a
    interferencia
signal-to-noise ratio razón de señal a ruido
signal trace trazo de señal
signal tracer seguidor de señal
signal tracing seguimiento de señal
signal transformation transformación de señal
signal transmission transmisión de señal
signal voltage tensión de señal
signal wave onda de señal
signal-wave envelope envolvente de onda de señal
signal winding arrollamiento de señal
signal window ventana de señal
signaling señalización
signaling battery batería de señalización
signaling bell campana de señalización
signaling channel canal de señalización
signaling circuit circuito de señalización
signaling code código de señalización
signaling device dispositivo de señalización
signaling equipment equipo de señalización
signaling frequency frecuencia de señalización
signaling lamp lámpara de señalización
signaling lead hilo de señalización

signaling light luz de señalización
signaling method método de señalización
signaling oscillator oscilador de señalización
signaling panel panel de señalización
signaling rate velocidad de señalización
signaling relay relé de señalización
signaling test prueba de señalización
signaling unit unidad de señalización
significance significación
significant code código significativo
significant condition estado significativo
significant condition of a modulation estado
    significativo de una modulación
significant digit dígito significativo
significant figure cifra significativa
significant instant instante significativo
significant instant of a modulation instante
    significativo de una modulación
significant interval intervalo significativo
silence cone cono de silencio
silencer silenciador
silent area área de silencio
silent interval intervalo de silencio
silent period periodo de silencio
silent zone zona de silencio
silicon silicio
silicon-boron photocell fotocélula de silicio-boro
silicon capacitor capacitor de silicio
silicon carbide carburo de silicio
silicon carbide rectifier rectificador de carburo de
    silicio
silicon carbide transistor transistor de carburo de
    silicio
silicon cell pila de silicio
silicon-controlled rectifier rectificador controlado de
    silicio
silicon-controlled switch conmutador controlado de
    silicio
silicon crystal cristal de silicio
silicon-crystal detector detector de cristal de silicio
silicon-crystal rectifier rectificador de cristal de
    silicio
silicon detector detector de silicio
silicon diode diodo de silicio
silicon dioxide dióxido de silicio
silicon double-base diode diodo de doble base de
    silicio
silicon epitaxial transistor transistor epitaxial de
    silicio
silicon gate puerta de silicio
silicon-junction diode diodo de unión de silicio
silicon monoxide monóxido de silicio
silicon on sapphire silicio sobre zafiro
silicon oxide óxido de silicio
silicon photocell fotocélula de silicio
silicon planar transistor transistor planar de silicio
silicon rectifier rectificador de silicio
silicon resistor resistor de silicio
silicon solar battery batería solar de silicio
silicon solar cell pila solar de silicio
silicon steel acero al silicio
silicon transistor transistor de silicio
silicon voltage regulator regulador de tensión de
    silicio
silicon wafer oblea de silicio
silicone silicona
silk-covered wire hilo recubierto de seda
silk-insulated cable cable aislado con seda
silk-insulated wire hilo aislado con seda
Silsbee effect efecto de Silsbee
Silsbee rule regla de Silsbee
silver plata
silver brazing soldadura fuerte con plata

silver-cadmium storage battery acumulador de plata-cadmio
silver-chloride cell pila de cloruro de plata
silver-mica capacitor capacitor de mica plateada
silver migration migración de plata
silver oxide óxido de plata
silver-oxide cell pila de óxido de plata
silver paint pintura de plata
silver-plated plateado
silver-plated aluminum aluminio plateado
silver-plated contact contacto plateado
silver-plated copper cobre plateado
silver-plating plateado
silver solder soldante de plata
silver-solder soldar con plata
silver-soldered soldado con plata
silver storage battery acumulador de plata
silver-to-silver contact contacto de plata a plata
silver-zinc primary cell pila primaria de plata-cinc
silver-zinc storage battery acumulador de plata-cinc
silvered plateado
silvered-ceramic capacitor capacitor de cerámica plateada
silvered-mica capacitor capacitor de mica plateada
silvered-mica dielectric dieléctrico de mica plateada
silvering plateado
silverstat silverstat
similar decimals decimales similares
simple arc arco simple
simple buffering almacenamiento intermedio simple
simple electrode electrodo simple
simple equation ecuación simple
simple harmonic current corriente armónica simple
simple harmonic electromotive force fuerza electromotriz armónica simple
simple harmonic motion moción armónica simple
simple harmonic wave onda armónica simple
simple image imagen simple
simple parameter parámetro simple
simple parity check comprobación de paridad simple
simple path trayectoria simple
simple periodic signal señal periódica simple
simple quad cuadrete simple
simple rectifier rectificador simple
simple rectifier circuit circuito rectificador simple
simple scanning exploración simple
simple signal señal simple
simple sound source fuente sonora puntual
simple stress estrés simple
simple tone tono puro
simple ventilation ventilación simple
simplex símplex
simplex apparatus aparato símplex
simplex channel canal símplex
simplex circuit circuito símplex
simplex communications link enlace de comunicaciones símplex
simplex line línea símplex
simplex mode modo símplex
simplex operation operación en símplex
simplex system sistema símplex
simplex telegraphy telegrafía símplex
simplex telephony telefonía símplex
simplex transmission transmisión símplex
simplicity simplicidad
simplified circuit circuito simplificado
simulate simular
simulated simulado
simulated job environment ambiente de trabajo simulado
simulated time tiempo simulado
simulation simulación
simulation model modelo de simulación

simulator simulador
simulator program programa simulador
simulator relay relé simulador
simulator routine rutina simuladora
simulator unit unidad simuladora
simulcast radiodifusión simultánea
simultaneity simultaneidad
simultaneous simultáneo
simultaneous access acceso simultáneo
simultaneous broadcast radiodifusión simultánea
simultaneous broadcasting radiodifusión simultánea
simultaneous carry acarreo simultáneo
simultaneous computer computadora simultánea
simultaneous contrast contraste simultáneo
simultaneous equations ecuaciones simultáneas
simultaneous frequency sharing compartimiento de frecuencia simultánea
simultaneous input-output entrada-salida simultánea
simultaneous lobing lobulación simultánea
simultaneous mode modo simultáneo
simultaneous multiplication multiplicación simultánea
simultaneous operations operaciones simultáneas
simultaneous playback reproducción simultánea
simultaneous-playback head cabeza de reproducción simultánea
simultaneous processing procesamiento simultáneo
simultaneous reception recepción simultánea
simultaneous recursion recursión simultánea
simultaneous scanning exploración simultánea
simultaneous system sistema simultáneo
simultaneous transmission transmisión simultánea
simultaneously simultáneamente
sine seno
sine current corriente sinusoidal
sine galvanometer galvanómetro de senos
sine law ley del seno
sine potentiometer potenciómetro sinusoidal
sine-squared pulse impulso de seno cuadrado
sine wave onda sinusoidal, onda senoidal
sine-wave alternator alternador de onda sinusoidal
sine-wave component componente de onda sinusoidal
sine-wave generator generador de ondas sinusoidales
sine-wave modulation modulación por onda sinusoidal
sine-wave oscillator oscilador de onda sinusoidal
sine-wave signal señal de onda sinusoidal
singing canto, silbido, zumbido
singing arc arco cantante
singing point punto de canto
singing suppressor supresor de canto
single sencillo, único
single access acceso único
single address dirección única
single-address code código de dirección única
single-address coding codificación de dirección única
single-address instruction instrucción de dirección única
single-address message mensaje de dirección única
single-anode magnetron magnetrón de ánodo único
single-anode rectifier rectificador de ánodo único
single-anode tank tubo de ánodo único
single-anode tube tubo de ánodo único
single-beam oscilloscope osciloscopio de haz único
single block bloque único
single-board computer computadora de panel único
single-board computer computadora de tablero único
single break ruptura única
single-break circuit-breaker cortacircuito de ruptura única

single-break switch conmutador de ruptura única
single-bus operation operación de bus único
single-button carbon microphone micrófono de
   carbón de cápsula única
single-button microphone micrófono de cápsula
   única
single-cable system sistema de cable único
single card tarjeta única
single-channel monocanal, de canal único
single-channel amplifier amplificador monocanal
single-channel codec codificador-descodificador
   monocanal
single-channel controller controlador monocanal
single-channel noise factor factor de ruido
   monocanal
single-channel receiver receptor monocanal
single-channel simplex símplex monocanal
single-chip codec codificador-descodificador en chip
   único
single circuit circuito único
single-circuit line línea de circuito único
single-circuit system sistema de circuito único
single-circuit transformer transformador de circuito
   único
single-coil filament filamento en espiral
single-coil relay relé de bobina única
single-component signal señal de componente único
single-conductor cable cable de conductor único
single contact contacto único
single-contact system sistema de contacto único
single control control único
single-core cable cable unipolar
single-core magnetic amplifier amplificador
   magnético de núcleo único
single-cotton-covered wire hilo recubierto con capa
   única de algodón
single crystal monocristal
single-crystal material material monocristalino
single-crystal semiconductor semiconductor
   monocristalino
single current corriente única
single-current system sistema de corriente única
single-current transmission transmisión por
   corriente única
single cycle ciclo único
single-dial control control de cuadrante único
single-diffused transistor transistor de difusión única
single diode diodo único
single element elemento único
single-element fuse fusible de elemento único
single-element rotary antenna antena rotatoria de
   elemento único
single-ended asimétrico
single-ended amplifier amplificador asimétrico
single-ended circuit circuito asimétrico
single-ended input entrada asimétrica
single-ended mixer mezclador asimétrico
single-ended output salida asimétrica
single-ended stage etapa asimétrica
single feeder alimentador único
single frequency frecuencia única
single-frequency amplifier amplificador de
   frecuencia única
single-frequency channel canal de frecuencia única
single-frequency duplex dúplex por frecuencia única
single-frequency operation operación en frecuencia
   única
single-frequency oscillator oscilador de frecuencia
   única
single-frequency pulsing pulsación de frecuencia
   única
single-frequency receiver receptor de frecuencia
   única

single-frequency signaling señalización con
   frecuencia única
single-frequency simplex símplex por frecuencia
   única
single-frequency sound sonido de frecuencia única
single-grid tube tubo de rejilla única
single-grid valve válvula de rejilla única
single-groove record disco monosurco
single hop salto único
single-hop path trayectoria de salto único
single-hop propagation propagación por salto único
single-hop radio relay radioenlace de salto único
single image imagen única
single-image response respuesta de imagen única
single junction unión única
single-junction transistor transistor de unión única
single layer capa única
single-layer coil bobina de capa única
single-layer winding arrollamiento de capa única
single-level address dirección de nivel único
single-level encoding codificación de nivel único
single-level storage almacenamiento de nivel único
single light luz única
single line línea única
single-line controller controlador de línea única
single-loop feedback retroalimentación de bucle
   único
single-loop memory memoria de bucle único
single-mode optical waveguide guíaondas óptico de
   modo único
single motion moción única
single-motion switch conmutador de moción única
single operation operación única
single-outage event evento de interrupción única
single pass pasada única
single-phase monofásico
single-phase alternator alternador monofásico
single-phase armature inducido monofásico
single-phase bridge rectifier rectificador en puente
   monofásico
single-phase circuit circuito monofásico
single-phase current corriente monofásica
single-phase full-wave bridge puente monofásico de
   onda completa
single-phase full-wave circuit circuito monofásico de
   onda completa
single-phase full-wave rectifier rectificador
   monofásico de onda completa
single-phase half-wave circuit circuito monofásico
   de media onda
single-phase induction motor motor de inducción
   monofásico
single-phase inverter inversor monofásico
single-phase machine máquina monofásica
single-phase motor motor monofásico
single-phase rectifier rectificador monofásico
single-phase system sistema monofásico
single-phase transformer transformador monofásico
single-point de punto único
single-point ground tierra de punto único
single-point grounding puesta a tierra de punto único
single polarity unipolar
single-polarity pulse impulso unipolar
single pole unipolar
single-pole double-throw switch conmutador
   unipolar de dos posiciones
single-pole fuse fusible unipolar
single-pole head cabeza unipolar
single-pole relay relé unipolar
single-pole single-throw switch conmutador unipolar
   de una posición
single-pole switch conmutador unipolar
single processor procesador único

single pulse impulso único
single-range de escala única
single scattering dispersión única
single-section filter filtro de sección única
single shot disparo único
single-shot multivibrator multivibrador monoestable
single sideband banda lateral única
single-sideband carrier system sistema de portadoras de banda lateral única
single-sideband communication comunicación por banda lateral única
single-sideband distortion distorsión de banda lateral única
single-sideband filter filtro de banda lateral única
single-sideband modulation modulación de banda lateral única
single-sideband receiver receptor de banda lateral única
single-sideband system sistema de banda lateral única
single-sideband transmission transmisión de banda lateral única
single-sideband transmitter transmisor de banda lateral única
single-sided de cara única, de una cara
single-sided disk disco de cara única
single signal señal única
single-signal receiver receptor de señal única
single-signal reception recepción por señal única
single-silk-covered wire hilo cubierto con capa única de seda
single-skip propagation propagación de salto único
single speed velocidad única
single stage etapa única
single-stage amplifier amplificador de etapa única
single station estación única
single step paso único
single-step debugging depuración de paso único
single-step mode modo de paso único
single-step operation operación de paso único
single sweep barrido único
single-sweep operation operación de barrido único
single-tank circuit-breaker cortacircuito de tanque único
single-tank switch conmutador de tanque único
single task tarea única
single-throw circuit-breaker cortacircuito de una posición
single-throw switch conmutador de una posición
single-tone keying manipulación de tono único
single track pista única
single-track recorder registrador de pista única
single-trip multivibrator multivibrador monoestable
single-tuned amplifier amplificador de sintonía única
single-tuned circuit circuito de sintonía única
single turn vuelta única
single-turn coil bobina de vuelta única
single-turn potentiometer potenciómetro de vuelta única
single-way connection conexión de una dirección
single winding arrollamiento sencillo
single wire hilo único
single-wire antenna antena de hilo único
single-wire circuit circuito de hilo único
single-wire-fed antenna antena de alimentación por hilo único
single-wire line línea de hilo único
single-wire transmission line línea de transmisión de hilo único
sink sumidero, sumidero de energía
sinter sínter, sinterizado, aglutinación
sintered sinterizado, aglutinado
sintered cathode cátodo sinterizado

sintered conductor conductor sinterizado
sintered magnetic material material magnético sinterizado
sintered metal metal sinterizado
sintered metallic powder polvo metálico sinterizado
sintered plate placa sinterizada
sintered-plate storage battery acumulador de placas sinterizadas
sintered powder polvo sinterizado
sintering sinterización, aglutinación
sinterization sinterización, aglutinación
sinterizing sinterización, aglutinación
sinusoid sinusoidal, senoidal, sinusoide
sinusoidal sinusoidal, senoidal
sinusoidal amplitude modulation modulación de amplitud sinusoidal
sinusoidal carrier wave onda portadora sinusoidal
sinusoidal component componente sinusoidal
sinusoidal current corriente sinusoidal
sinusoidal distribution distribución sinusoidal
sinusoidal electromagnetic wave onda electromagnética sinusoidal
sinusoidal electromotive force fuerza electromotriz sinusoidal
sinusoidal field campo sinusoidal
sinusoidal generator generador de ondas sinusoidales
sinusoidal input entrada sinusoidal
sinusoidal input signal señal de entrada sinusoidal
sinusoidal modulation modulación sinusoidal
sinusoidal oscillation oscilación sinusoidal
sinusoidal oscillator oscilador de onda sinusoidal
sinusoidal quantity magnitud sinusoidal
sinusoidal signal señal sinusoidal
sinusoidal-signal generator generador de señales sinusoidales
sinusoidal tone tono sinusoidal
sinusoidal vibration vibración sinusoidal
sinusoidal voltage tensión sinusoidal
sinusoidal wave onda sinusoidal
sinusoidal-wave generator generador de ondas sinusoidales
siren-tone generator generador de tono de sirena
site error error de emplazamiento
site interference interferencia por emplazamiento
six bit byte byte de seis bits
six-electrode tube tubo de seis electrodos
six-phase circuit circuito hexafásico
six-phase converter convertidor hexafásico
six-phase rectifier rectificador hexafásico
six-phase system sistema hexafásico
six-pole device dispositivo de seis polos
six-tube receiver receptor de seis tubos
sixteen-bit system sistema de dieciséis bits
skate patín, patinaje
skating patinaje
skating force fuerza de patinaje
skeletal esquelético
skeletal code código esquelético
skeleton esqueleto
skeleton bridge puente esquelético
skeleton-type assembly ensamblaje de tipo de esqueleto
skew sesgo, oblicuidad
skew arrangement arreglo oblicuo
skew factor factor de inclinación
skew ray rayo oblicuo
skewing sesgo, oblicuidad
skiatron esquiatrón
skin effect efecto superficial
skip salto
skip character carácter de salto
skip code código de salto

skip distance distancia de salto
skip effect efecto de salto
skip factor factor de salto
skip fading desvanecimiento por salto
skip instruction instrucción de salto
skip key tecla de salto
skip start comienzo de salto
skip stop parada de salto
skip test prueba de salto
skip zone zona de silencio
skirt selectivity selectividad de falda
Skrivanoff cell pila de Skrivanoff
sky factor factor de cielo
sky noise ruido celeste
sky wave onda espacial, onda celeste
sky wave correction corrección de onda espacial
skywave onda espacial, onda celeste
skywave communication comunicación por ondas espaciales
skywave correction corrección de onda espacial
skywave interference interferencia por ondas espaciales
skywave synchronization sincronización por onda espacial
skywave transmission transmisión por ondas espaciales
slab tajada gruesa, primera talla
slab coil bobina plana
sleeping sickness aumento gradual de corriente de fuga
sleetproof a prueba de aguanieve
sleeve manguito, manga
sleeve antenna antena de manguito
sleeve bearing cojinete de manguito
sleeve control control por tercer hilo
sleeve coupling acoplamiento de manguito
sleeve-dipole antenna antena dipolo de manguito
sleeve wire tercer hilo
sleeving manguito
slew rate velocidad de respuesta
slewing giro, giro rápido, giro paralelo
slewing rate velocidad de respuesta
slice tajada, plaquita
slicer rebanador, seccionador
slide-back voltmeter vóltmetro de oposición
slide coil bobina con cursor
slide contact contacto deslizante
slide-rule dial cuadrante de escala recta
slide-screw tuner sintonizador de tornillo deslizante
slide switch conmutador deslizante
slide-wire bridge puente de hilo
slide-wire potentiometer potenciómetro de hilo
slider contacto deslizante
sliding contact contacto deslizante
sliding-contact arm brazo de contacto deslizante
sliding-contact commutator conmutador de contacto deslizante
sliding-contact switch conmutador de contactos deslizantes
sliding electrode electrodo deslizante
sliding load carga deslizante
sliding plate placa deslizante
sliding short cortocircuito deslizante
sliding short-circuit cortocircuito deslizante
sliding voltage tensión deslizante
slinging wire hilo de suspensión
slip deslizamiento
slip clutch embrague deslizante
slip regulator regulador de deslizamiento
slip ring anillo rozante
slip-ring motor motor de anillos deslizantes
slip-ring rotor rotor de anillos deslizantes
slippage deslizamiento

slipping deslizamiento
slipping coupling acoplamiento deslizante
slope pendiente
slope attenuation atenuación de pendiente
slope compensation compensación de pendiente
slope compensator compensador de pendiente
slope conductance conductancia de pendiente
slope detection detección por pendiente
slope detector detector por pendiente
slope equalization ecualización de pendiente
slope equalizer ecualizador de pendiente
slope filter filtro de pendiente
slope ratio razón de pendientes
slope resistance resistencia diferencial
slot ranura, muesca
slot antenna antena de ranura
slot array red de ranuras
slot cell hoja aislante de ranura
slot coupling acoplamiento por ranura
slot effect efecto de ranura
slot-fed dipole dipolo ranurado
slot frequency frecuencia de ranura
slot group grupo de ranuras
slot insulation aislamiento de ranura
slot leakage fuga de ranura
slot number número de ranura
slot packing empaquetamiento de ranura
slot pitch paso de ranura
slot radiator radiador de ranura
slotted ranurado, con muescas
slotted armature inducido ranurado
slotted commutator conmutador ranurado
slotted core núcleo ranurado
slotted-cylinder antenna antena de cilindro ranurado
slotted-guide antenna antena de guía ranurada
slotted line línea ranurada
slotted-line recording system sistema de registro de línea ranurada
slotted rotor plate placa de rotor con muescas
slotted section sección ranurada
slotted waveguide guíaondas ranurado
slotted-waveguide antenna antena de guíaondas ranurado
slow-access storage almacenamiento de acceso lento
slow-acting relay relé de acción lenta
slow-action relay relé de acción lenta
slow-blow fuse fusible de fusión lenta
slow charge carga lenta
slow-closing de cierre lento
slow-closure device dispositivo de cierre lento
slow death muerte lenta
slow frequency drift deslizamiento de frecuencia lento
slow memory memoria lenta
slow-moving de movimiento lento
slow-operate de operación lenta
slow-operate relay relé de operación lenta
slow-operating de operación lenta
slow-operating relay relé de operación lenta
slow-release relay relé de desenganche lento
slow-releasing relay relé de desenganche lento
slow scan exploración lenta
slow-speed generator generador de velocidad lenta
slow-speed motor motor de velocidad lenta
slow-speed scan exploración de velocidad lenta
slow-speed starting arranque de velocidad lenta
slow storage almacenamiento lento
slow time tiempo lento
slow wave onda lenta
slow-wave circuit circuito de onda lenta
slug pedazo de metal, manguito de adaptación, blindaje de núcleo, núcleo
slug-tuned coil bobina de núcleo móvil

slug tuner sintonizador de núcleo
slug tuning sintonización por núcleo
slush compound compuesto antioxidante
small card tarjeta pequeña
small current corriente débil, corriente baja
small-current amplifier amplificador de corriente débil
small-current electronics electrónica de corriente débil
small ion ion pequeño
small-scale integration integración en pequeña escala
small signal señal pequeña, señal débil
small-signal analysis análisis de señal pequeña
small-signal bandwidth ancho de banda de señal pequeña
small-signal component componente de señal pequeña
small-signal current gain ganancia de corriente para señal pequeña
small-signal diode diodo para señales débiles
small-signal equivalent circuit circuito equivalente para señal pequeña
small-signal forward transadmittance transadmitancia directa para señal pequeña
small-signal input entrada para señal pequeña
small-signal power gain ganancia de potencia para señal pequeña
small-signal transconductance transconductancia para señal pequeña
small-signal transistor transistor para señal pequeña
smart card tarjeta inteligente
smart machine máquina inteligente
smart sensor sensor inteligente
smart terminal terminal inteligente
smartness inteligencia
smearing embarradura
Smith chart diagrama de Smith
smoke alarm alarma de humo
smoke-concentration indicator indicador de concentración de humo
smoke-density alarm alarma de densidad de humo
smoke detector detector de humo
smoke indicator indicador de humo
smokeproof a prueba de humo
smooth liso, uniforme
smooth line línea uniforme
smooth response respuesta uniforme
smooth traffic tráfico uniforme
smoothing aplanamiento, alisado
smoothing capacitor capacitor de aplanamiento
smoothing choke choque de aplanamiento
smoothing circuit circuito de aplanamiento
smoothing condenser condensador de aplanamiento
smoothing factor factor de aplanamiento
smoothing filter filtro de aplanamiento
smoothing resistor resistor de aplanamiento
snake culebra
snap-action de acción ultrarrápida
snap-action contact contacto de acción ultrarrápida
snap-action mechanism mecanismo de acción ultrarrápida
snap-action relay relé de acción ultrarrápida
snap-action switch conmutador de acción ultrarrápida
snap switch conmutador de acción ultrarrápida
snapshot dump vaciado selectivo
sneak current corriente parásita
sneak path trayectoria accidental
Snell's law ley de Snell
sniffer bobina exploradora
Snook rectifier rectificador de Snook
snow nieve
snow effect efecto de nieve

snow static estática por nieve
soaking temperature temperatura de remojo
soaking time tiempo de remojo
socket zócalo, tomacorriente, enchufe
socket adapter adaptador de zócalo
socket-outlet enchufe, base, zócalo
sodar sodar
sodium sodio
sodium silicate silicato de sodio
sodium-vapor lamp lámpara de vapor de sodio
sofar sofar
soft copy salida no impresa
soft-drawn wire hilo recocido
soft failure falla parcial
soft iron hierro blando, hierro dulce
soft-iron oscillograph oscilógrafo de hierro blando
soft solder soldante blando
soft soldering soldadura blanda
soft tube tubo blando
soft X-rays rayos X blandos
software programas, programática, software
software accuracy exactitud de programas
software bug error de programas
software command language lenguaje de mandos de programas
software consistency consistencia de programas
software control control de programas
software design diseño de programas
software development desarrollo de programas
software development process proceso de desarrollo de programas
software encryption codificación de programas
software error tolerance tolerancia de errores de programas
software features características de programas
software integrity integridad de programas
software library biblioteca de programas
software maintenance mantenimiento de programas
software model modelo de programas
software modularity modularidad de programas
software monitor monitor de programas
software network component componente de red de programas
software package paquete de programas
software project proyecto de programas
software-project management administración de proyecto de programas
software protection protección de programas
software quality calidad de programas
software reliability confiabilidad de programas
software-reliability management administración de confiabilidad de programas
software requirements requisitos de programas
software security seguridad de programas
software simulator simulador de programas
software specifications especificaciones de programas
software system sistema de programas
software test prueba de programas
solar solar
solar absorption index índice de absorción solar
solar activity actividad solar
solar-activity index índice de actividad solar
solar array red de células solares
solar battery batería solar
solar cell célula solar
solar constant constante solar
solar converter convertidor solar
solar cycle ciclo solar
solar energy energía solar
solar-energy collector colector de energía solar
solar-energy conversion conversión de energía solar
solar-energy converter convertidor de energía solar

solar-flare disturbance perturbación por erupción solar
solar laser láser solar
solar noise ruido solar
solar panel panel solar
solar power energía solar
solar-powered alimentado por energía solar
solar radiation radiación solar
solar-radiation simulator simulador de radiación solar
solar relay relé solar
solar simulator simulador solar
solar switch conmutador solar
solar wind viento solar
solder soldante, soldadura
solderability soldabilidad
solderable soldable
soldered joint junta soldada
soldering soldadura
soldering alloy aleación para soldar
soldering copper cobre de soldar
soldering gun pistola de soldar
soldering iron hierro de soldar, cautín
soldering paste pasta de soldar
soldering pliers pinzas de soldar
solderless sin soldadura
solderless breadboard montaje temporal sin soldadura
solderless connection conexión sin soldadura
solderless terminal terminal sin soldadura
solenoid solenoide
solenoid-actuated accionado por solenoide
solenoid brake freno de solenoide
solenoid braking frenado por solenoide
solenoid contactor contactor de solenoide
solenoid control control por solenoide
solenoid current corriente de solenoide
solenoid focusing enfoque por solenoide
solenoid-operated operado por solenoide
solenoid relay relé de solenoide
solenoid switch conmutador de solenoide
solenoid valve válvula de solenoide
solenoidal solenoidal
solenoidal field campo solenoidal
solid sólido
solid angle ángulo sólido
solid circuit circuito sólido
solid conductor conductor macizo, conductor sólido
solid-conductor cable cable de conductor macizo
solid contact contacto sólido
solid coupling acoplamiento sólido
solid-dielectric cable cable de dieléctrico sólido
solid electrolyte electrólito sólido
solid-electrolyte capacitor capacitor de electrólito sólido
solid enclosure caja sólida
solid error error sólido
solid ground tierra directa
solid pole polo sólido
solid-pole synchronous motor motor sincrónico de polo sólido
solid-state de estado sólido
solid-state battery batería de estado sólido
solid-state camera cámara de estado sólido
solid-state capacitor capacitor de estado sólido
solid-state chronometer cronómetro de estado sólido
solid-state circuit circuito de estado sólido
solid-state circuit-breaker cortacircuito de estado sólido
solid-state component componente de estado sólido
solid-state computer computadora de estado sólido
solid-state device dispositivo de estado sólido
solid-state diode diodo de estado sólido

solid-state dosimeter dosímetro de estado sólido
solid-state electronics electrónica de estado sólido
solid-state elements elementos de estado sólido
solid-state exciter excitador de estado sólido
solid-state integrated circuit circuito integrado de estado sólido
solid-state lamp lámpara de estado sólido
solid-state laser láser de estado sólido
solid-state logic lógica de estado sólido
solid-state maser máser de estado sólido
solid-state photosensor fotosensor de estado sólido
solid-state physics física del estado sólido
solid-state power supply fuente de alimentación de estado sólido
solid-state rectifier circuit circuito rectificador de estado sólido
solid-state relay relé de estado sólido
solid-state switch conmutador de estado sólido
solid-state thermometer termómetro de estado sólido
solid-state thyratron tiratrón de estado sólido
solid-state triode triodo de estado sólido
solid-state tuner sintonizador de estado sólido
solid-state voltmeter vóltmetro de estado sólido
solid tantalum capacitor capacitor de tantalio sólido
solid wire hilo macizo
solion solión
solistron solistrón
solubility solubilidad
solubilize solubilizar
soluble soluble
solute soluto
solution solución
solution conductivity conductividad de solución
solution pressure presión de solución
solvent solvente
solventless sin solvente
Sommerfeld equation ecuación de Sommerfeld
Sommerfeld formula fórmula de Sommerfeld
sonar sonar
sonar communication comunicación por sonar
sonar dome cúpula de sonar
sonar projector proyector de sonar
sonar receiver receptor de sonar
sonar signal señal de sonar
sonar-signal generator generador de señales de sonar
sonar transducer transductor de sonar
sonar transmitter transmisor de sonar
sonde sonda
sonic sónico
sonic altimeter altímetro sónico
sonic barrier barrera sónica
sonic boom estampido sónico
sonic delay line línea de retardo sónica
sonic depth finder buscador de profundidad acústico
sonic frequency frecuencia sónica
sonic sounding sondeo sónico
sonic speed velocidad sónica
sonics sónica
sonne sonne
sonobuoy sonoboya
sonogram sonograma
sonograph sonógrafo
sonoluminescence sonoluminiscencia
sonoluminescent sonoluminiscente
sonoptography sonoptografía
sophisticated electronics electrónica sofisticada
sorption process proceso de sorción
sort clasificar
sort brush escobilla de clasificación
sort control field campo de control de clasificación
sort field campo de clasificación
sort file archivo de clasificación

sort-merge clasificación-fusión
sort-merge program programa de
   clasificación-fusión
sort module módulo de clasificación
sort order orden de clasificación
sort pass pasada de clasificación
sort position posición de clasificación
sort program programa de clasificación
sort selection selección de clasificación
sort sequence secuencia de clasificación
sorter clasificador
sorter contact contacto de clasificación
sorter-reader clasificador-lectora
sorting clasificación
sorting brush escobilla de clasificación
sorting magnet electroimán de clasificación
sorting needle aguja de clasificación
sorting procedure procedimiento de clasificación
sorting program programa de clasificación
sorting routine rutina de clasificación
sorting string secuencia de clasificación
sorting suppression supresión de clasificación
sound sonido
sound-absorbent absorbente de sonido
sound absorber absorbente de sonido
sound-absorbing absorbente de sonido
sound-absorbing insulation aislamiento absorbente
   de sonido
sound absorption absorción de sonido
sound-absorption coefficient coeficiente de
   absorción de sonido
sound-absorption factor factor de absorción de
   sonido
sound amplification amplificación de sonido
sound amplifier amplificador de sonido
sound analyzer analizador de sonido
sound antenna antena de sonido
sound attenuation atenuación de sonido
sound-attenuation factor factor de atenuación de
   sonido
sound band banda sonora
sound barrier barrera de sonido
sound bars barras de interferencia por sonido
sound carrier portadora de sonido
sound-carrier amplifier amplificador de portadora
   de sonido
sound-carrier frequency frecuencia de portadora de
   sonido
sound-carrier monitor monitor de portadora de
   sonido
sound cell célula sonora
sound-cell microphone micrófono de célula sonora
sound chamber cámara de sonido
sound channel canal sonoro
sound circuit circuito de sonido
sound column columna sonora
sound-conductive conductor de sonido
sound control control de sonido
sound damping amortiguación de sonido
sound detection detección de sonido
sound-detection system sistema de detección de
   sonido
sound detector detector de sonido
sound diffuser difusor de sonido
sound-diffusing difusor de sonido
sound diplexer diplexor de sonido
sound directivity directividad de sonido
sound discriminator discriminador de sonido
sound dispersion dispersión de sonido
sound-dispersion pattern patrón de dispersión de
   sonido
sound distortion distorsión de sonido
sound-effects filter filtro de efectos de sonido

sound-emitting emisor de sonido
sound energy energía sonora
sound-energy density densidad de energía sonora
sound-energy flux flujo de energía sonora
sound equipment equipo de sonido
sound field campo sonoro
sound film película sonora
sound filter filtro sonoro
sound flux flujo sonoro
sound focusing enfoque de sonido
sound frequency frecuencia de sonido
sound generator generador de sonido
sound-hazard integrator integrador de peligro
   sonoro
sound image imagen de sonido
sound-insulated insonorizado, aislado del sonido
sound-insulating material material insonorizante
sound insulation insonorización
sound insulator insonorizador
sound intensity intensidad sonora
sound intensity level nivel de intensidad sonora
sound interference interferencia sonora
sound interference level nivel de interferencia
   sonora
sound intermediate frequency frecuencia intermedia
   de sonido
sound intermediate frequency amplifier
   amplificador de frecuencia intermedia de sonido
sound level nivel sonoro
sound-level calibrator calibrador de nivel sonoro
sound-level indicator indicador de nivel sonoro
sound-level meter medidor de nivel sonoro
sound-level scale escala de nivel sonoro
sound-measuring instrument instrumento de
   medición de sonido
sound-measuring system sistema de medición de
   sonido
sound meter medidor de sonido
sound mixer mezclador de sonidos
sound mixing mezcla de sonidos
sound-modulated wave onda modulada por sonido
sound modulation modulación por sonido
sound monitoring monitoreo de sonido
sound-on-film recording registro sonoro en película
sound-on-sound recording registro sobre registro
sound-operated relay relé operado por sonido
sound passband banda de paso de sonido
sound pickup captador de sonido, captación de
   sonido
sound power potencia sonora
sound-power density densidad de potencia sonora
sound-power level nivel de potencia sonora
sound-powered telephone teléfono alimentado por
   energía sonora
sound-powered telephone set aparato telefónico
   alimentado por energía sonora
sound pressure presión sonora
sound-pressure level nivel de presión sonora
sound-pressure measurement medida de presión
   sonora
sound probe sonda sonora
sound-proof insonoro, a prueba de sonido
sound radar radar sonoro
sound radiation radiación sonora
sound receiver receptor de sonido
sound reception recepción de sonido
sound recording registro de sonido
sound-recording equipment equipo de registro de
   sonido
sound-recording magnetic tape cinta magnética de
   registro de sonido, cinta magnetofónica
sound-recording system sistema de registro de
   sonido

sound-recording unit unidad de registro de sonido
sound reduction reducción sonora
sound-reduction factor factor de reducción sonora
sound reflection reflexión sonora
sound-reflection coefficient coeficiente de reflexión sonora
sound-reflection factor factor de reflexión sonora
sound reinforcement refuerzo de sonido
sound rejection rechazo de sonido
sound relay relé operado por sonido
sound reproducing reproducción de sonido
sound-reproducing system sistema de reproducción de sonido
sound reproduction reproducción de sonido
sound-reproduction system sistema de reproducción de sonido
sound scattering dispersión de sonido
sound section sección de sonido
sound signal señal de sonido
sound source fuente de sonido
sound spectrum espectro acústico
sound-spectrum analyzer analizador de espectro acústico
sound strength intensidad sonora
sound-survey meter medidor de características de sonido
sound system sistema de sonido
sound takeoff toma de sonido
sound tape cinta sonora
sound track pista de sonido
sound transmission transmisión de sonido
sound-transmission coefficient coeficiente de transmisión de sonido
sound-transmission factor factor de transmisión de sonido
sound trap trampa de sonido
sound unit unidad de sonido
sound vibration vibración sonora
sound volume volumen sonoro
sound wave onda sonora
sounder sonador
sounding board tablero sonoro
sounding buoy boya sonora
soundproof insonoro, a prueba de sonido
soundproofed insonorizado
soundproofing insonorización
source fuente, generador
source address dirección fuente
source block bloque fuente
source circuit circuito fuente
source code código fuente
source-code instruction instrucción de código fuente
source coding codificación fuente
source computer computadora fuente
source-data acquisition adquisición de datos de fuente
source-data automation automatización de datos de fuente
source-data card tarjeta de datos fuente
source database base de datos fuente
source disk disco fuente
source diskette disquete fuente
source document documento fuente
source efficiency eficiencia de fuente
source file archivo fuente
source follower seguidor de fuente
source impedance impedancia de fuente
source information información fuente
source language lenguaje fuente
source machine máquina fuente
source module módulo fuente
source of current fuente de corriente
source of energy fuente de energía

source of noise fuente de ruido
source program programa fuente
source recording registro fuente
source resistance resistencia de fuente
source segment segmento fuente
source statement sentencia fuente
source tape cinta fuente
south magnetic pole polo sur magnético
space espacio
space character carácter de espacio
space charge carga espacial
space-charge coupling acoplamiento por carga espacial
space-charge debunching desagrupamiento por carga espacial
space-charge density densidad de carga espacial
space-charge distortion distorsión de carga espacial
space-charge distribution distribución de carga espacial
space-charge effect efecto de carga espacial
space-charge field campo de carga espacial
space-charge filter filtro de carga espacial
space-charge generation generación de carga espacial
space-charge grid rejilla de carga espacial
space-charge layer capa de carga espacial
space-charge limitation limitación por carga espacial
space-charge-limited current corriente limitada por carga espacial
space-charge region región de carga espacial
space-charge repulsion repulsión de carga espacial
space-charge wave onda de carga espacial
space-charge wave propagation propagación de onda de carga espacial
space code código de espacio
space communication comunicación espacial
space current corriente espacial
space distribution distribución espacial
space diversity diversidad espacial
space-diversity reception recepción en diversidad espacial
space-diversity system sistema de diversidad espacial
space division división de espacio
space-division multiplex múltiplex por división de espacio
space factor factor de espacio
space group grupo espacial
space lattice retículo espacial
space parallax paralaje espacial
space pattern patrón de espacio
space probe sonda espacial
space radio radio espacial
space reflection reflexión espacial
space suppression supresión de espacio
space-time espacio-tiempo
space wave onda espacial
space-wave field campo de onda espacial
space-wave field strength intensidad de campo de onda espacial
spaced antennas antenas espaciadas
spacer espaciador, separador
spacing espaciado
spacing current corriente de espacio
spacing error error de separación
spacing frequency frecuencia de espacio
spacing impulse impulso de espacio
spacing of frequencies separación de frecuencias
spacing pulse impulso de espacio
spacing signal señal de espacio
spacing wave onda de espacio
spacistor espacistor
spaghetti espagueti

spallation espalación
span intervalo, espacio, vano
spare repuesto, reserva
spare cable cable de reserva
spare circuit circuito de reserva
spare component componente de repuesto
spare equipment equipo de repuesto
spare pair par de reserva
spare part parte de repuesto
spark chispa
spark absorber supresor de chispas
spark arrester supresor de chispas
spark blowing soplado de chispas
spark capacitor capacitor de chispas
spark coil bobina de chispa
spark condenser condensador de chispas
spark energy energía de chispa
spark extinguisher extintor de chispas
spark frequency frecuencia de chispa
spark gap distancia disruptiva, descargador de chispa, espinterómetro, explosor
spark-gap generator generador de chispas
spark-gap modulator modulador de chispas
spark-gap oscillator oscilador de chispas
spark-gap voltmeter vóltmetro de chispa
spark generator generador de chispas
spark killer supresor de chispas
spark lag retardo de chispa
spark length distancia disruptiva
spark micrometer micrómetro de chispas
spark-plug suppressor supresor de bujía
spark quencher extintor de chispas
spark quenching extinción de chispas
spark recorder registrador de chispa
spark-resistant resistente a chispas
spark sender emisor de chispa
spark spectrum espectro de chispa
spark-suppressing capacitor capacitor supresor de chispas
spark-suppressing filter filtro supresor de chispas
spark suppression supresión de chispas
spark suppressor supresor de chispas
spark transmitter transmisor de chispa
sparking chisporroteo
sparking distance distancia disruptiva
sparking potential potencial disruptivo
sparking voltage tensión disruptiva
sparkless sin chispas
sparkless commutation conmutación sin chispas
sparkover descarga disruptiva
sparkover test prueba de descarga disruptiva
sparkplug terminal borne de bujía
spatial coherence coherencia espacial
spatial distribution distribución espacial
spatial disturbance perturbación espacial
speaker altavoz
speaker damping amortiguamiento de altavoz
speaking arc arco parlante
special card tarjeta especial
special character carácter especial
special control control especial
special device dispositivo especial
special dial tone tono de marcar especial
special features características especiales
special function función especial
special-function key tecla de función especial
special link enlace especial
special-purpose calculator calculadora de uso especial
special-purpose character carácter de uso especial
special-purpose computer computadora de uso especial
special-purpose language lenguaje de uso especial

special-purpose logic lógica de uso especial
special-purpose motor motor de uso especial
special-purpose relay relé de uso especial
special register registro especial
special service servicio especial
special sign signo especial
special symbol símbolo especial
special variable variable especial
specialized especializado
specific específico
specific access acceso específico
specific acoustic impedance impedancia acústica específica
specific acoustic reactance reactancia acústica específica
specific acoustic resistance resistencia acústica específica
specific activity actividad específica
specific address dirección específica
specific addressing direccionamiento específico
specific capacity capacidad específica
specific code código específico
specific coding codificación específica
specific conductance conductancia específica
specific conductivity conductividad específica
specific consumption consumo específico
specific damping amortiguamiento específico
specific detectivity detectividad específica
specific dielectric strength rigidez dieléctrica específica
specific dispersivity dispersividad específica
specific electronic charge carga electrónica específica
specific emission emisión específica
specific energy energía específica
specific gravity gravedad específica
specific heat calor específico
specific humidity humedad específica
specific illumination iluminación específica
specific inductive capacitance capacitancia inductiva específica
specific inductive capacity capacidad inductiva específica
specific inertance inertancia específica
specific ionization ionización específica
specific irradiation irradiación específica
specific luminous intensity intensidad luminosa específica
specific magnetic resistance resistencia magnética específica
specific mode modo específico
specific output salida específica
specific permeability permeabilidad específica
specific photosensitivity fotosensibilidad específica
specific power potencia específica
specific pressure presión específica
specific program programa específico
specific refraction refracción específica
specific refractivity refractividad específica
specific reluctance reluctancia específica
specific resistance resistencia específica
specific rotation rotación específica
specific routine rutina específica
specific sound-energy flux flujo de energía sonora específico
specific speed velocidad específica
specific temperature temperatura específica
specific wattage wattaje específico
specification especificación
specification check comprobación de especificación
specification statement sentencia de especificación
specification verification verificación de especificación

specified achromatic lights luces acromáticas especificadas
spectral espectral
spectral absorption factor factor de absorción espectral
spectral analysis análisis espectral
spectral band banda espectral
spectral bandwidth ancho de banda espectral
spectral characteristic característica espectral
spectral color color espectral
spectral component componente espectral
spectral concentration concentración espectral
spectral correction corrección espectral
spectral density densidad espectral
spectral distribution distribución espectral
spectral emission emisión espectral
spectral emissivity emisividad espectral
spectral energy energía espectral
spectral energy distribution distribución de energía espectral
spectral filtering filtrado espectral
spectral index índice espectral
spectral irradiance irradiancia espectral
spectral irradiation irradiación espectral
spectral line línea espectral
spectral line width anchura de línea espectral
spectral locus lugar espectral
spectral luminance factor factor de luminancia espectral
spectral luminous efficiency eficiencia luminosa espectral
spectral luminous flux flujo luminoso espectral
spectral luminous gain ganancia luminosa espectral
spectral luminous intensity intensidad luminosa espectral
spectral output salida espectral
spectral purity pureza espectral
spectral quantum efficiency eficiencia cuántica espectral
spectral quantum yield rendimiento cuántico espectral
spectral radiance radiancia espectral
spectral radiant energy energía radiante espectral
spectral radiant flux flujo radiante espectral
spectral radiant gain ganancia radiante espectral
spectral radiant intensity intensidad radiante espectral
spectral reflectance reflectancia espectral
spectral reflection factor factor de reflexión espectral
spectral response respuesta espectral
spectral-response characteristic característica de respuesta espectral
spectral-response curve curva de respuesta espectral
spectral responsivity responsividad espectral
spectral scanning exploración espectral
spectral selectivity selectividad espectral
spectral sensitivity sensibilidad espectral
spectral-sensitivity characteristic característica de sensibilidad espectral
spectral series serie espectral
spectral temperature temperatura espectral
spectral transmission factor factor de transmisión espectral
spectral transmittance transmitancia espectral
spectral wavelength longitud de onda espectral
spectral width anchura espectral
spectral window ventana espectral
spectroanalysis espectroanálisis
spectrofluorometer espectrofluorómetro
spectrofluorometric espectrofluorométrico
spectrofluorometric analysis análisis espectrofluorométrico

spectrofluorometry espectrofluorometría
spectrogram espectrograma
spectrograph espectrógrafo
spectrographic espectrográfico
spectrography espectrografía
spectrometer espectrómetro
spectrometric espectrométrico
spectrometric instrument instrumento espectrométrico
spectrometry espectrometría
spectromicroscope espectromicroscopio
spectromicroscopic espectromicroscópico
spectrophotoelectric espectrofotoeléctrico
spectrophotography espectrofotografía
spectrophotometer espectrofotómetro
spectrophotometric espectrofotométrico
spectrophotometric analysis análisis espectrofotométrico
spectropolarimeter espectropolarímetro
spectropolarimetric espectropolarimétrico
spectroradiometer espectrorradiómetro
spectroscope espectroscopio
spectroscopic espectroscópico
spectroscopy espectroscopia
spectrum espectro
spectrum amplitude amplitud espectral
spectrum analysis análisis espectral
spectrum analyzer analizador espectral
spectrum congestion congestión espectral
spectrum conservation conservación espectral
spectrum crowding apiñamiento espectral
spectrum density densidad espectral
spectrum interval intervalo espectral
spectrum level nivel espectral
spectrum line línea espectral
spectrum locus lugar espectral
spectrum of frequencies espectro de frecuencias
spectrum recorder registrador de espectro
spectrum space intervalo espectral
spectrum utilization utilización espectral
spectrum width anchura espectral
specular especular
specular angle ángulo especular
specular density densidad especular
specular luminous reflectance reflectancia luminosa especular
specular reflectance reflectancia especular
specular reflection reflexión especular
specular reflector reflector especular
specular surface superficie especular
specular transmission transmisión especular
speech amplifier amplificador de voz
speech analyzer analizador de voz
speech band banda de voz
speech circuit circuito de conversación
speech clarifier clarificador de voz
speech clipper recortador de voz
speech compression compresión de voz
speech compressor compresor de voz
speech-controlled controlado por voz
speech current corriente de conversación
speech detector detector de voz
speech frequency frecuencia de voz
speech inversion inversión de voz
speech inverter inversor de voz
speech level nivel de conversación
speech limiter limitador de voz
speech-modulated modulado por voz
speech-noise ratio razón de conversación-ruido
speech-operated operado por voz
speech plus signaling telefonía con señalización
speech plus telegraphy telegrafía con señalización
speech power potencia de voz

speech quality calidad de voz
speech recognition reconocimiento de voz
speech recording registro de voz
speech reference signal señal de referencia de voz
speech scrambler codificador de voz
speech spectrum espectro de voz
speech synthesis síntesis de voz
speech synthesizer sintetizador de voz
speech test signal señal de prueba de voz
speech voltmeter vóltmetro de conversación
speed velocidad
speed adjusting ajuste de velocidad
speed adjustment ajuste de velocidad
speed change cambio de velocidad
speed-change control control de cambio de velocidad
speed changer cambiador de velocidad
speed constancy constancia de velocidad
speed control control de velocidad
speed-control mechanism mecanismo de control de velocidad
speed-control rheostat reóstato de control de velocidad
speed controller controlador de velocidad
speed governor regulador de velocidad
speed indicator indicador de velocidad
speed key manipulador de alta velocidad
speed limit velocidad límite
speed-limit indicator indicador de velocidad límite
speed limiter limitador de velocidad
speed-limiting device dispositivo limitador de velocidad
speed of light velocidad de la luz
speed of sound velocidad del sonido
speed of transmission velocidad de transmisión
speed range intervalo de velocidades
speed ratio razón de velocidades
speed recorder registrador de velocidad
speed-regulating rheostat reóstato de regulación de velocidad
speed regulation regulación de velocidad
speed-regulation characteristic característica de regulación de velocidad
speed regulator regulador de velocidad
speed selector selector de velocidad
speed switching conmutación de velocidad
speed test prueba de velocidad
speed variation variación de velocidad
speed variator variador de velocidad
speedup capacitor capacitor acelerador
sphere esfera
sphere gap explosor de esferas
sphere-gap voltmeter vóltmetro de explosor de esferas
sphere voltmeter vóltmetro de explosor de esferas
spherical esférico
spherical aberration aberración esférica
spherical aberration coefficient coeficiente de aberración esférica
spherical angle ángulo esférico
spherical antenna antena esférica
spherical array red esférica
spherical candlepower potencia lumínica esférica
spherical cavity cavidad esférica
spherical coordinates coordenadas esféricas
spherical degree grado esférico
spherical diffraction difracción esférica
spherical distance distancia esférica
spherical-earth factor factor de esfericidad de la tierra
spherical lens lente esférica
spherical mirror espejo esférico
spherical progressive wave onda progresiva esférica
spherical radiator radiador esférico

spherical reduction factor factor de reducción esférica
spherical reflector reflector esférico
spherical wave onda esférica
spherically esféricamente
spheroidal antenna antena esferoidal
spider araña
spiderweb antenna antena en telaraña
spiderweb coil bobina en telaraña
spike punta, punta de descarga
spike generator generador de puntas
spike potential potencial de punta
spike suppressor supresor de puntas
spike train tren de puntas
spillover desbordamiento
spin espín
spin echo eco de espín
spin paramagnetism paramagnetismo por espín
spin wave onda de espín
spin-wave resonance resonancia de onda de espín
spindle huso, eje, soporte
spinning electron electrón giratorio
spinthariscope espintariscopio
spiral espiral
spiral antenna antena en espiral
spiral coil bobina en espiral
spiral-coiled arrollado en espiral
spiral delay line línea de retardo en espiral
spiral distortion distorsión espiral
spiral-four cable cable de cuadretes en espiral
spiral grid rejilla en espiral
spiral-quad cuadrete en espiral
spiral scanning exploración en espiral
spiral sweep barrido en espiral
spiral trace trazo en espiral
spiral tuner sintonizador en espiral
spiral winding arrollamiento en espiral
splashproof a prueba de salpicaduras
splashproof enclosure caja a prueba de salpicaduras
splashproof machine máquina a prueba de salpicaduras
splashproof motor motor a prueba de salpicaduras
splatter emisiones espurias
splatter filter filtro contra emisiones espurias
splatter-suppressor circuit circuito supresor de emisiones espurias
splice empalme, unión
splicer empalmador
splicing chamber cámara de empalmes
splicing ear oreja de empalme
split anode ánodo dividido
split-anode magnetron magnetrón de ánodo dividido
split beam haz dividido
split brush escobilla dividida
split channel canal dividido
split-channel system sistema de canal dividido
split coil bobina dividida
split-conductor cable cable de conductores divididos
split-field motor motor de campo dividido
split fitting accesorio dividido
split image imagen dividida
split impulse impulso dividido
split-load circuit circuito de carga dividida
split-load phase inverter inversor de fase de carga dividida
split magnetron magnetrón dividido
split node nodo dividido
split phase fase dividida
split-phase circuit circuito de fase dividida
split-phase current corriente de fase dividida
split-phase motor motor de fase dividida
split picture imagen dividida
split projector proyector dividido

split pulse impulso dividido
split-rotor plate placa de rotor con muescas
split screen pantalla dividida
split selector selector dividido
split shield blindaje dividido
split-sound receiver receptor de sonido dividido
split-stator capacitor capacitor de estator dividido
split transducer transductor dividido
split winding arrollamiento dividido
split-winding loop bucle de arrollamiento dividido
splitter divisor
splitting división, corte
splitting time tiempo de corte
spontaneous emission emisión espontánea
spontaneous polarization polarización espontánea
spool carrete, bobina
spool insulator aislador de carrete
spool-wound resistor resistor arrollado en carrete
spooler almacenador temporal de entrada y salida
spooling almacenamiento temporal de entrada y salida
sporadic-E layer capa E esporádica
sporadic-E layer ionization ionización de capa E esporádica
sporadic ionization ionización esporádica
sporadic reflection reflexión esporádica
spot punto, punto luminoso, indicador luminoso
spot brightness brillo de punto
spot brilliance brillantez de punto
spot brilliancy brillantez de punto
spot check comprobación selectiva
spot cooling enfriamiento de punto
spot deflection deflexión de punto
spot diameter diámetro de punto
spot distortion distorsión de punto
spot frequency frecuencia puntual
spot height altura de punto
spot intensity intensidad de punto
spot jamming interferencia intencional de frecuencia puntual
spot motion moción de punto
spot noise ruido puntual
spot size tamaño de punto
spot speed velocidad de punto
spot welding soldadura por puntos
spot wobble oscilación de punto
spread spectrum espectro ensanchado
spread-spectrum communications comunicaciones por espectro ensanchado
spreader esparcidor, separador, propagador
spreading dispersión, ensanche
spreading current corriente de dispersión
spreading factor factor de dispersión
spreading loss pérdida por dispersión
spreading resistance resistencia de dispersión
spring contact contacto de resorte
spring switch conmutador de resorte
sprocket holes perforaciones de transporte
sprocket pulse impulso por perforaciones de transporte
spurious espurio
spurious count conteo espurio
spurious emission emisión espuria
spurious frequency frecuencia espuria
spurious harmonic armónica espuria
spurious-harmonic generation generación de armónicas espurias
spurious mode modo espurio
spurious modulation modulación espuria
spurious noise ruido espurio
spurious oscillation oscilación espuria
spurious output salida espuria
spurious pattern patrón espurio

spurious printing impresión espuria
spurious pulse impulso espurio
spurious radiation radiación espuria
spurious response respuesta espuria
spurious-response attenuation atenuación de respuestas espurias
spurious-response ratio razón de respuesta espuria
spurious shading sombra espuria
spurious sidebands bandas laterales espurias
spurious signal señal espuria
sputter deposición electrónica, evaporación catódica, chisporroteo
sputtering deposición electrónica, evaporación catódica, chisporroteo
square-law circuit circuito de ley cuadrática
square-law combiner combinador de ley cuadrática
square-law condenser condensador de ley cuadrática
square-law demodulator desmodulador de ley cuadrática
square-law detector detector de ley cuadrática
square-law response respuesta de ley cuadrática
square loop bucle cuadrado, antena de cuadro rectangular
square-loop antenna antena de cuadro rectangular
square pulse impulso cuadrado
square-pulse generator generador de impulsos cuadrados
square root raíz cuadrada
square signal señal cuadrada
square-signal generator generador de señales cuadradas
square tube tubo cuadrado
square wave onda cuadrada
square-wave amplifier amplificador de ondas cuadradas
square-wave component componente de ondas cuadradas
square-wave converter convertidor de ondas cuadradas
square-wave generator generador de ondas cuadradas
square-wave modulator modulador de onda cuadrada
square-wave oscillator oscilador de ondas cuadradas
square-wave output salida de onda cuadrada
square-wave response respuesta de onda cuadrada
square-wave response characteristic característica de respuesta de onda cuadrada
square-wave signal señal de onda cuadrada
square-wave test signal señal de prueba de onda cuadrada
square-wave testing pruebas con ondas cuadradas
square-wave voltage tensión de onda cuadrada
squaring amplifier amplificador cuadrador
squaring circuit circuito cuadrador
squawker altavoz de registro medio
squeal chillido
squeeze section sección compresible
squeeze track pista compresible
squeezeout aplastamiento
squegging autobloqueo
squegging oscillator oscilador de autobloqueo
squelch silenciador
squelch circuit circuito silenciador
squelch diode diodo silenciador
squelch level nivel silenciador
squelch system sistema silenciador
squelch triode triodo silenciador
squelch voltage tensión silenciadora
squelched silenciado
squelcher silenciador
squelching silenciamiento
squint estrabismo, ángulo de desviación

squirrel-cage antenna antena en jaula de ardilla
squirrel-cage induction motor motor de inducción en jaula de ardilla
squirrel-cage magnetron magnetrón en jaula de ardilla
squirrel-cage motor motor en jaula de ardilla
squirrel-cage rotor rotor en jaula de ardilla
squirrel-cage winding arrollamiento en jaula de ardilla
squitter disparo accidental
stabilitron estabilitrón
stability estabilidad
stability condition condición de estabilidad
stability factor factor de estabilidad
stability limit límite de estabilidad
stabilization estabilización
stabilization network red de estabilización
stabilization time tiempo de estabilización
stabilization voltage tensión de estabilización
stabilize estabilizar
stabilized estabilizado
stabilized amplifier amplificador estabilizado
stabilized feedback retroalimentación estabilizada
stabilized-gain amplifier amplificador de ganancia estabilizada
stabilized local oscillator oscilador local estabilizado
stabilized platform plataforma estabilizada
stabilized power supply fuente de alimentación estabilizada
stabilizer estabilizador
stabilizer mount montura estabilizadora
stabilizer tube tubo estabilizador
stabilizer unit unidad estabilizadora
stabilizing estabilizador
stabilizing amplifier amplificador estabilizador
stabilizing choke choque estabilizador
stabilizing circuit circuito estabilizador
stabilizing feedback retroalimentación estabilizadora
stabilizing network red estabilizadora
stabilizing potential potencial estabilizador
stabilizing resistor resistor estabilizador
stabilizing signal señal estabilizadora
stabilizing transistor transistor estabilizador
stabilizing tube tubo estabilizador
stabilizing voltage tensión estabilizadora
stabilizing winding arrollamiento estabilizador
stabistor estabistor
stable estable
stable circuit circuito estable
stable device dispositivo estable
stable element elemento estable
stable emitter emisor estable
stable orbit órbita estable
stable oscillation oscilación estable
stable platform plataforma estable
stable state estado estable
stable temperature temperatura estable
stack pila, pila de placas
stack control control de pila
stack manipulation manipulación de pila
stacked antenna antena de elementos apilados
stacked array red de antenas apiladas
stacked-dipole antenna antena de dipolos apilados
stacked dipoles dipolos apilados
stacked heads cabezas apiladas
stacked-job control control de trabajos apilados
stacked-job processing procesamiento de trabajos apilados
stacking apilamiento
stage etapa
stage-by-stage etapa por etapa
stage-by-stage elimination eliminación etapa por etapa

stage control control de etapa
stage efficiency eficiencia de etapa
stage gain ganancia por etapa
stage loss pérdida por etapa
stagger error por oscilación de punto, disposición en zigzag, tambaleo
stagger time tiempo de escalonamiento
stagger-tuned amplifier amplificador de sintonización escalonada
stagger-tuned circuits circuitos de sintonización escalonada
stagger tuning sintonización escalonada
staggered escalonado, alternado
staggered antenna antena escalonada
staggered arrangement arreglo escalonado
staggered array red escalonada
staggered heads cabezas escalonadas
staggered stereophonic tape cinta estereofónica escalonada
staggered-tuned amplifier amplificador de sintonización escalonada
staggered tuning sintonización escalonada
staggering escalonamiento, sintonización escalonada, tambaleo
staggering advantage ventaja por escalonamiento
stain spots manchas por exudación
stair-step wave onda en escalera
staircase escalera
staircase generator generador de onda en escalera
staircase signal señal en escalera
staircase wave onda en escalera
stalling torque par límite
stalo oscilador local estable
stamped circuit circuito estampado
stamped metal metal estampado
stamping estampado
stand-alone independiente, autónomo
stand-alone computer computadora autónoma
stand-alone dump vaciado autónomo
stand-alone modem módem autónomo
stand-alone program programa autónomo
stand-alone unit unidad autónoma
standard patrón, estándar, norma
standard absorption curve curva de absorción de referencia
standard ampere ampere patrón
standard antenna antena patrón
standard atmosphere atmósfera normal
standard band banda normal
standard broadcast band banda de radiodifusión normal
standard broadcast channel canal de radiodifusión normal
standard cable cable patrón
standard candle bujía patrón
standard capacitor capacitor patrón
standard cell pila patrón
standard chamber cámara patrón
standard color color patrón
standard color code código de colores patrón
standard compass compás magistral
standard component componente patrón
standard condenser condensador patrón
standard conductance conductancia patrón
standard connector conector patrón
standard crystal cristal patrón
standard current generator generador de corriente patrón
standard design diseño normal
standard deviation desviación estándar
standard electrode potential potencial de electrodo normal
standard equipment equipo normal

standard error error estándar
standard file archivo patrón
standard frequency frecuencia patrón
standard-frequency oscillator oscilador de frecuencia patrón
standard-frequency signal señal de frecuencia patrón
standard groove surco normal
standard hydrogen electrode electrodo de hidrógeno normal
standard inductor inductor patrón
standard instrument instrumento patrón
standard insulator aislador normal
standard interface interfaz patrón
standard ionization chamber cámara de ionización patrón
standard lamp lámpara patrón
standard loudspeaker altavoz patrón
standard microphone micrófono patrón
standard mismatcher desadaptador patrón
standard moisture humedad normal
standard noise factor factor de ruido normal
standard noise temperature temperatura de ruido normal
standard observer observador patrón
standard ohm ohm patrón
standard output level nivel de salida patrón
standard output load carga de salida patrón
standard pitch tono normal
standard potential potencial normal
standard program programa patrón
standard propagation propagación normal
standard pulse impulso patrón
standard radio atmosphere atmósfera de radio normal
standard radio horizon horizonte de radio normal
standard receiver receptor patrón
standard reference level nivel de referencia normal
standard reflector reflector normal
standard refraction refracción normal
standard resistance resistencia patrón
standard resistor resistor patrón
standard routine rutina patrón
standard signal señal patrón
standard-signal generator generador de señal patrón
standard-signal oscillator oscilador de señal patrón
standard solenoid solenoide patrón
standard solution solución normal
standard source fuente patrón
standard sphere gap explosor de esferas patrón
standard state estado normal
standard subroutine subrutina patrón
standard tape cinta patrón
standard television signal señal de televisión normal
standard temperature and pressure temperatura y presión normales
standard termination terminación patrón
standard test frequency frecuencia de prueba patrón
standard test tape cinta de prueba patrón
standard test tone tono de prueba patrón
standard tone tono normal
standard track pista normal
standard transmitter transmisor patrón
standard tuning frequency frecuencia de sintonización patrón
standard voltage tensión normal
standard-wave error error de onda patrón
standardization estandarización, normalización
standardize estandarizar, normalizar
standardized estandarizado, normalizado
standardized circuit circuito estandarizado
standardized component componente estandarizado
standardized pulses impulsos estandarizados

standardized television signal señal de televisión estandarizada
standby posición de espera, operación en espera
standby application aplicación de reserva
standby battery batería de reserva
standby block bloque de reserva
standby channel canal de reserva
standby circuit circuito de reserva
standby computer computadora de reserva
standby current corriente de reserva
standby equipment equipo de reserva
standby lighting alumbrado de reserva
standby line línea de reserva
standby operation operación de reserva
standby power energía de reserva
standby redundancy redundancia de reserva
standby register registro de reserva
standby repeater repetidor de reserva
standby station estación de reserva
standby time tiempo de reserva
standby transmitter transmisor de reserva
standby unit unidad de reserva
standing wave onda estacionaria
standing-wave amplifier amplificador de ondas estacionarias
standing-wave antenna antena de ondas estacionarias
standing-wave detector detector de ondas estacionarias
standing-wave distortion distorsión de ondas estacionarias
standing-wave indicator indicador de ondas estacionarias
standing-wave loss pérdida por ondas estacionarias
standing-wave loss factor factor de pérdidas por ondas estacionarias
standing-wave meter medidor de ondas estacionarias
standing-wave pattern patrón de ondas estacionarias
standing-wave ratio razón de ondas estacionarias
standing-wave ratio indicator indicador de razón de ondas estacionarias
standing-wave system sistema de ondas estacionarias
standoff insulator soporte aislante
star estrella
star circuit circuito en estrella
star-connected conectado en estrella
star-connected circuit circuito conectado en estrella
star connection conexión en estrella
star-delta starter arrancador estrella-delta
star-delta starting arranque estrella-delta
star-delta switch conmutador estrella-delta
star network red en estrella
star-quad cuadrete en estrella
star-quad cable cable de cuadretes en estrella
star rectifier rectificador en estrella
star-star connection conexión estrella-estrella
Stark broadening ensanchamiento de Stark
Stark effect efecto de Stark
start address dirección de arranque
start bit bit de arranque
start circuit circuito de arranque
start code código de arranque
start delay retardo de arranque
start element elemento de arranque
start impulse impulso de arranque
start lead hilo de inicio
start magnet electroimán de arranque
start pulse impulso de arranque
start-record signal señal de arranque de registro
start relay relé de arranque
start signal señal de arranque
start solenoid solenoide de arranque
start-stop character carácter de arranque-parada

start-stop code código de arranque-parada
start-stop control control de arranque-parada
start-stop device dispositivo de arranque-parada
start-stop multivibrator multivibrador de
　arranque-parada
start-stop push-button pulsador de arranque-parada
start-stop switch conmutador de arranque-parada
start-stop system sistema de arranque-parada
start-stop time tiempo de arranque-parada
start time tiempo de arranque
starter arrancador, cebador
starter anode ánodo de cebado
starter circuit circuito arrancador
starter coupling acoplamiento de arrancador
starter gap entrehierro de arrancador
starter rheostat reóstato de arranque
starter solenoid solenoide de arrancador
starter switch conmutador de arranque
starter voltage tensión de arrancador, tensión de
　cebado
starter voltage drop caída de tensión de arrancador
starting arranque, cebado
starting address dirección de inicio
starting anode ánodo de cebado
starting box caja de arranque
starting capacitor capacitor de arranque
starting circuit circuito de arranque
starting circuit-breaker cortacircuito de arranque
starting compensator compensador de arranque
starting compressor compresor de arranque
starting condenser condensador de arranque
starting current corriente de arranque
starting device dispositivo de cebado
starting electrode electrodo de cebado
starting element elemento de arranque
starting light luz de arranque
starting motor motor de arranque
starting power potencia de arranque
starting reactor reactor de arranque
starting relay relé de arranque
starting resistance resistencia de arranque
starting rheostat reóstato de arranque
starting signal señal de arranque
starting switch conmutador de arranque
starting torque par de arranque
starting value valor inicial
starting velocity velocidad de arranque
starting voltage tensión de arranque, tensión de
　cebado
starting winding arrollamiento de arranque
startup comienzo, arranque
startup disk disco de arranque
startup period periodo de arranque
startup procedure procedimiento de arranque
startup screen pantalla de arranque
starved amplifier amplificador subalimentado
starved tube tubo subalimentado
statampere estatampere
statcoulomb estatcoulomb
state estado
state code código de estado
state data datos de estado
state diagram diagrama de estados
state of charge estado de carga
state of equilibrium estado de equilibrio
state of polarization estado de polarización
state transition transición de estado
statement sentencia, instrucción, declaración
statement card tarjeta de sentencias
statement code código de sentencia
statement identifier identificador de sentencia
statement library biblioteca de sentencias
statfarad estatfarad

stathenry estathenry
static estática
static allocation asignación estática
static amplifier amplificador estático
static analysis análisis estático
static analyzer analizador estático
static balance equilibrio estático
static balancer equilibrador estático
static base current corriente de base estática
static base resistance resistencia de base estática
static base voltage tensión de base estática
static beam haz estático
static breeze brisa estática
static buffer allocation asignación de
　almacenamiento intermedio estática
static calibration calibración estática
static cathode current corriente de cátodo estática
static cathode resistance resistencia de cátodo
　estática
static cathode voltage tensión de cátodo estática
static characteristic característica estática
static characters caracteres estáticos
static charge carga estática
static check comprobación estática
static collector current corriente de colector estática
static collector resistance resistencia de colector
　estática
static collector voltage tensión de colector estática
static condenser condensador estático
static convergence convergencia estática
static converter convertidor estático
static decay decaimiento estático
static device dispositivo estático
static drain current corriente de consumo estática
static drain resistance resistencia de consumo
　estática
static drain voltage tensión de consumo estática
static dump vaciado estático
static electricity electricidad estática
static electrification electrificación estática
static electrode potential potencial de electrodo
　estático
static emitter current corriente de emisor estática
static emitter resistance resistencia de emisor
　estática
static emitter voltage tensión de emisor estática
static equilibrium equilibrio estático
static error error estático
static field campo estático
static flip-flop basculador estático
static focus foco estático
static frequency changer cambiador de frecuencia
　estático
static frequency converter convertidor de frecuencia
　estático
static frequency multiplier multiplicador de
　frecuencia estático
static friction fricción estática
static gain ganancia estática
static gate current corriente de puerta estática
static gate resistance resistencia de puerta estática
static gate voltage tensión de puerta estática
static grid current corriente de rejilla estática
static grid voltage tensión de rejilla estática
static hysteresis histéresis estática
static image imagen estática
static induced current corriente inducida estática
static induction inducción electrostática
static luminous sensitivity sensibilidad luminosa
　estática
static machine máquina electrostática
static magnetic cell célula magnética estática
static magnetic delay line línea de retardo magnética

estática
static magnetic field campo magnético estático
static magnetic storage almacenamiento magnético estático
static memory memoria estática
static-memory card tarjeta de memoria estática
static model modelo estático
static modulator modulador estático
static mutual conductance conductancia mutua estática
static overvoltage sobretensión estática
static parameter parámetro estático
static plate current corriente de placa estática
static plate resistance resistencia de placa estática
static plate voltage tensión de placa estática
static power converter convertidor de potencia estático
static pressure presión estática
static random-access memory memoria de acceso aleatorio estática
static rectifier rectificador estático
static register registro estático
static regulation regulación estática
static regulator regulador estático
static relay relé estático
static relocation relocalización estática
static resistance resistencia estática
static sensitivity sensibilidad estática
static skew sesgo estático
static source current fuente de corriente estática
static source resistance resistencia estática de fuente
static source voltage fuente de tensión estática
static storage almacenamiento estático
static storage allocation asignación de almacenamiento estático
static subroutine subrutina estática
static suppressor voltage tensión supresora estática
static switching conmutación estática
static test prueba estática
static torque par mínimo
static transconductance transconductancia estática
static transconductor transconductor estático
static transformer transformador estático
static value valor estático
static variable variable estática
static voltage regulator regulador de tensión estático
static voltage stabilizer estabilizador de tensión estático
static wave current corriente de onda estática
statical estático
statically estáticamente
statically balanced equilibrado estáticamente
statics estática, estáticos
station estación
station battery batería de estación
station code código de estación
station equipment equipo de estación
station line línea de estación
station selection selección de estación
station-selection code código de selección de estación
station test prueba de estación
station-to-station call llamada de estación a estación
stationary estacionario
stationary-anode tube tubo de ánodo estacionario
stationary antenna antena estacionaria
stationary battery batería estacionaria
stationary cell célula estacionaria
stationary coil bobina estacionaria
stationary contact contacto estacionario
stationary-contact assembly ensamblaje de contactos estacionarios
stationary field campo estacionario

stationary grid rejilla estacionaria
stationary magnetic field campo magnético estacionario
stationary node nodo estacionario
stationary rectifier rectificador estacionario
stationary satellite satélite estacionario
stationary sound wave onda sonora estacionaria
stationary state estado estacionario
stationary wave onda estacionaria
statistical estadístico
statistical quality control control de calidad estadístico
statitron estatitrón
statmho estatmho
statoersted estatoersted
statohm estatohm
stator estator
stator coil bobina de estator
stator current corriente de estator
stator-inductance starter arrancador inductivo de estator
stator plate placa de estator
stator-resistance starter arrancador resistivo de estator
statoscope estatoscopio
status estado
status analysis análisis de estado
status bits bits de estado
status change cambio de estado
status-change circuit circuito de cambio de estado
status-change pulse impulso de cambio de estado
status-change signal señal de cambio de estado
status code código de estado
status indicator indicador de estado
status line línea de estado
status message mensaje de estado
status register registro de estado
status word palabra de estado
statvolt estatvolt
statweber estatweber
steady current corriente estacionaria
steady state estado estacionario
steady-state component componente de estado estacionario
steady-state deviation desviación de estado estacionario
steady-state operation operación en estado estacionario
steady-state oscillation oscilación de estado estacionario
steady-state value valor de estado estacionario
steady-state vibration vibración de estado estacionario
steady voltage tensión estacionaria
steatite esteatita
steel acero
steel-tank rectifier rectificador de tanque de acero
steel wire hilo de acero
steerable orientable
steerable antenna antena orientable
steerable-beam antenna antena de haz orientable
steerable paraboloid paraboloide orientable
Stefan's law ley de Stefan
Stefan-Boltzmann constant constante de Stefan-Boltzmann
Stefan-Boltzmann law ley de Stefan-Boltzmann
Steinmetz coefficient coeficiente de Steinmetz
Steinmetz formula fórmula de Steinmetz
stellar estelar
stellar noise ruido estelar
stenode estenodo
stenode circuit circuito estenodo
step paso

step-back relay relé limitador de corriente
step-by-step action acción paso a paso
step-by-step automatic system sistema automático paso a paso
step-by-step control control paso a paso
step-by-step equipment equipo paso a paso
step-by-step excitation excitación paso a paso
step-by-step mode modo paso a paso
step-by-step operation operación paso a paso
step-by-step relay relé paso a paso
step-by-step selection selección paso a paso
step-by-step selector selector paso a paso
step-by-step switch conmutador paso a paso
step-by-step system sistema paso a paso
step-by-step test prueba paso a paso
step change incremento de paso
step counter contador de pasos
step-counter circuit circuito contador de pasos
step-counting circuit circuito contador de pasos
step-down ratio razón reductora
step-down transformer transformador reductor
step function función escalón
step-function generator generador de función escalón
step generator generador de pasos
step response respuesta al escalón
step switch conmutador de pasos
step-through operation operación paso a paso
step-up autotransformer autotransformador elevador
step-up connection conexión elevadora
step-up ratio razón elevadora
step-up transformer transformador elevador
step-up winding arrollamiento elevador
stepdown ratio razón reductora
stepdown transformer transformador reductor
stepped wave onda escalonada
stepper motor motor de avance paso a paso
stepping motor motor de avance paso a paso
stepping relay relé de avance paso a paso
stepping switch conmutador de avance paso a paso
steradian esterradian
Sterba array red de Sterba
stereo estéreo
stereo adapter adaptador estereofónico
stereo amplifier amplificador estereofónico
stereo broadcast radiodifusión estereofónica
stereo broadcasting radiodifusión estereofónica
stereo cartridge cápsula estereofónica
stereo channel canal estereofónico
stereo-channel separation separación de canales estereofónicos
stereo control unit unidad de control estereofónico
stereo decoder descodificador estereofónico
stereo discriminator discriminador estereofónico
stereo disk disco estereofónico
stereo head cabeza estereofónica
stereo headphones estereoaudífonos
stereo high-fidelity system sistema de alta fidelidad estereofónico
stereo image imagen estereofónica
stereo loudspeaker system sistema de altavoces estereofónico
stereo magnetic-tape recording registro en cinta magnética estereofónico
stereo microphone micrófono estereofónico
stereo multiplexing system sistema de multiplexión estereofónica
stereo phono cartridge cápsula fonocaptora estereofónica
stereo pickup fonocaptor estereofónico
stereo playback reproducción estereofónica
stereo radio broadcast radiodifusión estereofónica

stereo receiver receptor estereofónico
stereo reception recepción estereofónica
stereo record disco estereofónico
stereo recording registro estereofónico, grabación estereofónica
stereo reproduction reproducción estereofónica
stereo separation separación estereofónica
stereo-separation control control de separación estereofónica
stereo sound sonido estereofónico
stereo sound reproduction reproducción de sonido estereofónica
stereo sound system sistema de sonido estereofónico
stereo subcarrier subportadora estereofónica
stereo system sistema estereofónico
stereo tape cinta estereofónica
stereo tape recorder registrador en cinta estereofónica
stereo tape recording registro en cinta estereofónica
stereo tuner sintonizador estereofónico
stereocast radiodifusión estereofónica
stereocasting radiodifusión estereofónica
stereofluoroscopy estereofluoroscopia
stereographic estereográfico
stereomicroscope estereomicroscopio
stereophone estereófono
stereophonic estereofónico
stereophonic adapter adaptador estereofónico
stereophonic amplifier amplificador estereofónico
stereophonic broadcast radiodifusión estereofónica
stereophonic broadcasting radiodifusión estereofónica
stereophonic cartridge cartucho estereofónico
stereophonic channel canal estereofónico
stereophonic-channel separation separación de canales estereofónicos
stereophonic control unit unidad de control estereofónico
stereophonic decoder descodificador estereofónico
stereophonic discriminator discriminador estereofónico
stereophonic disk disco estereofónico
stereophonic head cabeza estereofónica
stereophonic headphones estereoaudífonos
stereophonic high-fidelity system sistema de alta fidelidad estereofónico
stereophonic image imagen estereofónica
stereophonic loudspeaker system sistema de altavoces estereofónico
stereophonic magnetic-tape recording registro en cinta magnética estereofónico
stereophonic microphone micrófono estereofónico
stereophonic multiplexing system sistema de multiplexión estereofónica
stereophonic phono cartridge cápsula fonocaptora estereofónica
stereophonic pickup fonocaptor estereofónico
stereophonic playback reproducción estereofónica
stereophonic radio broadcast radiodifusión estereofónica
stereophonic receiver receptor estereofónico
stereophonic reception recepción estereofónica
stereophonic record disco estereofónico
stereophonic recording registro estereofónico, grabación estereofónica
stereophonic reproduction reproducción estereofónica
stereophonic separation separación estereofónica
stereophonic-separation control control de separación estereofónica
stereophonic sound sonido estereofónico
stereophonic sound reproduction reproducción de sonido estereofónica

stereophonic sound system sistema de sonido estereofónico
stereophonic subcarrier subportadora estereofónica
stereophonic system sistema estereofónico
stereophonic tape cinta estereofónica
stereophonic tape recorder registrador en cinta estereofónica
stereophonic tape recording registro en cinta estereofónica
stereophonic tuner sintonizador estereofónico
stereophonically estereofónicamente
stereophonics estereofónica
stereophonous estereofónico
stereophony estereofonía
stereophotograph estereofotografía
stereophotography estereofotografía
stereoptics estereoóptica
stereoradar estereorradar
stereoradiographic estereorradiográfico
stereoradiography estereorradiografía
stereoradioscopy estereorradioscopía
stereoscillography estereooscilografía
stereoscope estereoscopio
stereoscopic estereoscópico
stereoscopic camera cámara estereoscópica
stereoscopic head cabeza estereoscópica
stereoscopic microscope microscopio estereoscópico
stereoscopic photography fotografía estereoscópica
stereoscopic prism prisma estereoscópica
stereoscopic system sistema estereoscópico
stereoscopic television televisión estereoscópica
stereoscopy estereoscopía
stereosonic estereosónico
stereospectrogram estereoespectrograma
stereotelemeter estereotelémetro
stereotelevision estereotelevisión
sterilizer esterilizador
stethoscope estetoscopio
stick circuit circuito de autorretención
stick relay relé de autorretención
sticking image imagen retenida
sticking voltage tensión de bloqueo
stiction fricción estática
still vista fija
stimulated emission emisión estimulada
stimulated transmission transmisión estimulada
stimulus estímulo
stochastic estocástico
stochastic path trayectoria estocástica
stochastic process proceso estocástico
stochastic simulation simulación estocástica
stoichiometric estequiométrico
stoichiometry estequiometría
Stokes' law ley de Stokes
stop parada, interrupción, tope
stop-band banda suprimida
stop-band ripple ondulación de banda suprimida
stop bit bit de parada
stop character carácter de parada
stop code código de parada
stop control control de parada
stop-control circuit circuito de control de parada
stop element elemento de parada
stop filter filtro supresor
stop instruction instrucción de parada
stop joint empalme de retención
stop key tecla de parada
stop lamp lámpara de parada
stop magnet electroimán de parada
stop pulse impulso de parada
stop-record signal señal de parada de registro
stop signal señal de parada
stop-start unit unidad de parada-arranque

stop time tiempo de parada
stopband banda suprimida
stopband ripple ondulación de banda suprimida
stopping capacitor capacitor de bloqueo
storage almacenamiento, memoria
storage access acceso al almacenamiento
storage access channel canal de acceso al almacenamiento
storage access method método de acceso al almacenamiento
storage access time tiempo de acceso al almacenamiento
storage allocation asignación de almacenamiento
storage allocation algorithm algoritmo de asignación de almacenamiento
storage area área de almacenamiento
storage assignment asignación de almacenamiento
storage battery batería de acumuladores
storage block bloque de almacenamiento
storage box caja de almacenamiento
storage buffer almacenador intermedio, memoria intermedia
storage camera tube tubo de cámara almacenador
storage capacity capacidad de almacenamiento
storage cell elemento de acumulador, celda de almacenamiento
storage circuit circuito de almacenamiento
storage-command pulse impulso de mando de acumulación
storage counter contador de acumulación
storage cycle ciclo de almacenamiento
storage-cycle time tiempo de ciclo de almacenamiento
storage density densidad de almacenamiento
storage descriptor descriptor de almacenamiento
storage device dispositivo de almacenamiento
storage drum tambor de almacenamiento
storage dump vaciado de almacenamiento
storage effect efecto de almacenamiento
storage efficiency eficiencia de almacenamiento
storage element elemento de almacenamiento
storage entry entrada de almacenamiento
storage exit salida de almacenamiento
storage factor factor de almacenamiento
storage field campo de almacenamiento
storage function función de almacenamiento
storage group grupo de almacenamiento
storage hierarchy jerarquía de almacenamiento
storage indicator indicador de almacenamiento
storage interference interferencia de almacenamiento
storage interleaving entrelazado de almacenamiento
storage keyboard teclado de almacenamiento
storage laser láser de almacenamiento
storage level nivel de almacenamiento
storage life vida de almacenamiento
storage load module módulo de carga de almacenamiento
storage location ubicación de almacenamiento
storage location selection selección de ubicación de almacenamiento
storage management administración de almacenamiento
storage map mapa de almacenamiento
storage mark marca de almacenamiento
storage medium medio de almacenamiento
storage module módulo de almacenamiento
storage oscilloscope osciloscopio de almacenamiento
storage position posición de almacenamiento
storage protection protección de almacenamiento
storage queue cola de almacenamiento
storage reconfiguration reconfiguración de almacenamiento
storage region región de almacenamiento

storage register registro de almacenamiento
storage relay relé de almacenamiento
storage requirement requisito de almacenamiento
storage ring anillo de almacenamiento
storage stack pila de almacenamiento
storage structure estructura de almacenamiento
storage surface superficie de almacenamiento
storage switch conmutador de almacenamiento
storage system sistema de almacenamiento
storage temperature temperatura de almacenamiento
storage test prueba de almacenamiento
storage time tiempo de almacenamiento
storage tube tubo de almacenamiento
storage unit unidad de almacenamiento
store almacenar, acumular
store controller controlador de almacenamiento
store dump vaciado de almacenamiento
store mark marca de almacenamiento
stored almacenado, acumulado
stored addition suma almacenada
stored base charge carga acumulada en la base
stored energy energía acumulada
stored-energy indicator indicador de energía
    acumulada
stored-energy welding soldadura por energía
    acumulada
stored logic lógica almacenada
stored-logic computer computadora de lógica
    almacenada
stored permanently permanentemente almacenado
stored program programa almacenado
stored-program computer computadora de
    programa almacenado
stored-program machine máquina de programa
    almacenado
stored-program processor procesador de programa
    almacenado
stored record registro almacenado
stored reference signal señal de referencia
    almacenada
stored routine rutina almacenada
storing almacenamiento
straight adapter adaptador en línea
straight amplifier amplificador directo
straight connector conector recto
straight dipole dipolo recto
straight filament filamento recto
straight-filament tube tubo de filamento recto
straight frequency frecuencia directa
straight horn bocina recta
straight-line capacitance capacitancia lineal
straight-line code código lineal
straight-line coding codificación lineal
straight-line control control lineal
straight-line detection detección lineal
straight-line detector detector lineal
straight-line frequency frecuencia lineal
straight-line path trayectoria lineal
straight-line programming programación lineal
straight-line radiator radiador lineal
straight-line wavelength longitud de onda lineal
straight receiver receptor directo
straight reception recepción directa
straightforward amplifier amplificador directo
strain esfuerzo, tensión, tirantez, deformación
strain gage deformímetro
strain-gage bridge puente deformimétrico
strain-gage transducer transductor deformimétrico
strain-gage transduction transducción
    deformimétrica
strain insulator aislador de tensión
strand hebra, hilo
stranded conductor conductor trenzado

stranded-conductor cable cable de conductores
    trenzados
stranded wire cable trenzado
strapping conexiones de puente, apareado
stratified estratificado
stratified discharge descarga estratificada
stratigraphy estratigrafía
stratosphere estratosfera
stray capacitance capacitancia parásita
stray capacity capacidad parásita
stray component componente parásito
stray coupling acoplamiento parásito
stray current corriente parásita
stray effect efecto parásito
stray emission emisión parásita
stray field campo parásito
stray impedance impedancia parásita
stray inductance inductancia parásita
stray light luz parásita
stray losses pérdidas por dispersión
stray magnetic field campo magnético parásito
stray radiation radiación parásita
stray reactance reactancia parásita
stray resistance resistencia parásita
streaking arrastre, falsa imagen
strength fuerza, intensidad
stress estrés, esfuerzo, tensión
stress analysis análisis de estrés
stress test prueba de estrés
stretch estiramiento
stretched string cuerda estirada
striction estricción
strike percutir, cebar
strike bath baño primario
strike deposit depósito primario
striking current corriente de cebado
striking device dispositivo de cebado
striking electrode electrodo de cebado
striking potential potencial de cebado
striking voltage tensión de cebado
string secuencia, serie, cadena de aisladores, cadena,
    cuerda
string break ruptura de secuencia
string control byte byte de control de secuencia
string electrometer electrómetro de cuerda
string galvanometer galvanómetro de cuerda
string handling manejo de secuencia
string language lenguaje de secuencia
string length longitud de secuencia
string manipulation manipulación de secuencias
string oscillograph oscilógrafo de cuerda
string oscilloscope osciloscopio de cuerda
string processing procesamiento de secuencia
string programming programación de secuencia
string variable variable de secuencia
strip regleta, franja, cinta, baño de eliminación
strip fuse fusible de cinta
stripper desforrador, baño de eliminación
stripper tank tanque de despojamiento
stripping desforrado, eliminación, lavado
stripping agent agente separador
strobe estroboscopio, lámpara estroboscópica, marca
    estroboscópica
strobe disk disco estroboscópico
strobe light lámpara estroboscópica
strobe marker marcador estroboscópico
strobe pulse impulso estroboscópico
strobing pulse impulso estroboscópico
strobing-pulse generator generador de impulsos
    estroboscópicos
stroboradiography estroborradiografía
stroboscope estroboscopio
stroboscope system sistema estroboscópico

stroboscopic estroboscópico
stroboscopic calibrating calibración estroboscópica
stroboscopic checking comprobación estroboscópica
stroboscopic disk disco estroboscópico
stroboscopic effect efecto estroboscópico
stroboscopic illumination iluminación estroboscópica
stroboscopic lamp lámpara estroboscópica
stroboscopic tachometer tacómetro estroboscópico
stroboscopic technique técnica estroboscópica
stroboscopic tube tubo estroboscópico
strobotron estrobotrón
strobotron lamp lámpara estrobotrón
stroke golpe, carrera
stroke speed frecuencia de exploración
strong coupling acoplamiento fuerte
strong electrolyte electrólito fuerte
strong focusing enfoque fuerte
strong signal señal fuerte
strong-signal area área de señal fuerte
strontium estroncio
Strowger selector selector de Strowger
Strowger switch conmutador de Strowger
Strowger system sistema de Strowger
structural failure falla estructural
structural instability inestabilidad estructural
structural resolution resolución estructural
structural stability estabilidad estructural
structural testing pruebas estructurales
structurally estructuralmente
structure estructura
structure conflict conflicto de estructura
structure diagram diagrama de estructura
structure factor factor de estructura
structured estructurado
structured design diseño estructurado
structured field campo estructurado
structured program programa estructurado
structured programming programación estructurada
structured programming language lenguaje de programación estructurado
stub adaptador, sección, brazo de reactancia
stub antenna antena corta
stub cable cable de conexión
studio-to-transmitter link enlace estudio-transmisor
stylus estilete, aguja
stylus alignment alineación de aguja
stylus drag arrastre de aguja
stylus force fuerza de aguja
stylus friction fricción de aguja
stylus pressure presión de aguja
stylus scratch ruido de aguja
stylus velocity velocidad de aguja
subaddress subdirección
subaqueous subacuático
subaqueous loudspeaker altavoz subacuático
subarea subárea
subarea address dirección de subárea
subarea node nodo de subárea
subassembly subensamblaje, subconjunto
subatomic subatómico
subatomic particle partícula subatómica
subaudible subaudible
subaudible tone tono subaudible
subaudio subaudio, infraacústico
subaudio band banda infraacústica
subaudio frequency frecuencia infraacústica
subaudio oscillator oscilador infraacústico
subaudio spectrum espectro infraacústico
subaudio telegraphy telegrafía infraacústica
subbase band subbanda base
subcarrier subportadora
subcarrier amplifier amplificador de subportadora

subcarrier band banda de subportadora
subcarrier channel canal de subportadora
subcarrier detector detector de subportadora
subcarrier discriminator discriminador de subportadora
subcarrier filter filtro de subportadora
subcarrier frequency frecuencia de subportadora
subcarrier frequency modulation modulación de frecuencia de subportadora
subcarrier generator generador de subportadora
subcarrier oscillator oscilador de subportadora
subcarrier reference signal señal de referencia de subportadora
subcarrier signal señal de subportadora
subcarrier wave onda subportadora
subcenter subcentro
subchannel subcanal
subchannel detector detector de subcanal
subchassis subchasis
subcommand submando
subcommutation subconmutación
subcomponent subcomponente
subcontrol subcontrol
subcontrol station estación de subcontrol
subcurrent subcorriente
subdirectory subdirectorio
subdistribution subdistribución
subdivide subdividir
subdivided capacitor capacitor subdividido
subdivided channel canal subdividido
subdivider subdivisor
subdivision subdivisión
subenvironment subambiente
subfeeder subalimentador
subfield subcampo
subfrequency subfrecuencia
subgroup subgrupo
subharmonic subarmónica, subarmónico
subharmonic detector detector subarmónico
subharmonic oscillation oscilación subarmónica
subharmonic oscillator oscilador subarmónico
subharmonic vibration vibración subarmónica
subjective photometer fotómetro subjetivo
sublayer subcapa
sublevel subnivel
submarine cable cable submarino
submarine line línea submarina
submarine repeater repetidor submarino
submarine telephone cable cable telefónico submarino
submenu submenú
submerged antenna antena sumergida
submerged arc welding soldadura por arco sumergido
submerged condenser condensador sumergido
submerged repeater repetidor sumergido
submersible sumergible
submersible enclosure caja sumergible
submersible fuse fusible sumergible
submersible machine máquina sumergible
submersible transformer transformador sumergible
submicroscopic submicroscópico
submillimeter submilimétrico
submillimeter wave onda submilimétrica
subminiature subminiatura
subminiature device dispositivo subminiatura
subminiature relay relé subminiatura
subminiature tube tubo subminiatura
subminiaturization subminiaturización
subminiaturize subminiaturizar
subminiaturized subminiaturizado
submodulator submodulador
submultiple submúltiplo

subnetwork subred
subnetwork configuration configuración de subred
subpanel subpanel
subparameter subparámetro
subpermanent subpermanente
subpermanent magnetism magnetismo
　　subpermanente
subprogram subprograma
subreflector subreflector
subrefraction subrefracción
subregion subregión
subregional subregional
subroutine subrutina
subroutine call llamada de subrutina
subroutine entry entrada de subrutina
subroutine instruction instrucción de subrutina
subsatellite subsatélite
subscheme subesquema
subscriber apparatus aparato de abonado
subscriber cable cable de abonado
subscriber connection conexión de abonado
subscriber installation instalación de abonado
subscriber line línea de abonado
subscriber meter contador de abonado
subscriber set aparato de abonado
subscriber station estación de abonado
subscriber telephone set aparato telefónico de
　　abonado
subset subconjunto, aparato de abonado
subsidiary conduit conducto subsidiario
subsonic subsónico
subsonic frequency frecuencia subsónica
substation subestación
substep subpaso
substitute character carácter de sustitución
substitute mode modo de sustitución
substitute track pista de sustitución
substitution sustitución
substitution capacitor capacitor de sustitución
substitution error error de sustitución
substitution inductor inductor de sustitución
substitution method método de sustitución
substitution resistor resistor de sustitución
substitution speaker altavoz de sustitución
substitution technique técnica de sustitución
substitution transformer transformador de
　　sustitución
substrate sustrato
substring subsecuencia
substructure subestructura
subsurface corrosion corrosión subsuperficial
subsynchronous subsincrónico
subsynchronous satellite satélite subsincrónico
subsystem subsistema
subsystem component componente de subsistema
subsystem identification identificación de subsistema
subsystem interface interfaz de subsistema
subsystem library biblioteca de subsistema
subtask subtarea
subterranean acoustical communication
　　comunicación acústica subterránea
subtotal subtotal
subtracter restador
subtracter-adder restador-sumador
subtraction resta, sustracción
subtraction circuit circuito de sustracción
subtractive sustractivo
subtractive color color sustractivo
subtractive color filter filtro de colores sustractivo
subtractive color system sistema de colores
　　sustractivo
subtractive filter filtro sustractivo
subtractive primaries primarios sustractivos

subtractive primary colors colores primarios
　　sustractivos
subtractive process proceso sustractivo
subtractor sustractor
subtrahend sustraendo
subtransient subtransitorio
subtransient electromotive force fuerza
　　electromotriz subtransitoria
subtransient reactance reactancia subtransitoria
subtransient time constant constante de tiempo
　　subtransitoria
subunit subunidad
subzone subzona
subzone center central de tránsito
successive sucesivo
sudden-change relay relé de cambio repentino
sudden death muerte repentina
sudden ionospheric disturbance perturbación
　　ionosférica repentina
Suhl effect efecto de Suhl
sulfate sulfato
sulfated battery batería sulfatada
sulfating sulfatación
sulfation sulfatación
sulfur azufre
sulfur hexafluoride hexafluoruro de azufre
sulfuric acid ácido sulfúrico
sum suma, adición
sum check comprobación por suma
sum frequency frecuencia de suma
sumcheck comprobación de suma
summary card tarjeta sumaria
summary counter contador sumario
summary punch perforadora sumaria
summary punching perforación sumaria
summary recorder registrador sumario
summation sumación, suma
summation bridge puente de suma
summation check comprobación por suma
summation instrument instrumento totalizador
summation meter medidor totalizador
summational sumacional
summator sumador
summer sumador
summing sumador
summing amplifier amplificador sumador
summing circuit circuito sumador
summing point punto sumador
sun battery batería solar
sun-powered alimentado por energía solar
sun-pump laser láser con bomba solar
sun radiation radiación solar
sun relay relé solar
sun switch conmutador solar
sunlight lamp lámpara de luz solar
sunlight-powered laser láser alimentado por luz
　　solar
sunspot mancha solar
sunspot cycle ciclo de manchas solares
sunspot disturbance perturbación por manchas
　　solares
sunspot interference interferencia por manchas
　　solares
sunspot noise ruido por manchas solares
super super
superacoustic superacústico
superaudible superaudible, ultraacústico
superaudio superaudio, ultraacústico
superaudio frequency frecuencia ultraacústica
superaudio telegraphy telegrafía ultraacústica
superbeta transistor transistor superbeta
supercommutation superconmutación
supercomputer supercomputadora

superconducting superconductor
superconducting cable cable superconductor
superconducting device dispositivo superconductor
superconducting electromagnet electroimán
    superconductor
superconducting generator generador
    superconductor
superconducting magnet electroimán
    superconductor
superconducting memory memoria superconductora
superconducting solenoid solenoide superconductor
superconductive superconductivo, superconductor
superconductivity superconductividad
superconductor superconductor
supercontrol supercontrol
supercontrol tube tubo de supercontrol
supercurrent supercorriente
superdirective superdirectivo
superdirective antenna antena superdirectiva
superdirectivity superdirectividad
superemitron superemitrón
superemitron camera cámara superemitrón
superficial superficial
superfine structure estructura superfina
superfluous superfluo
supergain superganancia
supergain antenna antena de superganancia
supergain array red de superganancia
supergrid superred
supergroup supergrupo
supergroup allocation asignación de supergrupos
supergroup band filter filtro de banda de
    supergrupos
supergroup carrier generator generador de
    portadora de supergrupo
supergroup control station estación de control de
    supergrupo
supergroup demodulator desmodulador de
    supergrupo
supergroup distribution frame repartidor de
    supergrupo
supergroup link enlace de supergrupos
supergroup modulating equipment equipo de
    modulación de supergrupo
supergroup modulator modulador de supergrupo
supergroup reference pilot piloto de referencia de
    supergrupo
supergroup section sección de supergrupo
supergroup subcontrol station estación de
    subcontrol de supergrupo
supergroup transfer point punto de transferencia de
    supergrupo
supergroup transfer station estación de
    transferencia de supergrupo
supergroup translation traslación de supergrupo
superheterodyne superheterodino
superheterodyne amplifier amplificador
    superheterodino
superheterodyne circuit circuito superheterodino
superheterodyne converter convertidor
    superheterodino
superheterodyne oscillator oscilador
    superheterodino
superheterodyne receiver receptor superheterodino
superheterodyne reception recepción
    superheterodina
superhigh frequency frecuencia superalta
superhigh-frequency radar radar de frecuencia
    superalta
supericonoscope supericonoscopio
superimposed superpuesto
superimposed circuit circuito superpuesto
superimposed coding codificación superpuesta

superimposed current corriente superpuesta
superimposed ringing llamada superpuesta
superluminous superluminoso
supermodulation supermodulación
superphantom circuit circuito superfantasma
superposed superpuesto
superposed circuit circuito superpuesto
superposition superposición
superposition effect efecto de superposición
superposition theorem teorema de la superposición
superpower superpotencia
superprecision superprecisión
superpressure superpresión
superpropagation superpropagación
superradiance superradiancia
superrefraction superrefracción
superrefractory superrefractario
superregeneration superregeneración
superregenerative superregenerativo
superregenerative circuit circuito superregenerativo
superregenerative detector detector
    superregenerativo
superregenerative receiver receptor
    superregenerativo
superregenerative reception recepción
    superregenerativa
supersaturate supersaturar
supersaturated supersaturado
supersaturated solution solución supersaturada
supersaturation supersaturación
supersensitive supersensible
supersensitive relay relé supersensible
supersonic supersónico
supersonic amplifier amplificador supersónico
supersonic beam haz supersónico
supersonic communication comunicación
    supersónica
supersonic detection detección supersónica
supersonic equipment equipo supersónico
supersonic flow flujo supersónico
supersonic frequency frecuencia supersónica
supersonic reception recepción supersónica
supersonic recording registro supersónico
supersonic signal señal supersónica
supersonic speed velocidad supersónica
supersonic wave onda supersónica
supersonics supersónica
supersynchronous supersincrónico
supersynchronous satellite satélite supersincrónico
supertelephone frequency frecuencia supertelefónica
superturnstile antenna antena de mariposa
supervisor supervisor
supervisory center centro supervisor
supervisory channel canal supervisor
supervisory console consola supervisora
supervisory control control supervisor
supervisory control system sistema de control
    supervisor
supervisory format formato supervisor
supervisory lamp lámpara supervisora
supervisory program programa supervisor
supervisory relay relé supervisor
supervisory routine rutina supervisora
supervisory signal señal supervisora
supervisory state estado supervisor
supervisory station estación supervisora
supervisory system sistema supervisor
supervisory terminal terminal supervisora
supervoltage supertensión
supplemental suplemental
supplementary suplementario
supplementary control control suplementario
supplementary equipment equipo suplementario

supplementary lighting alumbrado suplementario
supplementary maintenance mantenimiento suplementario
supplementary relay relé suplementario
supplier suministrador, abastecedor
supply fuente, alimentación, suministro, abasto
supply apparatus aparato de alimentación
supply battery batería de alimentación
supply circuit circuito de alimentación
supply current corriente de alimentación
supply equipment equipo de alimentación
supply lead conductor de alimentación
supply line línea de alimentación
supply mains red de alimentación
supply potential potencial de alimentación
supply power potencia de alimentación
supply rack bastidor de alimentación
supply section sección de alimentación
supply source fuente de alimentación
supply terminals terminales de alimentación
supply transformer transformador de alimentación
supply unit unidad de alimentación
supply voltage tensión de alimentación
support apoyo, soporte
support program programa de apoyo
support software programas de apoyo
support system sistema de apoyo
supporting mast mástil de soporte
suppress suprimir
suppressed suprimido
suppressed carrier portadora suprimida
suppressed-carrier modulation modulación con portadora suprimida
suppressed-carrier operation operación con portadora suprimida
suppressed-carrier system sistema con portadora suprimida
suppressed-carrier transmission transmisión con portadora suprimida
suppressed sideband banda lateral suprimida
suppressed-zero instrument instrumento de cero suprimido
suppressed-zero voltmeter vóltmetro de cero suprimido
suppression supresión, eliminación
suppression capacitor capacitor de supresión
suppression circuit circuito de supresión
suppression control control de supresión
suppression factor factor de supresión
suppression of carrier supresión de portadora
suppression pulse impulso de supresión
suppressor supresor
suppressor circuit circuito supresor
suppressor diode diodo supresor
suppressor grid rejilla supresora
suppressor-grid modulation modulación por rejilla supresora
suppressor pulse impulso supresor
surface superficie
surface-active agent agente tensoactivo
surface alloy aleación superficial
surface-alloy transistor transistor de aleación superficial
surface analyzer analizador de superficies
surface arc arco superficial
surface barrier barrera superficial
surface-barrier diffused transistor transistor de difusión de barrera superficial
surface-barrier diode diodo de barrera superficial
surface-barrier technique técnica de barrera superficial
surface-barrier transistor transistor de barrera superficial

surface charge carga superficial
surface-charge effect efecto de carga superficial
surface conductivity conductividad superficial
surface contact contacto de superficie
surface-contact rectifier rectificador de contacto por superficie
surface current corriente superficial
surface-current density densidad de corriente superficial
surface effect efecto superficial
surface energy energía superficial
surface impedance impedancia superficial
surface insulation aislamiento superficial
surface layer capa superficial
surface leakage fuga superficial
surface load carga superficial
surface migration migración superficial
surface noise ruido de superficie
surface passivation pasivación superficial
surface photoelectric effect efecto fotoeléctrico superficial
surface radar radar de superficie
surface recombination recombinación superficial
surface recombination rate velocidad de recombinación superficial
surface recombination velocity velocidad de recombinación superficial
surface reflection reflexión superficial
surface reflector reflector superficial
surface resistance resistencia superficial
surface resistivity resistividad superficial
surface superconductivity superconductividad superficial
surface switch conmutador de superficie
surface temperature temperatura superficial
surface tension tensión superficial
surface transfer impedance impedancia de transferencia superficial
surface treatment tratamiento superficial
surface wave onda de superficie
surface-wave amplifier amplificador de ondas de superficie
surface-wave antenna antena de ondas de superficie
surface-wave filter filtro de ondas de superficie
surface-wave transmission line línea de transmisión de ondas de superficie
surge sobretensión, sobretensión transitoria, pico transitorio, onda irruptiva
surge absorber absorbedor de picos transitorios
surge admittance admitancia ante picos transitorios
surge arrester supresor de picos transitorios
surge current sobrecorriente transitoria
surge electrode current corriente de electrodo anormal
surge generator generador de sobretensiones, generador de ondas de choque
surge impedance impedancia ante picos transitorios
surge limiting limitación de picos transitorios
surge protection protección contra picos transitorios
surge protector protector contra picos transitorios
surge recorder registrador de picos transitorios
surge relay relé de picos transitorios
surge suppressor supresor de picos transitorios
surge voltage sobretensión transitoria
surveillance vigilancia
surveillance radar radar de vigilancia
susceptance susceptancia
susceptance tube tubo de susceptancia
susceptibility susceptibilidad
susceptibility meter medidor de susceptibilidad
suspended call llamada suspendida
suspended load carga suspendida
suspension suspensión

suspension galvanometer galvanómetro de bobina móvil
suspension insulator aislador de suspensión
sustained interruption interrupción sostenida
sustained oscillation oscilación sostenida
sustained overvoltage sobretensión sostenida
sustained wave onda sostenida
swamping resistor resistor estabilizador
swap intercambio
swapping intercambio
swapping routine rutina de intercambio
sweep barrido, exploración
sweep accuracy exactitud de barrido
sweep amplifier amplificador de barrido
sweep circuit circuito de barrido
sweep delay retardo de barrido
sweep-delay circuit circuito de retardo de barrido
sweep duration duración de barrido
sweep excursion excursión de barrido
sweep expander expansor de barrido
sweep expansion expansión de barrido
sweep frequency frecuencia de barrido
sweep-frequency generator generador de frecuencia de barrido
sweep-frequency oscillator oscilador de frecuencia de barrido
sweep-frequency range alcance de frecuencias de barrido
sweep generator generador de barrido
sweep line línea de barrido
sweep lockout bloqueo de barrido
sweep magnification magnificación de barrido
sweep magnifier magnificador de barrido
sweep oscillator oscilador de barrido
sweep period periodo de barrido
sweep rate velocidad de barrido
sweep recovery time tiempo de recuperación de barrido
sweep reset restablecimiento de barrido
sweep reversal inversión de barrido
sweep signal señal de barrido
sweep-signal generator generador de señales de barrido
sweep speed velocidad de barrido
sweep test prueba de barrido
sweep time tiempo de barrido
sweep tone tono de barrido
sweep unit unidad de barrido
sweep velocity velocidad de barrido
sweep voltage tensión de barrido
sweep width anchura de barrido
sweep-width control control de anchura de barrido
sweeper generador de barrido
sweeping coil bobina de barrido
sweeping generator generador de barrido
sweeping speed velocidad de barrido
sweeping system sistema de barrido
swept-frequency oscillator oscilador con barrido de frecuencia
swept oscillator oscilador con barrido de frecuencia
swing oscilación, amplitud
swinging choke choque con entrehierro ajustable
Swiss-cheese packaging empaquetado en queso suizo
switch conmutador, interruptor
switch actuation actuación de conmutador
switch bay celda
switch code código de conmutador
switch current corriente de conmutador
switch-in conectar, poner en circuito
switch lever palanca de conmutador
switch-off desconectar, abrir el circuito
switch-on conectar, cerrar el circuito
switch-out desconectar, poner fuera del circuito

switch-panel panel de conmutadores
switch position posición de conmutador
switch rheostat reóstato de conmutador
switchbank banco de conmutadores
switchboard cuadro conmutador, cuadro de distribución, cuadro, tablero
switchbox caja de conmutador
switched beam haz conmutado
switched circuit circuito conmutado
switched connection conexión conmutada
switched line línea conmutada
switched network red conmutada
switched outlet tomacorriente conmutado
switched system sistema conmutado
switcher conmutador, interruptor
switchgear aparato de conexión, conmutador
switchgroup combinador
switchhook gancho conmutador
switching conmutación
switching accuracy exactitud de conmutación
switching action acción de conmutación
switching algebra álgebra de conmutación
switching apparatus aparato de conmutación
switching center centro de conmutación
switching characteristics características de conmutación
switching circuit circuito de conmutación
switching coefficient coeficiente de conmutación
switching constant constante de conmutación
switching control center centro de control de conmutación
switching control panel panel de control de conmutación
switching current corriente de conmutación
switching dependability confiabilidad de conmutación
switching design diseño de conmutación
switching desk pupitre de conmutación
switching device dispositivo de conmutación
switching diode diodo de conmutación
switching director director de conmutación
switching effect efecto de conmutación
switching element elemento de conmutación
switching equipment equipo de conmutación
switching flux flujo de conmutación
switching frequency frecuencia de conmutación
switching function función de conmutación
switching impulse impulso de conmutación
switching jack jack de conmutación
switching magnetomotive force fuerza magnetomotriz de conmutación
switching matrix matriz de conmutación
switching mode modo de conmutación
switching network red de conmutación
switching noise ruido de conmutación
switching-off desconexión, apertura de circuito
switching-on conexión, cierre de circuito
switching operation operación de conmutación
switching overvoltage sobretensión de conmutación
switching panel panel de conmutación
switching pilot piloto de conmutación
switching point punto de conmutación
switching pulse impulso de conmutación
switching rack bastidor de conmutación
switching rate frecuencia de conmutación
switching relay relé de conmutación
switching room sala de conmutación
switching section sección de conmutación
switching selector selector de conmutación
switching signal señal de conmutación
switching space espacio de conmutación
switching speed velocidad de conmutación
switching stage etapa de conmutación

switching station estación de conmutación
switching surge pico transitorio de conmutación
switching system sistema de conmutación
switching threshold umbral de conmutación
switching time tiempo de conmutación
switching tone tono de conmutación
switching transistor transistor de conmutación
switching tube tubo de conmutación
switching unit unidad de conmutación
switching valve válvula de conmutación
switching voltage tensión de conmutación
symbol símbolo
symbol analysis análisis de símbolos
symbol set conjunto de símbolos
symbolic simbólico
symbolic address dirección simbólica
symbolic addressing direccionamiento simbólico
symbolic analysis análisis simbólico
symbolic assembler ensamblador simbólico
symbolic assembly language lenguaje de ensamblaje simbólico
symbolic channel canal simbólico
symbolic character carácter simbólico
symbolic code código simbólico
symbolic coding codificación simbólica
symbolic data datos simbólicos
symbolic debugging depuración simbólica
symbolic diagram diagrama simbólico
symbolic execution ejecución simbólica
symbolic file archivo simbólico
symbolic image imagen simbólica
symbolic instruction instrucción simbólica
symbolic language lenguaje simbólico
symbolic logic lógica simbólica
symbolic machine máquina simbólica
symbolic model modelo simbólico
symbolic name nombre simbólico
symbolic notation notación simbólica
symbolic number número simbólico
symbolic parameter parámetro simbólico
symbolic program programa simbólico
symbolic programming programación simbólica
symbolic representation representación simbólica
symbolic unit unidad simbólica
symmetric simétrico
symmetrical simétrico
symmetrical alternating current corriente alterna simétrica
symmetrical alternating quantity magnitud alterna simétrica
symmetrical amplifier amplificador simétrico
symmetrical avalanche rectifier rectificador de avalancha simétrico
symmetrical cable cable simétrico
symmetrical channel canal simétrico
symmetrical circuit circuito simétrico
symmetrical component componente simétrico
symmetrical conductivity conductividad simétrica
symmetrical curve curva simétrica
symmetrical directional coupler acoplador direccional simétrico
symmetrical field-effect-transistor transistor de efecto de campo simétrico
symmetrical input entrada simétrica
symmetrical intensity distribution distribución de intensidad simétrica
symmetrical modulation voltage tensión de modulación simétrica
symmetrical multivibrator multivibrador simétrico
symmetrical network red simétrica
symmetrical output salida simétrica
symmetrical pair par simétrico
symmetrical processor procesador simétrico

symmetrical relay relé simétrico
symmetrical signal señal simétrica
symmetrical transducer transductor simétrico
symmetrical transistor transistor simétrico
symmetrical varistor varistor simétrico
symmetrical wave onda simétrica
symmetrically simétricamente
symmetry simetría
sympathetic simpático
sympathetic vibration vibración simpática
sync amplifier amplificador de sincronización
sync bits bits de sincronización
sync character carácter de sincronización
sync check comprobación de sincronización
sync compression compresión de señal de sincronización
sync generator generador de sincronización
sync level nivel de sincronización
sync pulse impulso de sincronización
sync-pulse generator generador de impulsos de sincronización
sync section sección de sincronización
sync separator separador de sincronización
sync-separator tube tubo separador de sincronización
sync signal señal de sincronización
sync-signal generator generador de señal de sincronización
sync takeoff punto de toma de impulsos de sincronización
sync-takeoff point punto de toma de impulsos de sincronización
synchro sincro, síncrono
synchro-control transformer sincrotransformador de control
synchro-differential receiver sincrorreceptor diferencial
synchro generator sincrogenerador
synchro motor sincromotor
synchro receiver sincrorreceptor
synchro repeater sincrorrepetidor
synchro system sincrosistema
synchro transmitter sincrotransmisor
synchrocyclotron sincrociclotrón
synchrodyne sincrodino
synchroflash sincrodestello
synchrogenerator sincrogenerador
synchronism sincronismo
synchronism indicator indicador de sincronismo
synchronization sincronización
synchronization apparatus aparato de sincronización
synchronization bits bits de sincronización
synchronization character carácter de sincronización
synchronization check comprobación de sincronización
synchronization control control de sincronización
synchronization error error de sincronización
synchronization indicator indicador de sincronización
synchronization pattern patrón de sincronización
synchronization pilot piloto de sincronización
synchronization point punto de sincronización
synchronization pulses impulsos de sincronización
synchronization signal señal de sincronización
synchronization time tiempo de sincronización
synchronize sincronizar
synchronized sincronizado
synchronized clamping fijación de nivel sincronizada
synchronized flash destello sincronizado
synchronized frequency frecuencia sincronizada

synchronized multivibrator multivibrador sincronizado
synchronized operation operación sincronizada
synchronized sweep barrido sincronizado
synchronized timing temporización sincronizada
synchronized transmitters transmisores sincronizados
synchronized wave onda sincronizada
synchronizer sincronizador
synchronizing sincronización
synchronizing amplifier amplificador de sincronización
synchronizing characters caracteres de sincronización
synchronizing circuit circuito de sincronización
synchronizing clock reloj de sincronización
synchronizing current corriente de sincronización
synchronizing generator generador de sincronización
synchronizing impulse impulso de sincronización
synchronizing inverter inversor de sincronización
synchronizing level nivel de sincronización
synchronizing pilot piloto de sincronización
synchronizing power potencia de sincronización
synchronizing pulse impulso de sincronización
synchronizing-pulse generator generador de impulsos de sincronización
synchronizing-pulse interval intervalo de impulsos de sincronización
synchronizing-pulse separation separación de impulsos de sincronización
synchronizing relay relé de sincronización
synchronizing separation separación de sincronización
synchronizing separator separador de sincronización
synchronizing signal señal de sincronización
synchronizing-signal amplitude amplitud de señal de sincronización
synchronizing-signal separation separación de señales de sincronización
synchronizing torque par de sincronización
synchronodyne sincronodino
synchronometer sincronómetro
synchronoscope sincronoscopio
synchronous sincrónico, síncrono
synchronous admittance admitancia sincrónica
synchronous alternator alternador sincrónico
synchronous-asynchronous motor motor sincrónico-asincrónico
synchronous booster converter convertidor sincrónico de autorregulación
synchronous capacitor capacitor sincrónico
synchronous chopper interruptor periódico sincrónico
synchronous clock reloj sincrónico
synchronous communication comunicación sincrónica
synchronous communications satellite satélite de comunicaciones sincrónicas
synchronous computer computadora sincrónica
synchronous condenser condensador sincrónico
synchronous converter convertidor sincrónico
synchronous coupling acoplamiento sincrónico
synchronous data datos sincrónicos
synchronous demodulator desmodulador sincrónico
synchronous detection detección sincrónica
synchronous detector detector sincrónico
synchronous device dispositivo sincrónico
synchronous electric clock reloj eléctrico sincrónico
synchronous electromotive force fuerza electromotriz sincrónica
synchronous flow flujo sincrónico
synchronous gate puerta sincrónica

synchronous generator generador sincrónico
synchronous ignitron ignitrón sincrónico
synchronous impedance impedancia sincrónica
synchronous induction motor motor de inducción sincrónico
synchronous inputs entradas sincrónicas
synchronous inverter inversor sincrónico
synchronous line adapter adaptador de línea sincrónico
synchronous logic lógica sincrónica
synchronous machine máquina sincrónica
synchronous mode modo sincrónico
synchronous motor motor sincrónico
synchronous network red sincrónica
synchronous operation operación sincrónica
synchronous orbit órbita sincrónica
synchronous processing procesamiento sincrónico
synchronous reactance reactancia sincrónica
synchronous rectifier rectificador sincrónico
synchronous repeater repetidor sincrónico
synchronous rotary converter convertidor rotativo sincrónico
synchronous rotary interrupter interruptor rotativo sincrónico
synchronous satellite satélite sincrónico
synchronous spark-gap descargador sincrónico
synchronous speed velocidad sincrónica
synchronous switch conmutador sincrónico
synchronous system sistema sincrónico
synchronous timer temporizador sincrónico
synchronous torque par sincrónico
synchronous transmission transmisión sincrónica
synchronous tuning sintonización sincrónica
synchronous vibrator vibrador sincrónico
synchronous voltage tensión sincrónica
synchronously sincrónicamente
synchrony sincronismo, sincronía
synchroscope sincroscopio
synchrotron sincrotrón
synchrotron magnet imán de sincrotrón
synergism sinergismo
synergistic sinérgico
syntax sintaxis
syntax error error de sintaxis
synthesis síntesis
synthesizer sintetizador
synthetic sintético
synthetic address dirección sintética
synthetic addressing direccionamiento sintético
synthetic crystal cristal sintético
synthetic display visualización sintética
synthetic distortion distorsión sintética
synthetic language lenguaje sintético
synthetic relationship relación sintética
synthetic resin resina sintética
syntonization sintonización
syntony sintonía
system sistema
system administrator administrador de sistema
system analyzer analizador de sistemas
system application aplicación de sistema
system architecture arquitectura de sistema
system architecture arquitectura de sistema
system area área de sistema
system bus bus de sistema
system check comprobación de sistema
system-check module módulo de comprobación de sistema
system command mando de sistema
system component componente de sistema
system configuration configuración de sistema
system console consola de sistema
system constant constante de sistema

system control control de sistema
system control area área de control de sistema
system control file archivo de control de sistema
system control panel panel de control de sistema
system control programming programación de
   control de sistema
system controller controlador de sistema
system data bus bus de datos de sistema
system definition definición de sistema
system delay time tiempo de retardo de sistema
system description descripción de sistema
system design diseño de sistema
system design language lenguaje de diseño de
   sistema
system design phase fase de diseño de sistema
system deviation desviación de sistema
system diagnostics diagnósticos de sistema
system directory directorio de sistema
system disk disco de sistema
system downtime tiempo muerto de sistema
system dump vaciado de sistema
system efficiency eficiencia de sistema
system element elemento de sistema
system error error de sistema
system evaluation evaluación de sistema
system file archivo de sistema
system flowchart organigrama de sistema
system format formato de sistema
system frequency frecuencia de sistema
system gain ganancia de sistema
system identification identificación de sistema
system input entrada de sistema
system input device dispositivo de entrada de
   sistema
system input unit unidad de entrada de sistema
system instruction instrucción de sistema
system integration integración de sistema
system integrity integridad de sistema
system interconnection interconexión de sistemas
system interface interfaz de sistema
system interface module módulo de interfaz de
   sistema
system interrupt interrupción de sistema
system key clave de sistema
system language lenguaje de sistema
system layout distribución de sistema
system load carga de sistema
system load factor factor de carga de sistema
system loader cargador de sistema
system logic lógica de sistema
system logical input device dispositivo de entrada
   lógica de sistema
system logical output device dispositivo de salida
   lógica de sistema
system macro macro de sistema
system macroinstruction macroinstrucción de
   sistema
system maintenance mantenimiento de sistema
system management administración de sistema
system mode modo de sistema
system model modelo de sistema
system modification modificación de sistema
system monitor monitor de sistema
system noise ruido de sistema
system of units sistema de unidades
system operator operador de sistema
system operator station estación de operador de
   sistema
system organization organización de sistema
system output salida de sistema
system output device dispositivo de salida de sistema
system output printer impresora de salida de
   sistema

system output tape cinta de salida de sistema
system output unit unidad de salida de sistema
system parameter parámetro de sistema
system partition partición de sistema
system printer impresora de sistema
system profile perfil de sistema
system program programa de sistema
system programming programación de sistema
system queue cola de sistema
system queue area área de colas de sistema
system recovery time tiempo de recuperación de
   sistema
system reference code código de referencia de
   sistema
system reliability confiabilidad de sistema
system requirements requisitos de sistema
system reserve reserva de sistema, reserva de red
system reset restablecimiento de sistema
system resident file archivo residente en el sistema
system resources recursos de sistema
system resources manager administrador de
   recursos de sistema
system restart reinicio de sistema
system service servicio de sistema
system service program programa de servicio de
   sistema
system software programas de sistema
system structure estructura de sistema
system support apoyo de sistema
system-support program programa de apoyo de
   sistema
system tape cinta de sistema
system task tarea de sistema
system test prueba de sistema
system testing pruebas de sistema
system time tiempo de sistema
system timer temporizador de sistema
system unit unidad de sistema
system verification verificación de sistema
system voltage tensión de sistema
systematic sistemático
systematic error error sistemático
systematic programming programación sistemática
systems analysis análisis de sistemas
systems flowchart organigrama de sistemas

T

T antenna antena en T
T circulator circulador en T
T connection conexión en T
T connector conector en T
T junction unión en T
T network red en T
table tabla, cuadro
table addressing direccionamiento de tablas
table card tarjeta de tabla
table code código de tabla
table element elemento de tabla
table function función de tabla
table item elemento de tabla
table look-up búsqueda en tabla
table look-up instruction instrucción de búsqueda en tabla
table reference character carácter de referencia de tabla
table-top device dispositivo de superficie de mesa
tabulate tabular
tabulating tabulación
tabulating equipment equipo de tabulación
tabulating machine máquina de tabulación
tabulation tabulación
tabulation character carácter de tabulación
tabulator tabulador
tacan tacán
tacan navigation system sistema de navegación tacán
tacan receiver receptor tacán
tacan transmitter transmisor tacán
tachogenerator tacogenerador
tachograph tacógrafo
tachometer tacómetro
tachometer generator generador tacométrico
tachometer pickup captor tacométrico
tachometric tacométrico
tachometric relay relé tacométrico
tachymeter taquímetro
tachymetric taquimétrico
tacitron tacitrón
tactile keyboard teclado táctil
tag etiqueta, terminal
tag reader lectora de etiquetas
tail cola
tail circuit circuito de cola
tail current corriente de cola
tail of pulse cola de impulso
tailing coleo, arrastre
take-up reel carrete receptor
talbot talbot
Talbot's law ley de Talbot
talk-listen switch conmutador de hablar-escuchar
talkback circuit circuito de intercomunicación
talkback speaker altavoz de intercomunicación
talking battery batería de conversación
talking terminal terminal de conversación
tally contar, cuadrar
tamper-proof a prueba de manipulación inexperta
tamper-resistant resistente a manipulación inexperta
Tanberg effect efecto de Tanberg
tandem tándem

tandem circuit circuito en tándem
tandem computers computadoras en tándem
tandem connection conexión en tándem
tandem exchange central tándem
tandem motor motor en tándem
tandem networks redes en tándem
tandem operation operación en tándem
tandem selection selección en tándem
tandem system sistema en tándem
tandem transistor transistores en tándem
tangent tangente
tangent galvanometer galvanómetro de tangentes
tangent ray rayo tangente
tangential tangencial
tangential focus foco tangencial
tangential incidence incidencia tangencial
tangential plane plano tangencial
tangential projection proyección tangencial
tangential sensitivity sensibilidad tangencial
tangential signal señal tangencial
tangential signal sensitivity sensibilidad de señal tangencial
tangential view vista tangencial
tangential wavepath trayectoria de onda tangéncial
tank tanque, cuba
tank circuit circuito tanque
tank coil bobina tanque
tank voltage tensión de tanque
tantalum tantalio
tantalum capacitor capacitor de tantalio
tantalum detector detector de tantalio
tantalum electrolytic capacitor capacitor electrolítico de tantalio
tantalum-foil electrolytic capacitor capacitor electrolítico de hojas de tantalio
tantalum lamp lámpara de tantalio
tantalum-nitride resistor resistor de nitruro de tantalio
tantalum rectifier rectificador de tantalio
tantalum thin-film circuit circuito de película delgada de tantalio
tap toma, derivación
tap box caja de tomas
tap changer conmutador de tomas
tap switch conmutador de tomas
tape cinta
tape advance avance de cinta
tape alternation alternación de cinta
tape backup unit unidad de reserva de cinta
tape-based basado en cintas
tape-bound limitado por cinta
tape cable cable en cinta
tape cartridge cartucho de cinta
tape-cartridge drive unidad de cartucho de cinta
tape cassette casete de cinta
tape channel canal de cinta
tape character carácter de cinta
tape characteristics características de cinta
tape checker comprobador de cintas
tape cleaner limpiador de cintas
tape code código de cinta
tape code checking comprobación de código de cinta
tape comparator comparador de cintas
tape connector conector de cinta
tape control control de cinta
tape-control unit unidad de control de cinta
tape-controlled controlado por cinta
tape-controlled card punch perforadora de tarjetas controlada por cinta
tape-controlled machine máquina controlada por cinta
tape-controlled transmitter transmisor controlado por cinta

tape conversion conversión de cinta
tape-conversion program programa de conversión de cinta
tape core núcleo de cinta
tape counter contador de cinta
tape data datos en cinta
tape data validation validación de datos de cinta
tape deck mecanismo de arrastre de cinta, unidad de cinta, magnetófono
tape-deck mechanism mecanismo de arrastre de cinta, mecanismo de magnetófono
tape density densidad de cinta
tape depressor depresor de cinta
tape drive mecanismo de arrastre de cinta, unidad de cinta
tape driven accionado por cinta
tape dump vaciado de cinta
tape duplication duplicación de cinta
tape encoder codificador de cinta
tape eraser borrador de cintas
tape error error de cinta
tape feed alimentación de cinta
tape-feed code código de alimentación de cinta
tape-feed switch conmutador de alimentación de cinta
tape feeding alimentación de cinta
tape file archivo en cinta
tape format formato de cinta
tape guide guía de cinta
tape handler manipulador de cinta
tape head cabeza de cinta
tape input entrada de cinta
tape label etiqueta de cinta
tape library biblioteca de cintas
tape lifter levantador de cinta
tape-limited limitado por cinta
tape load point punto de carga de cinta
tape loader cargador de cinta
tape loading carga de cinta
tape loop cinta sin fin
tape machine magnetófono, máquina de cinta magnética
tape magazine cartucho de cinta
tape management system sistema de administración de cintas
tape mark marca de cinta
tape memory memoria de cinta
tape monitor monitor de cinta
tape motion moción de cinta
tape movement movimiento de cinta
tape operating system sistema operativo de cinta
tape-oriented computer computadora orientada a cintas
tape parity paridad de cinta
tape path trayectoria de cinta
tape perforator perforadora de cinta
tape playback reproducción de cinta
tape-playback system sistema de reproducción de cinta
tape player reproductor de cinta
tape preparation preparación de cinta
tape printer impresora de cinta
tape processing procesamiento de cinta
tape programming programación de cinta
tape-programming system sistema de programación de cintas
tape punch perforadora de cinta
tape puncher perforadora de cinta
tape-punching machine máquina perforadora de cintas
tape-read unit unidad de lectura de cinta
tape reader lectora de cinta
tape-reading head cabeza de lectura de cinta

tape recorder magnetófono, registrador de cinta
tape recording registro en cinta
tape-recording unit unidad de registro en cinta
tape relay retransmisión por cinta
tape-relay center centro de retransmisión por cinta
tape-relay circuit circuito de retransmisión por cinta
tape-relay network red de retransmisión por cinta
tape-relay station estación de retransmisión por cinta
tape reperforator reperforadora de cinta
tape reproducer reproductor de cinta
tape resident system sistema residente en cinta
tape retransmission retransmisión por cinta
tape search búsqueda de cinta
tape skew sesgo de cinta
tape skip salto de cinta
tape sort clasificación de cinta
tape speed velocidad de cinta
tape-speed variation variación de velocidad de cinta
tape splicer empalmador de cinta
tape station estación de cinta
tape storage almacenamiento en cinta
tape synchronizer sincronizador de cintas
tape system sistema de cintas
tape-to-card de cinta a tarjeta
tape-to-card conversion conversión de cinta a tarjetas
tape-to-card converter convertidor de cinta a tarjetas
tape-to-head contact contacto de cinta a cabeza
tape-to-tape de cinta a cinta
tape-to-tape conversion conversión de cinta a cinta
tape-to-tape converter convertidor de cinta a cinta
tape track pista de cinta
tape transmitter transmisor de cinta
tape transport transporte de cinta, arrastre de cinta
tape-transport mechanism mecanismo de transporte de cinta, mecanismo de arrastre de cinta
tape-transport system sistema de transporte de cinta, sistema de arrastre de cinta
tape unit unidad de cinta
tape verification verificación de cinta
tape verifier verificadora de cinta
tape width anchura de cinta
tape-wound core núcleo de cinta arrollada
taper disminución progresiva, variación progresiva, guíaondas fusiforme, conicidad
tapered de disminución progresiva, de variación progresiva, ahusado
tapered attenuation atenuación progresiva
tapered distribution distribución progresiva
tapered illumination iluminación progresiva
tapered section sección progresiva
tapered transition transición progresiva
tapered waveguide guíaondas de sección variable
tapering disminución progresiva, variación progresiva, ahusamiento
tapped con tomas, con derivaciones
tapped battery batería con tomas
tapped coil bobina con tomas
tapped component componente con tomas
tapped control control con tomas
tapped delay line línea de retardo con tomas
tapped line línea con tomas
tapped resistor resistor con tomas
tapped rheostat reóstato con tomas
tapped transformer transformador con tomas
tapped winding arrollamiento con tomas
tapping point punto de toma, toma de regulación
target capacitance capacitancia de blanco
target computer computadora objeto
target configuration configuración objeto
target current corriente de blanco
target current amplifier amplificador de corriente de blanco

target cutoff voltage tensión de corte de blanco
target data set conjunto de datos objeto
target disk disco objeto
target diskette disquete objeto
target electrode electrodo de blanco
target language lenguaje objeto
target machine máquina objeto
target program programa objeto
target routine rutina objeto
target voltage tensión de blanco
task tarea
task control control de tareas
task-control block bloque de control de tareas
task-control character carácter de control de tareas
task-execution area área de ejecución de tareas
task identification identificación de tarea
task management administración de tareas
task queue cola de tareas
task set conjunto de tareas
task start iniciación de tarea
taut-band meter medidor de cinta tensa
Taylor connection conexión de Taylor
Taylor distribution distribución de Taylor
Taylor modulation modulación de Taylor
tearing desgarro
technetium tecnecio
technical técnico
technical information información técnica
technical specification especificación técnica
tecnetron tecnetrón
tee adapter adaptador en T
tee antenna antena en T
tee circuit circuito en T
tee connector conector en T
tee joint junta en T
tee junction unión en T
tee network red en T
tee switch conmutadores en T
teflon teflón
teflon-insulated aislado con teflón
teflon-insulated wire hilo aislado con teflón
teflon insulation aislamiento de teflón
telcothene telcoteno
telcothene-insulated aislado con telcoteno
telcothene-insulated wire hilo aislado con telcoteno
telcothene insulation aislamiento con telcoteno
teleammeter teleamperímetro
teleautograph teleautógrafo
teleautography teleautografía
telecamera telecámara
telecardiograph telecardiógrafo
telecast telerradiodifusión, teledifusión
telechrome telecromo
telecommunication telecomunicación
telecommunications access method método de
  acceso a telecomunicaciones
telecommunications cable cable de
  telecomunicaciones
telecommunications channel canal de
  telecomunicaciones
telecommunications circuit circuito de
  telecomunicaciones
telecommunications control unit unidad de control
  de telecomunicaciones
telecommunications exchange central de
  telecomunicaciones
telecommunications facility instalación de
  telecomunicaciones
telecommunications line línea de telecomunicaciones
telecommunications network red de
  telecomunicaciones
telecommunications relay relé de
  telecomunicaciones

telecommunications satellite satélite de
  telecomunicaciones
telecommunications service servicio de
  telecomunicaciones
telecommunications switching conmutación de
  telecomunicaciones
telecommunications system sistema de
  telecomunicaciones
telecommunications traffic tráfico de
  telecomunicaciones
telecompass telecompás
telecomputing telecomputación
teleconference teleconferencia
teleconnection teleconexión
telecontrol telecontrol
telecontrol equipment equipo de telecontrol
telecontrolled telecontrolado
telecontrolled substation substación telecontrolada
telecopier telecopiadora
telecounter telecontador
telediffusion teledifusión
telefacsimile telefacsímil
telegram telegrama
telegraph telégrafo
telegraph battery batería telegráfica
telegraph bias polarización telegráfica
telegraph center centro telegráfico
telegraph channel canal telegráfico
telegraph circuit circuito telegráfico
telegraph code código telegráfico
telegraph connection conexión telegráfica
telegraph demodulator desmodulador telegráfico
telegraph distortion distorsión telegráfica
telegraph distributor distribuidor telegráfico
telegraph electromagnet electroimán telegráfico
telegraph electronic relay relé electrónico
  telegráfico
telegraph emission emisión telegráfica
telegraph equipment equipo telegráfico
telegraph key manipulador telegráfico
telegraph level nivel telegráfico
telegraph line línea telegráfica
telegraph link enlace telegráfico
telegraph modem módem telegráfico
telegraph modulation modulación telegráfica
telegraph modulator modulador telegráfico
telegraph network red telegráfica
telegraph noise ruido telegráfico
telegraph pulse impulso telegráfico
telegraph receiver receptor telegráfico
telegraph recorder registrador telegráfico
telegraph relay relé telegráfico
telegraph repeater repetidor telegráfico
telegraph selector selector telegráfico
telegraph service servicio telegráfico
telegraph set aparato telegráfico
telegraph signal señal telegráfica
telegraph signal distortion distorsión de señal
  telegráfica
telegraph station estación telegráfica
telegraph system sistema telegráfico
telegraph terminal terminal telegráfica
telegraph transmission transmisión telegráfica
telegraph transmission speed velocidad de
  transmisión telegráfica
telegraph transmitter transmisor telegráfico
telegraphic telegráfico
telegraphic signal señal telegráfica
telegraphy telegrafía
teleindicator teleindicador
teleinformatics teleinformática
telematics telemática
telemechanism telemecanismo

telemeter telémetro, telemedidor
telemetering telemedida
telemetering antenna antena de telemedida
telemetering band banda de telemedida
telemetering channel canal de telemedida
telemetering circuit circuito de telemedida
telemetering coder codificador de telemedida
telemetering decoder descodificador de telemedida
telemetering demodulator desmodulador de telemedida
telemetering detector detector de telemedida
telemetering device dispositivo de telemedida
telemetering radiosonde radiosonda de telemedida
telemetering receiver receptor de telemedida
telemetering recorder registrador de telemedida
telemetering sampling muestreo de telemedida
telemetering system sistema de telemedida
telemetering transducer transductor de telemedida
telemetering transmitter transmisor de telemedida
telemetric telemétrico
telemetric receiver receptor telemétrico
telemetric transmitter transmisor telemétrico
telemetry telemetría
telemicroscopy telemicroscopía
telephone teléfono
telephone amplifier amplificador telefónico
telephone amplifying tube tubo amplificador telefónico
telephone-answering machine máquina de contestación telefónica
telephone-answering set aparato de contestación telefónica
telephone broadcasting radiodifusión telefónica
telephone cable cable telefónico
telephone call llamada telefónica
telephone call simulator simulador de llamadas telefónicas
telephone capacitor capacitor telefónico
telephone carrier portadora telefónica
telephone central office central telefónica
telephone channel canal telefónico
telephone circuit circuito telefónico
telephone communication comunicación telefónica
telephone connection conexión telefónica
telephone control panel panel de control telefónico
telephone current corriente telefónica
telephone dial disco telefónico
telephone earphone audífono telefónico
telephone equalization ecualización telefónica
telephone equipment equipo telefónico
telephone exchange central telefónica
telephone facilities instalaciones telefónicas
telephone feed circuit circuito de alimentación telefónica
telephone frequency frecuencia telefónica
telephone frequency characteristic característica de frecuencia telefónica
telephone handset microteléfono
telephone headset audífonos telefónicos
telephone induction coil bobina de inducción telefónica
telephone influence factor factor de influencia telefónica
telephone jack jack telefónico
telephone line línea telefónica
telephone link enlace telefónico
telephone modem módem telefónico
telephone modulation modulación telefónica
telephone network red telefónica
telephone noise ruido telefónico
telephone pickup captor telefónico
telephone plant planta telefónica
telephone plug clavija telefónica

telephone receiver receptor telefónico
telephone relay relé telefónico
telephone repeater repetidor telefónico
telephone ringer timbre telefónico
telephone service servicio telefónico
telephone set aparato telefónico
telephone sidetone captor telefónico
telephone silencer silenciador telefónico
telephone station estación telefónica
telephone switchboard cuadro conmutador telefónico
telephone switching conmutación telefónica
telephone system sistema telefónico
telephone terminal terminal telefónica
telephone traffic tráfico telefónico
telephone transmission transmisión telefónica
telephone transmitter transmisor telefónico
telephone wire hilo telefónico
telephonic telefónico
telephonograph telefonógrafo
telephonometer telefonómetro
telephonometry telefonometría
telephony telefonía
telephoto lens teleobjetivo
telephotograph telefotografía
telephotography telefotografía
telephotometer telefotómetro
teleprinter teleimpresora
teleprinter circuit circuito de teleimpresora
teleprinter connection conexión de teleimpresora
teleprinter interface interfaz de teleimpresora
teleprinter line línea de teleimpresora
teleprinter signal señal de teleimpresora
teleprinter system sistema de teleimpresora
teleprinter transmission transmisión de teleimpresora
teleprinter transmitter transmisor de teleimpresora
teleprinting teleimpresión
teleprocessing teleprocesamiento, teleproceso
teleprocessing network red de teleprocesamiento
teleprocessing system sistema de teleprocesamiento
teleprocessing terminal terminal de teleprocesamiento
teleprocessor teleprocesador
teleran telerán
teleran system sistema telerán
telereceiver telerreceptor, televisor
telereception telerrecepción, recepción de televisión
telerecorder telerregistrador, registrador de televisión
telerecording telerregistro, registro de televisión
telereference telerreferencia
teleregulation telerregulación
teleregulator telerregulador
telescope telescopio
telescopic telescópico
telescoping antenna antena telescópica
telescoping tower torre telescópica
telesignaling teleseñalización
teleswitch teleconmutador
teletape telecinta
teletext teletexto
telethermometer teletermómetro
telethermoscope teletermoscopio
teletransmission teletransmisión
teletype teletipo
teletype code código de teletipo
teletypewriter teleimpresora
teletypewriter channel canal de teleimpresora
teletypewriter circuit circuito de teleimpresora
teletypewriter exchange central de teleimpresoras
teletypewriter network red de teleimpresoras
televise televisar
televised image imagen televisada

televised picture imagen televisada
televising televisión
television televisión
television antenna antena de televisión
television band banda de televisión
television broadcast videodifusión
television broadcast station estación de
    videodifusión
television broadcasting videodifusión
television cable cable de televisión
television camera cámara de televisión
television-camera tube tubo de cámara de televisión
television channel canal de televisión
television circuit circuito de televisión
television connection conexión de televisión
television direct transmission transmisión directa de
    televisión
television distribution system sistema de
    distribución de televisión
television disturbance perturbación de televisión
television framing encuadre
television image imagen de televisión
television information información de televisión
television interference interferencia de televisión
television line línea de televisión
television link enlace de televisión
television magnetic tape cinta magnética para
    televisión
television master antenna antena colectiva de
    televisión
television network red de televisión
television pentode pentodo de televisión
television picture imagen de televisión
television picture tube tubo de imagen de televisión,
    cinescopio
television raster trama de televisión
television receiver receptor de televisión, televisor
television reception recepción de televisión
television relay retransmisión de televisión
television relay network red de retransmisión de
    televisión
television relay station estación de retransmisión de
    televisión
television relay system sistema de retransmisión de
    televisión
television repeater repetidora de televisión
television scanning exploración de televisión
television screen pantalla de televisión
television set aparato de televisión, televisor
television signal señal de televisión
television silencer silenciador de televisión
television sound sonido de televisión
television standard patrón de televisión
television station estación de televisión
television system converter convertidor de sistemas
    de televisión
television tape player reproductor de cintas
    magnéticas de televisión, magnetoscopio
television tape recorder grabadora de cintas
    magnéticas de televisión, magnetoscopio,
    videograbadora
television test card patrón de ajuste de televisión,
    mira
television transmission transmisión de televisión
television transmission standards patrones de
    transmisión de televisión
television transmitter transmisor de televisión
television tube tubo de televisión, cinescopio
television tuner sintonizador de televisión
televisor televisor
televoltmeter televóltmetro
telewattmeter telewáttmetro
telewriter teleautógrafo

telex télex
telex channel canal télex
telex circuit circuito télex
telex communication comunicación télex
telex connection conexión télex
telex link enlace télex
telex network red télex
telex station estación télex
telex system sistema télex
telex terminal terminal télex
telex transmission transmisión télex
telluric telúrico
telluric currents corrientes telúricas
telluric effects efectos telúricos
telluric-electricity electricidad telúrica
telluric field campo telúrico
tellurium telurio
temperature temperatura
temperature chamber cámara de temperatura
temperature class categoría de temperatura
temperature classification clasificación de
    temperatura
temperature coefficient coeficiente de temperatura
temperature coefficient of capacitance coeficiente
    de temperatura de capacitancia
temperature coefficient of capacity coeficiente de
    temperatura de capacidad
temperature coefficient of delay coeficiente de
    temperatura de retardo
temperature coefficient of electromotive force
    coeficiente de temperatura de fuerza electromotriz
temperature coefficient of frequency coeficiente de
    temperatura de frecuencia
temperature coefficient of permeability coeficiente
    de temperatura de permeabilidad
temperature coefficient of resistance coeficiente de
    temperatura de resistencia
temperature coefficient of resistivity coeficiente de
    temperatura de resistividad
temperature coefficient of sensitivity coeficiente de
    temperatura de sensibilidad
temperature coefficient of voltage drop coeficiente
    de temperatura de caída de tensión
temperature-compensated crystal oscillation
    oscilación de cristal con compensación de
    temperatura
temperature-compensated crystal oscillator
    oscilador de cristal con compensación de
    temperatura
temperature-compensated Zener diode diodo de
    Zener con compensación de temperatura
temperature-compensating capacitor capacitor
    compensador de temperatura
temperature-compensating component componente
    compensador de temperatura
temperature compensation compensación de
    temperatura
temperature control control de temperatura
temperature-control device dispositivo de control de
    temperatura
temperature-control system sistema de control de
    temperatura
temperature-controlled chamber cámara con
    temperatura controlada
temperature-controlled component componente con
    temperatura controlada
temperature-controlled crystal cristal con
    temperatura controlada
temperature-controlled crystal oscillator oscilador
    de cristal con temperatura controlada
temperature controller controlador de temperatura
temperature correction corrección de temperatura
temperature degree grado de temperatura

temperature detector detector de temperatura
temperature gradient gradiente de temperatura
temperature index índice de temperatura
temperature indicator indicador de temperatura
temperature interval intervalo de temperatura
temperature inversion inversión de temperatura
temperature-limited limitado por temperatura
temperature-limited diode diodo limitado por temperatura
temperature-limited emission emisión limitada por temperatura
temperature meter medidor de temperatura
temperature-regulating equipment equipo regulador de temperatura
temperature regulator regulador de temperatura
temperature relay relé de temperatura
temperature resistance coefficient coeficiente de resistencia de temperatura
temperature rise aumento de temperatura
temperature saturation saturación de temperatura
temperature scale escala de temperatura
temperature-sensitive sensible a la temperatura
temperature-sensitive resistor resistor sensible a la temperatura
temperature sensitivity sensibilidad a la temperatura
temperature sensor sensor de temperatura
temperature stability estabilidad de temperatura
temperature-to-current converter convertidor de temperatura a corriente
temperature-to-voltage converter convertidor de temperatura a tensión
temperature variation variación de temperatura
template patrón, plantilla
temporal temporal
temporal coherence coherencia temporal
temporal cohesion cohesión temporal
temporary temporal
temporary connection conexión temporal
temporary data datos temporales
temporary data set conjunto de datos temporal
temporary duty servicio temporal
temporary earth tierra temporal
temporary error error temporal
temporary file archivo temporal
temporary forced outage interrupción forzada temporal
temporary ground tierra temporal
temporary grounding puesta a tierra temporal
temporary installation instalación temporal
temporary interruption interrupción temporal
temporary magnet imán temporal
temporary magnetism magnetismo temporal
temporary memory memoria temporal
temporary overvoltage sobretensión temporal
temporary register registro temporal
temporary service servicio temporal
temporary storage almacenamiento temporal
temporary store almacenamiento temporal
ten-turn potentiometer potenciómetro de diez vueltas
tenebrescence tenebrescencia
tens complement complemento a diez
tension tensión
tera tera
terabyte terabyte
teracycle teraciclo
teraelectronvolt teraelectronvolt
terahertz terahertz
teraohm teraohm
terawatt terawatt
terbium terbio
terminal terminal, borna, borne
terminal address card tarjeta de dirección de terminal

terminal amplifier amplificador terminal
terminal area área terminal
terminal block bloque de terminales
terminal board tablero de terminales
terminal box caja de terminales
terminal chamber caja de terminales
terminal component componente terminal
terminal computer computadora terminal
terminal connector conector terminal
terminal control control de terminal
terminal controller controlador de terminal
terminal emulation emulación de terminal
terminal emulator emulador de terminal
terminal equipment equipo terminal
terminal exchange central terminal
terminal filter filtro terminal
terminal impedance impedancia terminal
terminal input buffer almacenador intermedio de entrada de terminal
terminal installation instalación terminal
terminal insulator aislador terminal
terminal interface interfaz de terminal
terminal network red terminal
terminal pair par de terminales
terminal plate placa de terminales
terminal point punto terminal
terminal pole poste terminal
terminal port puerto terminal
terminal repeater repetidor terminal
terminal room sala terminal
terminal screw tornillo de sujeción
terminal station estación terminal
terminal strip regleta de terminales
terminal unit unidad terminal
terminal voltage tensión terminal
terminate terminar
terminated terminado
terminated circuit circuito terminado
terminated line línea terminada
terminating terminación
terminating set conjunto de terminación
termination terminación
termination circuit circuito de terminación
termination plug clavija de terminación
termination unit unidad de terminación
terminator terminador
ternary ternario
ternary code código ternario
ternary logic lógica ternaria
ternary notation notación ternaria
ternary number system sistema numérico ternario
ternary relation relación ternaria
terrain echoes ecos de terreno
terrain error error por terreno
terrestrial terrestre
terrestrial current corriente terrestre
terrestrial magnetic field campo magnético terrestre
terrestrial magnetic poles polos magnéticos terrestres
terrestrial magnetism magnetismo terrestre
terrestrial microwave link enlace de microondas terrestre
terrestrial noise ruido terrestre
terrestrial radio noise ruido de radio terrestre
territorial broadcasting radiodifusión territorial
territorial transmission transmisión territorial
tertiary color color terciario
tertiary path trayectoria terciaria
tertiary winding arrollamiento terciario
tesla tesla
Tesla coil bobina de Tesla
Tesla current corriente de Tesla

Tesla transformer transformador de Tesla
test prueba
test access point punto de acceso de pruebas
test analysis análisis de pruebas
test bars barras de ajuste
test battery batería de pruebas
test bench banco de pruebas
test board tablero de pruebas
test box caja de pruebas
test brush escobilla de prueba
test bulb bombilla de prueba
test-busy signal señal de prueba en proceso
test cabinet gabinete de pruebas
test call llamada de prueba
test card tarjeta de prueba
test case caso de prueba
test-case generator generador de casos de prueba
test chamber cámara de pruebas
test chart patrón de pruebas
test circuit circuito de prueba
test clip pinza de pruebas
test condition condición de prueba
test connection conexión de prueba
test connector conector de prueba
test cord cordón de prueba
test coverage cobertura de prueba
test current corriente de prueba
test data datos de prueba
test-data generator generador de datos de prueba
test design diseño de prueba
test desk pupitre de pruebas
test duration duración de prueba
test equipment equipo de pruebas
test facility instalación de pruebas
test file archivo de pruebas
test frequency frecuencia de prueba
test handset microteléfono de prueba
test indicator indicador de prueba
test instruction instrucción de prueba
test instrument instrumento de prueba
test interval intervalo de prueba
test item artículo de prueba
test jack jack de prueba
test lamp lámpara de prueba
test lead conductor de prueba
test level nivel de prueba
test light luz de prueba
test line línea de pruebas
test load carga de prueba
test loop bucle de prueba
test message mensaje de prueba
test mode modo de prueba
test model modelo de prueba
test modulation modulación de prueba
test mount montura de prueba
test-oriented language lenguaje orientado a pruebas
test oscillator oscilador de pruebas
test pack paquete de prueba
test panel panel de pruebas
test pattern patrón de pruebas, imagen patrón
test phase fase de pruebas
test plug clavija de prueba
test point punto de prueba
test pole terminal de pruebas
test position posición de prueba
test probe sonda de prueba
test procedure procedimiento de prueba
test prod punta de prueba
test program programa de prueba
test rack bastidor de pruebas
test record disco de prueba
test relay relé de prueba
test requirement requisito de prueba

test routine rutina de pruebas
test run pasada de prueba
test section sección de prueba
test selector selector de prueba
test sequence secuencia de pruebas
test set equipo de pruebas
test signal señal de prueba
test-signal generator generador de señales de prueba
test software programas de pruebas
test specification especificación de prueba
test station estación de pruebas
test switch conmutador de prueba
test tape cinta de prueba
test terminal terminal de pruebas
test time tiempo de pruebas
test tone tono de prueba
test traffic tráfico de prueba
test transmission transmisión de prueba
test unit unidad de prueba
test validity validez de prueba
test voltage tensión de prueba
tester probador
testing area área de pruebas
testing cycle ciclo de pruebas
testing program programa de pruebas
testing routine rutina de pruebas
tetravalent tetravalente
tetrode tetrodo
tetrode field-effect-transistor transistor de efecto de campo tetrodo
tetrode oscillator oscilador tetrodo
tetrode transistor transistor tetrodo
text buffer almacenamiento intermedio de texto
text compression compresión de texto
text control control de texto
text mode modo de texto
text processing procesamiento de texto
thallium talio
thallium-activated activado con talio
thallium oxysulfide cell célula de oxisulfuro de talio
thalofide cell célula de oxisulfuro de talio
theoretical cutoff frequency frecuencia de corte teorética
theoretical electrical travel carrera eléctrica teorética
theory teoría
therm termia
thermal térmico, termal
thermal activation activación térmica
thermal aging envejecimiento térmico
thermal agitation agitación térmica
thermal-agitation noise ruido por agitación térmica
thermal alarm alarma térmica
thermal ammeter amperímetro térmico
thermal anemometer anemómetro térmico
thermal battery batería térmica
thermal capability capacidad térmica
thermal cell pila térmica
thermal circuit-breaker cortacircuito térmico
thermal coefficient coeficiente térmico
thermal compensation compensación térmica
thermal conduction conducción térmica
thermal conductivity conductividad térmica
thermal conductor conductor térmico
thermal convection convección térmica
thermal converter convertidor térmico
thermal cutout cortacircuito térmico
thermal delay relay relé de retardo térmico
thermal detector detector térmico
thermal diffusion difusión térmica
thermal diffusivity difusividad térmica
thermal drift deslizamiento térmico
thermal effect efecto térmico

thermal efficiency eficiencia térmica
thermal electromotive force fuerza electromotriz térmica
thermal emittance emitancia térmica
thermal endurance resistencia térmica
thermal energy energía térmica
thermal environment ambiente térmico
thermal equilibrium equilibrio térmico
thermal fluctuation fluctuación térmica
thermal flux flujo térmico
thermal generation generación térmica
thermal gradient gradiente térmico
thermal impedance impedancia térmica
thermal influence influencia térmica
thermal instability inestabilidad térmica
thermal instrument instrumento térmico
thermal insulation aislamiento térmico
thermal ionization ionización térmica
thermal junction unión térmica
thermal load carga térmica
thermal loss pérdida térmica
thermal meter medidor térmico
thermal microphone micrófono térmico
thermal neutron neutrón térmico
thermal noise ruido térmico
thermal-noise generator generador de ruido térmico
thermal overload sobrecarga térmica
thermal-overload capacity capacidad de sobrecarga térmica
thermal-overload protection protección contra sobrecarga térmica
thermal power potencia térmica
thermal printer impresora térmica
thermal probe sonda térmica
thermal protection protección térmica
thermal protector protector térmico
thermal radar radar térmico
thermal radiation radiación térmica
thermal radiator radiador térmico
thermal receiver receptor térmico
thermal recorder registrador térmico
thermal reflectance reflectancia térmica
thermal relay relé térmico
thermal release liberación térmica
thermal resistance resistencia térmica
thermal resistivity resistividad térmica
thermal resistor resistor térmico
thermal response time tiempo de respuesta térmica
thermal runaway avalancha térmica
thermal shock choque térmico
thermal stability estabilidad térmica
thermal switch conmutador térmico
thermal telephone receiver receptor telefónico térmico
thermal time constant constante de tiempo térmica
thermal time-delay relay relé de retardo de tiempo térmico
thermal transfer transferencia térmica
thermal tuner sintonizador térmico
thermal tuning sintonización térmica
thermal tuning sensitivity sensibilidad de sintonización térmica
thermal tuning system sistema de sintonización térmica
thermal tuning time tiempo de sintonización térmica
thermal unit unidad térmica
thermally térmicamente, termalmente
thermic térmico, termal
thermically térmicamente, termalmente
thermion termión
thermionic termoiónico
thermionic amplifier amplificador termoiónico
thermionic apparatus aparato termoiónico

thermionic cathode cátodo termoiónico
thermionic conversion conversión termoiónica
thermionic converter convertidor termoiónico
thermionic current corriente termoiónica
thermionic detector detector termoiónico
thermionic diode diodo termoiónico
thermionic effect efecto termoiónico
thermionic emission emisión termoiónica
thermionic-emission microscopy microscopía por emisión termoiónica
thermionic generator generador termoiónico
thermionic grid emission emisión de rejilla termoiónica
thermionic hollow cathode cátodo hueco termoiónico
thermionic instrument instrumento termoiónico
thermionic oscillator oscilador termoiónico
thermionic rectifier rectificador termoiónico
thermionic relay relé termoiónico
thermionic tube tubo termoiónico
thermionic valve válvula termoiónica
thermionic voltmeter vóltmetro termoiónico
thermionic work function función de trabajo termoiónico
thermionics termoiónica
thermistor termistor
thermistor bolometer bolómetro de termistor
thermistor bridge puente de termistores
thermistor mount montura de termistor
thermistor probe sonda de termistor
thermistor thermometer termómetro de termistor
thermistor thermostat termostato de termistor
thermoammeter termoamperímetro
thermocard termotarjeta
thermocatalytic termocatalítico
thermochemical termoquímico
thermocompensator termocompensador
thermoconducting termoconductor
thermoconducting plastic plástico termoconductor
thermoconductor termoconductor
thermocontact termocontacto
thermoconvection termoconvección
thermocouple termopar
thermocouple ammeter amperímetro de termopar
thermocouple bridge puente de termopar
thermocouple contact contacto de termopar
thermocouple converter convertidor de termopar
thermocouple galvanometer galvanómetro de termopar
thermocouple instrument instrumento de termopar
thermocouple meter medidor de termopar
thermocouple potentiometer potenciómetro de termopar
thermocouple pyrometer pirómetro de termopar
thermocouple thermometer termómetro de termopar
thermocutout termocortacircuito
thermodetector termodetector
thermodielectric termodieléctrico
thermodiffusion termodifusión
thermodilution termodilución
thermodynamic termodinámico
thermodynamic stability estabilidad termodinámica
thermodynamically termodinámicamente
thermodynamics termodinámica
thermoelastic termoelástico
thermoelectric termoeléctrico
thermoelectric arm brazo termoeléctrico
thermoelectric conversion conversión termoeléctrica
thermoelectric-conversion tube tubo de conversión termoeléctrica
thermoelectric cooler enfriador termoeléctrico
thermoelectric cooling enfriamiento termoeléctrico
thermoelectric couple par termoeléctrico

thermoelectric detector detector termoeléctrico
thermoelectric device dispositivo termoeléctrico
thermoelectric effect efecto termoeléctrico
thermoelectric element elemento termoeléctrico
thermoelectric generator generador termoeléctrico
thermoelectric heater calentador termoeléctrico
thermoelectric inversion inversión termoeléctrica
thermoelectric junction unión termoeléctrica
thermoelectric laws leyes termoeléctricas
thermoelectric material material termoeléctrico
thermoelectric plant central termoeléctrica
thermoelectric power potencia termoeléctrica
thermoelectric pyrometer pirómetro termoeléctrico
thermoelectric refrigeration refrigeración
  termoeléctrica
thermoelectric series serie termoeléctrica
thermoelectric solar cell pila solar termoeléctrica
thermoelectric thermometer termómetro
  termoeléctrico
thermoelectric voltage tensión termoeléctrica
thermoelectricity termoelectricidad
thermoelectromotive force fuerza
  termoelectromotriz
thermoelectron termoelectrón
thermoelectronic termoelectrónico
thermoelectronic emission emisión termoelectrónica
thermoelectronic rectifier rectificador
  termoelectrónico
thermoelement termoelemento
thermogalvanic termogalvánico
thermogalvanic cell célula termogalvánica
thermogalvanic corrosion corrosión termogalvánica
thermogalvanism termogalvanismo
thermogalvanometer termogalvanómetro
thermogenic termogénico
thermography termografía
thermoinsulation termoaislamiento
thermojunction termounión
thermoluminescence termoluminiscencia
thermoluminescent termoluminiscente
thermomagnetic termomagnético
thermomagnetic cooling enfriamiento
  termomagnético
thermomagnetic effect efecto termomagnético
thermometer termómetro
thermometer bridge puente termométrico
thermometer resistor resistor termométrico
thermometer scale escala termométrica
thermometric termométrico
thermomolecular termomolecular
thermonuclear reaction reacción termonuclear
thermopair termopar
thermophone termófono
thermophotochemistry termofotoquímica
thermophotovoltaic termofotovoltaico
thermopile termopila
thermoplastic termoplástico
thermoplastic material material termoplástico
thermoplastic recording registro termoplástico
thermoregulator termorregulador
thermorelay termorrelé
thermoresistance termorresistencia
thermosensitive termosensible
thermosensitive paper papel termosensible
thermosensitivity termosensibilidad
thermosetting termoendurecimiento
thermosetting material material termoendurecible
thermosetting plastic plástico termoendurecible
thermosetting resin resina termoendurecible
thermostability termoestabilidad
thermostable termoestable
thermostat termostato
thermostatic termostático

thermostatic alarm alarma termostática
thermostatic cutout cortacircuito termostático
thermostatic delay relay relé de retardo termostático
thermostatic relay relé termostático
thermostatic switch conmutador termostático
thermostatic valve válvula termostática
thermostatically termostáticamente
thermoswitch termoconmutador
thermotelluric current corriente termotelúrica
thermovariable resistor resistor termovariable
theta theta
theta polarization polarización theta
Thevenin's electrical theorem teorema eléctrico de
  Thevenin
Thevenin's theorem teorema de Thevenin
thick dipole dipolo grueso
thick film película gruesa
thick-film circuit circuito de película gruesa
thick-film component componente de película gruesa
thick-film hybrid circuit circuito híbrido de película
  gruesa
thick-film integrated circuit circuito integrado de
  película gruesa
thick-film passive circuit circuito pasivo de película
  gruesa
thick-film resistor resistor de película gruesa
thick-film technique técnica de películas gruesas
thick magnetic film película magnética gruesa
thin antenna antena delgada
thin film película delgada
thin-film capacitor capacitor de película delgada
thin-film circuit circuito de película delgada
thin-film component componente de película delgada
thin-film cryotron criotrón de película delgada
thin-film integrated circuit circuito integrado de
  película delgada
thin-film magnetoresistor magnetorresistor de
  película delgada
thin-film memory memoria de película delgada
thin-film microcircuit microcircuito de película
  delgada
thin-film microelectronic circuit circuito
  microelectrónico de película delgada
thin-film microelectronics microelectrónica de
  película delgada
thin-film processing procesamiento de película
  delgada
thin-film resistor resistor de película delgada
thin-film semiconductor semiconductor de película
  delgada
thin-film technique técnica de películas delgadas
thin-film transistor transistor de película delgada
thin-film waveguide guíaondas de película delgada
thin foil hoja delgada
thin layer capa delgada
thin linear antenna antena lineal delgada
thin magnetic film película magnética delgada
thin picture tube tubo de imagen delgado
thin plate placa delgada
thin-walled cable cable de pared delgada
thin-walled tube tubo de pared delgada
thin wire hilo delgado
third brush tercera escobilla
third channel tercer canal
third-generation de tercera generación
third-generation computer computadora de tercera
  generación
third-generation language lenguaje de tercera
  generación
third-generation tape cinta de tercera generación
third harmonic tercera armónica
third-harmonic crystal cristal de tercera armónica
third-harmonic distortion distorsión de tercera

armónica
third law of thermodynamics tercera ley de termodinámica
third-level address dirección de tercer nivel
third-level storage almacenamiento de tercer nivel
third rail tercer riel
thirty-two bit chip chip de treinta y dos bits
thirty-two bit computer computadora de treinta y dos bits
thirty-two bit system sistema de treinta y dos bits
Thomas resistor resistor de Thomas
Thomson bridge puente de Thomson
Thomson coefficient coeficiente de Thomson
Thomson cross-section sección transversal de Thomson
Thomson effect efecto de Thomson
Thomson electromotive force fuerza electromotriz de Thomson
Thomson heat calor de Thomson
Thomson meter medidor de Thomson
Thomson scattering dispersión de Thomson
Thomson voltage tensión de Thomson
thoriated toriado
thoriated emitter emisor toriado
thoriated filament filamento toriado
thoriated-tungsten cathode cátodo de tungsteno toriado
thoriated-tungsten filament filamento de tungsteno toriado
thorium torio
thread hilo
three-address de tres direcciones
three-address code código de tres direcciones
three-address computer computadora de tres direcciones
three-address instruction instrucción de tres direcciones
three-address system sistema de tres direcciones
three-bit byte byte de tres bits
three-cavity klystron klistrón de tres cavidades
three-channel loudspeaker system sistema de altavoces de tres canales
three-channel multiplex system sistema múltiplex de tres canales
three-channel stereo estéreo de tres canales
three-channel stereo system sistema estereofónico de tres canales
three-channel stereophonic system sistema estereofónico de tres canales
three-color beam haz de tres colores
three-color tube tubo de tres colores
three-conductor jack jack de tres conductores
three-conductor plug clavija de tres conductores
three-core cable cable trifilar
three-dimensional tridimensional
three-dimensional graphics gráficos tridimensionales
three-dimensional memory memoria tridimensional
three-dimensional picture imagen tridimensional
three-dimensional radar radar tridimensional
three-dimensional scanning exploración tridimensional
three-dimensional television televisión tridimensional
three-electrode tube tubo de tres electrodos
three-element antenna antena de tres elementos
three-element tube tubo de tres elementos
three-grid tube tubo de tres rejillas
three-grid valve válvula de tres rejillas
three-gun picture tube tubo de imagen de tres cañones
three-gun tube tubo de tres cañones
three-input adder sumador de tres entradas

three-input subtracter restador de tres entradas
three-junction transistor transistor de tres uniones
three-layer diode diodo de tres capas
three-level de tres niveles
three-level address dirección de tres niveles
three-level addressing direccionamiento de tres niveles
three-level laser láser de tres niveles
three-level maser máser de tres niveles
three-level subroutine subrutina de tres niveles
three-phase trifásico
three-phase alternator alternador trifásico
three-phase bridge rectifier rectificador en puente trifásico
three-phase circuit circuito trifásico
three-phase current corriente trifásica
three-phase generator generador trifásico
three-phase machine máquina trifásica
three-phase motor motor trifásico
three-phase power potencia trifásica
three-phase rectifier rectificador trifásico
three-phase system sistema trifásico
three-phase transformer transformador trifásico
three-phase voltage tensión trifásica
three-pin plug clavija de tres patillas
three-plus-one address dirección de tres más uno
three-pole tripolar
three-pole socket zócalo tripolar
three-pole switch conmutador tripolar
three-position de tres posiciones
three-position relay relé de tres posiciones
three-position switch conmutador de tres posiciones
three-quarter bridge rectificador en puente de tres cuartos
three-stage de tres etapas
three-stage amplifier amplificador de tres etapas
three-state logic lógica de tres estados
three-terminal de tres terminales
three-terminal capacitor capacitor de tres terminales
three-way loudspeaker system sistema de altavoces de tres canales
three-way switch conmutador de tres vías
three-way system sistema de tres canales
three-wire system sistema de tres hilos
threshold umbral
threshold amplifier amplificador de umbral
threshold circuit circuito de umbral
threshold component componente umbral
threshold control control de umbral
threshold current corriente umbral
threshold detector detector de umbral
threshold effect efecto de umbral
threshold energy energía umbral
threshold field campo umbral
threshold frequency frecuencia umbral
threshold gate puerta de umbral
threshold level nivel umbral
threshold linear amplifier amplificador lineal de umbral
threshold logic lógica de umbral
threshold of detection umbral de detección
threshold of luminescence umbral de luminiscencia
threshold of response umbral de respuesta
threshold of sensitivity umbral de sensibilidad
threshold signal señal umbral
threshold switch conmutador de umbral
threshold switching conmutación de umbral
threshold value valor umbral
threshold voltage tensión umbral
threshold wavelength longitud de onda umbral
throat microphone micrófono de garganta
throttle regulador
through channel canal de tránsito

through circuit circuito de tránsito
through joint empalmador
through level nivel de prueba no adaptado
through position posición de tránsito
through repeater repetidor directo
through trunk troncal de tránsito
throughput rendimiento total
throw carrera, tiro, vía
thulium tulio
thumbscrew tornillo de orejetas
thump golpeteo
Thury system sistema de Thury
Thury transmission system sistema de transmisión
   de Thury
thyratron tiratrón
thyratron amplifier amplificador tiratrónico
thyratron gate puerta tiratrónica
thyratron generator generador tiratrónico
thyratron inverter inversor tiratrónico
thyratron oscillator oscilador tiratrónico
thyratron timer temporizador tiratrónico
thyristor tiristor
thyrite tirita
thyrite resistor resistor de tirita
thyrite varistor varistor de tirita
tickler coil bobina de regeneración
tie amarre, ligadura
tie cable cable de enlace
tie feeder alimentador de enlace
tie point punto de conexión
tight coupling acoplamiento fuerte
tilt inclinación, distorsión de inclinación
tilt angle ángulo de inclinación
tilt control control de inclinación
tilt correction corrección de inclinación
tilt corrector corrector de inclinación
tilt stabilization estabilización de inclinación
tilt switch conmutador de inclinación
tilting inclinación
timbre timbre
time tiempo
time accuracy exactitud de tiempo
time alarm alarma de tiempo
time analyzer analizador de tiempo
time assignment asignación de tiempo
time axis eje de tiempo
time base base de tiempo
time-base circuit circuito de base de tiempo
time-base frequency frecuencia de base de tiempo
time-base oscillator oscilador de base de tiempo
time-base scanning exploración de base de tiempo
time-base signal señal de base de tiempo
time-base voltage tensión de base de tiempo
time calibration calibración de tiempo
time check comprobación de tiempo
time code código de tiempo
time comparator comparador de tiempo
time comparison comparación de tiempo
time constant constante de tiempo
time control control de tiempo
time-control pulse impulso de control de tiempo
time cycle ciclo de tiempo
time delay retardo de tiempo
time-delay circuit circuito de retardo de tiempo
time-delay relay relé de retardo de tiempo
time-delay switch conmutador de retardo de tiempo
time-dependent dependiente del tiempo
time-dependent process proceso dependiente del
   tiempo
time difference diferencia de tiempo
time discriminator discriminador de tiempo
time distortion distorsión de tiempo
time distribution distribución de tiempo

time-distribution analyzer analizador de distribución
   de tiempo
time-division multiplex múltiplex por división de
   tiempo
time-division multiplexing multiplexión por división
   de tiempo
time-division multiplier multiplicador por división
   de tiempo
time-division switching conmutación por división de
   tiempo
time-domain reflectometry reflectometría de
   dominio de tiempo
time duration duración de tiempo
time error error de tiempo
time factor factor de tiempo
time fuse fusible de tiempo
time gate puerta de tiempo
time interval intervalo de tiempo
time-interval counter contador de intervalos de
   tiempo
time-interval measurement medida de intervalos de
   tiempo
time-interval meter medidor de intervalos de tiempo
time-interval mode modo de intervalos de tiempo
time-interval selector selector de intervalos de
   tiempo
time jitter fluctuación de tiempo
time lag retardo de tiempo
time-lag fuse fusible de retardo de tiempo
time-lag relay relé de retardo de tiempo
time limit límite de tiempo
time mark marca de tiempo
time-mark generator generador de marcas de tiempo
time measurement medición de tiempo
time measurer medidor de tiempo
time meter medidor de tiempo, contador horario
time modulation modulación de tiempo
time multiplex múltiplex en tiempo
time multiplexing multiplexión en tiempo
time of operation tiempo de operación
time of response tiempo de respuesta
time of rise tiempo de subida
time of use tiempo de uso
time pattern modelo de tiempo
time rate régimen de descarga
time recorder registrador de tiempo
time recording registro de tiempo
time regulator regulador de tiempo
time relay relé de retardo de tiempo
time requirement requisito de tiempo
time response respuesta en función del tiempo
time scale escala de tiempo
time sharing compartimiento de tiempo
time-sharing console consola de tiempo compartido
time-sharing option opción de tiempo compartido
time-sharing priority prioridad de tiempo
   compartido
time-sharing system sistema de tiempo compartido
time-sharing terminal terminal de tiempo
   compartido
time-signal control control de señales de tiempo
time signals señales de tiempo
time slot intervalo de tiempo
time switch conmutador de tiempo
timed pulse impulso temporizado
timer temporizador, cronómetro
timing temporización, sincronización,
   cronometración
timing contact contacto de sincronización
timing control temporización, sincronización
timing device dispositivo de temporización,
   dispositivo de sincronización
timing element elemento de temporización

timing error error de temporización
timing marker marcador de sincronización
timing mechanism mecanismo de temporización
timing pulse impulso de temporización, impulso de sincronización
timing relay relé de temporización
timing resistor resistor de temporización
timing signal señal de sincronización
timing unit unidad de temporización
tin estaño
tin oxide óxido de estaño
tin-oxide resistor resistor de óxido de estaño
tin-plate estañar
tin-plated estañado
tinned wire hilo estañado
tinning estañado
tinsel oropel
tinsel conductor conductor de oropel
tinsel cord cordón de oropel
tint control control de tinte
tip cable cable de punta
tip jack jack de punta
tip wire hilo de punta
titanium titanio
titanium dioxide dióxido de titanio
toggle palanca acodada, palanca
toggle switch conmutador de palanca acodada
tolerable tolerable
tolerance tolerancia
tolerance field campo de tolerancia
toll cable cable interurbano
toll call llamada interurbana
toll center centro interurbano
toll circuit circuito interurbano
toll exchange central interurbana
toll line línea interurbana
toll network red interurbana
toll service servicio interurbano
toll switchboard cuadro conmutador interurbano
toll traffic tráfico interurbano
tomography tomografía
tonal tonal
tonal balance equilibrio tonal
tonal response respuesta tonal
tone tono
tone arm brazo de fonocaptor
tone burst ráfaga de tono
tone-burst generator generador de ráfagas de tono
tone channel canal de tono
tone control control de tono
tone-control circuit circuito de control de tono
tone converter convertidor de tono
tone correction corrección de tono
tone duration duración de tono
tone filter filtro de tono
tone frequency frecuencia de tono
tone generator generador de tono
tone keyer manipulador de tono
tone keying manipulación de tono
tone level nivel de tono
tone localizer localizador de tono
tone-modulated modulado por tono
tone-modulated wave onda modulada por tono
tone modulation modulación de tono
tone multiplex múltiplex de tonos
tone-operated operado por tono
tone oscillator oscilador de tono
tone range gama de tonos
tone receiver receptor de tono
tone switch conmutador de tono
tone telegraphy telegrafía por tonos
tone-wheel rueda fónica
tonewheel rueda fónica

top cap capacete
top channel canal superior
top-loaded antenna antena de capacidad terminal
topology topología
torn-tape relay retransmisión por cinta cortada
tornadotron tornadotrón
toroid toroide, bobina toroidal
toroidal coil bobina toroidal
toroidal core núcleo toroidal
toroidal element elemento toroidal
toroidal klystron klistrón toroidal
toroidal potentiometer potenciómetro toroidal
toroidal transformer transformador toroidal
toroidal winding arrollamiento toroidal
torque par, par motor, par de torsión
torque amplifier amplificador de par
torque control control de par
torque converter convertidor de par
torque measurer medidor de par
torque motor motor de par
torque synchro síncrono de par
torque-to-inertia ratio razón de par a inercia
torquemeter torsiómetro
torr torr
torsiometer torsiómetro
torsion torsión
torsion electrometer electrómetro de torsión
torsion galvanometer galvanómetro de torsión
torsion meter medidor de torsión
torsional torsional
torsionally torsionalmente
total amplitude amplitud total
total amplitude oscillation oscilación de amplitud total
total anode power input entrada de potencia anódica total
total attenuation coefficient coeficiente de atenuación total
total capacitance capacitancia total
total deviation desviación total
total distortion distorsión total
total earth conexión total a tierra
total efficiency eficiencia total
total electrode capacitance capacitancia de electrodo total
total emission emisión total
total emissivity emisividad total
total energy energía total
total field campo total
total filter filtro total
total functional resistance resistencia funcional total
total harmonic distortion distorsión armónica total
total heat calor total
total internal reflection reflexión interna total
total ionization ionización total
total losses pérdidas totales
total luminous flux flujo luminoso total
total multiplex signal señal múltiplex total
total noise level nivel de ruido total
total operation time tiempo de operación total
total polarization error error de polarización total
total pressure presión total
total radiation temperature temperatura de radiación total
total range intervalo total
total reflectance reflectancia total
total reflection reflexión total
total regulation regulación total
total switching time tiempo de conmutación total
total system downtime tiempo muerto de sistema total
total telegraph distortion distorsión telegráfica total
total time constant constante de tiempo total

total transition time tiempo de transición total
total transmittance transmitancia total
totally disconnected totalmente desconectado
totally enclosed totalmente encerrado
totally-enclosed apparatus aparato totalmente
    encerrado
totally-enclosed machine máquina totalmente
    encerrada
totally-enclosed motor motor totalmente encerrado
touch panel panel sensible al tacto
touch screen pantalla sensible al tacto
touch-sensitive sensible al tacto
touch-sensitive screen pantalla sensible al tacto
touch-tone calling llamada por tonos
touch-tone dialing marcación por tonos
touch-tone system sistema de llamada por tonos
touch-tone telephone teléfono de teclas de tonos
tourmaline turmalina
Touschek effect efecto de Touschek
tower torre
tower light luz de torre
tower lighting iluminación de torre
Townsend avalanche avalancha de Townsend
Townsend characteristic característica de Townsend
Townsend coefficient coeficiente de Townsend
Townsend criterion criterio de Townsend
Townsend discharge descarga de Townsend
Townsend ionization ionización de Townsend
Townsend ionization coefficient coeficiente de
    ionización de Townsend
trace traza, trazo
trace expansion expansión de trazo
trace interval intervalo de trazo
trace rotation rotación de trazo
trace time tiempo de trazo
trace width anchura de trazo
tracer trazador, rastreador
tracing trazado, rastreo
tracing equipment equipo de rastreo
tracing program programa de rastreo
tracing routine rutina de rastreo
track pista, surco
track circuit circuito de pista
track label etiqueta de pista
track number número de pista
track pitch paso entre pistas
track relay relé de pista
track selector selector de pista
track to track pista a pista
tracking seguimiento, rastreo, arrastre
tracking antenna antena de seguimiento
tracking beam haz de seguimiento
tracking characteristic característica de seguimiento
tracking circuit circuito de seguimiento
tracking error error de seguimiento
tracking filter filtro de seguimiento
tracking force fuerza vertical de aguja
tracking jitter fluctuación de seguimiento
tracking mechanism mecanismo de seguimiento
tracking receiver receptor de seguimiento
tracking signal señal de seguimiento
tracking station estación de seguimiento
tracking system sistema de seguimiento
tractor feed alimentación por tracción
tractor feeder alimentador de tracción
traffic tráfico
traffic analysis análisis de tráfico
traffic capacity capacidad de tráfico
traffic channel canal de tráfico
traffic circuit circuito de tráfico
traffic computer computadora de tráfico
traffic control control de tráfico
traffic diagram diagrama de tráfico

traffic distributor distribuidor de tráfico
traffic flow flujo de tráfico
traffic fluctuation fluctuación de tráfico
traffic frequency frecuencia de tráfico
traffic intensity intensidad de tráfico
traffic meter medidor de tráfico
traffic metering medición de tráfico
traffic pattern patrón de tráfico
traffic peak pico de tráfico
traffic point punto de tráfico
traffic record registro de tráfico
traffic recorder registrador de tráfico
traffic unit unidad de tráfico
traffic volume volumen de tráfico
traffic wave onda de tráfico
trailer record registro de arrastre
trailing blacks borde negro tras imagen blanca
trailing edge borde de salida
trailing whites borde blanco tras imagen negra
trailing zero cero a la derecha
train of waves tren de ondas
tramway tranvía
transaction transacción
transaction analysis análisis de transacción
transaction card tarjeta de transacción
transaction code código de transacción
transaction command mando de transacción
transaction data datos de transacción
transaction error error de transacción
transaction file archivo de transacción
transaction processing procesamiento de
    transacciones
transaction-processing system sistema de
    procesamiento de transacciones
transaction program programa de transacciones
transaction record registro de transacción
transaction recorder registrador de transacciones
transaction register registro de transacciones
transaction tape cinta de transacciones
transadmittance transadmitancia
transadmittance compression ratio razón de
    compresión de transadmitancia
transceiver transceptor
transcendental functions funciones transcendentales
transconductance transconductancia
transconductance amplifier amplificador de
    transconductancia
transconductance meter medidor de
    transconductancia
transconductor transconductor
transcribe transcribir
transcriber transcriptor
transcription transcripción
transdiode transdiodo
transduce transducir
transducer transductor
transducer amplifier amplificador de transductor
transducer-controlled controlado por transductor
transducer efficiency eficiencia de transductor
transducer gain ganancia de transductor
transducer head cabeza de transductor
transducer loss pérdida de transductor
transducer power gain ganancia de potencia de
    transductor
transducer power loss pérdida de potencia de
    transductor
transducer pulse delay retardo de impulso de
    transductor
transducing transducción
transduction transducción
transductor transductor
transductor amplifier amplificador de transductor
transductor-controlled controlado por transductor

transductor element　elemento de transductor
transductor-operated　operado por transductor
transfer　transferencia
transfer address　dirección de transferencia
transfer admittance　admitancia de transferencia
transfer bars　barras de transferencia
transfer card　tarjeta de transferencia
transfer characteristic　característica de transferencia
transfer check　comprobación de transferencia
transfer circuit　circuito de transferencia
transfer coefficient　coeficiente de transferencia
transfer command　mando de transferencia
transfer constant　constante de transferencia
transfer contact　contacto de transferencia
transfer control　control de transferencia
transfer current　corriente de transferencia
transfer-current ratio　razón de corrientes de transferencia
transfer effect　efecto de transferencia
transfer efficiency　eficiencia de transferencia
transfer electrode　electrodo de transferencia
transfer factor　factor de transferencia
transfer function　función de transferencia
transfer impedance　impedancia de transferencia
transfer instruction　instrucción de transferencia
transfer jack　jack de transferencia
transfer key　clave de transferencia
transfer loss　pérdida por transferencia
transfer mechanism　mecanismo de transferencia
transfer of control　transferencia de control
transfer-of-control card　tarjeta de transferencia de control
transfer operation　operación de transferencia
transfer option　opción de transferencia
transfer parameter　parámetro de transferencia
transfer position　posición de transferencia
transfer rate　velocidad de transferencia
transfer ratio　razón de transferencia
transfer relay　relé de transferencia
transfer signal　señal de transferencia
transfer standard　patrón de transferencia
transfer switch　conmutador de transferencia
transfer time　tiempo de transferencia
transfer trunk　enlace de transferencia
transfer unit　unidad de transferencia
transference　transferencia
transferred　transferido
transferred charge　carga transferida
transferred electron　electrón transferido
transferred-electron effect　efecto de electrón transferido
transferred-electron oscillator　oscilador de electrón transferido
transferred information　información transferida
transferred jitter　fluctuación transferida
transferred printed circuit　circuito impreso transferido
transferred voltage　tensión transferida
transfluxor　transfluxor
transform　transformar
transformation　transformación
transformation analysis　análisis de transformaciones
transformation ratio　razón de transformación
transformation resistance　resistencia de transformación
transformed　transformado
transformed conductance　conductancia transformada
transformer　transformador
transformer action　acción de transformador
transformer bridge　puente de transformador
transformer case　caja de transformador
transformer coil　bobina de transformador

transformer compound　compuesto para transformadores
transformer core　núcleo de transformador
transformer-correction factor　factor de corrección de transformador
transformer-coupled　acoplado por transformador
transformer-coupled amplifier　amplificador acoplado por transformador
transformer coupling　acoplamiento por transformador
transformer electromotive force　fuerza electromotriz de transformador
transformer feedback　retroalimentación por transformador
transformer input current　corriente de entrada de transformador
transformer input voltage　tensión de entrada de transformador
transformer loss　pérdida de transformador
transformer matching　adaptación por transformador
transformer noise　ruido de transformador
transformer oil　aceite de transformador
transformer output current　corriente de salida de transformador
transformer output voltage　tensión de salida de transformador
transformer utilization factor　factor de utilización de transformador
transformer voltage ratio　razón de tensiones de transformador
transformerless　sin transformador
transformerless amplifier　amplificador sin transformador
transformerless receiver　receptor sin transformador
transforming　transformador
transforming section　sección de transformación
transforming station　estación de transformación
transhorizon　transhorizonte
transhorizon link　enlace transhorizonte
transient　transitorio
transient analyzer　analizador de transitorios
transient area　área transitoria
transient command　mando transitorio
transient component　componente transitorio
transient condition　condición transitoria
transient current　corriente transitoria
transient data　datos transitorios
transient deviation　desviación transitoria
transient discharge　descarga transitoria
transient distortion　distorsión transitoria
transient effect　efecto transitorio
transient electric field　campo eléctrico transitorio
transient electromotive force　fuerza electromotriz transitoria
transient error　error transitorio
transient fault　falla transitoria
transient flux　flujo transitorio
transient forced outage　interrupción forzada transitoria
transient image　imagen transitoria
transient load　carga transitoria
transient magnetic field　campo magnético transitorio
transient motion　moción transitoria
transient oscillation　oscilación transitoria
transient period　periodo transitorio
transient phenomenon　fenómeno transitorio
transient potential　potencial transitorio
transient process　proceso transitorio
transient program　programa transitorio
transient reactance　reactancia transitoria
transient response　respuesta transitoria
transient routine　rutina transitoria
transient stability　estabilidad transitoria

transient state estado transitorio
transient time constant constante de tiempo transitoria
transient torque par transitorio
transient voltage tensión transitoria
transient wave onda transitoria
transimpedance transimpedancia
transistance transistancia
transistor transistor
transistor action acción de transistor
transistor-amplified amplificado por transistor
transistor amplifier amplificador de transistores
transistor analyzer analizador de transistores
transistor base base de transistor
transistor battery batería de transistores
transistor bias polarización de transistor
transistor calculator calculadora de transistores
transistor characteristic característica de transistor
transistor checker probador de transistores
transistor chip chip de transistor
transistor-controlled controlado por transistores
transistor-coupled logic lógica acoplada por transistores
transistor current meter medidor de corriente de transistores
transistor digital computer computadora digital de transistores
transistor-diode logic lógica transistor-diodo
transistor dissipation disipación de transistor
transistor logic circuit circuito lógico de transistor
transistor mount montura de transistor
transistor multivibrator multivibrador de transistores
transistor network red de transistores
transistor noise ruido de transistor
transistor-operated operado por transistores
transistor oscillator oscilador de transistores
transistor parameter parámetro de transistor
transistor pentode transistor pentodo
transistor power supply fuente de alimentación de transistores
transistor radio radio de transistores
transistor-resistor logic lógica transistor-resistor
transistor television set televisor de transistores
transistor tester probador de transistores
transistor-transistor logic lógica transistor-transistor
transistor transit time tiempo de tránsito de transistor
transistor triode transistor triodo
transistor voltmeter vóltmetro de transistores
transistorimeter transistorímetro
transistorization transistorización
transistorize transistorizar
transistorized transistorizado
transistorized amplifier amplificador transistorizado
transistorized cable cable transistorizado
transistorized circuitry circuitería transistorizada
transistorized direct-current motor motor de corriente continua transistorizado
transistorized discriminator discriminador transistorizado
transistorized filter filtro transistorizado
transistorized flip-flop basculador transistorizado
transistorized interphone interfono transistorizado
transistorized microphone micrófono transistorizado
transistorized radio radio transistorizado
transistorized relay relé transistorizado
transistorized television set televisor transistorizado
transistorized transmitter transmisor transistorizado
transistorized voltage regulator regulador de tensión transistorizado
transistorized voltmeter vóltmetro transistorizado
transistorizing transistorización

transit tránsito
transit administration administración de tránsito
transit angle ángulo de tránsito
transit call llamada de tránsito
transit center centro de tránsito
transit circuit circuito de tránsito
transit exchange central de tránsito
transit phase angle ángulo de fase de tránsito
transit telegram telegrama de tránsito
transit time tiempo de tránsito
transit-time mode modo de tiempo de tránsito
transit-time modulation modulación de tiempo de tránsito
transit-time oscillator oscilador de tiempo de tránsito
transit traffic tráfico de tránsito
transition transición
transition angle ángulo de transición
transition anode ánodo de transición
transition card tarjeta de transición
transition coil inductancia de paso
transition duration duración de transición
transition effect efecto de transición
transition element elemento de transición
transition energy energía de transición
transition factor factor de transición
transition frequency frecuencia de transición
transition layer capa de transición
transition load carga de transición
transition loss pérdida por transición
transition period periodo de transición
transition point punto de transición
transition pulse impulso de transición
transition rectifier rectificador de transición
transition region región de transición
transition resistance resistencia de transición
transition section sección de transición
transition temperature temperatura de transición
transition time tiempo de transición
transitional transicional
transitional coupling acoplamiento transicional
transitory transitorio
transitron transitrón
transitron oscillator oscilador transitrón
translating phase fase de traducción
translation traducción, traslación
translation loss pérdida por traducción
translator traductor
translucent translúcido
translucent body cuerpo translúcido
transmissibility transmisibilidad
transmissible transmisible
transmission transmisión, emisión
transmission accuracy exactitud de transmisión
transmission adapter adaptador de transmisión
transmission anomaly anomalía de transmisión
transmission area área de transmisión
transmission band banda de transmisión
transmission-band filter filtro de banda de transmisión
transmission bandwidth ancho de banda de transmisión
transmission block bloque de transmisión
transmission bridge puente de transmisión
transmission by cable transmisión por cable
transmission by line transmisión por línea
transmission by radio transmisión por radio
transmission by wire transmisión por hilo
transmission cable cable de transmisión
transmission center centro de transmisión
transmission chain cadena de transmisión
transmission channel canal de transmisión
transmission characteristics características de

transmisión
transmission code código de transmisión
transmission coefficient coeficiente de transmisión
transmission constant constante de transmisión
transmission control control de transmisión
transmission-control character carácter de control de transmisión
transmission-control unit unidad de control de transmisión
transmission controller controlador de transmisión
transmission curve curva de transmisión
transmission delay retardo de transmisión
transmission detector detector de transmisión
transmission diagram diagrama de transmisión
transmission direction dirección de transmisión
transmission distortion distorsión de transmisión
transmission efficiency eficiencia de transmisión
transmission equivalent equivalente de transmisión
transmission facility instalación de transmisión
transmission factor factor de transmisión
transmission fidelity fidelidad de transmisión
transmission frequency frecuencia de transmisión
transmission frequency meter frecuencímetro de transmisión
transmission gain ganancia de transmisión
transmission gate puerta de transmisión
transmission group grupo de transmisión
transmission identification identificación de transmisión
transmission impairment deterioro de transmisión
transmission interface interfaz de transmisión
transmission interruption interrupción de transmisión
transmission level nivel de transmisión
transmission limit límite de transmisión
transmission line línea de transmisión
transmission-line control control de línea de transmisión
transmission-line coupler acoplador de líneas de transmisión
transmission-line resonator resonador de línea de transmisión
transmission link enlace de transmisión
transmission loss pérdida de transmisión
transmission matrix matriz de transmisión
transmission measurement medida de transmisión
transmission medium medio de transmisión
transmission mode modo de transmisión
transmission modulation modulación de transmisión
transmission monitor monitor de transmisión
transmission network red de transmisión
transmission path trayectoria de transmisión
transmission performance calidad de transmisión
transmission plane plano de transmisión
transmission point punto de transmisión
transmission primaries primarios de transmisión
transmission priority prioridad de transmisión
transmission pulse impulso de transmisión
transmission quality calidad de transmisión
transmission range alcance de transmisión
transmission rate velocidad de transmisión
transmission ratio razón de transmisión
transmission regulator regulador de transmisión
transmission reliability confiabilidad de transmisión
transmission response respuesta de transmisión
transmission route ruta de transmisión
transmission security seguridad de transmisión
transmission sequence secuencia de transmisión
transmission speed velocidad de transmisión
transmission stability estabilidad de transmisión
transmission system sistema de transmisión
transmission test prueba de transmisión
transmission time tiempo de transmisión

transmission tower torre de transmisión
transmission unit unidad de transmisión
transmissive transmisivo
transmissivity transmisividad
transmissometer transmisómetro
transmit transmitir
transmit chain cadena de transmisión
transmit frequency frecuencia de transmisión
transmit gain ganancia de transmisión
transmit indicator indicador de transmisión
transmit-receive antenna antena de transmisión-recepción
transmit-receive relay relé de transmisión-recepción
transmit-receive switch conmutador de transmisión-recepción
transmit-receive tube tubo de transmisión-recepción
transmit signal señal de transmisión
transmittance transmitancia
transmittancy transmitancia
transmitted transmitido
transmitted band banda transmitida
transmitted information información transmitida
transmitted pulse impulso transmitido
transmitted sideband banda lateral transmitida
transmitted wave onda transmitida
transmitter transmisor, micrófono
transmitter distortion distorsión de transmisor
transmitter frequency frecuencia de transmisor
transmitter noise ruido de transmisor
transmitter power potencia de transmisor
transmitter range alcance de transmisor
transmitter-receiver transmisor-receptor
transmitter signal señal de transmisor
transmitting transmisor
transmitting antenna antena transmisora
transmitting apparatus aparato transmisor
transmitting branch rama transmisora
transmitting chain cadena transmisora
transmitting device dispositivo transmisor
transmitting equipment equipo transmisor
transmitting pair par transmisor
transmitting position posición transmisora
transmitting station estación transmisora
transmitting system sistema transmisor
transmitting terminal terminal transmisora
transmittivity transmisividad
transmodulation transmodulación
transoceanic transoceánico
transonic transónico
transparence transparencia
transparency transparencia
transparent transparente
transparent body cuerpo transparente
transparent plasma plasma transparente
transpolarizer transpolarizador
transponder transpondedor, respondedor
transponder beacon radiofaro respondedor
transponder dead time tiempo muerto de respondedor
transponder time delay tiempo de retardo de respondedor
transport transporte, arrastre
transport effects efectos de transporte
transport factor factor de transporte
transport mechanism mecanismo de transporte, mecanismo de arrastre
transport ratio razón de transporte
transportability transportabilidad
transportable transportable
transportable computer computadora transportable
transportable equipment equipo transportable
transportable station estación transportable
transportable substation subestación transportable

transportable transmitter transmisor transportable
transpose transponer
transposed pair par transpuesto
transposing transposición
transposition transposición
transposition interval intervalo de transposición
transposition section sección de transposición
transposition system sistema de transposición
transreceiver transceptor
transrectification transrectificación
transrectification characteristic característica de transrectificación
transrectification factor factor de transrectificación
transrectifier transrectificador
transresistance transresistencia
transresistance amplifier amplificador de transresistencia
transtrictor transtrictor
transuranium transuranio
transversal transversal
transverse transversal
transverse beam haz transversal
transverse check comprobación transversal
transverse current corriente transversal
transverse differential protection protección diferencial transversal
transverse drift deriva transversal
transverse effect efecto transversal
transverse electric and magnetic mode modo eléctrico y magnético transversal
transverse electric mode modo eléctrico transversal
transverse electric wave onda eléctrica transversal
transverse electromagnetic field campo electromagnético transversal
transverse electromagnetic mode modo electromagnético transversal
transverse electromagnetic wave onda electromagnética transversal
transverse field campo transversal
transverse interference interferencia transversal
transverse magnetic mode modo magnético transversal
transverse magnetic wave onda magnética transversal
transverse magnetization magnetización transversal
transverse mode modo transversal
transverse oscillation oscilación transversal
transverse overvoltage sobretensión transversal
transverse plate placa transversal
transverse recording registro transversal
transverse redundancy check comprobación de redundancia transversal
transverse scattering dispersión transversal
transverse septum septo transversal
transverse surge pico transitorio transversal
transverse voltage tensión transversal
transverse wave onda transversal
trap trampa
trap circuit circuito trampa
trapezoid trapecio
trapezoidal trapecial
trapezoidal distortion distorsión trapecial
trapezoidal modulation modulación trapecial
trapezoidal pattern patrón trapecial
trapezoidal pulse impulso trapecial
trapezoidal wave onda trapecial
trapped flux flujo atrapado
trapped radiation radiación atrapada
traveling detector detector móvil
traveling head cabeza móvil
traveling overvoltage sobretensión móvil
traveling probe sonda móvil
traveling wave onda progresiva

traveling-wave acoustic amplifier amplificador acústico de ondas progresivas
traveling-wave amplifier amplificador de ondas progresivas
traveling-wave antenna antena de ondas progresivas
traveling-wave interaction interacción de ondas progresivas
traveling-wave magnetron magnetrón de ondas progresivas
traveling-wave maser máser de ondas progresivas
traveling-wave oscillator oscilador de ondas progresivas
traveling-wave oscilloscope osciloscopio de ondas progresivas
traveling-wave parametric amplifier amplificador paramétrico de ondas progresivas
traveling-wave phototube fototubo de ondas progresivas
traveling-wave tube tubo de ondas progresivas
Travis discriminator discriminador de Travis
treble agudos
treble attenuator atenuador de agudos
treble boost refuerzo de agudos
treble compensation compensación de agudos
treble control control de agudos
treble correction corrección de agudos
treble loudspeaker altavoz de agudos
treble reproducer reproductor de agudos
treble speaker altavoz de agudos
trembler temblador
tri-tet oscillator oscilador tritet
triac triac
triad triada
triangle antenna antena triangular
triangular triangular
triangular code código triangular
triangular pulse impulso triangular
triangular random noise ruido aleatorio triangular
triangular wave onda triangular
triaxial cable cable triaxial
triaxial connector conector triaxial
triaxial loudspeaker altavoz triaxial
triaxial speaker altavoz triaxial
triboelectric triboeléctrico
triboelectric effect efecto triboeléctrico
triboelectric series serie triboeléctrica
triboelectricity triboelectricidad
triboluminescence triboluminiscencia
tributary circuit circuito tributario
tributary station estación tributaria
trichromatic tricromático
trichromatic coefficient coeficiente tricromático
trichromatic system sistema tricromático
trichromatic unit unidad tricromática
trickle charge carga de compensación
trickle charger cargador de compensación
trickle charging carga de compensación
trickle current corriente de compensación
triclinic system sistema triclínico
tricolor tricolor
tricolor beam haz tricolor
tricolor camera cámara tricolor
tricolor cathode-ray tube tubo de rayos catódicos tricolor
tricolor oscillograph oscilógrafo tricolor
tricolor picture tube tubo de imagen tricolor
tricolor tube tubo tricolor
tricon tricón
tridipole tridipolar
tridipole antenna antena tridipolar
tridirectional tridireccional
triductor triductor
trifurcation trifurcación

trigatron trigatrón
trigger disparador, gatillador, gatillo
trigger action acción de disparo
trigger button botón de disparo
trigger circuit circuito de disparo
trigger control control de disparo
trigger diode diodo de disparo
trigger effect efecto de disparo
trigger electrode electrodo de disparo
trigger grid rejilla de disparo
trigger level nivel de disparo
trigger point punto de disparo
trigger pulse impulso de disparo
trigger relay relé de disparo
trigger signal señal de disparo
trigger thyratron tiratrón de disparo
trigger tube tubo de disparo
triggered circuit circuito disparado
triggered sweep barrido disparado
triggering disparo, gatillado
triggering action acción de disparo
triggering circuit circuito de disparo
triggering electrode electrodo de disparo
triggering level nivel de disparo
triggering pulse impulso de disparo
triggering signal señal de disparo
triggering time tiempo de disparo
triggering voltage tensión de disparo
trigistor trigistor
trigonometric functions funciones trigonométricas
trigun tricañón
trigun picture tube tubo de imagen tricañón
trimmer ajuste fino, trímer
trimmer capacitor capacitor de ajuste
trimmer coil bobina de ajuste
trimmer condenser condensador de ajuste
trimmer inductor inductor de ajuste
trimmer potentiometer potenciómetro de ajuste
trimmer resistor resistor de ajuste
trimmer rheostat reóstato de ajuste
trimming ajuste, ajuste fino
trimming capacitor capacitor de ajuste
trimming condenser condensador de ajuste
trimming control control de ajuste
trimming potentiometer potenciómetro de ajuste
trimming rheostat reóstato de ajuste
trimming voltage tensión de ajuste
trinary number system sistema numérico trinario
trinistor trinistor
trinoscope trinoscopio
triode triodo
triode amplifier amplificador de triodos
triode-heptode triodo-heptodo
triode-heptode converter convertidor triodo-heptodo
triode-hexode triodo-hexodo
triode-hexode converter convertidor triodo-hexodo
triode-hexode mixer mezclador triodo-hexodo
triode oscillator oscilador triodo
triode-pentode triodo-pentodo
triode section sección triodo
triode switch conmutador triodo
triode transistor transistor triodo
trip coil bobina de disparo
trip current corriente de disparo
trip lever palanca de disparo
trip mechanism mecanismo de disparo
trip relay relé de disparo
trip value valor de disparo
trip voltage tensión de disparo
triple conversion triple conversión
triple-conversion receiver receptor de triple conversión
triple detection triple detección

triple-detection receiver receptor de triple detección
triple-diffused transistor transistor de triple difusión
triple diode triple diodo
triple-diode triode triodo triple diodo
triple-grid tube tubo de triple rejilla
triple precision triple precisión
triple-stub transformer transformador de triple adaptador
tripler triplicador
triplex triplex
triplex cable cable triplex
triplexer triplexor
triplexing triplexión
tripper disparador, desenganchador
tripping disparo, desenganche, desconexión, abertura
tripping coil bobina de disparo
tripping device dispositivo de disparo
tripping impulse impulso de disparo
tripping lever palanca de disparo
tripping mechanism mecanismo de disparo
tripping pulse impulso de disparo
tripping relay relé de disparo
triprocessor triprocesador
trisistor trisistor
tritium tritio
trochoidal mass analyzer analizador de masas trocoidal
trochotron trocotrón
troland troland
trolley trole
trolley base base de trole
trolley bus trolebús, omnibús de trole
trolley head cabeza de trole
trolley pivot pivote de trole
trolley shoe patín de contacto
trolley wheel ruedecilla de trole
trolley wire alambre de trole
tropical broadcasting radiodifusión tropical
tropical winding arrollamiento tropical
tropicalized tropicalizado
tropopause tropopausa
troposphere troposfera
tropospheric troposférico
tropospheric layer capa troposférica
tropospheric mode modo troposférico
tropospheric propagation propagación troposférica
tropospheric reflection reflexión troposférica
tropospheric refraction refracción troposférica
tropospheric scatter dispersión troposférica
tropospheric-scatter communication comunicación por dispersión troposférica
tropospheric-scatter propagation propagación por dispersión troposférica
tropospheric scattering dispersión troposférica
tropospheric wave onda troposférica
tropotron tropotrón
troubleshoot investigar fallas, depurar
troubleshooting investigación de fallas, depuración
true bearing rumbo real
true complement complemento de base
true ground tierra real
true ohm ohm real
true power potencia real
true resistance resistencia real
true time tiempo real
true watt watt real
truncate truncar
truncated truncado
truncated paraboloid paraboloide truncado
truncated picture imagen truncada
trunk tronco, troncal, enlace
trunk cable cable de enlace

trunk call llamada interurbana
trunk circuit circuito de enlace
trunk connection conexión interurbana
trunk dialing marcación interurbana
trunk exchange central interurbana
trunk feeder alimentador principal
trunk filter filtro de troncal
trunk group grupo de enlace
trunk line línea de enlace
trunk network red interurbana
trunk position posición interurbana
trunk route ruta de enlace
trunk switchboard cuadro conmutador interurbano
trunk zone zona interurbana
trunking enlace, enlazamiento
trunking diagram diagrama de enlaces
truth table tabla de verdad
tube tubo, envolvente
tube aging envejecimiento de tubos
tube amplifier amplificador de tubos
tube bridge puente de tubos
tube capacitances capacitancias de tubo
tube characteristic característica de tubo
tube checker probador de tubos
tube coefficient coeficiente de tubo
tube cooling enfriamiento de tubos
tube counter contador de tubo
tube electrode electrodo de tubo
tube electrometer electrómetro de tubo
tube element elemento de tubo
tube factor factor de tubo
tube failure falla de tubo
tube filament filamento de tubo
tube fuse fusible tubular
tube generator generador de tubo
tube heating time tiempo de calentamiento de tubo
tube interchangeability intercambialidad de tubos
tube kit juego de tubos
tube lifter levantatubos
tube mount montura de tubo
tube noise ruido de tubo
tube of force tubo de fuerza
tube oscillator oscilador de tubo
tube output salida de tubo
tube pin terminal de tubo
tube puller levantatubos, extractor de tubos
tube receiver receptor de tubos
tube relay relé de tubos
tube socket zócalo de tubo
tube tester probador de tubos
tube transmitter transmisor de tubo
tube voltage drop caída de tensión de tubo
tubeless sin tubos
tubeless fuse fusible sin tubo
tuberculation tuberculación
tubular busbar barra colectora tubular
tubular capacitor capacitor tubular
tubular condenser condensador tubular
tubular incandescent lamp lámpara incandescente tubular
tubular insulator aislador tubular
tubular lamp lámpara tubular
tubular probe sonda tubular
tubular waveguide guíaondas tubular
Tudor plate placa de Tudor
tumbler switch conmutador de volquete
tunable sintonizable
tunable band banda sintonizable
tunable cavity cavidad sintonizable
tunable filter filtro sintonizable
tunable magnetron magnetrón sintonizable
tunable maser máser sintonizable
tunable oscillator oscilador sintonizable

tunable probe sonda sintonizable
tunable receiver receptor sintonizable
tune sintonizar
tuned sintonizado
tuned amplifier amplificador sintonizado
tuned-anode circuit circuito de ánodo sintonizado
tuned-anode oscillator oscilador de ánodo sintonizado
tuned antenna antena sintonizada
tuned audio amplifier amplificador de audio sintonizado
tuned-base oscillator oscilador de base sintonizada
tuned cavity cavidad sintonizada
tuned circuit circuito sintonizado
tuned-collector oscillator oscilador de colector sintonizado
tuned coupler acoplador sintonizado
tuned detector detector sintonizado
tuned dipole dipolo sintonizado
tuned feeders alimentadores sintonizados
tuned filter filtro sintonizado
tuned-filter oscillator oscilador de filtro sintonizado
tuned-grid circuit circuito de rejilla sintonizada
tuned-grid impedance impedancia de rejilla sintonizada
tuned-grid oscillator oscilador de rejilla sintonizada
tuned harmonic ringing llamada armónica sintonizada
tuned line línea sintonizada
tuned-line amplifier amplificador de línea sintonizada
tuned load carga sintonizada
tuned-plate circuit circuito de placa sintonizada
tuned-plate oscillator oscilador de placa sintonizada
tuned radio-frequency radiofrecuencia sintonizada
tuned-radio-frequency amplifier amplificador de radiofrecuencia sintonizada
tuned-radio-frequency receiver receptor de radiofrecuencia sintonizada
tuned-radio-frequency reception recepción de radiofrecuencia sintonizada
tuned-radio-frequency transformer transformador de radiofrecuencia sintonizada
tuned reed lengüeta sintonizada
tuned relay relé sintonizado
tuned transformer transformador sintonizado
tuned voltmeter vóltmetro sintonizado
tuner sintonizador
tuner creep corrimiento de sintonizador
tungar bulb tubo tungar
tungar tube tubo tungar
tungsten tungsteno
tungsten arc arco de tungsteno
tungsten-arc lamp lámpara de arco de tungsteno
tungsten contact contacto de tungsteno
tungsten filament filamento de tungsteno
tungsten-filament lamp lámpara de filamento de tungsteno
tungsten lamp lámpara de tungsteno
tuning sintonización
tuning band banda de sintonización
tuning capacitor capacitor de sintonización
tuning cavity cavidad de sintonización
tuning check comprobación de sintonización
tuning circuit circuito de sintonización
tuning coil bobina de sintonización
tuning component componente de sintonización
tuning condenser condensador de sintonización
tuning constant constante de sintonización
tuning control control de sintonización
tuning core núcleo de sintonización
tuning creep corrimiento de sintonización
tuning curve curva de sintonización

tuning error error de sintonización
tuning fork diapasón
tuning-fork interrupter interruptor de diapasón
tuning-fork oscillator oscilador de diapasón
tuning-fork resonator resonador de diapasón
tuning frequency frecuencia de sintonización
tuning indicator indicador de sintonización
tuning-indicator tube tubo indicador de
  sintonización
tuning inductance inductancia de sintonización
tuning inductor inductor de sintonización
tuning meter medidor de sintonización
tuning probe sonda de sintonización
tuning range intervalo de sintonización
tuning ratio razón de sintonización
tuning screw tornillo de sintonización
tuning sensitivity sensibilidad de sintonización
tuning speed velocidad de sintonización
tuning strip regleta de sintonización
tuning stub adaptador de sintonización
tuning susceptance susceptancia de sintonización
tuning unit unidad de sintonización
tuning voltage tensión de sintonización
tuning wand varita de sintonización
tunnel cathode cátodo de efecto túnel
tunnel current corriente de efecto túnel
tunnel diode diodo de efecto túnel
tunnel-diode amplifier amplificador de diodo de
  efecto túnel
tunnel effect efecto túnel
tunnel rectifier rectificador de efecto túnel
tunnel resistor resistor de efecto túnel
tunnel triode triodo de efecto túnel
turboelectric turboeléctrico
turboexciter turboexcitador
turbogenerator turbogenerador
turbulence turbulencia
Turing machine máquina de Turing
turn vuelta
turn off apagar, desconectar
turn-off time tiempo de desconexión
turn on encender, conectar
turn-on time tiempo de conexión
turnaround time tiempo de reparación, tiempo de
  terminación, tiempo de inversión, tiempo de
  respuesta
turnoff time tiempo de desconexión
turnover frequency frecuencia de transición
turns factor factor de vueltas
turns ratio razón de vueltas
turnstile antenna antena cruzada
turntable plato giratorio
turntable rumble ronquido de plato giratorio
turret tuner sintonizador rotativo
tweeter altavoz de agudos
tweeter loudspeaker altavoz de agudos
twelve-channel group grupo de doce canales
twelve-phase rectifier rectificador de doce fases
twilight zone zona crepuscular
twin antennas antenas gemelas
twin cable cable gemelo
twin-cable system sistema de cables gemelos
twin contact doble contacto
twin-contact wire hilo de doble contacto
twin diode doble diodo
twin line línea bifilar plana
twin pentode doble pentodo
twin-T network red en doble T
twin triode doble triodo
twin wire cable bifilar
twinning apareamiento, maclación
twinplex twinplex
twist hélice

twisted joint junta torcida
twisted pair par torcido
twisted waveguide guíaondas torcido
twister cristal con efecto piezoeléctrico por torsión
twistor tuistor
two-address de dos direcciones
two-address code código de dos direcciones
two-address computer computadora de dos
  direcciones
two-address instruction instrucción de dos
  direcciones
two-bit byte byte de dos bits
two-channel amplifier amplificador de dos canales
two-channel loudspeaker system sistema de
  altavoces de dos canales
two-channel receiver receptor de dos canales
two-channel recorder registrador de dos canales
two-coil relay relé de dos arrollamientos
two-color cathode-ray-tube tubo de rayos catódicos
  de dos colores
two-condition modulation modulación bivalente
two-conductor cable cable de dos conductores
two-conductor plug clavija de dos conductores
two-dimensional bidimensional
two-dimensional circuit circuito bidimensional
two-dimensional memory memoria bidimensional
two-dimensional radar radar bidimensional
two-electrode tube tubo de dos electrodos, diodo
two-element antenna antena de dos elementos
two-element tube tubo de dos elementos
two-frequency de dos frecuencias
two-frequency duplex dúplex de dos frecuencias
two-layer de dos capas
two-level de dos niveles
two-level address dirección de dos niveles
two-level addressing direccionamiento de dos niveles
two-level laser láser de dos niveles
two-level maser máser de dos niveles
two-level subroutine subrutina de dos niveles
two-phase bifásico
two-phase alternator alternador bifásico
two-phase clock reloj bifásico
two-phase current corriente bifásica
two-phase modulation modulación bifásica
two-phase motor motor bifásico
two-phase short-circuit cortocircuito bifásico
two-phase system sistema bifásico
two-phase voltage tensión bifásica
two-phase winding arrollamiento bifásico
two-plus-one address dirección de dos más uno
two-point control control de dos puntos
two-pole de dos polos, de dos postes
two-pole switch conmutador de dos polos
two-port de dos puertos
two-port network red de dos puertos
two-position control control de dos posiciones
two-position relay relé de dos posiciones
two-position selector selector de dos posiciones
two-position switch conmutador de dos posiciones
two-position winding arrollamiento de dos pisos
two-range instrument instrumento de dos escalas
two-speed motor motor de dos velocidades
two-stage de dos etapas
two-stage amplifier amplificador de dos etapas
two-stage control control de dos etapas
two-stage process proceso de dos etapas
two-state device dispositivo de dos estados
two-state logic lógica de dos estados
two-step de dos pasos
two-step relay relé de dos pasos
two-terminal de dos terminales
two-terminal capacitor capacitor de dos terminales
two-terminal network red de dos terminales

two-throw switch conmutador de dos posiciones
two-tone de dos tonos
two-tone keying manipulación de dos tonos
two-tone modulation modulación de dos tonos
two-tone signal señal de dos tonos
two-tone system sistema de dos tonos
two-tone telegraph system sistema de telegrafía de
    dos tonos
two-track recording registro de dos pistas
two-way bidireccional
two-way amplifier amplificador bidireccional
two-way channel canal bidireccional
two-way circuit circuito bidireccional
two-way communication comunicación bidireccional
two-way conduction conducción bidireccional
two-way connection conexión bidireccional
two-way contact contacto bidireccional
two-way link enlace bidireccional
two-way loudspeaker system sistema de altavoces de
    dos canales
two-way mobile system sistema móvil bidireccional
two-way preamplifier preamplificador bidireccional
two-way radio radio bidireccional
two-way repeater repetidor bidireccional
two-way speaker system sistema de altavoces de dos
    canales
two-way switch conmutador bidireccional
two-way system sistema de dos canales, sistema
    bidireccional
two-way terminal terminal bidireccional
two-way traffic tráfico bidireccional
two-winding transformer transformador de dos
    arrollamientos
two-wire de dos hilos, bifilar
two-wire antenna antena bifilar
two-wire cable cable bifilar
two-wire circuit circuito bifilar
two-wire line línea bifilar
two-wire system sistema bifilar
two-wire termination terminación bifilar
two-wire winding arrollamiento bifilar
twos complement complemento a dos
Tyndall effect efecto de Tyndall
type-A display indicador visual tipo A
type-A facsimile facsímil tipo A
type-A wave onda tipo A
type-B display indicador visual tipo B
type-B facsimile facsímil tipo B
type-B wave onda tipo B
type-N semiconductor semiconductor tipo N
type of action tipo de acción
type of emission tipo de emisión
type of operation tipo de operación
type-P semiconductor semiconductor tipo P
typewriter máquina de escribir

ubitron ubitrón
Ulbricht sphere esfera de Ulbricht
ultimate capacity capacidad límite
ultimate load carga límite
ultimate ratio razón límite
ultimate sensitivity sensibilidad límite
ultor segundo ánodo
ultor anode segundo ánodo
ultra ultra
ultra-audible ultraaudible
ultra-audible frequency frecuencia ultraaudible
ultra-audion ultraaudión
ultra-audion circuit circuito ultraaudión
ultra-audion oscillator oscilador ultraaudión
ultra-high frequency frecuencia ultraalta
ultra-high frequency adapter adaptador de
    frecuencia ultraalta
ultra-high frequency antenna antena de frecuencia
    ultraalta
ultra-high frequency band banda de frecuencia
    ultraalta
ultra-high frequency capacitor capacitor de
    frecuencia ultraalta
ultra-high frequency converter convertidor de
    frecuencia ultraalta
ultra-high frequency diode diodo de frecuencia
    ultraalta
ultra-high frequency generator generador de
    frecuencias ultraaltas
ultra-high frequency link enlace de frecuencia
    ultraalta
ultra-high frequency loop antena de cuadro de
    frecuencia ultraalta
ultra-high frequency oscillator oscilador de
    frecuencia ultraalta
ultra-high frequency radar radar de frecuencia
    ultraalta
ultra-high frequency range gama de frecuencias
    ultraaltas
ultra-high frequency receiver receptor de frecuencia
    ultraalta
ultra-high frequency transistor transistor de
    frecuencia ultraalta
ultra-high frequency translator traductor de
    frecuencia ultraalta
ultra-high frequency transmitter transmisor de
    frecuencia ultraalta
ultra-high frequency tube tubo de frecuencia
    ultraalta
ultra-high frequency tuner sintonizador de
    frecuencia ultraalta
ultra-high resistance resistencia ultraalta
ultra-high speed velocidad ultraalta
ultra-high vacuum vacío ultraalto
ultra-high voltage tensión ultraalta
ultradyne ultradino
ultradyne reception recepción ultradina
ultradyne receptor receptor ultradino
ultrafast-recovery diode diodo de recuperación
    ultrarrápida
ultrafast switch conmutador ultrarrápido

ultrafiche ultraficha
ultrafine ultrafino
ultrahigh frequency frecuencia ultraalta
ultrahigh frequency adapter adaptador de frecuencia ultraalta
ultrahigh frequency antenna antena de frecuencia ultraalta
ultrahigh frequency band banda de frecuencia ultraalta
ultrahigh frequency capacitor capacitor de frecuencia ultraalta
ultrahigh frequency converter convertidor de frecuencia ultraalta
ultrahigh frequency diode diodo de frecuencia ultraalta
ultrahigh frequency generator generador de frecuencias ultraaltas
ultrahigh frequency link enlace de frecuencia ultraalta
ultrahigh frequency loop antena de cuadro de frecuencia ultraalta
ultrahigh frequency oscillator oscilador de frecuencia ultraalta
ultrahigh frequency radar radar de frecuencia ultraalta
ultrahigh frequency range gama de frecuencias ultraaltas
ultrahigh frequency receiver receptor de frecuencia ultraalta
ultrahigh frequency transistor transistor de frecuencia ultraalta
ultrahigh frequency translator traductor de frecuencia ultraalta
ultrahigh frequency transmitter transmisor de frecuencia ultraalta
ultrahigh frequency tube tubo de frecuencia ultraalta
ultrahigh frequency tuner sintonizador de frecuencia ultraalta
ultrahigh resistance resistencia ultraalta
ultrahigh speed velocidad ultraalta
ultrahigh vacuum vacío ultraalto
ultrahigh voltage tensión ultraalta
ultralinear ultralineal
ultralinear amplifier amplificador ultralineal
ultralinear circuit circuito ultralineal
ultralinear output stage etapa de salida ultralineal
ultralow ultrabaja
ultralow distortion distorsión ultrabaja
ultralow frequency frecuencia ultrabaja
ultramicrometer ultramicrómetro
ultramicroscope ultramicroscopio
ultramicrowave ultramicroonda
ultraminiature ultraminiatura
ultraphotic rays rayos ultrafóticos
ultraprecision ultraprecisión
ultrapure ultrapuro
ultrarapid ultrarrápido
ultrared ultrarrojo
ultrared rays rayos ultrarrojos
ultrasensitive ultrasensible
ultrasensitive instrument instrumento ultrasensible
ultrasensitive relay relé ultrasensible
ultrasensitive tube tubo ultrasensible
ultrashort ultracorto
ultrashort-wave antenna antena de ondas ultracortas
ultrashort-wave band banda de ondas ultracortas
ultrashort-wave broadcasting radiodifusión de ondas ultracortas
ultrashort-wave propagation propagación de ondas ultracortas
ultrashort-wave station estación de ondas ultracortas
ultrashort-wave transmitter transmisor de ondas ultracortas

ultrashort waves ondas ultracortas
ultrasonic ultrasónico
ultrasonic amplifier amplificador ultrasónico
ultrasonic bath baño ultrasónico
ultrasonic beam haz ultrasónico
ultrasonic bond enlace ultrasónico
ultrasonic brazing soldadura fuerte ultrasónica
ultrasonic cleaning limpieza ultrasónica
ultrasonic-cleaning bath baño de limpieza ultrasónica
ultrasonic-cleaning equipment equipo de limpieza ultrasónica
ultrasonic-cleaning system sistema de limpieza ultrasónica
ultrasonic communication comunicación ultrasónica
ultrasonic cutting corte ultrasónico
ultrasonic delay line línea de retardo ultrasónica
ultrasonic densitometer densitómetro ultrasónico
ultrasonic depth finder buscador de profundidad ultrasónico
ultrasonic detector detector ultrasónico
ultrasonic diffraction difracción ultrasónica
ultrasonic disintegrator desintegrador ultrasónico
ultrasonic dispersion dispersión ultrasónica
ultrasonic drill taladro ultrasónico
ultrasonic equipment equipo ultrasónico
ultrasonic filter filtro ultrasónico
ultrasonic flaw detection detección de imperfecciones ultrasónica
ultrasonic flaw detector detector de imperfecciones ultrasónico
ultrasonic frequency frecuencia ultrasónica
ultrasonic-frequency range gama de frecuencias ultrasónicas
ultrasonic generator generador ultrasónico
ultrasonic grating retículo ultrasónico
ultrasonic grating constant constante de retículo ultrasónico
ultrasonic heating calentamiento ultrasónico
ultrasonic image converter convertidor de imágenes ultrasónicas
ultrasonic inspection inspección ultrasónica
ultrasonic interferometry interferometría ultrasónica
ultrasonic level detector detector de nivel ultrasónico
ultrasonic light diffraction difracción de luz ultrasónica
ultrasonic light modulator modulador de luz ultrasónico
ultrasonic machining maquinado ultrasónico
ultrasonic piezoelectric transducer transductor piezoeléctrico ultrasónico
ultrasonic plating depósito ultrasónico
ultrasonic probe sonda ultrasónica
ultrasonic pulse impulso ultrasónico
ultrasonic receiver receptor ultrasónico
ultrasonic recording registro ultrasónico
ultrasonic relay relé ultrasónico
ultrasonic sensor sensor ultrasónico
ultrasonic signal señal ultrasónica
ultrasonic soldering soldadura por ultrasonidos
ultrasonic sounding sondeo ultrasónico
ultrasonic space grating retículo espacial ultrasónico
ultrasonic storage cell célula de almacenamiento ultrasónica
ultrasonic stroboscope estroboscopio ultrasónico
ultrasonic switch conmutador ultrasónico
ultrasonic system sistema ultrasónico
ultrasonic test prueba ultrasónica
ultrasonic tool herramienta ultrasónica
ultrasonic transducer transductor ultrasónico
ultrasonic vibration vibración ultrasónica

ultrasonic waves ondas ultrasónicas
ultrasonic welding soldadura por ultrasonidos
ultrasonically ultrasónicamente
ultrasonics ultrasónica
ultrasonography ultrasonografía
ultrasound ultrasonido
ultraspeed ultravelocidad
ultrastability ultraestabilidad
ultrastable ultraestable
ultrathin ultradelgado
ultraviolet ultravioleta
ultraviolet altimeter altímetro ultravioleta
ultraviolet component componente ultravioleta
ultraviolet crack detector detector de grietas ultravioleta
ultraviolet-excited excitado por ultravioleta
ultraviolet-induced inducido por ultravioleta
ultraviolet lamp lámpara ultravioleta
ultraviolet light luz ultravioleta
ultraviolet power potencia ultravioleta
ultraviolet radiation radiación ultravioleta
ultraviolet ray rayo ultravioleta
ultraviolet region región ultravioleta
ultraviolet spectroscopy espectroscopia ultravioleta
ultraviolet spectrum espectro ultravioleta
ultraviolet wave onda ultravioleta
ultraviolet wavelength longitud de onda ultravioleta
umbrella antenna antena en paraguas
Umklapp process proceso de Umklapp
unabsorbed field intensity intensidad de campo sin absorción
unabsorbed field strength intensidad de campo sin absorción
unacceptable condition condición inaceptable
unaddressable no direccionable
unaddressable storage almacenamiento no direccionable
unallocated no asignado
unamplified feedback retroalimentación sin amplificar
unary unario
unary operation operación unaria
unary operator operador unario
unassigned no asignado
unattended inatendido, desatendido
unattended mode modo inatendido
unattended operation operación inatendida
unattended repeater repetidor inatendido
unattended service servicio inatendido
unattended station estación inatendida
unattenuated no atenuado
unavailability indisponibilidad
unavailable indisponible
unavailable hours horas indisponibles
unbalance desequilibrio
unbalance factor factor de desequilibrio
unbalance ratio razón de desequilibrio
unbalance voltage tensión de desequilibrio
unbalanced desequilibrado
unbalanced antenna antena desequilibrada
unbalanced circuit circuito desequilibrado
unbalanced data link enlace de datos desequilibrado
unbalanced error error desequilibrado
unbalanced input entrada desequilibrada
unbalanced line línea desequilibrada
unbalanced load carga desequilibrada
unbalanced multivibrator multivibrador desequilibrado
unbalanced network red desequilibrada
unbalanced output salida desequilibrada
unbalanced phases fases desequilibradas
unbalanced system sistema desequilibrado
unbalanced transmission transmisión desequilibrada

unbalanced transmission line línea de transmisión desequilibrada
unbased tube tubo sin base
unbiased no polarizado
unbiased rectifier rectificador no polarizado
unblanking desbloqueo
unblanking interval intervalo de desbloqueo
unblanking pulse impulso de desbloqueo
unblanking time tiempo de desbloqueo
unblocked desbloqueado, no bloqueado
unblocking desbloqueo
unbypassed cathode resistor resistor de cátodo sin capacitor de desacoplamiento
uncalibrated no calibrado
uncalibrated unit unidad no calibrada
uncertainty incertidumbre
uncharged sin carga
uncoated filament filamento sin revestimiento
uncoded no codificado
uncompensated no compensado
uncompensated amplifier amplificador no compensado
uncompensated capacitor capacitor no compensado
uncompensated inductor inductor no compensado
uncompensated resistor resistor no compensado
uncompensated video amplifier videoamplificador no compensado
uncompensated volume control control de volumen no compensado
unconditional incondicional
unconditional assignment asignación incondicional
unconditional branch rama incondicional
unconditional control field campo de control incondicional
unconditional control transfer transferencia de control incondicional
unconditional jump salto incondicional
unconditional jump instruction instrucción de salto incondicional
unconditional stability estabilidad incondicional
unconditional statement sentencia incondicional
unconditional transfer transferencia incondicional
unconditional transfer instruction instrucción de transferencia incondicional
unconditional transfer of control transferencia incondicional de control
unconditionally incondicionalmente
uncontrolled no controlado
uncontrolled loop bucle no controlado
uncouple desacoplar, desconectar
uncoupled desacoplado, desconectado
uncoupling desacoplamiento, desconexión
undamped no amortiguado
undamped galvanometer galvanómetro no amortiguado
undamped meter medidor no amortiguado
undamped natural frequency frecuencia natural no amortiguada
undamped oscillation oscilación no amortiguada
undamped output circuit circuito de salida no amortiguado
undamped speaker enclosure caja de altavoz no amortiguada
undamped wave onda no amortiguada
undebugged no depurado
undefined indefinido
undelete recuperar archivos borrados
underbunching subagrupamiento
undercompensated subcompensado
undercompounded generator generador hipocompuesto
undercoupling subacoplamiento
undercurrent subcorriente

undercurrent relay relé de hipocorriente
underdamped subamortiguado
underdamping subamortiguamiento
underdriven subexcitado
underdriven unit unidad subexcitada
underexcited subexcitado
underexcited amplifier amplificador subexcitado
underflow desbordamiento negativo,
  subdesbordamiento
underflow error error por desbordamiento negativo
underflow indicator indicador de desbordamiento
  negativo
underfrequency subfrecuencia
underground subterráneo
underground antenna antena subterránea
underground cable cable subterráneo
underground circuit circuito subterráneo
underground communication comunicación
  subterránea
underground line línea subterránea
underground receiver receptor subterráneo
underground reception recepción subterránea
underground system sistema subterráneo
underground transmitter transmisor subterráneo
underlap falta de yuxtaposición
underlay capa base
underload circuit-breaker cortacircuito de carga
  mínima
underload relay relé de carga mínima
underload switch conmutador de carga mínima
undermodulation submodulación
underpower relay relé de mínima potencia
underpunch subperforación
underscan subexploración
undersea cable cable submarino
undershoot subimpulso
underswing suboscilación
undertuned subsintonizado
undervoltage subtensión, tensión insuficiente
undervoltage alarm alarma de subtensión
undervoltage circuit-breaker cortacircuito de
  tensión mínima
undervoltage protection protección contra
  subtensión
undervoltage relay relé de subtensión, relé de
  tensión mínima
undervoltage tripping desconexión por subtensión
underwater amplifier amplificador submarino
underwater antenna antena submarina
underwater link enlace submarino
underwater radar radar submarino
underwater repeater repetidor submarino
underwater signal señal submarina
underwater sounding sondeo submarino
underwater television televisión submarina
undesired response respuesta indeseada
undesired signal señal indeseada
undistorted sin distorsión
undistorted grid modulation modulación por rejilla
  sin distorsión
undistorted output salida sin distorsión
undistorted power potencia sin distorsión
undistorted power output salida de potencia sin
  distorsión
undistorted wave onda sin distorsión
undisturbed-one output salida de uno sin
  perturbación
undisturbed output signal señal de salida sin
  perturbación
undisturbed-zero output salida de cero sin
  perturbación
undoped no dopado
undulating ondulante

undulating current corriente ondulante
undulating voltage tensión ondulante
undulation ondulación
undulator ondulador
undulatory onduladorio
undulatory current corriente onduladoria
undulatory voltage tensión onduladoria
unequal alternation alternación desigual
unequal impulses impulsos desiguales
unerased no borrado
unexpected halt parada inesperada
unfiltered no filtrado
unformatted no formateado
unformatted diskette disquete no formateado
unformatted display visualización no formateada
unformatted image imagen no formateada
unformatted information información no formateada
unformatted input-output entrada-salida no
  formateada
unformatted mode modo no formateado
unformatted program programa no formateado
unformatted record registro no formateado
unfurlable antenna antena desplegable
ungrounded circuit circuito sin conexión a tierra
ungrounded system sistema sin conexión a tierra
unhook descolgar
uniaxial uniaxial
unibus canal único
uniconductor uniconductor
uniconductor cable cable uniconductor
unidirectional unidireccional
unidirectional action acción unidireccional
unidirectional antenna antena unidireccional
unidirectional array red unidireccional
unidirectional bus bus unidireccional
unidirectional circuit circuito unidireccional
unidirectional conductivity conductividad
  unidireccional
unidirectional coupler acoplador unidireccional
unidirectional current corriente unidireccional
unidirectional element elemento unidireccional
unidirectional hydrophone hidrófono unidireccional
unidirectional loudspeaker altavoz unidireccional
unidirectional microphone micrófono unidireccional
unidirectional network red unidireccional
unidirectional pattern patrón unidireccional
unidirectional pulse impulso unidireccional
unidirectional pulse train tren de impulsos
  unidireccional
unidirectional response respuesta unidireccional
unidirectional selector selector unidireccional
unidirectional speaker altavoz unidireccional
unidirectional transducer transductor unidireccional
unidirectional transmission transmisión
  unidireccional
unidirectional voltage tensión unidireccional
unidyne unidino
unidyne receiver receptor unidino
unidyne reception recepción unidina
unifilar unifilar
unifilar magnetometer magnetómetro unifilar
unifilar suspension suspensión unifilar
uniform uniforme
uniform beam haz uniforme
uniform charge carga uniforme
uniform current density densidad de corriente
  uniforme
uniform deposition deposición uniforme
uniform diffuse reflection reflexión difusa uniforme
uniform diffuser difusor uniforme
uniform electric field campo eléctrico uniforme
uniform field campo uniforme
uniform flow flujo uniforme

uniform frequency response respuesta de frecuencia uniforme
uniform illumination iluminación uniforme
uniform line línea uniforme
uniform linear array red lineal uniforme
uniform load carga uniforme
uniform magnetic field campo magnético uniforme
uniform plane wave onda plana uniforme
uniform precession precesión uniforme
uniform speed velocidad uniforme
uniform transmission line línea de transmisión uniforme
uniform velocity velocidad uniforme
uniform waveguide guíaondas uniforme
uniformity uniformidad
uniformity factor factor de uniformidad
uniformity ratio razón de uniformidad
unigrounded de conexión única a tierra
unijunction uniunión
unijunction transistor transistor uniunión
unilateral unilateral
unilateral amplifier amplificador unilateral
unilateral bearing rumbo unilateral
unilateral channel canal unilateral
unilateral conductivity conductividad unilateral
unilateral connection conexión unilateral
unilateral device dispositivo unilateral
unilateral element elemento unilateral
unilateral impedance impedancia unilateral
unilateral network red unilateral
unilateral relay relé unilateral
unilateral transducer transductor unilateral
unilateral transmission transmisión unilateral
unilateralization unilateralización
unimodal unimodal
unimode unimodo
uninsulated inaislado, no aislado
uninterruptable ininterrumpible
uninterruptable power supply fuente de alimentación ininterrumpible
uninterrupted ininterrumpido
uninterrupted duty servicio ininterrumpido
union unión
uniplex uniplex
unipolar unipolar
unipolar apparatus aparato unipolar
unipolar device dispositivo unipolar
unipolar electrode system sistema de electrodos unipolar
unipolar field-effect-transistor transistor de efecto de campo unipolar
unipolar induction inducción unipolar
unipolar input entrada unipolar
unipolar transistor transistor unipolar
unipolarity unipolaridad
unipole unipolo
unipotential unipotencial
unipotential cathode cátodo unipotencial
uniprocessor uniprocesador
uniprogramming uniprogramación
uniselector uniselector, selector de movimiento único
unit unidad
unit address dirección de unidad
unit amplifier amplificador unitario
unit area área unitaria
unit-area acoustic impedance impedancia acústica intrínseca
unit-area acoustic reactance reactancia acústica intrínseca
unit-area acoustic resistance resistencia acústica intrínseca
unit automatic exchange central automática unitaria
unit capacity factor factor de capacidad unitario

unit charge carga unitaria
unit check comprobación de unidad
unit control block bloque de control de unidad
unit device dispositivo unitario
unit dielectric strength rigidez dieléctrica específica
unit distance distancia unitaria
unit electric charge carga eléctrica unitaria
unit element elemento unitario
unit interval intervalo unitario
unit length longitud unidad
unit load carga unitaria
unit magnetic pole polo magnético unidad
unit of capacitance unidad de capacitancia
unit of current unidad de corriente
unit of electromotive force unidad de fuerza electromotriz
unit of information unidad de información
unit of light unidad de luz
unit of measurement unidad de medida
unit of resistance unidad de resistencia
unit of time unidad de tiempo
unit of transfer unidad de transferencia
unit pole polo unitario
unit pulse impulso unitario
unit record registro unitario
unit separation separación de unidades
unit separator separador de unidades
unit space espacio de unidad
unit-step function función escalón unitario
unit-step voltage tensión escalón unitario
unit string secuencia unitaria
unit substation subestación unitaria
unit tube tubo unitario
unitary code código unitario
uniterm system sistema de unitérminos
unitized construction construcción con subensamblajes
unitor unitor
unity coupling acoplamiento unitario
unity gain ganancia unitaria
unity-gain amplifier amplificador de ganancia unitaria
unity-gain antenna antena unitaria
unity power factor factor de potencia unitario
unity ratio razón unitaria
univalent univalente
universal universal
universal address dirección universal
universal antenna coupler acoplador de antena universal
universal bar barra universal
universal bridge puente universal
universal cable adapter adaptador de cable universal
universal character set conjunto de caracteres universal
universal controller controlador universal
universal coupler acoplador universal
universal coupling acoplamiento universal
universal device dispositivo universal
universal filter filtro universal
universal frequency counter contador de frecuencia universal
universal joint junta universal
universal measuring bridge puente de medida universal
universal motor motor universal
universal mounting montura universal
universal multimeter multímetro universal
universal output transformer transformador de salida universal
universal product code código de producto universal
universal receiver receptor universal
universal transformer transformador universal

universal transistor transistor universal
universal transmitter transmisor universal
universal Turing machine máquina de Turing universal
univibrator univibrador, multivibrador monoestable
unlike charges cargas desiguales
unlike poles polos desiguales
unlimited ilimitado
unload descargar
unloaded amplifier amplificador sin carga
unloaded antenna antena sin carga
unloaded battery batería sin carga
unloaded cable cable sin carga
unloaded generator generador sin carga
unloading descarga
unmatched elements elementos no adaptados
unmodified no modificado
unmodulated no modulado
unmodulated carrier portadora no modulada
unmodulated carrier wave onda portadora no modulada
unmodulated current corriente no modulada
unmodulated groove surco no modulado
unmodulated voltage tensión no modulada
unmodulated wave onda no modulada
unneutralized no neutralizado
unnumbered no numerado
unoccupied desocupado
unpack desagrupar
unperforated no perforado
unplanned outage interrupción no planificada
unpolarized light luz no polarizada
unpolarized plug clavija no polarizada
unpolarized radiation radiación no polarizada
unpolarized receptacle receptáculo no polarizado
unpolarized relay relé no polarizado
unpolarized socket zócalo no polarizado
unprocessable no procesable
unprocessed no procesado
unprogrammable no programable
unprogrammed no programado
unprotected sin protección
unprotected antenna antena sin protección
unpunched no perforado
unrecorded tape cinta sin registrar, cinta sin grabar
unrecoverable abend final anormal irrecuperable
unrecoverable error error irrecuperable
unreflected ray rayo sin reflexión
unreflected wave onda sin reflexión
unregulated no regulado
unregulated power supply fuente de alimentación no regulada
unsaturated no saturado
unsaturated core núcleo no saturado
unsaturated operation operación no saturada
unsaturated standard cell pila patrón no saturada
unsaturated tube tubo no saturado
unscheduled no programado
unscheduled down time tiempo muerto no programado
unscheduled interruption interrupción no programada
unscheduled maintenance mantenimiento no programado
unscheduled operation operación no programada
unscheduled outage interrupción no programada
unscheduled service servicio no programado
unscrambler descodificador
unsegmented no segmentado
unshielded cable cable no blindado
unshielded probe sonda no blindada
unsolder desoldar
unsorted no clasificado

unspecified no especificado
unstable inestable
unstable oscillation oscilación inestable
unstable region región inestable
unstable servo servo inestable
unstable state estado inestable
unsuppressed carrier portadora no suprimida
unswitched outlet tomacorriente no conmutada
unsymmetrical wave onda no simétrica
unterminated amplifier amplificador sin carga terminal
unterminated generator generador sin carga terminal
untuned no sintonizado
untuned amplifier amplificador no sintonizado
untuned antenna antena no sintonizada
untuned circuit circuito no sintonizado
untuned filter filtro no sintonizado
untuned line línea no sintonizada
untuned transformer transformador no sintonizado
untuned transmission line línea de transmisión no sintonizada
unverified no verificado
unwanted indeseado
unwanted emission emisión indeseada
unwanted modulation modulación indeseada
unwanted signal señal indeseada
unweighted no ponderado, sin compensación
unweighted decibels decibeles sin compensación
unweighted voltage tensión sin compensación
unwind desarrollar, desbobinar, codificar explícitamente
unwinding desarrollamiento, desbobinado, codificación explícita
up-conversion conversión ascendente
up-converter convertidor ascendente
up-time tiempo de funcionamiento
update actualizar
update operation operación de actualización
updated actualizado
updating actualización
updating operation operación de actualización
updating routine rutina de actualización
uplink señal de subida
uplink frequency frecuencia de señal de subida
upload cargar
upper atmosphere atmósfera superior
upper band banda superior
upper brush escobilla superior
upper disk disco superior
upper frequency limit límite de frecuencia superior
upper limit límite superior
upper sideband banda lateral superior
upper-sideband component componente de banda lateral superior
upper-sideband converter convertidor de banda lateral superior
upper storage almacenamiento superior
upset circuit circuito de cambio de estado
upset signal señal de cambio de estado
uptime tiempo de funcionamiento
upward compatibility compatibilidad ascendente
upward compression compresión ascendente
upward heterodyning heterodinaje ascendente
uranium uranio
urban exchange central urbana
urban satellite satélite urbano
urea plastic plástico de urea
urea resin resina de urea
ursigram ursigrama
usable accuracy exactitud utilizable
usable area área utilizable
usable bandwidth ancho de banda utilizable

usable frequency frecuencia utilizable
usable sensitivity sensibilidad utilizable
useful beam haz útil
useful field campo útil
useful frequency frecuencia útil
useful life vida útil
useful line línea útil
useful output salida útil
useful output power potencia de salida útil
useful power potencia útil
useful signal señal útil
user code código de usuario
user-controlled controlado por usuario
user-defined definido por usuario
user-friendly orientado al usuario
user identification identificación de usuario
user interface interfaz de usuario
user-oriented orientado al usuario
user-oriented language lenguaje orientado al usuario
user program programa de usuario
user routine rutina de usuario
user terminal terminal de usuario
utility factor factor de utilidad, factor de utilización
utility function función de apoyo operacional
utility outlet tomacorriente auxiliar
utility power potencia auxiliar
utility program programa de apoyo operacional
utility routine rutina de apoyo operacional
utility software programas de apoyo operacional
utilization utilización
utilization factor factor de utilización
utilization period periodo de utilización
utilization ratio razón de utilización
uvicon uvicón
uviol lamp lámpara de uviol

V antenna antena en V
V band banda V
V beam haz en V
V-beam radar radar de haz en V
vacant vacante
vacant code código vacante
vacant conductor conductor vacante
vacant level nivel vacante
vacant terminal terminal vacante
vacuum vacío
vacuum arc arco al vacío
vacuum-arc lamp lámpara de arco al vacío
vacuum arrester descargador de vacío
vacuum capacitor capacitor de vacío
vacuum cell célula de vacío
vacuum chamber cámara de vacío
vacuum condenser condensador al vacío
vacuum deposition deposición al vacío
vacuum diode diodo de vacío
vacuum evaporation evaporación en vacío
vacuum filter filtro de vacío
vacuum gage vacuómetro, medidor de vacío
vacuum impregnation impregnación en vacío
vacuum indicator indicador de vacío
vacuum lamp lámpara de vacío
vacuum leak fuga de vacío
vacuum-leak detector detector de fugas de vacío
vacuum level nivel de vacío
vacuum lightning arrester pararrayos de vacío
vacuum lightning protector pararrayos de vacío
vacuum-operated operado por vacío
vacuum photocell fotocélula de vacío
vacuum phototube fototubo de vacío
vacuum pump bomba de vacío
vacuum range alcance en vacío
vacuum rectifier rectificador de vacío
vacuum regulator regulador de vacío
vacuum relay relé de vacío
vacuum seal cierre de vacío
vacuum spectrograph espectrógrafo de vacío
vacuum switch conmutador de vacío
vacuum system sistema de vacío
vacuum tank tanque de vacío
vacuum test prueba en vacío
vacuum thermocouple termopar de vacío
vacuum-tight hermético al vacío
vacuum tube tubo de vacío
vacuum-tube amplifier amplificador de tubos de
    vacío
vacuum-tube bridge puente de tubos de vacío
vacuum-tube characteristics características de tubo
    de vacío
vacuum-tube detector detector de tubo de vacío
vacuum-tube electrometer electrómetro de tubo de
    vacío
vacuum-tube modulator modulador de tubos de
    vacío
vacuum-tube ohmmeter óhmetro de tubo de vacío
vacuum-tube oscillator oscilador de tubo de vacío
vacuum-tube receiver receptor de tubos de vacío
vacuum-tube rectifier rectificador de tubo de vacío

vacuum-tube transmitter transmisor de tubos de vacío
vacuum-tube voltmeter vóltmetro de tubo de vacío
vacuum valve válvula de vacío
vacuumtight hermético al vacío
valence valencia
valence band banda de valencia
valence bond enlace de valencia
valence electrons electrones de valencia
valence shell capa de valencia
valency valencia
valid válido
valid call llamada válida
valid memory address dirección de memoria válida
validate validar
validated validado
validation validación
validation process proceso de validación
validation test prueba de validación
validity validez
validity check comprobación de validez
validity checker comprobador de validez
validity checking comprobación de validez
valley current corriente de valle
valley point fondo de valle
valley-point current corriente de fondo de valle
valley voltage tensión de valle
valuator valuador
value valor
valve válvula
valve action acción de válvula
valve-actuated accionado por válvula
valve actuator actuador de válvula
valve adapter adaptador de válvula
valve adjustment ajuste de válvula
valve amplifier amplificador de válvulas
valve base base de válvula
valve box caja de válvulas
valve characteristic característica de válvula
valve computer computadora de válvulas
valve-controlled controlado por válvula
valve detector detector de válvula
valve effect efecto de válvula
valve element elemento de válvula
valve emission emisión de válvula
valve heater calentador de válvula
valve holder portaválvula
valve lifter levantaválvulas
valve module módulo de válvula
valve noise ruido de válvula
valve oscillator oscilador de válvula
valve pin terminal de válvula
valve plate placa de válvula
valve ratio razón de válvula
valve receiver receptor de válvulas
valve rectifier rectificador de válvula
valve socket zócalo de válvula
valve transconductance transconductancia de válvula
valve voltmeter vóltmetro de válvula
valveless sin válvulas
Van Allen belts cinturones de Van Allen
Van Allen radiation belts cinturones de radiación de Van Allen
Van de Graaff generator generador de Van de Graaff
Van der Pol oscillator oscilador de Van der Pol
Van der Waals forces fuerzas de Van der Waals
vanadium vanadio
vane paleta, aleta
vane-anode magnetron magnetrón de ánodo de paletas
vane attenuator atenuador de paleta longitudinal
vane magnetron magnetrón de paletas

vane-type instrument instrumento de tipo de paleta
vane wattmeter wáttmetro de paleta
vapor deposition deposición por vapor
vapor-filled lleno de vapor
vapor-free libre de vapor
vapor lamp lámpara de vapor
vapor plating depósito por vapor
vapor pressure presión de vapor
vaporization vaporización
vaporize vaporizar
vaporproof a prueba de vapores
vaportight hermético contra vapores
vapotron vapotrón
var var
varactor varactor
varactor amplifier amplificador de varactor
varactor frequency multiplier multiplicador de frecuencia de varactor
varhour varhora
varhour meter varhorímetro
variability variabilidad
variable variable
variable address dirección variable
variable antenna coupling acoplamiento de antena variable
variable area área variable
variable-area recording registro de área variable
variable-area track pista de área variable
variable attenuator atenuador variable
variable autotransformer autotransformador variable
variable bandwidth ancho de banda variable
variable beta beta variable
variable-beta transistor transistor de beta variable
variable bias polarización variable
variable-bias control control de polarización variable
variable block bloque variable
variable-block length longitud de bloque variable
variable capacitance capacitancia variable
variable-capacitance cartridge cartucho de capacitancia variable
variable-capacitance diode diodo de capacitancia variable
variable-capacitance pickup fonocaptor de capacitancia variable
variable-capacitance transducer transductor de capacitancia variable
variable capacitor capacitor variable
variable capacity capacidad variable
variable carrier portadora variable
variable-carrier modulator modulador de portadora variable
variable carrier wave onda portadora variable
variable command mando variable
variable condenser condensador variable
variable connector conector variable
variable coupling acoplamiento variable
variable-cutoff attenuator atenuador de corte variable
variable-cycle operation operación de ciclo variable
variable damping amortiguamiento variable
variable-damping control control de amortiguamiento variable
variable data datos variables
variable delay retardo variable
variable delay line línea de retardo variable
variable density densidad variable
variable-density recording registro de densidad variable
variable-density track pista de densidad variable
variable-depth sonar sonar de profundidad variable
variable duty servicio variable

variable efficiency eficiencia variable
variable-efficiency modulation modulación de eficiencia variable
variable erase borrado variable
variable-erase recording registro de borrado variable
variable field campo variable
variable field length longitud de campo variable
variable format formato variable
variable frequency frecuencia variable
variable-frequency oscillator oscilador de frecuencia variable
variable-frequency sensor sensor de frecuencia variable
variable-frequency wave onda de frecuencia variable
variable-gain amplifier amplificador de ganancia variable
variable-gain stage etapa de ganancia variable
variable impedance impedancia variable
variable-impedance tube tubo de impedancia variable
variable inductance inductancia variable
variable-inductance pickup fonocaptor de inductancia variable
variable-inductance transducer transductor de inductancia variable
variable-inductance tuning sintonización por inductancia variable
variable inductor inductor variable
variable input entrada variable
variable intensity intensidad variable
variable intermittent duty servicio intermitente variable
variable length longitud variable
variable-length field campo de longitud variable
variable-length instruction instrucción de longitud variable
variable-length record registro de longitud variable
variable-length sorting clasificación de longitud variable
variable logic lógica variable
variable losses pérdidas variables
variable monoenergetic emission emisión monoenergética variable
variable monoenergetic radiation radiación monoenergética variable
variable-mu pentode pentodo de mu variable
variable-mu tube tubo de mu variable
variable-mu valve válvula de mu variable
variable oscillator oscilador variable
variable output salida variable
variable-parameter amplifier amplificador de parámetros variables
variable-pitch recording registro de paso variable, grabación de paso variable
variable point punto variable, coma variable
variable potentiometer potenciómetro variable
variable pressure presión variable
variable queue cola variable
variable radio frequency radiofrecuencia variable
variable ratio razón variable
variable reactance reactancia variable
variable-reactance amplifier amplificador de reactancia variable
variable reactor reactor variable
variable record registro variable
variable record length longitud de registro variable
variable reluctance reluctancia variable
variable-reluctance cartridge cartucho de reluctancia variable
variable-reluctance microphone micrófono de reluctancia variable
variable-reluctance pickup fonocaptor de reluctancia variable

variable-reluctance stepping motor motor de avance paso a paso de reluctancia variable
variable-reluctance transducer transductor de reluctancia variable
variable resistance resistencia variable
variable-resistance pickup fonocaptor de resistencia variable
variable-resistance transducer transductor de resistencia variable
variable-resistance tuning sintonización de resistencia variable
variable resistor resistor variable
variable response respuesta variable
variable selectivity selectividad variable
variable-slope filter filtro de pendiente variable
variable-slope pulse impulso de pendiente variable
variable speed velocidad variable
variable-speed control control de velocidad variable
variable-speed device dispositivo de velocidad variable
variable-speed motor motor de velocidad variable
variable-speed scanning exploración de velocidad variable
variable-speed transmission transmisión de velocidad variable
variable sweep barrido variable
variable temporary duty servicio temporal variable
variable tension tensión variable
variable time tiempo variable
variable tone tono variable
variable-tone oscillator oscilador de tono variable
variable-torque motor motor de par variable
variable transformer transformador variable
variable tube tubo variable
variable valve válvula variable
variable voltage tensión variable
variable-voltage control control por tensión variable
variable-voltage generator generador de tensión variable
variable-voltage regulator regulador de tensión variable
variable-voltage stabilizer estabilizador de tensión variable
variable-voltage transformer transformador de tensión variable
variable width anchura variable
variable-width recording registro de anchura variable
variance variancia, variación
variation variación
variation factor factor de variación
variation of attenuation variación de atenuación
variational variacional
variator variador
varicap varicap
varindor varindor
variocoupler varioacoplador
variolosser atenuador variable
variometer variómetro
varioplex varioplex
varistor varistor
varistor capacitance capacitancia de varistor
varistor resistance resistencia de varistor
varistor voltage tensión de varistor
varmeter varmetro
varnished cable cable barnizado
varnished cambric cambray barnizado
varnished tape cinta barnizada
varying variable
varying amplitude amplitud variable
varying duty servicio variable
varying electrical field campo eléctrico variable

varying field campo variable
varying parameter parámetro variable
varying speed velocidad variable
varying-speed motor motor de velocidad variable
varying-voltage control control por tensión variable
vectograph vectógrafo
vector vector
vector addition suma vectorial
vector admittance admitancia vectorial
vector current corriente vectorial
vector diagram diagrama vectorial
vector electrocardiogram electrocardiograma
　　vectorial
vector field campo vectorial
vector function función vectorial
vector impedance impedancia vectorial
vector power potencia vectorial
vector power factor factor de potencia vectorial
vector product producto vectorial
vector quantity magnitud vectorial
vector sum suma vectorial
vector value valor vectorial
vectorial vectorial
vectorial field campo vectorial
vectorscope vectorscopio
vee antenna antena en V
vehicle vehículo
vehicular vehicular
vehicular antenna antena vehicular
velocimeter velocímetro
velocity velocidad
velocity antiresonance antirresonancia de velocidad
velocity constant constante de velocidad
velocity correction corrección de velocidad
velocity distribution distribución de velocidades
velocity error error de velocidad
velocity factor factor de velocidad
velocity filter filtro de velocidad
velocity fluctuation fluctuación de velocidad
velocity-fluctuation noise ruido de fluctuación de
　　velocidad
velocity hydrophone hidrófono de velocidad
velocity lag retardo de velocidad
velocity-lag error error de retardo de velocidad
velocity level nivel de velocidad
velocity microphone micrófono de velocidad
velocity-modulated amplifier amplificador de
　　modulación de velocidad
velocity-modulated oscillator oscilador de
　　modulación de velocidad
velocity-modulated tube tubo de modulación de
　　velocidad
velocity modulation modulación de velocidad
velocity-modulation amplifier amplificador de
　　modulación de velocidad
velocity-modulation generator generador de
　　modulación de velocidad
velocity-modulation oscillator oscilador de
　　modulación de velocidad
velocity of light velocidad de la luz
velocity of propagation velocidad de propagación
velocity of radio waves velocidad de ondas de radio
velocity of sound velocidad del sonido
velocity ratio razón de velocidades
velocity resonance resonancia de velocidad
velocity spectrograph espectrógrafo de velocidad
velocity transducer transductor de velocidad
velocity transformer transformador de velocidad
velocity variation variación de velocidad
velocity-variation amplifier amplificador de
　　variación de velocidad
velocity-variation oscillator oscilador de variación
　　de velocidad

venetian-blind antenna antena en persiana
venetian-blind effect efecto de persiana
Venn diagram diagrama de Venn
vent respiradero, abertura de ventilación
vented baffle bafle ventilado
vented-baffle loudspeaker altavoz de bafle ventilado
vented battery batería ventilada
vented enclosure caja ventilada
vented fuse fusible ventilado
ventilated ventilado
ventilated enclosure caja ventilada
ventilated motor motor ventilado
ventilation ventilación
ventricular fibrillation fibrilación ventricular
venturi venturi
verification verificación
verification mode modo de verificación
verification system sistema de verificación
verifier verificador
verify verificar
vernier vernier, nonio
vernier capacitor capacitor vernier
vernier coupling acoplamiento vernier
vernier dial cuadrante vernier
vernier resistor resistor vernier
vernier rheostat reóstato vernier
vernitel vernitel
versed cosine coseno verso
versed sine seno verso
versine seno verso
version versión
vertex vértice
vertex plate placa de vértice
vertical vertical
vertical advance avance vertical
vertical amplification amplificación vertical
vertical amplifier amplificador vertical
vertical amplitude amplitud vertical
vertical-amplitude control control de amplitud
　　vertical
vertical angle ángulo vertical
vertical antenna antena vertical
vertical axis eje vertical
vertical bar barra vertical
vertical beam haz vertical
vertical blanking extinción vertical
vertical-blanking pulse impulso de extinción vertical
vertical-blanking signal señal de extinción vertical
vertical centering centrado vertical
vertical-centering control control de centrado
　　vertical
vertical channel canal vertical
vertical compliance docilidad vertical
vertical component componente vertical
vertical convergence convergencia vertical
vertical-convergence amplifier amplificador de
　　convergencia vertical
vertical-convergence control control de
　　convergencia vertical
vertical coverage cobertura vertical
vertical-coverage pattern patrón de cobertura
　　vertical
vertical definition definición vertical
vertical deflection deflexión vertical
vertical-deflection amplifier amplificador de
　　deflexión vertical
vertical-deflection coils bobinas de deflexión vertical
vertical-deflection electrodes electrodos de deflexión
　　vertical
vertical-deflection generator generador de deflexión
　　vertical
vertical-deflection oscillator oscilador de deflexión
　　vertical

vertical-deflection plates placas de deflexión vertical
vertical dipole dipolo vertical
vertical dynamic convergence convergencia dinámica vertical
vertical dynamic focus enfoque dinámico vertical
vertical feed alimentación vertical
vertical field strength intensidad de campo vertical
vertical-field-strength diagram diagrama de intensidad de campo vertical
vertical flyback retorno vertical
vertical frequency frecuencia vertical
vertical frequency oscillator oscilador de frecuencia vertical
vertical frequency response respuesta de frecuencia vertical
vertical gain ganancia vertical
vertical-gain control control de ganancia vertical
vertical hold sincronismo vertical
vertical-hold control control de sincronismo vertical
vertical incidence incidencia vertical
vertical-incidence transmission transmisión de incidencia vertical
vertical input entrada vertical
vertical instability inestabilidad vertical
vertical interlace entrelazado vertical
vertical interval intervalo vertical
vertical-lateral recording grabación vertical-lateral, registro vertical-lateral
vertical linearity linealidad vertical
vertical-linearity control control de linealidad vertical
vertical microcode microcódigo vertical
vertical oscillator oscilador vertical
vertical output salida vertical
vertical-output regulator regulador de salida vertical
vertical-output stage etapa de salida vertical
vertical-output transformer transformador de salida vertical
vertical-output tube tubo de salida vertical
vertical parity check comprobación de paridad vertical
vertical pattern patrón vertical
vertical plane plano vertical
vertical polarization polarización vertical
vertical pulse impulso vertical
vertical quantity magnitud vertical
vertical radar radar vertical
vertical radiation radiación vertical
vertical radiation pattern patrón de radiación vertical
vertical radiator radiador vertical
vertical range alcance vertical
vertical recording grabación vertical, registro vertical
vertical redundancy redundancia vertical
vertical redundancy check comprobación de redundancia vertical
vertical resolution resolución vertical
vertical retrace retorno vertical
vertical-retrace blanking extinción de retorno vertical
vertical-retrace period periodo de retorno vertical
vertical-retrace time tiempo de retorno vertical
vertical scale escala vertical
vertical scan exploración vertical
vertical scanning exploración vertical
vertical-scanning frequency frecuencia de exploración vertical
vertical-scanning generator generador de exploración vertical
vertical-scanning oscillator oscilador de exploración vertical
vertical section sección vertical

vertical sensitivity sensibilidad vertical
vertical signal señal vertical
vertical-size control control de tamaño vertical
vertical-speed transducer transductor de velocidad vertical
vertical stylus force fuerza de aguja vertical
vertical sweep barrido vertical
vertical-sweep frequency frecuencia de barrido vertical
vertical-sweep generator generador de barrido vertical
vertical-sweep oscillator oscilador de barrido vertical
vertical-sweep transformer transformador de barrido vertical
vertical-sweep voltage tensión de barrido vertical
vertical switchboard cuadro conmutador vertical
vertical-sync pulse impulso de sincronismo vertical
vertical synchronism sincronismo vertical
vertical synchronization sincronización vertical
vertical-synchronizing pulse impulso de sincronismo vertical
vertical unipole unipolo vertical
vertical unipole antenna antena unipolar vertical
vertical wave onda vertical
vertical width control control de anchura vertical
vertically-polarized antenna antena polarizada verticalmente
vertically-polarized transmission transmisión polarizada verticalmente
vertically-polarized wave onda polarizada verticalmente
very high frequency muy alta frecuencia
very-high-frequency antenna antena de muy alta frecuencia
very-high-frequency band banda de muy alta frecuencia
very-high-frequency broadcasting radiodifusión de muy alta frecuencia
very-high-frequency channel canal de muy alta frecuencia
very-high-frequency direction-finder radiogoniómetro de muy alta frecuencia
very-high-frequency directional range radiofaro direccional de muy alta frecuencia
very-high-frequency ground station estación terrestre de muy alta frecuencia
very-high-frequency link enlace de muy alta frecuencia
very-high-frequency omnidirectional range radiofaro omnidireccional de muy alta frecuencia
very-high-frequency omnirange radiofaro omnidireccional de muy alta frecuencia
very-high-frequency omnirange system sistema de radiofaro omnidireccional de muy alta frecuencia
very-high-frequency oscillator oscilador de muy alta frecuencia
very-high-frequency radar radar de muy alta frecuencia
very-high-frequency radio beacon radiofaro de muy alta frecuencia
very-high-frequency radio wave onda de radio de muy alta frecuencia
very-high-frequency radiotelephone radioteléfono de muy alta frecuencia
very-high-frequency radiotelephony radiotelefonía de muy alta frecuencia
very-high-frequency receiver receptor de muy alta frecuencia
very-high-frequency television channel canal de televisión de muy alta frecuencia
very-high-frequency transmitter transmisor de muy alta frecuencia

very-high-frequency tube tubo de muy alta frecuencia
very high resistance muy alta resistencia
very-high-resistance voltmeter vóltmetro de muy alta resistencia
very high speed muy alta velocidad
very-high-speed integrated circuit circuito integrado de muy alta velocidad
very high tension muy alta tensión
very-large-scale integration integración a muy alta escala
very long range muy largo alcance
very-long-range radar radar de muy largo alcance
very low frequency muy baja frecuencia
very low resistance muy baja resistencia
very short range muy corto alcance
very-short-range radar radar de muy corto alcance
vestigial vestigial, residual
vestigial band banda residual
vestigial sideband banda lateral residual
vestigial-sideband filter filtro de banda lateral residual
vestigial-sideband modulation modulación de banda lateral residual
vestigial-sideband signal señal de banda lateral residual
vestigial-sideband transmission transmisión de banda lateral residual
vestigial-sideband transmitter transmisor de banda lateral residual
via circuit circuito de tránsito
via condition condición de tránsito
vibrating vibrante, oscilante
vibrating bell timbre
vibrating capacitor capacitor vibrante
vibrating-capacitor electrometer electrómetro de capacitor vibrante
vibrating coil bobina vibrante
vibrating condenser condensador vibrante
vibrating contact contacto vibrante
vibrating contactor contactor vibrante
vibrating converter convertidor vibrante
vibrating cylinder sensor sensor de cilindro vibrante
vibrating diaphragm diafragma vibrante
vibrating reed laminilla vibrante
vibrating-reed electrometer electrómetro de laminilla vibrante
vibrating-reed frequency meter frecuencímetro de laminillas vibrantes
vibrating-reed galvanometer galvanómetro de laminillas vibrantes
vibrating-reed instrument instrumento de laminillas vibrantes
vibrating-reed magnetometer magnetómetro de laminillas vibrantes
vibrating-reed meter medidor de laminilla vibrante
vibrating-reed oscillator oscilador de laminilla vibrante
vibrating-reed rectifier rectificador de laminilla vibrante
vibrating-reed relay relé de laminilla vibrante
vibrating relay relé vibrante
vibrating system sistema vibrante
vibrating-wire oscillator oscilador de hilo vibrante
vibrating-wire transducer transductor de hilo vibrante
vibration vibración, oscilación
vibration absorber absorbedor de vibraciones
vibration-absorbing material material absorbente de vibraciones
vibration analyzer analizador de vibraciones
vibration attenuation atenuación de vibraciones
vibration calibrator calibrador de vibraciones

vibration control control de vibraciones
vibration dampener amortiguador de vibraciones
vibration detection detección de vibraciones
vibration-detection system sistema de detección de vibraciones
vibration galvanometer galvanómetro de vibraciones
vibration generator generador de vibraciones
vibration isolator aislador de vibraciones
vibration meter medidor de vibraciones
vibration pickup captor de vibraciones
vibration-resistant resistente a las vibraciones
vibration sensitivity sensibilidad a las vibraciones
vibration test prueba de vibraciones
vibrational vibracional, vibratorio
vibrational resonance resonancia vibracional
vibrator vibrador
vibrator-type converter convertidor de tipo de vibrador
vibrator-type inverter inversor de tipo de vibrador
vibrator-type rectifier rectificador de tipo de vibrador
vibratory vibratorio
vibratory feeder alimentador vibratorio
vibratron vibratrón
vibrograph vibrógrafo
vibrometer vibrómetro
vibrotron vibrotrón
video video
video amplification amplificación de video, videoamplificación
video amplifier amplificador de video, videoamplificador
video attenuator atenuador de video, videoatenuador
video band banda de video, videobanda
video bandwidth ancho de banda de video
video buffer almacenamiento intermedio de video
video carrier portadora de video, videoportadora
video channel canal de video, videocanal
video circuit circuito de video
video connection conexión de video
video control control de video
video converter convertidor de video
video correction device dispositivo de corrección de video
video delay line línea de retardo de video
video detection detección de video, videodetección
video detector detector de video, videodetector
video-detector filter filtro de detector de video
video discrimination discriminación de video
video discriminator discriminador de video
video disk disco de video, videodisco
video-disk system sistema de discos de video
video display visualización de video
video display terminal terminal de visualización de video
video display unit unidad de visualización de video
video filter filtro de video
video frequency frecuencia de video, videofrecuencia
video-frequency amplification amplificación de videofrecuencia
video-frequency amplifier amplificador de videofrecuencia
video-frequency band banda de videofrecuencia
video-frequency component componente de videofrecuencia
video-frequency output salida de videofrecuencia
video-frequency signal señal de videofrecuencia
video-frequency voltage tensión de videofrecuencia
video gain ganancia de video
video gain control control de ganancia de video
video generator generador de video
video integration integración de video

video integrator integrador de video
video intermediate-frequency amplifier
     amplificador de frecuencia intermedia de video
video link enlace de video, videoenlace
video magnetic head cabeza magnética de video
video masking enmascaramiento de video
video mixer mezclador de video
video modulation modulación de video
video modulator modulador de video
video monitor monitor de video
video multimarker multimarcador de video
video noise ruido de video, videorruido
video noise level nivel de ruido de video
video output salida de video
video output amplifier amplificador de salida de
     video
video passband banda de paso de video
video pentode pentodo de video
video pickup captación de video, videocaptación
video player magnetoscopio, videograbadora,
     reproductor de cintas magnéticas de video
video power potencia de video
video preamplifier preamplificador de video
video pulse impulso de video
video pulse train tren de impulsos de video
video random-access memory memoria de acceso
     aleatorio de video
video receiver receptor de video, videorreceptor
video recording registro de video, grabación de
     video, videorregistro
video response respuesta de video
video screen pantalla de video
video signal señal de video, videoseñal
video-signal component componente de señal de
     video
video-signal monitor monitor de señal de video
video spectrum espectro de video
video stage etapa de video
video stretching alargamiento de video
video switcher conmutador de video
video switching conmutación de video
video synthesizer sintetizador de video
video tape videocinta, cinta de video
video-tape machine magnetoscopio, videograbadora,
     grabadora de cintas magnéticas de video
video-tape recorder magnetoscopio, videograbadora,
     grabadora de cintas magnéticas de video
video-tape recording grabación en cinta magnética
     de video, registro en cinta magnética de video
video-taping grabación en cinta magnética de video,
     registro en cinta magnética de video
video terminal terminal de video, videoterminal
video transmitter transmisor de video,
     videotransmisor
video tube tubo de video
video waveform forma de onda de video
videocassette recorder magnetoscopio,
     videograbadora, registrador de videocasete,
     grabadora de cintas magnéticas de video
videocast videodifusión
videocaster videodifusor
videodisk videodisco
videodisk recorder registrador de videodiscos
videonics videónica
videophone videófono
videotape videocinta, cinta de video
videotape machine magnetoscopio, videograbadora,
     grabadora de cintas magnéticas de video
videotape recorder magnetoscopio, videograbadora,
     grabadora de cintas magnéticas de video
videotape recording grabación en cinta magnética de
     video, registro en cinta magnética de video
videotex videotex

videotron videotrón
vidicon vidicón
vidicon camera cámara vidicón
vidicon system sistema vidicón
view-finder visor
viewfinder visor
viewfinder tube tubo de visor
viewing area área de visión
viewing mirror espejo de visión
viewing oscilloscope osciloscopio de visión
viewing screen pantalla de visión
viewing time tiempo de visión
Villard circuit circuito de Villard
Villari effect efecto de Villari
vinyl vinilo
vinyl-coated wire hilo revestido de vinilo
vinyl resin resina de vinilo
vinylidene chloride cloruro de vinilideno
violet ray rayo violeta
virtual virtual
virtual address dirección virtual
virtual-address area área de direcciones virtuales
virtual addressing direccionamiento virtual
virtual amperage amperaje efectivo
virtual anode ánodo virtual
virtual attribute atributo virtual
virtual card reader lectora de tarjetas virtual
virtual carrier portadora virtual
virtual cathode cátodo virtual
virtual circuit circuito virtual
virtual configuration configuración virtual
virtual current corriente virtual
virtual data datos virtuales
virtual disk disco virtual
virtual drive unidad virtual
virtual duration duración virtual
virtual field campo virtual
virtual height altura virtual
virtual image imagen virtual
virtual machine máquina virtual
virtual memory memoria virtual
virtual memory system sistema de memoria virtual
virtual memory technique técnica de memoria
     virtual
virtual mode modo virtual
virtual processing time tiempo de procesamiento
     virtual
virtual record registro virtual
virtual region región virtual
virtual relation relación virtual
virtual route ruta virtual
virtual scanner explorador virtual
virtual sequential access method método de acceso
     secuencial virtual
virtual state estado virtual
virtual storage almacenamiento virtual
virtual storage access method método de acceso al
     almacenamiento virtual
virtual storage address dirección de almacenamiento
     virtual
virtual storage management administración de
     almacenamiento virtual
virtual storage partition partición de
     almacenamiento virtual
virtual storage region región de almacenamiento
     virtual
virtual storage system sistema de almacenamiento
     virtual
virtual system sistema virtual
virtual terminal terminal virtual
virtual time tiempo virtual
virtual voltage tensión efectiva
virus virus

viscometer viscosímetro
viscosimeter viscosímetro
viscosity viscosidad
viscous-damped arm brazo con amortiguamiento
  viscoso
viscous damping amortiguamiento viscoso
viscous hysteresis histéresis viscosa
visibility visibilidad
visibility factor factor de visibilidad
visibility meter medidor de visibilidad
visible arc arco visible
visible light luz visible
visible radiation radiación visible
visible signal señal visible
visible spectrum espectro visible
vision carrier portadora de imagen
vision circuit circuito de imagen
vision frequency frecuencia de imagen
vision input entrada de imagen
vision modulation modulación de imagen
vision monitor monitor de imagen
vision receiver receptor de imagen
vision signal señal de imagen
vision transmitter transmisor de imagen
visual visual
visual alarm alarma visual
visual-aural radio range radiofaro direccional
  audiovisual
visual-aural range radiofaro direccional audiovisual
visual carrier frequency frecuencia portadora visual
visual check comprobación visual
visual communication comunicación visual
visual display visualizador, visualización
visual display terminal terminal de visualizador,
  terminal de visualización
visual display unit unidad de visualizador, unidad de
  visualización
visual indicator indicador visual
visual indicator tube tubo indicador visual
visual information información visual
visual monitoring monitoreo visual
visual photometer fotómetro visual
visual photometry fotometría visual
visual radio range radiofaro direccional visual
visual readout lectura visual
visual scanner explorador visual
visual signal señal visual
visual telegraphy telegrafía visual
visual telephony telefonía visual
visual terminal terminal visual
visual transmitter transmisor visual
visual-transmitter power potencia de transmisor
  visual
visual tuning sintonización visual
vital circuit circuito vital
vitreous vítreo
vitreous electricity electricidad vítrea
vocabulary vocabulario
vocoder vocoder
vodas vodas
voder voder
vogad vogad
voice-activated activado por voz
voice-actuated accionado por voz
voice-actuated modulator modulador accionado por
  voz
voice amplifier amplificador de voz
voice analyzer analizador de voz
voice band banda de voz
voice channel canal de voz
voice circuit circuito de voz
voice coder codificador de voz
voice coding codificación de voz

voice coil bobina móvil
voice communication comunicación vocal
voice-controlled relay relé controlado por voz
voice current corriente vocal
voice filter filtro de voz
voice frequency frecuencia de voz, frecuencia vocal
voice-frequency amplifier amplificador de
  frecuencias vocales
voice-frequency band banda de frecuencias vocales
voice-frequency carrier telegraphy telegrafía por
  portadora de frecuencia vocal
voice-frequency channel canal de frecuencia vocal
voice-frequency dialing marcación por frecuencia
  vocal
voice-frequency generator generador de frecuencias
  vocales
voice-frequency relay relé de frecuencia vocal
voice-frequency repeater repetidor de frecuencia
  vocal
voice-frequency ringing llamada de frecuencia vocal
voice-frequency signaling señalización por
  frecuencia vocal
voice-frequency system sistema de frecuencia vocal
voice-frequency telegraph channel canal telegráfico
  de frecuencia vocal
voice-frequency telegraph system sistema
  telegráfico de frecuencia vocal
voice-frequency telegraphy telegrafía de frecuencia
  vocal
voice-frequency telephony telefonía por frecuencia
  vocal
voice-frequency terminal terminal de frecuencia
  vocal
voice-grade channel canal de calidad vocal, canal de
  calidad telefónica
voice-grade circuit circuito de calidad vocal, circuito
  de calidad telefónica
voice-grade modem módem de calidad telefónica
voice modulation modulación vocal
voice-operated operado por voz
voice-operated device dispositivo operado por voz
voice-operated relay relé operado por voz
voice-operated telephone teléfono operado por voz
voice-operated transmission transmisión operada
  por voz
voice processing procesamiento de voz
voice-recognition device dispositivo de
  reconocimiento de voz
voice recorder registrador de voz
voice recording registro de voz
voice signal señal de voz
voice synthesizer sintetizador de voz
volatile volátil
volatile dynamic storage almacenamiento dinámico
  volátil
volatile file archivo volátil
volatile memory memoria volátil
volatile storage almacenamiento volátil
volatile store almacenamiento volátil
volt volt
volt-ammeter voltamperímetro
volt-ampere voltio-ampere, voltampere
volt-ampere-hour voltamperehora
volt-ampere-hour meter voltamperihorímetro
volt-ampere meter voltamperímetro
volt-ampere reactive voltampere reactivo
volt efficiency eficiencia de tensión
volt-hour volt-hora
volt-microammeter voltmicroamperímetro
volt-milliammeter voltmiliamperímetro
volt-ohm-ammeter voltohmamperímetro
volt-ohm-meter voltóhmetro
volt-ohm-milliammeter voltohmmiliamperímetro

volt-ohmmeter voltóhmetro
Volta's law ley de Volta
Volta's principle principio de Volta
Volta effect efecto de Volta
voltage voltaje, tensión
voltage-actuated accionado por tensión
voltage-actuated device dispositivo accionado por
 tensión
voltage adjuster ajustador de tensión
voltage adjustment ajuste de tensión
voltage amplification amplificación de tensión
voltage-amplification device dispositivo de
 amplificación de tensión
voltage amplifier amplificador de tensión
voltage-amplifier tube tubo amplificador de tensión
voltage-amplifying tube tubo amplificador de tensión
voltage antinode antinodo de tensión
voltage attenuation atenuación de tensión
voltage-balance relay relé accionado por diferencia
 de tensión
voltage bias polarización de tensión
voltage booster reforzador de tensión
voltage breakdown ruptura por tensión
voltage-breakdown test prueba de ruptura por
 tensión
voltage buildup acumulación de tensión
voltage calibration calibración de tensión
voltage calibrator calibrador de tensión
voltage-capacitance curve curva tensión-capacitancia
voltage circuit circuito de tensión
voltage class categoría de tensión
voltage coefficient coeficiente de tensión
voltage coefficient of capacitance coeficiente de
 tensión de capacitancia
voltage coefficient of resistance coeficiente de
 tensión de resistencia
voltage coil bobina de tensión
voltage comparator comparador de tensiones
voltage control control de tensión
voltage-controlled controlado por tensión
voltage-controlled amplifier amplificador controlado
 por tensión
voltage-controlled attenuator atenuador controlado
 por tensión
voltage-controlled blocking oscillator oscilador de
 bloqueo controlado por tensión
voltage-controlled capacitor capacitor controlado
 por tensión
voltage-controlled crystal oscillator oscilador de
 cristal controlado por tensión
voltage-controlled generator generador controlado
 por tensión
voltage-controlled magnetic amplifier amplificador
 magnético controlado por tensión
voltage-controlled oscillator oscilador controlado
 por tensión
voltage-controlled overcurrent relay relé de
 sobrecorriente controlado por tensión
voltage-controlled resistor resistor controlado por
 tensión
voltage converter convertidor de tensión
voltage corrector corrector de tensión
voltage crest cresta de tensión
voltage-current characteristic característica
 tensión-corriente
voltage-current curve curva tensión-corriente
voltage-current feedback retroalimentación
 tensión-corriente
voltage cutoff tensión de corte
voltage-dependent dependiente de la tensión
voltage-dependent capacitor capacitor dependiente
 de la tensión
voltage-dependent resistor resistor dependiente de la

 tensión
voltage detector detector de tensión
voltage-determined property propiedad determinada
 por tensión
voltage deviation desviación de tensión
voltage-difference detector detector de diferencia de
 tensión
voltage-directional relay relé polarizado
voltage discriminator discriminador de tensión
voltage distribution distribución de tensión
voltage divider divisor de tensión
voltage-divider circuit circuito divisor de tensión
voltage division división de tensión
voltage doubler doblador de tensión
voltage drop caída de tensión
voltage efficiency eficiencia de tensión
voltage endurance resistencia a la tensión
voltage-endurance test prueba de resistencia a la
 tensión
voltage factor factor de tensión
voltage-fed antenna antena alimentada en tensión
voltage feed alimentación de tensión
voltage feedback retroalimentación de tensión
voltage-feedback factor factor de retroalimentación
 de tensión
voltage fluctuation fluctuación de tensión
voltage-frequency curve curva tensión-frecuencia
voltage-frequency function función
 tensión-frecuencia
voltage gain ganancia de tensión
voltage generator generador de tensión
voltage gradient gradiente de tensión
voltage impulse impulso de tensión
voltage increment incremento de tensión
voltage indicator indicador de tensión
voltage influence influencia de tensión
voltage integrator integrador de tensión
voltage inverter inversor de tensión
voltage jump salto de tensión
voltage lag retardo de tensión
voltage lead adelanto de tensión
voltage level nivel de tensión
voltage-level difference diferencia de nivel de
 tensión
voltage limit límite de tensión
voltage limiter limitador de tensión
voltage-limiting device dispositivo limitador de
 tensión
voltage-limiting tube tubo limitador de tensión
voltage loop bucle de tensión
voltage loss pérdida de tensión
voltage maximum máximo de tensión
voltage measurement medida de tensión
voltage-measurement equipment equipo de medida
 de tensión
voltage meter medidor de tensión
voltage minimum mínimo de tensión
voltage multiplier multiplicador de tensión
voltage-multiplier circuit circuito multiplicador de
 tensión
voltage node nodo de tensión
voltage-operated operado por tensión
voltage overload sobrecarga de tensión
voltage peak pico de tensión
voltage probe sonda de tensión
voltage pulse impulso de tensión
voltage quadrupler cuadruplicador de tensión
voltage quintupler quintuplicador de tensión
voltage range intervalo de tensión
voltage-range multiplier multiplicador de alcance de
 tensión
voltage rating tensión nominal
voltage ratio razón de tensiones

voltage-ratio box caja de razones de tensión
voltage recovery recuperación de tensión
voltage recovery time tiempo de recuperación de tensión
voltage reducer reductor de tensión
voltage reference referencia de tensión
voltage-reference diode diodo de tensión de referencia
voltage-reference tube tubo de tensión de referencia
voltage reflection coefficient coeficiente de reflexión de tensión
voltage-regulating relay relé regulador de tensión
voltage-regulating transformer transformador regulador de tensión
voltage regulation regulación de tensión
voltage-regulation curve curva de regulación de tensión
voltage-regulation relay relé de regulación de tensión
voltage regulator regulador de tensión
voltage-regulator diode diodo regulador de tensión
voltage-regulator transformer transformador regulador de tensión
voltage-regulator tube tubo regulador de tensión
voltage relay relé de tensión
voltage resonance resonancia de tensión
voltage response respuesta de tensión
voltage rise subida de tensión
voltage saturation saturación de tensión
voltage selector selector de tensión
voltage sensing detección de tensión
voltage-sensing circuit circuito de detección de tensión
voltage-sensing relay relé de detección de tensión
voltage-sensitive bridge puente sensible a la tensión
voltage-sensitive capacitor capacitor sensible a la tensión
voltage-sensitive preamplifier preamplificador sensible a la tensión
voltage-sensitive resistor resistor sensible a la tensión
voltage sensitivity sensibilidad de tensión
voltage source fuente de tensión
voltage stabilization estabilización de tensión
voltage stabilizer estabilizador de tensión
voltage-stabilizer tube tubo estabilizador de tensión
voltage-stabilizing circuit circuito estabilizador de tensión
voltage-stabilizing diode diodo estabilizador de tensión
voltage-stabilizing tube tubo estabilizador de tensión
voltage standard patrón de tensión
voltage standing-wave ratio razón de ondas estacionarias de tensión
voltage supply fuente de tensión
voltage surge sobretensión
voltage-surge protector protector contra sobretensión
voltage-surge suppressor supresor de sobretensión
voltage swing oscilación de tensión
voltage test prueba de tensión
voltage to earth tensión respecto a tierra
voltage to ground tensión respecto a tierra
voltage to neutral tensión respecto a neutro
voltage transformer transformador de tensión
voltage trebler triplicador de tensión
voltage tripler triplicador de tensión
voltage tunable sintonizable por tensión
voltage-tunable magnetron magnetrón sintonizable por tensión
voltage-tunable oscillator oscilador sintonizable por tensión
voltage-tunable solistron solistrón sintonizable por tensión

voltage-tunable tube tubo sintonizable por tensión
voltage tuning sintonización por tensión
voltage-type telemeter telémetro de tipo de tensión
voltage unbalance desequilibrio de tensión
voltage-variable capacitor capacitor variable con la tensión
voltage variation variación de tensión
voltage vector vector de tensión
voltaic voltaico
voltaic battery batería voltaica
voltaic cell pila voltaica
voltaic couple par voltaico
voltaic pile pila voltaica
voltaism voltaísmo
voltameter voltámetro
voltametric voltamétrico
voltammeter voltamperímetro
voltampere voltampere
voltmeter vóltmetro, voltímetro
voltmeter-ammeter vóltmetro-amperímetro
voltmeter circuit circuito voltmétrico
voltmeter indicator indicador voltmétrico
voltmeter-millivoltmeter vóltmetro-milivóltmetro
voltmeter multiplier multiplicador de vóltmetro
voltmeter rectifier rectificador de vóltmetro
voltmeter sensitivity sensibilidad de vóltmetro
voltohmmeter voltóhmetro
volts per meter volts por metro
volume volumen
volume compensator compensador de volumen
volume compression compresión de volumen
volume compressor compresor de volumen
volume conductivity conductividad de volumen
volume control control de volumen
volume equivalent equivalente de referencia en volumen
volume expander expansor de volumen
volume expansion expansión de volumen
volume indicator indicador de volumen
volume level nivel de volumen
volume limiter limitador de volumen
volume-limiting amplifier amplificador limitador de volumen
volume meter medidor de volumen, indicador de volumen
volume range margen de volumen
volume recombination recombinación de volumen
volume recombination rate velocidad de recombinación de volumen
volume resistance resistencia de volumen
volume resistivity resistividad de volumen
volume unit unidad de volumen
volume-unit indicator indicador de unidades de volumen
volume-unit meter medidor de unidades de volumen
volume velocity velocidad de volumen
volumetric volumétrico
volumetric efficiency eficiencia volumétrica
volumetric radar radar volumétrico
von Hippel breakdown theory teoría de ruptura de von Hippel
vordac vordac
vortac vortac
vulcanized fiber fibra vulcanizada

W

W signal señal W
wafer oblea, plaquita
wafer socket base para obleas
wafer switch conmutador de obleas
Wagner earth tierra de Wagner
Wagner ground tierra de Wagner
Waidner-Burgess standard patrón de
    Waidner-Burgess
wait condition condición de espera
wait state estado de espera
wait time tiempo de espera
waiting condition condición de espera
waiting line línea de espera
waiting loop bucle de espera
waiting state estado de espera
waiting time tiempo de espera, tiempo de
    calentamiento
walkie-lookie videotransmisor portátil
walkie-talkie radioteléfono portátil
wall baffle bafle de pared
wall effect efecto de pared
wall energy energía de pared
wall housing caja de pared
wall mount montura de pared
wall-mounted montado en pared
wall outlet tomacorriente de pared
wall plug clavija de pared
wall receptacle receptáculo de pared
wall recombination recombinación de pared
wall switch conmutador de pared
wall telephone teléfono de pared
wall telephone set aparato telefónico de pared
Wallaston wire hilo de Wallaston
Wallman amplifier amplificador de Wallman
Walmsley antenna antena de Walmsley
wamoscope wamoscopio
wanted signal señal deseada
warble ululación, aullido
warble tone tono ululante
warble-tone generator generador de tono ululante
warm junction unión caliente
warm restart reiniciación caliente
warm start iniciación caliente
warm-up calentamiento
warm-up characteristic característica de
    calentamiento
warm-up interval intervalo de calentamiento
warm-up period periodo de calentamiento
warm-up time tiempo de calentamiento
warming-up calentamiento
warming-up characteristic característica de
    calentamiento
warming-up interval intervalo de calentamiento
warming-up period periodo de calentamiento
warming-up time tiempo de calentamiento
warmup calentamiento
warmup characteristic característica de
    calentamiento
warmup interval intervalo de calentamiento
warmup period periodo de calentamiento
warmup time tiempo de calentamiento

warning bell timbre de aviso, timbre de advertencia
warning circuit circuito de aviso
warning device dispositivo de aviso
warning indicator indicador de aviso
warning lamp lámpara de aviso
warning light luz de aviso
warning message mensaje de aviso
warning signal señal de aviso
warp alabeo
warpage alabeo
warped alabeado
warping alabeo
washer capacitor capacitor tipo arandela
washer resistor resistor tipo arandela
washer thermistor termistor tipo arandela
washer varistor varistor tipo arandela
waste instruction instrucción ficticia
water absorption absorción de agua
water-activated battery batería activada por agua
water adsorption adsorción de agua
water calorimeter calorímetro de agua
water capacitor capacitor de agua
water-cooled enfriado por agua
water-cooled klystron klistrón enfriado por agua
water-cooled laser láser enfriado por agua
water-cooled tube tubo enfriado por agua
water cooling enfriamiento por agua
water-cooling system sistema de enfriamiento por
    agua
water-flow alarm alarma de flujo de agua
water-flow control control de flujo de agua
water-flow switch conmutador de flujo de agua
water jacket cubierta de agua
water-level control control de nivel de agua
water-level gage indicador de nivel de agua
water load carga de agua
water power energía hidráulica
water-pressure alarm alarma de presión de agua
water-pressure control control de presión de agua
water-pressure gage indicador de presión de agua
water-pressure indicator indicador de presión de
    agua
water-pressure switch conmutador de presión de
    agua
water pump bomba de agua
water-repellent repelente al agua
water resistor resistor de agua
water rheostat reóstato de agua
water tester probador de agua
waterproof a prueba de agua, impermeable
waterproof apparatus aparato a prueba de agua
waterproof device dispositivo a prueba de agua
waterproof enclosure caja a prueba de agua
waterproof loudspeaker altavoz a prueba de agua
waterproof machine máquina a prueba de agua
watertight hermético al agua
watertight enclosure caja hermética al agua
watt watt
watt current corriente wattada
watt-decibel conversion conversión watts-decibeles
watt-hour watt-hora, watthora
watt-hour capacity capacidad en watthoras
watt-hour constant constante en watthoras
watt-hour efficiency eficiencia en watthoras
watt-hour energy energía en watthoras
watt-hour meter medidor de watthoras,
    watthorímetro
watt-second watt-segundo, wattsegundo
watt-second constant constante en wattsegundos
wattage wattaje
wattage rating wattaje nominal
watthour watt-hora, watthora
watthour capacity capacidad en watthoras

watthour constant constante en watthoras
watthour efficiency eficiencia en watthoras
watthour energy energía en watthoras
watthour meter medidor de watthoras, watthorímetro
wattless component componente deswattada
wattless current corriente deswattada
wattless power potencia deswattada
wattmeter wáttmetro
wattsecond watt-segundo, wattsegundo
wattsecond constant constante en wattsegundos
wave onda
wave absorption absorción de ondas
wave adapter adaptador de onda
wave aerial antena de ondas
wave amplitude amplitud de onda
wave analysis análisis de ondas
wave analyzer analizador de ondas
wave angle ángulo de onda
wave antenna antena de ondas
wave attenuation atenuación de onda
wave band banda de ondas, banda de frecuencias
wave beam haz de ondas
wave bounce rebote de ondas
wave clutter ecos parásitos de olas
wave collector colector de ondas
wave constant constante de onda
wave converter convertidor de onda
wave crest cresta de onda
wave current corriente ondulatoria
wave cycle ciclo de onda
wave detector detector de ondas
wave direction dirección de onda
wave distortion distorsión de onda
wave duct conducto de ondas
wave equation ecuación de onda
wave field campo ondulatorio
wave filter filtro de onda
wave form forma de onda
wave front frente de onda
wave function función de onda
wave generator generador de ondas
wave group grupo de ondas
wave-guide guíaondas, guía de ondas
wave heating calentamiento por ondas
wave impedance impedancia de onda
wave interference interferencia de ondas
wave-interference error error por interferencia de ondas
wave mechanics mecánica ondulatoria
wave meter ondámetro
wave motion moción ondulatoria
wave normal normal a la onda
wave number número de onda
wave packet paquete de ondas
wave parameter parámetro de onda
wave path trayectoria de ondas
wave period periodo de onda
wave phase fase de ondas
wave phenomenon fenómeno ondulatorio
wave polarization polarización de onda
wave propagation propagación de ondas
wave radiator radiador de ondas
wave range gama de ondas
wave receiver receptor de ondas
wave reception recepción de ondas
wave reflection reflexión de ondas
wave refraction refracción de ondas
wave shape forma de onda
wave tail cola de onda
wave telegraphy telegrafía por ondas
wave telephony telefonía por ondas
wave theory teoría ondulatoria

wave tilt inclinación de onda
wave train tren de ondas
wave trap trampa de ondas
wave trough valle de onda
wave vector vector de onda
wave velocity velocidad de onda
wave winding arrollamiento ondulado
waveband banda de ondas, banda de frecuencias
waveform forma de onda
waveform-amplitude distortion distorsión de amplitud de forma de onda
waveform analysis análisis de forma de onda
waveform analyzer analizador de forma de onda
waveform converter convertidor de forma de onda
waveform corrector corrector de forma de onda
waveform distortion distorsión de forma de onda
waveform error error de forma de onda
waveform generation generación de forma de onda
waveform generator generador de forma de onda
waveform influence influencia de forma de onda
waveform monitor monitor de forma de onda
waveform recorder registrador de forma de onda
waveform synthesizer sintetizador de forma de onda
waveform thermocouple termopar de forma de onda
wavefront frente de onda
waveguide guíaondas, guía de ondas
waveguide accelerator acelerador de guíaondas
waveguide adapter adaptador de guíaondas
waveguide apparatus aparato de guíaondas
waveguide attenuator atenuador de guíaondas
waveguide bend codo de guíaondas
waveguide calorimeter calorímetro de guíaondas
waveguide component componente de guíaondas
waveguide connector conector de guíaondas
waveguide converter convertidor de guíaondas
waveguide corner codo de guíaondas
waveguide coupling acoplamiento de guíaondas
waveguide critical dimension dimensión crítica de guíaondas
waveguide cutoff frequency frecuencia de corte de guíaondas
waveguide cutoff wavelength longitud de onda de corte de guíaondas
waveguide directional coupler acoplador direccional de guíaondas
waveguide dispersion dispersión de guíaondas
waveguide dummy load carga ficticia de guíaondas
waveguide elbow codo de guíaondas
waveguide filter filtro de guíaondas
waveguide flange brida de guíaondas
waveguide frequency meter frecuencímetro de guíaondas
waveguide gasket junta de guíaondas
waveguide impedance impedancia de guíaondas
waveguide iris iris de guíaondas
waveguide isolator aislador de guíaondas
waveguide joint junta de guíaondas
waveguide junction unión de guíaondas
waveguide lens lente de guíaondas
waveguide load carga de guíaondas
waveguide mode modo de guíaondas
waveguide-mode suppressor supresor de modo de guíaondas
waveguide phase shifter desfasador de guíaondas
waveguide post poste de guíaondas
waveguide probe sonda de guíaondas
waveguide propagation propagación por guíaondas
waveguide radiator radiador de guíaondas
waveguide reflector reflector de guíaondas
waveguide resonator resonador de guíaondas
waveguide scattering dispersión de guíaondas
waveguide seal cierre de guíaondas
waveguide section sección de guíaondas

waveguide shim junta de acoplamiento de guíaondas
waveguide shutter obturador de guíaondas
waveguide slug tuner sintonizador de manguito de guíaondas
waveguide stub adaptador de guíaondas
waveguide switch conmutador de guíaondas
waveguide system sistema de guíaondas
waveguide taper conicidad de guíaondas
waveguide tee unión en T de guíaondas
waveguide termination terminación de guíaondas
waveguide transformer transformador de guíaondas
waveguide tuner sintonizador de guíaondas
waveguide twist torcedura en guíaondas
waveguide wavelength longitud de onda de guíaondas
wavelength longitud de onda
wavelength band banda de longitudes de onda
wavelength calibration calibración de longitud de onda
wavelength constant constante de longitud de onda
wavelength range gama de longitudes de onda
wavelength shifter desplazador de longitud de onda
wavelength-to-frequency conversion conversión de longitud de onda a frecuencia
wavelength unit unidad de longitud de onda
wavemeter ondámetro
wavemeter dial cuadrante de ondámetro
waveshape forma de onda
waveshape analysis análisis de forma de onda
wavetrap trampa de ondas
wax cera
wax capacitor capacitor revestido de cera
wax-coated revestido de cera
wax-coated capacitor capacitor revestido de cera
wax-dipped capacitor capacitor inmerso en cera
wax-filled capacitor capacitor impregnado en cera
wax master original en cera
wax original original en cera
wax paper papel impregnado en cera
wax recording registro en cera
way point punto de referencia
weak battery batería débil
weak cell pila débil
weak color color débil
weak contrast contraste débil
weak coupling acoplamiento débil
weak current corriente débil
weak electrolyte electrólito débil
weak field campo débil
weak interaction interacción débil
weak magnet imán débil
weak signal señal débil
weak-signal detector detector de señal débil
weakly magnetic débilmente magnético
weakly magnetic material material débilmente magnético
weakly perturbed field campo débilmente perturbado
wearout deterioro completo
wearout failure falla por deterioro completo
wearout-failure period periodo de falla por deterioro completo
wearout point punto de deterioro completo
weather barrier barrera contra agentes atmosféricos
weather-protected protegido contra agentes atmosféricos
weather-protected machine máquina protegida contra agentes atmosféricos
weather-protected motor motor protegido contra agentes atmosféricos
weather protection protección contra agentes atmosféricos
weather radar radar meteorológico

weather-resistant resistente a agentes atmosféricos
weather satellite satélite meteorológico
weathering deterioro por agentes atmosféricos
weatherproof a prueba de agentes atmosféricos
weatherproof enclosure caja a prueba de agentes atmosféricos
weber weber
Weber's theory teoría de Weber
wedge cuña
wedge bonding unión en cuña
wedge filter filtro en cuña
Wehnelt cathode cátodo de Wehnelt
Wehnelt cylinder cilindro de Wehnelt
Wehnelt grid rejilla de Wehnelt
Wehnelt interrupter interruptor de Wehnelt
Wehnelt tube tubo de Wehnelt
Wehnelt voltage tensión de Wehnelt
weighted amplifier amplificador compensado
weighted code código ponderado
weighted current corriente ponderada
weighted current value valor de corriente ponderado
weighted distortion factor factor de distorsión ponderado
weighted level nivel ponderado
weighted noise ruido ponderado
weighted noise level nivel de ruido ponderado
weighted noise measurement medida de ruido ponderada
weighted sound level nivel de sonido ponderado
weighted value valor ponderado
weighted voltage value valor de tensión ponderado
weighting ponderación
weighting characteristic característica de ponderación
weighting factor factor de ponderación
weighting network red de ponderación
weighting resistor resistor de ponderación
Weir circuit circuito de Weir
Weiss constant constante de Weiss
Weiss theory of ferromagnetism teoría de Weiss del ferromagnetismo
Weissenberg method método de Weissenberg
weld soldadura
weld interval intervalo de soldadura
weld junction unión soldada
weld polarity polaridad de soldadura
weld time tiempo de soldadura
weld timer temporizador de soldadura
welded soldado
welded contact contacto soldado
welded joint junta soldada
welder soldadora
welding soldadura
welding control control de soldadura
welding-control circuit circuito de control de soldadura
welding current corriente de soldadura
welding cycle ciclo de soldadura
welding electrode electrodo de soldadura
welding point punto de soldadura
welding timer temporizador de soldadura
welding transformer transformador de soldadura
welding voltage tensión de soldadura
Wenner element elemento de Wenner
Wenner winding arrollamiento de Wenner
Wertheim effect efecto de Wertheim
Weston cell pila de Weston
Weston normal cell pila patrón de Weston
Weston standard cell pila patrón de Weston
wet battery batería húmeda
wet cell pila húmeda
wet contact contacto húmedo
wet electrolytic capacitor capacitor electrolítico

húmedo

wet flashover voltage tensión de contorneamiento en húmedo

wet-reed relay relé de laminillas húmedas

wet shelf life vida de estante húmeda

wet tantalum capacitor capacitor de tantalio húmedo

wetted-contact relay relé de contactos humedecidos

Wheatstone automatic system sistema automático de Wheatstone

Wheatstone automatic telegraphy telegrafía automática de Wheatstone

Wheatstone bridge puente de Wheatstone

Wheatstone instrument instrumento de Wheatstone

Wheatstone system sistema de Wheatstone

Wheatstone telegraphy telegrafía de Wheatstone

wheel printer impresora de rueda

whiffletree switch conmutador multiposición

whip antenna antena de látigo

Whippany effect efecto de Whippany

whisker bigote de gato, buscador

whistler silbido atmosférico

whistler mode modo de silbidos atmosféricos

whistler-mode propagation propagación por modo de silbidos atmosféricos

white adjustment ajuste del blanco

White circuit circuito de White

white compression compresión del blanco

white-dot pattern patrón de puntos blancos

white lamp lámpara blanca

white level nivel del blanco

white light luz blanca

white noise ruido blanco

white-noise generator generador de ruido blanco

white-noise record disco de ruido blanco

white peak pico del blanco

white radiation radiación blanca

white recording registro blanco

white saturation saturación del blanco

white signal señal blanca

white-to-black amplitude range margen de amplitud de blanco a negro

white-to-black frequency swing oscilación de frecuencia de blanco a negro

white transmission transmisión blanca

whole tone tono entero

wide-angle diffusion difusión de ángulo grande

wide-angle kinescope cinescopio de ángulo grande

wide band banda ancha

wide-band amplifier amplificador de banda ancha

wide-band antenna antena de banda ancha

wide-band axis eje de banda ancha

wide-band cable cable de banda ancha

wide-band carrier system sistema de portadora de banda ancha

wide-band channel canal de banda ancha

wide-band communications comunicaciones de banda ancha

wide-band communications system sistema de comunicaciones de banda ancha

wide-band dipole dipolo de banda ancha

wide-band generator generador de banda ancha

wide-band multichannel system sistema multicanal de banda ancha

wide-band noise ruido de banda ancha

wide-band noise signal señal de ruido de banda ancha

wide-band oscilloscope osciloscopio de banda ancha

wide-band power tube tubo de potencia de banda ancha

wide-band radio channel canal de radio de banda ancha

wide-band ratio razón de banda ancha

wide-band receiver receptor de banda ancha

wide-band repeater repetidor de banda ancha

wide-band signal señal de banda ancha

wide-band signal generator generador de señales de banda ancha

wide-band sweep barrido de banda ancha

wide-band transmitter transmisor de banda ancha

wide beam haz ancho

wide channel canal ancho

wide frequency band banda de frecuencias ancha

wide-open receiver receptor no sintonizado

wide passband banda de paso ancha

wide-range oscillator oscilador de gama ancha

wide-range reproduction reproducción de gama ancha

wide-range response respuesta de gama ancha

wideband banda ancha

wideband amplifier amplificador de banda ancha

wideband antenna antena de banda ancha

wideband axis eje de banda ancha

wideband cable cable de banda ancha

wideband carrier system sistema de portadora de banda ancha

wideband channel canal de banda ancha

wideband communications comunicaciones de banda ancha

wideband communications system sistema de comunicaciones de banda ancha

wideband dipole dipolo de banda ancha

wideband generator generador de banda ancha

wideband multichannel system sistema multicanal de banda ancha

wideband noise ruido de banda ancha

wideband noise signal señal de ruido de banda ancha

wideband oscilloscope osciloscopio de banda ancha

wideband power tube tubo de potencia de banda ancha

wideband radio channel canal de radio de banda ancha

wideband ratio razón de banda ancha

wideband receiver receptor de banda ancha

wideband repeater repetidor de banda ancha

wideband signal señal de banda ancha

wideband signal generator generador de señales de banda ancha

wideband sweep barrido de banda ancha

wideband transmitter transmisor de banda ancha

width coding codificación de anchura

width coil bobina de anchura

width control control de anchura

width potentiometer potenciómetro de anchura

Wiedemann effect efecto de Wiedemann

Wiedemann-Franz law ley de Wiedemann-Franz

Wiedemann-Franz ratio razón de Wiedemann-Franz

Wiegand effect efecto de Wiegand

Wien's displacement law ley de desplazamiento de Wien

Wien's first law primera ley de Wien

Wien's law ley de Wien

Wien's radiation law ley de radiación de Wien

Wien's second law segunda ley de Wien

Wien's third law tercera ley de Wien

Wien bridge puente de Wien

Wien-bridge distortion meter distorsiómetro de puente de Wien

Wien-bridge equivalent equivalente de puente de Wien

Wien-bridge filter filtro de puente de Wien

Wien-bridge frequency meter frecuencímetro de puente de Wien

Wien-bridge oscillator oscilador de puente de Wien

Wien capacitance bridge puente de capacitancias de Wien

Wien circuit circuito de Wien

Wien effect efecto de Wien
Wien inductance bridge puente de inductancias de Wien
Wien oscillator oscilador de Wien
willemite willemita
Williams tube tubo de Williams
Williamson amplifier amplificador de Williamson
Wilson chamber cámara de Wilson
Wilson cloud chamber cámara de niebla de Wilson
Wilson effect efecto de Wilson
Wilson electroscope electroscopio de Wilson
Wilson plate placa de Wilson
Wimshurst machine máquina de Wimshurst
wind charger cargador de viento
wind-driven battery charger cargador de baterías eólico
wind-driven generator generador eólico
winding arrollamiento, devanado, bobinado
winding coefficient coeficiente de arrollamiento
winding factor factor de arrollamiento
winding pitch paso de arrollamiento
winding ratio razón de transformación
winding wire hilo para arrollamientos
Windom antenna antena de Windom
window ventana, ventanilla
window counter tube tubo contador de ventana
window edge borde de ventana
window generator generador de ventana
window signal señal de ventana
window size tamaño de ventana
wipeout interferencia produciendo bloqueo total
wipeout area área de interferencia produciendo bloqueo total
wiper escobilla, contacto deslizante, frotador, cursor
wiping action acción deslizante
wiping contact contacto deslizante
wire alambre, hilo, conductor
wire antenna antena de hilo
wire broadcasting radiodifusión por hilo
wire-broadcasting system sistema de radiodifusión por hilo
wire brush escobilla de hilo
wire cable cable de hilo
wire communication comunicación por hilo
wire connection conexión por hilo
wire control control por hilo
wire dress arreglo óptimo de hilos
wire duct conducto de hilos
wire fuse fusible de hilo
wire gage calibre de hilos, calibrador de hilos
wire gauge calibre de hilos, calibrador de hilos
wire grid rejilla de hilo
wire insulation aislamiento de hilo
wire lead hilo de conexión
wire line línea de alambre
wire link enlace alámbrico
wire printer impresora de hilos
wire recorder registrador de hilo, grabadora de hilo
wire resistor resistor de hilo
wire splice empalme de hilos
wire stripper pelador de hilos
wire system sistema alámbrico
wire telegraphy telegrafía por hilo
wire telephony telefonía por hilo
wire-wound potentiometer potenciómetro bobinado
wire-wound resistor resistor bobinado
wire-wound rheostat reóstato bobinado
wire-wrap conexión arrollada
wire-wrap connection conexión arrollada
wire-wrap terminal terminal arrollada
wired control control alambrado
wired program programa alambrado
wired-program computer computadora de programa alambrado
wired radio radio por hilo
wired-radio receiver receptor de radio por hilo
wired-radio transmitter transmisor de radio por hilo
wireless inalámbrico, radiocomunicación
wireless communication comunicación inalámbrica
wireless device dispositivo inalámbrico
wireless microphone micrófono inalámbrico
wiretap dispositivo de intercepción de línea de comunicación, intercepción de línea de comunicación
wiretapping intercepción de línea de comunicación
wireway conducto de alambres
wirewound potentiometer potenciómetro bobinado
wirewound resistor resistor bobinado
wirewound rheostat reóstato bobinado
wiring alambrado, cableado, conexionado
wiring board tablero de conexiones
wiring capacitance capacitancia de alambrado
wiring connector conector de hilos
wiring diagram diagrama de alambrado
wiring harness arnés de conductores
wiring impedance impedancia de alambrado
wiring inductance inductancia de alambrado
wiring reactance reactancia de alambrado
wiring resistance resistencia de alambrado
withstand voltage tensión soportable sin descarga disruptiva
Witka circuit circuito de Witka
wobble modulation vobulación
wobbulation vobulación
wobbulator vobulador
Wollaston wire hilo de Wollaston
Wood's alloy aleación de Wood
Wood's lamp lámpara de Wood
Wood's metal metal de Wood
woofer altavoz de bajos
word-address format formato de dirección de palabra
word boundary límite de palabra
word format formato de palabra
word generator generador de palabras
word length longitud de palabra
word mark marca de palabra
word-oriented orientado a palabras
word processing procesamiento de palabras
word-processing program programa de procesamiento de palabras
word processor procesador de palabras
word recognizer reconocedor de palabras
word separator separador de palabras
word size tamaño de palabra
word time tiempo de palabra
word transfer transferencia de palabras
work area área de trabajo
work coil bobina de trabajo
work disk disco de trabajo
work electrodes electrodos de trabajo
work file archivo de trabajo
work flow flujo de trabajo
work-flow management administración de flujo de trabajo
work function función de trabajo
work load carga de trabajo
work register registro de trabajo
work signal señal de trabajo
work station estación de trabajo
work tape cinta de trabajo
working area área de trabajo
working circuit circuito de trabajo
working coil bobina de trabajo
working conditions condiciones de trabajo, régimen
working contact contacto de trabajo

working disk disco de trabajo
working file archivo de trabajo
working frequency frecuencia de trabajo
working load carga de trabajo
working memory memoria de trabajo
working parameter parámetro de trabajo
working range alcance útil
working routine rutina de trabajo
working space espacio de trabajo
working standard patrón de trabajo
working storage almacenamiento de trabajo
working surface superficie de trabajo
working tape cinta de trabajo
working value valor de trabajo
working voltage tensión de trabajo
working wave onda de trabajo
worst-case circuit analysis análisis de circuito en el peor caso
worst-case design diseño para operación normal aun en el peor caso
wound capacitor capacitor bobinado
wound resistor resistor bobinado
wound rotor rotor bobinado
wound-rotor induction motor motor de inducción de rotor bobinado
wound-rotor motor motor de rotor bobinado
wow gimoteo
wow and flutter gimoteo y centelleo
wow meter medidor de gimoteo
wrapping envoltura
Wratten filter filtro de Wratten
write escribir, registrar
write-access time tiempo de acceso de escritura
write authorization autorización de escritura
write control character carácter de control de escritura
write cycle ciclo de escritura
write enable habilitación de escritura
write error error de escritura
write head cabeza de escritura
write lockout bloqueo de escritura
write protect proteger contra escritura
write protected protegido contra escritura
write protection protección contra escritura
write pulse impulso de escritura
write-read head cabeza de escritura-lectura
write time tiempo de escritura
writing bar barra de impresión
writing current corriente de escritura
writing head cabeza de escritura
writing speed velocidad de escritura
Wullenweber antenna antena de Wullenweber
wye configuration configuración en estrella
wye connection conexión en estrella
wye junction unión en Y
wye network red en estrella

X address dirección X
X amplifier amplificador X
X and Z demodulation desmodulación X y Z
x-axis eje x
X band banda X
X-band oscillator oscilador de banda X
X-band radar radar en banda X
X bar barra X
X channel canal X
x coordinate coordenada x
X cut corte X
X-cut crystal cristal de corte X
X deflection deflexión X
X-deflection plate placa de deflexión X
X demodulation desmodulación X
X demodulator desmodulador X
X diode diodo X
X gain ganancia X
X line línea X
X meter medidor X
X operation operación X
X plate placa X
X position posición X
X punch perforación X
X radiation radiación X
X-ray rayo X
X-ray analysis análisis por rayos X
X-ray apparatus aparato de rayos X
X-ray crystallography cristalografía por rayos X
X-ray detecting device dispositivo detector de rayos X
X-ray diffraction difracción de rayos X
X-ray diffraction apparatus aparato de difracción de rayos X
X-ray diffraction camera cámara de difracción de rayos X
X-ray diffraction pattern patrón de difracción de rayos X
X-ray diffractometer difractómetro de rayos X
X-ray diffractometry difractometría de rayos X
X-ray excitation excitación por rayos X
X-ray film película de rayos X
X-ray fluorescence fluorescencia de rayos X
X-ray generator generador de rayos X
X-ray goniometer goniómetro de rayos X
X-ray hardness dureza de rayos X
X-ray inspection inspección por rayos X
X-ray laser láser de rayos X
X-ray machine máquina de rayos X
X-ray microanalyzer microanalizador de rayos X
X-ray microbeam microhaz de rayos X
X-ray microscope microscopio de rayos X
X-ray photograph fotografía por rayos X
X-ray spectrogoniometer espectrogoniómetro de rayos X
X-ray spectrogram espectrograma de rayos X
X-ray spectrograph espectrógrafo de rayos X
X-ray spectrometer espectrómetro de rayos X
X-ray spectrum espectro de rayos X
X-ray system sistema de rayos X
X-ray television televisión por rayos X

X-ray thickness gage medidor de espesores por rayos X
X-ray tube tubo de rayos X
X-ray unit unidad de rayos X
X unit unidad X
X wave onda X
x-y plotter trazador x-y
x-y plotting trazado x-y
x-y recorder registrador x-y
x-y recording registro x-y
xenon xenón
xenon flash lamp lámpara de destellos de xenón
xenon flash tube tubo de destellos de xenón
xenon rectifier rectificador de xenón
xenon tube tubo de xenón
xerographic xerográfico
xerographic printer impresora xerográfica
xerographic printing impresión xerográfica
xerography xerografía
xeroprinting xeroimpresión
xeroradiography xerorradiografía
XY-cut crystal cristal de corte XY

Y address dirección Y
Y amplifier amplificador Y
Y antenna antena en Y
y-axis eje y
Y bar barra Y
Y cable cable Y
Y channel canal Y
Y circulator circulador en estrella
Y component componente Y
Y-connected circuit circuito conectado en estrella
Y connection conexión en estrella
y coordinate coordenada y
Y cut corte Y
Y-cut crystal cristal de corte Y
Y deflection deflexión Y
Y-deflection plate placa de deflexión Y
Y gain ganancia Y
Y junction unión en Y
y line línea y
Y network red en Y
Y operation operación en Y
Y plate placa Y
Y position posición Y
Y punch perforación Y
Y signal señal Y
Y-Y connection conexión estrella-estrella
Yagi antenna antena de Yagi
yield rendimiento
yield map mapa de rendimiento
yoke yugo, culata
yoke coil bobina de yugo
yoke current corriente de yugo
yoke piece culata, yugo
yoke plug clavija de yugo
Young's modulus módulo de Young
ytterbium iterbio
yttria itria
yttrium itrio
yttrium aluminum garnet granate de itrio y aluminio
yttrium iron garnet granate de itrio e hierro
yttrium-iron-garnet device dispositivo de granate de itrio e hierro
yttrium-iron-garnet filter filtro de granate de itrio e hierro

Z

Z address dirección Z
Z amplifier amplificador Z
z-axis eje z
Z bar barra Z
Z channel canal Z
Z component componente Z
z coordinate coordenada z
Z cut corte Z
Z-cut crystal cristal de corte Z
Z deflection deflexión Z
Z demodulator desmodulador Z
Z gain ganancia Z
Z marker marcador Z
Z marker beacon radiobaliza Z
Z meter medidor de Z
Z-Y bridge puente Z-Y
Zamboni pile pila de Zamboni
Zeeman effect efecto de Zeeman
Zenely electroscope electroscopio de Zenely
Zener breakdown avalancha de Zener
Zener current corriente de Zener
Zener diode diodo de Zener
Zener-diode coupling element elemento de
    acoplamiento con diodo de Zener
Zener-diode regulator regulador de tensión con
    diodo de Zener
Zener-diode voltage regulation regulación de
    tensión con diodo de Zener
Zener-diode voltage regulator regulador de tensión
    con diodo de Zener
Zener effect efecto de Zener
Zener impedance impedancia de Zener
Zener region región de Zener
Zener voltage tensión de Zener
zeppelin antenna antena zepelín
zero-access memory memoria de tiempo de acceso
    cero
zero-access storage almacenamiento de tiempo de
    acceso cero
zero-address de dirección cero
zero-address computer computadora sin direcciones
zero-address instruction instrucción de dirección
    cero
zero adjust ajuste del cero
zero adjuster ajustador del cero
zero adjustment ajuste del cero
zero axis eje de cero
zero beat batido cero
zero-beat detection detección de batido cero
zero-beat detector detector de batido cero
zero-beat indicator indicador de batido cero
zero-beat method método de batido cero
zero-beat reception recepción por batido cero
zero bias polarización cero
zero-bias operation operación sin polarización
zero-bias tube tubo de polarización cero
zero bit bit cero
zero cancellation cancelación de ceros
zero capacitance capacitancia cero
zero cathode cátodo cero
zero-center ammeter amperímetro de cero central

zero-center meter medidor de cero central
zero-center microammeter microamperímetro de
    cero central
zero-center scale escala de cero central
zero-center voltmeter vóltmetro de cero central
zero charge carga cero
zero check comprobación del cero
zero complement complemento a cero
zero compression compresión de ceros
zero condition condición cero
zero constancy constancia del cero
zero crossing cruce del eje de cero
zero current corriente cero
zero-cut crystal cristal de corte cero
zero deflection deflexión cero
zero detection detección de cero
zero detector detector de cero
zero dispersion dispersión cero
zero drift deslizamiento del cero, deslizamiento cero
zero-drift error error por deslizamiento del cero
zero elimination eliminación de ceros
zero energy energía cero
zero error error cero, error por deslizamiento del
    cero
zero field emission emisión en campo uniforme
zero fill rellenar con ceros
zero frequency frecuencia cero
zero impedance impedancia cero
zero indication indicación de cero
zero indicator indicador de cero
zero inductance inductancia cero
zero input entrada cero
zero-input terminal terminal de puesta a cero
zero level nivel cero
zero-level address dirección de nivel cero
zero-level addressing direccionamiento de nivel cero
zero-level input entrada de nivel cero
zero-level output salida de nivel cero
zero-line stability estabilidad del ajuste de cero
zero load carga cero
zero-load meter medidor de carga cero
zero-load voltmeter vóltmetro de carga cero
zero-loss circuit circuito de pérdida cero
zero magnet imán para indicar cero
zero magnitude magnitud cero
zero method método de ajuste a cero
zero modulation modulación cero
zero output salida cero
zero-output terminal terminal de salida cero
zero-phase-sequence component componente
    homopolar
zero-phase-sequence protection protección
    homopolar
zero-phase-sequence relay relé homopolar
zero point punto cero
zero pole poste cero
zero position posición cero
zero potential potencial cero
zero power potencia cero
zero-power factor factor de potencia cero
zero-power resistance resistencia de disipación de
    potencia cero
zero punch perforación cero
zero reactance reactancia cero
zero reset restablecimiento a cero
zero-reset device dispositivo de restablecimiento a
    cero
zero resetting restablecimiento a cero
zero resistance resistencia cero
zero-sequence component componente homopolar
zero-sequence field impedance impedancia de
    campo homopolar
zero set ajuste del cero, puesta a cero

zero-setting device dispositivo de puesta a cero
zero shift deslizamiento del cero, deslizamiento cero
zero signal señal cero
zero-signal method método de señal cero
zero stability estabilidad del cero
zero state estado cero
zero static error error estático cero
zero subcarrier subportadora cero
zero-subcarrier chromaticity cromaticidad de subportadora cero
zero suppression supresión de ceros
zero temperature temperatura cero
zero test prueba de cero
zero time tiempo cero
zero-transmission-level point punto de nivel de transmisión cero
zero variation variación del cero, variación cero
zero voltage tensión cero
zeroing puesta a cero, ajuste del cero
zeta zeta
zeta potential potencial zeta
zigzag connection conexión en zigzag
zigzag rectifier rectificador en zigzag
zigzag reflections reflexiones en zigzag
zigzag winding arrollamiento en zigzag
zinc cinc
zinc-carbon cell pila de cinc-carbón
zinc oxide óxido de cinc
zinc-plating cincado
zinc standard cell pila patrón de cinc
zinc sulfide sulfuro de cinc
zinc telluride telururo de cinc
zincite cincita
zircon ceramic insulator aislador cerámico circónico
zirconia circonia
zirconium circonio
zirconium lamp lámpara de circonio
zone zona
zone bit bit de zona
zone blanking extinción de zona
zone boundary límite de zona
zone center centro de zona
zone control control de zona
zone digit dígito de zona
zone elimination eliminación de zona
zone leveling nivelación de zona
zone melting fusión de zona
zone of protection zona de protección
zone of silence zona de silencio
zone-position radar radar indicador de zona
zone punch perforación de zona
zone punching perforación de zona
zone purification purificación de zona
zone refining refinamiento de zona
zone selection selección de zona
zone selector selector de zona
zone suppression supresión de zona
zoom lens objetivo de distancia focal ajustable
zwitterion zwitterión

a prueba de agentes atmosféricos weatherproof
a prueba de agua waterproof
a prueba de aguanieve sleetproof
a prueba de atascamientos jamproof
a prueba de calor heatproof
a prueba de cortocircuitos short-circuitproof
a prueba de choques shockproof
a prueba de error errorproof
a prueba de fuego fireproof
a prueba de fugas leakproof
a prueba de gas gasproof
a prueba de goteo dripproof
a prueba de grasa greaseproof
a prueba de humedad humidityproof, moistureproof
a prueba de humo smokeproof
a prueba de impactos impactproof
a prueba de impericia foolproof
a prueba de inmersión immersionproof
a prueba de lluvia rainproof
a prueba de manipulación inexperta tamperproof
a prueba de polvo dustproof
a prueba de rayos lightningproof
a prueba de ruido noiseproof
a prueba de sacudidas shakeproof
a prueba de salpicaduras splashproof
a prueba de sonido soundproof
a prueba de vapores vaporproof
a prueba del ambiente environmentproof
ábaco abacus
abampere abampere
abampere por centímetro cuadrado abampere per
  square centimeter
abandonado abandoned
abandonar abandon
abandono abandonment
abastecedor supplier
abasto supply
abcoulomb abcoulomb
abcoulomb por centímetro cuadrado abcoulomb per
  square centimeter
aberración aberration
aberración axial axial aberration
aberración cromática chromatic aberration
aberración cromática lateral lateral chromatic
  aberration
aberración cromática longitudinal longitudinal
  chromatic aberration
aberración de colores color aberration
aberración esférica spherical aberration
aberración esférica longitudinal longitudinal
  spherical aberration
aberración focal focal aberration
aberración monocromática monochromatic
  aberration
abertura aperture, opening, hole, gap, tripping,
  break
abertura angular angular aperture
abertura auroral auroral opening
abertura automática automatic opening, automatic
  tripping
abertura de acoplamiento coupling aperture,

coupling hole
abertura de alimentación feed gap
abertura de bocina horn aperture
abertura de dispersión scattering aperture
abertura de exploración scanning aperture
abertura de haz beam hole
abertura de lente lens aperture
abertura de un archivo opening of a file
abertura de ventilación vent
abertura efectiva effective aperture
abertura falsa false tripping
abertura limitadora limiting aperture
abertura máxima efectiva maximum effective
  aperture
abertura numérica numerical aperture
abertura plana plane aperture
abertura relativa relative aperture, lens speed
abertura retardada delayed open
abertura selectiva selective opening
abfarad abfarad
abhenry abhenry
abierto open
abmho abmho
abocinado flaring
abohm abohm
abohm centímetro abohm centimeter
abortar abort
abrasión abrasion
abrazadera clamp, clip, saddle
abrazadera aislante insulating clamp
abrazadera de armadura armor clamp
abrazadera de cable cable clamp
abrazadera de montaje mounting clip
abrazadera de puesta a tierra grounding clamp
abrazadera de tierra ground clamp
abreviación abbreviation
abreviado abbreviated
abreviador abbreviator
abreviar abbreviate
abridor de puerta fotoeléctrico photoelectric door
  opener
abrillantador de tubo de imagen picture-tube
  brightener
abscisa abscissa
absiemens absiemens
absoluto absolute
absorbancia absorbance
absorbedor absorber
absorbedor de microondas microwave absorber
absorbedor de oscilaciones oscillation absorber
absorbedor de picos transitorios surge absorber
absorbedor de radiación radiation absorber
absorbedor de radiactividad radioactivity absorber
absorbedor de vibraciones vibration absorber
absorbedor neutro neutral absorber
absorbedor no selectivo nonselective absorber
absorbente absorbing
absorbente de sonido sound absorber
absorber absorb
absorbido absorbed
absorciometría absorptiometry
absorciómetro absorptiometer
absorción absorption
absorción acústica acoustic absorption
absorción anormal abnormal absorption
absorción atmosférica atmospheric absorption
absorción atómica atomic absorption
absorción auroral auroral absorption
absorción de agua water absorption
absorción de calor heat absorption
absorción de Compton Compton absorption
absorción de dígitos digit absorption
absorción de energía energy absorption

absorción de humedad moisture absorption
absorción de líquido liquid absorption
absorción de luz light absorption
absorción de ondas wave absorption
absorción de partículas cargadas absorption of charged particles
absorción de producción de pares pair production absorption
absorción de radiación radiation absorption
absorción de rayos gamma gamma-ray absorption
absorción de resonancia resonance absorption
absorción de Sabine Sabine absorption
absorción de saturación saturation absorption
absorción de sonido sound absorption
absorción de tierra ground absorption
absorción dieléctrica dielectric absorption
absorción eléctrica electrical absorption
absorción equivalente equivalent absorption
absorción fotoeléctrica photoelectric absorption
absorción fotónica photon absorption
absorción ionosférica ionospheric absorption
absorción paramagnética paramagnetic absorption
absorción por desviación deviation absorption
absorción selectiva selective absorption
absorción sin desviación nondeviated absorption
absortancia absorptance
absortancia interna internal absorptance
absortancia radiante radiant absorptance
absortividad absorptivity
absortividad acústica acoustic absorptivity
abstracción abstraction
abstracción de datos data abstraction
abtesla abtesla
abundancia abundance
abvolt abvolt
abvolt por centímetro abvolt per centimeter
abwatt abwatt
abweber abweber
acabado finishing
acaparamiento de corriente current hogging
acarreo cíclico end-around carry
acarreo de alta velocidad high-speed carry
acarreo parcial partial carry
acarreo simultáneo simultaneous carry
accesibilidad accessibility
accesible accessible
acceso access, port
acceso a datos secuencial sequential data access
acceso a memoria directo direct memory access
acceso al almacenamiento storage access
acceso al almacenamiento cíclico cyclic storage access
acceso aleatorio random access
acceso combinado combined access
acceso compartido shared access
acceso controlado controlled access
acceso de almacenamiento directo direct storage access
acceso de canal channel access
acceso de grupo group access
acceso de intervalo de control control interval access
acceso de lectura-escritura read-write access
acceso de número de línea line-number access
acceso de salida outgoing access
acceso dinámico dynamic access
acceso directo direct access
acceso directo direccionado addressed direct access
acceso dividido partitioned access
acceso en común joint access
acceso en paralelo parallel access
acceso en serie serial access
acceso específico specific access

acceso indexado indexed access
acceso inmediato immediate access
acceso instantáneo instantaneous access
acceso mínimo minimum access
acceso multinivel multilevel access
acceso múltiple multiple access
acceso múltiple preasignado preassigned multiple access
acceso obstaculizado access barred
acceso rápido fast access
acceso remoto remote access
acceso retardado delayed access
acceso secuencial sequential access
acceso secuencial aleatorio random sequential access
acceso secuencial indexado indexed sequential access
acceso simultáneo simultaneous access
acceso sólo de lectura read-only access
acceso único single access
accesorio accessory, fitting
accesorio de acceso access fitting
accesorio dividido split fitting
accesorios de cables cable accessories
acción action
acción correctiva corrective action
acción de amortiguamiento damping action
acción de bloqueo blocking action
acción de cierre closing action
acción de conmutación switching action
acción de contacto momentáneo momentary-contact action
acción de control control action
acción de control integral integral control action
acción de diodo diode action
acción de disparo triggering action
acción de memoria memory action
acción de muestreo sampling action
acción de palanca crowbar
acción de pistón piston action
acción de reposición reset action
acción de retardo delay action
acción de transformador transformer action
acción de transistor transistor action
acción de válvula valve action
acción derivada derivative action
acción deslizante wiping action
acción electrolítica electrolytic action
acción equilibradora balancing action
acción filtrante filtering action
acción flotante floating action
acción galvánica galvanic action
acción integral integral action
acción intermitente intermittent action
acción inversa reverse action
acción láser lasing
acción local local action
acción momentánea momentary action
acción multinivel multilevel action
acción multiposición multiposition action
acción paso a paso step-by-step action
acción permanente permanent action
acción positiva positive action
acción progresiva progressive action
acción proporcional proportional action
acción protectora protective action
acción protectora de puerta gate protective action
acción rápida rapid action
acción recíproca reciprocal action
acción rectificadora rectifying action
acción retardada delayed action
acción rotativa rotary action
acción todo o nada on-off action
acción unidireccional unidirectional action

accionado eléctricamente electrically driven
accionado electromagnéticamente
electromagnetically driven
accionado por cinta tape-driven
accionado por computadora computer-driven
accionado por mandos command-driven
accionado por menú menu-driven
accionado por motor motor-driven
accionado por palanca lever-actuated
accionado por parámetros parameter-driven
accionado por potencia power-actuated
accionado por presión pressure-actuated
accionado por pulsador pushbutton-actuated
accionado por solenoide solenoid-actuated
accionado por tensión voltage-actuated
accionado por válvula valve-actuated
accionado por voz voice-actuated
accionamiento directo direct drive
accionamiento eléctrico electric drive
accionamiento individual individual drive
accionamiento mecánico mechanical drive
aceite aislante insulating oil
aceite de castor castor oil
aceite de transformador transformer oil
aceite mineral mineral oil
aceite pasivado passivated oil
aceleración acceleration
aceleración angular angular acceleration
aceleración automática automatic acceleration
aceleración dinámica dynamic acceleration
aceleración lineal linear acceleration
aceleración negativa negative acceleration
aceleración postdeflexión post-deflection
acceleration
aceleración programada programmed acceleration
acelerado accelerated
acelerador accelerator
acelerador analógico electrónico electronic analog
accelerator
acelerador de Cockcroft-Walton Cockcroft-Walton
accelerator
acelerador de electrones accelerator of electrons
acelerador de guíaondas waveguide accelerator
acelerador de iones positivos positive-ion
accelerator
acelerador de Lawrence Lawrence accelerator
acelerador de partículas particle accelerator
acelerador de partículas pulsado pulsed particle
accelerator
acelerador de plasma plasma accelerator
acelerador electrónico electronic accelerator,
electron accelerator
acelerador electrónico lineal linear electron
accelerator
acelerador electrostático electrostatic accelerator
acelerador iónico lineal linear ion accelerator
acelerador lineal linear accelerator
acelerador lineal electrónico electron linear
accelerator
acelerador protónico proton accelerator
acelerar accelerate
acelerógrafo accelerograph
acelerómetro accelerometer
acelerómetro angular angular accelerometer
acelerómetro capacitivo capacitive accelerometer
acelerómetro de registro recording accelerometer
acelerómetro lineal linear accelerometer
acelerómetro piezoeléctrico piezoelectric
accelerometer
acentuación accentuation
acentuación armónica harmonic accentuation
acentuado accentuated
acentuador accentuator

acentuador armónico harmonic accentuator
aceptable acceptable
aceptable por máquina machine acceptable
aceptación de carga charge acceptance
aceptación de llamada call acceptance
aceptación de ruido noise acceptance
aceptante accepting
aceptar accept
aceptor acceptor
acercamiento de Boole Boolean approach
acercamiento heurístico heuristic approach
acero steel
acero al silicio silicon steel
acero aleado alloyed steel
acero de grano orientado grain-oriented steel
acero eléctrico electrical steel
acero inoxidable pasivo passive stainless steel
acero para imanes magnet steel
acero recubierto de cobre copper-clad steel
acetato acetate
acetato de celulosa cellulose acetate
acetato de polivinilo polyvinyl acetate
ácido acid
ácido bórico boric acid
ácido de batería battery acid
ácido sulfúrico sulfuric acid
acidómetro acidometer
acierto hit
acimut azimuth
acimut magnético magnetic azimuth
acimutal azimuthal
acometida service, service line, service entrance,
service connection
acomodación accommodation
acompañamiento de contacto contact follow
acondicionador de portadora carrier conditioner
acondicionador de señal signal conditioner
acondicionamiento conditioning
acondicionamiento de aire air conditioning
acondicionamiento de archivo file conditioning
acondicionamiento de circuito circuit conditioning
acondicionamiento de señal signal conditioning
acoplado directamente directly coupled
acoplado mecánicamente ganged
acoplado por resistencia resistance-coupled
acoplado por transformador transformer-coupled
acoplador coupler
acoplador acústico acoustic coupler
acoplador atenuante attenuating coupler
acoplador bidireccional bidirectional coupler
acoplador bidireccional resistivo resistive
bidirectional coupler
acoplador de acceso access coupler
acoplador de antena antenna coupler
acoplador de antena universal universal antenna
coupler
acoplador de audífono earphone coupler
acoplador de autopiloto autopilot coupler
acoplador de Cherenkov Cherenkov coupler
acoplador de Collins Collins coupler
acoplador de Cuccia Cuccia coupler
acoplador de datos data coupler
acoplador de guíaondas óptico optical waveguide
coupler
acoplador de líneas de transmisión
transmission-line coupler
acoplador de microondas microwave coupler
acoplador de potencia power coupler
acoplador de puente bridging coupler
acoplador de tubo electrónico electron-tube coupler
acoplador direccional directional coupler
acoplador direccional coaxial coaxial directional
coupler

acoplador direccional de guíaondas waveguide directional coupler
acoplador direccional multiagujero multihole directional coupler
acoplador direccional simétrico symmetrical directional coupler
acoplador eléctrico electrical coupler
acoplador electrónico electron coupler
acoplador flexible flexible coupler
acoplador flojo loose coupler
acoplador giromagnético gyromagnetic coupler
acoplador híbrido hybrid coupler
acoplador inverso reverse coupler
acoplador multiagujero multihole coupler
acoplador óptico optical coupler
acoplador optoeléctrico optoelectric coupler
acoplador plano plain coupling
acoplador rotativo rotary coupler, rotating coupler
acoplador sintonizado tuned coupler
acoplador unidireccional unidirectional coupler
acoplador universal universal coupler
acoplamiento coupling
acoplamiento acústico acoustic coupling
acoplamiento aislado insulated coupling
acoplamiento antiparalelo antiparallel coupling
acoplamiento capacitivo capacitive coupling
acoplamiento capacitivo equilibrado balanced capacitive coupling
acoplamiento cerrado close coupling
acoplamiento común common coupling
acoplamiento conductor conductive coupling
acoplamiento crítico critical coupling
acoplamiento curvo bend coupling
acoplamiento de antena antenna coupling
acoplamiento de antena variable variable antenna coupling
acoplamiento de arrancador starter coupling
acoplamiento de arrastre magnético magnetic-drag coupling
acoplamiento de cavidad cavity coupling
acoplamiento de control control coupling
acoplamiento de corriente alterna alternating-current coupling
acoplamiento de corriente continua direct-current coupling
acoplamiento de datos data coupling
acoplamiento de desenganche rápido quick-release coupling
acoplamiento de emergencia emergency coupling
acoplamiento de entrada-salida input-output coupling
acoplamiento de eslabón link coupling
acoplamiento de guíaondas waveguide coupling
acoplamiento de haz beam coupling
acoplamiento de inductancia mutua mutual-inductance coupling
acoplamiento de línea line coupling
acoplamiento de manguito sleeve coupling
acoplamiento de modos mode coupling
acoplamiento de paso de banda bandpass coupling
acoplamiento de programa program coupling
acoplamiento de reacción reaction coupling
acoplamiento de referencia reference coupling
acoplamiento débil weak coupling
acoplamiento deslizante slipping coupling
acoplamiento directo direct coupling
acoplamiento eléctrico electrical coupling
acoplamiento electromagnético electromagnetic coupling
acoplamiento electrónico electronic coupling, electron coupling
acoplamiento electrostático electrostatic coupling
acoplamiento en continua direct coupling

acoplamiento en paralelo paralleling
acoplamiento entre etapas interstage coupling
acoplamiento extensible extensible coupling
acoplamiento fibroóptico fiber-optic coupling
acoplamiento flexible flexible coupling
acoplamiento flojo loose coupling
acoplamiento fotónico photon coupling
acoplamiento fuerte tight coupling
acoplamiento híbrido hybrid coupling
acoplamiento indirecto indirect coupling
acoplamiento inductivo inductive coupling, flux linkage
acoplamiento integral integral coupling
acoplamiento interelectródico interelectrode coupling
acoplamiento intervalvular intervalve coupling
acoplamiento magnético magnetic coupling
acoplamiento mecánico ganging
acoplamiento mixto mixed coupling
acoplamiento mutuo mutual coupling
acoplamiento negativo negative coupling
acoplamiento óptico optical coupling
acoplamiento óptimo optimum coupling
acoplamiento parafásico paraphase coupling
acoplamiento parásito parasitic coupling, stray coupling
acoplamiento por bridas flange coupling
acoplamiento por capacitancia capacitance coupling
acoplamiento por capacitor común common-capacitor coupling
acoplamiento por carga espacial space-charge coupling
acoplamiento por cátodo cathode coupling
acoplamiento por circuito resonante resonant-circuit coupling
acoplamiento por componente común common-component coupling
acoplamiento por condensador condenser coupling
acoplamiento por choque choke coupling
acoplamiento por diafonía crosstalk coupling
acoplamiento por impedancia impedance coupling
acoplamiento por impedancia común common-impedance coupling
acoplamiento por inducción induction coupling
acoplamiento por inductor común common-inductor coupling
acoplamiento por ranura slot coupling
acoplamiento por resistencia resistance coupling
acoplamiento por resistencia-capacitancia resistance-capacitance coupling
acoplamiento por resistencia directa direct resistance coupling
acoplamiento por resistor común common-resistor coupling
acoplamiento por retroalimentación feedback coupling
acoplamiento por transformador transformer coupling
acoplamiento regenerativo regenerative coupling
acoplamiento resistivo resistive coupling
acoplamiento sincrónico synchronous coupling
acoplamiento sólido solid coupling
acoplamiento transicional transitional coupling
acoplamiento unitario unity coupling
acoplamiento universal universal coupling
acoplamiento variable variable coupling
acoplamiento vernier vernier coupling
acoplar mecánicamente gang
acorazado ironclad
acortamiento de impulsos pulse shortening
acromático achromatic
acrónimo acronym
actínico actinic

actinio actinium
actinismo actinism
actinodieléctrico actinodielectric
actinoelectricidad actinoelectricity
actinómetro actinometer
actitud attitude
activación activation
activación de bloque activation of block
activación de emisor activation of emitter
activación de revestimiento coating activation
activación de sesión session activation
activación directa direct activation
activación indirecta indirect activation
activación térmica thermal activation
activado activated, enabled
activado con talio thallium-activated
activado por voz voice-activated
activador activator
activar activate, enable
actividad activity
actividad beta beta activity
actividad catódica cathode activity
actividad de cristal crystal activity
actividad de entrada input activity
actividad de entrada-salida input-output activity
actividad de fondo background activity
actividad específica specific activity
actividad ficticia dummy activity
actividad fotoquímica photochemical activity
actividad geomagnética geomagnetic activity
actividad iónica ion activity
actividad magnética magnetic activity
actividad óptica optical activity
actividad piezoeléctrica piezoelectric activity
actividad residual residual activity
actividad solar solar activity
activo active, hot, live
actuación actuation
actuación de conmutador switch actuation
actuador actuator
actuador accionado por motor motor-driven
actuador
actuador auxiliar auxiliary actuator
actuador de enfasamiento phasing actuator
actuador de relé relay actuator
actuador de teclado keyboard actuator
actuador de válvula valve actuator
actuador electrostático electrostatic actuator
actuador lineal linear actuator
actuador rotativo rotary actuator
actualización updating
actualización de archivo file updating
actualización de bucle loop update
actualización de pantalla screen update
actualización diferida deferred update
actualización retardada delayed updating
actualizado updated
actualizar update
acumulación accumulation, buildup
acumulación de campo field accumulation
acumulación de huecos hole storage
acumulación de mensajes message accumulation
acumulación de óxido oxide buildup
acumulación de tensión voltage buildup
acumulado accumulated, stored
acumulador accumulator
acumulador alcalino alkaline storage battery,
alkaline accumulator
acumulador de calor eléctrico electric heat
accumulator
acumulador de Edison Edison storage battery,
Edison accumulator
acumulador de impulsos impulse accumulator

acumulador de níquel-cadmio nickel-cadmium
storage battery, nickel-cadmium accumulator
acumulador de níquel-hierro nickel-iron
accumulator
acumulador de placas sinterizadas sintered-plate
storage battery
acumulador de plata silver storage battery
acumulador de plata-cadmio silver-cadmium
storage battery
acumulador de plata-cinc silver-zinc storage battery
acumulador de plomo lead storage battery, lead
accumulator
acumulador de plomo-cinc lead-zinc storage battery
acumulador de tarjetas card stacker
acumular accumulate, store
acumulativo accumulative
acústica acoustics
acústica de microondas microwave acoustics
acústica de sala room acoustics
acústica equivalente equivalent acoustics
acústico acoustic
acustímetro acoustimeter
acustoeléctrico acoustoelectric
acustoelectrónica acoustoelectronics
acustoóptico acoustooptic
acutancia acutance
ad hoc ad hoc
adaptabilidad adaptability
adaptación adaptation, matching
adaptación a luz light adaptation
adaptación cromática chromatic adaptation
adaptación de antena antenna matching
adaptación de carga load matching
adaptación de impedancias impedance matching
adaptación por transformador transformer
matching
adaptador adapter, stub
adaptador coaxial coaxial stub
adaptador de antena antenna adapter
adaptador de archivos integrado integrated file
adapter
adaptador de audífono headphone adapter
adaptador de bastidor rack adapter
adaptador de bucle loop adapter
adaptador de cable cable adapter
adaptador de cable universal universal cable
adapter
adaptador de canal a canal channel-to-channel
adapter
adaptador de canales channel adapter
adaptador de cinta magnética magnetic-tape adapter
adaptador de clavija plug adapter
adaptador de comunicaciones communication
adapter
adaptador de comunicaciones integrado integrated
communications adapter
adaptador de cuarto de onda quarter-wave stub
adaptador de datos data adapter
adaptador de disco disk adapter
adaptador de enlace de datos data link adapter
adaptador de frecuencia frequency adapter
adaptador de frecuencia ultraalta ultra-high
frequency adapter
adaptador de gráficos graphics adapter
adaptador de gráficos en colores color graphics
adapter
adaptador de guíaondas waveguide adapter,
waveguide stub
adaptador de impedancia impedance adapter
adaptador de impulsor de generador generator
drive adapter
adaptador de interfaz interface adapter
adaptador de interfaz de periférico peripheral

interface adapter
adaptador de línea line adapter
adaptador de línea de comunicaciones communication line adapter
adaptador de línea sincrónico synchronous line adapter
adaptador de medidor de salida output-meter adapter
adaptador de micrófono microphone adapter
adaptador de onda wave adapter
adaptador de ondas cortas shortwave adapter
adaptador de resistencia resistance adapter
adaptador de sintonización tuning stub
adaptador de transmisión transmission adapter
adaptador de válvula valve adapter
adaptador de visualización display adapter
adaptador de visualizador display adapter
adaptador de visualizador monocromo monochrome display adapter
adaptador de zócalo socket adapter
adaptador en línea straight adapter
adaptador en T tee adapter
adaptador estereofónico stereophonic adapter
adaptador fonográfico phonograph adapter
adaptador gamma gamma match
adaptador múltiplex multiplex adapter
adaptador no disipativo nondissipative stub
adaptador panorámico panoramic adapter
adaptador resistivo resistive adapter
adaptante adapting
adaptar adapt
adaptividad adaptivity
adaptivo adaptive
adelantador de fase phase advancer
adelanto de corriente current lead
adelanto de tensión voltage lead
adherencia electromagnética electromagnetic adherence
adherido efectivamente effectively bonded
adhesión adhesion
adhesión de capa a capa layer-to-layer adhesion
adhesión electrostática electrostatic adhesion
adhesivo sensible a presión pressure-sensitive adhesive
adiabático adiabatic
adiactínico adiactinic
adición addition, sum
adición de doble precisión double-precision addition
adicional additional
adimensional dimensionless
aditamento attachment
aditivo additive
aditrón additron
administración de almacenamiento storage management
administración de almacenamiento virtual virtual storage management
administración de archivos file management
administración de base de datos database management
administración de base de datos de red network database management
administración de base de datos relacional relational database management
administración de cola queue management
administración de concurrencia concurrency management
administración de confiabilidad de programas software-reliability management
administración de datos data management
administración de enlaces link management
administración de flujo de trabajo work-flow management

administración de líneas line management
administración de memoria memory management
administración de procesamiento de datos data-processing management
administración de proyecto project management
administración de proyecto de programas software-project management
administración de recursos resource management
administración de recursos de información information resource management
administración de red network management
administración de red de comunicaciones communication network management
administración de sistema system management
administración de tareas task management
administración de trabajos job management
administración de tránsito transit administration
administrado por computadora computer managed
administrador de archivos file manager
administrador de asignación de memoria memory allocation manager
administrador de canal channel manager
administrador de cola queue manager
administrador de recursos de sistema system resources manager
administrador de red network administrator
administrador de registros record manager
administrador de sistema system administrator
admisión admittance
admitancia admittance
admitancia acústica acoustic admittance
admitancia ante picos transitorios surge admittance
admitancia de carga load admittance
admitancia de electrodo electrode admittance
admitancia de entrada input admittance
admitancia de entrada efectiva effective input admittance
admitancia de entrada en cortocircuito short-circuit input admittance
admitancia de espacio gap admittance
admitancia de espacio de circuito circuit-gap admittance
admitancia de espacio electrónica electronic-gap admittance
admitancia de retroalimentación feedback admittance
admitancia de retroalimentación en cortocircuito short-circuit feedback admittance
admitancia de salida output admittance
admitancia de salida efectiva effective output admittance
admitancia de salida en cortocircuito short-circuit output admittance
admitancia de transferencia transfer admittance
admitancia directa en cortocircuito short-circuit forward admittance
admitancia en circuito abierto open-circuit admittance
admitancia en cortocircuito short-circuit admittance
admitancia imagen image admittance
admitancia impulsiva impulsive admittance
admitancia indicial indicial admittance
admitancia mecánica mechanical admittance
admitancia normalizada normalized admittance
admitancia sin carga no-load admittance
admitancia sincrónica synchronous admittance
admitancia vectorial vector admittance
adosado back-to-back
adquisición acquisition
adquisición de datos data acquisition
adquisición de datos automática automatic data acquisition
adquisición de datos de fuente source-data

acquisition
adquisición de datos y control data-acquisition and control
adsorbente adsorbent
adsorción adsorption
adsorción de agua water adsorption
adsorción de gas gas adsorption
advertencia de aeronave aircraft warning
advertencia por radar radar warning
adyacencia adjacency
adyacente adjacent
aerodinámica aerodynamics
aerodinámico aerodynamic
aerofaro aerial beacon, aerophare
aerofase aerophase
aerofísica aerophysics
aerófono aerophone
aerogenerador aerogenerator
aerograma aerogram
aeromagnético aeromagnetic
aerometeorógrafo aerometeorograph
aerometría aerometry
aerómetro aerometer
aerosol aerosol
aferente afferent
afinidad electrónica electronic affinity
agente agent
agente antiestático antistatic agent
agente catalítico catalytic agent
agente de control control agent
agente de dopado doping agent
agente de extinción quenching agent
agente de suma addition agent
agente despolarizante depolarizing agent
agente extintor quench agent
agente impuro impurity agent
agente separador stripping agent
agente solicitante requesting agent
agente tensoactivo surface-active agent
agilidad de frecuencia frequency agility
agitación térmica thermal agitation
agitador agitator
agotamiento de memoria intermedia buffer depletion
agotar deplete
agregado aggregate
agregado de datos data aggregate
agregado de señales signal aggregate
agrupación de posiciones position grouping
agrupador buncher
agrupamiento bunching, grouping
agrupamiento de archivos file batching
agrupamiento de cortacircuito circuit-breaker grouping
agrupamiento de electrones electron bunching
agrupamiento de fases phase grouping
agrupamiento de operaciones operation grouping
agrupamiento en delta delta grouping
agrupamiento en fase phase bunching
agrupamiento ideal ideal bunching
agrupamiento óptimo optimum bunching
agrupamiento óptimo angular angle optimum bunching
agua pesada heavy water
agudeza sharpness
agudización peaking
agudo sharp
agudos treble
aguja needle, stylus
aguja de clasificación sorting needle
aguja de diamante diamond stylus
aguja de fibra fiber needle
aguja de grabación recording stylus

aguja de inmersión dipping needle
aguja de registro recording stylus
aguja de reproducción playback stylus
aguja de zafiro sapphire stylus
aguja eléctrica electric needle
aguja fonocaptora pickup needle
aguja fonográfica phonograph stylus, phonograph needle
aguja magnética magnetic needle
aguja móvil moving needle
agujero hole
agujero ciego knockout
agujero de circuito circuit hole
agujero de espiga pinhole
agujero de exploración scanning hole
agujero de montaje mounting hole
agujero de perforación punch hole
agujero diminuto pinhole
agujero piloto pilot hole
ahusado tapered
ahusamiento tapering
aire a aire air-to-air
aire a superficie air-to-surface
aire a tierra air-to-ground
aire ambiente ambient air
aislado insulated, isolated
aislado con plástico plastic-insulated
aislado con politeno polythene-insulated
aislado con teflón teflon-insulated
aislado con telcoteno telcothene-insulated
aislado del sonido sound-insulated
aislador insulator, isolator
aislador cerámico ceramic insulator
aislador cerámico circónico zircon ceramic insulator
aislador coaxial coaxial isolator
aislador de aceite oil insulator
aislador de antena antenna insulator
aislador de aparato apparatus insulator
aislador de avión airplane insulator
aislador de barra colectora busbar insulator
aislador de base base insulator
aislador de bayoneta bayonet insulator
aislador de botón knob insulator
aislador de cable cable insulator
aislador de carga load isolator
aislador de carga de ferrita ferrite load isolator
aislador de carrete spool insulator
aislador de electrodo electrode insulator
aislador de entrada leading-in insulator, bushing
aislador de ferrita ferrite isolator
aislador de fusible fuse insulator
aislador de guíaondas waveguide isolator
aislador de pocas pérdidas low-loss insulator
aislador de politeno polythene insulator
aislador de porcelana porcelain insulator
aislador de resonancia resonance isolator
aislador de rotación rotation isolator
aislador de sección section insulator
aislador de suspensión suspension insulator
aislador de tensión strain insulator
aislador de terminal pin insulator
aislador de vaina de cable cable-sheath insulator
aislador de vibraciones vibration isolator
aislador de vidrio glass insulator
aislador eléctrico electric insulator
aislador lateral lateral insulator
aislador metálico metallic insulator
aislador normal standard insulator
aislador óptico optical isolator
aislador pasamuros leading-through insulator, bushing insulator
aislador pasante feedthrough insulator
aislador terminal terminal insulator

aislador tubular tubular insulator
aislamiento insulation, isolation
aislamiento absorbente de sonido sound-absorbing insulation
aislamiento con plástico plastic insulation
aislamiento con telcoteno telcothene insulation
aislamiento de banda del inducido armature band insulation
aislamiento de barra bar insulation
aislamiento de barrera de gas gas barrier insulation
aislamiento de bobina coil insulation
aislamiento de caucho rubber insulation
aislamiento de colector commutator insulation
aislamiento de conductor conductor insulation
aislamiento de conexión connection insulation
aislamiento de desconexión off isolation
aislamiento de entrada-salida input-output isolation
aislamiento de falla fault isolation
aislamiento de fibra fiber insulation
aislamiento de fibra de vidrio fiberglass insulation
aislamiento de hilo wire insulation
aislamiento de junta joint insulation
aislamiento de microcircuitos microcircuit isolation
aislamiento de paraguta paragutta insulation
aislamiento de programa program isolation
aislamiento de ranura slot insulation
aislamiento de receptor receiver isolation
aislamiento de ruido noise insulation
aislamiento de teflón teflon insulation
aislamiento de tierra ground insulation
aislamiento defectuoso faulty insulation
aislamiento deteriorado impaired insulation
aislamiento dieléctrico dielectric isolation
aislamiento efectivo effective isolation
aislamiento elastomérico elastomeric insulation
aislamiento externo external insulation
aislamiento graduado graded insulation
aislamiento impregnado impregnated insulation
aislamiento impregnado sin escurrido mass-impregnated nondrained insulation
aislamiento impregnado y escurrido mass-impregnated and drained insulation
aislamiento interelectródico interelectrode insulation
aislamiento interno internal insulation
aislamiento lineal linear insulation
aislamiento paradiafónico near-end crosstalk isolation
aislamiento preimpregnado preimpregnated insulation
aislamiento superficial surface insulation
aislamiento telediafónico far-end crosstalk isolation
aislamiento térmico thermal insulation
aislante insulant
aislante fenólico phenolic insulant
aislar insulate, isolate
ajustable adjustable
ajustador de ancho de banda bandwidth adjuster
ajustador de fase phaser
ajustador de tensión voltage adjuster
ajustador del cero zero adjuster
ajustar adjust
ajuste adjustment, setting, scaling, trimming
ajuste compensador compensating adjustment
ajuste de acimut azimuth adjustment
ajuste de carácter character adjustment
ajuste de convergencia convergence adjustment
ajuste de factor de escala scale-factor adjustment
ajuste de fase phase adjustment, phasing
ajuste de ganancia gain adjustment
ajuste de línea automático automatic line adjust
ajuste de linealidad linearity adjustment
ajuste de linealidad horizontal horizontal-linearity adjustment

ajuste de precisión precision adjustment
ajuste de presión pressure adjustment
ajuste de pureza purity adjustment
ajuste de pureza de colores color-purity adjustment
ajuste de relé relay adjustment
ajuste de sensibilidad sensitivity adjustment
ajuste de tensión voltage adjustment
ajuste de válvula valve adjustment
ajuste de velocidad speed adjustment
ajuste de velocidad automático automatic speed adjustment
ajuste de volumen continuo ride gain
ajuste del blanco white adjustment
ajuste del cero zero adjustment, zero set, zeroing
ajuste final final adjustment
ajuste fino fine adjustment, trimming
ajuste fino de imagen picture fine adjustment
ajuste grueso coarse adjustment
ajuste horizontal horizontal adjustment
ajuste inicial setup adjustment, setup
ajuste mecánico mechanical adjustment
ajuste menor minor adjustment
ajuste Q Q adjustment
ajuste sin carga no-load adjustment
al vuelo on-the-fly
alabeado warped
alabeo warping, buckling
alambrado wiring
alambrado a mano hand wiring
alambrado atado laced wiring
alambrado de punto a punto point-to-point wiring
alambrado eléctrico electric wiring
alambrado impreso printed wiring
alambrado oculto concealed wiring
alambrado prefabricado prefabricated wiring
alambre wire
alambre de trole trolley wire
alambre desnudo bare wire
alambre trenzado braided wire
alargamiento de video video stretching
alargamiento permanente permanent stretch
alarma alarm
alarma audible audible alarm
alarma audible automática automatic audible alarm
alarma automática automatic alarm
alarma contra intrusión intrusion alarm
alarma contra ladrones acústica acoustic burglar alarm
alarma de batería battery alarm
alarma de congelación frost alarm
alarma de densidad de humo smoke-density alarm
alarma de falla de alimentación power-failure alarm
alarma de fallas fault alarm
alarma de flujo de agua water-flow alarm
alarma de flujo de combustible fuel-flow alarm
alarma de flujo de fluido fluid-flow alarm
alarma de flujo de gas gas-flow alarm
alarma de flujo de líquido liquid-flow alarm
alarma de fuego fire alarm
alarma de fusible fuse alarm
alarma de humo smoke alarm
alarma de humo fotoeléctrica photoelectric smoke alarm
alarma de incendios automática automatic fire alarm
alarma de línea abierta open-line alarm
alarma de llama flame alarm
alarma de monitoreo monitoring alarm
alarma de nivel de líquido liquid-level alarm
alarma de oficina office alarm
alarma de presión de aceite oil-pressure alarm
alarma de presión de agua water-pressure alarm
alarma de presión de gas gas-pressure alarm

alarma de presión de líquido   liquid-pressure alarm
alarma de proximidad   proximity alarm
alarma de repetidor   repeater alarm
alarma de sobremodulación   overmodulation alarm
alarma de subtensión   undervoltage alarm
alarma de tiempo   time alarm
alarma eléctrica   electric alarm
alarma fotoeléctrica   photoelectric alarm
alarma nocturna   night alarm
alarma remota   remote alarm
alarma secuencial   sequential alarm
alarma térmica   thermal alarm
alarma termostática   thermostatic alarm
alarma visual   visual alarm
albedo   albedo
albedógrafo   albedograph
alcalino   alkali
alcance   reach, range
alcance cuasióptico   quasi-optical range
alcance de detección   range of detection
alcance de eco   echo range
alcance de estación   range of station
alcance de exploración   search range
alcance de frecuencias de barrido   sweep-frequency
   range
alcance de instrumento   instrument range
alcance de línea visual   line-of-sight range
alcance de operación   operating range
alcance de operación nominal   rated operating range
alcance de radar   radar range
alcance de señal   signal range
alcance de servicio   service range
alcance de transmisión   transmission range
alcance de transmisor   transmitter range
alcance diurno   daylight range
alcance efectivo   effective range
alcance en vacío   vacuum range
alcance intermedio   intermediate range
alcance nocturno   night range
alcance normal   normal range
alcance óptico   optical range
alcance útil   working range
alcance vertical   vertical range
aleación   alloy
aleación antimagnética   antimagnetic alloy
aleación basada en galio   gallium-based alloy
aleación bimetálica   bimetal
aleación binaria   binary alloy
aleación de Heusler   Heusler alloys
aleación de Wood   Wood's alloy
aleación eutéctica   eutectic alloy
aleación fusible   fusible alloy
aleación hierro-cobalto   iron-cobalt alloy
aleación hierro-níquel   iron-nickel alloy
aleación intermetálica   intermetallic alloy
aleación magnética pulverizada   powdered magnetic
   alloy
aleación no corrosiva   noncorrosive alloy
aleación no magnética   nonmagnetic alloy
aleación para soldar   soldering alloy
aleación superficial   surface alloy
aleado   alloyed
aleatoriamente   randomly
aleatoriedad   randomness
aleatorización   randomization
aleatorizado   randomized
aleatorizar   randomize
aleta   vane, fin
aletas de enfriamiento   cooling fins
alfa   alpha
alfabético-numérico   alphabetic-numeric
alfabeto contiguo   contiguous alphabet
alfabeto fonético internacional   international

phonetic alphabet
alfafotográfico   alphaphotographic
alfageométrico   alphageometric
alfanumérico   alphanumeric
alfatrón   alphatron
alfiler   pin
álgebra binaria   binary algebra
álgebra de Boole   Boolean algebra
álgebra de circuitos   circuit algebra
álgebra de conmutación   switching algebra
algorítmico   algorithmic
algoritmo   algorithm
algoritmo criptográfico   cryptographic algorithm
algoritmo de asignación de almacenamiento   storage
   allocation algorithm
algoritmo de búsqueda   search algorithm
algoritmo de colocación   placement algorithm
algoritmo de paginación   paging algorithm
algoritmo iterativo   iterative algorithm
algoritmo programado   programmed algorithm
algoritmo recursivo   recursive algorithm
algoritmo secuencial   sequential algorithm
alias   alias
alimentación   feeding, feed
alimentación a mano   hand feeding
alimentación a presión   pressure-feed
alimentación acíclica   acyclic feeding
alimentación aleatoria   random feed
alimentación automática   automatic feed
alimentación boca abajo   face-down feed
alimentación central   center feed
alimentación cíclica   cyclic feed
alimentación común   common supply
alimentación de alta tensión   high-voltage feed
alimentación de alta velocidad   high-speed feed
alimentación de antena   antenna feed
alimentación de base   base feed
alimentación de Cassegrain   Cassegrain feed
alimentación de cinta   tape feed, ribbon feed
alimentación de corriente   current supply, current
   feed
alimentación de energía   power feed
alimentación de línea   line feed
alimentación de papel   paper feed
alimentación de placa   plate supply
alimentación de tarjetas   card feeding
alimentación de tensión   voltage feed
alimentación de unidad de lectura   read-unit feed
alimentación directiva   directive feed
alimentación en derivación   shunt feed
alimentación en paralelo   parallel feed
alimentación en serie   series feed, serial feed
alimentación errónea   misfeed
alimentación fija   fixed feed
alimentación frontal   front feed
alimentación fuera de centro   off-center feed
alimentación horizontal   horizontal feed
alimentación incremental   incremental feed
alimentación manual   manual feed
alimentación no calculada   brute supply
alimentación por el extremo   end feed
alimentación por espigas   pin feed
alimentación por fricción   friction feed
alimentación por red   mains supply
alimentación por tracción   tractor feed
alimentación posterior   rear feed
alimentación primaria   primary feed
alimentación secundaria   secondary feed
alimentación vertical   vertical feed
alimentado en cuadratura   quadrature-fed
alimentado en derivación   shunt-fed
alimentado por batería   battery powered, battery-fed
alimentado por cable   cable powered

alimentado por energía solar solar-powered
alimentador feeder
alimentador abierto open feeder
alimentador bifurcado bifurcated feeder
alimentador coaxial coaxial feeder
alimentador de alumbrado lighting feeder
alimentador de antena antenna feeder
alimentador de archivos file feed
alimentador de documentos automático automatic
　　document feeder
alimentador de enlace tie feeder
alimentador de entrada input feeder
alimentador de hojas sheet feeder
alimentador de línea line feeder
alimentador de línea aérea open-wire feeder
alimentador de llegada incoming feeder
alimentador de media onda half-wave feeder
alimentador de perforadora punch feed
alimentador de red network feeder
alimentador de retorno return feeder
alimentador de salida output feeder
alimentador de tarjetas card feed
alimentador de tracción tractor feeder
alimentador directo direct feeder
alimentador en paralelo parallel feeder
alimentador independiente independent feeder
alimentador múltiple multiple feeder
alimentador primario primary feeder
alimentador principal trunk feeder
alimentador radial radial feeder
alimentador resonante resonant feeder
alimentador rotativo rotating feeder
alimentador único single feeder
alimentador vibratorio vibratory feeder
alimentadores sintonizados tuned feeders
alineación alignment
alineación de acimut azimuth alignment
alineación de aguja stylus alignment
alineación de cabeza head alignment
alineación de coma decimal decimal point alignment
alineación de contactos contact alignment
alineación de cuadro frame alignment
alineación de etiqueta label alignment
alineación de haz beam alignment
alineación de límite boundary alignment
alineación decimal decimal alignment
alineación del nivel de negro black level alignment
alineación óptica boresighting
alineador de haz beam aligner
alineamiento de circuito de rejilla grid-circuit
　　alignment
alinear align
alisado smoothing
alma core
alma de cable cable core
almacén magazine
almacén de alimentación hopper, magazine
almacén de alimentación de entrada input hopper
almacén de alimentación de salida output hopper
almacén de alimentación de tarjetas card hopper
almacenado stored
almacenado permanentemente permanently stored
almacenado por computadora computer stored
almacenador borrable erasable storage
almacenador intermedio buffer, storage buffer
almacenador intermedio de entrada de terminal
　　terminal input buffer
**almacenador intermedio de imagen de memoria de
　　núcleos** core image buffer
almacenador intermedio de línea line buffer
almacenador intermedio de perforadora de tarjetas
　　card-punch buffer
almacenador temporal de entrada y salida spooler

almacenamiento storage
almacenamiento a largo plazo long-term storage
almacenamiento acústico acoustic storage
almacenamiento ascendente push-up storage
almacenamiento asociativo associative storage
almacenamiento automático automatic storage
almacenamiento auxiliar auxiliary storage
almacenamiento básico basic storage
almacenamiento central central storage
almacenamiento cíclico cyclic storage
almacenamiento circulatorio circulating storage
almacenamiento compartido shared storage
almacenamiento común common storage
almacenamiento criogénico cryogenic storage
almacenamiento de acceso aleatorio random-access
　　storage
almacenamiento de acceso directo direct-access
　　storage
almacenamiento de acceso en serie serial-access
　　storage
almacenamiento de acceso inmediato
　　immediate-access storage
almacenamiento de acceso lento slow-access storage
almacenamiento de acceso rápido rapid-access
　　storage
almacenamiento de acceso secuencial
　　sequential-access storage
almacenamiento de alta velocidad high-speed
　　storage
almacenamiento de archivo file storage
almacenamiento de baja velocidad low-speed
　　storage
almacenamiento de bits bit storage
almacenamiento de búsqueda en paralelo
　　parallel-search storage
almacenamiento de carga charge storage
almacenamiento de cinta magnética magnetic-tape
　　storage
almacenamiento de computadora computer storage
almacenamiento de computadora digital digital
　　computer storage
almacenamiento de control control storage
almacenamiento de controlador controller storage
almacenamiento de datos data storage
almacenamiento de datos archivados archival
　　storage
almacenamiento de desbordamiento overflow
　　storage
almacenamiento de desplazamiento descendente
　　push-down store
almacenamiento de diodo diode storage
almacenamiento de disco fijo fixed-disk storage
almacenamiento de disco magnético magnetic-disk
　　storage
almacenamiento de disco reusable reusable disk
　　storage
almacenamiento de energía energy storage
almacenamiento de energía magnética magnetic
　　energy storage
almacenamiento de entrada input storage
almacenamiento de entrada-salida input-output
　　storage
almacenamiento de factores factor storage
almacenamiento de formato format storage
almacenamiento de gran capacidad large-capacity
　　storage
almacenamiento de hilo magnético magnetic wire
　　storage
almacenamiento de impresión print storage
almacenamiento de información information storage
almacenamiento de instrucciones instruction storage
almacenamiento de lámparas de neón neon-bulb
　　storage

almacenamiento de línea de retardo delay-line storage
almacenamiento de línea de retardo magnetoestrictiva magnetostrictive delay-line storage
almacenamiento de máquina machine storage
almacenamiento de memoria memory storage
almacenamiento de memoria de acceso aleatorio random-access memory storage
almacenamiento de memoria intermedia intermediate memory storage
almacenamiento de mercurio mercury storage
almacenamiento de nivel único single-level storage
almacenamiento de núcleos core storage
almacenamiento de núcleos de ferrita ferrite-core storage
almacenamiento de núcleos de gran capacidad large-capacity core storage
almacenamiento de núcleos magnéticos magnetic-core storage
almacenamiento de película magnética magnetic-film storage
almacenamiento de portador de datos data carrier storage
almacenamiento de portadores carrier storage
almacenamiento de posición position storage
almacenamiento de primer nivel first-level storage
almacenamiento de programa residente resident program storage
almacenamiento de programas program storage
almacenamiento de rayos catódicos cathode-ray storage
almacenamiento de respaldo backup storage
almacenamiento de segundo nivel second-level storage
almacenamiento de señales signal storage
almacenamiento de tambor drum storage
almacenamiento de tambor magnético magnetic-drum storage
almacenamiento de tarjetas magnéticas magnetic card storage
almacenamiento de tercer nivel third-level storage
almacenamiento de tiempo de acceso cero zero-access storage
almacenamiento de trabajo working storage
almacenamiento de tubo de rayos catódicos cathode-ray tube storage
almacenamiento de uso general general-purpose storage
almacenamiento dedicado dedicated storage
almacenamiento dinámico dynamic storage
almacenamiento dinámico permanente permanent dynamic storage
almacenamiento dinámico protegido protected dynamic storage
almacenamiento dinámico volátil volatile dynamic storage
almacenamiento direccionado por contenido content-addressed storage
almacenamiento directo direct storage
almacenamiento eléctrico electric storage
almacenamiento electrónico electronic storage
almacenamiento electrostático electrostatic storage
almacenamiento en cinta tape storage
almacenamiento en condensador condenser storage
almacenamiento en disco disk storage
almacenamiento en línea on-line storage
almacenamiento en paralelo parallel storage
almacenamiento en serie serial storage
almacenamiento ensanchado expanded storage
almacenamiento estático static storage
almacenamiento externo external storage
almacenamiento fijo fixed storage

almacenamiento físico physical storage
almacenamiento físico asignado allocated physical storage
almacenamiento fotográfico photographic storage
almacenamiento fuera de línea off-line storage
almacenamiento general general storage
almacenamiento global global storage
almacenamiento holográfico holographic storage
almacenamiento inherente inherent storage
almacenamiento instantáneo instantaneous storage
almacenamiento intermedio buffer storage, intermediate storage, buffering
almacenamiento intermedio anticipatorio anticipatory buffering
almacenamiento intermedio bidireccional bidirectional buffer
almacenamiento intermedio de alta velocidad high-speed buffer
almacenamiento intermedio de impresión print buffer
almacenamiento intermedio de intercambio exchange buffering
almacenamiento intermedio de mensajes message buffering
almacenamiento intermedio de página page buffer
almacenamiento intermedio de pantalla screen buffer
almacenamiento intermedio de texto text buffer
almacenamiento intermedio de video video buffer
almacenamiento intermedio dinámico dynamic buffering
almacenamiento intermedio doble double buffering
almacenamiento intermedio para refrescar refresh buffer
almacenamiento intermedio simple simple buffering
almacenamiento interno internal storage
almacenamiento jerárquico hierarchical storage
almacenamiento láser laser storage
almacenamiento lento slow storage
almacenamiento libre free storage
almacenamiento local local storage
almacenamiento lógico logical storage
almacenamiento lógico asignado allocated logical storage
almacenamiento magnético magnetic storage
almacenamiento magnético estático static magnetic storage
almacenamiento manual manual storage
almacenamiento masivo mass storage
almacenamiento masivo central central mass storage
almacenamiento masivo inactivo inactive mass storage
almacenamiento matricial matrix store
almacenamiento monolítico monolithic storage
almacenamiento multinivel multilevel storage
almacenamiento no borrable nonerasable storage
almacenamiento no direccionable unaddressable storage
almacenamiento no volátil nonvolatile storage
almacenamiento permanente permanent storage
almacenamiento por capacitor capacitor storage
almacenamiento por categorías category storage
almacenamiento por semiconductor semiconductor storage
almacenamiento portátil portable storage
almacenamiento primario primary storage
almacenamiento principal main storage
almacenamiento protegido protected storage
almacenamiento real real storage
almacenamiento regenerativo regenerative storage
almacenamiento residente resident storage
almacenamiento secuencial sequential storage

almacenamiento secundario secondary storage
almacenamiento sólo de lectura read-only storage
almacenamiento sólo de lectura de capacitor capacitor read only storage
almacenamiento superior upper storage
almacenamiento temporal temporary storage
almacenamiento temporal de entrada y salida spooling
almacenamiento virtual virtual storage
almacenamiento volátil volatile storage
almacenar store, save
almohadilla de presión pressure pad
alocromático allochromatic
alotrópico allotropic
alta conductividad high conductivity
alta corriente high current
alta definición high definition
alta densidad high density
alta fidelidad high fidelity
alta frecuencia high frequency
alta impedancia high impedance
alta intensidad high intensity
alta presión high pressure
alta resistencia high resistance
alta resolución high resolution
alta sensibilidad high sensitivity
alta temperatura high temperature
alta tensión high voltage, high tension
alta velocidad high speed
alta velocidad de bits high bit rate
altamente sensible hair-trigger
altas mixtas mixed highs
altavoces múltiples multiple speakers
altavoz loudspeaker, speaker, driver
altavoz a prueba de agua waterproof loudspeaker
altavoz a prueba de explosiones explosionproof loudspeaker
altavoz aislado isolated speaker
altavoz amortiguado damped loudspeaker
altavoz biaxial biaxial loudspeaker
altavoz bidireccional bidirectional loudspeaker
altavoz coaxial coaxial loudspeaker
altavoz combinado combination speaker
altavoz complementario companion loudspeaker
altavoz compuesto composite loudspeaker
altavoz de agudos treble speaker, tweeter
altavoz de agudos electrostático electrostatic tweeter
altavoz de agudos piezoeléctrico piezoelectric tweeter
altavoz de aire comprimido compressed-air speaker
altavoz de alta frecuencia high-frequency driver, high-frequency speaker
altavoz de armadura magnética magnetic-armature loudspeaker
altavoz de armadura móvil moving-armature loudspeaker
altavoz de bafle ventilado vented-baffle loudspeaker
altavoz de baja frecuencia low-frequency driver, low-frequency loudspeaker
altavoz de bajos bass loudspeaker, woofer
altavoz de bobina móvil moving-coil loudspeaker
altavoz de bocina horn loudspeaker
altavoz de campo excitado excited-field loudspeaker
altavoz de capacitor capacitor loudspeaker
altavoz de cinta ribbon loudspeaker
altavoz de columna column speaker
altavoz de condensador condenser loudspeaker
altavoz de cono cone speaker
altavoz de control control loudspeaker
altavoz de cristal crystal loudspeaker
altavoz de dos conos dual-cone speaker
altavoz de extensión extension loudspeaker
altavoz de gama completa full-range speaker

altavoz de imán permanente permanent-magnet loudspeaker
altavoz de inducción induction loudspeaker
altavoz de intercomunicación talkback speaker
altavoz de laberinto labyrinth loudspeaker
altavoz de magnetoestricción magnetostriction loudspeaker
altavoz de monitoreo monitoring loudspeaker
altavoz de radiador directo direct-radiator loudspeaker
altavoz de registro medio midrange speaker, squawker
altavoz de sustitución substitution speaker
altavoz dinámico dynamic loudspeaker
altavoz dinámico de imán permanente permanent-magnet dynamic loudspeaker
altavoz electrodinámico electrodynamic loudspeaker
altavoz electromagnético electromagnetic loudspeaker
altavoz electrostático electrostatic loudspeaker
altavoz elíptico elliptic loudspeaker
altavoz en hongo mushroom loudspeaker
altavoz externo external loudspeaker
altavoz iónico ionic loudspeaker
altavoz magnético magnetic loudspeaker
altavoz magnetoestrictivo magnetostrictive loudspeaker
altavoz megafónico public-address speaker
altavoz-micrófono loudspeaker-microphone
altavoz monitor monitor loudspeaker
altavoz multicanal multichannel loudspeaker
altavoz multicelular multicellular loudspeaker
altavoz neumático pneumatic loudspeaker
altavoz ovalado oval loudspeaker
altavoz patrón standard loudspeaker
altavoz permamagnético permamagnetic speaker
altavoz piezoeléctrico piezoelectric loudspeaker
altavoz piloto pilot loudspeaker
altavoz plano flat loudspeaker
altavoz reflector de bajos bass-reflex speaker
altavoz reflejo reflex loudspeaker
altavoz subacuático subaqueous loudspeaker
altavoz triaxial triaxial loudspeaker
altavoz unidireccional unidirectional loudspeaker
alterable alterable
alterable eléctricamente electrically alterable
alteración alteration
alterar alter
alternación alternation
alternación de cinta tape alternation
alternación de fase phase alternation
alternación desigual unequal alternation
alternaciones iguales equal alternations
alternado staggered
alternador alternator
alternador autoexcitado self-excited alternator
alternador bifásico two-phase alternator
alternador de Alexanderson Alexanderson alternator
alternador de alta frecuencia high-frequency alternator
alternador de campo giratorio revolving-field alternator
alternador de Goldschmidt Goldschmidt alternator
alternador de Latour Latour alternator
alternador de onda sinusoidal sine-wave alternator
alternador de radiofrecuencia radio-frequency alternator
alternador de reacción reaction alternator
alternador heteropolar heteropolar alternator
alternador homopolar homopolar alternator
alternador inductor inductor alternator
alternador monofásico single-phase alternator

alternador multifásico multiphase alternator
alternador polifásico polyphase alternator
alternador sincrónico synchronous alternator
alternador trifásico three-phase alternator
alternativa alternative
alternativo alternate
altímetro altimeter
altímetro absoluto absolute altimeter
altímetro aneroide aneroid altimeter
altímetro barométrico barometric altimeter
altímetro de capacitancia capacitance altimeter
altímetro de radar radar altimeter
altímetro de rayos gamma gamma-ray altimeter
altímetro de reflexión reflection altimeter
altímetro electrónico electronic altimeter
altímetro láser laser altimeter
altímetro sónico sonic altimeter
altímetro ultravioleta ultraviolet altimeter
altitud altitude
altitud absoluta absolute altitude
altitud aparente apparent altitude
altitud cuadrantal quadrantal altitude
altitud por radar radar altitude
altitud por radio radio altitude
alto contraste high contrast
alto nivel high level
alto potencial high-potential
alto vacío high vacuum
altura angular angular height
altura de antena antenna height
altura de antena efectiva effective antenna height
altura de barrera barrier height
altura de carrete bobbin height
altura de imagen picture height
altura de impulso pulse height
altura de línea line height
altura de punto spot height
altura de radiación radiation height
altura efectiva effective height
altura libre headroom
altura real actual height
altura virtual virtual height
alumbrado lighting
alumbrado continuo continuous lighting
alumbrado de aproximación approach lighting
alumbrado de emergencia emergency lighting
alumbrado de instrumentos instrument lighting
alumbrado de reserva standby lighting
alumbrado difuso general general diffused lighting
alumbrado direccional directional lighting
alumbrado directo direct lighting
alumbrado eléctrico electric lighting
alumbrado general general lighting
alumbrado indirecto indirect lighting
alumbrado individual individual lighting
alumbrado local local lighting
alumbrado portátil portable lighting
alumbrado semidirecto semidirect lighting
alumbrado suplementario supplementary lighting
alúmina alumina
aluminio aluminum
aluminio plateado silver-plated aluminum
aluminizado aluminized
amalgama amalgam
amarre tie
ambiente ambient
ambiente compartido shared environment
ambiente controlado controlled environment
ambiente de acceso access environment
ambiente de aire air environment
ambiente de base de datos database environment
ambiente de información information environment
ambiente de laboratorio laboratory environment

ambiente de servicio service environment
ambiente de tierra ground environment
ambiente de trabajo simulado simulated job environment
ambiente electromagnético electromagnetic environment
ambiente inducido induced environment
ambiente interactivo interactive environment
ambiente magnetoiónico magnetoionic environment
ambiente mixto mixed environment
ambiente natural natural environment
ambiente operacional operational environment
ambiente operativo operating environment
ambiente peligroso hazardous environment
ambiente térmico thermal environment
ambigüedad ambiguity
ambigüedad de distancia range ambiguity
ambiguo ambiguous
ambipolar ambipolar
americio americium
amorfo amorphous
amortiguación de ruido noise damping
amortiguación de sonido sound damping
amortiguación magnetomecánica magnetomechanical damping
amortiguado buffered, damped, quenched
amortiguador buffer, quencher, damper, amortisseur, shock absorber
amortiguador común common buffer
amortiguador de diodo diode damper
amortiguador de fibra fiber buffer
amortiguador de vibraciones vibration dampener
amortiguador dinámico dynamic damper
amortiguador magnético magnetic damper
amortiguamiento buffering, damping
amortiguamiento acústico acoustic damping
amortiguamiento adiabático adiabatic damping
amortiguamiento aperiódico aperiodic damping
amortiguamiento crítico critical damping
amortiguamiento de altavoz loudspeaker damping
amortiguamiento de antena antenna damping
amortiguamiento de impulsos pulse damping
amortiguamiento de Landau Landau damping
amortiguamiento de oscilaciones oscillation damping
amortiguamiento de pérdida de antena antenna loss damping
amortiguamiento de radiación de antena antenna radiation damping
amortiguamiento eléctrico electric damping
amortiguamiento electromagnético electromagnetic damping
amortiguamiento específico specific damping
amortiguamiento exponencial exponential damping
amortiguamiento magnético magnetic damping
amortiguamiento mecánico mechanical damping
amortiguamiento óptimo optimum damping
amortiguamiento periódico periodic damping
amortiguamiento por fluido fluid damping
amortiguamiento reactivo reactive damping
amortiguamiento relativo relative damping
amortiguamiento variable variable damping
amortiguamiento viscoso viscous damping
amortiguar dampen
ampacidad ampacity
amperaje amperage
amperaje efectivo virtual amperage
ampere ampere
ampere absoluto absolute ampere
ampere efectivo effective ampere
ampere-hora ampere-hour
ampere internacional international ampere
ampere-minuto ampere-minute

ampere patrón standard ampere
ampere-pie ampere-foot
ampere por metro ampere per meter
amperevuelta ampere-turn
amperevuelta antagonista back ampere-turn
amperhorímetro quantity meter, ampere-hour meter
amperímetro ammeter
amperímetro aperiódico deadbeat ammeter
amperímetro de antena antenna ammeter
amperímetro de bobina móvil moving-coil ammeter
amperímetro de bolsillo pocket ammeter
amperímetro de cero central zero-center ammeter
amperímetro de fase phase ammeter
amperímetro de franja caliente hot-strip ammeter
amperímetro de hierro articulado hinged-iron ammeter
amperímetro de hierro móvil moving-iron ammeter
amperímetro de hilo caliente hot-wire ammeter
amperímetro de registro recording ammeter
amperímetro de termopar thermocouple ammeter
amperímetro dinamométrico dynamometer ammeter
amperímetro electromagnético electromagnetic ammeter
amperímetro multiescala multirange ammeter
amperímetro óptico optical ammeter
amperímetro polarizado polarized ammeter
amperímetro térmico thermal ammeter
ampliación ampliation
amplidino amplidyne
amplificación amplification
amplificación de audio audio amplification
amplificación de banda band amplification
amplificación de banda ancha broadband amplification
amplificación de corriente current amplification
amplificación de energía energy amplification
amplificación de frecuencia intermedia intermediate-frequency amplification
amplificación de gas gas amplification
amplificación de potencia power amplification
amplificación de radiofrecuencia radio-frequency amplification
amplificación de sonido sound amplification
amplificación de tensión voltage amplification
amplificación de tensión de modo común common-mode voltage amplification
amplificación de video video amplification
amplificación de videofrecuencia video-frequency amplification
amplificación electrónica electronic amplification
amplificación en contrafase push-pull amplification
amplificación horizontal horizontal amplification
amplificación interna internal amplification
amplificación lineal linear amplification
amplificación magnética magnetic amplification
amplificación óptica optical amplification
amplificación paramétrica parametric amplification
amplificación refleja reflex amplification
amplificación regenerativa regenerative amplification
amplificación total overall amplification
amplificación vertical vertical amplification
amplificado amplified
amplificado por transistor transistor-amplified
amplificador amplifier
amplificador AB AB amplifier
amplificador acoplado por cátodo cathode-coupled amplifier
amplificador acoplado por impedancia impedance-coupled amplifier
amplificador acoplado por resistencia resistance-coupled amplifier
amplificador acoplado por resistencia-capacitancia resistance-capacitance-coupled amplifier
amplificador acoplado por transformador transformer-coupled amplifier
amplificador acústico acoustic amplifier
amplificador acústico de ondas progresivas traveling-wave acoustic amplifier
amplificador acustoeléctrico acoustoelectric amplifier
amplificador aislado isolated amplifier
amplificador aislante isolating amplifier
amplificador alimentado en serie series-fed amplifier
amplificador analógico analog amplifier
amplificador armónico harmonic amplifier
amplificador asimétrico single-ended amplifier
amplificador atómico atomic amplifier
amplificador automodulado self-modulated amplifier
amplificador básico basic amplifier
amplificador bidireccional two-way amplifier
amplificador biestable bistable amplifier
amplificador bilateral bilateral amplifier
amplificador bipolar bipolar amplifier
amplificador capacitivo capacitive amplifier
amplificador cargado por el cátodo cathode-loaded amplifier
amplificador cascodo cascode amplifier
amplificador catódico cathodic amplifier
amplificador cerámico ceramic amplifier
amplificador clase A class A amplifier
amplificador clase AB class AB amplifier
amplificador clase B class B amplifier
amplificador compensado compensated amplifier, weighted amplifier
amplificador compresor compressor amplifier
amplificador computador computing amplifier
amplificador con ánodo a masa grounded-anode amplifier
amplificador con base a masa grounded-base amplifier
amplificador con cátodo a masa grounded-cathode amplifier
amplificador con colector a masa grounded-collector amplifier
amplificador con emisor a masa grounded-emitter amplifier
amplificador con filtro de muesca notch amplifier
amplificador con fuente a masa grounded-source amplifier
amplificador con placa a masa grounded-plate amplifier
amplificador con puerta a masa grounded-gate amplifier
amplificador con reacción retroactive amplifier
amplificador con rejilla a masa grounded-grid amplifier
amplificador conformador de señal signal-shaping amplifier
amplificador continuo continuous amplifier
amplificador controlado por corriente current-controlled amplifier
amplificador controlado por puerta gated amplifier
amplificador controlado por tensión voltage-controlled amplifier
amplificador cortocircuitable shortable amplifier
amplificador cromático chromatic amplifier
amplificador cuadrador squaring amplifier
amplificador de acoplamiento directo direct-coupled amplifier
amplificador de aislamiento isolation amplifier
amplificador de alta eficiencia high-efficiency amplifier
amplificador de alta fidelidad high-fidelity amplifier

amplificador de alta frecuencia high-frequency amplifier
amplificador de alta ganancia high-gain amplifier
amplificador de altavoz loudspeaker amplifier
amplificador de antena antenna amplifier, antenna booster
amplificador de aplanamiento máximo maximal flatness amplifier
amplificador de audífono headphone amplifier
amplificador de audio audio amplifier
amplificador de audio de dos canales dual-channel audio amplifier
amplificador de audio sintonizado tuned audio amplifier
amplificador de audiofrecuencia audio-frequency amplifier
amplificador de azul blue amplifier
amplificador de baja deriva low-drift amplifier
amplificador de baja distorsión low-distortion amplifier
amplificador de baja frecuencia low-frequency amplifier
amplificador de baja potencia low-power amplifier
amplificador de bajo ruido low-noise amplifier
amplificador de banda band amplifier
amplificador de banda ancha wideband amplifier, broadband amplifier
amplificador de banda angosta narrowband amplifier
amplificador de banda base baseband amplifier
amplificador de banda base de recepción receiving baseband amplifier
amplificador de banda de octava octave-band amplifier
amplificador de barrido sweep amplifier
amplificador de barrido horizontal horizontal-sweep amplifier
amplificador de base común common-base amplifier
amplificador de bloqueo lockout amplifier
amplificador de cabeza head amplifier
amplificador de cámara camera amplifier
amplificador de campos cruzados crossed-field amplifier
amplificador de capacitor capacitor amplifier
amplificador de captor fonográfico phonograph-pickup amplifier
amplificador de carga charge amplifier
amplificador de célula fotoeléctrica photoelectric-cell amplifier
amplificador de coincidencia coincidence amplifier
amplificador de colector común common-collector amplifier
amplificador de colores color amplifier
amplificador de comparación comparison amplifier
amplificador de componentes en paralelo parallel-component amplifier
amplificador de conexión en puente bridge-connected amplifier
amplificador de convergencia convergence amplifier
amplificador de convergencia vertical vertical-convergence amplifier
amplificador de corriente current amplifier
amplificador de corriente continua direct-current amplifier
amplificador de corriente continua lineal linear direct-current amplifier
amplificador de corriente continua logarítmico logarithmic direct-current amplifier
amplificador de corriente de blanco target current amplifier
amplificador de corriente débil small-current amplifier
amplificador de cristal crystal amplifier

amplificador de croma chroma amplifier
amplificador de crominancia chrominance amplifier
amplificador de cuatro hilos four-wire amplifier
amplificador de Darlington Darlington amplifier
amplificador de décadas decade amplifier
amplificador de deflexión vertical vertical-deflection amplifier
amplificador de desacentuación deemphasis amplifier
amplificador de diapasón fork amplifier
amplificador de diente de sierra sawtooth amplifier
amplificador de diferencia difference amplifier
amplificador de diodo diode amplifier
amplificador de diodo de efecto túnel tunnel-diode amplifier
amplificador de diodo de tipo de cavidad cavity-type diode amplifier
amplificador de distribución distribution amplifier
amplificador de distribución de impulsos pulse-distribution amplifier
amplificador de doble sintonía double-tuned amplifier
amplificador de Doherty Doherty amplifier
amplificador de dos canales two-channel amplifier
amplificador de dos etapas two-stage amplifier
amplificador de dos trayectorias double-stream amplifier
amplificador de efecto de Miller Miller-effect amplifier
amplificador de elementos en paralelo en contrafase push-pull-parallel amplifier
amplificador de enganche lock-in amplifier
amplificador de entrada diferencial differential-input amplifier
amplificador de error error amplifier
amplificador de etapa única single-stage amplifier
amplificador de exploración scanning amplifier
amplificador de explorador scanner amplifier
amplificador de extinción blanking amplifier
amplificador de fase phase amplifier
amplificador de fotocélula photocell amplifier
amplificador de frecuencia frequency amplifier
amplificador de frecuencia de señal signal frequency amplifier
amplificador de frecuencia fija fixed-frequency amplifier
amplificador de frecuencia intermedia intermediate-frequency amplifier
amplificador de frecuencia intermedia de sonido sound intermediate frequency amplifier
amplificador de frecuencia intermedia de video video intermediate-frequency amplifier
amplificador de frecuencia única single-frequency amplifier
amplificador de frecuencias vocales voice-frequency amplifier
amplificador de fuente común common-source amplifier
amplificador de gamma gamma amplifier
amplificador de ganancia constante constant-gain amplifier
amplificador de ganancia estabilizada stabilized-gain amplifier
amplificador de ganancia media medium-gain amplifier
amplificador de ganancia negativa negative-gain amplifier
amplificador de ganancia plana flat-gain amplifier
amplificador de ganancia variable variable-gain amplifier
amplificador de grupo group amplifier
amplificador de Holt Holt amplifier
amplificador de imagen image amplifier

amplificador de impresión print amplifier
amplificador de impulsos pulse amplifier
amplificador de impulsos de diodo diode pulse amplifier
amplificador de impulsos lineal linear pulse amplifier
amplificador de impulsos logarítmico logarithmic pulse amplifier
amplificador de impulsos magnético en paralelo parallel magnetic pulse amplifier
amplificador de instrumentación instrumentation amplifier
amplificador de instrumento instrument amplifier
amplificador de intensidad intensity amplifier
amplificador de interfono interphone amplifier
amplificador de interrupción periódica chopper amplifier
amplificador de lámpara lamp amplifier
amplificador de Lilienfeld Lilienfeld amplifier
amplificador de limitación clipping amplifier
amplificador de línea line amplifier
amplificador de línea de imagen picture line amplifier
amplificador de línea resonante resonant-line amplifier
amplificador de línea sintonizada tuned-line amplifier
amplificador de luminancia luminance amplifier
amplificador de luz light amplifier
amplificador de luz electrónico electronic light amplifier
amplificador de llamada calling amplifier
amplificador de masa negativa negative-mass amplifier
amplificador de medida measuring amplifier
amplificador de medidor meter amplifier
amplificador de micrófono microphone amplifier
amplificador de microondas molecular molecular microwave amplifier
amplificador de microteléfono handset amplifier
amplificador de Miller Miller amplifier
amplificador de modulación de velocidad velocity-modulation amplifier
amplificador de monitoreo monitoring amplifier
amplificador de motor motor amplifier
amplificador de muestreo y retención sample-and-hold amplifier
amplificador de neodimio neodymium amplifier
amplificador de ondas acústicas acoustic-wave amplifier
amplificador de ondas cuadradas square-wave amplifier
amplificador de ondas de superficie surface-wave amplifier
amplificador de ondas estacionarias standing-wave amplifier
amplificador de ondas milimétricas millimeter-wave amplifier
amplificador de ondas progresivas traveling-wave amplifier
amplificador de par torque amplifier
amplificador de parámetros variables variable-parameter amplifier
amplificador de paso de banda bandpass amplifier
amplificador de pico peak amplifier
amplificador de portadora carrier amplifier
amplificador de portadora de imagen picture-carrier amplifier
amplificador de portadora de sonido sound-carrier amplifier
amplificador de potencia power amplifier
amplificador de potencia básico basic power amplifier

amplificador de potencia de radiofrecuencia radio-frequency power amplifier
amplificador de potencia final final power amplifier
amplificador de potencia intermedio intermediate power amplifier
amplificador de potencia lineal linear power amplifier
amplificador de premodulación premodulation amplifier
amplificador de procesamiento processing amplifier
amplificador de programa program amplifier
amplificador de puente bridge amplifier, bridging amplifier
amplificador de puerta común common-gate amplifier
amplificador de radiofrecuencia radio-frequency amplifier
amplificador de radiofrecuencia sintonizada tuned-radio-frequency amplifier
amplificador de reactancia reactance amplifier
amplificador de reactancia variable variable-reactance amplifier
amplificador de reducción de ruido noise-reduction amplifier
amplificador de referencia reference amplifier
amplificador de refuerzo booster amplifier
amplificador de registro recording amplifier
amplificador de regrabación rerecording amplifier
amplificador de reproducción playback amplifier
amplificador de rerregistro rerecording amplifier
amplificador de resistencia negativa negative-resistance amplifier
amplificador de resonancia resonance amplifier
amplificador de respuesta logarítmica log-response amplifier
amplificador de retroalimentación feedback amplifier
amplificador de retroalimentación inversa inverse-feedback amplifier
amplificador de retroalimentación múltiple multiple-feedback amplifier
amplificador de retroalimentación negativa negative-feedback amplifier
amplificador de retroalimentación positiva positive-feedback amplifier
amplificador de rojo red amplifier
amplificador de ruido noise amplifier
amplificador de ruido directo direct noise amplifier
amplificador de salida output amplifier
amplificador de salida de video video output amplifier
amplificador de seguidor anódico anode-follower amplifier
amplificador de selenio selenium amplifier
amplificador de señal de imagen picture-signal amplifier
amplificador de servomotor servomotor amplifier
amplificador de sincronización synchronizing amplifier
amplificador de sintonía única single-tuned amplifier
amplificador de sintonización escalonada stagger-tuned amplifier
amplificador de sonido sound amplifier
amplificador de subportadora subcarrier amplifier
amplificador de tensión voltage amplifier
amplificador de tensión de control control-voltage amplifier
amplificador de transconductancia transconductance amplifier
amplificador de transductor transducer amplifier
amplificador de transistores transistor amplifier
amplificador de transistores complementarios

complementary-transistor amplifier
amplificador de transistores en contrafase
  push-pull transistor amplifier
amplificador de transmisión sending amplifier
amplificador de transresistencia transresistance
  amplifier
amplificador de tres etapas three-stage amplifier
amplificador de triodos triode amplifier
amplificador de tubos tube amplifier
amplificador de tubos de vacío vacuum-tube
  amplifier
amplificador de tubos electrónicos electron-tube
  amplifier
amplificador de tubos en paralelo parallel-tube
  amplifier
amplificador de umbral threshold amplifier
amplificador de válvulas valve amplifier
amplificador de varactor varactor amplifier
amplificador de variación de velocidad
  velocity-variation amplifier
amplificador de verde green amplifier
amplificador de video video amplifier
amplificador de videofrecuencia video-frequency
  amplifier
amplificador de vidrio glass amplifier
amplificador de voz voice amplifier
amplificador de Wallman Wallman amplifier
amplificador de Williamson Williamson amplifier
amplificador degenerado degenerate amplifier
amplificador degenerativo degenerative amplifier
amplificador desfasado out-of-phase amplifier
amplificador detector detector amplifier
amplificador dieléctrico dielectric amplifier
amplificador diferenciador differentiating amplifier
amplificador diferencial differential amplifier
amplificador diferencial de osciloscopio
  oscilloscope differential amplifier
amplificador diferencial operacional operational
  differential amplifier
amplificador dinámico dynamic amplifier
amplificador dinamoeléctrico dynamoelectric
  amplifier
amplificador directo straight amplifier
amplificador discontinuo discontinuous amplifier
amplificador discriminador discriminator amplifier
amplificador distribuido distributed amplifier
amplificador distribuidor distributing amplifier
amplificador electroacústico electroacoustic
  amplifier
amplificador electromecánico electromechanical
  amplifier
amplificador electrométrico electrometer amplifier
amplificador electrónico electronic amplifier
amplificador electrostático electrostatic amplifier
amplificador en cadena chain amplifier
amplificador en cascada cascaded amplifier
amplificador en celosía lattice amplifier
amplificador en contrafase push-pull amplifier
amplificador en contrafase-paralelo push-push
  amplifier
amplificador en contrafase quiescente quiescent
  push-pull amplifier
amplificador en cuadratura quadrature amplifier
amplificador en peine comb amplifier
amplificador enchufable plug-in amplifier
amplificador equilibrado balanced amplifier
amplificador estabilizado stabilized amplifier
amplificador estabilizado por interrupción
  periódica chopper-stabilized amplifier
amplificador estabilizador stabilizing amplifier
amplificador estático static amplifier
amplificador estereofónico stereophonic amplifier
amplificador excitador driver amplifier

amplificador exponencial exponential amplifier
amplificador ferrimagnético ferrimagnetic amplifier
amplificador ferroeléctrico ferroelectric amplifier
amplificador ferromagnético ferromagnetic
  amplifier
amplificador final final amplifier
amplificador fonográfico phonograph amplifier
amplificador fonónico phonon amplifier
amplificador fotoeléctrico photoelectric amplifier
amplificador fotoparamétrico photoparametric
  amplifier
amplificador gramofónico gramophone amplifier
amplificador horizontal horizontal amplifier
amplificador integrado integrated amplifier
amplificador integrador integrating amplifier
amplificador intermedio intermediate amplifier
amplificador invertido inverted amplifier
amplificador klistrón klystron amplifier
amplificador limitador limiting amplifier
amplificador limitador de baja distorsión
  low-distortion limiting amplifier
amplificador limitador de volumen volume-limiting
  amplifier
amplificador lineal linear amplifier
amplificador lineal de alta eficiencia high-efficiency
  linear amplifier
amplificador lineal de Doherty Doherty linear
  amplifier
amplificador lineal de umbral threshold linear
  amplifier
amplificador logarítmico logarithmic amplifier
amplificador magnético magnetic amplifier
amplificador magnético controlado por tensión
  voltage-controlled magnetic amplifier
amplificador magnético de núcleo único single-core
  magnetic amplifier
amplificador magnético en contrafase push-pull
  magnetic amplifier
amplificador magnético equilibrado balanced
  magnetic amplifier
amplificador magnético multietapa multistage
  magnetic amplifier
amplificador magnético rotativo rotating magnetic
  amplifier
amplificador máser maser amplifier
amplificador megafónico public-address amplifier
amplificador mezclador mixing amplifier
amplificador modulado modulated amplifier
amplificador modulado en placa plate-modulated
  amplifier
amplificador modulado por el cátodo
  cathode-modulated amplifier
amplificador modulado por rejilla grid-modulated
  amplifier
amplificador modulador modulating amplifier
amplificador molecular molecular amplifier
amplificador monitor monitor amplifier
amplificador monoaural monaural amplifier
amplificador monocanal single-channel amplifier
amplificador multibanda multiband amplifier
amplificador multicanal multichannel amplifier
amplificador multietapa multistage amplifier
amplificador multigama multirange amplifier
amplificador multiplicador multiplier amplifier
amplificador multiplicador de frecuencia
  frequency-multiplying amplifier
amplificador neutralizado neutralized amplifier
amplificador no compensado noncompensated
  amplifier, uncompensated amplifier
amplificador no lineal nonlinear amplifier
amplificador no saturable nonsaturable amplifier
amplificador no sintonizado untuned amplifier
amplificador operacional operational amplifier

amplificador óptico optical amplifier
amplificador optoelectrónico optoelectronic amplifier
amplificador parafásico paraphase amplifier
amplificador parafásico flotante floating paraphase amplifier
amplificador paramagnético paramagnetic amplifier
amplificador paramagnético distribuido distributed paramagnetic amplifier
amplificador paramétrico parametric amplifier
amplificador paramétrico de haz beam parametric amplifier
amplificador paramétrico de ondas progresivas traveling-wave parametric amplifier
amplificador paramétrico degenerado degenerate parametric amplifier
amplificador paramétrico fotodiódico photodiode parametric amplifier
amplificador piloto pilot amplifier
amplificador polarizado biased amplifier
amplificador preliminar preliminary amplifier
amplificador proporcional proportional amplifier
amplificador receptor receiving amplifier
amplificador recortador clipper amplifier
amplificador rectificador amplifier-rectifier
amplificador rectificador de línea line amplifier-rectifier
amplificador rectificador de ruido noise amplifier-rectifier
amplificador reflejo reflex amplifier
amplificador regenerativo regenerative amplifier
amplificador relé relay amplifier
amplificador remoto remote amplifier
amplificador rotativo rotary amplifier
amplificador selectivo selective amplifier
amplificador selectivo a la frecuencia frequency-selective amplifier
amplificador semiconductor semiconductor amplifier
amplificador sensible a la fase phase-sensitive amplifier
amplificador separador separating amplifier
amplificador simétrico symmetrical amplifier
amplificador sin carga unloaded amplifier
amplificador sin carga terminal unterminated amplifier
amplificador sin diafonía no-crosstalk amplifier
amplificador sin ruido no-noise amplifier
amplificador sin transformador transformerless amplifier
amplificador sintonizado tuned amplifier
amplificador sobrecargado overloaded amplifier
amplificador sobreexcitado overdriven amplifier
amplificador subalimentado starved amplifier
amplificador subexcitado underexcited amplifier
amplificador submarino underwater amplifier
amplificador sumador adding amplifier
amplificador superheterodino superheterodyne amplifier
amplificador supersónico supersonic amplifier
amplificador T en paralelo parallel-T amplifier
amplificador telefónico telephone amplifier
amplificador terminal terminal amplifier
amplificador termoiónico thermionic amplifier
amplificador tiratrónico thyratron amplifier
amplificador transistorizado transistorized amplifier
amplificador ultralineal ultralinear amplifier
amplificador ultrasónico ultrasonic amplifier
amplificador unilateral unilateral amplifier
amplificador unitario unit amplifier
amplificador vertical vertical amplifier
amplificador X X amplifier
amplificador Y Y amplifier

amplificador Z Z amplifier
amplificante amplifying
amplificar amplify
amplitrón amplitron
amplitud amplitude, swing
amplitud de campo field amplitude
amplitud de codo bend amplitude
amplitud de crominancia chrominance amplitude
amplitud de dispersión scattering amplitude
amplitud de eco echo amplitude
amplitud de imagen picture amplitude
amplitud de impulso absoluto medio average absolute pulse amplitude
amplitud de impulso de salida output-pulse amplitude
amplitud de impulso media average pulse amplitude
amplitud de impulsos pulse amplitude
amplitud de línea line amplitude
amplitud de onda wave amplitude
amplitud de ondulación ripple amplitude
amplitud de oscilación oscillation amplitude
amplitud de pico peak amplitude
amplitud de pico a pico peak-to-peak amplitude
amplitud de portadora carrier amplitude
amplitud de presión pressure amplitude
amplitud de probabilidad probability amplitude
amplitud de resonancia resonance amplitude
amplitud de ruido amplitude of noise
amplitud de señal signal amplitude
amplitud de señal de sincronización synchronizing-signal amplitude
amplitud de un impulso parásito pulse-spike amplitude
amplitud diferencial differential amplitude
amplitud eficaz root-mean-square amplitude
amplitud espectral spectrum amplitude
amplitud instantánea instantaneous amplitude
amplitud máxima maximum amplitude
amplitud máxima de portadora maximum carrier amplitude
amplitud media average amplitude
amplitud mínima minimum amplitude
amplitud pico de impulso peak pulse amplitude
amplitud pico de portadora peak carrier amplitude
amplitud total total amplitude
amplitud variable varying amplitude
amplitud vertical vertical amplitude
ampolla bulb
ampolla de vidrio glass bulb
ampolla metalizada metal-coated bulb
anacústico anacoustic
análisis analysis
análisis armónico harmonic analysis
análisis cualitativo qualitative analysis
análisis cuantitativo quantitative analysis
análisis de algoritmos algorithm analysis
análisis de altura de impulsos pulse-height analysis
análisis de banda de octava octave-band analysis
análisis de carga de red network load analysis
análisis de circuito circuit analysis
análisis de cristal crystal analysis
análisis de datos data analysis
análisis de diseño design analysis
análisis de distribución de amplitud amplitude distribution analysis
análisis de error error analysis
análisis de estado status analysis
análisis de estrés stress analysis
análisis de falla failure analysis
análisis de falla de circuito circuit-malfunction analysis
análisis de flujo de información information flow analysis

análisis de flujos de señales signal-flow analysis
análisis de forma de onda waveform analysis
análisis de Fourier Fourier analysis
análisis de frecuencias frequency analysis
análisis de imagen image analysis
análisis de información information analysis
análisis de ondas wave analysis
análisis de operación operation analysis
análisis de problema problem analysis
análisis de procedimientos procedure analysis
análisis de proceso process analysis
análisis de pruebas test analysis
análisis de red network analysis
análisis de requisitos requirements analysis
análisis de respuesta de frecuencia
    frequency-response analysis
análisis de ruido noise analysis
análisis de sensibilidad sensitivity analysis
análisis de señal pequeña small-signal analysis
análisis de señales fuertes large-signal analysis
análisis de símbolos symbol analysis
análisis de sistemas systems analysis
análisis de tiempo de impulsos pulse-time analysis
análisis de tráfico traffic analysis
análisis de transacción transaction analysis
análisis de transformaciones transformation analysis
análisis de trayectoria path analysis
análisis de volumen de errores error volume
    analysis
análisis dimensional dimensional analysis
análisis dinámico dynamic analysis
análisis espectral spectral analysis
análisis espectrofluorométrico spectrofluorometric
    analysis
análisis espectrofotométrico spectrophotometric
    analysis
análisis estático static analysis
análisis gráfico graphical analysis
análisis lógico logic analysis
análisis magnético magnetic analysis
análisis matemático mathematical analysis
análisis matricial matrix analysis
análisis modal modal analysis
análisis nodal nodal analysis
análisis numérico numerical analysis
análisis polarográfico polarographic analysis
análisis por absorción analysis by absorption
análisis por activación activation analysis
análisis por rayos positivos positive-ray analysis
análisis por rayos X X-ray analysis
análisis potenciométrico potentiometric analysis
análisis secuencial sequential analysis
análisis simbólico symbolic analysis
análisis sintáctico parsing
analítico analytical
analizador analyzer
analizador armónico harmonic analyzer
analizador armónico heterodino heterodyne
    harmonic analyzer
analizador de altura de impulsos pulse-height
    analyzer
analizador de amplitud amplitude analyzer
analizador de amplitud de impulsos pulse-amplitude
    analyzer
analizador de audio audio analyzer
analizador de banda angosta narrowband analyzer
analizador de canales channel analyzer
analizador de circuitos circuit analyzer
analizador de diseño design analyzer
analizador de distorsión distortion analyzer
analizador de distorsión de intermodulación
    intermodulation-distortion analyzer
analizador de distorsión de propagación

    propagation distortion analyzer
analizador de distribución de tiempo
    time-distribution analyzer
analizador de envolvente envelope analyzer
analizador de espectro acústico sound-spectrum
    analyzer
analizador de forma de onda waveform analyzer
analizador de impulsos pulse analyzer
analizador de impulsos multicanal multichannel
    pulse analyzer
analizador de interferencia interference analyzer
analizador de intermodulación intermodulation
    analyzer
analizador de léxico lexical analyzer
analizador de masa mass analyzer
analizador de masas trocoidal trochoidal mass
    analyzer
analizador de microcomputadora microcomputer
    analyzer
analizador de microprocesador microprocessor
    analyzer
analizador de ondas wave analyzer
analizador de ondas heterodino heterodyne wave
    analyzer
analizador de oxígeno oxygen analyzer
analizador de programa program analyzer
analizador de radioespectro radio spectrum analyzer
analizador de reacción magnético magnetic reaction
    analyzer
analizador de red network analyzer
analizador de ruido noise analyzer
analizador de ruido de banda de octava
    octave-band noise analyzer
analizador de sentencias de control control
    statement analyzer
analizador de señal signal analyzer
analizador de sistema de microprocesador
    microprocessor system analyzer
analizador de sistemas system analyzer
analizador de sonido sound analyzer
analizador de superficies surface analyzer
analizador de tiempo time analyzer
analizador de transistores transistor analyzer
analizador de transitorios transient analyzer
analizador de vibraciones vibration analyzer
analizador de voz voice analyzer
analizador diferencial differential analyzer
analizador diferencial digital digital differential
    analyzer
analizador diferencial electrónico electronic
    differential analyzer
analizador diferencial repetitivo repetitive
    differential analyzer
analizador digital digital analyzer
analizador dinámico dynamic analyzer
analizador espectral spectrum analyzer
analizador estático static analyzer
analizador lógico logic analyzer
analizador magnético magnetic analyzer
analizador multicanal multichannel analyzer
analizador panorámico panoramic analyzer
analizador rotacional rotational analyzer
analizador secuencial sequential analyzer
analizador sintáctico parser
analizar sintácticamente parse
analógico analog
analógico a digital analog-to-digital
analógico-digital analog-digital
análogo analogous
anarmónico anharmonic
anastigmático anastigmatic
ancla anchor
anclaje anchorage

anclaje de viento guy anchor
ancho de banda bandwidth
ancho de banda autorizado authorized bandwidth
ancho de banda básico basic bandwidth
ancho de banda constante constant bandwidth
ancho de banda crítico critical bandwidth
ancho de banda de amplificador amplifier
   bandwidth
ancho de banda de antena antenna bandwidth
ancho de banda de audiofrecuencia audio-frequency
   bandwidth
ancho de banda de canal channel bandwidth
ancho de banda de canal de crominancia
   chrominance-channel bandwidth
ancho de banda de entrada-salida input-output
   bandwidth
ancho de banda de espectro de un impulso pulse
   spectrum bandwidth
ancho de banda de facsímil facsimile bandwidth
ancho de banda de fase phase bandwidth
ancho de banda de fibra fiber bandwidth
ancho de banda de frecuencias frequency bandwidth
ancho de banda de impulso pulse bandwidth
ancho de banda de información information
   bandwidth
ancho de banda de modulación modulation
   bandwidth
ancho de banda de potencia power bandwidth
ancho de banda de radiofrecuencia radio-frequency
   bandwidth
ancho de banda de recepción receiving bandwidth
ancho de banda de receptor receiver bandwidth
ancho de banda de ruido noise bandwidth
ancho de banda de ruido aleatorio random-noise
   bandwidth
ancho de banda de ruido equivalente equivalent
   noise bandwidth
ancho de banda de señal pequeña small-signal
   bandwidth
ancho de banda de transmisión transmission
   bandwidth
ancho de banda de video video bandwidth
ancho de banda dispersivo dispersive bandwidth
ancho de banda efectivo effective bandwidth
ancho de banda en bucle abierto open-loop
   bandwidth
ancho de banda equivalente de ruido noise
   equivalent bandwidth
ancho de banda espectral spectral bandwidth
ancho de banda fibroóptica fiber-optic bandwidth
ancho de banda natural natural bandwidth
ancho de banda necesario necessary bandwidth
ancho de banda nominal nominal bandwidth
ancho de banda ocupado occupied bandwidth
ancho de banda óptico optical bandwidth
ancho de banda residual residual bandwidth
ancho de banda utilizable usable bandwidth
ancho de banda variable variable bandwidth
ancho de haz de señal de bajada downlink
   beamwidth
anchura angular angular width
anchura completa a media altura full width at half
   maximum
anchura de banda pasante passband width
anchura de barrido sweep width
anchura de canal channel width
anchura de cinta tape width
anchura de haz beamwidth
anchura de haz de antena antenna beamwidth
anchura de haz de media potencia half-power
   beamwidth
anchura de haz horizontal horizontal beamwidth
anchura de imagen picture width

anchura de impulso pulse width
anchura de línea line width
anchura de línea espectral spectral line width
anchura de línea nominal nominal line width
anchura de lóbulo lobe width
anchura de media potencia half-power width
anchura de modo mode width
anchura de surco groove width
anchura de trazo trace width
anchura del entrehierro gap width
anchura espectral spectral width
anchura lateral lateral width
anchura variable variable width
anecoico anechoic
aneléctrico anelectric
anelectrotono anelectrotonus
anemógrafo anemograph
anemómetro anemometer
anemómetro de hilo caliente hot-wire anemometer
anemómetro térmico thermal anemometer
anestesia eléctrica electrical anesthesia
angstrom angstrom
angular angular
ángulo angle
ángulo anódico anode angle
ángulo característico characteristic angle
ángulo circular circular angle
ángulo crítico critical angle
ángulo de abertura aperture angle
ángulo de aceptación acceptance angle
ángulo de acimut angle of azimuth
ángulo de agrupamiento bunching angle
ángulo de agrupamiento efectivo effective bunching
   angle
ángulo de arco arc angle
ángulo de arrastre drag angle
ángulo de ataque angle of attack
ángulo de avance angle of advance
ángulo de bipartición bipartition angle
ángulo de Bragg Bragg angle
ángulo de Brewster Brewster's angle
ángulo de campo angle of field
ángulo de carga load angle
ángulo de colimación angle of collimation
ángulo de conducción conduction angle
ángulo de conmutación commutating angle
ángulo de convergencia angle of convergence
ángulo de corrección de deriva drift-correction
   angle
ángulo de corte angle of cut
ángulo de declinación angle of declination
ángulo de deflexión deflection angle
ángulo de depresión angle of depression
ángulo de deriva drift angle
ángulo de descenso angle of descent
ángulo de desplazamiento displacement angle, shift
   angle
ángulo de desviación angle of deviation, squint
ángulo de diferencia de fase angle of phase
   difference
ángulo de difracción diffraction angle
ángulo de dirección direction angle
ángulo de directividad directivity angle
ángulo de disparo firing angle
ángulo de dispersión angle of dispersion, angle of
   scattering
ángulo de divergencia angle of divergence
ángulo de eje shaft angle
ángulo de eje óptico optical axis angle
ángulo de elevación elevation angle
ángulo de excentricidad offset angle
ángulo de exploración scanning angle
ángulo de extinción angle of extinction

ángulo de fase phase angle
ángulo de fase de dieléctrico dielectric phase angle
ángulo de fase de tránsito transit phase angle
ángulo de fase efectivo effective phase angle
ángulo de flujo angle of flow
ángulo de Hall Hall angle
ángulo de haz beam angle
ángulo de ignición angle of ignition
ángulo de impedancia impedance angle
ángulo de incidencia angle of incidence
ángulo de inclinación tilt angle
ángulo de inclinación de planeo glide-slope angle
ángulo de llegada angle of arrival
ángulo de onda wave angle
ángulo de operación operating angle
ángulo de orientación orientation angle
ángulo de pantalla screen angle
ángulo de pérdida dieléctrica dielectric loss angle
ángulo de pérdida magnética magnetic loss angle
ángulo de pérdidas loss angle
ángulo de planeo glide angle
ángulo de polarización angle of polarization
ángulo de posición position angle
ángulo de protección angle of protection
ángulo de radiación radiation angle
ángulo de recubrimiento overlap angle, overlap
ángulo de referencia reference angle
ángulo de reflexión angle of reflection
ángulo de refracción angle of refraction
ángulo de retardo delay angle, angle of lag
ángulo de retardo de fase phase-lag angle
ángulo de rotación rotation angle
ángulo de salida angle of departure
ángulo de solape angle of overlap
ángulo de sombra shadow angle
ángulo de subida angle of climb
ángulo de surco groove angle
ángulo de transición transition angle
ángulo de tránsito transit angle
ángulo de trayectoria de planeo glide-path angle
ángulo diferencial differential angle
ángulo eléctrico electrical angle
ángulo eléctrico efectivo effective electric angle
ángulo esférico spherical angle
ángulo especular specular angle
ángulo flujo de corriente angle of current flow
ángulo horizontal de desviación horizontal angle of deviation
ángulo indicado indicated angle
ángulo negativo negative angle
ángulo óptico optical angle
ángulo óptimo optimum angle
ángulo óptimo de incidencia optimum angle of incidence
ángulo óptimo de radiación optimum angle of radiation
ángulo positivo positive angle
ángulo rasante grazing angle
ángulo sólido solid angle
ángulo sólido de haz beam solid angle
ángulo vertical vertical angle
ángulo visual aparente apparent visual angle
anhistéresis anhysteresis
anidado nested
anidamiento nesting
anidar nest
anillo ring, loop, ring circuit
anillo colector collector ring
anillo de almacenamiento storage ring
anillo de conexión de rejilla grid connecting ring
anillo de distancia range ring
anillo de ferrita ferrite ring
anillo de Gramme Gramme ring

anillo de guarda guard ring
anillo de protección de archivos file protection ring
anillo de retención retaining ring
anillo de retención de colector commutator retaining ring
anillo de seguridad safety ring
anillo de sombra shading ring
anillo ferrorresonante ferroresonant ring
anillo híbrido hybrid ring
anillo intensificador intensifying ring
anillo luminoso luminous ring
anillo para cables cable ring
anillo rozante slip ring
anillos de Nobili Nobili's rings
anillos eléctricos electric rings
anión anion
aniseicón aniseikon
anisotropía cristalina crystalline anisotropy
anisotropía magnética magnetic anisotropy
anisotrópico anisotropic
anodal anodal
anódicamente anodically
anódico anodic
anodización anodization, anodizing
anodización de plasma plasma anodization
anodizado anodized
anodizar anodize
ánodo anode
ánodo a cátodo anode-to-cathode
ánodo acelerador accelerating anode
ánodo acelerador de haz beam accelerating anode
ánodo auxiliar auxiliary anode, keep-alive anode
ánodo carbonizado carbonized anode
ánodo colector collector anode
ánodo de carbón carbon anode
ánodo de cebado starter anode
ánodo de cobalto cobalt anode
ánodo de derivación bypass anode
ánodo de descarga relieving anode
ánodo de disco disk anode
ánodo de enfoque focusing anode
ánodo de excitación excitation anode
ánodo de grafito graphite anode
ánodo de magnesio magnesium anode
ánodo de transición transition anode
ánodo dividido split anode
ánodo en anillo ring anode
ánodo energizado energized anode
ánodo excitador exciting anode
ánodo final final anode
ánodo galvánico galvanic anode
ánodo hueco hollow anode
ánodo mantenedor holding anode
ánodo modulador modulating anode
ánodo postacelerador post-accelerating anode
ánodo principal main anode
ánodo reactivo reactive anode
ánodo rectificador rectifier anode
ánodo rotativo rotating anode
ánodo virtual virtual anode
anólito anolyte
anomalía anomaly
anomalía de transmisión transmission anomaly
anomalístico anomalistic
anómalo anomalous
anóptico anoptic
anormal abnormal
anotación annotation
anotrón anotron
antena antenna
antena abierta open antenna
antena activa active antenna
antena alimentada en tensión voltage-fed antenna

antena alimentada fuera de centro off-center-fed antenna
antena alimentada indirectamente indirectly fed antenna
antena alimentada por corriente current-fed antenna
antena alimentada por el extremo end-fed antenna
antena altamente direccional highly directional antenna
antena-amplificador antennafier
antena angular corner antenna
antena antidesvanecimiento antifading antenna
antena antiestática antistatic antenna
antena antiinterferencia antiinterference antenna
antena apantallada screened antenna
antena aperiódica aperiodic antenna
antena apoyada por cometa kite-supported antenna
antena armónica harmonic antenna
antena artificial artificial antenna
antena bicónica biconical antenna
antena bidireccional bidirectional antenna
antena bifilar two-wire antenna
antena bilateral bilateral antenna
antena buscadora homing antenna
antena capacitiva capacitive antenna
antena cargada loaded antenna
antena cargada por el centro center-loaded antenna
antena cilíndrica cylindrical antenna
antena circular circular antenna
antena coaxial coaxial antenna
antena colectiva collective antenna, master antenna
antena colectiva de televisión television master antenna
antena colineal collinear antenna
antena con conexión a tierra grounded antenna
antena con plano de tierra ground-plane antenna
antena con reflector reflector antenna
antena con reflector angular corner-reflector antenna
antena con reflector de Cassegrain Cassegrain reflector antenna
antena con reflector plano plane-reflector antenna
antena conforme conformal antenna
antena cónica conical antenna
antena corta stub antenna
antena cruzada turnstile antenna
antena cuadrantal quadrantal antenna
antena cúbica cubical antenna
antena de abertura aperture antenna
antena de adaptación en delta delta-matched antenna
antena de Adcock Adcock antenna
antena de aeronave aircraft antenna
antena de Alexanderson Alexanderson antenna
antena de Alford Alford antenna
antena de alimentación central center-feed antenna
antena de alimentación por hilo único single-wire-fed antenna
antena de alta frecuencia high-frequency antenna
antena de alta ganancia high-gain antenna
antena de aterrizaje por instrumentos instrument-landing antenna
antena de banda ancha wideband antenna
antena de barco shipboard antenna
antena de Bellini-Tosi Bellini-Tosi antenna
antena de Beverage Beverage antenna
antena de bocina horn antenna
antena de bocina bicónica biconical horn antenna
antena de bocina cónica conical-horn antenna
antena de bocina piramidal pyramidal horn antenna
antena de bocina rectangular rectangular-horn antenna
antena de bocina sectorial sectoral horn antenna
antena de Bruce Bruce antenna

antena de campo rotativo rotating-field antenna
antena de capacidad terminal top-loaded antenna
antena de capacitancia capacitor antenna
antena de carga de base base-loaded antenna
antena de Cassegrain Cassegrain antenna
antena de Chireix-Mesny Chireix-Mesny antenna
antena de Christiansen Christiansen antenna
antena de cilindro ranurado slotted-cylinder antenna
antena de corbata de lazo bow-tie antenna
antena de cosecante cosecant antenna
antena de cosecante cuadrada cosecant-squared antenna
antena de cuadrante quadrant antenna
antena de cuadro loop antenna, frame antenna
antena de cuadro apantallado shielded-loop antenna
antena de cuadro de Alford Alford loop antenna
antena de cuadro de frecuencia ultraalta ultra-high frequency loop antenna
antena de cuadro de media onda half-wave loop antenna
antena de cuadro de onda completa full-wave loop antenna
antena de cuadro equilibrada balanced loop antenna
antena de cuadro giratorio rotatable-loop antenna
antena de cuadro interior indoor loop antenna
antena de cuadro rectangular square-loop antenna
antena de cuadro rotativo rotating-loop antenna
antena de cuarto de onda quarter-wave antenna
antena de Cutler Cutler antenna
antena de dipolo doblado folded-dipole antenna
antena de dipolos apilados stacked-dipole antenna
antena de disco disk antenna
antena de doble diamante double-diamond antenna
antena de doble dipolo lazy-H antenna
antena de doble polarización dual-polarized antenna
antena de doblete doublet antenna
antena de doblete multihilo multiwire doublet antenna
antena de dos elementos two-element antenna
antena de elementos apilados stacked antenna
antena de emergencia emergency antenna
antena de espejo mirror antenna
antena de exploración scanning antenna
antena de ferrita ferrite antenna
antena de Franklin Franklin antenna
antena de frecuencia ultraalta ultra-high frequency antenna
antena de Fuchs Fuchs antenna
antena de globo balloon antenna
antena de guía ranurada slotted-guide antenna
antena de guíaondas ranurado slotted-waveguide antenna
antena de haz beam antenna
antena de haz conformado shaped-beam antenna
antena de haz de cosecante cuadrada cosecant-squared beam antenna
antena de haz en abanico fanned-beam antenna
antena de haz en cola de castor beavertail beam antenna
antena de haz en lápiz pencil-beam antenna
antena de haz orientable steerable-beam antenna
antena de haz rotativo rotary-beam antenna
antena de Hertz Hertz antenna
antena de hilo wire antenna
antena de hilo largo long-wire antenna
antena de hilo único single-wire antenna
antena de hilos múltiples multiple-wire antenna
antena de hilos paralelos parallel-wire antenna
antena de impedancia adaptada matched-impedance antenna
antena de interrogación interrogation antenna
antena de Kooman Kooman antenna

antena de Kraus Kraus antenna
antena de látigo whip antenna
antena de lente lens antenna
antena de lente de Fresnel Fresnel lens antenna
antena de lente de Luneberg Luneberg lens antenna
antena de lente dieléctrica dielectric-lens antenna
antena de lente geodésica geodesic lens antenna
antena de Lewis Lewis antenna
antena de lorán loran antenna
antena de manguito sleeve antenna
antena de Marconi Marconi antenna
antena de mariposa superturnstile antenna
antena de mástil mast antenna
antena de media onda half-wave antenna
antena de medida measuring antenna
antena de microondas microwave antenna
antena de monitoreo monitoring antenna
antena de monopolo doblado folded-monopole
    antenna
antena de muy alta frecuencia very-high-frequency
    antenna
antena de onda completa full-wave antenna
antena de onda con fugas leaky-wave antenna
antena de onda media medium-wave antenna
antena de ondas wave antenna
antena de ondas cortas short-wave antenna
antena de ondas de superficie surface-wave antenna
antena de ondas estacionarias standing-wave
    antenna
antena de ondas progresivas progressive-wave
    antenna
antena de ondas ultracortas ultrashort-wave antenna
antena de placa plana flat-plate antenna
antena de radar radar antenna
antena de radiación longitudinal end-fire antenna
antena de radiación transversal broadside antenna
antena de radio radio antenna
antena de radiobaliza marker antenna
antena de radiofaro radio-beacon antenna
antena de radiofaro direccional radio-range antenna
antena de ranura notch antenna, slot antenna
antena de red mains antenna
antena de referencia reference antenna
antena de reflector de bocina horn reflector antenna
antena de reflector parabólico parabolic-reflector
    antenna
antena de reflector paraboloidal
    paraboloidal-reflector antenna
antena de reflector plano billboard antenna
antena de reflexión angular corner-reflection
    antenna
antena de relé relay antenna
antena de salida output antenna
antena de satélite satellite antenna
antena de secciones múltiples multiple-bay antenna
antena de seguimiento tracking antenna
antena de seguimiento automático automatic
    tracking antenna
antena de sentido sense antenna
antena de sonido sound antenna
antena de superficie corrugada corrugated-surface
    antenna
antena de superganancia supergain antenna
antena de techo rooftop antenna
antena de telemedida telemetering antenna
antena de televisión television antenna
antena de televisión maestra master television
    antenna
antena de toda banda all-band antenna
antena de toda onda all-wave antenna
antena de tope plano flat-top antenna
antena de transmisión sending antenna
antena de transmisión-recepción transmit-receive

antenna
antena de tres elementos three-element antenna
antena de unidades múltiples multiple-unit antenna
antena de varilla rod antenna, loopstick antenna
antena de varilla de ferrita ferrite-rod antenna
antena de varilla dieléctrica dielectric-rod antenna
antena de Walmsley Walmsley antenna
antena de Windom Windom antenna
antena de Wullenweber Wullenweber antenna
antena de Yagi Yagi antenna
antena delgada thin antenna
antena desequilibrada unbalanced antenna
antena desplegable unfurlable antenna
antena dieléctrica dielectric antenna
antena dieléctrica de varillas polyrod antenna
antena dipolo dipole antenna
antena dipolo coaxial coaxial-dipole antenna
antena dipolo colineal collinear dipole antenna
antena dipolo de manguito sleeve-dipole antenna
antena direccional directional antenna
antena direccional de ondas cortas short-wave
    directional antenna
antena direccional de reflector plano bedspring
    antenna
antena direccional lanac lanac directional antenna
antena directiva directive antenna
antena directiva de elementos parásitos
    parasitic-element directive antenna
antena discocónica discone antenna
antena doblada bent antenna
antena elevada elevated antenna
antena en abanico fan antenna
antena en aleta fin antenna
antena en anillo ring antenna
antena en arpa harp antenna
antena en bobina coil antenna
antena en caja box antenna
antena en cruz cross antenna
antena en delta delta antenna
antena en embudo funnel antenna
antena en espacio libre free-space antenna
antena en espina de pescado fishbone antenna
antena en espiral spiral antenna
antena en forma de abanico fan-shaped antenna
antena en hoja plane antenna
antena en horquilla hairpin antenna
antena en jaula cage antenna
antena en jaula de ardilla squirrel-cage antenna
antena en mariposa bat-wing antenna
antena en orejas de conejo rabbit-ear antenna
antena en paraguas umbrella antenna
antena en peine comb antenna
antena en persiana venetian-blind antenna
antena en pino pine-tree antenna
antena en pirámide pyramid antenna
antena en prisma prism antenna
antena en queso cheese antenna
antena en rombo diamond antenna
antena en T T antenna
antena en telaraña spiderweb antenna
antena en tierra ground antenna
antena en trébol cloverleaf antenna
antena en V V antenna
antena en Y Y antenna
antena equilibrada balanced antenna
antena equilibradora balancing antenna
antena escalonada staggered antenna, echelon
    antenna
antena esférica spherical antenna
antena esferoidal spheroidal antenna
antena estacionaria stationary antenna
antena excitada driven antenna
antena exterior outdoor antenna

antena fantasma phantom antenna
antena ficticia dummy antenna, quiescent antenna, mute antenna
antena fija fixed antenna
antena fría cold antenna
antena giratoria revolving antenna, rotatable antenna
antena helicoidal helical antenna
antena hertziana Hertzian antenna
antena hiperdirectiva hyperdirective antenna
antena imagen image antenna
antena incorporada built-in antenna
antena independiente de frecuencia frequency-independent antenna
antena inferior bottom antenna
antena interior indoor antenna
antena isotrópica isotropic antenna
antena laminada laminated antenna
antena lineal linear antenna
antena lineal de alimentación central center-feed linear antenna
antena lineal delgada thin linear antenna
antena lineal seccionada sectionalized linear antenna
antena magnética magnetic antenna
antena metálica metal antenna
antena monitora monitor antenna
antena monopolar cónica conical monopole antenna
antena monopolo monopole antenna
antena móvil movable antenna, mobile antenna
antena multibanda multiband antenna
antena multibanda automática automatic multiband antenna
antena multicanal multichannel antenna
antena multielemento multielement antenna
antena multifrecuencia multifrequency antenna
antena multihaz multibeam antenna
antena multihilo multiwire antenna
antena múltiple multiple antenna
antena no direccional nondirectional antenna
antena no directiva nondirective antenna
antena no resonante nonresonant antenna
antena no sintonizada untuned antenna
antena omnidireccional omnidirectional antenna
antena omnidirectiva omnidirective antenna
antena operada remotamente remotely-operated antenna
antena orientable steerable antenna
antena orientable de unidades múltiples multiple-unit steerable antenna
antena parabólica parabolic antenna, dish antenna
antena parabólica de microondas microwave dish
antena paraboloidal paraboloidal antenna
antena parásita parasitic antenna
antena pasiva passive antenna
antena patrón standard antenna
antena periódica periodic antenna
antena periódica logarítmica logarithmic periodic antenna
antena periscópica periscopic antenna
antena planar planar antenna
antena polarizada verticalmente vertically-polarized antenna
antena portátil portable antenna
antena radiante radiating antenna
antena radiogoniométrica direction-finder antenna
antena receptora receiving antenna
antena receptora de cuarto de onda quarter-wave receiving antenna
antena reductora de ruido noise-reducing antenna
antena resonante resonant antenna
antena retrodirectiva retrodirective antenna
antena rómbica rhombic antenna
antena rómbica múltiple multiple rhombic antenna
antena rotativa rotary antenna

antena rotatoria de elemento único single-element rotary antenna
antena seccionada sectionalized antenna
antena sin carga unloaded antenna
antena sin pérdidas lossless antenna
antena sin protección unprotected antenna
antena sintonizada tuned antenna
antena submarina underwater antenna
antena subterránea underground antenna
antena sumergida submerged antenna
antena superdirectiva superdirective antenna
antena telescópica telescoping antenna
antena toroidal doughnut antenna
antena transmisora transmitting antenna
antena triangular triangle antenna
antena tridipolar tridipole antenna
antena unidireccional unidirectional antenna
antena unipolar vertical vertical unipole antenna
antena unitaria unity-gain antenna
antena vehicular vehicular antenna
antena vertical vertical antenna
antena vertical alimentada en derivación shunt-fed vertical antenna
antena vertical alimentada en serie series-fed vertical antenna
antena zepelín zeppelin antenna
antenas acopladas coupled antennas
antenas complementarias complementary antennas
antenas espaciadas spaced antennas
antenas gemelas twin antennas
antiamortiguamiento antidamping
anticátodo anticathode
anticiclotrón anticyclotron
anticipación anticipation
anticipatorio anticipatory
anticoincidencia anticoincidence
anticolisión anticollision
anticorrosión anticorrosion
anticorrosivo anticorrosive
anticuerpo de computadora computer antibody
antideflagrante flameproof
antidesvanecimiento antifading
antidispersión antiscatter
antiestático antistatic
antifase antiphase
antiferroeléctrico antiferroelectric
antiferromagnético antiferromagnetic
antiferromagnetismo antiferromagnetism
antifluctuación antihunt
antihistéresis antihysteresis
antiinducción antiinduction
antiinductivo antiinductive
antiinterferencia antiinterference
antilogaritmo antilogarithm
antílogo antilogous
antiluminiscente antiluminescent
antimagnético antimagnetic
antimateria antimatter
antimicrofónico antimicrophonic
antimonio antimony
antimoniuro de aluminio aluminum antimonide
antimoniuro de galio gallium antimonide
antimoniuro de indio indium antimonide
antineutrino antineutrino
antineutrón antineutron
antinodo antinode
antinodo de corriente current antinode
antinodo de tensión voltage antinode
antinucleón antinucleon
antioxidante antioxidant
antiparalelo antiparallel
antipartícula antiparticle
antípoda antipodal

antipolar antipolar
antipolo antipole
antiprotón antiproton
antiquark antiquark
antirradar antiradar
antirradiación antiradiation
antirreflexión antireflection
antirresonancia antiresonance
antirresonancia de desplazamiento displacement antiresonance
antirresonancia de velocidad velocity antiresonance
antirresonante antiresonant
antirruido antinoise
antivibrador antivibrator
anulador annulling
anular annular
anunciador annunciator
anunciador de llamada call announcer
añadir add on
añadir a una cola enqueue
años de servicio acumulados accumulated service years
apagado quenched
apagador quencher
apagar turn off, quench
apagón blackout
apantallado shielded
apantallamiento shielding
aparato apparatus, set, fixture
aparato a prueba de agua waterproof apparatus
aparato a prueba de explosiones explosionproof apparatus
aparato absoluto absolute apparatus
aparato analizador set analyzer
aparato antideflagrante flameproof apparatus
aparato bañado en aceite oil-immersed apparatus
aparato con cubierta metálica metal-enclosed apparatus
aparato de abonado subscriber set, subset
aparato de alimentación supply apparatus
aparato de alta tensión high-voltage apparatus
aparato de conexión switchgear
aparato de conmutación switching apparatus
aparato de contestación telefónica telephone-answering set
aparato de contraste de fase phase-contrast apparatus
aparato de control control apparatus
aparato de control eléctrico electric control apparatus
aparato de destellos flashing apparatus
aparato de difracción de rayos X X-ray diffraction apparatus
aparato de emisión-recepción automático automatic send-receive set
aparato de guíaondas waveguide apparatus
aparato de Morse Morse set
aparato de perforación punching apparatus
aparato de polarización polarization apparatus
aparato de rayos X X-ray apparatus
aparato de referencia reference apparatus
aparato de registro recording apparatus
aparato de reproducción reproducing apparatus
aparato de reserva reserve set
aparato de servicio service set
aparato de sincronización synchronization apparatus
aparato de televisión television set
aparato de transmisión sending set
aparato doméstico home appliance
aparato dúplex duplex apparatus
aparato eléctrico electric appliance, appliance
aparato eléctrico a prueba de choques shockproof electrical apparatus

aparato eléctrico asociado associated electrical apparatus
aparato eléctrico portátil portable appliance
aparato electroacústico electroacoustic apparatus
aparato electrodoméstico electric home appliance
aparato electromédico electromedical apparatus
aparato en derivación shunt apparatus
aparato exterior outdoor apparatus
aparato impresor printing apparatus
aparato integrador integrating apparatus
aparato interior indoor apparatus
aparato medidor de ruido noise-measuring set
aparato microtelefónico hand telephone set
aparato monitor monitor set
aparato monocromo monochrome set
aparato múltiple multiple apparatus
aparato múltiplex multiplex apparatus
aparato parcialmente encerrado partially enclosed apparatus
aparato portátil portable apparatus
aparato protegido protected apparatus
aparato receptor receiver set
aparato regulador regulating apparatus
aparato símplex simplex apparatus
aparato telefónico telephone set
aparato telefónico alimentado por energía sonora sound-powered telephone set
aparato telefónico combinado combined telephone set
aparato telefónico con dispositivo contra efectos locales antisidetone telephone set
aparato telefónico con efecto local sidetone telephone set
aparato telefónico de abonado subscriber telephone set
aparato telefónico de batería local local-battery telephone set
aparato telefónico de magneto magneto telephone set
aparato telefónico de pared wall telephone set
aparato telefónico manual manual telephone set
aparato telegráfico telegraph set
aparato termoiónico thermionic apparatus
aparato totalmente encerrado totally-enclosed apparatus
aparato transmisor transmitting apparatus
aparato unipolar unipolar apparatus
apareamiento matching, pairing, twinning
apareamiento de ecos echo matching
apareamiento de tarjetas card matching
aparente apparent
aperiódico aperiodic
apertura aperture
apertura de archivo file opening
apertura de circuito switching-off
apertura de observación observation aperture
apilamiento stacking
apilamiento de trabajos job stacking
apilamiento desplazado offset stacking
apiñamiento de caracteres character crowding
apiñamiento espectral spectrum crowding
aplanado flattened
aplanador de campo field flattener
aplanamiento flattening, smoothing
aplicabilidad applicability
aplicación application
aplicación de computadora computer application
aplicación de control control application
aplicación de control de tren automático automatic train control application
aplicación de red network application
aplicación de reserva standby application
aplicación de sistema system application

aplicación en línea on-line application
aplicación en tiempo real real-time application
aplicación fuera de línea off-line application
aplicación primaria primary application
aplicación retardada delayed application
aplicación secundaria secondary application
aplicaciones de gráficos graphics applications
aplicado applied
aplicador applicator
apocromático apochromatic
apocromatismo apochromatism
apogeo apogee
apostilbio apostilb
apoyo support
apoyo de sistema system support
aprendizaje basado en computadora
    computer-based learning
aprendizaje de máquina machine learning
aprendizaje por computadora computer learning
aprobado approved
aproximación approach, approximation
aproximación ciega blind approach
aproximación controlada desde tierra
    ground-controlled approach
aproximación de precisión precision approach
aproximación inicial initial approach
aproximación por instrumentos instrument
    approach
aproximación por radar radar approach
apuntamiento de haz beam pointing
aquavox aquavox
arandela aislante grommet
araña spider
araña interior inside spider
arbitraje arbitration
arbitraje de bus bus arbitration
arbitrario arbitrary
árbitro arbiter
árbol binario binary tree
árbol binario completo complete binary tree
árbol binario extendido extended binary tree
árbol de búsqueda search tree
árbol de decisión decision tree
arco arc
arco al vacío vacuum arc
arco auxiliar auxiliary arc
arco cantante singing arc
arco de carbón carbon arc
arco de cátodo frío cold-cathode arc
arco de contacto contact arc
arco de destello flash arc
arco de Duddell Duddell arc
arco de mercurio mercury arc
arco de Poulsen Poulsen arc
arco de seguridad safety arc
arco de tungsteno tungsten arc
arco eléctrico electrical arc
arco encerrado enclosed arc
arco menor minor arc
arco oscilante oscillating arc
arco parlante speaking arc
arco polar polar arc, pole arc
arco simple simple arc
arco superficial surface arc
arco visible visible arc
archivado archived
archival archival
archivo file
archivo abuelo grandfather file
archivo activo active file
archivo aleatorio random file
archivo auxiliar auxiliary file
archivo binario binary file

archivo central en línea on-line central file
archivo cerrado closed file, locked file
archivo clave key file
archivo combinado combined file
archivo compartido shared file
archivo completamente invertido fully inverted file
archivo comprimido compressed file
archivo compuesto composite file
archivo computarizado computerized file
archivo controlador controlling file
archivo conversacional conversational file
archivo de acceso aleatorio random-access file
archivo de actividad activity file
archivo de almacenamiento en disco disk storage
    file
archivo de almacenamiento masivo mass storage
    file
archivo de artículos item file
archivo de biblioteca library file
archivo de bloque fijo fixed block file
archivo de cambios change file
archivo de catálogo catalog file
archivo de clasificación sort file
archivo de cola queue file
archivo de computadora computer file
archivo de comunicaciones communication file
archivo de concordancia concordance file
archivo de configuración configuration file
archivo de continuación continuation file
archivo de control de sistema system control file
archivo de control de trabajos job control file
archivo de datos data file
archivo de datos externo external data file
archivo de datos gráficos graphics data file
archivo de datos internos internal data file
archivo de datos maestros master data file
archivo de datos privados private data file
archivo de destino destination file
archivo de detalle detail file
archivo de dirección address file
archivo de dirección de registro record address file
archivo de disco disk file
archivo de disco magnético magnetic-disk file
archivo de entrada input file
archivo de entrada de trabajos job-input file
archivo de entrada maestro master input file
archivo de entrada-salida input-output file
archivo de errores error file
archivo de formato format file
archivo de fusión merge file
archivo de imágenes de tarjeta card-image file
archivo de impresora printer file
archivo de información central central information
    file
archivo de informe report file
archivo de intercalación collation file
archivo de intercambio exchange file
archivo de lectura-escritura read-write file
archivo de mandos command file
archivo de mantenimiento maintenance file
archivo de máquina machine file
archivo de mensajes message file
archivo de película magnética magnetic-film file
archivo de programa program file
archivo de programas maestros master program file
archivo de pruebas test file
archivo de puntos de comprobación checkpoint file
archivo de referencia general general reference file
archivo de respaldo backup file
archivo de salida output file
archivo de salida maestro master output file
archivo de seguridad de recursos resource security
    file

archivo de serie de instrucciones batch file
archivo de sistema system file
archivo de tarjetas card file
archivo de tarjetas magnéticas magnetic card file
archivo de tarjetas perforadas punched-card file
archivo de trabajo working file
archivo de transacción transaction file
archivo de visualización display file
archivo directo direct file
archivo dividido partitioned file
archivo en cinta tape file
archivo en cinta magnética magnetic-tape file
archivo fijo fixed file
archivo físico physical file
archivo fuente source file
archivo inactivo inactive file
archivo indexado indexed file
archivo índice index file
archivo invertido inverted file
archivo lógico logic file
archivo maestro master file
archivo maestro activo active master file
archivo muerto dead file
archivo múltiple multiple file
archivo multivolumen multivolume file
archivo objeto object file
archivo parcialmente invertido partially inverted file
archivo patrón standard file
archivo plano flat file
archivo primario primary file
archivo principal main file
archivo privado private file
archivo relacional relational file
archivo relativo relative file
archivo residente resident file
archivo residente en el sistema system resident file
archivo reusable reusable file
archivo secuencial sequential file
archivo secuencial indexado indexed sequential file
archivo secundario secondary file
archivo simbólico symbolic file
archivo temporal temporary file
archivo temporal de tarea scratch file
archivo viciado corrupted file
archivo volátil volatile file
archivos de base de datos database files
archivos encadenados chained files
área area
área activa active area
área anódica anodic area
área automática automatic area
área brillante highlight
área coherente coherent area
área común common area
área constante constant area
área contigua contiguous area
área controlada controlled area
área crítica critical area
área de almacenaje save area
área de almacenamiento storage area
área de almacenamiento común common storage area
área de almacenamiento dinámica dynamic storage area
área de almacenamiento fija fixed-storage area
área de aproximación approach area
área de apuntes scratchpad area
área de búsqueda seek area
área de campo field area
área de captura capture area
área de cobertura coverage area
área de código code area

área de coherencia coherence area
área de cola de sistema local local system queue area
área de cola protegida protected queue area
área de colas de sistema system queue area
área de comunicaciones communication area
área de contacto contact area
área de control control area
área de control de sistema system control area
área de control por radar radar-control area
área de corrección patch area
área de datos data area
área de desbordamiento overflow area
área de desbordamiento de cilindro cylinder overflow area
área de desbordamiento independiente independent overflow area
área de desvanecimiento fading area
área de direcciones virtuales virtual-address area
área de dispersión scattering area
área de eco echo area
área de eco efectiva effective echo area
área de ejecución de tareas task-execution area
área de entrada input area
área de entrada-salida input-output area
área de etiqueta label area
área de exploración scanning area
área de flujo flow area
área de fondo background area
área de imagen image area
área de impresión print area
área de información de etiquetas label information area
área de instrucción instruction area
área de interferencia interference area
área de localización location area
área de mandos command area
área de memoria memory area
área de menú menu area
área de módulo de control control module area
área de operación operating area
área de paginación paging area
área de partición partition area
área de perforación punch area
área de perturbación mush area
área de prioridad foreground area
área de procesamiento processing area
área de programa program area
área de pruebas testing area
área de radar radar area
área de recepción reception area
área de registro recording area, record area, playing area
área de ruido noise area
área de salida output area
área de salida de datos gráficos graphic data output area
área de sección transversal cross-sectional area
área de sensación auditiva auditory sensation area
área de señal signal area
área de señal fuerte strong-signal area
área de señales clase A class A signal area
área de servicio service area
área de servicio común common service area
área de servicio intermitente intermittent-service area
área de servicio primario primary service area
área de servicio secundario secondary service area
área de silencio silent area
área de sistema system area
área de sombra shadow area
área de trabajo work area
área de trabajo de canal channel work area

área de transmisión transmission area
área de unión junction area
área de visión viewing area
área de visualización digital digital display area
área dinámica dynamic area
área dinámica paginable pageable dynamic area
área efectiva effective area
área efectiva de antena antenna effective area
área elemental elemental area
área en blanco blank area
área fija fixed area
área global global area
área inactiva inactive area
área libre clear area
área local local area
área luminosa luminous area
área marginal fringe area
área muerta dead area
área peligrosa hazardous area
área positiva positive area
área protegida protected area
área radiante radiating area
área reservada reserved area
área reubicable relocatable area
área secundaria secondary area
área terminal terminal area
área tipo N N-type area
área tipo P P-type area
área transitoria transient area
área unitaria unit area
área utilizable usable area
área variable variable area
área virtual compartida shared virtual area
argón argon
argumento argument
argumento de búsqueda de segmento segment
   search argument
argumento real actual argument
aritmética arithmetic
aritmética binaria binary arithmetic
aritmética de coma fija fixed-point arithmetic
aritmética de coma flotante floating-point arithmetic
aritmética de decimal fijo fixed-decimal arithmetic
aritmética de decimal flotante floating-decimal
   arithmetic
aritmética de precisión múltiple multiple-precision
   arithmetic
aritmética en serie serial arithmetic
aritmética interna internal arithmetic
aritmética paralela parallel arithmetic
aritmética parcial partial arithmetic
aritmético arithmetical
armadura armature, armor
armadura de cable cable armor
armadura de imán magnet armature, magnet
   keeper, keeper
armadura de relé relay armature
armadura lateral side armature
armadura magnética magnetic armature
armazón framework
armazón abierto open framework
armónica harmonic
armónica aural aural harmonic
armónica de audiofrecuencia audio-frequency
   harmonic
armónica de radiofrecuencia radio-frequency
   harmonic
armónica espuria spurious harmonic
armónica fundamental fundamental harmonic
armónica impar odd harmonic
armónica par even harmonic
armónico harmonic
arnés de conductores wiring harness

aro de brida flange ring
arqueo flashover
arquitectónico architectural
arquitectura architecture
arquitectura de computadora computer architecture
arquitectura de chip chip architecture
arquitectura de entrada-salida de
   microcomputadora microcomputer input-output
   architecture
arquitectura de microcomputadora microcomputer
   architecture
arquitectura de microprocesador microprocessor
   architecture
arquitectura de procesador en paralelo parallel
   processor architecture
arquitectura de programa program architecture
arquitectura de red network architecture
arquitectura de sistema system architecture
arquitectura de sistema abierto open-system
   architecture
arrancador starter
arrancador automático automatic starter
arrancador combinado combination starter
arrancador compensador compensator starter
arrancador de motor motor-starter
arrancador directo across-the-line starter
arrancador estrella-delta star-delta starter
arrancador inductivo de estator stator-inductance
   starter
arrancador magnético magnetic starter
arrancador resistivo de estator stator-resistance
   starter
arrancador secuencial sequential starter
arrancador semiautomático semiautomatic starter
arranque starting
arranque asincrónico asynchronous starting
arranque automático automatic starting
arranque de velocidad lenta slow-speed starting
arranque en frío cold boot
arranque estrella-delta star-delta starting
arranque frío cold start
arranque por autotransformador autotransformer
   starting
arranque reostático rheostatic starting
arrastramiento dragging
arrastre drag, carry, pulling, tracking
arrastre completo complete carry
arrastre de aguja stylus drag
arrastre de cinta tape transport
arrastre en cascada cascade carry
arrastre magnético magnetic creep, magnetic drag,
   magnetic drive
arrastre por borde rim drive
arreglado aleatoriamente randomly arranged
arreglo arrangement, layout
arreglo con retroalimentación negativa
   negative-feedback arrangement
arreglo de acceso a datos data access arrangement
arreglo de archivo file layout
arreglo de bastidor rack layout
arreglo de circuitos circuit arrangement
arreglo de contactos contact arrangement
arreglo de datos data layout
arreglo de equipo layout of equipment
arreglo de instrumentos instrument layout
arreglo de plataforma deck arrangement
arreglo de puertas gate array
arreglo de registros register arrangement
arreglo de resistencia-capacitancia
   resistance-capacitance arrangement
arreglo de respaldo backup arrangement
arreglo en circuito cerrado closed-circuit
   arrangement

arreglo en contrafase push-pull arrangement
arreglo en paralelo parallel arrangement
arreglo en serie series arrangement
arreglo en serie-paralelo series-parallel arrangement
arreglo escalonado staggered arrangement
arreglo oblicuo skew arrangement
arreglo óptimo de hilos wire dress
arriostramiento guying
arrollado en derivación shunt-wound
arrollado en espiral spiral-coiled
arrollado en serie series-wound
arrollamiento winding
arrollamiento a mano hand winding
arrollamiento aleatorio random winding
arrollamiento amortiguador damper winding, amortisseur winding
arrollamiento amplificador amplifying winding
arrollamiento antirresonante antiresonant winding
arrollamiento autoprotegido self-protected winding
arrollamiento auxiliar auxiliary winding
arrollamiento bifásico two-phase winding
arrollamiento bifilar bifilar winding
arrollamiento bipolar bipolar winding
arrollamiento cerrado reentrant winding
arrollamiento cilíndrico cylindrical winding
arrollamiento compensado compensated winding
arrollamiento compensador compensating winding
arrollamiento compensador polifásico polyphase compensating winding
arrollamiento compuesto compound winding
arrollamiento con derivación central center-tapped winding
arrollamiento con tomas tapped winding
arrollamiento concentrado concentrated winding
arrollamiento concéntrico concentric winding
arrollamiento cosenoidal cosine winding
arrollamiento chaprón chapron winding
arrollamiento de alta tensión high-voltage winding
arrollamiento de arranque starting winding
arrollamiento de Ayrton-Perry Ayrton-Perry winding
arrollamiento de borrado erasing winding
arrollamiento de campo field winding
arrollamiento de campo excitador exciter-field winding
arrollamiento de capa única single-layer winding
arrollamiento de cobre copper winding
arrollamiento de conmutación commutating winding
arrollamiento de control control winding
arrollamiento de dos pisos two-position winding
arrollamiento de entrada input winding
arrollamiento de excitación excitation winding
arrollamiento de fase phase winding
arrollamiento de filamento filament winding
arrollamiento de mantenimiento holding winding
arrollamiento de máquina machine winding
arrollamiento de oposición opposing winding
arrollamiento de paso corto short-pitch winding
arrollamiento de polarización bias winding
arrollamiento de potencia power winding
arrollamiento de puerta gate winding
arrollamiento de retroalimentación feedback winding
arrollamiento de salida output winding
arrollamiento de señal signal winding
arrollamiento de tensión fija fixed-voltage winding
arrollamiento de un piso one-position winding
arrollamiento de Wenner Wenner winding
arrollamiento del inducido armature winding
arrollamiento del rotor rotor winding
arrollamiento diferencial differential winding
arrollamiento distribuido distributed winding
arrollamiento dividido split winding

arrollamiento dúplex duplex winding
arrollamiento elevador step-up winding
arrollamiento en anillo ring winding
arrollamiento en barril barrel winding
arrollamiento en bobinas planas pie winding
arrollamiento en capas layer winding
arrollamiento en celosía lattice winding
arrollamiento en derivación shunt winding
arrollamiento en disco disk winding
arrollamiento en espiral spiral winding
arrollamiento en jaula de ardilla squirrel-cage winding
arrollamiento en panal honeycomb winding
arrollamiento en paralelo parallel winding
arrollamiento en rombo diamond winding
arrollamiento en serie series winding
arrollamiento en serie-paralelo series-parallel winding
arrollamiento en tambor drum winding
arrollamiento en zigzag zigzag winding
arrollamiento equipotencial equipotential winding
arrollamiento estabilizador stabilizing winding
arrollamiento exterior outer winding
arrollamiento imbricado lap winding
arrollamiento inductivo inductive winding
arrollamiento lateral lateral winding
arrollamiento multicapa multilayer winding
arrollamiento múltiple multiple winding
arrollamiento múltiplex multiplex winding
arrollamiento multiplicador multiplying winding
arrollamiento multitoma multitap winding
arrollamiento no inductivo noninductive winding
arrollamiento oblicuo oblique winding
arrollamiento ondulado wave winding
arrollamiento paralelo múltiple multiple parallel winding
arrollamiento polar pole winding
arrollamiento primario primary winding
arrollamiento principal main winding
arrollamiento ramificado branched winding
arrollamiento regulador regulating winding
arrollamiento seccionado sectionalized winding
arrollamiento secundario secondary winding
arrollamiento secundario auxiliar auxiliary secondary winding
arrollamiento sencillo single winding
arrollamiento sobre horma preformed winding
arrollamiento superpuesto banked winding
arrollamiento terciario tertiary winding
arrollamiento toroidal toroidal winding
arrollamiento tropical tropical winding
arrollamientos alternados sandwich windings
arrollamientos intercalados interleaved windings
arrollamientos opuestos opposed windings
arsénico arsenic
arseniuro de galio gallium arsenide
arseniuro de indio indium arsenide
artefacto artifact
articulación articulation
articulación de banda band articulation
artículo item
artículo binario binary item
artículo de base base item
artículo de coma flotante floating-point item
artículo de configuración de equipo físico hardware configuration item
artículo de datos data item
artículo de datos internos internal data item
artículo de formato format item
artículo de grupo group item
artículo de prueba test item
artículo decimal empaquetado packed decimal item
artículo elemental elementary item

artículo final end item
artículo no numérico nonnumeric item
artículo numérico numerical item
artificial artificial
artificialmente artificially
asbesto asbestos
ascendente ascending
asentamiento burn-in
asiento de clavija plug seat
asignable assignable
asignación assignment, allocation
asignación de almacenamiento storage allocation,
    storage assignment
asignación de almacenamiento dinámica dynamic
    storage allocation
asignación de almacenamiento estática static
    storage allocation
asignación de almacenamiento general
    general-storage assignment
asignación de almacenamiento intermedio estática
    static buffer allocation
asignación de bloque block allocation
asignación de canales channel assignment
asignación de confiabilidad reliability allocation
asignación de dirección address assignment
asignación de dispositivo ficticio dummy device
    assignment
asignación de espacio primaria primary space
    allocation
asignación de espacio secundario secondary space
    allocation
asignación de frecuencias frequency assignment,
    frequency allotment
asignación de grupos group allocation
asignación de memoria memory allocation
asignación de memoria dinámica dynamic memory
    allocation
asignación de núcleo core allocation
asignación de recursos resource allocation
asignación de recursos de computadora computer
    resource allocation
asignación de recursos dinámica dynamic resource
    allocation
asignación de supergrupos supergroup allocation
asignación de teclado key assignment
asignación de tiempo time assignment
asignación dinámica dynamic allocation
asignación directa direct allocation
asignación estática static allocation
asignación fuera de banda out-of-band assignment
asignación incondicional unconditional assignment
asignación múltiple multiple assignment
asignación primaria primary assignment
asignado assigned
asignar assign, allot
asignar frecuencias assign frequencies
asimetría asymmetry
asimétrico asymmetrical, single-ended
asincrónico asynchronous
asíncrono asynchronous
asindético asyndetic
asíntota asymptote
asintótico asymptotic
asistencia remota remote assistance
asistido assisted
asistido por computadora computer assisted
askarel askarel
asociación association
asociación de parámetros parameter association
asociado associated
asociativo associative
aspecto aspect
aspecto de señal signal aspect

asperezas asperities
astable astable
astático astatic
astatino astatine
astigmático astigmatic
astigmatismo astigmatism
astigmatismo anisotrópico anisotropic astigmatism
astriónica astrionics
astrocompás astrocompass
astrodinámica astrodynamics
astronomía de radar radar astronomy
astroseguidor astrotracker
ataque attack
atascamiento jam, jamming
atascamiento de tarjetas card jamming
atemperador attemperator
atención attention
atenuación attenuation
atenuación armónica harmonic attenuation
atenuación auroral auroral attenuation
atenuación compuesta composite attenuation
atenuación de acoplamiento directo direct-coupled
    attenuation
atenuación de banda band attenuation
atenuación de banda lateral sideband attenuation
atenuación de cable cable attenuation
atenuación de canal adyacente adjacent-channel
    attenuation
atenuación de corriente current attenuation
atenuación de diafonía crosstalk attenuation
atenuación de eco echo attenuation
atenuación de ecos parásitos clutter attenuation
atenuación de equilibrio balance attenuation
atenuación de modo diferencial differential-mode
    attenuation
atenuación de onda wave attenuation
atenuación de pendiente slope attenuation
atenuación de potencia power attenuation
atenuación de respuestas espurias
    spurious-response attenuation
atenuación de retroalimentación feedback
    attenuation
atenuación de segundo canal second-channel
    attenuation
atenuación de señal signal attenuation
atenuación de sombra shadow attenuation
atenuación de sonido sound attenuation
atenuación de tensión voltage attenuation
atenuación de trayectoria path attenuation
atenuación de vibraciones vibration attenuation
atenuación en espacio libre free-space attenuation
atenuación imagen image attenuation
atenuación infinita infinite attenuation
atenuación iterativa iterative attenuation
atenuación longitudinal longitudinal attenuation
atenuación paradiafónica near-end crosstalk
    attenuation
atenuación por filtro filter attenuation
atenuación por inserción insertion attenuation
atenuación por llama flame attenuation
atenuación por lluvia rain attenuation
atenuación progresiva tapered attenuation
atenuación relativa relative attenuation
atenuación relativa compleja complex relative
    attenuation
atenuación telediafónica far-end crosstalk
    attenuation
atenuación total overall attenuation
atenuado attenuated
atenuador attenuator
atenuador armónico harmonic attenuator
atenuador capacitivo capacitive attenuator
atenuador coaxial coaxial attenuator

atenuador controlado por tensión voltage-controlled attenuator
atenuador de absorción absorptive attenuator
atenuador de agudos treble attenuator
atenuador de aleta flap attenuator
atenuador de antena antenna attenuator
atenuador de bajos bass attenuator
atenuador de capacitancia mutua mutual-capacitance attenuator
atenuador de compensación de frecuencia frequency-compensated attenuator
atenuador de corte cutoff attenuator
atenuador de corte variable variable-cutoff attenuator
atenuador de cuadrante dial attenuator
atenuador de cuarto de onda quarter-wave attenuator
atenuador de décadas decade attenuator
atenuador de disco disk attenuator
atenuador de guíaondas waveguide attenuator
atenuador de guillotina guillotine attenuator
atenuador de impedancia adaptada matched-impedance attenuator
atenuador de inductancia mutua mutual-inductance attenuator
atenuador de interferencia interference attenuator
atenuador de línea line pad
atenuador de microondas microwave attenuator
atenuador de paleta longitudinal vane attenuator
atenuador de pistón piston attenuator
atenuador de potencial potential attenuator
atenuador de radiofrecuencia radio-frequency attenuator
atenuador de resistencia resistance attenuator
atenuador de salida output attenuator
atenuador de tipo de pistón piston type attenuator
atenuador de video video attenuator
atenuador decimal decimal attenuator
atenuador en escalera ladder attenuator
atenuador equilibrador balancing attenuator
atenuador fijo fixed attenuator, pad
atenuador logarítmico logarithmic attenuator
atenuador no reflejante nonreflecting attenuator
atenuador panorámico panoramic attenuator
atenuador reactivo reactive attenuator
atenuador resistivo resistive attenuator
atenuador rotativo rotary attenuator
atenuador variable variable attenuator
atenuador variable electrónico electronic variable attenuator
atenuador variable sin distorsión distortionless variable attenuator
atenuante attenuating
atenuar attenuate
atermancia athermancy
aterrizaje ciego blind landing
aterrizaje por instrumentos instrument landing
aterrizaje radioguiado glide-path landing
atmósfera atmosphere
atmósfera de hidrógeno hydrogen atmosphere
atmósfera de radio normal standard radio atmosphere
atmósfera normal standard atmosphere
atmósfera superior upper atmosphere
atmosférico atmospherical
atmosféricos atmospherics
atómica atomics
atómicamente atomically
atomicidad atomicity
atómico atomic
atomística atomistics
átomo atom
átomo aceptor acceptor atom

átomo de hidrógeno hydrogen atom
átomo de impureza impurity atom
átomo gramo gram-atom
atracción attraction
atracción de Coulomb Coulomb attraction
atracción eléctrica electrical attraction
atracción electromagnética electromagnetic attraction
atracción electrostática electrostatic attraction
atracción magnética magnetic attraction
atrapamiento de radiación radiation trapping
atributo attribute
atributo de archivo file attribute
atributo de archivo fijo fixed file attribute
atributo de campo field attribute
atributo de carácter character attribute
atributo de datos data attribute
atributo de enlace link attribute
atributo de entidad entity attribute
atributo de segmento segment attribute
atributo de visualización display attribute
atributo por omisión default attribute
atributo sólo de lectura read-only attribute
atributo virtual virtual attribute
attofarad attofarad
audar audar
audibilidad audibility
audible audible
audífono earphone, headphone, hearing aid
audífono de cabeza head receiver
audífono de cristal crystal earphone
audífono de inserción insert earphone
audífono de microteléfono handset earphone
audífono de monitoreo monitoring earphone
audífono electrostático electrostatic headphone
audífono monitor monitor earphone
audífono piezoeléctrico piezoelectric earphone
audífono telefónico telephone earphone
audífonos headset, headphones
audífonos de alta fidelidad high-fidelity headphones
audífonos telefónicos telephone headset
audio audio
audiocinta audiotape
audioconferencia audioconference
audiograma audiogram
audiograma de ruido noise audiogram
audiología audiology
audiólogo audiologist
audiometría audiometry
audiométrico audiometric
audiometrista audiometrist
audiómetro audiometer
audiómetro de ruido noise audiometer
audión audion
audiovisual audio-visual
auditivo auditory
auditoría audit
auditoría automatizada automated audit
auditoría de aplicación application audit
auditoría de código code audit
auditoría interna internal audit
aullador howler
aullido howl, warble
aullido acústico acoustic howl
aullido de audio audio howl
aumentador augmenter
aumentar augment
aumento magnification, rise, buildup, power
aumento de modulación modulation rise
aumento de resistencia a la izquierda left-hand taper
aumento de temperatura temperature rise
aumento de tensión absoluto absolute voltage rise

aumento exponencial exponential increase
aural aural
auricular earpiece
aurora aurora
auroral auroral
ausencia absence
ausente absent
austenítico austenitic
autenticación de red net authentication
autoabsorción self-absorption
autoadaptivo self-adapting
autoajustable self-adjustable
autoalarma autoalarm
autoalimentado self-powered
autoalimentador self-feeding
autoalineante self-aligning
autoamortiguamiento self-damping
autoapantallamiento self-screening
autoarrancador autostarter
autoarranque self-starting
autoblindaje self-shielding
autobloqueo squegging
autoborrado self-erasing
autocalentable self-heating
autocalentado self-heated
autocalibración self-calibration
autocalibrador self-calibrating
autocapacidad autocapacity
autocapacitancia self-capacitance
autocarga self-loading
autocentrador self-centering
autocerrador self-closing
autocodificador autocoder
autocolimador fotoeléctrico photoelectric
    autocollimator
autocompensado self-compensated
autocomprobación self-check, self-test
autocondensación autocondensation
autoconducción autoconduction
autoconsistente self-consistent
autoconstricción self-constriction
autocontrol autocontrol, self-control
autocontrolado self-controlled
autocorrección autocorrection
autocorrector self-correcting
autocorrelación autocorrelation
autodescarga self-discharge
autodiagnósitco self-diagnostic
autodino autodyne
autodispersión self-scattering
autoecualización self-equalizing
autoeléctrico autoelectric
autoelectrólisis autoelectrolysis
autoencendido self-ignition
autoenfriado self-cooled
autoenfriamiento self-cooling
autoenganchador self-latching
autoequilibrado self-balanced
autoequilibrador self-balancing
autoerección autoerection
autoestabilización self-stabilization
autoestabilizador self-stabilizing
autoexcitación autoexcitation, self-excitation
autoexcitación completa complete self-excitation
autoexcitado autoexcited
autoexcitador autoexciter
autoextintor self-extinguishing
autogenerativo self-generative
autoheterodino autoheterodyne
autoignición autoignition
autoimpedancia self-impedance
autoimpulso self-pulse
autoinducción self-induction

autoinductancia self-inductance
autoinductivo autoinductive, self-inductive
autoinductor self-inductor
autointerrumpido self-interrupted
autointerruptor self-interrupter
autoinversión self-reversal
autoionización autoionization
autolimitador self-limiting
autolubricación self-lubrication
autolubricante self-lubricating
autoluminoso self-luminous
automantenido self-maintained
automarcado autodial
autómata automaton
automática automatics
automáticamente automatically
automaticidad automaticity
automático automatic, self-operated
automatización automation
automatización de datos de fuente source-data
    automation
automatización de diseño design automation
automatización de microprocesador microprocessor
    automation
automatización de procesos process automation
automatización de producción production
    automation
automatizado automated
automatizar automate
automodificación self-modification
automodulado self-modulated
automonitor automonitor
autonavegante autonavigator
autónomo self-contained, stand-alone
autooperado self-operated
autoorganizador self-organizing
autoorientador self-orienting
autooscilación self-oscillation
autooscilador self-oscillator
autooscilador piloto pilot autooscillator
autooscilante self-oscillating
autopiloto autopilot
autopiloto electrónico electronic autopilot
autoplex autoplex
autopolarización autobias, self-bias
autopolarizado self-biased
autopolarizante self-biasing
autopropagación self-propagation
autopropagante self-propagating
autopropulsado self-propelled
autoprotección self-protection
autoprotegido self-protected
autoprueba self-test
autopulsante self-pulsing
autorización de escritura write authorization
autorizado authorized
autorradiografía autoradiography
autorrectificación self-rectification
autorrectificador self-rectifying
autorregistro self-recording
autorregulación self-regulating, self-governing
autorregulador self-governing
autorreposición self-reset, self-resetting
autorresistencia self-resistance
autorresonancia self-resonance
autorresonante self-resonant
autorrestablecimiento self-reset
autorrotación self-rotation
autosaturación self-saturation
autosaturante self-saturating
autosellador self-sealing
autosincrónico self-synchronous, autosynchronous
autosincronización self-synchronizing

autosíncrono autosyn
autosintonización self-tuning
autosintonizado self-tuned
autosoldadura self-soldering
autosostenido self-sustained
autotransductor autotransducter
autotransformador autotransformer
autotransformador de arranque autotransformer starter
autotransformador desfasador phase-shifting autotransformer
autotransformador elevador step-up autotransformer
autotransformador variable variable autotransformer
autoventilación self-ventilation
autoventilado self-ventilated
autoverificación autoverification
auxiliar auxiliary
avalancha avalanche
avalancha controlada controlled avalanche
avalancha de Townsend Townsend avalanche
avalancha de Zener Zener breakdown
avalancha electrónica electron avalanche
avalancha inversa reverse avalanche
avalancha iónica ion avalanche
avalancha térmica thermal runaway
avance lead
avance de cinta tape advance
avance de fase phase advance, phase leading
avance de registro record advance
avance horizontal horizontal advance
avance por cierres sucesivos jogging, inching
avance por saltos leapfrogging
avance rápido fast forward
avance vertical vertical advance
avería malfunction
avión radioguiado drone
aviso de mal funcionamiento malfunction alert
aviso temprano por microondas microwave early warning
axial axial
axialidad axiality
axiotrón axiotron
ayuda de diseño de pantalla screen design aid
ayuda de navegación navigation aid
ayuda de navegación autónoma self-contained navigation aid
ayuda en línea on-line help
ayuda navegacional navigational aid
azel azel
azimut azimuth
azogue quicksilver
azufre sulfur
azuza azuza

bactericida bactericidal
badajo clapper
bafle baffle
bafle circular circular baffle
bafle de altavoz loudspeaker baffle
bafle de laberinto labyrinth baffle
bafle de pared wall baffle
bafle direccional directional baffle
bafle infinito infinite baffle
bafle plano flat baffle
bafle reflejo reflex baffle
bafle ventilado vented baffle
baja actividad low activity
baja definición low definition
baja distorsión low distortion
baja energía low energy
baja frecuencia low frequency
baja impedancia low impedance
baja intensidad low intensity
baja potencia low power
baja presión low pressure
baja prioridad low priority
baja resistencia low resistance
baja resolución low resolution
baja temperatura low temperature
baja tensión low voltage, low tension
baja velocidad low speed
bajada drop, lead-in
bajo consumo low consumption, low drain
bajo nivel low level
bajo régimen low rate
bajo ruido low noise
bajo vacío low vacuum
bajo wattaje low-wattage
bajos bass
balance balance
balance de plasma plasma balance
balancín gimbal, rocker arm
balanza de Kelvin Kelvin balance
balasto ballast
balística ballistics
balística electrónica electron ballistics
baliza beacon
baliza luminosa light beacon
balizamiento beaconage
balún balun
balún de cuarto de onda quarter-wave balun
banco bank
banco de canales channel bank
banco de condensadores condenser bank
banco de conmutadores switchbank
banco de datos data bank
banco de lámparas lamp bank
banco de memoria memory bank
banco de programas program bank
banco de pruebas test bench
banco óptico optical bench
banda band
banda alta high band
banda ancha broadband, wideband
banda angosta narrowband, narrow bandwidth

banda anódica anode strap
banda atenuada attenuated band
banda baja low band
banda base baseband
banda base de recepción receiving baseband
banda básica basic band
banda característica characteristic band
banda ciudadana citizen's band
banda compartida shared band
banda crítica critical band
banda cruzada crossband
banda de absorción absorption band
banda de aficionados amateur band
banda de amplificación amplification band
banda de atenuación attenuation band
banda de atenuación por filtro filter attenuation band
banda de audio audio band
banda de audiofrecuencia audio-frequency band
banda de barrido de frecuencia frequency-sweep band
banda de bombeo pumping band
banda de comunicación communication band
banda de conducción conduction band
banda de conducción degenerada degenerate conduction band
banda de crominancia chrominance band
banda de décadas decade band
banda de diferencia difference band
banda de dispersión scatter band
banda de emisión emission band
banda de energía energy band
banda de energía de resonancia resonance energy band
banda de energía permitida allowed energy band
banda de energía prohibida forbidden energy band
banda de excitación excitation band
banda de exploración scanning band
banda de frecuencia de absorción absorption frequency band
banda de frecuencia ultraalta ultra-high frequency band
banda de frecuencias frequency band, waveband
banda de frecuencias ancha wide frequency band
banda de frecuencias de imagen picture frequency band
banda de frecuencias de radar radar frequency band
banda de frecuencias vocales voice-frequency band
banda de guarda guard band
banda de interferencia interference band
banda de longitudes de onda wavelength band
banda de lluvia rainband
banda de muy alta frecuencia very-high-frequency band
banda de octava octave band
banda de onda media medium-wave band
banda de ondas wave band
banda de ondas largas long-wave band
banda de ondas ultracortas ultrashort-wave band
banda de operación operating band
banda de paso ancha wide passband
banda de paso de filtro filter passband
banda de paso de sonido sound passband
banda de paso de video video passband
banda de radar radar band
banda de radiodifusión broadcast band
banda de radiodifusión de alta frecuencia high-frequency broadcast band
banda de radiodifusión normal standard broadcast band
banda de radiodifusión por modulación de frecuencia frequency-modulation broadcast band

banda de rechazo rejection band
banda de servicio service band
banda de sintonización tuning band
banda de subportadora subcarrier band
banda de supresión por filtro filter stop band
banda de telemedida telemetering band
banda de televisión television band
banda de tensión regulada band of regulated voltage
banda de transmisión transmission band
banda de transmisión de filtro filter transmission band
banda de un tercio de octava one-third-octave band
banda de valencia valence band
banda de video video band
banda de videofrecuencia video-frequency band
banda de voz voice band
banda decimétrica decimetric band
banda del inducido armature band
banda denaria denary band
banda efectiva effective band
banda electrónica electron band
banda espectral spectral band
banda exclusiva exclusive band
banda fundamental fundamental band
banda infraacústica subaudio band
banda lateral sideband
banda lateral asimétrica asymmetrical sideband
banda lateral cuasiúnica quasi-single sideband
banda lateral de manipulación keying sideband
banda lateral de modulación modulation sideband
banda lateral independiente independent sideband
banda lateral inferior lower sideband
banda lateral invertida inverted sideband
banda lateral parcialmente suprimida partially suppressed sideband
banda lateral principal main sideband
banda lateral residual vestigial sideband
banda lateral superior upper sideband
banda lateral suprimida suppressed sideband
banda lateral transmitida transmitted sideband
banda lateral única single sideband
banda libre clear band
banda llena filled band
banda magnética magnetic band
banda media medium band
banda métrica metric band
banda milimétrica millimeter band
banda muerta dead band
banda nominal nominal band
banda normal normal band, standard band
banda ocupada occupied band
banda parcialmente ocupada partially occupied band
banda pasante passband
banda permitida allowed band
banda perturbadora jammer band
banda prohibida forbidden band
banda proporcional proportional band
banda Q Q band
banda reducida reduced band
banda residual residual band, vestigial band
banda rotacional rotational band
banda sintonizable tunable band
banda sonora sound band
banda superior upper band
banda suprimida stopband
banda transmitida transmitted band
banda V V band
banda X X band
bandas laterales de colores color sidebands
bandas laterales espurias spurious sidebands
bandas relacionadas armónicamente harmonically related bands

bandeja de alimentación feed tray
bandeja de alimentación de papel paper-feed tray
bandeja de cables cable tray
bandeja de carga loading tray
bandeja de montaje mounting tray
bandeja de residuos de perforaciones chip tray
bandeja de tarjetas card tray
baño de aceite oil bath
baño de decapado pickling bath
baño de eliminación stripper, strip
baño de galvanoplastia plating bath
baño de limpieza ultrasónica ultrasonic-cleaning
  bath
baño de plomo lead bath
baño electrolítico electrobath
baño galvánico galvanic bath
baño primario strike bath
baño ultrasónico ultrasonic bath
baquelita bakelite
baquelizado bakelized
bario barium
barión baryon
barn barn
barniz aislante insulating varnish
baroconmutador baroswitch
barógrafo barograph
barométrico barometric
barómetro barometer
barómetro digital digital barometer
barorresistor baroresistor
barostato barostat
barotermógrafo barothermograph
barra bar, rod, shaft
barra amortiguadora amortisseur bar
barra colectora collecting bar, busbar, bus
barra colectora de potencia power busbar
barra colectora en anillo ring bus
barra colectora positiva positive busbar
barra colectora tubular tubular busbar
barra cortocircuitadora short-circuiting bar
barra cruzada crossbar
barra de anclaje anchor rod
barra de baja tensión low-voltage bar
barra de código code bar
barra de colores color bar
barra de conexión connection bar
barra de cuarzo quartz bar
barra de deriva drift bar
barra de descarga drop bar
barra de fase phase bar
barra de impresión writing bar
barra de zumbido hum bar
barra del inducido armature bar
barra distribuidora de potencia power-distributing
  bar
barra magnética magnetic bar
barra negativa negative bar
barra pararrayos lightning rod
barra Q Q bar
barra universal universal bar
barra vertical vertical bar
barra X X bar
barra Y Y bar
barra Z Z bar
barras apareadas paired bars
barras de ajuste test bars
barras de interferencia por sonido sound bars
barras de modulación modulation bars
barras de reserva reserve bars
barras de transferencia transfer bars
barras de zumbido horizontales horizontal-hum
  bars
barrera barrier

barrera colector-base collector-base barrier
barrera contra agentes atmosféricos weather
  barrier
barrera de energía de potencial de contacto
  contact-potential energy barrier
barrera de potencial potential barrier
barrera de potencial de contacto contact-potential
  barrier
barrera de sonido sound barrier
barrera de unión junction barrier
barrera magnética magnetic barrier
barrera PN PN barrier
barrera protectora protective barrier
barrera rectificadora rectifying barrier
barrera sónica sonic barrier
barrera superficial surface barrier
barrido sweep, scan, scanning
barrido calibrado calibrated sweep
barrido circular circular sweep
barrido cónico conical sweep
barrido controlado gated sweep
barrido cuasilineal quasi-linear sweep
barrido de banda ancha wideband sweep
barrido de campo field sweep
barrido de frecuencia frequency sweep
barrido de muesca notch sweep
barrido de pantalla screen sweep
barrido de precisión precision sweep
barrido directo forward sweep
barrido disparado triggered sweep
barrido en diente de sierra sawtooth sweep
barrido en espiral spiral sweep
barrido ensanchado expanded sweep
barrido exponencial exponential sweep
barrido helicoidal helical sweep
barrido horizontal horizontal sweep
barrido incremental incremental sweep
barrido lineal linear sweep
barrido magnético magnetic sweep
barrido mecánico mechanical sweep
barrido mixto mixed sweep
barrido no recurrente nonrecurrent sweep
barrido no repetitivo nonrepetitive sweep
barrido radial radial sweep
barrido recurrente recurrent sweep
barrido repetitivo repetitive sweep
barrido retardado delayed sweep
barrido retardante delaying sweep
barrido sincronizado synchronized sweep
barrido único single sweep
barrido variable variable sweep
barrido vertical vertical sweep
barrita de Flinders Flinders bar
basado en cintas tape-based
basado en computadora computer based
basado en sensores sensor-based
basal basal
basculador flip-flop
basculador acoplado de corriente alterna
  alternating-current-coupled flip-flop
basculador complementario complementary flip-flop
basculador controlado gated flip-flop
basculador de alta velocidad high-speed flip-flop
basculador de Eccles-Jordan Eccles-Jordan flip-flop
basculador de lámparas de neón neon-bulb flip-flop
basculador dinámico dynamic flip-flop
basculador electromecánico electromechanical
  flip-flop
basculador estático static flip-flop
basculador ferroeléctrico ferroelectric flip-flop
basculador ferrorresonante ferroresonant flip-flop
basculador magnético magnetic flip-flop
basculador NO-Y NAND flip-flop

basculador temporizado clocked flip-flop
basculador transistorizado transistorized flip-flop
base base, basis, radix, socket-outlet
base aislante insulating base
base binaria binary base
base de acetato acetate base
base de batería battery base
base de bayoneta bayonet base
base de clavija plug sleeve
base de coma flotante floating-point radix
base de complemento complement base
base de datos database
base de datos activa active database
base de datos archivados archival database
base de datos central central database
base de datos compartida shared database
base de datos de almacenamiento principal main
  storage database
base de datos de red network database
base de datos de referencia reference database
base de datos distribuida distributed database
base de datos en línea on-line database
base de datos física physical database
base de datos fuente source database
base de datos integrada integrated database
base de datos jerárquica hierarchical database
base de datos lógica logical database
base de datos numéricos numerical database
base de datos relacional relational database
base de enganche lock-in base
base de fusible fuse base
base de interfaz de línea line interface base
base de jack jack base
base de montaje temporal breadboard
base de poliéster polyester base
base de porcelana porcelain base
base de sistema numérico number system base
base de tiempo time base
base de tiempo lineal linear time base
base de transistor transistor base
base de trole trolley base
base de válvula valve base
base diheptal diheptal base
base loktal loktal base
base magnal magnal base
base mixta mixed base
base neperiana Naperian base
base no magnética nonmagnetic base
base noval noval base
base numérica number base
base octal octal base
base para obleas wafer socket
base pasiva passive base
base prefoco prefocus base
base tipo N N-type base
base tipo P P-type base
básico basic
bastidor rack, panel
bastidor de alimentación supply rack
bastidor de baterías battery rack
bastidor de bobinas repetidoras repeating-coil rack
bastidor de cables cable rack
bastidor de capacitores capacitor rack
bastidor de conmutación switching rack
bastidor de pruebas test rack
bastidor de relés relay rack
bastidor de repetidores repeater rack
basura garbage
batería battery
batería A A battery
batería activada por agua water-activated battery
batería alcalina alkaline battery
batería anódica anode battery

batería atómica atomic battery
batería B B battery
batería cargada charged battery
batería central central battery
batería compensadora buffer battery
batería común common battery
batería con tomas tapped battery
batería de acumuladores accumulator battery,
  storage battery
batería de aire air battery
batería de alimentación supply battery
batería de alta tensión high-voltage battery,
  high-tension battery
batería de audífono hearing-aid battery
batería de baja tensión low-voltage battery
batería de bloqueo blocking battery
batería de capacitores capacitor bank
batería de control control battery
batería de conversación talking battery
batería de Edison Edison battery
batería de emergencia emergency battery
batería de estación station battery
batería de estado sólido solid-state battery
batería de filamento filament battery
batería de imanes magnet battery
batería de línea line battery
batería de litio lithium battery
batería de llamada ringing battery
batería de mercurio mercury battery
batería de níquel-cadmio nickel-cadmium battery
batería de níquel-cadmio recargable rechargeable
  nickel-cadmium battery
batería de níquel-hierro nickel-iron battery
batería de pantalla screen battery
batería de pilas secas dry-cell battery, dry battery
batería de placa plate battery
batería de plomo lead battery
batería de plomo-ácido lead-acid battery
batería de plomo-antimonio lead-antimony battery
batería de plomo-calcio lead-calcium battery
batería de polarización bias battery
batería de polarización de rejilla grid-bias battery
batería de pruebas test battery
batería de refuerzo booster battery
batería de rejilla grid battery
batería de reserva reserve battery, standby battery
batería de respaldo backup battery
batería de señalización signaling battery
batería de transistores transistor battery
batería de unión junction battery
batería débil weak battery
batería ecualizadora equalizing battery
batería equilibradora balancing battery
batería estacionaria stationary battery
batería flotante floating battery
batería húmeda wet battery
batería local local battery
batería muerta dead battery
batería no recargable nonrechargeable battery
batería nuclear nuclear battery
batería polarizante polarizing battery
batería portátil portable battery
batería primaria primary battery
batería recargable rechargeable battery
batería secundaria secondary battery
batería secundaria alcalina alkaline secondary
  battery
batería silenciosa quiet battery
batería sin carga unloaded battery
batería solar solar battery
batería solar de silicio silicon solar battery
batería sulfatada sulfated battery
batería telegráfica telegraph battery

batería térmica thermal battery
batería ventilada vented battery
batería voltaica voltaic battery
batido beat, beating
batido cero zero beat
batido de portadora carrier beat
batidos interharmónicos interharmonic beats
batipitómetro bathypitometer
batitermógrafo bathythermograph
baudio baud
bazuca bazooka
bel bel
benito benito
bequerel becquerel
berilia beryllia
berquelio berkelium
beta beta
beta inversa inverse beta
beta variable variable beta
betatrón betatron
bevatrón bevatron
biaxial biaxial
biblioteca library
biblioteca de cintas tape library
biblioteca de cintas magnéticas magnetic-tape
　　library
biblioteca de datos data library
biblioteca de fases phase library
biblioteca de funciones function library
biblioteca de macroinstrucciones macro library
biblioteca de programas program library, software
　　library
biblioteca de sentencias statement library
biblioteca de sistema local local system library
biblioteca de subsistema subsystem library
biblioteca de trabajos job library
biblioteca de uso general general-purpose library
biblioteca maestra master library
biblioteca primaria primary library
biblioteca privada private library
bicóncavo biconcave
bicondicional biconditional
bicónico biconical
biconvexo biconvex
bidimensional two-dimensional
bidireccional bidirectional, two-way
biesférico bispherical
biestable bistable
biestático bistatic
bifásico biphase, two-phase
bifilar bifilar, two-wire
bifurcación bifurcation
bifurcado bifurcated
bigote de gato whisker
bilateral bilateral
bimetálico bimetallic
bimodal bimodal
bimorfo bimorphous
binario binary
binario contador counter binary
binario-decimal binary-decimal
binario fraccional fractional binary
binario natural natural binary
binario normal normal binary
binario ordinario ordinary binary
binario puro pure binary
binario regular regular binary
binaural binaural
binistor binistor
binocular binocular
binodo binode
binomio binomial
bioelectricidad bioelectricity

bioeléctrico bioelectric
bioelectrogénesis biooelectrogenesis
bioelectrónica bioelectronics
bioingeniería bioengineering
biónica bionics
biotelemetría biotelemetry
biotrón biotron
bipartición bipartition
bipolar bipolar
bipolar complejo complex bipolar
bipolaridad bipolarity
biquinario biquinary
birrefrigencia birefringence
birrejilla bigrid
bisector bisector
bisel bezel
bisimétrico bisymmetric
bisíncrono bysynchronous
bismuto bismuth
bisturí eléctrico electric bistoury
bit bit
bit cero zero bit
bit de arranque start bit
bit de cambio change bit
bit de comprobación check bit
bit de comprobación de redundancia
　　redundancy-check bit
bit de datos data bit
bit de encuadre framing bit
bit de guarda guard bit
bit de orden bajo low-order bit
bit de parada stop bit
bit de paridad parity bit
bit de paridad impar odd-parity bit
bit de referencia reference bit
bit de relleno filler bit
bit de signo sign bit
bit de zona zone bit
bit en serie serial bit
bit erróneo erroneous bit
bit indicador flag bit
bit más significativo most significant bit
bit menos significativo least significant bit
bit redundante redundant bit
bitónico bitonic
bits de estado status bits
bits de información information bits
bits de servicio service bits
bits de sincronización synchronization bits
bits por pulgada bits per inch
bits-segundo bits-second
bivalente bivalent
blanco acústico artificial artificial acoustic target
blanco de radar complejo complex radar target
blanco fantasma phantom target
blanco nominal nominal white
blanqueo bleaching
blindaje shield, shielding
blindaje de cable cable shield
blindaje de guarda guard shield
blindaje de núcleo slug
blindaje de plomo lead shielding
blindaje dividido split shield
blindaje entre etapas interstage shielding
blindaje interno internal shield
blindaje magnético magnetic shielding
blindaje no férrico nonferrous shield
blindaje no ferroso nonferrous shield
bloque block
bloque anidado nested block
bloque común common block
bloque corto short block
bloque de alimentación AB AB power pack

bloque de almacenamiento storage block
bloque de análisis analysis block
bloque de aplicación primaria primary application block
bloque de área de programa program area block
bloque de biblioteca library block
bloque de cabecera header block
bloque de código code block
bloque de contacto contact block
bloque de control control block
bloque de control de archivo file control block
bloque de control de canal channel control block
bloque de control de colas queue control block
bloque de control de conjunto de datos data set control block
bloque de control de datos data control block
bloque de control de dispositivo device control block
bloque de control de evento event control block
bloque de control de línea line-control block
bloque de control de página page-control block
bloque de control de procesos process-control block
bloque de control de tareas task-control block
bloque de control de trabajos job control block
bloque de control de unidad unit control block
bloque de cuarzo quartz block
bloque de datos data block
bloque de diodos diode pack
bloque de discos disk block
bloque de entrada entry block, input block
bloque de entrada-salida input-output block
bloque de imagen image block
bloque de memoria memory block
bloque de montaje mounting block
bloque de perforación punch block
bloque de procedimiento procedure block
bloque de programa program block
bloque de programa de canal channel program block
bloque de referencia reference block
bloque de registros record block
bloque de reserva standby block
bloque de salida output block
bloque de solicitud de interrupción interrupt request block
bloque de terminales terminal block
bloque de transmisión transmission block
bloque de unidad física physical unit block
bloque de unidad lógica logical unit block
bloque distribuidor distributing block
bloque electrónico funcional functional electronic block
bloque erróneo erroneous block
bloque fijo fixed block
bloque físico physical block
bloque fuente source block
bloque funcional functional block
bloque funcional molecular molecular functional block
bloque inactivo inactive block
bloque interno internal block
bloque lógico logical block
bloque magnetofónico motorboard
bloque permisivo permissive block
bloque primario primary block
bloque único single block
bloque variable variable block
bloqueado blocked
bloqueo blocking, locking, interlock, lockout, plugging
bloqueo absoluto absolute block
bloqueo automático de corriente alterna alternating-current automatic block

bloqueo automático de corriente continua direct-current automatic block
bloqueo de abertura aperture blockage
bloqueo de aproximación approach locking
bloqueo de barrido sweep lockout
bloqueo de escritura write lockout
bloqueo de línea line lockout
bloqueo de mensajes message blocking
bloqueo de rayo ray locking
bloqueo de rejilla grid blocking
bloqueo de teclado keyboard lockout, keyboard lockup
bloqueo interno internal blocking
bloqueo magnético magnetic locking
bloqueo permisivo absoluto absolute permissive block
bobina coil, roll, spool
bobina arrollada en capas layerwound coil
bobina captadora en horquilla hairpin pickup coil
bobina circular circular coil
bobina compensadora compensating coil
bobina con cursor slide coil
bobina con derivación central center-tapped coil
bobina con núcleo de hierro iron-core coil
bobina con tomas tapped coil
bobina correctora peaking coil
bobina correctora en derivación shunt peaking coil
bobina correctora en serie series-peaking coil
bobina de acoplamiento coupling coil
bobina de acoplamiento en horquilla hairpin coupling coil
bobina de ajuste trimmer coil
bobina de alta tensión high-tension coil
bobina de anchura width coil
bobina de antena antenna coil
bobina de antena con núcleo de ferrita ferrite-core antenna coil
bobina de arranque booster coil
bobina de barrido sweeping coil
bobina de campo field coil
bobina de campo en derivación shunt-field coil
bobina de campo magnético magnetic-field coil
bobina de capa única single-layer coil
bobina de captación pickup coil
bobina de carga loading coil
bobina de carga de antena antenna loading coil
bobina de cierre closing coil
bobina de contador current coil
bobina de convergencia convergence coil
bobina de corriente inversa reverse-current coil
bobina de chispa spark coil
bobina de choque choke coil
bobina de choque de modulador modulator choke coil
bobina de deflexión deflecting coil, deflection coil, deflector coil
bobina de deflexión electromagnética electromagnetic deflection coil
bobina de deflexión magnética magnetic deflecting coil
bobina de descarga discharge coil
bobina de desmagnetización degaussing coil
bobina de disparo tripping coil
bobina de dos tensiones dual-voltage coil
bobina de electroimán magnet coil
bobina de encuadre framing coil
bobina de enfoque focusing coil
bobina de enfoque magnético magnetic-focusing coil
bobina de equilibrio balance coil
bobina de estator stator coil
bobina de exploración probe coil, scanning coil, search coil
bobina de extinción quench coil, blowout coil

bobina de filamento filament coil
bobina de filtro filter coil
bobina de galvanómetro galvanometer coil
bobina de Helmholtz Helmholtz coil
bobina de ignición ignition coil
bobina de impedancia impedance coil
bobina de inducción induction coil
bobina de inducción contra efectos locales
　antisidetone induction coil
bobina de inducción de radiofrecuencia
　radio-frequency choke
bobina de inducción diferencial differential
　induction coil
bobina de inducción telefónica telephone induction
　coil
bobina de inductancia inductance coil, ballast
bobina de inductancia de antena aerial inductance
　coil
bobina de liberación release coil
bobina de Maxwell Maxwell coil
bobina de muestreo magnético magnetic test coil
bobina de núcleo de aire air-core coil
bobina de núcleo móvil slug-tuned coil
bobina de oscilador oscillator coil
bobina de placa plate coil
bobina de polarización magnética magnetic-biasing
　coil
bobina de Pupin Pupin coil
bobina de pureza purity coil
bobina de radiofrecuencia radio-frequency coil
bobina de reacción reaction coil
bobina de reactancia reactance coil
bobina de reactancia protectora protective
　reactance coil
bobina de reducción de zumbido hum-bucking coil
bobina de regeneración tickler coil
bobina de rejilla grid coil
bobina de relé relay coil
bobina de resistencia resistance coil
bobina de retardo retardation coil
bobina de retención retaining coil
bobina de retroalimentación feedback coil
bobina de Ruhmkorff Ruhmkorff coil
bobina de sintonización tuning coil
bobina de sintonización de antena antenna tuning
　coil
bobina de sombra shading coil
bobina de sostenimiento holding coil
bobina de suministro de batería battery-supply coil
bobina de tanque de placa plate tank coil
bobina de tensión voltage coil
bobina de Tesla Tesla coil
bobina de trabajo working coil, operating coil
bobina de transformador transformer coil
bobina de vuelta única single-turn coil
bobina de yugo yoke coil
bobina del inducido armature coil
bobina desviadora deflecting coil
bobina diferencial differential coil
bobina dividida split coil
bobina duolateral duolateral coil
bobina ecualizadora equalizing coil
bobina ecualizadora de tierra ground equalizer coil
bobina en cuadratura quadrature coil
bobina en derivación shunt coil
bobina en disco disk coil
bobina en espiral spiral coil
bobina en herradura horseshoe coil
bobina en hilado canastilla basket-weave coil
bobina en horquilla hairpin coil
bobina en panal honeycomb coil
bobina en serie series coil
bobina en telaraña spiderweb coil

bobina encapsulada encapsulated coil
bobina enchufable plug-in coil
bobina equilibradora balancing coil
bobina equilibradora anódica anode balancing coil
bobina equilibradora de fases phase balancing coil
bobina estacionaria stationary coil
bobina excitadora exciting coil, driver coil
bobina exploradora exploring coil, sniffer
bobina fantasma phantom coil
bobina ficticia dummy coil
bobina híbrida hybrid coil
bobina impregnada impregnated coil
bobina inductora inductor coil
bobina inductora de batería local local-battery
　inductor coil
bobina limitadora limiting coil
bobina limitadora de corriente current-limiting coil
bobina longitudinal longitudinal coil
bobina magnética magnetic coil
bobina magnetizante magnetizing coil
bobina mezcladora mixer coil
bobina moldeada molded coil
bobina móvil moving coil, voice coil
bobina móvil de altavoz loudspeaker voice coil
bobina multicapa multilayer coil
bobina multisección multisection coil
bobina mutua mutual coil
bobina neutralizadora de campo field-neutralizing
　coil
bobina neutralizante neutralizing coil
bobina no inductiva noninductive coil
bobina no lineal nonlinear coil
bobina oscilante oscillating coil
bobina plana flat coil
bobina poligonal polygonal coil
bobina primaria primary coil
bobina reguladora regulating coil
bobina repetidora repeating coil
bobina repetidora de circuito fantasma phantom
　circuit repeating coil
bobina repetidora fantasma phantom repeating coil
bobina rotativa rotating coil
bobina rotatoria rotor coil
bobina secundaria secondary coil
bobina selectora selector coil
bobina superpuesta backwound coil
bobina tanque tank coil
bobina térmica heat coil
bobina toroidal toroidal coil, toroid
bobina vibrante vibrating coil
bobinado winding
bobinas de deflexión horizontal
　horizontal-deflection coils
bobinas de deflexión vertical vertical-deflection
　coils
bobinas mutuamente acopladas mutually coupled
　coils
boca artificial artificial mouth
boca de bocina horn mouth
bocina horn, mouthpiece
bocina acústica acoustic horn
bocina circular circular horn
bocina compuesta compound horn
bocina con corrección de fase phase-corrected horn
bocina cónica conical horn
bocina curvada hoghorn
bocina de alimentación feed horn
bocina de Klipsch Klipsch horn
bocina de Potter Potter horn
bocina doblada folded horn
bocina eléctrica electric horn
bocina electromagnética electromagnetic horn
bocina exponencial exponential horn

bocina exponencial invertida inverted exponential horn
bocina hiperbólica hyperbolic horn
bocina logarítmica logarithmic horn
bocina multicelular multicellular horn
bocina multimodo multimode horn
bocina protectora protective horn
bocina recta straight horn
bocina rectangular rectangular horn
bocina sectorial sectoral horn
bola de avance advance ball
bolométrico bolometric
bolómetro bolometer
bolómetro coaxial coaxial bolometer
bolómetro de película film bolometer
bolómetro de termistor thermistor bolometer
bolómetro de tipo de filamento filament-type bolometer
bolus bolus
bomba pump
bomba de aceite oil pump
bomba de agua water pump
bomba de alimentación feed pump
bomba de calor heat pump
bomba de difusión diffusion pump
bomba de difusión de aceite oil-diffusion pump
bomba de difusión de mercurio mercury diffusion pump
bomba de mercurio mercury pump
bomba de vacío vacuum pump
bomba de vacío de alta presión high-pressure vacuum pump
bomba de vacío de baja presión low-pressure vacuum pump
bomba electromagnética electromagnetic pump
bomba electromagnética mecánica mechanical electromagnetic pump
bomba impelente force pump
bomba iónica ion pump
bomba magnética magnetic pump
bomba oscilante reciprocating pump
bombardeo bombardment
bombardeo catódico cathodic bombardment
bombardeo electrónico electron bombardment
bombeo pumping
bombeo en paralelo parallel pumping
bombeo óptico optical pumping
bombilla bulb
bombilla de lámpara lamp bulb
bombilla de neón neon bulb
bombilla de prueba test bulb
bombilla incandescente incandescent bulb
boquilla mouthpiece
borde border
borde ascendente rising edge
borde blanco ante imagen negra leading whites
borde blanco tras imagen negra trailing whites
borde de carácter character edge
borde de entrada leading edge
borde de guía guide edge
borde de imagen picture edge
borde de placa plate edge
borde de referencia reference edge
borde de salida trailing edge
borde de tarjeta card edge
borde de ventana window edge
borde delantero leading edge
borde descendente falling edge
borde luminoso luminous edge
borde negro ante imagen blanca leading blacks
borde negro tras imagen blanca trailing blacks
borna terminal
borne terminal, post

borne de bujía sparkplug terminal
bornita bornite
boro boron
borrable erasable
borrable eléctricamente electrically erasable
borrado erasure, deletion, blanking
borrado de archivo file deletion
borrado de campo field erasing
borrado de conjunto de datos data set deletion
borrado de datos data deletion
borrado de mensajes message delete
borrado por corriente alterna alternating-current erasing
borrado por corriente continua direct-current erasing
borrado por imán permanente permanent-magnet erase
borrado selectivo selective erase
borrado variable variable erase
borrador eraser
borrador de cinta magnética magnetic-tape eraser
borrador de cintas tape eraser
borrador masivo bulk eraser
borrar erase, delete, clear
borrar la memoria clear memory
borrar los registros de almacenamiento clear storage register
botella de Leyden Leyden bottle
botella magnética magnetic bottle
botón button, knob
botón contactor contactor button
botón de ajuste lateral lateral adjustment knob
botón de borrado erase button
botón de contacto contact button
botón de disparo trigger button
botón de micrófono microphone button
botón de oprimir para hablar press-to-talk button
botón de presión press button
botón de regulación setting knob
botón de reposición reset button
botón preajustable preset button
botón preajustado preset button
boya sonora sounding buoy
brazo arm
brazo activo active arm
brazo anódico anode arm
brazo auxiliar auxiliary arm
brazo con amortiguamiento viscoso viscous-damped arm
brazo de acceso access arm
brazo de contacto deslizante sliding-contact arm
brazo de fonocaptor pickup arm, tone arm
brazo de impedancia impedance arm
brazo de pivote pivot arm
brazo de portaescobillas brush holder arm
brazo de posicionamiento positioning arm
brazo de reactancia stub
brazo divisor dividing arm
brazo impresor printing arm
brazo inverso reverse arm
brazo termoeléctrico thermoelectric arm
brazos de razón ratio arms
bremsstrahlung bremsstrahlung
brida flange
brida de guía guide flange
brida de guíaondas waveguide flange
brillancímetro glossmeter
brillancímetro fotoeléctrico photoelectric glossmeter
brillantez brilliance
brillantez de imagen image brilliance
brillantez de punto spot brilliance
brillantez intrínseca intrinsic brilliance
brillo brightness, brilliance, gloss

brillo aparente apparent brightness
brillo bolométrico bolometric brightness
brillo de fondo background brightness
brillo de imagen picture brightness
brillo de pantalla screen brightness
brillo de punto spot brightness
brillo de superficie brightness of surface
brillo fotométrico photometric brightness
brillo intrínseco intrinsic brightness
brillo medio average brightness
brisa estática static breeze
bromo bromine
bronce bronze
bronce al berilio beryllium bronze
bronce fosforoso phosphor bronze
brújula compass
brújula maestra master compass
bruñido burnishing
bruñido a mano hand burnishing
bruñidor burnisher
bucle loop
bucle abierto open loop
bucle anidado nested loop
bucle básico basic loop
bucle cerrado closed loop
bucle con autorreposición self-resetting loop
bucle corriente current loop
bucle cuadrado square loop
bucle de acceso rápido rapid-access loop
bucle de acoplamiento coupling loop
bucle de acoplamiento de salida output coupling
    loop
bucle de alta ganancia high-gain loop
bucle de anidamiento nesting loop
bucle de arrollamiento dividido split-winding loop
bucle de cinta de papel paper-tape loop
bucle de conteo counting loop
bucle de control control loop
bucle de control de procesos process-control loop
bucle de control electrónico electronic control loop
bucle de control por retroalimentación feedback
    control loop
bucle de corriente current loop
bucle de enganche de fase phase-locked loop
bucle de espera waiting loop
bucle de grupo group loop
bucle de histéresis hysteresis loop
bucle de histéresis magnética magnetic hysteresis
    loop
bucle de línea line loop
bucle de línea aérea open-wire loop
bucle de medida measuring loop
bucle de modificación modification loop
bucle de proceso process loop
bucle de programa program loop
bucle de prueba test loop
bucle de recepción receiving loop
bucle de retroalimentación feedback loop
bucle de retroalimentación principal main feedback
    loop
bucle de sondeo polling loop
bucle de tensión voltage loop
bucle de tierra ground loop
bucle infinito infinite loop
bucle iterativo iterative loop
bucle local local loop
bucle menor minor loop
bucle no controlado uncontrolled loop
bucle principal major loop
bucle rectangular rectangular loop
bucle sin fin endless loop
bujía candle
bujía de Hefner Hefner candle

bujía decimal decimal candle
bujía internacional international candle
bujía patrón standard candle
bujía-pie foot-candle
burbujas magnéticas magnetic bubbles
bus bus
bus bidireccional bidirectional bus
bus de alta velocidad high-speed bus
bus de arbitraje arbitration bus
bus de cadena en margarita daisy chain bus
bus de comprobación check bus
bus de control control bus
bus de corriente continua direct-current bus
bus de datos data bus
bus de datos bidireccional bidirectional data bus
bus de datos de sistema system data bus
bus de datos óptico optical data bus
bus de dirección address bus
bus de entrada de datos data input bus
bus de entrada-salida input-output bus
bus de expansión expansion bus
bus de interfaz interface bus
bus de memoria memory bus
bus de puesta a tierra ground bus
bus de sistema system bus
bus de transferencia de datos data-transfer bus
bus de transferencia de dígitos digit-transfer bus
bus digital digital bus
bus en anillo ring bus
bus equilibrador balancing bus
bus interno internal bus
bus unidireccional unidirectional bus
buscador finder, searcher, whisker
buscador de fugas leakage finder
buscador de haz beam finder
buscador de línea line finder
buscador de llamada call finder
buscador de metales metal finder
buscador de polaridad polarity finder
buscador de profundidad acústico acoustic depth
    finder
buscador de profundidad eléctrico electric depth
    finder
buscador de profundidad ultrasónico ultrasonic
    depth finder
buscador de registro register finder
buscador de trabajos job seeker
buscador de uniones junction hunter
buscador distribuidor distributing finder
buscador secundario secondary finder
buscapolos pole finder
buscar seek
búsqueda search
búsqueda asociativa associative lookup
búsqueda automática automatic search
búsqueda binaria binary search
búsqueda cíclica cyclic search
búsqueda de archivo file search
búsqueda de área area search
búsqueda de base radix search
búsqueda de Boole Boolean search
búsqueda de cinta tape search
búsqueda de Fibonacci Fibonacci search
búsqueda de lista encadenada chained-list search
búsqueda de primera línea first line find
búsqueda dicotómica dichotomizing search
búsqueda en cadena chaining search
búsqueda en paralelo parallel search
búsqueda en serie serial search
búsqueda en tabla table look-up
búsqueda general general search
búsqueda global global search
búsqueda inversa reverse search, backward search

búsqueda lineal linear search
búsqueda logarítmica logarithmic search
búsqueda programada programmed search
búsqueda secuencial sequential search
búsqueda y reemplazo search and replace
búsqueda y reemplazo global global search and
 replace
byte byte
byte de atributo attribute byte
byte de atributo de pantalla screen attribute byte
byte de control control byte
byte de control de secuencia string control byte
byte de desplazamiento displacement byte, shift byte
byte de dos bits two-bit byte
byte de relleno padding byte
byte de respuesta extendido extended response byte
byte de seis bits six-bit byte
byte de siete bits seven-bit byte
byte de tres bits three-bit byte
byte indicador flag byte
bytes por pulgada bytes per inch

C

caballo de fuerza horsepower
caballo de fuerza fraccional fractional horsepower
caballo de fuerza-hora horsepower-hour
caballos de fuerza indicados indicated horsepower
cabecera header
cabecera de enlace link header
cabeza head
cabeza adaptadora adapter head
cabeza bipolar double-pole head
cabeza combinada combined head
cabeza de baja impedancia low-impedance head
cabeza de borrado erasing head, erase head
cabeza de borrado de corriente alterna
 alternating-current erasing head
cabeza de borrado de imán permanente
 permanent-magnet erasing head
cabeza de borrado por corriente continua
 direct-current erasing head
cabeza de borrado selectivo selective-erase head
cabeza de cable cable head
cabeza de cinta tape head
cabeza de cinta magnética magnetic-tape head
cabeza de contador de centelleos
 scintillation-counter head
cabeza de escritura write head
cabeza de escritura-lectura write-read head
cabeza de exploración scanning head
cabeza de grabación-lectura recording-playback
 head
cabeza de grabación longitudinal longitudinal
 recording head
cabeza de grabación-reproducción
 recording-reproducing head
cabeza de impresión print head
cabeza de lectura read head
cabeza de lectura de cinta tape-reading head
cabeza de lectura-escritura read-write head
cabeza de lectura-escritura combinada combined
 read-write head
cabeza de lectura magnética magnetic reading head
cabeza de manipulación keying head
cabeza de medida measuring head
cabeza de micrófono microphone head
cabeza de perforación punch head
cabeza de pista completa full-track head
cabeza de prelectura preread head
cabeza de proyección projection head
cabeza de proyector projector head
cabeza de radiofrecuencia radio-frequency head
cabeza de registro recording head
cabeza de registro en cinta magnética magnetic-tape
 recording head
cabeza de registro-lectura recording-playback head
cabeza de registro longitudinal longitudinal
 recording head
cabeza de registro magnético magnetic recording
 head
cabeza de registro mecánica mechanical recording
 head
cabeza de registro-reproducción
 recording-reproducing head

cabeza de reproducción playback head, reproducing head
cabeza de reproducción simultánea simultaneous-playback head
cabeza de transductor transducer head
cabeza de trazado plotting head
cabeza de trole trolley head
cabeza desalineada misaligned head
cabeza en anillo ring head
cabeza estereofónica stereophonic head
cabeza estereoscópica stereoscopic head
cabeza flotante floating head
cabeza fotométrica photometer head
cabeza impresora láser laser print head
cabeza lectora playback head
cabeza magnética magnetic head
cabeza magnética de video video magnetic head
cabeza magnetizada magnetized head
cabeza magnetizante magnetizing head
cabeza magnetodinámica magnetodynamic head
cabeza monitora monitor head
cabeza móvil flying head
cabeza multipista multitrack head
cabeza pirométrica pyrometer head
cabeza reproductora magnética magnetic reproducing head
cabeza rotativa rotating head
cabeza unipolar single-pole head
cabezal head assembly
cabezas apiladas stacked heads
cabezas escalonadas staggered heads
cable cable
cable a presión pressure cable
cable aéreo aerial cable, overhead cable
cable aéreo catenario catenary aerial cable
cable aislado insulated cable
cable aislado con plástico plastic-insulated cable
cable aislado con policloropreno polychloroprene-insulated cable
cable aislado con polietileno polyethylene-insulated cable
cable aislado con seda silk-insulated cable
cable aislado por aire air-insulated cable
cable aislado por papel paper-insulated cable
cable alimentador feeder cable
cable barnizado varnished cable
cable bifilar bifilar cable, two-wire cable
cable blindado armored cable
cable cargado loaded cable
cable coaxial coaxial cable
cable coaxial con espacio de aire air-space coax
cable coaxial de dieléctrico de aire air-dielectric coax
cable coaxial doble dual coaxial cable
cable coaxial flexible flexible coaxial cable
cable coaxial multicanal multichannel coaxial cable
cable coaxial perlado beaded coax
cable combinado combination cable
cable compuesto composite cable
cable con espacio de aire air-space cable
cable con suspensión continua lashed cable
cable concéntrico concentric cable
cable cubierto de caucho rubber-covered cable
cable de abonado subscriber cable
cable de acometida service cable, entrance cable
cable de alimentación power cable, feeding cable
cable de alimentación de baja tensión low-voltage power cable
cable de alta tensión high-tension cable
cable de antena antenna cable
cable de audio audio cable
cable de baja capacidad low-capacity cable

cable de bajada drop cable
cable de banda ancha wideband cable
cable de calentamiento heating cable
cable de cámara camera cable
cable de casa house cable
cable de cinta ribbon cable
cable de cobre copper cable
cable de comunicaciones communications cable
cable de conductor macizo solid-conductor cable
cable de conductor único single-conductor cable
cable de conductores de sector sector cable
cable de conductores divididos split-conductor cable
cable de conductores múltiples multiple-conductor cable
cable de conductores trenzados stranded-conductor cable
cable de conexión connecting cable, stub cable
cable de conexión flexible pigtail
cable de conexión temporal patch cable
cable de control control cable
cable de control de potencia power control cable
cable de corriente alterna alternating-current cable
cable de cuadrante dial cable
cable de cuadretes quadded cable
cable de cuadretes en espiral spiral-four cable
cable de cuadretes en estrella star-quad cable
cable de datos digitales digital-data cable
cable de derivación branch cable
cable de desmagnetización degaussing cable
cable de dieléctrico gaseoso gaseous-dielectric cable
cable de dieléctrico sólido solid-dielectric cable
cable de distribución distribution cable
cable de dos conductores two-conductor cable
cable de enlace tie cable, trunk cable
cable de entrada leading-in cable
cable de entrada de red mains cable
cable de extensión extension cable
cable de fibras ópticas fiber-optic cable
cable de gas gas-filled cable
cable de gran capacidad large-capacity cable
cable de hilo wire cable
cable de ignición ignition cable
cable de micrófono microphone cable
cable de núcleo de aire air-core cable
cable de pared delgada thin-walled cable
cable de pares paired cable
cable de pares gemelos multiple-twin cable
cable de pares múltiples multipair cable
cable de puente jumper cable
cable de puesta a tierra grounding cable
cable de punta tip cable
cable de radiofrecuencia radio-frequency cable
cable de reserva spare cable
cable de retardo delay cable
cable de retorno return cable
cable de Romex Romex cable
cable de telecomunicaciones telecommunications cable
cable de televisión television cable
cable de transmisión transmission cable
cable de unión junction cable
cable dedicado dedicated cable
cable desconectable detachable cable
cable deshilachado frayed cable
cable distribuidor distributing cable
cable dúplex duplex cable
cable eléctrico electric cable
cable en cinta tape cable
cable enterrado buried cable
cable esmaltado enameled cable
cable exterior outside cable
cable externo external cable
cable fibroóptico fiber-optic cable

cable flexible flexible cable
cable flexible plano flat flexible cable
cable forrado con yute jute-protected cable
cable gemelo twin cable
cable guía leader cable
cable hertziano Hertzian cable
cable ignífugo flame-resistant cable
cable interior inside cable
cable intermedio intermediate cable
cable interno internal cable
cable interurbano toll cable
cable inundado flooded cable
cable lateral lateral cable
cable local local cable
cable lleno de aceite oil-filled cable
cable multiconductor multiconductor cable
cable multifibra multifiber cable
cable multitrenza multistrand cable
cable no blindado unshielded cable
cable no cargado nonloaded cable
cable óptico optical cable
cable patrón standard cable
cable plano flat cable
cable plano de conductores conductor flat cable
cable plano flexible flexible flat cable
cable portador longitudinal longitudinal carrier
    cable
cable portador principal main carrier cable
cable preformado preformed cable
cable principal main cable
cable prolongador prolonger cable
cable protector protective cable
cable recubierto de plomo lead-covered cable
cable simétrico symmetrical cable
cable sin carga unloaded cable
cable sin pérdidas nonlossy cable
cable submarino submarine cable
cable subterráneo underground cable
cable superconductor superconducting cable
cable telefónico telephone cable
cable telefónico submarino submarine telephone
    cable
cable terrestre overland cable
cable transistorizado transistorized cable
cable trenzado braided cable
cable triaxial triaxial cable
cable trifilar three-core cable
cable triplex triplex cable
cable uniconductor uniconductor cable
cable unipolar single-core cable
cable Y Y cable
cableado cabling, wiring
cablegrama cablegram
cables agrupados bunched cables
cabrestante capstan
cadena antiferroeléctrica antiferroelectric chain
cadena binaria binary chain
cadena de aisladores string
cadena de amplificadores amplifier chain
cadena de cámara camera chain
cadena de datos data chain
cadena de equipo equipment chain
cadena de impresión print chain
cadena de mandos command chain
cadena de Markov Markov chain
cadena de películas film chain
cadena de recepción receive chain
cadena de relés relay chain
cadena de reproducción replay chain
cadena de transmisión transmit chain
cadena de transmisores chain of transmitters
cadena en margarita daisy chain
cadena potenciométrica potentiometer chain

cadena transmisora transmitting chain
cadmiado cadmium-plated
cadmio cadmium
caída fall, falling-off, drop
caída anódica anode fall
caída aparente apparent sag
caída catódica cathode fall
caída catódica anormal abnormal cathode fall
caída catódica normal normal cathode fall
caída de arco arc drop
caída de contacto contact drop
caída de electrodo electrode drop
caída de impulso pulse decay
caída de línea line drop
caída de potencia power drop
caída de potencial potential fall
caída de potencial anódico anode potential fall
caída de potencial catódico cathode potential fall
caída de presión pressure drop
caída de reactancia reactance drop
caída de tensión voltage drop
caída de tensión accesible accessible voltage drop
caída de tensión anódica anode voltage drop
caída de tensión de arrancador starter voltage drop
caída de tensión de encendedor ignitor voltage drop
caída de tensión de línea line-voltage drop
caída de tensión de tubo tube voltage drop
caída de tensión directa forward voltage drop
caída de tensión máxima permisible maximum
    permissible voltage drop
caída de tensión permisible permissible voltage drop
caída de tensión relativa relative voltage drop
caída de velocidad absoluta absolute speed drop
caída directa forward drop
caída directa de pico peak forward drop
caída óhmica ohmic drop
caída por impedancia impedance drop
caída por resistencia resistance drop
caída progresiva rolloff
caja enclosure, housing, cabinet, case, box
caja a prueba de agentes atmosféricos weatherproof
    enclosure
caja a prueba de agua waterproof enclosure
caja a prueba de explosiones explosionproof
    enclosure
caja a prueba de goteo dripproof enclosure
caja a prueba de salpicaduras splashproof enclosure
caja abierta por atrás open-back cabinet
caja acústica acoustic enclosure
caja aislada insulated enclosure
caja anecoica anechoic enclosure
caja azul blue box
caja contra transmisión-recepción
    antitransmit-receive box
caja de acometida service box
caja de almacenamiento storage box
caja de altavoz loudspeaker enclosure
caja de altavoz no amortiguada undamped speaker
    enclosure
caja de arranque starting box
caja de atenuación attenuation box
caja de bobinas coil box
caja de bobinas de carga loading-coil box
caja de bocina doblada folded-horn enclosure
caja de cable cable box
caja de capacitancias capacitance box
caja de capacitores capacitor box
caja de caracteres character box
caja de comprobación check box
caja de conexiones connection box
caja de conmutador switchbox
caja de control control box
caja de décadas decade box

caja de décadas de capacitancias decade capacitance box
caja de décadas de inductancias decade inductance box
caja de décadas de resistencias decade resistance box
caja de decisión decision box
caja de distribución distribution box
caja de distribución de potencia power-distribution box
caja de Doppler Doppler enclosure
caja de ecos echo box
caja de escobillas brush box
caja de fusibles fuse box
caja de imán magnet case
caja de inductancias inductance box
caja de inversión de fase phase-inversion cabinet
caja de jacks jack box
caja de junta joint box
caja de Klipsch Klipsch enclosure
caja de medidor meter case
caja de pared wall housing
caja de plástico plastic case
caja de pruebas test box
caja de puente post office box
caja de pulsadores push-button box
caja de razones ratio box
caja de razones de tensión voltage-ratio box
caja de receptor receiver cabinet
caja de relés relay box
caja de residuos de perforación chad box
caja de resistencias resistance box
caja de resonancia resonance box
caja de salida outlet box
caja de seccionamiento link box
caja de sustitución de resistencias resistance substitution box
caja de tarjetas card box
caja de terminal de cables cable terminal box
caja de terminales terminal box
caja de terminales antideflagrante flameproof terminal box
caja de tomas tap box
caja de transformador transformer case
caja de unión junction box
caja de uso general general-purpose enclosure
caja de válvulas valve box
caja distribuidora distributor box
caja distribuidora de impulsos pulse distributing box
caja exterior outdoor enclosure
caja hermética al agua watertight enclosure
caja hermética contra el polvo dusttight enclosure
caja moldeada molded case
caja negra black box
caja no continua noncontinuous enclosure
caja obscura de Salisbury Salisbury darkbox
caja protegida protected enclosure
caja reflectora de bajos bass-reflex enclosure
caja refleja reflex enclosure
caja registradora electrónica electronic cash register
caja sólida solid enclosure
caja sumergible submersible enclosure
caja ventilada vented enclosure
calcio calcium
calculadora calculator
calculadora automática automatic calculator
calculadora controlada por tarjeta card-controlled calculator
calculadora de Boole Boolean calculator
calculadora de razón ratio calculator
calculadora de red network calculator
calculadora de relés relay calculator

calculadora de tarjetas perforadas punched-card calculator
calculadora de transistores transistor calculator
calculadora de uso especial special-purpose calculator
calculadora eléctrica electric calculator
calculadora electrónica electronic calculator
calculadora impresora printing calculator
calculadora programable programmable calculator
calculadora remota remote calculator
calcular calculate
cálculo de Boole Boolean calculus
cálculo de coma fija fixed-point calculation
cálculo de coma flotante floating-point calculation
cálculo de grupo group calculation
cálculo de intensidad de campo field-strength calculation
cálculo diferencial differential calculus
cálculo en cadena chain calculation
cálculo integral integral calculus
cálculo iterativo iterative calculation
calculógrafo calculograph
calefacción eléctrica electric heating
calefacción radiante radiant heating
calefactor heater
calendario automático automatic calendar
calendario de conversión calendar of conversion
calentado eléctricamente electrically heated
calentador de aceite oil heater
calentador de inducción de tipo de núcleo core-type induction heater
calentador de inducción sin núcleo coreless induction heater
calentador de relé relay heater
calentador de resistencia resistance heater
calentador de válvula valve heater
calentador dieléctrico dielectric heater
calentador eléctrico electric heater
calentador electrónico electronic heater
calentador en obscuridad dark heater
calentador en serie series heater
calentador inductivo inductive heater
calentador por histéresis hysteresis heater
calentador por inducción induction heater
calentador por inducción de baja frecuencia low-frequency induction heater
calentador por inducción de dos frecuencias dual-frequency induction heater
calentador por inducción doméstico domestic induction heater
calentador por inmersión immersion heater
calentador radiante radiant heater
calentador termoeléctrico thermoelectric heater
calentamiento heating, warmup
calentamiento catódico paralelo parallel cathode heating
calentamiento de ánodo heating of anode
calentamiento de rejilla heating of grid
calentamiento dieléctrico dielectric heating
calentamiento diferencial differential heating
calentamiento dinámico dynamic heating
calentamiento directo direct heating
calentamiento electrónico electronic heating
calentamiento indirecto indirect heating
calentamiento inductivo inductive heating
calentamiento interno internal heating
calentamiento longitudinal longitudinal heating
calentamiento óhmico ohmic heating
calentamiento por alta frecuencia high-frequency heating
calentamiento por arco arc heating
calentamiento por corrientes parásitas eddy-current heating

calentamiento por haz electrónico electron-beam heating
calentamiento por histéresis hysteresis heating
calentamiento por impedancia impedance heating
calentamiento por inducción induction heating
calentamiento por inducción de baja frecuencia low-frequency induction heating
calentamiento por ondas wave heating
calentamiento por radiofrecuencia radio-frequency heating
calentamiento por resistencia resistance heating
calentamiento ultrasónico ultrasonic heating
calibración calibration, setting
calibración de canal channel calibration
calibración de frecuencia frequency calibration
calibración de instrumento instrument calibration
calibración de longitud de onda wavelength calibration
calibración de nivel de potencia power-level calibration
calibración de radar radar calibration
calibración de razón ratio calibration
calibración de tensión voltage calibration
calibración de tiempo time calibration
calibración dinámica dynamic calibration
calibración estática static calibration
calibración estroboscópica stroboscopic calibrating
calibración interna internal calibration
calibración lineal linear calibration
calibración relativa relative calibration
calibración secundaria secondary calibration
calibrado calibrated
calibrado por cristal crystal-calibrated
calibrador calibrator, gage
calibrador de cristal crystal calibrator
calibrador de cristal de cuarzo quartz-crystal calibrator
calibrador de dos frecuencias dual-frequency calibrator
calibrador de frecuencias frequency calibrator
calibrador de hilos wire gage
calibrador de medidor de combustible fuel-gage calibrator
calibrador de nivel sonoro sound-level calibrator
calibrador de tapón plug gage
calibrador de tensión voltage calibrator
calibrador de vibraciones vibration calibrator
calibrador marcador marker calibrator
calibraje gaging
calibrar calibrate, gage
calibre gage
calibre de alambres de Birmingham Birmingham wire gage
calibre de Brown y Sharpe Brown and Sharpe gage
calibre de hilos wire gage
calibre maestro master gage
calidad quality
calidad abstracta abstract quality
calidad de colores color quality
calidad de datos data quality
calidad de imagen picture quality
calidad de impresión print quality
calidad de programas software quality
calidad de transmisión transmission quality
calidad de voz speech quality
calidad dinámica dynamic quality
calidad media average quality
caliente hot
calificación qualification
calificación de equipo equipment qualification
californio californium
calomel calomel
calométrico calometric

calor de emisión heat of emission
calor de fusión heat of fusion
calor de Joule Joule heat
calor de Peltier Peltier heat
calor de reacción heat of reaction
calor de Thomson Thomson heat
calor de vaporización heat of vaporization
calor específico specific heat
calor específico de Debye Debye specific heat
calor infrarrojo infrared heat
calor luminoso luminous heat
calor por histéresis hysteresis heat
calor por resistencia resistance heat
calor radiante radiant heat
calor radiogénico radiogenic heat
calor reflejado reflected heat
calor total total heat
calorescencia calorescence
calorescente calorescent
caloría-gramo gram-calorie
calorimétrico calorimetric
calorímetro calorimeter
calorímetro de agua water calorimeter
calorímetro de guíaondas waveguide calorimeter
calorímetro de Joule Joule calorimeter
cama de tarjetas card bed
cámara ambiental environmental chamber
cámara anecoica anechoic chamber
cámara anecoica electromagnética electromagnetic anechoic chamber
cámara climática climate chamber
cámara con superorticón image-orthicon camera
cámara con temperatura controlada temperature-controlled chamber
cámara de calentamiento heating chamber
cámara de conductividad conductivity chamber
cámara de difracción diffraction camera
cámara de difracción de rayos X X-ray diffraction camera
cámara de difracción electrónica electron diffraction chamber
cámara de distribución distribution chamber
cámara de empalmes splicing chamber
cámara de estado sólido solid-state camera
cámara de expansión expansion chamber
cámara de explosión explosion chamber
cámara de figura de Lichtenberg Lichtenberg figure camera
cámara de iconoscopio iconoscope camera
cámara de infrarrojo infrared camera
cámara de ionización ionization chamber
cámara de ionización compensada compensated ionization chamber
cámara de ionización de aire libre open-air ionization chamber
cámara de ionización de conteo counting ionization chamber
cámara de ionización de impulsos pulse-ionization chamber
cámara de ionización de rejilla grid ionization chamber
cámara de ionización diferencial differential ionization chamber
cámara de ionización monitora monitor ionization chamber
cámara de ionización patrón standard ionization chamber
cámara de ionización proporcional proportional ionization chamber
cámara de niebla de Wilson Wilson cloud chamber
cámara de orticón orthicon camera
cámara de pruebas test chamber
cámara de pruebas ambiental environmental test

chamber
cámara de radar radar camera
cámara de rayos gamma gamma-ray camera
cámara de recolección electrónica electron
    collection chamber
cámara de reverberación reverberation chamber
cámara de Salisbury Salisbury chamber
cámara de Sievert Sievert chamber
cámara de sonido sound chamber
cámara de televisión television camera
cámara de temperatura temperature chamber
cámara de vacío vacuum chamber
cámara de Wilson Wilson chamber
cámara ecoica echo chamber
cámara electrónica electron camera
cámara estereoscópica stereoscopic camera
cámara fría cold chamber
cámara láser laser camera
cámara monoscópica monoscopic camera
cámara múltiple multiple chamber
cámara osciloscópica oscilloscope camera
cámara panorámica panoramic camera
cámara patrón standard chamber
cámara resonante resonant chamber
cámara superemitrón superemitron camera
cámara tricolor tricolor camera
cámara tricroma beam splitter camera
cámara vidicón vidicon camera
cambiador changer
cambiador de canal channel changer
cambiador de frecuencia frequency changer
cambiador de frecuencia de cristal
    crystal-frequency changer
cambiador de frecuencia de semiconductor
    semiconductor frequency changer
cambiador de frecuencia estático static frequency
    changer
cambiador de modo mode changer
cambiador de polarización polarization changer
cambiador de polos pole changer
cambiador de velocidad speed changer
cambio change, shift
cambio alfabético alphabetic shift
cambio colorimétrico colorimetric shift
cambio de amplitud amplitude change
cambio de canal channel shift
cambio de Compton Compton shift
cambio de control control change
cambio de control intermedio intermediate-control
    change
cambio de diámetro de guía de ondas flare angle
cambio de estado status change
cambio de fase phase change
cambio de frecuencia frequency change
cambio de frecuencia incremental incremental
    frequency change
cambio de modo mode change
cambio de presión pressure change
cambio de tipo de letras font change
cambio de velocidad speed change
cambio global global change
cambio mínimo detectable minimum detectable
    change
cambio primario primary change
cambray cambric
cambray barnizado varnished cambric
campana de señalización signaling bell
campana eléctrica electric bell
campana electromecánica electromechanical bell
campana electrónica electronic bell
campo field
campo acelerador accelerating field
campo alfabético alphabetic field

campo alfanumérico alphanumeric field
campo alterno alternating field
campo atenuado attenuated field
campo calculado calculated field
campo cinético motional field
campo clave key field
campo coercitivo coercive field
campo común common field
campo concatenado concatenated field
campo constante constant field
campo cristalino crystalline field
campo crítico critical field
campo crítico inferior lower critical field
campo cuantificado quantized field
campo de almacenamiento storage field
campo de antena antenna field
campo de arrastre drift field
campo de borrado erasing field
campo de búsqueda search field
campo de caracteres character field
campo de carga espacial space-charge field
campo de clase menor minor class field
campo de clasificación sort field
campo de código de operación operation-code field
campo de comentarios comment field
campo de comprobación check field
campo de conducción conduction field
campo de conmutación commutating field
campo de control control field
campo de control de acceso access-control field
campo de control de canal channel control field
campo de control de clasificación sort control field
campo de control forzado forced control field
campo de control incondicional unconditional
    control field
campo de control intermedio intermediate-control
    field
campo de control menor minor control field
campo de control normal normal control field
campo de control principal major control field
campo de corrección correction field
campo de corte cutoff field
campo de Coulomb Coulomb field
campo de cuenta de datos data count field
campo de datos data field
campo de datos de control control data field
campo de datos empaquetados packed data field
campo de deflexión deflecting field
campo de desbordamiento overflow field
campo de despolarización depolarization field
campo de destino destination field
campo de difracción diffraction field
campo de dirección address field
campo de dirección interna internal address field
campo de dispositivo device field
campo de enfoque focusing field
campo de enlace link field
campo de entrada input field
campo de estación de visualización display station
    field
campo de estado de canal channel status field
campo de exploración scanning field
campo de exploración impar odd scanning field
campo de flujo armónico harmonic flow field
campo de fuerza force field
campo de gráficos graphics field
campo de guía guide field
campo de Hall Hall field
campo de identificación de formato format
    identification field
campo de imagen image field
campo de inducción induction field
campo de información information field

campo de información de controlador controller information field
campo de instrucción instruction field
campo de interferencia interference field
campo de jacks jack field
campo de líneas impares odd-line field
campo de líneas pares even-line field
campo de longitud variable variable-length field
campo de Lorentz Lorentz field
campo de mandos command field
campo de mensajes message field
campo de microprograma microprogram field
campo de modo mode field
campo de motor motor field
campo de nivel de trabajos job-level field
campo de onda espacial space-wave field
campo de opciones option field
campo de operación operation field
campo de perforación punching field
campo de procesamiento de datos data-processing field
campo de radiación radiation field
campo de red array field
campo de referencia reference field
campo de referencia de campo field reference field
campo de respuesta extendido extended response field
campo de retrodispersión backscatter field
campo de ruido noise field
campo de salida output field
campo de secuencia sequence field
campo de selección selection field
campo de signo sign field
campo de sonido difuso diffuse sound field
campo de tarjeta card field
campo de tarjeta perforada punched-card field
campo de tolerancia tolerance field
campo de transmisión sending field
campo de visualización display field
campo de zumbido hum field
campo débil weak field
campo débilmente perturbado weakly perturbed field
campo derivado derived field
campo desmagnetizante demagnetizing field
campo difuso diffuse field, scattered field
campo distante distant field
campo eléctrico electric field
campo eléctrico cinético motional electric field
campo eléctrico de dipolo dipole electric field
campo eléctrico dinámico dynamic electric field
campo eléctrico oscilante oscillating electric field
campo eléctrico perturbado perturbed electric field
campo eléctrico transitorio transient electric field
campo eléctrico uniforme uniform electric field
campo eléctrico variable varying electric field
campo electromagnético electromagnetic field
campo electromagnético alterno alternating electromagnetic field
campo electromagnético cuantificado quantized electromagnetic field
campo electromagnético reflejado reflected electromagnetic field
campo electromagnético transversal transverse electromagnetic field
campo electrónico electronic field
campo electrostático electrostatic field
campo electrostático aleatorio random electrostatic field
campo elíptico elliptical field
campo empaquetado packed field
campo en cuadratura quadrature field
campo en derivación shunt field

campo entrelazado interlaced field
campo escalar scalar field
campo estacionario stationary field
campo estático static field
campo estructurado structured field
campo evanescente evanescent field
campo excitador exciting field
campo externo external field
campo fijo fixed field
campo giratorio revolving field
campo gravitacional gravitational field
campo individual individual field
campo intensificado intensified field
campo interferente interfering field
campo interno internal field
campo lejano far field
campo libre free field
campo libre de fuerza force-free field
campo magnético magnetic field
campo magnético circular circular magnetic field
campo magnético crítico critical magnetic field
campo magnético de constricción pinch magnetic field
campo magnético de dipolo dipole magnetic field
campo magnético de enfoque focusing magnetic field
campo magnético de la tierra earth's magnetic field
campo magnético dinámico dynamic magnetic field
campo magnético estacionario stationary magnetic field
campo magnético estático static magnetic field
campo magnético galáctico galactic magnetic field
campo magnético homogéneo homogeneous magnetic field
campo magnético longitudinal longitudinal magnetic field
campo magnético parásito stray magnetic field
campo magnético persistente persistent magnetic field
campo magnético perturbado perturbed magnetic field
campo magnético pulsado pulsed magnetic field
campo magnético pulsante pulsating magnetic field
campo magnético terrestre terrestrial magnetic field
campo magnético transitorio transient magnetic field
campo magnético uniforme uniform magnetic field
campo magnetizante magnetizing field
campo magnetostático magnetostatic field
campo móvil moving field
campo múltiple multiple field
campo no uniforme nonuniform field
campo nulo null field
campo numérico numerical field
campo numérico empaquetado packed numerical field
campo oculto hidden field
campo ondulatorio wave field
campo operando operand field
campo parásito stray field
campo perturbado perturbed field
campo plano flat field
campo polarizado linealmente linearly polarized field
campo por arrollamiento en serie series field
campo primario primary field
campo profundo deep field
campo protegido protected field
campo protegido programable programmable protected field
campo próximo near field
campo radiante radiant field
campo receptor receiving field

campo repetitivo repeating field
campo residual residual field
campo retardado retarded field
campo retardador retarding field
campo rotacional rotational field
campo rotativo rotary field
campo sinusoidal sinusoidal field
campo solenoidal solenoidal field
campo sonoro sound field
campo telúrico telluric field
campo total total field
campo transversal transverse field
campo umbral threshold field
campo uniforme uniform field
campo útil useful field
campo variable variable field
campo vectorial vectorial field
campo virtual virtual field
campos de líneas de potencia de corriente alterna
    alternating-current power-line fields
canal channel, bus
canal A A channel
canal adyacente adjacent channel
canal aleatorio random channel
canal alternativo alternate channel
canal analógico analog channel
canal ancho wide channel
canal arrendado leased channel
canal asignado allocated channel
canal asignado sin exclusividad on-call channel
canal autorizado authorized channel
canal B B channel
canal bidireccional two-way channel
canal binario binary channel
canal central derivado derived center channel
canal compartido shared channel
canal común common channel
canal de acceso al almacenamiento storage access
    channel
canal de acercamiento channel of approach
canal de alto rendimiento high-performance channel
canal de audio audio channel
canal de audio acompañante accompanying audio
    channel
canal de audio adyacente adjacent audio channel
canal de aviación aviation channel
canal de banda ancha wideband channel
canal de banda angosta narrowband channel
canal de banda base baseband channel
canal de cable cable channel
canal de calidad telefónica voice-grade channel
canal de calidad vocal voice-grade channel
canal de cinta tape channel
canal de cinta de papel paper-tape channel
canal de coincidencia coincidence channel
canal de colores color channel
canal de comunicación communication channel
canal de comunicaciones de datos data
    communications channel
canal de corriente portadora carrier-current channel
canal de crominancia chrominance channel
canal de cuatro hilos four-wire channel
canal de datos data channel
canal de desbordamiento overflow channel
canal de detección detection channel
canal de diferencia difference channel
canal de emergencia emergency channel
canal de encaminamiento routing channel
canal de enfasamiento phasing channel
canal de entrada input channel
canal de entrada-salida input-output channel
canal de entrada-salida de procesador processor
    input-output channel

canal de entrada-salida en serie serial input-output
    channel
canal de entrada-salida programable programmable
    input-output channel
canal de frecuencia intermedia
    intermediate-frequency channel
canal de frecuencia única single-frequency channel
canal de frecuencia vocal voice-frequency channel
canal de frecuencias frequency channel
canal de ida go channel
canal de ida y vuelta go-and-return channel
canal de imagen picture channel
canal de información information channel
canal de información multinivel multilevel
    information channel
canal de interfaz interface channel
canal de interfaz de periférico peripheral interface
    channel
canal de lectura reading channel
canal de lectura-escritura read-write channel
canal de luminancia luminance channel
canal de llamada calling channel
canal de llegada incoming channel
canal de mensajes message channel
canal de microondas microwave channel
canal de monitoreo monitoring channel
canal de muy alta frecuencia very-high-frequency
    channel
canal de operación operating channel
canal de papel paper channel
canal de portadora carrier channel
canal de programa program channel
canal de protección protection channel
canal de radio de banda ancha wideband radio
    channel
canal de radio de emergencia emergency radio
    channel
canal de radiodifusión broadcast channel
canal de radiodifusión normal standard broadcast
    channel
canal de radiodifusión por modulación de
    frecuencia frequency-modulation broadcast
    channel
canal de radiorrelé radio-relay channel
canal de redifusión rediffusion channel
canal de referencia reference channel
canal de registro recording channel
canal de reproducción playback channel,
    reproduction channel
canal de reserva standby channel
canal de retorno return channel
canal de salida output channel
canal de señal signal channel
canal de señalización signaling channel
canal de servicio service channel
canal de sonido adyacente adjacent sound channel
canal de sonido de microondas microwave sound
    channel
canal de subportadora subcarrier channel
canal de tarjetas card channel
canal de telecomunicaciones telecommunications
    channel
canal de teleimpresora teletypewriter channel
canal de telemedida telemetering channel
canal de televisión television channel
canal de televisión de muy alta frecuencia
    very-high-frequency television channel
canal de tono tone channel
canal de tono de audio audio-tone channel
canal de tráfico traffic channel
canal de tránsito through channel
canal de transmisión transmission channel
canal de video video channel

canal de voz voice channel
canal dedicado dedicated channel
canal desplazado offset channel
canal dividido split channel
canal dúplex duplex channel
canal en cuadratura quadrature channel
canal en tiempo real real-time channel
canal estereofónico stereophonic channel
canal fantasma phantom channel
canal fijo fixed channel
canal horizontal horizontal channel
canal impar odd channel
canal inferior bottom channel
canal inverso reverse channel, backward channel
canal invertido inverted channel
canal libre clear channel
canal local local channel
canal lógico logical channel
canal monocromo monochrome channel
canal múltiple multiple channel
canal múltiplex multiplex channel
canal normal normal channel
canal ómnibus omnibus channel
canal piloto pilot channel
canal principal principal channel
canal protector protective channel
canal Q Q channel
canal radiotelefónico radiotelephone channel
canal radiotelegráfico radiotelegraph channel
canal regional regional channel
canal regular regular channel
canal ruidoso noisy channel
canal secundario secondary channel
canal selector selector channel
canal semidúplex half-duplex channel
canal silencioso quiet channel
canal simbólico symbolic channel
canal simétrico symmetrical channel
canal símplex simplex channel
canal sonoro sound channel
canal subdividido subdivided channel
canal superior top channel
canal supervisor supervisory channel
canal telefónico telephone channel
canal telefónico de portadora carrier telephone channel
canal telegráfico telegraph channel
canal telegráfico de frecuencia vocal voice-frequency telegraph channel
canal télex telex channel
canal transversal header
canal único unibus
canal unidireccional one-way channel
canal unilateral unilateral channel
canal vacante idle channel
canal vertical vertical channel
canal X X channel
canal Y Y channel
canal Z Z channel
canaleta raceway
canalización channeling, ducting
canalización dúplex duplex channeling
cancelación cancellation
cancelación de ceros zero cancellation
cancelación de crominancia chrominance cancellation
cancelación de ecos echo cancellation
cancelación inmediata immediate cancel
cancelación por trayectorias múltiples multipath cancellation
candela candela
candoluminiscencia candoluminescence
canibalización cannibalization

cantidad quantity
cantidad adimensional dimensionless quantity
cantidad analógica analog quantity
cantidad binaria binary quantity
cantidad característica characteristic quantity
cantidad compleja complex quantity
cantidad de accionamiento actuating quantity
cantidad de base base quantity
cantidad de control control quantity
cantidad de desplazamiento de frecuencia frequency-shift amount
cantidad de electricidad quantity of electricity
cantidad de error amount of error
cantidad de iluminación quantity of illumination
cantidad de información quantity of information
cantidad de luz quantity of light
cantidad de modulación amount of modulation
cantidad de radiación quantity of radiation
cantidad de referencia reference quantity
cantidad digital digital quantity
cantidad discreta discrete quantity
cantidad exponencial exponential quantity
cantidad física physical quantity
cantidad horizontal horizontal quantity
cantidad nominal rated quantity
cantidad pulsante pulsating quantity
canto singing
cañón de azul blue gun
cañón de inundación flood gun
cañón de inyección injection gun
cañón de luz light gun
cañón de mantenimiento holding gun
cañón de Pierce Pierce gun
cañón de plasma plasma gun
cañón de rojo red gun
cañón de verde green gun
cañón doblado bent gun
cañón electrónico electron gun
cañón electrónico de flujo paralelo parallel-flow electron gun
cañón iónico ion gun
cañón pentodo pentode gun
cañón protónico proton gun
capa layer, shell
capa anódica anode layer
capa base underlay
capa catódica cathode layer
capa de agotamiento depletion layer
capa de Appleton Appleton layer
capa de barrera barrier layer
capa de bloqueo blocking layer
capa de carga espacial space-charge layer
capa de conducción conducting layer
capa de control de enlace de datos data link control layer
capa de difusión diffusion layer
capa de enlace de datos data link layer
capa de Heaviside Heaviside layer
capa de Heaviside-Kennelly Heaviside-Kennelly layer
capa de Kennelly Kennelly layer
capa de Kennelly-Heaviside Kennelly-Heaviside layer
capa de plasma plasma layer
capa de transición transition layer
capa de valencia valence shell
capa delgada thin layer
capa dieléctrica dielectric layer
capa doble de Helmholtz Helmholtz double layer
capa E esporádica sporadic-E layer
capa electrónica electron shell
capa epitaxial epitaxial layer
capa física physical layer

capa intermedia　intermediate layer
capa ionizada　ionized layer
capa magnética　magnetic shell
capa monoatómica　monatomic layer
capa monomolecular　monomolecular layer
capa muerta　dead layer
capa sensible　sensitive layer
capa superficial　surface layer
capa superior　higher layer
capa superpuesta　overlay
capa troposférica　tropospheric layer
capa única　single layer
capacete　top cap
capacidad　capacity, capability, capacitance
capacidad a tierra　earth capacity
capacidad aislante　insulating capacity
capacidad binaria　binary capacity
capacidad computacional　computational ability
capacidad concentrada　lumped capacity
capacidad de acceso　access capability
capacidad de almacenamiento　storage capacity
capacidad de almacenamiento de información
　information-storage capacity
capacidad de alta tensión　high-voltage capacity
capacidad de batería　battery capacity
capacidad de canal　channel capacity
capacidad de canal de banda base　baseband channel
　capacity
capacidad de carga　load capacity
capacidad de carga de corto tiempo　short-time
　rating
capacidad de carga neta　net load capability
capacidad de cierre　making capacity
capacidad de cierre nominal　rated making capacity
capacidad de circuito　circuit capacity
capacidad de computadora　computer capacity
capacidad de contador　counter capacity
capacidad de controlador programable
　programmable controller capability
capacidad de corriente　current-carrying capacity
capacidad de cortocircuito nominal　rated
　short-circuit capacity
capacidad de descarga　discharge capacity
capacidad de descarga de protector　arrester
　discharge capacity
capacidad de direccionamiento　addressing capacity
capacidad de disco　disk capacity
capacidad de emisión　emission capability
capacidad de energía　energy capacity
capacidad de entrada　input capacity
capacidad de llamadas　call capacity
capacidad de manejo de datos　data-handling
　capacity
capacidad de memoria　memory capacity
capacidad de modulación　modulation capability
capacidad de papel　paper capacity
capacidad de placa a rejilla　plate-to-grid capacity
capacidad de potencia　power capacity
capacidad de potencia de salida　power-output
　capability
capacidad de procesamiento　processing capability
capacidad de registro　register capacity
capacidad de rendimiento　performance capability
capacidad de resolución　resolving capability,
　resolution capability
capacidad de ruido　noise capability
capacidad de ruptura　breaking capacity
capacidad de salida　output capacity, fanout
capacidad de salida nominal　rated output capacity
capacidad de servicio　service capacity
capacidad de sobrecarga　overload capacity
capacidad de sobrecarga térmica　thermal-overload
　capacity

capacidad de tarjeta　card capacity
capacidad de tráfico　traffic capacity
capacidad de unidad de impresión　print-unit
　capacity
capacidad dieléctrica　dielectric capacity
capacidad digital　digital capacity
capacidad disponible　available capacity
capacidad disponible bruta　gross available capacity
capacidad distribuida　distributed capacity
capacidad efectiva　effective capacity
capacidad electrostática　electrostatic capacity
capacidad en ampere-horas　ampere-hour capacity
capacidad en derivación　shunt capacity
capacidad en metronios　metron capacity
capacidad en watthoras　watt-hour capacity
capacidad específica　specific capacity
capacidad extendida　extended capability
capacidad fibroóptica　fiber-optic capacity
capacidad física　physical capacity
capacidad inductiva　inductive capacity
capacidad inductiva específica　specific inductive
　capacity
capacidad infinita　infinite capacity
capacidad instalada　installed capacity
capacidad interelectródica　interelectrode capacity
capacidad interna　internal capacity
capacidad interruptora　interrupting capacity
capacidad limitada　limited capacity
capacidad límite　ultimate capacity
capacidad magnética　magnetic capacity
capacidad máxima　maximum capacity
capacidad máxima bruta　gross maximum capacity
capacidad máxima de potencia　power-handling
　capacity
capacidad mutua　mutual capacity
capacidad neta　net capability
capacidad nominal　rated capacity, rating
capacidad parásita　parasitic capacity, stray capacity
capacidad pico de emisión　peak emission capability
capacidad placa-cátodo　plate-cathode capacity
capacidad placa-filamento　plate-filament capacity
capacidad placa-rejilla　plate-grid capacity
capacidad posible　possible capacity
capacidad rejilla-filamento　grid-filament capacity
capacidad rejilla-pantalla　grid-screen capacity
capacidad térmica　thermal capability
capacidad térmica anódica　anode thermal capacity
capacidad variable　variable capacity
capacidad de entrada　fanin
capacímetro　capacimeter
capacitancia　capacitance
capacitancia acústica　acoustic capacitance
capacitancia ánodo-cátodo　anode-cathode
　capacitance
capacitancia calibrada　calibrated capacitance
capacitancia cátodo-ánodo　cathode-anode
　capacitance
capacitancia cátodo-filamento　cathode-filament
　capacitance
capacitancia cátodo-rejilla　cathode-grid capacitance
capacitancia cero　zero capacitance
capacitancia colector-base　collector-base
　capacitance
capacitancia de acoplamiento　coupling capacitance
capacitancia de alambrado　wiring capacitance
capacitancia de barrera　barrier capacitance
capacitancia de blanco　target capacitance
capacitancia de cable　cable capacitance
capacitancia de caja　case capacitance
capacitancia de capa de agotamiento　depletion-layer
　capacitance
capacitancia de capa de barrera　barrier-layer
　capacitance

capacitancia de carga   load capacitance
capacitancia de circuito   circuit capacitance
capacitancia de colector   collector capacitance
capacitancia de conexión   lead capacitance
capacitancia de difusión   diffusion capacitance
capacitancia de diodo   diode capacitance
capacitancia de discontinuidad   discontinuity capacitance
capacitancia de electrodo   electrode capacitance
capacitancia de electrodo total   total electrode capacitance
capacitancia de entrada   input capacitance
capacitancia de entrada diferencial   differential-input capacitance
capacitancia de entrada diferencial equivalente   equivalent differential input capacitance
capacitancia de entrada dinámica   dynamic input capacitance
capacitancia de entrada efectiva   effective input capacitance
capacitancia de entrada en cortocircuito   short-circuit input capacitance
capacitancia de espacio efectiva   effective gap capacitance
capacitancia de interfaz catódica   cathode-interface capacitance
capacitancia de mano   hand capacitance
capacitancia de Miller   Miller capacitance
capacitancia de placa   plate capacitance
capacitancia de placa a rejilla   plate-to-grid capacitance
capacitancia de polarización   polarization capacitance
capacitancia de puerta   gate capacitance
capacitancia de rejilla   grid capacitance
capacitancia de rejilla a cátodo   grid-to-cathode capacitance
capacitancia de rejilla a pantalla   grid-to-screen capacitance
capacitancia de rejilla a placa   grid-to-plate capacitance
capacitancia de retroalimentación   feedback capacitance
capacitancia de salida   output capacitance
capacitancia de salida efectiva   effective output capacitance
capacitancia de salida en cortocircuito   short-circuit output capacitance
capacitancia de sintonización de placa   plate tuning capacitance
capacitancia de tanque de placa   plate tank capacitance
capacitancia de tanque de rejilla   grid tank capacitance
capacitancia de transición de colector   collector transition capacitance
capacitancia de unión   junction capacitance
capacitancia de varistor   varistor capacitance
capacitancia del cuerpo   body capacitance
capacitancia diferencial   differential capacitance
capacitancia directa   direct capacitance
capacitancia distribuida   distributed capacitance
capacitancia efectiva   effective capacitance
capacitancia eléctrica   electrical capacitance
capacitancia electrostática   electrostatic capacitance
capacitancia en derivación   shunt capacitance
capacitancia en paralelo   parallel capacitance
capacitancia en paralelo efectiva   effective parallel capacitance
capacitancia en serie   series capacitance
capacitancia equilibrada   balanced capacitance
capacitancia equivalente   equivalent capacitance
capacitancia geométrica   geometric capacitance

capacitancia inductiva específica   specific inductive capacitance
capacitancia interelectródica   interelectrode capacitance
capacitancia interelectródica directa   direct interelectrode capacitance
capacitancia interelemento   interelement capacitance
capacitancia libre   free capacitance
capacitancia lineal   straight-line capacitance
capacitancia marginal   fringe capacitance
capacitancia mínima   minimum capacitance
capacitancia mutua   mutual capacitance
capacitancia natural   natural capacitance
capacitancia negativa   negative capacitance
capacitancia neta   net capacitance
capacitancia nominal   nominal capacitance
capacitancia parásita   parasitic capacitance, stray capacitance
capacitancia placa-cátodo   plate-cathode capacitance
capacitancia placa-filamento   plate-filament capacitance
capacitancia placa-pantalla   plate-screen capacitance
capacitancia placa-rejilla   plate-grid capacitance
capacitancia primaria   primary capacitance
capacitancia principal   main capacitance
capacitancia rejilla-ánodo   grid-anode capacitance
capacitancia rejilla-cátodo   grid-cathode capacitance
capacitancia rejilla-pantalla   grid-screen capacitance
capacitancia rejilla-placa   grid-plate capacitance
capacitancia residual   residual capacitance
capacitancia reversible   reversible capacitance
capacitancia reversible inicial   initial reversible capacitance
capacitancia secundaria   secondary capacitance
capacitancia total   total capacitance
capacitancia variable   variable capacitance
capacitancias de tubo   tube capacitances
capacitivo   capacitive
capacitor   capacitor
capacitor acelerador   speedup capacitor
capacitor aislante   isolating capacitor
capacitor ajustable   adjustable capacitor
capacitor amortiguador   buffer capacitor
capacitor automático   automatic capacitor
capacitor autorregenerativo   self-healing capacitor
capacitor bobinado   wound capacitor
capacitor calibrado   calibrated capacitor
capacitor cerámico   ceramic capacitor
capacitor cerámico de disco   disk ceramic capacitor
capacitor cerámico moldeado   molded ceramic capacitor
capacitor cilíndrico   cylindrical capacitor
capacitor coaxial   coaxial capacitor
capacitor compensador   compensating capacitor
capacitor compensador de temperatura   temperature-compensating capacitor
capacitor con fugas   leaky capacitor
capacitor concentrado   lumped capacitor
capacitor concéntrico   concentric capacitor
capacitor controlado por tensión   voltage-controlled capacitor
capacitor de aceite   oil capacitor
capacitor de aceite mineral   mineral-oil capacitor
capacitor de acoplamiento   coupling capacitor
capacitor de acoplamiento de entrada   input coupling capacitor
capacitor de acoplamiento de salida   output coupling capacitor
capacitor de acortamiento   shortening capacitor
capacitor de agua   water capacitor
capacitor de aire   air capacitor
capacitor de aire comprimido   compressed-air capacitor

capacitor de aislamiento isolation capacitor
capacitor de ajuste trimming capacitor
capacitor de almacenamiento de energía energy-storage capacitor
capacitor de altavoz loudspeaker capacitor
capacitor de antena antenna capacitor
capacitor de aplanamiento smoothing capacitor
capacitor de arranque starting capacitor
capacitor de arranque de motor motor-start capacitor
capacitor de bloqueo blocking capacitor
capacitor de bloqueo de salida output blocking capacitor
capacitor de cerámica plateada silvered-ceramic capacitor
capacitor de cilindro drum capacitor
capacitor de compensación padding capacitor
capacitor de composición composition capacitor
capacitor de conmutación commutating capacitor
capacitor de cristal crystal capacitor
capacitor de chispas spark capacitor
capacitor de décadas decade capacitor
capacitor de decrémetro decremeter capacitor
capacitor de derivación bypass capacitor
capacitor de desacoplamiento decoupling capacitor
capacitor de desacoplo anódico anode-bypass capacitor
capacitor de desacoplo de placa plate bypass capacitor
capacitor de desvío de cátodo cathode bypass capacitor
capacitor de dieléctrico cerámico ceramic dielectric capacitor
capacitor de dieléctrico de mica mica dielectric capacitor
capacitor de dieléctrico de vidrio glass dielectric capacitor
capacitor de dieléctrico gaseoso gas dielectric capacitor
capacitor de diodo diode capacitor
capacitor de disco disk capacitor
capacitor de dos terminales two-terminal capacitor
capacitor de electrólito sólido solid-electrolyte capacitor
capacitor de entrada input capacitor
capacitor de estado sólido solid-state capacitor
capacitor de estator dividido split-stator capacitor
capacitor de extinción quenching capacitor
capacitor de factor de potencia power-factor capacitor
capacitor de fijación de banda bandset capacitor
capacitor de filtro filter capacitor
capacitor de frecuencia ultraalta ultra-high frequency capacitor
capacitor de funcionamiento de motor motor-run capacitor
capacitor de gas gas capacitor
capacitor de guillotina guillotine capacitor
capacitor de hojas metálicas foil capacitor
capacitor de ley logarítmica logarithmic-law capacitor
capacitor de libro book capacitor
capacitor de mariposa butterfly capacitor
capacitor de mica mica capacitor
capacitor de mica moldeado molded mica capacitor
capacitor de mica plateada silvered-mica capacitor
capacitor de motor motor capacitor
capacitor de papel paper capacitor
capacitor de papel impregnado en aceite oil-impregnated paper capacitor
capacitor de papel metalizado metallized-paper capacitor
capacitor de papel moldeado molded paper capacitor

capacitor de parileno parylene capacitor
capacitor de película film capacitor
capacitor de película de óxido oxide-film capacitor
capacitor de película delgada thin-film capacitor
capacitor de placa a cátodo plate-to-cathode capacitor
capacitor de placa de vidrio glass plate capacitor
capacitor de placas paralelas parallel-plate capacitor
capacitor de plástico plastic capacitor
capacitor de policarbonato metalizado metallized-polycarbonate capacitor
capacitor de poliestireno polystyrene capacitor
capacitor de poliestirol polystyrol capacitor
capacitor de porcelana porcelain capacitor
capacitor de porcelana moldeada molded porcelain capacitor
capacitor de potencia power capacitor
capacitor de refuerzo boost capacitor
capacitor de rejilla grid capacitor
capacitor de salida output capacitor
capacitor de silicio silicon capacitor
capacitor de silicio metal-óxido metal-oxide-silicon capacitor
capacitor de sintonización tuning capacitor
capacitor de sintonización de ensanche de banda bandspread tuning capacitor
capacitor de supresión suppression capacitor
capacitor de sustitución substitution capacitor
capacitor de tantalio tantalum capacitor
capacitor de tantalio húmedo wet tantalum capacitor
capacitor de tantalio sólido solid tantalum capacitor
capacitor de tres terminales three-terminal capacitor
capacitor de tubo moldeado molded tube capacitor
capacitor de unidades múltiples multiple-unit capacitor
capacitor de unión junction capacitor
capacitor de vacío vacuum capacitor
capacitor de vidrio glass capacitor
capacitor de vidrio moldeado molded glass capacitor
capacitor dependiente de la tensión voltage-dependent capacitor
capacitor desfasador phase-shifting capacitor
capacitor dextrorso clockwise capacitor
capacitor diferencial differential capacitor
capacitor discreto discrete capacitor
capacitor electrolítico electrolytic capacitor
capacitor electrolítico de corriente alterna alternating-current electrolytic capacitor
capacitor electrolítico de hojas de tantalio tantalum-foil electrolytic capacitor
capacitor electrolítico de tantalio tantalum electrolytic capacitor
capacitor electrolítico húmedo wet electrolytic capacitor
capacitor electrolítico moldeado molded electrolytic capacitor
capacitor electrolítico no polarizado nonpolarized electrolytic capacitor
capacitor electrolítico polarizado polarized electrolytic capacitor
capacitor electrolítico seco dry electrolytic capacitor
capacitor electrostático electrostatic capacitor
capacitor en bañera bathtub capacitor
capacitor en cascada cascade capacitor
capacitor en chip chip capacitor
capacitor en derivación shunting capacitor
capacitor en paralelo parallel capacitor
capacitor en pomo doorknob capacitor
capacitor en serie series capacitor
capacitor encerrado enclosed capacitor
capacitor enchufable plug-in capacitor

capacitor entre etapas interstage capacitor
capacitor equilibrador balancing capacitor
capacitor externo external capacitor
capacitor ferroeléctrico ferroelectric capacitor
capacitor fijo fixed capacitor
capacitor hidromagnético hydromagnetic capacitor
capacitor ideal ideal capacitor
capacitor ideal no lineal nonlinear ideal capacitor
capacitor impregnado en aceite oil-impregnated
    capacitor
capacitor impregnado en cera wax-impregnated
    capacitor
capacitor impregnado en líquido liquid-impregnated
    capacitor
capacitor impreso printed capacitor
capacitor inductivo inductive capacitor
capacitor inmerso en cera wax-dipped capacitor
capacitor integrado integrated capacitor
capacitor integrador integrator capacitor
capacitor láser laser capacitor
capacitor líquido liquid capacitor
capacitor lleno de aceite oil-filled capacitor
capacitor lleno de líquido liquid-filled capacitor
capacitor metalizado metallized capacitor
capacitor moldeado molded capacitor
capacitor monolítico monolithic capacitor
capacitor multisección multisection capacitor
capacitor neutralizante neutralizing capacitor
capacitor no compensado uncompensated capacitor
capacitor no inductivo noninductive capacitor
capacitor no lineal nonlinear capacitor
capacitor NP0 NP0 capacitor
capacitor oscilador oscillator capacitor
capacitor pasante feedthrough capacitor
capacitor patrón standard capacitor
capacitor plano moldeado molded flat capacitor
capacitor polarizado polarized capacitor
capacitor primario primary capacitor
capacitor protector protective capacitor
capacitor resonante resonant capacitor
capacitor revestido de cera wax-coated capacitor
capacitor rotativo rotary capacitor
capacitor saturable saturable capacitor
capacitor secundario secondary capacitor
capacitor sensible a la tensión voltage-sensitive
    capacitor
capacitor separador separating capacitor
capacitor sincrónico synchronous capacitor
capacitor subdividido subdivided capacitor
capacitor supresor de chispas spark-suppressing
    capacitor
capacitor tándem gang capacitor
capacitor telefónico telephone capacitor
capacitor tipo arandela washer capacitor
capacitor tipo botón button capacitor
capacitor tipo N N-type capacitor
capacitor-transformador de resonancia resonance
    capacitor-transformer
capacitor tubular tubular capacitor
capacitor variable variable capacitor
capacitor variable con la tensión voltage-variable
    capacitor
capacitor variable eléctricamente electrically
    variable capacitor
capacitor vernier vernier capacitor
capacitor vibrante vibrating capacitor
capacitores acoplados mecánicamente ganged
    capacitors
capacitores conectados en paralelo
    parallel-connected capacitors
capacitores en paralelo-serie parallel-series
    capacitors
capacitores en serie-paralelo series-parallel

capacitors
capacitrón capacitron
capas ionosféricas ionospheric layers
capristor capristor
cápsula de carbón carbon button
cápsula de mercurio mercury capsule
cápsula de micrófono microphone cartridge
cápsula estereofónica stereo cartridge
cápsula fonocaptora phonograph cartridge, pickup
    cartridge
cápsula fonocaptora estereofónica stereophonic
    phonograph cartridge
cápsula fonocaptora fotoeléctrica photoelectric
    phonograph pickup
cápsula magnética magnetic cartridge
captación de sonido sound pickup
captación de video video pickup
captación directa direct pickup
captador de sonido sound pickup
captor pickup
captor de baja impedancia low-impedance pickup
captor de campo eléctrico electric-field pickup
captor de datos data pickup
captor de haz luminoso light-beam pickup
captor de presión pressure pickup
captor de temperatura abierta open temperature
    pickup
captor de vibraciones vibration pickup
captor fonográfico phonograph pickup
captor fotoeléctrico photoelectric pickup
captor tacométrico tachometer pickup
captor telefónico telephone pickup
captura capture
captura condicional conditional capture
captura de datos data capture
captura de huecos hole capture
captura de iones positivos positive-ion trapping
cara anódica anode face
cara catódica cathode face
cara de medidor meter face
cara de tarjeta card face
cara menor minor face
cara polar pole face
cara principal major face
carácter character
carácter adicional additional character
carácter alfabético alphabetic character
carácter alfanumérico alphanumeric character
carácter binario binary character
carácter clave key character
carácter codificado coded character
carácter codificado en binario binary-coded
    character
carácter de alimentación de formulario form-feed
    character
carácter de alimentación de línea line-feed
    character
carácter de arranque-parada start-stop character
carácter de atributo attribute character
carácter de bloque block character
carácter de bloque intermedio intermediate block
    character
carácter de borrado delete character
carácter de borrado de línea line deletion character
carácter de cambio change character
carácter de cambio de modo mode change character
carácter de cancelación cancel character
carácter de cinta tape character
carácter de código code character
carácter de comprobación check character
carácter de comprobación de redundancia
    redundancy-check character
carácter de comprobación de redundancia

longitudinal longitudinal redundancy check character
carácter de comprobación longitudinal longitudinal check character
carácter de comprobación por bloques block check character
carácter de control control character
carácter de control de comunicaciones communication control character
carácter de control de dispositivo device control character
carácter de control de enlace de datos data link control character
carácter de control de error error control character
carácter de control de escritura write control character
carácter de control de exactitud accuracy control character
carácter de control de impresión print-control character
carácter de control de llamadas call-control character
carácter de control de tareas task-control character
carácter de control de transmisión transmission-control character
carácter de datos data character
carácter de desplazamiento shift character
carácter de dígito binario binary digit character
carácter de edición editing character
carácter de error error character
carácter de escape escape character
carácter de escape de enlace link escape character
carácter de escape de enlace de datos data link escape character
carácter de espacio space character
carácter de extensión extension character
carácter de fin de mensaje end-of-message character
carácter de fin de texto end-of-text character
carácter de fin de transmisión end-of-transmission character
carácter de final de página page-end character
carácter de formato format character
carácter de función function character
carácter de índice index character
carácter de inserción insertion character
carácter de instrucción instruction character
carácter de intervalo gap character
carácter de línea line character
carácter de mando command character
carácter de no impresión nonprint character
carácter de parada stop character
carácter de protección protection character
carácter de reconocimiento acknowledgment character, recognition character
carácter de reconocimiento de mando command recognition character
carácter de reconocimiento negativo negative acknowledgment character
carácter de rechazo reject character
carácter de referencia de tabla table reference character
carácter de relación relation character
carácter de relleno filler character, idle character, padding character
carácter de repetición repeat character
carácter de retorno return character
carácter de salto skip character
carácter de selección de dispositivo device selection character
carácter de separación de archivos file separator character
carácter de signo sign character
carácter de sincronización synchronization character

carácter de sustitución substitute character
carácter de tabulación tabulation character
carácter de timbre bell character
carácter de tinta magnética magnetic ink character
carácter de tinta magnetizada magnetized ink character
carácter de uso especial special-purpose character
carácter en blanco blank character
carácter especial special character
carácter flotante floating character
carácter funcional functional character
carácter global global character
carácter gráfico graphic character
carácter ideográfico ideographic character
carácter ilegal illegal character
carácter impropio improper character
carácter inactivo inactive character
carácter magnético magnetic character
carácter más significativo most significant character
carácter matricial matrix character
carácter menos significativo least significant character
carácter no numérico nonnumeric character
carácter no válido invalid character
carácter nulo null character
carácter numérico numerical character
carácter operacional operational character
carácter óptico optical character
carácter primario primary character
carácter prohibido forbidden character
carácter redundante redundant character
carácter separador separator character
carácter separador de grupos group separator character
carácter simbólico symbolic character
caracteres de control de línea line-control characters
caracteres de direccionamiento de código code-directing characters
caracteres de sincronización synchronizing characters
caracteres de sondeo polling characters
caracteres estáticos static characters
caracteres por minuto characters per minute
caracteres por pulgada characters per inch
caracteres por segundo characters per second
característica characteristic, feature
característica amplitud-amplitud amplitude-amplitude characteristic
característica amplitud-frecuencia amplitude-frequency characteristic
característica anódica anode characteristic
característica ascendente rising characteristic
característica carga-frecuencia load-frequency characteristic
característica colorimétrica colorimetric characteristic
característica crítica critical characteristic
característica de amplitud amplitude characteristic
característica de ánodo a cátodo anode-to-cathode characteristic
característica de atenuación attenuation characteristic
característica de avalancha avalanche characteristic
característica de bloqueo blocking characteristic
característica de bucle loop feature
característica de caída progresiva rolloff characteristic
característica de calentamiento warmup characteristic
característica de capacitancia diferencial differential capacitance characteristic
característica de capacitancia reversible reversible capacitance characteristic

característica de carga load characteristic
característica de carga alterna alternating-charge characteristic
característica de carga media mean-charge characteristic
característica de comunicaciones de datos data communications feature
característica de control control characteristic
característica de control directo direct control feature
característica de corriente anódica anode current characteristic
característica de corriente constante constant-current characteristic
característica de corriente de rejilla grid-current characteristic
característica de corte cutoff characteristic
característica de datos data characteristic
característica de deflexión deflection characteristic
característica de diodo diode characteristic
característica de diseño design feature
característica de electrodo electrode characteristic
característica de emisión emission characteristic
característica de emisión secundaria secondary-emission characteristic
característica de entrada-salida input-output characteristic
característica de excitación de rejilla grid-drive characteristic
característica de fase phase characteristic
característica de fonocaptor pickup characteristic
característica de frecuencia frequency characteristic
característica de frecuencia de atenuación attenuation-frequency characteristic
característica de frecuencia plana flat frequency characteristic
característica de frecuencia telefónica telephone frequency characteristic
característica de frenaje braking characteristic
característica de fusible fuse characteristic
característica de impedancia impedance characteristic
característica de lectura playback characteristic
característica de meseta plateau characteristic
característica de modulación modulation characteristic
característica de operación operating characteristic
característica de operación operating feature
característica de persistencia persistence characteristic
característica de placa plate characteristic
característica de ponderación weighting characteristic
característica de propagación propagation characteristic
característica de protección contra daños por fallas failsafe feature
característica de registro recording characteristic
característica de regulación de velocidad speed-regulation characteristic
característica de rejilla grid characteristic
característica de rendimiento performance characteristic
característica de reproducción reproducing characteristic
característica de resistencia-capacitancia resistance-capacitance characteristic
característica de resonancia resonance characteristic
característica de respuesta response characteristic
característica de respuesta de fase phase-response characteristic
característica de respuesta de frecuencia frequency-response characteristic

característica de respuesta de frecuencia de área area frequency-response characteristic
característica de respuesta de onda cuadrada square-wave response characteristic
característica de respuesta espectral spectral-response characteristic
característica de respuesta plana flat-response characteristic
característica de retardo delay characteristic
característica de retardo de grupo group-delay characteristic
característica de retención retention characteristic
característica de saturación saturation characteristic
característica de secuencia sequence characteristic
característica de seguimiento tracking characteristic
característica de selectividad selectivity characteristic
característica de semitono halftone characteristic
característica de sensibilidad espectral spectral-sensitivity characteristic
característica de sobrecarga overload characteristic
característica de tipo de letras font characteristic
característica de Townsend Townsend characteristic
característica de transferencia transfer characteristic
característica de transferencia dinámica dynamic transfer characteristic
característica de transistor transistor characteristic
característica de transrectificación transrectification characteristic
característica de tubo tube characteristic
característica de válvula valve characteristic
característica descendente falling characteristic
característica dinámica dynamic characteristic
característica direccional directional characteristic
característica directa forward characteristic
característica en circuito abierto open-circuit characteristic
característica en cortocircuito short-circuit characteristic
característica espectral spectral characteristic
característica estática static characteristic
característica externa external characteristic
característica ganancia-frecuencia gain-frequency characteristic
característica incidental incidental characteristic
característica intrínseca intrinsic characteristic
característica inversa reverse characteristic
característica lineal linear characteristic
característica mutua mutual characteristic
característica no lineal nonlinear characteristic
característica nominal rated characteristic
característica operacional operational feature
característica rejilla-ánodo grid-anode characteristic
característica rejilla-placa grid-plate characteristic
característica sin carga no-load characteristic
característica tensión-corriente voltage-current characteristic
característica tensión-frecuencia lineal linear voltage-frequency characteristic
características de cinta tape characteristics
características de chip de microprocesador microprocessor chip characteristics
características de conmutación switching characteristics
características de operación operation characteristics
características de programas software features
características de señales fuertes large-signal characteristics
características de transmisión transmission characteristics
características de tubo de vacío vacuum-tube

characteristics
características eléctricas inversas inverse electrical
  characteristics
características especiales special features
características gamma gamma characteristics
características inversas inverse characteristics
caractrón caractron
carbón carbon
carbón granular granular carbon
carbón impregnado impregnated carbon
carbonizado carbonized
carbonizar carbonize
carborundo carborundum
carburo de silicio silicon carbide
carcinotrón carcinotron
cardiógrafo cardiograph
cardiograma cardiogram
cardioide cardioid
cardiotacómetro cardiotachometer
carga charge, charging, load, loading, burden
carga a masa charge-to-mass
carga activa active load
carga acumulada en la base stored base charge
carga acústica acoustic load
carga adaptada matched load
carga adelantada leading load
carga aleatoria random loading
carga anódica anode load
carga aparente apparent charge
carga artificial artificial load
carga atómica atomic charge
carga automática automatic load
carga auxiliar auxiliary burden
carga baja low load
carga base base load
carga básica basic load
carga capacitiva capacitive load
carga capacitiva de salida output capacitive loading
carga central center loading
carga cero zero load, zero charge
carga compensadora compensating charge
carga computacional computational load
carga conectada connected load
carga continua continuous load
carga controlable controllable load
carga crítica critical load
carga de actividad activity loading
carga de agua water load
carga de alto régimen high-rate charge
carga de alumbrado lighting load
carga de antena antenna loading
carga de aparato eléctrico appliance load
carga de arena sand load
carga de batería battery charge
carga de canal channel loading
carga de cinta tape loading
carga de circuito circuit load
carga de compensación trickle charge
carga de corriente constante constant-current charge
carga de corriente continua direct-current load
carga de diodo diode load
carga de electrón charge of electron
carga de entrada input loading
carga de frecuencia frequency loading
carga de guíaondas waveguide load
carga de haz electrónico electron-beam loading
carga de línea line load
carga de máquina machine load
carga de memoria de núcleos core load
carga de placa plate load
carga de placa óptima optimum plate load
carga de polarización polarization charge
carga de portadora carrier loading

carga de programa program load
carga de programa dinámica dynamic program
  loading
carga de programa inicial initial program load
carga de prueba test load
carga de radiofrecuencia radio-frequency load
carga de red network load
carga de refuerzo boost charge
carga de ruido noise load
carga de ruptura breaking load
carga de salida output load
carga de salida patrón standard output load
carga de servicio service load
carga de sistema system load
carga de tensión constante constant-voltage charge
carga de trabajo work load
carga de transición transition load
carga desequilibrada unbalanced load
carga deslizante sliding load
carga desplazada displaced charge
carga dinámica dynamic load
carga distribuida distributed load
carga ecualizadora equalizing charge
carga eficaz root-mean-square load
carga eléctrica electrical charge, electrical load
carga eléctrica unitaria unit electrical charge
carga electrónica electronic charge
carga electrónica específica specific electronic
  charge
carga electrostática electrostatic charge
carga elemental elemental charge
carga en derivación shunt loading
carga en serie series loading
carga equilibrada balanced load
carga equilibradora balancing load
carga espacial space charge
carga específica de electrón electron specific charge
carga estática static charge
carga excedente excess charge
carga externa external load
carga extraordinaria extraordinary load
carga ficticia dummy load
carga ficticia de guíaondas waveguide dummy load
carga flotante floating charge
carga frontal front loading
carga impulsiva impulsive load
carga inducida induced charge
carga inductiva inductive load
carga inicial initial charge
carga interna internal load
carga iónica ionic charge, ion charge
carga latente bound charge
carga lateral lateral load
carga lenta slow charge
carga libre free charge
carga límite limit load
carga líquida liquid load
carga liviana light load
carga magnética magnetic charge
carga máxima maximum load
carga máxima aplicada maximum applied load
carga máxima estimada estimated maximum load
carga mecánica mechanical load
carga media average charge
carga momentánea momentary load
carga móvil moving load
carga muerta dead load
carga negativa negative charge
carga neta net load
carga no disipativa nondissipative load
carga no inductiva noninductive load
carga no reactiva nonreactive load
carga no resonante nonresonant load

carga nominal nominal load, rated load
carga normal normal load
carga nuclear nuclear charge
carga óhmica ohmic load
carga óptima optimum load
carga permisible permissible load
carga por bobinas coil loading
carga por demanda demand charge
carga positiva positive charge
carga puntual point charge
carga química chemical load
carga rápida quick charge
carga reactiva reactive load
carga refleja reflex loading
carga remanente remanent charge
carga residual residual charge
carga resistiva resistive load
carga retardada lagging load
carga segura safe load
carga sintonizada tuned load
carga superficial surface charge
carga suspendida suspended load
carga térmica thermal load
carga transferida transferred charge
carga transitoria transient load
carga uniforme uniform charge, uniform load
carga unitaria unit charge
carga viva live load
cargado loaded
cargado positivamente positively charged
cargador charger, loader
cargador absoluto absolute loader
cargador automático automatic loader
cargador de batería de alto régimen high-rate
    battery charger
cargador de baterías battery charger
cargador de baterías eólico wind-driven battery
    charger
cargador de cinta tape loader
cargador de compensación trickle charger
cargador de memoria de acceso aleatorio
    random-access memory loader
cargador de programa program loader
cargador de programa inicial initial program loader
cargador de programa remoto remote program
    loader
cargador de programa reubicable relocatable
    program loader
cargador de rutina de carga bootstrap loader
cargador de sistema system loader
cargador de tarjetas card loader
cargador de uso general general-purpose loader
cargador de viento wind charger
cargador residente resident loader
cargador reubicador relocating loader
cargar upload
cargar y almacenar load-and-store
cargar y ejecutar load-and-go
cargas desiguales unlike charges
carillón electrónico electronic carillon
carrera travel, throw, stroke
carrera eléctrica teórica theoretical electrical
    travel
carrete reel, spool, bobbin
carrete de alimentación feed reel
carrete libre free reel
carrete receptor take-up reel
carril de contacto contact rail
carro automático automatic carriage
carro impresor printing carriage
cartucho cartridge, magazine
cartucho cerámico ceramic cartridge
cartucho de capacitancia variable

variable-capacitance cartridge
cartucho de cinta tape cartridge, tape magazine
cartucho de cinta magnética magnetic-tape cartridge
cartucho de cristal crystal cartridge
cartucho de datos data cartridge
cartucho de disco disk cartridge
cartucho de disco duro hard-disk cartridge
cartucho de fusible fuse cartridge
cartucho de micropelícula microfilm cartridge
cartucho de reemplazo replacement cartridge
cartucho de reluctancia variable variable-reluctance
    cartridge
cartucho de tipo de letras font cartridge
cartucho de vidrio glass cartridge
cartucho estereofónico stereophonic cartridge
cartucho fonográfico cerámico ceramic
    phonographic cartridge
cartucho maestro master cartridge
casa de sintonía doghouse
cascada cascade
casco shell
cascodo cascode
cascodo en contrafase push-pull cascode
casete cassette
casete de cinta tape cassette
casete magnético magnetic cassette
caso de prueba test case
casquete de rejilla grid cap
casquillo ferrule, cap, base
casquillo conectador hickey
casquillo de bayoneta bayonet cap
casquillo de lámpara lamp cap
casquillo de rosca Edison base
casquillo goliat mogul base
casquillo lateral side cap
castillo de plomo lead castle
catacústica catacoustics
catadióptrico catadioptric
cataforesis cataphoresis
catálisis catalysis
catalítico catalytic
catalizador catalyst
catamplificador cathamplifier
catautógrafo cathautograph
categoría de error error category
categoría de falla failure category
categoría de temperatura temperature class
categoría de tensión voltage class
categoría fundamental fundamental category
catelectrotono catelectrotonus
catetrón cathetron
catión cation
catódico cathodic
catodino cathodyne
cátodo cathode
cátodo activado activated cathode
cátodo calentado directamente directly heated
    cathode
cátodo calentado indirectamente indirectly heated
    cathode
cátodo caliente hot cathode
cátodo cero zero cathode
cátodo cilíndrico cylindrical cathode
cátodo compensado dispenser cathode
cátodo complejo complex cathode
cátodo de arco arc cathode
cátodo de baño de mercurio mercury-pool cathode
cátodo de bismuto bismuth cathode
cátodo de calentamiento iónico ionic-heated cathode
cátodo de descarga luminiscente glow-discharge
    cathode
cátodo de disco disk cathode
cátodo de efecto túnel tunnel cathode

cátodo de mercurio mercury cathode
cátodo de tungsteno toriado thoriated-tungsten cathode
cátodo de Wehnelt Wehnelt cathode
cátodo electrolítico electrolytic cathode
cátodo emisor brillante bright emitter cathode
cátodo equipotencial equipotential cathode
cátodo filamentario filamentary cathode
cátodo fotoeléctrico photoelectric cathode
cátodo fotoemisor photoemitter cathode
cátodo frío cold cathode
cátodo hueco hollow cathode
cátodo hueco de cesio cesium hollow cathode
cátodo hueco termoiónico thermionic hollow cathode
cátodo impregnado impregnated cathode
cátodo incandescente incandescent cathode
cátodo líquido liquid cathode, pool cathode
cátodo ovalado oval cathode
cátodo real actual cathode
cátodo rectangular rectangular cathode
cátodo rectificador rectifier cathode
cátodo revestido coated cathode
cátodo revestido de óxido oxide-coated cathode
cátodo sensible a la luz light-sensitive cathode
cátodo sinterizado sintered cathode
cátodo termoiónico thermionic cathode
cátodo unipotencial unipotential cathode
cátodo virtual virtual cathode
catodofluorescencia cathodofluorescence
catodofosforescencia cathodophosphorescence
catodógrafo cathodograph
catodoluminiscencia cathodoluminescence
católito catholyte
caucho rubber
caucho al plomo lead rubber
causa de falla failure cause
cautín soldering iron
cavidad cavity
cavidad cilíndrica cylindrical cavity
cavidad coaxial coaxial cavity
cavidad coaxial tabicada septate coaxial cavity
cavidad de célula cell cavity
cavidad de entrada input cavity
cavidad de microondas microwave cavity
cavidad de reacción reaction cavity
cavidad de referencia reference cavity
cavidad de salida output cavity
cavidad de sintonización tuning cavity
cavidad doblada folded cavity
cavidad dúplex duplex cavity
cavidad esférica spherical cavity
cavidad externa external cavity
cavidad láser laser cavity
cavidad óptica optical cavity
cavidad rectangular rectangular cavity
cavidad reentrante reentrant cavity
cavidad resonante resonating cavity
cavidad sintonizable tunable cavity
cavidad sintonizada tuned cavity
cavitación cavitation
cavitación pulsada pulsed cavitation
cebado priming, firing, starting
cebador primer, starter
cebar prime, strike
celda cell
celda activa active cell
celda acustoóptica acoustooptical cell
celda básica basic cell
celda binaria binary cell
celda corriente current cell
celda de almacenamiento storage cell
celda de batería battery cell

celda de carácter character cell
celda de datos data cell
celda de Golay Golay cell
celda de imagen image cell
celda de lectura read cell
celda de níquel-cadmio nickel-cadmium cell
celda de plomo lead cell
celda de plomo-ácido lead-acid cell
celda dinámica dynamic cell
celda electrolítica electrolytic cell
celda en blanco blank cell
celda líquida liquid cell
celda parcialmente seleccionada partially selected cell
celda piloto pilot cell
celda seleccionada selected cell
celosía lattice
célula cell
célula asimétrica asymmetrical cell
célula bimorfa bimorphous cell
célula bioquímica biochemical cell
célula conductora conductive cell
célula de almacenamiento ultrasónica ultrasonic storage cell
célula de barrera barrier cell
célula de calentamiento heating cell
célula de capa de barrera barrier-layer cell
célula de capacidad capacity cell
célula de carga load cell
célula de cesio cesium cell
célula de colores color cell
célula de conductividad conductivity cell
célula de fotooxidación photooxidation cell
célula de fotorreducción photoreduction cell
célula de fotounión photojunction cell
célula de Karlous Karlous cell
célula de Kerr Kerr cell
célula de memoria memory cell
célula de oxisulfuro de talio thallium oxysulfide cell
célula de reducción-oxidación redox cell
célula de selenio selenium cell
célula de sulfuro de cadmio cadmium-sulfide cell
célula de telururo de plomo lead telluride cell
célula de vacío vacuum cell
célula detectora detector cell
célula eléctrica electric cell
célula electroquímica electrochemical cell
célula estacionaria stationary cell
célula fotoconductiva photoconductive cell
célula fotoconductora photoconducting cell
célula fotoconductora de sulfuro de plomo lead sulfide photoconductive cell
célula fotoeléctrica photoelectric cell
célula fotoeléctrica de unión de germanio germanium junction photoelectric cell
célula fotoemisiva photoemissive cell
célula fotogalvánica photogalvanic cell
célula fotomultiplicadora photomultiplier cell
célula fotoquímica photochemical cell
célula fotorresistiva photoresistive cell
célula fotovoltaica photovoltaic cell
célula fotovoltaica de óxido de cobre copper-oxide photovoltaic cell
célula fotovoltaica de selenio selenium photovoltaic cell
célula helioeléctrica helioelectric cell
célula luminiscente luminescent cell
célula magnética magnetic cell
célula magnética estática static magnetic cell
célula pasabajos low-pass cell
célula primaria primary cell
célula rectificadora rectifier cell
célula solar solar cell

célula sonora sound cell
célula termogalvánica thermogalvanic cell
celuloide celluloid
centelleador scintillator
centelleante scintillating
centelleo scintillation, flutter
centelleo angular angular scintillation
centi centi
centímetro-gramo gram-centimeter
centipoise centipoise
centisegundo centisecond
centrado de imagen picture centering
centrado eléctrico electrical centering
centrado horizontal horizontal centering
centrado magnético magnetic centering
centrado mecánico mechanical centering
centrado por corriente continua direct-current
    centering
centrado por imán permanente permanent-magnet
    centering
centrado vertical vertical centering
central exchange, central
central adyacente adjacent exchange
central automática automatic exchange
central automática piloto pilot automatic exchange
central automática privada private automatic
    exchange
central automática privada conectada a red pública
    private automatic branch exchange
central automática rural rural automatic exchange
central automática unitaria unit automatic exchange
central controladora controlling exchange
central de agrupamiento grouping exchange
central de batería local local-battery exchange
central de control remoto remote-control exchange
central de radio radio central
central de salida outgoing exchange
central de telecomunicaciones telecommunications
    exchange
central de telecomunicaciones automática automatic
    telecommunications exchange
central de teleimpresoras teletypewriter exchange
central de tránsito transit exchange
central eléctrica generating plant
central eléctrica geotérmica geothermal power
    station
central generadora generating station
central hidroeléctrica hydroelectric power station
central interurbana toll exchange, trunk exchange
central local local exchange
central manual manual exchange
central manual privada private manual exchange
central manual privada conectada a red pública
    private manual branch exchange
central principal main exchange
central privada private exchange
central privada conectada a red pública private
    branch exchange
central satélite satellite exchange
central semiautomática semiautomatic exchange
central tándem tandem exchange
central telefónica telephone exchange
central telefónica privada private telephone
    exchange
central terminal terminal exchange
central termoeléctrica thermoelectric plant
central urbana urban exchange
centralógrafo centralograph
centrifugación centrifugation
centro acústico efectivo effective acoustic center
centro de agrupamiento grouping center
centro de banda band center, midband
centro de canal center of channel

centro de carga load center
centro de color color center
centro de computadoras computer center
centro de comunicaciones communication center
centro de conmutación switching center
centro de conmutación automático automatic
    switching center
centro de conmutación de mensajes message
    switching center
centro de conmutación internacional international
    switching center
centro de conmutadores automáticos automatic
    switch center
centro de control control center
centro de control de conmutación switching control
    center
centro de control de energía energy control center
centro de control de potencia power control center
centro de correspondencia correspondence center
centro de datos data center
centro de deflexión deflection center
centro de distribución distribution center
centro de fase phase center
centro de fase de red phase center of array
centro de filtro filter center
centro de gravedad center of gravity
centro de grupo group center
centro de haz center of beam
centro de impurezas impurity center
centro de información information center
centro de luz light center
centro de masa center of mass
centro de masa efectivo effective center of mass
centro de micrófono microphone center
centro de operaciones de red network operations
    center
centro de procesamiento processing center
centro de procesamiento de datos data-processing
    center
centro de procesamiento de datos electrónico
    electronic data processing center
centro de procesamiento de información
    information-processing center
centro de radiación center of radiation
centro de radiodifusión broadcast center
centro de recombinación recombination center
centro de referencia reference center
centro de retransmisión por cinta tape-relay center
centro de tránsito transit center
centro de transmisión transmission center
centro de zona zone center
centro efectivo effective center
centro eléctrico electrical center
centro intermedio intermediate center
centro interurbano toll center
centro óptico optical center
centro primario primary center
centro receptor receiving center
centro regional regional center
centro supervisor supervisory center
centro telegráfico telegraph center
cera wax
cera de carnauba carnauba wax
cera de ceresina ceresin wax
cerámica de pocas pérdidas low-loss ceramic
cerámica de zirconato-titanato de plomo lead
    zirconate-titanate ceramic
cerámica electroestrictiva electrostrictive ceramic
cerámica ferroeléctrica ferroelectric ceramic
cerámica piezoeléctrica piezoelectric ceramic
cerámica semiconductora semiconducting ceramic
cerca eléctrica electric fence
cerebro electrónico electronic brain

cerio cerium
cermet cermet
cero a la derecha trailing zero
cero a la izquierda leading zero
cero absoluto absolute zero
cero arbitrario arbitrary zero
cero automático automatic zero
cero binario binary zero
cero de máquina machine zero
cero definido defined zero
cero desplazado offset zero
cero eléctrico electrical zero
cero flotante floating zero
cero mecánico mechanical zero
cero positivo positive zero
cero real real zero
cerrador de puerta gate closer
cerradura eléctrica electric lock
cerradura electrónica electronic lock
cerrar el circuito switch-on
cerrojo latch
certificado de calibración calibration certificate
cesio cesium
cianuro de potasio potassium cyanide
cianuro libre free cyanide
cibernética cybernetics
ciclado cycling
cíclico cyclic
ciclo cycle, loop, run
ciclo abierto open cycle
ciclo abortado aborted cycle
ciclo abuelo grandfather cycle
ciclo anual annual cycle
ciclo básico basic cycle
ciclo cerrado closed cycle
ciclo de acceso access cycle
ciclo de alimentación feed cycle
ciclo de almacenamiento storage cycle
ciclo de arbitraje arbitration cycle
ciclo de atención attention cycle
ciclo de bombeo pumping cycle
ciclo de borrado erasing cycle
ciclo de bus bus cycle
ciclo de búsqueda search cycle
ciclo de calentamiento heating cycle
ciclo de calibración calibration cycle
ciclo de carácter character cycle
ciclo de carga loading cycle
ciclo de conexión y desconexión on-off cycle
ciclo de control control cycle
ciclo de control intermedio intermediate-control
    cycle
ciclo de copiado copy cycle
ciclo de datos data cycle
ciclo de dirección address cycle
ciclo de dirección secundaria secondary address
    cycle
ciclo de dirección solamente address-only cycle
ciclo de ejecución execution cycle
ciclo de ejecución de microcomputadora
    microcomputer execution cycle
ciclo de escritura write cycle
ciclo de escritura de bloque block write cycle
ciclo de exploración scanning cycle
ciclo de índice index cycle
ciclo de instalación installation cycle
ciclo de instrucción instruction cycle
ciclo de lectura read cycle
ciclo de lectura de bloque block read cycle
ciclo de lectura-escritura read-write cycle
ciclo de magnetización magnetization cycle
ciclo de manchas solares sunspot cycle
ciclo de manipulación keying cycle

ciclo de mantenimiento maintenance cycle
ciclo de máquina machine cycle
ciclo de memoria memory cycle
ciclo de modulación modulation cycle
ciclo de onda wave cycle
ciclo de operación operation cycle
ciclo de polarización polarization cycle
ciclo de procesamiento processing cycle
ciclo de procesamiento de computadora computer
    processing cycle
ciclo de procesamiento de datos data-processing
    cycle
ciclo de programa program cycle
ciclo de prueba de programa program test cycle
ciclo de pruebas testing cycle
ciclo de punto dot cycle
ciclo de Rayleigh Rayleigh cycle
ciclo de refresco refresh cycle
ciclo de relajación relaxation cycle
ciclo de reloj clock cycle
ciclo de salida exit cycle
ciclo de soldadura welding cycle
ciclo de tarjeta card cycle
ciclo de tiempo time cycle
ciclo de trabajo duty cycle
ciclo de trabajo de batería battery duty cycle
ciclo de trabajo de relé relay duty cycle
ciclo de vida life cycle
ciclo de visualización display cycle
ciclo directo direct cycle
ciclo ejecutivo executive cycle
ciclo fijo fixed cycle
ciclo intermedio intermediate cycle
ciclo magnético magnetic cycle
ciclo menor minor cycle
ciclo nulo null cycle
ciclo principal major cycle
ciclo reversible reversible cycle
ciclo solar solar cycle
ciclo único single cycle
ciclofón cyclophon
ciclograma cyclogram
cicloidotrón cycloidotron
ciclos por segundo cycles per second
ciclotrón cyclotron
ciclotrón de modulación de frecuencia
    frequency-modulated cyclotron
cierracircuito automático automatic circuit closer
cierre closedown
cierre closing, closure, make, seal, locking, lockout
cierre antes de abertura make-before-break
cierre de archivo file closing
cierre de botón de presión press-button lock
cierre de circuito circuit closing
cierre de contacto contact closure
cierre de guíaondas waveguide seal
cierre de vacío vacuum seal
cierre de vidrio glass seal
cierre hermético hermetic seal
cierre hermético de conductos duct sealing
cierre rápido quick closedown
cierre retardado delayed close, delayed make
cierre vidrio-metal glass-metal seal
cifra de calidad quality figure
cifra de rendimiento de radar radar performance
    figure
cifra de ruido noise figure
cifra de ruido de receptor receiver noise figure
cifra de ruido media average noise figure
cifra máxima peak figure
cifra significativa significant figure
cifrar encipher, encrypt
cilindro cylinder, drum

cilindro de caracteres character cylinder
cilindro de disco duro hard-disk cylinder
cilindro de Faraday Faraday cylinder
cilindro de información de etiquetas label
   information cylinder
cilindro de Wehnelt Wehnelt cylinder
cilindro electromagnético electromagnetic cylinder
cilindro hueco hollow cylinder
cinc zinc
cincado zinc-plating
cincita zincite
cinemático kinematic
cinematógrafo kinematograph
cinescópico kinescopic
cinescopio kinescope, television picture tube
cinescopio de ángulo grande wide-angle kinescope
cinescopio de colores color kinescope
cinético kinetic
cinetógrafo kinetograph
cinetoscopio kinetoscope
cinta tape, ribbon
cinta abuelo grandfather tape
cinta adhesiva adhesive tape
cinta aislante insulating tape
cinta barnizada varnished tape
cinta calibradora calibrating tape
cinta certificada certified tape
cinta codificada coded tape
cinta con perforación parcial chadless tape
cinta con residuos de perforación chadded tape
cinta de acetato acetate tape
cinta de acetato de celulosa cellulose acetate tape
cinta de alimentación de avance advance feed tape
cinta de alta densidad high-density tape
cinta de archivo de datos data file tape
cinta de archivo maestro master file tape
cinta de archivo múltiple multiple-file tape
cinta de archivo principal main file tape
cinta de archivos file tape
cinta de arrastre leader tape
cinta de baja densidad low-density tape
cinta de biblioteca library tape
cinta de biblioteca maestra master library tape
cinta de cambios change tape
cinta de casete cassette tape
cinta de caucho rubber tape
cinta de computadora computer tape
cinta de control control tape
cinta de control de papel paper-control tape
cinta de control numérico numerical control tape
cinta de cuatro pistas four-track tape
cinta de datos data tape
cinta de desbordamiento overflow tape
cinta de distribución distribution tape
cinta de dos caras double-faced tape
cinta de entrada input tape
cinta de errores error tape
cinta de facturación billing tape
cinta de fricción friction tape
cinta de goma combinada combination rubber tape
cinta de grabación record tape
cinta de impresora printer tape
cinta de instrucciones instruction tape
cinta de instrucciones maestra master instruction
   tape
cinta de intercalación collation tape
cinta de media pista half-track tape
cinta de memoria memory tape
cinta de origen originating tape
cinta de papel paper tape
cinta de perforación punch tape
cinta de problemas problem tape
cinta de programa program tape

cinta de programación programming tape
cinta de prueba test tape
cinta de prueba patrón standard test tape
cinta de puntos de comprobación checkpoint tape
cinta de referencia reference tape
cinta de registro recording tape
cinta de registro magnético magnetic recording tape
cinta de respaldo backup tape
cinta de salida output tape
cinta de salida de sistema system output tape
cinta de segunda generación second generation tape
cinta de sistema system tape
cinta de tercera generación third-generation tape
cinta de trabajo working tape
cinta de transacciones transaction tape
cinta de video video tape
cinta eléctrica electrical tape
cinta electrosensible electrosensitive tape
cinta en blanco blank tape
cinta en blanco raw tape
cinta estereofónica stereophonic tape
cinta estereofónica escalonada staggered
   stereophonic tape
cinta ferromagnética ferromagnetic tape
cinta fuente source tape
cinta grabada recorded tape
cinta homogénea homogeneous tape
cinta impregnada impregnated tape
cinta impresa printed tape
cinta llena filled tape
cinta maestra master tape
cinta magnética magnetic tape
cinta magnética de registro de sonido
   sound-recording magnetic tape
cinta magnética grabada recorded magnetic tape
cinta magnética para televisión television magnetic
   tape
cinta magnética registrada recorded magnetic tape
cinta magnetofónica sound-recording magnetic tape
cinta multiarchivo multifile tape
cinta multigeneracional multigenerational tape
cinta numérica numerical tape
cinta para grabación recording tape
cinta para registro recording tape
cinta patrón standard tape
cinta perforada perforated tape, punched tape
cinta piloto pilot tape
cinta pregrabada prerecorded tape
cinta registrada recorded tape
cinta revestida de óxido oxide-coated tape
cinta revestida de polvo magnético magnetic
   powder-coated tape
cinta semiconductora semiconducting tape
cinta sin fin tape loop
cinta sin grabar unrecorded tape
cinta sin registrar unrecorded tape
cinta sonora sound tape
cinta vacía empty tape
cinturón de radiación radiation belt
cinturones de radiación de Van Allen Van Allen
   radiation belts
cinturones de Van Allen Van Allen belts
circonia zirconia
circonio zirconium
circuital circuital
circuitería circuitry
circuitería de comprobación checking circuitry
circuitería de guarda guard circuitry
circuitería de interfaz interface circuitry
circuitería de refresco refresh circuitry
circuitería electrónica electronic circuitry
circuitería impresa printed circuitry
circuitería microelectrónica microelectronic

circuitry
circuitería transistorizada transistorized circuitry
circuito circuit
circuito a larga distancia long-distance circuit
circuito abierto open circuit
circuito absorbedor absorber circuit
circuito acentuador accentuator circuit
circuito aceptor acceptor circuit
circuito acoplado por resistor resistor-coupled circuit
circuito activo híbrido hybrid active circuit
circuito adaptador adapter circuit
circuito agrupado bunched circuit
circuito aislante isolating circuit
circuito ajustado adjusted circuit
circuito ajustador de línea line-adjusting circuit
circuito amortiguado losser circuit
circuito amortiguador damping circuit
circuito amplificador amplifying circuit
circuito amplificador de potencia power-amplifying circuit
circuito amplificador en contrafase push-pull amplifier circuit
circuito analógico analog circuit
circuito anódico anode circuit
circuito antena-tierra aerial-ground circuit
circuito anticoincidencia anticoincidence circuit
circuito antifluctuación antihunt circuit
circuito antirresonante antiresonant circuit
circuito apantallado screened circuit
circuito aperiódico aperiodic circuit
circuito aritmético arithmetic circuit
circuito arrancador starter circuit
circuito arrendado leased circuit
circuito asimétrico single-ended circuit
circuito asincrónico asynchronous circuit
circuito astable astable circuit
circuito autodino autodyne circuit
circuito autoelevador bootstrap circuit
circuito autoextintor self-extinguishing circuit
circuito autointerrumpido self-interrupted circuit
circuito automático automatic circuit
circuito autorresonante self-resonant circuit
circuito autosaturante self-saturating circuit
circuito auxiliar auxiliary circuit
circuito basculador flip-flop circuit
circuito básico basic circuit
circuito beta beta circuit
circuito bidimensional two-dimensional circuit
circuito bidireccional two-way circuit
circuito biestable bistable circuit
circuito bifilar two-wire circuit
circuito bifurcado forked circuit
circuito bilateral bilateral circuit
circuito binario binary circuit
circuito bioeléctrico bioelectric circuit
circuito bipolar bipolar circuit
circuito calefactor heater circuit
circuito calibrador calibrating circuit
circuito capacitivo capacitive circuit
circuito cargado loaded circuit
circuito cascodo cascode circuit
circuito cerrado closed circuit
circuito coaxial coaxial circuit
circuito codificador coder circuit, scrambler circuit
circuito colector-base collector-base circuit
circuito combinacional combinational circuit
circuito combinador combining circuit
circuito comparador comparator circuit
circuito compensador compensating circuit
circuito compensador de frecuencia frequency-compensating circuit
circuito complementario complementing circuit,

idling circuit
circuito completamente automático fully automatic circuit
circuito completo complete circuit
circuito compuesto compound circuit
circuito con base a masa grounded-base circuit
circuito con cátodo a masa grounded-cathode circuit
circuito con colector a masa grounded-collector circuit
circuito con emisor a masa grounded-emitter circuit
circuito con fuente a masa grounded-source circuit
circuito con neutralización cruzada cross-neutralized circuit
circuito con puerta a masa grounded-gate circuit
circuito con rejilla a masa grounded-grid circuit
circuito con retorno por tierra earth-return circuit
circuito concentrado lumped circuit
circuito conectado a tierra earthed circuit
circuito conectado en estrella star-connected circuit
circuito conmutado switched circuit
circuito conmutador biestable bistable switching circuit
circuito conmutador selectivo a la frecuencia frequency-selective switching circuit
circuito contador counter circuit
circuito contador de pasos step-counting circuit
circuito contador electrónico electronic counting circuit
circuito continuo continuous circuit
circuito contra atascamientos jam circuit
circuito contra ecos parásitos anticlutter circuit
circuito controlado controlled circuit
circuito controlador controlling circuit
circuito corrector corrector circuit
circuito corrector de frecuencia frequency-correcting circuit
circuito cuadrador squaring circuit
circuito cuádruplex quadruplex circuit
circuito cuasibiestable quasi-bistable circuit
circuito cuasimonoestable quasi-monostable circuit
circuito de absorción absorption circuit
circuito de acoplamiento coupling circuit
circuito de acoplamiento de antena antenna coupling circuit
circuito de adaptación applique circuit
circuito de adaptación de antena antenna-matching circuit
circuito de aire air circuit
circuito de alarma alarm circuit
circuito de alarma de fallas fault-alarm circuit
circuito de alarma de repetidor repeater-alarm circuit
circuito de alarma de sobretemperatura overtemperature alarm circuit
circuito de alimentación feed circuit, supply circuit
circuito de alimentación telefónica telephone feed circuit
circuito de almacenamiento storage circuit
circuito de alumbrado lighting circuit
circuito de ánodo sintonizado tuned-anode circuit
circuito de antena antenna circuit
circuito de aplanamiento smoothing circuit
circuito de aproximación approach circuit
circuito de arranque start circuit
circuito de audio audio circuit
circuito de aullador howler circuit
circuito de autorretención stick circuit
circuito de avance de fase phase-advance circuit
circuito de aviso warning circuit
circuito de baja energía low-energy circuit
circuito de baja potencia low-power circuit
circuito de bajada drop circuit
circuito de barrido sweep circuit

circuito de barrido de línea line sweep circuit
circuito de barrido horizontal horizontal-sweep circuit
circuito de base común common-base circuit
circuito de base de tiempo time-base circuit
circuito de batería battery circuit
circuito de bloqueo blocking circuit
circuito de bombeo pumping circuit
circuito de borrado erase circuit
circuito de bucle loop circuit
circuito de bus bus circuit
circuito de cable cable circuit
circuito de calidad telefónica voice-grade circuit
circuito de calidad vocal voice-grade circuit
circuito de cambio de estado status-change circuit
circuito de campo excitador exciter-field circuit
circuito de cancelación cancellation circuit
circuito de carga charging circuit, load circuit
circuito de carga de baterías battery charging circuit
circuito de carga dividida split-load circuit
circuito de cátodo común common-cathode circuit
circuito de centrado centering circuit
circuito de codificación coding circuit
circuito de coincidencia coincidence circuit
circuito de coincidencia de retardo delay coincidence circuit
circuito de cola tail circuit
circuito de colector común common-collector circuit
circuito de comparación de amplitudes amplitude-comparison circuit
circuito de compatibilidad compatibility circuit
circuito de compensación bucking circuit
circuito de comprobación checking circuit
circuito de comunicación communication circuit
circuito de conexión connecting circuit
circuito de conexión y desconexión on-off circuit
circuito de conferencia conference circuit
circuito de conmutación switching circuit
circuito de conmutación de impulsos pulse-switching circuit
circuito de conmutación electrónica electronic switching circuit
circuito de conmutación multiestable multistable switching circuit
circuito de constantes concentradas lumped-constant circuit
circuito de contacto contact circuit
circuito de conteo counting circuit
circuito de contestación answering circuit
circuito de contraderivación back-shunt circuit
circuito de control control circuit
circuito de control cruzado cross-control circuit
circuito de control de bucle cerrado closed-loop control circuit
circuito de control de ganancia diferencial differential gain control circuit
circuito de control de motor motor control circuit
circuito de control de parada stop-control circuit
circuito de control de soldadura welding-control circuit
circuito de control de tono tone-control circuit
circuito de control electrónico electronic control circuit
circuito de control por rejilla grid-control circuit
circuito de control remoto remote-control circuit
circuito de convergencia convergence circuit
circuito de convergencia de entrada fanin circuit
circuito de convergencia dinámica dynamic-convergence circuit
circuito de conversación speech circuit
circuito de cordón cord circuit
circuito de corriente current circuit
circuito de corriente alterna alternating-current circuit
circuito de corriente continua direct-current circuit
circuito de corriente de llamada ringing-current circuit
circuito de corta distancia short-distance circuit
circuito de croma chroma circuit
circuito de crominancia chrominance circuit
circuito de Crosby Crosby circuit
circuito de cruce crossover circuit
circuito de cuatro hilos four-wire circuit
circuito de chip chip circuit
circuito de datos data circuit
circuito de datos digitales digital-data circuit
circuito de decisión decision circuit
circuito de deflexión deflection circuit
circuito de derivación branch circuit
circuito de derivación de aparato eléctrico appliance branch circuit
circuito de derivación de uso general general-purpose branch circuit
circuito de derivación multihilo multiwire branch circuit
circuito de desacentuación deemphasis circuit
circuito de desacoplamiento decoupling circuit
circuito de descarga discharge circuit
circuito de descodificación decoding circuit
circuito de desfase phase-shift circuit
circuito de desmagnetización degaussing circuit
circuito de desvanecimiento fading circuit
circuito de detección de errores error-detection circuit
circuito de detección de tensión voltage-sensing circuit
circuito de disparo trigger circuit, firing circuit
circuito de disparo biestable bistable trigger circuit
circuito de disparo electrónico electronic trigger circuit
circuito de disparo monoestable monostable trigger circuit
circuito de divergencia de salida fanout circuit
circuito de doble sintonía double-tuned circuit
circuito de Eccles-Jordan Eccles-Jordan circuit
circuito de ecualización equalization circuit
circuito de efecto de Gunn Gunn-effect circuit
circuito de electroimán de perforación punch-magnet circuit
circuito de elementos distribuidos distributed-element circuit
circuito de elementos en paralelo en contrafase push-pull-parallel circuit
circuito de emergencia emergency circuit
circuito de emisor común common-emitter circuit
circuito de enclavamiento interlock circuit
circuito de enganche de fase phase-locked circuit
circuito de enganche horizontal horizontal-lock circuit
circuito de enlace link circuit, trunk circuit
circuito de entrada input circuit
circuito de entrada de modo común common-mode input circuit
circuito de entrada de puente bridge input circuit
circuito de entrada directa direct-input circuit
circuito de entretenimiento keep-alive circuit
circuito de escala de dos scale-of-two circuit
circuito de escala de ocho scale-of-eight circuit
circuito de estado sólido solid-state circuit
circuito de excitación drive circuit
circuito de exploración scanning circuit
circuito de extinción quenching circuit, blanking circuit
circuito de falla fault circuit
circuito de fase dividida split-phase circuit
circuito de fijación de nivel clamping circuit

circuito de fijación de nivel bidireccional
bidirectional clamping circuit
circuito de filamento filament circuit
circuito de filtro filter circuit
circuito de filtro pasabajos low-pass filter circuit
circuito de Flewelling Flewelling circuit
circuito de frenaje braking circuit
circuito de fuente común common-source circuit
circuito de fuente de alimentación power-supply
circuit
circuito de ganancia gain circuit
circuito de Goto Goto circuit
circuito de guarda guard circuit
circuito de Harkness Harkness circuit
circuito de Hazeltine Hazeltine circuit
circuito de hilo único single-wire circuit
circuito de hiperfrecuencias hyperfrequency circuit
circuito de ignición ignition circuit
circuito de igualdad equality circuit
circuito de imagen vision circuit
circuito de impulsos impulse circuit
circuito de impulsos pulse circuit
circuito de información information circuit
circuito de interacción interaction circuit
circuito de intercepción interception circuit
circuito de intercomunicación talkback circuit
circuito de interfaz interface circuit
circuito de interfaz de comunicaciones
communication interface circuit
circuito de inversión de fase phase-inversion circuit
circuito de ionización residual keep-alive circuit
circuito de lectura reading circuit
circuito de ley cuadrática square-law circuit
circuito de limitación clipping circuit
circuito de línea line circuit
circuito de línea coaxial coaxial-line circuit
circuito de línea compartida party-line circuit
circuito de línea de órdenes order-wire circuit
circuito de línea de paridad parity-line circuit
circuito de línea resonante resonant-line circuit
circuito de línea visual line-of-sight circuit
circuito de llamada calling circuit, ringing circuit
circuito de llegada incoming circuit
circuito de Loftin-White Loftin-White circuit
circuito de malla mesh circuit
circuito de manipulación keying circuit
circuito de mariposa butterfly circuit
circuito de medida measuring circuit
circuito de medidor meter circuit
circuito de Meissner Meissner circuit
circuito de memoria memory circuit
circuito de memoria de acceso aleatorio
random-access memory circuit
circuito de mensajes message circuit
circuito de Mesny Mesny circuit
circuito de micropotencia micropower circuit
circuito de Miller Miller circuit
circuito de monitoreo monitoring circuit
circuito de montaje temporal breadboard circuit
circuito de Morse Morse circuit
circuito de muestreo sampling circuit
circuito de muestreo y retención sample-and-hold
circuit
circuito de nanosegundo nanosecond circuit
circuito de negación negation circuit
circuito de neutralización por bobina
coil-neutralization circuit
circuito de no equivalencia nonequivalence circuit
circuito de onda lenta slow-wave circuit
circuito de operación operating circuit
circuito de origen originating circuit
circuito de oscilador de bloqueo blocking oscillator
circuit

circuito de parálisis paralysis circuit
circuito de paso de banda bandpass circuit
circuito de pedal pedal circuit
circuito de película delgada thin-film circuit
circuito de película delgada de tantalio tantalum
thin-film circuit
circuito de película delgada híbrido hybrid thin-film
circuit
circuito de película gruesa thick-film circuit
circuito de pérdida cero zero-loss circuit
circuito de pista track circuit
circuito de placa plate circuit
circuito de placa sintonizada tuned-plate circuit
circuito de polarización de rejilla grid-bias circuit
circuito de portadora carrier circuit
circuito de potencia power circuit
circuito de potencia de baja energía low-energy
power circuit
circuito de preparación set-up circuit
circuito de prioridad priority circuit
circuito de programa program circuit
circuito de prueba test circuit
circuito de puente degenerativo degenerative bridge
circuit
circuito de puerta gate circuit
circuito de puerta común common-gate circuit
circuito de puerta en paralelo parallel gate circuit
circuito de puerta lógica logical-gate circuit
circuito de puesta a tierra earthing circuit,
grounding circuit
circuito de punto a punto point-to-point circuit
circuito de radio radio circuit
circuito de radiocomunicación radiocommunication
circuit
circuito de radiocontrol radio-control circuit
circuito de radioenlace radio-link circuit
circuito de radiorrelé radio-relay circuit
circuito de reacción reaction circuit
circuito de receptáculo receptacle circuit
circuito de rectificadores en cascada cascade
rectifier circuit
circuito de rectificadores en paralelo
parallel-rectifier circuit
circuito de rechazo rejection circuit
circuito de referencia reference circuit
circuito de referencia hipotético hypothetical
reference circuit
circuito de regeneración regeneration circuit
circuito de registro recording circuit
circuito de rejilla grid circuit
circuito de rejilla común common-grid circuit
circuito de rejilla sintonizada tuned-grid circuit
circuito de relajación relaxation circuit
circuito de relajación biestable bistable relaxation
circuit
circuito de relé relay circuit
circuito de relé secuencial sequential relay circuit
circuito de repetidor repeater circuit
circuito de reposición reset circuit
circuito de reserva reserve circuit, standby circuit
circuito de resistencia-capacitancia
resistance-capacitance circuit
circuito de resistencia-inductancia
resistance-inductance circuit
circuito de resonancia resonance circuit
circuito de restauración restoration circuit
circuito de retardo delay circuit
circuito de retardo de barrido sweep-delay circuit
circuito de retardo de impulsos pulse-delay circuit
circuito de retardo de tiempo time-delay circuit
circuito de retardo por cable cable delay circuit
circuito de retención retention circuit
circuito de retorno return circuit

circuito de retorno a tierra ground-return circuit
circuito de retransmisión por cinta tape-relay circuit
circuito de retroalimentación feedback circuit
circuito de retroalimentación de corriente current feedback circuit
circuito de retroalimentación de desfase phase-shift feedback circuit
circuito de retroalimentación positiva positive-feedback circuit
circuito de salida outgoing circuit, output circuit
circuito de salida de receptor receiver output circuit
circuito de salida interno internal output circuit
circuito de salida no amortiguado undamped output circuit
circuito de Schmitt Schmitt circuit
circuito de seguimiento tracking circuit
circuito de selección selection circuit
circuito de selección de impulsos pulse-selection circuit
circuito de señal signal circuit
circuito de señalización signaling circuit
circuito de separación separation circuit
circuito de servicio service circuit
circuito de servicio continuo continuous-duty circuit
circuito de simetría complementaria complementary-symmetry circuit
circuito de sincronización synchronizing circuit
circuito de sincronización de colores color synchronizing circuit
circuito de sintonía única single-tuned circuit
circuito de sintonización tuning circuit
circuito de sonido sound circuit
circuito de suministro de batería battery-supply circuit
circuito de supresión suppression circuit
circuito de supresión de retorno retrace suppression circuit
circuito de supresión de ruido noise-suppression circuit
circuito de sustracción subtraction circuit
circuito de telecomunicación a larga distancia long-distance telecommunication circuit
circuito de telecomunicaciones telecommunications circuit
circuito de teleimpresora teleprinter circuit
circuito de telemedida telemetering circuit
circuito de televisión television circuit
circuito de temporización de impulsos pulse-timing circuit
circuito de tensión voltage circuit
circuito de tensión de referencia reference-voltage circuit
circuito de terminación termination circuit
circuito de tierra earth circuit, ground circuit
circuito de trabajo working circuit
circuito de tráfico traffic circuit
circuito de transferencia transfer circuit
circuito de tránsito transit circuit
circuito de tránsito directo direct-transit circuit
circuito de transmisión de programas program-transmission circuit
circuito de triodo con rejilla a masa grounded-grid triode circuit
circuito de umbral threshold circuit
circuito de unión junction circuit
circuito de unión local local-junction circuit
circuito de uso asignado allocated-use circuit
circuito de valor absoluto absolute value circuit
circuito de video video circuit
circuito de Villard Villard circuit
circuito de volante flywheel circuit
circuito de voz voice circuit

circuito de Weir Weir circuit
circuito de White White circuit
circuito de Wien Wien circuit
circuito de Witka Witka circuit
circuito dedicado dedicated circuit
circuito defectuoso faulty circuit
circuito derivado divided circuit
circuito descodificador decoder circuit
circuito desequilibrado unbalanced circuit
circuito desfasador phase-shifting circuit
circuito desmultiplexor demultiplexing circuit
circuito desvanecedor fader circuit
circuito detector detector circuit
circuito detector en contrafase push-pull detection circuit
circuito dieléctrico dielectric circuit
circuito diferenciador differentiating circuit
circuito digital digital circuit
circuito dinámico dynamic circuit
circuito directo direct circuit
circuito discreto discrete circuit
circuito discriminador discriminator circuit
circuito discriminador de desfase phase-shift discriminator circuit
circuito discriminador de razón ratio discriminator circuit
circuito disparado triggered circuit
circuito disparador de impulsos pulse trigger circuit
circuito distribuido distributed circuit
circuito divisor de fase phase-splitting circuit
circuito divisor de impulsos pulse-divider circuit
circuito divisor de tensión voltage-divider circuit
circuito doblador doubler circuit
circuito dúplex duplex circuit
circuito eléctrico electric circuit
circuito electrométrico electrometer circuit
circuito electrónico electronic circuit
circuito electrónico impreso printed electronic circuit
circuito elemental elementary circuit
circuito emisor-base emitter-base circuit
circuito emulador emulator circuit
circuito en anillo ring circuit
circuito en bucle looped circuit
circuito en común joint circuit
circuito en contrafase push-pull circuit
circuito en contrafase complementario complementary pushpull circuit
circuito en contrafase-paralelo push-push circuit
circuito en delta delta circuit
circuito en derivación shunt circuit
circuito en derivación-serie shunt-series circuit
circuito en estrella star circuit
circuito en oposición back-to-back circuit
circuito en paralelo parallel circuit
circuito en paralelo-serie parallel-series circuit
circuito en puente bridge circuit
circuito en reposo idle circuit
circuito en serie series circuit
circuito en serie equivalente equivalent series circuit
circuito en serie-paralelo series-parallel circuit
circuito en T tee circuit
circuito en tándem tandem circuit
circuito encapsulado encapsulated circuit, potted circuit
circuito encapsulado en resina resin-encapsulated circuit
circuito enchufable plug-in circuit
circuito endodino endodyne circuit
circuito equilibrado balanced circuit
circuito equilibrador balancing circuit
circuito equilibrador de ruido noise-balancing circuit

circuito equivalente equivalent circuit
circuito equivalente de señales fuertes large-signal equivalent circuit
circuito equivalente para señal pequeña small-signal equivalent circuit
circuito equivalente paralelo parallel equivalent circuit
circuito equivalente pasivo passive equivalent circuit
circuito escalador scaling circuit
circuito estabilizador stabilizing circuit
circuito estabilizador de tensión voltage-stabilizing circuit
circuito estable stable circuit
circuito estampado stamped circuit
circuito estandarizado standardized circuit
circuito estenodo stenode circuit
circuito excitador exciter circuit
circuito externo external circuit
circuito fantasma phantom circuit
circuito fantasma con retorno por tierra earth-return phantom circuit
circuito fantasma cuádruple quadruple phantom circuit
circuito ferrimagnético ferrimagnetic circuit
circuito ferrorresonante ferroresonant circuit
circuito fijo fixed circuit
circuito fijo permanente permanent fixed circuit
circuito filtrante filtering circuit
circuito físico physical circuit
circuito flotante floating circuit
circuito formador shaping circuit
circuito formador de impulsos pulse-forming circuit
circuito fotograbado photoengraved circuit
circuito fuente source circuit
circuito fundamental fundamental circuit
circuito generador de impulsos pulse-generating circuit
circuito grabado etched circuit
circuito helicoidal helix circuit
circuito hexafásico six-phase circuit
circuito híbrido hybrid circuit
circuito híbrido de película gruesa thick-film hybrid circuit
circuito hidráulico hydraulic circuit
circuito impreso printed circuit
circuito impreso de polvo compactado pressed powder printed circuit
circuito impreso electrónico electronic printed circuit
circuito impreso grabado etched printed circuit
circuito impreso por depósito plated printed circuit
circuito impreso sobre hoja metálica en relieve embossed-foil printed circuit
circuito impreso transferido transferred printed circuit
circuito indicador indicating circuit
circuito indicador de fallas fault-reporting circuit
circuito inductivo inductive circuit
circuito integrado integrated circuit
circuito integrado analógico analog integrated circuit
circuito integrado bipolar bipolar integrated circuit
circuito integrado compatible compatible integrated circuit
circuito integrado de estado sólido solid-state integrated circuit
circuito integrado de microondas microwave integrated circuit
circuito integrado de muy alta velocidad very-high-speed integrated circuit
circuito integrado de película delgada thin-film integrated circuit
circuito integrado de película gruesa thick-film integrated circuit
circuito integrado de películas film integrated circuit
circuito integrado digital digital integrated circuit
circuito integrado en gran escala large-scale integrated circuit
circuito integrado híbrido hybrid integrated circuit
circuito integrado lineal linear integrated circuit
circuito integrado magnético magnetic integrated circuit
circuito integrado monolítico monolithic integrated circuit
circuito integrado monolítico compatible compatible monolithic integrated circuit
circuito integrado multichip multichip integrated circuit
circuito integrado optoelectrónico optoelectronic integrated circuit
circuito integrado planar planar integrated circuit
circuito integrado plástico plastic integrated circuit
circuito integrado semiconductor semiconductor integrated circuit
circuito integrador integrating circuit
circuito intercambiador de calor heat-exchanger circuit
circuito interconectador interconnecting circuit
circuito intercontinental intercontinental circuit
circuito intermedio intermediate circuit
circuito intermitente gating circuit
circuito internacional international circuit
circuito internacional directo direct international circuit
circuito interrumpido interrupted circuit
circuito interurbano toll circuit
circuito intervalvular intervalve circuit
circuito lateral side circuit
circuito limitador limiting circuit
circuito limitador de amplitud amplitude-limiting circuit
circuito lineal linear circuit
circuito logarítmico logarithmic circuit
circuito lógico logic circuit
circuito lógico acoplado por emisor emitter-coupled logic circuit
circuito lógico de microenergía microenergy logic circuit
circuito lógico de microondas microwave logic circuit
circuito lógico de transistor transistor logic circuit
circuito lógico de transistores acoplado por emisor emitter-coupled transistor logic circuit
circuito lógico magnético magnetic logic circuit
circuito longitudinal longitudinal circuit
circuito magnético magnetic circuit
circuito magnético abierto open magnetic circuit
circuito magnético cerrado closed magnetic circuit
circuito matricial matrix circuit
circuito medido metered circuit
circuito metálico metallic circuit
circuito metálico equilibrado balanced metallic circuit
circuito métrico decimal decimal metric circuit
circuito mezclador mixing circuit
circuito mezclador de impulsos pulse-mixing circuit
circuito microelectrónico microelectronic circuit
circuito microelectrónico de película delgada thin-film microelectronic circuit
circuito miniaturizado miniaturized circuit
circuito mixto mixed circuit
circuito modulador modulating circuit
circuito modular modular circuit
circuito molecular molecular circuit
circuito monoestable monostable circuit

circuito monofásico single-phase circuit
circuito monofásico de media onda single-phase half-wave circuit
circuito monofásico de onda completa single-phase full-wave circuit
circuito monolítico monolithic circuit
circuito monolítico integrado integrated monolithic circuit
circuito mu mu circuit
circuito muerto dead circuit
circuito multicanal multichannel circuit
circuito multicapa multilayer circuit
circuito multichip multichip circuit
circuito multiestable multistable circuit
circuito multiestado multistate circuit
circuito multinivel multilevel circuit
circuito múltiple multiple circuit
circuito multiplicador de frecuencia frequency-multiplying circuit
circuito multiplicador de tensión voltage-multiplier circuit
circuito multipunto multipoint circuit
circuito multivibrador multivibrator circuit
circuito musical musical circuit
circuito mutuo mutual circuit
circuito neumático pneumatic circuit
circuito neutralizador de Hazeltine Hazeltine neutralizing circuit
circuito neutralizador de Rice Rice neutralizing circuit
circuito neutralizante neutralizing circuit
circuito neutro neutral circuit
circuito NO NOT circuit
circuito no esencial nonessential circuit
circuito no inductivo noninductive circuit
circuito no lineal nonlinear circuit
circuito NO-O NOR circuit
circuito no reactivo nonreactive circuit
circuito no sintonizado untuned circuit
circuito NO-Y NAND circuit
circuito NO-Y NOT-AND circuit
circuito normal normal circuit
circuito O OR circuit
circuito operado por portadora antirruido antinoise carrier-operated circuit
circuito óptico integrado integrated optical circuit
circuito oscilador oscillator circuit
circuito oscilante oscillating circuit
circuito oscilante de resonancia resonance oscillatory circuit
circuito pasivo passive circuit
circuito pasivo de película gruesa thick-film passive circuit
circuito periódico periodic circuit
circuito piloto pilot circuit
circuito pirométrico pyrometric circuit
circuito polar polar circuit
circuito polifásico polyphase circuit
circuito por línea terrestre landline circuit
circuito potenciométrico potentiometer circuit
circuito precableado prewired circuit
circuito prefabricado prefabricated circuit
circuito primario primary circuit
circuito principal main circuit
circuito privado private circuit, private wire
circuito procesado processed circuit
circuito productor de ruido noise-making circuit
circuito protector protective circuit
circuito puesto a masa grounded circuit
circuito puesto a tierra grounded circuit
circuito pulsante pulsing circuit
circuito quiescente quiescent circuit
circuito radial radial circuit

circuito radiante radiating circuit
circuito radioelectrónico radio-electronic circuit
circuito radiotelefónico radiotelephone circuit
circuito radiotelegráfico radiotelegraph circuit
circuito reactivo reactive circuit
circuito real real circuit
circuito receptor receiving circuit
circuito recortador clipper circuit
circuito recortador bidireccional bidirectional clipping circuit
circuito rectificador rectifying circuit
circuito rectificador de estado sólido solid-state rectifier circuit
circuito rectificador de onda completa full-wave rectifier circuit
circuito rectificador en derivación shunt rectifier circuit
circuito rectificador en puente bridge rectifier circuit
circuito rectificador en serie series rectifier circuit
circuito rectificador entrecruzado crisscross rectifier circuit
circuito rectificador múltiple multiple rectifier circuit
circuito rectificador polifásico polyphase rectifier circuit
circuito rectificador simple simple rectifier circuit
circuito rechazador rejector circuit
circuito redundante redundant circuit
circuito reentrante reentrant circuit
circuito reflejo reflex circuit
circuito regenerativo regenerative circuit
circuito regulado regulated circuit
circuito regulador regulating circuit
circuito repetidor de llamada calling repeater circuit
circuito resistivo resistive circuit
circuito resonador magnification circuit
circuito resonante resonant circuit
circuito resonante en paralelo parallel-resonant circuit
circuito resonante en serie series-resonant circuit
circuito restaurador de corriente continua direct-current restoring circuit
circuito reversible reversible circuit
circuito ruidoso noisy circuit
circuito seco dry circuit
circuito secuencial sequential circuit
circuito secundario secondary circuit
circuito seguidor follower circuit
circuito selector selector circuit
circuito sellado sealed circuit
circuito semidúplex semiduplex circuit
circuito sensor sensing circuit
circuito separador separator circuit
circuito silenciador muting circuit, squelch circuit
circuito silenciador de receptor receiver-muting circuit
circuito silencioso quiet circuit
circuito simétrico symmetrical circuit
circuito símplex simplex circuit
circuito simplificado simplified circuit
circuito sin conexión a tierra ungrounded circuit
circuito sintonizado tuned circuit
circuito sintonizado en serie series-tuned circuit
circuito sobrecargado overloaded circuit
circuito sólido solid circuit
circuito subterráneo underground circuit
circuito sumador adding circuit
circuito superfantasma superphantom circuit
circuito superheterodino superheterodyne circuit
circuito superheterodino de Armstrong Armstrong superheterodyne circuit
circuito superpuesto superimposed circuit

circuito superregenerativo superregenerative circuit
circuito superregenerativo de Armstrong
Armstrong superregenerative circuit
circuito supresor suppressor circuit, killer circuit
circuito supresor de colores color-killer circuit
circuito supresor de emisiones espurias
splatter-suppressor circuit
circuito tanque tank circuit
circuito telefónico telephone circuit
circuito telefónico comercial commercial telephone
circuit
circuito telefónico de portadora carrier telephone
circuit
circuito telefónico internacional international
telephone circuit
circuito telefónico por corriente portadora
carrier-current telephone circuit
circuito telegráfico telegraph circuit
circuito telegráfico fantasma phantom telegraph
circuit
circuito telegráfico internacional international
telegraph circuit
circuito télex telex circuit
circuito terminado terminated circuit
circuito trampa trap circuit
circuito transhorizonte beyond-the-horizon circuit
circuito tributario tributary circuit
circuito trifásico three-phase circuit
circuito ultraaudión ultra-audion circuit
circuito ultralineal ultralinear circuit
circuito único single circuit
circuito unidireccional unidirectional circuit
circuito virtual virtual circuit
circuito virtual permanente permanent virtual
circuit
circuito vital vital circuit
circuito vivo live circuit
circuito voltmétrico voltmeter circuit
circuito Y AND circuit
circuitos acoplados coupled circuits
circuitos asociados associated circuits
circuitos de computadora computer circuits
circuitos de comunicaciones por red network
communications circuits
circuitos de sintonización escalonada stagger-tuned
circuits
circuitos en tándem ganged circuits
circuitos isócronos isochronous circuits
circuitos magnéticos en serie series magnetic
circuits
circuitos mutuamente acoplados mutually coupled
circuits
circuitrón circuitron
circulación de electrólito circulation of electrolyte
circulador circulator
circulador de ferrita ferrite circulator
circulador de microondas microwave circulator
circulador de unión junction circulator
circulador en estrella Y circulator
circulador en T T circulator
circulante circulating
circulatorio circulating
círculo de calibración calibration circle
círculo de confusión circle of confusion
círculo de declinación circle of declination
círculo de espigas pin circle
círculo de guarda guard circle
círculo excéntrico eccentric circle
cirugía láser laser surgery
clarificador acústico acoustic clarifier
clarificador de voz speech clarifier
clase de aislamiento insulation class
clase de datos data class

clase de dispositivo device class
clase de exactitud accuracy class
clase de exactitud para medición accuracy class for
metering
clase de patrón pattern class
clase de servicio service class
clase de trabajo job class
clasificación rating, sorting
clasificación alfanumérica alphanumeric sorting
clasificación ascendente ascending sort
clasificación binaria binary sort
clasificación comparativa comparative sort
clasificación de aislamiento insulation rating
clasificación de banda muerta dead-band rating
clasificación de bloques block sort
clasificación de cálculos de dirección address
calculation sort
clasificación de carga continua continuous load
rating
clasificación de cinta tape sort
clasificación de clase de amplificador amplifier
class rating
clasificación de columna column sort
clasificación de corriente de servicio continuo
continuous-duty current rating
clasificación de dieléctrico dielectric rating
clasificación de disco disk sort
clasificación de distribución distribution sort
clasificación de documentos document sorting
clasificación de emergencia emergency rating
clasificación de entrada diferencial
differential-input rating
clasificación de exactitud accuracy rating
clasificación de formato format classification
clasificación de fusible fuse rating
clasificación de grupo group sorting
clasificación de inserción insertion sort
clasificación de intensidad sonora loudness rating
clasificación de intensidad sonora objetiva objective
loudness rating
clasificación de lista list sorting
clasificación de longitud variable variable-length
sorting
clasificación de medidor meter rating
clasificación de microcomputadora microcomputer
classification
clasificación de microprocesador microprocessor
classification
clasificación de ocho horas eight-hour rating
clasificación de ohms por volt ohms-per-volt rating
clasificación de patrones pattern classification
clasificación de relé relay rating
clasificación de resistencia a fuego fire-resistance
rating
clasificación de ruido noise rating
clasificación de seguridad safety class
clasificación de selección selection sort
clasificación de servicio service rating
clasificación de servicio continuo continuous-duty
rating
clasificación de servicio intermitente
intermittent-duty rating
clasificación de tabla de direcciones address table
sorting
clasificación de temperatura temperature
classification
clasificación descendente descending sort
clasificación digital digital sort
clasificación en cascada cascade sort
clasificación en serie serial sort
clasificación formal formal classification
clasificación-fusión sort-merge
clasificación generalizada generalized sort

clasificación interna internal sort
clasificación inversa backward sort
clasificación lineal linear sort
clasificación máxima maximum rating
clasificación momentánea momentary rating
clasificación multiarchivo multifile sorting
clasificación multinivel multilevel sort
clasificación nominal nominal rating
clasificación normal normal rating
clasificación numérica numerical sort
clasificación oscilante oscillating sort
clasificación periódica periodic rating
clasificación polifásica polyphase sort
clasificación por aguja needle sort
clasificación por base radix sort
clasificación por burbujas bubble sort
clasificación por lotes batch sort
clasificación principal major sort
clasificación rápida quick sort
clasificadora sorter
clasificadora de caracteres magnéticos magnetic character sorter
clasificadora de comprobación check sorter
clasificadora de documentos document sorter
clasificadora de tarjetas card sorter
clasificadora de tarjetas perforadas punched-card sorter
clasificadora fotoeléctrica photoelectric sorter
clasificadora-lectora sorter-reader
clasificar sort
clave key
clave compuesta composite key
clave criptográfica cryptographic key
clave de codificación de datos data encryption key
clave de código code key
clave de comprobación check key
clave de directorio directory key
clave de página page key
clave de registro record key
clave de sistema system key
clave de transferencia transfer key
clave empaquetada packed key
clave maestra criptográfica cryptographic master key
clave relativa relative key
clave remota remote key
clave secundaria secondary key
clavija plug
clavija abierta open plug
clavija adaptadora adapter plug
clavija banana banana plug
clavija coaxial coaxial plug
clavija concéntrica concentric plug
clavija cortocircuitadora shorting plug
clavija de audífono phone plug
clavija de circuito abierto open-circuit plug
clavija de conexión connecting plug, attachment plug
clavija de dos conductores two-conductor plug
clavija de infinidad infinity plug
clavija de Jones Jones plug
clavija de línea line plug
clavija de pared wall plug
clavija de patilla pin plug
clavija de prueba test plug
clavija de selector selector plug
clavija de teléfono phone plug
clavija de terminación termination plug
clavija de tres conductores three-conductor plug
clavija de tres patillas three-pin plug
clavija de yugo yoke plug
clavija desconectadora disconnecting plug
clavija fonográfica phono plug
clavija hembra female plug

clavija hermafrodítica hermaphroditic plug
clavija múltiple multiple plug
clavija no polarizada unpolarized plug
clavija no reversible nonreversible plug
clavija octal octal plug
clavija polarizada polarized plug
clavija sin cordón cordless plug
clavija telefónica telephone plug
clidonógrafo klydonograph
clidonograma klydonogram
clima adverso adverse weather
clima normal normal weather
climatómetro climatometer
cloro chlorine
cloruro de amonio ammonium chloride
cloruro de plomo lead chloride
cloruro de polivinilo polyvinyl chloride
cloruro de potasio potassium chloride
cloruro de vinilideno vinylidene chloride
coaxial coax
coaxil coax
cobalto cobalt
cobertura coverage
cobertura continua continuous coverage
cobertura de frecuencias frequency coverage
cobertura de prueba test coverage
cobertura de radar radar coverage
cobertura general general coverage
cobertura múltiple multiple coverage
cobertura nominal rated coverage
cobertura vertical vertical coverage
cobinotrón cobinotron
cobre copper
cobre al berilio beryllium copper
cobre de soldar soldering copper
cobre electrolítico electrolytic copper
cobre plateado silver-plated copper
cobre platinado platinum-clad copper
cobre recocido annealed copper
cociente multiplicador multiplier quotient
cocientímetro quotient meter
codán codan
codeposición codeposition
codificación coding, encoding
codificación absoluta absolute coding
codificación adaptiva adaptive coding
codificación aleatoria random coding
codificación alfabética alphabetic coding
codificación alfanumérica alphanumeric coding
codificación automática automatic coding
codificación básica basic coding
codificación binaria binary coding
codificación bipolar bipolar coding
codificación comprimida compressed encoding
codificación de anchura width coding
codificación de campo fijo fixed-field coding
codificación de colores color coding
codificación de datos data encoding
codificación de dirección única single-address coding
codificación de fase phase encoding
codificación de forma fija fixed-form coding
codificación de latencia mínima minimum latency coding
codificación de máquina machine coding
codificación de nivel único single-level encoding
codificación de programa program coding
codificación de retardo mínimo minimum delay coding
codificación de voz voice coding
codificación directa direct coding
codificación en línea inline coding
codificación específica specific coding

codificación forzada forced coding
codificación fuente source coding
codificación iterativa iterative coding
codificación lineal straight-line coding
codificación modular modular coding
codificación numérica numerical coding
codificación óptima optimum coding
codificación por bloques block coding
codificación por intervalos gap coding
codificación por muesca notch coding
codificación propia own coding
codificación relativa relative coding
codificación reubicable relocatable coding
codificación secuencial sequential coding
codificación simbólica symbolic coding
codificación sin pérdidas lossless encoding
codificación superpuesta superimposed coding
codificado coded
codificado en binario binary-coded
codificado en caracteres character coded
codificado por colores color coded
codificador coder, encoder, scrambler
codificador a-d a-d encoder
codificador de ángulo de eje shaft-angle encoder
codificador de cinta tape encoder
codificador de cinta magnética magnetic-tape encoder
codificador de colores color coder, color encoder
codificador de contacto contact encoder
codificador de cuantificación quantizing encoder
codificador de datos data encoder
codificador de fallas fault coder, fault encoder
codificador de impulsos pulse coder
codificador de posición de eje shaft-position encoder
codificador de tarjetas card encoder
codificador de teclado keyboard encoder
codificador de telemedida telemetering coder
codificador de voz voice coder, speech scrambler
codificador decimal decimal encoder
codificador-descodificador coder-decoder, codec
codificador-descodificador en chip único single-chip codec
codificador-descodificador monocanal single-channel codec
codificador fibroóptico fiber-optic scrambler
codificador rotativo rotary encoder
codificar encode
código code
código abierto open code
código absoluto absolute code
código abstracto abstract code
código alfabético alphabetic code
código alfanumérico alphanumeric code
código alfanumérico de ocho bits eight-bit alphanumeric code
código autocorrector self-correcting code
código automático automatic code
código binario binary code
código binario cíclico cyclic binary code
código binario-decimal binary-decimal code
código binario denso dense binary code
código binario en columnas column binary code
código binario en fila row binary code
código binario reflejado reflected binary code
código bipolar bipolar code
código biquinario biquinary code
código cíclico cyclic code
código comprobador de errores error-checking code
código con autocomplementación self-complementing code
código condicional conditional code
código continental continental code
código corrector de errores error-correcting code

código corto short code
código cuaternario quaternary code
código de acceso access code
código de acceso de operador operator access code
código de acceso mínimo minimum-access code
código de acción action code
código de alimentación de cinta tape-feed code
código de alimentación de línea line-feed code
código de área area code
código de arranque start code
código de arranque-parada start-stop code
código de atributo attribute code
código de autenticación de mensajes message authentication code
código de autocomprobación self-checking code
código de bajo nivel low-level code
código de barras bar code
código de barras ópticas optical bar code
código de Baudot Baudot code
código de bloque block code
código de bucle loop code
código de cable cable code
código de cadena chain code
código de canal vacante idle channel code
código de carácter character code
código de caracteres decimales codificado en binario binary-coded decimal character code
código de caracteres prohibidos forbidden character code
código de cinco canales five-channel code
código de cinco elementos five-element code
código de cinco niveles five-level code
código de cinco unidades five-unit code
código de cinta tape code
código de cinta de papel paper-tape code
código de cinta de perforación punch tape code
código de cinta magnética magnetic-tape code
código de cinta perforada perforated-tape code
código de colores color code
código de colores de resistores resistor color code
código de colores patrón standard color code
código de compilador compiler code
código de comprobación checking code
código de computadora computer code
código de condición condition code
código de conmutador switch code
código de contestación answerback code
código de control control code
código de control maestro master control code
código de control remoto remote-control code
código de conversión conversion code
código de corrección de errores error-correction code
código de cuatro direcciones four-address code
código de datos data code
código de datos alfabéticos alphabetic data code
código de datos alfanuméricos alphanumeric data code
código de datos de campo field data code
código de datos numéricos numerical data code
código de destino destination code
código de detección de errores error-detection code
código de dirección address code
código de dirección de llamada call direction code
código de dirección mnemónica mnemonic address code
código de dirección única single-address code
código de direccionamiento de llamada call-directing code
código de direcciones múltiples multiple-address code
código de dispositivo device code
código de dispositivo externo external device code

código de distancia mínima minimum distance code
código de dos direcciones two-address code
código de edición edit code
código de emisión emission code
código de encaminamiento routing code
código de ensamblaje assembly code
código de entrada condicional conditional entry code
código de error error code
código de escape escape code
código de espacio space code
código de estación station code
código de estado status code
código de exceso de cincuenta excess-fifty code
código de exceso de tres excess-three code
código de falla fault code
código de función function code
código de Gray Gray code
código de grupo group code
código de Hamming Hamming code
código de Hollerith Hollerith code
código de identificación de máquina machine identification code
código de identificación de mensajes message identification code
código de identificación de programa program identification code
código de impresión print code
código de impulsos pulse code
código de índice index code
código de información de máquina machine information code
código de instrucción instruction code
código de instrucción de entrada input instruction code
código de instrucción de máquina machine instruction code
código de instrucciones de computadora computer instruction code
código de intercambio interchange code
código de intercambio de datos data interchange code
código de interrupción interruption code
código de interrupción de máquina machine-interruption code
código de inversión de marcas alternativas alternate-mark inversion code
código de latencia mínima minimum latency code
código de lectura y perforación read-and-punch code
código de lenguaje language code
código de lenguaje de máquina machine-language code
código de línea line code
código de longitud de instrucción instruction length code
código de llamada selectiva selective-calling code
código de mando command code
código de máquina machine code
código de máquina absoluto absolute machine code
código de máquina reubicable relocatable machine code
código de modulación modulation code
código de Moore Moore code
código de Morse Morse code
código de Morse internacional international Morse code
código de Morse para cable cable Morse code
código de Morse por desplazamiento de frecuencia frequency-shift Morse code
código de nivel único one-level code
código de no impresión nonprint code
código de ocho niveles eight-level code

código de opción option code
código de operación operation code
código de operación aumentada augmented operation code
código de operación mnemónica mnemonic operation code
código de operador operator code
código de orden order code
código de parada stop code
código de perforación punch code
código de permutación permutation code
código de posición position code
código de privacidad privacy code
código de producto universal universal product code
código de razón constante constant-ratio code
código de razón fija fixed ratio code
código de recuperación retrieval code
código de redundancia cíclica cyclic redundancy code
código de referencia de sistema system reference code
código de registro record code
código de reproducción reproduction code
código de retardo mínimo minimum delay code
código de retención hold code
código de retorno return code
código de retorno primario primary return code
código de retroceso backspace code
código de salto skip code
código de selección selection code
código de selección de estación station-selection code
código de sentencia statement code
código de señalización signaling code
código de servicio service code
código de signo sign code
código de tabla table code
código de tarjeta card code
código de teletipo teletype code
código de terminación completion code
código de tiempo time code
código de trabajo job code
código de transacción transaction code
código de transmisión transmission code
código de tres direcciones three-address code
código de una dirección one-address code
código de usuario user code
código decimal decimal code
código decimal cíclico cyclic decimal code
código decimal codificado en binario binary-coded decimal code
código decimal de ocho bits eight-bit decimal code
código detector de errores error-detecting code
código directo direct code
código eléctrico electrical code
código en línea inline code
código en serie serial code
código ensamblador assembler code
código equilibrado balanced code
código específico specific code
código esquelético skeletal code
código falso false code
código fuente source code
código funcional functional code
código global global code
código ilegal illegal code
código indicador de programa program indicator code
código internacional international code
código interno internal code
código interpretador interpreter code
código interpretativo interpretive code
código lineal straight-line code

código magnético magnetic code
código mnemónico mnemonic code
código multifrecuencia multifrequency code
código nativo native code
código no válido invalid code
código numérico numerical code
código objeto object code
código oculto hidden code
código óptimo optimum code
código personal personal code
código ponderado weighted code
código predeterminado predetermined code
código privado private code
código Q Q code
código quibinario quibinary code
código quinario quinary code
código real actual code
código recurrente recurrent code
código redundante redundant code
código reentrante reentrant code
código reflejado reflected code
código reflejante reflective code
código relativo relative code
código reubicable relocatable code
código significativo significant code
código simbólico symbolic code
código telegráfico telegraph code
código ternario ternary code
código triangular triangular code
código unitario unitary code
código vacante vacant code
códigos de no reproducción nonreproducing codes
codistor codistor
codo bend
codo de guía guide elbow
codo de guíaondas waveguide elbow, waveguide
    corner, waveguide bend
coeficiente coefficient
coeficiente ambiental environmental coefficient
coeficiente de aberración esférica spherical
    aberration coefficient
coeficiente de absorción absorption coefficient
coeficiente de absorción atómica atomic absorption
    coefficient
coeficiente de absorción de energía energy
    absorption coefficient
coeficiente de absorción de masa mass absorption
    coefficient
coeficiente de absorción de reverberación
    reverberation absorption coefficient
coeficiente de absorción de sonido sound-absorption
    coefficient
coeficiente de absorción lineal linear absorption
    coefficient
coeficiente de absorción selectiva coefficient of
    selective absorption
coeficiente de acoplamiento coupling coefficient
coeficiente de acoplamiento de haz beam coupling
    coefficient
coeficiente de acoplamiento electromecánico
    electromechanical force factor
coeficiente de adhesión adhesion coefficient
coeficiente de amortiguamiento damping coefficient
coeficiente de ampere-vueltas ampere-turn
    coefficient
coeficiente de arrollamiento winding coefficient
coeficiente de atenuación attenuation coefficient
coeficiente de atenuación de dispersión scattering
    attenuation coefficient
coeficiente de atenuación de masa mass attenuation
    coefficient
coeficiente de atenuación de pares pair attenuation
    coefficient

coeficiente de atenuación fotoeléctrica photoelectric
    attenuation coefficient
coeficiente de atenuación imagen image attenuation
    coefficient
coeficiente de atenuación lineal linear attenuation
    coefficient
coeficiente de atenuación total total attenuation
    coefficient
coeficiente de autocorrelación autocorrelation
    coefficient
coeficiente de autoinducción coefficient of
    self-induction
coeficiente de captura capture coefficient
coeficiente de carga charging coefficient
coeficiente de coma flotante floating-point
    coefficient
coeficiente de condensación condensation coefficient
coeficiente de conmutación switching coefficient
coeficiente de correlación correlation coefficient
coeficiente de corriente de retorno return-current
    coefficient
coeficiente de cromaticidad chromaticity coefficient
coeficiente de decaimiento decay coefficient
coeficiente de deflexión deflection coefficient
coeficiente de densidad density coefficient
coeficiente de desfase phase-change coefficient
coeficiente de difusión diffusion coefficient
coeficiente de difusión lineal coefficient of linear
    diffusion
coeficiente de difusión radial radial-diffusion
    coefficient
coeficiente de directividad coefficient of directivity
coeficiente de dispersión dispersion coefficient,
    scattering coefficient
coeficiente de distorsión armónica coefficient of
    harmonic distortion
coeficiente de distorsión de fase phase-distortion
    coefficient
coeficiente de distorsión no lineal nonlinear
    distortion coefficient
coeficiente de distribución distribution coefficient
coeficiente de divergencia divergence coefficient
coeficiente de emisión secundaria
    secondary-emission coefficient
coeficiente de equivalencia de Potier Potier's
    coefficient of equivalence
coeficiente de error error coefficient
coeficiente de escala scale coefficient
coeficiente de excitación excitation coefficient
coeficiente de explotación coefficient of exploitation
coeficiente de fase acústica acoustic phase
    coefficient
coeficiente de fricción friction coefficient
coeficiente de fuga leakage coefficient
coeficiente de Hall Hall coefficient
coeficiente de histéresis hysteresis coefficient
coeficiente de inducción mutua coefficient of mutual
    induction
coeficiente de intermodulación intermodulation
    coefficient
coeficiente de ionización ionization coefficient
coeficiente de ionización de Townsend Townsend
    ionization coefficient
coeficiente de luminancia luminance coefficient
coeficiente de modulación modulation coefficient
coeficiente de ocupación coefficient of occupation
coeficiente de película film coefficient
coeficiente de Peltier Peltier coefficient
coeficiente de Peltier absoluto absolute Peltier
    coefficient
coeficiente de permeabilidad permeability
    coefficient
coeficiente de potencia power coefficient

coeficiente de presión pressure coefficient
coeficiente de propagación propagation coefficient
coeficiente de reciprocidad reciprocity coefficient
coeficiente de recombinación recombination coefficient
coeficiente de reflexión reflection coefficient
coeficiente de reflexión activa active reflection coefficient
coeficiente de reflexión de reverberación reverberation reflection coefficient
coeficiente de reflexión de tensión voltage reflection coefficient
coeficiente de reflexión sonora sound-reflection coefficient
coeficiente de refracción refractive coefficient
coeficiente de rendimiento coefficient of performance
coeficiente de resistencia resistance coefficient
coeficiente de resistencia de temperatura temperature resistance coefficient
coeficiente de resistencia negativa negative-resistance coefficient
coeficiente de retrodispersión backscattering coefficient
coeficiente de Sabine Sabine coefficient
coeficiente de sala room coefficient
coeficiente de salida output coefficient
coeficiente de Seebeck Seebeck coefficient
coeficiente de Seebeck absoluto absolute Seebeck coefficient
coeficiente de sensibilidad sensitivity coefficient
coeficiente de Steinmetz Steinmetz coefficient
coeficiente de temperatura temperature coefficient
coeficiente de temperatura de caída de tensión temperature coefficient of voltage drop
coeficiente de temperatura de capacidad temperature coefficient of capacity
coeficiente de temperatura de capacitancia temperature coefficient of capacitance
coeficiente de temperatura de frecuencia temperature coefficient of frequency
coeficiente de temperatura de fuerza electromotriz temperature coefficient of electromotive force
coeficiente de temperatura de ganancia gain temperature coefficient
coeficiente de temperatura de permeabilidad temperature coefficient of permeability
coeficiente de temperatura de resistencia temperature coefficient of resistance
coeficiente de temperatura de resistividad temperature coefficient of resistivity
coeficiente de temperatura de retardo temperature coefficient of delay
coeficiente de temperatura de sensibilidad temperature coefficient of sensitivity
coeficiente de temperatura negativo negative temperature coefficient
coeficiente de temperatura positivo positive temperature coefficient
coeficiente de tensión voltage coefficient
coeficiente de tensión de calefactor heater-voltage coefficient
coeficiente de tensión de capacitancia voltage coefficient of capacitance
coeficiente de tensión de resistencia voltage coefficient of resistance
coeficiente de Thomson Thomson coefficient
coeficiente de Townsend Townsend coefficient
coeficiente de transferencia transfer coefficient
coeficiente de transferencia de imagen image transfer coefficient
coeficiente de transmisión transmission coefficient
coeficiente de transmisión de reverberación reverberation transmission coefficient
coeficiente de transmisión de sonido sound-transmission coefficient
coeficiente de tubo tube coefficient
coeficiente de utilización coefficient of utilization
coeficiente de variación coefficient of variation
coeficiente dieléctrico dielectric coefficient
coeficiente diferencial differential coefficient
coeficiente luminoso luminous coefficient
coeficiente magnetorresistivo magnetoresistive coefficient
coeficiente osmótico osmotic coefficient
coeficiente reflejo reflex coefficient
coeficiente térmico thermal coefficient
coeficiente tricromático trichromatic coefficient
coeficientes de luminosidad luminosity coefficients
coercitividad coercivity
coercitividad intrínseca intrinsic coercivity
cogeneración cogeneration
coherencia coherence
coherencia de fase phase coherence
coherencia de modulación modulation coherence
coherencia espacial spatial coherence
coherencia temporal temporal coherence
coherente coherent
cohesión cohesion
cohesión lógica logical cohesion
cohesión secuencial sequential cohesion
cohesión temporal temporal cohesion
cohesor coherer
coincidencia coincidence
coincidencia aleatoria random coincidence
coincidencia de antena antenna coincidence
coincidencia de fases phase coincidence
coincidencia de impulsos pulse coincidence
coincidencia retardada delayed coincidence
cojinete bearing
cojinete autoalineable self-aligning bearing
cojinete de eje axle bearing
cojinete de joya jewel bearing
cojinete de manguito sleeve bearing
cola queue, tail
cola circular circular queue
cola de acceso directo direct-access queue
cola de actividad activity queue
cola de almacenamiento storage queue
cola de comunicaciones communication queue
cola de desplazamiento descendente push-down queue
cola de destino destination queue
cola de entrada input queue
cola de entrada de dispositivo device input queue
cola de espera de canales channel waiting queue
cola de impresión print queue
cola de impulso tail of pulse
cola de lectura adelantada read-ahead queue
cola de mensajes message queue
cola de mensajes de trabajo job message queue
cola de onda wave tail
cola de operaciones operations queue
cola de prioridad priority queue
cola de retención hold queue
cola de salida output queue
cola de sistema system queue
cola de tareas task queue
cola de trabajos job queue
cola de trabajos de entrada input job queue
cola de trabajos de salida output job queue
cola general general queue
cola no secuencial nonsequential queue
cola privada private queue
cola protegida protected queue
cola secuencial sequential queue

cola variable variable queue
colacionador collator
colacionador de tarjetas card collator
colacionar collate
colapez isinglass
colector collector, commutator
colector abierto open collector
colector de corriente current collector
colector de energía solar solar-energy collector
colector de ondas wave collector
colector de placa plana flat-plate collector
colector electrónico electron collector
colector tipo P P-type collector
coleo tailing
colgado on-hook
colgante pendant
colgante de acometida service drop
colidar colidar
colimación collimation
colimado collimated
colimador collimator
colimar collimate
colineal collinear
colisión electrónica electron collision
colisión iónica molecular molecular ion collision
colocar mal misplace
colodión collodion
color aditivo additive color
color cromático chromatic color
color de fondo background color
color de Planck Planckian color
color de referencia reference color
color débil weak color
color espectral spectral color
color espectral puro pure spectral color
color no saturado nonsaturated color
color patrón standard color
color saturado saturated color
color secundario secondary color
color sustractivo subtractive color
color terciario tertiary color
coloración de borde color edging
coloración extraña en bordes edging
colores complementarios complementary colors
colores primarios primary colors
colores primarios sustractivos subtractive primary colors
colorimetría colorimetry
colorimetría física physical colorimetry
colorimetría fotoeléctrica photoelectric colorimetry
colorimétrico colorimetric
colorímetro colorimeter
colorímetro aditivo additive colorimeter
colorímetro físico physical colorimeter
colorímetro fotoeléctrico photoelectric colorimeter
columbio columbium
columna column
columna binaria binary column
columna de aire air column
columna de continuación continuation column
columna de perforación punch column
columna de tarjeta card column
columna de visualización display column
columna en blanco blank column
columna positiva positive column
columna sonora sound column
columnas paralelas parallel columns
coma point
coma ajustable adjustable point
coma aritmética arithmetic point
coma base implícita assumed radix point
coma binaria binary point
coma binaria implícita implied binary point

coma de inserción binaria binary insertion point
coma de la base implícita implied radix point
coma de separación radix point
coma decimal decimal point, decimal comma
coma decimal implícita implied decimal point
coma fija fixed point
coma fija fraccional fractional fixed point
coma flotante floating point
coma hexadecimal hexadecimal point
coma octal octal point
coma variable variable point
comando command
combinación combination
combinación de bits bit combination
combinación de códigos code combination
combinación de contactos contact combination
combinación de contactos de relé relay-contact combination
combinación de perforaciones combination of punches
combinación prohibida forbidden combination
combinacional combinational
combinador combiner, switchgroup
combinador de canales channel combiner
combinador de ley cuadrática square-law combiner
combinador de potencia power switchgroup
combinador óptico optical combiner
combinatorio combinatorial
comienzo startup
comienzo de archivo beginning of file
comienzo de cinta beginning of tape
comienzo de programa program start
comienzo de salto skip start
compactación compaction, compacting, packing
compactación de datos data compaction
compactación de memoria memory compaction
compactación incremental incremental compaction
compactación local local compaction
comparación de amplitudes amplitude comparison
comparación de campos field comparison
comparación de colores color comparison
comparación de documentos document comparison
comparación de fase phase comparison
comparación de frecuencias frequency comparison
comparación de señales signal comparison
comparación de tiempo time comparison
comparación lógica logical comparison
comparador comparator
comparador analógico analog comparator
comparador cartográfico autoradar plot
comparador continuo continuous comparator
comparador de actividad de canal channel activity comparator
comparador de bobinas coil comparator
comparador de cintas tape comparator
comparador de colores color comparator
comparador de colores fotoeléctrico photoelectric color comparator
comparador de corrientes current comparator
comparador de dirección address comparator
comparador de diversidad diversity comparator
comparador de frecuencias frequency comparator
comparador de immitancias immitance comparator
comparador de impedancias impedance comparator
comparador de proyección projection comparator
comparador de reloj clock comparator
comparador de señales signal comparator
comparador de tensión analógico analog voltage comparator
comparador de tensiones voltage comparator
comparador de tiempo time comparator
comparador diferencial differential comparator
comparador digital digital comparator

comparador discreto discrete comparator
comparador lógico logic comparator
comparador óptico optical comparator
comparador polar polar comparator
comparar datos compare data
comparativo comparative
compartición sharing
compartimiento sharing
compartimiento auxiliar auxiliary compartment
compartimiento de archivos file sharing
compartimiento de control control compartment
compartimiento de cortacircuito circuit-breaker
  compartment
compartimiento de desbordamiento overflow
  bucket
compartimiento de frecuencia simultáneo
  simultaneous frequency sharing
compartimiento de frecuencias frequency sharing
compartimiento de programas program sharing
compartimiento de tiempo time sharing
compás compass
compás de aeronave aircraft compass
compás de inducción induction compass
compás giromagnético gyromagnetic compass
compás giroscópico gyroscopic compass
compás girostático gyrostatic compass
compás maestro master compass
compás magistral master compass
compás magnético magnetic compass
compás no magnético nonmagnetic compass
compás remoto remote compass
compatibilidad compatibility
compatibilidad ascendente upward compatibility
compatibilidad de diseños design compatibility
compatibilidad de disquete diskette compatibility
compatibilidad de equipos equipment compatibility
compatibilidad de periférico peripheral
  compatibility
compatibilidad de programas program compatibility
compatibilidad descendente downward compatibility
compatibilidad electromagnética electromagnetic
  compatibility
compatibilidad electromagnética entre sistemas
  intersystem electromagnetic compatibility
compatibilidad inversa reverse compatibility
compatible con computadora computer compatible
compensación compensation, offset
compensación bifásica biphase compensation
compensación catódica cathode compensation
compensación de abertura aperture compensation
compensación de agudos treble compensation
compensación de alta frecuencia high-frequency
  compensation
compensación de ambiente ambient compensation
compensación de atenuación attenuation
  compensation
compensación de baja frecuencia low-frequency
  compensation
compensación de bajos bass compensation
compensación de bajos automática automatic bass
  compensation
compensación de caída de reactancia
  reactance-drop compensation
compensación de carga automática automatic load
  compensation
compensación de corriente de reposo
  quiescent-current compensation
compensación de deslizamiento drift compensation
compensación de dopado doping compensation
compensación de fase phase compensation
compensación de frecuencia frequency
  compensation
compensación de intensidad sonora loudness

compensation
compensación de nivel automático automatic level
  compensation
compensación de paralaje parallax compensation
compensación de pendiente slope compensation
compensación de sombra shading compensation
compensación de temperatura temperature
  compensation
compensación en derivación shunt compensation
compensación en serie series compensation
compensación en serie-derivación series-shunt
  compensation
compensación por cable cable compensation
compensación térmica thermal compensation
compensador compensator
compensador acústico acoustic compensator
compensador de abertura aperture compensator
compensador de arranque starting compensator
compensador de atenuación attenuation
  compensator
compensador de bajos bass compensator
compensador de caída de línea line-drop
  compensator
compensador de corriente current compensator
compensador de corriente activa active-current
  compensator
compensador de corriente continua direct-current
  compensator
compensador de discos record compensator
compensador de fase phase compensator
compensador de impedancia impedance
  compensator
compensador de impedancia de filtro filter
  impedance compensator
compensador de intensidad sonora loudness
  compensator
compensador de nivel level compensator
compensador de pendiente slope compensator
compensador de polarización bias compensator
compensador de regrabación rerecording
  compensator
compensador de rerregistro rerecording
  compensator
compensador de temperatura magnético magnetic
  temperature compensator
compensador de volumen volume compensator
compensador magnético magnetic compensator
compensatorio compensatory
competencia de arbitraje arbitration contest
compilación compilation
compilación de programa program compilation
compilación separada separate compilation
compilador compiler
compilador científico scientific compiler
compilador conversacional conversational compiler
compilador cruzado cross compiler
compilador de alto nivel high-level compiler
compilador de control de procesos process-control
  compiler
compilador de microprocesador microprocessor
  compiler
compilador de programas program compiler
compilador de una pasada one-pass compiler
compilador de uno por uno one-for-one compiler
compilador incremental incremental compiler
compilador interactivo interactive compiler
compilador nativo native compiler
compilador por lotes batch compiler
compilador residente resident compiler
compilar compile
compilar y ejecutar compile-and-go
complejidad complexity
complejidad de dispositivo device complexity

complejo de coordinación coordination complex
complejo electromagnético electromagnetic complex
complementación complementation
complementación de datos data complementation
complementador complementer
complementario complementary
complemento complement
complemento a cero zero complement
complemento a diez tens complement
complemento a dos twos complement
complemento a nueve nines complement
complemento a uno ones complement
complemento de base radix complement
complemento de base menos uno radix-minus-one
    complement
complemento de cable cable complement
completamente automático fully automatic
completamente ecualizado fully equalized
completamente energizado fully energized
compolo compole
componente component
componente activo active component
componente activo de corriente active component of
    current
componente ajustable adjustable component
componente alterno alternating component
componente armónico harmonic component
componente biestable bistable component
componente binario binary component
componente capacitivo capacitive component
componente clase A class A component
componente compensador de fase
    phase-compensating component
componente compensador de temperatura
    temperature-compensating component
componente con temperatura controlada
    temperature-controlled component
componente con tomas tapped component
componente concentrado lumped component
componente cuadrático quadratic component
componente de acoplamiento coupling component
componente de audio audio component
componente de banda lateral sideband component
componente de banda lateral superior
    upper-sideband component
componente de base de datos database component
componente de brillo brightness component
componente de campo eléctrico electric-field
    component
componente de campo magnético magnetic-field
    component
componente de circuito circuit component
componente de color color component
componente de computadora computer component
componente de corriente alterna alternating-current
    component
componente de corriente continua direct-current
    component
componente de crominancia chrominance
    component
componente de datos data component
componente de desacoplamiento decoupling
    component
componente de equipo físico hardware component
componente de equipo integrado integrated
    equipment component
componente de estado estacionario steady-state
    component
componente de estado sólido solid-state component
componente de frecuencia frequency component
componente de frecuencia fundamental
    fundamental-frequency component
componente de frecuencia superior

higher-frequency component
componente de fuerza magnetomotriz
    magnetomotive force component
componente de guíaondas waveguide component
componente de imagen image component
componente de impedancia impedance component
componente de luminancia luminance component
componente de medida measurement component
componente de microcomputadora microcomputer
    component
componente de microondas no recíproco
    nonreciprocal microwave component
componente de microprocesador microprocessor
    component
componente de modulación de imagen
    picture-modulation component
componente de onda ordinaria ordinary-wave
    component
componente de onda sinusoidal sine-wave
    component
componente de ondas cuadradas square-wave
    component
componente de ondulación ripple component
componente de película delgada thin-film
    component
componente de película delgada discreto discrete
    thin-film component
componente de película gruesa thick-film
    component
componente de portadora carrier component
componente de potencia power component
componente de radio radio component
componente de radiofrecuencia radio-frequency
    component
componente de red de programas software network
    component
componente de repuesto spare component
componente de resistencia resistance component
componente de secuencia negativa
    negative-sequence component
componente de señal signal component
componente de señal de crominancia chrominance
    signal component
componente de señal de video video-signal
    component
componente de señal pequeña small-signal
    component
componente de señales fuertes large-signal
    component
componente de sintonización tuning component
componente de sistema system component
componente de subsistema subsystem component
componente de tono puro pure-tone component
componente de uso definido definite-purpose
    component
componente de uso general general-purpose
    component
componente de videofrecuencia video-frequency
    component
componente de visualización display component
componente desechable disposable component
componente desfasado out-of-phase component
componente deswattado wattless component
componente determinante de frecuencia
    frequency-determining component
componente directo direct component
componente discreto discrete component
componente distribuido distributed component
componente eléctrico electric component
componente electromagnético electromagnetic
    component
componente electrónico electronic component
componente electrostático electrostatic component

componente en cuadratura quadrature component
componente en fase in-phase component
componente encapsulado encapsulated component
componente encapsulado en plástico
   plastic-encapsulated component
componente encapsulado en resina
   resin-encapsulated component
componente enchufable plug-in component
componente energético energy component
componente enfriado por aire air-cooled component
componente espectral spectral component
componente estandarizado standardized component
componente extraño extraneous component
componente ferromagnético ferromagnetic
   component
componente fibroóptico fiber-optic component
componente ficticio dummy component
componente fijo fixed component
componente fotoconductor photoconductor
   component
componente funcional functional component
componente fundamental fundamental component
componente homopolar zero-phase-sequence
   component
componente horizontal horizontal component
componente ideal ideal component
componente imaginario imaginary component
componente impreso printed component
componente indirecto indirect component
componente inductivo inductive component
componente inherente inherent component
componente inmerso dipped component
componente integrado integrated component
componente lateral side component
componente longitudinal direct-axis component
componente magnético magnetic component
componente microminiatura microminiature
   component
componente miniatura miniature component
componente modular modular component
componente moldeado molded component
componente monocromático monochromatic
   component
componente multipuerto multiport component
componente neto net component
componente no recíproco nonreciprocal component
componente óhmico ohmic component
componente óhmico de tensión ohmic voltage
   component
componente ordinario ordinary component
componente parásito parasitic component, stray
   component
componente pasivo passive component
componente patrón standard component
componente periférico peripheral component
componente práctico practical component
componente quiescente quiescent component
componente reactivo reactive component
componente real real component
componente resistivo resistive component
componente rojo red component
componente simétrico symmetrical component
componente sinusoidal sinusoidal component
componente terminal terminal component
componente transitorio transient component
componente transversal quadrature-axis component
componente ultravioleta ultraviolet component
componente umbral threshold component
componente vertical vertical component
componente Y Y component
componente Z Z component
componentes apareados matched components
componentes de impedancia en paralelo

   parallel-impedance components
componentes de red de computadoras computer
   network components
componentes en paralelo parallel components
componentes externos outboard components
componentes independientes independent
   components
composición armónica harmonic composition
compresión compression
compresión ascendente upward compression
compresión de amplitud amplitude compression
compresión de ancho de banda bandwidth
   compression
compresión de archivo file compression
compresión de ceros zero compression
compresión de conjuntos de datos divididos
   partitioned data set compression
compresión de datos data compression
compresión de dígitos digit compression
compresión de imagen image compression
compresión de impulso pulse compression
compresión de modulación de amplitud
   amplitude-modulation compression
compresión de negro black compression
compresión de señal de sincronización sync
   compression
compresión de texto text compression
compresión de volumen volume compression
compresión de voz speech compression
compresión del blanco white compression
compresión descendente downward compression
compresión digital digital compression
compresión-expansión companding
compresión-expansión instantánea instantaneous
   companding
compresión lateral lateral compression
compresor compressor
compresor auxiliar auxiliary compressor
compresor de aire air compressor
compresor de arranque starting compressor
compresor de banda band compressor
compresor de fase phase compressor
compresor de volumen volume compressor
compresor de voz speech compressor
compresor electrodinámico electrodynamic
   compressor
compresor-expansor compander
compresor-expansor miniatura miniature
   compander
comprimir compress
comprobación check, proof
comprobación aleatoria random check
comprobación aritmética arithmetic check
comprobación automática automatic check
comprobación automática incorporada built-in
   automatic check
comprobación cíclica cyclic check
comprobación cruzada cross-checking
comprobación de alimentación feed check
comprobación de bit bit check
comprobación de calidad quality assurance
comprobación de canal channel check
comprobación de caracteres character check
comprobación de codificación coding check
comprobación de código de cinta tape code
   checking
comprobación de código falso false-code check
comprobación de combinación prohibida
   forbidden-combination check
comprobación de comparador comparator check
comprobación de confianza confidence check
comprobación de consistencia consistency check
comprobación de continuidad continuity check

comprobación de control de interfaz interface control check
comprobación de datos data check
comprobación de datos de lectura read data check
comprobación de desbordamiento overflow check
comprobación de dígito prohibido forbidden-digit check
comprobación de duplicación duplication check
comprobación de equilibrio balance check
comprobación de equipo físico hardware check
comprobación de error residual residual-error check
comprobación de errores error checking
comprobación de especificación specification check
comprobación de etiquetas label checking
comprobación de fidelidad fidelity check
comprobación de formato format check
comprobación de frecuencia frequency check
comprobación de frecuencia de oscilador local local-oscillator frequency check
comprobación de función function check
comprobación de ganancia gain check
comprobación de impresión print check
comprobación de lectura-escritura read-write check
comprobación de lectura tras escritura read-after-write check
comprobación de límite limit check
comprobación de máquina machine check
comprobación de margen range check
comprobación de módem modem check
comprobación de módulo module check
comprobación de paridad parity check
comprobación de paridad horizontal horizontal parity check
comprobación de paridad impar odd-parity check
comprobación de paridad lateral lateral parity check
comprobación de paridad longitudinal longitudinal parity check
comprobación de paridad par even parity check
comprobación de paridad por redundancia redundancy parity check
comprobación de paridad simple simple parity check
comprobación de paridad vertical vertical parity check
comprobación de perforación punch check
comprobación de problema problem check
comprobación de problemas dinámica dynamic problem checking
comprobación de procesador processor check
comprobación de programa program check
comprobación de redundancia redundancy check
comprobación de redundancia cíclica cyclic redundancy check
comprobación de redundancia lateral lateral redundancy check
comprobación de redundancia longitudinal longitudinal redundancy check
comprobación de redundancia transversal transverse redundancy check
comprobación de redundancia vertical vertical redundancy check
comprobación de reposición reset check
comprobación de residuo residue check
comprobación de secuencia sequence check
comprobación de selección selection check
comprobación de selección de dispositivo device selection check
comprobación de sincronización synchronization check
comprobación de sintonización tuning check
comprobación de sistema system check

comprobación de suma sumcheck
comprobación de tensión marginal marginal voltage check
comprobación de tiempo time check
comprobación de transferencia transfer check
comprobación de unidad unit check
comprobación de vaciado dump check
comprobación de validez validity check
comprobación del cero zero check
comprobación dinámica dynamic check
comprobación estática static check
comprobación estroboscópica stroboscopic checking
comprobación funcional functional check
comprobación incorporada built-in check
comprobación longitudinal longitudinal check
comprobación marginal marginal check
comprobación matemática mathematical check
comprobación módulo N modulo N check
comprobación periódica periodic check
comprobación por aguja needle check
comprobación por bloques block check
comprobación por bucle loop checking
comprobación por eco echo check
comprobación por suma sum check
comprobación posicional positional checking
comprobación programada programmed check
comprobación redundante redundant check
comprobación relacional relational checking
comprobación repetitiva repetitive check
comprobación selectiva spot check
comprobación transversal transverse check
comprobación visual visual check
comprobador de capacitores capacitor checker
comprobador de cintas tape checker
comprobador de condensador condenser checker
comprobador de convergencia convergence checker
comprobador de diodo diode checker
comprobador de ganancia gain checker
comprobador de linealidad linearity checker
comprobador de modulación modulation checker
comprobador de retorno flyback checker
comprobador de validez validity checker
compuerta gate
compuerta de decisión mayoritaria majority decision gate
compuesto compound
compuesto antioxidante slush compound
compuesto fotocrómico photochromic compound
compuesto intermetálico intermetallic compound
compuesto para sellado sealing compound
compuesto para transformadores transformer compound
computabilidad computability
computación computation, computing
computación automática automatic computing
computación implícita implicit computation
computación interactiva interactive computing
computación óptica optical computing
computación personal personal computing
computación remota remote computing
computacional computational
computadora computer
computadora activa active computer
computadora analógica analog computer
computadora analógica de corriente alterna alternating-current analog computer
computadora analógica de corriente continua direct-current analog computer
computadora analógica electrónica electronic analog computer
computadora analógica lineal linear analog computer
computadora analógica repetitiva repetitive analog

computer

computadora asincrónica asynchronous computer
computadora autoadaptiva self-adaptive computer
computadora automática automatic computer
computadora automática binaria binary automatic
computer
computadora autónoma stand-alone computer
computadora basada en sensores sensor-based
computer
computadora central central computer
computadora concurrente concurrent computer
computadora criogénica cryogenic computer
computadora de acceso múltiple multiple-access
computer
computadora de alta velocidad high-speed computer
computadora de automatización de producción
production automation computer
computadora de cinta magnética magnetic-tape
computer
computadora de coma flotante floating-point
computer
computadora de compilación compiling computer
computadora de comunicaciones communications
computer
computadora de contabilidad accounting computer
computadora de control control computer
computadora de control de procesos
process-control computer
computadora de cuarta generación fourth
generation computer
computadora de cuatro direcciones four-address
computer
computadora de dos direcciones two-address
computer
computadora de escritorio desktop computer
computadora de estado sólido solid-state computer
computadora de instrucciones fijas fixed-instruction
computer
computadora de línea de rumbo course-line
computer
computadora de lógica almacenada stored-logic
computer
computadora de memoria magnética magnetic
memory computer
computadora de memoria permanente
permanent-memory computer
computadora de navegación navigation computer
computadora de panel único single-board computer
computadora de primera generación first
generation computer
computadora de programa alambrado
wired-program computer
computadora de programa almacenado
stored-program computer
computadora de programa fijo fixed-program
computer
computadora de programación programming
computer
computadora de quinta generación fifth generation
computer
computadora de radiac radiac computer
computadora de relés relay computer
computadora de reserva standby computer
computadora de respaldo backup computer
computadora de secuencia arbitraria arbitrary
sequence computer
computadora de segunda generación second
generation computer
computadora de tablero único single-board
computer
computadora de tambor magnético magnetic-drum
computer
computadora de tarjetas perforadas punched-card

computer

computadora de teclado keyboard computer
computadora de tercera generación third-generation
computer
computadora de tráfico traffic computer
computadora de trayectoria de vuelo flight-path
computer
computadora de treinta y dos bits thirty-two bit
computer
computadora de tres direcciones three-address
computer
computadora de un chip one-chip computer
computadora de una dirección one-address
computer
computadora de uso especial special-purpose
computer
computadora de uso general general-purpose
computer
computadora de valor absoluto absolute value
computer
computadora de válvulas valve computer
computadora dedicada dedicated computer
computadora digital digital computer
computadora digital automática automatic digital
computer
computadora digital de transistores transistor
digital computer
computadora digital de uso general general-purpose
digital computer
computadora digital electrónica electronic digital
computer
computadora digital en serie serial digital computer
computadora digital paralela parallel digital
computer
computadora digital programable programmable
digital computer
computadora electromecánica electromechanical
computer
computadora electrónica electronic computer
computadora electrónica de tarjetas perforadas
punched-card electronic computer
computadora en línea on-line computer
computadora en paralelo parallel computer
computadora en serie serial computer
computadora en tiempo real real-time computer
computadora fuente source computer
computadora híbrida hybrid computer
computadora incremental incremental computer
computadora lineal linear computer
computadora maestra master computer
computadora mecánica mechanical computer
computadora microprogramable
microprogrammable computer
computadora multiprograma multiprogram
computer
computadora navegacional navigational computer
computadora neumática pneumatic computer
computadora no secuencial nonsequential computer
computadora numérica numerical computer
computadora objeto target computer
computadora orientada a caracteres
character-oriented computer
computadora orientada a cintas tape-oriented
computer
computadora orientada a comunicaciones
communications-oriented computer
computadora orientada a negocios business oriented
computer
computadora orientada a tarjetas card-oriented
computer
computadora periférica peripheral computer
computadora personal personal computer
computadora portátil portable computer

computadora primaria host computer
computadora principal main computer
computadora remota remote computer
computadora satélite satellite computer
computadora secuencial sequential computer
computadora simultánea simultaneous computer
computadora sin direcciones zero-address computer
computadora sincrónica synchronous computer
computadora terminal terminal computer
computadora transportable transportable computer
computadoras en tándem tandem computers
computar compute
computarizado computerized
cómputo de dirección address computation
cómputos en masa number crunching
comraz comraz
comunicación communication
comunicación a larga distancia long-distance communication
comunicación acústica acoustic communication
comunicación acústica subterránea subterranean acoustical communication
comunicación adaptiva adaptive communication
comunicación aire-tierra air-ground communication
comunicación asincrónica asynchronous communication
comunicación automatizada automated communication
comunicación bidireccional two-way communication
comunicación criptográfica cryptographic communication
comunicación de aire a aire air-to-air communication
comunicación de aire a tierra air-to-ground communication
comunicación de banda cruzada crossband communication
comunicación de barco a barco ship-to-ship communication
comunicación de barco a tierra ship-to-shore communication
comunicación de canal cruzado cross-channel communication
comunicación de datos data communication
comunicación de datos independiente independent data communication
comunicación de emergencia emergency communication
comunicación de fondo background communication
comunicación de paquetes packet communication
comunicación de punto a punto point-to-point communication
comunicación de seguridad safety communication
comunicación de socorro distress communication
comunicación de tierra a barco shore-to-ship communication
comunicación de tierra a tierra ground-to-ground communication
comunicación de tierra al aire ground-to-air communication
comunicación directa direct communication
comunicación dúplex duplex communication
comunicación eléctrica electric communication
comunicación en circuito cerrado closed-circuit communication
comunicación en serie serial communication
comunicación entre satélites intersatellite communication
comunicación espacial space communication
comunicación fibroóptica fiber-optic communication
comunicación impresa printed communication
comunicación inalámbrica wireless communication
comunicación indirecta indirect communication

comunicación múltiplex multiplex communication
comunicación por banda lateral única single-sideband communication
comunicación por cable cable communication
comunicación por corriente portadora carrier-current communication
comunicación por dispersión scattering communication
comunicación por dispersión troposférica tropospheric-scatter communication
comunicación por fibras ópticas fiber-optic communication
comunicación por haz láser laser-beam communication
comunicación por haz luminoso light-beam communication
comunicación por hilo wire communication
comunicación por infrarrojo infrared communication
comunicación por línea line communication
comunicación por línea de energía power-line communication
comunicación por módem modem communication
comunicación por ondas espaciales skywave communication
comunicación por portadora carrier communication
comunicación por saltos cortos short-skip communication
comunicación por satélite satellite communication
comunicación por sonar sonar communication
comunicación radiotelegráfica radiotelegraph communication
comunicación semidúplex semiduplex communication
comunicación sincrónica synchronous communication
comunicación subterránea underground communication
comunicación supersónica supersonic communication
comunicación telefónica telephone communication
comunicación télex telex communication
comunicación tierra-tierra ground-ground communication
comunicación transhorizonte beyond-the-horizon communication
comunicación ultrasónica ultrasonic communication
comunicación unidireccional one-way communication
comunicación visual visual communication
comunicación vocal voice communication
comunicaciones analógicas analog communications
comunicaciones de banda ancha wideband communications
comunicaciones digitales digital communications
comunicaciones electromagnéticas electromagnetic communications
comunicaciones en común joint communications
comunicaciones entre sistemas intersystem communications
comunicaciones móviles mobile communications
comunicaciones ópticas optical communications
comunicaciones por computadora computer communications
comunicaciones por espectro ensanchado spread-spectrum communications
comunicaciones por estelas meteóricas meteor-burst communications
con derivación central center tapped
con derivaciones tapped
con fugas leaky
con muescas slotted
con múltiples conexiones a tierra multigrounded

con puntas platinadas platinum-tipped
con tendencia a la obsolescencia obsolescence-prone
con tomas tapped
concatenación concatenation
concatenación de archivos file concatenation
concatenado concatenated
concatenar concatenate
concentración acústica acoustic concentration
concentración de datos data concentration
concentración de impurezas impurity concentration
concentración de iones de hidrógeno hydrogen-ion
    concentration
concentración de líneas line concentration
concentración de portadores carrier concentration
concentración electrónica electron concentration
concentración equivalente equivalent concentration
concentración espectral spectral concentration
concentración intrínseca intrinsic concentration
concentración iónica ion concentration
concentración molecular molecular concentration
concentración normal normal concentration
concentrador concentrator
concentrador de datos data concentrator
concentrador de datos de uso general
    general-purpose data concentrator
concentrador de líneas line concentrator
concentrador de mensajes message concentrator
concentrador de minicomputadora minicomputer
    concentrator
concentrador remoto remote concentrator
concentricidad concentricity
concéntrico concentric
concepto de caja negra black-box concept
concordancia concordance
concurrencia concurrency
concurrente concurrent
condensación condensation
condensación de archivo file condensation
condensador condenser, capacitor
condensador ajustable adjustable condenser
condensador al vacío vacuum condenser
condensador amortiguador buffer condenser
condensador antiinterferencia antiinterference
    condenser
condensador apocromático apochromatic condenser
condensador asincrónico asynchronous condenser
condensador calibrado calibrated condenser
condensador cerámico ceramic condenser
condensador con fugas leaky condenser
condensador de Abbe Abbe condenser
condensador de aceite oil condenser
condensador de acoplamiento coupling condenser
condensador de acoplamiento de antena antenna
    coupling condenser
condensador de acortamiento shortening condenser
condensador de ajuste trimming condenser
condensador de ajuste de altas frecuencias
    high-frequency trimmer
condensador de ajuste de oscilador oscillator
    trimmer
condensador de altavoz loudspeaker condenser
condensador de antena antenna condenser
condensador de aplanamiento smoothing condenser
condensador de arranque starting condenser
condensador de bloqueo blocking condenser
condensador de compensación padding condenser
condensador de chispas spark condenser
condensador de derivación bypass condenser
condensador de desacoplamiento decoupling
    condenser
condensador de desacoplo de placa plate bypass
    condenser
condensador de extinción quenching condenser

condensador de filtro filter condenser
condensador de ley cuadrática square-law
    condenser
condensador de Mansbridge Mansbridge condenser
condensador de mariposa butterfly condenser
condensador de mica mica condenser
condensador de motor motor condenser
condensador de papel paper condenser
condensador de placa plate condenser
condensador de policarbonato polycarbonate
    condenser
condensador de poliestirol polystyrol condenser
condensador de reacción reaction condenser
condensador de rejilla grid condenser
condensador de sintonización tuning condenser
condensador de sintonización de antena antenna
    tuning condenser
condensador de sintonización de placa plate tuning
    condenser
condensador de unidades múltiples multiple-unit
    condenser
condensador diferencial differential condenser
condensador electrolítico electrolytic condenser
condensador electrónico electronic condenser
condensador en bañera bathtub condenser
condensador en derivación shunt condenser
condensador en serie series condenser
condensador enchufable plug-in condenser
condensador equilibrador balancing condenser
condensador estático static condenser
condensador fijo fixed condenser
condensador hidromagnético hydromagnetic
    condenser
condensador mezclador mixer condenser
condensador neutralizante neutralizing condenser
condensador no inductivo noninductive condenser
condensador patrón standard condenser
condensador plano plane condenser
condensador rotativo rotary condenser
condensador sincrónico synchronous condenser
condensador sumergido submerged condenser
condensador tubular tubular condenser
condensador variable variable condenser
condensador vibrante vibrating condenser
condición anormal abnormal condition
condición atómica atomic condition
condición cero zero condition
condición combinada combined condition
condición compleja complex condition
condición compuesta compound condition
condición controlada controlled condition
condición de alarma alarm condition
condición de clase class condition
condición de colgado on-hook condition
condición de desbordamiento overflow condition
condición de descolgado off-hook condition
condición de disponibilidad ready condition
condición de entrada entry condition
condición de equilibrio equilibrium condition
condición de error error condition
condición de espera wait condition
condición de estabilidad stability condition
condición de estado de canal channel status
    condition
condición de exploración search condition
condición de fase phase condition
condición de frecuencia frequency condition
condición de grabación record condition
condición de límite boundary condition
condición de Lorentz Lorentz condition
condición de modulación modulation condition
condición de no listo not ready condition
condición de parada halt condition

condición de prueba test condition
condición de registro record condition
condición de relación relation condition
condición de reposo quiescent condition
condición de retención hold condition
condición de tensión degradada degraded voltage condition
condición de trabajo marking condition
condición de tránsito via condition
condición de trayectoria path condition
condición eléctrica peligrosa hazardous electrical condition
condición final final condition
condición habilitada enabled condition
condición inaceptable unacceptable condition
condición inicial initial condition
condición intermitente intermittent condition
condición limitadora limiting condition
condición necesaria necessary condition
condición negada negated condition
condición normal normal condition
condición transitoria transient condition
condición uno one condition
condicional conditional
condicionamiento de línea line conditioning
condiciones ambientales environmental conditions
condiciones de agua adversas adverse water conditions
condiciones de laboratorio laboratory conditions
condiciones de operación operating conditions
condiciones de operación de referencia reference operating conditions
condiciones de servicio service conditions
condiciones de trabajo working conditions
condiciones equilibradas balanced conditions
condiciones meteorológicas meteorological conditions
conducción conduction
conducción asimétrica asymmetrical conduction
conducción bidireccional two-way conduction
conducción de aire air conduction
conducción de electrones secundarios secondary-electron conduction
conducción directa forward conduction
conducción eléctrica electrical conduction
conducción electrolítica electrolytic conduction
conducción electrónica electronic conduction
conducción en obscuridad dark conduction
conducción extrínseca extrinsic conduction
conducción gaseosa gaseous conduction
conducción intrínseca intrinsic conduction
conducción inversa reverse conduction, inverse conduction
conducción iónica ionic conduction
conducción no metálica nonmetallic conduction
conducción ósea bone-conduction
conducción por agujeros defect conduction
conducción por avalancha avalanche conduction
conducción por electrones excedentes excess conduction
conducción por huecos hole conduction
conducción térmica thermal conduction
conducción tipo N N-type conduction
conducción tipo P P-type conduction
conducción unidireccional one-way conduction
conducta dinámica dynamic behavior
conductancia conductance
conductancia anódica anode conductance
conductancia colector-base collector-base conductance
conductancia de ánodo diferencial differential anode conductance
conductancia de conversión conversion conductance

conductancia de electrodo electrode conductance
conductancia de entrada input conductance
conductancia de fugas leak conductance
conductancia de pendiente slope conductance
conductancia de placa plate conductance
conductancia de rejilla grid conductance
conductancia de ruido equivalente equivalent noise conductance
conductancia de salida output conductance
conductancia directa forward conductance
conductancia en cortocircuito short-circuit conductance
conductancia equivalente equivalent conductance
conductancia específica specific conductance
conductancia extrínseca extrinsic conductance
conductancia inversa back conductance
conductancia máxima equivalente maximum equivalent conductance
conductancia molar molar conductance
conductancia molecular molecular conductance
conductancia mutua mutual conductance
conductancia mutua estática static mutual conductance
conductancia negativa negative conductance
conductancia patrón standard conductance
conductancia positiva positive conductance
conductancia relativa relative conductance
conductancia transformada transformed conductance
conductividad conductivity
conductividad acústica acoustic conductivity
conductividad asimétrica asymmetrical conductivity
conductividad bilateral bilateral conductivity
conductividad característica characteristic conductivity
conductividad compleja complex conductivity
conductividad de emisión secundaria secondary-emission conductivity
conductividad de equilibrio equilibrium conductivity
conductividad de revestimiento coating conductivity
conductividad de solución solution conductivity
conductividad de tierra ground conductivity
conductividad de volumen volume conductivity
conductividad efectiva effective conductivity
conductividad eléctrica electrical conductivity
conductividad electrolítica electrolytic conductivity
conductividad electrónica electronic conductivity
conductividad equivalente equivalent conductivity
conductividad específica specific conductivity
conductividad fotoeléctrica photoelectric conductivity
conductividad inducida de haz electrónico electron-beam induced conductivity
conductividad intrínseca intrinsic conductivity
conductividad magnética magnetic conductivity
conductividad molecular molecular conductivity
conductividad por bombardeo bombardment conductivity
conductividad por bombardeo electrónico ebiconductivity
conductividad simétrica symmetrical conductivity
conductividad superficial surface conductivity
conductividad térmica thermal conductivity
conductividad térmica electrónica electronic thermal conductivity
conductividad tipo N N-type conductivity
conductividad tipo P P-type conductivity
conductividad unidireccional unidirectional conductivity
conductividad unilateral unilateral conductivity
conducto duct, conduit
conducto atmosférico atmospheric duct
conducto de aire air duct

conducto de alambres wireway
conducto de arco arc chute
conducto de cable cable duct
conducto de goma flexible flexible rubber conduit
conducto de hilos wire duct
conducto de luz light pipe
conducto de metal flexible flexible metal conduit
conducto de ondas wave duct
conducto de radio atmosférico atmospheric radio duct
conducto de residuos de perforación chad duct
conducto de tierra ground conduit
conducto del inducido armature duct
conducto eléctrico raceway
conducto flexible flexible conduit
conducto magnetoiónico magnetoionic duct
conducto monolítico monolithic conduit
conducto múltiple multiple duct
conducto portacables cable duct
conducto subsidiario subsidiary conduit
conductómetro conductance meter
conductor conductor, lead, wire
conductor abierto open conductor
conductor acelerador accelerating conductor
conductor activo active conductor
conductor aéreo overhead conductor
conductor aislado insulated conductor
conductor anular annular conductor
conductor coaxial coaxial conductor
conductor compuesto composite conductor
conductor concéntrico concentric conductor
conductor conformado shaped conductor
conductor criogénico cryogenic conductor
conductor de acometida service conductor
conductor de alimentación power conductor, supply lead
conductor de aluminio aluminum conductor
conductor de bronce bronze conductor
conductor de cable cable conductor
conductor de comunicación communication conductor
conductor de contacto contact conductor
conductor de control control conductor
conductor de derivación branch conductor
conductor de entrada lead-in
conductor de fase phase conductor
conductor de hoja metálica foil conductor
conductor de línea line conductor
conductor de núcleo hueco hollow-core conductor
conductor de oropel tinsel conductor
conductor de pararrayos lightning conductor
conductor de prueba test lead
conductor de puesta a tierra grounding conductor
conductor de retorno return conductor
conductor de sonido sound-conductive
conductor de suministro de batería battery-supply conductor
conductor de tierra ground conductor
conductor desnudo bare conductor
conductor electrónico electron conductor
conductor exterior outer conductor
conductor fortuito fortuitous conductor
conductor hueco hollow conductor
conductor ideal ideal conductor
conductor imperfecto imperfect conductor
conductor impreso printed conductor
conductor influenciado influenced conductor
conductor interior inside conductor
conductor interno inner conductor
conductor lateral lateral conductor
conductor líquido liquid conductor
conductor macizo solid conductor
conductor móvil moving conductor

conductor múltiple multiple conductor
conductor negativo negative conductor
conductor neutro neutral conductor
conductor óptico optical conductor
conductor perfecto perfect conductor
conductor positivo positive conductor, positive lead
conductor principal main lead
conductor puesto a tierra grounded conductor
conductor radial radial conductor
conductor recubierto covered conductor
conductor resistivo resistive conductor
conductor segmental segmental conductor
conductor sinterizado sintered conductor
conductor sólido solid conductor
conductor térmico thermal conductor
conductor trenzado stranded conductor
conductor vacante vacant conductor
conductores conjugados conjugate conductors
conductores en derivación shunt leads
conectado a masa grounded
conectado a tierra grounded, earthed
conectado a tierra directamente directly grounded
conectado a tierra efectivamente effectively grounded
conectado a tierra indirectamente indirectly grounded
conectado a tierra permanentemente permanently earthed
conectado eléctricamente electrically connected
conectado en derivación shunt-connected
conectado en estrella star-connected
conectado en paralelo parallel-connected
conectado en serie series-connected
conectado permanentemente permanently connected
conectado por enlace link-attached
conectar connect, switch-on
conectar a tierra put to earth
conectividad connectivity
conector connector
conector a prueba de humedad moistureproof connector
conector adaptante adapting connector
conector anódico anode connector
conector coaxial coaxial connector
conector coaxial hermafrodita hermaphrodite coaxial connector
conector compatible compatible connector
conector de alimentación power connector
conector de antena antenna connector
conector de archivo file connector
conector de archivo interno internal file connector
conector de bridas flange connector
conector de cable cable connector
conector de cable flexible flexible-cable connector
conector de canales channel connector
conector de cinta tape connector
conector de circuitos impresos printed-circuit connector
conector de clavija plug connector
conector de fusible fuse connector
conector de guíaondas waveguide connector
conector de guíaondas óptico optical waveguide connector
conector de hilos wiring connector
conector de jack jack connector
conector de línea line connector
conector de presión pressure connector
conector de prueba test connector
conector de rompimiento muerto dead-break connector
conector de ruptura de carga load-break connector
conector de servicio service connector
conector eléctrico electric connector

conector en T T connector
conector enchufable plug-in connector
conector externo external connector
conector fijo fixed connector
conector flexible flexible connector
conector hermafrodita hermaphrodite connector
conector interno inner connector, internal connector
conector modular modular connector
conector multicontacto multicontact connector
conector multihilo multiwire connector
conector múltiple multiple connector
conector no reversible nonreversible connector
conector óptico optical connector
conector paralelo parallel connector
conector patrón standard connector
conector plano plain connector
conector recto straight connector
conector terminal terminal connector
conector triaxial triaxial connector
conector variable variable connector
conexión connection, hookup, attachment, lead
conexión a larga distancia long-distance connection
conexión a tierra ground connection, ground
conexión antiparalela antiparallel connection
conexión arrollada wire-wrap connection
conexión automática automatic connection
conexión bidireccional two-way connection
conexión capacitiva capacitive connection
conexión completa complete connection
conexión compuesta compound connection
conexión con base a masa grounded-base connection
conexión con cátodo a masa grounded-cathode connection
conexión con colector a masa grounded-collector connection
conexión con emisor a masa grounded-emitter connection
conexión conmutada switched connection
conexión cruzada cross-connection
conexión de abonado subscriber connection
conexión de antena antenna connection
conexión de cable cable connection
conexión de canal a canal channel-to-channel connection
conexión de clavija plug connection
conexión de clavija y jack plug-and-jack connection
conexión de colector común common-collector connection
conexión de conferencia conference connection
conexión de conmutador básica basic switch connection
conexión de corriente continua direct-current connection
conexión de datos data connection
conexión de derivación branch joint
conexión de derivación larga long-shunt connection
conexión de emisor común common-emitter connection
conexión de enlace link connection
conexión de espiga pin connection
conexión de filamento filament lead
conexión de Hoeppner Hoeppner connection
conexión de impedancia impedance bond
conexión de interfono interphone connection
conexión de Leblanc Leblanc connection
conexión de línea line connection
conexión de malla mesh connection
conexión de plomo lead connection
conexión de prueba test connection
conexión de puente jumper connection
conexión de puesta a tierra grounding connection
conexión de rejilla abierta open-grid connection
conexión de resina rosin connection

conexión de retorno return connection
conexión de retorno de rejilla grid-return lead
conexión de Scott Scott connection
conexión de servicio service connection
conexión de sobretensión overvoltage connection
conexión de Taylor Taylor connection
conexión de teleimpresora teleprinter connection
conexión de televisión television connection
conexión de tensión de línea line-voltage connection
conexión de tierra ground lead
conexión de una dirección single-way connection
conexión de video video connection
conexión defectuosa faulty connection
conexión ecualizadora equalizing connection
conexión eléctrica electric connection
conexión eléctrica de carriles rail bond
conexión elevadora step-up connection
conexión en anillo ring connection
conexión en base común common-base connection
conexión en bucle loop connection
conexión en cascada cascade connection
conexión en contrafase push-pull connection
conexión en delta delta connection
conexión en delta abierta open-delta connection
conexión en delta extendida extended delta connection
conexión en derivación shunt connection
conexión en estrella star connection, wye connection
conexión en oposición back-to-back connection
conexión en paralelo parallel connection, paralleling
conexión en puente bridging connection, bridge connection
conexión en serie serial connection, series connection
conexión en serie-paralelo series-parallel connection
conexión en T T connection
conexión en tándem tandem connection
conexión en zigzag zigzag connection
conexión enchufable plug-in connection
conexión equipotencial equipotential connection
conexión estrella-estrella star-star connection
conexión externa external connection
conexión física physical connection
conexión flexible flexible connection
conexión floja loose connection
conexión fonográfica phonograph connection
conexión fría cold connection
conexión general de aeronave aircraft bonding
conexión interfacial interfacial connection
conexión interna internal connection
conexión internacional international connection
conexión interurbana trunk connection
conexión inversora inverting connection
conexión local local connection
conexión matricial matrix connection
conexión mecánica mechanical connection
conexión múltiple multiple connection
conexión multipunto multipoint connection
conexión no inversora noninverting connection
conexión permanente permanent connection
conexión por circuito privado private-wire connection
conexión por hilo wire connection
conexión radial radial lead
conexión sin soldadura solderless connection
conexión telefónica telephone connection
conexión telegráfica telegraph connection
conexión télex telex connection
conexión temporal temporary connection, patch
conexión temporal de teléfono phone patch
conexión terminal end connection
conexión total a tierra total earth
conexión unidireccional one-way connection

conexión unilateral unilateral connection
conexionado wiring
confiabilidad reliability
confiabilidad alcanzada achieved reliability
confiabilidad de canal channel reliability
confiabilidad de circuito circuit reliability
confiabilidad de comunicación communication reliability
confiabilidad de conmutación switching dependability
confiabilidad de datos data reliability
confiabilidad de programas software reliability
confiabilidad de propagación propagation reliability
confiabilidad de señal signal reliability
confiabilidad de sistema system reliability
confiabilidad de transmisión transmission reliability
confiabilidad evaluada assessed reliability
confiabilidad numérica numerical reliability
confiabilidad observada observed reliability
confiabilidad operacional operational reliability
confiabilidad óptima optimum reliability
configurable configurable
configuración configuration
configuración binaria binary configuration
configuración con base a masa grounded-base configuration
configuración con colector a masa grounded-collector configuration
configuración con emisor a masa grounded-emitter configuration
configuración con rejilla a masa grounded-grid configuration
configuración cruzada cross configuration
configuración de bits bit configuration
configuración de circuito circuit configuration
configuración de circuitos impresos printed-circuit configuration
configuración de computadora computer configuration
configuración de controlador controller configuration
configuración de equipo físico hardware configuration
configuración de huecos hole configuration
configuración de máquina machine configuration
configuración de microprocesador microprocessor configuration
configuración de procesador processor configuration
configuración de programa program setup
configuración de red centralizada centralized network configuration
configuración de registro permisible allowable record configuration
configuración de sistema system configuration
configuración de sondeo paralelo parallel poll configuration
configuración de subred subnetwork configuration
configuración en bucle loop configuration
configuración en estrella wye configuration
configuración funcional functional configuration
configuración integrada integrated configuration
configuración iterativa iterative configuration
configuración lineal linear configuration
configuración mínima minimum configuration
configuración objeto target configuration
configuración virtual virtual configuration
configurar set-up
conflicto de acceso access conflict
conflicto de bus bus conflict
conflicto de estructura structure conflict
confocal confocal
conformidad conformance

congestión espectral spectrum congestion
congestión interna internal congestion
conicidad taper
conicidad de guíaondas waveguide taper
conjunto set
conjunto codificado coded set
conjunto codificado en binario binary-coded set
conjunto de archivos file set
conjunto de bastidor rack assembly
conjunto de bobinas coil set
conjunto de canales channel set
conjunto de caracteres character set
conjunto de caracteres alfabéticos alphabetic character set
conjunto de caracteres alfanuméricos alphanumeric character set
conjunto de caracteres alternativos alternate character set
conjunto de caracteres binarios binary character set
conjunto de caracteres codificado coded character set
conjunto de caracteres de biblioteca library character set
conjunto de caracteres de datos data character set
conjunto de caracteres de formato format character set
conjunto de caracteres de lenguaje language character set
conjunto de caracteres decimales codificado en binario binary-coded decimal character set
conjunto de caracteres extendido extended character set
conjunto de caracteres gráficos extendido extended graphic character set
conjunto de caracteres ideográficos ideographic character set
conjunto de caracteres numéricos numerical character set
conjunto de caracteres universal universal character set
conjunto de cartas genérico generic card set
conjunto de códigos code set
conjunto de datos data set
conjunto de datos catalogados cataloged data set
conjunto de datos concatenados concatenated data set
conjunto de datos consecutivos consecutive data set
conjunto de datos de entrada input data set
conjunto de datos de generación generation data set
conjunto de datos de impresión print data set
conjunto de datos de mensajes message data set
conjunto de datos de página page data set
conjunto de datos de panel panel data set
conjunto de datos de punto de comprobación checkpoint data set
conjunto de datos de radiodifusión broadcast data set
conjunto de datos de salida output data set
conjunto de datos de tipo de letras font data set
conjunto de datos dedicado dedicated data set
conjunto de datos directo direct data set
conjunto de datos divididos partitioned data set
conjunto de datos ficticios dummy data set
conjunto de datos generado generated data set
conjunto de datos gráficos graphic data set
conjunto de datos indexados indexed data set
conjunto de datos mixtos mixed data set
conjunto de datos objeto target data set
conjunto de datos relativos relative data set
conjunto de datos secundarios secondary data set
conjunto de datos temporal temporary data set
conjunto de discos disk pack
conjunto de elementos codificados coded element

set
conjunto de escobillas brush set
conjunto de etiquetas label set
conjunto de instrucciones instruction set
conjunto de instrucciones de computadora
  computer instruction set
conjunto de instrucciones de máquina machine
  instruction set
conjunto de instrucciones de microcódigo
  microcode instruction set
conjunto de instrucciones de microprocesador
  microprocessor instruction set
conjunto de instrucciones de microprograma
  microprogram instruction set
conjunto de montaje mounting assembly
conjunto de paneles de conexiones patch bay
conjunto de prueba de comunicaciones
  communications test set
conjunto de relés relay set
conjunto de secuencias sequence set
conjunto de símbolos symbol set
conjunto de símbolos gráficos graphics symbol set
conjunto de tareas task set
conjunto de tarjetas card set
conjunto de terminación terminating set
conjunto de tipo de letras font set
conjunto de trabajos job set
conjunto en cascada cascade set
conjunto híbrido hybrid set
conjunto magnético magnetic assembly
conjunto nulo null set
conjunto plano flat pack
conjuntor jack
conmutación switching, commutation
conmutación analógica analog switching
conmutación automática automatic switching
conmutación auxiliar auxiliary switching
conmutación bilateral bilateral switching
conmutación coaxial coaxial switching
conmutación complementaria complementary
  commutation
conmutación compuesta compound switching
conmutación de alcance range switching
conmutación de banda band switching
conmutación de circuito circuit switching
conmutación de colores color switching, sampling
conmutación de distancia range switching
conmutación de entrada-salida input-output
  switching
conmutación de escala range switching
conmutación de función alternativa alternate
  function switch
conmutación de haz beam switching
conmutación de impulsos pulse switching
conmutación de líneas line switching
conmutación de lóbulo lobe switching
conmutación de mensajes message switching
conmutación de modo de línea line-mode switching
conmutación de paquetes packet switching
conmutación de programas program switching
conmutación de Q Q switching
conmutación de telecomunicaciones
  telecommunications switching
conmutación de trabajo marking switching
conmutación de umbral threshold switching
conmutación de velocidad speed switching
conmutación de video video switching
conmutación digital digital switching
conmutación directa direct commutation
conmutación electrónica electronic switching
conmutación en tiempo real real-time switching
conmutación estática static switching
conmutación indirecta indirect commutation

conmutación inductiva inductive switching
conmutación intermedia intermediate switching
conmutación manual manual switching
conmutación momentánea momentary switching
conmutación múltiple multiple switching
conmutación no sincrónica nonsynchronous
  switching
conmutación por chorro de mercurio mercury-jet
  commutation
conmutación por división de tiempo time-division
  switching
conmutación por pulsador push-button switching
conmutación remota remote switching
conmutación secuencial sequential switching
conmutación semiautomática semiautomatic
  switching
conmutación sin cordones cordless switching
conmutación sin chispas sparkless commutation
conmutación sin puenteo nonbridging switching
conmutación telefónica telephone switching
conmutador switch, switcher, commutator, cutout
conmutador accionado por presión
  pressure-actuated switch
conmutador activado por luz light-activated switch
conmutador actuador actuator switch
conmutador aislante isolating switch
conmutador analógico analog switch
conmutador anticapacitivo anticapacitance switch
conmutador autoenganchador self-latching switch
conmutador automático automatic switch
conmutador auxiliar auxiliary switch
conmutador barométrico barometric switch
conmutador bidireccional two-way switch
conmutador bimetálico bimetallic switch
conmutador blindado metal-clad switch
conmutador buscador finder switch
conmutador centrífugo centrifugal switch
conmutador coaxial coaxial switch
conmutador colgante pendant switch
conmutador con cubierta metálica metal-enclosed
  switch
conmutador conectado por atrás back-connected
  switch
conmutador confirmante acknowledging switch
conmutador contra transmisión-recepción
  antitransmit-receive switch
conmutador controlado de silicio silicon-controlled
  switch
conmutador controlado por puerta gate-controlled
  switch
conmutador cortocircuitador short-circuiting switch
conmutador de abertura rápida quick-opening
  switch
conmutador de acción momentánea
  momentary-action switch
conmutador de acción rápida rapid-action switch
conmutador de acción ultrarrápida snap-action
  switch
conmutador de aceleración acceleration switch
conmutador de acometida service switch
conmutador de acoplamiento coupling switch
conmutador de adaptación de carga load-matching
  switch
conmutador de aire air switch
conmutador de alarma alarm switch
conmutador de alcance range switch
conmutador de alimentación power switch
conmutador de alimentación de cinta tape-feed
  switch
conmutador de alimentación principal main power
  switch
conmutador de almacenamiento storage switch
conmutador de almacenamiento manual manual

storage switch
conmutador de alta velocidad high-speed switch
conmutador de alumbrado lighting switch
conmutador de anillo ring switch
conmutador de antena antenna switch
conmutador de antena-tierra lightning switch
conmutador de arco arc switch
conmutador de arranque starter switch
conmutador de arranque-parada start-stop switch
conmutador de asignación alloter switch
conmutador de avance paso a paso stepping switch
conmutador de balancín rocker switch
conmutador de banda band switch
conmutador de banda base baseband switch
conmutador de barras cruzadas crossbar switch
conmutador de barras de código code-bar switch
conmutador de cadena chain switch
conmutador de campo field switch
conmutador de capacitor capacitor switch
conmutador de carga load switch, charge switch
conmutador de carga mínima underload switch
conmutador de cierre rápido quick-make switch
conmutador de cierre y abertura rápida
    quick-make-and-break switch
conmutador de circuito circuit switch
conmutador de circuito impreso printed-circuit
    switch
conmutador de clavija plug switch
conmutador de código de dirección address code
    switch
conmutador de conector connector switch
conmutador de conexión-desconexión on-off switch
conmutador de consola console switch
conmutador de contacto contact switch
conmutador de contacto a presión pressure-contact
    switch
conmutador de contacto de cuchilla knife-contact
    switch
conmutador de contacto deslizante sliding-contact
    commutator
conmutador de contacto momentáneo
    momentary-contact switch
conmutador de contactos deslizantes sliding-contact
    switch
conmutador de contactos múltiples multiple-contact
    switch
conmutador de control control switch
conmutador de control de impresión print-control
    switch
conmutador de control de motor motor control
    switch
conmutador de control principal main control
    switch
conmutador de corriente alterna alternating-current
    switch
conmutador de corte cutoff switch
conmutador de cuchilla knife switch
conmutador de chorro de mercurio mercury-jet
    switch
conmutador de datos automático automatic data
    switch
conmutador de descarga discharge switch
conmutador de desconexión disconnect switch
conmutador de diodo diode switch
conmutador de distancia range switch
conmutador de doble cierre double-make switch
conmutador de doble ruptura double-break switch
conmutador de dos polos two-pole switch
conmutador de dos posiciones two-position switch,
    two-throw switch
conmutador de emergencia emergency switch
conmutador de emisión-recepción send-receive
    switch

conmutador de encendido on-off switch
conmutador de enclavamiento interlock switch
conmutador de entrada entrance switch
conmutador de equilibrio de corriente
    current-balance switch
conmutador de escala scale switch, range switch
conmutador de estado sólido solid-state switch
conmutador de ferrita ferrite switch
conmutador de filamento filament switch
conmutador de flotador float switch
conmutador de flujo flow switch
conmutador de flujo de agua water-flow switch
conmutador de flujo de combustible fuel-flow
    switch
conmutador de flujo de fluido fluid-flow switch
conmutador de flujo de gas gas-flow switch
conmutador de flujo de líquido liquid-flow switch
conmutador de frente muerto dead-front switch
conmutador de función function switch
conmutador de gancho hook switch
conmutador de gravedad gravity switch
conmutador de guíaondas waveguide switch
conmutador de hablar-escuchar talk-listen switch
conmutador de ignición ignition switch
conmutador de impulsos pulse switch
conmutador de inclinación tilt switch
conmutador de inductancia inductance switch
conmutador de inercia inertia switch
conmutador de instrumento instrument switch
conmutador de interrupción breakpoint switch
conmutador de inversión de altavoces
    loudspeaker-reversal switch
conmutador de inversión de fase phase-reversal
    switch
conmutador de inversión de polaridad
    polarity-reversal switch
conmutador de laminilla reed switch
conmutador de laminillas secas dry-reed switch
conmutador de laminillas secas magnético magnetic
    dry-reed switch
conmutador de límite limit switch
conmutador de línea line switch
conmutador de línea primaria primary-line switch
conmutador de línea principal main line switch
conmutador de lóbulo lobe switch
conmutador de llamada call switch
conmutador de medidor meter switch
conmutador de membrana membrane switch
conmutador de mensajes message switch
conmutador de mercurio mercury switch
conmutador de mercurio magnético magnetic
    mercury switch
conmutador de moción única single-motion switch
conmutador de modo mode switch
conmutador de motor motor switch
conmutador de muestreo sampling switch
conmutador de nivel level switch
conmutador de nivel de fluido fluid-level switch
conmutador de nivel de líquido liquid-level switch
conmutador de obleas wafer switch
conmutador de oprimir para escuchar
    push-to-listen switch
conmutador de oprimir para hablar push-to-talk
    switch
conmutador de palanca bat-handle switch, lever
    switch
conmutador de palanca acodada toggle switch
conmutador de palanca en paleta paddle-handle
    switch
conmutador de panel panel switch
conmutador de panel frontal front-panel switch
conmutador de pared wall switch
conmutador de pasos step switch

conmutador de pasos rotativo rotary stepping switch
conmutador de pie foot switch
conmutador de polaridad polarity switch
conmutador de posición position switch
conmutador de preestablecimiento binario binary preset switch
conmutador de presión pressure switch
conmutador de presión de agua water-pressure switch
conmutador de presión de fluido fluid-pressure switch
conmutador de presión de líquido liquid-pressure switch
conmutador de presión diferencial pressure-differential switch
conmutador de privacidad privacy switch
conmutador de proximidad proximity switch
conmutador de prueba test switch
conmutador de puerta gate switch
conmutador de puesta a tierra grounding switch
conmutador de puesta a tierra de alta velocidad high-speed grounding switch
conmutador de pulsador push-button switch
conmutador de punto de cruce crosspoint switch
conmutador de reconexión reclosing switch
conmutador de reposición reset switch
conmutador de resorte spring switch
conmutador de retardo delay switch
conmutador de retardo de tiempo time-delay switch
conmutador de ruptura de carga load-break switch
conmutador de ruptura rápida quick-break switch
conmutador de ruptura única single-break switch
conmutador de secuencia sequence switch
conmutador de seguridad safety switch
conmutador de selección selection switch
conmutador de selección rotativo rotary selector switch
conmutador de semiconductor semiconductor switch
conmutador de sentido sense switch
conmutador de sobrerrecorrido overtravel switch
conmutador de solenoide solenoid switch
conmutador de Strowger Strowger switch
conmutador de superficie surface switch
conmutador de tambor drum switch
conmutador de tanque único single-tank switch
conmutador de tensión reducida reduced-voltage switch
conmutador de tiempo time switch
conmutador de tiempo automático automatic time switch
conmutador de tierra ground switch
conmutador de timbre bell switch
conmutador de tipo de laminilla reed-type switch
conmutador de tomas tap switch
conmutador de tono tone switch
conmutador de transferencia transfer switch
conmutador de transferencia automático automatic transfer switch
conmutador de transferencia de carga load transfer switch
conmutador de transferencia de emergencia emergency transfer switch
conmutador de transferencia de señal de elevador elevator signal-transfer switch
conmutador de transmisión-recepción transmit-receive switch
conmutador de trayectoria path switch
conmutador de tres posiciones three-position switch
conmutador de tres vías three-way switch
conmutador de umbral threshold switch
conmutador de una posición single-throw switch

conmutador de uso general general-use switch
conmutador de vacío vacuum switch
conmutador de video video switcher
conmutador de visualización display switch
conmutador de volquete tumbler switch
conmutador decimal-binario decimal-binary switch
conmutador deslizante slide switch
conmutador digital digital switch
conmutador electrolítico electrolytic switch
conmutador electromagnético electromagnetic switch
conmutador electromecánico electromechanical switch
conmutador electrónico electronic switch
conmutador electrónico de diodo diode electronic switch
conmutador electroquímico electrochemical switch
conmutador en aceite oil switch
conmutador en derivación shunt switch
conmutador en serie-paralelo series-parallel switch
conmutador encerrado enclosed switch
conmutador enlazado linked switch
conmutador estrella-delta star-delta switch
conmutador externo external switch
conmutador iluminado lighted switch
conmutador impreso printed switch
conmutador independiente independent switch
conmutador iniciador de fallas fault-initiating switch
conmutador integrador integrator switch
conmutador interruptor de carga load-interrupter switch
conmutador inversor reversing switch
conmutador iónico ionic switch
conmutador líquido liquid switch
conmutador maestro master switch
conmutador magnético magnetic switch
conmutador manual manual switch, manual cutout
conmutador motorizado motorized switch
conmutador multicontacto multicontact switch
conmutador múltiple multiple switch
conmutador multipolar multipole switch
conmutador multiposición multiposition switch
conmutador multipunto multipoint switch
conmutador neumático pneumatic switch
conmutador no sincrónico nonsynchronous switch
conmutador operado por luz light-operated switch
conmutador óptico optical switch
conmutador paso a paso step-by-step switch
conmutador pirométrico pyrometric switch
conmutador por luminiscencia glow switch
conmutador preajustable preset switch
conmutador preajustado preset switch
conmutador principal main switch
conmutador principal de batería battery main switch
conmutador programado programmed switch
conmutador químico chemical switch
conmutador ranurado slotted commutator
conmutador relé relay switch
conmutador rotativo rotary switch
conmutador secuencial sequential switch
conmutador selector selector switch
conmutador selector de programa program-selector switch
conmutador selector multiposición multiposition selector switch
conmutador sensible sensitive switch
conmutador silenciador muting switch
conmutador sin cordón cordless switch
conmutador sin cortocircuito nonshorting switch
conmutador sin rebotes debounced switch
conmutador sincrónico synchronous switch

conmutador solar solar switch
conmutador térmico thermal switch
conmutador térmico de enganche latching thermal switch
conmutador termosensible heat-sensitive switch
conmutador termostático thermostatic switch
conmutador termostático ajustable adjustable thermostatic switch
conmutador termostático ambiente ambient thermostatic switch
conmutador tetrapolar four-pole switch
conmutador triodo triode switch
conmutador tripolar three-pole switch
conmutador ultrarrápido ultrafast switch
conmutador ultrasónico ultrasonic switch
conmutador unidireccional one-way switch
conmutador unipolar single-pole switch
conmutador unipolar de dos posiciones single-pole double-throw switch
conmutador unipolar de una posición single-pole single-throw switch
conmutadores acoplados gang switch
conmutadores acoplados mecánicamente ganged switches
conmutadores en T tee switch
cono de altavoz loudspeaker cone
cono de compresión compression cone
cono de dispersión scattering cone
cono de protección cone of protection
cono de radiación radiation cone
cono de radiación nula cone of nulls
cono de recorrido largo long-travel cone
cono de silencio cone of silence
conoscopio conoscope
conservación de energía conservation of energy
conservación de paridad parity conservation
conservación de radiancia conservation of radiance
conservación espectral spectrum conservation
conservar save
consistencia de programas software consistency
consola console
consola compuesta composite console
consola de computadora computer console
consola de control control console
consola de control por computadora computer control console
consola de operación operating console
consola de operador operator console
consola de operador de red network-operator console
consola de radar radar console
consola de sistema system console
consola de sistema alternativo alternative system console
consola de tiempo compartido time-sharing console
consola de visualización display console
consola dúplex duplex console
consola maestra master console
consola monitora monitor console
consola múltiple multiple console
consola primaria primary console
consola receptora receiving console
consola remota remote console
consola secundaria secondary console
consola supervisora supervisory console
consolidación de archivos file consolidation
consonancia consonance
constancia de calibración calibration constancy
constancia de velocidad speed constancy
constancia del cero zero constancy
constantán constantan
constante constant
constante absoluta absolute constant

constante arbitraria arbitrary constant
constante aritmética arithmetic constant
constante automática automatic constant
constante característica characteristic constant
constante concentrada lumped constant
constante de abocinado flaring constant
constante de aceleración acceleration constant
constante de acoplamiento coupling constant
constante de acoplamiento de cuadripolo quadrupole coupling constant
constante de amortiguamiento damping constant
constante de apantallado screening constant
constante de archivo file constant
constante de atenuación attenuation constant
constante de atenuación acústica acoustic attenuation constant
constante de atenuación imagen image attenuation constant
constante de Avogadro Avogadro's constant
constante de bobina coil constant
constante de Boltzmann Boltzmann's constant
constante de calibración calibration constant
constante de capacitor capacitor constant
constante de caracteres character constant
constante de celda cell constant
constante de coma fija fixed-point constant
constante de coma flotante floating-point constant
constante de conmutación switching constant
constante de contador register constant
constante de datos data constant
constante de demanda demand constant
constante de densidad de radiación radiation-density constant
constante de desintegración decay constant
constante de detalle detail constant
constante de difusión diffusion constant
constante de dirección address constant
constante de disipación dissipation constant
constante de entrada entry constant
constante de error error constant
constante de error de aceleración acceleration error constant
constante de estructura fina fine-structure constant
constante de Faraday Faraday constant
constante de fase phase constant
constante de fase acústica acoustic phase constant
constante de fase imagen image phase constant
constante de frecuencia frequency constant
constante de fuerza force constant
constante de galvanómetro galvanometer constant
constante de Hall Hall constant
constante de Joule Joule constant
constante de línea line constant
constante de longitud de onda wavelength constant
constante de Nagaoka Nagaoka's constant
constante de onda wave constant
constante de oscilación oscillation constant
constante de Planck Planck's constant
constante de propagación propagation constant
constante de propagación axial axial propagation constant
constante de radiación radiation constant
constante de red network constant
constante de relocalización relocation constant
constante de resistencia-capacitancia resistance-capacitance constant
constante de retículo ultrasónico ultrasonic grating constant
constante de ruido noise constant
constante de sintonización tuning constant
constante de sistema system constant
constante de Stefan-Boltzmann Stefan-Boltzmann constant

constante de tiempo time constant
constante de tiempo aparente apparent time constant
constante de tiempo aperiódica aperiodic time constant
constante de tiempo corto fast-time constant
constante de tiempo de carga charge time constant
constante de tiempo de descarga discharge time constant
constante de tiempo de entrada input time constant
constante de tiempo de recuperación recovery-time constant
constante de tiempo de resistencia-capacitancia resistance-capacitance time constant
constante de tiempo de salida output time constant
constante de tiempo de temperatura ambiente ambient-temperature time constant
constante de tiempo mecánica mechanical time constant
constante de tiempo subtransitoria subtransient time constant
constante de tiempo térmica thermal time constant
constante de tiempo total total time constant
constante de tiempo transitoria transient time constant
constante de tierra ground constant
constante de transferencia transfer constant
constante de transferencia de imagen image transfer constant
constante de transferencia iterativa iterative transfer constant
constante de transmisión transmission constant
constante de velocidad velocity constant
constante de Weiss Weiss constant
constante dieléctrica dielectric constant
constante dieléctrica de tierra ground dielectric constant
constante dieléctrica efectiva effective dielectric constant
constante dieléctrica relativa relative dielectric constant
constante diferida deferred constant
constante distribuida distributed constant
constante eléctrica electrical constant
constante electromagnética electromagnetic constant
constante electrostática electrostatic constant
constante en watthoras watt-hour constant
constante en wattsegundos watt-second constant
constante figurativa figurative constant
constante fotoeléctrica photoelectric constant
constante gravitacional gravitational constant
constante hexadecimal hexadecimal constant
constante literal literal constant
constante lógica logical constant
constante magnética magnetic constant
constante numérica numerical constant
constante piezoeléctrica piezoelectric constant
constante real real constant
constante secundaria secondary constant
constante solar solar constant
constantes eléctricas lineales linear electrical constants
constreñimiento de rejilla grid constraint
constricción pinch
construcción a prueba de goteo dripproof construction
construcción con subensamblajes unitized construction
construcción condicional conditional construction
construcción de chasis abierto open-chassis construction
construcción de montaje temporal breadboard construction
construcción metalocerámica metal-ceramic construction
construcción modular modular construction
consulta inquiry
consulta aleatoria random inquiry
consulta en tiempo real real-time inquiry
consulta remota remote inquiry
consulta-respuesta inquiry-reply
consumo consumption, drain
consumo de corriente current consumption
consumo de corriente de calefactor heater-current consumption
consumo de corriente de filamento filament current consumption
consumo de energía energy consumption
consumo de energía en reposo quiescent power consumption
consumo de potencia power consumption
consumo específico specific consumption
consumo máximo específico maximum specific consumption
consumo nominal nominal consumption
contabilidad de mensajes automática automatic message accounting
contacto contact
contacto A A contact
contacto a presión pressure contact
contacto a tierra ground contact
contacto activo active contact
contacto bidireccional two-way contact
contacto bifurcado bifurcated contact
contacto cortocircuitador shorting contact
contacto de acción ultrarrápida snap-action contact
contacto de alarma alarm contact
contacto de alta presión high-pressure contact
contacto de bajo nivel low-level contact
contacto de bloqueo blocking contact
contacto de cabeza a cinta head-to-tape contact
contacto de carbón carbon contact
contacto de cierre make contact
contacto de cinta a cabeza tape-to-head contact
contacto de clasificación sorter contact
contacto de clavija plug contact
contacto de conexión connection contact
contacto de conmutación changeover contact
contacto de control control contact
contacto de doble ruptura double-break contact
contacto de dos posiciones double-throw contact
contacto de eje shaft contact
contacto de formación de arcos arcing contact
contacto de mínimo minimum contact
contacto de paso passing contact
contacto de plata a plata silver-to-silver contact
contacto de posición neutra midposition contact
contacto de potencia power contact
contacto de presión plunger
contacto de puente bridge contact
contacto de puesta a tierra grounding contact
contacto de punta point contact
contacto de reconexión reclosing contact
contacto de relé relay contact
contacto de reposición reset contact
contacto de reposo rest contact, back contact
contacto de resistencia resistance contact
contacto de resorte spring contact
contacto de ruptura break contact
contacto de salida output contact
contacto de sincronización timing contact
contacto de superficie surface contact
contacto de termopar thermocouple contact
contacto de trabajo working contact, front contact
contacto de transferencia transfer contact
contacto de tungsteno tungsten contact
contacto dependiente dependent contact

contacto deslizante sliding contact, slider, wiper
contacto eléctrico electrical contact
contacto eléctrico de circuito integrado integrated-circuit electrical contact
contacto enchufable plug-in contact
contacto estacionario stationary contact
contacto fijo fixed contact
contacto flexible flexible contact
contacto húmedo wet contact
contacto impreso printed contact
contacto independiente independent contact
contacto inferior bottom contact
contacto intermetálico intermetallic contact
contacto interno inner contact
contacto inverso reverse contact
contacto inversor reversing contact
contacto laminado laminated contact
contacto lateral side contact
contacto mantenido maintained contact
contacto microfónico microphonic contact
contacto momentáneo momentary contact
contacto móvil moving contact
contacto múltiple multiple contact
contacto normal normal contact
contacto normalmente abierto normally open contact
contacto normalmente cerrado normally closed contact
contacto óhmico ohmic contact
contacto pasivo passive contact
contacto plateado silver-plated contact
contacto por presión crimp contact
contacto positivo positive contact
contacto reforzado heavy-duty contact
contacto rodante rolling contact
contacto rodiado rhodium-plated contact
contacto sellado sealed contact
contacto semiconductor semiconductor contact
contacto sin puenteo nonbridging contact
contacto soldado welded contact
contacto sólido solid contact
contacto único single contact
contacto vibrante vibrating contact
contactor contactor
contactor de aeronave aircraft contactor
contactor de radiofrecuencia radio-frequency contactor
contactor de reposición reset contactor
contactor de ruptura rápida quick-break contactor
contactor de solenoide solenoid contactor
contactor electrónico electronic contactor
contactor inversor reversing contactor
contactor maestro master contactor
contactor magnético magnetic contactor
contactor mecánico mechanical contactor
contactor vibrante vibrating contactor
contactos autolimpiantes self-cleaning contacts
contactos auxiliares auxiliary contacts
contactos con solapamiento overlapping contacts
contactos de abrir antes de cerrar break-before-make contacts
contactos de baja capacitancia low-capacitance contacts
contactos de cerrar antes de abrir make-before-break contacts
contactos de doble cierre double-make contacts
contactos de exploración probe contacts
contactos de Kelvin Kelvin contacts
contactos de platino platinum contacts
contactos de ruptura múltiple multiple-break contacts
contactos dorados gold-plated contacts
contactos inferiores lower contacts

contactos instantáneos instantaneous contacts
contactos interruptores interrupter contacts
contactos principales main contacts
contactos retardados delayed contacts
contactos secos dry contacts
contactos tardíos late contacts
contador counter, meter, power meter, register
contador acumulante accumulating counter
contador básico basic counter
contador bidireccional bidirectional counter
contador binario binary counter
contador de abonado subscriber meter
contador de acumulación storage counter
contador de bucles loop counter
contador de caracteres character counter
contador de centelleos scintillation counter
contador de ciclos cycle counter
contador de cinta tape counter
contador de coincidencia coincidence counter
contador de control control counter
contador de cristal crystal counter
contador de décadas decade counter
contador de décadas electrónico electronic decade counter
contador de descargas discharge counter
contador de direcciones de memoria memory address counter
contador de direcciones de programa program address counter
contador de electricidad electricity meter
contador de eventos event counter
contador de flujo de gas gas-flow counter
contador de frecuencia frequency counter
contador de frecuencia universal universal frequency counter
contador de Geiger Geiger counter
contador de Geiger-Mueller Geiger-Mueller counter
contador de halógeno halogen counter
contador de impulsos pulse counter
contador de índice de ciclos cycle index counter
contador de instrucciones instruction counter
contador de intervalos de tiempo time-interval counter
contador de Johnson Johnson counter
contador de lectura-escritura read-write counter
contador de líneas line counter
contador de localizaciones location counter
contador de pasos step counter
contador de periodos period counter
contador de programa program counter
contador de punta point counter
contador de radiación radiation counter
contador de radiactividad radioactivity counter
contador de rayos gamma gamma-ray counter
contador de razón de frecuencias frequency-ratio counter
contador de reloj clock counter
contador de retardo delay counter
contador de secuencias sequence counter
contador de selección selection counter
contador de servicio service meter
contador de tarjetas card counter
contador de tubo tube counter
contador decimal scale-of-ten counter
contador digital digital counter
contador direccional directional counter
contador direccional fotoeléctrico photoelectric directional counter
contador eléctrico electric counter
contador electromecánico electromechanical counter
contador electrónico electronic counter
contador en anillo ring counter
contador en anillo de lámparas de neón neon-bulb

ring counter
contador ferrorresonante ferroresonant counter
contador fotoeléctrico photoelectric counter
contador horario time meter
contador índice index counter
contador iónico ion counter
contador lineal linear counter
contador multicanal multichannel counter
contador preajustable preset counter
contador preajustado preset counter
contador predeterminado predetermined counter
contador progresivo-regresivo forward-backward counter
contador proporcional proportional counter
contador quinario quinary counter
contador reversible reversible counter
contador sumario summary counter
contaminación contamination
contaminación atmosférica atmospheric contamination
contaminación catódica cathode contamination
contaminación de aire air contamination
contaminación de colores color contamination
contaminación de datos data contamination
contaminante contaminant
contar tally
contenido armónico harmonic content
contenido armónico relativo relative harmonic content
contenido de actividad activity content
contenido de humedad moisture content
contenido de información information content
contenido de información bruto gross information content
contenido de información medio mean information content
contenido de información neto net information content
conteo de coincidencia coincidence counting
conteo de impulsos pulse count
conteo espurio spurious count
contestación answering, answerback
contestación automática automatic answering
contiguo contiguous
continuamente ajustable continuously adjustable
continuidad continuity
contorneo flashover
contorno contour
contorno cilíndrico cylindrical contour
contorno de frecuencia frequency contour
contorno de Fresnel Fresnel contour
contorno de intensidad de campo field-intensity contour
contra ecos parásitos anticlutter
contra efectos locales antisidetone
contra interferencia intencional antijamming
contracción magnetoestrictiva magnetostrictive contraction
contracorriente reverse current
contraelectrodo counterelectrode
contragolpe kickback
contramedidas electrónicas activas active electronic countermeasures
contrapeso counterbalance, counterpoise
contrapolarización back bias
contrapolarización amplificada amplified back bias
contrapolarización automática automatic back bias
contrapresión backpressure
contrapuerta countergate
contraseña password
contraste contrast
contraste de brillo brightness contrast
contraste de detalle detail contrast

contraste de fase phase contrast
contraste de imagen image contrast
contraste de impresión print contrast
contraste de índice de refracción refractive index contrast
contraste de luminancia luminance contrast
contraste de modulación modulation contrast
contraste de señal signal contrast
contraste débil weak contrast
contraste equivalente equivalent contrast
contraste invertido reversed contrast
contraste simultáneo simultaneous contrast
contratensión countervoltage, back voltage
control adaptivo adaptive control
control alambrado wired control
control ambiental environmental control
control antipatinaje antiskating control
control asimétrico asymmetrical control
control astático astatic control
control autoalimentado self-powered control
control automático automatic control
control automático parcial partial automatic control
control autónomo self-contained control
control básico basic control
control bilateral bilateral control
control central central control
control centralizado centralized control
control compartido shared control
control común common control
control con retroalimentación lineal linear feedback control
control con tomas tapped control
control concéntrico concentric control
control concurrente concurrent control
control condicional conditional control
control continuo continuous control
control corrector peaking control
control cruzado cross control
control de absorción absorption control
control de acceso access control
control de acceso de medio medium access control
control de acción proporcional proportional-action control
control de acimut azimuth control
control de actitud attitude control
control de agudos treble control
control de ajuste trimming control
control de alimentación de papel paper-feed control
control de almacenamiento integrado integrated storage control
control de almacenamiento masivo mass storage control
control de altavoz loudspeaker control
control de altura height control
control de altura automática automatic height control
control de altura de imagen picture-height control
control de alumbrado fotoeléctrico photoelectric lighting control
control de amortiguamiento damping control
control de amortiguamiento variable variable-damping control
control de amplitud amplitude control
control de amplitud automático automatic amplitude control
control de amplitud de línea line-amplitude control
control de amplitud vertical vertical-amplitude control
control de ancho de banda bandwidth control
control de anchura width control
control de anchura de barrido sweep-width control
control de anchura de imagen picture-width control
control de anchura horizontal horizontal-width

control

control de anchura vertical vertical width control

control de aproximación approach control

control de aproximación automático automatic approach control

control de arco arc control

control de archivo file control

control de arranque-parada start-stop control

control de asimetría asymmetry control

control de audio audio control

control de bajos bass control

control de bajos automático automatic bass control

control de bordes edge control

control de brillantez brilliance control

control de brillantez de colores color brilliance control

control de brillo brightness control

control de brillo automático automatic brightness control

control de brillo de fondo background brightness control

control de brillo maestro master brightness control

control de bucle loop control

control de bucle cerrado closed-loop control

control de bucle fotoeléctrico photoelectric loop control

control de calidad quality control

control de calidad automático automatic quality control

control de calidad estadístico statistical quality control

control de cambio de velocidad speed-change control

control de campo field control

control de campo de generador generator field control

control de campo de motor motor-field control

control de canal de datos data channel control

control de cancelación cancellation control

control de carbón carbon control

control de carga load control

control de carga automático automatic load control

control de carga base base load control

control de carga de línea line-load control

control de carro carriage control

control de centrado centering control

control de centrado de imagen picture-centering control

control de centrado horizontal horizontal centering control

control de centrado vertical vertical-centering control

control de ceros a la izquierda left zero control

control de cinta tape control

control de colores color control

control de combustible por falla de llama flame-failure control

control de combustión automático automatic combustion control

control de comprobación de módem modem check control

control de comunicaciones communication control

control de conexión y desconexión on-off control

control de confiabilidad reliability control

control de configuración configuration control

control de conjunto de trabajos job set control

control de conmutación automático automatic switching control

control de conmutación de control control switching control

control de contraste contrast control

control de contraste automático automatic contrast control

control de contraste de imagen picture-contrast control

control de contraste manual manual contrast control

control de convergencia convergence control

control de convergencia de colores color-convergence control

control de convergencia horizontal horizontal-convergence control

control de convergencia vertical vertical-convergence control

control de copia copy control

control de corriente de entrada input-stream control

control de corriente de salida output stream control

control de croma chroma control

control de crominancia automático automatic chrominance control

control de cuadrante único single-dial control

control de cursor cursor control

control de datos data control

control de desmagnetización degaussing control

control de despacho económico economic dispatch control

control de desplazamiento shift control

control de desviación de frecuencia frequency-deviation control

control de dirección geográfica geographical address control

control de disparo trigger control

control de distancia range control

control de distribución distribution control

control de dos etapas two-stage control

control de dos posiciones two-position control

control de dos puntos two-point control

control de duración duration control

control de ecualización equalization control

control de ecualización de ganancia gain equalization control

control de electrodo electrode control

control de elevador elevator control

control de encuadre framing control

control de enfoque focusing control

control de enlace link control

control de enlace de datos data link control

control de enlace lógico logical link control

control de entrada-salida input-output control

control de equilibrio balance control

control de equilibrio de canales channel balance control

control de equilibrio de puente bridge balance control

control de error error control

control de etapa stage control

control de excitación drive control

control de excitación de campo field-excitation control

control de excitación horizontal horizontal-drive control

control de exploración scanning control

control de fase phase control

control de fase asimétrico asymmetrical phase control

control de fase automático automatic phase control

control de fase de convergencia convergence phase control

control de flujo flow control

control de flujo de agua water-flow control

control de flujo de combustible fuel-flow control

control de flujo de fluido fluid-flow control

control de flujo de gas gas-flow control

control de flujo de información information flow control

control de flujo de líquido liquid-flow control

control de flujo de trabajo job flow control

control de fondo background control
control de fondo automático automatic background control
control de formato format control
control de frecuencia frequency control
control de frecuencia automático automatic frequency control
control de frecuencia de campo field-frequency control
control de frecuencia fino fine frequency control
control de frecuencia incremental incremental frequency control
control de freno brake control
control de ganancia gain control
control de ganancia automático retardado delayed automatic gain control
control de ganancia automático automatic gain control
control de ganancia automático instantáneo instantaneous automatic gain control
control de ganancia automático silencioso quiet automatic gain control
control de ganancia contra ecos parásitos anticlutter gain control
control de ganancia de azul blue gain control
control de ganancia de crominancia chrominance gain control
control de ganancia de verde green gain control
control de ganancia de video video gain control
control de ganancia diferencial differential gain control
control de ganancia horizontal horizontal-gain control
control de ganancia maestro master gain control
control de ganancia remoto remote gain control
control de ganancia vertical vertical-gain control
control de generación automático automatic generation control
control de grupo group control
control de ignición ignition control
control de iluminación illumination control
control de imagen picture control
control de impresión print control
control de impresora printer control
control de inclinación tilt control
control de instrucciones instruction control
control de intensidad intensity control
control de intensidad sonora loudness control
control de interfaz interface control
control de interferencia interference control
control de limitación automático automatic limiting control
control de límite limit control
control de línea line control
control de línea de transmisión transmission-line control
control de linealidad linearity control
control de linealidad horizontal horizontal-linearity control
control de linealidad vertical vertical-linearity control
control de luz light control
control de llama flame control
control de llamadas call control
control de mandos command control
control de mantenimiento maintenance control
control de máquina machine control
control de máquina de microcomputadora microcomputer machine control
control de máquina dedicado dedicated machine control
control de mensajes message control
control de microprograma microprogram control

control de modulación automático automatic modulation control
control de motor motor control
control de motor de corriente continua direct-current motor control
control de muestreo sampling control
control de nivel level control
control de nivel automático automatic level control
control de nivel de agua water-level control
control de nivel de entrada input-level control
control de nivel de fluido fluid-level control
control de nivel de líquido liquid-level control
control de nivel de negro black level control
control de nivel de registro record-level control
control de operación operating control
control de operador extendido extended operator control
control de orientación orientation control
control de oscilación oscillation control
control de página page control
control de panel panel control
control de panel frontal front-panel control
control de pantalla screen control
control de papel paper control
control de par torque control
control de parada stop control
control de partición partition control
control de paso pitch control
control de pausa pause control
control de pila stack control
control de polarización bias control
control de polarización dinámico dynamic bias control
control de polarización variable variable-bias control
control de portadora carrier control
control de posición position control
control de posicionamiento horizontal horizontal-positioning control
control de potencia power control
control de presión de agua water-pressure control
control de presión de fluido fluid-pressure control
control de presión de gas gas-pressure control
control de presión de líquido liquid-pressure control
control de prioridad priority control
control de procedimiento procedure control
control de procesamiento de trabajos job-processing control
control de procesos process control
control de procesos críticos critical process control
control de producción production control
control de profundidad de página page depth control
control de programa program control
control de programa automático automatic program control
control de proyecto project control
control de prueba proof control
control de puente bridge control
control de puerta gate control
control de punto final end-point control
control de pureza purity control
control de razón ratio control
control de rebobinado rewind control
control de red network control, array control
control de regeneración regeneration control
control de registro register control
control de registro de datos data recording control
control de registro fotoeléctrico photoelectric register control
control de reposición reset control
control de retención hold control
control de rotación de imagen picture-rotation

control
control de ruido noise control
control de salida output control
control de saturación saturation control
control de secuencia sequence control
control de seguridad safety control
control de selección selection control
control de selectividad selectivity control
control de selectividad automático automatic
selectivity control
control de sensibilidad sensitivity control
control de sensibilidad automático automatic
sensitivity control
control de sensibilidad de ganancia gain sensitivity
control
control de señal signal control
control de señales de tiempo time-signal control
control de separación estereofónica
stereophonic-separation control
control de sesión session control
control de signo sign control
control de sincronismo horizontal horizontal-hold
control
control de sincronismo vertical vertical-hold control
control de sincronización synchronization control
control de sintonización tuning control
control de sintonización de ensanche de banda
bandspread tuning control
control de sintonización fina fine-tuning control
control de sistema system control
control de sistema ejecutivo executive system
control
control de soldadura welding control
control de sonido sound control
control de suma add control
control de supresión suppression control
control de tamaño de caracteres character size
control
control de tamaño vertical vertical-size control
control de tareas task control
control de temperatura temperature control
control de tensión voltage control
control de tensión ajustable adjustable voltage
control
control de tensión del inducido armature voltage
control
control de tensión independiente independent
voltage control
control de tensión variable ajustable adjustable
varying-voltage control
control de terminal terminal control
control de texto text control
control de tiempo time control
control de tiempo de arbitraje arbitration timing
control
control de tiempo de sensibilidad sensitivity-time
control
control de tinte tint control
control de tono tone control
control de tono por desfase phase-shift tone control
control de tonos de Baxandall Baxandall tone
control
control de trabajo job control
control de trabajos apilados stacked-job control
control de tráfico traffic control
control de tráfico aéreo air-traffic control
control de tráfico centralizado centralized traffic
control
control de tráfico de aeropuerto airport traffic
control
control de tráfico de área area traffic control
control de tráfico de entrada-salida input-output
traffic control

control de transferencia transfer control
control de transformador automático automatic
transformer control
control de transmisión transmission control
control de trayectoria path control
control de tren automático automatic train control
control de umbral threshold control
control de un punto one-point control
control de unidades múltiples multiple-unit control
control de velocidad speed control
control de velocidad de motor motor-speed control
control de velocidad variable variable-speed control
control de vibraciones vibration control
control de video video control
control de visualización display control
control de volumen volume control
control de volumen automático automatic volume
control
control de volumen automático instantáneo
instantaneous automatic volume control
control de volumen automático silencioso quiet
automatic volume control
control de volumen compensado compensated
volume control
control de volumen de altavoz loudspeaker volume
control
control de volumen manual manual volume control
control de volumen no compensado uncompensated
volume control
control de vuelo flight control
control de vuelo automático automatic flight control
control de zona zone control
control derivado derivative control
control desde tierra ground control
control desvanecedor fader control
control diferencial differential control
control digital digital control
control digital directo direct digital control
control dinámico dynamic control
control direccional directional control
control directo direct control
control ejecutivo executive control
control eléctrico electrical control
control electroneumático electropneumatic control
control electrónico electronic control
control en bucle abierto open-loop control
control en cascada cascade control
control en serie-paralelo series-parallel control
control en tándem gang control
control en tiempo real real-time control
control especial special control
control fino fine control
control flotante floating control
control fotoeléctrico photoelectric control
control gradual gradual control
control grueso coarse control
control indirecto indirect control
control individual individual control
control industrial industrial control
control inicial setup control
control inmediato immediate control
control integral integral control
control intermedio intermediate control
control intermitente intermittent control
control interno internal control
control-interrupción control-break
control intrínseco intrinsic control
control lineal linear control
control local local control
control maestro master control
control maestro de programa program master
control
control manual manual control

control menor minor control
control microprogramado microprogrammed control
control múltiple multiple control
control neumático pneumatic control
control normal normal control
control numérico numerical control
control numérico computarizado computerized numerical control
control numérico de computadoras computer numerical control
control numérico directo direct numerical control
control numérico distribuido distributed numerical control
control operacional operational control
control operado por motor motor-operated control
control operativo concurrente concurrent operating control
control panorámico panoramic control
control paso a paso step-by-step control
control periférico peripheral control
control permisivo permissive control
control por computadora computer control
control por corriente portadora carrier-current control
control por cristal crystal control
control por desfase phase-shift control
control por frecuencia constante constant-frequency control
control por hilo wire control
control por lotes batch control
control por pulsador push-button control
control por radar radar control
control por rejilla grid control
control por resistencia resistance control
control por retroalimentación feedback control
control por solenoide solenoid control
control por tensión variable varying-voltage control
control por tercer hilo sleeve control
control potenciométrico potentiometer control
control predictivo predictive control
control primario primary control
control principal main control
control programado programmed control
control proporcional proportional control
control relativo relative control
control remoto remote control
control remoto electromagnético electromagnetic remote control
control remoto multicanal multichannel remote control
control reostático rheostatic control
control residual residual control
control rotativo rotary control
control ruidoso noisy control
control secuencial sequential control
control secuencial dinámico dynamic sequential control
control semiajustable semiadjustable control
control semirremoto semiremote control
control supervisor supervisory control
control suplementario supplementary control
control suplementario de área area supplementary control
control único single control
controlable controllable
controlado automáticamente automatically controlled
controlado electrónicamente electronically controlled
controlado indirectamente indirectly controlled
controlado por cinta tape-controlled
controlado por computadora computer controlled

controlado por cuarzo quartz-controlled
controlado por diapasón fork-controlled
controlado por efecto de campo field-effect controlled
controlado por fase phase-controlled
controlado por programa program-controlled
controlado por puerta gate-controlled
controlado por pulsador push-button controlled
controlado por radar radar-controlled
controlado por relé relay-controlled
controlado por reverberación reverberation-controlled
controlado por tensión voltage-controlled
controlado por transductor transducer-controlled
controlado por transistores transistor-controlled
controlado por usuario user-controlled
controlado por válvula valve-controlled
controlado por voz speech-controlled
controlado remotamente remote-controlled
controlado reostáticamente rheostatically controlled
controlador controller, driver
controlador analógico analog controller
controlador automático automatic controller
controlador básico basic controller
controlador de acceso a memoria directo direct memory access controller
controlador de almacenamiento store controller
controlador de almacenamiento en disco disk storage controller
controlador de alumbrado fotoeléctrico photoelectric lighting controller
controlador de bus bus controller, bus driver
controlador de bus bidireccional bidirectional bus driver
controlador de bus de salida output bus driver
controlador de canales channel controller
controlador de carga load controller
controlador de cinta de papel paper-tape controller
controlador de cinta magnética magnetic-tape controller
controlador de circuitos circuit controller
controlador de comunicaciones communication controller
controlador de corriente alterna alternating-current controller
controlador de disco disk controller
controlador de disco duro hard-disk controller
controlador de dispositivo local local device controller
controlador de dos canales dual-channel controller
controlador de entrada-salida input-output controller
controlador de entrada-salida en serie serial input-output controller
controlador de grupo cluster controller
controlador de impresora printer controller
controlador de impresora de línea line printer controller
controlador de interfaz de red network interface controller
controlador de interrupción de prioridad priority interrupt controller
controlador de línea única single-line controller
controlador de memoria memory controller
controlador de microprocesador microprocessor controller
controlador de motor eléctrico electric motor controller
controlador de muestreo sampling controller
controlador de periférico peripheral driver
controlador de procesador processor controller
controlador de procesos process controller
controlador de programa program controller

controlador de puerto port controller
controlador de registro recording controller
controlador de respaldo backup controller
controlador de secuencia sequence controller
controlador de sistema system controller
controlador de tambor drum controller
controlador de temperatura temperature controller
controlador de terminal terminal controller
controlador de transferencia de datos data-transfer controller
controlador de transmisión transmission controller
controlador de uso definido definite-purpose controller
controlador de uso general general-purpose controller
controlador de velocidad speed controller
controlador digital digital controller
controlador eléctrico electric controller
controlador electroneumático electropneumatic controller
controlador electrónico electronic controller
controlador flotante floating controller
controlador indicador indicating controller
controlador integral integral controller
controlador inteligente intelligent controller
controlador maestro master controller
controlador magnético magnetic controller
controlador manual manual controller
controlador monocanal single-channel controller
controlador multilínea multiline controller
controlador piloto pilot controller
controlador preajustable preset controller
controlador preajustado preset controller
controlador programable programmable controller
controlador proporcional proportional controller
controlador semiautomático semiautomatic controller
controlador semimagnético semimagnetic controller
controlador universal universal controller
controles coaxiales coaxial controls
controles externos external control
convección convection
convección eléctrica electric convection
convección natural natural convection
convección térmica thermal convection
convectivo convective
convectrón convectron
convergencia convergence
convergencia de colores color convergence
convergencia de entrada fanin
convergencia de haces beam convergence
convergencia dinámica dynamic convergence
convergencia dinámica horizontal horizontal dynamic convergence
convergencia dinámica vertical vertical dynamic convergence
convergencia electrostática electrostatic convergence
convergencia estática static convergence
convergencia horizontal horizontal convergence
convergencia magnética magnetic convergence
convergencia radial radial convergence
convergencia vertical vertical convergence
convergente convergent, converging
conversación codificada scrambled speech
conversión conversion
conversión alfanumérica alphanumeric conversion
conversión analógica a digital analog-to-digital conversion
conversión aritmética arithmetic conversion
conversión ascendente up-conversion
conversión concurrente concurrent conversion
conversión de archivo file conversion

conversión de beta a alfa beta-to-alpha conversion
conversión de binario a decimal binary-to-decimal conversion
conversión de binario a hexadecimal binary-to-hexadecimal conversion
conversión de binario a octal binary-to-octal conversion
conversión de cinta tape conversion
conversión de cinta a cinta tape-to-tape conversion
conversión de cinta a tarjetas tape-to-card conversion
conversión de código code conversion
conversión de código de datos data code conversion
conversión de datos data conversion
conversión de decimal a binario decimal-to-binary conversion
conversión de decimal a hexadecimal decimal-to-hexadecimal conversion
conversión de digital a analógico digital-to-analog conversion
conversión de dirección address conversion
conversión de documento document conversion
conversión de energía energy conversion
conversión de energía electromagnética electromagnetic energy conversion
conversión de energía solar solar-energy conversion
conversión de exploración scanning conversion
conversión de formato format conversion
conversión de frecuencia frequency conversion
conversión de hexadecimal a decimal hexadecimal-to-decimal conversion
conversión de lógica negativa negative logic conversion
conversión de lógica positiva positive logic conversion
conversión de longitud de onda a frecuencia wavelength-to-frequency conversion
conversión de modo mode conversion
conversión de modo común common-mode conversion
conversión de octal a binario octal-to-binary conversion
conversión de octal a decimal octal-to-decimal conversion
conversión de procesos process conversion
conversión de programa program conversion
conversión de señal signal conversion
conversión de signo sign conversion
conversión de tarjeta a cinta card-to-tape conversion
conversión de tarjeta a disco card-to-disk conversion
conversión descendente down-conversion
conversión directa direct conversion
conversión en paralelo parallel conversion
conversión física physical conversion
conversión interna internal conversion
conversión paramétrica parametric conversion
conversión termoeléctrica thermoelectric conversion
conversión termoiónica thermionic conversion
conversión watts-decibeles watt-decibel conversion
convertidor converter
convertidor a-d a-d converter
convertidor analógico a digital analog-to-digital converter
convertidor analógico a frecuencia analog-to-frequency converter
convertidor analógico-digital analog-digital converter
convertidor anódico anode converter
convertidor ascendente up-converter
convertidor auxiliar auxiliary converter
convertidor binario binary converter

convertidor controlado por cristal crystal-controlled converter

convertidor corriente continua-corriente alterna direct-current-alternating-current converter

convertidor de alta frecuencia high-frequency converter

convertidor de arco arc converter

convertidor de arco de mercurio mercury-arc converter

convertidor de audio audio converter

convertidor de banda lateral inferior lower sideband converter

convertidor de banda lateral superior upper-sideband converter

convertidor de binario a decimal binary-to-decimal converter

convertidor de cinta a cinta tape-to-tape converter

convertidor de cinta a tarjetas tape-to-card converter

convertidor de cinta de papel a tarjetas paper-tape-to-card converter

convertidor de cinta magnética magnetic-tape converter

convertidor de código code converter

convertidor de colores color converter

convertidor de corriente alterna alternating-current converter

convertidor de corriente alterna directo direct alternating-current converter

convertidor de corriente continua direct-current converter

convertidor de datos data converter

convertidor de definición definition converter

convertidor de digital a analógico digital-to-analog converter

convertidor de diodo diode converter

convertidor de energía energy converter

convertidor de energía solar solar-energy converter

convertidor de entrada input converter

convertidor de equilibrio de línea line-balance converter

convertidor de exploración scanning converter

convertidor de fase phase converter

convertidor de fase rotativo rotary phase converter

convertidor de forma de onda waveform converter

convertidor de frecuencia frequency converter

convertidor de frecuencia a tensión frequency-to-voltage converter

convertidor de frecuencia de inducción induction frequency converter

convertidor de frecuencia estático static frequency converter

convertidor de frecuencia intermedia intermediate-frequency converter

convertidor de frecuencia paramétrico parametric frequency converter

convertidor de frecuencia ultraalta ultra-high frequency converter

convertidor de guíaondas waveguide converter

convertidor de imagen image converter

convertidor de imágenes de infrarrojo infrared image converter

convertidor de imágenes ultrasónicas ultrasonic image converter

convertidor de impedancia impedance converter

convertidor de impedancia generalizada generalized impedance converter

convertidor de impedancia negativa negative impedance converter

convertidor de interfaz interface converter

convertidor de interrupción periódica chopper converter

convertidor de lenguaje language converter

convertidor de línea line converter

convertidor de motor motor converter

convertidor de onda wave converter

convertidor de ondas cortas shortwave converter

convertidor de ondas cuadradas square-wave converter

convertidor de par torque converter

convertidor de par polifásico polyphase torque converter

convertidor de paralelo a serie parallel-to-serial converter

convertidor de potencia power converter

convertidor de potencia de semiconductor semiconductor power converter

convertidor de potencia electrónico electronic power converter

convertidor de potencia estático static power converter

convertidor de radiofrecuencia radio-frequency converter

convertidor de señal signal converter

convertidor de serie a paralelo serial-to-parallel converter

convertidor de sistemas de televisión television system converter

convertidor de tarjeta a cinta card-to-tape converter

convertidor de tarjeta a cinta magnética card-to-magnetic tape converter

convertidor de tarjeta a disco card-to-disk converter

convertidor de temperatura a corriente temperature-to-current converter

convertidor de temperatura a tensión temperature-to-voltage converter

convertidor de tensión voltage converter

convertidor de termopar thermocouple converter

convertidor de tipo de vibrador vibrator-type converter

convertidor de tono tone converter

convertidor de video video converter

convertidor descendente down-converter

convertidor digital digital converter

convertidor en cascada cascade converter

convertidor equilibrado balanced converter

convertidor estático static converter

convertidor ferroeléctrico ferroelectric converter

convertidor hexafásico six-phase converter

convertidor magnetohidrodinámico magnetohydrodynamic converter

convertidor modular modular converter

convertidor múltiple multiple converter

convertidor paramétrico parametric converter

convertidor pentarrejilla pentagrid converter

convertidor polifásico polyphase converter

convertidor reticular grating converter

convertidor rotativo rotary converter

convertidor rotativo sincrónico synchronous rotary converter

convertidor sincrónico synchronous converter

convertidor sincrónico de autorregulación synchronous booster converter

convertidor solar solar converter

convertidor superheterodino superheterodyne converter

convertidor térmico thermal converter

convertidor termoiónico thermionic converter

convertidor termoiónico de cesio cesium thermionic converter

convertidor triodo-heptodo triode-heptode converter

convertidor triodo-hexodo triode-hexode converter

convertidor vibrante vibrating converter

convertir convert

convexo-concavo convexo-concave

coordenada de cromaticidad chromaticity

coordinate
coordenada x x coordinate
coordenada y y coordinate
coordenada z z coordinate
coordenadas cartesianas Cartesian coordinates
coordenadas de Davisson Davisson coordinates
coordenadas esféricas spherical coordinates
coordenadas horizontales horizontal coordinates
coordenadas polares polar coordinates
coordenadas rectangulares rectangular coordinates
coordinación de aislamiento coordination of
    insulation
coordinación de frecuencias frequency coordination
coordinación de línea line coordination
coordinación de niveles level coordination
coordinación inductiva inductive coordination
copia de bloque block copy
copia de respaldo backup copy
copia heliográfica blueprint
copia impresa hard copy
copiadora electrostática electrostatic copier
copolarización copolarization
coprocesador coprocessor
coprocesador numérico numerical coprocessor
cordón cord
cordón a prueba de humedad moistureproof cord
cordón con dos clavijas double-ended cord
cordón de alimentación power cord, line cord
cordón de conexión connecting cord
cordón de conexión temporal patch cord
cordón de contestación answer cord
cordón de cuadrante dial cord
cordón de extensión extension cord
cordón de lámpara lamp cord
cordón de oropel tinsel cord
cordón de prueba test cord
cordón de puesta a tierra grounding cord
cordón eléctrico lamp cord
cordón flexible flexible cord
corona corona
corona continua continuous corona
corona electrostática electrostatic corona
correa de alimentación feeding belt
correa de transmisión drive belt
corrección correction
corrección acústica acoustic correction
corrección cromática chromatic correction
corrección de abertura aperture correction
corrección de agudos treble correction
corrección de bajos bass correction
corrección de cable cable correction
corrección de coincidencia coincidence correction
corrección de color color correction
corrección de convergencia convergence correction
corrección de deriva drift correction
corrección de distorsión distortion correction
corrección de error cuadrantal quadrantal error
    correction
corrección de errores error correction
corrección de errores automática automatic error
    correction
corrección de escala scale correction
corrección de factor de potencia power-factor
    correction
corrección de fase phase correction
corrección de frecuencia frequency correction
corrección de frecuencia automática automatic
    frequency correction
corrección de gamma gamma correction
corrección de impulso pulse correction
corrección de inclinación tilt correction
corrección de onda espacial skywave correction
corrección de paralaje parallax correction

corrección de ruido aleatorio random-noise
    correction
corrección de sombra shading correction
corrección de temperatura temperature correction
corrección de tiempo de resolución resolving-time
    correction
corrección de tiempo muerto dead-time correction
corrección de tono tone correction
corrección de velocidad velocity correction
corrección espectral spectral correction
corrección fuera de línea off-line correction
corrección geométrica geometric correction
corrección parcial partial correctness
correctivo corrective
corrector corrector
corrector de bajos bass corrector
corrector de cierre mínimo minimum-make
    corrector
corrector de distancia range corrector
corrector de errores error corrector
corrector de fase phase corrector
corrector de forma de onda waveform corrector
corrector de impedancia impedance corrector
corrector de impedancia de baja frecuencia
    low-frequency impedance corrector
corrector de impulso pulse corrector
corrector de inclinación tilt corrector
corrector de línea line corrector
corrector de retardo de fase phase-lag corrector
corrector de tensión voltage corrector
corrector de tiempo automático automatic time
    corrector
corrector de trama de color color-field corrector
corrector del haz de azul blue beam corrector
correlación correlation
correlación angular angular correlation
correlación cruzada cross-correlation
correlación de frecuencia instantánea instantaneous
    frequency correlation
correlación entre canales interchannel correlation
correlación no lineal nonlinear correlation
correlacionador correlator
correo computarizado computerized mail
correo electrónico electronic mail
corriente current
corriente activa active current
corriente adelantada leading current
corriente alterna alternating-current
corriente alterna-corriente continua
    alternating-current-direct-current
corriente alterna rectificada rectified alternating
    current
corriente alterna simétrica symmetrical alternating
    current
corriente anódica anode current
corriente aperiódica aperiodic current
corriente armónica harmonic current
corriente armónica simple simple harmonic current
corriente asimétrica asymmetrical current
corriente baja small current
corriente bidireccional bidirectional current
corriente bifásica biphase current, two-phase current
corriente capacitiva capacitive current
corriente catódica cathode current
corriente cero zero current
corriente circular circular current
corriente circulatoria circulating current
corriente compensadora compensating current
corriente compuesta composite current
corriente constante constant current
corriente continua direct-current, continuous current
corriente continua intermitente intermittent direct
    current

corriente continua interrumpida interrupted direct current
corriente continua nominal rated direct current
corriente continua permisible allowable continuous current
corriente continua proporcional proportional direct current
corriente continua pulsante pulsating direct current
corriente convencional conventional current
corriente correctora correcting current
corriente crítica critical current
corriente cruzada cross-current
corriente de absorción absorption current
corriente de acción action current
corriente de accionamiento actuating current
corriente de aire airflow
corriente de alimentación supply current, power current, feed current
corriente de alta frecuencia high-frequency current
corriente de alta tensión high-tension current
corriente de alto amperaje high-amperage current
corriente de amplificación amplification current
corriente de antena antenna current
corriente de arco arc current
corriente de arranque starting current
corriente de baja tensión low-voltage current
corriente de base base current
corriente de base de corriente alterna alternating-current base current
corriente de base de corriente continua direct-current base current
corriente de base dinámica dynamic base current
corriente de base estática static base current
corriente de batería battery current
corriente de bits bit stream
corriente de blanco target current
corriente de borrado erase current
corriente de calefactor heater current
corriente de calentamiento de filamento filament heating current
corriente de campo excitador exciter-field current
corriente de campo inducida induced field current
corriente de campo nominal rated field current
corriente de capacitancia capacitance current
corriente de carga load current, charging current
corriente de carga de línea line charging current
corriente de carga de salida output load current
corriente de cátodo de corriente alterna alternating-current cathode current
corriente de cátodo de corriente continua direct-current cathode current
corriente de cátodo dinámica dynamic cathode current
corriente de cátodo estática static cathode current
corriente de cebado striking current
corriente de centrado centering current
corriente de cierre making current
corriente de cola tail current
corriente de colector collector current
corriente de colector de corriente alterna alternating-current collector current
corriente de colector de corriente continua direct-current collector current
corriente de colector dinámica dynamic collector current
corriente de colector estática static collector current
corriente de compensación trickle current
corriente de conducción conduction current
corriente de conducción iónica ion conduction current
corriente de conducción magnética magnetic conduction current
corriente de conexión pull-in current, pullup current

corriente de conmutación switching current
corriente de conmutador switch current
corriente de constricción pinch current
corriente de consumo drain current
corriente de consumo de corriente continua direct-current drain current
corriente de consumo dinámica dynamic drain current
corriente de consumo estática static drain current
corriente de contactos nominal rated contact current
corriente de control control current
corriente de control crítica critical controlling current
corriente de control diferencial differential control current
corriente de control nominal nominal control current
corriente de controlador controller current
corriente de convección convection current
corriente de convección iónica ion convection current
corriente de conversación speech current
corriente de corona corona current
corriente de corte cutoff current
corriente de corte de colector collector cutoff current
corriente de corte de emisor emitter cutoff current
corriente de corto tiempo short-time current
corriente de cortocircuito short-circuit current
corriente de cortocircuito disponible available short-circuit current
corriente de cortocircuito nominal rated short-circuit current
corriente de cristal crystal current
corriente de D'Arsonval D'Arsonval current
corriente de datos data stream
corriente de deflexión deflecting current
corriente de demarcación demarcation current
corriente de derivación branch current, bypass current
corriente de desaccionamiento dropout current
corriente de descarga discharge current
corriente de descarga nominal nominal discharge current
corriente de desplazamiento displacement current, drift current
corriente de difusión diffusion current
corriente de dígito digit current
corriente de diodo diode current
corriente de disparo trip current
corriente de disparo de puerta gate trigger current
corriente de dispersión dispersion current
corriente de drenador de corriente alterna alternating-current drain current
corriente de drenaje bleeder current
corriente de efecto túnel tunnel current
corriente de eje shaft current
corriente de electrodo electrode current
corriente de electrodo anormal fault electrode current
corriente de electrodo inversa inverse electrode current
corriente de electrodo media average electrode current
corriente de emisión emission current
corriente de emisión de campo nulo field-free emission current
corriente de emisión de punto de flexión flexion point emission current
corriente de emisión de punto de inflexión inflection-point emission current
corriente de emisión primaria primary emission current

corriente de emisor emitter current
corriente de emisor de corriente alterna
alternating-current emitter current
corriente de emisor de corriente continua
direct-current emitter current
corriente de emisor dinámica dynamic emitter
current
corriente de emisor estática static emitter current
corriente de enfoque focusing current
corriente de enganche latching current
corriente de entrada input current
corriente de entrada de alto nivel high-level input
current
corriente de entrada de bajo nivel low-level input
current
corriente de entrada de corriente continua
direct-current input current
corriente de entrada de retroalimentación feedback
input current
corriente de entrada de trabajos job-input stream
corriente de entrada de transformador transformer
input current
corriente de error error current
corriente de escritura writing current
corriente de espacio spacing current
corriente de estator stator current
corriente de excitación excitation current, driving
current
corriente de excitación de campo field-excitation
current
corriente de falla fault current
corriente de fase phase current
corriente de fase dividida split-phase current
corriente de filamento filament current
corriente de fondo de valle valley-point current
corriente de fuente de corriente continua
direct-current source current
corriente de fuga leakage current
corriente de fuga a tierra earth-leakage current
corriente de fuga de corriente continua
direct-current leakage current
corriente de fuga de salida output leakage current
corriente de fusión fusing current, blowing current
corriente de fusión mínima minimum blowing
current
corriente de gas gas current
corriente de haz beam current
corriente de huecos hole current
corriente de impulsos pulse stream
corriente de interrupción de puerta gate turn-off
current
corriente de ionización ionization current
corriente de irrupción inrush current
corriente de Leduc Leduc current
corriente de liberación release current
corriente de límite de escala full-scale current
corriente de línea line current
corriente de línea de señal signal-line current
corriente de línea nominal rated line current
corriente de luminiscencia glow current
corriente de llamada ringing current
corriente de llegada incoming current
corriente de malla mesh current
corriente de máquina machine current
corriente de onda estática static wave current
corriente de ondulación ripple current
corriente de operación operating current
corriente de Oudin Oudin current
corriente de pantalla screen current
corriente de pérdidas loss current
corriente de placa plate current
corriente de placa de corriente alterna
alternating-current plate current

corriente de placa de corriente continua
direct-current plate current
corriente de placa dinámica dynamic plate current
corriente de placa estática static plate current
corriente de placa máxima de pico maximum peak
plate current
corriente de plasma plasma current
corriente de plena carga full-load current
corriente de polarización bias current, polarization
current
corriente de polarización de entrada input bias
current
corriente de polarización de referencia reference
bias current
corriente de portadora carrier current
corriente de preconducción preconduction current
corriente de preoscilación preoscillation current
corriente de protección protection current
corriente de prueba test current
corriente de puente bridge current
corriente de puerta gate current
corriente de puerta de corriente continua
direct-current gate current
corriente de puerta estática static gate current
corriente de puerta inversa reverse gate current
corriente de puesta en trabajo pickup current
corriente de punto de inflexión inflection-point
current
corriente de radiofrecuencia radio-frequency
current
corriente de rayos anódicos anode-ray current
corriente de rayos catódicos cathode-ray current
corriente de rayos positivos positive-ray current
corriente de recombinación recombination current
corriente de recuperación recovery current
corriente de referencia reference current
corriente de reinserción reinsertion current
corriente de rejilla grid current
corriente de rejilla crítica critical grid current
corriente de rejilla de control control-grid current
corriente de rejilla de corriente continua
direct-current grid current
corriente de rejilla de radiofrecuencia
radio-frequency grid current
corriente de rejilla estática static grid current
corriente de rejilla inversa inverse grid current
corriente de rejilla-pantalla screen-grid current
corriente de reposición reset current
corriente de reposo rest current
corriente de reserva standby current
corriente de resonador resonator current
corriente de retorno return current
corriente de ruido noise current
corriente de ruido de entrada input noise current
corriente de ruido equivalente equivalent noise
current
corriente de ruptura breakdown current, break
current
corriente de ruptura de aislamiento insulation
breakdown current
corriente de ruptura inversa reverse-breakdown
current
corriente de salida outgoing current, output current
corriente de salida de alto nivel high-level output
current
corriente de salida de bajo nivel low-level output
current
corriente de salida de señal signal output current
corriente de salida de trabajos job-output stream
corriente de salida de transformador transformer
output current
corriente de salida diferencial differential output
current

corriente de salida media average output current
corriente de salida nominal rated output current
corriente de saturación saturation current
corriente de saturación de colector collector saturation current
corriente de señal signal current
corriente de servicio service current
corriente de servicio continuo continuous-duty current
corriente de sincronización synchronizing current
corriente de sobrecarga overload current
corriente de soldadura welding current
corriente de solenoide solenoid current
corriente de sostenimiento hold current
corriente de Tesla Tesla current
corriente de tierra ground current
corriente de trabajo marking current
corriente de trabajos job stream
corriente de trabajos de entrada input job stream
corriente de trabajos de salida output job stream
corriente de transferencia transfer current
corriente de transición conductiva breakover current
corriente de valle valley current
corriente de velocidad inicial initial-velocity current
corriente de yugo yoke current
corriente de Zener Zener current
corriente débil small current, weak current
corriente desfasada out-of-phase current
corriente desmagnetizante demagnetizing current
corriente deswattada wattless current
corriente diacrítica diacritical current
corriente dieléctrica dielectric current
corriente diferencial differential current
corriente directa forward current
corriente directa de rectificador rectifier forward current
corriente directa instantánea instantaneous forward current
corriente directa pico recurrente peak recurrent forward current
corriente disponible available current
corriente distorsionada distorted current
corriente ecualizadora equalizing current
corriente efectiva effective current
corriente eficaz root-mean-square current
corriente eléctrica electrical current
corriente eléctrica interna internal electric current
corriente electrónica electron current, electronic current
corriente electrónica inicial initial electron current
corriente en anillo ring current
corriente en bucle loop current
corriente en circuito abierto open-circuit current
corriente en circuito cerrado closed-circuit current
corriente en cuadratura quadrature current
corriente en diente de sierra sawtooth current
corriente en espera idling current
corriente en fase in-phase current
corriente en obscuridad dark current
corriente en obscuridad de electrodo electrode dark current
corriente espacial space current
corriente estacionaria steady current
corriente excitadora exciting current
corriente farádica faradic current
corriente fluctuante fluctuating current
corriente fotoeléctrica photoelectric current
corriente fuerte power current
corriente galvánica galvanic current
corriente helicoidal helix current
corriente inducida induced current
corriente inducida directa direct induced current

corriente inducida estática static induced current
corriente inductora inducing current
corriente inicial initial drain
corriente instantánea instantaneous current
corriente intermitente intermittent current
corriente interruptora interrupting current
corriente inversa inverse current, back current
corriente inversa de electrodo electrode inverse current
corriente inversa de puerta gate reverse current
corriente inversa instantánea instantaneous reverse current
corriente iónica ion current, ionic current
corriente ionosférica ionospheric current
corriente limitada por carga espacial space-charge-limited current
corriente límite limiting current
corriente local local current
corriente longitudinal longitudinal current
corriente magnética magnetic current
corriente magnetizante magnetizing current
corriente magnetizante en cuadratura quadrature magnetizing current
corriente marginal marginal current
corriente máxima maximum current
corriente máxima de control maximum control current
corriente máxima de descarga maximum discharge current
corriente media average current
corriente medida measured current
corriente mínima minimum current
corriente mínima de fusión minimum fusing current
corriente mínima de operación minimum operate current
corriente modulada modulated current
corriente moduladora modulating current
corriente momentánea momentary current
corriente monofásica single-phase current
corriente multifásica multiphase current
corriente neta net current
corriente neutrónica neutron current
corriente no inducida noninduced current
corriente no modulada unmodulated current
corriente nominal nominal current, current rating
corriente nominal de fusible fuse current rating
corriente normal normal current
corriente nula null current
corriente ondulante undulating current
corriente ondulatoria wave current
corriente óptima optimum current
corriente oscilante oscillatory current
corriente parásita parasitic current, stray current, eddy current
corriente periódica periodic current
corriente persistente persistent current
corriente perturbadora equivalente equivalent disturbing current
corriente pico peak current
corriente pico de ánodo peak anode current
corriente pico de cátodo peak cathode current
corriente pico de descarga peak discharge current
corriente pico de electrodo peak electrode current
corriente pico de establecimiento peak making current
corriente pico de placa peak plate current
corriente polarizante polarizing current
corriente polifásica polyphase current
corriente ponderada weighted current
corriente por avalancha avalanche current
corriente primaria primary current
corriente primaria nominal rated primary current
corriente propia prospective current

corriente proporcional proportional current
corriente pulsante pulsating current
corriente pulsante multifrecuencia multifrequency
    pulsing current
corriente reactiva reactive current
corriente rectificada media average rectified current
corriente reflejada reflected current
corriente residual residual current
corriente resistiva resistive current
corriente retardada lagging current
corriente rotórica rotor current
corriente saturante saturating current
corriente secundaria secondary current
corriente secundaria nominal rated secondary
    current
corriente sin carga no-load current
corriente sin señal no-signal current
corriente sinusoidal sine current
corriente subsiguiente follow current
corriente superficial surface current
corriente superpuesta superimposed current
corriente telefónica telephone current
corriente térmica nominal rated thermal current
corriente termoiónica thermionic current
corriente termotelúrica thermotelluric current
corriente terrestre terrestrial current
corriente transitoria transient current
corriente transversal transverse current
corriente trifásica three-phase current
corriente umbral threshold current
corriente única single current
corriente unidireccional unidirectional current
corriente vectorial vector current
corriente virtual virtual current
corriente vocal voice current
corriente wattada watt current
corrientes coincidentes coincidence currents
corrientes de Foucault Foucault currents
corrientes de tierra earth currents
corrientes en contrafase push-pull currents
corrientes equilibradas balanced currents
corrientes metálicas metallic currents
corrientes opuestas opposed currents
corrientes telúricas telluric currents
corrimiento creepage
corrimiento adelantado de frecuencia frequency
    pulling
corrimiento atrasado de frecuencia frequency
    pushing
corrimiento de cable cable creepage
corrimiento de sintonización tuning creep
corrimiento de sintonizador tuner creep
corrimiento vertical roll
corrimiento vertical de imagen picture roll
corrosión corrosion
corrosión anódica anode corrosion
corrosión catódica cathode corrosion
corrosión de contacto contact corrosion
corrosión electrolítica electrolytic corrosion
corrosión electroquímica electrochemical corrosion
corrosión en hendiduras crevice corrosion
corrosión galvánica galvanic corrosion
corrosión interdendrítica interdendritic corrosion
corrosión intergranular intergranular corrosion
corrosión por almacenamiento shelf corrosion
corrosión por servicio service corrosion
corrosión química chemical corrosion
corrosión selectiva selective corrosion
corrosión subsuperficial subsurface corrosion
corrosión termogalvánica thermogalvanic corrosion
corrupción corruption
corrupción de datos data corruption
cortacircuito circuit-breaker, breaker, cutout, cutoff,

fuse
cortacircuito abierto open circuit-breaker
cortacircuito anódico anode circuit-breaker
cortacircuito automático automatic circuit-breaker
cortacircuito auxiliar auxiliary circuit-breaker
cortacircuito balístico ballistic circuit-breaker
cortacircuito bañado en aceite oil-immersed
    circuit-breaker
cortacircuito con autorreposición self-resetting
    circuit-breaker
cortacircuito de acción rápida fast-acting
    circuit-breaker
cortacircuito de aire air circuit-breaker
cortacircuito de alta velocidad high-speed
    circuit-breaker
cortacircuito de antena antenna circuit-breaker
cortacircuito de arranque starting circuit-breaker
cortacircuito de carga mínima underload
    circuit-breaker
cortacircuito de corriente alterna
    alternating-current circuit-breaker
cortacircuito de corriente continua direct-current
    circuit-breaker
cortacircuito de corriente inversa reverse-current
    circuit-breaker
cortacircuito de chorro de aire air-blast
    circuit-breaker
cortacircuito de dos posiciones double-throw
    circuit-breaker
cortacircuito de estado sólido solid-state
    circuit-breaker
cortacircuito de excitación de campo field
    circuit-breaker
cortacircuito de extinción magnética
    magnetic-blowout circuit-breaker
cortacircuito de fusible fuse cutout
cortacircuito de fusible abierto open-fuse cutout
cortacircuito de fusible en aceite oil-fuse cutout
cortacircuito de fusible restablecedor reclosing fuse
    cutout
cortacircuito de línea line circuit-breaker
cortacircuito de mínimo minimum circuit-breaker
cortacircuito de pulsador push-button
    circuit-breaker
cortacircuito de ruptura única single-break
    circuit-breaker
cortacircuito de salida outgoing circuit-breaker
cortacircuito de sección section circuit-breaker
cortacircuito de seguridad safety cutout
cortacircuito de sobrecarga overload circuit-breaker
cortacircuito de sobrecorriente overcurrent
    circuit-breaker
cortacircuito de sobretensión overvoltage
    circuit-breaker
cortacircuito de tanque único single-tank
    circuit-breaker
cortacircuito de tapón plug cutout
cortacircuito de tensión mínima undervoltage
    circuit-breaker
cortacircuito de una posición single-throw
    circuit-breaker
cortacircuito de uso definido definite-purpose
    circuit-breaker
cortacircuito de uso general general-purpose
    circuit-breaker
cortacircuito direccional directional circuit-breaker
cortacircuito en aceite oil circuit-breaker
cortacircuito enlazado linked circuit-breaker
cortacircuito exterior outdoor circuit-breaker
cortacircuito fusible fusible cutout
cortacircuito lleno de aceite oil-filled circuit-breaker
cortacircuito magnético magnetic circuit-breaker
cortacircuito magnetotérmico magnetothermal

circuit-breaker
cortacircuito manual manual cutout
cortacircuito multipolar multipole breaker
cortacircuito multiruptura multibreak
    circuit-breaker
cortacircuito principal main circuit-breaker
cortacircuito protector protective circuit-breaker
cortacircuito térmico thermal circuit-breaker
cortacircuito termostático thermostatic cutout
cortado en dados dicing
cortador cutter
cortadora de papel paper cutter
corte cutoff, cutting, cut
corte agudo sharp cutoff
corte de banda lateral sideband cutoff
corte de corriente de placa plate-current cutoff
corte de cristal crystal cut
corte de filtro filter cutoff
corte láser laser cutting
corte normal normal cut
corte paralelo parallel cut
corte remoto remote cutoff
corte ultrasónico ultrasonic cutting
corte X X cut
corte Y Y cut
corte Z Z cut
cortina radiante radiating curtain
cortina reflectora reflector curtain
corto alcance short range
corto plazo short term
cortocircuitado short-circuited, shorted
cortocircuitador short-circuiter, short-circuiting
cortocircuitador automático automatic
    short-circuiter
cortocircuito short-circuit, short
cortocircuito bifásico two-phase short-circuit
cortocircuito de alta resistencia high-resistance
    short
cortocircuito deslizante sliding short-circuit
cortocircuito directo direct short
cortocircuito total dead short
corutina coroutine
cosecante cosecant
cosecante de arco arc cosecant
cosecante hiperbólica hyperbolic cosecant
coseno cosine
coseno de arco arc cosine
coseno hiperbólico hyperbolic cosine
coseno verso versed cosine
cosmotrón cosmotron
costilla ridge
cosumando augend
cotangente cotangent
cotangente de arco arc cotangent
cotangente hiperbólica hyperbolic cotangent
coulomb coulomb
coulomb internacional international coulomb
coulómbmetro coulombmeter
coulómetro coulometer
cráter de arco arc crater
creación de base de datos database creation
creación no intencionada de bits drop-in
crecimiento de confiabilidad reliability growth
crecimiento dendrítico dendritic growth
crecimiento epitaxial epitaxial growth
crecimiento epitáxico epitaxic growth
crecimiento reotaxial rheotaxial growth
cresta crest, peak, pip
cresta de onda wave crest
cresta de tensión voltage crest
crioelectrónica cryoelectronics
criogenia cryogenics
criogénico cryogenic

criógeno cryogen
criómetro cryometer
criorresistencia cryoresistance
criosar cryosar
criosar compuesto compound cryosar
crioscópico cryoscopic
crioscopio cryoscope
criosistor cryosistor
criostato cryostat
criotrón cryotron
criotrón de película delgada thin-film cryotron
criotrónica cryotronics
criptográfico cryptographic
criptol kryptol
criptón krypton
cristal crystal
cristal antiferroeléctrico antiferroelectric crystal
cristal calibrador calibrator crystal
cristal cambiador de frecuencia frequency-changer
    crystal
cristal centelleador scintillator crystal
cristal centelleante scintillating crystal
cristal con efecto piezoeléctrico por torsión twister
cristal con temperatura controlada
    temperature-controlled crystal
cristal de alta frecuencia high-frequency crystal
cristal de bajo nivel low-level crystal
cristal de canal channel crystal
cristal de corte cero zero-cut crystal
cristal de corte paralelo parallel-cut crystal
cristal de corte X X-cut crystal
cristal de corte XY XY-cut crystal
cristal de corte Y Y-cut crystal
cristal de corte Z Z-cut crystal
cristal de cuarzo quartz crystal
cristal de galena galena crystal
cristal de germanio germanium crystal
cristal de roca flint glass
cristal de sal de Rochelle Rochelle-salt crystal
cristal de silicio silicon crystal
cristal de sobretono overtone crystal
cristal de tercera armónica third-harmonic crystal
cristal dieléctrico dielectric crystal
cristal en blanco blank
cristal ferroelástico ferroelastic crystal
cristal ferroeléctrico ferroelectric crystal
cristal fijo fixed crystal
cristal fotoconductor photoconducting crystal
cristal haloideo halide crystal
cristal ideal ideal crystal
cristal iónico ionic crystal
cristal líquido liquid crystal
cristal metálico metallic crystal
cristal mezclador mixer crystal
cristal modulador modulator crystal
cristal monoclínico monoclinic crystal
cristal negativo negative crystal
cristal nemático nematic crystal
cristal no polar nonpolar crystal
cristal oscilante oscillating crystal
cristal paraelástico paraelastic crystal
cristal paramagnético paramagnetic crystal
cristal patrón standard crystal
cristal perfecto perfect crystal
cristal piezoeléctrico piezoelectric crystal
cristal piezoeléctrico artificial artificial piezoelectric
    crystal
cristal piezoeléctrico enchufable plug-in
    piezoelectric crystal
cristal piezoeléctrico natural natural piezoelectric
    crystal
cristal positivo positive crystal
cristal rotativo rotating crystal

cristal semilla seed crystal
cristal sintético synthetic crystal
cristalografía crystallography
cristalografía electrónica electron crystallography
cristalografía neutrónica neutron crystallography
cristalografía por rayos X X-ray crystallography
cristalograma crystallogram
cristalomagnético crystallomagnetic
criterio de Barkhausen Barkhausen criterion
criterio de Barkhausen para osciladores
   Barkhausen criterion for oscillators
criterio de búsqueda search criterion
criterio de decisión decision criterion
criterio de Nyquist Nyquist criterion
criterio de Rayleigh Rayleigh criterion
criterio de Townsend Townsend criterion
criterios de aceptación acceptance criteria
criterios de falla failure criteria
croma chroma
cromado chrome-plated
cromaticidad chromaticity
cromaticidad de referencia reference chromaticity
cromaticidad de subportadora cero zero-subcarrier
   chromaticity
cromático chromatic
cromatografía de absorción absorption
   chromatography
cromatrón chromatron
cromel chromel
crominancia chrominance
crominancia de imagen picture chrominance
cromo chromium
cromoscopio chromoscope
cronistor chronistor
cronófero chronopher
cronofotómetro chronophotometer
cronógrafo chronograph
cronógrafo eléctrico electric chronograph
cronología absoluta absolute chronology
cronometración timing
cronometría de radar radar chronometry
cronómetro chronometer, timer
cronómetro de estado sólido solid-state chronometer
cronómetro de radar radar chronometer
cronómetro eléctrico electric chronometer
cronorrelé chronorelay
cronoscopio chronoscope
cronoscopio electrónico electronic chronoscope
cronotrón chronotron
cruce crossover
cruce aéreo aerial crossing
cruce de haz beam crossover
cruce del eje de cero zero crossing
cruce electrónico electronic crossover
cruzamiento frogging
cruzamiento de frecuencias frequency frogging
cuadradar quadradar
cuadrafonía quadraphony
cuadrafónico quadraphonic
cuadrantal quadrantal
cuadrante dial, quadrant
cuadrante de escala recta slide-rule dial
cuadrante de fijación de banda bandset dial
cuadrante de frecuencias frequency dial
cuadrante de lectura readout dial
cuadrante de ondámetro wavemeter dial
cuadrante de tambor drum dial
cuadrante iluminado por los bordes edge-lighted
   dial
cuadrante luminoso luminous dial
cuadrante vernier vernier dial
cuadratura quadrature
cuadratura de fase phase quadrature

cuadrete quad
cuadrete de dos cables de pares gemelos
   multiple-twin quad
cuadrete en espiral spiral-quad
cuadrete en estrella star-quad
cuadrete simple simple quad
cuadricorrelador quadricorrelator
cuadrícula raster
cuadrifónico quadraphonic
cuadripolo quadripole, quadrupole
cuadripolo axial axial quadrupole
cuadripolo de adaptación matching quadripole
cuadripolo linealizable linearizable quadripole
cuadro frame, switchboard, board, loop
cuadro apantallado shielded loop
cuadro conmutador switchboard
cuadro conmutador automático automatic
   switchboard
cuadro conmutador auxiliar auxiliary switchboard
cuadro conmutador de batería local local-battery
   switchboard
cuadro conmutador de distribución distribution
   switchboard
cuadro conmutador de doble lado cerrado dual
   switchboard
cuadro conmutador de emergencia emergency
   switchboard
cuadro conmutador dúplex duplex switchboard
cuadro conmutador encerrado enclosed
   switchboard
cuadro conmutador interurbano toll switchboard
cuadro conmutador manual manual switchboard
cuadro conmutador múltiple multiple switchboard
cuadro conmutador principal main switchboard
cuadro conmutador sin cordones cordless
   switchboard
cuadro conmutador telefónico telephone
   switchboard
cuadro conmutador vertical vertical switchboard
cuadro coplanar coplanar loop
cuadro de clavijas plug board
cuadro de distribución switchboard
cuadro de distribución de fuerza power switchboard
cuadro de Lecher Lecher frame
cuadro de página page frame
cuadro de resistencia resistance frame
cuadro de salida outward board
cuadro distribuidor distributing board
cuadro equilibrado balanced loop
cuadro giratorio rotatable loop
cuadro indicador indicator board
cuadro rotativo rotating loop
cuádruplex quadruplex
cuadruplexor quadruplexer
cuadruplicador quadrupler
cuadruplicador de frecuencia frequency quadrupler
cuadruplicador de tensión voltage quadrupler
cualidad quality
cualímetro qualimeter
cualitativo qualitative
cuantificación quantization
cuantificado quantized
cuantificador quantizer
cuantificar quantize
cuantímetro quantimeter
cuantitativamente quantitatively
cuantitativo quantitative
cuanto quantum
cuanto de energía quantum of energy
cuanto de luz quantum of light
cuanto de radiación quantum of radiation
cuantómetro quantometer
cuantos quanta

cuarto de canal quarter channel
cuarto de longitud de onda quarter wavelength
cuarto de onda quarter wave
cuarzo quartz
cuarzo fundido fused quartz
cuarzo piezoeléctrico piezoelectric quartz
cuarzo platinado platinized quartz
cuasar quasar
cuasi quasi
cuasialeatorio quasi-random
cuasiconductor quasi-conductor
cuasidieléctrico quasi-dielectric
cuasielástico quasi-elastic
cuasiestacionario quasi-stationary
cuasiinstrucción quasi-instruction
cuasilenguaje quasi-language
cuasinegativo quasi-negative
cuasióptico quasi-optical
cuasipositivo quasi-positive
cuaternario quaternary
cuatripartito quadripartite
cuatrivalente quadrivalent
cuba tank
cubículo de control control cubicle
cubículo de control electrónico electronic control
    cubicle
cubierta cover, sheath, jacket
cubierta a prueba de humedad moistureproof cover
cubierta aislante boot
cubierta contra polvo dust cover
cubierta de agua water jacket
cubierta de cable cable shackle
cubierta de electrones electron sheath
cubierta protectora protective cover
cubierto de caucho rubber-covered
cubrimiento blanketing
cuchilla de contacto contact blade
cuello neck
cuenta count
cuenta ambigua ambiguous count
cuenta de ciclos cycle count
cuenta de errores error count
cuenta de exploración raster count
cuenta de fondo background count
cuenta de registros record count
cuenta progresiva count up
cuenta regresiva countdown
cuerda chord
cuerda estirada stretched string
cuerno horn
cuerno polar pole horn
cuerpo de bucle loop body
cuerpo extraño foreign body
cuerpo gris gray body
cuerpo negro blackbody
cuerpo opaco opaque body
cuerpo translúcido translucent body
cuerpo transparente transparent body
culata yoke piece, yoke, heelpiece
culebra snake, fishing wire
cuña wedge
cúpula de enfoque focusing cup
cúpula de radar radar dome
cúpula de sonar sonar dome
curie curie
curio curium
curso de radio radio course
cursor cursor
cursor de potenciómetro potentiometer slider
cursor de visualización display cursor
cursor destructivo destructive cursor
cursor no destructivo nondestructive cursor
curtosis kurtosis

curva acampanada bell-shaped curve
curva anódica anode bend
curva aplanada flattened curve
curva característica characteristic curve
curva de absorción absorption curve
curva de absorción de referencia standard
    absorption curve
curva de activación activation curve
curva de actividad activity curve
curva de atenuación attenuation curve
curva de audibilidad audibility curve
curva de autoabsorción self-absorption curve
curva de Bode Bode curve
curva de Bragg Bragg curve
curva de calibración calibration curve
curva de carga load curve
curva de corriente constante constant-current curve
curva de desmagnetización demagnetization curve
curva de disminución decay curve
curva de dispersión scattering curve
curva de distorsión de frecuencia
    frequency-distortion curve
curva de distribución de frecuencias
    frequency-distribution curve
curva de distribución de intensidad intensity
    distribution curve
curva de distribución normal normal-distribution
    curve
curva de duración de carga load-duration curve
curva de ecualización equalization curve
curva de eficiencia efficiency curve
curva de emisión fotónica photon-emission curve
curva de energía potencial potential-energy curve
curva de error error curve
curva de excitabilidad excitability curve
curva de Gauss Gaussian curve
curva de histéresis hysteresis curve
curva de inducción normal normal-induction curve
curva de intensidad sonora equivalente
    equal-loudness curve
curva de Johnson Johnson curve
curva de magnetización magnetization curve
curva de magnetización inicial initial magnetization
    curve
curva de potencia power curve
curva de potencial potential curve
curva de probabilidad probability curve
curva de producto de energía energy-product curve
curva de propagación propagation curve
curva de reducción derating curve
curva de registro recording curve
curva de regulación regulation curve
curva de regulación de tensión voltage-regulation
    curve
curva de rendimiento performance curve
curva de resonancia resonance curve
curva de resonancia asimétrica asymmetrical
    resonance curve
curva de resonancia de doble pico double-hump
    resonance curve
curva de respuesta response curve
curva de respuesta de frecuencia
    frequency-response curve
curva de respuesta en cortocircuito short-circuit
    response curve
curva de respuesta espectral spectral-response
    curve
curva de respuesta global overall response curve
curva de respuesta plana flat-response curve
curva de respuesta selectiva selective response
    curve
curva de retardo delay curve
curva de salida output curve

curva de saturación saturation curve
curva de saturación de carga load saturation curve
curva de selectividad selectivity curve
curva de sintonización tuning curve
curva de transmisión transmission curve
curva dinámica dynamic curve
curva empírica empirical curve
curva en bañera bathtub curve
curva en mariposa butterfly curve
curva exponencial exponential curve
curva isocandela isocandela curve
curva logarítmica logarithmic curve
curva magnética magnetic curve
curva medida measured curve
curva normal normal curve
curva periódica periodic curve
curva polar polar curve
curva simétrica symmetrical curve
curva tensión-capacitancia voltage-capacitance
    curve
curva tensión-corriente voltage-current curve
curva tensión-frecuencia voltage-frequency curve
curvado bend
curvas Fletcher-Munson Fletcher-Munson curves

chaqueta jacket
chaqueta de cable cable jacket
chasis chassis
chasis de deflexión deflection chassis
chasis enchufable plug-in chassis
chasis experimental experimental chassis
chasquidos hash
chasquidos de manipulación key clicks
chillido squeal
chip chip
chip calculador calculator chip
chip cerámico ceramic chip
chip de arreglo de puertas gate-array chip
chip de memoria memory chip
chip de microcomputadora microcomputer chip
chip de microprocesador microprocessor chip
chip de música music chip
chip de procesador numérico numerical processor
    chip
chip de transistor transistor chip
chip de treinta y dos bits thirty-two bit chip
chip descodificador decoder chip
chip maestro master chip
chip monolítico monolithic chip
chip reversible flip chip
chip semiconductor semiconductor chip
chirrido chirp, chirping
chirridos de manipulación keying chirps
chispa spark
chispa amortiguada quenched spark
chispa caliente hot spark
chispa de cierre closing spark
chispa de disrupción disruption spark
chispa de ruptura break spark
chispa eléctrica electric spark
chispa retardada retarded spark
chispeo eléctrico electric sparking
chispómetro measuring spark-gap
chisporroteo sputter, sparking
choque choke, shock
choque acústico acoustic shock
choque aplicado applied shock
choque aversivo aversive shock
choque con entrehierro ajustable swinging choke
choque de alta frecuencia high-frequency choke
choque de aplanamiento smoothing choke
choque de baja frecuencia low-frequency choke
choque de cabeza head crash
choque de carga charging choke
choque de entrada input choke
choque de filamento filament choke
choque de filtro filter choke
choque de modulador modulator choke
choque de núcleo cerrado closed-core choke
choque de portadora carrier choke
choque de salida output choke
choque eléctrico electric shock
choque eléctrico secundario secondary electric
    shock
choque estabilizador stabilizing choke
choque mecánico mechanical shock

choque saturable saturable choke
choque térmico thermal shock
chorro jet
chorro electrónico electron jet

D

dado die
daño por cavitación cavitation damage
daraf daraf
datos data
datos absolutos absolute data
datos alfabéticos alphabetic data
datos alfanuméricos alphanumeric data
datos almacenados en computadora computer
  stored data
datos analógicos analog data
datos analógicos y digitales analog and digital data
datos apareados paired data
datos aproximados approximate data
datos aritméticos codificados coded arithmetic data
datos auxiliares auxiliary data
datos binarios binary data
datos binarios de coma fija fixed-point binary data
datos binarios empaquetados packed binary data
datos binarios fijos fixed binary data
datos característicos characteristic data
datos claves key data
datos codificados encoded data
datos complejos complex data
datos computarizados computerized data
datos constantes constant data
datos de adaptación adaptation data
datos de banda base baseband data
datos de bits numéricos numerical bit data
datos de Boole Boolean data
datos de campo field data
datos de caracteres character data
datos de coma fija fixed-point data
datos de coma flotante floating-point data
datos de computadora computer data
datos de confiabilidad reliability data
datos de control control data
datos de control de programa program-control data
datos de control intermedio intermediate-control
  data
datos de control menor minor control data
datos de control operacional operational control
  data
datos de control principal major control data
datos de controlador controller data
datos de decimal fijo fixed-decimal data
datos de desbordamiento overflow data
datos de encaminamiento routing data
datos de entrada input data
datos de estado state data
datos de falla failure data
datos de imagen image data
datos de imagen binarios binary image data
datos de línea de base base line data
datos de llegada inbound data
datos de mensajes message data
datos de operación operating data
datos de página page data
datos de programa program data
datos de prueba test data
datos de radar radar data
datos de salida output data

datos de transacción transaction data
datos decimales empaquetados packed decimal data
datos derivados derived data
datos digitales digital data
datos empaquetados packed data
datos en cinta tape data
datos en serie serial data
datos en tiempo real real-time data
datos externos external data
datos ficticios dummy data
datos fijos fixed data
datos globales global data
datos impresos printed data
datos incrementales incremental data
datos inmediatos immediate data
datos internos internal data
datos legibles por máquina machine-readable data
datos locales local data
datos maestros master data
datos masivos mass data
datos modificados modified data
datos muestreados sampled data
datos no pareados nonpaired data
datos nominales nominal data
datos nulos null data
datos numéricos numerical data
datos ordinales ordinal data
datos originales original data
datos permanentes permanent data
datos reales real data
datos relativos relative data
datos secundarios secondary data
datos simbólicos symbolic data
datos sin procesar raw data
datos sincrónicos synchronous data
datos temporales temporary data
datos transitorios transient data
datos variables variable data
datos virtuales virtual data
de abertura rápida quick-opening
de acceso rápido rapid-access
de acción rápida quick-acting
de acción ultrarrápida snap-action
de alto rendimiento high-performance
de calentamiento rápido quick-heating
de canal único single-channel
de cara única single-sided
de cierre lento slow-closing
de cierre rápido quick-make
de cinta a cinta tape-to-tape
de cinta a tarjeta tape-to-card
de conexión rápida quick-connecting
de conexión única a tierra ungrounded
de desconexión rápida quick-disconnect
de dirección cero zero-address
de disminución progresiva tapered
de doble uso dual use
de dos capas two-layer
de dos direcciones two-address
de dos estados bistate
de dos etapas two-stage
de dos frecuencias two-frequency
de dos hilos two-wire
de dos lados double-sided
de dos niveles two-level
de dos pasos two-step
de dos polos two-pole
de dos posiciones double-throw
de dos postes two-pole
de dos puertos two-port
de dos terminales two-terminal
de dos tonos two-tone
de escala única single-range

de estado sólido solid-state
de fibra óptica fiber-optic
de funciones múltiples multipurpose
de inserción rápida quick-insert
de largo alcance long-range
de llegada incoming
de montaje en bastidor rack-mount
de movimiento lento slow-moving
de operación lenta slow-operating
de origen originating
de pocas pérdidas low-loss
de posición positional
de punto único single-point
de recorrido largo long-throw
de ruptura rápida quick-break
de salida outgoing
de segundo orden second-order
de sólo recepción receive-only
de tercera generación third-generation
de toda onda all-wave
de todo clima all-weather
de tres direcciones three-address
de tres etapas three-stage
de tres niveles three-level
de tres posiciones three-position
de tres terminales three-terminal
de un lado one-sided
de un polo one-pole
de una cara single-sided
de una dimensión one-dimensional
de una dirección one-directional, one-way
de una etapa one-stage
de una fase one-phase
de una frecuencia one-frequency
de una señal one-signal
de uso general general purpose
de usos múltiples multiple-use
de variación progresiva tapered
de vida larga long-life
débilmente magnético weakly magnetic
debye debye
deca deca
década de capacitores capacitor decade
decaimiento decay
decaimiento anormal abnormal decay
decaimiento de ondas amortiguadas damped-wave
  decay
decaimiento dinámico dynamic decay
decaimiento estático static decay
decaimiento exponencial exponential decay
decalaje de escobillas brush shift
decapado pickling
decatrón dekatron
decca decca
decelerómetro decelerometer
deci deci
decibel decibel
decibeles ajustados adjusted decibels
decibeles sin compensación unweighted decibels
decibeles sobre ruido de referencia decibels above
  reference noise
decibeles sobre un kilowatt decibels above one
  kilowatt
decibeles sobre un volt decibels above one volt
decibeles sobre un watt decibels above one watt
decibelímetro decibel meter
decibelímetro de registro recording decibel meter
decibelio decibel
decigramo decigram
decila decile
decilitro deciliter
decimal decimal
decimal codificado coded decimal

decimal codificado en binario binary-coded decimal
decimal empaquetado packed decimal
decimal fijo fixed decimal
decimal flotante floating decimal
decimal periódico periodic decimal
decimal repetitivo repeating decimal
decimales similares similar decimals
decineper decineper
decisión lógica logic decision
declaración declaration, statement
declaración contextual contextual declaration
declaración implícita implicit declaration
declinación declination
declinación magnética magnetic declination
declinómetro declinometer
decómetro decometer
decremento decrement
decremento equivalente equivalent decrement
decremento lineal linear decrement
decremento logarítmico logarithmic decrement
decrémetro decremeter
decrémetro de Kolster Kolster decremeter
dectra dectra
dedo de contacto contact finger
defecto defect, bug
defecto de diseño design defect
defecto de límite boundary defect
defecto intermitente intermittent defect
defecto principal major defect
defectuoso defective, faulty
definición definition, resolution
definición aparente apparent definition
definición de archivo de comunicaciones communication file definition
definición de base de datos database definition
definición de bloque block definition
definición de campo field definition
definición de celda cell definition
definición de comparación comparison definition
definición de datos data definition
definición de formato format definition
definición de imagen picture definition
definición de mando command definition
definición de página page definition
definición de periodo period definition
definición de procedimiento procedure definition
definición de programa program definition
definición de sistema system definition
definición de sistema preconfigurado preconfigured system definition
definición de trabajo job definition
definición horizontal horizontal definition
definición vertical vertical definition
definido por usuario user-defined
deflector deflector
deflector de haz beam deflector
deflexión deflection
deflexión asimétrica asymmetrical deflection
deflexión cero zero deflection
deflexión completa full deflection
deflexión de aguja pointer deflection
deflexión de haz beam deflection
deflexión de límite de escala full-scale deflection
deflexión de punto spot deflection
deflexión de referencia reference deflection
deflexión electromagnética electromagnetic deflection
deflexión electrostática electrostatic deflection
deflexión horizontal horizontal deflection
deflexión iónica ion deflection
deflexión magnética magnetic deflection
deflexión máxima maximum deflection
deflexión pico peak deflection

deflexión residual residual deflection
deflexión vertical vertical deflection
deflexión X X deflection
deflexión Y Y deflection
deflexión Z Z deflection
deformación de surco groove deformation
deformación magnética magnetic strain
deformación permanente permanent deformation
deformímetro strain gage
deformímetro eléctrico electric-strain gage
degeneración degeneration
degeneración por calor heat aging
degeneración por emisor emitter degeneration
degenerativo degenerative
degradación degradation
degradación por ruido noise degradation
delga de colector commutator bar
delicuescente deliquescent
delimitador delimiter
delimitador de datos data delimiter
delineación de borde border delineation
delta delta
delta equivalente equivalent delta
deltic deltic
demanda acumulativa cumulative demand
demanda acumulativa continua continuous cumulative demand
demanda de información information demand
demanda de procesador processor demand
demanda directa direct demand
demanda instantánea instantaneous demand
demanda integrada integrated demand
demanda máxima maximum demand
demanda mínima minimum demand
demarcación demarcation
demostrador dinámico dynamic demonstrator
denario denary
dendrita dendrite
dendrítico dendritic
denominador denominator
densidad density
densidad de almacenamiento storage density
densidad de bits bit density
densidad de campo field density
densidad de caracteres character density
densidad de carga charge density
densidad de carga electrónica electron-charge density
densidad de carga espacial space-charge density
densidad de carga iónica ion-charge density
densidad de carga lineal linear charge density
densidad de carga magnética magnetic charge density
densidad de cinta tape density
densidad de componentes component density, packing density
densidad de componentes equivalente equivalent component density
densidad de conductores conductor density
densidad de corriente current density
densidad de corriente catódica cathode-current density
densidad de corriente de partículas particle current density
densidad de corriente directa forward current density
densidad de corriente eléctrica electric current density
densidad de corriente magnética magnetic current density
densidad de corriente superficial surface-current density
densidad de corriente uniforme uniform current

density
densidad de datos data density
densidad de desplazamiento eléctrico
    electric-displacement density
densidad de electrones density of electrons
densidad de energía energy density
densidad de energía magnética magnetic energy
    density
densidad de energía radiante radiant energy density
densidad de energía sonora sound-energy density
densidad de exploración scanning density
densidad de facsímil facsimile density
densidad de flujo flux density
densidad de flujo de desplazamiento displacement
    flux density
densidad de flujo de energía energy flux density
densidad de flujo de partículas particle-flux density
densidad de flujo de potencia power flux density
densidad de flujo de saturación saturation flux
    density
densidad de flujo dieléctrico dielectric flux density
densidad de flujo eléctrico electric-flux density
densidad de flujo intrínseco intrinsic flux density
densidad de flujo luminoso luminous flux density
densidad de flujo magnético magnetic flux density
densidad de flujo radiante radiant flux density
densidad de flujo remanente remanent flux density
densidad de flujo residual residual flux density
densidad de gas gas density
densidad de huecos hole density
densidad de impresión print density
densidad de impurezas impurity density
densidad de inducción induction density
densidad de información packing density,
    information density
densidad de ionización ionization density
densidad de línea line density
densidad de marcas mark density
densidad de partículas particle density
densidad de plasma plasma density
densidad de portadores carrier density
densidad de potencia power density
densidad de potencia sonora sound-power density
densidad de probabilidad probability density
densidad de radiación radiation density
densidad de registro recording density
densidad eléctrica electric density
densidad electrónica electron density
densidad en metronios metron density
densidad espectral spectral density
densidad especular specular density
densidad iónica ion density
densidad iónica lineal linear ion density
densidad lineal linear density
densidad luminosa luminous density
densidad magnética magnetic density
densidad máxima de flujo peak flux density
densidad molecular molecular density
densidad óptica optical density
densidad radiante radiant density
densidad superficial eléctrica electric surface
    density
densidad variable variable density
densidad volumétrica eléctrica electric volume
    density
densitómetro densitometer
densitómetro de proyección projection densitometer
densitómetro electrónico electronic densitometer
densitómetro fotoeléctrico photoelectric
    densitometer
densitómetro ultrasónico ultrasonic densitometer
dependencia angular angular dependence
dependencia de dispositivo device dependence

dependencia de frecuencia frequency dependence
dependencia funcional functional dependency
dependiente de configuración configuration
    dependent
dependiente de dispositivo device dependent
dependiente de frecuencia frequency-dependent
dependiente de la tensión voltage-dependent
dependiente de máquina machine-dependent
dependiente de memoria memory-dependent
dependiente de posición position-dependent
dependiente del tiempo time-dependent
deposición deposition
deposición al vacío vacuum deposition
deposición de aleación alloy deposition
deposición de película film deposition
deposición eléctrica electric deposition
deposición electroforética electrophoretic deposition
deposición electrolítica electrolytic deposition
deposición electrónica sputter
deposición epitaxial epitaxial deposition
deposición metálica metal deposition
deposición por vapor vapor deposition
deposición química chemical deposition
deposición uniforme uniform deposition
depositado plated
depósito plating, sediment
depósito catódico cathode deposit
depósito catódico flojo loose cathode deposit
depósito compuesto composite plate
depósito parcial partial plating
depósito por contacto contact plating
depósito por inmersión immersion plating
depósito por vapor vapor plating
depósito primario strike deposit
depósito radiactivo radioactive deposit
depósito semiautomático semiautomatic plating
depósito ultrasónico ultrasonic plating
depreciación por almacenamiento shelf depreciation
depresión pit
depresión de potencial potential trough
depresión eléctrica electrical depression
depresor de cinta tape depressor
depuración debugging, troubleshooting
depuración de interfaz interface debugging
depuración de microprograma microprogram
    debugging
depuración de paso único single-step debugging
depuración de programa program debugging
depuración de prototipo prototype debugging
depuración dinámica dynamic debugging
depuración en línea on-line debugging
depuración imperfecta imperfect debugging
depuración interactiva interactive debugging
depuración remota remote debugging
depuración simbólica symbolic debugging
depurar debug, troubleshoot
derecho de acceso access right
deriva drift
deriva lineal linear drift
deriva relativa relative drift
deriva térmica heat drift
deriva transversal transverse drift
derivación branch, branching, shunt, shunting, tap,
    bypass
derivación central center-tap
derivación de amperímetro ammeter shunt
derivación de Ayrton Ayrton shunt
derivación de galvanómetro galvanometer shunt
derivación de galvanómetro de Ayrton-Mather
    Ayrton-Mather galvanometer shunt
derivación de instrumento instrument shunt
derivación de medidor meter shunt
derivación de resistencia resistance shunt

derivación magnética magnetic shunt
derivación no inductiva noninductive shunt
derivación resonante resonant shunt
derivada derivative
derivada parcial partial derivative
derivado derivative
desaccionamiento dropout
desaceleración deceleration
desacentuación deemphasis
desacentuador deaccentuator
desacoplado uncoupled
desacoplador decoupler
desacoplamiento decoupling
desacoplamiento de pantalla screen decoupling
desacoplar decouple
desactivación deactivation
desactivación de sesión session deactivation
desactivación directa direct deactivation
desactivar deactivate
desactualización deupdating
desadaptación mismatch
desadaptación de impedancias impedance mismatch
desadaptador patrón standard mismatcher
desagrupamiento debunching
desagrupamiento por carga espacial space-charge
    debunching
desagrupar unpack
desalineación misalignment
desaparición progresiva fade out
desarrollar develop, unwind
desarrollo de aerosol aerosol development
desarrollo de circuitos circuit development
desarrollo de microprograma microprogram
    development
desarrollo de procesos process development
desarrollo de programa program development
desarrollo incremental incremental development
desasignación deallocation
desasignar deallocate
desatendido unattended
desaturación desaturation
desbloqueado unblocked
desbloquear deblock
desbloqueo unblocking
desbloqueo periódico gating
desbloqueo periódico de receptor receiver gating
desbobinado unwinding
desbobinar unwind
desbordamiento overflow
desbordamiento aritmético arithmetic overflow
desbordamiento característico characteristic
    overflow
desbordamiento de campo field overflow
desbordamiento de exponente exponent overflow
desbordamiento de memoria memory overflow
desbordamiento decimal decimal overflow
desbordamiento negativo underflow
desbordamiento negativo aritmético arithmetic
    underflow
desbordamiento negativo de exponente exponent
    underflow
desbordamiento negativo por característica
    characteristic underflow
desbordamiento progresivo progressive overflow
descarga discharge
descarga aperiódica aperiodic discharge
descarga atmosférica lightning strike
descarga automantenida self-maintained discharge
descarga autosostenida self-sustained discharge
descarga convectiva convective discharge
descarga de arco arc discharge
descarga de bajo régimen low-rate discharge
descarga de campo field discharge

descarga de cátodo frío cold-cathode discharge
descarga de corriente continua direct-current
    discharge
descarga de encendedor ignitor discharge
descarga de Geissler Geissler discharge
descarga de mantenimiento keep-alive discharge
descarga de Penning Penning discharge
descarga de rayo lightning stroke
descarga de Townsend Townsend discharge
descarga directa direct stroke
descarga disruptiva disruptive discharge, sparkover
descarga eléctrica electrical discharge
descarga electrostática electrostatic discharge
descarga en corona corona discharge
descarga en obscuridad dark discharge
descarga en penacho brush discharge
descarga estratificada stratified discharge
descarga gaseosa gas discharge
descarga impulsiva impulsive discharge
descarga indirecta indirect stroke
descarga láser laser burst
descarga luminiscente glow discharge, luminescent
    discharge
descarga luminiscente anormal abnormal glow
    discharge
descarga luminiscente de radiofrecuencia
    radio-frequency glow discharge
descarga luminiscente normal normal glow
    discharge
descarga luminosa luminous discharge
descarga oscilante oscillating discharge
descarga parcial partial discharge
descarga profunda deep discharge
descarga pulsada pulsed discharge
descarga residual residual discharge
descarga sin electrodos electrodeless discharge
descarga transitoria transient discharge
descargador discharger
descargador de chispa spark gap
descargador de chispa amortiguada quenched
    spark-gap
descargador de Lepel Lepel discharger
descargador de vacío vacuum arrester
descargador en diente de sierra sawtooth arrester
descargador rotativo rotary discharger, rotary
    spark-gap
descargador sincrónico synchronous spark-gap
descargar download
descifrable decodable
descifrado decryption
descifrar decipher, decrypt
descodificación decoding
descodificación de impulsos pulse decoding
descodificación en serie serial decoding
descodificación secuencial sequential decoding
descodificado decoded
descodificador decoder, unscrambler
descodificador activo active decoder
descodificador binario binary decoder
descodificador bipolar bipolar decoder
descodificador cuadrafónico quadraphonic decoder
descodificador de color color decoder
descodificador de dirección address decoder
descodificador de fallas fault decoder
descodificador de impulsos pulse decoder
descodificador de instrucciones instruction decoder
descodificador de operación operation decoder
descodificador de telemedida telemetering decoder
descodificador-desmultiplexor
    decoder-demultiplexer
descodificador estereofónico stereophonic decoder
descodificador pasivo passive decoder
descodificar decode

descolgado off-hook
descolgar unhook
descompilador decompiler
descompilar decompile
descomposición decomposition
descomposición de colores color breakup
descomposición electrolítica electrolytic decomposition
descomposición funcional functional decomposition
descomposición jerárquica hierarchical decomposition
descomposición modular modular decomposition
descompresión decompression
desconectable detachable
desconectado uncoupled
desconectador disconnecter
desconectar disconnect, uncouple, switch-off
desconector de protector arrester disconnector
desconexión disconnection, release, uncoupling, tripping, switching-off
desconexión automática automatic shutoff
desconexión por la última línea last-line release
desconexión por sobrecarga overload release
desconexión por subtensión undervoltage tripping
desconmutación decommutation
desconmutador decommutator
descrestación chopping
descrestador chopper
descripción alternativa alternate description
descripción de archivo file description
descripción de datos data description
descripción de diseño design description
descripción de formato format description
descripción de impresora printer description
descripción de línea line description
descripción de mensajes message description
descripción de problema problem description
descripción de programa program description
descripción de registro record description
descripción de sistema system description
descripción de trabajo job description
descriptor de almacenamiento storage descriptor
descriptor de base de datos database descriptor
descriptor de campo field descriptor
descriptor de control de partición partition control descriptor
descriptor de datos data descriptor
descriptor de formato de registro record format descriptor
descriptor de parámetro parameter descriptor
desecador dessicator
desecante desiccant
desechable expendable
desembrague clutch disengaging
desenergizado deenergized
desenergizar deenergize
desenfocado blurred
desenfoque defocusing, blurring
desenfoque de haz beam defocusing
desenfoque ionosférico ionospheric defocusing
desenfoque por deflexión deflection defocusing
desenganchador tripper
desenganche release, tripping
desenganche electromagnético electromagnetic release
desenganche instantáneo instantaneous release
desenganche retardado delayed release
desenganche selectivo selective release
desensamblador disassembler
desensamblar disassemble
desensibilización desensitization
desensibilizar desensitize
desequilibrado out of balance, unbalanced

desequilibrio unbalance, offset
desequilibrio de capacidad capacity unbalance
desequilibrio de capacitancias capacitance unbalance
desequilibrio de red network unbalance
desequilibrio de tensión voltage unbalance
desequilibrio diferencial differential unbalance
desequilibrio dinámico dynamic unbalance
desequilibrio residual residual unbalance
desequilibrio resistivo resistive unbalance
desexcitación deexcitation
desfasado out-of-phase
desfasador phase-shifter
desfasador de guíaondas waveguide phase-shifter
desfasador de resistencia-capacitancia resistance-capacitance phase-shifter
desfasador de resistencia-inductancia resistance-inductance phase-shifter
desfasador digital digital phase-shifter
desfasador direccional directional phase-shifter
desfasador electrostático electrostatic phase-shifter
desfasador rotativo rotary phase-shifter
desfasamiento outphasing
desfase phase shift, phase change, outphasing
desfase de inserción insertion phase shift
desfase progresivo progressive phase shift
desfibrilación defibrillation
desfibrilador defibrillator
desforrado stripping
desforrador stripper
desforrador de cables cable stripper
desgarre de imagen picture tearing
desgarro tearing
desgasificación degassing
deshumidificación dehumidification
deshumidificador dehumidifier
designación designation
designación de referencia reference designation
designación funcional functional designation
designador de canal channel designator
designador de registro register designator
designador de unidad drive designator
desinstalar deinstall
desintegración disintegration, decay
desintegración alfa alpha decay
desintegración atómica atomic disintegration
desintegración beta beta decay
desintegración catódica cathode disintegration
desintegración eléctrica electric disintegration
desintegración fotoeléctrica photoelectric disintegration
desintegración gamma gamma decay
desintegración intercristalina intercrystalline disintegration
desintegración natural natural disintegration
desintegrador disintegrator
desintegrador ultrasónico ultrasonic disintegrator
desintonización detuning
desintonizar detune
desionización deionization
desionizar deionize
deslaminación delamination
deslizamiento drift, shift, slip
deslizamiento a largo plazo long-term drift
deslizamiento aleatorio random drift
deslizamiento cero zero drift
deslizamiento de corriente de polarización bias current drift
deslizamiento de enfoque focusing drift
deslizamiento de frecuencia frequency drift
deslizamiento de frecuencia de oscilador oscillator drift
deslizamiento de frecuencia lento slow frequency

drift
deslizamiento de imagen image drift
deslizamiento de tensión drift voltage
deslizamiento del cero zero drift, zero shift
deslizamiento térmico thermal drift
deslizamiento vertical de imagen picture slip
desmagnetización demagnetization, degaussing
desmagnetización adiabática adiabatic demagnetization
desmagnetizador demagnetizer, degausser
desmagnetizador automático automatic degausser
desmagnetizador de cabeza head demagnetizer
desmagnetizante demagnetizing
desmagnetizar demagnetize, degauss
desmenuzado fritting
desmineralización demineralization
desmodulación demodulation
desmodulación analógica analog demodulation
desmodulación coherente coherent demodulation
desmodulación de amplitud amplitude demodulation
desmodulación de frecuencia frequency demodulation
desmodulación de portadora realzada enhanced-carrier demodulation
desmodulación magnética magnetic demodulation
desmodulación X X demodulation
desmodulación X y Z X and Z demodulation
desmodulado demodulated
desmodulador demodulator
desmodulador controlado gated demodulator
desmodulador de canal channel demodulator
desmodulador de canal de servicio service-channel demodulator
desmodulador de crominancia chrominance demodulator
desmodulador de diodo diode demodulator
desmodulador de doble diodo double-diode demodulator
desmodulador de enganche de fase phase-locked demodulator
desmodulador de envolvente envelope demodulator
desmodulador de grupo group demodulator
desmodulador de grupo maestro mastergroup demodulator
desmodulador de ley cuadrática square-law demodulator
desmodulador de luminancia luminance demodulator
desmodulador de producto product demodulator
desmodulador de subportadora de crominancia chrominance-subcarrier demodulator
desmodulador de supergrupo supergroup demodulator
desmodulador de telemedida telemetering demodulator
desmodulador equilibrado balanced demodulator
desmodulador Q Q demodulator
desmodulador sensible a la fase phase-sensitive demodulator
desmodulador sincrónico synchronous demodulator
desmodulador telegráfico telegraph demodulator
desmodulador X X demodulator
desmodulador Z Z demodulator
desmodularización demodularization
desmontable demountable
desmultiplexor demultiplexer
desmultiplicación de frecuencia frequency demultiplication
desocupado unoccupied
desoldar desolder
despachador de carga load dispatcher
despacho económico economic dispatch
despegue ciego blind takeoff

desplazador de frecuencia frequency shifter
desplazador de longitud de onda wavelength shifter
desplazamiento shift, shifting, displacement
desplazamiento a la derecha right shift
desplazamiento a la izquierda left shift
desplazamiento angular angular displacement
desplazamiento aritmético arithmetic shift
desplazamiento cíclico cyclic shift
desplazamiento circular circular shift
desplazamiento colorimétrico adaptivo adaptive colorimetric shift
desplazamiento de amplitud amplitude shift
desplazamiento de audiofrecuencia audio-frequency shift
desplazamiento de campo field displacement
desplazamiento de ciclos cycle shift
desplazamiento de color adaptivo adaptive color shift
desplazamiento de corriente de placa plate-current shift
desplazamiento de Doppler Doppler shift
desplazamiento de fase phase shift
desplazamiento de fase de rejilla grid phase shift
desplazamiento de fase diferencial differential phase shift
desplazamiento de frecuencia frequency shift, frequency displacement
desplazamiento de frecuencia de señal signal frequency shift
desplazamiento de frecuencia portadora carrier-frequency shift
desplazamiento de imagen image shift
desplazamiento de modo mode shift
desplazamiento de multiplicación multiplication shift
desplazamiento de partículas particle displacement
desplazamiento de polarización de rejilla grid-bias shift
desplazamiento de portadora carrier shift
desplazamiento del punto de operación operating-point shift
desplazamiento descendente pushdown
desplazamiento dieléctrico dielectric displacement
desplazamiento eléctrico electric displacement
desplazamiento en anillo ring shift
desplazamiento isotópico isotope shift
desplazamiento lateral lateral displacement
desplazamiento lógico logic shift
desplazamiento magnético magnetic displacement
desplazamiento no aritmético nonarithmetic shift
desplazamiento numérico numerical shift
desplazamiento positivo positive displacement
desplazamiento total resultant color shift
desplazar scroll
despolarización depolarization
despolarización resonante resonant depolarization
despolarizador depolarizer
despolarizador de dióxido de manganeso manganese-dioxide depolarizer
despolarizante depolarizing
destellador flasher
destellador de luz light flasher
destellador electrónico electronic flasher
destello flash, flare
destello de luz light flash
destello electrónico electronic flash
destello sincronizado synchronized flash
destellos periódicos periodic flashes
destino destination
destino lógico logical destination
destino secundario secondary destination
desvanecedor fader
desvanecimiento fading, fade

desvanecimiento a largo plazo long-term fading
desvanecimiento artificial artificial fading
desvanecimiento automático automatic fading
desvanecimiento de amplitud amplitude fading
desvanecimiento de imagen picture fading
desvanecimiento de largo periodo long-period fading
desvanecimiento de radio radio fadeout
desvanecimiento de señal signal fading
desvanecimiento plano flat fading
desvanecimiento por absorción absorption fading
desvanecimiento por polarización polarization fading
desvanecimiento por salto skip fading
desvanecimiento selectivo selective fading
desvanecimiento selectivo a la frecuencia frequency-selective fading
desviación deviation, deflection, shift
desviación angular angular deviation
desviación cuadrantal quadrantal deviation
desviación de acimut azimuth deviation
desviación de aguja needle deviation
desviación de capacidad capacity deviation
desviación de demanda demand deviation
desviación de estado estable absoluta absolute steady-state deviation
desviación de estado estacionario steady-state deviation
desviación de fase phase deviation
desviación de frecuencia frequency deviation
desviación de frecuencia nominal rated frequency deviation
desviación de haz beam bending
desviación de inclinación de planeo glide-slope deviation
desviación de línea de rumbo course-line deviation
desviación de modulación de frecuencia frequency-modulation deviation
desviación de portadora carrier deviation
desviación de potencia power deviation
desviación de sistema system deviation
desviación de sistema absoluto absolute system deviation
desviación de sistema nominal rated system deviation
desviación de tensión voltage deviation
desviación de trayectoria de vuelo flight-path deviation
desviación dinámica dynamic deviation
desviación estándar standard deviation
desviación lateral lateral deviation
desviación magnética magnetic deviation
desviación máxima maximum deviation
desviación máxima de frecuencia maximum frequency deviation
desviación máxima de sistema maximum system deviation
desviación media average deviation
desviación media cuadrática root-mean-square deviation
desviación paradiafónica near-end crosstalk deviation
desviación periódica periodic deviation
desviación pico de frecuencia peak frequency deviation
desviación radiogoniométrica direction-finder deviation
desviación residual residual deviation
desviación telediafónica far-end crosstalk deviation
desviación total total deviation
desviación transitoria transient deviation
desviación transitoria absoluta absolute transient deviation

desviador diverter
desviador de haz beam bender
desviador de papel paper deflector
desvío de cátodo cathode bypass
desviómetro deviation meter
detallado detailed
detalle de formato format detail
detalle de imagen image detail
detección detection, sensing
detección anódica anode detection
detección coherente coherent detection
detección de aguja point detection
detección de batido cero zero-beat detection
detección de borde border detection
detección de cambios change detection
detección de caracteres character sensing
detección de caracteres ópticos optical character sensing
detección de cero zero detection
detección de colisiones collision detection
detección de correlación correlation detection
detección de corriente de placa plate-current detection
detección de errores error detection
detección de errores automática automatic error detection
detección de errores remota remote error detection
detección de fallas fault detection
detección de fugas leak detection
detección de grietas magnética magnetic crack detection
detección de imperfecciones ultrasónica ultrasonic flaw detection
detección de lápiz óptico light-pen detection
detección de línea line detection
detección de luz coherente coherent-light detection
detección de marcas mark sensing
detección de placa plate detection
detección de potencia power detection
detección de producto product detection
detección de radar radar detection
detección de radiactividad radioactivity detection
detección de ruido noise sensing
detección de seudoerrores pseudoerror detection
detección de sonido sound detection
detección de tarjetas card sensing
detección de tensión voltage sensing
detección de vibraciones vibration detection
detección de video video detection
detección heterodina heterodyne detection
detección homodina homodyne detection
detección incoherente incoherent detection
detección infrarroja activa active infrared detection
detección inversa reverse detection
detección lineal linear detection
detección no lineal nonlinear detection
detección parabólica parabolic detection
detección pasiva passive detection
detección por diodo de germanio germanium-diode detection
detección por ecos echo detection
detección por pendiente slope detection
detección por rejilla grid detection
detección remota remote detection
detección retardada delayed detection
detección sincrónica synchronous detection
detección supersónica supersonic detection
detectabilidad detectability
detectable detectable
detectable por máquina machine-sensible
detectividad detectivity
detectividad específica specific detectivity
detectófono detectophone

detector detector
detector acústico acoustic detector
detector alfa alpha detector
detector amplificador amplifying detector
detector analógico analog detector
detector anódico anode detector
detector armónico harmonic detector
detector autodino autodyne detector
detector bolométrico bolometric detector
detector coaxial coaxial detector
detector coherente coherent detector
detector combinado combination detector
detector cuadrático quadratic detector
detector cuantitativo quantitative detector
detector de activación activation detector
detector de agujeros diminutos pinhole detector
detector de agujeros diminutos fotoeléctrico
　photoelectric pinhole detector
detector de alto nivel high-level detector
detector de altura de impulsos pulse-height detector
detector de amplitud amplitude detector
detector de anomalías magnéticas magnetic
　anomaly detector
detector de antena antenna detector
detector de audio audio detector
detector de banda angosta narrowband detector
detector de batido cero zero-beat detector
detector de Bradley Bradley detector
detector de calor heat detector
detector de cámara de ionización ionization
　chamber detector
detector de campo field detector
detector de campo retardador retarding-field
　detector
detector de captura electrónica electron capture
　detector
detector de carborundo carborundum detector
detector de centelleos scintillation detector
detector de cero zero detector
detector de circuito anódico anode circuit detector
detector de coincidencia coincidence detector
detector de colores color detector
detector de contacto contact detector
detector de conversión conversion detector
detector de corona corona detector
detector de correlación correlation detector
detector de corriente current detector
detector de cortocircuito short-circuit detector
detector de cristal crystal detector
detector de cristal de germanio germanium-crystal
　detector
detector de cristal de silicio silicon-crystal detector
detector de cristal fijo fixed-crystal detector
detector de croma chroma detector
detector de cuasipicos quasi-peak detector
detector de descarga discharge detector
detector de diferencia difference detector
detector de diferencia de tensión voltage-difference
　detector
detector de diodo diode detector
detector de diodo compensado compensated diode
　detector
detector de diodo de cristal crystal-diode detector
detector de diodo fijo fixed-diode detector
detector de doble sintonía double-tuned detector
detector de enganche de fase phase-locked detector
detector de envolvente envelope detector
detector de errores error detector
detector de excitación de rejilla grid-drive detector
detector de fallas fault detector
detector de fase phase detector
detector de fase de color color phase detector
detector de fase equilibrado balanced phase detector

detector de Foster-Seeley Foster-Seeley detector
detector de frecuencia frequency detector
detector de Fremodyne Fremodyne detector
detector de fuego fire detector
detector de fuga de rejilla grid-leak detector
detector de fugas leak detector
detector de fugas de vacío vacuum-leak detector
detector de galena galena detector
detector de gases gas detector
detector de gases de filamento caliente hot-filament
　gas detector
detector de Geiger-Mueller Geiger-Mueller detector
detector de grietas electromagnético
　electromagnetic crack detector
detector de grietas magnético magnetic crack
　detector
detector de grietas ultravioleta ultraviolet crack
　detector
detector de haz controlado gated-beam detector
detector de hilo caliente hot-wire detector
detector de hojas sheet detector
detector de humedad humidity detector
detector de humo smoke detector
detector de humo de bolómetro bolometer smoke
　detector
detector de humo de ionización ionization smoke
　detector
detector de humo fotoeléctrico photoelectric smoke
　detector
detector de imagen picture detector
detector de impedancia infinita infinite-impedance
　detector
detector de imperfecciones flaw detector
detector de imperfecciones fotoeléctrico
　photoelectric flaw detector
detector de imperfecciones magnético magnetic
　flaw detector
detector de imperfecciones ultrasónico ultrasonic
　flaw detector
detector de impulsos pulse detector
detector de incendios automático automatic fire
　detector
detector de infrarrojo infrared detector
detector de inversión de polaridad polarity-reversal
　detector
detector de ley cuadrática square-law detector
detector de líquido liquid detector
detector de llama flame detector
detector de llama fotoeléctrico photoelectric flame
　detector
detector de mentiras lie detector
detector de mineral mineral detector
detector de moción motion detector
detector de moción por infrarrojos infrared motion
　detector
detector de modulación de amplitud
　amplitude-modulation detector
detector de modulación de frecuencia
　frequency-modulation detector
detector de nivel level detector
detector de nivel ultrasónico ultrasonic level
　detector
detector de nota de batido beat note detector
detector de onda completa full-wave detector
detector de ondas wave detector
detector de ondas estacionarias standing-wave
　detector
detector de oscilaciones oscillation detector
detector de oscilador sincronizado locked-oscillator
　detector
detector de partículas particle detector
detector de pico peak detector
detector de picos de diodo diode peak detector

detector de placa plate detector
detector de polarización de rejilla grid-bias detector
detector de potencia power detector
detector de presión pressure detector
detector de producto product detector
detector de productos de diodo diode product detector
detector de proximidad proximity detector
detector de puente bridge detector
detector de puerta gate detector
detector de radar radar detector
detector de radiación radiation detector
detector de radiactividad radioactivity detector
detector de rayos lightning detector
detector de razón ratio detector
detector de ruido noise detector
detector de señal débil weak-signal detector
detector de seudoerrores pseudoerror detector
detector de silicio silicon detector
detector de sobretemperatura overtemperature detector
detector de sonido sound detector
detector de subcanal subchannel detector
detector de subportadora subcarrier detector
detector de tantalio tantalum detector
detector de telemedida telemetering detector
detector de temperatura temperature detector
detector de temperatura por resistencia resistance temperature detector
detector de tensión voltage detector
detector de termómetro de resistencia resistance-thermometer detector
detector de tierra ground detector
detector de tono piloto pilot-tone detector
detector de transmisión transmission detector
detector de tubo de vacío vacuum-tube detector
detector de umbral threshold detector
detector de válvula valve detector
detector de velocidad de subida rate-of-rise detector
detector de video video detector
detector de voz speech detector
detector electrolítico electrolytic detector
detector en contrafase push-pull detector
detector en cuadratura quadrature detector
detector en derivación shunt detector
detector en serie series detector
detector equilibrado balanced detector
detector fotoconductivo photoconductive detector
detector fotoeléctrico photoelectric detector
detector fotovoltaico photovoltaic detector
detector heterodino heterodyne detector
detector incoherente incoherent detector
detector infrarrojo enfriado cooled infrared detector
detector infrarrojo fotoconductivo photoconductive infrared detector
detector interno de temperatura internal detector of temperature
detector isotrópico isotropic detector
detector lineal linear detector
detector magnético magnetic detector
detector medio average detector
detector microfónico microphonic detector
detector móvil traveling detector
detector multielemento multielement detector
detector neumático pneumatic detector
detector no amplificador nonamplifying detector
detector no lineal nonlinear detector
detector no oscilante nonoscillating detector
detector no regenerativo nonregenerative detector
detector óptico optical detector
detector oscilante oscillating detector
detector pentarrejilla pentagrid detector
detector perikón perikon detector

detector piezoeléctrico piezoelectric detector
detector pirón pyron detector
detector polar polar detector
detector polarizado biased detector
detector por deriva de litio lithium-drifted detector
detector por pendiente slope detector
detector primario primary detector
detector químico chemical detector
detector radiac radiac detector
detector rectificador rectifying detector
detector regenerativo regenerative detector
detector selectivo selective detector
detector semiconductor semiconductor detector
detector sensible a la fase phase-sensitive detector
detector sensible a la luz light-sensitive detector
detector sensible a la posición position-sensitive detector
detector sincrónico synchronous detector
detector sintonizado tuned detector
detector subarmónico subharmonic detector
detector superregenerativo superregenerative detector
detector térmico thermal detector
detector termoeléctrico thermoelectric detector
detector termoiónico thermionic detector
detector ultrasónico ultrasonic detector
detención hold
deterioro deterioration, impairment
deterioro completo wearout
deterioro de rendimiento performance impairment
deterioro de transmisión transmission impairment
deterioro electroquímico electrochemical deterioration
deterioro galvánico galvanic deterioration
deterioro por agentes atmosféricos weathering
deterioro por transmisión de ruido noise transmission impairment
determinación de problema problem determination
determinación de sentido sense determination
determinante determinant
determinante de frecuencia frequency-determining
deuterio deuterium
devanado winding
dextrógiro clockwise
dextrorso clockwise
diac diac
diactor diactor
diafonía crosstalk
diafonía de interacción interaction crosstalk
diafonía entre antenas antenna crosstalk
diafonía entre canales interchannel crosstalk
diafonía lineal linear crosstalk
diafonía múltiple multiple crosstalk
diafotía cromática cross-color
diafragma diaphragm
diafragma capacitivo capacitive diaphragm
diafragma de adaptación matching diaphragm
diafragma de altavoz loudspeaker diaphragm
diafragma electrolítico electrolytic diaphragm
diafragma resonante resonant diaphragm
diafragma separador separator diaphragm
diafragma vibrante vibrating diaphragm
diagnóstico asistido por computadora computer-assisted diagnosis
diagnóstico de calidad quality diagnostic
diagnóstico de errores de programa program-error diagnosis
diagnóstico de problema problem diagnosis
diagnóstico ejecutivo executive diagnostic
diagnósticos de compilador compiler diagnostics
diagnósticos de errores error diagnostics
diagnósticos de sistema system diagnostics
diagnósticos en línea on-line diagnostics

diagnotor diagnotor
diagrama diagram, chart, layout, pattern
diagrama cardioide cardioid diagram
diagrama de acción action chart
diagrama de alambrado wiring diagram
diagrama de alambrado pictórico pictorial wiring diagram
diagrama de alineación alignment chart
diagrama de aplicación application diagram
diagrama de Applegate Applegate diagram
diagrama de banda de energía energy-band diagram
diagrama de Bode Bode diagram
diagrama de cableado cabling diagram
diagrama de caja box diagram
diagrama de circuito circuit diagram
diagrama de circuito esquemático schematic circuit diagram
diagrama de cobertura coverage diagram
diagrama de comunicaciones communication chart
diagrama de conexión connecting diagram
diagrama de conexiones connection diagram
diagrama de control control chart
diagrama de cromaticidad chromaticity diagram
diagrama de directividad directivity diagram
diagrama de directividad de antena antenna directivity diagram
diagrama de directividad horizontal horizontal-directivity diagram
diagrama de dispersión dispersion diagram
diagrama de ejecución run chart
diagrama de emisión emission chart
diagrama de enlaces trunking diagram
diagrama de espacio libre free-space diagram
diagrama de estados state diagram
diagrama de estructura structure diagram
diagrama de estructura de datos data-structure diagram
diagrama de fase phase diagram
diagrama de flujo flowchart
diagrama de flujo de datos data flowchart
diagrama de flujo de decisión decision flowchart
diagrama de flujos de señales signal flowchart
diagrama de fuerza force diagram
diagrama de Hull Hull diagram
diagrama de impedancia de carga load-impedance diagram
diagrama de instalación installation diagram
diagrama de intensidad de campo horizontal horizontal field-strength diagram
diagrama de intensidad de campo vertical vertical-field-strength diagram
diagrama de interconexiones interconnection diagram
diagrama de Laue Laue diagram
diagrama de lógica de equipo físico hardware logic diagram
diagrama de memoria memory diagram
diagrama de montaje mounting diagram
diagrama de niveles level diagram
diagrama de niveles de energía energy-level diagram
diagrama de numeración numbering diagram
diagrama de Nyquist Nyquist diagram
diagrama de perfil profile diagram
diagrama de potencial potential diagram
diagrama de Potier Potier's diagram
diagrama de radiación radiation diagram
diagrama de rendimiento performance chart
diagrama de Rieke Rieke diagram
diagrama de Rousseau Rousseau diagram
diagrama de Smith Smith chart
diagrama de tráfico traffic diagram
diagrama de transmisión transmission diagram

diagrama de uniones junction diagram
diagrama de Venn Venn diagram
diagrama direccional directional diagram
diagrama eléctrico electrical diagram
diagrama elemental elementary diagram
diagrama en bloques block diagram
diagrama en burbujas bubble chart
diagrama energético energy diagram
diagrama esquemático schematic diagram
diagrama esquemático de aplicación application schematic diagram
diagrama funcional functional diagram
diagrama ilustrativo illustrative diagram
diagrama lógico logic chart, logic diagram
diagrama lógico automatizado automated logic diagram
diagrama molecular molecular diagram
diagrama pictórico pictorial diagram
diagrama polar polar diagram
diagrama simbólico symbolic diagram
diagrama vectorial vector diagram
diagramático diagrammatic
dial dial
diamagnético diamagnetic
diamagnetismo diamagnetism
diamagnetómetro diamagnetometer
diametral diametral
diámetro catódico cathode diameter
diámetro de haz beam diameter
diámetro de punto spot diameter
diámetro de revestimiento cladding diameter
diámetro de surco groove diameter
diámetro del rotor rotor diameter
diámetro exterior outside diameter
diámetro interior inside diameter
diapasón diapason, tuning fork
diario de computadora computer log
diario de mantenimiento maintenance log
diario de máquina machine log
diatermia diathermy
diatermia de ondas largas long-wave diathermy
diatérmico diathermic
diatómico diatomic
dibit dibit
dibujo de líneas line drawing
dibujo esquemático schematic drawing
dibujo maestro master drawing
diccionario automático automatic dictionary
diccionario automatizado automated dictionary
diccionario de datos data dictionary
diccionario de datos activo active data dictionary
dicotomía dichotomy
dicroismo dichroism
dielectricidad dielectricity
dieléctrico dielectric
dieléctrico artificial artificial dielectric
dieléctrico cerámico ceramic dielectric
dieléctrico coaxial coaxial dielectric
dieléctrico de aceite oil dielectric
dieléctrico de mica plateada silvered-mica dielectric
dieléctrico de película química chemical-film dielectric
dieléctrico de politeno polythene dielectric
dieléctrico dispersivo dispersive dielectric
dieléctrico ferroeléctrico ferroelectric dielectric
dieléctrico gaseoso gaseous dielectric
dieléctrico ideal ideal dielectric
dieléctrico imperfecto imperfect dielectric
dieléctrico isotrópico isotropic dielectric
dieléctrico líquido liquid dielectric
dieléctrico no lineal nonlinear dielectric
dieléctrico perfecto perfect dielectric
dielectrómetro dielectrometer

diente de sierra sawtooth
diferencia angular angular difference
diferencia de fase phase difference
diferencia de fase angular angular phase difference
diferencia de frecuencias frequency difference
diferencia de fuerzas force differential
diferencia de nivel de potencia power-level difference
diferencia de nivel de tensión voltage-level difference
diferencia de potencia power difference
diferencia de potencial potential difference
diferencia de potencial de contacto contact-potential difference
diferencia de potencial eléctrico electric potential difference
diferencia de potencial magnético magnetic potential difference
diferencia de presión pressure difference
diferencia de retorno return difference
diferencia de temperatura media mean temperature difference
diferencia de tiempo time difference
diferencia lógica logic difference
diferenciación parcial partial differentiation
diferenciado differentiated
diferenciador differentiator
diferenciador analógico analog differentiator
diferenciador de resistencia-capacitancia resistance-capacitance differentiator
diferenciador electrónico electronic differentiator
diferencial differential
diferencial de tensión de entrada-salida input-output voltage differential
diferencial mecánico mechanical differential
diferenciar differentiate
difracción diffraction
difracción acústica acoustic diffraction
difracción costera shore effect
difracción de Bragg Bragg diffraction
difracción de cristal crystal diffraction
difracción de esquinas corner diffraction
difracción de Fraunhofer Fraunhofer diffraction
difracción de Fresnel Fresnel diffraction
difracción de luz light diffraction
difracción de luz ultrasónica ultrasonic light diffraction
difracción de rayos X X-ray diffraction
difracción electrónica electron diffraction
difracción esférica spherical diffraction
difracción óptica optical diffraction
difracción ultrasónica ultrasonic diffraction
difractógrafo diffractograph
difractometría de rayos X X-ray diffractometry
difractómetro diffractometer
difractómetro de rayos X X-ray diffractometer
difractor diffractor
difusión diffusion
difusión ambipolar ambipolar diffusion
difusión completa complete diffusion
difusión convectiva convective diffusion
difusión de ángulo angosto narrow-angle diffusion
difusión de ángulo grande wide-angle diffusion
difusión de Compton Compton diffusion
difusión de partículas particle diffusion
difusión de ruido noise diffusion
difusión electroquímica electrochemical diffusion
difusión gaseosa gaseous diffusion
difusión incompleta incomplete diffusion
difusión iónica ionic diffusion
difusión perfecta perfect diffusion
difusión planar planar diffusion
difusión por reflexión diffusion by reflection

difusión por transmisión diffusion by transmission
difusión radial radial diffusion
difusión selectiva selective diffusion
difusión térmica thermal diffusion
difusividad térmica thermal diffusivity
difuso diffuse, scattered
difusor de línea line diffuser
difusor de luz light diffuser
difusor de sonido sound diffuser
difusor neutro neutral diffuser
difusor no selectivo nonselective diffuser
difusor perfecto perfect diffuser
difusor selectivo selective diffuser
difusor uniforme uniform diffuser
digital digital
digitalizar digitize
digitización digitization
dígito digit
dígito aleatorio random digit
dígito binario binary digit
dígito codificado en binario binary-coded digit
dígito codificado en decimal decimal-coded digit
dígito de acarreo carry digit
dígito de arrastre carry digit
dígito de autocomprobación self-checking digit
dígito de código code digit
dígito de comprobación check digit
dígito de desbordamiento overflow digit
dígito de función function digit
dígito de guarda guard digit
dígito de intervalo gap digit
dígito de orden alto high-order digit
dígito de paridad parity digit
dígito de signo sign digit
dígito de zona zone digit
dígito decimal decimal digit
dígito decimal codificado coded-decimal digit
dígito decimal codificado en binario binary-coded decimal digit
dígito fraccional fractional digit
dígito hexadecimal hexadecimal digit
dígito más significativo most significant digit
dígito menos significativo least significant digit
dígito octal octal digit
dígito prohibido forbidden digit
dígito redundante redundant digit
dígito ruidoso noisy digit
dígito sexadecimal sexadecimal digit
dígito significativo significant digit
dígitos binarios equivalentes equivalent binary digits
digitrón digitron
dilatómetro dilatometer
dilución de color color dilution
dilución molecular molecular dilution
dilución normal normal dilution
dimensión absoluta absolute dimension
dimensión activa active dimension
dimensión crítica critical dimension
dimensión crítica de guíaondas waveguide critical dimension
dimensión incremental incremental dimension
dimensión no crítica noncritical dimension
dimensión normal normal dimension
dimensional dimensional
dina dyne
dina-centímetro dyne-centimeter
dinámetro dynameter
dinámica dynamics
dinámica de cristal crystal dynamics
dinámica de plasma plasma dynamics
dinámico dynamic
dínamo dynamo
dínamo de disco disk dynamo

dínamo en derivación shunt dynamo
dínamo en serie series dynamo
dinamoeléctrico dynamoelectric
dinamófono dynamophone
dinamómetro dynamometer
dinamómetro de absorción absorption dynamometer
dinamómetro eléctrico electric dynamometer
dinamotor dynamotor
dinatrón dynatron
dinistor dynistor
dinodo dynode
diodímetro diodemeter
diodo diode
diodo absorbedor absorber diode
diodo adherido con oro gold-bonded diode
diodo aislante isolating diode
diodo amortiguador damper diode
diodo amortiguador de impulsos pulse-damping diode
diodo amplificador amplifier diode
diodo capacitivo capacitive diode
diodo cerámico ceramic diode
diodo coaxial coaxial diode
diodo colector-base collector-base diode
diodo compensador compensating diode
diodo de acoplamiento coupling diode
diodo de aislamiento isolation diode
diodo de aleación alloy diode
diodo de alta velocidad high-speed diode
diodo de amortiguamiento damping diode
diodo de ánodo doble double-anode diode
diodo de arseniuro de galio gallium-arsenide diode
diodo de avalancha avalanche diode
diodo de avalancha controlada controlled-avalanche diode
diodo de barrera barrier diode
diodo de barrera de Schottky Schottky barrier diode
diodo de barrera intrínseca intrinsic-barrier diode
diodo de barrera superficial surface-barrier diode
diodo de capacitancia capacitance diode
diodo de capacitancia variable variable-capacitance diode
diodo de cartucho cartridge diode
diodo de cartucho reversible reversible cartridge diode
diodo de cátodo frío cold-cathode diode
diodo de cesio cesium diode
diodo de compensación offset diode
diodo de computadora computer diode
diodo de conmutación switching diode
diodo de contacto de punta point-contact diode
diodo de cristal crystal diode
diodo de cuatro capas four-layer diode
diodo de disparo trigger diode
diodo de disparo de puerta gate trigger diode
diodo de doble base double-base diode
diodo de doble base de silicio silicon double-base diode
diodo de efecto de campo field-effect diode
diodo de efecto túnel tunnel diode
diodo de estado sólido solid-state diode
diodo de estricción pinchoff diode
diodo de fijación de nivel clamping diode
diodo de fosfuro de galio gallium-phosphide diode
diodo de frecuencia ultraalta ultra-high frequency diode
diodo de gas gas diode
diodo de germanio germanium diode
diodo de Gunn Gunn diode
diodo de infrarrojo infrared diode
diodo de lectura read diode
diodo de medida measuring diode
diodo de óxido de cobre copper-oxide diode

diodo de óxido de níquel nickel-oxide diode
diodo de película metálica metal-film diode
diodo de plasma plasma diode
diodo de portadores activos hot-carrier diode
diodo de potencia power diode
diodo de protección de polaridad polarity-protection diode
diodo de reactancia reactance diode
diodo de recuperación efficiency diode
diodo de recuperación en derivación shunt efficiency diode
diodo de recuperación ultrarrápida ultrafast-recovery diode
diodo de referencia reference diode
diodo de refuerzo booster diode
diodo de resistencia negativa negative-resistance diode
diodo de ruido noise diode
diodo de ruido ideal ideal noise diode
diodo de Schockley Shockley diode
diodo de Schottky Schottky diode
diodo de selenio selenium diode
diodo de señales signal diode
diodo de silicio silicon diode
diodo de tensión de referencia voltage-reference diode
diodo de tres capas three-layer diode
diodo de unión junction diode
diodo de unión de aleación alloy-junction diode
diodo de unión de germanio germanium junction diode
diodo de unión de silicio silicon-junction diode
diodo de unión planar planar-junction diode
diodo de unión PN PN-junction diode
diodo de unión por crecimiento grown-junction diode
diodo de unión por fusión fused-junction diode
diodo de uso general general-purpose diode
diodo de vacío vacuum diode
diodo de vidrio glass diode
diodo de Zener Zener diode
diodo de Zener con compensación de temperatura temperature-compensated Zener diode
diodo detector detector diode
diodo detector de imagen picture-detector diode
diodo dieléctrico dielectric diode
diodo difuso diffused diode
diodo direccional directional diode
diodo dúplex duplex diode
diodo electroluminiscente electroluminescent diode
diodo emisor de luz light-emitting diode
diodo emisor de luz destellador flasher light-emitting diode
diodo entre etapas interstage diode
diodo epitaxial pasivado planar planar epitaxial passivated diode
diodo equivalente equivalent diode
diodo estabilizador de tensión voltage-stabilizing diode
diodo generador de ruido noise-generator diode
diodo-heptodo diode-heptode
diodo inverso backward diode
diodo láser laser diode
diodo limitado por temperatura temperature-limited diode
diodo limitador limiting diode
diodo lógico logic diode
diodo luminiscente de inyección injection luminescent diode
diodo mesa mesa diode
diodo mezclador mixer diode
diodo no conductor nonconducting diode
diodo oscilante oscillating diode

diodo para señales débiles small-signal diode
diodo paramétrico parametric diode
diodo-pentodo diode-pentode
diodo PIN PIN diode
diodo planar planar diode
diodo PNPN PNPN diode
diodo polarizado biased diode
diodo rápido fast diode
diodo rectificador rectifier diode
diodo regulador regulator diode
diodo regulador de tensión voltage-regulator diode
diodo restaurador de corriente continua
    direct-current restorer diode
diodo saturado saturated diode
diodo semiconductor semiconductor diode
diodo sensible a la luz light-sensitive diode
diodo silenciador squelch diode
diodo supresor suppressor diode
diodo supresor de arco arc-suppressor diode
diodo termoiónico thermionic diode
diodo-tetrodo diode-tetrode
diodo-triodo diode-triode
diodo único single diode
diodo varactor diode varactor
diodo varistor diode varistor
diodo X X diode
dioptoscopio dioptoscope
dioptral dioptral
dioptría diopter
dioptrómetro dioptrometer
diotrón diotron
dióxido de carbono agresivo aggressive carbon
    dioxide
dióxido de germanio germanium dioxide
dióxido de plomo lead dioxide
dióxido de selenio selenium dioxide
dióxido de silicio silicon dioxide
dióxido de titanio titanium dioxide
díplex diplex
díplex de cuatro frecuencias four-frequency diplex
diplexión diplexing
diplexor diplexer
diplexor de antena antenna diplexer
diplexor de muesca notch diplexer
diplexor de puente bridge diplexer
diplexor de sonido sound diplexer
dipolar dipolar
dipolarización dipolarization
dipolo dipole
dipolo corto short dipole
dipolo de banda ancha wideband dipole
dipolo de cuarto de onda quarter-wave dipole
dipolo de Hertz Hertz dipole
dipolo de media onda half-wave dipole
dipolo de referencia reference dipole
dipolo doblado folded dipole
dipolo eléctrico electric dipole
dipolo energizado energized dipole
dipolo grueso thick dipole
dipolo horizontal horizontal dipole
dipolo infinitesimal infinitesimal dipole
dipolo magnético magnetic dipole
dipolo multifrecuencia multifrequency dipole
dipolo pasivo passive dipole
dipolo ranurado slot-fed dipole
dipolo recto straight dipole
dipolo resonante resonant dipole
dipolo sintonizado tuned dipole
dipolo vertical vertical dipole
dipolos apilados stacked dipoles
dirección address, direction
dirección A A address
dirección absoluta absolute address

dirección alta high address
dirección aritmética arithmetic address
dirección cablegráfica cable address
dirección calculada calculated address
dirección codificada en binario binary-coded
    address
dirección conductora forward direction
dirección corriente current address
dirección de acumulador accumulator address
dirección de almacenamiento lógico logical storage
    address
dirección de almacenamiento real real storage
    address
dirección de almacenamiento virtual virtual storage
    address
dirección de arranque start address
dirección de base base address
dirección de bloque block address
dirección de byte byte address
dirección de byte relativo relative byte address
dirección de campo field direction
dirección de canal channel address
dirección de carácter character address
dirección de carga load address
dirección de celda cell address
dirección de código code address
dirección de componente component address
dirección de conducción conducting direction
dirección de cuatro más uno four-plus-one address
dirección de datos data address
dirección de derivación branch address
dirección de destino destination address
dirección de dispositivo device address
dirección de dispositivo física physical device
    address
dirección de dispositivo lógico logical device
    address
dirección de dos más uno two-plus-one address
dirección de dos niveles two-level address
dirección de elemento element address
dirección de enlace link address
dirección de entrada entry address
dirección de entrada-salida física physical
    input-output address
dirección de entrada-salida programada
    programmed input-output address
dirección de equipo físico hardware address
dirección de flujo flow direction
dirección de flujo de corriente direction of current
    flow
dirección de grupo group address
dirección de impresión print direction
dirección de inicio starting address
dirección de instrucción instruction address
dirección de línea base baseline direction
dirección de mando command address
dirección de máquina machine address
dirección de memoria memory address
dirección de memoria de entrada-salida
    input-output memory address
dirección de memoria válida valid memory address
dirección de n más uno n-plus-one address
dirección de nivel cero zero-level address
dirección de nivel secundario secondary-level
    address
dirección de nivel único single-level address
dirección de onda wave direction
dirección de pantalla screen address
dirección de polarización direction of polarization
dirección de primer nivel first-level address
dirección de procesador processor address
dirección de propagación direction of propagation
dirección de radiodifusión broadcast address

dirección de reanudación restart address
dirección de red network address
dirección de red principal main network address
dirección de referencia reference address
dirección de registro register address
dirección de reinicio restart address
dirección de retorno return address
dirección de rotación direction of rotation
dirección de segundo nivel second-level address
dirección de subárea subarea address
dirección de tercer nivel third-level address
dirección de transferencia transfer address
dirección de transmisión transmission direction
dirección de tres más uno three-plus-one address
dirección de tres niveles three-level address
dirección de unidad unit address
dirección de uno más uno one-plus-one address
dirección diferida deferred address
dirección dinámica dynamic address
dirección directa direct address
dirección efectiva effective address
dirección específica specific address
dirección explícita explicit address
dirección extendida extended address
dirección ficticia dummy address
dirección flotante floating address
dirección fuente source address
dirección funcional functional address
dirección generada generated address
dirección geográfica geographical address
dirección ilegal illegal address
dirección implícita implied address
dirección indexada indexed address
dirección indirecta indirect address
dirección individual individual address
dirección inmediata immediate address
dirección inversa inverse direction
dirección lógica logical address
dirección maestra master address
dirección mnemónica mnemonic address
dirección modal modal direction
dirección multinivel multilevel address
dirección múltiple multiple address
dirección no válida invalid address
dirección original original address
dirección primaria primary address
dirección Q Q address
dirección real real address
dirección regional regional address
dirección relativa relative address
dirección reubicable relocatable address
dirección secundaria secondary address
dirección simbólica symbolic address
dirección simbólica flotante floating symbolic
 address
dirección sintética synthetic address
dirección supuesta presumptive address
dirección única single address
dirección universal universal address
dirección variable variable address
dirección virtual virtual address
dirección X X address
dirección Y Y address
dirección Z Z address
direccionable addressable
direccionado addressed
direccional directional
direccionamiento addressing
direccionamiento absoluto absolute addressing
direccionamiento de archivo file addressing
direccionamiento de campo fijo fixed-field
 addressing
direccionamiento de dos niveles two-level

addressing
direccionamiento de grupo group addressing
direccionamiento de memoria memory addressing
direccionamiento de nivel cero zero-level addressing
direccionamiento de página page addressing
direccionamiento de registros register addressing
direccionamiento de segmento segment addressing
direccionamiento de segundo nivel second-level
 addressing
direccionamiento de tablas table addressing
direccionamiento de tres niveles three-level
 addressing
direccionamiento diferido deferred addressing
direccionamiento directo direct addressing
direccionamiento específico specific addressing
direccionamiento extendido extended addressing
direccionamiento geográfico geographical
 addressing
direccionamiento implícito implied addressing
direccionamiento indexado indexed addressing
direccionamiento indirecto indirect addressing
direccionamiento inmediato immediate addressing
direccionamiento multinivel multilevel addressing
direccionamiento relativo relative addressing
direccionamiento relativo en tiempo real real-time
 relative addressing
direccionamiento repetitivo repetitive addressing
direccionamiento simbólico symbolic addressing
direccionamiento sintético synthetic addressing
direccionamiento virtual virtual addressing
directividad directivity
directividad de antena antenna directivity
directividad de sonido sound directivity
directividad horizontal horizontal directivity
directividad normalizada normalized directivity
directividad parcial partial directivity
directivo directive
directo online
director de antena antenna director
director de conmutación switching director
director de dipolos doblados folded-dipole director
directorio de archivo file directory
directorio de datos data directory
directorio de segundo nivel second-level directory
directorio de sistema system directory
directorio raíz root directory
directriz directrix
dirigido por computadora computer directed
discado dialing
disco disk, record, dial
disco cerámico ceramic disk
disco compacto compact disk
disco de acetato acetate disk
disco de acetato de celulosa cellulose acetate disk
disco de arranque startup disk
disco de cabeza móvil moving-head disk
disco de cara única single-sided disk
disco de carga loading disk
disco de código de memoria memory code disk
disco de códigos code disk
disco de colores color disk
disco de Corbino Corbino disk
disco de datos data disk
disco de exploración scanning disk
disco de laca lacquer disk
disco de lentes lens disk
disco de memoria memory disk
disco de memoria de acceso aleatorio
 random-access memory disk
disco de Nipkow Nipkow disk
disco de nitrato de celulosa cellulose nitrate disk
disco de polietileno polyethylene disk
disco de programa program disk

disco de prueba test record
disco de Rayleigh Rayleigh disk
disco de registro magnético magnetic recording disk
disco de retén detent disk
disco de ruido blanco white-noise record
disco de sistema system disk
disco de surco fino fine-groove record
disco de trabajo working disk
disco de video video disk
disco duro hard disk
disco duro externo external hard disk
disco duro interno internal hard disk
disco estereofónico stereophonic record,
    stereophonic disk
disco estereofónico de frecuencia portadora
    carrier-frequency stereophonic disk
disco estroboscópico stroboscopic disk
disco fijo fixed disk
disco flexible floppy disk, flexible disk, diskette
disco fonográfico phonograph record, phonograph
    disk
disco fuente source disk
disco grabado recorded disk
disco gramofónico gramophone record
disco instantáneo instantaneous disk
disco laminado laminated disk
disco láser laser disk
disco láser óptico optical laser disk
disco magnético magnetic disk
disco magnético rotativo rotary magnetic disk
disco microsurco microgroove disk
disco minisurco minigroove record
disco monofónico monophonic record
disco monosurco single-groove record
disco multiespiral multispiral disk
disco no removible nonremovable disk
disco objeto target disk
disco óptico optical disk
disco para grabación recording disk
disco para registro recording disk
disco pregrabado prerecorded disk
disco rectificador rectifier disk
disco registrado recorded disk
disco removible removable disk
disco rígido rigid disk
disco rotativo rotating disk
disco sólo de lectura read-only disk
disco superior upper disk
disco telefónico telephone dial
disco virtual virtual disk
discontinuidad discontinuity
discontinuidad discreta discrete discontinuity
discontinuidad eléctrica electric discontinuity
discontinuidad lineal linear discontinuity
discontinuidad magnética magnetic discontinuity
discontinuo discontinuous, noncontinuous
discreto discrete
discriminación discrimination
discriminación de acimut azimuth discrimination
discriminación de antena antenna discrimination
discriminación de colores color discrimination
discriminación de distancia range discrimination
discriminación de filtro filter discrimination
discriminación de frecuencia frequency
    discrimination
discriminación de impulsos pulse discrimination
discriminación de polarización polarization
    discrimination
discriminación de polarización cruzada
    cross-polarization discrimination
discriminación de video video discrimination
discriminación direccional directional discrimination
discriminador discriminator

discriminador de altura de impulsos pulse-height
    discriminator
discriminador de amplitud amplitude discriminator
discriminador de amplitud de impulsos
    pulse-amplitude discriminator
discriminador de carga charge discriminator
discriminador de desfase phase-shift discriminator
discriminador de duración de impulsos
    pulse-duration discriminator
discriminador de fase phase discriminator
discriminador de Foster-Seeley Foster-Seeley
    discriminator
discriminador de frecuencia frequency
    discriminator
discriminador de impulsos pulse discriminator
discriminador de microondas microwave
    discriminator
discriminador de modulación de frecuencia
    frequency-modulation discriminator
discriminador de razón ratio discriminator
discriminador de retardo constante constant-delay
    discriminator
discriminador de sincronismo horizontal
    horizontal-sync discriminator
discriminador de sonido sound discriminator
discriminador de subportadora subcarrier
    discriminator
discriminador de tensión voltage discriminator
discriminador de tiempo time discriminator
discriminador de Travis Travis discriminator
discriminador de video video discriminator
discriminador diferencial differential discriminator
discriminador equilibrado balanced discriminator
discriminador estereofónico stereophonic
    discriminator
discriminador magnético magnetic discriminator
discriminador por conteo de impulsos pulse-count
    discriminator
discriminador transistorizado transistorized
    discriminator
discriminante discriminating
disección de imagen image dissection
disector dissector
disector de imagen image dissector
diseminación activa active dissemination
diseminación de datos data dissemination
diseñado a la medida custom designed
diseño design
diseño arquitectónico architectural design
diseño asistido por computadora computer-aided
    design
diseño certificado certified design
diseño conceptual conceptual design
diseño de abajo-arriba bottom-up design
diseño de archivo file design
diseño de base de datos database design
diseño de carácter character design
diseño de circuitos circuit design
diseño de circuitos impresos printed-circuit design
diseño de computadora computer design
diseño de conmutación switching design
diseño de datos data design
diseño de datos lógico logical data design
diseño de documento document design
diseño de equipo equipment design
diseño de exploración scanning design
diseño de implementación implementation design
diseño de lógica aleatoria random-logic design
diseño de macroinstrucción macroinstruction design
diseño de programa program design
diseño de prueba test design
diseño de sistema system design
diseño de tarjeta card design

diseño defectuoso faulty design
diseño detallado detailed design
diseño distribuido distributed design
diseño empírico empirical design
diseño estructurado structured design
diseño funcional functional design
diseño genérico generic design
diseño lógico logical design
diseño lógico de máquina machine logic design
diseño modular modular design
diseño normal standard design
diseño orientado a objetos object-oriented design
diseño para operación normal aun en el peor caso
    worst-case design
diseño preliminar preliminary design
disimétrico dissymmetrical
disipación dissipation
disipación anódica anode dissipation
disipación anormal abnormal dissipation
disipación de bobina coil dissipation
disipación de colector collector dissipation
disipación de electrodo electrode dissipation
disipación de energía en reposo quiescent power
    dissipation
disipación de pantalla screen dissipation
disipación de placa plate dissipation
disipación de potencia power dissipation
disipación de potencia de puerta gate power
    dissipation
disipación de rejilla grid dissipation
disipación de transistor transistor dissipation
disipación dieléctrica dielectric dissipation
disipación eléctrica electric dissipation
disipación máxima anódica maximum anode
    dissipation
disipación nominal rated dissipation
disipación térmica heat dissipation
disipador dissipator
disipador de potencia power-dissipating
disipador térmico heat dissipator, heat sink
disipativo dissipative
dislocación dislocation
disminución decay
disociación dissociation
disociación electrolítica electrolytic dissociation
disociación molecular molecular disassociation
disonancia dissonance
disparador trigger, tripper
disparador de línea line trigger
disparador de Schmitt Schmitt trigger
disparo triggering, tripping, firing
disparo accidental accidental triggering
disparo anódico anode firing
disparo anormal abnormal triggering
disparo automático automatic triggering
disparo de línea line triggering
disparo eléctrico electrical firing
disparo interdependiente intertripping
disparo interno internal triggering
disparo único single shot
dispersibilidad dispersibility
dispersímetro dispersion meter
dispersiómetro scatterometer
dispersión dispersion, scattering, scatter
dispersión acústica acoustic dispersion
dispersión aleatoria random scattering
dispersión anómala anomalous dispersion
dispersión atmosférica atmospheric scatter
dispersión atómica atomic scattering
dispersión cero zero dispersion
dispersión coherente coherent scattering
dispersión cromática chromatic dispersion
dispersión de Bragg Bragg scattering

dispersión de Compton Compton scattering
dispersión de corriente current spreading
dispersión de Debye-Jauncey Debye-Jauncey
    scattering
dispersión de electrones scattering of electrons
dispersión de frecuencias frequency spread
dispersión de guíaondas waveguide scattering
dispersión de haz beam spread
dispersión de impurezas impurity scattering
dispersión de luz light scattering
dispersión de Mie Mie scattering
dispersión de ondas de radio radio-wave scattering
dispersión de óxido oxide dispersion
dispersión de portadora carrier dispersion
dispersión de potencial potential scattering
dispersión de Raman Raman scattering
dispersión de Rayleigh Rayleigh scattering
dispersión de resonancia resonance scattering
dispersión de retardo delay dispersion
dispersión de sonido sound dispersion
dispersión de Thomson Thomson scattering
dispersión dieléctrica dielectric dispersion
dispersión directa forward scattering
dispersión eléctrica electric dispersion
dispersión en gran ángulo large-angle scattering
dispersión fibroóptica fiber-optic dispersion
dispersión fonónica phonon scattering
dispersión incoherente incoherent scattering
dispersión ionosférica ionospheric scatter
dispersión isotrópica isotropic scattering
dispersión lineal linear dispersion
dispersión lineal recíproca reciprocal linear
    dispersion
dispersión magnética magnetic leakage
dispersión meteórica meteor scatter
dispersión modulada modulated scattering
dispersión múltiple multiple scattering
dispersión negativa negative dispersion
dispersión no lineal nonlinear scattering
dispersión parcial partial dispersion
dispersión positiva positive dispersion
dispersión residual residual dispersion
dispersión rotativa rotary dispersion
dispersión transversal transverse scattering
dispersión troposférica tropospheric scattering
dispersión ultrasónica ultrasonic dispersion
dispersión única single scattering
dispersividad específica specific dispersivity
dispersivo dispersive
disperso scattered
disponibilidad availability
disponibilidad asintótica asymptotic availability
disponibilidad completa full availability
disponibilidad controlada controlled availability
disponibilidad de ciclos cycle availability
disponibilidad de registros de cilindro cylinder
    record availability
disponibilidad intrínseca intrinsic availability
disponibilidad limitada limited availability
disponibilidad operacional operational readiness
disponible available, ready
disposición disposition, layout, pattern, setting
disposición en zigzag stagger
dispositivo device
dispositivo a prueba de agua waterproof device
dispositivo a prueba de explosiones explosionproof
    device
dispositivo accionado por luz light-actuated device
dispositivo accionado por tensión voltage-actuated
    device
dispositivo acelerador o decelerador accelerating or
    decelerating device
dispositivo acoplado por cargas charge-coupled

device
dispositivo activo active device
dispositivo acústico acoustic device
dispositivo acustoóptico acoustooptic device
dispositivo aislado insulated device
dispositivo aislante insulating device
dispositivo alimentador de tarjetas card feed
  attachment
dispositivo amplificador de luz light-amplifying
  device
dispositivo amplificador de potencia
  power-amplifier device
dispositivo analógico analog device
dispositivo anticolisión anticollision device
dispositivo antifluctuación antihunt device
dispositivo antipatinaje antiskating device
dispositivo asimétrico asymmetrical device
dispositivo asincrónico asynchronous device
dispositivo autoalimentado self-powered device
dispositivo autónomo self-contained device
dispositivo autorrectificador self-rectifying device
dispositivo autosincrónico self-synchronous device
dispositivo auxiliar auxiliary device
dispositivo basculador flip-flop device
dispositivo básico basic device
dispositivo biestable bistable device
dispositivo binario binary device
dispositivo bipolar bipolar device
dispositivo comercial commercial device
dispositivo compartido shared device
dispositivo confirmante acknowledging device
dispositivo contra efectos locales antisidetone device
dispositivo contra interferencia intencional
  antijamming device
dispositivo corriente current device
dispositivo cortocircuitador short-circuiting device
dispositivo criogénico cryogenic device
dispositivo de acceso aleatorio random-access
  device
dispositivo de acceso directo direct-access device
dispositivo de accionamiento actuating device
dispositivo de acoplamiento en paralelo paralleling
  device
dispositivo de actuación actuation device
dispositivo de adaptación matching device
dispositivo de adaptación de impedancias
  impedance-matching device
dispositivo de agrupamiento bunching device
dispositivo de aislamiento isolation device
dispositivo de ajuste adjustment device
dispositivo de alarma alarm device
dispositivo de alarma de circuito abierto
  open-circuit alarm device
dispositivo de alarma de circuito cerrado
  closed-circuit alarm device
dispositivo de alivio de presión pressure-relief
  device
dispositivo de almacenamiento storage device
dispositivo de almacenamiento de acceso aleatorio
  random-access storage device
dispositivo de almacenamiento de acceso directo
  direct-access storage device
dispositivo de almacenamiento de datos data
  storage device
dispositivo de almacenamiento de energía
  energy-storage device
dispositivo de almacenamiento de imagen
  image-storage device
dispositivo de almacenamiento de información
  information-storage device
**dispositivo de almacenamiento de tambor
  magnético** magnetic-drum storage device
dispositivo de almacenamiento magnético magnetic

storage device
dispositivo de almacenamiento masivo bulk storage
  device
dispositivo de almacenamiento secuencial
  sequential-storage device
dispositivo de almacenamiento secundario
  secondary storage device
dispositivo de amortiguamiento externo external
  damping device
dispositivo de amplificación de tensión
  voltage-amplification device
dispositivo de antena aerial device
dispositivo de apilamiento desplazado
  offset-stacking device
dispositivo de arranque-parada start-stop device
dispositivo de autocontrol autocontrol device
dispositivo de avalancha avalanche device
dispositivo de avalancha controlada
  controlled-avalanche device
dispositivo de avance lento inching device
dispositivo de aviso warning device
dispositivo de aviso de colisiones collision warning
  device
dispositivo de banda ancha broadband device
dispositivo de bloqueo blocking device, locking
  device
dispositivo de bloqueo de aproximación approach
  locking device
dispositivo de borrar clearing device
dispositivo de búsqueda de archivos file search
  device
dispositivo de campos cruzados crossed-field device
dispositivo de caracteres character device
dispositivo de cebado striking device, starting device
dispositivo de cierre lento slow-closure device
dispositivo de cinco capas five-layer device
dispositivo de clasificación de grupo group sorting
  device
dispositivo de codificación de datos data encoding
  device
dispositivo de compensación padding device
dispositivo de comprobación checking device
dispositivo de comunicaciones communication
  device
dispositivo de conexión connecting device
dispositivo de conexión rápida quick-connecting
  device
dispositivo de conexión y desconexión on-off device
dispositivo de configuración rápida
  quick-configuration device
dispositivo de conmutación switching device
dispositivo de conmutación mecánica mechanical
  switching device
dispositivo de conmutación no mecánica
  nonmechanical switching device
dispositivo de contestación answering device
dispositivo de control control device
dispositivo de control de arco arc control device
dispositivo de control de comunicaciones
  communications control device
dispositivo de control de impresión print-control
  device
dispositivo de control de módem modem control
  device
dispositivo de control de temperatura
  temperature-control device
dispositivo de control permisivo permissive control
  device
dispositivo de conversión conversion device
dispositivo de conversión de energía energy
  conversion device
dispositivo de copia impresa hard-copy device
dispositivo de corrección de video video correction

device

dispositivo de corrientes parásitas eddy-current device

dispositivo de cuatro terminales four-terminal device

dispositivo de derivación bypass device

dispositivo de desacentuación deemphasis device

dispositivo de descarga discharge device

dispositivo de detección de marcas mark-sensing device

dispositivo de direcciones address device

dispositivo de disco disk device

dispositivo de disparo tripping device

dispositivo de dos estados two-state device

dispositivo de efecto de campo field-effect device

dispositivo de efecto de Hall Hall-effect device

dispositivo de enclavamiento interlocking device

dispositivo de enfoque focusing device

dispositivo de entrada input device

dispositivo de entrada de datos data input device

dispositivo de entrada de sistema system input device

dispositivo de entrada de trabajos job-input device

dispositivo de entrada gráfica graphic input device

dispositivo de entrada lógica logical input device

dispositivo de entrada lógica de sistema system logical input device

dispositivo de entrada-salida input-output device

dispositivo de entrada-salida de impresora printer input-output device

dispositivo de entrada-salida programable programmable input-output device

dispositivo de equipo físico hardware device

dispositivo de estado sólido solid-state device

dispositivo de exploración de campo field-scanning device

dispositivo de ferrita ferrite device

dispositivo de granate de itrio e hierro yttrium-iron-garnet device

dispositivo de imagen persistente persistent-image device

dispositivo de inductancia inductance device

dispositivo de interconexión interconnection device

dispositivo de interferencia intencional jamming device

dispositivo de interrupción interrupt device

dispositivo de inversión de tarjetas card reversing device

dispositivo de lectura readout device

dispositivo de lectura de lámpara de neón neon-lamp readout

dispositivo de línea line device

dispositivo de lógica aleatoria random-logic device

dispositivo de llamadas calling device

dispositivo de manejo handling device

dispositivo de manipulación keying device

dispositivo de mantenimiento de arco arc maintenance device

dispositivo de media altura half-height device

dispositivo de medida measuring device

dispositivo de medida de frecuencia frequency-measuring device

dispositivo de medida de potencia power measurement device

dispositivo de memoria memory device

dispositivo de memoria acoplado por cargas charge-coupled device memory

dispositivo de microcodificación microcoding device

dispositivo de monitoreo monitoring device

dispositivo de muestreo y retención sample-and-hold device

dispositivo de multiplicación multiplication device

dispositivo de numeración numbering device

dispositivo de observación observation device

dispositivo de operación operating device

dispositivo de paginación paging device

dispositivo de perforación perforation device

dispositivo de posicionamiento positioning device

dispositivo de potencia power device

dispositivo de precisión precision device

dispositivo de preparación de datos data preparation device

dispositivo de programa program device

dispositivo de programa alternativo alternate-program device

dispositivo de protección protection device

dispositivo de protección contra sobrecarga overload protective device

dispositivo de protección contra sobretensión overvoltage protection device

dispositivo de protección de archivos file protection device

dispositivo de protección de memoria memory protection device

dispositivo de puesta a cero zero-setting device

dispositivo de puesta a tierra grounding device

dispositivo de recepción díplex diplex receiving device

dispositivo de reconexión reclosing device

dispositivo de reconocimiento recognition device

dispositivo de reconocimiento de caracteres character recognition device

dispositivo de reconocimiento de voz voice-recognition device

dispositivo de reconocimiento óptico optical recognition device

dispositivo de red array device

dispositivo de registro de datos data recording device

dispositivo de reposición resetting device

dispositivo de resistencia negativa negative-resistance device

dispositivo de restablecimiento a cero zero-reset device

dispositivo de retardo delay device

dispositivo de retardo de dígito digit delay device

dispositivo de rotación de Faraday Faraday rotation device

dispositivo de salida output device

dispositivo de salida de sistema system output device

dispositivo de salida de trabajos job-output device

dispositivo de salida físico physical output device

dispositivo de salida gráfica graphic output device

dispositivo de salida lógica logical output device

dispositivo de salida lógica de sistema system logical output device

dispositivo de seguridad safety device

dispositivo de seis polos six-pole device

dispositivo de selección selecting device

dispositivo de señal audible audible signal device

dispositivo de señalización signaling device

dispositivo de servicio de interrupción interrupt service device

dispositivo de simetría complementaria complementary-symmetry device

dispositivo de sincronización timing device

dispositivo de sobrecorriente overcurrent device

dispositivo de superficie de mesa table-top device

dispositivo de teclado keyboard device

dispositivo de telemedida telemetering device

dispositivo de temporización timing device

dispositivo de transferencia de carga charge-transfer device

dispositivo de transferencia de frecuencia frequency transfer device

dispositivo de unión de ferrita ferrite junction device
dispositivo de valor absoluto absolute value device
dispositivo de velocidad variable variable-speed device
dispositivo de visualización display device
dispositivo de visualización alfanumérica alphanumeric display device
dispositivo de visualización con refresco refresh display device
dispositivo de visualización de caracteres character display device
dispositivo de visualización de exploración raster display device
dispositivo de visualización de exploración aleatoria random-scan display device
dispositivo de visualización electroluminiscente electroluminescent display device
dispositivo de visualización gráfica graphical display device
dispositivo de visualización monocroma monochrome display device
dispositivo de visualización panorámica panoramic display device
dispositivo dedicado dedicated device
dispositivo desconectable disconnectable device
dispositivo desconectador disconnecting device
dispositivo desfasador phase-shifting device
dispositivo detector de rayos X X-ray detecting device
dispositivo digital digital device
dispositivo dinámico dynamic device
dispositivo discreto discrete device
dispositivo divisor de fase phase-splitting device
dispositivo electroacústico electroacoustic device
dispositivo electroluminiscente coherente coherent electroluminescent device
dispositivo electromagnético electromagnetic device
dispositivo electromecánico electromechanical device
dispositivo electrónico electronic device
dispositivo electrónico fotosensible photosensitive electronic device
dispositivo electrotérmico electrothermic device
dispositivo emisor de tonos breves pager
dispositivo en línea on-line device
dispositivo enchufable plug-in device
dispositivo epitaxial epitaxial device
dispositivo especial special device
dispositivo estable stable device
dispositivo estático static device
dispositivo excitado separadamente separately excited device
dispositivo ficticio dummy device
dispositivo fotoeléctrico photoelectric device
dispositivo fotoelectrónico photoelectronic device
dispositivo fotosensible photosensitive device
dispositivo fuera de línea off-line device
dispositivo funcional functional device
dispositivo guiador homing device
dispositivo impresor printing device
dispositivo inalámbrico wireless device
dispositivo indicador indicating device
dispositivo indicador de carga charge indicating device
dispositivo indicador de llamadas call-indicating device
dispositivo libre clear device
dispositivo limitador limiting device
dispositivo limitador de corriente current-limiting device
dispositivo limitador de corriente de descarga discharge current limiting device

dispositivo limitador de corriente resistivo resistive current-limiting device
dispositivo limitador de picos peak-limiting device
dispositivo limitador de tensión voltage-limiting device
dispositivo limitador de velocidad speed-limiting device
dispositivo limpiador de cabeza head-cleaning device
dispositivo localizador locator device
dispositivo lógico logic device
dispositivo lógico programable programmable logic device
dispositivo magnético magnetic device
dispositivo mezclador mixing device
dispositivo microeléctrico microelectric device
dispositivo microelectrónico microelectronic device
dispositivo molecular molecular device
dispositivo monoestable monostable device
dispositivo multiabertura multiaperture device
dispositivo multibanda multiband device
dispositivo multietapa multistage device
dispositivo múltiplex multiplex device
dispositivo multiunión multijunction device
dispositivo no recíproco nonreciprocal device
dispositivo NPNP NPNP device
dispositivo opcional optional device
dispositivo operado a mano hand-operated device
dispositivo operado por voz voice-operated device
dispositivo optoelectrónico optoelectronic device
dispositivo paramétrico parametric device
dispositivo paramétrico no inversor noninverting parametric device
dispositivo pasivo passive device
dispositivo periférico peripheral device
dispositivo piezoeléctrico piezoelectric device
dispositivo PNPN PNPN device
dispositivo por difusión diffused device
dispositivo primario primary device
dispositivo protector protective device
dispositivo protector automático automatic protective device
dispositivo protector contra rayos lightning protective device
dispositivo protector contra sobrecorriente overcurrent protective device
dispositivo protector de fase phase-protective device
dispositivo recíproco reciprocal device
dispositivo regulador regulating device
dispositivo remoto remote device
dispositivo secuencial sequential device
dispositivo secundario secondary device
dispositivo semiconductor semiconductor device
dispositivo semiconductor de unidades múltiples multiple-unit semiconductor device
dispositivo sensible sensitive device
dispositivo sensor sensing device
dispositivo silenciador muting device
dispositivo sincrónico synchronous device
dispositivo sintonizador discriminador discriminator tuning device
dispositivo subminiatura subminiature device
dispositivo superconductor superconducting device
dispositivo térmico de aparato apparatus thermal device
dispositivo terminal end device
dispositivo termoeléctrico thermoelectric device
dispositivo transmisor transmitting device
dispositivo unilateral unilateral device
dispositivo unipolar unipolar device
dispositivo unitario unit device
dispositivo universal universal device
dispositivos agrupados clustered devices

dispositivos en oposición back-to-back devices
disprosio dysprosium
disquete diskette
disquete de alta densidad high-density diskette
disquete de distribución distribution diskette
disquete de doble densidad dual density diskette
disquete de dos lados double-sided diskette
disquete de instalación installation diskette
disquete de respaldo backup diskette
disquete en blanco blank diskette
disquete formateado formatted diskette
disquete fuente source diskette
disquete limpiador de cabeza head-cleaning diskette
disquete no formateado unformatted diskette
disquete objeto target diskette
disquete operativo operating diskette
disrupción disruption, breakdown
disrupción primaria primary breakdown
disruptivo disruptive
distancia range
distancia angular angular distance
distancia de coherencia coherence distance
distancia de coordinación coordination distance
distancia de correlación correlation distance
distancia de difusión diffusion distance
distancia de fuga leakage distance
distancia de Hamming Hamming distance
distancia de línea visual line-of-sight distance
distancia de protección protection distance
distancia de reacción reaction distance
distancia de recorrido de contacto contact travel
distancia de referencia reference distance
distancia de salto skip distance
distancia de separación separation distance
distancia disruptiva spark gap, sparking distance
distancia eléctrica electrical distance
distancia entre contactos abiertos clearance between open contacts
distancia entre puntas needle gap
distancia entre señales signal distance
distancia esférica spherical distance
distancia focal focal length
distancia hiperfocal hyperfocal distance
distancia modal modal distance
distancia radioóptica radio-optical distance
distancia unitaria unit distance
distintivo de llamada internacional international call sign
distorsiómetro distortion meter
distorsiómetro de puente de Wien Wien-bridge distortion meter
distorsión distortion
distorsión acústica acoustic distortion
distorsión amplitud-amplitud amplitude-amplitude distortion
distorsión amplitud-frecuencia amplitude-frequency distortion
distorsión armónica harmonic distortion
distorsión armónica de audiofrecuencia audio-frequency harmonic distortion
distorsión armónica total total harmonic distortion
distorsión asimétrica asymmetrical distortion
distorsión característica characteristic distortion
distorsión característica de línea line characteristic distortion
distorsión cíclica cyclic distortion
distorsión colorimétrica colorimetric distortion
distorsión cromática chromatic distortion
distorsión de abertura aperture distortion
distorsión de alimentador feeder distortion
distorsión de amplificador amplifier distortion
distorsión de amplitud amplitude distortion
distorsión de amplitud de forma de onda
　waveform-amplitude distortion
distorsión de anchura de impulso pulse-width distortion
distorsión de atenuación attenuation distortion
distorsión de audio audio distortion
distorsión de audiofrecuencia audio-frequency distortion
distorsión de banda lateral única single-sideband distortion
distorsión de brillo brightness distortion
distorsión de campo field distortion
distorsión de carácter character distortion
distorsión de carga espacial space-charge distortion
distorsión de circuito de salida output-circuit distortion
distorsión de cruce crossover distortion
distorsión de cuadro frame distortion
distorsión de cuantificación quantization distortion
distorsión de desviación deviation distortion
distorsión de envolvente envelope distortion
distorsión de exploración scanning distortion
distorsión de fase phase distortion
distorsión de forma form distortion
distorsión de forma de onda waveform distortion
distorsión de frecuencia frequency distortion
distorsión de frecuencia contra amplitud
　amplitude-versus-frequency distortion
distorsión de frecuencia de atenuación
　attenuation-frequency distortion
distorsión de frecuencia de líneas line-frequency distortion
distorsión de impulso pulse distortion
distorsión de inclinación tilt
distorsión de intermodulación intermodulation distortion
distorsión de intermodulación de radiofrecuencia
　radio-frequency intermodulation distortion
distorsión de línea line distortion
distorsión de linealidad linearity distortion
distorsión de modulación modulation distortion
distorsión de modulación de amplitud
　amplitude-modulation distortion
distorsión de onda wave distortion
distorsión de ondas estacionarias standing-wave distortion
distorsión de origen origin distortion
distorsión de polarización bias distortion
distorsión de punto spot distortion
distorsión de restitución restitution distortion
distorsión de retardo delay distortion
distorsión de retardo de grupo group-delay distortion
distorsión de segunda armónica second-harmonic distortion
distorsión de segundo orden second-order distortion
distorsión de señal signal distortion
distorsión de señal permisible permissible signal distortion
distorsión de señal telegráfica telegraph signal distortion
distorsión de servicio service distortion
distorsión de sobrecarga overload distortion
distorsión de sobreelongación overshoot distortion
distorsión de sonido sound distortion
distorsión de tercera armónica third-harmonic distortion
distorsión de tiempo time distortion
distorsión de transmisión transmission distortion
distorsión de transmisor transmitter distortion
distorsión diferencial differential distortion
distorsión en acerico pincushion distortion
distorsión en baja frecuencia low-frequency distortion

distorsión en barril barrel distortion
distorsión espiral spiral distortion
distorsión fase-amplitud phase-amplitude distortion
distorsión fibroóptica fiber-optic distortion
distorsión fortuita fortuitous distortion
distorsión geométrica geometric distortion
distorsión individual individual distortion
distorsión inharmónica inharmonic distortion
distorsión inherente inherent distortion
distorsión intramodal intramodal distortion
distorsión lineal linear distortion
distorsión máxima maximum distortion, peak distortion
distorsión multimodo multimode distortion
distorsión negativa negative distortion
distorsión no lineal nonlinear distortion
distorsión permisible allowable distortion
distorsión por armónicas impares odd-harmonic distortion
distorsión por desincronización jag
distorsión por histéresis hysteresis distortion
distorsión por imágenes superpuestas foldover
distorsión por movimientos de equipo judder
distorsión por retardo de envolvente envelope-delay distortion
distorsión por retardo de fase phase-delay distortion
distorsión por sobrerrecorrido overthrow distortion
distorsión por trayectorias múltiples multipath distortion
distorsión positiva positive distortion
distorsión sintética synthetic distortion
distorsión telegráfica telegraph distortion
distorsión telegráfica de polarización bias telegraph distortion
distorsión telegráfica total total telegraph distortion
distorsión total total distortion
distorsión transitoria transient distortion
distorsión trapecial trapezoidal distortion, keystone distortion
distorsión ultrabaja ultralow distortion
distorsionado distorted
distribución distribution, layout
distribución aleatoria random distribution
distribución angular angular distribution
distribución armónica harmonic distribution
distribución asimétrica asymmetrical distribution
distribución automática automatic distribution
distribución de amplitud amplitude distribution
distribución de amplitud de ruido noise amplitude distribution
distribución de buen clima fair weather distribution
distribución de campo field distribution
distribución de canales channel distribution
distribución de carga load distribution
distribución de carga espacial space-charge distribution
distribución de componentes component layout
distribución de corriente current distribution
distribución de corriente alterna alternating-current distribution
distribución de corriente continua direct-current distribution
distribución de densidad de amplitud amplitude density distribution
distribución de disco magnético magnetic-disk layout
distribución de energía energy distribution
distribución de energía espectral spectral energy distribution
distribución de energía espectral relativa relative spectral energy distribution
distribución de fallas failure distribution
distribución de fase phase distribution

distribución de flujo flux distribution
distribución de frecuencias frequency distribution
distribución de Gauss Gaussian distribution
distribución de impulsos pulse distribution
distribución de instrumentos instrument layout
distribución de intensidad intensity distribution
distribución de intensidad de campo field-strength distribution
distribución de intensidad simétrica symmetrical intensity distribution
distribución de memoria memory layout
distribución de montaje mounting layout
distribución de panel panel layout
distribución de potencia power distribution
distribución de potencial potential distribution
distribución de presión pressure distribution
distribución de Rayleigh Rayleigh distribution
distribución de señales signal distributing
distribución de sistema system layout
distribución de Taylor Taylor distribution
distribución de teclado keyboard layout
distribución de tensión voltage distribution
distribución de tiempo time distribution
distribución de velocidades velocity distribution
distribución en paralelo parallel distribution
distribución en todo clima all-weather distribution
distribución espacial spatial distribution, space distribution
distribución espectral spectral distribution
distribución gamma gamma distribution
distribución normal normal distribution
distribución progresiva tapered distribution
distribución secundaria secondary distribution
distribución sinusoidal sinusoidal distribution
distribuido distributed
distribuido aleatoriamente randomly distributed
distribuidor distributor
distribuidor de impulsos pulse distributor
distribuidor de llamada call distributor
distribuidor de llamadas automático automatic call distributor
distribuidor de potencia power distributor
distribuidor de repetidores repeater distribution frame
distribuidor de tráfico traffic distributor
distribuidor electrónico electronic distributor
distribuidor receptor receiving distributor
distribuidor telegráfico telegraph distributor
disyunción disjunction
disyuntor circuit-breaker, cutout
divergencia divergence
divergencia de frecuencia frequency divergence
divergencia de haz beam divergence
divergencia de salida fanout
divergente divergent
diversidad diversity
diversidad cuádruple quadruple diversity
diversidad de cargas load diversity
diversidad de frecuencia frequency diversity
diversidad de polarización polarization diversity
diversidad espacial space diversity
diversidad estacional seasonal diversity
división splitting
división binaria binary division
división capacitiva capacitive division
división de banda band splitting
división de carga load division
división de datos data division
división de ecos echo splitting
división de escala scale division
división de espacio space division
división de fase phase division
división de haz beam splitting

división de identificación identification division
división de intervalo de control control interval split
división de línea line splitting
división de regiones region partitioning
división de tensión voltage division
división de trabajo job dividing
división en paralelo parallel partitioning
división fija fixed partition
divisor divider, splitter
divisor analógico analog divider
divisor binario binary divider
divisor capacitivo capacitive divider
divisor de banda band splitter
divisor de colores color splitter
divisor de corriente current divider
divisor de décadas decade divider
divisor de drenaje bleeder divider
divisor de fase phase splitter
divisor de frecuencia frequency divider
divisor de frecuencia de impulsos pulse-frequency
    divider
divisor de haz beam splitter
divisor de línea line splitter
divisor de módem modem splitter
divisor de potencia power divider
divisor de potencial potential divider
divisor de resistencia-capacitancia
    resistance-capacitance divider
divisor de señal signal splitter
divisor de tensión voltage divider
divisor de tensión ajustable adjustable voltage
    divider
divisor de tensión capacitivo capacitive voltage
    divider
divisor de tensión de medida measurement voltage
    divider
divisor digital digital divider
divisor electrónico electronic divider
divisor por capacitancia capacitance divider
divisor potenciométrico potentiometer divider
divisor regenerativo regenerative divider
divisor resistivo resistive divider
doblador doubler
doblador de frecuencia frequency doubler
doblador de media onda half-wave doubler
doblador de onda completa full-wave doubler
doblador de tensión voltage doubler
doblador de tensión de media onda half-wave
    voltage doubler
doblador de tensión de onda completa full-wave
    voltage doubler
doblador de tensión en cascada cascade voltage
    doubler
doblamiento doubling
doblamiento de frecuencia frequency doubling
doble aislamiento double insulation
doble alimentación double feed
doble almacenamiento dual storage
doble armadura double armature
doble arrollamiento double winding
doble banda lateral double sideband
doble blindaje double shield
doble bus dual bus
doble capa double layer
doble capa eléctrica electric double layer
doble capacitor dual capacitor
doble codificación dual coding
doble conexión double connection
doble contacto twin contact
doble control dual control
doble controlador dual controller
doble conversación dual conversation
doble conversión double conversion

doble densidad double density
doble diodo double diode, dual diode, twin diode
doble diodo-doble triodo dual diode-dual triode
doble diversidad dual diversity
doble error double error
doble filamento double filament
doble fonocaptor dual pickup
doble imagen double image
doble impulso double pulse
doble intensidad dual intensity
doble limitador double limiter
doble longitud double length
doble modulación double modulation
doble pantalla double screen
doble pentodo dual pentode
doble perforación double punch
doble potenciómetro dual potentiometer
doble precisión double-precision
doble procesador dual processor
doble procesamiento double processing
doble puerta dual gate
doble reacción double reaction
doble red dual network
doble regeneración double regeneration
doble registro double register, dual recording
doble resistor dual resistor
doble respuesta double response
doble revestimiento double sheath
doble señal double signal
doble trazo dual trace
doble triodo double triode
doble velocidad dual speed
doblete doublet
doblete doblado folded doublet
doblete eléctrico electrical doublet
doblete horizontal horizontal doublet
doblete magnético magnetic doublet
doblete multihilo multiwire doublet
doblete radiante radiating doublet
dobletes apareados matched doublets
docilidad compliance
docilidad acústica acoustic compliance
docilidad lateral lateral compliance
docilidad mecánica mechanical compliance
docilidad negativa negative compliance
docilidad rectilínea rectilineal compliance
docilidad vertical vertical compliance
documentación documentation
documentación gráfica graphic documentation
documento de entrada input document
documento fuente source document
documento personal personal document
dominio domain
dominio de datos data domain
dominio eléctrico electric domain
dominio externo external domain
dominio ferroeléctrico ferroelectric domain
dominio ferromagnético ferromagnetic domain
dominio magnético magnetic domain
donador donor
dopado doping
dopado con oro gold doping
dopado de circuito integrado integrated-circuit
    doping
dopado de semiconductor semiconductor doping
dopante dopant
dopar dope
dorado gold plating
doran doran
dorso de tarjeta card back
dosímetro dosage meter
dosímetro de estado sólido solid-state dosimeter
dosímetro de película film dosimeter

dosímetro de vidrio glass dosimeter
dosímetro integrador integrating dosimeter
dosis absorbida absorbed dose
dosis acumulativa cumulative dose
dosis de radiación de equivalente en aire air equivalent radiation dose
drenaje eléctrico electrical drainage
drenaje eléctrico controlado controlled electric drainage
drenaje eléctrico directo direct electric drainage
drenaje eléctrico forzado forced electric drainage
drenaje eléctrico polarizado polarized electric drainage
drenaje forzado forced drainage
drenaje momentáneo momentary drainage
dualidad duality
ductilidad ductility
ductilímetro ductilimeter
duodecimal duodecimal
duodiodo duodiode
duodiodo-pentodo duodiode-pentode
duodiodo-tetrodo duodiode-tetrode
duodiodo-triodo duodiode-triode
duoplasmatrón duoplasmatron
duopolo duopole
duosexadecimal duosexadecimal
duotricenario duotricenary
duotriodo duotriode
dúplex duplex
dúplex completo full duplex
dúplex de doble canal double-channel duplex
dúplex de dos frecuencias two-frequency duplex
dúplex de puente bridge duplex
dúplex diferencial differential duplex
dúplex por frecuencia única single-frequency duplex
duplexión duplexing
duplexor duplexer
duplexor de antena antenna duplexer
duplexor de polarización circular circular-polarization duplexer
duplexor equilibrado balanced duplexer
duplicación automática automatic duplication
duplicación de cinta tape duplication
duplicación de disco disk duplication
duplicación de impulsos pulse duplication
duplicador duplicator
duración de barrido sweep duration
duración de centelleo scintillation duration
duración de compilación compile duration
duración de impulso pulse duration, pulse length
duración de impulso de salida output-pulse duration
duración de incremento repentino burst duration
duración de interrupción interruption duration, outage duration
duración de interrupción forzada forced outage duration
duración de llamada call duration
duración de media amplitud half-amplitude duration
duración de operación operation duration
duración de primera transición first transition duration
duración de prueba test duration
duración de respuesta response duration
duración de señal signal duration
duración de tiempo time duration
duración de tono tone duration
duración de transición transition duration
duración máxima maximum duration
duración virtual virtual duration
dural dural
duraluminio duralumin
dureza hardness
dureza de rayos X X-ray hardness

dureza magnética magnetic hardness
durómetro durometer, hardness tester

E

ebonita ebonite
ebullición boiling
eco echo
eco adelantado leading echo
eco artificial artificial echo
eco brillante bright echo
eco coherente coherent echo
eco de corriente current echo
eco de espín spin echo
eco de filtro filter echo
eco de larga duración long-duration echo
eco de radar radar echo
eco de reflexiones múltiples multiple-reflection echo
eco eléctrico electric echo
eco falso bogey
eco fijo fixed echo
eco lateral side echo
eco local local echo
eco meteórico meteor echo
eco múltiple multiple echo
eco musical musical echo
eco no identificado bogey
eco parásito parasitic echo, angel echo
eco permanente permanent echo
eco por lóbulos laterales side-lobe echo
eco por reflexión de espejo mirror-reflection echo
eco próximo near echo
eco reflejado back echo
eco terrestre ground return
ecoencefalografía echoencephalography
ecoencefalógrafo echoencephalograph
ecógrafo echograph
ecograma echogram
ecometría echometry
ecómetro echometer
ecómetro de impulsos pulse echometer
economía de combustible fuel economy
ecos de terreno terrain echoes
ecos parásitos clutter, hash
ecos parásitos de mar sea clutter
ecos parásitos de olas wave clutter
ecos parásitos de radar radar clutter
ecos parásitos de tierra ground clutter
ecos parásitos por lluvia rain clutter
ecos reflejos de mar sea return
ecoscopio echoscope
ecosonador echo sounder
ecosondeo echo sounding
ecuación característica characteristic equation
ecuación cuadrática quadratic equation
ecuación cuártica quartic equation
ecuación de Bjerknes Bjerknes equation
ecuación de Boole Boolean equation
ecuación de Bragg Bragg's equation
ecuación de campo field equation
ecuación de Child-Langmuir Child-Langmuir equation
ecuación de Clausius-Mosotti Clausius-Mosotti equation
ecuación de Clausius-Mosotti-Lorentz-Lorenz Clausius-Mosotti-Lorentz-Lorenz equation

ecuación de colores color equation
ecuación de Conwell-Weisskopf Conwell-Weisskopf equation
ecuación de de Broglie de Broglie's equation
ecuación de Debye Debye equation
ecuación de diodo diode equation
ecuación de equilibrio balance equation
ecuación de fuerza de Lorentz Lorentz force equation
ecuación de ganancia gain equation
ecuación de Kirchhoff Kirchhoff's equation
ecuación de Langmuir Langmuir's equation
ecuación de Laplace Laplace equation
ecuación de London London equation
ecuación de máquina machine equation
ecuación de Maxwell Maxwell's equation
ecuación de onda wave equation
ecuación de Planck-Einstein Planck-Einstein equation
ecuación de Poisson Poisson's equation
ecuación de radar radar equation
ecuación de radiofaro beacon equation
ecuación de Rayleigh-Jeans Rayleigh-Jeans equation
ecuación de Richardson Richardson's equation
ecuación de Richardson-Dushman Richardson-Dushman equation
ecuación de segundo grado second-degree equation
ecuación de Sommerfeld Sommerfeld equation
ecuación de superconductividad de London London superconductivity equation
ecuación diferencial differential equation
ecuación diferencial ordinaria ordinary differential equation
ecuación diferencial parcial partial differential equation
ecuación exponencial exponential equation
ecuación fotoeléctrica de Einstein Einstein photoelectric equation
ecuación lineal linear equation
ecuación logarítmica logarithmic equation
ecuación matricial matrix equation
ecuación simple simple equation
ecuaciones de Bloch Bloch equations
ecuaciones de Heaviside-Hertz Heaviside-Hertz equations
ecuaciones de Laue Laue equations
ecuaciones de mallas mesh equations
ecuaciones equivalentes equivalent equations
ecuaciones simultáneas simultaneous equations
ecuador magnético magnetic equator
ecualización equalization
ecualización de línea line equalization
ecualización de pendiente slope equalization
ecualización de pendiente lineal linear-slope equalization
ecualización de respuesta de frecuencia frequency-response equalization
ecualización telefónica telephone equalization
ecualizador equalizer
ecualizador de atenuación attenuation equalizer
ecualizador de cable cable equalizer
ecualizador de fase phase equalizer
ecualizador de ganancia gain equalizer
ecualizador de impulsos pulse equalizer
ecualizador de línea line equalizer
ecualizador de pendiente slope equalizer
ecualizador de potencia power equalizer
ecualizador de presión pressure equalizer
ecualizador de retardo delay equalizer
ecualizador de retardo de grupo group-delay equalizer
ecualizador en paralelo parallel equalizer
ecualizador en serie series equalizer

ecualizador fonográfico phonograph equalizer
ecualizador parabólico fijo fixed parabolic equalizer
edición editing
edificio apantallado shielded building
edificio auxiliar auxiliary building
editar edit
efapsis ephapse
efectividad de blindaje shield effectiveness
efectividad de filtro filter effectiveness
efectividad de interferencia intencional jamming effectiveness
efectividad relativa relative effectiveness
efecto actinoeléctrico actinoelectric effect
efecto acustoeléctrico acoustoelectric effect
efecto acustoóptico acoustooptic effect
efecto anódico anode effect
efecto anular annular effect
efecto asimétrico asymmetrical effect
efecto autoelectrónico autoelectronic effect
efecto binaural binaural effect
efecto cuerpo-antena body-antenna effect
efecto de almacenamiento storage effect
efecto de altura height effect
efecto de antena antenna effect
efecto de apantallamiento shielding effect
efecto de arrastre pulling effect
efecto de Auger Auger effect
efecto de avalancha avalanche effect
efecto de avión airplane effect
efecto de Barkhausen Barkhausen effect
efecto de Barnett Barnett effect
efecto de barómetro barometer effect
efecto de Becquerel Becquerel effect
efecto de Benedick Benedick effect
efecto de borde edge effect
efecto de Bridgman Bridgman effect
efecto de calentamiento heating effect
efecto de Calzecchi-Onesti Calzecchi-Onesti effect
efecto de campo field effect
efecto de canal channel effect
efecto de captura capture effect
efecto de carga de rejilla grid loading effect
efecto de carga espacial space-charge effect
efecto de carga superficial surface-charge effect
efecto de coincidencia coincidence effect
efecto de Compton Compton effect
efecto de conducto duct effect
efecto de conmutación switching effect
efecto de constricción pinch effect
efecto de contacto contact effect
efecto de Corbino Corbino effect
efecto de corona corona effect
efecto de corto tiempo short-time effect
efecto de Cotton-Mouton Cotton-Mouton effect
efecto de Damon Damon effect
efecto de Debye Debye effect
efecto de Dellinger Dellinger effect
efecto de Dember Dember effect
efecto de derivación shunting effect
efecto de desmagnetización demagnetization effect
efecto de Destriau Destriau effect
efecto de disparo trigger effect
efecto de doble pico double-hump effect
efecto de Doppler Doppler effect
efecto de eclipse eclipse effect
efecto de Edison Edison effect
efecto de Einstein-de Haas Einstein-de Haas effect
efecto de electrón transferido transferred-electron effect
efecto de enmascaramiento masking effect
efecto de espejo mirror effect
efecto de esquinas corner effect
efecto de Ettinghausen Ettinghausen effect

efecto de Faraday Faraday effect
efecto de fase phase effect
efecto de fatiga fatigue effect
efecto de flanqueo flanking effect
efecto de Fletcher-Munson Fletcher-Munson effect
efecto de flexión bending effect
efecto de fondo background effect
efecto de fotodifusión photodiffusion effect
efecto de Foucault Foucault effect
efecto de granalla shot effect
efecto de Gudden-Pohl Gudden-Pohl effect
efecto de Guillemin Guillemin effect
efecto de Gunn Gunn effect
efecto de Hall Hall effect
efecto de Hallwacks Hallwacks effect
efecto de haz beam effect
efecto de Hertz Hertz effect
efecto de imagen image effect
efecto de inducción induction effect
efecto de interfaz interface effect
efecto de ionización ionization effect
efecto de islote island effect
efecto de Johnson-Lark-Horowitz Johnson-Lark-Horowitz effect
efecto de Josephson Josephson effect
efecto de Joshi Joshi effect
efecto de Joule Joule effect
efecto de Kelvin Kelvin effect
efecto de Kendall Kendall effect
efecto de Kerr Kerr effect
efecto de Kundt Kundt effect
efecto de Larsen Larsen effect
efecto de latitud latitude effect
efecto de Leduc Leduc effect
efecto de Lilienfeld Lilienfeld effect
efecto de línea larga long-line effect
efecto de Lossev Lossev effect
efecto de Luxemberg Luxemberg effect
efecto de magnetoestricción magnetostriction effect
efecto de Malter Malter effect
efecto de mano hand effect
efecto de Marconi Marconi effect
efecto de Matteucci Matteucci effect
efecto de Meissner Meissner effect
efecto de Miller Miller effect
efecto de modulación de filamento filament modulation effect
efecto de montaña mountain effect
efecto de Mossbauer Mossbauer effect
efecto de motor motor effect
efecto de Nernst Nernst effect
efecto de Nernst-Ettinghausen Nernst-Ettinghausen effect
efecto de nieve snow effect
efecto de Ovshinsky Ovshinsky effect
efecto de Page Page effect
efecto de pantalla screen effect
efecto de pantalla magnética magnetic-screen effect
efecto de pared wall effect
efecto de parpadeo flicker effect
efecto de Paschen-Back Paschen-Back effect
efecto de Peltier Peltier effect
efecto de persiana venetian-blind effect
efecto de Pockels Pockels effect
efecto de polarización polarization effect
efecto de primer orden first-order effect
efecto de profundidad cinético kinetic depth effect
efecto de proximidad proximity effect
efecto de punta point effect
efecto de radiación radiation effect
efecto de Raman Raman effect
efecto de Ramsauer Ramsauer effect
efecto de ranura slot effect

efecto de reflexión reflection effect
efecto de resistencia negativa negative-resistance effect
efecto de resonancia resonance effect
efecto de retardo retardation effect
efecto de Richardson Richardson effect
efecto de Right-Leduc Right-Leduc effect
efecto de Sagnac Sagnac effect
efecto de salto skip effect
efecto de saturación saturation effect
efecto de Schottky Schottky effect
efecto de Seebeck Seebeck effect
efecto de segundo orden second-order effect
efecto de Silsbee Silsbee effect
efecto de sombra shadow effect
efecto de Stark Stark effect
efecto de Suhl Suhl effect
efecto de superposición superposition effect
efecto de talón heel effect
efecto de Tanberg Tanberg effect
efecto de terminación end effect
efecto de Thomson Thomson effect
efecto de tierra ground effect
efecto de Touschek Touschek effect
efecto de transferencia transfer effect
efecto de transición transition effect
efecto de transmisión de ruido noise transmission effect
efecto de trayectorias múltiples multipath effect
efecto de Tyndall Tyndall effect
efecto de umbral threshold effect
efecto de unión junction effect
efecto de válvula valve effect
efecto de Villari Villari effect
efecto de volante flywheel effect
efecto de Volta Volta effect
efecto de Wertheim Wertheim effect
efecto de Whippany Whippany effect
efecto de Wiedemann Wiedemann effect
efecto de Wiedemann directo direct Wiedemann effect
efecto de Wiedemann inverso inverse Wiedemann effect
efecto de Wiegand Wiegand effect
efecto de Wien Wien effect
efecto de Wilson Wilson effect
efecto de Zeeman Zeeman effect
efecto de Zeeman anómalo anomalous Zeeman effect
efecto de Zener Zener effect
efecto del cuerpo body effect
efecto del entrehierro gap effect
efecto diamagnético diamagnetic effect
efecto dinatrón dynatron effect
efecto direccional directional effect
efecto directo direct effect
efecto diurno daylight effect
efecto electrodérmico electrodermal effect
efecto electrofónico electrophonic effect
efecto electroóptico electrooptic effect
efecto electroóptico de Kerr Kerr electrooptical effect
efecto electrorresistivo electroresistive effect
efecto electrostático de largo alcance long-range electrostatic effect
efecto estacional seasonal effect
efecto estroboscópico stroboscopic effect
efecto fotoconductivo photoconductive effect
efecto fotoconductor photoconductor effect
efecto fotodieléctrico photodielectric effect
efecto fotoeléctrico photoelectric effect
efecto fotoeléctrico atómico atomic photoelectric effect

efecto fotoeléctrico inverso inverse photoelectric effect
efecto fotoeléctrico superficial surface photoelectric effect
efecto fotoelectromagnético photoelectromagnetic effect
efecto fotoemisivo photoemissive effect
efecto fotomagnético photomagnetic effect
efecto fotomagnetoeléctrico photomagnetoelectric effect
efecto fotonuclear photonuclear effect
efecto fotoquímico photochemical effect
efecto fotovoltaico photovoltaic effect
efecto galvanomagnético galvanomagnetic effect
efecto geoeléctrico geoelectric effect
efecto geomagnético geomagnetic effect
efecto giromagnético gyromagnetic effect
efecto giroscópico gyroscopic effect
efecto inductivo inductive effect
efecto inercial inertial effect
efecto integrador integrating effect
efecto isotópico isotope effect
efecto local sidetone effect
efecto longitudinal longitudinal effect
efecto magnetoelástico magnetoelastic effect
efecto magnetoeléctrico magnetoelectric effect
efecto magnetoestrictivo magnetostrictive effect
efecto magnetohidrodinámico magnetohydrodynamic effect
efecto magnetoóptico magnetooptical effect
efecto magnetoóptico de Kerr Kerr magnetooptical effect
efecto magnetorresistivo magnetoresistive effect
efecto magnetotermoeléctrico magnetothermoelectric effect
efecto magnetrón magnetron effect
efecto marginal fringe effect
efecto masivo bulk effect
efecto microfónico microphonic effect
efecto nocturno night effect
efecto parásito parasitic effect, stray effect
efecto piezoeléctrico piezoelectric effect
efecto piezoeléctrico converso converse piezoelectric effect
efecto piezoeléctrico directo direct piezoelectric effect
efecto piezoeléctrico inverso inverse piezoelectric effect
efecto piezorresistivo piezoresistive effect
efecto piroeléctrico pyroelectric effect
efecto piromagnético pyromagnetic effect
efecto plástico plastic effect
efecto posterior aftereffect
efecto posterior magnético magnetic aftereffect
efecto químico chemical effect
efecto secular secular effect
efecto seudoestereofónico pseudostereophonic effect
efecto superficial surface effect
efecto térmico thermal effect
efecto termoeléctrico thermoelectric effect
efecto termoiónico thermionic effect
efecto termomagnético thermomagnetic effect
efecto transitorio transient effect
efecto transversal transverse effect
efecto trapecial keystone effect
efecto triboeléctrico triboelectric effect
efecto túnel tunnel effect
efectos aurorales auroral effects
efectos de transporte transport effects
efectos del ambiente ambient effects
efectos telúricos telluric effects
eferente efferent
eficacia direccional front-to-back ratio

eficacia direccional de antena antenna front-to-back ratio
eficiencia efficiency
eficiencia absoluta absolute efficiency
eficiencia anódica anode efficiency
eficiencia aparente nominal rated apparent efficiency
eficiencia catódica cathode efficiency
eficiencia constante constant efficiency
eficiencia cuántica quantum efficiency
eficiencia cuántica espectral spectral quantum efficiency
eficiencia de abertura aperture efficiency
eficiencia de acoplamiento coupling efficiency
eficiencia de almacenamiento storage efficiency
eficiencia de altavoz loudspeaker efficiency
eficiencia de antena antenna efficiency
eficiencia de bomba pump efficiency
eficiencia de captura capture efficiency
eficiencia de carga charge efficiency
eficiencia de circuito circuit efficiency
eficiencia de circuito anódico anode circuit efficiency
eficiencia de circuito de carga load-circuit efficiency
eficiencia de circuito de placa plate-circuit efficiency
eficiencia de colector collector efficiency
eficiencia de computadora computer efficiency
eficiencia de cómputo computing efficiency
eficiencia de conteo counting efficiency
eficiencia de conversión conversion efficiency
eficiencia de conversión de centelleador scintillator conversion efficiency
eficiencia de conversión de centelleo scintillation conversion efficiency
eficiencia de corriente current efficiency
eficiencia de corrosión anódica anode corrosion efficiency
eficiencia de detección detection efficiency
eficiencia de difracción diffraction efficiency
eficiencia de ejecución execution efficiency
eficiencia de electrodo electrode efficiency
eficiencia de emisión emission efficiency
eficiencia de emisor emitter efficiency
eficiencia de estación generadora generating-station efficiency
eficiencia de etapa stage efficiency
eficiencia de fuente source efficiency
eficiencia de fuente de alimentación power-supply efficiency
eficiencia de generador generator efficiency
eficiencia de haz beam efficiency
eficiencia de iluminación de antena antenna illumination efficiency
eficiencia de inyección injection efficiency
eficiencia de luminaria luminaire efficiency
eficiencia de luminiscencia luminescence efficiency
eficiencia de motor motor efficiency
eficiencia de ocupación occupation efficiency
eficiencia de oscilación oscillation efficiency
eficiencia de panel panel efficiency
eficiencia de pantalla screen efficiency
eficiencia de placa plate efficiency
eficiencia de polarización polarization efficiency
eficiencia de potencia disponible available power efficiency
eficiencia de potencia total overall power efficiency
eficiencia de procesamiento processing efficiency
eficiencia de proyector projector efficiency
eficiencia de radiación radiation efficiency
eficiencia de recolección collection efficiency
eficiencia de rectificación rectification efficiency
eficiencia de salida output efficiency

eficiencia de sistema system efficiency
eficiencia de tensión voltage efficiency
eficiencia de tierra ground efficiency
eficiencia de transductor transducer efficiency
eficiencia de transferencia transfer efficiency
eficiencia de transmisión transmission efficiency
eficiencia de transmisión de corriente electrónica electron-stream transmission efficiency
eficiencia directiva directive efficiency
eficiencia efectiva effective efficiency
eficiencia eléctrica electrical efficiency
eficiencia eléctrica global overall electrical efficiency
eficiencia electrónica electronic efficiency
eficiencia electrotérmica electrothermal efficiency
eficiencia en ampere-horas ampere-hour efficiency
eficiencia en watthoras watt-hour efficiency
eficiencia energética energy efficiency
eficiencia fotoeléctrica photoelectric efficiency
eficiencia fundamental fundamental efficiency
eficiencia global overall efficiency
eficiencia ideal ideal efficiency
eficiencia luminosa luminous efficiency
eficiencia luminosa espectral spectral luminous efficiency
eficiencia luminosa relativa relative luminous efficiency
eficiencia mecánica mechanical efficiency
eficiencia radiante radiant efficiency
eficiencia real actual efficiency
eficiencia relativa relative efficiency
eficiencia térmica thermal efficiency
eficiencia total total efficiency
eficiencia variable variable efficiency
eficiencia volumétrica volumetric efficiency
efluvio effluve
einstenio einsteinium
eje axis, spindle, shaft
eje de abscisas axis of abscissas
eje de banda ancha wideband axis
eje de banda angosta narrowband axis
eje de cabrestante capstan shaft
eje de cero zero axis
eje de deflexión deflection axis
eje de exploración scanning axis
eje de haz beam axis
eje de imaginarias axis of imaginaries
eje de ordenadas axis of ordinates
eje de reales axis of reals
eje de referencia reference axis
eje de referencia de salida output reference axis
eje de salida output axis
eje de tiempo time axis
eje eléctrico electrical axis
eje en cuadratura quadrature axis
eje ferroeléctrico ferroelectric axis
eje flexible flexible shaft
eje horizontal horizontal axis
eje imaginario imaginary axis
eje magnético magnetic axis
eje mecánico mechanical axis
eje óptico optical axis
eje orientacional orientational axis
eje piezoeléctrico piezoelectric axis
eje polar polar axis
eje principal principal axis
eje real real axis
eje secundario secondary axis
eje vertical vertical axis
eje x x-axis
eje y y-axis
eje z z-axis
ejecución execution, run

ejecución concurrente concurrent execution
ejecución de programa program execution, program run
ejecución de programa normal normal program execution
ejecución de programas concurrentes concurrent program execution
ejecución en paralelo parallel run
ejecución incremental incremental execution
ejecución interpretativa interpretive execution
ejecución real actual execution
ejecución reversible reversible execution
ejecución simbólica symbolic execution
ejes de cristal crystal axes
ejes ortogonales orthogonal axes
elastancia elastance
elasticidad elasticity, compliance
elasticidad acústica acoustic elasticity
elasticidad eléctrica electric elasticity
elastividad elastivity
elastómero elastomer
elastorresistencia elastoresistance
ele ell
electra electra
electreto electret
electricidad electricity
electricidad atmosférica atmospheric electricity
electricidad atómica atomic electricity
electricidad de contacto contact electricity
electricidad dinámica dynamic electricity
electricidad estática static electricity
electricidad galvánica galvanic electricity
electricidad inducida por luz light-induced electricity
electricidad natural natural electricity
electricidad negativa negative electricity
electricidad nuclear nuclear electricity
electricidad por fricción frictional electricity
electricidad positiva positive electricity
electricidad telúrica telluric electricity
electricidad vítrea vitreous electricity
eléctrico electric
electrificación electrification
electrificación estática static electrification
electrificación negativa negative electrification
electrificación positiva positive electrification
electrificado electrified
electrificar electrify
electroacústica electroacoustics
electroacústico electroacoustic
electroafinidad electroaffinity
electroanálisis electroanalysis
electroanalizador electroanalyzer
electroanestesia electroanesthesia
electroatómico electroatomic
electrobalística electroballistics
electrobiología electrobiology
electrobioscopía electrobioscopy
electrocapilaridad electrocapillarity
electrocardiofonógrafo electrocardiophonograph
electrocardiografía electrocardiography
electrocardiógrafo electrocardiograph
electrocardiograma electrocardiogram
electrocardiograma vectorial vector electrocardiogram
electrocatálisis electrocatalysis
electrocauterización electrocautery
electrocibernética electrocybernetics
electrocinética electrokinetics
electrocinético electrokinetic
electrocinta electrotape
electrocirugía electrosurgery
electrocoagulación electrocoagulation

electrocorticograma electrocorticogram
electrocronómetro electrochronometer
electrocución electrocution
electrocultivo electroculture
electrocutar electrocute
electrochoque electroshock
electrochorro electrojet
electrodeposición electrodeposition, plating
electrodeposición selectiva selective electrodeposition
electrodepositado plated
electrodepósito electrodeposit
electrodérmico electrodermal
electrodermografía electrodermography
electrodermograma electrodermogram
electrodesecación electrodesiccation
electrodiagnóstico electrodiagnosis
electrodinámica electrodynamics
electrodinámica cuántica quantum electrodynamics
electrodinámico electrodynamic
electrodinamómetro electrodynamometer
electrodinamómetro de Siemens Siemens' electrodynamometer
electrodisolución electrodissolution
electrodo electrode
electrodo acelerador accelerating electrode
electrodo acelerador postdeflexión post-deflection accelerating electrode
electrodo activo active electrode
electrodo auxiliar auxiliary electrode
electrodo bipolar bipolar electrode
electrodo caliente hot electrode
electrodo colector collecting electrode
electrodo compuesto composite electrode
electrodo conectado al cuerpo body electrode
electrodo continuo continuous electrode
electrodo controlador controlling electrode
electrodo de aguja needle electrode
electrodo de autococción self-baking electrode
electrodo de baja impedancia low-impedance electrode
electrodo de base base electrode
electrodo de blanco target electrode
electrodo de cadmio cadmium electrode
electrodo de calomel calomel electrode
electrodo de calomel normal normal calomel electrode
electrodo de carbón carbon electrode
electrodo de cebado striking electrode, starting electrode
electrodo de comparación comparison electrode
electrodo de concentración concentration electrode
electrodo de contacto contact electrode
electrodo de contacto permanente hold electrode
electrodo de control control electrode
electrodo de control de rayo ray-control electrode
electrodo de convergencia convergence electrode
electrodo de corrimiento drift electrode
electrodo de deflexión deflecting electrode
electrodo de disipación dissipation electrode
electrodo de disparo trigger electrode
electrodo de encendido ignitor electrode
electrodo de enfoque focusing electrode
electrodo de entrada input electrode
electrodo de entretenimiento keep-alive electrode
electrodo de excitación keep-alive electrode
electrodo de grafito graphite electrode
electrodo de guarda guard electrode
electrodo de haz beam electrode
electrodo de hidrógeno hydrogen electrode
electrodo de hidrógeno normal standard hydrogen electrode
electrodo de hoja metálica foil electrode

electrodo de inmersión dipping electrode
electrodo de mosaico mosaic electrode
electrodo de Ostwald Ostwald electrode
electrodo de penetración penetration electrode
electrodo de placa plate electrode
electrodo de puerta gate electrode
electrodo de puesta a tierra grounding electrode
electrodo de punta point electrode
electrodo de quinhidrona quinhydrone electrode
electrodo de referencia reference electrode
electrodo de referencia difuso diffuse reference
   electrode
electrodo de referencia dispersivo dispersive
   reference electrode
electrodo de referencia no polarizable
   nonpolarizable reference electrode
electrodo de rejilla grid electrode
electrodo de salida output electrode
electrodo de salida de señal signal output electrode
electrodo de seguridad de llama flame safety
   electrode
electrodo de señal signal electrode
electrodo de soldadura welding electrode
electrodo de soldadura por arco arc-welding
   electrode
electrodo de soldadura por resistencia resistance
   welding electrode
electrodo de tierra ground electrode
electrodo de transferencia transfer electrode
electrodo de tubo tube electrode
electrodo de vidrio glass electrode
electrodo desacelerador decelerating electrode
electrodo deslizante sliding electrode
electrodo desnudo bare electrode
electrodo dispersivo dispersive electrode
electrodo emisor emitter electrode
electrodo en bigote de gato catwhisker
electrodo explorador exploring electrode
electrodo exterior outer electrode
electrodo frío cold electrode
electrodo frontal front electrode
electrodo hueco hollow electrode
electrodo inferior bottom electrode
electrodo inorgánico inorganic electrode
electrodo intensificador intensifier electrode
electrodo inyector injector electrode
electrodo metálico metal electrode
electrodo modulador modulating electrode
electrodo monopolar monopolar electrode
electrodo múltiple multiple electrode
electrodo negativo negative electrode
electrodo no continuo noncontinuous electrode
electrodo no polarizable nonpolarizable electrode
electrodo normal normal electrode
electrodo pasivo passive electrode
electrodo plano flat electrode
electrodo platinado platinum-clad electrode
electrodo positivo positive electrode
electrodo postacelerador post-accelerating electrode
electrodo posterior back electrode
electrodo principal main electrode
electrodo receptor receiving electrode
electrodo reflector reflector electrode
electrodo retardador retarding electrode
electrodo revestido coated electrode
electrodo simple simple electrode
electrodos coplanares coplanar electrodes
electrodos de deflexión horizontal
   horizontal-deflection electrodes
electrodos de deflexión vertical vertical-deflection
   electrodes
electrodos de guarda arcing horn
electrodos de trabajo work electrodes

electroencefalógrafo electroencephalograph
electroencefalograma electroencephalogram
electroencefaloscopio electroencephaloscope
electroenfriamiento electrocooling
electroerosión electroerosion
electroestatografía electrostatography
electroestenólisis electrostenolysis
electroestimulación electrostimulation
electroestricción electrostriction
electroestrictivo electrostrictive
electroextracción electroextraction
electrofílico electrophilic
electrofísica electrophysics
electrofisiología electrophysiology
electrofonía electrophony
electrofónico electrophonic
electrófono electrophone
electroforesis electrophoresis
electroforético electrophoretic
electroformación electroforming
electroformado electroformed
electroformar electroform
electróforo electrophorus
electrofotografía electrophotography
electrofotografía electrostática electrostatic
   electrophotography
electrofotográfico electrophotographic
electrofotómetro electrophotometer
electrogalvánico electrogalvanic
electrogalvanizado electrogalvanized
electrogalvanizar electrogalvanize
electrográfico electrographic
electrografito electrographite
electrógrafo electrograph
electrohidráulico electrohydraulic
electroimán electromagnet
electroimán acorazado ironclad electromagnet
electroimán anular annular electromagnet
electroimán apantallado shielded electromagnet
electroimán de arranque start magnet
electroimán de clasificación sorting magnet
electroimán de corriente alterna alternating-current
   electromagnet
electroimán de largo recorrido long-pull magnet
electroimán de liberación release magnet
electroimán de parada stop magnet
electroimán de perforación punch magnet
electroimán de recepción receive electromagnet
electroimán de relé relay magnet
electroimán elevador lifting magnet
electroimán polarizado polarized electromagnet
electroimán superconductor superconducting
   electromagnet
electroimán telegráfico telegraph electromagnet
electrólisis electrolysis
electrolítico electrolytic
electrólito electrolyte
electrólito de soda cáustica caustic-soda electrolyte
electrólito débil weak electrolyte
electrólito fuerte strong electrolyte
electrólito fundido fused electrolyte
electrólito impuro foul electrolyte
electrólito inorgánico inorganic electrolyte
electrólito sólido solid electrolyte
electrolizador electrolyzer
electrolizar electrolyze
electroluminiscencia electroluminescence
electroluminiscente electroluminescent
electromagnética electromagnetics
electromagnéticamente electromagnetically
electromagnético electromagnetic
electromagnetismo electromagnetism
electromagnetizador electromagnetizer

electromecánica electromechanics
electromecánico electromechanical
electrometalurgia electrometallurgy
electrometría electrometry
electrómetro electrometer
electrómetro absoluto absolute electrometer
electrómetro absoluto de Kelvin Kelvin absolute
    electrometer
electrómetro bifilar bifilar electrometer
electrómetro calibrador gage electrometer
electrómetro capilar capillary electrometer
electrómetro de balanza balance electrometer
electrómetro de capacitor vibrante
    vibrating-capacitor electrometer
electrómetro de cuadrantes quadrant electrometer
electrómetro de cuerda string electrometer
electrómetro de Dershem Dershem electrometer
electrómetro de Dolezalek Dolezalek electrometer
electrómetro de fibra de cuarzo quartz-fiber
    electrometer
electrómetro de filamento filament electrometer
electrómetro de hilo fiber electrometer
electrómetro de Hoffmann Hoffmann electrometer
electrómetro de Kelvin Kelvin electrometer
electrómetro de laminilla vibrante vibrating-reed
    electrometer
electrómetro de Lindemann Lindemann
    electrometer
electrómetro de torsión torsion electrometer
electrómetro de tubo tube electrometer
electrómetro de tubo de vacío vacuum-tube
    electrometer
electrómetro digital digital electrometer
electrómetro electrostático electrostatic electrometer
electrómetro mecánico mechanical electrometer
electrómetro múltiple multiple electrometer
electromigración electromigration
electromiografía electromyography
electromiográfico electromyographic
electromiógrafo electromyograph
electromiograma electromyogram
electromoción electromotion
electromotor electromotor
electromotriz electromotive
electrón electron
electrón de Auger Auger electron
electrón de conducción conduction electron
electrón de rechazo recoil electron
electrón desacelerado decelerated electron
electrón emitido emitted electron
electrón excedente excess electron
electrón ferromagnético ferromagnetic electron
electrón giratorio spinning electron
electrón libre free electron
electrón ligado bound electron
electrón negativo negative electron
electrón orbital orbital electron
electrón periférico peripheral electron
electrón pesado heavy electron
electrón planetario planetary electron
electrón positivo positive electron
electrón primario primary electron
electrón reflejado reflected electron
electrón secundario secondary electron
electrón transferido transferred electron
electrón-volt electron-volt
electronarcosis electronarcosis
electronegatividad electronegativity
electronegativo electronegative
electrones de valencia valence electrons
electrones dispersos scattered electrons
electrones monoenergéticos monoenergetic electrons
electroneumático electropneumatic

electrónica electronics
electrónica aeronáutica aeronautical electronics
electrónica analógica analog electronics
electrónica criogénica cryogenic electronics
electrónica cuántica quantum electronics
electrónica de aviación aviation electronics
electrónica de control de máquina machine control
    electronics
electrónica de corriente débil small-current
    electronics
electrónica de estado sólido solid-state electronics
electrónica de microsistemas microsystems
    electronics
electrónica digital digital electronics
electrónica doméstica domestic electronics
electrónica física physical electronics
electrónica industrial industrial electronics
electrónica integrada integrated electronics
electrónica molecular molecular electronics
electrónica navegacional navigational electronics
electrónica sofisticada sophisticated electronics
electrónicamente electronically
electrónico electronic
electrooculografía electrooculography
electrooculograma electrooculogram
electroóptica electrooptics
electroóptico electrooptical
electroósmosis electroosmosis
electroosmótico electroosmotic
electropolar electropolar
electropositivo electropositive
electropulido electropolishing
electroquímica electrochemistry
electroquímico electrochemical
electrorreducción electroreduction
electrorrefinado electrorefining
electrorrefinar electrorefine
electrorresistivo electroresistive
electrorretinograma electroretinogram
electroscopio electroscope
electroscopio apantallado shielded electroscope
electroscopio de Braun Braun electroscope
electroscopio de condensador condenser
    electroscope
electroscopio de fibra de cuarzo quartz-fiber
    electroscope
electroscopio de hoja metálica foil electroscope
electroscopio de hojas leaf electroscope
electroscopio de hojas de oro gold-leaf electroscope
electroscopio de Lauritsen Lauritsen electroscope
electroscopio de Wilson Wilson electroscope
electroscopio de Zenely Zenely electroscope
electrosección electrosection
electrosensible electrosensitive
electrosíntesis electrosynthesis
electrosoldadura electrowelding
electrostática electrostatics
electrostáticamente electrostatically
electrostático electrostatic
electrotaxia electrotaxis
electrotecnología electrotechnology
electrotelemedición electric telemetering
electrotelémetro electric telemeter
electrotelurógrafo electrotellurograph
electroterapéutica electrotherapeutics
electroterapia electrotherapy
electrotermia electrothermics
electrotérmico electrothermal
electrotitración electrotitration
electrotónico electrotonic
electrotono electrotonus
electrotropismo electrotropism
electrovalencia electrovalence

elemento element, item
elemento absoluto absolute element
elemento activo active element
elemento aperiódico aperiodic element
elemento aritmético arithmetic element
elemento astático astatic element
elemento base base element
elemento básico basic element
elemento biestable bistable element
elemento bilateral bilateral element
elemento bimetálico bimetallic element
elemento binario binary element
elemento catódico cathode element
elemento cerámico ceramic element
elemento colineal collinear element
elemento compensador compensating element
elemento computador computing element
elemento concentrado lumped element
elemento controlador controlling element
elemento criogénico cryogenic element
elemento de acoplamiento coupling element
elemento de acoplamiento con diodo de Zener
  Zener-diode coupling element
elemento de acumulador storage cell
elemento de almacenamiento storage element
elemento de almacenamiento binario binary storage
  element
elemento de antena antenna element
elemento de arranque starting element
elemento de batería battery element
elemento de bolómetro bolometer element
elemento de bucle loop element
elemento de calentamiento heating element
elemento de capacitor capacitor element
elemento de carácter character element
elemento de cinta ribbon element
elemento de circuito circuit element
elemento de cola queue element
elemento de conducción conducting element
elemento de conmutación switching element
elemento de contacto contact element
elemento de control control element
elemento de control básico basic control element
elemento de control final final control element
elemento de control fino fine control element
elemento de corte cutout element
elemento de datos data element
elemento de datos atómico atomic data element
elemento de datos de atributo attribute data element
elemento de datos elemental elementary data
  element
elemento de datos genérico generic data element
elemento de desplazamiento shifting element
elemento de detección primario primary detecting
  element
elemento de diseño design element
elemento de emergencia emergency cell
elemento de entrada input element
elemento de equivalencia equivalence element
elemento de excitación driver element
elemento de expansión electrotérmico
  electrothermal expansion element
elemento de identidad identity element
elemento de imagen image element
elemento de impresión print element
elemento de interrupción de programa program
  interruption element
elemento de lista list element
elemento de lógica binario binary logic element
elemento de lógica combinacional combinational
  logic element
elemento de lógica secuencial sequential logic
  element

elemento de mensaje message element
elemento de microcomputadora microcomputer
  element
elemento de modulación modulation element
elemento de monitoreo monitoring element
elemento de muestreo sampling element
elemento de negación negation element
elemento de no equivalencia nonequivalence element
elemento de parada stop element
elemento de paso pass element
elemento de procesamiento y control processing and
  control element
elemento de protección protection element
elemento de puerta gate element
elemento de radar radar element
elemento de red de antenas array element
elemento de regulación end cell
elemento de relé relay element
elemento de resistencia resistance element
elemento de resistencia negativa negative-resistance
  element
elemento de retardo delay element
elemento de retardo de dígito digit delay element
elemento de retroalimentación feedback element
elemento de señal signal element
elemento de sistema system element
elemento de tabla table element, table item
elemento de temporización timing element
elemento de trabajo marking element
elemento de transductor transductor element
elemento de transición transition element
elemento de tubo tube element
elemento de unidad de procesamiento central
  central processing unit element
elemento de válvula valve element
elemento de visualización display element
elemento de Wenner Wenner element
elemento desacelerador decelerating element
elemento detectable detectable element
elemento detector detecting element
elemento determinante de frecuencia
  frequency-determining element
elemento direccional directional element
elemento director director element
elemento discreto discrete element
elemento discriminante discriminating element
elemento distribuido distributed element
elemento eléctrico electrical element
elemento electrotérmico electrothermic element
elemento encapsulado encapsulated element
elemento estable stable element
elemento excitado driven element
elemento funcional functional element
elemento fusible fuse element
elemento helicoidal helical element
elemento impreso printed element
elemento impuro impurity element
elemento lineal linear element
elemento lógico logic element
elemento lógico compuesto compound logical
  element
elemento lógico digital digital logic element
elemento lógico magnético magnetic logic element
elemento luminoso luminous element
elemento maestro master element
elemento mayoritario majority element
elemento móvil moving element
elemento multihilo multiwire element
elemento NO NOT element
elemento no lineal nonlinear element
elemento no lineal pasivo passive nonlinear element
elemento NO-O NOR element
elemento NO-Y NAND element

elemento numérico numerical element
elemento O OR element
elemento optoelectrónico optoelectronic element
elemento paramagnético paramagnetic element
elemento parásito parasitic element
elemento pasivo passive element
elemento pentavalente pentavalent element
elemento portátil portable cell
elemento preajustable preset element
elemento preajustado preset element
elemento protector protecting element
elemento radiactivo radioactive element
elemento radiante radiating element
elemento radiante acústico acoustic radiating element
elemento rectificador rectifying element
elemento reflector reflective element
elemento regulador regulating element
elemento removible removable element
elemento resistente a la humedad moisture-resistant element
elemento resistivo resistive element
elemento resistor resistor element
elemento resonante resonant element
elemento rotativo rotating element
elemento secuencial sequential element
elemento seleccionable selectable element
elemento semiconductor semiconducting element
elemento sensible sensitive element
elemento sensor sensing element
elemento SI-ENTONCES IF-THEN element
elemento SI-Y-SOLO-SI IF-AND-ONLY-IF element
elemento termoeléctrico thermoelectric element
elemento toroidal toroidal element
elemento único single element
elemento unidireccional unidirectional element
elemento unilateral unilateral element
elemento unitario unit element
elemento Y AND element
elementos de código code elements
elementos de conmutación commutation elements
elementos de decisión decision elements
elementos de estado sólido solid-state elements
elementos de resolución resolution elements
elementos de señal coincidentes coincident signal elements
elementos en paralelo parallel elements
elementos en serie series elements
elementos no adaptados unmatched elements
elementos yuxtapuestos juxtaposed elements
elevador positive booster
elevador eléctrico electric elevator
elevador-reductor positive-negative booster, reversible booster
eliminación elimination, suppression, stripping
eliminación armónica harmonic elimination
eliminación de banda lateral sideband cutting
eliminación de bloque block delete
eliminación de ceros zero elimination
eliminación de portadora carrier elimination
eliminación de ruido noise elimination
eliminación de zona zone elimination
eliminación etapa por etapa stage-by-stage elimination
eliminador eliminator
eliminador armónico harmonic eliminator
eliminador de antena antenna eliminator
eliminador de Barkhausen Barkhausen eliminator
eliminador de batería battery eliminator
eliminador de eco echo eliminator
eliminador de ecos parásitos clutter eliminator
eliminador de interferencia interference eliminator
eliminador de módem modem eliminator

eliminador de parásitos parasitic eliminator
eliminador de ruido noise eliminator
eliminador heterodino heterodyne eliminator
eliminar delete
eliminar de una cola dequeue
elipse de Fresnel Fresnel ellipse
elipticidad ellipticity
elíptico elliptical
elongación elongation
elongación de línea line elongation
elongación magnética magnetic elongation
elongación permanente permanent elongation
emanación emanation
emanación radiactiva radioactive emanation
embalamiento runaway
embarradura smearing
embrague clutch
embrague de alimentación feed clutch
embrague de fricción magnética magnetic friction clutch
embrague deslizante slip clutch
embrague electromagnético electromagnetic clutch
embrague magnético magnetic clutch
embrague por histéresis hysteresis clutch
emergencia de capacidad capacity emergency
emisión emission, broadcast, transmission, sending
emisión aleatoria random emission
emisión armónica harmonic emission
emisión autoeléctrica autoelectric emission
emisión catódica cathode emission
emisión de banda ancha broadband emission
emisión de campo libre free-field emission
emisión de cátodo frío cold-cathode emission
emisión de filamento filament emission
emisión de impulsos pulse emission
emisión de iones positivos positive-ion emission
emisión de luz light emission
emisión de partículas particle emission
emisión de positrones positron emission
emisión de rejilla grid emission
emisión de rejilla primaria primary grid emission
emisión de rejilla secundaria secondary grid emission
emisión de rejilla termoiónica thermionic grid emission
emisión de ruido noise emission
emisión de Schottky Schottky emission
emisión de válvula valve emission
emisión distribuida distributed emission
emisión electrónica electronic emission
emisión en campo uniforme zero field emission
emisión específica specific emission
emisión espectral spectral emission
emisión espontánea spontaneous emission
emisión espontanea amplificada amplified spontaneous emission
emisión espuria spurious emission
emisión estimulada stimulated emission
emisión extraña extraneous emission
emisión fotoeléctrica photoelectric emission
emisión fotoeléctrica realzada por campo field-enhanced photoelectric emission
emisión fotónica photon emission
emisión fría cold emission
emisión imperfecta imperfect emission
emisión indeseada unwanted emission
emisión inducida induced emission
emisión inversa reverse emission, back emission
emisión iónica de campo field-ion emission
emisión lateral side emission
emisión limitada por temperatura temperature-limited emission
emisión modulada modulated emission

emisión monoenergética variable variable monoenergetic emission
emisión parásita parasitic emission, stray emission
emisión perfecta perfect emission
emisión perturbadora jammer emission
emisión por campo field emission
emisión primaria primary emission
emisión pulsada pulsed emission
emisión radiotelefónica radiotelephone emission
emisión-recepción automática automatic send-receive
emisión secundaria secondary emission
emisión secundaria realzada por campo field-enhanced secondary emission
emisión selectiva selective emission
emisión telegráfica telegraph emission
emisión termoelectrónica thermoelectronic emission
emisión termoiónica thermionic emission
emisión total total emission
emisiones espurias splatter
emisiones espurias de banda lateral sideband splatter
emisividad emissivity
emisividad direccional directional emissivity
emisividad espectral spectral emissivity
emisividad monocromática monochromatic emissivity
emisividad total total emissivity
emisivo emissive
emisor emitter, sender
emisor alfa alpha emitter
emisor beta beta emitter
emisor controlado controlled sender
emisor de caracteres character emitter
emisor de código code emitter
emisor de chispa spark sender
emisor de dígitos digit emitter
emisor de dígitos electrónico electronic digit emitter
emisor de dígitos selectivo selective digit emitter
emisor de electrones de cátodo frío cold-cathode electron emitter
emisor de impulsos impulse emitter
emisor de infrarrojo infrared emitter
emisor de luz light emitter
emisor de medio tiempo half-time emitter
emisor de óxido oxide emitter
emisor de óxido de magnesio magnesium-oxide emitter
emisor de partículas particle emitter
emisor de perforación punch emitter
emisor de platino-rutenio platinum-ruthenium emitter
emisor de positrones positron emitter
emisor de sonido sound-emitting
emisor débil dull emitter
emisor estable stable emitter
emisor gamma gamma emitter
emisor mayoritario majority emitter
emisor minoritario minority emitter
emisor puro pure emitter
emisor revestido de óxido oxide-coated emitter
emisor secundario secondary emitter
emisor selectivo selective emitter
emisor tipo N N-type emitter
emisor tipo P P-type emitter
emisor toriado thoriated emitter
emitancia emittance
emitancia luminosa luminous emittance
emitancia radiante radiant emittance
emitancia térmica thermal emittance
emitir emit
emitrón emitron
empalizada railings

empalmador splicer, jointer, through joint
empalmador de cables cable splicer
empalmador de cinta tape splicer
empalme splice, joint
empalme de aletas lug splice
empalme de cable cable splice
empalme de hilos wire splice
empalme de retención stop joint
empaque packing
empaque conductor conductive gasket
empaquetado packaging
empaquetado electrónico electronic packaging
empaquetado en queso suizo Swiss-cheese packaging
empaquetamiento packing
empaquetamiento de ranura slot packing
empírico empirical
empotrado embedded
emulación emulation
emulación de dispositivo device emulation
emulación de microprograma microprogram emulation
emulación de terminal terminal emulation
emulación de visualización display emulation
emulador emulator
emulador de circuito circuit emulator
emulador de terminal terminal emulator
emulador de terminal asincrónica asynchronous terminal emulator
emulador en circuito in-circuit emulator
emulador integrado integrated emulator
emular emulate
en blanco blank
en circuito in circuit
en contacto in contact
en derivación shunted
en el aire on the air
en estación on station
en fase in phase
en línea online, inline
en oposición in opposition
en paralelo in parallel, shunted
en reposo idle, quiescent
en serie in series, inline, serial
en serie aditiva series-aiding
en serie opositiva series-opposing
en serie por bits serial by bit
en serie por caracteres serial by character
en serie por palabras serial by word
en tándem ganged
encadenamiento chaining, linking, linkage
encadenamiento de bloque block chaining
encadenamiento de datos data chaining
encadenamiento de documentos document chaining
encadenamiento de flujo flux linkage
encadenamiento de mandos command chaining
encadenamiento de procedimientos procedure chaining
encadenamiento de programas program chaining
encadenamiento directo direct chaining
encaminamiento routing
encaminamiento alternativo alternate routing
encaminamiento de comunicaciones communication routing
encaminamiento de datos data routing
encaminamiento libre free routing
encaminamiento múltiple multiple routing
encapsulación encapsulation, potting
encapsulación plástica plastic encapsulation
encapsulación por inmersión dip encapsulation
encapsulado encapsulated
encapsulado en plástico plastic-encapsulated
encapsulado en resina resin-encapsulated

encapsulador encapsulating
encapsulante encapsulant
encefalografía encephalography
encefalograma encephalogram
encendedor igniter
encender turn on
encendido ignition
enclavamiento interlocking, interlock
enclavamiento automático automatic interlocking
enclavamiento de fase phase interlocking
enclavamiento de señal signal interlocking
enclavamiento direccional directional interlocking
enclavamiento directo direct interlocking
enclavamiento eléctrico electrical interlocking
enclavamiento electromecánico electromechanical
   interlocking
enclavamiento electroneumático electropneumatic
   interlocking
enclavamiento indirecto indirect interlocking
enclavamiento secuencial sequential interlocking
encogimiento shrinkage
encontrar y reemplazar find and replace
encuadre framing, television framing, frame
encuadre de imagen image framing
enchapado plating
enchufable plug-in, pluggable
enchufe plug, socket, socket-outlet
enderezador de espigas pin straightener
enderezador de patillas pin straightener
endodino endodyne
endoérgico endoergic
endorradiosonda endoradiosonde
endotérmico endothermic
endurecimiento de Cottrell Cottrell hardening
endurecimiento electroquímico electrochemical
   hardening
energético energetic
energía energy, power
energía activa active energy
energía acumulada stored energy
energía almacenada en volante flywheel-stored
   energy
energía aparente apparent energy
energía atómica atomic energy
energía característica characteristic energy
energía cero zero energy
energía cinética kinetic energy
energía de absorción absorption energy
energía de antena antenna energy
energía de chispa spark energy
energía de deformación magnética magnetic strain
   energy
energía de descarga discharge energy
energía de economía economy energy
energía de equilibrio equilibrium energy
energía de excitación excitation energy
energía de frecuencia fundamental
   fundamental-frequency energy
energía de haz beam energy
energía de histéresis hysteresis energy
energía de impulsos pulse energy
energía de ionización ionization energy
energía de ionización media mean ionization energy
energía de límite de banda band edge energy
energía de medida measurement energy
energía de pared wall energy
energía de partícula particle energy
energía de radiación radiation energy
energía de radiofrecuencia radio-frequency energy
energía de reacción reaction energy
energía de red lattice energy
energía de reserva standby power
energía de resonancia resonance energy

energía de salida output energy
energía de sistema neta net system energy
energía de transición transition energy
energía del entrehierro gap energy
energía efectiva effective energy
energía eléctrica electrical energy
energía eléctrica comercial commercial electric
   power
energía electrocinética electrokinetic energy
energía electromagnética electromagnetic energy
energía electromecánica electromechanical energy
energía electrónica media mean electron energy
energía electrostática electrostatic energy
energía en reposo rest energy
energía en watthoras watt-hour energy
energía específica specific energy
energía espectral spectral energy
energía fotónica photon energy
energía gravitacional gravitational energy
energía hidráulica water power
energía infrarroja infrared energy
energía interna internal energy
energía latente latent energy
energía luminosa luminous energy
energía magnética magnetic energy
energía magnetoelástica magnetoelastic energy
energía magnetohidrodinámica
   magnetohydrodynamic power
energía mecánica mechanical energy
energía media mean energy
energía neutrónica neutron energy
energía nuclear nuclear energy
energía orbital orbital energy
energía potencial potential energy
energía propia self-energy
energía química chemical energy
energía radiada radiated energy
energía radiante radiant energy
energía radiante espectral spectral radiant energy
energía reactiva reactive energy
energía reflejada reflected energy
energía relativa relative energy
energía residual residual energy
energía rotacional rotational energy
energía secundaria secondary energy
energía solar solar energy, solar power
energía sonora sound energy
energía superficial surface energy
energía térmica thermal energy
energía total total energy
energía umbral threshold energy
energizado energized
energizador energizer
energizar energize
enfasador phaser
enfasamiento phasing
enfasamiento de altavoz loudspeaker phasing
enfoque focus, focusing
enfoque automático automatic focusing
enfoque de haz electrónico electron-beam focusing
enfoque de sonido sound focusing
enfoque dinámico dynamic focusing
enfoque dinámico horizontal horizontal dynamic
   focus
enfoque dinámico vertical vertical dynamic focus
enfoque eléctrico electric focusing
enfoque electromagnético electromagnetic focusing
enfoque electrónico electron focusing
enfoque electrostático electrostatic focusing
enfoque fino fine focusing
enfoque fuerte strong focusing
enfoque iónico ionic focusing
enfoque ionosférico ionospheric focusing

enfoque magnético magnetic focusing
enfoque periódico periodic focusing
enfoque por gas gas focusing
enfoque por imanes permanentes permanent-magnet focusing
enfoque por solenoide solenoid focusing
enfoque postdeflexión post-deflection focus
enfriado por aceite oil-cooled
enfriado por agua water-cooled
enfriado por aire air cooled
enfriado por aire forzado forced-air cooled
enfriado por líquido liquid-cooled
enfriador termoeléctrico thermoelectric cooler
enfriamiento cooling
enfriamiento anódico anode cooling
enfriamiento de punto spot cooling
enfriamiento de tubos tube cooling
enfriamiento diferencial differential cooling
enfriamiento dinámico dynamic cooling
enfriamiento eléctrico electric cooling
enfriamiento electrostático electrostatic cooling
enfriamiento en circuito cerrado closed-circuit cooling
enfriamiento por agua water cooling
enfriamiento por aire air cooling
enfriamiento por aire forzado forced-air cooling
enfriamiento por conducción conduction cooling
enfriamiento por convección convection cooling
enfriamiento por líquido liquid cooling
enfriamiento por radiación radiation cooling
enfriamiento termoeléctrico thermoelectric cooling
enfriamiento termomagnético thermomagnetic cooling
enfriar rápidamente quench
enganche de fase phase lock
enganche horizontal horizontal lock
enganche magnético magnetic latching
engranaje gearing, gear
engranaje eléctrico electrical gearing
enlace link, linkage, bond, junction, trunk
enlace activo active link
enlace alámbrico wire link
enlace bidireccional two-way link
enlace compuesto composite link
enlace computador computing link
enlace cortocircuitador shorting link
enlace de anotaciones recording trunk
enlace de archivos file link
enlace de bordes edge linking
enlace de comunicaciones communication link
enlace de comunicaciones símplex simplex communications link
enlace de datos data link
enlace de datos desequilibrado unbalanced data link
enlace de datos digitales digital-data link
enlace de datos equilibrado balanced data link
enlace de desbordamiento overflow link
enlace de frecuencia ultraalta ultra-high frequency link
enlace de grupos group link
enlace de información information link
enlace de línea line link
enlace de línea digital digital line link
enlace de llegada incoming trunk
enlace de mandos command link
enlace de microondas terrestre terrestrial microwave link
enlace de modulación de frecuencia frequency-modulation link
enlace de muy alta frecuencia very-high-frequency link
enlace de programas program link
enlace de punto a punto point-to-point link

enlace de radiofrecuencia radio-frequency link
enlace de salida outgoing trunk
enlace de supergrupos supergroup link
enlace de televisión television link
enlace de transferencia transfer trunk
enlace de transmisión transmission link
enlace de valencia valence bond
enlace de video video link
enlace digital digital link
enlace dinámico dynamic link
enlace especial special link
enlace estudio-transmisor studio-to-transmitter link
enlace eutéctico eutectic bond
enlace físico physical link
enlace inactivo inactive link
enlace individual individual trunk
enlace iónico ionic bond
enlace local local trunk
enlace lógico logical link
enlace magnético magnetic linkage
enlace multicanal multichannel link
enlace múltiplex multiplex link
enlace multipunto multipoint link
enlace óptico optical link
enlace paralelo parallel link
enlace periférico peripheral link
enlace por cable cable link
enlace por microondas microwave link
enlace primario primary link
enlace radiotelefónico radiotelephone link
enlace submarino underwater link
enlace telefónico telephone link
enlace telegráfico telegraph link
enlace télex telex link
enlace transhorizonte transhorizon link
enlace ultrasónico ultrasonic bond
enlaces de datos ópticos optical data links
enlazado bonded, linked
enlazamiento link, linkage, bond, junction, trunking
enmascaramiento masking
enmascaramiento de falla fault masking
enmascaramiento de video video masking
ensamblado assembled
ensamblado de proceso process assembly
ensamblado en taller shop-assembled
ensamblador assembler
ensamblador absoluto absolute assembler
ensamblador básico basic assembler
ensamblador-compilador cruzado cross assembler-compiler
ensamblador cruzado cross assembler
ensamblador de equipo físico hardware assembler
ensamblador de lenguaje de microprocesador microprocessor language assembler
ensamblador de microcódigo microcode assembler
ensamblador de microprocesador microprocessor assembler
ensamblador de una pasada one-pass assembler
ensamblador de uno a uno one-to-one assembler
ensamblador de uno por uno one-for-one assembler
ensamblador-desensamblador de paquete packet assembler-disassembler
ensamblador eléctrico electrical assembler
ensamblador reubicable relocatable assembler
ensamblador reubicador relocating assembler
ensamblador simbólico symbolic assembler
ensamblaje assembly
ensamblaje condicional conditional assembly
ensamblaje de cabeza head assembly
ensamblaje de cabeza de registro recording head assembly
ensamblaje de cabeza enchufable plug-in head assembly

ensamblaje de cabeza grabadora recording head assembly
ensamblaje de cable cable assembly
ensamblaje de cable óptico optical cable assembly
ensamblaje de componentes component assembly
ensamblaje de conexión connection assembly
ensamblaje de conexión genérico generic connection assembly
ensamblaje de contacto estacionario stationary-contact assembly
ensamblaje de documento document assembly
ensamblaje de imán magnet assembly
ensamblaje de montaje mounting assembly
ensamblaje de programa program assembly
ensamblaje de puente de impedancia impedance-bridge assembly
ensamblaje de tipo de esqueleto skeleton-type assembly
ensamblaje en tándem gang assembly
ensamblaje múltiple multiple assembly
ensamblar assemble
ensamblar y ejecutar assemble-and-go
ensanchador de banda bandspreader
ensanchador de impulsos pulse stretcher
ensanchamiento de banda bandspreading
ensanchamiento de Doppler Doppler broadening
ensanchamiento de impulso pulse broadening
ensanchamiento de línea homogéneo homogeneous line broadening
ensanchamiento de Stark Stark broadening
ensanche broadening, spreading
ensanche de banda bandspread
ensanche de banda eléctrico electrical bandspread
ensanche de banda mecánico mechanical bandspread
ensanche de banda paralelo parallel bandspread
ensayo assay
entalla notch
entidad generalizada generalized entity
entrada entry, input
entrada acústica acoustic input
entrada aislada isolated input
entrada analógica analog input
entrada asimétrica asymmetrical input
entrada asincrónica asynchronous input
entrada binaria binary input
entrada bipolar bipolar input
entrada cero zero input
entrada coaxial coaxial input
entrada con guarda guarded input
entrada condicional conditional entry
entrada de almacenamiento storage entry
entrada de almacenamiento de factores factor storage entry
entrada de almacenamiento general general-storage entry
entrada de amplificador amplifier input
entrada de antena antenna input
entrada de antena equilibrada balanced antenna input
entrada de banda base baseband input
entrada de cable cable entry
entrada de canal channel entry
entrada de cinta tape input
entrada de cinta de papel paper-tape input
entrada de computadora computer input
entrada de computadora objeto object computer entry
entrada de conducto duct entrance
entrada de contacto contact input
entrada de contador counter entry
entrada de controlador programable programmable controller input

entrada de corriente en obscuridad equivalente equivalent dark-current input
entrada de datos data entry, data input
entrada de datos directa direct data entry
entrada de datos manual manual data input
entrada de descripción de archivo file description entry
entrada de descripción de datos data description entry
entrada de distribución distribution entry
entrada de gráficos graphics input
entrada de habilitación de chip chip enable input
entrada de imagen vision input
entrada de índice index entry
entrada de micrófono microphone input
entrada de modo diferencial differential-mode input
entrada de nivel cero zero-level input
entrada de orden order entry
entrada de placa plate input
entrada de potencia anódica anode power input
entrada de potencia anódica total total anode power input
entrada de potencia de circuito de carga load-circuit power input
entrada de potencia de placa plate power input
entrada de potencia principal main power input
entrada de proceso process entry
entrada de programa program entry
entrada de referencia reference input
entrada de ruido aleatorio random-noise input
entrada de ruido equivalente equivalent noise input
entrada de señal signal input
entrada de señal de imagen picture-signal input
entrada de señal fuerte large-signal input
entrada de sistema system input
entrada de subrutina subroutine entry
entrada de tarjetas card input
entrada desequilibrada unbalanced input
entrada diferencial differential input
entrada diferida deferred entry
entrada digital digital input, digital entry
entrada directa direct input
entrada en contrafase push-pull input
entrada en paralelo parallel entry
entrada en tiempo real real-time input
entrada equilibrada balanced input
entrada equivalente de ruido noise equivalent input
entrada ficticia dummy entry
entrada flotante floating input
entrada fonográfica phonograph input
entrada horizontal horizontal input
entrada ilegal illegal entry
entrada indirecta indirect entry
entrada inhibidora inhibiting input
entrada instrumental instrumental input
entrada inversa reverse entry
entrada inversora inverting input
entrada invertida inverted input
entrada libre clear input
entrada manual manual input
entrada no inversora noninverting input
entrada nominal nominal input, rated input
entrada para señal pequeña small-signal input
entrada por teclado keyboard entry
entrada positiva positive input
entrada primaria primary input
entrada-procesamiento-salida input-process-output
entrada requerida required input
entrada-salida input-output
entrada-salida analógica analog input-output
entrada-salida asimétrica asymmetrical input-output
entrada-salida controlada por programa program-controlled input-output

entrada-salida de página page input-output
entrada-salida en paralelo parallel input-output
entrada-salida en serie serial input-output
entrada-salida en tiempo real real-time input-output
entrada-salida formateada formatted input-output
entrada-salida gráfica graphic input-output
entrada-salida limitada limited input-output
entrada-salida no formateada unformatted
    input-output
entrada-salida programada programmed
    input-output
entrada-salida simultánea simultaneous input-output
entrada secundaria secondary input
entrada simétrica symmetrical input
entrada sinusoidal sinusoidal input
entrada unipolar unipolar input
entrada variable variable input
entrada vertical vertical input
entradas aleatorias random inputs
entradas de acción action entries
entradas sincrónicas synchronous inputs
entrar al sistema log in, log on
entrehierro gap, air gap, magnetic gap, pole gap
entrehierro de arrancador starter gap
entrehierro de cabeza head gap
entrehierro de contactos de relé relay-contact gap
entrehierro de imán magnet gap
entrehierro de registro recording gap
entrehierro de reproducción playback gap
entrehierro del inducido armature gap
entrehierro resonante resonant gap
entrelazado interlacing, interleaving
entrelazado de almacenamiento storage interleaving
entrelazado de frecuencias frequency interlace
entrelazado de impulsos pulse interleaving
entrelazado de líneas line interlacing
entrelazado de líneas impares odd-line interlace
entrelazado de puntos dot interlacing
entrelazado secuencial sequential interlace
entrelazado vertical vertical interlace
entrelazamiento interlacing
entrelazamiento de líneas impares odd-line
    interlacing
entrelazamiento progresivo progressive interlace
entrelazar interlace
entrenamiento basado en computadora
    computer-based training
entropía entropy
envainado sheathing
envejecimiento aging
envejecimiento de aislador aging of insulator
envejecimiento de componentes component aging
envejecimiento de tubos tube aging
envejecimiento magnético magnetic aging
envejecimiento térmico thermal aging
envenenamiento catódico cathode poisoning
envío de alarma alarm sending
envoltura wrapping
envoltura de núcleo core wrapping
envoltura final final wrap
envoltura interna inner jacket
envolvente envelope
envolvente de impulsos pulse envelope
envolvente de modulación modulation envelope
envolvente de onda de señal signal-wave envelope
envolvente de señal signal envelope
envolvente de vidrio glass envelope
envuelta housing
epitaxia epitaxy
epitaxial epitaxial
epitáxico epitaxic
épsilon epsilon
equifase equiphase

equilibrado balanced
equilibrado estáticamente statically balanced
equilibrador balancer
equilibrador de corriente continua direct-current
    balancer
equilibrador electrónico electronic balancer
equilibrador estático static balancer
equilibrio balance, equilibrium
equilibrio a cero null balance
equilibrio activo active balance
equilibrio de barrera barrier balance
equilibrio de canales channel balance
equilibrio de carga load balance
equilibrio de circuito circuit balance
equilibrio de colores color balance
equilibrio de corriente current balance
equilibrio de corriente continua direct-current
    balance
equilibrio de fases phase balance
equilibrio de línea line balance
equilibrio de presión pressure equilibrium
equilibrio de puente bridge balance
equilibrio de resistencias resistance balance
equilibrio dinámico dynamic equilibrium
equilibrio eléctrico electric balance
equilibrio electrónico electronic balance
equilibrio energético energy balance
equilibrio estático static equilibrium
equilibrio flotante floating balance
equilibrio híbrido hybrid balance
equilibrio longitudinal longitudinal balance
equilibrio magnético magnetic balance
equilibrio magnetohidrodinámico
    magnetohydrodynamic equilibrium
equilibrio reactivo reactive balance
equilibrio respecto a tierra balance to ground
equilibrio térmico thermal equilibrium
equilibrio tonal tonal balance
equipado con radar radar-equipped
equipo equipment, gear, device
equipo accesorio accessory equipment
equipo aerotransportado airborne equipment
equipo amplificador amplifying equipment
equipo asociado associated equipment
equipo automático automatic equipment
equipo automatizado automated equipment
equipo auxiliar auxiliary equipment
equipo compulsorio compulsory equipment
equipo computador computing equipment
equipo común common equipment
equipo con cubierta metálica metal-enclosed
    equipment
equipo contra efectos locales antisidetone equipment
equipo convencional conventional equipment
equipo criptográfico cryptographic equipment
equipo de acoplamiento en paralelo paralleling
    equipment
equipo de adquisición de datos data-acquisition
    equipment
equipo de aeronave aircraft equipment
equipo de alarma automática automatic alarm
    equipment
equipo de alimentación supply equipment
equipo de alta fidelidad high-fidelity equipment
equipo de alto rendimiento high-performance
    equipment
equipo de alumbrado lighting equipment
equipo de apoyo terrestre ground support equipment
equipo de barco shipboard equipment
equipo de barras cruzadas crossbar equipment
equipo de canal channel equipment
equipo de canalización channeling equipment
equipo de capacitor capacitor equipment

equipo de central exchange equipment
equipo de colimación collimation equipment
equipo de colores color equipment
equipo de comparación comparison equipment
equipo de comprobación checkout equipment
equipo de computadora computer equipment
equipo de comunicación por portadora carrier-communication equipment
equipo de comunicaciones de datos data communications equipment
equipo de comunicaciones móviles mobile communications equipment
equipo de conjunto de datos data set equipment
equipo de conmutación switching equipment
equipo de conmutación automático automatic switching equipment
equipo de contestación answering equipment
equipo de control control equipment
equipo de control de conmutación automático automatic switching control equipment
equipo de control de máquina automático automatic machine control equipment
equipo de control de procesos process-control equipment
equipo de control de transformador automático automatic transformer control equipment
equipo de control remoto remote-control equipment
equipo de controlador controller equipment
equipo de conversión conversion equipment
equipo de conversión de señal signal-conversion equipment
equipo de corriente alterna alternating-current equipment
equipo de corriente continua direct-current equipment
equipo de datos terrestre ground data equipment
equipo de detección de fugas de gas gas-leak detection equipment
equipo de detección de superficie de aeropuerto airport surface-detection equipment
equipo de dispersión directa forward-scatter equipment
equipo de drenaje drainage equipment
equipo de electroforesis electrophoresis equipment
equipo de emergencia emergency equipment
equipo de enfasamiento phasing equipment
equipo de ensamblaje automatizado automated assembly equipment
equipo de entrada input equipment
equipo de entrada-salida input-output equipment
equipo de estación station equipment
equipo de estación central central station equipment
equipo de excitación excitation equipment
equipo de exploración scanning equipment
equipo de facsímil facsimile equipment
equipo de guiado guidance equipment
equipo de Hollerith Hollerith equipment
equipo de indicación visual de fallas fault display equipment
equipo de interfaz de comunicaciones communications interface equipment
equipo de lectura readout equipment
equipo de lectura angular angular reading equipment
equipo de limpieza ultrasónica ultrasonic-cleaning equipment
equipo de línea line equipment
equipo de línea compartida party-line equipment
equipo de manejo de datos data-handling equipment
equipo de medida measuring equipment
equipo de medida de laboratorio laboratory measuring equipment
equipo de medida de pH pH measuring equipment

equipo de medida de tensión voltage-measurement equipment
equipo de microondas microwave equipment
equipo de modulación de grupo group modulation equipment
equipo de modulación de supergrupo supergroup modulating equipment
equipo de monitoreo automático automatic monitoring equipment
equipo de navegación inercial inertial navigation equipment
equipo de originación de colores color-origination equipment
equipo de portadora carrier equipment
equipo de preparación de datos data preparation equipment
equipo de privacidad privacy equipment
equipo de procesamiento de datos data-processing equipment
equipo de procesamiento de datos automático automatic data processing equipment
equipo de procesamiento de datos electrónico electronic data processing equipment
equipo de protección de línea line-protection equipment
equipo de prueba automático automatic test equipment
equipo de prueba incorporado built-in test equipment
equipo de prueba manual manual test equipment
equipo de prueba semiautomático semiautomatic test equipment
equipo de pruebas test equipment, test set
equipo de radar radar equipment, radar set
equipo de radar automático automatic radar equipment
equipo de radio radio equipment, radio set
equipo de radiorrelé radio-relay equipment
equipo de rastreo tracing equipment
equipo de recopilación de datos data collection equipment
equipo de registración de datos data-logging equipment
equipo de registro recording equipment
equipo de registro de sonido sound-recording equipment
equipo de repuesto spare equipment
equipo de reserva reserve equipment, standby equipment, backup
equipo de salida output equipment
equipo de señalización signaling equipment
equipo de servicio service equipment
equipo de soldadura por resistencia resistance welding equipment
equipo de sonido sound equipment
equipo de suministro eléctrico electric-supply equipment
equipo de tabulación tabulating equipment
equipo de tarjetas perforadas punched-card equipment
equipo de tecnología de información information technology equipment
equipo de telecontrol telecontrol equipment
equipo de terminal de datos data-terminal equipment
equipo de terminal de portadora carrier terminal equipment
equipo de tierra ground equipment
equipo de transferencia automático automatic transfer equipment
equipo de transmisión de datos data-transmission equipment
equipo de trayectoria de planeo glide-path

equipment
equipo digital digital equipment
equipo eléctrico electrical equipment
equipo eléctrico exterior outdoor electrical
    equipment
equipo eléctrico interior indoor electrical equipment
equipo electromédico electromedical equipment
equipo electrónico electronic equipment
equipo electrónico aerotransportado airborne
    electronic equipment
equipo electrónico industrial industrial electronic
    equipment
equipo electrónico marítimo marine electronic
    equipment
equipo electrónico miniaturizado miniaturized
    electronic equipment
equipo en línea on-line equipment
equipo fijo fixed equipment
equipo físico hardware
equipo físico adicional additional hardware
equipo físico compatible compatible hardware
equipo físico común common hardware
equipo físico de doble precisión double-precision
    hardware
equipo físico de entrada de gráficos graphics-input
    hardware
equipo físico de máquina machine hardware
equipo físico de montaje mounting hardware
equipo físico de salida de gráficos graphics-output
    hardware
equipo físico defectuoso faulty hardware
equipo flexible flexible equipment
equipo fuera de línea off-line equipment
equipo genérico generic equipment
equipo impresor printing equipment
equipo indicador de fallas fault-indicating equipment
equipo intermedio intermediate equipment
equipo manual manual equipment
equipo montado en bastidor rack-mounted
    equipment
equipo multicanal multichannel equipment
equipo múltiplex multiplex equipment
equipo normal standard equipment
equipo operable operable equipment
equipo operado por tarjetas card-operated
    equipment
equipo paso a paso step-by-step equipment
equipo periférico peripheral equipment
equipo periférico de control de procesos
    process-control peripheral equipment
equipo principal main equipment
equipo programable programmable equipment
equipo rebecca rebecca equipment
equipo rebecca-eureka rebecca-eureka equipment
equipo receptor receiving equipment, receiving set
equipo redundante redundant equipment
equipo regulador de temperatura
    temperature-regulating equipment
equipo sellable sealable equipment
equipo supersónico supersonic equipment
equipo suplementario supplementary equipment
equipo telefónico telephone equipment
equipo telegráfico telegraph equipment
equipo terminal terminal equipment
equipo terminal de canal channel terminal
    equipment
equipo transmisor transmitting equipment
equipo transportable transportable equipment
equipo ultrasónico ultrasonic equipment
equipotencial equipotential
equiseñal equisignal
equivalencia equivalence
equivalencia cuántica quantum equivalence

equivalencia lógica logical equivalence
equivalente equivalent
equivalente binario binary equivalent
equivalente de articulación articulation equivalent
equivalente de atenuación attenuation equivalent
equivalente de circuito circuit equivalent
equivalente de diodo diode equivalent
equivalente de Norton Norton's equivalent
equivalente de plomo lead equivalent
equivalente de presión de ruido noise pressure
    equivalent
equivalente de puente de Wien Wien-bridge
    equivalent
equivalente de referencia reference equivalent
equivalente de referencia en volumen volume
    equivalent
equivalente de repetición repetition equivalent
equivalente de transmisión transmission equivalent
equivalente decimal decimal equivalent
equivalente electroquímico electrochemical
    equivalent
equivalente-gramo gram-equivalent
equivalente mecánico de calor mechanical
    equivalent of heat
equivalente mecánico de luz mechanical equivalent
    of light
equivalente mínimo minimum equivalent
equivalente químico chemical equivalent
equivalente relativo relative equivalent
erbio erbium
erg erg
ergógrafo ergograph
ergómetro ergometer
erlang erlang
erosión erosion
erosión de contacto contact erosion
erosión eléctrica electrical erosion
erosión por chispa eléctrica electric-spark erosion
erosión química chemical erosion
errático erratic
erróneo erroneous
error error
error absoluto absolute error
error accidental accidental error
error acumulado accumulated error
error acumulativo cumulative error
error aleatorio random error
error aparente apparent error
error asignado assigned error
error catastrófico catastrophic error
error cero zero error
error compensador compensating error
error computacional computational error
error continuo continuous error
error cuadrantal quadrantal error
error de adaptación matching error
error de agente agent error
error de altura ionosférico ionospheric height error
error de ambigüedad ambiguity error
error de amplitud amplitude error
error de anchura de haz beamwidth error
error de archivo file error
error de bit bit error
error de bloque block error
error de bucle loop error
error de cadena chain error
error de calibración calibration error
error de carga loading error
error de cinta tape error
error de codificación coding error
error de coincidencia coincidence error
error de comparación comparison error
error de componente component error

error de conducción conduction error
error de control de área area control error
error de corriente current error
error de cuantificación quantization error
error de datos data error
error de desfase phase-displacement error
error de distancia range error
error de emplazamiento site error
error de equipo físico hardware error
error de escritura write error
error de espaciado residual residual spacing error
error de factor de escala scale-factor error
error de facturación billing error
error de fase phase error
error de flujo de gas gas-flow error
error de forma de onda waveform error
error de formato format error
error de frecuencia frequency error
error de impresión misprint
error de instrumento instrument error
error de lectura read error, readout error
error de lectura-escritura permanente permanent read-write error
error de límite de escala full-scale error
error de linealidad linearity error
error de máquina machine error
error de medida measurement error
error de medida negativo negative measurement error
error de medida positivo positive measurement error
error de memoria memory error
error de memoria sólo de lectura read-only memory error
error de no linealidad nonlinearity error
error de observación observation error
error de onda patrón standard-wave error
error de paralaje parallax error
error de paridad parity error
error de perforación punching error
error de polarización polarization error
error de polarización total total polarization error
error de precisión precision error
error de procedimiento procedural error
error de programa program error, program bug, bug
error de programación programming error
error de propagación propagation error
error de razón ratio error
error de redondeo rounding error
error de reflexión reflection error
error de registro registration error
error de repetibilidad repeatability error
error de resolución resolution error
error de retardo de fase phase-delay error
error de retardo de velocidad velocity-lag error
error de secuencia sequence error
error de seguimiento tracking error
error de separación spacing error
error de sincronización synchronization error
error de sintaxis syntax error
error de sintonización tuning error
error de sistema system error
error de sustitución substitution error
error de temporización timing error
error de tiempo time error
error de transacción transaction error
error de trayectorias múltiples multipath error
error de velocidad velocity error
error desequilibrado unbalanced error
error detectado detected error
error dinámico dynamic error
error electrostático electrostatic error

error equilibrado balanced error
error estándar standard error
error estático static error
error estático cero zero static error
error externo external error
error falso false error
error fatal fatal error
error fijo fixed error
error generado generated error
error geométrico geometric error
error heredado inherited error
error hiperbólico hyperbolic error
error inherente inherent error
error instrumental instrumental error
error intermitente intermittent error
error ionosférico ionospheric error
error irrecuperable unrecoverable error
error límite limiting error
error lógico logic error
error medio mean error
error múltiple multiple error
error no recuperable nonrecoverable error
error nocturno night error
error observable observable error
error octantal octantal error
error permanente permanent error
error por centelleo scintillation error
error por desbordamiento overflow error
error por desbordamiento negativo underflow error
error por deslizamiento del cero zero-drift error
error por duración de impulsos pulse-duration error
error por efecto nocturno night-effect error
error por fricción frictional error
error por histéresis hysteresis error
error por incremento repentino burst error
error por interferencia de ondas wave-interference error
error por oscilación de punto stagger
error por refracción refraction error
error por terreno terrain error
error posicional positional error
error preoperativo preoperative error
error probable probable error
error propagado propagated error
error recuperable recoverable error
error relativo relative error
error repetitivo repetitive error
error residual residual error
error resultante resulting error
error semántico semantic error
error semicircular semicircular error
error sensible al programa program-sensitive error
error sistemático systematic error
error sólido solid error
error temporal temporary error
error transitorio transient error
escala range, scale, scaler
escala absoluta absolute scale
escala acromática achromatic scale
escala binaria binary scale
escala calibrada calibrated scale
escala Celsius Celsius scale
escala centígrada centigrade scale
escala de calibración calibration scale
escala de cero central zero-center scale
escala de comparación comparison scale
escala de enfoque focusing scale
escala de espejo mirror scale
escala de frecuencias logarítmica logarithmic frequency scale
escala de grises gray scale
escala de Kelvin Kelvin scale
escala de legibilidad readability scale

escala de medidor meter scale
escala de nivel sonoro sound-level scale
escala de Pitágoras Pythagorean scale
escala de temperatura temperature scale
escala de temperatura absoluta absolute temperature scale
escala de tiempo time scale
escala de turbulencia scale of turbulence
escala decádica decade scaler
escala decibelimétrica decibel scale
escala ensanchada expanded scale
escala Fahrenheit Fahrenheit scale
escala horizontal horizontal scale
escala iluminada por los bordes edge-lighted scale
escala indicadora indicating scale
escala lineal linear scale
escala logarítmica logarithmic scale
escala no lineal nonlinear scale
escala Rankine Rankine scale
escala Reaumur Reaumur scale
escala termométrica thermometer scale
escala vertical vertical scale
escalamiento scaling
escalar scalar
escalera staircase
escalera eléctrica electric escalator
escalón echelon
escalón frontal front porch
escalón posterior back porch
escalonado staggered
escalonamiento staggering
escalonamiento de frecuencias frequency staggering
escandio scandium
escanistor scanistor
escansión scansion
escape escapement
escape de gas gas escape
escobilla brush, pickup, wiper
escobilla colectora feeder brush
escobilla compuesta compound brush
escobilla de carbón carbon brush
escobilla de carbón-grafito carbon-graphite brush
escobilla de clasificación sort brush
escobilla de cobre copper brush
escobilla de colector commutator brush
escobilla de control control brush
escobilla de grafito graphite brush
escobilla de hilo wire brush
escobilla de lectura reading brush
escobilla de lectura primaria primary reading brush
escobilla de lectura secundaria secondary reading brush
escobilla de línea line brush
escobilla de motor motor brush
escobilla de perforación punch brush
escobilla de prueba test brush
escobilla de secuencia sequence brush
escobilla dividida split brush
escobilla exterior outer brush
escobilla fija fixed brush
escobilla graduada graded brush
escobilla laminada laminated brush
escobilla metálica metal brush
escobilla metalizada metallized brush
escobilla miniatura miniature brush
escobilla positiva positive brush
escobilla primaria primary brush
escobilla secundaria secondary brush
escobilla superior upper brush
escobillas apareadas paired brushes
escribir write
esfera sphere
esfera de Poincare Poincare sphere

esfera de radiación radiation sphere
esfera de Ulbricht Ulbricht sphere
esféricamente spherically
esférico spherical
esfuerzo strain, stress
esfuerzo dieléctrico dielectric strain
esfuerzo dinámico dynamic strain
esfuerzo eléctrico electric strain
esfuerzo electrostático electrostatic strain
eslabón link
esmalte de porcelana porcelain enamel
espaciado spacing
espaciado angular angular spacing
espaciado de bobinas coil spacing
espaciado de bobinas de carga loading-coil spacing
espaciado de canales channel spacing
espaciado de caracteres character spacing
espaciado de electrodos electrode spacing
espaciado de frecuencias portadoras carrier-frequency spacing
espaciado de impulsos pulse spacing
espaciado de líneas line spacing
espaciado de placas plate spacing
espaciado de portadoras carrier spacing
espaciado entre caracteres intercharacter spacing
espaciado entre palabras interword spacing
espaciado incremental incremental spacing
espaciador spacer
espaciador de ebonita ebonite spacer
espaciador de jacks jack spacer
espaciamiento pitch
espacio gap, space
espacio adicional additional space
espacio aéreo controlado controlled airspace
espacio de aceleración acceleration space
espacio de aire air gap
espacio de archivo disponible available file space
espacio de bloques block gap
espacio de byte byte space
espacio de captación catcher space
espacio de carácter character space
espacio de colores color space
espacio de conmutación switching space
espacio de contacto contact gap
espacio de corrimiento drift space
espacio de datos data space
espacio de descarga discharge gap
espacio de dirección address space
espacio de dirección de procesador processor address space
espacio de dirección maestra master address space
espacio de direcciones de memoria memory address space
espacio de disco disk space
espacio de dispositivo device space
espacio de entrada input gap
espacio de Fresnel Fresnel clearance
espacio de índice index space
espacio de interacción interaction gap
espacio de jacks jack space
espacio de memoria memory space
espacio de mensajes message space
espacio de reflexión reflector space
espacio de salida output gap
espacio de trabajo working space
espacio de trabajo seguro safe working space
espacio de unidad unit space
espacio de visualización display space
espacio diferencial differential gap
espacio entre archivos file gap
espacio entre bloques interblock gap
espacio entre caracteres intercharacter space
espacio entre palabras interword gap

espacio entre polos clearance between poles
espacio entre registros interrecord gap
espacio entre varillas rod gap
espacio equipotencial equipotential space
espacio interelectródico interelectrode space
espacio libre free space
espacio libre distribuido distributed free space
espacio muerto dead space
espacio obscuro dark space
espacio obscuro anódico anode dark space
espacio obscuro catódico cathode dark space
espacio obscuro de Aston Aston dark space
espacio obscuro de Crookes Crookes dark space
espacio obscuro de Faraday Faraday dark space
espacio obscuro de Hittorf Hittorf dark space
espacio obscuro de Langmuir Langmuir dark space
espacio para trepar climb space
espacio protector protective gap
espacio requerido required space
espacio-tiempo space-time
espacios finales end spaces
espacistor spacistor
espagueti spaghetti
espalación spallation
esparcidor spreader
esparcimiento scatter, scattering
especializado specialized
especificación specification, entry
especificación conservadora conservative
    specification
especificación de condición condition entry
especificación de control control specification
especificación de datos data specification
especificación de desarrollo development
    specification
especificación de grupo cluster entry
especificación de interfaz interface specification
especificación de nivel de campo field level
    specification
especificación de producto product specification
especificación de programa program specification
especificación de prueba test specification
especificación de puntos de comprobación
    checkpoint entry
especificación de rendimiento performance
    specification
especificación de requisitos requirements
    specification
especificación eléctrica electrical specification
especificación formal formal specification
especificación funcional functional specification
especificación provisional provisional specification
especificación técnica technical specification
especificaciones ambientales environmental
    specifications
especificaciones de cabecera header specifications
especificaciones de cálculo calculation specifications
especificaciones de descripción de archivo file
    description specifications
especificaciones de descripción de campo field
    description specifications
especificaciones de descripción de datos data
    description specifications
especificaciones de diseño design specifications
especificaciones de entrada input specifications
especificaciones de procesamiento processing
    specifications
especificaciones de programas software
    specifications
especificaciones de salida output specifications
especificaciones de secuencia sequence
    specifications
especificaciones de tarjeta de control control card

specifications
específico specific
espectral spectral
espectro spectrum
espectro acústico sound spectrum
espectro atómico atomic spectrum
espectro audible audible spectrum
espectro continuo continuous spectrum
espectro cromático chromatic spectrum
espectro de absorción absorption spectrum
espectro de absorción de infrarrojo infrared
    absorption spectrum
espectro de amplitud amplitude spectrum
espectro de arco arc spectrum
espectro de audio audio spectrum
espectro de audiofrecuencia audio-frequency
    spectrum
espectro de bandas band spectrum
espectro de colores color spectrum
espectro de chispa spark spectrum
espectro de desvanecimiento fading spectrum
espectro de difracción diffraction spectrum
espectro de emisión emission spectrum
espectro de emisión fotónica photon-emission
    spectrum
espectro de energía constante equal-energy
    spectrum
espectro de fase phase spectrum
espectro de Fourier Fourier spectrum
espectro de frecuencias frequency spectrum
espectro de frecuencias de impulsos
    pulse-frequency spectrum
espectro de frecuencias electromagnético
    electromagnetic frequency spectrum
espectro de impulso pulse spectrum
espectro de interferencia interference spectrum
espectro de líneas line spectrum
espectro de luz light spectrum
espectro de masa mass spectrum
espectro de microondas microwave spectrum
espectro de onda media medium-wave spectrum
espectro de ondas cortas short-wave spectrum
espectro de ondas largas long-wave spectrum
espectro de polarización bias spectrum
espectro de potencia power spectrum
espectro de radio radio spectrum
espectro de radiofrecuencias radio-frequency
    spectrum
espectro de Raman Raman spectrum
espectro de rayos beta beta ray spectrum
espectro de rayos gamma gamma-ray spectrum
espectro de rayos X X-ray spectrum
espectro de resonancia resonance spectrum
espectro de resonancia magnética magnetic
    resonance spectrum
espectro de respuesta response spectrum
espectro de respuesta de banda ancha broadband
    response spectrum
espectro de respuesta de banda angosta
    narrowband response spectrum
espectro de rotación rotation spectrum
espectro de ruido noise spectrum
espectro de transferencia de carga charge-transfer
    spectrum
espectro de tren de impulsos pulse-train spectrum
espectro de video video spectrum
espectro de voz speech spectrum
espectro electromagnético electromagnetic spectrum
espectro energético energy spectrum
espectro ensanchado spread spectrum
espectro fotoeléctrico photoelectric spectrum
espectro infraacústico subaudio spectrum
espectro infrarrojo infrared spectrum

espectro luminoso luminous spectrum
espectro molecular molecular spectrum
espectro monocromático monochromatic spectrum
espectro óptico optical spectrum
espectro primario primary spectrum
espectro radiado radiated spectrum
espectro ultravioleta ultraviolet spectrum
espectro visible visible spectrum
espectroanálisis spectroanalysis
espectrofluorometría spectrofluorometry
espectrofluorométrico spectrofluorometric
espectrofluorómetro spectrofluorometer
espectrofotoeléctrico spectrophotoelectric
espectrofotografía spectrophotography
espectrofotométrico spectrophotometric
espectrofotómetro spectrophotometer
espectrofotómetro de absorción atómica atomic
      absorption spectrophotometer
espectrofotómetro de conteo de impulsos
      pulse-counting spectrophotometer
espectrofotómetro de llama flame
      spectrophotometer
espectrofotómetro de reflectancia reflectance
      spectrophotometer
espectrogoniómetro de rayos X X-ray
      spectrogoniometer
espectrografía spectrography
espectrográfico spectrographic
espectrógrafo spectrograph
espectrógrafo de cristal crystal spectrograph
espectrógrafo de frecuencia frequency spectrograph
espectrógrafo de masas mass spectrograph
espectrógrafo de rayos X X-ray spectrograph
espectrógrafo de vacío vacuum spectrograph
espectrógrafo de velocidad velocity spectrograph
espectrógrafo magnético magnetic spectrograph
espectrógrafo multicanal multichannel spectrograph
espectrograma spectrogram
espectrograma de rayos X X-ray spectrogram
espectrometría spectrometry
espectrometría de centelleos scintillation
      spectrometry
espectrometría de masa mass spectrometry
espectrométrico spectrometric
espectrómetro spectrometer
espectrómetro analizador analyzing spectrometer
espectrómetro de Bragg Bragg spectrometer
espectrómetro de centelleos scintillation
      spectrometer
espectrómetro de cristal crystal spectrometer
espectrómetro de helio helium spectrometer
espectrómetro de ionización ionization spectrometer
espectrómetro de llama flame spectrometer
espectrómetro de masa mass spectrometer
espectrómetro de masa de Aston Aston mass
      spectrometer
espectrómetro de masa de helio helium mass
      spectrometer
espectrómetro de radiofrecuencia radio-frequency
      spectrometer
espectrómetro de Raman Raman spectrometer
espectrómetro de rayos beta beta ray spectrometer
espectrómetro de rayos gamma gamma-ray
      spectrometer
espectrómetro de rayos X X-ray spectrometer
espectrómetro magnético magnetic spectrometer
espectrómetro neutrónico neutron spectrometer
espectrómetro óptico optical spectrometer
espectromicroscópico spectromicroscopic
espectromicroscopio spectromicroscope
espectropolarimétrico spectropolarimetric
espectropolarímetro spectropolarimeter
espectrorradiómetro spectroradiometer

espectroscopia spectroscopy
espectroscopia de fluorescencia fluorescence
      spectroscopy
espectroscopia de masa mass spectroscopy
espectroscopia de microondas microwave
      spectroscopy
espectroscopia de Raman Raman spectroscopy
espectroscopia gamma gamma spectroscopy
espectroscopia hertziana Hertzian spectroscopy
espectroscopia infrarroja infrared spectroscopy
espectroscopia por microondas molecular
      molecular microwave spectroscopy
espectroscopia ultravioleta ultraviolet spectroscopy
espectroscópico spectroscopic
espectroscopio spectroscope
espectroscopio de masa mass spectroscope
espectroscopio electrónico electronic spectroscope
especular specular
espejismo acústico acoustic mirage
espejismo auditivo auditory mirage
espejismo de radar radar mirage
espejo mirror
espejo de microondas microwave mirror
espejo de observación observation mirror
espejo de superficie reflectora frontal front-surface
      mirror
espejo de visión viewing mirror
espejo dicroico dichroic mirror
espejo dieléctrico dielectric mirror
espejo eléctrico electric mirror
espejo electromagnético electromagnetic mirror
espejo electrónico electronic mirror
espejo esférico spherical mirror
espejo magnético magnetic mirror
espejo móvil moving mirror
espejo óptico optical mirror
espejo parabólico parabolic mirror
espejo rotativo rotating mirror
espera activa active wait
espera de llamadas call waiting
espera de página page wait
espiga peg, pin, prong, dowel
espiga antideslizante drive pin
espín spin
espín de electrón electron spin
espín electrónico electronic spin
espinela ferromagnética ferromagnetic spinel
espintariscopio spinthariscope
espinterómetro spark gap
espinterómetro de chispa amortiguada quenched
      spark-gap
espiral spiral
espiral de entrada lead-in spiral
espiral de Roget Roget spiral
espurio spurious
esquelético skeletal
esqueleto skeleton
esquema de asignación allocation scheme
esquema de circuito circuit schematic
esquema de codificación coding scheme
esquema de control control scheme
esquema de montaje hookup
esquema de prioridad priority scheme
esquema de programa program scheme
esquema de red network chart
esquema interno internal schema
esquema lógico logical schema
esquemático schematic
esquiatrón skiatron
estabilidad stability
estabilidad a corto plazo short-term stability
estabilidad a largo plazo long-term stability
estabilidad absoluta absolute stability

estabilidad ambiental environmental stability
estabilidad computacional computational stability
estabilidad condicional conditional stability
estabilidad de alimentación feed stability
estabilidad de arco arc stability
estabilidad de circuito circuit stability
estabilidad de corto tiempo short-time stability
estabilidad de fase phase stability
estabilidad de frecuencia frequency stability
estabilidad de frecuencia portadora
    carrier-frequency stability
estabilidad de ganancia gain stability
estabilidad de nivel de negro black level stability
estabilidad de sistema de excitación excitation
    system stability
estabilidad de temperatura temperature stability
estabilidad de transmisión transmission stability
estabilidad del ajuste de cero zero-line stability
estabilidad del cero zero stability
estabilidad dimensional dimensional stability
estabilidad dinámica dynamic stability
estabilidad direccional dinámica dynamic
    directional stability
estabilidad en cortocircuito short-circuit stability
estabilidad estructural structural stability
estabilidad incondicional unconditional stability
estabilidad inherente inherent stability
estabilidad lateral lateral stability
estabilidad limitada limited stability
estabilidad longitudinal longitudinal stability
estabilidad marginal marginal stability
estabilidad mecánica mechanical stability
estabilidad natural natural stability
estabilidad neutra neutral stability
estabilidad térmica thermal stability
estabilidad termodinámica thermodynamic stability
estabilidad transitoria transient stability
estabilitrón stabilitron
estabilización stabilization, settling, burn-in
estabilización de acimut azimuth stabilization
estabilización de datos data stabilization
estabilización de deslizamiento drift stabilization
estabilización de emisor emitter stabilization
estabilización de frecuencia frequency stabilization
estabilización de inclinación tilt stabilization
estabilización de línea line stabilization
estabilización de oscilador oscillator stabilization
estabilización de plataforma platform stabilization
estabilización de polarización bias stabilization
estabilización de red network stabilization
estabilización de temperatura de diodo diode
    temperature stabilization
estabilización de tensión voltage stabilization
estabilización de tensión de carga load-voltage
    stabilization
estabilización de tensión de salida output voltage
    stabilization
estabilización por gradiente de gravedad gravity
    gradient stabilization
estabilización por interrupción periódica chopper
    stabilization
estabilización por retroalimentación feedback
    stabilization
estabilizado stabilized
estabilizado por resistencia resistance-stabilized
estabilizador stabilizer, ballast
estabilizador de carga load stabilizer
estabilizador de corona corona stabilizer
estabilizador de corriente current stabilizer
estabilizador de línea base baseline stabilizer
estabilizador de neón neon stabilizer
estabilizador de sistema de alimentación power
    system stabilizer

estabilizador de sistema de excitación excitation
    system stabilizer
estabilizador de tensión voltage stabilizer
estabilizador de tensión electrónico electronic
    voltage stabilizer
estabilizador de tensión estático static voltage
    stabilizer
estabilizador de tensión variable variable-voltage
    stabilizer
estabilizar stabilize
estabistor stabistor
estable stable
establecimiento settling
establecimiento de sesión session establishment
estación station
estación A A station
estación aceptante accepting station
estación activa active station
estación aeronáutica aeronautical station
estación autoalimentada self-powered station
estación automática automatic station
estación autónoma self-contained station
estación auxiliar auxiliary station
estación B B station
estación central central station
estación clase A class A station
estación clase B class B station
estación colgante pendant station
estación conectada por enlaces link-attached station
estación controlada remotamente remote-controlled
    station
estación costera coast station
estación de abonado subscriber station
estación de aeronave aircraft station
estación de alarmas central central alarm station
estación de altímetro altimeter station
estación de aterrizaje por instrumentos
    instrument-landing station
estación de barco ship station
estación de base base station
estación de bloque block station
estación de calentamiento heating station
estación de cinta tape station
estación de cinta magnética magnetic-tape station
estación de comparación comparison station
estación de comunicación de aerovía airway
    communication station
estación de conmutación switching station
estación de consulta inquiry station
estación de control control station
estación de control coaxial coaxial control station
estación de control de aeropuerto airport control
    station
estación de control de comunicaciones
    communications control station
estación de control de grupo group control station
estacíon de control de operador operator control
    station
estación de control de supergrupo supergroup
    control station
estación de datos data station
estación de datos múltiplex multiplex data station
estación de datos remota remote data station
estación de destino destination station
estación de distribución distribution station
estación de enlace link station
estación de enlace primario primary link station
estación de enlace secundario secondary link station
estación de entrada input station
estación de entrada de datos data input station
estación de escobillas brush station
estación de exploración óptica optical scanning
    station

estación de grupo cluster station
estación de impresión printing station
estación de información de trabajos job information station
estación de lectura reading station
estación de lectura adicional additional read station
estación de lectura de tarjetas card reading station
estación de lorán loran station
estación de medida measuring station
estación de microondas microwave station
estación de monitoreo monitoring station
estación de numeración en serie serial numbering station
estación de ondas ultracortas ultrashort-wave station
estación de operación operating station
estación de operador operator station
estación de operador de sistema system operator station
estación de origen originating station
estación de perforación punch station
estación de primera lectura first reading station
estación de procesamiento de datos data-processing station
estación de programa de control de red network control program station
estación de pruebas test station
estación de punto a punto point-to-point station
estación de radar radar station
estación de radar terrestre ground radar station
estación de radio aeronáutica aeronautical radio station
estación de radio de control aire-tierra air-ground control radio station
estación de radio de control de aeródromo aerodrome control radio station
estación de radio de superficie aeronáutica aeronautical ground radio station
estación de radiobaliza aeronáutica aeronautical marker-beacon station
estación de radiodifusión broadcast station
estación de radiodifusión de alta frecuencia high-frequency broadcast station
estación de radiodifusión de facsímil facsimile broadcast station
estación de radiodifusión de ondas largas long-wave broadcasting station
estación de radiodifusión internacional international broadcast station
estación de radiodifusión local local broadcast station
estación de radiodifusión marítima marine broadcast station
estación de radiodifusión por modulación de frecuencia frequency-modulation broadcast station
estación de radiofaro radio-beacon station
estación de radiofaro direccional radio-range station
estación de radiofaro marítimo marine radiobeacon station
estación de radiofaro omnidireccional omnidirectional range station
estación de radiolocalización radiolocation station
estación de radiorrelé radio-relay station
estación de recopilación de datos data collection station
estación de red network station
estación de referencia terrestre ground reference station
estación de relleno fill-in station
estación de reserva standby station
estación de respaldo backup station
estación de retransmisión de televisión television relay station
estación de retransmisión por cinta tape-relay station
estación de seguimiento tracking station
estación de servicio fijo aeronáutico aeronautical fixed service station
estación de subcontrol subcontrol station
estación de subcontrol de supergrupo supergroup subcontrol station
estación de suministro eléctrico electric-supply station
estación de superficie aeronáutica aeronautical ground station
estación de telecomunicación aeronáutica aeronautical telecommunication station
estación de televisión television station
estación de trabajo work station
estación de trabajo de función fija fixed-function work station
estación de trabajo local local work station
estación de trabajo maestra master workstation
estación de trabajo programable programmable work station
estación de trabajo remota remote work station
estación de transferencia de supergrupo supergroup transfer station
estación de transformación transforming station
estación de transmisión sending station
estación de trayectoria de planeo glide-path station
estación de unión junction station
estación de videodifusión television broadcast station
estación de visualización display station
estación de visualización de datos data display station
estación de visualización de mandos command display station
estación defectuosa faulty station
estación dependiente dependent station
estación emisora emitter station
estación equilibrada balanced station
estación experimental experimental station
estación fija fixed station
estación fija aeronáutica aeronautical fixed station
estación flotante floating station
estación generadora generating station
estación generadora hidroeléctrica hydroelectric generating station
estación guiadora homing station
estación hidroeléctrica hydroelectric station
estación inactiva inactive station
estación inatendida unattended station
estación interceptada intercepted station
estación intermedia intermediate station
estación local local station
estación maestra master station
estación meteorológica meteorological station
estación móvil mobile station
estación móvil de radionavegación radionavigation mobile station
estación móvil portátil portable mobile station
estación móvil terrestre land mobile station
estación pasiva passive station
estación portátil portable station
estación primaria primary station
estación principal main station
estación privada private station
estación pública public station
estación receptora receiving station
estación receptora privada private receiving station
estación receptora remota remote-receiver station
estación regulada regulated station
estación regular regular station
estación relé relay station

estación remota remote station
estación repetidora repeater station
estación repetidora automática automatic repeater station
estación repetidora de radio radio repeater station
estación retransmisora móvil mobile relay station
estación satélite satellite station
estación secundaria secondary station
estación semiautomática semiautomatic station
estación supervisora supervisory station
estación telefónica telephone station
estación telefónica privada private telephone station
estación telegráfica telegraph station
estación télex telex station
estación terminal terminal station
estación terrestre land station, ground station
estación terrestre de muy alta frecuencia very-high-frequency ground station
estación transmisora transmitting station
estación transmisora remota remote transmitter station
estación transportable transportable station
estación tributaria tributary station
estación única single station
estacional seasonal
estacionario stationary
estaciones de trabajo en grupo cluster workstations
estadística cuántica quantum statistics
estadístico statistical
estado state, status
estado absorbente absorbing state
estado activo active state
estado actual current state
estado anhiserético anhysteretic state
estado autorizado authorized state
estado cero zero state
estado corriente current state
estado cuántico quantum state
estado cuasiestable quasi-stable state
estado de agente agent status
estado de carga state of charge
estado de conexión on state
estado de configuración configuration state
estado de desconexión off state
estado de disponibilidad ready status
estado de dispositivo device status
estado de dispositivo externo external device status
estado de entrada input state
estado de entrada-salida input-output status
estado de equilibrio equilibrium state
estado de espera waiting state
estado de excitación excitation state
estado de interrupción outage state
estado de máquina machine state
estado de ocupado busy state
estado de plasma plasma state
estado de polarización state of polarization
estado de programa program status
estado de reposo resting state, quiescent state, idle state
estado de resonancia resonance state
estado de salida output state
estado de saturación saturation state
estado disponible available state
estado energético energy state
estado equilibrado balanced state
estado estable stable state
estado estacionario stationary state, steady state
estado final final state
estado fundamental ground state
estado inactivo inactive state
estado inestable unstable state
estado inicial initial state

estado intermedio intermediate state
estado interrumpible interruptable state
estado latente dormant state
estado lógico logic state
estado lógico interno internal logic state
estado magnético magnetic state
estado microscópico microscopic state
estado mixto mixed state
estado neutro neutral state
estado normal normal state, standard state
estado plástico plastic state
estado refractario absoluto absolute refractory state
estado significativo significant condition
estado significativo de una modulación significant condition of a modulation
estado supervisor supervisory state
estado transitorio transient state
estado uno one state
estado virtual virtual state
estampado stamping
estampido sónico sonic boom
estándar standard
estandarización standardization
estandarizado standardized
estandarizar standardize
estante shelf, rack, stand
estante de baterías battery stand
estante de llaves keyshelf
estante de montaje mounting shelf, mounting stand
estañado tin-plated
estañar tin-plate
estaño tin
estaño gris gray tin
estatampere statampere
estatcoulomb statcoulomb
estatfarad statfarad
estathenry stathenry
estática static
estática de precipitación precipitation static
estática estacional seasonal static
estática por nieve snow static
estáticamente statically
estático statical
estáticos statics
estatitrón statitron
estatmho statmho
estatoersted statoersted
estatohm statohm
estator stator
estatoscopio statoscope
estatvolt statvolt
estatweber statweber
esteatita steatite
estela de ionización meteórica meteor ionization trail
estela meteórica meteor trail
estelar stellar
estenodo stenode
estequiometría stoichiometry
estequiométrico stoichiometric
estéreo stereo
estéreo de tres canales three-channel stereo
estéreo por modulación de frecuencia frequency-modulation stereo
estereoaudífonos stereophonic headphones
estereoespectrograma stereospectrogram
estereofluoroscopia stereofluoroscopy
estereofonía stereophony
estereofonía multicanal multichannel stereophony
estereofónica stereophonics
estereofónicamente stereophonically
estereofónico stereophonic
estereófono stereophone

estereofotografía stereophotography
estereográfico stereographic
estereomicroscopio stereomicroscope
estereoóptica stereoptics
estereooscilografía stereoscillography
estereorradar stereoradar
estereorradiografía stereoradiography
estereorradiográfico stereoradiographic
estereorradioscopía stereoradioscopy
estereoscopía stereoscopy
estereoscópico stereoscopic
estereoscopio stereoscope
estereosónico stereosonic
estereotelémetro stereotelemeter
estereotelevisión stereotelevision
esterilizador sterilizer
esterradian steradian
estetoscopio stethoscope
estetoscopio electrónico electronic stethoscope
estilete stylus
estilete de fibra fiber stylus
estilo de conector connector style
estilo de línea line style
estimulación eléctrica electrical stimulation
estimulador cardíaco cardiac stimulator
estimulador electrónico electronic stimulator
estímulo stimulus
estímulo acromático achromatic stimulus
estímulo acumulante accumulating stimulus
estímulo acústico acoustic stimulus
estímulo de color color stimulus
estiramiento stretch
estiramiento muerto dead stretch
estocástico stochastic
estrabismo squint
estratificado stratified
estratigrafía stratigraphy
estratosfera stratosphere
estrella star
estrella del inducido armature spider
estrés stress
estrés de componentes component stress
estrés de impacto impact stress
estrés de reacción reaction stress
estrés dieléctrico dielectric stress
estrés eléctrico electrical stress
estrés electromagnético electromagnetic stress
estrés electrostático electrostatic stress
estrés interno internal stress
estrés magnético magnetic stress
estrés mecánico mechanical stress
estrés normal normal stress
estrés permisible allowable stress
estrés principal principal stress
estrés residual residual stress
estrés simple simple stress
estricción striction, pinchoff
estroborradiografía stroboradiography
estroboscópico stroboscopic
estroboscopio stroboscope, strobe
estroboscopio de lámpara de neón neon-bulb
    stroboscope
estroboscopio eléctrico electric stroboscope
estroboscopio electrónico electronic stroboscope
estroboscopio ultrasónico ultrasonic stroboscope
estrobotrón strobotron
estroncio strontium
estructura structure
estructura atómica atomic structure
estructura de almacenamiento storage structure
estructura de aplicación application structure
estructura de archivo file structure
estructura de banda de energía energy-band

structure
estructura de base de datos database structure
estructura de bloque block structure
estructura de bucle loop structure
estructura de bus bus structure
estructura de bus múltiple multiple bus structure
estructura de cadena en margarita daisy chain
    structure
estructura de control control structure
estructura de control condicional conditional
    control structure
estructura de control de bucle loop-control structure
estructura de control de secuencia sequence-control
    structure
estructura de cruce crossing structure
estructura de datos data structure
estructura de datos contiguos contiguous data
    structure
estructura de datos física physical data structure
estructura de datos gráficos graphic data structure
estructura de datos lineal linear data structure
estructura de datos lógica logical data structure
estructura de datos no lineal nonlinear data
    structure
estructura de datos secuencial sequential data
    structure
estructura de decisión decision structure
estructura de directorio directory structure
estructura de enfoque focusing structure
estructura de iteración iteration structure
estructura de línea de paridad parity-line structure
estructura de lista list structure
estructura de prioridades de bus bus priority
    structure
estructura de programa program structure
estructura de red network structure
estructura de rejilla grid structure
estructura de resistencia constante
    constant-resistance structure
estructura de secuencia sequence structure
estructura de selección selection structure
estructura de sistema system structure
estructura de soporte de cables cable supporting
    structure
estructura en celosía lattice structure
estructura fina fine structure
estructura física physical structure
estructura invertida inverted structure
estructura jerárquica hierarchical structure
estructura lineal linear structure
estructura lógica logical structure
estructura magnética periódica periodic magnetic
    structure
estructura menor minor structure
estructura molecular molecular structure
estructura relacional relational structure
estructura superfina superfine structure
estructurado structured
estructuralmente structurally
estudio anecoico anechoic studio
estudio ciego blind study
eta eta
etapa stage
etapa acoplada por impedancia impedance-coupled
    stage
etapa amplificadora de potencia power-amplifier
    stage
etapa amplificadora de radiofrecuencia
    radio-frequency amplifier stage
etapa amplificadora final final amplifier stage
etapa asimétrica single-ended stage
etapa binaria binary stage
etapa cambiadora de frecuencia frequency-changer

stage
etapa de amplificación amplification stage
etapa de amplificación modulada modulated amplification stage
etapa de antena antenna stage
etapa de conmutación switching stage
etapa de entrada input stage
etapa de exploración scanning stage
etapa de frecuencia intermedia intermediate-frequency stage
etapa de ganancia gain stage
etapa de ganancia variable variable-gain stage
etapa de memoria memory stage
etapa de multiplicación de frecuencia frequency multiplication stage
etapa de potencia power stage
etapa de preselección preselection stage
etapa de radiofrecuencia radio-frequency stage
etapa de salida output stage
etapa de salida horizontal horizontal-output stage
etapa de salida modulada modulated output stage
etapa de salida ultralineal ultralinear output stage
etapa de salida vertical vertical-output stage
etapa de selección selection stage
etapa de supresor de colores color-killer stage
etapa de transmisión sending stage
etapa de video video stage
etapa detectora detector stage
etapa diferencial differential stage
etapa discriminadora discriminator stage
etapa dobladora doubler stage
etapa dobladora de frecuencia frequency-doubler stage
etapa en cuadratura quadrature stage
etapa excitadora driver stage
etapa final final stage
etapa limitadora limiting stage
etapa limitadora de amplitud amplitude-limiter stage
etapa mezcladora mixer stage
etapa mezcladora en contrafase push-pull mixing stage
etapa modulada modulated stage
etapa moduladora modulator stage
etapa multiplicadora multiplier stage
etapa multiplicadora de frecuencia frequency-multiplier stage
etapa por etapa stage-by-stage
etapa preamplificadora preamplifier stage
etapa presintonizada pretuned stage
etapa separadora buffer stage
etapa única single stage
etiqueta label
etiqueta de archivo file label
etiqueta de cabecera header label
etiqueta de cinta tape label
etiqueta de conjunto de datos data set label
etiqueta de identificación identification label
etiqueta de pista track label
etiquetas futuras future labels
eureka eureka
europio europium
eutéctico eutectic
evacuación evacuation
evaluación de confiabilidad reliability evaluation
evaluación de daños damage assessment
evaluación de datos data evaluation
evaluación de programa program evaluation
evaluación de rendimiento performance evaluation
evaluación de sistema system evaluation
evaporación evaporation
evaporación catódica cathodic evaporation, sputtering

evaporación en vacío vacuum evaporation
evento aleatorio random event
evento de interrupción única single-outage event
evento de programa program event
evento ionizante ionizing event
evento ionizante inicial initial ionizing event
evento repetitivo repetitive event
exactitud accuracy
exactitud absoluta absolute accuracy
exactitud angular angular accuracy
exactitud de ajuste adjustment accuracy
exactitud de atenuación attenuation accuracy
exactitud de barrido sweep accuracy
exactitud de calibración calibration accuracy
exactitud de certificación certification accuracy
exactitud de conmutación switching accuracy
exactitud de control control accuracy
exactitud de frecuencia frequency accuracy
exactitud de instrumento instrument accuracy
exactitud de lectura readout accuracy, reading accuracy
exactitud de límite de escala full-scale accuracy
exactitud de medida measuring accuracy
exactitud de programas software accuracy
exactitud de referencia reference accuracy
exactitud de reloj clock accuracy
exactitud de rumbo bearing accuracy
exactitud de tiempo time accuracy
exactitud de transmisión transmission accuracy
exactitud dinámica dynamic accuracy
exactitud disponible available accuracy
exactitud media mean accuracy
exactitud nominal rated accuracy
exactitud relativa relative accuracy
exactitud utilizable usable accuracy
exacto accurate
examen microscópico microscopic examination
excentricidad eccentricity
excéntrico eccentric
excepción de desbordamiento overflow exception
excepción de direccionamiento addressing exception
excepción de operación operation exception
excepción de programa program exception
exceso de cincuenta excess fifty
excitabilidad excitability
excitabilidad eléctrica electrical excitability
excitación excitation
excitación armónica harmonic excitation
excitación compuesta compound excitation
excitación de antena antenna excitation
excitación de rejilla grid excitation
excitación diferencial differential excitation
excitación directa direct excitation
excitación en derivación shunt excitation
excitación en serie series excitation
excitación hipercompuesta overcompounded excitation
excitación independiente independent excitation
excitación indirecta indirect excitation
excitación múltiple multiple excitation
excitación natural natural excitation
excitación paramétrica parametric excitation
excitación paso a paso step-by-step excitation
excitación por choque shock excitation
excitación por impacto impact excitation
excitación por impulsos impulse excitation
excitación por rayos X X-ray excitation
excitación residual residual excitation
excitación separada separate excitation
excitado energized
excitado en derivación shunt-excited
excitado por ultravioleta ultraviolet-excited
excitado separadamente separately excited

excitador exciter, energizer, driver
excitador autoelevador bootstrap driver
excitador de alternador-rectificador
    alternator-rectifier exciter
excitador de conversión conversion exciter
excitador de Crosby Crosby exciter
excitador de estado sólido solid-state exciter
excitador de galvanómetro galvanometer driver
excitador de lámpara lamp driver
excitador de línea line driver
excitador de línea compartida party-line driver
excitador de modulador modulator driver
excitador de oscilador de bloqueo blocking
    oscillator driver
excitador de relé relay driver
excitador en serie series exciter
excitador maestro master driver
excitador piloto pilot exciter
excitador principal main exciter
excitancia radiante radiant excitance
excitar energize
excitón exciton
excitrón excitron
excursión excursion
excursión completa full excursion
excursión de barrido sweep excursion
excursión de frecuencia frequency excursion
excursión de frecuencia de pico a pico peak-to-peak
    frequency excursion
excursión de pico a pico peak-to-peak excursion
excursión de referencia reference excursion
excursión máxima maximum excursion
exotérmico exothermic
expansibilidad expansibility
expansible expandable
expansión expansion
expansión central center expansion
expansión de barrido sweep expansion
expansión de carácter character expansion
expansión de cociente quotient expansion
expansión de escala scale expansion
expansión de memoria memory expansion
expansión de trazo trace expansion
expansión de volumen volume expansion
expansión magnetoestrictiva magnetostrictive
    expansion
expansor expander
expansor de barrido sweep expander
expansor de haz beam expander
expansor de unidad de procesamiento central
    central processing unit expander
expansor de volumen volume expander
expectativa de vida life expectancy
experimental experimental
experimento experiment
exploración scan, scanning, sweep, search
exploración aleatoria random scan
exploración automática automatic scanning
exploración bilateral bilateral scanning
exploración circular circular scanning
exploración compensada compensated scan
exploración cónica conical scanning
exploración cuádruple quadruple scanning
exploración de alta velocidad high-velocity scanning
exploración de archivo file scan
exploración de baja velocidad low-velocity scanning
exploración de base de tiempo time-base scanning
exploración de campo field scanning
exploración de campo próximo near-field scanning
exploración de entrada-salida input-output scan
exploración de frecuencia frequency scanning
exploración de líneas line scanning
exploración de líneas electrónica electronic line
    scanning
exploración de mando command scan
exploración de marcas mark scanning
exploración de memoria memory scan
exploración de Palmer Palmer scan
exploración de película film scanning
exploración de pixel pixel scan
exploración de punto a punto point-to-point
    scanning
exploración de radar radar scan
exploración de sección section scanning
exploración de televisión television scanning
exploración de trama raster scan
exploración de velocidad lenta slow-speed scan
exploración de velocidad variable variable-speed
    scanning
exploración directa direct scanning
exploración eléctrica electrical scan
exploración electromagnética electromagnetic
    scanning
exploración electrónica electronic scanning
exploración electrostática electrostatic scanning
exploración en diente de sierra sawtooth scanning
exploración en espiral spiral scanning
exploración entrelazada interlaced scanning
exploración espectral spectral scanning
exploración ferroeléctrica ferroelectric scanning
exploración helicoidal helical scanning
exploración horizontal horizontal scanning
exploración indirecta indirect scanning
exploración inversa reverse scan
exploración isócrona isochronous scanning
exploración lenta slow scan
exploración lineal linear scanning
exploración magnética magnetic scan
exploración mecánica mechanical scanning
exploración múltiple multiple scanning
exploración no lineal nonlinear scanning
exploración no secuencial nonsequential scanning
exploración no sincronizada nonsynchronized
    scanning
exploración óptica optical scanning
exploración oscilante oscillatory scanning
exploración por punto móvil flying-spot scanning
exploración por puntos múltiples multiple-spot
    scanning
exploración por reacción reaction scanning
exploración por saltos de líneas line-jump scanning
exploración por sectores sector scanning
exploración posterior rear scanning
exploración progresiva progressive scanning
exploración rápida rapid scanning
exploración rectangular rectangular scan
exploración rectilínea rectilinear scanning
exploración retardada delayed scanning
exploración secuencial sequential scanning
exploración simple simple scanning
exploración simultánea simultaneous scanning
exploración tridimensional three-dimensional
    scanning
exploración vertical vertical scanning
explorador scanner, searcher
explorador de antena antenna scanner
explorador de cilindro drum scanner
explorador de código de barras bar code scanner
explorador de disco disk scanner
explorador de punto luminoso móvil light-spot
    scanner
explorador de punto móvil flying-spot scanner
explorador de radar radar scanner
explorador fotoeléctrico photoelectric scanner
explorador magnético magnetic explorer
explorador mecánico mechanical scanner

explorador óptico optical scanner
explorador óptico de código de barras bar code optical scanner
explorador principal main scanner
explorador rotativo rotating scanner
explorador secuencial sequential scanner
explorador virtual virtual scanner
explorador visual visual scanner
explosor spark gap
explosor de esferas sphere gap
explosor de esferas patrón standard sphere gap
exponencial exponential
exponente fraccional fractional exponent
exponente negativo negative exponent
exponente positivo positive exponent
exportar export
exposición a radiación radiation exposure
exposición aguda acute exposure
exposición bactericida bactericidal exposure
exposición radiante radiant exposure
exposímetro exposure meter
exposímetro fotográfico photographic exposure meter
expresión absoluta absolute expression
expresión algebraica algebraic expression
expresión aritmética arithmetic expression
expresión asintótica asymptotic expression
expresión condicional conditional expression
expresión de archivo file expression
expresión de Boole Boolean expression
expresión de caracteres character expression
expresión de operación operation expression
expresión escalar scalar expression
expresión lógica logical expression
expresión relacional relational expression
expresión reubicable relocatable expression
expresión SI IF expression
expuesto exposed
expulsor de campo magnético magnetic-field expeller
extensímetro extensimeter
extensímetro de resistencia resistance strain gage
extensión de distancia range extension
extensión de nombre de archivo file name extension
extensión de panel panel extension
extensión externa external extension
extensión por omisión default extension
extensómetro extensometer
extensómetro electroacústico electroacoustic extensometer
extensor de línea line stretcher
extensor de segmento segment extender
exterior outdoor, outside
extinción extinction, blanking, quenching, blowout
extinción de acimut azimuth blanking
extinción de arco arc extinction
extinción de chispas spark quenching
extinción de haz beam cutoff
extinción de línea line blanking
extinción de retorno horizontal horizontal-retrace blanking
extinción de retorno vertical vertical-retrace blanking
extinción de zona zone blanking
extinción durante retorno retrace blanking
extinción horizontal horizontal blanking
extinción magnética magnetic blowout
extinción recíproca reciprocal extinction
extinción vertical vertical blanking
extinguido extinguished, quenched
extinguir extinguish, quench
extintor extinguisher, quencher
extintor de chispas spark extinguisher, spark

quencher
extintor magnético magnetic blower
extracameral extracameral
extracción de cristal crystal pulling
extracción de papel paper extraction
extracción electrolítica electrowinning
extracción resonante resonant extraction
extracorriente extracurrent
extracto de base de datos database extract
extractor extractor
extractor de calor heat remover
extractor de lámparas lamp extractor
extractor de papel paper extractor
extractor de tarjetas card puller
extractor de tubos tube puller
extraer extract
extremo extreme
extremo de carga load end
extremo de transmisión sending end
extremo muerto dead end
extremo polar pole tip
extremo receptor receiving end
extremo sellado sealed end
extremos ambientales environmental extremes
extrínseco extrinsic
extrusión extrusion

F

fabricación de circuitos circuit fabrication
facilitación facilitation
facom facom
facsímil facsimile
facsímil cifrado enciphered facsimile
facsímil tipo A type-A facsimile
facsímil tipo B type-B facsimile
factor factor
factor armónico harmonic factor
factor balístico ballistic factor
factor constante constant factor
factor controlador controlling factor
factor de abocinado flare factor
factor de absorción de reverberación reverberation absorption factor
factor de absorción de sonido sound-absorption factor
factor de absorción espectral spectral absorption factor
factor de acción proporcional proportional-action factor
factor de aceleración acceleration factor
factor de acoplamiento coupling factor
factor de acoplamiento crítico critical coupling factor
factor de acoplamiento electromecánico electromechanical coupling factor
factor de actividad activity factor
factor de acumulación buildup factor
factor de acumulación de huecos hole-storage factor
factor de agrupamiento grouping factor
factor de aislamiento de ruido noise-insulation factor
factor de alcance reach factor
factor de almacenamiento storage factor
factor de amortiguamiento damping factor
factor de amplificación amplification factor
factor de amplificación de corriente current-amplification factor
factor de amplificación de gas gas-amplification factor
factor de amplificación de pantalla screen amplification factor
factor de amplificación geométrico geometric amplification factor
factor de amplificación inverso inverse amplification factor
factor de amplificación invertida inverted amplification factor
factor de amplitud amplitude factor
factor de antena antenna factor
factor de apantallado screening factor
factor de apantallamiento shielding factor
factor de aplanamiento smoothing factor
factor de aplicación application factor
factor de arrastre pulling figure
factor de arrollamiento winding factor
factor de asignación allocation factor
factor de atenuación attenuation factor
factor de atenuación de sonido sound-attenuation factor

factor de atenuación imagen image attenuation factor
factor de atenuación iterativa iterative attenuation factor
factor de aumento magnification factor
factor de bloqueo blocking factor
factor de bloqueo de salida output blocking factor
factor de brillo gloss factor
factor de calibración calibration factor
factor de calidad quality factor
factor de capacidad capacity factor
factor de capacidad unitario unit capacity factor
factor de captación pickup factor
factor de carga load factor, charge factor
factor de carga anual annual load factor
factor de carga catódica cathode loading factor
factor de carga de sistema system load factor
factor de carga instantáneo instantaneous load factor
factor de carga límite limit-load factor
factor de cielo sky factor
factor de circuito absorbedor absorber circuit factor
factor de coincidencia coincidence factor
factor de compresión compression factor
factor de comprobación checking factor
factor de confiabilidad reliability factor
factor de confianza confidence factor
factor de configuración configuration factor
factor de conmutación commutation factor
factor de convergencia convergence factor
factor de conversión conversion factor
factor de conversión de corriente alterna alternating-current conversion factor
factor de corrección correction factor
factor de corrección de antena antenna correction factor
factor de corrección de color color-correction factor
factor de corrección de medidor meter correction factor
factor de corrección de transformador transformer-correction factor
factor de cresta crest factor
factor de Debye-Waller Debye-Waller factor
factor de decaimiento decay factor
factor de decremento decrement factor
factor de deflexión deflection factor
factor de degradación degradation factor
factor de demanda demand factor
factor de desadaptación mismatch factor
factor de desequilibrio unbalance factor
factor de despolarización depolarization factor
factor de desviación deviation factor
factor de detectabilidad detectability factor
factor de directividad directivity factor
factor de discriminación discrimination factor
factor de diseño design factor
factor de disipación dissipation factor
factor de disipación de acoplamiento coupling-dissipation factor
factor de disipación dieléctrica dielectric dissipation factor
factor de dispersión scattering factor
factor de disponibilidad availability factor
factor de distorsión distortion factor
factor de distorsión de amplitud amplitude distortion factor
factor de distorsión de generador generator distortion factor
factor de distorsión no lineal nonlinear distortion factor
factor de distorsión ponderado weighted distortion factor
factor de distribución distribution factor

factor de diversidad diversity factor
factor de ductilidad ductility factor
factor de duplicación duplication factor
factor de efectividad de alumbrado lighting
    effectiveness factor
factor de eficiencia efficiency factor
factor de eficiencia de placa plate efficiency factor
factor de empaquetamiento packing factor
factor de entrada input factor
factor de entrelazado interlace factor
factor de envejecimiento aging factor
factor de escala scale factor, scaling factor
factor de escala de medidor meter scale factor
factor de esfericidad de la tierra spherical-earth
    factor
factor de espacio space factor
factor de estabilidad stability factor
factor de estructura structure factor
factor de fase phase factor
factor de fase imagen image phase factor
factor de filtro filter factor
factor de forma shape factor
factor de frecuencia máxima utilizable maximum
    usable frequency factor
factor de fricción friction factor
factor de fuerza force factor
factor de fuga leakage factor
factor de histéresis hysteresis factor
factor de impedancia impedance factor
factor de impulsos ausentes missing-pulse factor
factor de inclinación skew factor
factor de inducción induction factor
factor de inductancia inductance factor
factor de inductancia mutua mutual-inductance
    factor
factor de influencia factor of influence
factor de influencia telefónica telephone influence
    factor
factor de interacción interaction factor
factor de interferencia interference factor
factor de ionización ionization factor
factor de iteración iteration factor
factor de liberación release factor
factor de luminancia luminance factor
factor de luminancia espectral spectral luminance
    factor
factor de luminiscencia glow factor
factor de luminosidad luminosity factor
factor de luminosidad relativa relative luminosity
    factor
factor de luz solar daylight factor
factor de magnificación de bobina coil
    magnification factor
factor de mérito factor of merit
factor de modulación modulation factor
factor de modulación cruzada cross-modulation
    factor
factor de modulación de amplitud
    amplitude-modulation factor
factor de modulación de potencia power-modulation
    factor
factor de modulación negativa negative modulation
    factor
factor de modulación positiva positive modulation
    factor
factor de multiplicación multiplication factor
factor de multiplicación de gas gas multiplication
    factor
factor de multiplicación de lente lens multiplication
    factor
factor de ondulación ripple factor
factor de operación operation factor
factor de pantalla screen factor

factor de paso pitch factor
factor de penetración penetration factor
factor de pérdida de potencia power-loss factor
factor de pérdida dieléctrica dielectric loss factor
factor de pérdida magnética magnetic loss factor
factor de pérdidas loss factor
factor de pérdidas por ondas estacionarias
    standing-wave loss factor
factor de perturbación blur factor
factor de pico peak factor
factor de ponderación weighting factor
factor de potencia power factor
factor de potencia acústica acoustic power factor
factor de potencia cero zero-power factor
factor de potencia de aislamiento insulation power
    factor
factor de potencia de dieléctrico dielectric power
    factor
factor de potencia magnética magnetic power factor
factor de potencia nominal nominal power factor
factor de potencia radial radial power factor
factor de potencia unitario unity power factor
factor de potencia vectorial vector power factor
factor de propagación propagation factor
factor de propagación de cuadripolo quadripole
    propagation factor
factor de prueba proof factor
factor de radiancia radiance factor
factor de reactancia reactance factor
factor de recepción de polarización polarization
    receiving factor
factor de rectificación rectification factor
factor de red de antenas array factor
factor de reducción derating factor
factor de reducción de resistencia
    resistance-reduction factor
factor de reducción esférica spherical reduction
    factor
factor de reducción sonora sound-reduction factor
factor de redundancia redundancy factor
factor de reflectancia reflectance factor
factor de reflexión reflection factor
factor de reflexión de reverberación reverberation
    reflection factor
factor de reflexión difusa diffuse reflection factor
factor de reflexión espectral spectral reflection
    factor
factor de reflexión óptica optical-reflection factor
factor de reflexión regular regular reflection factor
factor de reflexión sonora sound-reflection factor
factor de regulación phase-control factor
factor de relocalización relocation factor
factor de reluctancia reluctance factor
factor de rendimiento performance factor
factor de repetición repetition factor
factor de retroalimentación feedback factor
factor de retroalimentación de tensión
    voltage-feedback factor
factor de retroalimentación negativa
    negative-feedback factor
factor de retrodispersión backscatter factor
factor de ruido noise factor
factor de ruido efectivo effective noise factor
factor de ruido excedente excess noise factor
factor de ruido global overall noise factor
factor de ruido medio average noise factor
factor de ruido monocanal single-channel noise
    factor
factor de ruido normal standard noise factor
factor de salida output factor
factor de salto skip factor
factor de saturación saturation factor
factor de seguridad safety factor

factor de seguridad para mantenimiento safety factor for holding
factor de seguridad para puesta en reposo safety factor for dropout
factor de seguridad para puesta en trabajo safety factor for pickup
factor de servicio service factor
factor de sobrecarga overload factor
factor de sobrecorriente overcurrent factor
factor de sobretensión magnification factor
factor de sombra shadow factor
factor de supresión suppression factor
factor de tambor drum factor
factor de tensión voltage factor
factor de tiempo time factor
factor de trabajo duty factor
factor de trabajo de impulsos pulse duty factor
factor de transferencia transfer factor
factor de transición transition factor
factor de transmisión transmission factor
factor de transmisión de reverberación reverberation transmission factor
factor de transmisión de sonido sound-transmission factor
factor de transmisión difusa diffuse transmission factor
factor de transmisión espectral spectral transmission factor
factor de transporte transport factor
factor de transrectificación transrectification factor
factor de tubo tube factor
factor de uniformidad uniformity factor
factor de utilidad utility factor
factor de utilización utilization factor
factor de utilización de transformador transformer utilization factor
factor de utilización secundaria secondary utilization factor
factor de variación variation factor
factor de velocidad velocity factor
factor de visibilidad visibility factor
factor de vueltas turns factor
factor desmagnetizante demagnetizing factor
factor dinámico dynamic factor
factor estacional seasonal factor
factor ganancia-ancho de banda gain-bandwidth factor
factor geométrico geometric factor
factor índice index factor
factor limitador limiting factor
factor magnetomecánico magnetomechanical factor
factor mu mu factor
factor multiplicador multiplier factor
factor primario primary factor
factor Q Q factor
factor reactivo reactive factor
factor reactivo aritmético arithmetic reactive factor
factores ambientales environmental factors
factores humanos human factors
factorial factorial
factorización factoring
falsa alarma false alarm
falso eco false echo
falta de yuxtaposición underlap
falla failure, fault, malfunction, breakdown
falla aleatoria random failure
falla catastrófica catastrophic failure
falla completa de disco disk crash
falla crítica critical failure
falla de aislamiento insulation failure, arcthrough
falla de alimentación power failure
falla de circuito circuit fault
falla de componente component failure

falla de conmutación commutation failure
falla de encendido misfire
falla de equipo equipment failure
falla de equipo físico hardware failure
falla de fusible fuse failure
falla de ignición ignition failure
falla de línea line fault
falla de máquina machine failure
falla de página page fault
falla de parte part failure
falla de pieza part failure
falla de programa program failure
falla de rectificación rectification failure
falla de red network failure, mains failure
falla de registro dropout
falla de retención retention failure
falla de tubo tube failure
falla electrónica electronic malfunction
falla estructural structural failure
falla gradual gradual failure
falla incipiente incipient failure
falla independiente independent failure
falla inducida induced failure
falla inicial initial failure
falla intermitente intermittent failure
falla invisible invisible failure
falla local local fault
falla marginal marginal fault
falla menor minor failure
falla no crítica noncritical failure
falla parcial partial failure
falla permanente permanent fault
falla por arco arc failure
falla por deterioro completo wearout failure
falla por impureza de modo moding
falla por ruido noise failure
falla primaria primary failure
falla principal major failure
falla que no impide funcionamiento failsoft
falla secundaria secondary failure
falla sensible al programa program-sensitive fault
falla transitoria transient fault
fallas equivalentes equivalent faults
fallo failure, fault, malfunction, breakdown
familia de colectores collector family
familia de curvas family of curves
familia de curvas de rejilla grid family of curves
familia de frecuencias family of frequencies
familia de partes parts family
familia de tipo de letras font family
fanotrón phanotron
fantasma ghost, phantom
fantasma adelantado leading ghost
fantasma durante retorno retrace ghost
fantasma positivo positive ghost
fantastrón phantastron
fantastrón acoplado por cátodo cathode-coupled phantastron
fantófono phantophone
farad farad
farad internacional international farad
faraday faraday
farádico faradic
faradímetro faradmeter
faradismo faradism
faradización faradization
faro beacon
faro aerotransportado airborne beacon
faro de navegación navigation beacon
faro de referencia landmark beacon
faro flotante floating beacon
faro navegacional navigational beacon
fase phase

fase abierta open phase
fase de adquisición acquisition phase
fase de análisis analysis phase
fase de caracteres character phase
fase de ciclo de vida life-cycle phase
fase de color color phase
fase de compilación compiling phase
fase de compresión compression phase
fase de concepto concept phase
fase de depuración debugging phase
fase de diseño design phase
fase de diseño de sistema system design phase
fase de ejecución execution phase
fase de ensamblaje assembling phase
fase de generación generation phase
fase de identificación identification phase
fase de imagen negativa negative picture phase
fase de imagen positiva positive picture phase
fase de implementación implementation phase
fase de ondas wave phase
fase de pruebas test phase
fase de recuperación recovery phase
fase de requisitos requirements phase
fase de resolución resolution phase
fase de respuesta reply phase
fase de señal signal phase
fase de solicitud request phase
fase de traducción translating phase
fase de transferencia de datos data-transfer phase
fase de un impulso pulse phase
fase diferencial differential phase
fase dividida split phase
fase imagen image phase
fase negativa negative phase
fase objeto object phase
fase opuesta opposite phase
fase paraeléctrica paraelectric phase
fase por fase phase-by-phase
fase positiva positive phase
fase principal main phase
fase retardada lagging phase
fases desequilibradas unbalanced phases
fasímetro phase meter
fasímetro electrónico electronic phase meter
fasitrón phasitron
fasor phasor
fatiga fatigue
fatiga de dieléctrico dielectric fatigue
fatiga de metal fatigue of metal
fatiga fotoeléctrica photoelectric fatigue
fatómetro fathometer
fax fax
fecha de desactivación deactivation date
fecha de instalación installation date
femto femto
femtoampere femtoampere
femtofarad femtofarad
femtovolt femtovolt
fenólico phenolic
fenómeno phenomenon
fenómeno aperiódico aperiodic phenomenon
fenómeno de difracción diffraction phenomenon
fenómeno de eco echo phenomenon
fenómeno de Gibbs Gibb's phenomenon
fenómeno de polarización polarization phenomenon
fenómeno físico physical phenomenon
fenómeno mecánico mechanical phenomenon
fenómeno ondulatorio wave phenomenon
fenómeno recurrente recurrent phenomenon
fenómeno repetitivo repetitive phenomenon
fenómeno transitorio transient phenomenon
fenómenos de descarga discharge phenomena
fermi fermi

fermio fermium
fernico fernico
ferrético ferretic
ferrielectricidad ferrielectricity
ferrieléctrico ferrielectric
ferrimagnético ferrimagnetic
ferrimagnetismo ferrimagnetism
ferristor ferristor
ferrita ferrite
ferrita orientada oriented ferrite
ferrítico ferritic
ferroacústico ferroacoustic
ferrocarril electrificado electrified railway
ferrodinámico ferrodynamic
ferroelástico ferroelastic
ferroelectricidad ferroelectricity
ferroeléctrico ferroelectric
ferroeléctrico cerámico ceramic ferroelectric
ferroeléctrico impropio improper ferroelectric
ferroespinela ferrospinel
ferroinductancia ferroinductance
ferroinductor ferroinductor
ferromagnética ferromagnetics
ferromagnético ferromagnetic
ferromagnetismo ferromagnetism
ferrómetro ferrometer
ferrorresonador ferroresonator
ferrorresonancia ferroresonance
ferrorresonante ferroresonant
ferroso ferrous
festoneado scalloping
fibra fiber
fibra de vidrio glass fiber
fibra multimodo multimode fiber
fibra óptica optical fiber
fibra vulcanizada vulcanized fiber
fibrilación fibrillation
fibrilación cardiaca heart fibrillation
fibrilación ventricular ventricular fibrillation
fibroóptica fiber optics
fibroóptica de infrarrojo infrared fiber optics
fibroóptico fiber-optic
fibroscopio fiberscope
ficticio fictitious, dummy
ficha card
fidelidad fidelity
fidelidad cromática chromatic fidelity
fidelidad de colores color fidelity
fidelidad de modulación modulation fidelity
fidelidad de transmisión transmission fidelity
fidelidad eléctrica electric fidelity
fidelidad geométrica geometric fidelity
figura de Lichtenberg Lichtenberg figure
figura de Lissajous Lissajous figure
fijación de imagen picture lock
fijación de nivel clamping, setup
fijación de nivel sincronizada synchronized
　　clamping
fijación de posición position fixing
fijación rígida rigid fastening
fijador fastener
fijador de nivel clamper
fila row
fila de caracteres character row
fila de comprobación check row
fila de dígitos digit row
fila de tarjetas card row
filamentario filamentary
filamento filament
filamento activado activated filament
filamento caliente hot filament
filamento carbonizado carbonized filament
filamento con derivación central center-tapped

filament
filamento de calentamiento rápido quick-heating filament
filamento de carbón carbon filament
filamento de cinta ribbon filament
filamento de emisión débil dull-emitter filament
filamento de plasma plasma filament
filamento de tubo tube filament
filamento de tungsteno tungsten filament
filamento de tungsteno toriado thoriated-tungsten filament
filamento doblado folded filament
filamento en bucle looped filament
filamento en espiral single-coil filament
filamento recto straight filament
filamento revestido coated filament
filamento revestido de óxido oxide-coated filament
filamento sin revestimiento uncoated filament
filamento toriado thoriated filament
fildistor fieldistor
fildistor de unión junction fieldistor
filtración filtration
filtración inherente inherent filtration
filtrado filtering
filtrado armónico harmonic filtering
filtrado espectral spectral filtering
filtrado no lineal nonlinear filtering
filtrado óptico optical filtering
filtro filter
filtro absorbente absorbing filter
filtro activo active filter
filtro acústico acoustic filter
filtro acústico pasabajos low-pass acoustic filter
filtro adaptado matched filter
filtro adicional added filter
filtro aditivo additive filter
filtro amortiguado damped filter
filtro armónico harmonic filter
filtro azul blue filter
filtro cerámico ceramic filter
filtro coaxial coaxial filter
filtro coloreado colored filter
filtro colorimétrico colorimetric filter
filtro compensador compensating filter
filtro compuesto compound filter
filtro contra emisiones espurias splatter filter
filtro contra ronquido rumble filter
filtro contra ruido mecánico rumble filter
filtro correctivo corrective filter
filtro de adaptación de impedancias impedance-matching filter
filtro de aislamiento isolation filter
filtro de altas high filter
filtro de aplanamiento smoothing filter
filtro de audiofrecuencia audio-frequency filter
filtro de bajos low filter
filtro de banda band filter
filtro de banda angosta narrowband filter
filtro de banda de frecuencia fija fixed-frequency band filter
filtro de banda de supergrupos supergroup band filter
filtro de banda de transmisión transmission-band filter
filtro de banda lateral residual vestigial-sideband filter
filtro de banda lateral única single-sideband filter
filtro de banda pasante passband filter
filtro de bandas laterales sideband filter
filtro de bloqueo blocking filter
filtro de bucle loop filter
filtro de Butterworth Butterworth filter
filtro de canal channel filter

filtro de capacitancia capacitance filter
filtro de capacitor capacitor filter
filtro de carga espacial space-charge filter
filtro de cavidad cavity filter
filtro de color aditivo additive color filter
filtro de color primario primary-color filter
filtro de colores color filter
filtro de colores sustractivo subtractive color filter
filtro de compensación compensation filter
filtro de constantes concentradas lumped-constant filter
filtro de corrección de color color-correction filter
filtro de corriente alterna alternating-current filter
filtro de corriente constante constant-current filter
filtro de corriente continua direct-current filter
filtro de corte agudo sharp-cutoff filter
filtro de cristal crystal filter
filtro de cuarto de onda quarter-wave filter
filtro de chasquidos hash filter
filtro de chasquidos de manipulación key-click filter
filtro de Chebyshev Chebyshev filter
filtro de choque de entrada choke-input filter
filtro de derivación bypass filter
filtro de desacoplamiento decoupling filter
filtro de detector de video video-detector filter
filtro de dígito digit filter
filtro de Doppler Doppler filter
filtro de ecos parásitos clutter filter
filtro de efectos de sonido sound-effects filter
filtro de eliminación de banda band-elimination filter
filtro de eliminación de portadora carrier-elimination filter
filtro de entrada input filter
filtro de entrada capacitiva capacitor-input filter
filtro de ferrita ferrite filter
filtro de frecuencias frequency filter
filtro de fuente de alimentación power-supply filter
filtro de función arbitraria arbitrary function filter
filtro de Gauss Gaussian filter
filtro de granate de itrio e hierro yttrium-iron-garnet filter
filtro de guíaondas waveguide filter
filtro de imagen picture filter
filtro de inductancia inductance filter
filtro de inductancia-capacitancia inductance-capacitance filter
filtro de interferencia interference filter
filtro de k constante constant-k filter
filtro de Kalman Kalman filter
filtro de línea line filter
filtro de línea de corriente alterna alternating-current line filter
filtro de línea de energía power-line filter
filtro de línea de estación auxiliar auxiliary station line filter
filtro de luz light filter
filtro de luz ambiente ambient-light filter
filtro de magnetoestricción magnetostriction filter
filtro de memoria coherente coherent memory filter
filtro de microondas microwave filter
filtro de modo en anillo ring-mode filter
filtro de modo resonante resonant-mode filter
filtro de modos mode filter
filtro de muesca notch filter
filtro de octava octave filter
filtro de onda wave filter
filtro de onda continua continuous-wave filter
filtro de ondas acústicas acoustic-wave filter
filtro de ondas compuesto composite wave filter
filtro de ondas de superficie surface-wave filter
filtro de ondas eléctricas electric-wave filter

filtro de ondas mecánico mechanical wave filter
filtro de ondulación ripple filter
filtro de paso de banda bandpass filter
filtro de paso de banda agudo sharp bandpass filter
filtro de paso de banda de recepción receiving
     bandpass filter
filtro de paso de canal channel pass filter
filtro de pendiente slope filter
filtro de pendiente variable variable-slope filter
filtro de pico peak filter
filtro de polarización bias filter
filtro de portadora carrier filter
filtro de potencia power filter
filtro de preénfasis preemphasis filter
filtro de puente de Wien Wien-bridge filter
filtro de radio radio filter
filtro de radiofrecuencia radio-frequency filter
filtro de radiorruido radio noise filter
filtro de ramificación branching filter
filtro de recepción receiving filter
filtro de rechazo rejection filter
filtro de rechazo de banda band-rejection filter
filtro de red network filter
filtro de red cristalina crystal-lattice filter
filtro de registro recording filter
filtro de resistencia-capacitancia
     resistance-capacitance filter
filtro de retroalimentación inversa inverse-feedback
     filter
filtro de rojo red filter
filtro de ruido noise filter
filtro de ruido de aguja scratch filter
filtro de salida output filter
filtro de sección pi pi-section filter
filtro de sección única single-section filter
filtro de seguimiento tracking filter
filtro de separación separation filter
filtro de separación direccional directional
     separation filter
filtro de subportadora subcarrier filter
filtro de supresión de banda band-suppression filter
filtro de supresión de diafonía crosstalk suppression
     filter
filtro de tono tone filter
filtro de transferencia de portadora carrier transfer
     filter
filtro de transmisión sending filter
filtro de troncal trunk filter
filtro de unión junction filter
filtro de vacío vacuum filter
filtro de velocidad velocity filter
filtro de verde green filter
filtro de video video filter
filtro de voz voice filter
filtro de Wratten Wratten filter
filtro dicroico dichroic filter
filtro digital digital filter
filtro direccional directional filter
filtro duplexor duplexer filter
filtro eléctrico electrical filter
filtro electromecánico electromechanical filter
filtro electrónico electronic filter
filtro electrostático electrostatic filter
filtro en anillo ring filter
filtro en celosía lattice filter
filtro en cuña wedge filter
filtro en derivación shunt filter
filtro en escalera ladder filter
filtro en peine comb filter
filtro en serie series filter
filtro equilibrado balanced filter
filtro formante formant filter
filtro giromagnético gyromagnetic filter

filtro graduado graded filter
filtro heterodino heterodyne filter
filtro ideal ideal filter
filtro inherente inherent filter
filtro integrador integrating filter
filtro iterativo iterative filter
filtro limitador limiting filter
filtro limitador de banda band-limiting filter
filtro lineal linear filter
filtro lineal aleatorio random linear filter
filtro magnetoestrictivo magnetostrictive filter
filtro mecánico mechanical filter
filtro monocromático monochromatic filter
filtro multicapa multilayer filter
filtro multimalla multimesh filter
filtro multisección multisection filter
filtro neutro neutral filter
filtro no calculado brute-force filter
filtro no sintonizado untuned filter
filtro óptico optical filter
filtro óptico sintonizable acústicamente acoustically
     tunable optical filter
filtro pasaaltos high-pass filter
filtro pasabajos low-pass filter
filtro pasabajos equilibrado balanced low-pass filter
filtro pasabanda electrónico electronic passband
     filter
filtro pasatodo all-pass filter
filtro periódico periodic filter
filtro piezoeléctrico piezoelectric filter
filtro polarizante polarizing filter
filtro polaroid polaroid filter
filtro primario primary filter
filtro rectificador rectifier filter
filtro resonante resonant filter
filtro secundario secondary filter
filtro selectivo selective filter
filtro separador separating filter
filtro sintonizable tunable filter
filtro sintonizado tuned filter
filtro sonoro sound filter
filtro supresor stop filter
filtro supresor de chispas spark-suppressing filter
filtro sustractivo subtractive filter
filtro terminal terminal filter
filtro total total filter
filtro transistorizado transistorized filter
filtro ultrasónico ultrasonic filter
filtro universal universal filter
filtrón filtron
fin de archivo end-of-file
fin de área end-of-area
fin de bloque end-of-block
fin de campo end-of-field
fin de cinta end-of-tape
fin de datos end-of-data
fin de dirección end-of-address
fin de línea end-of-line
fin de mensaje end-of-message
fin de procedimiento end-of-procedure
fin de programa end-of-program
fin de registro end-of-record
fin de sesión session end
fin de tarjeta end-of-card
fin de texto end-of-text
fin de trabajo end-of-job
fin de transmisión end-of-transmission
final anormal abnormal end, abend
final anormal irrecuperable unrecoverable abend
final anormal recuperable recoverable abend
final de tarea anormal abnormal end of task
finito finite
finsen finsen

finura de regulación notching ratio
física del estado sólido solid-state physics
física electrónica electron physics
fisión atómica atomic fission
fistrón phystron
flanco de impulso pulse edge
flanqueo flanking
flecha sag
flexibilidad flexibility
flexibilidad de frecuencia frequency flexibility
flexibilidad de procesamiento processing flexibility
flexibilidad operacional operational flexibility
flexible flexible
flexión flexion, bending
flexión atmosférica atmospheric bending
flexodo flexode
fluctuación fluctuation, jitter
fluctuación acumulada accumulated jitter
fluctuación de alineación alignment jitter
fluctuación de amplitud amplitude jitter
fluctuación de ángulo angle fluctuation
fluctuación de fase phase jitter
fluctuación de frecuencia frequency jitter
fluctuación de haz beam jitter
fluctuación de imagen picture jitter
fluctuación de impulsos pulse jitter
fluctuación de largo periodo long-period fluctuation
fluctuación de seguimiento tracking jitter
fluctuación de tensión voltage fluctuation
fluctuación de tiempo time jitter
fluctuación de tráfico traffic fluctuation
fluctuación de velocidad velocity fluctuation
fluctuación inducida induced jitter
fluctuación por zumbido hum jitter
fluctuación térmica thermal fluctuation
fluctuación transferida transferred jitter
fluctuación transitoria de facsímil facsimile
    transient
fluctuación transitoria oscilante oscillatory transient
fluctuaciones aleatorias random fluctuations
fluctuaciones transitorias de manipulación keying
    transients
fluctuante fluctuating
fluídica fluidics
fluido fluid
fluido amortiguador damping fluid
fluido de flotación flotation fluid
fluido eléctrico electric fluid
fluido magnético magnetic fluid
flujo flow, flux
flujo asincrónico asynchronous flow
flujo atrapado trapped flux
flujo axial axial flow
flujo bactericida bactericidal flux
flujo bidireccional bidirectional flow
flujo continuo continuous flow
flujo cruzado cross flux
flujo de aire air flow
flujo de bits bit stream
flujo de conmutación switching flux
flujo de control control flow, control flux
flujo de corriente current flow
flujo de datos data flow
flujo de electrones confinado confined electron flow
flujo de energía energy flux
flujo de energía geométrico geometric energy flux
flujo de energía sonora sound-energy flux
flujo de energía sonora específico specific
    sound-energy flux
flujo de fuga leakage flow, leakage flux
flujo de fuga de relé relay leakage flux
flujo de gas gas flow
flujo de huecos hole flow

flujo de inducción induction flux
flujo de información information flow
flujo de luz light flux
flujo de partículas particle flux
flujo de portadores primario primary carrier flow
flujo de potencia power flow, power flux
flujo de programa program flow
flujo de radiación radiation flux
flujo de red network flow
flujo de resonancia resonance flux
flujo de saturación saturation flux
flujo de señal signal flux
flujo de trabajo job flow, work flow
flujo de tráfico traffic flow
flujo dieléctrico dielectric flux
flujo eléctrico electric flux
flujo electromagnético electromagnetic flux
flujo electrónico electronic flow, electron flow
flujo electrostático electrostatic flux
flujo en cuadratura quadrature flux
flujo equilibrador balancing flux
flujo fotónico photon flux
flujo frío cold flow
flujo integrado integrated flux
flujo intrínseco intrinsic flux
flujo invertido reversed flow
flujo iónico ion flow
flujo lineal linear flow
flujo luminoso luminous flux
flujo luminoso espectral spectral luminous flux
flujo luminoso total total luminous flux
flujo magnético magnetic flux
flujo magnetodinámico magnetodynamic flow
flujo molecular molecular flow
flujo momentáneo momentary flow
flujo neutrónico neutron flux
flujo no corrosivo noncorrosive flux
flujo normal normal flow
flujo paralelo parallel flow
flujo polarizante polarizing flux
flujo preferencial preferential flow
flujo primario primary flow
flujo primario de portadores carrier primary flow
flujo radiante radiant flux
flujo radiante espectral spectral radiant flux
flujo reflejado reflected flow
flujo residual residual flux
flujo secundario secondary flow
flujo sincrónico synchronous flow
flujo sonoro sound flux
flujo supersónico supersonic flow
flujo térmico thermal flux
flujo transitorio transient flux
flujo uniforme uniform flow
flujógrafo fluxgraph
flujometría flowmetering
flujómetro flowmeter, fluxmeter
flujómetro acústico acoustic flowmeter
flujómetro de bismuto bismuth fluxmeter
flujómetro de efecto de Hall Hall-effect fluxmeter
flujómetro de hilo caliente hot-wire flowmeter
flujómetro de luz light fluxmeter
flujómetro de partículas particle fluxmeter
flujómetro electromagnético electromagnetic
    flowmeter
flujómetro electrónico electronic flowmeter
flujómetro magnético magnetic flowmeter
fluor fluorine
fluorescencia fluorescence
fluorescencia de rayos X X-ray fluorescence
fluorescencia de resonancia resonance fluorescence
fluorescencia por impacto impact fluorescence
fluorescente fluorescent

fluorimetría fluorimetry
fluorímetro fluorimeter
fluorofotómetro fluorophotometer
fluorometría fluorometry
fluorómetro fluorometer
fluoroscopia fluoroscopy
fluoroscópico fluoroscopic
fluoroscopio fluoroscope
focal focal
foco focus
foco estático static focus
foco fijo fixed focus
foco principal principal focus
foco real real focus
foco tangencial tangential focus
focómetro focometer
fondo background, faceplate
fondo ambiente ambient background
fondo de ruido noise background
fondo de valle valley point
fondo de visualización display background
fónica phonics
fonio phon
fonocaptor pickup
fonocaptor acústico acoustic pickup
fonocaptor cerámico ceramic pickup
fonocaptor de bobina móvil moving-coil pickup
fonocaptor de capacitancia variable variable-capacitance pickup
fonocaptor de carbón carbon pickup
fonocaptor de condensador condenser pickup
fonocaptor de contacto de carbón carbon-contact pickup
fonocaptor de cristal crystal pickup
fonocaptor de inductancia variable variable-inductance pickup
fonocaptor de modulación de frecuencia frequency-modulated pickup
fonocaptor de reluctancia variable variable-reluctance pickup
fonocaptor de resistencia variable variable-resistance pickup
fonocaptor dinámico dynamic pickup
fonocaptor electromagnético electromagnetic pickup
fonocaptor electrónico electronic pickup
fonocaptor electrostático capacitor pickup
fonocaptor estereofónico stereophonic pickup
fonocaptor magnético magnetic pickup, magnetic cartridge, magnetic pickup cartridge
fonocaptor piezoeléctrico piezoelectric pickup
fonocardiógrafo phonocardiograph
fonocardiograma phonocardiogram
fonodisco phonorecord
fonógrafo phonograph
fonógrafo acústico acoustic phonograph
fonógrafo mecánico mechanical phonograph
fonograma phonogram
fonoincisor cutter
fonoincisor con retroalimentación feedback cutter
fonometría phonometry
fonómetro phonometer
fonón phonon
fonón acústico acoustic phonon
fonorrecepción phonoreception
forma abreviada de mensaje abbreviated form of message
forma de bobina coil form
forma de cable cable form
forma de carácter character form
forma de complemento complement form
forma de entrada de datos data entry form
forma de envolvente de señal signal-envelope shape
forma de impulso pulse shape

forma de onda waveform, waveshape
forma de onda compleja complex waveform
forma de onda compuesta composite waveform
forma de onda de Gauss Gaussian waveform
forma de onda de impulso pulse waveform
forma de onda de línea line waveform
forma de onda de rejilla grid waveform
forma de onda de salida output waveform
forma de onda de video video waveform
forma de onda exponencial exponential waveform
forma de onda horizontal horizontal waveform
forma de onda parabólica parabolic waveform
forma de partículas particle shape
forma de respuesta response shape
forma de señal signal form
forma de surco groove shape
forma decimal decimal form
forma normalizada normalized form
forma numérica numerical form
formación formation, shaping
formación de arco en magnetrón magnetron arcing
formación de arcos arcing
formación de haz beam shaping
formación de impulsos pulse shaping
formación eléctrica electrical forming
formador shaper, former
formador de haz beam former
formador de impulsos pulse shaper
formaldehído formaldehyde
formante formant
formateado formatted
formato format
formato binario binary format
formato binario reubicable relocatable binary format
formato corto short format
formato de alto nivel high-level format
formato de archivo file format
formato de archivo de gráficos graphics file format
formato de bajo nivel low-level format
formato de bloque block format
formato de bloque fijo fixed block format
formato de celda cell format
formato de cinta tape format
formato de coma fija fixed-point format
formato de coma flotante floating-point format
formato de control control format
formato de datos data format
formato de datos de tarjeta card data format
formato de diagrama chart format
formato de dirección address format
formato de dirección de palabra word-address format
formato de entrada input format
formato de impresión print format
formato de instrucción instruction format
formato de instrucción corta short-instruction format
formato de intercambio de datos data interchange format
formato de mensajes message format
formato de orden order format
formato de palabra word format
formato de referencia reference format
formato de registro record format
formato de registro de datos data record format
formato de salida output format
formato de sistema system format
formato de sistema de disco disk system format
formato de sistema de tarjetas card system format
formato de tarjeta card format
formato de tarjeta perforada punched-card format
formato de visualización display format

formato decimal empaquetado packed decimal format
formato empaquetado packed format
formato enlazado linked format
formato fijo fixed format
formato físico physical format
formato general general format
formato global global format
formato interno internal format
formato largo long format
formato libre free format
formato lógico logical format
formato numérico numerical format
formato objeto object format
formato secuencial fijo fixed sequential format
formato supervisor supervisory format
formato variable variable format
fórmula cuadrática quadratic formula
fórmula de Bunet Bunet's formula
fórmula de Campbell Campbell's formula
fórmula de dispersión de Mott Mott scattering formula
fórmula de Esnault-Pelterie Esnault-Pelterie formula
fórmula de Johnson-Lark-Horowitz Johnson-Lark-Horowitz formula
fórmula de Landau Landau formula
fórmula de Langevin-Pauli Langevin-Pauli formula
fórmula de Nyquist Nyquist formula
fórmula de Preece Preece's formula
fórmula de radiación de Planck Planck's radiation formula
fórmula de reacción reaction formula
fórmula de Sommerfeld Sommerfeld formula
fórmula de Steinmetz Steinmetz formula
forro jacket, liner, lining
forsterita forsterite
forzado de campo field forcing
fosdic fosdic
fosforescencia phosphorescence
fosforescente phosphorescent
fósforo phosphorus, phosphor
fósforo de fluoruro de magnesio magnesium-flouride phosphor
fósforo de silicato de magnesio magnesium-silicate phosphor
fósforo de tungstato de magnesio magnesium-tungstate phosphor
fosforógeno phosphorogen
fosforoscopio phosphoroscope
fosforoso phosphorous
fosfuro de galio gallium phosphide
fosfuro de indio indium phosphide
foticón photicon
fotio phot
fotoactivo photoactive
fotoactuador photoactuator
fotoaislador photoisolator
fotocatalizador photocatalyst
fotocátodo photocathode
fotocátodo semitransparente semitransparent photocathode
fotocélula photocell, photoelectric cell
fotocélula autogeneradora self-generating photocell
fotocélula de cristal crystal photocell
fotocélula de emisión secundaria secondary-emission photocell
fotocélula de gas gas-filled photocell
fotocélula de germanio germanium photocell
fotocélula de óxido de cobre copper-oxide photocell
fotocélula de selenio selenium photocell
fotocélula de seleniuro de plomo lead selenide photocell

fotocélula de silicio silicon photocell
fotocélula de silicio-boro silicon-boron photocell
fotocélula de unión junction photocell
fotocélula de unión PN PN-junction photocell
fotocélula de unión por crecimiento grown-junction photocell
fotocélula de vacío vacuum photocell
fotocélula multiplicadora multiplier photocell
fotocélula rectificadora rectifier photocell
fotocoagulador láser laser photocoagulator
fotocolorímetro photocolorimeter
fotocomposición photocomposition
fotoconducción photoconduction
fotoconductancia photoconductance
fotoconductividad photoconductivity
fotoconductivo photoconductive
fotoconductor photoconductor
fotoconmutador photoswitch
fotocopiadora photocopier
fotocopiadora electrostática electrostatic photocopier
fotocorriente photocurrent
fotocrómico photochromic
fotodensitómetro photodensitometer
fotodescomposición photodecomposition
fotodesintegración photodisintegration
fotodestello photoflash
fotodetector photodetector
fotodieléctrico photodielectric
fotodifusión photodiffusion
fotodiodo photodiode
fotodiodo de avalancha avalanche photodiode
fotodiodo de contacto de punta point-contact photodiode
fotodiodo de emisión distribuida distributed-emission photodiode
fotodiodo de germanio germanium photodiode
fotodiodo de unión junction photodiode
fotodiodo de unión de germanio germanium junction photodiode
fotodiodo PIN PIN photodiode
fotodiodo planar planar photodiode
fotodisociación photodissociation
fotodispositivo photodevice
fotodosimetría photodosimetry
fotoefecto photoeffect
fotoelasticidad photoelasticity
fotoelasticimetría photoelasticimetry
fotoelasticímetro photoelasticimeter
fotoelástico photoelastic
fotoeléctricamente photoelectrically
fotoelectricidad photoelectricity
fotoeléctrico photoelectric
fotoelectroluminiscencia photoelectroluminescence
fotoelectromagnético photoelectromagnetic
fotoelectromotriz photoelectromotive
fotoelectrón photoelectron
fotoelectrónica photoelectronics
fotoelectrónico photoelectronic
fotoemisión photoemission
fotoemisivo photoemissive
fotoemisor photoemitter
fotoestimulación photostimulation
fotofabricación photofabrication
fotofluorescencia photofluorescence
fotofluorografía photofluorography
fotofluorógrafo photofluorograph
fotófono photophone
fotoforesis photophoresis
fotoformador photoformer
fotogalvánico photogalvanic
fotogenerador photogenerator
fotogoniómetro photogoniometer

fotograbado photoengraved
fotografía con infrarrojo infrared photography
fotografía de radar radar photograph
fotografía estereoscópica stereoscopic photography
fotografía por rayos X X-ray photograph
fotográfico photographic
fotograma photogram
fotointerruptor periódico photochopper
fotoionización photoionization
fotólisis photolysis
fotolitográfico photolithographic
fotoluminiscencia photoluminescence
fotomagnético photomagnetic
fotomagnetoeléctrico photomagnetoelectric
fotomáscara photomask
fotometría photometry
fotometría física physical photometry
fotometría fotoeléctrica photoelectric photometry
fotometría visual visual photometry
fotométrico photometric
fotómetro photometer, light meter
fotómetro de Bunsen Bunsen photometer
fotómetro de contraste contrast photometer
fotómetro de dispersión dispersion photometer
fotómetro de llama flame photometer
fotómetro de parpadeo flicker photometer
fotómetro de polarización polarization photometer
fotómetro de sombra shadow photometer
fotómetro digital digital photometer
fotómetro electrónico electronic photometer
fotómetro físico physical photometer
fotómetro fotoeléctrico photoelectric photometer
fotómetro integrador integrating photometer
fotómetro logarítmico logarithmic photometer
fotómetro objetivo objective photometer
fotómetro subjetivo subjective photometer
fotómetro visual visual photometer
fotomezclador photomixer
fotomicrografía photomicrography
fotomosaico photomosaic
fotomultiplicador photomultiplier
fotomultiplicador electrostático electrostatic
　photomultiplier
fotón photon
fotón óptico optical photon
fotonegativo photonegative
fotoneutrón photoneutron
fotonuclear photonuclear
fotoóptica photooptic
fotooxidación photooxidation
fotoparamétrico photoparametric
fotopico photo peak
fotópico photopic
fotopositivo photopositive
fotoprotón photoproton
fotoquímica photochemistry
fotoquímico photochemical
fotorreducción photoreduction
fotorrelé photorelay
fotorresistencia photoresistance
fotorresistente photoresistant
fotorresistividad photoresistivity
fotorresistivo photoresistive
fotorresistor photoresistor
fotoscopio photoscope
fotosensibilidad photosensitivity
fotosensibilidad específica specific photosensitivity
fotosensible photosensitive
fotosensor photosensor
fotosensor de estado sólido solid-state photosensor
fotosfera photosphere
fotosuperficie photosurface
fototelefonía phototelephony

fototelegrafía phototelegraphy
fototelegráfico phototelegraphic
fototelégrafo phototelegraph
fototelegrama phototelegram
fototemporizador phototimer
fototensión photovoltage
fototransductor phototransducer
fototransistor phototransistor
fototransistor de contacto de punta point-contact
　phototransistor
fototransistor de unión junction phototransistor
fototransistor PNP PNP phototransistor
fototubo phototube
fototubo de alto vacío high-vacuum phototube
fototubo de cesio cesium phototube
fototubo de gas gas phototube
fototubo de microondas microwave phototube
fototubo de ondas progresivas traveling-wave
　phototube
fototubo de vacío vacuum phototube
fototubo gaseoso gaseous phototube
fototubo multiplicador multiplier phototube
fototubo multiplicador de campos cruzados
　crossed-field multiplier phototube
fototubo multiplicador de microondas microwave
　multiplier phototube
fototubo multiplicador electrónico electron
　multiplier phototube
fotounión photojunction
fotoválvula photovalve
fotovaristor photovaristor
fotovoltaico photovoltaic
fotran photran
fracción de empaque packing fraction
fracción decimal decimal fraction
fragilización embrittlement
fragmentación fragmentation
fragmentación de archivo file fragmentation
fragmento fragment
francio francium
franja strip
franja de adaptación matching strip
franja de demarcación demarcation strip
franja magnética magnetic strip
franjas de colores marginales color fringing
franklin franklin
frasco de Dewar Dewar flask
frecuencia frequency
frecuencia absoluta absolute frequency
frecuencia acústica acoustic frequency
frecuencia adyacente adjacent frequency
frecuencia aire-tierra air-ground frequency
frecuencia ajustable adjustable frequency
frecuencia alternativa alternate frequency
frecuencia angular angular frequency
frecuencia antirresonante antiresonant frequency
frecuencia armónica harmonic frequency
frecuencia asignable assignable frequency
frecuencia asignada assigned frequency
frecuencia atómica atomic frequency
frecuencia audible audible frequency
frecuencia autorresonante self-resonant frequency
frecuencia básica basic frequency
frecuencia calibrada calibrated frequency
frecuencia característica characteristic frequency
frecuencia central center frequency
frecuencia central de canal channel center
　frequency
frecuencia cero zero frequency
frecuencia ciclotrónica cyclotron frequency
frecuencia circular circular frequency
frecuencia complementaria idler frequency
frecuencia crítica critical frequency

frecuencia cuadrantal quadrantal frequency
frecuencia de agrupamiento bunching frequency
frecuencia de ajuste adjustment frequency
frecuencia de antena natural natural antenna
    frequency
frecuencia de atenuación infinita infinite-attenuation
    frequency
frecuencia de audio audio frequency
frecuencia de banda base baseband frequency
frecuencia de banda lateral sideband frequency
frecuencia de barrera barrier frequency
frecuencia de barrido sweep frequency
frecuencia de barrido horizontal horizontal-sweep
    frequency
frecuencia de barrido vertical vertical-sweep
    frequency
frecuencia de base base frequency
frecuencia de base de tiempo time-base frequency
frecuencia de batido beat frequency
frecuencia de bombeo pumping frequency
frecuencia de borrado erasing frequency
frecuencia de campo field frequency
frecuencia de canal channel frequency
frecuencia de centro de banda midband frequency
frecuencia de codo bend frequency
frecuencia de coincidencia coincidence frequency
frecuencia de conmutación switching frequency
frecuencia de convergencia convergence frequency
frecuencia de corte cutoff frequency
frecuencia de corte alfa alpha cutoff frequency
frecuencia de corte beta beta cutoff frequency
frecuencia de corte de ganancia de potencia
    power-gain cutoff frequency
frecuencia de corte de guíaondas waveguide cutoff
    frequency
frecuencia de corte efectiva effective cutoff
    frequency
frecuencia de corte nominal nominal cutoff
    frequency
frecuencia de corte resistiva resistive cutoff
    frequency
frecuencia de corte teorética theoretical cutoff
    frequency
frecuencia de cristal crystal frequency
frecuencia de cruce crossover frequency
frecuencia de cuadro frame frequency
frecuencia de chispa spark frequency
frecuencia de Debye Debye frequency
frecuencia de destellos flashing rate
frecuencia de diapasón fork frequency
frecuencia de diferencia difference frequency
frecuencia de dispersión scattering frequency
frecuencia de Doppler Doppler frequency
frecuencia de Doppler diferencial differential
    Doppler frequency
frecuencia de eco echo frequency
frecuencia de emergencia emergency frequency
frecuencia de emergencia aeronáutica aeronautical
    emergency frequency
frecuencia de emisión emission frequency
frecuencia de entrada input frequency
frecuencia de espacio spacing frequency
frecuencia de esquina corner frequency
frecuencia de excitación excitation frequency
frecuencia de exploración scanning frequency
frecuencia de exploración de líneas line-scanning
    frequency
frecuencia de exploración horizontal
    horizontal-scanning frequency
frecuencia de exploración vertical vertical-scanning
    frequency
frecuencia de extinción quenching frequency
frecuencia de fusión fusion frequency

frecuencia de fusión de parpadeo flicker fusion
    frequency
frecuencia de ganancia cruzada gain-crossover
    frequency
frecuencia de grupo group frequency
frecuencia de iluminación rate of illumination
frecuencia de imagen picture frequency
frecuencia de impulsos pulse frequency
frecuencia de intermodulación intermodulation
    frequency
frecuencia de interrogación interrogation frequency
frecuencia de interrupciones interruption frequency
frecuencia de interruptor interrupter frequency
frecuencia de Larmor Larmor frequency
frecuencia de límite de escala full-scale frequency
frecuencia de línea de energía power-line frequency
frecuencia de líneas line frequency
frecuencia de líneas de exploración scanning-line
    frequency
frecuencia de líneas de imagen picture line
    frequency
frecuencia de líneas horizontales horizontal-line
    frequency
frecuencia de llamada calling frequency, ringing
    frequency
frecuencia de manipulación keying frequency
frecuencia de medida measuring frequency
frecuencia de modulación modulation frequency
frecuencia de muestreo sampling frequency
frecuencia de muestreo de colores color sampling
    rate
frecuencia de Nyquist Nyquist frequency
frecuencia de ocurrencia frequency of occurrence
frecuencia de onda media medium-wave frequency
frecuencia de ondulación ripple frequency
frecuencia de operación operating frequency
frecuencia de oscilación oscillation frequency
frecuencia de oscilador oscillator frequency
frecuencia de parpadeo flicker frequency
frecuencia de penetración penetration frequency
frecuencia de plasma plasma frequency
frecuencia de polarización bias frequency
frecuencia de portadora carrier frequency
frecuencia de portadora de canal channel-carrier
    frequency
frecuencia de portadora de imagen picture-carrier
    frequency
frecuencia de portadora de sonido sound-carrier
    frequency
frecuencia de portadora media mean carrier
    frequency
frecuencia de precesión de Larmor Larmor
    precession frequency
frecuencia de proyección projection frequency
frecuencia de prueba test frequency
frecuencia de prueba patrón standard test frequency
frecuencia de pulsación pulsation frequency
frecuencia de puntos dot frequency
frecuencia de radio aire-tierra air-ground radio
    frequency
frecuencia de radio de aire a tierra air-to-ground
    radio frequency
frecuencia de ranura slot frequency
frecuencia de recepción receive frequency
frecuencia de recurrencia recurrence frequency
frecuencia de red power frequency, mains frequency
frecuencia de referencia reference frequency
frecuencia de refresco refresh rate
frecuencia de relajación relaxation frequency
frecuencia de reloj clock frequency
frecuencia de reloj maestro master clock frequency
frecuencia de repetición repetition frequency,
    repetition rate

frecuencia de repetición básica basic repetition frequency
frecuencia de repetición de cuadro frame repetition rate
frecuencia de repetición de imagen picture repetition rate
frecuencia de repetición de impulsos pulse repetition frequency, pulse repetition rate
frecuencia de reposo resting frequency
frecuencia de resonancia resonance frequency
frecuencia de respuesta response frequency
frecuencia de salida output frequency
frecuencia de salida nominal rated output frequency
frecuencia de señal signal frequency
frecuencia de señal de bajada downlink frequency
frecuencia de señal de subida uplink frequency
frecuencia de señalización signaling frequency
frecuencia de sintonización tuning frequency
frecuencia de sintonización patrón standard tuning frequency
frecuencia de sistema system frequency
frecuencia de sistema nominal rated system frequency
frecuencia de socorro distress frequency
frecuencia de sonido sound frequency
frecuencia de subportadora subcarrier frequency
frecuencia de subportadora de colores color-subcarrier frequency
frecuencia de suma sum frequency
frecuencia de supresión quench frequency
frecuencia de tono tone frequency
frecuencia de trabajo working frequency
frecuencia de trabajo común common working frequency
frecuencia de trabajo óptima optimum working frequency
frecuencia de tráfico traffic frequency
frecuencia de tráfico óptima optimum traffic frequency
frecuencia de transición transition frequency
frecuencia de transmisión transmission frequency
frecuencia de transmisor transmitter frequency
frecuencia de video video frequency
frecuencia de voz voice frequency
frecuencia de zumbido hum frequency
frecuencia desplazada offset frequency
frecuencia determinada por cristal crystal-determined frequency
frecuencia directa straight frequency
frecuencia discreta discrete frequency
frecuencia espuria spurious frequency
frecuencia extremadamente alta extremely high frequency
frecuencia extremadamente baja extremely low frequency
frecuencia fija fixed frequency
frecuencia forzada forced frequency
frecuencia fundamental fundamental frequency
frecuencia giromagnética gyromagnetic frequency
frecuencia heterodina heterodyne frequency
frecuencia horizontal horizontal frequency
frecuencia imagen image frequency
frecuencia inaudible inaudible frequency
frecuencia inferior lower frequency
frecuencia infinita infinite frequency
frecuencia infraacústica subaudio frequency
frecuencia infrabaja infralow frequency
frecuencia infrasónica infrasonic frequency
frecuencia instantánea instantaneous frequency
frecuencia interferente interfering frequency
frecuencia intermedia intermediate frequency
frecuencia intermedia de imagen picture intermediate frequency

frecuencia intermedia de sonido sound intermediate frequency
frecuencia inversa backward frequency
frecuencia lateral side frequency
frecuencia limitante por absorción absorption limiting frequency
frecuencia límite limiting frequency
frecuencia lineal straight-line frequency
frecuencia marcadora marker frequency
frecuencia máxima de banda base maximum baseband frequency
frecuencia máxima de manipulación maximum keying frequency
frecuencia máxima de modulación maximum modulation frequency
frecuencia máxima observada maximum observed frequency
frecuencia máxima utilizable maximum usable frequency
frecuencia media medium frequency
frecuencia mínima de muestreo minimum sampling frequency
frecuencia moduladora modulating frequency
frecuencia monocroma monochrome frequency
frecuencia musical musical frequency
frecuencia natural natural frequency
frecuencia natural amortiguada damped natural frequency
frecuencia natural no amortiguada undamped natural frequency
frecuencia no armónica nonharmonic frequency
frecuencia nocturna night frequency
frecuencia nominal nominal frequency, rated frequency
frecuencia normal normal frequency
frecuencia normalizada normalized frequency
frecuencia nula null frequency
frecuencia óptima optimum frequency
frecuencia pasante passing frequency
frecuencia patrón standard frequency
frecuencia penetrante penetrating frequency
frecuencia piloto pilot frequency
frecuencia piloto de línea line-pilot frequency
frecuencia portadora angular angular carrier frequency
frecuencia portadora de imagen picture tone
frecuencia portadora visual visual carrier frequency
frecuencia presintonizada pretuned frequency
frecuencia programada scheduled frequency
frecuencia propia free-running frequency
frecuencia puntual spot frequency
frecuencia regulada regulated frequency
frecuencia relativa relative frequency
frecuencia resonante resonant frequency
frecuencia resonante de antena antenna resonant frequency
frecuencia resonante de bajos bass-resonant frequency
frecuencia resonante de cuarto de onda quarter-wave resonant frequency
frecuencia resonante natural natural resonant frequency
frecuencia secundaria secondary frequency
frecuencia seleccionada selected frequency
frecuencia sincronizada synchronized frequency
frecuencia sónica sonic frequency
frecuencia subsónica subsonic frequency
frecuencia superalta superhigh frequency
frecuencia supersónica supersonic frequency
frecuencia supertelefónica supertelephone frequency
frecuencia telefónica telephone frequency
frecuencia ultraacústica superaudio frequency
frecuencia ultraalta ultra-high frequency

frecuencia ultraaudible ultra-audible frequency
frecuencia ultrabaja ultralow frequency
frecuencia ultrasónica ultrasonic frequency
frecuencia umbral threshold frequency
frecuencia única single frequency
frecuencia útil useful frequency
frecuencia útil más baja lowest useful frequency
frecuencia utilizable usable frequency
frecuencia utilizable más baja lowest usable frequency
frecuencia variable variable frequency
frecuencia vertical vertical frequency
frecuencia vocal voice frequency
frecuencias de media potencia half-power frequencies
frecuencias de registro recording frequencies
frecuencímetro frequency meter
frecuencímetro analógico analog frequency meter
frecuencímetro contador de impulsos pulse-counter frequency meter
frecuencímetro de absorción absorption frequency meter
frecuencímetro de aguja pointer frequency meter
frecuencímetro de cavidad cavity frequency meter
frecuencímetro de Frahm Frahm frequency meter
frecuencímetro de guíaondas waveguide frequency meter
frecuencímetro de impedancia impedance frequency-meter
frecuencímetro de laminillas vibrantes vibrating-reed frequency meter
frecuencímetro de lectura directa direct-reading frequency meter
frecuencímetro de línea coaxial coaxial-line frequency meter
frecuencímetro de precisión precision frequency meter
frecuencímetro de puente de Wien Wien-bridge frequency meter
frecuencímetro de referencia reference frequency-meter
frecuencímetro de registro recording frequency-meter
frecuencímetro de resonancia resonance frequency-meter
frecuencímetro de tipo heterodino heterodyne-type frequency meter
frecuencímetro de transmisión transmission frequency meter
frecuencímetro diferencial differential frequency meter
frecuencímetro digital digital frequency meter
frecuencímetro electromecánico electromechanical frequency meter
frecuencímetro electrónico electronic frequency meter
frecuencímetro electroóptico electrooptical frequency meter
frecuencímetro heterodino heterodyne frequency meter
frecuencímetro integrador integrating frequency meter
frenado braking
frenado magnético magnetic braking
frenado por contracorriente plugging
frenado por resistencia resistance braking
frenado por solenoide solenoid braking
frenado regenerativo regenerative braking
frenado reostático rheostatic braking
frenaje braking
frenaje dinámico dynamic braking
frenaje eléctrico electric braking
frenaje electrodinámico electrodynamic braking

frenaje electromagnético electromagnetic braking
frenaje por corrientes parásitas eddy-current braking
freno de cortocircuito short-circuit brake
freno de rejilla grid stopper
freno de solenoide solenoid brake
freno eléctrico electric brake
freno electromagnético electromagnetic brake
freno electromecánico electromechanical brake
freno electroneumático electropneumatic brake
freno magnético magnetic brake
freno por histéresis hysteresis brake
frente de fase phase front
frente de onda wavefront
frente muerto dead front
freón freon
fresnel fresnel
fricción friction
fricción de aguja stylus friction, needle drag
fricción de arranque breakaway friction
fricción estática static friction
fricción magnética magnetic friction
fricción molecular molecular friction
fricción seca dry friction
fritado fritting
fritado de relé relay fritting
frotador wiper
fuente source, supply
fuente A A supply
fuente aceptable acceptable source
fuente anódica anode supply
fuente B B supply
fuente de alimentación power source, power supply
fuente de alimentación A A power supply
fuente de alimentación alternativa alternate power source
fuente de alimentación anódica anode power supply
fuente de alimentación autónoma self-contained power supply
fuente de alimentación auxiliar auxiliary power supply
fuente de alimentación B B power supply
fuente de alimentación de alta resolución high-resolution power supply
fuente de alimentación de alta tensión high-voltage power supply
fuente de alimentación de batería battery power supply
fuente de alimentación de calefactor heater power supply
fuente de alimentación de corriente alterna alternating-current power supply
fuente de alimentación de corriente constante constant-current power supply
fuente de alimentación de corriente continua direct-current power supply
fuente de alimentación de emergencia emergency power supply
fuente de alimentación de estado sólido solid-state power supply
fuente de alimentación de filamento filament power supply
fuente de alimentación de laboratorio laboratory power supply
fuente de alimentación de línea line power supply
fuente de alimentación de oscilador oscillator power supply
fuente de alimentación de periférico peripheral power supply
fuente de alimentación de placa plate power supply
fuente de alimentación de radiofrecuencia radio-frequency power supply
fuente de alimentación de referencia reference

power supply

fuente de alimentación de retorno flyback power supply

fuente de alimentación de transistores transistor power supply

fuente de alimentación electrónica electronic power supply

fuente de alimentación estabilizada stabilized power supply

fuente de alimentación externa external power supply

fuente de alimentación ininterrumpible uninterruptable power supply

fuente de alimentación interna internal power supply

fuente de alimentación moduladora modulating power supply

fuente de alimentación no regulada unregulated power supply

fuente de alimentación operacional operational power supply

fuente de alimentación preferida preferred power supply

fuente de alimentación regulada regulated power supply

fuente de alimentación regulada electrónicamente electronically regulated power supply

fuente de alimentación secundaria secondary power supply

fuente de alimentación sin interrupción no-break power supply

fuente de batería común common-battery supply

fuente de calor heat source

fuente de carga charge source

fuente de corriente source of current

fuente de corriente alterna alternating-current source

fuente de corriente constante constant-current source

fuente de corriente continua direct-current source

fuente de corriente de salida output source current

fuente de corriente estática static source current

fuente de datos data source

fuente de electrones electron source

fuente de energía source of energy

fuente de energía centralizada centralized power source

fuente de energía constante equal-energy source

fuente de energía sin interrupción no-break power source

fuente de imagen picture source

fuente de información information source

fuente de información discreta discrete information source

fuente de infrarrojo infrared source

fuente de línea line source

fuente de luz light source

fuente de luz de diodo diode light source

fuente de luz de unión junction light source

fuente de luz primaria primary light source

fuente de luz pulsada pulsed light source

fuente de luz secundaria secondary light source

fuente de mensajes message source

fuente de microtensiones microvolter

fuente de ondas portadoras carrier-wave supply

fuente de polarización bias source

fuente de portadoras carrier supply

fuente de puente bridge source

fuente de rayos gamma gamma-ray source

fuente de referencia reference source

fuente de ruido noise source

fuente de señales signal source

fuente de sonido sound source

fuente de tensión voltage source, voltage supply

fuente de tensión alterna alternating-current source voltage

fuente de tensión constante constant-voltage source

fuente de tensión de referencia reference-voltage source

fuente de tensión de rejilla grid voltage supply

fuente de tensión estática static source voltage

fuente de tensión ideal ideal voltage source

fuente distante distant source

fuente emisora emitting source

fuente equivalente equivalent source

fuente extendida extended source

fuente flotante floating source

fuente incandescente incandescent source

fuente iónica ion source

fuente isotrópica isotropic source

fuente local local source

fuente patrón standard source

fuente plana plane source

fuente polifásica equilibrada balanced polyphase source

fuente primaria primary source

fuente pulsada pulsed source

fuente puntual point source

fuente radiactiva radioactive source

fuente regulada regulated supply

fuente secundaria secondary source

fuente sonora puntual simple sound source

fuentes de Huygens Huygens' sources

fuera de banda out of band

fuera de circuito off-circuit

fuera de contacto out of contact

fuera de equilibrio out of balance

fuera de escala off-scale

fuera de frecuencia off-frequency

fuera de línea off-line

fuera de servicio out of service, out of order

fuera de sincronismo out of step

fuera de sintonía off tune

fuera de uso out of use

fuera del aire off the air

fuera del límite off-limit

fuerza force, strength

fuerza aceleradora accelerating force

fuerza bruta brute force

fuerza centrífuga centrifugal force

fuerza centrípeta centripetal force

fuerza coercitiva coercive force

fuerza coercitiva intrínseca intrinsic coercive force

fuerza contraelectromotriz counterelectromotive force

fuerza de accionamiento actuating force

fuerza de adherencia bond strength

fuerza de aguja stylus force

fuerza de aguja vertical vertical stylus force

fuerza de arranque breakaway force

fuerza de contacto contact force

fuerza de Coulomb Coulomb force

fuerza de deflexión deflecting force

fuerza de inducción induction force

fuerza de largo alcance long-range force

fuerza de liberación release force

fuerza de Lorentz Lorentz force

fuerza de patinaje skating force

fuerza de restauración restoring force

fuerza desmagnetizante demagnetizing force

fuerza disruptiva disruptive force

fuerza efectiva effective force

fuerza electromotriz transitoria transient electromotive force

fuerza eléctrica electric force

fuerza electrodinámica electrodynamic force

fuerza electromagnética electromagnetic force
fuerza electromotriz electromotive force
fuerza electromotriz alterna alternating electromotive force
fuerza electromotriz armónica simple simple harmonic electromotive force
fuerza electromotriz de contacto contact electromotive force
fuerza electromotriz de Peltier Peltier electromotive force
fuerza electromotriz de Potier Potier's electromotive force
fuerza electromotriz de reposo rest electromotive force
fuerza electromotriz de Seebeck Seebeck electromotive force
fuerza electromotriz de Thomson Thomson electromotive force
fuerza electromotriz de transformador transformer electromotive force
fuerza electromotriz directa direct electromotive force
fuerza electromotriz inducida induced electromotive force
fuerza electromotriz longitudinal longitudinal electromotive force
fuerza electromotriz oscilante oscillatory electromotive force
fuerza electromotriz sincrónica synchronous electromotive force
fuerza electromotriz sinusoidal sinusoidal electromotive force
fuerza electromotriz subtransitoria subtransient electromotive force
fuerza electromotriz térmica thermal electromotive force
fuerza equilibradora balancing force
fuerza fotoelectromotriz photoelectromotive force
fuerza impulsiva impulsive force
fuerza iónica ionic strength
fuerza lateral side thrust
fuerza ligante binding force
fuerza magnética magnetic force
fuerza magnetizante magnetizing force
fuerza magnetizante máxima peak magnetizing force
fuerza magnetomotriz magnetomotive force
fuerza magnetomotriz de conmutación switching magnetomotive force
fuerza mecánica mechanical force
fuerza media mean force
fuerza motriz motive power
fuerza nuclear nuclear force
fuerza repulsiva repelling force
fuerza retardadora retarding force
fuerza separadora de contacto contact separating force
fuerza termoelectromotriz thermoelectromotive force
fuerza vertical de aguja tracking force
fuerzas de Van der Waals Van der Waals forces
fuerzas ligantes iónicas ionic binding forces
fuerzas ligantes metálicas metallic binding forces
fuerzas opuestas opposed forces
fuga leak, leakage
fuga a tierra ground leak, earth leakage
fuga calibrada calibrated leak
fuga de capacitor capacitor leakage
fuga de condensador condenser leakage
fuga de corriente continua direct-current leakage
fuga de datos data leakage
fuga de flujo flux leakage
fuga de flujo magnético magnetic flux leakage

fuga de gas gas leak
fuga de gases outgassing
fuga de línea line leakage
fuga de portadora carrier leak
fuga de radiofrecuencia radio-frequency leak
fuga de ranura slot leakage
fuga de rejilla grid leak
fuga de vacío vacuum leak
fuga detectable detectable leak
fuga eléctrica electric leakage
fuga inversa inverse leakage
fuga mínima detectable minimum detectable leak
fuga superficial surface leakage
fulguración fulguration
función function
función agregada aggregate function
función algebraica algebraic function
función alterna alternating function
función analógica analog function
función aperiódica aperiodic function
función aritmética arithmetic function
función armónica harmonic function
función automática automatic function
función auxiliar auxiliary function
función beta beta function
función binaria binary function
función circular circular function
función combinacional combinational function
función compleja complex function
función complementaria complementary function
función continua continuous function
función de acumulador accumulator function
función de alarma alarm function
función de almacenamiento storage function
función de ambigüedad ambiguity function
función de apoyo operacional utility function
función de arco arc function
función de autocorrelación autocorrelation function
función de Bessel Bessel function
función de Boole Boolean function
función de borrado delete function
función de Brillouin Brillouin function
función de Butterworth Butterworth function
función de circuito circuit function
función de codificación algebraica algebraic coding function
función de componente component function
función de conmutación switching function
función de consola console function
función de control control function
función de control de acceso access-control function
función de control de microprograma microprogram control function
función de control dinámico dynamic control function
función de controlador controller function
función de convergencia convergence function
función de copiado copy function
función de correlación correlation function
función de densidad de Rayleigh Rayleigh density function
función de distribución distribution function
función de distribución de Fermi-Dirac Fermi-Dirac distribution function
función de enrutamiento intermedia intermediate routing function
función de ganancia gain function
función de Gauss Gaussian function
función de immitancia immittance function
función de impedancia impedance function
función de impulsos impulse function
función de interpolación interpolation function
función de Langevin Langevin function

función de localización de línea line-finding function
función de lógica combinacional combinational logic function
función de luminancia luminance function
función de máquina machine function
función de microcomputadora microcomputer function
función de omisión omit function
función de onda wave function
función de pérdidas loss function
función de programa program function
función de protocolo protocol function
función de reactancia reactance function
función de recuperación recovery function
función de red network function
función de repetición repeat function
función de respuesta response function
función de respuesta de banda base baseband response function
función de respuesta de frecuencia frequency-response function
función de seguridad safety function
función de tabla table function
función de trabajo job function
función de trabajo termoiónico thermionic work function
función de transferencia transfer function
función de transferencia de archivo file transfer function
función de transferencia de banda base baseband transfer function
función de transferencia de bloques block transfer function
función de transferencia de bucle loop-transfer function
función de transferencia de error error transfer function
función de transferencia de red network transfer function
función de transferencia de retroalimentación feedback transfer function
función de transferencia directa forward transfer function
función de transferencia en bucle abierto open-loop transfer function
función derivada derived function
función discontinua discontinuous function
función distribuida distributed function
función elíptica elliptical function
función escalar scalar function
función escalón step function
función escalón unitario unit-step function
función especial special function
función explícita explicit function
función exponencial exponential function
función externa external function
función fija fixed function
función fuera de línea off-line function
función gamma gamma function
función generadora generating function
función hiperbólica hyperbolic function
función implícita implicit function
función incorporada built-in function
función intrínseca intrinsic function
función inversa inverse function
función lineal linear function
función lógica logic function
función lógica biestable bistable logic function
función lógica secuencial sequential logical function
función manual manual function
función matemática mathematical function
función miscelánea miscellaneous function
función negativa negative function

función no lineal nonlinear function
función pasatodo all-pass function
función periódica periodic function
función positiva positive function
función potencial potential function
función predefinida predefined function
función primaria primary function
función recursiva recursive function
función tensión-frecuencia voltage-frequency function
función trabajo work function
función trigonométrica inversa inverse trigonometric function
función vectorial vector function
función Y AND function
funcional functional
funcionamiento performance, operation
funcionamiento adecuado adequate performance
funcionamiento en paralelo parallel running
funciones básicas basic functions
funciones de Bloch Bloch functions
funciones transcendentales transcendental functions
funciones trigonométricas trigonometric functions
functor functor
fundamental fundamental
fundido blown, fused, cast
fundido a presión die cast
fundir blow
fusible fuse, cutout
fusible a prueba de explosiones explosionproof fuse
fusible abierto open fuse
fusible bipolar bipolar fuse
fusible con alarma grasshopper fuse
fusible conectado por atrás back-connected fuse
fusible de acometida service fuse
fusible de cartucho cartridge fuse
fusible de cartucho de vidrio glass-cartridge fuse
fusible de cinta strip fuse
fusible de cuchilla knife fuse
fusible de diodo diode fuse
fusible de elemento único single-element fuse
fusible de expulsión expulsion fuse
fusible de fusión lenta slow-blow fuse
fusible de hilo wire fuse
fusible de hilo descubierto link fuse
fusible de instrumento instrument fuse
fusible de línea line fuse
fusible de potencial potential fuse
fusible de proximidad proximity fuse
fusible de puente bridge fuse
fusible de retardo de tiempo time-lag fuse
fusible de ruptura rápida quick-break fuse
fusible de seguridad safety fuse
fusible de tapón plug fuse
fusible de tetracloruro de carbono carbon tetrachloride fuse
fusible de tiempo time fuse
fusible de un solo uso one-time fuse
fusible defectuoso defective fuse
fusible desconectador disconnecting fuse
fusible encerrado enclosed fuse
fusible enchufable plug-in fuse
fusible ficticio dummy fuse
fusible fundido blown fuse
fusible indicador indicating fuse
fusible limitador de corriente current-limiting fuse
fusible líquido liquid fuse
fusible multipolar multipolar fuse
fusible no intercambiable noninterchangeable fuse
fusible piloto pilot fuse
fusible protector protective fuse
fusible renovable renewable fuse
fusible secundario secondary fuse

fusible sin retardo nondelay fuse
fusible sin tubo tubeless fuse
fusible sumergible submersible fuse
fusible tubular tube fuse
fusible unipolar single-pole fuse
fusible ventilado vented fuse
fusión fusion, blowing, blowout, merge
fusión atómica atomic fusion
fusión-clasificación merge-sort
fusión de zona zone melting
fusión equilibrada balanced merge

G

gabinete cabinet
gabinete de pruebas test cabinet
gadolinio gadolinium
gafas de protección goggles
galena galena
galga piezoeléctrica piezoelectric gage
galio gallium
galvánico galvanic
galvanismo galvanism
galvanización galvanization
galvanizar galvanize
galvanomagnético galvanomagnetic
galvanometría galvanometry
galvanométrico galvanometric
galvanómetro galvanometer
galvanómetro absoluto absolute galvanometer
galvanómetro amortiguado damped galvanometer
galvanómetro aperiódico deadbeat galvanometer
galvanómetro astático astatic galvanometer
galvanómetro balístico ballistic galvanometer
galvanómetro de aguja pointer galvanometer
galvanómetro de aguja móvil moving-needle
  galvanometer
galvanómetro de bobina móvil moving-coil
  galvanometer
galvanómetro de Broca Broca galvanometer
galvanómetro de bucle loop galvanometer
galvanómetro de cuerda string galvanometer
galvanómetro de chorro de tinta ink-jet
  galvanometer
galvanómetro de D'Arsonval D'Arsonval
  galvanometer
galvanómetro de espejo mirror galvanometer
galvanómetro de haz luminoso light-beam
  galvanometer
galvanómetro de hilo de Einthoven Einthoven string
  galvanometer
galvanómetro de imán móvil moving-magnet
  galvanometer
galvanómetro de imán permanente
  permanent-magnet galvanometer
galvanómetro de laminillas vibrantes vibrating-reed
  galvanometer
galvanómetro de potencial potential galvanometer
galvanómetro de registro recording galvanometer
galvanómetro de senos sine galvanometer
galvanómetro de tangentes tangent galvanometer
galvanómetro de termopar thermocouple
  galvanometer
galvanómetro de torsión torsion galvanometer
galvanómetro de vibraciones vibration galvanometer
galvanómetro diferencial differential galvanometer
galvanómetro electrostático electrostatic
  galvanometer
galvanómetro integrador integrating galvanometer
galvanómetro no amortiguado undamped
  galvanometer
galvanómetro oscilográfico oscillograph
  galvanometer
galvanómetro reflector reflecting galvanometer
galvanómetro sobreamortiguado overdamped

galvanometer
galvanoplasteado plated
galvanoplastia galvanoplasty, electroplating, plating
galvanoplastia mecánica mechanical electroplating
galvanoplastia por corriente modulada
modulated-current electroplating
galvanoplastia rápida flash plating
galvanoplastia semiautomática semiautomatic
electroplating
galvanoplastiado electroplated
galvanoplastiar electroplate
galvanoscopio galvanoscope
galvanotaxia galvanotaxis
galvanoterapia galvanotherapy
galvanotropismo galvanotropism
galvatrón galvatron
gama gamut, range
gama de brillo range of brightness
gama de colores color range
gama de frecuencias range of frequencies
gama de frecuencias de modulación
modulation-frequency range
gama de frecuencias de operación
operating-frequency range
gama de frecuencias ópticas optical frequency range
gama de frecuencias ultraaltas ultra-high frequency
range
gama de frecuencias ultrasónicas
ultrasonic-frequency range
gama de longitudes de onda wavelength range
gama de medida measurement range
gama de microondas microwave range
gama de ondas wave range
gama de ondas cortas shortwave range
gama de potencia power range
gama de resistencias resistance range
gama de respuesta response range
gama de temperaturas range of temperature
gama de tintes hue range
gama de tonos tone range
gama de valores range of values
gamma gamma
gammámetro gammameter
ganancia gain
ganancia absoluta absolute gain
ganancia absoluta de antena absolute gain of
antenna
ganancia ajustable adjustable gain
ganancia compuesta composite gain
ganancia de amplificador amplifier gain
ganancia de antena antenna gain
ganancia de bucle loop gain
ganancia de bucle cerrado closed-loop gain
ganancia de bucle neta net loop gain
ganancia de conversión conversion gain
ganancia de conversión disponible available
conversion gain
ganancia de corriente current gain
ganancia de corriente para señal pequeña
small-signal current gain
ganancia de energía energy gain
ganancia de energía neta net energy gain
ganancia de intensidad sonora loudness gain
ganancia de modo común common-mode gain
ganancia de modo diferencial differential-mode gain
ganancia de potencia power gain
ganancia de potencia de señal fuerte large-signal
power gain
ganancia de potencia de transductor transducer
power gain
ganancia de potencia disponible available power
gain
ganancia de potencia en antena antenna power gain

ganancia de potencia en espacio libre free-space
power gain
ganancia de potencia para señal pequeña
small-signal power gain
ganancia de potencia relativa relative power gain
ganancia de potencia total overall power gain
ganancia de puenteo bridging gain
ganancia de reforzador booster gain
ganancia de repetidor repeater gain
ganancia de señal signal gain
ganancia de sistema system gain
ganancia de tensión voltage gain
ganancia de tensión de bucle cerrado closed-loop
voltage gain
ganancia de tensión de conversión conversion
voltage gain
ganancia de tensión de modo común common-mode
voltage gain
ganancia de tensión diferencial differential voltage
gain
ganancia de tensión en bucle abierto open-loop
voltage gain
ganancia de tensión en circuito abierto open-circuit
voltage gain
ganancia de transductor transducer gain
ganancia de transmisión transmission gain
ganancia de video video gain
ganancia diferencial differential gain
ganancia dinámica dynamic gain
ganancia direccional directional gain
ganancia directa forward gain
ganancia directiva directive gain
ganancia directiva relativa relative directive gain
ganancia disponible available gain
ganancia en bucle abierto open-loop gain
ganancia en radiofrecuencia radio-frequency gain
ganancia estática static gain
ganancia fraccional fractional gain
ganancia homocromática homochromatic gain
ganancia horizontal horizontal gain
ganancia incremental incremental gain
ganancia isotrópica isotropic gain
ganancia luminosa espectral spectral luminous gain
ganancia máxima de potencia disponible maximum
available power gain
ganancia máxima disponible maximum available
gain
ganancia negativa negative gain
ganancia neta net gain
ganancia no recíproca nonreciprocal gain
ganancia omnidireccional omnidirectional gain
ganancia operacional operational gain
ganancia parcial partial gain
ganancia parcial relativa relative partial gain
ganancia-pérdida gain-loss
ganancia plana flat gain
ganancia por altura height gain
ganancia por diversidad diversity gain
ganancia por etapa stage gain
ganancia por inserción insertion gain
ganancia por obstáculo obstacle gain
ganancia por reflexión reflection gain
ganancia proporcional proportional gain
ganancia radiante radiant gain
ganancia radiante espectral spectral radiant gain
ganancia relativa relative gain
ganancia total overall gain
ganancia unitaria unity gain
ganancia vertical vertical gain
ganancia X X gain
ganancia Y Y gain
ganancia Z Z gain
gananciómetro gain-measuring set

gancho conmutador switchhook
gancho PN PN hook
garganta de bocina horn throat
gas gas
gas de dopado doping gas
gas dieléctrico dielectric gas
gas electrolítico electrolytic gas
gas electrónico electronic gas
gas ideal ideal gas
gas inerte inert gas
gas ionizado ionized gas
gas noble noble gas
gas ocluido occluded gas
gas perfecto perfect gas
gas raro rare gas
gas residual residual gas
gaseoso gaseous
gasificación gassing
gaston gaston
gatillado triggering
gatillador trigger
gatillo trigger
gausaje gaussage
gausiómetro gaussmeter
gausitrón gausitron
gauss gauss
Gee Gee
gel gel
genemotor genemotor
generación generation
generación bruta gross generation
generación de armónicas espurias
    spurious-harmonic generation
generación de barrido circular circular sweep
    generation
generación de caracteres character generation
generación de carga espacial space-charge
    generation
generación de dirección address generation
generación de energía magnetohidrodinámica
    magnetohydrodynamic power generation
generación de forma de onda waveform generation
generación de gas gassing
generación de haz electrónico electron-beam
    fabrication
generación de imagen en tiempo real real-time
    image generation
generación de impulsos pulse generation
generación de impulsos rectangulares
    rectangular-pulse generation
generación de informe report generation
generación de microondas microwave generation
generación de números aleatorios random-number
    generation
generación de programa program generation
generación disponible available generation
generación máxima maximum generation
generación máxima bruta gross maximum
    generation
generación neta net generation
generación real actual generation
generación real bruta gross actual generation
generación térmica thermal generation
generado por computadora computer generated
generado por datos data generated
generador generator, source
generador abierto open generator
generador acústico acoustic generator
generador armónico harmonic generator
generador armónico klistrón klystron harmonic
    generator
generador arrollado en derivación shunt-wound
    generator

generador arrollado en serie series-wound generator
generador autoexcitado self-excited generator
generador auxiliar auxiliary generator
generador bipolar bipolar generator
generador compuesto compound generator
generador controlado por cuarzo quartz-controlled
    generator
generador controlado por tensión voltage-controlled
    generator
generador de aplicación application generator
generador de arco iris rainbow generator
generador de audio audio generator
generador de banda ancha wideband generator
generador de barras bar generator
generador de barras de colores color-bar generator
generador de barrido sweep generator
generador de barrido horizontal horizontal-sweep
    generator
generador de barrido vertical vertical-sweep
    generator
generador de borrado erase generator
generador de campo giratorio revolving-field
    generator
generador de caracteres character generator
generador de carga de baterías battery charging
    generator
generador de casos de prueba test-case generator
generador de casos de prueba automatizado
    automated test case generator
generador de circuitos impresos printed-circuit
    generator
generador de código code generator
generador de colores color generator
generador de compilador compiler generator
generador de conjunto de trabajos job set generator
generador de control control generator
generador de correa belt generator
generador de corriente current generator
generador de corriente alterna alternating-current
    generator
generador de corriente constante constant-current
    generator
generador de corriente continua direct-current
    generator
generador de corriente de ruido noise-current
    generator
generador de corriente patrón standard current
    generator
generador de corrientes múltiples multiple-current
    generator
generador de chispas spark generator
generador de datos data generator
generador de datos de prueba test-data generator
generador de datos de prueba automatizado
    automated test data generator
generador de deflexión deflection generator
generador de deflexión vertical vertical-deflection
    generator
generador de desmagnetización degaussing
    generator
generador de destellos flash generator
generador de disco disk generator
generador de doble arrollamiento double winding
    generator
generador de doble corriente double-current
    generator
generador de Dobrowolsky Dobrowolsky generator
generador de energía magnetohidrodinámico
    magnetohydrodynamic power generator
generador de exploración scanning generator
generador de exploración horizontal
    horizontal-scanning generator
generador de exploración vertical vertical-scanning

generator
generador de extinción blanking generator
generador de forma de onda waveform generator
generador de frecuencia de barrido
sweep-frequency generator
generador de frecuencia de batido beat-frequency
generator
generador de frecuencia de línea line-frequency
generator
generador de frecuencia piloto pilot-frequency
generator
generador de frecuencias frequency generator
generador de frecuencias ultraaltas ultra-high
frequency generator
generador de frecuencias vocales voice-frequency
generator
generador de función escalón step-function
generator
generador de función fija fixed-function generator
generador de función natural natural function
generator
generador de funciones function generator
generador de funciones de diodo diode function
generator
generador de funciones de uso general
general-purpose function generator
generador de fusión merge generator
generador de Hall Hall generator
generador de haz electrónico electron-beam
generator
generador de imagen picture generator
generador de imagen patrón pattern generator
generador de imán permanente permanent-magnet
generator
generador de impulsos pulse generator
generador de impulsos cuadrados square-pulse
generator
generador de impulsos de desbloqueo gate
generator
generador de impulsos de reloj clock pulse
generator
generador de impulsos de sincronización
synchronizing-pulse generator
generador de impulsos estroboscópicos
strobing-pulse generator
generador de impulsos fotoeléctrico photoelectric
pulse generator
generador de impulsos retardados delayed-pulse
generator
generador de inducción induction generator
generador de inducido giratorio revolving-armature
generator
generador de informes report generator
generador de interferencia interference generator
generador de iones negativos negative-ion generator
generador de línea line generator
generador de mandos command generator
generador de mano hand generator
generador de marcas de distancia range-marker
generator
generador de marcas de tiempo time-mark
generator
generador de marcas múltiples multiple-marker
generator
generador de Marx Marx generator
generador de modulación de velocidad
velocity-modulation generator
generador de números aleatorios random-number
generator
generador de onda en escalera staircase generator
generador de onda rectangular rectangular-wave
generator
generador de ondas wave generator

generador de ondas cuadradas square-wave
generator
generador de ondas de choque surge generator
generador de ondas en diente de sierra sawtooth
generator
generador de ondas sinusoidales sine-wave
generator
generador de ondulación ripple generator
generador de oscilaciones oscillation generator
generador de palabras word generator
generador de pantalla screen generator
generador de pasos step generator
generador de patrón de linealidad linearity pattern
generator
generador de patrones de exploración raster pattern
generator
generador de portadora de canal channel-carrier
generator
generador de portadora de grupo group carrier
generator
generador de portadora de supergrupo supergroup
carrier generator
generador de programas program generator
generador de pruebas automatizado automated test
generator
generador de puente bridge generator
generador de puntas spike generator
generador de puntos dot generator
generador de radiofrecuencia radio-frequency
generator
generador de ráfagas de tono tone-burst generator
generador de rampa ramp generator
generador de rayos X X-ray generator
generador de referencia reference generator
generador de referencias cruzadas cross reference
generator
generador de rejilla grating generator
generador de rejilla negativa negative-grid
generator
generador de reloj clock generator
generador de resistencia-capacitancia
resistance-capacitance generator
generador de ruido noise generator
generador de ruido aleatorio random-noise
generator
generador de ruido blanco white-noise generator
generador de ruido de Gauss Gaussian-noise
generator
generador de ruido de microondas microwave noise
generator
generador de ruido térmico thermal-noise generator
generador de señal de audio audio signal generator
generador de señal de reloj clock signal generator
generador de señal de ruido noise signal generator
generador de señal de sincronización sync-signal
generator
generador de señal modulada en amplitud
amplitude-modulated signal generator
generador de señal patrón standard-signal generator
generador de señales signal generator
generador de señales aleatorias random signal
generator
generador de señales cuadradas square-signal
generator
generador de señales de banda ancha wideband
signal generator
generador de señales de barrido sweep-signal
generator
generador de señales de microondas microwave
signal generator
generador de señales de prueba test-signal
generator
generador de señales de radiofrecuencia

radio-frequency signal generator
generador de señales de sonar sonar-signal generator
generador de señales moduladas modulated signal generator
generador de señales sinusoidales sinusoidal-signal generator
generador de sincronización synchronizing generator
generador de sobretensiones surge generator
generador de sonido sound generator
generador de subportadora subcarrier generator
generador de tensión voltage generator
generador de tensión constante constant-voltage generator
generador de tensión de ruido noise-voltage generator
generador de tensión ficticio fictitious voltage generator
generador de tensión fija fixed-voltage generator
generador de tensión variable variable-voltage generator
generador de toda onda all-wave generator
generador de tono tone generator
generador de tono constante constant-tone generator
generador de tono de sirena siren-tone generator
generador de tono piloto pilot-tone generator
generador de tono ululante warble-tone generator
generador de trama de imagen crosshatch generator
generador de trenes de impulsos pulse-train generator
generador de tubo tube generator
generador de tubo electrónico electron-tube generator
generador de tubo gaseoso gaseous-tube generator
generador de Van de Graaff Van de Graaff generator
generador de velocidad de bauds baud rate generator
generador de velocidad lenta slow-speed generator
generador de ventana window generator
generador de vibraciones vibration generator
generador de video video generator
generador diferencial differential generator
generador eléctrico electric generator
generador eléctrico rotativo rotating electric generator
generador electrónico electronic generator
generador electrostático electrostatic generator
generador electrostático rotativo rotary electrostatic generator
generador en derivación shunt generator
generador en serie series generator
generador eólico wind-driven generator
generador heterodino multifrecuencia multifrequency heterodyne generator
generador heteropolar heteropolar generator
generador hipercompuesto overcompounded generator
generador hipocompuesto undercompounded generator
generador homopolar homopolar generator
generador klistrón klystron generator
generador magnetoeléctrico magnetoelectric generator
generador magnetohidrodinámico magnetohydrodynamic generator
generador magnetoplasmadinámico magnetoplasmadynamic generator
generador marcador marker generator
generador modulado en amplitud amplitude-modulated generator
generador monoscópico monoscopic generator

generador-motor generator-motor
generador multifásico multiphase generator
generador multipolar multipolar generator
generador polifásico polyphase generator
generador por efecto de Hall Hall-effect generator
generador portátil portable generator
generador principal main generator
generador puro pure generator
generador rotativo rotary generator
generador selsyn selsyn generator
generador sin carga unloaded generator
generador sin carga terminal unterminated generator
generador sincrónico synchronous generator
generador síncrono polifásico polyphase synchronous generator
generador superconductor superconducting generator
generador tacométrico tachometer generator
generador termoeléctrico thermoelectric generator
generador termoiónico thermionic generator
generador tiratrónico thyratron generator
generador trifásico three-phase generator
generador ultrasónico ultrasonic generator
generador-verificador de paridad parity generator-checker
generadores conectados en paralelo parallel-connected generators
generalizado generalized
generar generate
genérico generic
geoalerta geoalert
geocéntrico geocentric
geodésico geodesic
geodímetro geodimeter
geoelectricidad geoelectricity
geoeléctrico geoelectric
geomagnético geomagnetic
geomagnetismo geomagnetism
geometría geometry
geometría de redes network geometry
geotérmico geothermal
geotropismo geotropism
germanio germanium
germanio de alta pureza high-purity germanium
germanio dopado con oro gold-doped germanium
germanio tipo N N-type germanium
germanio tipo P P-type germanium
giga giga
gigabit gigabit
gigabyte gigabyte
gigaciclo gigacycle
gigaelectronvolt gigaelectronvolt
gigahertz gigahertz
gigámetro gigameter
gigaohm gigaohm
gigawatt gigawatt
gilbert gilbert
gilberts por centímetro gilberts per centimeter
gimoteo wow
gimoteo y centelleo wow and flutter
girador gyrator
girador de microondas microwave gyrator
giratorio rotatable
giro paralelo slewing
giro rápido slewing
girocompás gyrocompass
girocompás maestro master gyrocompass
girocompás magistral master gyrocompass
girofrecuencia gyrofrequency
giromagnético gyromagnetic
giropiloto gyropilot
giroscópico gyroscopic

giroscopio gyroscope, gyro
giroscopio de inducción magnética
  magnetic-induction gyroscope
giroscopio de velocidades angulares rate gyroscope
giroscopio direccional directional gyroscope
giroscopio electrostático electrostatic gyroscope
giroscopio libre free gyro
giroscopio magnetohidrodinámico
  magnetohydrodynamic gyroscope
girostático gyrostatic
giróstato gyrostat
girotrón gyrotron
girotrópico gyrotropic
global global
glosario automatizado automated glossary
glucinio glucinium
golpe stroke
golpeteo thump
goma rubber
goma dura hard rubber
goniofotómetro goniophotometer
goniógrafo goniograph
goniometría goniometry
goniómetro goniometer
goniómetro de rayos X X-ray goniometer
goniómetro magnético magnetic direction-finder
goniómetro reflector reflecting goniometer
googol googol
googolplex googolplex
grabación recording
grabación a velocidad constante constant-velocity
  recording
grabación de paso variable variable-pitch recording
grabación de video video recording
grabación de video electrónica electronic video
  recording
grabación en cinta magnética de video video-tape
  recording
grabación estereofónica stereo recording
grabación por modulación de frecuencia
  frequency-modulation recording
grabación química etching
grabación vertical vertical recording
grabación vertical-lateral vertical-lateral recording
grabado scribing
grabador recorder, cutter
grabador magnético magnetic cutter
grabador químico etchant
grabadora de casete cassette recorder
grabadora de cintas magnéticas de televisión
  television tape recorder
grabadora de cintas magnéticas de video
  videocassette recorder, video-tape recorder
grabadora de hilo wire recorder
grabadora de velocidad constante constant-velocity
  recorder
grabadora de video electrónica electronic video
  recorder
grabadora portátil portable recorder
gradación gradation, grading
gradiente gradient
gradiente alterno alternating gradient
gradiente barométrico barometric gradient
gradiente de calor heat gradient
gradiente de calor de Joule Joule heat gradient
gradiente de campo field gradient
gradiente de campo magnético magnetic-field
  gradient
gradiente de concentración concentration gradient
gradiente de contraste contrast gradient
gradiente de energía energy gradient
gradiente de flujo flux gradient
gradiente de índice de refracción refractive index
  gradient
gradiente de potencial potential gradient
gradiente de presión pressure gradient
gradiente de temperatura temperature gradient
gradiente de tensión voltage gradient
gradiente térmico thermal gradient
gradientímetro gradient meter
gradiómetro gradiometer
grado degree, grade
grado de amortiguamiento degree of damping
grado de asimetría degree of asymmetry
grado de coherencia degree of coherence
grado de disociación degree of dissociation
grado de distorsión degree of distortion
grado de exactitud degree of accuracy
grado de reacción reaction degree
grado de rectificación de corriente degree of
  current rectification
grado de rectificación de tensión degree of voltage
  rectification
grado de servicio grade of service
grado de temperatura temperature degree
grado de vacío degree of vacuum
grado de vibración degree of vibration
grado eléctrico electrical degree
grado esférico spherical degree
grados de libertad degrees of freedom
graduado graduated, graded
gradual gradual
gráfica graph, plot
gráfica de barras bar graph
gráfica de Bode Bode plot
gráfica logarítmica logarithmic graph
graficador plotter
gráfico graphic
gráfico de llamadas call graph
graficón graphecon
gráficos graphics
gráficos codificados coded graphics
gráficos de alta resolución high-resolution graphics
gráficos de baja resolución low-resolution graphics
gráficos de caracteres character graphics
gráficos de computadora computer graphics
gráficos de exploración raster graphics
gráficos de información information graphics
gráficos en colores color graphics
gráficos interactivos interactive graphics
gráficos no codificados noncoded graphics
gráficos pasivos passive graphics
gráficos tridimensionales three-dimensional graphics
grafitización graphitization
grafito graphite
grafito coloidal colloidal graphite
grafófono graphophone
gramo gram
gramófono gramophone
gran capacidad large capacity
granalla de carbón carbon granules
granate garnet
granate de itrio e hierro yttrium iron garnet
granate de itrio y aluminio yttrium aluminum garnet
grano grain
granosidad graininess
granular granular
granulosidad granularity
grapa de cable cable hanger
gravedad gravity
gravedad específica specific gravity
gridistor gridistor
grieta de imán permanente permanent-magnet crack
grosor de capa muerta dead-layer thickness
grosor de revestimiento coating thickness
grosor óptico optical thickness

grupo group, cluster
grupo básico basic group
grupo binario binary group
grupo coherente coherent bundle
grupo convertidor motor-generator set
grupo de acceso de proceso process access group
grupo de almacenamiento storage group
grupo de base base group, base cluster
grupo de canales channel group
grupo de capacitores capacitor group
grupo de caracteres character group
grupo de carga load group
grupo de circuitos circuit group
grupo de códigos code group
grupo de conductores harness
grupo de conjunto de datos data set group
grupo de conmutación commutating group
grupo de control control group
grupo de datos data group
grupo de datos de generación generation data group
grupo de doce canales twelve-channel group
grupo de enlace link group, trunk group
grupo de enlace de alto uso high-usage trunk group
grupo de etiquetas label group
grupo de identificación identification group
grupo de impulsos pulse group
grupo de línea line group
grupo de llegada incoming group
grupo de ondas wave group
grupo de portadoras carrier group
grupo de ranuras slot group
grupo de relés relay group
grupo de reserva reserve group
grupo de salida outgoing group
grupo de transmisión transmission group
grupo de uniones junction group
grupo de visualización display group
grupo espacial space group
grupo fantasma phantom group
grupo fundamental fundamental group
grupo iónico ion cluster
grupo lógico logical group
grupo maestro mastergroup
grupo primario primary group, primary cluster
grupo puntual point group
grupo quinario quinary group
grupo repetitivo repeating group
grupo secundario secondary group
guantes aislantes insulating gloves
guantes de caucho al plomo lead-rubber gloves
guarda de memoria memory guard
guía guide, leader
guía circular circular guide
guía de alimentación de papel paper-feed guide
guía de cinta tape guide
guía de flujo flux guide
guía de luz light guide
guía de ondas waveguide
guía de papel paper guide
guía de tarjetas card guide
guía dieléctrica dielectric guide
guiado guidance, homing
guiado activo active homing
guiado direccional directional homing
guiado inercial inertial guidance
guiado por haz beam-rider guidance
guiado por infrarrojo infrared homing
guiado por lorán loran guidance
guiado por radar radar homing
guiado semiactivo semiactive guidance
guianza guidance
guíaondas waveguide
guíaondas acanalado ridge waveguide

guíaondas adaptado matched waveguide
guíaondas cerrado closed waveguide
guíaondas cilíndrico cylindrical waveguide
guíaondas circular circular waveguide
guíaondas con fugas leaky waveguide
guíaondas con pérdidas lossy waveguide
guíaondas de corte cutoff waveguide
guíaondas de haz beam waveguide
guíaondas de película delgada thin-film waveguide
guíaondas de placas paralelas parallel-plate waveguide
guíaondas de planos paralelos parallel-plane waveguide
guíaondas de referencia reference waveguide
guíaondas de sección variable tapered waveguide
guíaondas de varilla dieléctrica dielectric-rod waveguide
guíaondas dieléctrico planar planar dielectric waveguide
guíaondas dieléctrico dielectric waveguide
guíaondas elíptico elliptical waveguide
guíaondas evanescente evanescent waveguide
guíaondas flexible flexible waveguide
guíaondas fusiforme taper
guíaondas helicoidal helix waveguide
guíaondas multimodo multimode waveguide
guíaondas óptico optical waveguide
guíaondas óptico de modo único single-mode optical waveguide
guíaondas periódico periodic waveguide
guíaondas radiante radiating guide
guíaondas ranurado slotted waveguide
guíaondas rectangular rectangular waveguide
guíaondas sin reflexión reflectionless waveguide
guíaondas tabicado septate waveguide
guíaondas torcido twisted waveguide
guíaondas tubular tubular waveguide
guíaondas uniforme uniform waveguide
gutapercha gutta-percha

habilidad ability
habilitación de bus de datos data bus enable
habilitación de chip chip enable
habilitación de escritura write enable
habilitación de salida output enable
habilitado enabled
habilitar enable
hacer ininteligible garble
hafnio hafnium
halación halation
halo halo
halógeno halogen
haluro halide
hardware hardware
hartley hartley
haz beam
haz ancho wide beam
haz catódico cathode beam
haz ciclotrónico cyclotron beam
haz circular circular beam
haz concentrado concentrated beam
haz confinado confined beam
haz conformado shaped beam
haz cónico conical beam
haz conmutado switched beam
haz controlado gated beam
haz convergente convergent beam
haz de aterrizaje landing beam
haz de azul blue beam
haz de comprobación check beam
haz de cosecante cuadrada cosecant-squared beam
haz de electrones beam of electrons
haz de electrones agrupados bunched electron beam
haz de exploración scanning beam
haz de Gauss Gaussian beam
haz de microondas microwave beam
haz de neutrones beam of neutrons
haz de ondas wave beam
haz de plasma plasma beam
haz de radar radar beam
haz de radiación radiation beam
haz de radio radio beam
haz de radiofaro direccional radio-range beam
haz de rayos ray beam
haz de retorno return beam
haz de seguimiento tracking beam
haz de sostenimiento holding beam
haz de tres colores three-color beam
haz direccional directional beam
haz disperso scattered beam
haz divergente divergent beam
haz dividido split beam
haz electrónico electron beam
haz electrónico de alta velocidad high-speed electron beam
haz electrónico de baja velocidad low-speed electron beam
haz electrónico modulado modulated electron beam
haz electrónico multivelocidad multivelocity electron beam
haz en abanico fanned beam

haz en cola de castor beavertail beam
haz en lápiz pencil beam
haz en V V beam
haz estático static beam
haz filiforme pencil beam
haz hertziano Hertzian beam
haz horizontal horizontal beam
haz hueco hollow beam
haz incidente incident beam
haz iónico ion beam
haz iónico modulado modulated ion beam
haz iónico molecular molecular ion beam
haz iónico monoenergético monoenergetic ion beam
haz láser laser beam
haz localizador localizer beam
haz luminoso light beam
haz luminoso modulado modulated light beam
haz modulado modulated beam
haz molecular molecular beam
haz monoenergético monoenergetic beam
haz móvil moving beam
haz neutrónico neutron beam
haz paralelo parallel beam
haz periódico periodic beam
haz polarizado polarized beam
haz posterior back beam
haz principal main beam
haz protónico polarizado polarized proton beam
haz pulsado pulsed beam
haz radial radial beam
haz reentrante reentrant beam
haz reflejado reflected beam
haz rotativo rotary beam
haz sellado sealed beam
haz supersónico supersonic beam
haz transversal transverse beam
haz tricolor tricolor beam
haz ultrasónico ultrasonic beam
haz uniforme uniform beam
haz útil useful beam
haz vertical vertical beam
hebra strand
hecto hecto
hectoampere hectoampere
hectométrico hectometric
hectómetro hectometer
hectowatt hectowatt
hecho a la medida custom made
hélice helix, twist
helicono helicone
helio helium
heliónica helionics
heliostato heliostat
helitrón helitron
hemimórfico hemimorphic
hemisférico hemispherical
henry henry
henry internacional international henry
heptodo heptode
heptodo cambiador de frecuencia frequency-changer heptode
heptodo de doble control dual-control heptode
hermético hermetic, airtight
hermético al agua watertight
hermético al vacío vacuumtight
hermético contra aceite oiltight
hermético contra el goteo driptight
hermético contra el polvo dusttight
hermético contra gas gastight
hermético contra la lluvia raintight
hermético contra vapores vaportight
herramienta aislada insulated tool
herramienta aislante insulating tool

herramienta de alineación alignment tool
herramienta de diseño automatizado automated design tool
herramienta eléctrica electric tool
herramienta ultrasónica ultrasonic tool
herramientas de verificación automatizadas automated verification tools
herramientas programadas automáticamente automatically programmed tools
hertz hertz
hesitación hesitation
heterocromático heterochromatic
heterodinaje heterodyning
heterodinaje ascendente upward heterodyning
heterodinaje descendente downward heterodyning
heterodinaje óptico optical heterodyning
heterodino heterodyne
heterogeneidad heterogeneity
heterogéneo heterogenous
heteropolar heteropolar
heurístico heuristic
hexadecimal hexadecimal
hexafluoruro de azufre sulfur hexafluoride
hexodo hexode
hexodo mezclador mixer hexode
híbrido hybrid
híbrido coaxial coaxial hybrid
hidroacústica hydroacoustics
hidroacústico hydroacoustic
hidrocinético hydrokinetic
hidrodinámico hydrodynamic
hidroeléctrico hydroelectric
hidrófono hydrophone
hidrófono de bobina móvil moving-coil hydrophone
hidrófono de carbón carbon hydrophone
hidrófono de conductor móvil moving-conductor hydrophone
hidrófono de cristal crystal hydrophone
hidrófono de línea line hydrophone
hidrófono de magnetoestricción magnetostriction hydrophone
hidrófono de presión pressure hydrophone
hidrófono de velocidad velocity hydrophone
hidrófono direccional directional hydrophone
hidrófono electrocinético electrokinetic hydrophone
hidrófono omnidireccional omnidirectional hydrophone
hidrófono piezoeléctrico piezoelectric hydrophone
hidrófono unidireccional unidirectional hydrophone
hidrófugo moistureproof
hidrógeno hydrogen
hidrógeno pesado heavy hydrogen
hidrólisis hydrolysis
hidromagnética hydromagnetics
hidromagnético hydromagnetic
hidrómetro hydrometer
hierro iron
hierro blando soft iron
hierro de soldar soldering iron
hierro de soldar eléctrico electric soldering iron
hierro dulce soft iron
hierro electrolítico electrolytic iron
hierro galvanizado galvanized iron
hierro móvil moving iron
hierro pasivado passivated iron
hierro pasivo passive iron
higristor hygristor
higrógrafo hygrograph
higrómetro hygrometer
higrómetro de cabello hair hygrometer
higrómetro eléctrico electric hygrometer
higrómetro electrónico electronic hygrometer
higrostato hygrostat

higrotermógrafo hygrothermograph
hilo wire, thread, strand
hilo aéreo open wire
hilo aislado insulated wire
hilo aislado con seda silk-insulated wire
hilo aislado con teflón teflon-insulated wire
hilo aislado con telcoteno telcothene-insulated wire
hilo aislado por papel paper-insulated wire
hilo antideflagrante flameproof wire
hilo apantallado shielded wire
hilo bimetálico bimetallic wire
hilo con capa magnética magnetic plated wire
hilo cubierto con capa única de seda single-silk-covered wire
hilo cubierto de caucho rubber-covered wire
hilo de acero steel wire
hilo de acometida drop wire
hilo de alimentación feed wire
hilo de antena antenna wire
hilo de aparato fixture wire
hilo de aparato eléctrico appliance wire
hilo de audio audio wire
hilo de avance advance wire
hilo de batería battery wire
hilo de cobre desnudo bare copper wire
hilo de cobre recocido annealed copper wire
hilo de conexión connecting wire, hookup wire, lead, wire lead
hilo de contacto contact wire
hilo de doble contacto twin-contact wire
hilo de entrada lead-in wire
hilo de equilibrio balance wire
hilo de guarda guard wire
hilo de inicio start lead
hilo de lectura read wire
hilo de línea line wire
hilo de Litzendraht Litzendraht wire
hilo de manganina manganin wire
hilo de medidor meter wire
hilo de platino platinum wire
hilo de puente jumper wire
hilo de punta tip wire
hilo de registro magnético magnetic recording wire
hilo de rejilla grid wire
hilo de resistencia resistance wire
hilo de retorno return wire
hilo de señalización signaling lead
hilo de suspensión slinging wire
hilo de tierra ground wire
hilo de timbre bell wire
hilo de Wollaston Wollaston wire
hilo delgado thin wire
hilo desnudo bare wire
hilo dieléctrico dielectric wire
hilo en blanco blank wire
hilo esmaltado enameled wire
hilo estañado tinned wire
hilo estirado en duro hard-drawn wire
hilo externo outside lead
hilo flexible flexible lead
hilo fusible fusible wire, fuse wire
hilo interior inside lead
hilo litz litz wire
hilo macizo solid wire
hilo magnético magnetic wire
hilo marcador marker thread
hilo multifilamento multifilament wire
hilo multitrenza multistrand wire
hilo muy fino hair wire
hilo negativo negative wire
hilo neutro neutral wire
hilo oscilante oscillating wire
hilo para arrollamientos winding wire

hilo para electroimanes magnet wire
hilo piloto pilot wire
hilo plano flat wire
hilo positivo positive wire
hilo precortado precut wire
hilo radial radial wire
hilo recocido annealed wire
hilo recocido en negro black annealed wire
hilo rectangular rectangular wire
hilo recubierto covered wire
hilo recubierto con capa única de algodón
   single-cotton-covered wire
hilo recubierto de cobre copper-clad wire
hilo recubierto de papel paper-covered wire
hilo recubierto de plomo lead-covered wire
hilo recubierto de seda silk-covered wire
hilo resistivo desnudo bare resistance wire
hilo revestido coated wire
hilo revestido de vinilo vinyl-coated wire
hilo telefónico telephone wire
hilo único single wire
hilo vivo live wire
hilos de Lecher Lecher wires
hiperacústica hyperacoustics
hiperconvertidor paramétrico parametric-up
   converter
hiperfrecuencia hyperfrequency
hiperfrecuencias microwave frequencies
hiperón hyperon
hiperpolarización hyperpolarization
hipersónico hypersonic
hipersonido hypersound
hipótesis hypothesis
hipsograma hypsogram
hipsómetro hypsometer
histeresígrafo hysteresigraph
histeresigrama hysteresigram
histeresímetro hysteresimeter
histéresis hysteresis
histéresis de frecuencia frequency hysteresis
histéresis de sintonización electrónica electronic
   tuning hysteresis
histéresis dieléctrica dielectric hysteresis
histéresis eléctrica electric hysteresis
histéresis electrostática electrostatic hysteresis
histéresis estática static hysteresis
histéresis magnética magnetic hysteresis
histéresis magnetomecánica magnetomechanical
   hysteresis
histéresis rotacional rotational hysteresis
histéresis viscosa viscous hysteresis
histeresiscopio hysteresiscope
histograma histogram
hodoscopio hodoscope
hoja aislante insulating sheet
hoja aislante de ranura slot cell
hoja de codificación coding sheet
hoja de códigos code sheet
hoja de control control sheet
hoja de datos data sheet
hoja de oro gold leaf
hoja de plástico plastic sheet
hoja delgada thin foil
hoja eléctrica electrical sheet
hoja exterior outside foil
hoja laminada laminated sheet
hoja magnética magnetic sheet
hoja metálica metal foil, foil
hoja plana flat sheet
holmio holmium
holocámara holocamera
holografía holography
holográfico holographic

holograma hologram
homeostasis homeostasis
homocromático homochromatic
homodino homodyne
homogeneidad homogeneity
homogéneo homogeneous
homomorfismo homomorphism
homopolar homopolar
hora cargada busy hour
hora ocupada busy hour
hora ocupada de circuito circuit busy hour
hora pico busy hour, rush hour
hora pico de circuito circuit busy hour
hora pico de grupo de circuitos circuit-group busy
   hour
horario schedule
horas de consumo máximo peak hours
horas de interrupción forzada forced outage hours
horas de preparación setup hours
horas de servicio service hours
horas disponibles available hours
horas equivalentes equivalent hours
horas indisponibles unavailable hours
horizontal horizontal
horizontalmente horizontally
horizonte horizon
horizonte aparente apparent horizon
horizonte artificial artificial horizon
horizonte de radar radar horizon
horizonte de radio normal standard radio horizon
horizonte intermedio intermediate horizon
horizonte natural natural horizon
horizonte óptico optical horizon
horizonte real real horizon
horno de alta frecuencia high-frequency furnace
horno de arco arc furnace
horno de arco eléctrico electric arc furnace
horno de baja frecuencia low-frequency furnace
horno de baño de plomo lead-bath furnace
horno de Heroult Heroult furnace
horno de inducción induction furnace
horno de microondas microwave oven
horno de rayos catódicos cathode-ray furnace
horno de resistencia resistance furnace
horno eléctrico electric furnace
horno para cristal crystal oven
horno por inducción de baja frecuencia
   low-frequency induction furnace
hoyo pit
hueco hole
hueco positivo positive hole
humedad humidity, moisture
humedad absoluta absolute humidity
humedad ambiente ambient humidity
humedad atmosférica atmospheric humidity
humedad crítica critical humidity
humedad específica specific humidity
humedad normal standard moisture
humedad relativa relative humidity
húmedo moist, humid
humidificación humidification
humidímetro humidity meter, moisture meter
huso spindle

I

iconoscópico iconoscopic
iconoscopio iconoscope
iconoscopio de Farnsworth Farnsworth iconoscope
iconoscopio de imagen image iconoscope
iconotrón iconotron
ideal ideal
idealizado idealized
identidad de suma addition identity
identificación identification
identificación automática automatic identification
identificación de archivo file identification
identificación de circuito circuit identification
identificación de configuración asignada allocated configuration identification
identificación de conjunto de datos data set identification
identificación de correlación correlation identification
identificación de dirección address identification
identificación de grupo group identification
identificación de interrupción interrupt identification
identificación de mensajes message identification
identificación de nodo node identification
identificación de número automática automatic number identification
identificación de patrón pattern identification
identificación de problema problem identification
identificación de programa program identification
identificación de puntos point identification
identificación de sesión local local session identification
identificación de sistema system identification
identificación de subsistema subsystem identification
identificación de tarea task identification
identificación de transmisión transmission identification
identificación de usuario user identification
identificación falsa false identification
identificación por radar radar identification
identificado identified
identificador identifier
identificador de archivo file identifier
identificador de caracteres character identifier
identificador de comunicación communication identifier
identificador de etiqueta label identifier
identificador de función function identifier
identificador de llamadas call identifier
identificador de mensaje message identifier
identificador de nodo node identifier
identificador de partición partition identifier
identificador de señal signal identifier
identificador de sentencia statement identifier
ideográfico ideographic
idiocromático idiochromatic
idioeléctrico idioelectric
ignición ignition
ignición apantallada screened ignition
ignición blindada shielded ignition
ignición electrónica electronic ignition

ignición piezoeléctrica piezoelectric ignition
ignición por bobina coil ignition
ignición por magneto magneto ignition
ignífugo flame-resistant
ignitor igniter
ignitrón ignitron
ignitrón sincrónico synchronous ignitron
igualación equalization
igualación adaptiva adaptive equalization
igualación de abertura aperture equalization
igualación de atenuación attenuation equalization
ilimitado unlimited
iluminación illumination
iluminación de abertura aperture illumination
iluminación de cebado priming illumination
iluminación de panel panel illumination
iluminación de pantalla screen illumination
iluminación de torre tower lighting
iluminación directa direct illumination
iluminación específica specific illumination
iluminación estroboscópica stroboscopic illumination
iluminación indirecta indirect illumination
iluminación mínima minimum illumination
iluminación por los bordes edge lighting
iluminación progresiva tapered illumination
iluminación uniforme uniform illumination
iluminado illuminated
iluminado por los bordes edge-lighted
iluminador de infrarrojo infrared illuminator
iluminador monocromático monochromatic illuminator
iluminancia illuminance
iluminante illuminant
iluminar illuminate
iluminómetro illuminometer
imagen image, picture
imagen binaria binary image
imagen codificada coded image
imagen de alta definición high-definition image
imagen de alto contraste high-contrast image
imagen de audio audio image
imagen de baja definición low-definition image
imagen de borde edge image
imagen de configuración configuration image
imagen de espejo mirror image
imagen de fondo background image
imagen de memoria de núcleos core image
imagen de pantalla screen image
imagen de prueba de definición resolution test chart, resolution chart, resolution pattern
imagen de radar radar image
imagen de sonido sound image
imagen de tarjeta card image
imagen de televisión television image, television picture
imagen de tono continuo continuous tone image
imagen de visualización display image
imagen digital digital image
imagen dividida split image
imagen dura hard image
imagen eléctrica electric image
imagen electrónica electron image
imagen electrónica latente latent electronic image
imagen electrostática electrostatic image
imagen estática static image
imagen estereofónica stereophonic image
imagen explorada scanned image
imagen fantasma ghost image, ghost, echo image
imagen fantasma negativa negative ghost
imagen formateada formatted image
imagen fotográfica photographic image
imagen inversa reverse image

imagen invertida reversed image
imagen latente latent image
imagen maestra master image
imagen monocroma monochrome image
imagen monocromática monochromatic picture
imagen múltiple multiple image
imagen negativa negative picture
imagen nítida sharp image
imagen no formateada unformatted image
imagen óptica optical image
imagen patrón test pattern
imagen persistente afterimage
imagen persistente negativa negative afterimage
imagen policroma polychrome picture
imagen positiva positive picture
imagen real real image
imagen retenida retained image, image burn
imagen simbólica symbolic image
imagen simple simple image
imagen televisada televised image, televised picture
imagen transitoria transient image
imagen tridimensional three-dimensional picture
imagen truncada truncated picture
imagen única single image
imagen virtual virtual image
imágenes en escala de grises gray scale images
imágenes yuxtapuestas juxtaposed images
imán magnet
imán amortiguador damping magnet
imán cerámico ceramic magnet
imán cilíndrico cylindrical magnet
imán circular circular magnet
imán compensador compensating magnet
imán de altavoz loudspeaker magnet
imán de arrastre drag magnet
imán de barra bar magnet
imán de bloqueo blocking magnet
imán de borrado erase magnet
imán de campo field magnet
imán de cilindro drum magnet
imán de coincidencia coincidence magnet
imán de control control magnet
imán de convergencia convergence magnet
imán de convergencia de haces beam magnet
imán de corrección lateral lateral-correction magnet
imán de encuadre framing magnet
imán de enfoque focusing magnet
imán de extinción blowout magnet
imán de frenaje braking magnet
imán de hierro iron magnet
imán de parrilla grate magnet
imán de placa plate magnet
imán de platino-cobalto platinum-cobalt magnet
imán de posicionamiento positioning magnet
imán de pureza purity magnet
imán de pureza de colores color-purity magnet
imán de retención holding magnet
imán de sincrotrón synchrotron magnet
imán de trampa iónica ion-trap magnet
imán débil weak magnet
imán en anillo ring magnet
imán en cilindro cylinder magnet
imán en herradura horseshoe magnet
imán en varilla rod magnet
imán homopolar homopolar magnet
imán laminado laminated magnet
imán lateral lateral magnet
imán móvil moving magnet
imán natural natural magnet
imán neutralizador de campo field-neutralizing magnet
imán para indicar cero zero magnet
imán permanente permanent magnet

imán permanente de platino-cobalto platinum-cobalt permanent magnet
imán permanente periódico periodic permanent magnet
imán retardador retarding magnet
imán rotativo rotating magnet
imán selector selector magnet
imán temporal temporary magnet
imitación imitation
immitancia immittance
impacto ambiental environmental impact
impacto electrónico electron impact
impedancia impedance
impedancia acoplada coupled impedance
impedancia activa active impedance
impedancia acústica acoustic impedance
impedancia acústica característica characteristic acoustic impedance
impedancia acústica específica specific acoustic impedance
impedancia acústica intrínseca unit-area acoustic impedance
impedancia adaptada matched impedance
impedancia amortiguada damped impedance
impedancia anódica anode impedance
impedancia ante picos transitorios surge impedance
impedancia asincrónica asynchronous impedance
impedancia bilateral bilateral impedance
impedancia bilateral lineal linear bilateral impedance
impedancia bloqueada blocked impedance
impedancia característica characteristic impedance
impedancia característica en espacio libre free-space characteristic impedance
impedancia cargada loaded impedance
impedancia catódica cathode impedance
impedancia cero zero impedance
impedancia cinética motional impedance
impedancia compleja complex impedance
impedancia común common impedance
impedancia concentrada lumped impedance
impedancia de acoplamiento coupling impedance
impedancia de adaptación matching impedance
impedancia de alambrado wiring impedance
impedancia de alimentación de antena antenna feed impedance
impedancia de altavoz loudspeaker impedance
impedancia de antena antenna impedance
impedancia de antena nominal nominal antenna impedance
impedancia de aparato eléctrico appliance impedance
impedancia de avalancha avalanche impedance
impedancia de banda base baseband impedance
impedancia de base base impedance
impedancia de campo homopolar homopolar field impedance
impedancia de carga load impedance
impedancia de carga anódica anode load impedance
impedancia de carga de placa plate load impedance
impedancia de carga óptima optimum load impedance
impedancia de cavidad cavity impedance
impedancia de circuito de rejilla grid-circuit impedance
impedancia de conmutación commutating impedance
impedancia de derivación branch impedance
impedancia de diodo diode impedance
impedancia de electrodo electrode impedance
impedancia de emisor emitter impedance
impedancia de entrada input impedance
impedancia de entrada de bucle cerrado

closed-loop input impedance
impedancia de entrada de modo común
common-mode input impedance
impedancia de entrada de rejilla grid-input
impedance
impedancia de entrada diferencial differential-input
impedance
impedancia de entrada diferencial equivalente
equivalent differential input impedance
impedancia de entrada efectiva effective input
impedance
impedancia de entrada en bucle abierto open-loop
input impedance
impedancia de entrada interna internal input
impedance
impedancia de equilibrio balance impedance
impedancia de etapa excitadora driver impedance
impedancia de fuente source impedance
impedancia de guíaondas waveguide impedance
impedancia de interacción interaction impedance
impedancia de interfaz catódica cathode-interface
impedance
impedancia de línea line impedance
impedancia de máquina machine impedance
impedancia de onda wave impedance
impedancia de onda característica characteristic
wave impedance
impedancia de onda de guía guide-wave impedance
impedancia de placa plate impedance
impedancia de placa a placa plate-to-plate
impedance
impedancia de placa dinámica dynamic plate
impedance
impedancia de puerta gate impedance
impedancia de radiación radiation impedance
impedancia de rejilla grid impedance
impedancia de rejilla sintonizada tuned-grid
impedance
impedancia de relé relay impedance
impedancia de retroalimentación feedback
impedance
impedancia de revestimiento catódico
cathode-coating impedance
impedancia de salida output impedance
impedancia de salida de bucle cerrado closed-loop
output impedance
impedancia de salida dinámica dynamic output
impedance
impedancia de salida efectiva effective output
impedance
impedancia de salida en bucle abierto open-loop
output impedance
impedancia de salida interna internal output
impedance
impedancia de secuencia de fase negativa negative
phase-sequence impedance
impedancia de secuencia positiva positive-sequence
impedance
impedancia de tierra earth impedance
impedancia de transferencia transfer impedance
impedancia de transferencia superficial surface
transfer impedance
impedancia de Zener Zener impedance
impedancia del extremo de transmisión sending-end
impedance
impedancia diferencial differential impedance
impedancia dinámica dynamic impedance
impedancia en circuito abierto open-circuit
impedance
impedancia en cortocircuito short-circuit impedance
impedancia en derivación shunt impedance
impedancia en paralelo parallel impedance
impedancia en serie series impedance

impedancia equilibrada balanced impedance
impedancia equilibradora balancing impedance
impedancia equivalente equivalent impedance
impedancia externa external impedance
impedancia generalizada generalized impedance
impedancia imagen image impedance
impedancia incremental incremental impedance
impedancia inductiva inductive impedance
impedancia infinita infinite impedance
impedancia interna internal impedance
impedancia intrínseca intrinsic impedance
impedancia inversa inverse impedance
impedancia iterativa iterative impedance
impedancia libre free impedance
impedancia longitudinal longitudinal impedance
impedancia mecánica mechanical impedance
impedancia mecánica angular angular mechanical
impedance
impedancia mutua mutual impedance
impedancia negativa negative impedance
impedancia neta net impedance
impedancia nominal nominal impedance, rated
impedance
impedancia normal normal impedance
impedancia normalizada normalized impedance
impedancia operacional operational impedance
impedancia parásita stray impedance
impedancia primaria primary impedance
impedancia puntual point impedance
impedancia reflejada reflected impedance
impedancia residual residual impedance
impedancia resistiva resistive impedance
impedancia secundaria secondary impedance
impedancia sincrónica synchronous impedance
impedancia sincrónica longitudinal direct-axis
synchronous impedance
impedancia sincrónica transversal quadrature-axis
synchronous impedance
impedancia subtransitoria longitudinal direct-axis
subtransient impedance
impedancia subtransitoria transversal
quadrature-axis subtransient impedance
impedancia superficial surface impedance
impedancia térmica thermal impedance
impedancia terminal terminal impedance
impedancia transitoria longitudinal direct-axis
transient impedance
impedancia transitoria transversal quadrature-axis
transient impedance
impedancia unilateral unilateral impedance
impedancia variable variable impedance
impedancia vectorial vector impedance
impedancias conjugadas conjugate impedances
impedancias desadaptadas mismatched impedances
impedancias recíprocas reciprocal impedances
impedancímetro impedance meter
impedor impedor
imperfección imperfection, flaw, blemish
imperfección de cristal crystal imperfection
imperfección de red lattice imperfection
imperfecto imperfect
impermeable waterproof
implantación iónica ion implantation
implementación implementation
implícito implicit
implosión implosion
importación de datos data import
importar import
imprecisión imprecision
impregnación impregnation
impregnación en vacío vacuum impregnation
impregnación por inmersión dip impregnation
impregnado impregnated

impregnado de plástico plastic-impregnated
impregnado en aceite oil-impregnated
impregnante impregnant
impresión printing, printout
impresión al vuelo on-the-fly printing
impresión bidireccional bidirectional printing
impresión de archivo file print
impresión de datos data printout
impresión de fondo background printing
impresión de grupo group printing
impresión de líneas múltiples multiple-line printing
impresión de memoria memory printout
impresión de resumen de alarmas alarm summary
     printout
impresión de tarjetas card printing
impresión de tiempo transcurrido elapsed-time
     printout
impresión diferida deferred printing
impresión dinámica dynamic printout
impresión dúplex duplex printing
impresión en paralelo parallel printing
impresión en tiempo real real-time printout
impresión espuria spurious printing
impresión magnética magnetic printing
impresión por línea line printing
impresión por matrices matrix printing
impresión por pasadas múltiples multiple-pass
     printing
impresión xerográfica xerographic printing
impreso printed
impresor de barras bar printer
impresor de caracteres character printer
impresora printer
impresora al vuelo on-the-fly printer
impresora bidireccional bidirectional printer
impresora de alta velocidad high-speed printer
impresora de banda band printer
impresora de cinta tape printer
impresora de colores color printer
impresora de consola console printer
impresora de chorro de tinta ink-jet printer
impresora de gráficos graphics printer
impresora de hilos wire printer
impresora de impacto de caracteres character
     impact printer
impresora de línea line printer
impresora de matriz de puntos dot-matrix printer
impresora de micropelícula microfilm printer
impresora de página page printer
impresora de referencia reference printer
impresora de rueda wheel printer
impresora de rueda en margarita daisy wheel
     printer
impresora de salida output printer
impresora de salida de sistema system output
     printer
impresora de sistema system printer
impresora de tambor drum printer
impresora electrosensible electrosensitive printer
impresora electrostática electrostatic printer
impresora electrotérmica electrothermal printer
impresora en cadena chain printer
impresora en paralelo parallel printer
impresora en serie serial printer
impresora inteligente intelligent printer
impresora láser laser printer
impresora magnética magnetic printer
impresora óptica optical printer
impresora por impactos impact printer
impresora rápida rapid printer
impresora térmica thermal printer
impresora xerográfica xerographic printer
imprimir print

improductivo nonproductive
impulsión impulsion
impulsivo impulsive
impulso impulse, pulse, ping, pip, flash
impulso acústico acoustic impulse
impulso agudo sharp pulse
impulso ancho broad impulse
impulso aserrado serrated pulse
impulso bidireccional bidirectional pulse
impulso bipolar bipolar pulse
impulso completo complete impulse
impulso compuesto composite pulse
impulso continuo continuous pulse
impulso corto short pulse
impulso cuadrado square pulse
impulso de arranque start pulse
impulso de bloqueo blocking pulse
impulso de cambio de estado status-change pulse
impulso de campo field pulse
impulso de canal channel pulse
impulso de cierre make pulse
impulso de codificación coding pulse
impulso de conmutación switching pulse
impulso de control de tiempo time-control pulse
impulso de corriente current pulse
impulso de corriente alterna alternating-current
     pulse
impulso de corriente de corta duración
     short-duration current pulse
impulso de corta duración short-duration pulse
impulso de definición set pulse
impulso de desbloqueo unblanking pulse
impulso de desplazamiento shift pulse
impulso de disco dial pulse
impulso de disparo trigger pulse, tripping pulse
impulso de disposición set pulse
impulso de eco echo pulse
impulso de encuadre framing pulse
impulso de entrada input pulse
impulso de escritura write pulse
impulso de escritura parcial partial-write pulse
impulso de espacio spacing pulse
impulso de excitación driving pulse
impulso de extinción blanking pulse
impulso de extinción horizontal horizontal-blanking
     pulse
impulso de extinción vertical vertical-blanking pulse
impulso de frecuencia portadora carrier-frequency
     pulse
impulso de Gauss Gaussian pulse
impulso de habilitación enable pulse
impulso de información information pulse
impulso de inhibición inhibit pulse
impulso de interferencia interference pulse
impulso de interrogación interrogation pulse
impulso de ionización ionization pulse
impulso de lectura read pulse
impulso de lectura parcial partial-read pulse
impulso de línea line pulse
impulso de luz light impulse
impulso de mando command pulse, drive pulse, gate
impulso de mando de acumulación
     storage-command pulse
impulso de mando parcial partial drive pulse
impulso de no impresión nonprint impulse
impulso de nube cloud pulse
impulso de parada stop pulse
impulso de pendiente variable variable-slope pulse
impulso de potencia power pulse
impulso de preparación set-up impulse
impulso de puerta gate pulse
impulso de radar radar pip, radar pulse
impulso de radiofrecuencia radio-frequency pulse

impulso de reloj clock pulse
impulso de reposición reset pulse
impulso de respuesta reply pulse
impulso de retorno de línea line-flyback pulse
impulso de ruido noise pulse
impulso de ruptura break impulse
impulso de salida output pulse
impulso de selección gating pulse
impulso de señal signal pulse
impulso de seno cuadrado sine-squared pulse
impulso de sincronismo horizontal horizontal-sync
  pulse
impulso de sincronismo vertical
  vertical-synchronizing pulse
impulso de sincronismo vertical aserrado serrated
  vertical synchronizing pulse
impulso de sincronización synchronizing pulse,
  timing pulse
impulso de sincronización de imagen
  picture-synchronizing pulse
impulso de sincronización maestro master
  synchronization pulse
impulso de sobrecarga overload pulse
impulso de supresión suppression pulse
impulso de temporización timing pulse
impulso de tensión voltage pulse
impulso de tensión de corta duración short-duration
  voltage pulse
impulso de trabajo marking pulse
impulso de transición transition pulse
impulso de transmisión transmission pulse
impulso de video video pulse
impulso decimal decimal pulse
impulso derivado derived pulse
impulso directo direct pulse
impulso discreto discrete pulse
impulso dividido split pulse
impulso eléctrico electrical impulse
impulso eléctrico discreto discrete electric impulse
impulso electromagnético electromagnetic pulse
impulso electrónico electronic pulse
impulso en diente de sierra sawtooth pulse
impulso espurio spurious pulse
impulso estroboscópico strobe pulse
impulso fantasma ghost pulse
impulso habilitante enabling pulse
impulso horizontal horizontal pulse
impulso incidente incident pulse
impulso incremental incremental impulse
impulso inducido afterpulse
impulso intensificador de línea line-intensifying
  pulse
impulso interrogante interrogating pulse
impulso inverso reverse impulse
impulso magnético magnetic impulse
impulso marcador marker pip, marker pulse
impulso modulado modulated pulse
impulso multifrecuencia multifrequency pulse
impulso negativo negative pulse
impulso no recurrente nonrecurring pulse
impulso parabólico parabolic pulse
impulso parásito pulse spike
impulso patrón standard pulse
impulso piloto pilot pulse
impulso por perforaciones de transporte sprocket
  pulse
impulso positivo positive pulse
impulso principal main bang
impulso rápido rapid pulse
impulso rectangular rectangular pulse
impulso residual residual impulse
impulso seccionado chopped impulse
impulso selector selector pulse

impulso sincronizador de línea line-synchronizing
  pulse
impulso supresor suppressor pulse
impulso telegráfico telegraph pulse
impulso temporizado timed pulse
impulso transmitido transmitted pulse
impulso trapecial trapezoidal pulse
impulso triangular triangular pulse
impulso ultrasónico ultrasonic pulse
impulso único single pulse
impulso unidireccional unidirectional pulse
impulso unipolar single-polarity pulse
impulso unitario unit pulse
impulso vertical vertical pulse
impulso vertical aserrado serrated vertical pulse
impulsor magnético magnetic impulser
impulsos coherentes coherent pulses
impulsos de corriente en obscuridad dark-current
  pulses
impulsos de llamada ringing impulses
impulsos de sincronización synchronization pulses
impulsos desiguales unequal impulses
impulsos ecualizadores equalizing pulses
impulsos estandarizados standardized pulses
impulsos generados electromagnéticamente
  electromagnetically generated pulses
impulsos por segundo pulses per second
impureza impurity, killer
impureza aceptora acceptor impurity
impureza de color color impurity
impureza de cristal crystal impurity
impureza donadora donor impurity
inactivo inactive, quiescent
inaislado uninsulated
inalámbrico wireless
inatendido unattended
inaudible nonaudible
incandescencia incandescence
incandescente incandescent
incertidumbre uncertainty
incertidumbre de entrada input uncertainty
incertidumbre de medida measurement uncertainty
incertidumbre relativa relative uncertainty
incidencia incidence
incidencia aleatoria random incidence
incidencia axial axial incidence
incidencia oblicua oblique incidence
incidencia tangencial tangential incidence
incidencia vertical vertical incidence
incidental incidental
incidente incident
inclinación inclination, tilt
inclinación de onda wave tilt
inclinación de planeo glide slope
inclinación del techo de impulso pulse tilt
inclinación magnética magnetic inclination
inclinómetro inclinometer
inclusión inclusion
incoherencia incoherency
incoherente incoherent, noncoherent
incombustible fireproof
incompletamente incompletely
incompleto incomplete
incondicional unconditional
incondicionalmente unconditionally
inconsistencia inconsistency
inconsistente inconsistent
incorporado built-in, self-contained
incremental incremental
incremento increment
incremento de carácter character increment
incremento de paso step change
incremento de tensión voltage increment

incremento digital digital increment
incremento entre caracteres intercharacter
    increment
incremento finito finite increment
incremento repentino burst
incremento repentino de errores error burst
indefinido undefined
indentificador de conexión connection identifier
independencia de datos data independence
independencia de datos física physical data
    independence
independencia de dispositivo device independence
independencia de máquina machine independence
independencia de plataforma platform independence
independiente stand-alone
independiente de configuración configuration
    independent
independiente de dispositivo device independent
independiente de frecuencia frequency-independent
independiente de máquina machine independent
indeseado unwanted
indexación indexing
indexación asignada assigned indexing
indexación de asignación assignment indexing
indexación de tarjetas card indexing
indexado indexed
indicación audible audible indication
indicación de alarma remota remote alarm
    indication
indicación de cero zero indication
indicación de fallas fault reporting
indicación de grupo group indication
indicación de medidor meter indication
indicación de señal signal indication
indicación de sentido sense finding
indicación de sobrecarga overload indication
indicación de tierra ground indication
indicación negativa negative indication
indicación remota remote indication
indicación visual de atención attention display
indicación visual de fallas fault display
indicado indicated
indicador indicator, pointer, display
indicador acimutal automático omnibearing
    indicator
indicador analógico analog indicator
indicador automático automatic indicator
indicador azel azel indicator
indicador de acción action indicator
indicador de aceleración acceleration indicator
indicador de aguja needle pointer
indicador de agujas cruzadas crossed-pointer
    indicator
indicador de aislamiento insulation indicator
indicador de alarma alarm indicator
indicador de almacenamiento storage indicator
indicador de aproximación approach indicator
indicador de aviso warning indicator
indicador de batido beat indicator
indicador de batido cero zero-beat indicator
indicador de bloque block indicator
indicador de cabina audible audible cab indicator
indicador de calibración calibration indicator
indicador de calidad de canal channel quality
    indicator
indicador de campo field indicator
indicador de canal channel indicator
indicador de cancelación cancel indicator
indicador de cantidad amount indicator
indicador de carga charge indicator
indicador de carga de batería battery charge
    indicator
indicador de celda corriente current cell indicator

indicador de cero zero indicator
indicador de cinta rota broken tape indicator
indicador de columna column indicator
indicador de comprobación check indicator
indicador de comprobación de máquina
    machine-check indicator
indicador de comprobación de signo sign-check
    indicator
indicador de concentración de humo
    smoke-concentration indicator
indicador de condición condition indicator
indicador de contacto a tierra ground-contact
    indicator
indicador de contaminación de aire air
    contamination indicator
indicador de continuación continuation indicator
indicador de corriente de rejilla grid-current
    indicator
indicador de cortocircuito short indicator
indicador de datos data pointer
indicador de demanda máxima maximum-demand
    indicator
indicador de desbordamiento overflow indicator
indicador de desbordamiento negativo underflow
    indicator
indicador de descarga discharge indicator
indicador de desfase phase-difference indicator
indicador de desplazamiento de portadora carrier
    shift indicator
indicador de destino destination indicator
indicador de desviación de línea de rumbo
    course-line deviation indicator
indicador de desviación de trayectoria de vuelo
    flight-path deviation indicator
indicador de dirección direction indicator, address
    indicator
indicador de dirección de aterrizaje landing
    direction indicator
indicador de dirección magnético magnetic
    direction-indicator
indicador de distancia-altura range-height indicator
indicador de distancia-amplitud range-amplitude
    display
indicador de distancia-rumbo range-bearing display
indicador de ejecución run indicator
indicador de encaminamiento routing indicator
indicador de energía acumulada stored-energy
    indicator
indicador de enlace link indicator
indicador de envolvente de radiofrecuencia
    radio-frequency envelope indicator
indicador de equilibrio balance indicator
indicador de error permanente permanent error
    indicator
indicador de errores error indicator
indicador de estado status indicator
indicador de factor de potencia power-factor
    indicator
indicador de falla failure indicator, fault indicator
indicador de falla de alimentación power-failure
    indicator
indicador de fase phase indicator
indicador de fin de archivo end-of-file indicator
indicador de fin de página end-of-page indicator
indicador de flujo de combustible fuel-flow
    indicator
indicador de flujo de fluido fluid-flow indicator
indicador de flujo de gas gas-flow indicator
indicador de flujo de líquido liquid-flow indicator
indicador de frecuencia frequency indicator
indicador de frecuencia de circuito resonante
    resonant-circuit frequency indicator
indicador de fugas leakage indicator

indicador de fugas de gas gas-leak indicator
indicador de haz luminoso light-beam pointer
indicador de humedad humidity indicator, moisture
indicador
indicador de humo smoke indicator
indicador de lámpara de neón neon-bulb indicator
indicador de lectura readout indicator
indicador de lectura directa direct-reading indicator
indicador de lectura remota remote-reading
indicador
indicador de localización location indicator
indicador de lorán loran indicator
indicador de luz de neón neon-light indicator
indicador de llamadas call indicator
indicador de mensaje message indicator
indicador de moción motion indicator
indicador de modo mode indicator
indicador de neón neon indicator
indicador de neutralización neutralization indicator
indicador de nivel level indicator
indicador de nivel de aceite oil-level gage
indicador de nivel de agua water-level gage
indicador de nivel de control control level indicator
indicador de nivel de fluido fluid-level indicator
indicador de nivel de grabación recording-level
indicador
indicador de nivel de líquido liquid-level indicator
indicador de nivel de potencia power-level indicator
indicador de nivel de programa program level
indicador
indicador de nivel de registro recording-level
indicador
indicador de nivel de señal signal-level indicator
indicador de nivel sonoro sound-level indicator
indicador de nivel superior higher-level indicator
indicador de ondas estacionarias standing-wave
indicador
indicador de operación operation indicator
indicador de orientación orientation indicator
indicador de parada halt indicator
indicador de pH pH indicator
indicador de pico peak indicator
indicador de picos de lámpara de neón neon-bulb
peak indicator
indicador de polaridad polarity indicator
indicador de polos pole indicator
indicador de posición position indicator
indicador de posición de archivo file position
indicador
indicador de posición de eje shaft-position indicator
indicador de posición electrónico electronic position
indicador
indicador de posición en el aire air-position
indicador
indicador de posición en el plano plan-position
indicador
indicador de posición ensanchado expanded position
indicador
indicador de posición relativa ground-position
indicador
indicador de potencia electrotérmico electrothermic
power indicator
indicador de presión pressure indicator
indicador de presión de aceite oil-pressure gage
indicador de presión de agua water-pressure gage
indicador de presión de combustible fuel-pressure
indicador
indicador de presión de fluido fluid-pressure
indicador
indicador de presión de gas gas-pressure indicator
indicador de presión de líquido liquid-pressure
indicador
indicador de prioridad priority indicator

indicador de profundidad depth indicator
indicador de prueba test indicator
indicador de puente bridge indicator
indicador de radar radar indicator
indicador de radiación radiation indicator
indicador de radiocompás radio-compass indicator
indicador de radiofrecuencia radio-frequency
indicador
indicador de rayos catódicos cathode-ray indicator
indicador de razón de ondas estacionarias
standing-wave ratio indicator
indicador de reacción reaction indicator
indicador de referencia reference indicator
indicador de resonancia resonance indicator
indicador de rumbo course indicator
indicador de salida output indicator
indicador de secuencia de fases phase-sequence
indicador
indicador de sentido sense finder
indicador de sincronismo synchronism indicator
indicador de sincronización synchronization
indicador
indicador de sintonía por sombra shadow-tuning
indicador
indicador de sintonización tuning indicator
indicador de sintonización de rayos catódicos
cathode-ray tuning indicator
indicador de sistema de aterrizaje por instrumentos
instrument-landing system indicator
indicador de sobrecarga overload indicator
indicador de sobremodulación overmodulation
indicador
indicador de sobremodulación de lámpara de neón
neon-bulb overmodulation indicator
indicador de sobretemperatura overtemperature
indicador
indicador de sobretensión de lámpara de neón
neon-bulb overvoltage indicator
indicador de tecla de mandos command key
indicador
indicador de temperatura temperature indicator
indicador de temperatura digital digital temperature
indicador
indicador de tensión voltage indicator
indicador de tiempo transcurrido elapsed-time
indicador
indicador de tierra ground indicator
indicador de tipo de mano hand-type pointer
indicador de transmisión transmit indicator
indicador de trayectoria de planeo glide-path
indicador
indicador de unidades de volumen volume-unit
indicador
indicador de vacío vacuum indicator
indicador de valor medio mean value indicator
indicador de velocidad speed indicator
indicador de velocidad del aire airspeed indicator
indicador de velocidad límite speed-limit indicator
indicador de volumen volume indicator
indicador de volumen de lámparas de neón
neon-bulb volume indicator
indicador de vuelo flight indicator
indicador de zona de comunicaciones
communications zone indicator
indicador digital digital indicator
indicador externo external indicator
indicador Gee Gee indicator
indicador giroeléctrico de actitud attitude
gyroelectric indicator
indicador lógico logical indicator
indicador luminoso spot
indicador magnético de radiocompás radio-compass
magnetic indicator

indicador manual manual indicator
indicador negativo negative indicator
indicador óptico optical indicator
indicador panorámico panoramic indicator
indicador piezoeléctrico piezoelectric indicator
indicador por bajada drop indicator
indicador positivo positive indicator
indicador potenciométrico potentiometer indicator
indicador radiológico radiological indicator
indicador radiomagnético radio magnetic indicator
indicador visual display, visual indicator
indicador visual A A display
indicador visual B B display
indicador visual complejo complex display
indicador visual conmutado chopped display
indicador visual de medidor meter display
indicador visual tipo A type-A display
indicador visual tipo B type-B display
indicador voltmétrico voltmeter indicator
indicar read
índice index
índice activo active index
índice alternativo alternate index
índice automático automatic index
índice de absorción solar solar absorption index
índice de actividad solar solar-activity index
índice de agrupamiento clustering index
índice de alimentación feed index
índice de calidad quality index
índice de ciclos cycle index
índice de concordancia de color percibido color
  rendering index
índice de confiabilidad reliability index
índice de corrimiento atrasado pushing figure
índice de diafonía crosstalk index
índice de direccionalidad directional gain
índice de directividad directivity index
índice de dosis absorbida absorbed dose index
índice de duración de interrupción
  interruption-duration index
índice de enfoque focusing index
índice de fragmentación fragmentation index
índice de frecuencia de interrupciones
  interruption-frequency index
índice de grupo group index
índice de líneas de código code line index
índice de local room index
índice de modulación modulation index
índice de no linealidad nonlinearity index
índice de pérdidas loss index
índice de permutación permutation index
índice de polarización polarization index
índice de refracción refractive index
índice de refracción modificado modified refractive
  index
índice de refracción relativo relative refractive
  index
índice de regulación setting index
índice de rendimiento performance index
índice de ruido noise index
índice de temperatura temperature index
índice de temperatura relativa relative temperature
  index
índice de utilización de canal channel-utilization
  index
índice espectral spectral index
índice fino fine index
índice maestro master index
índice numérico numerical index
índice secundario secondary index
indicial indicial
indio indium
indirectamente indirectly

indirecto indirect
indisponibilidad unavailability
indisponible unavailable
inducción induction
inducción característica characteristic induction
inducción crítica critical induction
inducción de campo eléctrico electric-field induction
inducción de campo electromagnético
  electromagnetic field induction
inducción de campo magnético magnetic-field
  induction
inducción de saturación saturation induction
inducción eléctrica electric induction
inducción electromagnética electromagnetic
  induction
inducción electrónica electronic induction
inducción electrostática electrostatic induction
inducción férrica ferric induction
inducción homopolar homopolar induction
inducción incremental incremental induction
inducción intrínseca intrinsic induction
inducción magnética magnetic induction
inducción magnética residual residual magnetic
  induction
inducción magnetoeléctrica magnetoelectric
  induction
inducción mutua mutual induction
inducción normal normal induction
inducción nuclear nuclear induction
inducción parásita parasitic induction
inducción peristáltica peristaltic induction
inducción remanente remanent induction
inducción residual residual induction
inducción unipolar unipolar induction
inducido armature
inducido artificialmente artificially induced
inducido de alambrado impreso printed-wiring
  armature
inducido de generador generator armature
inducido de Gramme Gramme armature
inducido de medidor meter armature
inducido de motor motor armature
inducido en anillo ring armature
inducido fijo fixed armature
inducido giratorio revolving armature
inducido monofásico single-phase armature
inducido por ruido noise-induced
inducido por ultravioleta ultraviolet-induced
inducido ranurado slotted armature
inductancia inductance
inductancia acústica acoustic inductance
inductancia aparente apparent inductance
inductancia-capacitancia inductance-capacitance
inductancia-capacitancia-resistencia
  inductance-capacitance-resistance
inductancia cero zero inductance
inductancia concentrada concentrated inductance
inductancia crítica critical inductance
inductancia de alambrado wiring inductance
inductancia de antena antenna inductance
inductancia de carga loading inductance
inductancia de circuito oscilante oscillating-circuit
  inductance
inductancia de compensación padding inductance
inductancia de conexiones lead inductance
inductancia de etapa excitadora driver inductance
inductancia de fuga leakage inductance
inductancia de núcleo de aire air-core inductance
inductancia de paso transition coil
inductancia de saturación saturation inductance
inductancia de sintonización tuning inductance
inductancia de sintonización de antena antenna
  tuning inductance

inductancia de sintonización de placa plate tuning inductance
inductancia de tanque de placa plate tank inductance
inductancia de tanque de rejilla grid tank inductance
inductancia distribuida distributed inductance
inductancia efectiva effective inductance
inductancia en circuito abierto open-circuit inductance
inductancia en cortocircuito short-circuit inductance
inductancia en paralelo parallel inductance
inductancia en serie series inductance
inductancia en serie efectiva effective series inductance
inductancia equivalente equivalent inductance
inductancia incremental incremental inductance
inductancia magnetizante magnetizing inductance
inductancia máxima equivalente maximum equivalent inductance
inductancia mutua mutual inductance
inductancia natural natural inductance
inductancia negativa negative inductance
inductancia neta net inductance
inductancia operacional operational inductance
inductancia parásita parasitic inductance, stray inductance
inductancia primaria primary inductance
inductancia pura pure inductance
inductancia residual residual inductance
inductancia secundaria secondary inductance
inductancia variable variable inductance
inductancímetro inductance meter
inductivamente inductively
inductividad inductivity
inductivo inductive
inductófono inductophone
inductómetro inductometer
inductómetro variable de Brooks Brooks variable inductometer
inductor inductor
inductor bifilar bifilar inductor
inductor cilíndrico cylindrical inductor
inductor con derivación central center-tapped inductor
inductor concentrado lumped inductor
inductor continuamente ajustable continuously-adjustable inductor
inductor de ajuste trimmer inductor
inductor de control control inductor
inductor de décadas decade inductor
inductor de filtro filter inductor
inductor de núcleo cerrado closed-core inductor
inductor de sintonización tuning inductor
inductor de sustitución substitution inductor
inductor de tierra earth inductor
inductor discreto discrete inductor
inductor fijo fixed inductor
inductor impreso printed inductor
inductor moldeado molded inductor
inductor mutuo mutual inductor
inductor no compensado uncompensated inductor
inductor no lineal nonlinear inductor
inductor patrón standard inductor
inductor regulador regulating inductor
inductor variable variable inductor
inductor variable eléctricamente electrically variable inductor
inductores ecualizadores de tierra ground equalizer inductors
inductores en paralelo parallel inductors
inductores en serie series inductors
inductores en serie-paralelo series-parallel inductors

inercia inertia
inercia eléctrica electrical inertia
inercia electromagnética electromagnetic inertia
inercial inertial
inertancia inertance
inertancia acústica acoustic inertance
inertancia específica specific inertance
inestabilidad instability
inestabilidad a largo plazo long-term instability
inestabilidad de frecuencia residual residual frequency instability
inestabilidad estructural structural instability
inestabilidad horizontal horizontal instability
inestabilidad magnética magnetic instability
inestabilidad magnetohidrodinámica magnetohydrodynamic instability
inestabilidad mecánica mechanical instability
inestabilidad térmica thermal instability
inestabilidad vertical vertical instability
inestable unstable
inexactitud inaccuracy
inexacto inaccurate
infección infection
infiltración infiltration
infinitesimal infinitesimal
infinito infinite
inflamabilidad flammability
inflamable flammable
influencia influence
influencia de campo externo external field influence
influencia de forma de onda waveform influence
influencia de operación operation influence
influencia de proximidad proximity influence
influencia de tensión voltage influence
influencia eléctrica electric influence
influencia electrostática electrostatic influence
influencia inductiva inductive influence
influencia por frecuencia frequency influence
influencia térmica thermal influence
información information
información aeronáutica aeronautical information
información alfanumérica alphanumeric information
información analógica analog information
información binaria binary information
información de avance advance information
información de colores color information
información de control control information
información de control de trabajos job control information
información de controlador controller information
información de cromaticidad chromaticity information
información de crominancia chrominance information
información de imagen picture information
información de luminancia luminance information
información de máquina machine information
información de radar radar information
información de retroalimentación feedback information
información de secuencia sequence information
información de televisión television information
información de verde green information
información de vuelo flight information
información detectable por máquina machine-sensible information
información digital digital information
información en línea on-line information
información formateada formatted information
información fuente source information
información impresa printed information
información legible por máquina machine-readable information

información maestra master information
información meteorológica meteorological
    information
información miscelánea miscellaneous information
información monocroma monochrome information
información mutua mutual information
información mutua media average mutual
    information
información no formateada unformatted information
información numérica numerical information
información preliminar preliminary information
información registrada recorded information
información selectiva selective information
información técnica technical information
información transferida transferred information
información transmitida transmitted information
información visual visual information
informática informatics
informe report
informe computarizado computerized report
informe de errores error report
informe de excepciones exception report
informe de facturación billing report
informe de mal funcionamiento malfunction report
infra infra
infraacústico infraacoustic
infradino infradyne
infrarrojo infrared
infrarrojo lejano far infrared
infrarrojo próximo near infrared
infrasónica infrasonics
infrasónico infrasonic
infrasonido infrasound
ingeniería asistida por computadora
    computer-aided engineering
ingeniería de computadoras computer engineering
ingeniería humana human engineering
ingeniero de circuitos circuit engineer
ingeniero de computadoras computer engineer
ingeniero electricista electrical engineer
ingrediente activo active ingredient
inherente inherent
inhibición inhibition
inhibición acústica acoustic inhibition
inhibición auditiva auditory inhibition
inhibidor inhibitor
inhibidor de decapado pickling inhibitor
inhibidor de reacción reaction inhibitor
inhibir inhibit
inhomogéneo inhomogeneous
iniciación caliente warm start
iniciación de interrupción outage initiation
iniciación de paso de trabajo job step initiation
iniciación de puntos de comprobación checkpoint
    start
iniciación de sesión session initiation
iniciación de tarea task start
iniciación eléctrica electrical initiation
iniciación fría cold start
iniciador de impulsos pulse initiator
iniciador de programa program initiator
inicialización initialization
inicialización de bucle loop initialization
inicialización de campo field initialization
inicialización de trabajo job initialization
inicialización primaria primary initialization
inicializar initialize
iniciar initiate
iniciar una sesión log in, log on
inicio de impulso pulse start
inicio de sesión session initiation
inicio rápido quick start
ininflamable nonflammable

ininteligible garbled
ininterrumpible uninterruptable
ininterrumpido uninterrupted
inmediato immediate
inmersión dipping
inmunidad immunity
inmunidad al ruido noise immunity
inmunidad al ruido de corriente alterna
    alternating-current noise immunity
inoperante inoperative, dummy
inorgánico inorganic
inseparable nonseparable
inserción insertion
inserción de datos data insertion
inserción de repetidor repeater insertion
insonorización soundproofing, sound insulation
insonorizado soundproofed, sound-insulated
insonorizador sound insulator
insonoro soundproof
inspección inspection
inspección asistida por computadora
    computer-aided inspection
inspección de diseño design inspection
inspección de requisitos requirements inspection
inspección fotoeléctrica photoelectric inspection
inspección magnetoscópica magnetoscopic
    inspection
inspección por rayos X X-ray inspection
inspección selectiva selective inspection
inspección ultrasónica ultrasonic inspection
inspectoscopio inspectoscope
instalación installation
instalación abreviada abbreviated installation
instalación automática automatic installation
instalación de abonado subscriber installation
instalación de capacitor capacitor installation
instalación de dispersión directa forward-scatter
    installation
instalación de inclinación de planeo glide-slope
    facility
instalación de instrumento instrument installation
instalación de mandos command installation
instalación de pruebas test facility
instalación de radar radar installation
instalación de reserva reserve installation
instalación de retransmisión retransmission
    installation
instalación de telecomunicaciones
    telecommunications facility
instalación de transmisión transmission facility
instalación dúplex diferencial differential duplex
    installation
instalación eléctrica electrical installation
instalación expuesta exposed installation
instalación interior indoor installation
instalación normal normal install
instalación semiautomática semiautomatic
    installation
instalación temporal temporary installation
instalación terminal terminal installation
instalaciones facilities
instalaciones de alumbrado lighting facilities
instalaciones de comunicaciones communications
    facilities
instalaciones de radio radio facilities
instalaciones radiotelefónicas radiotelephone
    facilities
instalaciones telefónicas telephone facilities
instantáneamente instantaneously
instantáneo instantaneous
instante instant
instante significativo significant instant
instante significativo de una modulación significant

instant of a modulation
instrucción instruction, statement
instrucción absoluta absolute instruction
instrucción alternativa alternate instruction
instrucción aritmética arithmetic instruction
instrucción asistida por computadora computer-assisted instruction
instrucción basada en computadora computer-based instruction
instrucción básica basic instruction
instrucción completa complete instruction
instrucción condicional conditional instruction
instrucción corriente current instruction
instrucción corta short instruction
instrucción de búsqueda fetch instruction
instrucción de búsqueda en tabla table look-up instruction
instrucción de carga load instruction
instrucción de código de máquina machine-code instruction
instrucción de código fuente source-code instruction
instrucción de comparar compare instruction
instrucción de computadora computer instruction
instrucción de control control instruction
instrucción de control de procedimiento procedure control instruction
instrucción de cuatro direcciones four-address instruction
instrucción de datos data instruction
instrucción de decisión decision instruction
instrucción de derivación branch instruction
instrucción de desplazamiento shift instruction
instrucción de desplazamiento de acumulador accumulator shift instruction
instrucción de dirección cero zero-address instruction
instrucción de dirección única single-address instruction
instrucción de direcciones múltiples multiple-address instruction
instrucción de dos direcciones two-address instruction
instrucción de ejecución execution instruction
instrucción de enlace linkage instruction
instrucción de ensamblador assembler instruction
instrucción de entrada entry instruction
instrucción de entrada-salida input-output instruction
instrucción de entrada-salida programada programmed input-output instruction
instrucción de extracción extract instruction
instrucción de indicador de comprobación check indicator instruction
instrucción de interrupción breakpoint instruction
instrucción de interrupción condicional conditional breakpoint instruction
instrucción de lenguaje de ensamblaje assembly language instruction
instrucción de longitud variable variable-length instruction
instrucción de llamada call instruction
instrucción de máquina machine instruction
instrucción de microcódigo microcode instruction
instrucción de microprocesador microprocessor instruction
instrucción de multiplicar-dividir multiply-divide instruction
instrucción de no impresión nonprint instruction
instrucción de no operación no-operation instruction
instrucción de parada stop instruction, halt instruction
instrucción de parada condicional conditional stop instruction

instrucción de parada de programa program stop instruction
instrucción de parada opcional optional stop instruction
instrucción de pausa pause instruction
instrucción de perforación punching instruction
instrucción de preparación housekeeping instruction
instrucción de procesador processor instruction
instrucción de programa program instruction
instrucción de prueba test instruction
instrucción de rama condicional conditional branch instruction
instrucción de reanudación restart instruction
instrucción de referencia reference instruction
instrucción de reinicio restart instruction
instrucción de repetición repetition instruction
instrucción de retorno return instruction
instrucción de salto jump instruction
instrucción de salto de acumulador accumulator jump instruction
instrucción de salto incondicional unconditional jump instruction
instrucción de selección selection instruction
instrucción de sistema system instruction
instrucción de subrutina subroutine instruction
instrucción de suma add instruction
instrucción de transferencia transfer instruction
instrucción de transferencia de acumulador accumulator transfer instruction
instrucción de transferencia de control control transfer instruction
instrucción de transferencia incondicional unconditional transfer instruction
instrucción de transporte interno internal transport instruction
instrucción de tres direcciones three-address instruction
instrucción de una dirección one-address instruction
instrucción de uno más uno one-plus-one instruction
instrucción de visualización display instruction
instrucción directa direct instruction
instrucción efectiva effective instruction
instrucción ejecutiva executive instruction
instrucción ficticia dummy instruction
instrucción fija fixed instruction
instrucción ilegal illegal instruction
instrucción indirecta indirect instruction
instrucción inmediata immediate instruction
instrucción lógica logic instruction
instrucción microprogramable microprogrammable instruction
instrucción multidirección multiaddress instruction
instrucción no privilegiada nonprivileged instruction
instrucción no válida invalid instruction
instrucción nula null instruction
instrucción primaria primary instruction
instrucción privilegiada privileged instruction
instrucción programada programmed instruction
instrucción real actual instruction
instrucción relativa relative instruction
instrucción SI IF instruction
instrucción simbólica symbolic instruction
instrucción sin dirección addressless instruction
instrucción supuesta presumptive instruction
instrucciones condensadas condensed instructions
instrucciones de instalación installation instructions
instrucciones de mantenimiento maintenance instructions
instrucciones de máquina de microprograma microprogram machine instructions
instrucciones de operación operating instructions
instrucciones de programación programming instructions

instrucciones iniciales initial instructions
instrumentación instrumentation
instrumentación de análisis analysis instrumentation
instrumentación de aviación aviation instrumentation
instrumentación electrónica electronic instrumentation
instrumentación gráfica graphic instrumentation
instrumentación meteorológica meteorological instrumentation
instrumental instrumental
instrumento instrument
instrumento ajustable adjustable instrument
instrumento amortiguado magnéticamente magnetically damped instrument
instrumento aperiódico aperiodic instrument
instrumento autocalibrador self-calibrating instrument
instrumento autónomo self-contained instrument
instrumento autorregistrador self-recording instrument
instrumento autosincrónico self-synchronous instrument
instrumento bilateral bilateral instrument
instrumento bimetálico bimetallic instrument
instrumento bolométrico bolometric instrument
instrumento calibrado calibrated instrument
instrumento con autocomprobación self-checking instrument
instrumento con rectificador rectifier instrument
instrumento controlador controlling instrument
instrumento de aguja pointer instrument
instrumento de aterrizaje landing instrument
instrumento de aviación aviation instrument
instrumento de bobina móvil moving-coil instrument
instrumento de cero suprimido suppressed-zero instrument
instrumento de dos escalas two-range instrument
instrumento de espejo mirror instrument
instrumento de haz electrónico electron-beam instrument
instrumento de hierro articulado hinged-iron instrument
instrumento de hierro móvil moving-iron instrument
instrumento de hilo caliente hot-wire instrument
instrumento de imán móvil moving-magnet instrument
instrumento de imán permanente permanent-magnet instrument
instrumento de inducción induction instrument
instrumento de ionización ionization instrument
instrumento de laminillas vibrantes vibrating-reed instrument
instrumento de lectura directa direct-reading instrument
instrumento de medición astático astatic measuring instrument
instrumento de medición de sonido sound-measuring instrument
instrumento de medida measuring instrument
instrumento de medida con pinza hook-on instrument
instrumento de microcomputadora microcomputer instrument
instrumento de microprocesador microprocessor instrument
instrumento de navegación navigation instrument
instrumento de precisión precision instrument
instrumento de prueba test instrument
instrumento de referencia reference instrument
instrumento de registro recording instrument
instrumento de registro de accionamiento directo direct-acting recording instrument
instrumento de registro directo direct-recording instrument
instrumento de resistencia resistance instrument
instrumento de termopar thermocouple instrument
instrumento de tipo de paleta vane-type instrument
instrumento de tono fijo fixed-pitch instrument
instrumento de vuelo flight instrument
instrumento de Wheatstone Wheatstone instrument
instrumento diferencial differential instrument
instrumento digital de lectura directa direct-reading digital instrument
instrumento dinamométrico dynamometer instrument
instrumento eléctrico electrical instrument
instrumento electrodinámico electrodynamic instrument
instrumento electromagnético electromagnetic instrument
instrumento electrónico electronic instrument
instrumento electrostático electrostatic instrument
instrumento electrotérmico electrothermal instrument
instrumento espectrométrico spectrometric instrument
instrumento ferrodinámico ferrodynamic instrument
instrumento flotante floating instrument
instrumento iluminado illuminated instrument
instrumento indicador indicating instrument
instrumento integrador integrating instrument
instrumento multiescala multirange instrument
instrumento navegacional navigational instrument
instrumento patrón standard instrument
instrumento radiológico radiological instrument
instrumento térmico thermal instrument
instrumento terminal end instrument
instrumento termoiónico thermionic instrument
instrumento totalizador summation instrument
instrumento ultrasensible ultrasensitive instrument
integración integration
integración a muy alta escala very-large-scale integration
integración analógica analog integration
integración coherente coherent integration
integración de escala media medium-scale integration
integración de impulsos pulse integration
integración de señales signal integration
integración de sistema system integration
integración de video video integration
integración en gran escala large-scale integration
integración en pequeña escala small-scale integration
integración no coherente noncoherent integration
integración numérica numerical integration
integración parcial partial integration
integrado integrated
integrador integrator
integrador analógico analog integrator
integrador de corriente current integrator
integrador de ganancia gain integrator
integrador de impulsos pulse integrator
integrador de Miller Miller integrator
integrador de peligro sonoro sound-hazard integrator
integrador de tensión voltage integrator
integrador de video video integrator
integrador digital digital integrator
integrador eléctrico electrical integrator
integrador electrónico electronic integrator
integrador fotométrico photometric integrator
integrador incremental incremental integrator
integrador limitado limited integrator

integral integral
integral de línea line integral
integral múltiple multiple integral
integrando integrand
integrar integrate
integridad integrity
integridad de aplicación application integrity
integridad de base de datos database integrity
integridad de datos data integrity
integridad de programas software integrity
integridad de sistema system integrity
inteligencia artificial artificial intelligence
inteligencia de máquina machine intelligence
inteligencia de microprocesador microprocessor
    intelligence
inteligencia distribuida distributed intelligence
inteligencia electrónica electronic intelligence
inteligibilidad intelligibility
inteligible intelligible
intensidad intensity, strength
intensidad acústica acoustic intensity
intensidad armónica harmonic intensity
intensidad de campo field intensity, field strength
intensidad de campo de interferencia interference
    field strength
intensidad de campo de onda espacial space-wave
    field strength
intensidad de campo de radio radio field intensity
intensidad de campo de ruido noise field intensity
intensidad de campo efectiva effective field intensity
intensidad de campo eléctrico electric-field strength
intensidad de campo eléctrico de corriente alterna
    alternating-current electrical field strength
intensidad de campo electromagnético
    electromagnetic field intensity
intensidad de campo electrostático electrostatic field
    intensity
intensidad de campo en espacio libre free-space
    field intensity
intensidad de campo horizontal horizontal field
    strength
intensidad de campo incidente incident field
    intensity
intensidad de campo magnético magnetic-field
    intensity
intensidad de campo sin absorción unabsorbed field
    intensity
intensidad de campo vertical vertical field strength
intensidad de color color intensity
intensidad de corriente current strength, current
    intensity
intensidad de haz electrónico electron-beam
    intensity
intensidad de iluminación intensity of illumination
intensidad de magnetización intensity of
    magnetization
intensidad de polo magnético magnetic pole strength
intensidad de punto spot intensity
intensidad de radiación radiation intensity
intensidad de recepción receiving intensity
intensidad de referencia reference intensity
intensidad de señal signal intensity, signal strength
intensidad de señal de retorno return-signal
    intensity
intensidad de señal máxima maximal signal strength
intensidad de tráfico traffic intensity
intensidad horizontal media mean horizontal
    intensity
intensidad luminosa luminous intensity
intensidad luminosa específica specific luminous
    intensity
intensidad luminosa espectral spectral luminous
    intensity

intensidad magnética magnetic intensity
intensidad máxima de campo peak field strength
intensidad pico peak intensity
intensidad radiante radiant intensity
intensidad radiante espectral spectral radiant
    intensity
intensidad relativa relative intensity
intensidad sonora sound intensity, loudness
intensidad sonora aparente overall loudness
intensidad sonora equivalente equivalent loudness
intensidad variable variable intensity
intensificación intensification
intensificación armónica harmonic intensification
intensificación de armónicas impares odd-harmonic
    intensification
intensificación de imagen image intensification
intensificado intensified
intensificador intensifier
intensificador armónico harmonic intensifier
intensificador de eco echo intensifier
intensificador de imagen image intensifier
intensificador de imagen fluoroscópica fluoroscopic
    image intensifier
intensímetro intensimeter
intensitómetro intensitometer
intento attempt
intento de llamada call attempt
intento de llamada abandonado abandoned call
    attempt
intentos de llamada por hora call attempts per hour
interacción interaction
interacción conversacional conversational
    interaction
interacción cuantificada quantized interaction
interacción de campos cruzados crossed-field
    interaction
interacción de encendedor ignitor interaction
interacción de impurezas impurity interaction
interacción de ondas progresivas traveling-wave
    interaction
interacción débil weak interaction
interacción electromagnética electromagnetic
    interaction
interacción mutua mutual interaction
interacciones de Coulomb Coulomb interactions
interacciones de Coulomb entre partículas particle
    Coulomb interactions
interacoplamiento cross-coupling
interactivo interactive
intercalación interleaving, collation
intercalación de paridad odd-even interleaving
intercaladora collator
intercalar interleave, collate
intercambiable interchangeable
intercambiador de calor heat exchanger
intercambialidad interchangeability
intercambialidad de tubos tube interchangeability
intercambialidad eléctrica electrical
    interchangeability
intercambialidad mecánica mechanical
    interchangeability
intercambio exchange, swap
intercambio de calor heat exchange
intercambio de datos data exchange
intercambio de datos básicos basic data exchange
intercambio de energía energy exchange
intercambio de impulsos de sincronización
    handshaking
intercambio de información information interchange
intercambio de memoria memory exchange
intercambio de mensajes message exchange
intercambio iónico ion exchange
intercambio químico chemical exchange

intercepción automática automatic intercept
intercepción de línea de comunicación wiretapping
intercepción de mensajes message interception
intercepción miscelánea miscellaneous interception
intercomunicación intercommunication
intercomunicador intercommunicator, intercom
intercomunicador de oprimir para hablar
   press-to-talk intercom
interconectado interconnected
interconectador interconnector
interconectar interconnect
interconexión interconnection
interconexión de frecuencia intermedia
   intermediate-frequency interconnection
interconexión de sistemas system interconnection
interconexión eléctrica electric interconnection
interconexión progresiva progressive
   interconnection
interconexión regional regional interconnection
interconstruído embedded
intercristalino intercrystalline
interdigital interdigital
interfacial interfacial
interfaz interface
interfaz computadora-televisión computer-television
   interface
interfaz de bus bus interface
interfaz de cable coaxial coaxial-cable interface
interfaz de casete cassette interface
interfaz de codificador encoder interface
interfaz de comunicaciones communication interface
interfaz de comunicaciones de datos data
   communications interface
interfaz de comunicaciones programable
   programmable communications interface
interfaz de conector connector interface
interfaz de copia impresa hard-copy interface
interfaz de disco disk interface
interfaz de enlace de paquetes de datos packet data
   link interface
interfaz de entrada-salida input-output interface
interfaz de expansión expansion interface
interfaz de grupo cluster interface
interfaz de intercambio de impulsos de
   sincronización handshake interface
interfaz de máquina machine interface
interfaz de memoria de microprocesador
   microprocessor memory interface
interfaz de microcontrolador microcontroller
   interface
interfaz de minicomputadora minicomputer
   interface
interfaz de modo común common-mode interface
interfaz de nivel de paquete packet level interface
interfaz de operación operating interface
interfaz de periférico programable programmable
   peripheral interface
interfaz de programa program interface
interfaz de red network interface
interfaz de sistema system interface
interfaz de subsistema subsystem interface
interfaz de teleimpresora teleprinter interface
interfaz de terminal terminal interface
interfaz de transmisión transmission interface
interfaz de unidad de conexión attachment unit
   interface
interfaz de uso general general-purpose interface
interfaz de usuario user interface
interfaz digital digital interface
interfaz en serie serial interface
interfaz en tiempo real real-time interface
interfaz física physical interface
interfaz funcional functional interface

interfaz híbrida hybrid interface
interfaz mecánica mechanical interface
interfaz múltiplex multiplex interface
interfaz paralela parallel interface
interfaz patrón standard interface
interferencia interference
interferencia aleatoria random interference
interferencia armónica harmonic interference
interferencia armónica de frecuencia intermedia
   intermediate-frequency harmonic interference
interferencia armónica de oscilador oscillator
   harmonic interference
interferencia atmosférica atmospheric interference
interferencia auroral auroral interference
interferencia cuántica quantum interference
interferencia cuasiimpulsiva quasi-impulsive
   interference
interferencia dañina harmful interference
interferencia de almacenamiento storage
   interference
interferencia de banda ancha broadband
   interference
interferencia de banda angosta narrowband
   interference
interferencia de banda lateral sideband interference
interferencia de Barkhausen Barkhausen
   interference
interferencia de bloqueo blocking interference
interferencia de campo magnético magnetic-field
   interference
interferencia de canal adyacente adjacent-channel
   interference
interferencia de canal alternativo alternate-channel
   interference
interferencia de canal común common-channel
   interference
interferencia de frecuencia intermedia
   intermediate-frequency interference
interferencia de intermodulación intermodulation
   interference
interferencia de modo común common-mode
   interference
interferencia de modo diferencial differential-mode
   interference
interferencia de ondas wave interference
interferencia de oscilador oscillator interference
interferencia de radiodifusión broadcast
   interference
interferencia de radiofrecuencia radio-frequency
   interference
interferencia de ruido noise interference
interferencia de segundo canal second-channel
   interference
interferencia de televisión television interference
interferencia de zumbido hum interference
interferencia destructiva destructive interference
interferencia eléctrica electrical interference
interferencia electromagnética electromagnetic
   interference
interferencia electrónica electronic interference
interferencia entre canales interchannel interference
interferencia entre símbolos intersymbol
   interference
interferencia heterodina heterodyne interference
interferencia imagen image interference
interferencia inductiva inductive interference
interferencia industrial industrial interference
interferencia inherente inherent interference
interferencia intencional jamming
interferencia intencional activa active jamming
interferencia intencional de ángulo angle jamming
interferencia intencional de frecuencia puntual spot
   jamming

interferencia intencional de radar radar jamming
interferencia intencional electrónica electronic jamming
interferencia intencional multibanda barrage jamming
interferencia intencional selectiva selective jamming
interferencia intermitente intermittent interference
interferencia longitudinal longitudinal interference
interferencia mutua mutual interference
interferencia natural natural interference
interferencia permisible permissible interference
interferencia polarizante polarizing interference
interferencia por diatermia diathermy interference
interferencia por eco echo interference
interferencia por emplazamiento site interference
interferencia por ignición ignition interference
interferencia por manchas solares sunspot interference
interferencia por ondas espaciales skywave interference
interferencia produciendo bloqueo total wipeout
interferencia radiada radiated interference
interferencia selectiva selective interference
interferencia sonora sound interference
interferencia transversal transverse interference
interferente interfering
interferometría de microondas microwave interferometry
interferometría ultrasónica ultrasonic interferometry
interferómetro interferometer
interferómetro acústico acoustic interferometer
interferómetro de espejo mirror interferometer
interferómetro de Fabry-Perot Fabry-Perot interferometer
interferómetro de Michelson Michelson interferometer
interferómetro de microondas microwave interferometer
interferómetro de polarización polarization interferometer
interferómetro multimodo multimode interferometer
interfono interphone
interfono transistorizado transistorized interphone
intermetálico intermetallic
intermitente intermittent
intermodulación intermodulation
intermodulación de haces beam intermodulation
intermodular intermodulate
intermolecular intermolecular
interno internal
interpolación interpolation
interpolación lineal linear interpolation
interpolador interpolator
interpolar interpolate
interpolo interpole
interportadora intercarrier
interpretador interpreter
interpretador de lenguaje language interpreter
interpretador de tarjetas card interpreter
interpretador de tarjetas perforadas punched-card interpreter
interpretativo interpretive
interrelación interfix
interresonancia interresonance
interrogación interrogation
interrogación por impulsos pulse interrogation
interrogador interrogator
interrogador-respondedor interrogator-responder
interrumpible interruptable
interrumpido interrupted
interrumpir interrupt
interrupción interruption, chopping, outage, break

interrupción automática automatic interruption, automatic outage
interrupción condicional conditional breakpoint
interrupción controlada por programa program-controlled interruption
interrupción corta short break
interrupción de alimentación power outage
interrupción de cadena en margarita daisy chain interrupt
interrupción de circuito circuit dropout
interrupción de comprobación de máquina machine-check interruption
interrupción de contingencia contingency interrupt
interrupción de corriente alterna alternating-current interruption
interrupción de entrada-salida input-output interruption
interrupción de equipo físico hardware interrupt
interrupción de error error interrupt
interrupción de línea de base base line break
interrupción de máquina machine interruption
interrupción de multiprogramación multiprogramming interrupt
interrupción de paridad parity interrupt
interrupción de periférico peripheral interrupt
interrupción de prioridad priority interrupt
interrupción de procesador processor interrupt
interrupción de proceso process interrupt
interrupción de programa program interruption
interrupción de programa automática automatic program interrupt
interrupción de rutina de control control routine interrupt
interrupción de servicio outage
interrupción de sistema system interrupt
interrupción de transmisión transmission interruption
interrupción dependiente de procesador processor-dependent interrupt
interrupción externa external interruption
interrupción forzada forced interruption, forced outage
interrupción forzada temporal temporary forced outage
interrupción forzada transitoria transient forced outage
interrupción interna internal interruption
interrupción involuntaria involuntary interrupt
interrupción momentánea momentary interruption
interrupción múltiple multiple interrupt
interrupción no planificada unplanned outage
interrupción no programada unscheduled interruption, unscheduled outage
interrupción planificada planned outage
interrupción planificada extendida extended planned outage
interrupción por error de procesador processor-error interrupt
interrupción por error de programa program-error interrupt
interrupción por falla de alimentación power-failure interrupt
interrupción programada scheduled interruption, scheduled outage
interrupción secundaria secondary outage
interrupción sostenida sustained interruption
interrupción temporal temporary interruption
interruptor interrupter, switch, cutout, breaker
interruptor accionado por pérdida de conexión a tierra ground-fault interrupter
interruptor de circuito circuit interrupter
interruptor de contacto contact breaker
interruptor de corriente current interrupter

interruptor de diapasón tuning-fork interrupter
interruptor de Foucault Foucault interrupter
interruptor de línea line interrupter
interruptor de Wehnelt Wehnelt interrupter
interruptor electrolítico electrolytic interrupter
interruptor periódico chopper
interruptor periódico de impulsos pulse chopper
interruptor periódico de instrumento instrument chopper
interruptor periódico de luz light chopper
interruptor periódico fotoconductivo photoconductive chopper
interruptor periódico sincrónico synchronous chopper
interruptor rotativo rotary interrupter
interruptor rotativo sincrónico synchronous rotary interrupter
intersección intersection
intervalo interval, range, span, gap
intervalo audible audible range
intervalo corto short interval
intervalo de amplitud amplitude range
intervalo de antena antenna range
intervalo de audiofrecuencia audio-frequency range
intervalo de barrido de frecuencia frequency-sweep range
intervalo de bloqueo blocking interval
intervalo de calentamiento warm-up interval
intervalo de calibración calibration interval
intervalo de caracteres character interval
intervalo de conducción conduction interval
intervalo de confianza confidence interval
intervalo de conmutación commutation interval
intervalo de conteo counting interval
intervalo de contestación answering interval
intervalo de contraste contrast range
intervalo de control control interval
intervalo de corriente básico basic current range
intervalo de corriente de referencia reference current range
intervalo de demanda demand interval
intervalo de desbloqueo unblanking interval
intervalo de enganche lock-in range
intervalo de errores error range
intervalo de escala scale range
intervalo de extinción blanking interval
intervalo de extinción de línea line-blanking interval
intervalo de frecuencia portadora carrier-frequency range
intervalo de frecuencias frequency range
intervalo de guarda guard interval
intervalo de impedancias impedance range
intervalo de impulsos pulse interval
intervalo de impulsos de sincronización synchronizing-pulse interval
intervalo de manipulación keying interval
intervalo de mantenimiento maintenance interval
intervalo de medida measuring range
intervalo de muestreo sampling interval
intervalo de Nyquist Nyquist interval
intervalo de operación operate interval
intervalo de orientación orientation range
intervalo de prueba test interval
intervalo de reconexión reclosing interval
intervalo de recuperación inversa reverse recovery interval
intervalo de reflexión reflection interval
intervalo de repetición de impulsos pulse repetition interval
intervalo de reposición reset interval
intervalo de retorno return interval
intervalo de silencio silent interval
intervalo de sintonización tuning range

intervalo de sintonización electrónica electronic tuning range
intervalo de soldadura weld interval
intervalo de sondeo polling interval
intervalo de temperatura temperature interval
intervalo de temperatura ambiente ambient-temperature range
intervalo de temperatura de operación operating-temperature range
intervalo de temperatura de operación ambiente ambient operating-temperature range
intervalo de temperatura intrínseca intrinsic temperature range
intervalo de tensión voltage range
intervalo de tensión básica basic voltage range
intervalo de tensión de modo común common-mode voltage range
intervalo de tiempo time interval
intervalo de trabajo marking interval
intervalo de transposición transposition interval
intervalo de trazo trace interval
intervalo de velocidades speed range
intervalo diferencial differential gap
intervalo entre dígitos interdigit interval
intervalo espectral spectrum interval
intervalo mínimo minimum interval
intervalo muerto dead interval
intervalo principal main gap
intervalo significativo significant interval
intervalo total total range
intervalo unitario unit interval
intervalo vertical vertical interval
intervalómetro intervalometer
intrínseco intrinsic
invar invar
inventario de datos data inventory
inversamente inversely
inversión inversion, reversal
inversión automática automatic reversing
inversión de barrido sweep reversal
inversión de canales channel reversal
inversión de corriente reversal of current
inversión de dígito alternativo alternate digit inversion
inversión de ejes axis inversion
inversión de fase phase inversion, phase reversal
inversión de imagen picture inversion
inversión de marca codificada coded mark inversion
inversión de polaridad polarity reversing
inversión de polarización polarization reversal
inversión de polos polarity reversing
inversión de secuencia de fases phase-sequence reversal
inversión de señal signal inversion
inversión de temperatura temperature inversion
inversión de voz speech inversion
inversión lateral lateral inversion
inversión magnética magnetic reversal
inversión termoeléctrica thermoelectric inversion
inverso inverse
inversor inverter, reverser
inversor de arco de mercurio mercury-arc inverter
inversor de corriente current reverser
inversor de corriente continua direct-current inverter
inversor de fase phase inverter
inversor de fase acústico acoustic phase inverter
inversor de fase de carga dividida split-load phase inverter
inversor de fase de Schmitt Schmitt phase inverter
inversor de frecuencia frequency inverter
inversor de impedancia impedance inverter
inversor de impulsos pulse inverter

inversor de interferencia interference inverter
inversor de polaridad polarity reverser
inversor de polos pole reverser
inversor de potencia power inverter
inversor de relajación relaxation inverter
inversor de señal signal inverter
inversor de sincronización synchronizing inverter
inversor de tensión voltage inverter
inversor de tipo de vibrador vibrator-type inverter
inversor de voz speech inverter
inversor electrónico electronic inverter
inversor magnético magnetic inverter
inversor monofásico single-phase inverter
inversor parafásico paraphase inverter
inversor parafásico flotante floating paraphase
    inverter
inversor rotativo rotary inverter
inversor sincrónico synchronous inverter
inversor tiratrónico thyratron inverter
invertido inverted, reversed
invertir invert, revert
investigación básica basic research
inyección injection
inyección catódica cathode injection
inyección de electrones electron injection
inyección de huecos hole injection
inyección de papel paper injection
inyección de portadora luminosa light-carrier
    injection
inyección de portadores carrier injection
inyección de portadores minoritarios
    minority-carrier injection
inyección de señal signal injection
inyección por rejilla de control control-grid
    injection
inyección por rejilla exterior outer-grid injection
inyector injector
inyector de electrones electron injector
inyector de huecos hole injector
inyector de papel paper injector
inyector de señal signal injector
inyector-extractor de papel paper injector-extractor
ion ion
ion de impureza impurity ion
ion de Langevin Langevin ion
ion negativo negative ion
ion pequeño small ion
ion positivo positive ion
ion primario primary ion
ion residual residual ion
ion secundario secondary ion
iónico ionic
ionización ionization
ionización acumulativa cumulative ionization
ionización artificial artificial ionization
ionización atmosférica atmospheric ionization
ionización de campo field ionization
ionización de capa E esporádica sporadic-E layer
    ionization
ionización de Townsend Townsend ionization
ionización electrolítica electrolytic ionization
ionización específica specific ionization
ionización esporádica sporadic ionization
ionización lineal linear ionization
ionización meteórica meteoric ionization
ionización mínima minimum ionization
ionización múltiple multiple ionization
ionización por impacto impact ionization
ionización por radiación radiation ionization
ionización primaria primary ionization
ionización proporcional proportional ionization
ionización residual residual ionization
ionización secundaria secondary ionization

ionización térmica thermal ionization
ionización total total ionization
ionizado ionized
ionizado positivamente positively ionized
ionizante ionizing
ionizar ionize
ionófono ionophone
ionógeno ionogen
ionograma ionogram
ionómetro ionometer
ionómetro de registro recording ionometer
ionósfera ionosphere
ionosférico ionospheric
ionosonda ionosonde
ir a go to
iraser iraser
iridio iridium
iridiscencia iridescence
iris iris
iris de guíaondas waveguide iris
iris resonante resonant iris
irradiación irradiation
irradiación específica specific irradiation
irradiación espectral spectral irradiation
irradiado irradiated
irradiancia irradiance
irradiancia espectral spectral irradiance
irradiar irradiate
irrecuperable nonrecoverable
irregular irregular
irregularidad irregularity
irregularidad cíclica cyclic irregularity
irregularidad de impedancia impedance irregularity
irregularidades ionosféricas ionospheric
    irregularities
irreversible irreversible
isocromático isochromatic
isocrona isochrone
isocronismo isochronism
isócrono isochronous
isodosis isodose
isoeléctrico isoelectric
isogono isogonal
isolantita isolantite
isolito isolith
isolux isolux
isomagnético isomagnetic
isómero isomer
isoplanar isoplanar
isotérmico isothermal
isotópico isotopic
isótopo isotope
isótopo pesado heavy isotope
isótopo radiactivo radioactive isotope
isotrón isotron
isotrónica isotronics
isotrópico isotropic
ítem item
iteración iteration
iterativo iterative
iterbio ytterbium
itria yttria
itrio yttrium

J

jack jack
jack banana banana jack
jack coaxial coaxial jack
jack concéntrico concentric jack
jack de audífono phone jack
jack de circuito abierto open-circuit jack
jack de conmutación switching jack
jack de contestación answering jack
jack de lámpara lamp jack
jack de línea line jack
jack de micrófono microphone jack
jack de patilla pin jack
jack de prueba test jack
jack de punta tip jack
jack de ruptura break jack
jack de salida output jack
jack de teléfono phone jack
jack de transferencia transfer jack
jack de tres conductores three-conductor jack
jack fonográfico phono jack
jack local local jack
jack múltiple multiple jack
jack telefónico telephone jack
jaula cage
jaula de Faraday Faraday cage
jerarquía hierarchy
jerarquía de almacenamiento storage hierarchy
jerarquía de datos data hierarchy
jerarquía de gráficos graphics hierarchy
jerarquía de memoria memory hierarchy
jerárquico hierarchical
jirafa boom
joule joule
joule internacional international joule
juego kit
juego de adaptadores adapter kit
juego de alineación alignment kit
juego de tubos tube kit
junta joint, gasket
junta aislante insulating joint
junta blindada shielded joint
junta de acoplamiento de guíaondas waveguide
   shim
junta de alta resistencia high-resistance joint
junta de bridas flange joint
junta de cable cable joint
junta de choque choke joint
junta de gas gas-filled joint
junta de guíaondas waveguide joint
junta de resina rosin joint
junta de solape lap joint
junta en T tee joint
junta hermética seal
junta hermética de cerámica a metal
   ceramic-to-metal seal
junta mecánica mechanical joint
junta multifibra multifiber joint
junta múltiple multiple joint
junta rotativa rotary joint
junta soldada soldered joint
junta torcida twisted joint

K

L

kalirotrón kallirotron
kappa kappa
kenopliotrón kenopliotron
kenotrón kenotron
keraunófono keraunophone
kilo kilo
kiloampere kiloampere
kilobaud kilobaud
kilobit kilobit
kilobyte kilobyte
kilocaloría kilocalorie
kilociclo kilocycle
kilocurie kilocurie
kiloelectrónvolt kiloelectronvolt
kilogauss kilogauss
kilogramo kilogram
kilohenry kilohenry
kilohertz kilohertz
kilolínea kiloline
kilolumen kilolumen
kilomaxwell kilomaxwell
kilométrico kilometric
kilómetro kilometer
kilooersted kilooersted
kiloohm kiloohm
kilorroentgen kiloroentgen
kilosegundo kilosecond
kilovar kilovar
kilovar-hora kilovar-hour
kilovolt kilovolt
kilovolt-ampere kilovolt-ampere
kilovolt-ampere reactivo reactive kilovolt-ampere
kilovóltmetro kilovoltmeter
kilowatt kilowatt
kilowatt-hora kilowatt-hour
klistrón klystron
klistrón de ánodo modulador modulating-anode klystron
klistrón de haces múltiples multiple-beam klystron
klistrón de potencia power klystron
klistrón de radar radar klystron
klistrón de tres cavidades three-cavity klystron
klistrón enfriado por agua water-cooled klystron
klistrón mezclador mixer klystron
klistrón multiplicador de frecuencia frequency-multiplier klystron
klistrón oscilador oscillator klystron
klistrón oscilante oscillating klystron
klistrón pulsado pulsed klystron
klistrón reflejo reflex klystron
klistrón toroidal toroidal klystron
kovar kovar
krarupización krarupization

laberinto acústico acoustic labyrinth
lábil labile
laboratorio laboratory
laboratorio de calibración calibration laboratory
ladar ladar
lado de baja frecuencia low-frequency side
lado de bobina coil side
lado de línea line side
lado de sección section side
lado negativo negative side
lado positivo positive side
lambda lambda
lambert lambert
lambert-pie foot-lambert
lámina metálica foil
laminación lamination
laminaciones recocidas annealed laminations
laminado laminated
laminilla reed
laminilla vibrante vibrating reed
laminografía laminography
lámpara lamp
lámpara bactericida bactericidal lamp
lámpara blanca white lamp
lámpara con reflector reflector lamp
lámpara de alarma alarm lamp
lámpara de arco arc lamp
lámpara de arco abierto open-arc lamp
lámpara de arco al vacío vacuum-arc lamp
lámpara de arco concentrado concentrated-arc lamp
lámpara de arco de carbón carbon-arc lamp
lámpara de arco de llama flame arc lamp
lámpara de arco de proyección projector arc lamp
lámpara de arco de tungsteno tungsten-arc lamp
lámpara de aviso warning lamp
lámpara de cadmio cadmium lamp
lámpara de carbón carbon lamp
lámpara de cátodo caliente hot-cathode lamp
lámpara de cátodo frío cold-cathode lamp
lámpara de circonio zirconium lamp
lámpara de circuito impreso printed-circuit lamp
lámpara de comparación comparison lamp
lámpara de contestación answer lamp
lámpara de cráter crater lamp
lámpara de cuadrante dial lamp
lámpara de cuarzo quartz lamp
lámpara de cuarzo fundido fused-quartz lamp
lámpara de cuarzo-halógeno quartz-halogen lamp
lámpara de cuarzo-yodo quartz-iodine lamp
lámpara de descarga discharge lamp
lámpara de descarga de alta intensidad high-intensity discharge lamp
lámpara de descarga de cátodo caliente hot-cathode discharge lamp
lámpara de descarga de cátodo frío cold-cathode discharge lamp
lámpara de descarga eléctrica electric-discharge lamp
lámpara de descarga gaseosa gas-discharge lamp
lámpara de descarga luminiscente glow-discharge lamp

lámpara de descarga luminosa luminous-discharge lamp
lámpara de destello flash lamp, flashbulb
lámpara de destello electrónico electronic flash lamp
lámpara de destellos flashing lamp
lámpara de destellos de xenón xenon flash lamp
lámpara de destellos repetitiva repeating flash tube
lámpara de diodo diode lamp
lámpara de encendido rápido rapid-start lamp
lámpara de estado sólido solid-state lamp
lámpara de exploración scanning lamp
lámpara de filamento filament lamp
lámpara de filamento de carbón carbon-filament lamp
lámpara de filamento de gas gas-filled filament lamp
lámpara de filamento de tungsteno tungsten-filament lamp
lámpara de filamento eléctrico electric filament lamp
lámpara de filamento incandescente incandescent filament lamp
lámpara de filamento metálico metal-filament lamp
lámpara de fotodestellos photoflash lamp
lámpara de galvanómetro galvanometer lamp
lámpara de gas gas-filled lamp
lámpara de Hefner Hefner lamp
lámpara de helio helium lamp
lámpara de hidrógeno hydrogen lamp
lámpara de infrarrojos infrared lamp
lámpara de instrumentos instrument lamp
lámpara de lectura readout lamp
lámpara de línea line lamp
lámpara de luz mixta mixed-light lamp
lámpara de luz solar sunlight lamp
lámpara de llama flame lamp
lámpara de mano hand lamp
lámpara de medidor meter lamp
lámpara de mercurio mercury lamp
lámpara de mercurio de baja presión low-pressure mercury lamp
lámpara de microdestellos microflash lamp
lámpara de Moore Moore lamp
lámpara de navegación navigation lamp
lámpara de neón neon lamp
lámpara de Nernst Nernst lamp
lámpara de nitrógeno nitrogen lamp
lámpara de panel panel lamp
lámpara de parada stop lamp
lámpara de precisión precision lamp
lámpara de proyección projection lamp
lámpara de prueba test lamp
lámpara de rayos catódicos cathode-ray lamp
lámpara de referencia reference lamp
lámpara de registro recording lamp
lámpara de repetidor repeater lamp
lámpara de resistencia resistance lamp
lámpara de resonancia resonance lamp
lámpara de seguridad safety lamp
lámpara de señal signal lamp
lámpara de señalización signaling lamp
lámpara de sodio de alta presión high-pressure sodium lamp
lámpara de sodio de baja presión low-pressure sodium lamp
lámpara de tantalio tantalum lamp
lámpara de tungsteno tungsten lamp
lámpara de uviol uviol lamp
lámpara de vacío vacuum lamp
lámpara de vapor vapor lamp
lámpara de vapor de cesio cesium-vapor lamp
lámpara de vapor de mercurio mercury-vapor lamp

lámpara de vapor de sodio sodium-vapor lamp
lámpara de vapor metálico metal-vapor lamp
lámpara de Wood Wood's lamp
lámpara desnuda bare lamp
lámpara eléctrica electric lamp
lámpara eléctrica incandescente incandescent electric lamp
lámpara electroluminiscente electroluminescent lamp
lámpara en hongo mushroom lamp
lámpara en serie series lamp
lámpara enchufable plug-in lamp
lámpara estabilizadora ballast lamp
lámpara estroboscópica stroboscopic lamp, strobe light
lámpara estrobotrón strobotron lamp
lámpara excitadora exciter lamp
lámpara expuesta exposed lamp
lámpara fluorescente fluorescent lamp
lámpara fluorescente de baja temperatura low-temperature fluorescent lamp
lámpara fluorescente de bajo brillo low-brightness fluorescent lamp
lámpara fluorescente de encendido rápido rapid-start fluorescent lamp
lámpara fluorescente de mercurio mercury fluorescent lamp
lámpara incandescente incandescent lamp
lámpara incandescente tubular tubular incandescent lamp
lámpara indicadora indicator lamp
lámpara indicadora de dirección direction indicator lamp
lámpara indicadora de sobrecarga overload indicator lamp
lámpara lineal linear lamp
lámpara luminiscente glow lamp
lámpara luminiscente de argón argon glow lamp
lámpara luminiscente de neón neon glow lamp
lámpara mateada frosted lamp
lámpara metalizada metallized lamp
lámpara microminiatura microminiature lamp
lámpara miniatura miniature lamp
lámpara normal normal lamp
lámpara opalina opal lamp
lámpara parpadeante blinking lamp
lámpara patrón standard lamp
lámpara permanente permanent lamp
lámpara piloto pilot lamp
lámpara piroeléctrica pyroelectric lamp
lámpara portátil portable lamp
lámpara prefoco prefocus lamp
lámpara protectora protecting lamp
lámpara supervisora supervisory lamp
lámpara tubular tubular lamp
lámpara ultravioleta ultraviolet lamp
lanac lanac
lantanio lanthanum
lanzamiento launching
lapidado lapping
lápiz conductor conductive pencil
lápiz óptico light pen
lápiz óptico de gráficos graphics light pen
lapso lapse
larga distancia long distance
larga duración long duration
laringófono laryngophone
láser laser
láser alimentado por luz solar sunlight-powered laser
láser con bomba solar sun-pump laser
láser de almacenamiento storage laser
láser de argón argon laser

láser de arseniuro de galio gallium-arsenide laser
láser de cristal crystal laser
láser de cuatro niveles four-level laser
láser de diodo diode laser
láser de dióxido de carbono carbon dioxide laser
láser de Doppler Doppler laser
láser de dos niveles two-level laser
láser de estado sólido solid-state laser
láser de gas gas laser
láser de impulsos gigantes giant-pulse laser
láser de inyección injection laser
láser de luz verde green laser
láser de neodimio neodymium laser
láser de neón-helio neon-helium laser
láser de onda continua continuous-wave laser
láser de rayos X X-ray laser
láser de rubí ruby laser
láser de rubí pulsado pulsed ruby laser
láser de salida pulsada pulsed-output laser
láser de semiconductor semiconductor laser
láser de tres niveles three-level laser
láser de unión junction laser
láser de vidrio de neodimio neodymium-glass laser
láser enfriado por agua water-cooled laser
láser fonónico phonon laser
láser helio-neón helium-neon laser
láser iónico ion laser
láser iónico de criptón krypton ion laser
láser líquido liquid laser
láser magnetoóptico magnetooptical laser
láser multimodo multimode laser
láser pulsado pulsed laser
láser solar solar laser
latencia latency
latencia de interrupción interrupt latency
latencia media average latency
latencia mínima minimum latency
latente latent
lateral lateral
latón brass
laurencio lawrencium
lavado stripping
lazo loop
lazo de goteo drip loop
lazo equilibrado balanced loop
lector reader, pickup, player
lectora reader, pickup, player
lectora alfanumérica alphanumeric reader
lectora-clasificadora reader-sorter
lectora-clasificadora de código de barras bar code
    reader-sorter
lectora de alta velocidad high-speed reader
lectora de caracteres character reader
lectora de caracteres de tinta magnética magnetic
    ink character reader
lectora de caracteres fotoeléctrica photoelectric
    character reader
lectora de caracteres magnéticos magnetic character
    reader
lectora de caracteres óptica optical character reader
lectora de cinta tape reader
lectora de cinta de papel paper-tape reader
lectora de cinta fotoeléctrica photoelectric tape
    reader
lectora de cinta magnética magnetic-tape reader
lectora de cinta perforada perforated-tape reader
lectora de códigos code reader
lectora de códigos de barras ópticas optical bar
    code reader
lectora de documentos document reader
lectora de documentos óptica optical document
    reader
lectora de entrada input reader

lectora de escritura handwriting reader
lectora de etiquetas tag reader
lectora de mandos command reader
lectora de marcas mark reader
lectora de marcas óptica optical mark reader
lectora de microfichas microfiche reader
lectora de microformas microform reader
lectora de páginas page reader
lectora de película film reader
lectora de tarjetas card reader
lectora de tarjetas de identificación
    identification-card reader
lectora de tarjetas fotoeléctrica photoelectric card
    reader
lectora de tarjetas magnéticas magnetic card reader
lectora de tarjetas perforadas punched-card reader
lectora de tarjetas virtual virtual card reader
lectora en serie serial reader
lectora electrónica electronic reader
lectora fotoeléctrica photoelectric reader
lectora fotográfica photo reader
lectora-impresora reader-printer
lectora interna internal reader
lectora-interpretadora reader-interpreter
lectora magnética magnetic reader
lectora óptica optical reader
lectora óptica alfanumérica alphanumeric optical
    reader
lectora residente resident reader
lectura readout, reading, sensing, playback
lectura absoluta absolute readout
lectura acústica acoustic reading
lectura alfanumérica alphanumeric readout
lectura de aguja pointer reading
lectura de caracteres character reading
lectura de comprobación check reading
lectura de doble impulso double-pulse reading
lectura de instrumento instrument reading
lectura de instrumento media mean instrument
    reading
lectura de límite de escala full-scale reading
lectura de marcas mark reading
lectura de marcas óptica optical mark reading
lectura de medidor meter reading, meter readout
lectura de pantalla screen read
lectura de registro register reading
lectura defectuosa faulty reading
lectura destructiva destructive reading
lectura digital digital readout, digital reading
lectura directa direct reading
lectura dispersa scatter read
lectura en línea inline readout
lectura en paralelo parallel reading
lectura en serie serial reading
lectura-escritura read-write
lectura inversa reverse reading
lectura magnética magnetic readout
lectura no destructiva nondestructive reading
lectura numérica numerical readout
lectura parcial partial read
lectura por espejo mirror reading
lectura regenerativa regenerative reading
lectura remota remote reading, remote readout
lectura sin carga no-load reading
lectura visual visual readout
lechada grout
legibilidad readability
legible readable
legible por máquina machine readable
lenguaje language
lenguaje absoluto absolute language
lenguaje algebraico algebraic language
lenguaje algorítmico algorithmic language

lenguaje artificial artificial language
lenguaje básico basic language
lenguaje binario binary language
lenguaje científico scientific language
lenguaje comercial commercial language
lenguaje común common language
lenguaje conversacional conversational language
lenguaje de alto nivel high-level language
lenguaje de bajo nivel low-level language
lenguaje de compilador compiler language
lenguaje de computadora computer language
lenguaje de control control language
lenguaje de control de operación operation control language
lenguaje de control de procesos process-control language
lenguaje de control de programación lineal linear programming control language
lenguaje de control de trabajos job control language
lenguaje de control remoto remote-control language
lenguaje de conversión de formato format-conversion language
lenguaje de cuarta generación fourth generation language
lenguaje de definición de datos data definition language
lenguaje de desarrollo de aplicaciones application development language
lenguaje de descripción de datos data description language
lenguaje de descripción de equipo físico hardware description language
lenguaje de descripción de página page description language
lenguaje de diseño design language
lenguaje de diseño de equipo físico hardware design language
lenguaje de diseño de programa program design language
lenguaje de diseño de sistema system design language
lenguaje de ensamblaje assembly language
lenguaje de ensamblaje de microprograma microprogram assembly language
lenguaje de ensamblaje simbólico symbolic assembly language
lenguaje de entrada input language
lenguaje de equipo físico hardware language
lenguaje de generador generator language
lenguaje de gráficos graphics language
lenguaje de macroensamblaje macro-assembly language
lenguaje de mandos command language
lenguaje de mandos de base de datos database command language
lenguaje de mandos de programas software command language
lenguaje de manipulación de datos data manipulation language
lenguaje de máquina machine language
lenguaje de máquina común common machine language
lenguaje de nivel superior higher-level language
lenguaje de orden alto high-order language
lenguaje de orden superior higher-order language
lenguaje de primera generación first generation language
lenguaje de problemas problem language
lenguaje de procedimientos procedural language
lenguaje de procesamiento de listas list-processing language
lenguaje de producción production language
lenguaje de programa program language

lenguaje de programación programming language
lenguaje de programación concurrente concurrent programming language
lenguaje de programación de alto nivel high-level programming language
lenguaje de programación de aplicaciones application programming language
lenguaje de programación de bajo nivel low-level programming language
lenguaje de programación de computadora computer programming language
lenguaje de programación estructurado structured programming language
lenguaje de programación lógica logic programming language
lenguaje de quinta generación fifth generation language
lenguaje de referencia reference language
lenguaje de salida output language
lenguaje de secuencia string language
lenguaje de segunda generación second generation language
lenguaje de sistema system language
lenguaje de sistema de computación remota remote-computing system language
lenguaje de sistema de computadora remoto remote computer system language
lenguaje de solución de problemas problem-solving language
lenguaje de tercera generación third-generation language
lenguaje de transferencia de registros register-transfer language
lenguaje de uso especial special-purpose language
lenguaje declarativo declarative language
lenguaje efectivo effective language
lenguaje ejecutivo executive language
lenguaje ensamblador assembler language
lenguaje formal formal language
lenguaje fuente source language
lenguaje funcional functional language
lenguaje independiente de computadoras computer independent language
lenguaje independiente de máquina machine-independent language
lenguaje interactivo interactive language
lenguaje intermedio intermediate language
lenguaje interpretativo interpretive language
lenguaje mnemónico mnemonic language
lenguaje objeto target language
lenguaje orientado a aplicaciones application-oriented language
lenguaje orientado a computadoras computer-oriented language
lenguaje orientado a máquina machine-oriented language
lenguaje orientado a objetos object-oriented language
lenguaje orientado a problemas problem-oriented language
lenguaje orientado a procedimientos procedure-oriented language
lenguaje orientado a procesos process-oriented language
lenguaje orientado a pruebas test-oriented language
lenguaje orientado al usuario user-oriented language
lenguaje relacional relational language
lenguaje simbólico symbolic language
lenguaje sintético synthetic language
lengüeta reed
lengüeta de conexión lug
lengüeta sintonizada tuned reed

lente lens
lente acromática achromatic lens
lente acústica acoustic lens
lente anastigmática anastigmatic lens
lente aplanadora de campo field-flattener lens
lente compuesta compound lens
lente convergente converging lens
lente cromática chromatic lens
lente de abertura aperture lens
lente de antena antenna lens
lente de antena metálica metallic antenna lens
lente de cuarzo quartz lens
lente de enfoque focusing lens
lente de Fresnel Fresnel lens
lente de guíaondas waveguide lens
lente de Luneberg Luneberg lens
lente de microondas microwave lens
lente de placas paralelas parallel-plate lens
lente de proyección projection lens
lente de retardo delay lens
lente dieléctrica dielectric lens
lente dieléctrica artificial artificial dielectric lens
lente divergente diverging lens
lente electromagnética electromagnetic lens
lente electrónica electron lens
lente electrostática electrostatic lens
lente esférica spherical lens
lente magnética magnetic lens
lente metálica metallic lens
lente óptica optical lens
lenticular lenticular
letra de código code letter
letra de unidad drive letter
letras de llamada call letters
levantador lifter
levantador de cinta tape lifter
levantamiento geoeléctrico geoelectric survey
levantatubos tube puller, tube lifter
levantaválvulas valve lifter
léxico automatizado automated lexicon
ley A A-law
ley de Ampere Ampere's law
ley de atracción electrostática law of electrostatic
  attraction
ley de Beer Beer's law
ley de Biot-Savart Biot-Savart Law
ley de Bragg Bragg's law
ley de Brewster Brewster's law
ley de cargas law of charges
ley de cargas eléctricas law of electric charges
ley de Child Child's law
ley de compresión-expansión companding law
ley de corrientes de Kirchhoff Kirchhoff's current
  law
ley de corrientes inducidas law of induced current
ley de Coulomb Coulomb's law
ley de Curie Curie's law
ley de Curie-Weiss Curie-Weiss law
ley de desplazamiento de Wien Wien's displacement
  law
ley de distribución normal law of normal
  distribution
ley de Einstein Einstein's law
ley de emisión del coseno cosine emission law
ley de Faraday Faraday's law
ley de Faraday de inducción Faraday's law of
  induction
ley de Fleming-Kennelly Fleming-Kennelly law
ley de Gauss Gauss' law
ley de Gudden-Pohl Gudden-Pohl law
ley de Hooke Hooke's law
ley de inducción law of induction
ley de inducción electromagnética law of

electromagnetic induction
ley de Jacob Jacob's law
ley de Joule Joule's law
ley de Kelvin Kelvin's law
ley de Kennelly Kennelly's law
ley de Kirchhoff Kirchhoff's law
ley de Kundt Kundt's law
ley de Lambert Lambert's law
ley de Lambert de iluminación Lambert's law of
  illumination
ley de Langmuir Langmuir's law
ley de Laplace Laplace's law
ley de lentes law of lenses
ley de Lenz Lenz's law
ley de Lorentz Lorentz law
ley de Lorentz-Lorentz Lorentz-Lorentz law
ley de magnetismo law of magnetism
ley de Maxwell Maxwell's law
ley de Moseley Moseley's law
ley de Neumann Neumann's law
ley de Ohm Ohm's law
ley de Paschen Paschen's law
ley de Pfeffer law of Pfeffer
ley de Planck Planck's law
ley de promedios law of averages
ley de radiación law of radiation
ley de radiación de Kirchhoff Kirchhoff's radiation
  law
ley de radiación de Planck Planck's radiation law
ley de radiación de Wien Wien's radiation law
ley de Rayleigh Rayleigh law
ley de Rayleigh-Jeans Rayleigh-Jeans law
ley de reflexión law of reflection
ley de refracción law of refraction
ley de repulsión electrostática law of electrostatic
  repulsion
ley de Sabine Sabine's law
ley de sistemas electromagnéticos law of
  electromagnetic systems
ley de Snell Snell's law
ley de Stefan Stefan's law
ley de Stefan-Boltzmann Stefan-Boltzmann law
ley de Stokes Stokes' law
ley de Talbot Talbot's law
ley de tensiones de Kirchhoff Kirchhoff's voltage
  law
ley de Volta Volta's law
ley de Wiedemann-Franz Wiedemann-Franz law
ley de Wien Wien's law
ley del coseno cosine law
ley del coseno de Lambert Lambert's cosine law
ley del seno sine law
ley inversa de cuadrados law of inverse squares
leyes de Faraday de electrólisis Faraday's laws of
  electrolysis
leyes de moción de Newton Newton's laws of
  motion
leyes termoeléctricas thermoelectric laws
liberación release
liberación de papel paper release
liberación forzada forced release
liberación térmica thermal release
liberar la línea clear the line
libra-pie foot-pound
libra-pulgada inch-pound
libras por pulgada cuadrada pounds per square inch
libre free, idle
libre de bloqueos lockout-free
libre de mantenimiento maintenance-free
libre de paralaje parallax-free
libre de pérdidas loss-free
libre de ruido noise-free
libre de vapor vapor-free

libre de zumbido hum-free
licencia avanzada advanced license
lidar lidar
ligador binder
ligadura tie
limen limen
limen de diferencia difference limen
limitación limitation, limiting, clipping
limitación de amplitud amplitude limiting
limitación de base base limiting
limitación de carga automática automatic load limitation
limitación de corriente current limiting
limitación de corte cutoff limiting
limitación de diseño design limitation
limitación de impulsos pulse limiting
limitación de modulación modulation limiting
limitación de picos peak limiting
limitación de picos de potencia power-peak limitation
limitación de picos transitorios surge limiting
limitación de ruido noise limiting
limitación de ruido automática automatic noise limiting
limitación de salida output limiting
limitación finita finite clipping
limitación por carga espacial space-charge limitation
limitación por emisión emission limitation
limitación por rejilla grid limiting
limitación por saturación saturation limiting
limitado limited, bound
limitado por cinta tape-bound
limitado por computadora computer-limited
limitado por entrada input-bound
limitado por entrada-salida input-output limited
limitado por impresión print-bound
limitado por impresora printer-bound
limitado por periférico peripheral-bound
limitado por procesador processor-bound
limitado por procesador central central-processor limited
limitado por proceso process-bound
limitado por salida output-bound
limitado por temperatura temperature-limited
limitador limiter
limitador automático automatic limiter
limitador de amplitud amplitude limiter
limitador de audio audio limiter
limitador de banda angosta narrowband limiter
limitador de base base limiter
limitador de corriente current limiter
limitador de corriente automático automatic current limiter
limitador de demanda demand limiter
limitador de diodo diode limiter
limitador de doble diodo double-diode limiter
limitador de ferrita ferrite limiter
limitador de modulación modulation limiter
limitador de modulación de frecuencia frequency-modulation limiter
limitador de picos peak limiter
limitador de picos automático automatic peak limiter
limitador de picos de audio audio peak limiter
limitador de picos de audiofrecuencia audio-frequency peak limiter
limitador de potencia power limiter
limitador de puente bridge limiter
limitador de rejilla grid limiter
limitador de ruido noise limiter
limitador de ruido automático automatic noise limiter

limitador de ruido de diodos diode noise limiter
limitador de ruido dinámico dynamic noise limiter
limitador de salida output limiter
limitador de Schmitt Schmitt limiter
limitador de señal signal limiter
limitador de tensión voltage limiter
limitador de velocidad speed limiter
limitador de velocidad de subida rate-of-rise limiter
limitador de volumen volume limiter
limitador de voz speech limiter
limitador dinámico dynamic limiter
limitador en cascada cascade limiter
limitador en derivación shunt limiter
limitador en paralelo parallel limiter
limitador en serie series limiter
limitador ferrimagnético ferrimagnetic limiter
limitador giromagnético gyromagnetic limiter
limitador inverso inverse limiter
limitador pasivo passive limiter
limitador recortador clipper limiter
limitador separador buffer limiter
límite limit, boundary
límite característico characteristic boundary
límite catódico cathode border
límite cuántico quantum limit
límite de banda band edge
límite de calentamiento heating limit
límite de carácter character boundary
límite de cilindro cylinder boundary
límite de corriente current limit
límite de detección detection limit
límite de durabilidad endurance limit
límite de error limit of error
límite de estabilidad stability limit
límite de estabilidad natural natural stability limit
límite de exploración scanning limit
límite de fatiga fatigue limit
límite de frecuencia frequency limit
límite de frecuencia superior upper frequency limit
límite de nitidez sharpness limit
límite de palabra word boundary
límite de potencia power limit
límite de presión pressure limit
límite de resolución limit of resolution
límite de salida output limit
límite de sobrerrecorrido overtravel limit
límite de temperatura de operación operating-temperature limit
límite de tensión voltage limit
límite de tensión de emergencia emergency voltage limit
límite de tensión normal normal voltage limit
límite de tiempo time limit
límite de transmisión transmission limit
límite de zona zone boundary
límite difuso diffuse boundary
límite elástico elastic limit
límite inferior lower limit
límite intergranular grain boundary
límite mecánico mechanical limit
límite PN PN boundary
límite superior upper limit
límites de capacidad capacity limits
límites de confianza confidence limits
límites de diseño design limits
límites de regulación setting range
limpiador de cintas tape cleaner
limpieza alcalina alkaline cleaning
limpieza anódica anode cleaning
limpieza catódica cathode cleaning
limpieza de archivo file cleanup
limpieza de gas gas cleanup
limpieza electrolítica electrolytic cleaning

limpieza ultrasónica ultrasonic cleaning
linac linac
línea line
línea abierta open line
línea abierta equilibrada balanced open line
línea aclínica aclinic line
línea activa active line
línea acústica acoustic line
línea adaptada matched line
línea aérea overhead line, open-wire line, overhead
    wire
línea agónica agonic line
línea aislada por aire air-insulated line
línea alimentadora feeder line
línea antirresonante antiresonant line
línea apantallada shielded line
línea arrendada leased line
línea artificial artificial line
línea artificial equilibradora duplex artificial line
línea auxiliar auxiliary line
línea base baseline
línea base asignada allocated baseline
línea base de referencia reference baseline
línea bifilar two-wire line
línea bifilar plana twin line
línea bipolar bipolar line
línea cargada busy line, loaded line
línea central centerline
línea coaxial coaxial line
línea compartida shared line, party line
línea con fugas leaky line
línea con múltiples conexiones a tierra
    multigrounded line
línea con pérdidas lossy line
línea con tomas tapped line
línea concéntrica concentric line
línea conmutada switched line
línea corriente current line
línea cortocircuitada short-circuited line
línea de abonado subscriber line
línea de absorción absorption line
línea de acceso access line
línea de adaptación matching stub
línea de aire air line
línea de alambre wire line
línea de alimentación supply line, feed line, power
    lead
línea de alta tensión high-tension line
línea de alta velocidad high-speed line
línea de altavoz loudspeaker line
línea de audio audio line
línea de bajada drop line
línea de barrido sweep line
línea de base base line
línea de bus bus line
línea de cable aéreo aerial cable line
línea de campo field line
línea de campo de control control field line
línea de campo eléctrico electric-field line
línea de campo magnético magnetic-field line
línea de carga load line
línea de centros line of centers
línea de circuito único single-circuit line
línea de codificación coding line
línea de código line of code
línea de comentarios comment line
línea de comunicación line of communication
línea de comunicación de audio audio
    communication line
línea de comunicaciones de datos data
    communications line
línea de concentración concentration line
línea de contacto contact line

línea de contacto aéreo aerial contact line
línea de contacto catenaria catenary contact line
línea de contestación answering line
línea de continuación continuation line
línea de control control line
línea de control secuencial sequential-control line
línea de convergencia convergence line
línea de conversión de datos data conversion line
línea de corriente alterna alternating-current line
línea de cuarto de longitud de onda
    quarter-wavelength line
línea de cuarto de onda quarter-wave line
línea de datos data line
línea de depuración debugging line
línea de derivación branch line, bypass line
línea de desbordamiento overflow line
línea de disipación dissipation line
línea de disipación de potencia constante
    constant-power dissipation line
línea de distribución distribution line
línea de energía power line
línea de enfasamiento phasing line
línea de enlace trunk line
línea de entrada input line
línea de entrada de interrupción interrupt input line
línea de escala scale line
línea de espacio de aire air-gap line
línea de espera waiting line
línea de estación station line
línea de estado status line
línea de excepción exception line
línea de exploración scanning line
línea de extensión extension line
línea de extensión artificial artificial extension line
línea de flujo flow line, line of flux
línea de formato format line
línea de fuente de alimentación power-supply line
línea de fuente de alimentación auxiliar auxiliary
    power supply line
línea de fuerza line of force
línea de fuerza electrostática electrostatic line of
    force
línea de fuerza magnética magnetic line of force
línea de guía guide line
línea de Guillemin Guillemin line
línea de hidrógeno hydrogen line
línea de hilo único single-wire line
línea de hilos paralelos parallel-wire line
línea de impedancia constante constant-impedance
    line
línea de impresión print line
línea de impulsos pulse line
línea de inducción line of induction
línea de información information line
línea de interconexión interconnection line
línea de llamada calling line
línea de llegada incoming line
línea de mandos command line
línea de media onda half-wave line
línea de mensajes message line
línea de nivel level line
línea de omisión omit line
línea de operación operating line
línea de órdenes order wire
línea de placas paralelas pillbox line
línea de pocas pérdidas low-loss line
línea de portadora carrier line
línea de posición position line
línea de programa program line
línea de propagación line of propagation
línea de pruebas test line
línea de punto a punto point-to-point line
línea de radiofrecuencia radio-frequency line

línea de Rayleigh Rayleigh line
línea de referencia reference line
línea de referencia de magnitud magnitude reference line
línea de reserva standby line
línea de resonancia resonance line
línea de retardo delay line
línea de retardo acústica acoustic delay line
línea de retardo acústica magnetoestrictiva magnetostrictive acoustic delay line
línea de retardo amplificadora amplifying delay line
línea de retardo artificial artificial delay line
línea de retardo con tomas tapped delay line
línea de retardo concentrada lumped delay line
línea de retardo de cuarzo quartz delay line
línea de retardo de difracción diffraction delay line
línea de retardo de mercurio mercury delay line
línea de retardo de microondas microwave delay line
línea de retardo de níquel nickel delay line
línea de retardo de video video delay line
línea de retardo eléctrica electric delay line
línea de retardo electromagnética electromagnetic delay line
línea de retardo en espiral spiral delay line
línea de retardo equivalente equivalent delay line
línea de retardo interdigital interdigital delay line
línea de retardo longitudinal longitudinal delay line
línea de retardo magnética magnetic delay line
línea de retardo magnética estática static magnetic delay line
línea de retardo magnetoestrictiva magnetostrictive delay line
línea de retardo sónica sonic delay line
línea de retardo ultrasónica ultrasonic delay line
línea de retardo variable variable delay line
línea de retorno return line, retrace line, flyback line
línea de retorno horizontal horizontal-retrace line
línea de rumbo course line
línea de salida outgoing line, output line
línea de salto jump line
línea de señal signal line
línea de señales bidireccional bidirectional signal line
línea de servicio service line
línea de telecomunicaciones telecommunications line
línea de teleimpresora teleprinter line
línea de televisión television line
línea de tensión constante constant-voltage line
línea de transmisión transmission line
línea de transmisión abierta open transmission line
línea de transmisión adaptada matched transmission line
línea de transmisión aérea open-wire transmission line
línea de transmisión apantallada shielded transmission line
línea de transmisión artificial artificial transmission line
línea de transmisión coaxial coaxial transmission line
línea de transmisión concéntrica concentric transmission line
línea de transmisión de cuarto de onda quarter-wave transmission line
línea de transmisión de datos data-transmission line
línea de transmisión de energía power transmission line
línea de transmisión de haz beam transmission line
línea de transmisión de hilo único single-wire transmission line
línea de transmisión de media onda half-wave transmission line
línea de transmisión de microondas microwave transmission line
línea de transmisión de ondas de superficie surface-wave transmission line
línea de transmisión de radiofrecuencia radio-frequency transmission line
línea de transmisión desequilibrada unbalanced transmission line
línea de transmisión equilibrada balanced transmission line
línea de transmisión exponencial exponential transmission line
línea de transmisión finita finite transmission line
línea de transmisión helicoidal helical transmission line
línea de transmisión impresa printed transmission line
línea de transmisión infinita infinite transmission line
línea de transmisión no resonante nonresonant transmission line
línea de transmisión no sintonizada untuned transmission line
línea de transmisión planar planar transmission line
línea de transmisión presurizada pressurized transmission line
línea de transmisión radial radial transmission line
línea de transmisión resonante resonant transmission line
línea de transmisión semirresonante semiresonant transmission line
línea de transmisión sin reflexión reflectionless transmission line
línea de transmisión uniforme uniform transmission line
línea de unión junction line
línea de visualizador display line
línea dedicada dedicated line
línea defectuosa faulty line
línea desequilibrada unbalanced line
línea directa direct line
línea disponible available line
línea dúplex duplex line
línea duplicada duplicate line
línea ecualizadora equalizing line
línea eléctrica electric line
línea eléctrica de alta confiabilidad high-reliability electric line
línea en blanco blank line
línea en circuito abierto open-circuited line
línea en derivación shunt line
línea energizada hot line
línea equilibrada balanced line
línea equilibradora balancing line
línea equipotencial equipotential line
línea espectral spectral line, spectrum line
línea exponencial exponential line
línea externa external line
línea fantasma phantom line
línea ficticia dummy line
línea fina hairline
línea física physical line
línea formadora de impulsos pulse-forming line
línea helicoidal helical line
línea imaginaria imaginary line
línea impar odd-numbered line
línea individual individual line
línea infinita infinite line
línea interna internal line
línea internacional international line
línea interurbana toll line
línea isoclínica isoclinic line
línea isodinámica isodynamic line

línea isogónica isogonic line
línea isolux isolux line
línea isomagnética isomagnetic line
línea larga long line
línea libre free line, idle line
línea llamada called line
línea local local line
línea lógica logical line
línea magnética magnetic line
línea medida metered line
línea monitora monitor line
línea muerta dead line
línea multipunto multipoint line
línea negativa negative line
línea neutra neutral line
línea no cargada nonloaded line
línea no resonante nonresonant line
línea no sintonizada untuned line
línea nula null line
línea oculta hidden line
línea ocupada busy line
línea par even line
línea periódica periodic line
línea piloto pilot line
línea positiva positive line
línea primaria primary line
línea principal main line
línea privada private line
línea pública public line
línea radial radial line
línea ranurada slotted line
línea real real line
línea remota remote line
línea resonante resonant line
línea resonante de cuarto de onda quarter-wave resonant line
línea resonante múltiple multiple-resonant line
línea ruidosa noisy line
línea rural rural line
línea secundaria secondary line
línea semirresonante semiresonant line
línea símplex simplex line
línea sin pérdidas lossless line
línea sintonizada tuned line
línea submarina submarine line
línea subterránea underground line
línea telefónica telephone line
línea telefónica equilibrada balanced telephone line
línea telefónica internacional international telephone line
línea telegráfica telegraph line
línea terminada terminated line
línea terrestre landline
línea única single line
línea uniforme uniform line
línea útil useful line
línea visual line-of-sight
línea X X line
línea Y Y line
lineal linear, lineal
linealidad linearity
linealidad de exploración scanning linearity
linealidad de fase phase linearity
linealidad de frecuencia frequency linearity
linealidad de ganancia gain linearity
linealidad de imagen picture linearity
linealidad de modulación modulation linearity
linealidad de nivel de ganancia gain-level linearity
linealidad de respuesta response linearity
linealidad de señal linearity of signal
linealidad dependiente dependent linearity
linealidad horizontal horizontal linearity
linealidad normal normal linearity

linealidad vertical vertical linearity
linealizable linearizable
linealizar linearize
linealmente linearly
líneas de código code lines
líneas de comunicaciones communication lines
líneas de emisión emission lines
líneas de flujo flux lines
líneas de flujo eléctrico electric-flux lines
líneas de fuerza eléctrica electric lines of force
líneas de Kikuchi Kikuchi lines
líneas de Lecher Lecher lines
líneas de suministro eléctrico electric-supply lines
líneas inactivas inactive lines
líneas por minuto lines per minute
líneas principales mains
linotrón linotron
linterna flashlight
líquido liquid
líquido criogénico cryogenic liquid
líquido ionizado ionized liquid
líquido nemático nematic liquid
liso smooth
lista list
lista alfabética alphabetic list
lista bloqueada blocked list
lista circular circular list
lista controlada controlled list
lista de bibliotecas library list
lista de carga load list
lista de comprobación checklist
lista de datos data list
lista de desplazamiento descendente push-down list
lista de distribución distribution list
lista de errores error list
lista de llamadas calling list
lista de mandos command list
lista de parámetros parameter list
lista de parámetros de comunicaciones communication parameters list
lista de prioridad priority list
lista de puesta en cola queuing list
lista de sondeo polling list
lista de visualización display list
lista en cadena chain list
lista encadenada chained list
lista enlazada linked list
lista lineal linear list
lista lineal enlazada linked linear list
lista mixta mixed list
lista secuencial sequential list
listado listing
listado de ensamblaje assembly listing
listo ready
litio lithium
litro liter
lobulación lobing
lobulación simultánea simultaneous lobing
lóbulo lobe
lóbulo de antena antenna lobe
lóbulo de coma coma lobe
lóbulo de radiación radiation lobe
lóbulo direccional directional lobe
lóbulo lateral side lobe
lóbulo menor minor lobe
lóbulo posterior back lobe
lóbulo principal main lobe
lóbulo secundario secondary lobe
localización location, localization
localización aislada isolated location
localización de base base location
localización de bit bit location
localización de byte efectiva effective byte location

localización de fallas fault location
localización de línea line finding
localización de memoria memory location
localización de recepción receiving location
localización por eco echo-location
localización protegida protected location
localización selectiva selective localization
localizado localized
localizador locator, localizer
localizador de comparación de fase
    phase-comparison localizer
localizador de fallas fault finder
localizador de fugas leak locator
localizador de haz luminoso light-beam localizer
localizador de pista runway localizer
localizador de tono tone localizer
localizador de trayectoria de planeo glide-path
    localizer
localizador electrónico electronic locator
localizador equiseñal equisignal localizer
locomotora eléctrica electric locomotive
locomotora eléctrica de corriente alterna
    alternating-current electric locomotive
locomotora eléctrica de corriente continua
    direct-current electric locomotive
lodar lodar
logarítmicamente logarithmically
logarítmico logarithmic
logaritmo logarithm, log
logaritmo común common logarithm
logaritmo hiperbólico hyperbolic logarithm
logaritmo natural natural logarithm
logaritmo negativo negative logarithm
logaritmo neperiano Naperian logarithm
lógica logic
lógica acoplada por transistores transistor-coupled
    logic
lógica activa active logic
lógica aleatoria random logic
lógica almacenada stored logic
lógica auxiliar ancillary logic
lógica binaria binary logic
lógica binaria en paralelo parallel binary logic
lógica combinacional combinational logic
lógica combinatoria combinatorial logic
lógica compartida shared logic
lógica compuesta compound logic
lógica común common logic
lógica de alta velocidad high-speed logic
lógica de alto umbral high-threshold logic
lógica de baja velocidad low-speed logic
lógica de bajo nivel low-level logic
lógica de Boole Boolean logic
lógica de circuitos circuit logic
lógica de concurrencia concurrency logic
lógica de control control logic
lógica de control ejecutivo executive control logic
lógica de diodos diode logic
lógica de dos estados two-state logic
lógica de ejecución de instrucción instruction
    execution logic
lógica de entrada lineal linear input logic
lógica de estado sólido solid-state logic
lógica de fase phase logic
lógica de instrucción de exploración scanning
    instruction logic
lógica de interfaz de tablero de conectores de
    circuitos impresos backplane interface logic
lógica de lámparas de neón neon-bulb logic
lógica de máquina machine logic
lógica de n niveles n-level logic
lógica de programa program logic
lógica de resistor-capacitor-transistor
    resistor-capacitor-transistor logic
lógica de resistor-transistor resistor-transistor logic
lógica de Schottky Schottky logic
lógica de sistema system logic
lógica de sondeo paralelo parallel poll logic
lógica de transistores de acoplamiento directo
    direct-coupled transistor logic
lógica de tres estados three-state logic
lógica de umbral threshold logic
lógica de uso especial special-purpose logic
lógica deductiva deductive logic
lógica digital digital logic
lógica digital programable programmable digital
    logic
lógica diodo-transistor diode-transistor logic
lógica en modo de corriente current-mode logic
lógica fija fixed logic
lógica formal formal logic
lógica inductiva inductive logic
lógica matemática mathematical logic
lógica mayoritaria majority logic
lógica monoestable one-shot logic
lógica negativa negative logic
lógica neumática pneumatic logic
lógica NO NOT logic
lógica no saturada nonsaturated logic
lógica O OR logic
lógica positiva positive logic
lógica programable programmable logic
lógica programada programmed logic
lógica secuencial sequential logic
lógica simbólica symbolic logic
lógica sincrónica synchronous logic
lógica ternaria ternary logic
lógica transistor-diodo transistor-diode logic
lógica transistor-resistor transistor-resistor logic
lógica transistor-transistor transistor-transistor logic
lógica variable variable logic
lógico logical
logómetro ratiometer
logonio logon
loktal loktal
longevidad longevity
longevidad ante flexiones flex life
longevidad de retención retention longevity
longitud length
longitud activa active length
longitud angular angular length
longitud de antena efectiva effective antenna length
longitud de arco arc length
longitud de bloque block length
longitud de bloque fijo fixed block length
longitud de bloque variable variable-block length
longitud de cable length of cable
longitud de campo field length
longitud de campo variable variable field length
longitud de coherencia coherence length
longitud de comprobación check length
longitud de Debye Debye length
longitud de difusión diffusion length
longitud de equilibrio equilibrium length
longitud de escala scale length
longitud de fase phase length
longitud de inserción insertion length
longitud de instrucción instruction length
longitud de Landau Landau length
longitud de línea line length
longitud de línea de exploración scanning-line
    length
longitud de meseta plateau length
longitud de onda wavelength
longitud de onda complementaria complementary
    wavelength

longitud de onda crítica critical wavelength
longitud de onda de absorción crítica critical absorption wavelength
longitud de onda de calibración calibration wavelength
longitud de onda de corte cutoff wavelength
longitud de onda de corte de guíaondas waveguide cutoff wavelength
longitud de onda de emisión emission wavelength
longitud de onda de guíaondas waveguide wavelength
longitud de onda de oscilador oscillator wavelength
longitud de onda de prioridad priority wavelength
longitud de onda dominante dominant wavelength
longitud de onda efectiva effective wavelength
longitud de onda eléctrica electrical wavelength
longitud de onda en espacio libre free-space wavelength
longitud de onda en la guía guide wavelength
longitud de onda espectral spectral wavelength
longitud de onda fundamental fundamental wavelength
longitud de onda lineal straight-line wavelength
longitud de onda media medium wavelength
longitud de onda natural natural wavelength
longitud de onda operacional operational wavelength
longitud de onda ultravioleta ultraviolet wavelength
longitud de onda umbral threshold wavelength
longitud de palabra word length
longitud de palabra de máquina machine-word length
longitud de palabra fija fixed word length
longitud de paquete packet length
longitud de recorrido length of travel
longitud de registro record length
longitud de registro variable variable record length
longitud de ruptura length of break
longitud de secuencia string length
longitud de trayectoria path length
longitud de trayectoria óptica optical path length
longitud del entrehierro gap length
longitud eléctrica electrical length
longitud fija fixed length
longitud focal posterior back focal length
longitud máxima de línea maximum line length
longitud mínima de onda minimum wavelength
longitud unidad unit length
longitud variable variable length
longitudinal longitudinal
lorac lorac
lorán loran
lorán de baja frecuencia low-frequency loran
lote de trabajos job batch
luces acromáticas especificadas specified achromatic lights
luces de ángulo de aproximación angle of approach lights
luces de aproximación approach lights
luces de canal channel lights
luces de límites range lights
luces de posición position lights
lugar acromático achromatic locus
lugar de dígito digit place
lugar de perforación hole site
lugar de Planck Planckian locus
lugar de procesamiento processing site
lugar decimal decimal place
lugar en una cola queue place
lugar espectral spectrum locus, spectral locus
lugar ideal ideal site
lumen lumen
lumen-hora lumen-hour

lumen-metro lumen-meter
lumen-segundo lumen-second
lumens por watt lumen per watt
luminancia luminance
luminancia de pantalla screen luminance
luminancia media average luminance
luminancia relativa relative luminance
luminaria luminaire
luminaria empotrada recessed luminaire
luminiscencia luminescence, glow
luminiscencia anódica anode glow
luminiscencia anormal abnormal glow
luminiscencia azul blue glow
luminiscencia catódica cathode luminescence
luminiscencia de campo equivalente equivalent field luminescence
luminiscencia eléctrica electric glow
luminiscencia fosforescente phosphorescent glow
luminiscencia negativa negative glow
luminiscencia normal normal glow
luminiscencia persistente afterglow
luminiscencia positiva positive glow
luminiscente luminescent
luminóforo luminophor
luminosidad luminosity
luminosidad catódica cathode glow
luminosidad relativa relative luminosity
luminoso luminous
lumistor lumistor
lutecio lutetium
lux lux
lux-segundo lux-second
luxómetro luxmeter, illumination meter
luz light
luz aeronáutica aeronautical light
luz alterna alternating light
luz ámbar amber light
luz ambiente ambient light
luz anticolisión anticollision light
luz blanca white light
luz coherente coherent light
luz de aeronave aircraft light
luz de alta intensidad high-intensity light
luz de anclaje anchor light
luz de arranque starting light
luz de aterrizaje landing light
luz de aviso warning light
luz de baja intensidad low-intensity light
luz de base base light
luz de clasificación classification light
luz de código code light
luz de control control light
luz de cuadrante dial light
luz de destellos flashing light
luz de dirección de aterrizaje landing direction light
luz de error error light
luz de falla failure light
luz de freno brake light
luz de funcionamiento running light
luz de fusible fuse light
luz de identificación identification light
luz de instrumentos instrument light
luz de límite boundary light
luz de navegación navigation light
luz de neón neon light
luz de obstrucción obstruction light
luz de ocultación occulting light
luz de panel panel light
luz de prueba test light
luz de respaldo backup light
luz de retorno return light
luz de señal signal light
luz de señalización signaling light

luz de superficie aeronáutica aeronautical ground
    light
luz de tablero de instrumentos instrument panel
    light
luz de torre tower light
luz de visualizador display light
luz difusa scattered light
luz directa direct light
luz eléctrica electric light
luz fija fixed light
luz fría cold light
luz giratoria revolving light
luz incidente incident light
luz incoherente incoherent light
luz indicadora indicator light
luz indirecta indirect light
luz infrarroja infrared light
luz intermitente intermittent light
luz lineal linear light
luz marcadora marker light
luz modulada modulated light
luz monocromática monochromatic light
luz no polarizada unpolarized light
luz parásita stray light
luz parpadeante blinking light
luz periódica periodic light
luz piloto pilot light
luz polarizada polarized light
luz polarizada en un plano plane-polarized light
luz posterior de señal signal back light
luz pulsada pulsed light
luz puntual point light
luz química chemical light
luz reflejada reflected light
luz refractada refracted light
luz rítmica rhythmic light
luz semiindirecta semiindirect lighting
luz ultravioleta ultraviolet light
luz única single light
luz visible visible light

llamada call, ringing
llamada a larga distancia long-distance call
llamada armónica harmonic ringing
llamada armónica sintonizada tuned harmonic
    ringing
llamada audible audible call
llamada automática automatic call
llamada codificada coded call
llamada completada completed call
llamada condicional conditional call
llamada contestada answered call
llamada de asistencia assistance call
llamada de conferencia conference call
llamada de corriente alterna-corriente continua
    alternating-current-direct-current ringing
llamada de dirección abreviada abbreviated address
    calling
llamada de entrada entry call
llamada de estación a estación station-to-station call
llamada de frecuencia vocal voice-frequency ringing
llamada de macro macro call
llamada de potencia power ringing
llamada de procedimiento procedure call
llamada de prueba test call
llamada de salida outgoing call
llamada de socorro distress call
llamada de subrutina subroutine call
llamada de tránsito transit call
llamada de una unidad one-unit call
llamada directa direct call
llamada en código code ringing
llamada externa external call
llamada inmediata immediate ringing
llamada internacional international call
llamada interrumpida interrupted ringing
llamada interurbana toll call, trunk call
llamada local local call
llamada manual manual ringing, ringdown
llamada múltiple multiple call
llamada nacional nationwide call
llamada perdida lost call
llamada por magneto magneto ringing
llamada por tonos touch-tone calling
llamada repetida repeated call
llamada retardada delayed call
llamada selectiva selective calling
llamada selectiva selective ringing
llamada selectiva a la frecuencia frequency-selective
    ringing
llamada semiautomática semiautomatic ringing
llamada semiselectiva semiselective ringing
llamada sin llave keyless ringing
llamada superpuesta superimposed ringing
llamada suspendida suspended call
llamada telefónica telephone call
llamada válida valid call
llamador ringer
llamar por nombre call by name
llamar por referencia call by reference
llamar por valor call by value
llamar por variable call by variable

llave key
llave de descarga discharge key
llave de monitoreo monitoring key
llave de palanca lever key
llave real actual key
lleno de vapor vapor-filled

maclación twinning
macro macro
macro de acceso access macro
macro de llamada calling macro
macro de sistema system macro
macro de teclado keyboard macro
macrocodificación macrocoding
macrocódigo macrocode
macrodefinición macrodefinition
macrodefinición de biblioteca library
    macrodefinition
macroensamblador macroassembler
macroensamblador residente resident
    macroassembler
macroensamblaje macroassembly
macroexpansión macroexpansion
macrogeneración macrogeneration
macrogenerador macrogenerator
macroinstrucción macroinstruction, macro
macroinstrucción de salida exit macroinstruction
macroinstrucción de sistema system
    macroinstruction
macroinstrucción declarativa declarative
    macroinstruction
macroinstrucción exterior outer macroinstruction
macroinstrucción funcional functional
    macroinstruction
macroinstrucción imperativa imperative
    macroinstruction
macroinstrucción interna inner macroinstruction
macrolenguaje macrolanguage
macromando macrocommand
macroonda macrowave
macrooperación macrooperation
macroorganigrama macroflowchart
macroprocesador macroprocessor
macroprocesamiento macroprocessing
macroprograma macroprogram
macroprogramación macroprogramming
macroscópico macroscopic
macrosentencia macrostatement
macrosónica macrosonics
madera dura hard wood
madistor madistor
magnalio magnalium
magnesio magnesium
magnética magnetics
magnéticamente magnetically
magnético magnetic
magnetismo magnetism
magnetismo libre free magnetism
magnetismo natural natural magnetism
magnetismo residual residual magnetism
magnetismo subpermanente subpermanent
    magnetism
magnetismo temporal temporary magnetism
magnetismo terrestre terrestrial magnetism
magnetita magnetite, magnetic iron ore
magnetizabilidad magnetizability
magnetizable magnetizable
magnetización magnetization

magnetización anhisterética anhysteretic magnetization
magnetización cíclica cyclic magnetization
magnetización de saturación saturation magnetization
magnetización ideal ideal magnetization
magnetización longitudinal longitudinal magnetization
magnetización normal normal magnetization
magnetización perpendicular perpendicular magnetization
magnetización por impulsos pulse magnetization
magnetización por inducción magnetization by induction
magnetización remanente remanent magnetization
magnetización residual residual magnetization
magnetización transversal transverse magnetization
magnetizado magnetized
magnetizado cíclicamente cyclically magnetized
magnetizador magnetizer
magnetizador de impulso de semiciclo half-cycle impulse magnetizer
magnetizador de semiciclo half-cycle magnetizer
magnetizante magnetizing
magnetizar magnetize
magneto magneto
magneto de ignición ignition magneto
magnetoacústica magnetoacoustics
magnetoacústico magnetoacoustic
magnetocardiógrafo magnetocardiograph
magnetocardiograma magnetocardiogram
magnetoconductividad magnetoconductivity
magnetocristalino magnetocrystalline
magnetodinámico magnetodynamic
magnetodiodo magnetodiode
magnetoelástico magnetoelastic
magnetoeléctrico magnetoelectric
magnetoestricción magnetostriction
magnetoestricción de saturación saturation magnetostriction
magnetoestricción lineal linear magnetostriction
magnetoestricción positiva positive magnetostriction
magnetoestrictivo magnetostrictive
magnetoestrictor magnetostrictor
magnetofluidodinámica magnetofluidynamics
magnetofluidodinámico magnetofluidynamic
magnetofluidomecánica magnetofluidmechanics
magnetofluidomecánico magnetofluidmechanic
magnetófono cassette recorder
magnetófono magnetophone, magnetic-tape cassette recorder, cassette recorder, tape recorder
magnetógrafo magnetograph
magnetograma magnetogram
magnetohidrodinámica magnetohydrodynamics
magnetohidrodinámico magnetohydrodynamic
magnetoiónica magnetoionics
magnetoiónico magnetoionic
magnetomecánico magnetomechanical
magnetómetro magnetometer
magnetómetro de haz electrónico electron-beam magnetometer
magnetómetro de imán móvil moving-magnet magnetometer
magnetómetro de impedancia impedance magnetometer
magnetómetro de Kew Kew magnetometer
magnetómetro de laminillas vibrantes vibrating-reed magnetometer
magnetómetro de magnetoestricción magnetostriction magnetometer
magnetómetro de núcleo saturable saturable-core magnetometer
magnetómetro de resistencia resistance magnetometer

magnetómetro generador generating magnetometer
magnetómetro protónico proton magnetometer
magnetómetro unifilar unifilar magnetometer
magnetomotor permanent-magnet motor
magnetomotriz magnetomotive
magnetón magneton
magnetón de Bohr Bohr magneton
magnetón nuclear nuclear magneton
magnetoóptica magnetooptics
magnetoóptico magnetooptical
magnetopausa magnetopause
magnetoplasmadinámica magnetoplasmadynamics
magnetoplasmadinámico magnetoplasmadynamic
magnetoquímica magnetochemistry
magnetor magnettor
magnetorresistencia magnetoresistance
magnetorresistividad magnetoresistivity
magnetorresistivo magnetoresistive
magnetorresistor magnetoresistor
magnetorresistor de película delgada thin-film magnetoresistor
magnetoscopía magnetoscopy
magnetoscópico magnetoscopic
magnetoscopio magnetoscope, video-tape recorder, videocassette recorder, television tape recorder
magnetosfera magnetosphere
magnetostática magnetostatics
magnetostático magnetostatic
magnetostricción magnetostriction
magnetotérmico magnetothermal
magnetotermoelasticidad magnetothermoelasticity
magnetotermoelástico magnetothermoelastic
magnetotermoeléctrico magnetothermoelectric
magnetotorsión magnetotorsion
magnetotransistor magnetotransistor
magnetrón magnetron
magnetrón de ánodo de paletas vane-anode magnetron
magnetrón de ánodo dividido split-anode magnetron
magnetrón de ánodo neutro neutral-anode magnetron
magnetrón de ánodo único single-anode magnetron
magnetrón de capacitor resonante resonant-capacitor magnetron
magnetrón de cavidad resonante resonant-cavity magnetron
magnetrón de cavidades cavity magnetron
magnetrón de cavidades múltiples multiple-cavity magnetron
magnetrón de cilindros coaxiales coaxial-cylinder magnetron
magnetrón de ondas continuas continuous-wave magnetron
magnetrón de ondas inversas backward-wave magnetron
magnetrón de ondas progresivas traveling-wave magnetron
magnetrón de paletas vane magnetron
magnetrón de placas terminales end-plate magnetron
magnetrón de resistencia negativa negative-resistance magnetron
magnetrón de sol naciente rising-sun magnetron
magnetrón dividido split magnetron
magnetrón en jaula de ardilla squirrel-cage magnetron
magnetrón integrado packaged magnetron
magnetrón interdigital interdigital magnetron
magnetrón lineal linear magnetron
magnetrón multicavidad multicavity magnetron
magnetrón multisegmento multisegment magnetron
magnetrón pulsado pulsed magnetron

magnetrón sintonizable tunable magnetron
magnetrón sintonizable por tensión voltage-tunable
  magnetron
magnificación magnification
magnificación de barrido sweep magnification
magnificación de potencia power magnification
magnificación electrónica electronic magnification
magnificación lineal linear magnification
magnificación longitudinal longitudinal
  magnification
magnificación por gas gas magnification
magnificador de barrido sweep magnifier
magnistor magnistor
magnitud magnitude, quantity
magnitud absoluta absolute magnitude
magnitud alterna simétrica symmetrical alternating
  quantity
magnitud aparente apparent magnitude
magnitud cero zero magnitude
magnitud de corriente current magnitude
magnitud de descarga aparente apparent discharge
  magnitude
magnitud de incremento repentino burst magnitude
magnitud de ráfaga absoluta media average
  absolute burst magnitude
magnitud dinámica dynamic quantity
magnitud eficaz root-mean-square magnitude
magnitud escalar scalar quantity
magnitud física physical magnitude
magnitud generalizada generalized quantity
magnitud periódica periodic quantity
magnitud relativa relative magnitude
magnitud sinusoidal sinusoidal quantity
magnitud vectorial vector quantity
magnitud vertical vertical quantity
mal funcionamiento malfunction
mala definición bad definition
maleabilidad malleability
malla mesh
malla de campo field mesh
malla de colector collector mesh
malla de rejilla grid mesh
mancha electródica luminosa luminous electrode
  spot
mancha iónica ion spot, ion burn
mancha obscura dark spot
mancha solar sunspot
mancha superluminosa flare spot
manchas por exudación stain spots
mandato command
mandil de caucho al plomo lead-rubber apron
mando command
mando absoluto absolute command
mando anidado nested command
mando convertido converted command
mando de canal channel command
mando de control control command
mando de edición edit command
mando de entrada-salida de canal channel
  input-output command
mando de impresora printer command
mando de lectura read command
mando de línea line command
mando de no operación no-operation command
mando de operador operator command
mando de preparación housekeeping command
mando de procedimiento procedure command
mando de reposición reset command
mando de sistema system command
mando de transacción transaction command
mando de transferencia transfer command
mando de visualización display command
mando directo positive control

mando empotrado embedded command
mando externo external command
mando inmediato immediate command
mando interno internal command
mando manual manual command
mando no empotrado nonembedded command
mando regular regular command
mando relativo relative command
mando transitorio transient command
mando variable variable command
mandril magnético magnetic chuck
manejador de papel paper handler
manejo de datos data handling
manejo de información information handling
manejo de interrupción interrupt handling
manejo de mensajes message handling
manejo de papel paper handling
manejo de patrón pattern handling
manejo de secuencia string handling
manejo de tarjetas card handling
manga sleeve
manganeso manganese
manganina manganin
manguito sleeve
manguito aislante insulating sleeve, boot
manguito catódico cathode sleeve
manguito de adaptación slug
manguito de colector commutator sleeve
manguito de empalme jointing sleeve, ferrule
manguito de plomo lead sleeve
manguito de porcelana porcelain sleeve
manguito protector protective sleeve
manipulación manipulation, keying
manipulación algebraica algebraic manipulation
manipulación de base de datos database
  manipulation
manipulación de bits bit manipulation
manipulación de bytes byte manipulation
manipulación de datos data manipulation
manipulación de dos tonos two-tone keying
manipulación de frecuencia frequency keying
manipulación de pila stack manipulation
manipulación de secuencias string manipulation
manipulación de tono tone keying
manipulación de tono único single-tone keying
manipulación electrónica electronic keying
manipulación por bloqueo de rejilla grid-block
  keying
manipulación por cátodo cathode keying
manipulación por derivación central center-tap
  keying
manipulación por desfase phase-shift keying
manipulación por desplazamiento de frecuencia
  frequency-shift keying
manipulación por placa plate keying
manipulación por rejilla grid keying
manipulación primaria primary keying
manipulación remota remote manipulation
manipulación todo o nada on-off keying
manipulador manipulator, handler, key, hand key
manipulador de alta velocidad speed key
manipulador de archivos file handler
manipulador de bloques block handler
manipulador de cinta tape handler
manipulador de cinta perforada punched-tape
  handler
manipulador de comprobación de máquina
  machine-check handler
manipulador de control remoto remote-control
  manipulator
manipulador de mensajes message handler
manipulador de teclado keyboard keyer
manipulador de tono tone keyer

manipulador electromecánico electromechanical manipulator
manipulador electrónico electronic key
manipulador inversor reversing key
manipulador remoto remote manipulator
manipulador telegráfico telegraph key
manipular manipulate
mano artificial artificial hand
manómetro manometer
manómetro piezoeléctrico piezoelectric pressure gage
mantenimiento maintenance
mantenimiento adaptivo adaptive maintenance
mantenimiento automático automatic maintenance
mantenimiento coactivo coactive maintenance
mantenimiento correctivo corrective maintenance
mantenimiento de archivos file maintenance
mantenimiento de arco arc maintenance
mantenimiento de datos data maintenance
mantenimiento de emergencia emergency maintenance
mantenimiento de equipo maintenance of equipment
mantenimiento de línea line maintenance
mantenimiento de programa program maintenance
mantenimiento de seguridad security maintenance
mantenimiento de servicio maintenance of service
mantenimiento de sistema system maintenance
mantenimiento diferido deferred maintenance
mantenimiento inmediato immediate maintenance
mantenimiento intermedio intermediate maintenance
mantenimiento mínimo minimum maintenance
mantenimiento no programado unscheduled maintenance
mantenimiento operacional operational maintenance
mantenimiento preventivo preventive maintenance
mantenimiento programado scheduled maintenance
mantenimiento remoto remote maintenance
mantenimiento rutinario routine maintenance
mantenimiento suplementario supplementary maintenance
manual manual
manualmente manually
manufactura asistida por computadora computer-aided manufacturing
manufactura integrada por computadora computer integrated manufacturing
mapa de almacenamiento storage map
mapa de archivo file map
mapa de bits bit map
mapa de caracteres character map
mapa de carga load map
mapa de emisión emission map
mapa de entrada-salida input-output map
mapa de Karnaugh Karnaugh map
mapa de memoria memory map
mapa de rendimiento yield map
mapa físico physical map
mapa lógico logic map
máquina a prueba de agua waterproof machine
máquina a prueba de gas gasproof machine
máquina a prueba de goteo dripproof machine
máquina a prueba de salpicaduras splashproof machine
máquina abierta open machine
máquina abstracta abstract machine
máquina acíclica acyclic machine
máquina analógica analog machine
máquina antideflagrante flameproof machine
máquina arrollada en derivación shunt-wound machine
máquina asincrónica asynchronous machine
máquina autoexcitada self-excited machine
máquina autoventilada self-ventilated machine

máquina bipolar bipolar machine
máquina clasificadora de tarjetas card sorting machine
máquina computadora computing machine
máquina controlada por cinta tape-controlled machine
máquina de abrasión abrasion machine
máquina de accionamiento directo direct-drive machine
máquina de base de datos database machine
máquina de cajero automatizado automated teller machine
máquina de cinta magnética tape machine
máquina de código code machine
máquina de colector de corriente alterna alternating-current commutator machine
máquina de contabilidad accounting machine
máquina de contabilidad eléctrica electric accounting machine
máquina de contabilidad electrónica electronic accounting machine
máquina de contestación answering machine
máquina de contestación telefónica telephone-answering machine
máquina de control control machine
máquina de desvanecimiento fading machine
máquina de direccionamiento addressing machine
máquina de emergencia emergency machine
máquina de ensamblaje assembly machine
máquina de escribir typewriter
máquina de facsímil facsimile machine
máquina de inducción induction machine
máquina de inducción polifásica polyphase induction machine
máquina de influencia influence machine
máquina de lectura reading machine
máquina de numeración numbering machine
máquina de oficina business machine
máquina de perforación punching machine
máquina de procesamiento de datos data-processing machine
máquina de procesamiento de datos electrónico electronic data processing machine
máquina de procesamiento de información information-processing machine
máquina de programa almacenado stored-program machine
máquina de rayos X X-ray machine
máquina de Scherbius Scherbius machine
máquina de tabulación tabulating machine
máquina de tarjetas card machine
máquina de tarjetas perforadas punched-card machine
máquina de trazado plotting machine
máquina de Turing Turing machine
máquina de Turing universal universal Turing machine
máquina de ventilación forzada forced-ventilated machine
máquina de Wimshurst Wimshurst machine
máquina eléctrica electric machine
máquina electrostática electrostatic machine
máquina excitada separadamente separately excited machine
máquina fuente source machine
máquina heteropolar heteropolar machine
máquina hidroeléctrica hydroelectric machine
máquina horizontal horizontal machine
máquina inteligente smart machine
máquina magnetoeléctrica permanent-magnet machine
máquina monofásica single-phase machine
máquina multipolar multipolar machine

máquina numérica numerical machine
máquina objeto target machine
máquina orientada al operador operator-oriented machine
máquina perforadora perforating machine
máquina perforadora de cintas tape-punching machine
máquina polifásica polyphase machine
máquina principal host machine
máquina protegida protected machine
máquina protegida contra agentes atmosféricos weather-protected machine
máquina rotativa rotary machine
máquina secuencial sequential machine
máquina semiatendida semiattended machine
máquina semiautomática semiautomatic machine
máquina semiprotegida semiprotected machine
máquina simbólica symbolic machine
máquina sin escobillas brushless machine
máquina sincrónica synchronous machine
máquina sincrónica sin escobillas brushless synchronous machine
máquina sumadora adding machine
máquina sumergible submersible machine
máquina totalmente encerrada totally-enclosed machine
máquina trifásica three-phase machine
máquina ventilada externamente externally ventilated machine
máquina virtual virtual machine
máquina virtual de acceso múltiple multiple-access virtual machine
maquinado machining
maquinado por chispa eléctrica electric-spark machining
maquinado por descargas eléctricas electrical-discharge machining
maquinado por haz electrónico electron-beam machining
maquinado ultrasónico ultrasonic machining
maquinaria de computadora computer machinery
maquinaria dinamoeléctrica dynamoelectric machinery
marca mark, marker
marca calibradora calibrating marker
marca continua continuous mark
marca de almacenamiento storage mark
marca de archivo file mark
marca de bloque block mark
marca de cinta tape mark
marca de control control mark
marca de distancia distance mark
marca de escala scale mark
marca de fin end mark
marca de fin de archivo end-of-file mark
marca de fin de campo end-of-field mark
marca de fin de cinta end-of-tape mark
marca de fin de datos end-of-data mark
marca de grupo group mark
marca de identificación identification mark
marca de índice index mark
marca de palabra word mark
marca de párrafo paragraph mark
marca de referencia reference mark
marca de registro register mark
marca de segmento segment mark
marca de tambor drum mark
marca de tiempo time mark
marca estroboscópica strobe
marca óptica optical mark
marca sensora sensing mark
marcación dialing
marcación a larga distancia long-distance dialing

marcación abreviada abbreviated dialing
marcación de distancia distance dialing
marcación de distancia directa direct-distance dialing
marcación de distancia directa internacional international direct-distance dialing
marcación de distancia internacional international distance dialing
marcación en bucle loop dialing
marcación interurbana trunk dialing
marcación manual manual dialing
marcación por corriente alterna alternating-current dialing
marcación por frecuencia vocal voice-frequency dialing
marcación por impulsos pulse dialing
marcación por tonos touch-tone dialing
marcación rotativa rotary dialing
marcaciones cruzadas cross bearings
marcado dialing
marcador marker, dialer
marcador central central marker
marcador de absorción absorption marker
marcador de altura height marker
marcador de calibración calibration marker
marcador de campo field marker
marcador de cristal crystal marker
marcador de distancia range marker
marcador de grupo group marker
marcador de límite boundary marker
marcador de radar radar marker
marcador de sincronización timing marker
marcador decimal decimal marker
marcador en abanico fan marker
marcador estroboscópico strobe marker
marcador Z Z marker
marcapasos pacemaker
marcapasos cardíaco cardiac pacemaker
marcar dial, dial-up
marco frame, framework
marco de montaje mounting frame
marco de potencia power frame
margen margin, range
margen de amplitud de blanco a negro white-to-black amplitude range
margen de bucle loop margin
margen de capacidad capability margin
margen de conmutación margin of commutation
margen de corriente current margin
margen de desvanecimiento fading margin
margen de diseño design margin
margen de enfoque focusing range
margen de estabilidad margin of stability
margen de estabilidad de Duerdoth Duerdoth's stability margin
margen de fase phase margin
margen de ganancia gain margin
margen de guía guide margin
margen de potencia margin of power
margen de ruido noise margin
margen de ruido de corriente continua direct-current noise margin
margen de seguridad safety margin
margen de sintonización mecánica mechanical tuning range
margen de sobrecarga overload margin
margen de tensión de operación operating-voltage range
margen de tolerancia margin of allowance
margen de volumen volume range
margen dinámico dynamic range
margen efectivo effective margin
margen neto net margin

margen nominal　nominal margin
margen normal　normal margin
margen protector　protective margin
marginal　marginal
martillo　hammer
martillo de impresión　print hammer
martillo impresor　printing hammer
masa　ground
masa activa　active mass
masa acústica　acoustic mass
masa atómica　atomic mass
masa de aire　air mass
masa de chasis　chassis ground
masa de electrón　electron mass
masa electromagnética　electromagnetic mass
masa en reposo de electrón　electron rest mass
masa mecánica　mechanical mass
masa molecular　molecular mass
máscara　mask
máscara de abertura　aperture mask
máscara de edición　editing mask
máscara de papel　paper mask
máscara de plástico　plastic mask
máscara de programa　program mask
máscara de sombra　shadow mask
máser　maser
máser de capacitor resonante　resonant-capacitor
　maser
máser de cavidad　cavity maser
máser de cavidad óptica　optical cavity maser
máser de dos niveles　two-level maser
máser de estado sólido　solid-state maser
máser de gas　gas maser
máser de granate　garnet maser
máser de hidrógeno　hydrogen maser
máser de infrarrojo　infrared maser
máser de infrarrojo lejano　far infrared maser
máser de ondas progresivas　traveling-wave maser
máser de rubí　ruby maser
máser de rubí pulsado　pulsed ruby maser
máser de semiconductor　semiconductor maser
máser de tres niveles　three-level maser
máser fonónico　phonon maser
máser óptico　optical maser
máser óptico de semiconductor　semiconductor
　optical maser
máser pulsado　pulsed maser
máser sintonizable　tunable maser
mástil　mast, boom
mástil de antena　radio mast
mástil de antena arriostrado　guyed antenna mast
mástil de soporte　supporting mast
mástil rotativo　rotary mast
materia　matter
material a prueba de fuego　fireproof material
material absorbente　absorbing material
material absorbente de vibraciones
　vibration-absorbing material
material aceptor　acceptor material
material activo　active material
material acústico　acoustic material
material aislante　insulating material
material aislante eléctrico　electrical insulating
　material
material amortiguador　damping material
material antiferroeléctrico　antiferroelectric material
material antiferromagnético　antiferromagnetic
　material
material antirruido　antinoise material
material centelleador　scintillator material
material centelleante　scintillating material
material con resistencia　resistance material
material conductor　conducting material

material contaminado　contaminated material
material cristalino　crystalline material
material de alta energía　high-energy material
material de base　base material
material de bolus　bolus material
material de contactos　contact material
material de encapsulación　potting material
material de pantalla　screen material
material de pocas pérdidas　low-loss material
material débilmente magnético　weakly magnetic
　material
material delicuescente　deliquescent material
material diamagnético　diamagnetic material
material dieléctrico　dielectric material
material dieléctrico de pocas pérdidas　low-loss
　dielectric material
material disipativo　dissipative material
material emisivo　emissive material
material encapsulador　encapsulating material
material fenólico　phenolic material
material ferrimagnético　ferrimagnetic material
material ferroeléctrico　ferroelectric material
material ferromagnético　ferromagnetic material
material fluorescente　fluorescent material
material fotoconductivo　photoconductive material
material fotoeléctrico　photoelectric material
material fotoemisivo　photoemissive material
material fotorresistivo　photoresistive material
material fotosensible　photosensitive material
material fotovoltaico　photovoltaic material
material fuertemente magnético　hard magnetic
　material
material giromagnético　gyromagnetic material
material higroscópico　hygroscopic material
material impuro　impurity material
material insonorizante　sound-insulating material
material luminiscente　luminescent material
material magnético　magnetic material
material magnético sinterizado　sintered magnetic
　material
material monocristalino　monocrystalline material
material no conductor　nonconducting material
material para imanes permanentes
　permanent-magnet material
material para sellado　sealing material
material paramagnético　paramagnetic material
material peligroso　hazardous material
material piroeléctrico　pyroelectric material
material policristalino　polycrystalline material
material protector　protective material
material radiactivo　radioactive material
material radiotransparente　radiotransparent
　material
material retrorreflejante　retroreflecting material
material semiconductor　semiconductor material
material sensible a la luz　light-sensitive material
material termoeléctrico　thermoelectric material
material termoendurecible　thermosetting material
material termoplástico　thermoplastic material
material tipo N　N-type material
material tipo P　P-type material
materiales de chip　chip materials
matiz　hue
matriz　matrix
matriz de acceso　access matrix
matriz de adaptación　adaptation matrix
matriz de admitancia　admittance matrix
matriz de carácter　character matrix
matriz de conmutación　switching matrix
matriz de correlación　correlation matrix
matriz de datos　data matrix
matriz de diodos　diode matrix
matriz de dispersión　dispersion matrix

matriz de immitancia immittance matrix
matriz de núcleos core matrix
matriz de puntos dot matrix
matriz de transmisión transmission matrix
matriz electrónica electronic matrix
matriz negativa negative matrix
matriz nula null matrix
matriz operando operand matrix
matriz positiva positive matrix
mavar mavar
maxicomputadora mainframe computer
máximo maximum, ceiling
máximo de tensión voltage maximum
máximos y mínimos maxima and minima
maxwell maxwell
maxwell-vuelta maxwell-turn
mayoría de resistencia a la derecha right-hand taper
mecánica mechanics
mecánica cuántica quantum mechanics
mecánica ondulatoria wave mechanics
mecánicamente mechanically
mecánico mechanical
mecanismo mechanism, movement
mecanismo amortiguador damping mechanism
mecanismo autoorientador self-orienting mechanism
mecanismo corrector correcting mechanism
mecanismo de acceso access mechanism
mecanismo de acción ultrarrápida snap-action
  mechanism
mecanismo de accionamiento actuating mechanism
mecanismo de acoplamiento coupling mechanism
mecanismo de alimentación feed mechanism
mecanismo de alimentación de cinta ribbon-feed
  mechanism
mecanismo de alimentación de papel paper-feed
  mechanism
mecanismo de altavoz loudspeaker mechanism
mecanismo de arrastre transport mechanism
mecanismo de arrastre de carretes reel-to-reel tape
  deck
mecanismo de arrastre de cinta tape-deck
  mechanism, tape deck, tape-transport mechanism,
  tape drive
mecanismo de arrastre de cinta magnética
  magnetic-tape deck, magnetic-tape drive
mecanismo de avance de papel paper advance
  mechanism
mecanismo de bloqueo lockout mechanism
mecanismo de bobina móvil moving-coil mechanism
mecanismo de contacto contact mechanism
mecanismo de conteo counting mechanism
mecanismo de control control mechanism
mecanismo de control de acceso access-control
  mechanism
mecanismo de control de motor motor control
  mechanism
mecanismo de control de velocidad speed-control
  mechanism
mecanismo de D'Arsonval D'Arsonval movement
mecanismo de decisión decision mechanism
mecanismo de disparo tripping mechanism
mecanismo de envejecimiento aging mechanism
mecanismo de falla failure mechanism
mecanismo de impresión print mechanism
mecanismo de liberación release mechanism
mecanismo de magnetófono tape-deck mechanism
mecanismo de medida measurement mechanism
mecanismo de medidor meter mechanism
mecanismo de memoria memory mechanism
mecanismo de operación operating mechanism
mecanismo de perforación perforating mechanism,
  punching mechanism
mecanismo de posicionamiento positioning
  mechanism

mecanismo de registro recording mechanism
mecanismo de retardo retardation mechanism
mecanismo de retén detent mechanism
mecanismo de retroceso backspace mechanism
mecanismo de seguimiento tracking mechanism
mecanismo de seguridad safety mechanism
mecanismo de seguridad electromagnético
  electromagnetic safety mechanism
mecanismo de selección selecting mechanism
mecanismo de teclado keyboard mechanism
mecanismo de temporización timing mechanism
mecanismo de transferencia transfer mechanism
mecanismo de transporte transport mechanism
mecanismo de transporte de cinta tape-transport
  mechanism
mecanismo eléctrico electric mechanism
mecanismo equilibrador de ancho de banda
  bandwidth balancing mechanism
mecanismo generador de señales signal-generator
  mechanism
mecanismo impresor printing mechanism
mecanismo integrador integrating mechanism
mecanismo para mantener cerrado hold-closed
  mechanism
mecanoeléctrico mechanoelectric
mecanoelectrónico mechanoelectronic
media mean
media aritmética arithmetic mean
media armónica harmonic mean
media carga half-load
media geométrica geometric mean
media logarítmica logarithmic mean
media onda half-wave
media palabra half-word
media tuerca half-nut
media vida half-life
mediana median
medible measurable
medible directamente directly measurable
medición measurement, metering, gaging
medición de tiempo time measurement
medición de tráfico traffic metering
medición de velocidad por radar radar speed
  measurement
medición múltiple multiple metering
medida measurement
medida absoluta absolute measurement
medida absoluta de corriente absolute measurement
  of current
medida absoluta de tensión absolute measurement of
  voltage
medida balística ballistic measurement
medida calibrada calibrated measurement
medida cuantitativa quantitative measurement
medida de aislamiento insulation measurement
medida de atenuación attenuation measurement
medida de campo field measurement
medida de conductancia mutua mutual-conductance
  measurement
medida de contador counter measurement
medida de corriente current measurement
medida de corriente alterna alternating-current
  measurement
medida de desfase phase-angle measurement
medida de desplazamiento displacement
  measurement
medida de diafonía crosstalk measurement
medida de distancia range measurement
medida de distancia electrónica electronic distance
  measurement
medida de distorsión de fase phase-distortion
  measurement

medida de elevación automática automatic elevation measurement
medida de enmascaramiento masking measurement
medida de entrada diferencial differential-input measurement
medida de entrada flotante floating-input measurement
medida de fase phase measurement
medida de frecuencia frequency measurement
medida de ganancia gain measurement
medida de impedancia impedance measurement
medida de información measure of information
medida de intensidad de campo field-intensity measurement
medida de interferencia interference measurement
medida de intervalos de tiempo time-interval measurement
medida de ionización ionization measurement
medida de laboratorio laboratory measurement
medida de periodos period measurement
medida de potencia power measurement
medida de presión sonora sound-pressure measurement
medida de radiación radiation measurement
medida de razón de frecuencias frequency-ratio measurement
medida de rendimiento performance measurement
medida de resistencia resistance measurement
medida de resistividad resistivity measurement
medida de respuesta de frecuencia frequency-response measurement
medida de ruido noise measurement
medida de ruido ponderada weighted noise measurement
medida de temperatura sin contacto noncontact temperature measurement
medida de tensión voltage measurement
medida de transmisión transmission measurement
medida directa direct measurement
medida eléctrica electrical measurement
medida electrónica electronic measurement
medida electroquímica electrochemical measurement
medida en bucle abierto open-loop measurement
medida física physical measurement
medida indirecta indirect measurement
medida lineal linear measure
medida magnética magnetic measurement
medida objetiva objective measurement
medida por comparación comparison measurement
medida por ondas métricas metric-wave measurement
medida punto por punto point-by-point measurement
medida sensible a la frecuencia frequency-sensitive measurement
medido remotamente remote-metered
medidor meter, gage
medidor amortiguado damped meter
medidor analógico analog meter
medidor aperiódico deadbeat meter
medidor calibrado calibrated meter
medidor con cero central center-zero meter
medidor de actividad activity meter
medidor de alto vacío high-vacuum gage
medidor de ampere-horas ampere-hour meter
medidor de aplausos applause meter
medidor de audiofrecuencias audio-frequency meter
medidor de bobina móvil movable-coil meter
medidor de bolsillo pocket meter
medidor de brillo glossmeter
medidor de campo field meter
medidor de campo eléctrico generador generating electric field meter

medidor de cantidad quantity meter
medidor de capacidad capacity meter
medidor de capacitancia capacitance meter
medidor de capacitancia digital digital capacitance meter
medidor de características de sonido sound-survey meter
medidor de carga cero zero-load meter
medidor de centelleo scintillation meter
medidor de cero null meter
medidor de cero central zero-center meter
medidor de cinta tensa taut-band meter
medidor de coeficiente de reflexión reflection-coefficient meter
medidor de combustible fuel gage
medidor de Compton Compton meter
medidor de computadora computer meter
medidor de conductancia mutua mutual-conductance meter
medidor de conductividad conductivity meter
medidor de conductividad térmica heat-conductivity gage
medidor de contacto contact-making meter
medidor de contaminación de aire air contamination meter
medidor de corriente current meter
medidor de corriente alterna alternating-current meter
medidor de corriente de diodo diode current meter
medidor de corriente de placa plate-current meter
medidor de corriente de transistores transistor current meter
medidor de corriente electrolítico electrolytic current meter
medidor de corriente electrónica electronic current meter
medidor de cristal crystal meter
medidor de chispa measuring spark-gap
medidor de D'Arsonval D'Arsonval meter
medidor de demanda demand meter
medidor de demanda acumulativa cumulative-demand meter
medidor de demanda con retardo lagged-demand meter
medidor de demanda de registro recording demand meter
medidor de demanda integrada integrated-demand meter
medidor de desbordamiento overflow meter
medidor de desplazamiento displacement meter
medidor de desviación deviation meter
medidor de desviación de frecuencia frequency-deviation meter
medidor de desviación de portadora carrier-deviation meter
medidor de desviación de trayectoria de vuelo flight-path deviation meter
medidor de diafonía crosstalk meter
medidor de distorsión armónica harmonic-distortion meter
medidor de distorsión de modulación cruzada cross-modulation distortion meter
medidor de distribución de amplitud amplitude distribution meter
medidor de emisión emission meter
medidor de energía activa active energy meter
medidor de escala ensanchada expanded-scale meter
medidor de escala logarítmica logarithmic-scale meter
medidor de espesores por rayos X X-ray thickness gage
medidor de excedente excess meter
medidor de factor de calidad quality-factor meter

medidor de factor de potencia power-factor meter
medidor de factor de ruido noise-factor meter
medidor de factor reactivo reactive-factor meter
medidor de flujo de combustible fuel-flow gage
medidor de flujo de fluido fluid-flow gage
medidor de flujo de gas gas-flow gage
medidor de flujo de líquido liquid-flow gage
medidor de frecuencia frequency measurer
medidor de funciones múltiples multiple-purpose
   meter
medidor de ganancia gain meter
medidor de gimoteo wow meter
medidor de gradiente gradient meter
medidor de gráfica de barras bar-graph meter
medidor de grosor magnético magnetic thickness
   measurer
medidor de haz luminoso light-beam meter
medidor de hierro móvil moving-iron meter
medidor de hilo caliente hot-wire meter
medidor de horas hour meter
medidor de imán magnet meter
medidor de imán permanente permanent-magnet
   meter
medidor de impedancia acústica acoustic impedance
   meter
medidor de impedancia de tipo de puente
   bridge-type impedance meter
medidor de inducción induction meter
medidor de inductancia-capacitancia
   inductance-capacitance meter
medidor de intensidad de campo field-intensity
   meter, field-strength meter
medidor de intensidad de campo de radio radio
   field strength meter
medidor de intensidad de campo eléctrico
   electric-field strength meter
medidor de intensidad de campo eléctrico de
   corriente alterna alternating-current electrical
   field strength meter
medidor de intensidad de ruido noise-intensity
   meter
medidor de intensidad de señal signal-strength
   meter
medidor de intermodulación intermodulation meter
medidor de intervalo interval meter
medidor de intervalo de impulsos pulse-interval
   meter
medidor de intervalos de tiempo time-interval meter
medidor de ionización ionization gage
medidor de ionización de cátodo frío cold-cathode
   ionization gage
medidor de ionización de Penning Penning
   ionization gage
medidor de laminilla vibrante vibrating-reed meter
medidor de lectura directa direct-reading meter
medidor de llamadas call meter
medidor de McLeod McLeod gage
medidor de modulación modulation meter
medidor de monitoreo monitoring meter
medidor de motor motor meter
medidor de nivel level gage
medidor de nivel de audio audio-level meter
medidor de nivel de fluido fluid-level meter
medidor de nivel de líquido liquid-level meter
medidor de nivel de ruido noise-level meter
medidor de nivel sonoro sound-level meter
medidor de ocupación de grupo group occupancy
   meter
medidor de ondas estacionarias standing-wave
   meter
medidor de paleta de hierro iron-vane meter
medidor de paleta magnética magnetic-vane meter
medidor de paleta móvil moving-vane meter

medidor de panel panel meter
medidor de panel digital digital panel meter
medidor de par torque measurer
medidor de péndulo pendulum meter
medidor de pérdidas loss meter
medidor de periodos period meter
medidor de pH pH meter
medidor de Phillips Phillips gage
medidor de pico a pico peak-to-peak meter
medidor de Pirani Pirani gage
medidor de placa plate meter
medidor de porcentaje de modulación
   percent-modulation meter
medidor de posición position meter
medidor de potencia power meter
medidor de potencia activa active power meter
medidor de potencia aparente apparent power meter
medidor de potencia bolométrico bolometric power
   meter
medidor de potencia calométrico calometric power
   meter
medidor de potencia de salida output-power meter
medidor de potencia de tipo de puente bridge-type
   power meter
medidor de potencia digital digital power meter
medidor de potencia electrotérmico electrothermic
   power meter
medidor de potencia reflejada reflected-power
   meter
medidor de presión pressure meter
medidor de presión de combustible fuel-pressure
   meter
medidor de presión de fluido fluid-pressure meter
medidor de presión de gas gas-pressure meter
medidor de presión de líquido liquid-pressure meter
medidor de Q Q meter
medidor de radiofrecuencia radio-frequency meter
medidor de razón ratio meter
medidor de razón de frecuencias frequency-ratio
   meter
medidor de registro recording meter
medidor de resonancia resonance meter
medidor de ruido noise meter
medidor de ruido de circuito circuit-noise meter
medidor de ruido objetivo objective noise meter
medidor de salida output meter
medidor de sintonización tuning meter
medidor de sonido sound meter
medidor de susceptibilidad susceptibility meter
medidor de tarifas múltiples multiple-tariff meter
medidor de temperatura temperature meter
medidor de temperatura de color color temperature
   meter
medidor de tensión voltage meter
medidor de termopar thermocouple meter
medidor de Thomson Thomson meter
medidor de tiempo time meter, time measurer
medidor de tiempo de funcionamiento running-time
   meter
medidor de tiempo de reacción reaction-time meter
medidor de tiempo transcurrido elapsed-time meter
medidor de tipo de aguja pointer-type meter
medidor de tipo de espejo mirror-type meter
medidor de tipo de puente bridge-type meter
medidor de torsión torsion meter
medidor de tráfico traffic meter
medidor de transconductancia transconductance
   meter
medidor de una dirección one-directional meter
medidor de unidades de volumen volume-unit meter
medidor de vacío vacuum gage
medidor de vibraciones vibration meter
medidor de visibilidad visibility meter

medidor de volumen volume meter
medidor de watthoras watt-hour meter
medidor de Z Z meter
medidor doble dual meter
medidor eléctrico electric meter
medidor electrolítico electrolytic meter
medidor electromagnético de par electromagnetic
  torquemeter
medidor en joules joulemeter
medidor enchufable plug-in meter
medidor integrador integrating meter
medidor logarítmico logarithmic meter
medidor monitor monitor meter
medidor multiescala multirange meter
medidor no amortiguado undamped meter
medidor no electrónico nonelectronic meter
medidor no oscilante nonblinking meter
medidor oscilante oscillating meter
medidor polarizado polarized meter
medidor polifásico polyphase meter
medidor remoto remote meter
medidor segmental segmental meter
medidor sellado sealed meter
medidor térmico thermal meter
medidor totalizador summation meter
medidor X X meter
medio medium
medio absorbente absorbing medium
medio aleatorio random medium
medio anisotrópico anisotropic medium
medio con pérdidas lossy medium
medio conductor conductive medium
medio controlado controlled medium
medio de almacenamiento storage medium
medio de almacenamiento de datos data storage
  medium
medio de almacenamiento de información
  information-storage medium
medio de almacenamiento magnético magnetic
  storage medium
medio de almacenamiento secundario secondary
  storage medium
medio de comunicación communication medium
medio de datos data medium
medio de datos automatizado automated data
  medium
medio de datos en paralelo parallel data medium
medio de datos portátil portable data medium
medio de datos prerregistrado prerecorded data
  medium
medio de enfriamiento cooling medium
medio de entrada input medium
medio de entrada-salida input-output medium
medio de información information medium
medio de láser activo active laser medium
medio de parámetros parameter medium
medio de presentación presentation medium
medio de procesamiento processing medium
medio de propagación propagation medium
medio de registro recording medium
medio de registro de datos data recording medium
medio de registro magnético magnetic recording
  medium
medio de registro no magnético nonmagnetic
  recording medium
medio de retardo delay medium
medio de retorno return medium
medio de salida output medium
medio de transmisión transmission medium
medio dispersivo dispersive medium
medio en blanco blank medium
medio gaseoso gaseous medium
medio láser laser medium

medio legible por máquina machine-readable
  medium
medio magnético magnetic medium
medio magnetoiónico magnetoionic medium
medio multiplicador multiplying medium
medio ultilizable por máquina machine-usable
  medium
medio vacío empty medium
medios de almacenamiento removibles removable
  storage media
mega mega
megabar megabar
megabit megabit
megabyte megabyte
megaciclo megacycle
megaciclos por segundo megacycles per second
megacurie megacurie
megadina megadyne
megaelectrón-volt megaelectron-volt
megáfono megaphone
megahertz megahertz
megalumen megalumen
megampere megampere
megarroentgen megaroentgen
megatonelada megaton
megatrón megatron
megavar megavar
megavolt megavolt
megavolt-ampere megavolt-ampere
megawatt megawatt
megawatt-hora megawatt-hour
megohm megohm
megohm-farads megohm-farads
megohm-microfarads megohm-microfarads
megóhmetro megohmmeter
mejora de diseño design improvement
mejora de factor de potencia power-factor
  improvement
mel mel
memistor memistor
memoria memory, storage
memoria activa active memory
memoria acústica acoustic memory
memoria alterable alterable memory
memoria asincrónica asynchronous memory
memoria asociativa associative memory
memoria asociativa magnética magnetic associative
  memory
memoria auxiliar auxiliary memory
memoria bidimensional two-dimensional memory
memoria borrable erasable memory
memoria central central memory
memoria cíclica cyclic memory
memoria circulatoria circulating memory
memoria compartida shared memory
memoria común common pool
memoria continua continuous memory
memoria controlada controlled memory
memoria criogénica cryogenic memory
memoria cuasidinámica quasi-dynamic memory
memoria de acceso aleatorio random-access
  memory
memoria de acceso aleatorio de video video
  random-access memory
memoria de acceso aleatorio de visualización
  display random access memory
memoria de acceso aleatorio dinámica dynamic
  random access memory
memoria de acceso aleatorio estática static
  random-access memory
memoria de acceso cuasialeatorio quasi-random
  access memory
memoria de acceso en serie serial-access memory

memoria de acceso rápido rapid-access memory
memoria de acceso secuencial sequential-access memory
memoria de agujas needle memory
memoria de alta velocidad high-speed memory
memoria de apuntes scratchpad memory
memoria de base base memory
memoria de bucle único single-loop memory
memoria de burbujas bubble memory
memoria de burbujas magnéticas magnetic bubble memory
memoria de búsqueda search memory
memoria de búsqueda en paralelo parallel-search memory
memoria de cinta tape memory
memoria de cinta de papel paper-tape memory
memoria de computadora computer memory
memoria de computadora electrónica electronic computer memory
memoria de control control memory
memoria de disco disk memory
memoria de disco magnético magnetic-disk memory
memoria de ferrita ferrite memory
memoria de lámparas de neón neon-bulb memory
memoria de lectura-escritura read-write memory
memoria de lectura y escritura read-and-write memory
memoria de línea de retardo delay-line memory
memoria de mercurio mercury memory
memoria de nivel único one-level memory
memoria de núcleos core memory
memoria de núcleos de ferrita ferrite-core memory
memoria de núcleos de ferrita lítica lithium ferrite-core memory
memoria de núcleos magnéticos magnetic-core memory
memoria de película delgada thin-film memory
memoria de película magnética magnetic-film memory
memoria de procedimientos procedure memory
memoria de tambor drum memory
memoria de tambor magnético magnetic-drum memory
memoria de tiempo de acceso cero zero-access memory
memoria de trabajo working memory
memoria de tubo de rayos catódicos cathode-ray tube memory
memoria dinámica dynamic memory
memoria direccionable addressable memory
memoria direccionada addressed memory
memoria disponible available memory
memoria electrónica electronic memory
memoria electrostática electrostatic memory
memoria en cinta magnética magnetic-tape memory
memoria en paralelo parallel memory
memoria en serie serial memory
memoria entrelazada interlaced memory
memoria estática static memory
memoria expandida expanded memory
memoria extendida extended memory
memoria externa external memory
memoria fija fixed memory
memoria fotoóptica photooptic memory
memoria holográfica holographic memory
memoria inherente inherent memory
memoria inmediata cache memory
memoria inmediata de disco disk cache
memoria inmediata de microprocesador microprocessor cache memory
memoria intercalada interleaved memory
memoria intermedia buffer memory, buffer, intermediate memory

memoria intermedia de comunicaciones communications buffer
memoria intermedia de comunicaciones de datos data communications buffer
memoria intermedia de datos data buffer
memoria intermedia de disco disk buffer
memoria intermedia de entrada input buffer
memoria intermedia de entrada-salida input-output buffer
memoria intermedia de red network buffer
memoria intermedia de registro de salida output register buffer
memoria intermedia de salida output buffer
memoria intermedia de teclado keyboard buffer
memoria intermedia de visualización display buffer
memoria intermedia elástica elastic buffer
memoria intermedia periférica peripheral buffer
memoria interna internal memory
memoria láser laser memory
memoria lenta slow memory
memoria local local memory
memoria magnética magnetic memory
memoria magnética multidimensional multidimensional magnetic memory
memoria masiva mass memory
memoria matricial matrix memory
memoria multicinta multitape memory
memoria multipuerto multiport memory
memoria no destructiva nondestructive memory
memoria no volátil nonvolatile memory
memoria óptica optical memory
memoria para refrescar refresh memory
memoria permanente permanent memory
memoria primaria primary memory
memoria principal main memory
memoria programable programmable memory
memoria rápida rapid memory
memoria real real memory
memoria regenerativa regenerative memory
memoria rotativa rotating memory
memoria secuencial sequential memory
memoria secundaria secondary memory
memoria semiconductora semiconductor memory
memoria seudodinámica pseudodynamic memory
memoria sólo de lectura read-only memory
memoria sólo de lectura programable programmable read-only memory
memoria superconductora superconducting memory
memoria temporal temporary memory
memoria tridimensional three-dimensional memory
memoria virtual virtual memory
memoria volátil volatile memory
memoscopio memoscope
memotrón memotron
mendelevio mendelevium
menor que less than
menos significativo least significant
mensaje message
mensaje administrativo aeronáutico aeronautical administrative message
mensaje de alto nivel high-level message
mensaje de aviso warning message
mensaje de consulta inquiry message
mensaje de dirección de grupo group address message
mensaje de dirección única single-address message
mensaje de direcciones múltiples multiple-address message
mensaje de entrada input message
mensaje de error error message
mensaje de estado status message
mensaje de excepción exception message
mensaje de formato fijo fixed-format message

mensaje de primer nivel first-level message
mensaje de procedimiento procedure message
mensaje de programa a programa
    program-to-program message
mensaje de prueba test message
mensaje de salida outgoing message
mensaje de segundo nivel second-level message
mensaje de terminación completion message
mensaje de una dirección one-address message
mensaje emitido broadcast message
mensaje físico physical message
mensaje funcional functional message
mensaje ininteligible garbled message
mensaje lógico logical message
mensaje multidirección multiaddress message
mensaje predefinido predefined message
mensaje sin texto nontext message
mensuración mensuration
mensurando measurand
menú menu
menú de mandos command menu
menú de mandos principal main command menu
menú de selección selection menu
menú principal main menu
mercurio mercury
meridiano magnético magnetic meridian
mesa de control control table
mesa de sacudidas shake table
mesa vibratoria shake table
meseta de potencial plateau of potential
mesócrono mesochronous
metacompilador metacompiler
metadata metadata
metadina metadyne
metal metal
metal alcalino alkali metal
metal alcalinotérreo alkaline-earth metal
metal con resistencia resistance metal
metal de base base metal
metal de Muntz Muntz metal
metal de relleno filler metal
metal de Wood Wood's metal
metal depositado deposited metal
metal estampado stamped metal
metal magnetorresistivo magnetoresistive metal
metal monel monel metal
metal no férrico nonferrous metal
metal no ferroso nonferrous metal
metal noble noble metal
metal pesado heavy metal
metal pulverizado powdered metal
metal resistente a la corrosión corrosion-resistant
    metal
metal sinterizado sintered metal
metalenguaje metalanguage
metalización metallization, plating
metalización electrolítica electrolytic metallization
metalización galvánica galvanic metallization
metalizado metallized, plated
metalizar metallize
metalocerámico metal-ceramic
metaloide metalloid
metalurgia de fibras fiber metallurgy
metámero metamer
metascopio metascope
meteórico meteoric
meteorógrafo meteorograph
meteorológico meteorological
método A-B A-B method
método de acceso access method
método de acceso a telecomunicaciones
    telecommunications access method
método de acceso al almacenamiento storage access

method
método de acceso al almacenamiento virtual virtual
    storage access method
método de acceso aleatorio random-access method
método de acceso básico basic access method
método de acceso de base de datos database access
    method
método de acceso de gráficos graphics access
    method
método de acceso de terminal remota
    remote-terminal access method
método de acceso directo direct-access method
método de acceso directo jerárquico hierarchical
    direct-access method
método de acceso dividido partitioned access method
método de acceso por colas queue access method
método de acceso residente resident access method
método de acceso secuencial sequential-access
    method
método de acceso secuencial generalizado
    generalized sequential access method
método de acceso secuencial indexado indexed
    sequential access method
método de acceso secuencial jerárquico hierarchical
    sequential-access method
método de acceso secuencial relativo relative
    sequential access method
método de acceso secuencial virtual virtual
    sequential access method
método de acceso secundario secondary access
    method
método de ajuste a cero zero method
método de anillo ring method
método de Barnett Barnett method
método de batido beat method
método de batido cero zero-beat method
método de cero null method
método de control control method
método de Czochralski Czochralski method
método de chasquidos click method
método de Debye-Scherrer Debye-Scherrer method
método de dopado doping method
método de enfasamiento phasing method
método de equilibrio equilibrium method
método de exploración scanning method
método de filtro pasaaltos high-pass filter method
método de interportadora intercarrier method
método de medida absoluta absolute method of
    measurement
método de operación method of operation
método de oposición opposition method
método de polarización biasing method
método de Potier Potier's method
método de prueba alternativo alternative test
    method
método de radioeco radio-echo method
método de rayos refractados refracted-ray method
método de recálculo recalculation method
método de reciprocidad reciprocity method
método de reflexión de Fresnel Fresnel-reflection
    method
método de resonancia resonance method
método de Schlieren Schlieren method
método de señal cero zero-signal method
método de señalización signaling method
método de sustitución substitution method
método de trayectoria crítica critical path method
método de variación de frecuencia
    frequency-variation method
método de Weissenberg Weissenberg method
método del cristal oscilante oscillating-crystal
    method
método deltic deltic method

método directo direct method
método empírico empirical method
método heterodino heterodyne method
método heterostático heterostatic method
método idiostático idiostatic method
método magnetoacústico magnetoacoustic method
método operacional operational method
método piloto pilot method
método potenciométrico potentiometer method
metodología de desarrollo development methodology
métrico metric
metro meter
metro-ampere meter-ampere
metro-bujía meter-candle
metro-kilogramo-segundo meter-kilogram-second
metrología metrology
metronio metron
metrónomo metronome
mezcla mixture, mix
mezcla aditiva additive mixture
mezcla de colores color mixture
mezcla de impulsos pulse mixing
mezcla de instrucciones instruction mix
mezcla de sonidos sound mixing
mezcla despolarizante depolarizing mix
mezcla gaseosa gaseous mixture
mezclador mixer
mezclador asimétrico single-ended mixer
mezclador de audio audio mixer
mezclador de canales channel mixer
mezclador de coincidencia coincidence mixer
mezclador de colores color mixer
mezclador de cristal crystal mixer
mezclador de crominancia chrominance mixer
mezclador de diodo diode mixer
mezclador de diodo de cristal crystal-diode mixer
mezclador de imagen picture mixer
mezclador de microondas microwave mixer
mezclador de señales signal mixer
mezclador de sonidos sound mixer
mezclador de triodo con rejilla a masa grounded-grid triode mixer
mezclador de video video mixer
mezclador en contrafase push-pull mixer
mezclador equilibrado balanced mixer
mezclador heterodino heterodyne mixer
mezclador pasivo passive mixer
mezclador pentarrejilla pentagrid mixer
mezclador triodo-hexodo triode-hexode mixer
mho mho
mica mica, isinglass
mica adherida con vidrio glass-bonded mica
micanita micanite
micarta micarta
micro micro
microampere microampere
microamperímetro microammeter
microamperímetro de cero central zero-center microammeter
microamperímetro electrónico electronic microammeter
microanálisis por sonda electrónica electron-probe microanalysis
microanalizador microanalyzer
microanalizador de rayos X X-ray microanalyzer
microanalizador electrónico electron microanalyzer
microarquitectura microarchitecture
microbalanza microscale
microbanda microband, microstrip
microbar microbar
microbarn microbarn
microbarógrafo microbarograph

microcanal microchannel
microciclo microcycle
microcircuitería microcircuitry
microcircuito microcircuit
microcircuito basado en cerámica ceramic-based microcircuit
microcircuito de componentes discretos discrete-component microcircuit
microcircuito de película delgada thin-film microcircuit
microcircuito híbrido hybrid microcircuit
microcodificación microcoding
microcódigo microcode
microcódigo vertical vertical microcode
microcomponente microcomponent
microcomputadora microcomputer
microcomputadora de escritorio desktop microcomputer
microconmutador microswitch
microcontrol microcontrol
microcontrolador microcontroller
microcristal microcrystal
microcurie microcurie
microchip microchip
microchispa microspark
microdensitómetro microdensitometer
microdestello microflash
microdisco microdisk
microdisco flexible microfloppy disk
microdisquete microdiskette
microeléctrico microelectric
microelectrodo microelectrode
microelectroforesis microelectrophoresis
microelectrólisis microelectrolysis
microelectrónica microelectronics
microelectrónica de película delgada thin-film microelectronics
microelectrónico microelectronic
microelectroscopio microelectroscope
microelemento microelement
microenergía microenergy
microenganche microlock
microensamblador microassembler
microescala microscale
microespaciamiento microspacing
microespectrofotómetro microspectrophotometer
microestructura microstructure
microestructura híbrida hybrid microstructure
microfarad microfarad
microfaradímetro microfaradmeter
microfenómeno microphenomenon
microficha microfiche
microfísica microphysics
microfonía microphony
microfónica microphonics
microfónico microphonic
microfonismo microphonism
micrófono microphone
micrófono antirruido antinoise microphone
micrófono astático astatic microphone
micrófono bidireccional bidirectional microphone
micrófono cardioide cardioid microphone
micrófono cerámico ceramic microphone
micrófono combinado combination microphone
micrófono con elemento cerámico ceramic element microphone
micrófono de armadura magnética magnetic-armature microphone
micrófono de bobina móvil moving-coil microphone
micrófono de botón button microphone
micrófono de campo libre free-field microphone
micrófono de cancelación de ruido noise-cancelling microphone

micrófono de cápsula única single-button microphone
micrófono de captación pickup microphone
micrófono de carbón carbon microphone
micrófono de carbón de cápsula única single-button carbon microphone
micrófono de carbón en contrafase push-pull carbon microphone
micrófono de célula sonora sound-cell microphone
micrófono de cinta ribbon microphone
micrófono de condensador condenser microphone
micrófono de condensador de alta frecuencia high-frequency condenser microphone
micrófono de conductor móvil moving-conductor microphone
micrófono de contacto contact microphone
micrófono de cristal crystal microphone
micrófono de cristal de sal de Rochelle Rochelle-salt crystal microphone
micrófono de descarga luminiscente glow-discharge microphone
micrófono de desfase phase-shift microphone
micrófono de doble botón double-button microphone
micrófono de electreto electret microphone
micrófono de garganta throat microphone
micrófono de gradiente gradient microphone
micrófono de gradiente de presión pressure-gradient microphone
micrófono de granalla de carbón carbon-granules microphone
micrófono de hilo caliente hot-wire microphone
micrófono de incidencia aleatoria random-incidence microphone
micrófono de labio lip microphone
micrófono de línea line microphone
micrófono de llama flame microphone
micrófono de magnetoestricción magnetostriction microphone
micrófono de medida measuring microphone
micrófono de oprimir para hablar press-to-talk microphone
micrófono de presión pressure microphone
micrófono de pulsador push-button microphone
micrófono de reflector parabólico parabolic-reflector microphone
micrófono de reluctancia variable variable-reluctance microphone
micrófono de solapa lapel microphone
micrófono de velocidad velocity microphone
micrófono diferencial differential microphone
micrófono dinámico dynamic microphone
micrófono direccional directional microphone
micrófono electrodinámico electrodynamic microphone
micrófono electromagnético electromagnetic microphone
micrófono electrónico electronic microphone
micrófono electrostático electrostatic microphone
micrófono en contrafase push-pull microphone
micrófono estereofónico stereophonic microphone
micrófono granular granular microphone
micrófono inalámbrico wireless microphone
micrófono inductivo inductive microphone
micrófono inductor inductor microphone
micrófono iónico ionic microphone
micrófono magnético magnetic microphone
micrófono magnético de baja impedancia low-impedance magnetic microphone
micrófono magnetoestrictivo magnetostrictive microphone
micrófono mezclador mixing microphone
micrófono miniatura miniature microphone
micrófono no direccional nondirectional microphone

micrófono omnidireccional omnidirectional microphone
micrófono para hablar de cerca close-talk microphone
micrófono parabólico parabolic microphone
micrófono patrón standard microphone
micrófono piezoeléctrico piezoelectric microphone
micrófono polidireccional polydirectional microphone
micrófono semiconductor semiconductor microphone
micrófono semidireccional semidirectional microphone
micrófono sonda probe microphone
micrófono térmico thermal microphone
micrófono transistorizado transistorized microphone
micrófono unidireccional unidirectional microphone
microfonógrafo microphonograph
micrófonos apareados matched microphones
microfonoscopio microphonoscope
microforma microform
microfotografía microphotography
microfotómetro microphotometer
microfuga microleak
microfunción microfunction
microgalvanómetro microgalvanometer
microgauss microgauss
micrografía micrography
micrográficos micrographics
microgramo microgram
microhaz microbeam
microhaz de rayos X X-ray microbeam
microhenry microhenry
microhm microhm
microimagen microimage
microinstrucción microinstruction
microinstrucción horizontal horizontal microinstruction
microlámpara microlamp
microlitro microliter
micrológica micrologic
micromalla micromesh
micromanipulador micromanipulator
micromaquinado micromachining
micrometría micrometry
micrométrico micrometric
micrómetro micrometer
micrómetro de chispas spark micrometer
micrómetro electrónico electronic micrometer
micromho micromho
micromhómetro micromhometer
micromicro micromicro
microminiatura microminiature
micromódulo micromodule
micrón micron
micrónica micronics
microoblea microwafer
microóhmetro microhmmeter
microonda microwave
microoperación microoperation
microoscilógrafo microoscillograph
micropelícula microfilm
micropotencia micropower
microprocesador microprocessor
microprocesador de chip chip microprocessor
microprocesador multilínea multiline microprocessor
microprocesamiento microprocessing
microprograma microprogram
microprograma de control control microprogram
microprogramable microprogrammable
microprogramación microprogramming
microprogramación de microcontrol microcontrol

microprogramming
microprogramado microprogrammed
micropulgada microinch
micropulsación micropulsation
microrradar microradar
microrradiografía microradiography
microrradiómetro microradiometer
microrreflectómetro microreflectometer
microrroentgen microroentgen
microrrutina microroutine
microscopía microscopy
microscopía de barrido scanning microscopy
microscopía electrónica electron microscopy
microscopía electrónica de barrido scanning
    electron microscopy
microscopía por emisión termoiónica
    thermionic-emission microscopy
microscópico microscopic
microscopio microscope
microscopio de emisión de campo field-emission
    microscope
microscopio de infrarrojo infrared microscope
microscopio de medida measuring microscope
microscopio de punto móvil flying-spot microscope
microscopio de rayos X X-ray microscope
microscopio electrónico electron microscope,
    electronic microscope
microscopio electrónico de barrido scanning
    electron microscope
microscopio electrónico electrostático electrostatic
    electron microscope
microscopio electrónico magnético magnetic
    electron microscope
microscopio electrostático electrostatic microscope
microscopio estereoscópico stereoscopic microscope
microscopio fotoeléctrico photoelectric microscope
microscopio iónico de campo field-ion microscope
microscopio magnético magnetic microscope
microscopio protónico proton microscope
microsegundo microsecond
microsegundo-luz light-microsecond
microsiemens microsiemens
microsin microsyn
microsistema microsystem
microsonda microprobe
microsonda electrónica electron microprobe
microsurco microgroove
microtarjeta microcard
microteléfono microtelephone, telephone handset,
    handset
microteléfono de prueba test handset
microtensión microvoltage
microtrón microtron
microvolt microvolt
microvóltmetro microvoltmeter
microvóltmetro electrónico electronic
    microvoltmeter
microvolts por metro microvolts per meter
microwatt microwatt
microwattaje microwattage
microwáttmetro microwattmeter
midicomputadora midicomputer
migración migration
migración atómica atomic migration
migración de datos data migration
migración de plata silver migration
migración electroquímica electrochemical migration
migración iónica ion migration
migración superficial surface migration
mil circular circular mil
mili milli
miliampere milliampere
miliampere-segundo milliampere-second

miliamperímetro milliammeter
miliamperímetro de rejilla grid milliammeter
miliamperímetro electrónico electronic
    milliammeter
milibar millibar
milibarn millibarn
milicurie millicurie
milifarad millifarad
milifot milliphot
miligauss milligauss
miligramo milligram
milihenry millihenry
mililambert millilambert
mililitro milliliter
milimaxwell millimaxwell
milimétrico millimetric
milímetro millimeter
milimicro millimicro
milimol millimole
miliohm milliohm
milióhmetro milliohmmeter
milirradián milliradian
milirroentgen milliroentgen
milirutherford millirutherford
milisegundo millisecond
milivolt millivolt
milivolt-ampere millivolt-ampere
milivóltmetro millivoltmeter
milivóltmetro electrónico electronic millivoltmeter
milivolts por metro millivolts per meter
miliwatt milliwatt
miliwatt digital digital milliwatt
miliwáttmetro milliwattmeter
millones de instrucciones por segundo millions of
    instructions per second
mina acústica acoustic mine
mineral mineral
miniatura miniature
miniaturización miniaturization
miniaturizado miniaturized
miniaturizar miniaturize
minicalculadora minicalculator
minicomputadora minicomputer
minicomputadora personal personal minicomputer
miniconmutador miniswitch
minidisco minidisk
minidisco flexible minifloppy
miniensamblador miniassembler
minimización minimization
mínimo minimum
mínimo de potencial potential minimum
mínimo de tensión voltage minimum
miniperiférico miniperipheral
minipista minitrack
miniscopio miniscope
minómetro minometer
minuendo minuend
minuto minute
minuto de arco arc minute
mioelectricidad myoelectricity
mioeléctrico myoelectric
mira television test card, sight
miriámetro myriameter
misil balístico ballistic missile
mistor mistor
mnemónico mnemonic
moción motion
moción aparente apparent motion
moción armónica harmonic motion
moción armónica simple simple harmonic motion
moción de cinta tape motion
moción de deriva drift motion
moción de electrón electron motion

moción de punto spot motion
moción inversa reverse motion
moción lateral lateral motion
moción libre free motion
moción ondulatoria wave motion
moción rotativa rotary motion
moción transitoria transient motion
moción única single motion
modal modal
modelo model
modelo a escala scale model
modelo analógico analog model
modelo de antena antenna model
modelo de cadena de Markov absorbente absorbing Markov chain model
modelo de caja negra black-box model
modelo de circuito circuit model
modelo de confiabilidad reliability model
modelo de datos data model
modelo de datos internos internal data model
modelo de datos lógico logical data model
modelo de disponibilidad availability model
modelo de entrada-salida input-output model
modelo de error error model
modelo de montaje temporal breadboard model
modelo de predicción de errores error prediction model
modelo de producción production model
modelo de programas software model
modelo de propagación empírico empirical propagation model
modelo de prueba test model
modelo de red network model
modelo de simulación simulation model
modelo de sistema system model
modelo de tiempo time pattern
modelo descriptivo descriptive model
modelo dinámico dynamic model
modelo discreto discrete model
modelo estático static model
modelo experimental experimental model
modelo gráfico graphical model
modelo interno internal model
modelo jerárquico hierarchical model
modelo matemático mathematical model
modelo natural natural model
modelo numérico numerical model
modelo piloto pilot model
modelo relacional relational model
modelo representacional representational model
modelo simbólico symbolic model
módem modem
módem asincrónico asynchronous modem
módem autónomo stand-alone modem
módem de banda base baseband modem
módem de calidad telefónica voice-grade modem
módem de datos data modem
módem de grupo group modem
módem de grupo de base basegroup modem
módem de microprocesador microprocessor modem
módem de servicio service modem
módem externo external modem
módem integrado integrated modem
módem interno internal modem
módem nulo null modem
módem telefónico telephone modem
módem telegráfico telegraph modem
moderador moderator
modificación modification
modificación de bucle loop modification
modificación de caracteres gráficos graphic character modification
modificación de copia copy modification

modificación de datos data modification
modificación de dirección address modification
modificación de instrucción instruction modification
modificación de programa program modification
modificación de sistema system modification
modificado modified
modificador modifier
modificador de carácter character modifier
modificador de dirección address modifier
modificador de fase phase modifier
modificador de instrucción instruction modifier
modificar modify
modo mode
modo abierto open mode
modo acústico acoustic mode
modo alternativo alternate mode
modo aritmético arithmetic mode
modo asincrónico asynchronous mode
modo automático de operación automatic mode of operation
modo axial axial mode
modo básico basic mode
modo binario binary mode
modo binario en columnas column binary mode
modo calculador calculator mode
modo complementario complementing mode
modo comprimido compressed mode
modo común common mode
modo continuo continuous mode
modo conversacional conversational mode
modo de acceso access mode
modo de acceso de archivo file access mode
modo de acceso de lectura-escritura read-write access mode
modo de acceso directo direct-access mode
modo de agotamiento depletion mode
modo de agotamiento-enriquecimiento depletion-enhancement mode
modo de ajuste de líneas adjust line mode
modo de anticipación anticipation mode
modo de bloque block mode
modo de bloqueo blocking mode
modo de carácter character mode
modo de carga load mode
modo de compatibilidad compatibility mode
modo de computar compute mode
modo de conmutación switching mode
modo de consulta inquiry mode
modo de control control mode
modo de control básico basic control mode
modo de control de red network control mode
modo de control matemático mathematical control mode
modo de control por computadora computer control mode
modo de conversión conversion mode
modo de datos data mode
modo de desplazamiento numérico numerical shift mode
modo de direccionamiento addressing mode
modo de direccionamiento de memoria memory addressing mode
modo de direccionamiento de microcomputadora microcomputer addressing mode
modo de direccionamiento relativo relative addressing mode
modo de eco echo mode
modo de edición edit mode
modo de emisión emission mode
modo de enfoque-desenfoque focus-defocus mode
modo de entrada input mode, enter mode
modo de entrada-salida input-output mode
modo de equilibrio equilibrium mode

modo de excitación excitation mode
modo de falla failure mode
modo de formato format mode
modo de función function mode
modo de gráficos graphics mode
modo de guíaondas waveguide mode
modo de imagen image mode
modo de imagen real real image mode
modo de impulsos pulse mode
modo de inserción insert mode
modo de interrupción interrupt mode
modo de interrupción de mandos command interrupt mode
modo de intervalo interval mode
modo de intervalos de tiempo time-interval mode
modo de línea line mode
modo de mandos command mode
modo de mantenimiento maintenance mode, hold mode
modo de mensajes múltiples multiple message mode
modo de multiplexor de bloque block multiplexer mode
modo de multisistema multisystem mode
modo de operación operation mode
modo de operación de entrada-salida input-output operation mode
modo de orden superior higher-order mode
modo de oscilación oscillation mode
modo de oscilador oscillator mode
modo de página page mode
modo de pantalla completa full-screen mode
modo de parada halt mode
modo de paso único single-step mode
modo de prioridad priority mode
modo de procesamiento processing mode
modo de procesamiento por lotes batch processing mode
modo de programa program mode
modo de propagación propagation mode
modo de prueba test mode
modo de radiación radiation mode
modo de recepción reception mode
modo de reflexión reflection mode
modo de registro record mode
modo de resonador resonator mode
modo de resonancia resonance mode
modo de respuesta response mode
modo de respuesta de línea line response mode
modo de respuesta inmediata immediate-response mode
modo de respuesta normal normal response mode
modo de revestimiento cladding mode
modo de salida output mode
modo de servicio service mode
modo de silbidos atmosféricos whistler mode
modo de sistema system mode
modo de solicitud inmediata immediate-request mode
modo de suma add mode
modo de sustitución substitute mode
modo de teclado keyboard mode
modo de texto text mode
modo de tiempo de tránsito transit-time mode
modo de transferencia de datos data-transfer mode
modo de transmisión transmission mode
modo de verificación verification mode
modo de vibración mode of vibration
modo de visualización display mode
modo de visualización gráfica graphic display mode
modo degenerado degenerate mode
modo dividido partitioned mode
modo dominante dominant mode
modo eléctrico electric mode

modo eléctrico circular circular electric mode
modo eléctrico transversal transverse electric mode
modo eléctrico y magnético transversal transverse electric and magnetic mode
modo electromagnético transversal transverse electromagnetic mode
modo en línea on-line mode
modo en tiempo real real-time mode
modo específico specific mode
modo espurio spurious mode
modo evanescente evanescent mode
modo excluyente exclusive mode
modo expandido expanded mode
modo fantasma ghost mode
modo fuera de eje off-axis mode
modo fuera de línea off-line mode
modo fundamental fundamental mode
modo genérico generic mode
modo híbrido hybrid mode
modo inatendido unattended mode
modo independiente independent mode
modo indirecto indirect mode
modo interactivo interactive mode
modo local local mode
modo longitudinal longitudinal mode
modo maestro master mode
modo magnético transversal transverse magnetic mode
modo manual manual mode
modo monitor monitor mode
modo múltiplex multiplex mode
modo nativo native mode
modo natural natural mode
modo no formateado unformatted mode
modo no operacional nonoperational mode
modo normal normal mode
modo normal de radiación normal mode of radiation
modo normal de vibración normal mode of vibration
modo operacional operational mode
modo óptico optical mode
modo ortogonal orthogonal mode
modo pasivo passive mode
modo paso a paso step-by-step mode
modo pi pi mode
modo polarizado linealmente linearly polarized mode
modo principal principal mode
modo privilegiado privileged mode
modo programado programmed mode
modo protegido protected mode
modo real real mode
modo resonante resonant mode
modo ruidoso noisy mode
modo seccionado chopped mode
modo seleccionado selected mode
modo selector selector mode
modo símplex simplex mode
modo simultáneo simultaneous mode
modo sincrónico synchronous mode
modo transversal transverse mode
modo troposférico tropospheric mode
modo virtual virtual mode
modor modor
modulación modulation
modulación angular angular modulation
modulación anódica anode modulation
modulación bifásica two-phase modulation
modulación binaria binary modulation
modulación bivalente two-condition modulation
modulación característica characteristic modulation
modulación cero zero modulation
modulación clase A class A modulation

modulación clase B class B modulation
modulación completa complete modulation
modulación compuesta compound modulation
modulación con portadora suprimida
suppressed-carrier modulation
modulación controlada por portadora
carrier-controlled modulation
modulación cruzada cross-modulation
modulación cruzada externa external
cross-modulation
modulación de acceso múltiple multiple-access
modulation
modulación de alta eficiencia high-efficiency
modulation
modulación de alto nivel high-level modulation
modulación de amplitud amplitude-modulation
modulación de amplitud asimétrica asymmetrical
amplitude modulation
modulación de amplitud de impulsos
pulse-amplitude modulation
modulación de amplitud en cuadratura quadrature
amplitude modulation
modulación de amplitud incidental incidental
amplitude modulation
modulación de amplitud negativa negative
amplitude modulation
modulación de amplitud positiva positive amplitude
modulation
modulación de amplitud residual residual amplitude
modulation
modulación de amplitud sinusoidal sinusoidal
amplitude modulation
modulación de ángulo angle modulation
modulación de Armstrong Armstrong modulation
modulación de baja potencia low-power modulation
modulación de bajo nivel low-level modulation
modulación de banda lateral residual
vestigial-sideband modulation
modulación de banda lateral única single-sideband
modulation
modulación de bobina de choque choke-coil
modulation
modulación de brillantez brilliance modulation
modulación de canal channel modulation
modulación de colector collector modulation
modulación de conductividad conductivity
modulation
modulación de corriente anódica anode current
modulation
modulación de corriente constante constant-current
modulation
modulación de corriente de conducción
conduction-current modulation
modulación de corriente de placa plate-current
modulation
modulación de densidad density modulation
modulación de doble banda lateral double-sideband
modulation
modulación de dos tonos two-tone modulation
modulación de eficiencia variable
variable-efficiency modulation
modulación de espaciado de impulsos pulse-spacing
modulation
modulación de fase phase modulation
modulación de frecuencia frequency-modulation
modulación de frecuencia de banda angosta
narrowband frequency-modulation
modulación de frecuencia de impulsos
pulse-frequency modulation
modulación de frecuencia de subportadora
subcarrier frequency modulation
modulación de frecuencia directa direct frequency
modulation

modulación de frecuencia incidental incidental
frequency modulation
modulación de frecuencia indirecta indirect
frequency modulation
modulación de frecuencia-modulación de fase
frequency-modulation-phase-modulation
modulación de frecuencia-modulación de frecuencia
frequency-modulation-frequency-modulation
modulación de frecuencia positiva positive
frequency modulation
modulación de frecuencia residual residual
frequency modulation
modulación de grupo group modulation
modulación de grupo primario primary-group
modulation
modulación de haz beam modulation
modulación de Heising Heising modulation
modulación de imagen picture modulation
modulación de imagen negativa negative picture
modulation
modulación de imagen positiva positive picture
modulation
modulación de impulsos pulse modulation
modulación de impulsos codificados en delta delta
pulse-code modulation
modulación de impulsos por cátodo cathode pulse
modulation
modulación de impulsos por placa plate pulse
modulation
modulación de impulsos rectangulares
rectangular-pulse modulation
modulación de intensidad intensity modulation
modulación de intervalo de impulsos pulse-interval
modulation
modulación de luz light modulation
modulación de luz negativa negative light
modulation
modulación de pantalla screen modulation
modulación de pendiente de impulso pulse-slope
modulation
modulación de pérdidas loss modulation
modulación de picos positivos positive-peak
modulation
modulación de placa plate modulation
modulación de polarización bias modulation
modulación de portadora carrier modulation
modulación de portadora controlada
controlled-carrier modulation
modulación de portadora dividida divided-carrier
modulation
modulación de portadora flotante floating-carrier
modulation
modulación de portadora quiescente
quiescent-carrier modulation
modulación de potencia power modulation
modulación de prueba test modulation
modulación de referencia reference modulation
modulación de reflexión reflection modulation
modulación de señal de imagen picture-signal
modulation
modulación de Taylor Taylor modulation
modulación de tiempo time modulation
modulación de tiempo de tránsito transit-time
modulation
modulación de tono tone modulation
modulación de tono por diapasón fork tone
modulation
modulación de transmisión transmission modulation
modulación de velocidad velocity modulation
modulación de video video modulation
modulación de voz de banda angosta narrowband
voice modulation
modulación de zumbido hum modulation

modulación delta-sigma delta-sigma modulation
modulación descendente downward modulation
modulación diferencial differential modulation
modulación eléctrica electric modulation
modulación electrónica electronic modulation
modulación en anchura de impulsos pulse-width modulation
modulación en cuadratura quadrature modulation
modulación en delta delta modulation
modulación en delta de alta información high-information delta modulation
modulación en serie series modulation
modulación equilibrada balanced modulation
modulación espuria spurious modulation
modulación exterior outer modulation
modulación externa external modulation
modulación fiel faithful modulation
modulación incorrecta incorrect modulation
modulación indeseada unwanted modulation
modulación interna internal modulation
modulación isócrona isochronous modulation
modulación lineal linear modulation
modulación mixta mixed modulation
modulación multidimensional multidimensional modulation
modulación múltiple multiple modulation
modulación múltiplex multiplex modulation
modulación negativa negative modulation
modulación paramétrica parametric modulation
modulación perfecta perfect modulation
modulación placa-pantalla plate-screen modulation
modulación por absorción absorption modulation
modulación por autoimpulso self-pulse modulation
modulación por cátodo cathode modulation
modulación por conteo de impulsos pulse-count modulation
modulación por corriente de convección convection-current modulation
modulación por derivación central center-tap modulation
modulación por desfasamiento outphasing modulation
modulación por desfase phase-shift modulation
modulación por desplazamiento de frecuencia frequency-shift modulation
modulación por duración de impulsos pulse-duration modulation
modulación por efecto de campo field-effect modulation
modulación por impulsos codificados pulse-code modulation
modulación por impulsos cuantificados quantized pulse modulation
modulación por impulsos de rejilla grid-pulse modulation
modulación por inversión de fase phase-inversion modulation
modulación por onda sinusoidal sine-wave modulation
modulación por polarización polarization modulation
modulación por polarización de rejilla grid-bias modulation
modulación por posición de impulsos pulse-position modulation
modulación por rejilla grid modulation
modulación por rejilla de alta eficiencia high-efficiency grid modulation
modulación por rejilla de bajo nivel low-level grid modulation
modulación por rejilla de control control-grid modulation
modulación por rejilla-pantalla screen-grid modulation

modulación por rejilla sin distorsión undistorted grid modulation
modulación por rejilla supresora suppressor-grid modulation
modulación por sonido sound modulation
modulación por tiempo de impulsos pulse-time modulation
modulación positiva positive modulation
modulación residual residual modulation
modulación sin distorsión distortionless modulation
modulación sinusoidal sinusoidal modulation
modulación telefónica telephone modulation
modulación telegráfica telegraph modulation
modulación trapecial trapezoidal modulation
modulación vocal voice modulation
modulado modulated
modulado en amplitud amplitude modulated
modulado en fase phase-modulated
modulado en placa plate-modulated
modulado por impulsos pulse-modulated
modulado por tono tone-modulated
modulado por voz speech-modulated
modulador modulator
modulador accionado por voz voice-actuated modulator
modulador activo active modulator
modulador-amplificador modulator-amplifier
modulador cambiador de frecuencia frequency-changer modulator
modulador clase A class A modulator
modulador clase AB class AB modulator
modulador clase B class B modulator
modulador de amplitud amplitude modulator
modulador de bobina móvil movable-coil modulator
modulador de canal channel modulator
modulador de canal de servicio service-channel modulator
modulador de contacto contact modulator
modulador de crominancia chrominance modulator
modulador de chirrido chirp modulator
modulador de chispas spark-gap modulator
modulador de desplazamiento de audiofrecuencia audio-frequency-shift modulator
modulador de diodo diode modulator
modulador de fase phase modulator
modulador de fase magnético magnetic phase modulator
modulador de frecuencia frequency modulator
modulador de grupo group modulator
modulador de grupo maestro mastergroup modulator
modulador de Hall Hall modulator
modulador de impulsos pulse modulator
modulador de impulsos de radar radar-pulse modulator
modulador de inducción induction modulator
modulador de luz light modulator
modulador de luz ultrasónico ultrasonic light modulator
modulador de medida measuring modulator
modulador de onda cuadrada square-wave modulator
modulador de óxido de cobre copper-oxide modulator
modulador de portadora controlada controlled-carrier modulator
modulador de portadora variable variable-carrier modulator
modulador de producto product modulator
modulador de radar radar modulator
modulador de reactancia reactance modulator
modulador de subportadora de crominancia

chrominance-subcarrier modulator
modulador de supergrupo supergroup modulator
modulador de tipo de medidor meter-type modulator
modulador de tubo de inductancia inductance-tube modulator
modulador de tubo de reactancia reactance tube modulator
modulador de tubos de vacío vacuum-tube modulator
modulador de video video modulator
modulador-desmodulador modulator-demodulator
modulador electromecánico electromechanical modulator
modulador en anillo ring modulator
modulador en contrafase push-pull modulator
modulador en derivación shunt modulator
modulador en serie series modulator
modulador equilibrado balanced modulator
modulador estático static modulator
modulador galvanométrico galvanometer modulator
modulador heterodino heterodyne modulator
modulador lineal linear modulator
modulador magnético magnetic modulator
modulador mecánico mechanical modulator
modulador paramétrico parametric modulator
modulador pasivo passive modulator
modulador por efecto de Hall Hall-effect modulator
modulador por rejilla grid modulator
modulador pulsante pulsing modulator
modulador regenerativo regenerative modulator
modulador serrasoide serrasoid modulator
modulador telegráfico telegraph modulator
modular modulate
modularidad modularity
modularidad de programas software modularity
modularización modularization
módulo module, modulus
módulo compuesto composite module
módulo de adquisición de datos data-acquisition module
módulo de almacenamiento storage module
módulo de almacenamiento en disco disk storage module
módulo de bloqueo lockout module
módulo de carga load module
módulo de carga absoluta absolute load module
módulo de carga de almacenamiento storage load module
módulo de carga residente resident load module
módulo de clasificación sort module
módulo de comprobación de sistema system-check module
módulo de control control module
módulo de control de entrada-salida input-output control module
módulo de controlador driver module
módulo de convertidor de entrada input-converter module
módulo de datos data module
módulo de detección de contactos contact sense module
módulo de elasticidad modulus of elasticity
módulo de entrada input module
módulo de entrada-salida aislado isolated input-output module
módulo de evaluación de procesador processor-evaluation module
módulo de expansión de memoria memory expansion module
módulo de interfaz interface module
módulo de interfaz de periférico peripheral interface module

módulo de interfaz de procesador processor interface module
módulo de interfaz de sistema system interface module
módulo de interrupción de prioridad priority interrupt module
módulo de memoria memory module
módulo de procesador processor module
módulo de programa program module
módulo de programación programming module
módulo de refracción refractive modulus
módulo de reloj clock module
módulo de reloj de tiempo real real-time clock module
módulo de salida output module
módulo de salida de control control output module
módulo de salida digital digital output module
módulo de salida digital aislado isolated digital output module
módulo de servicios central central services module
módulo de tarjetas card module
módulo de temporización de microcomputadora microcomputer timing module
módulo de transferencia de registros register-transfer module
módulo de válvula valve module
módulo de visualización display module
módulo de Young Young's modulus
módulo encapsulado encapsulated module
módulo enchufable plug-in module
módulo ficticio dummy module
módulo fotovoltaico photovoltaic module
módulo fuente source module
módulo funcional functional module
módulo habilitado enabled module
módulo local fibroóptico fiber-optic local module
módulo lógico digital digital logic module
módulo microelectrónico microelectronic module
módulo multifunción multifunction module
módulo objeto object module
módulo residente resident module
módulo reubicable relocatable module
modulómetro modulometer
módulos inactivos inactive modules
molalidad molality
molaridad molarity
molde mold
moldeado molded
moldeado en la pieza molded-in
molécula molecule
molécula-gramo gram-molecule
molécula polar polar molecule
molecular molecular
moleculoy moleculoy
molibdeno molybdenum
momento moment
momento angular angular momentum
momento dipolar dipole moment
momento dipolar magnético magnetic dipole moment
momento eléctrico electric moment
momento magnético magnetic moment
momento magnético electrónico electron magnetic moment
momento magnético nuclear nuclear magnetic moment
momento multipolar multipole moment
monádico monadic
moniscopio moniscope
monitor monitor, screen
monitor activo active monitor
monitor básico basic monitor
monitor cardíaco cardiac monitor

monitor de aire continuo continuous air monitor
monitor de área area monitor
monitor de cámara camera monitor
monitor de cinta tape monitor
monitor de circuito abierto open-circuit monitor
monitor de colores color monitor
monitor de colores compuesto composite color
    monitor
monitor de comunicaciones communications monitor
monitor de condición atmosférica atmospheric
    condition monitor
monitor de contaminación de aire air contamination
    monitor
monitor de control de cámara camera control
    monitor
monitor de control maestro master control monitor
monitor de desviación deviation monitor
monitor de ejecución de programa
    program-execution monitor
monitor de equipo físico hardware monitor
monitor de fase phase monitor
monitor de fondo background monitor
monitor de forma de onda waveform monitor
monitor de frecuencia frequency monitor
monitor de imagen image monitor
monitor de línea de datos data line monitor
monitor de línea de energía power-line monitor
monitor de localización location monitor
monitor de manipulación keying monitor
monitor de microprocesador microprocessor
    monitor
monitor de modulación modulation monitor
monitor de multisincronización multisync monitor
monitor de onda continua continuous-wave monitor
monitor de operaciones operation monitor
monitor de portadora de imagen picture-carrier
    monitor
monitor de portadora de sonido sound-carrier
    monitor
monitor de potencia power monitor
monitor de programa program monitor
monitor de radar radar monitor
monitor de radiación radiation monitor
monitor de radiación directa direct radiation
    monitor
monitor de radiactividad radioactivity monitor
monitor de radio radio monitor
monitor de radiofaro direccional radio-range
    monitor
monitor de radiofrecuencia radio-frequency monitor
monitor de registro maestro master recording
    monitor
monitor de rendimiento performance monitor
monitor de salida output monitor
monitor de señal de video video-signal monitor
monitor de sistema system monitor
monitor de subportadora de colores
    color-subcarrier monitor
monitor de transmisión transmission monitor
monitor de video video monitor
monitor de visualización display monitor
monitor digital digital monitor
monitor en tiempo real real-time monitor
monitor lógico logical monitor
monitor maestro master monitor
monitor monocromo monochrome monitor
monitor panorámico panoramic monitor
monitor radiológico radiological monitor
monitor remoto remote monitor
monitoreo monitoring
monitoreo automático de sistema automatic
    monitoring of system
monitoreo de área area monitoring

monitoreo de fase phase monitoring
monitoreo de frecuencia frequency monitoring
monitoreo de nivel level monitoring
monitoreo de procesos process monitoring
monitoreo de sonido sound monitoring
monitoreo en tiempo real real-time monitoring
monitoreo periódico periodic monitoring
monitoreo por radar radar monitoring
monitoreo secuencial sequential monitoring
monitoreo visual visual monitoring
monitorización monitoring
monoanódico monoanodic
monoatómico monatomic
monoaural monaural
monoaxial monaxial
monocanal single-channel
monocapa monolayer
monocélula monocell
monocinético monokinetic
monoclínico monoclinic
monocristal monocrystal
monocristalización monocrystallization
monocromador monochromator
monocromador de cuarzo quartz monochromator
monocromaticidad monochromaticity
monocromático monochromatic
monocromo monochrome
monodo monode
monoenergético monoenergetic
monoespaciamiento monospacing
monoestable monostable
monofásico single-phase
monofier monofier
monofonía monophony
monofónico monophonic
monoimpulso monopulse
monoimpulso comparador de amplitudes
    amplitude-comparison monopulse
monolítico monolithic
monometálico monometallic
monomio monomial
monomolecular monomolecular
monopieza monopiece
monopolar monopolar
monopolo monopole
monopolo de cuarto de onda quarter-wave
    monopole
monopolo doblado folded monopole
monopolo magnético magnetic monopole
monoprogramación monoprogramming
monoscópico monoscopic
monoscopio monoscope
monostático monostatic
monotonicidad monotonicity
monotrón monotron
monovalente monovalent
monóxido de silicio silicon monoxide
montado en bastidor rack-mounted
montado en pared wall-mounted
montado sobre pivote pivot-mounted
montaje mounting, setup
montaje de instrumentos instrument mounting
montaje de panel panel mounting
montaje de pruebas breadboard
montaje de relé relay mounting
montaje en bastidor rack mounting
montaje preliminar electrónico electronic
    breadboard
montaje temporal electrónico electronic breadboard
montaje temporal sin soldadura solderless
    breadboard
montar set-up
montura mount

montura de bayoneta bayonet mount
montura de bolómetro bolometer mount
montura de bolómetro coaxial coaxial-bolometer
  mount
montura de frente muerto dead-front mounting
montura de jacks jack mounting
montura de pared wall mount
montura de prueba test mount
montura de techo roof mount
montura de termistor thermistor mount
montura de transistor transistor mount
montura de tubo tube mount
montura estabilizadora stabilizer mount
montura flexible flexible mounting
montura universal universal mounting
mosaico mosaic
motor motor
motor a prueba de goteo dripproof motor
motor a prueba de salpicaduras splashproof motor
motor abierto open motor
motor adherido bonded motor
motor arrollado en derivación shunt-wound motor
motor arrollado en serie series-wound motor
motor asincrónico asynchronous motor
motor autoventilado self-ventilated motor
motor auxiliar auxiliary motor
motor bifásico two-phase motor
motor compensado compensated motor
motor compuesto compound motor
motor compuesto diferencial differential compound
  motor
motor con autoarranque self-starting motor
motor con engranaje reductor gear motor
motor criogénico cryogenic motor
motor de accionamiento directo direct-drive motor
motor de aguja point machine
motor de alimentación feed motor
motor de altavoz loudspeaker motor
motor de anillos deslizantes slip-ring motor
motor de arranque starting motor
motor de arranque con resistencia resistance-start
  motor
motor de arranque por repulsión repulsion-start
  motor
motor de avance paso a paso stepping motor
motor de avance paso a paso de reluctancia
  variable variable-reluctance stepping motor
motor de bobina móvil moving-coil motor
motor de campo dividido split-field motor
motor de circuito impreso printed-circuit motor
motor de colector commutator motor
motor de colector de corriente alterna
  alternating-current commutator motor
motor de colector polifásico polyphase commutator
  motor
motor de condensador capacitor motor
motor de corriente alterna alternating-current motor
motor de corriente alterna-corriente continua
  alternating-current-direct-current motor
motor de corriente constante constant-current motor
motor de corriente continua direct-current motor
motor de corriente continua transistorizado
  transistorized direct-current motor
motor de dos velocidades two-speed motor
motor de exploración rápida rapid-scanning motor
motor de fase dividida split-phase motor
motor de histéresis hysteresis motor
motor de imán magnet motor
motor de impresión print engine
motor de inducción induction motor
motor de inducción de arranque por repulsión
  repulsion-start induction motor
motor de inducción de colector commutator

induction motor
motor de inducción de rotor bobinado wound-rotor
  induction motor
motor de inducción de uso general general-purpose
  induction motor
motor de inducción en jaula de ardilla squirrel-cage
  induction motor
motor de inducción lineal linear induction motor
motor de inducción monofásico single-phase
  induction motor
motor de inducción multivelocidad multispeed
  induction motor
motor de inducción polifásico polyphase induction
  motor
motor de inducción sincrónico synchronous
  induction motor
motor de par torque motor
motor de par constante constant-torque motor
motor de par de accionamiento directo direct-drive
  torque motor
motor de par variable variable-torque motor
motor de polo sombreado shaded-pole motor
motor de polos distribuidos distributed-pole motor
motor de potencia constante constant-power motor
motor de reacción reaction motor
motor de reluctancia reluctance motor
motor de repulsión repulsion motor
motor de repulsión-inducción repulsion-induction
  motor
motor de rotor bobinado wound-rotor motor
motor de Schrage Schrage motor
motor de taza de arrastre drag-cup motor
motor de uso definido definite-purpose motor
motor de uso especial special-purpose motor
motor de uso general general-purpose motor
motor de velocidad ajustable adjustable-speed motor
motor de velocidad constante constant-speed motor
motor de velocidad constante ajustable adjustable
  constant-speed motor
motor de velocidad lenta slow-speed motor
motor de velocidad variable variable-speed motor
motor de velocidad variable ajustable adjustable
  varying-speed motor
motor de velocidades múltiples multiple-speed
  motor
motor diferencial differential motor
motor eléctrico electric motor
motor en derivación shunt motor
motor en derivación polifásico polyphase shunt
  motor
motor en jaula de ardilla squirrel-cage motor
motor en serie series motor
motor en tándem tandem motor
motor fónico phonic motor
motor-generador motor-generator
motor hermético hermetic motor
motor lineal linear motor
motor logarítmico logarithmic motor
motor monofásico single-phase motor
motor multipolar multipolar motor
motor paramétrico parametric motor
motor piromagnético pyromagnetic motor
motor plano pancake motor
motor polifásico polyphase motor
motor primario prime mover
motor protegido contra agentes atmosféricos
  weather-protected motor
motor reversible reversible motor
motor selsyn selsyn motor
motor semiprotegido semiprotected motor
motor sincrónico synchronous motor
motor sincrónico-asincrónico
  synchronous-asynchronous motor

motor sincrónico con autoarranque self-starting synchronous motor
motor sincrónico de polo sólido solid-pole synchronous motor
motor síncrono polifásico polyphase synchronous motor
motor totalmente encerrado totally-enclosed motor
motor trifásico three-phase motor
motor universal universal motor
motor ventilado ventilated motor
motorizado motorized
móvil movable, mobile
movilidad mobility
movilidad de deslizamiento drift mobility
movilidad de Hall Hall mobility
movilidad de huecos hole mobility
movilidad de portadores carrier mobility
movilidad electrónica electronic mobility
movilidad intrínseca intrinsic mobility
movilidad iónica ionic mobility
movimiento movement
movimiento browniano brownian movement
movimiento de bloque block movement
movimiento de cinta tape movement
movimiento de cursor cursor movement
movimiento de datos data movement
movimiento de imagen lateral lateral image movement
movimiento incremental incremental movement
movimiento natural natural movement
mu mu
muaré moire
muerte lenta slow death
muerte repentina sudden death
muesca notch, slot
muesca de rechazo rejection notch
muestra sample
muestra aleatoria random sample
muestra analógica analog sample
muestra de comprobación check sample
muestra instantánea instantaneous sample
muestreo sampling
muestreo aleatorio random sampling
muestreo analógico analog sampling
muestreo de bits bit sampling
muestreo de corriente current sensing
muestreo de telemedida telemetering sampling
muestreo discreto discrete sampling
muestreo instantáneo instantaneous sampling
muestreo secuencial sequential sampling
muestreo y retención sample and hold
multiabertura multiaperture
multiacceso multiaccess
multiacoplador multicoupler
multiacoplador de antena antenna multicoupler
multiagujero multihole
multialambre multiwire
multiánodo multianode
multiar multiar
multiarchivo multifile
multibanda multiband
multibucle multiloop
multicanal multichannel
multicañón multigun
multicapa multilayer
multicátodo multicathode
multicavidad multicavity
multicelular multicellular
multiciclo multicycle
multicinta multitape
multicircuito multicircuit
multicomponente multicomponent
multicomputadora multicomputer

multicón multicon
multiconductor multiconductor
multiconmutación multiswitching
multiconmutador multiswitch
multicontacto multicontact
multicorriente multicurrent
multicromo multichrome
multichip multichip
multidimensional multidimensional
multidirección multiaddress
multidireccional multidirectional
multielectrodo multielectrode
multielemento multielement
multiemisor multiemitter
multienlace multilink
multiescala multirange
multiescalonado multistep
multiestable multistable
multiestación multistation
multiestado multistate
multietapa multistage
multifásico multiphase
multifibra multifiber
multifilamento multifilament
multiflujo multiflow
multifocal multifocal
multifrecuencia multifrequency
multifunción multifunction
multigama multirange
multigrupo multigroup
multihaz multibeam
multihilo multiwire
multihoja multileaf
multilínea multiline
multimalla multimesh
multimarcador de video video multimarker
multímetro multimeter
multímetro digital digital multimeter
multímetro electrónico electronic multimeter
multímetro universal universal multimeter
multimodal multimodal
multimodo multimode
multinivel multilevel
multioperacional multioperational
multipactor multipactor
multipar multipair
multipasada multipass
multiperforación multihole, multipunching, gang punching
multipista multitrack
múltiple multiple
múltiple de comprobación check multiple
múltiple parcial partial multiple
múltiplex multiplex
múltiplex asincrónico asynchronous multiplex
múltiplex de tonos tone multiplex
múltiplex digital digital multiplex
múltiplex en tiempo time multiplex
múltiplex por división de espacio space-division multiplex
múltiplex por división de fase phase-division multiplex
múltiplex por división de frecuencia frequency-division multiplex
múltiplex por división de tiempo time-division multiplex
múltiplex por frecuencia frequency multiplex
múltiplex por modos de impulsos pulse-mode multiplex
múltiplex por modulación de impulsos pulse-modulation multiplex
multiplexión multiplexing
multiplexión de bytes byte multiplexing

multiplexión de canales channel multiplexing
multiplexión de radio radio multiplexing
multiplexión digital digital multiplexing
multiplexión en tiempo time multiplexing
multiplexión óptica optical multiplexing
multiplexión por división de frecuencia
    frequency-division multiplexing
multiplexión por división de tiempo time-division
    multiplexing
multiplexor multiplexer
multiplexor analógico analog multiplexer
multiplexor asincrónico asynchronous multiplexer
multiplexor de alto nivel high-level multiplexer
multiplexor de bloque block multiplexer
multiplexor de canal de datos data channel
    multiplexer
multiplexor de datos data multiplexer
multiplexor de impedancia constante
    constant-impedance multiplexer
multiplexor de módem modem multiplexer
multiplexor de primer orden first-order multiplexer
multiplexor de salida output multiplexer
multiplexor diferencial differential multiplexer
multiplexor óptico optical multiplexer
multiplicación de carga charge multiplication
multiplicación de colector collector multiplication
multiplicación de frecuencia frequency
    multiplication
multiplicación de gas gas multiplication
multiplicación de patrones pattern multiplication
multiplicación electrónica electron multiplication
multiplicación lógica logic multiplication
multiplicación simultánea simultaneous
    multiplication
multiplicador multiplier
multiplicador analógico analog multiplier
multiplicador analógico de Hall Hall analog
    multiplier
multiplicador constante constant multiplier
multiplicador de alcance range multiplier
multiplicador de alcance de tensión voltage-range
    multiplier
multiplicador de campo hiperbólico hyperbolic-field
    multiplier
multiplicador de campos cruzados crossed-field
    multiplier
multiplicador de coincidencia coincidence multiplier
multiplicador de electrones secundarios
    secondary-electron multiplier
multiplicador de emisión secundaria
    secondary-emission multiplier
multiplicador de escala scale multiplier
multiplicador de Farnsworth Farnsworth multiplier
multiplicador de fase phase multiplier
multiplicador de fotorresistor photoresistor
    multiplier
multiplicador de frecuencia frequency multiplier
multiplicador de frecuencia de varactor varactor
    frequency multiplier
multiplicador de frecuencia estático static frequency
    multiplier
multiplicador de frecuencia klistrón klystron
    frequency multiplier
multiplicador de frecuencia por reactancia
    reactance frequency multiplier
multiplicador de Hall Hall multiplier
multiplicador de haz circular circular-beam
    multiplier
multiplicador de instrumento instrument multiplier
multiplicador de modulación de frecuencia y
    modulación de amplitud
    frequency-modulation-amplitude-modulation
    multiplier

multiplicador de probabilidad probability multiplier
multiplicador de Q Q multiplier
multiplicador de tensión voltage multiplier
multiplicador de vóltmetro voltmeter multiplier
multiplicador digital digital multiplier
multiplicador dinamométrico dynamometer
    multiplier
multiplicador electrónico electronic multiplier,
    electron multiplier
multiplicador electrónico de impulsos pulse
    electronic multiplier
multiplicador electrónico magnético magnetic
    electron multiplier
multiplicador fijo fixed multiplier
multiplicador fotoeléctrico photoelectric multiplier
multiplicador instantáneo instantaneous multiplier
multiplicador orbital orbital multiplier
multiplicador por división de tiempo time-division
    multiplier
multiplicador por efecto de Hall Hall-effect
    multiplier
multiplicando multiplicand
múltiplo integral integral multiple
multipolar multipole, multipolar
multiposición multiposition
multiprecisión multiprecision
multiprobador multitester
multiprocesador multiprocessor
multiprocesamiento multiprocessing
multiproceso multiprocess
multiprograma multiprogram
multiprogramación multiprogramming
multipuerta multiport
multipuerto multiport
multipunto multipoint
multisalida multioutput
multisección multisection
multisecuencia multisequence
multisegmento multisegment
multisensor multisensor
multisistema multisystem
multitarea multitask
multitensión multivoltage
multiterminal multiterminal
multitomacorriente multioutlet
multitono multitone
multitrazo multitrace
multitrenza multistrand
multitrón multitron
multitubo multitube
multiunidad multiunit
multiunión multijunction
multiusuario multiuser
multivalente multivalent
multivariable multivariable
multivelocidad multispeed
multivibrador multivibrator
multivibrador acoplado por cátodo cathode-coupled
    multivibrator
multivibrador asimétrico asymmetrical
    multivibrator
multivibrador astable astable multivibrator
multivibrador biestable bistable multivibrator
multivibrador de acoplamiento de placa
    plate-coupled multivibrator
multivibrador de acoplamiento electrónico
    electron-coupled multivibrator
multivibrador de arranque-parada start-stop
    multivibrator
multivibrador de ciclo único one-cycle multivibrator
multivibrador de Eccles-Jordan Eccles-Jordan
    multivibrator
multivibrador de lámparas de neón neon-bulb

multivibrator
multivibrador de Potter Potter multivibrator
multivibrador de retardo delay multivibrator
multivibrador de transistores transistor
   multivibrator
multivibrador desequilibrado unbalanced
   multivibrator
multivibrador divisor de frecuencia
   frequency-dividing multivibrator
multivibrador equilibrado balanced multivibrator
multivibrador horizontal horizontal multivibrator
multivibrador interacoplado cross-coupled
   multivibrator
multivibrador libre free-running multivibrator
multivibrador monoestable monostable
   multivibrator, one-shot multivibrator, univibrator
multivibrador simétrico symmetrical multivibrator
multivibrador sincronizado synchronized
   multivibrator
multivolumen multivolume
multivuelta multiturn
mumetal mumetal
murmullo babble
muro posterior backwall
música electrónica electronic music
mutilado mutilated
mutuamente mutually
muy alta frecuencia very high frequency
muy alta resistencia very high resistance
muy alta tensión very high tension
muy alta velocidad very high speed
muy baja frecuencia very low frequency
muy baja resistencia very low resistance
muy corto alcance very short range
muy largo alcance very long range

N

nano nano
nanoalmacenamiento nanostore
nanoampere nanoampere
nanocircuito nanocircuit
nanocomputadora nanocomputer
nanofarad nanofarad
nanohenry nanohenry
nanoinstrucción nanoinstruction
nanosegundo nanosecond
nanovolt nanovolt
nanovóltmetro nanovoltmeter
nanowatt nanowatt
nanowáttmetro nanowattmeter
navaglobo navaglobe
navar navar
navegación a la estima dead-reckoning navigation
navegación ciega blind navigation
navegación con satélite satellite navigation
navegación de aproximación approach navigation
navegación de largo alcance long-range navigation
navegación decca decca navigation
navegación electrónica electronic navigation
navegación hiperbólica hyperbolic navigation
navegación por radar radar navigation
navegación rho-theta rho-theta navigation
navegador de Doppler Doppler navigator
neblina metálica metal fog
nefelometría nephelometry
nefelómetro nephelometer
negación negation
negación alternativa alternative denial
negado negated
negador negator
negar negate
negativamente negatively
negativo negative
negativo común common negative
negativo metálico metal negative
negativo original original master, metal negative,
   metal master
negatrón negatron
negro de humo lampblack
negro nominal nominal black
negro tras blanco black after white
nemático nematic
neodimio neodymium
neón neon
neopreno neoprene
neotrón neotron
neper neper
neptunio neptunium
nesistor nesistor
neumático pneumatic
neuristor neuristor
neuroelectricidad neuroelectricity
neutralidad eléctrica electrical neutrality
neutralización neutralization
neutralización anódica anode neutralization
neutralización catódica cathode neutralization
neutralización cruzada cross-neutralization
neutralización de carga charge neutralization

neutralización de Hazeltine Hazeltine neutralization
neutralización de placa plate neutralization
neutralización de rejilla grid neutralization
neutralización de Rice Rice neutralization
neutralización entrecruzada crisscross
    neutralization
neutralización inductiva inductive neutralization
neutralización por bobina coil neutralization
neutralización por conexión cruzada
    cross-connected neutralization
neutralizado neutralized
neutralizador de carga charge neutralizer
neutralizante neutralizing
neutralizar neutralize
neutro neutral
neutro aislado insulated neutral
neutro conectado a tierra earthed neutral
neutro flotante floating neutral
neutrodino neutrodyne
neutrón neutron
neutrón de baja energía low-energy neutron
neutrón frío cold neutron
neutrón térmico thermal neutron
neutrónica neutronics
neutrónico neutronic
newton newton
nexo nexus
nibble nibble
nicromo nichrome
nieve snow
nilón nylon
niobio niobium
níquel nickel
níquel-cadmio nickel-cadmium
níquel-hierro nickel-iron
niquelado nickel-plating
nit nit
nitidez sharpness
nitidez de imagen image sharpness
nítido sharp
nitrato de celulosa cellulose nitrate
nitrocelulosa nitrocellulose
nitrógeno nitrogen
nivel level
nivel absoluto absolute level
nivel aceptor acceptor level
nivel ambiente ambient level
nivel cero zero level
nivel compuesto composite level
nivel cuántico quantum level
nivel de acceso access level
nivel de acceso físico physical access level
nivel de acceso lógico logical access level
nivel de actividad activity level
nivel de aislamiento insulation level, isolation level
nivel de almacenamiento storage level
nivel de amplitud amplitude level
nivel de bajada drop level
nivel de base base level
nivel de borrado blank level
nivel de calibración calibration level
nivel de calidad aceptable acceptable quality level
nivel de campo field level
nivel de código code level
nivel de color color level
nivel de confianza confidence level
nivel de congelación freezing level
nivel de control control level
nivel de conversación speech level
nivel de corriente absoluto absolute current level
nivel de corriente continua direct-current level
nivel de corriente relativo relative current level
nivel de cuantificación quantization level

nivel de cuasipico quasi-peak level
nivel de datos data level
nivel de decisión decision level
nivel de demarcación demarcation level
nivel de descarga parcial terminal aceptable
    acceptable terminal partial discharge level
nivel de diafonía crosstalk level
nivel de direccionamiento addressing level
nivel de diseño design level
nivel de disparo triggering level
nivel de documentación documentation level
nivel de dopado doping level
nivel de eco echo level
nivel de energía energy level
nivel de energía atómica atomic energy level
nivel de energía molecular molecular energy level
nivel de energía normal normal energy level
nivel de enlace de datos data link level
nivel de enmascaramiento masking level
nivel de entrada entry level, input level
nivel de extinción blanking level
nivel de Fermi Fermi level
nivel de flujo de luz light-flux level
nivel de fondo background level
nivel de fusión merge level
nivel de imagen medio average picture level
nivel de impedancia impedance level
nivel de impurezas impurity level
nivel de índice index level
nivel de integración integration level
nivel de intensidad intensity level
nivel de intensidad sonora sound intensity level,
    loudness level
nivel de interferencia interference level
nivel de interferencia sonora sound interference
    level
nivel de intermodulación intermodulation level
nivel de interrupción interrupt level
nivel de interrupción de prioridad priority interrupt
    level
nivel de limitación clipping level
nivel de línea line level
nivel de mantenimiento maintenance level
nivel de modificación modification level
nivel de modulación modulation level
nivel de modularidad modularity level
nivel de negro black level
nivel de operación operate level
nivel de operación normal normal operating level
nivel de paquete packet level
nivel de pedestal pedestal level
nivel de perturbación blur level
nivel de portadora carrier level
nivel de portadora medio average carrier level
nivel de potencia power level
nivel de potencia absoluta absolute power level
nivel de potencia alta crítico critical high-power
    level
nivel de potencia relativa relative power level
nivel de potencia sonora sound-power level
nivel de presión pressure level
nivel de presión de banda band pressure level
nivel de presión de banda de octava octave-band
    pressure level
nivel de presión de octava octave pressure level
nivel de presión sonora sound-pressure level
nivel de presión sonora de banda band sound
    pressure level
nivel de prioridad priority level
nivel de procedimiento procedure level
nivel de procesamiento processing level
nivel de procesamiento de inicialización de trabajo
    job initialization processing level

nivel de procesamiento general general processing level
nivel de programa program level
nivel de protección contra sobretensión overvoltage protection level
nivel de protocolo protocol level
nivel de prueba test level
nivel de prueba no adaptado through level
nivel de radiación radiation level
nivel de radiorruido radio noise level
nivel de recepción reception level
nivel de redundancia de línea line redundancy level
nivel de referencia reference level
nivel de referencia de amplitud amplitude reference level
nivel de referencia de visualización display reference level
nivel de referencia normal standard reference level
nivel de registro recording level
nivel de registro normal normal recording level
nivel de resonancia resonance level
nivel de ruido noise level
nivel de ruido de circuito circuit-noise level
nivel de ruido de fondo background noise level
nivel de ruido de fondo energizado aceptable acceptable energized background noise level
nivel de ruido de modulación de amplitud amplitude-modulation noise level
nivel de ruido de portadora carrier noise level
nivel de ruido de video video noise level
nivel de ruido global overall noise level
nivel de ruido ponderado weighted noise level
nivel de ruido total total noise level
nivel de salida output level
nivel de salida patrón standard output level
nivel de saturación saturation level
nivel de seguridad security level
nivel de selección selection level
nivel de sensibilidad sensitivity level
nivel de señal signal level
nivel de señal de facsímil facsimile-signal level
nivel de señal mínimo minimum signal level
nivel de servicio service level
nivel de sesión session level
nivel de silenciamiento quieting level
nivel de sincronización synchronizing level
nivel de sobrecarga overload level
nivel de sonido diurno-nocturno day-night sound level
nivel de sonido ponderado weighted sound level
nivel de tasa de fallas failure rate level
nivel de tensión voltage level
nivel de tensión absoluto absolute voltage level
nivel de tensión relativo relative voltage level
nivel de tono tone level
nivel de transición de código code transition level
nivel de transmisión transmission level
nivel de transmisión relativo relative transmission level
nivel de vacío vacuum level
nivel de velocidad velocity level
nivel de volumen volume level
nivel de voz medio average voice level
nivel de zumbido hum level
nivel del blanco white level
nivel del blanco de referencia reference white level
nivel del gris gray level
nivel del negro picture black
nivel del negro de referencia reference black level
nivel discreto discrete level
nivel donador donor level
nivel espectral spectrum level
nivel fundamental ground level

nivel lógico logic level
nivel luminoso luminous level
nivel máximo de registro maximum record level
nivel máximo de señal maximum signal level
nivel medio average level
nivel molecular molecular level
nivel muerto dead level
nivel musical musical level
nivel nominal nominal level
nivel pico peak level
nivel pico de señal peak signal level
nivel ponderado weighted level
nivel relativo relative level
nivel secundario secondary level
nivel seguro de soltar safe let-go level
nivel silenciador squelch level
nivel sobre el umbral level above threshold
nivel sonoro sound level
nivel superior higher level
nivel telegráfico telegraph level
nivel umbral threshold level
nivel vacante vacant level
nivel vacío empty level
nivelación leveling
nivelación de zona zone leveling
nivelador de potencia power leveler
no aislado noninsulated
no ajustable nonadjustable
no amortiguado undamped
no amplificador nonamplifying
no aritmético nonarithmetic
no asignado unassigned
no atenuado unattenuated
no audible nonaudible
no automantenido nonself-maintained
no automático nonautomatic
no autosincronizante nonself-synchronizing
no bloqueado unblocked
no borrable nonerasable
no borrado unerased
no calibrado uncalibrated
no cancelable noncancellable
no cargado nonloaded
no clasificado unsorted
no codificado uncoded
no coherente noncoherent
no compensado uncompensated
no conductor nonconductor
no continuo noncontinuous
no controlado uncontrolled
no corrosivo noncorrosive
no cristalino noncrystalline
no crítico noncritical
no depurado undebugged
no desmodulante nondemodulating
no destructivo nondestructive
no direccionable nonaddressable
no direccional nondirectional
no directivo nondirective
no disipativo nondissipative
no dispersivo nondispersive
no dopado undoped
no ejecutable nonexecutable
no elástico nonelastic
no eléctrico nonelectrical
no electrificado nonelectrified
no electrólito nonelectrolyte
no electrónico nonelectronic
no empotrado nonembedded
no en blanco nonblank
no entrelazado noninterlaced
no equivalente nonequivalent
no esencial nonessential

no especificado unspecified
no específico nonspecific
no especular nonspecular
no estacionario nonstationary
no estructural nonstructural
no férrico nonferrous
no ferroso nonferrous
no filtrado unfiltered
no formateado unformatted
no higroscópico nonhygroscopic
no homogéneo nonhomogeneous
no inducido noninduced
no inductivo noninductive
no inflamable nonflammable
no intercambiable noninterchangeable
no iónico nonionic
no ionizante nonionizing
no lineal nonlinear
no linealidad nonlinearity
no linealidad de amplificador amplifier nonlinearity
no linealidad de fase phase nonlinearity
no linealidad diferencial differential nonlinearity
no listo not ready
no local nonlocal
no localizado nonlocalized
no magnético nonmagnetic
no matemático nonmathematical
no mecánico nonmechanical
no metal nonmetal
no metálico nonmetallic
no modificado unmodified
no modulado unmodulated
no neutralizado unneutralized
no numerado unnumbered
no numérico nonnumeric
NO-O excluyente exclusive NOR
no óhmico nonohmic
no operable nonoperable
no operacional nonoperational
no operativo nonoperating
no oscilante nonoscillating
no oscilatorio nonoscillatory
no paralelo nonparallel
no paramétrico nonparametric
no pareado nonpaired
no perforado unperforated
no periódico nonperiodic
no planar nonplanar
no polar nonpolar
no polarizable nonpolarizable
no polarizado nonpolarized, unbiased
no ponderado unweighted
no ponteante nonbridging
no privilegiado nonprivileged
no procesable unprocessable
no procesado unprocessed
no productivo nonproductive
no programable unprogrammable
no programado unprogrammed
no radiactivo nonradioactive
no reactivo nonreactive
no recargable nonrechargeable
no reciprocidad nonreciprocity
no recíproco nonreciprocal
no recuperable nonrecoverable
no recurrente nonrecurrent
no reflejante nonreflecting
no regenerativo nonregenerative
no regulado unregulated
no removible nonremovable
no renovable nonrenewable
no repetitivo nonrepetitive
no residente nonresident

no resonante nonresonating
no reversible nonreversible
no saturable nonsaturable
no saturado nonsaturated
no secuencial nonsequential
no segmentado unsegmented
no selectivo nonselective
no separable nonseparable
no simétrico nonsymmetrical
no simultáneo nonsimultaneous
no sincrónico nonsynchronous
no sincronizado nonsynchronized
no sintonizado untuned
no sinusoidal nonsinusoidal
no solapante nonoverlapping
no térmico nonthermal
no uniforme nonuniform
no ventilado nonventilated
no verificado unverified
no volátil nonvolatile
nobelio nobelium
noble noble
noctovisión noctovision
nodal nodal
nodo node, branch point
nodo ascendente ascending node
nodo de corriente current node
nodo de enrutamiento intermedio intermediate routing node
nodo de presión node of pressure
nodo de procesamiento de datos data-processing node
nodo de red network node
nodo de subárea subarea node
nodo de tensión voltage node
nodo dependiente dependent node
nodo dividido split node
nodo estacionario stationary node
nodo inactivo inactive node
nodo intermedio intermediate node
nodo maestro master node
nodo menor minor node
nodo parcial partial node
nodo periférico peripheral node
nodo principal major node
nódulos nodules
nombre alternativo alternate name
nombre de acceso access name
nombre de archivo file name
nombre de biblioteca library name
nombre de campo field name
nombre de clase class name
nombre de condición condition name
nombre de conjunto de datos data set name
nombre de correlación correlation name
nombre de datos data name
nombre de definición de datos data definition name
nombre de diseño de letra genérico generic font name
nombre de dispositivo device name
nombre de entrada entry name
nombre de grupo group name
nombre de lenguaje language name
nombre de mando command name
nombre de modo mode name
nombre de procedimiento procedure name
nombre de programa program name
nombre de registro record name
nombre de sección section name
nombre de trabajo job name
nombre de trayectoria path name
nombre de unidad lógica logical unit name
nombre genérico generic name

nombre global global name
nombre mnemónico mnemonic name
nombre nulo null name
nombre simbólico symbolic name
nomenclatura nomenclature
nominal nominal
nonio vernier
nonio de frecuencia frequency vernier
nonio de sintonización fina fine-tuning vernier
norma standard
normal normal
normal a la onda wave normal
normalización normalization, standardization
normalización de colores color standardization
normalizado normalized, standardized
normalizar normalize, standardize
normalmente normally
normalmente abierto normally open
normalmente cerrado normally closed
norte magnético magnetic north
nota de batido beat note
notación notation
notación alfanumérica alphanumeric notation
notación binaria binary notation
notación biquinaria biquinary notation
notación científica scientific notation
notación codificada coded notation
notación codificada en binario binary-coded
  notation
notación compleja complex notation
notación de base base notation, radix notation
notación de base fija fixed-base notation
notación de base mixta mixed-base notation
notación de coma fija fixed-point notation
notación de coma flotante floating-point notation
notación de Lukasiewicz Lukasiewicz notation
notación decimal decimal notation
notación decimal codificada coded-decimal notation
notación decimal codificada en binario
  binary-coded decimal notation
notación duodecimal duodecimal notation
notación hexadecimal hexadecimal notation
notación infija infix notation
notación matricial matrix notation
notación mixta mixed notation
notación mnemónica mnemonic notation
notación numérica number notation
notación octal octal notation
notación octal codificada en binario binary-coded
  octal notation
notación por prefijos prefix notation
notación posicional positional notation
notación sexadecimal sexadecimal notation
notación simbólica symbolic notation
notación ternaria ternary notation
noval noval
novar novar
noy noy
nu nu
nube electrónica electron cloud
nuclear nuclear
núcleo nucleus, core
núcleo abierto open core
núcleo binario binary core
núcleo cerámico ceramic core
núcleo cerrado closed core
núcleo de abertura aperture core
núcleo de carrete bobbin core
núcleo de cinta tape core
núcleo de cinta arrollada tape-wound core
núcleo de cinta magnética magnetic-tape core
núcleo de electroimán magnet core
núcleo de ferrita ferrite core

núcleo de hierro iron core
núcleo de hierro laminado laminated iron core
núcleo de hierro pulverizado powdered-iron core
núcleo de inserción insert core
núcleo de memoria memory core
núcleo de polvo dust core
núcleo de polvo magnético magnetic powder core
núcleo de relé relay core
núcleo de resistor resistor core
núcleo de sintonización tuning core
núcleo de transformador transformer core
núcleo del inducido armature core
núcleo dieléctrico dielectric core
núcleo en E E core
núcleo ferromagnético ferromagnetic core
núcleo ferromagnético laminado laminated
  ferromagnetic core
núcleo hueco hollow core
núcleo laminado laminated core
núcleo magnético magnetic core
núcleo magnético saturable saturable magnetic core
núcleo no saturado unsaturated core
núcleo ranurado slotted core
núcleo saturable saturable core
núcleo saturado saturated core
núcleo toroidal toroidal core
nucleónica nucleonics
nulo null
numeración numeration, numbering
numeración automática automatic numbering
numeración consecutiva consecutive numbering
numeración de base radix numeration
numeración de base fija fixed-radix numeration
numeración en serie serial numbering
numeración nacional national numbering
numerado en serie serially numbered
numerador numerator
numeral numeral
numeral binario binary numeral
numeralización numeralization
numéricamente numerically
numérico numerical
número aleatorio random number
número atómico atomic number
número binario binary number
número biquinario biquinary number
número codificado coded number
número complejo complex number
número cuántico quantum number
número de autocomprobación self-checking number
número de banda band number
número de banda de frecuencias frequency-band
  number
número de base base number, radix number
número de base mixta mixed-base number
número de bloque relativo relative block number
número de capas number of layers
número de coma fija fixed-point number
número de coma flotante floating-point number
número de comprobación check number
número de conexión connection number
número de control control number
número de designación designation number
número de dispositivo device number
número de doble precisión double-precision number
número de equipo equipment number
número de Fibonacci Fibonacci number
número de Fresnel Fresnel number
número de generación generation number
número de grupo group number
número de grupo lógico logical group number
número de identificación de circuito circuit
  identification number

número de identificación personal personal identification number
número de impulsos pulse number
número de línea line number
número de líneas de exploración number of scanning lines
número de longitud múltiple multiple-length number
número de llamada call number
número de Lorentz Lorentz number
número de masa mass number
número de modo mode number
número de nivel level number
número de onda wave number
número de operación operation number
número de oscilación oscillation number
número de página page number
número de panel panel number
número de pista track number
número de ranura slot number
número de Rayleigh Rayleigh number
número de referencia reference number
número de registro record number
número de registro relativo relative-record number
número de sección section number
número de secuencia sequence number
número de secuencia de archivo file sequence number
número de segmento segment number
número de serie de bloque block serial number
número de serie de circuito circuit serial number
número de trabajo job number
número de unidad drive number
número de unidad lógica logical unit number
número decimal codificado en binario binary-coded decimal number
número decimal repetitivo repeating decimal number
número en serie serial number
número entero integral number
número hexadecimal hexadecimal number
número imaginario imaginary number
número internacional international number
número irracional irrational number
número lógico logical number
número mixto mixed number
número nacional national number
número natural natural number
número negativo negative number
número normal normal number
número octal octal number
número positivo positive number
número racional rational number
número real real number
número redondeado rounded number
número sexadecimal sexadecimal number
número simbólico symbolic number
números preferidos preferred numbers
números seudoaleatorios pseudorandom numbers
nutación nutation
nuvistor nuvistor

O excluyente exclusive OR
objetivo objective lens
objetivo de distancia focal ajustable zoom lens
objeto global global object
objeto interno internal object
oblea wafer
oblea de germanio germanium wafer
oblea de microcircuitos microcircuit wafer
oblea de microelemento microelement wafer
oblea de silicio silicon wafer
oblicuidad obliquity, skew, skewing
oblicuo oblique
oboe oboe
obscurecimiento darkening
observable observable
observación observation
observación de servicio service observation
observacional observational
observador patrón standard observer
obsolescencia obsolescence
obsoleto obsolete
obturador shutter
obturador de guíaondas waveguide shutter
obturador de Kerr Kerr shutter
obturador electroóptico de Kerr Kerr electrooptical shutter
octal octal
octal codificado en binario binary-coded octal
octava octave
octeto octet
octodo octode
octonario octonary
ocular eyepiece
ocultación occultation
ocupación occupation
ocupación media mean occupancy
ocurrencia occurrence
odógrafo odograph
odómetro odometer
oersted oersted
oficina automatizada automated office
oficina central local local central office
oficina central manual manual central office
ofitrón ophitron
ohm ohm
ohm absoluto absolute ohm
ohm acústico acoustic ohm
ohm-centímetro ohm-centimeter
ohm internacional international ohm
ohm-milla ohm-mile
ohm patrón standard ohm
ohm real true ohm
ohm recíproco reciprocal ohm
óhmetro ohmmeter
óhmetro de tubo de vacío vacuum-tube ohmmeter
ohmiaje ohmage
óhmico ohmic
ohms por cuadrado ohms per square
ohms por volt ohms per volt
oído artificial artificial ear
oído artificial de referencia reference artificial ear

ojo artificial artificial eye
ojo eléctrico electric eye
ojo electrónico electronic eye
ojo fotoeléctrico photoelectric eye
ojo mágico magic eye
omega omega
omegatrón omegatron
ómnibus eléctrico electric bus
ómnibus de trole trolley bus
omnidireccional omnidirectional
omnidirectivo omnidirective
omnígrafo omnigraph
onda wave
onda absorbida absorbed wave
onda acústica acoustic wave
onda amortiguada damped wave
onda amortiguada manipulada keyed damped wave
onda armónica harmonic wave
onda armónica simple simple harmonic wave
onda asimétrica asymmetrical wave
onda beta beta wave
onda característica characteristic wave
onda celeste skywave
onda ciclotrónica cyclotron wave
onda cilíndrica cylindrical wave
onda compartida shared wave
onda compensadora compensating wave
onda complementaria complementary wave
onda componente magnetoiónica magnetoionic wave component
onda con polarización dextrorsa clockwise-polarized wave
onda con polarización levógira left-hand polarized wave
onda cónica conical wave
onda continua continuous wave
onda continua de corriente alterna alternating-current continuous wave
onda continua interrumpida interrupted continuous wave
onda continua manipulada keyed continuous wave
onda continua modulada modulated continuous wave
onda corta shortwave
onda cosenoidal cosine wave
onda cuadrada square wave
onda cuasicuadrada quasi-square wave
onda cuasióptica quasi-optical wave
onda cuasirectangular quasi-rectangular wave
onda de Alfven Alfven wave
onda de canal channel wave
onda de carga espacial space-charge wave
onda de compresión compression wave
onda de contestación answering wave
onda de corriente current wave
onda de choque shock wave
onda de choque completa full-impulse wave
onda de choque magnetohidrodinámica magnetohydrodynamic shock wave
onda de doble impulso double-pulse wave
onda de doble pico double-hump wave
onda de espacio spacing wave
onda de espín spin wave
onda de excitación excitation wave
onda de fase phase wave
onda de frecuencia variable variable-frequency wave
onda de impulsos impulse wave
onda de incidencia oblicua oblique-incidence wave
onda de Lamb Lamb wave
onda de límite boundary wave
onda de llamada calling wave
onda de manipulación keying wave

onda de modulación modulation wave
onda de modulación de fase phase-modulation wave
onda de partículas particle wave
onda de plasma plasma wave
onda de polarización elíptica elliptically polarized wave
onda de radar radar wave
onda de radio radio wave
onda de radio atmosférica atmospheric radio wave
onda de radio de muy alta frecuencia very-high-frequency radio wave
onda de Rayleigh Rayleigh wave
onda de recepción receive wave
onda de retorno return wave
onda de ruido noise wave
onda de salida outgoing wave, output wave
onda de señal signal wave
onda de socorro distress wave
onda de superficie surface wave
onda de superficie de Rayleigh Rayleigh surface wave
onda de trabajo working wave, marking wave
onda de tráfico traffic wave
onda difractada diffracted wave
onda directa direct wave
onda dispersa scattered wave
onda diurna diurnal wave
onda divergente diverging wave
onda dominante dominant wave
onda elástica elastic wave
onda eléctrica electric wave
onda eléctrica circular circular electric wave
onda eléctrica transversal transverse electric wave
onda electromagnética guiada guided electromagnetic wave
onda electromagnética híbrida hybrid electromagnetic wave
onda electromagnética periódica periodic electromagnetic wave
onda electromagnética plana plane electromagnetic wave
onda electromagnética sinusoidal sinusoidal electromagnetic wave
onda electromagnética transversal transverse electromagnetic wave
onda electrónica electron wave
onda electrostática electrostatic wave
onda electrotónica electrotonic wave
onda en diente de sierra sawtooth wave
onda en diente de sierra lineal linear sawtooth waveform
onda en escalera staircase wave
onda escalonada stepped wave
onda esférica spherical wave
onda espacial space wave, skywave
onda estacionaria standing wave, stationary wave
onda evanescente evanescent wave
onda extraordinaria extraordinary wave
onda fundamental fundamental wave
onda gravitacional gravity wave
onda gravitacional acústica acoustic-gravity wave
onda guiada guided wave
onda hectométrica hectometric wave
onda híbrida hybrid wave
onda hidromagnética hydromagnetic wave
onda horizontal horizontal wave
onda incidente incident wave
onda indirecta indirect wave
onda interferente interfering wave
onda inversa reverse wave
onda ionosférica ionospheric wave
onda irruptiva surge
onda kilométrica kilometric wave

onda lambda lambda wave
onda larga long wave
onda lateral lateral wave
onda lenta slow wave
onda longitudinal longitudinal wave
onda luminosa light wave
onda magnética circular circular magnetic wave
onda magnética transversal transverse magnetic wave
onda magnetoacústica magnetoacoustic wave
onda magnetohidrodinámica magnetohydrodynamic wave
onda manipulada keyed wave
onda marítima maritime wave
onda media medium wave
onda milimétrica millimetric wave
onda modulada modulated wave
onda modulada en amplitud amplitude-modulated wave
onda modulada en fase phase-modulated wave
onda modulada por impulsos pulse-modulated wave
onda modulada por sonido sound-modulated wave
onda modulada por tono tone-modulated wave
onda moduladora modulating wave
onda móvil moving wave
onda no amortiguada undamped wave
onda no modulada unmodulated wave
onda no simétrica nonsymmetrical wave
onda no sinusoidal nonsinusoidal wave
onda ordinaria ordinary wave
onda periódica periodic wave
onda periódica compleja complex periodic wave
onda piloto pilot wave
onda plana plane wave
onda plana uniforme uniform plane wave
onda plástica plastic wave
onda polarizada polarized wave
onda polarizada a la derecha right-hand polarized wave
onda polarizada en un plano plane-polarized wave
onda polarizada horizontalmente horizontally-polarized wave
onda polarizada linealmente linearly polarized wave
onda polarizada sinistrórsum counterclockwise-polarized wave
onda polarizada verticalmente vertically-polarized wave
onda portadora carrier wave
onda portadora modulada modulated carrier wave
onda portadora no modulada unmodulated carrier wave
onda portadora sinusoidal sinusoidal carrier wave
onda portadora variable variable carrier wave
onda positiva positive wave
onda posterior back wave
onda progresiva progressive wave
onda progresiva esférica spherical progressive wave
onda progresiva libre free progressive wave
onda progresiva plana plane-progressive wave
onda pulsante pulsating wave
onda pura pure wave
onda Q Q wave
onda radioeléctrica radioelectric wave
onda rápida fast wave
onda rectangular rectangular wave
onda reflejada reflected wave, echo wave
onda reflejada por tierra ground-reflected wave
onda refractada refracted wave
onda restaurada restored wave
onda rotacional rotational wave
onda senoidal sine wave
onda simétrica symmetrical wave
onda sin distorsión undistorted wave

onda sin reflexión unreflected wave
onda sincronizada synchronized wave
onda sinusoidal sine wave, sinusoidal wave
onda sinusoidal distorsionada distorted sine wave
onda sinusoidal equivalente equivalent sine wave
onda sonora sound wave
onda sonora directa direct sound wave
onda sonora estacionaria stationary sound wave
onda sonora polarizada en un plano plane-polarized sound wave
onda sonora polarizada linealmente linearly polarized sound wave
onda sostenida sustained wave
onda submilimétrica submillimeter wave
onda subportadora subcarrier wave
onda supersónica supersonic wave
onda térmica heat wave
onda terrestre ground wave
onda tipo A type-A wave
onda tipo B type-B wave
onda transitoria transient wave
onda transmitida transmitted wave
onda transversal transverse wave
onda trapecial trapezoidal wave
onda triangular triangular wave
onda troposférica tropospheric wave
onda ultravioleta ultraviolet wave
onda vertical vertical wave
onda X X wave
ondámetro wavemeter
ondámetro absoluto absolute wavemeter
ondámetro coaxial coaxial wavemeter
ondámetro de absorción absorption wavemeter
ondámetro de cavidad cavity wavemeter
ondámetro de cavidad resonante resonant-cavity wavemeter
ondámetro de varilla rod wavemeter
ondámetro dinámico dynamic wavemeter
ondámetro heterodino heterodyne wavemeter
ondas aéreas airwaves
ondas centimétricas centimetric waves
ondas cerebrales brain waves
ondas complejas complex waves
ondas continuas puras pure continuous waves
ondas de de Broglie de Broglie waves
ondas de hiperfrecuencia hyperfrequency waves
ondas de radio galácticas galactic radio waves
ondas decamétricas decametric waves
ondas decimétricas decimetric waves
ondas delta delta waves
ondas electromagnéticas electromagnetic waves
ondas hertzianas Hertzian waves
ondas incoherentes incoherent waves
ondas infrarrojas infrared waves
ondas interrumpidas interrupted waves
ondas inversas backward waves
ondas métricas metric waves
ondas miriamétricas myriametric waves
ondas no dispersivas nondispersive waves
ondas regresivas backward waves
ondas ultracortas ultrashort waves
ondas ultrasónicas ultrasonic waves
ondógrafo ondograph
ondoscopio ondoscope
ondulación undulation, ripple
ondulación de banda pasante passband ripple
ondulación de banda suprimida stop-band ripple
ondulación de colector commutator ripple
ondulación de corriente current ripple
ondulación periódica periodic undulation
ondulación residual residual ripple
ondulador undulator, ondulator
onduladorio undulatory

ondulante undulating
opacidad opacity
opacímetro opacimeter
opacímetro fotoeléctrico photoelectric opacimeter
opaco opaque
opción de borrado de mensajes message delete option
opción de exploración de memoria memory scan option
opción de protección de memoria memory protection option
opción de tiempo compartido time-sharing option
opción de transferencia transfer option
opción de visualización display option
opción implícita assumed option
opción por omisión default option
opcional optional
opciones de compilador compiler options
opdar opdar
operable operable
operable externamente externally operable
operación operation
operación a mano hand operation
operación algebraica algebraic operation
operación alterna alternating operation
operación alternativa alternate operation
operación anticoincidencia anticoincidence operation
operación aritmética arithmetic operation
operación aritmética-lógica arithmetic logical operation
operación aritmética binaria binary arithmetic operation
operación aritmética decimal decimal arithmetic operation
operación asincrónica asynchronous operation
operación asistida assisted operation
operación automática automatic operation
operación auxiliar auxiliary operation
operación bicondicional bicondicional operation
operación bidireccional bidirectional operation
operación biestable bistable operation
operación bifurcada forked working
operación binaria binary operation
operación bipolar bipolar operation
operación clase A class A operation
operación clase B class B operation
operación complementaria complementary operation
operación completa complete operation
operación completamente automática fully automatic operation
operación con portadora suprimida suppressed-carrier operation
operación concurrente concurrent operation
operación condicional conditional operation
operación confiable dependable operation
operación consecutiva consecutive operation
operación continua continuous operation
operación conversacional conversational operation
operación de actualización update operation
operación de arbitraje arbitration operation
operación de banda cruzada crossband operation
operación de barrido único single-sweep operation
operación de bloque block operation
operación de Boole Boolean operation
operación de Boole binaria binary Boolean operation
operación de bucle loop operation
operación de bucle continuo continuous-loop operation
operación de bus bus operation
operación de bus único single-bus operation
operación de cálculo calculating operation
operación de cálculo medio average calculation

operation
operación de canal adyacente adjacent-channel operation
operación de canal común common-channel operation
operación de carga loading operation
operación de ciclo fijo fixed-cycle operation
operación de ciclo variable variable-cycle operation
operación de coma fija fixed-point operation
operación de coma flotante floating-point operation
operación de comparación comparison operation
operación de computadora computer operation
operación de conexión y desconexión on-off operation
operación de conmutación switching operation
operación de consulta-respuesta inquiry-response operation
operación de conteo counting operation
operación de control control operation
operación de copiado copy operation
operación de cristal crystal operation
operación de cuantificación quantizing operation
operación de desbordamiento overflow operation
operación de desconexión y conexión off-on operation
operación de entrada-salida input-output operation
operación de entrada-salida programada programmed input-output operation
operación de equivalencia equivalence operation
operación de identidad identity operation
operación de impresora printer operation
operación de impulsos pulse operation
operación de impulsos coherentes coherent-pulse operation
operación de intercalación collation operation
operación de interfaz de entrada-salida input-output interface operation
operación de interrupción interrupt operation
operación de lápiz óptico light-pen operation
operación de línea compartida party-line operation
operación de longitud fija fixed-length operation
operación de máquina machine operation
operación de medidor de corriente current-meter operation
operación de memoria de burbujas bubble memory operation
operación de modo de paquetes packet mode operation
operación de multiplicación multiply operation
operación de multiprocesamiento multiprocessing operation
operación de no equivalencia nonequivalence operation
operación de no identidad nonidentity operation
operación de obtención get operation
operación de oscilador sincronizado locked-oscillator operation
operación de paso único single-step operation
operación de preparación housekeeping operation
operación de relé relay operation
operación de relé incorrecta incorrect relay operation
operación de reserva standby operation
operación de salto jump operation
operación de señales fuertes large-signal operation
operación de suma add operation
operación de transferencia transfer operation
operación de transferencia de datos data-transfer operation
operación de trayectoria continua continuous path operation
operación de un paso one-step operation
operación de unidad de procesamiento central

central processing unit operation
operación de validación de datos data validation operation
operación declarativa declarative operation
operación degradada degraded operation
operación destructiva destructive operation
operación diádica dyadic operation
operación díplex diplex operation
operación direccional directional operation
operación directa direct operation
operación doble dual operation
operación eléctrica electrical operation
operación en bucle looping
operación en contrafase push-pull operation
operación en dúplex duplex operation
operación en espera standby, idling
operación en estado estacionario steady-state operation
operación en frecuencia única single-frequency operation
operación en grupo group operation
operación en línea on-line operation
operación en modo manual manual mode operation
operación en paralelo parallel operation
operación en serie serial operation, series operation
operación en símplex simplex operation
operación en tándem tandem operation
operación en tiempo real real-time operation
operación en Y Y operation
operación falsa false operation
operación fuera de línea off-line operation
operación fundamental fundamental operation
operación global global operation
operación ilegal illegal operation
operación inatendida unattended operation
operación incorrecta incorrect operation
operación independiente independent operation
operación indirecta indirect operation
operación interactiva interactive operation
operación intermitente intermittent operation
operación iterativa iterative operation
operación limitada por computadora computer-limited operation
operación limitada por distorsión distortion-limited operation
operación limitadora limiting operation
operación lógica logic operation
operación manual manual operation
operación matricial matrix operation
operación mecánica mechanical operation
operación monádica monadic operation
operación multicanal multichannel operation
operación multimodo multimode operation
operación múltiplex multiplex operation
operación multitarea multitask operation
operación necesaria necessary operation
operación no automática nonautomatic operation
operación no destructiva nondestructive operation
operación NO-O NOR operation
operación no programada unscheduled operation
operación no saturada unsaturated operation
operación no sincrónica nonsynchronous operation
operación NO-Y NAND operation
operación NO-Y NOT-AND operation
operación normal normal operation
operación O OR operation
operación O incluyente inclusive OR operation
operación paso a paso step-by-step operation
operación periférica peripheral operation
operación por control remoto remote-control operation
operación por frecuencias agrupadas grouped-frequency operation

operación por pulsador push-button operation
operación por señales signal operation
operación por servomotor power operation
operación por voz automática automatic voice operation
operación preliminar de prueba dry run
operación privilegiada privileged operation
operación programada scheduled operation
operación protegida contra daños por fallas failsafe operation
operación remota remote operation
operación repetitiva repetitive operation
operación ruidosa noisy operation
operación saturada saturated operation
operación secuencial sequential operation
operación secuencial automática automatic sequential operation
operación semiautomática semiautomatic operation
operación semidúplex semiduplex operation
operación SI-ENTONCES IF-THEN operation
operación SI-Y-SOLO-SI IF-AND-ONLY-IF operation
operación sin carga no-load operation
operación sin espera no-wait operation
operación sin polarización zero-bias operation
operación sin ruido noiseless operation
operación sincrónica synchronous operation
operación sincronizada synchronized operation
operación unaria unary operation
operación única single operation
operación X X operation
operacional operational
operaciones por minuto operations per minute
operaciones simultáneas simultaneous operations
operado directamente directly operated
operado eléctricamente electrically operated
operado manualmente manually operated
operado por batería battery-operated
operado por motor motor-operated
operado por presión pressure-operated
operado por pulsador push-button operated
operado por radar radar-operated
operado por relé relay-operated
operado por ruido noise-operated
operado por señales signal-operated
operado por solenoide solenoid-operated
operado por teclado keyboard operated
operado por tensión voltage-operated
operado por tono tone-operated
operado por transductor transductor-operated
operado por transistores transistor-operated
operado por vacío vacuum-operated
operado por voz voice-operated
operado remotamente remotely operated
operado secuencialmente sequentially operated
operador operator
operador A A operator
operador autorizado authorized operator
operador B B operator
operador binario binary operator
operador complejo complex operator
operador complementario complementary operator
operador compuesto composite operator
operador condicional conditional operator
operador de Boole Boolean operator
operador de Boole monádico monadic Boolean operator
operador de caracteres character operator
operador de comparación comparison operator
operador de computadora computer operator
operador de control control operator
operador de dominio domain operator
operador de nodo node operator

operador de programa program operator
operador de red network operator
operador de sistema system operator
operador diádico dyadic operator
operador infijo infix operator
operador lógico logic operator
operador monádico monadic operator
operador no lineal nonlinear operator
operador numérico numerical operator
operador O OR operator
operador unario unary operator
operando operand
operando binario binary operand
operando inmediato immediate operand
operando posicional positional operand
operandos literales literal operands
oposición opposition, bucking
oposición de fase phase opposition
óptica optics
óptica de fibras fiber optics
óptica de infrarrojo infrared optics
óptica de proyección projection optics
óptica de Schmidt Schmidt optics
óptica electrónica electronic optics
óptica equivalente equivalent optics
óptica física physical optics
óptica geométrica geometric optics
óptica parageométrica parageometrical optics
ópticamente optically
ópticamente acoplado optically coupled
ópticamente activo optically active
ópticamente plano optically flat
óptico optic
optimización optimization
optimización de disco disk optimizing
optimización no lineal nonlinear optimization
optimizado optimized
optimizador optimizer
optimizador de disco disk optimizer
optimizar optimize
óptimo optimum, optimal
optoacoplador optocoupler
optoaislador optoisolator
optoeléctrico optoelectric
optoelectrónica optoelectronics
optoelectrónico optoelectronic
optófono optophone
órbita orbit
órbita de Larmor Larmor orbit
órbita de satélite satellite orbit
órbita ecuatorial equatorial orbit
órbita electrónica electronic orbit, electron orbit
órbita estable stable orbit
órbita geoestacionaria geostationary orbit
órbita polar polar orbit
órbita retrógrada retrograde orbit
órbita sincrónica synchronous orbit
orbital orbital
orden order, command
orden alto high order
orden aplicativo applicative order
orden bajo low order
orden de clasificación sort order
orden de entrada-salida input-output order
orden de interferencia order of interference
orden de lógica order of logic
orden de llamadas order of calls
orden de magnitud order of magnitude
orden de muestreo sampling order
orden de prioridad priority order
orden de reflexión order of reflection
orden descendente descending order
orden inicial initial order

orden inverso reverse order
orden relativo relative order
ordenación arrangement
ordenada ordinate
ordenador computer
ordinario ordinary
oreja de empalme splicing ear
organigrama flowchart, flow diagram
organigrama de proceso process flowchart
organigrama de programa program flowchart
organigrama de programación programming
  flowchart
organigrama de sistema system flowchart
organigrama de sistemas systems flowchart
organigrama lógico logic flow chart
organización consecutiva consecutive organization
organización de almacenamiento físico physical
  storage organization
organización de archivo file organization
organización de base de datos database organization
organización de conjunto de datos data set
  organization
organización de datos data organization
organización de disco disk organization
organización de memoria memory organization
organización de memoria de bits bit memory
  organization
organización de multiprocesamiento
  multiprocessing organization
organización de sistema system organization
organización de sistemas de microprocesador
  microprocessor system organization
organización del procesador central central
  processor organization
organización directa jerárquica hierarchical direct
  organization
organización dividida partitioned organization
organización indexada indexed organization
organización relativa relative organization
organización secuencial sequential organization
organización secuencial indexada indexed sequential
  organization
organización secuencial jerárquica hierarchical
  sequential organization
orientación orientation, attitude
orientación de página page orientation
orientación de partículas particle orientation
orientación horizontal landscape orientation
orientación preferida preferred orientation
orientación vertical portrait orientation
orientacional orientational
orientado oriented
orientado a caracteres character-oriented
orientado a palabras word-oriented
orientado a trabajos job-oriented
orientado al usuario user-oriented, user-friendly
orificio orifice
origen origin
origen absoluto absolute origin
origen ensamblado assembled origin
originación de datos data origination
original en cera wax original, wax master
original en laca lacquer original
original metálico metal master
orioscopio orioscope
oro gold
oropel tinsel
orticón orthicon
orticón de imagen image orthicon
orticonoscopio orthiconoscope
ortocódigo orthocode
ortogonalidad orthogonality
ortonúcleo orthocore

ortorrómbico orthorhombic
osciductor osciducer
oscilación oscillation, swing, vibration
oscilación acústica acoustic oscillation
oscilación amortiguada damped oscillation
oscilación anormal abnormal oscillation
oscilación armónica harmonic oscillation
oscilación continua continuous oscillation
oscilación controlada por cuarzo quartz-controlled
    oscillation
oscilación de amplitud total total amplitude
    oscillation
oscilación de arco arc oscillation
oscilación de Barkhausen Barkhausen oscillation
oscilación de Barkhausen-Kurz Barkhausen-Kurz
    oscillation
**oscilación de cristal con compensación de
    temperatura** temperature-compensated crystal
    oscillation
oscilación de encendedor ignitor oscillation
oscilación de estado estacionario steady-state
    oscillation
oscilación de fase phase swinging
oscilación de frecuencia frequency swing
oscilación de frecuencia de blanco a negro
    white-to-black frequency swing
oscilación de pico a pico peak-to-peak swing
oscilación de plasma plasma oscillation
oscilación de portadora carrier swing
oscilación de punto spot wobble
oscilación de rejilla grid swing
oscilación de servomecanismo servo oscillation
oscilación de tensión voltage swing
oscilación de tensión de salida output voltage swing
oscilación electrónica electronic oscillation
oscilación en diente de sierra sawtooth oscillation
oscilación espuria spurious oscillation
oscilación estable stable oscillation
oscilación fundamental fundamental swing
oscilación hidromagnética hydromagnetic oscillation
oscilación inestable unstable oscillation
oscilación intermitente intermittent oscillation
oscilación lógica logic swing
oscilación longitudinal longitudinal oscillation
oscilación máxima de frecuencia maximum
    frequency swing
oscilación modulada modulated oscillation
oscilación modulada en fase phase-modulated
    oscillation
oscilación multietapa multistage oscillation
oscilación natural natural oscillation
oscilación no amortiguada undamped oscillation
oscilación parásita parasitic oscillation
oscilación periódica periodic oscillation
oscilación pulsada pulsed oscillation
oscilación pura pure oscillation
oscilación radial radial oscillation
oscilación resonante resonant oscillation
oscilación sinusoidal sinusoidal oscillation
oscilación sostenida sustained oscillation
oscilación subarmónica subharmonic oscillation
oscilación transitoria transient oscillation
oscilación transversal transverse oscillation
oscilaciones amortiguadas ringing
oscilaciones autosostenidas self-sustained
    oscillations
oscilaciones coherentes coherent oscillations
oscilaciones de Gunn Gunn oscillations
oscilaciones eléctricas electric oscillations
oscilaciones eléctricas amortiguadas damped
    electric oscillations
oscilaciones forzadas forced oscillations
oscilaciones incoherentes incoherent oscillations

oscilaciones libres free oscillations
oscilaciones no armónicas nonharmonic oscillations
oscilaciones persistentes persistent oscillations
oscilaciones transitorias ringing
oscilador oscillator
oscilador acoplado por cátodo cathode-coupled
    oscillator
oscilador alimentado en derivación shunt-fed
    oscillator
oscilador alimentado en serie series-fed oscillator
oscilador anarmónico anharmonic oscillator
oscilador armónico harmonic oscillator
oscilador autocontrolado self-controlled oscillator
oscilador autodino autodyne oscillator
oscilador autoexcitado self-excited oscillator
oscilador autoextintor self-quenching oscillator
oscilador automodulado self-modulated oscillator
oscilador bloqueado blocked oscillator
oscilador cambiador de frecuencia
    frequency-changer oscillator
oscilador cautivo captive oscillator
oscilador coherente coherent oscillator
oscilador con barrido de frecuencia
    swept-frequency oscillator
oscilador con rejilla a masa grounded-grid oscillator
oscilador con tubo de gas gas-tube oscillator
oscilador controlado gated oscillator
oscilador controlado por cristal crystal-controlled
    oscillator
oscilador controlado por luz light-controlled
    oscillator
oscilador controlado por tensión voltage-controlled
    oscillator
oscilador-convertidor oscillator-converter
oscilador de acoplamiento electrónico
    electron-coupled oscillator
oscilador de alta frecuencia high-frequency
    oscillator
oscilador de ánodo sintonizado tuned-anode
    oscillator
oscilador de arco arc oscillator
oscilador de Armstrong Armstrong oscillator
oscilador de audio audio oscillator
oscilador de audiofrecuencia audio-frequency
    oscillator
oscilador de autobloqueo squegging oscillator
oscilador de banda de octava octave-band oscillator
oscilador de banda X X-band oscillator
oscilador de Barkhausen Barkhausen oscillator
oscilador de Barkhausen-Kurz Barkhausen-Kurz
    oscillator
oscilador de barrido sweep oscillator
oscilador de barrido vertical vertical-sweep
    oscillator
oscilador de base de tiempo time-base oscillator
oscilador de base sintonizada tuned-base oscillator
oscilador de batido beat oscillator
oscilador de bloqueo blocking oscillator
oscilador de bloqueo autopulsante self-pulsing
    blocking oscillator
oscilador de bloqueo controlado por tensión
    voltage-controlled blocking oscillator
oscilador de bloqueo monoestable monostable
    blocking oscillator
oscilador de borrado erase oscillator
oscilador de Butler Butler oscillator
oscilador de calibración calibration oscillator
oscilador de campo retardador retarding-field
    oscillator
oscilador de cavidades cavity oscillator
oscilador de Clapp Clapp oscillator
oscilador de Clapp-Gouriet Clapp-Gouriet oscillator
oscilador de colector sintonizado tuned-collector

oscillator
oscilador de colores color oscillator
oscilador de Colpitts Colpitts oscillator
oscilador de componentes en paralelo
parallel-component oscillator
oscilador de cristal crystal oscillator
**oscilador de cristal con compensación de
temperatura** temperature-compensated crystal
oscillator
oscilador de cristal con temperatura controlada
temperature-controlled crystal oscillator
oscilador de cristal controlado por tensión
voltage-controlled crystal oscillator
oscilador de cristal de cuarzo quartz-crystal
oscillator
oscilador de cristal de Reinartz Reinartz crystal
oscillator
oscilador de cristal multicanal multichannel crystal
oscillator
oscilador de croma chroma oscillator
oscilador de cuarzo quartz oscillator
oscilador de chispas spark-gap oscillator
oscilador de décadas decade oscillator
oscilador de deflexión deflection oscillator
oscilador de deflexión horizontal
horizontal-deflection oscillator
oscilador de deflexión vertical vertical-deflection
oscillator
oscilador de descodificación decoding oscillator
oscilador de desfase phase-shift oscillator
oscilador de diapasón tuning-fork oscillator
oscilador de diodo diode oscillator
oscilador de Dow Dow oscillator
oscilador de efecto de Gunn Gunn-effect oscillator
oscilador de electrón transferido
transferred-electron oscillator
oscilador de elementos en paralelo en contrafase
push-pull-parallel oscillator
oscilador de enganche de fase phase-locked
oscillator
oscilador de exploración search oscillator
oscilador de exploración de líneas line-scanning
oscillator
oscilador de exploración vertical vertical-scanning
oscillator
oscilador de extinción quenching oscillator
oscilador de filtro sintonizado tuned-filter oscillator
oscilador de Franklin Franklin oscillator
oscilador de frecuencia de barrido sweep-frequency
oscillator
oscilador de frecuencia de batido beat-frequency
oscillator
oscilador de frecuencia estable frequency-stable
oscillator
oscilador de frecuencia fija fixed-frequency
oscillator
oscilador de frecuencia patrón standard-frequency
oscillator
oscilador de frecuencia ultraalta ultra-high
frequency oscillator
oscilador de frecuencia única single-frequency
oscillator
oscilador de frecuencia variable variable-frequency
oscillator
oscilador de frecuencia vertical vertical frequency
oscillator
oscilador de gama ancha wide-range oscillator
oscilador de Gill-Morrell Gill-Morrell oscillator
oscilador de Gunn Gunn oscillator
oscilador de Hartley Hartley oscillator
oscilador de haz molecular molecular-beam
oscillator
oscilador de Heil Heil oscillator

oscilador de Hertz Hertz oscillator
oscilador de hilo vibrante vibrating-wire oscillator
oscilador de impedancia negativa negative
impedance oscillator
oscilador de impulsos pulse oscillator
oscilador de inducción induction oscillator
oscilador de King King oscillator
oscilador de Kozanowski Kozanowski oscillator
oscilador de laboratorio laboratory oscillator
oscilador de laminilla vibrante vibrating-reed
oscillator
oscilador de lámpara de neón neon-bulb oscillator
oscilador de Lampkin Lampkin oscillator
oscilador de Lecher Lecher oscillator
oscilador de línea coaxial coaxial-line oscillator
oscilador de línea concéntrica concentric-line
oscillator
oscilador de línea resonante resonant-line oscillator
oscilador de luz coherente coherent-light oscillator
oscilador de magnetoestricción magnetostriction
oscillator
oscilador de mariposa butterfly oscillator
oscilador de Meacham Meacham oscillator
oscilador de Meissner Meissner oscillator
oscilador de micrófono microphone oscillator
oscilador de microondas microwave oscillator
oscilador de Miller Miller oscillator
oscilador de modulación de velocidad
velocity-modulation oscillator
oscilador de muy alta frecuencia
very-high-frequency oscillator
oscilador de neón neon oscillator
oscilador de núcleo saturable saturable-core
oscillator
oscilador de onda continua continuous-wave
oscillator
oscilador de onda sinusoidal sine-wave oscillator
oscilador de ondas cortas short-wave oscillator
oscilador de ondas cuadradas square-wave
oscillator
oscilador de ondas inversas backward-wave
oscillator
oscilador de ondas inversas lineal linear
backward-wave oscillator
oscilador de ondas progresivas traveling-wave
oscillator
oscilador de Pierce Pierce oscillator
oscilador de placa sintonizada tuned-plate oscillator
oscilador de placas paralelas parallel-plate oscillator
oscilador de polarización bias oscillator
oscilador de portadora carrier oscillator
oscilador de portadora de canal channel-carrier
oscillator
oscilador de potencia power oscillator
oscilador de Potter Potter oscillator
oscilador de pruebas test oscillator
oscilador de puente bridge oscillator
oscilador de puente de Wien Wien-bridge oscillator
oscilador de radiofrecuencia radio-frequency
oscillator
oscilador de red en paralelo parallel-network
oscillator
oscilador de referencia reference oscillator
oscilador de rejilla negativa negative-grid oscillator
oscilador de rejilla positiva positive-grid oscillator
oscilador de rejilla sintonizada tuned-grid oscillator
oscilador de relajación relaxation oscillator
oscilador de reloj clock oscillator
oscilador de resistencia-capacitancia
resistance-capacitance oscillator
oscilador de resistencia negativa negative-resistance
oscillator
oscilador de retroalimentación feedback oscillator

oscilador de Scott Scott oscillator
oscilador de señal patrón standard-signal oscillator
oscilador de señalización signaling oscillator
oscilador de servicio service oscillator
oscilador de sobretono overtone oscillator
oscilador de subportadora subcarrier oscillator
oscilador de subportadora de colores
  color-subcarrier oscillator
oscilador de subportadora de crominancia
  chrominance-subcarrier oscillator
oscilador de tiempo de tránsito transit-time
  oscillator
oscilador de tipo de puente bridge-type oscillator
oscilador de toda onda all-wave oscillator
oscilador de tono tone oscillator
oscilador de tono variable variable-tone oscillator
oscilador de transconductancia negativa
  negative-transconductance oscillator
oscilador de transistores transistor oscillator
oscilador de transmisión sending oscillator
oscilador de tubo tube oscillator
oscilador de tubo de vacío vacuum-tube oscillator
oscilador de tubo tipo faro lighthouse-tube oscillator
oscilador de válvula valve oscillator
oscilador de Van der Pol Van der Pol oscillator
oscilador de variación de velocidad
  velocity-variation oscillator
oscilador de varillas paralelas parallel-rod oscillator
oscilador de Wien Wien oscillator
oscilador desplazado offset oscillator
oscilador-detector oscillator-detector
oscilador dinatrón dynatron oscillator
oscilador-doblador oscillator-doubler
oscilador electromecánico electromechanical
  oscillator
oscilador electrónico electron oscillator
oscilador en anillo ring oscillator
oscilador en contrafase push-pull oscillator
oscilador equilibrado balanced oscillator
oscilador estabilizado por línea line-stabilized
  oscillator
oscilador estabilizado por resistencia
  resistance-stabilized oscillator
oscilador excitado por choque shock-excited
  oscillator
oscilador fijo fixed oscillator
oscilador fonográfico phonograph oscillator
oscilador helitrón helitron oscillator
oscilador hertziano Hertzian oscillator
oscilador heterodino heterodyne oscillator
oscilador hidrodinámico hydrodynamic oscillator
oscilador horizontal horizontal oscillator
oscilador infraacústico subaudio oscillator
oscilador kalitrón kallitron oscillator
oscilador klistrón klystron oscillator
oscilador lábil labile oscillator
oscilador lineal linear oscillator
oscilador local local oscillator
oscilador local de dos señales double local oscillator
oscilador local estabilizado stabilized local oscillator
oscilador local estable stalo
oscilador maestro master oscillator
oscilador maestro-amplificador de potencia master
  oscillator-power amplifier
oscilador maestro controlado por cristal
  crystal-controlled master oscillator
oscilador maestro de cuarzo quartz master oscillator
oscilador magnetoestrictivo magnetostrictive
  oscillator
oscilador magnetrón magnetron oscillator
oscilador mezclador mixer oscillator
oscilador-mezclador-detector
  oscillator-mixer-detector

oscilador modulado modulated oscillator
oscilador modulado por impulsos pulse-modulated
  oscillator
oscilador molecular molecular oscillator
oscilador-multiplicador oscillator-multiplier
oscilador paramétrico parametric oscillator
oscilador piezoeléctrico piezoelectric oscillator
oscilador piloto pilot oscillator
oscilador polifásico polyphase oscillator
oscilador pulsado pulsed oscillator
oscilador reentrante reentrant oscillator
oscilador reflejo reflex oscillator
oscilador sincronizado locked oscillator
oscilador sintonizable tunable oscillator
oscilador sintonizable por tensión voltage-tunable
  oscillator
oscilador sintonizado eléctricamente electrically
  tuned oscillator
oscilador sintonizado por ferrita ferrite-tuned
  oscillator
oscilador sintonizado por resistencia
  resistance-tuned oscillator
oscilador sobrecargado overloaded oscillator
oscilador subarmónico subharmonic oscillator
oscilador subarmónico de enganche de fase
  phase-locked subharmonic oscillator
oscilador subarmónico paramétrico parametric
  subharmonic oscillator
oscilador superheterodino superheterodyne
  oscillator
oscilador T en paralelo parallel-T oscillator
oscilador termoiónico thermionic oscillator
oscilador tetrodo tetrode oscillator
oscilador tiratrónico thyratron oscillator
oscilador transitrón transitron oscillator
oscilador triodo triode oscillator
oscilador tritet tri-tet oscillator
oscilador ultraaudión ultra-audion oscillator
oscilador variable variable oscillator
oscilador vertical vertical oscillator
osciladores acoplados coupled oscillators
oscilante oscillating, oscillatory, vibrating
oscilar oscillate
oscilistor oscillistor
oscilografía oscillography
oscilografía de alta velocidad high-speed
  oscillography
oscilográfico oscillographic
oscilógrafo oscillograph
oscilógrafo bifilar bifilar oscillograph
oscilógrafo de cuerda string oscillograph
oscilógrafo de chorro de líquido liquid-jet
  oscillograph
oscilógrafo de doble haz double-beam oscillograph
oscilógrafo de espejo mirror oscillograph
oscilógrafo de espejo móvil moving-mirror
  oscillograph
oscilógrafo de galvanómetro de espejo
  mirror-galvanometer oscillograph
oscilógrafo de haz luminoso light-beam oscillograph
oscilógrafo de hierro blando soft-iron oscillograph
oscilógrafo de rayos catódicos cathode-ray
  oscillograph
oscilógrafo electromagnético electromagnetic
  oscillograph
oscilógrafo mecánico mechanical oscillograph
oscilógrafo multicanal multichannel oscillograph
oscilógrafo tricolor tricolor oscillograph
oscilograma oscillogram
oscilómetro oscillometer
osciloscópico oscilloscopic
osciloscopio oscilloscope
osciloscopio clase A class A oscilloscope

osciloscopio clase B class B oscilloscope
osciloscopio de almacenamiento storage oscilloscope
osciloscopio de alta velocidad high-speed
    oscilloscope
osciloscopio de banda ancha wideband oscilloscope
osciloscopio de corriente continua direct-current
    oscilloscope
osciloscopio de cuerda string oscilloscope
osciloscopio de doble haz double-beam oscilloscope
osciloscopio de doble trazo dual-trace oscilloscope
osciloscopio de dos canales dual-channel
    oscilloscope
osciloscopio de haz único single-beam oscilloscope
osciloscopio de laboratorio laboratory oscilloscope
osciloscopio de monitoreo monitoring oscilloscope
osciloscopio de muestreo sampling oscilloscope
osciloscopio de ondas progresivas traveling-wave
    oscilloscope
osciloscopio de rayo luminoso light-ray oscilloscope
osciloscopio de rayos catódicos cathode-ray
    oscilloscope
osciloscopio de rayos catódicos multihaz multibeam
    cathode-ray oscilloscope
osciloscopio de visión viewing oscilloscope
osciloscopio electromagnético electromagnetic
    oscilloscope
osciloscopio electromecánico electromechanical
    oscilloscope
osciloscopio multibanda multiband oscilloscope
osciloscopio multicanal multiple-channel
    oscilloscope
osciloscopio multihaz multibeam oscilloscope
osmio osmium
osmiridio osmiridium
ósmosis osmosis
osmótico osmotic
osófono osophone
osteófono osteophone
óvalo auroral auroral oval
oxicorte flame cutting
oxicorte por arco oxy-arc cutting
oxidación oxidation
oxidación anódica anodic oxidation
oxidación electrolítica electrolytic oxidation
oxidación electroquímica electrochemical oxidation
oxidación interna internal oxidation
oxidación preferencial preferential oxidation
oxidante oxidant
oxidar oxidize
óxido oxide
óxido de bario-estroncio barium-strontium oxide
óxido de berilio beryllium oxide
óxido de cinc zinc oxide
óxido de cromo pasivo passive chromium oxide
óxido de estaño tin oxide
óxido de hierro iron oxide
óxido de hierro magnético magnetic iron oxide
óxido de plata silver oxide
óxido de plomo lead oxide
óxido de silicio silicon oxide
óxido férrico ferric oxide
óxido férrico gamma gamma ferric oxide
óxido ferroso ferrous oxide
óxido magnético magnetic oxide
oxígeno oxygen
oxímetro oximeter
ozono ozone

P

pacor pacor
padar padar
pader padder
pader de baja frecuencia low-frequency padder
pader de oscilador oscillator padder
página activa active page
página de memoria memory page
página física physical page
página lógica logical page
página parcial partial page
paginable pageable
paginación pagination, paging
paginación anticipatoria anticipatory paging
paginación automática automatic pagination
palabra binaria binary word
palabra clave keyword
palabra clave en contexto keyword in context
palabra clave fuera de contexto keyword out of
    context
palabra de código code word
palabra de comprobación check word
palabra de computadora computer word
palabra de comunicaciones communications word
palabra de control control word
palabra de control de canal channel control word
palabra de control de datos data control word
palabra de control de formato format control word
palabra de dirección de canal channel address word
palabra de edición edit word
palabra de enlace link word
palabra de estado status word
palabra de estado de canal channel status word
palabra de estado de procesador processor status
    word
palabra de estado de programa program status
    word
palabra de fin de registro end-of-record word
palabra de información information word
palabra de instrucción instruction word
palabra de longitud fija fixed-length word
palabra de mando de canal channel command word
palabra de máquina machine word
palabra de memoria memory word
palabra de parámetro parameter word
palabra de secuencia alfabética alphabetic string
    word
palabra en serie serial word
palabra identificadora identifier word
palabra índice index word
palabra numérica numerical word
palabra parcial partial word
palabra reservada reserved word
palabras de control de acceso access-control words
palabras de datos data words
paladio palladium
palanca toggle
palanca acodada toggle
palanca de código code lever
palanca de conmutador switch lever
palanca de disparo tripping lever
palanca de reposición reset lever

palanca de tarjetas card lever
paleomagnetismo paleomagnetism
paleta vane, paddle
paleta móvil moving vane
paleta resistiva resistive vane
palomas pigeons
palpador feeler
pancromático panchromatic
panel panel
panel acústico resonante resonant acoustic panel
panel combinador combiner panel
panel convertidor de imagen image-converter panel
panel de ajuste de frecuencia frequency adjustment
   panel
panel de alimentación power-supply panel
panel de ayuda help panel
panel de bastidor rack panel
panel de carga charging panel
panel de circuitos impresos printed-circuit panel
panel de conexiones patching panel
panel de conmutación switching panel
panel de conmutación de programas
   program-switching panel
panel de conmutadores switch-panel
panel de consola de operador operator console
   panel
panel de control control panel
panel de control de aterrizaje por instrumentos
   instrument-landing control panel
panel de control de circuito circuit control panel
panel de control de conmutación switching control
   panel
panel de control de mantenimiento maintenance
   control panel
panel de control de microcomputadora
   microcomputer control panel
panel de control de operador operator control panel
panel de control de potencia power control panel
panel de control de procesos process-control panel
panel de control de programación programming
   control panel
panel de control de programador programmer
   control panel
panel de control de sistema system control panel
panel de control fijo fixed control panel
panel de control intercambiable interchangeable
   control panel
panel de control maestro master control panel
panel de control manual manual control panel
panel de control por computadora computer control
   panel
panel de control telefónico telephone control panel
panel de distribución distribution panel
panel de distribución de potencia power-distribution
   panel
panel de distribución de potencia eléctrica electric
   power distribution panel
panel de entrada de datos data entry panel
panel de equipo equipment panel
panel de filtro filter panel
panel de frente muerto dead-front panel
panel de fusibles fuse panel
panel de gas gas panel
panel de identificación identification panel
panel de indicación visual de fallas fault display
   panel
panel de jacks jack panel
panel de lámparas lamp panel
panel de luces indicadoras indicator-light panel
panel de mantenimiento maintenance panel
panel de medidor meter panel
panel de monitoreo monitoring panel
panel de operación operating panel

panel de pantalla completa full-screen panel
panel de pruebas test panel
panel de relés relay panel
panel de señalización signaling panel
panel de separación separation panel
panel de servicio service panel
panel de uniones junction panel
panel de uniones modular modular junction panel
panel de visualización display panel
panel de visualización de plasma plasma display
   panel
panel distribuidor distributing panel
panel electroluminiscente electroluminescent panel
panel ficticio dummy panel
panel fonográfico phono panel
panel fotocrómico photochromic panel
panel fotovoltaico photovoltaic panel
panel frontal front panel
panel gráfico graphic panel
panel híbrido hybrid panel
panel indicador indicator panel
panel indicador de fallas fault-indicator panel
panel intensificador de imagen image-intensifier
   panel
panel interno inner panel
panel lateral side panel
panel monitor monitor panel
panel perforado perforated board
panel sensible al tacto touch panel
panel solar solar panel
panorámico panoramic
pantalla screen, display, shield
pantalla acústica baffle
pantalla aluminizada aluminized screen
pantalla anticorona corona shield
pantalla borrada clear screen
pantalla catódica cathode screen
pantalla completa full screen
pantalla cuadriculada graph screen
pantalla de arranque startup screen
pantalla de colores color screen
pantalla de desviación baffle
pantalla de enfoque focusing screen
pantalla de entrada entry screen
pantalla de Faraday Faraday screen
pantalla de imagen de colores color picture screen
pantalla de larga persistencia long-persistence
   screen
pantalla de lectura read screen
pantalla de menú menu screen
pantalla de persistencia persistence screen
pantalla de radar radar screen
pantalla de rayos catódicos cathode-ray screen
pantalla de retardo delay screen
pantalla de televisión television screen
pantalla de tubo de rayos catódicos cathode-ray
   tube screen
pantalla de video video screen
pantalla de visión viewing screen
pantalla de visualización display screen
pantalla dividida split screen
pantalla eléctrica electric screen
pantalla electroluminiscente electroluminescent
   screen
pantalla electromagnética electromagnetic screen
pantalla electrostática electrostatic screen
pantalla electrostática de Faraday Faraday
   electrostatic screen
pantalla fluorescente fluorescent screen
pantalla fluoroscópica fluoroscopic screen
pantalla fosforescente phosphorescent screen
pantalla fotosensible photosensitive screen
pantalla intensificadora intensifying screen

pantalla luminiscente luminescent screen
pantalla luminosa luminous screen
pantalla magnética magnetic screen
pantalla metalizada metallized screen
pantalla mosaico mosaic screen
pantalla opaca opaque screen
pantalla osciloscópica oscilloscope screen
pantalla pasiva passive screen
pantalla plana flat screen
pantalla protectora protective screen
pantalla reflectora reflecting screen
pantalla sensible al tacto touch-sensitive screen
pantalla térmica heat shield
pantalla terminal end shield
pantografía pantography
pantógrafo pantograph
papel aceitado oiled paper
papel cristal glassine
papel de pescado fishpaper
papel electrosensible electrosensitive paper
papel fosforescente phosphorescent paper
papel impregnado en cera wax paper
papel kraft kraft paper
papel termosensible thermosensitive paper
paquete package, pack, packet
paquete de entrada input deck
paquete de fibras fiber bundle
paquete de fibras ópticas optical-fiber bundle
paquete de ondas wave packet
paquete de programas program package
paquete de prueba test pack
paquete fibroóptico fiber-optic bundle
par pair, torque
par acelerador accelerating torque
par amortiguador damping torque
par apantallado shielded pair, screened pair
par apareado matched pair
par asincrónico asynchronous torque
par astático astatic pair
par cargado loaded pair
par coaxial coaxial pair
par complementario complementary pair
par concéntrico concentric pair
par de aceleración acceleration torque
par de antenas antenna pair
par de arranque starting torque
par de carga load torque
par de deflexión deflecting torque
par de electrones electron pair
par de frenaje braking torque
par de Goto Goto pair
par de histéresis hysteresis torque
par de impulsos pulse pair
par de impulsos de referencia reference pulse pair
par de medidor meter torque
par de recepción receiving pair
par de reserva reserve pair
par de restauración restoring torque
par de retención holding torque
par de sincronización synchronizing torque
par de terminales terminal pair
par de torsión torque
par de transistores complementarios complementary-transistor pair
par electrón-hueco electron-hole pair
par equilibrado balanced pair
par galvánico galvanic couple
par hueco-electrón hole-electron pair
par límite stalling torque
par magnético magnetic couple
par máximo peak torque
par mínimo static torque
par motor motor torque, torque

par negativo negative torque
par no cargado nonloaded pair
par nominal rated torque
par retardador retarding torque
par simétrico symmetrical pair
par sincrónico synchronous torque
par termoeléctrico thermoelectric couple
par torcido twisted pair
par transitorio transient torque
par transmisor transmitting pair
par transpuesto transposed pair
par voltaico voltaic couple
parábola parabola
parábola de corte cutoff parabola
parabólico parabolic
paraboloidal paraboloidal
paraboloide paraboloid
paraboloide orientable steerable paraboloid
paraboloide truncado truncated paraboloid
parada stop, halt
parada absoluta absolute stop
parada automática automatic stop
parada codificada coded stop
parada con señal auditiva hoot stop
parada condicional conditional stop
parada de control control stop
parada de emergencia emergency stop
parada de impulso pulse stop
parada de máquina machine halt
parada de programa program stop
parada de salto skip stop
parada dinámica dynamic stop
parada imprevista hangup
parada inesperada unexpected halt
parada no programada nonprogrammed halt
parada opcional optional stop
parada ordenada orderly halt
parada por error error stop
parada programada programmed halt
paradiafonía near-end crosstalk
paraeléctrico paraelectric
parafina paraffin
paraguta paragutta
paralaje parallax
paralaje espacial space parallax
paralelo parallel
paralelo-serie parallel-serial
paramagnético paramagnetic
paramagnetismo paramagnetism
paramagnetismo por espín spin paramagnetism
paramétrico parametric
parámetro parameter
parámetro activo active parameter
parámetro adimensional dimensionless parameter
parámetro con variación lineal linear varying parameter
parámetro concentrado lumped parameter
parámetro controlado controlled parameter
parámetro de adaptación adaptation parameter
parámetro de ajuste adjustment parameter
parámetro de archivo file parameter
parámetro de cable cable parameter
parámetro de circuito circuit parameter
parámetro de códigos code parameter
parámetro de dispositivo device parameter
parámetro de generación generation parameter
parámetro de modulación modulation parameter
parámetro de onda wave parameter
parámetro de palabra clave keyword parameter
parámetro de programa program parameter
parámetro de red network parameter
parámetro de referencia reference parameter
parámetro de señal signal parameter

parámetro de sesión session parameter
parámetro de sistema system parameter
parámetro de trabajo working parameter
parámetro de transferencia transfer parameter
parámetro de transistor transistor parameter
parámetro descriptivo descriptive parameter
parámetro dinámico dynamic parameter
parámetro distribuido distributed parameter
parámetro eléctrico electrical parameter
parámetro en circuito abierto open-circuit parameter
parámetro en cortocircuito short-circuit parameter
parámetro estático static parameter
parámetro ficticio dummy parameter
parámetro fijo fixed parameter
parámetro formal formal parameter
parámetro global global parameter
parámetro híbrido hybrid parameter
parámetro matemático mathematical parameter
parámetro no lineal nonlinear parameter
parámetro por omisión default parameter
parámetro posicional positional parameter
parámetro preajustable preset parameter
parámetro preajustado preset parameter
parámetro real actual parameter
parámetro secundario secondary parameter
parámetro simbólico symbolic parameter
parámetro simple simple parameter
parámetro variable varying parameter
parametrón parametron
parámetros de admitancia admittance parameters
parámetros de comunicaciones communication parameters
parámetros de imagen image parameters
parámetros de línea line parameters
parámetros de operación operating parameters
parámetros eléctricos lineales linear electrical parameters
paramistor paramistor
pararrayos lightning arrester, arrester
pararrayos con tubo de gas gas-tube lightning arrester
pararrayos de cuchilla knife lightning arrester
pararrayos de cuernos horn arrester
pararrayos de vacío vacuum lightning arrester
parásito parasitic
paraxial paraxial
parcial partial
parcialmente partially
pared de Bloch Bloch wall
pared de surco groove wall
pared de vidrio glass wall
pareja acoplada por el cátodo long-tailed pair
pareo de frecuencias frequency pairing
parfocal parfocal
paridad parity
paridad de bit bit parity
paridad de bloque block parity
paridad de caracteres character parity
paridad de cinta tape parity
paridad de cinta magnética magnetic-tape parity
paridad de memoria memory parity
paridad de tambor drum parity
paridad impar odd parity
paridad lateral lateral parity
paridad longitudinal longitudinal parity
paridad par even parity
parileno parylene
parpadeo flicker, flickering, blinking, glint
parpadeo de carga load flicker
parpadeo de colores color flicker
parpadeo de cromaticidad chromaticity flicker
parpadeo de línea line flicker

parpadeo de luminancia luminance flicker
parque de antenas antenna park
parrilla magnética magnetic grate
parte part
parte básica basic part
parte de dirección address part
parte de metal vivo live-metal part
parte de operación operation part
parte de reemplazo replacement part
parte de repuesto spare part
parte defectuosa faulty part
parte discreta discrete part
parte electrónica electronic part
parte fraccional fractional part
parte funcional functional part
parte imaginaria imaginary part
parte inferior de tarjeta perforada lower curtate
parte llamada called party
parte plana flat top
parte presentada displayed part
parte real real part
parte renovable renewable part
parte superior de impulso pulse top
parte viva live part
partes con conexión a tierra grounded parts
partes estructurales asociadas associated structural parts
partición partition, partitioning
partición compartida shared partition
partición de almacenamiento principal main storage partition
partición de almacenamiento virtual virtual storage partition
partición de disco disk partition
partición de sistema system partition
partición dinámica dynamic partition
partición interactiva interactive partition
partición paginable pageable partition
partícula particle
partícula alfa alpha particle
partícula beta beta particle
partícula cargada charged particle
partícula coloidal colloidal particle
partícula de alta energía high-energy particle
partícula de alta velocidad high-speed particle
partícula de baja energía low-energy particle
partícula de campo field particle
partícula elemental elementary particle
partícula magnética magnetic particle
partícula no ionizante nonionizing particle
partícula subatómica subatomic particle
pasada pass, run
pasada de clasificación sort pass
pasada de fusión merge pass
pasada de máquina machine pass, machine run
pasada de prueba test run
pasada única single pass
pasamuros bushing
pasante feedthrough
pascal pascal
pasivación passivation
pasivación anódica anodic passivation
pasivación con vidrio glassivation
pasivación electroquímica electrochemical passivation
pasivación química chemical passivation
pasivación superficial surface passivation
pasivado passivated
pasivar passivate
pasividad passivity
pasividad química chemical passivity
pasivo passive
paso pitch, step

paso de alimentación feed pitch
paso de arrollamiento winding pitch
paso de banda bandpass
paso de encaminamiento routing step
paso de exploración scanning pitch
paso de procedimiento procedure step
paso de procesamiento processing step
paso de procesamiento de datos data-processing
   step
paso de programa program step
paso de ranura slot pitch
paso de rejilla grid pitch
paso de trabajo job step
paso entre filas row pitch
paso entre pistas track pitch
paso parcial partial pitch
paso polar pole pitch
paso único single step
pasta paste
pasta de soldar soldering paste
pasta no corrosiva noncorrosive paste
patilla pin, post, leg
patilla de guía guide pin
patilla de rejilla grid pin
patín skate, collector shoe, shoe
patín de contacto trolley shoe
pátina patina
patinaje skating
patrón pattern, standard
patrón cardioide cardioid pattern
patrón conductor conductive pattern
patrón de ajuste de televisión television test card
patrón de antena antenna pattern
patrón de arco iris rainbow pattern
patrón de barras bar pattern
patrón de barras de colores color-bar pattern
patrón de bits bit pattern
patrón de calentamiento heating pattern
patrón de calibración calibration standard
patrón de campo field pattern
patrón de campo lejano far-field pattern
patrón de campo próximo near-field pattern
patrón de caracteres character pattern
patrón de cobertura vertical vertical-coverage
   pattern
patrón de códigos code pattern
patrón de comparación comparison standard
patrón de cosecante cuadrada cosecant-squared
   pattern
patrón de datos data pattern
patrón de definición definition pattern
patrón de desfase phase-shift standard
patrón de diferencia difference pattern
patrón de difracción diffraction pattern
patrón de difracción de campo lejano far-field
   diffraction pattern
patrón de difracción de campo próximo near-field
   diffraction pattern
patrón de difracción de Fresnel Fresnel diffraction
   pattern
patrón de difracción de rayos X X-ray diffraction
   pattern
patrón de directividad directivity pattern
patrón de dispersión dispersion pattern
patrón de dispersión de sonido sound-dispersion
   pattern
patrón de distribución distribution pattern
patrón de encaminamiento routing pattern
patrón de espacio space pattern
patrón de espacio libre free-space pattern
patrón de exploración scanning pattern
patrón de fase phase pattern
patrón de flujo flux pattern

patrón de frecuencia frequency standard
patrón de frecuencia de haz molecular
   molecular-beam frequency standard
patrón de frecuencia molecular molecular
   frequency standard
patrón de frecuencia primaria primary frequency
   standard
patrón de frecuencia secundaria secondary
   frequency standard
patrón de Fresnel Fresnel pattern
patrón de haz beam pattern
patrón de huecos hole pattern
patrón de imagen image pattern
patrón de inductancia inductance standard
patrón de intensidad de campo field-intensity
   pattern
patrón de interconexiones interconnection pattern
patrón de interfaz interface standard
patrón de interferencia interference pattern
patrón de laboratorio laboratory standard
patrón de Laue Laue pattern
patrón de lenguaje language standard
patrón de líneas de imagen picture line standard
patrón de Lissajous Lissajous pattern
patrón de luz light pattern, light standard
patrón de medida measurement standard
patrón de nomenclatura nomenclature standard
patrón de ondas estacionarias standing-wave pattern
patrón de polarización polarization pattern
patrón de polvo powder pattern
patrón de potencia power pattern
patrón de prueba de barras bar test pattern
patrón de prueba de barras de colores color-bar
   test pattern
patrón de pruebas test pattern, test chart
patrón de puntos dot pattern
patrón de puntos blancos white-dot pattern
patrón de radiación radiation pattern
patrón de radiación de campo lejano far-field
   radiation pattern
patrón de radiación de campo próximo near-field
   radiation pattern
patrón de radiación en espacio libre free-space
   radiation pattern
patrón de radiación horizontal horizontal radiation
   pattern
patrón de radiación vertical vertical radiation
   pattern
patrón de radiactividad radioactivity standard
patrón de referencia reference standard
patrón de referencia básico basic reference standard
patrón de referencia primario primary reference
   standard
patrón de rendimiento performance pattern
patrón de resistencia resistance standard
patrón de respuesta direccional directional response
   pattern
patrón de rueda dentada gear-wheel pattern
patrón de sincronización synchronization pattern
patrón de televisión television standard
patrón de tensión voltage standard
patrón de trabajo working standard
patrón de tráfico traffic pattern
patrón de trama de imagen crosshatch pattern
patrón de transferencia transfer standard
patrón de Waidner-Burgess Waidner-Burgess
   standard
patrón direccional directional pattern
patrón doblado folded pattern
patrón en espina de pescado herringbone pattern
patrón espurio spurious pattern
patrón horizontal horizontal pattern
patrón internacional international standard

patrón matriz master pattern
patrón nacional national standard
patrón normalizado normalized pattern
patrón óptico optical pattern
patrón primario primary pattern, primary standard
patrón radial radial pattern
patrón relativo relative pattern
patrón reticulado graticule pattern
patrón secundario secondary pattern, secondary standard
patrón trapecial trapezoidal pattern
patrón unidireccional unidirectional pattern
patrón vertical vertical pattern
patrones de control control standards
patrones de facsímil facsimile standards
patrones de procesamiento de datos data-processing standards
patrones de televisión a colores color-television standards
patrones de transmisión de televisión television transmission standards
pausa pause
pausa de manipulación keying pause
pausa mínima minimum pause
pedestal pedestal
pedestal de extinción blanking pedestal
pelador de hilos wire stripper
película film
película criogénica cryogenic film
película de base base film
película de contacto contact film
película de óxido oxide film
película de poliéster polyester film
película de rayos X X-ray film
película delgada thin film
película dieléctrica dielectric film
película emisora de luz light-emitting film
película epitaxial epitaxial film
película gruesa thick film
película magnética magnetic film
película magnética delgada thin magnetic film
película magnética gruesa thick magnetic film
película metálica metal film
película monomolecular monomolecular film
película pancromática panchromatic film
película pasiva passive film
película radiográfica radiographic film
película reotaxial rheotaxial film
película sensible a la luz light-sensitive film
película sonora sound film
peligro danger, hazard
peligro de radiación radiation danger
peligros ambientales environmental hazards
peligroso hazardous
pelusa de tarjetas card fluff
pellizco pinch
pendiente slope
pendiente de impulso pulse slope
pendiente de meseta plateau slope
pendiente relativa de meseta plateau relative slope
penduleo hunting
péndulo pendulum
péndulo de Helmholtz Helmholtz pendulum
penetrabilidad penetrability
penetración de cable cable penetration
penetración de campo magnético magnetic-field penetration
penetrante penetrating
penetrómetro penetrometer
penetrón penetron
pentarrejilla pentagrid
pentatrón pentatron
pentodo pentode

pentodo de amplificador amplifier pentode
pentodo de corte agudo sharp-cutoff pentode
pentodo de mu variable variable-mu pentode
pentodo de potencia power pentode
pentodo de potencia lineal linear power pentode
pentodo de salida output pentode
pentodo de salida de línea line-output pentode
pentodo de televisión television pentode
pentodo de video video pentode
perceptrón perceptron
percutir strike
pérdida loss
pérdida accidental accidental loss
pérdida de absorción absorptive loss
pérdida de acimut azimuth loss
pérdida de antena antenna loss
pérdida de arco arc loss
pérdida de bucle loop loss
pérdida de bucle de recepción receiving-loop loss
pérdida de cable cable loss
pérdida de caída de arco arc-drop loss
pérdida de calor heat loss
pérdida de calor de Joule Joule heat loss
pérdida de campo débil low-field loss
pérdida de conexión a tierra ground fault
pérdida de contacto contact miss
pérdida de conversión conversion loss
pérdida de cuantificación quantizing loss
pérdida de desviación angular angular deviation loss
pérdida de eco neta net echo loss
pérdida de energía energy loss
pérdida de energía indirecta indirect energy loss
pérdida de energía irreversible irreversible energy loss
pérdida de flujo irreversible irreversible flux loss
pérdida de grabación recording loss
pérdida de información loss of information
pérdida de inserción característica characteristic insertion loss
pérdida de lectura playback loss
pérdida de línea line loss
pérdida de potencia power loss
pérdida de potencia aparente apparent power loss
pérdida de potencia de rejilla grid power loss
pérdida de potencia de transductor transducer power loss
pérdida de potencia directa forward power loss
pérdida de potencia externa external power loss
pérdida de potencia interna internal power loss
pérdida de presión pressure loss
pérdida de propagación propagation loss
pérdida de puenteo bridging loss
pérdida de registro recording loss
pérdida de reproducción reproduction loss
pérdida de retorno return loss
pérdida de sensibilidad sensitivity loss
pérdida de señal signal loss
pérdida de tensión voltage loss
pérdida de transductor transducer loss
pérdida de transformador transformer loss
pérdida de transmisión transmission loss
pérdida de transmisión básica basic transmission loss
pérdida de trayectoria path loss
pérdida de unión junction loss
pérdida de visualización display loss
pérdida dieléctrica dielectric loss
pérdida directa forward loss
pérdida eléctrica electric loss
pérdida en el hierro iron loss
pérdida en espacio libre free-space loss
pérdida magnética magnetic loss

pérdida mecánica mechanical loss
pérdida neta net loss
pérdida neta mínima minimum net loss
pérdida neta mínima de trabajo minimum working net loss
pérdida no recíproca nonreciprocal loss
pérdida óhmica ohmic loss
pérdida plana flat loss
pérdida por absorción absorption loss
pérdida por absorción acústica acoustic absorption loss
pérdida por acoplamiento coupling loss
pérdida por aislador insulator loss
pérdida por alimentador feeder loss
pérdida por calor heat loss
pérdida por coincidencia coincidence loss
pérdida por conductor conductor loss
pérdida por conversión de modo mode conversion loss
pérdida por corrientes parásitas eddy-current loss
pérdida por desadaptación mismatch loss
pérdida por desalineación angular angular misalignment loss
pérdida por desbordamiento overflow loss
pérdida por desvanecimiento fading loss
pérdida por difracción diffraction loss
pérdida por dispersión spreading loss
pérdida por divergencia divergence loss
pérdida por efecto de Joule Joule loss
pérdida por efecto rasante grazing loss
pérdida por entrehierro gap loss
pérdida por etapa stage loss
pérdida por exploración scanning loss
pérdida por fluctuación fluctuation loss
pérdida por fricción frictional loss
pérdida por histéresis hysteresis loss
pérdida por histéresis dieléctrica dielectric hysteresis loss
pérdida por histéresis incremental incremental hysteresis loss
pérdida por histéresis magnética magnetic hysteresis loss
pérdida por inserción insertion loss
pérdida por interacción interaction loss
pérdida por ionización ionization loss
pérdida por ondas estacionarias standing-wave loss
pérdida por radiación radiation loss
pérdida por reflexión reflection loss
pérdida por refracción refraction loss
pérdida por resistencia resistance loss
pérdida por separación separation loss
pérdida por sombra shadow loss
pérdida por traducción translation loss
pérdida por transferencia transfer loss
pérdida por transición transition loss
pérdida proporcional proportional loss
pérdida residual residual loss
pérdida resistiva resistive loss
pérdida supuesta presumptive loss
pérdida térmica thermal loss
pérdida total overall loss
pérdidas de circuito circuit loss
pérdidas de cruce crossover loss
pérdidas en el cobre copper loss
pérdidas en el núcleo core loss
pérdidas por carga load losses
pérdidas por corona corona loss
pérdidas por diafonía crosstalk loss
pérdidas por dispersión stray losses
pérdidas por excitación excitation losses
pérdidas por inducción induction loss
pérdidas por reflexiones múltiples multiple-reflection losses

pérdidas sin carga no-load losses
pérdidas totales total losses
pérdidas variables variable losses
perditancia leakance
perfil de carga load profile
perfil de conjunto de datos data set profile
perfil de índice index profile
perfil de índice de refracción refractive index profile
perfil de índice graduado graded-index profile
perfil de potencial potential profile
perfil de programa program profile
perfil de refractividad refractivity profile
perfil de sistema system profile
perfil de trayectoria path profile
perfil lateral lateral profile
perfil longitudinal longitudinal profile
perfil parabólico parabolic profile
perfilado shaping
perfilómetro perfilometer
perfilómetro electrónico electronic profilometer
perforación perforation, punch, hole
perforación cero zero punch
perforación de alimentación feed hole
perforación de alimentación por espigas pin feed hole
perforación de cambio de página page-change hole
perforación de comprobación check punch
perforación de control control hole
perforación de designación designation hole
perforación de función function punch
perforación de zona zone punch
perforación desalineada off-punch
perforación en cadena lacing
perforación en paralelo parallel punching
perforación en serie serial punching
perforación entre posiciones usuales interstage punching
perforación longitudinal longitudinal perforation
perforación múltiple multiple punching
perforación numérica numeric punch
perforación sumaria summary punching
perforación X X punch
perforación Y Y punch
perforaciones de código code holes
perforaciones de transporte sprocket holes
perforado perforated, punched
perforado en los bordes edge perforated
perforador perforator
perforador-comprobador perforator-checker
perforador de recepción receiving perforator
perforador impresor printing perforator
perforador manual manual perforator
perforadora puncher, punch, keypunch
perforadora calculadora electrónica electronic calculating punch
perforadora de alta velocidad high-speed punch
perforadora de cinta tape puncher
perforadora de cinta de papel paper-tape punch
perforadora de columnas column punch
perforadora de mano hand punch
perforadora de salida output punch
perforadora de tarjetas card punch
perforadora de tarjetas controlada por cinta tape-controlled card punch
perforadora de tarjetas controlada por tarjeta card-controlled tape punch
perforadora de teclado keyboard punch
perforadora multiplicadora multiplying punch
perforadora sumaria summary punch
perforar punch
periférico peripheral
periférico de entrada input peripheral

periférico de salida output peripheral
periférico remoto remote peripheral
periódicamente periodically
periodicidad periodicity
periodicidad de llamada ringing periodicity
periódico periodic
periodo period
periodo actual current period
periodo anomalístico anomalistic period
periodo corriente current period
periodo corto short period
periodo de abertura break period
periodo de acción action period
periodo de arranque startup period
periodo de barrido sweep period
periodo de bloqueo blocking period
periodo de calentamiento warm-up period
periodo de campo field period
periodo de caracteres character period
periodo de carga charging period
periodo de carga aumentada on-peak period
periodo de carga reducida off-peak period
periodo de conducción conducting period
periodo de conexión on period
periodo de conmutación commutating period
periodo de cuadro frame period
periodo de depuración debugging period
periodo de desconexión off period
periodo de dígito digit period
periodo de enfriamiento cooling period
periodo de estabilización burn-in period
periodo de evaluación evaluation period
periodo de excitación period of excitation
periodo de exploración scanning period
periodo de exposición exposure period
periodo de extinción blanking period
periodo de falla por deterioro completo
    wearout-failure period
periodo de fallas normal normal failure period
periodo de fallas tempranas early-failure period
periodo de imagen picture period
periodo de impulsos impulse period
periodo de línea line period
periodo de muestreo sampling period
periodo de onda wave period
periodo de operación operating period
periodo de oscilación period of oscillation
periodo de pulsación pulsation period
periodo de recurrencia recurrence period
periodo de regeneración regeneration period
periodo de reintento retry period
periodo de repetición de impulsos pulse repetition
    period
periodo de reposo idle period
periodo de retención retention period
periodo de retorno return period
periodo de retorno vertical vertical-retrace period
periodo de reverberación reverberation period
periodo de servicio service period
periodo de silencio silent period
periodo de trabajo period of duty
periodo de tráfico máximo peak period
periodo de transición transition period
periodo de utilización utilization period
periodo de validez period of validity
periodo inverso reverse period
periodo latente latent period
periodo muerto dead period
periodo natural natural period
periodo orbital orbital period
periodo recurrente recurrent period
periodo transitorio transient period
periscópico periscopic

peristáltico peristaltic
permaloy permalloy
permanencia permanence
permanente permanent
permanentemente permanently
permanentemente almacenado stored permanently
permatrón permatron
permeabilidad permeability
permeabilidad absoluta absolute permeability
permeabilidad aparente apparent permeability
permeabilidad compleja complex permeability
permeabilidad de amplitud amplitude permeability
permeabilidad diferencial differential permeability
permeabilidad en espacio libre free-space
    permeability
permeabilidad equivalente equivalent permeability
permeabilidad específica specific permeability
permeabilidad ideal ideal permeability
permeabilidad incremental incremental permeability
permeabilidad inicial initial permeability
permeabilidad intrínseca intrinsic permeability
permeabilidad magnética magnetic permeability
permeabilidad máxima maximum permeability
permeabilidad normal normal permeability
permeabilidad relativa relative permeability
permeabilidad relativa efectiva effective relative
    permeability
permeabilidad reversible reversible permeability
permeable permeable
permeámetro permeameter
permeancia permeance
permendur permendur
perminvar perminvar
permisible permissible
permisivo permissive
permiso de acceso access permission
permitividad permittivity
permitividad absoluta absolute permittivity
permitividad compleja complex permittivity
permitividad dieléctrica dielectric permittivity
permitividad diferencial differential permittivity
permitividad relativa relative permittivity
permitividad reversible reversible permittivity
permutación permutation
perno pin
peróxido de plomo peroxide of lead
perpendicular perpendicular
perpetuo perpetual
perrera doghouse
persistencia persistence
persistencia de bajas hangover
persistencia de imagen lag
persistente persistent
persistor persistor
persistrón persistron
perturbación disturbance, perturbation
perturbación aleatoria random disturbance
perturbación de campo field disturbance
perturbación de propagación propagation
    disturbance
perturbación de radar radar disturbance
perturbación de radio radio disturbance
perturbación de televisión television disturbance
perturbación eléctrica atmosférica atmospheric
    electric disturbance
perturbación electromagnética electromagnetic
    disturbance
perturbación espacial spatial disturbance
perturbación intencional jamming
perturbación ionosférica ionospheric disturbance
perturbación ionosférica repentina sudden
    ionospheric disturbance
perturbación magnética magnetic disturbance

perturbación momentánea momentary disturbance
perturbación parásita parasitic disturbance
perturbación periódica periodic disturbance
perturbación por erupción solar solar-flare disturbance
perturbación por manchas solares sunspot disturbance
perturbado perturbed
perturbador jammer
perveancia perveance
pescante davit
peso atómico atomic weight
peso atómico-gramo gram-atomic weight
peso molecular molecular weight
peso sofométrico psophometric weight
peta peta
petición de salida output request
pez griega rosin
pH pH
phi phi
pi pi
picaduras pits
pico peak
pico absoluto absolute peak
pico de audio audio peak
pico de blanco peak white
pico de modulación modulation peak
pico de modulación negativa negative modulation peak
pico de negro peak black
pico de potencia power peak
pico de potencial potential peak
pico de resonancia resonance peak
pico de ruido noise spike
pico de señal signal peak
pico de tensión voltage peak
pico de tráfico traffic peak
pico fotoeléctrico photoelectric peak
pico negativo negative peak
pico positivo positive peak
pico repetitivo repetitive peak
pico transitorio surge
pico transitorio de conmutación switching surge
pico transitorio transversal transverse surge
picoampere picoampere
picoamperímetro picoammeter
picoamperímetro electrónico electronic picoammeter
picocomputadora picocomputer
picocoulomb picocoulomb
picocurie picocurie
picofarad picofarad
picohenry picohenry
picosegundo picosecond
picovolt picovolt
picovóltmetro electrónico electronic picovoltmeter
picowatt picowatt
pictórico pictorial
piedra imán lodestone
pieza part
pieza polar pole piece
piezo piezo
piezocristal piezocrystal
piezodieléctrico piezodielectric
piezoelectricidad piezoelectricity
piezoelectricidad directa direct piezoelectricity
piezoelectricidad indirecta indirect piezoelectricity
piezoeléctrico piezoelectric
piezoelemento cerámico ceramic piezoelement
piezómetro piezometer
piezorresistencia piezoresistance
piezorresistividad piezoresistivity
piezorresistivo piezoresistive

piezotérmico piezothermal
piezotropismo piezotropism
pigmento fotosensible photosensitive pigment
pila cell, battery, pile, stack
pila alcalina alkaline cell
pila calibrada calibrated cell
pila contraelectromotriz counterelectromotive cell
pila Daniell Daniell cell
pila de ácido acid cell
pila de aire air cell
pila de aire-cinc air-zinc cell
pila de almacenamiento storage stack
pila de bicromato bichromate cell
pila de Bunsen Bunsen cell
pila de cadmio cadmium cell
pila de carbón-cinc carbon-zinc cell
pila de circuito abierto open-circuit cell
pila de circuito cerrado closed-circuit cell
pila de Clark Clark cell
pila de cloruro de plata silver-chloride cell
pila de combustible fuel cell
pila de combustible de ciclo cerrado closed-cycle fuel cell
pila de concentración concentration cell
pila de consumo de carbón carbon-consuming cell
pila de desplazamiento displacement cell
pila de Edison Edison cell
pila de electrólito fundido fused-electrolyte cell
pila de electrólito único one-fluid cell
pila de energía energy cell
pila de gas gas cell
pila de gravedad gravity cell
pila de Grove Grove cell
pila de Leclanche Leclanche cell
pila de litio lithium cell
pila de magnesio magnesium cell
pila de mercurio mercury cell
pila de óxido de mercurio mercury oxide cell
pila de óxido de plata silver-oxide cell
pila de placas stack
pila de polarización biasing cell
pila de polarización de rejilla grid-bias cell
pila de prometio promethium cell
pila de rectificadores rectifier stack
pila de reserva reserve cell
pila de sal de amoniaco sal ammoniac cell
pila de silicio silicon cell
pila de Skrivanoff Skrivanoff cell
pila de soda cáustica caustic-soda cell
pila de trabajos job stack
pila de Weston Weston cell
pila de Zamboni Zamboni pile
pila débil weak cell
pila galvánica galvanic cell
pila húmeda wet cell
pila Lalande Lalande cell
pila local local cell
pila patrón standard cell
pila patrón de cadmio cadmium standard cell
pila patrón de cinc zinc standard cell
pila patrón de Weston Weston standard cell
pila patrón no saturada unsaturated standard cell
pila primaria de dióxido de plomo lead dioxide primary cell
pila primaria de plata-cinc silver-zinc primary cell
pila reversible reversible cell
pila seca dry cell
pila secundaria secondary cell
pila solar de silicio silicon solar cell
pila solar termoeléctrica thermoelectric solar cell
pila térmica thermal cell
pila voltaica voltaic cell
pilar de cables cable pillar

piloto pilot
piloto automático automatic pilot
piloto de conmutación switching pilot
piloto de línea line pilot
piloto de llamada ringing pilot
piloto de referencia reference pilot
piloto de referencia de supergrupo supergroup reference pilot
piloto de sincronización synchronization pilot
piloto regulador regulating pilot
pintura conductora conducting paint
pintura de plata silver paint
pinza clip
pinza aislada insulated clip
pinza cocodrilo crocodile clip
pinza de batería battery clip
pinza de contacto contact clip
pinza de Fahnestock Fahnestock clip
pinza de fusible fuse clip
pinza de pruebas test clip
pinza de puesta a tierra ground clip
pinza de rejilla grid clip
pinzas de soldar soldering pliers
pip pip
piramidal pyramidal
pirheliómetro pyrheliometer
pirita de hierro iron pyrite
piroconductividad pyroconductivity
piroelectricidad pyroelectricity
piroeléctrico pyroelectric
piroelectrólisis pyroelectrolysis
pirólisis pyrolysis
piromagnético pyromagnetic
piromagnetismo pyromagnetism
pirométrico pyrometric
pirómetro pyrometer
pirómetro de célula fotoeléctrica photoelectric-cell pyrometer
pirómetro de radiación radiation pyrometer
pirómetro de resistencia resistance pyrometer
pirómetro de termopar thermocouple pyrometer
pirómetro eléctrico electric pyrometer
pirómetro fotoeléctrico photoelectric pyrometer
pirómetro óptico optical pyrometer
pirómetro termoeléctrico thermoelectric pyrometer
pista track
pista a pista track to track
pista compresible squeeze track
pista de alimentación feed track
pista de área variable variable-area track
pista de auditoría audit trail
pista de cinta tape track
pista de control control track
pista de datos data track
pista de densidad variable variable-density track
pista de desbordamiento overflow track
pista de disco flexible floppy-disk track
pista de entrada input track
pista de inserción insertion track
pista de lectura reading track
pista de papel paper track
pista de perforación punching track
pista de registro recording track
pista de sonido sound track
pista de sonido magnética magnetic sound track
pista de sonido múltiple multiple sound track
pista de sonido óptica optical sound track
pista de sustitución substitute track
pista de tarjetas card track
pista de transferencia de dígitos digit-transfer track
pista defectuosa defective track, bad track
pista electrificada electrified track
pista llena full track

pista magnética magnetic track
pista normal standard track
pista óptica optical track
pista primaria primary track
pista relativa relative track
pista única single track
pistas independientes independent tracks
pistola de soldar soldering gun
pistón piston, plunger
pistón de contacto contact piston
pistón de choque choke piston
pistonófono pistonphone
pivote pivot
pivote de trole trolley pivot
pixel pixel
pixel de borde edge pixel
placa plate, board
placa acorazada ironclad plate
placa adaptadora adapter plate
placa anódica anode plate
placa antirremanente residual plate
placa carbonizada carbonized plate
placa colectora collector plate
placa de abertura aperture plate
placa de adaptación dieléctrica dielectric matching plate
placa de aleación alloy plate
placa de apoyo backing plate
placa de base base plate
placa de capacitor capacitor plate
placa de conexión connection plate
placa de cristal crystal plate
placa de cuarto de onda quarter-wave plate
placa de cuarzo quartz plate
placa de cubierta cover plate
placa de deflexión deflector plate
placa de deflexión X X-deflection plate
placa de deflexión Y Y-deflection plate
placa de estator stator plate
placa de Faure Faure plate
placa de formación de haz beam-forming plate
placa de guarda guard plate
placa de guía guide plate
placa de Hall Hall plate
placa de media onda half-wave plate
placa de memoria magnética magnetic memory plate
placa de mica mica plate
placa de montaje mounting plate
placa de Plante Plante plate
placa de presión pressure plate
placa de retención retaining plate
placa de rosetas rosette plate
placa de rotor con muescas slotted rotor plate
placa de señal signal plate
placa de terminales terminal plate
placa de tierra ground plate, grounding plate
placa de Tudor Tudor plate
placa de válvula valve plate
placa de vértice vertex plate
placa de vidrio glass plate
placa de Wilson Wilson plate
placa decorativa escutcheon
placa delgada thin plate
placa deslizante sliding plate
placa frontal faceplate
placa frontal de medidor meter faceplate
placa helicoidal helical plate
placa holográfica holographic plate
placa indicadora rating plate
placa lateral sideplate
placa madre motherboard
placa móvil rotor plate

placa negativa negative plate
placa pancromática panchromatic plate
placa para calentar hotplate
placa plana flat plate
placa positiva positive plate
placa posterior backplate
placa protectora protecting plate
placa rectangular rectangular plate
placa sensible sensitive plate
placa sinterizada sintered plate
placa terminal end plate
placa terminal de núcleo core end plate
placa transversal transverse plate
placa X X plate
placa Y Y plate
placas de Chladni Chladni's plates
placas de deflexión horizontal horizontal-deflection
    plates
placas de deflexión vertical vertical-deflection plates
plaga púrpura purple plague
plan de aplicación application plan
plan de mantenimiento maintenance plan
plan de muestreo de aceptación acceptance
    sampling plan
plan de numeración numbering plan
plan de numeración nacional national numbering
    plan
planar planar
planigrafía planigraphy
planígrafo planigraph
planímetro planimeter
plano plane
plano cartesiano Cartesian plane
plano complejo complex plane
plano de abertura aperture plane
plano de almacenamiento de núcleos magnéticos
    magnetic-core storage plane
plano de colores color plane
plano de conducción conducting plane
plano de convergencia convergence plane
plano de deflexión deflection plane
plano de dígito digit plane
plano de incidencia plane of incidence
plano de núcleos core plane
plano de núcleos magnéticos magnetic-core plane
plano de polarización plane of polarization
plano de propagación plane of propagation
plano de proyección plane of projection
plano de prueba proof plane
plano de referencia reference plane
plano de reflexión reflection plane
plano de rotación plane of rotation
plano de tierra ground plane
plano de transmisión transmission plane
plano de unión junction plane
plano de vibración plane of vibration
plano focal focal plane
plano neutro neutral plane
plano nodal nodal plane
plano principal principal plane
plano sagital sagittal plane
plano tangencial tangential plane
plano vertical vertical plane
planocóncavo planoconcave
planoconvexo planoconvex
planta eléctrica electric plant, generating plant,
    power plant, power station
planta telefónica telephone plant
plantilla template
plaquita wafer, slice
plasma plasma
plasma opaco opaque plasma
plasma rotativo rotating plasma

plasma transparente transparent plasma
plasmático plasmatic
plasmatrón plasmatron
plasmoide plasmoid
plasmón plasmon
plasmotrón plasmotron
plasticidad plasticity
plástico plastic
plástico conductor conductive plastic
plástico de poliéster polyester plastic
plástico de urea urea plastic
plástico fenólico phenolic plastic
plástico reforzado reinforced plastic
plástico termoconductor thermoconducting plastic
plástico termoendurecible thermosetting plastic
plastificante plasticizer
plata silver
plata alemana German silver
plata níquel nickel silver
plataforma platform, deck
plataforma de bobinas coil brace
plataforma de capacitores capacitor platform
plataforma de carga loading platform
plataforma de conectividad connectivity platform
plataforma de radar aerotransportada airborne
    radar platform
plataforma estabilizada stabilized platform
plataforma estable stable platform
plataforma giroestabilizada gyrostabilized platform
plateado silver-plated, silvered
platina platen
platinado platinum-plated, platinized
platinar platinum-plate
platinita platinite
platinizar platinize
platino platinum
platinodo platinode
platinoide platinoid
platinotrón platinotron
plato giratorio turntable
plato rotativo rotating turntable
plena carga full load, full rating
plena potencia full power
pliodinatrón pliodynatron
pliotrón pliotron
plomo lead
pluma de registro recording pen
pluma sensible a presión pressure-sensitive pen
plumbicón plumbicon
plutonio plutonium
población de portadores carrier population
poder de absorción equivalente equivalent
    absorbing power
poder de definición resolving power
poder de penetración penetrating power
poder de penetración de radiación penetrating
    power of radiation
poder de refracción refractive power
poder de resolución resolving power
poder dispersivo dispersive power
poder emisivo emissive power
poder emisor emitting power
poder multiplicador multiplying power
poise poise
polar polar
polaridad polarity
polaridad automática automatic polarity
polaridad de altavoz loudspeaker polarity
polaridad de conexiones lead polarity
polaridad de deflexión deflection polarity
polaridad de imán magnet polarity
polaridad de señal de imagen picture-signal polarity
polaridad de soldadura weld polarity

polaridad eléctrica electrical polarity
polaridad inversa reverse polarity
polaridad invertida reversed polarity
polaridad magnética magnetic polarity
polaridad negativa negative polarity
polaridad positiva positive polarity
polarimetría polarimetry
polarímetro polarimeter
polarímetro de Laurent Laurent polarimeter
polariscopio polariscope
polarizabilidad polarizability
polarizabilidad iónica ionic polarizability
polarizabilidad magnética magnetic polarizability
polarizable polarizable
polarización polarization, bias
polarización anódica anodic polarization
polarización automática automatic bias
polarización base base bias
polarización bolométrica bolometric bias
polarización catódica cathodic polarization
polarización cero zero bias
polarización circular circular polarization
polarización cruzada cross-polarization
polarización de activación activation polarization
polarización de alta frecuencia high-frequency bias
polarización de antena antenna polarization
polarización de batería battery bias
polarización de calefactor heater bias
polarización de contacto contact bias
polarización de corriente current bias
polarización de corriente alterna alternating-current
    bias
polarización de corte cutoff bias
polarización de detector detector bias
polarización de diodo diode bias
polarización de electrodo electrode bias
polarización de emisor emitter bias
polarización de frecuencia frequency bias
polarización de línea line bias
polarización de onda wave polarization
polarización de operación operating bias
polarización de puerta automática automatic gate
    bias
polarización de recepción receiving polarization
polarización de rejilla grid bias
polarización de rejilla automática automatic grid
    bias
polarización de rejilla de control control-grid bias
polarización de rejilla de corriente continua
    direct-current grid bias
polarización de rejilla directa direct grid bias
polarización de rejilla negativa negative-grid bias
polarización de relé relay bias
polarización de retardo de rectificador rectifier
    delay bias
polarización de saturación saturation polarization
polarización de señal signal bias
polarización de tensión voltage bias
polarización de trabajo marking bias
polarización de transistor transistor bias
polarización deseada desired polarization
polarización dieléctrica dielectric polarization
polarización directa forward bias
polarización eléctrica electric polarization, electric
    bias
polarización eléctrica de relé relay electric bias
polarización electrolítica electrolytic polarization
polarización electromagnética electromagnetic bias
polarización electroquímica electrochemical
    polarization
polarización elíptica elliptical polarization
polarización equilibrada de detector detector
    balanced bias

polarización espontánea spontaneous polarization
polarización fija fixed bias
polarización horizontal horizontal polarization
polarización imperfecta imperfect polarization
polarización inducida induced polarization
polarización inversa inverse bias
polarización limitadora limiting polarization
polarización lineal linear polarization
polarización magnética magnetic polarization,
    magnetic biasing
polarización magnética de corriente alterna
    alternating-current magnetic bias
polarización magnética de corriente continua
    direct-current magnetic biasing
polarización magnética de relé relay magnetic bias
polarización magnética por corriente alterna
    alternating-current magnetic biasing
polarización magnetoiónica magnetoionic
    polarization
polarización mecánica mechanical polarization,
    mechanical bias
polarización mecánica de relé relay mechanical bias
polarización mixta mixed polarization
polarización negativa negative bias
polarización normal normal polarization
polarización óptima optimum bias
polarización perfecta perfect polarization
polarización perpendicular perpendicular
    polarization
polarización plana plane polarization
polarización por concentración concentration
    polarization
polarización por fuga de rejilla grid-leak bias
polarización por potencial de contacto
    contact-potential bias
polarización positiva positive bias
polarización protectora protective bias
polarización remanente remanent polarization
polarización selectiva selective polarization
polarización telegráfica telegraph bias
polarización theta theta polarization
polarización variable variable bias
polarización vertical vertical polarization
polarizado polarized, biased
polarizado en un plano plane-polarized
polarizado linealmente linearly polarized
polarizado magnéticamente magnetically biased
polarizado negativamente negatively biased
polarizador polarizer
polarizador de infrarrojo infrared polarizer
polarizante polarizing
polarizar polarize
polarografía polarography
polarográfico polarographic
polarógrafo polarograph
polarograma polarogram
polaroid polaroid
polea magnética magnetic pulley
polianódico polyanode
poliatómico polyatomic
policarbonato polycarbonate
policlopreno polychloroprene
policromador polychromator
policromático polychromatic
policromo polychrome
polidireccional polydirectional
polielectrodo polyelectrode
polielectrólito polyelectrolyte
polienergético polyenergetic
poliéster polyester
poliestireno polystyrene
poliestirol polystyrol
polietileno polyethylene

polifásico polyphase
polifotal polyphotal
polifoto polyphote
poligonal polygonal
polígrafo polygraph
polihierro polyiron
polilínea polyline
polimérico polymeric
polimerización polymerization
polimerizar polymerize
polímero polymer
polímero ferroeléctrico ferroelectric polymer
poliodo polyode
poliolefina polyolefin
polióptico polyoptic
poliplexor polyplexer
polipropileno polypropylene
polisilicio polysilicon
politeno polythene
politetrafluoretileno polytetrafluoroethylene
poliuretano polyurethane
polivalencia polyvalence
polivalente polyvalent
polivinilo polyvinyl
polo pole
polo análogo analogous pole
polo antílogo antilogous pole
polo de campo field pole
polo de conmutación commutating pole
polo distribuido distributed pole
polo laminado laminated pole
polo magnético magnetic pole
polo magnético libre free magnetic pole
polo magnético unidad unit magnetic pole
polo negativo negative pole
polo norte north pole
polo norte magnético north magnetic pole
polo opuesto opposite pole
polo positivo positive pole
polo principal main pole
polo rotativo rotating pole
polo saliente salient pole
polo sólido solid pole
polo sombreado shaded pole
polo sur magnético south magnetic pole
polo unitario unit pole
polonio polonium
polos consecuentes consequent poles
polos de entrada-salida input-output poles
polos desiguales unlike poles
polos magnéticos terrestres terrestrial magnetic
    poles
polos similares like poles
polvo compactado pressed powder
polvo de carbón carbon dust
polvo de hierro iron dust
polvo fotosensible photosensitive powder
polvo magnético magnetic powder
polvo metálico sinterizado sintered metallic powder
polvo sinterizado sintered powder
ponderación weighting
poner en circuito switch-in
poner fuera del circuito switch-out
porcelana porcelain
porcentaje percentage, percent
porcentaje armónico harmonic percentage
porcentaje de absorción percent absorption
porcentaje de acoplamiento percentage coupling
porcentaje de apantallamiento shield percentage
porcentaje de distorsión percent distortion
porcentaje de distorsión armónica percent harmonic
    distortion
porcentaje de distorsión de intermodulación

    intermodulation-distortion percentage
porcentaje de error percentage error
porcentaje de incertidumbre percentage uncertainty
porcentaje de modulación percent modulation
porcentaje de modulación de haz beam modulation
    percentage
porcentaje de ondulación ripple percentage
porcentaje de pérdidas percentage loss
porcentaje de retroalimentación feedback
    percentage
porcentaje de sincronización percent
    synchronization
porcentaje efectivo de modulación effective
    percentage of modulation
porcentaje máximo de modulación maximum
    percentage modulation
porción de acción action portion
porción de dirección address portion
porción en cuadratura quadrature portion
porción no residente nonresident portion
poroso porous
portaaguja needle holder
portabilidad portability
portacristal crystal holder
portachasis chassis holder
portachip chip carrier
portadiodo diode holder
portador carrier
portador activo hot carrier
portador adyacente adjacent carrier
portador de carga charge carrier
portador de corriente current carrier
portador de datos data carrier
portador en equilibrio equilibrium carrier
portador excedente excess carrier
portador libre free carrier
portador mayoritario majority carrier
portador minoritario minority carrier
portador minoritario excedente excess minority
    carrier
portadora carrier
portadora aumentada augmented carrier
portadora auxiliar auxiliary carrier
portadora coherente coherent carrier
portadora completa full carrier
portadora condicionada conditioned carrier
portadora continua continuous carrier
portadora controlada controlled carrier
portadora controlada por cuarzo quartz-controlled
    carrier
portadora de amplitud constante constant-amplitude
    carrier
portadora de canal channel carrier
portadora de colores color carrier
portadora de crominancia chrominance carrier
portadora de frecuencia intermedia
    intermediate-frequency carrier
portadora de frecuencia piloto pilot-frequency
    carrier
portadora de grupo group carrier
portadora de imagen image carrier
portadora de impulsos pulse carrier
portadora de modulación de frecuencia
    frequency-modulation carrier
portadora de radiofrecuencia radio-frequency
    carrier
portadora de sonido sound carrier
portadora de video video carrier
portadora de video adyacente adjacent video carrier
portadora en fase in-phase carrier
portadora en línea line carrier
portadora en línea de energía power-line carrier
portadora flotante floating carrier

portadora fundamental fundamental carrier
portadora incrementada exalted carrier
portadora manipulada keyed carrier
portadora modulada modulated carrier
portadora modulada en fase phase-modulated
    carrier
portadora multicanal multichannel carrier
portadora no modulada unmodulated carrier
portadora no suprimida unsuppressed carrier
portadora piloto pilot carrier
portadora principal main carrier
portadora pulsada pulsed carrier
portadora quiescente quiescent carrier
portadora reducida reduced carrier
portadora suprimida suppressed carrier
portadora telefónica telephone carrier
portadora variable variable carrier
portadora virtual virtual carrier
portaelectrodo electrode holder
portaescobillas brush holder
portafusible fuseholder
portaimán magnet carrier
portalámpara lampholder, lamp socket
portapastilla pellet holder
portátil portable
portaválvula valve holder
posición position, setting
posición A A position
posición abierta open position
posición actual current position
posición angular angular position
posición aparente apparent position
posición astronómica astronomical position
posición auxiliar ancillary position
posición B B position
posición binaria binary position
posición cero zero position
posición conectada connected position
posición controladora controlling position
posición corriente current position
posición de almacenamiento storage position
posición de base de datos database position
posición de bit bit position
posición de caracteres character position
posición de centro exacto dead-center position
posición de cierre make position
posición de clasificación sort position
posición de código code position
posición de comprobación check position
posición de concentración concentration position
posición de conexión on position
posición de conmutador switch position
posición de contador counter position
posición de control control position
posición de control maestro master control position
posición de desbordamiento overflow position
posición de desconexión off position
posición de dígito digit position
posición de dígito decimal decimal digit position
posición de eje shaft position
posición de espera standby
posición de guarda guard position
posición de haz beam position
posición de impresión print position
posición de impulso pulse position
posición de índice index position
posición de liberación releasing position
posición de montaje mounting position
posición de operación operating position
posición de orden alto high-order position
posición de orden bajo low-order position
posición de perforación punch position
posición de programa program position

posición de prueba test position
posición de recepción receiving position
posición de registro recording position
posición de reposición reset position
posición de reposo rest position, key-up position
posición de salida outgoing position
posición de transferencia transfer position
posición de tránsito through position
posición de visualización display position
posición del signo sign position
posición desconectada disconnected position
posición diferencial differential position
posición fija fixed position
posición final end position
posición implícita assumed position
posición inicial initial position
posición interurbana trunk position
posición inversa reverse position
posición manual manual position
posición neutra neutral position
posición normal normal position
posición numérica numerical position
posición transmisora transmitting position
posición X X position
posición Y Y position
posicionador positioner
posicionador de chips chip positioner
posicional positional
posicionamiento positioning
posicionamiento de cabeza head positioning
posicionamiento de colector collector positioning
posicionamiento por corriente continua
    direct-current positioning
posicionamiento proporcional proportional
    positioning
posicionamiento relativo relative positioning
posicionamiento rotativo rotary positioning
posiciones agrupadas grouped positions
posiciones de llegada incoming positions
positivamente positively
positivo positive
positrón positron
postaceleración post acceleration
poste post, pole
poste arriostrado guyed pole
poste cero zero pole
poste de cable cable post
poste de guíaondas waveguide post
poste de línea line post
poste de unión junction pole
poste terminal terminal pole
posteco post-echo
postecualización post-equalization
postprocesador post-processor
postprocesamiento post-processing
postsincronización post-synchronization
potasio potassium
potencia power
potencia absoluta absolute power
potencia absorbida absorbed power
potencia activa active power
potencia activa media average active power
potencia acústica acoustic power
potencia acústica instantánea instantaneous acoustic
    power
potencia acústica instantánea instantaneous sound
    power
potencia amplificadora amplifying power
potencia anódica anode power
potencia aparente apparent power
potencia aparente base base apparent power
potencia aparente nominal nominal apparent power
potencia aplicada applied power

potencia atómica atomic power
potencia auxiliar auxiliary power
potencia calorimétrica calorimetric power
potencia cero zero power
potencia computacional computational power
potencia continua continuous power
potencia de alimentación supply power
potencia de amplificación amplification power
potencia de amplificador amplifier power
potencia de antena aerial power
potencia de arranque starting power
potencia de audio audio power
potencia de bandas laterales sideband power
potencia de calefactor heater power
potencia de calentamiento heating power
potencia de carga load power
potencia de computadora computer power
potencia de cómputo computing power
potencia de conversión disponible available
  conversion power
potencia de corriente alterna alternating-current
  power
potencia de corriente continua direct-current power
potencia de desaccionamiento dropout power
potencia de drenaje bleeder power
potencia de eco echo power
potencia de economía economy power
potencia de emisión emission power
potencia de entrada input power
potencia de entrada anódica anode input power
potencia de entrada de audio audio input power
potencia de entrada de corriente continua
  direct-current input power
potencia de entrada de placa plate input power
potencia de entrada nominal nominal input power,
  rated input power
potencia de excitación driving power
potencia de excitación de rejilla grid-driving power
potencia de filamento filament power
potencia de fuga leakage power
potencia de fuga armónica harmonic leakage power
potencia de haz beam power
potencia de impulso pulse power
potencia de interferencia interference power
potencia de línea line power
potencia de magnificación magnifying power
potencia de memoria memory power
potencia de onda sinusoidal equivalente equivalent
  sine wave power
potencia de operación operating power
potencia de placa plate power
potencia de plena carga full-load power
potencia de polarización bias power
potencia de portadora carrier power
potencia de procesamiento processing power
potencia de radiación radiation power
potencia de radiofrecuencia radio-frequency power
potencia de reacción reaction power
potencia de reserva de corriente alterna
  alternating-current standby power
potencia de ruido noise power
potencia de salida power output, output power
potencia de salida aparente apparent output power
potencia de salida de audio audio output power
potencia de salida nominal rated output power
potencia de salida útil useful output power
potencia de señal signal power
potencia de sincronización synchronizing power
potencia de transmisor transmitter power
potencia de transmisor visual visual-transmitter
  power
potencia de video video power
potencia de voz speech power

potencia deswattada wattless power
potencia disponible available power
potencia efectiva effective power
potencia efectiva mínima lowest effective power
potencia eficaz root-mean-square rating
potencia eléctrica electrical power
potencia en antena antenna power
potencia en espera idling power
potencia equivalente de ruido noise equivalent
  power
potencia específica specific power
potencia excitadora exciting power
potencia ficticia fictitious power
potencia fluctuante fluctuating power
potencia hidroeléctrica hydroelectric power
potencia incidente incident power
potencia indicada indicated power
potencia instantánea instantaneous power
potencia inversa reverse power
potencia lumínica aparente apparent candlepower
potencia lumínica de bujía candlepower
potencia lumínica esférica spherical candlepower
potencia luminosa luminous power
potencia magnética magnetic power
potencia máxima maximum power
potencia máxima continua maximum continuous
  power
potencia máxima de entrada segura maximum safe
  input power
potencia máxima media maximum average power
potencia máxima utilizable maximum usable power
potencia mecánica mechanical power
potencia media average power
potencia mínima de disparo minimum firing power
potencia mínima detectable minimum detectable
  power
potencia modulada modulated power
potencia moduladora modulating power
potencia musical music power
potencia neta net power
potencia nominal nominal power, nominal power
  rating, power rating, nameplate rating
potencia óptica optical power
potencia pico peak power
potencia pico de envolvente peak envelope power
potencia pico de impulso peak pulse power
potencia pico de radiación peak radiated power
potencia pico de radiofrecuencia peak
  radiofrequency output
potencia pico de salida peak power output
potencia pico de señal peak signal power
potencia pico instantánea instantaneous peak power
potencia polifásica polyphase power
potencia primaria primary power
potencia pulsada pulsed power
potencia radiada radiated power
potencia radiada aparente apparent radiated power
potencia radiada efectiva effective radiated power
potencia radiante radiant power
potencia reactiva reactive power
potencia real real power
potencia recibida received power
potencia reducida reduced power
potencia reflejada reflected power
potencia relativa relative power
potencia rotatoria rotatory power
potencia secundaria secondary power
potencia sin distorsión undistorted power
potencia sofométrica psophometric power
potencia sonora sound power
potencia térmica thermal power
potencia termoeléctrica thermoelectric power
potencia trifásica three-phase power

potencia ultravioleta ultraviolet power
potencia útil useful power
potencia vectorial vector power
potencia vocal instantánea instantaneous speech power
potencial potential
potencial absoluto absolute potential
potencial acelerador accelerating potential
potencial activo active potential
potencial anódico anode potential
potencial avanzado advanced potential
potencial cero zero potential
potencial crítico critical potential
potencial de acción action potential
potencial de aceleración acceleration potential
potencial de alimentación supply potential
potencial de aparición appearance potential
potencial de atracción attractive potential
potencial de barrera barrier potential
potencial de base base potential
potencial de cebado striking potential
potencial de centrifugación centrifugation potential
potencial de contacto contact potential
potencial de contacto de rejilla grid contact potential
potencial de corriente electrónica electron-stream potential
potencial de corte cutoff potential
potencial de Coulomb Coulomb potential
potencial de deflexión deflecting potential
potencial de deformación deformation potential
potencial de demarcación demarcation potential
potencial de descarga discharge potential
potencial de descomposición decomposition potential
potencial de desionización deionization potential
potencial de electrodo electrode potential
potencial de electrodo absoluto absolute electrode potential
potencial de electrodo de equilibrio equilibrium electrode potential
potencial de electrodo dinámico dynamic electrode potential
potencial de electrodo estático static electrode potential
potencial de electrodo normal standard electrode potential
potencial de equilibrio equilibrium potential
potencial de excitación excitation potential
potencial de extinción extinction potential
potencial de guarda guard potential
potencial de ignición ignition potential
potencial de ionización ionization potential
potencial de límite boundary potential
potencial de luminiscencia glow potential
potencial de membrana membrane potential
potencial de oxidación-reducción oxidation-reduction potential
potencial de pantalla screen potential
potencial de placa plate potential
potencial de polarización polarization potential, biasing potential
potencial de polielectrodo mixto mixed polyelectrode potential
potencial de punta spike potential
potencial de radiación radiation potential
potencial de reacción de equilibrio equilibrium reaction potential
potencial de reducción-oxidación redox potential
potencial de rejilla grid potential
potencial de rejilla inverso inverse grid potential
potencial de rejilla-pantalla screen-grid potential
potencial de reposo resting potential
potencial de sedimentación sedimentation potential
potencial de segundo ánodo second-anode potential

potencial de tierra ground potential, earth potential
potencial de unión de líquidos liquid-junction potential
potencial disruptivo sparking potential
potencial eléctrico electric potential
potencial electrocinético electrokinetic potential
potencial electroforético electrophoretic potential
potencial electrolítico electrolytic potential
potencial electromiográfico electromyographic potential
potencial electronegativo electronegative potential
potencial electroosmótico electroosmotic potential
potencial electropositivo electropositive potential
potencial electrostático electrostatic potential
potencial en circuito abierto open-circuit potential
potencial en reposo rest potential
potencial estabilizador stabilizing potential
potencial flotante floating potential
potencial inducido por antena antenna-induced potential
potencial magnético magnetic potential
potencial negativo negative potential
potencial normal standard potential
potencial positivo positive potential
potencial propagado propagated potential
potencial retardado delayed potential
potencial transitorio transient potential
potencial vectorial magnético magnetic vector potential
potencial zeta zeta potential
potenciales de Debye Debye potentials
potenciométricamente potentiometrically
potenciométrico potentiometric
potenciómetro potentiometer
potenciómetro automático automatic potentiometer
potenciómetro bobinado wire-wound potentiometer
potenciómetro capacitivo capacitive potentiometer
potenciómetro con derivación central center-tapped potentiometer
potenciómetro de ajuste trimmer potentiometer
potenciómetro de anchura width potentiometer
potenciómetro de carbón carbon potentiometer
potenciómetro de centrado centering potentiometer
potenciómetro de contraste contrast potentiometer
potenciómetro de control control potentiometer
potenciómetro de diez vueltas ten-turn potentiometer
potenciómetro de entrada input potentiometer
potenciómetro de equilibrio a cero null-balance potentiometer
potenciómetro de hilo slide-wire potentiometer
potenciómetro de imagen picture potentiometer
potenciómetro de inducción induction potentiometer
potenciómetro de medida measuring potentiometer
potenciómetro de microfricción microfriction potentiometer
potenciómetro de milivolts millivolt potentiometer
potenciómetro de pasos fijos fixed-step potentiometer
potenciómetro de precisión precision potentiometer
potenciómetro de rejilla grid potentiometer
potenciómetro de termopar thermocouple potentiometer
potenciómetro de vuelta única single-turn potentiometer
potenciómetro digital digital potentiometer
potenciómetro electrónico electronic potentiometer
potenciómetro fotoeléctrico photoelectric potentiometer
potenciómetro graduado graded potentiometer
potenciómetro helicoidal helical potentiometer
potenciómetro lineal linear potentiometer
potenciómetro logarítmico logarithmic potentiometer

potenciómetro magnético magnetic potentiometer
potenciómetro manual manual potentiometer
potenciómetro multicanal multichannel potentiometer
potenciómetro multiplicador multiplier potentiometer
potenciómetro multivuelta multiturn potentiometer
potenciómetro no lineal nonlinear potentiometer
potenciómetro panorámico panoramic potentiometer
potenciómetro paramétrico parametric potentiometer
potenciómetro sinusoidal sine potentiometer
potenciómetro toroidal toroidal potentiometer
potenciómetro variable variable potentiometer
potenciómetros acoplados mecánicamente ganged potentiometers
potenciostato potentiostat
poundal poundal
pozo de Gauss Gaussian well
pragilbert pragilbert
pragilberts por weber pragilberts per weber
praoersted praoersted
praseodimio praseodymium
preajustable preset
preajustado preset
preajuste manual manual preset
prealmacenamiento prestorage
prealmacenar prestore
preamplificación preamplification
preamplificador preamplifier
preamplificador bidireccional two-way preamplifier
preamplificador compensado compensated preamplifier
preamplificador de antena antenna preamplifier
preamplificador de cámara camera preamplifier
preamplificador de instrumento instrument preamplifier
preamplificador de micrófono microphone preamplifier
preamplificador de reproducción playback preamplifier
preamplificador de video video preamplifier
preamplificador ecualizador equalizing preamplifier
preamplificador integrador integrating preamplifier
preamplificador mezclador mixer preamplifier
preamplificador monoaural monaural preamplifier
preamplificador multigama multirange preamplifier
preamplificador sensible a la tensión voltage-sensitive preamplifier
preanálisis preanalysis
preasignación preallocation
preasignado preassigned
precableado prewired
precalentador preheater
precalentamiento preheating
precalentar preheat
precalibrado precalibrated
precargado preloaded
precargar preload
precesión precession
precesión de Larmor Larmor precession
precesión uniforme uniform precession
precipitabilidad precipitability
precipitable precipitable
precipitación precipitation
precipitación de partículas electrostática electrostatic particle precipitation
precipitación eléctrica electric precipitation
precipitación electrostática electrostatic precipitation
precipitador precipitator
precipitador de polvo dust precipitator
precipitador de polvo eléctrico electric dust precipitator

precipitador de polvo electrónico electronic dust precipitator
precipitador eléctrico electric precipitator
precipitador electrónico electronic precipitator
precipitador electrostático electrostatic precipitator
precipitrón precipitron
precisión precision, accuracy
precisión completa full precision
precisión de control control precision
precisión engañosa misleading precision
precisión extendida extended precision
precisión falsa false precision
precisión múltiple multiple precision
preciso precise, accurate
precodificado precoded
precompilación precompilation
precompilador precompiler
precomprobación prechecking
precondición precondition
precondicionamiento preconditioning
preconducción preconduction
preconfigurado preconfigured
precontrol precontrol
precorrección precorrection
precorrección de fase phase precorrection
precortado precut
precursor precursor
predefinido predefined
predetección predetection
predeterminado predetermined
predicción adaptiva adaptive prediction
predicción de errores error prediction
predicción de fallas failure prediction
predicción ionosférica ionospheric prediction
predicción lineal linear prediction
predisociación predissociation
predistorsión predistortion
predistorsión de imagen picture predistortion
preeco preecho
preecualización preequalization
preejecución preexecution
preénfasis preemphasis
preenfocado prefocused
preenfoque prefocusing
preensamblado preassembly
prefabricación prefabrication
prefabricado prefabricated
preferencial preferential
preferido preferred
prefijo prefix
prefijo de bloque block prefix
prefijo de borrar clearing prefix
prefijo multiplicador multiplier prefix
preformado preformed
pregrabado prerecorded
pregrupo pregroup
pregunta query
preignición preignition
preimpregnado preimpregnated
preimpregnar preimpregnate
preimpresión preprinting
preimpreso preprinted
preindicación preindication
preionización preionization
preionizante preionizing
preliminar preliminary
premodulación premodulation
prenumeración prenumbering
preoscilación preoscillation
preparación setup
preparación de cinta tape preparation
preparación de datos data preparation
preparación de programa program setup

preparar set-up
preperforado prepunched
preperforar prepunch
preprocesado preprocessed
preprocesador preprocessor
preprocesamiento preprocessing
preprocesar preprocess
preproducción preproduction
preprogramado preprogrammed
prerregistrado prerecorded
prerregistrar prerecord
prerrevestimiento precoating
presductor presducer
preselección preselection
preseleccionado preselected
preselector preselector
preselector de radiofrecuencia radio-frequency
     preselector
presentación presentation
presentación panorámica panoramic presentation
presentación visual display
presintonizable pretunable
presintonizado pretuned
presintonizar pretune
presión pressure
presión absoluta absolute pressure
presión activa active pressure
presión acústica acoustic pressure
presión acústica instantánea instantaneous sound
     pressure
presión ambiente ambient pressure
presión atmosférica atmospheric pressure
presión barométrica barometric pressure
presión de aceite oil pressure
presión de aguja needle pressure
presión de banda band pressure
presión de contacto contact pressure
presión de contacto final final contact pressure
presión de diseño design pressure
presión de gas gas pressure
presión de ionización ionization pressure
presión de plasma plasma pressure
presión de prueba proof pressure
presión de radiación radiation pressure
presión de radiación acústica acoustic radiation
     pressure
presión de referencia reference pressure
presión de ruido noise pressure
presión de solución solution pressure
presión de vapor vapor pressure
presión desactuante deactuating pressure
presión diferencial differential pressure
presión ecualizadora equalizing pressure
presión eléctrica electric pressure
presión electromagnética electromagnetic pressure
presión electrostática electrostatic pressure
presión específica specific pressure
presión estática static pressure
presión hidrodinámica hydrodynamic pressure
presión hidrostática hydrostatic pressure
presión interna nominal rated internal pressure
presión magnética magnetic pressure
presión máxima acústica maximum sound pressure
presión máxima de operación maximum operating
     pressure
presión nominal rated pressure
presión normal normal pressure
presión osmótica osmotic pressure
presión parcial partial pressure
presión positiva positive pressure
presión relativa relative pressure
presión sonora sound pressure
presión sonora efectiva effective sound pressure

presión sonora eficaz root-mean-square sound
     pressure
presión sonora excedente excess sound pressure
presión sonora máxima peak sound pressure
presión total total pressure
presión variable variable pressure
presostato pressurestat
presurización pressurization
presurizado pressurized
presurizar pressurize
pretersónico pretersonic
primario primary
primario de crominancia chrominance primary
primario de crominancia fina fine-chrominance
     primary
primario de luminancia luminance primary
primarios aditivos additive primaries
primarios de receptor receiver primaries
primarios de transmisión transmission primaries
primarios sustractivos subtractive primaries
primer ánodo first anode
primer detector first detector
primer dígito first digit
primer dinodo first dynode
primer oscilador local first local oscillator
primer selector first selector
primera armónica first harmonic
primera descarga de Townsend first Townsend
     discharge
primera etapa first stage
primera etapa de audio first audio stage
primera lectura first reading
primera ley de Kirchhoff Kirchhoff's first law
primera ley de termodinámica first law of
     thermodynamics
primera ley de Wien Wien's first law
primera región de Fresnel first Fresnel region
primera talla slab
primera tarjeta first card
primera transacción first transaction
primera transición first transition
primera zona de Fresnel first Fresnel zone
primero en entrar-primero en salir first-in-first-out
principio de Boltzmann Boltzmann's principle
principio de equivalencia cuántica quantum
     equivalence principle
principio de Fermat Fermat's principle
principio de Hamilton Hamilton's principle
principio de Huygens Huygens' principle
principio de incertidumbre de Heisenberg
     Heisenberg uncertainty principle
principio de reciprocidad reciprocity principle
principio de Volta Volta's principle
prioridad priority
prioridad absoluta absolute priority
prioridad de acceso access priority
prioridad de dispositivo device priority
prioridad de ejecución de máquina machine
     execution priority
prioridad de interrupción interrupt priority
prioridad de mensajes message priority
prioridad de multiprogramación multiprogramming
     priority
prioridad de programa program priority
prioridad de salida output priority
prioridad de selección selection priority
prioridad de tiempo compartido time-sharing
     priority
prioridad de trabajo job priority
prioridad de transmisión transmission priority
prioridad externa external priority
prioridad normal normal priority
prisma prism

prisma de cuarzo quartz prism
prisma de Nicol Nicol prism
prisma estereoscópica stereoscopic prism
prisma inversor reversing prism
prisma ocular ocular prism
prisma polarizante polarizing prism
prisma rotativo rotating prism
prismático prismatic
privacidad de datos data privacy
privilegio de archivo file privilege
probabilidad probability
probabilidad de adquisición acquisition probability
probabilidad de captura capture probability
probabilidad de colisión probability of collision
probabilidad de detección detection probability
probabilidad de error de bit bit error probability
probabilidad de error de facturación billing error
    probability
probabilidad de excitación excitation probability
probabilidad de falla probability of failure
probabilidad de falsa alarma false-alarm probability
probabilidad de ionización probability of ionization
probabilidad de operación falsa false-operation
    probability
probabilidad de penetración penetration probability
probabilidad empírica empirical probability
probador tester
probador de agua water tester
probador de capacitores capacitor tester
probador de cinta magnética magnetic-tape tester
probador de circuito de neón neon circuit tester
probador de circuito de rejilla grid-circuit tester
probador de circuitos circuit tester
probador de continuidad continuity tester
probador de cristal crystal tester
probador de diodo diode tester
probador de diodo dinámico dynamic diode tester
probador de emisión emission tester
probador de funciones múltiples multipurpose tester
probador de fusible fuse tester
probador de imanes magnet tester
probador de inducidos growler
probador de intoxicación intoxication tester
probador de micrófonos microphone tester
probador de módem modem tester
probador de neón neon tester
probador de resistores resistor tester
probador de tarjetas card tester
probador de transistores transistor tester
probador de transistores dinámico dynamic
    transistor tester
probador de tubos tube tester
probador de tubos dinámico dynamic tube tester
probador de uso general general-purpose tester
probador en circuito in-circuit tester
probador no destructivo nondestructive tester
probador portátil portable tester
problema de comprobación check problem
problema de prueba de referencia benchmark
    problem
procedimiento procedure
procedimiento anidado nested procedure
procedimiento asincrónico asynchronous procedure
procedimiento catalogado cataloged procedure
procedimiento de acceso access procedure
procedimiento de acceso de enlace link-access
    procedure
procedimiento de actualización local local update
    procedure
procedimiento de análisis de mantenimiento
    maintenance analysis procedure
procedimiento de análisis sintáctico parsing
    procedure

procedimiento de arranque startup procedure
procedimiento de búsqueda search procedure
procedimiento de calibración calibration procedure
procedimiento de carga loading procedure
procedimiento de clasificación sorting procedure
procedimiento de complemento complement
    procedure
procedimiento de contingencia contingency
    procedure
procedimiento de control control procedure
procedimiento de control de acceso de medio
    medium access control procedure
procedimiento de control de enlace lógico logical
    link control procedure
procedimiento de control de llamadas call-control
    procedure
procedimiento de decisión decision procedure
procedimiento de entrada input procedure
procedimiento de entrada-salida input-output
    procedure
procedimiento de exploración search procedure
procedimiento de función function procedure
procedimiento de mandos command procedure
procedimiento de medida measurement procedure
procedimiento de operación operating procedure
procedimiento de preparación setup procedure
procedimiento de prueba test procedure
procedimiento de reconstrucción de archivo file
    reconstruction procedure
procedimiento de recuperación recovery procedure
procedimiento de reducción de ruido noise
    abatement procedure
procedimiento de reserva fallback procedure
procedimiento de respaldo backup procedure
procedimiento de salida output procedure
procedimiento de verificación de instalación
    installation verification procedure
procedimiento externo external procedure
procedimiento inicial initial procedure
procedimiento interno internal procedure
procedimiento operacional operational procedure
procedimiento recursivo recursive procedure
procedimientos encadenados chained procedures
procesable por máquina machine processable
procesador processor
procesador auxiliar auxiliary processor
procesador central central processor
procesador de acceso de mantenimiento
    maintenance access processor
procesador de almacenamiento de control control
    storage processor
procesador de almacenamiento principal main
    storage processor
procesador de alta velocidad high-speed processor
procesador de archivo de mandos command file
    processor
procesador de asignación de recursos resource
    allocation processor
procesador de bloque block processor
procesador de coma flotante floating-point
    processor
procesador de comunicaciones communication
    processor
procesador de comunicaciones de datos data
    communications processor
procesador de control de red network control
    processor
procesador de control de trabajos job control
    processor
procesador de corriente de trabajos job stream
    processor
procesador de datos data processor
procesador de datos digitales digital-data processor

procesador de datos electrónico electronic data processor
procesador de dispositivo device processor
procesador de enlace de paquetes de datos packet data link processor
procesador de entrada-salida input-output processor
procesador de gráficos graphics processor
procesador de imagen image processor
procesador de imagen de exploración raster image processor
procesador de interfaz interface processor
procesador de lenguaje language processor
procesador de lenguaje de ensamblaje assembly language processor
procesador de mandos command processor
procesador de mantenimiento maintenance processor
procesador de palabras word processor
procesador de programa almacenado stored-program processor
procesador de secuencias sequence processor
procesador de señal signal processor
procesador de teclado keyboard processor
procesador de trabajo gráfico graphic job processor
procesador de unidad de procesamiento central central processing unit processor
procesador de uso general general-purpose processor
procesador digital digital processor
procesador en paralelo parallel processor
procesador global global processor
procesador local local processor
procesador maestro master processor
procesador microprogramable microprogrammable processor
procesador microprogramado microprogrammed processor
procesador numérico numerical processor
procesador periférico peripheral processor
procesador posterior back-end processor
procesador principal host processor
procesador redundante redundant processor
procesador remoto remote processor
procesador satélite satellite processor
procesador simétrico symmetrical processor
procesador único single processor
procesamiento processing
procesamiento aleatorio random processing
procesamiento asincrónico asynchronous processing
procesamiento automático automatic processing
procesamiento centralizado centralized processing
procesamiento concurrente concurrent processing
procesamiento consecutivo consecutive processing
procesamiento continuo continuous processing
procesamiento conversacional conversational processing
procesamiento de archivos file processing
procesamiento de base de datos database processing
procesamiento de campo por campo field-by-field processing
procesamiento de cinta tape processing
procesamiento de computadora computer processing
procesamiento de comunicaciones communications processing
procesamiento de comunicaciones de datos data communications processing
procesamiento de consulta inquiry processing
procesamiento de datos data processing
procesamiento de datos automático automatic data processing
procesamiento de datos automatizado automated data processing

procesamiento de datos centralizado centralized data processing
procesamiento de datos de acceso remoto remote-access data processing
procesamiento de datos descentralizado decentralized data processing
procesamiento de datos distribuido distributed data processing
procesamiento de datos electrónico electronic data processing
procesamiento de datos en línea on-line data processing
procesamiento de datos gráficos graphic data processing
procesamiento de datos integrado integrated data processing
procesamiento de datos mecánico mechanical data processing
procesamiento de datos no numérico nonnumeric data processing
procesamiento de datos remoto remote data processing
procesamiento de documentos document processing
procesamiento de entrada de bloques block input processing
procesamiento de etiquetas label processing
procesamiento de fondo background processing
procesamiento de grupo group processing
procesamiento de imagen image processing
procesamiento de imágenes por computadora computer image processing
procesamiento de información information processing
procesamiento de información automático automatic information processing
procesamiento de información electrónico electronic information processing
procesamiento de lenguaje automatizado automated language processing
procesamiento de listas list processing
procesamiento de llamadas call processing
procesamiento de mandos command processing
procesamiento de mensajes message processing
procesamiento de orden order processing
procesamiento de palabras word processing
procesamiento de pantalla completa full-screen processing
procesamiento de película delgada thin-film processing
procesamiento de prioridad foreground processing
procesamiento de secuencia string processing
procesamiento de secuencias múltiples multiple string processing
procesamiento de señal signal processing
procesamiento de señal coherente coherent signal processing
procesamiento de señales digitales digital signal processing
procesamiento de texto text processing
procesamiento de trabajo gráfico graphic job processing
procesamiento de trabajo remoto remote job processing
procesamiento de trabajos job processing
procesamiento de trabajos apilados stacked-job processing
procesamiento de trabajos múltiples multiple-job processing
procesamiento de transacciones transaction processing
procesamiento de voz voice processing
procesamiento descentralizado decentralized processing

procesamiento diferido deferred processing
procesamiento directo direct processing
procesamiento distribuido distributed processing
procesamiento en línea on-line processing
procesamiento en paralelo parallel processing
procesamiento en serie serial processing
procesamiento en tiempo real real-time processing
procesamiento fuera de línea off-line processing
procesamiento inmediato immediate processing
procesamiento interactivo interactive processing
procesamiento interno internal processing
procesamiento inverso backward processing
procesamiento modular modular processing
procesamiento multitarea multitask processing
procesamiento numérico numerical processing
procesamiento por lotes batch processing
procesamiento por lotes en línea on-line batch
    processing
procesamiento por lotes en tiempo real real-time
    batch processing
procesamiento por lotes secuencial sequential batch
    processing
procesamiento por prioridad priority processing
procesamiento primario primary processing
procesamiento remoto remote processing
procesamiento secuencial sequential processing
procesamiento simultáneo simultaneous processing
procesamiento sincrónico synchronous processing
proceso process
proceso adiabático adiabatic process
proceso aditivo additive process
proceso de administración de red network
    management process
proceso de carga loading process
proceso de codificación coding process
proceso de conversión conversion process
proceso de conversión fotoeléctrica photoelectric
    conversion process
proceso de Cottrell Cottrell process
proceso de crecimiento epitaxial epitaxial-growth
    process
proceso de desarrollo de programas software
    development process
proceso de difusión diffusion process
proceso de dos etapas two-stage process
proceso de ensamblaje assembly process
proceso de entrada input process
proceso de fabricación por lotes batch fabrication
    process
proceso de grabación recording process
proceso de limpieza cleanup process
proceso de memoria memory process
proceso de muestreo sampling process
proceso de refusión meltback process
proceso de registro recording process
proceso de salida output process
proceso de sorción sorption process
proceso de Umklapp Umklapp process
proceso de validación validation process
proceso dependiente del tiempo time-dependent
    process
proceso electrofotográfico electrophotographic
    process
proceso electrolítico irreversible irreversible
    electrolytic process
proceso electrolítico reversible reversible
    electrolytic process
proceso electrostático electrostatic process
proceso en línea on-line process
proceso epitaxial epitaxial process
proceso estocástico stochastic process
proceso fotolitográfico photolithographic process
proceso irreversible irreversible process

proceso iterativo iterative process
proceso magnético reversible reversible magnetic
    process
proceso multietapa multistage process
proceso múltiple multiple process
proceso natural natural process
proceso planar planar process
proceso predefinido predefined process
proceso recursivo recursive process
proceso reversible reversible process
proceso sustractivo subtractive process
proceso transitorio transient process
procesos concurrentes concurrent processes
procesos interactivos interactive processes
producción production
producción asistida por computadora
    computer-aided production
producción de energía energy production
producción de energía media mean energy
    production
producción de pares pair production
producción en masa mass production
producción piloto pilot production
producción por lotes batch production
producido mecánicamente mechanically produced
producto de energía energy product
producto electrónico electronic product
producto escalar scalar product
producto ganancia-ancho de banda gain-bandwidth
    product
producto lógico logic product
producto parcial partial product
producto vectorial vector product
productor armónico harmonic producer
productor de ruido noise producer
productos de intermodulación intermodulation
    products
profundidad de calentamiento heating depth
profundidad de campo field depth
profundidad de corte depth of cut
profundidad de descarga depth of discharge
profundidad de desvanecimiento fading depth
profundidad de modulación modulation depth
profundidad de penetración penetration depth
profundidad de pixel pixel depth
profundidad de surco groove depth
profundidad del entrehierro gap depth
profundidad óptica optical depth
programa program, schedule
programa absoluto absolute program
programa accionado por mandos command-driven
    program
programa accionado por mensajes message-driven
    program
programa accionado por menú menu-driven
    program
programa activo active program
programa alambrado wired program
programa almacenado stored program
programa alternativo alternate program
programa autoadaptivo self-adapting program
programa automático automatic program
programa autónomo stand-alone program
programa autorizado authorized program
programa auxiliar auxiliary program
programa basado en caracteres character-based
    program
programa codificado coded program
programa compilado separadamente separately
    compiled program
programa común common program
programa cruzado cross program
programa de acceso a datos data access program

programa de administración de archivos file management program
programa de administración de base de datos database management program
programa de análisis selectivo selective trace program
programa de aplicación application program
programa de aplicación primaria primary application program
programa de apoyo support program
programa de apoyo de sistema system-support program
programa de apoyo operacional utility program
programa de aprendizaje learning program
programa de biblioteca library program
programa de bucle cerrado closed-loop program
programa de búsqueda fetch program
programa de canal channel program
programa de carga load program
programa de cinta perforada punched-tape program
programa de clasificación sorting program
programa de clasificación de disco disk sort program
programa de clasificación-fusión sort-merge program
programa de clasificación-fusión generalizado generalized sort-merge program
programa de compilación compiling program
programa de compilador compiler program
programa de comprobación checking program
programa de computadora computer program
programa de comunicaciones communications program
programa de conjunto de trabajos job set program
programa de consulta inquiry program
programa de control control program
programa de control de comunicaciones communications control program
programa de control de entrada-salida input-output control program
programa de control de interrupción de comunicaciones communication interrupt control program
programa de control de mandos command control program
programa de control de mensajes message control program
programa de control de red network control program
programa de control de red local local network control program
programa de control de red remoto remote network control program
programa de control de trabajos job control program
programa de control ejecutivo executive control program
programa de control primario primary-control program
programa de control residente resident control program
programa de control residual residual-control program
programa de conversión conversion program
programa de conversión de archivo file conversion program
programa de conversión de cinta tape-conversion program
programa de conversión de lenguaje language conversion program
programa de corrección correction program
programa de definición de panel panel definition program

programa de demostración demonstration program
programa de depuración debugging program
programa de depuración de microprocesador microprocessor debugging program
programa de depuración en tiempo real real-time debugging program
programa de desarrollo de productos product development program
programa de diagnóstico diagnostic program
programa de diagnóstico fuera de línea off-line diagnostic program
programa de diseño design program
programa de distribución de página page layout program
programa de edición edit program
programa de emulador integrado integrated emulator program
programa de ensamblador cruzado cross assembler program
programa de ensamblaje assembly program
programa de entrada input program
programa de entrada de rutina de carga bootstrap input program
programa de fondo background program
programa de función function program
programa de generación generation program
programa de gráficos graphics program
programa de inicialización de disco disk initialization program
programa de instalación installation program
programa de interrupción breakpoint program
programa de lenguaje de control control language program
programa de lenguaje de ensamblaje assembly language program
programa de lista de direcciones mailing list program
programa de localización de fallas fault-location program
programa de llamada calling program
programa de macroensamblaje macro-assembly program
programa de macros preprocesado preprocessed macro program
programa de mantenimiento maintenance schedule, maintenance program
programa de mantenimiento de biblioteca library maintenance program
programa de máquina machine program
programa de monitoreo monitoring program
programa de periférico peripheral program
programa de preparación housekeeping program
programa de prioridad priority program
programa de procesamiento processing program
programa de procesamiento de aplicación application processing program
programa de procesamiento de mandos command processing program
programa de procesamiento de mensajes message processing program
programa de procesamiento de palabras word-processing program
programa de procesamiento por lotes batch processing program
programa de producción production program
programa de prueba test program
programa de prueba de referencia benchmark program
programa de prueba en línea on-line test program
programa de pruebas testing program
programa de rastreo tracing program
programa de recubrimiento overlay program
programa de recuperación recovery program

programa de recuperación automática automatic recovery program
programa de recuperación de final anormal abend recovery program
programa de relocalización dinámica dynamic relocation program
programa de reserva de disco duro hard-disk backup program
programa de salida output program, exit program
programa de seguridad security program
programa de servicio service program
programa de servicio de sistema system service program
programa de sistema system program
programa de solución de problemas problem-solving program
programa de sondeo polling program
programa de tarjetas card program
programa de traducción de lenguaje language translation program
programa de transacciones transaction program
programa de uso general general-purpose program
programa de usuario user program
programa de vaciado de memoria de núcleos core dump program
programa de visualización gráfica graphic display program
programa dependiente de computadora computer dependent program
programa ejecutable executable program
programa ejecutivo executive program
programa emulador emulator program
programa en bucle loop program
programa en línea on-line program
programa encadenado chained program
programa ensamblador assembler program
programa específico specific program
programa estructurado structured program
programa exterior outer program
programa fijo fixed program
programa formateado formatted program
programa fuente source program
programa generador generator program
programa general general program
programa heurístico heuristic program
programa inactivo inactive program
programa incompleto incomplete program
programa integrado integrated program
programa interactivo interactive program
programa interno inner program
programa interpretador interpreter program
programa interpretativo interpretive program
programa iterativo iterative program
programa lineal linear program
programa local local program
programa llamado called program
programa macrogenerador macro-generating program
programa maestro master program
programa miniensamblador miniassembler program
programa modular modular program
programa monitor monitor program
programa multifásico multiphase program
programa no formateado unformatted program
programa no residente nonresident program
programa objeto target program
programa operacional operational program
programa operativo de uso general general-purpose operating program
programa óptimo optimum program
programa patrón standard program
programa por omisión default program
programa post mortem post-mortem program

programa primario primary program
programa principal main program
programa reentrable reenterable program
programa reentrante reentrant program
programa residente resident program
programa residente en memoria memory resident program
programa reubicable relocatable program
programa reusable reusable program
programa secuencial sequential program
programa secundario secondary program
programa segmentado segmented program
programa simbólico symbolic program
programa simulador simulator program
programa solicitante requesting program
programa supervisor supervisory program
programa supervisor de procesos process supervisory program
programa transitorio transient program
programable programmable
programable externamente externally programmable
programación programming
programación absoluta absolute programming
programación automática automatic programming
programación concurrente concurrent programming
programación convencional conventional programming
programación conversacional conversational programming
programación cruzada cross programming
programación cuadrática quadratic programming
programación de acceso aleatorio random-access programming
programación de acceso mínimo minimum-access programming
programación de cinta tape programming
programación de computadora computer programming
programación de control de sistema system control programming
programación de control integral integral control programming
programación de diagnóstico diagnostic programming
programación de latencia mínima minimum latency programming
programación de lenguaje de máquina machine-language programming
programación de máquina machine programming
programación de prioridades priority scheduling
programación de retardo delay programming
programación de retardo mínimo minimum delay programming
programación de secuencia string programming
programación de sistema system programming
programación dinámica dynamic programming
programación discreta discrete programming
programación en serie serial programming
programación estructurada structured programming
programación funcional functional programming
programación heurística heuristic programming
programación incidental incidental programming
programación incremental incremental programming
programación lineal linear programming
programación lógica logic programming
programación matemática mathematical programming
programación modular modular programming
programación múltiple multiple programming
programación no lineal nonlinear programming
programación no numérica nonnumeric

programming
programación operacional operational programming
programación óptima optimum programming
programación orientada a archivos file-oriented programming
programación paralela parallel programming
programación paramétrica parametric programming
programación por tarjetas perforadas punched-card programming
programación secuencial sequential programming
programación simbólica symbolic programming
programación sistemática systematic programming
programado programmed
programado externamente externally programmed
programado por tarjetas card-programmed
programador programmer
programador de computadora computer programmer
programador de memoria sólo de lectura read-only memory programmer
programador de prioridades priority scheduler
programador de secuencias sequence programmer
programador de tambor drum programmer
programas software, programs
programas accionados por menú menu-driven software
programas compatibles compatible software
programas comunes common software
programas concurrentes concurrent programs
programas críticos critical software
programas de aplicación application software
programas de apoyo support software
programas de apoyo operacional utility software
programas de biblioteca library software
programas de comunicaciones communications software
programas de control control software
programas de control de errores error control software
programas de controlador driver software
programas de diagnóstico diagnostic software
programas de gráficos graphics software
programas de pruebas test software
programas de sistema system software
programas defectuosos faulty software
programas en línea on-line software
programas inalterables firmware
programas integrados integrated software
programática programmatics, software
progresión progression
progresión aritmética arithmetic progression
progresión armónica harmonic progression
progresión geométrica geometric progression
progresivo progressive
prolongador de línea line lengthener
prometio promethium
propagación propagation
propagación acústica acoustic propagation
propagación anómala anomalous propagation
propagación anormal abnormal propagation
propagación auroral auroral propagation
propagación cuasióptica quasi-optical propagation
propagación de onda de carga espacial space-charge wave propagation
propagación de ondas wave propagation
propagación de ondas de radio radio-wave propagation
propagación de ondas largas long-wave propagation
propagación de ondas métricas metric-wave propagation
propagación de ondas ultracortas ultrashort-wave propagation
propagación de salto único single-skip propagation

propagación electromagnética electromagnetic propagation
propagación en espacio libre free-space propagation
propagación guiada guided propagation
propagación ionosférica ionospheric propagation
propagación normal normal propagation
propagación por difracción diffraction propagation
propagación por dispersión scatter propagation
propagación por dispersión directa forward-scatter propagation
propagación por dispersión troposférica tropospheric-scatter propagation
propagación por guíaondas waveguide propagation
propagación por modo de silbidos atmosféricos whistler-mode propagation
propagación por salto único single-hop propagation
propagación por saltos múltiples multihop propagation
propagación por trayectorias múltiples multipath propagation
propagación transhorizonte beyond-the-horizon propagation
propagación troposférica tropospheric propagation
propagado propagated
propagador spreader
propenso a errores error-prone
propiedad determinada por tensión voltage-determined property
propiedad extrínseca extrinsic property
propiedad física physical property
propiedad generalizada generalized property
propiedad intrínseca intrinsic property
propiedades de caracteres character properties
propiedades de circuito circuit properties
propiedades directivas directive properties
proporción proportion
proporcionado proportionate
proporcional proportional
proporcionalidad proportionality
proporcionamiento proportioning
propulsión eléctrica electric propulsion
protactinio protactinium
protección protection
protección anódica anodic protection
protección automática automatic protection
protección básica basic protection
protección catódica cathodic protection
protección contra agentes atmosféricos weather protection
protección contra búsqueda fetch protection
protección contra copia copy protection
protección contra corriente inversa reverse-current protection
protección contra cortocircuitos short-circuit protection
protección contra descarga directa direct-stroke protection
protección contra descarga indirecta indirect-stroke protection
protección contra escritura write protection
protección contra fase abierta open-phase protection
protección contra frenado de contramarcha antiplugging protection
protección contra fuego fire protection
protección contra humedad humidity protection
protección contra inversión de fase phase-reversal protection
protección contra lectura read protection
protección contra pérdida de sincronismo out-of-step protection
protección contra picos transitorios surge protection
protección contra polaridad inversa

reverse-polarity protection
protección contra rayos lightning protection
protección contra sobrecarga overload protection
protección contra sobrecarga térmica
thermal-overload protection
protección contra sobrecargas automática
automatic overload protection
protección contra sobrecorriente overcurrent
protection
protección contra sobretemperatura
overtemperature protection
protección contra sobretensión overvoltage
protection
protección contra sobretensiones incorporada
built-in overvoltage protection
protección contra sobrevelocidad overspeed
protection
protección contra subtensión undervoltage
protection
protección cruzada cross-protection
protección de almacenamiento storage protection
protección de archivos file protection
protección de avalancha avalanche protection
protección de bloque block protection
protección de campo field protection
protección de cátodo cathode protection
protección de celda cell protection
protección de circuito circuit protection
protección de contacto contact protection
protección de contraseña password protection
protección de corriente direccional directional
current protection
protección de cortocircuito automática automatic
short-circuit protection
protección de datos data protection
protección de descarga de campo field discharge
protection
protección de distancia distance protection
protección de entrada input protection
protección de falla de campo field-failure protection
protección de frecuencia frequency protection
protección de impedancia impedance protection
protección de lectura-escritura read-write protection
protección de memoria memory protection
protección de página page protection
protección de potencia power protection
protección de potencia direccional directional
power protection
protección de privacidad privacy protection
protección de programa program protection
protección de reactancia reactance protection
protección de reserva reserve protection
protección de resistencia resistance protection
protección de respaldo backup protection
protección de sobrecorriente direccional directional
overcurrent protection
protección de tierra ground protection
protección diferencial differential protection
protección diferencial longitudinal longitudinal
differential protection
protección diferencial transversal transverse
differential protection
protección homopolar zero-phase-sequence
protection
protección individual individual protection
protección por comparación de fase
phase-comparison protection
protección por corriente portadora carrier-current
protection
protección por hilo piloto pilot-wire protection
protección por piloto pilot protection
protección por relé de tierra ground relay protection
protección primaria primary protection

protección principal main protection
protección selectiva selective protection
protección térmica thermal protection
protector protector, arrester
protector con espacios gap arrester
protector contra picos transitorios surge protector
protector contra sobretensión voltage-surge
protector
protector de aluminio aluminum arrester
protector de células de aluminio aluminum-cell
arrester
protector de contacto contact protector
protector de medidor meter protector
protector de pantalla screen saver
protector de red network protector
protector de sobretensión de espacio de aire air-gap
surge protector
protector exterior outdoor arrester
protector interior indoor arrester
protector químico chemical protector
protector térmico thermal protector
proteger contra escritura write protect
proteger contra lectura read-protect
protegido protected
protegido contra agentes atmosféricos
weather-protected
protegido contra daños por fallas failsafe
protegido contra escritura write protected
protegido por fusible fused
protocolo protocol
protocolo de acceso a datos data access protocol
protocolo de acceso seguro assured access protocol
protocolo de cambio de dirección change direction
protocol
protocolo de comunicaciones communications
protocol
protocolo de control de enlace de datos data link
control protocol
protocolo de control de enlace lógico logical link
control protocol
protocolo de enlace link protocol
protocolo de enlace de datos data link protocol
protocolo de mantenimiento de archivos file
maintenance protocol
protocolo de transferencia de archivo file transfer
protocol
protocolo orientado a caracteres character-oriented
protocol
protón proton
protón negativo negative proton
prototipo prototype
prototipo de macro macro prototype
protuberancia bulge
proustita proustite
provisional provisional
proximidad proximity
proyección axial axial projection
proyección frontal front projection
proyección lateral lateral projection
proyección oblicua oblique projection
proyección ortogonal orthogonal projection
proyección tangencial tangential projection
proyecto de programas software project
proyector projector
proyector cinematográfico motion-picture projector
proyector de sonar sonar projector
proyector de tubo de rayos catódicos cathode-ray
tube projector
proyector dividido split projector
prueba test
prueba A-B A-B test
prueba acelerada accelerated test
prueba aceptada accepted test

prueba ambiental environmental test
prueba audible audible test
prueba automática automatic test
prueba beta beta test
prueba biaxial biaxial test
prueba comparativa comparative test
prueba condicional conditional test
prueba continua continuous test
prueba cualitativa qualitative test
prueba cuantitativa quantitative test
prueba de abocinado flaring test
prueba de aceptación acceptance test
prueba de actividad de filamento filament activity test
prueba de aislamiento insulation test
prueba de almacenamiento storage test
prueba de alta tensión high-voltage test
prueba de alto potencial high-potential test
prueba de aprobación approval test
prueba de articulación articulation test
prueba de bajada drop test
prueba de banco bench test
prueba de barrido sweep test
prueba de batería battery test
prueba de bucle abierto open-loop test
prueba de bucle de Murray Murray loop test
prueba de cable cable test
prueba de calibración calibration test
prueba de capacidad capacity test
prueba de cero zero test
prueba de circuito circuit test
prueba de circuito abierto open-circuit test
prueba de comparación comparison test
prueba de compatibilidad compatibility test
prueba de confiabilidad reliability test
prueba de confiabilidad de campo field-reliability test
prueba de confianza confidence test
prueba de conformidad conformance test
prueba de construcción construction test
prueba de continuidad continuity test
prueba de corbata de lazo bow-tie test
prueba de corriente current test
prueba de corriente alterna alternating-current test
prueba de corriente continua continuous-current test
prueba de corriente de cierre making-current test
prueba de choque shock test
prueba de descarga disruptiva sparkover test
prueba de destellos flashing test
prueba de diagnóstico diagnostic test
prueba de durabilidad endurance test
prueba de duración life test
prueba de duración acelerada accelerated life test
prueba de eficiencia efficiency test
prueba de ejecución en paralelo parallel run test
prueba de energía de descarga discharge-energy test
prueba de equilibrio balance test
prueba de equipo equipment test
prueba de estación station test
prueba de estrés stress test
prueba de exactitud accuracy test
prueba de formato format test
prueba de frenaje braking test
prueba de freno brake test
prueba de fugas leakage test
prueba de impacto impact test
prueba de impulsos impulse test
prueba de instalación installation test
prueba de integración integration test
prueba de intervalo ambiental aceptable acceptable-environmental-range test

prueba de inyección de señal signal-injection test
prueba de laboratorio laboratory test
prueba de límites limit test
prueba de localización de fallas fault-location test
prueba de llamada calling test
prueba de manipulación keying test
prueba de mantenimiento maintenance test
prueba de mantenimiento preventivo preventive maintenance test
prueba de McProud McProud test
prueba de memoria de acceso aleatorio random-access memory test
prueba de micrófono microphone test
prueba de módulo module test
prueba de nivel de ruido noise-level test
prueba de ocupado busy test
prueba de oscilación oscillation test
prueba de parámetro parameter test
prueba de potencia musical music power test
prueba de potencia nominal power rating test
prueba de potencial aplicado applied-potential test
prueba de presión pressure test
prueba de procedimiento procedural test
prueba de producción production test
prueba de programa program test
prueba de propagación propagation test
prueba de prototipo prototype test
prueba de radiación dinámica dynamic radiation test
prueba de receptor receiver test
prueba de referencia benchmark
prueba de regresión regression testing
prueba de rendimiento performance test
prueba de resistencia resistance test
prueba de resistencia a la tensión voltage-endurance test
prueba de resistencia óhmica ohmic resistance test
prueba de respuesta de frecuencia frequency-response test
prueba de rocío salino salt-spray test
prueba de ruido noise test
prueba de ruptura de carga load-break test
prueba de ruptura por tensión voltage-breakdown test
prueba de sacudidas shake test
prueba de salida output test
prueba de salto skip test
prueba de secuencia de fases phase-sequence test
prueba de sensibilidad sensitivity test
prueba de señalización signaling test
prueba de servicio service test
prueba de servicio acelerado accelerated service test
prueba de sistema system test
prueba de sobretensión overvoltage test
prueba de sobretensión controlada controlled overvoltage test
prueba de sobrevelocidad overspeed test
prueba de taller shop test
prueba de tensión voltage test
prueba de tensión aplicada applied voltage test
prueba de tiempo de alta tensión high-voltage time test
prueba de transmisión transmission test
prueba de trayectoria path test
prueba de validación validation test
prueba de velocidad speed test
prueba de vibraciones vibration test
prueba de vuelo flight test
prueba destructiva destructive test
prueba dieléctrica dielectric test
prueba dinámica dynamic test
prueba dinamométrica dynamometer test
prueba directa direct test

prueba en bucle loop test
prueba en cortocircuito short-circuit test
prueba en línea on-line test
prueba en vacío vacuum test
prueba estática static test
prueba formal formal test
prueba fuera de línea off-line test
prueba funcional functional test
prueba incorporada built-in test
prueba inicial initial test
prueba instantánea flash test
prueba intermitente intermittent test
prueba interna internal test
prueba marginal marginal test
prueba multifrecuencia multifrequency test
prueba no destructiva nondestructive test
prueba operacional operational test
prueba pasiva passive test
prueba paso a paso step-by-step test
prueba periódica periodic test
prueba piloto pilot test
prueba por saltos leapfrog test
prueba práctica field test
prueba Q Q test
prueba remota remote test
prueba reproducible reproducible test
prueba retardada delayed test
prueba rutinaria routine test
prueba ultrasónica ultrasonic test
pruebas con ondas cuadradas square-wave testing
pruebas de certificación certification tests
pruebas de diseño design tests
pruebas de sistema system testing
pruebas en oposición back-to-back testing
pruebas estructurales structural testing
pruebas por lotes batch testing
puente bridge, jumper
puente acústico acoustic bridge
puente anódico anode bridge
puente autoequilibrado self-balanced bridge
puente calibrador calibrating bridge
puente combinado combination bridge
puente completo full bridge
puente conjugado conjugate bridge
puente de alimentación feeding bridge
puente de Anderson Anderson bridge
puente de banda base baseband bridge
puente de bolómetro bolometer bridge
puente de brazos de razón ratio-arm bridge
puente de Campbell Campbell bridge
puente de Campbell-Colpitts Campbell-Colpitts bridge
puente de canal de servicio service-channel bridge
puente de capacidad capacity bridge
puente de capacitancias capacitance bridge
puente de capacitancias de Wien Wien capacitance bridge
puente de Carey-Foster Carey-Foster bridge
puente de cero null bridge
puente de comparación comparison bridge
puente de conductividad conductivity bridge
puente de corriente alterna alternating-current bridge
puente de cuatro elementos four-element bridge
puente de décadas decade bridge
puente de DeSauty DeSauty's bridge
puente de desfase phase-shift bridge
puente de distorsión distortion bridge
puente de frecuencias frequency bridge
puente de Graetz Graetz bridge
puente de Hay Hay bridge
puente de Heaviside-Campbell Heaviside-Campbell bridge

puente de hilo slide-wire bridge
puente de immitancias immittance bridge
puente de impedancia impedance bridge
puente de inducción induction bridge
puente de inductancia-capacitancia inductance-capacitance bridge
puente de inductancia-capacitancia-resistencia inductance-capacitance-resistance bridge
puente de inductancia mutua mutual-inductance bridge
puente de inductancia mutua de Carey-Foster Carey-Foster mutual inductance bridge
puente de inductancia mutua de Heaviside Heaviside mutual-inductance bridge
puente de inductancia mutua de Maxwell Maxwell mutual-inductance bridge
puente de inductancias inductance bridge
puente de inductancias de Maxwell Maxwell inductance bridge
puente de inductancias de Wien Wien inductance bridge
puente de Kelvin Kelvin bridge
puente de Lecher Lecher bridge
puente de Maxwell Maxwell bridge
puente de medida measuring bridge
puente de medida universal universal measuring bridge
puente de Miller Miller bridge
puente de Nernst Nernst bridge
puente de Owen Owen bridge
puente de radiofrecuencia radio-frequency bridge
puente de red network bridge
puente de registro recording bridge
puente de resistencia-capacitancia resistance-capacitance bridge
puente de resistencia en paralelo parallel-resistance bridge
puente de resistencia-inductancia resistance-inductance bridge
puente de resistencias resistance bridge
puente de resonancia resonance bridge
puente de resonancia en serie series-resonance bridge
puente de retroalimentación feedback bridge
puente de Robinson Robinson bridge
puente de Schering Schering bridge
puente de suma summation bridge
puente de termistores thermistor bridge
puente de termómetro de resistencia resistance-thermometer bridge
puente de termopar thermocouple bridge
puente de Thomson Thomson bridge
puente de transformador transformer bridge
puente de transmisión transmission bridge
puente de tubos tube bridge
puente de tubos de vacío vacuum-tube bridge
puente de uso general general-purpose bridge
puente de Wheatstone Wheatstone bridge
puente de Wheatstone resistivo resistive Wheatstone bridge
puente de Wien Wien bridge
puente deformimétrico strain-gage bridge
puente desequilibrado off-balance bridge
puente desfasador phase-shifting bridge
puente divisor divider bridge
puente doble double bridge
puente doble de Kelvin Kelvin double bridge
puente eléctrico electric bridge
puente eliminador de modulación modulation-eliminator bridge
puente en anillo ring bridge
puente equilibrado balanced bridge
puente esquelético skeleton bridge

puente magnético magnetic bridge
puente monofásico de onda completa single-phase full-wave bridge
puente no lineal nonlinear bridge
puente rectificador rectifier bridge
puente sensible a la frecuencia frequency-sensitive bridge
puente sensible a la tensión voltage-sensitive bridge
puente termométrico thermometer bridge
puente universal universal bridge
puente Z-Y Z-Y bridge
puenteo bridging
puerta gate, port
puerta A y no B A and not B gate
puerta analógica analog gate
puerta anticoincidencia anticoincidence gate
puerta bicondicional biconditional gate
puerta coherente coherent gate
puerta de amplitud amplitude gate
puerta de arco arc gate
puerta de bloqueo lockout gate
puerta de coincidencia coincidence gate
puerta de color color gate
puerta de cruce crossover gate
puerta de decisión decision gate
puerta de diodo diode gate
puerta de elementos de decisión decision element gate
puerta de entrada gate of entry
puerta de excepción except gate
puerta de flujo flux gate
puerta de información information gate
puerta de inhibición inhibit gate
puerta de lámpara de neón neon-bulb gate
puerta de muestreo sampling gate
puerta de negación negation gate
puerta de Phillips Phillips gate
puerta de señal signal gate
puerta de silicio silicon gate
puerta de sincronización cromática burst gate
puerta de tiempo time gate
puerta de transmisión transmission gate
puerta de umbral threshold gate
puerta de un elemento one-element gate
puerta electrónica electronic gate
puerta expansible expandable gate
puerta flotante floating gate
puerta habilitante enabling gate
puerta lineal linear gate
puerta lógica logic gate
puerta magnética magnetic gate
puerta NO NOT gate
puerta NO-O NOR gate
puerta NO-Y NAND gate
puerta O OR gate
puerta posterior back gate
puerta SI-Y-SOLO-SI IF-AND-ONLY-IF gate
puerta sincrónica synchronous gate
puerta tiratrónica thyratron gate
puerta uno one gate
puerta Y AND gate
puerto port
puerto de comunicaciones communication port
puerto de entrada-salida input-output port
puerto de entrada-salida en serie serial input-output port
puerto de impresora printer port
puerto de juegos game port
puerto de memoria memory port
puerto de salida output port
puerto serie serial port
puerto terminal terminal port
puesta a cero zeroing, zero set, reset, resetting

puesta a masa grounding
puesta a tierra grounding, earthing
puesta a tierra de punto único single-point grounding
puesta a tierra protectora protective grounding
puesta a tierra temporal temporary grounding
puesta en cola queuing
puesta en espera camp-on
puesto a tierra por resistencia resistance-grounded
puesto de bloqueo block post
puesto de mando power signal box
pulido electrolítico electrolytic polishing
pulsación pulsing, pulsation, beat
pulsación aleatoria random pulsing
pulsación anódica anode pulsing
pulsación de frecuencia frequency pulsing
pulsación de frecuencia única single-frequency pulsing
pulsación de línea line pulsing
pulsación de retardo delay pulsing
pulsación eléctrica electric pulsation
pulsación periódica periodic pulsation
pulsación por rejilla grid pulsing
pulsación rítmica rhythmic pulsing
pulsado pulsed
pulsador push-button, plunger, pulser
pulsador de arranque-parada start-stop push-button
pulsador de línea line pulser
pulsador iluminado illuminated push-button
pulsador momentáneo momentary push-button
pulsante pulsating
pulsar pulsate
pulsatorio pulsating
pulso pulse
pulso de dígito digit pulse
pulverización catódica cathode sputtering
pulverizado powdered
pulvimetalurgia powder metallurgy
punta peak, point, spike
punta de acción action spike
punta de cable cable end
punta de carbón carbon tip
punta de contacto prod
punta de descarga spike
punta de flecha arrowhead
punta de prueba test prod
punto point, spot, dot, notch
punto caliente hot spot
punto catódico cathode spot
punto cero zero point
punto ciego blind spot
punto ciego de radar radar blind spot
punto colector collector dot
punto de acceso de pruebas test access point
punto de alimentación feed point
punto de anclaje anchor point
punto de aplicación application point
punto de calor heat spot
punto de canto singing point
punto de captura capture spot
punto de carga load point
punto de carga de cinta tape load point
punto de código code point
punto de color color dot
punto de comprobación checkpoint
punto de comunicación point of communication
punto de conexión point of connection, tie point
punto de congelación freezing point
punto de conmutación switching point
punto de contacto point of contact
punto de control control point
punto de control de unidad física physical unit control point

punto de corte cutoff point
punto de cruce crossover point, crossover, crosspoint
punto de Curie Curie point
punto de Curie ferroeléctrico ferroelectric Curie point
punto de destrucción destruction point
punto de deterioro completo wearout point
punto de disparo trigger point, firing point
punto de distribución de cable cable distribution point
punto de embrague clutch point
punto de entrada entry point
punto de entrada asincrónico asynchronous entry point
punto de entrada primario primary entry point
punto de entrada secundario secondary entry point
punto de exploración scanning point, scanning spot
punto de fase point of phase
punto de fusión melting point, blowing point
punto de ignición ignition point
punto de imagen picture dot
punto de inflamabilidad flash point
punto de inflexión inflection point
punto de inserción insertion point
punto de intercambio interchange point
punto de interrupción breakpoint
punto de interrupción de datos data breakpoint
punto de inversión inversion point
punto de limitación clipping point
punto de media potencia half-power point
punto de medida measuring point
punto de multiplicación multiplication point
punto de nivel de transmisión cero zero-transmission-level point
punto de operación operating point
punto de operación de corriente continua direct-current operating point
punto de origen point of origin
punto de polarización bias point
punto de potencia power point
punto de prueba test point
punto de prueba de aguja needle test point
punto de puesta a tierra grounding point
punto de reanudación restart point
punto de recalescencia recalescent point
punto de recuperación recovery point
punto de recuperación de mensajes message recovery point
punto de reejecución rerun point
punto de reentrada reentry point
punto de referencia reference point
punto de referencia de caracteres character reference point
punto de registro recording spot
punto de regulación setting point
punto de reinicio restart point
punto de reposo quiescent point
punto de retorno return point
punto de rocío dew point
punto de ruptura breakpoint
punto de salida exit point
punto de saturación saturation point
punto de sincronización synchronization point
punto de sobrecarga overload point
punto de soldadura welding point
punto de toma tapping point
punto de toma de impulsos de sincronización sync-takeoff point
punto de tráfico traffic point
punto de transferencia de grupo group transfer point
punto de transferencia de supergrupo supergroup

transfer point
punto de transición transition point
punto de transición conductiva breakover point
punto de transmisión transmission point
punto de unión junction point
punto de vaciado dump point
punto de venta point of sale
punto decimal real actual decimal point
punto direccionable addressable point
punto fijado set point
punto final end point
punto focal focal point
punto fosforescente phosphor dot
punto frío cold spot
punto índice index point
punto isoeléctrico isoelectric point
punto luminoso light spot
punto modal modal point
punto móvil flying spot
punto muerto dead spot
punto negro black spot
punto neutro neutral point
punto nodal nodal point
punto nulo null point
punto objeto object point
punto obscuro dark spot
punto pico peak point
punto por punto point-by-point
punto principal principal point
punto Q Q point
punto sumador summing point
punto terminal terminal point
punto variable variable point
pupila artificial artificial pupil
pupitre de conmutación switching desk
pupitre de control control desk
pupitre de control maestro master control desk
pupitre de monitoreo monitoring desk
pupitre de observación observation desk
pupitre de pruebas test desk
pupitre monitor monitor desk
pureza purity
pureza colorimétrica colorimetric purity
pureza de colores color purity
pureza de excitación excitation purity
pureza de modo mode purity
pureza de señal signal purity
pureza espectral spectral purity
purificación purification
purificación de datos data purification
purificación de electrólito purification of electrolyte
purificación de zona zone purification
purificador purifier
purificar purify

Q

quark quark
quemadura burn, burnout
quemadura de trama raster burn
quiescente quiescent
quimógrafo kymograph
quinario quinary
quintuplete quintuplet
quintuplicador quintupler
quintuplicador de frecuencia frequency quintupler
quintuplicador de tensión voltage quintupler

R

rabo de cerdo pigtail
racional rational
racon racon
rad rad
radan radan
radar radar
radar aerotransportado airborne radar
radar anticolisión anticollision radar
radar basado en tierra ground-based radar
radar bidimensional two-dimensional radar
radar biestático bistatic radar
radar de adquisición acquisition radar
radar de adquisición y rastreo acquisition and tracking radar
radar de aeronave aircraft radar
radar de aeropuerto airport radar
radar de alcance medio medium-range radar
radar de alta definición high-definition radar
radar de alta frecuencia high-frequency radar
radar de alta resolución high-resolution radar
radar de aproximación de precisión precision approach radar
radar de aviso de colisiones collision warning radar
radar de búsqueda aerotransportado airborne search radar
radar de campo de aviación airfield radar
radar de comparación de fase phase-comparison radar
radar de control control radar
radar de control de aeródromo aerodrome control radar
radar de control de aproximación approach control radar
radar de control de área area control radar
radar de control de puerto harbor control radar
radar de control de tráfico aéreo air-traffic control radar
radar de corto alcance short-range radar
radar de chirrido chirp radar
radar de detección detection radar
radar de doble haz double-beam radar
radar de Doppler Doppler radar
radar de Doppler de impulsos pulse Doppler radar
radar de Doppler láser laser Doppler radar
radar de exploración search radar
radar de frecuencia superalta superhigh-frequency radar
radar de frecuencia ultraalta ultra-high frequency radar
radar de haz en V V-beam radar
radar de impulsos pulse radar
radar de impulsos coherentes coherent-pulse radar
radar de información general general-information radar
radar de instrumentación instrumentation radar
radar de intercepción aerotransportado airborne intercept radar
radar de largo alcance long-range radar
radar de luz coherente coherent-light radar
radar de mano hand radar
radar de modulación de frecuencia

frequency-modulation radar
radar de monoimpulsos monopulse radar
radar de muy alta frecuencia very-high-frequency
    radar
radar de muy corto alcance very-short-range radar
radar de muy largo alcance very-long-range radar
radar de navegación navigation radar
radar de onda continua continuous-wave radar
radar de ondas largas long-wave radar
radar de precisión precision radar
radar de puntería aiming radar
radar de superficie surface radar
radar de uso general general-purpose radar
radar de vigilancia surveillance radar
radar de vigilancia aérea air surveillance radar
radar de vigilancia aerotransportado airborne
    surveillance radar
radar de vigilancia de aeropuerto airport
    surveillance radar
radar de vigilancia de rutas aéreas air-route
    surveillance radar
radar de vigilancia terrestre ground surveillance
    radar
radar decca decca radar
radar en banda X X-band radar
radar en diversidad diversity radar
radar indicador de zona zone-position radar
radar láser laser radar
radar maestro master radar
radar meteorológico weather radar
radar meteorológico aerotransportado airborne
    weather radar
radar meteorológico de colores color weather radar
radar milimétrico millimetric radar
radar monostático monostatic radar
radar navegacional navigational radar
radar óptico optical radar
radar panorámico panoramic radar
radar pasivo passive radar
radar primario primary radar
radar secundario secondary radar
radar sonoro sound radar
radar submarino underwater radar
radar telealtímetro height-finder radar
radar térmico thermal radar
radar transhorizonte over-the-horizon radar
radar tridimensional three-dimensional radar
radar vertical vertical radar
radar volumétrico volumetric radar
radariscopio radarscope
radarizado radarized
radecón radechon
radiac radiac
radiación radiation
radiación acústica acoustic radiation
radiación alfa alpha radiation
radiación ambiente ambient radiation
radiación anormal abnormal radiation
radiación atómica atomic radiation
radiación atrapada trapped radiation
radiación bactericida bactericidal radiation
radiación beta beta radiation
radiación blanca white radiation
radiación característica characteristic radiation
radiación ciclotrónica cyclotron radiation
radiación coherente coherent radiation
radiación compleja complex radiation
radiación continua continuous radiation
radiación cósmica cosmic radiation
radiación dañina harmful radiation
radiación de antena antenna radiation
radiación de bombeo pumping radiation
radiación de cavidad cavity radiation

radiación de Cerenkov Cerenkov radiation
radiación de cuadripolo quadripole radiation
radiación de cuerpo negro blackbody radiation
radiación de fondo background radiation
radiación de fuga leakage radiation
radiación de ondas posteriores back-wave radiation
radiación de oscilador oscillator radiation
radiación de oscilador local local-oscillator radiation
radiación de plasma plasma radiation
radiación de radar radar radiation
radiación de radiofrecuencia radio-frequency
    radiation
radiación de receptor receiver radiation
radiación de resonancia resonance radiation
radiación directa direct radiation
radiación directiva directive radiation
radiación efectiva effective radiation
radiación eléctrica electric radiation
radiación electromagnética electromagnetic
    radiation
radiación espuria spurious radiation
radiación fluorescente fluorescent radiation
radiación fotónica photon radiation
radiación fuera de banda out-of-band radiation
radiación gamma gamma radiation
radiación hertziana Hertzian radiation
radiación heterogénea heterogenous radiation
radiación homogénea homogeneous radiation
radiación incoherente incoherent radiation
radiación infrarroja infrared radiation
radiación infrarroja polarizada polarized infrared
    radiation
radiación integrada integrated radiation
radiación interior indoor radiation
radiación ionizante ionizing radiation
radiación monocromática monochromatic radiation
radiación monoenergética monoenergetic radiation
radiación monoenergética variable variable
    monoenergetic radiation
radiación natural natural radiation
radiación no coherente noncoherent radiation
radiación no ionizante nonionizing radiation
radiación no polarizada unpolarized radiation
radiación no térmica nonthermal radiation
radiación normal normal radiation
radiación parásita stray radiation
radiación penetrante penetrating radiation
radiación polarizada polarized radiation
radiación policromática polychromatic radiation
radiación posterior back radiation
radiación primaria primary radiation
radiación secundaria secondary radiation
radiación selectiva selective radiation
radiación solar solar radiation
radiación sonora sound radiation
radiación térmica thermal radiation
radiación ultravioleta ultraviolet radiation
radiación vertical vertical radiation
radiación visible visible radiation
radiación X X radiation
radiactividad radioactivity
radiactividad aérea airborne radioactivity
radiactividad artificial artificial radioactivity
radiactividad atmosférica atmospheric radioactivity
radiactividad inducida induced radioactivity
radiactividad residual residual radioactivity
radiactivo radioactive
radiado radiated
radiador radiator
radiador acústico acoustic radiator
radiador anisotrópico anisotropic radiator
radiador anódico anode radiator
radiador completo full radiator

radiador de antena antenna radiator
radiador de bocina circular circular-horn radiator
radiador de bocina piramidal pyramidal horn radiator
radiador de bocina rectangular rectangular-horn radiator
radiador de calor heat radiator
radiador de cuarto de onda quarter-wave radiator
radiador de electrodo electrode radiator
radiador de guíaondas waveguide radiator
radiador de Hertz Hertz radiator
radiador de Lambert Lambertian radiator
radiador de media onda half-wave radiator
radiador de ondas wave radiator
radiador de Planck Planckian radiator
radiador de ranura slot radiator
radiador dieléctrico dielectric radiator
radiador dipolo dipole radiator
radiador eléctrico electric radiator
radiador esférico spherical radiator
radiador incidental incidental radiator
radiador isotrópico isotropic radiator
radiador lineal straight-line radiator
radiador no selectivo nonselective radiator
radiador pasivo passive radiator
radiador primario primary radiator
radiador secundario secondary radiator
radiador selectivo selective radiator
radiador térmico thermal radiator
radiador vertical vertical radiator
radiador vertical doblado folded vertical radiator
radial radial
radián radian
radián eléctrico electrical radian
radián hiperbólico hyperbolic radian
radiancia radiance
radiancia básica basic radiance
radiancia espectral spectral radiance
radiante radiating
radiar radiate
radiática radiatics
radiativo radiative
radio radio, radium, radius
radio bidireccional two-way radio
radio de aficionados amateur radio
radio de Bohr Bohr radius
radio de cristal crystal radio
radio de electrón clásico classical electron radius
radio de referencia reference radius
radio de transistores transistor radio
radio espacial space radio
radio marítimo marine radio
radio por hilo wired radio
radio portátil portable radio
radio transistorizado transistorized radio
radio unidireccional one-way radio
radioactivo radioactive
radioacústica radioacoustics
radioacústico radioacoustic
radioaltímetro radio altimeter
radioamplificación radio amplification
radioastronomía radio astronomy
radioastronómico radio-astronomical
radioatenuación radio attenuation
radioatmósfera radio atmosphere
radioayuda radio aid
radiobaliza marker beacon, marker, beacon
radiobaliza de límite boundary marker beacon
radiobaliza de posición radio marker beacon
radiobaliza en abanico fan-marker beacon
radiobaliza exterior outer marker beacon
radiobaliza intermedia middle marker beacon
radiobaliza interna inner marker beacon

radiobaliza Z Z marker beacon
radioblindaje radio shield
radioboya radio buoy
radiocanal radio channel
radiocompás radio compass
radiocompás automático automatic radio compass
radiocompás de una frecuencia one-frequency radiocompass
radiocomunicación radiocommunication
radiocomunicación de aviación aviation radiocommunication
radioconductor radioconductor
radiocontrol radiocontrol
radiocontrolado radio-controlled
radiocristalografía radiocrystallography
radiocronómetro radio chronometer
radiodespacho radio-dispatching
radiodetección radiodetection
radiodifusión broadcasting, radiobroadcasting, broadcast, radiobroadcast, radiodiffusion
radiodifusión a larga distancia long-distance broadcasting
radiodifusión de alta frecuencia high-frequency broadcasting
radiodifusión de colores color broadcasting
radiodifusión de modulación de amplitud amplitude-modulation broadcasting
radiodifusión de muy alta frecuencia very-high-frequency broadcasting
radiodifusión de ondas ultracortas ultrashort-wave broadcasting
radiodifusión en cadena chain broadcasting
radiodifusión estereofónica stereophonic broadcasting, stereophonic broadcast, stereocasting, stereocast
radiodifusión por canal común cochannel broadcasting
radiodifusión por frecuencia común common-frequency broadcasting
radiodifusión por línea line broadcasting, wire broadcasting, line radio
radiodifusión por modulación de frecuencia frequency-modulated broadcasting
radiodifusión por ondas métricas metric-wave broadcasting
radiodifusión simultánea simultaneous broadcasting, simulcast
radiodifusión telefónica telephone broadcasting
radiodifusión territorial territorial broadcasting
radiodifusión tropical tropical broadcasting
radiodifusor broadcaster
radiodistribución radiodistribution
radioeco radio echo
radioelectricidad radioelectricity
radioeléctrico radioelectric
radioelectrónica radio-electronics
radioelemento radioelement
radioemanación radio emanation
radioemisión radio emission
radioemisor radio sender
radioenlace radio link
radioenlace de línea visual line-of-sight radio link
radioenlace de salto único single-hop radio relay
radioenlace multicanal multichannel radio link, multichannel radio relay
radioenlace por microondas microwave radio link, microwave radio relay, microwave relay link
radioenlace por modulación de impulsos pulse-modulation radio link
radioespectro radio spectrum
radioespectroscopio radio spectroscope
radioestación radio station
radioestrella radio star

radiofacsímil radiofacsimile
radiofaro radiobeacon, beacon
radiofaro aeronáutico aeronautical beacon
radiofaro de aeródromo aerodrome beacon
radiofaro de aeropuerto airport beacon
radiofaro de aerovía airway beacon
radiofaro de aproximación approach beacon
radiofaro de aproximación ciega blind approach beacon
radiofaro de aterrizaje landing beacon
radiofaro de código code beacon
radiofaro de corto alcance short-range radiobeacon
radiofaro de destellos flashing beacon
radiofaro de emergencia emergency beacon
radiofaro de identificación identification beacon
radiofaro de infrarrojo infrared beacon
radiofaro de luz de aproximación approach-light beacon
radiofaro de microondas microwave beacon
radiofaro de muy alta frecuencia very-high-frequency radio beacon
radiofaro de peligro hazard beacon
radiofaro de peligro de aeródromo aerodrome hazard beacon
radiofaro de peligro de aeropuerto airport danger beacon
radiofaro de pista de aeropuerto airport runway beacon
radiofaro de radar radar beacon
radiofaro de seguridad de control de tráfico aéreo air-traffic control safety beacon
radiofaro direccional radio-range beacon, homing beacon, radio range
radiofaro direccional audiovisual visual-aural radio range
radiofaro direccional aural aural radio range
radiofaro direccional de Adcock Adcock radio range
radiofaro direccional de cuatro vías four-course radio range
radiofaro direccional de muy alta frecuencia very-high-frequency directional range
radiofaro direccional equiseñal equisignal radio range beacon
radiofaro direccional visual visual radio range
radiofaro directivo directive beacon
radiofaro equiseñal equisignal beacon
radiofaro fantasma ghost beacon
radiofaro fijo fixed beacon
radiofaro giratorio revolving radiobeacon
radiofaro indicador de rumbo course-indicating beacon
radiofaro lanac lanac beacon
radiofaro localizador de pista runway localizer beacon
radiofaro marítimo marine radiobeacon
radiofaro no direccional nondirectional radio beacon
radiofaro omnidireccional omnidirectional radio beacon, omnidirectional beacon, omnidirectional radio range, omnirange
radiofaro omnidireccional de muy alta frecuencia very-high-frequency omnirange
radiofaro pasivo passive radio beacon
radiofaro pulsado pulsed beacon
radiofaro rebecca-eureka rebecca-eureka beacon
radiofaro respondedor responder beacon
radiofaro rotativo rotating radio beacon
radiofonía radiophony
radiofónico radiophonic
radiófono radiophone
radiofonógrafo radiophonograph
radiofoto radiophoto
radiofotografía radiophotography

radiofotograma radiophotogram
radiofotoluminiscencia radiophotoluminescence
radiofrecuencia radio frequency
radiofrecuencia sintonizada tuned radio-frequency
radiofrecuencia variable variable radio frequency
radiofuente radio source
radiofuente cuasiestelar quasi-stellar radio source
radiogalaxia radio galaxy
radiogénico radiogenic
radiogoniometría radiogoniometry, radio direction-finding
radiogoniómetro radiogoniometer, radio direction-finder, direction finder
radiogoniómetro automático automatic direction-finder
radiogoniómetro de Adcock Adcock direction-finder
radiogoniómetro de alta frecuencia high-frequency direction finder
radiogoniómetro de baja frecuencia low-frequency direction finder
radiogoniómetro de Bellini-Tosi Bellini-Tosi direction-finder
radiogoniómetro de cuadro loop direction finder, frame direction-finder
radiogoniómetro de cuadro compensado compensated-loop direction-finder
radiogoniómetro de cuadro rotativo rotating-loop direction-finder
radiogoniómetro de microondas microwave direction finder
radiogoniómetro de muy alta frecuencia very-high-frequency direction-finder
radiogoniómetro de rayos catódicos cathode-ray direction-finder
radiogoniómetro de Robinson Robinson direction-finder
radiogoniómetro de Robinson-Adcock Robinson-Adcock direction-finder
radiogoniómetro fijo fixed direction-finder
radiogoniómetro manual manual direction-finder
radiogoniómetro rotativo rotating direction-finder
radiografía radiography
radiografía industrial industrial radiography
radiografía instantánea flash radiography
radiográficamente radiographically
radiográfico radiographic
radiograma radiogram
radioguiado radio guidance, radio homing
radiohorizonte radio horizon
radiointerferencia radio interference
radiointerferómetro radiointerferometer
radioisótopo radioisotope
radiolocalización radiolocation
radiolocalización por Doppler radio Doppler
radiolocalizador radiolocator
radiología radiology
radiológico radiological
radiolucencia radiolucency
radiolucente radiolucent
radioluminiscencia radioluminescence
radiomalla radio mesh
radiometalografía radiometallography
radiometeorógrafo radiometeorograph
radiometeorología radiometeorology
radiometría radiometry
radiometría de microondas microwave radiometry
radiométrico radiometric
radiómetro radiometer
radiómetro acústico acoustic radiometer
radiómetro de Crookes Crookes radiometer
radiómetro de Dicke Dicke radiometer
radiómetro de microondas microwave radiometer
radiómetro de Nichols Nichols radiometer

radiómetro neto net radiometer
radiomicrófono radio microphone
radiomicrómetro de Boys Boys radiomicrometer
radionavegación radionavigation
radionavegación aeronáutica aeronautical radionavigation
radiónica radionics
radioopacidad radioopacity
radioopaco radiopaque
radiooperado radio-operated
radioportadora radio carrier
radiopropagación radio propagation
radioprospección radio prospecting
radioproximidad radio proximity
radiorrecepción radio reception
radiorreceptor radio receiver
radiorreflector radio reflector
radiorregistrador radio recorder
radiorrelé radio relay
radiorresistencia radioresistance
radiorresonancia radioresonance
radiorruido radio noise
radioscopía radioscopy
radioscopio radioscope
radiosensibilidad radiosensitivity
radiosensible radiosensitive
radioseñal radio signal
radiosistema radio system
radiosonda radioprobe, radiosonde
radiosonda de telemedida telemetering radiosonde
radiosondeo radiosounding
radiosónico radiosonic
radiotelecomunicación radiotelecommunication
radiotelefonía radiotelephony
radiotelefonía de muy alta frecuencia very-high-frequency radiotelephony
radiotelefónico radiotelephonic
radioteléfono radiotelephone
radioteléfono de muy alta frecuencia very-high-frequency radiotelephone
radioteléfono portátil walkie-talkie, handie-talkie
radiotelegrafía radiotelegraphy
radiotelegráfico radiotelegraphic
radiotelégrafo radiotelegraph
radiotelegrama radio telegram
radioteleimpresora radioteleprinter
radiotelemedición radiotelemetering
radiotelemetría radiotelemetry
radiotelescopio radio telescope
radioteletipo radioteletype
radiotélex radiotelex
radiotermia radiothermics
radiotransceptor radio transceiver
radiotransmisión radio transmission
radiotransmisión díplex diplex radio transmission
radiotransmisión multidireccional multidirectional radio transmission
radiotransmisor radio transmitter
radiotransmisor automático automatic radio transmitter
radiotransmisor multicanal multichannel radio transmitter
radiotransparente radiotransparent
radiotrazador radiotracer
radioviento rawin
radomo radome
radón radon
radux radux
ráfaga burst
ráfaga de color color burst
ráfaga de tono tone burst
ráfaga láser laser burst
raíz root

raíz cuadrada square root
raíz cuadrada de media de cuadrados root mean square
rama branch, leg
rama acústica acoustic branch
rama auxiliar auxiliary branch
rama condicional conditional branch
rama de impedancia impedance branch
rama inactiva inactive leg
rama incondicional unconditional branch
rama transmisora transmitting branch
ramac ramac
ramark ramark
ramas conjugadas conjugate branches
ramificación branching
rampa ramp
ranura slot
ranura de acoplamiento coupling slot
ranura de cabeza head slot
ranura de disquete de entrada-salida input-output diskette slot
ranura de entrada-salida input-output slot
ranura de expansión expansion slot
ranura de exploración scanning slot
ranura de filtro filter slot
ranura de periférico peripheral slot
ranura de tarjetas card slot
ranura longitudinal longitudinal slot
ranura polarizante polarizing slot
ranura posicionadora keyway
ranura radial radial slot
ranura radiante radiating slot
ranurado slotted
ranuras paralelas parallel slots
rapcon rapcon
rapidez rapidity
rapidez de decaimiento decay rate
rapidez de modulación modulation rate
rarefacción rarefaction
rarefactor getter
rarificado rarefied
ras flush
ráser raser
rasguño scratch
rastreador tracer
rastreo tracing, tracking, roaming
rastreo de llamada call trace
ratón mouse
ratón en serie serial mouse
raya dash, scratch
rayadura scratch
raydist raydist
rayl rayl
rayo ray, lightning
rayo axial axial ray
rayo central central ray
rayo con fugas leaky ray
rayo cósmico duro hard cosmic ray
rayo de luz ray of light
rayo de revestimiento cladding ray
rayo directo direct ray
rayo extraordinario extraordinary ray
rayo guiado guided ray
rayo incidente incident ray
rayo indirecto indirect ray
rayo inicial initial ray
rayo iónico ion ray
rayo ionosférico ionospheric ray
rayo luminoso light ray
rayo luminoso modulado modulated light ray
rayo meridional meridional ray
rayo monocromático monochromatic ray
rayo oblicuo skew ray

rayo ordinario ordinary ray
rayo paraxial paraxial ray
rayo positivo positive ray
rayo principal principal ray
rayo reflejado reflected ray
rayo refractado refracted ray
rayo Roentgen Roentgen ray
rayo sin reflexión unreflected ray
rayo tangente tangent ray
rayo ultravioleta ultraviolet ray
rayo violeta violet ray
rayo X X-ray
rayos actínicos actinic rays
rayos alfa alpha rays
rayos anódicos anode rays
rayos beta beta rays
rayos canales canal rays
rayos catódicos cathode-rays
rayos colimados collimated rays
rayos convergentes converging rays
rayos cósmicos cosmic rays
rayos de Lenard Lenard rays
rayos delta delta rays
rayos gamma gamma rays
rayos infrarrojos infrared rays
rayos límite limit rays
rayos penetrantes penetrating rays
rayos retrógrados retrograde rays
rayos secundarios secondary rays
rayos ultrafóticos ultraphotic rays
rayos ultrarrojos ultrared rays
rayos X blandos soft X-rays
rayos X continuos continuous X-rays
rayos X duros hard X-rays
rayos X fluorescentes fluorescent X-rays
rayos X polarizados polarized X-rays
rayos X secundarios secondary X-rays
razón ratio, rate
razón armónica harmonic ratio
razón atómica atomic ratio
razón axial axial ratio
razón de abertura aperture ratio
razón de absorción diferencial differential
    absorption ratio
razón de aciertos hit ratio
razón de acoplamiento coupling ratio
razón de actividad activity ratio
razón de adaptación de impedancias
    impedance-matching ratio
razón de amortiguamiento damping ratio
razón de amplitud amplitude ratio
razón de ancho de banda bandwidth ratio
razón de armadura de relé relay-armature ratio
razón de aspecto aspect ratio
razón de atenuación attenuation ratio
razón de banda ancha wideband ratio
razón de bloqueo blocking ratio
razón de brillos brightness ratio
razón de campo field ratio
razón de cancelación cancellation ratio
razón de capacitancias capacitance ratio
razón de captación pickup ratio
razón de captura capture ratio
razón de cavidad cavity ratio
razón de cierre-abertura make-and-break ratio
razón de compresión compression ratio
razón de compresión de transadmitancia
    transadmittance compression ratio
razón de conducción conduction ratio
razón de conexión-desconexión on-off ratio
razón de contador register ratio
razón de contraste contrast ratio
razón de contraste de impresión print contrast ratio

razón de control control ratio
razón de convergencia convergence ratio
razón de conversación-ruido speech-noise ratio
razón de conversión conversion ratio
razón de corriente de excitación excitation current
    ratio
razón de corriente de transferencia directa forward
    transfer current ratio
razón de corriente primaria primary-current ratio
razón de corrientes current ratio
razón de corrientes de transferencia
    transfer-current ratio
razón de cortocircuito short-circuit ratio
razón de definición resolution ratio
razón de desaccionamiento dropout ratio
razón de desequilibrio unbalance ratio
razón de desviación deviation ratio
razón de disponibilidad availability ratio
razón de disponibilidad de equipo físico hardware
    availability ratio
razón de distorsión de intermodulación
    intermodulation-distortion ratio
razón de durabilidad endurance ratio
razón de emisión secundaria secondary-emission
    ratio
razón de energía energy ratio
razón de engranaje gear ratio
razón de equilibrio balance ratio
razón de equilibrio de corriente current-balance
    ratio
razón de errores error ratio
razón de errores a bits bit-error ratio
razón de errores de caracteres character-error ratio
razón de errores residuales residual-error ratio
razón de escala scaling ratio
razón de escape escape ratio
razón de expansión expansion ratio
razón de fallas failure ratio
razón de frecuencias frequency ratio
razón de ganancia gain ratio
razón de ganancia de conversión conversion-gain
    ratio
razón de imagen picture ratio
razón de impedancias impedance ratio
razón de impulso pulse ratio
razón de interferencia de frecuencia intermedia
    intermediate-frequency interference ratio
razón de interferencia imagen image-interference
    ratio
razón de inversión inversion ratio
razón de luminancia luminance ratio
razón de magnificación ratio of magnification
razón de mezcla de gases gas admixture ratio
razón de modulación modulation ratio
razón de ondas estacionarias standing-wave ratio
razón de ondas estacionarias de tensión voltage
    standing-wave ratio
razón de ondulación ripple ratio
razón de operación operating ratio
razón de óxido oxide ratio
razón de par a inercia torque-to-inertia ratio
razón de paso pitch ratio
razón de pendientes slope ratio
razón de polarización polarization ratio
razón de portadora a ruido carrier-to-noise ratio
razón de potencia power ratio
razón de potencias de ruido noise-power ratio
razón de propagación propagation ratio
razón de propagación de fase phase-propagation
    ratio
razón de protección protection ratio
razón de puente bridge ratio
razón de reactancia-resistencia reactance-resistance

ratio
razón de rectificación rectification ratio
razón de recuperación recovery ratio
razón de rechazo de modo común common-mode rejection ratio
razón de reproducción reproduction ratio
razón de resistencias resistance ratio
razón de respuesta de frecuencia intermedia intermediate-frequency response ratio
razón de respuesta espuria spurious-response ratio
razón de retorno return ratio
razón de retroalimentación feedback ratio
razón de ruido noise ratio
razón de salida uno a salida cero one-to-zero ratio
razón de selección selection ratio
razón de sensibilidad sensitivity ratio
razón de señal a diafonía signal-to-crosstalk ratio
razón de señal a distorsión signal-to-distortion ratio
razón de señal a imagen signal-to-image ratio
razón de señal a interferencia signal-to-interference ratio
razón de señal a ruido signal-to-noise ratio
razón de señal a ruido disponible available signal-to-noise ratio
razón de señal-ruido signal-noise ratio
razón de señal visual a exploración blip-scan ratio
razón de sintonización tuning ratio
razón de supresión de amplitud amplitude suppression ratio
razón de tensiones voltage ratio
razón de tensiones de transformador transformer voltage ratio
razón de tiempo de impulsos pulse-time ratio
razón de trabajo duty ratio
razón de transferencia transfer ratio
razón de transferencia de bucle loop-transfer ratio
razón de transferencia de corriente directa forward current-transfer ratio
razón de transformación transformation ratio
razón de transmisión transmission ratio
razón de transporte transport ratio
razón de uniformidad uniformity ratio
razón de uno a uno one-to-one ratio
razón de utilidad serviceability ratio
razón de utilización utilization ratio
razón de vacío gas ratio
razón de válvula valve ratio
razón de velocidades velocity ratio
razón de vueltas turns ratio
razón de Wiedemann-Franz Wiedemann-Franz ratio
razón dimensional dimensional ratio
razón dinámica dynamic ratio
razón directa direct ratio
razón elevadora step-up ratio
razón entre canales ratio between channels
razón error-bloques block-error ratio
razón escalar scalar ratio
razón fija fixed ratio
razón giromagnética gyromagnetic ratio
razón imagen image ratio
razón inversa inverse ratio
razón límite ultimate ratio
razón magnetomecánica magnetomechanical ratio
razón magnetorresistiva magnetoresistive ratio
razón marca-espacio mark-space ratio
razón nominal nominal ratio
razón numérica numerical ratio
razón reductora step-down ratio
razón señal-imagen signal-image ratio
razón unitaria unity ratio
razón variable variable ratio
reacción reaction
reacción anódica anodic reaction

reacción catódica cathodic reaction
reacción del inducido armature reaction
reacción electrodérmica electrodermal reaction
reacción electromagnética electromagnetic reaction
reacción en cadena chain reaction
reacción endotérmica endothermic reaction
reacción exotérmica exothermic reaction
reacción fotoquímica photochemical reaction
reacción inversa reverse reaction
reacción negativa negative reaction
reacción normal normal reaction
reacción nuclear nuclear reaction
reacción polar polar reaction
reacción prohibida forbidden reaction
reacción secundaria secondary reaction
reacción termonuclear thermonuclear reaction
reactancia reactance
reactancia acústica acoustic reactance
reactancia acústica específica specific acoustic reactance
reactancia acústica intrínseca unit-area acoustic reactance
reactancia asincrónica asynchronous reactance
reactancia auxiliar ballast
reactancia capacitiva capacitive reactance
reactancia cero zero reactance
reactancia combinada combined reactance
reactancia de alambrado wiring reactance
reactancia de antena antenna reactance
reactancia de capacidad capacity reactance
reactancia de conmutación commutating reactance
reactancia de electrodo electrode reactance
reactancia de fuga leakage reactance
reactancia de línea line reactance
reactancia de masa mass reactance
reactancia de polarización polarization reactance
reactancia de Potier Potier's reactance
reactancia de salida output reactance
reactancia de saturación saturation reactance
reactancia de secuencia de fase negativa negative phase-sequence reactance
reactancia de secuencia de fase positiva positive phase-sequence reactance
reactancia de secuencia negativa negative-sequence reactance
reactancia del inducido armature reactance
reactancia dieléctrica dielectric reactance
reactancia efectiva effective reactance
reactancia en circuito abierto open-circuit reactance
reactancia en cortocircuito short-circuit reactance
reactancia en cuadratura quadrature reactance
reactancia equivalente equivalent reactance
reactancia inductiva inductive reactance
reactancia limitadora de potencia power-limiting reactance
reactancia mecánica mechanical reactance
reactancia mutua mutual reactance
reactancia negativa negative reactance
reactancia neta net reactance
reactancia parásita stray reactance
reactancia rotórica rotor reactance
reactancia sincrónica synchronous reactance
reactancia sincrónica longitudinal direct-axis synchronous reactance
reactancia subtransitoria subtransient reactance
reactancia subtransitoria longitudinal direct-axis subtransient reactance
reactancia transitoria transient reactance
reactancia transitoria longitudinal direct-axis transient reactance
reactancia variable variable reactance
reactancímetro reactance meter
reactatrón reactatron

reactivación reactivation
reactivación de filamento filament reactivation
reactivador reactivator
reactivar reactivate
reactividad reactivity
reactivo reactive
reactor reactor
reactor atómico atomic reactor
reactor de arranque starting reactor
reactor de haz beam reactor
reactor de núcleo saturable saturable-core reactor
reactor de potencia power reactor
reactor en derivación shunt reactor
reactor en paralelo anódico anode paralleling reactor
reactor en serie series reactor
reactor saturable saturable reactor
reactor saturable de corriente alterna alternating-current saturable reactor
reactor variable variable reactor
reajustar readjust
realambrable rewireable
realambrar rewire
realzado de bordes edge enhancement
realzado de contraste contrast enhancement
realzado de imagen image enhancement
realzar enhance
reanudar resume, restart
rearrollar rewind
reasignación reallocation
reasignar reallocate
rebanada de bits bit slice
rebanador slicer
rebobinado rewinding
rebobinador rewinder
rebobinar rewind
rebose overflow
rebote bounce
rebote de contacto contact bounce
rebote de contacto de relé relay-contact bounce
rebote de ondas wave bounce
rebote lunar moon bounce
recalada homing
recalcular recalculate
recálculo recalculation
recalescencia recalescence
recalibrado recalibrated
recalibrar recalibrate
recarga recharge
recargable rechargeable, reloadable
recargar reload, recharge
recepción reception
recepción a larga distancia long-distance reception
recepción aural aural reception
recepción autodina autodyne reception
recepción ciega blind reception
recepción con diagrama en ocho figure-of-eight reception
recepción de datos data reception
recepción de fotografía photograph reception
recepción de imagen picture reception
recepción de largo alcance long-range reception
recepción de modulación de frecuencia frequency-modulation reception
recepción de ondas wave reception
recepción de portadora incrementada exalted-carrier reception
recepción de radiofaro beacon reception
recepción de radiofrecuencia sintonizada tuned-radio-frequency reception
recepción de sonido sound reception
recepción de televisión television reception
recepción díplex diplex reception

recepción direccional directional reception
recepción directa straight reception
recepción en cardioide cardioid reception
recepción en diversidad diversity reception
recepción en diversidad espacial space-diversity reception
recepción en doble diversidad dual-diversity reception
recepción endodina endodyne reception
recepción estereofónica stereophonic reception
recepción fotográfica photographic reception
recepción heterodina heterodyne reception
recepción homodina homodyne reception
recepción infradina infradyne reception
recepción local local reception
recepción monocroma monochrome reception
recepción monofónica monophonic reception
recepción múltiple multiple reception
recepción panorámica panoramic reception
recepción por batido cero zero-beat reception
recepción por señal única single-signal reception
recepción por trayectorias múltiples multipath reception
recepción radiotelegráfica radiotelegraph reception
recepción regenerativa regenerative reception
recepción simultánea simultaneous reception
recepción subterránea underground reception
recepción superheterodina superheterodyne reception
recepción superregenerativa superregenerative reception
recepción supersónica supersonic reception
recepción telefónica phone reception
recepción ultradina ultradyne reception
recepción unidina unidyne reception
receptáculo receptacle
receptáculo accesorio accessory receptacle
receptáculo coaxial coaxial receptacle
receptáculo concéntrico concentric receptacle
receptáculo de conveniencia convenience receptacle
receptáculo de pared wall receptacle
receptáculo embutido flush receptacle
receptáculo empotrado flush receptacle
receptáculo no polarizado unpolarized receptacle
receptáculo polarizado polarized receptacle
receptividad receptivity
receptor receiver, receptor
receptor acústico acoustic receiver
receptor aerotransportado airborne receiver
receptor altamente sensible highly sensitive receiver
receptor autodino autodyne receiver
receptor auxiliar auxiliary receiver
receptor bipolar bipolar receiver
receptor comercial commercial receiver
receptor controlado por cristal crystal-controlled receiver
receptor de alta fidelidad high-fidelity receiver
receptor de audífono headphone receiver
receptor de audio de cristal crystal audio receiver
receptor de avión airplane receiver
receptor de bajo ruido low-noise receiver
receptor de banda ancha wideband receiver
receptor de banda angosta narrowband receiver
receptor de banda lateral única single-sideband receiver
receptor de base base receiver
receptor de batería battery receiver
receptor de cobertura general general-coverage receiver
receptor de códigos code receiver
receptor de colores color receiver
receptor de comparación comparison receiver
receptor de comunicaciones communications

receiver

receptor de conducción ósea bone-conduction receiver

receptor de conductor móvil moving-conductor receiver

receptor de control remoto remote-control receiver

receptor de conversión directa direct-conversion receiver

receptor de corriente alterna alternating-current receiver

receptor de corriente alterna-corriente continua alternating-current-direct-current receiver

receptor de corriente continua direct-current receiver

receptor de corriente portadora carrier-current receiver

receptor de corte cutoff receiver

receptor de cristal crystal receiver, crystal set

receptor de cuasipico quasi-peak receiver

receptor de datos data receiver

receptor de datos coordenados coordinate data receiver

receptor de diversidad cuádruple quadruple-diversity receiver

receptor de doble conversión double-conversion receiver

receptor de doble señal double-signal receiver

receptor de dos canales two-channel receiver

receptor de emergencia emergency receiver

receptor de enganche de fase phase-locked receiver

receptor de exploración scanning receiver

receptor de exploración automática automatic scanning receiver

receptor de facsímil facsimile receiver

receptor de frecuencia fija fixed-frequency receiver

receptor de frecuencia ultraalta ultra-high frequency receiver

receptor de frecuencia única single-frequency receiver

receptor de haz luminoso light-beam receiver

receptor de hierro móvil moving-iron receiver

receptor de imagen picture receiver

receptor de impulsos pulse receiver

receptor de inclinación de planeo glide-slope receiver

receptor de infrarrojo infrared receiver

receptor de interportadora intercarrier receiver

receptor de línea line receiver

receptor de lorán loran receiver

receptor de mano hand receiver

receptor de medida measuring receiver

receptor de microondas microwave receiver

receptor de modulación de amplitud-modulación de frecuencia amplitude-modulation-frequency-modulation receiver

receptor de modulación de frecuencia frequency-modulation receiver

receptor de monitoreo monitoring receiver

receptor de Morse Morse receiver

receptor de muy alta frecuencia very-high-frequency receiver

receptor de navegación navigation receiver

receptor de ondas wave receiver

receptor de ondas cortas shortwave receiver

receptor de portadora incrementada exalted-carrier receiver

receptor de proyección projection receiver

receptor de radar radar receiver

receptor de radio por hilo wired-radio receiver

receptor de radiobaliza marker-beacon receiver

receptor de radiodifusión broadcast receiver

receptor de radiofaro radio-beacon receiver, beacon

receiver

receptor de radiofrecuencia sintonizada tuned-radio-frequency receiver

receptor de radiosonda radiosonde receiver

receptor de seguimiento tracking receiver

receptor de seis tubos six-tube receiver

receptor de señal única single-signal receiver

receptor de señales signal receiver

receptor de señales de toda onda all-wave signal receiver

receptor de socorro distress receiver

receptor de sonar sonar receiver

receptor de sonido sound receiver

receptor de sonido dividido split-sound receiver

receptor de sonido maestro master sound receiver

receptor de tambor drum receiver

receptor de tarjetas card receiver

receptor de telemedida telemetering receiver

receptor de televisión television receiver

receptor de televisión a colores color-television receiver

receptor de televisión de proyección projection television receiver

receptor de tipo de correlación correlation-type receiver

receptor de toda onda all-wave receiver

receptor de tono tone receiver

receptor de trayectoria de planeo glide-path receiver

receptor de triple conversión triple-conversion receiver

receptor de triple detección triple-detection receiver

receptor de tubos tube receiver

receptor de tubos de vacío vacuum-tube receiver

receptor de válvulas valve receiver

receptor de video video receiver

receptor de video de cristal crystal video receiver

receptor diferencial differential receiver

receptor directo straight receiver

receptor electromagnético electromagnetic receiver

receptor electrostático electrostatic receiver

receptor en diversidad diversity receiver

receptor en doble diversidad dual-diversity receiver

receptor estereofónico stereophonic receiver

receptor-excitador receiver-exciter

receptor físico physical receptor

receptor fotoeléctrico photoelectric receiver

receptor Gee Gee receiver

receptor heterodino heterodyne receiver

receptor homodino homodyne receiver

receptor infradino infradyne receiver

receptor lineal-logarítmico lin-log receiver

receptor localizador localizer receiver

receptor logarítmico logarithmic receiver

receptor mixto mixed receiver

receptor-modulador receiver-modulator

receptor monitor monitor receiver

receptor monocanal single-channel receiver

receptor monocromo monochrome receiver

receptor monofónico monophonic receiver

receptor móvil mobile receiver

receptor multibanda multiband receiver

receptor multicanal multichannel receiver

receptor multipatrón multistandard receiver

receptor neumático pneumatic receptor

receptor neutrodino neutrodyne receiver

receptor no regenerativo nonregenerative receiver

receptor no sincrónico nonsynchronous receiver

receptor no sintonizado wide-open receiver

receptor panorámico panoramic receiver

receptor paramétrico parametric receiver

receptor patrón standard receiver

receptor piezoeléctrico piezoelectric receiver

receptor portátil portable receiver
receptor presintonizado pretuned receiver
receptor principal main receiver
receptor que radia señales blooper
receptor radiotelegráfico radiotelegraph receiver
receptor regenerativo regenerative receiver
receptor remoto remote receiver
receptor selectivo selective receiver
receptor sensible sensitive receiver
receptor sin transformador transformerless receiver
receptor sintonizable tunable receiver
receptor subterráneo underground receiver
receptor superheterodino superheterodyne receiver
receptor superregenerativo superregenerative
receiver
receptor tacán tacan receiver
receptor telefónico telephone receiver
receptor telefónico térmico thermal telephone
receiver
receptor telegráfico telegraph receiver
receptor telemétrico telemetric receiver
receptor térmico thermal receiver
receptor-transmisor receiver-transmitter
receptor ultradino ultradyne receptor
receptor ultrasónico ultrasonic receiver
receptor unidino unidyne receiver
receptor universal universal receiver
reciclado recycling
recierre reclosing
reciprocación reciprocation
reciprocidad reciprocity
recíproco reciprocal
reclasificación resorting
reclasificar resort
recocer anneal
recocido annealing
recocido eléctrico electric annealing
recocido gaseoso gaseous annealing
recodificación recoding
recodificar recode
recolección electrónica electron collection
recolector de polvo dust collector
recombinación recombination
recombinación de pared wall recombination
recombinación de volumen volume recombination
recombinación superficial surface recombination
recompilar recompile
reconfiguración reconfiguration
reconfiguración de almacenamiento storage
reconfiguration
reconfiguración de dispositivo dinámico dynamic
device reconfiguration
reconfiguración dinámica dynamic reconfiguration
reconmutación reswitching
reconmutar reswitch
reconocedor de palabras word recognizer
reconocible recognizable
reconocible por máquina machine recognizable
reconocimiento recognition, acknowledgment
reconocimiento de caracteres character recognition
reconocimiento de caracteres de tinta magnética
magnetic ink character recognition
reconocimiento de caracteres magnéticos magnetic
character recognition
reconocimiento de caracteres ópticos optical
character recognition
reconocimiento de marcas óptico optical mark
recognition
reconocimiento de patrón pattern recognition
reconocimiento de volumen automático automatic
volume recognition
reconocimiento de voz speech recognition
reconstitución de archivo file reconstitution

reconstrucción reconstruction
reconstrucción de archivo file reconstruction
reconstrucción de imagen image reconstruction,
picture reconstruction
reconstructor de imagen image reconstructor
reconstructor espacial dinámico dynamic spatial
reconstructor
reconversión reconversion
reconvertir reconvert
recopilación de datos data gathering
recopilación de información information gathering
recopilador de datos data collector
recorrido travel, run, path
recorrido de armadura armature travel
recorrido de armadura de relé relay-armature travel
recorrido de frecuencias frequency run
recorrido libre medio mean free path
recorrido medio mean path
recortador clipper, chopper
recortador de diodo diode clipper
recortador de impulsos pulse clipper
recortador de nivel level clipper
recortador de picos peak clipper
recortador de picos positivos positive-peak clipper
recortador de ruido noise clipper
recortador de voz speech clipper
recortamiento clipping
recortamiento de impulsos pulse clipping
recortamiento duro hard clipping
recorte clipping
recorte de picos peak clipping
recorte de ruido noise clipping
recorte lineal linear clipping
rectificable rectifiable
rectificación rectification
rectificación de capa de agotamiento depletion-layer
rectification
rectificación de media onda half-wave rectification
rectificación de onda completa full-wave
rectification
rectificación de potencia power rectification
rectificación de señal signal rectification
rectificación farádica faradic rectification
rectificación lineal linear rectification
rectificación mecánica mechanical rectification
rectificación por diodo diode rectification
rectificación por fugas de rejilla leaky-grid
rectification
rectificación por rejilla grid rectification
rectificado rectified
rectificador rectifier
rectificador complementario complementary
rectifier
rectificador controlado controlled rectifier
rectificador controlado de silicio silicon-controlled
rectifier
rectificador controlado por fase phase-controlled
rectifier
rectificador controlado por semiconductor
semiconductor-controlled rectifier
rectificador de alta potencia high-power rectifier
rectificador de alta tensión high-voltage rectifier
rectificador de alto vacío high-vacuum rectifier
rectificador de ampolla de vidrio glass-bulb rectifier
rectificador de ánodo único single-anode rectifier
rectificador de arco arc rectifier
rectificador de arco de mercurio mercury-arc
rectifier
rectificador de avalancha avalanche rectifier
rectificador de avalancha controlada
controlled-avalanche rectifier
rectificador de avalancha simétrico symmetrical
avalanche rectifier

rectificador de baja tensión low-voltage rectifier
rectificador de baño de mercurio mercury-pool rectifier
rectificador de carburo de silicio silicon carbide rectifier
rectificador de carga de baterías battery charging rectifier
rectificador de cátodo caliente hot-cathode rectifier
rectificador de cátodo frío cold-cathode rectifier
rectificador de cátodo líquido pool rectifier
rectificador de conmutación commutating rectifier
rectificador de contacto contact rectifier
rectificador de contacto de punta point-contact rectifier
rectificador de contacto por superficie surface-contact rectifier
rectificador de contacto seco dry-contact rectifier
rectificador de control control rectifier
rectificador de control por rejilla grid-controlled rectifier
rectificador de corriente current rectifier
rectificador de cristal crystal rectifier
rectificador de cristal de silicio silicon-crystal rectifier
rectificador de cristal de sulfuro de cadmio cadmium-sulfide crystal rectifier
rectificador de descarga de arco arc-discharge rectifier
rectificador de descarga luminiscente glow-discharge rectifier
rectificador de diodo diode rectifier
rectificador de dirección direction rectifier
rectificador de disco disk rectifier
rectificador de disco seco dry-disk rectifier
rectificador de discos metálicos metallic-disk rectifier
rectificador de doble diodo double-diode rectifier
rectificador de doce fases twelve-phase rectifier
rectificador de efecto túnel tunnel rectifier
rectificador de enfoque focus rectifier
rectificador de gas gas-filled rectifier
rectificador de germanio germanium rectifier
rectificador de Gratz Gratz rectifier
rectificador de ignición ignition rectifier
rectificador de laminilla vibrante vibrating-reed rectifier
rectificador de magnesio-sulfuro de cobre magnesium-copper-sulfide rectifier
rectificador de media onda half-wave rectifier
rectificador de media onda bifásica biphase half-wave rectifier
rectificador de medidor meter rectifier
rectificador de mercurio mercury rectifier
rectificador de onda completa full-wave rectifier
rectificador de óxido oxide rectifier
rectificador de óxido de cobre copper-oxide rectifier
rectificador de pico a pico peak-to-peak rectifier
rectificador de pico negativo negative-peak rectifier
rectificador de placa metálica metal-plate rectifier
rectificador de placa seca dry-plate rectifier
rectificador de polaridad invertida reversed-polarity rectifier
rectificador de potencia power rectifier
rectificador de potencia de germanio germanium power rectifier
rectificador de retorno flyback rectifier
rectificador de ruido noise rectifier
rectificador de selenio selenium rectifier
rectificador de semiconductor semiconductor rectifier
rectificador de señales signal rectifier
rectificador de silicio silicon rectifier
rectificador de Snook Snook rectifier

rectificador de sulfuro de cobre copper-sulfide rectifier
rectificador de tanque de acero steel-tank rectifier
rectificador de tantalio tantalum rectifier
rectificador de tensión constante constant-voltage rectifier
rectificador de tipo de vibrador vibrator-type rectifier
rectificador de transición transition rectifier
rectificador de tubo de vacío vacuum-tube rectifier
rectificador de tubo electrónico electronic tube rectifier
rectificador de unión junction rectifier
rectificador de unión PN PN-junction rectifier
rectificador de unión por difusión diffused-junction rectifier
rectificador de vacío vacuum rectifier
rectificador de válvula valve rectifier
rectificador de vapor de cesio cesium-vapor rectifier
rectificador de vapor de mercurio mercury-vapor rectifier
rectificador de video de cristal crystal video rectifier
rectificador de vóltmetro voltmeter rectifier
rectificador de xenón xenon rectifier
rectificador desmodulador demodulator rectifier
rectificador discriminador de fase phase-discriminating rectifier
rectificador electrolítico electrolytic rectifier
rectificador electromecánico electromechanical rectifier
rectificador electrónico electronic rectifier
rectificador electroquímico electrochemical rectifier
rectificador en estrella star rectifier
rectificador en paralelo parallel rectifier
rectificador en puente bridge rectifier
rectificador en puente de tres cuartos three-quarter bridge
rectificador en puente monofásico single-phase bridge rectifier
rectificador en puente trifásico three-phase bridge rectifier
rectificador en zigzag zigzag rectifier
rectificador estacionario stationary rectifier
rectificador estático static rectifier
rectificador gaseoso gaseous rectifier
rectificador hexafásico six-phase rectifier
rectificador ideal ideal rectifier
rectificador lineal linear rectifier
rectificador líquido liquid rectifier
rectificador magnetrón magnetron rectifier
rectificador mecánico mechanical rectifier
rectificador metálico metallic rectifier
rectificador metálico básico basic metallic rectifier
rectificador mezclador mixer rectifier
rectificador monoanódico monoanodic rectifier
rectificador monofásico single-phase rectifier
rectificador monofásico de onda completa single-phase full-wave rectifier
rectificador multiánodo multianode rectifier
rectificador multipactor multipactor rectifier
rectificador múltiple multiple rectifier
rectificador no polarizado unbiased rectifier
rectificador PN PN rectifier
rectificador polarizado biased rectifier
rectificador polianódico polyanode rectifier
rectificador polifásico polyphase rectifier
rectificador por retroalimentación de diodo diode feedback rectifier
rectificador químico chemical rectifier
rectificador rotativo rotating rectifier
rectificador seco dry rectifier
rectificador sellado sealed rectifier

rectificador simple simple rectifier
rectificador sin bomba pumpless rectifier
rectificador sincrónico synchronous rectifier
rectificador termoelectrónico thermoelectronic rectifier
rectificador termoiónico thermionic rectifier
rectificador trifásico three-phase rectifier
rectificar rectify
rectigón rectigon
rectilíneo rectilinear
recubierto de plomo lead-covered
recubrimiento overlapping, overlay, covering, coating
recubrimiento metálico interior metal backing
recuperable retrievable
recuperación recovery, retrieval
recuperación alternativa alternate recovery
recuperación de alto nivel high-level recovery
recuperación de archivo file recovery
recuperación de archivo inversa backward file recovery
recuperación de asignación allocation recovery
recuperación de datos data retrieval
recuperación de documentos document retrieval
recuperación de equipo físico hardware recovery
recuperación de error de equipo físico hardware error recovery
recuperación de errores error recovery
recuperación de información information retrieval
recuperación de mensajes message retrieval
recuperación de portadora carrier recovery
recuperación de puntos de comprobación checkpoint recovery
recuperación de señal signal recovery
recuperación de tensión voltage recovery
recuperación directa forward recovery
recuperación falsa false retrieval
recuperación tras falla failure recovery
recuperar retrieve
recuperar archivos borrados undelete
recurrencia recurrence
recurrente recurrent
recursión recursion
recursión simultánea simultaneous recursion
recursivo recursive
recurso de datos data resource
recurso de red network resource
recursos compartidos shared resources
recursos de capacidad capacity resources
recursos de equipo físico hardware resources
recursos de sistema system resources
rechazador rejector
rechazo rejection
rechazo de banda band rejection, bandstop
rechazo de frecuencia frequency rejection
rechazo de imagen image rejection
rechazo de modo común common-mode rejection
rechazo de modo diferencial differential-mode rejection
rechazo de modo normal normal-mode rejection
rechazo de ruido noise rejection
rechazo de sonido sound rejection
rechazo de tensión de modo común common-mode voltage rejection
red network, mains, array, lattice
red abierta open network
red activa active network
red activa-pasiva active-passive network
red aislada isolated network
red analógica analog network
red anuladora annulling network
red apiñada close-spaced array
red arrendada leased network

red básica basic network
red bifásica quarter-phase network
red bilateral bilateral network
red binomia binomial array
red centralizada centralized network
red cerrada closed array
red cilíndrica cylindrical array
red circular circular array
red colineal collinear array
red combinada combined network
red combinadora combining network
red compensadora compensating network
red compensadora de fase phase-compensating network
red completamente conectada fully connected network
red compresora compressor network
red común common network
red conformadora de señal signal-shaping network
red conforme conformal array
red cónica conical array
red conmutada switched network
red continua continuous network
red contra interferencia intencional antijamming array
red controlada por computadora computer controlled network
red convertidora de frecuencia frequency-converting network
red cooperativa cooperative network
red correctiva corrective network, shaping network
red correctora de fase phase-correcting network
red cristalina crystal lattice
red de abertura circular circular-aperture array
red de acceso múltiple multiple-access network
red de acoplamiento coupling network
red de adaptación matching network
red de adaptación de carga load-matching network
red de adaptación de impedancias impedance-matching network
red de agrupamiento grouping network
red de aislamiento isolation network
red de alimentación supply mains
red de anillos ringed network
red de antenas antenna array, array
red de antenas activas active antenna array
red de antenas apiladas stacked array
red de antenas binomia binomial antenna array
red de antenas lineal linear antenna array
red de antenas omnidireccionales omnidirectional antenna array
red de área area network
red de área local local area network
red de atenuación attenuation network
red de bus bus network
red de cable cable network
red de cables coaxiales coaxial-cable network
red de celdas cell array
red de células solares solar array
red de Chebyshev Chebyshev array
red de circuitos network of circuits
red de circuitos integrados integrated-circuit array
red de codificación secuencial sequential-coding network
red de compensación de fase phase-compensation network
red de computadoras computer network
red de computadoras centralizada centralized computer network
red de computadoras descentralizada decentralized computer network
red de cómputo computing network
red de comunicaciones communications network

red de comunicaciones de datos data communications network
red de conductancia constante constant-conductance network
red de conmutación switching network
red de conmutación de paquetes packet switching network
red de control de trayectoria path control network
red de control distribuida distributed control network
red de correo electrónico electronic mail network
red de cuatro terminales four-terminal network
red de datos data network
red de datos de conmutación de paquetes packet switching data network
red de desacentuación deemphasis network
red de desacoplamiento decoupling network
red de descodificación decoding network
red de desfase phase-shift network
red de diodos diode array
red de dipolos doblados folded-dipole array
red de distribución distribution network, grid
red de dos puertos two-port network
red de dos terminales two-terminal network
red de energía power network
red de enfasamiento phasing network
red de estabilización stabilization network
red de facsímil facsimile network
red de fase mínima minimum-phase network
red de filtros filter network
red de Hall Hall network
red de información information network
red de k constante constant-k network
red de línea visual line-of-sight network
red de líneas network of lines
red de malla mesh network
red de mandos command network
red de memoria memory array
red de muesca notch network
red de multisistema multisystem network
red de núcleos core array
red de parámetros con variación lineal linear varying-parameter network
red de parámetros distribuidos distributed-parameter network
red de ponderación weighting network
red de precisión precision network
red de preénfasis preemphasis network
red de procesamiento de datos distribuida distributed data processing network
red de protección cradle-guard
red de punto a punto point-to-point network
red de radar radar network
red de radiación longitudinal end-fire array
red de radiación transversal broadside array
red de radio radio network
red de radiorrelés radio-relay network
red de ramificación branching network
red de ranuras slot array
red de resistencia-capacitancia resistance-capacitance network
red de resistencia constante constant-resistance network
red de resistores resistor network
red de retardo delay network
red de retardo de impulsos pulse-delay network
red de retransmisión de televisión television relay network
red de retransmisión por cinta tape-relay network
red de salida output network
red de Sterba Sterba array
red de superganancia supergain array
red de telecomunicaciones telecommunications network

red de telecomunicaciones fija aeronáutica aeronautical fixed telecommunications network
red de teleimpresoras teletypewriter network
red de teleprocesamiento teleprocessing network
red de televisión television network
red de transistores transistor network
red de transmisión transmission network
red de uniones junction network
red desequilibrada unbalanced network
red desfasadora phase-shifting network
red determinante de frecuencia frequency-determining network
red diferenciadora differentiating network
red digital automática automatic digital network
red digital integrada integrated digital network
red direccional directional array
red disimétrica dissymmetrical network
red distribuida distributed network
red divisora dividing network
red divisora de altavoz loudspeaker dividing network
red divisora de frecuencia frequency-dividing network
red ecualizadora equalizing network
red ecualizadora de fase phase-equalizing network
red eléctrica electric network, mains
red eléctrica activa active electric network
red eléctrica pasiva passive electric network
red en anillo ring network
red en bucle loop network
red en cascada cascade network
red en celosía lattice network
red en cortina curtain array
red en cuadratura quadrature network
red en delta delta network
red en doble T twin-T network
red en escalera ladder network
red en estrella star network
red en fase phased array
red en pi pi network
red en pino pine-tree array
red en puente bridge network
red en serie-derivación series-shunt network
red en serie-paralelo series-parallel network
red en T T network
red en T puenteada bridged-T network
red en Y Y network
red equilibrada balanced network
red equilibradora balancing network
red equilibradora de fases phase balancing network
red equivalente equivalent network
red escalonada staggered array
red esférica spherical array
red estabilizadora stabilizing network
red extendida extended network
red fantasma phantom network
red filtrante filtering network
red formadora de impulsos pulse-forming network
red híbrida hybrid network
red integradora integrating network
red interconectada interconnected network
red interurbana toll network, trunk network
red inversa inverse network
red jerárquica hierarchical network
red libre free network
red lineal linear array, linear network
red lineal uniforme uniform linear array
red local local network
red lógica logic network
red matricial matrix network
red molecular molecular lattice
red multimalla multimesh network

red multipuerto multiport network
red multipunto multipoint network
red nacional nationwide network
red no disipativa nondissipative network
red no lineal nonlinear network
red no planar nonplanar network
red no sincrónica nonsynchronous network
red parásita parasitic array
red parcial partial network
red pasatodo all-pass network
red pasiva passive network
red planar planar array
red potenciométrica potentiometer network
red primaria primary network
red principal main network
red privada private network
red pública public network
red radial radial network
red receptora receiving net
red rectangular rectangular array
red recurrente recurrent network
red resistiva resistive network
red rural rural network
red secundaria secondary network
red selectiva selective network
red simétrica symmetrical network
red sin pérdidas lossless network
red sincrónica synchronous network
red sumadora adding network
red T en paralelo parallel-T network
red telefónica telephone network
red telefónica privada private telephone network
red telefónica pública public telephone network
red telegráfica telegraph network
red télex telex network
red terminal terminal network
red unidireccional unidirectional network,
   unidirectional array
red unilateral unilateral network
redes en tándem tandem networks
redes recíprocas reciprocal networks
redifusión rediffusion
redirigir redirect
redistribución redistribution
redistribución de área area redistribution
redistribución de energía energy redistribution
redondeamiento rounding
redondear round-off
redondear hacia arriba round-up
redondeo rounding
reducción reduction
reducción armónica harmonic reduction
reducción de articulación articulation reduction
reducción de contraste contrast reduction
reducción de chip chip reduction
reducción de datos data reduction
reducción de datos en línea on-line data reduction
reducción de ganancia gain reduction
reducción de ganancia de acimut azimuth gain
   reduction
reducción de resistencia resistance reduction
reducción de ruido noise reduction
reducción de sensibilidad sensitivity decrease
reducción de tensión de línea brownout
reducción de zumbido hum bucking
reducción electrolítica electrolytic reduction
reducción electroquímica electrochemical reduction
reducción exponencial exponential decrease
reducción óptica optical reduction
reducción química chemical reduction
reducción sonora sound reduction
reducción telediafónica far-end crosstalk reduction
reducible reducible

reducir reduce
reducir a escala scale down
reductor reducer
reductor armónico harmonic reducer
reductor de intensidad dimmer
reductor de intensidad de lámpara lamp dimmer
reductor de intensidad de luz light dimmer
reductor de ruido noise reducer
reductor de tensión voltage reducer
redundancia redundancy
redundancia activa active redundancy
redundancia cíclica cyclic redundancy
redundancia de comprobación check redundancy
redundancia de datos data redundancy
redundancia de línea line redundancy
redundancia de red network redundancy
redundancia de reserva standby redundancy
redundancia homogénea homogeneous redundancy
redundancia longitudinal longitudinal redundancy
redundancia relativa relative redundancy
redundancia vertical vertical redundancy
redundante redundant
reejecución rerun
reemisión rebroadcasting, rebroadcast
reemisión de radio radio relay
reemplazo replacement
reemplazo global global replace
reencaminamiento rerouting
reencaminar reroute
reencendido reignition, restriking
reensamblaje reassembly
reensamblar reassemble
reentrable reenterable
reentrada reentry
reentrante reentrant
reentrar reenter
reequilibrar rebalance
reescribir rewrite
reestructuración dinámica dynamic restructuring
reexplorar rescan
referencia reference
referencia ambigua ambiguous reference
referencia circular circular reference
referencia coherente coherent reference
referencia cruzada cross reference
referencia de archivo file reference
referencia de celda cell reference
referencia de celda absoluta absolute cell reference
referencia de celda relativa relative cell reference
referencia de corriente continua direct-current
   reference
referencia de dirección address reference
referencia de entrada input reference
referencia de frecuencia interna internal frequency
   reference
referencia de línea line reference
referencia de mandos command reference
referencia de portadora de colores color-carrier
   reference
referencia de portadora de crominancia
   chrominance-carrier reference
referencia de subportadora de colores
   color-subcarrier reference
referencia de subportadora de crominancia
   chrominance-subcarrier reference
referencia de tensión voltage reference
referencia incluyente inclusive reference
referencia inversa backward reference
referencia piloto pilot reference
referencia posterior back reference
refinado electrolítico electrolytic refining
refinamiento de zona zone refining
reflectancia reflectance

reflectancia aparente apparent reflectance
reflectancia bicónica biconical reflectance
reflectancia bidireccional bidirectional reflectance
reflectancia bihemisférica bihemispherical reflectance
reflectancia de fondo background reflectance
reflectancia espectral spectral reflectance
reflectancia especular specular reflectance
reflectancia luminosa luminous reflectance
reflectancia luminosa especular specular luminous reflectance
reflectancia regular regular reflectance
reflectancia térmica thermal reflectance
reflectancia total total reflectance
reflectividad reflectivity
reflectividad acústica acoustic reflectivity
reflectividad de ecos parásitos clutter reflectivity
reflectividad difusa diffuse reflectivity
reflectividad monostática monostatic reflectivity
reflectógrafo reflectograph
reflectograma reflectogram
reflectometría de dominio de tiempo time-domain reflectometry
reflectómetro reflection meter, reflectometer
reflectómetro fotoeléctrico photoelectric reflectometer
reflectómetro híbrido hybrid reflectometer
reflector reflector
reflector angular corner reflector
reflector angular pasivo passive corner reflector
reflector asimétrico asymmetrical reflector
reflector cilíndrico cylindrical reflector
reflector complejo complex reflector
reflector con corrección de fase phase-corrected reflector
reflector de antena antenna reflector
reflector de célula fotoeléctrica photoelectric-cell reflector
reflector de dipolo doblado folded-dipole reflector
reflector de guíaondas waveguide reflector
reflector de porcelana porcelain reflector
reflector de radar radar reflector
reflector de rejilla grating reflector
reflector de varillas rod reflector
reflector dieléctrico multicapa multilayer dielectric reflector
reflector en cortina curtain reflector
reflector esférico spherical reflector
reflector especular specular reflector
reflector magnético magnetic reflector
reflector monostático monostatic reflector
reflector normal standard reflector
reflector parabólico parabolic reflector
reflector paraboloidal paraboloidal reflector
reflector parásito parasitic reflector
reflector pasivo passive reflector
reflector pasivo codificado coded passive reflector
reflector plano plane reflector
reflector principal main reflector
reflector retrodirectivo retrodirective reflector
reflector superficial surface reflector
reflejado reflected
reflexión reflection
reflexión acústica acoustic reflection
reflexión angular corner reflection
reflexión atmosférica atmospheric reflection
reflexión auroral auroral reflection
reflexión de Bragg Bragg reflection
reflexión de espejo mirror reflection
reflexión de Fresnel Fresnel reflection
reflexión de impulsos pulse reflection
reflexión de línea line reflection
reflexión de luz light reflection

reflexión de ondas wave reflection
reflexión de radar radar reflection
reflexión difusa diffuse reflection
reflexión difusa uniforme uniform diffuse reflection
reflexión espacial space reflection
reflexión especular specular reflection
reflexión esporádica sporadic reflection
reflexión interna internal reflection
reflexión interna total total internal reflection
reflexión ionosférica ionospheric reflection
reflexión mixta mixed reflection
reflexión no especular nonspecular reflection
reflexión óptica optical reflection
reflexión prohibida forbidden reflection
reflexión regular regular reflection
reflexión selectiva selective reflection
reflexión sonora sound reflection
reflexión superficial surface reflection
reflexión terrestre ground reflection
reflexión total total reflection
reflexión troposférica tropospheric reflection
reflexiones anormales abnormal reflections
reflexiones de estelas meteóricas meteor-trail reflections
reflexiones en zigzag zigzag reflections
reformación reforming
reformado reforming
reforzador booster
reforzador de bajos bass booster
reforzador de impedancia negativa negative impedance booster
reforzador de radiofrecuencia radio-frequency booster
reforzador de señal signal booster
reforzador de tensión voltage booster
refracción refraction
refracción acústica acoustic refraction
refracción atmosférica atmospheric refraction
refracción costera coastal refraction
refracción de flujo flux refraction
refracción de ondas wave refraction
refracción difusa diffuse refraction
refracción específica specific refraction
refracción normal standard refraction
refracción regular regular refraction
refracción troposférica tropospheric refraction
refractado refracted
refractancia refractance
refractario refractory
refractividad refractivity
refractividad específica specific refractivity
refractométrico refractometric
refractómetro refractometer
refractómetro de microondas microwave refractometer
refractor refractor
refractor asimétrico asymmetrical refractor
refrescar refresh
refresco de memoria de acceso aleatorio random-access memory refresh
refrigeración termoeléctrica thermoelectric refrigeration
refrigerante refrigerant, coolant
refrigerante hermético hermetic refrigerant
refuerzo boost
refuerzo de agudos treble boost
refuerzo de altas high boost
refuerzo de bajos bass boost
refuerzo de sonido sound reinforcement
regeneración regeneration
regeneración acústica acoustic regeneration
regeneración catódica cathode regeneration
regeneración de electrólito regeneration of

electrolyte
regeneración de imagen image regeneration
regeneración de impulsos pulse regeneration
regeneración de pantalla screen regeneration
regeneración de señal signal regeneration
regenerador regenerator
regenerador de impulsos pulse regenerator
regenerador de señal signal regenerator
regenerador de subportadora de crominancia
chrominance-subcarrier regenerator
regenerar regenerate
regenerativo regenerative
régimen rating, rate, working conditions
régimen continuo continuous rating
régimen de descarga discharge rate
régimen de descarga por hora hour-discharge rate
régimen de terminación de carga finishing rate
régimen máximo continuo maximum continuous
rating
régimen nominal rating
región activa active region
región anódica anode region
región catódica cathode region
región de agotamiento depletion region
región de almacenamiento storage region
región de almacenamiento principal main storage
region
región de almacenamiento virtual virtual storage
region
región de antena antenna region
región de barrera barrier region
región de base base region
región de Bragg Bragg region
región de campo lejano far-field region
región de campo próximo near-field region
región de carga espacial space-charge region
región de Chapman Chapman region
región de comunicación communication region
región de confusión confusion region
región de control control region
región de corte cutoff region
región de cruce crossover region
región de descarga luminiscente glow-discharge
region
región de difracción diffraction region
región de fondo background region
región de Fraunhofer Fraunhofer region
región de Fresnel Fresnel region
región de Geiger Geiger region
región de Geiger-Mueller Geiger-Mueller region
región de histéresis hysteresis region
región de infrarrojo próximo near infrared region
región de linealidad linearity region
región de microondas microwave region
región de operación region of operation
región de procesamiento de mensajes message
processing region
región de Raman-Nath Raman-Nath region
región de rejilla negativa negative-grid region
región de resistencia negativa negative-resistance
region
región de saturación saturation region
región de Schumann Schumann region
región de sombra shadow region
región de transición transition region
región de ultravioleta próxima near ultraviolet
region
región de Zener Zener region
región inactiva inactive region
región inestable unstable region
región intrínseca intrinsic region
región ionosférica ionospheric region
región negativa negative region

región obscura catódica cathode dark region
región óhmica ohmic region
región paginable pageable region
región paraeléctrica paraelectric region
región prohibida forbidden region
región proporcional proportional region
región próxima near region
región semiconductora semiconducting region
región ultravioleta ultraviolet region
región virtual virtual region
regional regional
registrable recordable
registración de datos data logging
registrador recorder
registrador acústico acoustic recorder
registrador analógico analog recorder
registrador autoequilibrador self-balancing recorder
registrador autorregulador self-regulating recorder
registrador cinescópico kinescope recorder
registrador continuo continuous recorder
registrador controlador controlling recorder
registrador de accionamiento directo direct-acting
recorder
registrador de carretes reel-to-reel tape recorder
registrador de casete cassette recorder
registrador de cinta tape recorder
registrador de cinta perforada punched-tape
recorder
registrador de cuadrante dial recorder
registrador de chispa spark recorder
registrador de datos data recorder
registrador de demanda máxima maximum-demand
recorder
registrador de dirección de base base address
recorder
registrador de disco disk recorder
registrador de doble trazo dual-trace recorder
registrador de dos canales two-channel recorder
registrador de electreto electret recorder
registrador de entrada input recorder
registrador de espectro spectrum recorder
registrador de facsímil facsimile recorder
registrador de fase phase recorder
registrador de forma de onda waveform recorder
registrador de haz luminoso light-beam recorder
registrador de hilo wire recorder
registrador de impulsos pulse recorder
registrador de intensidad de campo field-strength
recorder
registrador de media pista half-track recorder
registrador de nivel level recorder
registrador de nivel de líquido liquid-level recorder
registrador de pH pH recorder
registrador de picos transitorios surge recorder
registrador de pista única single-track recorder
registrador de pluma pen recorder
registrador de pluma caliente hot-pen recorder
registrador de radiosonda radiosonde recorder
registrador de respuesta de frecuencia
frequency-response recorder
registrador de respuesta rápida quick-response
recorder
registrador de sector angosto narrow-sector
recorder
registrador de señal de radar radar signal recorder
registrador de sonido fotográfico photographic
sound recorder
registrador de sonido óptico optical sound recorder
registrador de tambor drum recorder
registrador de telemedida telemetering recorder
registrador de temperatura electrónico electronic
temperature recorder
registrador de tiempo time recorder

registrador de tinta ink recorder
registrador de trabajos job recorder
registrador de tráfico traffic recorder
registrador de transacciones transaction recorder
registrador de vapor de tinta ink-vapor recorder, ink-mist recorder
registrador de velocidad speed recorder
registrador de video electrónico electronic video recorder
registrador de videocasete videocassette recorder
registrador de videodiscos videodisk recorder
registrador de voz voice recorder
registrador de vuelo flight recorder
registrador digital digital recorder
registrador digital incremental incremental digital recorder
registrador electrolítico electrolytic recorder
registrador electromecánico electromechanical recorder
registrador electrotérmico electrothermal recorder
registrador en cinta estereofónica stereophonic tape recorder
registrador en cinta magnética magnetic-tape recorder
registrador en película film recorder
registrador fonográfico phonograph recorder
registrador fotoeléctrico photoelectric recorder
registrador fotográfico photographic recorder
registrador galvanométrico galvanometer recorder
registrador gráfico graphic recorder
registrador helicoidal helix recorder
registrador holográfico holographic recorder
registrador magnético magnetic recorder
registrador mecánico mechanical recorder
registrador monoaural monaural recorder
registrador monofónico monophonic recorder
registrador multitrazo multitrace recorder
registrador no sincrónico nonsynchronous recorder
registrador oscilográfico oscillographic recorder
registrador portátil portable recorder
registrador sumario summary recorder
registrador telegráfico telegraph recorder
registrador térmico thermal recorder
registrador x-y x-y recorder
registrar write
registro record, register, recording, log
registro activo active record
registro actual current record
registro almacenado stored record
registro analógico analog record
registro aritmético arithmetic register
registro auxiliar auxiliary register
registro B B register
registro blanco white recording
registro bloqueado blocked record
registro cerrado locked record
registro cinescópico kinescopic recording
registro circulatorio circulating register
registro corriente current record
registro cuádruple quadruple register
registro de acceso a datos data access register
registro de activación activation record
registro de acumulador accumulator register
registro de almacenamiento storage register
registro de almacenamiento asociativo associative storage register
registro de almacenamiento electrónico electronic storage register
registro de almacenamiento intermedio de memoria memory buffer register
registro de alta fidelidad high-fidelity recording
registro de amplitud constante constant-amplitude recording

registro de anchura variable variable-width recording
registro de archivo de longitud fija fixed-length file record
registro de archivo maestro master file record
registro de área variable variable-area recording
registro de arrastre carry register
registro de base base register
registro de base de datos database record
registro de base de datos física physical database record
registro de borrado deletion record
registro de borrado variable variable-erase recording
registro de cabecera header record
registro de cambios change record
registro de cociente multiplicador multiplier-quotient register
registro de código de operación operation-code register
registro de colores registration of colors
registro de coma flotante floating-point register
registro de comprobación check register
registro de comunicaciones communications recording
registro de control control register, control record
registro de control de acceso access-control register
registro de control de secuencia sequence-control register
registro de cosumando augend register
registro de datos data record, data register, data recording
registro de datos de memoria memory data register
registro de datos externo external data record
registro de datos interno internal data record
registro de demanda demand register
registro de demanda acumulativa cumulative-demand register
registro de densidad variable variable-density recording
registro de desbordamiento overflow register
registro de desplazamiento shift register
registro de desplazamiento ferrorresonante ferroresonant shift register
registro de desplazamiento magnético magnetic shift register
registro de destino destination register
registro de detalle detail record
registro de dirección address register
registro de dirección de base base address register
registro de dirección de instrucción instruction address register
registro de direcciones de memoria memory address register
registro de direcciones de programa program address register
registro de doble densidad double density recording
registro de dos pistas two-track recording
registro de eco echo record
registro de enlace link register
registro de enlace de cadena chain link record
registro de entrada input register
registro de entrada manual manual input register
registro de entrada-salida input-output register
registro de errores error record
registro de estado status register
registro de estado de programa program status register
registro de etiqueta label record
registro de extensión extension register
registro de fallas failure logging
registro de formato format record
registro de frecuencia frequency record

registro de imagen picture record
registro de impresora printer record
registro de impulsos impulse recording
registro de instrucción instruction register
registro de instrucción de control control-instruction register
registro de intensidad de campo field-strength recording
registro de interrupción interrupt register
registro de línea de retardo delay-line register
registro de longitud fija fixed-length record
registro de longitud variable variable-length record
registro de llegada incoming record
registro de memoria memory register
registro de memoria intermedia buffer register
registro de mensajes message logging
registro de microprocesador microprocessor register
registro de modulación de frecuencia frequency-modulation recording
registro de núcleos magnéticos magnetic-core register
registro de paso variable variable-pitch recording
registro de pista completa full-track recording
registro de precisión precision recording
registro de problemas problem register
registro de programa program register
registro de puntos de comprobación checkpoint register
registro de radiofrecuencia radio-frequency record
registro de receptor receiver register
registro de referencia reference recording
registro de relleno padding record
registro de reserva standby register
registro de salida output register
registro de secuencias sequence register
registro de señal signal recording
registro de señal de radar radar signal recording
registro de sonido sound recording
registro de sonido fotográfico photographic sound recording
registro de sonido magnético magnetic sound recording
registro de sonido óptico optical sound recording
registro de suma addition record
registro de sumando addend register
registro de tamaño fijo fixed-size record
registro de tarjeta card record
registro de tiempo time recording
registro de trabajo work register
registro de tráfico traffic record
registro de transacción transaction record
registro de transferencia de datos data-transfer register
registro de uso general general-purpose register
registro de video video recording
registro de video electrónico electronic video recording
registro de visualización display register
registro de voz speech recording, voice recording
registro dedicado dedicated register
registro desfasado out-of-phase recording
registro digital digital recording
registro direccionable addressable register
registro directo direct recording
registro duplicado duplicate record
registro eléctrico electric recording
registro electrográfico electrographic recording
registro electrolítico electrolytic recording
registro electromecánico electromechanical recording
registro electrónico electronic recording, electronic register

registro electroquímico electrochemical recording
registro electrosensible electrosensitive recording
registro electrostático electrostatic recording
registro electrotérmico electrothermal recording
registro en cera wax recording
registro en cinta tape recording
registro en cinta estereofónica stereophonic tape recording
registro en cinta magnética de video video-tape recording, video-taping, magnetic-tape recording
registro en cinta magnética estereofónico stereophonic magnetic-tape recording
registro en cuatro pistas four-track recording
registro en disco disk recording
registro en fase in-phase recording
registro en laca lacquer recording
registro en película film recording
registro en pistas múltiples multiple-track recording
registro encadenado chained record
registro especial special register
registro estático static register
registro estereofónico stereo recording
registro estereofónico monosurco monogroove stereophonic recording
registro externo external register
registro físico physical record
registro fluorográfico fluorographic recording
registro fonográfico phonograph recording
registro formateado formatted record
registro fotográfico photographic recording
registro fotosensible photosensitive recording
registro fuente source recording
registro general general register
registro horizontal horizontal recording
registro inactivo inactive record
registro incidente incident record
registro índice index register
registro instantáneo instantaneous recording
registro interno internal record
registro lateral lateral recording
registro lógico logical record
registro maestro master record
registro magnético magnetic recording
registro mecánico mechanical register
registro modificador modifier register
registro multipista multitrack recording
registro multiplicador multiplier register
registro multitrazo multitrace recording
registro negro black recording
registro no formateado unformatted record
registro numérico numerical recording
registro óptico optical recording
registro oscilográfico oscillographic record
registro osciloscópico oscilloscope recording
registro permanente permanent record
registro perpendicular perpendicular recording
registro por desfase phase-difference recording
registro por haz electrónico electron-beam recording
registro por modulación de fase phase-modulation recording
registro por modulación de impulsos pulse-modulation recording
registro por surcos en relieve embossed-groove recording
registro primario primary register
registro principal principal register
registro raíz root record
registro redundante redundant recording
registro relativo relative record
registro remoto remote recording
registro reposicionable resettable register
registro sin retorno a cero nonreturn-to-zero

recording
registro sin ruido noiseless recording
registro sobre registro sound-on-sound recording
registro sonoro en película sound-on-film recording
registro supersónico supersonic recording
registro temporal temporary register
registro termoplástico thermoplastic recording
registro transversal transverse recording
registro ultrasónico ultrasonic recording
registro unitario unit record
registro variable variable record
registro vertical vertical recording
registro vertical-lateral vertical-lateral recording
registro virtual virtual record
registro x-y x-y recording
registros agrupados grouped records
regla de Ampere Ampere's rule
regla de decisión decision rule
regla de Fleming Fleming's rule
regla de generadores de Fleming Fleming's generator rule
regla de la mano hand rule
regla de la mano derecha right-hand rule
regla de la mano derecha de Fleming Fleming's right-hand rule
regla de la mano izquierda left-hand rule
regla de la mano izquierda de Fleming Fleming's left-hand rule
regla de Maxwell Maxwell's rule
regla de ruido de Nyquist Nyquist noise rule
regla de Silsbee Silsbee rule
reglas de compartimiento sharing rules
reglas de prioridad priority rules
regleta strip
regleta de contacto contact strip
regleta de jacks jack strip
regleta de montaje mounting strip
regleta de resistencia resistance strip
regleta de resistores resistor strip
regleta de sintonización tuning strip
regleta de terminales terminal strip
regrabación rerecording, dubbing
regrabar rerecord
regulación regulation
regulación automática automatic regulation
regulación automática lineal linear automatic regulation
regulación de amplitud de portadora carrier-amplitude regulation
regulación de carga load regulation
regulación de corriente current regulation
regulación de corriente constante constant-current regulation
regulación de corriente de batería battery-current regulation
regulación de frecuencia frequency regulation
regulación de frecuencia de línea line-frequency regulation
regulación de línea line regulation
regulación de nivel level regulation
regulación de nivel automática automatic level regulation
regulación de potencia reactiva idle-power regulation
regulación de presión pressure regulation
regulación de tensión voltage regulation
regulación de tensión con diodo de Zener Zener-diode voltage regulation
regulación de tensión constante constant-voltage regulation
regulación de tensión de salida output voltage regulation
regulación de tensión electrónica electronic voltage

regulation
regulación de tensión ferrorresonante ferroresonant voltage regulation
regulación de velocidad speed regulation
regulación dinámica dynamic regulation
regulación electrónica electronic regulation
regulación estática static regulation
regulación manual manual regulation
regulación por resistencia resistance regulation
regulación relativa relative regulation
regulación total total regulation
regulado automáticamente automatically regulated
regulado electrónicamente electronically regulated
regulador regulator, governor
regulador automático automatic regulator
regulador de accionamiento progresivo forward-acting regulator
regulador de alta tensión high-voltage regulator
regulador de alta velocidad high-speed regulator
regulador de banda base baseband regulator
regulador de bobina móvil moving-coil regulator
regulador de campo field regulator
regulador de campo de motor motor-field regulator
regulador de carga load regulator
regulador de corriente current regulator
regulador de corriente automático automatic current regulator
regulador de corriente constante constant-current regulator
regulador de corriente de enfoque focus current regulator
regulador de corriente de filamento filament current regulator
regulador de deslizamiento slip regulator
regulador de factor de potencia power-factor regulator
regulador de fase phase regulator
regulador de frecuencia frequency regulator
regulador de ganancia gain regulator
regulador de generador generator regulator
regulador de inducción induction regulator
regulador de lámpara lamp regulator
regulador de placas de carbón carbon-pile regulator
regulador de potencia power regulator
regulador de potencial constante constant-potential regulator
regulador de presión pressure regulator
regulador de registro recording regulator
regulador de retroalimentación feedback regulator
regulador de salida output regulator
regulador de salida vertical vertical-output regulator
regulador de sensibilidad sensitivity regulator
regulador de temperatura temperature regulator
regulador de temperatura electrónico electronic temperature regulator
regulador de tensión voltage regulator
regulador de tensión automático automatic voltage regulator
regulador de tensión con diodo de Zener Zener-diode voltage regulator
regulador de tensión constante constant-voltage regulator
regulador de tensión de alimentador feeder voltage regulator
regulador de tensión de descarga luminiscente glow-discharge voltage regulator
regulador de tensión de lámpara de neón neon-bulb voltage regulator
regulador de tensión de línea line-voltage regulator
regulador de tensión de placa plate-voltage regulator
regulador de tensión de placas de carbón carbon-pile voltage regulator

regulador de tensión de red mains voltage regulator
regulador de tensión de silicio silicon voltage regulator
regulador de tensión electrónico electronic voltage regulator
regulador de tensión estático static voltage regulator
regulador de tensión ferrorresonante ferroresonant voltage regulator
regulador de tensión gaseosa gaseous voltage regulator
regulador de tensión por inducción induction voltage regulator
regulador de tensión transistorizado transistorized voltage regulator
regulador de tensión variable variable-voltage regulator
regulador de tiempo time regulator
regulador de transmisión transmission regulator
regulador de vacío vacuum regulator
regulador de velocidad speed regulator
regulador diferencial differential regulator
regulador dinámico dynamic regulator
regulador eléctrico electric governor
regulador electroneumático electropneumatic regulator
regulador electrónico electronic regulator
regulador en derivación shunt regulator
regulador en serie series regulator
regulador estático static regulator
regulador lineal linear regulator
regulador maestro master regulator
regulador magnético magnetic regulator
regulador piloto pilot regulator
regulador primario primary regulator
regulador reostático rheostatic regulator
reignición reignition
reiniciación caliente warm restart
reiniciación condicional conditional restart
reiniciación de puntos de comprobación checkpoint restart
reiniciación diferida deferred restart
reiniciación fría cold restart
reiniciación normal normal restart
reinicialización reinitialization
reinicializar reinitialize
reiniciar restart, reinitiate, reinitialize
reinicio de sistema system restart
reinserción reinsertion
reinserción de corriente continua direct-current reinsertion
reintento de mando command retry
rejilla grid, grating, grille
rejilla abierta open grid
rejilla aceleradora accelerator grid
rejilla agrupadora buncher grid
rejilla aislada insulated grid
rejilla antidispersión antiscatter grid
rejilla antihistéresis antihysteresis grid
rejilla autopolarizada self-biased grid
rejilla bloqueada blocked grid
rejilla catódica cathode grid
rejilla circular circular grating
rejilla colectora collector grid
rejilla concéntrica concentric grid
rejilla de apantallamiento shield grid
rejilla de barrera barrier grid
rejilla de carga espacial space-charge grid
rejilla de control control grid
rejilla de corte remoto remote-cutoff grid
rejilla de cuadro frame grid
rejilla de desionización deionization grid
rejilla de difracción diffraction grating
rejilla de disparo trigger grid

rejilla de enfoque focusing grid
rejilla de hilo wire grid
rejilla de inyección injection grid
rejilla de inyección de señales signal-injection grid
rejilla de Lysholm Lysholm grid
rejilla de malla fina fine-mesh grid
rejilla de mando driving grid
rejilla de referencia reference grid
rejilla de resonador resonator grid
rejilla de resonador de entrada input-resonator grid
rejilla de Wehnelt Wehnelt grid
rejilla dorada gold-plated grid
rejilla en cuadratura quadrature grid
rejilla en espiral spiral grid
rejilla escalonada echelon grating
rejilla estacionaria stationary grid
rejilla exterior outer grid
rejilla flotante floating grid
rejilla flotante libre free-floating grid
rejilla helicoidal helical grid
rejilla inyectora injector grid
rejilla libre free grid
rejilla moduladora modulator grid
rejilla móvil moving grid
rejilla negativa negative grid
rejilla osciladora oscillator grid
rejilla oscilante reciprocating grid
rejilla ovalada oval grid
rejilla-pantalla screen grid
rejilla polarizada polarized grid
rejilla positiva positive grid
rejilla reflectora reflecting grating
rejilla reguladora regulating grid
rejilla rotativa rotating grating
rejilla supresora suppressor grid
rel rel
relación relation, ratio
relación alcance-energía range-energy relation
relación aritmética arithmetic relation
relación binaria binary relation
relación corriente-tensión current-voltage relation
relación de Bridgman Bridgman relation
relación de caracteres character relation
relación de control control relationship
relación de fase phase relationship
relación de reciprocidad reciprocity relation
relación derivada derived relation
relación directa direct relation
relación lógica logical relation
relación O OR relationship
relación sintética synthetic relationship
relación ternaria ternary relation
relación virtual virtual relation
relación Y AND relationship
relacional relational
relajación relaxation
relajación dieléctrica dielectric relaxation
relativo relative
relé relay
relé abierto open relay
relé accionado por diferencia de tensión voltage-balance relay
relé accionado por gas gas-actuated relay
relé accionado por motor motor-driven relay
relé acelerador accelerating relay
relé acelerador de campo field accelerating relay
relé acelerador de campo de generador generator field accelerating relay
relé alimentador feeder relay
relé anunciador annunciator relay
relé autoenganchador self-latching relay
relé auxiliar auxiliary relay
relé bidireccional bidirectional relay

relé biestable bistable relay
relé bimetálico bimetallic relay
relé binario binary relay
relé buscador de cero null-seeking relay
relé centrífugo centrifugal relay
relé cerámico ceramic relay
relé coaxial coaxial relay
relé con autorreposición self-resetting relay
relé con memoria memory relay
relé controlado por voz voice-controlled relay
relé cortocircuitador shorting relay
relé de acción lenta slow-action relay
relé de acción mecánica mechanical-action relay
relé de acción rápida fast-acting relay
relé de acción rotativa rotary-action relay
relé de acción ultrarrápida snap-action relay
relé de alarma alarm relay
relé de almacenamiento storage relay
relé de alta confiabilidad high-reliability relay
relé de alta impedancia high-impedance relay
relé de alta tensión high-voltage relay
relé de alta velocidad high-speed relay
relé de altavoz loudspeaker relay
relé de alumbrado de aproximación
    approach-lighting relay
relé de amortiguador dashpot relay
relé de antena antenna relay
relé de anunciación annunciation relay
relé de aplicación de campo field application relay
relé de armadura armature relay
relé de armadura lateral side-armature relay
relé de arranque start relay
relé de arranque de motor motor-start relay
relé de autorretención stick relay
relé de avance lento inching relay
relé de avance paso a paso stepping relay
relé de barra colectora busbar relay
relé de base octal octal-base relay
relé de bloqueo blocking relay, locking relay
relé de bobina móvil moving-coil relay
relé de bobina única single-coil relay
relé de bobinas múltiples multiple-coil relay
relé de cambio repentino sudden-change relay
relé de campo field relay
relé de campo en derivación shunt-field relay
relé de capacitancia capacitance relay
relé de carga de baterías battery charging relay
relé de carga mínima underload relay
relé de cierre closing relay
relé de circuito de placa plate-circuit relay
relé de circuito impreso printed-circuit relay
relé de cociente quotient relay
relé de conductancia conductance relay
relé de conexión connecting relay
relé de conmutación switching relay
relé de contactos de mercurio mercury-contact relay
relé de contactos humedecidos wetted-contact relay
relé de contactos humedecidos por mercurio
    mercury-wetted relay
relé de contactos múltiples multiple-contact relay
relé de conteo counting relay
relé de control control relay
relé de control de potencia power control relay
relé de control intermedio intermediate-control relay
relé de control magnético magnetic control relay
relé de conversión binario-decimal binary-decimal
    conversion relay
relé de corriente current relay
relé de corriente alterna alternating-current relay
relé de corriente continua direct-current relay
relé de corriente de placa plate-current relay
relé de corriente inversa inverse current relay
relé de corriente máxima maximum current relay

relé de corriente mínima de corriente continua
    direct-current undercurrent relay
relé de corriente portadora carrier-current relay
relé de corte cutoff relay
relé de desconexión de carga load-disconnect relay
relé de desenganche lento slow-release relay
relé de desenganche rápido fast-release relay
relé de desplazamiento de mercurio mercury
    displacement relay
relé de detección de tensión voltage-sensing relay
relé de diferencia de impulsos impulse-difference
    relay
relé de dirección de potencia power-direction relay
relé de disco móvil movable-disk relay
relé de disparo trigger relay, trip relay
relé de distancia distance relay
relé de distribución distribution relay
relé de dos arrollamientos two-coil relay
relé de dos bobinas dual-coil relay
relé de dos pasos two-step relay
relé de dos posiciones two-position relay
relé de emisión-recepción send-receive relay
relé de enclavamiento interlocking relay, latching
    relay
relé de enganche magnético magnetic latching relay
relé de equilibrio balance relay
relé de equilibrio de corriente current-balance relay
relé de equilibrio de fases phase-balance relay
relé de estado sólido solid-state relay
relé de factor de potencia power-factor relay
relé de falla fault relay
relé de falla de campo field-failure relay
relé de fase abierta open-phase relay
relé de flujo flow relay
relé de forzado de campo field forcing relay
relé de frecuencia frequency relay
relé de frecuencia de campo polarizado polarized
    field frequency relay
relé de frecuencia vocal voice-frequency relay
relé de función function relay
relé de funcionamiento rápido fast-operate relay
relé de gas gas relay
relé de guarda guard relay
relé de hilo caliente hot-wire relay
relé de hipocorriente undercurrent relay
relé de imán permanente permanent-magnet relay
relé de impedancia impedance relay
relé de impulsos impulse relay
relé de inducción induction relay
relé de inercia inertia relay
relé de instrumento instrument relay
relé de inversión de fase phase-reversal relay
relé de Kipp Kipp relay
relé de laminilla de punto de cruce crosspoint reed
    relay
relé de laminilla vibrante vibrating-reed relay
relé de laminillas reed relay
relé de laminillas húmedas wet-reed relay
relé de laminillas humedecidas por mercurio
    mercury-wetted reed relay
relé de laminillas secas dry-reed relay
relé de línea line relay
relé de llamada calling relay, ringing relay
relé de manipulación keying relay
relé de medida measuring relay
relé de mercurio mercury relay
relé de microondas microwave relay
relé de mínima potencia underpower relay
relé de Morse Morse relay
relé de operación lenta slow-operate relay
relé de pasos rotativo rotary stepping relay
relé de perforación punch relay
relé de picos transitorios surge relay

relé de pista track relay
relé de placa plate relay
relé de polo sombreado shaded-pole relay
relé de potencia power relay
relé de potencia activa active power relay
relé de potencia direccional directional power relay
relé de potencia inversa reverse-power relay
relé de potencia reactiva reactive-power relay
relé de presión pressure relay
relé de presión de gas gas-pressure relay
relé de producto product relay
relé de protección de puesta a tierra protection ground relay
relé de protección diferencial differential protective relay
relé de prueba test relay
relé de puesta a tierra earthing relay
relé de pulsador push-button relay
relé de radar radar relay
relé de reactancia reactance relay
relé de recepción receiving relay
relé de reconexión reclosing relay
relé de reconexión de corriente alterna alternating-current reclosing relay
relé de rectificador rectifier relay
relé de red network relay
relé de registro record relay
relé de regulación de tensión voltage-regulation relay
relé de reposición homing relay
relé de reposición eléctrica electric reset relay
relé de reposición manual manual-reset relay
relé de resistencia resistance relay
relé de respaldo backup relay
relé de retardo delay relay
relé de retardo de tiempo time-delay relay, time relay
relé de retardo de tiempo térmico thermal time-delay relay
relé de retardo inverso inverse time-lag relay
relé de retardo térmico thermal delay relay
relé de retardo termostático thermostatic delay relay
relé de retención holding relay, lock-up relay
relé de rotación de fase phase-rotation relay
relé de secuencia sequence relay
relé de secuencia de fase negativa negative phase-sequence relay
relé de secuencia de fase positiva positive phase-sequence relay
relé de secuencia de fases phase-sequence relay
relé de selenio selenium relay
relé de semiconductor semiconductor relay
relé de señal signal relay
relé de señalización signaling relay
relé de servicio intermitente intermittent-duty relay
relé de sincronización synchronizing relay
relé de sobrecarga overload relay
relé de sobrecarga momentánea momentary-overload relay
relé de sobrecorriente overcurrent relay
relé de sobrecorriente controlado por tensión voltage-controlled overcurrent relay
relé de sobrecorriente de corriente continua direct-current overcurrent relay
relé de sobrecorriente de tiempo de corriente alterna alternating-current time overcurrent relay
relé de sobrecorriente direccional directional overcurrent relay
relé de sobrecorriente direccional de corriente alterna alternating-current directional overcurrent relay
relé de sobrecorriente instantánea instantaneous overcurrent relay

relé de sobrefrecuencia overfrequency relay
relé de sobrepotencia overpower relay
relé de sobretensión overvoltage relay
relé de sobretensión de corriente continua direct-current overvoltage relay
relé de solenoide solenoid relay
relé de solenoide rotativo rotary solenoid relay
relé de subtensión undervoltage relay
relé de suma y resta add-and-subtract relay
relé de suministro de batería battery-supply relay
relé de supresión de ruido noise-suppression relay
relé de tasa de cambio rate-of-change relay
relé de telecomunicaciones telecommunications relay
relé de temperatura temperature relay
relé de temporización timing relay
relé de tensión voltage relay
relé de tensión de secuencia de fases phase-sequence voltage relay
relé de tensión mínima undervoltage relay
relé de tensión mínima de corriente continua direct-current undervoltage relay
relé de tiempo definido definite-time relay
relé de tiempo magnético magnetic time relay
relé de tierra ground relay
relé de tipo de instrumento instrument-type relay
relé de tipo de medidor meter-type relay
relé de tipo de motor motor-type relay
relé de tipo plano flat-type relay
relé de transferencia transfer relay
relé de transferencia de potencia power transfer relay
relé de transmisión-recepción transmit-receive relay
relé de tres posiciones three-position relay
relé de trinquete ratchet relay
relé de tubos tube relay
relé de uso especial special-purpose relay
relé de uso general general-purpose relay
relé de vacío vacuum relay
relé desacelerador decelerating relay
relé desacelerador de campo de generador generator field decelerating relay
relé destellador flasher relay
relé diferencial differential relay
relé diferencial polarizado polarized differential relay
relé direccional directional relay
relé direccional de potencia power-directional relay
relé discriminante discriminating relay
relé electrodinámico electrodynamic relay
relé electroestrictivo electrostrictive relay
relé electromagnético electromagnetic relay
relé electromecánico electromechanical relay
relé electroneumático electropneumatic relay
relé electrónico electronic relay
relé electrónico telegráfico telegraph electronic relay
relé electrostático electrostatic relay
relé electrotérmico electrothermal relay
relé en serie series relay
relé encerrado enclosed relay
relé enchufable plug-in relay
relé estable en posición central center-stable relay
relé estático static relay
relé excitador exciter relay
relé ferrodinámico ferrodynamic relay
relé fotoeléctrico photoelectric relay
relé fotoelectrónico photoelectronic relay
relé galvanométrico galvanometer relay
relé homopolar zero-phase-sequence relay
relé indicador indicating relay
relé instantáneo instantaneous relay
relé integrador integrating relay
relé libre de zumbido hum-free relay
relé limitador de corriente step-back relay

relé lógico logic relay
relé maestro master relay
relé maestro de red network master relay
relé magnético magnetic relay
relé magnético sensible sensitive magnetic relay
relé magnetoeléctrico magnetoelectric relay
relé magnetoestrictivo magnetostrictive relay
relé marginal marginal relay
relé mecánico mechanical relay
relé medidor meter relay
relé modulador modulator relay
relé multibobina multicoil relay
relé multiposición multiposition relay
relé neumático pneumatic relay
relé neutro neutral relay
relé no polar nonpolar relay
relé no polarizado nonpolarized relay
relé operacional operational relay
relé operado por capacidad capacity-operated relay
relé operado por corriente continua direct-current
     operated relay
relé operado por luz light-operated relay
relé operado por portadora carrier-operated relay
relé operado por sonido sound-operated relay
relé operado por voz voice-operated relay
relé óptico optical relay
relé optoelectrónico optoelectronic relay
relé parcialmente energizado partially energized
     relay
relé pasivo passive relay
relé paso a paso step-by-step relay
relé permisivo permissive relay
relé piloto pilot relay
relé polarizado polarized relay, biased relay
relé primario primary relay
relé protector protective relay
relé protector de campo field protective relay
relé pulsante pulsing relay
relé reforzado heavy-duty relay
relé regulador regulating relay
relé regulador de tensión voltage-regulating relay
relé repetidor repeating relay
relé resistente a choques shock-resistant relay
relé resonante resonant relay
relé retardado delayed relay
relé rotativo rotary relay
relé secuencial sequential relay
relé secundario secondary relay
relé selectivo selective relay
relé selectivo a la frecuencia frequency-selective
     relay
relé selector selector relay
relé sellado sealed relay
relé sellado herméticamente hermetically-sealed
     relay
relé sensible sensitive relay
relé sensible a la frecuencia frequency-sensitive
     relay
relé sensible a la polaridad polarity-sensitive relay
relé simétrico symmetrical relay
relé simulador simulator relay
relé sin polarización no-bias relay
relé sintonizado tuned relay
relé solar solar relay
relé subminiatura subminiature relay
relé supersensible supersensitive relay
relé supervisor supervisory relay
relé suplementario supplementary relay
relé tacométrico tachometric relay
relé telefónico telephone relay
relé telegráfico telegraph relay
relé temporizado mecánicamente mechanically
     timed relay

relé térmico thermal relay
relé termoiónico thermionic relay
relé termostático thermostatic relay
relé todo o nada all-or-nothing relay
relé transistorizado transistorized relay
relé ultrasensible ultrasensitive relay
relé ultrasónico ultrasonic relay
relé unilateral unilateral relay
relé unipolar single-pole relay
relé vibrante vibrating relay
relevador relay
relocalización relocation
relocalización de programa program relocation
relocalización dinámica dynamic relocation
relocalización estática static relocation
reloj atómico atomic clock
reloj bifásico two-phase clock
reloj de cuarzo quartz clock
reloj de impulsos pulse clock
reloj de sincronización synchronizing clock
reloj de tiempo real real-time clock
reloj digital digital clock
reloj eléctrico electric clock, electric watch
reloj eléctrico sincrónico synchronous electric clock
reloj electromagnético electromagnetic clock
reloj electrónico electronic clock, electronic watch
reloj interno internal clock
reloj local local clock
reloj maestro master clock
reloj nuclear nuclear clock
reloj principal principal clock
reloj programable programmable clock
reloj secundario secondary clock
reloj sincrónico synchronous clock
reluctancia reluctance
reluctancia del entrehierro gap reluctance
reluctancia específica specific reluctance
reluctancia magnética magnetic reluctance
reluctancia variable variable reluctance
reluctividad reluctivity
reluctivo reluctive
reluctómetro reluctometer
rellenar con ceros zero fill
relleno filler, filling
relleno de cable cable filler
relleno de espacios gap filling
relleno de fusible fuse filler
relleno de rejilla grid filling
remagnetizador remagnetizer
remanencia remanence
remanencia de imagen burn
remanencia magnética magnetic remanence
remanente remanent
remodulación remodulation
remodulador remodulator
remolino magnético magnetic whirl
remotamente remotely
remoto remote
remover remove
removible removable
rendimiento yield, performance, efficiency
rendimiento cuántico quantum yield
rendimiento cuántico espectral spectral quantum
     yield
rendimiento de máquina machine performance
rendimiento en presión pressure sensitivity
rendimiento fotoeléctrico photoelectric yield
rendimiento óptimo optimum performance
rendimiento total throughput
renio rhenium
renovable renewable
reoelectricidad rheoelectricity
reoeléctrico rheoelectric

reoestricción rheostriction
reóforo rheophore
reográfico rheographic
reógrafo rheograph
reorganización reorganization
reorganización concurrente concurrent reorganization
reorganización de base de datos database reorganization
reostáticamente rheostatically
reostático rheostatic
reóstato rheostat
reóstato bobinado wire-wound rheostat
reóstato con tomas tapped rheostat
reóstato de agua water rheostat
reóstato de ajuste trimmer rheostat
reóstato de arranque starter rheostat
reóstato de campo field rheostat
reóstato de carbón carbon rheostat
reóstato de carga load rheostat
reóstato de carga ficticia dummy-load rheostat
reóstato de conmutador switch rheostat
reóstato de control de velocidad speed-control rheostat
reóstato de excitación excitation rheostat
reóstato de filamento filament rheostat
reóstato de placa frontal faceplate rheostat
reóstato de regulación de velocidad speed-regulating rheostat
reóstato de rejilla grid rheostat
reóstato en anillo ring rheostat
reóstato exponencial exponential rheostat
reóstato líquido liquid rheostat
reóstato potenciométrico potentiometer rheostat
reóstato rotórico rotor rheostat
reóstato vernier vernier rheostat
reóstatos acoplados mecánicamente ganged rheostats
reotaxial rheotaxial
reotomo rheotome
reotrópico rheotropic
rep rep
reparabilidad repairability
reparación repair
reparación de emergencia emergency repair
reparación principal major repair
repartidor de grupos group distribution frame
repartidor de supergrupo supergroup distribution frame
repartidor intermedio intermediate distributing frame
repartidor principal main distributing frame
repeledor repeller
repelente al agua water-repellent
reperforación repunching
reperforador reperforator
reperforador de recepción receiving reperforator
reperforador impresor printing reperforator
reperforadora de cinta tape reperforator
reperforar repunch
repetibilidad repeatability
repetible repeatable
repetición repetition
repetición automática automatic repetition
repetición de impulsos pulse repetition
repetición de programa program repeat
repetidamente repeatedly
repetido repeated
repetidor repeater
repetidor activo active repeater
repetidor amplificador amplifying repeater
repetidor analógico analog repeater
repetidor automático automatic repeater

repetidor bidireccional two-way repeater
repetidor de banda band repeater
repetidor de banda ancha wideband repeater
repetidor de conferencia conference repeater
repetidor de cuatro hilos four-wire repeater
repetidor de frecuencia de impulsos pulse-frequency repeater
repetidor de frecuencia vocal voice-frequency repeater
repetidor de impedancia negativa negative impedance repeater
repetidor de impulsos pulse repeater
repetidor de impulsos de radar radar-pulse repeater
repetidor de línea line repeater
repetidor de llamada ringing repeater
repetidor de microondas microwave repeater
repetidor de modulación de frecuencia frequency-modulation repeater
repetidor de portadora carrier repeater
repetidor de programa program repeater
repetidor de radar radar repeater
repetidor de radio radio repeater
repetidor de reserva standby repeater
repetidor de resistencia negativa negative-resistance repeater
repetidor desmodulador back-to-back repeater
repetidor digital digital repeater
repetidor directo through repeater
repetidor discriminante discriminating repeater
repetidor electromecánico electromechanical repeater
repetidor electrónico electronic repeater
repetidor en contrafase push-pull repeater
repetidor giroscópico electrónico electronic gyroscopic repeater
repetidor heterodino heterodyne repeater
repetidor inatendido unattended repeater
repetidor intermedio intermediate repeater
repetidor klistrón klystron repeater
repetidor local local repeater
repetidor multicanal multichannel repeater
repetidor óptico optical repeater
repetidor pasivo passive repeater
repetidor principal main repeater
repetidor regenerativo regenerative repeater
repetidor retardado delayed repeater
repetidor rotativo rotary repeater
repetidor semiactivo semiactive repeater
repetidor semidúplex half-duplex repeater
repetidor sincrónico synchronous repeater
repetidor submarino submarine repeater
repetidor sumergido submerged repeater
repetidor telefónico telephone repeater
repetidor telegráfico telegraph repeater
repetidor terminal terminal repeater
repetidor unidireccional one-way repeater
repetidora de televisión television repeater
repetitivo repetitive
reponer restore
reporte report
reposición reset, resetting
reposición automática automatic reset
reposición de equipo físico hardware reset
reposición eléctrica electrical reset
reposición externa external reset
reposición instantánea instantaneous reset
reposición manual manual reset
reposicionabilidad resettability
reposicionable resettable
representación analógica analog representation
representación binaria binary representation
representación binaria incremental incremental binary representation

representación codificada coded representation
representación de coma fija fixed-point representation
representación de coma flotante floating-point representation
representación de datos data representation
representación de datos internos internal data representation
representación de valor absoluto absolute value representation
representación decimal decimal representation
representación digital digital representation
representación discreta discrete representation
representación incremental incremental representation
representación incremental binaria binary incremental representation
representación lineal linear representation
representación lógica logical representation
representación nula null representation
representación numérica numerical representation
representación posicional positional representation
representación simbólica symbolic representation
representativo representative
reprocesamiento reprocessing
reproducción reproduction, playback
reproducción de cinta tape playback
reproducción de datos data playback
reproducción de gama ancha wide-range reproduction
reproducción de imagen picture reproduction
reproducción de sonido sound reproduction
reproducción de sonido estereofónica stereophonic sound reproduction
reproducción directa direct playback
reproducción eléctrica electrical reproduction
reproducción estereofónica stereophonic playback, stereophonic reproduction
reproducción fiel faithful reproduction
reproducción fonográfica phonograph reproduction
reproducción óptica optical playback
reproducción simultánea simultaneous playback
reproducibilidad reproducibility
reproducible reproducible
reproducir reproduce
reproductor player, reproducer
reproductor acumulante accumulating reproducer
reproductor de agudos treble reproducer
reproductor de carretes reel-to-reel tape recorder
reproductor de cinta tape reproducer, tape player
reproductor de cinta de papel paper-tape reproducer
reproductor de cinta magnética magnetic-tape reproducer
reproductor de cintas magnéticas de televisión television tape player
reproductor de cintas magnéticas de video video player
reproductor de facsímil facsimile reproducer
reproductor de imagen image reproducer
reproductor de película film reproducer
reproductor de registro magnético magnetic recording reproducer
reproductor de sonido fotográfico photographic sound reproducer
reproductor de sonido óptico optical sound reproducer
reproductor de tarjetas card reproducer
reproductor dinámico dynamic reproducer
reproductor fonográfico phonograph reproducer
reproductor lateral lateral reproducer
reproductor magnético magnetic reproducer
reproductor mecánico mechanical reproducer

reprogramación reprogramming
reprogramar reprogram
repuesta de estado de entrada-salida input-output status response
repuesto spare
repulsión repulsion
repulsión de carga espacial space-charge repulsion
repulsión de Coulomb Coulomb repulsion
repulsión eléctrica electrical repulsion
repulsión electromagnética electromagnetic repulsion
repulsión magnética magnetic repulsion
repulsión molecular molecular repulsion
repulsividad repulsiveness
repulsivo repulsive
requisito requirement
requisito de almacenamiento storage requirement
requisito de diseño design requirement
requisito de implementación implementation requirement
requisito de información information requirement
requisito de interfaz interface requirement
requisito de prueba test requirement
requisito de rendimiento performance requirement
requisito de servicio service requirement
requisito de tiempo time requirement
requisito físico physical requirement
requisito funcional functional requirement
requisitos de alimentación power requirements
requisitos de procesamiento processing requirements
requisitos de programas software requirements
requisitos de sistema system requirements
rerradiación reradiation
rerradiar reradiate
rerradiodifusión rebroadcasting, rebroadcast
rerregistrar rerecord
rerregistro rerecording, dubbing
reserva reserve, backup, spare
reserva completa full backup
reserva de estación generadora generating-station reserve
reserva de gas gas reservoir
reserva de ignición ignition reserve
reserva de operación operating reserve
reserva de potencia power reserve
reserva de red system reserve
reserva de sistema system reserve
reserva degradada degraded backup
reserva eléctrica electrical reserve
reserva global global backup
reserva incremental incremental backup
reserva instalada installed reserve
reserva requerida required reserve
reservado reserved
residente resident
residente en memoria memory resident
residente en memoria de núcleos core memory resident
residente permanentemente permanently resident
residual residual
residuo garbage
residuo anódico anode scrap
residuo de ruido noise residue
residuo eléctrico electric residue
residuos de perforación chad
resiliencia resilience
resina resin, rosin
resina acrílica acrylic resin
resina de policlorotrifluoroetileno polychlorotrifluoroethylene resin
resina de politetrafluoretileno polytetrafluoroethylene resin

resina de urea urea resin
resina de vinilo vinyl resin
resina epóxica epoxy resin
resina fenólica phenolic resin
resina fluorocarbúrica fluorocarbon resin
resina sintética synthetic resin
resina termoendurecible thermosetting resin
resinoso resinous
resintonización retuning
resintonizar retune
resistencia resistance, resistor
resistencia a impactos impact strength
resistencia a interferencia intencional jamming resistance
resistencia a la abrasión abrasion resistance
resistencia a la fatiga fatigue resistance
resistencia a la humedad moisture resistance
resistencia a la tensión voltage endurance
resistencia a pérdidas loss resistance
resistencia a tierra resistance to earth
resistencia aceleradora accelerating resistance
resistencia acústica acoustic resistance
resistencia acústica específica specific acoustic resistance
resistencia acústica intrínseca unit-area acoustic resistance
resistencia al calor heat resistance
resistencia amortiguadora damping resistance
resistencia anódica anode resistance
resistencia asincrónica asynchronous resistance
resistencia bifilar bifilar resistance
resistencia bloqueada blocked resistance
resistencia caliente hot resistance
resistencia-capacitancia-inductancia resistance-capacitance-inductance
resistencia cero zero resistance
resistencia crítica critical resistance
resistencia de acoplamiento coupling resistance
resistencia de aislamiento insulation resistance
resistencia de alambrado wiring resistance
resistencia de amortiguamiento crítico externa external critical damping resistance
resistencia de ánodo diferencial differential anode resistance
resistencia de antena antenna resistance
resistencia de antena efectiva effective antenna resistance
resistencia de arco arc resistance
resistencia de arranque starting resistance
resistencia de barrera barrier resistance
resistencia de base base resistance
resistencia de base de corriente alterna alternating-current base resistance
resistencia de base de corriente continua direct-current base resistance
resistencia de base dinámica dynamic base resistance
resistencia de base estática static base resistance
resistencia de bloqueo blocking resistance
resistencia de bobina coil resistance
resistencia de bobina de relé relay-coil resistance
resistencia de Bronson Bronson resistance
resistencia de bucle loop resistance
resistencia de campo field resistance
resistencia de capa de barrera barrier-layer resistance
resistencia de carga load resistance
resistencia de carga de diodo diode load resistance
resistencia de carga de placa plate load resistance
resistencia de carga de rejilla grid-load resistance
resistencia de carga efectiva effective load resistance
resistencia de cátodo de corriente alterna alternating-current cathode resistance
resistencia de cátodo de corriente continua direct-current cathode resistance
resistencia de cátodo dinámica dynamic cathode resistance
resistencia de cátodo estática static cathode resistance
resistencia de cebado crítica critical build-up resistance
resistencia de colector collector resistance
resistencia de colector de corriente alterna alternating-current collector resistance
resistencia de colector de corriente continua direct-current collector resistance
resistencia de colector dinámica dynamic collector resistance
resistencia de colector estática static collector resistance
resistencia de conmutación commutating resistance
resistencia de consumo de corriente continua direct-current drain resistance
resistencia de consumo dinámica dynamic drain resistance
resistencia de consumo estática static drain resistance
resistencia de contacto contact resistance
resistencia de contacto dinámica dynamic contact resistance
resistencia de corriente alterna alternating-current resistance
resistencia de desacoplamiento decoupling resistance
resistencia de disipación de potencia cero zero-power resistance
resistencia de dispersión spreading resistance
resistencia de dispersión de base base spreading resistance
resistencia de drenador de corriente alterna alternating-current drain resistance
resistencia de drenaje bleeder resistance
resistencia de electrodo electrode resistance
resistencia de emisor emitter resistance
resistencia de emisor de corriente alterna alternating-current emitter resistance
resistencia de emisor de corriente continua direct-current emitter resistance
resistencia de emisor dinámica dynamic emitter resistance
resistencia de emisor estática static emitter resistance
resistencia de entrada input resistance
resistencia de entrada diferencial differential-input resistance
resistencia de entrada diferencial equivalente equivalent differential input resistance
resistencia de entrada interna internal input resistance
resistencia de etapa excitadora driver resistance
resistencia de extinción quenching resistance
resistencia de fase phase resistance
resistencia de filamento filament resistance
resistencia de filamento caliente filament hot resistance
resistencia de filamento frío filament cold resistance
resistencia de fuente source resistance
resistencia de fuente de corriente continua direct-current source resistance
resistencia de fuga leakage resistance
resistencia de fuga de encendedor ignitor leakage resistance
resistencia de fuga de rejilla grid-leak resistance
resistencia de inserción insertion resistance
resistencia de instrumento instrument resistance

resistencia de interfaz interface resistance
resistencia de interfaz catódica cathode-interface
   resistance
resistencia de ionización ionization resistance
resistencia de Koch Koch resistance
resistencia de lámpara lamp resistance
resistencia de malla mesh resistance
resistencia de medidor meter resistance
resistencia de pantalla screen resistance
resistencia de placa plate resistance
resistencia de placa de corriente alterna
   alternating-current plate resistance
resistencia de placa de corriente continua
   direct-current plate resistance
resistencia de placa dinámica dynamic plate
   resistance
resistencia de placa estática static plate resistance
resistencia de platino platinum resistance
resistencia de polarización polarization resistance,
   biasing resistance
resistencia de puerta de corriente continua
   direct-current gate resistance
resistencia de puerta estática static gate resistance
resistencia de radiación radiation resistance
resistencia de radiación de antena aerial radiation
   resistance
resistencia de rejilla grid resistance
resistencia de resonancia resonance resistance
resistencia de retroalimentación feedback resistance
resistencia de ruido equivalente equivalent noise
   resistance
resistencia de ruptura breakdown resistance
resistencia de salida output resistance
resistencia de salida en bucle abierto open-loop
   output resistance
resistencia de salida interna internal output
   resistance
resistencia de saturación saturation resistance
resistencia de secuencia de fase negativa negative
   phase-sequence resistance
resistencia de secuencia de fase positiva positive
   phase-sequence resistance
resistencia de secuencia negativa negative-sequence
   resistance
resistencia de secuencia positiva positive-sequence
   resistance
resistencia de señal signal resistance
resistencia de tierra ground resistance, earth
   resistance
resistencia de transformación transformation
   resistance
resistencia de transición transition resistance
resistencia de unión junction resistance
resistencia de varistor varistor resistance
resistencia de volumen volume resistance
resistencia del inducido armature resistance
resistencia dieléctrica dielectric resistance
resistencia diferencial differential resistance
resistencia diferencial anódica anode differential
   resistance
resistencia dinámica dynamic resistance
resistencia directa forward resistance
resistencia distribuida distributed resistance
resistencia efectiva effective resistance
resistencia eléctrica electrical resistance
resistencia en alta frecuencia high-frequency
   resistance
resistencia en circuito abierto open-circuit
   resistance
resistencia en corriente continua direct-current
   resistance
resistencia en cortocircuito short-circuit resistance
resistencia en derivación shunt resistance

resistencia en obscuridad dark resistance
resistencia en paralelo parallel resistance
resistencia en paralelo efectiva effective parallel
   resistance
resistencia en radiofrecuencia radio-frequency
   resistance
resistencia en serie series resistance
resistencia en serie efectiva effective series
   resistance
resistencia en serie equivalente equivalent series
   resistance
resistencia entre bases interbase resistance
resistencia equilibradora balancing resistance
resistencia equivalente equivalent resistance
resistencia específica specific resistance
resistencia estabilizadora ballast resistance
resistencia estática static resistance
resistencia estática de fuente static source resistance
resistencia externa external resistance
resistencia fría cold resistance
resistencia funcional total total functional resistance
resistencia incremental incremental resistance
resistencia inductiva inductive resistance
resistencia interna internal resistance
resistencia interna aparente apparent internal
   resistance
resistencia intrínseca intrinsic resistance
resistencia inversa inverse resistance, back
   resistance
resistencia lineal linear resistance
resistencia magnética magnetic resistance
resistencia magnética específica specific magnetic
   resistance
resistencia magnetotérmica magnetothermal
   resistance
resistencia mecánica mechanical resistance
resistencia mínima minimum resistance
resistencia mínima absoluta absolute minimum
   resistance
resistencia multiplicadora multiplier resistance
resistencia mutua mutual resistance
resistencia negativa negative resistance
resistencia neta net resistance
resistencia no inductiva noninductive resistance
resistencia no lineal nonlinear resistance
resistencia no reactiva nonreactive resistance
resistencia nominal nominal resistance
resistencia óhmica ohmic resistance
resistencia parásita parasitic resistance, stray
   resistance
resistencia pasiva passive resistance
resistencia patrón standard resistance
resistencia positiva positive resistance
resistencia primaria primary resistance
resistencia protectora protective resistance
resistencia pura pure resistance
resistencia real true resistance
resistencia reflejada reflected resistance
resistencia residual residual resistance
resistencia rotórica rotor resistance
resistencia secundaria secondary resistance
resistencia superficial surface resistance
resistencia térmica thermal resistance
resistencia térmica efectiva effective thermal
   resistance
resistencia ultraalta ultra-high resistance
resistencia variable variable resistance
resistente resistant
resistente a agentes atmosféricos weather-resistant
resistente a chispas spark-resistant
resistente a choques shock-resistant
resistente a la corrosión corrosion-resistant
resistente a la herrumbre rust-resistant

resistente a la humedad moisture-resistant
resistente a la oxidación oxidation-resistant
resistente a las vibraciones vibration-resistant
resistente a radiación radiation-resistant
resistente al ácido acid-resistant
resistente al fuego fire-resistant
resistente al polvo dust resistant
resistividad resistivity
resistividad de base base resistivity
resistividad de corriente continua direct-current resistivity
resistividad de masa mass resistivity
resistividad de tierra earth resistivity
resistividad de volumen volume resistivity
resistividad efectiva effective resistivity
resistividad eléctrica electrical resistivity
resistividad intrínseca intrinsic resistivity
resistividad magnética magnetic resistivity
resistividad molecular molecular resistivity
resistividad residual residual resistivity
resistividad superficial surface resistivity
resistividad térmica thermal resistivity
resistivo resistive
resistor resistor
resistor aislado insulated resistor
resistor aislante isolating resistor
resistor ajustable adjustable resistor
resistor amortiguador damping resistor
resistor arrollado en carrete spool-wound resistor
resistor bifilar bifilar resistor
resistor bobinado wound resistor, wire-wound resistor
resistor catódico cathode resistor
resistor cerámico ceramic resistor
resistor cermet cermet resistor
resistor compensador compensating resistor
resistor con derivación central center-tapped resistor
resistor con impurezas compensadas compensated-impurity resistor
resistor con tomas tapped resistor
resistor concentrado lumped resistor
resistor controlado por tensión voltage-controlled resistor
resistor corrector peaking resistor
resistor de acoplamiento coupling resistor
resistor de adaptación matching resistor
resistor de agua water resistor
resistor de aislamiento isolation resistor
resistor de ajuste trimmer resistor
resistor de aplanamiento smoothing resistor
resistor de autopolarización self-bias resistor
resistor de banco de lámparas lamp-bank resistor
resistor de base base resistor
resistor de bucle loop resistor
resistor de caída dropping resistor
resistor de caída en serie series dropping resistor
resistor de calentamiento heating resistor
resistor de campo field resistor
resistor de carbón carbon resistor
resistor de carbón aislado insulated carbon resistor
resistor de carbón craqueado cracked-carbon resistor
resistor de carbón depositado deposited carbon resistor
resistor de carga load resistor, loading resistor, charging resistor
resistor de carga de baterías battery charging resistor
resistor de carga de diodo diode load resistor
resistor de carga de rejilla grid-load resistor
resistor de cátodo sin capacitor de desacoplamiento unbypassed cathode resistor

resistor de cinta adhesiva adhesive-tape resistor
resistor de compensación compensation resistor
resistor de composición composition resistor
resistor de composición fija fixed composition resistor
resistor de conexión temporal patching resistor
resistor de cordón de alimentación line-cord resistor
resistor de cristal crystal resistor
resistor de décadas decade resistor
resistor de derivación central ajustable adjustable center-tap resistor
resistor de desacoplamiento decoupling resistor
resistor de descarga discharge resistor, discharging resistor
resistor de diodo diode resistor
resistor de disco disk resistor
resistor de drenaje bleeder resistor
resistor de efecto túnel tunnel resistor
resistor de entrada input resistor
resistor de extinción quenching resistor
resistor de frenaje braking resistor
resistor de freno de rejilla grid stopper resistor
resistor de fuga de rejilla grid-leak resistor
resistor de grafito graphite resistor
resistor de hilo wire resistor
resistor de línea de grafito graphite-line resistor
resistor de metal-óxido metal-oxide resistor
resistor de muestreo de corriente current-sensing resistor
resistor de nitruro de tantalio tantalum-nitride resistor
resistor de óxido de estaño tin-oxide resistor
resistor de pantalla screen resistor
resistor de pastilla pellet resistor
resistor de película film resistor
resistor de película de carbón carbon-film resistor
resistor de película de metal-óxido metal-oxide film resistor
resistor de película delgada thin-film resistor
resistor de película gruesa thick-film resistor
resistor de película metálica metallic-film resistor
resistor de placa plate resistor
resistor de polarización bias resistor
resistor de polarización catódica cathode-bias resistor
resistor de polarización de rejilla grid-bias resistor
resistor de ponderación weighting resistor
resistor de potencia power resistor
resistor de precisión precision resistor
resistor de rejilla grid resistor
resistor de silicio silicon resistor
resistor de sustitución substitution resistor
resistor de temporización timing resistor
resistor de Thomas Thomas resistor
resistor de tipo de disco disk-type resistor
resistor de tipo plano flat-type resistor
resistor de tirita thyrite resistor
resistor de vidrio glass resistor
resistor degenerativo degenerative resistor
resistor dependiente de la tensión voltage-dependent resistor
resistor dependiente de luz light-dependent resistor
resistor discreto discrete resistor
resistor disipador de potencia power-dissipating resistor
resistor ecualizador equalizing resistor
resistor electrolítico electrolytic resistor
resistor electrónico electronic resistor
resistor en cilindro drum resistor
resistor en chip chip resistor
resistor en derivación shunt resistor
resistor enchufable plug-in resistor

resistor estabilizador stabilizing resistor, ballast resistor
resistor ficticio dummy resistor
resistor fijo fixed resistor
resistor flexible flexible resistor
resistor flotante floating resistor
resistor fusible fusible resistor, resistor fuse
resistor impreso printed resistor
resistor incorporado built-in resistor
resistor indicador de carga load-indicating resistor
resistor inductivo inductive resistor
resistor integrado integrated resistor
resistor limitador limiting resistor
resistor limitador de corriente current-limiting resistor
resistor lineal linear resistor
resistor líquido liquid resistor
resistor metálico metallic resistor
resistor metalizado metallized resistor
resistor moldeado molded resistor
resistor negativo negative resistor
resistor no aislado noninsulated resistor
resistor no compensado uncompensated resistor
resistor no inductivo noninductive resistor
resistor no lineal nonlinear resistor
resistor patrón standard resistor
resistor por difusión diffused resistor
resistor positivo positive resistor
resistor potenciométrico potentiometer-type resistor
resistor protector protective resistor
resistor químico chemical resistor
resistor reductor de tensión de pantalla screen dropping resistor
resistor sensible a la luz light-sensitive resistor
resistor sensible a la temperatura temperature-sensitive resistor
resistor sensible a la tensión voltage-sensitive resistor
resistor térmico thermal resistor
resistor termométrico thermometer resistor
resistor termosensible heat-sensitive resistor
resistor termovariable thermovariable resistor
resistor tipo arandela washer resistor
resistor-transistor resistor-transistor
resistor variable variable resistor
resistor variable eléctricamente electrically variable resistor
resistor vernier vernier resistor
resistores en paralelo resistors in parallel, parallel resistors
resistores en paralelo-serie resistors in parallel-series
resistores en serie resistors in series, series resistors
resistores en serie-paralelo resistors in series-parallel
resnatrón resnatron
resolución resolution
resolución angular angular resolution
resolución de acimut azimuth resolution
resolución de altura de impulsos pulse-height resolution
resolución de cuadrante dial resolution
resolución de distancia distance resolution
resolución de energía energy resolution
resolución de exploración scanning resolution
resolución de fase phase resolution
resolución de frecuencia frequency resolution
resolución de gráficos graphics resolution
resolución de imagen picture resolution
resolución de impulsos pulse resolution
resolución de lectura readout resolution
resolución de radar radar resolution
resolución de rayos gamma gamma-ray resolution

resolución de rumbo bearing resolution
resolución de visualización display resolution
resolución de visualización gráfica graphic display resolution
resolución estructural structural resolution
resolución horizontal horizontal resolution
resolución infinita infinite resolution
resolución límite limiting resolution
resolución óptica optical resolution
resolución vertical vertical resolution
resolutor resolver
resolutor eléctrico electrical resolver
resolvedor resolver
resolvedor angular angular resolver
resolvedor de ecuaciones equation solver
resolver resolve
resonador resonator
resonador abierto open resonator
resonador acústico acoustic resonator
resonador agrupador buncher
resonador anódico anode resonator
resonador anular annular resonator
resonador bivalvo shell circuit
resonador compuesto compound resonator
resonador concéntrico concentric resonator
resonador confocal confocal resonator
resonador de cavidad cavity resonator
resonador de cavidad cilíndrica cylindrical-cavity resonator
resonador de cavidad de entrada buncher resonator
resonador de cavidad de microondas microwave cavity resonator
resonador de cristal crystal resonator
resonador de cristal de cuarzo quartz-crystal resonator
resonador de cuarzo quartz resonator
resonador de diapasón tuning-fork resonator
resonador de entrada input resonator
resonador de guíaondas waveguide resonator
resonador de haz beam resonator
resonador de Helmholtz Helmholtz resonator
resonador de hilos paralelos parallel-wire resonator
resonador de línea coaxial coaxial-line resonator
resonador de línea de transmisión transmission-line resonator
resonador de mariposa butterfly resonator
resonador de microondas microwave resonator
resonador de Oudin Oudin resonator
resonador de placa de cuarzo quartz-plate resonator
resonador de salida output resonator
resonador de sol naciente rising-sun resonator
resonador de tipo de cavidad cavity-type resonator
resonador interdigital interdigital resonator
resonador magnetoestrictivo magnetostrictive resonator
resonador multimodo multimode resonator
resonador piezoeléctrico piezoelectric resonator
resonancia resonance
resonancia acústica acoustic resonance
resonancia antiferromagnética antiferromagnetic resonance
resonancia armónica harmonic resonance
resonancia atómica atomic resonance
resonancia de amplitud amplitude resonance
resonancia de antena antenna resonance
resonancia de caja enclosure resonance
resonancia de cavidad cavity resonance
resonancia de cono cone resonance
resonancia de cuarto de onda quarter-wave resonance
resonancia de doble pico double-hump resonance
resonancia de fase phase resonance
resonancia de la tierra earth resonance

resonancia de onda de espín spin-wave resonance
resonancia de plasma plasma resonance
resonancia de precesión precession resonance
resonancia de presión pressure resonance
resonancia de sala room resonance
resonancia de tensión voltage resonance
resonancia de velocidad velocity resonance
resonancia del espín electrónico electron spin resonance
resonancia electroacústica electroacoustic resonance
resonancia electrónica electron resonance
resonancia en paralelo parallel resonance
resonancia en serie series resonance
resonancia ferromagnética ferromagnetic resonance
resonancia fundamental fundamental resonance
resonancia giromagnética gyromagnetic resonance
resonancia inversa inverse resonance
resonancia longitudinal longitudinal resonance
resonancia magnética magnetic resonance
resonancia magnética electrónica electron magnetic resonance
resonancia magnética nuclear nuclear magnetic resonance
resonancia magnética protónica proton magnetic resonance
resonancia microfónica microphonic resonance
resonancia múltiple multiple resonance
resonancia natural natural resonance
resonancia óptica optical resonance
resonancia paramagnética paramagnetic resonance
resonancia paramagnética electrónica electron paramagnetic resonance
resonancia paramagnética electrónica electronic paramagnetic resonance
resonancia paramétrica parametric resonance
resonancia periódica periodic resonance
resonancia selectiva selective resonance
resonancia vibracional vibrational resonance
resonante resonating, resonant
resonar resonate
resonistor resonistor
resorte de retención retaining spring
respaldado por computadora computer backed
respaldo backup, backing
respaldo de impresora printer backup
respaldo en línea on-line backup
respaldo local local backup
respaldo remoto remote backup
respiradero vent
resplandor glare
resplandor incomodante discomfort glare
resplandor indirecto indirect glare
resplandor reflejado reflected glare
respondedor responder, transponder, replier
respondedor aerotransportado airborne replier
respondedor de radar radar responder
respondedor de radar aerotransportado airborne radar responder
responsividad responsivity
responsividad espectral spectral responsivity
respuesta response, replay
respuesta a presión pressure response
respuesta acústica de campo libre free-field acoustic response
respuesta al escalón step response
respuesta aleatoria random response
respuesta amplitud-frecuencia amplitude-frequency response
respuesta ancha broad response
respuesta armónica harmonic response
respuesta axial axial response
respuesta comprimida compressed response
respuesta de actividad activity response

respuesta de amplificador amplifier response
respuesta de amplitud amplitude response
respuesta de amplitud lineal linear amplitude response
respuesta de audio audio response
respuesta de audiofrecuencia audio-frequency response
respuesta de banda angosta narrowband response
respuesta de banda pasante passband response
respuesta de campo difuso diffuse-field response
respuesta de campo libre free-field response
respuesta de color color response
respuesta de corriente constante constant-current response
respuesta de corriente de campo libre free-field current response
respuesta de doble pico double-hump response
respuesta de excepción exception response
respuesta de excitación excitation response
respuesta de fase phase response
respuesta de fondo background response
respuesta de fotocátodo photocathode response
respuesta de frecuencia frequency response
respuesta de frecuencia acústica acoustic frequency response
respuesta de frecuencia de banda base baseband frequency response
respuesta de frecuencia de plena potencia full-power frequency response
respuesta de frecuencia horizontal horizontal frequency response
respuesta de frecuencia plana flat frequency response
respuesta de frecuencia uniforme uniform frequency response
respuesta de frecuencia vertical vertical frequency response
respuesta de frenaje braking response
respuesta de gama ancha wide-range response
respuesta de Gauss Gaussian response
respuesta de imagen única single-image response
respuesta de impulsos pulse response
respuesta de incidencia aleatoria random-incidence response
respuesta de la piel galvánica galvanic skin response
respuesta de ley cuadrática square-law response
respuesta de llamada call response
respuesta de onda cuadrada square-wave response
respuesta de paso de banda bandpass response
respuesta de potencia disponible available power response
respuesta de radar radar response
respuesta de receptor receiver response
respuesta de reverberación reverberation response
respuesta de tensión voltage response
respuesta de tensión de campo libre free-field voltage response
respuesta de tensión de sistema de excitación excitation system voltage response
respuesta de transmisión transmission response
respuesta de video video response
respuesta dinámica dynamic response
respuesta direccional directional response
respuesta en alta frecuencia high-frequency response
respuesta en baja frecuencia low-frequency response
respuesta en cortocircuito short-circuit response
respuesta en función del tiempo time response
respuesta en presión pressure response
respuesta en radiofrecuencia radio-frequency response
respuesta espectral spectral response

respuesta espectral absoluta absolute spectral response
respuesta espectral de fotocátodo absoluta absolute photocathode spectral response
respuesta espuria spurious response
respuesta extraña extraneous response
respuesta forzada forced response
respuesta fotovoltaica photovoltaic response
respuesta geoeléctrica geoelectric response
respuesta imagen image response
respuesta indeseada undesired response
respuesta indicial indicial response
respuesta inmediata immediate response
respuesta lineal linear response
respuesta logarítmica logarithmic response
respuesta máxima peak response
respuesta negativa negative response
respuesta no lineal nonlinear response
respuesta no óhmica nonohmic response
respuesta normal normal response
respuesta normalizada normalized response
respuesta óhmica ohmic response
respuesta parcial partial response
respuesta plana flat response
respuesta positiva positive response
respuesta propagada propagated response
respuesta proporcional proportional response
respuesta rápida rapid response
respuesta relativa relative response
respuesta residual residual response
respuesta retardada delayed response
respuesta tonal tonal response
respuesta transitoria transient response
respuesta unidireccional unidirectional response
respuesta uniforme smooth response
respuesta variable variable response
resta subtraction
restablecimiento resetting, reset
restablecimiento a cero zero resetting
restablecimiento de barrido sweep reset
restablecimiento de sistema system reset
restador subtracter
restador analógico analog subtracter
restador completo en paralelo parallel full-subtracter
restador completo en serie serial full-subtracter
restador de tres entradas three-input subtracter
restador de un dígito one-digit subtracter
restador digital digital subtracter
restador electrónico electronic subtracter
restador-sumador subtracter-adder
restauración restoration
restauración de carga load restoration
restauración de corriente continua direct-current restoration
restauración de imagen image restoration
restauración de línea base baseline restoration
restaurador restorer
restaurador de azul blue restorer
restaurador de canales channel restorer
restaurador de corriente continua direct-current restorer
restaurador de verde green restorer
restaurar restore
restitución restitution
resultado intermedio intermediate result
resultado parcial partial result
resultante resultant
retardador retarder
retardante delaying
retardo delay, lag
retardo absoluto absolute delay
retardo acústico acoustic delay

retardo de altitud altitude delay
retardo de arranque start delay
retardo de barrido sweep delay
retardo de código code delay
retardo de conexión on-delay
retardo de contestación answering delay
retardo de corriente current lag
retardo de chispa spark lag
retardo de desconexión off-delay
retardo de destello flash delay
retardo de envolvente envelope delay
retardo de fase phase delay, phase lag
retardo de filamento filament lag
retardo de frecuencia central center-frequency delay
retardo de grupo group delay
retardo de ignición ignition delay
retardo de impulso de transductor transducer pulse delay
retardo de impulsos pulse delay
retardo de liberación release delay
retardo de línea line delay
retardo de línea base baseline delay
retardo de memoria intermedia buffer delay
retardo de modo diferencial differential-mode delay
retardo de operación operate delay, operate lag
retardo de propagación propagation delay
retardo de radiofaro beacon delay
retardo de restitución restitution delay
retardo de señal signal delay
retardo de sondeo polling delay
retardo de tensión voltage lag
retardo de tiempo time delay, time lag
retardo de tiempo de señal signal time delay
retardo de tiempo definido definite-time delay
retardo de tono de marcar dial-tone delay
retardo de transmisión transmission delay
retardo de velocidad velocity lag
retardo diferencial differential delay
retardo eléctrico electric delay
retardo externo external delay
retardo magnético magnetic delay
retardo medio average delay
retardo por distorsión distortion delay
retardo por trayectorias múltiples multipath delay
retardo relativo relative delay
retardo variable variable delay
retén detent
retención retention, hold
retención automática automatic hold
retención de capacidad capacity retention
retención de carga charge retention, charge holding
retención de corriente constante constant-current retention
retención de imagen image retention
retención en memoria memory retention
retención en reposo lock-down
retención en trabajo lockup
retención manual manual hold
retentividad retentivity
reticulado reticulated
retículo reticle, grating, graticule
retículo acústico acoustic grating
retículo de tipo de letras font reticle
retículo espacial space lattice
retículo espacial ultrasónico ultrasonic space grating
retículo radial radial grating
retículo ultrasónico ultrasonic grating
retorno return, retrace, flyback, kickback
retorno a cero return to zero
retorno a tierra ground return
retorno aislado insulated return
retorno común common return

retorno de carro carriage return
retorno de filamento filament return
retorno de línea line return
retorno de rejilla grid return
retorno horizontal horizontal retrace, horizontal flyback
retorno por carril rail return
retorno por tierra earth return
retorno terrestre land return
retorno vertical vertical retrace, vertical flyback
retractable retractable
retransmisión retransmission, relay
retransmisión automática automatic retransmission
retransmisión de mensajes message retransmission
retransmisión de televisión television relay
retransmisión manual manual retransmission
retransmisión por cinta tape retransmission
retransmisión por cinta cortada torn-tape relay
retransmisión por líneas terrestres landline relay
retransmisor retransmitter
retransmisor automático automatic retransmitter
retransmisor de cinta perforada perforated-tape retransmitter
retrazo retrace
retroacción retroaction, feedback
retroacoplamiento back coupling
retroactivamente retroactively
retroactivo retroactive
retroalimentación feedback
retroalimentación acústica acoustic feedback
retroalimentación capacitiva capacitive feedback
retroalimentación correctiva corrective feedback
retroalimentación corriente-tensión current-voltage feedback
retroalimentación de arrastre carry feedback
retroalimentación de audiofrecuencia audio-frequency feedback
retroalimentación de bucle único single-loop feedback
retroalimentación de capacitancia capacitance feedback
retroalimentación de corriente current feedback
retroalimentación de información information feedback
retroalimentación de mensajes message feedback
retroalimentación de modulación de frecuencia frequency-modulation feedback
retroalimentación de posición position feedback
retroalimentación de salida output feedback
retroalimentación de señal de interrupción interrupt signal feedback
retroalimentación de tensión voltage feedback
retroalimentación de tensión negativa negative voltage feedback
retroalimentación degenerativa degenerative feedback
retroalimentación en derivación shunt feedback
retroalimentación en fase in-phase feedback
retroalimentación en puente bridge feedback
retroalimentación en serie series feedback
retroalimentación estabilizada stabilized feedback
retroalimentación estabilizadora stabilizing feedback
retroalimentación externa external feedback
retroalimentación global overall feedback
retroalimentación inductiva inductive feedback
retroalimentación inherente inherent feedback
retroalimentación inversa inverse feedback
retroalimentación local local feedback
retroalimentación magnética magnetic feedback
retroalimentación medida measured feedback
retroalimentación multietapa multistage feedback
retroalimentación negativa negative feedback

retroalimentación por transformador transformer feedback
retroalimentación positiva positive feedback
retroalimentación primaria primary feedback
retroalimentación regenerativa regenerative feedback
retroalimentación sin amplificar unamplified feedback
retroalimentación tensión-corriente voltage-current feedback
retroarco arcback
retroceso rápido fast reverse
retrodirectivo retrodirective
retrodispersión backscatter, backscattering, retrodispersion
retrodispersión indirecta indirect backscatter
retroemisión backemission
retrógrado retrograde
retroimpulso backswing
retromodificación retrofit
retropolarización back bias
retrorreflector retroreflector
retrorreflejante retroreflecting
retrorreflexión retroreflection
reubicable relocatable
reubicación relocation
reubicar relocate
reusable reusable
reverberación reverberation, boom
reverberador reverberator
reverberante reverberant
reverberar reverberate
reverberatorio reverberatory
reversibilidad reversibility
reversible reversible
reversión reversion
revertir revert
revestido de cera wax-coated
revestido de óxido oxide-coated
revestido de plomo lead-coated
revestidor liner
revestimiento coat, coating, sheath, sheathing, liner, lining
revestimiento antirreflexión antireflection coating
revestimiento catódico cathode coating
revestimiento conductor conductive coating
revestimiento de bobina coil serving
revestimiento de oro gold coating
revestimiento de óxido oxide coating
revestimiento de rodio rhodium coating
revestimiento dieléctrico dielectric coating
revestimiento exterior outer coating
revestimiento homogéneo homogeneous cladding
revestimiento magnético magnetic coating
revestimiento por inmersión dip coating
revestimiento primario primary coating
revestimiento protector protective coating
revestimiento sensible sensitive lining
revisión de alarma alarm checking
revisión de diseño design review
revisión de requisitos requirements review
revolución revolution
rho rho
riesgo de choque eléctrico shock hazard
riesgo de radiación radiation hazard
rigidez dieléctrica dielectric rigidity, dielectric strength
rigidez dieléctrica específica specific dielectric strength
rigidez eléctrica electric rigidity, electric strength
rigidez eléctrica intrínseca intrinsic electric strength
rigidez magnética magnetic rigidity
rígido rigid

riómetro riometer
riostra guy wire
riotrón ryotron
rítmico rhythmic
ritmo de crecimiento rate of growth
ritmo de cuenta de fondo background count rate
ritmo de interrogación rate of interrogation
ritmo real actual rhythm
robo de ciclos cycle stealing
robot robot
robótica robotics
robótico robotic
robotización robotization
robustez robustness
rodamiento rolling
rodiado rhodium-plated
rodillo platen
rodillo de alimentación de papel paper-feed roller
rodillo de presión pressure roller
rodio rhodium
roentgen roentgen
roentgenografía roentgenography
roentgenograma roentgenogram
roentgenómetro roentgenometer
roentgenoscopía roentgenoscopy
roentgenoscopio roentgenoscope
rojo-verde-azul red-green-blue
rollo roll
rómbico rhombic
rompimiento muerto dead break
ronquido rumble
ronquido de plato giratorio turntable rumble
rosca izquierda left-hand thread
roseta rosette
rosetón rosette
rotación rotation
rotación de dominio domain rotation
rotación de Faraday Faraday rotation
rotación de Faraday magnetoiónica magnetoionic Faraday rotation
rotación de fase phase rotation
rotación de imagen picture rotation
rotación de trazo trace rotation
rotación específica specific rotation
rotación magnética magnetic rotation
rotación magnetoóptica magnetooptical rotation
rotación molecular molecular rotation
rotación reversible reversible rotation
rotacional rotational
rotador rotator
rotador de Faraday Faraday rotator
rotador de ferrita ferrite rotator
rotámetro rotameter
rotar rotate
rotativo rotary, rotatory, rotating
rotatomotriz rotatomotive
rotatorio rotary, rotatory, rotating
rotor rotor
rotor bobinado wound rotor
rotor de alta impedancia high-impedance rotor
rotor de alta reactancia high-reactance rotor
rotor de alta resistencia high-resistance rotor
rotor de anillos deslizantes slip-ring rotor
rotor de ranuras radiales radial-slot rotor
rotor de volante flywheel rotor
rotor en jaula de ardilla squirrel-cage rotor
rozamiento friction
rubidio rubidium
rueda de caracteres character wheel
rueda de espejos mirror wheel
rueda de impresión print wheel
rueda de transferencia idler wheel
rueda en margarita daisy wheel

rueda fónica phonic wheel, tonewheel
ruedecilla de trole trolley wheel
ruido noise
ruido acústico acoustic noise
ruido aéreo airborne noise
ruido aleatorio random noise
ruido aleatorio de Gauss Gaussian random noise
ruido aleatorio triangular triangular random noise
ruido ambiente ambient noise
ruido atmosférico atmospheric noise
ruido audible audible noise
ruido básico basic noise
ruido blanco white noise
ruido celeste sky noise
ruido cíclico cyclic noise
ruido continuo continuous noise
ruido cósmico cosmic noise
ruido cuántico quantum noise
ruido cuasiimpulsivo quasi-impulsive noise
ruido de absorción atmosférica atmospheric absorption noise
ruido de aguja needle noise, needle scratch
ruido de amplificador amplifier noise
ruido de amplificador de radiofrecuencia radio-frequency amplifier noise
ruido de amplitud amplitude noise
ruido de ángulo angle noise
ruido de antena antenna noise
ruido de aparato set noise
ruido de audiofrecuencia audio-frequency noise
ruido de avalancha avalanche noise
ruido de banda ancha wideband noise
ruido de canal vacante idle channel noise
ruido de canoa motorboating
ruido de cavitación cavitation noise
ruido de circuito circuit noise
ruido de conmutación switching noise
ruido de contacto contact noise
ruido de convertidor converter noise
ruido de corriente current noise
ruido de corriente alterna alternating-current noise
ruido de corriente continua direct-current noise
ruido de cuantificación quantizing noise
ruido de diafonía crosstalk noise
ruido de efecto de granalla shot-effect noise
ruido de equipo equipment noise
ruido de fase phase noise
ruido de fluctuación fluctuation noise
ruido de fluctuación de velocidad velocity-fluctuation noise
ruido de fondo background noise
ruido de fondo eléctrico electric background noise
ruido de Gauss Gaussian noise
ruido de generador generator noise
ruido de granalla shot noise
ruido de haz beam noise
ruido de ignición ignition noise
ruido de impulsos pulse noise
ruido de impulsos extraños extraneous impulse noise
ruido de inducción induction noise
ruido de intermodulación intermodulation noise
ruido de Jansky Jansky noise
ruido de Johnson Johnson noise
ruido de línea line noise
ruido de mezclador mixer noise
ruido de micrófono microphone noise
ruido de modo común common-mode noise
ruido de modo normal normal-mode noise
ruido de modulación modulation noise
ruido de modulación cruzada cross-modulation noise
ruido de modulación de amplitud

amplitude-modulation noise
ruido de modulación de frecuencia
  frequency-modulation noise
ruido de motor motorboating, rumble
ruido de palomitas de maíz popcorn noise
ruido de parpadeo flicker noise
ruido de partición partition noise
ruido de portadora carrier noise
ruido de propagación path distortion noise
ruido de radar radar noise
ruido de radio radio noise
ruido de radio ambiental environmental radio noise
ruido de radio ambiente ambient radio noise
ruido de radio de banda ancha broadband radio
  noise
ruido de radio de banda angosta narrowband radio
  noise
ruido de radio de modo diferencial
  differential-mode radio noise
ruido de radio galáctico galactic radio noise
ruido de radio natural natural radio noise
ruido de radio radiado radiated radio noise
ruido de radio terrestre terrestrial radio noise
ruido de receptor receiver noise
ruido de rectificador rectifier noise
ruido de red mains noise
ruido de referencia reference noise
ruido de resistencia resistance noise
ruido de retardo de grupo group-delay noise
ruido de sala room noise
ruido de saturación saturation noise
ruido de Schottky Schottky noise
ruido de seguimiento angular angle tracking noise
ruido de sistema system noise
ruido de superficie surface noise
ruido de transformador transformer noise
ruido de transistor transistor noise
ruido de transmisor transmitter noise
ruido de tubo tube noise
ruido de válvula valve noise
ruido de video video noise
ruido eléctrico electrical noise
ruido eléctrico aleatorio random electrical noise
ruido eléctrico de banda ancha broadband electrical
  noise
ruido electromagnético electromagnetic noise
ruido en cascada cascade noise
ruido entre estaciones interstation noise
ruido errático erratic noise
ruido espurio spurious noise
ruido estelar stellar noise
ruido excedente excess noise
ruido externo external noise
ruido extraño extraneous noise
ruido galáctico galactic noise
ruido generado generated noise
ruido impulsivo impulsive noise
ruido inducido induced noise
ruido inherente inherent noise
ruido interno internal noise
ruido intrínseco intrinsic noise
ruido iónico ion noise
ruido local local noise
ruido mecánico mechanical noise, rumble
ruido mecánico extraño extraneous mechanical noise
ruido medio average noise
ruido metálico metallic noise
ruido microfónico microphonic noise
ruido modal modal noise
ruido natural natural noise
ruido parásito parasitic noise
ruido periódico periodic noise
ruido ponderado weighted noise

ruido por agitación térmica thermal-agitation noise
ruido por flexión flexural noise
ruido por gas gas noise
ruido por impacto impact noise
ruido por ionización ionization noise
ruido por manchas solares sunspot noise
ruido por retardo de alimentador feeder delay noise
ruido puntual spot noise
ruido radiado radiated noise
ruido residual residual noise
ruido rosado pink noise
ruido seudoaleatorio pseudorandom noise
ruido sofométrico psophometric noise
ruido solar solar noise
ruido telefónico telephone noise
ruido telegráfico telegraph noise
ruido térmico thermal noise
ruido terrestre terrestrial noise
ruidoso noisy
rumbatrón rhumbatron
rumbo bearing, heading, course
rumbo angular angular bearing
rumbo aparente apparent bearing
rumbo de círculo máximo great circle bearing
rumbo falso false course
rumbo indicado indicated bearing
rumbo magnético magnetic bearing, magnetic
  heading
rumbo múltiple multiple course
rumbo por compás compass bearing
rumbo por radar radar bearing, radar heading
rumbo por radiofaro omnidireccional omnibearing
rumbo real true bearing
rumbo recíproco reciprocal bearing
rumbo relativo relative bearing
rumbo unilateral unilateral bearing
ruptor breaker
ruptura break, breaking, breakdown, rupture
ruptura de aislamiento insulation breakdown
ruptura de gas gas breakdown
ruptura de línea line break
ruptura de secuencia sequence break
ruptura destructiva destructive breakdown
ruptura dieléctrica dielectric breakdown
ruptura eléctrica electrical breakdown
ruptura inversa reverse breakdown
ruptura múltiple multiple break
ruptura por avalancha avalanche breakdown
ruptura por fatiga fatigue breakage
ruptura por tensión voltage breakdown
ruptura retardada delayed break
ruptura secundaria secondary breakdown
ruptura única single break
ruta route, path
ruta alternativa alternate route
ruta auxiliar auxiliary route
ruta de cable cable route
ruta de desbordamiento overflow route
ruta de emergencia emergency route
ruta de enlace trunk route
ruta de microondas microwave route
ruta de radiorrelés radio-relay route
ruta de transmisión transmission route
ruta explícita explicit route
ruta extendida extended route
ruta extendida primaria primary extended route
ruta mixta mixed route
ruta normal normal route
ruta primaria primary route
ruta principal main route
ruta secundaria secondary route
ruta terrestre overland route
ruta virtual virtual route

rutenio ruthenium
rutherford rutherford
rutina routine
rutina abierta open routine
rutina algorítmica algorithmic routine
rutina almacenada stored routine
rutina alternativa alternate routine
rutina automática automatic routine
rutina auxiliar auxiliary routine
rutina cerrada closed routine
rutina compartida shared routine
rutina completa complete routine
rutina condensadora condensing routine
rutina correctora de errores error-correcting routine
rutina de acceso mínimo minimum-access routine
rutina de acción action routine
rutina de actualización updating routine
rutina de análisis analysis routine
rutina de análisis de aislamiento de falla fault isolation analysis routine
rutina de análisis de error error analysis routine
rutina de apertura de archivo file-opening routine
rutina de aplicación application routine
rutina de apoyo operacional utility routine
rutina de autocarga self-loading routine
rutina de autocomprobación self-checking routine
rutina de ayuda en depuración debugging aid routine
rutina de biblioteca library routine
rutina de búsqueda fetch routine
rutina de carga bootstrap routine, bootstrap, loading routine
rutina de carga de sistema operativo de disco disk operating system booting
rutina de carga fría cold boot
rutina de cargador loader routine
rutina de clasificación sorting routine
rutina de coma flotante floating-point routine
rutina de compilación compiling routine
rutina de comprobación checking routine
rutina de comprobación de errores error checking routine
rutina de comprobación de secuencia sequence-checking routine
rutina de computadora computer routine
rutina de contabilidad accounting routine
rutina de continuar tras desastre disaster continue routine
rutina de control control routine
rutina de control maestro master control routine
rutina de conversión conversion routine
rutina de corrección patch routine
rutina de corrección de errores error-correction routine
rutina de depuración debugging routine
rutina de descodificación decoding routine
rutina de detección de errores error-detection routine
rutina de diagnóstico diagnostic routine
rutina de edición edit routine
rutina de ejecución execution routine
rutina de ensamblaje assembly routine
rutina de entrada input routine
rutina de entrada-salida input-output routine
rutina de error error routine
rutina de estado de canal channel status routine
rutina de fin de archivo end-of-file routine
rutina de fin de cinta end-of-tape routine
rutina de fusión merge routine
rutina de generación generation routine
rutina de inserción directa direct-insert routine
rutina de instalación installation routine

rutina de intercambio swapping routine
rutina de intercepción de impresión print intercept routine
rutina de interfaz interface routine
rutina de justificación justification routine
rutina de latencia mínima minimum latency routine
rutina de llamada call routine
rutina de mal funcionamiento malfunction routine
rutina de manejo de bloques block handling routine
rutina de manipulación de archivo file-handling routine
rutina de mantenimiento maintenance routine
rutina de mensajes message routine
rutina de normalización normalization routine
rutina de preparación housekeeping routine
rutina de procesamiento de etiquetas label processing routine
rutina de procesamiento de interrupción interrupt processing routine
rutina de prueba de referencia benchmark routine
rutina de pruebas testing routine
rutina de puntos de comprobación checkpoint routine
rutina de rastreo tracing routine
rutina de reanudación restart routine
rutina de recuperación recovery routine
rutina de recuperación de errores error recovery routine
rutina de redondeo rounding routine
rutina de reejecución rerun routine
rutina de reinicio restart routine
rutina de retorno return routine
rutina de salida exit routine, output routine
rutina de salto jump routine
rutina de servicio service routine
rutina de servicio de interrupción interrupt service routine
rutina de sistema ejecutivo executive system routine
rutina de sondeo polling routine
rutina de tiempo de ejecución run time routine
rutina de trabajo working routine
rutina de usuario user routine
rutina de vaciado de memoria de núcleos core dump routine
rutina detectora de errores error-detecting routine
rutina ejecutiva executive routine
rutina específica specific routine
rutina fija fixed routine
rutina fotográfica photographic routine
rutina generadora generating routine
rutina general general routine
rutina generalizada generalized routine
rutina heurística heuristic routine
rutina incompleta incomplete routine
rutina incorporada built-in routine
rutina interpretadora interpreter routine
rutina interpretativa interpretive routine
rutina iterativa iterative routine
rutina maestra master routine
rutina monitora monitor routine
rutina no residente nonresident routine
rutina objeto target routine
rutina óptima optimum routine
rutina patrón standard routine
rutina post mortem post-mortem routine
rutina principal main routine
rutina recursiva recursive routine
rutina reentrable reenterable routine
rutina reentrante reentrant routine
rutina residente resident routine
rutina reusable reusable routine
rutina simuladora simulator routine
rutina supervisora supervisory routine

rutina transitoria transient routine

S

sabine sabin
sabor eléctrico electrical taste
sabor galvánico galvanic taste
sacafusibles fuse puller
sacudida shock
sacudidor shaker
sal conductora conducting salt
sal de amoniaco sal ammoniac
sal de Rochelle Rochelle salt
sal de Seignette Seignette salt
sala anecoica anechoic room, dead room
sala de computadoras computer room
sala de conmutación switching room
sala de control control room
sala de control central central control room
sala de control maestro master control room
sala de control principal main control room
sala de instrumentos instrument room
sala de máquinas machine room
sala de reverberación reverberation room
sala limpia clean room
sala terminal terminal room
salida output, exit
salida acústica acoustic output
salida aleatoria random output
salida analógica analog output
salida anormal abnormal exit
salida asimétrica asymmetrical output
salida asincrónica asynchronous exit
salida binaria binary output
salida cero zero output
salida controlada controlled output
salida de almacenamiento storage exit
salida de alta impedancia high-impedance output
salida de amplificador amplifier output
salida de audio audio output
salida de bloque block output
salida de canales mixtos mixed-channel output
salida de cero sin perturbación undisturbed-zero output
salida de computadora computer output
salida de contador counter exit
salida de control control output
salida de copia impresa hard-copy output
salida de corriente current output
salida de corriente alterna alternating-current outlet
salida de datos data output
salida de frecuencia de batido beat-frequency output
salida de gráficos graphics output
salida de imagen picture output
salida de impresora printer output
salida de lectura read output
salida de línea line output
salida de nivel cero zero-level output
salida de onda cuadrada square-wave output
salida de página page exit
salida de portadora carrier output
salida de potencia de audio audio power output
salida de potencia de placa plate power output
salida de potencia de portadora carrier power output

salida de potencia de radiofrecuencia radio-frequency power output
salida de potencia instantánea instantaneous power output
salida de potencia media average power output
salida de potencia normal normal power output
salida de potencia radiada radiated power output
salida de potencia sin distorsión undistorted power output
salida de programa program exit
salida de radiofrecuencia radio-frequency output
salida de receptor receiver output
salida de ruido noise output
salida de selección parcial partial-select output
salida de señal signal output
salida de sistema system output
salida de suma add output
salida de tarjetas card output
salida de trabajos job output
salida de tubo tube output
salida de uno sin perturbación undisturbed-one output
salida de video video output
salida de videofrecuencia video-frequency output
salida desequilibrada unbalanced output
salida diferida deferred exit
salida digital digital output
salida directa direct output
salida discreta discrete output
salida en paralelo parallel output
salida en serie serial output
salida en tiempo real real-time output
salida equilibrada balanced output
salida específica specific output
salida espectral spectral output
salida espuria spurious output
salida fantasma phantom output
salida filtrada filtered output
salida flotante floating output
salida gráfica graphic output
salida luminosa luminous output
salida máxima maximum output
salida máxima de potencia maximum power output
salida máxima de potencia de señal maximum signal power output
salida máxima de potencia media maximum average power output
salida máxima de potencia sin distorsión maximum undistorted power output
salida máxima sin distorsión maximum undistorted output
salida máxima útil maximum useful output
salida momentánea momentary output
salida no impresa soft copy
salida nominal nominal output, rated output
salida pulsada pulsed output
salida rectificada rectified output
salida requerida required output
salida retardada delayed output
salida secuencial sequential output
salida simétrica symmetrical output
salida sin distorsión undistorted output
salida sin percances graceful exit
salida uno one output
salida útil useful output
salida variable variable output
salida vertical vertical output
salir del sistema log off, log out
salir del sistema log out
salto jump, skip
salto condicional conditional jump
salto corto short skip
salto cuántico quantum jump

salto de arco arcover
salto de arco de aislador insulator arcover
salto de cinta tape skip
salto de energía energy gap
salto de modo mode jump
salto de página page skip
salto de papel paper skip
salto de tensión voltage jump
salto de tensión negativo negative voltage jump
salto horizontal de imagen picture weave
salto incondicional unconditional jump
salto por desbordamiento overflow skip
salto único single hop
salto vertical de imagen picture jump
saltos de Barkhausen Barkhausen jumps
samario samarium
sanafán sanafant
sanatrón sanatron
sangrado de tinta ink bleed
satélite satellite
satélite activo active satellite
satélite artificial artificial satellite
satélite de comunicaciones communications satellite
satélite de comunicaciones activo active communications satellite
satélite de comunicaciones pasivo passive communications satellite
satélite de comunicaciones sincrónicas synchronous communications satellite
satélite de navegación navigation satellite
satélite de telecomunicaciones telecommunications satellite
satélite estacionario stationary satellite
satélite geodésico geodesic satellite
satélite geoestacionario geostationary satellite
satélite meteorológico weather satellite
satélite navegacional navigational satellite
satélite no sincrónico nonsynchronous satellite
satélite pasivo passive satellite
satélite reflector reflector satellite
satélite relé relay satellite
satélite repetidor retardado delayed repeater satellite
satélite sincrónico synchronous satellite
satélite subsincrónico subsynchronous satellite
satélite supersincrónico supersynchronous satellite
satélite urbano urban satellite
saturable saturable
saturación saturation
saturación anódica anode saturation
saturación de blanco white saturation
saturación de color color saturation
saturación de corriente current saturation
saturación de corriente de placa plate-current saturation
saturación de emisión emission saturation
saturación de filamento filament saturation
saturación de negro black saturation
saturación de núcleo core saturation
saturación de placa plate saturation
saturación de temperatura temperature saturation
saturación de tensión voltage saturation
saturación magnética magnetic saturation
saturado saturated
saturador saturator
saturante saturating
saturar saturate
secam secam
secante secant
secante de arco arc secant
secante hiperbólica hyperbolic secant
sección section
sección aislada insulated section

sección aritmética arithmetic section
sección compresible squeeze section
sección controladora controlling section
sección cortocircuitadora short-circuiting section
sección crítica critical section
sección de adaptación matching section
sección de agrupamiento bunching section
sección de alimentación supply section
sección de archivo file section
sección de cable coaxial coaxial-cable section
sección de computadoras computer section
sección de configuración configuration section
sección de conmutación switching section
sección de control control section
sección de control común common control section
sección de control de datos data control section
sección de control de microprograma
    microprogram control section
sección de control ficticia dummy control section
sección de control global global control section
sección de controlador controller section
sección de depuración debugging section
sección de entrada input section
sección de entrada-salida input-output section
sección de filtro filter section
sección de grupo group section
sección de grupo maestro mastergroup section
sección de guíaondas waveguide section
sección de línea line section
sección de línea digital digital line section
sección de modulación modulation section
sección de perforación punching section
sección de procesamiento processing section
sección de prueba test section
sección de radio radio section
sección de repetición principal main repeating
    section
sección de reserva reserve section
sección de salida output section
sección de separación phase break
sección de sincronización sync section
sección de sonido sound section
sección de supergrupo supergroup section
sección de tipo de letras font section
sección de transformación transforming section
sección de transición transition section
sección de transposición transposition section
sección digital digital section
sección eficaz de absorción absorption cross section
sección eficaz de radar radar cross-section
sección en celosía lattice section
sección excitada driven section
sección ficticia dummy section
sección funcional functional section
sección intermedia intermediate section
sección pentodo pentode section
sección pi pi section
sección principal main section
sección progresiva tapered section
sección ranurada slotted section
sección terminal end section
sección transversal cross-section
sección transversal biestática bistatic cross-section
sección transversal de antena antenna cross-section
sección transversal de dispersión scattering
    cross-section
sección transversal de retrodispersión
    backscattering cross-section
sección transversal de Thomson Thomson
    cross-section
sección triodo triode section
sección vertical vertical section
seccionado chopping

seccionador sectionalizer, section switch,
    disconnecting switch, switch, chopper
seccionador de línea automático automatic line
    sectionalizer
seccionador exterior outdoor disconnecting switch
seccionador primario primary disconnecting switch
seccionador principal main disconnect switch
seccional sectional
seccionamiento sectioning
seco dry
sector sector
sector alternativo alternative sector
sector ciego blind sector
sector de control de biblioteca library control sector
sector de disco disk sector
sector de disco flexible floppy-disk sector
sector de exploración scanning sector
sector de inclinación de planeo glide-slope sector
sector de linealidad linearity sector
sector de trayectoria de planeo glide-path sector
sector defectuoso bad sector
sector encadenado chained sector
sector equiseñal equisignal sector
sector localizador localizer sector
sector muerto dead sector
sectorial sectoral
secuencia sequence, string
secuencia aleatoria random sequence
secuencia alfabética alphabetic string
secuencia binaria binary sequence
secuencia codificada coded sequence
secuencia de abortar abort sequence
secuencia de bits bit string
secuencia de búsqueda search string
secuencia de campo field string
secuencia de caracteres character string
secuencia de caracteres nulos null character string
secuencia de clasificación sort sequence
secuencia de codificación coding sequence
secuencia de colores color sequence
secuencia de conteo counting sequence
secuencia de control control sequence
secuencia de datos mixtos mixed data string
secuencia de dígitos binarios binary digit string
secuencia de ejecución execution sequence
secuencia de elementos binarios binary element
    string
secuencia de entrada entry sequence
secuencia de exploración scanning sequence
secuencia de fases phase sequence
secuencia de identificación identification sequence
secuencia de impulsos pulse sequence
secuencia de intercalación collation sequence
secuencia de longitud fija fixed-length string
secuencia de llamada calling sequence
secuencia de microinstrucciones microinstruction
    sequence
secuencia de muestreo de colores color sampling
    sequence
secuencia de números aleatorios random-number
    sequence
secuencia de operaciones operation sequence
secuencia de prioridades priority sequence
secuencia de procesamiento processing sequence
secuencia de procesamiento de datos
    data-processing sequence
secuencia de procesamiento secundario secondary
    processing sequence
secuencia de programa program sequence
secuencia de proseguir go-ahead sequence
secuencia de pruebas test sequence
secuencia de sentencias sequence of statements
secuencia de trabajo job sequence

secuencia de transmisión transmission sequence
secuencia de visualización primaria primary display sequence
secuencia decimal codificada coded-decimal sequence
secuencia descendiente descending sequence
secuencia en línea inline sequence
secuencia flotante floating string
secuencia forzada forced sequence
secuencia gráfica graphic string
secuencia iterativa iterative sequence
secuencia jerárquica hierarchical sequence
secuencia natural natural sequence
secuencia normal normal sequence
secuencia nula null string
secuencia numérica numeric sequence
secuencia opuesta opposite sequence
secuencia predeterminada predetermined sequence
secuencia primaria primary sequence
secuencia selectiva selective sequence
secuencia unitaria unit string
secuencia vacía empty string
secuenciador sequencer
secuenciador de control control sequencer
secuencial sequential
secuencialmente sequentially
secuenciamiento automático automatic sequencing
secuenciamiento de paquetes packet sequencing
secundario secondary
sedimentación sedimentation
sedimento sediment
segmentación segmentation
segmentación de imagen image segmentation
segmentación de programa program segmentation
segmentado segmented
segmental segmental
segmento segment
segmento compartido shared segment
segmento común common segment
segmento cortocircuitador shorting segment
segmento de base de datos database segment
segmento de cabecera header segment
segmento de cable cable segment
segmento de cable coaxial coaxial-cable segment
segmento de capacitor capacitor segment
segmento de códigos code segment
segmento de colector commutator segment
segmento de datos data segment
segmento de gráficos graphics segment
segmento de indexación indexing segment
segmento de línea line segment
segmento de página page segment
segmento de programa program segment
segmento de recubrimiento overlay segment
segmento de visualización display segment
segmento dependiente dependent segment
segmento detectable detectable segment
segmento en blanco blank segment
segmento extendido extended segment
segmento físico physical segment
segmento fuente source segment
segmento indexado indexed segment
segmento lógico logical segment
segmento primario primary segment
segmento principal main segment
segmento secundario secondary segment
segmentos incluyentes inclusive segments
seguido por radar radar-tracked
seguidor follower
seguidor anódico anode follower
seguidor catódico cathode follower
seguidor de emisor emitter follower
seguidor de fuente source follower

seguidor de señal signal tracer
seguidor de trayectoria automática automatic track follower
seguimiento tracking
seguimiento angular angle tracking
seguimiento automático automatic tracking
seguimiento ayudado aided tracking
seguimiento de Doppler Doppler tracking
seguimiento de fase phase tracking
seguimiento de señal signal tracing
segunda armónica second harmonic
segunda descarga de Townsend second Townsend discharge
segunda generación second generation
segunda ley de Kirchhoff Kirchhoff's second law
segunda ley de termodinámica second law of thermodynamics
segunda ley de Wien Wien's second law
segundo second
segundo ánodo second anode, ultor
segundo de arco arc second
segundo detector second detector
segundo grupo second group
segundo internacional international second
segundo sin error error-free second
seguridad security, safety
seguridad de acceso a recursos resource access security
seguridad de archivos file security
seguridad de base de datos database security
seguridad de computadoras computer security
seguridad de conjunto de datos data set security
seguridad de datos data security
seguridad de programa program security
seguridad de recursos resource security
seguridad de red network security
seguridad de sistema de procesamiento de datos data-processing system security
seguridad de transmisión transmission security
seguridad multinivel multilevel security
selcal selcal
selección selection
selección a larga distancia long-distance selection
selección abreviada abbreviated selection
selección alternativa alternate selection
selección de amplitud amplitude selection
selección de campo field selection
selección de canal channel entry
selección de canal de entrada-salida input-output channel selection
selección de clasificación sort selection
selección de corriente alterna alternating-current selection
selección de chip chip selection
selección de datos data selection
selección de dígitos múltiples multiple digit selection
selección de dirección address selection
selección de estación station selection
selección de formato format selection
selección de frecuencia frequency selection
selección de grupo group selection
selección de impresión print selection
selección de impulsos pulse selection
selección de línea line selection
selección de línea automática automatic line selection
selección de programa program selection
selección de registro register select
selección de ruta automática automatic route selection
selección de señal gating
selección de ubicación de almacenamiento storage location selection

selección de zona zone selection
selección directa direct selection
selección en tándem tandem selection
selección manual manual selection
selección múltiple multiple selection
selección numérica numerical selection
selección paso a paso step-by-step selection
selección por corrientes coincidentes
    coincident-current selection
selección por pulsador push-button selection
seleccionable selectable
seleccionado selected
seleccionar select
selectancia selectance
selectivamente selectively
selectividad selectivity
selectividad de canal adyacente adjacent-channel
    selectivity
selectividad de falda skirt selectivity
selectividad de filtro filter selectivity
selectividad de frecuencia frequency selectivity
selectividad de frecuencia intermedia
    intermediate-frequency selectivity
selectividad de radiofrecuencia radio-frequency
    selectivity
selectividad de receptor receiver selectivity
selectividad direccional directional selectivity
selectividad efectiva effective selectivity
selectividad espectral spectral selectivity
selectividad global overall selectivity
selectividad polifásica polyphase selectivity
selectividad variable variable selectivity
selectividad y desensibilización de canal adyacente
    adjacent-channel selectivity and desensitization
selectivo selective
selector selector
selector anticoincidencia anticoincidence selector
selector de acceso access selector
selector de alcance range selector
selector de altura de impulsos pulse-height selector
selector de amplitud amplitude selector
selector de amplitud de impulsos pulse-amplitude
    selector
selector de banda band selector
selector de bandas laterales sideband selector
selector de barras cruzadas crossbar selector
selector de canales channel selector
selector de código code selector
selector de coincidencia coincidence selector
selector de conmutación switching selector
selector de corrientes por pulsador push-button
    current selector
selector de cuadrante dial selector
selector de datos data selector
selector de dígito A A-digit selector
selector de dígitos digit selector
selector de distancia range selector
selector de diversidad diversity selector
selector de doble control dual-control selector
selector de dos posiciones two-position selector
selector de eco echo selector
selector de energía energy selector
selector de escala scale selector
selector de formato format selector
selector de frecuencias frequency selector
selector de grupo group selector
selector de grupo de línea line group selector
selector de grupo final final group selector
selector de impulsos pulse selector
selector de intervalos de tiempo time-interval
    selector
selector de línea line selector
selector de llegada incoming selector

selector de modo mode selector
selector de movimiento único uniselector
selector de oficina office selector
selector de perforación punch selector
selector de pista track selector
selector de potencia power selector
selector de programa program selector
selector de prueba test selector
selector de puerto port selector
selector de registro register selector
selector de rumbo course selector
selector de salida outgoing selector
selector de señales signal selector
selector de Strowger Strowger selector
selector de tensión voltage selector
selector de trayectoria path selector
selector de velocidad speed selector
selector de zona zone selector
selector discriminante discriminating selector
selector dividido split selector
selector final final selector
selector-multiplexor de datos data
    selector-multiplexer
selector paso a paso step-by-step selector
selector primario primary selector
selector repetidor repeating selector
selector telegráfico telegraph selector
selector unidireccional unidirectional selector
selectrón selectron
selenio selenium
seleniuro de plomo lead selenide
selsyn selsyn
selsyn diferencial differential selsyn
sellado sealing
sellado ambientalmente environmentally sealed
sellado herméticamente hermetically sealed
sellado polióptico polyoptic sealing
semáforo semaphore
semáforo de dirección direction semaphore
semántica semantics
semántico semantic
semator semator
semiacarreo half-carry
semiactivo semiactive
semiajustable semiadjustable
semianchura half-width
semiautomático semiautomatic
semicelda half-cell
semicelda de calomel calomel half-cell
semicelda de quinhidrona quinhydrone half-cell
semicelda de vidrio glass half-cell
semicerrado semienclosed
semiciclo half-cycle, semicycle
semiciclo de reposición resetting half-cycle
semicompilado semicompiled
semiconductible semiconductible
semiconductor semiconductor
semiconductor aceptante accepting semiconductor
semiconductor aceptor acceptor semiconductor
semiconductor amorfo amorphous semiconductor
semiconductor bipolar bipolar semiconductor
semiconductor compensado compensated
    semiconductor
semiconductor de impurezas impurity
    semiconductor
semiconductor de película delgada thin-film
    semiconductor
semiconductor degenerado degenerate
    semiconductor
semiconductor donador donor semiconductor
semiconductor dopado doped semiconductor
semiconductor elemental elemental semiconductor
semiconductor extrínseco extrinsic semiconductor

semiconductor fundido fused semiconductor
semiconductor intrínseco intrinsic semiconductor
semiconductor iónico ionic semiconductor
semiconductor líquido liquid semiconductor
semiconductor metal-óxido
    metal-oxide-semiconductor
semiconductor metal-óxido complementario
    complementary metal-oxide semiconductor
semiconductor mixto mixed semiconductor
semiconductor monocristalino single-crystal
    semiconductor
semiconductor orgánico organic semiconductor
semiconductor tipo N N-type semiconductor
semiconductor tipo P P-type semiconductor
semiconductorizado semiconductorized
semicontinuo semicontinuous
semicristalino semicrystalline
semidipolo half-dipole
semidireccional semidirectional
semidirecto semidirect
semidúplex half-duplex, semiduplex
semiencerrado semienclosed
semiestable semistable
semifrecuencia half-frequency
semilogarítmico semilogarithmic
semimecánico semimechanical
semimetal semimetal
semimetálico semimetallic
semiperiodo half-period
semipila half-cell
semiportátil semiportable
semiproporcional semiproportional
semiprotegido semiprotected
semipuente half-bridge
semirresonante semiresonant
semirrestador half-subtracter
semirrestador en paralelo parallel half-subtracter
semirrestador en serie serial half-subtracter
semirrígido semirigid
semiselectivo semiselective
semisuma half-add
semisumador half-adder
semisumador en paralelo parallel half-adder
semisumador en serie serial half-adder
semitono halftone, semitone
semitransparente semitransparent
semiuniversal semiuniversal
senario senary
sencillo single
seno sine
seno de arco arc sine
seno hiperbólico hyperbolic sine
seno verso versed sine, versine
senoidal sinusoid
sensibilidad sensitivity, sensibility
sensibilidad a la temperatura temperature
    sensitivity
sensibilidad a las vibraciones vibration sensitivity
sensibilidad a presión pressure sensitivity
sensibilidad aleatoria random sensitivity
sensibilidad anódica anode sensitivity
sensibilidad axial axial sensitivity
sensibilidad de aceleración angular
    angular-acceleration sensitivity
sensibilidad de campo difuso diffuse-field sensitivity
sensibilidad de canal channel sensitivity
sensibilidad de colores color sensitivity
sensibilidad de contraste contrast sensitivity
sensibilidad de contraste relativa relative contrast
    sensitivity
sensibilidad de corriente current sensitivity
sensibilidad de corriente de campo libre free-field
    current sensitivity

sensibilidad de deflexión deflection sensitivity
sensibilidad de deflexión electrostática electrostatic
    deflection sensitivity
sensibilidad de desviación deviation sensitivity
sensibilidad de desviación angular angular deviation
    sensitivity
sensibilidad de desviación máxima
    maximum-deviation sensitivity
sensibilidad de entrada input sensitivity
sensibilidad de exploración scanning sensitivity
sensibilidad de fotocátodo photocathode sensitivity
sensibilidad de frecuencia frequency sensitivity
sensibilidad de galvanómetro galvanometer
    sensitivity
sensibilidad de instrumento instrument sensitivity
sensibilidad de límite de escala full-scale sensitivity
sensibilidad de medidor meter sensitivity
sensibilidad de operación operating sensitivity
sensibilidad de potencia power sensibility
sensibilidad de receptor receiver sensitivity
sensibilidad de referencia reference sensitivity
sensibilidad de relé relay sensitivity
sensibilidad de resolución resolution sensitivity
sensibilidad de rumbo bearing sensitivity
sensibilidad de señal tangencial tangential signal
    sensitivity
sensibilidad de silenciamiento quieting sensitivity
sensibilidad de sintonización tuning sensitivity
sensibilidad de sintonización electrónica electronic
    tuning sensitivity
sensibilidad de sintonización térmica thermal tuning
    sensitivity
sensibilidad de tensión voltage sensitivity
sensibilidad de tensión de campo libre free-field
    voltage sensitivity
sensibilidad de velocidad angular angular velocity
    sensitivity
sensibilidad de vibración angular angular vibration
    sensitivity
sensibilidad de vóltmetro voltmeter sensitivity
sensibilidad dinámica dynamic sensitivity
sensibilidad eléctrica electrical sensitivity
sensibilidad espectral spectral sensitivity
sensibilidad estática static sensitivity
sensibilidad fotoeléctrica photoelectric sensitivity
sensibilidad horizontal horizontal sensitivity
sensibilidad incremental incremental sensitivity
sensibilidad límite ultimate sensitivity
sensibilidad local local sensitivity
sensibilidad luminosa luminous sensitivity
sensibilidad luminosa catódica cathode luminous
    sensitivity
sensibilidad luminosa dinámica dynamic luminous
    sensitivity
sensibilidad luminosa estática static luminous
    sensitivity
sensibilidad magnética magnetic sensitivity
sensibilidad máxima maximum sensitivity
sensibilidad máxima de receptor receiver maximum
    sensitivity
sensibilidad monocromática monochromatic
    sensitivity
sensibilidad radiante radiant sensitivity
sensibilidad radiogoniométrica direction-finder
    sensitivity
sensibilidad relativa relative sensibility
sensibilidad tangencial tangential sensitivity
sensibilidad utilizable usable sensitivity
sensibilidad vertical vertical sensitivity
sensibilización sensitization, sensitizing
sensibilizado sensitized
sensibilizador sensitizer
sensibilizar sensitize

sensible sensitive
sensible a la fase phase-sensitive
sensible a la frecuencia frequency-sensitive
sensible a la luz light-sensitive
sensible a la posición position-sensitive
sensible a la temperatura temperature-sensitive
sensible a presión pressure sensitive
sensible a radiación radiation-sensitive
sensible al calor heat-sensitive
sensible al tacto touch-sensitive
sensitómetro sensitometer
sensor sensor
sensor aerotransportado airborne sensor
sensor de arco arc sensor
sensor de cilindro vibrante vibrating cylinder sensor
sensor de cristal crystal sensor
sensor de fin de cinta end-of-tape sensor
sensor de frecuencia variable variable-frequency
  sensor
sensor de gas gas sensor
sensor de hilo de resistencia resistance-wire sensor
sensor de humedad humidity sensor
sensor de imagen image sensor
sensor de impulsos pulse sensor
sensor de intrusión intrusion sensor
sensor de luz light sensor
sensor de moción rotativo rotary-motion sensor
sensor de nivel de líquido liquid-level sensor
sensor de papel paper sensor
sensor de perturbación de campo field disturbance
  sensor
sensor de posición position sensor
sensor de radiación radiation sensor
sensor de radiosonda radiosonde sensor
sensor de selección de frecuencia
  frequency-selection sensor
sensor de temperatura temperature sensor
sensor de temperatura resistivo resistive
  temperature sensor
sensor inteligente smart sensor
sensor iónico ion sensor
sensor magnético remoto remote magnetic sensor
sensor piezoeléctrico piezoelectric sensor
sensor sensible a la luz light-sensitive sensor
sensor ultrasónico ultrasonic sensor
sentencia statement
sentencia anormal abnormal statement
sentencia compuesta compound statement
sentencia condicional conditional statement
sentencia de asignación assignment statement
sentencia de comentarios comment statement
sentencia de condición de error error condition
  statement
sentencia de control control statement
sentencia de control de bucle loop-control statement
sentencia de control de operación operation control
  statement
sentencia de control de trabajos job control
  statement
sentencia de control ejecutivo executive control
  statement
sentencia de datos data statement
sentencia de definición de datos data definition
  statement
sentencia de depuración debugging statement
sentencia de edición edit statement
sentencia de entrada-salida input-output statement
sentencia de especificación specification statement
sentencia de impresión print statement
sentencia de instrucción instruction statement
sentencia de lenguaje language statement
sentencia de mando command statement
sentencia de manejo de patrón pattern-handling

statement
sentencia de parámetro parameter statement
sentencia de preprocesador preprocessor statement
sentencia de procedimiento procedure statement
sentencia de programa program statement
sentencia de programación programming statement
sentencia de segundo nivel second-level statement
sentencia de trabajo job statement
sentencia de transferencia de control control
  transfer statement
sentencia de un problema problem statement
sentencia declarativa declarative statement
sentencia fuente source statement
sentencia imperativa imperative statement
sentencia incondicional unconditional statement
sentencia modelo model statement
sentencia no ejecutable nonexecutable statement
sentencia nula null statement
sentido sense
sentido de portadora carrier sense
señal signal, sign
señal abierta clear signal
señal acústica acoustic signal
señal agregada aggregate signal
señal agregada equilibrada balanced aggregate
  signal
señal aleatoria random signal
señal alfabética alphabetic signal
señal alta activa active-high signal
señal amortiguadora damping signal
señal amplificada amplified signal
señal analógica analog signal
señal audible audible signal
señal audible continua continuous audible signal
señal aural aural signal
señal automática automatic signal
señal baja activa active-low signal
señal bidireccional bidirectional signal
señal binaria binary signal
señal bipolar bipolar signal
señal blanca white signal
señal calibrada calibrated signal
señal cero zero signal
señal codificada coded signal
señal coherente coherent signal
señal compleja complex signal
señal compuesta compound signal
señal condicional conditional signal
señal continua continuous signal
señal controlada gated signal
señal correctora correcting signal
señal cuadrada square signal
señal cuantificada quantized signal
señal cuasialeatoria quasi-random signal
señal cuasianalógica quasi-analog signal
señal de accionamiento actuating signal
señal de aceptación de llamada call-accepted signal
señal de alarma alarm signal
señal de alarma de fuego fire-alarm signal
señal de alarma de incendios manual manual
  fire-alarm signal
señal de alineación de cuadro frame-alignment
  signal
señal de alta fidelidad high-fidelity signal
señal de altitud altitude signal
señal de aproximación approach signal
señal de arranque start signal
señal de arranque de registro start-record signal
señal de atención attention signal
señal de aterrizaje landing signal
señal de audio audio signal
señal de audiofrecuencia audio-frequency signal
señal de aviso warning signal

señal de aviso de falla de alimentación
power-failure warning signal
señal de azul blue signal
señal de bajada downlink
señal de bajo nivel low-level signal
señal de banda ancha broadband signal, wideband
signal
señal de banda base baseband signal
señal de banda lateral residual vestigial-sideband
signal
señal de barrido sweep signal
señal de base de tiempo time-base signal
señal de bloqueo blocking signal
señal de bobina coil signal
señal de borrado erase signal
señal de brillo brightness signal
señal de cabina automática automatic cab signal
señal de calibración calibration signal
señal de cámara camera signal
señal de cambio de estado status-change signal
señal de cancelación canceling signal
señal de carácter character signal
señal de codificación coding signal
señal de código de Morse Morse code signal
señal de colgado on-hook signal
señal de color de portadora carrier color signal
señal de color primario primary-color signal
señal de colores color signal
señal de colores compuesta composite color signal
señal de compensación compensation signal
señal de compensación de sombra
shading-compensation signal
señal de componente único single-component signal
señal de conmutación switching signal
señal de contestación answer signal
señal de contraste de impresión print contrast signal
señal de control control signal
señal de control cíclica cyclic control signal
señal de control por retroalimentación feedback
control signal
señal de control remoto remote-control signal
señal de convergencia dinámica
dynamic-convergence signal
señal de cromaticidad chromaticity signal
señal de crominancia chrominance signal
señal de crominancia de portadora carrier
chrominance signal
señal de crominancia Q Q chrominance signal
señal de cruce de ferrocarril highway crossing
signal
señal de datos data signal
señal de descodificación decoding signal
señal de desconexión disconnect signal
señal de destellos flashing signal
señal de diferencia difference signal
señal de diferencia de color color-difference signal
señal de dirección address signal
señal de directividad directivity signal
señal de disco disk signal
señal de disparo trigger signal
señal de distancia range signal
señal de diversidad diversity signal
señal de doble rebote double-bounce signal
señal de Doppler Doppler signal
señal de dos tonos two-tone signal
señal de eco echo signal
señal de encuadre framing signal
señal de enfasamiento phasing signal
señal de entrada input signal
señal de entrada al bucle loop-input signal
señal de entrada de modo común common-mode
input signal
señal de entrada de referencia reference input

signal
señal de entrada diferencial differential-input signal
señal de entrada sinusoidal sinusoidal input signal
señal de error error signal
señal de error de bucle loop-error signal
señal de espacio spacing signal
señal de excitación driving signal
señal de extinción blanking signal
señal de extinción horizontal horizontal-blanking
signal
señal de extinción vertical vertical-blanking signal
señal de facsímil facsimile signal
señal de falla fault signal
señal de fin de bloque end-of-block signal
señal de frecuencia intermedia
intermediate-frequency signal
señal de frecuencia patrón standard-frequency
signal
señal de función function signal
señal de grupo ocupado group-busy signal
señal de guarda guard signal
señal de haz beam signal
señal de imagen picture signal
señal de imagen borrada blanked picture signal
señal de imagen de colores color picture signal
señal de impulsos pulse signal
señal de indicación de alarma alarm indication
signal
señal de inhibición inhibit signal
señal de intensidad mínima minimum-strength
signal
señal de interrogación interrogation signal
señal de interrupción interrupt signal
señal de inversión de marcas alternativas
alternate-mark inversion signal
señal de invitación a marcar proceed-to-dial signal
señal de invitación a transmitir proceed-to-transmit
signal
señal de lectura readout signal
señal de liberación release signal
señal de línea line signal, dial tone
señal de lorán loran signal
señal de luminancia luminance signal
señal de luminancia constante constant-luminance
signal
señal de llamada call signal, ringing signal, calling
signal, ringback signal, ringback
señal de llamada colectiva collective-call sign
señal de mando command signal
señal de medida measuring signal
señal de micrófono microphone signal
señal de microondas microwave signal
señal de modo común common-mode signal
señal de modo diferencial differential-mode signal
señal de modulación modulation signal
señal de Morse Morse signal
señal de nivel único one-level signal
señal de normalización normalization signal
señal de ocupado audible audible busy signal
señal de onda cuadrada square-wave signal
señal de onda sinusoidal sine-wave signal
señal de parada stop signal
señal de parada absoluta absolute stop signal
señal de parada de registro stop-record signal
señal de peligro danger signal
señal de polaridad negativa negative-polarity signal
señal de polaridad positiva positive-polarity signal
señal de portadora carrier signal
señal de prioridad priority signal
señal de programa program signal
señal de progreso de llamada call progress signal
señal de prueba test signal
señal de prueba de onda cuadrada square-wave test

signal
señal de prueba de voz speech test signal
señal de prueba en proceso test-busy signal
señal de puerta gate signal
señal de punto obscuro dark-spot signal
señal de radar radar signal
señal de radiofrecuencia radio-frequency signal
señal de reconocimiento acknowledgment signal
señal de referencia reference signal
señal de referencia almacenada stored reference
signal
señal de referencia de onda continua
continuous-wave reference signal
señal de referencia de subportadora subcarrier
reference signal
señal de referencia de voz speech reference signal
señal de referencia digital digital reference signal
señal de rejilla grid signal
señal de reloj clock signal
señal de respuesta response signal
señal de retorno return signal
señal de retorno de carro carriage return signal
señal de retorno en bucle loop-return signal
señal de retroalimentación feedback signal
señal de retroalimentación de bucle loop-feedback
signal
señal de rojo red signal
señal de ruido noise signal
señal de ruido de banda ancha wideband noise
signal
señal de ruido eléctrico electric-noise signal
señal de salida outgoing signal, output signal
señal de salida de bucle loop-output signal
señal de salida de lectura read output signal
señal de salida sin perturbación undisturbed output
signal
señal de salida única one-output signal
señal de saturación saturation signal
señal de secuencia sequence signal
señal de seguimiento tracking signal
señal de seguridad safety signal
señal de selección selection signal
señal de selección de canal channel select signal
señal de sincronización synchronization signal,
synchronizing signal
señal de sincronización compuesta composite
synchronization signal
señal de sincronización de colores color
synchronizing signal
señal de socorro distress signal
señal de solicitud de llamada call request signal
señal de sonar sonar signal
señal de sonido sound signal
señal de subida uplink
señal de subportadora subcarrier signal
señal de subportadora en cuadratura de fase
quadrature-phase subcarrier signal
señal de supresión de haz blackout signal
señal de tablero board signal
señal de teleimpresora teleprinter signal
señal de televisión television signal
señal de televisión a colores color-television signal
señal de televisión compuesta composite television
signal
señal de televisión estandarizada standardized
television signal
señal de televisión normal standard television signal
señal de trabajo work signal, marking signal
señal de transferencia transfer signal
señal de transmisión transmit signal
señal de transmisor transmitter signal
señal de ventana window signal
señal de verde green signal

señal de video video signal
señal de videofrecuencia video-frequency signal
señal de voz voice signal
señal débil weak signal, small signal
señal defectuosa faulty signal
señal deseada desired signal
señal desfasada out-of-phase signal
señal desmodulada demodulated signal
señal diferencial differential signal
señal diferencial de bucle loop-difference signal
señal digital digital signal
señal directa forward signal
señal discreta discrete signal
señal distorsionada distorted signal
señal eléctrica electric signal
señal eléctrica interferente interfering electric signal
señal electromagnética electromagnetic signal
señal electroneumática electropneumatic signal
señal en cuadratura quadrature signal
señal en escalera staircase signal
señal en fase in-phase signal
señal engañosa misleading signal
señal espuria spurious signal
señal estabilizadora stabilizing signal
señal externa external signal
señal extraña extraneous signal
señal falsa false signal
señal fantasma phantom signal, ghost signal
señal fija fixed signal
señal fotoeléctrica photoelectric signal
señal fuera de banda out-of-band signal
señal fuera de frecuencia off-frequency signal
señal fuerte strong signal, large signal
señal grande large signal
señal habilitante enabling signal
señal horizontal horizontal signal
señal imagen image signal
señal indeseada undesired signal
señal inhibidora inhibiting signal
señal inicial initial signal
señal inmediata immediate signal
señal interferente interfering signal
señal intermitente intermittent signal
señal inversa backward signal
señal legible readable signal
señal libre de fluctuaciones jitter-free signal
señal limitada limited signal
señal lógica logic signal
señal longitudinal longitudinal signal
señal luminosa luminous signal
señal marcadora marker signal
señal medida measured signal
señal mínima detectable minimum detectable signal
señal mínima discernible minimum discernible
signal
señal modulada en amplitud amplitude-modulated
signal
señal moduladora modulating signal
señal monocroma monochrome signal
señal multicanal multichannel signal
señal multicomponente multicomponent signal
señal múltiplex total total multiplex signal
señal negra black signal
señal no coherente noncoherent signal
señal octonaria octonary signal
señal parásita parasitic signal
señal patrón standard signal
señal pequeña small signal
señal periódica periodic signal
señal periódica simple simple periodic signal
señal permanente permanent signal
señal pulsada pulsed signal
señal Q Q signal

señal radiada radiated signal
señal radiada efectiva effective signal radiated
señal reconocible recognizable signal
señal rectangular rectangular signal
señal réctificada rectified signal
señal reflejada reflected signal
señal repetida repeated signal
señal saturante saturating signal
señal seccionada chopped signal
señal secuencial sequential signal
señal semiautomática semiautomatic signal
señal simétrica symmetrical signal
señal simple simple signal
señal sincronizadora de cuadro
    frame-synchronizing signal
señal sincronizadora de línea line-synchronizing
    signal
señal sinusoidal sinusoidal signal
señal submarina underwater signal
señal supersónica supersonic signal
señal supervisora supervisory signal
señal tangencial tangential signal
señal telegráfica telegraphic signal
señal transhorizonte over-the-horizon signal
señal ultrasónica ultrasonic signal
señal umbral threshold signal
señal única single signal
señal útil useful signal
señal vertical vertical signal
señal visible visible signal
señal visual visual signal, blip
señal visual de eco echo blip
señal W W signal
señal Y Y signal
señales de enclavamiento interlocking signals
señales de radio galácticas galaxy radio signals
señales de tiempo time signals
señales mioeléctricas myoelectric signals
señales por estelas meteóricas meteor-burst signals
señalización signaling
señalización automática automatic signaling
señalización con frecuencia única single-frequency
    signaling
señalización de baja frecuencia low-frequency
    signaling
señalización de circuito abierto open-circuit
    signaling
señalización de circuito cerrado closed-circuit
    signaling
señalización de fallas fault signaling
señalización de portadora carrier signaling
señalización de potencia power signaling
señalización directa forward signaling
señalización electroneumática electropneumatic
    signaling
señalización en bucle loop signaling
señalización ferroviaria railway signaling
señalización fuera de banda out-of-band signaling
señalización multifrecuencia multifrequency
    signaling
señalización por batería central central-battery
    signaling
señalización por bucle de corriente continua
    direct-current loop signaling
señalización por cambios de frecuencia
    frequency-change signaling
señalización por corriente continua direct-current
    signaling
señalización por desplazamiento de frecuencia
    frequency-shift signaling
señalización por frecuencia vocal voice-frequency
    signaling
señalización protectora protective signaling

señalización remota remote signaling
señalizador ringer
señalizador de paridad parity flag
separabilidad separability
separable separable
separación separation, gap, spacing
separación aleatoria random separation
separación angular angular separation
separación aparente apparent separation
separación con guiones automática automatic
    hyphenation
separación de amplitud amplitude separation
separación de canales channel separation
separación de canales estereofónicos
    stereophonic-channel separation
separación de colores color separation
separación de contactos contact separation
separación de contactos de relé relay-contact
    separation
separación de frecuencias separation of frequencies,
    spacing of frequencies
separación de impulsos pulse separation
separación de impulsos de sincronización
    synchronizing-pulse separation
separación de modos mode separation
separación de picos peak separation
separación de señales de sincronización
    synchronizing-signal separation
separación de sincronización synchronizing
    separation
separación de unidades unit separation
separación electromagnética electromagnetic
    separation
separación electrostática electrostatic separation
separación energética energy gap
separación entre frecuencias frequency separation
separación equivalente equivalent separation
separación estereofónica stereophonic separation
separación fija fixed spacing
separación magnética magnetic separation
separadamente separately
separado separate
separador separator, spacer, buffer
separador de amplitud amplitude separator
separador de archivo file separator
separador de argumentos argument separator
separador de bloques block separator
separador de cable cable separator
separador de campo field separator
separador de columnas column separator
separador de ferrita ferrite separator
separador de frecuencias frequency separator
separador de grupos group separator
separador de impulsos impulse separator
separador de información information separator
separador de isótopos magnético magnetic isotope
    separator
separador de palabras word separator
separador de registros record separator
separador de señales signal separator
separador de sincronización synchronizing
    separator
separador de tambor magnético magnetic-drum
    separator
separador de unidades unit separator
separador dieléctrico dielectric separator
separador electromagnético electromagnetic
    separator
separador electrónico de frecuencias electronic
    crossover
separador electrostático electrostatic separator
separador magnético magnetic separator
septenario septanary

septendecimal septendecimal
septeto septet
septo septum
septo transversal transverse septum
serdes serdes
serial serial
serializador-deserializador serializer-deserializer
serie series, string, bank
serie armónica harmonic series
serie armónica de tonos harmonic series of tones
serie característica characteristic series
serie de calefactores heater string
serie de Fibonacci Fibonacci series
serie de Fourier Fourier series
serie de Maclaurin Maclaurin series
serie de potencia power series
serie-derivación series-shunt
serie divergente divergent series
serie dominante dominant series
serie electromotriz electromotive series
serie electroquímica electrochemical series
serie electrostática electrostatic series
serie espectral spectral series
serie exponencial exponential series
serie finita finite series
serie galvánica galvanic series
serie hiperarmónica hyperharmonic series
serie homogénea homogeneous series
serie infinita infinite series
serie logarítmica logarithmic series
serie-paralelo serial-parallel
serie termoeléctrica thermoelectric series
serie triboeléctrica triboelectric series
serrodino serrodyne
servicio service, duty
servicio aeromóvil aeromobile service
servicio automático automatic service
servicio comercial continuo continuous commercial service
servicio compartido shared service
servicio continuo continuous service, continuous duty
servicio de advertencia de aeronave aircraft warning service
servicio de aplicación application service
servicio de computadoras computer services
servicio de control de aeródromo aerodrome control service
servicio de control de aproximación approach control service
servicio de corto tiempo short-time duty
servicio de emergencia emergency service
servicio de información information service
servicio de información en línea on-line information service
servicio de larga distancia automático automatic long-distance service
servicio de mantenimiento maintenance service
servicio de radio aeronáutico aeronautical radio service
servicio de radio móvil mobile radio service
servicio de radiodifusión broadcast service
servicio de radiodifusión aeronáutico aeronautical broadcasting service
servicio de radionavegación aeronáutico aeronautical radionavigation service
servicio de radionavegación marítima maritime radionavigation service
servicio de red network service
servicio de seguridad safety service
servicio de sistema system service
servicio de telecomunicación aeronáutica aeronautical telecommunication service

servicio de telecomunicaciones telecommunications service
servicio de telecomunicaciones fijo aeronáutico aeronautical fixed telecommunications service
servicio de una hora one-hour duty
servicio dedicado dedicated service
servicio duplicado duplicate service
servicio especial special service
servicio experimental experimental service
servicio fijo fixed service
servicio fijo aeronáutico aeronautical fixed service
servicio general general service
servicio inatendido unattended service
servicio ininterrumpido uninterrupted duty
servicio intermitente intermittent service, intermittent duty
servicio intermitente variable variable intermittent duty
servicio interurbano toll service
servicio local local service
servicio medido measured service, metered service
servicio mixto mixed service
servicio móvil mobile service
servicio móvil aéreo air mobile service
servicio móvil aeronáutico aeronautical mobile service
servicio móvil terrestre land mobile service
servicio no esencial nonessential service
servicio no programado unscheduled service
servicio nominal nominal service, rated duty
servicio periódico periodic service, periodic duty
servicio periódico intermitente intermittent periodic service, intermittent periodic duty
servicio por circuito privado private-wire service
servicio preventivo preventive service
servicio privado private service
servicio programado scheduled service
servicio rápido rapid service
servicio semidúplex half-duplex service
servicio sin conexiones connectionless service
servicio sin espera no-delay service
servicio telefónico telephone service
servicio telefónico internacional international telephone service
servicio telegráfico telegraph service
servicio temporal temporary service, temporary duty
servicio temporal variable variable temporary duty
servicio variable variable duty
servicios de comunicaciones communication services
servidor server
servidor de archivo file server
servidor de archivos dedicado dedicated file server
servidor de base de datos database server
servidor de comunicaciones communication server
servidor de disco disk server
servidor de impresora printer server
servidor de red network server
servo de control proporcional proportional-control servo
servo inestable unstable servo
servo regulador regulating servo
servoaccionado servodriven, servoactuated
servoactuador servoactuator
servoaltímetro servoaltimeter
servoamplificador servoamplifier
servoanalizador servoanalyzer
servoayuda servoassistance
servoayudado servoassisted
servobucle servoloop
servocilindro servocylinder
servocircuito servocircuit
servocontrol servocontrol
servocontrolado servocontrolled

servocontrolador servocontroller
servointegrador servointegrator
servomanipulador servomanipulator
servomecánica servomecanics
servomecanismo servomechanism
servomecanismo de bucle abierto open-loop servomechanism
servomecanismo electrohidráulico electrohydraulic servomechanism
servomecanismo posicional positional servomechanism
servomodulador servomodulator
servomotor servomotor
servomotor de piloto automático automatic-pilot servomotor
servomotor electroneumático electropneumatic servomotor
servomotor multiplicador multiplying servo
servomotorizado servomotorized
servomultiplicador servomultiplier
servooperado servooperated
servopotenciómetro servopotentiometer
servorregulador servoregulator, servogovernor
servosistema servosystem
servosistema de enganche de fase phase-locked servosystem
servosonda servoprobe
servoválvula servovalve
servoválvula electrohidráulica electrohydraulic servovalve
sesgo skew, skewing, bias
sesgo de cinta tape skew
sesgo estático static skew
sesión session
sesión de consulta inquiry session
sesiones en paralelo parallel sessions
seudoaleatorio pseudorandom
seudoángulo de Brewster pseudo-Brewster angle
seudobinario pseudobinary
seudocódigo pseudocode
seudoerror pseudoerror
seudoestereofónico pseudostereophonic
seudoinstrucción pseudoinstruction
seudolenguaje pseudolanguage
seudooperación pseudooperation
seudoportadora pseudocarrier
seudorregistro pseudoregister
seudorreloj pseudoclock
seudorruido pseudonoise
seudotemporizador pseudotimer
seudotexto pseudotext
seudovariable pseudovariable
sexadecimal sexadecimal
sexagesimal sexagesimal
sextante electrónico electronic sextant
sexteto sextet
shorán shoran
shunt shunt
shuntado shunted
SI-ENTONCES IF-THEN
SI-Y-SOLO-SI IF-AND-ONLY-IF
siembra de errores error seeding
siembra de fallas fault seeding
siemens siemens
sievert sievert
sigma sigma
sigmatrón sigmatron
significación significance
signo sign, signal
signo algebraico algebraic sign
signo característico characteristic sign
signo de cambio change sign
signo de código code sign

signo de identificación identification sign
signo especial special sign
silbato confirmante acknowledging whistle
silbato de Galton Galton whistle
silbido whistling, singing, birdie
silbido atmosférico whistler
silbido heterodino heterodyne whistle
silenciador silencer, squelcher, squelch
silenciador de audio audio squelch
silenciador de ruido noise silencer, noise squelch
silenciador de televisión television silencer
silenciador selectivo selective squelch
silenciador telefónico telephone silencer
silenciamiento quieting
silenciamiento silencing, quieting, squelching, muting
silenciamiento de receptor receiver muting
silencio de radio internacional international radio silence
silicato de sodio sodium silicate
silicio silicon
silicio metal-óxido metal-oxide-silicon
silicio sobre zafiro silicon on sapphire
silicio tipo N N-type silicon
silicio tipo P P-type silicon
silicona silicone
silverstat silverstat
simbólico symbolic
símbolo symbol
símbolo abstracto abstract symbol
símbolo compuesto composite symbol
símbolo de código code symbol
símbolo de componente component symbol
símbolo de comprobación check symbol
símbolo de edición editing symbol
símbolo de entrada entry symbol
símbolo de entrada-salida input-output symbol
símbolo de escape lógico logical escape symbol
símbolo de fin de línea line-end symbol
símbolo de localización location symbol
símbolo de procesamiento processing symbol
símbolo de secuencia sequence symbol
símbolo de variable global global variable symbol
símbolo decimal decimal symbol
símbolo especial special symbol
símbolo esquemático schematic symbol
símbolo externo external symbol
símbolo funcional functional symbol
símbolo gráfico graphic symbol
símbolo lógico logic symbol
símbolo matemático mathematical symbol
símbolo mnemónico mnemonic symbol
símbolos programados programmed symbols
simetría symmetry
simetría aritmética arithmetic symmetry
simetría complementaria complementary symmetry
simetría geométrica geometric symmetry
simétricamente symmetrically
simétrico symmetrical
símplex simplex
símplex de doble canal double-channel simplex
símplex monocanal single-channel simplex
símplex por frecuencia única single-frequency simplex
simplicidad simplicity
simplificación de circuito circuit simplification
simulación simulation
simulación académica academic simulation
simulación ambiental environmental simulation
simulación analógica analog simulation
simulación basada en actividad activity-based simulation
simulación continua continuous simulation

simulación de laboratorio laboratory simulation
simulación digital digital simulation
simulación discreta discrete simulation
simulación en tiempo real real-time simulation
simulación estocástica stochastic simulation
simulación híbrida hybrid simulation
simulación incorporada built-in simulation
simulación matemática mathematical simulation
simulación por computadora computer simulation
simulado simulated
simulado por computadora computer simulated
simulador simulator
simulador analógico electrónico electronic analog
   simulator
simulador de computadora computer simulator
simulador de ensamblador de microprocesador
   microprocessor assembler simulator
simulador de fase phase simulator
simulador de llamadas telefónicas telephone call
   simulator
simulador de memoria sólo de lectura read-only
   memory simulator
simulador de microcontrolador microcontroller
   simulator
simulador de modo de equilibrio equilibrium mode
   simulator
simulador de programa program simulator
simulador de radiación solar solar-radiation
   simulator
simulador de vuelo flight simulator
simulador incorporado built-in simulator
simulador interactivo interactive simulator
simulador solar solar simulator
simular simulate
simultáneamente simultaneously
simultaneidad simultaneity
simultáneo simultaneous
sin almacenamiento nonstorage
sin carga uncharged
sin compensación unweighted
sin conexiones connectionless
sin contactos contactless
sin cordón cordless
sin cortocircuito nonshorting
sin chispas sparkless
sin desviación nondeviated
sin dirección addressless
sin distorsión distortionless
sin ecos echoless
sin error error-free
sin errores de programa bug free
sin escobillas brushless
sin espacios gapless
sin formato formatless
sin memoria intermedia nonbuffered
sin núcleo coreless
sin obsolescencia obsolescence-free
sin paridad no parity
sin pérdidas lossless
sin perforaciones punchless
sin protección unprotected
sin puenteo nonbridging
sin reflexión reflectionless
sin reposición nonreset
sin retardo nondelay
sin retorno a cero nonreturn-to-zero
sin ruido noiseless
sin soldadura solderless
sin solvente solventless
sin transformador transformerless
sin tubos tubeless
sin válvulas valveless
sincro synchro

sincro diferencial differential synchro
sincrociclotrón synchrocyclotron
sincrodestello synchroflash
sincrodino synchrodyne
sincrogenerador synchrogenerator
sincromotor synchro motor
sincronía synchrony
sincrónicamente synchronously
sincrónico synchronous
sincronismo synchronism
sincronismo horizontal horizontal hold
sincronismo vertical vertical hold
sincronización synchronization, synchronizing,
   timing
sincronización contradireccional contradirectional
   timing
sincronización de bloques block synchronization
sincronización de colores color synchronization
sincronización de cuadro frame synchronization
sincronización de imagen picture synchronization
sincronización de líneas line synchronization
sincronización de módem modem synchronization
sincronización de oscilador oscillator
   synchronization
sincronización de oscilador de bloqueo blocking
   oscillator synchronization
sincronización de portadora carrier synchronization
sincronización de programas program
   synchronization
sincronización de puerta gate timing
sincronización de volante flywheel synchronization
sincronización horizontal horizontal synchronization
sincronización mutua mutual synchronization
sincronización por onda espacial skywave
   synchronization
sincronización positiva positive synchronization
sincronización secuencial sequential synchronization
sincronización vertical vertical synchronization
sincronizado synchronized
sincronizado en fase phase-synchronized
sincronizador synchronizer
sincronizador automático automatic synchronizer
sincronizador de canales channel synchronizer
sincronizador de ciclos cycle timer
sincronizador de cintas tape synchronizer
sincronizador de colores color synchronizer
sincronizador de datos data synchronizer
sincronizador de radar radar synchronizer
sincronizador maestro master synchronizer
sincronizar synchronize
síncrono synchronous
síncrono de par torque synchro
sincronodino synchronodyne
sincronómetro synchronometer
sincronoscopio synchronoscope
sincrorreceptor synchro receiver
sincrorreceptor diferencial synchro-differential
   receiver
sincrorrepetidor synchro repeater
sincroscopio synchroscope
sincrosistema synchro system
sincrotransformador de control synchro-control
   transformer
sincrotransmisor synchro transmitter
sincrotrón synchrotron
sincrotrón electrónico electron synchrotron
sincrotrón protónico proton synchrotron
sinérgico synergistic
sinergismo synergism
sinistrórsum counterclockwise
sintaxis syntax
sínter sinter
sinterización sinterization, sintering

sinterizado sintered
síntesis synthesis
síntesis de circuitos circuit synthesis
síntesis de imagen picture synthesis
síntesis de programa program synthesis
síntesis de red network synthesis
síntesis de voz speech synthesis
sintético synthetic
sintetizador synthesizer
sintetizador de forma de onda waveform synthesizer
sintetizador de frecuencias frequency synthesizer
sintetizador de frecuencias electrónico electronic frequency synthesizer
sintetizador de impulsos pulse synthesizer
sintetizador de señales signal synthesizer
sintetizador de video video synthesizer
sintetizador de voz voice synthesizer
sintetizador directo direct synthesizer
sintetizador musical music synthesizer
síntoma de falla fault symptom
sintonía syntony
sintonizable tunable
sintonizable por tensión voltage tunable
sintonización tuning, syntonization
sintonización ancha broad tuning
sintonización automática automatic tuning
sintonización capacitiva capacitive tuning
sintonización controlada por cristal crystal-controlled tuning
sintonización crítica critical tuning
sintonización de antena antenna tuning
sintonización de antena en paralelo parallel antenna tuning
sintonización de banda ancha broadband tuning
sintonización de oscilador oscillator tuning
sintonización de paso de banda bandpass tuning
sintonización de placa plate tuning
sintonización de receptor receiver tuning
sintonización de resistencia variable variable-resistance tuning
sintonización de volante flywheel tuning
sintonización eléctrica electric tuning
sintonización electrónica electronic tuning
sintonización en derivación shunt tuning
sintonización en línea inline tuning
sintonización en tándem ganged tuning
sintonización escalonada staggered tuning
sintonización fina fine tuning
sintonización gruesa coarse tuning
sintonización incremental incremental tuning
sintonización inductiva inductive tuning
sintonización magnética magnetic tuning
sintonización manual manual tuning
sintonización mecánica mechanical tuning
sintonización múltiple multiple tuning
sintonización plana flat tuning
sintonización por accionamiento directo direct-drive tuning
sintonización por circuito doble double-circuit tuning
sintonización por inductancia variable variable-inductance tuning
sintonización por núcleo slug tuning
sintonización por ojo mágico magic-eye tuning
sintonización por permeabilidad permeability tuning
sintonización por pulsador push-button tuning
sintonización por rejilla grid tuning
sintonización por reluctancia reluctance tuning
sintonización por resistencia resistance tuning
sintonización por resistencia-capacitancia resistance-capacitance tuning
sintonización por tensión voltage tuning

sintonización por varillas paralelas parallel-rod tuning
sintonización remota remote tuning
sintonización selectiva selective tuning
sintonización silenciosa quiet tuning
sintonización sincrónica synchronous tuning
sintonización térmica thermal tuning
sintonización visual visual tuning
sintonizado tuned
sintonizado en paralelo parallel-tuned
sintonizado por resistencia resistance-tuned
sintonizador tuner
sintonizador básico basic tuner
sintonizador cascodo cascode tuner
sintonizador coaxial coaxial tuner
sintonizador continuo continuous tuner
sintonizador de antena antenna tuner
sintonizador de cuarto de onda quarter-wave tuner
sintonizador de estado sólido solid-state tuner
sintonizador de frecuencia ultraalta ultra-high frequency tuner
sintonizador de guíaondas waveguide tuner
sintonizador de líneas resonantes resonant-line tuner
sintonizador de manguito de guíaondas waveguide slug tuner
sintonizador de modulación de amplitud-modulación de frecuencia amplitude-modulation-frequency-modulation tuner
sintonizador de modulación de frecuencia frequency-modulation tuner
sintonizador de modulación de frecuencia y modulación de amplitud frequency-modulation-amplitude-modulation tuner
sintonizador de núcleo slug tuner
sintonizador de pulsador push-button tuner
sintonizador de radio radio tuner
sintonizador de radiofrecuencia multicanal multichannel radio-frequency tuner
sintonizador de televisión television tuner
sintonizador de tornillo deslizante slide-screw tuner
sintonizador discriminador discriminator tuner
sintonizador en espiral spiral tuner
sintonizador estereofónico stereophonic tuner
sintonizador múltiple multiple tuner
sintonizador rotativo turret tuner
sintonizador térmico thermal tuner
sintonizar tune
sinusoidal sinusoidal
sinusoide sinusoid
siseo hiss
siseo auroral auroral hiss
siseo de micrófono microphone hiss
sismófono seismophone
sismógrafo seismograph
sismógrafo de inducción induction seismograph
sismógrafo electromagnético electromagnetic seismograph
sismógrafo piezoeléctrico piezoelectric seismograph
sismograma seismogram
sismómetro seismometer
sismoscopio seismoscope
sistema system
sistema a demanda on-demand system
sistema abierto open system
sistema absoluto absolute system
sistema absoluto de unidades absolute system of units
sistema activo active system
sistema acústico acoustic system
sistema adaptivo adaptive system
sistema adjunto adjoint system
sistema aislado insulated system

sistema alámbrico wire system
sistema alfa alpha system
sistema analógico analog system
sistema anticolisiones collision-avoidance system
sistema anunciador annunciator system
sistema armónico amortiguado damped harmonic
system
sistema audiovisual audio-visual system
sistema automático automatic system
sistema automático de relés all-relay system
sistema automático de Wheatstone Wheatstone
automatic system
sistema automático paso a paso step-by-step
automatic system
sistema basado en el compilador compiler-based
system
sistema biamplificador biamplifier system
sistema bidireccional two-way system
sistema bifásico two-phase system
sistema bifilar two-wire system
sistema binario binary system
sistema cegesimal centimeter-gram-second system
sistema centralizado centralized system
sistema cerrado closed system
sistema codificador absoluto absolute encoder
system
sistema coherente coherent system
sistema colorimétrico colorimetric system
sistema compacto compact system
sistema compatible compatible system
sistema común common system
sistema con conexión a tierra grounded system
sistema con neutro aislado isolated-neutral system
sistema con portadora suprimida suppressed-carrier
system
sistema con retorno por tierra earth-return system
sistema con retroalimentación feedback system
sistema conectado connected system
sistema conmutado switched system
sistema continuo continuous system
sistema controlado controlled system
sistema controlador controlling system
sistema conversacional conversational system
sistema convertidor converter system
sistema corrector de errores error-correcting system
sistema cristalino crystal system
sistema crítico critical system
sistema cuadrafónico quadraphonic system
sistema cuádruple quadruple system
sistema cuádruplex quadruplex system
sistema cuantificado quantized system
sistema de acceso en serie serial-access system
sistema de acceso múltiple multiple-access system
sistema de accionamiento actuating system
sistema de Adcock Adcock system
sistema de administración management system
sistema de administración de archivos file
management system
sistema de administración de base de datos
database management system
sistema de administración de base de datos de red
network database management system
**sistema de administración de base de datos
jerárquica** hierarchical database management
system
sistema de administración de cintas tape
management system
sistema de administración de datos data
management system
sistema de administración de información
information-management system
sistema de administración de proyectos project
management system

sistema de adquisición de datos data-acquisition
system
sistema de adquisición y control acquisition and
control system
sistema de aislamiento insulation system
sistema de aislamiento eléctrico electrical insulation
system
sistema de alarma de circuito abierto open-circuit
alarm system
sistema de alarma de circuito cerrado closed-circuit
alarm system
sistema de alarma de fallas fault alarm system
sistema de alarma de fuego fire-alarm system
sistema de alarma de humo automático automatic
smoke alarm system
sistema de alarma de incendio automática
automatic fire-alarm system
sistema de alarma integrado integrated alarm
system
sistema de alimentación power system, feed system
sistema de alimentación aislado isolated power
system
sistema de alimentación en paralelo parallel-feed
system
sistema de alimentación en serie series-feed system
sistema de alimentación fotovoltaico photovoltaic
power system
sistema de almacenamiento storage system
sistema de almacenamiento de datos magnético
magnetic data-storage system
sistema de almacenamiento de energía
energy-storage system
sistema de almacenamiento virtual virtual storage
system
sistema de alta fidelidad estereofónico stereophonic
high-fidelity system
sistema de alta tensión high-voltage system
sistema de altavoces loudspeaker system
sistema de altavoces de dos canales two-way
loudspeaker system
sistema de altavoces de tres canales three-way
loudspeaker system
sistema de altavoces estereofónico stereophonic
loudspeaker system
sistema de altavoces múltiples multiple-loudspeaker
system
sistema de alternación de fase por línea
phase-alternation line system
sistema de alumbrado lighting system
sistema de alumbrado de aproximación
approach-lighting system
sistema de alumbrado de aproximación de precisión
precision approach lighting system
sistema de antena antenna system
sistema de antena colectiva master antenna system
sistema de antena integrado integrated antenna
system
sistema de antenas adaptivo adaptive antenna
system
sistema de antenas equilibrado balanced antenna
system
sistema de apoyo support system
sistema de aproximación controlada desde tierra
ground-controlled approach system
sistema de aproximación por instrumentos
instrument-approach system
sistema de archivos compartidos shared files system
sistema de archivos jerárquico hierarchical file
system
sistema de arranque-parada start-stop system
sistema de arrastre de cinta tape-transport system
sistema de atenuación attenuator system
sistema de aterrizaje óptico optical landing system

sistema de aterrizaje por instrumentos instrument-landing system

sistema de aterrizaje por instrumentos de Lorentz Lorentz instrument-landing system

sistema de aterrizajes radioguiado glide-path landing system

sistema de audio audio system

sistema de automatización de producción production automation system

sistema de avance rápido fast forward system

sistema de baja tensión low-voltage system

sistema de banda ancha broadband system

sistema de banda lateral única single-sideband system

sistema de bandeja de cables cable tray system

sistema de barras cruzadas crossbar system

sistema de barrido sweeping system

sistema de base de datos database system

sistema de batería común common-battery system

sistema de batería local local-battery system

sistema de Batten Batten system

sistema de Baudot Baudot system

sistema de bloqueo blocking system, block system

sistema de bloqueo automático automatic block system

sistema de bloqueo de receptor receiver lockout system

sistema de bloqueo manual manual block system

sistema de bucle abierto open-loop system

sistema de bucle cerrado closed-loop system

sistema de bus bus system

sistema de bus abierto open-bus system

sistema de bus cerrado closed bus system

sistema de cable guía leader-cable system

sistema de cable único single-cable system

sistema de cables cable system

sistema de cables gemelos twin-cable system

sistema de calefacción radiante radiant heating system

sistema de calentamiento heating system

sistema de calentamiento integrado integrated heating system

sistema de canal dividido split-channel system

sistema de cebado priming system

sistema de ciclo cerrado closed-cycle system

sistema de cinco hilos five wire system

sistema de cinta magnética magnetic-tape system

sistema de cintas tape system

sistema de circuito doble double-circuit system

sistema de circuito único single-circuit system

sistema de codificación coding system

sistema de codificación multidimensional multidimensional coding system

sistema de códigos code system

sistema de colores color system

sistema de colores aditivos additive color system

sistema de colores de Munsell Munsell color system

sistema de colores sustractivo subtractive color system

sistema de coma fija fixed-point system

sistema de coma flotante floating-point system

sistema de compilador compiler system

sistema de computación remota remote-computing system

sistema de computadora computer system

sistema de computadora auxiliar auxiliary computer system

sistema de computadora de procesos process computer system

sistema de computadora en línea on-line computer system

sistema de computadora híbrido hybrid computer system

sistema de computadora integrado integrated computer system

sistema de computadora portátil portable computer system

sistema de computadora remoto remote computer system

sistema de computadoras dúplex duplex computer system

sistema de computadoras múltiples multiple computer system

sistema de comunicación de información information communication system

sistema de comunicación por impulsos pulse communication system

sistema de comunicaciones communication system

sistema de comunicaciones de banda ancha wideband communications system

sistema de comunicaciones de datos data communications system

sistema de comunicaciones digitales digital communications system

sistema de comunicaciones móviles mobile communications system

sistema de comunicaciones por computadora computer-communications system

sistema de conducto duct system

sistema de conductores conductor system

sistema de conductos conduit system

sistema de conmutación switching system

sistema de conmutación automático automatic switching system

sistema de conmutación de barras cruzadas crossbar switching system

sistema de conmutación de mensajes message switching system

sistema de conmutación de retardo delay switching system

sistema de conmutación electromecánico electromechanical switching system

sistema de conmutación electrónica electronic switching system

sistema de conmutación semiautomática semiautomatic switching system

sistema de contacto único single-contact system

sistema de conteo de impulsos pulse-counting system

sistema de control control system

sistema de control activo active control system

sistema de control adaptivo adaptive control system

sistema de control aerotransportado airborne control system

sistema de control automático automatic control system

sistema de control con retroalimentación lineal linear feedback control system

sistema de control concurrente concurrent control system

sistema de control continuo continuous-control system

sistema de control de adquisición de datos data-acquisition control system

sistema de control de bucle cerrado closed-loop control system

sistema de control de entrada-salida input-output control system

sistema de control de exactitud accuracy control system

sistema de control de excitación excitation control system

sistema de control de iluminación fotoeléctrico photoelectric illumination control system

sistema de control de microcomputadora microcomputer control system

sistema de control de muestreo sampling control system
sistema de control de posición position-control system
sistema de control de posicionamiento positioning control system
sistema de control de procesos process-control system
sistema de control de proyectos project control system
sistema de control de sensores sensor-control system
sistema de control de temperatura temperature-control system
sistema de control de vuelo automático automatic flight control system
sistema de control digital digital control system
sistema de control ejecutivo executive control system
sistema de control en bucle abierto open-loop control system
sistema de control en cascada cascade control system
sistema de control en tiempo real real-time control system
sistema de control interno internal control system
sistema de control manual manual control system
sistema de control numérico numerical control system
sistema de control por retroalimentación feedback control system
sistema de control proporcional proportional-control system
sistema de control supervisor supervisory control system
sistema de conversión en paralelo parallel conversion system
sistema de coordenadas coordinate system
sistema de coordenadas cartesianas Cartesian coordinate system
sistema de corriente alterna alternating-current system
sistema de corriente constante constant-current system
sistema de corriente única single-current system
sistema de cuatro fases four-phase system
sistema de cuatro hilos four-wire system
sistema de cuatro hilos equivalente equivalent four-wire system
sistema de chips chip system
sistema de datos data system
sistema de datos digitales digital-data system
sistema de deflexión deflection system
sistema de deflexión de haz electrónico electron-beam deflection system
sistema de depuración interactiva interactive debugging system
sistema de desarrollo development system
sistema de desarrollo de aplicaciones application development system
sistema de desarrollo de microprocesador microprocessor development system
sistema de desarrollo de programas program-development system
sistema de despacho automático automatic dispatching system
sistema de destino destination system
sistema de desviación deviation system
sistema de detección de sonido sound-detection system
sistema de detección de vibraciones vibration-detection system
sistema de dieciséis bits sixteen-bit system

sistema de direccionamiento addressing system
sistema de disco disk system
sistema de disco duro hard-disk system
sistema de discos de video video-disk system
sistema de discos flexibles floppy-disk system
sistema de distribución distribution system
sistema de distribución de datos data distribution system
sistema de distribución de Edison Edison distribution system
sistema de distribución de televisión television distribution system
sistema de distribución primario primary distribution system
sistema de distribución secundario secondary distribution system
sistema de diversidad cuádruple quadruple-diversity system
sistema de diversidad espacial space-diversity system
sistema de doble banda lateral double-sideband system
sistema de doble bus double bus system
sistema de doble haz dual-beam system
sistema de doble precisión double-precision system
sistema de doble procesador dual processor system
sistema de doble procesamiento double processing system
sistema de Doppler Doppler system
sistema de dos canales two-way system
sistema de dos tonos two-tone system
sistema de ecosondeo echo sounding system
sistema de electrodos bipolares bipolar electrode system
sistema de electrodos coaxiales coaxial electrode system
sistema de electrodos concéntricos concentric electrode system
sistema de electrodos monopolar monopolar electrode system
sistema de electrodos unipolar unipolar electrode system
sistema de emergencia emergency system
sistema de encaminamiento de datos data routing system
sistema de enclavamiento interlocking system
sistema de enfoque focusing system
sistema de enfriamiento cooling system
sistema de enfriamiento de intercambiador de calor heat-exchanger cooling system
sistema de enfriamiento por agua water-cooling system
sistema de enlace por microondas microwave-link system
sistema de ensamblaje assembly system
sistema de entrada input system
sistema de entrada-salida input-output system
sistema de entrada-salida básico basic input-output system
sistema de equilibrio a cero null-balance system
sistema de equipos equipment system
sistema de excitación excitation system
sistema de excitación de alta velocidad high-speed excitation system
sistema de exploración scanning system
sistema de facsímil facsimile system
sistema de frecuencia vocal voice-frequency system
sistema de frenaje braking system
sistema de Gauss Gaussian system
sistema de generación de programas program generation system
sistema de Giorgi Giorgi system
sistema de gráficos graphics system

sistema de gráficos interactivos interactive graphics system

sistema de guía acústico acoustic homing system

sistema de guiado guidance system

sistema de guíaondas waveguide system

sistema de Halstead Halstead system

sistema de haz beam system

sistema de Heaviside-Lorentz Heaviside-Lorentz system

sistema de identificación identification system

sistema de ignición ignition system

sistema de Ilgner Ilgner system

sistema de imanes magnet system

sistema de impedancia constante constant-impedance system

sistema de impulsos inversos revertive-pulse system

sistema de indexación indexing system

sistema de inducción induction system

sistema de información information system

sistema de información compartida shared information system

sistema de información de computadoras computer information system

sistema de información en tiempo real real-time information system

sistema de intercomunicación intercommunication system

sistema de interfaz interface system

sistema de interfaz de proceso process-interface system

sistema de interfono interphone system

sistema de interportadora intercarrier system

sistema de interrupción interrupt system

sistema de interrupción de prioridad priority interrupt system

sistema de interrupción flexible flexible interruption system

sistema de Karlous Karlous system

sistema de Kramer Kramer system

sistema de laboratorio laboratory system

sistema de lápiz óptico light-pen system

sistema de Leblanc Leblanc system

sistema de lectura read system

sistema de lentes de Karlous Karlous lens system

sistema de limpieza ultrasónica ultrasonic-cleaning system

sistema de línea equilibrada balanced line system

sistema de línea visual line-of-sight system

sistema de líneas terrestres landline system

sistema de lógica binario binary logic system

sistema de lógica secuencial sequential logic system

sistema de luminancia constante constant-luminance system

sistema de llamada por tonos touch-tone system

sistema de manejo de datos data-handling system

sistema de manejo de información information-handling system

sistema de medición de sonido sound-measuring system

sistema de medida measuring system

sistema de memoria memory system

sistema de memoria de acceso aleatorio random-access memory system

sistema de memoria en serie serial-memory system

sistema de memoria virtual virtual memory system

sistema de mensuración mensuration system

sistema de microcomputadora microcomputer system

sistema de microondas microwave system

sistema de microprocesador microprocessor system

sistema de microprocesamiento microprocessing system

sistema de minicomputadora minicomputer system

sistema de modulación de frecuencia de Armstrong Armstrong frequency-modulation system

sistema de monitoreo monitoring system

sistema de monitoreo de control remoto remote-control monitoring system

sistema de monitores múltiples multiple monitor system

sistema de muestreo sampling system

sistema de multiplexión estereofónica stereophonic multiplexing system

sistema de multiplexión óptica optical multiplexing system

sistema de multiplexión por división de frecuencia frequency-division multiplexing system

sistema de multiplexión por portadora carrier multiplexing system

sistema de multiplexión remoto remote multiplexing system

sistema de multiprocesador interactivo interactive multiprocessor system

sistema de multiprocesamiento multiprocessing system

sistema de multiprogramación multiprogramming system

sistema de Munsell Munsell system

sistema de navegación de alto rendimiento high-performance navigation system

sistema de navegación de corto alcance short-range navigation system

sistema de navegación de Doppler Doppler navigation system

sistema de navegación de largo alcance long-range navigation system

sistema de navegación decca decca navigation system

sistema de navegación hiperbólica hyperbolic navigation system

sistema de navegación inercial inertial navigation system

sistema de navegación por impulsos pulse navigation system

sistema de navegación por radar radar navigation system

sistema de navegación rho-theta rho-theta navigation system

sistema de navegación tacán tacan navigation system

sistema de numeración numeration system

sistema de numeración binario binary numeration system

sistema de numeración biquinario biquinary numeration system

sistema de numeración de base radix numeration system

sistema de numeración de base fija fixed-radix numbering system

sistema de numeración de base mixta mixed-base numeration system

sistema de numeración hexadecimal hexadecimal numeration system

sistema de numeración octal octal numeration system

sistema de ondas estacionarias standing-wave system

sistema de operación de línea de mandos command line operating system

sistema de operación de tarjetas card operating system

sistema de paginación paging system

sistema de panel panel system

sistema de parámetros fijos fixed-parameter system

sistema de pie-libra-segundo foot-pound-second system

sistema de pilas de combustible fuel-cell system
sistema de portadora carrier system
sistema de portadora de banda ancha wideband carrier system
sistema de portadora flotante floating-carrier system
sistema de portadora quiescente quiescent-carrier system
sistema de portadoras de banda lateral única single-sideband carrier system
sistema de privacidad privacy system
sistema de procesamiento processing system
sistema de procesamiento de datos data-processing system
sistema de procesamiento de datos automático automatic data processing system
sistema de procesamiento de datos electrónico electronic data processing system
sistema de procesamiento de información information-processing system
sistema de procesamiento de transacciones transaction-processing system
sistema de procesamiento distribuido distributed processing system
sistema de procesamiento en paralelo parallel processing system
sistema de programación programming system
sistema de programación cruzada cross programming system
sistema de programación de cintas tape-programming system
sistema de programación de disco disk programming system
sistema de programación lineal linear programming system
sistema de programación matemática mathematical programming system
sistema de programación orientado a máquina machine-oriented programming system
sistema de propagación por saltos múltiples multihop propagation system
sistema de propulsión eléctrica electric propulsion system
sistema de protección protection system
sistema de protección diferencial differential protective system
sistema de prueba de programa program test system
sistema de prueba en línea on-line test system
sistema de Pupin Pupin system
sistema de radar radar system
sistema de radar en cadena chain radar system
sistema de radiación radiation system
sistema de radiodespacho radio-dispatching system
sistema de radiodifusión estereofónica múltiplex multiplex stereophonic broadcasting system
sistema de radiodifusión por hilo wire-broadcasting system
sistema de radiodistribución radiodistribution system
sistema de radioenlace radio-link system
sistema de radioenlace multicanal multichannel radio relay system
sistema de radioenlace por microondas microwave radio relay system
sistema de radiofaro de aproximación ciega blind approach beacon system
sistema de radiofaro omnidireccional de muy alta frecuencia very-high-frequency omnirange system
sistema de radiofaros radio-beacon system
sistema de radionavegación radionavigation system
sistema de radiorrelé radio-relay system

sistema de rayos X X-ray system
sistema de reacción reaction system
sistema de realzado de imagen image-enhancement system
sistema de recepción de datos data-reception system
sistema de recopilación de datos data gathering system
sistema de rectificación central central rectifier system
sistema de recuperación de datos data retrieval system
sistema de recuperación de información information-retrieval system
sistema de reducción de datos data-reduction system
sistema de reducción-oxidación redox system
sistema de referencia reference system
sistema de refrigeración en cascada cascade refrigeration system
sistema de registro de línea ranurada slotted-line recording system
sistema de registro de longitud fija fixed-length record system
sistema de registro de sonido sound-recording system
sistema de registro multipista multitrack recording system
sistema de regrabación rerecording system
sistema de relés relay system
sistema de representación representation system
sistema de representación de coma fija fixed-point representation system
sistema de representación de coma flotante floating-point representation system
sistema de representación numérica number representation system
sistema de representación posicional positional representation system
sistema de reproducción playback system
sistema de reproducción de cinta tape-playback system
sistema de reproducción de sonido sound-reproduction system
sistema de rerregistro rerecording system
sistema de reserva de batería battery backup system
sistema de respaldo backup system
sistema de respaldo de biblioteca library backup system
sistema de retardo de tambor magnético magnetic-drum delay system
sistema de retorno aislado insulated-return system
sistema de retransmisión de televisión television relay system
sistema de retroalimentación cuasilineal quasi-linear feedback system
sistema de retroalimentación de bucle múltiple multiple-loop feedback system
sistema de retroalimentación de decisión decision feedback system
sistema de retroalimentación de información information-feedback system
sistema de reverberación reverberation system
sistema de rotación de fase phase-rotation system
sistema de satélite de acceso múltiple multiple-access satellite system
sistema de satélites satellite system
sistema de Scherbius Scherbius system
sistema de Schmidt Schmidt system
sistema de Scott Scott system
sistema de secuencia de campos field-sequential system
sistema de secuencia de líneas line-sequential system
sistema de seguimiento tracking system
sistema de seguimiento activo active tracking system

sistema de seguimiento de onda continua
continuous-wave tracking system
sistema de seguimiento de radar avanzado
advanced radar tracking system
sistema de seguimiento pasivo passive tracking
system
sistema de seguimiento por comparación de fase
phase-comparison tracking system
sistema de seguimiento por radar radar tracking
system
sistema de segundo nivel second-level system
sistema de seguridad security system
sistema de señal de bloqueo automático automatic
block signal system
sistema de señal de cabina automática automatic
cab signal system
sistema de señalización por corriente alterna
alternating-current signaling system
sistema de servicio fijo fixed-service system
sistema de silenciamiento quieting system
sistema de sincronización de colores
color-synchronization system
sistema de sintonización por accionamiento directo
direct-drive tuning system
sistema de sintonización por pulsador push-button
tuning system
sistema de sintonización térmica thermal tuning
system
sistema de sonido sound system
sistema de sonido de cuatro canales four-channel
sound system
sistema de sonido estereofónico stereophonic sound
system
sistema de sonido magnético magnetic sound system
sistema de sonido monoaural monaural sound
system
sistema de sonido monofónico monophonic sound
system
sistema de sonido por interportadora intercarrier
sound system
sistema de soporte de decisiones decision-support
system
sistema de soporte dinámico dynamic support
system
sistema de Strowger Strowger system
sistema de tarjetas card system
sistema de telecomunicaciones telecommunications
system
sistema de telecomunicaciones automático
automatic telecommunications system
sistema de telecomunicaciones de control directo
automático automatic direct-control
telecommunications system
sistema de telecomunicaciones manual manual
telecommunications system
sistema de telegrafía de dos tonos two-tone
telegraph system
sistema de teleimpresora teleprinter system
sistema de telemedida telemetering system
sistema de teleprocesamiento teleprocessing system
sistema de televisión comunitaria community
television system
sistema de tensión constante constant-voltage system
sistema de tensión extra alta extra-high voltage
system
sistema de tensión media medium-voltage system
sistema de terminación digital digital termination
system
sistema de Thury Thury system
sistema de tiempo compartido time-sharing system
sistema de tierra ground system
sistema de tierra de antena antenna ground system
sistema de tierra radial radial ground system

sistema de transmisión transmission system
sistema de transmisión acústica acoustic
transmission system
sistema de transmisión de datos data-transmission
system
sistema de transmisión de información
information-transmission system
sistema de transmisión de placas paralelas
parallel-plate transmission system
sistema de transmisión de Thury Thury
transmission system
sistema de transmisión múltiplex multiplex
transmission system
sistema de transmisión por impulsos pulse
transmission system
sistema de transporte de cinta tape-transport system
sistema de transposición transposition system
sistema de treinta y dos bits thirty-two bit system
sistema de tres canales three-way system
sistema de tres direcciones three-address system
sistema de tres hilos three-wire system
sistema de unidad de cartucho cartridge drive
system
sistema de unidad de disco disk drive system
sistema de unidades system of units
sistema de unitérminos uniterm system
sistema de uso general general-purpose system
sistema de usuarios múltiples multiple-user system
sistema de vacío vacuum system
sistema de verificación verification system
sistema de verificación automatizado automated
verification system
sistema de videodisco óptico láser laser optical
videodisk system
sistema de visualización de datos data display
system
sistema de visualización gráfica graphic display
system
sistema de Wheatstone Wheatstone system
sistema decca decca system
sistema decimal decimal system
sistema decimal codificado en binario binary-coded
decimal system
sistema dectra dectra system
sistema dedicado dedicated system
sistema dependiente de códigos code dependent
system
sistema desequilibrado unbalanced system
sistema detector de errores error-detecting system
sistema detector de fallas fault-detecting system
sistema digital en tiempo real real-time digital
system
sistema dimensional dimensional system
sistema díplex diplex system
sistema director director system
sistema disipativo dissipative system
sistema distribuido distributed system
sistema distribuido central central distributed
system
sistema dúplex duplex system
sistema dúplex completo full-duplex system
sistema dúplex diferencial differential duplex system
sistema ejecutivo compartido shared executive
system
sistema ejecutivo de multiprogramación
multiprogramming executive system
sistema eléctrico electrical system
sistema eléctrico internacional international
electrical system
sistema electrohidráulico electrohydraulic system
sistema electromagnético electromagnetic system
sistema electromagnético de Heaviside-Lorentz
Heaviside-Lorentz electromagnetic system

sistema electromagnético práctico practical electromagnetic system
sistema electrostático electrostatic system
sistema empotrado embedded system
sistema en anillo ring system
sistema en bucle loop system
sistema en diversidad diversity system
sistema en doble diversidad dual-diversity system
sistema en línea on-line system
sistema en paralelo parallel system
sistema en serie series system
sistema en tándem tandem system
sistema en tiempo real real-time system
sistema equilibrado balanced system
sistema equilibrador de ruido noise-balancing system
sistema estereofónico stereophonic system
sistema estereofónico de Minter Minter stereophonic system
sistema estereofónico de tres canales three-channel stereophonic system
sistema estereofónico multicanal multichannel stereophonic system
sistema estereoscópico stereoscopic system
sistema estroboscópico stroboscope system
sistema extintor de fuego fire-extinguisher system
sistema fotoeléctrico photoelectric system
sistema fotoeléctrico de haz modulado modulated-beam photoelectric system
sistema fuera de línea off-line system
sistema Gee Gee system
sistema hexadecimal hexadecimal system
sistema hexafásico six-phase system
sistema híbrido hybrid system
sistema idealizado idealized system
sistema incremental incremental system
sistema independiente de códigos code independent system
sistema indicador de fallas fault-reporting system
sistema integrado integrated system, packaged system
sistema interactivo interactive system
sistema interconectado interconnected system
sistema intermedio intermediate system
sistema internacional international system
Sistema Internacional de Unidades International System of Units
sistema láser laser system
sistema lineal linear system
sistema lógico programable programmable logic system
sistema magnético magnetic system
sistema magnético astático astatic magnetic system
sistema magnético multipista multitrack magnetic system
sistema magnetohidrodinámico magnetohydrodynamic system
sistema manual manual system
sistema megafónico public-address system
sistema métrico metric system
sistema metro-kilogramo-segundo meter-kilogram-second system
sistema modular modular system
sistema monitor monitor system
sistema monoclínico monoclinic system
sistema monofásico single-phase system
sistema monofónico monophonic system
sistema móvil mobile system
sistema móvil bidireccional two-way mobile system
sistema móvil unidireccional one-way mobile system
sistema multiacceso multiaccess system
sistema multicanal multichannel system
sistema multicanal de banda ancha wideband

multichannel system
sistema multicomputadora multicomputer system
sistema multidimensional multidimensional system
sistema multifásico multiphase system
sistema multifrecuencia multifrequency system
sistema múltiple multiple system
sistema múltiplex multiplex system
sistema múltiplex de tres canales three-channel multiplex system
sistema multiplicador multiplying system
sistema multiprocesador multiprocessor system
sistema multitarea multitasking system
sistema multiunidad multiunit system
sistema multiusuario multiuser system
sistema no entrelazado noninterlaced system
sistema no lineal nonlinear system
sistema numérico number system
sistema numérico binario binary number system
sistema numérico decimal decimal number system
sistema numérico duodecimal duodecimal number system
sistema numérico hexadecimal hexadecimal number system
sistema numérico octal octal number system
sistema numérico posicional positional number system
sistema numérico sexadecimal sexadecimal number system
sistema numérico sexagesimal sexagesimal number system
sistema numérico ternario ternary number system
sistema numérico trinario trinary number system
sistema oboe oboe system
sistema octal octal system
sistema octonario octonary system
sistema operativo operating system
sistema operativo básico basic operating system
sistema operativo de cinta tape operating system
sistema operativo de disco disk operating system
sistema operativo de disco de microcomputadora microcomputer disk operating system
sistema operativo de procesos múltiples multiple process operating system
sistema operativo de red network operating system
sistema operativo de uso general general-purpose operating system
sistema operativo en tiempo real real-time operating system
sistema operativo extendido extended operating system
sistema operativo multitarea multitasking operating system
sistema óptico optical system
sistema óptico de microondas microwave optical system
sistema óptico de Schmidt Schmidt optical system
sistema óptico perfecto perfect optical system
sistema ortorrómbico orthorhombic system
sistema oscilante equivalente equivalent oscillating system
sistema para evitar colisiones collision-avoidance system
sistema pasivo passive system
sistema paso a paso step-by-step system
sistema piloto pilot system
sistema polifásico polyphase system
sistema polifásico equilibrado balanced polyphase system
sistema por magneto magneto system
sistema práctico practical system
sistema principal main system, host system
sistema protector protective system
sistema pulsante pulsing system

sistema que permite funcionamiento tras falla failsoft system
sistema racionalizado rationalized system
sistema radial radial system
sistema rebecca rebecca system
sistema rebecca-eureka rebecca-eureka system
sistema receptor receiving system
sistema redundante redundant system
sistema regulador regulating system
sistema residente resident system
sistema residente en cinta tape resident system
sistema residente en disco disk resident system
sistema rho-theta rho-theta system
sistema robótico robotic system
sistema rotativo rotary system
sistema secuencial sequential system
sistema selectivo selective system
sistema selsyn selsyn system
sistema semiautomático semiautomatic system
sistema semidúplex half-duplex system
sistema semimecánico semimechanical system
sistema silenciador muting system, squelch system
sistema símplex simplex system
sistema simultáneo simultaneous system
sistema sin bloqueos nonblocking system
sistema sin conexión a tierra ungrounded system
sistema sin interrupción no-break system
sistema sincrónico synchronous system
sistema subterráneo underground system
sistema supervisor supervisory system
sistema telefónico telephone system
sistema telefónico manual manual telephone system
sistema telefónico móvil mobile telephone system
sistema telefónico semiautomático semiautomatic telephone system
sistema telegráfico telegraph system
sistema telegráfico de frecuencia vocal voice-frequency telegraph system
sistema telegráfico ómnibus omnibus telegraph system
sistema telegráfico por desfase phase-shift telegraph system
sistema telerán teleran system
sistema télex telex system
sistema transmisor transmitting system
sistema triclínico triclinic system
sistema tricromático trichromatic system
sistema trifásico three-phase system
sistema ultrasónico ultrasonic system
sistema vibrante vibrating system
sistema vidicón vidicon system
sistema virtual virtual system
sistemas compartidos shared systems
sistemas eléctricos esenciales essential electrical systems
sistemas eléctricos interactivos interactive electrical systems
sistemático systematic
sobreacoplado overcoupled
sobreacoplamiento overcoupling
sobreagrupamiento overbunching
sobrealcance overshoot
sobreamortiguado overdamped
sobreamortiguamiento overdamping
sobrecarga overload
sobrecarga de canal channel overload
sobrecarga de circuito circuit overload
sobrecarga de corto tiempo short-time overload
sobrecarga de información information overload
sobrecarga de operación operating overload
sobrecarga de tensión voltage overload
sobrecarga momentánea momentary overload
sobrecarga normal normal overload

sobrecarga térmica thermal overload
sobrecargado overloaded
sobrecargar overcharge
sobrecarrera eléctrica electrical overtravel
sobrecarrera mecánica mechanical overtravel
sobrecorriente overcurrent, excess current, current surge
sobrecorriente instantánea instantaneous overcurrent
sobrecorriente transitoria surge current
sobrecorte overcutting
sobrediseñar overdesign
sobredisparo overshoot
sobreelongación overshoot
sobreescribir overwrite
sobreexcitación overexcitation, overdrive
sobreexcitado overexcited, overdriven
sobreexploración overscanning
sobreexponer overexpose
sobreexposición overexposure
sobreflujo overflux
sobrefrecuencia overfrequency
sobreimpresión overprinting
sobreimprimir overprint
sobreimpulso overshoot
sobremodulación overmodulation
sobremodular overmodulate
sobreoscilación overswing
sobreperforación overpunching
sobrepolarización overbias
sobrepotencia overpower
sobrepotencial overpotential
sobrepresión overpressure
sobrerrecorrido overtravel, overthrow
sobresaturación de colores color overload
sobretemperatura overtemperature
sobretensión overvoltage, voltage surge, surge
sobretensión de conmutación switching overvoltage
sobretensión de contador counter overvoltage
sobretensión de línea line surge
sobretensión electrolítica electrolytic excess voltage
sobretensión electrónica electronic surge
sobretensión estática static overvoltage
sobretensión inductiva inductive kick
sobretensión longitudinal longitudinal overvoltage
sobretensión momentánea momentary surge
sobretensión móvil traveling overvoltage
sobretensión óhmica ohmic overvoltage
sobretensión oscilante oscillatory surge
sobretensión por rayo lightning surge
sobretensión sostenida sustained overvoltage
sobretensión temporal temporary overvoltage
sobretensión transitoria surge voltage, surge
sobretensión transversal transverse overvoltage
sobretono overtone
sobrevelocidad overspeed
sodar sodar
sodio sodium
sofar sofar
sofométricamente psophometrically
sofométrico psophometric
sofómetro psophometer
software software
solapamiento overlapping
solapamiento de señales signal overlapping
solape overlap
solape de banda band overlap
solape de entrada-salida input-output overlap
solape de frecuencias frequency overlap
solape de procesamiento processing overlap
solar solar
soldabilidad solderability
soldable solderable

soldado welded
soldado con bronce bronzewelded
soldado con láser laser-welded
soldado con plata silver-soldered
soldadora welder
soldadora de puntos múltiples multiple-spot welder
soldadora multipunto multipoint welder
soldadura solder, soldering, weld, welding
soldadura a tope butt welding
soldadura blanda soft soldering
soldadura capacitiva capacitive welding
soldadura con bronce bronzewelding
soldadura con núcleo de resina rosin-core solder
soldadura continua continuous welding
soldadura eléctrica electrical welding
soldadura en serie series welding
soldadura fría cold weld
soldadura fuerte brazing
soldadura fuerte con plata silver brazing
soldadura fuerte eléctrica electric brazing
soldadura fuerte por inducción induction brazing
soldadura fuerte por resistencia resistance brazing
soldadura fuerte ultrasónica ultrasonic brazing
soldadura láser laser welding
soldadura manual manual welding
soldadura percusiva percussive welding
soldadura por aire caliente hot-air soldering
soldadura por arco arc welding
soldadura por arco eléctrico electric arc welding
soldadura por arco manual manual arc welding
soldadura por arco metálico metal-arc welding
soldadura por arco sumergido submerged arc
　welding
soldadura por cortocircuito short-circuit welding
soldadura por electrodos múltiples
　multiple-electrode welding
soldadura por energía acumulada stored-energy
　welding
soldadura por fusión fusion welding
soldadura por haz electrónico electron-beam
　welding
soldadura por inducción induction welding
soldadura por inmersión dip soldering
soldadura por ola flow soldering
soldadura por percusión percussion welding
soldadura por presión pressure welding
soldadura por presión fría cold pressure welding
soldadura por pulsaciones pulsation welding
soldadura por puntos spot welding
soldadura por puntos múltiples multiple-spot
　welding
soldadura por resistencia resistance soldering
soldadura por ultrasonidos ultrasonic soldering
soldante solder
soldante blando soft solder
soldante de plata silver solder
soldante duro hard solder
soldante en pasta paste solder
solenoidal solenoidal
solenoide solenoid
solenoide de arrancador starter solenoid
solenoide de arranque start solenoid
solenoide patrón standard solenoid
solenoide rotativo rotary solenoid
solenoide superconductor superconducting solenoid
solicitud request
solicitud de activación de sesión session activation
　request
solicitud de canal channel request
solicitud de desactivación de sesión session
　deactivation request
solicitud de distribución distribution request
solicitud de entrada input request

solicitud de establecimiento de sesión session
　establishment request
solicitud de información information request
solicitud de iniciación de sesión session initiation
　request
solicitud de interrupción interrupt request
solicitud de terminación de sesión session
　termination request
solicitud inmediata immediate request
solicitud primaria primary request
solicitud secundaria secondary request
sólido solid
solión solion
solistrón solistron
solistrón sintonizable por tensión voltage-tunable
　solistron
sólo de lectura read-only
solubilidad solubility
solubilizar solubilize
soluble soluble
solución solution
solución anódica anodic solution
solución de decapado pickling solution
solución de galvanoplastia plating solution
solución de problemas problem solving
solución electrolítica electrolytic solution
solución gráfica graphic solution
solución molar molar solution
solución normal normal solution, standard solution
solución saturada saturated solution
solución supersaturada supersaturated solution
soluto solute
solvente solvent
sombra shading
sombra de radar radar shadow
sombra espuria spurious shading
sombreado shading
sonador sounder
sonar sonar
sonar activo active sonar
sonar de exploración scanning sonar
sonar de profundidad variable variable-depth sonar
sonar pasivo passive sonar
sonda probe, sonde
sonda accesoria accessory probe
sonda aislada insulated probe
sonda apantallada shielded probe
sonda capacitiva capacitive probe
sonda de acoplamiento coupling probe
sonda de aguja needle probe
sonda de alta resistencia high-resistance probe
sonda de alta tensión high-voltage probe
sonda de baja capacitancia low-capacitance probe
sonda de captación pickup probe
sonda de corriente current probe
sonda de corriente continua direct-current probe
sonda de cristal crystal probe
sonda de descarga discharge probe
sonda de diodo diode probe
sonda de diodo de cristal crystal-diode probe
sonda de efecto de Hall Hall-effect probe
sonda de exploración search probe
sonda de funciones múltiples multipurpose probe
sonda de guíaondas waveguide probe
sonda de Hall Hall probe
sonda de haz iónico ion-beam probe
sonda de pico peak probe
sonda de potencial potential probe
sonda de prueba test probe
sonda de punta de aguja needle-tip probe
sonda de radiofrecuencia radio-frequency probe
sonda de rayos gamma gamma-ray probe
sonda de resistividad resistivity probe

sonda de sintonización tuning probe
sonda de tensión voltage probe
sonda de termistor thermistor probe
sonda desmoduladora demodulator probe
sonda detectora detector probe
sonda divisora divider probe
sonda eléctrica electric probe
sonda electrónica electron probe
sonda equilibrada balanced probe
sonda espacial space probe
sonda fibroóptica fiber-optic probe
sonda flotante floating probe
sonda fotoeléctrica photoelectric probe
sonda indicadora indicator probe
sonda lógica logic probe
sonda magnética magnetic probe
sonda móvil moving probe
sonda multiplicadora multiplier probe
sonda no blindada unshielded probe
sonda óptica optical probe
sonda piezoeléctrica piezoelectric probe
sonda pirométrica pyrometer probe
sonda rectificadora rectifier probe
sonda sintonizable tunable probe
sonda sonora sound probe
sonda térmica thermal probe
sonda tubular tubular probe
sonda ultrasónica ultrasonic probe
sondeo probing, sounding, polling
sondeo aleatorio random probing
sondeo con incidencia oblicua oblique-incidence
   probing
sondeo paralelo parallel polling
sondeo sónico sonic sounding
sondeo submarino underwater sounding
sondeo ultrasónico ultrasonic sounding
sónica sonics
sónico sonic
sonido sound
sonido aerodinámico aerodynamic sound
sonido audible audible sound
sonido binaural binaural sound
sonido cuadrafónico quadraphonic sound
sonido de aguja needle chatter
sonido de dos canales dual-channel sound
sonido de enmascaramiento masking sound
sonido de fondo background sound
sonido de frecuencia única single-frequency sound
sonido de televisión television sound
sonido de timbre ringing
sonido difuso diffused sound
sonido digital digital sound
sonido equilibrado balanced sound
sonido estereofónico stereophonic sound
sonido incidente incident sound
sonido panorámico panoramic sound
sonido puro pure sound
sonido reverberante reverberant sound
sonne sonne
sonoboya sonobuoy
sonógrafo sonograph
sonograma sonogram
sonoluminiscencia sonoluminescence
sonoluminiscente sonoluminescent
sonoptografía sonoptography
sonoridad loudness
soplado de chispas spark blowing
soplador blower
soplete de plasma de radiofrecuencia
   radio-frequency plasma torch
soplo blowing
soporte support, bracket, rack, bearing
soporte aislante standoff insulator

soporte anódico anode support
soporte de aislador insulator bracket
soporte de aplicación application support
soporte de baterías battery rack
soporte de bobina coil form
soporte de cable cable bracket
soporte de decisión decision support
soporte de electrodo electrode support
soporte de electrodos en D dee line
soporte de medidor meter support
soporte de montaje mounting bracket
soporte de portaescobillas brush holder support
soporte de programa program support
soporte perlado beaded support
soporte rígido rigid support
soporte vertical drop bracket
sostenimiento hold
subacoplamiento undercoupling
subacuático subaqueous
subagrupamiento underbunching
subalimentador subfeeder
subambiente subenvironment
subamortiguado underdamped
subamortiguamiento underdamping
subárea subarea
subarmónica subharmonic
subarmónico subharmonic
subatómico subatomic
subaudible subaudible
subaudio subaudio
subbanda base subbase band
subbanda de frecuencia frequency subband
subcampo subfield
subcanal subchannel
subcapa sublayer
subcentro subcenter
subcompensado undercompensated
subcomponente subcomponent
subconjunto subset
subconjunto de caracteres character subset
subconjunto de caracteres numéricos numerical
   character subset
subconjunto de lenguaje language subset
subconjunto digital digital subset
subconjunto gráfico graphic subset
subconmutación subcommutation
subcontrol subcontrol
subconvertidor paramétrico parametric-down
   converter
subcorriente subcurrent
subchasis subchassis
subdesbordamiento underflow
subdirección subaddress
subdirectorio subdirectory
subdistribución subdistribution
subdividir subdivide
subdivisión subdivision
subdivisión de bloque blockette
subdivisor subdivider
subensamblaje subassembly
subesquema subscheme
subestación substation
subestación aislada por gas gas-insulated substation
subestación automática automatic substation
subestación de acumuladores accumulator
   substation
subestación de distribución distribution substation
subestación de rectificadores rectifier substation
subestación de unidad móvil mobile unit substation
subestación exterior outdoor substation
subestación móvil mobile substation
subestación semiautomática semiautomatic
   substation

subestación telecontrolada telecontrolled substation
subestación transportable transportable substation
subestación unitaria unit substation
subestructura substructure
subexcitado underexcited, underdriven
subexploración underscan
subfrecuencia subfrequency
subgrupo subgroup
subgrupo de portadoras carrier subgroup
subida de tensión voltage rise
subida de tensión relativa relative voltage rise
subida de tensión resonante resonant-voltage rise
subida de velocidad absoluta absolute speed rise
subida rápida rapid rise
subimpulso undershoot
sublenguaje de base de datos database sublanguage
sublenguaje de datos data sublanguage
submando subcommand
submenú submenu
submicroscópico submicroscopic
submilimétrico submillimeter
subminiatura subminiature
subminiaturización subminiaturization
subminiaturizado subminiaturized
subminiaturizar subminiaturize
submodulación undermodulation
submodulador submodulator
submúltiplo submultiple
subnivel sublevel
suboscilación underswing
subpanel subpanel
subparámetro subparameter
subpaso substep
subperforación underpunch
subpermanente subpermanent
subportadora subcarrier
subportadora cero zero subcarrier
subportadora de audio audio subcarrier
subportadora de colores color subcarrier
subportadora de colores modulada modulated color
   subcarrier
subportadora de crominancia chrominance
   subcarrier
subportadora estereofónica stereophonic subcarrier
subportadora intermedia intermediate subcarrier
subportadora piloto pilot subcarrier
subprograma subprogram
subprograma de función function subprogram
subprograma de procedimientos procedure
   subprogram
subred subnetwork
subreflector subreflector
subrefracción subrefraction
subregión subregion
subregional subregional
subrutina subroutine
subrutina abierta open subroutine
subrutina anidada nested subroutine
subrutina cerrada closed subroutine
subrutina de biblioteca library subroutine
subrutina de división division subroutine
subrutina de dos niveles two-level subroutine
subrutina de inserción directa direct-insert
   subroutine
subrutina de llamada call subroutine
subrutina de memoria de acceso aleatorio
   random-access memory subroutine
subrutina de nivel de interrupción interrupt level
   subroutine
subrutina de nivel único one-level subroutine
subrutina de segundo nivel second-level subroutine
subrutina de tres niveles three-level subroutine
subrutina de vaciado dump subroutine

subrutina dinámica dynamic subroutine
subrutina en línea inline subroutine
subrutina enlazada linked subroutine
subrutina estática static subroutine
subrutina insertada inserted subroutine
subrutina matemática mathematical subroutine
subrutina microprogramada microprogrammed
   subroutine
subrutina paramétrica parametric subroutine
subrutina patrón standard subroutine
subrutina recursiva recursive subroutine
subrutina reentrante reentrant subroutine
subrutina reubicable relocatable subroutine
subrutina reusable reusable subroutine
subsatélite subsatellite
subsecuencia substring
subsincrónico subsynchronous
subsintonizado undertuned
subsistema subsystem
subsistema central central subsystem
subsistema de almacenamiento eléctrico electric
   storage subsystem
subsistema de almacenamiento en disco disk
   storage subsystem
subsistema de cinta magnética magnetic-tape
   subsystem
subsistema de disco disk subsystem
subsistema de entrada input subsystem
subsistema de entrada de trabajos job-entry
   subsystem
subsistema de lenguaje language subsystem
subsistema de programador programmer subsystem
subsistema de red array subsystem
subsistema de salida output subsystem
subsistema funcional functional subsystem
subsistema interactivo interactive subsystem
subsistema periférico peripheral subsystem
subsónico subsonic
substrato activo active substrate
substrato anisotrópico anisotropic substrate
subtarea subtask
subtensión undervoltage
subterráneo underground
subtotal subtotal
subtransitorio subtransient
subunidad subunit
subzona subzone
sucesivo successive
sueño eléctrico electric sleep
sujeción clamping
sujetador fastener, holder
sujetador de cordón cord fastener
sujetador de papel paper holder
sujetador magnético magnetic holder
sulfatación sulfation
sulfato sulfate
sulfato de plomo lead sulfate
sulfuro de cinc zinc sulfide
sulfuro de plomo lead sulfide
suma sum, addition
suma algebraica algebraic sum
suma almacenada stored addition
suma aritmética arithmetic sum
suma de comprobación check sum
suma de frecuencias frequency sum
suma destructiva destructive addition
suma en paralelo parallel addition
suma en serie serial addition
suma falsa false add
suma lógica logical sum
suma no destructiva nondestructive addition
suma parcial partial sum
suma repetitiva repetitive addition

suma vectorial vector addition
sumación summation
sumacional summational
sumador adder, summer
sumador-acumulador adder-accumulator
sumador algebraico algebraic adder
sumador analógico analog adder
sumador binario binary adder
sumador completo full adder
sumador completo en paralelo parallel full-adder
sumador completo en serie serial full-adder
sumador de marcas marker adder
sumador de tres entradas three-input adder
sumador de un dígito one-digit adder
sumador digital en serie serial digital adder
sumador electrónico electronic adder
sumador en paralelo parallel adder
sumador en serie serial adder
sumador inversor inverting adder
sumador inversor analógico analog inverting adder
sumador lineal linear adder
sumador-restador adder-subtracter
sumador-restador analógico analog adder-subtracter
sumando addend
sumergible submersible
sumidero sink
sumidero de energía sink
suministrador supplier
suministro supply
suministro de batería battery supply
suministro eléctrico electric supply
super super
superacústico superacoustic
superaudible superaudible
superaudio superaudio
supercomputadora supercomputer
superconductividad superconductivity
superconductividad superficial surface superconductivity
superconductivo superconductive
superconductor superconductor
superconductor duro hard superconductor
superconmutación supercommutation
supercontrol supercontrol
supercorriente supercurrent
superdirectividad superdirectivity
superdirectivo superdirective
superemitrón superemitron
superficial superficial
superficie surface
superficie aleatoria random surface
superficie de almacenamiento storage surface
superficie de calentamiento heating surface
superficie de colector commutator surface
superficie de comparación comparison surface
superficie de contacto contact surface
superficie de convergencia convergence surface
superficie de dispersión scattering surface
superficie de Lambert Lambertian surface
superficie de nivel level surface
superficie de recolección electrónica electron collection surface
superficie de referencia reference surface
superficie de registro recording surface
superficie de trabajo working surface
superficie de visualización display surface
superficie en piel de naranja orange peel
superficie equifase equiphase surface
superficie equipotencial equipotential surface
superficie equiseñal equisignal surface
superficie especular specular surface
superficie fotoconductora photoconducting surface
superficie inferior bottom surface

superficie magnética magnetic surface
superficie mate matte surface
superficie oculta hidden surface
superficie pasivada passivated surface
superficie radiante radiating surface
superficie reflectora reflective surface
superficie retrorreflejante retroreflecting surface
superficie sensibilizada sensitized surface
superficie sensible a la luz light-sensitive surface
superficies paralelas parallel surfaces
superfluo superfluous
superganancia supergain
supergrupo supergroup
supergrupo básico basic supergroup
supergrupo de canales channel supergroup
superheterodino superheterodyne
superheterodino de doble conversión double-conversion superheterodyne
supericonoscopio supericonoscope
superimposición de colores color registration
superluminoso superluminous
supermodulación supermodulation
superorticón image orthicon
superposición superposition, overlay
superposición de impulsos pulse superposition
superpotencia superpower
superprecisión superprecision
superpresión superpressure
superpropagación superpropagation
superpuesto superimposed
superradiancia superradiance
superred supergrid
superrefracción superrefraction
superrefractario superrefractory
superregeneración superregeneration
superregenerativo superregenerative
supersaturación supersaturation
supersaturado supersaturated
supersaturar supersaturate
supersensible supersensitive
supersincrónico supersynchronous
supersónica supersonics
supersónico supersonic
supertensión supervoltage
supervisión de fallas fault supervision
supervisor supervisor
suplemental supplemental
suplementario supplementary
supresión suppression
supresión armónica harmonic suppression
supresión completa complete suppression
supresión de arco arc suppression
supresión de bajos bass suppression
supresión de banda band suppression
supresión de banda lateral sideband suppression
supresión de ceros zero suppression
supresión de ceros a la izquierda leading zero suppression
supresión de clasificación sorting suppression
supresión de chispas spark suppression
supresión de ecos echo suppression
supresión de ecos parásitos clutter suppression
supresión de espacio space suppression
supresión de haz beam suppression, beam extinction
supresión de impresión print suppression
supresión de interferencia interference suppression
supresión de lóbulos laterales sidelobe suppression
supresión de lóbulos secundarios secondary-lobe suppression
supresión de modulación modulation suppression
supresión de perforación punch suppression
supresión de portadora carrier suppression
supresión de puerta gate suppression

supresión de radiointerferencia radio interference suppression
supresión de ruido noise suppression
supresión de ruido automática automatic noise suppression
supresión de ruido de interportadora intercarrier noise suppression
supresión de ruido entre estaciones interstation noise suppression
supresión de zona zone suppression
supresión en cuadratura quadrature suppression
supresión fundamental fundamental suppression
supresión incompleta incomplete suppression
supresor suppressor, killer
supresor armónico harmonic suppressor
supresor de bujía spark-plug suppressor
supresor de campo field suppressor
supresor de canto singing suppressor
supresor de colores color killer
supresor de chispas spark suppressor, spark arrester, spark killer
supresor de eco echo suppressor
supresor de eco completo full echo suppressor
supresor de eco intermedio intermediate echo suppressor
supresor de interferencia interference suppressor
supresor de modo mode suppressor
supresor de modo de guíaondas waveguide-mode suppressor
supresor de onda inversa inverse suppressor
supresor de parásitos parasitic suppressor
supresor de picos transitorios surge suppressor
supresor de puntas spike suppressor
supresor de radiofrecuencia radio-frequency suppressor
supresor de reacción reaction suppressor
supresor de rejilla grid suppressor
supresor de ruido noise supressor, noise killer
supresor de ruido dinámico dynamic noise suppressor
supresor de ruido eléctrico electric-noise suppressor
supresor de ruido entre canales interchannel noise suppressor
supresor de sobretensión overvoltage suppressor
supresor de velocidad de subida rate-of-rise suppressor
supresor eléctrico electrical suppressor
suprimido suppressed
suprimir suppress
sur magnético magnetic south
surco groove, track
surco cerrado locked groove
surco de entrada lead-in groove
surco de salida lead-out groove
surco entre registros leadover groove
surco excéntrico eccentric groove
surco fino fine groove
surco no modulado unmodulated groove
surco normal standard groove
surco rápido fast groove
surcos por gas gas grooves
susceptancia susceptance
susceptancia de electrodo electrode susceptance
susceptancia de sintonización tuning susceptance
susceptibilidad susceptibility
susceptibilidad antiferromagnética antiferromagnetic susceptibility
susceptibilidad diamagnética diamagnetic susceptibility
susceptibilidad dieléctrica dielectric susceptibility
susceptibilidad eléctrica electric susceptibility
susceptibilidad electromagnética electromagnetic susceptibility

susceptibilidad inicial initial susceptibility
susceptibilidad magnética magnetic susceptibility
susceptibilidad reversible reversible susceptibility
suspensión suspension
suspensión acústica acoustic suspension
suspensión bifilar bifilar suspension
suspensión catenaria catenary suspension
suspensión continua lashing
suspensión de recorrido largo long-throw suspension
suspensión dinámica dynamic suspension
suspensión rígida rigid suspension
suspensión unifilar unifilar suspension
sustancia amorfa amorphous substance
sustancia antiferroeléctrica antiferroelectric substance
sustancia antiferromagnética antiferromagnetic substance
sustancia diamagnética diamagnetic substance
sustancia ferromagnética ferromagnetic substance
sustancia paramagnética paramagnetic substance
sustancia peligrosa hazardous substance
sustitución substitution
sustitución de dirección address substitution
sustitución de parámetro parameter substitution
sustitución directa direct substitution
sustracción subtraction
sustractivo subtractive
sustractor subtractor
sustraendo subtrahend
sustrato substrate
sustrato de zafiro sapphire substrate
sustrato pasivo passive substrate

T

T híbrida hybrid tee
T mágica magic tee
tabique septum
tabla table
tabla de archivos file table
tabla de arreglo de caracteres character arrangement table
tabla de asignación allocation table
tabla de asignación de archivo file allocation table
tabla de asignación de disco disk allocation table
tabla de asignación de periféricos peripheral assignment table
tabla de cabecera header table
tabla de caracteres character table
tabla de códigos code table
tabla de colores color table
tabla de configuración de red network configuration table
tabla de control de archivo file control table
tabla de control de nivel level control table
tabla de control de partición partition control table
tabla de control de trabajos job control table
tabla de conversión conversion table
tabla de decisión decision table
tabla de descripción de datos data description table
tabla de entrada-salida input-output table
tabla de estado de canal channel status table
tabla de frecuencias frequency table
tabla de funciones function table
tabla de imagen image table
tabla de niveles level table
tabla de recuperación de programa program recovery table
tabla de referencia de programa program reference table
tabla de secuencia sequence table
tabla de trabajos job table
tabla de verdad truth table
tabla periódica periodic table
tablero board, switchboard, panel
tablero A A board
tablero acelerador accelerator board
tablero de alambrado impreso printed-wiring board
tablero de anuncios electrónico electronic bulletin board
tablero de base baseboard
tablero de carga charging board
tablero de circuitos circuit board
tablero de circuitos impresos printed-circuit board
tablero de componentes impresos printed-component board
tablero de conectores de circuitos impresos backplane
tablero de conexiones patchboard, wiring board
tablero de control control board
tablero de control maestro master control board
tablero de entrada-salida input-output board
tablero de expansión expansion board
tablero de extensión extension board
tablero de fusibles fuse board
tablero de instrumentos instrument board, instrument panel
tablero de lectura read board
tablero de memoria memory board
tablero de montaje mounting board
tablero de motor motorboard
tablero de pruebas test board
tablero de pulsadores push-button board
tablero de reloj-calendario clock-calendar board
tablero de terminales terminal board
tablero direccionado addressed board
tablero electrónico electronic board
tablero expansor expander board
tablero lógico logic board
tablero matriz motherboard
tablero multicapa multilayer board
tablero multifunción multifunction board
tablero sonoro sounding board
tabulación tabulation
tabulación de base base tabulation
tabulación decimal decimal tabulation
tabulación inversa back tabulation
tabulador tabulator
tabular tabulate
taburete aislante insulating stool
tacán tacan
tacitrón tacitron
tacogenerador tachogenerator
tacógrafo tachograph
tacométrico tachometric
tacómetro tachometer
tacómetro de arrastre magnético magnetic-drag tachometer
tacómetro de corriente continua direct-current tachometer
tacómetro de corrientes parásitas eddy-current tachometer
tacómetro eléctrico electric tachometer
tacómetro electrónico electronic tachometer
tacómetro estroboscópico stroboscopic tachometer
tacómetro magnético magnetic tachometer
tacómetro óptico optical tachometer
tajada slice
tajada gruesa slab
taladro láser laser drill
taladro ultrasónico ultrasonic drill
talbot talbot
talio thallium
tamaño de archivo file size
tamaño de base de datos database size
tamaño de bloque block size
tamaño de cartucho cartridge size
tamaño de conjunto de datos data set size
tamaño de contacto contact size
tamaño de chip chip size
tamaño de imagen picture size
tamaño de memoria memory size
tamaño de muestra sample size
tamaño de palabra word size
tamaño de pantalla screen size
tamaño de partículas particle size
tamaño de punto spot size
tamaño de registro record size
tamaño de ventana window size
tamaño máximo de archivo maximum file size
tambaleo stagger, staggering
tambor drum
tambor de almacenamiento storage drum
tambor de cable cable drum
tambor de espejos mirror drum
tambor de exploración scanning drum
tambor de impresión print drum
tambor de lentes lens drum
tambor de memoria memory drum

tambor de programa program drum
tambor magnético magnetic drum
tambor rotativo rotary drum
tándem tandem
tangencial tangential
tangente tangent
tangente de arco arc tangent
tangente de pérdidas loss tangent
tangente hiperbólica hyperbolic tangent
tanque tank
tanque de despojamiento stripper tank
tanque de galvanoplastia plating tank
tanque de liberación liberator tank
tanque de mercurio mercury tank
tanque de placa plate tank
tanque de rejilla grid tank
tanque de vacío vacuum tank
tanque electrolítico electrolytic tank
tantalio tantalum
tapa lid
tapa de medidor meter cover
tapete aislante insulating mat
tapete de tierra ground mat
taquimétrico tachymetric
taquímetro tachymeter
tarea task, job
tarea abortada aborted job
tarea de fondo background job
tarea de lectura reading task
tarea de paso de trabajo job step task
tarea de prioridad foreground task
tarea de procesamiento de datos data-processing task
tarea de sistema system task
tarea dependiente dependent job
tarea inmediata immediate task
tarea no productiva nonproductive task
tarea opcional optional task
tarea primaria primary task
tarea principal main task, main job
tarea productiva productive task
tarea secundaria secondary task
tarea única single task
tareas concurrentes concurrent tasks, concurrent jobs
tarifa de llamada call rate
tarjeta card
tarjeta a cinta card-to-tape
tarjeta a disco card-to-disk
tarjeta a tarjeta card-to-card
tarjeta binaria binary card
tarjeta binaria en columnas column binary card
tarjeta compuesta composite card
tarjeta con franja magnética magnetic strip card
tarjeta de alambrado impreso printed-wiring card
tarjeta de archivo file card
tarjeta de búsqueda search card
tarjeta de cabecera header card
tarjeta de carga load card
tarjeta de circuitos circuit card
tarjeta de circuitos impresos printed-circuit card
tarjeta de circuitos integrados integrated-circuit card
tarjeta de comprobación check card
tarjeta de comunicaciones communication card
tarjeta de continuación continuation card
tarjeta de control control card
tarjeta de control de fin de trabajo end-of-job control card
tarjeta de control de programa program-control card
tarjeta de control de trabajos job control card
tarjeta de controlador controller card

tarjeta de datos data card
tarjeta de datos fuente source-data card
tarjeta de dirección de terminal terminal address card
tarjeta de entrada input card
tarjeta de entrada-salida en paralelo parallel input-output card
tarjeta de entrada analógica analog input card
tarjeta de error error card
tarjeta de expansión expansion card
tarjeta de fin de trabajo end-of-job card
tarjeta de gráficos graphics card
tarjeta de identificación identification card
tarjeta de índice index card
tarjeta de instrucción instruction card
tarjeta de interfaz interface card
tarjeta de interfaz de red network interface card
tarjeta de interfaz en serie serial-interface card
tarjeta de llamada de agenda agendum call card
tarjeta de memoria de acceso aleatorio random-access memory card
tarjeta de memoria estática static-memory card
tarjeta de microcanal microchannel card
tarjeta de microcomputadora microcomputer card
tarjeta de microprocesador microprocessor card
tarjeta de noventa columnas ninety-column card
tarjeta de noventa y seis columnas ninety six-column card
tarjeta de orden order card
tarjeta de papel paper card
tarjeta de parámetro parameter card
tarjeta de procesamiento de datos data-processing card
tarjeta de programa program card
tarjeta de prueba test card
tarjeta de receptor receiver card
tarjeta de referencia reference card
tarjeta de registro de control control record card
tarjeta de rutina de carga bootstrap card
tarjeta de salida analógica analog output card
tarjeta de selección selection card
tarjeta de sentencias statement card
tarjeta de tabla table card
tarjeta de trabajo job card
tarjeta de transacción transaction card
tarjeta de transferencia transfer card
tarjeta de transferencia de control transfer-of-control card
tarjeta de transición transition card
tarjeta de usos múltiples multiple-use card
tarjeta doble double card
tarjeta en blanco blank card
tarjeta especial special card
tarjeta inteligente smart card
tarjeta lógica logic card
tarjeta maestra master card
tarjeta maestra preperforada prepunched master card
tarjeta magnética magnetic card
tarjeta muescada notched card
tarjeta omitida omitted card
tarjeta pequeña small card
tarjeta perforable punch card
tarjeta perforada punched card
tarjeta perforada en los bordes edge-punched card
tarjeta perforada en los márgenes margin-punched card
tarjeta piloto pilot card
tarjeta preperforada prepunched card
tarjeta primaria primary card
tarjeta sumaria summary card
tarjeta única single card
tarjetas continuas continuous cards

tarjetas por minuto cards per minute
tasa rate
tasa angular angular rate
tasa de alimentación feed rate
tasa de borrado erasing rate
tasa de cambio rate of change
tasa de conteo counting rate
tasa de demanda demand rate
tasa de deslizamiento drift rate
tasa de deslizamiento aleatorio random drift rate
tasa de deslizamiento insensible a la aceleración
acceleration-insensitive drift rate
tasa de deslizamiento sensible a la aceleración
acceleration-sensitive drift rate
tasa de desvanecimientos fading rate
tasa de emisión de rayos gamma gamma-ray
emission rate
tasa de errores error rate
tasa de errores de bits bit error rate
tasa de errores de caracteres character-error rate
tasa de errores de elementos element error rate
tasa de errores de perforación keypunch error rate
tasa de errores en bloques block error rate
tasa de errores residuales residual-error rate
tasa de fallas failure rate
tasa de fallas de componentes component failure
rate
tasa de fallas de línea line failure rate
tasa de fallas por demanda demand failure rate
tasa de interrupciones outage rate
tasa de interrupciones programadas scheduled
outage rate
tasa de llamadas call rate
tasa de reposición reset rate
tasa nominal nominal rate
taza de arrastre drag cup
tecla key
tecla alternativa alternate key
tecla de acceso de programa program access key
tecla de alimentación de formulario form-feed key
tecla de alimentación de línea line-feed key
tecla de avance de programa program advance key
tecla de ayuda help key
tecla de borrado erasing key
tecla de cambio shift key
tecla de cancelación cancel key
tecla de carácter character key
tecla de carga load key
tecla de control control key
tecla de control de teclado keyboard control key
tecla de corrección correction key
tecla de cursor cursor key
tecla de dígito digit key
tecla de escape escape key
tecla de función function key
tecla de función alternativa alternate function key
tecla de función de mandos command function key
tecla de función de programa program function key
tecla de función de teclado keyboard function key
tecla de función de tubo de rayos catódicos
cathode-ray tube function key
tecla de función especial special-function key
tecla de función programable programmable
function key
tecla de impresión printing key
tecla de liberación release key
tecla de línea line key
tecla de mandos command key
tecla de parada stop key
tecla de paro de alimentación de formulario
form-feed stop key
tecla de perforación punching key
tecla de repetición repeat key

tecla de repetición automática automatic repeat key
tecla de reposición reset key
tecla de retroceso backspace key
tecla de salto skip key
tecla de señal de error error signal key
tecla no válida invalid key
teclado keyboard
teclado alfabético alphabetic keyboard
teclado alfanumérico alphanumeric keyboard
teclado bloqueado locked-up keyboard
teclado complementario companion keyboard
teclado de almacenamiento storage keyboard
teclado de consola console keyboard
teclado de entrada input keyboard
teclado de entrada de datos data entry keyboard
teclado de membrana membrane keyboard
teclado desconectable detachable keyboard
teclado digital digital keyboard
teclado intermedio intermediate keyboard
teclado magnético magnetic keyboard
teclado motorizado motorized keyboard
teclado numérico numerical keypad, numerical
keyboard
teclado principal main keyboard
teclado programado programmed keyboard
teclado sensible a presión pressure-sensitive
keyboard
teclado táctil tactile keyboard
tecleado keying
tecleado equilibrado back-shunt keying
tecnecio technetium
tecnetrón tecnetron
técnica bolométrica bolometric technique
técnica criptográfica cryptographic technique
técnica de acceso access technique
técnica de almacenamiento de datos data storage
technique
técnica de asignación allocation technique
técnica de barrera superficial surface-barrier
technique
técnica de bloques de construcción building-block
technique
técnica de control control technique
técnica de Czochralski Czochralski technique
técnica de desarrollo development technique
técnica de enganche lock-in technique
técnica de ensamblaje assembly technique
técnica de espacio de aire air-gap technique
técnica de indexación indexing technique
técnica de intercambio iónico ion-exchange
technique
técnica de medición metering technique
técnica de medida measuring technique
técnica de memoria virtual virtual memory
technique
técnica de paginación paging technique
técnica de películas delgadas thin-film technique
técnica de películas gruesas thick-film technique
técnica de procesamiento epitaxial epitaxial
processing technique
técnica de retrodispersión backscattering technique
técnica de sustitución substitution technique
técnica de zona flotante floating-zone technique
técnica electrónica electronic technique
técnica estroboscópica stroboscopic technique
técnica metalocerámica metal-ceramic technique
técnica modular modular technique
técnica molecular molecular technique
técnica operacional operational technique
técnicas de microprogramación microprogramming
techniques
técnico technical
técnico de computadora computer technician

tecnología de información information technology
tecnología de procesamiento de datos
    data-processing technology
tecnología monolítica monolithic technology
techo de impulso pulse flat
teflón teflon
tela carbonizada carbonized cloth
telcoteno telcothene
teleamperímetro teleammeter
teleautografía teleautography
teleautógrafo teleautograph
telecámara telecamera
telecardiógrafo telecardiograph
telecinta teletape
telecompás telecompass
telecomputación telecomputing
telecomunicación telecommunication
telecomunicación eléctrica electric
    telecommunication
telecomunicación óptica optical telecommunication
telecomunicación por ondas portadoras
    carrier-wave telecommunication
telecomunicaciones aeronáuticas aeronautical
    telecommunications
teleconexión teleconnection
teleconferencia teleconference
teleconmutador teleswitch
telecontador telecounter
telecontrol telecontrol
telecontrolado telecontrolled
telecopiadora telecopier
telecromo telechrome
telediafonía far-end crosstalk
teledifusión telediffusion, telecast
telefacsímil telefacsimile
telefonía telephony
telefonía de portadora quiescente quiescent-carrier
    telephony
telefonía múltiplex multiplex telephony
telefonía por frecuencia vocal voice-frequency
    telephony
telefonía por hilo wire telephony
telefonía por onda luminosa light-wave telephony
telefonía por ondas wave telephony
telefonía por portadora carrier telephony
telefonía símplex simplex telephony
telefonía visual visual telephony
telefónico telephonic
teléfono telephone, phone
teléfono alimentado por energía sonora
    sound-powered telephone
teléfono celular cellular telephone
teléfono colgante pendant telephone
teléfono con efecto local sidetone telephone
teléfono de columna deskstand telephone
teléfono de disco dial telephone
teléfono de líneas múltiples multiple-line telephone
teléfono de magneto magneto telephone
teléfono de pared wall telephone
teléfono de teclas de tonos touch-tone telephone
teléfono manual manual telephone
teléfono operado por voz voice-operated telephone
teléfono público public telephone
telefonógrafo telephonograph
telefonometría telephonometry
telefonómetro telephonometer
telefotografía telephotography
telefotómetro telephotometer
telegrafía telegraphy
telegrafía alfabética alphabetic telegraphy
telegrafía armónica harmonic telegraphy
telegrafía automática automatic telegraphy
telegrafía automática de Wheatstone Wheatstone

automatic telegraphy
telegrafía con señalización speech plus telegraphy
telegrafía cuádruplex quadruplex telegraphy
telegrafía de frecuencia vocal voice-frequency
    telegraphy
telegrafía de Morse Morse telegraphy
telegrafía de puntos y rayas dot-and-dash telegraphy
telegrafía de rayas y puntos dash-dot telegraphy
telegrafía de todo o nada on-off telegraphy
telegrafía de Wheatstone Wheatstone telegraphy
telegrafía eléctrica electric telegraphy
telegrafía electroquímica electrochemical telegraphy
telegrafía en facsímil facsimile telegraphy
telegrafía escalonada echelon telegraphy
telegrafía infraacústica subaudio telegraphy
telegrafía manual manual telegraphy
telegrafía modulada modulated telegraphy
telegrafía mosaico mosaic telegraphy
telegrafía multicanal multichannel telegraphy
telegrafía múltiplex multiplex telegraphy
telegrafía por corriente continua direct-current
    telegraphy
telegrafía por corriente portadora carrier-current
    telegraphy
telegrafía por desplazamiento de frecuencia
    frequency-shift telegraphy
telegrafía por hilo wire telegraphy
telegrafía por ondas wave telegraphy
telegrafía por portadora carrier telegraphy
telegrafía por portadora de frecuencia vocal
    voice-frequency carrier telegraphy
telegrafía por tonos tone telegraphy
telegrafía símplex simplex telegraphy
telegrafía ultraacústica superaudio telegraphy
telegrafía visual visual telegraphy
telegráfico telegraphic
telégrafo telegraph
telégrafo acústico acoustic telegraph
telégrafo corrector de errores error-correcting
    telegraph
telégrafo de aguja needle telegraph
telégrafo de Morse Morse telegraph
telégrafo impresor printing telegraph
telegrama telegram
telegrama de tránsito transit telegram
teleimpresión teleprinting
teleimpresora teleprinter
teleimpresora de página page teleprinter
teleindicador teleindicator
teleinformática teleinformatics
telemandado remote-controlled
telemando remote control
telemática telematics
telemecanismo telemechanism
telemedición analógica analog telemetering
telemedición móvil mobile telemetering
telemedida telemetering
telemedido remote-metered
telemedidor telemeter
telemedidor de tipo de frecuencia frequency-type
    telemeter
telemedidor digital digital telemeter
telemetría telemetry
telemétrico telemetric
telémetro telemeter
telémetro de tipo de tensión voltage-type telemeter
telemicroscopía telemicroscopy
teleobjetivo telephoto lens
teleprocesador teleprocessor
teleprocesamiento teleprocessing
teleproceso teleprocessing
telerán teleran
telerradiodifusión telecast

telerrecepción telereception
telerreceptor telereceiver
telerreferencia telereference
telerregistrador telerecorder
telerregistro telerecording
telerregulación teleregulation
telerregulador teleregulator
telescópico telescopic
telescopio telescope
telescopio de radar radar telescope
telescopio electrónico electron telescope
telescopio reflector reflecting telescope
teleseñalización telesignaling
teletermómetro telethermometer
teletermoscopio telethermoscope
teletexto teletext
teletipo teletype
teletipo multicanal multichannel teletype
teletipo receptor receiving teletype
teletransmisión teletransmission
televisar televise
televisión television
televisión a colores color television
televisión a colores secuencial sequential color television
televisión de alta definición high-definition television
televisión de baja definición low-definition television
televisión de banda angosta narrowband television
televisión de proyección projection television
televisión digital digital television
televisión en colores color television
televisión estereoscópica stereoscopic television
televisión monocroma monochrome television
televisión multicanal multichannel television
televisión por antena comunitaria community-antenna television
televisión por cable cable television
televisión por circuito cerrado closed-circuit television
televisión por rayos X X-ray television
televisión submarina underwater television
televisión tridimensional three-dimensional television
televisor television receiver, television set, televisor
televisor de transistores transistor television set
televisor transistorizado transistorized television set
televóltmetro televoltmeter
telewáttmetro telewattmeter
télex telex
télex internacional international telex
telúrico telluric
telurio tellurium
telururo de cinc zinc telluride
telururo de plomo lead telluride
temblador trembler
temperatura temperature
temperatura absoluta absolute temperature
temperatura ambiental environmental temperature
temperatura ambiente ambient temperature, room temperature
temperatura ambiente base base ambient temperature
temperatura ambiente de operación operating ambient temperature
temperatura ambiente límite limiting ambient temperature
temperatura característica characteristic temperature
temperatura cero zero temperature
temperatura criogénica cryogenic temperature
temperatura crítica critical temperature

temperatura de aire ambiente ambient air temperature
temperatura de almacenamiento storage temperature
temperatura de antena antenna temperature
temperatura de brillo brightness temperature
temperatura de caja case temperature
temperatura de calefactor heater temperature
temperatura de color color temperature
temperatura de condensación condensation temperature
temperatura de conductor conductor temperature
temperatura de cuerpo body temperature
temperatura de Curie Curie temperature
temperatura de Curie-Weiss Curie-Weiss temperature
temperatura de Curie ferroeléctrica ferroelectric Curie temperature
temperatura de Debye Debye temperature
temperatura de drenaje bleeder temperature
temperatura de luminancia luminance temperature
temperatura de operación operating temperature
temperatura de operación ambiente ambient operating temperature
temperatura de radiación radiation temperature
temperatura de radiación total total radiation temperature
temperatura de radiancia radiance temperature
temperatura de referencia reference temperature
temperatura de remojo soaking temperature
temperatura de ruido noise temperature
temperatura de ruido efectiva effective noise temperature
temperatura de ruido equivalente equivalent noise temperature
temperatura de ruido media average noise temperature
temperatura de ruido normal standard noise temperature
temperatura de sala room temperature
temperatura de transición transition temperature
temperatura de unión junction temperature
temperatura de vaina sheath temperature
temperatura del aire interior media average inside air temperature
temperatura del punto más caliente hottest-spot temperature
temperatura efectiva effective temperature
temperatura electrónica electronic temperature
temperatura específica specific temperature
temperatura espectral spectral temperature
temperatura estable stable temperature
temperatura externa external temperature
temperatura intrínseca intrinsic temperature
temperatura límite limiting temperature
temperatura máxima de exposición maximum exposure temperature
temperatura media mean temperature
temperatura medida measured temperature
temperatura negativa negative temperature
temperatura normal normal temperature
temperatura observable observable temperature
temperatura superficial surface temperature
temperatura y presión normales standard temperature and pressure
templar quench
temporal temporary
temporización timing
temporización de impulsos pulse timing
temporización sincronizada synchronized timing
temporizador timer
temporizador de bus bus timer
temporizador de cierre make timer

temporizador de intervalo interval timer
temporizador de intervalos programable programmable interval timer
temporizador de programa program timer
temporizador de reciclado recycle timer
temporizador de reposición reset timer
temporizador de retardo delay timer
temporizador de secuencia sequence timer
temporizador de sistema system timer
temporizador de soldadura weld timer
temporizador eléctrico electric timer
temporizador electromecánico electromechanical timer
temporizador electrónico electronic timer
temporizador fotoeléctrico photoelectric timer
temporizador integrador integrating timer
temporizador monoestable monostable timer
temporizador neumático pneumatic timer
temporizador secuencial sequential timer
temporizador sin reposición eléctrica nonreset timer
temporizador sincrónico synchronous timer
temporizador tiratrónico thyratron timer
tenazas de fusibles fuse tongs
tendido de cable cable run
tenebrescencia tenebrescence
tensión voltage, tension, potential, strain, pressure
tensión A A voltage
tensión absorbida absorbed voltage
tensión aceleradora accelerating voltage
tensión aceleradora de haz beam accelerating voltage
tensión aceleradora iónica ion accelerating voltage
tensión activa active voltage
tensión adelantada leading voltage
tensión ajustable adjustable voltage
tensión alterna alternating voltage
tensión alterna básica basic alternating voltage
tensión anódica anode voltage
tensión anódica característica characteristic anode voltage
tensión anódica inversa inverse anode voltage
tensión aplicada applied voltage
tensión B B voltage
tensión bifásica two-phase voltage
tensión catódica cathode voltage
tensión cero zero voltage
tensión compensadora compensating voltage
tensión compuesta composite voltage
tensión concentrada lumped voltage
tensión constante constant voltage
tensión continua direct voltage
tensión controlada controlled voltage
tensión crítica critical voltage
tensión crítica de relé relay critical voltage
tensión crítica disruptiva disruptive critical voltage
tensión cuántica quantum voltage
tensión de accionamiento actuating voltage
tensión de aceleración acceleration voltage
tensión de acometida service voltage
tensión de agotamiento depletion voltage
tensión de aislamiento insulation voltage
tensión de ajuste trimming voltage
tensión de alimentación supply voltage
tensión de alimentación anódica anode supply voltage
tensión de alimentación de corriente continua direct-current supply voltage
tensión de alimentación de placa plate supply voltage
tensión de alimentación máxima absoluta absolute maximum supply voltage
tensión de ánodo a cátodo anode-to-cathode voltage
tensión de ánodo crítica critical anode voltage

tensión de ánodo directa forward anode voltage
tensión de arco arc voltage
tensión de arqueo en seco dry flashover voltage
tensión de arrancador starter voltage
tensión de arranque starting voltage
tensión de audio audio voltage
tensión de audiofrecuencia audio-frequency voltage
tensión de baño bath voltage
tensión de barrera barrier voltage
tensión de barrido sweep voltage
tensión de barrido lineal linear sweep voltage
tensión de barrido vertical vertical-sweep voltage
tensión de base base voltage
tensión de base de corriente alterna alternating-current base voltage
tensión de base de corriente continua direct-current base voltage
tensión de base de tiempo time-base voltage
tensión de base dinámica dynamic base voltage
tensión de base estática static base voltage
tensión de batería battery voltage
tensión de batería nominal nominal battery voltage
tensión de blanco target voltage
tensión de bloqueo blocking voltage
tensión de bombeo pumping voltage
tensión de borrado erasing voltage
tensión de caída de arco arc-drop voltage
tensión de calefactor heater voltage
tensión de calibración calibration voltage
tensión de campo field voltage
tensión de campo nominal rated field voltage
tensión de capa a capa layer-to-layer voltage
tensión de capacitancia diferencial differential capacitance voltage
tensión de carga charge voltage, charging voltage, load voltage
tensión de cátodo de corriente alterna alternating-current cathode voltage
tensión de cátodo de corriente continua direct-current cathode voltage
tensión de cátodo dinámica dynamic cathode voltage
tensión de cátodo estática static cathode voltage
tensión de cebado striking voltage, starting voltage, firing voltage
tensión de cebado baja low-striking voltage
tensión de celda cell voltage
tensión de cierre sealing voltage
tensión de circuito circuit voltage
tensión de circuito de control control-circuit voltage
tensión de circuito nominal nominal circuit voltage, rated circuit voltage
tensión de colector collector voltage
tensión de colector de corriente alterna alternating-current collector voltage
tensión de colector de corriente continua direct-current collector voltage
tensión de colector dinámica dynamic collector voltage
tensión de colector estática static collector voltage
tensión de comienzo de descarga discharge inception voltage
tensión de compensación bucking voltage
tensión de conexión pull-in voltage
tensión de conmutación switching voltage
tensión de consumo de corriente continua direct-current drain voltage
tensión de consumo dinámica dynamic drain voltage
tensión de consumo estática static drain voltage
tensión de contacto contact voltage
tensión de contacto de campo field contact voltage
tensión de contorneamiento flashover voltage
tensión de contorneamiento en húmedo wet flashover voltage

tensión de control control voltage
tensión de control compuesta composite controlling voltage
tensión de control de ganancia automático automatic gain control voltage
tensión de control diferencial differential control voltage
tensión de convergencia dinámica dynamic-convergence voltage
tensión de corrección interna internal correction voltage
tensión de corriente alterna alternating-current voltage
tensión de corriente continua direct-current voltage
tensión de corte cutoff voltage
tensión de corte de blanco target cutoff voltage
tensión de corte de rejilla grid cutoff voltage
tensión de cresta crest voltage
tensión de deflexión deflection voltage
tensión de deposición deposition voltage
tensión de derivación branch voltage
tensión de desaccionamiento dropout voltage
tensión de descarga discharge voltage
tensión de descarga disruptiva disruptive discharge voltage
tensión de descomposición decomposition voltage
tensión de desequilibrio offset voltage
tensión de desequilibrio de entrada input offset voltage
tensión de desintegración disintegration voltage
tensión de desionización deionization voltage
tensión de destello flashing voltage
tensión de diodo equivalente equivalent diode voltage
tensión de diseño design voltage
tensión de disparo triggering voltage, trip voltage
tensión de disparo de puerta gate trigger voltage
tensión de drenador de corriente alterna alternating-current drain voltage
tensión de eco echo voltage
tensión de ecualización equalization voltage
tensión de electrodo electrode voltage
tensión de emisor emitter voltage
tensión de emisor de corriente alterna alternating-current emitter voltage
tensión de emisor de corriente continua direct-current emitter voltage
tensión de emisor dinámica dynamic emitter voltage
tensión de emisor estática static emitter voltage
tensión de enfoque focusing voltage
tensión de enmascaramiento masking voltage
tensión de entrada input voltage
tensión de entrada de transformador transformer input voltage
tensión de entrada diferencial differential-input voltage
tensión de equilibrio equilibrium voltage
tensión de equilibrio de fases phase-balance voltage
tensión de error error voltage
tensión de error de entrada input error voltage
tensión de error de salida output error voltage
tensión de escobillas brush voltage
tensión de estabilización stabilization voltage
tensión de estado de conexión on-state voltage
tensión de estricción pinchoff voltage
tensión de excitación excitation voltage, driving voltage
tensión de exploración scanning voltage
tensión de extinción extinction voltage, quenching voltage
tensión de extinción de corona corona extinction voltage
tensión de extinción de descarga discharge

extinction voltage
tensión de fase phase voltage
tensión de filamento filament voltage
tensión de fluctuación fluctuation voltage
tensión de formación formation voltage
tensión de fuente de alimentación anódica anode power supply voltage
tensión de fuente de corriente continua direct-current source voltage
tensión de fuga leakage voltage
tensión de funcionamiento running voltage
tensión de Hall Hall voltage
tensión de histéresis hysteresis voltage
tensión de ignición ignition voltage
tensión de impedancia impedance voltage
tensión de impulso impulse voltage
tensión de impulso seccionado chopped-impulse voltage
tensión de integración integration voltage
tensión de interrupción de puerta gate turn-off voltage
tensión de ionización ionization voltage
tensión de ionización de fondo background ionization voltage
tensión de línea line voltage
tensión de línea de corriente alterna alternating-current line voltage
tensión de línea nominal rated line voltage
tensión de luminiscencia glow voltage
tensión de malla mesh voltage
tensión de máquina machine voltage
tensión de modo común common-mode voltage
tensión de modo diferencial differential-mode voltage
tensión de modulación simétrica symmetrical modulation voltage
tensión de nodo node voltage
tensión de onda cuadrada square-wave voltage
tensión de ondulación ripple voltage
tensión de operación operating voltage
tensión de oposición opposing voltage
tensión de pantalla screen voltage
tensión de penetración penetration voltage
tensión de perforación puncture voltage
tensión de pico a pico peak-to-peak voltage
tensión de pico inversa inverse peak voltage
tensión de placa plate voltage
tensión de placa a filamento plate-to-filament voltage
tensión de placa de corriente alterna alternating-current plate voltage
tensión de placa de corriente continua direct-current plate voltage
tensión de placa dinámica dynamic plate voltage
tensión de placa equivalente equivalent plate voltage
tensión de placa estática static plate voltage
tensión de plena carga full-load voltage
tensión de polarización bias voltage
tensión de polarización polarization voltage, bias voltage
tensión de polarización de rejilla grid-bias voltage, grid polarization voltage
tensión de portadora carrier voltage
tensión de postaceleración post-acceleration voltage
tensión de prueba test voltage
tensión de puerta gate voltage
tensión de puerta de corriente alterna alternating-current gate voltage
tensión de puerta de corriente continua direct-current gate voltage
tensión de puerta dinámica dynamic gate voltage
tensión de puerta estática static gate voltage
tensión de puerta inversa reverse gate voltage

tensión de puesta en trabajo pickup voltage
tensión de punto de inflexión inflection-point voltage
tensión de reactancia reactance voltage
tensión de recuperación recovery voltage
tensión de rechazo de modo común common-mode rejection voltage
tensión de reencendido restriking voltage
tensión de referencia reference voltage
tensión de referencia interna internal reference voltage
tensión de referencia preseleccionada preselected reference voltage
tensión de reflector reflector voltage
tensión de reinserción reinsertion voltage
tensión de rejilla grid voltage
tensión de rejilla crítica critical grid voltage
tensión de rejilla de corriente alterna alternating-current grid voltage
tensión de rejilla de corriente continua direct-current grid voltage
tensión de rejilla dinámica dynamic grid voltage
tensión de rejilla equivalente equivalent grid voltage
tensión de rejilla estática static grid voltage
tensión de rejilla inversa inverse grid voltage
tensión de rejilla negativa negative-grid voltage
tensión de rejilla-pantalla screen-grid voltage
tensión de reposición reset voltage
tensión de retorno return voltage
tensión de retroalimentación feedback voltage
tensión de ruido noise voltage
tensión de ruido de entrada input noise voltage
tensión de ruptura breakdown voltage
tensión de ruptura anódica anode breakdown voltage
tensión de ruptura asintótica asymptotic breakdown voltage
tensión de ruptura de corriente alterna alternating-current breakdown voltage
tensión de ruptura de corriente continua direct-current breakdown voltage
tensión de ruptura dieléctrica dielectric breakdown voltage
tensión de ruptura eléctrica electric breakdown voltage
tensión de ruptura en circuito abierto open-circuit breakdown voltage
tensión de ruptura en cortocircuito short-circuit breakdown voltage
tensión de ruptura inversa reverse-breakdown voltage
tensión de salida output voltage
tensión de salida de transformador transformer output voltage
tensión de salida nominal rated output voltage
tensión de salto de arco arcover voltage
tensión de saturación saturation voltage
tensión de saturación de emisor emitter saturation voltage
tensión de señal signal voltage
tensión de servicio service voltage
tensión de sintonización tuning voltage
tensión de sintonización de corriente continua direct-current tuning voltage
tensión de sistema system voltage
tensión de sistema nominal nominal system voltage, rated system voltage
tensión de soldadura welding voltage
tensión de sostenimiento keep-alive voltage
tensión de supresión de haz blackout voltage
tensión de tanque tank voltage
tensión de tanque de placa plate tank voltage
tensión de tanque de rejilla grid tank voltage

tensión de terminal asimétrica asymmetrical terminal voltage
tensión de Thomson Thomson voltage
tensión de trabajo working voltage
tensión de trabajo nominal nominal working voltage, rated working voltage
tensión de trabajo segura safe working voltage
tensión de transición conductiva breakover voltage
tensión de transición conductiva directa forward breakover voltage
tensión de valle valley voltage
tensión de varistor varistor voltage
tensión de video de azul blue video voltage
tensión de videofrecuencia video-frequency voltage
tensión de Wehnelt Wehnelt voltage
tensión de Zener Zener voltage
tensión de zumbido hum voltage
tensión del inducido armature voltage
tensión del video de rojo red video voltage
tensión del video de verde green video voltage
tensión desfasada out-of-phase voltage
tensión deslizante sliding voltage
tensión desmodulada demodulated voltage
tensión desplazadora de cero falso antistickoff voltage
tensión diametral diametral voltage
tensión diferencial differential voltage
tensión directa forward voltage
tensión directa instantánea instantaneous forward voltage
tensión disruptiva disruptive voltage
tensión ecualizadora equalizing voltage
tensión efectiva effective voltage
tensión eficaz root-mean-square voltage
tensión eléctrica electric voltage
tensión emisor-base emitter-base voltage
tensión en circuito abierto open-circuit voltage
tensión en circuito cerrado closed-circuit voltage
tensión en cortocircuito short-circuit voltage
tensión en cuadratura quadrature voltage
tensión en delta delta voltage
tensión en diente de sierra sawtooth voltage
tensión en espera idling voltage
tensión en estado de desconexión off-state voltage
tensión en fase in-phase voltage
tensión entre bases interbase voltage
tensión entre fases line-to-line voltage
tensión escalón unitario unit-step voltage
tensión estabilizadora stabilizing voltage
tensión estabilizante de frecuencia frequency-stabilizing voltage
tensión estacionaria steady voltage
tensión excitadora exciting voltage
tensión extra alta extra-high voltage
tensión final final voltage
tensión fluctuante fluctuating voltage
tensión fotoeléctrica photoelectric voltage
tensión generada generated voltage
tensión inducida induced voltage
tensión inicial initial voltage
tensión inicial de corona corona inception voltage
tensión instantánea instantaneous voltage
tensión insuficiente undervoltage
tensión interfásica interphase tension
tensión interferente interfering voltage
tensión inversa inverse voltage
tensión inversa de ionización flashback voltage
tensión inversa de pico repetitiva repetitive peak inverse voltage
tensión inversa inicial initial inverse voltage
tensión inversa instantánea instantaneous reverse voltage
tensión inversa máxima de pico maximum peak

inverse voltage

tensión longitudinal longitudinal voltage

tensión máxima maximum voltage, ceiling voltage

tensión máxima de diseño maximum design voltage

tensión máxima de operación maximum operating voltage

tensión máxima de salida maximum output voltage

tensión máxima de sistema maximum system voltage

tensión máxima excitadora exciter ceiling voltage

tensión media average voltage

tensión media de electrodo average voltage of electrode

tensión mínima de arranque minimum starting voltage

tensión mínima disruptiva minimum flashover voltage

tensión modulada linealmente linearly modulated voltage

tensión modulante pico peak modulating voltage

tensión neta net voltage

tensión neutralizante neutralizing voltage

tensión no modulada unmodulated voltage

tensión nominal nominal voltage, rated voltage

tensión nominal de sistema de excitación excitation system rated voltage

tensión normal standard voltage

tensión nula null voltage

tensión onduladoria undulatory voltage

tensión ondulante undulating voltage

tensión oscilante oscillating voltage

tensión permisible allowable voltage

tensión perturbadora equivalente equivalent disturbing voltage

tensión pico bloqueada peak blocked voltage

tensión pico de ánodo peak anode voltage

tensión pico de ánodo directo peak forward anode voltage

tensión pico de placa peak plate voltage

tensión pico inversa peak inverse voltage

tensión pico inversa de ánodo peak inverse anode voltage

tensión polarizante polarizing voltage

tensión polarizante de corriente continua direct-current polarizing voltage

tensión positiva positive voltage

tensión primaria primary voltage

tensión primaria nominal rated primary voltage

tensión pulsante pulsating voltage

tensión reactiva reactive voltage

tensión rectangular rectangular voltage

tensión rectificada rectified voltage

tensión reducida reduced voltage

tensión regulada regulated voltage

tensión residual residual voltage

tensión resistiva resistive voltage

tensión respecto a neutro voltage to neutral

tensión respecto a tierra voltage to ground, voltage to earth

tensión resultante resultant voltage

tensión secundaria secondary voltage

tensión secundaria nominal rated secondary voltage

tensión silenciadora squelch voltage

tensión sin carga no-load voltage

tensión sin compensación unweighted voltage

tensión sincrónica synchronous voltage

tensión sinusoidal sinusoidal voltage

tensión sofométrica psophometric voltage

tensión soportable sin descarga disruptiva withstand voltage

tensión subtransitoria longitudinal direct-axis subtransient voltage

tensión superficial surface tension

tensión supresora de corriente continua direct-current suppressor voltage

tensión supresora estática static suppressor voltage

tensión terminal terminal voltage

tensión termoeléctrica thermoelectric voltage

tensión transferida transferred voltage

tensión transitoria transient voltage

tensión transitoria longitudinal direct-axis transient voltage

tensión transversal transverse voltage

tensión trifásica three-phase voltage

tensión ultraalta ultra-high voltage

tensión umbral threshold voltage

tensión unidireccional unidirectional voltage

tensión variable variable voltage

tensiones en contrafase push-pull voltages

tensiones equilibradas balanced voltages

teorema binomio binomial theorem

teorema de Ampere Ampere's theorem

teorema de Ampere-Laplace Ampere-Laplace theorem

teorema de Bloch Bloch theorem

teorema de Carnot Carnot theorem

teorema de compensación compensation theorem

teorema de Foster Foster theorem

teorema de Gauss Gauss' theorem

teorema de la superposición superposition theorem

teorema de Norton Norton's theorem

teorema de Nyquist Nyquist theorem

teorema de Rayleigh-Carson Rayleigh-Carson theorem

teorema de reciprocidad reciprocity theorem

teorema de reciprocidad de Rayleigh Rayleigh reciprocity theorem

teorema de Shannon Shannon's theorem

teorema de Thevenin Thevenin's theorem

teorema de transferencia de potencia power transfer theorem

teorema de transferencia máxima de potencia maximum power transfer theorem

teorema eléctrico de Thevenin Thevenin's electrical theorem

teoría theory

teoría atómica atomic theory

teoría cuántica quantum theory

teoría cuántica del magnetismo quantum theory of magnetism

teoría de autómatas automata theory

teoría de autómatas abstracta abstract automata theory

teoría de banda band theory

teoría de circuitos circuit theory

teoría de colas queuing theory

teoría de decisión decision theory

teoría de diamagnetismo de Langevin Langevin theory of diamagnetism

teoría de difusión diffusion theory

teoría de la luz electromagnética electromagnetic theory of light

teoría de redes network theory

teoría de ruptura de von Hippel von Hippel breakdown theory

teoría de Weber Weber's theory

teoría de Weiss del ferromagnetismo Weiss theory of ferromagnetism

teoría del muestreo sampling theory

teoría molecular del magnetismo molecular theory of magnetism

teoría ondulatoria wave theory

tera tera

terabyte terabyte

teraciclo teracycle

teraelectronvolt teraelectronvolt

terahertz terahertz
teraohm teraohm
terawatt terawatt
terbio terbium
tercer canal third channel
tercer hilo sleeve wire
tercer riel third rail
tercera armónica third harmonic
tercera escobilla third brush
tercera ley de termodinámica third law of
   thermodynamics
tercera ley de Wien Wien's third law
termal thermal
termalmente thermally
termia therm
térmicamente thermally
térmico thermal
terminación termination
terminación adaptada matched termination
terminación anormal abnormal termination
terminación bifilar two-wire termination
terminación con pérdidas lossy termination
terminación de aparato apparatus termination
terminación de ciclo cycle termination
terminación de circuito abierto open-circuit
   termination
terminación de guíaondas waveguide termination
terminación de guíaondas óptico optical waveguide
   termination
terminación de resistencia resistance termination
terminación de sesión session termination
terminación en cuarto de onda quarter-wave
   termination
terminación equilibrada balanced termination
terminación exterior outdoor termination
terminación interior indoor termination
terminación parcial partial completion
terminación patrón standard termination
terminación sin pérdidas lossless termination
terminado terminated
terminador terminator
terminador de cables cable terminator
terminal terminal, prong, pin
terminal accesible accessible terminal
terminal anódica anode terminal
terminal arrollada wire-wrap terminal
terminal asincrónica asynchronous terminal
terminal banana banana pin
terminal bidireccional bidirectional pin
terminal catódica cathode terminal
terminal central central terminal
terminal coaxial coaxial terminal
terminal conectada por enlaces link-attached
   terminal
terminal conversacional conversational terminal
terminal de adquisición de datos data-acquisition
   terminal
terminal de alineación alignment pin
terminal de alta velocidad high-speed terminal
terminal de audio audio terminal
terminal de baja velocidad low-speed terminal
terminal de base base terminal, base pin
terminal de cables cable terminal
terminal de carga loading terminal
terminal de clavija banana banana plug terminal
terminal de computadora computer terminal
terminal de comunicaciones communications
   terminal
terminal de comunicaciones de datos data
   communications terminal
terminal de conexión connection terminal
terminal de consulta inquiry terminal
terminal de consulta-respuesta inquiry-response

terminal
terminal de contacto contact pin
terminal de control control terminal
terminal de conversación talking terminal
terminal de copia impresa hard-copy terminal
terminal de cordón cord terminal
terminal de datos data terminal
terminal de datos múltiplex multiplex data terminal
terminal de datos remota remote data terminal
terminal de definición set terminal
terminal de desconexión rápida quick-disconnect
   terminal
terminal de destino destination terminal
terminal de detección sense terminal
terminal de disposición set terminal
terminal de edición editing terminal
terminal de entrada input terminal
terminal de entrada única one-input terminal
terminal de facsímil facsimile terminal
terminal de fase phase terminal
terminal de frecuencia vocal voice-frequency
   terminal
terminal de función fija fixed-function terminal
terminal de guarda guard terminal
terminal de línea line terminal
terminal de memoria de acceso aleatorio
   random-access memory terminal
terminal de memoria sólo de lectura read-only
   memory terminal
terminal de microcomputadora microcomputer
   terminal
terminal de microondas microwave terminal
terminal de microprocesador microprocessor
   terminal
terminal de minicomputadora minicomputer
   terminal
terminal de modo de paquetes packet mode terminal
terminal de onda portadora carrier-wave terminal
terminal de origen originating terminal
terminal de portadora carrier terminal
terminal de procesamiento de datos data-processing
   terminal
terminal de pruebas test terminal
terminal de puerta gate terminal
terminal de puesta a cero zero-input terminal
terminal de puesta a tierra grounding terminal
terminal de recopilación de datos data collection
   terminal
terminal de red network terminal
terminal de reposición reset terminal
terminal de salida output terminal
terminal de salida cero zero-output terminal
terminal de salida gráfica graphic output terminal
terminal de salida única one-output terminal
terminal de sujeción binding post
terminal de teclado keyboard terminal
terminal de teleprocesamiento teleprocessing
   terminal
terminal de tiempo compartido time-sharing
   terminal
terminal de tierra ground terminal, earth terminal
terminal de tornillo screw terminal
terminal de transmisión de datos data-transmission
   terminal
terminal de tubo tube pin
terminal de tubo de rayos catódicos cathode-ray
   tube terminal
terminal de usuario user terminal
terminal de válvula valve pin
terminal de video video terminal
terminal de visualización visual display terminal
terminal de visualización de consulta inquiry
   display terminal

terminal de visualización de video video display terminal

terminal de visualización gráfica graphic display terminal

terminal de visualizador display terminal

terminal del inducido armature terminal

terminal eléctrica electric terminal

terminal en diversidad diversity terminal

terminal en paralelo parallel terminal

terminal en serie serial terminal

terminal en tiempo real remota remote real-time terminal

terminal física physical terminal

terminal gráfica graphic terminal

terminal inteligente intelligent terminal

terminal interactiva interactive terminal

terminal local local terminal

terminal lógica logical terminal

terminal maestra master terminal

terminal monitora monitor terminal

terminal multicanal multichannel terminal

terminal multifunción multifunction terminal

terminal múltiple multiple terminal

terminal negativa negative terminal

terminal neutra neutral terminal

terminal orientada a trabajos job-oriented terminal

terminal pasante feedthrough terminal

terminal positiva positive terminal

terminal programable programmable terminal

terminal receptora receiving terminal

terminal remota remote terminal

terminal seleccionada por programa program-selected terminal

terminal sin soldadura solderless terminal

terminal solicitante requesting terminal

terminal supervisora supervisory terminal

terminal telefónica telephone terminal

terminal telefónica de onda portadora carrier-wave telephone terminal

terminal telegráfica telegraph terminal

terminal télex telex terminal

terminal transmisora transmitting terminal

terminal vacante vacant terminal

terminal virtual virtual terminal

terminal visual visual terminal

terminales axiales axial leads

terminales de alimentación supply terminals

terminales de antena antenna terminals

terminales de corriente current terminals

terminales de Hall Hall terminals

terminales de medida measurement terminals

terminar terminate

terminar una sesión log off, log out

término absoluto absolute term

término dominante dominant term

termión thermion

termistor thermistor

termistor autocalentable self-heating thermistor

termistor calentado directamente directly heated thermistor

termistor calentado indirectamente indirectly heated thermistor

termistor de cuenta bead thermistor

termistor en disco disk thermistor

termistor sonda probe thermistor

termistor tipo arandela washer thermistor

termoaislamiento thermoinsulation, heat insulation

termoaislante heat-insulating

termoamperímetro thermoammeter

termocatalítico thermocatalytic

termocompensador thermocompensator

termoconductor thermoconductor

termoconmutador thermoswitch

termocontacto thermocontact

termoconvección thermoconvection

termocortacircuito thermocutout

termodetector thermodetector

termodieléctrico thermodielectric

termodifusión thermodiffusion

termodilución thermodilution

termodinámica thermodynamics

termodinámicamente thermodynamically

termodinámico thermodynamic

termoelástico thermoelastic

termoelectricidad thermoelectricity

termoeléctrico thermoelectric

termoelectrón thermoelectron

termoelectrónico thermoelectronic

termoelemento thermoelement

termoendurecimiento thermosetting

termoestabilidad thermostability

termoestable thermostable

termófono thermophone

termofotoquímica thermophotochemistry

termofotovoltaico thermophotovoltaic

termogalvánico thermogalvanic

termogalvanismo thermogalvanism

termogalvanómetro thermogalvanometer

termogénico thermogenic

termografía thermography

termoiónica thermionics

termoiónico thermionic

termoluminiscencia thermoluminescence

termoluminiscente thermoluminescent

termomagnético thermomagnetic

termometría de resistencia resistance thermometry

termométrico thermometric

termómetro thermometer

termómetro bimetálico bimetallic thermometer

termómetro de cuarzo quartz thermometer

termómetro de estado sólido solid-state thermometer

termómetro de lectura remota remote-reading thermometer

termómetro de radiación radiation thermometer

termómetro de resistencia resistance thermometer

termómetro de resistencia de platino platinum-resistance thermometer

termómetro de termistor thermistor thermometer

termómetro de termopar thermocouple thermometer

termómetro digital digital thermometer

termómetro eléctrico electric thermometer

termómetro electrónico electronic thermometer

termómetro óptico optical thermometer

termómetro termoeléctrico thermoelectric thermometer

termomolecular thermomolecular

termopar thermocouple, thermopair

termopar autocalentado self-heated thermocouple

termopar calentado directamente directly heated thermocouple

termopar calentado indirectamente indirectly heated thermocouple

termopar cobre-constantán copper-constantan thermocouple

termopar de alto vacío high-vacuum thermocouple

termopar de bismuto bismuth thermocouple

termopar de forma de onda waveform thermocouple

termopar de plasma plasma thermocouple

termopar de vacío vacuum thermocouple

termopar hierro-constantán iron-constantan thermocouple

termopar platino-telurio platinum-tellurium thermocouple

termopar sonda probe thermocouple

termopila thermopile

termoplástico thermoplastic
termoquímico thermochemical
termorregulador thermoregulator
termorregulador electrónico electronic
   thermoregulator
termorrelé thermorelay
termorresistencia thermoresistance
termosensibilidad thermosensitivity
termosensible thermosensitive, heat-sensitive
termostáticamente thermostatically
termostático thermostatic
termostato thermostat
termostato bimetálico bimetallic thermostat
termostato de inmersión immersion thermostat
termostato de mercurio mercury thermostat
termostato de seguridad safety thermostat
termostato de termistor thermistor thermostat
termostato eléctrico electric thermostat
termotarjeta thermocard
termounión thermojunction
ternario ternary
terrestre terrestrial
tesauro automatizado automated thesaurus
tesla tesla
tetrapolar four-pole
tetravalente tetravalent
tetrodo tetrode
tetrodo de cristal crystal tetrode
tetrodo de doble base double-base tetrode
tetrodo de efecto de campo field-effect tetrode
tetrodo de formación de haz beam-forming tetrode
tetrodo de gas gas tetrode
tetrodo de haz beam tetrode
tetrodo de potencia power tetrode
tetrodo de uniones junction tetrode
tetrodo PNP PNP tetrode
tetrodo transistor de contactos de punta
   point-contact transistor tetrode
texto complejo complex text
texto de gráficos graphics text
texto estandarizado almacenado boilerplate text
theta theta
tiempo time
tiempo activo on-time
tiempo atómico atomic time
tiempo cero zero time
tiempo corto de respuesta short response time
tiempo corto de subida short rise time
tiempo de abertura aperture time, opening time,
   break time
tiempo de acceso access time
tiempo de acceso a disco disk access time
tiempo de acceso al almacenamiento storage access
   time
tiempo de acceso aleatorio random-access time
tiempo de acceso de búsqueda seek access time
tiempo de acceso de escritura write-access time
tiempo de acceso de lectura read access time
tiempo de acceso medio average access time
tiempo de acción derivada rate time
tiempo de accionamiento actuating time
tiempo de aceleración accelerating time
tiempo de activación activation time
tiempo de actuación actuation time
tiempo de actuación de relé relay actuation time
tiempo de actuación efectivo effective actuation time
tiempo de actuación final final actuation time
tiempo de adquisición acquisition time
tiempo de agrupamiento bunching time
tiempo de almacenamiento storage time
tiempo de almacenamiento de diodo diode storage
   time
tiempo de almacenamiento de portadores carrier

   storage time
tiempo de análisis analysis time
tiempo de arranque start time
tiempo de arranque-parada start-stop time
tiempo de arrastre carry time
tiempo de asentamiento seating time
tiempo de ataque attack time
tiempo de avance lead time
tiempo de barrido sweep time
tiempo de bit bit time
tiempo de borde delantero de impulso leading-edge
   pulse time
tiempo de borrado erasing time
tiempo de búsqueda seek time
tiempo de caída fall time
tiempo de caída de impulso pulse fall time
tiempo de cálculo calculating time
tiempo de cálculo representativo representative
   calculating time
tiempo de calentamiento warm-up time, heating time
tiempo de calentamiento catódico cathode heating
   time
tiempo de calentamiento controlado controlled
   warm-up time
tiempo de calentamiento de tubo tube heating time
tiempo de carga charging time, load time
tiempo de ciclo cycle time
tiempo de ciclo de almacenamiento storage-cycle
   time
tiempo de ciclo de lectura read-cycle time
tiempo de ciclo de lectura-escritura read-write cycle
   time
tiempo de ciclo de visualización display cycle time
tiempo de cierre closing time, make time
tiempo de coherencia coherence time
tiempo de cola medio average queue time
tiempo de compilación compilation time
tiempo de comprobación checkout time
tiempo de comprobación de código code checking
   time
tiempo de computadora computer time
tiempo de conexión connect time, turn-on time
tiempo de conmutación switching time
tiempo de conmutación total total switching time
tiempo de conteo counting time
tiempo de contestación answering time
tiempo de conversión conversion time
tiempo de corrección correction time
tiempo de corte splitting time
tiempo de cruce crossover time
tiempo de decaimiento decay time
tiempo de decaimiento de centelleo scintillation
   decay time
tiempo de decaimiento de señal signal decay time
tiempo de deposición deposition time
tiempo de desaccionamiento dropout time
tiempo de desaceleración deceleration time
tiempo de desarrollo development time
tiempo de desarrollo de programas
   program-development time
tiempo de desbloqueo unblanking time
tiempo de desconexión turn-off time, off-time
tiempo de desionización deionization time
tiempo de difusión diffusion time
tiempo de dígito digit time
tiempo de disparo triggering time
tiempo de duración duration time
tiempo de ejecución execution time, run time
tiempo de electrificación electrification time
tiempo de encendido de encendedor ignitor firing
   time
tiempo de enfasamiento phasing time
tiempo de enfriamiento cooling time

tiempo de ensamblaje assembly time
tiempo de entrada entry time
tiempo de escalonamiento stagger time
tiempo de escritura write time
tiempo de espera waiting time
tiempo de espera medio mean waiting time
tiempo de estabilización stabilization time
tiempo de establecimiento settling time
tiempo de expansión expansion time
tiempo de exploración scanning time
tiempo de exposición exposure time
tiempo de extinción blanking time
tiempo de falla fault time
tiempo de fluctuación fluctuation time
tiempo de formación de arcos arcing time
tiempo de funcionamiento running time, up-time
tiempo de fusión melting time
tiempo de fusión de fusible fuse melting time
tiempo de impulso pulse time
tiempo de impulso medio mean pulse time
tiempo de inactividad off-time
tiempo de instalación installation time
tiempo de instrucción instruction time
tiempo de interrupción outage time
tiempo de ionización ionization time
tiempo de latencia latency time
tiempo de latencia de memoria memory latency
 time
tiempo de lectura read time
tiempo de liberación release time
tiempo de limitación clipping time
tiempo de mantenimiento maintenance time
tiempo de mantenimiento activo active maintenance
 time
tiempo de mantenimiento correctivo corrective
 maintenance time
tiempo de mantenimiento de emergencia emergency
 maintenance time
tiempo de mantenimiento preventivo preventive
 maintenance time
tiempo de mantenimiento preventivo activo active
 preventive maintenance time
tiempo de mantenimiento rutinario routine
 maintenance time
tiempo de máquina machine time
tiempo de medidor meter time
tiempo de montura set-up time
tiempo de movimiento de datos data movement time
tiempo de multiplicación multiplication time
tiempo de observación observation time
tiempo de ocupado busy time
tiempo de operación operation time
tiempo de operación medio average operation time
tiempo de operación total total operation time
tiempo de oscilación oscillation time
tiempo de oscilación parásita ring time
tiempo de palabra word time
tiempo de parada stop time
tiempo de parada medio mean stopping time
tiempo de parálisis paralysis time
tiempo de paso pass time
tiempo de posicionamiento positioning time
tiempo de precalentamiento preheating time
tiempo de precalentamiento catódico cathode
 preheating time
tiempo de preensamblado preassembly time
tiempo de preparación set-up time
tiempo de procesamiento processing time
tiempo de procesamiento virtual virtual processing
 time
tiempo de proceso process time
tiempo de producción production time
tiempo de programa program time

tiempo de propagación propagation time
tiempo de prueba de programa program test time
tiempo de pruebas test time
tiempo de reacción reaction time
tiempo de readquisición reacquisition time
tiempo de recolección electrónica electron collection
 time
tiempo de recolección iónica ion-collection time
tiempo de reconexión reclosing time
tiempo de recuperación recovery time
tiempo de recuperación de barrido sweep recovery
 time
tiempo de recuperación de diodo diode recovery
 time
tiempo de recuperación de fase phase recovery time
tiempo de recuperación de media amplitud
 half-amplitude recovery time
tiempo de recuperación de puerta gate recovery
 time
tiempo de recuperación de relé relay recovery time
tiempo de recuperación de sistema system recovery
 time
tiempo de recuperación de tensión voltage recovery
 time
tiempo de recuperación directa forward recovery
 time
tiempo de recuperación inversa reverse recovery
 time
tiempo de recuperación tras sobrecarga overload
 recovery time
tiempo de referencia reference time
tiempo de registro recording time
tiempo de relajación relaxation time
tiempo de relajación radiativa radiative relaxation
 time
tiempo de reloj clock time
tiempo de remojo soaking time
tiempo de reparación repair time
tiempo de reparación activo active repair time
tiempo de reparación medio mean repair time
tiempo de repetición de impulsos pulse repetition
 time
tiempo de reposición resetting time
tiempo de reposo idle time
tiempo de reproducción playing time
tiempo de reserva standby time
tiempo de resolución resolution time
tiempo de respuesta response time
tiempo de respuesta de interrupción interrupt
 response time
tiempo de respuesta de receptor receiver response
 time
tiempo de respuesta térmica thermal response time
tiempo de restauración restoring time
tiempo de retardo delay time
tiempo de retardo controlado por puerta
 gate-controlled delay time
tiempo de retardo de fase phase-delay time
tiempo de retardo de grupo group-delay time
tiempo de retardo de medida measuring delay time
tiempo de retardo de propagación propagation
 delay time
tiempo de retardo de respondedor transponder time
 delay
tiempo de retardo de sistema system delay time
tiempo de retardo medio mean delay time
tiempo de retención retention time, holding time
tiempo de retención de bits bit retention time
tiempo de retorno return time, retrace time
tiempo de retorno vertical vertical-retrace time
tiempo de reverberación reverberation time
tiempo de selección de canal channel select time
tiempo de servicio service time

tiempo de sincronización synchronization time
tiempo de sintonización térmica thermal tuning time
tiempo de sistema system time
tiempo de sobrecarga overload time
tiempo de soldadura weld time
tiempo de subida rise time
tiempo de subida controlado por puerta gate-controlled rise time
tiempo de subida de centelleo scintillation rise time
tiempo de subida de impulso pulse rise time
tiempo de subida de señal signal rise time
tiempo de suma add time
tiempo de suma y resta add-subtract time
tiempo de transferencia transfer time
tiempo de transferencia de procesador processor transfer time
tiempo de transición transition time
tiempo de transición total total transition time
tiempo de tránsito transit time
tiempo de tránsito de electrón electron transit time
tiempo de tránsito de transistor transistor transit time
tiempo de tránsito interelectródico interelectrode transit time
tiempo de transmisión transmission time
tiempo de trazo trace time
tiempo de unidad de procesamiento central central processing unit time
tiempo de uso time of use
tiempo de visión viewing time
tiempo de visualización display time
tiempo debatible debatable time
tiempo definido definite time
tiempo disponible available time
tiempo efectivo effective time
tiempo entre dígitos interdigit time
tiempo entre falsas alarmas false-alarm time
tiempo equivalente equivalent time
tiempo focal focal time
tiempo fuera de servicio out-of-service time
tiempo inactivo inactive time
tiempo incidental incidental time
tiempo ineficaz ineffective time
tiempo lento slow time
tiempo máximo de acceso maximum access time
tiempo máximo de almacenamiento maximum storage time
tiempo máximo de retención maximum retention time
tiempo medio entre fallas mean time between failures
tiempo medio entre interrupciones mean time between outages
tiempo medio hasta falla mean time to failure
tiempo medio hasta la primera falla mean time to first failure
tiempo medio hasta reparación mean time to repair
tiempo mínimo de ejecución minimum run time
tiempo mínimo utilizable minimum usable time
tiempo muerto down time, dead time
tiempo muerto acumulado accumulated down time
tiempo muerto administrativo administrative downtime
tiempo muerto aparente apparent dead time
tiempo muerto de respondedor transponder dead time
tiempo muerto de sistema system downtime
tiempo muerto de sistema total total system downtime
tiempo muerto efectivo effective dead time
tiempo muerto medio mean down time
tiempo muerto no programado unscheduled down time

tiempo muerto programado scheduled down time
tiempo operable operable time
tiempo operacional operational time
tiempo rápido fast time
tiempo real real time
tiempo requerido required time
tiempo simulado simulated time
tiempo transcurrido elapsed time
tiempo variable variable time
tiempo virtual virtual time
tierra ground, earth
tierra a tierra ground-to-ground
tierra accidental accidental ground, ground fault
tierra alcalina alkaline earth
tierra artificial artificial ground, artificial earth
tierra común common ground
tierra de amplificador amplifier ground
tierra de equipo equipment ground
tierra de protector arrester ground
tierra de punto único single-point ground
tierra de seguridad safety ground
tierra de Wagner Wagner ground
tierra directa direct ground
tierra ficticia fictitious earth
tierra general general ground
tierra indirecta indirect ground
tierra neutra neutral ground
tierra perfecta dead ground
tierra por falla fault ground
tierra principal main ground
tierra radial radial ground
tierra rara rare earth
tierra real true ground
tierra temporal temporary ground
timbre timbre, ringer, bell
timbre vibrating bell
timbre de advertencia warning bell
timbre de alarma alarm bell
timbre de aviso warning bell
timbre de batería battery bell
timbre de corriente continua direct-current bell
timbre piloto pilot bell
timbre polarizado polarized ringer
timbre telefónico telephone ringer
timbre telefónico armónico harmonic telephone ringer
tinta magnética magnetic ink
tinte hue
tinte de contraste contrast hue
tinte intrínseco intrinsic hue
tinte primario primary hue
tipo atómico atomic type
tipo compuesto composite type
tipo de acceso access type
tipo de acción type of action
tipo de archivo file type
tipo de cable cable type
tipo de carácter character type
tipo de datos data type
tipo de datos abstracto abstract data type
tipo de dispositivo device type
tipo de emisión emission type
tipo de entidad entity type
tipo de formato format type
tipo de freno brake type
tipo de interrupción outage type
tipo de letra codificado coded font
tipo de letra de impresora printer font
tipo de letra de pantalla screen font
tipo de letra incorporada built-in font
tipo de letras font
tipo de línea line type
tipo de longitud de campo field length type

tipo de memoria memory type
tipo de nodo node type
tipo de operación type of operation
tipo de pantalla screen type
tipo de registro record type
tipo de segmento segment type
tipo de tarjeta card type
tipo de visualización display type
tipo discreto discrete type
tipo lógico logical type
tipos de tubos preferidos preferred tube types
tira de barrera barrier strip
tirantez strain
tiras metálicas antirradar chaff
tiratrón thyratron
tiratrón autoextintor self-extinguishing thyratron
tiratrón controlado por fase phase-controlled thyratron
tiratrón de corriente continua direct-current thyratron
tiratrón de disparo trigger thyratron
tiratrón de estado sólido solid-state thyratron
tiratrón de gas gas-filled thyratron
tiratrón de hidrógeno hydrogen thyratron
tiratrón de vapor de mercurio mercury-vapor thyratron
tiratrón miniatura miniature thyratron
tiristor thyristor
tiristor bidireccional bidirectional thyristor
tiristor bidireccional controlado gated bidirectional thyristor
tiristor epitaxial epitaxial thyristor
tiristor triódico bidireccional bidirectional triode thyristor
tiristor triodo de bloqueo inverso reverse-blocking triode-thyristor
tirita thyrite
tiro throw
titanato de bario barium titanate
titanio titanium
titilación flutter
titilación auroral auroral flutter
titilación de aeronave aircraft flutter
titilación de aeroplano aeroplane flutter
titilación de avión airplane flutter
titilación diferencial differential flutter
tocadiscos record player
todo eléctrico all-electric
todo electrónico all-electronic
todo o nada all-or-nothing
tolerable tolerable
tolerancia tolerance, allowance
tolerancia a fallas fault tolerance
tolerancia a fluctuaciones jitter tolerance
tolerancia absoluta absolute tolerance
tolerancia armónica harmonic tolerance
tolerancia de ajuste adjustment tolerance
tolerancia de desvanecimiento fading allowance
tolerancia de distorsión distortion tolerance
tolerancia de errores error tolerance
tolerancia de errores de programas software error tolerance
tolerancia de fabricación fabrication tolerance
tolerancia de factor de escala scale-factor tolerance
tolerancia de frecuencia frequency tolerance
tolerancia de manufactura manufacturing tolerance
tolerancia de visualización display tolerance
tolerancia nominal nominal tolerance
tolerancia permisible allowable tolerance
tolerante a fallas fault tolerant
toma tap, outlet, intake
toma de acoplamiento coupling tap
toma de aire air intake

toma de bobina coil tap
toma de corriente outlet, current tap
toma de corriente con conexión a tierra ground outlet
toma de corriente eléctrica electrical outlet
toma de corriente empotrada flush outlet
toma de corriente múltiple multioutlet
toma de potencia power takeoff
toma de receptáculo receptacle outlet
toma de regulación tapping point
toma de sonido sound takeoff
toma media half tap
tomacorriente outlet, power outlet, power receptacle, socket, plug
tomacorriente accesorio accessory outlet
tomacorriente auxiliar utility outlet
tomacorriente con conexión a tierra grounded outlet
tomacorriente conmutado switched outlet
tomacorriente de alumbrado lighting outlet
tomacorriente de aparato eléctrico appliance outlet
tomacorriente de clavija plug-in outlet
tomacorriente de conveniencia convenience outlet
tomacorriente de pared wall outlet
tomacorriente eléctrico electrical outlet
tomacorriente embutido flush outlet
tomacorriente no conmutada unswitched outlet
tomacorriente puesto a tierra safety outlet
tomografía tomography
tomografía asistida por computadora computer-assisted tomography
tomografía axial computarizada computerized axial tomography
tomografía computarizada computed tomography
tonal tonal
tono tone, pitch
tono absoluto absolute pitch
tono adicional additional tone
tono audible audible tone
tono breve beep
tono combinado combination tone
tono complejo complex tone
tono continuo continuous tone
tono de alarma alarm tone
tono de alarma de fallas fault alarm tone
tono de audio fijo fixed audio tone
tono de audiofrecuencia audio-frequency tone
tono de barrido sweep tone
tono de batido beat tone
tono de conmutación switching tone
tono de diferencia difference tone
tono de fallas fault tone
tono de grupo ocupado group-busy tone
tono de guarda guard tone
tono de línea ocupada line busy tone
tono de llamada ringing tone
tono de marcar dial tone
tono de marcar bloqueado blocked dial tone
tono de marcar especial special dial tone
tono de ocupado busy tone
tono de ocupado de canal channel busy tone
tono de orden order tone
tono de prueba test tone
tono de prueba patrón standard test tone
tono de referencia reference tone
tono de trabajo marking tone
tono discreto discrete tone
tono entero whole tone
tono fijo fixed pitch
tono fundamental fundamental tone
tono indicador de fallas fault-reporting tone
tono intermitente intermittent tone
tono local sidetone
tono musical musical tone

tono normal standard tone, standard pitch
tono piloto pilot tone
tono puro pure tone
tono resultante resultant tone
tono secundario secondary tone
tono sinusoidal sinusoidal tone
tono subaudible subaudible tone
tono ululante warble tone
tono variable variable tone
tope antirremanente residual stop
tope plano flat top
tope posterior back stop
topología topology
topología de redes network topology
torcedura en guíaondas waveguide twist
torcido adicional additional twisting
toriado thoriated
torio thorium
tormenta de radiorruido radio noise storm
tormenta de ruido noise storm
tormenta eléctrica electric storm
tormenta ionosférica ionospheric storm
tormenta magnética magnetic storm
tornadotrón tornadotron
tornillo de ajuste fino fine-adjustment screw
tornillo de avance lead screw
tornillo de guía lead screw
tornillo de orejetas thumbscrew
tornillo de sintonización tuning screw
tornillo de sintonización capacitiva capacitive
   tuning screw
tornillo de sujeción binding screw, terminal screw
toroide toroid
toroide de ferrita ferrite toroid
torr torr
torre tower
torre angular angle tower
torre arriostrada guyed tower
torre de antena radio tower
torre de comunicaciones communication tower
torre de control control tower
torre de control de aeródromo aerodrome control
   tower
torre de control de tráfico de aeropuerto airport
   traffic control tower
torre de microondas microwave tower
torre de radar radar tower
torre de transmisión transmission tower
torre radiante radiating tower
torre telescópica telescoping tower
torre terminal dead-end tower
torsiómetro torsiometer, torquemeter
torsiómetro electromagnético electromagnetic
   torquemeter
torsión torsion
torsional torsional
torsionalmente torsionally
total acumulado accumulated total
total de comprobación check total, hash total
total de control control total
total de prueba proof total
total intermedio intermediate total
total menor minor total
total progresivo progressive total
totalmente desconectado totally disconnected
totalmente encerrado totally enclosed
toxímetro de anoxemia anoxemia toximeter
trabajo job
trabajo de operación operating duty
trabajo en línea on-line work
trabajo interactivo interactive job
trabajo sin carga no-load work
tracción eléctrica electric traction

tractor de papel paper tractor
traducción translation
traducción algorítmica algorithmic translation
traducción de código code translation
traducción de datos data translation
traducción de dirección address translation
traducción de lenguaje language translation
traducción de líneas line translation
traducción de programa program translation
traducción de programa de canal channel program
   translation
traductor translator
traductor de códigos code translator
traductor de entrada input translator
traductor de frecuencia ultraalta ultra-high
   frequency translator
traductor de lenguaje language translator
traductor de nivel level translator
traductor de uno a uno one-to-one translator
tráfico traffic
tráfico A A traffic
tráfico a larga distancia long-distance traffic
tráfico artificial artificial traffic
tráfico bidireccional two-way traffic
tráfico de comunicaciones communication traffic
tráfico de datos data traffic
tráfico de desbordamiento overflow traffic
tráfico de información information traffic
tráfico de llegada incoming traffic
tráfico de prueba test traffic
tráfico de salida outgoing traffic
tráfico de seguridad safety traffic
tráfico de telecomunicaciones telecommunications
   traffic
tráfico de tránsito transit traffic
tráfico intercontinental intercontinental traffic
tráfico internacional international traffic
tráfico interno internal traffic
tráfico interno de llegada incoming internal traffic
tráfico interurbano toll traffic
tráfico liviano light traffic
tráfico medio average traffic
tráfico mixto mixed traffic
tráfico telefónico telephone traffic
tráfico uniforme smooth traffic
trama de color color field
trama de exploración raster grid
trama de televisión television raster
tramo section
trampa trap
trampa de absorción absorption trap
trampa de huecos hole trap
trampa de humedad moisture trap
trampa de interferencia interference trap
trampa de ondas wavetrap
trampa de polaridad polarity trap
trampa de sonido sound trap
trampa iónica ion trap
transacción transaction
transacción de borrado delete transaction
transacción de dirección solamente address-only
   transaction
transacción de escritura de bloque block write
   transaction
transacción de lectura de bloque block read
   transaction
transacción de suma add transaction
transacción local local transaction
transacción no recuperable nonrecoverable
   transaction
transacción nula null transaction
transadmitancia transadmittance
transadmitancia de haz beam transadmittance

transadmitancia directa forward transadmittance
transadmitancia directa para señal pequeña
small-signal forward transadmittance
transadmitancia en cortocircuito short-circuit
transadmittance
transadmitancia interelectródica interelectrode
transadmittance
transadmitancia inversa reverse transadmittance
transceptor transceiver, transreceiver
transceptor fibroóptico fiber-optic transceiver
transceptor magnetoestrictivo magnetostrictive
transceiver
transconductancia transconductance
transconductancia de conversión conversion
transconductance
transconductancia de pantalla screen
transconductance
transconductancia de rejilla a placa grid-to-plate
transconductance
transconductancia de válvula valve
transconductance
transconductancia directa forward transconductance
transconductancia estática static transconductance
transconductancia extrínseca extrinsic
transconductance
transconductancia interelectródica interelectrode
transconductance
transconductancia intrínseca intrinsic
transconductance
transconductancia negativa negative
transconductance
transconductancia para señal pequeña small-signal
transconductance
transconductancia refleja reflex transconductance
transconductancia rejilla-placa grid-plate
transconductance
transconductor transconductor
transconductor estático static transconductor
transcribir transcribe
transcripción transcription
transcripción cinescópica kinescope transcription
transcripción de datos data transcription
transcripción eléctrica electrical transcription
transcripción lateral lateral transcription
transcriptor transcriber
transdiodo transdiode
transducción transduction
transducción deformimétrica strain-gage
transduction
transducción reluctiva reluctive transduction
transducción resistiva resistive transduction
transducir transduce
transductor transducer, pickup
transductor activo active transducer
transductor acústico acoustic transducer
transductor autogenerador self-generating
transducer
transductor bidireccional bidirectional transducer
transductor bilateral bilateral transducer
transductor capacitivo capacitive transducer
transductor cerámico ceramic transducer
transductor cinematográfico motion-picture pickup
transductor compuesto composite transducer
transductor de acoplamiento en paralelo parallel
transducer
transductor de acoplamiento en serie series
transductor
transductor de capacitancia variable
variable-capacitance transducer
transductor de capacitor capacitor transducer
transductor de cinta ribbon transducer
transductor de conducción ósea bone-conduction
transducer

transductor de control de posición position-control
transducer
transductor de conversión conversion transducer
transductor de conversión armónico harmonic
conversion transducer
transductor de conversión heterodino heterodyne
conversion transducer
transductor de corriente alterna alternating-current
transducer
transductor de corriente continua direct-current
transducer
transductor de cristal crystal transducer
transductor de cristal cerámico ceramic crystal
transducer
transductor de datos data transducer
transductor de desplazamiento displacement
transducer
transductor de diodo diode transducer
transductor de gama completa full-range transducer
transductor de hilo caliente hot-wire transducer
transductor de hilo vibrante vibrating-wire
transducer
transductor de inductancia diferencial
differential-inductance transducer
transductor de inductancia mutua
mutual-inductance transducer
transductor de inductancia variable
variable-inductance transducer
transductor de ionización ionization transducer
transductor de línea line transducer
transductor de medida measuring transductor
transductor de moción lineal linear motion
transducer
transductor de película magnetoestrictiva
magnetostrictive film transducer
transductor de posición position transducer
transductor de posición digital absoluto absolute
digital position transducer
transductor de presión pressure transducer
transductor de presión absoluta absolute pressure
transducer
transductor de presión diferencial
differential-pressure transducer
transductor de reluctancia variable
variable-reluctance transducer
transductor de resistencia resistance transducer
transductor de resistencia variable
variable-resistance transducer
transductor de sonar sonar transducer
transductor de telemedida telemetering transducer
transductor de velocidad velocity transducer
transductor de velocidad vertical vertical-speed
transducer
transductor deformimétrico strain-gage transducer
transductor diferencial differential transducer
transductor digital digital transducer
transductor disimétrico dissymmetrical transducer
transductor dividido split transducer
transductor eléctrico electrical transducer
transductor eléctrico acústico acoustic electric
transducer
transductor electroacústico electroacoustic
transducer
transductor electroestrictivo electrostrictive
transducer
transductor electromagnético electromagnetic
transducer
transductor electromecánico electromechanical
transducer
transductor electroquímico electrochemical
transducer
transductor electrostático electrostatic transducer
transductor fotoeléctrico photoelectric transducer

transductor hidroacústico hydroacoustic transducer
transductor ideal ideal transducer
transductor inductivo inductive transducer
transductor lineal linear transducer
transductor magnético magnetic transducer
transductor magnetoeléctrico magnetoelectric
    transducer
transductor magnetoestrictivo magnetostrictive
    transducer
transductor mecanoeléctrico mechanoelectric
    transducer
transductor oscilante oscillating transducer
transductor pasatodo all-pass transducer
transductor pasivo passive transducer
transductor piezoeléctrico piezoelectric transducer
transductor piezoeléctrico ultrasónico ultrasonic
    piezoelectric transducer
transductor piroeléctrico pyroelectric transducer
transductor potenciométrico potentiometric
    transducer
transductor radiactivo radioactive transducer
transductor recíproco reciprocal transducer
transductor resistivo resistive transducer
transductor reversible reversible transducer
transductor rotativo digital digital rotary transducer
transductor simétrico symmetrical transducer
transductor ultrasónico ultrasonic transducer
transductor unidireccional unidirectional transducer
transductor unilateral unilateral transducer
transferencia transfer, transference
transferencia automática automatic transfer
transferencia condicional conditional transfer
transferencia condicional de control conditional
    transfer of control
transferencia de acceso a memoria directo direct
    memory access transfer
transferencia de archivo file transfer
transferencia de bloque block transfer
transferencia de calor heat transfer
transferencia de campo field transfer
transferencia de capa a capa layer-to-layer transfer
transferencia de carga charge transfer
transferencia de control control transfer
transferencia de control condicional conditional
    control transfer
transferencia de control de programa
    program-control transfer
transferencia de control incondicional unconditional
    control transfer
transferencia de datos data transfer
transferencia de datos asincrónica asynchronous
    data transfer
transferencia de datos de alta velocidad high-speed
    data transfer
transferencia de datos instantánea instantaneous
    data transfer
transferencia de datos lógica logical data transfer
transferencia de datos programada programmed
    data transfer
transferencia de dígitos digit transfer
transferencia de dirección address transfer
transferencia de energía energy transfer
transferencia de energía lineal linear energy
    transfer
transferencia de imagen image transfer
transferencia de información information transfer
transferencia de mandos command transfer
transferencia de palabras word transfer
transferencia de potencia power transfer
transferencia de registro register transfer
transferencia de señal de capa a capa layer-to-layer
    signal transfer
transferencia de tensión accidental accidental

voltage transfer
transferencia directa forward transfer
transferencia electrónica electron transfer
transferencia en paralelo parallel transfer
transferencia en serie serial transfer
transferencia incondicional unconditional transfer
transferencia incondicional de control unconditional
    transfer of control
transferencia magnética magnetic transfer,
    print-through
transferencia periférica peripheral transfer
transferencia por cable cable transfer
transferencia radial radial transfer
transferencia secuencial sequential transfer
transferencia térmica thermal transfer
transferido transferred
transferir download
transfluxor transfluxor
transformación transformation
transformación beta beta transformation
transformación de coordenadas de color
    color-coordinate transformation
transformación de datos data transformation
transformación de energía energy transformation
transformación de fase phase transformation
transformación de frecuencia frequency
    transformation
transformación de impedancia impedance
    transformation
transformación de Laplace Laplace transform
transformación de Norton Norton transformation
transformación de potencia power transformation
transformación de señal signal transformation
transformado transformed
transformador transformer
transformador acorazado shell-type transformer
transformador aislado por aceite oil-insulated
    transformer
transformador aislado por gas gas-insulated
    transformer
transformador aislante isolating transformer,
    insulating transformer
transformador ajustable adjustable transformer
transformador ajustador de fase giratorio rotatable
    phase-adjusting transformer
transformador antifluctuación antihunt transformer
transformador armónico harmonic transformer
transformador autoenfriado self-cooled transformer
transformador auxiliar auxiliary transformer
transformador auxiliar de retorno booster
    transformer
transformador bañado en aceite oil-immersed
    transformer
transformador bifilar bifilar transformer
transformador con derivación central center-tapped
    transformer
transformador con núcleo de hierro iron-core
    transformer
transformador con tomas tapped transformer
transformador continuamente ajustable
    continuously-adjustable transformer
transformador criogénico cryogenic transformer
transformador de acoplamiento coupling
    transformer
transformador de acoplamiento controlado
    controlled-coupling transformer
transformador de acoplamiento de entrada input
    coupling transformer
transformador de acoplamiento de salida output
    coupling transformer
transformador de adaptación matching transformer
transformador de adaptación de altavoz
    loudspeaker matching transformer

transformador de adaptación de impedancias
impedance-matching transformer
transformador de adaptación de línea line-matching
transformer
transformador de adaptación en delta delta
matching transformer
transformador de aislamiento isolation transformer
transformador de alimentación power-supply
transformer
transformador de alta frecuencia high-frequency
transformer
transformador de alta reactancia high-reactance
transformer
transformador de alta tensión high-voltage
transformer
transformador de alumbrado lighting transformer
transformador de audio audio transformer
transformador de audiofrecuencia audio-frequency
transformer
transformador de baja tensión low-voltage
transformer
transformador de barras cruzadas crossbar
transformer
transformador de barrido horizontal
horizontal-sweep transformer
transformador de barrido vertical vertical-sweep
transformer
transformador de bloqueo blocking transformer
transformador de calefactor heater transformer
transformador de circuito de control control-circuit
transformer
transformador de circuito único single-circuit
transformer
transformador de control control transformer
transformador de conversión conversion
transformer
transformador de corriente current transformer
transformador de corriente compensado
compensated current transformer
transformador de corriente constante
constant-current transformer
transformador de corriente continua direct-current
transformer
transformador de distribución distribution
transformer
transformador de dos arrollamientos two-winding
transformer
transformador de enfasamiento phasing transformer
transformador de entrada input transformer
transformador de etapa excitadora driver
transformer
transformador de excitación drive transformer
transformador de fase phase transformer
transformador de fase aislada phase-isolated
transformer
transformador de filamento filament transformer
transformador de frecuencia frequency transformer
transformador de frecuencia intermedia
intermediate-frequency transformer
transformador de generador de impulsos peaking
transformer
transformador de guíaondas waveguide transformer
transformador de ignición ignition transformer
transformador de imagen picture transformer
transformador de impedancia impedance
transformer
transformador de impedancia de antena antenna
impedance transformer
transformador de impulsos pulse transformer
transformador de Joly Joly transformer
transformador de línea line transformer
transformador de medidor instrument transformer
transformador de micrófono microphone

transformer
transformador de modo mode transformer
transformador de modulación modulation
transformer
transformador de núcleo core transformer
transformador de núcleo cerrado closed-core
transformer
transformador de núcleo de aire air-core
transformer
transformador de oscilación oscillation transformer
transformador de placa plate transformer
transformador de potencia power transformer
transformador de potencia auxiliar auxiliary power
transformer
transformador de potencial potential transformer
transformador de preamplificación preamplification
transformer
transformador de puente bridge transformer
transformador de puesta a tierra grounding
transformer
transformador de radiofrecuencia radio-frequency
transformer
transformador de radiofrecuencia sintonizada
tuned-radio-frequency transformer
transformador de razones múltiples multiple-ratio
transformer
transformador de rectificador rectifier transformer
transformador de rejilla grid transformer
transformador de retorno flyback transformer
transformador de salida output transformer
transformador de salida de línea line-output
transformer
transformador de salida horizontal
horizontal-output transformer
transformador de salida universal universal output
transformer
transformador de salida vertical vertical-output
transformer
transformador de Scott Scott transformer
transformador de simétrico a asimétrico
balanced-to-unbalanced transformer
transformador de soldadura welding transformer
transformador de sustitución substitution
transformer
transformador de tensión voltage transformer
transformador de tensión constante
constant-voltage transformer
transformador de tensión por capacitor capacitor
voltage transformer
transformador de tensión variable variable-voltage
transformer
transformador de Tesla Tesla transformer
transformador de timbre bell transformer
transformador de tipo aislador insulator-type
transformer
transformador de tipo seco dry-type transformer
transformador de triple adaptador triple-stub
transformer
transformador de uno a uno one-to-one transformer
transformador de uso general general-purpose
transformer
transformador de velocidad velocity transformer
transformador desfasador phase-shifting
transformer
transformador diferencial differential transformer
transformador diferencial lineal linear differential
transformer
transformador discriminador discriminator
transformer
transformador distribuidor de potencia
power-distributing transformer
transformador electrónico electronic transformer
transformador elevador step-up transformer

transformador en contrafase push-pull transformer
transformador en cuadratura quadrature transformer
transformador en cuarto de onda quarter-wave transformer
transformador en derivación shunt transformer
transformador en serie series transformer
transformador enchufable plug-in transformer
transformador enfriado por aceite oil-cooled transformer
transformador entre etapas interstage transformer
transformador equilibrado balanced transformer
transformador equilibrador balancer transformer
transformador estático static transformer
transformador exterior outside transformer, outdoor transformer
transformador giratorio rotatable transformer
transformador híbrido hybrid transformer
transformador ideal ideal transformer
transformador incorporado built-in transformer
transformador independiente independent transformer
transformador interfásico interphase transformer
transformador interior inside transformer, indoor transformer
transformador lineal linear transformer
transformador lleno de aceite oil-filled transformer
transformador lleno de líquido liquid-filled transformer
transformador microfónico microphonic transformer
transformador monofásico single-phase transformer
transformador múltiple multiple transformer
transformador multiplicador de fase phase-multiplying transformer
transformador multiplicador de frecuencia frequency-multiplying transformer
transformador multitoma multitap transformer
transformador no sintonizado untuned transformer
transformador oscilante oscillating transformer
transformador pasamuros bushing transformer
transformador perfecto perfect transformer
transformador polifásico polyphase transformer
transformador principal main transformer
transformador reductor stepdown transformer
transformador regulador regulating transformer
transformador regulador de tensión voltage-regulating transformer
transformador rotativo rotary transformer
transformador saturable saturable transformer
transformador seco dry transformer
transformador simétrico-asimétrico balanced-unbalanced transformer
transformador sintonizado tuned transformer
transformador sobreacoplado overcoupled transformer
transformador sonda probe transformer
transformador sumergible submersible transformer
transformador toroidal toroidal transformer
transformador trifásico three-phase transformer
transformador universal universal transformer
transformador variable variable transformer
transformadores en paralelo banked transformers
transformar transform
transhorizonte transhorizon
transición transition
transición automática automatic crossover
transición conductiva breakover
transición conductiva directa forward breakover
transición cuántica quantum transition
transición de estado state transition
transición de fase phase transition
transición de fase difusa diffuse phase transition

transición de puente bridge transition
transición de puente equilibrado balanced bridge transition
transición en circuito abierto open-circuit transition
transición negativa negative transition
transición por derivación shunt transition
transición positiva positive transition
transición progresiva tapered transition
transicional transitional
transimpedancia transimpedance
transimpedancia directa forward transimpedance
transimpedancia inversa reverse transimpedance
transistancia transistance
transistor transistor
transistor aleado alloyed transistor
transistor anular annular transistor
transistor bidireccional bidirectional transistor
transistor bipolar bipolar transistor
transistor coaxial coaxial transistor
transistor compuesto compound transistor
transistor de aleación alloy transistor
transistor de aleación difusa diffused-alloy transistor
transistor de aleación superficial surface-alloy transistor
transistor de alta velocidad high-speed transistor
transistor de audiofrecuencia audio-frequency transistor
transistor de avalancha avalanche transistor
transistor de barrera barrier transistor
transistor de barrera intrínseca intrinsic-barrier transistor
transistor de barrera superficial surface-barrier transistor
transistor de base aleada bonded-barrier transistor
transistor de base difusa diffused-base transistor
transistor de base epitaxial epitaxial base transistor
transistor de base graduada graded-base transistor
transistor de base homotaxial hometaxial-base transistor
transistor de base metálica metal-base transistor
transistor de beta variable variable-beta transistor
transistor de campo de arrastre drift-field transistor
transistor de campo interno drift transistor
transistor de capa de agotamiento depletion-layer transistor
transistor de capa superpuesta overlay transistor
transistor de capas de difusión grown-diffused transistor
transistor de carburo de silicio silicon carbide transistor
transistor de colector difuso electroquímico electrochemical diffused collector transistor
transistor de conmutación switching transistor
transistor de conmutación de alta velocidad high-speed switching transistor
transistor de contactos de punta point-contact transistor
transistor de cuatro capas four-layer transistor
transistor de difusión diffusion transistor
transistor de difusión de aleación alloy-diffused transistor
transistor de difusión de barrera superficial surface-barrier diffused transistor
transistor de difusión doble double-diffused transistor
transistor de difusión microaleado microalloy diffused transistor
transistor de difusión única single-diffused transistor
transistor de doble base double-base transistor
transistor de doble dopado double-doped transistor
transistor de doble emisor dual-emitter transistor

transistor de doble superficie double-surface transistor
transistor de efecto de campo field-effect transistor
transistor de efecto de campo de agotamiento depletion field-effect-transistor
transistor de efecto de campo de unión junction field-effect-transistor
transistor de efecto de campo epitaxial epitaxial field-effect-transistor
transistor de efecto de campo fotoeléctrico photoelectric field-effect-transistor
transistor de efecto de campo multicanal multichannel field-effect transistor
transistor de efecto de campo simétrico symmetrical field-effect-transistor
transistor de efecto de campo tetrodo tetrode field-effect-transistor
transistor de efecto de campo unipolar unipolar field-effect-transistor
transistor de frecuencia ultraalta ultra-high frequency transistor
transistor de gancho hook transistor
transistor de germanio germanium transistor
transistor de indio indium transistor
transistor de interrupción periódica chopper transistor
transistor de meseta de crecimiento epitaxial epitaxial-growth mesa transistor
transistor de meseta epitaxial epitaxial mesa transistor
transistor de microondas microwave transistor
transistor de modulación de conductividad conductivity-modulation transistor
transistor de oscilador oscillator transistor
transistor de película delgada thin-film transistor
transistor de potencia power transistor
transistor de puerta resonante resonant-gate transistor
transistor de punta y uniones point-junction transistor
transistor de puntas point transistor
transistor de radiofrecuencia radio-frequency transistor
transistor de reactancia reactance transistor
transistor de refusión meltback transistor
transistor de región intrínseca intrinsic-region transistor
transistor de salida output transistor
transistor de semiconductor metal-óxido metal-oxide-semiconductor transistor
transistor de silicio silicon transistor
transistor de silicio planar planar silicon transistor
transistor de tres uniones three-junction transistor
transistor de triple difusión triple-diffused transistor
transistor de unión de aleación alloy-junction transistor
transistor de unión de aleación de germanio germanium alloy-junction transistor
transistor de unión de doble base double-base junction transistor
transistor de unión flotante floating-junction transistor
transistor de unión graduada graded-junction transistor
transistor de unión NP NP junction transistor
transistor de unión PN PN-junction transistor
transistor de unión PNP PNP junction transistor
transistor de unión por crecimiento grown-junction transistor
transistor de unión por difusión diffused-junction transistor
transistor de unión por difusión PNP PNP diffused-junction transistor

transistor de unión por fusión fused-junction transistor
transistor de unión única single-junction transistor
transistor de uniones junction transistor
transistor de uniones NPIN NPIN junction transistor
transistor de uniones NPN NPN junction transistor
transistor de uniones planares planar-junction transistor
transistor de uso general general-purpose transistor
transistor diodo diode transistor
transistor electroóptico electrooptical transistor
transistor en derivación shunt transistor
transistor enfriado por aire air-cooled transistor
transistor epitaxial epitaxial transistor
transistor epitaxial de silicio silicon epitaxial transistor
transistor epitaxial pasivado planar planar epitaxial passivated transistor
transistor epitaxial planar planar epitaxial transistor
transistor estabilizador stabilizing transistor
transistor mesa mesa transistor
transistor microaleado microalloy transistor
transistor moldeado molded transistor
transistor multiemisor multiemitter transistor
transistor NPIN NPIN transistor
transistor NPIP NPIP transistor
transistor NPN NPN transistor
transistor NPN de germanio germanium NPN transistor
transistor NPNP NPNP transistor
transistor optoelectrónico optoelectronic transistor
transistor para señal pequeña small-signal transistor
transistor pasivado passivated transistor
transistor pentodo pentode transistor
transistor pentodo de efecto de campo pentode field-effect transistor
transistor planar planar transistor
transistor planar de silicio silicon planar transistor
transistor planar epitaxial epitaxial planar transistor
transistor planar por difusión diffused-planar transistor
transistor PNIN PNIN transistor
transistor PNIP PNIP transistor
transistor PNP PNP transistor
transistor PNPN PNPN transistor
transistor por crecimiento variable rate-grown transistor
transistor por difusión diffused transistor
transistor simétrico symmetrical transistor
transistor superbeta superbeta transistor
transistor tetrodo tetrode transistor
transistor tipo N N-type transistor
transistor triodo triode transistor
transistor unipolar unipolar transistor
transistor uniunión unijunction transistor
transistor universal universal transistor
transistores acoplados coupled transistors
transistores complementarios complementary transistors
transistores en tándem tandem transistor
transistorímetro transistorimeter
transistorización transistorization
transistorizado transistorized
transistorizar transistorize
tránsito transit
tránsito automático automatic transit
tránsito directo direct transit
transitorio transitory, transient
transitrón transitron
translúcido translucent
transmisibilidad transmissibility
transmisible transmissible
transmisión transmission, sending

transmisión a larga distancia long-distance transmission
transmisión acústica acoustic transmission
transmisión analógica analog transmission
transmisión asincrónica asynchronous transmission
transmisión automática automatic transmission
transmisión bipolar bipolar transmission
transmisión bisíncrona bysynchronous transmission
transmisión blanca white transmission
transmisión ciega blind transmission
transmisión colectiva collective transmission
transmisión con bloqueo por voz antivoice-operated transmission
transmisión con incidencia oblicua oblique-incidence transmission
transmisión con portadora suprimida suppressed-carrier transmission
transmisión de baja velocidad low-speed transmission
transmisión de banda ancha broadband transmission
transmisión de banda lateral asimétrica asymmetrical sideband transmission
transmisión de banda lateral residual vestigial-sideband transmission
transmisión de banda lateral única single-sideband transmission
transmisión de caracteres paralela parallel character transmission
transmisión de cinta perforada perforated-tape transmission
transmisión de colores color transmission
transmisión de corriente alterna alternating-current transmission
transmisión de corriente continua direct-current transmission
transmisión de corriente de bits bit stream transmission
transmisión de datos data transmission
transmisión de datos asincrónica asynchronous data transmission
transmisión de datos automática automatic data transmission
transmisión de datos digitales digital-data transmission
transmisión de datos directa direct data transmission
transmisión de doble banda lateral double-sideband transmission
transmisión de energía power transmission
transmisión de facsímil facsimile transmission
transmisión de flujo de bits bit stream transmission
transmisión de fotografía photograph transmission
transmisión de haz beam transmission
transmisión de imagen image transmission
transmisión de imagen de corriente continua direct-current picture transmission
transmisión de impulsos pulse transmission
transmisión de incidencia vertical vertical-incidence transmission
transmisión de información information transmission
transmisión de información digital digital information transmission
transmisión de luminancia constante constant-luminance transmission
transmisión de mensajes message transmission
transmisión de paquetes packet transmission
transmisión de portadora analógica analog carrier transmission
transmisión de portadora controlada controlled-carrier transmission
transmisión de portadora quiescente quiescent-carrier transmission

transmisión de programa program transmission
transmisión de prueba test transmission
transmisión de punto a punto point-to-point transmission
transmisión de radiodifusión broadcast transmission
transmisión de radiofacsímil radio-facsimile transmission
transmisión de rumbo bearing transmission
transmisión de señal signal transmission
transmisión de sonido sound transmission
transmisión de teleimpresora teleprinter transmission
transmisión de televisión television transmission
transmisión de velocidad variable variable-speed transmission
transmisión desequilibrada unbalanced transmission
transmisión difusa diffuse transmission
transmisión digital digital transmission
transmisión díplex diplex transmission
transmisión direccionada addressed transmission
transmisión direccional directional transmission
transmisión directa direct transmission
transmisión directa de televisión television direct transmission
transmisión dúplex duplex transmission
transmisión efectiva effective transmission
transmisión en bucle loop transmission
transmisión en espacio libre free-space transmission
transmisión en paralelo parallel transmission
transmisión en serie serial transmission
transmisión en tiempo real real-time transmission
transmisión especular specular transmission
transmisión estimulada stimulated transmission
transmisión física physical transmission
transmisión fotográfica photographic transmission
transmisión híbrida hybrid transmission
transmisión individual individual transmission
transmisión isócrona isochronous transmission
transmisión local local transmission
transmisión manual manual transmission
transmisión mixta mixed transmission
transmisión modulada en amplitud amplitude-modulated transmission
transmisión modulada por impulsos pulse-modulated transmission
transmisión monocroma monochrome transmission
transmisión múltiple multiple transmission
transmisión múltiplex multiplex transmission
transmisión negativa negative transmission
transmisión negra black transmission
transmisión neutra neutral transmission
transmisión no simultánea nonsimultaneous transmission
transmisión operada por voz voice-operated transmission
transmisión óptica optical transmission
transmisión polar polar transmission
transmisión polarizada verticalmente vertically-polarized transmission
transmisión por cable transmission by cable
transmisión por cable coaxial de banda ancha broadband coaxial cable transmission
transmisión por correa belt drive
transmisión por corriente única single-current transmission
transmisión por desplazamiento de frecuencia frequency-shift transmission
transmisión por dispersión scatter transmission
transmisión por hilo transmission by wire
transmisión por incrementos repentinos burst transmission
transmisión por línea line transmission
transmisión por ondas espaciales skywave

transmission
transmisión por ondas portadoras carrier-wave transmission
transmisión por portadora carrier transmission
transmisión por radio transmission by radio
transmisión por saltos múltiples multiskip transmission, multihop transmission
transmisión por satélite satellite transmission
transmisión por trayectorias múltiples multipath transmission
transmisión positiva positive transmission
transmisión regular regular transmission
transmisión secuencial sequential transmission
transmisión selectiva selective transmission
transmisión semidúplex half-duplex transmission
transmisión símplex simplex transmission
transmisión simultánea simultaneous transmission
transmisión sincrónica synchronous transmission
transmisión telefónica telephone transmission
transmisión telegráfica telegraph transmission
transmisión telegráfica de alta velocidad high-speed telegraph transmission
transmisión télex telex transmission
transmisión territorial territorial transmission
transmisión transhorizonte over-the-horizon transmission
transmisión unidireccional unidirectional transmission
transmisión unilateral unilateral transmission
transmisividad transmissivity, transmittivity
transmisividad acústica acoustic transmissivity
transmisividad atmosférica atmospheric transmissivity
transmisivo transmissive
transmisómetro transmissometer
transmisor transmitter
transmisor aerotransportado airborne transmitter
transmisor aural aural transmitter
transmisor autoexcitado self-excited transmitter
transmisor automático automatic transmitter
transmisor auxiliar auxiliary transmitter
transmisor controlado por cinta tape-controlled transmitter
transmisor controlado por cristal crystal-controlled transmitter
transmisor de aeronave aircraft transmitter
transmisor de alta frecuencia high-frequency transmitter
transmisor de alternador alternator transmitter
transmisor de arco arc transmitter
transmisor de audio audio transmitter
transmisor de avión airplane transmitter
transmisor de baja potencia low-power transmitter
transmisor de banda ancha wideband transmitter
transmisor de banda lateral residual vestigial-sideband transmitter
transmisor de banda lateral única single-sideband transmitter
transmisor de carbón carbon transmitter
transmisor de cinta tape transmitter
transmisor de códigos code transmitter
transmisor de colores color transmitter
transmisor de control remoto remote-control transmitter
transmisor de corriente portadora carrier-current transmitter
transmisor de chispa spark transmitter
transmisor de datos data transmitter
transmisor de datos coordenados coordinate data transmitter
transmisor de distancia range transmitter
transmisor de doble banda lateral double-sideband transmitter

transmisor de emergencia emergency transmitter
transmisor de enlace link transmitter
transmisor de facsímil facsimile transmitter
transmisor de fallas fault transmitter
transmisor de frecuencia fija fixed-frequency transmitter
transmisor de frecuencia media medium-frequency transmitter
transmisor de frecuencia ultraalta ultra-high frequency transmitter
transmisor de haz luminoso light-beam transmitter
transmisor de imagen picture transmitter
transmisor de impulsos pulse transmitter
transmisor de inclinación de planeo glide-slope transmitter
transmisor de infrarrojo infrared transmitter
transmisor de lorán loran transmitter
transmisor de micrófono microphone transmitter
transmisor de microondas microwave transmitter
transmisor de modulación de amplitud-modulación de frecuencia amplitude-modulation-frequency-modulation transmitter
transmisor de modulación de fase phase-modulation transmitter
transmisor de modulación de frecuencia frequency-modulated transmitter
transmisor de muy alta frecuencia very-high-frequency transmitter
transmisor de numeración numbering transmitter
transmisor de onda media medium-wave transmitter
transmisor de ondas continuas continuous-wave transmitter
transmisor de ondas cortas shortwave transmitter
transmisor de ondas ultracortas ultrashort-wave transmitter
transmisor de programas program transmitter
transmisor de radar radar transmitter
transmisor de radio de emergencia emergency radio transmitter
transmisor de radio por hilo wired-radio transmitter
transmisor de radiodifusión broadcast transmitter
transmisor de radiodifusión de alta frecuencia high-frequency broadcast transmitter
transmisor de radioenlace radio-link transmitter
transmisor de radiofaro beacon transmitter
transmisor de radiofaro direccional radio-range transmitter
transmisor de radiosonda radiosonde transmitter
transmisor de reserva standby transmitter
transmisor de rumbo bearing transmitter
transmisor de sonar sonar transmitter
transmisor de tambor drum transmitter
transmisor de teclado keyboard transmitter
transmisor de teleimpresora teleprinter transmitter
transmisor de telemedida telemetering transmitter
transmisor de televisión television transmitter
transmisor de trayectoria de planeo glide-path transmitter
transmisor de tubo tube transmitter
transmisor de tubos de vacío vacuum-tube transmitter
transmisor de video video transmitter
transmisor diferencial differential transmitter
transmisor direccional directional transmitter
transmisor electromagnético electromagnetic transmitter
transmisor estabilizado por cristal crystal-stabilized transmitter
transmisor fijo fixed transmitter
transmisor localizador localizer transmitter
transmisor maestro master transmitter
transmisor modulado en amplitud

amplitude-modulated transmitter
transmisor modulado en fase phase-modulated transmitter
transmisor móvil mobile transmitter
transmisor multicanal multichannel transmitter
transmisor multietapa multistage transmitter
transmisor multifrecuencia multifrequency transmitter
transmisor múltiple multiple transmitter
transmisor múltiplex multiplex transmitter
transmisor omnidireccional omnidirectional transmitter
transmisor patrón standard transmitter
transmisor portátil portable transmitter
transmisor principal main transmitter
transmisor pulsado pulsed transmitter
transmisor radiotelefónico radiotelephone transmitter
transmisor radiotelegráfico radiotelegraph transmitter
transmisor-receptor transmitter-receiver
transmisor relé relay transmitter
transmisor repetidor repeater transmitter
transmisor satélite satellite transmitter
transmisor secundario secondary transmitter
transmisor selsyn selsyn transmitter
transmisor sincronizado en fase phase-synchronized transmitter
transmisor subterráneo underground transmitter
transmisor tacán tacan transmitter
transmisor telefónico telephone transmitter
transmisor telegráfico telegraph transmitter
transmisor telemétrico telemetric transmitter
transmisor transistorizado transistorized transmitter
transmisor transportable transportable transmitter
transmisor universal universal transmitter
transmisor visual visual transmitter
transmisores sincronizados synchronized transmitters
transmitancia transmittance
transmitancia bicónica biconical transmittance
transmitancia bidireccional bidirectional transmittance
transmitancia bihemisférica bihemispherical transmittance
transmitancia de bucle loop transmittance
transmitancia de trayectoria path transmittance
transmitancia espectral spectral transmittance
transmitancia interna internal transmittance
transmitancia luminosa luminous transmittance
transmitancia regular regular transmittance
transmitancia total total transmittance
transmitido transmitted
transmitir transmit
transmodulación transmodulation
transoceánico transoceanic
transónico transonic
transparencia transparency
transparente transparent
transpolarizador transpolarizer
transpondedor transponder
transpondedor coherente coherent transponder
transpondedor de banda cruzada crossband transponder
transpondedor de radar radar transponder
transponer transpose
transportabilidad transportability
transportable transportable
transporte transport
transporte cíclico end-around carry
transporte de cinta tape transport
transporte de portadores carrier transport
transposición transposition

transposición de Roebel Roebel transposition
transrectificación transrectification
transrectificador transrectifier
transresistencia transresistance
transtrictor transtrictor
transuranio transuranium
transversal transversal
tranvía tramway
trapecial trapezoidal
trapecio trapezoid
traslación de frecuencia frequency translation
traslación de supergrupo supergroup translation
traslado de llamadas call forwarding
traspaso handover
traspaso de canal channel handover
tratamiento acústico acoustic treatment
tratamiento superficial surface treatment
tratamiento térmico heat treatment
trayectoria abierta open path
trayectoria accidental sneak path
trayectoria autorizada authorized path
trayectoria crítica critical path
trayectoria de acceso access path
trayectoria de acceso compartido shared-access path
trayectoria de aproximación approach path
trayectoria de arco arc path
trayectoria de ascensor hoistway
trayectoria de aterrizaje landing path
trayectoria de cinta tape path
trayectoria de conducción eléctrica electric conducting path
trayectoria de control control path
trayectoria de datos data path
trayectoria de eco echo path
trayectoria de enlace lógico logical link path
trayectoria de exploración scan path
trayectoria de flujo flow path
trayectoria de fuga leakage path
trayectoria de información information path
trayectoria de ionización ionization path
trayectoria de lectura read path
trayectoria de línea visual line-of-sight path
trayectoria de onda tangencial tangential wavepath
trayectoria de ondas wave path
trayectoria de papel paper path
trayectoria de planeo glide path
trayectoria de progresión forward path
trayectoria de propagación propagation path
trayectoria de radiopropagación radio-propagation path
trayectoria de recubrimiento overlay path
trayectoria de red network path
trayectoria de retorno return path
trayectoria de retroalimentación feedback path
trayectoria de salto único single-hop path
trayectoria de señal signal path
trayectoria de tarjetas card path
trayectoria de transmisión transmission path
trayectoria de transmisión de microondas microwave transmission path
trayectoria de vuelo flight path
trayectoria digital digital path
trayectoria electrónica electron trajectory
trayectoria en espacio libre free-space path
trayectoria estocástica stochastic path
trayectoria helicoidal helical path
trayectoria libre free path
trayectoria libre molecular molecular free path
trayectoria lineal straight-line path
trayectoria magnética magnetic path
trayectoria múltiple multiple path
trayectoria óptica optical path

trayectoria perforada punched path
trayectoria principal principal path
trayectoria rasante grazing path
trayectoria reversible reversible path
trayectoria simple simple path
trayectoria terciaria tertiary path
trayectoria terrestre land path
trayectorias atmosféricas atmospheric paths
traza trace
trazado plot, plotting, tracing, layout
trazado de circuito circuit layout
trazado de diagrama chart layout
trazado de gráficos graphics plotting
trazado de línea line plot
trazado x-y x-y plotting
trazador plotter, tracer
trazador de alta velocidad high-speed plotter
trazador de datos data plotter
trazador de gráficos graph plotter
trazador de pluma pen plotter
trazador electrónico electronic plotter
trazador electrostático electrostatic plotter
trazador en línea on-line plotter
trazador incremental incremental plotter
trazador radiactivo radioactive tracer
trazador x-y x-y plotter
trazo trace, scan
trazo circular circular trace
trazo de impulso pulse trace
trazo de registro recording trace
trazo de retorno return trace
trazo de señal signal trace
trazo en espiral spiral trace
trazo osciloscópico oscilloscope trace
tren de impulsos pulse train
tren de impulsos de video video pulse train
tren de impulsos periódico periodic pulse train
tren de impulsos unidireccional unidirectional pulse train
tren de ondas wave train
tren de puntas spike train
trenza braid
triac triac
triada triad
triada de colores color triad
triangular triangular
triángulo de colores color triangle
triángulo de impedancia impedance triangle
triboelectricidad triboelectricity
triboeléctrico triboelectric
triboluminiscencia triboluminescence
tricañón trigun
tricolor tricolor
tricón tricon
tricromático trichromatic
tridimensional three-dimensional
tridipolar tridipole
tridireccional tridirectional
triductor triductor
trifásico three-phase
trifurcación trifurcation
trigatrón trigatron
trigistor trigistor
trímer trimmer
trinistor trinistor
trinoscopio trinoscope
triodo triode
triodo cambiador de frecuencia frequency-changer triode
triodo con rejilla a masa grounded-grid triode
triodo de alta impedancia high-impedance triode
triodo de alta potencia high-power triode
triodo de baja impedancia low-impedance triode

triodo de cristal crystal triode
triodo de efecto túnel tunnel triode
triodo de estado sólido solid-state triode
triodo de gas gas triode
triodo de impedancia media medium-impedance triode
triodo de neón neon triode
triodo de planos paralelos parallel-plane triode
triodo de potencia power triode
triodo de salida output triode
triodo electrométrico electrometer triode
triodo-heptodo triode-heptode
triodo-hexodo triode-hexode
triodo oscilador oscillator triode
triodo-pentodo triode-pentode
triodo planar planar triode
triodo regulador regulator triode
triodo silenciador squelch triode
triodo triple diodo triple-diode triode
triple conversión triple conversion
triple detección triple detection
triple diodo triple diode
triple precisión triple precision
triplex triplex
triplexión triplexing
triplexor triplexer
triplicación de frecuencia frequency trebling
triplicador tripler
triplicador de frecuencia frequency tripler
triplicador de tensión voltage tripler
tripolar three-pole
triprocesador triprocessor
trisistor trisistor
tritio tritium
trocotrón trochotron
troland troland
trole trolley
trolebús trolley bus
troncal trunk
troncal de alto uso high-usage trunk
troncal de salida outward trunk
troncal de tránsito through trunk
tronco trunk
tropicalizado tropicalized
tropopausa tropopause
troposfera troposphere
troposférico tropospheric
tropotrón tropotron
troquel die
truncado truncated
truncar truncate
tuberculación tuberculation
tubería con costura seamed tubing
tubería de cobre copper tubing
tubería flexible flexible tubing
tubería sin costura seamless tubing
tubería termocontráctil heat-shrink tubing
tubo tube, pipe
tubo aluminizado aluminized tube
tubo amortiguador damping tube
tubo amplificador amplifying tube, amplifier tube
tubo amplificador de potencia power-amplifier tube
tubo amplificador de tensión voltage-amplifying tube
tubo amplificador electrónico electronic amplifying tube
tubo amplificador final final amplifier tube
tubo amplificador telefónico telephone amplifying tube
tubo autoprotegido self-protected tube
tubo autorrectificador self-rectifying tube
tubo bañado en aceite oil-immersed tube
tubo blando soft tube

tubo calentado indirectamente indirectly heated tube
tubo cerámico ceramic tube
tubo cinescópico kinescope tube
tubo combinado combination tube
tubo compuesto composite tube
tubo con bombeo continuo pumped tube
tubo con cátodo de calentamiento iónico
    ionic-heated cathode tube
tubo con cátodo revestido de óxido oxide-cathode
    tube
tubo con costura seamed tube
tubo con fugas leaky tube
tubo con gas residual gassy tube
tubo con pantalla metalizada metallized-screen tube
tubo contactor contactor tube
tubo contador counter tube
tubo contador autoextintor self-quenched counter
    tube
tubo contador autoextintor de halógeno
    halogen-quenched counter tube
tubo contador de cátodo frío cold-cathode counter
    tube
tubo contador de flujo de gas gas-flow counter tube
tubo contador de flujo de líquido liquid-flow
    counter tube
tubo contador de gas gas-filled counter tube
tubo contador de Geiger-Mueller Geiger-Mueller
    counter tube
tubo contador de punta point counter tube
tubo contador de radiación radiation-counter tube
tubo contador de transferencia luminiscente
    glow-transfer counter tube
tubo contador de ventana window counter tube
tubo contador multicátodo multicathode counter
    tube
tubo contador para líquidos liquid counter tube
tubo contador proporcional proportional counter
    tube
tubo contra transmisión-recepción
    antitransmit-receive tube
tubo convertidor converter tube
tubo convertidor de exploración scanning-converter
    tube
tubo convertidor de imagen image-converter tube
tubo convertidor pentarrejilla pentagrid converter
    tube
tubo cuadrado square tube
tubo de almacenamiento storage tube
tubo de almacenamiento de cámara camera storage
    tube
tubo de almacenamiento de carga charge-storage
    tube
tubo de almacenamiento de computadora computer
    storage tube
tubo de almacenamiento de imagen image-storage
    tube
tubo de almacenamiento de información
    information-storage tube
tubo de almacenamiento de señales eléctricas
    electric signal storage tube
tubo de almacenamiento de visualización
    display-storage tube
tubo de almacenamiento electrónico electronic
    storage tube
tubo de almacenamiento fotoelectrónico
    photoelectronic storage tube
tubo de almacenamiento por rayos catódicos
    cathode-ray storage tube
tubo de alta confiabilidad high-reliability tube
tubo de alta gamma high-gamma tube
tubo de alta mu high-mu tube
tubo de alta potencia high-power tube
tubo de alto vacío high-vacuum tube

tubo de amplificación amplification tube
tubo de ánodo estacionario stationary-anode tube
tubo de ánodo móvil movable-anode tube
tubo de ánodo rotativo rotating-anode tube
tubo de ánodo único single-anode tube
tubo de arco arc tube
tubo de argón argon tube
tubo de atenuación attenuator tube
tubo de baja capacitancia low-capacitance tube
tubo de bajo mu low-mu tube
tubo de bajo ruido low-noise tube
tubo de baquelita bakelite tube
tubo de Barkhausen Barkhausen tube
tubo de base octal octal-base tube
tubo de batería battery tube
tubo de bellota acorn tube
tubo de bloqueo lockout tube
tubo de Braun Braun tube
tubo de cámara camera tube
tubo de cámara almacenador storage camera tube
tubo de cámara de televisión television-camera tube
tubo de cámara fotoconductivo photoconductive
    camera tube
tubo de cámara fotoemisivo photoemissive camera
    tube
tubo de campo field tube
tubo de campo retardador retarding-field tube
tubo de cátodo caliente hot-cathode tube
tubo de cátodo de baño de mercurio mercury-pool
    cathode tube
tubo de cátodo de mercurio con rejilla grid-pool
    tube
tubo de cátodo frío cold-cathode tube
tubo de cátodo hueco hollow-cathode tube
tubo de cátodo líquido pool-cathode tube
tubo de Chaoul Chaoul tube
tubo de cinco electrodos five-electrode tube
tubo de cinco elementos five-element tube
tubo de conmutación switching tube
tubo de conmutación de haz beam-switching tube
tubo de cono metálico metal-cone tube
tubo de contacto contact tube
tubo de control control tube
tubo de control por rejilla grid-controlled tube
tubo de conversión termoeléctrica
    thermoelectric-conversion tube
tubo de Coolidge Coolidge tube
tubo de corte cutoff tube
tubo de corte agudo sharp-cutoff tube
tubo de corte extendido extended cutoff tube
tubo de corte remoto remote-cutoff tube
tubo de crominancia chrominance tube
tubo de Crookes Crookes tube
tubo de cuatro electrodos four-electrode tube
tubo de cuatro elementos four-element tube
tubo de deflexión de haz beam-deflection tube
tubo de deflexión electromagnética electromagnetic
    deflection tube
tubo de deflexión electrostática electrostatic
    deflection tube
tubo de descarga discharge tube
tubo de descarga de arco arc-discharge tube
tubo de descarga de criptón krypton discharge tube
tubo de descarga de deuterio deuterium discharge
    tube
tubo de descarga eléctrica electric-discharge tube
tubo de descarga electrónica electron discharge tube
tubo de descarga gaseosa gas-discharge tube
tubo de descarga horizontal horizontal-discharge
    tube
tubo de descarga luminiscente glow-discharge tube
tubo de descarga secuencial sequential discharge
    tube

tubo de destellos flash tube
tubo de destellos de xenón xenon flash tube
tubo de destellos electrónicos electronic flash tube
tubo de destellos láser laser flash tube
tubo de disparo trigger tube
tubo de dos electrodos two-electrode tube
tubo de dos elementos two-element tube
tubo de ebonita ebonite tube
tubo de efecto de campo field-effect tube
tubo de emisión secundaria secondary-emission tube
tubo de enfoque focusing tube
tubo de enfoque de línea line-focus tube
tubo de enganche lock-in tube
tubo de entrada input tube
tubo de etapa excitadora driver tube
tubo de Faraday Faraday tube
tubo de filamento brillante bright filament tube
tubo de filamento recto straight-filament tube
tubo de flujo flux tube
tubo de fotodestellos photoflash tube
tubo de frecuencia ultraalta ultra-high frequency
   tube
tubo de fuerza tube of force
tubo de funciones múltiples multipurpose tube
tubo de gas gas tube
tubo de gas de cátodo frío cold-cathode gas tube
tubo de gas raro rare-gas tube
tubo de Geiger Geiger tube
tubo de Geiger-Mueller Geiger-Mueller tube
tubo de Geissler Geissler tube
tubo de gran ángulo large-angle tube
tubo de guarda guard tube
tubo de haz beam tube
tubo de haz conformado shaped-beam tube
tubo de haz controlado gated-beam tube
tubo de haz de cesio cesium-beam tube
tubo de haz electrónico electron-beam tube
tubo de haz lineal linear beam tube
tubo de haz orbital orbital-beam tube
tubo de haz radial radial-beam tube
tubo de Heil Heil tube
tubo de Herschel-Quincke Herschel-Quincke tube
tubo de hiperfrecuencias hyperfrequency tube
tubo de Hittorf Hittorf tube
tubo de Holtz Holtz tube
tubo de imagen picture tube, image tube
tubo de imagen autoenfocador self-focusing picture
   tube
tubo de imagen cromático chromatic picture tube
tubo de imagen de colores color picture tube
tubo de imagen de televisión television picture tube
tubo de imagen de televisión a colores
   color-television picture tube
tubo de imagen de tres cañones three-gun picture
   tube
tubo de imagen delgado thin picture tube
tubo de imagen electrónica electron image tube
tubo de imagen electrostático electrostatic picture
   tube
tubo de imagen en cascada cascade image tube
tubo de imagen en color de reflexión reflection
   color tube
tubo de imagen monocroma monochrome picture
   tube
tubo de imagen tricañón trigun picture tube
tubo de imagen tricolor tricolor picture tube
tubo de impedancia variable variable-impedance
   tube
tubo de inductancia inductance tube
tubo de inserción insert tube
tubo de Kundt Kundt tube
tubo de Lawrence Lawrence tube
tubo de Lenard Lenard tube

tubo de Lilienfeld Lilienfeld tube
tubo de línea coaxial coaxial-line tube
tubo de manipulación keying tube
tubo de McNally McNally tube
tubo de memoria memory tube
tubo de memoria electrónico electronic memory
   tube
tubo de memoria electrostático electrostatic
   memory tube
tubo de microondas microwave tube
tubo de modulación de velocidad
   velocity-modulated tube
tubo de monitoreo monitoring tube
tubo de mu variable variable-mu tube
tubo de Muller Muller tube
tubo de muy alta frecuencia very-high-frequency
   tube
tubo de neón neon tube
tubo de ondas electrónicas electron-wave tube
tubo de ondas inversas backward-wave tube
tubo de ondas milimétricas millimeter-wave tube
tubo de ondas progresivas traveling-wave tube
tubo de oscilador de rejilla positiva positive-grid
   oscillator tube
tubo de oscilador local local-oscillator tube
tubo de oscilógrafo oscillograph tube
tubo de osciloscopio oscilloscope tube
tubo de osciloscopio de rayos catódicos cathode-ray
   oscilloscope tube
tubo de pantalla absorbente dark-trace tube
tubo de pantalla luminiscente luminescent-screen
   tube
tubo de pared delgada thin-walled tube
tubo de Pitot Pitot tube
tubo de polarización cero zero-bias tube
tubo de porcelana porcelain tube
tubo de postaceleración post-acceleration tube
tubo de potencia power tube
tubo de potencia de banda ancha wideband power
   tube
tubo de potencia de haces beam power tube
tubo de potencia de salida power-output tube
tubo de proyección projection tube
tubo de puerta gate tube
tubo de punto móvil flying-spot tube
tubo de radio radio tube
tubo de radiofrecuencia radio-frequency tube
tubo de rayos catódicos cathode-ray tube
tubo de rayos catódicos aluminizado aluminized
   cathode-ray tube
tubo de rayos catódicos de doble haz double-beam
   cathode ray tube
tubo de rayos catódicos de dos colores two-color
   cathode-ray-tube
tubo de rayos catódicos de proyección projection
   cathode-ray tube
tubo de rayos catódicos electromagnético
   electromagnetic cathode-ray-tube
tubo de rayos catódicos electrostático electrostatic
   cathode-ray tube
tubo de rayos catódicos multicañón multigun
   cathode-ray tube
tubo de rayos catódicos tricolor tricolor cathode-ray
   tube
tubo de rayos límite grenz tube
tubo de rayos X X-ray tube
tubo de rayos X blindado shielded X-ray tube
tubo de rayos X de cátodo caliente hot-cathode
   X-ray tube
tubo de rayos X de Coolidge Coolidge X-ray tube
tubo de rayos X de gas gas X-ray tube
tubo de rayos X multietapa multistage X-ray tube
tubo de reactancia reactance tube

tubo de reemplazo replacement tube
tubo de referencia reference tube
tubo de rejilla de cuadro frame-grid tube
tubo de rejilla luminiscente grid-glow tube
tubo de rejilla única single-grid tube
tubo de rejillas alineadas aligned-grid tube
tubo de rejillas coplanares coplanar-grid tube
tubo de rejillas múltiples multiple-grid tube
tubo de resistencia resistance tube
tubo de resonancia resonance tube
tubo de salida output tube
tubo de salida horizontal horizontal-output tube
tubo de salida vertical vertical-output tube
tubo de Schuler Schuler tube
tubo de seis electrodos six-electrode tube
tubo de Shepherd Shepherd tube
tubo de supercontrol supercontrol tube
tubo de susceptancia susceptance tube
tubo de televisión television tube
tubo de televisión de rayos católicos cathode-ray television tube
tubo de tensión de referencia voltage-reference tube
tubo de tipo de célula cell-type tube
tubo de tipo de filamento filament-type tube
tubo de transconductancia negativa negative-transconductance tube
tubo de transmisión-recepción transmit-receive tube
tubo de trazo obscuro dark-trace tube
tubo de tres cañones three-gun tube
tubo de tres colores three-color tube
tubo de tres electrodos three-electrode tube
tubo de tres elementos three-element tube
tubo de tres rejillas three-grid tube
tubo de triple rejilla triple-grid tube
tubo de unidades múltiples multiple-unit tube
tubo de unión junction tube
tubo de vacío vacuum tube
tubo de vapor de mercurio mercury-vapor tube
tubo de vida larga long-life tube
tubo de video video tube
tubo de vidrio glass tube
tubo de visión directa direct-viewing tube
tubo de visor viewfinder tube
tubo de visualizador display tube
tubo de Wehnelt Wehnelt tube
tubo de Williams Williams tube
tubo de xenón xenon tube
tubo decatrón dekatron tube
tubo desmodulador demodulator tube
tubo desmontable demountable tube
tubo detector detector tube
tubo detector de Farnsworth Farnsworth detector tube
tubo diodo diode tube
tubo disector dissector tube
tubo disector de imagen image-dissector tube
tubo dúplex duplex tube
tubo duro hard tube
tubo electromagnético electromagnetic tube
tubo electrométrico electrometer tube
tubo electrónico electronic tube, electron tube
tubo electrónico de cesio cesium electron tube
tubo en cascada cascade tube
tubo en cuadratura quadrature tube
tubo en pomo doorknob tube
tubo enfriado por aceite oil-cooled tube
tubo enfriado por agua water-cooled tube
tubo enfriado por aire air-cooled tube
tubo estabilizador stabilizer tube, ballast tube
tubo estabilizador de tensión voltage-stabilizer tube
tubo estroboscópico stroboscopic tube
tubo faro disk-seal tube
tubo fijador de nivel clamp tube

tubo final final tube
tubo flexible flexible tube, flexible pipe
tubo fluorescente fluorescent tube
tubo fotoeléctrico photoelectric tube
tubo fotoemisivo photoemissive tube
tubo fotoluminiscente photoglow tube
tubo fotomultiplicador photomultiplier tube
tubo gaseoso gaseous tube
tubo generador de ruido noise-generator tube
tubo indicador indicator tube
tubo indicador católico electron-ray indicator tube
tubo indicador de neón neon indicator tube
tubo indicador de sintonización tuning-indicator tube
tubo indicador visual visual indicator tube
tubo industrial industrial tube
tubo intensificador de imagen image-intensifier tube
tubo interdigital interdigital tube
tubo inversor de fase phase-inverter tube
tubo invertido inverted tube
tubo limitador limiting tube
tubo limitador de tensión voltage-limiting tube
tubo luminiscente glow tube
tubo luminiscente modulador modulator glow tube
tubo magnetrón magnetron tube
tubo medidor de ionización ionization-gage tube
tubo metálico metal tube
tubo mezclador mixing tube
tubo microfónico microphonic tube
tubo microminiatura microminiature tube
tubo miniatura miniature tube, bantam tube
tubo modulador modulator tube
tubo multiánodo multianode tube
tubo multicañón multigun tube
tubo multicátodo multicathode tube
tubo multielectrodo multielectrode tube
tubo multielemento multielement tube
tubo multietapa multistage tube
tubo multihaz multibeam tube
tubo múltiple multiple tube
tubo multiplicador multiplier tube
tubo multiplicador de haz beam multiplier tube
tubo multiplicador electrónico electron multiplier tube
tubo multiplicador fotoeléctrico photoelectric electron-multiplier tube
tubo multirrejilla multigrid tube
tubo multiunidad multiunit tube
tubo no saturado unsaturated tube
tubo noval noval tube
tubo octal octal tube
tubo octodo octode tube
tubo oscilador oscillator tube
tubo pentarrejilla pentagrid tube
tubo piloto pilot tube
tubo pirométrico pyrometer tube
tubo planar planar tube
tubo plano flat tube
tubo preamplificador preamplifier tube
tubo protector protector tube
tubo recortador clipper tube
tubo rectangular rectangular tube
tubo rectificador rectifier tube
tubo rectificador de onda completa full-wave rectifier tube
tubo regulador regulator tube
tubo regulador de corriente current-regulator tube
tubo regulador de tensión voltage-regulator tube
tubo relé relay tube
tubo repetidor repeater tube
tubo selector selector tube
tubo sellado sealed tube
tubo sensible a la luz light-sensitive tube

tubo separador separator tube
tubo separador de sincronización sync-separator tube
tubo sin base unbased tube
tubo sin costura seamless tube
tubo sin electrodos electrodeless tube
tubo sintonizable por tensión voltage-tunable tube
tubo subalimentado starved tube
tubo subminiatura subminiature tube
tubo subminiatura en línea inline subminiature tube
tubo supresor de colores color-killer tube
tubo termoiónico thermionic tube
tubo tipo faro lighthouse tube
tubo tipo lápiz pencil tube
tubo tricolor tricolor tube
tubo tungar tungar tube
tubo ultrasensible ultrasensitive tube
tubo unitario unit tube
tubo variable variable tube
tubos electrónicos apareados matched electron tubes
tuistor twistor
tulio thulium
túnel de corrimiento drift tunnel
tungsteno tungsten
turbina de extracción automática automatic extraction turbine
turbina de reacción reaction turbine
turboeléctrico turboelectric
turboexcitador turboexciter
turbogenerador turbogenerator
turbulencia turbulence
turbulencia de microondas microwave turbulence
turmalina tourmaline
twinplex twinplex

U

ubicación location
ubicación de almacenamiento storage location
ubicador de haz beam locater
ubicador de metales metal locator
ubitrón ubitron
ultra ultra
ultraacústico superaudible
ultraaudible ultra-audible
ultraaudión ultra-audion
ultrabaja ultralow
ultracorto ultrashort
ultradelgado ultrathin
ultradino ultradyne
ultraestabilidad ultrastability
ultraestable ultrastable
ultraficha ultrafiche
ultrafino ultrafine
ultralineal ultralinear
ultramicrómetro ultramicrometer
ultramicroonda ultramicrowave
ultramicroscopio ultramicroscope
ultraminiatura ultraminiature
ultraprecisión ultraprecision
ultrapuro ultrapure
ultrarrápido ultrarapid
ultrarrojo ultrared
ultrasensible ultrasensitive
ultrasónica ultrasonics
ultrasónica de hiperfrecuencias microwave ultrasonics
ultrasónicamente ultrasonically
ultrasónico ultrasonic
ultrasonido ultrasound
ultrasonografía ultrasonography
ultravelocidad ultraspeed
ultravioleta ultraviolet
ultravioleta lejano far ultraviolet
ululación warble
umbral threshold, porch
umbral absoluto absolute threshold
umbral anterior front porch
umbral de conmutación switching threshold
umbral de corriente continua direct-current threshold
umbral de decisión decision threshold
umbral de detección detection threshold
umbral de emisión emission threshold
umbral de frecuencia frequency threshold
umbral de Geiger Geiger threshold
umbral de Geiger-Mueller Geiger-Mueller threshold
umbral de interferencia interference threshold
umbral de luminancia luminance threshold
umbral de luminancia absoluta absolute luminance threshold
umbral de luminiscencia luminescence threshold
umbral de mejora improvement threshold
umbral de modulación modulation threshold
umbral de parpadeo flicker threshold
umbral de respuesta threshold of response
umbral de sensibilidad threshold of sensitivity
umbral de señal signal threshold

umbral diferencial differential threshold
umbral diferencial de luminancia
    luminance-difference threshold
umbral fotoeléctrico photoelectric threshold
umbral logarítmico logarithmic threshold
umbral normal normal threshold
unario unary
uniaxial uniaxial
único single
uniconductor uniconductor
unidad unit, drive
unidad absoluta absolute unit
unidad accesoria accessory unit
unidad adaptadora de datos data adapter unit
unidad amplificadora amplifier unit
unidad amplificadora de potencia power-amplifier
    unit
unidad angstrom angstrom unit
unidad anticoincidencia anticoincidence unit
unidad aritmética arithmetic unit
unidad aritmética-lógica arithmetic logic unit
unidad aritmética paralela parallel arithmetic unit
unidad aritmética y lógica arithmetic and logic unit
unidad astronómica astronomical unit
unidad autónoma stand-alone unit
unidad biestable bistable unit
unidad binaria binary unit
unidad cambiadora de frecuencia
    frequency-changer unit
unidad central central unit, mainframe
unidad certificada certified unit
unidad combinadora combining unit
unidad convertidora converter unit
unidad corriente current drive
unidad de acceso access unit
unidad de acceso directo direct-access unit
unidad de acoplamiento coupling unit
unidad de acoplamiento de antena antenna coupling
    unit
unidad de aislador de aparato apparatus insulator
    unit
unidad de alimentación power-supply unit, supply
    unit, power pack
unidad de almacenamiento storage unit
unidad de almacenamiento auxiliar auxiliary
    storage unit
unidad de almacenamiento de disco magnético
    magnetic-disk storage unit
unidad de almacenamiento de disco removible
    removable-disk storage unit
unidad de almacenamiento de núcleos core storage
    unit
unidad de almacenamiento de núcleos magnéticos
    magnetic-core storage unit
unidad de almacenamiento de programas program
    storage unit
unidad de almacenamiento de tambor magnético
    magnetic-drum storage unit
unidad de almacenamiento en disco disk storage
    unit
unidad de almacenamiento magnético magnetic
    storage unit
unidad de almacenamiento masivo bulk storage unit
unidad de alta frecuencia high-frequency unit
unidad de archivo file unit
unidad de aritmética en serie serial-arithmetic unit
unidad de barrido sweep unit
unidad de base base unit
unidad de bolómetro bolometer unit
unidad de cabezas múltiples head stack
unidad de cálculo calculating unit
unidad de calentamiento heating unit
unidad de canal channel unit

unidad de capacitancia unit of capacitance
unidad de capacitor capacitor unit
unidad de capacitor-resistor capacitor-resistor unit
unidad de carga loading unit, charging unit
unidad de cartucho cartridge drive
unidad de cartucho de cinta tape-cartridge drive
unidad de casete cassette drive, cassette unit
unidad de cinta tape drive, tape deck, tape unit
unidad de cinta magnética magnetic-tape drive,
    magnetic-tape deck, magnetic-tape unit
unidad de color primario primary-color unit
unidad de comparación comparing unit
unidad de compilación compilation unit
unidad de conmutación switching unit
unidad de conmutación auxiliar auxiliary switching
    unit
unidad de consulta inquiry unit
unidad de contacto contact unit
unidad de contestación answerback unit
unidad de control control unit
unidad de control central central control unit
unidad de control de cámara camera control unit
unidad de control de canal channel control unit
unidad de control de cinta tape-control unit
unidad de control de cinta magnética magnetic-tape
    control unit
unidad de control de comunicaciones
    communication control unit
unidad de control de comunicaciones de datos data
    communications control unit
unidad de control de datos data control unit
unidad de control de disco disk control unit
unidad de control de dispositivo device control unit
unidad de control de entrada-salida input-output
    control unit
unidad de control de grupo cluster control unit
unidad de control de impresora printer-control unit
unidad de control de instrucciones instruction
    control unit
unidad de control de línea line-control unit
unidad de control de microprograma microprogram
    control unit
unidad de control de motor motor control unit
unidad de control de potencia power control unit
unidad de control de programa program-control
    unit
unidad de control de secuencia sequence-control
    unit
unidad de control de telecomunicaciones
    telecommunications control unit
unidad de control de transmisión
    transmission-control unit
unidad de control de transmisión de datos
    data-transmission control unit
unidad de control estereofónico stereophonic
    control unit
unidad de control periférico peripheral control unit
unidad de control principal main control unit
unidad de control remoto remote-control unit
unidad de controlador controller unit
unidad de controlador de interrupción de prioridad
    priority interrupt controller unit
unidad de conversión de energía energy conversion
    unit
unidad de corriente unit of current
unidad de corriente de centrado centering current
    unit
unidad de cristal crystal unit
unidad de datos data unit
unidad de Debye Debye unit
unidad de desplazamiento shift unit
unidad de diafonía crosstalk unit
unidad de diagnóstico diagnostic unit

unidad de disco disk drive, disk unit
unidad de disco de cabeza móvil movable-head disk unit
unidad de disco fijo fixed disk drive
unidad de disco flexible flexible disk drive, floppy-disk drive
unidad de disco magnético magnetic-disk unit
unidad de diseño design unit
unidad de disquete diskette drive
unidad de distancia range unit
unidad de distancia astronómica astronomical unit of distance
unidad de distribución de antena antenna distribution unit
unidad de distribución de potencia power-distribution unit
unidad de distribución de refrigerante coolant distribution unit
unidad de documento document drive
unidad de drenaje drainage unit
unidad de ejecución run unit
unidad de elementos de código code element unit
unidad de emisión-recepción automática automatic send-receive unit
unidad de energía energy unit
unidad de energía atómica atomic unit of energy
unidad de energía sin interrupción no-break power unit
unidad de ensamblaje assembly unit
unidad de entrada input unit
unidad de entrada de sistema system input unit
unidad de entrada manual manual input unit
unidad de entrada-salida input-output unit
unidad de equipo equipment unit
unidad de expansión expansion unit
unidad de exploración scanning unit, raster unit
unidad de extensión de distancia range-extension unit
unidad de falla failure unit
unidad de fuerza electromotriz unit of electromotive force
unidad de fusible fuse unit
unidad de fusible de expulsión expulsion fuse unit
unidad de fusible limitadora de corriente current-limiting fuse unit
unidad de fusible líquido liquid fuse unit
unidad de fusible no renovable nonrenewable fuse unit
unidad de grabación recording unit
unidad de Hefner Hefner unit
unidad de identidad identity unit
unidad de iluminación illumination unit
unidad de impresión print unit
unidad de información unit of information
unidad de información básica basic information unit
unidad de información de trayectoria path information unit
unidad de intercambio de mensajes message exchange unit
unidad de interfaz interface unit
unidad de interfaz de computadora computer interface unit
unidad de interferencia interference unit
unidad de lectura read unit
unidad de lectura de cinta tape-read unit
unidad de línea line unit
unidad de llamada automática automatic calling unit
unidad de longitud de onda wavelength unit
unidad de luz unit of light
unidad de marcación automática automatic dialing unit
unidad de masa mass unit
unidad de masa atómica atomic mass unit

unidad de media altura half-height drive
unidad de medida unit of measurement
unidad de memoria memory unit
unidad de memoria de línea de retardo delay-line memory unit
unidad de memoria de núcleos de ferrita ferrite-core memory unit
unidad de memoria intermedia buffer unit
unidad de microprocesamiento microprocessing unit
unidad de motor motor unit
unidad de núcleos magnéticos magnetic-core unit
unidad de operación operating unit
unidad de origen originating unit
unidad de parada-arranque stop-start unit
unidad de perforación punching unit
unidad de perforadora de tarjetas card-punch unit
unidad de peso atómico atomic weight unit
unidad de potencia power unit
unidad de procesador processor unit
unidad de procesamiento processing unit
unidad de procesamiento central central processing unit
unidad de procesamiento central de microcomputadora microcomputer central processing unit
unidad de procesamiento gráfico graphic processing unit
unidad de procesamiento primario primary-processing unit
unidad de producción production unit
unidad de programa program unit
unidad de programa genérica generic program unit
unidad de programación programming unit
unidad de prueba test unit
unidad de prueba portátil portable test unit
unidad de radar radar unit
unidad de rayos X X-ray unit
unidad de reemplazo replacement unit
unidad de registro recording unit
unidad de registro de sonido sound-recording unit
unidad de registro en cinta tape-recording unit
unidad de registro en cinta magnética magnetic-tape recording unit
unidad de registro en disco disk recording unit
unidad de reproducción playback unit
unidad de reserva standby unit
unidad de reserva de cinta tape backup unit
unidad de resistencia unit of resistance
unidad de resistor-capacitor resistor-capacitor unit
unidad de respaldo backup unit
unidad de respuesta de audio audio response unit
unidad de retardo delay unit
unidad de retardo de bucle loop-delay unit
unidad de retardo de destello flash-delay unit
unidad de retardo lineal linear delay unit
unidad de reverberación reverberation unit
unidad de ruido noise unit
unidad de salida output unit
unidad de salida de sistema system output unit
unidad de secuencia sequence unit
unidad de segmentación segmentation unit
unidad de señal de reconocimiento acknowledgment signal unit
unidad de señalización signaling unit
unidad de servicio service unit
unidad de sintonización tuning unit
unidad de sintonización de antena antenna tuning unit
unidad de sistema system unit
unidad de solicitud request unit
unidad de sonido sound unit
unidad de tambor magnético magnetic-drum unit
unidad de teclado a cinta key-to-tape unit

unidad de teclado a disco key-to-disk unit
unidad de temporización timing unit
unidad de terminación termination unit
unidad de terminal de datos data-terminal unit
unidad de tiempo unit of time
unidad de tráfico traffic unit
unidad de transferencia transfer unit
unidad de transferencia unit of transfer
unidad de transferencia de frecuencia frequency transfer unit
unidad de transmisión transmission unit
unidad de transmisión básica basic transmission unit
unidad de transmisión de datos data-transmission unit
unidad de transmisión de rumbo bearing transmission unit
unidad de visualización display unit
unidad de visualización visual display unit
unidad de visualización de análisis de datos data analysis display unit
unidad de visualización de radar radar display unit
unidad de visualización de registros register display unit
unidad de visualización de video video display unit
unidad de visualización digital digital display unit
unidad de visualización directa direct display unit
unidad de visualización gráfica graphic display unit
unidad de visualización remota remote-display unit
unidad de visualizador visual display unit
unidad de volumen volume unit
unidad de zona muerta dead zone unit
unidad del procesador central central processor unit
unidad descodificadora decoder unit
unidad desfasadora phase-shifting unit
unidad discreta discrete unit
unidad eléctrica electrical unit
unidad electromagnética electromagnetic unit
unidad electrostática electrostatic unit
unidad electrotérmica electrothermic unit
unidad en línea on-line unit
unidad enchufable plug-in unit
unidad equilibradora balancing unit
unidad estabilizadora stabilizer unit
unidad excitadora exciter unit
unidad exploradora scanner unit
unidad fija fixed drive
unidad física physical unit, physical drive
unidad física de periférico peripheral physical unit
unidad fotométrica photometer unit
unidad fuera de línea off-line unit
unidad funcional functional unit
unidad generadora generating unit
unidad genérica generic unit
unidad impresora printing unit
unidad impresora en cadena chain printer unit
unidad indicadora indicator unit
unidad indicadora de falla local local fault display unit
unidad integradora integrating unit
unidad intercambiable interchangeable unit
unidad internacional international unit
unidad inversora de fase phase-reversing unit
unidad lectora de tarjetas card reader unit
unidad logarítmica logarithmic unit
unidad lógica logical unit
unidad lógica periférica peripheral logical unit
unidad lógica primaria primary logical unit
unidad lógica secundaria secondary logical unit
unidad maestra master unit
unidad matricial matrix unit
unidad monitora monitor unit
unidad montada en bastidor rack-mounted unit

unidad motriz motive unit
unidad móvil mobile unit
unidad móvil autónoma self-contained mobile unit
unidad móvil portátil portable mobile unit
unidad multiplicadora multiplier unit
unidad no calibrada uncalibrated unit
unidad operacional operational unit
unidad osciladora oscillator unit
unidad periférica peripheral unit
unidad por omisión default drive
unidad portátil portable unit
unidad prefabricada prefabricated unit
unidad principal main unit
unidad racionalizada rationalized unit
unidad receptora receiving unit
unidad rectificadora rectifier unit
unidad reguladora regulating unit
unidad removible removable drive
unidad repetidora repeater unit
unidad reservada reserved unit
unidad selectora de canales channel selector unit
unidad sensora sensing unit
unidad simbólica symbolic unit
unidad simuladora simulator unit
unidad sobreexcitada overdriven unit
unidad subexcitada underdriven unit
unidad térmica thermal unit
unidad térmica británica British thermal unit
unidad terminal terminal unit
unidad terminal central central terminal unit
unidad tricromática trichromatic unit
unidad virtual virtual drive
unidad visualizadora de tubo de rayos catódicos cathode-ray tube display unit
unidad X X unit
unidades de Giorgi Giorgi units
unidades de Heaviside Heaviside units
unidades de máquina machine units
unidades derivadas derived units
unidades eléctricas absolutas absolute electrical units
unidades eléctricas internacionales international electrical units
unidades eléctricas prácticas practical electrical units
unidades estructurales de memoria memory structural units
unidades fundamentales fundamental units
unidades prácticas practical units
unidimensional one-dimensional
unidino unidyne
unidireccional unidirectional, one-way
unifilar unifilar
uniforme uniform
uniformidad uniformity
uniformidad de campo field uniformity
unilateral unilateral
unilateralización unilateralization
unimodal unimodal
unimodo unimode
unión union, junction, splice
unión abrupta abrupt junction
unión adaptada matched junction
unión caliente hot junction
unión con polarización inversa reverse-biased junction
unión de aleación alloy junction
unión de colector collector junction
unión de contacto de punta point-contact junction
unión de emisor emitter junction
unión de germanio tipo P P-type germanium junction
unión de guía guide junction

unión de guíaondas waveguide junction
unión de Peltier Peltier junction
unión de salida outgoing junction
unión dopada doped junction
unión emisor-base emitter-base junction
unión en cascada cascade junction
unión en cuña wedge bonding
unión en T tee junction
unión en T de guíaondas waveguide tee
unión en T paralelo shunt-tee junction
unión en T serie series tee junction
unión en Y Y junction, wye junction
unión epitaxial epitaxial junction
unión flotante floating junction
unión fotosensible photosensitive junction
unión graduada graded junction
unión híbrida hybrid junction
unión ideal ideal junction
unión líquida liquid junction
unión local local junction
unión n-n n-n junction
unión normal normal joint
unión NP NP junction
unión planar planar junction
unión PN PN junction
unión PN de germanio germanium PN junction
unión PNP PNP junction
unión por crecimiento grown junction
unión por crecimiento variable rate-grown junction
unión por difusión diffused junction
unión por fusión fused junction
unión PP PP junction
unión rectificadora rectifier junction
unión semiconductor-metal semiconductor-metal junction
unión semiconductora semiconductor junction
unión soldada weld junction
unión térmica thermal junction
unión terminada en jack jack-ended junction
unión termoeléctrica thermoelectric junction
unión única single junction
uniplex uniplex
unipolar unipolar
unipolaridad unipolarity
unipolo unipole
unipolo vertical vertical unipole
unipotencial unipotential
uniprocesador uniprocessor
uniprogramación uniprogramming
uniselector uniselector
unitor unitor
uniunión unijunction
univalente univalent
universal universal
univibrador univibrator
uno binario binary one
uranio uranium
urgencia de reparación repair urgency
ursigrama ursigram
uso compartido shared use
uso en común joint use
uso exclusivo exclusive use
uso final end use
usuario autorizado authorized user
usuario final end user
utilización utilization
utilización de máquina machine utilization
utilización espectral spectrum utilization
uvicón uvicon

vacante idle
vacante vacant
vaciado dump, dumping
vaciado autónomo stand-alone dump
vaciado condicional conditional dump
vaciado de almacenamiento storage dump
vaciado de cambios change dump
vaciado de cinta tape dump
vaciado de corriente continua direct-current dump
vaciado de disco disk dump
vaciado de errores error dump
vaciado de memoria memory dump
vaciado de memoria de núcleos core dump
vaciado de memoria diferencial differential memory dump
vaciado de pantalla screen dump
vaciado de puntos de comprobación checkpoint dump
vaciado de sistema system dump
vaciado de tambor drum dump
vaciado diferencial differential dump
vaciado dinámico dynamic dump
vaciado ejecutivo executive dump
vaciado estático static dump
vaciado formateado formatted dump
vaciado post mortem post-mortem dump
vaciado programado programmed dump
vaciado selectivo selective dump
vaciado y reiniciación dump and restart
vacío vacuum
vacío ultraalto ultra-high vacuum
vacuómetro vacuum gage
vacuómetro de ionización ionization vacuum gage
vaina sheath
vaina anódica anode sheath
vaina de aluminio aluminum sheath
vaina de cable cable sheath
vaina de electrones electron sheath
vaina de iones positivos positive-ion sheath
vaina de plasma plasma sheath
vaina de plomo lead sheath
vaina iónica ion sheath
valencia valence
valencia positiva positive valence
validación validation
validación cruzada cross validation
validación de datos data validation
validación de datos de cinta tape data validation
validación de datos de entrada input data validation
validación de entrada input validation
validación de programa program validation
validado validated
validar validate
validez validity
validez de datos data validity
validez de prueba test validity
válido valid
valor value
valor absoluto absolute value
valor absoluto medio average absolute value
valor acumulado accumulated value

valor aproximado approximate value
valor asignado assigned value
valor calculado calculated value
valor de acceso access value
valor de base base value
valor de Boole Boolean value
valor de código code value
valor de comparación comparison value
valor de condición condition value
valor de corriente ponderado weighted current value
valor de cresta crest value
valor de cuasipico quasi-peak value
valor de datos data value
valor de decisión decision value
valor de desprendimiento dropout value
valor de disparo trip value
valor de entrada entry value
valor de estado estacionario steady-state value
valor de fondo de escala full scale
valor de límite de escala full-scale value
valor de Munsell Munsell value
valor de pH pH value
valor de pico a pico peak-to-peak value
valor de potencial potential value
valor de puesta en trabajo pickup value
valor de recuperación recovery value
valor de referencia reference value
valor de reposo quiescent value
valor de resistencia resistance value
valor de retorno return value
valor de saturación saturation value
valor de tensión instantáneo instantaneous voltage value
valor de tensión ponderado weighted voltage value
valor de trabajo working value
valor deseado desired value
valor efectivo effective value
valor eficaz root-mean-square value
valor estático static value
valor fijado set value
valor final final value
valor fuera de tolerancia out-of-tolerance value
valor ideal ideal value
valor indicado indicated value
valor inicial initial value
valor inicial por omisión default initial value
valor inmediato immediate value
valor instantáneo instantaneous value
valor límite boundary value, limiting value
valor lógico logical value
valor máximo maximum value
valor máximo absoluto absolute maximum rating
valor medido measured value
valor medio average value
valor medio de potencia average value of power
valor nominal nominal value, rated value, rating
valor nulo null value
valor numérico numerical value
valor óhmico ohmic value
valor pico peak value
valor ponderado weighted value
valor por omisión default value
valor positivo positive value
valor principal principal value
valor probable probable value
valor real real value
valor rectificado rectified value
valor registrado recorded value
valor relativo relative value
valor reproducible reproducible value
valor residual residual value
valor seguro safe value

valor umbral threshold value
valor vectorial vector value
valores preferidos preferred values
valores preferidos de componentes preferred values of components
valuador valuator
válvula valve
válvula actuadora actuator valve
válvula amplificadora amplifying valve
válvula calentada indirectamente indirectly heated valve
válvula con bombeo continuo pumped valve
válvula con cátodo revestido de óxido oxide-cathode valve
válvula de accionamiento directo direct-acting valve
válvula de admisión de gas gas admittance valve
válvula de aguja needle valve
válvula de alimentación feed valve
válvula de alta mu high-mu valve
válvula de alta potencia high-power valve
válvula de alto vacío high-vacuum valve
válvula de aplicación application valve
válvula de bridas flanged valve
válvula de conmutación switching valve
válvula de cuatro electrodos four-electrode valve
válvula de deflexión deflection valve
válvula de desbordamiento overflow valve
válvula de descarga discharge valve
válvula de Fleming Fleming valve
válvula de gas gas valve
válvula de mariposa butterfly valve
válvula de mu variable variable-mu valve
válvula de Nodon Nodon valve
válvula de placa plate valve
válvula de puerta gate valve
válvula de radio radio valve
válvula de reactancia reactance valve
válvula de rejilla única single-grid valve
válvula de rejillas múltiples multiple-grid valve
válvula de solenoide solenoid valve
válvula de tres rejillas three-grid valve
válvula de unidades múltiples multiple-unit valve
válvula de vacío vacuum valve
válvula de vapor de mercurio mercury-vapor valve
válvula electrolítica electrolytic valve
válvula electromecánica electromechanical valve
válvula electroneumática electropneumatic valve
válvula electrónica electronic valve
válvula electroquímica electrochemical valve
válvula en hongo mushroom valve
válvula estabilizadora ballast valve
válvula fotoeléctrica photoelectric valve
válvula inversora de fase phase-inverter valve
válvula iónica ionic valve
válvula maestra master valve
válvula metálica metal valve
válvula mezcladora mixer valve
válvula miniatura miniature valve
válvula moduladora modulator valve
válvula multielectrodo multielectrode valve
válvula operada eléctricamente electrically operated valve
válvula osciladora oscillator valve
válvula preamplificadora preamplifying valve
válvula rectificadora rectifying valve
válvula rectificadora de potencia power rectifying valve
válvula termoiónica thermionic valve
válvula termostática thermostatic valve
válvula variable variable valve
valle de impulso pulse valley
valle de onda wave trough
vanadio vanadium

vapor de mercurio mercury vapor
vapor metálico metal vapor
vapor residual residual vapor
vapor saturado saturated vapor
vaporización vaporization
vaporizar vaporize
vapotrón vapotron
var var
varactor varactor
varactor de arseniuro de galio gallium-arsenide varactor
varhora varhour
varhorímetro varhour meter
variabilidad variability
variable variable
variable aleatoria random variable
variable analógica analog variable
variable asignada allocated variable
variable automática automatic variable
variable binaria binary variable
variable compleja complex variable
variable condicional conditional variable
variable continua continuous variable
variable controlada controlled variable
variable controlada indirectamente indirectly controlled variable
variable cruzada crossed variable
variable de archivo file variable
variable de Boole Boolean variable
variable de carácter character variable
variable de control de bucle loop-control variable
variable de entrada entry variable
variable de lenguaje de control control language variable
variable de máquina machine variable
variable de programa program variable
variable de referencia reference variable
variable de retardo lag variable
variable de salida output variable
variable de secuencia string variable
variable dependiente dependent variable
variable discreta discrete variable
variable escalar scalar variable
variable especial special variable
variable estática static variable
variable ficticia dummy variable
variable fija fixed variable
variable global global variable
variable independiente independent variable
variable interna internal variable
variable local local variable
variable lógica logical variable
variable medida measured variable
variable real real variable
variación variation, variance
variación aleatoria random variation
variación angular angular variation
variación aperiódica aperiodic variation
variación cero zero variation
variación de amplitud amplitude variation
variación de atenuación variation of attenuation
variación de campo field variation
variación de frecuencia frequency variation
variación de fuente de alimentación power-supply variation
variación de impedancia impedance variation
variación de potencial potential variation
variación de resistencia resistance variation
variación de temperatura temperature variation
variación de tensión voltage variation
variación de tensión absoluta absolute voltage variation
variación de tensión cíclica cyclic voltage variation

variación de tensión relativa relative voltage variation
variación de umbral automática automatic threshold variation
variación de velocidad speed variation
variación de velocidad absoluta absolute speed variation
variación de velocidad de cinta tape-speed variation
variación de velocidad relativa relative speed variation
variación del cero zero variation
variación diurna diurnal variation
variación lineal linear taper
variación magnética magnetic variation
variación paramétrica parametric variation
variación repentina longitudinal longitudinal surge
variación secular secular variation
variacional variational
variaciones de tensión de línea line-voltage variations
variador variator
variador de velocidad speed variator
variancia variance
variante de conector connector variant
varicap varicap
varilla rod, bar
varilla de adaptación matching pillar, matching pole
varilla de ferrita ferrite rod
varilla de guía guide rod
varilla de tierra grounding rod, ground rod
varilla dieléctrica dielectric rod
varilla magnética magnetic rod
varilla oscilante oscillating rod
varindor varindor
varioacoplador variocoupler
variómetro variometer
varioplex varioplex
varistor varistor
varistor de cilindro drum varistor
varistor de efecto de campo field-effect varistor
varistor de metal-óxido metal-oxide varistor
varistor de tirita thyrite varistor
varistor en disco disk varistor
varistor simétrico symmetrical varistor
varistor tipo arandela washer varistor
varita de sintonización tuning wand
varita óptica optical wand
varmetro varmeter
vaso poroso porous pot
vectógrafo vectograph
vector vector
vector de activación de programa program activation vector
vector de atenuación attenuation vector
vector de campo eléctrico electric-field vector
vector de corriente current vector
vector de Hertz Hertz vector
vector de onda wave vector
vector de Poynting Poynting vector
vector de tensión voltage vector
vector eléctrico electric vector
vector hertziano Hertzian vector
vector impedancia impedance vector
vector magnético magnetic vector
vector Q Q vector
vector unitario de polarización polarization unit vector
vectores de campo acoplados coupled field vectors
vectorial vectorial
vectorscopio vectorscope
vehicular vehicular
vehículo vehicle
velocidad velocity, speed, rate

velocidad ajustada adjusted speed
velocidad angular angular velocity
velocidad base base speed
velocidad ciega blind speed
velocidad completa full speed
velocidad controlada electrónicamente electronically controlled speed
velocidad controlada por regulador governor-controlled speed
velocidad crítica critical speed
velocidad de acimut azimuth rate
velocidad de aguja stylus velocity
velocidad de Alfven Alfven velocity
velocidad de aproximación approach speed
velocidad de arranque starting velocity
velocidad de arrastre drift speed
velocidad de barrido sweep speed
velocidad de bauds baud rate
velocidad de bits bit rate
velocidad de bits digital digital bit rate
velocidad de bits equivalente equivalent bit rate
velocidad de bits fibroóptica fiber-optic bit rate
velocidad de borrado erasing speed
velocidad de caída rate of fall
velocidad de cálculo calculating speed
velocidad de calentamiento catódico cathode heating rate
velocidad de carga charging rate
velocidad de cebado priming speed
velocidad de cebado crítica critical build-up speed
velocidad de cinta tape speed
velocidad de códigos code speed
velocidad de cómputo computing speed
velocidad de conmutación switching speed
velocidad de conversión conversion rate
velocidad de corrección correction rate
velocidad de datos data rate
velocidad de deposición deposition speed
velocidad de desionización deionization rate
velocidad de deslizamiento drift velocity
velocidad de difusión rate of diffusion
velocidad de disco disk speed
velocidad de disipación dissipation rate
velocidad de electrón electron velocity
velocidad de emisión emission velocity
velocidad de entrada input speed
velocidad de escritura writing speed
velocidad de exploración scanning speed, scanning rate
velocidad de extinción rate of decay
velocidad de fase phase velocity
velocidad de flujo flow rate
velocidad de flujo electrónico electron drift velocity
velocidad de generación generation rate
velocidad de grupo group velocity
velocidad de haz beam velocity
velocidad de impresión printing speed
velocidad de impresión efectiva effective printing rate
velocidad de impulsos impulse speed
velocidad de información information rate
velocidad de ionización ionization rate
velocidad de la luz speed of light
velocidad de lectura reading speed, reading rate
velocidad de línea line speed
velocidad de manipulación keying rate
velocidad de medidor meter speed
velocidad de modulación rate of modulation
velocidad de muestreo sampling speed
velocidad de muestreo de canal channel sampling rate
velocidad de Nyquist Nyquist rate
velocidad de onda wave velocity

velocidad de ondas de radio velocity of radio waves
velocidad de operación operating speed
velocidad de paginación paging rate
velocidad de papel paper speed
velocidad de partículas particle velocity
velocidad de película film speed
velocidad de penetración penetration rate
velocidad de perforación punching rate
velocidad de perforación de tarjetas card-punching rate
velocidad de portadores carrier velocity
velocidad de procesamiento processing speed
velocidad de programación programming speed
velocidad de propagación propagation velocity
velocidad de punto spot speed
velocidad de reacción rate of reaction
velocidad de rebobinado rewinding speed
velocidad de recombinación recombination rate
velocidad de recombinación de volumen volume recombination rate
velocidad de recombinación superficial surface recombination rate
velocidad de recuperación recovery rate
velocidad de registro recording speed
velocidad de reloj clock speed
velocidad de repetición básica basic repetition rate
velocidad de reproducción reproduction speed
velocidad de respuesta response rate, response speed
velocidad de retorno flyback velocity
velocidad de salida output speed
velocidad de señalización signaling rate
velocidad de sintonización tuning speed
velocidad de subida rate of rise
velocidad de surco groove speed
velocidad de tambor drum speed
velocidad de tarjetas card speed
velocidad de titilación flutter rate
velocidad de transferencia transfer rate
velocidad de transferencia de bits bit transfer rate
velocidad de transferencia de caracteres character transfer rate
velocidad de transferencia de datos data-transfer rate
velocidad de transferencia de datos efectiva effective data transfer rate
velocidad de transferencia de datos media average data transfer rate
velocidad de transferencia de datos real actual data transfer rate
velocidad de transferencia de energía energy transfer rate
velocidad de transferencia de mensajes message transfer rate
velocidad de transferencia efectiva effective transfer rate
velocidad de transferencia media average transfer rate
velocidad de transferencia real actual transfer rate
velocidad de transformación media average transformation rate
velocidad de transmisión transmission rate, transmission speed
velocidad de transmisión de datos data transmission rate
velocidad de transmisión efectiva effective transmission rate
velocidad de transmisión media average transmission rate
velocidad de transmisión telegráfica telegraph transmission speed
velocidad de trazado plotting speed
velocidad de volumen volume velocity

velocidad del sonido speed of sound
velocidad efectiva effective speed
velocidad eficaz root-mean-square velocity
velocidad electrónica media mean electron velocity
velocidad en bauds agregada aggregate baud rate
velocidad equilibradora balancing speed
velocidad específica specific speed
velocidad hipersónica hypersonic speed
velocidad inicial initial velocity
velocidad libre free speed
velocidad límite speed limit
velocidad máxima maximum speed, peak speed
velocidad máxima de respuesta maximum response speed
velocidad media average speed
velocidad molecular molecular velocity
velocidad nominal nominal speed, rated speed
velocidad nominal de motor rated engine speed
velocidad normal normal speed
velocidad óptica optical speed
velocidad orbital orbital velocity
velocidad pico de transferencia peak transfer rate
velocidad pico de transferencia de datos peak data transfer rate
velocidad reducida reduced speed
velocidad relativa relative speed, relative velocity
velocidad relativa a la tierra ground speed
velocidad rotacional rotational velocity
velocidad sin carga no-load speed
velocidad sincrónica synchronous speed
velocidad sónica sonic speed
velocidad supersónica supersonic speed
velocidad ultraalta ultra-high speed
velocidad única single speed
velocidad uniforme uniform speed, uniform velocity
velocidad variable variable speed
velocímetro velocimeter
ventaja por escalonamiento staggering advantage
ventaja sobre ruido noise advantage
ventana window
ventana abierta open window
ventana activa active window
ventana capacitiva capacitive window
ventana de comprobación check window
ventana de infrarrojo infrared window
ventana de muestreo sampling window
ventana de radio radio window
ventana de radio atmosférica atmospheric radio window
ventana de salida output window
ventana de seguridad safety window
ventana de señal signal window
ventana de visualización display window
ventana eléctrica electrical window
ventana espectral spectral window
ventana inactiva inactive window
ventana inductiva inductive window
ventana resonante resonant window
ventanilla window
ventanilla de inspección inspection window
ventanilla de observación observation window
ventilación ventilation
ventilación por aire air ventilation
ventilación simple simple ventilation
ventilado ventilated
ventilador de emergencia emergency fan
ventilador de enfriamiento cooling fan
ventilador de motor motor ventilator
ventilador en hongo mushroom ventilator
venturi venturi
verificación verification
verificación de cinta tape verification
verificación de cinta de papel paper-tape verification

verificación de copia copy check
verificación de datos data verification
verificación de diseño design verification
verificación de especificación specification verification
verificación de ocupado busy verification
verificación de rendimiento performance verification
verificación de requisitos requirements verification
verificación de sistema system verification
verificación de tarjetas card verification
verificador verifier
verificador automático automatic verifier
verificador de cinta de papel paper-tape verifier
verificador de tarjetas card verifier
verificador de tarjetas perforadas punched-card verifier
verificador numérico numerical verifier
verificadora de cinta tape verifier
verificar verify
vernier vernier
vernitel vernitel
versión version
versión de archivo file version
versión de base base version
versión de red network version
vertical vertical
vertical aparente apparent vertical
vértice vertex
vestigial vestigial
vía path, route
vibración vibration, dithering, chatter
vibración aleatoria random vibration
vibración amortiguada damped vibration
vibración cuasiarmónica quasi-harmonic vibration
vibración de contacto contact chatter
vibración de contacto de relé relay-contact chatter
vibración de estado estacionario steady-state vibration
vibración de resonancia resonance vibration
vibración libre free vibration
vibración lineal linear vibration
vibración magnetoestrictiva magnetostrictive vibration
vibración molecular molecular vibration
vibración no periódica nonperiodic vibration
vibración normal normal vibration
vibración periódica periodic vibration
vibración simpática sympathetic vibration
vibración sinusoidal sinusoidal vibration
vibración sonora sound vibration
vibración subarmónica subharmonic vibration
vibración ultrasónica ultrasonic vibration
vibracional vibrational
vibrador vibrator, shaker
vibrador asincrónico asynchronous vibrator
vibrador autorrectificador self-rectifying vibrator
vibrador de cristal crystal vibrator
vibrador de cuarzo quartz vibrator
vibrador de media onda half-wave vibrator
vibrador de onda completa full-wave vibrator
vibrador de placa plate vibrator
vibrador electromagnético electromagnetic vibrator
vibrador no sincrónico nonsynchronous vibrator
vibrador piezoeléctrico piezoelectric vibrator
vibrador sincrónico synchronous vibrator
vibrador ultrasónico magnetoestrictivo magnetostrictive ultrasonic vibrator
vibrante vibrating
vibratorio vibratory, vibrational
vibratrón vibratron
vibrógrafo vibrograph
vibrómetro vibrometer

vibrotrón vibrotron
vida calificada qualified life
vida de almacenamiento storage life
vida de almacenamiento seco dry shelf life
vida de batería battery life
vida de carga load life
vida de ciclos cycle life
vida de diseño design life
vida de equipo equipment life
vida de estante húmeda wet shelf life
vida de operación operating life
vida de portadores carrier lifetime
vida de servicio service life
vida en almacenamiento shelf life
vida estimada estimated life
vida instalada installed life
vida mecánica mechanical life
vida media average life
vida nominal rated life
vida útil useful life
video video
video bipolar bipolar video
video coherente coherent video
video comprimido compressed video
video compuesto composite video
video inverso inverse video
videoamplificación video amplification
videoamplificador video amplifier
videoamplificador compensado compensated video
    amplifier
videoamplificador no compensado uncompensated
    video amplifier
videoatenuador video attenuator
videobanda video band
videocanal video channel
videocaptación video pickup
videocinta video tape
videodetección video detection
videodetector video detector
videodifusión television broadcasting, television
    broadcast, videocast
videodifusor videocaster
videodisco videodisk
videoenlace video link
videófono videophone
videofrecuencia video frequency
videograbadora video-tape recorder, video-tape
    machine, videocassette recorder
videónica videonics
videoportadora video carrier
videorreceptor video receiver
videorregistro video recording
videorruido video noise
videoseñal video signal
videoseñal compuesta composite video signal
videoseñal de colores color video signal
videoseñal de salida output video signal
videoseñal monocroma monochrome video signal
videoseñales de crominancia chrominance video
    signals
videoterminal video terminal
videotex videotex
videotransmisor video transmitter
videotransmisor portátil walkie-lookie
videotrón videotron
vidicón vidicon
vidrio glass
vidrio de Lindemann Lindemann glass
vidrio de plomo lead glass
vidrio de protección protection glass
vidrio de seguridad safety glass
vidrio eléctrico electrical glass
vidrio fotocrómico photochromic glass

vidrio protector protective glass
vidrio resistente al calor heat-resistant glass
viento eléctrico electric wind
viento solar solar wind
vientre de corriente current loop
vigilancia surveillance
vinilo vinyl
violación bipolar bipolar violation
violación de código code violation
violación de código bipolar bipolar code violation
violación de paridad parity violation
virtual virtual
virus virus
virus de computadora computer virus
viscosidad viscosity
viscosidad dieléctrica dielectric viscosity
viscosidad magnética magnetic viscosity
viscosímetro viscosimeter
viscosímetro electromagnético electromagnetic
    viscometer
visibilidad visibility
visibilidad de visualización display visibility
visión directa direct vision
visión escotópica scotopic vision
visor viewfinder
visor eléctrico electrical boresight
vista fija still
vista oblicua oblique view
vista tangencial tangential view
visual visual
visualización display, visual display
visualización alternativa alternate display
visualización de caracteres character display
visualización de consola console display
visualización de datos data display
visualización de datos de radar radar data display
visualización de gráficos graphics display
visualización de indicador de distancia-altura
    range-height indicator display
visualización de información information display
visualización de información digital digital
    information display
visualización de mandos command display
visualización de menú menu display
visualización de microinstrucciones
    microinstruction display
visualización de microprograma microprogram
    display
visualización de sector sector display
visualización de video video display
visualización descentrada off-center display
visualización digital digital display
visualización directa direct display
visualización ensanchada expanded display
visualización formateada formatted display
visualización gráfica graphic display
visualización no formateada unformatted display
visualización numérica numerical display
visualización orientada a caracteres
    character-oriented display
visualización oscilográfica oscillographic display
visualización panorámica panoramic display
visualización preprocesada preprocessed display
visualización remota remote display
visualización sintética synthetic display
visualizador display, display device, visual display
visualizador analógico analog display
visualizador borrado clear display
visualizador cinescópico kinescope display
visualizador de centro abierto open-center display
visualizador de cristal líquido liquid-crystal display
visualizador de cristal nemático nematic-crystal
    display

visualizador de descarga gaseosa gas-discharge display
visualizador de línea line display
visualizador de matriz de puntos dot-matrix display
visualizador de panel plano flat-panel display
visualizador de plasma plasma display
visualizador de plasma gaseoso gas-plasma display
visualizador de radar radar display
visualizador de rayos catódicos cathode-ray display
visualizador electrocrómico electrochromic display
visualizador electroluminiscente electroluminescent display
visualizador incremental incremental display
visualizador interactivo interactive display
visualizador osciloscópico oscilloscopic display
visualizador sin almacenamiento nonstorage display
vítreo vitreous
vivo live, alive
vobulación wobble modulation, wobbulation
vobulador wobbulator
vocabulario vocabulary
vocoder vocoder
vodas vodas
voder voder
vogad vogad
volátil volatile
volatilidad de datos data volatility
volcado dump, dumping
volcado de disco disk dump
volt volt
volt-ampere volt-ampere
volt-ampere reactivo reactive volt-ampere
volt-hora volt-hour
volt internacional international volt
voltaico voltaic
voltaísmo voltaism
voltaje voltage
voltamétrico voltametric
voltámetro voltameter
voltampere volt-ampere
voltampere reactivo volt-ampere reactive
voltamperehora volt-ampere-hour
voltamperihorímetro volt-ampere-hour meter
voltamperímetro volt-ampere meter, volt-ammeter
vóltmetro voltmeter
vóltmetro-amperímetro voltmeter-ammeter
vóltmetro bipolar bipolar voltmeter
vóltmetro de alta impedancia high-impedance voltmeter
vóltmetro de alta resistencia high-resistance voltmeter
vóltmetro de audiofrecuencia audio-frequency voltmeter
vóltmetro de bobina móvil moving-coil voltmeter
vóltmetro de carga cero zero-load voltmeter
vóltmetro de cero central zero-center voltmeter
vóltmetro de cero suprimido suppressed-zero voltmeter
vóltmetro de contacto contact voltmeter
vóltmetro de conversación speech voltmeter
vóltmetro de corriente continua direct-current voltmeter
vóltmetro de cresta crest voltmeter
vóltmetro de cristal crystal voltmeter
vóltmetro de chispa spark-gap voltmeter
vóltmetro de desfase phase-angle voltmeter
vóltmetro de diodo diode voltmeter
vóltmetro de efecto corona corona voltmeter
vóltmetro de estado sólido solid-state voltmeter
vóltmetro de explosor de esferas sphere-gap voltmeter
vóltmetro de fase phase voltmeter
vóltmetro de hierro móvil moving-iron voltmeter

vóltmetro de Kelvin Kelvin voltmeter
vóltmetro de muestreo sampling voltmeter
vóltmetro de muy alta resistencia very-high-resistance voltmeter
vóltmetro de oposición slide-back voltmeter
vóltmetro de pico peak voltmeter
vóltmetro de pico a pico peak-to-peak voltmeter
vóltmetro de pico negativo negative-peak voltmeter
vóltmetro de pico positivo positive-peak voltmeter
vóltmetro de picos de diodo diode peak voltmeter
vóltmetro de rayos catódicos cathode-ray voltmeter
vóltmetro de registro recording voltmeter
vóltmetro de salida output voltmeter
vóltmetro de transistores transistor voltmeter
vóltmetro de tubo de vacío vacuum-tube voltmeter
vóltmetro de válvula valve voltmeter
vóltmetro diferencial differential voltmeter
vóltmetro digital digital voltmeter
vóltmetro digital de registro recording digital voltmeter
vóltmetro dinamométrico dynamometer voltmeter
vóltmetro electrónico electronic voltmeter
vóltmetro electrónico equilibrado balanced electronic voltmeter
vóltmetro electrostático electrostatic voltmeter
vóltmetro generador generating voltmeter
vóltmetro logarítmico logarithmic voltmeter
vóltmetro-milivóltmetro voltmeter-millivoltmeter
vóltmetro multiescala multirange voltmeter
vóltmetro portátil portable voltmeter
vóltmetro potenciométrico potentiometric voltmeter
vóltmetro-potenciómetro digital digital voltmeter-potentiometer
vóltmetro rectificador-amplificador rectifier-amplifier voltmeter
vóltmetro rotativo rotary voltmeter
vóltmetro segmental segmental voltmeter
vóltmetro selectivo a la frecuencia frequency-selective voltmeter
vóltmetro sintonizado tuned voltmeter
vóltmetro termoiónico thermionic voltmeter
vóltmetro transistorizado transistorized voltmeter
voltmicroamperímetro volt-microammeter
voltmiliamperímetro volt-milliammeter
voltohmamperímetro volt-ohm-ammeter
voltóhmetro volt-ohmmeter, volt-ohm-meter
voltohmmiliamperímetro volt-ohm-milliammeter
volts por metro volts per meter
volumen volume, bulk
volumen de almacenamiento masivo mass storage volume
volumen de disco disk volume
volumen de gas gas volume
volumen de referencia reference volume
volumen de tráfico traffic volume
volumen de uso general general-use volume
volumen global overall volume
volumen primario primary volume
volumen sonoro sound volume
volumétrico volumetric
vordac vordac
vortac vortac
voz artificial artificial voice
vuelco dump
vuelo ciego blind flight
vuelo controlado controlled flight
vuelo por instrumentos instrument flight
vuelta turn
vuelta única single turn
vueltas secundarias secondary turns

wamoscopio wamoscope
watt watt
watt-hora watt-hour
watt internacional international watt
watt real true watt
watt-segundo watt-second
wattaje wattage
wattaje de calefactor heater wattage
wattaje de carga load wattage
wattaje de plena carga full-load wattage
wattaje específico specific wattage
wattaje máximo maximum wattage
wattaje nominal wattage rating
watthora watt-hour
watthorímetro watt-hour meter, kilowatt-hour meter
wáttmetro wattmeter
wáttmetro coaxial coaxial wattmeter
wáttmetro de inducción induction wattmeter
wáttmetro de paleta vane wattmeter
wáttmetro de registro recording wattmeter
wáttmetro digital digital wattmeter
wáttmetro dinamométrico dynamometer wattmeter
wáttmetro electrónico electronic wattmeter
wáttmetro electrostático electrostatic wattmeter
wáttmetro fotoeléctrico photoelectric wattmeter
wáttmetro indicador indicating wattmeter
watts aparentes apparent watts
wattsegundo watt-second
weber weber
willemita willemite

xenón xenon
xerografía xerography
xerográfico xerographic
xeroimpresión xeroprinting
xerorradiografía xeroradiography

Y

Z

yodo iodine
yugo yoke, yoke piece
yugo de deflexión deflecting yoke
yugo de exploración scanning yoke
yugo de imán magnet yoke
yugo electromagnético electromagnetic yoke
yugo magnético magnetic yoke
yute jute

zapata shoe
zapata polar pole shoe
zeta zeta
zirconato-titanato de plomo lead zirconate-titanate
zócalo socket, socket-outlet
zócalo de bayoneta bayonet socket
zócalo de cristal crystal socket
zócalo de porcelana porcelain socket
zócalo de tubo tube socket
zócalo de válvula valve socket
zócalo duodecal duodecal socket
zócalo loktal loktal socket
zócalo magnal magnal socket
zócalo no polarizado unpolarized socket
zócalo octal octal socket
zócalo tripolar three-pole socket
zona zone
zona anacústica anacoustic zone
zona auroral auroral zone
zona ciega blind zone
zona crepuscular twilight zone
zona de aproximación approach zone
zona de aterrizaje landing zone
zona de Brillouin Brillouin zone
zona de carga loading zone
zona de central exchange area
zona de control control zone
zona de convergencia convergence zone
zona de difracción diffraction zone
zona de distribución distribution zone
zona de Fresnel Fresnel zone
zona de inducción induction zone
zona de peligro de radiación radiation danger zone
zona de protección zone of protection
zona de radiación radiation zone
zona de silencio silent zone
zona de sombra shadow zone
zona equifase equiphase zone
zona equiseñal equisignal zone
zona flotante floating zone
zona global global zone
zona hiperacústica hyperacoustical zone
zona interurbana trunk zone
zona lejana far zone
zona muerta dead zone
zona neutra neutral zone
zona obscura obscure zone
zona plana flat zone
zona primaria primary zone
zona prohibida forbidden zone
zona protegida protected zone
zona próxima near zone
zumbador buzzer
zumbador eléctrico electric buzzer
zumbador electrónico electronic buzzer
zumbido hum, buzz
zumbido anódico anode hum
zumbido catódico cathode hum
zumbido de calefactor heater hum
zumbido de corriente alterna alternating-current hum

zumbido de filamento filament hum
zumbido de fuente de alimentación power-supply
  hum
zumbido de micrófono microphone hum
zumbido de portadora carrier hum
zumbido de red mains hum
zumbido de relé relay hum
zumbido electrónico electronic hum
zumbido inducido induced hum
zumbido inductivo inductive hum
zumbido magnético magnetic hum
zumbido residual residual hum
zwitterión zwitterion